Microelectronic Circuit Design, 6th Edition

深入理解微电子电路设计

电子元器件、数字电路、模拟电路原理及应用

原书第6版

理查德·C. 耶格（Richard C. Jaeger）
[美] 特拉维斯·N. 布莱洛克（Travis N. Blalock） 著
本杰明·J. 布莱洛克（Benjamin J. Blalock）

朱前成 朱秀华 编译

清华大学出版社
北京

内 容 简 介

全书共分为 15 章,系统论述微电子电路的基本知识及其应用,涵盖固态电子学与器件、模拟电路、运算放大器和反馈三部分知识体系。固态电子学与器件部分主要介绍电子学的基本原理、固态电子学基础、二极管的 i-v 特性及晶体管的 SPICE 模型等内容,给出电路设计中常用的最差情况分析、蒙特卡洛分析等主要分析方法。模拟电路部分从放大器入手,详细介绍相关概念、二端口模型、反馈放大器频率响应、小信号建模、单晶体管放大器、差分放大器、反馈放大器及振荡器等。运算放大器和反馈部分着重讲解运算放大器的相关概念、非线性运算放大器和反馈放大器的稳定性、运算放大器的应用、差分放大器和运算放大器的设计、模拟集成电路设计的技术及晶体管反馈放大器与振荡器等。

书中给出大量的设计实例及练习供读者学习与实践,可以作为电子信息类、电气类专业本科生或研究生的专业教材或参考书,也可以作为从事固态电子学与器件、模拟电路设计或开发,以及与运算放大器相关的工程技术人员的参考书。

北京市版权局著作权合同登记号　图字: 01-2023-0119

All Rights reserved. No part of this publication may be reproduced or transmitted in any form or by any means, electronic or mechanical, including without limitation photocopying, recording, taping, or any database, information or retrieval system, without the prior written permission of the publisher.

This authorized Bilingual edition is published by Tsinghua University Press Limited in arrangement with McGraw-Hill Education (Singapore) Pte. Ltd.. This edition is authorized for sale in the People's Republic of China only, excluding Hong Kong, Macao SAR and Taiwan.

Translation Copyright © 2024 by McGraw-Hill Education (Singapore) Pte. Ltd and Tsinghua University Press Limited.

版权所有。未经出版人事先书面许可,对本出版物的任何部分不得以任何方式或途径复制或传播,包括但不限于复印、录制、录音,或通过任何数据库、信息或可检索的系统。

此双语版本经授权仅限在中华人民共和国境内(不包括香港特别行政区、澳门特别行政区和台湾)销售。

翻译版权©2024 由麦格劳-希尔教育(新加坡)有限公司与清华大学出版社有限公司所有。

本书封面贴有 McGraw-Hill Education 公司防伪标签,无标签者不得销售。
版权所有,侵权必究。举报: 010-62782989, beiqinquan@tup.tsinghua.edu.cn。

图书在版编目(CIP)数据

深入理解微电子电路设计: 电子元器件、数字电路、模拟电路原理及应用: 原书第 6 版/(美)理查德·C. 耶格(Richard C. Jaeger), (美)特拉维斯·N. 布莱洛克(Travis N. Blalock), (美)本杰明·J. 布莱洛克(Benjamin J. Blalock)著; 朱前成,朱秀华编译. —北京: 清华大学出版社, 2024.7
(清华开发者书库)
书名原文: Microelectronic Circuit Design, 6th Edition
ISBN 978-7-302-65819-1

Ⅰ.①深… Ⅱ.①理…②特…③本…④朱…⑤朱… Ⅲ.①超大规模集成电路-电路设计 Ⅳ.①TN470.2

中国国家版本馆 CIP 数据核字(2024)第 058991 号

责任编辑: 曾　珊
封面设计: 李召霞
责任校对: 李建庄
责任印制: 宋　林

出版发行: 清华大学出版社
　　网　　址: https://www.tup.com.cn, https://www.wqxuetang.com
　　地　　址: 北京清华大学学研大厦 A 座　　邮　编: 100084
　　社 总 机: 010-83470000　　邮　购: 010-62786544
　　投稿与读者服务: 010-62776969, c-service@tup.tsinghua.edu.cn
　　质量反馈: 010-62772015, zhiliang@tup.tsinghua.edu.cn
　　课件下载: https://www.tup.com.cn, 010-83470236
印 装 者: 三河市龙大印装有限公司
经　　销: 全国新华书店
开　　本: 203mm×260mm　　印　张: 75.75　　字　数: 2323 千字
版　　次: 2024 年 8 月第 1 版　　印　次: 2024 年 8 月第 1 次印刷
印　　数: 1~2000
定　　价: 199.00 元

产品编号: 097418-01

译 者 序

随着集成电路工艺的不断发展，微电子电路的性能、设计与分析方法都发生了变化。目前，集成电路制造工艺的特征尺寸越来越小，使得集成电路的集成度越来越高，集成电路正朝着小型化、低功耗、系统集成方向发展。本书系统论述了微电子电路的基本知识及其应用，涵盖了固态电子学与器件、模拟电路、运算放大器和反馈三部分知识体系。通过本书的学习，读者可以对现代电子设计的基本技术、模拟电路和数字电路及分立电路和集成电路进行全面了解。

本书是微电子电路设计领域的一部大作，作者具有丰富的业界设计经验，而且本书经过不断修订，已经成为微电子电路设计领域中的权威教材及工具书。书中涉及内容广泛，将数字电路或者模拟电路内容单独拿出来都可以自成体系，以供读者学习。

本书全面讲述了微电子电路的基础知识及其应用技术，书中没有简单罗列各种器件或者电路，而是关注于让读者理解器件或电路的基本概念、设计方法和仿真验证手段，从全局上把握微电子电路的发展、现状及主要技术。本书覆盖了固态电子学、半导体器件、模拟电路及运算放大器领域的主要内容，可以帮助读者更好地理解和把握微电子电路的设计方法和设计理念。本书强调微电子电路的设计与分析，适合作为高校电子信息、电气工程、计算机及工程技术类相关专业的教材，也可以作为工程技术人员的参考书。本书的主要特色有：

- **注重方法**。本书从工程应用角度定义了9步问题求解方法，书中大量设计实例都是采用该方法进行求解，掌握该方法对于解决电子电路问题及工程问题都会有益处。
- **注重实践**。本书注重理论与实践相结合，每章都提供了大量的设计实例及课后练习，以及在线的习题解答。
- **注重仿真**。本书以计算机作为辅助工具，利用MATLAB、电子表格或者计算机高级语言来开发设计项目，许多电路设计都提供了SPICE仿真模型，便于读者对所设计电路进行性能上的模拟验证，易于读者理解和掌握。

需要特别说明的是，本书中电路元器件及各种量的表示形式遵照英文原书。在本书的编译过程中，我们力求忠实于原著，但由于编译者技术和编译水平有限，书中难免存在各种不足，敬请读者批评指正。

编译者

2024 年 4 月

前　　言

本书全面讲解了现代模拟电子电路设计中的基本技术。通过本书的学习,读者可以对固态电子学、半导体器件、模拟电路、运算放大器和反馈、集成电路技术等有深入的理解。尽管大多数读者可能不会从事集成电路设计的相关工作,但对于集成电路结构的深入理解,有助于读者从系统设计的角度加深认识,从而消除系统设计中的隐患,增强集成电路的可靠性。

本书的创作完成得益于作者在精密数字设计领域多年的丰富工作经历以及多年的教学总结。书中涉及内容广泛,读者可以根据需要选择适当的内容用于两个学期或者连续三个学期的学习。

为了减少纸质图书的篇幅,第 6 版中涉及数字电路的章节已作为补充章节,收录在电子书形式的教科书版本中。

本书说明

本书第 6 版对书中内容进行了更新,强化和改进了一些概念的讲解,更利于读者学习和掌握。

对第一部分的 5 个章节进行了重组。其中,第 3 章讲述二极管相关内容之后,在第 4 章讨论"双极型晶体管",第 5 章具体介绍"场效应晶体管",这部分内容将在全书的第二部分和第三部分继续探讨。

同时,本书至少增加了 30% 的习题;修订和扩展了流行的 Electronic-in-Action 功能,包括 IEEE 社团、SPICE 的历史发展、身体传感器网络、JONES 混频器、高级 CMOS 技术、数/模转换及模/数转换。

本书具有鲜明的特点,可以归纳如下:
(1) 所有实例均采用了结构性问题求解方法。
(2) 每章都提供了相关的电子应用案例。
(3) 每章开始都介绍了与本章内容相关的电子学领域技术的重要发展历程。
(4) 重点强调设计要点,给出了大量的实际电路设计案例。
(5) 本书设计实例中充分利用了 SPICE 仿真软件。
(6) 在 SPICE 中整合了器件模型。
(7) 整合了网站素材。
(8) 增加了大量习题。
(9) 可以从 McGraw-Hill 获得最新的资源包。

第二部分由第 6~9 章组成,首先概述放大器的一般特性,然后讲述小信号建模和单晶体管放大器,以及放大器。

第三部分的前 3 章聚焦于理想运算放大器和非理想运算放大器,包括反馈及放大器的稳定性;后 3 章集中介绍模拟集成电路的设计及设计技术。

设计

在工程师培训中,设计仍然是一个较难的课题。本书定义了非常清晰的问题求解方法,利用该方法可以加深学生对于设计相关问题的理解能力。书中提供的设计实例有助于建立对设计流程的了解。

本书模拟电路部分一直强调采用设计模拟决策的方法。在任何适合的情况下,都在标准混合 π 模型

表示的基础上将放大器特性表达式进行了简化。例如,在绝大多数图书中放大器的电压增益表达式只能写为$|A_v| = g_m R_L$,而隐藏了电源电压作为基本设计变量这一事实。本书对此表达式进行了改进,将双极型晶体管的电压增益近似为$g_m R_L \cong 10 V_{CC}$,或将FET电路的电压增益近似为$g_m R_L \cong V_{DD}$,明确地揭示了放大器设计与电源供电电压选择的关系,为共发射极放大器和共源极放大器的电压增益提供了一种简单的一阶设计估算方法。双极型放大器的增益优势也显而易见。只要有可能,书中经常会给出近似估算此类性能的技巧和方法。双极型电路和FET电路之间的比较和设计权衡贯穿了第三部分内容。

在第1章结尾处介绍了最差情况分析和蒙特卡洛分析技术。传统的本科生课程并不会包含这些内容,但这却是在面临较多的元件容限和差异情况下进行电路设计时需要具备的重要技能,在本书的例子和习题中对采用标准元件和给定元件容限的电路利用该技术都进行了讨论。

习题与指导

在每章的最后都给出专门的设计习题、计算机操作习题和SPICE习题。设计习题用 💡 表示,计算机操作习题用 💻 表示,SPICE习题用 Ⓢ 表示。习题对书中的内容而言十分重要,更难或需要花费更多时间的习题用 * 和 ** 表示。此外,书中所有图标都制成了PowerPoint文件,可以从Connect提供的教师资源部分获得,也可从网站上进行检索。从中还可以找到制作成PowerPoint文件的指导和注意事项。

致谢

感谢对本书编写及筹备做出贡献的工作者。我们的学生在对原稿的润色上提供了极大的帮助,并尽力完成了原稿的多次修订。一直以来,我们的系领导——奥本大学的J.D.Irwin和Mark Nelms,以及弗吉尼亚大学的J.C.Lach,高度支持员工努力写出更高水平的教材。

感谢以下审阅人员:明尼苏达大学德卢斯分校的Stanley Burns、北卡罗来纳农工州立大学的Numan Dogan、波特兰州立大学的Melinda Holtzman、加利福尼亚州立大学北岭分校的Bradley Jackson、新泽西理工学院的Serhiy Levkov和田纳西大学诺克斯维尔分校的Jayne Wu(Jie Wu)。

感谢J.F.Pierce和T.J.Paulus的课堂练习"电子应用"给我们带来的灵感。Blalock教授多年前就跟随Pierce教授学习有关电子学内容,至今仍盛赞他们早已绝版的教材中所采用的诸多分析技术。

那些熟悉Don Pederson教授的"Yellow Peril"的人会在整本书中看到它的影响力。在Jaeger教授成为佛罗里达大学Art Brodersen教授的学生之后不久,他很幸运地获得了Pederson的书,并从头到尾进行了仔细研究。

最后,感谢McGraw-Hill团队的支持,包括环球出版社的Raghothaman Srinivasan、产品开发员Vincent Bradshaw、市场经理Nick McFadden、项目经理Jane Mohr。

在写作本书的过程中,我们尽力将自身在模拟和数字设计领域的业界背景与多年的课堂经验融合在一起,希望能获得一定程度的成功。欢迎大家提出建设性意见和建议。

Richard C.Jaeger

奥本大学

Travis N.Blalock

弗吉尼亚大学

Benjamin J.Blalock

田纳西大学诺克斯维尔分校

PREFACE

Through study of this text, the reader will develop a comprehensive understanding of the basic techniques of modern analog electronic circuit design. Even though most readers may not ultimately be engaged in the design of integrated circuits (ICs) themselves, a thorough understanding of the internal circuit structure of ICs is prerequisite to avoiding many pitfalls that prevent the effective and reliable application of integrated circuits in system design.

The writing integrates the authors' extensive industrial backgrounds in precision analog and digital design with their many years of experience in the classroom. A broad spectrum of topics is included, and material can easily be selected to satisfy either a two-semester or three-quarter sequence in electronics.

In order to reduce the length, cost, and weight of the text, the digital electronics chapters from earlier editions have been included as supplemental chapters in the e-book version of the textbook that is available in Connect.

IN THIS EDITION

This edition continues to update the material to achieve improved readability and accessibility to the student. In addition to general material updates, a number of specific changes have been included.

The five chapters of Part One have been reorganized to improve material flow. Chapter 4, "Bipolar Junction Transistors" now follows directly after the diode chapter, and "Field-Effect Transistors" becomes Chapter 5. A new low-power, low-voltage, and weak inversion thread begins in Part One. Chapter 5 specifically introduces the behavior and modeling of the FET in the moderate and weak inversion regions, and this thread continues throughout Parts Two and Three.

Other important elements include:

- At least 30 percent revised or new problems.
- Updated PowerPoint slides are available from the authors at www.JaegerBlalock.com or Connect.
- Popular digital features can be found through McGraw Hill Education's Connect platform, details of which can be found later in the Preface.
- The structured problem-solving approach continues throughout the examples.
- Popular Electronics in Action features have been revised and expanded to include IEEE Societies, Historical Development of SPICE, Body Sensor Networks, Jones Mixer, Advanced CMOS Technology, Fully Differential Amplifiers, and DACs and ADCs to name a few.

Chapter openers enhance the reader's understanding of historical developments in electronics. Design notes highlight important ideas that the circuit designer should remember. The Internet is viewed as an integral extension of the text.

Features of the book are outlined below.

- The Structured Problem-Solving Approach is used throughout the examples.
- Electronics in Action features in each chapter.
- Chapter openers highlighting developments in the field of electronics.
- Design Notes and emphasis on practical circuit design.
- Broad use of SPICE throughout the text, examples, and problems.
- Integrated treatment of device modeling in SPICE.
- Numerous Exercises, Examples, and Design Examples.
- Large number of problems.
- Integrated web materials.

Part Two consists of Chapters 6 through 9 and begins with an overview of general amplifier characteristics, followed by small-signal modeling of transistors and comprehensive discussion of classical single-stage amplifier design including frequency response.

The first three chapters of Part Three focus on ideal and nonideal operational amplifiers, including feedback and amplifier stability. The last three chapters concentrate on analog integrated circuit design and design techniques.

DESIGN

Design remains a difficult issue in educating engineers. The use of the well-defined problem-solving methodology presented in this text can significantly enhance the students ability to understand issues related to design. The design examples assist in building an understanding of the design process.

Methods for making design estimates and decisions are stressed throughout the analog portion of the text. Expressions for amplifier behavior are simplified beyond the standard hybrid-pi model expressions whenever appropriate. For example, the expression for the voltage gain of an amplifier in most texts is simply written as $|A_v| = g_m R_L$, which tends to hide the power supply voltage as the fundamental design variable. Rewriting this expression in approximate form as $g_m R_L \cong 10 V_{CC}$ for the BJT, or $g_m R_L \cong V_{DD}$ for the FET, explicitly displays the dependence of amplifier design on the choice of power supply voltage and provides a simple first-order design estimate for the voltage gain of the common-emitter and common-source amplifiers. The gain advantage of the BJT stage is also clear. These approximation techniques and methods for performance estimation are included as often as possible. Comparisons and design tradeoffs between the properties of BJTs and FETs are included throughout Part Three.

Worst-case and Monte-Carlo analysis techniques are introduced at the end of the first chapter. These are not topics traditionally included in undergraduate courses. However, the ability to design circuits in the face of wide component tolerances and variations is a key component of electronic circuit design, and the design of circuits using standard components and tolerance assignment are discussed in examples and included in many problems.

PROBLEMS AND INSTRUCTOR SUPPORT

Specific design problems, computer problems, and SPICE problems are included at the end of each chapter. Design problems are indicated by 💡, computer problems are indicated by 💻, and SPICE problems are indicated by 🌀. The problems are keyed to the topics in the text with the more difficult or time-consuming problems indicated by * and **. An Instructor's Manual containing solutions to all the problems is available to instructors from the authors. In addition, the graphs and figures are available as PowerPoint files and can be retrieved on the Instructor's Resources section of Connect, along with various web materials referenced in the textbook for students. Instructor notes are available as PowerPoint slides.

To access the Instructor Resources through Connect, you must first contact your McGraw Hill Learning Technology Representative to obtain a password. If you do not know your McGraw Hill representative, please go to www.mhhe.com/rep, to find your representative.

Once you have your password, please go to connect.mheducation.com, and log in. Click on the course for which you are using *Microelectronic Circuit Design, 6e*. If you have not added a course, click "Add Course," and select "Engineering-Electrical and Computer" from the drop-down menu. Select this textbook and click "Next."

Once you have added the course, click on the "Library" link, and then click "Instructor Resources."

ACKNOWLEDGMENTS

We want to thank the large number of people who have had an impact on the material in this text and on its preparation. Our students have helped immensely in polishing the manuscript and have managed to survive the many revisions of the manuscript. Our department heads, J. D. Irwin and Mark Nelms of Auburn University, N. Sidiropoulos of the University of Virginia and Gregory Peterson of the University of Tennessee, have always been highly supportive of faculty efforts to develop improved texts.

We want to thank all reviewers, including the following:

Stanley Burns	*University of Minnesota, Duluth*
Numan Dogan	*North Carolina Agricultural and Technical State University*
Melinda Holtzman	*Portland State University*
Bradley Jackson	*California State University, Northridge*
Serhiy Levkov	*New Jersey Institute of Technology*
Jayne Wu (Jie Wu)	*The University of Tennessee, Knoxville*

We are also thankful for inspiration from the classic text *Applied Electronics* by J. F. Pierce and T. J. Paulus. Professor Travis Blalock Learned Electronics from Professor Pierce many years ago and still appreciates many of the analytical techniques employed in their long out-of-print text.

Those familiar with Professor Don Pederson's "Yellow Peril" will see its influence throughout this text. Shortly after Professor Jaeger became Professor Art

Brodersen's student at the University of Florida, he was fortunate to be given a copy of Pederson's book to study from cover to cover.

Finally, we want to thank the team at McGraw Hill, including Theresa Collins and Erin Kamm, Product Developers; Jane Mohr, Content Project Manager; Lisa Granger, Marketing Manager; and Sadika Rehman, Full-Service Project Manager.

In developing this text, we have attempted to integrate our industrial backgrounds in analog and digital design with many years of experience in the classroom. We hope we have at least succeeded to some extent. Constructive suggestions and comments will be appreciated.

Richard C. Jaeger
Auburn University

Travis N. Blalock
University of Virginia

Benjamin J. Blalock
University of Tennessee, Knoxville

CHAPTER-BY-CHAPTER SUMMARY

PART ONE—SOLID-STATE ELECTRONICS AND DEVICES

Chapter 1 provides a historical perspective on the field of electronics beginning with vacuum tubes and advancing to Tera-scale integration and its impact on the global economy. Chapter 1 also provides a classification of electronic signals and a review of some important tools from network analysis, including the ideal operational amplifier. Because developing a good problem-solving methodology is of such import to an engineer's career, the comprehensive Structured Problem Solving Approach is used to help students develop their problem solving skills. The structured approach is discussed in detail in the first chapter and used in the subsequent examples in the text. Component tolerances and variations play an extremely important role in practical circuit design, and Chapter 1 closes with introductions to tolerances, temperature coefficients, worst-case design, and Monte Carlo analysis.

Chapter 2 discusses semiconductor materials including the covalent-bond and energy-band models of semiconductors. The chapter includes material on intrinsic carrier density, electron and hole populations, n- and p-type material, and impurity doping. Mobility, resistivity, and carrier transport by both drift and diffusion are included as topics. Velocity saturation is discussed, as well as an introductory discussion of microelectronic fabrication.

Chapter 3 introduces the structure and i-v characteristics of solid-state diodes. Discussions of Schottky diodes, variable capacitance diodes, photo-diodes, solar cells, and LEDs are also included. This chapter introduces the concepts of device modeling and the use of different levels of modeling to achieve various approximations to reality. The SPICE model for the diode is discussed. The concepts of bias, operating point, and load-line are all introduced, and iterative mathematical solutions are also used to find the operating point with MATLAB and spreadsheets. Diode applications in rectifiers are discussed in detail and a discussion of the dynamic switching characteristics of diodes is also presented.

Chapter 4 introduces the bipolar junction transistor and presents a heuristic development of the transport (simplified Gummel-Poon) model of the BJT based upon superposition. The various regions of operation are discussed in detail. Common-emitter and common-base current gains are defined, and base transit-time, diffusion capacitance, and cutoff frequency are all discussed. Bipolar technology and physical structure are introduced. The four-resistor bias circuit is discussed in detail. The SPICE model for the BJT and SPICE model parameters are also discussed in Chapter 4.

Chapter 5 discusses MOS and junction field-effect transistors, starting with a qualitative description of the MOS capacitor. Models are developed for the FET i-v characteristics, and a complete discussion of the regions of operation of the device is presented. Body effect is included. MOS transistor performance limits—including scaling, cut-off frequency, and subthreshold conduction—are discussed as well as basic Λ-based layout methods. Biasing circuits and load-line analysis are presented. The concept of velocity saturation from Chapter 2 is reinforced with the addition of the unified MOS model of Rabaey and Chandrakasan to Chapter 5. FET SPICE models and model parameters are discussed in Chapter 5. In the 6th edition, the discussion of moderate and weak inversion is expanded, and a low voltage/weak inversion thread continues through the rest of the text.

PART TWO—ANALOG ELECTRONICS

Chapter 6 provides a succinct introduction to analog electronics. The concepts of voltage gain, current gain, and power gain are developed using two-port circuit models. Much care has been taken to be consistent in the use of the notation that defines these quantities as well as in the use of dc, ac, and total signal notation throughout the book. Bode plots are reviewed and amplifiers are classified by frequency response. MATLAB is utilized as a tool for producing Bode plots. SPICE simulation using built-in SPICE models is introduced.

Chapter 7 begins the general discussion of linear amplification using the BJT and FET as C-E and C-S amplifiers. Biasing for linear operation and the concept of small-signal modeling are both introduced, and small-signal models of the diode, BJT, and FET are all developed. The limits for small-signal operation are all carefully defined. The use of coupling and bypass capacitors and inductors to separate the ac and dc designs is explored. The important $10V_{CC}$ and V_{DD} design estimates for the voltage gain of the C-E and C-S amplifiers are introduced, and the role of the transistor's intrinsic gain in bounding circuit performance is discussed. The role of Q-point design on power dissipation and signal range is also introduced.

Chapter 8 proceeds with an in-depth comparison of the characteristics of single-transistor amplifiers, including small-signal amplitude limitations. Appropriate points for signal injection and extraction are identified, and amplifiers are classified as inverting amplifiers (C-E, C-S), noninverting amplifiers (C-B, C-G), and followers (C-C, C-D). The treatment of MOS and bipolar devices is merged from Chapter 8 on, and design tradeoffs between the use of the BJT and the FET in amplifier circuits is an important thread that is followed through all of Part Two. A detailed discussion of the design of coupling and bypass capacitors and the role of these capacitors in controlling the low frequency response of amplifiers appears in this chapter.

Chapter 9 discusses the frequency response of analog circuits. The behavior of each of the three categories of single-stage amplifiers (C-E/C-S, C-B/C-G, and C-C/C-D) is discussed in detail, and BJT behavior is contrasted with that of the FET. The frequency response of the transistor is discussed, and the high frequency, small-signal models are developed for both the BJT and FET. Miller multiplication is used to obtain estimates of the lower and upper cutoff frequencies of complex multistage amplifiers. Gain-bandwidth products and gain-bandwidth tradeoffs in design are discussed. Cascode amplifier frequency response, and tuned amplifiers are included in this chapter. The important short-circuit and open-circuit time-constant techniques for estimating the dominant low- and high-frequency poles are covered in detail.

Because of the renaissance and pervasive use of RF circuits, Chapter 9 includes an introductory section on RF amplifiers, including shunt peaked and tuned amplifiers. A discussion of gate resistance in FETs mirrors that of base resistance in the BJT. The discussion of the impact of the frequency-dependent current gain of the FET includes both the input and output impedances of the source follower configuration. Material on mixers includes passive and active single- and double-balanced mixers and the widely used Jones Mixer.

PART THREE—OPERATIONAL AMPLIFIERS AND FEEDBACK

Chapter 10 reviews classic ideal operational amplifier circuits that include the inverting, noninverting, summing, and difference amplifiers as well as the integrator, differentiator, and low-pass and high-pass filters.

Chapter 11 focuses on a comprehensive discussion of the characteristics and limitations of real operational amplifiers, including the effects of finite gain and input resistance, nonzero output resistance, input offset voltage, input bias and offset currents, output voltage and current limits, finite bandwidth, and common-mode rejection. A consistent loop-gain analysis approach is used to study the four classic feedback configurations, and Blackman's theorem is utilized to find input and output resistances of closed-loop amplifiers. The important successive voltage and current injection technique for finding loop-gain is included in Chapter 11. Stability of first-, second-, and third-order systems is discussed, and the concepts of phase and gain margin are introduced. Relationships between Nyquist and Bode techniques are explicitly discussed. A section concerning the relationship between phase margin and time domain response is included. The macro model concept is introduced and the discussion of SPICE simulation of op-amp circuits using various levels of models continues in Chapter 11.

Chapter 12 covers a wide range of operational amplifier applications that include multistage amplifiers, the instrumentation amplifier, and continuous time and discrete time active filters. Cascade amplifiers are investigated including a discussion of the bandwidth of multistage amplifiers. An introduction to D/A and A/D converters appears in this chapter. The Barkhausen criterion for oscillation are presented and followed by a discussion of op-amp-based sinusoidal oscillators. High frequency oscillators are discussed in Chapter 15. Nonlinear circuits applications including rectifiers, Schmitt triggers, and multivibrators conclude the material in Chapter 12.

Chapter 13 explores the design of multistage direct coupled amplifiers. An evolutionary approach to multistage op amp design is used. MOS and bipolar differential amplifiers are first introduced. Subsequent addition of a second gain stage and then an output stage convert the differential amplifiers into simple op amps. Class A, B, and AB operations are defined. Electronic current sources are designed and used for biasing of the basic operational amplifiers. Discussion of important FET-BJT design tradeoffs are included wherever appropriate. Additional low voltage/weak inversion problems have been added to Chapters 13, 14, and 15.

Chapter 14 introduces techniques that are of particular import in integrated circuit design. A variety of current mirror circuits are introduced and applied in bias circuits and as active loads in operational amplifiers. A wealth of circuits and analog design techniques are explored through the detailed analysis of the classic 741 operational amplifier. The Brokaw bandgap reference and Gilbert analog multiplier as well as the MOS weak inversion reference are introduced in Chapter 14.

Chapter 15 presents detailed examples of feedback as applied to transistor amplifier circuits. The loop-gain analysis approach introduced in Chapter 11 is used to find the closed-loop gain of various amplifiers, and Blackman's theorem is utilized to find input and output resistances of closed-loop amplifiers.

Amplifier stability is also discussed in Chapter 15, and Nyquist diagrams and Bode plots (with MATLAB) are used to explore the phase and gain margin of amplifiers. Basic single-pole op-amp compensation is discussed, and the unity gain-bandwidth product is related to amplifier slew rate. Design of op-amp compensation to achieve a desired phase margin is presented. The discussion of transistor oscillator circuits includes the classic Colpitts, Hartley, and negative G_m configurations. Crystal oscillators, ring oscillators and a discussion of positive feedback in flip-flops are also included.

The Digital Electronics chapters from the fifth edition are now included as supplemental chapters in the e-book version of this text, which is available to users of this edition through Connect.

Four Appendices include tables of standard component values (Appendix A), summary of the device models and sample SPICE parameters (Appendix B), review of two-port networks (Appendix C), and Physical Constants and Transistor Model Summary (Appendix D). Data sheets for representative solid-state devices and operational amplifiers are available via the Internet. A table in Appendix C helps relate various two-port parameters that often appear in specification sheets to the FET and BJT model parameters that appear in the text.

目　录

PART ONE
SOLID-STATE ELECTRONICS AND DEVICES

CHAPTER 1
INTRODUCTION TO ELECTRONICS　3

- 1.1 A Brief History of Electronics: From Vacuum Tubes to Giga-Scale Integration　6
- 1.2 Classification of Electronic Signals　10
  - 1.2.1 Digital Signals　10
  - 1.2.2 Analog Signals　11
  - 1.2.3 A/D and D/A Converters—Bridging the Analog and Digital Domains　12
- 1.3 Notational Conventions　14
- 1.4 Problem-Solving Approach　15
- 1.5 Important Concepts from Circuit Theory　17
  - 1.5.1 Voltage and Current Division　17
  - 1.5.2 Thévenin and Norton Circuit Representations　18
- 1.6 Frequency Spectrum of Electronic Signals　23
- 1.7 Amplifiers　24
  - 1.7.1 Ideal Operational Amplifiers　25
  - 1.7.2 Amplifier Frequency Response　28
- 1.8 Element Variations in Circuit Design　28
  - 1.8.1 Mathematical Modeling of Tolerances　29
  - 1.8.2 Worst-Case Analysis　29
  - 1.8.3 Monte Carlo Analysis　31
  - 1.8.4 Temperature Coefficients　34
- 1.9 Numeric Precision　36
- Summary　36
- Key Terms　37
- References　38
- Additional Reading　38
- Problems　39

第一部分
固态电子学与器件

第 1 章
电子学简介

- 1.1 电子学发展简史：从真空管到吉规模集成电路
- 1.2 电信号的分类
  - 1.2.1 数字信号
  - 1.2.2 模拟信号
  - 1.2.3 A/D 和 D/A 转换器——模拟与数字信号的桥梁
- 1.3 符号约定
- 1.4 问题求解的方法
- 1.5 电路理论的主要概念
  - 1.5.1 分压和分流
  - 1.5.2 戴维南定理和诺顿定理
- 1.6 电信号的频谱
- 1.7 放大器
  - 1.7.1 理想运算放大器
  - 1.7.2 放大器频率响应
- 1.8 电路设计中元件参数的变化
  - 1.8.1 容差的数学模型
  - 1.8.2 最差情况分析
  - 1.8.3 蒙特卡洛分析
  - 1.8.4 温度系数
- 1.9 数值精度
- 小结
- 关键词
- 参考文献
- 补充阅读
- 习题

CHAPTER 2
SOLID-STATE ELECTRONICS   46

2.1 Solid-State Electronic Materials   50
2.2 Covalent Bond Model   51
2.3 Drift Currents and Mobility in Semiconductors   54
     2.3.1 Drift Currents   54
     2.3.2 Mobility   55
     2.3.3 Velocity Saturation   55
2.4 Resistivity of Intrinsic Silicon   56
2.5 Impurities in Semiconductors   57
     2.5.1 Donor Impurities in Silicon   58
     2.5.2 Acceptor Impurities in Silicon   58
2.6 Electron and Hole Concentrations in Doped Semiconductors   58
     2.6.1 n-Type Material($N_D>N_A$)   59
     2.6.2 p-Type Material($N_A>N_D$)   60
2.7 Mobility and Resistivity in Doped Semiconductors   61
2.8 Diffusion Currents   65
2.9 Total Current   66
2.10 Energy Band Model   67
     2.10.1 Electron—Hole Pair Generation in an Intrinsic Semiconductor   67
     2.10.2 Energy Band Model for a Doped Semiconductor   68
     2.10.3 Compensated Semiconductors   68
2.11 Overview of Integrated Circuit Fabrication   70
Summary   73
Key Terms   74
References   75
Additional Reading   75
Problems   75

CHAPTER 3
SOLID-STATE DIODES AND DIODE CIRCUITS   81

3.1 The pn Junction Diode   84
     3.1.1 pn Junction Electrostatics   84
     3.1.2 Internal Diode Currents   88
3.2 The $i\text{-}v$ Characteristics of the Diode   89
3.3 The Diode Equation：A Mathematical Model for

第 2 章
固态电子学

2.1 固态电子材料
2.2 共价键模型
2.3 半导体中的漂移电流和迁移率
     2.3.1 漂移电流
     2.3.2 迁移率
     2.3.3 速度饱和
2.4 本征硅的电阻率
2.5 半导体中的杂质
     2.5.1 硅中的施主杂质
     2.5.2 硅中的受主杂质
2.6 掺杂半导体中的电子和空穴浓度
     2.6.1 n 型材料（$N_D>N_A$）
     2.6.2 p 型材料（$N_A>N_D$）
2.7 掺杂半导体中的迁移率和电阻率
2.8 扩散电流
2.9 总电流
2.10 能带模型
     2.10.1 本征半导体中电子空穴对的产生
     2.10.2 掺杂半导体的能带模型
     2.10.3 补偿半导体
2.11 集成电路制造综述
小结
关键词
参考文献
补充阅读
习题

第 3 章
固态二极管和二极管电路

3.1 pn 结二极管
     3.1.1 pn 结静电学
     3.1.2 二极管内部电流
3.2 二极管的 $i\text{-}v$ 特性
3.3 二极管方程：二极管的数学模型

		the Diode	91		
3.4		Diode Characteristics under Reverse, Zero, and Forward Bias	94	3.4	二极管特性之反偏、零偏和正偏
	3.4.1	Reverse Bias	94	3.4.1	反偏
	3.4.2	Zero Bias	94	3.4.2	零偏
	3.4.3	Forward Bias	95	3.4.3	正偏
3.5		Diode Temperature Coefficient	97	3.5	二极管的温度系数
3.6		Diodes under Reverse Bias	97	3.6	反偏下的二极管
	3.6.1	Saturation Current in Real Diodes	98	3.6.1	实际二极管的饱和电流
	3.6.2	Reverse Breakdown	100	3.6.2	反向击穿
	3.6.3	Diode Model for the Breakdown Region	101	3.6.3	击穿区的二极管模型
3.7		pn Junction Capacitance	101	3.7	pn 结电容
	3.7.1	Reverse Bias	101	3.7.1	反偏
	3.7.2	Forward Bias	102	3.7.2	正偏
3.8		Schottky Barrier Diode	104	3.8	肖特基势垒二极管
3.9		Diode SPICE Model and Layout	104	3.9	二极管的 SPICE 模型及版图
	3.9.1	Diode Layout	105	3.9.1	二极管的版图
3.10		Diode Circuit Analysis	106	3.10	二极管电路分析
	3.10.1	Load-Line Analysis	107	3.10.1	负载线分析法
	3.10.2	Analysis Using the Mathematical Model for the Diode	108	3.10.2	二极管数学模型分析法
	3.10.3	The Ideal Diode Model	112	3.10.3	理想二极管模型
	3.10.4	Constant Voltage Drop Model	114	3.10.4	恒压降模型
	3.10.5	Model Comparison and Discussion	115	3.10.5	模型比较与讨论
3.11		Multiple-Diode Circuits	116	3.11	多二极管电路
3.12		Analysis of Diodes Operating in the Breakdown Region	119	3.12	二极管工作在击穿区域的分析
	3.12.1	Load-Line Analysis	119	3.12.1	负载线分析
	3.12.2	Analysis with the Piecewise Linear Model	119	3.12.2	分段线性模型分析
	3.12.3	Voltage Regulation	120	3.12.3	稳压器
	3.12.4	Analysis Including Zener Resistance	121	3.12.4	包含齐纳电阻的电路分析
	3.12.5	Line and Load Regulation	122	3.12.5	线性调整率和负载调整率
3.13		Half-Wave Rectifier Circuits	123	3.13	半波整流电路
	3.13.1	Half-Wave Rectifier with Resistor Load	123	3.13.1	带负载电阻的半波整流器
	3.13.2	Rectifier Filter Capacitor	124	3.13.2	整流滤波电容
	3.13.3	Half-Wave Rectifier with RC Load	125	3.13.3	带 RC 负载的半波整流器
	3.13.4	Ripple Voltage and Conduction Interval	126	3.13.4	纹波电压和导通期
	3.13.5	Diode Current	128	3.13.5	二极管电流
	3.13.6	Surge Current	130	3.13.6	浪涌电流
	3.13.7	Peak-Inverse-Voltage (PIV) Rating	130	3.13.7	额定峰值反向电压
	3.13.8	Diode Power Dissipation	130	3.13.8	二极管功耗

	3.13.9	Half-Wave Rectifier with Negative Output Voltage	131	3.13.9	输出负电压的半波整流器
3.14	Full-Wave Rectifier Circuits		133	3.14	全波整流电路
	3.14.1	Full-Wave Rectifier with Negative Output Voltage	134	3.14.1	输出负电压的全波整流器
3.15	Full-Wave Bridge Rectification		134	3.15	全波桥式整流
3.16	Rectifier Comparison and Design Tradeoffs		135	3.16	整流器的比较和折中设计
3.17	Dynamic Switching Behavior of the Diode		139	3.17	二极管的动态开关行为
3.18	Photo Diodes, Solar Cells, and Light-Emitting Diodes		140	3.18	光电二极管、太阳能电池和发光二极管
	3.18.1	Photo Diodes and Photodetectors	140	3.18.1	光电二极管和光探测器
	3.18.2	Power Generation from Solar Cells	141	3.18.2	太阳能电池
	3.18.3	Light-Emitting Diodes (LEDs)	142	3.18.3	发光二极管(LED)
Summary			143	小结	
Key Terms			144	关键词	
Reference			145	参考文献	
Additional Reading			145	扩展阅读	
Problems			145	习题	

CHAPTER 4
BIPOLAR JUNCTION TRANSISTORS   158

第 4 章
双极型晶体管

4.1	Physical Structure of the Bipolar Transistor		161	4.1	双极型晶体管的物理结构
4.2	The Transport Model for the *npn* Transistor		162	4.2	*npn* 晶体管的传输模型
	4.2.1	Forward Characteristics	163	4.2.1	正向特性
	4.2.2	Reverse Characteristics	165	4.2.2	反向特性
	4.2.3	The Complete Transport Model Equations for Arbitrary Bias Conditions	166	4.2.3	任意偏置条件下晶体管传输模型方程
4.3	The *pnp* Transistor		168	4.3	*pnp* 晶体管
4.4	Equivalent Circuit Representations for the Transport Models		170	4.4	晶体管传输模型的等效电路
4.5	The *i-v* Characteristics of the Bipolar Transistor		171	4.5	双极型晶体管的 *i-v* 特性
	4.5.1	Output Characteristics	171	4.5.1	输出特性
	4.5.2	Transfer Characteristics	172	4.5.2	传输特性
4.6	The Operating Regions of the Bipolar Transistor		173	4.6	双极型晶体管的工作区
4.7	Transport Model Simplifications		174	4.7	传输模型的简化
	4.7.1	Simplified Model for the Cutoff Region	174	4.7.1	截止区的简化模型
	4.7.2	Model Simplifications for the Forward-Active Region	176	4.7.2	正向有源区的模型简化
	4.7.3	Diodes in Bipolar Integrated Circuits	182	4.7.3	双极型集成电路中的二极管
	4.7.4	Simplified Model for the Reverse-Active Region	183	4.7.4	反向有源区的简化模型

4.7.5	Modeling Operation in the Saturation Region	186	4.7.5 饱和区模型
4.8	Nonideal Behavior of the Bipolar Transistor	189	4.8 双极型晶体管的非理想特性
4.8.1	Junction Breakdown Voltages	189	4.8.1 结击穿电压
4.8.2	Minority-Carrier Transport in the Base Region	189	4.8.2 基区的少数载流子传输
4.8.3	Base Transit Time	190	4.8.3 基区传输时间
4.8.4	Diffusion Capacitance	192	4.8.4 扩散电容
4.8.5	Frequency Dependence of the Common-Emitter Current Gain	193	4.8.5 共发电流增益对频率的依赖性
4.8.6	The Early Effect and Early Voltage	193	4.8.6 Early 效应和 Early 电压
4.8.7	Modeling the Early Effect	194	4.8.7 Early 效应的建模
4.8.8	Origin of the Early Effect	194	4.8.8 Early 效应的产生原因
4.9	Transconductance	195	4.9 跨导
4.10	Bipolar Technology and SPICE Model	196	4.10 双极型工艺与 SPICE 模型
4.10.1	Qualitative Description	196	4.10.1 定量描述
4.10.2	SPICE Model Equations	197	4.10.2 SPICE 模型方程
4.10.3	High-Performance Bipolar Transistors	199	4.10.3 高性能双极型晶体管
4.11	Practical Bias Circuits for the BJT	199	4.11 BJT 的实际偏置电路
4.11.1	Four-Resistor Bias Network	201	4.11.1 四电阻偏置网络
4.11.2	Design Objectives for the Four-Resistor Bias Network	203	4.11.2 四电阻偏置网络的设计目标
4.11.3	Iterative Analysis of the Four-Resistor Bias Circuit	207	4.11.3 四电阻偏置电路的迭代分析
4.12	Tolerances in Bias Circuits	208	4.12 偏置电路的容差
4.12.1	Worst-Case Analysis	208	4.12.1 最差情况分析
4.12.2	Monte Carlo Analysis	210	4.12.2 蒙特卡洛分析
Summary		213	小结
Key Terms		215	关键词
References		216	参考文献
Additional Readings		216	补充阅读
Problems		216	习题

CHAPTER 5
FIELD-EFFECT TRANSISTORS 228

第 5 章
场效应晶体管

5.1	Characteristics of the MOS Capacitor	231	5.1 MOS 电容特性
5.1.1	Accumulation Region	232	5.1.1 积累区
5.1.2	Depletion Region	233	5.1.2 耗尽区
5.1.3	Inversion Region	233	5.1.3 反型区
5.2	The NMOS Transistor	233	5.2 NMOS 晶体管
5.2.1	Qualitative i-v Behavior of the NMOS Transistor	234	5.2.1 NMOS 晶体管的 i-v 特性的定性描述
5.2.2	Triode Region Characteristics of the NMOS Transistor	235	5.2.2 NMOS 晶体管的线性区特性
5.2.3	On Resistance	238	5.2.3 导通电阻

5.2.4	Transconductance	239
5.2.5	Saturation of the *i-v* Characteristics	240
5.2.6	Mathematical Model in the Saturation (Pinch-Off) Region	241
5.2.7	Transconductance in Saturation	242
5.2.8	Channel-Length Modulation	242
5.2.9	Transfer Characteristics and Depletion-Mode MOSFETs	243
5.2.10	Body Effect or Substrate Sensitivity	245
5.3	PMOS Transistors	246
5.4	MOSFET Circuit Symbols	248
5.5	MOS Transistor Symmetry	249
5.5.1	The One-Transistor Dram Cell	249
5.5.2	Data Storage in the 1-T Cell	250
5.5.3	Reading Data from the 1-T Cell	251
5.6	CMOS Technology	254
5.6.1	CMOS Voltage Transfer Characteristics	256
5.7	CMOS Latchup	258
5.8	Capacitances in MOS Transistors	260
5.8.1	NMOS Transistor Capacitances in the Triode Region	260
5.8.2	Capacitances in the Saturation Region	263
5.8.3	Capacitances in Cutoff	263
5.9	MOSFET Modeling in SPICE	263
5.10	MOS Transistor Scaling	265
5.10.1	Drain Current	266
5.10.2	Gate Capacitance	266
5.10.3	Circuit and Power Densities	266
5.10.4	Power-Delay Product	267
5.10.5	Cutoff Frequency	267
5.10.6	High Field Limitations	268
5.10.7	The Unified MOS Transistor Model Including High Field Limitations	269
5.10.8	Subthreshold Conduction	270
5.11	All Region Modeling	271
5.11.1	Interpolation Model	271
5.11.2	Interpolation Model in the Saturation Region	271
5.11.3	Transconductance Efficiency	272
5.12	MOS Transistor Fabrication and Layout Design Rules	274
5.12.1	Minimum Feature Size and Alignment	

5.2.4 跨导
5.2.5 *i-v* 特性的饱和
5.2.6 饱和(夹断)区的数学模型
5.2.7 饱和跨导
5.2.8 沟道长度调制
5.2.9 传输特性及耗尽型 MOSFET
5.2.10 体效应或衬底灵敏度
5.3 PMOS 晶体管
5.4 MOSFET 电路符号
5.5 MOS 晶体管对称性
5.5.1 单晶体管 Dram 单元
5.5.2 1-T 细胞中的数据存储
5.5.3 从 1-T 细胞读取数据
5.6 CMOS 技术
5.6.1 CMOS 电压传输特性
5.7 CMOS 锁存器
5.8 MOS 晶体管电容
5.8.1 NMOS 晶体管的线性区电容
5.8.2 饱和区电容
5.8.3 截止区电容
5.9 SPICE 中的 MOSFET 建模
5.10 MOS 晶体管的等比例缩放
5.10.1 漏极电流
5.10.2 栅极电容
5.10.3 电流和功率密度
5.10.4 功耗-延迟积
5.10.5 截止频率
5.10.6 大电场限制
5.10.7 包含高场限制的统一 MOS 晶体管模型
5.10.8 亚阈值导通
5.11 全区域建模
5.11.1 插值模型
5.11.2 饱和区域的插值模型
5.11.3 跨导效率
5.12 MOS 晶体管的制造工艺及版图设计规则
5.12.1 最小特征尺寸和对准

	Tolerance	274		容差	
5.12.2	MOS Transistor Layout	275	5.12.2	MOS 晶体管的版图	
5.12.3	CMOS Inverter Layout	277	5.12.3	CMOS 逆变器布局	
5.13	Advanced CMOS Technologies	278	5.13	先进 CMOS 技术	
5.14	Biasing the NMOS Field-Effect Transistor	281	5.14	NMOS 场效应晶体管的偏置	
5.14.1	Why Do We Need Bias?	281	5.14.1	为什么需要偏置	
5.14.2	Four-Resistor Biasing	283	5.14.2	四电阻偏置	
5.14.3	Constant Gate-Source Voltage Bias	287	5.14.3	恒定栅-源电压偏置	
5.14.4	Graphical Analysis for the Q-Point	287	5.14.4	Q 点的图形分析	
5.14.5	Analysis Including Body Effect	288	5.14.5	包含体效应的分析	
5.14.6	Analysis Using the Unified Model	290	5.14.6	使用统一模型进行分析	
5.14.7	NMOS Circuit Analysis Comparisons	292	5.14.7	NMOS 电路分析比较	
5.14.8	Two-Resistor Bias	292	5.14.8	双电阻器偏置	
5.15	Biasing the PMOS Field-Effect Transistor	292	5.15	PMOS 场效应晶体管的偏置	
5.16	Biasing the CMOS Inverter as an Amplifier	295	5.16	偏置 CMOS 反相器作为放大器	
5.17	The CMOS Transmission Gate	296	5.17	CMOS 传输门	
5.18	The Junction Field-Effect Transistor (JFET)	298	5.18	结型场效应晶体管（JFET）	
5.18.1	The JFET with Bias Applied	299	5.18.1	偏压下的 JFET	
5.18.2	JFET Channel with Drain-Source Bias	299	5.18.2	漏源偏置下的 JFET 沟道	
5.18.3	n-Channel JFET i-v Characteristics	301	5.18.3	n 沟道 JFET 的 i-v 特性	
5.18.4	The p-Channel JFET	302	5.18.4	p 沟道 JFET	
5.18.5	Circuit Symbols and JFET Model Summary	302	5.18.5	JFET 的电路符号和模型小结	
5.18.6	JFET Capacitances	303	5.18.6	JFET 电容	
5.19	JFET Modeling in SPICE	304	5.19	JFET 的 SPICE 模型	
5.20	Biasing the JFET and Depletion-Mode MOSFET	305	5.20	JFET 和耗尽型 MOSFET 的偏置	
Summary		307	小结		
Key Terms		309	关键词		
References		310	参考文献		
Additional Readings		311	补充阅读		
Problems		312	习题		

PART TWO
ANALOG ELECTRONICS

第二部分
模拟电路

CHAPTER 6
INTRODUCTION TO AMPLIFIERS   331

第 6 章
放大器简介

6.1	An Example of an Analog Electronic System	334	6.1	模拟电子系统示例	
6.2	Amplification	335	6.2	放大作用	
6.2.1	Voltage Gain	336	6.2.1	电压增益	
6.2.2	Current Gain	337	6.2.2	电流增益	

6.2.3	Power Gain	337	6.2.3 功率增益
6.2.4	Location of the Amplifier	337	6.2.4 放大器的位置
6.2.5	The Decibel Scale	338	6.2.5 分贝
6.3	Two-Port Models for Amplifiers	341	6.3 放大器的二端口模型
6.3.1	The g-Parameters	341	6.3.1 g 参数
6.4	Mismatched Source and Load Resistances	345	6.4 源和负载电阻的失配
6.5	The Differential Amplifier	348	6.5 差分放大器
6.5.1	Differential Amplifier Voltage Transfer Characteristic	349	6.5.1 差分放大器的电压传输特性
6.5.2	Voltage Gain	349	6.5.2 电压增益
6.6	Distortion in Amplifiers	351	6.6 放大器的失真
6.7	Differential Amplifier Model	352	6.7 差分放大器模型
6.8	Amplifier Frequency Response	354	6.8 放大器的频率响应
6.8.1	Bode Plots	354	6.8.1 伯德图
6.8.2	The Low-Pass Amplifier	355	6.8.2 低通放大器
6.8.3	The High-Pass Amplifier	358	6.8.3 高通放大器
6.8.4	Band-Pass Amplifiers	361	6.8.4 带通放大器
Summary		364	小结
Key Terms		365	关键词
References		365	参考文献
Additional Reading		365	补充阅读
Problems		365	习题

CHAPTER 7
THE TRANSISTOR AS AN AMPLIFIER 374

第 7 章
晶体管放大器

7.1	The Transistor as an Amplifier	377	7.1 晶体管放大器
7.1.1	The BJT Amplifier	378	7.1.1 BJT 放大器
7.1.2	The MOSFET Amplifier	379	7.1.2 MOSFET 放大器
7.2	Coupling and Bypass Capacitors	380	7.2 耦合电容和旁路电容
7.3	Circuit Analysis Using dc and ac Equivalent Circuits	382	7.3 用直流和交流等效电路进行电路分析
7.3.1	Menu for dc and ac Analysis	382	7.3.1 直流和交流分析步骤
7.4	Introduction to Small-Signal Modeling	386	7.4 小信号模型简介
7.4.1	Graphical Interpretation of the Small-Signal Behavior of the Diode	386	7.4.1 二极管小信号行为的图形解释
7.4.2	Small-Signal Modeling of the Diode	387	7.4.2 二极管的小信号建模
7.5	Small-Signal Models for Bipolar Junction Transistors	389	7.5 双极型晶体管的小信号模型
7.5.1	The Hybrid-Pi Model	391	7.5.1 混合 π 模型
7.5.2	Graphical Interpretation of the Transconductance	392	7.5.2 图解跨导

7.5.3	Small-Signal Current Gain	392
7.5.4	The Intrinsic Voltage Gain of the BJT	393
7.5.5	Equivalent Forms of the Small-Signal Model	394
7.5.6	Simplified Hybrid-Pi Model	395
7.5.7	Definition of a Small Signal for the Bipolar Transistor	395
7.5.8	Small-Signal Model for the *pnp* Transistor	397
7.5.9	ac Analysis versus Transient Analysis in SPICE	398
7.6	The Common-Emitter (C-E) Amplifier	398
7.6.1	Terminal Voltage Gain	398
7.6.2	Input Resistance	400
7.6.3	Signal Source Voltage Gain	400
7.7	Important Limits and Model Simplifications	400
7.7.1	A Design Guide for the Common-Emitter Amplifier	401
7.7.2	Upper Bound on the Common-Emitter Gain	402
7.7.3	Small-Signal Limit for the Common-Emitter Amplifier	402
7.8	Small-Signal Models for Field-Effect Transistors	405
7.8.1	Small-Signal Model for the MOSFET	405
7.8.2	Intrinsic Voltage Gain of the MOSFET	407
7.8.3	Definition of Small-Signal Operation for the MOSFET	408
7.8.4	Body Effect in the Four-Terminal MOSFET	409
7.8.5	Small-Signal Model for the PMOS Transistor	410
7.8.6	Small-Signal Model for MOS Transistors in Weak Inversion	411
7.8.7	Small-Signal Model for the Junction Field-Effect Transistor	411
7.9	Summary and Comparison of the Small-Signal Models of the BJT and FET	412
7.10	The Common-Source (C-S) Amplifier	415
7.10.1	Common-Source Terminal Voltage Gain	416
7.10.2	Signal Source Voltage Gain for the Common-Source Amplifier	416

7.5.3	小信号电流增益
7.5.4	BJT 的固有电压增益
7.5.5	小信号模型的等效形式
7.5.6	简化的混合 π 模型
7.5.7	双极型晶体管的小信号定义
7.5.8	*pnp* 晶体管的小信号模型
7.5.9	用 SPICE 进行交流分析和瞬态分析的对比
7.6	共射极放大器
7.6.1	端电压增益
7.6.2	输入电阻
7.6.3	信号源电压增益
7.7	重要限制及模型简化
7.7.1	共射极放大器的设计指导
7.7.2	共射极增益的上限
7.7.3	共射极放大器的小信号限制
7.8	场效应晶体管的小信号模型
7.8.1	MOSFET 的小信号模型
7.8.2	MOSFET 的本征电压增益
7.8.3	MOSFET 小信号工作的定义
7.8.4	四端 MOSFET 中的体效应
7.8.5	PMOS 晶体管的小信号模型
7.8.6	弱反型区(亚阈值区)模式下 MOS 管的小信号模型
7.8.7	结型场效应晶体管的小信号模型
7.9	BJT 和 FET 小信号模型的小结与对比
7.10	共源极放大器
7.10.1	共源极端电压增益
7.10.2	共源极放大器的信号源电压增益

7.10.3	A Design Guide for the Common-Source Amplifier	417	
7.10.4	Small-Signal Limit for the Common-Source Amplifier	418	
7.10.5	Input Resistances of the Common-Emitter and Common-Source Amplifiers	420	
7.10.6	Common-Emitter and Common-Source Output Resistances	422	
7.10.7	Comparison of the Three Amplifier Examples	428	
7.11	Common-Emitter and Common-Source Amplifier Summary	429	
7.11.1	Guidelines for Neglecting the Transistor Output Resistance	429	
7.12	Amplifier Power and Signal Range	430	
7.12.1	Power Dissipation	430	
7.12.2	Signal Range	431	

Summary 434
Key Terms 435
Reference 435
Problems 436

7.10.3 共源极放大器的设计指导
7.10.4 共源极放大器的小信号限制
7.10.5 共射极放大器和共源极放大器的输入电阻
7.10.6 共射极和共源极的输出电阻
7.10.7 三个放大器实例的比较
7.11 共射极放大器和共源极放大器小结
7.11.1 可忽略晶体管输出电阻的指南
7.12 放大器功率和信号范围
7.12.1 功耗
7.12.2 信号范围
小结
关键词
参考文献
习题

CHAPTER 8
TRANSISTOR AMPLIFIER BUILDING BLOCKS 449

8.1	Amplifier Classification	452	
8.1.1	Signal Injection and Extraction—the BJT	452	
8.1.2	Signal Injection and Extraction—the FET	453	
8.1.3	Common-Emitter (C-E) and Common-Source (C-S) Amplifiers	454	
8.1.4	Common-Collector (C-C) and Common-Drain (C-D) Topologies	455	
8.1.5	Common-Base (C-B) and Common-Gate (C-G) Amplifiers	457	
8.1.6	Small-Signal Model Review	458	
8.2	Inverting Amplifiers—Common-Emitter and Common-Source Circuits	458	
8.2.1	The Common-Emitter (C-E) Amplifier	458	
8.2.2	Common-Emitter Example Comparison	470	
8.2.3	The Common-Source Amplifier	471	
8.2.4	Small-Signal Limit for the Common-		

第 8 章
采用单晶体管放大器构建块

8.1 放大器分类
　8.1.1 双极型晶体管的信号注入和抽取
　8.1.2 场效应管的信号注入和抽取
　8.1.3 共发射极(C-E)和共源极(C-S)放大器
　8.1.4 共集电极(C-C)和共漏极(C-D)拓扑图
　8.1.5 共基极(C-B)和共栅极(C-G)放大器
　8.1.6 小信号模型回顾
8.2 反相放大器——共射极和共源极放大器电路
　8.2.1 共发射极(C-E)放大器
　8.2.2 共发射极实例的比较
　8.2.3 共源极放大器
　8.2.4 共源极放大器的小信号范围

		Source Amplifier		474		8.2.5	共发射极和共源极放大器
	8.2.5	Common-Emitter and Common-Source Amplifier Characteristics		478			特性
	8.2.6	C-E/C-S Amplifier Summary		479		8.2.6	C-E/C-S 放大器小结
	8.2.7	Equivalent Transistor Representation of the Generalized C-E/C-S Transistor		479		8.2.7	通用 C-E/C-S 晶体管的等效晶体管表示
8.3	Follower Circuits—Common-Collector and Common-Drain Amplifiers			480	8.3	跟随器电路——共集电极和共漏极放大器	
	8.3.1	Terminal Voltage Gain		482		8.3.1	端电压增益
	8.3.2	Input Resistance		483		8.3.2	输入电阻
	8.3.3	Signal Source Voltage Gain		483		8.3.3	信号源电压增益
	8.3.4	Follower Signal Range		484		8.3.4	跟随器信号范围
	8.3.5	Follower Output Resistance		484		8.3.5	跟随器的输出电阻
	8.3.6	Current Gain		486		8.3.6	电流增益
	8.3.7	C-C/C-D Amplifier Summary		486		8.3.7	C-C/C-D 放大器小结
8.4	Noninverting Amplifiers—Common-Base and Common-Gate Circuits			490	8.4	同相放大器——共基极和共栅极电路	
	8.4.1	Terminal Voltage Gain and Input Resistance		492		8.4.1	端电压增益和输入电阻
	8.4.2	Signal Source Voltage Gain		493		8.4.2	信号源电压增益
	8.4.3	Input Signal Range		494		8.4.3	输入信号范围
	8.4.4	Resistance at the Collector and Drain Terminals		494		8.4.4	集电极和漏极端的电阻
	8.4.5	Current Gain		495		8.4.5	电流增益
	8.4.6	Overall Input and Output Resistances for the Noninverting Amplifiers		495		8.4.6	同相放大器的总体输入和输出电阻
	8.4.7	C-B/C-G Amplifier Summary		499		8.4.7	C-B/C-G 放大器小结
8.5	Amplifier Prototype Review and Comparison			500	8.5	放大器原型回顾和比较	
	8.5.1	The BJT Amplifiers		500		8.5.1	双极型晶体管放大器
	8.5.2	The FET Amplifiers		502		8.5.2	FET 放大器
8.6	Common-Source Amplifiers Using MOS Transistor Loads			505	8.6	采用 MOS 反相器的共源极放大器	
	8.6.1	Voltage Gain Estimate		505		8.6.1	电压增益估算
	8.6.2	Detailed Analysis		506		8.6.2	详细分析
	8.6.3	Alternative Loads		507		8.6.3	其他可选负载
	8.6.4	Input and Output Resistances		508		8.6.4	输入和输出电阻
8.7	Coupling and Bypass Capacitor Design			511	8.7	耦合和旁路电容设计	
	8.7.1	Common-Emitter and Common-Source Amplifiers		511		8.7.1	共发射极和共源极放大器
	8.7.2	Common-Collector and Common-Drain Amplifiers		515		8.7.2	共集电极和共漏极放大器

8.7.3	Common-Base and Common-Gate Amplifiers	518	8.7.3	共基极和共栅极放大器
8.7.4	Setting Lower Cutoff Frequency f_L	521	8.7.4	设置下限截止频率 f_L
8.8	Amplifier Design Examples	522	8.8	放大器设计实例
8.8.1	Monte Carlo Evaluation of the Common-Base Amplifier Design	531	8.8.1	共基极放大器设计的蒙特卡洛分析
8.9	Multistage ac-Coupled Amplifiers	536	8.9	多级交流耦合放大器
8.9.1	A Three-Stage ac-Coupled Amplifier	537	8.9.1	三级交流耦合放大器
8.9.2	Voltage Gain	539	8.9.2	电压增益
8.9.3	Input Resistance	540	8.9.3	输入电阻
8.9.4	Signal Source Voltage Gain	540	8.9.4	信号源的电压增益
8.9.5	Output Resistance	541	8.9.5	输出电阻
8.9.6	Current and Power Gain	542	8.9.6	电流和功率增益
8.9.7	Input Signal Range	542	8.9.7	输入信号范围
8.9.8	Estimating the Lower Cutoff Frequency of the Multistage Amplifier	546	8.9.8	估算多级放大器的截止频率下限
8.10	Introduction to dc-Coupled Amplifiers	546	8.10	直流耦合放大器简介
8.10.1	A dc-Coupled Three-Stage Amplifier	548	8.10.1	直流耦合三级放大器
8.10.2	Two Transistor dc-Coupled Amplifiers	549	8.10.2	双晶体管直流耦合放大器
Summary		551	小结	
Key Terms		553	关键词	
Additional Reading		553	扩展阅读	
Problems		553	习题	

CHAPTER 9
AMPLIFIER FREQUENCY RESPONSE 571

第 9 章
放大器频率响应

9.1	Amplifier Frequency Response	574	9.1	放大器频率响应
9.1.1	Low-Frequency Response	575	9.1.1	低频响应
9.1.2	Estimating ω_L in the Absence of a Dominant Pole	575	9.1.2	缺少主极点情况下估算 ω_L
9.1.3	High-Frequency Response	578	9.1.3	高频响应
9.1.4	Estimating ω_H in the Absence of a Dominant Pole	578	9.1.4	缺少主极点情况下估算 ω_H
9.2	Direct Determination of the Low-Frequency Poles and Zeros—the Common-Source Amplifier	579	9.2	直接确定低频极点和零点——共源放大器
9.3	Estimation of ω_L Using the Short-Circuit Time-Constant Method	584	9.3	用短路时间常数法估算 ω_L 的值
9.3.1	Estimate of ω_L for the Common-Emitter Amplifier	585	9.3.1	估算共发射极放大器的 ω_L
9.3.2	Estimate of ω_L for the Common-Source Amplifier	589	9.3.2	估算共源极放大器的 ω_L

	9.3.3	Estimate of ω_L for the Common-Base Amplifier	590	9.3.3	估算共基极放大器的ω_L
	9.3.4	Estimate of ω_L for the Common-Gate Amplifier	591	9.3.4	估算共栅极放大器的ω_L
	9.3.5	Estimate of ω_L for the Common-Collector Amplifier	592	9.3.5	估算共集电极放大器的ω_L
	9.3.6	Estimate of ω_L for the Common-Drain Amplifier	592	9.3.6	估算共漏极放大器的ω_L
9.4	Transistor Models at High Frequencies		593	9.4	高频晶体管模型
	9.4.1	Frequency-Dependent Hybrid-Pi Model for the Bipolar Transistor	593	9.4.1	双极型晶体管与频率相关的混合π模型
	9.4.2	Modeling C_π and C_μ in SPICE	594	9.4.2	在SPICE中对C_π和C_μ建模
	9.4.3	Unity-Gain Frequency f_T	594	9.4.3	单位增益频率f_T
	9.4.4	High-Frequency Model for the FET	597	9.4.4	FET的高频模型
	9.4.5	Modeling C_{GS} and C_{GD} in SPICE	598	9.4.5	运用SPICE为C_{GS}和C_{GD}建模
	9.4.6	Channel Length Dependence of f_T	598	9.4.6	f_T与沟道长度的关系
	9.4.7	Limitations of the High-Frequency Models	600	9.4.7	高频模型的局限性
9.5	Base and Gate Resistances in the Small-Signal Models		600	9.5	混合π模型中的基区电阻
	9.5.1	Effect of Base and Gate Resistances on Midband Amplifiers	601	9.5.1	基区电阻对中频放大器的影响
9.6	High-Frequency Common-Emitter and Common-Source Amplifier Analysis		602	9.6	共发射极和共源极放大器的高频响应
	9.6.1	The Miller Effect	604	9.6.1	密勒效应
	9.6.2	Common-Emitter and Common-Source Amplifier High-Frequency Response	606	9.6.2	共发射极和共源极放大器的高频响应
	9.6.3	Direct Analysis of the Common-Emitter Transfer Characteristic	608	9.6.3	共发射极放大器传输特性的直接分析
	9.6.4	Poles of the Common-Emitter Amplifier	609	9.6.4	共发射极放大器的极点
	9.6.5	Dominant Pole for the Common-Source Amplifier	612	9.6.5	共源极放大器的主极点
	9.6.6	Estimation of ω_H Using the Open-Circuit Time-Constant Method	614	9.6.6	用开路时间常数法估算ω_H
	9.6.7	Common-Source Amplifier with Source Degeneration Resistance	615	9.6.7	包含源极衰减电阻的共源放大器
	9.6.8	Poles of the Common-Emitter with Emitter Degeneration Resistance	617	9.6.8	包含发射极衰减电阻的共发射极放大器的极点
9.7	Common-Base and Common-Gate Amplifier High-Frequency Response		620	9.7	共基极和共栅极放大器的高频响应
9.8	Common-Collector and Common-Drain Amplifier High-Frequency Response		622	9.8	共集电极和共漏极放大器的高频响应

9.9	Single-Stage Amplifier High-Frequency Response Summary	625
	9.9.1 Amplifier Gain-Bandwidth (GBW) Limitations	625
9.10	Frequency Response of Multistage Amplifiers	626
	9.10.1 Differential Amplifier	626
	9.10.2 The Common-Collector/Common-Base Cascade	628
	9.10.3 High-Frequency Response of the Cascode Amplifier	629
	9.10.4 Cutoff Frequency for the Current Mirror	630
	9.10.5 Three-Stage Amplifier Example	631
9.11	Introduction to Radio Frequency Circuits	639
	9.11.1 Radio Frequency Amplifiers	640
	9.11.2 The Shunt-Peaked Amplifier	640
	9.11.3 Single-Tuned Amplifier	642
	9.11.4 Use of a Tapped Inductor—the Auto Transformer	644
	9.11.5 Multiple Tuned Circuits—Synchronous and Stagger Tuning	646
	9.11.6 Common-Source Amplifier with Inductive Degeneration	647
9.12	Mixers and Balanced Modulators	651
	9.12.1 Introduction to Mixer Operation	651
	9.12.2 A Single-Balanced Mixer	652
	9.12.3 The Differential Pair as a Single-Balanced Mixer	653
	9.12.4 A Double-Balanced Mixer	655
	9.12.5 The Jones Mixer—a Double-Balanced Mixer/Modulator	657
Summary		661
Key Terms		662
References		662
Problems		663

PART THREE
OPERATIONAL AMPLIFIERS AND FEEDBACK

CHAPTER 10
IDEAL OPERATIONAL AMPLIFIERS   679

10.1 Ideal Differential and Operational Amplifiers   681

	10.1.1	Assumptions for Ideal Operational Amplifier Analysis	682
10.2		Analysis of Circuits Containing Ideal Operational Amplifiers	682
	10.2.1	The Inverting Amplifier	683
	10.2.2	The Transresistance Amplifier—a Current-to-Voltage Converter	686
	10.2.3	The Noninverting Amplifier	688
	10.2.4	The Unity-Gain Buffer, or Voltage Follower	690
	10.2.5	The Summing Amplifier	693
	10.2.6	The Difference Amplifier	695
10.3		Frequency Dependent Feedback	697
	10.3.1	An Active Low-Pass Filter	698
	10.3.2	An Active High-Pass Filter	701
	10.3.3	The Integrator	702
	10.3.4	The Differentiator	706
Summary			706
Key Terms			707
References			708
Additional Reading			708
Problems			708

CHAPTER 11
NONIDEAL OPERATIONAL AMPLIFIERS AND FEEDBACK AMPLIFIER STABILITY 715

11.1		Classic Feedback Systems	718
	11.1.1	Closed-Loop Gain Analysis	719
	11.1.2	Gain Error	719
11.2		Analysis of Circuits Containing Nonideal Operational Amplifiers	720
	11.2.1	Finite Open-Loop Gain	720
	11.2.2	Nonzero Output Resistance	723
	11.2.3	Finite Input Resistance	727
	11.2.4	Summary of Nonideal Inverting and Noninverting Amplifiers	731
11.3		Series and Shunt Feedback Circuits	732
	11.3.1	Feedback Amplifier Categories	732
	11.3.2	Voltage Amplifiers—Series-Shunt Feedback	733
	11.3.3	Transimpedance Amplifiers—	

	10.1.1	理想运算放大器分析中的假设
10.2		理想运算放大器电路的分析
	10.2.1	反相放大器
	10.2.2	互阻放大器——电流/电压转换器
	10.2.3	同相放大器
	10.2.4	单位增益缓冲器或电压跟随器
	10.2.5	求和放大器
	10.2.6	差分放大器
10.3		反馈放大器的频率特性
	10.3.1	有源低通滤波器
	10.3.2	有源高通滤波器
	10.3.3	积分器
	10.3.4	微分器
小结		
关键词		
参考文献		
补充阅读		
习题		

第 11 章
非线性运算放大器和反馈放大器的稳定性

11.1		经典反馈系统
	11.1.1	闭环增益分析
	11.1.2	增益误差
11.2		含有非理想运算放大器的电路分析
	11.2.1	有限开环增益
	11.2.2	非零输出电阻
	11.2.3	有限输入电阻
	11.2.4	非理想反相和同相放大器小结
11.3		串联反馈和并联反馈电路
	11.3.1	反馈放大器类型
	11.3.2	电压放大器——电压串联反馈
	11.3.3	跨阻放大器——电压并联

		Shunt-Shunt Feedback	733		反馈
	11.3.4	Current Amplifiers—Shunt-Series Feedback	733	11.3.4	电流放大器——电流并联反馈
	11.3.5	Transconductance Amplifiers—Series-Series Feedback	733	11.3.5	跨导放大器——电流串联反馈
11.4	Unified Approach to Feedback Amplifier Gain Calculation		733	11.4	反馈放大器计算的统一方法
	11.4.1	Closed-Loop Gain Analysis	734	11.4.1	闭环增益分析
	11.4.2	Resistance Calculations Using Blackman's Theorem	734	11.4.2	利用 Blackman 理论计算电阻
11.5	Series-Shunt Feedback—Voltage Amplifiers		734	11.5	电压串联反馈放大器——电压放大器
	11.5.1	Closed-Loop Gain Calculation	735	11.5.1	闭环增益计算
	11.5.2	Input Resistance Calculations	735	11.5.2	输入电阻计算
	11.5.3	Output Resistance Calculations	736	11.5.3	输出电阻计算
	11.5.4	Series-Shunt Feedback Amplifier Summary	737	11.5.4	电压串联反馈放大器小结
11.6	Shunt-Shunt Feedback—Transresistance Amplifiers		741	11.6	电压并联反馈放大器——跨阻放大器
	11.6.1	Closed-Loop Gain Calculation	741	11.6.1	闭环增益分析
	11.6.2	Input Resistance Calculations	742	11.6.2	输入电阻计算
	11.6.3	Output Resistance Calculations	742	11.6.3	输出电阻计算
	11.6.4	Shunt-Shunt Feedback Amplifier Summary	743	11.6.4	电压并联反馈放大器小结
11.7	Series-Series Feedback—Transconductance Amplifiers		746	11.7	电流串联反馈放大器——跨导放大器
	11.7.1	Closed-Loop Gain Calculation	747	11.7.1	闭环增益计算
	11.7.2	Input Resistance Calculation	747	11.7.2	输入电阻计算
	11.7.3	Output Resistance Calculation	748	11.7.3	输出电阻计算
	11.7.4	Series-Series Feedback Amplifier Summary	748	11.7.4	电流串联反馈放大器小结
11.8	Shunt-Series Feedback—Current Amplifiers		750	11.8	电流并联反馈放大器——电流放大器
	11.8.1	Closed-Loop Gain Calculation	751	11.8.1	闭环增益计算
	11.8.2	Input Resistance Calculation	751	11.8.2	输入电阻计算
	11.8.3	Output Resistance Calculation	752	11.8.3	输出电阻计算
	11.8.4	Shunt-Series Feedback Amplifier Summary	752	11.8.4	电流并联反馈放大器总结
11.9	Finding the Loop Gain Using Successive Voltage and Current Injection		755	11.9	使用持续电压和电流注入法计算回路增益
	11.9.1	Simplifications	758	11.9.1	简化
11.10	Distortion Reduction Through the Use of Feedback		758	11.10	利用反馈减小失真
11.11	DC Error Sources and Output Range			11.11	直流误差源和输出摆幅

	Limitations	759
11.11.1	Input-Offset Voltage	759
11.11.2	Offset-Voltage Adjustment	761
11.11.3	Input-Bias and Offset Currents	762
11.11.4	Output Voltage and Current Limits	764

11.12 Common-Mode Rejection and Input Resistance 767
- 11.12.1 Finite Common-Mode Rejection Ratio 767
- 11.12.2 Why Is CMRR Important? 768
- 11.12.3 Voltage-Follower Gain Error due to CMRR 771
- 11.12.4 Common-Mode Input Resistance 774
- 11.12.5 An Alternate Interpretation of CMRR 775
- 11.12.6 Power Supply Rejection Ratio 775

11.13 Frequency Response and Bandwidth of Operational Amplifiers 777
- 11.13.1 Frequency Response of the Noninverting Amplifier 779
- 11.13.2 Inverting Amplifier Frequency Response 782
- 11.13.3 Using Feedback to Control Frequency Response 784
- 11.13.4 Large-Signal Limitations—Slew Rate and Full-Power Bandwidth 786
- 11.13.5 Macro Model for Operational Amplifier Frequency Response 787
- 11.13.6 Complete Op Amp Macro Models in SPICE 788
- 11.13.7 Examples of Commercial General-Purpose Operational Amplifiers 788

11.14 Stability of Feedback Amplifiers 789
- 11.14.1 The Nyquist Plot 789
- 11.14.2 First-Order Systems 790
- 11.14.3 Second-Order Systems and Phase Margin 791
- 11.14.4 Step Response and Phase Margin 792
- 11.14.5 Third-Order Systems and Gain Margin 795
- 11.14.6 Determining Stability from the Bode Plot 796

Summary 800

Key Terms	802	关键词	
References	802	参考文献	
Problems	803	习题	

CHAPTER 12
OPERATIONAL AMPLIFIER APPLICATIONS 817

第 12 章
运算放大器应用

12.1 Cascaded Amplifiers	820	12.1 级联放大器
12.1.1 Two-Port Representations	820	12.1.1 二端口表示
12.1.2 Amplifier Terminology Review	822	12.1.2 放大器专有名词回顾
12.1.3 Frequency Response of Cascaded Amplifiers	825	12.1.3 级联放大器的频率响应
12.2 The Instrumentation Amplifier	833	12.2 仪表放大器
12.3 Active Filters	836	12.3 有源滤波器
12.3.1 Low-Pass Filter	836	12.3.1 低通滤波器
12.3.2 A High-Pass Filter with Gain	840	12.3.2 带增益的高通滤波器
12.3.3 Band-Pass Filter	842	12.3.3 带通滤波器
12.3.4 Sensitivity	844	12.3.4 灵敏度
12.3.5 Magnitude and Frequency Scaling	845	12.3.5 幅值和频率缩放
12.4 Switched-Capacitor Circuits	846	12.4 开关电容电路
12.4.1 A Switched-Capacitor Integrator	846	12.4.1 开关-电容积分器
12.4.2 Noninverting SC Integrator	848	12.4.2 同相 SC 积分器
12.4.3 Switched-Capacitor Filters	850	12.4.3 开关电容滤波器
12.5 Digital-to-Analog Conversion	853	12.5 数/模转换
12.5.1 D/A Converter Fundamentals	853	12.5.1 数/模转换器基础
12.5.2 D/A Converter Errors	854	12.5.2 数/模转换器误差
12.5.3 Digital-to-Analog Converter Circuits	856	12.5.3 数/模转换电路
12.6 Analog-to-Digital Conversion	860	12.6 模/数转换
12.6.1 A/D Converter Fundamentals	861	12.6.1 模/数转换器基础
12.6.2 Analog-to-Digital Converter Errors	862	12.6.2 模/数转换器误差
12.6.3 Basic A/D Conversion Techniques	863	12.6.3 基本模/数转换技术
12.7 Oscillators	874	12.7 振荡器
12.7.1 The Barkhausen Criteria for Oscillation	874	12.7.1 振荡的巴克豪森准则
12.7.2 Oscillators Employing Frequency-Selective RC Networks	875	12.7.2 带频率选择 RC 网络的振荡器
12.8 Nonlinear Circuit Applications	879	12.8 非线性电路的应用
12.8.1 A Precision Half-Wave Rectifier	879	12.8.1 精密半波整流器
12.8.2 Nonsaturating Precision-Rectifier Circuit	880	12.8.2 非饱和的精准整流电路
12.9 Circuits Using Positive Feedback	882	12.9 正反馈电路
12.9.1 The Comparator and Schmitt Trigger	882	12.9.1 比较器和施密特触发器

12.9.2	The Astable Multivibrator	884
12.9.3	The Monostable Multivibrator or One Shot	885
Summary		889
Key Terms		891
Additional Reading		892
Problems		892

CHAPTER 13
DIFFERENTIAL AMPLIFIERS AND OPERATIONAL AMPLIFIER DESIGN 906

- 13.1 Differential Amplifiers 909
  - 13.1.1 Bipolar and MOS Differential Amplifiers 909
  - 13.1.2 dc Analysis of the Bipolar Differential Amplifier 910
  - 13.1.3 Transfer Characteristic for the Bipolar Differential Amplifier 912
  - 13.1.4 ac Analysis of the Bipolar Differential Amplifier 913
  - 13.1.5 Differential-Mode Gain and Input and Output Resistances 914
  - 13.1.6 Common-Mode Gain and Input Resistance 916
  - 13.1.7 Common-Mode Rejection Ratio (CMRR) 918
  - 13.1.8 Analysis Using Differential—and Common-Mode Half-Circuits 919
  - 13.1.9 Biasing with Electronic Current Sources 922
  - 13.1.10 Modeling the Electronic Current Source in SPICE 923
  - 13.1.11 dc Analysis of the MOSFET Differential Amplifier 923
  - 13.1.12 Differential-Mode Input Signals 926
  - 13.1.13 Small-Signal Transfer Characteristic for the MOS Differential Amplifier 927
  - 13.1.14 Common-Mode Input Signals 927
  - 13.1.15 Model for Differential Pairs 928
- 13.2 Evolution to Basic Operational Amplifiers 932
  - 13.2.1 A Two-Stage Prototype for an

	Operational Amplifier	933
13.2.2	Improving the Op Amp Voltage Gain	938
13.2.3	Darlington Pairs	939
13.2.4	Output Resistance Reduction	940
13.2.5	A CMOS Operational Amplifier Prototype	944
13.2.6	BiCMOS Amplifiers	946
13.2.7	All Transistor Implementations	946
13.3	Output Stages	948
13.3.1	The Source Follower—a Class-A Output Stage	948
13.3.2	Efficiency of Class-A Amplifiers	949
13.3.3	Class-B Push-Pull Output Stage	950
13.3.4	Class-AB Amplifiers	952
13.3.5	Class-AB Output Stages for Operational Amplifiers	953
13.3.6	Short-Circuit Protection	953
13.3.7	Transformer Coupling	955
13.4	Electronic Current Sources	958
13.4.1	Single-Transistor Current Sources	959
13.4.2	Figure of Merit for Current Sources	959
13.4.3	Higher Output Resistance Sources	960
13.4.4	Current Source Design Examples	961
Summary		969
Key Terms		970
References		970
Additional Reading		971
Problems		971

CHAPTER 14
ANALOG INTEGRATED CIRCUIT DESIGN TECHNIQUES 989

14.1	Circuit Element Matching	992
14.2	Current Mirrors	993
14.2.1	dc Analysis of the MOS Transistor Current Mirror	994
14.2.2	Changing the MOS Mirror Ratio	996
14.2.3	dc Analysis of the Bipolar Transistor Current Mirror	997
14.2.4	Altering the BJT Current Mirror Ratio	999
14.2.5	Multiple Current Sources	1000
14.2.6	Buffered Current Mirror	1001

13.2.2	提高运算放大器的电压增益	
13.2.3	达林顿对	
13.2.4	减小输出电阻	
13.2.5	CMOS 运算放大器原型	
13.2.6	BiCMOS 放大器	
13.2.7	全晶体管实现电路	
13.3	输出级	
13.3.1	源极跟随器——A 类输出级	
13.3.2	A 类放大器的效率	
13.3.3	B 类推挽输出级	
13.3.4	AB 类放大器	
13.3.5	运算放大器的 AB 类输出级	
13.3.6	短路保护	
13.3.7	变压器耦合	
13.4	电子电流源	
13.4.1	单晶体管电流源	
13.4.2	电路源的品质因数	
13.4.3	高输出电阻电流源	
13.4.4	电流源设计实例	

小结
关键词
参考文献
补充阅读
习题

第 14 章
模拟集成电路设计技术

14.1	电路元件匹配	
14.2	电流镜	
14.2.1	MOS 晶体管电流镜的直流分析	
14.2.2	改变 MOS 镜像比率	
14.2.3	双极型晶体管电流镜的直流分析	
14.2.4	改变 BJT 电流镜的镜像比率	
14.2.5	多级电流源	
14.2.6	缓冲电流镜	

	14.2.7	Output Resistance of the Current Mirrors	1002	14.2.7	电流镜像的输出阻抗
	14.2.8	Two-Port Model for the Current Mirror	1003	14.2.8	电流镜的二端口模型
	14.2.9	The Widlar Current Source	1005	14.2.9	Widlar 电流源
	14.2.10	The MOS Version of the Widlar Source	1008	14.2.10	MOS 管 Widlar 电流源
	14.2.11	MOS Widlar Source in Weak Inversion	1008	14.2.11	弱反转中的 MOS Widlar 源
14.3	High-Output-Resistance Current Mirrors		1009	14.3	高输出电阻电流镜
	14.3.1	The Wilson Current Sources	1010	14.3.1	Wilson 电流源
	14.3.2	Output Resistance of the Wilson Source	1011	14.3.2	Wilson 电流源的输出电阻
	14.3.3	Cascode Current Sources	1012	14.3.3	Cascode 电流源
	14.3.4	Output Resistance of the Cascode Sources	1013	14.3.4	Cascode 电流源的输出电阻
	14.3.5	Regulated Cascode Current Source	1014	14.3.5	可调 Cascode 电流源
	14.3.6	Current Mirror Summary	1015	14.3.6	电流镜小结
14.4	Reference Current Generation		1018	14.4	参考电流的产生
14.5	Supply-Independent Biasing		1019	14.5	与电源电压无关的偏置
	14.5.1	A V_{BE}-Based Reference	1019	14.5.1	基于 V_{BE} 的参考源
	14.5.2	The Widlar Source	1019	14.5.2	Widlar 电流源
	14.5.3	Power-Supply-Independent Bias Cell	1020	14.5.3	与电源电压无关的偏置单元
	14.5.4	A Supply-Independent MOS Reference Cell	1021	14.5.4	与电源电压无关的 MOS 参考单元
14.6	The Bandgap Reference		1023	14.6	带隙基准源
14.7	The Current Mirror as an Active Load		1027	14.7	电流镜作为有源负载
	14.7.1	CMOS Differential Amplifier with Active Load	1027	14.7.1	带有源负载的 CMOS 差分放大器
	14.7.2	Bipolar Differential Amplifier with Active Load	1034	14.7.2	带有源负载的双极差分放大器
14.8	Active Loads in Operational Amplifiers		1038	14.8	运算放大器中的源负载
	14.8.1	CMOS Op-Amp Voltage Gain	1038	14.8.1	CMOS 运算放大器电压增益
	14.8.2	dc Design Considerations	1039	14.8.2	直流设计注意事项
	14.8.3	Bipolar Operational Amplifiers	1041	14.8.3	双极型运算放大器
	14.8.4	Input Stage Breakdown	1042	14.8.4	输入级击穿
14.9	The μA741 Operational Amplifier		1043	14.9	μA741 运算放大器
	14.9.1	Overall Circuit Operation	1043	14.9.1	电路总体工作原理
	14.9.2	Bias Circuitry	1044	14.9.2	偏置电路
	14.9.3	dc Analysis of the 741 Input Stage	1045	14.9.3	μA741 输入级的直流分析
	14.9.4	ac Analysis of the 741 Input Stage	1048	14.9.4	μA741 输入级的交流分析

14.9.5	Voltage Gain of the Complete Amplifier	1049	14.9.5	整体放大器的电压增益	
14.9.6	The 741 Output Stage	1053	14.9.6	μA741 的输出级	
14.9.7	Output Resistance	1055	14.9.7	输出阻抗	
14.9.8	Short-Circuit Protection	1055	14.9.8	短路保护电路	
14.9.9	Summary of the μA741 Operational Amplifier Characteristics	1055	14.9.9	μA741 运算放大器特性小结	
14.10	The Gilbert Analog Multiplier	1056	14.10	Gilbert 模拟乘法器	
Summary		1058	小结		
Key Terms		1059	关键词		
References		1060	参考文献		
Additional Readings		1060	补充阅读		
Problems		1060	习题		

CHAPTER 15
TRANSISTOR FEEDBACK AMPLIFIERS AND OSCILLATORS   1077

第 15 章
晶体管反馈放大器与振荡器

15.1	Basic Feedback System Review	1080	15.1	基本反馈系统回顾	
	15.1.1 Closed-Loop Gain	1080		15.1.1 闭环增益	
	15.1.2 Closed-Loop Impedances	1081		15.1.2 闭环阻抗	
	15.1.3 Feedback Effects	1081		15.1.3 反馈的作用	
15.2	Feedback Amplifier Analysis at Midband	1083	15.2	反馈放大器的中频分析	
	15.2.1 Closed-Loop Gain	1083		15.2.1 闭环增益	
	15.2.2 Input Resistance	1084		15.2.2 输入电阻	
	15.2.3 Output Resistance	1084		15.2.3 输出电阻	
	15.2.4 Offset Voltage Calculation	1085		15.2.4 偏移电压计算	
15.3	Feedback Amplifier Circuit Examples	1086	15.3	反馈放大电路举例	
	15.3.1 Series-Shunt Feedback—Voltage Amplifiers	1086		15.3.1 串-并反馈（电压串联反馈）——电压放大器	
	15.3.2 Differential Input Series-Shunt Voltage Amplifier	1091		15.3.2 差分输入串-并电压放大器	
	15.3.3 Shunt-Shunt Feedback—Transresistance Amplifiers	1094		15.3.3 并-并反馈（电压并联反馈）——跨阻放大器	
	15.3.4 Series-Series Feedback—Transconductance Amplifiers	1100		15.3.4 串-串反馈（电流串联反馈）——跨导放大器	
	15.3.5 Shunt-Series Feedback—Current Amplifiers	1103		15.3.5 并-串反馈（电流并联反馈）——电流放大器	
15.4	Review of Feedback Amplifier Stability	1106	15.4	反馈放大器稳定性回顾	
	15.4.1 Closed-Loop Response of the Uncompensated Amplifier	1107		15.4.1 未补偿放大器的闭环响应	
	15.4.2 Phase Margin	1108		15.4.2 相位裕度	

15.4.3	Higher-Order Effects	1112
15.4.4	Response of the Compensated Amplifier	1113
15.4.5	Small-Signal Limitations	1115

15.5 Single-Pole Operational Amplifier Compensation 1115
- 15.5.1 Three-Stage Op-Amp Analysis 1116
- 15.5.2 Transmission Zeros in FET Op Amps 1118
- 15.5.3 Bipolar Amplifier Compensation 1119
- 15.5.4 Slew Rate of the Operational Amplifier 1120
- 15.5.5 Relationships between Slew Rate and Gain-Bandwidth Product 1121

15.6 High-Frequency Oscillators 1130
- 15.6.1 The Colpitts Oscillator 1131
- 15.6.2 The Hartley Oscillator 1132
- 15.6.3 Amplitude Stabilization in *LC* Oscillators 1133
- 15.6.4 Negative Resistance in Oscillators 1133
- 15.6.5 Negative G_m Oscillator 1134
- 15.6.6 Crystal Oscillators 1136
- 15.6.7 Ring Oscillators 1139
- 15.6.8 Positive Feedback and Latchup 1140

Summary 1143
Key Terms 1145
Additional Readings 1145
Problems 1145

APPENDICES 1159

- **A** Standard Discrete Component Values 1159
- **B** Solid-State Device Models and SPICE Simulation Parameters 1162
- **C** Two-Port Review (Section 6.3) 1167
- **D** Physical Constants and Transistor Model Summary 1170

PART ONE
SOLID-STATE ELECTRONICS AND DEVICES

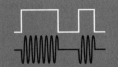

CHAPTER 1
INTRODUCTION TO ELECTRONICS

CHAPTER 2
SOLID-STATE ELECTRONICS

CHAPTER 3
SOLID-STATE DIODES AND DIODE CIRCUITS

CHAPTER 4
BIPOLAR JUNCTION TRANSISTORS

CHAPTER 5
FIELD-EFFECT TRANSISTORS

CHAPTER 1
INTRODUCTION TO ELECTRONICS

第1章　电子学简介

本章提纲

1.1　电子学发展简史：从真空管到吉规模集成电路

1.2　电信号的分类

1.3　符号约定

1.4　问题求解的方法

1.5　电路理论的主要概念

1.6　电信号的频谱

1.7　放大器

1.8　电路设计中元件参数的变化

1.9　数值精度

本章目标

- 了解电子学的发展简史；
- 了解集成电路技术的开拓性进展；
- 了解电信号的分类；
- 理解电路理论中重要的符号约定及主要概念；
- 了解电路分析中的容差分析方法；
- 了解本书中解决问题的方法。

本章导读

　　本章对电子学的发展历史进行了概述。20世纪中叶Pickard发明了晶体二极管探测器标志着电子时代的到来，接着Fleming发明了真空二极管，Deforest发明了真空三极管。

　　第二次世界大战是电子学飞速发展的时期。这个时期，Bardeen、Brattain和Shockley在1947年发明双极型晶体管，Kilby和Noyce、Moore几乎同时发明了集成电路。同时，集成电路很快实现了产业化，自20世纪60年代中期以来，无论是以存储器密度、微处理器晶体管数量还是以最小特征尺寸衡量，集成电路的复杂度都呈指数式发展。集成电路经历了小规模集成电路、中等规模集成电路、大规模集成电路、超大规模集成电路、甚大规模集成电路等阶段的发展，现已进入吉规模集成电路(giga-scale integration, GSI)时代。

模拟信号和数字信号是电路设计中主要处理的两类信号。模拟电信号可以在有限的电压电流范围内任意取值，而数字信号只能取有限个分立值。最常见的数字信号是二进制信号，用两个分立值来表示。模拟和数字世界联系的桥梁是数/模(D/A)转换器和模/数(A/D)转换器。数/模转换器把数字信息转换成模拟电压或电流，而模/数转换器把输入的模拟电压或电流转换成数字输出。

本章也给出了复杂信号处理的概要，介绍了傅里叶定理在复杂信号处理中的关键作用，可以用正弦信号的线性组合表示复杂信号。用线性放大器对这些信号进行模拟信号处理，可以改变信号的幅值和相位。线性放大器只能改变各频率成分的幅值和相位，而不改变信号的频率组成。

放大器是电路应用中的重要器件，本章对于放大器的类型进行了归纳介绍。根据频率响应特性，可将放大器分为低通、高通、带通、带阻、全通放大器这几种类型。常把用于放大某一频率范围内信号的放大器称为滤波器。

本章最为重要的一点是作者归纳了一套解决电路问题的方法步骤。针对复杂电路问题的求解，作者将问题分为几个步骤，并且本书给出的所有例题都采用这种解决问题的方法。作者归纳的解决问题的步骤如下：

(1) 尽可能清晰地将问题描述出来。
(2) 列出已知信息和数据。
(3) 找出解决问题所必需的未知条件。
(4) 列出自己的假设，以及在分析过程中还可能发现的新假设。
(5) 在许多可能的备选方案中寻找问题的解决方案。
(6) 为找到问题的解决方案进行详细分析。
(7) 结果检验。问题是否得到解决？数据分析是否正确？是否找到了所有的未知条件？所提假设是否成立？所得的结果是否能通过简单的一致性检查？
(8) 结果评价。方案是否现实？是否成立？如果不成立，重复步骤（4）～（7），直至获得满意的解决方案。
(9) 计算机辅助分析。验证所得到的结果是否满足问题的需求。

本章在探索解决问题方法的同时，对于问题求解中常用的初值分析、容差分析、最差情况分析以及蒙特卡洛法分析等都给出了说明。分析元器件容差对电路性能影响的方法包括最差情况分析和统计学的蒙特卡洛分析。大部分电路分析程序能够确定温度对绝大多数元器件的影响。在最差情况分析中，令所有元器件同时取极限值，对于电路行为的预测往往过于悲观。蒙特卡洛法分析电路的大量随机情况，以形成对电路性能统计分布的实际估测。SPICE等电路分析软件包可以提供蒙特卡洛分析选项。

Chapter Outline

1.1 A Brief History of Electronics: From Vacuum Tubes to Giga-Scale Integration
1.2 Classification of Electronic Signals
1.3 Notational Conventions
1.4 Problem-Solving Approach
1.5 Important Concepts from Circuit Theory
1.6 Frequency Spectrum of Electronic Signals
1.7 Amplifiers
1.8 Element Variations in Circuit Design
1.9 Numeric Precision
Summary
Key Terms
References
Additional Reading
Problems

Chapter Goals

- Present a brief history of electronics
- Quantify the explosive development of integrated circuit technology
- Discuss initial classification of electronic signals
- Review important notational conventions and concepts from circuit theory
- Introduce methods for including tolerances in circuit analysis
- Present the problem-solving approach used in this text

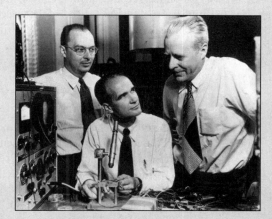

Figure 1.1 John Bardeen, William Shockley, and Walter Brattain in Brattain's laboratory in 1948.
Reprinted with permission of Alcatel-Lucent USA Inc.

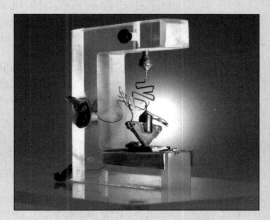

Figure 1.2 The first germanium bipolar transistor.
Reprinted with permission of Alcatel-Lucent USA Inc.

November 2022 is the 75th anniversary of the 1947 discovery of the bipolar transistor by John Bardeen and Walter Brattain at Bell Laboratories, a seminal event that marked the beginning of the semiconductor age (see Figs. 1.1 and 1.2). The invention of the transistor and the subsequent development of microelectronics have done more to shape the modern era than any other event. The transistor and microelectronics have reshaped how business is transacted, machines are designed, information moves, wars are fought, people interact, and countless other areas of our lives.

This textbook develops the basic operating principles and design techniques governing the behavior of the devices and circuits that form the backbone of much of the infrastructure of our modern world. This knowledge will enable students who aspire to design and create the next generation of this technological revolution to build a solid foundation for more advanced design courses. In addition, students who expect to work in some other technology area will learn material that will help them understand microelectronics, a technology that will continue to have impact on how their chosen field develops. This understanding will enable them to fully exploit microelectronics in their own technology area. Now let us return to our short history of the transistor.

After the discovery of the transistor, it was but a few months until William Shockley developed a theory that described the operation of the bipolar junction transistor.

Only 10 years later, in 1956, Bardeen, Brattain, and Shockley received the Nobel Prize in physics for the discovery of the transistor.

In June 1948 Bell Laboratories held a major press conference to announce the discovery. In 1952 Bell Laboratories, operating under legal consent decrees, made licenses for the transistor available for the modest fee of $25,000 plus future royalty payments. About this time, Gordon Teal, another member of the solid-state group, left Bell Laboratories to work on the transistor at Geophysical Services, Inc., which subsequently became Texas Instruments (TI). There he made the first silicon transistors, and TI marketed the first all-transistor radio. Another early licensee of the transistor was Tokyo Tsushin Kogyo, which became the Sony Company in 1955. Sony subsequently sold a transistor radio with a marketing strategy based on the idea that everyone could now have a personal radio; thus was launched the consumer market for transistors. A very interesting account of these and other developments can be found in [1, 2] and their references.

Activity in electronics began more than a century ago with the first radio transmissions in 1895 by Marconi, and these experiments were followed after only a few years by the invention of the first electronic amplifying device, the triode vacuum tube. In this period, electronics—loosely defined as the design and application of electron devices—has had such a significant impact on our lives that we often overlook just how pervasive electronics has really become. One measure of the degree of this impact can be found in the gross domestic product (GDP) of the world. In 2020 the world GDP was approximately U.S. $90 trillion, and of this total more than 15 percent was directly traceable to electronics [3–5].

We commonly encounter electronics in the form of cellular phones, radios, televisions, and audio equipment, but electronics can be found even in seemingly mundane appliances such as vacuum cleaners, washing machines, and refrigerators. Wherever one looks in industry, electronics is found. The corporate world obviously depends heavily on data processing systems to manage its operations. In fact, it is hard to see how the computer industry could have evolved without the use of its own products. In addition, the design process depends ever more heavily on computer-aided design (CAD) systems, and manufacturing relies on electronic systems for process control—in petroleum refining, automobile tire production, food processing, power generation, and so on.

1.1 A BRIEF HISTORY OF ELECTRONICS: FROM VACUUM TUBES TO GIGA-SCALE INTEGRATION

Because most of us have grown up with electronic products all around us, we often lose perspective of how far the industry has come in a relatively short time. At the beginning of the twentieth century, there were no commercial electron devices, and transistors were not invented until the late 1940s! Explosive growth was triggered by first the commercial availability of the bipolar transistor in the late 1950s, and then the realization of the integrated circuit (IC) in 1961. Since that time, signal processing using electron devices and electronic technology has become a pervasive force in our lives.

Table 1.1 lists a number of important milestones in the evolution of the field of electronics. The Age of Electronics began in the early 1900s with the invention of the first electronic two-terminal devices, called **diodes.** The **vacuum diode,** or diode **vacuum tube,** was invented by Fleming in 1904; in 1906 Pickard created a diode by forming a point contact to a silicon crystal. (Our study of electron devices begins with the introduction of the solid-state diode in Chapter 3.)

Deforest's invention of the three-element vacuum tube known as the **triode** was an extremely important milestone. The addition of a third element to a diode enabled electronic amplification to take place with good isolation between the input and output ports of the device.

TABLE 1.1
Milestones in Electronics

YEAR	EVENT
1874	Ferdinand Braun invents the solid-state rectifier.
1884	American Institute of Electrical Engineers (AIEE) formed.
1895	Marconi makes first radio transmissions.
1904	Fleming invents diode vacuum tube—Age of Electronics begins.
1906	Pickard creates solid-state point-contact diode (silicon).
1906	Deforest invents triode vacuum tube (audion).
1910–1911	"Reliable" tubes fabricated.
1912	Institute of Radio Engineers (IRE) founded.
1907–1927	First radio circuits developed from diodes and triodes.
1920	Armstrong invents super heterodyne receiver.
1925	TV demonstrated.
1925	Lilienfeld files patent application on the field-effect device.
1927–1936	Multigrid tubes developed.
1933	Armstrong invents FM modulation.
1935	Heil receives British patent on a field-effect device.
1940	Radar developed during World War II—TV in limited use.
1947	Bardeen, Brattain, and Shockley at Bell Laboratories invent bipolar transistors.
1950	First demonstration of color TV.
1952	Shockley describes the unipolar field-effect transistor.
1952	Commercial production of silicon bipolar transistors begins at Texas Instruments.
1952	Ian Ross and George Dacey demonstrate the junction field-effect transistor.
1956	Bardeen, Brattain, and Shockley receive Nobel Prize for invention of bipolar transistors.
1958	Integrated circuit developed simultaneously by Kilby at Texas Instruments and Noyce and Moore at Fairchild Semiconductor.
1961	First commercial digital IC available from Fairchild Semiconductor.
1963	AIEE and IRE merge to become the Institute of Electrical and Electronic Engineers (IEEE)
1967	First semiconductor RAM (64 bits) discussed at the IEEE International Solid-State Circuits Conference (ISSCC).
1968	First commercial IC operational amplifier—the µA709—introduced by Fairchild Semiconductor.
1970	One-transistor dynamic memory cell invented by Dennard at IBM.
1970	Low-loss optical fiber invented.
1971	4004 microprocessor introduced by Intel.
1972	First 8-bit microprocessor—the 8008—introduced by Intel.
1973	Martin Cooper demonstrated a prototype of Motorola's handheld mobile phone.
1974	First commercial 1-kilobit memory chip developed.
1974	8080 microprocessor introduced.
1978	First 16-bit microprocessor developed.
1984	Megabit memory chip introduced.
1985	Flash memory introduced at ISSCC.
1987	Erbium doped, laser-pumped optical fiber amplifiers demonstrated.
1995	Experimental gigabit memory chip presented at the IEEE ISSCC.
2000	Alferov, Kilby, and Kromer share the Nobel Prize in physics for optoelectronics, invention of the integrated circuit, and heterostructure devices, respectively.
2007	Fert and Grünberg share the Nobel Prize in physics for the discovery of giant magnetoresistance.
2009	Kao shares one-half of the 2009 Nobel Prize in physics for fiber optic communication using light with Boyle and Smith for invention of the Charge-Coupled Device (CCD).
2010	Geim and Novoaelov share the Nobel Prize in physics for groundbreaking experiments regarding the two-dimensional material graphene.
2014	Akasaki, Amano, and Nakamura share the Nobel Prize in physics for the invention of efficient blue light-emitting diodes, which has enabled bright and energy-saving white light sources.
2018	Ten billion transistor integrated circuit chip presented at ISSCC.
2019	Goodenough, Whittingham, and Yoshino share the Nobel Prize in chemistry for the development of lithium-ion batteries.

Silicon-based three-element devices now form the basis of virtually all electronic systems. Fabrication of tubes that could be used reliably in circuits followed the invention of the triode by a few years and enabled rapid circuit innovation. Amplifiers and oscillators were developed that significantly improved radio transmission and reception. Armstrong invented the super heterodyne receiver in 1920 and FM modulation in 1933. Electronics developed rapidly during World War II, with great advances in the field of radio communications and the development of radar. Although first demonstrated in 1930, television did not begin to come into widespread use until the 1950s.

An important event in electronics occurred in 1947, when John Bardeen, Walter Brattain, and William Shockley at Bell Telephone Laboratories invented the **bipolar transistor.**[1] Although field-effect devices had actually been conceived by Lilienfeld in 1925, Heil in 1935, and Shockley in 1952 [2], the technology to produce such devices on a commercial basis did not yet exist. Bipolar devices, however, were rapidly commercialized.

Then in 1958, the nearly simultaneous invention of the **integrated circuit (IC)** by Kilby at Texas Instruments and Noyce and Moore at Fairchild Semiconductor produced a new technology that would profoundly change our lives. The miniaturization achievable through IC technology made available complex electronic functions with high performance at low cost. The attendant characteristics of high reliability, low power, and small physical size and weight were additional important advantages.

In 2000, Jack St. Clair Kilby received a share of the Nobel Prize for the invention of the integrated circuit. In the mind of the authors, this was an exceptional event as it represented one of the first awards to an electronic technologist.

Most of us have had some experience with personal computers, and nowhere is the impact of the integrated circuit more evident than in the area of digital electronics. For example, 4-gigabit (Gb) dynamic memory chips, similar to those in Fig. 1.3(c), contain more than 4 billion transistors. A 128-Gb flash memory chip stores 2 or 3 bits per memory cell using multilevel storage techniques and has more than 17 billion transistors in the memory array alone, not counting address decoding and sensing circuitry. Creating this much memory using individual vacuum tubes [depicted in Fig. 1.3(a)] or even discrete transistors [shown in Fig. 1.3(b)] would be almost inconceivable (see Prob. 1.9).

Levels of Integration

The dramatic progress of integrated circuit miniaturization is shown graphically in Figs. 1.4 and 1.5. The complexities of memory chips and microprocessors have grown exponentially with time. In over four decades since 1970, the number of transistors on a microprocessor chip has increased by a factor of 10 million as depicted in Fig. 1.4. Similarly, memory density has grown by a factor of more than 10 million from a 64-bit chip in 1968 to the announcement of 32-Gb chip production in 2018.

Since the commercial introduction of the integrated circuit, these increases in density have been achieved through a continued reduction in the minimum line width, or **minimum feature size,** that can be defined on the surface of the integrated circuit (see Fig. 1.5). Today most corporate semiconductor laboratories around the world are actively working on deep submicron processes with feature sizes below 10 nm—less than one five-thousandth the diameter of a human hair.

As the miniaturization process has continued, a series of commonly used abbreviations has evolved to characterize the various levels of integration. Prior to the invention of the integrated circuit, electronic systems were implemented in discrete form. Early ICs, with fewer than

---

[1] The term **transistor** is said to have originated as a contraction of "transfer resistor," based on the voltage-controlled resistance of the characteristics of the MOS transistor.

1.1 A Brief History of Electronics: From Vacuum Tubes to Giga-Scale Integration

Figure 1.3 Comparison of (a) vacuum tubes, (b) individual transistors, (c) integrated circuits in dual-in-line packages (DIPs), and (d) ICs in surface mount packages.
Source: (a) Courtesy of ARRL Handbook for Radio Amateurs, 1992; (b, c, and d) Richard Jaeger

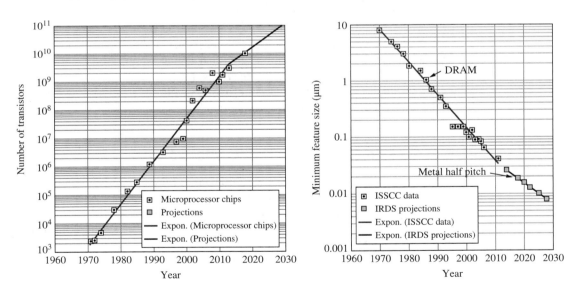

Figure 1.4 Microprocessor complexity versus time.

Figure 1.5 DRAM feature size versus year.

100 components, were characterized as **small-scale integration,** or **SSI.** As density increased, circuits became identified as **medium-scale integration (MSI,** 100–1000 components/chip), **large-scale integration (LSI,** 10^3–10^4 components/chip), and **very-large-scale integration (VLSI,** 10^4–10^9 components/chip). Today discussions focus on **giga-scale integration (GSI,** above 10^9 components/chip) and beyond.

ELECTRONICS IN ACTION

Cellular Phone Evolution

The impact of technology scaling is ever present in our daily lives. One example appears visually in the pictures of cellular phone evolution below. Early mobile phones were often large and had to be carried in a relatively large pouch (hence the term "bag phone"). The next generation of analog phones could easily fit in your hand, but they had poor battery life caused by their analog communications technology. Implementations of fourth- and fifth-generation digital cellular technology are considerably smaller and have much longer battery life. As IC density increased, additional functions such as high-function cameras, GPS, Bluetooth, and Wifi were integrated with the digital phone.

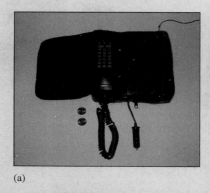

(a)   (b)   (c)

A decade of cellular phone evolution: (a) early Uniden "bag phone," (b) Nokia analog phone, and (c) Apple iPhone.
Source: (a and b) Richard Jaeger; (c) Yalcin Sonat/Shutterstock

Cell phones also represent excellent examples of the application of **mixed-signal** integrated circuits that contain both analog and digital circuitry on the same chip. ICs in the cell phone contain analog radio-frequency receiver and transmitter circuitry, analog-to-digital and digital-to-analog converters, CMOS logic and memory, power conversion circuits, imaging chips, accelerometers, and more.

1.2 CLASSIFICATION OF ELECTRONIC SIGNALS

The signals that electronic devices are designed to process can be classified into two broad categories: analog and digital. **Analog signals** can take on a continuous range of values, and thus represent continuously varying quantities; purely **digital signals** can appear at only one of several discrete levels. Examples of these types of signals are described in more detail in the next two subsections, along with the concepts of digital-to-analog and analog-to-digital conversion, which make possible the interface between the two systems.

1.2.1 DIGITAL SIGNALS

When we speak of digital electronics, we are most often referring to electronic processing of **binary digital signals,** or signals that can take on only one of two discrete amplitude levels as illustrated in Fig. 1.6. The status of binary systems can be represented by two symbols: a logical 1 is assigned to represent one level, and a logical 0 is assigned to the second level.[2] The two

---

[2] This assignment facilitates the use of Boolean algebra, reviewed in Chapter S6 of the eBook.

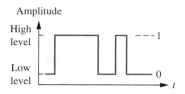

Figure 1.6 A time-varying binary digital signal.

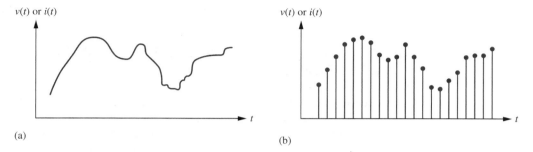

Figure 1.7 (a) A continuous analog signal; (b) sampled data version of signal in (a).

logic states generally correspond to two separate voltages—V_H and V_L—representing the high and low amplitude levels, and a number of voltage ranges are in common use. Although $V_H = 5$ V and $V_L = 0$ V represented the primary standard for many years, these have given way to lower voltage levels because of power consumption and semiconductor device limitations. Systems employing $V_H = 3.3$, down to 1 V or less with $V_L = 0$ V, are now used in many types of electronics.

However, binary voltage levels can also be negative or even bipolar. One high-performance logic family called ECL uses $V_H = -0.8$ V and $V_L = -2.0$ V, and the early standard RS-422 and RS-232 communication links between a small computer and its peripherals used $V_H = +12$ V and $V_L = -12$ V. In addition, the time-varying binary signal in Fig. 1.6 could equally well represent the amplitude of a current or that of an optical signal being transmitted down a fiber in an optical digital communication system. Recent USB and similar standards returned to the use of a single positive supply voltage.

Detailed discussion of logic circuits that were included in earlier editions can now be found in Chapters S6–S9 of the e-book. These include PMOS, NMOS, and CMOS logic,[3] which use field-effect transistors, and the TTL and ECL families, which are based on bipolar transistors.

1.2.2 ANALOG SIGNALS

Although quantities such as electronic charge and electron spin or the position of a switch are discrete, much of the physical world is really analog in nature. Our senses of vision, hearing, smell, taste, and touch are all analog processes. Analog signals directly represent variables such as temperature, humidity, pressure, light intensity, or sound—all of which may take on any value, typically within some finite range. In practice, classification of digital and analog signals is largely one of perception. If we look at a digital signal similar to the one in Fig. 1.6 with an oscilloscope, we find that it actually makes a continuous transition between the high and low levels. The signal cannot make truly abrupt transitions between two levels. Designers of high-speed digital systems soon realize that they are really dealing with analog signals. The time-varying voltage or current plotted in Fig. 1.7(a) could be the electrical representation of temperature, flow rate, or pressure versus time, or the continuous audio output from a microphone. Some analog transducers

---

[3] For now, let us accept these initials as proper names without further definition. The details of each of these circuits are developed in Chapters S6–S9 of the eBook.

produce output *voltages* in the range of 0 to 5 or 0 to 10 V, whereas others are designed to produce an output *current* that ranges between 4 and 20 mA. At the other extreme, signals detected by a radio antenna can be as small as a fraction of a microvolt.

To process the information contained in these analog signals, electronic circuits are used to selectively modify the amplitude, phase, and frequency content of the signals. In addition, significant increases in the voltage, current, and power level of the signal are usually needed. All these modifications to the signal characteristics are achieved using various forms of amplifiers, and Parts Two and Three of this text provide an in-depth discussion of the analysis and design of a wide range of amplifiers using operational amplifiers and bipolar and field-effect transistors.

1.2.3 A/D AND D/A CONVERTERS—BRIDGING THE ANALOG AND DIGITAL DOMAINS

For analog and digital systems to be able to operate together, we must be able to convert signals from analog to digital form and vice versa. We sample the input signal at various points in time as in Fig. 1.7(b) and convert or quantize its amplitude into a digital representation. The quantized value can be represented in binary form or can be a decimal representation as given by the display on a digital multimeter. The electronic circuits that perform these translations are called analog-to-digital (A/D) and digital-to-analog (D/A) converters.

Digital-to-Analog Conversion

The **digital-to-analog converter,** often referred to as a **D/A converter** or **DAC,** provides an interface between the digital signals of computer systems and the continuous signals of the analog world. The D/A converter takes digital information, most often in binary form, as input and generates an output voltage or current that may be used for electronic control or analog information display. In the DAC in Fig. 1.8(a), an n-bit binary input word $(b_1, b_2, \cdots, b_n)$ is treated as a binary fraction and multiplied by a full-scale reference voltage V_{FS} to set the output of the D/A converter. The behavior of the DAC can be expressed mathematically as

$$v_O = (b_1 2^{-1} + b_2 2^{-2} + \cdots + b_n 2^{-n}) V_{FS} \qquad \text{for } b_i \in \{1, 0\} \tag{1.1}$$

Examples of typical values of the full-scale voltage V_{FS} are 1, 2, 5, 5.12, 10, and 10.24 V. The smallest voltage change that can occur at the output takes place when the **least significant bit** b_n, or **LSB,** in the digital word changes from a 0 to a 1. This minimum voltage change is given by

$$V_{LSB} = 2^{-n} V_{FS} \tag{1.2}$$

At the other extreme, b_1 is referred to as the **most significant bit, or MSB,** and has a weight of one-half V_{FS}.

The **resolution of a converter** is typically specified in terms of the number of digital bits (i.e., 8-, 10-, 12-, 14- or 16-bit resolution, and so on).

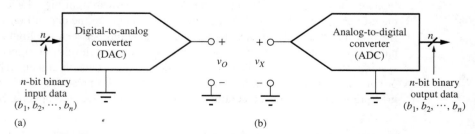

Figure 1.8 Block diagram representation for (a) a D/A converter and (b) an A/D converter.

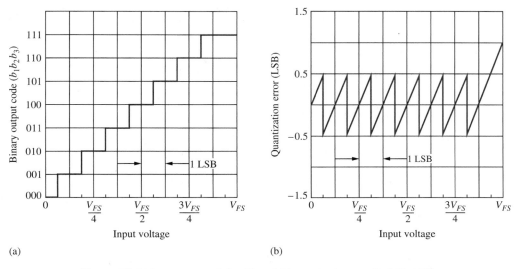

Figure 1.9 (a) Input–output relationship and (b) quantization error for 3-bit ADC.

EXERCISE: A 10-bit D/A converter has $V_{FS} = 5.12$ V. What is the output voltage for a binary input code of (1100010001)? What is V_{LSB}? What is the size of the MSB?

ANSWERS: 3.925 V; 5 mV; 2.56 V

Analog-to-Digital Conversion

The **analog-to-digital converter (A/D converter** or **ADC)** is used to transform analog information in electrical form into digital data. The ADC in Fig. 1.8(b) takes an unknown continuous analog input signal, usually a voltage v_X, and converts it into an n-bit binary number that can be easily manipulated by a computer. The n-bit number is a binary fraction representing the ratio between the unknown input voltage v_X and the converter's full-scale voltage V_{FS}.

For example, the input–output relationship for an ideal 3-bit A/D converter is shown in Fig. 1.9(a). As the input increases from zero to full scale, the output digital code word stair-steps from 000 to 111.[4] The output code is constant for an input voltage range equal to 1 LSB of the ADC. Thus, as the input voltage increases, the output code first underestimates and then overestimates the input voltage. This error, called **quantization error,** is plotted against input voltage in Fig. 1.9(b).

For a given output code, we know only that the value of the input voltage lies somewhere within a 1-LSB quantization interval. For example, if the output code of the 3-bit ADC is 100, corresponding to a voltage $V_{FS}/2$, then the input voltage can be anywhere between $\frac{7}{16}V_{FS}$ and $\frac{9}{16}V_{FS}$, a range of $V_{FS}/8$ V or 1 LSB. From a mathematical point of view, the ADC circuitry in Fig. 1.8(b) picks the values of the bits in the binary word to minimize the magnitude of the quantization error v_e between the unknown input voltage v_X and the nearest quantized voltage level:

$$v_e = |v_X - (b_1 2^{-1} + b_2 2^{-2} + \cdots + b_n 2^{-n})V_{FS}| \tag{1.3}$$

---

[4] The binary point is understood to be to the immediate left of the digits of the code word. As the code word stair-steps from 000 to 111, the binary fraction steps from 0.000 to 0.111.

> **EXERCISE:** (a) An 8-bit A/D converter has $V_{FS} = 5$ V. What is the digital output code word for an input of 1.2 V? What is the voltage range corresponding to 1 LSB of the converter? (b) Repeat for $V_{FS} = 5.12$ V.
>
> **ANSWERS:** 00111101; 19.5 mV; 00111100; 20.0 mV

1.3 NOTATIONAL CONVENTIONS

In many circuits we will be dealing with both dc and time-varying values of voltages and currents. The following standard notation will be used to keep track of the various components of an electrical signal. Total quantities will be represented by lowercase letters with capital subscripts, such as v_T and i_T in Eq. (1.4). The dc components are represented by capital letters with capital subscripts as, for example, V_{DC} and I_{DC} in Eq. (1.4); changes or variations from the dc value are represented by signal components v_{sig} and i_{sig}. Total quantities are then given by

$$v_T = V_{DC} + v_{\text{sig}} \quad \text{or} \quad i_T = I_{DC} + i_{\text{sig}} \tag{1.4}$$

As examples, the total base-emitter voltage v_{BE} of a transistor and the total drain current i_D of a field-effect transistor are written as

$$v_{BE} = V_{BE} + v_{be} \quad \text{and} \quad i_D = I_D + i_d \tag{1.5}$$

Unless otherwise indicated, the equations describing a given network will be written assuming a consistent set of units: volts, amperes, and ohms. For example, the equation $5 \text{ V} = (10{,}000 \text{ }\Omega)I_1 + 0.6$ V may be written as $5 = 10{,}000 I_1 + 0.6$.

The fourth upper/lowercase combination, such as V_{be} or I_d, is reserved for the amplitude of a sinusoidal signal's phasor representation as defined in Sec. 1.5.

> **EXERCISE:** Suppose the voltage at a circuit node is described by
>
> $$v_A = (5 \sin 2000\pi t + 4 + 3 \cos 1000\pi t) \text{ V}$$
>
> What are the expressions for V_A and v_a?
>
> **ANSWERS:** $V_A = 4$ V; $v_a = (5 \sin 2000\pi t + 3 \cos 1000\pi t)$ V

Resistance and Conductance Representations

In the circuits throughout this text, resistors will be indicated symbolically as R_x or r_x, and the values will be expressed in Ω, kΩ, MΩ, and so on. During analysis, however, it may be more convenient to work in terms of conductance with the following convention:

$$G_x = \frac{1}{R_x} \quad \text{and} \quad g_\pi = \frac{1}{r_\pi} \tag{1.6}$$

For example, conductance G_x always represents the reciprocal of the value of R_x, and g_π represents the reciprocal of r_π. The values next to a resistor symbol will always be expressed in terms of resistance (Ω, kΩ, MΩ).

Dependent Sources

In electronics, **dependent** (or **controlled**) **sources** are used extensively. Four types of dependent sources are summarized in Fig. 1.10, in which the standard diamond shape is used for controlled sources. The **voltage-controlled current source (VCCS), current-controlled current source (CCCS),** and **voltage-controlled voltage source (VCVS)** are used routinely in this text to model transistors and amplifiers or to simplify more complex circuits. Only the **current-controlled voltage source (CCVS)** sees limited use.

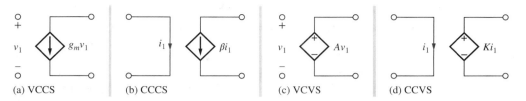

Figure 1.10 Controlled sources: (a) voltage-controlled current source (VCCS); (b) current-controlled current source (CCCS); (c) voltage-controlled voltage source (VCVS); (d) current-controlled voltage source (CCVS).

1.4 PROBLEM-SOLVING APPROACH

Solving problems is a centerpiece of an engineer's activity. As engineers, we use our creativity to find new solutions to problems that are presented to us. A well-defined **problem-solving approach** is essential. The examples in this text highlight an approach that can be used in all facets of your career, as a student and as an engineer in industry. The method is outlined in the following nine steps:

1. State the **problem** as clearly as possible.
2. List the **known information and given data.**
3. Define the **unknowns** that must be found to solve the problem.
4. List your **assumptions.** You may discover additional assumptions as the analysis progresses.
5. Develop an **approach** from a group of possible alternatives.
6. Perform an **analysis** to find a solution to the problem. As part of the analysis, be sure to draw the circuit and label the variables.
7. **Check the results.** Has the problem been solved? Is the math correct? Have all the unknowns been found? Have the assumptions been satisfied? Do the results satisfy simple consistency checks?
8. **Evaluate the solution.** Is the solution realistic? Can it be built? If not, repeat steps 4–7 until a satisfactory solution is obtained.
9. **Computer-aided analysis.** SPICE and other computer tools are highly useful to check the results and to see if the solution satisfies the problem requirements. Compare the computer results to your hand results.

To begin solving a problem, we must try to understand its details. The first four steps, which attempt to clearly define the problem, can be the most important part of the solution process. Time spent in understanding, clarifying, and defining the problem can save much time and frustration.

The first step is to write down a statement of the problem. The original problem description may be quite vague; we must try to understand the problem as well as, or even better than, the individual who posed the problem. As part of this focus on understanding the problem, we list the information that is known and unknown. Problem-solving errors can often be traced to imprecise definition of the unknown quantities. For example, it is very important for analysis to draw the circuit properly and to clearly label voltages and currents on our circuit diagrams.

Often there are more unknowns than constraints, and we need engineering judgment to reach a solution. Part of our task in studying electronics is to build up the background for selecting between various alternatives. Along the way, we often need to make approximations and assumptions that simplify the problem or form the basis of the chosen approach. It is important to state these assumptions, so that we can be sure to check their validity at the end. Throughout this text you will encounter opportunities to make assumptions. Most often, you should make assumptions that simplify your computational effort yet still achieve useful results.

The exposition of the known information, unknowns, and assumptions helps us not only to better understand the problem but also to think about various alternative solutions. We must choose the approach that appears to have the best chance of solving the problem. There may be more than one satisfactory approach. Each person will view the problem somewhat differently, and the approach that is clearest to one individual may not be the best for another. Pick the one that seems best to you. As part of defining the approach, be sure to think about what computational tools are available to assist in the solution, including MATLAB®, Mathcad®, spreadsheets, SPICE, and your calculator.

Once the problem and approach are defined as clearly as possible, then we can perform any analysis required and solve the problem. After the analysis is completed we need to check the results. A number of questions should be resolved. First, have all the unknowns been found? Do the results make sense? Are they consistent with each other? Are the results consistent with assumptions used in developing the approach to the problem?

Then we need to evaluate the solution. Are the results viable? For example, are the voltage, current, and power levels reasonable? Can the circuit be realized with reasonable yield with real components? Will the circuit continue to function within specifications in the face of significant component variations? Is the cost of the circuit within specifications? If the solution is not satisfactory, we need to modify our approach and assumptions and attempt a new solution. An iterative solution is often required to meet the specifications in realistic design situations. SPICE and other computer tools are highly useful for checking results and ensuring that the solution satisfies the problem requirements.

The solutions to the examples in this text have been structured following the problem-solving approach introduced here. Although some examples may appear trivial, the power of the structured approach increases as the problem becomes more complex.

WHAT ARE REASONABLE NUMBERS?

Part of our "check of results" should be to decide if the answer is "reasonable" and makes sense. Over time we must build up an understanding of what numbers are reasonable. Most solid-state devices that we will encounter are designed to operate from voltages ranging from a battery voltage of less than 1 V on the low end to no more than 40–50 V[5] at the high end. Typical power supply voltages will range from less than 1 V to 20 V or so, and typical resistance values encountered will range from a few ohms up to many GΩ.

Based on our knowledge of dc circuits, we should expect that the voltages in our circuits not exceed the power supply voltages. For example, if a circuit is operating from +8 and −5-V supplies, all of our calculated dc voltages should be between −5 and +8 V. In addition, the peak-to-peak amplitude of an ac signal should not exceed 13 V, the difference of the two supply voltages. With a 10-V supply, the maximum current that can go through a 100-Ω resistor is 100 mA; the current through a 10-MΩ resistor can be no more than 1 μA. Thus we should remember the following "rules" to check our results:

1. With few exceptions, the dc voltages in our circuits will not exceed the power supply voltages. The peak-to-peak amplitude of an ac signal should not exceed the difference of the power supply voltages.
2. The currents in our circuits will range from microamperes to no more than a hundred milliamperes or so.
3. So if a calculation yields a voltage exceeding that of the power supply, or a current of greater than 1 A, you need to check your work.

---

[5] The primary exception is in the area of power electronics, where one encounters much larger voltages and currents than the ones mentioned here.

1.5 IMPORTANT CONCEPTS FROM CIRCUIT THEORY

Analysis and design of electronic circuits make continuous use of a number of important techniques from basic network theory. Circuits are most often analyzed using a combination of **Kirchhoff's voltage law,** abbreviated **KVL,** and **Kirchhoff's current law,** abbreviated **KCL.** Occasionally, the solution relies on systematic application of **nodal** or **mesh analysis.** **Thévenin** and **Norton circuit transformations** are often used to help simplify circuits, and the notions of voltage and current division are extremely useful. Models of active devices invariably involve dependent sources, as mentioned in the last section, and we need to be familiar with dependent sources in all forms. Amplifier analysis also uses two-port network theory. A review of two-port networks is deferred until the introductory discussion of amplifiers in Chapter 6. If the reader feels uncomfortable with any of the concepts just mentioned, this is a good time for review. To help, a brief review of these important circuit techniques follows.

1.5.1 VOLTAGE AND CURRENT DIVISION

Voltage and current division are highly useful circuit analysis techniques that can be derived directly from basic circuit theory. They are both used routinely throughout this text, and it is very important to be sure to understand the conditions for which each technique is valid! Examples of both methods are provided next.

Voltage division is demonstrated by the circuit in Fig. 1.11(a) in which the voltages v_1 and v_2 can be expressed as

$$v_1 = i_i R_1 \quad \text{and} \quad v_2 = i_i R_2 \tag{1.7}$$

Applying KVL to the single loop,

$$v_i = v_1 + v_2 = i_i(R_1 + R_2) \quad \text{and} \quad i_i = \frac{v_i}{R_1 + R_2} \tag{1.8}$$

Combining Eqs. (1.7) and (1.8) yields the basic voltage division formula:

$$v_1 = v_i \frac{R_1}{R_1 + R_2} \quad \text{and} \quad v_2 = v_i \frac{R_2}{R_1 + R_2} \tag{1.9}$$

For the resistor values in Fig. 1.11(a),

$$v_1 = 10\text{ V} \frac{8\text{ k}\Omega}{8\text{ k}\Omega + 2\text{ k}\Omega} = 8.00\text{ V} \quad \text{and} \quad v_2 = 10\text{ V} \frac{2\text{ k}\Omega}{8\text{ k}\Omega + 2\text{ k}\Omega} = 2.00\text{ V} \tag{1.10}$$

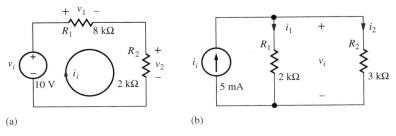

Figure 1.11 (a) A resistive voltage divider; (b) current division in a simple network.

DESIGN NOTE

VOLTAGE DIVIDER RESTRICTIONS

Note that the voltage divider relationships in Eq. (1.9) can be applied only when the current through the two resistor branches is the same. Also, note that the formulas are correct if the resistances are replaced by complex impedances and the voltages are represented as **phasors**.

$$\mathbf{V_1} = \mathbf{V_i}\frac{Z_1}{Z_1 + Z_2} \quad \text{and} \quad \mathbf{V_2} = \mathbf{V_i}\frac{Z_2}{Z_1 + Z_2}$$

Current division is also very useful. Let us find the currents i_1 and i_2 in the circuit in Fig. 1.11(b). Using KCL at the single node,

$$i_i = i_1 + i_2 \quad \text{where } i_1 = \frac{v_i}{R_1} \text{ and } i_2 = \frac{v_i}{R_2} \quad (1.11)$$

and solving for v_S yields

$$v_i = i_i \frac{1}{\frac{1}{R_1} + \frac{1}{R_2}} = i_i \frac{R_1 R_2}{R_1 + R_2} = i_i(R_1 \| R_2) \quad (1.12)$$

in which the notation $R_1 \| R_2$ represents the parallel combination of resistors R_1 and R_2. Combining Eqs. (1.11) and (1.12) yields the current division formulas:

$$i_1 = i_i \frac{R_2}{R_1 + R_2} \quad \text{and} \quad i_2 = i_i \frac{R_1}{R_1 + R_2} \quad (1.13)$$

For the values in Fig. 1.11(b),

$$i_1 = 5 \text{ mA} \frac{3 \text{ k}\Omega}{2 \text{ k}\Omega + 3 \text{ k}\Omega} = 3.00 \text{ mA} \quad i_2 = 5 \text{ mA} \frac{2 \text{ k}\Omega}{2 \text{ k}\Omega + 3 \text{ k}\Omega} = 2.00 \text{ mA}$$

DESIGN NOTE

CURRENT DIVIDER RESTRICTIONS

It is important to note that the same voltage must appear across both resistors in order for the current division expressions in Eq. (1.13) to be valid. Here again, the formulas are correct if the resistances are replaced by complex impedances and the currents are represented as **phasors**.

$$\mathbf{I_1} = \mathbf{I_i}\frac{Z_2}{Z_1 + Z_2} \quad \text{and} \quad \mathbf{I_2} = \mathbf{I_i}\frac{Z_1}{Z_1 + Z_2}$$

1.5.2 THÉVENIN AND NORTON CIRCUIT REPRESENTATIONS

Let us now review the method for finding **Thévenin and Norton equivalent circuits,** including a dependent source; the circuit in Fig. 1.12(a) serves as our illustration. Because the linear network in the dashed box has only two terminals, it can be represented by either the Thévenin or Norton equivalent circuits in Fig. 1.12(b) and (c). The work of Thévenin and Norton permits us to reduce complex circuits to a single source and equivalent resistance. We illustrate these two important techniques with the next four examples.

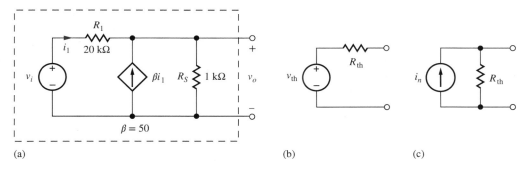

Figure 1.12 (a) Two-terminal circuit and its (b) Thévenin and (c) Norton equivalents.

EXAMPLE 1.1 THÉVENIN EQUIVALENT CIRCUIT

Let's practice finding the Thévenin and Norton equivalent circuits for the network in Fig. 1.12(a).

PROBLEM Find the Thévenin and Norton equivalent representations for the circuit in Fig. 1.12(a).

SOLUTION **Known Information and Given Data:** Circuit topology and values appear in Fig. 1.12(a).

Unknowns: Thévenin equivalent voltage v_{th}, Thévenin equivalent resistance R_{th}, and Norton equivalent current i_n.

Approach: Voltage source v_{th} is defined as the open-circuit voltage at the terminals of the circuit. R_{th} is the equivalent resistance at the terminals of the circuit terminals with all **independent sources** set to zero. Source i_n represents the short-circuit current available at the output terminals and is equal to v_{th}/R_{th}.

Assumptions: None

Analysis: We will first find the value of v_{th}, then R_{th} and finally i_n. Open-circuit voltage v_{th} can be found by applying KCL at the output terminals where the notational convention for conductance from Sec. 1.3 ($G_S = 1/R_S$) has been applied.

$$\beta i_1 = \frac{v_o - v_i}{R_1} + \frac{v_o}{R_S} = G_1(v_o - v_i) + G_S v_o \qquad (1.14)$$

Current i_1 is given by

$$i_1 = G_1(v_i - v_o) \qquad (1.15)$$

Substituting Eq. (1.15) into Eq. (1.14) and combining terms yields

$$G_1(\beta + 1)v_i = [G_1(\beta + 1) + G_S]v_o \qquad (1.16)$$

The Thévenin equivalent output voltage is then found to be

$$v_{th} = \frac{G_1(\beta + 1)}{[G_1(\beta + 1) + G_S]} v_i = \frac{(\beta + 1)R_S}{[(\beta + 1)R_S + R_1]} v_i \qquad (1.17)$$

where the second relationship was found by multiplying numerator and denominator by $(R_1 R_S)$. For the values in this problem,

$$v_o = \frac{(50 + 1)1 \text{ k}\Omega}{[(50 + 1)1 \text{ k}\Omega + 20 \text{ k}\Omega]} v_i = 0.718 v_i \quad \text{and} \quad v_{th} = 0.718 v_i \qquad (1.18)$$

R_{th} represents the equivalent resistance present at the output terminals with all independent sources set to zero. To find the **Thévenin equivalent resistance** R_{th}, we first set the independent sources in the network to zero. Remember, however, that **dependent sources must remain active.** A test voltage or current source is then applied to the network terminals and the corresponding current or voltage calculated. In Fig. 1.13 v_i is set to zero (i.e., replaced by a short circuit), voltage source v_x is applied to the network, and the current i_x must be determined so that

$$R_{th} = \frac{v_x}{i_x} \tag{1.19}$$

can be calculated.

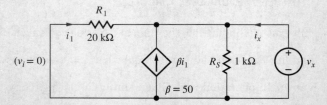

Figure 1.13 A test source v_x is applied to the network to find R_{th}.

$$i_x = -i_1 - \beta i_1 + G_S v_x \quad \text{in which } i_1 = -G_1 v_x \tag{1.20}$$

Combining and simplifying these two expressions yield

$$i_x = [(\beta + 1)G_1 + G_S]v_x \quad \text{and} \quad R_{th} = \frac{v_x}{i_x} = \frac{1}{(\beta + 1)G_1 + G_S} \tag{1.21}$$

The denominator of Eq. (1.21) represents the sum of two conductances, which corresponds to the parallel combination of two resistances. Therefore, Eq. (1.21) can be rewritten as

$$R_{th} = \frac{1}{(\beta+1)G_1 + G_S} = \frac{R_S \dfrac{R_1}{(\beta+1)}}{R_S + \dfrac{R_1}{(\beta+1)}} = R_S \left\| \dfrac{R_1}{(\beta+1)} \right. \tag{1.22}$$

For the values in this example,

$$R_{th} = R_S \left\| \dfrac{R_1}{(\beta+1)} \right. = 1 \text{ k}\Omega \left\| \dfrac{20 \text{ k}\Omega}{(50+1)} \right. = 1 \text{ k}\Omega \| 392 \text{ } \Omega = 282 \text{ } \Omega \tag{1.23}$$

Norton source i_n represents the short-circuit current available from the original network. Since we already have the Thévenin equivalent circuit, we can use it to easily find the value of i_n.

$$i_n = \frac{v_{th}}{R_{th}} = \frac{0.718 v_i}{282 \text{ } \Omega} = 2.55 \times 10^{-3} v_i$$

The Thévenin and Norton equivalent circuits for Fig. 1.12 calculated in the previous example appear for comparison in Fig. 1.14.

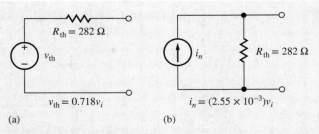

Figure 1.14 Completed (a) Thévenin and (b) Norton equivalent circuits for the two-terminal network in Fig. 1.12(a).

Check of Results: We have found the three unknowns required. A recheck of the calculations indicates they are done correctly. The value of v_{th} is the same order of magnitude as v_i, so its value should not be unusually large or small. The value of R_{th} is less than 1 kΩ, which seems reasonable, since we should not expect the resistance to exceed the value of R_S that appears in parallel with the output terminals. We can double-check everything by directly calculating i_n from the original circuit. If we short the output terminals in Fig. 1.12, we find the short-circuit current (see Ex. 1.2) to be $i_n = (\beta + 1)v_i/R_1 = 2.55 \times 10^{-3} v_i$ and in agreement with the other method.

EXAMPLE 1.2 NORTON EQUIVALENT CIRCUIT

Practice finding the Norton equivalent circuit for a network containing a dependent source.

PROBLEM Find the Norton equivalent (Fig. 1.12(c)) for the circuit in Fig. 1.12(a).

SOLUTION **Known Information and Given Data:** Circuit topology and circuit values appear in Fig. 1.12(a). The value of R_{th} was calculated in the previous example.

Unknowns: Norton equivalent current i_n.

Approach: The Norton equivalent current is found by determining the current coming out of the network when a short circuit is applied to the terminals.

Assumptions: None.

Analysis: For the circuit in Fig. 1.15, the output current will be

$$i_n = i_1 + \beta i_1 \quad \text{and} \quad i_1 = v_i/R_1 \tag{1.24}$$

since the short circuit across the output forces the current through R_S to be 0. Combining the two expressions in Eq. (1.24) yields

$$i_n = (\beta + 1)G_1 v_i = \frac{(\beta + 1)}{R_1} v_i \tag{1.25}$$

or

$$i_n = \frac{(50 + 1)}{20 \text{ k}\Omega} v_i = \frac{v_i}{392 \text{ }\Omega} = (2.55 \text{ mS}) v_i \tag{1.26}$$

The resistance in the Norton equivalent circuit also equals R_{th} found in Eq. (1.23).

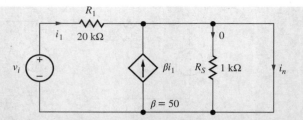

Figure 1.15 Circuit for determining short-circuit output current.

Check of Results: We have found the Norton equivalent current. Note that $v_{th} = i_n R_{th}$ and this result can be used to check the calculations: $i_n R_{th} = (2.55 \text{ mS}) v_s (282 \text{ }\Omega) = 0.719 v_s$, which agrees within round-off error with the previous example.

ELECTRONICS IN ACTION

Player Characteristics

The headphone amplifier in a personal music player represents an everyday example of a basic audio amplifier. The traditional audio band spans the frequencies from 20 Hz to 20 kHz, a range that extends beyond the hearing capability of most individuals at both the upper and lower ends.

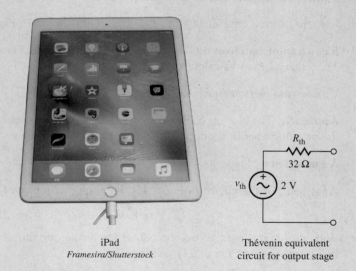

iPad
Framesira/Shutterstock

Thévenin equivalent circuit for output stage

The characteristics of the Apple iPad in the accompanying figure are representative of a high-quality audio output stage in an MP3 player or a computer sound card. The output can be represented by a Thévenin equivalent circuit with $v_{th} = 2$ V and $R_{th} = 32$ ohms, and the output stage is designed to deliver a power of approximately 15 mW into each channel of a headphone with a matched impedance of 32 ohms. The output power is approximately constant over the 20 Hz–20 kHz frequency range. At the lower and upper cutoff frequencies, f_L and f_H, the output power will be reduced by 3 dB, a factor of 2.

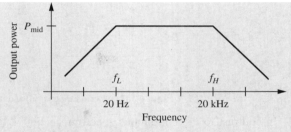

Power versus frequency for an audio amplifier

The distortion characteristics of the amplifier are also important, and this is an area that often distinguishes one sound card or MP3 player from another. A good audio system will have a total harmonic distortion (THD) specification of less than 0.1 percent at full power.

1.6 FREQUENCY SPECTRUM OF ELECTRONIC SIGNALS

Fourier analysis and the **Fourier series** represent extremely powerful tools in electrical engineering. Results from Fourier theory show that complicated signals are actually composed of a continuum of sinusoidal components, each having a distinct amplitude, frequency, and phase. The **frequency spectrum** of a signal presents the amplitude and phase of the components of the signal versus frequency.

Nonrepetitive signals have continuous spectra with signals that may occupy a broad range of frequencies. For example, the amplitude spectrum of a television signal measured during a small time interval is depicted in Fig. 1.16. The TV video signal is designed to occupy the frequency range from 0 to 4.5 MHz.[6] Other types of signals occupy different regions of the frequency spectrum. Table 1.2 identifies the frequency ranges associated with various categories of common signals.

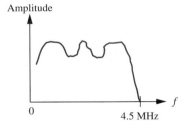

Figure 1.16 Spectrum of a TV signal.

TABLE 1.2
Frequencies Associated with Common Signals

CATEGORY	FREQUENCY RANGE
Audible sounds	20 Hz–20 kHz
Baseband video (TV) signal	0–4.5 MHz
AM radio broadcasting	0.54–1.6 MHz
High-frequency radio communications	1.6–54 MHz
VHF television (Channels 2–6)	54–88 MHz
FM radio broadcasting	88–108 MHz
VHF radio communication	108–174 MHz
VHF television (Channels 7–13)	174–216 MHz
Maritime and government communications	216–450 MHz
Business communications	450–470 MHz
UHF television (Channels 14–69)	470–806 MHz
Fixed and mobile communications including allocations for analog and digital cellular telephones, personal communications, and other wireless devices, 3G/4G/LTE/5G	0.60–5.0 GHz
Satellite television	3.7–4.2 GHz
Wireless devices	5.0–5.5 GHz
Industrial, scientific, and medical (ISM) bands	6 MHz–246 MHz
Automotive radarband	76–81 GHz
Radio Astronomy	73 GHz

[6] This signal is combined with a much higher carrier frequency prior to transmission.

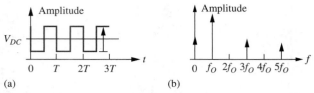

Figure 1.17 A periodic signal (a) and its amplitude spectrum (b).

In contrast to the continuous spectrum in Fig. 1.16, Fourier series analysis shows that *any periodic* signal, such as the square wave of Fig. 1.17, contains spectral components only at discrete frequencies[7] that are related directly to the period of the signal. For example, the square wave of Fig. 1.17 having an amplitude V_O and period T can be represented by the Fourier series

$$v(t) = V_{DC} + \frac{2V_O}{\pi}\left(\sin\omega_o t + \frac{1}{3}\sin 3\omega_o t + \frac{1}{5}\sin 5\omega_o t + \cdots\right) \quad (1.27)$$

in which $\omega_o = 2\pi/T$ (rad/s) is the **fundamental radian frequency** of the square wave. We refer to $f_o = 1/T$ (Hz) as the **fundamental frequency** of the signal, and the frequency components at $2f_o, 3f_o, 4f_o, \ldots$ are called the second, third, fourth, and so on **harmonic frequencies.**

1.7 AMPLIFIERS

The characteristics of analog signals are most often manipulated using linear amplifiers that affect the amplitude and/or phase of the signal without changing its frequency. Although a complex signal may have many individual components, as just described in Sec. 1.6, linearity permits us to use the **superposition principle** to treat each component individually.

For example, suppose the amplifier with voltage gain A in Fig. 1.18(a) is fed a sinusoidal input signal component v_i with amplitude V_i, frequency ω_i, and phase ϕ:

$$v_i = V_i \sin(\omega_i t + \phi) \quad (1.28)$$

Then, if the amplifier is linear, the output corresponding to this signal component will also be a sinusoidal signal at the same frequency but with a different amplitude and phase:

$$v_o = V_o \sin(\omega_i t + \phi + \theta) \quad (1.29)$$

Using phasor notation, the input and output signals would be represented as

$$\mathbf{V_i} = V_i \angle \phi \quad \text{and} \quad \mathbf{V_o} = V_o \angle (\phi + \theta) \quad (1.30)$$

The **voltage gain** of the amplifier is defined in terms of these phasors:

$$A = \frac{\mathbf{V_o}}{\mathbf{V_i}} = \frac{V_o \angle (\phi + \theta)}{V_i \angle \phi} = \frac{V_o}{V_i} \angle \theta \quad (1.31)$$

This amplifier has a voltage gain with magnitude equal to V_o/V_i and a phase shift of θ. In general, both the magnitude and phase of the voltage gain will be a function of frequency. Note that amplifiers also often provide current gain and power gain as well as voltage gain, but these concepts will not be explored further until Chapter 6.

The curves in Fig. 1.19 represent the input and output voltage waveforms for an inverting amplifier with $A_v = -5$ and $v_i = 1 \sin 2000\pi t$ V. Both the factor of five increase in signal amplitude and the 180° phase shift (multiplication by −1) are apparent in the graph.

At this point, a note regarding the phase angle is needed. In Eqs. (1.28) and (1.29), ωt, ϕ, and θ must have the same units. With ωt normally expressed in radians, ϕ should also be in radians. However, in electrical engineering texts, ϕ is often expressed in degrees. We must be

---

[7] There are an infinite number of components, however.

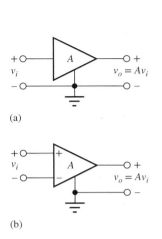
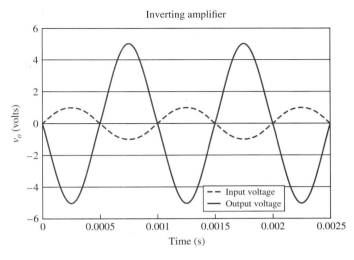

Figure 1.18 (a) Symbol for amplifier with single input and voltage gain A; (b) differential amplifier having two inputs and gain A.

Figure 1.19 Voltage waveforms of input v_i and output v_o for an amplifier with gain $A_v = -5$ and $v_i = 1 \sin 2000\pi t$ V.

aware of this mixed system of units and remember to convert degrees to radians before making any numeric calculations.

> **EXERCISE:** The input and output voltages of an amplifier are expressed as
>
> $$v_i = 0.001 \sin(2000\pi t) \text{ V} \quad \text{and} \quad v_o = -5\cos(2000\pi t + 25°) \text{ V}$$
>
> in which v_i and v_o are specified in volts when t is in seconds. What are $\mathbf{V}_i$, $\mathbf{V}_o$, and the voltage gain of the amplifier?
>
> **ANSWERS:** $0.001\angle 0°$; $5\angle -65°$; $5000\angle -65°$

1.7.1 IDEAL OPERATIONAL AMPLIFIERS

The **operational amplifier**, "op amp" for short, is a fundamental building block in electronic design and is discussed in most introductory circuit courses. A brief review of the ideal op amp is provided here; an in-depth study of the properties of ideal and nonideal op amps and the circuits used to build the op amp itself are the subjects of Chapters 10–14. Although it is impossible to realize the **ideal operational amplifier**, its use allows us to quickly understand the basic behavior to be expected from a given circuit and serves as a fundamental building block in circuit design.

From our basic circuit courses, we may recall that op amps are differential (or difference) amplifiers that respond to the signal voltage that appears between the + and − input terminals of the amplifier depicted in Fig. 1.18(b). Ideal op amps are assumed to have infinite **voltage gain** and infinite **input resistance**, and these properties lead to two special assumptions that are used to analyze circuits containing ideal op amps:

1. The voltage difference across the input terminals is zero; that is, $v_- = v_+$.
2. Both input currents are zero.

Applying the Assumptions—The Inverting Amplifier

The classic **inverting amplifier** circuit will be used to refresh our memory of the analysis of circuits employing op amps. The inverting amplifier is built by grounding the positive input of the operational amplifier and connecting resistors R_1 and R_2, called the **feedback network**, between the inverting input and the signal source and amplifier output node, respectively, as in

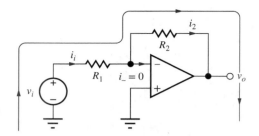

Figure 1.20 Inverting amplifier using op amp.

Fig. 1.20. Note that the ideal op amp is represented by a triangular amplifier symbol without a gain A indicated.

Our goal is to determine the voltage gain A_v of the overall amplifier, and to find A_v, we must find a relationship between $\mathbf{v}_i$ and $\mathbf{v}_o$. One approach is to write an equation for the single loop shown in Fig. 1.20:

$$\mathbf{v}_i - \mathbf{i}_i R_1 - \mathbf{i}_2 R_2 - \mathbf{v}_o = 0 \tag{1.32}$$

Now we need to express $\mathbf{i}_i$ and $\mathbf{i}_2$ in terms of $\mathbf{v}_i$ and $\mathbf{v}_o$. By applying KCL at the inverting input to the amplifier, we see that $\mathbf{i}_2$ must equal $\mathbf{i}_i$ because Assumption 2 states that $\mathbf{i}_-$ must be zero:

$$\mathbf{i}_i = \mathbf{i}_2 \tag{1.33}$$

Current $\mathbf{i}_i$ can be written in terms of $\mathbf{v}_i$ as

$$\mathbf{i}_i = \frac{\mathbf{v}_i - \mathbf{v}_-}{R_1} \tag{1.34}$$

where $\mathbf{v}_-$ is the voltage at the inverting input (negative input) of the op amp. But Assumption 1 states that the input voltage between the op amp terminals must be zero, so $\mathbf{v}_-$ must be zero because the positive input is grounded. Therefore

$$\mathbf{i}_i = \frac{\mathbf{v}_i}{R_1} \tag{1.35}$$

Combining Eqs. (1.32)–(1.35), the voltage gain is given by

$$A_v = \frac{\mathbf{v}_o}{\mathbf{v}_i} = -\frac{R_2}{R_1} \tag{1.36}$$

Referring to Eq. (1.36), we should note several things. The voltage gain is negative, indicative of an inverting amplifier with a 180° phase shift between its input and output signals. In addition, the magnitude of the gain can be greater than or equal to 1 if $R_2 \geq R_1$ (the most common case), but it can also be less than 1 for $R_2 < R_1$.

In the amplifier circuit in Fig. 1.20, the inverting-input terminal of the operational amplifier is at ground potential, 0 V, and is referred to as a **virtual ground.** The ideal operational amplifier adjusts its output to whatever voltage is necessary to force v_- to be zero. (In practical circuits, the output levels will be limited by the power supply voltages.)

DESIGN NOTE — VIRTUAL GROUND IN OP AMP CIRCUITS

Although the inverting input represents a virtual ground, it is *not* connected directly to ground (there is no direct dc path for current to reach ground). Shorting this terminal to ground for analysis purposes is a common mistake that must be avoided.

EXERCISE: Suppose $R_2 = 100$ kΩ. What value of R_1 gives a gain of −5?

ANSWER: 20 kΩ

ELECTRONICS IN ACTION

A Familiar Electronic System—The Cellular Phone

A cellular phone contains many radios operating over a multitude of frequencies including those of Bluetooth, cellular, GPS, ISM, and Wifi, to name a few (see Table 1.2). In addition, a "world" phone must operate on cellular frequency allocations that are different in North America, Europe, and Asia, for example.

The accompanying block diagram represents a typical mixed-signal (analog and digital) radio-frequency (RF) receiver that uses a number of amplifiers as well as digital technology. The signal from the antenna can be very small, often in the microvolt range. The signal's amplitude and power level are increased sequentially by the RF, intermediate frequency (IF), and audio amplifiers. The power available from the antenna may amount to only picowatts, whereas at the output, the amplifier may be providing a few tens of milliwatts to a headset or pair of earpods, or a high-power audio system could be delivering a 100-W or larger audio signal to its speaker system.

The local oscillator and frequency synthesizer tune the receiver to the desired frequency or channel and represent special applications of amplifiers and digital hardware. The mixer circuit actually changes the frequency of the incoming signal ($f_M = f_{RF} \pm f_{OSC}$) and is an application of a multiplier whose design draws heavily on linear amplifier and/or digital multiplexing circuit concepts. Finally, the demodulator/detector may be implemented using a combination of analog and digital circuitry employing A/D and D/A converters.

Chapters throughout this text provide in-depth exploration of the design techniques used in linear amplifiers and oscillators as well as the foundation needed to understand more complex circuits such as mixers, modulators, detectors, and A/D and D/A converters.

Examples of Several Radios in a Cellular Phone

TYPE	CAPABILITY
Bluetooth	Receive Rx/Transmit Tx
Cellular	Rx/Tx
GPS	Rx
Wifi (ISM)	Rx/Tx

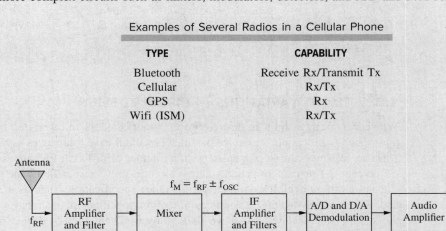

Block diagram for a basic radio receiver.

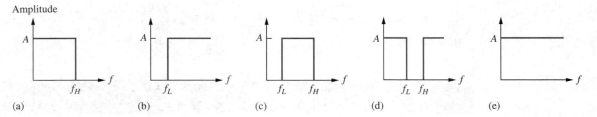

Figure 1.21 Ideal amplifier frequency responses: (a) low-pass, (b) high-pass, (c) band-pass, (d) band-reject, and (e) all-pass characteristics.

1.7.2 AMPLIFIER FREQUENCY RESPONSE

In addition to modifying the voltage, current, and/or power level of a given signal, amplifiers are often designed to selectively process signals of different frequency ranges. Amplifiers are classified into a number of categories based on their frequency response; five possible categories are shown in Fig. 1.21. The **low-pass amplifier,** Fig. 1.21(a), passes all signals below some upper cutoff frequency f_H, whereas the **high-pass amplifier,** Fig. 1.21(b), amplifies all signals above the lower cutoff frequency f_L. The **band-pass amplifier** passes all signals between the two cutoff frequencies f_L and f_H, as in Fig. 1.21(c). The **band-reject amplifier** in Fig. 1.21(d) rejects all signals having frequencies lying between f_L and f_H. Finally, the **all-pass amplifier** in Fig. 1.21(e) amplifies signals at any frequency. The all-pass amplifier is actually used to tailor the phase of the signal rather than its amplitude. Circuits that are designed to amplify specific ranges of signal frequencies are often referred to as **filters.**

EXERCISE: (a) The band-pass amplifier in Fig. 1.21(c) has $f_L = 1.5$ kHz, $f_H = 2.5$ kHz, and $A = 10$. If the input voltage is given by

$$v_i = [0.5 \sin(2000\pi t) + \sin(4000\pi t) + 1.5 \sin(6000\pi t)] \text{ V}$$

what is the output voltage of the amplifier? (b) Suppose the same input signal is applied to the low-pass amplifier in Fig. 1.21(a), which has $A = 5$ and $f_H = 1.75$ kHz. What is the output voltage?

ANSWERS: $10.0 \sin 4000\pi t$ V; $2.50 \sin 2000\pi t$ V

1.8 ELEMENT VARIATIONS IN CIRCUIT DESIGN

Whether a circuit is built in discrete form or fabricated as an integrated circuit, the passive components and semiconductor device parameters will all have **tolerances** associated with their values. Discrete resistors can be purchased with a number of different tolerances including ± 10 percent, ± 5 percent, ± 1 percent, or better, whereas resistors in ICs can exhibit wide variations (± 30 percent). Capacitors often exhibit asymmetrical tolerance specifications such as $+20$ percent$/-50$ percent, and power supply voltage tolerances are often specified in the range of 1 to 10 percent. For the semiconductor devices that we shall study in Chapters 3–5, device parameters may vary by 30 percent or more.

In addition to this initial value uncertainty due to tolerances, the values of the circuit components and parameters will vary with temperature and circuit age. It is important to understand the effect of these element changes on our circuits and to be able to design circuits that will continue to operate correctly in the face of such element variations. We will explore two analysis approaches, worst-case analysis and Monte Carlo analysis, that can help quantify the effects of tolerances on circuit performance.

1.8.1 MATHEMATICAL MODELING OF TOLERANCES

A mathematical model for symmetrical parameter variations is

$$P_{nom}(1 - \varepsilon) \leq P \leq P_{nom}(1 + \varepsilon) \qquad (1.37)$$

in which P_{nom} is the nominal specification for the parameter such as the resistor value or independent source value, and ε is the fractional tolerance for the component. For example, a resistor R with **nominal value** of 10 kΩ and a 5 percent tolerance could exhibit a resistance anywhere in the following range:

$$10{,}000 \, \Omega(1 - 0.05) \leq R \leq 10{,}000 \, \Omega(1 + 0.05)$$

or

$$9500 \, \Omega \leq R \leq 10{,}500 \, \Omega$$

> **EXERCISE:** A 39-kΩ resistor has a 10 percent tolerance. What is the range of resistor values corresponding to this resistor? Repeat for a 3.6-kΩ resistor with a 1 percent tolerance.
>
> **ANSWERS:** $35.1 \leq R \leq 42.9$ kΩ; $3.56 \leq R \leq 3.64$ kΩ

1.8.2 WORST-CASE ANALYSIS

Worst-case analysis is often used to ensure that a design will function under a given set of component variations. Worst-case analysis is performed by choosing values of the various components that make a desired variable (such as voltage, current, power, gain, or bandwidth) as large and as small as possible. These two limits are usually found by analyzing a circuit with the values of the various circuit elements pushed to their extremes. Although a worst-case design is often too conservative and represents "overdesign," it is important to understand the technique and its limitations. An easy way to explore worst-case analysis is with an example.

EXAMPLE 1.3 **WORST-CASE ANALYSIS**

Here we apply worst-case analysis to a simple voltage divider circuit.

PROBLEM Find the nominal and worst-case values (highest and lowest) of output voltage V_O and source current I_I for the voltage divider circuit of Fig. 1.22.

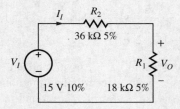

Figure 1.22 Resistor voltage divider circuit with tolerances.

SOLUTION **Known Information and Given Data:** We have been given the voltage divider circuit in Fig. 1.22; the 15-V source V_I has a 10 percent tolerance; resistor R_1 has a nominal value of 18 kΩ with a 5 percent tolerance; resistor R_2 has a nominal value of 36 kΩ with a 5 percent tolerance. Expressions for V_O and I_I are

$$V_O = V_I \frac{R_1}{R_1 + R_2} \quad \text{and} \quad I_I = \frac{V_I}{R_1 + R_2} \qquad (1.38)$$

Unknowns: V_O^{nom}, V_O^{max}, V_O^{min}, I_I^{nom}, I_I^{max}, I_I^{min}

Approach: Find the nominal values of V_O and I_I with all circuit elements set to their nominal (ideal) values. Find the worst-case values by selecting the individual voltage and resistance values that force V_O and I_I to their extremes. Note that the values selected for the various circuit elements to produce V_O^{max} will most likely differ from those that produce I_I^{max}, and so on.

Assumptions: None.

Analysis:
(a) Nominal Values
The nominal value of voltage V_O is found using the nominal values for all the parameters:

$$V_O^{nom} = V_I^{nom} \frac{R_1^{nom}}{R_1^{nom} + R_2^{nom}} = 15\,\text{V} \frac{18\,\text{k}\Omega}{18\,\text{k}\Omega + 36\,\text{k}\Omega} = 5\,\text{V} \tag{1.39}$$

Similarly, the nominal value of source current I_I is

$$I_I^{nom} = \frac{V_S^{nom}}{R_1^{nom} + R_2^{nom}} = \frac{15\,\text{V}}{18\,\text{k}\Omega + 36\,\text{k}\Omega} = 278\,\mu\text{A} \tag{1.40}$$

(b) Worst-Case Limits
Now let us find the **worst-case values (the largest and smallest possible values)** of voltage V_O and current I_I that can occur for the given set of element tolerances. First, the values of the components will be selected to make V_O as large as possible. However, it may not always be obvious at first to which extreme to adjust the individual component values. Rewriting Eq. (1.38) for voltage V_O will help:

$$V_O = V_I \frac{R_1}{R_1 + R_2} = \frac{V_I}{1 + R_2/R_1} \tag{1.41}$$

In order to make V_O as large as possible, the numerator of Eq. (1.41) should be large and the denominator small. Therefore, V_I and R_1 should be chosen to be as large as possible and R_2 as small as possible. Conversely, in order to make V_O as small as possible, V_I and R_1 must be small and R_2 must be large. Using this approach, the maximum and minimum values of V_O are

$$V_O^{max} = \frac{15\,\text{V}(1.1)}{1 + \frac{36\,\text{k}\Omega(0.95)}{18\,\text{k}\Omega(1.05)}} = 5.87\,\text{V} \quad \text{and} \quad V_O^{min} = \frac{15\,\text{V}(.90)}{1 + \frac{36\,\text{k}\Omega(1.05)}{18\,\text{k}\Omega(0.95)}} = 4.20\,\text{V} \tag{1.42}$$

The maximum value of V_O is 17 percent greater than the nominal value of 5 V, and the minimum value is 16 percent below the nominal value.

The worst-case values of I_I are found in a similar manner but require different choices for the values of the resistors:

$$I_I^{max} = \frac{V_I^{max}}{R_1^{min} + R_2^{min}} = \frac{15\,\text{V}(1.1)}{18\,\text{k}\Omega(0.95) + 36\,\text{k}\Omega(0.95)} = 322\,\mu\text{A}$$

$$I_I^{min} = \frac{V_I^{min}}{R_1^{max} + R_2^{max}} = \frac{15\,\text{V}(0.9)}{18\,\text{k}\Omega(1.05) + 36\,\text{k}\Omega(1.05)} = 238\,\mu\text{A} \tag{1.43}$$

The maximum of I_I is 16 percent greater than the nominal value, and the minimum value is 14 percent less than nominal.

Check of Results: The nominal and worst-case values have been determined and range 14 to 17 percent above and below the nominal values. We have three circuit elements that are varying, and the sum of the three tolerances is 20 percent. Our worst-case values differ from the nominal case by somewhat less than this amount, so the results appear reasonable.

EXERCISE: Find the nominal and worst-case values of the power delivered by source V_I in Fig. 1.22.

ANSWERS: 4.17 mW, 3.21 mW, 5.31 mW

DESIGN NOTE

BE WARY OF WORST-CASE DESIGN

In a real circuit, the parameters will be randomly distributed between the limits, and it is unlikely that the various components will all reach their extremes at the same time. Thus the worst-case analysis technique will overestimate (often badly) the extremes of circuit behavior, and a design based on worst-case analysis usually represents an unnecessary overdesign that is more costly than necessary to achieve the specifications with satisfactory yield. A better, although more complex, approach is to attack the problem statistically using Monte Carlo analysis. It is also difficult to see how to adjust values to achieve the worst-case situation in complex circuits. However, if every circuit must work no matter what, worst-case analysis may be appropriate.

1.8.3 MONTE CARLO ANALYSIS

Monte Carlo analysis uses randomly selected versions of a given circuit to predict its behavior from a statistical basis. For Monte Carlo analysis, a value for each of the elements in the circuit is selected at random from the possible distributions of parameters, and the circuit is then analyzed using the randomly selected element values. Many such randomly selected realizations ("cases" or "instances") of the circuit are generated, and the statistical behavior of the circuit is built up from analysis of the many test cases. Obviously, this is a good use of the computer. Before proceeding, we need to refresh our memory concerning a few results from probability and random variables.

Uniformly Distributed Parameters

In this section, the variable parameters will be assumed to be uniformly distributed between the two extremes. In other words, the probability that any given value of the parameter will occur is the same. In fact, when the parameter tolerance expression in Eq. (1.37) was first encountered, most of us probably visualized it in terms of a uniform distribution as depicted by the probability density function $p(r)$ for a uniformly distributed resistor r represented graphically in Fig. 1.23(a). The probability that a resistor value lies between r and $(r + dr)$ is equal to $p(r) dr$. The total probability P must equal unity, so

$$P = \int_{-\infty}^{+\infty} p(r)\,dr = 1 \tag{1.44}$$

Using this equation with the uniform probability density of Fig. 1.23(a) yields $p(r) = \frac{1}{2\varepsilon R_{\text{nom}}}$ as indicated in the figure.

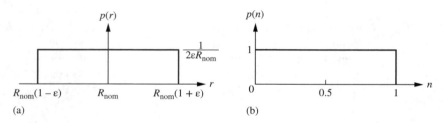

Figure 1.23 (a) Probability density function for a uniformly distributed resistor; (b) probability density function for a random variable uniformly distributed between 0 and 1.

Monte Carlo analysis can be readily implemented with a spreadsheet, MATLAB®, Mathcad®, or other computer program using the **uniform random number generators** that are built into the software. Successive calls to these random number generators produce a sequence of pseudo-**random numbers** that are uniformly distributed between 0 and 1 with a mean of 0.5 as in Fig. 1.23(b).

For example, the Excel® spreadsheet contains the function called RAND() (used with a null argument), whereas MATLAB uses rand,[8] and Mathcad uses rnd(1). These functions generate random numbers with the distribution in Fig. 1.23(b). Other software products contain random number generators with similar names. In order to use RAND() to generate the distribution in Fig. 1.23(a), the mean must be centered at R_{nom} and the width of the distribution set to $(2\varepsilon) \times R_{nom}$:

$$R = R_{nom}(1 + 2\varepsilon(\text{RAND}() - 0.5)) \tag{1.45}$$

Now let us see how we use Eq. (1.45) in implementing a Monte Carlo analysis.

EXAMPLE 1.4 MONTE CARLO ANALYSIS

Now we will apply Monte Carlo analysis to the voltage divider circuit.

PROBLEM Perform a Monte Carlo analysis of the circuit in Fig. 1.22. Find the mean, standard deviation, and largest and smallest values for V_O, I_I, and the power delivered from the source.

SOLUTION **Known Information and Given Data:** The voltage divider circuit appears in Fig. 1.22. The 15-V source V_I has a 10 percent tolerance, resistor R_1 has a nominal value of 18 kΩ with a 5 percent tolerance, and resistor R_2 has a nominal value of 36 kΩ with a 5 percent tolerance. Expressions for V_O, I_I, and P_I are

$$V_O = V_I \frac{R_1}{R_1 + R_2} \qquad I_I = \frac{V_I}{R_1 + R_2} \qquad P_I = V_I I_I$$

Unknowns: The mean, standard deviation, and largest and smallest values for V_O, I_I, and P_I.

Approach: To perform a Monte Carlo analysis of the circuit in Fig. 1.22, we assign randomly selected values to V_I, R_1, and R_2 and then use the values to determine V_O and I_I. Using Eq. (1.45) with the tolerances specified in Fig. 1.22, the power supply and resistor values are represented as

1. $V_I = 15(1 + 0.2(\text{RAND}() - 0.5))$
2. $R_1 = 18{,}000(1 + 0.1(\text{RAND}() - 0.5))$ \qquad (1.46)
3. $R_2 = 36{,}000(1 + 0.1(\text{RAND}() - 0.5))$

---

[8] In MATLAB, rand generates a single random number, rand(n) is an $n \times n$ matrix of random numbers, and rand(n, m) is an $n \times m$ matrix of random numbers. In Mathcad, rnd(x) returns a number uniformly distributed between 0 and x.

Note that each variable must invoke a separate call of the function RAND() so that the random values will be independently selected. The random elements in Eq. (1.46) are then used to evaluate the equations that characterize the circuit, including the power delivered from the source:

$$4. \quad V_O = V_I \frac{R_1}{R_1 + R_2}$$

$$5. \quad I_I = \frac{V_s}{R_1 + R_2} \quad (1.47)$$

$$6. \quad P_I = V_I I_I$$

This example will utilize a spreadsheet. However, any number of computer tools could be used: MATLAB®, Mathcad®, C++, SPICE, or the like.

Assumptions: The parameters are uniformly distributed between their means. A 100-case analysis will be performed.

Analysis: The spreadsheet used in this analysis appears in Table 1.3. Equation sets (1.46) and (1.47) are entered into the first row of the spreadsheet, and then that row may be copied into as many additional rows as the number of statistical cases that are desired. The analysis is automatically repeated for the random selections to build up the statistical distributions, with each row representing one analysis of the circuit. At the bottom of the columns, the mean, standard deviation, and minimum and maximum values can all be calculated using built-in spreadsheet functions, and the overall spreadsheet data can be used to build histograms for the circuit performance. A portion of the spreadsheet output for 100 cases of the circuit of Fig. 1.22 is shown in Table 1.3.

TABLE 1.3

	V_I (V)	R_1 (Ω)	R_2 (Ω)	V_O (V)	I_I (A)	P (W)
TOLERANCE	10.00%	5.00%	5.00%			
Case 1	15.94	17,248	35,542	5.21	3.02E−04	4.81E−03
2	14.90	18,791	35,981	5.11	2.72E−04	4.05E−03
3	14.69	18,300	36,725	4.89	2.67E−04	3.92E−03
4	16.34	18,149	36,394	5.44	3.00E−04	4.90E−03
5	14.31	17,436	37,409	4.55	2.61E−04	3.74E−03
...						
95	16.34	17,323	36,722	5.24	3.02E−04	4.94E−03
96	16.38	17,800	35,455	5.47	3.08E−04	5.04E−03
97	15.99	17,102	35,208	5.23	3.06E−04	4.89E−03
98	14.06	18,277	35,655	4.76	2.61E−04	3.66E−03
99	13.87	17,392	37,778	4.37	2.51E−04	3.49E−03
Case 100	15.52	18,401	34,780	5.37	2.92E−04	4.53E−03
Avg	14.88	17,998	36,004	4.96	2.76E−04	4.12E−03
Nom.	15.00	18,000	36,000	5.00	2.78E−04	4.17E−03
Stdev	0.86	476	976	0.30	1.73E−05	4.90E−04
Max	16.46	18,881	37,778	5.70	3.10E−04	5.04E−03
WC-Max	16.50	18,900	37,800	5.87	3.22E−04	—
Min	13.52	17,102	34,201	4.37	2.42E−04	3.29E−03
WC-Min	13.50	17,100	34,200	4.20	2.38E−04	—

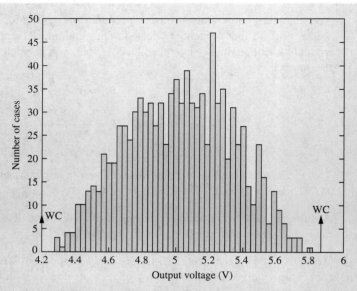

Figure 1.24 Histogram of a 1000-case simulation.

Check of Results: The average values for V_O and I_I are 4.96 V and 276 µA, respectively, which are close to the values originally estimated from the nominal circuit elements. The averages will more closely approach the nominal values as the number of cases used in the analysis is increased. The standard deviations are 0.30 V and 17.3 µA, respectively.

A histogram (generated with MATLAB® hist(x, n)) of the results of a 1000-case simulation of the output voltage in the same problem appears in Fig. 1.24. Note that the overall distribution is becoming Gaussian in shape with the peak in the center near the mean value. The worst-case values calculated earlier are several standard deviations from the mean and lie outside the minimum and maximum values that occurred even in this 1000-case Monte Carlo analysis.

Some implementations of the SPICE circuit analysis program, PSPICE® for example, actually contain a Monte Carlo option in which a full circuit simulation is automatically performed for any number of randomly selected test cases. These programs, which provide a powerful tool for much more complex statistical analysis than is possible by hand, can perform statistical estimates of delay, frequency response, and the like for circuits with many elements.

1.8.4 TEMPERATURE COEFFICIENTS

In the real world, all physical circuit elements change value as the temperature changes. Our circuit designs must continue to operate properly as the temperature changes. For example, the temperature range for commercial products is typically 0 to 70°C, whereas the standard military temperature range is −55 to +85°C. Other environments, such as the engine compartment of an automobile, can be even more extreme.

Mathematical Model
The basic mathematical model for incorporating element variation with temperature is

$$P = P_{\text{nom}}(1 + \alpha_1 \Delta T + \alpha_2 \Delta T^2) \qquad \text{with} \qquad \Delta T = T - T_{\text{nom}} \qquad (1.48)$$

Coefficients α_1 and α_2 represent the first- and second-order[9] temperature coefficients, and ΔT represents the difference between the actual temperature T and the temperature at which the nominal value is specified:

$$P = P_{\text{nom}} \quad \text{for} \quad T = T_{\text{nom}} \qquad (1.49)$$

Common values for the magnitude of α_1 range from 0 to plus or minus several thousand parts per million per degree Celsius (1000 ppm/°C = 0.1%/°C). For example, nichrome resistors are highly stable and can exhibit a **temperature coefficient of resistance** (**TCR** $= \alpha_1$) of only 50 ppm/°C. In contrast, diffused resistors in integrated circuits may have α_1 as large as several thousand ppm/°C. Most elements will also exhibit some curvature in their characteristics as a function of temperature, and α_2 will be nonzero, although small. We will neglect α_2 unless otherwise stated.

SPICE Model

Most SPICE programs contain models for the temperature dependencies of many circuit elements. For example, the temperature-dependent SPICE model for the resistor is equivalent to that given in Eq. (1.48):

$$R(T) = R(\text{TNOM}) * [1 + \text{TC1} * (T - \text{TNOM}) + \text{TC2} * (T - \text{TNOM})^2] \qquad (1.50)$$

in which the SPICE parameters are defined as follows:

TNOM = temperature at which the nominal resistor value is measured
T = temperature at which the simulation is performed
TC1 = first-order temperature coefficient
TC2 = second-order temperature coefficient

EXAMPLE 1.5   **TCR ANALYSIS**

Find the value of a resistor at various temperatures.

PROBLEM A diffused resistor has a nominal value of 10 kΩ at a temperature of 25°C and has a TCR of +2000 ppm/°C. Find its resistance at 40 and 75°C.

SOLUTION **Known Information and Given Data:** The resistor's nominal value is 10 kΩ at $T = 25°C$. The TCR is 2000 ppm/°C.

Unknowns: The resistor values at 40 and 75°C.

Approach: Use the known values to evaluate Eq. (1.48).

Assumptions: Based on the TCR statement, $\alpha_1 = 2000$ ppm/°C and $\alpha_2 = 0$.

Analysis: The TCR of +2000 ppm/°C corresponds to

$$\alpha_1 = \frac{2 \times 10^3}{10^6} \frac{1}{°C} = 2 \times 10^{-3}/°C$$

The resistor value at 40°C would be

$$R = 10 \text{ k}\Omega \left[1 + \frac{2 \times 10^{-3}}{°C}(40 - 25)°C\right] = 10.3 \text{ k}\Omega$$

---

[9] Higher-order temperature dependencies can also be included.

and at 75°C the value would be

$$R = 10 \text{ k}\Omega \left[1 + \frac{2 \times 10^{-3}}{°C} (75 - 25)°C \right] = 11.0 \text{ k}\Omega$$

Check of Results: 2000 ppm/°C corresponds to 0.2%/°C or 20 Ω/°C for the 10-kΩ resistor. A 15°C temperature change should shift the resistor value by 300 Ω, whereas a 50°C change should change the value by 1000 Ω. Thus the answers appear correct.

> **EXERCISE:** What will the resistor value in Ex. 1.5 be for $T = -55°C$ and $T = +85°C$?
>
> **ANSWERS:** 8.40 kΩ; 11.2 kΩ

1.9 NUMERIC PRECISION

Many numeric calculations will be performed throughout this book. Keep in mind that the circuits being designed can all be built in discrete form in the laboratory or can be implemented as integrated circuits. In designing circuits, we will be dealing with components that have tolerances ranging from less than ±1 percent to greater than ±50 percent, and calculating results to a precision of more than three significant digits represents a meaningless exercise except in very limited circumstances. Thus, the results in this text are consistently represented with three significant digits: 2.03 mA, 5.72 V, 0.0436 μA, and so on. For example, see the answers in Eqs. (1.18), (1.23), and so on.

SUMMARY

- The age of electronics began in the early 1900s with Pickard's creation of the crystal diode detector, Fleming's invention of the diode vacuum tube, and then Deforest's development of the triode vacuum tube. Since that time, the electronics industry has grown to account for as much as 15 percent of the world's gross domestic product.
- The real catalysts for the explosive growth of electronics occurred following World War II. The first was the invention of the bipolar transistor by Bardeen, Brattain, and Shockley in 1947; the second was the simultaneous invention of the integrated circuit by Kilby and by Noyce and Moore in 1958.
- Integrated circuits quickly became a commercial reality, and the complexity, whether measured in memory density (bits/chip), microprocessor transistor count, or minimum feature size, has changed exponentially since the mid-1960s. We are now in an era of giga-scale integration (GSI), having already put lower levels of integration—SSI, MSI, LSI, and VLSI—behind us.
- Electronic circuit design deals with two major categories of signals. Analog electrical signals may take on any value within some finite range of voltage or current. Digital signals, however, can take on only a finite set of discrete levels. The most common digital signals are binary signals, which are represented by two discrete levels.
- Bridging between the analog and digital worlds are the digital-to-analog and analog-to-digital conversion circuits (DAC and ADC, respectively). The DAC converts digital information into an analog voltage or current, whereas the ADC creates a digital number at its output that is proportional to an analog input voltage or current.

- A number of fundamental network techniques are used to analyze circuits throughout this text. These include current and voltage division, Thévenin and Norton equivalent circuits, and nodal and loop analysis. Be sure you understand these methods!
- Fourier demonstrated that complex signals can be represented as a linear combination of sinusoidal signals. Analog signal processing is applied to these signals using linear amplifiers; these modify the amplitude and phase of analog signals.
- Amplifiers are often classified by their frequency response into low-pass, high-pass, band-pass, band-reject, and all-pass categories. Electronic circuits that are designed to amplify specific ranges of signal frequencies are usually referred to as filters.
- Solving problems is one focal point of an engineer's career. A well-defined problem-solving approach is essential, and to this end, a structured problem-solving approach has been introduced in this chapter as outlined in these nine steps. Throughout the rest of this text, the examples will follow this problem-solving approach:
  1. State the **problem** as clearly as possible.
  2. List the **known information and given data.**
  3. Define the **unknowns** that must be found to solve the problem.
  4. List your **assumptions.** You may discover additional assumptions as the analysis progresses.
  5. Select and develop an **approach** from a list of possible alternatives.
  6. Perform an **analysis** to find a solution to the problem.
  7. **Check the results.** Is the math correct? Have all the unknowns been found? Do the results satisfy simple consistency checks?
  8. **Evaluate the solution.** Is the solution realistic? Can it be built? If not, repeat steps 4–7 until a satisfactory solution is obtained.
  9. Use **computer-aided analysis** to check the results and to see if the solution satisfies the problem requirements.
- Our circuit designs will be implemented using real components whose initial values differ from those of the design and that change with time and temperature. Techniques for analyzing the influence of element tolerances on circuit performance include the worst-case analysis and statistical Monte Carlo analysis methods. Circuit analysis programs include the ability to specify temperature dependencies for most circuit elements.
- In worst-case analysis, element values are simultaneously pushed to their extremes, and the resulting predictions of circuit behavior are often overly pessimistic.
- The Monte Carlo method analyzes a large number of randomly selected versions of a circuit to build up a realistic estimate of the statistical distribution of circuit performance. Random number generators in high-level computer languages, spreadsheets, Mathcad®, or MATLAB® can be used to randomly select element values for use in Monte Carlo analysis. Some circuit analysis packages such as PSPICE® provide a Monte Carlo analysis option as part of the program.

KEY TERMS

All-pass amplifier
Analog signal
Analog-to-digital converter (A/D converter or ADC)
Band-pass amplifier
Band-reject amplifier
Binary digital signal
Bipolar transistor
Current-controlled current source (CCCS)
Current-controlled voltage source (CCVS)
Current division
Dependent (or controlled) source
Digital signal
Digital-to-analog converter (D/A converter or DAC)
Diode

Feedback network
Filters
Fourier analysis
Fourier series
Frequency spectrum
Fundamental frequency
Fundamental radian frequency
Giga-scale integration (GSI)
Harmonic frequency
High-pass amplifier
Ideal operational amplifier
Integrated circuit (IC)
Input resistance
Inverting amplifier
Kirchhoff's current law (KCL)
Kirchhoff's voltage law (KVL)
Large-scale integration (LSI)
Least significant bit (LSB)
Low-pass amplifier
Medium-scale integration (MSI)
Mesh analysis
Minimum feature size
Mixed signal
Monte Carlo analysis
Most significant bit (MSB)
Nodal analysis
Nominal value

Norton circuit transformation
Norton equivalent circuit
Operational amplifier (op amp)
Phasor
Problem-solving approach
Quantization error
Random numbers
Resolution of a converter
Small-scale integration (SSI)
Superposition principle
Temperature coefficient of resistance (TCR)
Thévenin circuit transformation
Thévenin equivalent circuit
Thévenin equivalent resistance
Tolerance
Transistor
Triode
Uniform random number generator
Vacuum diode
Vacuum tube
Very-large-scale integration (VLSI)
Virtual ground
Voltage-controlled current source (VCCS)
Voltage-controlled voltage source (VCVS)
Voltage division
Voltage gain
Worst-case analysis

REFERENCES

1. W. F. Brinkman, D. E. Haggan, and W. W. Troutman, "A History of the Invention of the Transistor and Where It Will Lead Us," *IEEE Journal of Solid-State Circuits,* vol. 32, no. 12, pp. 1858–65, December 1997.
2. www.pbs.org/transistor/index.html
3. *CIA Factbook,* www.cia.gov
4. *Fortune* Global 500, www.fortune.com/global500
5. *Fortune* 500, www.fortune.com/fortune500

ADDITIONAL READING

Commemorative Supplement to the Digest of Technical Papers, 1993 IEEE International Solid-State Circuits Conference Digest, vol. 36, February 1993.

Digest of Technical Papers of the IEEE Custom Integrated International Circuits Conference, ieee-cicc.org, yearly.

Digest of Technical Papers of the IEEE International Electronic Devices Meeting, December of each year.

Digest of Technical Papers of the IEEE International Solid-State Circuits Conference, February of each year.

Digest of Technical Papers of the IEEE International Symposia on VLSI Technology and Circuits, June of each year.

Electronics, Special Commemorative Issue, April 17, 1980.

Frequency allocations: www.fcc.gov

Garratt, G. R. M. *The Early History of Radio from Faraday to Marconi.* London: Institution of Electrical Engineers (IEE), 1994.

IEDM: www.ieee-iedm.org
IEEE: www.ieee.org
IEEE Solid-State Circuits Society: www.sscs.org
International Technology Roadmap for Semiconductors: www.itrs2.net
ISSCC: www.isscc.org
J. T. Wallmark, "The Field-Effect Transistor—An Old Device with New Promise," *IEEE Spectrum,* March 1964.
2017 and 2018 IRDS™ Roadmaps: https://irds.ieee.org/editions

PROBLEMS

1.1 A Brief History of Electronics: From Vacuum Tubes to Giga-Scale Integration

1.1. Make a list of 20 items in your environment that contain electronics. A PC and its peripherals are considered one item. (Do not confuse electromechanical timers, common in clothes dryers or the switch in a simple thermostat, with electronic circuits.)

1.2. The upper line in Fig. 1.4 is described by $N = (2.512 \times 10^9) \times 10^{(\text{Year}-2012)/4.343}$. Based upon a straight-line projection, estimate the number of transistors in a complex IC chip in the year 2025.

1.3. The lower line in Fig. 1.4 is described by $N = 1327 \times 10^{(\text{Year}-1970)/6.52}$. Based on a straight-line projection of this figure, what will be the number of transistors in a microprocessor in the year 2020?

1.4. The change in memory density with time can be described by $B = 19.97 \times 10^{0.1977(\text{Year}-1960)}$. If a straight-line projection is made using this equation, what will be the number of memory bits/chip in the year 2026?

1.5. (a) How many years does it take for the number of transistors to increase by a factor of 2, based on the equation in Prob. 1.3? (b) By a factor of 10?

1.6. (a) How many years does it take for memory chip density to increase by a factor of 2, based on the equation in Prob. 1.4? (b) By a factor of 10?

1.7. Repeat Prob. 1.5 using the equation in Prob. 1.2.

1.8. If you make a straight-line projection from the upper curve in Fig. 1.5, what will be the minimum feature size in integrated circuits in the year 2026? The curve can be described by $F = 8.00 \times 10^{-0.05806(\text{Year}-1970)}$ μm. Do you think this is possible? Why or why not?

1.9. The filament of a small vacuum tube uses a power of approximately 1.5 W. Suppose that 17.2 billion of these tubes are used to build the equivalent of a 16 Gb memory. (a) How much power is required for this memory? (b) If this power is supplied from a 220-V ac source, what is the current required by this memory? (c) If the vacuum tube occupies a volume of 80 cm^3, what would be the volume occupied by the tubes in a 16-Gb memory?

1.2 Classification of Electronic Signals

1.10. Classify each of the following as an analog or digital quantity: (a) status of a light switch, (b) status of a thermostat, (c) water pressure, (d) gas tank level, (e) bank overdraft status, (f) light bulb intensity, (g) stereo volume, (h) full or empty cup, (i) room temperature, (j) TV channel selection, and (k) tire pressure.

1.11. An 8-bit A/D converter has $V_{FS} = 2.5$ V. What is the value of the voltage corresponding to the LSB? If the input voltage is 1.40 V, what is the binary output code of the converter?

1.12. A 10-bit D/A converter has a full-scale voltage of 1.0 V. What is the voltage corresponding to the LSB? What is the output voltage if the binary input code is equal to (0101100110)?

1.13. A 12-bit D/A converter has a full-scale voltage of 5.12 V. What is the voltage corresponding to the LSB? To the MSB? What is the output voltage if the binary input code is equal to (100100101001)?

1.14. A 15-bit A/D converter has $V_{FS} = 10.24$ V. What is the value of the LSB? If the input voltage is 6.89 V, what is the binary output code of the converter?

1.15. (a) A digital multimeter is being designed to have a readout with a range of 0 to 10000. How many bits will be required in its A/D converter? (b) Repeat for six decimal digits.

1.16. A 14-bit ADC has $V_{FS} = 2.56$ V and the output code is (10101110111010). What is the size of the LSB for the converter? What range of input voltages corresponds to the ADC output code?

1.3 Notational Conventions

1.17. If $i_B = 0.003(2.5 + 0.25 \cos 1000t)$ A, what are I_B and i_b?

1.18. If $v_{GS} = (1.5 + 0.5u(t-1) + 0.2 \cos 2000\pi t)$ V, what are V_{GS} and v_{gs}? [$u(t)$ is the unit step function.]

1.19. If $V_{CE} = 2$ V and $v_{ce} = (5 \cos 5000t)$ V, write the expression for v_{CE}.

1.20. If $V_{DS} = 2.5$ V and $v_{ds} = (0.2 \sin 2500t + 4 \sin 1000t)$ V, write the expression for v_{DS}.

1.5 Important Concepts from Circuit Theory

1.21. Use voltage and current division to find V_1, V_2, I_2, and I_3 in the circuit in Fig. P1.21 if $V = 1$ V, $R_1 = 24$ kΩ, $R_2 = 30$ kΩ, and $R_3 = 13$ kΩ.

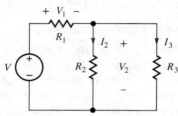

Figure P1.21

1.22. Use voltage and current division to find V_1, V_2, I_2, and I_3 in the circuit in Fig. P1.21 if $V = 8$ V, $R_1 = 30$ kΩ, $R_2 = 24$ kΩ, and $R_3 = 18$ kΩ.

1.23. Use current and voltage division to find I_1, I_2, and V_3 in the circuit in Fig. P1.23 if $I = 250$ μA, $R_1 = 150$ kΩ, $R_2 = 68$ kΩ, and $R_3 = 82$ kΩ.

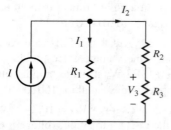

Figure P1.23

1.24. Use current and voltage division to find I_1, I_2, and V_3 in the circuit in Fig. P1.23 if $I = 4$ mA, $R_1 = 9.1$ kΩ, $R_2 = 3.9$ kΩ, and $R_3 = 5.6$ kΩ.

1.25. (a) Remove R_3 from the circuit in Fig. P1.21 and find the Thévenin equivalent circuit for the remaining network. Then use the Thévenin circuit to find the current in R_3. (b) Repeat using the Norton equivalent circuit.

1.26. (a) Remove R_3 from the circuit in Fig. P1.23 and find the Norton equivalent circuit for the remaining network. Then use the Norton circuit to find voltage V_3. (b) Repeat using the Thévenin equivalent circuit.

1.27. Find the Thévenin equivalent representation of the circuit in Fig. P1.27 if $g_m = 0.025$ S and $R_1 = 10$ kΩ.

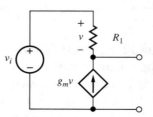

Figure P1.27

1.28. Find the Norton equivalent representation of the circuit in Fig. P1.27 if $g_m = 0.002$ S and $R_1 = 11$ kΩ.

1.29. Find the Norton equivalent representation of the circuit in Fig. P1.29(a) if $\beta = 100$, $R_1 = 47$ kΩ, and $R_2 = 100$ kΩ. (b) Repeat for the circuit in Fig. P1.29(b).

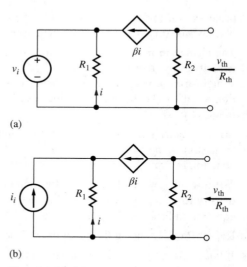

(a)

(b)

Figure P1.29

1.30. (a) Find the Thévenin equivalent representation of the circuit in Fig. P1.29(a) if $\beta = 120$, $R_1 = 75$ kΩ, and $R_2 = 75$ kΩ. (b) Repeat for Fig. P1.27(b).

1.31. (a) What is the resistance presented to source v_i by the circuit in Fig. P1.29(a) if $\beta = 75$, $R_1 = 100$ kΩ, and $R_2 = 39$ kΩ? (b) Repeat for Fig. P1.29(b).

1.32. Find the Thévenin equivalent representation of the circuit in Fig. P1.32 if $g_m = .0025$ S, $R_1 = 200$ kΩ, and $R_2 = 2$ MΩ.

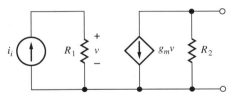

Figure P1.32

1.33. (a) What is the equivalent resistance between terminals A and B in Fig. P1.33? (b) What is the equivalent resistance between terminals C and D? (c) What is the equivalent resistance between terminals E and F? (d) Between terminals B and D? (e) Between A and E?

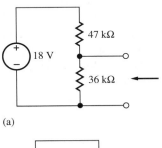

Figure P1.33

1.34. (a) Find the Thévenin equivalent circuit for the network in Fig. P1.34(a). (b) What is the Norton equivalent circuit? (c) Find the Thévenin equivalent circuit for the network in Fig. P1.34(b). (d) What is the Norton equivalent circuit?

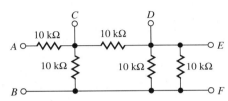

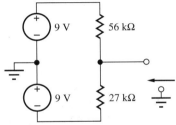

Figure P1.34

1.35. Resistor R_L is connected between the terminals of a one-port (two-terminal) circuit. Two measurements are made. If $R_L = 2$ kΩ, the voltage across R_L is 2 V. If $R_L = 4$ kΩ, the voltage across R_L is 3 V. (a) What is the Thévenin equivalent circuit for the one-port [as in Fig. 1.12(b)]? (b) What is the Norton equivalent circuit for the one-port [as in Fig. 1.12(c)]?

1.36. (a) Assume that the voltage source in Fig. P1.34(b) is the correct value, but there is some problem with the connections in the rest of the circuit. What are the maximum and minimum possible voltages at the output of the circuit? (b) Repeat for the circuit in Fig. P1.34(a).

1.6 Frequency Spectrum of Electronic Signals

1.37. A signal voltage is expressed as $v(t) = (4 \sin 4000\pi t + 2.6 \cos 2000\pi t)$ V. Draw a graph of the amplitude spectrum for $v(t)$ similar to the one in Fig. 1.17(b).

*1.38. (a) Voltage $v_1 = 4 \sin 20{,}000\pi t$ is multiplied by voltage $v_2 = 2 \sin 2000\pi t$. Draw a graph of the amplitude spectrum for $v = v_1 \times v_2$ similar to the one in Fig. 1.17(b). (Note. In electronics the product is often called *mixing* because it produces a signal that contains output frequencies that are not in the input signal but depend directly on the input frequencies.)

1.7 Amplifiers

1.39. The input and output voltages of an amplifier are expressed as $v_i = 10^{-4} \sin(2 \times 10^7 \pi t)$ V and $v_o = 2 \sin(2 \times 10^7 \pi t + 43°)$ V. What are the magnitude and phase of the voltage gain of the amplifier?

*1.40. The input and output voltages of an amplifier are expressed as

$$v_i = [10^{-3} \sin(5000\pi t) + 2 \times 10^{-3} \sin(3000\pi t)] \text{ V}$$

and

$$v_o = [10^{-2} \sin(3000\pi t - 45°) + 10^{-1} \sin(5000\pi t - 12°)] \text{ V}$$

(a) What are the magnitude and phase of the voltage gain of the amplifier at a frequency of 2500 Hz? (b) At 1500 Hz?

1.41. What is the voltage gain of the amplifier in Fig. 1.20 if (a) $R_1 = 5.6$ kΩ and $R_2 = 560$ kΩ? (b) For $R_1 = 18$ kΩ and $R_2 = 240$ kΩ? (c) For $R_1 = 2$ kΩ and $R_2 = 75$ kΩ? (d) Write an expression for the output voltage $v_o(t)$ of the circuit in Fig. 1.20 if $R_1 = 910$ Ω, $R_2 = 8.2$ kΩ, and $v_i(t) = (0.01 \sin 750\pi t)$ V. (e) Write an expression for the current $i_s(t)$.

1.42. (a) Find an expression for the voltage gain $A_v = v_o/v_i$ for the amplifier in Fig. P1.42(a). (b) Find an expression for the voltage gain $A_v = v_o/v_i$ for the amplifier in Fig. P1.42(b).

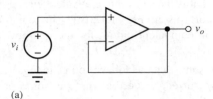

(a)

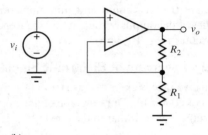

(b)

Figure P1.42

1.43. Write an expression for the output voltage $v_o(t)$ of the circuit in Fig. P1.43 if $R_1 = 10$ kΩ, $R_2 = 10$ kΩ, $R_3 = 51$ kΩ, $v_1(t) = (0.01 \sin 3770t)$ V, and $v_2(t) = (0.05 \sin 10{,}000t)$ V. Write an expression for the voltage appearing at the inverting input (v_-).

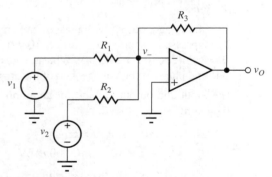

Figure P1.43

1.44. The circuit in Fig. P1.44 can be used as a simple 3-bit digital-to-analog converter (DAC). The individual bits of the binary input word ($b_1b_2b_3$) are used to control the position of the switches, with the resistor connected to 0 V if $b_i = 0$ and connected to V_{REF} if $b_i = 1$. (a) What is the output voltage for the DAC as shown with input data of (011) if $V_{\text{REF}} = 1.0$ V? (b) Suppose the input data change to (100). What will be the new output voltage? (c) Make a table giving the output voltages for all eight possible input data combinations.

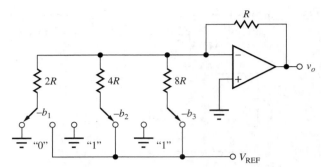

Figure P1.44

Amplifier Frequency Response

1.45. An amplifier has a voltage gain of 7 for frequencies below 6000 Hz, and zero gain for frequencies above 6000 Hz. Classify this amplifier.

1.46. An amplifier has a voltage gain of 20 for frequencies above 10 kHz, and zero gain for frequencies below 10 kHz. Classify this amplifier.

1.47. An amplifier has a voltage gain of zero for frequencies below 2000 Hz, and zero gain for frequencies above 5000 Hz. In between these two frequencies the amplifier has a gain of 20. Classify this amplifier.

1.48. The amplifier in Prob. 1.45 has an input signal given by $v_s(t) = (3 \sin 2000\pi t + 2 \cos 8000\pi t + 2 \cos 15{,}000\pi t)$ V. Write an expression for the output voltage of the amplifier.

1.49. The amplifier in Prob. 1.46 has an input signal given by $v_s(t) = (0.8 \sin 2500\pi t + 0.5 \cos 8000\pi t + 0.75 \cos 12{,}000\pi t)$ V. Write an expression for the output voltage of the amplifier.

1.50. The amplifier in Prob. 1.47 has an input signal given by $v_s(t) = (0.6 \sin 2500\pi t + 0.5 \cos 8000\pi t + 0.6 \cos 12{,}000\pi t)$ V. Write an expression for the output voltage of the amplifier.

1.51. An amplifier has an input signal that can be represented as

$$v(t) = \frac{4}{\pi}\left(\sin \omega_o t + \frac{1}{3}\sin 3\omega_o t + \frac{1}{5}\sin 5\omega_o t\right) \text{ V}$$

where $f_o = 1000$ Hz
(a) Use MATLAB to plot the signal for $0 \le t \le 5$ ms.
(b) The signal $v(t)$ is amplified by an amplifier that provides a voltage gain of 5 at all frequencies. Plot the output voltage for this amplifier for $0 \le t \le 5$ ms.
(c) A second amplifier has a voltage gain of 5 for frequencies below 2000 Hz but zero gain for frequencies above 2000 Hz. Plot the output voltage

for this amplifier for $0 \leq t \leq 5$ ms. (d) A third amplifier has a gain of 5 at 1000 Hz, a gain of 3 at 3000 Hz, and a gain of 1 at 5000 Hz. Plot the output voltage for this amplifier for $0 \leq t \leq 5$ ms.

1.8 Element Variations in Circuit Design

1.52. (a) A 4.7-kΩ resistor is purchased with a tolerance of 1 percent. What is the possible range of values for this resistor? (b) Repeat for a 5 percent tolerance. (c) Repeat for a 10 percent tolerance.

1.53. A 10,000 µF capacitor has an asymmetric tolerance specification of $+20\%/-50\%$. What is the possible range of values for this capacitor?

1.54. The power supply voltage for a circuit must vary by no more than 50 mV from its nominal value of 1.8 V. What is its tolerance specification?

1.55. An 8200-Ω resistor is purchased with a tolerance of 10 percent. It is measured with an ohmmeter and found to have a value of 7905 Ω. Is this resistor within its specification limits? Explain your answer.

1.56. (a) The output voltage of a 5-V power supply is measured to be 5.30 V. The power supply has a 5 percent tolerance specification. Is the supply operating within its specification limits? Explain your answer. (b) The voltmeter that was used to make the measurement has a 1.5 percent tolerance. Does that change your answer? Explain.

1.57. A resistor is measured and found to have a value of 6066 Ω at 0°C and 6562 Ω at 100°C. What are the temperature coefficient and nominal value for the resistor? Assume $T_{NOM} = 27°C$.

1.58. A resistor has a value of 10 kΩ at 30°C, a tolerance of 5 percent, and a TCR of 2200 ppm/°C. What is the possible range of values of this resistor at a temperature of 75°C?

1.59. Find the worst-case values of I_1, I_2, and V_3 for the circuit in Prob. 1.23 if the resistor tolerances are 5 percent and the current source tolerance is 2 percent.

1.60. Find the worst-case values of V_1, I_2, and I_3 for the circuit in Prob. 1.21 if the resistor tolerances are 5 percent and the voltage source tolerance is 5 percent.

1.61. Find the worst-case values for the Thévenin equivalent resistance for the circuit in Prob. 1.27 if the resistor tolerance is 10 percent and the tolerance on g_m is also 20 percent.

1.62. Perform a 200-case Monte Carlo analysis for the circuit in Prob. 1.59 and compare the results to the worst-case calculations.

1.63. Perform a 200-case Monte Carlo analysis for the circuit in Prob. 1.60 and compare the results to the worst-case calculations.

1.64. The tolerance on each of the elements in Fig. P1.34(b) is 5 percent. (a) What are the maximum and minimum values of the Thévenin equivalent voltage of the circuit? (b) Repeat for the Norton equivalent current. (c) Repeat for the value of R_{th}.

1.65. Repeat Prob. 1.64 for the circuit in Fig. P1.34(a).

1.9 Numeric Precision

1.66. (a) Express the following numbers to three significant digits of precision: 3.2947, 0.995171, -6.1551. (b) To four significant digits. (c) Check these answers using your calculator.

1.67. (a) What is the voltage developed by a current of 1.763 mA in a resistor of 20.70 kΩ? Express the answer with three significant digits. (b) Express the answer with two significant digits. (c) Repeat for $I = 102.1$ µA and $R = 97.80$ kΩ.

1.68. Calculate the power delivered to 32-Ω headphones using the Thévenin equivalent circuit for the music player discussed on page 20 for a frequency of 10 kHz.

术语对照

All-pass amplifier	全通放大器
Analog signal	模拟信号
Analog-to-digital converter (A/D converter or ADC)	模/数转换器
Band-pass amplifier	带通放大器
Band-reject amplifier	带阻放大器
Binary digital signal	二进制数字信号
Bipolar transistor	双极型晶体管
Current-controlled current source(CCCS)	电流控制电流源(CCCS)
Current-controlled voltage source(CCVS)	电流控制电压源(CCVS)
Current division	分流
D/A converter	D/A转换器(DAC)
Dependent(or controlled) source	受控源
Digital signal	数字信号
Digital-to-analog converter (D/A converter or DAC)	数/模转换器
Diode	二极管
Distortion characteristics	失真特性
Feedback network	反馈网络
Filters	滤波器
Fourier analysis	傅里叶分析
Fourier series	傅里叶级数
Frequency spectrum	频谱
Fundamental frequency	基频
Fundamental radian frequency	基本角频率
Giga-scale integration(GSI)	巨大规模集成电路(或吉规模集成电路)
Harmonic frequency	谐波频率
High-pass amplifier	高通放大器
Ideal operational amplifier	理想运算放大器
Independent source	独立源
Integrated circuit(IC)	集成电路
Input resistance	输入电阻
Inverting amplifier	反相放大器
Kirchhoff's current law(KCL)	基尔霍夫电流定律
Kirchhoff's voltage law(KVL)	基尔霍夫电压定律
Large-scale integration(LSI)	大规模集成
Least significant bit(LSB)	最低有效位
Low-pass amplifier	低通放大器
Medium-scale integration(MSI)	中规模集成
Mesh analysis	网孔分析法
Minimum feature size	最小特征尺寸
Monte Carlo analysis	蒙特卡洛分析

Most significant bit(MSB)	最高有效位
Nodal analysis	节点分析
Nominal value	额定值
Nominal resistor value	额定电阻值
Norton circuit transformation	诺顿电路变换
Norton equivalent circuit	诺顿等效电路
Operational amplifier(op amp)	运算放大器(op amp)
Phasor	相量
Problem-solving approach	解题方法
Quantization error	量化误差
Random numbers	随机数
Resolution of the converter	转换器分辨率
Small-scale integration(SSI)	小规模集成
Superposition principle	叠加原理
Temperature coefficient	温度系数
Temperature coefficient of resistance (TCR)	电阻温度系数
Temperature dependent SPICE model	温度相关性SPICE模型
Thévenin circuit transformation	戴维南电路变换
Thévenin equivalent circuit	戴维南等效电路
Thévenin equivalent resistance	戴维南等效电阻
Tolerance	容差
Total harmonic distortion	总谐波失真
Transistor	晶体管
Triode	三极管
Ultra-large-scale integration(ULSI)	甚大规模集成
Uniform random number generator	均匀随机数生成器
Vacuum diode	真空二极管
Vacuum tube	真空管
Very-large-scale integration(VLSI)	超大规模集成
Virtual ground	虚地
Voltage-controlled current source(VCCS)	压控电流源
Voltage-controlled voltage source(VCVS)	压控电压源
Voltage division	分压
Voltage gain	电压增益
Worst-case analysis	最差情况分析
Worst-case values	最差情况值

CHAPTER 2
SOLID-STATE ELECTRONICS

第2章 固态电子学

本章提纲

2.1 固态电子材料

2.2 共价键模型

2.3 半导体中的漂移电流和迁移率

2.4 本征硅的电阻率

2.5 半导体中的杂质

2.6 掺杂半导体中的电子和空穴浓度

2.7 掺杂半导体中的迁移率和电阻率

2.8 扩散电流

2.9 总电流

2.10 能带模型

2.11 集成电路制造综述

本章目标

- 了解半导体特性，探索通过控制半导体特性来制造电子器件的方法；
- 了解半导体及导体的电阻率、绝缘特性；
- 了解半导体的共价键模型和能带模型；
- 理解带隙能量和本征载流子浓度的概念；
- 理解半导体中的两种带电载流子——电子和空穴；
- 理解半导体中的受主杂质和施主杂质；
- 掌握杂质掺杂法以控制电子数量和空穴数量；
- 理解半导体中的漂移电流和扩散电流的含义；
- 理解弱电场下的迁移率和速度饱和概念；
- 理解迁移率与掺杂水平的无关性；
- 了解集成电路的主要制造工艺。

本章导读

本章主要研究了固态电子学相关的半导体材料以及半导体材料的电学特性——半导体电路设计的基础。本章从器件物理所需的固态电子材料以及相关基本概念入手，深入浅出地讲解了半导体特性、半导体模型、半导体掺杂、迁移率以及速度饱和等概念，并简要介绍了集成电路的制造工艺。

本章首先给出了物质的三种存在形式，即非结晶体、多晶体和单晶体，并分别介绍了它们的特点。非结晶体的原子排列是随机的，无规则的；多晶材料由小晶体组成，晶体材料在整个宏观晶体中呈现出原子间高度规则的导电结构。与之相应的是三种类型的电子材料：绝缘体、导体和半导体。绝缘体的电阻率大于$10^5\ \Omega \cdot cm$，导体的电阻率低于$10^{-3}\ \Omega \cdot cm$，半导体的电阻率介于绝缘体和半导体之间。硅（Si）是现代工业中最重要的半导体材料，主要用于超大规模集成电路（VLSI）的制造。复合半导体材料一般由砷化镓和磷化铟两种元素组成，是光电应用中最重要的材料，如发光二极管、激光器和光电探测器等。

接着，本章从半导体材料的性能着手，指出半导体的实用性主要来自其晶体的周期性，并引入了半导体中两个重要的模型概念：共价键模型和能带模型。这两个模型受温度影响很大，当温度低至接近0K时，半导体中的所有共价键都是完好的，半导体不导电。温度升高时，半导体吸收能量，使得一部分共价键断开。断开共价键所需能量称为带隙能量E_G。共价键断开时，产生两个带电载流子：电子和空穴。电子带电量为$-q$，可以在导带自由移动；空穴带电量为$+q$，可以在价带自由移动。

对于半导体材料，本章进一步给出了纯净物质及半导体材料的定义。纯净物质指的是本征材料，电子密度n和空穴密度p均等于本征载流子密度n_i，室温时硅的n_i约为$10^{10}/cm^3$。用杂质原子替换少量晶体中的电子，就可以大大改变空穴电子浓度。硅是第Ⅴ列元素，最外层有四个电子，与最邻近的四个原子形成共价键。而杂质元素的原子（元素周期表中Ⅲ族或Ⅴ族的元素）最外层电子数为三个或五个。第Ⅴ列元素如P、As、Sb最外层电子数比硅多一个，在硅中起到施主元素的作用，可以直接增加导电带电子。第Ⅲ列元素，例如B，最外层只有三个电子，在价带上产生一个自由空穴。

对于掺杂的浓度进一步分为施主和受主杂质的浓度，分别用N_D和N_A表示。如果n大于p，则称半导体是n型半导体，电子是多数载流子，空穴是少数载流子。如果p大于n，则称半导体是p型半导体，此时空穴是多数载流子，电子是少数载流子。

本章进一步从电子和空穴的移动概念引申出扩散电流及扩散系数相关概念。电子和空穴移动形成的电流分别称为电子电流和空穴电流，二者均由漂移电流和扩散电流两部分组成。漂移电流是载流子在外加电场下运动产生的；漂移电流与电子迁移率和空穴迁移率（分别用μ_n和μ_p表示）成正比。扩散电流产生的原因是存在电子或空穴浓度梯度，其大小正比于电子和空穴的扩散系数（分别用D_n和D_p表示）。扩散系数和迁移率的关系可以用爱因斯坦关系式$D/\mu=kT/q$表示，掺杂半导体会破坏晶体晶格的周期性，并且随着杂质掺杂浓度的增加，迁移率和扩散率都会单调降低。

本章最后给出了复杂固态器件的制造工艺，包括氧化、光刻、刻蚀、离子注入、扩散、蒸发、溅射、化学气相沉积（CVD）和外延层生长等。制造高性能固态器件和高集成度电路的核心是通过掺入杂质改变导电类型，以及控制空穴和电子的浓度。制造集成电路，需要在硅表面有选择性地形成局部n型区和p型区。注入扩散过程中，用二氧化硅、氮化硅、多晶硅、光刻胶和其他材料覆盖晶圆表面的某些区域，防止杂质原子污染原材料。掩模板是利用计算机辅助设计和光摄影递减技术制作出来的，其上有保护层上所开窗口区域的图形。使用高分辨率的光刻技术，可将掩模板上的图形转移到晶圆表面。

Chapter Outline

2.1 Solid-State Electronic Materials
2.2 Covalent Bond Model
2.3 Drift Currents and Mobility in Semiconductors
2.4 Resistivity of Intrinsic Silicon
2.5 Impurities in Semiconductors
2.6 Electron and Hole Concentrations in Doped Semiconductors
2.7 Mobility and Resistivity in Doped Semiconductors
2.8 Diffusion Currents
2.9 Total Current
2.10 Energy Band Model
2.11 Overview of Integrated Circuit Fabrication
Summary
Key Terms
References
Additional Reading
Problems

Chapter Goals

- Explore the characteristics of semiconductors and discover how engineers control semiconductor properties to fabricate electronic devices
- Characterize resistivity and insulators, semiconductors, and conductors
- Develop the covalent bond and energy band models for semiconductors
- Understand the concepts of bandgap energy and intrinsic carrier concentration
- Explore the behavior of the two charge carriers in semiconductors—electrons and holes
- Discuss acceptor and donor impurities in semiconductors
- Learn to control the electron and hole populations using impurity doping
- Understand drift and diffusion currents in semiconductors
- Explore the concepts of low-field mobility and velocity saturation
- Discuss the dependence of mobility on doping level
- Explore basic IC fabrication processes

Jack St. Clair Kilby.
Courtesy of Texas Instruments

The Kilby integrated circuit.
Courtesy of Texas Instruments

Jack Kilby from Texas Instruments Inc. and Gordon Moore and Robert Noyce from Fairchild Semiconductor pioneered the nearly simultaneous development of the integrated circuit in the late 1950s. After years of litigation, the basic integrated circuit patents of Jack Kilby and Texas Instruments were upheld, and also finally recognized in Japan in 1994. Gorden E. Moore, Robert Noyce, and Andrew S. Grove founded the Intel Corporation in 1968. Kilby shared the 2000 Nobel Prize in physics for invention of the integrated circuit.

Andy Grove, Robert Noyce, and Gordon Moore with Intel 8080 processor rubylith in 1978.
Courtesy of Intel Corporation

As discussed in Chapter 1, the evolution of solid-state materials and the subsequent development of the technology for integrated circuit fabrication have revolutionized electronics and made possible the modern information and technological revolution. Using silicon as well as other crystalline semiconductor materials, we can now fabricate integrated circuits (ICs) that have billions of electronic components on a single 20 mm × 20 mm die. Most of us have some familiarity with the very high-speed microprocessor and memory components that form the building blocks for personal computers and workstations. Consider for a moment the content of a 4-Gb memory chip. The memory array alone on this chip will contain more than 4×10^9 transistors and 4×10^9 capacitors—more than 8 billion electronic components on a single die!

Our ability to build such phenomenal electronic system components is based on a detailed understanding of solid-state physics as well as on development of fabrication processes necessary to turn the theory into a manufacturable reality. Integrated circuit manufacturing is an excellent example of a process requiring a broad understanding of many disciplines. IC fabrication requires knowledge of physics, chemistry, electrical engineering, mechanical engineering, materials engineering, and metallurgy, to mention just a few disciplines. The breadth of understanding required is a challenge, but it makes the field of solid-state electronics an extremely exciting and vibrant area of specialization.

It is possible to explore the behavior of electronic circuits from a "black box" perspective, simply trusting a set of equations that model the terminal voltage and current characteristics of each of the electronic devices. However, understanding the underlying behavior of the devices leads a designer to develop an intuition that extends beyond the simplified models of a black box approach. Building our devices and circuits from fundamentals enables us to understand the limitations and appropriate uses of particular devices and circuits. This is especially true when we experimentally observe deviations from our model predictions. One goal of this chapter is to develop a basic understanding of the underlying operational principles of semiconductor devices that enables us to place our simplified models in the appropriate context.

The material in this chapter provides the background necessary for understanding the behavior of the solid-state devices presented in subsequent chapters. We begin our study of solid-state electronics by exploring the characteristics of crystalline materials, with an emphasis on silicon, the most commercially important semiconductor. We look at electrical conductivity and resistivity and discuss the mechanisms of electronic conduction. The technique of impurity doping is discussed, along with its use in controlling conductivity and resistivity type.

2.1 SOLID-STATE ELECTRONIC MATERIALS

Electronic materials generally can be divided into three categories: **insulators, conductors,** and **semiconductors.** The primary parameter used to distinguish among these materials is the **resistivity** ρ, with units of $\Omega \cdot cm$. As indicated in Table 2.1, insulators have resistivities greater than $10^5 \; \Omega \cdot cm$, whereas conductors have resistivities below $10^{-3} \; \Omega \cdot cm$. For example, diamond, one of the highest quality insulators, has a very large resistivity, $10^{16} \; \Omega \cdot cm$. On the other hand, pure copper, a good conductor, has a resistivity of only $3 \times 10^{-6} \; \Omega \cdot cm$. Semiconductors occupy the full range of resistivities between the insulator and conductor boundaries; moreover, the resistivity can be controlled by adding various impurity atoms to the semiconductor crystal.

Elemental semiconductors are formed from a single type of atom (column IV of the periodic table of elements; see Table 2.2), whereas **compound semiconductors** can be formed from combinations of elements from columns III and V or columns II and VI. These later materials are often referred to as III–V (3–5) or II–VI (2–6) compound semiconductors. Table 2.3 presents some of the most useful possibilities. There are also ternary materials such as mercury cadmium telluride, gallium aluminum arsenide, gallium indium arsenide, and gallium indium phosphide.

Historically, germanium was one of the first semiconductors to be used. However, it was rapidly supplanted by silicon, which today is the most important semiconductor material. Silicon has a wider bandgap energy,[1] enabling it to be used in higher-temperature applications than germanium, and oxidation forms a stable insulating oxide on silicon, giving silicon significant processing advantages over germanium during fabrication of ICs. In addition to silicon, gallium arsenide, indium phosphide, silicon carbide, and silicon germanium are commonly encountered today, although germanium is still used in some limited applications. Silicon germanium has emerged as an important material over the last decade or so, and silicon germanium technology has been used to achieve record high frequency performance in silicon-based bipolar transistors.

The compound semiconductor materials gallium arsenide (GaAs) and indium phosphide (InP) are the most important materials for optoelectronic applications, including light-emitting diodes (LEDs), lasers, and photodetectors.

TABLE 2.1
Electrical Classification of Solid Materials

MATERIALS	RESISTIVITY ρ ($\Omega \cdot cm$)
Insulators	$10^5 < \rho$
Semiconductors	$10^{-3} < \rho < 10^5$
Conductors	$\rho < 10^{-3}$

---

[1] The meaning of bandgap energy is discussed in detail in Secs. 2.2 and 2.10.

TABLE 2.2
Portion of the Periodic Table, Including the Most Important Semiconductor Elements (shaded)

IIB	IIIA	IVA	VA	VIA
	5 10.811 **B** Boron	6 12.01115 **C** Carbon	7 14.0067 **N** Nitrogen	8 15.9994 **O** Oxygen
	13 26.9815 **Al** Aluminum	14 28.086 **Si** Silicon	15 30.9738 **P** Phosphorus	16 32.064 **S** Sulfur
30 65.37 **Zn** Zinc	31 69.72 **Ga** Gallium	32 72.59 **Ge** Germanium	33 74.922 **As** Arsenic	34 78.96 **Se** Selenium
48 112.40 **Cd** Cadmium	49 114.82 **In** Indium	50 118.69 **Sn** Tin	51 121.75 **Sb** Antimony	52 127.60 **Te** Tellurium
80 200.59 **Hg** Mercury	81 204.37 **Tl** Thallium	82 207.19 **Pb** Lead	83 208.980 **Bi** Bismuth	84 (210) **Po** Polonium

TABLE 2.3
Semiconductor Materials

SEMICONDUCTOR	BANDGAP ENERGY E_G (eV)
Carbon (diamond)	5.47
Silicon	1.12
Germanium	0.66
Tin	0.082
Gallium arsenide	1.42
Gallium nitride	3.49
Indium phosphide	1.35
Boron nitride	7.50
Silicon carbide	3.26
Silicon germanium	1.10
Cadmium selenide	1.70

Many research laboratories are exploring the formation of diamond, boron nitride, silicon carbide, and silicon germanium materials. Diamond and boron nitride are excellent insulators at room temperature, but they, as well as silicon carbide, can be used as semiconductors at much higher temperatures (600°C). Adding a small percentage (<10 percent) of germanium to silicon has been shown to offer improved device performance in a process compatible with normal silicon processing. Because of its performance advantages, this SiGe technology [1] has rapidly been introduced into fabrication of devices for RF (radio-frequency) applications, particularly for use in the telecommunications marketplace.

EXERCISE: What are the chemical symbols for antimony, arsenic, aluminum, boron, gallium, germanium, indium, phosphorus, and silicon?

ANSWERS: Sb, As, Al, B, Ga, Ge, In, P, Si

2.2 COVALENT BOND MODEL

Atoms can bond together in **amorphous, polycrystalline,** or **single-crystal** forms. Amorphous materials have a disordered structure, whereas polycrystalline material consists of a large number of small crystallites. Most of the useful properties of semiconductors, however, occur in high-purity, single-crystal material. Silicon—column IV in the periodic table—has four **electrons** in the outer shell. Single-crystal material is formed by the covalent bonding of each silicon atom with its four nearest neighbors in a highly regular three-dimensional array of atoms, as shown in Fig. 2.1. Much of the behavior we discuss can be visualized using the simplified two-dimensional **covalent bond model** of Fig. 2.2.

At temperatures approaching absolute zero, all the electrons reside in the covalent bonds shared between the atoms in the array, with no electrons free for conduction. The outer shells of the silicon atoms are full, and the material behaves as an insulator. As the temperature increases, thermal energy

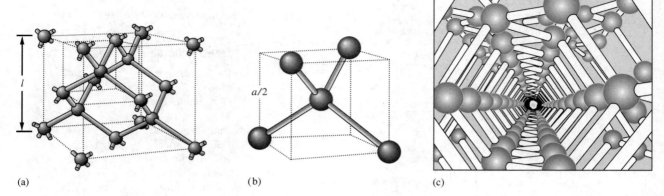

Figure 2.1 Silicon crystal lattice structure. (a) Diamond lattice unit cell. The cube side length $l = 0.543$ nm. (b) Enlarged top corner of the diamond lattice, showing the four nearest neighbors bonding within the structure.
Source: (c) *From S. M. Sze,* Semiconductor Devices: Physics and Technology. *Copyright © 1985 John Wiley & Sons. Adapted.*

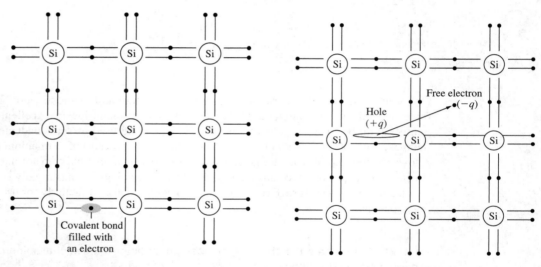

Figure 2.2 Two-dimensional silicon lattice with shared covalent bonds. At temperatures approaching absolute zero, 0 K, all bonds are filled, and the outer shells of the silicon atoms are completely full.

Figure 2.3 An electron–hole pair is generated whenever a covalent bond is broken.

is added to the crystal and some bonds break, freeing a small number of electrons for conduction, as in Fig. 2.3. The density of these free electrons is equal to the **intrinsic carrier density** n_i (cm^{-3}), which is determined by material properties and temperature:

$$n_i^2 = BT^3 \exp\left(-\frac{E_G}{kT}\right) \quad \text{cm}^{-6} \tag{2.1}$$

where E_G = semiconductor bandgap energy in eV (electron volts)

k = Boltzmann's constant, 8.62×10^{-5} eV/K

T = absolute temperature, K

B = material-dependent parameter, 1.08×10^{31} K^{-3} · cm^{-6} for Si

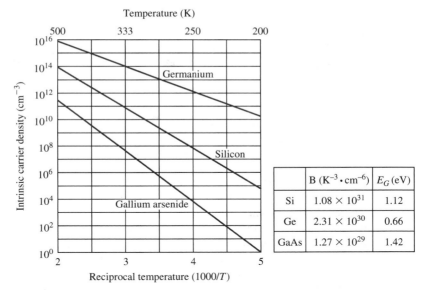

Figure 2.4 Intrinsic carrier density versus temperature from Eq. (2.1).

Bandgap energy E_G is the minimum energy needed to break a covalent bond in the semiconductor crystal, thus freeing electrons for conduction. Table 2.3 lists values of the bandgap energy for various semiconductors.

The *density of conduction (or free) electrons* is represented by the symbol **n** (electrons/cm³), and for **intrinsic material** $n = n_i$. The term *intrinsic* refers to the generic properties of pure material. Although n_i is an intrinsic property of each semiconductor, it is extremely temperature-dependent for all materials. Figure 2.4 has examples of the strong variation of intrinsic carrier density with temperature for germanium, silicon, and gallium arsenide.

EXAMPLE 2.1 **INTRINSIC CARRIER CONCENTRATION**

Calculate the theoretical value of n_i in silicon at room temperature.

PROBLEM Calculate the value of n_i in silicon at room temperature (300 K).

SOLUTION **Known Information and Given Data:** Equation (2.1) defines n_i, B, and k. $E_G = 1.12$ eV from Table 2.3.

Unknowns: Intrinsic carrier concentration n_i.

Approach: Calculate n_i by evaluating Eq. (2.1).

Assumptions: $T = 300$ K at room temperature.

Analysis:

$$n_i^2 = 1.08 \times 10^{31} \ (\text{K}^{-3} \cdot \text{cm}^{-6})(300 \ \text{K})^3 \exp\left[\frac{-1.12 \ \text{eV}}{(8.62 \times 10^{-5} \ \text{eV/K})(300 \ \text{K})}\right]$$

$$n_i^2 = 4.52 \times 10^{19}/\text{cm}^6 \quad \text{or} \quad n_i = 6.73 \times 10^9/\text{cm}^3$$

Check of Results: The desired unknown has been found, and the value agrees with the results graphed in Fig. 2.4.

Discussion: For simplicity, in subsequent calculations we use $n_i = 10^{10}/\text{cm}^3$ as the room temperature value of n_i for silicon. The density of silicon atoms in the crystal lattice is approximately $5 \times 10^{22}/\text{cm}^3$. We see from this example that only one bond in approximately 10^{13} is broken at room temperature.

EXERCISE: Calculate the value of n_i in germanium at a temperature of 300 K.

ANSWER: $2.27 \times 10^{13}/\text{cm}^3$

A second charge carrier is actually formed when the covalent bond in Fig. 2.3 is broken. As an electron, which has charge $-q$ equal to -1.602×10^{-19} C, moves away from the covalent bond, it leaves behind a **vacancy** in the bond structure in the vicinity of its parent silicon atom. The vacancy is left with an effective charge of $+q$. An electron from an adjacent bond can fill this vacancy, creating a new vacancy in another position. This process allows the vacancy to move through the crystal. The moving vacancy behaves just as a particle with charge $+q$ and is called a **hole. Hole density** is represented by the symbol p (holes/cm^3).

As already described, two charged particles are created for each bond that is broken: one electron and one hole. For intrinsic silicon, $n = n_i = p$, and the product of the electron and hole concentrations is

$$pn = n_i^2 \tag{2.2}$$

The **pn product** is given by Eq. (2.2) whenever a semiconductor is in **thermal equilibrium.** (This very important result is used later.) In thermal equilibrium, material properties are dependent only on the temperature T, with no other form of stimulus applied. Equation (2.2) does not apply to semiconductors operating in the presence of an external stimulus such as an applied voltage or current or an optical excitation.

EXERCISE: Calculate the intrinsic carrier density in silicon at 50 K and 325 K. On the average, what is the length of one side of the cube of silicon that is needed to find one electron and one hole at $T = 50$ K?

ANSWERS: $4.34 \times 10^{-39}/\text{cm}^3$; $4.01 \times 10^{10}/\text{cm}^3$; 6.13×10^{10} m

---

2.3 DRIFT CURRENTS AND MOBILITY IN SEMICONDUCTORS

2.3.1 DRIFT CURRENTS

Electrical resistivity ρ and its reciprocal, **conductivity** σ, characterize current flow in a material when an electric field is applied. Charged particles move or *drift* in response to the electric field, and the resulting current is called *drift current*. The **drift current density** j is defined as

$$j = Q\mathbf{v} \quad (\text{C/cm}^3)(\text{cm/s}) = \text{A/cm}^2 \tag{2.3}$$

where j = current density,[2] the charge in coulombs moving through an area of unit cross section

Q = charge density[2], the charge in a unit volume

$\mathbf{v}$ = velocity of charge in an electric field

---

[2] Note that "density" has different meanings based on the context. Current density involves a cross-sectional area, whereas charge density is a volumetric quantity.

In order to find the charge density, we explore the structure of silicon using both the covalent bond model and (later) the energy band model for semiconductors. Next, we relate the velocity of the charge carriers to the applied electric field.

2.3.2 MOBILITY

We know from electromagnetics that charged particles move in response to an applied electric field. This movement is termed **drift,** and the resulting current flow is known as **drift current.** Positive charges drift in the same direction as the electric field, whereas negative charges drift in a direction opposed to the electric field. At low fields carrier drift velocity **v** (cm/s) is proportional to the electric field **E** (V/cm); the constant of proportionality is called the **mobility** μ:

$$\mathbf{v}_n = -\mu_n \mathbf{E} \quad \text{and} \quad \mathbf{v}_p = \mu_p \mathbf{E} \tag{2.4}$$

where $\mathbf{v}_n$ = velocity of electrons (cm/s)
$\mathbf{v}_p$ = velocity of holes (cm/s)
μ_n = **electron mobility,** 1420 cm^2/V · s in intrinsic Si
μ_p = **hole mobility,** 470 cm^2/V · s in intrinsic Si

Conceptually, holes are localized to move through the covalent bond structure, but electrons are free to move about the crystal. Thus, one might expect hole mobility to be less than electron mobility.

> **EXERCISE:** Calculate the velocity of a hole in an electric field of 10 V/cm. What is the electron velocity in an electric field of 1000 V/cm? The voltage across a resistor is 1 V, and the length of the resistor is 2 μm. What is the electric field in the resistor?
>
> **ANSWERS:** 4.70 × 10^3 cm/s; −1.42 × 10^6 cm/s; 5.00 × 10^3 V/cm

2.3.3 VELOCITY SATURATION

From physics, we know that the velocity of carriers cannot increase indefinitely, certainly not beyond the speed of light. In silicon, for example, the linear velocity-field relationship assumed in Eq. (2.4) is valid only for fields below approximately 5000 V/cm or 0.5 V/μm. As the electric field increases above this value, the velocity of both holes and electrons begins to saturate, as indicated in Fig. 2.5. At low fields, the slope of the characteristic represents the mobility, as defined by Eq. (2.4). For fields above approximately 3 × 10^4 V/cm in silicon, carrier velocity approaches the **saturated drift velocity** v_{sat}. For electrons and holes in silicon, v_{sat} is approximately 10^7 cm/s. The velocity saturation phenomenon ultimately places an upper limit on the frequency response of solid-state devices.

> **EXERCISE:** (a) What are the maximum drift velocities for electrons and holes in germanium. What are the low field mobilities? (b) What is the maximum drift velocity for electrons in gallium arsenide? What is the electron mobility in gallium arsenide?
>
> **ANSWERS:** 6 × 10^6 cm/s, 4300 cm^2/V · s, 2100 cm^2/V · s; 2 × 10^7 cm/s, 8500 cm^2/V · s

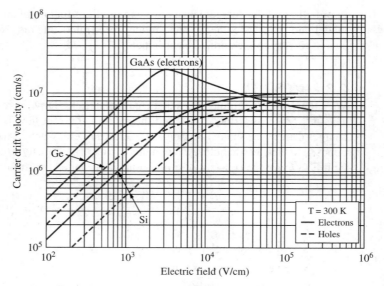

Figure 2.5 Carrier velocity versus electric field in semiconductors at 300 K.
Source: Semiconductor Devices: Physics and Technology by S. M. Sze © 1985.

2.4 RESISTIVITY OF INTRINSIC SILICON

We are now in a position to calculate the electron and hole drift current densities j_n^{drift} and j_p^{drift}. For simplicity, we assume a one-dimensional current and avoid the vector notation of Eqs. (2.3) and (2.4):

$$\begin{aligned} j_n^{\text{drift}} &= Q_n v_n = (-qn)(-\mu_n E) = qn\mu_n E \quad \text{A/cm}^2 \\ j_p^{\text{drift}} &= Q_p v_p = (+qp)(+\mu_p E) = qp\mu_p E \quad \text{A/cm}^2 \end{aligned} \qquad (2.5)$$

in which $Q_n = (-qn)$ and $Q_p = (+qp)$ represent the charge densities (C/cm^3) of electrons and holes, respectively. The total drift current density is then given by

$$j_T^{\text{drift}} = j_n + j_p = q(n\mu_n + p\mu_p)E = \sigma E \qquad (2.6)$$

This equation defines σ, the **electrical conductivity**:

$$\sigma = q(n\mu_n + p\mu_p) \quad (\Omega \cdot \text{cm})^{-1} \qquad (2.7)$$

Resistivity ρ is the reciprocal of conductivity:

$$\rho = \frac{1}{\sigma} \quad (\Omega \cdot \text{cm}) \qquad (2.8)$$

The unit of resistivity, the Ohm-cm, may seem strange to many of us, but from Eq. (2.6), ρ represents the ratio of electric field to drift current density. The resistivity unit is therefore

$$\rho = \frac{E}{j_T^{\text{drift}}} \quad \text{and} \quad \frac{\text{V/cm}}{\text{A/cm}^2} = \Omega \cdot \text{cm} \qquad (2.9)$$

EXAMPLE 2.2   RESISTIVITY OF INTRINSIC SILICON

Here we determine if intrinsic silicon is an insulator, semiconductor, or conductor at room temperature by calculating its resistivity.

PROBLEM Find the resistivity of intrinsic silicon at room temperature and classify it as an insulator, semiconductor, or conductor.

SOLUTION **Known Information and Given Data:** The room temperature mobilities for intrinsic silicon were given right after Eq. (2.4). For intrinsic silicon, the electron and hole densities are both equal to n_i.

Unknowns: Resistivity ρ and classification.

Approach: The conductivity and resistivity can be found using Eqs. (2.7) and (2.8), respectively. The results are then compared to the definitions in Table 2.1.

Assumptions: Temperature is unspecified; assume "room temperature" with $n_i = 10^{10}/\text{cm}^3$.

Analysis: For intrinsic silicon, the charge density of electrons is given by $Q_n = -qn_i$, whereas the charge density for holes is $Q_p = +qn_i$. Substituting the given values into Eq. (2.7) yields

$$\sigma = (1.60 \times 10^{-19})[(10^{10})(1420) + (10^{10})(470)] \quad (\text{C})(\text{cm}^{-3})(\text{cm}^2/\text{V} \cdot \text{s})$$
$$= 3.02 \times 10^{-6} \ (\Omega \cdot \text{cm})^{-1}$$

The resistivity ρ is equal to the reciprocal of the conductivity, so for intrinsic silicon

$$\rho = \frac{1}{\sigma} = 3.31 \times 10^5 \ \Omega \cdot \text{cm}$$

From Table 2.1, we see that intrinsic silicon can be characterized as an insulator, albeit near the low end of the insulator resistivity range.

Check of Results: The resistivity has been found, and intrinsic silicon is a poor insulator.

EXERCISE: Find the resistivity of intrinsic silicon at 400 K and classify it as an insulator, semiconductor, or conductor. Use the mobility values from Ex. 2.2.

ANSWERS: 1420 $\Omega \cdot$ cm, semiconductor

EXERCISE: Calculate the resistivity of intrinsic silicon at 50 K if the electron mobility is 6500 cm^2/V $\cdot$ s and the hole mobility is 2000 cm^2/V $\cdot$ s. Classify the material.

ANSWER: 1.69×10^{53} $\Omega \cdot$ cm, insulator

2.5 IMPURITIES IN SEMICONDUCTORS

The real advantages of semiconductors emerge when **impurities** are added to the material in minute but well-controlled amounts. This process is called **impurity doping**, or just **doping**, and the material that results is termed a **doped semiconductor**. Impurity doping enables us to change the resistivity

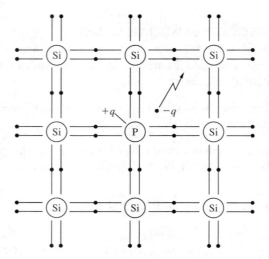

Figure 2.6 An extra electron is available from a phosphorus donor atom.

over a very wide range and to determine whether the electron or hole population controls the resistivity of the material. The following discussion focuses on silicon, although the concepts of impurity doping apply equally well to other materials. The impurities that we use with silicon are from columns III and V of the periodic table.

2.5.1 DONOR IMPURITIES IN SILICON

Donor impurities in silicon are from column V, having five valence electrons in the outer shell. The most commonly used elements are phosphorus, arsenic, and antimony. When a donor atom replaces a silicon atom in the crystal lattice, as shown in Fig. 2.6, four of the five outer shell electrons fill the covalent bond structure; it then takes very little thermal energy to free the extra electron for conduction. At room temperature, essentially every donor atom contributes (donates) an electron for conduction. Each donor atom that becomes ionized by giving up an electron will have a net charge of $+q$ and represents an immobile fixed charge in the crystal lattice.

2.5.2 ACCEPTOR IMPURITIES IN SILICON

Acceptor impurities in silicon are from column III and have one less electron than silicon in the outer shell. The primary acceptor impurity is boron, which is shown in place of a silicon atom in the lattice in Fig. 2.7. Because boron has only three electrons in its shell, a vacancy exists in the bond structure, and it is easy for a nearby electron to move into this vacancy, creating another vacancy in the bond structure. This mobile vacancy represents a hole that can move through the lattice, as illustrated in Fig. 2.7(a) and (b), and the hole may simply be visualized as a particle with a charge of $+q$. Each impurity atom that becomes ionized by accepting an electron has a net charge of $-q$ and is immobile in the lattice, as in Fig. 2.7.

2.6 ELECTRON AND HOLE CONCENTRATIONS IN DOPED SEMICONDUCTORS

We now discover how to calculate the **electron** and **hole concentrations** in a semiconductor containing donor and acceptor impurities. In doped material, the electron and hole concentrations are no longer equal. If $n > p$, the material is called **n-type,** and if $p > n$, the material is referred to as **p-type.** The carrier with the larger population is called the **majority carrier,** and the carrier with the smaller population is termed the **minority carrier.**

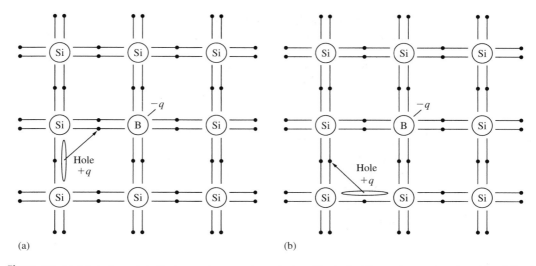

Figure 2.7 (a) A hole is created after boron atom accepts an electron. The ionized boron atom represents an immobilized charge of $-q$. The vacancy in the silicon bond structure represents a mobile hole with charge $+q$. (b) Mobile hole moving through the silicon lattice.

To make detailed calculations of electron and hole densities, we need to keep track of the donor and acceptor impurity concentrations:

N_D = **donor impurity concentration**   atoms/cm^3
N_A = **acceptor impurity concentration**   atoms/cm^3

Two additional pieces of information are needed. First, the semiconductor material must remain charge neutral, which requires that the sum of the total positive charge and negative charge be zero. Ionized donors and holes represent positive charge, whereas ionized acceptors and electrons carry negative charge. Thus **charge neutrality** requires

$$q(N_D + p - N_A - n) = 0 \qquad (2.10)$$

Second, the product of the electron and hole concentrations in intrinsic material was given in Eq. (2.2) as $pn = n_i^2$. It can be shown theoretically that $pn = n_i^2$ even for doped semiconductors in thermal equilibrium, and Eq. (2.2) is valid for a very wide range of doping concentrations.

2.6.1 *n*-TYPE MATERIAL ($N_D > N_A$)

Solving Eq. (2.2) for p and substituting into Eq. (2.10) yields a quadratic equation for n:

$$n^2 - (N_D - N_A)n - n_i^2 = 0$$

Now solving for n,

$$n = \frac{(N_D - N_A) + \sqrt{(N_D - N_A)^2 + 4n_i^2}}{2} \quad \text{and} \quad p = \frac{n_i^2}{n} \qquad (2.11)$$

In practical situations $(N_D - N_A) \gg 2n_i$, and n is given approximately by $n \cong (N_D - N_A)$. The formulas in Eq. (2.11) should be used for $N_D > N_A$.

2.6.2 p-TYPE MATERIAL ($N_A > N_D$)

For the case of $N_A > N_D$, we substitute for n in Eq. (2.10) and use the quadratic formula to solve for p:

$$p = \frac{(N_A - N_D) + \sqrt{(N_A - N_D)^2 + 4n_i^2}}{2} \quad \text{and} \quad n = \frac{n_i^2}{p} \qquad (2.12)$$

Again, the usual case is $(N_A - N_D) \gg 2n_i$, and p is given approximately by $p \cong (N_A - N_D)$. Equation (2.12) should be used for $N_A > N_D$.

Because of practical process-control limitations, impurity densities that can be introduced into the silicon lattice range from approximately 10^{14} to 10^{21} atoms/cm³. Thus, N_A and N_D normally will be much greater than the intrinsic carrier concentration in silicon at room temperature. From the preceding approximate expressions, we see that the majority carrier density is set directly by the net impurity concentration: $p \cong (N_A - N_D)$ for $N_A > N_D$ or $n \cong (N_D - N_A)$ for $N_D > N_A$.

DESIGN NOTE   PRACTICAL DOPING LEVELS

In both n- and p-type semiconductors, the majority carrier concentrations are established "at the factory" by the engineer's choice of N_A and N_D and are independent of temperature over a wide range. In contrast, the minority carrier concentrations, although small, are proportional to n_i^2 and highly temperature dependent. For practical doping levels,

$$\text{For } n\text{-type } (N_D > N_A): \quad n \cong N_D - N_A \quad p = \frac{n_i^2}{N_D - N_A}$$

$$\text{For } p\text{-type } (N_A > N_D): \quad p \cong N_A - N_D \quad n = \frac{n_i^2}{N_A - N_D}$$

Typical values of doping fall in this range:

$$10^{14}/\text{cm}^3 \leq |N_A - N_D| \leq 10^{21}/\text{cm}^3$$

EXAMPLE 2.3   ELECTRON AND HOLE CONCENTRATIONS

Calculate the electron and hole concentrations in a silicon sample containing both acceptor and donor impurities.

PROBLEM Find the type and electron and hole concentrations in a silicon sample at room temperature if it is doped with a boron concentration of $10^{16}/\text{cm}^3$ and a phosphorus concentration of $2 \times 10^{15}/\text{cm}^3$.

SOLUTION **Known Information and Given Data:** Boron and phosphorus doping concentrations and room temperature operation are specified.

Unknowns: Electron and hole concentrations (n and p).

Approach: Identify the donor and acceptor impurity concentrations and use their values to find n and p with Eq. (2.11) or Eq. (2.12), as appropriate.

Assumptions: At room temperature, $n_i = 10^{10}/\text{cm}^3$.

Analysis: Using Table 2.2 we find that boron is an acceptor impurity and phosphorus is a donor impurity. Therefore

$$N_A = 10^{16}/\text{cm}^3 \quad \text{and} \quad N_D = 2 \times 10^{15}/\text{cm}^3$$

Since $N_A > N_D$, the material is p-type, and we have $(N_A - N_D) = 8 \times 10^{15}/\text{cm}^3$. For $n_i = 10^{10}/\text{cm}^3$, $(N_A - N_D) \gg 2n_i$, and we can use the simplified form of Eq. (2.12):

$$p \cong (N_A - N_D) = 8.00 \times 10^{15} \text{ holes/cm}^3$$

$$n = \frac{n_i^2}{p} = \frac{10^{20}/\text{cm}^6}{8.00 \times 10^{15}/\text{cm}^3} = 1.25 \times 10^4 \text{ electrons/cm}^3$$

Check of Results: We have found the electron and hole concentrations. We can double check the pn product: $pn = 10^{20}/\text{cm}^6$, which is correct.

EXERCISE: Find the type and electron and hole concentrations in a silicon sample at a temperature of 400 K if it is doped with a boron concentration of $10^{16}/\text{cm}^3$ and a phosphorus concentration of $2 \times 10^{15}/\text{cm}^3$.

ANSWERS: $8.00 \times 10^{15}/\text{cm}^3$; $6.75 \times 10^8/\text{cm}^3$; n-type.

EXERCISE: Silicon is doped with an antimony concentration of $2 \times 10^{16}/\text{cm}^3$. Is antimony a donor or acceptor impurity? Find the electron and hole concentrations at 300 K. Is this material n- or p-type?

ANSWERS: Donor; $2 \times 10^{16}/\text{cm}^3$; $5 \times 10^3/\text{cm}^3$; n-type.

One might ask why we care about the minority carriers if they are so small in number. Indeed, we find shortly that semiconductor resistivity is controlled by the majority carrier concentration, and in Chapter 4 we find that field-effect transistors (FETs) are also majority carrier devices. However, the characteristics of diodes and bipolar junction transistors, discussed in Chapters 3 and 5, respectively, depend strongly on the minority carrier populations. Thus, to be able to design a variety of solid-state devices, we must understand how to manipulate both the majority and minority carrier concentrations.

2.7 MOBILITY AND RESISTIVITY IN DOPED SEMICONDUCTORS

The introduction of impurities into a semiconductor such as silicon actually degrades the mobility of the carriers in the material. Impurity atoms have slightly different sizes than the silicon atoms that they replace and hence disrupt the periodicity of the lattice. In addition, the impurity atoms are ionized and represent regions of localized charge that were not present in the original crystal. Both these effects cause the electrons and holes to scatter as they move through the semiconductor and reduce the mobility of the carriers in the crystal.

Figure 2.8 shows the dependence of mobility on the *total* impurity doping density $N_T = (N_A + N_D)$ in silicon. We see that mobility drops rapidly as the doping level in the crystal increases. Mobility in heavily doped material can be more than an order of magnitude less than that in lightly doped material. On the other hand, doping vastly increases the density of majority carriers in the semiconductor material and thus has a dramatic effect on resistivity that overcomes the influence of decreased mobility.

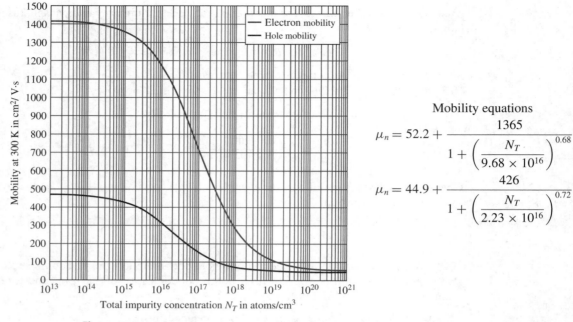

Figure 2.8 Dependence of electron and hole mobility on total impurity concentration in silicon at 300 K.

> **EXERCISE:** What are the electron and hole mobilities in a silicon sample with an acceptor impurity density of $10^{16}/cm^3$? Repeat for a donor density of $3 \times 10^{17}/cm^3$.
>
> **ANSWERS:** 1180 cm²/V·s; 318 cm²/V·s; 484 cm²/V·s; 102 cm²/V·s
>
> **EXERCISE:** What are the electron and hole mobilities in a silicon sample with an acceptor impurity density of $4 \times 10^{16}/cm^3$ and a donor impurity density of $6 \times 10^{16}/cm^3$?
>
> **ANSWERS:** 727 cm²/V·s, 153 cm²/V·s

Remember that impurity doping also determines whether the material is *n*- or *p*-type, and simplified expressions can be used to calculate the conductivity of most extrinsic material. Note that $\mu_n n \gg \mu_p p$ in the expression for σ in Ex. 2.4. For doping levels normally encountered, this inequality will be true for *n*-type material, and $\mu_p p \gg \mu_n n$ will be valid for *p*-type material. The majority carrier concentration controls the conductivity of the material so that

$$\sigma \cong q\mu_n n \cong q\mu_n(N_D - N_A) \qquad \text{for } n\text{-type material}$$
$$\sigma \cong q\mu_p p \cong q\mu_p(N_A - N_D) \qquad \text{for } p\text{-type material}$$
(2.13)

We now explore the relationship between doping and resistivity with an example.

EXAMPLE 2.4 RESISTIVITY CALCULATION OF DOPED SILICON

This example contrasts the resistivity of doped silicon to that of pure silicon.

PROBLEM Calculate the resistivity of silicon doped with a donor density $N_D = 4 \times 10^{15}/cm^3$. What is the material type? Classify the sample as an insulator, semiconductor, or conductor.

SOLUTION **Known Information and Given Data:** $N_D = 4 \times 10^{15}/\text{cm}^3$.

Unknowns: Resistivity ρ, which also requires us to find the hole and electron concentrations (p and n) and mobilities (μ_p and μ_n); material type.

Approach: Use the doping concentration to find n and p and μ_n and μ_p; substitute these values into the expression for σ.

Assumptions: Since N_A is not mentioned, assume $N_A = 0$. Assume room temperature with $n_i = 10^{10}/\text{cm}^3$.

Analysis: In this case, $N_D > N_A$ and much much greater than n_i, so

$$n = N_D = 4 \times 10^{15} \text{ electrons/cm}^3$$

$$p = \frac{n_i^2}{n} = 10^{20}/4 \times 10^{15} = 2.5 \times 10^4 \text{ holes/cm}^3$$

Because $n > p$, the silicon is n-type material. From the equations in Fig. 2.8, the electron and hole mobilities for an impurity concentration of $2 \times 10^{15}/\text{cm}^3$ are

$$\mu_n = 1280 \text{ cm}^2/\text{V} \cdot \text{s} \qquad \mu_p = 375 \text{ cm}^2/\text{V} \cdot \text{s}$$

The conductivity and resistivity are now found to be

$$\sigma = 1.6 \times 10^{19}[(1280)(4 \times 10^{15}) + (375)(2.5 \times 10^4)] = 0.817 \ (\Omega \cdot \text{cm})^{-1}$$

and

$$\rho = 1/\sigma = 1.22 \ \Omega \cdot \text{cm}$$

This silicon sample is a semiconductor.

Check of Results: We have found the required unknowns.

Discussion: Comparing these results to those for intrinsic silicon, we note that the introduction of a minute fraction of impurities into the silicon lattice has changed the resistivity by 5 orders of magnitude, changing the material in fact from an insulator to a midrange semiconductor. Based upon this observation, it is not unreasonable to assume that additional doping can change silicon into a conductor (see the exercise following Ex. 2.5). Note that the doping level in this example represents a replacement of less than 10^{-5} percent of the atoms in the silicon crystal.

EXAMPLE 2.5 **WAFER DOPING—AN ITERATIVE CALCULATION**

Solutions to many engineering problems require iterative calculations as well as the integration of mathematical and graphical information.

PROBLEM An n-type silicon wafer has a resistivity of $0.025 \ \Omega \cdot \text{cm}$. What is the donor concentration N_D?

SOLUTION **Known Information and Given Data:** The wafer is n-type silicon; resistivity is $0.025 \ \Omega \cdot \text{cm}$.

Unknowns: Doping concentration N_D required to achieve the desired resistivity.

Approach: For this problem, an iterative trial-and-error solution is necessary. Because the resistivity is low, it should be safe to assume that

$$\sigma = q\mu_n n = q\mu_n N_D \qquad \text{and} \qquad \mu_n N_D = \frac{\sigma}{q}$$

We know that μ_n is a function of the doping concentration N_D, but the functional dependence may be available only in graphical form. This is an example of a type of problem often encountered in engineering. The solution requires an iterative trial-and-error approach. To solve the problem, we need to establish a logical progression of steps in which the choice of one parameter enables us to evaluate other parameters that lead to the solution. One method for this problem is

1. Choose a value of N_D.
2. Calculate electron mobility μ_n using the equations in Fig. 2.8.
3. Calculate $\mu_n N_D$.
4. If $\mu_n N_D$ is not correct, go back to step 1.

Obviously, we hope we can make educated choices that will lead to convergence of the process after a few trials.

Assumptions: Assume the wafer contains only donor impurities.

Analysis: For this problem,

$$\frac{\sigma}{q} = (0.025 \times 1.60 \times 10^{-19})^{-1} = 2.50 \times 10^{20} \ (\text{V} \cdot \text{s} \cdot \text{cm})^{-1}$$

Choosing a first guess of $N_D = 1 \times 10^{18}/\text{cm}^3$:

TRIAL	N_D (cm^{-3})	μ_n (cm^2/V · s)	$\mu_n N_D$ (V · s · cm)$^{-1}$
1	1.00E + 18	2.84E + 02	2.84E + 20
2	8.81E + 17	3.01E + 02	2.65E + 20
3	8.31E + 17	3.09E + 02	2.57E + 20
4	8.09E + 17	3.13E + 02	2.53E + 20
5	7.99E + 17	3.15E + 02	2.51E + 20
6	7.95E + 17	3.15E + 02	2.51E + 20
7	7.92E + 17	3.16E + 02	2.50E + 20

After six iterations, we find $N_D = 7.92 \times 10^{17}$ donor atoms/cm^3.

Check of Results: We have found the only unknown. $N_D = 7.92 \times 10^{17}/\text{cm}^3$ is in the range of practically achievable doping. See the Design Note in Sec. 2.6. Double checking the results. For $N_D = 7.92 \times 10^{17}/\text{cm}^3$, the calculated mobility is 316 cm^2/V·s, in agreement with our iterative analysis.

EXERCISE: What is the minimum value of donor doping required to convert silicon to a conductor at room temperature? What is the resistivity?

ANSWER: $9.68 \times 10^{19}/\text{cm}^3$ with $\mu_n = 64.5$ cm^2/V · s, 0.001 Ω · cm

EXERCISE: Silicon is doped with a phosphorus concentration of $2 \times 10^{16}/\text{cm}^3$. What are N_A and N_D? What are the electron and hole mobilities? What are the mobilities if boron in a concentration of $3 \times 10^{16}/\text{cm}^3$ is added to the silicon? What are the resistivities?

ANSWERS: $N_A = 0/\text{cm}^3$; $N_D = 2 \times 10^{16}/\text{cm}^3$; $\mu_n = 1070$ cm^2/V · s, $\mu_p = 266$ cm^2/V · s; $\mu_n = 886$ cm^2/V · s; $\mu_p = 198$ cm^2/V · s; 0.292 Ω · cm; 3.16 Ω · cm

EXERCISE: Silicon is doped with a boron concentration of $4 \times 10^{18}/cm^3$. Is boron a donor or acceptor impurity? Find the electron and hole concentrations at 300 K. Is this material *n*-type or *p*-type? Find the electron and hole mobilities. What is the resistivity of the material?

ANSWERS: Acceptor; $n = 25/cm^3$, $p = 4 \times 10^{18}/cm^3$; *p*-type; $\mu_n = 153$ cm²/V·s and $\mu_p = 54.8$ cm²/V·s; 0.0285 Ω·cm

EXERCISE: Silicon is doped with an indium concentration of $7 \times 10^{19}/cm^3$. Is indium a donor or acceptor impurity? Find the electron and hole concentrations, the electron and hole mobilities, and the resistivity of this silicon material at 300 K. Is this material *n*- or *p*-type?

ANSWERS: Acceptor; $n = 1.4/cm^3$, $p = 7 \times 10^{19}/cm^3$; $\mu_n = 67.5$ cm²/V·s and $\mu_p = 46.2$ cm²/V·s; $\rho = 0.00193$ Ω·cm; *p*-type

2.8 DIFFUSION CURRENTS

As already described, the electron and hole populations in a semiconductor in thermal equilibrium are controlled by the impurity doping concentrations N_A and N_D. Up to this point we have tacitly assumed that the doping is uniform in the semiconductor, but this need not be the case. Changes in doping are encountered often in semiconductors, and there will be gradients in the electron and hole concentrations. Gradients in these free carrier densities give rise to a second current flow mechanism, called **diffusion**. The free carriers tend to move (diffuse) from regions of high concentration to regions of low concentration in much the same way as a puff of smoke in one corner of a room rapidly spreads throughout the entire room.

A simple one-dimensional gradient in the electron or hole density is shown in Fig. 2.9. The gradient in this figure is positive in the $+x$ direction, but the carriers diffuse in the $-x$ direction, from high to low concentration. Thus the **diffusion current** densities are proportional to the negative of the carrier gradient:

$$j_p^{\text{diff}} = (+q)D_p\left(-\frac{\partial p}{\partial x}\right) = -qD_p\frac{\partial p}{\partial x}$$
$$j_n^{\text{diff}} = (-q)D_n\left(-\frac{\partial n}{\partial x}\right) = +qD_n\frac{\partial n}{\partial x}$$

A/cm² (2.14)

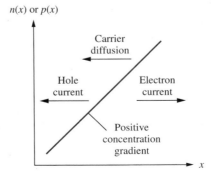

Figure 2.9 Carrier diffusion in the presence of a concentration gradient.

The proportionality constants D_p and D_n are the **hole** and **electron diffusivities,** with units cm²/s. Diffusivity and mobility are related by **Einstein's relationship:**

$$\frac{D_n}{\mu_n} = \frac{kT}{q} = \frac{D_p}{\mu_p} \qquad (2.15)$$

The quantity $(kT/q = V_T)$ is called the **thermal voltage V_T**, and its value is approximately 0.026 V at room temperature. We encounter the parameter V_T in several different contexts throughout this book. Typical values of the diffusivities (also referred to as the **diffusion coefficients**) in silicon are in the range 2 to 35 cm²/s for electrons and 1 to 15 cm²/s for holes at room temperature.

> **EXERCISE:** Calculate the value of the thermal voltage V_T for $T = 50$ K, 77 K, 300 K, and 400 K.
>
> **ANSWERS:** 4.3 mV; 6.63 mV; 25.8 mV; 34.5 mV

DESIGN NOTE **THERMAL VOLTAGE V_T**

$$V_T = kT/q = 0.0258 \text{ V at 300 K}$$

> **EXERCISE:** What are the maximum values of the room temperature values (300 K) of the diffusion coefficients for electrons and holes in silicon based on the mobilities in Fig. 2.8?
>
> **ANSWERS:** Using $V_T = 25.8$ mV; 36.6 cm²/s, 12.1 cm²/s
>
> **EXERCISE:** An electron gradient of $+10^{16}$/(cm³ · μm) exists in a semiconductor. What is the diffusion current density at room temperature if the electron diffusivity is 20 cm²/s? Repeat for a hole gradient of $+10^{20}$/cm⁴ with $D_p = 4$ cm²/s.
>
> **ANSWERS:** +320 A/cm²; −64 A/cm²

2.9 TOTAL CURRENT

Generally, currents in a semiconductor have both drift and diffusion components. The total electron and hole current densities j_n^T and j_p^T can be found by adding the corresponding drift and diffusion components from Eqs. (2.5) and (2.14):

$$j_n^T = q\mu_n n E + q D_n \frac{\partial n}{\partial x} \quad \text{and} \quad j_p^T = q\mu_p p E - q D_p \frac{\partial p}{\partial x} \qquad (2.16)$$

Using Einstein's relationship from Eq. (2.15), Eq. (2.16) can be rewritten as

$$j_n^T = q\mu_n n \left(E + V_T \frac{1}{n} \frac{\partial n}{\partial x} \right) \quad \text{and} \quad j_p^T = q\mu_p p \left(E - V_T \frac{1}{p} \frac{\partial p}{\partial x} \right) \qquad (2.17)$$

Equation (2.16) or (2.17) combined with Gauss' law

$$\nabla \cdot (\varepsilon E) = Q \quad (2.18)$$

where ε permittivity (F/cm), E = electric field (V/cm), and Q = charge density (C/cm^3) gives us a powerful mathematics approach for analyzing the behavior of semiconductors and forms the basis for many of the results presented in later chapters.

2.10 ENERGY BAND MODEL

This section discusses the **energy band model** for a semiconductor, which provides a useful alternative view of the electron–hole creation process and the control of carrier concentrations by impurities. Quantum mechanics predicts that the highly regular crystalline structure of a semiconductor produces periodic quantized ranges of allowed and disallowed energy states for the electrons surrounding the atoms in the crystal. Figure 2.10 is a conceptual picture of this band structure in the semiconductor, in which the regions labeled **conduction band** and **valence band** represent allowed energy states for electrons. Energy E_V corresponds to the top edge of the valence band and represents the highest permissible energy for a valence electron. Energy E_C corresponds to the bottom edge of the conduction band and represents the lowest available energy level in the conduction band. Although these bands are shown as continuums in Fig. 2.10, they actually consist of a very large number of closely spaced, discrete energy levels.

Electrons are not permitted to assume values of energy lying between E_C and E_V. The difference between E_C and E_V is called the *bandgap energy* E_G:

$$E_G = E_C - E_V \quad (2.19)$$

Table 2.3 listed examples of the bandgap energy for a number of semiconductors.

2.10.1 ELECTRON—HOLE PAIR GENERATION IN AN INTRINSIC SEMICONDUCTOR

In silicon at very low temperatures (≈ 0 K), the valence band states are completely filled with electrons, and the conduction band states are completely empty, as shown in Fig. 2.11. The semiconductor

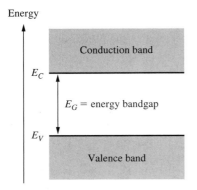

Figure 2.10 Energy band model for a semiconductor with bandgap E_G.

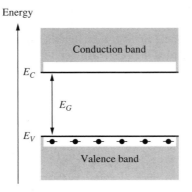

Figure 2.11 Semiconductor at 0 K with filled valence band and empty conduction band. This figure corresponds to the bond model in Fig. 2.2.

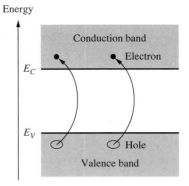

Figure 2.12 Creation of electron–hole pairs by thermal excitation across the energy bandgap. This figure corresponds to the bond model of Fig. 2.3.

in this situation does not conduct current when an electric field is applied. There are no free electrons in the conduction band, and no holes exist in the completely filled valence band to support current flow. The band model of Fig. 2.11 corresponds directly to the completely filled bond model of Fig. 2.2.

As temperature rises above 0 K, thermal energy is added to the crystal. A few electrons gain the energy required to surmount the energy bandgap and jump from the valence band into the conduction band, as shown in Fig. 2.12. Each electron that jumps the bandgap creates an electron–hole pair. This **electron–hole pair generation** situation corresponds directly to that presented in Fig. 2.3.

2.10.2 ENERGY BAND MODEL FOR A DOPED SEMICONDUCTOR

Figures 2.13 to 2.15 present the band model for **extrinsic material** containing donor and/or acceptor atoms. In Fig. 2.13, a concentration N_D of donor atoms has been added to the semiconductor. The donor atoms introduce new localized energy levels within the bandgap at a **donor energy level E_D** near the conduction band edge. The value of $(E_C - E_D)$ for phosphorus is approximately 0.045 eV, so it takes very little thermal energy to promote the extra electrons from the donor sites into the conduction band. The density of conduction-band states is so high that the probability of finding an electron in a donor state is practically zero, except for heavily doped material (large N_D) or at very low temperature. Thus at room temperature, essentially all the available donor electrons are free for conduction. Figure 2.13 corresponds to the bond model of Fig. 2.6.

In Fig. 2.14, a concentration N_A of acceptor atoms has been added to the semiconductor. The acceptor atoms introduce energy levels within the bandgap at the **acceptor energy level E_A** near the valence band edge. The value of $(E_A - E_V)$ for boron is approximately 0.044 eV, and it takes very little thermal energy to promote electrons from the valence band into the acceptor energy levels. At room temperature, essentially all the available acceptor sites are filled, and each promoted electron creates a hole that is free for conduction. Figure 2.14 corresponds to the bond model of Fig. 2.7.

2.10.3 COMPENSATED SEMICONDUCTORS

The situation for a **compensated semiconductor,** one containing both acceptor and donor impurities, is depicted in Fig. 2.15 for the case in which there are more donor atoms than acceptor atoms. Electrons seek the lowest energy states available, and they fall from donor sites, filling all the available acceptor sites. The remaining free electron population is given by $n = (N_D - N_A)$.

The energy band model just discussed represents a conceptual model that is complementary to the covalent bond model of Sec. 2.2. Together they help us visualize the processes involved in creating holes and electrons in doped semiconductors.

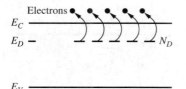

Figure 2.13 Donor level with activation energy $(E_C - E_D)$. This figure corresponds to the bond model of Fig. 2.6.

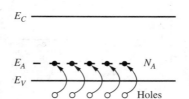

Figure 2.14 Acceptor level with activation energy $(E_A - E_V)$. This figure corresponds to the bond model of Fig. 2.7(b).

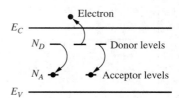

Figure 2.15 Compensated semiconductor containing both donor and acceptor atoms with $N_D > N_A$.

ELECTRONICS IN ACTION

CCD Cameras

Modern astronomy is highly dependent on microelectronics for both the collection and analysis of astronomical data. Tremendous advancements in astronomy have been made possible by the combination of electronic image capture and computer analysis of the acquired images.

In the case of optical telescopes, the Charge-Coupled Device (CCD) camera converts photons to electrical signals that are then formed into a computer image. Like other photo-detector circuits, the CCD captures electrons that are generated when incident photons interact with the semiconductor material and create hole–electron pairs as in Fig. 2.12. A two-dimensional array of millions of CCD cells is formed on a single chip, similar to the one shown below. CCD imagers are especially important to astronomers because of their very high sensitivity and low electronic noise.

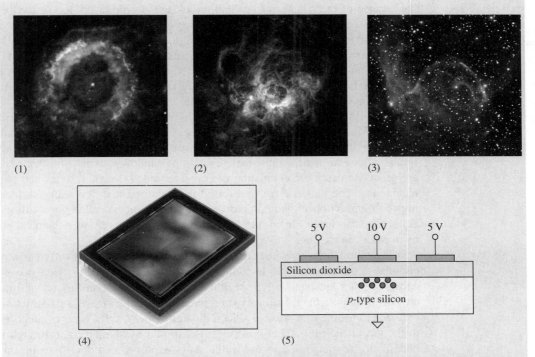

Source: (1) NGC6369: The Little Ghost Nebula. Credit: Hubble Heritage Team, NASA; (2) NGC604: Giant Stellar Nursery. Credit: H. Yang (UIUC), HST, NASA; (3) NGC2359: Thors Helmet. Credits: Christine and David Smith, Steve Mandel, Adam Block (KPNO Visitor Program), NOAO, AURA, NSF. (4) The chip pictured above is a 33 megapixel Dalsa CCD image sensor © *Teledyne DALSA. Reprinted by permission.*

A simplified view of a CCD cell is shown here. A group of electrons have accumulated under the middle electrode due to the higher voltage present. The electrons are held within the semiconductor by the combination of the insulating silicon-dioxide layer and the fields created by the electrodes. The more incident light, the more electrons are captured. To read the charge out of the cell, the electrode voltages are manipulated to move the charge from electrode to electrode until it is converted to a voltage at the edge of the imaging array. The astronomical images were acquired with CCD cameras located on the Hubble Space Telescope.

2.11 OVERVIEW OF INTEGRATED CIRCUIT FABRICATION

Before we leave this chapter, we explore how an engineer uses selective control of semiconductor doping to form a simple electronic device. We do this by studying the basic fabrication steps utilized to fabricate a solid-state diode. These ideas help us understand the characteristics of many electronic devices that depend strongly on the physical structure of the device.

Complex solid-state devices and circuits are fabricated through the repeated application of a number of basic IC processing steps including oxidation, photolithography, etching, ion implantation, diffusion, evaporation, sputtering, chemical vapor deposition, and epitaxial growth. **Silicon dioxide** (SiO_2) layers are formed by heating silicon wafers to a high temperature (1000 to 1200°C) in the presence of pure oxygen or water vapor. This process is called **oxidation.** Thin layers of metal films are deposited through **evaporation** by heating the metal to its melting point in a vacuum. In contrast, both conducting metal films and insulators can be deposited through a process called **sputtering,** which uses physical ion bombardment to effect transfer of atoms from a source target to the wafer surface.

Thin films of polysilicon, silicon dioxide, and **silicon nitride** can all be formed through **chemical vapor deposition** (CVD), in which the material is precipitated from a gaseous mixture directly onto the surface of the silicon wafer. Shallow n- and p-type layers are formed by **ion implantation,** where the wafer is bombarded by high-energy (10-keV to 5-MeV) acceptor or donor impurity atoms generated by a high-voltage particle accelerator. A greater depth of the impurity layers can be achieved by **diffusion** of the impurities at high temperatures, typically 1000 to 1200°C, in either an inert or oxidizing environment. Bipolar processes, as well as some CMOS processes, employ the **epitaxial growth** technique to form thin high-quality layers of crystalline silicon on top of the wafer. The epitaxial layer replicates the crystal structure of the original silicon substrate.

To build integrated circuits, localized n- and p-type regions must be formed selectively in the silicon surface. Silicon dioxide, silicon nitride, **polysilicon, photoresist,** and other materials can all be used to block out areas of the wafer surface to prevent penetration of impurity atoms during implantation and/or diffusion. **Masks** containing window patterns to be opened in the protective layers are produced using a combination of computer-aided design systems and photographic reduction techniques. The patterns are transferred from the mask to the wafer surface through the use of high-resolution optical photographic techniques, a process called **photolithography.** The windows defined by the masks are cut through the protective layers by wet-chemical **etching** using acids or by dry-plasma etching.

The fabrication steps just outlined can be combined in many different ways to form integrated circuits. A simple example is contained in Figs. 2.16 and 2.17. Here we wish to form a diode consisting of a localized p-type region diffused into an n-type silicon substrate. Metallic contacts are needed to both the n- and p-type regions. In Fig. 2.17(a), a 500-μm-thick silicon wafer has been oxidized to form a thin layer of silicon dioxide (1 μm), and a layer of photoresist has been applied to the top of the SiO_2. The photoresist is exposed by shining light through a mask that contains patterns to be transferred to the wafer. After exposure and development, this photoresist (called positive resist) has an opening where it was exposed, as in Fig. 2.17(b). Next, the oxide is etched away using the photoresist as a barrier layer, leaving a window through both the photoresist and oxide layers, as in Fig. 2.17(c). Acceptor impurities are now implanted into the silicon through the window, but are blocked everywhere else by the barrier formed by the photoresist and oxide layers. After photoresist removal, a localized p-type region exists in the silicon below the window in the SiO_2, as in Fig. 2.17(d). The p-type region will typically extend from a few tenths of a micron to at most a few microns below the silicon surface.

Oxide is regrown on the wafer surface and coated with a new layer of photoresist, as indicated in Fig. 2.17(e). Contact windows are exposed through a second mask. The structure in Fig. 2.17(f) results following completion of the photolithography step and subsequent etching of the contact windows in the oxide. Contacts will be made to both the n-type substrate and the p-type region through these openings. Next, photoresist is removed, and an aluminum layer is evaporated onto the silicon wafer

2.11 Overview of Integrated Circuit Fabrication

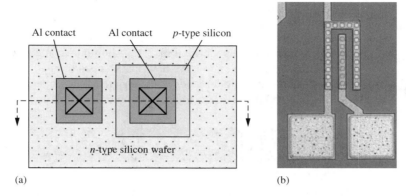

(a) (b)

Figure 2.16 (a) Top view of the *pn* diode structure formed by fabrication steps in Fig. 2.17. (b) Photomicrograph of an actual diode with multiple contacts to the *p*-type region and an *n*-type collar surrounding the *p* diffusion.

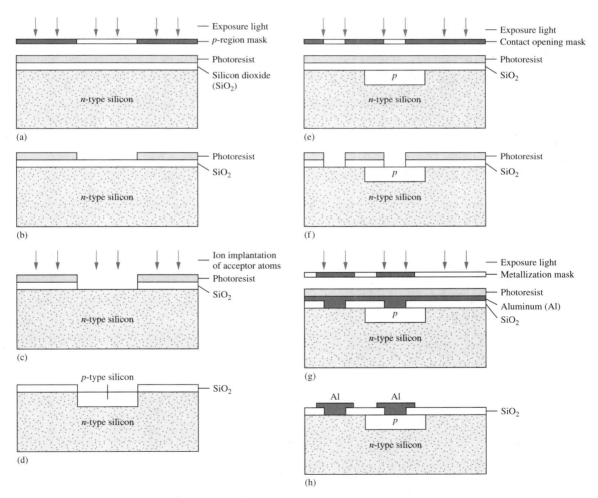

Figure 2.17 Silicon wafer (a) at first mask exposure step, (b) after exposure and development of photoresist, (c) following etching of silicon dioxide, and (d) after implantation/diffusion of acceptor impurity and resist removal. (e) Exposure of contact opening mask (f) after resist development and etching of contact openings. (g) Exposure of metal mask. (h) Final structure after etching of aluminum and resist removal.

which is once again coated with photoresist as in Fig. 2.17(g). A third mask and photolithography step are used to transfer the desired metallization pattern to the wafer surface, and then the aluminum is etched away wherever it is not coated with photoresist. The completed structure appears in Fig. 2.17(h) and corresponds to the top view in Fig. 2.16. Aluminum contacts have been made to both the n-type substrate and the p-type region. We have just stepped through the fabrication of our first solid-state device—a pn junction diode! Study of the characteristics, operation, and application of diodes is the topic of Chapter 3. Figure 2.16(b) is a photomicrograph of an actual diode.

ELECTRONICS IN ACTION

Lab-on-a-chip

The photo below illustrates the integration of silicon microelectronic circuits, microfluidics, and a printed circuit board to realize a nanoliter DNA analysis device. DNA fluid samples are introduced at one end of the device, metered into nanoliter sized droplets, and propelled along a fluidic channel where the sample is mixed with other materials, heated, and optically stimulated. Integrated optical detectors are used to measure the resulting fluorescence for detection of target genetic bio-materials.

Devices such as the one below are revolutionizing health-care by improving our understanding of disease and disease mechanisms, enabling rapid diagnostics and providing for the screening of large numbers of potential treatments in a low-cost fashion. Bioengineering and in particular the application of microelectronics to health-care and life sciences is a rapidly growing and exciting field.

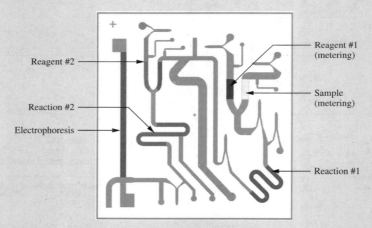

Courtesy of Mark A. Burns, University of Michigan

Courtesy of Mark A. Burns, University of Michigan

SUMMARY

- Materials are found in three primary forms: amorphous, polycrystalline, and crystalline. An amorphous material is a totally disordered or random material that shows no short range order. In polycrystalline material, large numbers of small crystallites can be identified. A crystalline material exhibits a highly regular bonding structure among the atoms over the entire macroscopic crystal.
- Electronic materials can be separated into three classifications based on their electrical resistivity. Insulators have resistivities above 10^5 $\Omega \cdot$ cm, whereas conductors have resistivities below 10^{-3} $\Omega \cdot$ cm. Between these two extremes lie semiconductor materials.
- Today's most important semiconductor is silicon (Si), which is used for fabrication of very-large-scale-integrated (VLSI) circuits. Two compound semiconductor materials, gallium arsenide (GaAs) and indium phosphide (InP), are the most important materials for optoelectronic applications including light-emitting diodes (LEDs), lasers, and photodetectors.
- The highly useful properties of semiconductors arise from the periodic nature of crystalline material, and two conceptual models for these semiconductors were introduced: the covalent bond model and the energy band model.
- At very low temperatures approaching 0 K, all the covalent bonds in a semiconductor crystal will be intact and the material will actually be an insulator. As temperature is raised, the added thermal energy causes a small number of covalent bonds to break. The amount of energy required to break a covalent bond is equal to the bandgap energy E_G.
- When a covalent bond is broken, two charge carriers are produced: an electron, with charge $-q$, that is free to move about the conduction band; and a hole, with charge $+q$, that is free to move through the valence band.
- Pure material is referred to as intrinsic material, and the electron density n and hole density p in an intrinsic material are both equal to the intrinsic carrier density n_i, which is approximately equal to 10^{10} carriers/cm^3 in silicon at room temperature. In a material in thermal equilibrium, the product of the electron and hole concentrations is a constant: $pn = n_i^2$.
- The hole and electron concentrations can be significantly altered by replacing small numbers of atoms in the original crystal with impurity atoms. Silicon, a column IV element, has four electrons in its outer shell and forms covalent bonds with its four nearest neighbors in the crystal. In contrast, the impurity elements (from columns III and V of the periodic table) have either three or five electrons in their outer shells.
- In silicon, column V elements such as phosphorus, arsenic, and antimony, with an extra electron in the outer shell, act as donors and add electrons directly to the conduction band. A column III element such as boron has only three outer shell electrons and creates a free hole in the valence band.
- Donor and acceptor impurity densities are usually represented by N_D and N_A, respectively.
- If n exceeds p, the semiconductor is referred to as n-type material, and electrons are the majority carriers and holes are the minority carriers. If p exceeds n, the semiconductor is referred to as p-type material, and holes become the majority carriers and electrons, the minority carriers.
- Electron and hole currents each have two components: a drift current and a diffusion current.
- Drift current is the result of carrier motion caused by an applied electric field. Drift currents are proportional to the electron and hole mobilities (μ_n and μ_p, respectively).
- Diffusion currents arise from gradients in the electron or hole concentrations. The magnitudes of the diffusion currents are proportional to the electron and hole diffusivities (D_n and D_p, respectively).
- Diffusivity and mobility are related by the Einstein relationship: $D/\mu = kT/q$. Doping the semiconductor disrupts the periodicity of the crystal lattice, and the mobility—and hence diffusivity—both decrease monotonically as the impurity doping concentration is increased.
- The expression kT/q has units of voltage and is often referred to as the thermal voltage V_T. At room temperature, the value of V_T is approximately 26 mV.

- The ability to add impurities to change the conductivity type and to control hole and electron concentrations is at the heart of our ability to fabricate high-performance solid-state devices and high-density integrated circuits. In the next several chapters, we see how this capability is used to form diodes, field-effect transistors (FETs), and bipolar junction transistors (BJTs).
- Complex solid-state devices and circuits are fabricated through the repeated application of a number of basic IC processing steps, including oxidation, photolithography, etching, ion implantation, diffusion, evaporation, sputtering, chemical vapor deposition (CVD), and epitaxial growth.
- To build integrated circuits, localized n- and p-type regions must be formed selectively in the silicon surface. Silicon dioxide, silicon nitride, polysilicon, photoresist, and other materials can all be used to block out areas of the wafer surface to prevent penetration of impurity atoms during implantation and/or diffusion. Masks containing window patterns to be opened in the protective layers are produced using a combination of computer-aided design systems and photographic reduction techniques. The patterns are transferred from the mask to the wafer surface through the use of high-resolution photolithography.

KEY TERMS

Acceptor energy level
Acceptor impurities
Acceptor impurity concentration
Amorphous material
Bandgap energy
Charge neutrality
Chemical vapor deposition
Compensated semiconductor
Compound semiconductor
Conduction band
Conductivity
Conductor
Covalent bond model
Diffusion
Diffusion coefficients
Diffusion current
Donor energy level
Donor impurities
Donor impurity concentration
Doped semiconductor
Doping
Drift current
Einstein's relationship
Electrical conductivity
Electron
Electron concentration
Electron diffusivity
Electron–hole pair generation
Electron mobility
Elemental semiconductor
Energy band model
Epitaxial growth
Etching
Evaporation
Extrinsic material

Hole
Hole concentration
Hole density
Hole diffusivity
Hole mobility
Impurities
Impurity doping
Insulator
Intrinsic carrier density
Intrinsic material
Ion implantation
Majority carrier
Mask
Minority carrier
Mobility
n-type material
Oxidation
p-type material
Photolithography
Photoresist
pn product
Polycrystalline material
Polysilicon
Resistivity
Saturated drift velocity
Semiconductor
Silicon dioxide
Silicon nitride
Single-crystal material
Sputtering
Thermal equilibrium
Thermal voltage
Vacancy
Valence band

REFERENCES

1. J. D. Cressler, "Re-Engineering Silicon: SiGe Heterojunction Bipolar Technology," *IEEE Spectrum,* pp. 49–55, March 1995.

ADDITIONAL READING

Campbell, S. A. *Fabrication Engineering at the Micro- and Nanoscale,* 4th ed. Oxford University Press, New York: 2012.
Jaeger, R. C. *Introduction to Microelectronic Fabrication,* 2d ed. Prentice-Hall, Reading, MA: 2001.
Pierret, R. F. *Semiconductor Fundamentals*, 2d ed. Prentice-Hall, Reading, MA: 1988.
Sze, S. M. and Ng, K. K. *Physics of Semiconductor Devices,* Wiley, New York: 2006.
Yang, E. S. *Microelectronic Devices.* McGraw-Hill, New York: 1988.

PROBLEMS

2.1 Solid-State Electronic Materials

2.1. Pure aluminum has a resistivity of $2.82\ \mu\Omega \cdot$ cm. Based on its resistivity, should aluminum be classified as an insulator, semiconductor, or conductor?

2.2. The resistivity of silicon dioxide is $10^{15}\ \Omega \cdot$ cm. Is this material a conductor, semiconductor, or insulator?

2.3. An aluminum interconnection line in an integrated circuit can be operated with a current density up to 10 MA/cm². If the line is 5 μm wide and 1 μm high, what is the maximum current permitted in the line?

2.2 Covalent Bond Model

2.4. An aluminum interconnection line runs diagonally from one corner of an 18 mm × 18 mm silicon integrated circuit die to the other corner. (a) What is the resistance of this line if it is 1 μm thick and 5 μm wide? (b) Repeat for a 0.5 μm thick line. The resistivity of pure aluminum is $2.82\ \mu\Omega \cdot$ cm.

2.5. Copper interconnections have been introduced into state-of-the-art ICs because of its lower resistivity. Repeat Prob. 2.4 for pure copper with a resistivity of $1.66\ \mu\Omega \cdot$ cm.

2.6. Use Eq. (2.1) to calculate the actual temperature that corresponds to the value $n_i = 10^{10}$/cm³ in silicon.

2.7. Calculate the intrinsic carrier densities in silicon and germanium at (a) 77 K, (b) 300 K, and (c) 450 K. Use the information from the table in Fig. 2.4.

2.8. (a) At what temperature will $n_i = 10^{15}$/cm³ in silicon? (b) If the donor doping is 10^{15}/cm³, what are the electron and hole populations at the temperature in part (a)? (c) What would be the electron and hole populations at room temperature?

2.9. Calculate the intrinsic carrier density in gallium arsenide at (a) 300 K, (b) 100 K, (c) 450 K. Use the information from the table in Fig. 2.4.

2.3 Drift Currents and Mobility in Semiconductors

2.10. Electrons and holes are moving in a uniform, one-dimensional electric field $E = -2000$ V/cm. The electrons and holes have mobilities of 700 and 250 cm²/V · s, respectively. What are the electron and hole velocities? If $n = 10^{17}$/cm³ and $p = 10^3$/cm³, what are the electron and hole current densities?

2.11. The maximum drift velocities of electrons and holes in silicon are approximately 10^7 cm/s. What are the electron and hole current densities if $n = 10^{18}$/cm³ and $p = 10^2$/cm³? What is the total current density? If the sample has a cross section of 1 μm × 25 μm, what is the maximum current?

2.12. The maximum drift velocity of electrons in silicon is 10^7 cm/s. If the silicon has a charge density of 0.4 C/cm³, what is the maximum current density in the material?

2.13. A current density of +2500 A/cm² exists in a semiconductor having a charge density of 0.01 C/cm³. What are the carrier velocities?

2.14. A silicon sample is supporting an electric field of −1500 V/cm, and the mobilities of electrons and holes are 1000 and 400 cm²/V · s, respectively. What are the electron and hole velocities?

If $p = 10^{17}/cm^3$ and $n = 10^3/cm^3$, what are the electron and hole current densities?

2.15. (a) A voltage of 5 V is applied across a 5-μm-long region of silicon. What is the electric field? (b) Suppose the maximum field allowed in silicon is 10^5 V/cm. How large a voltage can be applied to the 10-μm region?

2.4 Resistivity of Intrinsic Silicon

2.16. At what temperature will intrinsic silicon become an insulator, based on the definitions in Table 2.1? Assume that $\mu_n = 1800$ cm²/V · s and $\mu_p = 700$ cm²/V · s.

2.17. At what temperature will intrinsic silicon become a conductor based on the definitions in Table 2.1? Assume that $\mu_n = 120$ cm²/V · s and $\mu_p = 60$ cm²/V · s. (Note that silicon melts at 1430 K.)

2.5 Impurities in Semiconductors

2.18. Draw a two-dimensional conceptual picture (similar to Fig. 2.6) of the silicon lattice containing one donor atom and one acceptor atom in adjacent lattice positions. Are there any free electrons or holes?

2.19. Crystalline germanium has a lattice similar to that of silicon. (a) What are the possible donor atoms in Ge based on Table 2.2? (b) What are the possible acceptor atoms in Ge based on Table 2.2?

2.20. GaAs is composed of equal numbers of atoms of gallium and arsenic in a lattice similar to that of silicon. (a) Suppose a silicon atom replaces a gallium atom in the lattice. Do you expect the silicon atom to behave as a donor or acceptor impurity? Why? (b) Suppose a silicon atom replaces an arsenic atom in the lattice. Do you expect the silicon atom to behave as a donor or acceptor impurity? Why?

2.21. InP is composed of equal atoms of indium and phosphorus in a lattice similar to that of silicon. (a) Suppose a germanium atom replaces an indium atom in the lattice. Do you expect the germanium atom to behave as a donor or acceptor impurity? Why? (b) Suppose a germanium atom replaces a phosphorus atom in the lattice. Do you expect the germanium atom to behave as a donor or acceptor impurity? Explain.

2.22. A current density of 5000 A/cm² exists in a 0.02 Ω · cm n-type silicon sample. What is the electric field needed to support this drift current density?

2.23. Silicon is doped with 10^{16} boron atoms/cm³. How many boron atoms will be in a silicon region that is 180 μm long, 2 μm wide, and 0.5 μm deep?

2.6 Electron and Hole Concentrations in Doped Semiconductors

2.24. Silicon is doped with 7×10^{18} boron atoms/cm³. (a) Is this n- or p-type silicon? (b) What are the hole and electron concentrations at room temperature? (c) What are the hole and electron concentrations at 200 K?

2.25. Silicon is doped with 3×10^{17} arsenic atoms/cm³. (a) Is this n- or p-type silicon? (b) What are the hole and electron concentrations at room temperature? (c) What are the hole and electron concentrations at 250 K?

2.26. Silicon is doped with 3×10^{18} arsenic atoms/cm³ and 8×10^{18} boron atoms/cm³. (a) Is this n- or p-type silicon? (b) What are the hole and electron concentrations at room temperature?

2.27. Silicon is doped with 6×10^{17} boron atoms/cm³ and 2×10^{17} phosphorus atoms/cm³ (a) Is this n- or p-type silicon? (b) What are the hole and electron concentrations at room temperature?

2.28. Suppose a semiconductor has $N_A = 2 \times 10^{17}/cm^3$, $N_D = 3 \times 10^{17}/cm^3$, and $n_i = 10^{17}/cm^3$. What are the electron and hole concentrations?

2.29. Suppose a semiconductor has $N_D = 10^{16}/cm^3$, $N_A = 5 \times 10^{16}/cm^3$, and $n_i = 10^{11}/cm^3$. What are the electron and hole concentrations?

2.7 Mobility and Resistivity in Doped Semiconductors

2.30. Silicon is doped with a donor concentration of $5 \times 10^{16}/cm^3$. Find the electron and hole concentrations, the electron and hole mobilities, and the resistivity of this silicon material at 300 K. Is this material n- or p-type?

2.31. Silicon is doped with an acceptor concentration of $2.5 \times 10^{18}/cm^3$. Find the electron and hole concentrations, the electron and hole mobilities, and the resistivity of this silicon material at 300 K. Is this material n- or p-type?

2.32. Silicon is doped with an indium concentration of $8 \times 10^{19}/cm^3$. Is indium a donor or acceptor impurity? Find the electron and hole concentrations, the electron and hole mobilities, and the resistivity of this silicon material at 300 K. Is this material n- or p-type?

2.33. A silicon wafer is uniformly doped with 4.5×10^{16} phosphorus atoms/cm^3 and 5.5×10^{16} boron atoms/cm^3. Find the electron and hole concentrations, the electron and hole mobilities, and the resistivity of this silicon material at 300 K. Is this material n- or p-type?

2.34. Repeat Ex. 2.5 for p-type silicon. Assume that the silicon contains only acceptor impurities. What is the acceptor concentration N_A?

*2.35. A p-type silicon wafer has a resistivity of $0.5\ \Omega \cdot$ cm. It is known that silicon contains only acceptor impurities. What is the acceptor concentration N_A?

*2.36. It is conceptually possible to produce extrinsic silicon with a higher resistivity than that of intrinsic silicon. How would this occur?

*2.37. n-type silicon wafers with a resistivity of $3.0\ \Omega \cdot$ cm are needed for integrated circuit fabrication. What donor concentration N_D is required in the wafers? Assume $N_A = 0$.

2.38. A silicon sample is doped with 5.0×10^{19} donor atoms/cm^3 and 5.0×10^{19} acceptor atoms/cm^3. (a) What is its resistivity? (b) Is this an insulator, conductor, or semiconductor? (c) Is this intrinsic material? Explain your answers.

2.39. (a) What is the minimum donor doping required to convert silicon into a conductor based on the definitions in Table 2.1? (b) What is the minimum acceptor doping required to convert silicon into a conductor?

*2.40. Measurements of a silicon wafer indicate that it is p-type with a resistivity of $1\ \Omega \cdot$ cm. It is also known that it contains only boron impurities. (a) What additional acceptor concentration must be added to the sample to change its resistivity to $0.25\ \Omega \cdot$ cm? (b) What concentration of donors would have to be added to the original sample to change the resistivity to $0.25\ \Omega \cdot$ cm? Would the resulting material be classified as n- or p-type silicon?

*2.41. A silicon wafer has a background concentration of 1×10^{16} boron atoms/cm^3. (a) Determine the conductivity of the wafer. (b) What concentration of phosphorus atoms must be added to the wafer to make the conductivity equal to $4.5\ (\Omega \cdot \text{cm})^{-1}$?

*2.42. A silicon wafer has a doping concentration of 1×10^{16} phosphorus atoms/cm^3. (a) Determine the conductivity of the wafer. (b) What concentration of boron atoms must be added to the wafer to make the conductivity equal to $5.0\ (\Omega \cdot \text{cm})^{-1}$?

2.8 Diffusion Currents

2.43. Make a table of the values of thermal voltage V_T for $T = 50$ K, 75 K, 100 K, 150 K, 200 K, 250 K, 300 K, 350 K, and 400 K.

2.44. The electron concentration in a region of silicon is shown in Fig. P2.44. (a) If the electron mobility is $350\ \text{cm}^2/\text{V} \cdot \text{s}$ and the width $W_B = 0.25\ \mu$m, determine the electron diffusion current density. Assume room temperature. (b) Plot the electron velocity for $0 \le x \le W_B$.

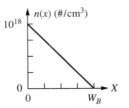

Figure P2.44

2.45. Suppose the hole concentration in silicon sample is described mathematically by

$$p(x) = 10^5 + 10^{19} \exp\left(-\frac{x}{L_p}\right) \text{ holes/cm}^3,\ x \ge 0$$

in which L_p is known as the diffusion length for holes and is equal to 2.0 μm. Find the diffusion current density for holes as a function of distance for $x \ge 0$ if $D_p = 15\ \text{cm}^2/\text{s}$. What is the diffusion current at $x = 0$ if the cross-sectional area is 10 μm^2?

2.9 Total Current

*2.46. A 5-μm-long block of p-type silicon has an acceptor doping profile given by $N_A(x) = 10^{14} + 10^{18} \exp(-10^4 x)$, where x is measured in cm. Use Eq. (2.17) to demonstrate that the material must have a nonzero internal electric field E. What is the value of E at $x = 0$ and $x = 5\ \mu$m? (*Hint:* In thermal equilibrium, the total electron and total hole currents must each be zero.)

2.47. Figure P2.47 gives the electron and hole concentrations in a 2-μm-wide region of silicon. In addition, there is a constant electric field of 25 V/cm present in the sample. What is the total current density at $x = 0$? What are the individual drift and diffusion components of the hole and electron current densities at $x = 1.0\ \mu$m? Assume that the electron and hole mobilities are 350 and 150 cm^2/V $\cdot$ s, respectively.

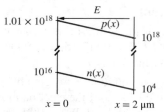

Figure P2.47

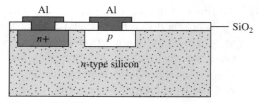

Figure P2.51

2.10 Energy Band Model

2.48. Draw a figure similar to Fig. 2.15 for the case $N_A > N_D$ in which there are two acceptor atoms for each donor atom.

*2.49. Electron–hole pairs can be created by means other than the thermal activation process as described in Figs. 2.3 and 2.12. For example, energy may be added to electrons through optical means by shining light on the sample. If enough optical energy is absorbed, electrons can jump the energy bandgap, creating electron–hole pairs. What is the maximum wavelength of light that we should expect silicon to be able to absorb? (*Hint:* Remember from physics that energy E is related to wavelength λ by $E = hc/\lambda$ in which Planck's constant $h = 6.626 \times 10^{-34}$ J·s and the velocity of light $c = 3 \times 10^{10}$ cm/s.)

2.11 Overview of Integrated Circuit Fabrication

2.50. Draw the cross section for a *pn* diode similar to that in Fig. 2.17(h) if the fabrication process utilizes a *p*-type substrate in place of the *n*-type substrate depicted in Fig. 2.17.

2.51. To ensure that a good ohmic contact is formed between aluminum and *n*-type silicon, an additional doping step is added to the diode in Fig. 2.17(h) to place an *n*+ region beneath the left-hand contact as in Fig. P2.51. Where might this step go in the process flow in Fig. 2.17? Draw a top and side view of a mask that could be used in the process.

Miscellaneous

*2.52. Single crystal silicon consists of three-dimensional arrays of the basic unit cell in Fig. 2.1(a). (a) How many atoms are in each unit cell? (b) What is the volume of the unit cell in cm^3? (c) Show that the atomic density of silicon is 5×10^{22} atoms/cm^3. (d) The density of silicon is 2.33 g/cm^3. What is the mass of one unit cell? (e) Based on your calculations here, what is the mass of a proton? Assume that protons and neutrons have the same mass and that electrons are much much lighter. Is your answer reasonable? Explain.

2.53. Silicon carbide (SiC) comes in several crystalline forms: 3H SiC, 4H SiC, and 6H SiC. (a) Find the bandgap of these three forms of SiC. (b) Find the electron and hole mobilities for these forms of SiC.

2.54. Gallium nitride (GaN) comes in several crystalline forms. Zinc blend has a face-centered cubic lattice, whereas wurtzite has a hexagonal lattice. (a) Find the bandgap of the zinc blend form. (b) Find the bandgap of the wurtzite form. (c) Find the electron and hole mobilities for these forms of GaN.

2.55. Tin (Sn) is in column IV of the periodic table (Table 2.2). Why is it not very useful as a semiconductor material? (*Hint:* Calculate the intrinsic carrier concentration in tin at room temperature.)

2.56. Access the Internet and identify at least four II–VI compound materials that are useful in electronic applications. Identify chemical symbols for the materials and their application areas.

术语对照

Acceptor energy level	受主能级
Acceptor impurities	受主杂质
Acceptor impurity concentration	受主掺杂浓度
Amorphous material	非晶材料
Bandgap energy	带隙能量
Charge neutrality	电中性
Chemical vapor deposition	化学气相沉积
Compensated semiconductor	补偿半导体
Compound semiconductor	化合物半导体
Conduction band	导带
Conductivity	电导率
Conductor	导体
Covalent bond model	共价键模型
Diffusion	扩散
Diffusion coefficients	扩散系数
Diffusion current	扩散电流
Donor energy level	施主能级
Donor impurities	施主杂质
Donor impurity concentration	施主掺杂浓度
Doped semiconductor	掺杂半导体
Doping	掺杂
Drift	漂移
Drift current	漂移电流
Einstein's relationship	爱因斯坦关系式
Electrical conductivity	电导率
Electron	电子
Electron concentration	电子浓度
Electron diffusivity	电子扩散
Electron-hole pair generation	电子-空穴对的产生
Electron mobility	电子迁移率
Elemental semiconductor	元素半导体
Energy band model	能带模型
Epitaxial growth	外延生长
Etching	刻蚀
Evaporation	蒸发
Extrinsic material	非本征半导体材料
Hole	空穴
Hole concentration	空穴浓度
Hole density	空穴密度
Hole diffusivity	空穴扩散

Hole mobility	空穴迁移率
Impurities	杂质
Impurity doping	杂质掺杂
Insulator	绝缘体
Intrinsic carrier density	本征载流子密度
Intrinsic material	本征材料
Ion implantation	离子注入
Majority carrier	多数载流子
Mask	掩模板
Minority carrier	少数载流子
Mobility	迁移率
n-type material	n型材料
Oxidation	氧化
p-type material	p型材料
Photolithography	光刻
Photoresist	光刻胶
pn product	pn结
Polycrystalline material	多晶材料
Polysilicon	多晶硅
Resistivity	电阻率
Saturated drift velocity	饱和漂移速度
Semiconductor	半导体
Silicon dioxide	二氧化硅
Silicon nitride	氮化硅
Single-crystal material	单晶材料
Sputtering	溅射
Thermal equilibrium	热平衡
Thermal voltage	热电压
Vacancy	真空
Valence band	价带

CHAPTER 3
SOLID-STATE DIODES AND DIODE CIRCUITS

第3章　固态二极管和二极管电路

本章提纲

3.1　pn结二极管

3.2　二极管的i-v特性

3.3　二极管方程：二极管的数学模型

3.4　二极管特性之反偏、零偏和正偏

3.5　二极管的温度系数

3.6　反偏下的二极管

3.7　pn结电容

3.8　肖特基势垒二极管

3.9　二极管的SPICE模型及版图

3.10　二极管电路分析

3.11　多二极管电路

3.12　二极管工作在击穿区域的分析

3.13　半波整流电路

3.14　全波整流电路

3.15　全波桥式整流

3.16　整流器的比较和折中设计

3.17　二极管的动态开关行为

3.18　光电二极管、太阳能电池和发光二极管

本章目标

- 理解二极管结构及基本版图；
- 了解pn结的电学特性；
- 熟悉各种二极管模型，包括数学模型、理想模型及恒定压降模型；
- 理解二极管的SPICE描述及二极管的模型参数；
- 熟悉二极管的工作区，包括正向偏置、反向偏置以及反向击穿；
- 能在电路分析中应用不同模型；
- 熟悉不同类型二极管，包括齐纳二极管、变容二极管、肖特基势垒二极管、太阳能电池和发光二极管（LED）；
- 熟悉pn结二极管的动态开关行为；

- 熟悉二极管整流器；
- 熟悉二极管电路SPICE。

本章导读

　　本章主要研究了固态二极管相关的原理、特性、模型及电路应用。首先从pn结入手，介绍pn结的电学特性。pn结二极管是p型半导体和n型半导体的紧密接触形成的。在pn结二极管中，冶金结附近存在较大的浓度梯度，由此产生较大的电子和空穴扩散电流。在零偏条件下，二极管两端以及空间电荷区中均没有电流存在。空间电荷区将产生内建电势和内部电场，而内部电场将导致电子和空穴电流的漂移电流，正好与扩散电流相抵消。

　　接下来，研究了二极管两端施加电压时的i-v特性以及二极管模型方程。在二极管两端施加电压时，二极管内部结区域的平衡遭到破坏，产生电流的传导。根据二极管的i-v特性可以通过二极管方程实现精确建模。根据二极管两端加载电压方向的不同，分为二极管反偏和正偏。反偏条件下二极管电流等于$-I_S$，这个值非常小。正偏条件下，电流可以很大，二极管的压降为0.6~0.7V。室温下，二极管电压每变化60mV可以引起10倍的二极管电流变化；室温下，硅二极管的温度系数为-1.8mV/℃。

　　击穿现象是二极管电路中的常见现象。本章对击穿现象的定义、预防及利用都进行了介绍。如果二极管两端的反向偏压过大，内部电场会使二极管发生击穿，包括齐纳击穿和雪崩击穿。工作在击穿区的二极管具有基本固定的压降，必须严格限制二极管上的电流，否则很容易烧毁器件。工作在击穿区的齐纳二极管可用于稳压器电路设计。电压调整率和负载调整率分别用于表征输出电压随输入电压和输出电流的变化。

　　二极管电容对于二极管的性能具有很大的影响。本章给出了二极管电容产生的原因以及影响。如果二极管两端电压发生变化，则在空间电荷区附近存储的电荷量也会随之变化，这意味着二极管模型中应包含电容。二极管反向偏置时，电容与外加电压的平方根成反比；正向偏置时，电容与工作电流和二极管传输时间成正比。由于这一电容的存在，使得二极管不能立即截止或导通，在截止时会形成一个电荷存储延迟。

　　本章重点给出了二极管电路计算和模拟的主要方法，包括迭代法、负载线分析法、恒压降法、SPICE电路分析法等。在电路计算中，迭代法主要针对直接使用二极管方程时的情况。负载线分析法、理想二极管模型以及恒压降法常被用于二极管电路的简化分析。SPICE电路分析程序包含了精确描述理想和非理想二极管特性的内建模型，能够方便地分析含有二极管的电路特性。

　　在二极管的应用方面，本章重点讲解了半波、全波以及全波桥式整流电路。整流电路主要将交流电压转换为直流电压。在电源电路中使用的整流器必须能够经受大的周期峰值电流，以及刚上电时的浪涌电流。在整流电路中，滤波电容的设计将决定纹波电压和二极管的导通角。由于存在内部电容，二极管的导通和截止必须经过电容充放电，因而不能瞬间完成。导通时间通常都是很小的，但截止过程则要慢得多，它必须将二极管中的存储电荷都转移掉，这样就产生一个存储延迟τ_S。在存储延迟期间，可能出现较大的反向电流。

　　本章最后研究了pn结发光和检测光的能力，讨论了光电二极管、太阳能电池和发光二极管的基本特性。

Chapter Outline

3.1 The *pn* Junction Diode
3.2 The *i-v* Characteristics of the Diode
3.3 The Diode Equation: A Mathematical Model for the Diode
3.4 Diode Characteristics under Reverse, Zero, and Forward Bias
3.5 Diode Temperature Coefficient
3.6 Diodes under Reverse Bias
3.7 *pn* Junction Capacitance
3.8 Schottky Barrier Diode
3.9 Diode SPICE Model and Layout
3.10 Diode Circuit Analysis
3.11 Multiple-Diode Circuits
3.12 Analysis of Diodes Operating in the Breakdown Region
3.13 Half-Wave Rectifier Circuits
3.14 Full-Wave Rectifier Circuits
3.15 Full-Wave Bridge Rectification
3.16 Rectifier Comparison and Design Tradeoffs
3.17 Dynamic Switching Behavior of the Diode
3.18 Photo Diodes, Solar Cells, and Light-Emitting Diodes
Summary
Key Terms
Reference
Additional Reading
Problems

Chapter Goals

- Understand diode structure and basic layout
- Develop electrostatics of the *pn* junction
- Explore various diode models including the mathematical model, the ideal diode model, and the constant voltage drop model
- Understand the SPICE representation and model parameters for the diode
- Define regions of operation of the diode, including forward and reverse bias and reverse breakdown
- Apply the various types of models in circuit analysis
- Explore different types of diodes including Zener, variable capacitance, and Schottky barrier diodes as well as solar cells and light emitting diodes (LEDs)
- Discuss the dynamic switching behavior of the *pn* junction diode
- Explore diode rectifiers
- Practice simulating diode circuits using SPICE

Photograph of an assortment of diodes

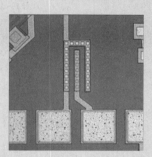

Fabricated diode

The first electronic circuit element that we explore is the solid-state *pn* junction diode. The diode is an extremely important device in its own right with many important applications including ac-dc power conversion (rectification), solar power generation, and high-frequency mixers for RF communications. In addition, the *pn* junction diode is a fundamental building block for other solid-state devices. In later chapters, we will find that two closely coupled diodes are used to form the bipolar junction transistor (BJT), and two diodes form an integral part of the metal-oxide-semiconductor field-effect transistor (MOSFET) and the junction field-effect transistor (JFET). Gaining an understanding of diode characteristics is prerequisite to understanding the behavior of the field-effect and bipolar transistors that are used to realize both digital logic circuits and analog amplifiers.

The *pn* junction diode is formed by fabricating adjoining regions of *p*-type and *n*-type semiconductor material. Another type of diode, called the Schottky barrier diode, is formed by a non-ohmic contact between a metal such as aluminum, palladium, or platinum and an *n*-type or *p*-type semiconductor. Both types of solid-state diodes are discussed in this chapter. The vacuum diode, which was used before the advent of semiconductor diodes, still finds application in very high voltage situations.

Chapter 3 Solid-State Diodes and Diode Circuits

The *pn* junction diode is a nonlinear element, and for many of us, this will be our first encounter with a nonlinear device. The diode is a two-terminal circuit element similar to a resistor, but its *i-v* characteristic, the relationship between the current through the element and the voltage across the element, is not a straight line. This nonlinear behavior allows electronic circuits to be designed to provide many useful operations, including rectification, mixing (a form of multiplication), and wave shaping. Diodes can also be used to perform elementary logic operations such as the AND and OR functions.

This chapter begins with a basic discussion of the structure and behavior of the *pn* junction diode and its terminal characteristics. Next is an introduction to the concept of modeling, and several different models for the diode are introduced and used to analyze the behavior of diode circuits. We begin to develop the intuition needed to make choices between models of various complexities in order to simplify electronic circuit analysis and design. Diode circuits are then explored, including the detailed application of the diode in rectifier circuits. The characteristics of Zener diodes, photo diodes, solar cells, and light-emitting diodes are also discussed.

3.1 THE *pn* JUNCTION DIODE

The **pn junction diode** is formed by fabrication of a *p*-type semiconductor region in intimate contact with an *n*-type semiconductor region, as illustrated in Fig. 3.1. The diode is constructed using the impurity doping process discussed in the last section of Chapter 2.

An actual diode can be formed by starting with an *n*-type wafer with doping N_D and selectively converting a portion of the wafer to *p*-type by adding acceptor impurities with $N_A > N_D$. The point at which the material changes from *p*-type to *n*-type is called the metallurgical junction. The *p*-type region is also referred to as the **anode** of the diode, and the *n*-type region is called the **cathode** of the diode.

Figure 3.2 gives the circuit symbol for the diode, with the left-hand end corresponding to the *p*-type region of the diode and the right-hand side corresponding to the *n*-type region. We will see shortly that the "arrow" points in the direction of positive current in the diode.

3.1.1 *pn* JUNCTION ELECTROSTATICS

Consider a *pn* junction diode similar to Fig. 3.1 having $N_A = 10^{17}/\text{cm}^3$ on the *p*-type side and $N_D = 10^{16}/\text{cm}^3$ on the *n*-type side. The hole and electron concentrations on the two sides of the junction will be

$$\begin{aligned} p\text{-type side:} \quad & p_p = 10^{17} \text{ holes/cm}^3 \quad & n_p = 10^3 \text{ electrons/cm}^3 \\ n\text{-type side:} \quad & p_n = 10^4 \text{ holes/cm}^3 \quad & n_n = 10^{16} \text{ electrons/cm}^3 \end{aligned} \quad (3.1)$$

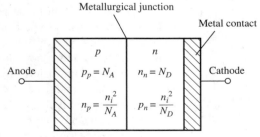

Figure 3.1 Basic *pn* junction diode.   **Figure 3.2** Diode circuit symbol.

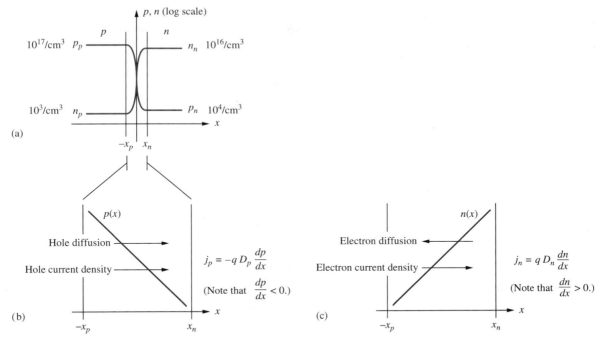

Figure 3.3 (a) Carrier concentrations; (b) hole diffusion current in the space charge region; (c) electron diffusion current in the space charge region.

As shown in Fig. 3.3(a), a very large concentration of holes exists on the *p*-type side of the metallurgical junction, whereas a much smaller hole concentration exists on the *n*-type side. Likewise, there is a very large concentration of electrons on the *n*-type side of the junction and a very low concentration on the *p*-type side.

From our knowledge of diffusion from Chapter 2, we know that mobile holes will diffuse from the region of high concentration on the *p*-type side toward the region of low concentration on the *n*-type side and that mobile electrons will diffuse from the *n*-type side to the *p*-type side, as in Fig. 3.3(b) and (c). If the diffusion processes were to continue unabated, there would eventually be a uniform concentration of holes and electrons throughout the entire semiconductor region, and the *pn* junction would cease to exist. Note that the two diffusion current densities are both directed in the positive *x* direction, but this is inconsistent with zero current in the open-circuited terminals of the diode.

A second, competing process must be established to balance the diffusion current. The competing mechanism is a drift current, as discussed in Chapter 2, and its origin can be understood by focusing on the region in the vicinity of the **metallurgical junction** shown in Fig. 3.4. As mobile holes move out of the *p*-type material, they leave behind immobile negatively charged acceptor atoms. Correspondingly, mobile electrons leave behind immobile ionized donor atoms with a localized positive charge. A **space charge region (SCR),** depleted of mobile carriers, develops in the region immediately around the metallurgical junction. This region is also often called the **depletion region,** or **depletion layer.**

From electromagnetics, we know that a region of space charge ρ_c (C/cm^3) will be accompanied by an electric field E measured in V/cm through Gauss' law

$$\nabla \cdot E = \frac{\rho_c}{\varepsilon_s} \tag{3.2}$$

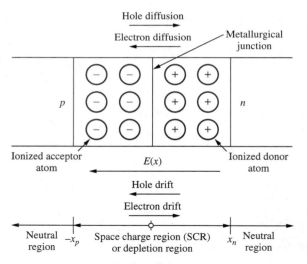

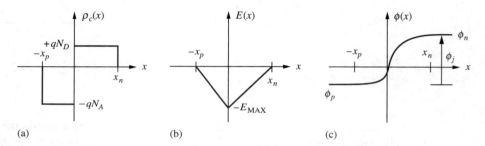

Figure 3.4 Space charge region formation near the metallurgical junction.

Figure 3.5 (a) Charge density (C/cm³), (b) electric field (V/cm), and (c) electrostatic potential (V) in the space charge region of a *pn* junction.

written assuming a constant semiconductor permittivity ε_s (F/cm). In one dimension, Eq. (3.2) can be rearranged to give

$$E(x) = \frac{1}{\varepsilon_s} \int \rho_c(x)\, dx \tag{3.3}$$

Figure 3.5 illustrates the space charge and electric field in the diode for the case of uniform (constant) doping on both sides of the junction. As illustrated in Fig. 3.5(a), the value of the space charge density on the *p*-type side will be $-qN_A$ and will extend from the metallurgical junction at $x = 0$ to $-x_p$, whereas that on the *n*-type side will be $+qN_D$ and will extend from 0 to $+x_n$. The overall diode must be charge neutral, so

$$qN_A x_p = qN_D x_n \tag{3.4}$$

The electric field is proportional to the integral of the space charge density and will be zero in the (charge) neutral regions outside of the depletion region. Using this zero-field boundary condition yields the triangular electric field distribution in Fig. 3.5(b).

Figure 3.5(c) represents the integral of the electric field and shows that a **built-in potential** or **junction potential** ϕ_j exists across the *pn* junction space charge region according to

$$\phi_j = -\int E(x)\, dx \quad \text{V} \tag{3.5}$$

ϕ_j represents the difference in the internal chemical potentials between the n and p sides of the diode, and it can be shown [1] to be given by

$$\phi_j = V_T \ln\left(\frac{N_A N_D}{n_i^2}\right) \qquad (3.6)$$

where the **thermal voltage** $V_T = kT/q$ was originally defined in Chapter 2.

Equations (3.3) to (3.6) can be used to determine the total width of the depletion region w_{do} in terms of the built-in potential:

$$w_{do} = (x_n + x_p) = \sqrt{\frac{2\varepsilon_s}{q}\left(\frac{1}{N_A} + \frac{1}{N_D}\right)\phi_j} \quad \text{cm} \qquad (3.7)$$

From Eq. (3.7), we see that the doping on the more lightly doped side of the junction will be the most important in determining the **depletion-layer width.**

EXAMPLE 3.1 DIODE SPACE CHARGE REGION WIDTH

When diodes are actually fabricated, the doping levels on opposite sides of the pn junction tend to be quite asymmetric, and the resulting depletion layer tends to extend primarily on one side of the junction and is referred to as a "one-sided" step junction or one-sided abrupt junction. The pn junction that we analyze provides an example of the magnitudes of the distances involved in such a pn junction.

PROBLEM Calculate the built-in potential and depletion-region width for a silicon diode with $N_A = 10^{17}/\text{cm}^3$ on the p-type side and $N_D = 10^{20}/\text{cm}^3$ on the n-type side.

SOLUTION **Known Information and Given Data:** On the p-type side, $N_A = 10^{17}/\text{cm}^3$; on the n-type side, $N_D = 10^{20}/\text{cm}^3$. Theory describing the pn junction is given by Eqs. (3.4) through (3.7).

Unknowns: Built-in potential ϕ_j and depletion-region width w_{do}

Approach: Find the built-in potential using Eq. (3.6); use ϕ_j to calculate w_{do} in Eq. (3.7).

Assumptions: The diode operates at room temperature with $V_T = 0.026$ V. There are only donor impurities on the n-type side and acceptor impurities on the p-type side of the junction. The doping levels are constant on each side of the junction.

Analysis: The built-in potential is given by

$$\phi_j = V_T \ln\left(\frac{N_A N_D}{n_i^2}\right) = (0.026 \text{ V}) \ln\left[\frac{(10^{17}/\text{cm}^3)(10^{20}/\text{cm}^3)}{(10^{20}/\text{cm}^6)}\right] = 1.018 \text{ V}$$

For silicon, $\varepsilon_s = 11.7\varepsilon_o$, where $\varepsilon_o = 8.85 \times 10^{-14}$ F/cm represents the permittivity of free space.

$$w_{do} = \sqrt{\frac{2\varepsilon_s}{q}\left(\frac{1}{N_A} + \frac{1}{N_D}\right)\phi_j}$$

$$w_{do} = \sqrt{\frac{2 \cdot 11.7 \cdot (8.85 \times 10^{-14} \text{ F/cm})}{1.60 \times 10^{-19} \text{ C}}\left(\frac{1}{10^{17}/\text{cm}^3} + \frac{1}{10^{20}/\text{cm}^3}\right) 1.018 \text{ V}} = 0.115 \text{ μm}$$

Check of Results: The built-in potential should be less than the bandgap of the material. For silicon the bandgap is approximately 1.12 V (see Table 2.3), so ϕ_j appears reasonable. The depletion-layer width seems quite small, but a double check of the numbers indicates that the calculation is correct.

Discussion: The numbers in this example are fairly typical of a *pn* junction diode. For the normal doping levels encountered in solid-state diodes, the built-in potential ranges between 0.5 V and 1.0 V, and the total depletion-layer width w_{do} can range from a fraction of 1 μm in heavily doped diodes to tens of microns in lightly doped diodes.

EXERCISE: Calculate the built-in potential and depletion-region width for a silicon diode if N_A is increased to 2×10^{18}/cm^3 on the *p*-type side and $N_D = 10^{20}$/cm^3 on the *n*-type side.

ANSWERS: 1.05 V; 0.0263 μm

3.1.2 INTERNAL DIODE CURRENTS

Remember that the electric field E points in the direction that a positive carrier will move, so electrons drift toward the positive x direction and holes drift in the negative x direction in Fig. 3.4. Because the terminal currents must be zero, a dynamic equilibrium is established in the junction region. Hole diffusion is precisely balanced by hole drift, and electron diffusion is exactly balanced by electron drift. This balance is stated mathematically in Eq. (3.8), in which the total hole and electron current densities must each be identically zero:

$$j_n^T = qn\mu_n E + qD_n \frac{\partial n}{\partial x} = 0 \quad \text{and} \quad j_p^T = qp\mu_p E - qD_p \frac{\partial p}{\partial x} = 0 \quad \text{A/cm}^2 \quad (3.8)$$

The difference in potential in Fig. 3.5(c) represents a barrier to both hole and electron flow across the junction. When a voltage is applied to the diode, the potential barrier is modified, and the delicate balances in Eq. (3.8) are disturbed, resulting in a current in the diode terminals.

EXAMPLE 3.2   DIODE ELECTRIC FIELD AND SPACE-CHARGE REGION EXTENTS

Now we find the value of the electric field in the diode and the size of the individual depletion layers on either side of the *pn* junction.

PROBLEM Find x_n, x_p, and E_{MAX} for the diode in Ex. 3.1.

SOLUTION **Known Information and Given Data:** On the *p*-type side, $N_A = 10^{17}$/cm^3; on the *n*-type side, $N_D = 10^{20}$/cm^3. Theory describing the *pn* junction is given by Eqs. (3.4) through (3.7). From Ex. 3.1, $\phi_j = 0.979$ V and $w_{do} = 0.113$ μm.

Unknowns: x_n, x_p, and E_{MAX}

Approach: Use Eqs. (3.4) and (3.7) to find x_n and x_p; use Eq. (3.5) to find E_{MAX}.

Assumptions: Room temperature operation

Analysis: Using Eq. (3.4), we can write

$$w_{do} = x_n + x_p = x_n \left(1 + \frac{N_D}{N_A}\right) \quad \text{and} \quad w_{do} = x_n + x_p = x_p \left(1 + \frac{N_A}{N_D}\right)$$

Solving for x_n and x_p gives

$$x_n = \frac{w_{do}}{\left(1 + \dfrac{N_D}{N_A}\right)} = \frac{0.113 \ \mu\text{m}}{\left(1 + \dfrac{10^{20}/\text{cm}^3}{10^{17}/\text{cm}^3}\right)} = 1.13 \times 10^{-4} \ \mu\text{m}$$

and

$$x_p = \frac{w_{do}}{\left(1 + \dfrac{N_A}{N_D}\right)} = \frac{0.113 \ \mu\text{m}}{\left(1 + \dfrac{10^{17}/\text{cm}^3}{10^{20}/\text{cm}^3}\right)} = 0.113 \ \mu\text{m}$$

Equation (3.5) indicates that the built-in potential is equal to the area under the triangle in Fig. 3.5(b). The height of the triangle is $(-E_{\text{MAX}})$ and the base of the triangle is $x_n + x_p = w_{do}$:

$$\phi_j = \frac{1}{2} E_{\text{MAX}} w_{do} \quad \text{and} \quad E_{\text{MAX}} = \frac{2\phi_j}{w_{do}} = \frac{2(0.979 \ \text{V})}{0.113 \ \mu\text{m}} = 173 \ \text{kV/cm}$$

Check of Results: From Eqs. (3.3) and (3.4), E_{MAX} can also be found from the doping levels and depletion-layer widths on each side of the junction. The equation in the next exercise can be used as a check of the answer.

EXERCISE: Using Eq. (3.3) and Fig. 3.5(a) and (b), show that the maximum field is given by

$$E_{\text{MAX}} = \frac{qN_A x_p}{\varepsilon_s} = \frac{qN_D x_n}{\varepsilon_s}$$

Use this formula to find E_{MAX}.

ANSWER: 175 kV/cm

EXERCISE: Calculate E_{MAX}, x_p, and x_n for a silicon diode if $N_A = 2 \times 10^{18}/\text{cm}^3$ on the *p*-type side and $N_D = 10^{20}/\text{cm}^3$ on the *n*-type side. Use $\phi_j = 1.05$ V and $w_{do} = 0.0263 \ \mu\text{m}$.

ANSWERS: 798 kV/cm; $5.16 \times 10^{-4} \ \mu\text{m}$; 0.0258 μm

---

3.2 THE *i-v* CHARACTERISTICS OF THE DIODE

The diode is the electronic equivalent of a mechanical check valve—it permits current to flow in one direction in a circuit, but prevents movement of current in the opposite direction. We will find that this nonlinear behavior has many useful applications in electronic circuit design. To understand this phenomenon, we explore the relationship between the current in the diode and the voltage applied to the diode. This information, called the *i-v* characteristic of the diode, is first presented graphically and then mathematically in this section and Sec. 3.3.

The current in the diode is determined by the voltage applied across the diode terminals, and the diode is shown with a voltage applied in Fig. 3.6. Voltage v_D represents the voltage applied between the diode terminals; i_D is the current through the diode. The neutral regions of the diode represent a low resistance to current, and essentially all the external applied voltage is dropped across the space charge region.

The applied voltage disturbs the balance between the drift and diffusion currents at the junction specified in the two expressions in Eq. (3.8). A positive applied voltage reduces the potential barrier for electrons and holes, as in Fig. 3.7, and current easily crosses the junction. A negative voltage

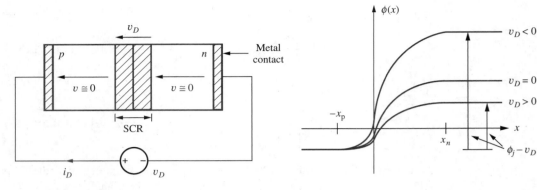

Figure 3.6 Diode with external applied voltage v_D.

Figure 3.7 Electrostatic junction potential for different applied voltages.

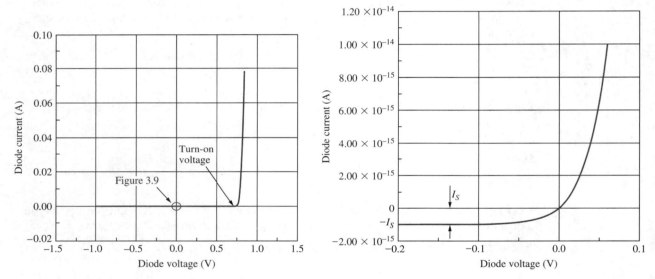

Figure 3.8 Graph of the i-v characteristics of a pn junction diode.

Figure 3.9 Diode behavior near the origin with $I_S = 10^{-15}$ A and $n = 1$.

increases the potential barrier, and although the balance in Eq. (3.8) is disturbed, the increased barrier results in a very small current.

The most important details of the diode i-v characteristic appear in Fig. 3.8. The diode characteristic is definitely not linear. For voltages less than zero, the diode is essentially nonconducting, with $i_D \cong 0$. As the voltage increases above zero, the current remains nearly zero until the voltage v_D exceeds approximately 0.5 to 0.7 V. At this point, the diode current increases rapidly, and the voltage across the diode becomes almost independent of current. The voltage required to bring the diode into significant conduction is often called either the **turn-on** or **cut-in voltage** of the diode.

Figure 3.9 is an enlargement of the region around the origin in Fig. 3.8. We see that the i-v characteristic passes through the origin; the current is zero when the applied voltage is zero. For negative voltages the current is not actually zero but reaches a limiting value labeled as $-I_S$ for voltages less than -0.1 V. I_S is called the **reverse saturation current**, or just **saturation current**, of the diode.

3.3 THE DIODE EQUATION: A MATHEMATICAL MODEL FOR THE DIODE

When performing both hand and computer analysis of circuits containing diodes, it is very helpful to have a mathematical representation, or model, for the i-v characteristics depicted in Figs. 3.8 and 3.9. In fact, solid-state device theory has been used to formulate a mathematical expression that agrees amazingly well with the measured i-v characteristics of the pn junction diode. We study this extremely important formula called the **diode equation** in this section.

A positive voltage v_D is applied to the diode in Fig. 3.10; in the figure the diode is represented by its circuit symbol from Fig. 3.2. Although we will not attempt to do so here, Eq. (3.8) can be solved for the hole and electron concentrations and the terminal current in the diode as a function of the voltage v_D across the diode. The resulting diode equation, given in Eq. (3.9), provides a **mathematical model** for the i-v characteristics of the diode:

$$i_D = I_S \left[\exp\left(\frac{qv_D}{nkT}\right) - 1 \right] = I_S \left[\exp\left(\frac{v_D}{nV_T}\right) - 1 \right] \quad (3.9)$$

where I_S = reverse saturation current of diode (A) T = absolute temperature (K)
        v_D = voltage applied to diode (V) n = nonideality factor (dimensionless)
        q = electronic charge (1.60×10^{-19} C) $V_T = kT/q$ = thermal voltage (V)
        k = Boltzmann's constant (1.38×10^{-23} J/K)

The total current through the diode is i_D, and the voltage drop across the diode terminals is v_D. Positive directions for the terminal voltage and current are indicated in Fig. 3.10. V_T is the thermal voltage encountered previously in Chapter 2 and will be assumed equal to 0.025 V at room temperature. I_S is the (reverse) saturation current of the diode encountered in Fig. 3.9, and n is a dimensionless parameter discussed in more detail shortly. The saturation current is typically in the range

$$10^{-18} \text{ A} \leq I_S \leq 10^{-9} \text{ A} \quad (3.10)$$

From device physics, it can be shown that the diode saturation current is proportional to n_i^2, where n_i is the density of electrons and holes in intrinsic semiconductor material. After reviewing Eq. (2.1) in Chapter 2, we realize that I_S will be strongly dependent on temperature. Detailed discussion of this temperature dependence is in Sec. 3.5.

Parameter n is termed the **nonideality factor.** For most silicon diodes, n is in the range 1.0 to 1.1, although it approaches a value of 2 in diodes operating at high current densities. From this point on, we assume that $n = 1$ unless otherwise indicated, and the diode equation will be written as

$$i_D = I_S \left[\exp\left(\frac{v_D}{V_T}\right) - 1 \right] \quad (3.11)$$

It is difficult to distinguish small variations in the value of n from an uncertainty in our knowledge in

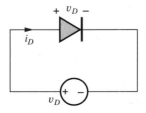

Figure 3.10 Diode with applied voltage v_D.

the absolute temperature. This is one reason that we will assume that $n = 1$ in this text. The problem can be investigated further by working on the next exercise.

> **EXERCISE:** For $n = 1$ and $T = 300$ K, $n(KT/q) = 25.9$ mV. Verify this calculation. Now, suppose $n = 1.03$. What temperature gives the same value for nV_T?
>
> **ANSWER:** 291 K

The mathematical model in Eq. (3.11) provides a highly accurate prediction of the i-v characteristics of the pn junction diode. The model is useful for understanding the detailed behavior of diodes and also forms the heart of the diode model in the SPICE circuit simulation program. It provides a basis for understanding the i-v characteristics of the bipolar transistor in Chapter 5.

DESIGN NOTE

> The static i-v characteristics of the diode are well-characterized by three parameters: saturation current I_S, temperature via the thermal voltage V_T, and nonideality factor n.
>
> $$i_D = I_S \left[\exp\left(\frac{v_D}{nV_T}\right) - 1 \right]$$

EXAMPLE 3.3 DIODE VOLTAGE AND CURRENT CALCULATIONS

In this example, we calculate some typical values of diode voltages for several different current levels and types of diodes.

PROBLEM (a) Find the diode voltage for a silicon diode with $I_S = 0.1$ fA operating at room temperature at a current of 300 μA. What is the diode voltage if $I_S = 10$ fA? What is the diode voltage if the current increases to 1 mA?
(b) Find the diode voltage for a silicon power diode with $I_S = 10$ nA and $n = 2$ operating at room temperature at a current of 10 A.
(c) A silicon diode is operating with a temperature of 50°C and the diode voltage is measured to be 0.736 V at a current of 2.50 mA. What is the saturation current of the diode?

SOLUTION (a) Known Information and Given Data: The diode currents are given and the saturation current parameter I_S is specified.

Unknowns: Diode voltage at each of the operating currents

Approach: Solve Eq. (3.9) for the diode voltage and evaluate the expression at each operating current.

Assumptions: At room temperature, we will use $V_T = 0.025$ V $= 1/40$ V; assume $n = 1$, since it is not specified otherwise; assume dc operation: $i_D = I_D$ and $v_D = V_D$.

Analysis: Solving Eq. (3.9) for V_D with $I_D = 0.1$ fA yields

$$V_D = nV_T \ln\left(1 + \frac{I_D}{I_S}\right) = 1(0.025 \text{ V}) \ln\left(1 + \frac{3 \times 10^{-4} \text{ A}}{10^{-16} \text{ A}}\right) = 0.718 \text{ V}$$

For $I_S = 10$ fA:

$$V_D = nV_T \ln\left(1 + \frac{I_D}{I_S}\right) = 1(0.025 \text{ V}) \ln\left(1 + \frac{3 \times 10^{-4} \text{ A}}{10^{-14} \text{ A}}\right) = 0.603 \text{ V}$$

For $I_D = 1$ mA with $I_S = 0.1$ fA:

$$V_D = nV_T \ln\left(1 + \frac{I_D}{I_S}\right) = 1(0.025 \text{ V}) \ln\left(1 + \frac{10^{-3} \text{ A}}{10^{-16} \text{ A}}\right) = 0.748 \text{ V}$$

Check of Results: The diode voltages are all between 0.5 V and 1.0 V and are reasonable (the diode voltage should not exceed the bandgap for $n = 1$).

SOLUTION (b) **Known Information and Given Data:** The diode current is given and the values of the saturation current parameter I_S and n are both specified.

Unknowns: Diode voltage at the operating current

Approach: Solve Eq. (3.9) for the diode voltage and evaluate the resulting expression.

Assumptions: At room temperature, we will use $V_T = 0.025$ V $= 1/40$ V.

Analysis: The diode voltage will be

$$V_D = nV_T \ln\left(1 + \frac{I_D}{I_S}\right) = 2(0.025 \text{ V}) \ln\left(1 + \frac{10 \text{ A}}{10^{-8} \text{ A}}\right) = 1.04 \text{ V}$$

Check of Results: Based on the comment at the end of part (a) and realizing that $n = 2$, voltages between 1 V and 2 V are reasonable for power diodes operating at high currents.

SOLUTION (c) **Known Information and Given Data:** The diode current is 2.50 mA and voltage is 0.736 V. The diode is operating at a temperature of 50°C.

Unknowns: Diode saturation current I_S

Approach: Solve Eq. (3.9) for the saturation current and evaluate the resulting expression. The value of the thermal voltage V_T will need to be calculated for $T = 50°$C.

Assumptions: The value of n is unspecified, so assume $n = 1$.

Analysis: Converting $T = 50°$C to Kelvins, $T = (273 + 50)$ K $= 323$ K, and

$$V_T = \frac{kT}{q} = \frac{(1.38 \times 10^{-23} \text{ J/K})(323 \text{ K})}{1.60 \times 10^{-19} \, °\text{C}} = 27.9 \text{ mV}$$

Solving Eq. (3.9) for I_S yields

$$I_S = \frac{I_D}{\exp\left(\frac{V_D}{nV_T}\right) - 1} = \frac{2.5 \text{ mA}}{\exp\left(\frac{0.736 \text{ V}}{0.0279 \text{ V}}\right) - 1} = 8.74 \times 10^{-15} \text{ A} = 8.74 \text{ fA}$$

Check of Results: The saturation current is within the range of typical values specified in Eq. (3.10).

> **EXERCISE:** A diode has a reverse saturation current of 40 fA. Calculate i_D for diode voltages of 0.55 and 0.7 V. What is the diode voltage if $i_D = 6$ mA?
>
> **ANSWERS:** 143 µA; 57.9 mA; 0.643 V

3.4 DIODE CHARACTERISTICS UNDER REVERSE, ZERO, AND FORWARD BIAS

When a dc voltage or current is applied to an electronic device, we say that we are providing a dc bias or simply a **bias** to the device. As we develop our electronics expertise, choosing the bias will be important to all of the circuits that we analyze and design. We will find that bias determines device characteristics, power dissipation, voltage and current limitations, and other important circuit parameters such as impedance levels and voltage gain. For a diode, there are three important bias conditions. **Reverse bias** and **forward bias** correspond to $v_D < 0$ V and $v_D > 0$ V, respectively. The **zero bias** condition, with $v_D = 0$ V, represents the boundary between the forward and reverse bias regions. When the diode is operating with reverse bias, we consider the diode "off" or nonconducting because the current is very small ($i_D = -I_S$). For forward bias, the diode is usually in a highly conducting state and is considered "on."

3.4.1 REVERSE BIAS

For $v_D < 0$, the diode is said to be operating under reverse bias. Only a very small reverse leakage current, approximately equal to I_S, flows through the diode. This current is small enough that we usually think of the diode as being in the nonconducting or off state when it is reverse-biased. For example, suppose that a dc voltage $V = -4V_T = -0.1$ V is applied to the diode terminals so that $v_D = -0.1$ V. Substituting this value into Eq. (3.11) gives

$$i_D = I_S \left[\exp\left(\frac{v_D}{V_T}\right) - 1 \right] = I_S[\overset{\text{negligible}}{\exp(-4)} - 1] \approx -I_S \quad (3.12)$$

because $\exp(-4) = 0.018$. For a reverse bias greater than $4V_T$, that is, $v_D \leq -4V_T = -0.1$ V, the exponential term $\exp(v_D/V_T)$ is much less than 1, and the diode current will be approximately equal to $-I_S$, a very small current. The current I_S was identified in Fig. 3.9.

> **EXERCISE:** A diode has a reverse saturation current of 5 fA. Calculate i_D for diode voltages of -0.04 V and -2 V (see Sec. 3.6).
>
> **ANSWERS:** -3.99 fA; -5 fA

The situation depicted in Fig. 3.9 and Eq. (3.12) actually represents an idealized picture of the diode. In a real diode, the reverse leakage current is several orders of magnitude larger than I_S due to the generation of electron–hole pairs within the depletion region. In addition, i_D does not saturate but increases gradually with reverse bias as the width of the depletion layer increases with reverse bias. (See Sec. 3.6.1).

3.4.2 ZERO BIAS

Although it may seem to be a trivial result, it is important to remember that the i-v characteristic of the diode passes through the origin. For zero bias with $v_D = 0$, we find $i_D = 0$. Just as for a resistor, there must be a voltage across the diode terminals in order for a nonzero current to exist.

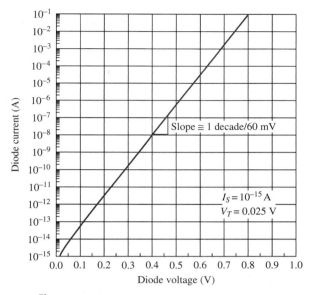

Figure 3.11 Diode i-v characteristic on semilog scale.

3.4.3 FORWARD BIAS

For the case $v_D > 0$, the diode is said to be operating under forward bias, and a large current can be present in the diode. Suppose that a voltage $v_D \geq +4V_T = +0.1$ V is applied to the diode terminals. The exponential term $\exp(v_D/V_T)$ is now much greater than 1, and Eq. (3.9) reduces to

$$i_D = I_S \left[\exp\left(\frac{v_D}{V_T}\right) - \overset{\text{negligible}}{\cancel{1}} \right] \cong I_S \exp\left(\frac{v_D}{V_T}\right) \qquad (3.13)$$

The diode current grows exponentially with applied voltage for a forward bias greater than approximately $4V_T$.

The diode i-v characteristic for forward voltages is redrawn in semilogarithmic form in Fig. 3.11. The straight line behavior predicted by Eq. (3.13) for voltages $v_D \geq 4V_T$ is apparent. A slight curvature can be observed near the origin, where the -1 term in Eq. (3.13) is no longer negligible. The slope of the graph in the exponential region is very important. Only a 60-mV increase in the forward voltage is required to increase the diode current by a factor of 10. This is the reason for the almost vertical increase in current noted in Fig. 3.8 for voltages above the turn-on voltage.

EXAMPLE 3.4   **DIODE VOLTAGE CHANGE VERSUS CURRENT**

The slope of the diode i-v characteristic is an important number for circuit designers to remember.

PROBLEM   Use Eq. (3.13) to accurately calculate the voltage change required to increase the diode current by a factor of 10.

SOLUTION   **Known Information and Given Data:** The current changes by a factor of 10.

Unknowns: The diode voltage change corresponding to a one decade change in current; the saturation current has not been given.

Approach: Form an expression for the ratio of two diode currents using the diode equation. The saturation current will cancel out and is not needed.

Assumptions: Room temperature operation with $V_T = 25.0$ mV; assume $I_D \gg I_S$.

Analysis: Let

$$i_{D1} = I_S \exp\left(\frac{v_{D1}}{V_T}\right) \quad \text{and} \quad i_{D2} = I_S \exp\left(\frac{v_{D2}}{V_T}\right)$$

Taking the ratio of the two currents and setting it equal to 10 yields

$$\frac{i_{D2}}{i_{D1}} = \exp\left(\frac{v_{D2} - v_{D1}}{V_T}\right) = \exp\left(\frac{\Delta v_D}{V_T}\right) = 10 \quad \text{and} \quad \Delta v_D = V_T \ln 10 = 2.3 V_T$$

Therefore $\Delta V_D = 2.3 V_T = 57.5$ mV (or approximately 60 mV) at room temperature.

Check of Results: The result is consistent with the logarithmic plot in Fig. 3.11. The diode voltage changes approximately 60 mV for each decade change in forward current.

EXERCISE: A diode has a saturation current of 2 fA. (a) What is the diode voltage at a diode current of 40 μA (assume $V_T = 25.0$ mV)? Repeat for a diode current of 400 μA. What is the difference in the two diode voltages? (b) Repeat for $V_T = 25.9$ mV.

ANSWERS: 0.593 V, 0.651 V, 57.6 mV; 0.614 V, 0.674 V, 59.6 mV

DESIGN NOTE

The diode voltage changes by approximately *60 mV per decade* change in diode current. Sixty mV/decade often plays an important role in our thinking about the design of circuits containing both diodes and bipolar transistors and is a good number to remember.

Figure 3.12 compares the characteristics of three diodes with different values of saturation current. The saturation current of diode A is 10 times larger than that of diode B, and the saturation current of diode B is 10 times that of diode C. The spacing between each pair of curves is

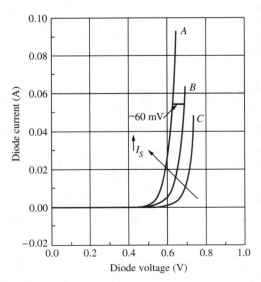

Figure 3.12 Diode characteristics for three different reverse saturation currents: (a) 10^{-12} A, (b) 10^{-13} A, and (c) 10^{-14} A.

approximately 60 mV. If the saturation current of the diode is reduced by a factor of 10, then the diode voltage must increase by approximately 60 mV to reach the same operating current level. Figure 3.12 also shows the relatively low sensitivity of the forward diode voltage to changes in the parameter I_S. For a fixed diode current, a change of two orders of magnitude in I_S results in a diode voltage change of only 120 mV.

3.5 DIODE TEMPERATURE COEFFICIENT

Another important number to keep in mind is the temperature coefficient associated with the diode voltage v_D. Solving Eq. (3.11) for the diode voltage under forward bias

$$v_D = V_T \ln\left(\frac{i_D}{I_S} + 1\right) = \frac{kT}{q} \ln\left(\frac{i_D}{I_S} + 1\right) \cong \frac{kT}{q} \ln\left(\frac{i_D}{I_S}\right) \quad \text{V} \quad (3.14)$$

and taking the derivative with respect to temperature yields

$$\frac{dv_D}{dT} = \frac{k}{q} \ln\left(\frac{i_D}{I_S}\right) - \frac{kT}{q}\frac{1}{I_S}\frac{dI_S}{dT} = \frac{v_D}{T} - V_T\frac{1}{I_S}\frac{dI_S}{dT} = \frac{v_D - V_{GO} - 3V_T}{T} \quad \text{V/K} \quad (3.15)$$

where it is assumed that $i_D \gg I_S$ and $I_S \propto n_i^2$. In the numerator of Eq. (3.15), v_D represents the diode voltage, V_{GO} is the voltage corresponding to the silicon bandgap energy at 0 K ($V_{GO} = E_G/q$), and V_T is the thermal potential. The last two terms result from the temperature dependence of n_i^2 as defined by Eq. (2.2). Evaluating the terms in Eq. (3.15) for a silicon diode with $v_D = 0.65$ V, $E_G = 1.12$ eV, and $V_T = 0.025$ V yields

$$\frac{dv_D}{dT} = \frac{(0.65 - 1.12 - 0.075) \text{ V}}{300 \text{ K}} = -1.82 \text{ mV/K} \quad (3.16)$$

DESIGN NOTE

The forward voltage of the diode decreases as temperature increases, and the diode exhibits a temperature coefficient of approximately -1.8 mV/°C at room temperature.

EXERCISE: (a) Verify Eq. (3.15) using the expression for n_i^2 from Eq. (2.1). (b) A silicon diode is operating at $T = 300$ K, with $i_D = 1$ mA, and $v_D = 0.680$ V. Use the result from Eq. (3.16) to estimate the diode voltage at 275 K and at 350 K.

ANSWERS: 0.726 V; 0.589 V

3.6 DIODES UNDER REVERSE BIAS

We must be aware of several other phenomena that occur in diodes operated under reverse bias. As depicted in Fig. 3.13, the reverse voltage v_R applied across the diode terminals is dropped across the space charge region and adds directly to the built-in potential of the junction:

$$v_j = \phi_j + v_R \quad \text{for } v_R > 0 \quad (3.17)$$

The increased voltage results in a larger internal electric field that must be supported by additional charge in the depletion layer, as defined by Eqs. (3.2) to (3.5). Using Eq. (3.7) with the voltage

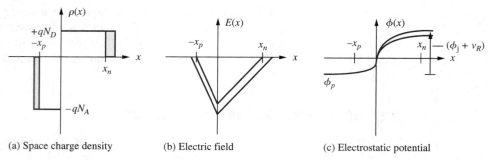

Figure 3.13 The *pn* junction diode under reverse bias.

from Eq. (3.17), the general expression for the depletion-layer width w_d for an applied reverse-bias voltage v_R becomes

$$w_d = (x_n + x_p) = \sqrt{\frac{2\varepsilon_s}{q}\left(\frac{1}{N_A} + \frac{1}{N_D}\right)(\phi_j + v_R)}$$

or

$$w_d = w_{do}\sqrt{1 + \frac{v_R}{\phi_j}} \quad \text{where } w_{do} = \sqrt{\frac{2\varepsilon_s}{q}\left(\frac{1}{N_A} + \frac{1}{N_D}\right)\phi_j} \quad (3.18)$$

The width of the space charge region increases approximately in proportion to the square root of the applied voltage.

> **EXERCISE:** The diode in Ex. 3.1 had a zero-bias depletion-layer width of 0.113 μm and a built-in voltage of 0.979 V. What will be the depletion-layer width for a 10-V reverse bias? What is the new value of E_{MAX}?
>
> **ANSWERS:** 0.378 μm; 581 kV/cm

3.6.1 SATURATION CURRENT IN REAL DIODES

The reverse saturation current actually results from the thermal generation of hole–electron pairs in the depletion region that surrounds the *pn* junction and is therefore proportional to the volume of the depletion region. Since the depletion-layer width increases with reverse bias, as described by Eq. (3.18), the reverse current does not truly saturate, as depicted in Fig. 3.9 and Eq. (3.9). Instead, there is gradual increase in reverse current as the magnitude of the reverse bias voltage is increased.

$$I_S = I_{SO}\sqrt{1 + \frac{v_R}{\phi_j}} \quad (3.19)$$

Under forward bias, the depletion-layer width changes very little, and $I_S = I_{SO}$ for forward bias.

> **EXERCISE:** A diode has $I_{SO} = 10$ fA and a built-in voltage of 0.8 V. What is I_S for a reverse bias of 10 V?
>
> **ANSWER:** 36.7 fA.

ELECTRONICS IN ACTION

The PTAT Voltage and Electronic Thermometry

The well-defined temperature dependence of the diode voltage discussed in Secs. 3.3 to 3.5 is actually used as the basis for most digital thermometers. We can build a simple electronic thermometer based on the circuit shown here in which two identical diodes are biased by current sources I_1 and I_2.

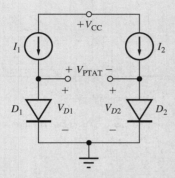
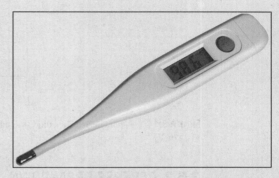

Digital thermometer: © D. Hurst/Alamy RF.

If we calculate the difference between the diode voltages using Eq. (3.14), we discover a voltage that is directly **proportional to absolute temperature** (PTAT), referred to as the PTAT voltage or V_{PTAT}:

$$V_{PTAT} = V_{D1} - V_{D2} = V_T \ln\left(\frac{I_{D1}}{I_S}\right) - V_T \ln\left(\frac{I_{D2}}{I_S}\right) = V_T \ln\left(\frac{I_{D1}}{I_{D2}}\right) = \frac{kT}{q} \ln\left(\frac{I_{D1}}{I_{D2}}\right)$$

The PTAT voltage has a temperature coefficient given by

$$\frac{dV_{PTAT}}{dT} = \frac{k}{q} \ln\left(\frac{I_{D1}}{I_{D2}}\right) = \frac{V_{PTAT}}{T}$$

By using two diodes, the temperature dependence of I_S has been eliminated from the equation. For example, suppose $T = 295$ K, $I_{D1} = 250$ μA, and $I_{D2} = 50$ μA. Then $V_{PTAT} = 40.9$ mV with a temperature coefficient of $+0.139$ mV/K.

This simple but elegant PTAT voltage circuit forms the heart of most of today's highly accurate electronic thermometers as depicted in the block diagram here. The analog PTAT voltage is amplified and then converted to a digital representation by an A/D converter. The digital output is scaled and offset to properly represent either the Fahrenheit or Celsius temperature scales and appears on an alphanumeric display. The scaling and offset shift can also be done in analog form prior to the A/D conversion operation.

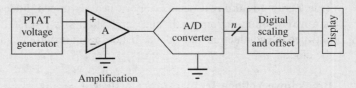

Block diagram of a digital thermometer.

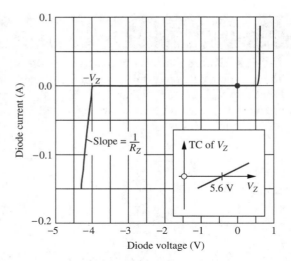

Figure 3.14 *i-v* characteristic of a diode including the reverse-breakdown region. The inset shows the temperature coefficient (TC) of V_Z.

3.6.2 REVERSE BREAKDOWN

As the reverse voltage increases, the electric field within the device grows, and the diode eventually enters the **breakdown region.** The onset of the breakdown process is fairly abrupt, and the current increases rapidly for any further increase in the applied voltage, as shown in the *i-v* characteristic of Fig. 3.14.

The magnitude of the voltage at which breakdown occurs is called the **breakdown voltage** V_Z of the diode and is typically in the range $2 \text{ V} \leq V_Z \leq 2000 \text{ V}$. The value of V_Z is determined primarily by the doping level on the more lightly doped side of the *pn* junction, but the heavier the doping, the smaller the breakdown voltage of the diode.

Two separate breakdown mechanisms have been identified: *avalanche breakdown* and *Zener breakdown*. These are discussed in the following two sections.

Avalanche Breakdown

Silicon diodes with breakdown voltages greater than approximately 5.6 V enter breakdown through a mechanism called **avalanche breakdown.** As the width of the depletion layer increases under reverse bias, the electric field increases, as indicated in Fig. 3.13. Free carriers in the depletion region are accelerated by this electric field, and as the carriers move through the depletion region, they collide with the fixed atoms. At some point, the electric field and the width of the space charge region become large enough that some carriers gain energy sufficient to break covalent bonds upon impact, thereby creating electron–hole pairs. The new carriers created can also accelerate and create additional electron–hole pairs through this **impact-ionization process,** as illustrated in Fig. 3.15.

Zener Breakdown

True **Zener breakdown** occurs only in heavily doped diodes. The high doping results in a very narrow depletion-region width, and application of a reverse bias causes carriers to tunnel directly between the conduction and valence bands, again resulting in a rapidly increasing reverse current in the diode.

Breakdown Voltage Temperature Coefficient

We can differentiate between the two types of breakdown because the breakdown voltages associated with the two mechanisms exhibit opposite temperature coefficients (TC). In avalanche breakdown,

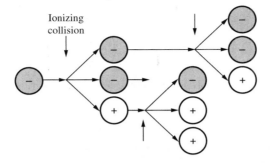
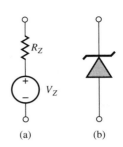

Figure 3.15 The avalanche breakdown process. (Note that the positive and negative charge carriers will actually be moving in opposite directions in the electric field in the depletion region.)

Figure 3.16 (a) Model for reverse-breakdown region of diode. (b) Zener diode symbol.

V_Z increases with temperature; in Zener breakdown, V_Z decreases with temperature. For silicon diodes, a zero temperature coefficient is achieved at approximately 5.6 V. The avalanche breakdown mechanism dominates in diodes that exhibit breakdown voltages of more than 5.6 V, whereas diodes with breakdown voltages below 5.6 V enter breakdown via the Zener mechanism.

3.6.3 DIODE MODEL FOR THE BREAKDOWN REGION
In breakdown, the diode can be modeled by a voltage source of value V_Z in series with resistor R_Z, which sets the slope of the i-v characteristic in the breakdown region, as indicated in Fig. 3.14. The value of R_Z is normally small ($R_Z \leq 100 \, \Omega$), and the reverse current flowing in the diode must be limited by the external circuit or the diode will be destroyed.

From the i-v characteristic in Fig. 3.14 and the model in Fig. 3.16, we see that the voltage across the diode is almost constant, independent of current, in the reverse-breakdown region. Some diodes are actually designed to be operated in **reverse breakdown.** These diodes are called **Zener diodes**[1] and have the special circuit symbol given in Fig. 3.16(b). Links to data sheets for a series of Zener diode can be found on the MCD website.

3.7 *pn* JUNCTION CAPACITANCE
Forward- and reverse-biased diodes have a capacitance associated with the *pn* junction. This capacitance is important under dynamic signal conditions because it prevents the voltage across the diode from changing instantaneously. The capacitance is referred to as the *pn* junction capacitance or the **depletion-layer capacitance.**

3.7.1 REVERSE BIAS
Under reverse bias, w_d increases beyond its zero-bias value, as expressed by Eq. (3.18), and hence the amount of charge in the depletion region also increases. Since the charge in the diode is changing with voltage, a capacitance results. Using Eqs. (3.4) and (3.7), the total space charge on the *n*-side

---

[1] The term *Zener diode* is typically used to refer to diodes that breakdown by either the Zener or avalanche mechanism.

of the diode is given by

$$Q_n = qN_D x_n A = q\left(\frac{N_A N_D}{N_A + N_D}\right) w_d A \quad \text{C} \qquad (3.20)$$

where A is the cross-sectional area of the diode and w_d is described by Eq. (3.18). The capacitance of the reverse-biased pn junction is given by

$$C_j = \frac{dQ_n}{dv_R} = \frac{C_{jo} A}{\sqrt{1 + \frac{v_R}{\phi_j}}} \quad \text{where } C_{jo} = \frac{\varepsilon_s}{w_{do}} \quad \text{F/cm}^2 \qquad (3.21)$$

in which C_{jo} represents the **zero-bias junction capacitance** per unit area of the diode.

Equation (3.21) shows that the capacitance of the diode changes with applied voltage. The capacitance decreases as the reverse bias increases, exhibiting an inverse square root relationship. This voltage-controlled capacitance can be very useful in certain electronic circuits. Diodes can be designed with impurity profiles (called *hyper-abrupt profiles*) specifically optimized for operation as voltage-controlled capacitors. As for the case of Zener diodes, a special symbol exists for the variable capacitance diode, as shown in Fig. 3.17. Remember that this diode is designed to be operated under reverse bias, but it conducts in the forward direction. Links to data sheets for a series of variable capacitance diodes can be found on the MCD website.

Figure 3.17 Circuit symbol for the variable capacitance diode (varactor).

> **EXERCISE:** What is the value of C_{jo} for the diode in Ex. 3.1? What is the zero bias value of C_j if the diode junction area is 100 μm × 125 μm? What is the capacitance at a reverse bias of 5 V?
>
> **ANSWERS:** 91.7 nf/cm²; 11.5 pF; 4.64 pF

3.7.2 FORWARD BIAS

When the diode is operating under forward bias, additional charge is stored in the neutral regions near the edges of the space charge region. The amount of charge Q_D stored in the diode is proportional to the diode current:

$$Q_D = i_D \tau_T \quad \text{C} \qquad (3.22)$$

The proportionality constant τ_T is called the diode **transit time** and ranges from 10^{-15} s to more than 10^{-6} s (1 fs to 1 μs) depending on the size and type of diode. Because we know that i_D is dependent on the diode voltage through the diode equation, there is an additional capacitance, the **diffusion capacitance** C_D, associated with the forward region of operation:

$$C_D = \frac{dQ_D}{dv_D} = \frac{(i_D + I_S)\tau_T}{V_T} \cong \frac{i_D \tau_T}{V_T} \quad \text{F} \qquad (3.23)$$

in which V_T is the thermal voltage. The diffusion capacitance is proportional to current and can become quite large at high currents.

ELECTRONICS IN ACTION

The SPICE Circuit Simulation Program An IEEE Global History Network Milestone[2]

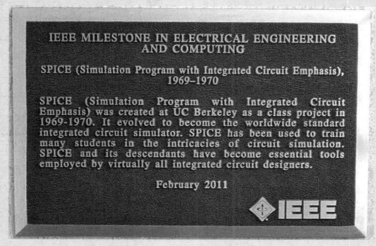

Courtesy of David Hodges

"SPICE (Simulation Program with Integrated Circuit Emphasis) was created at UC Berkeley as a class project in 1969–1970. It evolved to become the worldwide standard integrated circuit simulator. SPICE has been used to train many students in the intricacies of circuit simulation. SPICE and its descendants have become essential tools employed by virtually all integrated circuit designers.

SPICE was the first computer program for simulating the performance of integrated circuits that was readily available to undergraduate students for study of integrated circuit design. Hundreds of graduates from UC Berkeley and other universities became the backbone of the engineering workforce that moved the United States to microelectronics industry leadership in the 1970s. Graduates of Berkeley became leaders of today's largest firms delivering design automation capabilities for advanced microelectronics."

Above is a photograph of the SPICE Commemorative Plaque that may be found just inside the main entrance to Cory Hall, the Electrical Engineering Building at the University of California, Berkeley, CA. Cory Hall is the building where SPICE was developed.

Professor Donald O. Pederson guided the students that developed the SPICE program, and he was awarded the 1998 IEEE Medal of Honor "For creation of the SPICE program universally used for the computer aided design of circuits." Further information can be found in the *IEEE Solid-State Circuits Magazine*, "SPICE Commemorative Issue," vol. 3, no. 2, Spring 2011, and on the IEEE Global History Network website.

Professor Don Pederson, "Father of SPICE", IEEE Medal of Honor Recipient:
Courtesy of David Hodges

---

[2] Quoted from the SPICE Circuit Simulation Program Milestone of the IEEE Global History Network: http://www.ieeeghn.org/wiki/index.php/Milestones:SPICE_Circuit_Simulation_Program Milestone photograph courtesy of Professor David A. Hodges, used with permission.

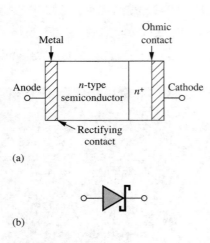

Figure 3.18 (a) Schottky barrier diode structure; (b) Schottky diode symbol.

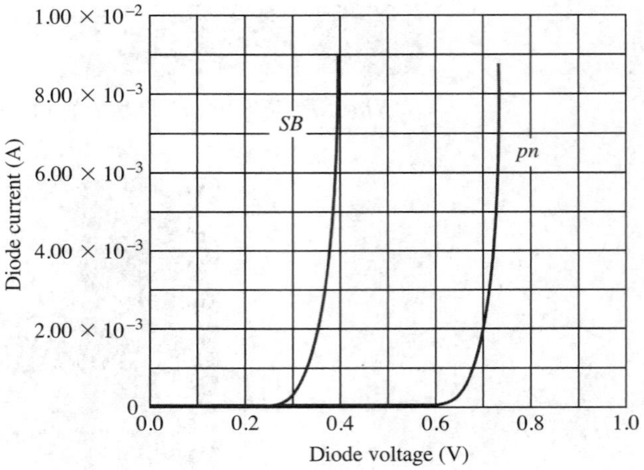

Figure 3.19 Comparison of *pn* junction (*pn*) and Schottky barrier diode (*SB*) *i*-*v* characteristics.

EXERCISE: A diode has a transit time of 10 ns. What is the diffusion capacitance of the diode for currents of 10 µA, 0.8 mA, and 50 mA at room temperature?

ANSWERS: 4 pF; 320 pF; 20 nF

3.8 SCHOTTKY BARRIER DIODE

In a p^+n junction diode, the *p*-side is a highly doped region (a conductor), and one might wonder if it could be replaced with a metallic layer. That is in fact the case, and in the **Schottky barrier diode,** one of the semiconductor regions of the *pn* junction diode is replaced by a non-ohmic rectifying metal contact, as indicated in Fig. 3.18. It is easiest to form a Schottky contact to *n*-type silicon, and for this case the metal region becomes the diode anode. An n^+ region is added to ensure that the cathode contact is ohmic. The symbol for the Schottky barrier diode appears in Fig. 3.18(b).

The Schottky diode turns on at a much lower voltage than its *pn*-junction counterpart, as indicated in Fig. 3.19. It also has significantly reduced internal charge storage under forward bias. We encounter an important use of the Schottky diode in bipolar logic circuits in Chapter 9. Schottky diodes also find important applications in high-power rectifier circuits and fast switching applications.

3.9 DIODE SPICE MODEL AND LAYOUT

The circuit in Fig. 3.20 represents the diode model that is included in SPICE programs. Resistance R_S represents the inevitable series resistance that always accompanies fabrication of, and making contacts to, a real device structure. The current source represents the ideal exponential behavior of the diode as described by Eq. 3.12 and **SPICE parameters IS, N,** and V_T. The model equation for i_D also includes a second term, not shown here, that models the effects of carrier generation in the space charge region in a manner similar to Eq. (3.19).

The capacitor specification includes the depletion-layer capacitance for the reverse-bias region modeled by **SPICE parameters CJO, VJ, and M,** as well as the diffusion capacitance associated

3.9 Diode SPICE Model and Layout

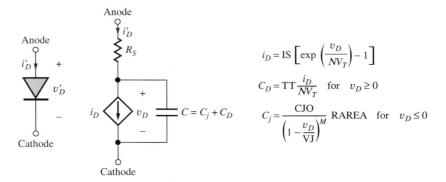

Figure 3.20 Diode equivalent circuit and simplified versions of the model equations used in SPICE programs.

with the junction under forward bias and defined by N and the transit-time parameter TT. In SPICE, the "junction grading coefficient" is an adjustable parameter. Using the typical value of $M = 0.5$ results in Eq. (3.21).

 EXERCISE: Find the default values of the seven parameters in Table 3.1 for the SPICE program that you use in class. Compare to the values in Table 3.1.

TABLE 3.1
SPICE Diode Parameter Equivalences

PARAMETER	OUR TEXT	SPICE	TYPICAL DEFAULT VALUES
Saturation current	I_S	IS	10 fA
Ohmic series resistance	R_S	RS	0 Ω
Ideality factor or emission coefficient	n	N	1
Transit time	τ_T	TT	0 s
Zero-bias junction capacitance for a unit area diode RAREA = 1	$C_{jo} \cdot A$	CJO	0 F
Built-in potential	ϕ_j	VJ	1 V
Junction grading coefficient	—	M	0.5
Relative junction area	—	RAREA	1

3.9.1 DIODE LAYOUT

Figure 3.21(a) shows the layout of a simple diode fabricated by forming a *p*-type diffusion in an *n*-type silicon wafer, as outlined in Chapter 2. This diode has a long rectangular *p*-type diffusion to increase the value of I_S, which is proportional to the junction area. Multiple contacts are formed to the *p*-type anode, and the *p*-region is surrounded by a collar of contacts to the *n*-type region. Both these sets of contacts are used to minimize the value of the extrinsic series resistance R_S of the diode, as included in the model in Fig. 3.20. Identical contacts are used so that they all tend to etch open at the same time during the fabrication process. The use of multiple identical contacts also facilitates calculation of the overall contact resistance. Heavily doped *n*-type regions are placed under the *n*-region contacts to ensure formation of an ohmic contact and prevent formation of a Schottky barrier diode.

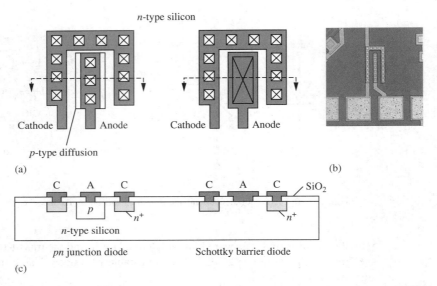

Figure 3.21 (a) Layout of a *pn* junction diode and a Schottky diode (b) *pn* junction diode photograph (c) Cross-section of the two diodes (See top view of diode in Chapter 3 opener.)

A conceptual drawing of a metal-semiconductor or Schottky diode also appears in Fig. 3.21(a) in which the aluminum metallization acts as the anode of the diode and the *n*-type semiconductor is the diode cathode. Careful attention to processing details is needed to form a diode rather than just an ohmic contact.

3.10 DIODE CIRCUIT ANALYSIS

We now begin our analysis of circuits containing diodes and introduce simplified circuit models for the diode. Figure 3.22 presents a series circuit containing a voltage source, resistor, and diode. Note that V and R may represent the Thévenin equivalent of a more complicated two-terminal network. Also note the notational change in Fig. 3.22. In the circuits that we analyze in the next few sections, the applied voltage and resulting diode voltage and current will all be dc quantities. (Recall that the dc components of the total quantities i_D and v_D are indicated by I_D and V_D, respectively.)

One common objective of diode circuit analysis is to find the **quiescent operating point** (Q-point), or **bias point,** for the diode. The Q-point consists of the dc current and voltage (I_D, V_D) that define the point of operation on the diode's i-v characteristic. We start the analysis by writing the loop equation for the circuit of Fig. 3.22:

$$V = I_D R + V_D \qquad (3.24)$$

Equation (3.24) represents a constraint placed on the diode operating point by the circuit elements. The diode i-v characteristic in Fig. 3.8 represents the allowed values of I_D and V_D as determined by the solid-state diode itself. Simultaneous solution of these two sets of constraints defines the Q-point.

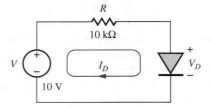

Figure 3.22 Diode circuit containing a voltage source and resistor.

3.10 Diode Circuit Analysis

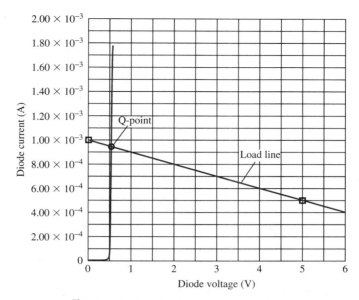

Figure 3.23 Diode i-v characteristic and load line.

We explore several methods for determining the solution to Eq. (3.24), including graphical analysis and the use of models of varying complexity for the diode. These techniques will include

- Graphical analysis using the load-line technique.
- Analysis with the mathematical model for the diode.
- Simplified analysis with an ideal diode model.
- Simplified analysis using the constant voltage drop model.

3.10.1 LOAD-LINE ANALYSIS

In some cases, the i-v characteristic of the solid-state device may be available only in graphical form, as in Fig. 3.23. We can then use a graphical approach (**load-line analysis**) to find the simultaneous solution of Eq. (3.24) with the graphical characteristic. Equation (3.24) defines the **load line** for the diode. The Q-point can be found by plotting the graph of the load line on the i-v characteristic for the diode. The intersection of the two curves represents the quiescent operating point, or Q-point, for the diode.

EXAMPLE 3.5   **LOAD-LINE ANALYSIS**

The graphical load-line approach is an important concept for visualizing the behavior of diode circuits as well as for estimating the actual Q-point.

PROBLEM Use load-line analysis to find the Q-point for the diode circuit in Fig. 3.22 using the i-v characteristic in Fig. 3.23.

SOLUTION **Known Information and Given Data:** The diode i-v characteristic is presented graphically in Fig. 3.23. Diode circuit is given in Fig. 3.22 with $V = 10$ V and $R = 10$ kΩ.

Unknowns: Diode Q-point (I_D, V_D).

Approach: Write the load-line equation and find two points on the load line that can be plotted on the graph in Fig. 3.23. The Q-point is at the intersection of the load line with the diode i-v characteristic.

Assumptions: Diode temperature corresponds to the temperature at which the graph in Fig. 3.23 was measured.

Analysis: Using the values from Fig. 3.22, Eq. (3.24) can be rewritten as

$$10 = 10^4 I_D + V_D \tag{3.25}$$

Two points are needed to define the line. The simplest choices are

$$I_D = (10\text{ V}/10\text{ k}\Omega) = 1\text{ mA} \quad \text{for} \quad V_D = 0 \quad \text{and} \quad V_D = 10\text{ V} \quad \text{for} \quad I_D = 0$$

Unfortunately, the second point is not in the range of the graph presented in Fig. 3.23, but we are free to choose any point that satisfies Eq. (3.25). Let's pick $V_D = 5$ V:

$$I_D = (10 - 5)\text{V}/10^4\text{ }\Omega = 0.5\text{ mA} \quad \text{for } V_D = 5$$

These points and the resulting load line are plotted in Fig. 3.23. The Q-point is given by the intersection of the load line and the diode characteristic:

$$\text{Q-point} = (0.95\text{ mA}, 0.6\text{ V})$$

Check of Results: We can double check our result by substituting the diode voltage found from the graph into Eq. (3.25) and calculating I_D. Using $V_D = 0.6$ V in Eq. (3.25) yields an improved estimate for the Q-point: (0.94 mA, 0.6 V). [We could also substitute 0.95 mA into Eq. (3.25) and calculate V_D.]

Discussion: Note that the values determined graphically are not quite on the load line since they do not precisely satisfy the load-line equation. This is a result of the limited precision that we can obtain by reading the graph.

EXERCISE: Repeat the load-line analysis if $V = 5$ V and $R = 5$ kΩ.

ANSWERS: (0.88 mA, 0.6 V)

EXERCISE: Use SPICE to find the Q-point for the circuit in Fig. 3.22. Use the default values of parameters in your SPICE program.

ANSWERS: (935 μA, 0.653 V) for $I_S = 10$ fA and $T = 300$ K

3.10.2 ANALYSIS USING THE MATHEMATICAL MODEL FOR THE DIODE

We can use our mathematical model for the diode to approach the solution of Eq. (3.25) more directly. The particular diode characteristic in Fig. 3.23 is represented quite accurately by diode Eq. (3.11), with $I_S = 10^{-13}$ A, $n = 1$, and $V_T = 0.025$ V:

$$I_D = I_S \left[\exp\left(\frac{V_D}{V_T}\right) - 1 \right] = 10^{-13}[\exp(40 V_D) - 1] \tag{3.26}$$

Eliminating I_D by substituting Eq. (3.26) into Eq. (3.25) yields

$$10 = 10^4 \cdot 10^{-13}[\exp(40 V_D) - 1] + V_D \tag{3.27}$$

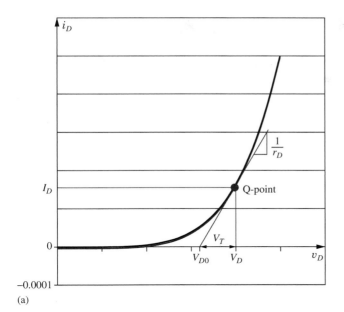
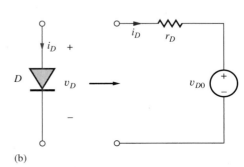

Figure 3.24 (a) Diode behavior around the Q-point; (b) linear model for the diode at the Q-point.

The expression in Eq. (3.27) is called a *transcendental equation* and does not have a closed-form analytical solution, so we settle for a numerical answer to the problem.

One approach to finding a numerical solution to Eq. (3.27) is through simple trial and error. We can guess a value of V_D and see if it satisfies Eq. (3.27). Based on the result, a new guess can be formulated and Eq. (3.27) evaluated again. The human brain is quite good at finding a sequence of values that will converge to the desired solution.

On the other hand, it is often preferable to use a computer to find the solution to Eq. (3.27), particularly if we need to find the answer to several different problems or parameter sets. The computer, however, requires a much more well-defined iteration strategy than brute force trial and error.

We can develop an iterative solution method for the diode circuit in Fig. 3.22 by creating a linear model for the diode equation in the vicinity of the diode Q-point as depicted in Fig. 3.24(a). First we find the slope of the diode characteristic at the operating point:

$$g_D = \left.\frac{\partial i_D}{\partial v_D}\right|_{Q-Pt} = \frac{I_S}{V_T} \exp\left(\frac{V_D}{V_T}\right) = \frac{I_D + I_S}{V_T} \cong \frac{I_D}{V_T} \quad \text{and} \quad r_D = \frac{1}{g_D} = \frac{V_T}{I_D} \quad (3.28)$$

Slope g_D is called the diode conductance, and its reciprocal r_D is termed the diode resistance. Next we can use the slope to find the x-axis intercept point V_{D0}:

$$V_{D0} = V_D - I_D r_D = V_D - V_T \quad (3.29)$$

V_{D0} and r_D represent a two-element linear circuit model for the diode as in Fig. 3.24(b), and this circuit model replaces the diode in the single loop circuit in Fig. 3.25.

Now we can use an iterative process to find the Q-point of the diode in the circuit.

1. Pick a starting guess for I_D.
2. Calculate the diode voltage using $V_D = V_T \ln\left(1 + \frac{I_D}{I_S}\right)$.
3. Calculate the values of V_{D0} and r_D.
4. Calculate a new estimate for I_D from the circuit in Fig. 3.25(b): $I_D = \frac{V - V_{D0}}{R + r_D}$.
5. Repeat steps 2 to 4 until convergence is obtained.

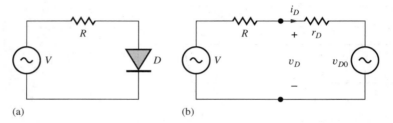

Figure 3.25 (a) Diode circuit; (b) circuit with two-element diode model.

TABLE 3.2
Example of Iterative Analysis

I_D (A)	V_D (V)	R_D (Ω)	V_{D0} (V)
1.0000E−03	0.5756	25.80	0.5498
9.4258E−04	0.5742	27.37	0.5484
9.4258E−04	0.5742	27.37	0.5484

Table 3.2 presents the results of performing the above iteration process using a spreadsheet. The diode current and voltage converge rapidly in only three iterations.

Note that one can achieve answers to an almost arbitrary precision using the numerical approach. However, in most real circuit situations, we will not have an accurate value for the saturation current of the diode, and there will be significant tolerances associated with the sources and passive components in the circuit. For example, the saturation current specification for a given diode type may vary by factors ranging from 10:1 to as much as 100:1. In addition, resistors commonly have ±5 percent or ±10 percent tolerances, and we do not know the exact operating temperature of the diode (remember the −1.8 mV/K temperature coefficient) or the precise value of the parameter n. Hence, it does not make sense to try to obtain answers with a precision of more than a few significant digits.

An alternative to the use of a spreadsheet is to write a simple program using a high-level language. The solution to Eq. (3.28) also can be found using the "solver" routines in many calculators, which use iteration procedures more sophisticated than that just described. MATLAB also provides the function fzero, which will calculate the zeros of a function as outlined in Ex. 3.6.

EXERCISE: An alternative expression (another transcendental equation) for the basic diode circuit can be found by eliminating V_D in Eq. (3.25) using Eq. (3.14). Show that the result is

$$10 = 10^4 I_D + 0.025 \ln\left(1 + \frac{I_D}{I_S}\right)$$

EXAMPLE 3.6 SOLUTION OF THE DIODE EQUATION USING MATLAB

MATLAB is one example of a computer tool that can be used to find the solution to transcendental equations.

PROBLEM Use MATLAB to find the solution to Eq. (3.27).

SOLUTION **Known Information and Given Data:** Diode circuit in Fig. 3.22 with $V = 10$ V, $R = 10$ kΩ, $I_S = 10^{-13}$ A, $n = 1$, and $V_T = 0.025$ V.

Unknowns: Diode voltage V_D.

Approach: Create a MATLAB "M-File" describing Eq. (3.27). Execute the program to find the diode voltage.

Assumptions: Room temperature operation with $V_T = 1/40$ V.

Analysis: First, create an M-File for the function 'diode':

$$\text{function xd} = \text{diode(vd)}$$
$$\text{xd} = 10 - (10^\wedge(-9)) * (\exp(40 * \text{vd}) - 1) - \text{vd}$$

Then find the solution near 1 V:

$$\text{fzero('diode', 1)}$$

Answer: 0.5742 V.

Check of Results: The diode voltage is positive and in the range of 0.5 to 0.8 V, which is expected for a diode. Substituting this value of voltage into the diode equation yields a current of 0.944 mA. This answer appears reasonable since we know that the diode current cannot exceed $10 \text{ V}/10 \text{ k}\Omega = 1.0$ mA, which is the maximum current available from the circuit [i.e., if the diode were replaced with a short circuit ($V_D = 0$), the current in the circuit would be 1 mA]. See Sec. 3.10.3.

EXERCISE: Use the MATLAB to find the solution to

$$10 = 10^4 I_D + 0.025 \ln\left(1 + \frac{I_D}{I_S}\right) \quad \text{for } I_S = 10^{-13} \text{ A}$$

ANSWER: 942.6 μA.

EXAMPLE 3.7   EFFECT OF DEVICE TOLERANCES ON DIODE Q-POINTS

Let us now see how sensitive our Q-point results are to the exact value of the diode saturation current.

PROBLEM Suppose that there is a tolerance on the value of the saturation current such that the value is given by

$$I_S^{\text{nom}} = 10^{-15} \text{ A} \quad \text{and} \quad 2 \times 10^{-16} \text{ A} \leq I_S \leq 5 \times 10^{-15} \text{ A}$$

Find the nominal, smallest, and largest values of the diode voltage and current in the circuit in Fig. 3.22.

SOLUTION **Known Information and Given Data:** The nominal and worst-case values of saturation current are given as well as the circuit values in Fig. 3.22.

Unknowns: Nominal and worst-case values for the diode Q-point: (I_D, V_D)

Approach: Use MATLAB or the solver on our calculator to find the diode voltages and then the currents for the nominal and worst-case values of I_S. Note from Eq. (3.24) that the maximum value of diode voltage corresponds to minimum current and vice versa.

Assumptions: Room temperature operation with $V_T = 0.025$ V. The voltage and resistance in the circuit do not have tolerances associated with them.

Analysis: For the nominal case, Eq. (3.28) becomes
$$f = 10 - 10^4(10^{-15})[\exp(40V_D) - 1] - V_D$$
for which the solver yields
$$V_D^{\text{nom}} = 0.689 \text{ V} \quad \text{and} \quad I_D^{\text{nom}} = \frac{(10 - 0.689) \text{ V}}{10^4 \text{ }\Omega} = 0.931 \text{ mA}$$
For the minimum I_S case, Eq. (3.28) is
$$f = 10 - 10^4(2 \times 10^{-16})[\exp(40V_D) - 1] - V_D$$
and the solver yields
$$V_D^{\text{max}} = 0.729 \text{ V} \quad \text{and} \quad I_D^{\text{min}} = \frac{(10 - 0.729) \text{ V}}{10^4 \text{ }\Omega} = 0.927 \text{ mA}$$
Finally, for the maximum value of I_S, Eq. (3.28) becomes
$$f = 10 - 10^4(5 \times 10^{-15})[\exp(40V_D) - 1] - V_D$$
and the solver gives
$$V_D^{\text{min}} = 0.649 \text{ V} \quad \text{and} \quad I_D^{\text{max}} = \frac{(10 - 0.649) \text{ V}}{10^4 \text{ }\Omega} = 0.935 \text{ mA}$$

Check of Results: The diode voltages are positive and in the range of 0.5 to 0.8 V which is expected for a diode. The diode currents are all less than the short circuit current available from the voltage source (10 V/10 kΩ = 1.0 mA).

Discussion: Note that even though the diode saturation current in this circuit changes by a factor of 5:1 in either direction, the current changes by less than $\pm 0.5\%$. As long as the driving voltage in the circuit is much larger than the diode voltage, the current should be relatively insensitive to changes in the diode voltage or the diode saturation current.

EXERCISE: Find V_D and I_D if the upper limit on I_S is increased to 10^{-14} A.

ANSWERS: 0.6316 V; 0.9368 A

EXERCISE: Use the Solver function in your calculator to find the solution to
$$10 = 10^4 I_D + 0.025 \ln\left(1 + \frac{I_D}{I_S}\right) \quad \text{for } I_S = 10^{-13} \text{ A} \quad \text{and} \quad I_S = 10^{-15} \text{ A}$$

ANSWERS: 0.9426 mA; 0.9311 mA

3.10.3 THE IDEAL DIODE MODEL

Graphical load-line analysis provides insight into the operation of the diode circuit of Fig. 3.22, and the mathematical model can be used to provide more accurate solutions to the load-line problem. The next method discussed provides simplified solutions to the diode circuit of Fig. 3.22 by introducing simplified diode circuit models of varying complexity.

The diode, as described by its i-v characteristic in Fig. 3.8 or by Eq. (3.11), is obviously a nonlinear device. However, most, if not all, of the circuit analysis that we have learned in electrical engineering thus far assumed that the circuits were composed of linear elements. To use this wealth of analysis techniques, we will use **piecewise linear** approximations to the diode characteristic.

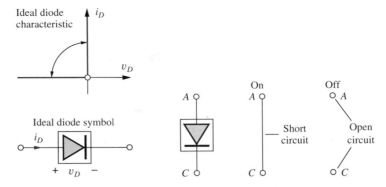

Figure 3.26 (a) Ideal diode i-v characteristics and circuit symbol; (b) circuit models for on and off states of the ideal diode.

The **ideal diode model** is the simplest model for the diode. The i-v characteristic for the **ideal diode** in Fig. 3.26 consists of two straight-line segments. If the diode is conducting a forward or positive current (forward-biased), then the voltage across the diode is zero. If the diode is reverse-biased, with $v_D < 0$, then the current through the diode is zero. These conditions can be stated mathematically as

$$v_D = 0 \quad \text{for } i_D > 0 \quad \text{and} \quad i_D = 0 \quad \text{for } v_D \le 0$$

The special symbol in Fig. 3.26 is used to represent the ideal diode in circuit diagrams.

We can now think of the diode as having two states. The diode is either conducting in the *on* state, or nonconducting and *off*. For circuit analysis, we use the models in Fig. 3.26(b) for the two states. If the diode is on, then it is modeled by a "short" circuit, a wire. For the off state, the diode is modeled by an "open" circuit, no connection.

Analysis Using the Ideal Diode Model

Let us now analyze the circuit of Fig. 3.22 assuming that the diode can be modeled by the ideal diode of Fig. 3.26(b). The diode has two possible states, and our analysis of diode circuits proceeds as follows:

1. Select a model for the diode.
2. Identify the anode and cathode of the diode and label the diode voltage v_D and current i_D.
3. Make an (educated) guess concerning the region of operation of the diode based on the circuit configuration.
4. Analyze the circuit using the diode model appropriate for the assumption in step 3.
5. Check the results to see if they are consistent with the assumptions.

For this analysis, we select the ideal diode model. The diode in the original circuit is replaced by the ideal diode, as in Fig. 3.27(b). Next we must guess the state of the diode. Because the voltage source appears to be trying to force a positive current through the diode, our first guess will be to

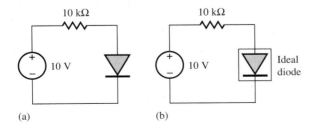

Figure 3.27 (a) Original diode circuit; (b) circuit modeled by an ideal diode.

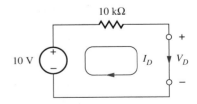

Figure 3.28 Ideal diode replaced with its model for the on state.

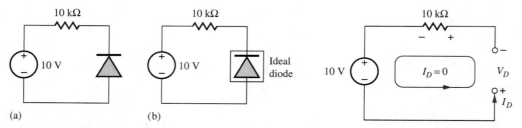

Figure 3.29 (a) Circuit with reverse-biased diode; (b) circuit modeled by ideal diode.

Figure 3.30 Ideal diode replaced with its model for the off region.

assume that the diode is on. The ideal diode of Fig. 3.27(b) is replaced by its piecewise linear model for the on region in Fig. 3.28, and the diode current is given by

$$I_D = \frac{(10 - 0) \text{ V}}{10 \text{ k}\Omega} = 1.00 \text{ mA}$$

The current $I_D \geq 0$, which is consistent with the assumption that the diode is on. Based on the ideal diode model, we find that the diode is forward-biased and operating with a current of 1 mA. The Q-point is therefore equal to (1 mA, 0 V).

Analysis of a Circuit Containing a Reverse-Biased Diode

A second circuit example in which the diode terminals have been reversed appears in Fig. 3.29; the ideal diode model is again used to model the diode [Fig. 3.29(b)]. The voltage source now appears to be trying to force a current backward through the diode. Because the diode cannot conduct in this direction, we assume the diode is off. The ideal diode of Fig. 3.29(b) is replaced by the open circuit model for the off region, as in Fig. 3.30.

Writing the loop equation for this case,

$$10 + V_D + 10^4 I_D = 0$$

Because $I_D = 0$, $V_D = -10$ V. The calculated diode voltage is negative, which is consistent with the starting assumption that the diode is off. The analysis shows that the diode in the circuit of Fig. 3.29 is indeed reverse-biased. The Q-point is $(0, -10$ V$)$.

Although these two problems may seem rather simple, the complexity of diode circuit analysis increases rapidly as the number of diodes increases. If the circuit has N diodes, then the number of possible states is 2^N. A circuit with 10 diodes has 1024 different possible circuits that could be analyzed! Only through practice can we develop the intuition needed to avoid analysis of many incorrect cases. We analyze more complex circuits shortly, but first let's look at a slightly better piecewise linear model for the diode.

3.10.4 CONSTANT VOLTAGE DROP MODEL

We know from our earlier discussion that there is a small, nearly constant voltage across the forward-biased diode. The ideal diode model ignores the presence of this voltage. However, the piecewise linear model for the diode can be improved by adding a constant voltage V_{on} in series with the ideal diode, as shown in Fig. 3.31(b). This is the **constant voltage drop (CVD) model**. V_{on} offsets the i-v characteristic of the ideal diode, as indicated in Fig. 3.31(c). The piecewise linear models for the two states become a voltage source V_{on} for the on state and an open circuit for the off state. We now have

$$v_D = V_{\text{on}} \quad \text{for } i_D > 0 \quad \text{and} \quad i_D = 0 \quad \text{for } v_D \leq V_{\text{on}}$$

We may consider the ideal diode model to be the special case of the constant voltage drop model for which $V_{\text{on}} = 0$. From the i-v characteristics presented in Fig. 3.8, we see that a reasonable choice for V_{on} is 0.6 to 0.7 V. We use a voltage of 0.6 V as the turn-on voltage for our diode circuit analysis.

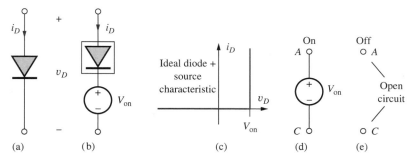

Figure 3.31 Constant voltage drop model for diode: (a) actual diode; (b) ideal diode plus voltage source V_{on}; (c) composite i-v characteristic; (d) CVD model for the on state; (e) model for the off state.

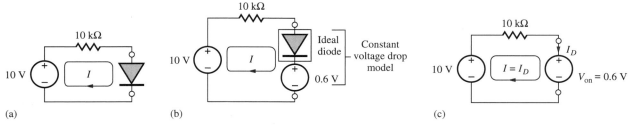

Figure 3.32 Diode circuit analysis using constant voltage drop model: (a) original diode circuit; (b) circuit with diode replaced by the constant voltage drop model; (c) circuit with ideal diode replaced by the piecewise linear model.

Diode Analysis with the Constant Voltage Drop Model

Let us analyze the diode circuit from Fig. 3.22 using the CVD model for the diode. The diode in Fig. 3.32(a) is replaced by its CVD model in Fig. 3.32(b). The 10-V source once again appears to be forward biasing the diode, so assume that the ideal diode is on, resulting in the simplified circuit in Fig. 3.32(c). The diode current is given by

$$I_D = \frac{(10 - V_{on})\text{ V}}{10 \text{ k}\Omega} = \frac{(10 - 0.6)\text{ V}}{10 \text{ k}\Omega} = 0.940 \text{ mA} \tag{3.30}$$

which is slightly smaller than that predicted by the ideal diode model but quite close to the exact result found earlier in Ex. 3.6.

3.10.5 MODEL COMPARISON AND DISCUSSION

We have analyzed the circuit of Fig. 3.22 using four different approaches; the various results appear in Table 3.3. All four sets of predicted voltages and currents are quite similar. Even the simple ideal diode model only overestimates the current by less than 10 percent compared to the mathematical model. We see that the current is quite insensitive to the actual choice of diode voltage. This is a result of the exponential dependence of the diode current on voltage as well as the large source voltage (10 V) in this particular circuit.

TABLE 3.3
Comparison of Diode Circuit Analysis Results

ANALYSIS TECHNIQUE	DIODE CURRENT	DIODE VOLTAGE
Load-line analysis	0.94 mA	0.6 V
Mathematical model	0.942 mA	0.547 V
Ideal diode model	1.00 mA	0 V
Constant voltage drop model	0.940 mA	0.600 V

116   Chapter 3   Solid-State Diodes and Diode Circuits

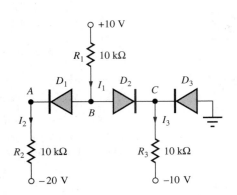

Figure 3.33 Example of a circuit containing three diodes.

TABLE 3.4
Possible Diode States for Circuit in Fig. 3.33

D_1	D_2	D_3
Off	Off	Off
Off	Off	On
Off	On	Off
Off	On	On
On	Off	Off
On	Off	On
On	On	Off
On	On	On

Rewriting Eq. (3.30),

$$I_D = \frac{10 - V_{\text{on}}}{10\ \text{k}\Omega} = \frac{10\ \text{V}}{10\ \text{k}\Omega}\left(1 - \frac{V_{\text{on}}}{10}\right) = (1.00\ \text{mA})\left(1 - \frac{V_{\text{on}}}{10}\right) \quad (3.31)$$

we see that the value of I_D is approximately 1 mA for $V_{\text{on}} \ll 10$ V. Variations in V_{on} have only a small effect on the result. However, the situation would be significantly different if the source voltage were only 1 V for example (see Prob. 3.62).

3.11 MULTIPLE-DIODE CIRCUITS

The load-line technique is applicable only to single-diode circuits, and the mathematical model, or numerical iteration technique, becomes much more complex for circuits with more than one nonlinear element. In fact, the SPICE electronic circuit simulation program referred to throughout this book is designed to provide numerical solutions to just such complex problems. However, we also need to be able to perform hand analysis to predict the operation of multidiode circuits as well as to build our understanding and intuition for diode circuit operation. In this section we discuss the use of the simplified diode models for hand analysis of more complicated diode circuits.

As the complexity of diode circuits increases, we must rely on our intuition to eliminate unreasonable solution choices. Even so, analysis of diode circuits may require several iterations. Intuition can only be developed over time by working problems, and here we analyze a circuit containing three diodes.

Figure 3.33 is an example of a circuit with several diodes. In the analysis of this circuit, we will use the CVD model to improve the accuracy of our hand calculations.

EXAMPLE 3.8   **ANALYSIS OF A CIRCUIT CONTAINING THREE DIODES**

Now we will attempt to find the solution for a three-diode circuit. Our analysis will employ the CVD model.

PROBLEM Find the Q-points for the three diodes in Fig. 3.33. Use the constant voltage drop model for the diodes.

SOLUTION **Known Information and Given Data:** Circuit topology and element values in Fig. 3.33

3.11 Multiple-Diode Circuits

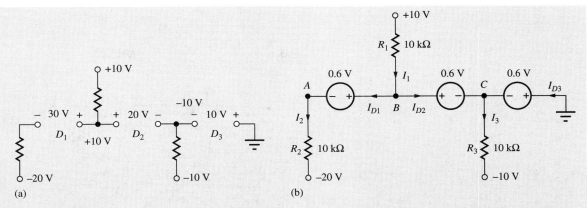

Figure 3.34 (a) Three diode circuit model with all diodes off; (b) circuit model for circuit of Fig. 3.33 with all diodes on.

Unknowns: $(I_{D1}, V_{D1}), (I_{D2}, V_{D2}), (I_{D3}, V_{D3})$

Approach: With three diodes, there are the eight On/Off combinations indicated in Table 3.4. A common method that we often use to find a starting point is to consider the circuit with all the diodes in the off state as in Fig 3.34(a). Here we see that the circuit tends to produce large forward biases across D_1, D_2, and D_3. So our second step will be to assume that all the diodes are on.

Assumptions: Use the constant voltage drop model with $V_{on} = 0.6$ V.

Analysis: The circuit is redrawn using the CVD diode models in Fig. 3.34(b). Here we skipped the step of physically drawing the circuit with the ideal diode symbols but instead incorporated the piecewise linear models directly into the figure. Working from right to left, we see that the voltages at nodes C, B, and A are given by

$$V_C = -0.6 \text{ V} \qquad V_B = -0.6 + 0.6 = 0 \text{ V} \qquad V_A = 0 - 0.6 = -0.6 \text{ V}$$

With the node voltages specified, it is easy to find the current through each resistor:

$$I_1 = \frac{10 - 0}{10} \frac{\text{V}}{\text{k}\Omega} = 1 \text{ mA} \qquad I_2 = \frac{-0.6 - (-20)}{10} \frac{\text{V}}{\text{k}\Omega} = 1.94 \text{ mA}$$

$$I_3 = \frac{-0.6 - (-10)}{10} \frac{\text{V}}{\text{k}\Omega} = 0.94 \text{ mA} \qquad (3.32)$$

Using Kirchhoff's current law, we also have

$$I_2 = I_{D1} \qquad I_1 = I_{D1} + I_{D2} \qquad I_3 = I_{D2} + I_{D3} \qquad (3.33)$$

Combining Eqs. (3.32) and (3.33) yields the three diode currents:

$$I_{D1} = 1.94 \text{ mA} > 0 \checkmark \qquad I_{D2} = -0.94 \text{ mA} < 0 \times \qquad I_{D3} = 1.86 \text{ mA} > 0 \checkmark \qquad (3.34)$$

Check of Results: I_{D1} and I_{D3} are greater than zero and therefore consistent with the original assumptions. However, I_{D2}, which is less than zero, represents a contradiction. So we must try again.

SECOND ITERATION For our second attempt, let us assume D_1 and D_3 are on and D_2 is off, as in Fig. 3.35(a). We now have

$$+10 - 10{,}000I_1 - 0.6 - 10{,}000I_2 + 20 = 0 \qquad \text{with } I_1 = I_{D1} = I_2 \qquad (3.35)$$

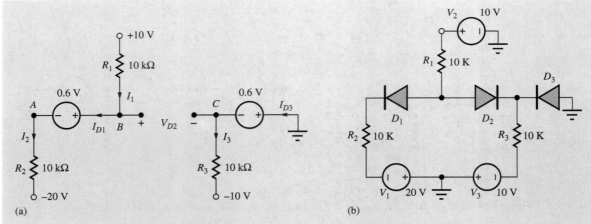

Figure 3.35 (a) Circuit with diodes D_1 and D_3 on and D_2 off; (b) Circuit for SPICE simulation.

which yields

$$I_{D1} = \frac{29.4 \text{ V}}{20 \text{ k}\Omega} = 1.47 \text{ mA} > 0 \checkmark$$

Also

$$I_{D3} = I_3 = \frac{-0.6 - (-10)}{10} \frac{\text{V}}{\text{k}\Omega} = 0.940 \text{ mA} > 0 \checkmark$$

The voltage across diode D_2 is given by

$$V_{D2} = 10 - 10{,}000 I_1 - (-0.6) = 10 - 14.7 + 0.6 = -4.10 \text{ V} < 0 \checkmark$$

Check of Results: I_{D1}, I_{D3}, and V_{D2} are now all consistent with the circuit assumptions, so the Q-points for the circuit are

$$D_1: (1.47 \text{ mA}, 0.6 \text{ V}) \qquad D_2: (0 \text{ mA}, -4.10 \text{ V}) \qquad D_3: (0.940 \text{ mA}, 0.6 \text{ V})$$

Discussion: The Q-point values that we would have obtained using the ideal diode model are (see Prob. 3.73):

$$D_1: (1.50 \text{ mA}, 0 \text{ V}) \qquad D_2: (0 \text{ mA}, -5.00 \text{ V}) \qquad D_3: (1.00 \text{ mA}, 0 \text{ V})$$

The values of I_{D1} and I_{D3} differ by less than 6 percent. However, the reverse-bias voltage on D_2 differs by 20 percent. This shows the difference that the choice of models can make. The results from the circuit using the CVD model should be a more accurate estimate of how the circuit will actually perform than would result from the ideal diode case. Remember, however, that these calculations are both just approximations based on our models for the actual behavior of the real diode circuit.

Computer-Aided Analysis: SPICE analysis yields the following Q-points for the circuit in Fig. 3.35(b): (1.47 mA, 0.665 V), (−4.02 pA, −4.01 V), (0.935 mA, 0.653 V). Device parameter and Q-point information are found directly using the SHOW and SHOWMOD commands in SPICE. Or, voltmeters and ammeters (zero-valued current and voltage sources) can be inserted in the circuit in some implementations of SPICE. Note that the −4 pA current in D_2 is much larger than the reverse saturation current of the diode (IS defaults to 10 fA), and results from a more complete SPICE model in the author's version of SPICE.

> **EXERCISE:** Find the Q-points for the three diodes in Fig. 3.33 if R_1 is changed to 2.5 kΩ.
>
> **ANSWERS:** (2.13 mA, 0.6 V); (1.13 mA, 0.6 V); (0 mA, −1.27 V)
>
> **EXERCISE:** Use SPICE to calculate the Q-points of the diodes in the previous exercise. Use $I_S = 1$ fA.
>
> **ANSWERS:** (2.12 mA, 0.734 V); (1.12 mA, 0.718 V); (0 mA, −1.19 V)

3.12 ANALYSIS OF DIODES OPERATING IN THE BREAKDOWN REGION

Reverse breakdown is actually a highly useful region of operation for the diode. The reverse breakdown voltage is nearly independent of current and can be used as either a voltage regulator or voltage reference. Thus, it is important to understand the analysis of diodes operating in reverse breakdown.

Figure 3.36 is a single-loop circuit containing a 20-V source supplying current to a Zener diode with a reverse breakdown voltage of 5 V. The voltage source has a polarity that will tend to reverse-bias the diode. Because the source voltage exceeds the Zener voltage rating of the diode, $V_Z = 5$ V, we should expect the diode to be operating in its breakdown region.

3.12.1 LOAD-LINE ANALYSIS

The i-v characteristic for this Zener diode is given in Fig. 3.37, and load-line analysis can be used to find the Q-point for the diode, independent of the region of operation. The normal polarities for I_D and V_D are indicated in Fig. 3.36, and the loop equation is

$$-20 = V_D + 5000 I_D \tag{3.36}$$

In order to draw the load line, we choose two points on the graph:

$$V_D = 0, \ I_D = -4 \text{ mA} \quad \text{and} \quad V_D = -5 \text{ V}, \ I_D = -3 \text{ mA}$$

In this case the load line intersects the diode characteristic at a Q-point in the breakdown region: (−2.9 mA, −5.2 V).

3.12.2 ANALYSIS WITH THE PIECEWISE LINEAR MODEL

The assumption of reverse breakdown requires that the diode current I_D be less than zero or that the Zener current $I_Z = -I_D > 0$. We will analyze the circuit with the piecewise linear model and test this condition to see if it is consistent with the reverse-breakdown assumption.

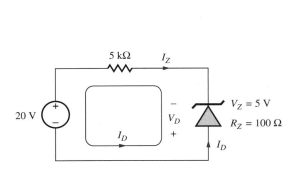

Figure 3.36 Circuit containing a Zener diode with $V_Z = 5$ V and $R_Z = 100$ Ω.

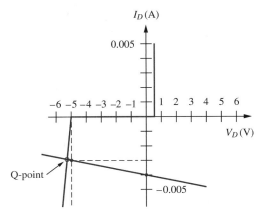

Figure 3.37 Load line for Zener diode.

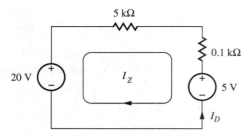
Figure 3.38 Circuit with piecewise linear model for Zener diode. Note that the diode model is valid only in the breakdown region of the characteristic.

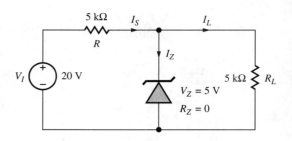
Figure 3.39 Zener diode voltage regulator circuit.

In Fig. 3.38, the Zener diode has been replaced with its piecewise linear model from Fig. 3.16 in Sec. 3.6, with $V_Z = 5$ V and $R_Z = 100$ Ω. Writing the loop equation this time in terms of I_Z:

$$20 - 5100 I_Z - 5 = 0 \quad \text{or} \quad I_Z = \frac{(20-5) \text{ V}}{5100 \text{ Ω}} = 2.94 \text{ mA} \quad (3.37)$$

Because I_Z is greater than zero ($I_D < 0$), the solution is consistent with our assumption of Zener breakdown operation.

It is worth noting that diodes have three possible states when the breakdown region is included, (on, off, and reverse breakdown), further increasing analysis complexity.

3.12.3 VOLTAGE REGULATION

A useful application of the Zener diode is as a **voltage regulator,** as shown in the circuit of Fig. 3.39. The function of the Zener diode is to maintain a constant voltage across load resistor R_L. As long as the diode is operating in reverse breakdown, a voltage of approximately V_Z will appear across R_L. To ensure that the diode is operating in the Zener breakdown region, we must have $I_Z > 0$.

The circuit of Fig. 3.39 has been redrawn in Fig. 3.40 with the model for the Zener diode, with $R_Z = 0$. Using nodal analysis, the Zener current is expressed by $I_Z = I_I - I_L$. The currents I_I and I_L are equal to

$$I_I = \frac{V_I - V_Z}{R} = \frac{(20-5) \text{ V}}{5 \text{ kΩ}} = 3 \text{ mA} \quad \text{and} \quad I_L = \frac{V_Z}{R_L} = \frac{5 \text{ V}}{5 \text{ kΩ}} = 1 \text{ mA} \quad (3.38)$$

resulting in a Zener current $I_Z = 2$ mA. $I_Z > 0$, which is again consistent with our assumptions. If the calculated value of I_Z were less than zero, then the Zener diode no longer controls the voltage across R_L, and the voltage regulator is said to have "dropped out of regulation."

For proper regulation to take place, the Zener current must be positive:

$$I_Z = I_I - I_L = \frac{V_I - V_Z}{R} - \frac{V_Z}{R_L} \geq 0 \quad (3.39)$$

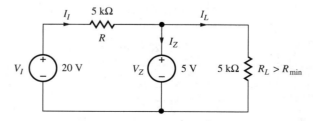
Figure 3.40 Circuit with a constant voltage model for the Zener diode.

Solving for R_L yields a lower bound on the value of load resistance for which the Zener diode will continue to act as a voltage regulator.

$$R_L > \frac{R}{\left(\frac{V_S}{V_Z} - 1\right)} = R_{\min} \quad (3.40)$$

EXERCISE: What is the value of $R_{\min}$ for the Zener voltage regulator circuit in Figs. 3.39 and 3.40? What is the output voltage for $R_L = 1\ \text{k}\Omega$? For $R_L = 2\ \text{k}\Omega$?

ANSWERS: 1.67 kΩ; 3.33 V; 5.00 V

3.12.4 ANALYSIS INCLUDING ZENER RESISTANCE

The voltage regulator circuit in Fig. 3.39 has been redrawn in Fig. 3.41 and now includes a nonzero Zener resistance R_Z. The output voltage is now a function of the current I_Z through the Zener diode. For small values of R_Z, however, the change in output voltage will be small.

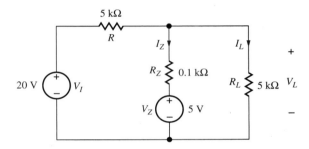

Figure 3.41 Zener diode regulator circuit, including Zener resistance.

EXAMPLE 3.9   **DC ANALYSIS OF A ZENER DIODE REGULATOR CIRCUIT**

Find the operating point for a Zener-diode-based voltage regulator circuit.

PROBLEM Find the output voltage and Zener diode current for the Zener diode regulator in Figs. 3.39 and 3.41 if $R_Z = 100\ \Omega$ and $V_Z = 5$ V.

SOLUTION **Known Information and Given Data:** Zener diode regulator circuit as modeled in Fig. 3.41 with $V_I = 20$ V, $R = 5\ \text{k}\Omega$, $R_Z = 0.1\ \text{k}\Omega$, and $V_Z = 5$ V

Unknowns: V_L, I_Z

Approach: The circuit contains a single unknown node voltage V_L, and a nodal equation can be written to find the voltage. Once V_L is found, I_Z can be determined using Ohm's law.

Assumptions: Use the piecewise linear model for the diode as drawn in Fig. 3.41.

Analysis: Writing the nodal equation for V_L yields

$$\frac{V_L - 20\ \text{V}}{5000\ \Omega} + \frac{V_L - 5\ \text{V}}{100\ \Omega} + \frac{V_L}{5000\ \Omega} = 0$$

Multiplying the equation by 5000 Ω and collecting terms gives

$$52V_L = 270 \text{ V} \quad \text{and} \quad V_L = 5.19 \text{ V}$$

The Zener diode current is equal to

$$I_Z = \frac{V_L - 5 \text{ V}}{100 \text{ }\Omega} = \frac{5.19 \text{ V} - 5 \text{ V}}{100 \text{ }\Omega} = 1.90 \text{ mA} > 0$$

Check of Results: $I_Z > 0$ confirms operation in reverse breakdown. We see that the output voltage of the regulator is slightly higher than for the case with $R_Z = 0$, and the Zener diode current is reduced slightly. Both changes are consistent with the addition of R_Z to the circuit.

Computer-Aided Analysis: We can use SPICE to simulate the Zener circuit if we specify the breakdown voltage using SPICE parameters BV, IBV, and RS. BV sets the breakdown voltage, and IBV represents the current at breakdown. Setting BV = 5 V and RS = 100 Ω and letting IBV default to 1 mA yields $V_L = 5.21$ V and $I_Z = 1.92$ mA, which agree well with our hand calculations. A transfer function analysis from V_S to V_L gives yields a sensitivity of 21 mV/V and an output resistance of 108 Ω. The meaning of these numbers is discussed in the next section.

> **EXERCISE:** Find V_L, I_Z, and the Zener power dissipation in Fig. 3.41 if $R = 1$ kΩ.
>
> **ANSWERS:** 6.25 V; 12.5 mA; 78.1 mW

3.12.5 LINE AND LOAD REGULATION

Two important parameters characterizing a voltage regulator circuit are **line regulation** and **load regulation.** Line regulation characterizes how sensitive the output voltage is to input voltage changes and is expressed as V/V or as a percentage. Load regulation characterizes how sensitive the output voltage is to changes in the load current withdrawn from the regulator and has the units of Ohms.

$$\text{Line regulation} = \frac{dV_L}{dV_I} \quad \text{and} \quad \text{Load regulation} = \frac{dV_L}{dI_L} \quad (3.41)$$

We can find expressions for these quantities from a straightforward analysis of the circuit in Fig. 3.41 similar to that in Ex. 3.9:

$$\frac{V_L - V_I}{R} + \frac{V_L - V_Z}{R_Z} + I_L = 0 \quad (3.42)$$

For a fixed load current, we find the line regulation as

$$\text{Line regulation} = \frac{R_Z}{R + R_Z} \quad (3.43)$$

and for changes in I_L as

$$\text{Load regulation} = -(R_Z \| R) \quad (3.44)$$

The load regulation should be recognized as the Thévenin equivalent resistance looking back into the regulator from the load terminals.

> **EXERCISE:** What are the values of the load and line regulation for the circuit in Fig. 3.41?
>
> **ANSWERS:** 19.6 mV/V; 98.0 Ω. Note that these are close to the SPICE results in Ex. 3.9.

3.13 HALF-WAVE RECTIFIER CIRCUITS

Rectifiers represent an application of diodes that we encounter frequently every day, but they may not be recognized as such. The basic **rectifier circuit** converts an ac voltage to a pulsating dc voltage. An LC or RC filter is then added to eliminate the ac components of the waveform and produce a nearly constant dc voltage output. Virtually every electronic device that is plugged into the wall utilizes a rectifier circuit to convert the 120-V, 60-Hz ac power line source to the various dc voltages required to operate electronic devices such as personal computers, audio systems, radio receivers, televisions, and the like. All of our battery chargers and "wall-warts" contain rectifiers. As a matter of fact, the vast majority of electronic circuits are powered by a dc source, usually based on some form of rectifier.

This section explores half-wave rectifier circuits with capacitor filters that form the basis for many dc power supplies. Up to this point, we have looked at only steady-state dc circuits in which the diode remained in one of its three possible states (on, off, or reverse breakdown). Now, however, the diode state will be changing with time, and a given piecewise linear model for the circuit will be valid for only a certain time interval.

3.13.1 HALF-WAVE RECTIFIER WITH RESISTOR LOAD

A single diode is used to form the **half-wave rectifier circuit** in Fig. 3.42. A sinusoidal voltage source $v_I = V_P \sin \omega t$ is connected to the series combination of diode D_1 and load resistor R. During the first half of the cycle, for which $v_I > 0$, the source forces a current through diode D_1 in the forward direction, and D_1 will be on. During the second half of the cycle, $v_I < 0$. Because a negative current cannot exist in the diode (unless it is in breakdown), it turns off. These two states are modeled in Fig. 3.43 using the ideal diode model.

When the diode is on, voltage source v_S is connected directly to the output and $v_O = v_I$. When the diode is off, the current in the resistor is zero, and the output voltage is zero. The input and output voltage waveforms are shown in Fig. 3.44(b), and the resulting current is called pulsating

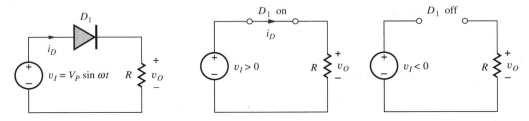

Figure 3.42 Half-wave rectifier circuit.

Figure 3.43 Ideal diode models for the two half-wave rectifier states.

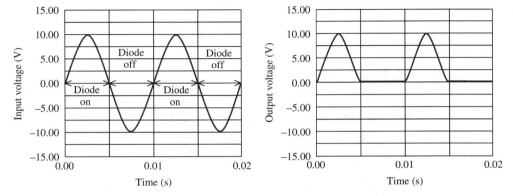

Figure 3.44 Sinusoidal input voltage v_S and pulsating dc output voltage v_O for the half-wave rectifier circuit.

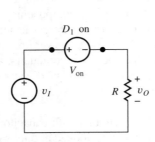

Figure 3.45 CVD model for the rectifier on state.

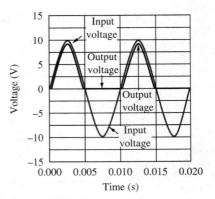

Figure 3.46 Half-wave rectifier output voltage with $V_P = 10$ V and $V_{on} = 0.7$ V.

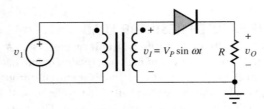

Figure 3.47 Transformer-driven half-wave rectifier.

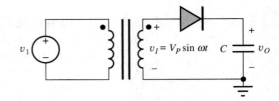

Figure 3.48 Rectifier with capacitor load (peak detector).

direct current. In this circuit, the diode is conducting 50 percent of the time and is off 50 percent of the time.

In some cases, the forward voltage drop across the diode can be important. Figure 3.45 shows the circuit model for the on-state using the CVD model. For this case, the output voltage is one diode-drop smaller than the input voltage during the conduction interval:

$$v_O = (V_P \sin \omega t) - V_{on} \qquad \text{for } V_P \sin \omega t \geq V_{on} \qquad (3.45)$$

The output voltage remains zero during the off-state interval. The input and output waveforms for the half-wave rectifier, including the effect of V_{on}, are shown in Fig. 3.46 for $V_P = 10$ V and $V_{on} = 0.7$ V.

In many applications, a transformer is used to convert from the 120-V ac, 60-Hz voltage available from the power line to the desired ac voltage level, as in Fig. 3.47. The transformer can step the voltage up or down depending on the application; it also enhances safety by providing isolation from the power line. From circuit theory we know that the output of an ideal transformer can be represented by an ideal voltage source, and we use this knowledge to simplify the representation of subsequent rectifier circuit diagrams.

The unfiltered output of the half-wave rectifier in Fig. 3.42 or 3.47 is not suitable for operation of most electronic circuits because constant power supply voltages are required to establish proper bias for the electronic devices. A **filter capacitor** (or more complex circuit) can be added to filter the output of the circuit in Fig. 3.47 to remove the time-varying components from the waveform.

3.13.2 RECTIFIER FILTER CAPACITOR

To understand operation of the rectifier filter, we first consider operation of the **peak-detector** circuit in Fig. 3.48. This circuit is similar to that in Fig. 3.47 except that the resistor is replaced with a capacitor C that is initially discharged $[v_O(0) = 0]$.

Models for the circuit with the diode in the on and off states are in Fig. 3.49, and the input and output voltage waveforms associated with this circuit are in Fig. 3.50. As the input voltage starts to

3.13 Half-Wave Rectifier Circuits

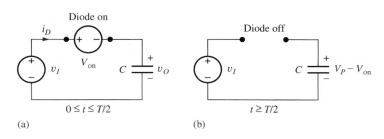

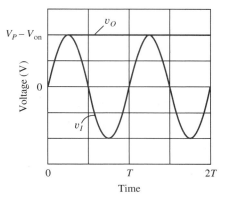

Figure 3.49 Peak-detector circuit models (constant voltage drop model). (a) The diode is on for $0 \le t \le T/2$. (b) The diode is off for $t > T/2$.

Figure 3.50 Input and output waveforms for the peak-detector circuit.

rise, the diode turns on and connects the capacitor to the source. The capacitor voltage equals the input voltage minus the voltage drop across the diode.

At the peak of the input voltage waveform, the current through the diode tries to reverse direction because $i_D = C[d(v_I - V_{on})/dt] < 0$, the diode cuts off, and the capacitor is disconnected from the rest of the circuit. There is no circuit path to discharge the capacitor, so the voltage on the capacitor remains constant. Because the amplitude of the input voltage source v_I can never exceed V_P, the capacitor remains disconnected from v_I for $t > T/2$. Thus, the capacitor in the circuit in Fig. 3.48 charges up to a voltage one diode-drop below the peak of the input waveform and then remains constant, thereby producing a dc output voltage

$$V_{dc} = V_P - V_{on} \tag{3.46}$$

3.13.3 HALF-WAVE RECTIFIER WITH RC LOAD

To make use of this output voltage, a load must be connected to the circuit as represented by the resistor R in Fig. 3.51. Now there is a path available to discharge the capacitor during the time the diode is not conducting. Models for the conducting and nonconducting time intervals are shown in Fig. 3.52; the waveforms for the circuit are shown in Fig. 3.53. The capacitor is again assumed to be

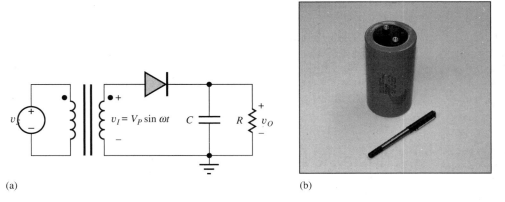

Figure 3.51 (a) Half-wave rectifier circuit with filter capacitor; (b) a-175,000-μF, 15-V filter capacitor. Capacitance tolerance is −10 percent, +75 percent.

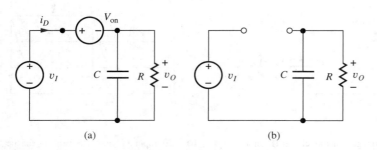

Figure 3.52 Half-wave rectifier circuit models. (a) Diode on. (b) Diode off.

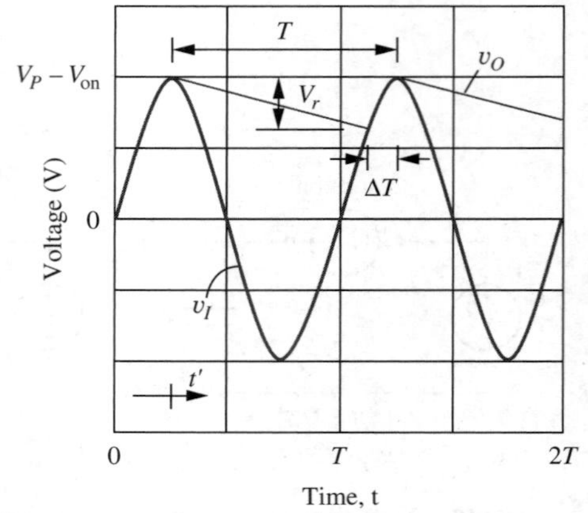

Figure 3.53 Input and output voltage waveforms for the half-wave rectifier circuit ($RC \gg T$).

initially discharged and the time constant RC is assumed to be $\gg T$. During the first quarter cycle, the diode conducts, and the capacitor is rapidly charged toward the peak value of the input voltage source. The diode cuts off at the peak of v_I, and the capacitor voltage then discharges exponentially through the resistor R, as governed by the circuit in Fig. 3.52(b). The discharge continues until the voltage $v_I - v_{on}$ exceeds the output voltage v_O, which occurs near the peak of the next cycle. The process is then repeated once every cycle.

3.13.4 RIPPLE VOLTAGE AND CONDUCTION INTERVAL

The output voltage is no longer constant as in the ideal peak-detector circuit but has a **ripple voltage** V_r. In addition, the diode only conducts for a short time ΔT during each cycle. This time ΔT is called the **conduction interval,** and its angular equivalent is the **conduction angle θ_c** where $\theta_c = \omega \Delta T$. The variables ΔT, θ_c, and V_r are important values related to dc power supply design, and we will now develop expressions for these parameters.

During the discharge period, the voltage across the capacitor is described by

$$v_o(t') = (V_P - V_{on}) \exp\left(-\frac{t'}{RC}\right) \qquad \text{for } t' = \left(t - \frac{T}{4}\right) \geq 0 \qquad (3.47)$$

We have referenced the t' time axis to $t = T/4$ to simplify the equation. The ripple voltage V_r is given by

$$V_r = (V_P - V_{on}) - v_o(t') = (V_P - V_{on})\left[1 - \exp\left(-\frac{T - \Delta T}{RC}\right)\right] \qquad (3.48)$$

A small value of V_r is desired in most power supply designs; a small value requires RC to be much greater than $T - \Delta T$. Using $\exp(-x) \cong 1 - x$ for small x results in an approximate expression for the ripple voltage:

$$V_r \cong (V_P - V_{on})\frac{T}{RC}\left(1 - \frac{\Delta T}{T}\right) \qquad (3.49)$$

A small ripple voltage also means $\Delta T \ll T$, and the final simplified expression for the ripple voltage becomes

$$V_r \cong \frac{(V_P - V_{on})}{R}\frac{T}{C} = I_{dc}\frac{T}{C} \qquad (3.50)$$

where

$$I_{dc} = \frac{V_P - V_{on}}{R} \qquad (3.51)$$

The approximation of the exponential used in Eqs. (3.49) and (3.50) is equivalent to assuming that the capacitor is being discharged by a constant current so that the discharge waveform is a straight line. The ripple voltage V_R can be considered to be determined by an equivalent dc current discharging the capacitor C for a time period T (that is, $\Delta V = (I_{dc}/C)\,T$).

Approximate expressions can also be obtained for conduction angle θ_C and conduction interval ΔT. At time $t = \frac{5}{4}T - \Delta T$, the input voltage just exceeds the output voltage, and the diode is conducting. Therefore, $\theta = \omega t = 5\pi/2 - \theta_C$ and

$$V_p \sin\left(\frac{5}{2}\pi - \theta_C\right) - V_{on} = (V_P - V_{on}) - V_r \qquad (3.52)$$

Remembering that $\sin(5\pi/2 - \theta_C) = \cos\theta_C$, we can simplify the above expression to

$$\cos\theta_C = 1 - \frac{V_r}{V_P} \qquad (3.53)$$

For small values of θ_C, $\cos\theta_C \cong 1 - \theta_C^2/2$. Solving for the conduction angle and conduction interval gives

$$\theta_C = \sqrt{\frac{2V_r}{V_P}} \quad \text{and} \quad \Delta T = \frac{\theta_C}{\omega} = \frac{1}{\omega}\sqrt{\frac{2V_r}{V_P}} \qquad (3.54)$$

EXAMPLE 3.10 HALF-WAVE RECTIFIER ANALYSIS

Here we see an illustration of numerical results for a half-wave rectifier with a capacitive filter.

PROBLEM Find the value of the dc output voltage, dc output current, ripple voltage, conduction interval, and conduction angle for a half-wave rectifier driven from a transformer having a secondary voltage of 12.6 V_{rms} (60 Hz) with $R = 15\ \Omega$ and $C = 25{,}000\ \mu F$. Assume the diode on-voltage $V_{on} = 1$ V.

SOLUTION **Known Information and Given Data:** Half-wave rectifier circuit with RC load as depicted in Fig. 3.51; transformer secondary voltage is 12.6 V_{rms}, operating frequency is 60 Hz, $R = 15\ \Omega$, and $C = 25{,}000\ \mu F$.

Unknowns: dc output voltage V_{dc}, output current I_{dc}, ripple voltage V_r, conduction interval ΔT, conduction angle θ_C

Approach: Given data can be used directly to evaluate Eqs. (3.46), (3.50), (3.51), and (3.54).

Assumptions: Diode on-voltage is 1 V. Remember that the derived results assume the ripple voltage is much less than the dc output voltage ($V_r \ll V_{dc}$) and the conduction interval is much less than the period of the ac signal ($\Delta T \ll T$).

Analysis: The ideal dc output voltage in the absence of ripple is given by Eq. (3.46):

$$V_{dc} = V_P - V_{on} = \left(12.6\sqrt{2} - 1\right)\ \text{V} = 16.8\ \text{V}$$

The nominal dc current delivered by the supply is

$$I_{dc} = \frac{V_P - V_{on}}{R} = \frac{16.8 \text{ V}}{15 \, \Omega} = 1.12 \text{ A}$$

The ripple voltage is calculated using Eq. (3.50) with the discharge interval $T = 1/60$ s:

$$V_r \cong I_{dC}\frac{T}{C} = 1.12 \text{ A} \frac{\frac{1}{60}\text{ s}}{2.5 \times 10^{-2}\text{ F}} = 0.747 \text{ V}$$

The conduction angle is calculated using Eq. (3.54)

$$\theta_c = \omega \Delta T = \sqrt{\frac{2V_r}{V_P}} = \sqrt{\frac{2 \cdot 0.75}{17.8}} = 0.290 \text{ rad or } 16.6°$$

and the conduction interval is

$$\Delta T = \frac{\theta_c}{\omega} = \frac{\theta_c}{2\pi f} = \frac{0.29}{120\pi} = 0.769 \text{ ms}$$

Check of Results: The ripple voltage represents 4.4 percent of the dc output voltage. Thus the assumption that the voltage is approximately constant is justified. The conduction time is 0.769 ms out of a total period $T = 16.7$ ms, and the assumption that $\Delta T \ll T$ is also satisfied.

Discussion: From this example, we see that even a 1-A power supply requires a significant filter capacitance C to maintain a low ripple percentage. In this case, $C = 0.025\text{F} = 25,000$ μF.

EXERCISE: Find the value of the dc output voltage, dc output current, ripple voltage, conduction interval, and conduction angle for a half-wave rectifier that is being supplied from a transformer having a secondary voltage of 6.3 V_{rms} (60 Hz) with $R = 0.5$ Ω and $C = 500{,}000$ μF. Assume the diode on voltage $V_{on} = 1$ V.

ANSWERS: 7.91 V; 15.8 A; 0.527 V; 0.912 ms; 19.7°

EXERCISE: What are the values of the dc output voltage and dc output current for a half-wave rectifier that is being supplied from a transformer having a secondary voltage of 10 V_{rms} (60 Hz) and a 2-Ω load resistor? Assume the diode on voltage $V_{on} = 1$ V. What value of filter capacitance is required to have a ripple voltage of no more than 0.1 V? What is the conduction angle?

ANSWERS: 13.1 V; 6.57 A; 1.10 F; 6.82°

3.13.5 DIODE CURRENT

In rectifier circuits, a nonzero current is present in the diode for only a very small fraction of the period T, yet an almost constant dc current is flowing out of the filter capacitor to the load. The total charge lost from the capacitor during each cycle must be replenished by the current through the diode during the short conduction interval ΔT, which leads to high peak diode currents. Figure 3.54 shows

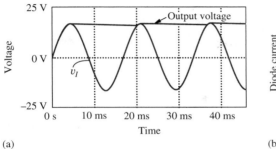
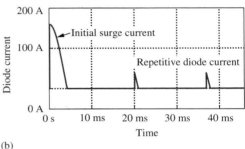

Figure 3.54 SPICE simulation of the half-wave rectifier circuit: (a) voltage waveforms; (b) diode current.

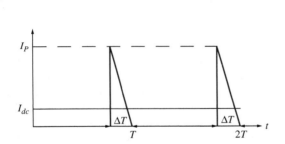

Figure 3.55 Triangular approximation to diode current pulse.

Figure 3.56 Peak reverse voltage across the diode in a half-wave rectifier.

the results of SPICE simulation of the diode current. The repetitive current pulse can be modeled approximately by a triangle of height I_P and width ΔT, as in Fig. 3.55.

Equating the charge supplied through the diode during the conduction interval to the charge lost from the filter capacitor during the complete period yields

$$Q = I_P \frac{\Delta T}{2} = I_{dc} T \quad \text{or} \quad \boxed{I_P = I_{dc} \frac{2T}{\Delta T}} \tag{3.55}$$

Here we remember that the integral of current over time represents charge Q. Therefore the charge supplied by the triangular current pulse in Fig. 3.55 is given by the area of the triangle, $I_P \Delta T / 2$.

For Ex. 3.10, the peak diode current would be

$$I_P = 1.12 \frac{2 \cdot 16.7}{0.769} = 48.6 \text{ A} \tag{3.56}$$

which agrees well with the simulation results in Fig. 3.55. The diode must be built to handle these high peak currents, which occur over and over. This high peak current is also the reason for the relatively large choice of V_{on} used in Ex. 3.10. (See Prob. 3.82.)

> **EXERCISE:** (a) What is the forward voltage of a diode operating at a current of 48.6 A at 300 K if $I_S = 10^{-15}$ A? (b) At 50 C?
>
> **ANSWERS:** 0.994 V; 1.07 V

3.13.6 SURGE CURRENT

When the power supply is first turned on, the capacitor is completely discharged, and there will be an even larger current through the diode, as is visible in Fig. 3.54. During the first quarter cycle, the current through the diode is given approximately by

$$i_d(t) = i_c(t) \cong C \left[\frac{d}{dt} V_P \sin \omega t \right] = \omega C V_P \cos \omega t \qquad (3.57)$$

The peak value of this initial **surge current** occurs at $t = 0^+$ and is given by

$$I_{SC} = \omega C V_P = 2\pi (60 \text{ Hz})(0.025 \text{ F})(17.8 \text{ V}) = 168 \text{ A}$$

Using the numbers from Ex. 3.10 yields an initial surge current of almost 170 A! This value, again, agrees well with the simulation results in Fig. 3.54. If the input signal v_I does not happen to be crossing through zero when the power supply is turned on, the situation can be even worse, and rectifier diodes selected for power supply applications must be capable of withstanding very large surge currents as well as the large repetitive current pulses required each cycle.

In most practical circuits, the surge current will be large but cannot actually reach the values predicted by Eq. (3.57) because of series resistances in the circuit that we have neglected. The rectifier diode itself will have an internal series resistance (review the SPICE model in Sec. 3.9 for example), and the transformer will have resistances associated with both the primary and secondary windings. A total series resistance in the secondary of only a few tenths of an ohm will significantly reduce both the surge current and peak repetitive current in the circuit. In addition, the large time constant associated with the series resistance and filter capacitance causes the rectifier output to take many cycles to reach its steady-state voltage. (See SPICE simulation problems at the end of this chapter.)

3.13.7 PEAK-INVERSE-VOLTAGE (PIV) RATING

We must also be concerned about the breakdown voltage rating of the diodes used in rectifier circuits. This breakdown voltage is called the **peak-inverse-voltage (PIV)** rating of the rectifier diode. The worst-case situation for the half-wave rectifier is depicted in Fig. 3.56 in which it is assumed that the ripple voltage V_r is very small. When the diode is off, as in Fig. 3.52(b), the largest reverse bias across the diode is equal to $V_{dc} - v_I$ which occurs when v_I reaches its negative peak of $-V_P$. The diode must therefore be able to withstand a reverse bias of at least

$$\text{PIV} \geq V_{dc} - v_I^{\min} = V_P - V_{on} - (-V_P) = 2V_P - V_{on} \cong 2V_P \qquad (3.58)$$

From Eq. (3.58), we see that diodes used in the half-wave rectifier circuit must have a PIV rating equal to twice the peak voltage supplied by the source v_I. The PIV value corresponds to the minimum value of Zener breakdown voltage for the rectifier diode. A safety margin of at least 25 to 50 percent is usually specified for the diode PIV rating in power supply designs.

3.13.8 DIODE POWER DISSIPATION

In high-current power supply applications, the power dissipation in the rectifier diodes can become significant. The average power dissipation in the diode is defined by

$$P_D = \frac{1}{T} \int_0^T v_D(t) i_D(t) \, dt \qquad (3.59)$$

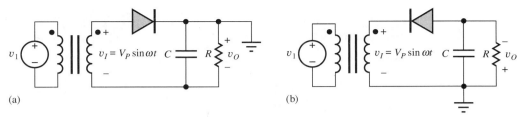

Figure 3.57 Half-wave rectifier circuits that develop negative output voltages.

This expression can be simplified by assuming that the voltage across the diode is approximately constant at $v_D(t) = V_{on}$ and by using the triangular approximation to the diode current $i_D(t)$ shown in Fig. 3.55. Eq. (3.59) becomes

$$P_D = \frac{1}{T}\int_0^T V_{on} i_D(t)\,dt = \frac{V_{on}}{T}\int_{T-\Delta T}^T i_D(t)\,dt = V_{on}\frac{I_P}{2}\frac{\Delta T}{T} = V_{on}I_{dc} \qquad (3.60)$$

Using Eq. (3.55) we see that the power dissipation is equivalent to the constant dc output current multiplied by the on-voltage of the diode. For the half-wave rectifier example, $P_D = (1\text{ V})(1.1\text{ A}) = 1.1$ W. This rectifier diode would probably need a heat sink to maintain its temperature at a reasonable level. Note that the average current through the diode is I_{dc}.

Another source of power dissipation is caused by resistive loss within the diode. Diodes have a small internal series resistance R_S, and the average power dissipation in this resistance can be calculated using

$$P_D = \frac{1}{T}\int_0^T i_D^2(t)R_S\,dt \qquad (3.61)$$

Evaluation of this integral (left for Prob. 3.87) for the triangular current wave form in Fig. 3.55 yields

$$P_D = \frac{1}{3}I_P^2 R_S \frac{\Delta T}{T} = \frac{4}{3}\frac{T}{\Delta T}I_{dc}^2 R_S \qquad (3.62)$$

Using the number from the rectifier example with $R_S = 0.20\ \Omega$ yields $P_D = 7.3$ W! This is significantly greater than the component of power dissipation caused by the diode on-voltage calculated using Eq. (3.60). The component of power dissipation described by Eq. (3.62) can be reduced by minimizing the peak current I_P through the use of the minimum required size of filter capacitor or by using the full-wave rectifier circuits, which are discussed in Sec. 3.14.

3.13.9 HALF-WAVE RECTIFIER WITH NEGATIVE OUTPUT VOLTAGE

The circuit of Fig. 3.51 can also be used to produce a negative output voltage if the top rather than the bottom of the capacitor is grounded, as depicted in Fig. 3.57(a) or by reversing the direction of the diode in the original circuit as in Fig. 3.57(b). These two circuits are equivalent. In the circuit in Fig. 3.57(b), the diode conducts on the negative half cycle of the transformer voltage v_I, and the dc output voltage is $V_{dc} = -(V_P - V_{on})$.

ELECTRONICS IN ACTION

AM Demodulation

The waveform for a 100 percent amplitude modulated (AM) signal is shown in the figure below and described mathematically by $v_{AM} = 2 \sin \omega_C t (1 + \sin \omega_M t)$ V in which ω_C is the carrier frequency ($f_C = 50$ kHz) and ω_M is the modulating frequency ($f_M = 5$ kHz). The envelope of

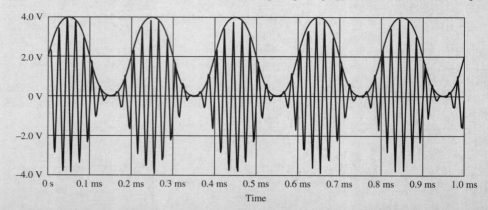

the AM signal contains the information being transmitted, and the envelope can be recovered from the signal using a simple half-wave rectifier. In the SPICE circuit below, the signal to be demodulated is applied as the input signal to the rectifier, and the rectifier, and the $R_2 C_1$ time

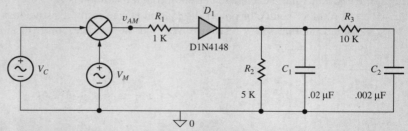

constant is set to filter out the carrier frequency but follow the signal's envelope. Additional filtering is provided by the low-pass filter formed by R_3 and C_2. SPICE simulation results appear below along with the results of a Fourier analysis of the demodulated signal. The plots of v_{C1} and v_{C2} represent the voltages across capacitors C_1 and C_2 respectively.

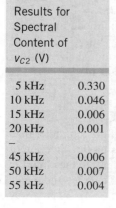

SPICE Results for Spectral Content of v_{C2} (V)

5 kHz	0.330
10 kHz	0.046
15 kHz	0.006
20 kHz	0.001
—	
45 kHz	0.006
50 kHz	0.007
55 kHz	0.004

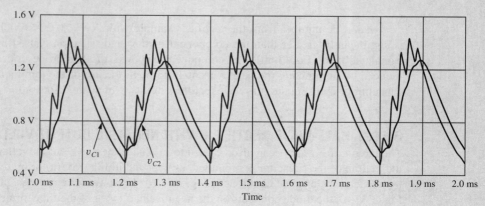

3.14 FULL-WAVE RECTIFIER CIRCUITS

Full-wave rectifier circuits cut the capacitor discharge time in half and offer the advantage of requiring only one-half the filter capacitance to achieve a given ripple voltage. The full-wave rectifier circuit in Fig. 3.58 uses a **center-tapped transformer** to generate two voltages that have equal amplitudes but are 180 degrees out of phase. With voltage v_I applied to the anode of D_1, and $-v_I$ applied to the anode of D_2, the two diodes form a pair of half-wave rectifiers operating on alternate half cycles of the input waveform. Proper phasing is indicated by the dots on the two halves of the transformer.

For $v_I > 0$, D_1 will be functioning as a half-wave rectifier, and D_2 will be off, as indicated in Fig. 3.59. The current exits the upper terminal of the transformer, goes through diode D_1, through the RC load, and returns back into the center tap of the transformer.

For $v_I < 0$, D_1 will be off, and D_2 will be functioning as a half-wave rectifier as indicated in Fig. 3.60. During this portion of the cycle, the current path leaves the bottom terminal of the transformer, goes through D_2, down through the RC load, and again returns into the transformer center tap. The current direction in the load is the same during both halves of the cycle; one-half of the transformer is utilized during each half cycle.

The load, consisting of the filter capacitor C and load resistor R, now receives two current pulses per cycle, and the capacitor discharge time is reduced to less than $T/2$, as indicated in the graph in Fig. 3.61. An analysis similar to that for the half-wave rectifier yields the same formulas for dc output voltage, ripple voltage, and ΔT, except that the discharge interval is $T/2$ rather than T. For a given capacitor value, the ripple voltage is one-half as large, and the conduction interval and peak

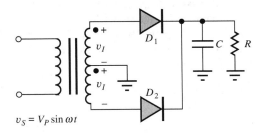

Figure 3.58 Full-wave rectifier circuit using two diodes and a center-tapped transformer. This circuit produces a positive output voltage.

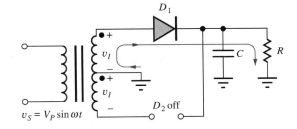

Figure 3.59 Equivalent circuit for $v_I > 0$.

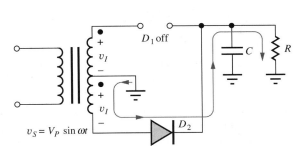

Figure 3.60 Equivalent circuit for $v_I < 0$.

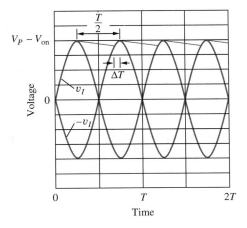

Figure 3.61 Voltage waveforms for the full-wave rectifier.

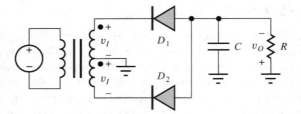

Figure 3.62 Full-wave rectifier with negative output voltage.

current are reduced. The peak-inverse-voltage waveform for each diode is similar to the one shown in Fig. 3.56 for the half-wave rectifier, with the result that the PIV rating of each diode is the same as in the half-wave rectifier.

3.14.1 FULL-WAVE RECTIFIER WITH NEGATIVE OUTPUT VOLTAGE

By reversing the polarity of the diodes, as in Fig. 3.62, a full-wave rectifier circuit with a negative output voltage is realized. Other aspects of the circuit remain the same as the previous full-wave rectifiers with positive output voltages.

3.15 FULL-WAVE BRIDGE RECTIFICATION

The requirement for a center-tapped transformer in the full-wave rectifier can be eliminated through the use of two additional diodes in the **full-wave bridge rectifier circuit** configuration shown in Fig. 3.63. For $v_I > 0$, D_2 and D_4 will be on and D_1 and D_3 will be off, as indicated in Fig. 3.64. Current exits the top of the transformer, goes through D_2 into the RC load, and returns to the transformer through D_4. The full transformer voltage, now minus two diode voltage drops, appears across the load capacitor yielding a dc output voltage

$$V_{dc} = V_P - 2V_{on} \qquad (3.63)$$

The peak voltage at node 1, which represents the maximum reverse voltage appearing across D_1, is equal to $(V_P - V_{on})$. Similarly, the peak reverse voltage across diode D_3 is $(V_P - 2V_{on}) - (-V_{on}) = (V_P - V_{on})$.

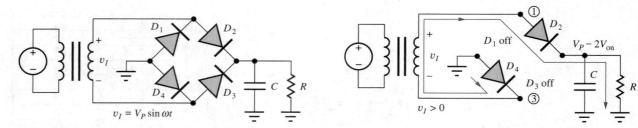

Figure 3.63 Full-wave bridge rectifier circuit with positive output voltage.

Figure 3.64 Full-wave bridge rectifier circuit for $v_I > 0$.

For $v_I < 0$, D_1 and D_3 will be on and D_2 and D_4 will be off, as depicted in Fig. 3.65. Current leaves the bottom of the transformer, goes through D_3 into the RC load, and back through D_1 to the transformer. The full transformer voltage is again being utilized. The peak voltage at node 3 is now equal to $(V_P - V_{on})$ and is the maximum reverse voltage appearing across D_4. Similarly, the peak reverse voltage across diode D_2 is $(V_P - 2V_{on}) - (-V_{on}) = (V_P - V_{on})$.

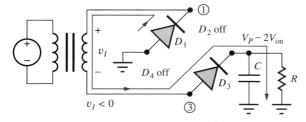

Figure 3.65 Full-wave bridge rectifier circuit for $v_I < 0$.

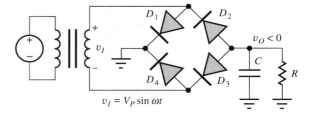

Figure 3.66 Full-wave bridge rectifier circuit with $v_O < 0$.

From the analysis of the two half cycles, we see that each diode must have a PIV rating given by

$$\text{PIV} = V_P - V_{\text{on}} \cong V_P \tag{3.64}$$

As with the previous rectifier circuits, a negative output voltage can be generated by reversing the direction of the diodes, as in the circuit in Fig. 3.66.

3.16 RECTIFIER COMPARISON AND DESIGN TRADEOFFS

Tables 3.5 and 3.6 summarize the characteristics of the half-wave, full-wave, and full-wave bridge rectifiers introduced in Secs. 3.13 to 3.15. The filter capacitor often represents a significant economic factor in terms of cost, size, and weight in the design of **rectifier circuits.** For a given ripple voltage, the value of the filter capacitor required in the full-wave rectifier is one-half that for the half-wave rectifier.

The reduction in peak current in the full-wave rectifier can significantly reduce heat dissipation in the diodes. The addition of the second diode and the use of a center-tapped transformer represent additional expenses that offset some of the advantage. However, the benefits of full-wave rectification usually outweigh the minor increase in circuit complexity.

The bridge rectifier eliminates the need for the center-tapped transformer, and the PIV rating of the diodes is reduced, which can be particularly important in high-voltage circuits. The cost of the extra diodes is usually negligible, particularly since four-diode bridge rectifiers can be purchased in single-component form.

TABLE 3.5
Rectifier Equation Summary

HALF-WAVE RECTIFIER	FULL-WAVE RECTIFIER	FULL-WAVE BRIDGE RECTIFIER
$V_{dc} = V_P - V_{\text{on}} \quad I_{dc} = \dfrac{(V_P - V_{\text{on}})}{R}$	$V_{dc} = V_P - V_{\text{on}} \quad I_{dc} = \dfrac{(V_P - V_{\text{on}})}{R}$	$V_{dc} = V_P - 2V_{\text{on}} \quad I_{dc} = \dfrac{(V_P - 2V_{\text{on}})}{R}$
$V_r = \dfrac{(V_P - V_{\text{on}})}{R} \dfrac{T}{C} = I_{dc} \dfrac{T}{C}$	$V_r = \dfrac{(V_P - V_{\text{on}})}{R} \dfrac{T}{2C} = I_{dc} \dfrac{T}{2C}$	$V_r = \dfrac{(V_P - 2V_{\text{on}})}{R} \dfrac{T}{2C} = I_{dc} \dfrac{T}{2C}$
$\Delta T = \dfrac{1}{\omega}\sqrt{\dfrac{2V_r}{V_P}} \quad \theta_c = \omega \Delta T$	$\Delta T = \dfrac{1}{\omega}\sqrt{\dfrac{2V_r}{V_P}} \quad \theta_c = \omega \Delta T$	$\Delta T = \dfrac{1}{\omega}\sqrt{\dfrac{2V_r}{V_P}} \quad \theta_c = \omega \Delta T$
$I_P = I_{dc} \dfrac{2T}{\Delta T} \quad \text{PIV} = 2V_P$	$I_P = I_{dc} \dfrac{T}{\Delta T} \quad \text{PIV} = 2V_P$	$I_P = I_{dc} \dfrac{T}{\Delta T} \quad \text{PIV} = V_P$

TABLE 3.6
Comparison of Rectifiers with Capacitive Filters

RECTIFIER PARAMETER	HALF-WAVE RECTIFIER	FULL-WAVE RECTIFIER	FULL-WAVE BRIDGE RECTIFIER
Filter capacitor	C	$\dfrac{C}{2}$	$\dfrac{C}{2}$
PIV rating	$2V_P$	$2V_P$	V_P
Peak diode current (constant V_r)	Highest I_P	Reduced $\dfrac{I_P}{2}$	Reduced $\dfrac{I_P}{2}$
Surge current	Highest	Reduced ($\propto C$)	Reduced ($\propto C$)
Comments	Least complexity	Smaller capacitor Requires center-tapped transformer Two diodes	Smaller capacitor Four diodes No center tap on transformer

DESIGN EXAMPLE 3.11

RECTIFIER DESIGN

Now we will use our rectifier theory to design a rectifier circuit that will provide a specified output voltage and ripple voltage.

PROBLEM Design a rectifier to provide a dc output voltage of 15 V with no more than 1 percent ripple at a load current of 2 A.

SOLUTION **Known Information and Given Data:** $V_{dc} = 15$ V, $V_r < 0.15$ V, $I_{dc} = 2$ A

Unknowns: Circuit topology, transformer voltage, filter capacitor, diode PIV rating, diode repetitive current rating, diode surge current rating.

Approach: Use given data to evaluate rectifier circuit equations. Let us choose a full-wave bridge topology that requires a smaller value of filter capacitance, a smaller diode PIV voltage, and no center tap in the transformer.

Assumptions: Assume diode on-voltage is 1 V. The ripple voltage is much less than the dc output voltage ($V_r \ll V_{dc}$), and the conduction interval should be much less than the period of the ac signal ($\Delta T \ll T$).

Analysis: The required transformer voltage is

$$V = \frac{V_P}{\sqrt{2}} = \frac{V_{dc} + 2V_{\text{on}}}{\sqrt{2}} = \frac{15 + 2}{\sqrt{2}} \text{ V} = 12.0 \text{ V}_{\text{rms}}$$

The filter capacitor is found using the ripple voltage, output current, and discharge interval:

$$C = I_{dc}\left(\frac{T/2}{V_r}\right) = 2\,\text{A}\left(\frac{1}{120}\,\text{s}\right)\left(\frac{1}{0.15\,\text{V}}\right) = 0.111\,\text{F}$$

To find I_P, the conduction time is calculated using Eq. (3.54)

$$\Delta T = \frac{1}{\omega}\sqrt{\frac{2V_r}{V_P}} = \frac{1}{120\pi}\sqrt{\frac{2(0.15)\text{ V}}{17\text{ V}}} = 0.352 \text{ ms}$$

and the peak repetitive current is found to be

$$I_P = I_{dc}\left(\frac{2}{\Delta T}\right)\left(\frac{T}{2}\right) = 2\text{ A}\frac{(1/60)\text{ s}}{0.352 \text{ ms}} = 94.7 \text{ A}$$

The surge current estimate is

$$I_{\text{surge}} = \omega C V_P = 120\pi(0.111)(17) = 711 \text{ A}$$

The minimum diode PIV is $V_P = 17$ V. A choice with a safety margin would be PIV > 20 V. The repetitive current rating should be 95 A with a surge current rating of 710 A. Note that both of these calculations overestimate the magnitude of the currents because we have neglected series resistance of the transformer and diode. The minimum filter capacitor needs to be 111,000 μF. Assuming a tolerance of -30 percent, a nominal filter capacitance of 160,000 μF would be required.

Check of Results: The ripple voltage is designed to be 1 percent of the dc output voltage. Thus the assumption that the voltage is approximately constant is justified. The conduction time is 0.352 ms out of a total period $T = 16.7$ ms. Thus the assumption that $\Delta T \ll T$ is satisfied.

Computer-Aided Analysis: This design example represents an excellent place where simulation can be used to explore the magnitude of the diode currents and improve the design so that we don't over-specify the rectifier diodes. A SPICE simulation with $R_S = 0.1$ Ω, $n = 2$, $I_S = 1$ μA, and a transformer series resistance of 0.1 Ω yields a number of unexpected results: $I_P = 11$ A, $I_{\text{surge}} = 70$ A, and $V_{dc} = 13$ V! The surge current and peak repetitive current are both reduced by almost an order of magnitude compared to our hand calculations! In addition the output voltage is lower than expected. If we think further, a peak current of 11 A will cause a peak voltage drop of 2.2 V across the total series resistance of 0.2 Ω, so it should not be surprising that the output voltage is 2 V lower than originally expected. The series resistances actually help to reduce the stress on the diodes. The time constant of the series resistance and the filter capacitor is 0.44 s, so the circuit takes many cycles to reach the steady-state output voltage.

EXERCISE: Repeat the rectifier design assuming the use of a half-wave rectifier.

ANSWERS: $V = 11.3$ V_{rms}; $C = 222,000$ μF; $I_P = 184$ A; $I_{SC} = 1340$ A.

ELECTRONICS IN ACTION

Power Cubes and Cell Phone Chargers

We actually encounter the unfiltered transformer driven half-wave rectifier circuit depicted in Fig. 3.47 frequently in our everyday lives in the form of "power cubes" and battery chargers for many portable electronic devices. An example is shown in the accompanying figure. The power cube contains only a small transformer and rectifier diode. The transformer is wound with small wire and has a significant resistance in both the primary and secondary windings. In the transformer in the photograph, the primary resistance is 600 Ω and the secondary resistance is 15 Ω, and these resistances actually help provide protection from failure of the transformer windings. Load resistance R in Fig. 3.51 represents the actual electronic device that is receiving power from the power cube and may often be a rechargable battery. In some cases, a filter capacitor may be included as part of the circuit that forms the load for the power cube.

Part (c) of the figure below shows a much more complex device used for recharging the batteries in a cell phone. The simplified schematic in part (c) utilizes a full-wave bridge rectifier with filter capacitor connected directly to the ac line. The rectifier's high voltage output is filtered by capacitor C_1 and feeds a switching regulator consisting of a switch, the transformer driving a half-wave rectifier with pi-filter (D_5, C_2, L, and C_3), and a feedback circuit that controls the output voltage by modulating the duty cycle of the switch. The transformer steps down the voltage and provides isolation from the high voltage ac line input. Diode D_6 and R clamp the inductor voltage when the switch opens. The feedback signal path is isolated from the input using an optical isolator. (See Electronics in Action in Chapter 5 for discussion of an optical isolator.) Note the wide range of input voltages accomodated by the circuit. Thus, most international voltage standards can be accommodated by one adopter.

(a)

(b)

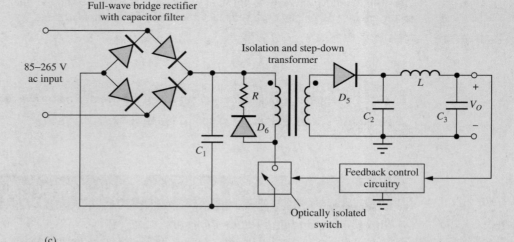

(c)

(a) Inside a simple power cube; (b) cell phone charger; (c) simplified schematic for the cell phone charger.

3.17 DYNAMIC SWITCHING BEHAVIOR OF THE DIODE

Up to this point, we have tacitly assumed that diodes can turn on and off instantaneously. However, an unusual phenomenon characterizes the dynamic switching behavior of the *pn* junction diode. SPICE simulation is used to illustrate the switching of the diode in the circuit in Fig. 3.67, in which diode D_1 is being driven from voltage source v_I through resistor R_1.

The source is zero for $t < 0$. At $t = 0$, the source voltage rapidly switches to $+1.5$ V, forcing a current into the diode to turn it on. The voltage remains constant until $t = 7.5$ ns. At this point the source switches to -1.5 V in order to turn the diode back off.

The simulation results are presented in Fig. 3.68. Following the voltage source change at $t = 0+$, the current increases rapidly, but the internal capacitance of the diode prevents the diode voltage from changing instantaneously. The current actually overshoots its final value and then decreases as the diode turns on and the diode voltage increases to approximately 0.7 V. At any given time, the current flowing into the diode is given by

$$i_D(t) = \frac{v_1(t) - v_D(t)}{0.75 \text{ k}\Omega} \tag{3.65}$$

The initial peak of the current occurs when v_I reaches 1.5 V and v_D is still nearly zero:

$$i_{D\max} = \frac{1.5 \text{ V}}{0.75 \text{ k}\Omega} = 2.0 \text{ mA} \tag{3.66}$$

After the diode voltage reaches its final value with $V_{\text{on}} \approx 0.7$ V, the current stabilizes at a forward current I_F of

$$I_F = \frac{1.5 - 0.7}{0.75 \text{ k}\Omega} = 1.1 \text{ mA} \tag{3.67}$$

At $t = 7.5$ ns, the input source rapidly changes polarity to -1.5 V, and a surprising thing happens. The diode current also rapidly reverses direction and is much greater than the reverse saturation current of the diode! The diode does not turn off immediately. In fact, the diode actually

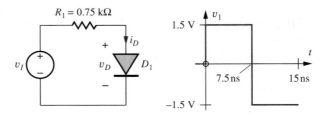

Figure 3.67 Circuit used to explore diode-switching behavior.

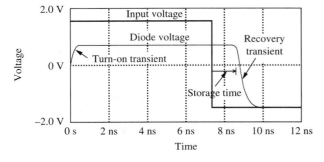
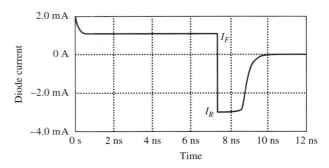

Figure 3.68 SPICE simulation results for the diode circuit in Fig. 3.67. (The diode transit time is equal to 5 ns.)

remains forward-biased by the charge stored in the diode, with $v_D = V_{on}$, even though the current has changed direction! The reverse current I_R is equal to

$$I_R = \frac{-1.5 - 0.7}{0.75 \text{ k}\Omega} = -2.9 \text{ mA} \quad (3.68)$$

The current remains at -2.9 mA for a period of time called the diode **storage time** τ_S, during which the internal charge stored in the diode is removed. Once the stored charge has been removed, the voltage across the diode begins to drop and charges toward the final value of -1.5 V. The current in the diode drops rapidly to zero as the diode voltage begins to fall.

The turn-on time and recovery time are determined primarily by the charging and discharging of the nonlinear depletion-layer capacitance C_j through the resistance R_S. The storage time is determined by the diffusion capacitance and diode transit time defined in Eq. (3.22) and by the values of the forward and reverse currents I_F and I_R:

$$\tau_S = \tau_T \ln\left[1 - \frac{I_F}{I_R}\right] = 5 \ln\left[1 - \frac{1.1 \text{ mA}}{-2.9 \text{ mA}}\right] \text{ ns} = 1.6 \text{ ns} \quad (3.69)$$

The SPICE simulation results in Fig. 3.68 agree well with this value.

Always remember that solid-state devices do not turn off instantaneously. The unusual storage time behavior of the diode is an excellent example of the switching delays that occur in *pn* junction devices in which carrier flow is dominated by the minority-carrier diffusion process. This behavior is not present in field-effect transistors, in which current flow is dominated by majority-carrier drift.

3.18 PHOTO DIODES, SOLAR CELLS, AND LIGHT-EMITTING DIODES

Several other important applications of diodes include photo detectors in communication systems, solar cells for generating electric power, and light-emitting diodes (LEDs). These applications all rely on the solid-state diode's ability to detect on produce optical emissions.

3.18.1 PHOTO DIODES AND PHOTODETECTORS

If the depletion region of a *pn* junction diode is illuminated with light of sufficiently high energy, the photons can cause electrons to jump the semiconductor bandgap, creating electron–hole pairs. For photon absorption to occur, the incident photons must have an energy E_p that exceeds the bandgap of the semiconductor:

$$E_p = h\nu = \frac{hc}{\lambda} \geq E_G \quad (3.70)$$

where $h =$ Planck's constant (6.626×10^{-34} J · s) $\lambda =$ wavelength of optical illumination
$\nu =$ frequency of optical illumination $c =$ velocity of light (3×10^8 m/s)

The *i-v* characteristic of a diode with and without illumination is shown in Fig. 3.69. The original diode characteristic is shifted vertically downward by the photon-generated current. Photon absorption creates an additional current crossing the *pn* junction that can be modeled by a current source i_{PH} in parallel with the *pn* junction diode, as shown in Fig. 3.70.

Based on this model, we see that the incident optical signal can be converted to an electrical signal voltage using the simple **photodetector circuit** in Fig. 3.71. The diode is reverse-biased to enhance the width and electric field in the depletion region. The photon-generated current i_{PH} will flow through resistor R and produce an output signal voltage given by

$$v_o = i_{PH} R \quad (3.71)$$

3.18 Photo Diodes, Solar Cells, and Light-Emitting Diodes   141

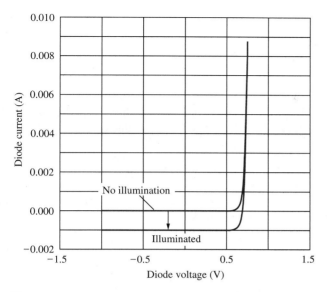

Figure 3.69 Diode i-v characteristic with and without optical illumination.

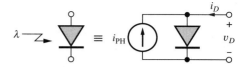

Figure 3.70 Model for optically illuminated diode. i_{PH} represents the current generated by absorption of photons in the vicinity of the pn junction.

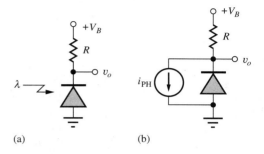

Figure 3.71 Basic photodetector circuit (a) and model (b).

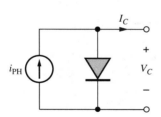

Figure 3.72 pn Diode under steady-state illumination as a solar cell.

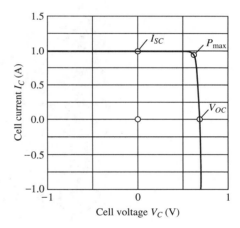

Figure 3.73 Terminal characteristics for a pn junction solar cell.

In optical fiber communication systems, the amplitude of the incident light is modulated by rapidly changing digital data, and i_{PH} includes a time-varying signal component. The time-varying signal voltage at v_o is fed to additional electronic circuits to demodulate the signal and recover the original data that were transmitted down the optical fiber.

3.18.2 POWER GENERATION FROM SOLAR CELLS

In **solar cell** applications, the optical illumination is constant, and a dc current I_{PH} is generated. The goal is to extract power from the cell, and the i-v characteristics of solar cells are usually plotted in terms of the cell current I_C and cell voltage V_C, as defined in Fig. 3.72.

The i-v characteristic of the pn junction used for solar cell applications is plotted in terms of these terminal variables in Fig. 3.73. Also indicated on the graph are the short-circuit current I_{SC}, the open-circuit voltage V_{OC}, and the maximum power point $P_{\max}$. I_{SC} represents the maximum current available from the cell, and V_{OC} is the voltage across the open-circuited cell when all the photo current is flowing into the internal pn junction. For the solar cell to supply power to an external circuit, the product $I_C \times V_C$ must be positive, corresponding to the first quadrant of the characteristic. An attempt is made to operate the cell near the point of maximum output power $P_{\max}$.

ELECTRONICS IN ACTION

Solar Energy

The photograph below depicts the Long Island Solar Farm installation on the Brookhaven National Laboratory (BNL) site in the center of Long Island, New York. The installation, consisting of 164,312 solar panels utilizing crystalline silicon technology, is capable of a generating a peak power of 32 MW with an estimated annual energy output of 44 million kilowatt-hours, enough to power an estimated 4500 homes for a year. The project was a collaboration between the Department of Energy, BP Solar, and the Long Island Power Authority and became operational near the end of 2011. The Long Island Power Authority purchases 100 percent of the power generated by the installation that is estimated to offset production of approximately 30,000 metric tons of CO_2 per year as well as significant amounts of other pollutants.

Long Island Solar Farm Installation. The main BNL campus is at the upper center of the picture.
Courtesy Brookhaven National Laboratory.

3.18.3 LIGHT-EMITTING DIODES (LEDs)

Light-emitting diodes, or **LEDs,** rely on the annihilation of electrons and holes through recombination rather than on the generation of carriers, as in the case of the photo diode. When a hole and electron recombine, an energy equal to the bandgap of the semiconductor can be released in the form of a photon. This recombination process is present in the forward-biased pn junction diode. In silicon, the recombination process actually involves the interaction of photons and lattice vibrations called phonons, so the optical emission process in silicon is not nearly as efficient as that in the III–V compound semiconductor GaAs or the ternary materials such as $GaIn_{1-x}As_x$ and $GaIn_{1-x}P_x$. LEDs in these compound semiconductor materials provide visible illumination, and the color of the output can be controlled by varying the fraction x of arsenic or phosphorus in the material which changes this bandgap energy.

SUMMARY

In this chapter we investigated the detailed behavior of the solid-state diode.

- A *pn* junction diode is created when *p*-type and *n*-type semiconductor regions are formed in intimate contact with each other. In the *pn* diode, large concentration gradients exist in the vicinity of the metallurgical junction, giving rise to large electron and hole diffusion currents.

- Under zero bias, no current can exist at the diode terminals, and a space charge region forms in the vicinity of the *pn* junction. The region of space charge results in both a built-in potential and an internal electric field, and the electric field produces electron and hole drift currents that exactly cancel the corresponding components of diffusion current.

- When a voltage is applied to the diode, the balance in the junction region is disturbed, and the diode conducts a current. The resulting i-v characteristics of the diode are accurately modeled by the diode equation:

$$i_D = I_S \left[\exp\left(\frac{v_D}{nV_T}\right) - 1 \right]$$

  where I_S = reverse saturation current of the diode

  n = nonideality factor (approximately 1)

  $V_T = kT/q$ = thermal voltage (0.025 V at room temperature)

- Under reverse bias, the diode current equals $-I_S$, a very small current.

- For forward bias, however, large currents are possible, and the diode presents an almost constant voltage drop of 0.6 to 0.7 V.

- At room temperature, an order of magnitude change in diode current requires a change of less than 60 mV in the diode voltage. At room temperature, the silicon diode voltage exhibits a temperature coefficient of approximately -1.8 mV/°C.

- One must also be aware of the reverse-breakdown phenomenon that is not included in the diode equation. If too large a reverse voltage is applied to the diode, the internal electric field becomes so large that the diode enters the breakdown region, either through Zener breakdown or avalanche breakdown. In the breakdown region, the diode again represents an almost fixed voltage drop, and the current must be limited by the external circuit or the diode can easily be destroyed.

- Diodes called Zener diodes are designed to operate in breakdown and can be used in simple voltage regulator circuits. Line regulation and load regulation characterize the change in output voltage of a power supply due to changes in input voltage and output current, respectively.

- As the voltage across the diode changes, the charge stored in the vicinity of the space charge region of the diode changes, and a complete diode model must include a capacitance. Under reverse bias, the capacitance varies inversely with the square root of the applied voltage. Under forward bias, the capacitance is proportional to the operating current and the diode transit time. These capacitances prevent the diode from turning on and off instantaneously and cause a storage time delay during turn-off.

- Direct use of the nonlinear diode equation in circuit calculations usually requires iterative numeric techniques. Several methods for simplifying the analysis of diode circuits were discussed, including the graphical load-line method and use of the ideal diode and constant voltage drop models.

- SPICE circuit analysis programs include a comprehensive built-in model for the diode that accurately reproduces both the ideal and nonideal characteristics of the diode and is useful for exploring the detailed behavior of circuits containing diodes.

- Important applications of diodes include half-wave, full-wave, and full-wave bridge rectifier circuits used to convert from ac to dc voltages in power supplies. Simple power supply circuits use capacitive filters, and the design of the filter capacitor determines power supply ripple voltage and diode conduction angle. Diodes used as rectifiers in power supplies must be able to withstand large peak repetitive currents as well as surge currents when the power supplies are first turned on. The reverse-breakdown voltage of rectifier diodes is referred to as the peak-inverse-voltage, or PIV, rating of the diode.

- Real diodes cannot turn on or off instantaneously because the internal capacitances of the diodes must be charged and discharged. The turn-on time is usually quite short, but diodes that have been conducting turn off much less abruptly. It takes time to remove stored charge within the diode, and this time delay is characterized by storage time τ_s. During the storage time, it is possible for large reverse currents to occur in the diode.

- Finally, the ability of the *pn* junction device to generate and detect light was discussed, and the basic characteristics of photo diodes, solar cells, and light-emitting diodes were presented.

KEY TERMS

Anode
Avalanche breakdown
Bias current and voltage
Breakdown region
Breakdown voltage
Built-in potential (or voltage)
Cathode
Center-tapped transformer
Conduction angle
Conduction interval
Constant voltage drop (CVD) model
Cut-in voltage
Depletion layer
Depletion-layer capacitance
Depletion-layer width
Depletion region
Diffusion capacitance
Diode equation
Diode SPICE parameters (IS, RS, N, TT, CJO, VJ, M)
Filter capacitor
Forward bias
Full-wave bridge rectifier circuit
Full-wave rectifier circuit
Half-wave rectifier circuit
Ideal diode
Ideal diode model
Impact-ionization process
Junction capacitance
Junction potential
Light-emitting diode (LED)
Line regulation
Load line
Load-line analysis
Load regulation
Mathematical model
Metallurgical junction
Nonideality factor (n)
Peak detector
Peak inverse voltage (PIV)
Photodetector circuit
Piecewise linear model
pn junction diode
Q-point
Quiescent operating point
Rectifier circuits
Reverse bias
Reverse breakdown
Reverse saturation current (I_S)
Ripple current
Ripple voltage
Saturation current
Schottky barrier diode
Solar cell
Space charge region (SCR)
Storage time
Surge current
Thermal voltage (V_T)
Transit time
Turn-on voltage
Voltage regulator
Voltage transfer characteristic (VTC)
Zener breakdown
Zener diode
Zero bias
Zero-bias junction capacitance

REFERENCE

1. G. W. Neudeck, *The PN Junction Diode,* 2d ed. Pearson Education, Upper Saddle River, NJ: 1989.

ADDITIONAL READING

PSPICE, ORCAD, now owned by Cadence Design Systems, San Jose, CA.
LTspice available from Linear Technology Corp.
Tina-TI SPICE-based analog simulation program available from Texas Instruments.
T. Quarles, A. R. Newton, D. O. Pederson, and A. Sangiovanni-Vincentelli, *SPICE3 Version 3f3 User's Manual.* UC Berkeley: May 1993.
A. S. Sedra, and K. C. Smith. *Microelectronic Circuits.* 8th ed. Oxford University Press, New York: 2020.

PROBLEMS

(Use $V_T = 25.9$ mV as the default value in the problem set.)

3.1 The *pn* Junction Diode

3.1. A diode is doped with $N_A = 10^{18}/\text{cm}^3$ on the *p*-type side and $N_D = 10^{19}/\text{cm}^3$ on the *n*-type side. (a) What is the depletion-layer width w_{do}? (b) What are the values of x_p and x_n? (c) What is the value of the built-in potential of the junction? (d) What is the value of E_{MAX}? Use Eq. (3.3) and Fig. 3.5.

3.2. A diode is doped with $N_A = 10^{15}/\text{cm}^3$ on the *p*-type side and $N_D = 10^{18}/\text{cm}^3$ on the *n*-type side. (a) What are the values of p_p, p_n, n_p, and n_n? (b) What are the depletion-region width w_{do} and built-in voltage?

3.3. Repeat Prob. 3.2 for a diode with $N_A = 10^{15}/\text{cm}^3$ on the *p*-type side and $N_D = 10^{20}/\text{cm}^3$ on the *n*-type side.

3.4. Repeat Prob. 3.2 for a diode with $N_A = 10^{16}/\text{cm}^3$ on the *p*-type side and $N_D = 5 \times 10^{19}/\text{cm}^3$ on the *n*-type side.

3.5. A diode has $w_{do} = 1$ µm and $\phi_j = 0.75$ V. (a) What reverse bias is required to double the depletion-layer width? (b) What is the depletion region width if a reverse bias of 10 V is applied to the diode?

3.6. A diode has $w_{do} = 0.4$ µm and $\phi_j = 0.85$ V. (a) What reverse bias is required to triple the depletion-layer width? (b) What is the depletion region width if a reverse bias of 7 V is applied to the diode?

3.7. Suppose a drift current density of 5000 A/cm^2 exists in the neutral region on the *p*-type side of a diode that has a resistivity of 0.5 Ω · cm. What is the electric field needed to support this drift current density?

3.8. Suppose a drift current density of 2000 A/cm^2 exists in the neutral region on the *n*-type side of a diode that has a resistivity of 2.5 Ω · cm. What is the electric field needed to support this drift current density?

3.9. The maximum velocity of carriers in silicon is approximately 10^7 cm/s. (a) What is the maximum drift current density that can be supported in a region of *p*-type silicon with a doping of $5 \times 10^{17}/\text{cm}^3$? (b) Repeat for a region of *n*-type silicon with a doping of $4 \times 10^{15}/\text{cm}^3$?

**3.10. Suppose that $N_A(x) = N_o \exp(-x/L)$ in a region of silicon extending from $x = 0$ to $x = 10$ µm, where N_o is a constant. Assume that $p(x) = N_A(x)$. Assuming that j_p must be zero in thermal equilibrium, show that a built-in electric field must exist and find its value for $L = 2$ µm and $N_o = 10^{18}/\text{cm}^3$.

3.11. What carrier gradient is needed to generate a diffusion current density of $j_n = 1500$ A/cm^2 if $\mu_n = 600$ cm^2/V · s?

3.12. Use the solver routine in your calculator to find the solution to Eq. (3.25) for $I_S = 5 \times 10^{-16}$ A.

3.13. Use a spreadsheet to iteratively find the solution to Eq. (3.25) for $I_S = 10^{-14}$ A.

3.14. (a) Use MATLAB or MATHCAD to find the solution to Eq. (3.25) for $I_S = 10^{-12}$ A. (b) Repeat for $I_S = 10^{-15}$ A.

3.2–3.4 The i-v Characteristics of the Diode; the Diode Equation: A Mathematical Model for the Diode; and Diode Characteristics under Reverse, Zero, and Forward Bias

3.15. To what temperature does $V_T = 0.025$ V actually correspond? What is the value of V_T for temperatures of $-55°C$, $0°C$, and $+125°C$?

3.16. (a) Plot a graph of the diode equation similar to Fig. 3.8 for a diode with $I_S = 10^{-12}$ A and $n = 1$. (b) Repeat for $n = 2$. (c) Repeat (a) for $I_S = 10^{-15}$ A.

3.17. A diode has $n = 1.06$ at $T = 320$ K. What is the value of $n \cdot V_T$? What temperature would give the same value of $n \cdot V_T$ if $n = 1.00$?

3.18. Plot the diode current for a diode with $I_{SO} = 15$ fA and $\phi_j = 0.75$ V for -10 V $\leq v_D \leq 0$ V using Eq. (3.19).

*3.19. What are the values of I_S and n for the diode in the graph in Fig. P3.19? Assume $V_T = 0.0259$ V.

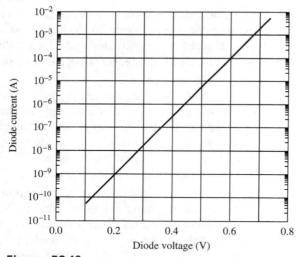

Figure P3.19

3.20. A diode has $I_S = 10^{-17}$ A and $n = 1.05$. (a) What is the diode voltage if the diode current is 70 µA? (b) What is the diode voltage if the diode current is 5 µA? (c) What is the diode current for $v_D = 0$ V? (d) What is the diode current for $v_D = -0.075$ V? (e) What is the diode current for $v_D = -5$ V?

3.21. A diode has $I_S = 5$ aA and $n = 1$. (a) What is the diode voltage if the diode current is 500 µA?

(b) What is the diode voltage if the diode current is 10 µA? (c) What is the diode current for $v_D = 0$ V? (d) What is the diode current for $v_D = -0.06$ V? (e) What is the diode current for $v_D = -4$ V?

3.22. A diode has $I_S = 0.2$ fA and $n = 1$. (a) What is the diode current if the diode voltage is 0.675 V? (b) What will be the diode voltage if the current increases by a factor of 3?

3.23. A diode has $I_S = 10^{-10}$ A and $n = 2$. (a) What is the diode voltage if the diode current is 25 A? (b) What is the diode voltage if the diode current is 75 A?

3.24. A diode is operating with $i_D = 2$ mA and $v_D = 0.82$ V. (a) What is I_S if $n = 1$? (b) What is the diode current for $v_D = -5$ V?

3.25. A diode is operating with $i_D = 300$ µA and $v_D = 0.75$ V. (a) What is I_S if $n = 1.07$? (b) What is the diode current for $v_D = -3$ V?

3.26. The saturation current for diodes with the same part number may vary widely. Suppose it is known that 10^{-14} A $\leq I_S \leq 10^{-12}$ A. What is the range of forward voltages that may be exhibited by the diode if it is biased with $i_D = 2$ mA?

3.27. A diode is biased by a 0.9-V dc source, and its current is found to be 100 µA at $T = 35°C$. (a) At what temperature will the current double? (b) At what temperature will the current be 50 µA?

**3.28. The i-v characteristic for a diode has been measured under carefully controlled temperature conditions ($T = 307$ K), and the data are in Table P3.28. Use a spreadsheet or MATLAB to find the values of I_S and n that provide the best fit of the diode equation to the measurements in the least-squares sense. [That is, find the values of I_S and n that minimize the function

TABLE P3.28
Diode i-v Measurements

DIODE VOLTAGE	DIODE CURRENT
0.500	6.591×10^{-7}
0.550	3.647×10^{-6}
0.600	2.158×10^{-5}
0.650	1.780×10^{-4}
0.675	3.601×10^{-4}
0.700	8.963×10^{-4}
0.725	2.335×10^{-3}
0.750	6.035×10^{-3}
0.775	1.316×10^{-2}

$M = \sum_{m=1}^{n}(i_D^m - I_{Dm})^2$, where i_D is the diode equation from Eq. (3.1) and I_{Dm} are the measured data.] For your values of I_S and n, what is the minimum value of $M = \sum_{m=1}^{n}(i_D^m - I_{Dm})^2$?

3.5 Diode Temperature Coefficient

3.29. What is the value of V_T for temperatures of $-40°C$, $0°C$, and $+50°C$?

3.30. A diode has $I_S = 10^{-16}$ A and $n = 1.05$. (a) What is the diode voltage if the diode current is 250 μA at $T = 25°C$? (b) What is the diode voltage at $T = 85°C$? Assume the diode voltage temperature coefficient is -2 mV/K at 55°C.

3.31. A diode has $I_S = 20$ fA and $n = 1$. (a) What is the diode voltage if the diode current is 100 μA at $T = 25°C$? (b) What is the diode voltage at $T = 50°C$? Assume the diode voltage temperature coefficient is -1.8 mV/K at 0°C.

*3.32. The temperature dependence of I_S is described approximately by

$$I_S = CT^3 \exp\left(-\frac{E_G}{kT}\right)$$

What is the diode voltage temperature coefficient based on this expression and Eq. (3.15) if $E_G = 1.12$ eV, $V_D = 0.7$ V, and $T = 315$ K?

3.33. The saturation current of a silicon diode is described by the expression in Prob. 3.32. (a) What temperature change will cause I_S to double at $T = 300$ K? (b) To increase by 20 times? (c) To decrease by 100 times?

3.6 Diodes under Reverse Bias

3.34. A diode has $w_{do} = 1.5$ μm and $\phi_j = 0.8$ V. (a) What is the depletion layer width for $V_R = 5$ V? (b) For $V_D = -10$ V?

3.35. A diode has a doping of $N_D = 10^{20}/cm^3$ on the n-type side and $N_A = 10^{18}/cm^3$ on the p-type side. What are the values of w_{do} and ϕ_j? What is the value of w_d at a reverse bias of 5 V? At 25 V?

3.36. A diode has a doping of $N_D = 10^{15}/cm^3$ on the n-type side and $N_A = 10^{17}/cm^3$ on the p-type side. What are the values of w_{do} and ϕ_j? What is the value of w_d at a reverse bias of 10 V? At 100 V?

*3.37. A diode has $w_{do} = 2$ μm and $\phi_j = 0.65$ V. If the diode breaks down when the internal electric field reaches 300 kV/cm, what is the breakdown voltage of the diode?

*3.38. Silicon breaks down when the internal electric field exceeds 300 kV/cm. At what reverse bias do you expect the diode of Prob. 3.5 to break down?

3.39. What are the breakdown voltage V_Z and Zener resistance R_Z of the diode depicted in Fig. P3.39?

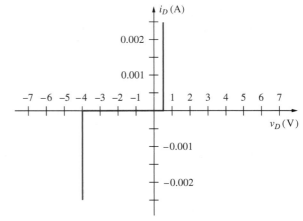

Figure P3.39

**3.40. A diode is fabricated with $N_A \gg N_D$. What value of doping is required on the lightly doped side to achieve a reverse-breakdown voltage of 750 V if the semiconductor material breaks down at a field of 300 kV/cm?

3.41. The diode in Fig. 3.36 is described by the i-v characteristics in Fig. P3.39. (a) Draw the load line and find the Q-point for a 10-V source and a 5-kΩ resistor. (b) Repeat for a -4-V source and a 2-kΩ resistor. (c) Repeat for a -8-V source and a 4-kΩ resistor. (d) Repeat for a 6-V source and a 3-kΩ resistor.

3.7 pn Junction Capacitance

3.42. What is the zero-bias junction capacitance/cm² for a diode with $N_A = 10^{18}/cm^3$ on the p-type side and $N_D = 10^{20}/cm^3$ on the n-type side? What is the diode capacitance with a 3-V reverse bias if the diode area is 0.05 cm²?

3.43. What is the zero-bias junction capacitance per cm² for a diode with $N_A = 10^{18}/cm^3$ on the p-type side and $N_D = 10^{15}/cm^3$ on the n-type side. What is the diode capacitance with a 9 V reverse bias if the diode area is 0.02 cm²?

3.44. A diode is operating at a current of 250 μA. (a) What is the diffusion capacitance if the diode transit time is 100 ps? (b) How much charge is stored in the diode? (c) Repeat for $i_D = 3$ mA.

3.45. A diode is operating at a current of 2 A. (a) What is the diffusion capacitance if the diode transit time is 10 ns? (b) How much charge is stored in the diode? (c) Repeat for $i_D = 100$ mA.

3.46. A square pn junction diode is 5 mm on a side. The p-type side has a doping concentration of $10^{19}/cm^3$ and the n-type side has a doping concentration of $10^{16}/cm^3$. (a) What is the zero-bias capacitance of the diode? What is the capacitance at a reverse bias of 4 V? (b) Repeat for an area of 10^4 µm².

3.47. A variable capacitance diode with $C_{jo} = 39$ pF and $\phi_j = 0.80$ V is used to tune a resonant LC circuit as shown in Fig. P3.47. The impedance of the RFC (radio frequency choke) can be considered infinite. What are the resonant frequencies ($f_o = \frac{1}{2\pi\sqrt{LC}}$) for $V_{DC} = 1$ V and $V_{DC} = 9$ V?

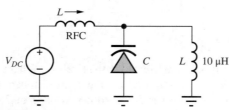

Figure P3.47

3.8 Schottky Barrier Diode

3.48. A Schottky barrier diode is modeled by the diode equation in Eq. (3.11) with $I_S = 5 \times 10^{-11}$ A. (a) What is the diode voltage at a current of 5 mA? (b) What would be the voltage of a pn junction diode with $I_S = 10^{-14}$ A operating at the same current?

3.49. Suppose a Schottky barrier diode can be modeled by the diode equation in Eq. (3.11) with $I_S = 5 \times 10^{-7}$ A. (a) What is the diode voltage at a current of 40 A? (b) What would be the voltage of a pn junction diode with $I_S = 10^{-15}$ A and $n = 2$?

3.9 Diode SPICE Model and Layout

3.50. (a) A diode has $I_S = 5 \times 10^{-16}$ A and $R_S = 15$ Ω and is operating at a current of 1 mA at room temperature. What are the values of V_D and V'_D? (b) Repeat for $R_S = 200$ Ω.

3.51. A pn diode has a resistivity of 1 Ω·cm on the p-type side and 0.02 Ω·cm on the n-type side. What is the value of R_S for this diode if the cross-sectional area of the diode is 0.01 cm² and the lengths of the p- and n-sides of the diode are each 250 µm?

*3.52. A diode fabrication process has a specific contact resistance of 10 Ω·µm². If the contacts are each 1 µm × 1 µm in size, what are the total contact resistances associated with the anode and cathode contacts to the diode in Fig. 3.21(a)?

3.10 Diode Circuit Analysis

3.53. (a) Plot the load line and find the Q-point for the diode circuit in Fig. P3.53 if $V = 5$ V and $R = 2$ kΩ. Use the i-v characteristic in Fig. P3.39. (b) Repeat for $V = -6$ V and $R = 3$ kΩ. (c) Repeat for $V = -3$ V and $R = 3$ kΩ.

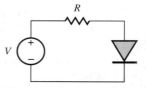

Figure P3.53

3.54. (a) Plot the load line and find the Q-point for the diode circuit in Fig. P3.53 if $V = 10$ V and $R = 5$ kΩ. Use the i-v characteristic in Fig. P3.39. (b) Repeat for $V = -10$ V and $R = 5$ kΩ. (c) Repeat for $V = -2$ V and $R = 2$ kΩ.

3.55. Simulate the circuit in Prob. 3.54 with SPICE and compare the results to those in Prob. 3.54. Use $I_S = 10^{-15}$ A.

3.56. Use the i-v characteristic in Fig. P3.39. (a) Plot the load line and find the Q-point for the diode circuit in Fig. P3.54 if $V = 6$ V and $R = 4$ kΩ. (b) For $V = -6$ V and $R = 3$ kΩ. (c) For $V = -3$ V and $R = 3$ kΩ. (d) For $V = 12$ V and $R = 8$ kΩ. (e) For $V = -25$ V and $R = 10$ kΩ.

Iterative Analysis and the Mathematical Model

3.57. Use direct trial and error to find the solution to the diode circuit in Fig. 3.22 using Eq. (3.27).

3.58. Repeat the iterative procedure used in the spreadsheet in Table 3.2 for initial guesses of 1 µA, 5 mA, and 5 A and 0 A. How many iterations are required for each case? Did any problem arise? If so, what is the source of the problem?

3.59. A diode has $I_S = 0.1$ fA and is operating at $T = 300$ K. (a) What are the values of V_{DO} and r_D if $I_D = 100$ µA? (b) If $I_D = 2.5$ mA? (c) If $I_D = 25$ mA?

3.60. (a) Use the iterative procedure in the spreadsheet in Table 3.2 to find the diode current and voltage for

the circuit in Fig. 3.22 if $V = 2.5$ V and $R = 3.3$ kΩ.
(b) Repeat for $V = 7.5$ V and $R = 13$ kΩ.

3.61. (a) Use the iterative procedure in the spreadsheet in Table 3.2 to find the diode current and voltage for the circuit in Fig. 3.22 if $V = 1$ V and $R = 12$ kΩ.
(b) Repeat for $V = 3$ V and $R = 6.8$ kΩ.

3.62. Use MATLAB or MATHCAD to numerically find the Q-point for the circuit in Fig. 3.22 using the equation in the exercise on page 100.

Ideal Diode and Constant Voltage Drop Models

*3.63. Find the Q-point for the circuit in Fig. 3.22 using the same four methods as in Sec. 3.10 if the voltage source is 1 V. Compare the answers in a manner similar to Table 3.3.

3.64. Find the Q-point for the diode in Fig. P3.64 using (a) the ideal diode model and (b) the constant voltage drop model with $V_{on} = 0.6$ V. (c) Discuss the results. Which answer do you feel is most correct? (d) Use iterative analysis to find the actual Q-point if $I_S = 0.1$ fA.

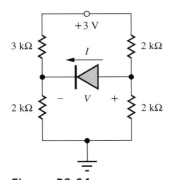

Figure P3.64

3.65. Simulate the circuit of Fig. P3.64 and find the diode Q-point. Compare the results to those in Prob. 3.64.

3.66. (a) Find the worst-case values of the Q-point current for the diode in Fig. P3.64 using the ideal diode model if the resistors all have 10 percent tolerances. (b) Repeat using the CVD model with $V_{on} = 0.6$ V.

3.67. (a) Find I and V in the four circuits in Fig. P3.67 using the ideal diode model. (b) Repeat using the constant voltage drop model with $V_{on} = 0.65$ V.

3.68. (a) Find I and V in the four circuits in Fig. P3.67 using the ideal diode model if the resistor values are changed to 68 kΩ. (b) Repeat using the constant voltage drop model with $V_{on} = 0.6$ V.

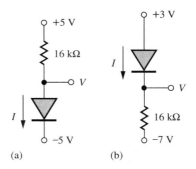

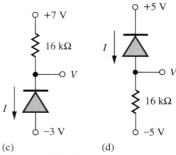

Figure P3.67

3.11 Multiple Diode Circuits

3.69. Find the Q-points for the diodes in the four circuits in Fig. P3.69 using (a) the ideal diode model and (b) the constant voltage drop model with $V_{on} = 0.75$ V.

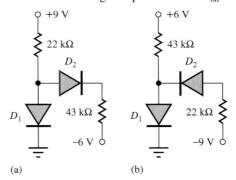

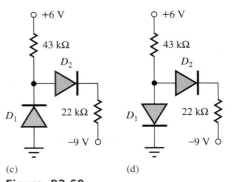

Figure P3.69

150 Chapter 3 Solid-State Diodes and Diode Circuits

3.70. Repeat Prob. 3.69 if the direction of diode D_1 is reversed in each of the four circuits.

3.71. Repeat Prob. 3.69 if the direction of diode D_2 is reversed in each of the four circuits.

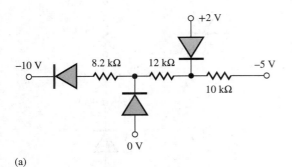

(a)

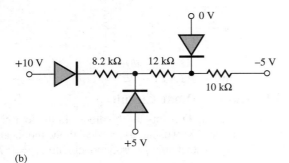

(b)

3.72. Find the Q-points for the diodes in the four circuits in Fig. P3.69 if the values of all the resistors are changed to 15 kΩ using (a) the ideal diode model and (b) the constant voltage drop model with $V_{on} = 0.60$ V.

3.73. Find the Q-point for the diodes in the circuits in Fig. P3.73 using the ideal diode model.

3.74. Find the Q-point for the diodes in the circuits in Fig. P3.73 using the constant voltage drop model with $V_{on} = 0.65$ V.

3.75. Reverse the direction of the middle diode in each of the circuits in Fig. P3.69. (a) Redraw the four circuits. Find the new Q-points using (b) the ideal diode model (c) the CVD model with $V_{on} = 0.75$ V.

3.76. Simulate the diode circuits in Fig. P3.73 and compare your results to those in Prob. 3.73.

3.77. Verify that the values presented in Ex. 3.8 using the ideal diode model are correct.

3.78. Simulate the circuit in Fig. 3.33 and compare to the results in Ex. 3.8.

3.12 Analysis of Diodes Operating in the Breakdown Region

3.79. Draw the load line for the circuit in Fig. P3.79 on the characteristics in Fig. P3.39 and find the Q-point.

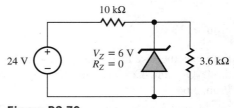

Figure P3.79

3.80. (a) Find the Q-point for the Zener diode in Fig. P3.79. (b) Repeat if $R_Z = 100$ Ω. (c) The Zener diode is accidentally put in the circuit upside down. Redraw the circuit and find the new Q-point.

3.81. What is maximum load current I_L that can be drawn from the Zener regulator in Fig. P3.81 if it is to maintain a regulated output? What is the minimum value of R_L that can be used and still have a regulated output voltage?

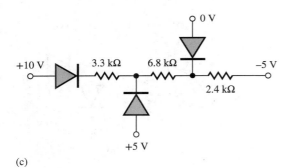

(c)

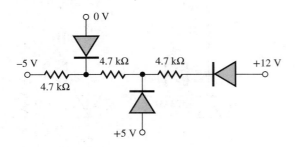

(d)

Figure P3.73

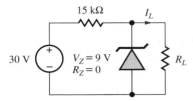

Figure P3.81

3.82. What is power dissipation in the Zener diode in Fig. P3.81 for (a) $R_L = 2$ kΩ (b) $R_L = 4.7$ kΩ (c) $R_L = 15$ kΩ (d) $R_L = \infty$?

3.83. Load resistor R_L in Fig. P3.81 is 12 kΩ. What are the nominal and worst-case values of Zener diode current and power dissipation if the power supply voltage, Zener breakdown voltage and resistors all have 5 percent tolerances?

3.84. What is power dissipation in the Zener diode in Fig. P3.84 for (a) $R_L = 150$ Ω? (b) $R_L = \infty$? (c) $R_L = 50$ Ω?

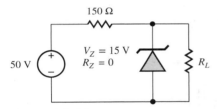

Figure P3.84

3.85. In the lab, the Zener diode in Fig. P3.84 is accidentally put in the circuit upside down. Redraw the circuit and find the new Q-point.

3.86. Load resistor R_L in Fig. P3.84 is 100 Ω. What are the nominal and worst-case values of Zener diode current and power dissipation if the power supply voltage, Zener breakdown voltage, and resistors all have 10 percent tolerances?

3.13 Half-Wave Rectifier Circuits

3.87. A power diode has a reverse saturation current of 10^{-9} A and $n = 1.6$. What is the forward voltage drop at the peak current of 48.6 A that was calculated in the Exercise in Sec. 3.13.5?

3.88. A power diode has a reverse saturation current of 10^{-8} A and $n = 2$. What is the forward voltage drop at the peak current of 100 A? What is the power dissipation in the diode in a half-wave rectifier application operating at 60 Hz if the series resistance is 0.01 Ω and the conduction time is 1 ms?

*3.89. (a) Use a spreadsheet or MATLAB or write a computer program to find the numeric solution to the conduction angle equation for a 60 Hz half-wave rectifier circuit that uses a filter capacitance of 100,000 μF. The circuit is designed to provide 5 V at 5 A. {That is, solve $[(V_P - V_{on})\exp(-t/RC) = V_P \cos \omega t - V_{on}]$. Be careful! There are an infinite number of solutions to this equation. Be sure your algorithm finds the desired answer to the problem.} Assume $V_{on} = 1$ V. (b) Compare to calculations using Eq. (3.57).

3.90. What is the actual average value (the dc value) of the rectifier output voltage for the waveform in Fig. P3.90 if V_r is 10 percent of $V_P - V_{on} = 18$ V?

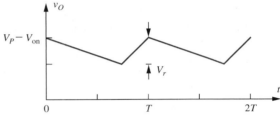

Figure P3.90

3.91. Draw the voltage waveforms, similar to those in Fig. 3.53, for the negative output rectifier in Fig. 3.57(b).

*3.92. Show that evaluation of Eq. (3.61) will yield the result in Eq. (3.62).

3.93. The half-wave rectifier in Fig. P3.93 is operating at a frequency of 60 Hz, and the rms value of the transformer output voltage v_I is 12.6 V $\pm$ 10%. What are the nominal and worst case values of the dc output voltage V_O if the diode voltage drop is 1 V?

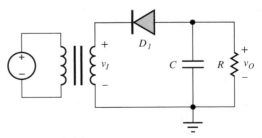

Figure P3.93

3.94. The half-wave rectifier in Fig. P3.93 is operating at a frequency of 60 Hz, and the rms value of the transformer output voltage is 6.3 V. (a) What is the value of the dc output voltage V_O if the diode voltage drop is 1 V? (b) What is the minimum value of C required to maintain the ripple voltage to less than 0.25 V if $R = 0.5\ \Omega$? (c) What is the PIV rating of the diode in this circuit? (d) What is the surge current when power is first applied? (e) What is the amplitude of the repetitive current in the diode?

3.95. Simulate the behavior of the half-wave rectifier in Fig. P3.93 for $v_I = 10\sin 120\pi t$, $R = 0.025\ \Omega$ and $C = 0.5$ F. (Use IS $= 10^{-10}$ A, RS $= 0$, and RELTOL $= 10^{-6}$.) Compare the simulated values of dc output voltage, ripple voltage, and peak diode current to hand calculations. Repeat simulation with $R_S = 0.02\ \Omega$.

3.96. (a) Repeat Prob. 3.94 for a frequency of 400 Hz. (b) Repeat Prob. 3.94 for a frequency of 70 kHz.

3.97. A 3.3-V, 50-A dc power supply is to be designed with a ripple of less than 1.5 percent. Assume that a half-wave rectifier circuit (60 Hz) with a capacitor filter is used. (a) What is the size of the filter capacitor C? (b) What is the PIV rating for the diode? (c) What is the rms value of the transformer voltage needed for the rectifier? (d) What is the value of the peak repetitive diode current in the diode? (e) What is the surge current at $t = 0^+$?

3.98. A 2500-V, 3-A, dc power supply is to be designed with a ripple voltage ≤ 0.5 percent. Assume that a half-wave rectifier circuit (60 Hz) with a capacitor filter is used. (a) What is the size of the filter capacitor C? (b) What is the minimum PIV rating for the diode? (c) What is the rms value of the transformer voltage needed for the rectifier? (d) What is the peak value of the repetitive current in the diode? (e) What is the surge current at $t = 0^+$?

*3.99. Draw the voltage waveforms at nodes v_O and v_1 for the "voltage-doubler" circuit in Fig. P3.99 for the first two cycles of the input sine wave. What is the steady-state output voltage if $V_P = 20$ V?

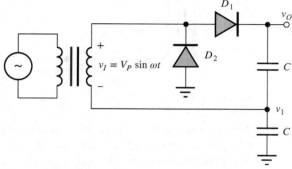

Figure P3.99

3.100. Simulate the voltage-doubler rectifier circuit in Fig. P3.99 for $C = 500\ \mu$F and $v_I = 1500\sin 2\pi(60)t$ with a load resistance of $R_L = 3000\ \Omega$ added between v_O and ground. Calculate the ripple voltage and compare to the simulation.

3.101. Estimate the maximum surge current in a half-wave rectifier with a transformer having an rms secondary voltage of 50 V and a secondary resistance of 0.25 Ω. Assume the filter capacitance is 0.5 F and a frequency of 60 Hz.

3.14 Full-Wave Rectifier Circuits

3.102. The full-wave rectifier in Fig. P3.102 is operating at a frequency of 60 Hz, and the rms value of the transformer output voltage is 18 V. (a) What is the value of the dc output voltage if the diode voltage drop is 1 V? (b) What is the minimum value of C required to maintain the ripple voltage to less than 0.25 V if $R = 0.5\ \Omega$? (c) What is the PIV rating of the diode in this circuit? (d) What is the surge current when power is first applied? (e) What is the amplitude of the repetitive current in the diode?

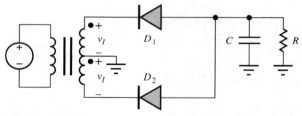

Figure P3.102

3.103. Repeat Prob. 3.102 if the rms value of the transformer output voltage v_I is 15 V.

3.104. A 60-Hz full-wave rectifier is built with a transformer having an rms secondary voltage of 20 V and filter capacitance $C = 150{,}000\ \mu$F. What is the largest current that can be supplied by the rectifier circuit if the ripple must be less than 0.3 V?

3.105. Simulate the behavior of the full-wave rectifier in Fig. P3.102 for $R = 3\ \Omega$ and $C = 22{,}000\ \mu$F. Assume that the rms value of v_I is 10.0 V and the frequency is 400 Hz. (Use IS $= 10^{-10}$ A, RS $= 0$, and RELTOL $= 10^{-6}$.) Compare the simulated values of dc output voltage, ripple voltage, and peak diode current to hand calculations. Repeat simulation with $R_S = 0.25$.

3.106. Repeat Prob. 3.97 for a full-wave rectifier circuit.

3.107. Repeat Prob. 3.98 for a full-wave rectifier circuit.

*3.108. The full-wave rectifier circuit in Fig. P3.108(a) was designed to have a maximum ripple of approximately 1 V, but it is not operating properly. The measured waveforms at the three nodes in the circuit are shown in Fig. P3.108(b). What is wrong with the circuit?

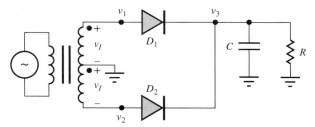

Figure P3.108(a)

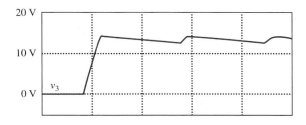

Figure P3.108(b) Waveforms for the circuit in Fig. P3.108(a).

3.109. For the Zener regulated power supply in Fig. P3.109, the rms value of v_I is 15 V, the operating frequency is 60 Hz, $R = 100\ \Omega$, $C = 1000\ \mu\text{F}$, the on-voltage of diodes D_1 and D_2 is 0.75 V, and the Zener voltage of diode D_3 is 15 V. (a) What type of rectifier is used in this power supply circuit? (b) What is the dc voltage at V_1? (c) What is the dc output voltage V_O? (d) What is the magnitude of the ripple voltage at V_1? (e) What is the minimum PIV rating for the rectifier diodes? (f) Draw a new version of the circuit that will produce an output voltage of -15 V.

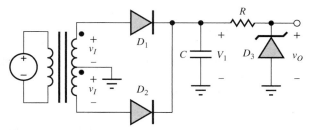

Figure P3.109

3.15 Full-Wave Bridge Rectification

3.110. Repeat Prob. 3.102 for a full-wave bridge rectifier circuit. Draw the circuit.

3.111. Repeat Prob. 3.97 for a full-wave bridge rectifier circuit. Draw the circuit.

3.112. Repeat Prob. 3.98 for a full-wave bridge rectifier circuit. Draw the circuit.

*3.113. What are the dc output voltages V_1 and V_2 for the rectifier circuit in Fig. P3.113 if $v_I = 40 \sin 377t$ and $C = 20{,}000\ \mu\text{F}$?

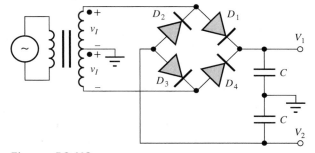

Figure P3.113

3.114. (a) Draw a graph of the output voltage waveform for the bridge rectifier in Fig. 3.63. (b) Diode D_3 fails and becomes an open circuit in Fig. 3.63. Draw a graph of the output waveform. (c) Repeat if diode D_4 fails as an open circuit. (d) Repeat if diode D_4 fails as a short circuit.

3.115. (a) Draw a graph of the output voltage waveform for the bridge rectifier in Fig. 3.66. (b) Diode D_2 fails and becomes an open circuit in Fig. 3.66. Draw a graph of the output waveform. (c) Repeat if diode D_1 fails as an open circuit. (d) Repeat if diode D_1 fails as a short circuit.

3.116. (a) The bridge rectifier design in Example 3.11 requires a 12-V_{rms} transformer, a 111,000-μF capacitor, and a 94.7-A diode bridge with a PIV rating of 20 V. Go to an electronic parts supplier on the web and find components that meet these requirements as well as the cost of these components. (b) Repeat for the half-wave rectifier case.

3.117. Simulate the rectifier circuit in Fig. P3.113 for $C = 100$ mF and $v_I = 40 \sin 2\pi(60)t$ with a 500-Ω load connected between each output and ground.

3.118. Repeat Prob. 3.104 if the full-wave bridge circuit is used instead of the rectifier in Fig. P3.102. Draw the circuit!

3.16 Rectifier Comparison and Design Tradeoffs

3.119. A 3.3-V, 30-A dc power supply is to be designed to have a ripple voltage of no more than 10 mV. Compare the pros and cons of implementing this power supply with half-wave, full-wave, and full-wave bridge rectifiers.

3.120. A 200-V, 0.5-A dc power supply is to be designed with less than a 2 percent ripple voltage. Compare the pros and cons of implementing this power supply with half-wave, full-wave, and full-wave bridge rectifiers.

3.121. A 3000-V, 1-A dc power supply is to be designed with less than a 4 percent ripple voltage. Compare the pros and cons of implementing this power supply with half-wave, full-wave, and full-wave bridge rectifiers.

3.17 Dynamic Switching Behavior of the Diode

*3.122. (a) Calculate the current at $t = 0^+$ in the circuit in Fig. P3.122. (b) Calculate I_F, I_R, and the storage time expected when the diode is switched off if $\tau_T = 8$ ns.

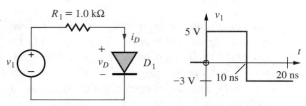

Figure P3.122

3.123. Reverse the direction of the diode in Fig. P3.122 and rework Prob. 3.122.

3.124. (a) Simulate the switching behavior of the circuit in Fig. P3.122. (b) Compare the simulation results to the hand calculations in Prob. 3.122.

*3.125. (a) Calculate the current at $t = 0^+$ in the circuit in Fig. P3.122 if R_1 is changed to 5 Ω. (b) Calculate I_F, I_R, and the storage time expected when the diode is switched off at $t = 10$ µs if $\tau_T = 250$ ns.

**3.126. The simulation results presented in Fig. 3.68 were performed with the diode transit time $\tau_T = 5$ ns. (a) Repeat the simulation of the diode circuit in Fig. P3.126(a) with the diode transit time changed to $\tau_T = 50$ ns. Does the storage time that you observe change in proportion to the value of τ_T in your simulation? Discuss. (b) Repeat the simulation with the input voltage changed to the one in Fig. P3.126(b), in which it is assumed that v_1 has been at 1.5 V for a long time, and compare the results to those obtained in (a). What is the reason for the difference between the results in (a) and (b)?

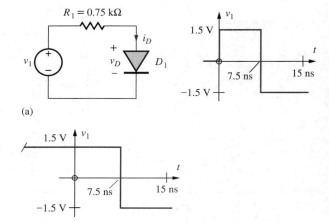

Figure P3.126

3.18 Photo Diodes, Solar Cells, and LEDs

*3.127. The output of a diode used as a solar cell is given by

$$I_C = 1 - 10^{-14}[\exp(40V_C) - 1] \text{ amperes}$$

What operating point corresponds to P_{max}? What is P_{max}? What are the values of I_{SC} and V_{OC}?

*3.128. Three diodes are connected in series to increase the output voltage of a solar cell. The individual outputs of the three diodes are given by

$$I_{C1} = 1.06 - 10^{-15}[\exp(40V_{C1}) - 1] \text{ A}$$
$$I_{C2} = 1.00 - 10^{-15}[\exp(40V_{C2}) - 1] \text{ A}$$
$$I_{C3} = 0.94 - 10^{-15}[\exp(40V_{C3}) - 1] \text{ A}$$

(a) What are the values of I_{SC} and V_{OC} for the series connected cell? (b) What is the value of P_{max}?

3.129. Write an expression for the total photo current i_{PH} for a diode having dc plus signal current components.

**3.130. The bandgaps of silicon and gallium arsenide are 1.12 eV and 1.42 eV, respectively. What are the wavelengths of light that you would expect to be emitted from these devices based on direct recombination of holes and electrons? To what "colors" of light do these wavelengths correspond?

**3.131. Repeat Prob. 3.130 for Ge, GaN, InP, InAs, BN, SiC, and CdSe.

3.132. A sinusoidal optical signal is incident on the photo diode in Fig. P3.132 and generates a sinusoidal current in the diode at a frequency of 100 kHz and an amplitude of 5 mA. (a) Write an expression for the op amp output voltage v_O if the diode has $V_{OC} = 0.65$ V, $I_{SC} = 5$ mA, $R = 200$ Ω, and $V_{BIAS} = 0$. (b) Repeat if $V_{BIAS} = 1$ V.

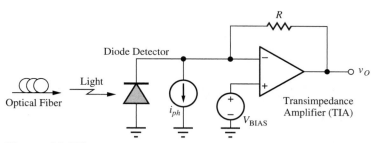

Figure P3.132

术语对照

Anode	阳极
Avalanche breakdown	雪崩击穿
Bias current and voltage	偏置电流和电压
Breakdown region	故障区域
Breakdown voltage	击穿电压
Built-in potential(or voltage)	内置电位（或电压）
Cathode	阴极
Center-tapped transformer	中心抽头变压器
Conduction angle	导通角
Conduction interval	传导时间间隔
Constant voltage drop(CVD) mode	恒压降（CVD）模型
Cut-in voltage	接通电压
Depletion layer	耗尽层
Depletion-layer capacitance	耗尽层电容
Depletion-layer width	耗尽层宽度
Depletion region	耗尽区
Diffusion capacitance	扩散电容
Diode equation	二极管方程
Diode SPICE parameters(IS, RS,N,TT,CJO, VJ,M)	二极管SPICE参数
Filter capacitor	滤波电容器
Forward bias	正向偏压
Full-wave bridge rectifier circuit	电桥式整流电路
Full-wave rectifier circuit	全波整流电路
Half-wave rectifier circuit	半波整流电路
Ideal diode	理想二极管
Ideal diode model	理想二极管模型
impact-ionization process	碰撞电离过程
Junction capacitance	结电容
Junction potential	结面电位
Light-emitting diode(LED)	发光二极管（LED）
Line regulation	电源调整率
Load line	载重线

Load-line analysis	载重线分析
Load regulation	负载调整率
Mathematical model	数学模型
Metallurgical junction	冶金结
Nonideality factor(n)	非理想性因素（n）
Peak detector	峰值检波器
Peak inverse voltage(PIV)	峰值反向电压（PIV）
Photodetector circuit	光电探测器电路
Piecewise linear model	分段线性模型
pn junction diode	pn结二极管
proportional to absolute temperature(PTAT)	与绝对温度（PTAT）成正比
Q-point	Q点
Quiescent operating point	静态操作点
Rectifier circuits	整流电路
Reverse bias	反向偏压
Reverse breakdown	反向击穿
Reverse saturation current(I_S)	反向饱和电流（I_S）
Ripple current	波纹电流
Ripple voltage	波纹电压
Saturation current	饱和电流
Schottky barrier diode	肖特基势垒二极管
Solar cell	太阳能电池
Space charge region(SCR)	空间电荷区域（SCR）
storage time	存储时间
Surge current	冲击电流
Thermal voltage(V_T)	热电压（V_T）
transit time	通过时间
Turn-on voltage	开启电压
Voltage regulator	稳压器
Voltage transfer characteristic(VTC)	电压传递特性（VTC）
Zener breakdown	齐纳击穿
Zener diode	齐纳二极管
Zero bias	零偏压
Zero-bias junction capacitance	零偏压结电容

CHAPTER 4
BIPOLAR JUNCTION TRANSISTORS

第4章 双极型晶体管

本章提纲
4.1 双极型晶体管的物理结构
4.2 *npn*晶体管的传输模型
4.3 *pnp*晶体管
4.4 晶体管传输模型的等效电路
4.5 双极型晶体管的i-v特性
4.6 双极型晶体管的工作区
4.7 传输模型的简化
4.8 双极型晶体管的非理想特性
4.9 跨导
4.10 双极型工艺与SPICE模型
4.11 BJT的实际偏置电路
4.12 偏置电路的容差

本章目标
- 理解双极型晶体管（BJT）的物理结构；
- 理解双极型晶体管的行为及载流子通过基区传输的重要性；
- 理解双极型晶体管及其特性；
- 掌握*npn*晶体管及*pnp*晶体管的区别；
- 理解双极型器件的传输模型；
- 理解BJT的四个工作区；
- 学会每个工作区的简化模型；
- 理解Early效应的来源及模型；
- 了解双极型晶体管的SPICE模型描述；
- 了解偏置电路的最差情况分析及蒙特卡洛（Monte Carlo）分析。

本章导读

  本章重点介绍双极型晶体管的相关概念，包括*npn*晶体管以及*pnp*晶体管，比较全面地讲解了双极型晶体管的物理结构、传输模型、等效电路、输出特性、SPICE模拟以及偏置电路等内容，既介绍双极型晶体管的发展由来、电路结构，又介绍相关的模型计算及设计，重点部分以设计实例的形式给出了设计与计算的方法。

  双极型晶体管于20世纪40年代末由贝尔电话实验室的Bardeen、Brattain和Shockley发明，并成为第一个商业上成功的三端固态器件，三人也因发明晶体管于1956年获得诺贝尔物理学奖。双极型晶体管的成功面市主要取决于其结构的特点，该晶体管的有源区位于半导体材料表面以下，使得晶体管的工作不受表面特性和清洁度的制约。因此，最初制作双极型晶体管要比制作MOS管更容易。20世纪50年代后期双极型晶体管开始商业化。20世纪60年代初，出现了第一个集成电路，即电阻-晶体管逻辑门，以及由若干晶体管和电阻构成的运算放大器。

目前，双极型晶体管在分立电路和集成电路设计中仍然广泛使用，特别是在高速、高精度电路中双极型晶体管依然是优先选择的器件，如运算放大器、A/D和D/A转换器以及无线通信产品，锗硅BJT的工作效率在所有硅晶体管中是最高的。

双极型晶体管的物理结构包含三个掺杂半导体区，由p型和n型交替半导体材料组成的三层夹层构成，可以制作成npn或pnp两种形式，形成晶体管的发射极将载流子注入基极。大多数载流子横穿基区，并由集电极收集。没有完全穿过基极区域的载流子在基极端子中产生小电流。

本章详细介绍了双极型晶体管的i-v特性以及简化传输模型，给出了表征双极型晶体管传输模型的三个独立参数，即饱和电流I_S、正向和反向共发射极电流增益β_F和β_R。β_F非常大，范围是20~500，它表征了BJT的强大的电流放大能力。实际制造中的限制导致双极型晶体管结构具有固有的不对称性，且β_R比β_F小得多，典型值为0~10。

双极型晶体管中，由于每个pn结都有正偏和反偏两种状态，因此一共有四种可能的工作区。本章对双极型晶体管的四个工作区进行了定义，并详细介绍了适合于各工作区的简化模型。根据施加到基极-发射极和基极-集电极的偏置电压，决定了晶体管具有的四个工作区，即截止区、正向有源区、反向有源区和饱和区。截止和饱和区经常用于开关和逻辑电路中。晶体管截止时，等效为断开的开关，当处于饱和区时等效为闭合的开关。正向有源区的双极型晶体管能够提供很高的电压和电流增益，可用于模拟信号的放大。双极型晶体管的i-v特性通常以图示的形式表示输出特性和传输特性。输出特性主要表示I_C与V_{CE}的关系，传输特性主要表示I_C与V_{BE}或V_{EB}的关系。

双极型晶体管的特性在许多方面与理想数学模型存在偏差，在晶体管的使用和选择过程中需要考虑许多条件及限制，本章对相关的反向击穿电压、扩散电容、最高频率Early效应（Early effect）等都做了详细的讲解。

为了得到双极型晶体管的综合仿真模型，除了运用晶体管的物理结构相关知识外，还需要晶体管传输模型表达式并进行相关实验。本章在分析双极型晶体管工艺的基础上给出了双极型晶体管的SPICE模型及其简化，给出了SPICE表达式中常用的参数。理解SPICE内部模型，有助于判断与分析SPICE仿真结果与仿真器件的特定应用是否一致。

双极型晶体管的偏置电路设计与分析是研究双极型晶体管的重要内容。本章研究了双极型晶体管的实际偏置电路，对四电阻网络的设计做了深入研究。偏置的目标是建立已知的静态工作点或Q点，表示晶体管最初所处的工作区。对于双极型晶体管而言，npn晶体管的Q点由集电极电流的直流量I_C和集电极-发射极电压V_{CE}表示；pnp晶体管的Q点则由集电极电流的直流量I_C和发射极-集电极电压V_{EC}表示。四电阻偏置电路可以很好地控制Q点，是能够保持晶体管Q点稳定的最好的电路之一。

本章详细讲解了四电阻偏置的电路、设计目标，提供了用于偏置双极型晶体管的四电阻偏置电路和二电阻偏置电路实例，并给出了四电阻偏置电路设计中的迭代分析方法。

无论电路是在实验室制作还是批量的集成电路制造，电路元件都存在参数取值的容差。分立电阻容差可以为10%、5%或者1%，而集成电路中电阻的容差会更大（±30%）。电源电压的容差一般为5%~10%。电路元件的容差会对偏置电路设计和分析产生很大的影响，本章重点给出了两种分析元件容差对电路影响的方法，即最差情况分析和统计蒙特卡洛分析，并给出了分析实例。在最差情况分析中，将元件的参数值取极限值，得到的结果往往很差。蒙特卡洛方法分析了大量随机选择的电路，以建立电路性能统计分布的实际估计。计算机高级语言、电子表格或者MATLAB中的随机数据发生器可以为蒙特卡洛分析提供随机的元件数值。SPICE中的一些电路分析软件包还提供蒙特卡洛分析选项。

Chapter Outline

4.1 Physical Structure of the Bipolar Transistor
4.2 The Transport Model for the *npn* Transistor
4.3 The *pnp* Transistor
4.4 Equivalent Circuit Representations for the Transport Models
4.5 The *i-v* Characteristics of the Bipolar Transistor
4.6 The Operating Regions of the Bipolar Transistor
4.7 Transport Model Simplifications
4.8 Nonideal Behavior of the Bipolar Transistor
4.9 Transconductance
4.10 Bipolar Technology and SPICE Model
4.11 Practical Bias Circuits for the BJT
4.12 Tolerances in Bias Circuits
Summary
Key Terms
References
Additional Readings
Problems

Chapter Goals

- Explore the physical structure of the bipolar transistor
- Understand bipolar transistor action and the importance of carrier transport across the base region
- Study the terminal characteristics of the BJT
- Explore the differences between *npn* and *pnp* transistors
- Develop the transport model for the bipolar device
- Define the four regions of operation of the BJT
- Explore model simplifications for each region of operation
- Understand the origin and modeling of the Early effect
- Present the SPICE model for the bipolar transistor
- Provide examples of worst-case and Monte Carlo analysis of bias circuits

November 2017 is the 70th anniversary of the discovery of the bipolar transistor by John Bardeen and Walter Brattain at Bell Laboratories. In a matter of a few months, William Shockley managed to develop a theory describing the operation of the bipolar junction transistor. Only a few years later in 1956, Bardeen, Brattain, and Shockley received the Nobel Prize in Physics for the discovery of the transistor.

John Bardeen, William Shockley, and Walter Brattain in Brattain's Laboratory in 1948.
Reprinted with permission of Alacatel-Lucent USA Inc.

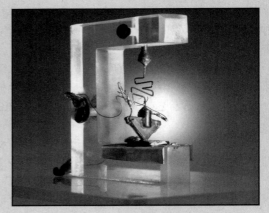

The first germanium bipolar transistor
Reprinted with permission of Alacatel-Lucent USA Inc.

In June 1948, Bell Laboratories held a major press conference to announce the discovery (which of course went essentially unnoticed by the public). Later in 1952, Bell Laboratories, operating under legal consent decrees, made licenses for the transistor available for the modest fee of $25,000 plus future royalty payments. About this time, Gordon Teal, another member of the solid-state group, left Bell Laboratories to work on the transistor at Geophysical Services Inc., which subsequently became Texas Instruments (TI). There he made the first silicon transistors, and

TI marketed the first all transistor radio. Another of the early licensees of the transistor was Tokyo Tsushin Kogyo which became the Sony Company in 1955. Sony subsequently sold a transistor radio with a marketing strategy based upon the idea that everyone could now have their own personal radio; thus was launched the consumer market for transistors. A very interesting account of these and other developments can be found in [1, 2] and their references.

Following its invention and demonstration in the late 1940s by Bardeen, Brattain, and Shockley at Bell Laboratories, the **bipolar junction transistor,** or **BJT,** became the first commercially successful three-terminal solid-state device. Its commercial success was based on its structure in which the active base region of the transistor is below the surface of the semiconductor material, making it much less dependent on surface properties and cleanliness. Thus, it was initially easier to manufacture BJTs than MOS transistors, and commercial bipolar transistors were available in the late 1950s. The first integrated circuits, resistor-transistor logic gates and operational amplifiers, consisting of a few transistors and resistors appeared in the early 1960s.

While the FET has become the dominant device technology in modern integrated circuits, bipolar transistors are still widely used in both discrete and integrated circuit design. In particular, the BJT is still the preferred device in many applications that require high speed and/or high precision. Typical of these application areas are circuits for the growing families of wireless computing and communication products, and silicon-germanium (SiGe) BJTs offer the highest operating frequencies of any silicon transistor.

The bipolar transistor is composed of a sandwich of three doped semiconductor regions and comes in two forms: the *npn* transistor and the *pnp* transistor. Performance of the bipolar transistor is dominated by *minority-carrier* transport via diffusion and drift in the central region of the transistor. Because carrier mobility and diffusivity are higher for electrons than holes, the *npn* transistor is an inherently higher-performance device than the *pnp* transistor. In Part III of this book, we will learn that the bipolar transistor typically offers a much higher voltage gain capability than the FET. On the other hand, the BJT input resistance is much lower, because a current must be supplied to the control electrode.

Our study of the BJT begins with a discussion of the *npn* transistor, followed by a discussion of the *pnp* device. The **transport model,** a simplified version of the Gummel-Poon model, is developed and used as our mathematical model for the behavior of the BJT. Four regions of operation of the BJT are defined and simplified models developed for each region. Examples of circuits that can be used to bias the bipolar transistor are presented. The chapter closes with a discussion of the worst-case and Monte Carlo analyses of the effects of tolerances on bias circuits.

4.1 PHYSICAL STRUCTURE OF THE BIPOLAR TRANSISTOR

The bipolar transistor structure consists of three alternating layers of *n*- and *p*-type semiconductor material. These layers are referred to as the **emitter (E), base (B),** and **collector (C).** Either an ***npn*** or a ***pnp* transistor** can be fabricated. The behavior of the device can be seen from the simplified cross section of the *npn* transistor in Fig. 4.1(a). During normal operation, a majority of the current enters the collector terminal, crosses the base region, and exits from the emitter terminal. A small current also enters the base terminal, crosses the base-emitter junction of the transistor, and exits the emitter.

The most important part of the bipolar transistor is the active base region between the dashed lines directly beneath the heavily doped ($n+$) emitter. Carrier transport in this region dominates the i-v characteristics of the BJT. Figure 4.1(b) illustrates the rather complex physical structure actually used to realize an *npn* transistor in integrated circuit form. Most of the structure in Fig. 4.1(b) is required to fabricate the external contacts to the collector, base, and emitter regions and to isolate one bipolar transistor from another. In the *npn* structure shown, collector current i_C and base current i_B enter the

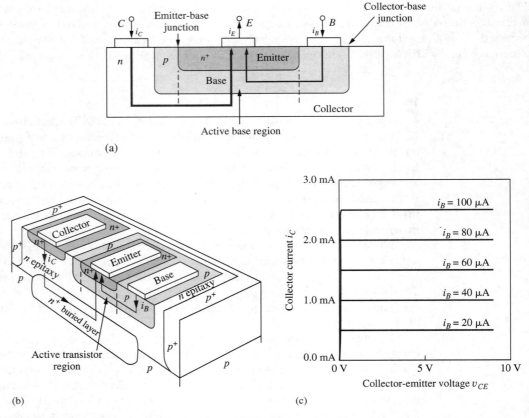

Figure 4.1 (a) Simplified cross section of an *npn* transistor with currents that occur during "normal" operation; (b) three-dimensional view of an integrated *npn* bipolar junction transistor; (c) output characteristics of an *npn* transistor.

collector (C) and base (B) terminals of the transistor, and emitter current i_E exits from the emitter (E) terminal. An example of the **output characteristics** of the bipolar transistor appears in Fig. 4.1(c), which plots collector current i_C versus collector-emitter voltage v_{CE} with base current as a parameter. The characteristics exhibit an appearance very similar to the output characteristics of the field-effect transistor. We find that a primary difference, however, is that a significant dc current must be supplied to the base of the device, whereas the dc gate current of the FET is zero. In the sections that follow, a mathematical model is developed for these *i-v* characteristics for both *npn* and *pnp* transistors.

4.2 THE TRANSPORT MODEL FOR THE *npn* TRANSISTOR

Figure 4.2 represents a conceptual model for the active region of the *npn* bipolar junction transistor structure. At first glance, the BJT appears to simply be two *pn* junctions connected back to back. However, the central region (the base) is very thin (0.1 to 100 μm), and the close proximity of the two junctions leads to coupling between the two diodes. This coupling is the essence of the bipolar device. The lower *n*-type region (the emitter) injects electrons into the *p*-type base region of the device. Almost all these injected electrons travel across the narrow base region and are removed (or collected) by the upper *n*-type region (the collector).

The three terminal currents are the **collector current i_C**, the **emitter current i_E**, and the **base current i_B**. The base-emitter voltage v_{BE} and the base-collector voltage v_{BC} applied to the two *pn* junctions in Fig. 4.2 determine the magnitude of these three currents in the bipolar transistor and

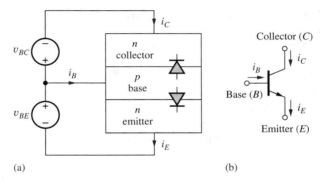

Figure 4.2 (a) Idealized *npn* transistor structure for a general-bias condition; (b) circuit symbol for the *npn* transistor.

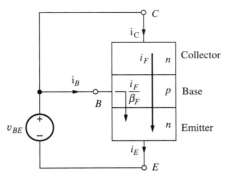

Figure 4.3 *npn* transistor with v_{BE} applied and $v_{BC} = 0$.

TABLE 4.1
Common-Emitter and Common-Base Current Gain Comparison

α_F or α_R	$\beta_F = \dfrac{\alpha_F}{1-\alpha_F}$ or $\beta_R = \dfrac{\alpha_R}{1-\alpha_R}$
0.1	0.11
0.5	1
0.9	9
0.95	19
0.99	99
0.998	499

are defined as positive when they forward-bias their respective *pn* junctions. The arrows indicate the directions of positive current in most *npn* circuit applications. The circuit symbol for the *npn* transistor appears in Fig. 4.2(b). The arrow part of the symbol identifies the emitter terminal and indicates that dc current normally exits the emitter of the *npn* transistor.

4.2.1 FORWARD CHARACTERISTICS

To facilitate both hand and computer analysis, we need to construct a mathematical model that closely matches the behavior of the transistor, and equations that describe the static *i-v* characteristics of the device can be constructed by summing currents within the transistor structure.[1] In Fig. 4.3, an arbitrary voltage v_{BE} is applied to the base-emitter junction, and the voltage applied to the base-collector junction is set to zero. The base-emitter voltage establishes emitter current i_E, which equals the total current crossing the base-emitter junction. This current is composed of two components. The largest portion, the **forward-transport current** i_F, enters the collector, travels completely across the very narrow base region, and exits the emitter terminal. The collector current i_C is equal to i_F, which has the form of an ideal diode current

$$i_C = i_F = I_S \left[\exp\left(\frac{v_{BE}}{V_T}\right) - 1 \right] \qquad (4.1)$$

---

[1] The differential equations that describe the internal physics of the BJT are linear second-order differential equations. These equations are linear in terms of the hole and electron concentrations; the currents are directly related to these carrier concentrations. Thus, superposition can be used with respect to the currents flowing in the device.

Parameter I_S is the **transistor saturation current**—that is, the saturation current of the bipolar transistor. I_S is proportional to the cross-sectional area of the active base region of the transistor, and can have a wide range of values:

$$10^{-18} \text{ A} \leq I_S \leq 10^{-9} \text{ A}$$

In Eq. (4.1), V_T should be recognized as the thermal voltage introduced in Chapter 2 and given by $V_T = kT/q \cong 0.025$ V at room temperature.

In addition to i_F, a second, much smaller component of current crosses the base-emitter junction. This current forms the base current i_B of the transistor, and it is directly proportional to i_F:

$$i_B = \frac{i_F}{\beta_F} = \frac{I_S}{\beta_F} \left[\exp\left(\frac{v_{BE}}{V_T}\right) - 1 \right] \qquad (4.2)$$

Parameter β_F is called the **forward** (or **normal**[2]) **common-emitter current gain.** Its value typically falls in the range

$$10 \leq \beta_F \leq 500$$

Emitter current i_E can be calculated by treating the transistor as a super node for which

$$i_C + i_B = i_E \qquad (4.3)$$

Adding Eqs. (4.1) and (4.2) together yields

$$i_E = \left(I_S + \frac{I_S}{\beta_F} \right) \left[\exp\left(\frac{v_{BE}}{V_T}\right) - 1 \right] \qquad (4.4)$$

which can be rewritten as

$$i_E = I_S \left(\frac{\beta_F + 1}{\beta_F}\right) \left[\exp\left(\frac{v_{BE}}{V_T}\right) - 1 \right] = \frac{I_S}{\alpha_F} \left[\exp\left(\frac{v_{BE}}{V_T}\right) - 1 \right] \qquad (4.5)$$

The parameter α_F is called the **forward** (or **normal**[3]) **common-base current gain,** and its value typically falls in the range

$$0.95 \leq \alpha_F < 1.0$$

The parameters α_F and β_F are related by

$$\alpha_F = \frac{\beta_F}{\beta_F + 1} \quad \text{or} \quad \beta_F = \frac{\alpha_F}{1 - \alpha_F} \qquad (4.6)$$

Equations (4.1), (4.2), and (4.5) express the fundamental physics-based characteristics of the bipolar transistor. The three terminal currents are all exponentially dependent on the base-emitter voltage of the transistor. This is a much stronger nonlinear dependence than the square-law behavior of the FET.

For the bias conditions in Fig. 4.3, the transistor is actually operating in a region of high current gain, called the **forward-active region**[4] of operation, which is discussed more fully in Sec. 4.9. Three extremely useful auxiliary relationships are valid in the forward-active region. The first two

---

[2] β_N is sometimes used to represent the normal common-emitter current gain.

[3] α_N is sometimes used to represent the normal common-base current gain.

[4] Four regions of operation are fully defined in Sec. 4.6.

can be found from the ratio of the collector and base current in Eqs. (4.1) and (4.2):

$$\frac{i_C}{i_B} = \beta_F \quad \text{or} \quad i_C = \beta_F i_B \quad \text{and} \quad i_E = (\beta_F + 1)i_B \tag{4.7}$$

using Eq. (4.3). The third relationship is found from the ratio of the collector and emitter currents in Eqs. (4.1) and (4.5):

$$\frac{i_C}{i_E} = \alpha_F \quad \text{or} \quad i_C = \alpha_F i_E \tag{4.8}$$

Equation (4.7) expresses important and useful properties of the bipolar transistor: The transistor "amplifies" (magnifies) its base current by the factor β_F. Because the current gain $\beta_F \gg 1$, injection of a small current into the base of the transistor produces a much larger current in both the collector and the emitter terminals. Equation (4.8) indicates that the collector and emitter currents are almost identical because $\alpha_F \cong 1$.

4.2.2 REVERSE CHARACTERISTICS

Now consider the transistor in Fig. 4.4, in which voltage v_{BC} is applied to the base-collector junction, and the base-emitter junction is zero-biased. The base-collector voltage establishes the collector current i_C, now crossing the base-collector junction. The largest portion of the collector current, the reverse-transport current i_R, enters the emitter, travels completely across the narrow base region, and exits the collector terminal. Current i_R has a form identical to i_F:

$$i_R = I_S \left[\exp\left(\frac{v_{BC}}{V_T}\right) - 1 \right] \quad \text{and} \quad i_E = -i_R \tag{4.9}$$

except the controlling voltage is now v_{BC}.

In this case, a fraction of the current i_R must also be supplied as base current through the base terminal:

$$i_B = \frac{i_R}{\beta_R} = \frac{I_S}{\beta_R} \left[\exp\left(\frac{v_{BC}}{V_T}\right) - 1 \right] \tag{4.10}$$

Parameter β_R is called the **reverse (or inverse[5]) common-emitter current gain.**

In Chapter 5, we discovered that the FET was an inherently symmetric device. For the bipolar transistor, Eqs. (4.1) and (4.9) show the symmetry that is inherent in the current that traverses the base region of the bipolar transistor. However, the impurity doping levels of the emitter and collector regions of the BJT structure are quite asymmetric, and this fact causes the base currents in the

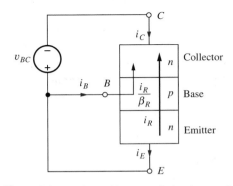

Figure 4.4 Transistor with v_{BC} applied and $v_{BE} = 0$.

---

[5] β_I is sometimes used to represent the inverse common-emitter current gain.

forward and reverse modes to be significantly different. For typical BJTs, $0 < \beta_R \leq 10$ whereas $10 \leq \beta_F \leq 500$.

The collector current in Fig. 4.4 can be found by combining the base and emitter currents, as was done to obtain Eq. (4.5):

$$i_C = -\frac{I_S}{\alpha_R}\left[\exp\left(\frac{v_{BC}}{V_T}\right) - 1\right] \tag{4.11}$$

in which the parameter α_R is called the **reverse** (or **inverse**[6]) **common-base current gain**:

$$\alpha_R = \frac{\beta_R}{\beta_R + 1} \quad \text{or} \quad \beta_R = \frac{\alpha_R}{1 - \alpha_R} \tag{4.12}$$

Typical values of α_R fall in the range

$$0 < \alpha_R \leq 0.95$$

Values of the common-base current gain α and the common-emitter current gain β are compared in Table 4.1 on page 218. Because α_F is typically greater than 0.95, β_F can be quite large. Values ranging from 10 to 500 are quite common for β_F, although it is possible to fabricate special-purpose transistors[7] with β_F as high as 5000. In contrast, α_R is typically less than 0.5, which results in values of β_R of less than 1.

> **EXERCISE:** (a) What values of β correspond to $\alpha = 0.970, 0.993, 0.250$? (b) What values of α correspond to $\beta = 40, 200, 3$?
>
> **ANSWERS:** (a) 32.3; 142; 0.333 (b) 0.976; 0.995; 0.750

4.2.3 THE COMPLETE TRANSPORT MODEL EQUATIONS FOR ARBITRARY BIAS CONDITIONS

Combining the expressions for the two collector, emitter, and base currents from Eqs. (4.1) and (4.11), (4.4) and (4.9), and (4.2) and (4.10) yields expressions for the total collector, emitter, and base currents for the *npn* transistor that are valid for the completely general-bias voltage situation in Fig. 4.2:

$$i_C = I_S\left[\exp\left(\frac{v_{BE}}{V_T}\right) - \exp\left(\frac{v_{BC}}{V_T}\right)\right] - \frac{I_S}{\beta_R}\left[\exp\left(\frac{v_{BC}}{V_T}\right) - 1\right]$$

$$i_E = I_S\left[\exp\left(\frac{v_{BE}}{V_T}\right) - \exp\left(\frac{v_{BC}}{V_T}\right)\right] + \frac{I_S}{\beta_F}\left[\exp\left(\frac{v_{BE}}{V_T}\right) - 1\right] \tag{4.13}$$

$$i_B = \frac{I_S}{\beta_F}\left[\exp\left(\frac{v_{BE}}{V_T}\right) - 1\right] + \frac{I_S}{\beta_R}\left[\exp\left(\frac{v_{BC}}{V_T}\right) - 1\right]$$

From this equation set, we see that three parameters are required to characterize an individual BJT: I_S, β_F, and β_R. (Remember that temperature is also an important parameter because $V_T = kT/q$.)

The first term in both the emitter and collector current expressions in Eqs. (4.13) is

$$i_T = I_S\left[\exp\left(\frac{v_{BE}}{V_T}\right) - \exp\left(\frac{v_{BC}}{V_T}\right)\right] \tag{4.14}$$

which represents the current being transported completely across the base region of the transistor. Equation (4.14) demonstrates the symmetry that exists between the base-emitter and base-collector voltages in establishing the dominant current in the bipolar transistor.

---

[6] α_I is sometimes used to represent the inverse common-base current gain.

[7] These devices are often called "super-beta" transistors.

Equation (4.13) actually represents a simplified version of the more complex **Gummel-Poon model** [3, 4] and form the heart of the BJT model used in the SPICE simulation program. The full Gummel-Poon model accurately describes the characteristics of BJTs over a wide range of operating conditions, and it has largely supplanted its predecessor, the **Ebers-Moll model** [5] (see Prob. 4.23).

EXAMPLE 4.1   TRANSPORT MODEL CALCULATIONS

The advantage of the full transport model is that it can be used to estimate the currents in the bipolar transistor for any given set of bias voltages.

PROBLEM Use the transport model equations to find the terminal voltages and currents in the circuit in Fig. 4.5 in which an *npn* transistor is biased by two dc voltage sources.

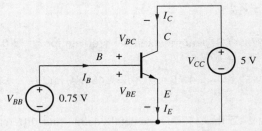

Figure 4.5 *npn* transistor circuit example: $I_S = 10^{-16}$ A, $\beta_F = 50$, $\beta_R = 1$.

SOLUTION   **Known Information and Given Data:** The *npn* transistor in Fig. 4.5 is biased by two dc sources $V_{BB} = 0.75$ V and $V_{CC} = 5.0$ V. The transistor parameters are $I_S = 10^{-16}$ A, $\beta_F = 50$, and $\beta_R = 1$.

Unknowns: Junction bias voltages V_{BE} and V_{BC}; emitter current I_E, collector current I_C, base current I_B

Approach: Determine V_{BE} and V_{BC} from the circuit. Use these voltages and the transistor parameters to calculate the currents using Eq. (4.13).

Assumptions: The transistor is modeled by the transport equations and is operating at room temperature with $V_T = 25.0$ mV.

Analysis: In this circuit, the base emitter voltage V_{BE} is set directly by source V_{BB}, and the base collector voltage is the difference between V_{BB} and V_{CC}:

$$V_{BE} = V_{BB} = 0.75 \text{ V}$$
$$V_{BC} = V_{BB} - V_{CC} = 0.75 \text{ V} - 5.00 \text{ V} = -4.25 \text{ V}$$

Substituting these voltages into Eqs. (4.13) along with the transistor parameters yields

$$I_C = 10^{-16} \text{ A} \left[\exp\left(\frac{0.75 \text{ V}}{0.025 \text{ V}}\right) - \overset{0}{\cancel{\exp\left(\frac{-4.75 \text{ V}}{0.025 \text{ V}}\right)}} \right] - \frac{10^{-16}}{1} \text{ A} \left[\overset{0}{\cancel{\exp\left(\frac{-4.75 \text{ V}}{0.025 \text{ V}}\right)}} - 1 \right]$$

$$I_E = 10^{-16} \text{ A} \left[\exp\left(\frac{0.75 \text{ V}}{0.025 \text{ V}}\right) - \overset{0}{\cancel{\exp\left(\frac{-4.75 \text{ V}}{0.025 \text{ V}}\right)}} \right] + \frac{10^{-16}}{50} \text{ A} \left[\exp\left(\frac{0.75 \text{ V}}{0.025 \text{ V}}\right) - 1 \right]$$

$$I_B = \frac{10^{-16}}{50} \text{ A} \left[\exp\left(\frac{0.75 \text{ V}}{0.025 \text{ V}}\right) - 1 \right] + \frac{10^{-16}}{1} \text{ A} \left[\overset{0}{\cancel{\exp\left(\frac{-4.75 \text{ V}}{0.025 \text{ V}}\right)}} - 1 \right]$$

and evaluating these expressions gives

$$I_C = 1.07 \text{ mA} \qquad I_E = 1.09 \text{ mA} \qquad I_B = 21.4 \text{ μA}$$

Check of Results: The sum of the collector and base currents equals the emitter current as required by KCL for the transistor treated as a super node. Also, the terminal currents range from microamperes to milliamperes, which are reasonable for most transistors.

Discussion: Note that the collector-base junction in Fig. 4.5 is reverse-biased, so the terms containing V_{BC} become negligibly small. In this example, the transistor is biased in the forward-active region of operation for which

$$\beta_F = \frac{I_C}{I_B} = \frac{1.07 \text{ mA}}{0.0214 \text{ mA}} = 50 \qquad \text{and} \qquad \alpha_F = \frac{I_C}{I_E} = \frac{1.07 \text{ mA}}{1.09 \text{ mA}} = 0.982$$

EXERCISE: Repeat the example problem for $I_S = 10^{-15}$ A, $\beta_F = 100$, $\beta_R = 0.50$, $V_{BE} = 0.70$ V, and $V_{CC} = 10$ V.

ANSWERS: $I_C = 1.45$ mA, $I_E = 1.46$ mA, and $I_B = 14.5$ μA

In Secs. 4.5 to 4.11 we completely define four different regions of operation of the transistor and find simplified models for each region. First, however, let us develop the transport model for the *pnp* transistor in a manner similar to that for the *npn* transistor.

4.3 THE *pnp* TRANSISTOR

In Chapter 5, we found we could make either NMOS or PMOS transistors by simply interchanging the *n*- and *p*-type regions in the device structure. One might expect the same to be true of bipolar transistors, and we can indeed fabricate *pnp* transistors as well as *npn* transistors.

The *pnp* transistor is fabricated by reversing the layers of the transistor, as diagrammed in Fig. 4.6. The transistor has been drawn with the emitter at the top of the diagram, as it appears in most circuit diagrams throughout this book. The arrows again indicate the normal directions of positive current in the *pnp* transistor in most circuit applications. The voltages applied to the two *pn* junctions are the emitter-base voltage v_{EB} and the collector-base voltage v_{CB}. These voltages are again positive when they forward-bias their respective *pn* junctions. Collector current i_C and base current i_B exit the transistor terminals, and the emitter current i_E enters the device. The circuit symbol for the *pnp*

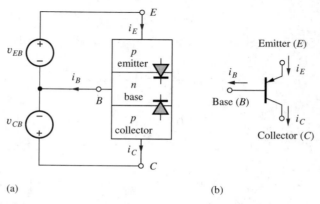

Figure 4.6 (a) Idealized *pnp* transistor structure for a general-bias condition; (b) circuit symbol for the *pnp* transistor.

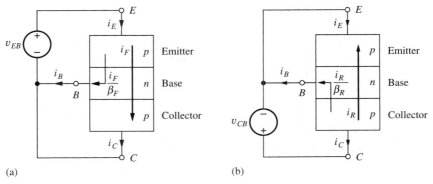

Figure 4.7 (a) *pnp* transistor with v_{EB} applied and $v_{CB} = 0$; (b) *pnp* transistor with v_{CB} applied and $v_{EB} = 0$.

transistor appears in Fig. 4.6(b). The arrow identifies the emitter of the *pnp* transistor and points in the direction of normal positive-emitter current.

Equations that describe the static *i-v* characteristics of the *pnp* transistor can be constructed by summing currents within the structure just as for the *npn* transistor. In Fig. 4.7(a), voltage v_{EB} is applied to the emitter-base junction, and the collector-base voltage is set to zero. The emitter-base voltage establishes forward-transport current i_F that traverses the narrow base region and base current i_B that crosses the emitter-base junction of the transistor:

$$i_C = i_F = I_S \left[\exp\left(\frac{v_{EB}}{V_T}\right) - 1 \right] \qquad i_B = \frac{i_F}{\beta_F} = \frac{I_S}{\beta_F} \left[\exp\left(\frac{v_{EB}}{V_T}\right) - 1 \right]$$

and (4.15)

$$i_E = i_C + i_B = I_S \left(1 + \frac{1}{\beta_F}\right) \left[\exp\left(\frac{v_{EB}}{V_T}\right) - 1 \right]$$

In Fig. 4.7(b), a voltage v_{CB} is applied to the collector-base junction, and the emitter-base junction is zero-biased. The collector-base voltage establishes the reverse-transport current i_R and base current i_B:

$$-i_E = i_R = I_S \left[\exp\left(\frac{v_{CB}}{V_T}\right) - 1 \right] \qquad i_B = \frac{i_R}{\beta_R} = \frac{I_S}{\beta_R} \left[\exp\left(\frac{v_{CB}}{V_T}\right) - 1 \right]$$

and (4.16)

$$i_C = -I_S \left(1 + \frac{1}{\beta_R}\right) \left[\exp\left(\frac{v_{CB}}{V_T}\right) - 1 \right]$$

where the collector current is given by $i_C = i_E - i_B$.

For the general-bias voltage situation in Fig. 4.6, Eqs. (4.15) and (4.16) are combined to give the total collector, emitter, and base currents of the *pnp* transistor:

$$i_C = I_S \left[\exp\left(\frac{v_{EB}}{V_T}\right) - \exp\left(\frac{v_{CB}}{V_T}\right) \right] - \frac{I_S}{\beta_R} \left[\exp\left(\frac{v_{CB}}{V_T}\right) - 1 \right]$$

$$i_E = I_S \left[\exp\left(\frac{v_{EB}}{V_T}\right) - \exp\left(\frac{v_{CB}}{V_T}\right) \right] + \frac{I_S}{\beta_F} \left[\exp\left(\frac{v_{EB}}{V_T}\right) - 1 \right] \qquad (4.17)$$

$$i_B = \frac{I_S}{\beta_F} \left[\exp\left(\frac{v_{EB}}{V_T}\right) - 1 \right] + \frac{I_S}{\beta_R} \left[\exp\left(\frac{v_{CB}}{V_T}\right) - 1 \right]$$

These equations represent the simplified Gummel-Poon or transport model equations for the *pnp* transistor and can be used to relate the terminal voltages and currents of the *pnp* transistor for any general-bias condition. Note that these equations are identical to those for the *npn* transistor except that v_{EB} and v_{CB} replace v_{BE} and v_{BC}, respectively, and are a result of our careful choice for the direction of positive currents in Figs. 4.2 and 4.6.

> **EXERCISE:** Find I_C, I_E, and I_B for a *pnp* transistor if $I_S = 10^{-16}$ A, $\beta_F = 75$, $\beta_R = 0.40$, $V_{EB} = 0.75$ V, and $V_{CB} = +0.70$ V.
>
> **ANSWERS:** $I_C = 0.563$ mA, $I_E = 0.938$ mA, $I_B = 0.376$ mA

4.4 EQUIVALENT CIRCUIT REPRESENTATIONS FOR THE TRANSPORT MODELS

For circuit simulation, as well as hand analysis purposes, the transport model equations for the *npn* and *pnp* transistors can be represented by the equivalent circuits shown in Fig. 4.8(a) and (b), respectively. In the *npn* model in Fig. 4.8(a), the total transport current i_T traversing the base is determined by I_S, v_{BE}, and v_{BC}, and is modeled by the current source i_T:

$$i_T = i_F - i_R = I_S \left[\exp\left(\frac{v_{BE}}{V_T}\right) - \exp\left(\frac{v_{BC}}{V_T}\right) \right] \tag{4.18}$$

The diode currents correspond directly to the two components of the base current:

$$i_B = \frac{I_S}{\beta_F} \left[\exp\left(\frac{v_{BE}}{V_T}\right) - 1 \right] + \frac{I_S}{\beta_R} \left[\exp\left(\frac{v_{BC}}{V_T}\right) - 1 \right] \tag{4.19}$$

Directly analogous arguments hold for the circuit elements in the *pnp* circuit model of Fig. 4.8(b).

> **EXERCISE:** Find I_T if $I_S = 10^{-15}$ A, $V_{BE} = 0.75$ V, and $V_{BC} = -2.0$ V.
>
> **ANSWER:** 10.7 mA
>
> **EXERCISE:** Find the dc transport current I_T for the transistor in Example 4.1.
>
> **ANSWER:** $I_T = 1.07$ mA

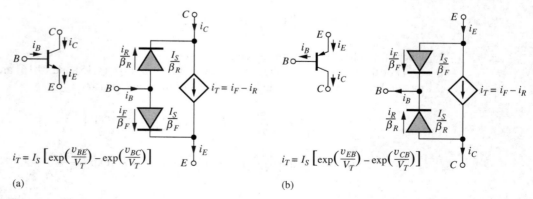

Figure 4.8 (a) Transport model equivalent circuit for the *npn* transistor; (b) transport model equivalent circuit for the *pnp* transistor.

ANOTHER LOOK AT THE FORWARD-ACTIVE REGION

As discussed in Sec. 4.4, the strict definition of the forward-active region requires the collector-base junction to be zero- or reverse-biased. However, in low-power, low-voltage circuits, every tenth of a volt is important. Here we look more closely at the behavior of the transistor by rearranging the transport current expression for the *npn* collector current from Eq. (4.18):

$$I_C = I_S\left(e^{\frac{v_{BE}}{V_T}} - e^{\frac{v_{BC}}{V_T}}\right) = I_S e^{\frac{v_{BE}}{V_T}}\left(1 - e^{\frac{v_{BC}-v_{BE}}{V_T}}\right) \tag{4.20}$$

If we are willing to accept a 1 percent error in the expected value of collector current, then we require

$$e^{\frac{v_{BC}-v_{BE}}{V_T}} \leq 0.01 \quad \text{or} \quad v_{BE} - v_{BC} \geq -V_T \ln(0.01) \tag{4.21}$$

Noting that $(v_{BE} - v_{BC})$ is equal to the transistor's collector-emitter voltage v_{CE} and using 25.8 mV for V_T, we find v_{CE} must exceed only 120 mV rather than the full base-emitter voltage in the range of 0.5–0.7 V! For circuits operating from 1.5 V or less, this difference is significant! The same result can be derived for the minimum value of v_{EC} of the *pnp* transistor. Various circuit examples appear throughout later chapters.

> **EXERCISE:** What value of V_{CE} is required for the collector current error in Eq. (4.20) to be less than 0.1 percent?
>
> **ANSWER:** 179 mV. (Note that this is another example of 60 mV/decade of current change.)

4.5 THE *i-v* CHARACTERISTICS OF THE BIPOLAR TRANSISTOR

Two complementary views of the *i-v* behavior of the BJT are represented by the device's **output characteristic** and its **transfer characteristic**. (We will see similar characteristics for the FETs presented in Chapter 5.) The output characteristics represent the relationship between the collector current and collector-emitter or collector-base voltage of the transistor, whereas the transfer characteristic relates the collector current to the base-emitter voltage. A knowledge of both *i-v* characteristics is basic to understanding the overall behavior of the bipolar transistor.

4.5.1 OUTPUT CHARACTERISTICS

Circuits for measuring or simulating the **common-emitter output characteristics** are shown in Fig. 4.9. In these circuits, the base of the transistor is driven by a constant current source, and the output characteristics represent a graph of i_C vs. v_{CE} for the *npn* transistor (or i_C vs. v_{EC} for the *pnp*) with base current i_B as a parameter. Note that the Q-point (I_C, V_{CE}) or (I_C, V_{EC}) locates the BJT operating point on the output characteristics.

First, consider the *npn* transistor operating with $v_{CE} \geq 0$, represented by the first quadrant of the graph in Fig. 4.10. For $i_B = 0$, the transistor is nonconducting or cut off. As i_B increases above 0, i_C also increases. For $v_{CE} \geq v_{BE}$, the *npn* transistor is in the forward-active region, and collector current is independent of v_{CE} and equal to $\beta_F i_B$. Remember, it was demonstrated earlier that $i_C \cong \beta_F i_B$ in the forward-active region. For $v_{CE} \leq v_{BE}$, the transistor enters the **saturation region** of operation in which the total voltage between the collector and emitter terminals of the transistor is small.

It is important to note that the saturation region of the BJT does not correspond to the saturation region of the FET that will be defined in Chapter 5. The **forward-active region** (or just **active region**) of the BJT corresponds to the saturation region of the FET. When we begin our discussion of amplifiers in Parts Two and Three, we will simply apply the term active region to both devices. The active region is the region most often used in transistor implementations of amplifiers.

172 Chapter 4 Bipolar Junction Transistors

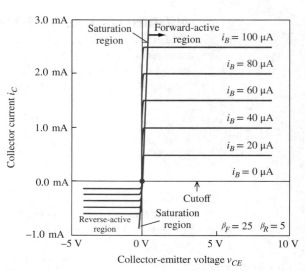

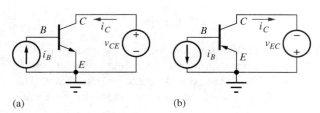

Figure 4.9 Circuits for determining common-emitter output characteristics: (a) *npn* transistor, (b) *pnp* transistor.

Figure 4.10 Common-emitter output characteristics for the bipolar transistor (i_C vs. v_{CE} for the *npn* transistor or i_C vs. v_{EC} for the *pnp* transistor).

In the third quadrant for $v_{CE} \leq 0$, the roles of the collector and emitter reverse. For $v_{BE} \leq v_{CE} \leq 0$, the transistor remains in saturation. For $v_{CE} \leq v_{BE}$, the transistor enters the **reverse-active region,** in which the *i-v* characteristics again become independent of v_{CE}, and now $i_C \cong -(\beta_R + 1)i_B$. The reverse-active region curves have been plotted for a relatively large value of reverse common-emitter current gain, $\beta_R = 5$, to enhance their visibility. As noted earlier, the reverse-current gain β_R is often less than 1.

Using the polarities defined in Fig. 4.9(b) for the *pnp* transistor, the output characteristics will appear exactly the same as in Fig. 4.10, except that the horizontal axis will be the voltage v_{EC} rather than v_{CE}. Remember that $i_B > 0$ and $i_C > 0$ correspond to currents exiting the base and collector terminals of the *pnp* transistor.

4.5.2 TRANSFER CHARACTERISTICS

The **common-emitter transfer characteristic** of the BJT defines the relationship between the collector current and the base-emitter voltage of the transistor. An example of the transfer characteristic for an *npn* transistor is shown in graphical form in Fig. 4.11, with both linear and semilog scales for

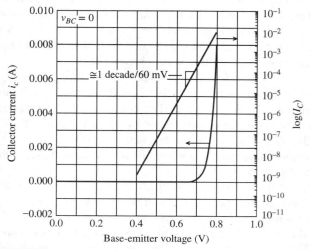

Figure 4.11 BJT transfer characteristic in the forward-active region.

the particular case of $v_{BC} = 0$. The transfer characteristic is virtually identical to that of a *pn* junction diode. This behavior can also be expressed mathematically by setting $v_{BC} = 0$ in the collector-current expression in Eq. (4.13):

$$i_C = I_S \left[\exp\left(\frac{v_{BE}}{V_T}\right) - 1 \right] \quad \text{for } v_{BC} = 0 \tag{4.22}$$

Because of the exponential relationship in Eq. (4.22), the semilog plot exhibits the same slope as that for a *pn* junction diode. Only a 60-mV change in v_{BE} is required to change the collector current by a factor of 10, and for a fixed collector current, the base-emitter voltage of the silicon BJT will exhibit a -1.8-mV/°C temperature coefficient, just as for the silicon diode (see Sec. 3.5).

> **EXERCISE:** What base-emitter voltage V_{BE} corresponds to $I_C = 100$ µA in an *npn* transistor at room temperature if $I_S = 10^{-16}$ A? For $I_C = 1$ mA?
>
> **ANSWERS:** 0.691 V; 0.748 V

4.6 THE OPERATING REGIONS OF THE BIPOLAR TRANSISTOR

In the bipolar transistor, each *pn* junction may be independently forward-biased or reverse-biased, so there are four possible regions of operation, as defined in Table 4.2. The operating point establishes the region of operation of the transistor and can be defined by any two of the four terminal voltages or currents. The characteristics of the transistor are quite different for each of the four regions of operation, and in order to simplify our circuit analysis task, we need to be able to make an educated guess as to the region of operation of the BJT.

When both junctions are reverse-biased, the transistor is essentially nonconducting or *cut off* (**cutoff region**) and can be considered an open switch. If both junctions are forward-biased, the transistor is operating in the **saturation region** and appears as a closed switch. Cutoff and saturation are most often used to represent the two states in binary logic circuits implemented with BJTs. For example, switching between these two operating regions occurs in the bipolar transistor-transistor logic circuits. This information can be found online through Connect.

In the **forward-active region** (also called the **normal-active region** or just **active region**), in which the base-emitter junction is forward-biased and the base-collector junction is reverse-biased, the BJT can provide high current, voltage, and power gains. The forward-active region is most often

TABLE 4.2
Regions of Operation of the Bipolar Transistor

BASE-EMITTER JUNCTION	BASE-COLLECTOR JUNCTION	
	Reverse Bias	Forward Bias
Forward Bias	Forward-active region (Normal-active region) (Good amplifier)	Saturation region* (Closed switch)
Reverse Bias	Cutoff region (Open switch)	Reverse-active region (Inverse-active region) (Poor amplifier)

*It is important to note that the saturation region of the bipolar transistor does *not* correspond to the saturation region of the FET. This unfortunate use of terms is historical in nature and something we just have to accept.

used to achieve high-quality amplification. In addition, in the fastest form of bipolar logic, called emitter-coupled logic, the transistors switch between the cutoff and the forward-active regions.

In the **reverse-active region** (or **inverse-active region**), the base-emitter junction is reverse-biased and the base-collector junction is forward-biased. In this region, the transistor exhibits low current gain, and the reverse-active region is not often used. However, we will see an important application of the reverse-active region in transistor-transistor logic circuits in Chapter 9. Reverse operation of the bipolar transistor has also found use in analog-switching applications.

The transport model equations describe the behavior of the bipolar transistor for any combination of terminal voltages and currents. However, the complete sets of equations in (4.13) and (4.17) are quite imposing. In subsequent sections, bias conditions specific to each of the four regions of operation will be used to obtain simplified sets of relationships that are valid for the individual regions. The Q-point for the BJT is (I_C, V_{CE}) for the *npn* transistor and (I_C, V_{EC}) for the *pnp*.

> **EXERCISE:** What is the region of operation of (a) an *npn* transistor with $V_{BE} = 0.75$ V and $V_{BC} = -0.70$ V? (b) A *pnp* transistor with $V_{CB} = 0.70$ V and $V_{EB} = 0.75$ V?
>
> **ANSWERS:** Forward-active region; saturation region

4.7 TRANSPORT MODEL SIMPLIFICATIONS

The complete sets of transport model equations developed in Secs. 4.2 and 4.3 describe the behavior of the *npn* and *pnp* transistors for any pair of terminal voltages and currents, and these equations are indeed the basis for the models used in SPICE circuit simulation. However, the full sets of equations are quite complex, so now we will explore simplifications that can be used to reduce the complexity of the model descriptions for each of the four different regions of operation identified in Table 4.2.

4.7.1 SIMPLIFIED MODEL FOR THE CUTOFF REGION

The easiest region to understand is the cutoff region, in which both junctions are reverse-biased. For an *npn* transistor, the cutoff region requires $v_{BE} \leq 0$ and $v_{BC} \leq 0$. Let us further assume that

$$v_{BE} < -4\frac{kT}{q} \quad \text{and} \quad v_{BC} < -4\frac{kT}{q} \quad \text{where} \quad -4\frac{kT}{q} = -0.1 \text{V}$$

These two conditions allow us to neglect the exponential terms in Eq. (4.13), yielding the following simplified equations for the *npn* terminal currents in cutoff:

$$i_C = I_S\left[\cancel{\exp\left(\frac{v_{BE}}{V_T}\right)}^0 - \cancel{\exp\left(\frac{v_{BC}}{V_T}\right)}^0\right] - \frac{I_S}{\beta_R}\left[\cancel{\exp\left(\frac{v_{BC}}{V_T}\right)}^0 - 1\right]$$

$$i_E = I_S\left[\cancel{\exp\left(\frac{v_{BE}}{V_T}\right)}^0 - \cancel{\exp\left(\frac{v_{BC}}{V_T}\right)}^0\right] + \frac{I_S}{\beta_F}\left[\cancel{\exp\left(\frac{v_{BE}}{V_T}\right)}^0 - 1\right] \quad (4.23)$$

$$i_B = \frac{I_S}{\beta_F}\left[\cancel{\exp\left(\frac{v_{BE}}{V_T}\right)}^0 - 1\right] + \frac{I_S}{\beta_R}\left[\cancel{\exp\left(\frac{v_{BC}}{V_T}\right)}^0 - 1\right]$$

or

$$i_C = +\frac{I_S}{\beta_R} \qquad i_E = -\frac{I_S}{\beta_F} \qquad i_B = -\frac{I_S}{\beta_F} - \frac{I_S}{\beta_R}$$

In cutoff, the three terminal currents—i_C, i_E, and i_B—are all constant and smaller than the saturation current I_S of the transistor. The simplified model for this situation is shown in Fig. 4.12(b).

4.7 Transport Model Simplifications

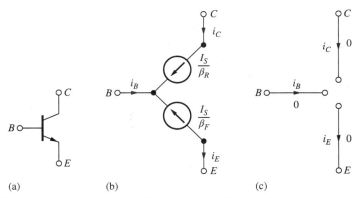

Figure 4.12 Modeling the *npn* transistor in cutoff: (a) *npn* transistor, (b) constant leakage current model, (c) open-circuit model.

In cutoff, only very small leakage currents appear in the three transistor terminals. In most cases, these currents are negligibly small and can be assumed to be zero. Therefore, we usually think of the transistor operating in the cutoff region as being "off" with essentially zero terminal currents, as indicated by the three-terminal open-circuit model in Fig. 4.12(c). The cutoff region represents an open switch and is used as one of the two states required for binary logic circuits.

EXAMPLE 4.2   A BJT BIASED IN CUTOFF

Cutoff represents the "off state" in switching applications, so an understanding of the magnitudes of the currents involved is important. In this example, we explore how closely the "off state" approaches zero.

PROBLEM Figure 4.13 is an example of a circuit in which the transistor is biased in the cutoff region. Estimate the currents using the simplified model in Fig. 4.12, and compare to calculations using the full transport model.

SOLUTION **Known Information and Given Data:** From the figure, $I_S = 10^{-16}$ A, $\alpha_F = 0.95$, $\alpha_R = 0.25$, $V_{BE} = 0$ V, $V_{BC} = -5$ V

Unknowns: I_C, I_B, I_E

Approach: First analyze the circuit using the simplified model of Fig. 4.12. Then, compare the results to calculations using the voltages to simplify the transport equations.

Assumptions: $V_{BE} = 0$ V, so the "diode" terms containing V_{BE} are equal to 0. $V_{BC} = -5$ V, which is much less than $-4kT/q = -100$ mV, so the transport model equations can be simplified.

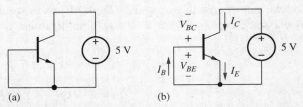

Figure 4.13 (a) *npn* transistor bias in the cutoff region (for calculations, use $I_S = 10^{-16}$ A, $\alpha_F = 0.95$, $\alpha_R = 0.25$); (b) normal current directions.

Analysis: The voltages $V_{BE} = 0$ and $V_{BC} = -5$ V are consistent with the definition of the cutoff region. If we use the open-circuit model in Fig. 4.12(c), the currents I_C, I_E, and I_B are all predicted to be zero.

To obtain a more exact estimate of the currents, we use the transport model equations. For the circuit in Fig. 4.13, the base-emitter voltage is exactly zero, and $V_{BC} \ll 0$. Therefore, Eq. (4.13) reduces to

$$I_C = I_S\left(1 + \frac{1}{\beta_R}\right) = \frac{I_S}{\alpha_R} = \frac{10^{-16}\,\text{A}}{0.25} = 4 \times 10^{-16}\,\text{A} = 0.4\,\text{fA}$$

$$I_E = I_S = 10^{-16}\,\text{A} = 1\,\text{fA} \quad \text{and} \quad I_B = -\frac{I_S}{\beta_R} = -\frac{10^{-16}\,\text{A}}{\frac{1}{3}} = -3 \times 10^{-16}\,\text{A} = 3\,\text{fA}$$

The calculated currents in the terminals are very small but nonzero. Note, in particular, that the base current is not zero and that small currents exit both the emitter and base terminals of the transistor.

Check of Results: As a check on our results, we see that Kirchhoff's current law is satisfied for the transistor treated as a super node: $i_C + i_B = i_E$.

Discussion: The voltages $V_{BE} = 0$ and $V_{BC} = -5$ V are consistent with the definition of the cutoff region. Thus, we expect the currents to be negligibly small. Here again we see an example of the use of different levels of modeling to achieve different degrees of precision in the answer $[(I_C, I_E, I_B) = (0, 0, 0)$ or $(4 \times 10^{-16}\,\text{A}, 10^{-16}\,\text{A}, -3 \times 10^{-16}\,\text{A})]$.

EXERCISE: Calculate the values of the currents in the circuit in Fig. 4.13(a) if the value of the voltage source is changed to 10 V and (b) if the base-emitter voltage is set to −3 V using a second voltage source.

ANSWERS: (a) No change; (b) 0.300 fA, 5.26 aA, −0.305 fA

4.7.2 MODEL SIMPLIFICATIONS FOR THE FORWARD-ACTIVE REGION

Arguably the most important region of operation of the BJT is the forward-active region, in which the emitter-base junction is forward-biased and the collector-base junction is reverse-biased. In this region, the transistor can exhibit high voltage and current gains and is useful for analog amplification. From Table 4.2, we see that the forward-active region of an *npn* transistor corresponds to $v_{BE} \geq 0$ and $v_{BC} \leq 0$. In most cases, the forward-active region will have

$$v_{BE} > 4\frac{kT}{q} = 0.1\,\text{V} \quad \text{and} \quad v_{BC} < -4\frac{kT}{q} = -0.1\,\text{V}$$

and we can assume that $\exp(-v_{BC}/V_T) \ll 1$ just as we did in simplifying Eq. set (4.23). We can also assume $\exp(v_{BE}/V_T) \gg 1$. These simplifications yield

$$i_C = I_S \exp\left(\frac{v_{BE}}{V_T}\right) + \frac{I_S}{\beta_R}$$

$$i_E = \frac{I_S}{\alpha_F} \exp\left(\frac{v_{BE}}{V_T}\right) + \frac{I_S}{\beta_F} \quad (4.24)$$

$$i_B = \frac{I_S}{\beta_F} \exp\left(\frac{v_{BE}}{V_T}\right) - \frac{I_S}{\beta_F} - \frac{I_S}{\beta_R}$$

4.7 Transport Model Simplifications

The exponential term in each of these expressions is usually huge compared to the other terms. By neglecting the small terms, we find the most useful simplifications of the BJT model for the forward-active region:

$$i_C = I_S \exp\left(\frac{v_{BE}}{V_T}\right) \qquad i_E = \frac{I_S}{\alpha_F} \exp\left(\frac{v_{BE}}{V_T}\right) \qquad i_B = \frac{I_S}{\beta_F} \exp\left(\frac{v_{BE}}{V_T}\right) \qquad (4.25)$$

In these equations, the fundamental, exponential relationship between all the terminal currents and the base-emitter voltage v_{BE} is once again clear. In the forward-active region, the terminal currents all have the form of diode currents in which the controlling voltage is the base-emitter junction potential. It is also important to note that the currents are all independent of the base-collector voltage v_{BC}. The collector current i_C can be modeled as a voltage-controlled current source that is controlled by the base-emitter voltage and independent of the collector voltage.

By taking ratios of the terminal currents in Eq. (4.25), two important auxiliary relationships for the forward-active region are found, and observing that $i_E = i_C + i_B$ yields a third important result:

$$i_C = \alpha_F i_E \quad \text{and} \quad i_C = \beta_F i_B \qquad i_E = (\beta_F + 1) i_B \qquad (4.26)$$

The results from Eq. (4.26) are placed in a circuit context in the next two examples from Fig. 4.14.

DESIGN NOTE

FORWARD-ACTIVE REGION

Operating points in the forward-active region are normally used for linear amplifiers. Our dc model for the forward-active region is quite simple:

$$I_C = \beta_F I_B \quad \text{and} \quad I_E = (\beta_F + 1) I_B \quad \text{with} \quad V_{BE} \cong 0.7 \text{ V}.$$

Forward-active operation requires $V_{BE} > 0$ and $V_{CE} \geq V_{BE}$.

EXAMPLE 4.3 FORWARD-ACTIVE REGION OPERATION WITH EMITTER CURRENT BIAS

Current sources are widely utilized for biasing in circuit design, and such a source is used to set the Q-point current in the transistor in Fig. 4.14(a).

PROBLEM Find the emitter, base and collector currents, and base-emitter voltage for the transistor biased by a current source in Fig. 4.14(a).

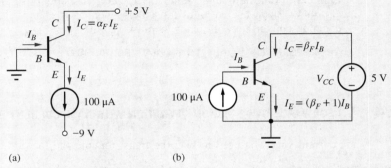

Figure 4.14 Two *npn* transistors operating in the forward-active region ($I_S = 10^{-16}$ A and $\alpha_F = 0.95$ are assumed for the example calculations).

SOLUTION **Known Information and Given Data:** An *npn* transistor biased by the circuit in Fig. 4.14(a) with $I_S = 10^{-16}$ A and $\alpha_F = 0.95$. From the circuit, $V_{BC} = V_B - V_C = -5$ V and $I_E = +100$ μA.

Unknowns: I_C, I_B, V_{BE}

Approach: Show that the transistor is in the forward-active region of operation and use Eqs. (4.25) and (4.24) to find the unknown currents and voltage.

Assumptions: Room temperature operation with $V_T = 25.9$ mV

Analysis: From the circuit, we observe that the emitter current is forced by the current source to be $I_E = +100$ μA, and the current source will forward-bias the base-emitter diode. Study of the mathematical model in Eq. (4.13) also confirms that the base-emitter voltage must be positive (forward bias) in order for the emitter current to be positive. Thus, we have $V_{BE} > 0$ and $V_{BC} < 0$, which correspond to the forward-active region of operation for the *npn* transistor.

The base and collector currents can be found using Eq. (4.26) with $I_E = 100$ μA:

$$I_C = \alpha_F I_E = 0.95 \cdot 100 \text{ μA} = 95 \text{ μA}$$

Solving for β_F gives $\quad \beta_F = \dfrac{\alpha_F}{1 - \alpha_F} = \dfrac{0.95}{1 - .95} = 19 \quad \beta_F + 1 = 20$

and

$$I_B = \frac{I_E}{\beta_F + 1} = \frac{100 \text{ μA}}{20} = 5 \text{ μA}$$

The base-emitter voltage is found from the emitter current expression in Eq. (4.25):

$$V_{BE} = V_T \ln \frac{\alpha_F I_E}{I_S} = (0.0259 \text{ V}) \ln \frac{0.95(10^{-4} \text{ A})}{10^{-16} \text{ A}} = 0.715 \text{ V}$$

Check of Results: As a check on our results, we see that Kirchhoff's current law is satisfied for the transistor treated as a super node: $i_C + i_B = i_E$. Also we can check V_{BE} using both the collector and base current expressions in Eq. (4.25).

Discussion: We see that most of the current being forced or "pulled" out of the emitter by the current source comes directly through the transistor from the collector. This is the common-base mode in which $i_C = \alpha_F i_E$ with $\alpha_F \cong 1$.

EXERCISE: Calculate the values of the currents and base-emitter voltage in the circuit in Fig. 4.14(a) if (a) the value of the voltage source is changed to 10 V and (b) the transistor's common-emitter current gain is increased to 50.

ANSWERS: (a) No change; (b) 100 μA, 1.96 μA, 98.0 μA, 0.715 V

EXAMPLE **4.4** **FORWARD-ACTIVE REGION OPERATION WITH BASE CURRENT BIAS**

A current source is used to bias the transistor into the forward-active region in Fig. 4.14(b).

PROBLEM Find the emitter, base and collector currents, and base-emitter and base-collector voltages for the transistor biased by the base current source in Fig. 4.14(b).

SOLUTION **Known Information and Given Data:** An *npn* transistor biased by the circuit in Fig. 4.14(b) with $I_S = 10^{-16}$ A and $\alpha_F = 0.95$. From the circuit, $V_C = +5$ V and $I_B = +100$ µA.

Unknowns: I_C, I_B, V_{BE}, V_{BC}

Approach: Show that the transistor is in the forward-active region of operation and use Eqs. (4.25) and (4.26) to find the unknown currents and voltage.

Assumptions: Room temperature operation with $V_T = 25.9$ mV

Analysis: In the circuit in Fig. 4.14(b), base current I_B is now forced to equal 100 µA by the ideal current source. This current enters the base and will exit the emitter, forward-biasing the base-emitter junction. From the mathematical model in Eq. (4.13), we see that positive base current can occur for positive V_{BE} and positive V_{BC}. However, we have $V_{BC} = V_B - V_C = V_{BE} - V_C$. Since the base-emitter diode voltage will be approximately 0.7 V, and $V_C = 5$ V, V_{BC} will be negative (e.g., $V_{BC} \cong 0.7 - 5.0 = -4.3$ V). Thus we have $V_{BE} > 0$ and $V_{BC} < 0$, which corresponds to the forward-active region of operation for the *npn* transistor, and the collector and emitter currents can be found using Eq. (4.26) with $I_B = 100$ µA:

$$I_C = \beta_F I_B = 19 \cdot 100 \text{ µA} = 1.90 \text{ mA}$$

$$I_E = (\beta_F + 1) I_B = 20 \cdot 100 \text{ µA} = 2.00 \text{ mA}$$

The base-emitter voltage can be found from the collector current expression in Eq. (4.25):

$$V_{BE} = V_T \ln \frac{I_C}{I_S} = (0.0259 \text{ V}) \ln \frac{1.9 \times 10^{-3} \text{ A}}{10^{-16} \text{ A}} = 0.792 \text{ V}$$

$$V_{BC} = V_B - V_C = V_{BE} - V_C = 0.792 - 5 = -4.21 \text{ V}$$

Check of Results: As a check on our results, we see that Kirchhoff's current law is satisfied for the transistor treated as a super node: $i_C + i_B = i_E$. Also we can check the value of V_{BE} using either the emitter or base current expressions in Eq. (4.25). The calculated values of V_{BE} and V_{BC} correspond to forward-active region operation.

Discussion: A large amplification of the current takes place when the current source is injected into the base terminal in Fig. 4.14(b) in contrast to the situation when the source is connected to the emitter terminal in Fig. 4.14(a).

EXERCISE: Calculate the values of the currents and base-emitter voltage in the circuit in Fig. 4.14(b) if (a) the value of the voltage source is changed to 10 V and (b) the transistor's common-emitter current gain is increased to 50.

ANSWERS: (a) No change; (b) 5.00 mA, 100 µA, 5.10 mA, 0.817 V

EXERCISE: What is the minimum value of V_{CC} that corresponds to forward-active region bias in Fig. 4.14(b)?

ANSWER: $V_{CC} = V_{BE} = 0.792$ V

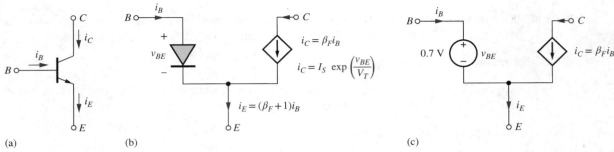

Figure 4.15 (a) *npn* transistor; (b) simplified model for the forward-active region; (c) further simplification for the forward-active region using the CVD model for the diode.

As illustrated in Examples 4.3 and 4.4, Eq. (4.26) can often be used to greatly simplify the analysis of circuits operating in the forward-active region. However, remember this caveat well: **The results in Eq. (4.26) are valid *only* for the forward-active region of operation!**

The BJT is often considered a current-controlled device. However, from Eq. (4.25), we see that the fundamental physics-based behavior of the BJT in the forward-active region is that of a (nonlinear) voltage-controlled current source. The base current should be considered as an unwanted defect current that must be supplied to the base in order for the transistor to operate. In an ideal BJT, β_F would be infinite, the base current would be zero, and the collector and emitter currents would be identical, just as we will find for the FET currents in Chapter 5. Unfortunately, it is impossible to fabricate such a BJT.

Equation (4.25) leads to the simplified circuit model for the forward-active region shown in Fig. 4.15. The current in the base-emitter diode is amplified by the common-emitter current gain β_F and appears in the collector terminal. However, remember that the base and collector currents are exponentially related to the base-emitter voltage. Because the base-emitter diode is forward-biased in the forward-active region, the transistor model of Fig. 4.15(b) can be further simplified to that of Fig. 4.15(c), in which the diode is replaced by its constant voltage drop (CVD) model, in this case $V_{BE} = 0.7$ V. The dc base and emitter voltages differ by the 0.7-V diode voltage drop in the forward-active region.

EXAMPLE 4.5   FORWARD-ACTIVE REGION BIAS USING TWO POWER SUPPLIES

Analog circuits frequently operate from a pair of positive and negative power supplies so that bipolar input and output signals can easily be accommodated. The circuit in Fig. 4.16 provides one possible circuit configuration in which resistor R and -9-V source replace the current source utilized in Fig. 4.14(a). Collector resistor R_C has been added to reduce the collector-emitter voltage.

PROBLEM   Find the Q-point for the transistor in the circuit in Fig. 4.16.

SOLUTION   **Known Information and Given Data:** *npn* transistor in the circuit in Fig. 4.16(a) with $\beta_F = 50$ and $\beta_R = 1$

Unknowns: Q-point (I_C, V_{CE})

Approach: In this circuit, the base-collector junction will tend to be reverse-biased by the 9-V source. The combination of the resistor and the -9-V source will force a current out of the emitter and forward-bias the base-emitter junction. Thus, the transistor appears to be biased in the forward-active region of operation.

4.7 Transport Model Simplifications

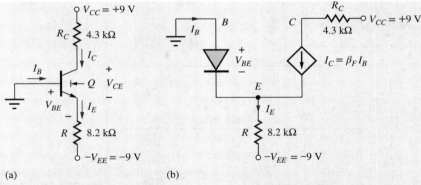

Figure 4.16 (a) *npn* Transistor circuit (assume $\beta_F = 50$ and $\beta_R = 1$); (b) simplified model for the forward-active region.

Assumptions: Assume forward-active region operation; since we do not know the saturation current, assume $V_{BE} = 0.7$ V; use the simplified model for the forward-active region to analyze the circuit as in Fig. 4.16(b).

Analysis: The currents can now be found by using KVL around the base-emitter loop:

$$V_{BE} + 8200 I_E - V_{EE} = 0 \quad \text{or} \quad V_T \ln\left(1 + \frac{I_C}{I_S}\right) + 8200 \frac{I_C}{\alpha_F} - V_{EE} = 0$$

For $V_{BE} = 0.7$ V, $0.7 + 8200 I_E - 9 = 0$ or $I_E = \dfrac{8.3 \text{ V}}{8200 \text{ }\Omega} = 1.01$ mA

At the emitter node, $I_E = (\beta_F + 1) I_B$, so

$$I_B = \frac{1.02 \text{ mA}}{50 + 1} = 19.8 \text{ }\mu\text{A} \quad \text{and} \quad I_C = \beta_F I_B = 0.990 \text{ mA}$$

The collector-emitter voltage is equal to

$$V_{CE} = V_{CC} - I_C R_C - (-V_{BE}) = 9 - .990 \text{ mA}(4.3 \text{ k}\Omega) + 0.7 = 5.44 \text{ V}$$

The Q-point is (0.990 mA, 5.44 V).

Check of Results: We see that KVL is satisfied around the output loop containing the collector-emitter voltage: $+9 - V_{RC} - V_{CE} - V_R - (-9) = 9 - 4.3 - 5.4 - 8.3 + 9 = 0$. We must check the forward-active region assumption $V_{CE} = 5.4$ V which is greater than $V_{BE} = 0.7$ V. Also, the currents are all positive and $I_C + I_B = I_E$.

Discussion: In this circuit, the combination of the resistor and the -9-V source replace the current source that was used to bias the transistor in Fig. 4.14(a).

Computer-Aided Analysis: SPICE contains a built-in model for the bipolar transistor that will be discussed in detail in Sec. 4.10. SPICE simulation with the default *npn* transistor model yields a Q-point that agrees well with our hand analysis: (0.993 mA, 5.50 V).

Solving the transcendental equation for I_C using MATLAB$^{(R)}$ or WolframAlpha$^{(R)}$ with $I_S = 0.1$ fA and $V_T = 25.9$ mV yields $(I_C, V_{CE}) = (0.9905$ mA, 5.55 V).

EXERCISE: (a) Find the Q-point in Ex. 4.5 if resistor R is changed to 5.6 kΩ. (b) What value of R is required to set the current to approximately 100 μA in the original circuit?

ANSWERS: (a) (1.45 mA, 3.5 V); (b) 82 kΩ.

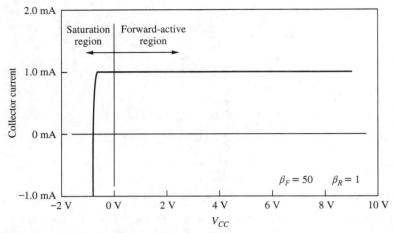

Figure 4.17 Simulation of output characteristics of circuit of Fig. 4.16(a).

Figure 4.18 Diode-connected transistor.

Figure 4.17 displays the results of simulation of the collector current of the transistor in Fig. 4.16 versus the supply voltage V_{CC}. For $V_{CC} > 0$, the collector-base junction will be reverse-biased, and the transistor will be in the forward-active region. In this region, the circuit behaves essentially as a 1-mA ideal current source in which the output current is independent of V_{CC}. Note that the circuit actually behaves as a current source for V_{CC} down to approximately -0.5 V. By the definitions in Table 4.2, the transistor enters saturation for $V_{CC} < 0$, but the transistor does not actually enter heavy saturation until the base-collector junction begins to conduct for $V_{BC} \geq +0.5$ V.

> **EXERCISE:** Find the three terminal currents in the transistor in Fig. 4.16 if the 8.2 kΩ resistor value is changed to 5.6 kΩ.
>
> **ANSWERS:** 1.48 mA, 29.1 μA, 1.45 mA
>
> **EXERCISE:** What are the actual values of V_{BE} and V_{CE} for the transistor in Fig. 4.16(a) if $I_S = 5 \times 10^{-16}$ A? (Note that an iterative solution is necessary.)
>
> **ANSWERS:** 0.933 V, 5.48 V

4.7.3 DIODES IN BIPOLAR INTEGRATED CIRCUITS

In integrated circuits, we often want the characteristics of a diode to match those of the BJT as closely as possible. In addition, it takes about the same amount of area to fabricate a diode as a full bipolar transistor. For these reasons, a diode is usually formed by connecting the base and collector terminals of a bipolar transistor, as shown in Fig. 4.18. This connection forces $v_{BC} = 0$.

Using the transport model equations for BJT with this boundary condition yields an expression for the terminal current of the "diode":

$$i_D = (i_C + i_B) = \left(I_S + \frac{I_S}{\beta_F}\right)\left[\exp\left(\frac{v_{BE}}{V_T}\right) - 1\right] = \frac{I_S}{\alpha_F}\left[\exp\left(\frac{v_D}{V_T}\right) - 1\right] \quad (4.27)$$

The terminal current has an i-v characteristic corresponding to that of a diode with a reverse saturation current that is determined by the BJT parameters. This technique is often used in both analog and digital circuit design; we will see many examples of its use in the analog designs in Part Two.

EXERCISE: What is the equivalent saturation current of the diode in Fig 4.18 if the transistor is described by $I_S = 2 \times 10^{-14}$ A and $\alpha_F = 0.95$?

ANSWER: 21.1 fA

ELECTRONICS IN ACTION

The Bipolar Transistor PTAT Cell

The diode version of the PTAT cell that generates an output voltage **proportional to absolute temperature** was introduced back in Chapter 3. We can also easily implement the PTAT cell using two bipolar transistors as shown in the figure here in which two identical bipolar transistors are biased in the forward-active region by current sources with a 10:1 current ratio.

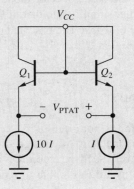

Logo © Auburn University. Photo courtesy of Richard Jaeger

The PTAT voltage is given by

$$V_{\text{PTAT}} = V_{E2} - V_{E1} = (V_{CC} - V_{BE2}) - (V_{CC} - V_{BE1}) = V_{BE1} - V_{BE2}$$

$$V_{\text{PTAT}} = V_T \ln\left(\frac{10I}{I_S}\right) - V_T \ln\left(\frac{I}{I_S}\right) = \frac{kT}{q}\ln(10) \quad \text{and} \quad \frac{dV_{\text{PTAT}}}{dT} = \frac{198\ \mu V}{°K}$$

The bipolar PTAT cell is a circuit commonly used in electronic thermometry and temperature compensation in more complex circuitry.

4.7.4 SIMPLIFIED MODEL FOR THE REVERSE-ACTIVE REGION

In the reverse-active region, also called the inverse-active region, the roles of the emitter and collector terminals are reversed. The base-collector diode is forward-biased and the base-emitter junction is reverse-biased, and we can assume that $\exp(v_{BE}/V_T) \ll 1$ for $v_{BE} < -0.1$ V just as we did in simplifying Eq. set (4.23). Applying this approximation to Eq. (4.13) and neglecting the -1 terms relative to the exponential terms yields the simplified equations for the reverse-active region:

$$i_C = -\frac{I_S}{\alpha_R}\exp\left(\frac{v_{BC}}{V_T}\right) \qquad i_E = -I_S\exp\left(\frac{v_{BC}}{V_T}\right) \qquad i_B = \frac{I_S}{\beta_R}\exp\left(\frac{v_{BC}}{V_T}\right) \qquad (4.28)$$

Ratios of these equations yield $i_E = -\beta_R i_B$ and $i_E = \alpha_R i_C$.

Equation (4.28) leads to the simplified circuit model for the reverse-active region shown in Fig. 4.19. The base current in the base-collector diode is amplified by the reverse common-emitter current gain β_R and enters the emitter terminal.

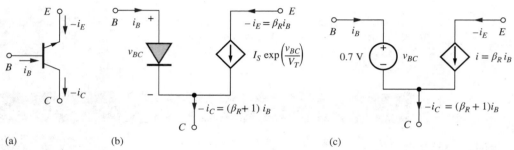

Figure 4.19 (a) *npn* transistor in the reverse-active region; (b) simplified circuit model for the reverse-active region; (c) further simplification in the reverse-active region using the CVD model for the diode.

In the reverse-active region, the base-collector diode is now forward-biased, and the transistor model of Fig. 4.19(b) can be further simplified to that of Fig. 4.19(c), in which the diode is replaced by its CVD model with a voltage of 0.7 V. The base and collector voltages differ only by one 0.7-V diode drop in the reverse-active region.

EXAMPLE 4.6   REVERSE-ACTIVE REGION ANALYSIS

Although the reverse-active region is not often used, one does encounter it fairly frequently in the laboratory. If the transistor is inadvertently plugged in upside down, for example, the transistor will be operating in the reverse-active region. On the surface, the circuit will seem to be working but not very well. It is useful to be able to recognize when this error has occurred.

PROBLEM The collector and emitter terminals of the *npn* transistor in Fig. 4.16 have been interchanged in the circuit in Fig. 4.20 (perhaps the transistor was plugged into the circuit backwards by accident). Find the new Q-point for the transistor in the circuit in Fig. 4.20.

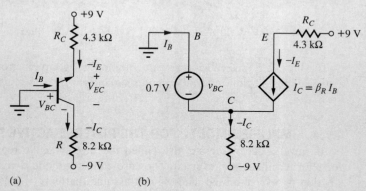

Figure 4.20 (a) Circuit of Fig. 4.16 with *npn* transistor orientation reversed; (b) circuit simplification using the model for the reverse-active region (analysis of the circuit uses $\beta_F = 50$ and $\beta_R = 1$).

SOLUTION **Known Information and Given Data:** *npn* transistor in the circuit in Fig. 4.20 with $\beta_F = 50$ and $\beta_R = 1$

Unknowns: Q-point (I_C, V_{CE})

Approach: In this circuit, the base-emitter junction is reverse-biased by the 9-V source ($V_{BE} = V_B - V_E = -9$ V). The combination of the 8.2-kΩ resistor and the -9-V source will pull a current *out of* the collector and forward-bias the base-collector junction. Thus, the transistor appears to be biased in the reverse-active region of operation.

Assumptions: Assume reverse-active region operation; since we do not know the saturation current, assume $V_{BC} = 0.7$ V; use the simplified model for the reverse-active region to analyze the circuit as in Fig. 4.20(b).

Analysis: The current exiting from the collector $(-I_C)$ is now equal to

$$(-I_C) = \frac{-0.7 \text{ V} - (-9 \text{ V})}{8200 \, \Omega} = 1.01 \text{ mA}$$

The current through the 8.2-kΩ resistor is unchanged compared to that in Fig. 4.16. However, significant differences exist in the currents in the base terminal and the +9-V source. At the collector node, $(-I_C) = (\beta_R + 1)I_B$, and at the emitter, $(-I_E) = \beta_R I_B$:

$$I_B = \frac{1.01 \text{ mA}}{2} = 0.505 \text{ mA} \quad \text{and} \quad -I_E = (1) I_B = 0.505 \text{ mA}$$

$$V_{EC} = 9 - 4300(.505 \text{ mA}) - (-0.7 \text{ V}) = 7.5 \text{ V}$$

Check of Results: We see that KVL is satisfied around the output loop containing the collector-emitter voltage: $+9 - V_{CE} - V_R - (-9) = 9 - 9.7 - 8.3 + 9 = 0$. Also, $I_C + I_B = I_E$, and the calculated current directions are all consistent with the assumption of reverse-active region operation. Finally $V_{EB} = 9 - 43 \text{ k}\Omega \, (0.505 \text{ mA}) = 6.8$ V. $V_{EB} > 0$ V, and the reverse active assumption is correct.

Discussion: Note that the base current is much larger than expected, whereas the current entering the upper terminal of the device is much smaller than would be expected if the transistor were in the circuit as originally drawn in Fig. 4.16. These significant differences in current often lead to unexpected shifts in voltage levels at the base and collector terminals of the transistor in more complicated circuits.

Computer-Aided Design: The built-in SPICE model is valid for any operating region, and simulation with the default model gives results very similar to hand calculations.

DESIGN NOTE

REVERSE-ACTIVE REGION CHARACTERISTICS

Note that the currents for reverse-active region operation are usually very different from those found for forward-active region operation in Fig. 4.16. These drastic differences are often useful in debugging circuits that we have built in the lab and can be used to discover transistors that have been improperly inserted into a circuit breadboard.

EXERCISE: Find the three terminal currents in the transistor in Fig. 4.20 if resistor R is changed to 5.6 kΩ.

ANSWERS: 1.48 mA, 0.741 mA, 0.741 mA

4.7.5 MODELING OPERATION IN THE SATURATION REGION

The fourth and final region of operation is called the saturation region. In this region, both junctions are forward-biased, and the transistor typically operates with a small voltage between collector and emitter terminals. In the saturation region, the dc value of v_{CE} is called the **saturation voltage** of the transistor: v_{CESAT} for the *npn* transistor or v_{ECSAT} for the *pnp* transistor.

In order to determine v_{CESAT}, we assume that both junctions are forward-biased so that i_C and i_B from Eq. (4.13) can be approximated as

$$i_C = I_S \exp\left(\frac{v_{BE}}{V_T}\right) - \frac{I_S}{\alpha_R} \exp\left(\frac{v_{BC}}{V_T}\right)$$
$$i_B = \frac{I_S}{\beta_F} \exp\left(\frac{v_{BE}}{V_T}\right) + \frac{I_S}{\beta_R} \exp\left(\frac{v_{BC}}{V_T}\right) \tag{4.29}$$

Simultaneous solution of these equations using $\beta_R = \alpha_R/(1 - \alpha_R)$ yields expressions for the base-emitter and base-collector voltages:

$$v_{BE} = V_T \ln \frac{i_B + (1 - \alpha_R) i_C}{I_S \left[\frac{1}{\beta_F} + (1 - \alpha_R)\right]} \quad \text{and} \quad v_{BC} = V_T \ln \frac{i_B - \frac{i_C}{\beta_F}}{I_S \left[\frac{1}{\alpha_R}\right]\left[\frac{1}{\beta_F} + (1 - \alpha_R)\right]} \tag{4.30}$$

By applying KVL to the transistor in Fig. 4.21, we find that the collector-emitter voltage of the transistor is $v_{CE} = v_{BE} - v_{BC}$, and substituting the results from Eq. (4.30) into this equation yields an expression for the saturation voltage of the *npn* transistor:

$$v_{CESAT} = V_T \ln \left[\left(\frac{1}{\alpha_R}\right) \frac{1 + \frac{i_C}{(\beta_R + 1) i_B}}{1 - \frac{i_C}{\beta_F i_B}}\right] \quad \text{for } i_B > \frac{i_C}{\beta_F} \tag{4.31}$$

This equation is important and highly useful in the design of saturated digital switching circuits. For a given value of collector current, Eq. (4.31) can be used to determine the base current required to achieve a desired value of v_{CESAT}.

Note that Eq. (4.31) is valid only for $i_B > i_C/\beta_F$. This is an auxiliary condition that can be used to define saturation region operation. The ratio i_C/β_F represents the base current needed to maintain transistor operation in the forward-active region. If the base current exceeds the value needed for forward-active region operation, the transistor will enter saturation. The actual value of i_C/i_B is often called the **forced beta** β_{FOR} of the transistor, where $\beta_{FOR} \leq \beta_F$.

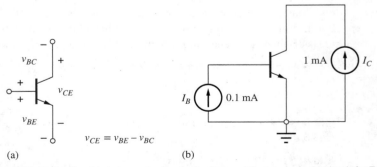

Figure 4.21 (a) Relationship between the terminal voltages of the transistor; (b) circuit for Ex. 4.8.

EXAMPLE 4.7 SATURATION VOLTAGE CALCULATION

The BJT saturation voltage is important in many switching applications including logic circuits and power supplies. Here we find an example of the value of the saturation voltage for a forced beta of 10.

PROBLEM Calculate the saturation voltage for an *npn* transistor with $I_C = 1$ mA, $I_B = 0.1$ mA, $\beta_F = 50$, and $\beta_R = 1$.

SOLUTION **Known Information and Given Data:** An *npn* transistor is operating with $I_C = 1$ mA, $I_B = 0.1$ mA, $\beta_F = 50$, and $\beta_R = 1$.

Unknowns: Collector-emitter voltage of the transistor.

Approach: Because $I_C/I_B = 10 < \beta_F$, the transistor will indeed be saturated. Therefore we can use Eq. (4.31) to find the saturation voltage.

Assumptions: Room temperature operation with $V_T = 0.0259$ V

Analysis: Using $\alpha_R = \beta_R/(\beta_R + 1) = 0.5$ and $I_C/I_B = 10$ yields

$$v_{\text{CESAT}} = (0.025 \text{ V}) \ln\left[\left(\frac{1}{0.5}\right) \frac{1 + \dfrac{1 \text{ mA}}{2(0.1 \text{ mA})}}{1 - \dfrac{1 \text{ mA}}{50(0.1 \text{ mA})}}\right] = 0.070 \text{ V}$$

Check of Results: A small, nearly zero, value of saturation voltage is expected; thus the calculated value appears reasonable.

Discussion: We see that the value of V_{CE} in this example is indeed quite small. However, it is nonzero even for $i_C = 0$ [see Prob. 4.56]! It is impossible to force the forward voltages across both *pn* junctions to be exactly equal, which is a consequence of the asymmetric values of the forward and reverse current gains. The existence of this small voltage "offset" is an important difference between the BJT and the MOSFET.

Computer-Aided Analysis: We can simulate the situation in this example by driving the base of the BJT with one current source and the collector with a second. (This is one of the few circuit situations in which we can force a current into the collector using a current source.) SPICE yields $V_{\text{CESAT}} = 0.070$ V. The default temperature in SPICE is 27°C, and the slight difference in V_T accounts for the difference between SPICE result and our hand calculations.

EXERCISE: What is the saturation voltage in Ex. 4.7 if the base current is reduced to 40 μA?

ANSWER: 103 mV.

EXERCISE: Use Eq. (4.30) to find V_{BESAT} and V_{BCSAT} for the transistor in Ex. 4.7 if $I_S = 10^{-15}$ A.

ANSWERS: 0.719 V, 0.650 V.

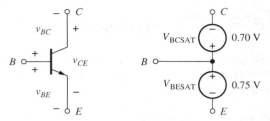

Figure 4.22 Simplified model for the *npn* transistor in saturation.

Figure 4.22 shows the simplified model for the transistor in saturation in which the two diodes are assumed to be forward-biased and replaced by their respective on-voltages. The forward voltages of both diodes are normally higher in saturation than in the forward-active region, as indicated in the figure by $V_{BESAT} = 0.75$ V and $V_{BCSAT} = 0.7$ V. In this case, V_{CESAT} is 50 mV. In saturation, the terminal currents are determined by the external circuit elements; no simplifying relationships exist between i_C, i_B, and i_E other than $i_C + i_B = i_E$.

ELECTRONICS IN ACTION

Optical Isolators

The optical isolator drawn schematically here represents a highly useful circuit that behaves much like a single transistor, but provides a very high breakdown voltage and low capacitance between its input and output terminals. Input current i_{IN} drives a light emitting diode (LED) whose output illuminates the base region of an *npn* transistor. Energy lost by the photons creates hole–electron pairs in the base of the *npn*. The holes represent base current that is then amplified by the current gain β_F of the transistor, whereas the electrons simply become part of the collector current.

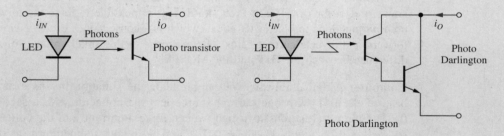

Photo Darlington

The output characteristics of the optical isolator are very similar to those of a BJT operating in the active region in Fig. 4.10. However, the conversion of photons to hole–electron pairs is not very efficient in silicon, and the current transfer ratio, $\beta_F = i_O/i_{IN}$, of the optical isolator is often only around unity. The "Darlington connection" of two transistors (see Sec. 13.2.3) is often used to improve the overall current gain of the isolator. In this case, the output current is increased by the current gain of the second transistor.

The dc isolation provided by such devices can exceed a thousand volts and is limited primarily by the spacing of the pins and the characteristics of the circuit board that the isolator is mounted upon. AC isolation is limited to the low picofarad range by stray capacitance between the input and outputs pins.

4.8 NONIDEAL BEHAVIOR OF THE BIPOLAR TRANSISTOR

As with all devices, the BJT characteristics deviate from our ideal mathematical models in a number of ways. The emitter-base and collector-base diodes that form the bipolar transistor have finite reverse breakdown voltages (see Sec. 3.6.2) that we must carefully consider when choosing a transistor or the power supplies for our circuits. There are also capacitances associated with each of the diodes, and these capacitances place limitations on the high-frequency response of the transistor. In addition, we know that holes and electrons in semiconductor materials have finite velocities. Thus, it takes time for the carriers to move from the emitter to the collector, and this time delay places an additional limit on the upper frequency of operation of the bipolar transistor. Finally, the output characteristics of the BJT exhibit a dependence on collector-emitter voltage similar to the channel-length modulation effect that occurs in the MOS transistor (Sec. 5.2.7). This section considers each of these limitations in more detail.

4.8.1 JUNCTION BREAKDOWN VOLTAGES

The bipolar transistor is formed from two back-to-back diodes, each of which has a Zener breakdown voltage associated with it. If the reverse voltage across either *pn* junction is too large, the corresponding diode will break down. In the transistor structure in Fig. 4.1, the emitter region is the most heavily doped region and the collector is the most lightly doped region. These doping differences lead to a relatively low breakdown voltage for the base-emitter diode, typically in the range of 3 to 10 V. On the other hand, the collector-base diode can be designed to break down at much larger voltages. Transistors can be fabricated with collector-base breakdown voltages as high as several hundred volts.[8]

Transistors must be selected with breakdown voltages commensurate with the reverse voltages that will be encountered in the circuit. In the forward-active region, for example, the collector-base junction is operated under reverse bias and must not break down. In the cutoff region, both junctions are reverse-biased, and the relatively low breakdown voltage of the emitter-base junction must not be exceeded.

4.8.2 MINORITY-CARRIER TRANSPORT IN THE BASE REGION

Current in the BJT is predominantly determined by the transport of *minority carriers* across the base region. In the *npn* transistor in Fig. 4.23, transport current i_T results from the diffusion of minority carriers—electrons in the *npn* transistor or holes in the *pnp*—across the base. Base current i_B is composed of hole injection back into the emitter and collector, as well as a small additional current I_{REC} needed to replenish holes lost to recombination with electrons in the base. These three components of base current are shown in Fig. 4.23(a).

An expression for the transport current i_T can be developed using our knowledge of carrier diffusion and the values of base-emitter and base-collector voltages. It can be shown from device physics (beyond the scope of this text) that the voltages applied to the base-emitter and base-collector junctions define the minority-carrier concentrations at the two ends of the base region through these relationships:

$$n(0) = n_{bo} \exp\left(\frac{v_{BE}}{V_T}\right) \quad \text{and} \quad n(W_B) = n_{bo} \exp\left(\frac{v_{BC}}{V_T}\right) \quad (4.32)$$

in which **n_{bo}** is the **equilibrium electron density** in the *p*-type base region.

The two junction voltages establish a minority-carrier concentration gradient across the base region, as illustrated in Fig. 4.23(b). For a narrow base, the minority-carrier density decreases

---

[8] Specially designed power transistors may have breakdown voltages in the 1000-V range.

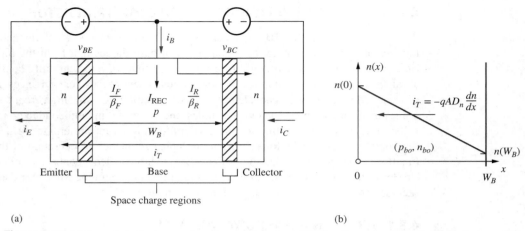

Figure 4.23 (a) Currents in the base region of an *npn* transistor; (b) minority-carrier concentration in the base of the *npn* transistor.

linearly across the base, and the diffusion current in the base can be calculated using the diffusion current expression in Eq. (2.14):

$$i_T = -qAD_n\frac{dn}{dx} = +qAD_n\frac{n_{bo}}{W_B}\left[\exp\left(\frac{v_{BE}}{V_T}\right) - \exp\left(\frac{v_{BC}}{V_T}\right)\right] \quad (4.33)$$

where A = cross-sectional area of base region and W_B = **base width.** Because the carrier gradient is negative, electron current i_T is directed in the negative x direction, exiting the emitter terminal (positive i_T).

Comparing Eqs. (4.33) and (4.18) yields a value for the bipolar transistor saturation current I_S:

$$I_S = qAD_n\frac{n_{bo}}{W_B} = \frac{qAD_n n_i^2}{N_{AB} W_B} \quad (4.34a)$$

where N_{AB} = doping concentration in base of transistor, n_i = intrinsic-carrier concentration (10^{10}/cm^3), and $n_{bo} = n_i^2/N_{AB}$ using Eq. (2.12).

The corresponding expression for the saturation current of the *pnp* transistor is

$$I_S = qAD_p\frac{p_{bo}}{W_B} = \frac{qAD_p n_i^2}{N_{DB} W_B} \quad (4.34b)$$

Remembering from Chapter 2 that mobility μ, and hence diffusivity $D = (kT/q)\mu$ (cm^2/s), is larger for electrons than holes ($\mu_n > \mu_p$), we see from Eqs. (4.34) that the *npn* transistor will conduct a higher current than the *pnp* transistor for a given set of applied voltages.

> **EXERCISE:** (a) What is the value of D_n at room temperature if μ_n = 500 cm^2/V · s? (b) What is I_S for a transistor with A = 50 μm^2, W = 1 μm, D_n = 12.5 cm^2/s, and N_{AB} = 10^{18}/cm^3?
>
> **ANSWERS:** 13.0 cm^2/s; 10^{-18} A

4.8.3 BASE TRANSIT TIME

To turn on the bipolar transistor, minority-carrier charge must be introduced into the base to establish the carrier gradient in Fig. 4.23(b). The **forward transit time** τ_F represents the time constant associated with storing the required charge Q in the base region and is defined by

$$\tau_F = \frac{Q}{I_T} \quad (4.35)$$

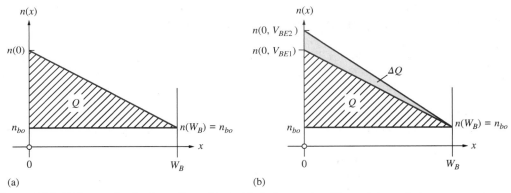

Figure 4.24 (a) Excess minority charge Q stored in the bipolar base region; (b) stored charge Q changes as v_{BE} changes.

Figure 4.24 depicts the situation in the neutral base region of an *npn* transistor operating in the forward-active region with $v_{BE} > 0$ and $v_{BC} = 0$. The area under the triangle represents the excess minority charge Q that must be stored in the base to support the diffusion current. For the dimensions in Fig. 4.24 and using Eq. (4.32)

$$Q = qA[n(0) - n_{bo}]\frac{W_B}{2} = qAn_{bo}\left[\exp\left(\frac{v_{BE}}{V_T}\right) - 1\right]\frac{W_B}{2} \qquad (4.36)$$

For the conditions in Fig. 4.24(a),

$$i_T = \frac{qAD_n}{W_B}n_{bo}\left[\exp\left(\frac{v_{BE}}{V_T}\right) - 1\right] \qquad (4.37)$$

Substituting Eqs. (4.36) and (4.37) into Eq. (4.35), the forward transit time for the *npn* transistor is found to be

$$\tau_F = \frac{W_B^2}{2D_n} = \frac{W_B^2}{2V_T\mu_n} \qquad (4.38a)$$

The corresponding expression for the transit time of the *pnp* transistor is

$$\tau_F = \frac{W_B^2}{2D_p} = \frac{W_B^2}{2V_T\mu_p} \qquad (4.38b)$$

The base transit time can be viewed as the average time required for a carrier emitted by the emitter to arrive at the collector. Hence, one would not expect the transistor to be able to reproduce frequencies with periods that are less than the transit time, and the base transit time in Eq. (4.38) places an upper limit on the useful operating frequency f of the transistor:

$$f \leq \frac{1}{2\pi\tau_F} \qquad (4.39)$$

From Eq. (4.38), we see that the transit time is inversely proportional to the minority-carrier mobility in the base, and the difference between electron and hole mobility leads to an inherent frequency and speed advantage for the *npn* transistor. Thus, an *npn* transistor may be expected to be 2 to 2.5 times as fast as a *pnp* transistor for a given geometry and doping. Equation (4.38) also indicates the importance of shrinking the base width W_B of the transistor as much as possible. Early transistors had base widths of 10 μm or more, whereas the base width of transistors in research laboratories today is 0.1 μm (100 nm) and below.

EXAMPLE 4.8   SATURATION CURRENT AND TRANSIT TIME

Device physics has provided us with expressions that can be used to estimate transistor saturation current and transit time based on a knowledge of physical constants and structural device information. Here we find representative values of I_S and τ_F for a bipolar transistor.

PROBLEM Find the saturation current and base transit time for an *npn* transistor with a 100 μm × 100 μm emitter region, a base doping of $10^{17}/\text{cm}^3$, and a base width of 1 μm. Assume $\mu_n = 500 \text{ cm}^2/\text{V} \cdot \text{s}$.

SOLUTION **Known Information and Given Data:** Emitter area = 100 μm × 100 μm, $N_{AB} = 10^{17}/\text{cm}^3$, $W_B = 1$ μm, $\mu_n = 500 \text{ cm}^2/\text{V} \cdot \text{s}$

Unknowns: Saturation current I_S; transit time τ_F

Approach: Evaluate Eqs. (4.35) and (4.39) using the given data.

Assumptions: Room temperature operation with $V_T = 0.0259$ V and $n_i = 10^{10}/\text{cm}^3$

Analysis: Using Eq. (4.34) for I_S:

$$I_S = \frac{qAD_n n_i^2}{N_{AB} W_B} = \frac{(1.6 \times 10^{-19} \text{ C})(10^{-2} \text{ cm})^2 \left(0.0259 \text{ V} \times 500 \frac{\text{cm}^2}{\text{V} \cdot \text{s}}\right)\left(\frac{10^{20}}{\text{cm}^6}\right)}{\left(\frac{10^{17}}{\text{cm}^3}\right)(10^{-4} \text{ cm})} = 2.07 \times 10^{-15} \text{ A}$$

in which $D_n = (kT/q)\mu_n$ has been used [remember Eq. (2.15)].
Using Eq. (4.38)

$$\tau_F = \frac{W_B^2}{2V_T \mu_n} = \frac{(10^{-4} \text{ cm})^2}{2(0.0259 \text{ V})\left(500 \frac{\text{cm}^2}{\text{V} \cdot \text{s}}\right)} = 3.86 \times 10^{-10} \text{ s}$$

Check of Results: The calculations appear correct, and the value of I_S is within the range given in Sec. 4.2.

Discussion: Operation of this particular transistor is limited to frequencies below $f = 1/(2\pi\tau_F) = 400$ MHz.

4.8.4 DIFFUSION CAPACITANCE

Capacitances are circuit elements that limit the high-frequency performance of both MOS and bipolar devices. For the base-emitter voltage and hence the collector current in the BJT to change, the charge stored in the base region also must change, as illustrated in Fig. 4.24(b). This change in charge with v_{BE} can be modeled by a capacitance C_D, called the **diffusion capacitance**, placed in parallel with the forward-biased base-emitter diode as defined by

$$C_D = \left.\frac{dQ}{dv_{BE}}\right|_{Q\text{-point}} = \frac{1}{V_T}\frac{qAn_{bo}W_B}{2}\exp\left(\frac{V_{BE}}{V_T}\right) \quad (4.40)$$

This equation can be rewritten as

$$C_D = \frac{1}{V_T}\left[\frac{qAD_n n_{bo}}{W_B}\exp\left(\frac{V_{BE}}{V_T}\right)\right]\left(\frac{W_B^2}{2D_n}\right) \cong \frac{I_T}{V_T}\tau_F \quad (4.41)$$

Because the transport current actually represents the collector current in the forward-active region, the expression for the diffusion capacitance is normally written as

$$C_D = \frac{I_C}{V_T}\tau_F \qquad (4.42)$$

From Eq. (4.42), we see that the diffusion capacitance C_D is directly proportional to current and inversely proportional to temperature T. For example, a BJT operating at a current of 1 mA with $\tau_F = 4 \times 10^{-10}$ s has a diffusion capacitance of

$$C_D = \frac{I_C}{V_T}\tau_F = \frac{10^{-3}\text{ A}}{0.0259\text{ V}}(3.86 \times 10^{-10}\text{ s}) = 14.9 \times 10^{-12}\text{ F} = 14.9\text{ pF}$$

This is a substantial capacitance, but it can be even larger if the transistor is operating at significantly higher currents.

EXERCISE: Calculate the value of the diffusion capacitance for a power transistor operating at a current of 10 A and a temperature of 100°C if $\tau_F = 4$ nS.

ANSWER: 1.24 μF—a significant capacitance!

4.8.5 FREQUENCY DEPENDENCE OF THE COMMON-EMITTER CURRENT GAIN

The forward-biased diffusion and reverse-biased *pn* junction capacitances of the bipolar transistor cause the current gain of the transistor to be frequency-dependent. An example of this dependence is given in Fig. 4.25. At low frequencies, the current gain has a constant value β_F, but as frequency increases, the current gain begins to decrease. The **unity-gain frequency** f_T is defined to be the frequency at which the magnitude of the current gain is equal to 1. The behavior in the graph is described mathematically by

$$\beta(f) = \frac{\beta_F}{\sqrt{1 + \left(\frac{f}{f_\beta}\right)^2}} \qquad (4.43)$$

where $f_\beta = f_T/\beta_F$ is the **β-cutoff frequency**. For the transistor in Fig. 4.25, $\beta_F = 125$ and $f_T = 300$ MHz.

EXERCISE: What is the β-cutoff frequency for the transistor in Fig. 4.25?

ANSWER: 2.4 MHz

4.8.6 THE EARLY EFFECT AND EARLY VOLTAGE

In the transistor output characteristics in Fig. 4.10, the current is saturated at a constant value in the forward-active region. However, in a real transistor, there is actually a positive slope to the characteristics, as shown in Fig. 4.26. The collector current is not truly independent of v_{CE}.

It has been observed experimentally that when the output characteristic curves are extrapolated back to the point of zero collector current, the curves all intersect at approximately a common point, $v_{CE} = -V_A$. This phenomenon is called the **Early effect** [6], and the voltage V_A is called the **Early voltage** after James Early from Bell Laboratories, who first identified the source of the behavior. A relatively small value of Early voltage (14 V) has been used in Fig. 4.26 to exaggerate the characteristics. Values for the Early voltage more typically fall in the range

$$10\text{ V} \leq V_A \leq 200\text{ V}$$

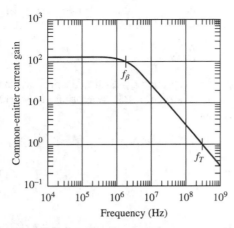
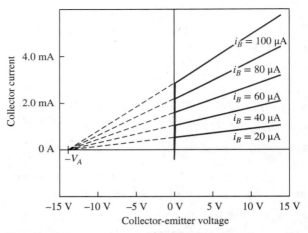

Figure 4.25 Magnitude of the common-emitter current gain β vs. frequency.

Figure 4.26 Transistor output characteristics identifying the Early voltage V_A.

4.8.7 MODELING THE EARLY EFFECT

The dependence of the collector current on collector-emitter voltage is easily included in the simplified mathematical model for the forward-active region of the BJT by modifying Eqs. (4.25) as follows:

$$i_C = I_S\left[\exp\left(\frac{v_{BE}}{V_T}\right)\right]\left[1 + \frac{v_{CE}}{V_A}\right]$$

$$\beta_F = \beta_{FO}\left[1 + \frac{v_{CE}}{V_A}\right] \qquad (4.44)$$

$$i_B = \frac{I_S}{\beta_{FO}}\left[\exp\left(\frac{v_{BE}}{V_T}\right)\right]$$

β_{FO} represents the value of β_F extrapolated to $V_{CE} = 0$. In these expressions, the collector current and current gain now have the same dependence on v_{CE}, but the base current remains independent of v_{CE}. This result assumes that the current gain is determined by back injection into the emitter [7]. This is consistent with Fig. 4.26, in which the separation of the constant-base-current curves in the forward-active region increases as v_{CE} increases, indicating that the current gain β_F is increasing with v_{CE}.

> **EXERCISE:** A transistor has $I_S = 10^{-15}$ A, $\beta_{FO} = 75$, and $V_A = 50$ V and is operating with $V_{BE} = 0.7$ V and $V_{CE} = 10$ V. What are I_B, β_F, and I_C? What would be β_F and I_C if $V_A = \infty$?
>
> **ANSWERS:** 7.29 µA, 90, 0.656 mA; 75, 0.547 mA

4.8.8 ORIGIN OF THE EARLY EFFECT

Modulation of the base width W_B of the transistor by the collector-base voltage is the cause of the Early effect. As the reverse bias across the collector-base junction increases, the width of the collector-base depletion layer increases, and width W_B of the base decreases. This mechanism, termed **base-width modulation,** is depicted in Fig. 4.27, in which the collector-base space charge region width is shown for two different values of collector-base voltage corresponding to effective base widths of W_B and W_B'. Equation (4.33) demonstrated that collector current is inversely proportional to the base width W_B, so a decrease in W_B results in an increase in transport current i_T. This decrease in W_B as V_{CB} increases is the cause of the Early effect.

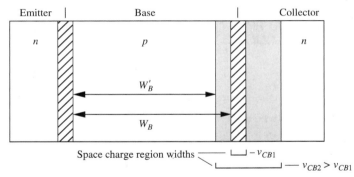

Figure 4.27 Base-width modulation, or Early effect.

The Early effect reduces the output resistance of the bipolar transistor and places an important limit on the amplification factor of the BJT. These limitations are discussed in detail in Part Two, Chapter 7.

4.9 TRANSCONDUCTANCE

One of the most important parameters for all electronic devices is **transconductance** g_m, which characterizes how the output current of the device changes in response to a change in its input voltage. For the bipolar transistor, g_m relates changes in i_C to changes in v_{BE} as defined by

$$g_m = \left.\frac{di_C}{dv_{BE}}\right|_{Q\text{-point}} \tag{4.45}$$

For Q-points in the forward-active region, Eq. (4.45) can be evaluated using the collector-current expression from Eq. (4.25):

$$g_m = \left.\frac{d}{dv_{BE}}\left\{I_S \exp\left(\frac{v_{BE}}{V_T}\right)\right\}\right|_{Q\text{-point}} = \frac{1}{V_T} I_S \exp\left(\frac{V_{BE}}{V_T}\right) = \frac{I_C}{V_T} \cong 40\, I_C \tag{4.46}$$

Equation (4.46) represents the fundamental relationship for the transconductance of the bipolar transistor, in which we find g_m is directly proportional to collector current. This is an important result that is used many times in bipolar circuit design. It is worth noting that the expression for the transit time defined in Eq. (4.42) can be rewritten as

$$\tau_F = \frac{C_D}{g_m} \quad \text{or} \quad C_D = g_m \tau_F \tag{4.47}$$

DESIGN NOTE

BIPOLAR TRANSCONDUCTANCE AND TRANSCONDUCTANCE EFFICIENCY

$$g_m = \frac{I_C}{V_T} \cong 40\, I_C \quad \text{and} \quad \frac{g_m}{I_C} = \frac{1}{V_T} \cong 40$$

Transconductance g_m, defined above in Eq. (4.45), is one of the most important parameters for all electronic devices. Transconductance efficiency g_m/I_C characterizes how much transconductance is achieved per unit of bias current and is a parameter utilized to compare different electronic devices. The bipolar device has a comparatively large value of g_m/I_C, approaching a value of 40 at room temperature. Note that transconductance efficiency is independent of operating current. In the next chapter we will discover that this is not true for the MOS transistor.

DESIGN NOTE

TRANSIT TIME

$$\tau_F = \frac{C_D}{g_m}$$

Transit time τ_F places an upper limit on the frequency response of the bipolar device.

EXERCISE: What is the value of the BJT transconductance g_m at $I_C = 100$ µA and $I_C = 1$ mA? Assume $1/V_T = 40$ V^{-1}. What is the value of the diffusion capacitance for each of these currents if the base transit time is 25 ps? What is the actual transconductance efficiency for these currents at $T = 300$ K?

ANSWERS: 4 mS; 40 mS; 0.1 pF; 1.0 pF; 38.6 V^{-1}; 38.6 V^{-1}

4.10 BIPOLAR TECHNOLOGY AND SPICE MODEL

In order to create a comprehensive simulation model of the bipolar transistor, our knowledge of the physical structure of the transistor is coupled with the transport model expressions and experimental observations. We typically start with a circuit representation of our mathematical model that describes the intrinsic behavior of the transistor, and then add additional elements to model parasitic effects introduced by the actual physical structure. Remember, in any case, that our SPICE models represent only lumped element equivalent circuits for the distributed structure that we actually fabricate.

Although we will seldom use the equations that make up the simulation model in hand calculations, awareness and understanding of the equations can help when SPICE generates unexpected results. This can happen when we attempt to use a device in an unusual way, or the simulator may produce a circuit result that does not fit within our understanding of the device behavior. Understanding the internal model in SPICE will help us interpret whether our knowledge of the device is wrong or if the simulation has some built-in assumptions that may not be consistent with a particular application of the device.

4.10.1 QUALITATIVE DESCRIPTION

A detailed cross section of the classic *npn* structure from Fig. 4.1 is given in Fig. 4.28(a), and the corresponding SPICE circuit model appears in Fig. 4.28(b). Circuit elements i_C, i_B, C_{BE}, and C_{BC} describe the intrinsic transistor behavior that we have discussed thus far. Current source i_C represents the current transported across the base from collector to emitter, and current source i_B models the total base current of the transistor. **Base-emitter** and **base-collector capacitances** C_{BE} and C_{BC} include models for the diffusion capacitances and the junction capacitances associated with the base-emitter and base-collector diodes.

Additional circuit elements are added to account for nonideal characteristics of the real transistor. The physical structure has a large-area *pn* junction that isolates the collector from the substrate of the transistor and separates one transistor from the next. The primary components related to this junction are diode current i_S and substrate capacitance C_{JS}. Base resistance R_B accounts for the resistance between the external base contact and the intrinsic base region of the transistor. Similarly, collector current must pass through R_C on its way to the active region of the collector-base junction, and R_E models any extrinsic emitter resistance present in the device.

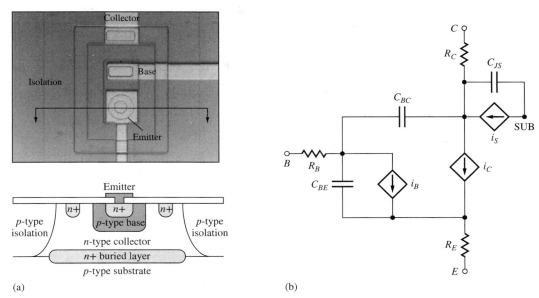

Figure 4.28 (a) Top view and cross section of a junction-isolated transistor; (b) SPICE model for the *npn* transistor.

4.10.2 SPICE MODEL EQUATIONS

The SPICE models are comprehensive but quite complex. Even the model equations presented below represent simplified versions of the actual models. Table 4.3 defines the SPICE parameters that are used in these expressions. More complete descriptions can be found in [8].

The collector and base currents are given by

$$i_C = \frac{(i_F - i_R)}{\text{KBQ}} - \frac{i_R}{\text{BR}} - i_{RG} \quad \text{and} \quad i_B = \frac{i_F}{\text{BF}} + \frac{i_R}{\text{BR}} + i_{FG} + i_{RG}$$

in which the forward and reverse components of the transport current are

$$i_F = \text{IS} \cdot \left[\exp\left(\frac{v_{BE}}{\text{NF} \cdot V_T}\right) - 1\right] \quad \text{and} \quad i_R = \text{IS} \cdot \left[\exp\left(\frac{v_{BC}}{\text{NR} \cdot V_T}\right) - 1\right] \quad (4.48)$$

Base current i_B includes two added terms to model additional space-charge region currents associated with the base-emitter and base-collector junctions:

$$i_{FG} = \text{ISE} \cdot \left[\exp\left(\frac{v_{BE}}{\text{NE} \cdot V_T}\right) - 1\right] \quad \text{and} \quad i_{RG} = \text{ISC} \cdot \left[\exp\left(\frac{v_{BC}}{\text{NC} \cdot V_T}\right) - 1\right]$$

Another new addition is the KBQ term that includes voltages VAF and VAR to model the Early effect in both the forward and reverse modes, as well as "knee current" parameters IKF and IKR that model current gain fall-off at high operating currents.

$$\text{KBQ} = \left(\frac{1}{2}\right) \frac{1 + \left[1 + 4\left(\frac{i_F}{\text{IKF}} + \frac{i_R}{\text{IKR}}\right)\right]^{NK}}{1 + \frac{v_{CB}}{\text{VAF}} + \frac{v_{EB}}{\text{VAR}}}$$

Note as well that the Early effect is cast in terms of v_{BC} rather than v_{CE} as we have used in Eq. (4.44).

The substrate junction current is expressed as

$$i_S = \text{ISS} \cdot \left[\exp\left(\frac{v_{\text{SUB-C}}}{\text{NS} \cdot V_T}\right) - 1\right]$$

TABLE 4.3
Bipolar Device Parameters for Circuit Simulation (npn/pnp)

PARAMETER	NAME	DEFAULT	TYPICAL VALUES
Saturation current	IS	10^{-16} A	3×10^{-17} A
Forward current gain	BF	100	100
Forward emission coefficient	NF	1	1.03
Forward Early voltage	VAF	∞	75 V
Forward knee current	IKF	∞	0.05 A
Reverse knee current	IKR	∞	0.01 A
Reverse current gain	BR	1	0.5
Reverse emission coefficient	NR	1	1.05
Base resistance	RB	0	250 Ω
Collector resistance	RC	0	50 Ω
Emitter resistance	RE	0	1 Ω
Forward transit time	TF	0	0.15 ns
Reverse transit time	TR	0	15 ns
Base-emitter leakage saturation current	ISE	0	1 pA
Base-emitter leakage emission coefficient	NE	1.5	1.4
Base-emitter junction capacitance	CJE	0	0.5 pF
Base-emitter junction potential	PHIE	0.8 V	0.8 V
Base-emitter grading coefficient	ME	0.5	0.5
Base-collector leakage saturation current	ISC	0	1 pA
Base-collector leakage emission coefficient	NC	1.5	1.4
Base-collector junction capacitance	CJC	0	1 pF
Base-collector junction potential	PHIC	0.75 V	0.7 V
Base-collector grading coefficient	MC	0.33	0.33
Substrate saturation current	ISS	0	1 fA
Substrate emission coefficient	NS	1	1
Collector-substrate junction capacitance	CJS	0	3 pF
Collector-substrate junction potential	VJS	0.75 V	0.75 V
Collector-substrate grading coefficient	MJS	0	0.5

The three device capacitances in Fig. 4.28(b) are represented by

$$C_{BE} = \frac{i_F}{\text{NE} \cdot V_T}\text{TF} + \frac{\text{CJE}}{\left(1 - \frac{v_{BE}}{\text{PHIE}}\right)^{\text{MJE}}} \quad \text{and} \quad C_{BC} = \frac{i_R}{\text{NC} \cdot V_T}\text{TR} + \frac{\text{CJC}}{\left(1 - \frac{v_{BC}}{\text{PHIC}}\right)^{\text{MJC}}}$$

$$C_{JS} = \frac{\text{CJS}}{\left(1 + \frac{v_{\text{SUB-C}}}{\text{VJS}}\right)^{\text{MJS}}} \tag{4.49}$$

C_{BE} and C_{BC} consist of two terms representing the diffusion capacitance (modeled by TF and NE or TR and NC) and depletion-region capacitance (modeled by CJE, PHIE, and MJE or CJC, PHIC, and MJC). The substrate diode is normally reverse biased, so it is modeled by just the depletion-layer capacitance (CJS, VJS, and MJS). The base, collector, and emitter series resistances are RB, RC, and RE, respectively.

The SPICE model for the *pnp* transistor is similar to that presented in Fig. 4.28(b) except for reversal of the current sources and of the positive polarity for the transistor currents and voltages.

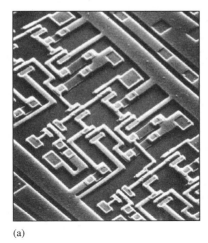
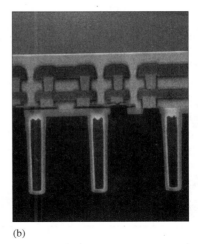

(a)   (b)

Figure 4.29 (a) Top view of a high-performance trench-isolated integrated circuit; (b) cross section of a high-performance trench-isolated bipolar transistor.
Cressler, J. D. "Reengineering Silicon: SiGe Heterojunction Bipolar Technology." *IEEE Spectrum* 32, no. 3 (1995): 49–55. Copyright ©1995 IEEE. Reprinted with permission.

4.10.3 HIGH-PERFORMANCE BIPOLAR TRANSISTORS

Modern transistors designed for high-speed switching and analog RF applications use combinations of sophisticated shallow and deep trench isolation processes to reduce the device capacitances and minimize the transit times. These devices typically utilize polysilicon emitters, have extremely narrow bases, and may incorporate SiGe base regions. A layout and cross section of a very high frequency, trench-isolated SiGe bipolar transistor appear in Fig. 4.29. In the research laboratory, SiGe transistors have already exhibited cutoff frequencies in excess of 500 GHz and are predicted to exceed 1 THz.

> **EXERCISE:** A bipolar transistor has a current gain of 80, a collector current of 350 μA for V_{BE} = 0.68 V, and an Early voltage of 70 V. What are the values of SPICE parameters BF, IS, and VAF? Assume $T = 27°C$ and $V_{BC} = 0$ V.
>
> **ANSWERS:** 80, 1.39 fA, 70 V

4.11 PRACTICAL BIAS CIRCUITS FOR THE BJT

The goal of biasing is to establish a known **quiescent operating point,** or **Q-point** that represents the initial operating region of the transistor. In the bipolar transistor, the Q-point is represented by the dc values of the collector-current and collector-emitter voltage (I_C, V_{CE}) for the *npn* transistor, or emitter-collector voltage (I_C, V_{EC}) for the *pnp*.

Logic gates and linear amplifiers use very different operating points. For example, the circuit in Fig. 4.30(a) can be used as either a logic inverter or a linear amplifier depending upon our choice of operating points. The voltage transfer characteristic (VTC) for the circuit appears in Fig. 4.31(a), and the corresponding output characteristics and load line[9] appear in Fig. 4.31(b). For low values of v_{BE}, the transistor is nearly cut off, and the output voltage is 5 V, corresponding to a binary "1" in a logic applications. As v_{BE} increases above 0.6 V, the output drops quickly and reaches its "on-state" voltage of 0.18 V for v_{BE} greater than 0.8 V. The BJT is now operating in its saturation region, and the small "on-voltage" would correspond to a "0" in binary logic. These two logic states are also shown on the transistor output characteristics in Fig. 4.31(b).

---

[9] See Section 3.10.1.

200 Chapter 4 Bipolar Junction Transistors

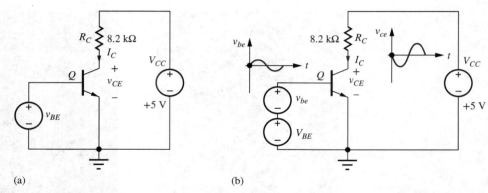

Figure 4.30 (a) Circuit for a logic inverter; (b) the same transistor used as a linear amplifier.

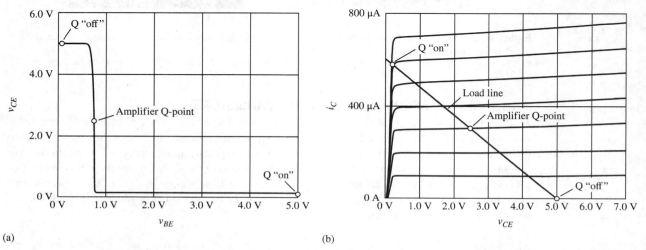

Figure 4.31 (a) Voltage transfer characteristic (VTC) with quiescent operating points (Q-points) corresponding to an "on-switch," an amplifier, and an "off switch"; (b) the same three operating points located on the transistor output characteristics.

When the transistor is "on," it conducts a substantial current, and v_{CE} falls to 0.18 V. When the transistor is off, v_{CE} equals 5 V.[10]

For amplifier applications, the Q-point is located in the region of high slope (high gain) of the voltage transfer characteristic, also indicated in Fig. 4.31(a). At this operating point, the transistor is operating in the forward-active region, the region in which high voltage, current, and/or power gain can be achieved. To establish this Q-point, a **dc bias** V_{BE} is applied to the base as in Fig. 4.30(b), and a small ac signal v_{be} is added to vary the base voltage around the bias value.[11] The variation in total base-emitter voltage v_{BE} causes the collector current to change, and an amplified replica of the ac input voltage appears at the collector with a phase shift of 180°. Our study of the design of transistor amplifiers begins in Chapter 7 of this book.

In Secs. 4.6 to 4.10, we presented simplified models for the four operating regions of the BJT. In general, we will not explicitly insert the simplified circuit models for the transistor into the circuit but instead will use the mathematical relationships that were derived for the specific operating region of interest. For example, in the forward-active region, the results $V_{BE} = 0.7$ V and $I_C = \beta_F I_B$ will be utilized to directly simplify the circuit analysis.

---

[10] Detailed design of logic gates are included in the Instructor Resources and can be accessed through Connect.

[11] Remember $v_{BE} = V_{BE} + v_{be}$.

In the dc biasing examples that follow, the Early voltage is assumed to be infinite. In general, including the Early voltage in bias circuit calculations substantially increases the complexity of the analysis but typically changes the results by less than 10 percent. In most cases, the tolerances on the values of resistors and independent sources will be 5 to 10 percent, and the transistor current-gain β_F may vary by a factor of 4:1 to 10:1. For example, the current gain of a transistor may be specified to be a minimum of 50 with a typical value of 100 but no upper bound specified. These tolerances will swamp out any error due to neglect of the Early voltage. Thus, basic hand design will be done ignoring the Early effect, and if more precision is needed, the calculations can be refined through SPICE analysis.

4.11.1 FOUR-RESISTOR BIAS NETWORK

Because of the BJT's exponential relationship between current and voltage and its strong dependence on temperature T, the constant V_{BE} form of biasing utilized in Fig. 4.30 does not represent a practical technique. One of the best circuits for stabilizing the Q-point of a transistor is the four-resistor bias network in Fig. 4.32. R_1 and R_2 form a resistive voltage divider across the power supplies (12 V and 0 V) and attempt to establish a fixed voltage at the base of transistor Q_1. R_E and R_C are used to define the emitter current and collector-emitter voltage of the transistor.

Our goal is to find the Q-point of the transistor: (I_C, V_{CE}). The first steps in analysis of the circuit in Fig. 4.32(a) are to split the power supply into two equal voltages, as in Fig. 4.32(b),

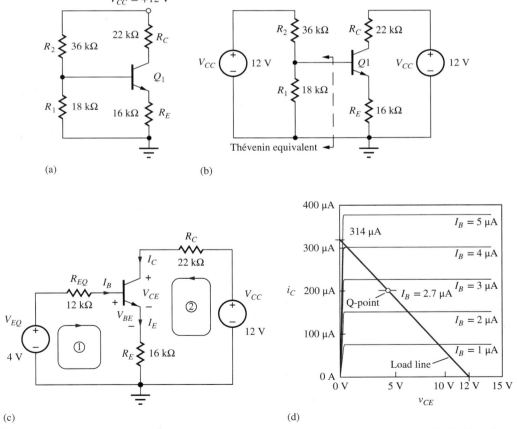

Figure 4.32 (a) The four-resistor bias network (assume $\beta_F = 75$ for analysis); (b) four-resistor bias circuit with replicated sources; (c) Thévenin simplification of the four-resistor bias network; (d) load line for the four-resistor bias circuit.

and then to simplify the circuit by replacing the base-bias network by its Thévenin equivalent circuit, as shown in Fig. 4.32(c). V_{EQ} and R_{EQ} are given by

$$V_{EQ} = V_{CC} \frac{R_1}{R_1 + R_2} \qquad R_{EQ} = \frac{R_1 R_2}{R_1 + R_2} \qquad (4.50)$$

For the values in Fig. 4.32(c), $V_{EQ} = 4$ V and $R_{EQ} = 12$ kΩ.

Detailed analysis begins by assuming a region of operation in order to simplify the BJT model equations. Because the most common region of operation for this bias circuit is the forward-active region, we will assume it to be the region of operation and neglect the Early voltage term. Using Kirchhoff's voltage law around loop 1:

$$V_{EQ} = I_B R_{EQ} + V_{BE} + I_E R_E = I_B R_{EQ} + V_{BE} + (\beta_F + 1) I_B R_E \qquad (4.51)$$

Solving for I_B yields

$$I_B = \frac{V_{EQ} - V_{BE}}{R_{EQ} + (\beta_F + 1) R_E} \qquad \text{where} \qquad V_{BE} = V_T \ln\left(\frac{I_B}{I_S/\beta_F} + 1\right) \qquad (4.52)$$

Unfortunately, combining these expressions yields a transcendental equation. However, if we assume an approximate value of V_{BE}, then we can find the collector and emitter currents using our auxillary relationships $I_C = \beta_F I_B$ and $I_E = (\beta_F + 1) I_B$:

$$I_C = \frac{V_{EQ} - V_{BE}}{\dfrac{R_{EQ}}{\beta_F} + \dfrac{(\beta_F + 1)}{\beta_F} R_E} \qquad \text{and} \qquad I_E = \frac{V_{EQ} - V_{BE}}{\dfrac{R_{EQ}}{(\beta_F + 1)} + R_E} \qquad (4.53)$$

For large current gain ($\beta_F \gg 1$), Eqs. (4.52) and (4.53) simplify to

$$I_E \cong I_C \cong \frac{V_{EQ} - V_{BE}}{\dfrac{R_{EQ}}{\beta_F} + R_E} \qquad \text{with} \qquad I_B \cong \frac{V_{EQ} - V_{BE}}{R_{EQ} + \beta_F R_E} \qquad (4.54)$$

Now that I_C is known, we can use loop 2 to find collector-emitter voltage V_{CE}:

$$V_{CE} = V_{CC} - I_C R_C - I_E R_E = V_{CC} - I_C\left(R_C + \frac{R_E}{\alpha_F}\right) \qquad (4.55)$$

since $I_E = I_C/\alpha_F$. Normally $\alpha_F \cong 1$, and Eq. (4.55) can be simplified to

$$V_{CE} \cong V_{CC} - I_C(R_C + R_E) \qquad (4.56)$$

For the circuit in Fig. 4.32, we are assuming forward-active region operation with $V_{BE} = 0.7$ V, and the Q-point values (I_C, V_{CE}) are

$$I_C \cong \frac{V_{EQ} - V_{BE}}{\dfrac{R_{EQ}}{\beta_F} + R_E} = \frac{(4 - 0.7)\text{V}}{\dfrac{12\ \text{k}\Omega}{75} + 16\ \text{k}\Omega} = 204\ \mu\text{A} \qquad \text{with} \qquad I_B = \frac{204\ \mu\text{A}}{75} = 2.72\ \mu\text{A}$$

$$V_{CE} \cong V_{CC} - I_C(R_C + R_E) = 12 - 2.04\ \mu\text{A}(22\ \text{k}\Omega + 16\ \text{k}\Omega) = 4.25\ \text{V}$$

A more precise estimate using Eqs. (4.53) and (4.55) gives a Q-point of (202 μA, 4.30 V). Since we don't know the actual value of V_{BE}, and haven't considered any tolerances, the approximate expressions give excellent engineering results.

All the calculated currents are greater than zero, and using the result in Eq. (4.55), $V_{BC} = V_{BE} - V_{CE} = 0.7 - 4.32 = -3.62$ V. Thus, the base-collector junction is reverse-biased, and the assumption of forward-active region operation was correct. The Q-point resulting from our analysis is (204 µA, 4.25 V).

Before leaving this bias example, let us draw the load line for the circuit and locate the Q-point on the output characteristics. The load-line equation for this circuit already appeared as Eq. (4.55):

$$V_{CE} = V_{CC} - \left(R_C + \frac{R_E}{\alpha_F}\right)I_C = 12 - 38{,}200 I_C \tag{4.57}$$

Two points are needed to plot the load line. Choosing $I_C = 0$ yields $V_{CE} = 12$ V, and picking $V_{CE} = 0$ yields $I_C = 314$ µA. The resulting load line is plotted on the transistor common-emitter output characteristics in Fig. 4.32(d). The base current was already found to be 2.7 µA, and the intersection of the $I_B = 2.7$-µA characteristic with the load line defines the Q-point. In this case we must estimate the location of the $I_B = 2.7$-µA curve that corresponds to $I_C = 202$ µA.

EXERCISE: Find the values of I_B, I_C, I_E, and V_{CE} using the exact expressions in Eqs. (4.52), (4.53), and (4.55).

ANSWERS: 2.69 µA, 202 µA, 204 µA, 4.28 V

EXERCISE: Find the Q-point for the circuit in Fig. 4.32(d) if $R_1 = 180$ kΩ and $R_2 = 360$ kΩ.

ANSWER: (185 µA, 4.93 V)

DESIGN NOTE

Good engineering approximations for the Q-point in the four-resistor bias circuit for the bipolar transistor are:

$$I_C \cong \frac{V_{EQ} - V_{BE}}{\dfrac{R_{EQ}}{\beta_F} + R_E} \cong \frac{V_{EQ} - V_{BE}}{R_E} \quad \text{and} \quad V_{CE} \cong V_{CC} - I_C(R_C + R_E)$$

4.11.2 DESIGN OBJECTIVES FOR THE FOUR-RESISTOR BIAS NETWORK

Now that we have analyzed a circuit involving the four-resistor bias network, let us explore the design objectives of this bias technique through further simplification of the expression for the collector and emitter currents in Eq. (4.54) by assuming that $R_{EQ}/\beta_F \ll R_E$. Then

$$I_E \cong I_C \cong \frac{V_{EQ} - V_{BE}}{R_E} \tag{4.58}$$

The value of the Thévenin equivalent resistance R_{EQ} is normally designed to be small enough to neglect the voltage drop caused by the base current flowing through R_{EQ}. Under these conditions, I_C and I_E are set by the combination of V_{EQ}, V_{BE}, and R_E. In addition, V_{EQ} is normally designed to be large enough that small variations in the assumed value of V_{BE} will not materially affect the value of I_E.

In the original bias circuit reproduced in Fig. 4.33, the assumption that the voltage drop $I_B R_{EQ} \ll (V_{EQ} - V_{BE})$ is equivalent to assuming $I_B \ll I_2$ so that $I_1 \cong I_2$. For this case, the base current of Q_1 does not disturb the voltage divider action of R_1 and R_2. Using the approximate expression in Eq. (4.58) estimates the emitter current in the circuit in Fig. 4.32 to be

$$I_C \cong I_E \cong \frac{4 \text{ V} - 0.7 \text{ V}}{16{,}000 \text{ Ω}} = 206 \text{ µA}$$

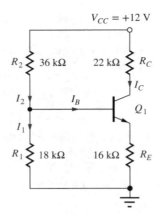

Figure 4.33 Currents in the base-bias network.

which is essentially the same as the result that was calculated using the more exact expression. This is the result that should be achieved with a proper bias network design. If the Q-point is independent of I_B, it will also be independent of current gain β (a poorly controlled transistor parameter). The emitter current will then be approximately the same for a transistor with a current gain of 50 or 500.

Generally, a very large number of possible combinations of R_1 and R_2 will yield the desired value of V_{EQ}. An additional constraint is needed to finalize the design choice. A useful choice is to limit the current used in the base-voltage-divider network by choosing $I_2 \leq I_C/5$. This choice ensures that the power dissipated in bias resistors R_1 and R_2 is less than 20 percent of the total quiescent power consumed by the circuit and at the same time ensures that $I_2 \gg I_B$ for $\beta \geq 50$.

Resistor R_E adds feedback to the circuit that tends to stabilize the bias point. If the emitter current increases slightly, then the voltage across R_E increases and decreases the base-emitter voltage, which tends to reduce the value of the emitter current. This effect is a form of negative feedback that will be studied in detail in Parts Two and Three of this book.

EXERCISE: Show that choosing $I_2 = I_C/5$ is equivalent to setting $I_2 = 10\ I_B$ when $\beta_F = 50$.

EXERCISE: Find the Q-point for the circuit in Fig. 4.32(a) if β_F is 500.

ANSWER: (206 µA, 4.18 V).

DESIGN EXAMPLE 4.9 FOUR-RESISTOR BIAS DESIGN

Here we explore the design of the network most commonly utilized to bias the BJT—the four-resistor bias circuit.

PROBLEM Design a four-resistor bias circuit to give a Q-point of (750 µA, 5 V) using a 15-V supply with an *npn* transistor having a minimum current gain of 100.

SOLUTION **Known Information and Given Data:** The bias circuit in Fig. 4.33 with $V_{CC} = 15$ V; the *npn* transistor has $\beta_F = 100$, $I_C = 750$ µA, and $V_{CE} = 5$ V.

Unknowns: Base voltage V_B, voltages across resistors R_E and R_C; values for R_1, R_2, R_C, and R_E

Approach: First, partition V_{CC} between the collector-emitter voltage of the transistor and the voltage drops across R_C and R_E. Next, choose currents I_1 and I_2 for the base-bias network. Finally, use the assigned voltages and currents to calculate the unknown resistor values.

Assumptions: The transistor is to operate in the forward-active region. The base-emitter voltage of the transistor is 0.7 V. The Early voltage is infinite.

Analysis: To calculate values for the resistors, we must know the voltage across the emitter and collector resistors and the voltage V_B. V_{CE} is designed to be 5 V. One common choice is to divide the remaining power supply voltage $(V_{CC} - V_{CE}) = 10$ V equally between R_E and R_C. Thus, $V_E = 5$ V and $V_C = 5 + V_{CE} = 10$ V. The values of R_C and R_E are then given by

$$R_C = \frac{V_{CC} - V_C}{I_C} = \frac{5 \text{ V}}{750 \text{ μA}} = 6.67 \text{ kΩ} \quad \text{and} \quad R_E = \frac{V_E}{I_E} = \frac{5 \text{ V}}{758 \text{ μA}} = 6.60 \text{ kΩ}$$

The base voltage is given by $V_B = V_E + V_{BE} = 5.7$ V. For forward-active region operation, we know that $I_B = I_C/\beta_F = 750 \text{ μA}/100 = 7.5 \text{ μA}$. Now choosing $I_2 = 10 I_B$, we have $I_2 = 75$ μA, $I_1 = 9 I_B = 67.5$ μA, and R_1 and R_2 can be determined:

$$R_1 = \frac{V_B}{9 I_B} = \frac{5.7 \text{ V}}{67.5 \text{ μA}} = 84.4 \text{ kΩ} \quad R_2 = \frac{V_{CC} - V_B}{10 I_B} = \frac{15 - 5.7 \text{ V}}{75 \text{ μA}} = 124 \text{ kΩ} \quad (4.59)$$

Check of Results: We have $V_{BE} = 0.7$ V and $V_{BC} = 5.7 - 10 = -4.3$ V, which are consistent with the forward-active region assumption.

Discussion: The values calculated above should yield a Q-point very close to the design goals. However, if we were going to build this circuit in the laboratory, we must use standard values for the resistors. In order to complete the design, we refer to the table of resistor values in Appendix A. There we find that the closest available values are $R_1 = 82$ kΩ, $R_2 = 120$ kΩ, $R_E = 6.8$ kΩ, and $R_C = 6.8$ kΩ.

Computer-Aided Analysis: SPICE can now be used as a tool to check our design. The final design using these values appears in Fig. 4.34 for which SPICE (with IS $= 2 \times 10^{-15}$ A) predicts the Q-point to be (734 μA, 4.97 V), with $V_{BE} = 0.65$ V. We neglected the Early effect in our hand calculations, but SPICE represents an easy way to check this assumption. If we set VAF $= 75$ V in SPICE, keeping the other parameters the same, the new Q-point is (737 μA, 4.93 V). Clearly, the changes caused by the Early effect are negligible.

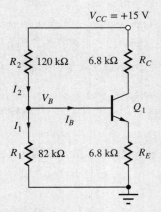

Figure 4.34 Final bias circuit design for a Q-point of (750 μA, 5 V).

EXERCISE: Redesign the four-resistor bias circuit to yield $I_C = 75$ μA and $V_{CE} = 5$ V.

ANSWERS: (66.7 kΩ, 66.0 kΩ, 844 kΩ, 1.24 MΩ) → (68 kΩ, 68 kΩ, 820 kΩ, 1.20 MΩ)

Note the scaling that has occurred. We reduced the current by a factor of 10 by increasing the value of each resistor by the same amount.

DESIGN NOTE — FOUR-RESISTOR BIAS DESIGN

1. Choose the Thévenin equivalent base voltage V_{EQ}: $\quad \dfrac{V_{CC}}{4} \leq V_{EQ} \leq \dfrac{V_{CC}}{2}$

2. Select R_1 to set $I_1 = 9I_B$: $\quad R_1 = \dfrac{V_{EQ}}{9I_B}$

3. Select R_2 to set $I_2 = 10I_B$: $\quad R_2 = \dfrac{V_{CC} - V_{EQ}}{10I_B}$

4. R_E is determined by V_{EQ} and the desired collector current: $\quad R_E \cong \dfrac{V_{EQ} - V_{BE}}{I_C}$

5. R_C is determined by the desired collector-emitter voltage: $\quad R_C \cong \dfrac{V_{CC} - V_{CE}}{I_C} - R_E$

EXAMPLE 4.10   TWO-RESISTOR BIASING

In this example, we explore an alternative bias circuit that requires only two resistors and apply it to biasing a *pnp* transistor. (A similar circuit can also be used for *npn* biasing.)

PROBLEM Find the Q-point for the *pnp* transistor in the two-resistor bias circuit in Fig. 4.35. Assume $\beta_F = 50$.

SOLUTION **Known Information and Given Data:** Two-resistor bias circuit in Fig. 4.35 with a *pnp* transistor with $\beta_F = 50$

Unknowns: I_C, V_{CE}.

Approach: Assume a region of operation and analyze the circuit to determine the Q-point; check answer to see if it is consistent with the assumptions.

Assumptions: Forward-active region operation with $V_{EB} = 0.7$ V and $V_A = \infty$

Analysis: The voltages and currents are first carefully labeled as in Fig. 4.35. To find the Q-point, an equation is written involving V_{EB}, I_B, and I_C:

$$9 = V_{EB} + 18{,}000 I_B + 1000(I_C + I_B) \qquad (4.60)$$

Applying the assumption of forward-active region operation with $\beta_F = 50$ and $V_{EB} = 0.7$ V

$$9 = 0.7 + 18{,}000 I_B + 1000(51) I_B \qquad (4.61)$$

and

$$I_B = \dfrac{9 \text{ V} - 0.7 \text{ V}}{69{,}000 \ \Omega} = 120 \ \mu\text{A} \qquad I_C = 50 I_B = 6.01 \text{ mA} \qquad (4.62)$$

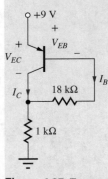

Figure 4.35 Two-resistor bias circuit with a *pnp* transistor.

The emitter-collector voltage is given by

$$V_{EC} = 9 - 1000(I_C + I_B) = 2.88 \text{ V} \quad \text{and} \quad V_{BC} = 2.18 \text{ V} \quad (4.63)$$

The Q-point is $(I_C, V_{EC}) = (6.01 \text{ mA}, 2.88 \text{ V})$.

Check of Results: Because I_B, I_C, and V_{BC} are all greater than zero, the assumption of forward-active region operation is valid, and the Q-point is correct.

Computer-Aided Analysis: For this circuit, SPICE simulation yields (6.04 mA, 2.95 V), which agrees with the Q-point found from our hand calculations.

EXERCISE: What is the Q-point if the 18-kΩ resistor is increased to 36 kΩ?

ANSWER: (4.77 mA, 4.13 V)

DISCUSSION: This circuit utilizes negative feedback similar to that in the four-resistor bias circuit. Suppose the collector current increases by a small amount. Then the voltage across the collector resistor increases, and the base current will decrease, thereby decreasing the collector current.

EXERCISE: Draw the two-resistor bias circuit (a "mirror image" of Fig. 4.35) that would be used to bias an *npn* transistor from a single +9-V supply using the same two resistor values as in Fig. 4.35.

ANSWER: See circuit topology in Fig. P4.100.

The bias circuit examples that have been presented in this section have only scratched the surface of the possible techniques that can be used to bias *npn* and *pnp* transistors. However, the analysis techniques have illustrated the basic approaches that need to be followed in order to determine the Q-point of any bias circuit.

4.11.3 ITERATIVE ANALYSIS OF THE FOUR-RESISTOR BIAS CIRCUIT

To find I_C in the circuit in Fig. 4.32, we need to find a solution to the following pair of equations:

$$I_C = \frac{V_{EQ} - V_{BE}}{\frac{R_{EQ}}{\beta_F} + \frac{(\beta_F + 1)}{\beta_F} R_E} \quad \text{where} \quad V_{BE} = V_T \ln\left(\frac{I_C}{I_S} + 1\right) \quad (4.64)$$

In the analysis presented in Sec. 4.11, we avoided the problems associated with solving the resulting transcendental equation by assuming that we knew an approximate value for V_{BE}. However, we can find a numerical solution to these two equations with a simple iterative process using a spreadsheet, "fzero" in MATLAB®, or WolframAlpha® for example.

1. Guess a value for V_{BE}.
2. Calculate the corresponding value of I_C using $I_C = \dfrac{V_{EQ} - V_{BE}}{\dfrac{R_{EQ}}{\beta_F} + \dfrac{(\beta_F + 1)}{\beta_F} R_E}$.
3. Update the estimate for V_{BE} as $V'_{BE} = V_T \ln\left(\dfrac{I_C}{I_S} + 1\right)$.
4. Repeat steps 2 and 3 until convergence is obtained.

Table 4.4 presents the results of this iterative method showing convergence in only three iterations. This rapid convergence occurs because of the very steep nature of the $I_C - V_{BE}$ characteristic.

TABLE 4.4
BJT Iterative Bias Solution $I_S = 10^{-15}$ A, $V_T = 25$ mV (See Fig. 4.32)

V_{BE} (V)	I_C (A)	V'_{BE} (V)
0.7000	2.015E–04	0.6507
0.6507	2.046E–04	0.6511
0.6511	2.045E–04	0.6511

One might ask if this result is better than the one obtained earlier in Sec. 4.11.1. As in most cases, the results are only as good as the input data. Here we need to accurately know the values of saturation current I_S and temperature T in order to calculate V_{BE}. In the earlier solution we simply estimated V_{BE}. In reality, we seldom will know exact values of either I_S or T, so we most often are just satisfied with a direct estimate for V_{BE}. In addition, we haven't considered tolerances discussed in the next section.

EXERCISE: Repeat the iterative analysis above to find the values of I_C and V_{BE} if $V_T = 25.9$ mV.

ANSWERS: 203.1 μA, 0.6744 V

4.12 TOLERANCES IN BIAS CIRCUITS

When a circuit is actually built in discrete form in the laboratory or fabricated as an integrated circuit, the components and device parameters all have tolerances associated with their values. Discrete resistors can easily be purchased with 10 percent, 5 percent, or 1 percent tolerances, whereas typical resistors in ICs can exhibit even wider variations (± 30 percent). Power supply voltage tolerances are often 5 to 10 percent.

For a given bipolar transistor type, parameters such as current gain may cover a range of 5:1 to 10:1, or may be specified with only a nominal value and lower bound. The BJT (or diode) saturation current may vary by a factor varying from 10:1 to 100:1, and the Early voltage may vary by ± 20 percent. In FET circuits, the values of threshold voltage and the transconductance parameter can vary widely, and in op-amp circuits all the op-amp parameters (e.g., open-loop gain, input resistance, output resistance, input bias current, unity gain frequency, and the like) typically exhibit wide specification ranges.

In addition to these initial value uncertainties, the values of the circuit components and parameters change as temperature changes and the circuit ages. It is important to understand the effect of these variations on our circuits and be able to design circuits that will continue to operate correctly in the face of these element variations. Worst-case analysis and Monte Carlo analysis, introduced in Chapter 1, are two approaches that can be used to quantify the effects of tolerances on circuit performance.

4.12.1 WORST-CASE ANALYSIS

Worst-case analysis is often used to ensure that a design will function under an expected set of component variations. In Q-point analysis, for example, the values of components are simultaneously pushed to their various extremes in order to determine the worst possible range of Q-point values. Unfortunately, a design based on worst-case analysis is usually an unnecessary overdesign and economically undesirable, but it is important to understand the technique and its limitations.

EXAMPLE 4.11 WORST-CASE ANALYSIS OF THE FOUR-RESISTOR BIAS NETWORK

Now we explore the application of worst-case analysis to the four-resistor bias network with a given set of tolerances assigned to the elements. In Ex. 4.12, the bounds generated by the worst-case analysis will be compared to a statistical sample of the possible network realizations using Monte Carlo analysis.

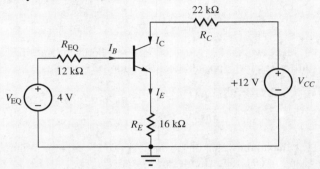

Figure 4.36 Simplified four-resistor bias circuit of Fig. 4.32(c) assuming nominal element values.

PROBLEM Find the worst-case values of I_C and V_{CE} for the transistor circuit in Fig. 4.36 that is the simplified version of the four-resistor bias circuit in Fig. 4.32. Assume that the 12-V power supply has a 5 percent tolerance and the resistors have 10 percent tolerances. Also, assume that the transistor current gain has a nominal value of 75 with a 50 percent tolerance.

SOLUTION **Known Information and Given Data:** Simplified version of the four-resistor bias circuit in Fig. 4.36; 5 percent tolerance on V_{CC}; 10 percent tolerance for each resistor; current $\beta_{FO} = 75$ with a 50 percent tolerance

Unknowns: Minimum and maximum values of I_C and V_{CE}

Approach: Find the worst-case values of V_{EQ} and R_{EQ}; use the results to find the extreme values of the base and collector current; use the collector current values to find the worst-case values of collector-emitter voltage.

Assumptions: To simplify the analysis, assume that the voltage drop in R_{EQ} can be neglected and β_F is large so that I_C is given by

$$I_C \cong I_E = \frac{V_{EQ} - V_{BE}}{R_E} \quad (4.65)$$

Assume V_{BE} is fixed at 0.7 V.

Analysis: To make I_C as large as possible, V_{EQ} should be at its maximum extreme and R_E should be a minimum value. To make I_C as small as possible, V_{EQ} should be minimum and R_E should be a maximum value. Variations in V_{BE} are assumed to be negligible but could also be included if desired.

The extremes of R_E are 0.9×16 kΩ = 14.4 kΩ, and 1.1×16 kΩ = 17.6 kΩ. The extreme values of V_{EQ} are somewhat more complicated:

$$V_{EQ} = V_{CC}\frac{R_1}{R_1 + R_2} = \frac{V_{CC}}{1 + \frac{R_2}{R_1}} \quad (4.66)$$

To make V_{EQ} as large as possible, the numerator of Eq. (4.66) should be large and the denominator small. Therefore, V_{CC} and R_1 must be as large as possible and R_2 as small as possible.

Conversely, to make V_{EQ} as small as possible, V_{CC} and R_1 must be small and R_2 must be large. Using this approach, the maximum and minimum values of V_{EQ} are

$$V_{EQ}^{\max} = \frac{12 \text{ V}(1.05)}{1 + \dfrac{36 \text{ k}\Omega(0.9)}{18 \text{ k}\Omega(1.1)}} = 4.78 \text{ V} \quad \text{and} \quad V_{EQ}^{\min} = \frac{12 \text{ V}(.95)}{1 + \dfrac{36 \text{ k}\Omega(1.1)}{18 \text{ k}\Omega(0.9)}} = 3.31 \text{ V}$$

Substituting these values in Eq. (4.62) gives the following extremes for I_C:

$$I_C^{\max} = \frac{4.78 \text{ V} - 0.7 \text{ V}}{14{,}400 \text{ }\Omega} = 283 \text{ }\mu\text{A} \quad \text{and} \quad I_C^{\min} = \frac{3.31 \text{ V} - 0.7 \text{ V}}{17{,}600 \text{ }\Omega} = 148 \text{ }\mu\text{A}$$

The worst-case range of V_{CE} will be calculated in a similar manner, but we must be careful to watch for possible cancellation of variables:

$$V_{CE} = V_{CC} - I_C R_C - I_E R_E \cong V_{CC} - I_C R_C - \frac{V_{EQ} - V_{BE}}{R_E} R_E \quad (4.67)$$

$$V_{CE} \cong V_{CC} - I_C R_C - V_{EQ} + V_{BE}$$

The maximum value of V_{CE} in Eq. (4.67) occurs for minimum I_C and minimum R_C and vice versa. Using (4.67), the extremes of V_{CE} are

$$V_{CE}^{\max} \cong 12 \text{ V}(1.05) - (148 \text{ }\mu\text{A})(22 \text{ k}\Omega \times 0.9) - 3.31 \text{ V} + 0.7 \text{ V} = 7.06 \text{ V} \checkmark$$

$$V_{CE}^{\min} \cong 12 \text{ V}(0.95) - (283 \text{ }\mu\text{A})(22 \text{ k}\Omega \times 1.1) - 4.78 \text{ V} + 0.7 \text{ V} = 0.471 \text{ Saturated!}$$

Check of Results: The transistor remains in the forward-active region for the upper extreme, but the transistor saturates (weakly) at the lower extreme. Because the forward-active region assumption is violated in the latter case, the calculated values of V_{CE} and I_C would not actually be correct for this case.

Discussion: Note that the worst-case values of I_C differ by a factor of almost 2:1! The maximum I_C is 38 percent greater than the nominal value of 210 µA, and the minimum value is 37 percent below the nominal value. The failure of the bias circuit to maintain the transistor in the desired region of operation for the worst-case values is evident. Note also, that a 2:1 ratio of currents only changes V_{BE} by 18 mV and changes the results by a small amount.

4.12.2 MONTE CARLO ANALYSIS

In a real circuit, the parameters will have some statistical distribution, and it is unlikely that the various components will all reach their extremes at the same time. Thus, the worst-case analysis technique will overestimate (often badly) the extremes of circuit behavior. A better approach is to attack the problem statistically using the method of Monte Carlo analysis.

As discussed in Chapter 1, **Monte Carlo analysis** uses randomly selected versions of a given circuit to predict its behavior from a statistical basis. For Monte Carlo analysis, values for each parameter in the circuit are selected at random from the possible distributions of parameters, and the circuit is then analyzed using the randomly selected element values. Many random parameter sets are generated, and the statistical behavior of the circuit is built up from analysis of the many test cases.

In Ex. 4.12, an Excel spreadsheet[12] will be used to perform a Monte Carlo analysis of the four-resistor bias circuit. As discussed in Chapter 1, Excel contains the function RAND( ), which generates random numbers uniformly distributed between 0 and 1, but for Monte Carlo analysis, the mean must be centered on R_{nom} and the width of the distribution set to $(2\varepsilon) \times R_{\text{nom}}$:

$$R = R_{\text{nom}}[1 + 2\varepsilon(\text{RAND}(\ ) - 0.5)] \quad (4.68)$$

---

[12] Similar results are easily found with MATLAB®.

EXAMPLE 4.12   MONTE CARLO ANALYSIS OF THE FOUR-RESISTOR BIAS NETWORK

Now, let us compare the worst-case results from Ex. 4.11 to a statistical sample of 500 randomly generated realizations of the transistor embedded in the four-resistor bias network.

PROBLEM Perform a Monte Carlo analysis to determine statistical distributions for the collector current and collector-emitter voltage for the four-resistor circuit in Figs. 4.32 and 4.36 with a 5 percent tolerance on V_{CC}, 10 percent tolerances for each resistor and a 50 percent tolerance on the current gain $\beta_{FO} = 75$.

SOLUTION **Known Information and Given Data:** Circuit in Fig. 4.32(a) as simplified in Fig. 4.36; 5 percent tolerance on the 12-V power supply V_{CC}; 10 percent tolerance on each resistor; current $\beta_{FO} = 75$ with a 50 percent tolerance

Unknowns: Statistical distributions of I_C and V_{CE}

Approach: To perform a Monte Carlo analysis of the circuit in Fig. 4.32, random values are assigned to V_{CC}, R_1, R_2, R_C, R_E, and β_F and then used to determine I_C and V_{CE}. A spreadsheet is used to make the repetitive calculations.

Assumptions: V_{BE} is fixed at 0.7 V. Random values are statistically independent of each other.

Computer-Aided Analysis: Using the tolerances from the worst-case analysis, the power supply, resistors, and current gain are represented as

$$
\begin{aligned}
&1. \quad V_{CC} = 12(1 + 0.1(\text{RAND}(\ ) - 0.5)) \\
&2. \quad R_1 = 18{,}000(1 + 0.2(\text{RAND}(\ ) - 0.5)) \\
&3. \quad R_2 = 36{,}000(1 + 0.2(\text{RAND}(\ ) - 0.5)) \\
&4. \quad R_E = 16{,}000(1 + 0.2(\text{RAND}(\ ) - 0.5)) \\
&5. \quad R_C = 22{,}000(1 + 0.2(\text{RAND}(\ ) - 0.5)) \\
&6. \quad \beta_F = 75(1 + (\text{RAND}(\ ) - 0.5))
\end{aligned}
\quad (4.69)
$$

Remember, each variable evaluation must invoke a separate call of the function RAND( ) so that the random values will be independent of each other.

In the spreadsheet results presented in Fig. 4.37, the random elements in Eq. (4.69) are used to evaluate the equations that characterize the bias circuit:

$$
\begin{aligned}
&7. \quad V_{EQ} = V_{CC}\frac{R_1}{R_1 + R_2} \qquad &10. \quad I_C = \beta_F I_B \\
&8. \quad R_{EQ} = \frac{R_1 R_2}{R_1 + R_2} \qquad &11. \quad I_E = \frac{I_C}{\alpha_F} \\
&9. \quad I_B = \frac{V_{EQ} - V_{BE}}{R_{EQ} + (\beta_F + 1)R_E} \qquad &12. \quad V_{CE} = V_{CC} - I_C R_C - I_E R_E
\end{aligned}
\quad (4.70)
$$

Because the computer is doing the work, the complete expressions rather than the approximate relations for the various calculations are used in Eq. (4.70).[13] Once Eqs. (4.69) and (4.70) have been entered into one row of the spreadsheet, that row can be copied into as many additional rows as the number of statistical cases that are desired. The analysis is automatically repeated for the random selections to build up the statistical distributions, with each row representing one analysis of the circuit. At the end of the columns, the mean and standard deviation can be

---

[13] Note that V_{BE} could also be treated as a random variable.

Monte Carlo Spreadsheet

Case #	V_{CC} (1)	R_1 (2)	R_2 (3)	R_E (4)	R_C (5)	β_F (6)	V_{EQ} (7)	R_{EQ} (8)	I_B (9)	I_C (10)	V_{CE} (12)
1	12.277	16827	38577	15780	23257	67.46	3.729	11716	2.87E-06	1.93E-04	4.687
2	12.202	18188	32588	15304	23586	46.60	4.371	11673	5.09E-06	2.37E-04	2.891
3	11.526	16648	35643	14627	20682	110.73	3.669	11348	1.87E-06	2.07E-04	4.206
4	11.658	17354	33589	14639	22243	44.24	3.971	11442	5.00E-06	2.21E-04	3.420
5	11.932	19035	32886	16295	20863	62.34	4.374	12056	3.61E-06	2.25E-04	3.500
6	11.857	18706	32615	15563	21064	60.63	4.322	11888	3.83E-06	2.32E-04	3.286
7	11.669	18984	39463	17566	21034	42.86	3.790	12818	4.07E-06	1.75E-04	4.859
8	12.222	19291	37736	15285	22938	63.76	4.135	12765	3.53E-06	2.25E-04	3.577
9	11.601	17589	34032	17334	23098	103.07	3.953	11596	1.85E-06	1.90E-04	3.873
10	11.533	17514	33895	17333	19869	71.28	3.929	11547	2.63E-06	1.88E-04	4.505
11	11.436	19333	34160	15107	22593	68.20	4.133	12346	3.34E-06	2.28E-04	2.797
12	11.962	18810	33999	15545	22035	53.69	4.261	12110	4.25E-06	2.28E-04	3.330
13	11.801	19610	37917	14559	21544	109.65	4.023	12925	2.11E-06	2.31E-04	3.426
14	12.401	17947	34286	15952	21086	107.84	4.261	11780	2.09E-06	2.26E-04	4.002
15	11.894	16209	35321	17321	23940	45.00	3.741	11111	3.89E-06	1.75E-04	4.607
16	12.329	16209	37873	16662	23658	112.01	3.695	11351	1.63E-06	1.83E-04	4.923
17	11.685	19070	35267	15966	21864	64.85	4.101	12377	3.29E-06	2.13E-04	3.559
18	11.456	18096	37476	15529	20141	91.14	3.730	12203	2.17E-06	1.98E-04	4.370
19	12.527	18752	38261	15186	21556	69.26	4.120	12584	3.26E-06	2.26E-04	4.180
20	12.489	17705	36467	17325	20587	83.95	4.082	11919	2.35E-06	1.97E-04	4.979
21	11.436	18773	34697	16949	21848	65.26	4.015	12182	3.01E-06	1.96E-04	3.768
22	11.549	16830	38578	16736	19942	109.22	3.508	11718	1.57E-06	1.71E-04	5.247
23	11.733	16959	39116	15944	21413	62.82	3.548	11830	2.86E-06	1.80E-04	4.965
24	11.738	18486	35520	17526	20455	70.65	4.018	12158	2.70E-06	1.90E-04	4.457
25	11.679	18908	38236	15160	21191	103.12	3.864	12652	2.05E-06	2.12E-04	3.958
Mean	11.848	18014	35102	15973	21863	67.30	4.024	11885	3.44E-06	2.09E-04	3.880
Std. Dev.	0.296	958	2596	1108	1309	23.14	0.264	520	1.14E-06	2.18E-05	0.657

(X) = Equation number in text

Figure 4.37 Example of a Monte Carlo analysis using a spreadsheet.

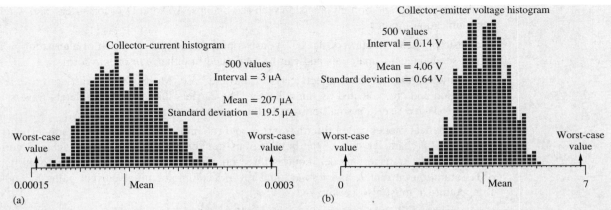

Figure 4.38 (a) Collector-current histogram; (b) collector-emitter voltage histogram.

calculated using built-in spreadsheet functions, and the overall spreadsheet data can be used to build histograms for the circuit performance.

An example of a portion of the spreadsheet output for 25 cases of the circuit in Fig. 4.36 is shown in Fig. 4.37, whereas the full results of the analysis of 500 cases of the four-resistor bias circuit are given in the histograms for I_C and V_{CE} in Fig. 4.38. The mean values for I_C and V_{CE} are 207 μA and 4.06 V, respectively, which are close to the values originally estimated from the nominal circuit elements. The standard deviations are 19.6 μA and 0.64 V, respectively.

Check of Results and Discussion: The worst-case calculations from Sec. 4.12.1 are indicated by the arrows in the figures. It can be seen that the worst-case values of V_{CE} lie well beyond the edges of the statistical distribution, and that saturation does not actually occur for the worst statistical case evaluated. If the Q-point distribution results in the histograms in Fig. 4.38 were not sufficient to meet the design criteria, the parameter tolerances could be changed and the Monte Carlo simulation redone. For example, if too large a fraction of the circuits failed to be within some specified limits, the tolerances could be tightened by specifying more expensive, higher accuracy resistors.

Some implementations of the SPICE circuit analysis program actually contain a Monte Carlo option in which a full circuit simulation is automatically performed for any number of randomly selected test cases. These programs are a powerful tool for performing much more complex statistical analysis than is possible by hand. Using these programs, statistical estimates of delay, frequency response, and so on of circuits with large numbers of transistors can be performed. The impact of temperature changes is also studied with Monte Carlo analyses.

SUMMARY

- The bipolar junction transistor (BJT) was invented in the late 1940s at the Bell Telephone Laboratories by Bardeen, Brattain, and Shockley and became the first commercially successful three-terminal solid-state device.
- Although the field-effect transistor (FET) has become the dominant device technology in modern integrated circuits, bipolar transistors are still widely used in both discrete and integrated circuit design. In particular, the BJT is still the preferred device in many applications that

require high speed and/or high precision such as op-amps, A/D and D/A converters, and wireless communication products.

- The basic physical structure of the BJT consists of a three-layer sandwich of alternating *p*- and *n*-type semiconductor materials and can be fabricated in either *npn* or *pnp* form.

- The emitter of the transistor injects carriers into the base. Most of these carriers traverse the base region and are collected by the collector. The carriers that do not completely traverse the base region give rise to a small current in the base terminal.

- A mathematical model called the transport model (a simplified Gummel-Poon model) characterizes the *i-v* characteristics of the bipolar transistor for general terminal voltage and current conditions. The transport model requires three unique parameters to characterize a particular BJT: saturation current I_S and forward and reverse common-emitter current gains β_F and β_R. Temperature T must also be known.

- β_F is a relatively large number, ranging from 10 to 500, and characterizes the significant current amplification capability of the BJT. Practical fabrication limitations cause the bipolar transistor structure to be inherently asymmetric, and the value of β_R is much smaller than β_F, typically between 0 and 10.

- SPICE circuit analysis programs contain an all-region built-in model for the transistor that is an extension of the transport model.

- Four regions of operation—cutoff, forward-active, reverse-active, and saturation—were identified for the BJT based on the bias voltages applied to the base-emitter and base-collector junctions. The transport model can be simplified for each individual region of operation.

- The cutoff and saturation regions are most often used in switching applications and logic circuits. In cutoff, the transistor approximates an open switch, whereas in saturation, the transistor represents a closed switch. The saturated bipolar transistor has a small voltage, the collector-emitter saturation voltage V_{CESAT}, between its collector and emitter terminals, even when operating with zero collector current.

- In the forward-active region, the bipolar transistor can provide high voltage and current gain for amplification of analog signals. The reverse-active region finds limited use in some analog- and digital-switching applications.

- The *i-v* characteristics of the bipolar transistor are often presented graphically in the form of the output characteristics, i_C versus v_{CE} or v_{CB}, and the transfer characteristics, i_C versus v_{BE} or v_{EB}.

- In the forward-active region, the collector current increases slightly as the collector-emitter voltage increases. The origin of this effect is base-width modulation, known as the Early effect, and it can be included in the model for the forward-active region through addition of the parameter called the Early voltage V_A.

- The collector current of the bipolar transistor is determined by minority-carrier diffusion across the base of the transistor, and expressions were developed that relate the saturation current and base transit time of the transistor to physical device parameters. The base width plays a crucial role in determining the base transit time and the high-frequency operating limits of the transistor.

- Minority-carrier charge is stored in the base of the transistor during its operation, and changes in this stored charge with applied voltage result in diffusion capacitances being associated with forward-biased junctions. The value of the diffusion capacitance is proportional to the collector current I_C.

- Capacitances of the bipolar transistor cause the current gain to be frequency-dependent. At the beta-cutoff frequency f_β, the current gain has fallen to 71 percent of its low frequency value, whereas the value of the current gain is only 1 at the unity-gain frequency f_T.

- The transconductance g_m of the bipolar transistor in the forward-active region relates differential changes in collector current and base-emitter voltage and was shown to be directly proportional to the dc collector current I_C.
- Transconductance efficiency g_m/I_C is used to compare various types of electronic devices. For the BJT, $g_m/I_C = 1/V_T$, approximately 40 at room temperature.
- Design of the four-resistor network was investigated in detail. The four-resistor bias circuit provides highly stable control of the Q-point and is the most important bias circuit for discrete design. A simple two-resistor bias circuit was also discussed.
- Techniques for analyzing the influence of element tolerances on circuit performance include the worst-case analysis and statistical Monte Carlo analysis methods. In worst-case analysis, element values are simultaneously pushed to their extremes, and the resulting predictions of circuit behavior are often overly pessimistic. The Monte Carlo method analyzes a large number of randomly selected versions of a circuit to build up a realistic estimate of the statistical distribution of circuit performance. Random number generators in high-level computer languages, spreadsheets, or MATLAB can be used to randomly select element values for use in the Monte Carlo analysis. Some circuit analysis packages such as PSPICE provide a Monte Carlo analysis option as part of the program.

KEY TERMS

Active region
Base
Base current
Base width
Base-collector capacitance
Base-collector voltage
Base-emitter capacitance
Base-emitter voltage
Base-width modulation
β-cutoff frequency f_β
Bipolar junction transistor (BJT)
Collector
Collector-base voltage
Collector current
Common-emitter output characteristic
Common-emitter transfer characteristic
Cutoff region
dc bias
Diffusion capacitance
Early effect
Early voltage V_A
Ebers-Moll model
Emitter
Emitter-base voltage
Emitter current
Equilibrium electron density
Forced beta
Forward-active region
Forward common-emitter current gain β_F

Forward common-base current gain α_F
Forward transit time τ_F
Forward-transport current
Gummel-Poon model
Inverse-active region
Inverse common-emitter current gain
Inverse common-base current gain
Monte Carlo analysis
Normal-active region
Normal common-emitter current gain
Normal common-base current gain
npn transistor
Output characteristic
pnp transistor
Quiescent operating point
Q-point
Reverse-active region
Reverse common-base current gain α_R
Reverse common-emitter current gain β_R
Saturation region
Saturation voltage
SPICE model parameters BF, IS, VAF
Transconductance
Transfer characteristic
Transport current i_T
Transistor saturation current
Transport model
Unity-gain frequency f_T
Worst-case analysis

REFERENCES

1. William F. Brinkman, "The transistor: 50 glorious years and where we are going," *IEEE International Solid-State Circuits Conference Digest,* vol. 40, pp. 22–26, February 1997.
2. William F. Brinkman, Douglas E. Haggan, and William W. Troutman, "A history of the invention of the transistor and where it will lead us," *IEEE Journal of Solid-State Circuits,* vol. 32, pp. 1858–1865, December 1997.
3. H. K. Gummel and H. C. Poon, "A compact bipolar transistor model," *ISSCC Digest of Technical Papers,* pp. 78, 79, 146, February 1970.
4. H. K. Gummel, "A charge control relation for bipolar transistors," *Bell System Technical Journal,* January 1970.
5. J. J. Ebers and J. L. Moll, "Large signal behavior of junction transistors," *Proc. IRE.,* pp. 1761–1772, December 1954.
6. J. M. Early, "Effects of space-charge layer widening in junction transistors," *Proc. IRE.,* pp. 1401–1406, November 1952.
7. J. D. Cressler and G. Niu, *Silicon-Germanium Heterojunction Bipolar Transistors,* Artech House, Boston: 2003 (ISBN 1-58053-361-2).
8. B. M. Wilamowski and R. C. Jaeger, *Computerized Circuit Analysis Using SPICE Programs,* McGraw-Hill, New York: 1997.

ADDITIONAL READINGS

J. D. Cressler (Editor), *Silicon Heterostructure Handbook—Materials, Fabrication, Devices, Circuits, and Applications of SiGe and Si Strained-Layer Epitaxy,* CRC Press, Taylor & Francis Group, Boca Raton, FL: 2006 (ISBN 0-8493-3559-0).

R. C. Jaeger and A. J. Brodersen, "Self consistent bipolar transistor models for computer simulation," *Solid-State Electronics*, vol. 21, no. 10, pp. 1269–1272, October 1978.

PROBLEMS

If not otherwise specified, use $I_S = 10^{-16}$ A, $V_A = 50$ V, $\beta_F = 100$, $\beta_R = 1$, $V_{BE} = 0.70$ V, and $V_T = 25.9$ mV.

4.1 Physical Structure of the Bipolar Transistor

4.1. Figure P4.1 is a cross section of an *npn* bipolar transistor similar to that in Fig. 4.1. Indicate the letter (A to G) that identifies the base contact, collector contact, emitter contact, *n*-type emitter region, *n*-type collector region, and the active or intrinsic transistor region.

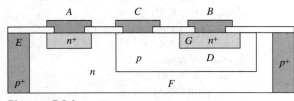

Figure P4.1

4.2 The Transport Model for the *npn* Transistor

4.2. (a) Label the collector, base, and emitter terminals of the transistor in the circuit in Fig. P4.2. (b) Label the base-emitter and base-collector voltages, V_{BE} and V_{BC}, respectively. (c) If $V = 0.710$ V, $I_C = 275$ μA, and $I_B = 2.5$ μA, find the values of I_S, β_F, and β_R for the transistor if $\alpha_R = 0.53$.

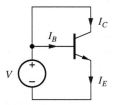

Figure P4.2

4.3. (a) Label the collector, base, and emitter terminals of the transistor in the circuit in Fig. P4.3. (b) Label the base-emitter and base-collector voltages, V_{BE}

and V_{BC}, and the positive directions of the collector, base, and emitter currents. (c) If $V = 0.710$ V, $I_E = -275$ µA, and $I_B = 140$ µA, find the values of I_S, β_F, and β_R for the transistor if $\alpha_F = 0.975$.

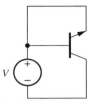

Figure P4.3

4.4. Fill in the missing entries in Table P4.1

TABLE P4.1

α	β
	0.150
0.400	
0.825	
	10.0
0.973	
	250
	1000
0.9998	

4.5. (a) Find the current I_{CBS} in Fig. P4.5(a). (Use the parameters specified at the beginning of the problem set.) (b) Find the current I_{CBO} and the voltage V_{BE} in Fig. P4.5(b).

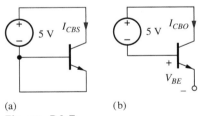

(a) (b)

Figure P4.5

4.6. For the transistor in Fig. P4.6, $I_S = 5 \times 10^{-16}$ A, $\beta_F = 100$, and $\beta_R = 0.25$. (a) Label the collector, base, and emitter terminals of the transistor. (b) What is the transistor type? (c) Label the base-emitter and base-collector voltages, V_{BE} and V_{BC}, respectively, and label the normal directions for I_E, I_C, and I_B. (d) What is the relationship between V_{BE} and V_{BC}? (e) Write the simplified form of the transport model equations that apply to this particular circuit configuration. Write an expression for I_E/I_B. Write an expression for I_E/I_C. (f) Find the values of I_E, I_C, I_B, V_{BC}, and V_{BE}.

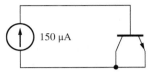

Figure P4.6

4.7. For the transistor in Fig. P4.7, $I_S = 6 \times 10^{-16}$ A, $\beta_F = 120$, and $\beta_R = 0.40$. (a) Label the collector, base, and emitter terminals of the transistor. (b) What is the transistor type? (c) Label the base-emitter and base-collector voltages, V_{BE} and V_{BC}, and the normal directions for I_E, I_C, and I_B. (d) Find the values of I_E, I_C, I_B, V_{BC}, and V_{BE} if $I = 175$ µA.

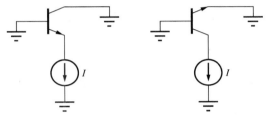

Figure P4.7   **Figure P4.8**

4.8. For the transistor in Fig. P4.8, $I_S = 6 \times 10^{-16}$ A, $\beta_F = 120$, and $\beta_R = 0.40$. (a) Label the collector, base, and emitter terminals of the transistor. (b) What is the transistor type? (c) Label the base-emitter and base-collector voltages, V_{BE} and V_{BC}, and label the normal directions for I_E, I_C, and I_B. (d) Find the values of I_E, I_C, I_B, V_{BC}, and V_{BE} if $I = 175$ µA.

4.9. The *npn* transistor is connected in a "diode" configuration in Fig. P4.9(a). Use the transport model equations to show that the *i-v* characteristics of this connection are similar to those of a diode as defined by Eq. (3.11). What is the reverse saturation current of this "diode" if $I_S = 4 \times 10^{-15}$ A, $\beta_F = 100$, and $\beta_R = 0.33$?

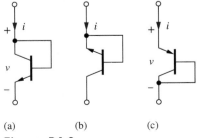

(a)   (b)   (c)

Figure P4.9

4.10. The *npn* transistor is connected in an alternate "diode" configuration in Fig. P4.9(b). Use the transport model equations to show that the *i-v* characteristics of this connection are similar to those of a diode as defined by Eq. (3.11). What is the reverse saturation current of this "diode" if $I_S = 5 \times 10^{-16}$ A, $\beta_F = 70$, and $\beta_R = 3$?

4.11. Calculate i_T for an *npn* transistor with $I_S = 10^{-15}$ A for (a) $V_{BE} = 0.735$ V and $V_{BC} = -3$ V and (b) $V_{BC} = 0.735$ V and $V_{BE} = -3$ V.

4.12. Calculate i_T for an *npn* transistor with $I_S = 10^{-16}$ A for (a) $V_{BE} = 0.685$ V and $V_{BC} = -3$ V and (b) $V_{BC} = 0.685$ V and $V_{BE} = -3$ V.

4.3 The *pnp* Transistor

4.13. Figure P4.13 is a cross section of a *pnp* bipolar transistor similar to the *npn* transistor in Fig. 4.1. Indicate the letter (*A* to *G*) that represents the base contact, collector contact, emitter contact, *p*-type emitter region, *p*-type collector region, and the active or intrinsic transistor region.

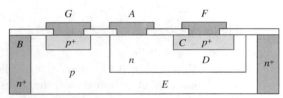

Figure P4.13

4.14. For the transistor in Fig. P4.14(a), $I_S = 6 \times 10^{-16}$ A, $\alpha_F = 0.985$, and $\alpha_R = 0.33$. (a) What type of transistor is in this circuit? (b) Label the collector, base, and emitter terminals of the transistor. (c) Label the emitter-base and collector-base voltages, and label the normal direction for I_E, I_C, and I_B. (d) Write the simplified form of the transport model equations that apply to this particular circuit configuration. Write an expression for I_E/I_C. Write an expression for I_E/I_B. (e) Find the values of I_E, I_C, I_B, β_F, β_R, V_{EB}, and V_{CB}.

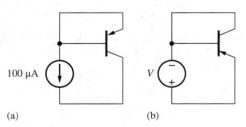

Figure P4.14

4.15. (a) Label the collector, base, and emitter terminals of the transistor in the circuit in Fig. P4.14(b). (b) Label the emitter-base and collector-base voltages, V_{EB} and V_{CB}, and the normal directions for I_E, I_C, and I_B. (c) If $V = 0.640$ V, $I_C = 300$ µA, and $I_B = 4$ µA, find the values of I_S, β_F, and β_R for the transistor if $\alpha_R = 0.25$.

4.16. Repeat Prob. 4.9 for the "diode-connected" *pnp* transistor in Fig. P4.9(c).

4.17. For the transistor in Fig. P4.17, $I_S = 4 \times 10^{-16}$ A, $\beta_F = 90$, and $\beta_R = 4$. (a) Label the collector, base, and emitter terminals of the transistor. (b) What is the transistor type? (c) Label the emitter-base and collector-base voltages, and label the normal direction for I_E, I_C, and I_B. (d) Write the simplified form of the transport model equations that apply to this particular circuit configuration. Write an expression for I_E/I_B. Write an expression for I_E/I_C. (e) Find the values of I_E, I_C, I_B, V_{CB}, and V_{EB}.

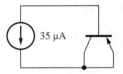

Figure P4.17

4.18. For the transistor in Fig. P4.18(a), $I_S = 2.5 \times 10^{-16}$ A, $\beta_F = 100$, and $\beta_R = 5$. (a) Label the collector, base, and emitter terminals of the transistor. (b) What is the transistor type? (c) Label the emitter-base and collector-base voltages, V_{EB} and V_{CB}, and the normal directions for I_E, I_C, and I_B. (d) Find the values of I_E, I_C, I_B, V_{CB}, and V_{EB} if $I = 250$ µA.

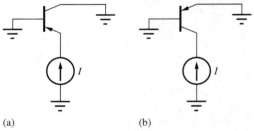

Figure P4.18

4.19. For the transistor in Fig. P4.18(b), $I_S = 2.5 \times 10^{-16}$ A, $\beta_F = 75$, and $\beta_R = 1$. (a) Label the collector, base, and emitter terminals of the transistor. (b) What is the transistor type? (c) Label the emitter-base and collector-base voltages, V_{EB} and V_{CB}, and label the normal directions for I_E, I_C, and I_B. (d) Find the values of I_E, I_C, I_B, V_{CB}, and V_{EB} if $I = 200$ µA.

4.20. Calculate i_T for a *pnp* transistor with $I_S = 6 \times 10^{-16}$ A for (a) $V_{EB} = 0.70$ V and $V_{CB} = -3$ V and (b) $V_{CB} = 0.67$ V and $V_{EB} = -3$ V.

4.4 Equivalent Circuit Representations for the Transport Models

4.21. Calculate the values of i_T and the two diode currents for the equivalent circuit in Fig. 4.8(a) for an *npn* transistor with $I_S = 2.5 \times 10^{-16}$ A, $\beta_F = 80$, and $\beta_R = 2$ for (a) $V_{BE} = 0.75$ V and $V_{BC} = -3$ V and (b) $V_{BC} = 0.75$ V and $V_{BE} = -3$ V.

4.22. Calculate the values of i_T and the two diode currents for the equivalent circuit in Fig. 4.8(b) for a *pnp* transistor with $I_S = 4 \times 10^{-15}$ A, $\beta_F = 60$, and $\beta_R = 3$ for (a) $V_{EB} = 0.65$ V and $V_{CB} = -3$ V and (b) $V_{CB} = 0.65$ V and $V_{EB} = -3$ V.

4.23. The Ebers-Moll model was one of the first mathematical models used to describe the characteristics of the bipolar transistor. Show that the *npn* transport model equations can be transformed into the Ebers-Moll equations below. [*Hint:* Add and subtract 1 from the collector and emitter current expressions in Eqs. (4.13).]

$$i_E = I_{ES}\left[\exp\left(\frac{v_{BE}}{V_T}\right) - 1\right] - \alpha_R I_{CS}\left[\exp\left(\frac{v_{BC}}{V_T}\right) - 1\right]$$

$$i_C = \alpha_F I_{ES}\left[\exp\left(\frac{v_{BE}}{V_T}\right) - 1\right] - I_{CS}\left[\exp\left(\frac{v_{BC}}{V_T}\right) - 1\right]$$

$$i_B = (1 - \alpha_F)I_{ES}\left[\exp\left(\frac{v_{BE}}{V_T}\right) - 1\right] + (1 - \alpha_R)I_{CS}\left[\exp\left(\frac{v_{BC}}{V_T}\right) - 1\right]$$

$$\alpha_F I_{ES} = \alpha_R I_{CS}$$

4.24. What are the values of α_F, α_R, I_{ES}, and I_{CS} for an *npn* transistor with $I_S = 4 \times 10^{-15}$ A, $\beta_F = 100$, and $\beta_R = 1.5$? Show that $\alpha_F I_{ES} = \alpha_R I_{CS}$.

4.25. The Ebers-Moll model was one of the first mathematical models used to describe the characteristics of the bipolar transistor. Show that the *pnp* transport model equations can be transformed into the Ebers-Moll equations that follow. [*Hint:* Add and subtract 1 from the collector and emitter current expressions in Eqs. (4.17).]

4.5 The *i-v* Characteristics of the Bipolar Transistor

*4.26. The common-emitter output characteristics for an *npn* transistor are given in Fig. P4.26. What are the values of β_F at (a) $I_C = 5$ mA and $V_{CE} = 5$ V? (b) $I_C = 7$ mA and $V_{CE} = 7.5$ V? (c) $I_C = 10$ mA and $V_{CE} = 14$ V? (d) $I_B = 100$ µA and $V_{CE} = 10$ V? (e) $I_B = 40$ µA and $V_{CE} = 5$ V?

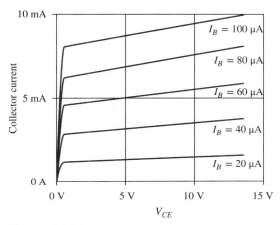

Figure P4.26

4.27. Plot the common-emitter output characteristics for an *npn* transistor having $I_S = 1$ fA, $\beta_{FO} = 75$, and $V_A = 50$ V for six equally spaced base current steps ranging from 0 to 250 µA and V_{CE} ranging from 0 to 10 V.

4.28. Use SPICE to plot the common-emitter output characteristics for the *npn* transistor in Prob. 4.27.

4.29. Plot the common-emitter output characteristics for a *pnp* transistor having $I_S = 1$ fA, $\beta_{FO} = 75$, and $V_A = 50$ V for six equally spaced base current steps ranging from 0 to 200 µA and V_{EC} ranging from 0 to 10 V.

4.30. Use SPICE to plot the common-emitter output characteristics for the *pnp* transistor in Prob. 4.29.

$$i_E = I_{ES}\left[\exp\left(\frac{v_{EB}}{V_T}\right) - 1\right] - \alpha_R I_{CS}\left[\exp\left(\frac{v_{CB}}{V_T}\right) - 1\right]$$

$$i_C = \alpha_F I_{ES}\left[\exp\left(\frac{v_{EB}}{V_T}\right) - 1\right] - I_{CS}\left[\exp\left(\frac{v_{CB}}{V_T}\right) - 1\right]$$

$$i_B = (1 - \alpha_F)I_{ES}\left[\exp\left(\frac{v_{EB}}{V_T}\right) - 1\right] + (1 - \alpha_R)I_{CS}\left[\exp\left(\frac{v_{CB}}{V_T}\right) - 1\right]$$

$$\alpha_F I_{ES} = \alpha_R I_{CS}$$

4.31. What is the reciprocal of the slope (in mV/decade) of the logarithmic transfer characteristic for an *npn* transistor in the common-emitter configuration at a temperature of (a) 200 K, (b) 250 K, (c) 300 K, and (d) 350 K?

Junction Breakdown Voltages

*4.32. In the circuits in Fig. P4.9, the Zener breakdown voltages of the collector-base and emitter-base junctions of the transistors are 50 V and 6 V, respectively. What is the Zener breakdown voltage for each "diode" connected transistor configuration?

4.33. In the circuits in Fig. P4.33, the Zener breakdown voltages of the collector-base and emitter-base junctions of the *npn* transistors are 40 V and 6.3 V, respectively. What is the current in the resistor in each circuit? (*Hint:* The equivalent circuits for the transport model equations may help in visualizing the circuit.)

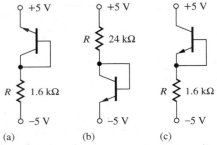

Figure P4.33

4.34. An *npn* transistor is biased as indicated in Fig. 4.9(a). What is the largest value of V_{CE} that can be applied without junction breakdown if the breakdown voltages of the collector-base and emitter-base junctions of the *npn* transistors are 60 V and 5 V, respectively?

*4.35. (a) For the circuit in Fig. P4.35, what is the maximum value of I according to the transport model equations if $I_S = 1 \times 10^{-16}$ A, $\beta_F = 50$, and $\beta_R = 0.5$? (b) Suppose that $I = 1$ mA. What happens to the transistor? (*Hint:* The equivalent circuits for the transport model equations may help in visualizing the circuit.)

Figure P4.35

4.6 The Operating Regions of the Bipolar Transistor

4.36. Indicate the region of operation in the following table for an *npn* transistor biased with the indicated voltages.

BASE-EMITTER VOLTAGE	BASE-COLLECTOR VOLTAGE	
	0.7 V	−5.0 V
−5.0 V		
0.7 V		

4.37. (a) What are the regions of operation for the transistors in Fig. P4.9? (b) In Fig. P4.44(a)? (c) In Fig. P4.47? (d) In Fig. P4.60?

4.38. (a) What is the region of operation for the transistor in Fig. P4.5(a)? (b) In Fig. P4.5(b)?

4.39. (a) What is the region of operation for the transistor in Fig. P4.6? (b) In Fig. P4.7? (c) In Fig. P4.8?

4.40. Indicate the region of operation in the following table for a *pnp* transistor biased with the indicated voltages.

EMITTER-BASE VOLTAGE	COLLECTOR-BASE VOLTAGE	
	0.7 V	−0.65 V
0.7 V		
−0.65 V		

4.41. (a) What is the region of operation for the transistor in Fig. P4.2? (b) In Fig. P4.3?

4.42. (a) What is the region of operation for the transistor in Fig. P4.14(a)? (b) In Fig. P4.14(b)?

4.43. (a) What is the region of operation for the transistor in Fig. P4.17? (b) In Fig. P4.18(a)? (c) In Fig. P4.18(b)?

4.7 Transport Model Simplifications
Cutoff Region

4.44. (a) What are the three terminal currents I_E, I_B, and I_C in the transistor in Fig. P4.44(a) if $I_S = 2 \times 10^{-16}$ A, $\beta_F = 80$, and $\beta_R = 2$? (b) Repeat for Fig. P4.44(b).

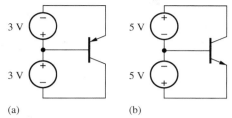

Figure P4.44

**4.45. An *npn* transistor with $I_S = 5 \times 10^{-16}$ A, $\alpha_F = 0.95$, and $\alpha_R = 0.5$ is operating with $V_{BE} = 0.3$ V and $V_{BC} = -5$ V. This transistor is not truly operating in the region defined to be cutoff, but we still say the transistor is off. Why? Use the transport model equations to justify your answer. In what region is the transistor actually operating according to our definitions?

Forward-Active Region

4.46. What are the values of β_F and I_S for the transistor in Fig. P4.46?

4.47. What are the values of β_F and I_S for the transistor in Fig. P4.47?

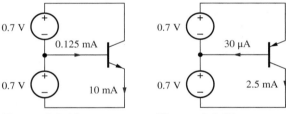

Figure P4.46 **Figure P4.47**

4.48. What are the emitter, base, and collector currents in the circuit in Fig. 4.16 if $V_{EE} = 3.3$ V, $R = 47$ kΩ, and $\beta_F = 80$?

**4.49. A transistor has $f_T = 500$ MHz and $\beta_F = 75$. (a) What is the β-cutoff frequency f_β of this transistor? (b) Use Eq. (4.43) to find an expression for the frequency dependence of α_F—that is, $\alpha_F(f)$. [*Hint*: Write an expression for $\beta(s)$.] What is the α-cutoff frequency for this transistor?

*4.50. (a) Start with the transport model equations for the *pnp* transistor, Eq. (4.17), and construct the simplified version of the *pnp* equations that apply to the forward-active region [similar to Eq. (4.25)]. (b) Draw the simplified model for the *pnp* transistor similar to the *npn* version in Fig. 4.19(c).

Reverse-Active Region

4.51. What are the values of β_R and I_S for the transistor in Fig. P4.51?

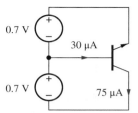

Figure P4.51

4.52. What are the values of β_R and I_S for the transistor in Fig. P4.52?

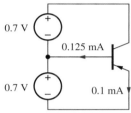

Figure P4.52

4.53. Find the emitter, base, and collector currents in the circuit in Fig. 4.20 if the negative power supply is -3.3 V, $R = 56$ kΩ, and $\beta_R = 0.75$.

Saturation Region

4.54. What is the saturation voltage of an *npn* transistor operating with $I_C = 1$ mA and $I_B = 1$ mA if $\beta_F = 75$ and $\beta_R = 3$? What is the forced β of this transistor? What is the value of V_{BE} if $I_S = 10^{-15}$ A?

4.55. Derive an expression for the saturation voltage V_{ECSAT} of the *pnp* transistor in a manner similar to that used to derive Eq. (4.31).

4.56. (a) What is the collector-emitter voltage for the transistor in Fig. P4.56(a) if $I_S = 7 \times 10^{-16}$ A, $\alpha_F = 0.983$, and $\alpha_R = 0.5$? (b) What is the emitter-collector voltage for the transistor in Fig. P4.56(b) for the same transistor parameters?

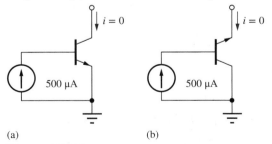

Figure P4.56

4.57. Repeat Prob. 4.56 for $\alpha_F = 0.95$ and $\alpha_R = 0.33$.

4.58. (a) What base current is required to achieve a saturation voltage of $V_{CESAT} = 0.1$ V in an *npn* power transistor that is operating with a collector current of 20 A if $\beta_F = 20$ and $\beta_R = 0.9$? What is the forced β of this transistor? (b) Repeat for $V_{CESAT} = 0.04$ V.

**4.59. An *npn* transistor with $I_S = 1 \times 10^{-16}$ A, $\alpha_F = 0.975$, and $\alpha_R = 0.5$ is operating with $V_{BE} = 0.70$ V and $V_{BC} = 0.50$ V. By definition, this transistor is operating in the saturation region. However, in the discussion of Fig. 4.17 it was noted that this transistor actually behaves as if it is still in the forward-active region even though $V_{BC} > 0$. Why? Use the transport model equations to justify your answer.

4.60. The current I in both circuits in Fig. P4.60 is 200 µA. Find the value of V_{BE} for both circuits if $I_S = 4 \times 10^{-16}$ A, $\beta_F = 50$, and $\beta_R = 0.5$. What is V_{CESAT} in Fig. P4.60(b)?

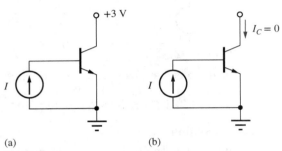

Figure P4.60

Diodes in Bipolar Integrated Circuits

4.61. Derive the result in Eq. (4.27) by applying the circuit constraints to the transport equations.

4.62. What is the reverse saturation current of the diode in Fig. 4.18 if the transistor is described by $I_S = 3 \times 10^{-15}$ A, $\alpha_R = 0.20$, and $\alpha_F = 0.98$?

4.63. The two transistors in Fig. P4.63 are identical. What is the collector current of Q_2 if $I = 25$ µA and $\beta_F = 60$?

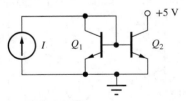

Figure P4.63

4.8 Nonideal Behavior of the Bipolar Transistor

4.64. Calculate the diffusion capacitance of a bipolar transistor with a forward transit time $\tau_F = 50$ ps for collector currents of (a) 2 µA, (b) 200 µA, (c) 20 mA.

4.65. (a) What is the forward transit time τ_F for an *npn* transistor with a base width $W_B = 0.5$ µm and a base doping of $10^{18}/cm^3$? (b) Repeat the calculation for a *pnp* transistor.

4.66. What is the diffusion capacitance for an *npn* transistor with $\tau_F = 10$ ps if it is operating at 300 K with a collector currents of 1 µA, 1 mA, and 10 mA?

4.67. A transistor has $f_T = 750$ MHz and $f_\beta = 10$ MHz. What is the dc current gain of the transistor? What is the current gain of the transistor at 50 MHz? At 250 MHz?

4.68. A transistor has a dc current gain of 200 and a current gain of 10 at 75 MHz. What are the unity-gain and beta-cutoff frequencies of the transistor?

4.69. (a) An *npn* transistor is needed that will operate at a frequency of at least 5 GHz. What base width is required for the transistor if the base doping is $5 \times 10^{18}/cm^3$? (b) Repeat for 10 GHz.

4.70. What is the saturation current for a transistor with a base doping of $6 \times 10^{18}/cm^3$, a base width of 0.18 µm, and a cross-sectional area of 17 µm^2?

The Early Effect and Early Voltage

4.71. An *npn* transistor is operating in the forward-active region with a base current of 3 µA. It is found that $I_C = 225$ µA for $V_{CE} = 5$ V and $I_C = 265$ µA for $V_{CE} = 10$ V. What are the values of β_{FO} and V_A for this transistor?

4.72. An *npn* transistor with $I_S = 5 \times 10^{-16}$ A, $\beta_F = 100$, and $V_A = 65$ V is biased in the forward-active region with $V_{BE} = 0.72$ V and $V_{CE} = 10$ V. (a) What is the collector current I_C? (b) What would be the collector current I_C if $V_A = \infty$? (c) What is the ratio of the two answers in parts (a) and (b)?

4.73. The common-emitter output characteristics for an *npn* transistor are given in Fig. P4.26. What are the values of β_{FO} and V_A for this transistor?

4.74. (a) Recalculate the currents in the transistor in Fig. 4.14 if $I_S = 5 \times 10^{-16}$ A, $\beta_{FO} = 19$, and $V_A = 50$ V. What is V_{BE}? (b) What was V_{BE} for $V_A = \infty$?

4.75. Recalculate the currents in the transistor in Fig. 4.16 if $\beta_{FO} = 50$ and $V_A = 50$ V.

4.76. Repeat Prob. 4.63 if $V_A = 50$ V and $V_{BE} = 0.7$ V.

4.9 Transconductance

4.77. What is the transconductance of an *npn* transistor operating at 350 K and a collector current of (a) 10 µA, (b) 100 µA, (c) 1 mA, and (d) 10 mA? (e) Repeat for a *pnp* transistor.

4.78. (a) What collector current is required for an *npn* transistor to have a transconductance of 25 mS at a temperature of 320 K? (b) Repeat for a *pnp* transistor. (c) Repeat parts (a) and (b) for a transconductance of 40 µS.

4.79. Calculate the transconductance for the transistor in Fig. P4.26 for a Q-point for $I_B = 60$ µA and $V_{CE} = 5$ V for temperatures of −55 C, 30 C, and 125 C.

4.80. Calculate the transconductance efficiency for the transistor in Fig. P4.26 for a Q-point for $I_B = 80$ µA and $V_{CE} = 10$ V for temperatures of −55 C, 30 C, and 125 C.

4.10 Bipolar Technology and SPICE Model

4.81. (a) Find the default values of the following parameters for the generic *npn* transistor in the version of SPICE that you use in class: IS, BF, BR, VAF, VAR, TF, TR, NF, NE, RB, RC, RE, ISE, ISC, ISS, IKF, IKR, CJE, CJC. (*Note:* The values in Table P4.1 may not agree exactly with your version of SPICE.) (b) Repeat for the generic *pnp* transistor.

4.82. A SPICE model for a bipolar transistor has a forward knee current IKF = 10 mA and NK = 0.5. How much does the KBQ factor reduce the collector current of the transistor in the forward-active region if i_F is (a) 1 mA? (b) 10 mA? (c) 50 mA?

4.83. Plot a graph of KBQ versus i_F for an *npn* transistor with IKF = 40 mA and NK = 0.5. Assume forward-active region operation with VAF = ∞.

4.11 Practical Bias Circuits for the BJT

Four-Resistor Biasing

4.84. (a) Find the Q-point for the circuit in Fig. P4.84(a). Assume that $\beta_F = 50$ and $V_{BE} = 0.7$ V. (b) Repeat the calculation if all the resistor values are decreased by a factor of 5. (c) Repeat if all the resistor values are increased by a factor of 5. (d) Find the Q-point in part (a) using the numerical iteration method if $I_S = 0.5$ fA and $V_T = 25.8$ mV.

4.85. (a) Calculate the worst-case values of the Q-point for Fig. P4.84(a) if the resistor tolerances are all 5 percent. (b) Repeat if the tolerances on β_F are ±20 percent/±50 percent (no resistor tolerances).

4.86. (a) Scale the resistor values in the circuits in Fig. P4.84 to increase the currents by a factor of 5. Pick the nearest 5 percent standard values from Table A.2. (b) Repeat to reduce the currents in Fig. P4.84 by a factor of 3.

4.87. (a) Find the Q-point for the circuit in Fig. P4.84(a) if the 27-kΩ resistor is replaced with a 33-kΩ resistor. Assume that $\beta_F = 75$. (b) Calculate the worst-case values of the Q-point if resistor tolerances are all 10 percent. (c) Calculate the worst-case values of the Q-point if the tolerances on β_F are +20 percent/−50 percent.

4.88. (a) Find the Q-point for the circuit in Fig. P4.84(b). Assume $\beta_F = 50$ and $V_{BE} = 0.7$ V. (b) Repeat if all the resistor values are decreased by a factor of 5. (c) Repeat if all the resistor values are increased by a factor of 5. (d) Find the Q-point in part (a) using the numerical iteration method if $I_S = 0.4$ fA and $V_T = 25.8$ mV.

4.89. (a) Find the Q-point for the circuit in Fig. P4.84(b) if the 27-kΩ resistor is replaced with a 43-kΩ resistor. Assume $\beta_F = 75$ and $V_{BE} = 0.7$ V. (b) Repeat if all the resistor values are decreased by a factor of 5. (c) Repeat if all the resistor values are increased by a factor of 5. (d) Find the Q-point in part (a) using the numerical iteration method if $I_S = 1$ fA and $V_T = 25.8$ mV.

4.90. (a) Calculate the worst-case values of the Q-point for Problem 4.85(a) if the resistor tolerances are all 10 percent. (b) Repeat if the tolerances on β_F are +20 percent/−50 percent (no resistor tolerances).

4.91. (a) Simulate the circuits in Fig. P4.84 and compare the SPICE results to your hand calculations of the Q-point. Use $I_S = 1 \times 10^{-16}$ A, $\beta_F = 50$, $\beta_R = 0.25$, and $V_A = \infty$. (b) Repeat for $V_A = 60$ V. (c) Repeat (a) for the circuit in Fig. 4.32(c). (d) Repeat (b) for the circuit in Fig. 4.32(c).

4.92. Find the Q-point in the circuit in Fig. 4.32 if $R_1 = 120$ kΩ, $R_2 = 270$ kΩ, $R_E = 100$ kΩ, $R_C = 150$ kΩ, $\beta_F = 100$, and the positive power supply voltage is 10 V.

4.93. (a) Find the Q-point in the circuit in Fig. 4.32 if $R_1 = 6.8$ kΩ, $R_2 = 13$ kΩ, $R_C = 5.1$ kΩ, $R_E = 7.5$ kΩ, $\beta_F = 100$, and the positive power supply voltage is 15 V. (b) Find the worst-case values of the Q-point if resistor tolerances are all 5 percent.

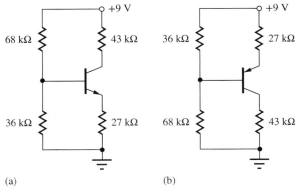

(a) (b)

Figure P4.84

4.94. (a) Design a four-resistor bias network for an *npn* transistor to give $I_C = 12$ µA and $V_{CE} = 6$ V if $V_{CC} = 18$ V and $\beta_F = 75$. (b) Replace your exact values with the nearest values from the resistor table in Appendix C and find the resulting Q-point. (c) Find the worst-case values of the Q-point in (b) if the resistor tolerances are all 5 percent.

4.95. (a) Design a four-resistor bias network for an *npn* transistor to give $I_C = 1$ mA, $V_{CE} = 5$ V, and $V_E = 3$ V if $V_{CC} = 12$ V and $\beta_F = 100$. (b) Replace your exact values with the nearest values from the resistor table in Appendix C and find the resulting Q-point. (c) Find the worst-case values of the Q-point in (b) if the resistor tolerances are all 10 percent.

4.96. (a) Design a four-resistor bias network for a *pnp* transistor to give $I_C = 750$ µA, $V_{EC} = 2$ V, and $V_E = 1$ V if $V_{CC} = 5$ V and $\beta_F = 60$. (b) Replace your exact values with the nearest values from the resistor table in Appendix C and find the resulting Q-point. (c) Find the worst-case values of the Q-point in part (b) if the resistor tolerances are all 5 percent.

4.97. (a) Design a four-resistor bias network for a *pnp* transistor to give $I_C = 9$ mA and $V_{EC} = 5$ V if $V_{RE} = 1$ V, $V_{CC} = -15$ V, and $\beta_F = 50$. (b) Replace your exact values with the nearest values from the resistor table in Appendix C and find the resulting Q-point.

Load Line Analysis

*4.98. Find the Q-point for the circuit in Fig. P4.98 using the graphical load-line approach. Use the characteristics in Fig. P4.26.

*4.99. Find the Q-point for the circuit in Fig. P4.99 using the graphical load-line approach. Use the characteristics in Fig. P4.26, assuming that the graph is a plot of i_C vs. v_{EC} rather than i_C vs. v_{CE}.

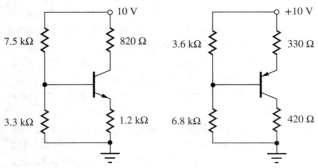

Figure P4.98   Figure P4.99

Bias Circuits and Applications

4.100. Find the Q-point for the circuit in Fig. P4.100 for (a) $\beta_F = 40$, (b) $\beta_F = 120$, (c) $\beta_F = 250$, (d) $\beta_F = \infty$. (e) Find the Q-point in part (a) using the numerical iteration method if $I_S = 0.5$ fA and $V_T = 25.8$ mV. (f) Find the Q-point in part (c) using the numerical iteration method if $I_S = 0.5$ fA and $V_T = 25.8$ mV.

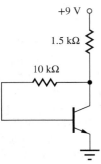

Figure P4.100

4.101. (a) Find the worst-case values of the Q-point in Prob. 4.100(a) if the resistor tolerances are 10 percent. (b) Repeat for Prob. 4.100(b). (c) Repeat for Prob. 4.100(c).

4.102. (a) Design the bias circuit in Fig. P4.102 to give a Q-point of $I_C = 10$ mA and $V_{EC} = 3$ V if the transistor current gain $\beta_F = 60$. (b) What is the Q-point if the current gain of the transistor is actually 40? (c) Find the worst-case values of the Q-point in part (a) if resistor tolerances are all 5 percent.

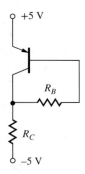

Figure P4.102

4.103. (a) Design the bias circuit in Fig. P4.103 to give a Q-point of $I_C = 20$ µA and $V_{CE} = 0.90$ V if the transistor current gain is $\beta_F = 50$ and $V_{BE} = 0.65$ V. (b) What is the Q-point if the current gain of the transistor is actually 125? (c) Find the worst-case values of the Q-point in part (a) if resistor tolerances are all 5 percent.

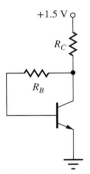

Figure P4.103

Bias Circuit Applications

4.104. The Zener diode in Fig. P4.104 has $V_Z = 6$ V and $R_Z = 100\ \Omega$. What is the output voltage if $I_L = 20$ mA? Use $I_S = 1 \times 10^{-16}$ A, $\beta_F = 50$, and $\beta_R = 0.5$ to find a precise answer.

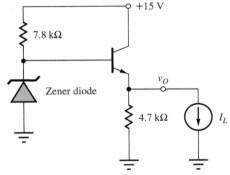

Figure P4.104

4.105. (a) Find the Line Regulation for the circuit in the previous problem (see Section 3.12.5). (b) Find the load regulation for the circuit in the previous problem.

*4.106. Create a model for the Zener diode and simulate the circuit in Prob. P4.105. Compare the SPICE results to your hand calculations. Use $I_S = 1 \times 10^{-16}$ A, $\beta_F = 50$, and $\beta_R = 0.5$.

**4.107. The circuit in Fig. P4.107 has $V_{EQ} = 7$ V and $R_{EQ} = 100\ \Omega$. What is the output resistance R_o of this circuit for $i_L = 20$ mA if R_o is defined as $R_o = -dv_O/di_L$? Assume $\beta_F = 50$.

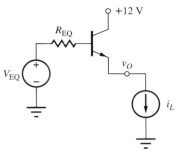

Figure P4.107

4.108. What is the output voltage v_O in Fig. P4.108 if the op-amp is ideal? What are the values of the base and emitter currents and the total current supplied by the 15-V source? Assume $\beta_F = 60$. What is the op-amp output voltage?

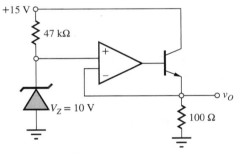

Figure P4.108

4.109. What is the output voltage v_O in Fig. P4.109 if the op-amp is ideal? What are the values of the base and emitter currents and the total current supplied by the 15-V source? Assume $\beta_F = 40$. What is the op-amp output voltage?

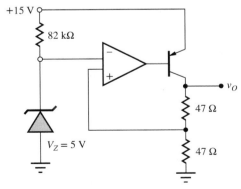

Figure P4.109

术语对照

Active region	有源区
Base	基极
Base current	基极电流
Base width	基区宽度
Base-collector capacitance	基极-集电区电容
Base-emitter capacitance	基极-发射区电容
Base-width modulation	基区宽度调制
β-cutoff frequency f_β	β 截止频率 f_β
Bipolar junction transistor(BJT)	双极型晶体管（BJT）
Collector	集电极
Collector current	集电极电流
Common-base output characteristic	共基极输出特性
Common-emitter output characteristic	共发射极输出特性
Common-emitter transfer characteristic	共发射极转移特性
Cutoff region	截止区
Dc bias	直流偏置
Diffusion capacitance	扩散电容
Early effect	Early 效应
Early voltage V_A	Early 电压 V_A
Ebers-Moll model	EM 模型
Emitter	发射极
Emitter current	发射极电流
Equilibrium electron density	平衡电子密度
Transfer characteristic	传输特性
Forward-active region	正向有源区
Forward common-emitter current gain β_F	正向共发射极电流增益 β_F
Forward common-base current gain	正向共基极电流增益
Forward transit time τ_F	正向传输时间 τ_F
Forward transport current	正向传输电流
Gummel-Poon model	GP 模型
Inverse-active region	反向有源区
Inverse common-emitter current gain	反向共发射极电流增益
Inverse common-base current gain	反向共基极电流增益

Monte Carlo analysis	蒙特卡洛分析
Normal- active region	正向有源区
Normal common-emitter current gain	共发射极电流增益
Normal common-base current gain	共基极电流增益
npn transistor	*npn*晶体管
Output characteristic	输出特性
pnp transistor	*pnp*晶体管
Quiescent operating point	静态工作点
Q-point	Q点
Reverse-active region	反向有源区
Reverse common-emitter current gain	反向共发射极电流增益
Reverse common-base current gain α_R	反向共基极电流增益 α_R
Reverse common-base current gain β_R	反向共发射极电流增益 β_R
Saturation region	饱和区
Saturation voltage	饱和电压
SPICE model parameters BF、IS、VAF	SPICE模型参数BF、IS、VAF
Transconductance	跨导
Transfer characteristic	转移特性
Transistor saturation current	晶体管饱和电流增益
Transport model	传输模型
Transistor Unity-gain frequency	电流增益频率
Unity-gain frequency f_T	单位增益频率 f_T
Worst-case analysis	最差情况分析

CHAPTER 5
FIELD-EFFECT TRANSISTORS

第5章 场效应晶体管

本章提纲
5.1 MOS电容特性
5.2 NMOS晶体管
5.3 PMOS晶体管
5.4 MOSFET电路符号
5.5 MOS晶体管对称性
5.6 CMOS技术
5.7 CMOS锁存器
5.8 MOS晶体管电容
5.9 SPICE中的MOSFET建模
5.10 MOS晶体管的等比例缩放
5.11 全区域建模
5.12 MOS晶体管的制造工艺及版图设计规则
5.13 先进CMOS技术
5.14 NMOS场效应晶体管的偏置
5.15 PMOS场效应晶体管的偏置
5.16 偏置CMOS反相器作为放大器
5.17 CMOS传输门
5.18 结型场效应晶体管（JFET）
5.19 JFET的SPICE模型
5.20 JFET和耗尽型MOSFET的偏置

本章目标
- 能定性研究MOS场效应晶体管的工作原理；
- 能定义并研究场效应管工作的截止区、线性区（三极管区）及饱和区；
- 能研究MOSFET中电流-电压（i-v）特性的数学模型；
- 了解电子器件的输出特性及传输特性；
- 能分类比较NMOS场效应晶体管和PMOS场效应晶体管增强和耗尽模型特性；
- 学会电路原理图中场效应管的电路符号；
- 学会晶体管工作在不同工作区的偏置电路；
- 学会MOS晶体管和电路的基本结构及掩模板版图设计；
- 学会MOS器件按比例缩小的概念；
- 了解三端器件和四端器件的特性；
- 理解MOSFET中电容的来源；
- 了解SPICE中的场效应管建模。

本章导读

  第一个MOSFET（金属氧化物半导体场效应晶体管）在20世纪50年代末制造成功，随后人们花费了将近十年的时间开发出一条可靠的MOS商业制造流程。由于PMOS场效应晶体管制作工艺较为简单，IC产业中首先使用的是PMOS场效应晶体管，第一台微处理器就是利用PMOS工艺制作出来的。20世纪60年代末，人们对制作工艺流程有了进一步的理解，并能更好地控制整个工艺流程，NMOS场效应晶体管开始大量使用，并迅速取代了PMOS场效应晶体管。由于NMOS场效应晶体管的迁移率大于PMOS场效应晶体管，使得NMOS器件有更好的电路性能。20世纪80年代中期，功耗成为一大难题，互补型MOS器件（CMOS，同时采用NMOS场效应晶体管和PMOS场效应晶体管）以其低功耗特点迅速占领市场，虽然CMOS器件存在制作复杂、成本较高的缺点，但其发展十分迅速，现在CMOS技术已经成为电子产业的主导技术。

  本章重点讨论了MOSFET和结型场效应晶体管（junction field-effect transistor，JFET）两种类型场效应晶体管（FET）的结构和i-v特性曲线。MOSFET是工业上应用最成功的固态器件，它是构成高集成度超大规模集成电路（包括微处理器和存储器）的基本单元。JFET电学基础是pn结，在模拟电路和射频电路设计中广泛应用。

  MOSFET的核心是MOS电容，该电容通过绝缘氧化层将金属栅电极和半导体隔离而形成。栅电势直接控制栅下半导体区域中的载流子浓度；MOS电容有三个工作区：积累区、耗尽区、反型区。MOSFET是在MOS电容的半导体区域中添加两个pn结形成的。这两个pn结分别作为MOS场效应晶体管的源端和漏端，为MOSFET的沟道区提供载流子。源结和漏结必须总是保持反偏，从而将沟道和衬底隔离开来。MOS场效应晶体管的沟道可以是n型或是p型，分别对应于NMOS场效应晶体管和PMOS场效应晶体管。此外，MOSFET还可以制作成增强型或者耗尽型。对于增强型器件，其栅-源电压必须要超出阈值电压，以保证源漏之间导电沟道的建立。对于耗尽型器件，在其制造过程中已内建沟道，对其施加栅压的作用是抑制导通。

  JFET利用pn结来控制导通沟道区的电阻。栅-源电压用来调制栅与沟道之间耗尽层的宽度，从而改变沟道区的宽度。JFET也分为n型沟道和p型沟道两种，根据其结构特点，JFET本质上是一种耗尽型器件，通常出现在运算放大器和射频电路设计等模拟电路的应用中。

  本章从定性和定量两个方面讨论MOSFET和JFET的i-v特性及数学模型，并研究不同类型晶体管的区别，随后介绍晶体管在不同工作区的偏置技术。MOSFET实际上是一个四端器件，其阈值电压与晶体管的源衬电压有关。JFET是一个夹断电压恒定的三端器件，其重要参数包括：饱和电流I_{DSS}、夹断电压V_P以及沟道长度调制系数λ。

  本章给出了多种偏置电路的例子，针对各种类型的MOSFET使用数学模型求解其静态工作点（Q点）。Q点表示漏电流和漏-源电压的直流值（I_D，V_{DS}）。晶体管的i-v特性通常以图形化的形式展现，包括表示i_D-v_{DS}关系的输出特性曲线和表示i_D-v_{GS}关系的传输特性曲线。本章讨论了用负载线法和迭代法求解Q点的例子。四电阻偏置是分立设计中最重要的偏置电路，能够提供稳定的工作点。

  本章还讨论了MOS管的栅极-源极、栅极-漏极、漏极-衬底、源极-衬底以及栅极-衬底电容，介绍了栅-源电容和栅-衬电容的Meyer模型。这里所有的电容都是晶体管端口电压的非线性函数。JFET的电容主要来源于反偏的栅极-沟道结，同样表现出与晶体管端电压非线性的关系。

  在SPICE电路分析程序中含有MOSFET和JFET的复杂模型。这些模型中包含了很多电路元件和参数，尽可能对晶体管的实际性能进行建模。

  本章最后介绍了恒电场等比例缩小理论，该理论为MOS器件合理最小化提供了基本框架。在该理论下晶体管密度增大，但功率密度保持不变，故电路的延迟将按照α缩小，功耗——延迟积按α的立方缩小。晶体管截止频率f_T表示晶体管能够提供放大功能的最高频率，其将按照α等比例缩放。在小尺寸器件中电场变得很高，当电场达到10kV/cm以上后，载流子速度趋于饱和。当器件缩小到很小的尺寸，亚阈值电流就逐渐重要起来。

  虽然双极型晶体管先于FET投入使用，但FET更易于理解，而且也是迄今为止实际生产中最重要的器件。

Chapter Outline

5.1 Characteristics of the MOS Capacitor
5.2 The NMOS Transistor
5.3 PMOS Transistors
5.4 MOSFET Circuit Symbols
5.5 MOS Transistor Symmetry
5.6 CMOS Technology
5.7 CMOS Latchup
5.8 Capacitances in MOS Transistors
5.9 MOSFET Modeling in SPICE
5.10 MOS Transistor Scaling
5.11 All Region Modeling
5.12 MOS Transistor Fabrication and Layout Design Rules
5.13 Advanced CMOS Technologies
5.14 Biasing the NMOS Field-Effect Transistor
5.15 Biasing the PMOS Field-Effect Transistor
5.16 Biasing the CMOS Inverter as an Amplifier
5.17 The CMOS Transmission Gate
5.18 The Junction Field-Effect Transistor (JFET)
5.19 JFET Modeling in SPICE
5.20 Biasing the JFET and Depletion-Mode MOSFET
Summary
Key Terms
References
Additional Readings
Problems

Chapter Goals

- Develop a qualitative understanding of the operation of the MOS field-effect transistor
- Define and explore FET characteristics in the cutoff, triode, and saturation regions of operation
- Develop mathematical models for the current-voltage (i-v) characteristics of MOSFETs and JFETs
- Discuss the output and transfer characteristic descriptions of MOS devices
- Explore the voltage transfer characteristics of CMOS inverters
- Explore layout of the CMOS inverter
- Understand the problem of "latchup" in CMOS technology
- Catalog and contrast the characteristics of both NMOS and PMOS enhancement-mode and depletion-mode FETs
- Explore complementary MOS (CMOS) technology and circuits
- Learn the symbols used to represent FETs in circuit schematics
- Investigate circuits used to bias the transistors into various regions of operation
- Learn the basic structure and mask layout for MOS transistors and circuits
- Explore the concept of MOS device scaling
- Contrast three- and four-terminal device behavior
- Understand sources of capacitance in MOSFETs
- Explore FET Modeling in SPICE

In this chapter we begin to explore the field-effect transistor or FET. The FET has emerged as the dominant device in modern integrated circuits and is present in the vast majority of semiconductor products produced today. The ability to dramatically shrink the size of the FET device has made possible handheld computational power unimagined just 20 years ago.

As noted in Chapter 1, various versions of the field-effect device were conceived by Lilienfeld in 1928, Heil in 1935, and Shockley in 1952, well before the technology to produce such devices existed. The first successful metal-oxide-semiconductor field-effect transistors, or MOSFETs, were fabricated in the late 1950s, but it took nearly a decade to develop reliable commercial fabrication processes for MOS devices. Because of fabrication-related difficulties, MOSFETs with a p-type conducting region, PMOS devices, were the first to be commercially available in IC form, and the first microprocessors were built using PMOS processes. By the late 1960s, understanding and control of fabrication processes had improved to the point that devices with an n-type conducting region, NMOS transistors, could be reliably fabricated in large numbers, and NMOS rapidly supplanted PMOS technology because the improved mobility of the NMOS device translated directly into higher circuit performance. By the mid-1980s, power had become a severe problem, and the low-power characteristics of Complementary MOS (CMOS) caused a rapid shift to that technology even though it was a more complex and costly process. Today CMOS technology, which utilizes both NMOS and PMOS transistors, is the dominant technology in the electronics industry.

An additional type of FET, the junction field-effect transistor or JFET, is based upon a pn junction structure and is typically found in analog applications including the design of op amps and RF circuits.

5.1 Characteristics of the MOS Capacitor

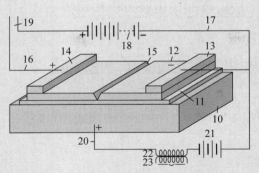

Drawing from the 1928 Lilienfeld patent [1]

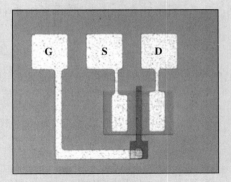

Top view of a simple MOSFET

Chapter 5 explores the characteristics of the **metal-oxide-semiconductor field-effect transistor (MOSFET)** that is without doubt the most commercially successful solid-state device. It is the primary component in high-density VLSI chips, including microprocessors and memories. A second type of FET, the **junction field-effect transistor (JFET)**, is based on a *pn* junction structure and finds application particularly in analog and RF circuit design.

P-**channel MOS (PMOS) transistors** were the first MOS devices to be successfully fabricated in large-scale integrated (LSI) circuits. Early microprocessor chips used PMOS technology. Greater performance was later obtained with the commercial introduction of ***n*-channel MOS (NMOS)** technology, using both enhancement-mode and ion-implanted depletion-mode devices.

This chapter discusses the qualitative and quantitative i-v behavior of FETs and investigates the differences between the various types of transistors. Techniques for biasing the transistors in various regions of operation are also presented.

Early integrated circuit chips contained only a few transistors, whereas today, the International Technology Roadmap for Semiconductors (ITRS [2]) projects the existence of chips with greater than 100 billion transistors by the year 2026! This phenomenal increase in transistor density has been the force behind the explosive growth of the electronics industry outlined in Chapter 1 that has been driven by our ability to reduce (scale) the dimensions of the transistor without compromising its operating characteristics.

5.1 CHARACTERISTICS OF THE MOS CAPACITOR

At the heart of the MOSFET is the **MOS capacitor** structure depicted in Fig. 5.1. Understanding the qualitative behavior of this capacitor provides a basis for understanding operation of the MOSFET. The MOS capacitor is used to induce charge at the interface between the semiconductor and oxide. The top electrode of the MOS capacitor is formed of a low-resistivity material, typically aluminum or heavily doped polysilicon (polycrystalline silicon). We refer to this electrode as the **gate (*G*)** for reasons that become apparent shortly. A thin insulating layer, typically silicon dioxide, isolates the gate from the substrate or body—the semiconductor region that acts as the second electrode of the capacitor. Silicon dioxide is a stable, high-quality electrical insulator readily formed by thermal oxidation of the silicon substrate. The ability to form this stable high-quality insulator is one of the basic reasons that silicon is the dominant semiconductor material today. The semiconductor region may be *n*- or *p*-type. A *p*-type substrate is depicted in Fig. 5.1.

Chapter 5 Field-Effect Transistors

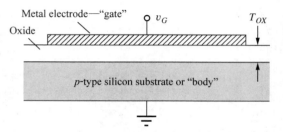

Figure 5.1 MOS capacitor structure on p-type silicon.

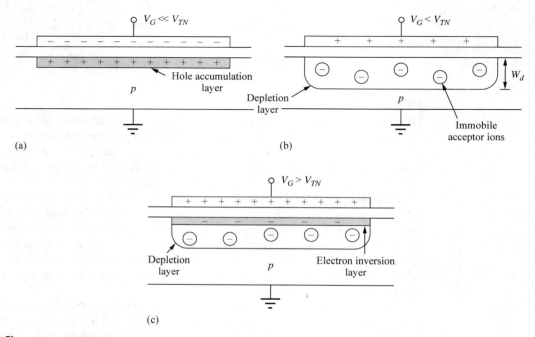

Figure 5.2 MOS capacitor operating in (a) accumulation, (b) depletion, and (c) inversion. Parameter V_{TN} in the figure is called the threshold voltage and represents the voltage required to just begin formation of the inversion layer.

The semiconductor forming the bottom electrode of the capacitor typically has a substantial resistivity and a limited supply of holes and electrons. Because the semiconductor can therefore be depleted of carriers, as discussed in Chapter 2, the capacitance of this structure is a nonlinear function of voltage. Figure 5.2 shows the conditions in the region of the substrate immediately below the gate electrode for three different bias conditions: accumulation, depletion, and inversion.

5.1.1 ACCUMULATION REGION

The situation for a large negative bias on the gate with respect to the substrate is indicated in Fig. 5.2(a). The large negative charge on the metallic gate is balanced by positively charged holes attracted to the silicon-silicon dioxide interface directly below the gate. For the bias condition shown, the hole density at the surface exceeds that which is present in the original p-type substrate, and the surface is said to be operating in the **accumulation region** or just in **accumulation.** This majority carrier accumulation layer is extremely shallow, effectively existing as a charge sheet directly below the gate.

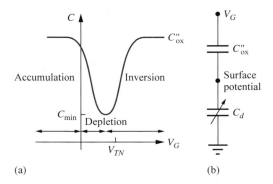

Figure 5.3 (a) Low frequency capacitance-voltage (C-V) characteristics for a MOS capacitor on a p-type substrate; (b) series capacitance model for the C-V characteristic.

5.1.2 DEPLETION REGION

Now consider the situation as the gate voltage is slowly increased. First, holes are repelled from the surface. Eventually, the hole density near the surface is reduced below the majority-carrier level set by the substrate doping level, as depicted in Fig. 5.2(b). This condition is called **depletion** and the region, the **depletion region.** The region beneath the metal electrode is depleted of free carriers in much the same way as the depletion region that exists near the metallurgical junction of the *pn* junction diode. In Fig. 5.2(b), positive charge on the gate electrode is balanced by the negative charge of the ionized acceptor atoms in the depletion layer. The depletion-region width w_d can range from a fraction of a micron to tens of microns, depending on the applied voltage and substrate doping levels.

5.1.3 INVERSION REGION

As the voltage on the top electrode increases further, electrons are attracted to the surface. At some particular voltage level, the electron density at the surface exceeds the hole density. At this voltage, the surface has inverted from the *p*-type polarity of the original substrate to an *n*-type **inversion layer,** or **inversion region,** directly underneath the top plate as indicated in Fig. 5.2(c). This inversion region is an extremely shallow layer, existing as a charge sheet directly below the gate. In the MOS capacitor, the high density of electrons in the inversion layer is supplied by the electron–hole generation process within the depletion layer.

The positive charge on the gate is balanced by the combination of negative charge in the inversion layer plus negative ionic acceptor charge in the depletion layer. The voltage at which the surface inversion layer just forms plays an extremely important role in field-effect transistors and is called the **threshold voltage** V_{TN}.

Figure 5.3 depicts the variation of the capacitance of the NMOS structure with gate voltage. At voltages well below threshold, the surface is in accumulation, corresponding to Fig. 5.2(a), and the capacitance is high and determined by the **oxide thickness.** This *gate oxide capacitance* per unit area C''_{ox} is the parallel plate capacitor formed between the gate and the accumulation layer with a silicon dioxide dielectric. As the gate voltage increases, the surface depletion layer forms as in Fig. 5.2(b), the effective separation of the capacitor plates increases, and the capacitance decreases. The total capacitance can be modeled as the series combination of the fixed oxide capacitance C''_{ox} and the voltage-dependent depletion-layer capacitance C_d, as in Fig. 5.3(b). The inversion layer forms at the surface as V_G exceeds threshold voltage V_{TN}, as in Fig. 5.2(c), and the capacitance rapidly increases back to the parallel plate capacitor value determined by the oxide layer thickness.

5.2 THE NMOS TRANSISTOR

A MOSFET is formed by adding two heavily doped *n*-type (n^+) diffusions to the cross section of Fig. 5.1, resulting in the structure in Fig. 5.4. The diffusions provide a supply of electrons that can

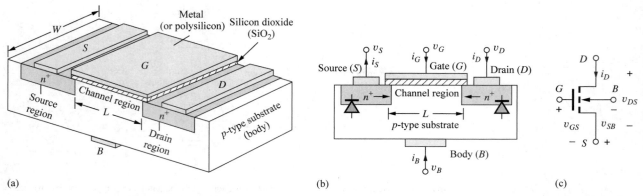

Figure 5.4 (a) NMOS transistor structure, (b) cross section, and (c) circuit symbol for the four-terminal NMOSFET.

readily move under the gate as well as terminals that can be used to apply a voltage and cause a current in the channel region of the transistor.

Figure 5.4 shows a three-dimensional view, cross section, and circuit symbol of an ***n*-channel MOSFET,** usually called an **NMOS transistor,** or **NMOSFET.** The central region of the NMOSFET is the MOS capacitor discussed in Sec. 5.1, and the top electrode of the capacitor is called the gate. The two heavily doped *n*-type regions (n^+ regions), called the **source** *S* and **drain** *D*, are formed in the *p*-type substrate and nearly aligned with the edge of the gate. The source and drain provide a supply of carriers so that the inversion layer can rapidly form in response to the gate voltage. The substrate of the NMOS transistor represents a fourth device terminal and is referred to synonymously as the **substrate terminal,** or the **body terminal** *B*.

The terminal voltages and currents for the NMOS device are defined in Fig. 5.4(b) and (c). Drain current i_D, source current i_S, gate current i_G, and body current i_B are all defined, with the positive direction of each current indicated for an NMOS transistor. The important terminal voltages are the gate-source voltage $v_{GS} = v_G - v_S$, the drain-source voltage $v_{DS} = v_D - v_S$, and the source-bulk voltage $v_{SB} = v_S - v_B$. These voltages are all positive during normal operation of the NMOSFET.

Note that the source and drain regions form *pn* junctions with the substrate. These two junctions are kept reverse-biased at all times to provide isolation between the junctions and the substrate as well as between adjacent MOS transistors. Thus, the bulk voltage must be less than or equal to the voltages applied to the source and drain terminals to ensure that these *pn* junctions are properly reverse-biased.

The semiconductor region between the source and drain regions directly below the gate is called the **channel region** of the FET, and two dimensions of critical import are defined in Fig. 5.4. *L* represents the **channel length,** which is measured in the direction of current in the channel. *W* is the **channel width,** which is measured perpendicular to the direction of current. In this and later chapters we will find that choosing the values for *W* and *L* is an important aspect of the digital and analog IC designer's task.

5.2.1 QUALITATIVE *i-v* BEHAVIOR OF THE NMOS TRANSISTOR

Before attempting to derive an expression for the current-voltage characteristic of the NMOS transistor, let us try to develop a qualitative understanding of what we might expect by referring to Fig. 5.5. In the figure, the source, drain, and body of the NMOSFET are all grounded.

For a dc gate-source voltage, $v_{GS} = V_{GS}$, well below threshold voltage V_{TN}, as in Fig. 5.5(a), back-to-back *pn* junctions exist between the source and drain, and only a small leakage current can flow between these two terminals. For V_{GS} near but still below threshold, a depletion region forms beneath the gate and merges with the depletion regions of the source and drain, as indicated in Fig. 5.5(b). The depletion region is devoid of free carriers, so a current still does not appear between the source and drain. Finally, when the gate-channel voltage exceeds the threshold voltage V_{TN}, as

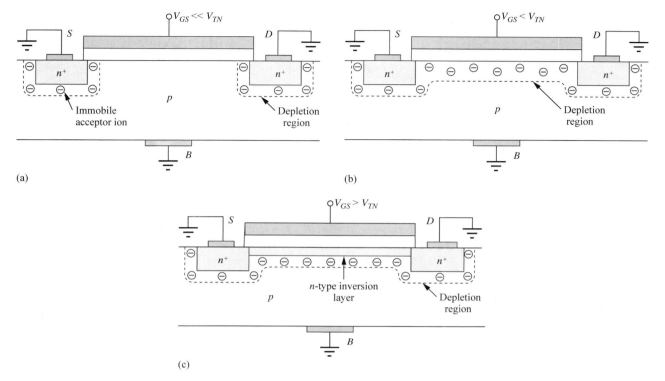

Figure 5.5 (a) $V_{GS} \ll V_{TN}$; (b) $V_{GS} < V_{TN}$; (c) $V_{GS} > V_{TN}$.

in Fig. 5.5(c), electrons flow in from the source and drain to form an inversion layer that connects the n^+ source region to the n^+ drain. A resistive connection, the channel, now exists between the source and drain terminals.

If a positive voltage is applied between the drain and source terminals, electrons in the channel inversion layer will drift in the electric field, creating a current in the terminals. Positive current in the NMOS transistor enters the drain terminal, travels down the channel, and exits the source terminal, as indicated by the polarities in Fig. 5.4(b). The gate terminal is insulated from the channel; thus, there is no dc gate current, and $i_G = 0$. The drain-bulk and source-bulk (and induced channel-to-bulk) pn junctions must be reverse-biased at all times to ensure that only a small reverse-bias leakage current exists in these diodes. This current is usually negligible with respect to the channel current i_D and is neglected. Thus we assume that $i_B = 0$.

In the device in Fig. 5.5, a channel must be induced by the applied gate voltage for conduction to occur. The gate voltage "enhances" the conductivity of the channel; this type of MOSFET is termed an **enhancement-mode device.** Later in this chapter we identify an additional type of MOSFET called a **depletion-mode device.** In Sec. 5.2.2, we develop a mathematical model for the current in the terminals of the NMOS device in terms of the applied voltages.

5.2.2 TRIODE[1] REGION CHARACTERISTICS OF THE NMOS TRANSISTOR

An expression for the drain current i_D can be developed by considering the transport of charge in the channel in Fig. 5.6, which is depicted for a small value of v_{DS}. We saw in Sec. 5.2.1 that both i_G and i_B are zero. Therefore, the current entering the drain in Fig. 5.6 must be equal to the current

---

[1] This region of operation is also referred to as the "linear region." We will use triode region to avoid confusion with the concept of linear amplification introduced later in the text.

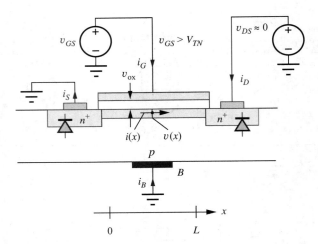

Figure 5.6 Model for determining i-v characteristics of the NMOS transistor.

leaving the source:

$$i_S = i_D \tag{5.1}$$

The electron charge per unit length (a line charge — C/cm) at any point in the channel is given by

$$Q' = -WC_{ox}''(v_{ox} - V_{TN}) \qquad \text{C/cm for } v_{ox} \geq V_{TN} \tag{5.2}$$

where $C_{ox}'' = \varepsilon_{ox}/T_{ox}$, the oxide capacitance per unit area (F/cm²)

ε_{ox} = oxide permittivity (F/cm) T_{ox} = oxide thickness (cm)

For silicon dioxide, $\varepsilon_{ox} = 3.9\varepsilon_o$, where $\varepsilon_o = 8.854 \times 10^{-14}$ F/cm.

The voltage v_{ox} represents the voltage across the oxide and will be a function of position in the channel:

$$v_{ox} = v_{GS} - v(x) \tag{5.3}$$

where $v(x)$ is the voltage at any point x in the channel referred to the source. Note that v_{ox} must exceed V_{TN} for an inversion layer to exist, so Q' will be zero until $v_{ox} > V_{TN}$. At the source end of the channel, $v_{ox} = v_{GS}$, and it decreases to $v_{ox} = v_{GS} - v_{DS}$ at the drain end of the channel.

The electron drift current at any point in the channel is given by the product of the charge per unit length times the velocity v_x:

$$i(x) = Q'(x)v_x(x) \tag{5.4}$$

The charge Q' is represented by Eq. (5.2), and the velocity v_x of electrons in the channel is determined by the electron mobility and the transverse electric field in the channel:

$$i(x) = Q'v_x = [-WC_{ox}''(v_{ox} - V_{TN})](-\mu_n E_x) \tag{5.5}$$

The transverse field is equal to the negative of the spatial derivative of the voltage in the channel

$$E_x = -\frac{dv(x)}{dx} \tag{5.6}$$

Combining Eqs. (5.3) to (5.6) yields an expression for the current at any point in the channel:

$$i(x) = -\mu_n C_{ox}'' W[v_{GS} - v(x) - V_{TN}]\frac{dv(x)}{dx} \tag{5.7}$$

We know the voltages applied to the device terminals are $v(0) = 0$ and $v(L) = v_{DS}$, and we can integrate Eq. (5.7) between 0 and L:

$$\int_0^L i(x)\,dx = -\int_0^{v_{DS}} \mu_n C''_{ox} W [v_{GS} - v(x) - V_{TN}]\,dv(x) \tag{5.8}$$

Because there is no mechanism to lose current as it goes down the channel, the current must be equal to the same value i_D at every point x in the channel, $i(x) = i_D$, and Eq. (5.8) finally yields

$$i_D = \mu_n C''_{ox} \frac{W}{L}\left(v_{GS} - V_{TN} - \frac{v_{DS}}{2}\right) v_{DS} \tag{5.9}$$

The value of $\mu_n C''_{ox}$ is fixed for a given technology and cannot be changed by the circuit designer. For circuit analysis and design purposes, Eq. (5.9) is therefore most often written as

$$i_D = K'_n \frac{W}{L}\left(v_{GS} - V_{TN} - \frac{v_{DS}}{2}\right) v_{DS} \quad \text{or just} \quad i_D = K_n \left(v_{GS} - V_{TN} - \frac{v_{DS}}{2}\right) v_{DS} \tag{5.10}$$

where $K_n = K'_n W/L$ and $K'_n = \mu_n C''_{ox}$. Parameters K_n and K'_n are called **transconductance parameters** and both have units of A/V^2.

Equation (5.10) represents the classic expression for the drain-source current for the NMOS transistor in its **linear region** or **triode region** of operation, in which a resistive channel directly connects the source and drain. This resistive connection will exist as long as the voltage across the oxide exceeds the threshold voltage at every point in the channel:

$$v_{GS} - v(x) \geq V_{TN} \qquad \text{for } 0 \leq x \leq L \tag{5.11}$$

The voltage in the channel is maximum at the drain end where $v(L) = v_{DS}$. Thus, Eqs. (5.9) and (5.10) are valid as long as

$$v_{GS} - v_{DS} \geq V_{TN} \qquad \text{or} \qquad v_{GS} - V_{TN} \geq v_{DS} \tag{5.12}$$

Recapitulating for the triode region,

$$i_D = K'_n \frac{W}{L}\left(v_{GS} - V_{TN} - \frac{v_{DS}}{2}\right) v_{DS} \quad \text{for} \quad v_{GS} - V_{TN} \geq v_{DS} \geq 0 \quad \text{and} \quad K'_n = \mu_n C''_{ox} \tag{5.13}$$

Equation (5.13) is used frequently in the rest of this text. Commit it to memory!

Some additional insight into the mathematical model can be gained by regrouping the terms in Eq. (5.13):

$$i_D = \left[C''_{ox} W \left(v_{GS} - V_{TN} - \frac{v_{DS}}{2}\right)\right]\left(\mu_n \frac{v_{DS}}{L}\right) \tag{5.14}$$

For small drain-source voltages, the first term in brackets represents the average charge per unit length in the channel because the average channel voltage $\overline{v(x)} = v_{DS}/2$. The second term represents the drift velocity in the channel, where the average electric field is equal to the total voltage v_{DS} across the channel divided by the channel length L.

We should note that the term *triode region* is used because the drain current of the FET depends on the drain voltage of the transistor, and this behavior is similar to that of the electronic vacuum triode that appeared many decades earlier (see Table 1.2 — Milestones in Electronics).

Note also that the **Quiescent operating point** or **Q-point** for the FET is typically expressed as (I_D, V_{DS}) although it can be any two of the three values I_D, V_{DS}, or V_{GS}.

> **EXERCISE:** Calculate K'_n for a transistor with $\mu_n = 500$ cm²/v·s and $T_{ox} = 25$ nm. Repeat for $T_{ox} = 5$ nm.
>
> **ANSWERS:** 69.1 µA/V²; 345 µA/V²
>
> **EXERCISE:** An NMOS transistor has $K'_n = 50$ µA/V². What is the value of K_n if $W = 20$ µm, $L = 1$ µm? If $W = 60$ µm, $L = 3$ µm? If $W = 10$ µm, $L = 0.25$ µm?
>
> **ANSWERS:** 1000 µA/V²; 1000 µA/V²; 2000 µA/V²
>
> **EXERCISE:** Calculate the drain current in an NMOS transistor for $V_{GS} = 0, 1$ V, 2 V, and 3 V, with $V_{DS} = 0.1$ V, if $W = 10$ µm, $L = 1$ µm, $V_{TN} = 1.5$ V, and $K'_n = 25$ µA/V². What is the value of K_n?
>
> **ANSWERS:** 0; 0; 11.3 µA; 36.3 µA; 250 µA/V²

5.2.3 ON RESISTANCE

The i-v characteristics in the triode region generated from Eq. (5.13) are drawn in Fig. 5.7 for the case of $V_{TN} = 1$ V and $K_n = 250$ µA/V². The curves in Fig. 5.7 represent a portion of the common-source **output characteristics** for the NMOS device. The output characteristics for the MOSFET are graphs of drain current i_D as a function of drain-source voltage v_{DS}. A family of curves is generated, with each curve corresponding to a different value of gate-source voltage v_{GS}. The output characteristics in Fig. 5.7 appear to be a family of nearly straight lines, hence the alternate name **linear region** (of operation). However, some curvature can be noted in the characteristics, particularly for $V_{GS} = 2$ V.

Let us explore the triode region behavior in more detail using Eq. (5.9). For small drain-source voltages such that $v_{DS}/2 \ll v_{GS} - V_{TN}$, Eq. (5.9) can be reduced to

$$i_D \cong \mu_n C''_{ox} \frac{W}{L}(v_{GS} - V_{TN})v_{DS} \tag{5.15}$$

in which the current i_D through the MOSFET is directly proportional to the voltage v_{DS} across the MOSFET. The FET behaves much like a resistor connected between the drain and source terminals, but the resistor value can be controlled by the gate-source voltage. It has been said that this voltage-controlled resistance behavior originally gave rise to the name transistor, a contraction of "transfer-resistor."

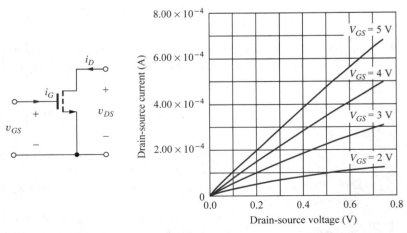

Figure 5.7 NMOS i-v characteristics in the triode region ($V_{SB} = 0$). A three-terminal NMOS circuit symbol is often used when $v_{SB} = 0$.

The resistance of the FET in the triode region near the origin, called the **on-resistance** R_{on}, is defined in Eq. (5.16) and can be found by taking the derivative of Eq. (5.13):

$$R_{on} = \left[\frac{\partial i_D}{\partial v_{DS}}\bigg|_{v_{DS}\to 0}\right]^{-1}_{Q\text{-}pt} = \frac{1}{K'_n \frac{W}{L}(V_{GS} - V_{TN} - V_{DS})}\bigg|_{V_{DS}\to 0} = \frac{1}{K'_n \frac{W}{L}(V_{GS} - V_{TN})} \quad (5.16)$$

We will find that the value of R_{on} plays a very important role in the operation of MOS logic circuits in Chapters 6 to 8. Note that R_{on} is also equal to the ratio v_{DS}/i_D from Eq. (5.15).

Near the origin, the i-v curves are indeed straight lines. However, curvature develops as the assumption $v_{DS} \ll v_{GS} - V_{TN}$ starts to be violated. For the lowest curve in Fig. 5.7, $V_{GS} - V_{TN} = 2 - 1 = 1$ V, and we should expect linear behavior only for values of v_{DS} below 0.1 to 0.2 V. On the other hand, the curve for $V_{GS} = 5$ V exhibits quasi-linear behavior throughout most of the range of Fig. 5.7. Note that a three-terminal NMOS circuit symbol is often used (see Figs. 5.7 and 5.8) when the bulk terminal is connected to the source terminal forcing $v_{SB} = 0$.

> **EXERCISE:** Calculate the on-resistance of an NMOS transistor for $V_{GS} = 2$ V and $V_{GS} = 5$ V if $V_{TN} = 1$ V and $K_n = 250$ μA/V². What value of V_{GS} is required for an on-resistance of 2 kΩ?
>
> **ANSWERS:** 4 kΩ; 1 kΩ; 3v

5.2.4 TRANSCONDUCTANCE

An important characteristic of transistors is the **transconductance** given the symbol g_m. The transconductance of the MOS devices relates the change in drain current to a change in gate-source voltage. For the linear region:

$$g_m = \frac{i_D}{v_{GS}}\bigg|_{Q\text{-}pt} = K_n V_{DS} = \frac{I_D}{V_{GS} - V_{TN} - V_{DS}/2} \quad (5.17)$$

where we have taken the derivative of Eq. (5.15) and evaluated the result at the Q-point. We encounter g_m frequently in electronics, particularly during our study of analog circuit design. The larger the

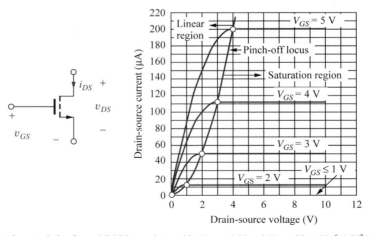

Figure 5.8 Output characteristics for an NMOS transistor with $V_{TN} = 1$ V and $K_n = 25 \times 10^{-6}$ A/V² ($v_{SB} = 0$). A three-terminal NMOS circuit symbol is used when $v_{SB} = 0$.

device transconductance, the more gain we can expect from an amplifier that utilizes the transistor. It is interesting to note that g_m is the reciprocal of the on-resistance defined in Eq. (5.16).

> **EXERCISE:** Find the drain current and transconductance for an NMOS transistor operating with $V_{GS} = 2.5$ V, $V_{TN} = 1$ V, $V_{DS} = 0.28$ V, and $K_n = 1$ mA/V².
>
> **ANSWERS:** 0.344 mA; 0.250 mS

5.2.5 SATURATION OF THE *i-v* CHARACTERISTICS

As discussed, Eq. (5.13) is valid as long as the resistive channel region directly connects the source to the drain. However, an unexpected phenomenon occurs in the MOSFET as the drain voltage increases above the triode region limit in Eq. (5.13). The current does not continue to increase, but instead saturates at an almost constant value. This unusual behavior is depicted in the *i-v* characteristics in Fig. 5.8 for several fixed gate-source voltages.

We can try to understand the origin of the current saturation by studying the device cross sections in Fig. 5.9. In Fig. 5.9(a), the MOSFET is operating in the triode region with $v_{DS} < v_{GS} - V_{TN}$, as discussed previously. In Fig. 5.9(b), the value of v_{DS} has increased to $v_{DS} = v_{GS} - V_{TN}$, for which the channel just disappears at the drain. Figure 5.9(c) shows the channel for an even larger value of v_{DS}. The channel region has disappeared, or *pinched off*, before reaching the drain end of the channel, and the resistive channel region is no longer in contact with the drain. At first glance, one may be inclined to expect that the current should become zero in the MOSFET; however, this is not the case.

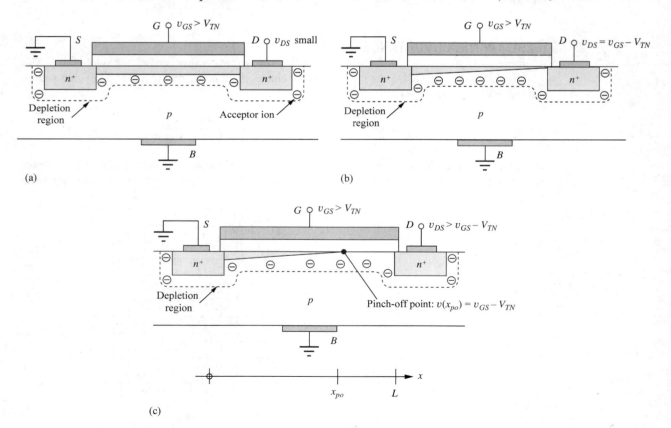

Figure 5.9 (a) MOSFET in the linear region; (b) MOSFET with channel just pinched off at the drain; (c) channel pinch-off for $v_{DS} > v_{GS} - V_{TN}$.

As depicted in Fig. 5.9(c), the voltage at the **pinch-off point** in the channel is always equal to

$$v_{GS} - v(x_{po}) = V_{TN} \quad \text{or} \quad v(x_{po}) = v_{GS} - V_{TN}$$

There is still a voltage equal to $v_{GS} - V_{TN}$ across the inverted portion of the channel, and electrons will be drifting down the channel from left to right. When the electrons reach the pinch-off point, they are injected into the depleted region between the end of the channel and the drain, and the electric field in the depletion region then sweeps these electrons on to the drain. Once the channel has reached pinch-off, the voltage drop across the inverted channel region is constant; hence, the drain current becomes constant and independent of drain-source voltage.

This region of operation of the MOSFET is often referred to as either the **saturation region** or the **pinch-off region** of operation. However, we will learn a different meaning for saturation when we discuss bipolar transistors in the next chapter. On the other hand, operation beyond pinchoff is the regime that we most often use for analog amplification, and in Part III we will use the term **active region** to refer to this region for both MOS and bipolar devices.

5.2.6 MATHEMATICAL MODEL IN THE SATURATION (PINCH-OFF) REGION

Now let us find an expression for the MOSFET drain current in the pinched-off channel. The drain-source voltage just needed to pinch off the channel at the drain is $v_{DS} = v_{GS} - V_{TN}$, and substituting this value into Eq. (5.13) yields an expression for the NMOS current in the saturation region of operation:

$$i_D = \frac{K'_n}{2} \frac{W}{L} (v_{GS} - V_{TN})^2 \quad \text{for } v_{DS} \geq (v_{GS} - V_{TN}) \geq 0 \qquad (5.18)$$

This is the classic square-law expression for the drain-source current for the n-channel MOSFET operating in pinch-off. The current depends on the square of $v_{GS} - V_{TN}$ but is now independent of the drain-source voltage v_{DS}. Equation (5.18) is *also used* frequently in the rest of this text. Be sure to commit it to memory!

The value of v_{DS} for which the transistor saturates is given the special name v_{DSAT} defined by

$$v_{DSAT} = v_{GS} - V_{TN} \qquad (5.19)$$

and v_{DSAT} is referred to as the **saturation voltage**, or **pinch-off voltage**, of the MOSFET. Equation (5.18) can be interpreted in a manner similar to that of Eq. (5.14):

$$i_D = \left(C''_{ox} W \frac{v_{GS} - V_{TN}}{2} \right) \left(\mu_n \frac{v_{GS} - V_{TN}}{L} \right) \qquad (5.20)$$

The inverted channel region has a voltage of $v_{GS} - V_{TN}$ across it, as depicted in Fig. 5.9(c). Thus, the first term represents the magnitude of the average electron charge in the inversion layer, and the second term is the magnitude of the velocity of electrons in an electric field equal to $(v_{GS} - V_{TN})/L$.

An example of the overall output characteristics for an NMOS transistor with $V_{TN} = 1$ V and $K_n = 25$ μA/V^2 appeared in Fig. 5.8, in which the locus of pinch-off points is determined by $v_{DS} = v_{DSAT}$. To the left of the **pinch-off locus**, the transistor is operating in the triode region, and it is operating in the saturation region for operating points to the right of the locus. For $v_{GS} \leq V_{TN} = 1$ V, the transistor is cut off, and the drain current is zero. As the gate voltage is increased in the saturation region, the curves spread out due to the square-law nature of Eq. (5.18).

Figure 5.10 gives an individual output characteristic for $V_{GS} = 3$ V, showing the behavior of the individual triode and saturation region equations. The triode region expression given in Eq. (5.13) is represented by the inverted parabola in Fig. 5.10. Note that it does not represent a valid model for the i-v behavior for $V_{DS} > V_{GS} - V_{TN} = 2$ V for this particular device. Note also that the maximum drain voltage must never exceed the Zener breakdown voltage of the drain-substrate pn junction diode.

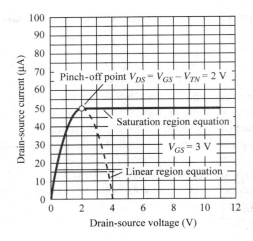

Figure 5.10 Output characteristic showing intersection of the triode (linear) region and saturation region equations at the pinch-off point.

EXERCISE: Calculate the drain current for an NMOS transistor operating with $V_{GS} = 5$ V and $V_{DS} = 10$ V if $V_{TN} = 1$ V and $K_n = 1$ mA/V². What is the W/L ratio of this device if $K'_n = 40$ μA/V²? What is W if $L = 0.35$ μm?

ANSWERS: 8.00 mA; 25/1; 8.75 μm

5.2.7 TRANSCONDUCTANCE IN SATURATION

Transconductance g_m was defined in sec. 5.2.4 and relates the change in drain current to a change in gate-source voltage. For the saturation region:

$$g_m = \left. \frac{di_D}{dv_{GS}} \right|_{Q\text{-}pt} = K_n(V_{GS} - V_{TN}) = \frac{2I_D}{V_{GS} - V_{TN}} \qquad (5.21)$$

where we have taken the derivative of Eq. (5.18) and evaluated the result at the Q-point. The larger the device transconductance, the more gain we can expect from an amplifier that utilizes the transistor. It is interesting to note that the value of g_m in saturation is approximately twice that in the linear region.

EXERCISE: Find the drain current and transconductance for an NMOS transistor operating with $V_{GS} = 2.5$ V, $V_{TN} = 1$ V, $V_{DS} = 0.28$ V and $K_n = 1$ mA/V².

ANSWERS: 1.13 mA; 1.5 mS

5.2.8 CHANNEL-LENGTH MODULATION

The output characteristics of the device in Fig. 5.8 indicate that the drain current is constant once the device enters the saturation region of operation. However, this is not quite true. Rather, the $i\text{-}v$ curves have a small positive slope, as indicated in Fig. 5.11(a). The drain current increases slightly as the drain-source voltage increases. The increase in drain current visible in Fig. 5.11 is the result of a phenomenon called **channel-length modulation** that can be understood by referring to Fig. 5.11(b), in which the channel region of the NMOS transistor is depicted for the case of $v_{DS} > v_{DSAT}$. The channel pinches off before it makes contact with the drain. Thus, the actual length of the resistive channel is given by $L = L_M - \Delta L$, where L_M is the spacing between the edges of the source and drain diffusions. As v_{DS} increases above v_{DSAT}, the length of the depleted channel region ΔL also increases, and the effective value of L decreases. Therefore, the value of L in the denominator of Eq. (5.18) actually has a slight inverse dependence on v_{DS}, leading to an increase in drain current increases as v_{DS} increases. The expression in Eq. (5.18) can be heuristically modified to include this drain-voltage dependence as

$$i_D = \frac{K'_n}{2} \frac{W}{L} (v_{GS} - V_{TN})^2 (1 + \lambda v_{DS}) \tag{5.22}$$

in which λ is called the **channel-length modulation parameter.** The value of λ is dependent on the channel length, and typical values are $0 \text{ V}^{-1} \leq \lambda \leq 0.2 \text{ V}^{-1}$. In Fig. 5.11, λ is approximately 0.01 V^{-1}, which yields a 10 percent increase in drain current for a drain-source voltage change of 10 V.

> **EXERCISE:** Calculate the drain current for an NMOS transistor operating with $V_{GS} = 5$ V and $V_{DS} = 10$ V if $V_{TN} = 1$ V, $K_n = 1$ mA/V^2, and $\lambda = 0.02$ V^{-1}. What is I_D for $\lambda = 0$?
>
> **ANSWERS:** 9.60 mA; 8.00 mA
>
> **EXERCISE:** Calculate the drain current for the NMOS transistor in Fig. 4.11 operating with $V_{GS} = 4$ V and $V_{DS} = 5$ V if $V_{TN} = 1$ V, $K_n = 25$ μA/V^2, and $\lambda = 0.01$ V^{-1}. Repeat for $V_{GS} = 5$ V and $V_{DS} = 10$ V.
>
> **ANSWERS:** 118 μA; 220 μA

5.2.9 TRANSFER CHARACTERISTICS AND DEPLETION-MODE MOSFETS

The output characteristics in Figs. 5.7, 5.8, and 5.11 represent our first look at graphical representations of the i-v characteristics of the transistor. The output characteristics plot drain current versus drain-source voltage for fixed values of the gate-source voltage. The second commonly used graphical format, called the **transfer characteristic,** plots drain current versus gate-source voltage for a fixed drain-source voltage. An example of this form of characteristic is given in Fig. 5.12 for two

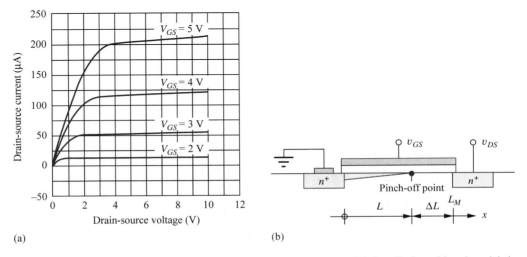

Figure 5.11 (a) Output characteristics including the effects of channel-length modulation; (b) channel-length modulation.

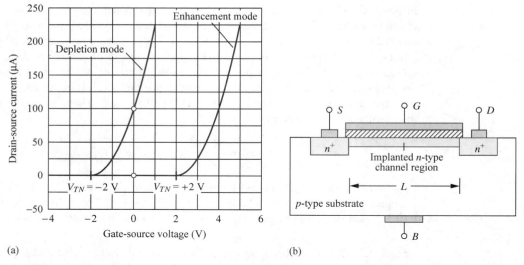

Figure 5.12 (a) Transfer characteristics for enhancement-mode and depletion-mode NMOS transistors operating in the saturation region; (b) cross section of a depletion-mode NMOS transistor.

NMOS transistors in the pinch-off region. Up to now, we have been assuming that the threshold voltage of the NMOS transistor is positive, as in the right-hand curve in Fig. 5.12. This curve corresponds to an enhancement-mode device with $V_{TN} = +2$ V. Here we can clearly see the turn-on of the transistor as v_{GS} increases. The device is off (nonconducting) for $v_{GS} \le V_{TN}$, and it starts to conduct as v_{GS} exceeds V_{TN}. The curvature reflects the square-law behavior of the transistor in the saturation region as described by Eq. (5.18).

However, it is also possible to fabricate NMOS transistors with values of $V_{TN} \le 0$. These transistors are called **depletion-mode MOSFETs,** and the transfer characteristic for such a device with $V_{TN} = -2$ V is depicted in the left-hand curve in Fig. 5.12(a). Note that a nonzero drain current exists in the depletion-mode MOSFET for $v_{GS} = 0$; a negative value of v_{GS} is required to turn the device off.

The cross section of the structure of a depletion-mode NMOSFET is shown in Fig. 5.12(b). A process called *ion implantation* is used to form a built-in n-type channel in the device so that the source and drain are connected through the resistive channel region. A negative voltage must be applied to the gate to deplete the n-type channel region and eliminate the current path between the source and drain (hence the name depletion-mode device). In Chapter 6 we will see that the ion-implanted depletion-mode device played an important role in the evolution of MOS logic circuits. The addition of the depletion-mode MOSFET to NMOS technology provided substantial performance improvement, and it was a rapidly accepted change in technology in the mid-1970s.

EXERCISE: Calculate the drain current for the NMOS depletion-mode transistor in Fig. 4.12 for $V_{GS} = 0$ V if $K_n = 50$ μA/V². Assume the transistor is in the pinch-off region. What value of V_{GS} is required to achieve the same current in the enhancement-mode transistor in the same figure?

ANSWERS: 100 μA; 4 V

EXERCISE: Calculate the drain current for the NMOS depletion-mode transistor in Fig. 4.12 for $V_{GS} = +1$ V if $K_n = 50$ μA/V². Assume the transistor is in the pinch-off region.

ANSWER: 225 μA

5.2.10 BODY EFFECT OR SUBSTRATE SENSITIVITY

Thus far, it has been assumed that the source-bulk voltage v_{SB} is zero. With $v_{SB} = 0$, the MOSFET behaves as if it were a three-terminal device. However, we find many circuits, particularly in ICs, in which the bulk and source of the MOSFET must be connected to different voltages so that $v_{SB} \neq 0$. A nonzero value of v_{SB} affects the i-v characteristics of the MOSFET by changing the value of the threshold voltage. This effect is called **substrate sensitivity,** or **body effect,** and can be modeled by

$$V_{TN} = V_{TO} + \gamma \left(\sqrt{v_{SB} + 2\phi_F} - \sqrt{2\phi_F} \right) \tag{5.23}$$

where V_{TO} = **zero-substrate-bias value for** V_{TN} (V)

γ = **body-effect parameter** ($\sqrt{V}$)

$2\phi_F$ = **surface potential parameter** (V)

Parameter γ determines the intensity of the body effect, and its value is set by the relative sizes of the oxide and depletion-layer capacitances C''_{ox} and C_d in Fig. 5.3. The surface potential represents the approximate voltage across the depletion layer at the onset of inversion. For typical NMOS transistors, $-5 \text{ V} \leq V_{TO} \leq +5 \text{ V}$, $0 \leq \gamma \leq 3\sqrt{V}$, and $0.3 \text{ V} \leq 2\phi_F \leq 1 \text{ V}$.

We use $2\phi_F = 0.6$ V throughout the rest of this text, and Eq. (5.23) will be represented as

$$V_{TN} = V_{TO} + \gamma \left(\sqrt{v_{SB} + 0.6} - \sqrt{0.6} \right) \tag{5.24}$$

Figure 5.13 plots an example of the threshold-voltage variation with source-bulk voltage for an NMOS transistor, with $V_{TO} = 1$ V and $\gamma = 0.75\sqrt{V}$. We see that $V_{TN} = V_{TO} = 1$ V for $v_{SB} = 0$ V, but the value of V_{TN} more than doubles for $v_{SB} = 5$ V. In Chapter 6, we will see that this behavior can have a significant impact on the design of MOS logic circuits.

DESIGN NOTE

The mathematical model for the NMOS transistor in its various regions of operation is summarized in the equation set below and should be committed to memory!

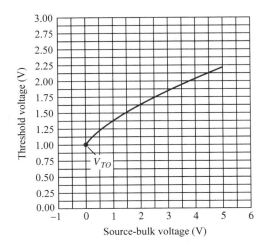

Figure 5.13 Threshold variation with source-bulk voltage for an NMOS transistor, with $V_{TO} = 1$ V, $2\phi_F = 0.6$ V, and $\gamma = 0.75\sqrt{V}$.

NMOS TRANSISTOR MATHEMATICAL MODEL SUMMARY

Equations (5.25) through (5.29) represent the complete model for the i-v behavior of the NMOS transistor.

For all regions,

$$K_n = K'_n \frac{W}{L} \qquad K'_n = \mu_n C''_{ox} \qquad i_G = 0 \qquad i_B = 0 \qquad (5.25)$$

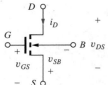

NMOS transistor

Cutoff region:
$$i_D = 0 \qquad \text{for } v_{GS} \leq V_{TN} \qquad (5.26)$$

Triode region:
$$i_D = K_n \left(v_{GS} - V_{TN} - \frac{v_{DS}}{2} \right) v_{DS} \qquad \text{for } v_{GS} - V_{TN} \geq v_{DS} \geq 0 \qquad (5.27)$$

Saturation region:
$$i_D = \frac{K_n}{2}(v_{GS} - V_{TN})^2 (1 + \lambda v_{DS}) \qquad \text{for } v_{DS} \geq (v_{GS} - V_{TN}) \geq 0 \qquad (5.28)$$

Threshold voltage:
$$V_{TN} = V_{TO} + \gamma \left(\sqrt{v_{SB} + 2\phi_F} - \sqrt{2\phi_F} \right) \qquad (5.29)$$

$V_{TN} > 0$ for enhancement-mode NMOS transistors. Depletion-mode NMOS devices can also be fabricated, and $V_{TN} \leq 0$ for these transistors.

EXERCISE: Calculate the threshold voltage for the MOSFET of Fig. 5.13 for source-bulk voltages of 0 V, 1.5 V, and 3 V.

ANSWERS: 1.00 V; 1.51 V; 1.84 V

EXERCISE: What is the region of operation and drain current of an NMOS transistor having $V_{TN} = 1$ V, $K_n = 1$ mA/V^2, and $\lambda = 0.02$ V^{-1} for (a) $V_{GS} = 0$ V, $V_{DS} = 1$ V; (b) $V_{GS} = 2$ V, $V_{DS} = 0.5$ V; (c) $V_{GS} = 2$ V, $V_{DS} = 2$ V?

ANSWERS: (a) cutoff, 0 A; (b) triode, 375 μA; (c) saturation, 520 μA

5.3 PMOS TRANSISTORS

MOS transistors with p-type channels (PMOS transistors) can also easily be fabricated. In fact, as mentioned earlier, the first commercial MOS transistors and integrated circuits used PMOS devices because it was easier to control the fabrication process for PMOS technology. The PMOS device is built by forming p-type source and drain regions in an n-type substrate, as depicted in the device cross section in Fig. 5.14(a).

The qualitative behavior of the transistor is essentially the same as that of an NMOS device except that the normal voltage and current polarities are reversed. The normal directions of current in the **PMOS transistor** are indicated in Fig. 5.14. A negative voltage on the gate relative to the source ($v_{GS} < 0$) is required to attract holes and create a p-type inversion layer in the channel region. To initiate conduction in the enhancement-mode PMOS transistor, the gate-source voltage must be more negative than the threshold voltage of the p-channel device, denoted by V_{TP}. To keep the source-substrate and drain-substrate junctions reverse-biased, v_{SB} and v_{DB} must also be less than zero. This requirement is satisfied by $v_{DS} \leq 0$.

An example of the output characteristics for an enhancement-mode PMOS transistor is given in Fig. 5.14(b). For $v_{GS} \geq V_{TP} = -1$ V, the transistor is off. For more negative values of v_{GS}, the drain current increases in magnitude. The PMOS device is in the triode region for small values of V_{DS}, and the saturation of the characteristics is apparent at larger V_{DS}. The curves look just like those for

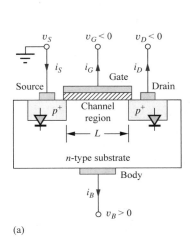
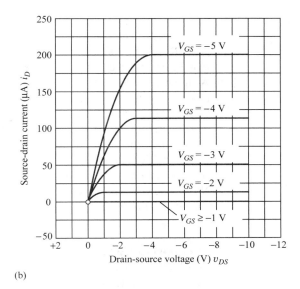

Figure 5.14 (a) Cross section of an enhancement-mode PMOS transistor; (b) output characteristics for a PMOS transistor with $V_{TP} = -1$ V.

the NMOS device except for sign changes on the values of v_{GS} and v_{DS}. This is a result of assigning the positive current direction to current exiting from the drain terminal of the PMOS transistor.

> **DESIGN NOTE**
>
> The mathematical model for the PMOS transistor in its various regions of operation is summarized in the equation set below and should be committed to memory!

PMOS TRANSISTOR MATHEMATICAL MODEL SUMMARY

Equations (5.30) through (5.34) represent the complete model for the i-v behavior of the PMOS transistor.

PMOS transistor

For all regions

$$K_p = K'_p \frac{W}{L} \qquad K'_p = \mu_p C''_{ox} \qquad i_G = 0 \qquad i_B = 0 \qquad (5.30)$$

Cutoff region:

$$i_D = 0 \qquad \text{for } V_{GS} \geq V_{TP} \qquad (5.31)$$

Triode region:

$$i_D = K_p \left(v_{GS} - V_{TP} - \frac{v_{DS}}{2} \right) v_{DS} \qquad \text{for } 0 \leq |v_{DS}| \leq |v_{GS} - V_{TP}| \qquad (5.32)$$

Saturation region:

$$i_D = \frac{K_p}{2} (v_{GS} - V_{TP})^2 (1 + \lambda |v_{DS}|) \qquad \text{for } |v_{DS}| \geq |v_{GS} - V_{TP}| \geq 0 \qquad (5.33)$$

Threshold voltage:

$$V_{TP} = V_{TO} - \gamma (\sqrt{v_{BS} + 2\phi_F} - \sqrt{2\phi_F}) \qquad (5.34)$$

For the enhancement-mode PMOS transistor, $V_{TP} < 0$. Depletion-mode PMOS devices can also be fabricated; $V_{TP} \geq 0$ for these devices.

Various authors have different ways of writing the equations that describe the PMOS transistor. Our choice attempts to avoid as many confusing minus signs as possible. The drain-current expressions for the PMOS transistor are written in similar form to those for the NMOS transistor except that the drain-current direction is reversed and the values of v_{GS} and v_{DS} are now negative quantities. A sign must still be changed in the expressions, however. The parameter γ is normally specified as a positive value for both n- and p-channel devices, and a positive bulk-source potential will cause the PMOS threshold voltage to become more negative.

An important parametric difference appears in the expressions for K_p and K_n. In the PMOS device, the charge carriers in the channel are holes, so current is proportional to hole mobility μ_p. Hole mobility is typically only 40 percent of the electron mobility, so for a given size device and set of voltage bias conditions, the PMOS device will conduct only 40 percent of the current of the NMOS device! Higher current capability leads to higher frequency operation in both digital and analog circuits. Thus, NMOS devices are preferred over PMOS devices in many applications.

EXERCISE: What is the region of operation and drain current of a PMOS transistor having $V_{TP} = -1$ V, $K_p = 0.4$ mA/V^2, and $\lambda = 0.02$ V^{-1} for (a) $V_{GS} = 0$ V, $V_{DS} = -1$ V; (b) $V_{GS} = -2$ V, $V_{DS} = -0.5$ V; (c) $V_{GS} = -2$ V, $V_{DS} = -2$ V?

ANSWERS: (a) cutoff, 0 A; (b) triode, 150 µA; (c) saturation, 208 µA

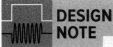
DESIGN NOTE

Electron mobility is typically 2.5 times that of the hole mobility, so for a given size device and set of voltage bias conditions, the NMOS device will conduct 2.5 times the current of the PMOS device.

5.4 MOSFET CIRCUIT SYMBOLS

Standard circuit symbols for four different types of MOSFETs are given in Fig. 5.15: (a) NMOS enhancement-mode, (b) PMOS enhancement-mode, (c) NMOS depletion-mode, and (d) PMOS depletion-mode transistors. The four terminals of the MOSFET are identified as source (S), drain (D), gate (G), and bulk (B). The arrow on the **bulk terminal** indicates the polarity of the bulk-drain, bulk-source, and bulk-channel pn junction diodes; the arrow points inward for an NMOS device and outward for the PMOS transistor. Enhancement-mode devices are indicated by the dashed line in the channel region, whereas depletion-mode devices have a solid line, indicating the existence of the built-in channel. The gap between the gate and channel represents the insulating oxide region. Table 5.1 summarizes the threshold-voltage values for the four types of NMOS and PMOS transistors.

In many circuit applications, the MOSFET substrate terminal is connected to its source. The shorthand notation in Fig. 5.15(e) and (f) is often used to represent these three-terminal MOSFETs. The arrow identifies the source terminal and points in the direction of normal positive current.

To further add to the confusing array of symbols that the circuit designer must deal with, a number of additional symbols are used in other texts and reference books and in papers in technical journals. The wide diversity of symbols is unfortunate, but it is a fact of life that circuit designers must accept. For example, if one tires of drawing the dashed line for the enhancement-mode device as well as the substrate arrow, one arrives at the NMOS transistor symbol in Fig. 5.15(g); the channel line is then thickened to represent the NMOS depletion-mode device, as in Fig. 5.15(h). In a similar vein,

TABLE 5.1
Categories of MOS transistors

	NMOS DEVICE	PMOS DEVICE
Enhancement-mode	$V_{TN} > 0$	$V_{TP} < 0$
Depletion-mode	$V_{TN} \leq 0$	$V_{TP} \geq 0$

the symbol in Fig. 5.15(i) represents the enhancement-mode PMOS transistor, and the corresponding depletion-mode PMOS device appears in Fig. 5.15(j). In the last two symbols, the circles represent a carry-over from logic design and are meant to indicate the logical inversion operation. We explore this more fully in Part II of this book. The symbols in Fig. 5.15(g) and (i) commonly appear in books discussing VLSI logic design.

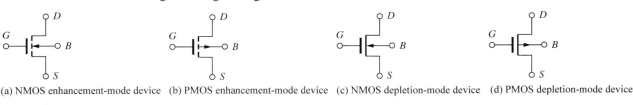

(a) NMOS enhancement-mode device (b) PMOS enhancement-mode device (c) NMOS depletion-mode device (d) PMOS depletion-mode device

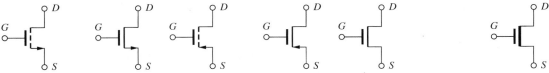

(e) Three-terminal NMOS transistors (f) Three-terminal PMOS transistors (g) Shorthand notation—NMOS enhancement-mode device (h) Shorthand notation—NMOS depletion-mode device

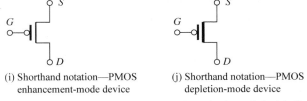

(i) Shorthand notation—PMOS enhancement-mode device (j) Shorthand notation—PMOS depletion-mode device

Figure 5.15 (a)–(f) IEEE Standard MOS transistor circuit symbols; (g)–(j) other commonly used symbols.

5.5 MOS TRANSISTOR SYMMETRY

The symmetry of MOS devices should be noted in the cross sections of Figs. 5.4 and 5.14. The terminal that is acting as the drain is actually determined by the applied potentials. Current can traverse the channel in either direction, depending on the applied voltages. For NMOS transistors, the n^+ region that is at the highest voltage will be the drain, and the one at the lowest voltage will be the source. For the PMOS transistor, the p^+ region at the lowest voltage will be the drain, and the one at the highest voltage will be the source. This symmetry is highly useful in certain applications, particularly dynamic random-access memory (DRAM) circuits, switched capacitor circuits, and analog signal switches.

The FET symmetry is in contrast to the bipolar transistor in which the structure of the BJT in Chapter 4 is inherently asymmetric. However, it is possible to make lateral BJTs with symmetric characteristics, although typically with much poorer current gain and frequency response.

5.5.1 THE ONE-TRANSISTOR DRAM CELL

The heart of high-density memory chips is the dynamic random-access memory (DRAM) cell that is based upon a basic one-transistor/one-capacitor circuit termed the 1-T cell. A 32-gigabit

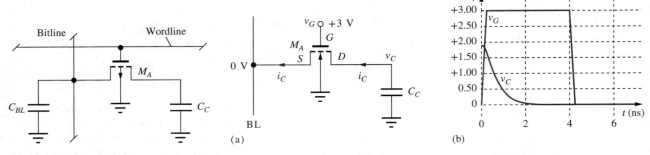

Figure 5.16 One-transistor (1-T) storage cell in which binary data is represented by the presence or absence of charge on C_C.

Figure 5.17 (a) Writing a 0 into the 1-T cell; (b) waveform during WRITE operation.

DRAM chip contains more than 32 billion transistors and 32 billion capacitors!

In the 1-T cell in Fig. 5.16, data is stored as the presence or absence of charge on **cell capacitor** C_C. Because leakage currents exist in the drain-bulk and source-bulk junctions of the transistor and in the transistor channel, the information stored on C_C is eventually corrupted. To prevent this loss of information, the state of the cell is periodically read and then written back into the cell to reestablish the desired cell voltages. This operation is referred to as the **refresh operation.** Each storage cell in a DRAM typically must be refreshed every few milliseconds.

In the next two sections, we shall explore the write and read operations of the DRAM cell and see how they depend upon the symmetrical behavior of the MOS transistor. Current may transverse the channel in either direction depending upon the data to be stored in the cell or the contents of the data read from the cell.

5.5.2 DATA STORAGE IN THE 1-T CELL

In the analysis that follows, a 0 will be represented by 0 V on capacitor C_C, and a 1 will be represented by a high level on C_C. These data are written into the 1-T cell by placing the desired voltage level on the "bitline" and then turning on access transistor M_A.

Storing a 0

Consider first the situation for storing a 0 in the cell, as in Fig. 5.17. In this case, the bitline is held at 0 V, and the bitline terminal of the MOSFET acts as the source of the FET. The gate is raised to $V_{DD} = 3$ V. If the cell voltage is already zero, then the drain-source voltage of the MOSFET is zero, and the current is zero. If the cell contains a 1 as in Fig. 5.17(b) with $v_C > 0$, then the MOSFET completely discharges C_C, also yielding $v_C = 0$.

The cell voltage waveform resulting from writing a zero into a cell containing a one is given in Fig. 5.18(b). The initial capacitor voltage, calculated in the next section, is rapidly discharged by the access transistor.

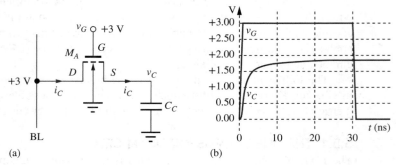

Figure 5.18 (a) Conditions for writing a 1 into the 1-T cell; (b) waveform during WRITE operation.

> **EXERCISE:** (a) What is the cell current i_C in Fig. 5.16 just after access transistor M_A is turned on if $V_C = 1.9$ V, $K_n = 60$ µA/V^2, and $V_{TO} = 0.7$ V? (b) Estimate the fall time of the voltage on the capacitor using Eqs. (S.15) and (S.20), with $C_C = 50$ fF. (c) Recalculate the current in (a) using the unified model with $V_{SAT} = 1$ V.
>
> **ANSWERS:** 154 µA; 1.34 ns; 108 µA

Storing a 1

Now consider the case of writing a 1 into the 1-T cell in Fig. 5.18. The bitline is first set to V_{DD} (3 V), and the wordline is then raised to V_{DD}. The bitline terminal of M_A acts as its drain, and the cell capacitance terminal acts as the FET source. Because $V_{DS} = V_{GS}$, and M_A is an enhancement mode device, M_A will operate in the saturation region. If a full 1 level already exists in the cell, then the current is zero in M_A. However, if V_C is less than a full 1 level, current through M_A will charge up the capacitor to a potential one threshold voltage below the gate voltage.

We see from this analysis that the voltage levels corresponding to 0 and 1 in the 1-T cell are 0 V and $V_G - V_{TN}$. The threshold voltage must be evaluated for a source-bulk voltage equal to V_C:

$$V_C = V_G - V_{TN} = V_G - \left[V_{TO} + \gamma\left(\sqrt{V_C + 2\phi_F} - \sqrt{2\phi_F}\right)\right] \qquad (5.35)$$

Once again we see the important use of the bidirectional characteristics of the MOSFET. Charge must be able to flow in both directions through the transistor in order to write the desired data into the cell. To read the data, current must also be able to change directions.

The waveform for writing a one into the 1-T cell appears in Fig. 5.18(b). Note the relatively long time required to reach final value. However, access transistor M_A can be turned off at the 10-ns point without significant loss in cell voltage.

> **EXERCISE:** Find the cell voltage V_C if $V_{DD} = 3$ V, $V_{TO} = 0.7$ V, $\gamma = 0.5$ √V, and $2\phi_F = 0.6$ V. What is V_C if $\gamma = 0$?
>
> **ANSWERS:** 1.89 V; 2.3 V
>
> **EXERCISE:** If a cell is in a 1 state, how many electrons are stored on the cell capacitor if $C_C = 25$ fF?
>
> **ANSWER:** 2.95×10^5 electrons

The results in the preceding exercises are typical of the situation for the 1 level in the cell. A significant part of the power supply voltage is lost because of the threshold voltage of the MOSFET, and the body effect has an important role in further reducing the cell voltage for the 1 state. If there were no body effect in the first exercise, then V_C would increase to 2.3 V.

5.5.3 READING DATA FROM THE 1-T CELL

To read the information from the 1-T cell, the bitline is first precharged (**bitline precharge**) to a known voltage, typically V_{DD} or one-half V_{DD}. The access transistor is then turned on, and the cell capacitance is connected to the bitline through M_A. A phenomenon called **charge sharing** occurs. The total charge, originally stored separately on the **bitline capacitance** C_{BL} and cell capacitance C_C, is shared between the two capacitors following the switch closure, and the voltage on the bitline changes slightly. The magnitude and sign of the change are related to the stored information.

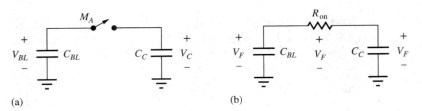

Figure 5.19 Model for charge sharing between the 1-T storage cell capacitance and the bitline capacitance: (a) circuit model before activation of access transistor; (b) circuit following closure of access switch.

Detailed behavior of data readout can be understood using the circuit model in Fig. 5.19. Before access transistor M_A is turned on, the switch is open, and the total initial charge Q_I on the two capacitors is

$$Q_I = C_{BL}V_{BL} + C_C V_C \tag{5.36}$$

After access transistor M_A is activated, corresponding to closing the switch, current through the on-resistance of M_A equalizes the voltage on the two capacitors. The final value of the stored charge Q_F is given by

$$Q_F = (C_{BL} + C_C)V_F \tag{5.37}$$

Because no mechanism for charge loss exists, Q_F must equal Q_I. Equating Eqs. (5.36) and (5.37) and solving for V_F yields

$$V_F = \frac{C_{BL}V_{BL} + C_C V_C}{C_{BL} + C_C} \tag{5.38}$$

The signal to be detected is the change in the voltage on the bitline from its initial precharged value:

$$\Delta V = V_F - V_{BL} = \frac{C_C}{C_{BL} + C_C}(V_C - V_{BL}) = \frac{(V_C - V_{BL})}{\frac{C_{BL}}{C_C} + 1} \tag{5.39}$$

Equation (5.39) can be used to guide our selection of the precharge voltage. If V_{BL} is set midway between the 1 and 0 levels, then ΔV will be positive if a 1 is stored in the cell and negative if a 0 is stored. Study of Eq. (5.39) also shows that the signal voltage ΔV can be quite small. Equal and opposite voltages are exactly what we desire for driving sense amplifiers with differential inputs. If there are 128 rows in our memory array, then there will be 128 access transistors connected to the bitline, and the ratio of bitline capacitance to cell capacitance can be quite large. If we assume that $C_{BL} \gg C_C$, Eq. (5.38) shows that the final voltage on the bitline and cell is

$$V_F = \frac{C_{BL}V_{BL} + C_C V_C}{C_{BL} + C_C} \cong V_{BL} \tag{5.40}$$

Thus, the content of the cell is destroyed during the process of reading the data from the cell—the 1-T cell is a cell with destructive readout. To restore the original contents following a read operation, the data must be written back into the cell.

Except for the case of an ideal switch, charge transfer cannot occur instantaneously. If the on-resistance were constant, the voltages and currents in the circuit in Fig. 5.19 would change exponentially with a time constant τ determined by R_{on} and the series combination of C_{BL} and C_C:

$$\tau = R_{on}\frac{C_C C_{BL}}{C_C + C_{BL}} \cong R_{on}C_C \quad \text{for } C_{BL} \gg C_C \tag{5.41}$$

EXERCISE: (a) Find the change in bitline voltage for a memory array in which $C_{BL} = 49\, C_C$ if the bitline is precharged midway between the voltages corresponding to a 1 and a 0. Assume that 0 V corresponds to a 0 and 1.9 V corresponds to a 1. (b) What is τ if $R_{on} = 5\ \text{k}\Omega$ and $C_C = 25\ \text{fF}$?

ANSWERS: +19.0 mV, −19.0 mV; 0.125 ns

The preceding exercise reinforces the fact that the voltage change that must be detected by a "sense" amplifier for a 1-T cell is quite small. Designing such an amplifier to rapidly detect this small change is one of the major challenges of DRAM design. Also, note that the **charge transfer** occurs rapidly.

MOS DEVICE SYMMETRY

The MOS transistor terminal that is acting as the drain is actually determined by the applied potentials. Current can traverse the channel in either direction, depending on the applied voltages.

ELECTRONICS IN ACTION

MOS Memory

For many years, high-density memory has served as the IC industry's vehicle for driving technology to ever smaller dimensions. In the mid-1960s, the first random-access memory (RAM) chip using MOS technology [3] was discussed at the IEEE International Solid-State Circuits Conference (ISSCC) [4], and in 1974 the first commercial 1024-bit (1-Kb) memory was introduced [5]. By 2000, experimental 1-gigabit (Gb) chips had been described at the ISSCC, and, as this edition is being written in 2021, 32-Gb memory chips are commercially available.

Robert H. Dennard, inventor of the 1-transistor DRAM cell
Business Wire/Handout/Getty Images

Fujio Masuoka Flash Memory Inventor
2014 IEEE

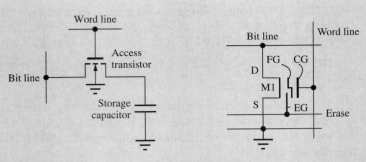

1-T DRAM cell.

Flash Memory Cell Schematic [7]. Floating Gate (FG), Control gate (CG), Erase Gate (EG)

The circuit that made these incredible memory chips possible is called the **one-transistor dynamic RAM cell** or 1-T DRAM. This elegant circuit, which requires only one transistor and one capacitor to store a single bit of information [6], was invented in 1966 by Robert H. Dennard of the IBM Thomas J. Watson Research Center. In this circuit, patented in 1968, the binary information is temporarily stored as a charge on the capacitor, and the data must be periodically refreshed in order to prevent information loss. In addition, the process of reading the data out of most DRAM circuits destroys the information, and the data must be put back into memory as part of the read operation. At the time of the invention, Dennard and a few of his colleagues were probably the only ones that believed the circuit could be made to actually work. Today, there are arguably more 1-T DRAM bits in the world than any other electronic circuit.

More recently, ever-increasing numbers of low-power, handheld electronic devices have employed a class of memory devices known as Flash Memory, invented by Dr. Fujio Masuoka [7, 8] of Tokyo Shibaura Electric Co. Ltd., now Toshiba Corporation. Cell phones maintain their user data in flash memory chips, digital cameras store pictures and digital audio players store music in flash memory, and many computer users carry high-density flash memories in their pockets. All of these memory chips are "non-volatile" maintaining the stored information without power but can be both written or read when power is applied. In the past decade, these devices have become broadly used to replace hard disk drives in laptop and desktop computers, and flash memory has now eclipsed DRAM in the number of bits shipped per year [9].

The flash memory cell can be fabricated by inserting two additional gates in the MOS structure between the original gate and the channel. In operation, the floating gate can be charged or discharged creating shifts in the threshold voltage of the transistor. The erase gate assists in discharging the floating gate.

5.6 CMOS TECHNOLOGY

CMOS (complementary MOS) requires a fabrication technology that can produce both NMOS and PMOS transistors together on one IC substrate, and the basic IC structure used to accomplish this task is shown in Fig. 5.20. In this cross section, NMOS transistors are shown fabricated in a normal manner in a *p*-type silicon substrate. A lightly doped *n*-type region, called the *n*-well, is formed in the *p*-type substrate, and PMOS transistors are then fabricated in the *n*-well region, which becomes the body of the PMOS device. Note that a large-area diode exists between the *n*-well and *p*-type substrate. This *pn* diode must always be kept reverse-biased by proper connection of V_{DD} and V_{SS} (e.g., 2.5 V and 0 V).

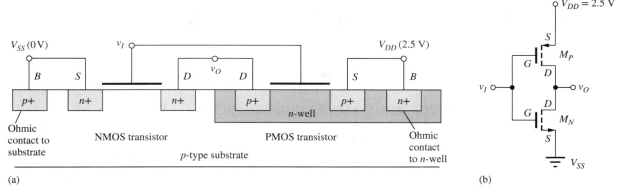

Figure 5.20 (a) *n*-well CMOS structure for forming both NMOS and PMOS transistors in a single silicon substrate; (b) basic CMOS circuit that can be used as an inverting amplifier or digital logic inverter.

The connections between the transistors needed to form the basic CMOS circuit are also indicated in Fig. 5.20 and correspond to the circuit schematic for in Fig. 5.20(b). The circuit consists of one NMOS and one PMOS transistor. In the CMOS circuit the source of the PMOS transistor is connected to V_{DD}, the source of the NMOS transistor is connected to V_{SS} (0 V in this case), the drain terminals of the two MOSFETs are connected together to form the output node, and the two gate terminals are connected together to form the input node. The substrates of both the NMOS and PMOS transistors are connected to their respective sources, so body effect is eliminated in both devices. The structure and circuit in Fig. 5.20 can be operated as either an analog amplifier or as a logic inverter.

Before we explore the design of the CMOS circuit, we need parameters for the CMOS devices, as given in Table 5.2. In CMOS technology, the transistors are normally designed to have equal and opposite threshold voltages—for example, $V_{TON} = +0.6$ V and $V_{TOP} = -0.6$ V. Remember that $K'_n = \mu_n C''_{ox}$ and $K'_p = \mu_p C''_{ox}$ and that hole mobility in the channel of the PMOSFET is approximately 40 percent of the electron mobility in the NMOSFET channel. The values in Table 5.2 reflect this difference because the value of K' for the NMOS device is shown as 2.5 times that for the PMOS transistor. Processing differences are also reflected in the different values for γ and $2\phi_F$ in the two types of transistors.

Remember that the threshold voltage of the NMOS transistor is denoted by V_{TN} and that of the PMOS transistor by V_{TP}:

$$V_{TN} = V_{TON} + \gamma_N\left(\sqrt{V_{SBN} + 2\phi_{FN}} - \sqrt{2\phi_{FN}}\right)$$

and (5.42)

$$V_{TP} = V_{TOP} - \gamma_P\left(\sqrt{V_{BSP} + 2\phi_{FP}} - \sqrt{2\phi_{FP}}\right)$$

For $v_{SBN} = 0$, $V_{TN} = V_{TON}$, and for $v_{BSP} = 0$, $V_{TP} = V_{TOP}$.

TABLE 5.2
CMOS Transistor Parameters

	NMOS DEVICE	PMOS DEVICE
V_{TO}	0.6 V	−0.6 V
γ	0.50 $\sqrt{V}$	0.75 $\sqrt{V}$
$2\phi_F$	0.60 V	0.70 V
K'	100 μA/V^2	40 μA/V^2

EXERCISES: (a) What are the values of K_p and K_n for transistors with $W/L = 20/1$? (b) An NMOS transistor has $V_{SB} = 2.5$ V. What is the value of V_{TN}? (c) A PMOS transistor has $V_{BS} = 2.5$ V. What is the value of V_{TP}?

ANSWERS: (a) 800 µA/V², 2 mA/V²; (b) 1.09 V; (c) −1.31 V

5.6.1 CMOS VOLTAGE TRANSFER CHARACTERISTICS

Figure 5.21 shows the result of simulation of the voltage transfer characteristic (VTC) of a **symmetrical CMOS circuit,** designed with $K_P = K_N$. The VTC can be divided into five different regions, as shown in the figure and summarized in Table 5.3. For an input voltage less than $V_{TN} = 0.6$ V in region 1, the NMOS transistor is off, and the output is maintained at $V_O = 2.5$ V by the PMOS device, which is on.

Similarly, for an input voltage greater than $(V_{DD} - |V_{TP}|)$ (1.9 V) in Region 5, the PMOS device is off, and the output is maintained at $V_O = 0$ V by the NMOS transistor that is on. In Region 2, the NMOS transistor is saturated and the PMOS transistor is in the triode region. For the input voltage near $V_{DD}/2$ (Region 3), both transistors are operating in the saturation region. The boundary between Regions 2 and 3 is defined by the boundary between the saturation and triode regions of operation for the PMOS transistor. Saturation of the PMOS device requires $|v_{DS}| \geq |v_{GS} - V_{TP}|$:

$$(2.5 - v_O) \geq (2.5 - v_I) - 0.6 \quad \text{or} \quad v_O \leq v_I + 0.6 \tag{5.43}$$

In a similar manner, the boundary between Regions 3 and 4 is defined by saturation of the NMOS device:

$$v_{DS} \geq v_{GS} - V_{TN} \quad \text{or} \quad v_O \geq v_I - 0.6 \tag{5.44}$$

In Region 4, the voltages place the NMOS transistor in the triode region, and the PMOS transistor remains saturated.

The CMOS structure and circuit in Figs. 5.20 and 5.21 can be operated as either an analog inverting amplifier, or logic inverter. The mode of operation is dependent upon the choice of the

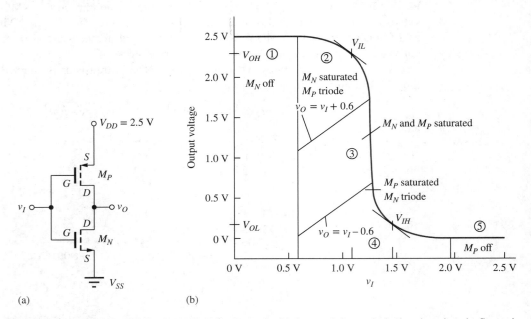

Figure 5.21 (a) Basic CMOS circuit. (b) CMOS voltage transfer characteristic may be broken down into the five regions outlined in Table 5.3.

TABLE 5.3
Regions of Operation of Transistors in a Symmetrical CMOS Circuit

REGION	INPUT VOLTAGE v_I	OUTPUT VOLTAGE v_O	NMOS TRANSISTOR	PMOS TRANSISTOR
1	$v_I \leq V_{TN}$	$V_H = V_{DD}$	Cutoff	Triode
2	$V_{TN} < v_I \leq v_O + V_{TP}$	High	Saturation	Triode
3	$v_I \cong V_{DD}/2$	$V_{DD}/2$	Saturation	Saturation
4	$v_O + V_{TN} < v_I \leq (V_{DD} - |V_{TP}|)$	Low	Triode	Saturation
5	$v_I \geq (V_{DD} - |V_{TP}|)$	$V_L = 0$	Triode	Cutoff

circuit's quiescent operating point (Q-point). In this and subsequent chapters, we focus on the CMOS amplifier characteristics, whereas detailed discussion of basic CMOS logic design can be found in the supplemental chapters in the e-book.[5]

The largest voltage gain (i.e., $A_v = dv_O/dv_I$) is achieved in Region 3 where both transistors are saturated, and this region is normally used for amplifier operation. In order to have the largest signal voltage swing at the output, the quiescent output voltage is typically set $V_O = V_{DD}/2$, or 1.25 V for this particular case. Voltages V_{IL} and V_{IH} correspond to the points at which the magnitude of the voltage gain has actually fallen to unity, and the circuit is no longer very useful as an amplifier.

EXERCISE: Suppose $v_I = 1$ V for the CMOS circuit in Fig. 5.21. (a) What is the range of values of v_O for which M_N is saturated and M_P is in the triode region? (b) For which values are both transistors saturated? (c) For which values is M_P saturated and M_N in the triode region?

ANSWERS: (1.6 V $\leq v_O \leq$ 2.5 V); (0.4 V $\leq v_O \leq$ 1.6 V); (0 V $\leq v_O \leq$ 0.4 V)

EXERCISE: The $(W/L)_N$ of M_N in Fig. 5.21 is 10/1. What is the value of $(W/L)_P$ required to form a symmetrical circuit?

ANSWER: 25/1

Figure 5.22 shows the results of simulation of the voltage transfer characteristics for a CMOS circuit with a symmetrical design ($K_p = K_n$) for several values of V_{DD}. Note that the output voltage levels V_H and V_L are always determined by the two power supplies. As the input voltage rises from 0 to V_{DD}, the output remains constant for $v_I < V_{TN}$ and $v_I > (V_{DD} - |V_{TP}|)$. For this symmetrical design case, the transition between V_H and V_L is centered at $v_I = V_{DD}/2$. The straight line on the graph represents $v_O = v_I$, which occurs for $v_I = V_{DD}/2$ for the symmetrical circuit.

If $K_p \neq K_n$, then the transition shifts away from $V_{DD}/2$. To simplify notation, a parameter K_R is defined: $K_R = K_n/K_p$. K_R represents the relative current drive capability between the NMOS and PMOS devices in the circuit. Voltage transfer characteristics for circuits with $K_R = 5$, 1, and 0.2 are shown in Fig. 5.23. For $K_R > 1$, the NMOS current drive capability exceeds that of the PMOS transistor, so the transition region shifts to $v_I < V_{DD}/2$. Conversely, for $K_R < 1$, PMOS current drive capability is greater than that of the NMOS device, and the transition region occurs for $v_I > V_{DD}/2$. Note that the red dot indicates the typical Q-point for the symmetrical amplifier design.

---

[5] As a logic inverter, the input is switched between two input logic levels, $V_L = 0$ V and $V_H = 2.5$ V, which then force the output to reach high and low logic levels V_H and V_L, respectively.

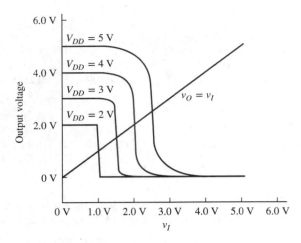

Figure 5.22 Voltage transfer characteristics for a symmetrical CMOS circuit ($K_R = 1$) with $V_{DD} = 5$ V, 4 V, 3 V, and 2 V.

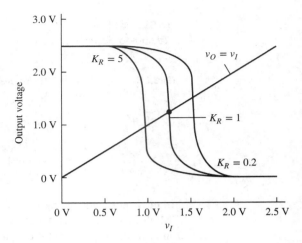

Figure 5.23 CMOS voltage transfer characteristics for $K_R = 5$, 1, and 0.2 for $V_{DD} = 2.5$ V. $K_R = K_n/K_p$.

As will be discussed shortly, FETs do not actually turn off abruptly as indicated in Eq. (5.9), but conduct small currents for gate-source voltages below threshold. This characteristic enables a CMOS circuit to function at very low supply voltages. In fact, it has been shown that the minimum supply voltage for operation of CMOS is only $[2V_T \ln(2)]$ V [10, 11]. At room temperature, this voltage is less than 40 mV!

EXERCISE: Equate the expressions for the drain currents of M_N and M_P to show that $v_O = v_I$ occurs for a voltage equal to $V_{DD}/2$ in a symmetrical circuit. What voltage corresponds to $v_O = v_I$ in an inverter with $K_R = 10$ and $V_{DD} = 4$ V? For $K_R = 0.1$ and $V_{DD} = 4$ V?

ANSWERS: 1.27 V; 2.73 V

5.7 CMOS LATCHUP

The basic CMOS structure has a potentially destructive failure mechanism called **latchup** that was a major concern in early implementations and helped delay broad adoption of the technology. By the mid-1980s, technological solutions were developed that effectively suppress the latchup phenomenon. However, it is important to understand the source of the problem that arises from the complex nature of the CMOS integrated structure that produces **parasitic bipolar transistors.** These bipolar transistors are normally off but can conduct under some transient fault conditions.

In the cross section of the CMOS structure in Fig. 5.24(a), a *pnp* transistor is formed by the source region of the PMOS transistor, the *n*-well, and the *p*-type substrate, whereas a lateral *npn* transistor is formed by the source region of the NMOS transistor, the *p*-type substrate, and the *n*-well. The physical structure connects the *npn* and *pnp* transistors together in the "lumped" equivalent circuit shown in Fig. 5.24(b). R_n and R_p model the series resistances existing between the external power supply connections and the internal base terminals of the bipolar transistors.

If the currents in R_n and R_p in the circuit model in Fig. 5.24(b) are zero, then the base-emitter voltages of both bipolar transistors are zero, and both devices are off. The total supply

5.7 CMOS Latchup

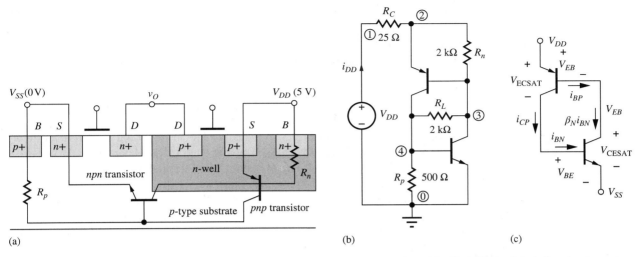

Figure 5.24 (a) CMOS structure with parasitic *npn* and *pnp* transistors identified; (b) circuit including shunting resistors R_n and R_p for SPICE simulation. See text for description of R_L; (c) regenerative structure formed by parasitic *npn* and *pnp* transistors.

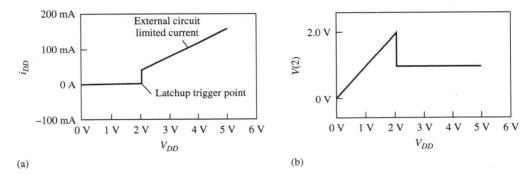

Figure 5.25 SPICE simulation of latchup in the circuit of Fig. 7.34(a): (a) current from V_{DD}; (b) voltage at node 2.

voltage ($V_{DD} - V_{SS}$) appears across the reverse-biased well-to-substrate junction that forms the collector-base junction of the two parasitic bipolar transistors. However, if a current should develop in the base of either the *npn* or the *pnp* transistor, latchup can be triggered, and high currents can destroy the structure. One source of current is the leakage across the large area *pn*-junction formed between the *n*-well and substrate. Another source is a transient that momentarily forward biases the drain substrate diode of the one of the MOS transistors. Once latchup is triggered, the current is limited only by the external circuit components.

The problem can be more fully understood by referring to Fig. 5.24(c), in which R_n and R_p have been neglected for the moment. Suppose a base current i_{BN} begins to flow in the base of the *npn* transistor. This base current is amplified by the *npn* current gain β_N and must be supplied from the base of the *pnp* transistor. The *pnp* base current is then amplified further by the current gain β_P of the *pnp* transistor, yielding a collector current equal to

$$i_{CP} = \beta_P i_{BP} = \beta_P (\beta_N i_{BN}) \tag{5.45}$$

However, the *pnp* collector current i_{CP} is also equal to the *npn* base current i_{BN}. If the product of the two current gains $\beta_P \beta_N$ exceeds unity, then all the currents will grow without bound. This situation is called *latchup*. Once the circuit has entered the latchup state, both transistors enter their low impedance state, and the voltage across the structure collapses to one diode drop plus one emitter-collector voltage:

$$V = V_{EB} + V_{\text{CESAT}} = V_{BE} + V_{\text{ECSAT}} \qquad (5.46)$$

Shunting resistors R_n and R_p shown in Fig. 5.24 actually play an important role in determining the latchup conditions in a real CMOS structure. As mentioned before, latchup would not occur in an ideal structure for which $R_n = 0 = R_p$, and modern CMOS technology uses special substrates and processing to minimize the values of these two resistors.

The results of SPICE simulation of the behavior of the circuit in Fig. 5.24(b) for representative circuit elements are presented in Fig. 5.25. Resistor R_L is added to the circuit to provide a leakage path across the collector-base junctions to initiate the latchup phenomenon in the simulation. Prior to latchup in Fig. 5.25, all currents are very small, and the voltage at node 2 is directly proportional to the input voltage V_{DD}. At the point that latchup is triggered, the voltage across the CMOS structure collapses to approximately 0.8 V, and the current increases abruptly to $(V_{DD} - 0.8)/R_C$. The current level is limited only by the external circuit component values. Large currents cause high power dissipation that can rapidly destroy most CMOS structures.

Under normal operating conditions, latchup does not occur. However, if a fault or transient occurs that causes one of the source or drain diffusions to momentarily exceed the power supply voltage levels, then latchup can be triggered. Ionizing radiation or intense optical illumination are two other possible sources of latchup initiation.

Note that this section actually introduced another form of modeling. Figure 5.24(a) is a cross section of a complex three-dimensional distributed structure, whereas Figs. 5.24(b) and (c) are attempts to represent or model this complex structure using a simplified network of discrete transistors and resistors. Note, too, that Fig. 5.24(b) is only a crude model of the real situation, so significant deviations between model predictions and actual measurements should not be surprising. It is easy to forget that circuit schematics generally represent only idealized models for the behavior of highly complex circuits.

5.8 CAPACITANCES IN MOS TRANSISTORS

Every electronic device has internal capacitances that limit the high-frequency performance of the particular device. In logic applications, these capacitances limit the switching speed of the circuits, and in amplifiers, the capacitances limit the frequency at which useful amplification can be obtained. Thus knowledge of the origin and modeling of these capacitances is quite important, and an introductory discussion of the capacitances of the MOS transistor appears in this section.

5.8.1 NMOS TRANSISTOR CAPACITANCES IN THE TRIODE REGION

Figure 5.26(a) shows the various capacitances associated with the MOS field-effect transistor operating in the triode region, in which the resistive channel region connects the source and drain. A simple model for these capacitances was presented by Meyer [12]. The total gate-channel capacitance C_{GC} is equal to the product of the **gate-channel capacitance** per unit area C''_{ox} (F/m^2) and the area of the gate:

$$C_{GC} = C''_{ox} WL \qquad (5.47)$$

ELECTRONICS IN ACTION

CMOS Camera on a Chip

Earlier in this text we examined the CCD image sensor widely used in astronomy. Although the CCD imager produces very high quality images, it requires an expensive specialized manufacturing process, complex control circuitry, and consumes a substantial amount of power. In the early 1990s, designers began developing techniques to integrate photo-detection circuitry onto inexpensive mainstream digital CMOS processes. In 1993, Dr. Eric Fossum's group at the Jet Propulsion Laboratory announced a CMOS digital camera on a chip. Since that time, many companies have designed camera chips that are based on mainstream CMOS processes, allowing the merging of many camera functions onto a single chip.

Pictured here is a photo of such a chip from Teledyne DALSA.[6] The device produces full color images and has 33 million pixels in the imaging array.

33 Megapixel Dalsa CCD image sensor
Teledyne DALSA. Reprinted by permission.

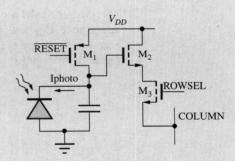

A basic photo diode pixel architecture.

A typical photodiode-based imaging pixel is also shown above. After asserting the $\overline{\text{RESET}}$ signal, the storage capacitor is fully charged to V_{DD} through transistor M_1. The reset signal is then removed, and light incident on the photodiode generates a photo current that discharges the capacitor. Different light intensities produce different voltages on the capacitor at the end of the light integration time. To read the stored value, the row select (ROWSEL) signal is asserted, and the capacitor voltage is driven onto the COLUMN bus via transistors M_2 and M_3.

In many designs random variations in the device characteristics will cause variations in the signal produced by each pixel for the same intensity of incident light. To correct for many of these variations, a technique known as *correlated double sampling* is used. After the signal level is read from a pixel, the pixel is reset and then read again to acquire a baseline signal. The baseline signal is subtracted from the desired signal, thereby removing the non-uniformities and noise sources which are common to both of the acquired signals.

Chips like this one are now common in digital cameras and digital camcorders. These common and inexpensive portable devices are enabled by the integration of analog photo-sensitive pixel structures with mainstream CMOS processes.

---

[6] The chip pictured above is a 33 Megapixel Dalsa CCD image sensor. The image is courtesy of the Teledyne DALSA.

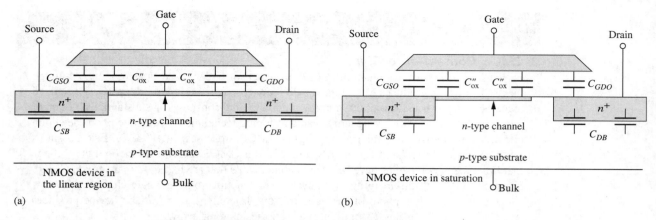

Figure 5.26 (a) NMOS capacitances in the linear region; (b) NMOS capacitances in the active region.

In the Meyer model for the triode region, C_{GC} is partitioned into two equal parts. The **gate-source capacitance** C_{GS} and the **gate-drain capacitance** C_{GD} each consists of one-half of the gate-channel capacitance plus the **overlap capacitances** C_{GSO} and C_{GDO} associated with the gate-source or gate-drain regions:

$$C_{GS} = \frac{C_{GC}}{2} + C_{GSO}W = C''_{ox}\frac{WL}{2} + C_{GSO}W$$
$$C_{GD} = \frac{C_{GC}}{2} + C_{GDO}W = C''_{ox}\frac{WL}{2} + C_{GDO}W \quad (5.48)$$

The overlap capacitances arise from two sources. First, the gate is actually not perfectly aligned to the edges of the source and drain diffusion but overlaps the diffusions somewhat. In addition, fringing fields between the gate and the source and drain regions contribute to the values of C_{GSO} and C_{GDO}.

The **gate-source** and **gate-drain overlap capacitances** C_{GSO} and C_{GDO} are normally specified as oxide **capacitances per unit width** (F/m). Note that C_{GS} and C_{GD} each have a component that is proportional to the area of the gate and one proportional to the width of the gate.

The capacitances of the reverse-biased pn junctions, indicated by the **source-body** and **drain-body capacitances** C_{SB} and C_{DB}, respectively, exist between the source and drain diffusions and the substrate of the MOSFET. Each capacitance consists of a component proportional to the junction bottom area of the source (A_S) or drain (A_D) region and a sidewall component that is proportional to the perimeter of the source (P_S) or drain (P_D) junction region:

$$C_{SB} = C_j A_S + C_{jsw} P_S \qquad C_{DB} = C_j A_D + C_{jsw} P_D \quad (5.49)$$

Here C_j is called the junction bottom capacitance per unit area (F/m^2), and C_{jsw} is the junction sidewall capacitance per unit length. C_{SB} and C_{DB} will be present regardless of the region of operation. Note that the junction capacitances are voltage dependent [see Eq. (3.21)].

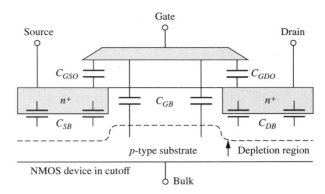

Figure 5.27 NMOS capacitances in the cutoff region.

5.8.2 CAPACITANCES IN THE SATURATION REGION

In the saturation region of operation, depicted in Fig. 5.26(b), the portion of the channel beyond the pinch-off point disappears. The Meyer models for the values of C_{GS} and C_{GD} become

$$C_{GS} = \tfrac{2}{3} C_{GC} + C_{GSO} W \quad \text{and} \quad C_{GD} = C_{GDO} W \quad (5.50)$$

in which C_{GS} now contains two-thirds of C_{GC}, but only the overlap capacitance contributes to C_{GD}. Now C_{GD} is directly proportional to W, whereas C_{GS} retains a component dependent on $W \times L$.

5.8.3 CAPACITANCES IN CUTOFF

In the cutoff region of operation, depicted in Fig. 5.27, the conducting channel region is gone. The values of C_{GS} and C_{GD} now contain only the overlap capacitances:

$$C_{GS} = C_{GSO} W \quad \text{and} \quad C_{GD} = C_{GDO} W \quad (5.51)$$

In the cutoff region, a small capacitance C_{GB} appears between the gate and bulk terminal, as indicated in Fig. 5.27.

$$C_{GB} = C_{GBO} L \quad (5.52)$$

in which C_{GBO} is the gate-bulk **capacitance per unit length.**

It should be clear from Eqs. (5.47) to (5.52) that MOSFET capacitances depend on the region of operation of the transistor and are nonlinear functions of the voltages applied to the terminals of the device. In subsequent chapters we analyze the impact of these capacitances on the behavior of digital and analog circuits. Complete models for these nonlinear capacitances are included in circuit simulation programs such as SPICE, and circuit simulation is an excellent tool for exploring the detailed impact of these capacitances on circuit performance.

> **EXERCISE:** Calculate C_{GS} and C_{GD} for a transistor operating in the triode and saturation regions if $C''_{ox} = 200\ \mu\text{F/m}^2$, $C_{GSO} = C_{GDO} = 300\ \text{pF/m}$, $L = 0.5\ \mu\text{m}$, and $W = 5\ \mu\text{m}$.
>
> **ANSWERS:** 1.75 fF, 1.75 fF; 1.83 fF, 1.5 fF

5.9 MOSFET MODELING IN SPICE

The SPICE circuit analysis program is used to simulate more complicated circuits and to make much more detailed calculations than we can perform by hand analysis. The circuit representation for the MOSFET model that is implemented in SPICE is given in Fig. 5.28, and as we can

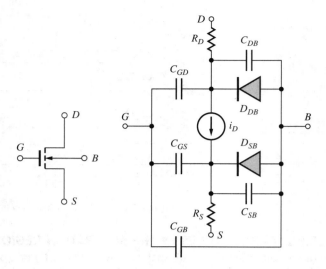

Figure 5.28 SPICE model for the NMOS transistor.

observe, the model uses quite a number of circuit elements in an attempt to accurately represent the characteristics of a real MOSFET. For example, small resistances R_S and R_D appear in series with the external MOSFET source and drain terminals, and diodes are included between the source and drain regions and the substrate. The need for the power of the computer is clear here. It would be virtually impossible for us to use this sophisticated model in our hand calculations.

Many different MOSFET models [13] of varying complexity are built into various versions of the SPICE simulation program, and they are denoted by "Level=Model_Number." The levels each have a unique mathematical formulation for **current source** i_D and for the various device capacitances. The model we have studied in this chapter is the most basic model and is referred to as the Level-1 model (LEVEL=1). Largely because of a lack of standard parameter usage at the time SPICE was first written, as well as the limitations of the programming languages originally used, the parameter names that appear in the models often differ from those used in this text and throughout the literature. The LEVEL=1 model is coded into SPICE using the following formulas, which are similar to those we have already studied.

Table 5.4 contains the equivalences of the **SPICE model** parameters and our equations summarized in Sec. 5.2. Typical and default values of the SPICE model parameters can be found in Table 5.4. A similar model is used for the PMOS transistor, but the polarities of the voltages and currents, and the directions of the diodes, are reversed.

Triode region: $\quad i_D = \text{KP} \dfrac{W}{L}\left(v_{GS} - \text{VT} - \dfrac{v_{DS}}{2}\right)v_{DS}(1 + \text{LAMBDA} \cdot v_{DS})$

Saturation region: $\quad i_D = \dfrac{\text{KP}}{2} \dfrac{W}{L}(v_{GS} - \text{VT})^2(1 + \text{LAMBDA} \cdot v_{DS})$ \hfill (5.53)

Threshold voltage: $\text{VT} = \text{VTO} + \gamma\left(\sqrt{v_{SB} + \text{PHI}} - \sqrt{\text{PHI}}\right)$

Notice that the SPICE level-1 description includes the addition of channel-length modulation to the triode region expression. Addition of the channel-length modulation term eliminates a discontinuity between the triode and saturation regions. Also, be sure not to confuse SPICE threshold voltage **VT** with thermal voltage V_T.

TABLE 5.4
SPICE Parameter Equivalences

PARAMETER	OUR TEXT	SPICE	DEFAULT
Transconductance	K'_n or K'_p	KP	20 µA/V^2
Threshold voltage	V_{TN} or V_{TP}	VT	—
Zero-bias threshold voltage	V_{TO}	VTO	1V
Surface potential	$2\phi_F$	PHI	0.6 V
Body effect	γ	GAMMA	0
Channel length modulation	λ	LAMBDA	0
Mobility	μ_n or μ_p	UO	600 cm^2/V·s
Gate-drain capacitance per unit width	C_{GDO}	CGDO	0
Gate-source capacitance per unit width	C_{GSO}	CGSO	0
Gate-bulk capacitance per unit length	C_{GBO}	CGBO	0
Junction bottom capacitance per unit area	C_J	CJ	0
Grading coefficient	MJ	MJ	0.5 V$^{0.5}$
Sidewall capacitance	C_{JSW}	CJSW	0
Sidewall grading coefficient	MJSW	MJSW	0.5 V$^{0.5}$
Oxide thickness	T_{ox}	TOX	100 nm
Junction saturation current	I_S	IS	10 fA
Built-in potential	ϕ_j	PB	0.8 V
Ohmic drain resistance	—	RD	0
Ohmic source resistance	—	RS	0

The junction capacitances are modeled in SPICE by a generalized form of the capacitance expression in Eq. (3.21)

$$C_J = \frac{CJO}{\left(1 + \frac{v_R}{PB}\right)^{MJ}} \quad \text{and} \quad C_{JSW} = \frac{CJSWO}{\left(1 + \frac{v_R}{PB}\right)^{MJSW}} \quad (5.54)$$

in which v_R is the reverse bias across the *pn* junction.

EXERCISE: What are the values of SPICE model parameters **KP, LAMBDA, VTO, PHI,** *W*, and *L* for a transistor with the following characteristics: $V_{TN} = 1$ V, $K_n = 150$ µA/V^2, $W = 1.5$ µm, $L = 0.25$ µm, $\lambda = 0.0133$ V^{-1}, and $2\phi_F = 0.6$ V?

ANSWERS: 150 µA/V^2; 0.0133 V^{-1}; 1 V; 0.6 V; 1.5 µm; 0.25 µm (specified in SPICE as 150U; 0.0133; 1; 0.6; 1.5U; 0.25U)

5.10 MOS TRANSISTOR SCALING

In Chapter 1, we discussed the phenomenal increase in integrated circuit density and complexity. These changes have been driven by our ability to aggressively scale the physical dimensions of the MOS transistor. A theoretical framework for MOSFET miniaturization was first provided by Dennard, Gaensslen, Kuhn, and Yu [14, 15]. The basic tenant of the theory is to require that the electrical fields be maintained constant within the device as the geometry is changed. Thus, if a physical dimension is reduced by a factor of α, then the voltage applied across that dimension must also be decreased by the same factor.

5.10.1 DRAIN CURRENT

These rules are applied to the transconductance parameter and triode region drain current expressions for the MOSFET in Eq. (5.55) in which the three physical dimensions W, L, and T_{ox} are all reduced by the factor α, and each of the voltages including the threshold voltage is reduced by the same factor.

$$K_n^* = \mu_n \frac{\varepsilon_{ox}}{T_{ox}/\alpha} \frac{W/\alpha}{L/\alpha} = \alpha \mu_n \frac{\varepsilon_{ox}}{T_{ox}} \frac{W}{L} = \alpha K_n$$

$$i_D^* = \mu_n \frac{\varepsilon_{ox}}{T_{ox}/\alpha} \frac{W/\alpha}{L/\alpha} \left(\frac{v_{GS}}{\alpha} - \frac{V_{TN}}{\alpha} - \frac{v_{DS}}{2\alpha}\right) \frac{v_{DS}}{\alpha} = \frac{i_D}{\alpha}$$
(5.55)

We see that scaled transconductance parameter K_n^* is increased by the scale factor α, whereas the scaled drain current is reduced from the original value by the scale factor.

5.10.2 GATE CAPACITANCE

In a similar manner, the total gate-channel capacitance of the device is also found to be reduced by α:

$$C_{GC}^* = (C_{ox}'')^* W^* L^* = \frac{\varepsilon_{ox}}{T_{ox}/\alpha} \frac{W/\alpha}{L/\alpha} = \frac{C_{GC}}{\alpha}$$
(5.56)

It can be demonstrated[7] that the delay of logic gates is limited by the transistor's ability to charge and discharge the capacitance associated with the circuit. Based on $i = C\,dv/dt$, an estimate of the delay of a scaled logic circuit is

$$\tau^* = C_{GC}^* \frac{\Delta V^*}{i_D^*} = \frac{C_{GC}}{\alpha} \frac{\Delta V/\alpha}{i_D/\alpha} = \frac{\tau}{\alpha}$$
(5.57)

We find that circuit delay is also improved (reduced) by the scale factor α.

5.10.3 CIRCUIT AND POWER DENSITIES

As we scale down the dimensions by α, the number of circuits in a given area will increase by a factor of α^2. An important concern in scaling is therefore what happens to the power per circuit, and hence the power per unit area (power density) as dimensions are reduced. The total power supplied to a transistor circuit will be equal to the product of the supply voltage and the transistor drain current:

$$P^* = V_{DD}^* i_D^* = \left(\frac{V_{DD}}{\alpha}\right)\left(\frac{i_D}{\alpha}\right) = \frac{P}{\alpha^2}$$

and
(5.58)

$$\frac{P^*}{A^*} = \frac{P^*}{W^* L^*} = \frac{P/\alpha^2}{(W/\alpha)(L/\alpha)} = \frac{P}{WL} = \frac{P}{A}$$

The result in Eq. (5.58) is extremely important. It indicates that the power per unit area remains constant if a technology is properly scaled. Even though we are increasing the number of circuits by α^2, the total power for a given size integrated circuit die will remain constant. Violation of the **scaling theory** over many years, by maintaining a constant 5-V power supply as dimensions were reduced, led to unmanagable power levels in integrated circuits. The power problem was finally resolved by changing from NMOS to CMOS technology, and then by reducing the power supply voltages. Today, power supply voltages in advanced technologies are 1 V or less!

---

[7] See Chapter S6 in the e-book.

TABLE 5.5
Constant Electric Field Scaling Results

PERFORMANCE MEASURE	SCALE FACTOR	PERFORMANCE MEASURE	SCALE FACTOR
Area/circuit	$1/\alpha^2$	Circuit delay	$1/\alpha$
Transconductance parameter	α	Power/circuit	$1/\alpha^2$
Current	$1/\alpha$	Power/unit area (power density)	1
Capacitance	$1/\alpha$	Power-delay product (PDP)	$1/\alpha^3$

5.10.4 POWER-DELAY PRODUCT

A useful figure of merit for comparing logic families is the **power-delay product (PDP),** which is discussed in detail in Chapters 6 to 9 of the 5th edition.[8] The product of power and delay time represents energy, and the power-delay product represents a measure of the energy required to perform a simple logic operation.

$$\text{PDP}^* = P^* \tau^* = \frac{P}{\alpha^2} \frac{\tau}{\alpha} = \frac{\text{PDP}}{\alpha^3} \quad (5.59)$$

The PDP figure of merit shows the full power of technology scaling. The power-delay product is reduced by the cube of the scaling factor.

Each new generation of lithography technology corresponds to a scale factor $\alpha = \sqrt{2}$. Therefore each new technology generation increases the potential number of circuits per chip by a factor of 2 and improves the PDP by a factor of almost 3. Table 5.5 summarizes the performance changes achieved with **constant electric field scaling.**

EXERCISE: A MOS technology is scaled from a 0.18-μm feature size to 10 nm. What is the increase in the number of circuits/cm^2? What is the improvement in the power-delay product?

ANSWERS: 324 times; 583 times.

EXERCISE: Suppose that the voltages are not scaled as the dimensions are reduced by a factor of α? How does the drain current of the transistor change? How do the power/circuit and power density scale?

ANSWERS: $I_D^* = \alpha I_D; P^* = \alpha P; P^*/A^* = \alpha^3 P/A!$

5.10.5 CUTOFF FREQUENCY

The ratio of transconductance g_m to gate-channel capacitance C_{GC} represents the highest useful frequency of operation of the transistor, and this ratio is called the **cutoff frequency f_T** of the device. The cutoff frequency represents the highest frequency at which the transistor can provide amplification. We can find f_T for the MOSFET by combining Eqs. (5.22) and (5.47):

$$f_T = \frac{1}{2\pi} \frac{g_m}{C_{GC}} = \frac{1}{2\pi} \frac{\mu_n}{L^2} (V_{GS} - V_{TN}) \quad (5.60)$$

Here we see clearly the advantage of scaling the channel length of MOSFET. The cutoff frequency improves with the square of the reduction in channel length.

---

[8] See Chapters S6–S9 of the e-book.

> **EXERCISE:** (a) A MOSFET has a mobility of 500 cm²/V · s and channel length of 0.25 μm. What is its cutoff frequency if the gate voltage exceeds the threshold voltage by 1 V? (b) Repeat for a channel length of 20 nm.
>
> **ANSWERS:** (a) 127 GHz; (b) 4.97 THz

5.10.6 HIGH FIELD LIMITATIONS

Unfortunately the assumptions underlying constant-field scaling have often been violated due to a number of factors. For many years, the supply voltage was maintained constant at a standard level of 5 V, while the dimensions of the transistor were reduced, thus increasing the electric fields within the MOSFET. Increasing the electric field in the device can reduce long-term reliability and ultimately lead to breakdown of the gate oxide or *pn* junction.

High fields directly affect MOS transistor mobility in two ways. The first effect is a reduction in the mobility of the MOS transistor due to increasing carrier scattering at the channel oxide interface. The second effect of high electric fields is to cause a breakdown of the linear mobility–field relationship as discussed in Chapter 2. At low fields, carrier velocity is directly proportional to electric field, as assumed in Eq. (5.5), but for fields exceeding approximately 10^5 V/cm, the carriers reach a maximum velocity of approximately 10^7 cm/s called the saturation velocity v_{SAT} (see Fig. 2.5). Both mobility reduction and velocity saturation tend to linearize the drain current expressions for the MOSFET. The results of these effects can be incorporated into the drain current model for the MOSFET as indicated in Eq. (5.61) by limiting the drain-source voltage to V_{SAT}, the voltage at which saturation velocity v_{SAT} is reached.

$$i_D = \mu_n C''_{ox} \frac{W}{L}\left(v_{GS} - V_{TN} - \frac{V_{SAT}}{2}\right)V_{SAT} = K_n\left(v_{GS} - V_{TN} - \frac{V_{SAT}}{2}\right)V_{SAT} \quad \text{where} \quad V_{SAT} = \frac{v_{SAT} L}{\mu_n} \tag{5.61}$$

The use of V_{SAT} represents a piecewise linear approximation to the velocity-field characteristic as in Fig. 5.29.

Note that i_D becomes a linear function of v_{GS} in the *velocity limited region*, and the transconductance becomes independent of operating point but with only 50% of the value in the saturation region. [See Eq. (5.22).]

$$g_m = \left.\frac{\partial i_D}{\partial v_{GS}}\right|_{Q-Pt} = K_n V_{SAT} = \frac{I_D}{V_{GS} - V_{TN} - \frac{V_{SAT}}{2}} \tag{5.62}$$

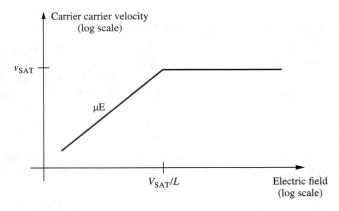

Figure 5.29 Piecewise linear approximation of velocity-field characteristic.

> **EXERCISE:** A MOSFET has a channel length of 1 μm. What value of V_{DS} will cause the electrons to reach saturation velocity? Repeat for a channel length of 0.1 μm. Repeat for $L = 15$ nm.
>
> **ANSWERS:** 10 V, 1 V, 0.15 V

5.10.7 THE UNIFIED MOS TRANSISTOR MODEL, INCLUDING HIGH FIELD LIMITATIONS

Rabaey, Chandrakasan, and Nikolic [16] have combined the equations describing linear, saturation, and high field regions of operation of the MOSFET into the single description in Eq. (5.62) commonly referred to as the *Unified Model*.

$$i_D = K'_n \frac{W}{L}\left(v_{GS} - V_{TN} - \frac{V_{MIN}}{2}\right) V_{MIN} (1 + \lambda V_{DS}) \text{ where}$$

$$V_{MIN} = \min\{(V_{GS} - V_{TN}), V_{DS}, V_{SAT}\} \tag{5.63}$$

In order to only have one equation to remember, the $(1 + \lambda V_{DS})$ term has been included for all three regions of operation (see the SPICE model equations in Sec. 5.10). Although this is not strictly correct for in the linear region, the error introduced is very small for the typically small values of V_{DS} encountered in the linear region. Remember also, that this is a simplified expression designed for hand calculations and is only represents an approximation of the $i - v$ characteristics of real transistors.

A graph of the output characteristics generated by the Unified Model appears in Fig. 5.30. As V_{GS} increases from zero, we see the normal quadratic behavior in the saturation region. However, as for $(V_{GS} - V_{TN}) > V_{SAT}$, the transistor enters velocity saturation and the spacing between the curves becomes constant for constant steps in V_{GS}.

> **EXERCISE:** Calculate the value of the drain current for the transistor in Fig. 5.30 at $V_{GS} = V_{DS} = 4$ V. What would be the current in the transistor if $V_{SAT} = 20$ V?
>
> **ANSWERS:** 607.5 μA; 661.5 μA

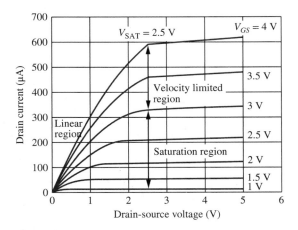

Figure 5.30 An example of an MOS transistor output characteristics produced by the Unified Model with $K_n = 100$ μA/V², $V_{TN} = 0.5$ V, $V_{SAT} = 2.5$ V, and $\lambda = 0.02$/V.

UNIFIED MATHEMATICAL MODEL SUMMARY

NMOS transistor:

$$K_n = K'_n \frac{W}{L} \quad K'_n = \mu_n C''_{ox} \quad i_G = 0 \quad i_B = 0$$

$$V_{TN} = V_{T0} + \gamma\left(\sqrt{v_{SB} + 2\phi_F} - \sqrt{2\phi_F}\right)$$

$$i_D = 0 \quad \text{for} \quad v_{GS} \leq V_{TN} \tag{5.64}$$

$$i_D = K_n\left(v_{GS} - V_{TN} - \frac{V_{\text{MIN}}}{2}\right) V_{\text{MIN}}(1 + \lambda V_{DS}) \quad \text{for} \quad V_{GS} > V_{TN}$$

$$V_{\text{MIN}} = \min\{(V_{GS} - V_{TN}), V_{DS}, V_{\text{SAT}}\}$$

PMOS transistor:

$$K_p = K'_p \frac{W}{L} \quad K'_p = \mu_p C''_{ox} \quad i_G = 0 \quad i_B = 0$$

$$V_{TP} = V_{T0} - \gamma\left(\sqrt{v_{SB} + 2\phi_F} - \sqrt{2\phi_F}\right)$$

$$i_D = 0 \quad \text{for} \quad v_{GS} \geq V_{TP} \tag{5.65}$$

$$i_D = K_p\left(v_{GS} - V_{TP} - \frac{V_{\text{MIN}}}{2}\right) V_{\text{MIN}}(1 + \lambda V_{DS}) \quad \text{for} \quad V_{GS} < V_{TP}$$

$$V_{\text{MIN}} = \max\{(V_{GS} - V_{TP}), V_{DS}, V_{\text{SAT}}\}$$

5.10.8 SUBTHRESHOLD CONDUCTION

In our discussion of the MOSFET thus far, we have assumed that the transistor turns off abruptly as the gate-source voltage drops below the threshold voltage. In reality, this is not the case. Although the resistive channel disappears, conduction does not actually cease.

Looking carefully at the channel region near the surface of the NMOS transistor structure in Fig. 5.31, we observe that the n^+ source, p substrate, and n^+ drain form a lateral *npn* bipolar transistor! The n^+ source becomes the emitter, the p-region is the base, and the n^+ drain is the collector. Electrons still enter the p-region from the source and then diffuse to the drain, just as in the BJT.

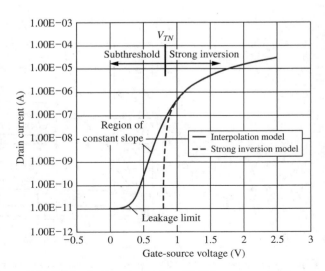

Figure 5.31 Subthreshold conduction in an NMOS transistor with $V_{TN} = 0.8$ V.

The current transported across the base of the "BJT" can be written using Eq. (5.66) as

$$I_D = I_{DO}\left[\exp\left(\frac{v_{GS} - v_{TN}}{nV_T}\right)\right]\left[1 - \exp\left(-\frac{v_{DS}}{V_T}\right)\right] \cong I_{DO}\exp\left(\frac{v_{GS} - v_{TN}}{nV_T}\right) \quad (5.66)$$

where $I_{DO} = 2nV_T^2 K_n$ and saturation operation can be assumed for $v_{DS} > 3V_T$, as in the approximate result. I_{DO} characterizes the FET current in the subthreshold region. Parameter n accounts for the division of the gate voltage between the oxide capacitance and that of the depletion-layer capacitance (see Fig. 5.3) in the absence of the channel.

As depicted in Fig. 5.31, the drain current decreases exponentially for values of v_{GS} less than V_{TN} (referred to as the subthreshold region), as indicated by the region of constant slope in the graph. At very low gate-source voltages, the drain current ultimately becomes limited by small currents such as drain-bulk diode leakage. The graph also depicts the curve for the basic square-law model described previously by Eq. (5.19) for $v_{GS} \geq V_{TN}$.

A measure of the rate of turn off of the MOSFET in the subthreshold region is specified as the reciprocal of the slope $(1/S)$ in mV/decade of current change. Typical values range from 60 to 120 mV/decade. The value of n depends on the relative magnitudes of C''_{ox} and C_d in Fig. 5.3(b), and its value typically ranges between 1 and 2. Note that $S = 2.3nV_T$.

5.11 ALL REGION MODELING

A simple theoretical model does not exist that combines strong inversion through subthreshold behavior. For design and hand calculations, a heuristic interpolation model has been developed as discussed below. Very sophisticated versions of the interpolation model, such as the EKV model [17], are used for simulation purposes.

5.11.1 INTERPOLATION MODEL

The **Interpolation Model** in Eq. (5.67) is a "simple" mathematical model that combines the basic models for strong inversion and subthreshold in a form that becomes the strong inversion model (Eq. 5.14) for $V_{GS} \gg V_{TN}$ and the subthreshold model (Eq. 5.66) for $V_{GS} \ll V_{TN}$ as well as providing a good fit in between [18, 19].

$$I_D = I_{DO}\left\{\left[\ln\left(1 + \exp\left(\frac{V_{GS} - V_{TN}}{2nV_T}\right)\right)\right]^2 - \left[\ln\left(1 + \exp\left(\frac{V_{GS} - V_{TN} - nV_{DS}}{2nV_T}\right)\right)\right]^2\right\}$$

$$I_{DO} = 2nV_T^2 K_n \quad (5.67)$$

There is a small discrepancy in the current limits in saturation. However, the model provides an engineering approximation that is good for design, and the error is typically less than the tolerances associated with a given process technology.

5.11.2 INTERPOLATION MODEL IN THE SATURATION REGION

If V_{DS} is greater than 200–300 mV in the subthreshold region, then the second exponential term in Eq. (5.67) can be neglected, and the drain current becomes independent of the drain voltage (i.e., the MOSFET drain current is saturated). The overall drain current expression can be simplified to

$$I_D \cong I_{DO}\left[\ln\left(1 + \exp\left(\frac{V_{GS} - V_{TN}}{2nV_T}\right)\right)\right]^2 \cong I_{DO}\exp\left(\frac{V_{GS} - V_{TN}}{nV_T}\right) \quad (5.68)$$

The limits can be found using the following approximations: for $V_{GS} \gg V_{TN}$, the logarithmic term has the form $\ln(\exp(x)) = x$, whereas for $V_{GS} \ll V_{TN}$, the term has the form $\ln(1 + y) \cong y$. The graph in Fig. 5.31 was generated using Eq. (5.68) with addition of a fixed leakage current. We will base our designs below strong inversion in the rest of this text on Eq. (5.68).

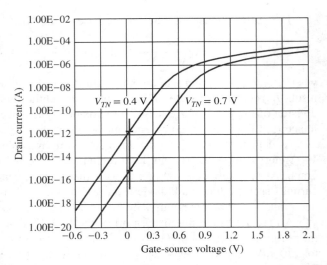

Figure 5.32 Graph of the interpolation model in Eq. (5.68) for two values of threshold voltage showing the change in drain current at zero gate-source voltage ($K_n = 25$ µA/V², $n = 1.5$, and $V_T = 25.8$ mV).

From Eq. (5.55), we see that the threshold voltage of the transistor should be reduced as the dimensions are reduced. However, the subthreshold region does not scale properly, and the curve in Fig. 5.32 tends to shift horizontally as V_{TN} is decreased. The reduced threshold increases the drain current in "off" devices, which ultimately limits data storage time in the dynamic memory cells and plays an important role in limiting battery life in low-power portable devices.

5.11.3 TRANSCONDUCTANCE EFFICIENCY

Much of today's state-of-the-art circuit design focuses on achieving low-power design at low voltages. One of the advantages for FETs is a significant improvement in the transconductance efficiency of the transistor when operating below strong inversion. Using Eq. (5.68), we quickly find that the transconductance efficiency in subthreshold becomes $1/nV_T$ approaching that of the BJT.

Figure 5.33 plots the FET transconductance efficiency versus V_{GS} for an NMOS transistor with $V_{TN} = 0.5$ V. In the solid curve for gate-source voltages above one volt in strong inversion, g_m/I_D falls below three, but as V_{GS} decreases, the efficiency increases and approaches a value of 28 well below threshold.[9] The dashed current includes a small constant leakage current (1 pA), which causes the efficiency to drop rapidly at very low drain currents.

Figure 5.34 shows the transconductance efficiency obtained from measurement characterization of an NMOS transistor on the commercially available MOSFET array, the ALD1106 (https://www.aldinc.com/pdf/ALD1106.pdf). Here the transconductance efficiency is shown as a function of the FET's levels of inversion, weak, moderate, and strong. Borrowing nomenclature from the previously mentioned EKV model, *inversion coefficient*, *IC*, is used to quantify the level of inversion. For weak inversion, $IC \leq 0.1$, $0.1 < IC < 10$ for moderate inversion, and $IC \geq 10$ for strong inversion. Equation (5.69) from the EKV model explicitly provides transconductance efficiency

---

[9] In practice, "long channel" CMOS technology ($L \geq 0.25$ µm) typically operates with $(V_{GS} - V_{TN}) \geq 250$ mV for strong inversion. This is helpful for recognizing the strong inversion bias condition. The quantity $(V_{GS} - V_{TN})$ is sometimes referred to as the "gate overdrive" voltage.

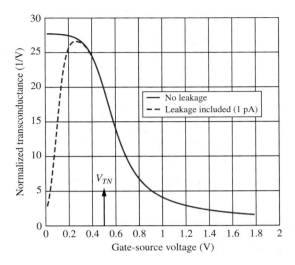

Figure 5.33 Interpolation model calculations for transconductance efficiency vs. gate-source voltage for an NMOS transistor [19] with (dashed curve) and without low-level (solid curve) leakage currents for $V_{TN} = 0.5$ V.

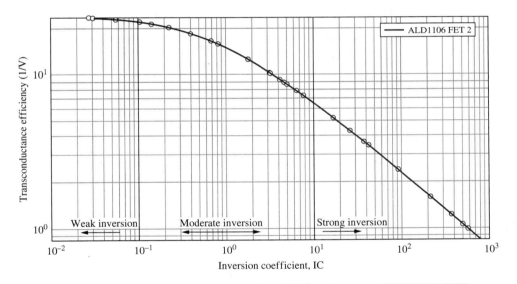

Figure 5.34 Measured transconductance efficiency for an NMOS transistor on an ALD1106 MOSFET array.

as a function of *IC*. Though not explicitly shown in Fig. 5.34, moderate inversion can span two decades of drain current.

$$\frac{g_m}{I_D} = \frac{1}{nV_T}\left[\frac{1}{\sqrt{IC + 0.25} + 0.5}\right] \quad \text{where} \quad IC = \frac{I_D}{I_{D0}} \quad (5.69)$$

The difference in FET properties between weak inversion and strong inversion are based in the fundamental conduction mechanisms for these operational modes. For strong inversion, the FET is a drift current device (discussed earlier in this chapter). In weak inversion, the FET is a diffusion current device, analogous to the BJT. Hence their similar transconductance efficiency.

Today's low-voltage low-power circuits often require operation of FETs in a range spanning from below strong inversion down into weak inversion. Throughout the rest of the text we will periodically explore operation of circuits utilizing FETs operating in saturation below strong inversion.

> **EXERCISE:** (a) What is the leakage current in the device in Fig. 5.31 for $V_{GS} = 0.25$ V? (b) Suppose the transistor in Fig. 5.31 had $V_{TN} = 0.5$ V. What will be the leakage current for $V_{GS} = 0$ V? (c) A memory chip uses 10^9 of the transistors in part (b). What is the total leakage current if $V_{GS} = 0$ V for all the transistors?
>
> **ANSWERS:** (a) $\cong 10^{-18}$ A; (b) 3×10^{-15} A; (c) 3 µA
>
> **EXERCISE:** Find the values of I_{DO} and n for the graph in Fig. 5.31. Use the data for drain currents of 10 nA and 100 pA. Assume $V_{TN} = 0.5$ V.
>
> **ANSWER:** $\cong 16.7$ µA, 2.1
>
> **EXERCISE:** Calculate the subthreshold slope for the graph in Fig. 5.31 and express the value in mV/decade current.
>
> **ANSWER:** 54.3 mV/decade
>
> **EXERCISE:** What is the value of inversion coefficient IC when the MOS gate-source voltage equals the threshold voltage?
>
> **ANSWER:** 0.48
>
> **EXERCISE:** In the interpolation model, assume $I_{DO} = 100$ µA, $V_{GS} = 0.4$ V, $V_{TN} = 0.4$ V, and $n = 1.5$. (a) What value of V_{DS} is required for the right-hand term to be only 5 percent of the left-hand term in the drain current Eq. 5.70. What is the value of I_D? (b) Repeat for $V_{GS} = 0.2$ V.
>
> **ANSWERS:** 71.4 mV, 22.8 µA; 61.9 mV, 0.251 µA

5.12 MOS TRANSISTOR FABRICATION AND LAYOUT DESIGN RULES[10]

In addition to choosing the circuit topology, the MOS integrated circuit designer must pick the values of the W/L ratios of the transistors and develop a layout for the circuit that ensures that it will achieve the performance specifications. Design of the layout of transistors and circuits in integrated form is constrained by a set of rules termed the **design rules** or **ground rules.** These rules are technology specific and specify minimum sizes, spacings, and overlaps for the various shapes that define transistors. The sets of rules are different for MOS and bipolar processes, for MOS processes designed specifically for logic and memory, and even for similar processes from different companies.

5.12.1 MINIMUM FEATURE SIZE AND ALIGNMENT TOLERANCE

Basic layout can be defined around a **minimum feature size** F, which represents the width of the smallest line or space that can be reliably transferred to the surface of a wafer using a given generation of lithographic manufacturing tools. (See Fig. 1.5(b).) To produce a basic set of ground

---

[10] Richard C. Jaeger, *Introduction to Microelectronic Fabrication: Volume 5 of Modular Series on Solid State Devices,* 2nd edition, © 2002. Electronically reproduced by permission of Pearson Education, Inc., Upper Saddle River, New Jersey.

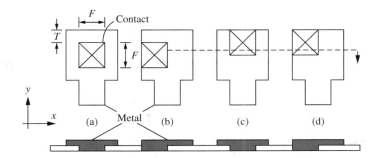

Figure 5.35 Misalignment of a metal pattern over a contact opening: (a) desired alignment, (b) one possible worst-case misalignment in the x direction, (c) one possible worst-case misalignment in the y direction, and (d) misalignment in both directions.

rules, we must also know the maximum misalignment which can occur between two mask levels during fabrication. For example, Fig. 5.35(a) shows the nominal position of a metal line aligned over a contact window (indicated by the box with an × in it). The metal overlaps the contact window by at least one **alignment tolerance** T in all directions. During the fabrication process, alignment will not be perfect, and the actual structure may exhibit misalignment in the x or y directions or both. Figure 5.35(b) through (d) show the result of one possible set of worst-case alignments of the patterns in the x, y, and both directions simultaneously. Our set of design rules assume that T is the same in both directions. Transistors designed with our ground rules may fail to operate properly if the misalignment exceeds tolerance T.

5.12.2 MOS TRANSISTOR LAYOUT

Figure 5.36 outlines the process and mask sequence used to fabricate a basic planar polysilicon-gate transistor. The first mask defines the active area, or thin oxide region of the transistor, and

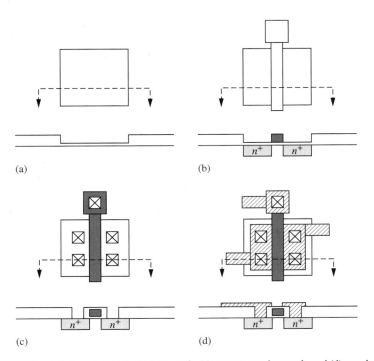

Figure 5.36 (a) Active area mask, (b) gate mask, (c) contact opening mask, and (d) metal mask.

the second mask defines the polysilicon gate of the transistor. The channel region of the transistor is actually produced by the intersection of these first two mask layers; the source and/or drain regions are formed wherever the active layer (mask 1) is *not* covered by the gate layer (mask 2). The third and fourth masks delineate the contact openings and the metal pattern. The overall mask sequence is

Active area mask	Mask 1
Polysilicon-gate mask	Mask 2—align to mask 1
Contact window mask	Mask 3—align to mask 2
Metal mask	Mask 4—align to mask 3

The alignment sequence must be specified to properly account for alignment tolerances in the ground rules. In this particular example, each mask is aligned to the one used in the preceding step, but this is not always the case.

We will now explore a set of design rules similar in concept to those developed by Mead and Conway [20]. These ground rules were designed to permit easy translation of a design from one generation of technology to another by simply changing the size of one parameter Λ. To achieve this goal, the rules are quite forgiving in terms of the mask-to-mask alignment tolerance.

A composite set of rules for a transistor is shown graphically in Fig. 5.37 in which the minimum feature size $F = 2\Lambda$ and the alignment tolerance $T = F/2 = \Lambda$. (Parameter Λ could be 0.5, 0.25, or 0.1 μm, for example.) Note that an alignment tolerance equal to one-half the minimum feature size is a very forgiving alignment tolerance.

For the transistor in Fig. 5.37, all linewidths and spaces must be a minimum feature size of 2Λ. Square contacts are a minimum feature size of 2Λ in each dimension. To ensure that the

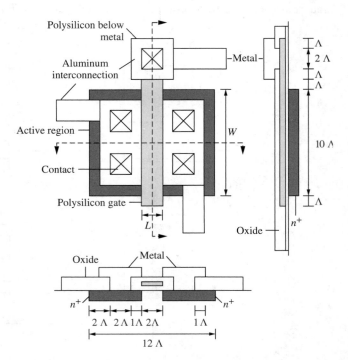

Figure 5.37 Composite top view and cross sections of a transistor with $W/L = 5/1$ demonstrating a basic set of ground rules.

metal completely covers the contact for worst-case misalignment, a 1 Λ border of metal is required around the contact region. The polysilicon gate must overlap the edge of the active area and the contact openings by 1 Λ. However, because of the potential for tolerance accumulation during successive misalignments of masks 2 and 3, the contacts must be inside the edges of the active area by 2 Λ.

The transistor in Fig. 5.37 has a W/L ratio of 10 Λ/2 Λ or 5/1, and the total active area is 120 Λ². Thus the active channel region represents approximately 17 percent of the total area of the transistor. Note that the polysilicon gate defines the edges of the source and/or drain regions and results in "self-alignment" of the edges of the gate to the edges of the channel region. Self-alignment of the gate to the channel reduces the size of the transistor and minimizes the "overlap capacitances" associated with the transistor.

EXERCISE: (a) What is the active area of the transistor in Fig. 5.37 if Λ = 0.125 µm? What are the values of W and L for the transistor? What is the area of the transistor gate region? How many of these transistors could be packed together on a 10 mm × 10 mm integrated circuit die if the active areas of the individual transistors must be spaced apart by a minimum of 4 Λ? (b) Repeat for Λ = 20 nm.

ANSWERS: (a) 1.88 µm²; 1.25 µm, 0.25 µm; 0.31 µm²; 28.6 × 10⁶
(b) 0.048 µm²; 200 nm, 40 nm; 8.0 × 10⁻³ µm²; 1.30 × 10⁹

5.12.3 CMOS INVERTER LAYOUT

Two possible layouts for CMOS inverters appear in Fig. 5.38. In the left-hand layout, the lower transistor is the NMOS device in the p-type substrate, whereas the upper transistor is the PMOS device, which is within the boundary of the n-well. The polysilicon gates of the two transistors are connected together to form the input v_I, and the drain diffusions of the two transistors are connected together by the metallization to form output v_O. The sources of the NMOS and PMOS

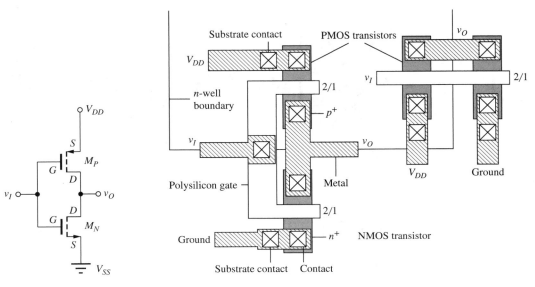

Figure 5.38 Layout of two CMOS inverters.

devices are at the bottom and top of the drawing, respectively, and are connected to the ground and power supply buss metallization. Each device has a local substrate contact connected to the source of the transistor. In this particular layout, each transistor has a W/L of 2/1. In the right-hand layout, the two transistors appear side by side with a common polysilicon gate running through both. Various options for the design of the W/L ratios of the transistors are discussed in the rest of this chapter.

5.13 ADVANCED CMOS TECHNOLOGIES

As semiconductor technology is scaled to smaller and smaller dimensions, leakage current in the subthreshold region becomes more and more of a problem, particularly for logic circuits. When the gate-source voltage is zero, the transistor is expected to be off (nonconducting), but we know from Fig. 5.31 that there is a small nonzero leakage current.

As mentioned earlier, the threshold voltage must be reduced as the transistor is scaled to smaller dimensions, and this substantially increases the leakage current as depicted in Fig. 5.33. For $V_{GS} = 0$, the drain currents are 0.7 fA for $V_{TN} = 0.7$ V and 1.7 pA with $V_{TN} = 0.4$ V, an increase of a factor of more than 2000! These numbers sound small but become large when an IC die contains 10 or 100 billion transistors.

The leakage problem occurs because the gate sits on the top of the transistor structure in the planar device and is simply unable to quench the channel "deep" in semiconductor. Since the beginning of the last decade (2010 and beyond), the new non-planar structures in Fig. 5.39 have been created that gain better control of the device by having the gate surround the channel. The FinFET has been in use below the 28-nm technology node, whereas the stacked nanosheet transistor appears to be a prime candidate below 5- to 7-nm geometries, albeit many fabrication issues remain to be overcome.

The accompanying Electronics in Action box describes the FinFET structure and layout in detail. Note that the gate surrounds the channel on three sides, and the fins are all aligned in the same direction (Fig. 5.39) due to the difficulty of achieving uniform FETs during fabrication.

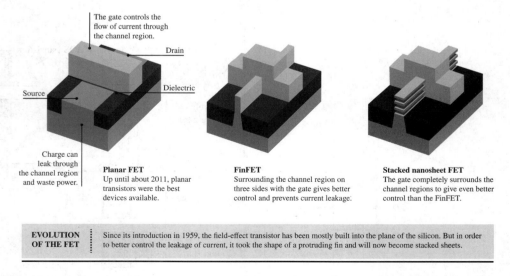

Figure 5.39 FET evolution from the classic planar structure to the three-dimensional FinFET and Nanosheet transistors [21].

Republished with permission of The Institute of Electrical and Electronics Engineers, Incorporated (IEEE), from The Last Silicon Transistor, P. Ye, T. Ernst and M. V. Khare, IEEE Spectrum Volume: 56, Issue: 8, Aug. 2019, permission conveyed through Copyright Clearance Center, Inc.

ELECTRONICS IN ACTION

FinFET Technologies

CMOS processes aimed at the 22-nm technology node and below are using some form of the "FinFET" transistor depicted in Fig. 1(a). In this idealized drawing of Intel's Tri-gate structure, a vertical silicon fin is created on the surface of the silicon wafer and surrounded by an insulating oxide region.[11] Holes or electrons will move laterally along the fin from source to drain. The metallic gate, TiN for example, is fabricated as a stripe perpendicular to the channel fin, and the channel length L of the device is equal to the width of this gate stripe. The TiN gate surrounds the channel region of the device, effectively placing three gates (i.e., Tri-gate) in parallel yielding a gate width given approximately by $W = 2H_{fin} + W_{fin}$. The gate and channel are separated by a high dielectric constant ("high-k") gate material to provide large C''_{ox}. Contacts are subsequently made to the source and drain ends of the fin as well as to the gate. Substrate contact is provided where the fin extends down through the oxide. The oxides in Fig. 1 are formed using a process called shallow trench isolation or STI that was originally developed to provide isolation in planar CMOS technologies and represents another important feature of advanced CMOS processes.

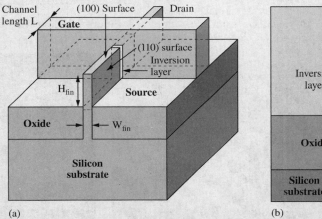
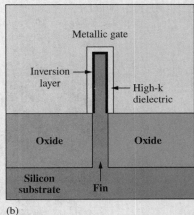

Figure 1 (a) Tri-gate Transistor structure; (b) cross-section in gate perpendicular to the channel. *Auth, Chris. "22 nm fully depleted Tri-gate CMOS transistors," IEEE CICC Digest, pp. 1–6, 2012. https://ieeexplore.ieee.org/document/6330657.*[12]

Figure 2 is a photomicrograph of an actual fin that is approximately 8 nm wide and 34 nm high. The fin structure provides high transconductance with low parasitic capacitances. The surrounding gate quenches the channel more effectively than in a planar device thereby minimizing off-state (subthreshold) leakage current, a significant problem in large microprocessors (see Sec. S7.4, online only). This structure also minimizes or eliminates the latch-up phenomenon.

Multi-fin structures, as in Fig. 3, can be used to increase the current carrying capacity of the device by connecting multiple fins in parallel, or this structure can be used as three transistors with a common gate signal. Figure 4 shows the realization of an array of six-transistor (6T) static memory cells in Tri-gate technology.[13] The layout demonstrates the use of multiple gates on a

---

[11] Think of a shark fin sticking up out of the water.
[12] Adapted from C. Auth, "22 nm fully depleted Tri-gate CMOS transistors," *2012 IEEE CICC Digest*, pp. 1–6. © IEEE.
[13] E. Karl et al., "A 4.6 Ghz 162 Mb SRAM design in 22 nm Tri-gate CMOS technology with integrated read and write assist circuitry," *IEEE JSSC*, vol. 48, no. 1, pp. 150–158, January 2013.

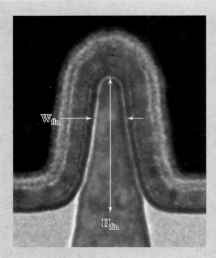

Figure 2 PMOS Fin — $W_{Fin} = 8$ nm and $H_{Fin} = 34$ nm.[10]
Adapted from C. Auth, "22 nm fully depleted Tri-gate CMOS transistors," 2012 IEEE CICC Digest, pp. 1–6. ©IEEE. Used with permission.

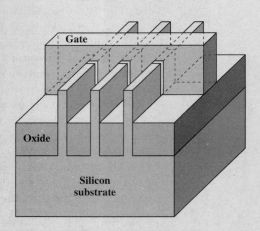

Figure 3 Multi-fin Tri-gate structure © *IEEE Adapted.*[10]

Figure 4 6-T SRAM cell in Tri-gate technology. *Adapted from C. Auth, "22 nm fully depleted Tri-gate CMOS transistors," 2012 IEEE CICC Digest, pp. 1–6. ©IEEE. Used with permission.*

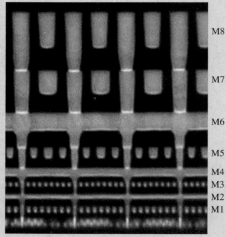

Figure 5 Eight-level metallization in an advanced CMOS process. *Courtesy of Intel Corporation.*

single fin as well as the connections between the ends of fins and gates to implement the various logic functions required by the SRAM circuit (see Chapter S8, e-book only).

Because of the extreme complexity of state-of-the-art microprocessor chips, advanced CMOS processes require the use of many levels of metallization in order to be able to successfully connect billions of transistors. Figure 5 shows the eight-levels of metallization above the transistor level as utilized in the Tri-gate technology discussed here.[14]

---

[14] C. Auth et al., "A 22 nm high performance and low-power CMOS technology featuring fully-depleted Tri-Gate transistors, self-aligned contacts and high density MIM capacitors," *2012 VLSI Technology Symposium Digest*, pp. 131–132, June 2012.

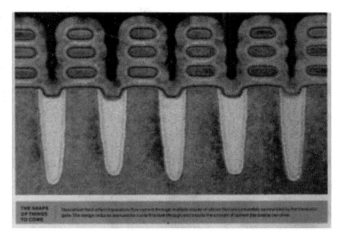

Figure 5.40 Nanosheet transistors.
Republished with permission of The Institute of Electrical and Electronics Engineers, Incorporated (IEEE), from The Last Silicon Transistor, P. Ye, T. Ernst and M. V. Khare, IEEE Spectrum Volume: 56, Issue: 8, Aug. 2019, permission conveyed through Copyright Clearance Center, Inc.

As mentioned earlier, the stacked nanosheet transistor in Fig. 5.39 appears to be a prime technology candidate for use below 5–7 nm. In the multi-layer nanosheet transistors in Figs. 5.39 and 5.40, the gates of the various sheets completely surround the channel regions in the individual layers and are connected directly together as part of the fabrication. The nanosheet structure achieves maximum control of the channel region minimizing the subthreshold leakage currents and must be duplicated for both NMOS and PMOS transistors for CMOS designs.

5.14 BIASING THE NMOS FIELD-EFFECT TRANSISTOR

The MOS circuit designer has the flexibility to choose the circuit topology and W/L ratios of the devices in the circuit, and to a lesser extent, the voltages applied to the devices. As designers, we need to develop a mental catalog of useful circuit configurations, and we begin by looking at several basic circuits for biasing the MOSFET.

5.14.1 WHY DO WE NEED BIAS?

We have found that the MOSFET has three regions of operation: cutoff, triode, and saturation (four, if one includes velocity saturation). For circuit applications, we want to establish a well-defined **quiescent operating point,** or **Q-point,** for the MOSFET in a particular region of operation. The Q-point for the MOSFET is represented by the dc values (I_D, V_{DS}) that locate the operating point on the MOSFET output characteristics. [In reality, we need the three values (I_D, V_{DS}, V_{GS}), but two are enough to calculate the third if we know the region of operation of the device.]

For binary logic circuits, the transistor acts as an "on-off" switch, and the Q-point is set to be in either the cutoff region ("off") or the triode region ("on"). For example, let us explore the circuit in Fig. 5.41(a) that can be used as either a logic inverter or a linear amplifier depending upon our choice of operating points. The voltage transfer characteristic (VTC) for the circuit appears in Fig. 5.42(a). For low values of v_{GS}, the MOSFET is off, and the output voltage is 5 V, corresponding to a binary "1" in a logic application. As v_{GS} increases, the output begins to drop and finally reaches its "on-state" voltage of 0.65 V for $v_{GS} = 5$ V. This voltage would correspond to a "0" in binary logic. These two logic states are also shown on the transistor output characteristics in Fig. 5.42(b). When the transistor is "on," it conducts a substantial current, and v_{DS} falls to 0.65 V. When the transistor is off, v_{DS} equals 5 V. The design of logic gates is discussed in detail in the supplemental chapters in the e-book.

282 Chapter 5 Field-Effect Transistors

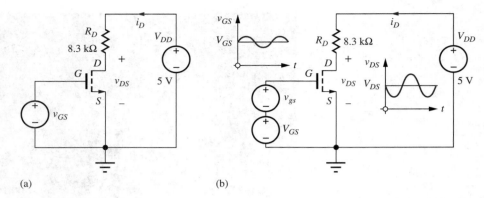

Figure 5.41 (a) Circuit for a logic inverter; (b) the same transistor used as a linear amplifier.

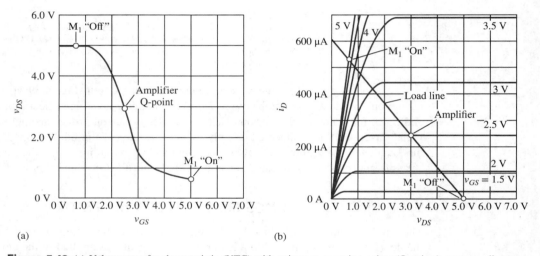

Figure 5.42 (a) Voltage transfer characteristic (VTC) with quiescent operating points (Q-points) corresponding to an "on-switch," an amplifier, and an "off-switch"; (b) the same three operating points located on the transistor output characteristics.

For amplifier applications, the Q-point is located in the region of high slope (high gain) near the center of the voltage transfer characteristic, also indicated in Fig. 5.42(a). At this operating point, the transistor is operating in saturation, the region in which high voltage, current, and/or power gain can be achieved. To establish this Q-point, a **dc bias** V_{GS} is applied to the gate as in Fig. 5.41(b), and a small ac signal v_{gs} is added to vary the gate voltage around the bias value.[15] The variation in total gate-source voltage v_{GS} causes the drain current to change, and an amplified replica of the ac input voltage appears at the drain. Our study of the design of transistor amplifiers begins in Chapter 7.

The straight line connecting the Q-points in Fig. 5.42(b) is the *load line* that was first encountered in Chapters 3 and 4. The dc load line plots the permissible values of I_D and V_{DS} as determined by the external circuit. In this case, the load line equation is given by

$$V_{DD} = I_D R_D + V_{DS}$$

---

[15] Remember $v_{GS} = V_{GS} + v_{gs}$.

For hand analysis and design of Q-points, channel-length modulation is usually ignored by assuming $\lambda = 0$. A review of Fig. 5.11 indicates that including λ changes the drain current by less than 10 percent. Generally, we do not know the values of transistor parameters to this accuracy, and the tolerances on both discrete or integrated circuit elements may be as large as 30 to 50 percent. If you explore some transistor specification sheets, you will discover parameters that have a 4 or 5 to 1 spread in values. You will also find parameters with only a minimum or maximum value specified. Thus, neglecting λ will not significantly affect the validity of our analysis. Also, many bias circuits involve feedback which further reduces the influence of λ. On the other hand, in Part Two we will see that λ can play an extremely important role in limiting the voltage gain of analog amplifier circuits, and the effect of λ must often be included in the analysis of these circuits.

To analyze circuits containing MOSFETs, we must first assume a region of operation, just as we did to analyze diode circuits in Chapter 3. The bias circuits that follow will most often be used to place the transistor Q-point in the saturation region, and by examining Eq. (5.30) with $\lambda = 0$, we see that we must know the gate-source voltage V_{GS} to calculate the drain current I_D. Then, once we know I_D, we can find V_{DS} from the constraints of Kirchhoff's voltage law. Thus our most frequently used analysis approach will be to first find V_{GS} and then to use its value to find the value of I_D. I_D will then be used to calculate V_{DS}.

Menu for Bias Analysis

1. Assume a region of operation (most often either the saturation or velocity-limited regions).
2. Use circuit analysis to find V_{GS}.
3. Use V_{GS} to calculate I_D, and I_D to determine V_{DS}.
4. Check the validity of the operating region assumptions.
5. Change assumptions and analyze again if necessary.

DESIGN NOTE   **SATURATION BY CONNECTION!**

When making bias calculations for analysis or design, it is useful to remember that an NMOS *enhancement-mode* device that is operating with $V_{DS} = V_{GS}$ will always be in the saturation region. The same is true for an enhancement-mode PMOS transistor with $V_{SD} = V_{SG}$.

To demonstrate this result, it is easiest to keep the signs straight by considering an NMOS device with dc bias. For saturation, it is required that

$$V_{DS} \geq V_{GS} - V_{TN}$$

But if $V_{DS} = V_{GS}$, this condition becomes

$$V_{DS} \geq V_{DS} - V_{TN} \quad \text{or} \quad V_{TN} \geq 0$$

which is always true if V_{TN} is a positive number. $V_{TN} > 0$ corresponds to an NMOS enhancement-mode device. Thus an enhancement-mode device operating with $V_{DS} = V_{GS}$ is always saturated! Similar arguments hold true for enhancement-mode PMOS devices.

5.14.2 FOUR-RESISTOR BIASING

The circuit in Fig. 5.41(b) provides a fixed gate-source bias voltage to the transistor. Theoretically, this works fine. However, in practice the values of K_n, V_{TN}, and λ for the MOSFET will not be known with high precision and the Q-point is not well-controlled. In addition, we must be concerned about resistor and power supply tolerances (you may wish to review Sec. 1.8) as well as component value drift with both time and temperature in an actual circuit. The four-resistor bias circuit discussed in this section uses negative feedback to provide a well-stabilized Q-point. We will encounter circuits employing negative feedback throughout the rest of this text.

EXAMPLE 5.1 FOUR-RESISTOR BIAS CIRCUIT

The most general and important bias method that we will encounter is the **four-resistor bias circuit** in Fig. 5.43(a) that is same circuit topology as that discussed for the bipolar transistor in Chapter 4, Sec. 4.7. In this case, resistor R_S helps stabilize the MOSFET Q-point in the face of many types of circuit parameter variations. This bias circuit is actually a form of *feedback circuit*, which will be studied in great detail in Parts Two and Three of this text. Also observe that a single voltage source V_{DD} is used to supply both the gate-bias voltage and the drain current. The four-resistor bias circuit is most often used to place the transistor in the saturation region of operation for use as an amplifier for analog signals.

PROBLEM Find the Q-point = (I_D, V_{DS}) for the MOSFET in the four-resistor bias circuit in Fig. 5.43.

SOLUTION **Known Information and Given Data:** Circuit schematic in Fig. 5.43 with $V_{DD} = 10$ V, $R_1 = 1$ MΩ, $R_2 = 1.5$ MΩ, $R_D = 75$ kΩ, $R_S = 39$ kΩ, $K_n = 25$ μA/V^2, and $V_{TN} = 1$ V.

Unknowns: Q-point = (I_D, V_{DS}), V_{GS}, and region of operation

Approach: We can find the Q-point using the mathematical model for the NMOS transistor. We assume a region of operation, determine the Q-point, and check to see if the resulting Q-point is consistent with the assumed region of operation.

Assumptions: The first step in our Q-point analysis of the equivalent circuit in Fig. 5.43 is to assume that the transistor is saturated (remember to use $\lambda = 0$):

$$I_D = \frac{K_n}{2}(V_{GS} - V_{TN})^2 \tag{5.70}$$

Since V_{SAT} is not specified we will assume there are no velocity limits. Also, $I_G = 0 = I_B$. Using the $\lambda = 0$ assumption simplifies the mathematics because I_D is then modeled as being independent of V_{DS}.

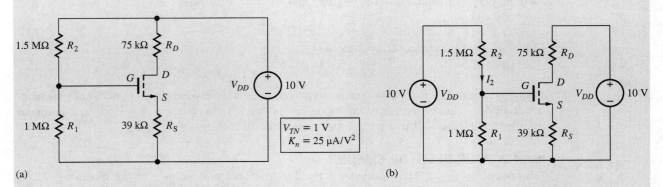

Figure 5.43 (a) Four-resistor bias network for a MOSFET; (b) equivalent circuit with replicated sources.

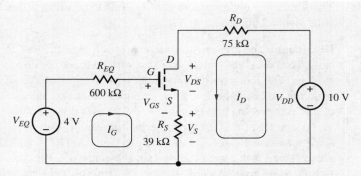

Figure 5.44 Equivalent circuit for the four-resistor bias network.

Analysis: To find I_D, the gate-source voltage must be determined, and we begin by simplifying the circuit. In the equivalent circuit in Fig. 5.43(b), the voltage source V_{DD} has been split into two equal-valued sources. After the Thévenin transformation is applied to this circuit, the resulting equivalent circuit is given in Fig. 5.44 in which the variables have been clearly labeled. This is the final circuit to be analyzed. We recognize that the gate-bias voltage is determined by V_{EQ} and R_{EQ}.

Detailed analysis begins by writing the input loop equation containing V_{GS}:

$$V_{EQ} = I_G R_{EQ} + V_{GS} + (I_G + I_D)R_S \quad \text{or} \quad V_{EQ} = V_{GS} + I_D R_S \tag{5.71}$$

because we know that $I_G = 0$. Substituting Eq. (5.70) into Eq. (5.71) yields

$$V_{EQ} = V_{GS} + \frac{K_n R_s}{2}(V_{GS} - V_{TN})^2 \tag{5.72}$$

and solving for V_{GS} using the quadratic equation yields

$$V_{GS} = V_{TN} + \frac{1}{K_n R_S}\left(\sqrt{1 + 2K_n R_S(V_{EQ} - V_{TN})} - 1\right) \tag{5.73}$$

Substitution of this result back into Eq. (5.70) gives us I_D:

$$I_D = \frac{1}{2K_n R_S^2}\left(\sqrt{1 + 2K_n R_S(V_{EQ} - V_{TN})} - 1\right)^2 \tag{5.74}$$

The second part of the Q-point, V_{DS}, can now be determined by writing the "output" loop equation including the drain-source terminals of the device.

$$V_{DD} = I_D R_D + V_{DS} + (I_G + I_D)R_S \quad \text{or} \quad V_{DS} = V_{DD} - I_D(R_D + R_S) \tag{5.75}$$

Equation (5.75) has been simplified since we know $I_G = 0$.

For the specific values in Fig. 5.44 with $V_{TN} = 1$, $K_n = 25\,\mu\text{A/V}^2$, $V_{EQ} = 4$ V and $V_{DD} = 10$ V, the Q-point values are

$$I_D = \frac{1}{2(25 \times 10^{-6})(39 \times 10^3)^2}\left(\sqrt{1 + 2(25 \times 10^{-6})(39 \times 10^3)(4 - 1)} - 1\right)^2 = 34.4\,\mu\text{A}$$

$$V_{DS} = 10 - 34.4\,\mu\text{A}(75\,\text{k}\Omega + 39\,\text{k}\Omega) = 6.08\,\text{V}$$

Check of Results: Checking the saturation region assumption for $V_{DS} = 6.08$ V, we have

$$V_{GS} - V_{TN} = \frac{1}{K_n R_S}\left(\sqrt{1 + 2K_n R_S(V_{EQ} - V_{TN})} - 1\right) = 1.66\,\text{V} \quad \text{so} \quad V_{DS} > (V_{GS} - V_{TN}) \checkmark$$

The saturation region assumption is consistent with the resulting Q-point: (34.4 µA, 6.08 V) with $V_{GS} = 2.66$ V.

Discussion: The four-resistor bias circuit is one of the best for biasing transistors in discrete circuits. The bias point is well stabilized with respect to device parameter variations and temperature changes. The four-resistor bias circuit is most often used to place the transistor in the saturation region of operation for use as an amplifier for analog signals, and as mentioned at the beginning of this example, the bias circuit in Fig. 5.43 uses negative feedback to stabilize the operating point. The operation of this feedback mechanism can be viewed in the following manner. Suppose for some reason that I_D begins to increase. Equation (5.71) indicates that an increase in I_D must be accompanied by a decrease in V_{GS} since V_{EQ} is fixed. But, this decrease in V_{GS} will tend to restore I_D back to its original value [see Eq. (5.70)]. This is negative feedback in action!

Note that this circuit uses the three-terminal representation for the MOSFET, in which it is assumed that the bulk terminal is tied to the source. If the bulk terminal is instead grounded, the analysis becomes more complex because the threshold voltage is then a function of the voltage developed at the source terminal of the device. This case will be investigated in more detail in Ex. 5.2. Let us now use the computer to explore the impact of neglecting λ in our hand analysis.

Computer-Aided Analysis: If we use SPICE to simulate the circuit using a LEVEL = 1 model and the parameters from our hand analysis (KP = 25 µA/V^2 and VTO = 1 V), we get exactly the same Q-point (34.4 µA, 6.08 V). If we add LAMBDA = 0.02 V^{-1}, SPICE yields a new Q-point of (35.9 µA, 5.91 V). The Q-point values change by less than 5 percent, a value that is well below our uncertainty in the device parameter and resistor values in a real situation.

EXERCISE: Use the quadratic equation to derive Eq. (5.76) and then verify the result given in Eq. (5.76)

EXERCISE: Suppose K_n increases to 30 µA/V^2 for the transistor in Fig. 5.44. What is the new Q-point for the circuit?

ANSWER: (36.8 µA, 5.81 V)

EXERCISE: Suppose V_{TN} changes from 1 V to 1.5 V for the MOSFET in Fig. 5.44. What is the new Q-point for the circuit?

ANSWER: (26.7 µA, 6.96 V)

EXERCISE: Find the Q-point in the circuit in Fig. 5.46 if R_S is changed to 62 kΩ.

ANSWER: (25.4 µA, 6.52 V)

DESIGN NOTE

The Q-point values (I_D, V_{DS}) for the MOS transistor operating in the saturation region using the four-resistor bias network are

$$I_D = \frac{1}{2K_n R_S^2}\left(\sqrt{1 + 2K_n R_S(V_{EQ} - V_{TN})} - 1\right)^2 \quad \text{and} \quad V_{DS} = V_{DD} - I_D(R_D + R_S)$$

where V_{EQ} is the Thévenin equivalent voltage between the gate terminal and ground.

EXERCISE: Show that the Q-point in the circuit in Fig. 5.43 for $R_1 = 1\,\text{M}\Omega$, $R_2 = 1.5\,\text{M}\Omega$, $R_S = 1.8\,\text{k}\Omega$, and $R_D = 39\,\text{k}\Omega$ is (99.5 µA, 5.94 V).

EXERCISE: Find the Q-point in the circuit in Fig. 5.43 for $R_1 = 1.5\,\text{M}\Omega$, $R_2 = 1\,\text{M}\Omega$, $R_S = 22\,\text{k}\Omega$, and $R_D = 18\,\text{k}\Omega$.

ANSWER: (99.2 µA, 6.03 V)

EXERCISE: Redesign the values of R_1 and R_2 to set the bias current to 2 µA while maintaining $V_{EQ} = 6\,\text{V}$. What is the value of R_{EQ}?

ANSWERS: 3 MΩ, 2 MΩ, 1.2 MΩ

DESIGN NOTE — GATE VOLTAGE DIVIDER DESIGN

Resistors R_1 and R_2 in Fig. 5.43 are required to set the value of V_{EQ}, but the current in the resistors does not contribute directly to operation of the transistor. Thus we would like to minimize the current "lost" through R_1 and R_2. The sum $(R_1 + R_2)$ sets the current in the gate bias resistors. As a rule of thumb, $R_1 + R_2$ is usually chosen to limit the current to no more than a few percent of the value of the drain current. The value of current I_2 is 4 percent of the drain current $I_2 = 10\,\text{V}/(1\,\text{M}\Omega + 1.5\,\text{M}\Omega) = 4\,\mu\text{A}$.

5.14.3 CONSTANT GATE-SOURCE VOLTAGE BIAS

The circuit in Fig. 5.41(a) represents a special case of the four-resistor bias circuit in which $R_S = 0$. For this case, Eqs. (5.72), (5.73), and (5.75) become

$$V_{GS} = V_{EQ} \qquad i_D = \frac{K_n}{2}(V_{EQ} - V_{TN})^2 \qquad V_{DS} = V_{DD} - I_D R_D \qquad (5.76)$$

The problem with this type of bias is immediately apparent. The drain current is highly dependent upon transistor parameters K_n and V_{TN} that typically have a broad spread in their specified values. Thus the Q-point, and possibly the operating region, will be poorly controlled, so this type of bias is seldom utilized.

5.14.4 GRAPHICAL ANALYSIS FOR THE Q-POINT

The Q-point for the four-resistor bias circuit in Figs. 5.43 and 5.44 can also be found graphically with a load line method similar to the one used for diode circuits in Sec. 3.10 or the BJT in Sec. 4.11. The graphical approach helps us visualize the operating point of the device and its location relative to the boundaries of the various operating regions of the transistor. For this circuit, however, the gate-source voltage of the MOSFET depends upon the drain current, and we must use both the output and transfer characteristics of the transistor to find the Q-point.

Load Line and Bias Line Analysis

The *Load Line* relates V_{DS} to I_D. For the four-resistor bias circuit, V_{DS} is equal to the power supply voltage minus the voltage drops across the drain and source resistors:

$$V_{DS} = V_{DD} - I_D(R_D + R_S) \qquad (5.77)$$

In this circuit, V_{GS} is also dependent upon I_D. Since gate current I_G is 0, the value of V_{GS} is equal to gate bias voltage V_{EQ} minus the voltage drop across the source resistor:

$$V_{GS} = V_{EQ} - I_D R_S \qquad (5.78)$$

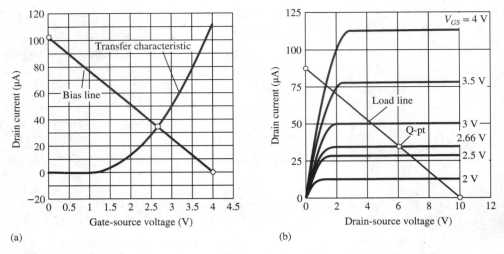

Figure 5.45 (a) Transfer characteristics and bias line; (b) output characteristics, load line, and Q-point.

This expression is referred to as the *Bias Line*. To find the Q-point, we first plot the bias line on the transfer characteristic for the transistor to find the value of V_{GS}. Then we plot the load line on the output characteristics and locate the Q-point based upon the value of V_{GS}. For the values in Fig. 5.46, the bias and load line expressions are

$$V_{GS} = 4 - 3.90 \times 10^4 I_D \quad \text{and} \quad V_{DS} = 10 - 1.14 \times 10^5 I_D \quad (5.79)$$

The bias line is plotted on the transfer characteristic (I_D vs. V_{GS}) in Fig. 5.45(a) just as for the diode circuits in Sec. 3.10. Two points are required to establish the bias line. For $I_D = 0$, $V_{GS} = 4$ V, and for $V_{GS} = 0$, $I_D = 103$ μA. The bias line is plotted on the transfer characteristics, and the intersection of the two curves is the value of V_{GS}. In this case, $V_{GS} = 2.66$ V.

Similarly, we need two points to plot the load line: for $I_D = 0$, $V_{DS} = 10$ V, and for $V_{DS} = 0$, $I_D = 87.7$ μA. The load line is plotted on the output characteristics in Fig. 5.45(b). The $V_{GS} = 2.66$ V point is estimated by interpolation between the 2.5 V and 3 V curves. Reading the Q-point from the graph yields $V_{DS} = 6$ V and $I_D = 35$ μA. These Q-point values are close to the more precise values found using the mathematical model.

5.14.5 ANALYSIS INCLUDING BODY EFFECT

In integrated circuits, MOS transistors are frequently operated with separate connections to the source and bulk terminals resulting in a nonzero source-bulk voltage V_{SB} as depicted in Fig. 5.46. In this circuit, $V_{SB} = I_D R_S$, and hence threshold voltage V_{TN} is not fixed, but varies as the drain current changes. This complicates the analysis and alters the Q-point from the one found in Sec. 5.15.2 (34.4 μA, 6.08 V). Example 5.2 explores the impact of the connection in Fig. 5.46.

EXAMPLE 5.2 ANALYSIS INCLUDING BODY EFFECT

The NMOS transistor in Fig. 5.44 was connected as a three-terminal device. This example explores how the Q-point is altered when the substrate is connected as shown in Fig. 5.46.

5.14 Biasing the NMOS Field-Effect Transistor

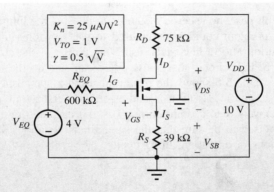

Figure 5.46 MOSFET with redesigned bias circuit.

PROBLEM Find the Q-point = (I_D, V_{DS}) for the MOSFET in the four-resistor biasing circuit in Fig. 5.46 including the influence of body effect on the transistor threshold.

SOLUTION **Known Information and Given Data:** The circuit schematic in Fig. 5.46 with $V_{EQ} = 4$ V, $R_{EQ} = 600$ kΩ, $R_S = 39$ kΩ, $R_D = 75$ kΩ, $K_n = 25$ µA/V^2, $V_{TO} = 1$ V, and $\gamma = 0.5$ V^{-1}

Unknowns: I_D, V_{DS}, V_{GS}, V_{BS}, V_{TN}, and region of operation

Approach: In this case, the source-bulk voltage, $V_{SB} = I_S R_S = I_D R_S$, is no longer zero, and we must solve the following set of equations:

$$V_{GS} = V_{EQ} - I_D R_S \qquad V_{SB} = I_D R_S$$
$$V_{TN} = V_{TO} + \gamma \left(\sqrt{V_{SB} + 2\phi_F} - \sqrt{2\phi_F} \right) \qquad (5.80)$$
$$I_D = \frac{K_n}{2}(V_{GS} - V_{TN})^2$$

Although it may be possible to solve these equations analytically, it will be more expedient to find the Q-point by iteration using the computer with a spreadsheet, MATLAB, MATHCAD, or with a calculator.

Assumptions: Saturation region operation with $I_G = 0$, $I_B = 0$, and $2\phi_F = 0.6$ V

Analysis: Using the assumptions and values in Fig. 5.45, Eq. set (5.80) becomes

$$V_{GS} = 4 - 39{,}000 I_D \qquad V_{SB} = 39{,}000 I_D$$
$$V_{TN} = 1 + 0.5\left(\sqrt{V_{SB} + 0.6} - \sqrt{0.6}\right) \qquad I'_D = \frac{25 \times 10^{-6}}{2}(V_{GS} - V_{TN})^2 \qquad (5.81)$$

and the drain-source voltage is found from

$$V_{DS} = V_{DD} - I_D(R_D + R_S) = 10 - 114{,}000 I_D \qquad (5.82)$$

The expressions in Eq. (5.81) have been arranged in a logical order for an iterative solution:

1. Estimate the value of I_D.
2. Use I_D to calculate the values of V_{GS} and V_{SB}.
3. Calculate the resulting value of V_{TN} using V_{SB}.
4. Calculate I'_D using the results of steps 1 to 3, and compare to the original estimate for I_D.
5. If the calculated value of I'_D is not equal to the original estimate for I_D, then go back to step 1.

In this case, no specific method for choosing the improved estimate for I_D is provided (although the problem could be structured to use Newton's method), but it is easy to converge to the solution after a few trials, using the power of the computer to do the calculations. (Note that the SPICE circuit analysis program can also do the job for us.)

Table 5.6 shows the results of using a spreadsheet to iteratively find the solution to Eqs. (5.81) and (5.82) by trial and error. The first iteration sequence used by the author is shown; it converges to a drain current of 30.0 μA and drain-source voltage of 6.58 V. Care must be exercised to be sure that the spreadsheet equations are properly formulated to account for all regions of operation. In particular, $I_D = 0$ if $V_{GS} < V_{TN}$.

TABLE 5.6
Four-Resistor Bias Iteration

I_D	V_{SB}	V_{GS}	V_{TN}	I'_D	V_{DS}
4.750E-05	1.85E+00	2.15E+00	1.40E+00	7.065E-06	9.19E+00
5.000E-05	1.95E+00	2.05E+00	1.41E+00	5.102E-06	9.42E+00
4.000E-05	1.56E+00	2.44E+00	1.35E+00	1.492E-05	8.30E+00
3.000E-05	1.17E+00	2.83E+00	1.28E+00	3.011E-05	6.57E+00
2.995E-05	1.17E+00	2.83E+00	1.28E+00	3.020E-05	6.56E+00
3.010E-05	1.17E+00	2.83E+00	1.28E+00	2.993E-05	6.59E+00
3.005E-05	1.17E+00	2.83E+00	1.28E+00	3.002E-05	6.58E+00
3.004E-05	1.17E+00	2.83E+00	1.28E+00	3.004E-05	6.58E+00

Check of Results: For this design, we now have

$$V_{DS} = 6.58 \text{ V}, V_{GS} - V_{TN} = 1.55 \text{ V} \quad \text{and} \quad V_{DS} > (V_{GS} - V_{TN}) \checkmark$$

The saturation region assumption is consistent with the solution, and the Q-point is (30.0 μA, 6.58 V).

Discussion: Now that the analysis is complete, we see that the presence of body effect in the circuit has caused the threshold voltage to increase from 1 V to 1.28 V and the drain current to decrease by approximately 13 percent from 34.4 μA to 30.0 μA.

EXERCISE: Find the new drain current in the circuit in Fig. 5.45 if $\gamma = 0.75\sqrt{\text{V}}$.

ANSWER: 28.2 μA

Examples 5.1 and 5.2 have demonstrated the techniques that we need to analyze most of the circuits we will encounter. The four-resistor and two-resistor bias circuits (see Probs. 5.115–5.150) are most often encountered in discrete design, whereas current sources and current mirrors, introduced in Chapter 13, find extensive application in integrated circuit design.

5.14.6 ANALYSIS USING THE UNIFIED MODEL
The Unified Model includes saturation of the carrier velocity, and this section presents an example of analysis of the MOSFET operating in the velocity-limited region.

EXAMPLE 5.3   ANALYSIS USING THE UNIFIED MODEL

This example finds the Q-point for the four-resistor bias circuit from Fig. 5.44 using the Unified Model with $V_{SAT} = 1$ V.

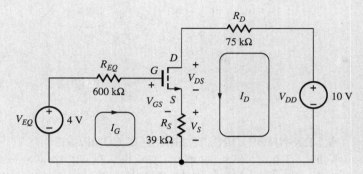

PROBLEM Find the Q-point (I_D, V_{DS}) for the MOSFET in the circuit in Fig 5.44 using the Unified Model with $V_{SAT} = 1$ V.

SOLUTION **Known Information and Given Data:** The circuit schematic in Fig. 5.44 with $V_{EQ} = 4$ V, $R_{EQ} = 600$ kΩ, $R_S = 39$ kΩ, $R_D = 75$ kΩ, $K_n = 25$ μA/V^2, and $V_{TN} = 1$ V, and $V_{SAT} = 1$ V.

Unknowns: Q-point (I_D, V_{DS}), V_{GS}, and the region of operation

Approach: In this case, we need to solve this set of equations:

$$I_D = K_n\left(V_{GS} - V_{TN} - \frac{V_{MIN}}{2}\right) V_{MIN} \quad \text{with} \quad V_{MIN} = \min\{V_{GS} - V_{TN}, V_{DS}, V_{SAT}\}$$

and $V_{GS} = V_{EQ} - I_D R_S$

Assumptions: Velocity-limited operation with $V_{GS} - V_{TN} > V_{SAT}$ so $V_{MIN} = V_{SAT}$.

Analysis: Using the values for the transistor and circuits yields

$$I_D = 25 \times 10^{-6}\left(V_{GS} - 1 - \frac{1}{2}\right)(1) \quad \text{and} \quad V_{GS} = 4 - 3.9 \times 10^4 I_D$$

Combining these equations and solving for I_D gives

$$I_D = \frac{2.5}{7.9 \times 10^4} = 31.7 \ \mu A$$

The drain-source voltage is found as

$$V_{DS} = V_{DD} - I_D(R_D + R_S) = 10 \text{ V} - 31.7 \ \mu A(114 \text{ k}\Omega) = 6.39 \text{ V}$$

and the Q-point is (31.7 μA, 6.39 V).

Check of Results: $V_{GS} - V_{TN} = 1.76$ V and $V_{DS} = 6.39$ V, so $V_{MIN} = 1$ V is correct.

Discussion: The Q-point is also similar to that found earlier for the saturation region of operation. The most important factor responsible for the relatively small change is the negative feedback provided by source resistor R_S that helps stabilize the Q-point against variations in transistor characteristics.

TABLE 5.7
Comparison of NMOS Model Results

MODEL	I_D (μA)	V_{DS} (V)	V_{GS} (V)
Basic 4R bias	34.4	6.08	2.66
4R Bias with $\lambda = 0.04$/V	37.1	5.77	2.55
4R + body effect	30.0	6.58	2.83
Unified Model	31.7	6.39	2.76
Monte Carlo analysis with 10% tolerances	(29.9, 40.0)	(5.68, 6.62)	(2.55, 2.79)
Average	34.1	6.10	2.65

5.14.7 NMOS CIRCUIT ANALYSIS COMPARISONS

Table 5.7 provides a comparison of the Q-point results for the previous three models as well as an additional Monte Carlo analysis of the basic four-resistor bias circuit with all tolerances of 10 percent, and an analysis with $\lambda = 0.04$/V. The worst-case deviations from the mean are approximately 10 percent. The tolerances on the resistors and transistor parameters cause larger differences than those between the models, as indicated by the Monte Carlo analysis in the last two rows of the table, so we tend to use the basic transistor model whenever possible and utilize circuit simulation when more complete modeling is desired.

5.14.8 TWO-RESISTOR BIAS

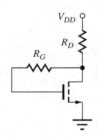

Figure 5.47 Basic two-resistor bias circuit.

A simple **feedback bias** circuit employing only two resistors is shown in Fig. 5.47. R_D and the transistor work together to set the Q-point. When the circuit is used as an amplifier, R_G provides isolation between the signal on the gate and that on the drain.

Since the gate current is zero, gate-source voltage V_{GS} equals drain-source voltage V_{DS}, and the transistor is saturated.

$$V_{GS} = V_{DD} - I_D R_D \quad \text{or} \quad V_{GS} = V_{DD} - \frac{K_n R_D}{2}(V_{GS} - V_{TN})^2 \quad (5.83)$$

Solving for V_{GS} yields

$$V_{GS} = V_{TN} - \frac{1}{K_n R_D} \pm \sqrt{\left(V_{TN} - \frac{1}{K_n R_D}\right)^2 + \frac{2V_{DD}}{K_n R_D} - V_{TN}^2} \quad (5.84)$$

and

$$I_D = \frac{K_n}{2}(V_{GS} - V_{TN})^2 \quad \text{and} \quad V_{DS} = V_{GS} \quad (5.85)$$

EXERCISE: Find the Q-point for the NMOS transistor in the two-resistor bias circuit if $V_{DD} = 10$ V, $K_n = 50$ μA, $V_{TN} = 1$ V, $R_D = 75$ kΩ and $R_G = 100$ kΩ.

ANSWER: $(I_D, V_{DS}) = (94.1$ μA, 2.94 V$)$. Checking: $V_{DS} = 10 - I_D R_D = 2.94$ V.

5.15 BIASING THE PMOS FIELD-EFFECT TRANSISTOR

CMOS technology, which uses a combination of NMOS and PMOS transistors, is the dominant IC technology in use today, and it is thus very important to know how to bias both types of devices. PMOS bias techniques mirror those used in the previous NMOS bias examples.

In the circuits that follow, you will observe that the source of the PMOS transistor will be consistently drawn at the top of the device since the source of the PMOS device is normally connected to a potential that is higher than the drain. This is in contrast to the NMOS transistor in which the drain is connected to a more positive voltage than the source. The PMOS model equations were summarized in Sec. 5.3. Remember that the drain current I_D is positive when coming out of the drain terminal of the PMOS device, and the values of V_{GS} and V_{DS} will be negative.

EXAMPLE 5.4 FOUR-RESISTOR BIAS FOR THE PMOS FET

The four-resistor bias circuit in Fig. 5.48 functions in a manner similar to that used for the NMOS device in Ex. 5.3. In the circuit in Fig. 5.48(a), a single voltage source V_{DD} is used to supply both the gate-bias voltage and the source-drain current. R_1 and R_2 form the gate voltage divider circuit. R_S sets the source/drain current, and R_D determines the source-drain voltage.

PROBLEM Find the quiescent operating point Q-point (I_D, V_{DS}) for the PMOS transistor in the four-resistor bias circuit in Fig. 5.48.

SOLUTION **Known Information and Given Data:** Circuit schematic in Fig. 5.48 with $V_{DD} = 10$ V, $R_1 = 1$ MΩ, $R_2 = 1.5$ MΩ, $R_D = 75$ kΩ, $R_S = 39$ kΩ, $K_P = 25$ μA/V^2, $V_{TP} = -1$ V, and $I_G = 0$

Unknowns: I_D, V_{DS}, V_{GS}, and the region of operation

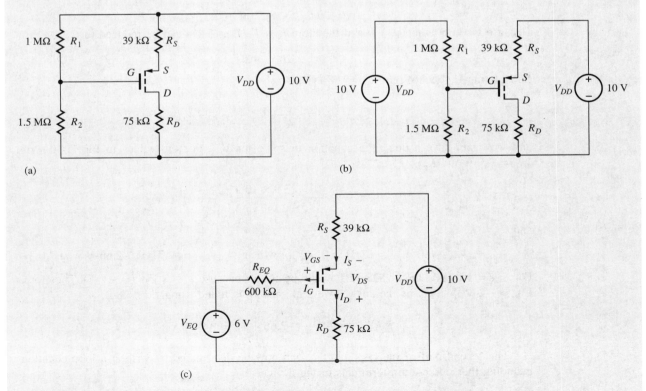

Figure 5.48 Four-resistor bias for a PMOS transistor.

Approach: We can find the Q-point using the mathematical model for the PMOS transistor. We assume a region of operation, determine the Q-point, and check to see if the Q-point is consistent with the assumed region of operation. First find the value of V_{GS}; use V_{GS} to find I_D; use I_D to find V_{DS}.

Assumptions: Assume that the transistor is operating in the saturation region (once again, remember to use $\lambda = 0$)

$$I_D = \frac{K_p}{2}(V_{GS} - V_{TP})^2 \tag{5.86}$$

Analysis: We begin by simplifying the circuit. In the equivalent circuit in Fig. 5.48(b), the voltage source has been split into two equal-valued sources, and in Fig. 5.48(c), the gate-bias circuit is replaced by its Thévenin equivalent

$$V_{EQ} = 10\text{ V}\frac{1.5\text{ M}\Omega}{1\text{ M}\Omega + 1.5\text{ M}\Omega} = 6\text{ V} \quad \text{and} \quad R_{EQ} = 1\text{ M}\Omega \parallel 1.5\text{ M}\Omega = 600\text{ k}\Omega$$

Figure 5.48(c) represents the final circuit to be analyzed (be sure to label the variables). Note that this circuit uses the three-terminal representation for the MOSFET, in which it is assumed that the bulk terminal is tied to the source. If the bulk terminal were connected to V_{DD}, the analysis would be similar to that used in Ex. 5.2 because the threshold voltage would then be a function of the voltage developed at the source terminal of the device.

To find I_D, the gate-source voltage must be determined, and we write the input loop equation containing V_{GS}:

$$V_{DD} = I_S R_S - V_{GS} + I_G R_G + V_{EQ} \tag{5.87}$$

Because we know that $I_G = 0$ and therefore $I_S = I_D$, Eq. (5.87) can be reduced to

$$V_{DD} - V_{EQ} = I_D R_S - V_{GS} \tag{5.88}$$

Substituting Eq. (5.86) into Eq. (5.88) yields

$$V_{DD} - V_{EQ} = \frac{K_p R_S}{2}(V_{GS} - V_{TP})^2 - V_{GS} \tag{5.89}$$

and we again have a quadratic equation to solve for V_{GS}. For the values in Fig. 5.50 with $V_{TP} = -1$ V and $K_p = 25$ µA/V^2

$$10 - 6 = \frac{(25 \times 10^{-6})(3.9 \times 10^4)}{2}(V_{GS} + 1)^2 - V_{GS}$$

and

$$V_{GS}^2 - 0.051\,V_{GS} - 7.21 = 0 \quad \text{for which} \quad V_{GS} = +2.71\text{ V}, -2.66\text{ V}$$

For $V_{GS} = +2.71$ V, the PMOS FET would be cut off because $V_{GS} > V_{TP} (= -1\text{ V})$. Therefore, $V_{GS} = -2.66$ V must be the answer we seek, and I_D is found using Eq. (5.86):

$$I_D = \frac{25 \times 10^{-6}}{2}(-2.66 + 1)^2 = 34.4\text{ µA}$$

The second part of the Q-point, V_{DS}, can now be determined by writing a loop equation including the source-drain terminals of the device:

$$V_{DD} = I_S R_S - V_{DS} + I_D R_D \quad \text{or} \quad V_{DD} = I_D(R_S + R_D) - V_{DS} \tag{5.90}$$

Eq. (5.90) has been simplified since we know that $I_S = I_D$. Substituting the values from the circuit gives

$$10 \text{ V} = (34.4 \text{ µA})(39 \text{ k}\Omega + 75 \text{ k}\Omega) - V_{DS} \quad \text{or} \quad V_{DS} = -6.08 \text{ V}$$

Check of Results: We have

$$V_{DS} = -6.08 \text{ V} \quad \text{and} \quad V_{GS} - V_{TP} = -2.66 \text{ V} + 1 \text{ V} = -1.66 \text{ V}$$

and $|V_{DS}| > |V_{GS} - V_{TP}|$. Therefore the saturation region assumption is consistent with the resulting Q-point (34.4 µA, −6.08 V) with $V_{GS} = -2.66$ V.

Evaluation and Discussion: As mentioned in Ex. 5.1, the bias circuit in Fig. 5.48 uses negative feedback to stabilize the operating point. Suppose I_D begins to increase. Since V_{EQ} is fixed, an increase in I_D will cause a decrease in the magnitude of V_{GS} [see Eq. (5.88)], and this decrease will tend to restore I_D back to its original value.

EXERCISE: Find the Q-point in the circuit in Fig. 5.48 if R_S is changed to 62 kΩ.

ANSWER: (25.4 µA, −6.52 V)

EXERCISE: (a) Use SPICE to find the Q-point in the circuit in Fig. 5.46. (b) Repeat if R_S is changed to 62 kΩ. (c) Repeat parts (a) and (b) with $\lambda = 0.02$.

ANSWERS: (a) (34.4 µA, −6.08 V); (b) (25.4 µA, −6.52 V); (c) (35.9 µA, −5.91 V), (26.3 µA, −6.39 V)

5.16 BIASING THE CMOS INVERTER AS AN AMPLIFIER

In the output and transfer characteristics in Figs. 5.49(b) and 5.44, it was noted that the Q-point for an amplifier should be chosen in the high gain region of the voltage transfer characteristics (i.e., region of high slope). Choosing a Q-point that is midway between the power supplies permits maximum signal swing at the output.

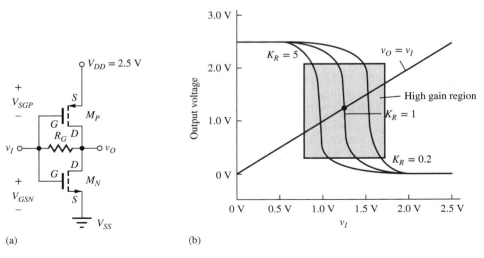

Figure 5.49 (a) CMOS inverting amplifier biased with single resistor R_G; (b) CMOS voltage transfer characteristic indicating the high gain region for three values of K_R.

The basic CMOS amplifier in Fig. 5.49 is biased with the single resistor R_G, again a form of feedback biasing. In order to find the Q-point, we utilize the constraints on the circuit as in Eq. (5.91) knowing that the gate currents are zero, and the transistors are both saturated by "connection."

$$I_{DN} = I_{DP} \quad \text{and} \quad V_{DD} = V_{SGP} + V_{GSN} \qquad (5.91)$$

Equating the drain currents with $K_R = K_n/K_p$,

$$\frac{K_n}{2}(V_{GSN} - V_{TN})^2 = \frac{K_p}{2}(V_{SGP} + V_{TP})^2 \quad \text{or} \quad \sqrt{K_R}(V_{GSN} - V_{TN}) = (V_{SGP} + V_{TP}) \qquad (5.92)$$

yields

$$V_{GSN} = \frac{V_{DD} + \sqrt{K_R}\,V_{TN} + V_{TP}}{1 + \sqrt{K_R}} \quad \text{and} \quad V_{SGP} = V_{DD} - V_{GSN} \qquad (5.93)$$

The Q-points become (I_{DN}, V_{DSN}) and (I_{DP}, V_{SDP}).

EXERCISE: What are the Q-point values for the following values of K_R using $K_n = 50\ \mu A/V^2$, $V_{TN} = 0.6$ V and $V_{TP} = -0.6$ V and $V_{DD} = 2.5$ V: (a) 1; (b) 5; (c) 0.2? Compare your results with Fig. 5.49.

ANSWERS: Both (106 μA, 1.25 V); (4.00 μA, 1.00 V), (4.00 μA, 1.50 V); (20.3 μA, 1.50 V), (20.3 μA, 1.00 V); The voltages agree.

EXERCISE: Derive an expression for V_{SGP} similar to that for V_{GSN}.

ANSWER: $V_{SGP} = \dfrac{\sqrt{K_R}(V_{DD} - V_{TN}) - V_{TP}}{1 + \sqrt{K_R}}$.

5.17 THE CMOS TRANSMISSION GATE

The **CMOS transmission gate** in Fig. 5.50 is a circuit useful in both analog and digital design. The circuit consists of NMOS and PMOS transistors, with source and drain terminals connected in parallel and gate terminals driven by opposite phase logic signals indicated by A and $\overline{A}$. The transmission gate is used so often that it is given the special circuit symbol shown in Fig. 5.50(c). For $A = 0$, both transistors are off, and the transmission gate represents an open circuit.

When the transmission gate is in the conducting state ($A = 1$), the input and output terminals are connected together through the parallel combination of the on-resistances of the two transistors, and the transmission gate represents a *bidirectional* resistive connection between the input and output terminals. The individual on-resistances R_{onp} and R_{onn}, as well as the equivalent on-resistance R_{EQ}, all vary as a function of the input voltage v_I, as shown in Fig. 5.51. The value of R_{EQ} is equal to the parallel combination of R_{onp} and R_{onn}

$$R_{EQ} = \frac{R_{onp} R_{onn}}{R_{onp} + R_{onn}} \qquad (5.94)$$

where R_{on} was discussed in Sec. 5.2.3.

ELECTRONICS IN ACTION

Thermal Inkjet Printers

Inkjet printers have moved from a few niche applications in the 1960s to a widespread, mainstream consumer presence. Thermal inkjet technology was invented in 1979 at Hewlett-Packard Laboratories. Since that time, inkjet technology has evolved to the point where modern thermal inkjet printers deliver 10 to 20 picoliter droplets at rates of several kHz. Integration of the ink handling structures with microelectronics has been an important component of this evolution. Early versions of thermal inkjet printers had drive electronics that were separate from the ink delivery devices. Through the use of MEMS (micro-electro-mechanical system) technology, it has been possible to combine MOS transistors onto the same substrate with the ink handling structures.

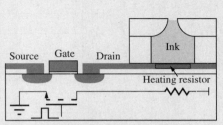

Simplified diagram of thermal inkjet structure integrated with MOS drive transistors. A voltage pulse on the gate causes I^2R heating in the resistor.

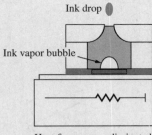

Heat from power dissipated in the resistor vaporizes a small amount of ink causing the ejection of an ink droplet out of the nozzle.

1994–2006 Hewlett-Packard Company. All Rights Reserved.

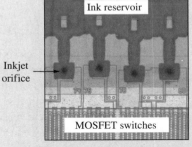

Photomicrograph of inkjet print head
Courtesy of Charles D. Ellis

This diagram is a simplified illustration of a merged thermal inkjet system. A MOSFET transistor is located in the left segment of the silicon substrate. A metal layer connects the drain of the transistor to the thin-film resistive heating material directly under the ink cavity. When the gate of the transistor is driven with a voltage pulse, current passes through the resistor leading to a rapid heating of the ink in the cavity. The temperature of the ink in contact with the resistor increases until a small portion of the ink vaporizes. The vapor bubble forces an ink drop to be ejected from the nozzle at the top of the ink cavity and onto the paper. (In practice, the drops are directed down onto the paper.) At the end of the gate drive pulse, the resistor cools and the vapor bubble collapses, allowing more ink to be drawn into the cavity from an ink reservoir.

Due to the high densities and resolutions made possible by the merging of control and drive electronics with the printing structures, inkjet printers are now capable of generating photo-quality images at reasonable costs. As we will see throughout this text, making high-technology affordable and widely available is a common trait of microelectronics-based systems. This is a clever example of how sensors and/or actuators can be integrated with electronic components in semiconductor technology.

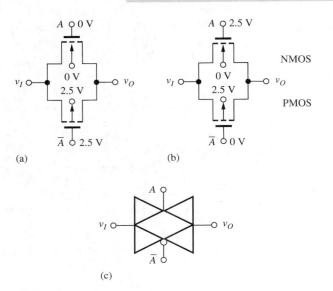

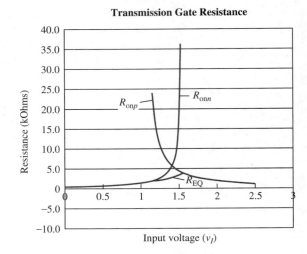

Figure 5.50 CMOS transmission gate in (a) off state and (b) on state; (c) special circuit symbol for the transmission gate.

Figure 5.51 On-resistance of a transmission gate versus input voltage v_I including body effect using the values from Table 5.2 and $(W/L)_N = (W/L)_P = 10/1$. The maximum value of R_{EQ} is approximately 4 kΩ.

5.18 THE JUNCTION FIELD-EFFECT TRANSISTOR (JFET)

Another type of field-effect transistor can be formed without the need for an insulating oxide by using pn junctions, as illustrated in Fig. 5.52. This device, the **junction field-effect transistor, or JFET,** consists of an n-type block of semiconductor material and two pn junctions that form the gate. Although less prevalent than MOSFETs, JFETs have many applications in both integrated and discrete circuit design, particularly in analog and RF and applications. In integrated circuits, JFETs are most often found in BiFET processes, which combine bipolar transistors with

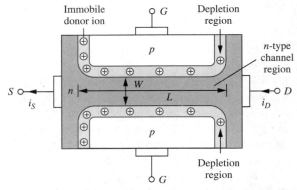

Figure 5.52 Basic n-channel JFET structure and important dimensions. (Note that for clarity the depletion layer in the p-type material is not indicated in the figure.)

JFETs. The JFET provides a device with much lower input current and much higher input impedance than that typically achieved with the bipolar transistor.

In the **n-channel JFET,** current again enters the channel region at the drain and exits from the source. The resistance of the channel region is controlled by changing the physical width of the channel through modulation of the depletion layers that surround the *pn* junctions between the gate and the channel (see Secs. 3.1 and 3.6). In its triode region, the JFET can be thought of as simply a voltage-controlled resistor with its channel resistance determined by

$$R_{CH} = \frac{\rho}{t}\frac{L}{W} \tag{5.95}$$

where ρ = resistivity of the channel region

L = channel length

W = width of channel between the *pn* junction depletions regions

t = depth of channel into the page

When a voltage is applied between the drain and source, the channel resistance determines the current.

With no bias applied, as in Fig. 5.52, a resistive channel region exists connecting the drain and source. Application of a reverse bias to the gate-channel diodes will cause the depletion layers to widen, reducing the channel width and decreasing the current. Thus, the JFET is inherently a depletion-mode device—a voltage must be applied to the gate to turn the device off.

The JFET in Fig. 5.52 is drawn assuming one-sided step junctions ($N_A \gg N_D$) between the gate and channel in which the depletion layers extend only into the channel region of the device (see Secs. 3.1 and 3.6). Note how an understanding of the physics of the *pn* junction is used to create the JFET.

5.18.1 THE JFET WITH BIAS APPLIED

Figure 5.53(a) shows a JFET with 0 V on the drain and source and with the gate voltage $v_{GS} = 0$. The channel width is W. During normal operation, a reverse bias must be maintained across the *pn* junctions to provide isolation between the gate and channel. This reverse bias requires $v_{GS} \leq 0$ V.

In Fig. 5.53(b), v_{GS} has decreased to a negative value, and the depletion layers have increased in width. The width of the channel has now decreased, with $W' < W$, increasing the resistance of the channel region; see Eq. (5.95). Because the gate-source junction is reverse-biased, the gate current will equal the reverse saturation current of the *pn* junction, normally a very small value, and we will assume that $i_G \cong 0$.

For more negative values of v_{GS}, the channel width continues to decrease, increasing the resistance of the channel region. Finally, the condition in Fig. 5.53(c) is reached for $v_{GS} = V_P$, the pinch-off voltage; V_P is the (negative) value of gate-source voltage for which the conducting channel region completely disappears. The channel becomes pinched-off as the depletion regions from the two *pn* junctions merge at the center of the channel. At this point, the resistance of the channel region has become infinitely large. Further increases in v_{GS} do not substantially affect the internal appearance of the device in Fig. 5.54(c). However, v_{GS} must not exceed the reverse breakdown voltage of the gate-channel junction.

5.18.2 JFET CHANNEL WITH DRAIN-SOURCE BIAS

Figure 5.54(a) to (c) shows conditions in the JFET for increasing values of drain-source voltage v_{DS} and a fixed value of v_{GS}. For a small value of v_{DS}, as in Fig. 5.54(a), the resistive channel connects the source and drain, the JFET is operating in its triode region, and the drain current will be dependent on the drain-source voltage v_{DS}. Assuming $i_G = 0$, the current entering the drain must exit from the source, as in the MOSFET. Note, however, that the reverse bias across the gate-channel junction is larger at the drain end of the channel than at the source end, and so the depletion layer is wider at the drain end of the device than at the source end. For increasing

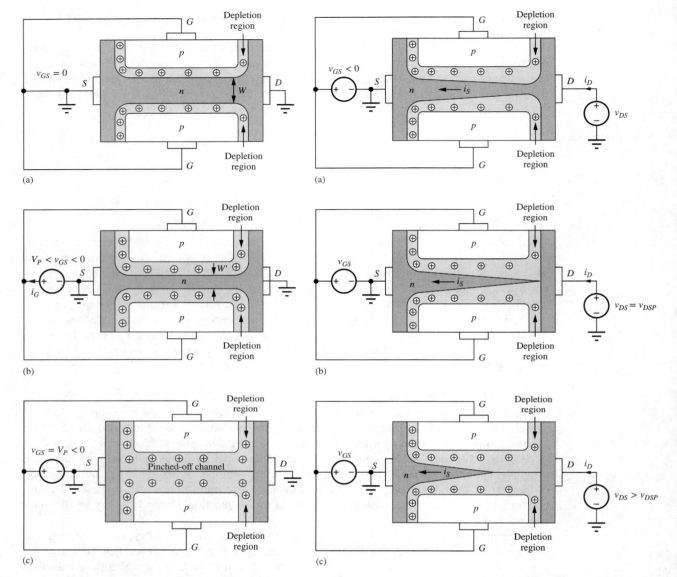

Figure 5.53 (a) JFET with zero gate-source bias; (b) JFET with negative gate-source voltage that is less negative than the pinch-off voltage V_P. (note $W' < W$); (c) JFET at pinchoff with $v_{GS} = V_P$.

Figure 5.54 (a) JFET with small drain source; (b) JFET with channel just at pinchoff with $v_{DS} = v_{DSP}$; (c) JFET with v_{DS} greater voltage than v_{DSP}.

values of v_{DS}, the depletion layer at the drain becomes wider and wider until the channel pinches off near the drain, as in Fig. 5.54(b). Pinchoff first occurs for

$$v_{GS} - v_{DSP} = V_P \quad \text{or} \quad v_{DSP} = v_{GS} - V_P \qquad (5.96)$$

in which v_{DSP} is the value of drain voltage required to just pinch off the channel. Once the JFET channel pinches-off, the drain current saturates, just as for the MOSFET. Electrons are accelerated down the channel, injected into the depletion region, and swept on to the drain by the electric field.

Figure 5.54(c) shows the situation for an even larger value of v_{DS}. The pinch-off point moves toward the source, shortening the length of the resistive channel region. Thus, the JFET suffers from channel-length modulation in a manner similar to that of the MOSFET.

5.18.3 n-CHANNEL JFET i-v CHARACTERISTICS

Since the structure of the JFET is considerably different from the MOSFET, it is quite surprising that the i-v characteristics are virtually identical. We will rely on this similarity and not try to derive the JFET equations here. However, although mathematically equivalent, the equations for the JFET are usually written in a form slightly different from those of the MOSFET. We can develop this form starting with the saturation region expression for a MOSFET, in which the threshold voltage V_{TN} is replaced with the pinch-off voltage V_P:

$$i_D = \frac{K_n}{2}(v_{GS} - V_P)^2 = \frac{K_n}{2}(-V_P)^2 \left(1 - \frac{v_{GS}}{V_P}\right)^2 \quad \text{or} \quad i_D = I_{DSS}\left(1 - \frac{v_{GS}}{V_P}\right)^2 \quad (5.97)$$

in which the parameter I_{DSS} is defined by

$$I_{DSS} = \frac{K_n}{2} V_P^2 \quad \text{or} \quad K_n = \frac{2 I_{DSS}}{V_P^2} \quad (5.98)$$

I_{DSS} is the value of the drain current for $v_{GS} = 0$, and represents the maximum current in the JFET under normal operating conditions, since the gate diode should be kept reverse-biased with $v_{GS} \leq 0$. The pinch-off voltage V_P typically ranges from 0 to -25 V, and the value of I_{DSS} can range from 10 μA to more than 10 A, depending upon the value of W and L in K_n.

If we include channel-length modulation, the expression for the drain current in pinchoff (saturation) becomes

$$I_D = I_{DSS}\left(1 - \frac{v_{GS}}{V_P}\right)^2 (1 + \lambda v_{DS}) \quad \text{for} \quad v_{DS} \geq v_{GS} - V_P \geq 0 \quad (5.99)$$

The transfer characteristic for a JFET operating in pinchoff, based on Eq. (5.99), is shown in Fig. 5.55, and the overall output characteristics for an n-channel JFET are reproduced in Fig. 5.56 with $\lambda = 0$. We see that the drain current decreases from a maximum of I_{DSS} toward zero as v_{GS} ranges from zero to the negative pinch-off voltage V_P.

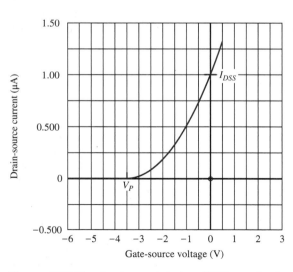

Figure 5.55 Transfer characteristic for a JFET operating in pinchoff with $I_{DSS} = 1$ mA and $V_P = -3.5$ V.

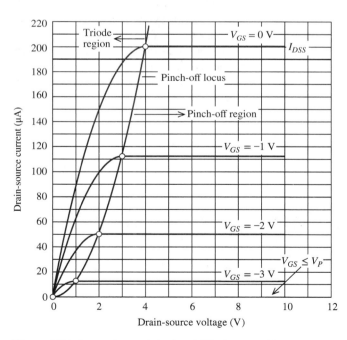

Figure 5.56 Output characteristics for a JFET with $I_{DSS} = 200$ μA and $V_P = -4$ V.

The triode region of the device is also apparent in Fig. 5.55 for $v_{DS} \leq v_{GS} - V_P$. We can obtain an expression for the triode region of the JFET using the equation for the MOSFET triode region. Substituting for K_n and V_{TN} in Eq. (5.29) yields

$$i_D = \frac{2I_{DSS}}{V_P^2}\left(v_{GS} - V_P - \frac{v_{DS}}{2}\right)v_{DS} \quad \text{for } v_{GS} \geq V_P \quad \text{and} \quad v_{GS} - V_P \geq v_{DS} \geq 0 \quad (5.100)$$

Equations (5.99) and (5.100) represent our mathematical model for the n-channel JFET.

> **EXERCISE:** (a) Calculate the current for the JFET in Fig. 5.54 for $V_{GS} = -2$ V and $V_{DS} = 3$ V. What is the minimum drain voltage required to pinch off the JFET? (b) Repeat for $V_{GS} = -1$ V and $V_{DS} = 6$ V. (c) Repeat for $V_{GS} = -2$ V and $V_{DS} = 0.5$ V.
>
> **ANSWERS:** (a) 184 µA, 1.5 V; (b) 510 µA, 2.5 V; (c) 51.0 µA, 1.5 V
>
> **EXERCISE:** (a) Calculate the current for the JFET in Fig. 5.55 for $V_{GS} = -2$ V and $V_{DS} = 0.5$ V. (b) Repeat for $V_{GS} = -1$ V and $V_{DS} = 6$ V.
>
> **ANSWERS:** (a) 21.9 µA; (b) 113 µA

5.18.4 THE p-CHANNEL JFET

A *p*-channel version of the JFET can be fabricated by reversing the polarities of the *n*- and *p*-type regions in Fig. 5.52, as depicted in Fig. 5.57. As for the PMOS FET, the direction of current in the channel is opposite to that of the *n*-channel device, and the signs of the operating bias voltages will be reversed.

5.18.5 CIRCUIT SYMBOLS AND JFET MODEL SUMMARY

The circuit symbols and terminal voltages and currents for *n*-channel and *p*-channel JFETs are presented in Fig. 5.58. The arrow identifies the polarity of the gate-channel diode. The JFET structures in Figs. 5.52 and 5.57 are inherently symmetric, as were those of the MOSFET, and the source and drain are actually determined by the voltages in the circuit in which the JFET is used. However, there are various more complex JFET structures that are no longer symmetrical, and the arrow that indicates the gate-channel junction is often offset to indicate the preferred source terminal of the device.

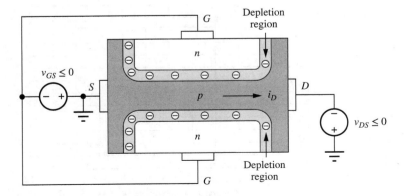

Figure 5.57 *p*-channel JFET with bias voltages.

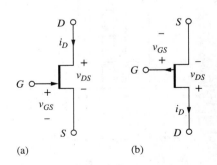

Figure 5.58 (a) *n*-channel and (b) *p*-channel JFET circuit symbols.

A summary of the mathematical models for the *n*-channel and *p*-channel JFETs follows. Because the JFET is a three-terminal device, the pinch-off voltage is independent of the terminal voltages.

---

n-CHANNEL JFET

For all regions
$$i_G = 0 \quad \text{for} \quad v_{GS} \leq 0 \quad (5.101)$$

Cutoff region:
$$i_D = 0 \quad \text{for} \quad v_{GS} \leq V_P \quad (V_P < 0) \quad (5.102)$$

Triode region:
$$i_D = \frac{2I_{DSS}}{V_P^2}\left(v_{GS} - V_P - \frac{v_{DS}}{2}\right)v_{DS} \quad \text{for} \quad v_{GS} \geq V_P \quad \text{and} \quad v_{GS} - V_P \geq v_{DS} \geq 0 \quad (5.103)$$

Pinch-off region:
$$i_D = I_{DSS}\left(1 - \frac{v_{GS}}{V_P}\right)^2 (1 + \lambda v_{DS}) \quad \text{for} \quad v_{DS} \geq v_{GS} - V_P \geq 0 \quad (5.104)$$

---

p-CHANNEL JFET

For all regions
$$i_G = 0 \quad \text{for} \quad v_{GS} \geq 0 \quad (5.105)$$

Cutoff region:
$$i_D = 0 \quad \text{for} \quad v_{GS} \geq V_P \quad (V_P > 0) \quad (5.106)$$

Triode region:
$$i_D = \frac{2I_{DSS}}{V_P^2}\left(v_{GS} - V_P - \frac{v_{DS}}{2}\right)v_{DS} \quad \text{for} \quad v_{GS} \leq V_P \quad \text{and} \quad |v_{GS} - V_P| \geq |v_{DS}| \geq 0 \quad (5.107)$$

Pinch-off region:
$$i_D = I_{DSS}\left(1 - \frac{v_{GS}}{V_P}\right)^2 (1 + \lambda |v_{DS}|) \quad \text{for} \quad |v_{DS}| \geq |v_{GS} - V_P| \geq 0 \quad (5.108)$$

Overall, JFETs behave in a manner very similar to that of depletion-mode MOSFETs, and the JFET is biased in the same way as a depletion-mode MOSFET. In addition, most circuit designs must ensure that the gate-channel diode remains reverse-biased. This is not a concern for the MOSFET. In certain circumstances, however, forward bias of the JFET diode can actually be used to advantage. For instance, we know that a silicon diode can be forward-biased by up to 0.4 to 0.5 V without significant conduction. In other applications, the gate diode can be used as a built-in diode clamp, and in some oscillator circuits, forward conduction of the gate diode is used to help stabilize the amplitude of the oscillation. This effect is explored in more detail during the discussion of oscillator circuits in Chapter 15.

5.18.6 JFET CAPACITANCES

The gate-source and gate-drain capacitances of the JFET, C_{GS} and C_{GD}, are determined by the depletion-layer capacitances of the reverse-biased *pn* junctions forming the gate of the transistor and will exhibit a bias dependence similar to that described by Eq. (3.21) in Chapter 3.

EXERCISE: (a) Calculate the drain current for a p-channel JFET described by I_{DSS} = 2.5 mA and V_P = 4 V and operating with V_{GS} = 3 V and V_{DS} = −3 V. What is the minimum drain-source voltage required to pinch off the JFET? (b) Repeat for V_{GS} = 1 V and V_{DS} = −6 V. (c) Repeat for V_{GS} = 2 V and V_{DS} = −0.5 V.

ANSWERS: (a) 156 µA, −1.00 V; (b) 1.41 mA, −3.00 V; (c) 273 µA, −2.00 V

5.19 JFET MODELING IN SPICE

The circuit representation for the basic JFET model that is implemented in SPICE is given in Fig. 5.59. As for the MOSFET, the JFET model contains a number of additional parameters in an attempt to accurately represent the real device characteristics. Small resistances R_S and R_D appear in series with the JFET source and drain terminals, diodes are included between the gate and internal source and drain terminals, and device capacitances are included in the model.

The model for i_D is an adaptation of the MOSFET model and uses some of the parameter names and formulas from the MOSFET as can be observed in Eq. (5.109).

$$\text{Triode region:} \quad i_D = 2 \cdot \text{BETA}\left(v_{GS} - \text{VTO} - \frac{v_{DS}}{2}\right) v_{DS}(1 + \text{LAMBDA} \cdot v_{DS})$$
$$\text{for} \quad v_{GS} - \text{VTO} \geq v_{DS} \geq 0 \quad (5.109)$$

$$\text{Pinch-off region:} \quad i_D = \text{BETA}(v_{GS} - \text{VTO})^2(1 + \text{LAMBDA} \cdot v_{DS})$$
$$\text{for} \quad v_{DS} \geq v_{GS} - \text{VTO} \geq 0$$

The transconductance parameter BETA is related to the JFET parameters by

$$\text{BETA} = \frac{I_{DSS}}{V_P^2} \quad (5.110)$$

The SPICE description adds a channel-length modulation term to the triode region expression. An additional quirk is that the value of VTO is always specified as a positive number for both n- and p-channel JFETS. Table 5.8 contains the equivalences of the SPICE model parameters and our equations summarized at the end of the previous section. Typical and default values of the SPICE model parameters can also be found in Table 5.8. For more detail see [13].

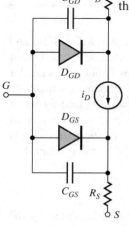

Figure 5.59 SPICE model for the n-channel JFET.

TABLE 5.8
SPICE JFET Parameter Equivalences

PARAMETER	OUR TEXT	SPICE	DEFAULT
Transconductance	—	BETA	100 µA/V²
Zero-bias drain current	I_{DSS}	—	—
Pinch-off voltage	V_P	VTO	−2 V
Channel length modulation	λ	LAMBDA	0
Zero-bias gate-drain capacitance	C_{GD}	CGD	0
Zero-bias gate-source capacitance	C_{GS}	CGS	0
Gate-bulk capacitance per unit width	C_{GBO}	CGBO	0
Ohmic drain resistance	—	RD	0
Ohmic source resistance	—	RS	0
Gate diode saturation current	I_S	IS	10 fA

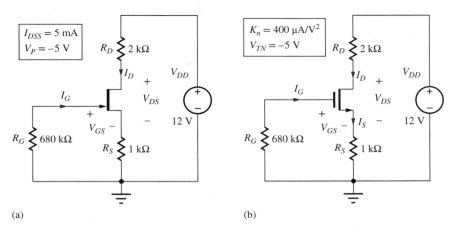

Figure 5.60 Bias circuits for (a) *n*-channel JFET and (b) depletion-mode MOSFET.

EXERCISE: An *n*-channel JFET is described by $I_{DSS} = 2.5$ mA, $V_P = -2$ V, and $\lambda = 0.025$ V^{-1}. What are the values of BETA, VTO, and LAMBDA for this transistor?

ANSWERS: 625 µA; 2 V; 0.025 V^{-1}

EXERCISE: A *p*-channel JFET is described by $I_{DSS} = 5$ mA, $V_P = 2$ V, and $\lambda = 0.02$ V^{-1}. What are the values of BETA, VTO, and LAMBDA for this transistor?

ANSWERS: 1.25 mA; 2 V; 0.02 V^{-1}

5.20 BIASING THE JFET AND DEPLETION-MODE MOSFET

The basic bias circuit for an *n*-channel JFET or depletion-mode MOSFET appears in Fig. 5.60. Because depletion-mode transistors conduct for $v_{GS} = 0$, a separate gate bias voltage is not required, and the bias circuit requires one less resistor than the four-resistor bias circuit discussed earlier in this chapter. In the circuits in Fig. 5.60, the value of R_S will set the source and drain currents, and the sum of R_S and R_D will determine the drain-source voltage. R_G is used to provide a dc connection between the gate and ground while maintaining a high resistance path for ac signal voltages that may be applied to the gate (in amplifier applications, for example). In some cases, even R_G may be omitted.

EXAMPLE 5.5   **BIASING THE JFET AND DEPLETION-MODE MOSFET**

Biasing of JFETs and depletion-mode MOSFETS is very similar, and this example presents a set of bias calculations for the two devices.

PROBLEM Find the quiescent operating point for the circuit in Fig. 5.60(a).

SOLUTION **Known Information and Given Data:** Circuit topology in Fig. 5.60(a) with $V_{DD} = 12$ V, $R_S = 1$ kΩ, $R_D = 2$ kΩ, $R_G = 680$ kΩ, $I_{DSS} = 5$ mA, and $V_P = -5$ V

Unknowns: V_{GS}, I_D, V_{DS}

Approach: Analyze the input loop to find V_{GS}. Use V_{GS} to find I_D, and I_D to determine V_{DS}.

Assumptions: The JFET is pinched-off, the gate-channel junction is reverse biased, and the reverse leakage current of the gate is negligible.

Analysis: Write the input loop equation including V_{GS}:

$$I_G R_G + V_{GS} + I_S R_S = 0 \quad \text{or} \quad V_{GS} = -I_D R_S \tag{5.111}$$

Equation (5.111) was simplified since $I_G = 0$ and $I_S = I_D$. By assuming the JFET is in the pinch-off region and using Eq. (5.97), Eq. (5.111) becomes

$$V_{GS} = -I_{DSS} R_S \left(1 - \frac{V_{GS}}{V_P}\right)^2 \tag{5.112}$$

Substituting in the circuit and transistor values into Eq. (5.112) yields

$$V_{GS} = -(5 \times 10^{-3}\,\text{A})(1000\,\Omega)\left(1 - \frac{V_{GS}}{-5\,\text{V}}\right)^2 \quad \text{or} \quad V_{GS}^2 + 15 V_{GS} + 25 = 0 \tag{5.113}$$

which has the roots -1.91 and -13.1 V. The second value is more negative than the pinch-off voltage of -5 V, so the transistor would be cutoff for this value of V_{GS}. Therefore $V_{GS} = -1.91$ V, and the drain and source currents are

$$I_D = I_S = \frac{1.91\,\text{V}}{1\,\text{k}\Omega} = 1.91\,\text{mA}$$

The drain-source voltage is found by writing the output loop equation:

$$V_{DD} = I_D R_D + V_{DS} + I_S R_S \tag{5.114}$$

which can be rearranged to yield

$$V_{DS} = V_{DD} - I_D(R_D + R_S) = 12 - (1.91\,\text{mA})(3\,\text{k}\Omega) = 6.27\,\text{V}$$

Check of Results: Our analysis yields

$$V_{GS} - V_P = -1.91\,\text{V} - (-5\,\text{V}) = +3.09\,\text{V} \quad \text{and} \quad V_{DS} = 6.27\,\text{V}$$

Because V_{DS} exceeds $(V_{GS} - V_P)$, the device is pinched off. In addition, the gate-source junction is reverse biased by 1.91 V. So, the JFET Q-point is (1.91 mA, 6.27 V).

Discussion: Because depletion-mode transistors conduct for $v_{GS} = 0$, a separate gate bias voltage is not required, and the bias circuit requires one less resistor than the four-resistor bias circuit discussed earlier in this chapter. The circuitry for biasing depletion-mode MOSFETs is identical as indicated in Fig. 5.60(b)—see the exercises after this example.

Computer-Aided Analysis: SPICE analysis yields the same Q-point as our hand calculations. If we add $\lambda = 0.02\,\text{V}^{-1}$, the Q-point shifts to (2.10 mA, 5.98 V).

EXERCISE: What are the values of VTO, BETA, and LAMBDA used in the simulation in the last example?

ANSWERS: -5 V; 0.2 mA; 0.02 V^{-1}

EXERCISE: Show that the expression for the gate-source voltage of the MOSFET in Fig. 5.60(b) is identical to Eq. (5.89). Find the Q-point for the MOSFET and show that it is the same as that for the JFET.

EXERCISE: What is the Q-point for the JFET in Fig. 5.60(a) if $V_{DD} = 9$ V?

ANSWER: (1.91 mA, 3.27 V)

EXERCISE: Find the Q-point in the circuit in Fig. 5.60(a) if R_S is changed to 2 kΩ.

ANSWER: (1.25 mA, 7.00 V)

EXERCISE: (a) Suppose the gate diode of the JFET in Fig. 5.60(a) has a reverse saturation current of 10 nA. Since the diode is reverse biased, $I_G = -10$ nA. What is the voltage at the gate terminal of the transistor? [See Eq. (5.116).] What is the new value of V_{GS}? What will be the new Q-point of the JFET? (b) Repeat if the saturation current is 1 μA.

ANSWERS: (a) +6.80 mV, −1.91 V, (1.91 mA, 6.27 V); (b) 0.680 V, −2.22 V, (1.54 mA, 7.36 V)

SUMMARY

- This chapter discussed the structures and *i-v* characteristics of two types of field-effect transistors (FETs): the metal-oxide-semiconductor FET, or MOSFET, and the junction FET, or JFET.
- At the heart of the MOSFET is the MOS capacitor, formed by a metallic gate electrode insulated from the semiconductor by an insulating oxide layer. The potential on the gate controls the carrier concentration in the semiconductor region directly beneath the gate; three regions of operation of the MOS capacitor were identified: accumulation, depletion, and inversion.
- A MOSFET is formed when two *pn* junctions are added to the semiconductor region of the MOS capacitor. The junctions act as the source and drain terminals of the MOS transistor and provide a ready supply of carriers for the channel region of the MOSFET. The source and drain junctions must be kept reverse-biased at all times in order to isolate the channel from the substrate.
- MOS transistors can be fabricated with either *n*- or *p*-type channel regions and are referred to as NMOS or PMOS transistors, respectively. In addition, MOSFETs can be fabricated as either enhancement-mode or depletion-mode devices.
- For an *enhancement-mode device,* a gate-source voltage exceeding the threshold voltage must be applied to the transistor to establish a conducting channel between source and drain.
- In the *depletion-mode device,* a channel is built into the device during its fabrication, and a voltage must be applied to the transistor's gate to quench conduction.
- The JFET uses *pn* junctions to control the resistance of the conducting channel region. The gate-source voltage modulates the width of the depletion layers surrounding the gate-channel junctions and thereby changes the width of the channel region. A JFET can be fabricated with either *n*- or *p*-type channel regions, but because of its structure, the JFET is inherently a depletion-mode device.
- Both the MOSFET and JFET are symmetrical devices. The source and drain terminals of the device are actually determined by the voltages applied to the terminals. For a given geometry and set of voltages, the *n*-channel transistor will conduct two to three times the current of the *p*-channel device because of the difference between the electron and hole mobilities in the channel.

- Although structurally different, the *i-v* characteristics of MOSFETs and JFETs are very similar, and each type of FET has three basic regions of operation.

- In the subthreshold region (e.g., $V_{GS} < V_{TN}$), the channel does not abruptly disappear! Instead, the drain-source current decreases exponentially and eventually reaches a very small value limited by *pn* junction leakage currents.

- In *cutoff*, a channel does not exist, and the terminal currents are zero.

- In the *triode region* of operation, the drain current in the FET depends on both the gate-source and drain-source voltages of the transistor. For small values of drain-source voltage, the transistor exhibits an almost linear relationship between its drain current and drain-source voltage. In the triode region, the FET can be used as a voltage-controlled resistor, in which the on-resistance of the transistor is controlled by the gate-source voltage of the transistor. Because of this behavior, the name *transistor* was developed as a contraction of "transfer resistor."

- For values of drain-source voltage exceeding the pinch-off voltage, the drain current of the FET becomes almost independent of the drain-source voltage. In this region, referred to variously as the *pinch-off* region, the *saturation region*, or the *active region*, the drain-source current exhibits a square-law dependence on the voltage applied between the gate and source terminals. Variations in drain-source voltage do cause small changes in drain current in saturation due to channel-length modulation.

- As transistors are scaled to very small dimensions, the electric field in the channel becomes high and the carriers reach saturation velocity. For this case, the transistor is operating in the *velocity-limited region*.

- Mathematical models for the *i-v* characteristics of both MOSFETs and JFETs were presented. The MOSFET is actually a four-terminal device and has a threshold voltage that depends on the source-body voltage of the transistor.

- Key parameters for the MOSFET include the transconductance parameters K_n or K_p, the zero-bias threshold voltage V_{TO}, body effect parameter γ, and channel-length modulation parameter λ as well as the width W and length L of the channel.

- The *Unified Model* introduces parameter V_{SAT} to model the velocity-limited region of operation and defines parameter V_{MIN} which replaces V_{DS} in the basic mathematical model of the transistor. $V_{MIN} = \min\{V_{GS} - V_{TN}, V_{DS}, V_{SAT}\}$.

- The "all-region" *interpolation model* provides a continuous, although approximate, representation of the drain current from strong inversion deep into the subthreshold region and is highly useful in circuit design and simulation.

- The JFET was modeled as a three-terminal device with constant pinch-off voltage. Key parameters for the JFET include saturation current I_{DSS}, pinch-off voltage V_P, and channel-length modulation parameter λ.

- CMOS (complementary MOS) integrated circuit technology enables fabrication of both NMOS and PMOS transistors on the same chip and has arguably become today's most widely utilized semiconductor technology. CMOS facilitates the design of low-power, low-voltage circuits for mobile devices.

- A variety of examples of bias circuits were presented, and the mathematical model was used to find the quiescent operating point, or Q-point, for various types of MOSFETs. The Q-point represents the dc values of drain current and drain-source voltage: (I_D, V_{DS}).

- The *i-v* characteristics are often displayed graphically in the form of either the output characteristics, that plot i_D versus v_{DS}, or the transfer characteristics, that graph i_D versus v_{GS}. Examples of finding the Q-point using graphical load-line and iterative numerical analyses were discussed.

- The most important bias circuit in discrete design is the four-resistor circuit, which yields a well-stabilized operating point.
- The gate-source, gate-drain, drain-bulk, source-bulk, and gate-bulk capacitances of MOS transistors were discussed, and the Meyer model for the gate-source and gate-bulk capacitances was introduced. All the capacitances are nonlinear functions of the terminal voltages of the transistor. The capacitances of the JFET are determined by the capacitance of the reverse-biased gate-channel junctions and also exhibit a nonlinear dependence on the terminal voltages of the transistor.
- Complex models for MOSFETs and JFETs are built into SPICE circuit analysis programs. These models contain many circuit elements and parameters to attempt to model the true behavior of the transistor as closely as possible.
- Part of the IC designer's job often includes layout of the transistors based on a set of technology-specific ground rules that define minimum feature dimensions and spaces between features.
- Constant electric field scaling provides a framework for proper miniaturization of MOS devices in which the power density remains constant as the transistor density increases. In this case, circuit delay improves directly with the scale factor α, whereas the power-delay product improves with the cube of α.
- The cutoff frequency f_T of the transistor represents the highest frequency at which the transistor can provide amplification. Cutoff frequency f_T improves directly with the scale factor.
- The electric fields in small devices can become very high, and the carrier velocity tends to saturate at fields above 10 kV/cm. Subthreshold leakage current becomes increasingly important as devices are scaled to small dimension.
- *Static power dissipation:* For low-power applications, particularly where battery life is important, leakage current from subthreshold conduction and the reverse-biased wells and drain-substrate junctions can become an important source of power dissipation. This leakage current places a lower bound on the power required to operate a CMOS circuit.
- *The CMOS transmission gate:* A bidirectional circuit element, the CMOS transmission gate that utilizes the parallel connection of an NMOS and a PMOS, transistor, was introduced. When the transmission gate is on, it provides a low-resistance connection between its input and output terminals over the entire input voltage range. We will find the transmission gate used in circuit implementations of both the D latch and the master-slave D flip-flop. It has also been widely used in analog multiplexers.
- *Latchup:* An important potential failure mechanism in CMOS is the phenomenon called latchup, which is caused by the existence of parasitic *npn* and *pnp* bipolar transistors in the CMOS structure. A lumped circuit model for latchup was developed and used to simulate the latchup behavior of a CMOS inverter. Special substrates and IC processing are used to minimize the possibility of latchup in modern CMOS technologies.

KEY TERMS

Accumulation
Accumulation region
Active region
Alignment tolerance T
Back-gate
Bitline capacitance
Bitline precharge
Body effect

Body-effect parameter γ
Body terminal (B)
Bulk terminal (B)
C_{GS}, C_{GD}, C_{GB}, C_{DB}, C_{SB}, C''_{ox}, C_{GDO}, C_{GSO}
Capacitance per unit length
Capacitance per unit width
Cell capacitor
Channel length L

Channel-length modulation
Channel-length modulation parameter λ
Channel region
Channel width W
Charge sharing
Charge transfer
CMOS transmission gate
Complementary MOS (CMOS)
Constant electric field scaling
Current source
Cutoff frequency
dc bias
Depletion
Depletion-mode device
Depletion-mode MOSFETs
Depletion region
Design rules
Drain-bulk capacitance
Drain (D)
Enhancement-mode device
Feedback bias
Field-effect transistor (FET)
Four-resistor bias
Gate (G)
Gate-channel capacitance C_{GC}
Gate-drain capacitance
Gate-drain capacitance C_{GD}
Gate-source capacitance
Gate-source capacitance C_{GS}
Ground rules
High field limitations
Interpolation model
Inversion layer
Inversion region
Junction field-effect transistor (JFET)
KP
K_n, K_p
K'_n, K'_p
LAMBDA, λ
Latchup
Linear region
Metal-oxide-semiconductor field-effect transistor (MOSFET)
Minimum feature size F

MOS capacitor
n-channel JFET
n-channel MOS (NMOS)
n-channel MOSFET
NMOSFET
NMOS transistor
One-transistor dynamic cell
On-resistance (R_{on})
Output characteristics
Overlap capacitance
Oxide thickness
Parasitic bipolar transistors
p-channel JFET
p-channel MOS (PMOS)
PHI
Pinch-off locus
Pinch-off point
Pinch-off region
Pinch-off voltage
PMOS transistor
Power delay product (PDP)
Quiescent operating point
Q-point
Refresh operation
Saturation region
Saturation voltage
Scaling theory
SPICE model
Source-bulk capacitance
Source (S)
Substrate sensitivity
Substrate terminal
Surface potential parameter $2\phi_F$
Subthreshold region
Symmetrical CMOS circuit
Threshold voltage V_{TN}, V_{TP}
Transconductance g_m
Transconductance parameter—K'_n, K'_p, KP, K_n, K_p
Transfer characteristic
Triode region
V_{TN}, V_{TP}, VT, VTO
Zero-substrate-bias value for V_{TN}

REFERENCES

1. F. M. Wanlass and C. T. Sah, "Nanowatt logic using field-effect metal-oxide-semiconductor triodes," *IEEE International Solid-State Circuits Conference Digest,* vol. VI, pp. 32–33, February 1963.

2. IDRS, *International Roadmap for Devices and Systems*, IDRS.ieee.org. The IDRS replaces the older ITRS, *International Technology Road Map for Semiconductors*, public.itrs.net
3. J. Wood and R. G. Ball, "The use of insulated-gate field-effect transistors in digital storage systems," *ISSCC Digest of Technical Papers*, vol. 8, pp. 82–83, February 1965.
4. ISSCC, *Digest of Technical Papers of the IEEE International Solid-State Circuits Conference* (ISSCC), February of each year.
5. W. M. Regitz and J. A. Karp, "A three-transistor cell, 1024-bit, 500 ns MOS RAM," *ISSCC Digest of Technical Papers*, vol. 13, pp. 36–39, February 1970.
6. Robert H. Dennard; U.S. patent 3,387,286 assigned to the IBM Corporation.
7. F. Masuoka, U.S. Patent 4,437,174 March 1984, assigned to Toshiba Corp.
8. F. Masuoka, U.S. Patents 3,825,945, 4,115,795, and 4,5437,174 assigned to Toshiba Corp.
9. F. Masuoka, "Great encounters leading me to the inventions of flash memories and surrounding gate transistor technology, *IEEE Solid-State Circuits Magazine*, vol. 5, no. 4, pp. 10–20, Fall 2013. Also see three companion articles by Koji Sakui and Ken Takeuchi.
10. J. D. Meindl and J. A. Davis, "The fundamental limit on binary switching energy for terascale integration (TSI)," *IEEE Journal of Solid-State Circuits,* vol. 35, no. 10, pp. 1515–1516, October 2000.
11. R. M. Swanson and J. D. Meindl, "Ion-implanted complementary MOS transistors in low-voltage circuits," *IEEE Journal of Solid-State Circuits,* vol. SC-7, no. 2, pp. 146–153, April 1972.
12. J. E. Meyer, "MOS models and circuit simulations," *RCA Review,* vol. 32, pp. 42–63, March 1971.
13. B. M. Wilamowski and R. C. Jaeger, *Computerized Circuit Analysis Using SPICE Programs,* McGraw-Hill, New York: 1997.
14. R. H. Dennard, F. H. Gaensslen, L. Kuhn, and H. N. Yu, "Design of micron MOS switching devices," *IEEE IEDM Digest,* pp. 168–171, December 1972.
15. R. H. Dennard, F. H. Gaensslen, H-N. Yu, V. L. Rideout, E. Bassous, and A. R. LeBlanc, "Design of ion-implanted MOSFET's with very small physical dimensions," *IEEE J. Solid-State Circuits,* vol. SC-9, no. 5, pp. 256–268, October 1974.
16. J. M. Rabaey, A. Chandrakasan, and B. Nikolic, *Digital Integrated Circuits*. 2nd ed., Prentice Hall, New Jersey: 2005.
17. C. C. Enz, F. Krummenacher, and E. A. Vittoz, "An analytical MOS transistor model valid in all regions of operation and dedicated to low-voltage and low-current applications," *Analog Integrated Circuits and Signal Processing*, vol. 8, pp. 83–114, 1995.
18. Y. Tsividis, K. Suyama, and K. Vavelidis, "A simple 'Reconciliation' MOSFET model valid in all regions," *Electronics Letters*, vol. 31, pp. 506–508, 1995.
19. Yannis Tsividis and Colin McAndrew, *Operation and Modeling of the MOS Transistor*, 3rd ed., Oxford University Press, 2011.
20. Carver Mead and Lynn Conway, *Introduction to VLSI Systems,* Addison Wesley, Reading, Massachusetts: 1980.
21. P. Ye, T. Ernst, and M. V. Khare, "The last silicon transistor," *IEEE Spectrum*, pp. 30–35, August 2019.

ADDITIONAL READINGS

S. Aritome, "NAND flash innovations," *IEEE Solid-State Circuits Magazine*, vol. 5, no. 4, pp. 21–29, Fall 2013.

Digest of the IEEE International Electron Devices Meeting (IEDM), December of each year.

U.S. Patent 1,900,018. Also see 1,745,175 and 1,877,140.

PROBLEMS

Use the parameters in Table 5.9 as needed in the problems here.

TABLE 5.9
MOS Transistor Parameters

	NMOS DEVICE	PMOS DEVICE
V_{TO}	+0.75 V	−0.75 V
γ	$0.75\sqrt{V}$	$0.5\sqrt{V}$
$2\phi_F$	0.6 V	0.6 V
K'	100 µA/V²	40 µA/V²

$\varepsilon_{ox} = 3.9\varepsilon_o$ and $\varepsilon_s = 11.7\varepsilon_o$ where $\varepsilon_o = 8.854 \times 10^{-14}$ F/cm

5.1 Characteristics of the MOS Capacitor

5.1. (a) The MOS capacitor in Fig. 5.1 has $V_{TN} = 0.7$ V and $V_G = 2$ V. To what region of operation does this bias condition correspond? (b) Repeat for $V_G = -2$ V. (c) Repeat for $V_G = 0.5$ V.

5.2. Calculate the capacitance of an MOS capacitor with an oxide thickness T_{ox} of (a) 50 nm, (b) 25 nm, (c) 10 nm, and (d) 5 nm.

5.3. The minimum value of the depletion-layer capacitance can be estimated using an expression similar to Eq. (3.18): $C_d = \varepsilon_S/x_d$ in which the depletion-layer width is $x_d \cong \sqrt{\frac{2\varepsilon_S}{qN_B}(0.75 \text{ V})}$ and N_B is the substrate doping. Estimate C_d for $N_B = 10^{15}$/cm³. Repeat for $N_B = 10^{17}$/cm³.

5.2 The NMOS Transistor

Triode (Linear) Region Characteristics

5.4. Calculate K'_n for an NMOS transistor with $\mu_n = 500$ cm²/V · s for an oxide thickness of (a) 40 nm, (b) 20 nm, (c) 10 nm, (d) 5 nm, and (e) 2 nm.

5.5. (a) What is the charge density (C/cm²) in the channel if the oxide thickness is 17.5 nm and the oxide voltage exceeds the threshold voltage by 1.2 V? (b) Repeat for a 6-nm oxide and a bias 0.75 V above threshold.

5.6. (a) What is the electron velocity in the channel if $\mu_n = 525$ cm²/V · s and the electric field is 3000 V/cm? (b) Repeat for $\mu_n = 400$ cm²/V · s with a field of 1500 V/cm.

5.7. An NMOS transistor has $K'_n = 160$ µA/V². What is the value of K_n if $W = 60$ µm, $L = 3$ µm? If $W = 10$ µm, $L = 0.25$ µm? If $W = 3$ µm, $L = 40$ nm?

5.8. Calculate the drain current in an NMOS transistor for $V_{GS} = 0, 1$ V, 2 V, and 3 V, with $V_{DS} = 0.3$ V, if $W = 6$ µm, $L = 0.5$ µm, $V_{TN} = 0.75$ V, and $K'_n = 200$ µA/V². What is the value of K_n?

5.9. Calculate the drain current in an NMOS transistor for $V_{GS} = 0, 1$ V, 2 V, and 3 V, with $V_{DS} = 0.1$ V, if $W = 10$ µm, $L = 0.2$ µm, $V_{TN} = 0.8$ V, and $K'_n = 250$ µA/V². What is the value of K_n?

5.10. Equation (5.2) indicates that the charge/unit · length in the channel of a pinched-off transistor decreases as one proceeds from source to drain. However, our text argued that the current entering the drain terminal is equal to the current exiting from the source terminal. How can a constant current exist everywhere in the channel between the drain and source terminals if the first statement is indeed true?

5.11. Identify the source, drain, gate, and bulk terminals and find the current I in the transistors in Fig. P5.11. Assume $V_{TN} = 0.6$ V.

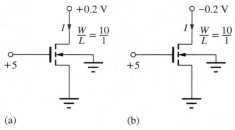

(a)   (b)

Figure P5.11

5.12. (a) What is the current in the transistor in Fig. P5.11(a) if the 0.2 V is changed to 0.5 V? Assume $V_{TN} = 0.5$ V. (b) Repeat if the gate voltage is changed to 3 V and the other voltage remains at 0.2 V?

5.13. (a) What is the current in the transistor in Fig. P5.11(b) if −0.2 V is changed to −0.5 V? Assume $V_{TN} = 0.70$ V. (b) If the gate voltage is changed to 3 V and the upper terminal voltage is replaced by −1 V?

5.14. (a) Design a transistor (choose W) to have $K_n = 4$ mA/V² if $L = 0.12$ µm. (See Table 5.9.) (b) Repeat for $K_n = 800$ µA/V².

On Resistance

5.15. What is the on-resistance of an NMOS transistor with $W/L = 100/1$ if $V_{GS} = 5$ V and $V_{TN} = 0.65$ V? (b) If $V_{GS} = 2.5$ V and $V_{TN} = 0.50$ V? (See Table 5.9.)

5.16. (a) What is the W/L ratio required for an NMOS transistor to have an on-resistance of 500 Ω when $V_{GS} = 5$ V and $V_{SB} = 0$? (b) Repeat for $V_{GS} = 3.3$ V. (c) Repeat for 50 Ω.

5.17. Suppose that an NMOS transistor must conduct a current $I_D = 25$ A with $V_{DS} \leq 0.1$ V when it is on. What is the maximum on-resistance of the transistor? If $V_G = 5$ V is used to turn on the transistor and $V_{TN} = 2$ V, what is the minimum value of K_n required to achieve the required on-resistance?

Saturation of the i-v Characteristics

*5.18. The output characteristics for an NMOS transistor are given in Fig. P5.18. What are the values of K_n and V_{TN} for this transistor? Is this an enhancement-mode or depletion-mode transistor? What is W/L for this device?

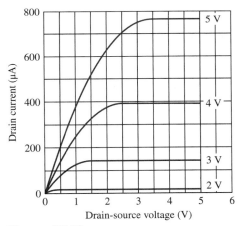

Figure P5.18

5.19. Add the $V_{GS} = 3.5$ V and $V_{GS} = 4.5$ V curves to the i-v characteristic of Fig. P5.18. What are the values of i_{DSAT} and v_{DSAT} for these new curves?

5.20. Calculate the drain current in an NMOS transistor for $V_{GS} = 0$, 1 V, 2 V, and 3 V, with $V_{DS} = 3.3$ V, if $W = 5$ μm, $L = 0.5$ μm, $V_{TN} = 1$ V, and $K'_n = 375$ μA/V². What is the value of K_n? Check the saturation region assumption.

5.21. Calculate the drain current in an NMOS transistor for $V_{GS} = 0$, 1 V, 2 V, and 3 V, with $V_{DS} = 4$ V, if $W = 10$ μm, $L = 1$ μm, $V_{TN} = 1.5$ V, and $K'_N = 200$ μA/V². What is the value of K_n? Check the saturation region assumption.

Regions of Operation

5.22. Find the region of operation and drain current in an NMOS transistor with $K'_n = 200$ μA/V², $W/L = 10/1$, $V_{TN} = 0.75$ V and (a) $V_{GS} = 2$ V and $V_{DS} = 2.5$ V, (b) $V_{GS} = 2$ V and $V_{DS} = 0.2$ V and (c) $V_{GS} = 0$ V and $V_{DS} = 4$ V. (d) repeat for $K'_n = 300$ μA/V².

5.23. Identify the region of operation of an NMOS transistor with $K_n = 250$ μA/V² and $V_{TN} = 1$ V for (a) $V_{GS} = 5$ V and $V_{DS} = 6$ V, (b) $V_{GS} = 0$ V and $V_{DS} = 6$ V, (c) $V_{GS} = 2$ V and $V_{DS} = 2$ V, (d) $V_{GS} = 1.5$ V and $V_{DS} = 0.5$, (e) $V_{GS} = 2$ V and $V_{DS} = -0.5$ V, and (f) $V_{GS} = 3$ V and $V_{DS} = -6$ V.

5.24. Identify the region of operation of an NMOS transistor with $K_n = 400$ μA/V² and $V_{TN} = 0.7$ V for (a) $V_{GS} = 3.3$ V and $V_{DS} = 3.3$ V, (b) $V_{GS} = 0$ V and $V_{DS} = 3.3$ V, (c) $V_{GS} = 2$ V and $V_{DS} = 2$ V, (d) $V_{GS} = 1.5$ V and $V_{DS} = 0.5$, (e) $V_{GS} = 2$ V and $V_{DS} = -0.5$ V, and (f) $V_{GS} = 3$ V and $V_{DS} = -3$ V.

5.25. (a) Identify the source, drain, gate, and bulk terminals for each of the transistors in the circuit in Fig. P5.25(a). Assume $V_{DD} > 0$. (b) Repeat for the circuit in Fig. P5.25(b).

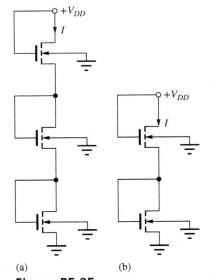

Figure P5.25

5.26. (a) Identify the source, drain, gate, and bulk terminals for the transistor in the circuit in Fig. P5.26.

Assume $V_{DD} > 0$. (b) Repeat for $V_{DD} < 0$. (c) An issue occurs with operation of the circuit in Fig. P5.26 with $V_{DD} < 0$. What is the problem?

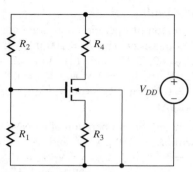

Figure P5.26

Transconductance

5.27. Calculate the transconductance for an NMOS transistor for $V_{GS} = 2$ V and 3.3 V, with $V_{DS} = 3.3$ V, if $W = 20$ µm, $L = 1$ µm, $V_{TN} = 0.7$ V, and $K'_n = 250$ µA/V^2. Check the saturation region assumption. What is the value of g_m/I_D?

5.28. (a) Estimate the transconductance for the transistor in Fig. P5.18 for $V_{GS} = 4$ V and $V_{DS} = 4$ V. (b) Repeat for $V_{GS} = 3$ V and $V_{DS} = 4.5$ V. (c) What are the values of g_m/I_D?

5.29. What is the transconductance of the MOSFET in Prob. 5.27 with $V_{GS} = 2$ V and 3.3 V with $V_{DS} = 1$ V? What is the value of g_m/I_D?

Channel-Length Modulation

5.30. (a) Calculate the drain current in an NMOS transistor if $K_n = 225$ µA/V^2, $V_{TN} = 0.70$ V, $\lambda = 0.02$ V^{-1}, $V_{GS} = 5$ V, and $V_{DS} = 6$ V. (b) Repeat assuming $\lambda = 0$.

5.31. (a) Calculate the drain current in an NMOS transistor if $K_n = 450$ µA/V^2, $V_{TN} = 1$ V, $\lambda = 0.03$ V^{-1}, $V_{GS} = 4$ V, and $V_{DS} = 5$ V. (b) Repeat assuming $\lambda = 0$.

5.32. (a) Find the drain current for the transistor in Fig. P5.32 if $\lambda = 0$. (b) Repeat if $\lambda = 0.05$ V^{-1}. (c) Repeat part (a) if the W/L ratio is changed to 30/1.

5.33. (a) Find the drain current for the transistor in Fig. P5.32 if $\lambda = 0$ and the W/L ratio is changed to 20/1. (b) Repeat if $\lambda = 0.025$ V^{-1}.

5.34. (a) Find the current I in Fig. P5.34 if $V_{DD} = 10$ V and $\lambda = 0$. Both transistors have $W/L = 10/1$.

(b) What is the current if both transistors have $W/L = 20/1$. (c) Repeat part (a) for $\lambda = 0.05$ V^{-1}.

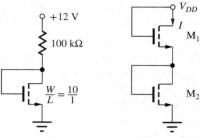

Figure P5.32 **Figure P5.34**

5.35. (a) Find the currents in the two transistors in Fig. P5.34 if $(W/L)_1 = 25/1$, $(W/L)_2 = 12.5/1$ and $\lambda = 0$ for both transistors. (b) Repeat part (a) if $\lambda = 0.04/V$ for both transistors.

5.36. (a) Find the currents in the two transistors in Fig. P5.34 if $(W/L)_1 = 10/1$, $(W/L)_2 = 30/1$, and $\lambda = 0$ for both transistors. (b) Repeat for $(W/L)_2 = 30/1$ and $(W/L)_1 = 10/1$. (c) Repeat part (a) if $\lambda = 0.25/V$ for both transistors.

Transfer Characteristics and the Depletion-Mode MOSFET

5.37. (a) Calculate the drain current in an NMOS transistor if $K_n = 250$ µA/V^2, $V_{TN} = -3$ V, $\lambda = 0$, $V_{GS} = 0$ V, and $V_{DS} = 6$ V. (b) Repeat assuming $\lambda = 0.025$ V^{-1}.

5.38. An NMOS depletion-mode transistor is operating with $V_{DS} = V_{GS} > 0$. What is the region of operation for this device?

5.39. (a) Calculate the drain current in an NMOS transistor if $K_n = 250$ µA/V^2, $V_{TN} = -2.5$ V, $\lambda = 0$, $V_{GS} = 5$ V, and $V_{DS} = 6$ V. (b) Repeat assuming $\lambda = 0.04$ V^{-1}.

5.40. (a) Find the Q-point for the transistor in Fig. P5.40(a) if $V_{TN} = -2.2$ V. (b) Repeat for $R = 50$ kΩ and $W/L = 20/1$. (c) Repeat parts (a) & (b) for Fig. 5.40(b).

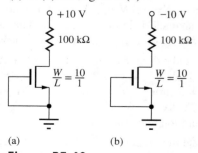

Figure P5.40

Body Effect or Substrate Sensitivity

5.41. Repeat Problem 5.20 for $V_{SB} = 1.3$ V with the values from Table 5.9.

5.42. Repeat Prob. 5.21 for $V_{SB} = 1.75$ V with the values from Table 5.9.

5.43. (a) An NMOS transistor with $W/L = 7.5/1$ has $V_{TO} = 1$ V, $2\phi_F = 0.6$ V, and $\gamma = 0.7$ $\sqrt{V}$. The transistor is operating with $V_{SB} = 3$ V, $V_{GS} = 2.5$ V, and $V_{DS} = 4$ V. What is the drain current in the transistor? (b) Repeat for $V_{DS} = 0.5$ V.

5.44. An NMOS transistor with $W/L = 13.7/1$ has $V_{TO} = 1.5$ V, $2\phi_F = 0.75$ V, and $\gamma = 0.55$ $\sqrt{V}$. The transistor is operating with $V_{SB} = 4$ V, $V_{GS} = 2$ V, and $V_{DS} = 5$ V. What is the drain current in the transistor? (b) Repeat for $V_{DS} = 0.5$ V.

5.45. A depletion-mode NMOS transistor has $V_{TO} = -1.65$ V, $2\phi_F = 0.75$ V, and $\gamma = 1.5$ $\sqrt{V}$. What source-bulk voltage is required to change this transistor into an enhancement-mode device with a threshold voltage of $+0.70$ V?

*5.46. The measured body-effect characteristic for an NMOS transistor is given in Table 5.10. What are the best values of V_{TO}, γ, and $2\phi_F$ (in the least-squares sense—see Prob. 3.31) for this transistor?

TABLE 5.10

V_{SB} (V)	V_{TN} (V)
0	0.710
0.5	0.912
1.0	1.092
1.5	1.232
2.0	1.377
2.5	1.506
3.0	1.604
3.5	1.724
4.0	1.822
4.5	1.904
5.0	2.005

5.3 PMOS Transistors

5.47. Calculate K'_p for a PMOS transistor with $\mu_p = 200$ cm^2/V · s for an oxide thickness of (a) 50 nm, (b) 20 nm, (c) 10 nm, and (d) 5 nm.

*5.48. The output characteristics for a PMOS transistor are given in Fig. P5.48. What are the values of K_p and V_{TP} for this transistor? Is this an enhancement-mode or depletion-mode transistor?

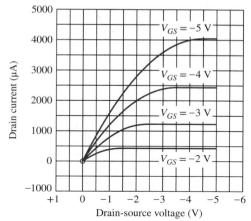

Figure P5.48

5.49. Add the $V_{GS} = -3.5$ V and $V_{GS} = -4.5$ V curves to the i-v characteristic of Fig. P5.48. What are the values of i_{DSAT} and v_{DSAT} for these new curves?

5.50. Find the region of operation and drain current in a PMOS transistor with $W/L = 20/1$ for $V_{BS} = 0$ V and (a) $V_{GS} = -1.1$ V and $V_{DS} = -0.2$ V and (b) $V_{GS} = -1.3$ V and $V_{DS} = -0.2$ V. (c) Repeat parts (a) and (b) for $V_{BS} = 1$ V.

5.51. (a) Calculate the on-resistance for a PMOS transistor having $W/L = 250/1$ and operating with $V_{GS} = -5$ V and $V_{TP} = -0.75$ V. (b) Repeat for a similar NMOS transistor with $V_{GS} = 5$ V and $V_{TN} = 0.75$ V. (c) What W/L ratio is required for the PMOS transistor to have the same R_{on} as the NMOS transistor in (b)?

5.52. (a) What is the W/L ratio required for an PMOS transistor to have an on-resistance of 1 kΩ when $V_{GS} = -5$ V and $V_{BS} = 0$? Assume $V_{TP} = -0.70$ V. (b) Repeat for an NMOS transistor with $V_{GS} = +5$ V and $V_{BS} = 0$. Assume $V_{TN} = 0.70$ V.

5.53. (a) What is the W/L ratio required for a PMOS transistor to have an on-resistance of 15 Ω when $V_{GS} = -5$ V and $V_{SB} = 0$? Assume $V_{TP} = -0.70$ V. (b) Repeat for an NMOS transistor with $V_{GS} = +5$ V and $V_{BS} = 0$. Assume $V_{TN} = 0.70$ V.

5.54. (a) Identify the source, drain, gate, and bulk terminals for the transistors in the two circuits in Fig. P5.54(a). Assume $V_{DD} = 10$ V. (b) Repeat for Fig. P5.54(b).

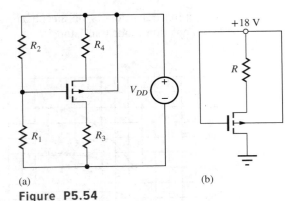

Figure P5.54

5.55. What is the on-resistance and voltage V_O for the parallel combination of the NMOS ($W/L = 14/1$) and PMOS ($W/L = 35/1$) transistors in Fig. P5.55 for $V_{IN} = 0$ V? (b) For $V_{IN} = 5$ V? This circuit is called a transmission-gate.

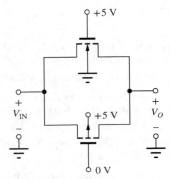

Figure P5.55

5.56. Suppose a PMOS transistor must conduct a current $I_D = 0.75$ A with $V_{SD} \le 0.1$ V when it is on. What is the maximum on-resistance? If $V_G = 0$ V is used to turn on the transistor with $V_S = 10$ V and $V_{TP} = -2$ V, what is the minimum value of K_p required to achieve the required on-resistance?

5.57. A PMOS transistor is operating with $V_{BS} = 0$ V, $V_{GS} = -1.5$ V, and $V_{DS} = -0.5$ V. What are the region of operation and drain current in this device if $W/L = 35/1$?

5.58. A PMOS transistor is operating with $V_{BS} = 4$ V, $V_{GS} = -1.5$ V, and $V_{DS} = -4$ V. What are the region of operation and drain current in this device if $W/L = 40/1$?

5.4 MOSFET Circuit Symbols

5.59. The PMOS transistor in Fig. P5.54(a) is conducting current. Is $V_{TP} > 0$ or $V_{TP} < 0$ for this transistor? Based on this value of V_{TP}, what type transistor is in the circuit? Is the proper symbol used in this circuit for this transistor? If not, what symbol should be used?

5.60. The PMOS transistor in Fig. P5.54(b) is conducting current. Is $V_{TP} > 0$ or $V_{TP} < 0$ for this transistor? Based on this value of V_{TP}, what type transistor is in the circuit? Is the proper symbol used in this circuit for this transistor? If not, what symbol should be used?

5.61. (a) Redraw the circuits in Fig. P5.54(a) with a three-terminal PMOS transistor with its body connected to its source. (b) Repeat for Fig. P5.54(b).

5.62. Redraw the circuit in Fig. 5.43 with a four-terminal NMOS transistor with its body connected to -3 V.

5.63. Redraw the circuit in Fig. 5.44 with a four-terminal NMOS transistor with its body connected to -5 V.

5.5 MOS Transistor Symmetry

5.64. The 1-T cell in Fig. P5.64 uses a bitline voltage of 2.5 V and a wordline voltage of 2.5 V. (a) What are the cell voltages stored on C_C for a 1 and 0 if $V_{TO} = 0.6$ V, $\gamma = 0.5\sqrt{V}$, and $2\varphi_F = 0.6$ V? (b) What would be the minimum wordline voltage needed in order for the cell voltage to reach 2.5 V for a 1?

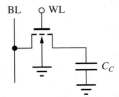

Figure P5.64

5.65. Repeat Prob. 5.64 if the bitline and wordline voltages are 1.8 V.

*5.66. Substrate leakage currents usually tend to destroy only one of the two possible states in the 1-T cell. For the circuit in Fig. P5.64, which level is the most sensitive to leakage currents and why?

5.67. Find an expression for the energy that is lost during the charge redistribution for reading out the data in the 1-T memory cell? (a) How much energy is lost if $V_C = 1.9$ V, $V_{BL} = 1$ V, and $C_C = 25$ fF. State your assumptions. (b) Suppose a 256 Mb memory using these cells is refreshed every 5 ms. What is the average power consumed by the charge redistribution operation?

*5.68. The gate-source and drain-source capacitances of the MOSFET in Fig. P5.64 are each 100 fF, and $C_C = 75$ fF. The bitline and wordline have been stable at 2.5 V for a long time. The wordline signal is shown in Fig. P5.68. What is the voltage stored on C_C before the wordline drops? Estimate the drop in voltage on the C_C due to coupling of the wordline signal through the gate-source capacitance. Use $V_{TO} = 0.70$ V, $\gamma = 0.5\ \sqrt{V}$, and $2\varphi_F = 0.6$ V.

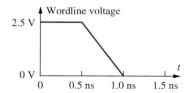

Figure P5.68

5.69. A 1-T cell has $C_C = 50$ fF and $C_{BL} = 7.5$ pF. (a) If the bitlines are precharged to 2.5 V, and the cell voltage is 0 V, what is the change in bitline voltage ΔV following cell access? (b) What is the final voltage in the cell?

5.70. A 1-T cell memory can be fabricated using PMOS transistors shown in Fig. P8.70. (a) What are the voltages stored on the capacitor corresponding to logic 0 and 1 levels for a technology using $V_{DD} = 3.3$ V? (b) For $V_{DD} = 2.5$ V? (c) For $V_{DD} = 1.8$ V?

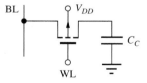

Figure P5.70

Use $K'_n = 100\ \mu A/V^2$, $K'_p = 40\ \mu A/V^2$, $V_{TN} = 0.6$ V, and $V_{TP} = -0.6$ V unless otherwise indicated. *For simulation purposes, use the values in Appendix B.*

5.6 CMOS Technology

5.71. Calculate the values of K'_n and K'_p for NMOS and PMOS transistors with a gate oxide thickness of 10 nm and 5 nm. Assume that $\mu_n = 500\ cm^2/V \cdot s$, $\mu_p = 200\ cm^2/V \cdot s$, and the relative permittivity of the gate oxide is 3.9. ($\varepsilon_0 = 8.854 \times 10^{-14}$ F/cm).

5.72. Draw a cross section similar to that in Fig. 5.20 for a CMOS process that uses a p-well instead of an n-well. Show the connections for a CMOS inverter, and draw an annotated version of the corresponding circuit schematic. (*Hint:* Start with an n-type substrate and interchange all the n- and p-type regions.)

*5.73. (a) The n-well in a CMOS process covers an area of 1 cm × 0.5 cm, and the junction saturation current density is 500 pA/cm². What is the total leakage current of the reverse-biased well? (b) Suppose the drain and source regions of the NMOS and PMOS transistors are 2 μm × 5 μm, and the saturation current density of the junctions is 100 pA/cm². If the chip has 20 million inverters, what is the total leakage current due to the reverse-biased junctions when $v_O = 2.5$ V? Assume $V_{DD} = 2.5$ V and $V_{SS} = 0$ V. (c) When $v_O = 0$ V?

*5.74. A particular interconnection between two circuits on an IC chip runs one-half the distance across a 10-mm-wide die. If the line is 1 μm wide and the oxide ($\varepsilon_r = 3.9$, $\varepsilon_0 = 8.854 \times 10^{-14}$ F/cm) beneath the line is 1 μm thick, what is the total capacitance of this line, assuming that the capacitance is three times that predicted by the parallel plate capacitance formula? Assume that the silicon beneath the oxide represents a conducting ground plane.

5.75. Simulate the VTC for a CMOS inverter with $K_n = 2.5 K_p$. Find the input voltage for which $v_O = v_I$ and compare to the value calculated by hand. Use $V_{DD} = 2.5$ V.

5.76. (a) What bias voltage is required at V_I to set $V_O = 1.65$ V in the circuit in Fig. P5.76 (b) What are the currents in the two transistors at the bias voltage in part (a)?

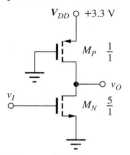

Figure P5.76

5.77. The outputs of two CMOS inverters are accidentally tied together, as shown in Fig. P5.77. What is the voltage at the common output node if the NMOS and PMOS transistors have W/L ratios of 20/1 and 40/1, respectively? What is the current in the circuit?

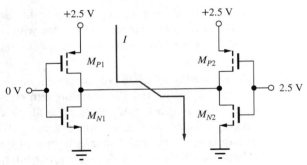

Figure P5.77

5.7 CMOS Latchup

5.78. Simulate CMOS latchup using the circuit in Fig. 5.24(b) and plot graphs of the voltages at nodes 2, 3, and 4 as well as the current supplied by V_{DD}. Discuss the behavior of the voltages and identify important voltage levels, current levels, and slopes on the graphs.

5.79. Repeat Prob. 5.78 if the values of R_n and R_p are reduced by a factor of 10.

5.80. Draw the cross section and equivalent circuit, similar to Fig. 5.24, for a *p*-well CMOS technology.

5.8 Capacitances in MOS Transistors

5.81. Calculate C''_{ox} and C_{GC} for an MOS transistor with $W = 25$ μm and $L = 0.25$ μm with an oxide thickness of (a) 50 nm, (b) 20 nm, (c) 10 nm, and (d) 5 nm.

5.82. Calculate C''_{ox} and C_{GC} for an MOS transistor with $W = 10$ μm and $L = 0.5$ μm with an oxide thickness of (a) 25 nm, (b) 10 nm, and (c) 3 nm.

5.83. In a certain MOSFET, the value of C'_{OL} can be calculated using an effective overlap distance of 0.5 μm. What is the value of C'_{OL} for an oxide thickness of 10 nm.

5.84. What are the values of C_{GS} and C_{GD} for a transistor with $C''_{ox} = 1.4 \times 10^{-3}$ F/m² and $C'_{OL} = 5 \times 10^{-9}$ F/m if $W = 10$ μm and $L = 1$ μm operating in (a) the triode region, (b) the saturation region, and (c) cutoff?

5.85. A large-power MOSFET has an effective gate area of 60×10^6 μm². What is the value of C_{GC} if T_{ox} is 100 nm?

5.86. (a) Find C_{GS} and C_{GD} for the transistor in Fig. 5.37 for the triode region if $\Lambda = 0.4$ μm, $T_{ox} = 150$ nm, and $C_{GSO} = C_{GDO} = 20$ pF/m. (b) Repeat for the saturation region. (c) Repeat for the cutoff region.

5.87. (a) Repeat Prob. 5.86 for a transistor similar to Fig. 5.37 but with $W/L = 10/1$. (b) With $W/L = 100/1$. Assume $L = 1$ μm.

5.88. Find C_{SB} and C_{DB} for the transistor in Fig. 5.37 if $\Lambda = 0.5$ μm, the substrate doping is 10^{16}/cm³, the source and drain doping is 10^{20}/cm³, and $C_{JSW} = C_J \times (5 \times 10^{-4}$/cm$)$.

5.9 MOSFET Modeling in SPICE

5.89. (a) What are the values of SPICE model parameters KP, LAMBDA, VTO, W, and L for the transistor in Fig. 5.7 if $K'_n = 50$ μA/V² and $L = 0.5$ μm? (b) Repeat L for the transistor in Fig. 5.8 if $K'_n = 10$ μA/V² and $L = 0.6$ μm?

5.90. What are the values of SPICE model parameters KP, LAMBDA, VTO, PHI, W, and L for a transistor with the following characteristics: $V_{TN} = 0.7$ V, $K_n = 175$ μA/V², $W = 5$ μm, $L = 0.25$ μm, $\lambda = 0.02$ V^{-1}, and $2\phi_F = 0.8$ V?

5.91. (a) What are the values of SPICE model parameters KP, LAMBDA, VTO, W, and L for the transistor in Fig. 5.14 if $K'_p = 10$ μA/V² and $L = 0.5$ μm? (b) Repeat for the transistor in Fig. 5.42(b) if $K'_n = 25$ μA/V² and $L = 0.6$ μm?

5.92. (a) What are the values of SPICE model parameters VTO, PHI, and GAMMA for the transistor in Fig. 5.13? (b) Repeat for the transistor in Prob. 5.44.

5.10 MOS Transistor Scaling

5.93. (a) A transistor has $T_{ox} = 40$ nm, $V_{TN} = 1$ V, $\mu_n = 500$ cm²/V·s, $L = 2$ μm, and $W = 20$ μm. What are K_n and the saturated value of i_D for this transistor if $V_{GS} = 3.75$ V? (b) The technology is scaled down by a factor of 2. What are the new values of $T_{ox}, W, L, V_{TN}, V_{GS}, K_n$, and i_D?

5.94. (a) A transistor has an oxide thickness of 20 nm with $L = 1$ μm and $W = 30$ μm. What is C_{GC} for this transistor? (b) The technology is scaled down by a factor of 2. What are the new values of T_{ox}, W, L, and C_{GC}?

5.95. Show that the cutoff frequency of a PMOS device is given by $f_T = \frac{1}{2\pi}\frac{\mu_p}{L^2}|V_{GS} - V_{TP}|$.

5.96. (a) An NMOS device has $\mu_n = 450$ cm²/V·s. What is the cutoff frequency for $L = 1$ μm if the transistor is biased at 1 V above threshold? What would be the cutoff frequency of a similar PMOS device if $\mu_p = 0.4\,\mu_n$? (b) Repeat for $L = 0.1$ μm.

5.97. An NMOS transistor has $T_{ox} = 80$ nm, $\mu_n = 400$ cm^2/V·s, $L = 0.1$ μm, $W = 2$ μm, and $V_{GS} - V_{TN} = 2$ V. (a) What is the saturation region current predicted by Eq. (5.17)? (b) What is the saturation current predicted by Eq. (5.65) if we assume $v_{SAT} = 10^7$ cm/s?

5.98. Repeat Prob. 5.23 if the transistor has $V_{SAT} = 1.8$ V.

5.99. Repeat Prob. 5.24 if the transistor has $V_{SAT} = 2.5$ V.

5.100. Repeat Prob. 5.30 if the transistor has $V_{SAT} = 3$ V.

5.101. Repeat Prob. 5.31 if the transistor has $V_{SAT} = 2.5$ V.

5.102. The NMOS transistor in Fig. 5.31 is biased with $V_{GS} = 0$ V. What is the drain current? (b) What is the drain current if the threshold voltage is reduced to 0.5 V?

5.11 All Region Modeling

5.103. Plot the output characteristics (i_D vs. v_{DS}) for an NMOSFET for v_{DS} ranging from 0 to 0.75 V and ($v_{GS} - V_{TN}$) = 0, −10 mV, −20 mV, −30 mV, −40 mV, and −50 mV, assuming $I_{DO} = 25$ nA, $n = 1.3$, and $V_T = 25.8$ mV.

5.104. An NMOS transistor has $K_n = 250$ μA/V^2, $V_{TN} = 0.5$ V, and $n = 1.67$. (a) Given a value of V_{GS} at the Q-point, how can you tell if the transistor is operating in the subthreshold region of operation? (b) Given a value of Q-point current I_D, how can you tell if the transistor is operating in the subthreshold region of operation?

5.105. (a) What is the maximum value of slope S for an FET operating in the subthreshold region at room temperature (300 K)? (b) Repeat for $n = 1.5$. (c) Repeat part (b) for liquid nitrogen temperature (77 K). (d) Repeat part (b) for liquid helium temperature (4 K).

5.106. Equations (5.70) and (5.71) are written for an NMOS transistor. Write similar equations that describe the drain current for the PMOS transistor in terms of V_{SG}, V_{SD}, and V_{TP}.

5.107. (a) Use the interpolation model in Eq. (5.71) to find the current in the FET in Fig. P5.107 if $I_{D0} = 40$ nA, $V_{TN} = 0.5$ V and $V_{DD} = 0.5$ V. (b) Repeat for $V_{DD} = 0.4$ V. (c) Repeat for $V_{DD} = 0.6$ V.

Figure P5.107

5.108. (a) Use the interpolation model in Eq. (5.71) to find the currents in the FETs in Fig. P5.108 if $I_{D0} = 60$ nA, $V_{TN} = 0.5$ V, $\gamma = 0$, and $V_{DD} = 1$ V. (b) Repeat for $V_{DD} = 0.7$ V.

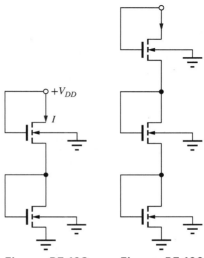

Figure P5.108   Figure P5.109

5.109. Use the interpolation model in Eq. (5.71) to find the currents in the FETs in Fig. P5.109 if $I_{D0} = 50$ nA, $V_{TN} = 0.5$ V, $\gamma = 0$, and $V_{DD} = 1.5$ V. (b) Repeat for $V_{DD} = 1.2$ V.

5.12 MOS Transistor Fabrication and Layout Design Rules

5.110. Layout a transistor with $W/L = 10/1$ similar to Fig. 5.37. What fraction of the total area does the channel represent?

5.111. Layout a transistor with $W/L = 5/1$ similar to Fig. 5.37 using $T = F = 2\,\Lambda$. What fraction of the total area does the channel represent?

5.112. Layout a transistor with $W/L = 5/1$ similar to Fig. 5.37 but change the alignment so that masks 2, 3, and 4 are all aligned to mask 1. What fraction of the total area does the channel represent?

5.113. Layout a transistor with $W/L = 5/1$ similar to Fig. 5.37 but change the alignment so that mask 3 is aligned to mask 1. What fraction of the total area does the channel represent?

5.13 Advanced CMOS Technologies

5.114. What is the W/L of the transistor in Fig. 1 of the EIA on page 263 if the channel is 20 nm long? Use the fin dimensions from Fig. 2 on page 264.

5.14 Biasing the NMOS Field-Effect Transistor

Four-Resistor Biasing

5.115. (a) Find the Q-point for the transistor in Fig. P5.115 for $R_1 = 120$ kΩ, $R_2 = 240$ kΩ, $R_3 = 24$ kΩ, $R_4 = 13$ kΩ, and $V_{DD} = 10$ V. Assume that $V_{TO} = 1$ V, $\gamma = 0$, and $W/L = 7/1$. (b) Repeat for $W/L = 14/1$.

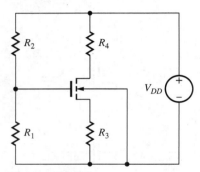

Figure P5.115

5.116. (a) Repeat Prob. 5.115(a) if all resistor values are increased by a factor of 10. (b) Repeat Prob. 5.115(a) if all resistor values are reduced by a factor of 10 and $W/L = 20/1$. (c) Repeat (b) for $W/L = 60/1$.

5.117. Repeat Prob. 5.115 with $V_{DD} = 12$ V.

5.118. Find the Q-point for the transistor in Fig. P5.115 for $R_1 = 220$ kΩ, $R_2 = 430$ kΩ, $R_3 = 51$ kΩ, $R_4 = 22$ kΩ, and $V_{DD} = 12$ V. Assume that $V_{TO} = 1$ V, $\gamma = 0$, and $W/L = 5/1$. (b) Repeat for $W/L = 15/1$.

5.119. Use SPICE to simulate the circuit in Prob. 5.115 and compare the results to hand calculations.

5.120. Use SPICE to simulate the circuit in Prob. 5.117 and compare the results to hand calculations.

5.121. Repeat Prob. 5.118(a) with all resistor values reduced by a factor of 5.

5.122. Use SPICE to simulate the circuit in Prob. 5.121 and compare the results to hand calculations.

5.123. Use SPICE to simulate the circuit in Prob. 5.118 and compare the results to hand calculations.

5.124. The drain current in the circuit in Fig. 5.43 was found to be 34.4 µA. The gate bias circuit in the example could have been designed with many different choices for resistors R_1 and R_2. Some possibilities for (R_1, R_2) are (2 kΩ, 3 kΩ), (10 kΩ, 15 kΩ), (200 kΩ, 300 kΩ), and (1.2 MΩ, 1.8 MΩ). Which of these choices would be the best and why?

*5.125. Suppose the design of Ex. 5.1 is implemented with $V_{EQ} = 4$ V, $R_S = 22$ kΩ, and $R_D = 43$ kΩ. (a) What would be the Q-point if $K_n = 35$ µA/V^2? (b) If $K_n = 25$ µA/V^2 but $V_{TN} = 0.75$ V?

5.126. (a) Simulate the circuit in Ex. 5.1 and compare the results to the calculations. (b) Repeat for the circuit design in Ex. 5.2.

5.127. Design a four-resistor bias network for an NMOS transistor to give a Q-point of (250 µA, 4.5 V) with $V_{DD} = 9$ V and $R_{EQ} \cong 250$ kΩ. Use the parameters from Table 5.9.

5.128. Design a four-resistor bias network for an NMOS transistor to give a Q-point of (100 µA, 6 V) with $V_{DD} = 12$ V and $R_{EQ} \cong 250$ kΩ. Use the parameters from Table 5.9.

5.129. Design a four-resistor bias network for an NMOS transistor to give a Q-point of (500 µA, 5 V) with $V_{DD} = 12$ V and $R_{EQ} \cong 600$ kΩ. Use the parameters from Table 5.9.

Load Line Analysis

5.130. Draw the load line for the circuit in Fig. P5.130 on the output characteristics in Fig. P5.18 and locate the Q-point. Assume $V_{DD} = 1 + 4$ V. What is the operating region of the transistor?

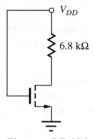

Figure P5.130

5.131. Draw the load line for the circuit in Fig. P5.130 on the output characteristics in Fig. P5.18 and locate the Q-point. Assume $V_{DD} = +5$ V and the resistor is changed to 8.3 kΩ. What is the operating region of the transistor?

5.132. Draw the load line for the circuit in Fig. P5.132 on the output characteristics in Fig. P5.18 and locate the Q-point. Assume $V_{DD} = +6$ V. What is the operating region of the transistor?

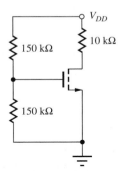

Figure P5.132

5.133. Draw the load line for the circuit in Fig. P5.132 on the output characteristics in Fig. P5.18 and locate the Q-point. Assume $V_{DD} = +8$ V. What is the operating region of the transistor?

Depletion-Mode Devices

5.134. What is the Q-point of the transistor in Fig. P5.115 if $R_1 = 2$ MΩ, $R_2 = \infty$, $R_3 = 10$ kΩ, $R_4 = 5$ kΩ, and $V_{DD} = 12$ V for $V_{TN} = -4.5$ V and $K_n = 2$ mA/V^2?

5.135. What is the Q-point of the transistor in Fig. P5.115 if $R_1 = 470$ kΩ, $R_2 = \infty$, $R_3 = 27$ kΩ, $R_4 = 51$ kΩ, and $V_{DD} = 8$ V for $V_{TN} = -3.5$ V and $K_n = 600$ µA/V^2?

5.136. Design a bias network for a depletion-mode NMOS transistor to give a Q-point of (250 µA, 7.5 V) with $V_{DD} = 12$ V if $V_{TN} = -4$ V and $K_n = 1$ mA/V^2.

*5.137. Design a bias network for a depletion-mode NMOS transistor to give a Q-point of (2 mA, 6 V) with $V_{DD} = 10$ V if $V_{TN} = -2.5$ V and $K_n = 250$ µA/V^2. (*Hint:* You may wish to consider the four-resistor bias network.)

Two-Resistor Biasing

The two-resistor bias circuit represents a simple alternative strategy for biasing the MOS transistor.

5.138. (a) Find the Q-point for the transistor in the circuit in Fig. P5.138(a) if $V_{DD} = +10$ V. (b) Repeat for the circuit in Fig. P5.138(b).

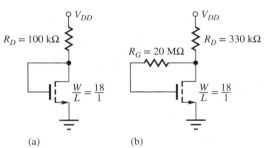

Figure P5.138

5.139. (a) Find the Q-point for the transistor in the circuit in Fig. P5.138(a) if $V_{DD} = +9$ V and W/L is changed to 80? (b) Repeat for the circuit in Fig. P5.138(b).

5.140. (a) Find the Q-point for the transistor in the circuit in Fig. P5.138(b) if $V_{DD} = 9$ V and the 330 kΩ resistor is increased to 430 kΩ. (b) Repeat if the 20 MΩ resistor is reduced to 2 MΩ.

Body Effect

5.141. Find the solution to Eq. set (5.83) using MATLAB. (b) Repeat for $\gamma = 0.75\ \sqrt{V}$.

5.142. Find the solution to Eq. set (5.83) using a spreadsheet if $\gamma = 0.75\ \sqrt{V}$. (b) Repeat for $\gamma = 1.25\ \sqrt{V}$.

5.143. Redesign the values of R_S and R_D in the circuit in Ex. 5.2 to compensate for the body effect and restore the Q-point to its original value (34 µA, 6.1 V).

5.144. Find the Q-point for the transistor in Fig. P5.115 for $R_1 = 100$ kΩ, $R_2 = 220$ kΩ, $R_3 = 24$ kΩ, $R_4 = 12$ kΩ, and $V_{DD} = 10$ V. Assume that $V_{TO} = 1$ V, $\gamma = 0.6\ \sqrt{V}$, and $W/L = 5/1$.

*5.145. (a) Repeat Prob. 5.144 with $\gamma = 0.75\ \sqrt{V}$. (b) Repeat Prob. 5.144 with $R_4 = 24$ kΩ.

5.146. (a) Use SPICE to simulate the circuit in Prob. 5.145(a) and compare the results to hand calculations. (b) Repeat for Prob. 5.145(b).

5.147. Simulate the circuit in Prob. 5.115 using (a) $\gamma = 0$ and (b) $\gamma = 0.5$ V$^{-0.5}$ and $2\phi_F = 0.6$ V and compare the results. Does our neglect of body effect in hand calculations appear to be justified?

5.148. Simulate the circuit in Prob. 5.116(a) using (a) $\gamma = 0$ and (b) $\gamma = 0.5$ V$^{-0.5}$ and $2\phi_F = 0.6$ V and compare the results. Does our neglect of body effect in hand calculations appear to be justified?

5.149. Simulate the circuit in Prob. 5.116(b) and (c) using (a) $\gamma = 0$ and (b) $\gamma = 0.5$ V$^{-0.5}$ and $2\phi_F = 0.6$ V and compare the results. Does our neglect of body effect in hand calculations appear to be justified?

5.150. Simulate the circuit in Prob. 5.117 using (a) $\gamma = 0$ and (b) $\gamma = 0.5$ V$^{-0.5}$ and $2\phi_F = 0.6$ V and compare the results. Does our neglect of body effect in hand calculations appear to be justified?

General Bias Problems

5.151. (a) Find the current I in Fig. P5.151 if $V_{DD} = 3$ V assuming that $\gamma = 0$, $V_{TO} = 1$ V, and the transistors both have $W/L = 20/1$. (b) Repeat for $V_{DD} = 5$ V. *(c) Repeat part (a) with $\gamma = 0.5 \sqrt{V}$.

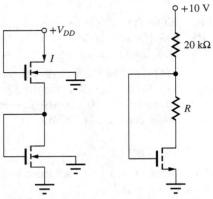

Figure P5.151   **Figure P5.152**

5.152. Find the Q-point for the transistor in Fig. P5.152 if $R = 10$ kΩ, $V_{TO} = 1$ V, and $W/L = 4/1$.

5.153. Find the Q-point for the transistor in Fig. P5.152 if $R = 20$ kΩ, $V_{TO} = 1$ V, and $W/L = 2/1$.

**5.154. (a) Find the current I in Fig. P5.154 assuming that $\gamma = 0$ and $W/L = 25/1$ for each transistor. (b) Repeat part (a) for $W/L = 40/1$. **(c) Repeat part (a) with $\gamma = 0.5 \sqrt{V}$.

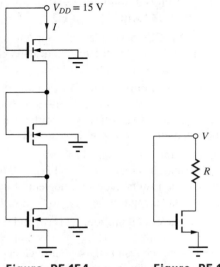

Figure P5.154   **Figure P5.156**

5.155. (a) Simulate the circuit in Fig. P5.154 using SPICE and compare the results to those of Prob. 5.158(a). (b) Repeat for Prob. 5.154(b). **(c) Repeat for Prob. 5.156(c).

5.156. What value of W/L is required to set $V_{DS} = 0.25$ V in the circuit in Fig. P5.156 if $V = 2.5$ V and $R = 160$ kΩ?

5.157. What value of W/L is required to set $V_{DS} = 0.50$ V in the circuit in Fig. P5.156 if $V = 2.5$ V and $R = 68$ kΩ?

NMOS Subthreshold Conduction

5.158. (a) Use the interpolation model in Eq. (5.71) to find the current in the FET in Fig. P5.138(a) if $I_{D0} = 40$ nA, $V_{TN} = 0.5$ V, $n = 1.6$, and $V_{DD} = 0.5$ V. (b) Repeat for $V_{DD} = 0.4$ V. (b) Repeat for $V_{DD} = 0.6$ V.

5.159. (a) Use the interpolation model in Eq. (5.71) to find the currents in the FETs in Fig. P5.151 if $I_{D0} = 60$ nA, $V_{TN} = 0.5$ V, $\gamma = 0$, $n = 1.7$, and $V_{DD} = 1$ V. (b) Repeat for $V_{DD} = 1$ V with $\gamma = 0.5$ V$^{0.5}$. (c) Repeat for $V_{DD} = 0.8$ V with $\gamma = 0.5$ V$^{0.5}$.

5.160. (a) Use the interpolation model in Eq. (5.71) to find the currents in the FETs in Fig. P5.154 if $I_{D0} = 50$ nA, $V_{TN} = 0.5$ V, $\gamma = 0$, $n = 1.5$, and $V_{DD} = 1.2$ V. (b) Repeat for $V_{DD} = 1.2$ V with $\gamma = 0.5$ V$^{0.5}$. (c) Repeat for $V_{DD} = 0.9$ V with $\gamma = 0.5$ V$^{0.5}$.

5.161. (a) Find the current in the circuit in Fig. P5.138(a) if the supply voltage is 1.5 V, $V_{TN} = 0.6$ V, $R_D = 10$ MΩ, $K_n = 25$ μA/V^2, $n = 1.5$, and $T = 300$ K. (b) Repeat for the circuit in Fig. P5.138(b) if $R_D = 10$ MΩ and $R_G = 10$ MΩ.

5.162. Find the current in the circuit in Fig. P5.151 if the supply voltage is 1.2 V, $V_{TN} = 0.6$ V, $\gamma = 0$, $K'_n = 25$ μA/V^2, $n = 1.5$, $T = 300$ K and $W/L = 20/1$.

5.163. Use SPICE to plot the i-v characteristic for the circuit in Fig. P5.196 for $0 \leq V \leq 15$ V if the JFETs have $I_{DSS1} = 200$ μA, $V_{P1} = -2$ V, $I_{DSS2} = 500$ μA, and $V_{P2} = -4$ V.

5.164. The circuit in Fig. P5.164 is a voltage regulator utilizing an ideal op amp. (a) Find the output voltage of the circuit if the Zener diode voltage is 5 V. (b) What are the current in the Zener diode and the drain current in the NMOS transistor? (c) What is the op amp output voltage if the MOSFET has $V_{TN} = 1.25$ V and $K_n = 150$ mA/V^2?

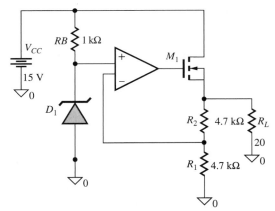

Figure P5.164

5.15 Biasing the PMOS Field-Effect Transistor

5.165. (a) Find the Q-point for the transistor in Fig. P5.165(a) if $V_{DD} = -10$ V, $R = 75$ kΩ, and $W/L = 1/1$. (b) Find the Q-point for the transistor in Fig. P5.165(b) if $V_{DD} = -10$ V, $R = 75$ kΩ, and $W/L = 1/1$.

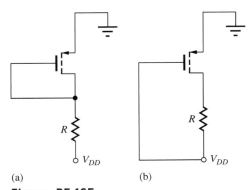

Figure P5.165

5.166. Simulate the circuits in Prob. 5.165 with $V_{DD} = -10$ V and compare the Q-point results to hand calculations.

**5.167. (a) Find the current I in Fig. P5.167 assuming that $\gamma = 0$ and $W/L = 30/1$ for each transistor. (b) Repeat part (a) for $W/L = 55/1$. **(c) Repeat part (a) with $\gamma = 0.5 \sqrt{V}$.

*5.168. (a) Find current I and voltage V_O in Fig. P5.168(a) if $W/L = 20/1$ for both transistors and $V_{DD} = 10$ V. (b) What is the current if $W/L = 80/1$? (c) Repeat for the circuit in Fig. P5.168(b).

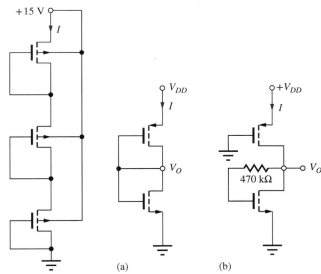

Figure P5.167   Figure P5.168

*5.169. (a) Simulate the circuit in Prob. 5.167(a) and compare the results to those of Prob. 5.167(a). (b) Repeat for Prob. 5.167(b). (c) Repeat for Prob. 5.167(c).

5.170. Draw the load line for the circuit in Fig. P5.170 on the output characteristics in Fig. P5.48 and locate the Q-point. What is the operating region of the transistor?

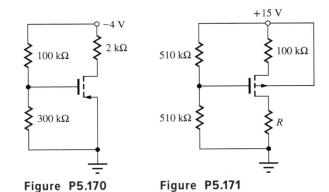

Figure P5.170   Figure P5.171

5.171. (a) Find the Q-point for the transistor in Fig. P5.171 if $R = 50$ kΩ. Assume that $\gamma = 0$ and $W/L = 20/1$. (b) What is the permissible range of values for R if the transistor is to remain in the saturation region?

5.172. Simulate the circuit of Prob. 5.171(a) and find the Q-point. Compare the results to hand calculations.

*5.173. (a) Find the Q-point for the transistor in Fig. P5.171 if $R = 43$ kΩ. Assume that $\gamma = 0.5 \sqrt{V}$ and $W/L = 20/1$. (b) What is the permissible range of values for R if the transistor is to remain in the saturation region?

324 Chapter 5 Field-Effect Transistors

5.174. Simulate the circuit of Prob. 5.173(a) and find the Q-point. Compare the results to hand calculations.

5.175. (a) Find the Q-point for the transistor in Fig. P5.175 if $V_{DD} = 14$ V, $R = 100$ kΩ, $W/L = 10/1$, and $\gamma = 0$. (b) Repeat for $\gamma = 1 \sqrt{V}$.

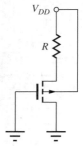

Figure P5.175

5.176. Find the Q-point current for the transistor in Fig. P5.171 if all resistors are reduced by a factor of 2. Assume saturation region operation. What value of R is needed to set $V_{DS} = -5$ V. Assume that $\gamma = 0$ and $W/L = 40/1$.

5.177. Repeat Prob. 5.176 if $\gamma = 0.5 \sqrt{V}$ and $W/L = 40/1$.

5.178. (a) Design a four-resistor bias network for a PMOS transistor to give a Q-point of (1 mA, −3 V) with $V_{DD} = -9$ V and $R_{EQ} \geq 1$ MΩ. Use the parameters from Table 5.9. (b) Repeat for an NMOS transistor with $V_{DS} = +3$ V and $V_{DD} = +9$ V.

5.179. (a) Design a four-resistor bias network for a PMOS transistor to give a Q-point of (500 μA, −5 V) with $V_{DD} = -15$ V and $R_{EQ} \geq 100$ kΩ. Use the parameters from Table 5.9. (b) Repeat for an NMOS transistor with $V_{DS} = +6$ V and $V_{DD} = +15$ V.

5.180. (a) Find the Q-point for the transistor in Fig. P5.180 if $V_{TO} = +4$ V, $\gamma = 0$, and $W/L = 10/1$. (b) Repeat if $\gamma = 0.25 \sqrt{V}$.

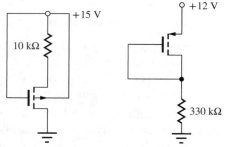

Figure P5.180 Figure P5.181

5.181. (a) Find the Q-point for the transistor in Fig. P5.181 if $V_{TO} = -1$ V and $W/L = 10/1$. (b) Repeat for $V_{TO} = -3$ V and $W/L = 30/1$.

5.182. What is the Q-point for each transistor in Fig. P5.182?

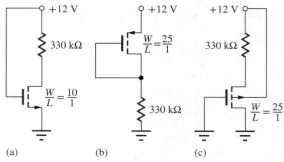

Figure P5.182

5.183. The circuit in Fig. P5.183 is a voltage regulator utilizing an ideal op amp. (a) Find the output voltage of the circuit if the Zener diode voltage is 5 V. (b) What are the current in the Zener diode and the drain current in the PMOS transistor? (c) What is the op amp output voltage if the MOSFET has $V_{TP} = -1.5$ V and $K_n = 50$ mA/V²?

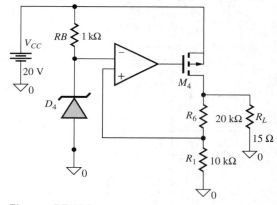

Figure P5.183

5.16 Biasing the CMOS Inverter as an Amplifier

5.184. (a) What is the value of I_{D0} at room temperature for a 10/1 NMOS FET based upon the parameters in Table 5.9 with $n = 1.6$. (b) Repeat part (a) for a 25/1 PMOS FET with $n = 1.7$. (c) What W/L ratio will give the PMOS transistor the same value of I_{D0} as the NMOS device?

5.185. (a) Find the Q-points of the transistors in the CMOS amplifier in Fig. P5.185 if the supply voltage is 1.5 V, $V_{TN} = 0.6$ V, $V_{TP} = -0.6$ V, $R_G = 1$ MΩ, $K_n = K_p = 50$ μA/V², $n = 1.5$, and $T = 300$ K. (b) Repeat for the circuit for $R_G = 5.1$ MΩ. (c) Repeat (a) for $V_{DD} = 3.3$ V.

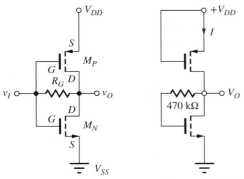

Figure P5.185 **Figure P5.186**

5.186. (a) Find the Q-points of the transistors in the CMOS amplifier in Fig. P5.185 if the supply voltage is 3.3 V, $V_{TN} = 0.6$ V, $V_{TP} = -0.6$ V, $R_G = 1$ MΩ, $K_n = 55$ µA/V^2, $K_p = 45$ µA/V^2, $n = 1.5$, and $T = 300$ K. (b) Repeat for $V_{DD} = 1.5$ V.

5.187. (a) Find the Q-points of the transistors in the CMOS amplifier in Fig. P5.185 if the supply voltage is 3.3 V, $V_{TN} = 0.575$ V, $V_{TP} = -0.625$ V, $R_G = 1$ MΩ, $K_n = K_p = 50$ µA/V^2, $n = 1.5$, and $T = 300$ K. (b) Repeat for $V_{DD} = 1.5$ V.

5.188. Find the Q-points of the transistors in the CMOS amplifier in Fig. P5.185 if the supply voltage is 1.5 V, $V_{TN} = 0.6$ V, $V_{TP} = -0.6$ V, $R_G = 2$ MΩ, $K_n = K_p = 50$ µA/V^2, $n = 1.55$, for the NMOS device, $n = 1.7$ for the PMOS device and $T = 300$ K.

5.189. (a) Find the Q-points of the transistors in the CMOS amplifier in Fig. P5.185 if the supply voltage is 1.2 V, $V_{TN} = 0.6$ V, $V_{TP} = -0.6$ V, $R_G = 1$ MΩ, $K_n = K_p = 50$ µA/V^2, $n = 1.5$, and $T = 300$ K. (b) Repeat for $V_{DD} = 1.0$ V. (c) Repeat for $V_{DD} = 0.8$ V.

5.190. During testing, an engineer accidentally shorts the output of the CMOS amplifier in Fig. P5.185 to V_{DD}. What are the values of the two drain currents if the supply voltage is 1.5 V, $V_{TN} = 0.6$ V, $V_{TP} = -0.6$ V, $R_G = 1$ MΩ, $K_n = K_p = 50$ µA/V^2, $n = 1.5$, and $T = 300$ K? (b) Repeat if the output of the amplifier is shorted to ground and $R_G = 5.1$ MΩ. (c) Repeat for $V_{DD} = 2.5$ V.

5.191. (a) Find the Q-points of the transistors in the CMOS amplifier in Fig. P5.186 if supply voltage V_{DD} is 1.5 V, $V_{TN} = 0.5$ V, $V_{TP} = -0.5$ V, $R_G = 4.7$ MΩ, $K_n = K_p = 50$ µA/V^2, $n = 1.65$, and $T = 300$ K. Note that the PMOS transistor is operating in subthreshold. (b) Repeat part (a) for

the circuit if the resistor is changed to 2 MΩ. (c) Repeat part (a) for $V_{DD} = 1.0$ V.

5.17 The CMOS Transmission Gate

5.192. (a) Calculate the on-resistance of an NMOS transistor with $W/L = 20/1$ for $V_{GS} = 2.5$ V, $V_{SB} = 0$ V, and $V_{DS} = 0$ V. (b) Calculate the on-resistance of a PMOS transistor with $W/L = 20/1$ for $V_{SG} = 2.5$ V, $V_{SB} = 0$ V, and $V_{SD} = 0$ V. (c) What do we mean when we say that a transistor is "on" even though I_D and $V_{DS} = 0$?

5.193. Calculate the maximum and minimum values of the equivalent on-resistance for the transmission gate in Fig. 5.58.

5.194. (a) What is the largest value of the on-resistance of a transmission gate with $W/L = 10/1$ for both transistors if the input voltage range is $0 \le v_I \le 1$ V and the power supply is 2.5 V? At what input voltage does it occur? (b) Repeat for $0 \le v_I \le 2.5$ V.

5.195. A certain analog multiplexer application requires the equivalent on-resistance R_{EQ} of a transmission gate to always be less than 250 Ω for $0 \le v_I \le 2.5$ V. What are the minimum values of W/L for the NMOS and PMOS transistors if $V_{TON} = 0.75$ V, $V_{TOP} = -0.75$ V, $\gamma = 0.5 \sqrt{V}$, $2\phi_F = 0.6$ V, $K'_p = 40$ µA/V^2, and $K'_n = 100$ µA/V^2?

5.196. (a) What are the voltages at the nodes in the pass-transistor networks in Fig. P5.196. For NMOS transistors, use $V_{TO} = 0.70$ V, $\gamma = 0.6 \sqrt{V}$, and $2\phi_F = 0.6$ V. For PMOS transistors, $V_{TO} = -0.70$ V and $\gamma = 0.5 \sqrt{V}$. (b) What would be the voltages if transmission gates were used in place of each transistor?

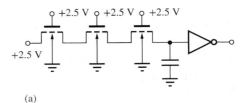

(a)

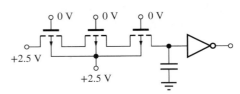

(b)

Figure P5.196

5.18 The Junction Field-Effect Transistor (JFET)

5.197. The JFET in Fig. P5.197 has $I_{DSS} = 500$ μA and $V_P = -3$ V. Find the Q-point for the JFET for (a) $R = 0$ and $V = 5$ V, (b) $R = 0$ and $V = 0.25$ V, and (c) $R = 8.2$ kΩ and $V = 5$ V.

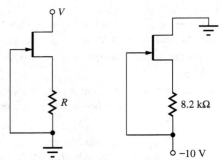

Figure P5.197 **Figure P5.198**

5.198. Find the Q-point for the JFET in Fig. P5.198 if $I_{DSS} = 5$ mA and $V_P = -5$ V.

5.199. Find the on-resistance of the JFET in Fig. P5.199 if $I_{DSS} = 1$ mA and $V_P = -5$ V. Repeat for $I_{DSS} = 100$ μA and $V_P = -2$ V.

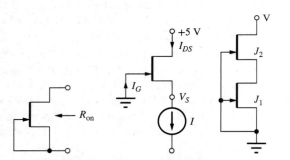

Figure P5.199 **Figure P5.200** **Figure P5.201**

5.200. The JFET in Fig. P5.200 has $I_{DSS} = 1$ mA and $V_P = -4$ V. Find I_D, I_G, and V_S for the JFET if (a) $I = 0.5$ mA and (b) $I = 2$ mA.

*5.201. The JFETs in Fig. P5.201 have $I_{DSS1} = 200$ μA, $V_{P1} = -2$ V, $I_{DSS2} = 500$ μA, and $V_{P2} = -4$ V. (a) Find the Q-point for the two JFETs if $V = 9$ V. (b) What is the minimum value of V that will ensure that both J_1 and J_2 are in pinch-off?

5.202. Simulate the circuit in Prob. 5.201(a) and compare the results to hand calculations.

*5.203. The JFETs in Fig. P5.203 have $I_{DSS} = 200$ μA and $V_P = +2$ V. (a) Find the Q-point for the two JFETs if $R = 10$ kΩ and $V = 15$ V. (b) What is the minimum value of V that will ensure that both J_1 and J_2 are in pinch-off if $R = 10$ kΩ?

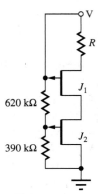

Figure P5.203

5.204. Simulate the circuit in Prob. 5.203(a) and compare the results to hand calculations.

5.205. (a) The JFET in Fig. P5.205(a) has $I_{DSS} = 250$ μA and $V_P = -2$ V. Find the Q-point for the JFET. (b) The JFET in Fig. P5.205(b) has $I_{DSS} = 250$ μA and $V_P = +2$ V. Find the Q-point for the JFET.

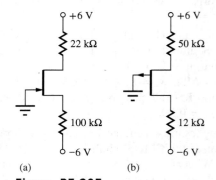

(a)   (b)

Figure P5.205

5.206. Simulate the circuit in Prob. 5.205(a) and compare the results to hand calculations. (b) Repeat for Prob. 5.205(b).

术语对照

Accumulation	积累
Accumulation region	积累区
Active region	有源区
Alignment tolerance T	对准容限 T
Body effect	体效应
Body-effect parameter λ	体效应系数 λ
Body terminal(B)	体端
Bulk terminal(B)	衬底端
C_{GS}	栅-源电容
C_{GD}	栅-漏电容
C_{GB}	栅衬电容,栅极对衬底的电容
C_{DB}	漏衬电容,漏极对衬底的电容
C_{SB}	源-体电容,源极对衬底的电
C_{OX}''	容单位面积的栅极-沟道电容
C_{GDO}	栅-漏交叠电容
C_{GSO}	栅-源交叠电容
Capacitance per unit width	单位宽度电容
Channel length L	沟道长度 L
Channel-length modulation	沟道长度调制
Channel-length modulation parameter λ	沟道长度调制参数 λ
Channel region	沟道区
Channel width W	沟道宽度 W
Constant electric field scaling	恒定电场缩放
Current sink	电流沉
Current source	电流源
Cutoff frequency	截止频率
Depletion	耗尽
Depletion-mode device	耗尽型装置
Depletion-mode MOSFETs	耗尽型MOSFET晶体管
Depletion region	耗尽区
Design rules	设计规则
Drain(D)	漏极（D）
Electronic current source	电流源
Enhancement-mode device	增强型器件
Field-effect transistor(FET)	场效应晶体管（FET）
Four-resistor bias	四电阻偏置
Gate(G)	栅极（G）
Gate-channel capacitance C_{GC}	栅-沟道电容 C_{GC}
Gate-drain capacitance C_{GD}	栅-漏电容 C_{GD}
Gate-source capacitance C_{GS}	栅-源电容 C_{GS}
Ground rules	接地规则
High field limitation	高场限制
Inversion layer	反相层
Inversion region	反相区

English	中文
K_P	SPICE中的跨导参数
K'_n, K'_P	跨导参数
LAMBDA, λ	沟道调制系数，λ
Triode region	线性区
Metal-oxide-semiconductor field-effect transistor(MOSFET)	金属-氧化物半导体场效应晶体管
Minimum feature size F	最小特征尺寸 F
Mirror ratio	镜像比
MOS capacitor	MOS电容
n-channel MOS(NMOS)	n沟道MOS（NMOS）
n-channel MOSFET	n沟道MOSFET
n-channel transistor	n沟道晶体管
NMOSFET	NMOS场效应管
NMOS transistor	NMOS晶体管
On-resistance (R_{ON})	导通电阻（R_{ON}）
Output characteristics	输出特性
Output resistance	输出电阻
Overlap capacitance	交叠电容
Oxide thickness	氧化层厚度
p-channel MOS(PMOS)	p沟道MOS（PMOS）
PHI	黄金分割
Pinch-off locus	夹断轨迹
Pinch-off point	夹断点
Pinch-off region	夹断区
PMOS transistor	PMOS晶体管
Power delay product	功耗-延迟积
Quiescent operating point	准静态工作点
Q-point	Q点
Saturation region	饱和区
Saturation voltage	饱和电压
Scaling theory	按比例缩小理论
Small-signal output resistance	小信号输出电阻
SPICE models	SPICE 模型
Source(S)	源极（S）
Substrate sensitivity	衬底灵敏度
Substrate terminal	衬底端
Surface potential parameter $2\phi_F$	表面电势参数 $2\phi_F$
Subthreshold region	亚阈值区
Threshold voltage V_{TN}, V_{TP}	阈值电压 V_{TN}, V_{TP}
Transconductance g_m	跨导 g_m
Transconductance parameter—K'_n, K'_P, K_P	跨导参数 K'_n, K'_P, K_P
Transfer characteristic	传输特性
Triode region	线性区
V_{TN}, V_{TP}	零偏阈值电压参数
V_T, V_{TO}	SPICE中的零偏阈值电压参数
Zero-substrate-bias value for V_{TN}	V_{TN}的零衬底偏置值

PART TWO
ANALOG ELECTRONICS

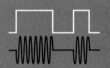

CHAPTER 6
INTRODUCTION TO AMPLIFIERS

CHAPTER 7
THE TRANSISTOR AS AN AMPLIFIER

CHAPTER 8
TRANSISTOR AMPLIFIER BUILDING BLOCKS

CHAPTER 9
AMPLIFIER FREQUENCY RESPONSE

CHAPTER 6
INTRODUCTION TO AMPLIFIERS

第6章　放大器简介

本章提纲

6.1　模拟电子系统示例
6.2　放大作用
6.3　放大器的二端口模型
6.4　源和负载电阻的失配
6.5　差分放大器
6.6　放大器的失真
6.7　差分放大器模型
6.8　放大器的频率响应

本章目标

- 理解电压增益、电流增益和功率增益；
- 理解用分贝方式表示增益；
- 理解输入电阻和输出电阻；
- 理解传输函数和伯德图；
- 理解低通和高通放大器；
- 理解截止频率和带宽；
- 二端口网络；
- 理解线性放大器的偏压；
- 理解放大器的失真；
- 了解用SPICE分析交流和传输性能；
- 了解用于确定通用放大器电路增益、输入阻抗和输出阻抗的方法。

本章导读

　　本章开始讲解模拟信号电路，对模拟信号的形式、作用及处理等进行了说明。模拟电路技术是电子产品中的关键技术，通常以模/数（A/D）转换器和数/模（D/A）转换器实现计算机与模拟信号的采集、转换与处理。每天，世界都被日益丰富的各种通信形式连接。光纤系统、有线调制解调器、数字用户线和无线通信技术等，都依赖放大器产生并检测信号非常微弱的传输信息。温度、湿度、压力、速度、光强、声音等都是模拟信号，这些信号反映了一定的物理量，可以在一定连续的范围内任意取值。在电子学中，这些信号可以用测量压力、温度或流速的传感器输出，也可以是麦克风或立体声放大器的音频信号。这些信息的特征通常用线性放大器来操控。这种放大器可以改变信号的幅度或相位，而不影响其频谱特性。

　　本章首先给出了放大作用的概念，介绍了电压增益、电流增益、功率增益以及分贝等概念。各种增益表达式通常涉及的数值都比较大，习惯上用分贝(decibel, dB)来表示电压、电流和功率增益的值。分贝数是以10

为底的算术功率比值的对数的10倍，分贝的加法和减法对应对数的乘法和除法。由于功率均与电压和电流的二次方成正比，因此在A_{vdB}和A_{idB}的表达式中会出现因子20。

放大器系统的建模往往十分复杂，借助于二端口网络可方便地对复杂系统中的放大器进行建模，可以采用相对简单的二端口网络来替代相对复杂的电路。本章给出了二端口网络的数学模型，其中采用g参数作为描述放大器的重点参数。对于g参数的计算本章给出了详细的计算实例。

运算放大器(operational amplifier)，简称运放(OP amp)，是模拟电路设计的一个基本构件模块。本节重点研究了运算放大器电路及其特性，其中重点分析了运算放大器基本电路的应用，包括反相和同相放大器、求和放大器、积分器和基本滤波器。运算放大器是差分放大器的一种形式。差分放大器对两个输入信号之差产生响应，是一类非常有用的电路。放大器的电压增益A用于描述输入信号的变化量与输出信号的变化量之间的关系，用放大器VTC曲线的斜率来定义，根据放大器输入和输出的相位不同，放大器又分为同相放大器(noninverting amplifier)和反相放大器(inverting amplifier)。

为了使放大器表现为线性工作，放大器必须合理偏置，以确保它在其线性区内工作。放大器偏置点，即Q点的选择会同时影响放大器的增益和实现线性放大的输入信号范围。偏置点选择不当，可能会使放大器工作在非线性区，导致信号失真。失真是将信号中不需要的谐波与需要的谐波进行比较，由总谐波失真(total harmonic distortion, THD）表示信号失真。

Chapter Outline

6.1 An Example of an Analog Electronic System
6.2 Amplification
6.3 Two-Port Models for Amplifiers
6.4 Mismatched Source and Load Resistances
6.5 The Differential Amplifier
6.6 Distortion in Amplifiers
6.7 Differential Amplifier Model
6.8 Amplifier Frequency Response
Summary
Key Terms
References
Additional Reading
Problems

Chapter Goals

Chapter 6 begins our study of the circuits used for analog electronic signal processing. We will develop an understanding of concepts related to linear amplification and methods of characterizing amplifiers.

- Voltage gain, current gain, and power gain
- Gain conversion to decibel representation
- Input resistance and output resistance
- Transfer functions and Bode plots
- Low-pass, high-pass, and band-pass amplifiers
- Cutoff frequencies and bandwidth
- Two-port models
- Distortion in amplifiers
- Use of ac and transfer function (TF) analyses in SPICE
- Techniques used to determine voltage gain, input resistance, and output resistance of general amplifier circuits

Lee deForest and the Audion.
Bettmann/Getty Images

Invention of the Audion tube by Lee DeForest in 1906 was a milestone event in electronics as it represented the first device that provided amplification with reasonable isolation between the input and output [1–4]. Amplifiers today, most often in solid-state integrated circuit form, play a key role in the multitude of electronic devices that we encounter in our daily activities, even in devices that we often think are digital in nature. Examples include 4G and 5G cell phones, disk drives, digital audio players, global positioning systems, and RF systems such as Bluetooth and WiFi. All these devices utilize amplifiers to transform very small analog signals to levels where they can be reliably converted into digital form. Analog circuit technology also lies at the heart of the interface between the analog and digital portions of these devices in the form of analog-to-digital (A/D) and digital-to-analog (D/A) converters. Every day the world is becoming connected through an increasing variety of communications links. Optical fiber systems, cable modems, digital subscriber lines, and wireless communications technologies rely on amplifiers to both generate and then detect extremely small signals containing the transmitted information.

Much of the information about the world around us, such as temperature, humidity, pressure, velocity, light intensity, sound, and so on, is "analog" in nature, may take on any value within some continuous range, and can be represented by the analog signal in Fig. 6.1. In electrical form, these signals may be the output of transducers that measure pressure, temperature, or flow rate, or the audio signal from a microphone or audio amplifier. The characteristics of these signals are most often manipulated using linear amplifiers, which change the amplitude and/or phase of a signal.

Chapter 6 Introduction to Amplifiers

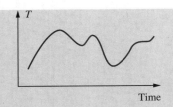

Figure 6.1 Temperature versus time—a continuous analog signal.

Today we recognize it is frequently advantageous to do as much signal processing as possible in the digital domain. Because of noise and dynamic range considerations, most A/D and D/A converters have full-scale ranges of 1 to 5 volts, whereas signals in sensor, transducer, communications, and many other applications are typically at much lower levels. For example, temperature sensor outputs may be less than 1 mV/°C, and cell phones and satellite radios require sensitivities in the microvolt range. Thus, we require the use of amplifiers to increase the voltage, current, and/or power levels of these signals. At the same time, amplifiers are used to limit (filter) the frequency content of the signals.

Chapter 6 discusses important amplifier parameters including voltage-, current-, and power-gain, input and output resistance, and frequency response. Chapter 7 introduces the use of coupling and bypass capacitors to differentiate the ac and dc characteristics of amplifiers. Small-signal modeling is introduced including development of small-signal models for the diode and BJT and MOS transistors.

Chapter 8 explores the detailed characteristics of single-transistor amplifiers including the common-emitter, common-base and common-collector BJT amplifiers and the common-source, common-gate and common-drain MOS amplifiers. Low frequency response of these amplifiers is discussed in detail, Chapter 9 explores amplifier overall frequency response including short-circuit and open-circuit time constant methods for estimating lower and upper cutoff frequencies of amplifiers. Discussions of A/D and D/A conversion will wait until Chapter 12.

6.1 AN EXAMPLE OF AN ANALOG ELECTRONIC SYSTEM

We begin exploring some of the uses for analog amplifiers by examining a familiar electronic system, a wireless headset, shown schematically in Fig. 6.2. This figure is representative of a wide range of radio frequency (RF) wireless speaker systems. At the receiving antenna are **VHF**,[1] **UHF**, and microwave radio signals in the 1–5 GHz range that contain the information for at least two channels of stereo music.[2] In our headset receiver, these signals may have amplitudes as small as 1 μV, whereas the audio amplifiers develop the voltage and current necessary to deliver 25 mW of power to 24-Ω earphones in the 50- to 15,000-Hz audio frequency range.

This receiver is a complex analog system that provides many forms of analog signal processing, some linear and some nonlinear (see Table 6.1). For example, the amplitude of the signal

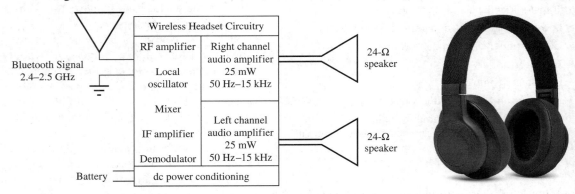

Figure 6.2 Wireless headset.
Photoongraphy/Shutterstock

---

[1] The radio spectrum is traditionally divided into different frequency bands: RF, or radio frequency (0.5–50 MHz); VHF, or very high frequency (50–150 MHz); UHF, or ultra high frequencies (150–1000 MHz); and so on. Today, however, RF is commonly used to refer to the whole radio spectrum from 0.5 MHz to 10 GHz and higher. (See Sec. 1.6.)

[2] A satellite or cell phone radio receiver is very similar except the input frequencies range from 1 to 5 GHz. Bluetooth and WiFi both oprerate in the 2.4 GHz ISM frequency band (see Table 1.3).

TABLE 6.1
Wireless Headphone Set

LINEAR CIRCUIT FUNCTIONS	NONLINEAR CIRCUIT FUNCTIONS
Radio frequency amplification	dc power supply (rectification)
Audio frequency amplification	Frequency conversion (mixing)
Frequency selection (tuning)	Detection/demodulation
Antenna impedance matching	Frequency synthesis (tuning)
Tailoring audio frequency response	Digital-to-analog conversion
Local oscillator	
Mixer	

must be increased at **radio** and **audio frequencies** (**RF** and **AF**, respectively). Large overall voltage, current, and power gains are required to go from the very small signal received from the antenna to the audio signal delivered to the speaker. The input of the receiver is often designed to match the low impedance of the antenna.

In addition, we usually want only one channel to be heard at a time. The desired signal must be selected from the multitude of signals appearing at the antenna, and the receiver requires circuits with high frequency selectivity at its input. An adjustable frequency signal source, called the *local oscillator*, is also needed to tune the receiver. The electronic implementations of all these functions are based on **linear amplifiers.**

In most receivers, the incoming signal frequency is changed, through a process called *mixing*, to a lower **intermediate frequency (IF),**[3] where the audio information can be readily separated from the RF carrier through a process called *demodulation*. Mixing and demodulation are two basic examples of analog signal processing that employ multiplication of signals. But even these circuits are based on linear amplifier designs. Finally, the dc voltages needed to power the system are obtained from the battery via voltage regulators.

6.2 AMPLIFICATION

Linear amplifiers are an extremely important class of circuits, and Part Two discusses various aspects of their analysis and design in detail. As an introduction to amplification, let us concentrate on one of the channels of the audio portion of the headset in Fig. 6.3. In this figure, the input to the stereo amplifier channel is represented by the Thévenin equivalent source v_i and source resistor R_I. The headphone at the output is modeled by a 24-Ω resistor.

Based on Fourier analysis, we know that a complex periodic signal v_i can be represented as the sum of many individual sine waves:

$$v_i = \sum_{j=1}^{\infty} V_j \sin(\omega_j t + \phi_j) \tag{6.1}$$

where V_j = amplitude of jth component of signal, ω_j is the radian frequency, and ϕ_j is the phase.

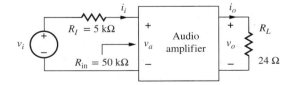

Figure 6.3 Amplifier channel from a set of wireless headphones.

---

[3] Common IF frequencies are 11.7 MHz, 455 kHz, and 262 kHz.

If the amplifier is linear, the principle of superposition applies, so that each signal component can be treated individually and the results summed to find the complete signal. For simplicity in our analysis, we will consider only the ith component of the signal, with frequency ω_i and amplitude V_i:

$$v_i = V_i \sin \omega_i t \tag{6.2}$$

For this example, we assume $V_i = 0.001$ V, 1 mV; because this signal serves as our reference input, we can assume $\phi_i = 0$ without loss of generality.

The output of the linear amplifier is a sinusoidal signal at the same frequency but with a different amplitude V_o and phase θ:

$$v_o = V_o \sin(\omega_i t + \theta) \tag{6.3}$$

The amplifier output power is

$$P_o = \left(\frac{V_o}{\sqrt{2}}\right)^2 \frac{1}{R_L} \tag{6.4}$$

where the quantity $V_o/\sqrt{2}$ represents the rms value of the sinusoidal voltage signal. For an amplifier delivering 25 mW to the 24-Ω load, the amplitude of the output voltage is

$$V_o = \sqrt{2 P_o R_L} = \sqrt{2 \times 0.025 \text{ W} \times 24\ \Omega} = 1.10 \text{ V}^{\,4}$$

This output power level also requires an output current

$$i_o = I_o \sin(\omega_i t + \theta) \tag{6.5}$$

with an amplitude

$$I_o = \frac{V_o}{R_L} = \frac{1.10 \text{ V}}{24\ \Omega} = 45.8 \text{ mA}$$

Note that because the load element is a resistor, i_o and v_o have the same phase (θ).

6.2.1 VOLTAGE GAIN

For sinusoidal signals, the **voltage gain** A_v of an amplifier is defined in terms of the **phasor representations** of the input and output voltages. Using $\sin \omega t = \text{Im}[e^{j\omega t}]$ as our reference, the phasor representation of v_i is $\mathbf{v}_i = V_i \angle 0°$ and that for v_O is $\mathbf{v}_o = V_o \angle \theta$. Similarly, $\mathbf{i}_i = I_i \angle 0°$ and $\mathbf{i}_o = I_o \angle \theta$. The overall voltage gain is then expressed by the phasor ratio:

$$|A_v| = \frac{V_o}{V_a} \quad \text{and} \quad \angle A_v = \theta \tag{6.6}$$

For the audio amplifier itself in Fig. 6.3, the magnitude of the required voltage gain is

$$|A_v| = \frac{V_o}{V_a} = \frac{1.10 \text{ V}}{0.909 \text{ mV}} = 1210 \quad \text{where} \quad V_a = \frac{R_{\text{in}}}{R_I + R_{\text{in}}} V_i = \frac{50 \text{ k}\Omega}{55 \text{ k}\Omega} 1 \text{ mV} = 0.909 \text{ mV}$$

accounts for the loss of signal between **source resistance R_I** and **input resistance R_{in}**.

We will find that the amplifier building blocks studied in the next several chapters have either $\theta = 0°$ or $\theta = 180°$ for frequencies in the "midband" range of the amplifier. (Midband will be defined later in Sec. 6.8.4.)

In addition, achieving this level of voltage gain usually requires several stages of amplification. Be sure to note that the magnitude of the gain is defined by the amplitudes of the signals and is a constant; it is *not* a function of time! For the rest of this section, we concentrate on the magnitudes of the gains, saving a more detailed consideration of amplifier phase for Sec. 6.8.

---

[4] Note that this voltage level is compatible with typical lithium-ion battery voltages.

6.2.2 CURRENT GAIN

The audio amplifier in our example requires a substantial increase in current level as well. The input current is determined by R_I and R_{in} of the amplifier. When we write the input current as $i_i = I_i \sin \omega_i t$, the amplitude of the current is

$$I_i = \frac{V_i}{R_I + R_{in}} = \frac{10^{-3} \text{ V}}{5 \text{ k}\Omega + 50 \text{ k}\Omega} = 1.82 \times 10^{-8} \text{ A} = 18.2 \text{ nA} \tag{6.7}$$

Phase $\phi = 0$ because the circuit is purely resistive.

The **current gain** is defined as the ratio of the phasor representations of i_o and i_i:

$$A_i = \frac{i_o}{i_i} = \frac{I_o \angle \theta}{I_i \angle 0} = \frac{I_o}{I_i} \angle \theta \tag{6.8}$$

The magnitude of the overall current gain is equal to the ratio of the amplitudes of the output and input currents:

$$|A_i| = \frac{I_o}{I_i} = \frac{45.6 \times 10^{-3} \text{ A}}{1.82 \times 10^{-8} \text{ A}} = 2.51 \times 10^6$$

This level of current gain also requires several stages of amplification.

6.2.3 POWER GAIN

The power delivered to the amplifier input is quite small, whereas the power delivered to the speaker is substantial. Thus, the amplifier also exhibits a very large power gain. **Power gain A_P** is defined as the ratio of the output power P_o delivered to the load, to the power P_i delivered from the source:

$$A_P = \frac{P_o}{P_i} = \frac{\frac{V_o}{\sqrt{2}} \frac{I_o}{\sqrt{2}}}{\frac{V_i}{\sqrt{2}} \frac{I_i}{\sqrt{2}}} = \frac{V_o I_o}{V_i I_i} = |A_v||A_i| \tag{6.9}$$

Note from Eq. (6.9) that either rms or peak values of voltage and current may be used to define power gain as long as the choice is applied consistently at the input and output of the amplifier. (This is also true for A_v and A_i.) For our ongoing example, we find the power gain to be a very large number:

$$A_P = \frac{1.10 \times 0.0456}{0.909 \times 10^{-3} \times 1.82 \times 10^{-8}} = 3.03 \times 10^9$$

> **EXERCISE:** (a) Verify that $|A_P| = |A_v||A_i|$. (b) An amplifier must deliver 20 W to a 16-Ω speaker. The sinusoidal input signal source can be represented as a 5-mV source in series with a 10-kΩ resistor. If the input resistance of the amplifier is 20 kΩ, what are the voltage, current, and power gains required of the overall amplifier?
>
> **ANSWERS:** 7590, 9.49×10^6, 7.20×10^{10}

6.2.4 LOCATION OF THE AMPLIFIER

Locating the boundaries of a given amplifier can be confusing and often depends upon the context of the discussion. In Fig. 6.3, the input and output voltages of the "audio amplifier" were chosen to be v_a and v_o. In later chapters we will refer to this as the "terminal" voltage gain. However, the overall amplifier voltage gain from source to load can be defined as

$$A_v = \frac{v_o}{v_i} \tag{6.10}$$

Note that the gain of the audio amplifier must be greater than that of the gain desired for the overall amplifier due to the voltage loss caused by voltage division between source resistance R_I and the audio amplifier input resistance R_{in}. The magnitude of the gain of the overall amplifier is

$$A_v = \frac{V_o}{V_i} = \frac{1.10 \text{ V}}{1.00 \text{ mV}} = 1100 \tag{6.11}$$

6.2.5 THE DECIBEL SCALE

The various gain expressions often involve some rather large numbers, and it is customary to express the values of voltage, current, and power gain in terms of the **decibel**, or **dB** (one-tenth of a Bel):

$$A_{PdB} = 10 \log A_P \qquad A_{vdB} = 20 \log |A_v| \qquad A_{idB} = 20 \log |A_i| \tag{6.12}$$

The number of decibels is 10 times the base 10 logarithm of the arithmetic power ratio, and decibels are added and subtracted just like logarithms to represent multiplication and division. Because power is proportional to the square of both voltage and current, a factor of 20 appears in the expressions for A_{vdB} and A_{idB}.

Table 6.2 has a number of useful examples. From this table, we can see that an increase in voltage or current gain by a factor of 10 corresponds to a change of 20 dB, whereas a factor of 10 increase in power gain corresponds to a change of 10 dB. A factor of 2 corresponds to a 6-dB change in voltage or current gain or a 3-dB change in power gain. In the chapters that follow, the various gains routinely are expressed interchangeably in terms of arithmetic values or dB, so it is important to become comfortable with the conversions in Eq. (6.12) and Table 6.2.

EXERCISE: Express the voltage gain, current gain, and power gain in the exercise at the end of Sec. 6.2.3 in dB.

ANSWERS: 77.6 dB, 140 dB, 108 dB.

EXERCISE: Express the voltage gain, current gain, and power gain of the overall amplifier in Fig. 6.3 in dB.

ANSWERS: 61.7 dB, 128 dB, 94.8 dB.

EXERCISE: Express the voltage gain of the audio amplifier from Section 6.2.2 in dB.

ANSWER: 61.6 dB.

TABLE 6.2
Expressing Gain in Decibels

	|GAIN|	A_{vdB} or A_{idB}	A_{PdB}
	1000	60 dB	30 dB
	500	54 dB	27 dB
	300	50 dB	25 dB
$A_{vdB} = 20 \log |A_v|$	100	40 dB	20 dB
$A_{idB} = 20 \log |A_i|$	20	26 dB	13 dB
$A_{PdB} = 10 \log A_P$	10	20 dB	10 dB
	$\sqrt{10} = 3.16$	10 dB	5 dB
	2	6 dB	3 dB
	1	0 dB	0 dB
	0.5	−6 dB	−3 dB
	0.1	−20 dB	−10 dB

EXAMPLE 6.1   IMPEDANCE LEVEL TRANSFORMATION

Let's explore another example of what an amplifier might do for us. Suppose we have a signal $v_i = 0.1 \sin 2000\pi t$ volts from some transducer (e.g., a microphone in a computer or the output of a digital-to-analog converter) with a Thévenin equivalent source resistance R_I of 2 kΩ, and we'd like to listen to that signal with a 32-Ω ear bud, represented by the load R_L, as depicted by the circuit model in Fig. 6.4. Unfortunately, only a very small fraction of the transducer voltage will get to the load because of the large impedance mismatch between the 2-kΩ source resistance and the 32-Ω load resistance.

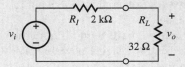

Figure 6.4 Circuit model for transducer connected directly to a load.

Since we are dealing with a resistive network, the output voltage will also be a sine wave with the same phase as the input, $v_o = V_o \sin 2000\pi t$ volts, and the amplitude of the output voltage is found using voltage division:

$$V_o = V_i \frac{R_L}{R_I + R_L} = 0.1\,\text{V}\left(\frac{32\,\Omega}{2032\,\Omega}\right) = 1.58\,\text{mV} \tag{6.13}$$

The signal is reduced by a factor of almost 100 and will probably be inaudible. We can use an amplifier to solve this problem as depicted in Fig. 6.5. Here we are using a **two-port model** for the amplifier consisting of an input resistance R_{in}, a voltage gain A, and an **output resistance** R_{out}.

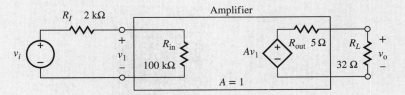

Figure 6.5 Circuit model with amplifier inserted in the network.

Let us assume (arbitrarily for the moment) that $R_{\text{in}} = 100\,\text{k}\Omega$, $A = 1$ (0 dB) and $R_{\text{out}} = 5\,\Omega$, and recalculate the output voltage using voltage division:

$$V_o = AV_1 \frac{R_L}{R_{\text{out}} + R_L} \quad \text{and} \quad V_1 = V_i \frac{R_{\text{in}}}{R_I + R_{\text{in}}} \tag{6.14}$$

Combing and evaluating these expressions yields

$$V_o = A_v\left(V_i \frac{R_{\text{in}}}{R_I + R_{\text{in}}}\right)\left(\frac{R_L}{R_{\text{out}} + R_L}\right) = 1(0.1\,\text{V})\left(\frac{100\,\text{k}\Omega}{102\,\text{k}\Omega}\right)\left(\frac{32\,\Omega}{37\,\Omega}\right) = 84.8\,\text{mV} \tag{6.15}$$

and the actual output signal is $v_o = 84.8 \sin 2000\pi t$ mV. Now we have succeeded in applying about 85 percent of the signal to the desired load, the earphone. The power delivered to the earphone is still fairly small:

$$P_o = \frac{V_o^2}{2R_L} = \frac{(84.8\,\text{mV})^2}{2(32\,\Omega)} = 0.112\,\text{mW} \tag{6.16}$$

If we would like to increase the power in the earphone, we can increase the voltage gain of the amplifier. Suppose we increase the internal gain of the amplifier by 26 dB and see what happens.

We must convert A from dB, and then repeat the calculations:

$$A = 10^{\frac{26}{20}} = 20.0$$

$$V_o = 20(0.1\text{ V})\left(\frac{100\text{ k}\Omega}{102\text{ k}\Omega}\right)\left(\frac{32\text{ }\Omega}{37\text{ }\Omega}\right) = 1.709\text{ V} \qquad (6.17)$$

$$P_o = \frac{(1.70\text{ V})^2}{2(32\text{ }\Omega)} = 45.2\text{ mW}$$

Now we have a substantial audio signal in our earphone (possibly near the specification limit of the earphone).

In Ex. 6.1, we have used an amplifier to provide an impedance level transformation as well as increasing the signal power applied to the ear bud. The amplifier "buffers" the signal source from the low impedance (32-Ω) load. These are only two of the many uses of amplifiers. One of the most common additional applications is to tailor the frequency response of the signal. In this case the amplifier circuitry becomes a *filter*.

ELECTRONICS IN ACTION

Player Characteristics

The headphone amplifier in a personal music player represents an everyday example of a basic audio amplifier. The traditional audio band spans the frequencies from 20 Hz to 20 kHz, a range that extends beyond the hearing capability of most individuals at both the upper and lower ends.

Apple iPhone
*MollieGPhoto/
Shutterstock*

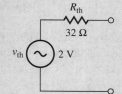

Thévenin equivalent
circuit for output stage

The characteristics of the Apple iPod in the accompanying figure are representative of a high-quality audio output stage in a digital audio player or a computer sound card. The output can be represented by a Thévenin equivalent circuit with $v_{th} = 2$ V and $R_{th} = 32$ ohms, and the output stage is designed to deliver a power of approximately 15 mW into each channel of a headphone with a matched impedance of 32 ohms. The output power is approximately constant over the 20 Hz to 20 kHz audio frequency range.

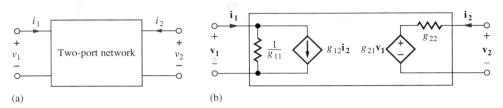

Figure 6.6 (a) Two-port network representation; (b) two port g-parameter representation.

6.3 TWO-PORT MODELS FOR AMPLIFIERS

The simple three-element model for the amplifier in Fig. 6.5, which we introduced in Ex. 6.1, is referred to as a **two-port network** or just a **two-port** in electrical circuits texts and is very useful for modeling the behavior of amplifiers in complex systems. We can use the two-port to provide a relatively simple representation of a much more complicated circuit. Thus, the two-port helps us hide or encapsulate the complexity of the circuit so we can more easily manage the overall analysis and design. One important limitation must be remembered, however. The two-ports we use are linear network models, and are valid under small-signal conditions that will be fully discussed in Chapter 13.

From network theory, we know that two-port networks can be represented in terms of **two-port parameters:** the g-, h-, y-, z-, s-, and $abcd$-parameters. Note in these two-port representations that (v_1, i_1) and (v_2, i_2) represent the signal components of the voltages and currents at the two ports of the network. We will focus on the g-parameter description. Other parameter sets are discussed in Appendix C.

6.3.1 THE g-PARAMETERS

The **g-parameter** description is one of the most commonly used two-port representations for a voltage amplifier:

$$\mathbf{i_1} = g_{11}\mathbf{v_1} + g_{12}\mathbf{i_2}$$
$$\mathbf{v_2} = g_{21}\mathbf{v_1} + g_{22}\mathbf{i_2}$$
(6.18)

Figure 6.6(b) is a network representation of these equations.

The g-parameters are determined from a given network using a combination of **open-circuit** ($i = 0$) and **short-circuit** ($v = 0$) **termination** conditions by applying these parameter definitions:

$$g_{11} = \frac{\mathbf{i_1}}{\mathbf{v_1}}\bigg|_{\mathbf{i_2}=0} = \text{open-circuit input conductance}$$

$$g_{12} = \frac{\mathbf{i_1}}{\mathbf{i_2}}\bigg|_{\mathbf{v_1}=0} = \text{reverse \textbf{short-circuit current gain}}$$

$$g_{21} = \frac{\mathbf{v_2}}{\mathbf{v_1}}\bigg|_{\mathbf{i_2}=0} = \text{forward \textbf{open-circuit voltage gain}}$$

$$g_{22} = \frac{\mathbf{v_2}}{\mathbf{i_2}}\bigg|_{\mathbf{v_1}=0} = \text{\textbf{short-circuit output resistance}}$$

(6.19)

Unfortunately, the classic g-parameter notation doesn't provide much support for our intuition, so the more descriptive representation that was used in Ex. 6.1 is described by Eq. (6.20) and Fig. 6.7.

$$\mathbf{v_1} = \mathbf{i_1} R_{\text{in}}$$
$$\mathbf{v_2} = A\mathbf{v_1} + \mathbf{i_2} R_{\text{out}}$$
(6.20)

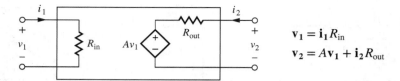

Figure 6.7 Simplified two-port with more intuitive notation, and "g_{12}" = 0.

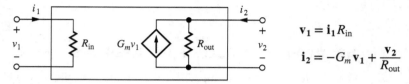

Figure 6.8 Norton transformation of the circuit in Fig. 6.7 in which $G_m = \dfrac{A}{R_{out}}$.

R_{in} represents the input resistance to the amplifier, A is the voltage gain when there is no external load on the amplifier, and R_{out} is the output resistance of the amplifier. In a normal amplifier design, we desire the forward gain (g_{21}) to be much larger than the reverse gain (g_{12}), that is, $g_{21} \gg g_{12}$, and Eq. (6.20) and Fig. 6.7 show the simplified two-port representation in which the reverse gain g_{12} is assumed to be zero. Figure 6.8 presents an alternate two-port representation that we shall encounter frequently in our transistor circuits. In this equivalent circuit, the output port components have been found using Norton's theorem that yields $G_m = A/R_{out}$.

EXAMPLE 6.2 FINDING A SET OF g-PARAMETERS

This example calculates a set of g-parameters for a network containing a dependent current source. We encounter this type of circuit often in analog circuit analysis and design because our models for both bipolar and field-effect transistors contain dependent current sources.

PROBLEM Find the g-parameters for the circuit shown here. Include g_{12} for completeness, and compare it to g_{21}.

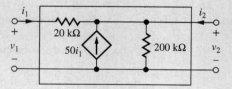

SOLUTION **Known Information and Given Data:** Circuit as given in the problem statement including element values; g-parameter definitions in Eq. (6.19)

Unknowns: Values of the four g-parameters

Approach: Apply the boundary conditions specified for each g-parameter and use circuit analysis to find the values of the four parameters. Note that each set of boundary conditions applies to two parameters.

Assumptions: None

Analysis—g_{11} and g_{21}: Looking at the definitions of the g-parameters,

$$G_{in} = g_{11} = \left.\frac{i_1}{v_1}\right|_{i_2=0} \quad \text{and} \quad A = g_{21} = \left.\frac{v_2}{v_1}\right|_{i_2=0}$$

we see that g_{11} and g_{21} use the same boundary conditions. We apply voltage v_1 to the input port, and the output port is open circuited (i.e., i_2 is set to zero), as in the figure here.

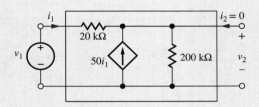

g_{11}: Writing an equation around the input loop and applying KCL at the output node yields

$$\mathbf{v_1} = (2 \times 10^4)\mathbf{i_1} + (\mathbf{i_1} + 50\mathbf{i_1})(200 \text{ k}\Omega)$$

$$G_{in} = \frac{i_1}{v_1} = \frac{1}{2 \times 10^4 \,\Omega + 51(200 \text{ k}\Omega)} = \frac{1}{10.2 \text{ M}\Omega} = 9.79 \times 10^{-8} \text{ S} \quad \text{and} \quad R_{in} = 10.2 \text{ M}\Omega$$

g_{21}: Since the external port current i_2 is zero, the voltage v_2 is given by

$$\mathbf{v_2} = (\mathbf{i_1} + 50\,i_1)(200 \text{ k}\Omega) = \mathbf{i_1}(51)(200 \text{ k}\Omega)$$

and i_1 can be related to v_1 using g_{11}:

$$\mathbf{v_2} = (g_{11}\mathbf{v_1})(51)(200 \text{ k}\Omega)$$

$$A = \frac{\mathbf{v_2}}{\mathbf{v_1}} = g_{11}(51)(200 \text{ k}\Omega) = (9.79 \times 10^{-8} \text{ S})(51)(200 \text{ k}\Omega) = +0.998$$

Analysis—g_{12} and g_{22}: Looking again at the definitions of the g-parameters, we see that g_{12} and g_{22} use the same boundary condition.

$$g_{12} = \left.\frac{i_1}{i_2}\right|_{v_1=0} \quad \text{and} \quad R_{out} = g_{22} = \left.\frac{v_2}{i_2}\right|_{v_1=0}$$

A current source i_2 is applied to the output port, and the input port is short-circuited (i.e., v_1 is set to zero) as shown in this figure:

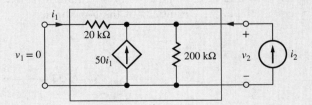

g_{22}: With $v_1 = 0$, we see that the network is just a single-node circuit. Writing a nodal equation for v_2 yields

$$(\mathbf{i_2} + 50\mathbf{i_1}) = \frac{v_2}{200 \text{ k}\Omega} + \frac{v_2}{20 \text{ k}\Omega}$$

But, i_1 can be written directly in terms of v_2 as $i_1 = -v_2/20\text{ k}\Omega$. Combining these two equations yields the short-circuit output resistance g_{22}:

$$i_2 = \frac{v_2}{200\text{ k}\Omega} + \frac{v_2}{20\text{ k}\Omega} + 50\frac{v_2}{20\text{ k}\Omega} \quad \text{and} \quad R_{\text{out}} = \frac{v_2}{i_2} = \frac{1}{\frac{1}{200\text{ k}\Omega} + \frac{51}{20\text{ k}\Omega}} = 391\text{ }\Omega$$

g_{12}: The reverse short-circuit current gain g_{12} can be found using the preceding results:

$$i_1 = -\frac{v_2}{20\text{ k}\Omega} = -\frac{R_{\text{out}} i_2}{20\text{ k}\Omega} \quad \text{and} \quad g_{12} = \frac{i_1}{i_2} = -\frac{391\text{ }\Omega}{20\text{ k}\Omega} = -0.0196$$

The final g-parameter equations for the network are

$$i_1 = 9.79 \times 10^{-8} v_1 - 1.96 \times 10^{-2} i_2$$
$$v_2 = 0.998 v_1 + 3.91 \times 10^2 i_2$$

Check of Results: The results are confirmed below using SPICE.

Discussion: Note that the values of $R_{\text{in}} = 10.2\text{ M}\Omega$ and $R_{\text{out}} = 391\text{ }\Omega$ differ greatly from any of the resistor values in the network. This is a result of the action of the dependent current source and is an important effect that we will see throughout the analysis of analog transistor circuits. Here we see that g_{12} is indeed small and that $g_{12} \ll g_{21}$. We will make use of this observation when we study feedback. The simplified mathematical and two-port models for the circuit with g_{12} neglected become

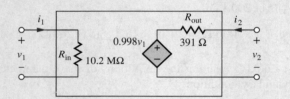

$$v_1 = (10.2\text{ M}\Omega) i_1$$
$$v_2 = 0.998 v_1 + (391\text{ }\Omega) i_2$$
$$R_{\text{in}} = 10.2\text{ M}\Omega$$
$$R_{\text{out}} = 391\text{ }\Omega$$
$$A = 0.998$$

Computer-Aided Analysis: Numerical values for two-port parameters can easily be found using the **transfer function** (TF) analysis capability of SPICE. In order to find the g-parameters for the circuit in this example, we drive the network with voltage source V1 at the input and current source I2 at the output, as in the figure here. These choices correspond to the boundary conditions in the definitions of the g-parameters.

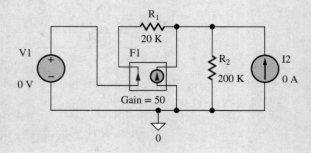

Both independent sources are assigned zero values. The TF analysis calculates how variables change in response to changes in an independent source. Therefore, a starting point of zero is fine. The zero value sources directly satisfy the boundary conditions required to calculate the g-parameters.

Two TF analyses are used—one to find g_{11} and g_{21} and a second to find g_{12} and g_{22}. The first analysis requests calculation of the transfer function from source V1 to the voltage at the output node, and SPICE will calculate three quantities: the value of the transfer function, resistance at the input source node, and resistance at the output node. The SPICE results are transfer function = 0.998, input resistance = 10.2 MΩ, and output resistance = 391 Ω. Parameter g_{21} is the open-circuit voltage gain, which agrees with the hand calculations, and g_{11} is the input conductance equal to the reciprocal of 10.2 MΩ, again in agreement with our hand calculations.

The second analysis requests the transfer function from source I2 to the current in source V1. The results from SPICE are transfer function = 0.0196 and input resistance = 391 Ω. In this case, the output resistance (at V1) cannot be calculated because V1 represents a short at the input. Note that parameter g_{12} is the negative of the TF value. The sign difference arises from the passive sign convention assumed by SPICE in which positive current is directed downward through source V1. Parameter g_{22} is the 391-Ω resistance presented to source I2, which is the "input resistance" in this calculation. We find precise agreement with our hand calculations. It is important to remember that the SPICE TF analysis is a form of dc analysis and should not be used in networks containing capacitors and inductors!

EXERCISE: Find the g-parameters for the circuit in Ex. 6.2 if the 200-kΩ resistor is replaced with a 50-kΩ resistor, and the dependent source is changed to $75i_1$.

ANSWERS: 2.62×10^{-7} S; 0.995, −0.0131; 262 Ω

EXERCISE: Confirm your calculations in the previous exercise using SPICE.

DESIGN NOTE

Remember, the transfer function analysis in SPICE is a dc analysis and should not be used in circuits containing capacitors and inductors!

6.4 MISMATCHED SOURCE AND LOAD RESISTANCES

In introductory circuit theory, the maximum power transfer theorem is usually discussed. Maximum power transfer occurs when the source and load resistances are matched (equal in value). In most amplifier applications, however, the opposite situation is desired. A completely mismatched condition is often used at both the input and output ports of the amplifier, with the exception of RF systems and power amplifiers.

To understand the statement above, let's further consider the voltage amplifier in Fig. 6.9, which has the same structure as the one in Ex. 6.1. The input to the two-port is a Thévenin equivalent representation of the input source, and the output is connected to a load represented by resistor R_L.

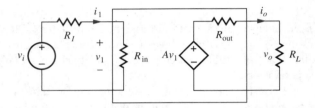

Figure 6.9 Two-port representation of an amplifier with source and load connected.

To find the overall voltage gain, voltage division is applied to each loop:

$$\mathbf{v_o} = A\mathbf{v_1}\frac{R_L}{R_{out} + R_L} \quad \text{and} \quad \mathbf{v_1} = \mathbf{v_i}\frac{R_{in}}{R_I + R_{in}} \quad (6.21)$$

Combining these two equations yields an expression for the magnitude of the voltage gain A_v:

$$|A_v| = \frac{V_o}{V_i} = A\frac{R_{in}}{R_I + R_{in}}\frac{R_L}{R_{out} + R_L} \quad (6.22)$$

To achieve maximum voltage gain, the resistors should satisfy $R_{in} \gg R_I$ and $R_{out} \ll R_L$. For this case,

$$|A_v| \cong A \quad (6.23)$$

The situation described by these two equations is a totally mismatched condition at both the input and the output ports. An **ideal voltage amplifier** satisfies the conditions above by having $R_{in} = \infty$ and $R_{out} = 0$.

The magnitude of the current gain of the amplifier in Fig. 6.9 can be expressed as

$$|A_i| = \frac{I_o}{I_1} = \frac{\frac{V_o}{R_L}}{\frac{V_i}{R_I + R_{in}}} = \frac{V_o}{V_i}\frac{R_I + R_{in}}{R_L} \quad \text{or} \quad |A_i| = |A_v|\frac{R_I + R_{in}}{R_L} \quad (6.24)$$

EXERCISE: What is the current gain of an ideal voltage amplifier?

ANSWER: ∞

EXERCISE: Write an expression for the power gain of the amplifier in Fig. 6.9 in terms of the voltage gain.

ANSWER: $A_P = A_v^2 \frac{R_I + R_{in}}{R_L}$

EXERCISE: Suppose the audio amplifier in Fig. 6.3 can be modeled by $R_{in} = 50$ kΩ and $R_{out} = 0.5$ Ω. What value of open-circuit gain A is required to achieve an output power of 100 W if $v_i = 0.001 \sin 2000\pi t$? How much power is being dissipated in R_{out}? What is the current gain?

ANSWERS: 46,800 (93.4 dB); 6.25 W; 2.75×10^8 (169 dB)

EXERCISE: Repeat the preceding exercise if the input and output ports are matched to the source and load respectively (that is, $R_{in} = 5$ kΩ and $R_{out} = 8$ Ω). (It should become clear why we don't design R_{out} to match the load resistance.)

ANSWERS: 160,000 (104 dB); 100 W; 5×10^7 (153 dB)

ELECTRONICS IN ACTION

Laptop Computer Touchpad

An essential element of a graphical user interface is a pointing device. This was clear to Douglas Engelbart in the late 1960s during his experimentation with graphical computer interfaces. In order to provide the feeling that a user was directly manipulating objects on the screen, Engelbart invented the computer mouse in 1968. It did not move into the computing mainstream until the introduction of the Apple Macintosh in 1984. As integrated circuit technology advanced and made possible the creation of laptop computers, it became necessary to develop a pointing device that was contained within the form factor of the laptop computer but maintained the "connective" feel between the user and the computer interface. Trackballs were used in early machines, but they didn't allow the intuitive x-y hand displacement feedback of the mouse. Trackballs were also prone to accumulation of dirt and other debris, which reduced their robustness.

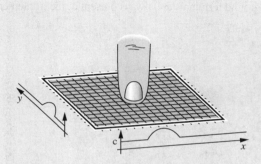

Sergio Azenha/Alamy Stock Photo

Touch screens were available in the early 1980s, but they required non-robust resistive membranes and/or expensive fabrication techniques. In the early 1990s, Synaptics Corporation developed the capacitive sensing touchpad. A simplified drawing of a capacitive touchpad is shown above. A thin insulating surface covers an x-y grid of wires. When a finger is placed on or near the surface, the capacitance of the wires directly underneath is changed. By measuring the capacitance between the wires and ground, it is possible to detect the presence of an object. If the capacitance measurement is performed with each of the wires in sequence, a capacitance versus position profile is developed. Calculating the centroid of the broad profile allows the system to form a precise indication of finger position over the touchpad.

The measurement itself can be done in a number of ways. The capacitance could be part of a tuned circuit to control the frequency of an oscillator. One could drive the capacitance with a sinusoid current and measure the peak-to-peak value of the resulting voltage. Or, as is the case with most touchpads, a step voltage is driven onto the wire and the resulting charging current is integrated. The magnitude of the integral is proportional to capacitance. Once again, integrated circuit technology made the device practical and inexpensive. A significant number of wires is required to achieve adequate resolution. If implemented with discrete components, the switches, signal routing, and signal processing would be large and expensive. A single mixed-signal CMOS integrated circuit, integrating precision analog circuits and digital processing, was designed to provide all of the necessary functionality, as well as to provide a digital interface that is easily incorporated into a computer. Bridging the gap between real world analog information and digital computers is an important and recurring theme in analog microelectronics.

6.5 THE DIFFERENTIAL AMPLIFIER

The **differential amplifier** is a form of amplifier that responds to the difference of two input signals (and hence is sometimes referred to as a difference amplifier) and represents an extremely useful class of circuits (see Fig. 6.10). For example, they are used as error amplifiers in almost all electronic feedback and control systems, and operational amplifiers themselves are in fact very high performance versions of the differential amplifier. Thus, we set the stage for our study of op amps in Part Three by exploring the characteristics of the basic differential amplifier shown in schematic form in Fig. 6.11. In addition, high-performance analog ICs will typically have many differential amplifiers in the circuitry.

The amplifier has two inputs where signals v_+ and v_- are connected, and a single output v_o, all referenced to the common (ground) terminal between the two power supplies V_{CC} and V_{EE}. In most applications, $V_{CC} \geq 0$ and $-V_{EE} \leq 0$, and the voltages are often symmetric—that is, ±1.5 V, ±5 V, ±12 V, ±15 V, ±18 V, ±22 V, and so on. These power supply voltages limit the output voltage range: $-V_{EE} \leq v_O \leq V_{CC}$. For simplicity, the amplifier is most often drawn without explicitly showing either the power supplies, as in Fig. 6.12(a), or the ground connection, as in Fig. 6.12(b)—but we must remember that the power and ground terminals are always present in the implementation of a real circuit.

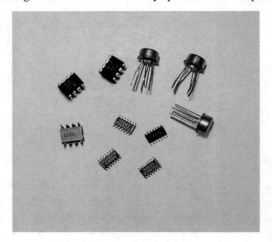

Figure 6.10 Discrete operational amplifiers, a class of high performance differential amplifiers.
Richard Jaeger

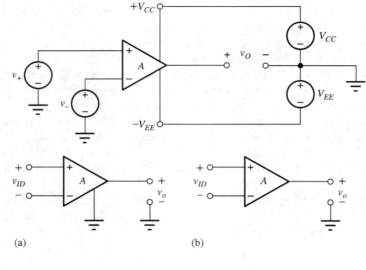

Figure 6.11 The differential amplifier, including power supplies.

Figure 6.12 (a) Amplifier without power supplies explicitly included; (b) differential amplifier with implied ground connection.

6.5.1 DIFFERENTIAL AMPLIFIER VOLTAGE TRANSFER CHARACTERISTIC

The **voltage transfer characteristic** or **VTC** for a differential amplifier biased by power supplies V_{CC} and $-V_{EE}$ is shown in Fig. 6.13. The VTC graphs the total output voltage v_O versus the total differential input voltage v_{ID}. In this particular case, V_{CC} and $-V_{EE}$ are symmetrical 10-V supplies, and thus the output voltage is restricted to $-10\text{ V} \leq v_O \leq +10\text{ V}$.

Because of the power supply limits, we see that the input–output relationship is linear over only a limited region of the characteristic. Using our standard notation introduced in Chapter 1, the total input voltage v_{ID} is represented as the sum of two components:

$$v_{ID} = V_{ID} + v_{id} \tag{6.25}$$

in which V_{ID} represents the dc value of v_{ID}, and v_{id} is the signal component of the input voltage. Similarly, the total output voltage is represented by

$$v_O = V_O + v_o \tag{6.26}$$

in which V_O represents the dc value of the output voltage, and v_o is the signal component of the output voltage. For the amplifier to provide linear amplification of the signal v_{id}, the total input signal must be biased by the dc voltage V_{ID} into the central high-slope region of the characteristic.

6.5.2 VOLTAGE GAIN

The voltage gain A of an amplifier describes the relation between changes in the input signal and changes in the output signal and is defined by the *slope* of the amplifier's VTC, evaluated for an input voltage equal to the dc bias voltage V_{ID}:

$$A = \left.\frac{\partial v_O}{\partial v_{ID}}\right|_{v_{ID}=V_{ID}} \tag{6.27}$$

For the VTC in Fig. 6.13 with $V_{ID} = 1$ V,

$$A = \frac{10 - 0}{1.5 - 0.5}\left(\frac{V}{V}\right) = +10 \quad \text{or} \quad A_{vdB} = 20\log(10) = 20\text{ dB} \tag{6.28}$$

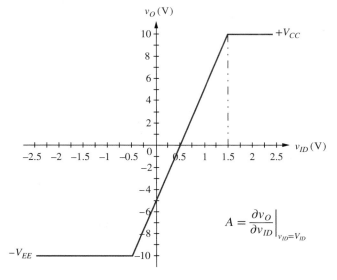

Figure 6.13 Voltage transfer characteristic (VTC) for a differential amplifier with $V_{CC} = 10$ V, $-V_{EE} = -10$ V and gain $A = +10$ (20 dB). Gain A is equal to the slope of the VTC.

Note that the gain is **not** equal to the ratio of the total output voltage to the total input voltage! For example, for $v_{ID} = +1$ V,

$$\frac{v_O}{v_{ID}} = \frac{5}{1} = +5 \neq A \tag{6.29}$$

In this amplifier, the slope of the VTC in Fig. 6.13 is everywhere ≥ 0, so the amplifier input and output are in phase; this amplifier is a **noninverting amplifier.** If the slope had been negative, then the input and output signals would be 180° out of phase, and the amplifier would be characterized as an **inverting amplifier.**

Offset Voltage

The amplifier in Fig. 6.13 is said to have an **input offset voltage V_{OS}** defined as the dc input voltage necessary to force the output to zero volts. In this case, $V_{OS} = +0.5$ V. Offset voltage is one of the non-ideal characteristics of an operational amplifier, and we will discuss it further until Chapter 11.

Signal Amplification

A graphical representation of the VTC with a sinusoidal input signal and sinusoidal output signal appears in Fig. 6.14 in which v_{ID1} and v_{O1} are given by

$$v_{ID1} = 1 + 0.25 \sin 2000\pi t \text{ volts} \quad \text{and} \quad v_{O1} = 5 + 2.5 \sin 2000\pi t \text{ volts} \tag{6.30}$$

Notice also that there is a limited range of input voltage for which the amplifier will behave in a linear manner. For an input bias of 1 V as in Eq. (6.28) and Fig. 6.14, the maximum input voltage signal amplitude must be less than 0.5 V, which corresponds to a maximum output signal amplitude of 5 V.

If the ac input signal exceeds 0.5 V, then the top part of the output signal will be clipped off. Figure 6.15 presents the results of SPICE simulation of the amplifier VTC in Fig. 6.14 with the input signal in Eq. (6.28) and $v_{ID2} = 1 + 1.5 \sin 2000\pi t$ volts. As the value of v_{ID2} exceeds 1.5 V, the output stays constant at +10 V. Any further increase in input voltage results in no change in the output voltage! The voltage gain in this region is zero because the slope of the VTC is 0.

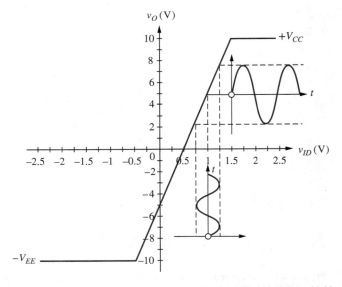

Figure 6.14 Graphical interpretation voltage transfer characteristic (VTC) with a sinusoidal input signal applied.

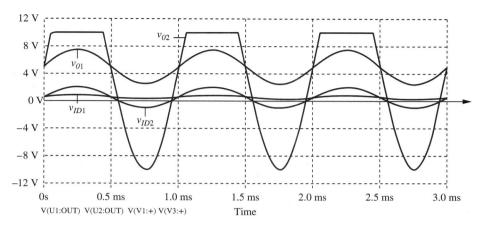

Figure 6.15 SPICE simulation results for the amplifier VTC in Fig. 6.14 for two input signals: $v_{ID1} = 1 + 0.25 \sin 2000\pi t$ volts and $v_{ID2} = 1 + 1.5 \sin 2000\pi t$ volts. The amplifier is operating in a linear manner for input one, but is driven into nonlinear operation by input signal two.

> **EXERCISE:** (a) What input bias point should be chosen for the amplifier in Fig. 6.13 to provide the maximum possible linear input signal magnitude? What are the maximum input and output signal amplitudes? (b) What is the voltage gain if the amplifier input is biased at $V_{ID} = -1.0$ V?
>
> **ANSWERS:** 0.5 V, $|v_i| \leq 1.0$ V; 10 V; 0
>
> **EXERCISE:** Write an expression for $v_O(t)$ for the amplifier in Fig. 6.13 if $v_{ID}(t) = (0.25 + 0.75 \sin 1000\pi t)$ V. What dc bias appears as part of the output voltage?
>
> **ANSWERS:** $(-2.5 + 7.5 \sin 1000\pi t)$ V; -2.5 V

6.6 DISTORTION IN AMPLIFIERS

As mentioned in Sec. 6.5, if the input signal is too large, then the output waveform will be significantly distorted since the gain for positive values of the input signal will be different from the gain for negative values. In Fig. 6.15, the top of the largest waveform appears "flattened," and there is a slope discontinuity in the waveform.

A measure of the distortion in such a signal is given by its **total harmonic distortion (THD),** which compares the undesired harmonic content of a signal to the desired component. If we expand the Fourier series representation for a signal $v(t)$, we have

$$v(t) = \underset{\text{dc}}{V_O} + \underset{\substack{\text{desired} \\ \text{output}}}{V_1 \sin(\omega_o t + \phi_1)} + \underset{\substack{\text{2nd harmonic} \\ \text{distortion}}}{V_2 \sin(2\omega_o t + \phi_2)} + \underset{\substack{\text{3rd harmonic} \\ \text{distortion}}}{V_3 \sin(3\omega_o t + \phi_3)} + \cdots \quad (6.31)$$

The signal at frequency ω_o is the desired output that has the same frequency as the input signal. The terms at $2\omega_o$, $3\omega_o$, etc. represent second-, third-, and higher-order harmonic distortion. The percent THD is defined by

$$\text{THD\%} = 100\% \times \frac{\sqrt{\sum_{2}^{\infty} V_n^2}}{V_1} \quad (6.32)$$

The numerator of this expression combines the amplitudes of the individual distortion terms in rms form, whereas the denominator contains only the desired component. Normally, only the first few terms are important in the numerator. For example, SPICE Fourier analysis yields this representation for the distorted signal in Fig. 6.15,

$$v(t) = 2.46 + 10.6 \sin(2000\pi t) + 2.67 \sin(4000\pi t + 90°)$$
$$+ 0.886 \sin(6000\pi t) + 0.177 \sin(8000\pi t + 90°) + 0.372 \sin(10000\pi t)$$

for which the total distortion is approximately

$$\text{THD} \cong 100\% \times \frac{\sqrt{2.67^2 + 0.886^2 + 0.177^2 + 0.372^2}}{3} = 26.8\%$$

This value of THD represents a large amount of distortion, which is clearly visible in Fig. 6.15. Good distortion levels are well below 1 percent and are not readily apparent to the eye.

EXERCISE: Use MATLAB or Mathcad to plot both the distorted output signal in Fig. 6.15 and its reconstruction described by v(t) above.

ANSWER: wt = 2000∗pi∗linspace(0,.002,1024);
 v = min(10,(5+15∗sin(wt));
 f = 2.46+10.6∗sin(wt)+2.67∗cos(2∗wt)+0.866∗sin(3∗wt)+0.177∗cos(4∗wt)
  +0.372∗sin(5∗wt);
 plot(wt,v,wt,f)

(Note the close match between the two curves with only a few components of the series.)

EXERCISE: Use MATLAB to find the Fourier representation for v(t).

ANSWER: wt = 2000∗pi∗linspace(0,.001,512);
 v = min(10,(5+15∗sin(wt));
 s = fft(v)/512;
 mag=sqrt(s.∗conj(s))
 mag(1:10)

(Note that the fft function in MATLAB generates the coefficients for the complex Fourier series.)

6.7 DIFFERENTIAL AMPLIFIER MODEL

For purposes of signal analysis, the differential amplifier can be represented by its input resistance R_{id}, output resistance R_o, and controlled voltage source Av_{id}, as in Fig. 6.16. This is the two-port representation introduced to Sec. 6.3.

$$\begin{aligned} A &= \text{voltage gain (open-circuit voltage gain)} \\ v_{id} &= (v_+ - v_-) = \text{differential input signal voltage} \\ R_{id} &= \text{amplifier input resistance} \\ R_o &= \text{amplifier output resistance} \end{aligned} \quad (6.33)$$

The signal voltage developed at the output of the amplifier is in phase with the voltage applied to the + input terminal and 180° out of phase with the signal applied to the − input terminal. The v_+ and v_- terminals are therefore referred to as the **noninverting input** and **inverting input**, respectively.

In a typical application, the amplifier is driven by a signal source having a Thévenin equivalent voltage v_i and resistance R_I and is connected to a load represented by the resistor R_L, as in Fig. 6.17. For this simple circuit, the input voltage v_{id} and the output voltage can be written in

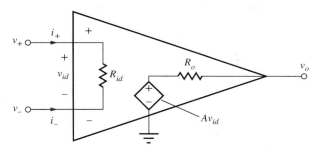

Figure 6.16 Differential amplifier.

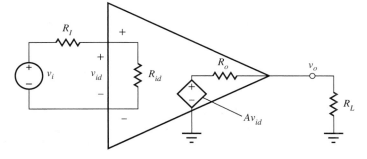

Figure 6.17 Amplifier with source and load attached.

terms of the circuit elements as

$$\mathbf{v_{id}} = \mathbf{v_i}\frac{R_{id}}{R_{id} + R_I} \quad \text{and} \quad \mathbf{v_o} = A\mathbf{v_{id}}\frac{R_L}{R_o + R_L} \quad (6.34)$$

Combining Eq. (6.34) yields an expression for the overall voltage gain of the amplifier circuit in Fig. 6.17 for arbitrary values of R_I and R_L:

$$A_v = \frac{\mathbf{v_o}}{\mathbf{v_i}} = A\frac{R_{id}}{R_I + R_{id}}\frac{R_L}{R_o + R_L} \quad (6.35)$$

Differential amplifier circuits are most often **dc-coupled amplifiers,** and the signals v_o and v_i may in fact have a dc component that represents a dc shift of the input away from the initial operating point of the circuit (the Q-point). The op amp that you probably studied in basic circuits courses amplifies not only the ac components of the signal but also this dc component. We must remember that the ratio needed to find A_v, as indicated in Eq. (6.35), is determined by the amplitude and phase of the individual signal components and is not a time-varying quantity, but $\omega = 0$ is a valid signal frequency! Recall from Chapter 1 that v_i, v_o, i_2 and so on represent our signal voltages and currents and are generally functions of time: $v_i(t), v_o(t), i_2(t)$. But whenever we do algebraic calculations of voltage gain, current gain, input resistance, output resistance, and so on, we must use phasor representations of the individual signal components in our calculations: $\mathbf{v_i}$, $\mathbf{v_o}$, $\mathbf{i_2}$. Signals $v_i(t), v_o(t), i_2(t)$ and so on may be composed of many individual signal components, one of which may be a dc shift away from the Q-point value. Many of the amplifiers that we shall initially study will have only one input and are equivalent to a differential amplifier with one input grounded. These amplifiers can be either inverting or non-inverting amplifiers.

EXAMPLE 6.3   VOLTAGE GAIN ANALYSIS

Find the overall gain of a differential amplifier including the effects of load and source resistance.

PROBLEM Calculate the voltage gain for an amplifier with the following parameters: $A = 100$, $R_{id} = 100$ kΩ, and $R_o = 100$ Ω, with $R_I = 10$ kΩ and $R_L = 1000$ Ω. Express the result in dB.

SOLUTION **Known Information and Given Data:** $A = 100$, $R_{id} = 100 \text{ k}\Omega$, $R_o = 100 \text{ }\Omega$, $R_I = 10 \text{ k}\Omega$, and $R_L = 1000 \text{ }\Omega$

Unknown: Voltage gain A_v.

Approach: Evaluate the expression in Eq. (6.35). Convert answer to dB.

Assumptions: None.

Analysis: Using Eq. (6.35),

$$A_v = 100\left(\frac{100 \text{ k}\Omega}{10 \text{ k}\Omega + 100 \text{ k}\Omega}\right)\left(\frac{1000 \text{ }\Omega}{100 \text{ }\Omega + 1000 \text{ }\Omega}\right) = 82.6$$

$$A_{vdB} = 20 \log |A_v| = 20 \log |82.6| = 38.3 \text{ dB}$$

Check of Results: We have found the only unknown requested.

Discussion: The amplifier's internal voltage gain capability is $A = 100$, but an overall gain of only 82.6 is being realized because a portion of the signal source voltage ($\cong 9$ percent) is being dropped across R_I, and part of the internal amplifier voltage (Av_{id}) (also $\cong 9$ percent) is being lost across R_o.

Computer-Aided Analysis: The SPICE circuit is shown here, and a transfer function analysis from source VI to the output node is used to characterize the amplifier in this example.

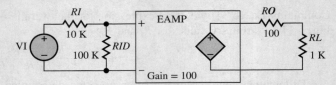

The SPICE results are transfer function = 82.6, input resistance = 110 kΩ, and output resistance = 90.9 Ω. A_v equals the value of the transfer function, the resistance at the terminals of VI is the input resistance, and the output resistance represents the total resistance at the output node. The voltage gain agrees with our hand analysis.

6.8 AMPLIFIER FREQUENCY RESPONSE

General amplifier transfer functions can be quite complicated, having many poles and zeros, but their overall behavior can be broken down into a number of categories including low-pass, high-pass, and band-pass amplifiers to name a few. In the next several sections we will review Bode plots for low-pass, high-pass and band-pass amplifiers. Other types of transfer functions were discussed in Chapter 1 and will be discussed further in subsequent chapters.

6.8.1 BODE PLOTS

When we explore the characteristics of amplifiers, we are usually interested in the behavior of the transfer function for physical frequencies ω—that is, for $s = j\omega$, and the transfer function can then be represented in polar form by its **magnitude** $|A_v(j\omega)|$ and **phase angle** $\angle A_v(j\omega)$, which are both functions of frequency:

$$A_v(j\omega) = |A_v(j\omega)| \angle A_v(j\omega) \tag{6.36}$$

6.8 Amplifier Frequency Response

It is often convenient to display this information separately in a graphical form called a **Bode plot**. The Bode plot displays the magnitude of the transfer function in dB and the phase in degrees (or radians) versus a logarithmic frequency scale. Bode plots for low-pass, high-pass and band-pass amplifiers are discussed in the next several sections.

6.8.2 THE LOW-PASS AMPLIFIER

Circuits that amplify signals over a range of frequencies including dc are an extremely important class of circuits and are referred to as low-pass amplifiers. For instance, most operational amplifiers are inherently low-pass amplifiers. The simplest low-pass amplifier circuit is described by the **single-pole**[5] transfer function

$$A_v(s) = \frac{A_o \omega_H}{s + \omega_H} = \frac{A_o}{1 + \dfrac{s}{\omega_H}} \quad (6.37)$$

in which A_o is the low-frequency gain and ω_H represents the cutoff frequency of this low-pass amplifier. Let us first explore the behavior of the magnitude of $A_v(s)$ and then look at the phase response.

Magnitude Response

Substituting $s = j\omega$ into Eq. (6.37) and finding the magnitude of the function $A_v(j\omega)$ yields

$$|A_v(j\omega)| = \left|\frac{A_o \omega_H}{j\omega + \omega_H}\right| = \frac{|A_o \omega_H|}{\sqrt{\omega^2 + \omega_H^2}} \quad (6.38)$$

The Bode magnitude plot is given in terms of dB:

$$|A_v(j\omega)|_{dB} = 20 \log |A_o \omega_H| - 20 \log \sqrt{\omega^2 + \omega_H^2} \quad (6.39)$$

For a given set of numeric values, Eq. (6.39) can be easily evaluated and plotted using a package such as MATLAB or a spreadsheet, and results in the graph in Fig. 6.18.

For the general case, the graph is conveniently plotted in terms of its asymptotic behavior at low and high frequencies. For low frequencies, $\omega \ll \omega_H$, the magnitude is approximately constant:

$$\left.\frac{A_o \omega_H}{\sqrt{\omega^2 + \omega_H^2}}\right|_{\omega \ll \omega_H} \cong \frac{A_o \omega_H}{\sqrt{\omega_H^2}} = A_o \quad \text{or} \quad (20 \log A_o) \text{ dB} \quad (6.40)$$

At frequencies well below ω_H, the gain of the amplifier is constant and equal to A_o, which corresponds to the horizontal asymptote in Fig. 6.18. Signals at frequencies below ω_H are amplified by the gain A_o. In fact, the gain of this amplifier is constant down to dc ($\omega = 0$)!

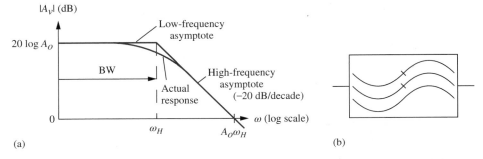

Figure 6.18 (a) Low-pass amplifier: BW $= \omega_H$ and GBW $= A_o \omega_H$; (b) low-pass filter symbol.

---

[5] A general low-pass amplifier may have many poles. The single-pole version is the simplest approximation to the ideal low-pass characteristic described in Chapter 1.

However, as ω exceeds ω_H, the gain of the amplifier begins to decrease (high-frequency roll-off). For sufficiently high frequencies, $\omega \gg \omega_H$, the magnitude can be approximated by

$$\left.\frac{A_o \omega_H}{\sqrt{\omega^2 + \omega_H^2}}\right|_{\omega \gg \omega_H} \cong \frac{A_o \omega_H}{\sqrt{\omega^2}} = \frac{A_o \omega_H}{\omega} \tag{6.41}$$

and converting Eq. (6.40) to dB yields

$$|A_v(j\omega)|_{dB} \cong \left(20 \log A_o - 20 \log \frac{\omega}{\omega_H}\right) \text{ dB} \tag{6.42}$$

For frequencies much greater than ω_H, the transfer function decreases at a rate of 20 dB per decade increase in frequency, as indicated by the high-frequency asymptote in Fig. 6.18. Obviously, ω_H plays an important role in characterizing the amplifier; this critical frequency is called the **upper-cutoff frequency** of the amplifier. At $\omega = \omega_H$, the gain of the amplifier is

$$|A_v(j\omega_H)| = \frac{A_o \omega_H}{\sqrt{\omega_H^2 + \omega_H^2}} = \frac{A_o}{\sqrt{2}} \quad \text{or} \quad [(20 \log A_o) - 3] \text{ dB} \tag{6.43}$$

and ω_H is sometimes referred to as the **upper -3-dB frequency** of the amplifier. ω_H is also often termed the **upper half-power point** of the amplifier because the output power of the amplifier, which is proportional to the square of the voltage, is reduced by a factor of 2 at $\omega = \omega_H$. Note that when the expressions for the two asymptotes given in Eqs. (6.40) and (6.41) are equated, they intersect precisely at $\omega = \omega_H$.

Bandwidth

The gain of the amplifier in Fig. 6.18 is approximately uniform (it varies by less than 3 dB) for all frequencies below ω_H. This is called a **low-pass amplifier**. The **bandwidth (BW)** of an amplifier is defined by the range of frequencies in which the amplification is approximately constant; it is expressed in either radians/second or Hz. For the low-pass amplifier,

$$\text{BW} = \omega_H \text{ (rad/s)} \quad \text{or} \quad \text{BW} = f_H = \frac{\omega_H}{2\pi} \text{ Hz} \tag{6.44}$$

Gain-Bandwidth Product

The **gain-bandwidth product** is often used as a figure-of-merit for comparing amplifiers, and for a low-pass amplifier it is simply the product of the low-frequency gain and the bandwidth of the amplifier:

$$\text{GBW} = A_o \omega_H \tag{6.45}$$

For a single-pole, low-pass characteristic, the GBW also represents the **unity-gain frequency** ω_T of the amplifier, the frequency at which the magnitude of the gain becomes 1 or 0 dB. We can find ω_T using Eq. (6.41) for $\omega \gg \omega_H$:

$$|A_v(j\omega_T)| = 1 \quad \text{or} \quad \frac{A_o \omega_H}{\omega_T} = 1 \quad \text{and} \quad \omega_T = A_o \omega_H \tag{6.46}$$

> **EXERCISE:** Find the low-frequency gain, cutoff frequency, bandwidth, and gain-bandwidth product of the low-pass amplifier with the following transfer function:
>
> $$A_v(s) = -\frac{2\pi \times 10^6}{(s + 5000\pi)}$$
>
> **ANSWERS:** -400, 2.5 kHz, 2.5 kHz, 1 MHz

Phase Response

The phase behavior versus frequency is also of interest in many applications and later will be found to be of great importance to the stability of feedback amplifiers. Again substituting $s = j\omega$ in Eq. (6.36), the phase response of the low-pass amplifier is found to be

$$\angle A_v(j\omega) = \angle \frac{A_o}{1+j\frac{\omega}{\omega_H}} = \angle A_o - \tan^{-1}\left(\frac{\omega}{\omega_H}\right) \qquad (6.47)$$

The phase angle of A_o is $0°$ if A_o is positive and $180°$ if A_o is negative.

TABLE 6.3
Inverse Tangent

ω	$\tan^{-1}\frac{\omega}{\omega_C}$
$0.01\,\omega_C$	$0.057°$
$0.1\,\omega_C$	$5.7°$
ω_C	$45°$
$10\,\omega_C$	$84.3°$
$100\,\omega_C$	$89.4°$

Figure 6.19 Phase versus normalized frequency (ω/ω_C) resulting from a single inverse tangent term $+\tan^{-1}(\omega/\omega_C)$. The straight-line approximation is also given.

The frequency-dependent phase term associated with each pole or zero in a transfer function involves the evaluation of the inverse tangent function, as in Eq. (6.47). Important values appear in Table 6.3, and a graph of the complete inverse tangent function is given in Fig. 6.19. At the pole or zero frequency indicated by critical frequency ω_C, the magnitude of the phase shift is $45°$. One decade below ω_C, the phase is $5.7°$, and one decade above ω_C, the phase is $84.3°$. Two decades away from ω_C, the phase approaches its asymptotic limits of $0°$ and $90°$. Note that the phase response can also be approximated by the three straight-line segments in Fig. 6.19, in a manner similar to the asymptotes of the magnitude response.

The phase of more complex transfer functions with multiple poles and zeros is simply given by the appropriate sum and differences of inverse tangent functions. However, they are most easily evaluated with a computer or calculator.

EXAMPLE 6.4   **THE *RC* LOW-PASS FILTER**

An *RC* low-pass network is a simple but important passive circuit that we will encounter in the upcoming chapters.

PROBLEM Find the voltage transfer function $V_o(s)/V_i(s)$ for the low-pass network in the figure below.

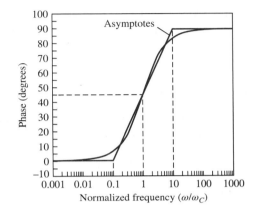

$$V_o(s) = V_i(s)\,\frac{\dfrac{R_2/sC}{R_2 + 1/sC}}{R_1 + \dfrac{R_2/sC}{R_2 + 1/sC}}$$

$$A_v(s) = \frac{V_o(s)}{V_i(s)} = \left(\frac{R_2}{R_1+R_2}\right)\left(\frac{1}{1+\dfrac{s}{\omega_H}}\right)$$

SOLUTION **Known Information and Given Data:** Circuit as given above in the problem statement.

Unknowns: Voltage transfer function $V_o(s)/V_i(s)$

Approach: Find the transfer function by applying voltage division in the frequency domain (s domain). Remember that the impedance of the capacitor is $1/sC$.

Assumptions: None

Analysis: Direct application of voltage division yields the equations next to the schematic, where the upper cutoff frequency is

$$\omega_H = \frac{1}{(R_1 \| R_2)C}$$

Check of Results: For $s \ll \omega_H$, the gain through the network is $R_2/(R_1 + R_2)$, which is correct.

Discussion: The cutoff frequency ω_H occurs at the frequency for which the reactance of the capacitor equals the parallel combination of resistors R_1 and R_2: $1/\omega_H C = R_1 \| R_2$. $R_1 \| R_2$ represents the Thévenin equivalent resistance present at the capacitor terminals. For $\omega \ll \omega_H$, the impedance of the capacitor is negligible with respect to the resistance in the circuit.

EXERCISE: What is the cutoff frequency for the low-pass circuit in Ex. 6.4 if $R_1 = 1$ kΩ, $R_2 = 100$ kΩ, and $C = 200$ pF?

ANSWER: 804 kHz

6.8.3 THE HIGH-PASS AMPLIFIER

A second basic single-pole transfer function is the high-pass characteristic, which includes a single pole plus a zero at the origin. We most often find this function combined with the low-pass function to form **band-pass amplifiers.** In fact a true high-pass characteristic is impossible to obtain since we will see that it requires infinite bandwidth. The best we can hope to do is approximate the high-pass characteristic over some finite range of frequencies.

The transfer function for an ideal single-pole **high-pass amplifier** can be written as

$$A_v(s) = \frac{A_o s}{s + \omega_L} = \frac{A_o}{1 + \frac{\omega_L}{s}} \qquad (6.48)$$

and for $s = j\omega$ the magnitude of Eq. (6.48) is

$$|A_v(j\omega)| = \left|\frac{A_o j\omega}{j\omega + \omega_L}\right| = \frac{A_o \omega}{\sqrt{\omega^2 + \omega_L^2}} = \frac{A_o}{\sqrt{1 + \left(\frac{\omega_L}{\omega}\right)^2}} \qquad (6.49)$$

The Bode magnitude plot for this function is depicted in Fig. 6.20. In this case, the gain of the amplifier is constant for all frequencies above the **lower-cutoff frequency** ω_L. At frequencies high enough to satisfy $\omega \gg \omega_L$, the magnitude can be approximated by

$$\left.\frac{A_o \omega}{\sqrt{\omega^2 + \omega_L^2}}\right|_{\omega \gg \omega_L} \cong \frac{A_o \omega}{\sqrt{\omega^2}} = A_o \qquad \text{or} \qquad (20 \log A_o) \text{ dB} \qquad (6.50)$$

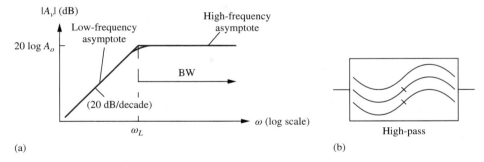

Figure 6.20 (a) High-pass amplifier; (b) high-pass filter symbol.

For ω exceeding ω_L, the gain is constant at gain A_o. At frequencies well below ω_L,

$$\left.\frac{A_o\omega}{\sqrt{\omega^2 + \omega_L^2}}\right|_{\omega \ll \omega_L} \cong \frac{A_o\omega}{\sqrt{\omega_L^2}} = \frac{A_o\omega}{\omega_L} \qquad (6.51)$$

Converting Eq. (6.51) to dB yields

$$|A_v(j\omega)| \cong (20 \log A_o) + 20 \log \frac{\omega}{\omega_L} \qquad (6.52)$$

At frequencies below ω_L, the gain increases at a rate of 20 dB per decade increase in frequency. At **critical frequency** $\omega = \omega_L$,

$$|A_v(j\omega_L)| = \frac{A_o\omega_L}{\sqrt{\omega_L^2 + \omega_L^2}} = \frac{A_o}{\sqrt{2}} \quad \text{or} \quad [(20 \log A_o) - 3] \text{ dB} \qquad (6.53)$$

The gain is again 3 dB below its midband value. Besides being called the lower-cutoff frequency, ω_L is referred to as the **lower −3-dB frequency** or the **lower half-power point.**

The high-pass amplifier provides approximately uniform gain at all frequencies above ω_L, and its bandwidth is therefore infinite:

$$\text{BW} = \infty - \omega_L = \infty \qquad (6.54)$$

The phase dependence of the high-pass amplifier is found by evaluating the phase of $A_v(j\omega)$ from Eq. (6.48):

$$\text{and} \quad \angle A_v(j\omega) = \angle \frac{A_o j\omega}{j\omega + \omega_L} = \angle A_o + 90° - \tan^{-1}\left(\frac{\omega}{\omega_L}\right) \qquad (6.55)$$

This phase expression is similar to that of the low-pass amplifier, except for a +90° shift due to the s term in the numerator.

EXERCISE: Find the midband gain, cutoff frequency, and bandwidth of the amplifier with this transfer function:

$$A_v(s) = \frac{250s}{(s + 250\pi)}$$

ANSWERS: 250; 125 Hz; ∞

EXERCISE: Use MATLAB to produce a Bode plot of this transfer function.

ANSWER: w = logspace(1,5,100)
bode([250 0],[1 250*pi],w)

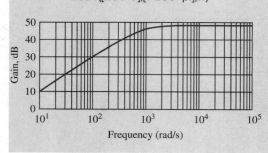
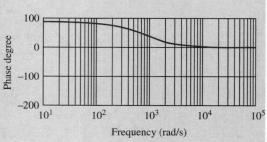

EXAMPLE 6.5 THE *RC* HIGH-PASS FILTER

The *RC* high-pass network is another important passive circuit that we will encounter in the upcoming chapters.

PROBLEM Find the voltage transfer function $V_o(s)/V_i(s)$ for the high-pass network in the figure below.

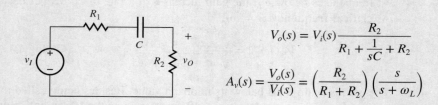

$$V_o(s) = V_i(s)\frac{R_2}{R_1 + \frac{1}{sC} + R_2}$$

$$A_v(s) = \frac{V_o(s)}{V_i(s)} = \left(\frac{R_2}{R_1 + R_2}\right)\left(\frac{s}{s + \omega_L}\right)$$

SOLUTION **Known Information and Given Data:** Circuit as given in the problem statement.

Unknowns: Voltage transfer function V_o/V_i

Approach: Find the transfer function by applying voltage division in the frequency domain (*s* domain). Remember that the impedance of the capacitor is $1/sC$.

Assumptions: None

Analysis: Direct application of voltage division yields the equation next to the circuit schematic, where the lower cutoff frequency is

$$\omega_L = \frac{1}{(R_1 + R_2)C}$$

Check of Results: For $\omega \gg \omega_L$, the gain through the network is $R_2/(R_1 + R_2)$, which is correct.

Discussion: Cutoff frequency ω_L occurs at the frequency for which the reactance of the capacitor equals the sum of resistors R_1 and R_2, which represents the Thévenin equivalent resistance at the capacitor terminals: $1/\omega_L C = R_1 + R_2$. For $\omega \gg \omega_L$, the impedance of the capacitor is negligible with respect to the resistance in the circuit.

EXERCISE: What is the cutoff frequency for the high-pass circuit in Ex. 6.5 if $R_1 = 1\ \text{k}\Omega$, $R_2 = 100\ \text{k}\Omega$, and $C = 0.1\ \mu\text{F}$?

ANSWER: 15.8 Hz

6.8.4 BAND-PASS AMPLIFIERS

Many amplifiers combine low-pass and high-pass characteristics to form a bandpass amplifier as shown in the graph in Fig. 6.21. For example, audio amplifiers are often designed to only pass frequencies in the 20 Hz–20 kHz range. In this case, the lower and upper cutoff frequencies f_L and f_H would be set to 20 Hz and 20 kHz, respectively. The region of constant gain between ω_L and ω_H is referred to as midband.

The transfer function for a basic **band-pass amplifier** can be constructed from the product of the low-pass and high-pass transfer functions from Eqs. (6.37) and (6.48):

$$A_v(s) = \frac{A_o s \omega_H}{(s + \omega_L)(s + \omega_H)} = A_o \frac{s}{(s + \omega_L)} \frac{1}{\left(\frac{s}{\omega_H} + 1\right)} \quad (6.56)$$

The **midband** range of frequencies is defined by $\omega_L \leq \omega \leq \omega_H$, for which

$$|A_v(j\omega)| \cong A_o \quad (6.57)$$

A_o represents the gain in this midband region and is called the **midband gain**: $A_{\text{mid}} = A_o$.

The mathematical expression for the magnitude of $A_v(j\omega)$ is

$$|A_v(j\omega)| = \left|\frac{A_o j\omega\, \omega_H}{(j\omega + \omega_L)(j\omega + \omega_H)}\right| = \frac{A_o \omega\, \omega_H}{\sqrt{(\omega^2 + \omega_L^2)(\omega^2 + \omega_H^2)}} \quad (6.58)$$

or

$$|A_v(j\omega)| = \frac{A_{\text{mid}}}{\sqrt{\left(1 + \frac{\omega_L^2}{\omega^2}\right)\left(1 + \frac{\omega^2}{\omega_H^2}\right)}} \quad \text{where } A_{\text{mid}} = A_o \quad (6.59)$$

The expression in Eq. (6.59) has been written in a form that exposes the gain in the midband region. At both ω_L and ω_H, it is easy to show, assuming $\omega_L \ll \omega_H$, that

$$|A_v(j\omega_L)| = |Av(j\omega_{HJ})| = \frac{A_o}{\sqrt{2}} \quad \text{or} \quad [(20 \log A_o) - 3]\ \text{dB} \quad (6.60)$$

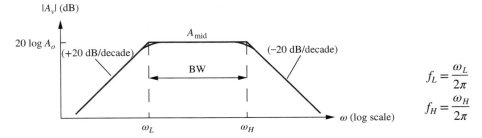

Figure 6.21 Band-pass amplifier.

The gain is 3 dB below the midband gain at both critical frequencies. The region of approximately uniform gain (that is, the region of less than 3 dB variation) extends from ω_L to ω_H (f_L to f_H), and hence the bandwidth of the band-pass amplifier is

$$\text{BW} = f_H - f_L = \frac{\omega_H - \omega_L}{2\pi} \tag{6.61}$$

Evaluating the phase of $A_v(j\omega)$ from Eq. (6.56),

$$\angle A_v(j\omega) = \angle A_o + 90° - \tan^{-1}\left(\frac{\omega}{\omega_L}\right) - \tan^{-1}\left(\frac{\omega}{\omega_H}\right) \tag{6.62}$$

An example of this phase response is in the next exercise.

EXAMPLE 6.6 TRANSFER FUNCTION EVALUATION

This example is included to help refresh our memory on calculations involving complex numbers.

PROBLEM Find the magnitude and phase of this voltage transfer function for $\omega = 0$ and $\omega = 3$ rad/s.

$$A_v(s) = 50\frac{s^2 + 4}{s^2 + 2s + 2}$$

SOLUTION **Known Information and Given Data:** Transfer function describing the voltage gain

Unknowns: Magnitude and phase of $A_v(j0)$ and $A_v(j3)$

Approach: Substitute $s = j\omega$ into the expression for $A_v(s)$ and simplify. Substitute $\omega = 0$ and $\omega = 3$ into the resulting expressions. Then find magnitude and phase of the resulting complex numbers.

Assumptions: We remember how to do arithmetic with complex numbers.

Analysis: Inserting $s = j\omega$ into $A_v(s)$ and rearranging yields

$$A_v(j\omega) = 50\frac{4 - \omega^2}{(2 - \omega^2) + 2j\omega}$$

The magnitude and phase of this expression are

$$|A_v(j\omega)| = 50\frac{|4 - \omega^2|}{\sqrt{(2 - \omega^2)^2 + 4\omega^2}} \quad \text{and} \quad \angle A_v(j\omega) = \angle(4 - \omega^2) - \tan^{-1}\left(\frac{2\omega}{2 - \omega^2}\right)$$

Substituting $\omega = 0$ gives

$$|A_v(j0)| = \frac{200}{\sqrt{4}} = 100 \quad \text{or} \quad 40.0 \text{ dB}$$

$$\angle A_v(j0) = \angle(200) - \tan^{-1}(0) = 0°$$

Substituting $\omega = 3$ gives

$$|A_v(j3)| = \frac{250}{\sqrt{49 + 36}} = 27.1 \quad \text{or} \quad 28.7 \text{ dB}$$

$$\angle A_v(j\omega) = \angle(250) - \tan^{-1}\left(\frac{-6}{7}\right) = 0° - (-40.6) = 40.6°$$

Check of Results: We can easily check the results using MATLAB or a calculator. With MATLAB,

$$h = \text{freqs}([50\ 0\ 200], [1\ 2\ 2], [0\ 3]);$$
$$\text{abs}(h)$$
$$\text{angle }(h) * 180/\text{pi}$$

The results confirm the preceding analysis.

EXERCISE: Find the magnitude and phase of the voltage gain in Ex. 6.6 for $\omega = 1$ rad/s and $\omega = 5$ rad/s.

ANSWERS: 36.5 dB, −63.4°; 32.4 dB, 23.5°

EXERCISE: Find the magnitude and phase of the following transfer function for $\omega = 0.95$, 1.0, and 1.10.

$$A_v(s) = 20\,\frac{s^2 + 1}{s^2 + 0.1s + 1}$$

ANSWERS: 14.3, −44.3°; 0, −90°; 17.7, +27.6°

EXERCISE: Make a Bode plot of the following $A_v(s)$ using MATLAB: $A_v(s) = -\dfrac{2\pi \times 10^6}{(s + 5000\pi)}$

ANSWER: w = logspace(2, 6, 100)
bode(2*pi*1e6,[1 5000*pi],w)

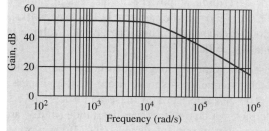
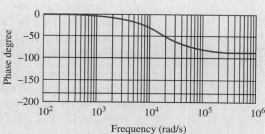

EXERCISE: Find the midband gain, lower- and upper-cutoff frequencies, and bandwidth of the amplifier with the following transfer function:

$$A_v(s) = -\frac{2 \times 10^7 s}{(s + 100)(s + 50000)}$$

ANSWERS: 52 dB; 15.9 Hz; 7.96 kHz; 7.94 kHz

EXERCISE: Write an expression for the phase of the transfer function above. What is the phase shift for w = 0, 100, 50000, and ∞?

ANSWERS: $\angle A_v(j\omega) = -90° - \tan^{-1}\left(\dfrac{\omega}{100}\right) - \tan^{-1}\left(\dfrac{\omega}{50000}\right)$; −90°, −135°, −225°, −270°

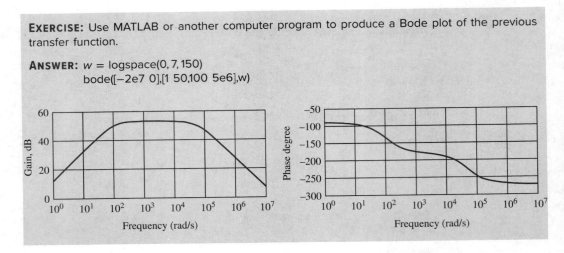

EXERCISE: Use MATLAB or another computer program to produce a Bode plot of the previous transfer function.

ANSWER: w = logspace(0, 7, 150)
bode([−2e7 0],[1 50,100 5e6],w)

SUMMARY

- This chapter introduced important characteristics of linear amplifiers and explored simplified models for the amplifiers. Voltage gain A_v, current gain A_i, power gain A_P, input resistance, and output resistance were all defined. Gains are expressed in terms of the phasor representations of sinusoidal signals or as transfer functions using Laplace transforms. The magnitudes of the gains are often expressed in terms of the logarithmic decibel or dB scale.

- Linear amplifiers can be conveniently modeled using two-port representations. The g-parameters are of particular interest for describing amplifiers in this text. In most of the amplifiers we consider, the 1–2 parameter (i.e., g_{12}) will be neglected. These networks were recast in terms of input resistance R_{in}, output resistance R_{out}, and open-circuit voltage gain A. Ideal voltage amplifiers have $R_{in} = \infty$ and $R_{out} = 0$.

- Linear amplifiers can be used to tailor the magnitude and/or phase of sinusoidal signals and are often characterized by their frequency response. Low-pass, high-pass and band-pass characteristics were discussed. The characteristics of these amplifiers are conveniently displayed in graphical form as a Bode plot, which presents the magnitude (in dB) and phase (in degrees) of a transfer function versus a logarithmic frequency scale. Bode plots can be created easily using MATLAB.

- In an amplifier, the midband gain A_{mid} represents the maximum gain of the amplifier. At the upper- and lower-cutoff frequencies—f_H and f_L, respectively—the voltage gain is equal to $A_{mid}/\sqrt{2}$ and is 3 dB below its midband value ($20 \log |A_{mid}|$). The bandwidth of the amplifier extends from f_L to f_H and is defined as $BW = f_H - f_L$.

- Amplifier circuits that are designed to tailor the frequency response of a signal are often referred to as filters.

- An amplifier must be properly biased to ensure that it operates in its linear region. The choice of bias point of the amplifier, its Q-point, can affect both the gain of the amplifier and the size of the input signal range for which linear amplification will occur. Improper choice of bias point can lead to nonlinear operation of an amplifier and distortion of the signal. One measure of linearity of a signal is its percent total harmonic distortion (THD).

KEY TERMS

Audio frequency (AF)
Band-pass amplifier
Bandwidth
Bode plot
Critical frequency
Current gain (A_i)
dc-coupled amplifier
Decibel (dB)
Differential amplifier
g-parameters
High-pass amplifier
Input offset voltage (V_{os})
Input resistance (R_{in})
Intermediate frequency (IF)
Inverting amplifier
Inverting input
Linear amplifier
Lower-cutoff frequency
Lower half-power point
Lower −3-dB frequency
Low-pass amplifier
Magnitude
Midband gain
Noninverting input

Open-circuit input conductance
Open-circuit termination
Open-circuit voltage gain
Output resistance
Phase angle
Phasor representation
Power gain (A_P)
Q-point
Short-circuit current gain
Short-circuit output resistance
Short-circuit termination
Single-pole frequency response
Source resistance
Total harmonic distortion (THD)
Transfer function
Two-port model
Two-port network
Unity-gain frequency
Upper-cutoff frequency
Upper half-power point
Upper −3-dB frequency
VHF
Voltage amplifier
Voltage gain

REFERENCES

1. T. Lewis, *Empire on the Air: The Men Who Made Radio*, Harper Collins: 1991.
2. J. A. Hijiya, *Lee de Forest and the Fatherhood of Radio*, Lehigh University Press: 1992.
3. T. H. Lee, "A Non Linear History of Radio," Chapter 1 in *The Design of CMOS Radio-Frequency Integrated Circuits*, Cambridge University Press: 1998.
4. National Geographic Society, *Those Inventive Americans*, (Editor and Publisher, 1971). pp. 182–187 (Lee de Forest by H. J. Lewis).

ADDITIONAL READING

P. R. Gray, P. J. Hurst, S. H. Lewis, and R. G. Meyer, *Analysis and Design of Analog Integrated Circuits*, Fifth Edition, John Wiley and Sons, New York, NY: 2009.
"Radio Spectrum," *Wikipedia, The Free Encyclopedia*, https://en.wikipedia.org/w/index.php?title=Radio_spectrum&oldid=946313909 (accessed April 27, 2020).

PROBLEMS

6.1 An Example of an Analog Electronic System

6.1. In addition to those given in the introduction, list 15 physical variables in your everyday life that can be represented as continuous analog signals.

6.2 Amplification

6.2. Convert the following to decibels: (a) voltage gains of 120, −60, 50,000, −100,000, 0.90; (b) current gains of 600, 3000, -10^6, 200,000,

0.95; (c) power gains of 2×10^9, 4×10^5, 6×10^8, 10^{10}.

6.3. Express the voltage, current, and power gains in Ex. 6.1 in dB.

6.4. Convert the following gains from dB to decimal values: 125, 36, −24, 3000, 1000000, 0.87, .001, −.02.

6.5. For what value of voltage gain A_v does $A_v = 20 \log(A_v)$?

6.6. Suppose the input and output voltages of an amplifier are given by

$$v_I = 1 \sin(1000\pi t) + 0.333 \sin(3000\pi t) + 0.200 \sin(5000\pi t) \text{ V}$$

and

$$v_O = 2 \sin\left(1000\pi t + \frac{\pi}{6}\right) + \sin\left(3000\pi t + \frac{\pi}{6}\right) + \sin\left(5000\pi t + \frac{\pi}{6}\right) \text{ V}$$

(a) Plot the input and output voltage waveforms of v_I and v_O for $0 \le t \le 5$ ms. (b) What are the amplitudes, frequencies, and phases of the individual signal components in v_I? (c) What are the amplitudes, frequencies, and phases of the individual signal components in v_O? (d) What are the voltage gains at the three frequencies? (e) Is this a linear amplifier?

6.7. What are the voltage gain, current gain, and power gain required of the amplifier in Fig. 6.3 if $V_i = 3.0$ mV and the desired output power is 25 W?

6.8. What are the voltage gain, current gain, and power gain required of the amplifier in Fig. 6.3 in dB if $V_i = 10$ mV, $R_I = 2$ kΩ, and the output power is 30 W?

6.9. The output of a PC sound card was set to be a 1-kHz sine wave with an amplitude of 1 V using MATLAB. The outputs were monitored with an oscilloscope and ac voltmeter. (a) For the left channel, the rms value of the open-circuit output voltage at 1 kHz was measured to be 0.760 V, and it dropped to 0.740 V with a 1040-Ω load resistor attached. Draw the Thévenin equivalent circuit representation for the left output of the sound card (i.e., What are v_{th} and R_{th}?). (b) For the right channel, the rms value of the open-circuit output voltage at 1 kHz was measured to be 0.768 V, and it dropped to 0.721 V with a 430-Ω load resistor attached. Draw the Thévenin equivalent circuit representation for the output of the right channel. (c) What were the values of the measured amplitudes of the two open-circuit output voltages? What percent error was observed between the actual voltage and the desired voltage as defined by MATLAB? (d) Go to the lab and determine the Thévenin equivalent output voltage and resistance for the sound card in your laptop PC.

6.10. Suppose that each output channel of a computer's sound card can be represented by a 1-V ac source in series with a 32-Ω resistor. Each channel of the amplifier in the external speakers has an input resistance of 20 kΩ, and must deliver 10 W into an 8-Ω speaker. (a) What are the voltage gain, current gain, and power gain required of the amplifier? (b) What would be a reasonable dc power supply voltage for this amplifier?

6.11. What is the minimum battery voltage required for the amplifier in Fig. 6.3 for an output power of 50 mW?

6.12. The amplifier in a battery-powered device is being designed to deliver 50 mW to a set of headphones. The impedance of the headphones can be chosen to be 8 Ω, 32 Ω, or 1000 Ω. Calculate the voltage and current required to deliver 50 mW to each of these resistances. Which resistance seems the best choice for a battery powered application?

6.3 Two-Port Models for Amplifiers

6.13. (a) Calculate the g-parameters for the circuit in Fig. P6.13. Compare g_{12} and g_{21}. (b) Repeat for the h-parameters (see Appendix C).

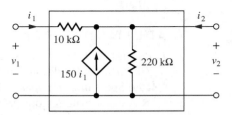

Figure P6.13

6.14. (a) Use SPICE transfer function analysis to find the g-parameters for the circuit in Fig. P6.13. (b) Repeat for h-parameters.

6.15. (a) Calculate the g-parameters for the circuit in Fig. P6.15. Compare g_{21} and g_{12}. (b) Repeat for z-parameters. (c) Confirm your results in parts (a) and (b) using SPICE.

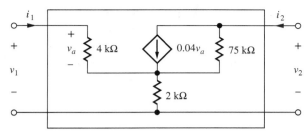

Figure P6.15

6.16. (a) Calculate the g-parameters for the circuit in Fig. P6.16. Compare g_{12} and g_{21}. (b) Repeat for y-parameters.

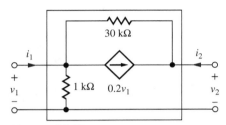

Figure P6.16

6.17. (a) Use SPICE transfer function analysis to find the g-parameters for the circuit in Fig. P6.16. (b) Repeat for the h-parameters. (c) Repeat for the z-parameters.

6.4 Mismatched Source and Load Resistances

6.18. An amplifier connected in the circuit in Fig. P6.18 has the two-port parameters listed below, with $R_I = 5$ kΩ and $R_L = 8$ Ω. (a) Find the overall voltage gain A_v, current gain A_i, and power gain A_P for the amplifier and express the results in dB. (b) What is the amplitude V_i of the sinusoidal input signal v_i needed to deliver 1 W to the 8-Ω load resistor? (c) How much power is dissipated in the amplifier when 1 W is delivered to the load resistor?

Input resistance $R_{in} = 1$ MΩ
Output resistance $R_{out} = 0.5$ Ω
$A = 56$ dB
$v_i = V_i \sin \omega t$

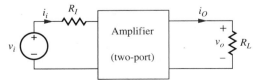

Figure P6.18

6.19. Suppose that the amplifier of Fig. P6.18 has been designed to match the source and load resistances with the parameters below. (a) What is the amplitude of the input signal v_i needed to deliver 1 W to the 16-Ω load resistor? (b) How much power is dissipated in the amplifier when 1 W is delivered to the load resistor?

Input resistance $R_{in} = 1$ kΩ
Output resistance $R_{out} = 16$ Ω
$A = 56$ db

6.20. The headphone amplifier in a battery-powered device has an output resistance of 32 Ω and is designed to deliver 0.1 W to the headphones. If the resistance of the headphones is 16 Ω, calculate the voltage and current required from the dependent voltage source (A in our model) to deliver 0.1 W to the headphones. How much power is delivered from dependent source? How much power is lost in the output resistance?

6.21. Repeat Prob. 6.20 if the headphones have a resistance of 500 Ω.

6.22. For the circuit in Fig. 6.9, $R_I = 1$ kΩ, $R_L = 16$ Ω, and $A = -2000$. What values of R_{in} and R_{out} will produce maximum power in the load resistor R_L? What is the maximum power that can be delivered to R_L if v_i is a sine wave with an amplitude of 20 mV? What is the power gain of this amplifier?

6.23. For the circuit in Fig. 6.9, $R_I = 2$ kΩ, $R_{in} = 20$ kΩ, $R_{out} = 60$ Ω, and $R_L = 2$ kΩ. What value of A is required to produce a voltage gain of 70 dB if the amplifier is to be an inverting amplifier ($\theta = 180°$)?

6.24. The circuit in Fig. P6.24 represents a two-port model for a current amplifier. Write expressions for input current i_1, output current i_o and the current gain $A_i = I_o/I_s$. What values of R_{in} and R_{out} provide maximum magnitude for the current gain?

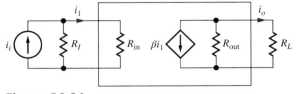

Figure P6.24

6.25. For the circuit in Fig. P6.24, $R_I = 100$ kΩ, $R_L = 10$ kΩ, and $\beta = 400$. What values of R_{in} and R_{out} will produce maximum power in the load resistor R_L? What is the maximum power that can be delivered to R_L if i_i is a sine wave with an amplitude of 1 μA? What is the power gain of this amplifier?

6.26. For the circuit in Fig. P6.24, $R_I = 100$ kΩ, $R_{in} = 20$ kΩ, $R_{out} = 300$ kΩ, and $R_L = 56$ kΩ. What value of β is required to produce a current gain of 150?

*6.27. For the circuit in Fig. 6.9, show that

$$A_{PdB} = A_{vdB} - 10 \log\left(\frac{R_L}{R_I + R_{in}}\right)$$

and

$$A_{PdB} = A_{idB} + 10 \log\left(\frac{R_L}{R_I + R_{in}}\right)$$

6.28. Two amplifiers are connected in series, or cascaded, in the circuit in Fig. P6.28. If $R_I = 1$ kΩ, $R_{in} = 5$ kΩ, $R_{out} = 500$ Ω, $R_L = 500$ Ω, and $A = -120$, what are the voltage gain, current gain, and power gain of the overall amplifier? Express your answers in dB.

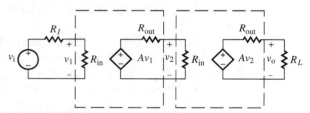

Figure P6.28

6.5 The Differential Amplifier

6.29. The circuit inside the box in Fig. P6.29 contains only resistors and diodes. The terminal V_O is connected to some point in the circuit inside the box. (a) Is the largest possible value of V_O most nearly 0 V, −9 V, +6 V, or +15 V? Why? (b) Is the smallest possible value of V_O most nearly 0 V, −9 V, +6 V, or +15 V? Why? (c) Repeat parts (a) and (b) for supply voltages of 0 and −9 V.

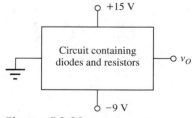

Figure P6.29

6.30. (a) The input voltage applied to the amplifier in Fig. P6.30 is $v_I = V_B + V_M \sin 1000t$. What is the voltage gain of the amplifier for small values of V_M if $V_B = 0.6$ V? What is the maximum value of V_M that can be used and still have an undistorted sinusoidal signal at v_O? (b) Write expressions for $v_I(t)$ and $v_O(t)$. (c) What is the gain for $V_B = 0.7$ V?

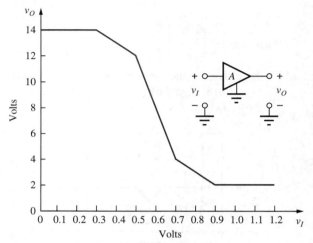

Figure P6.30

6.31. (a) Repeat Prob. 6.30 for $V_B = 0.8$ V. (b) For $V_B = 0.2$ V.

6.32. (a) Repeat Prob. 6.30 for $V_B = 0.5$ V. (b) For $V_B = 1.1$ V.

*6.33. The input voltage applied to the amplifier in Fig. P6.30 is $v_I = (0.6 + 0.1 \sin 1000t)$ V. (a) Write expressions for the output voltage. (b) Draw a graph of two cycles of the output voltage. (c) Calculate the first five spectral components of this signal. You may use MATLAB or other computer analysis tools.

6.34. The input voltage applied to the amplifier in Fig. P6.30 is $v_I = (0.5 + 0.1 \sin 1000t)$ V. (a) Write expressions for the output voltage. (b) Draw a graph of two cycles of the output voltage. (c) Calculate the first five spectral components of this signal. You may use MATLAB or other computer analysis tools.

6.6 Distortion in Amplifiers

6.35. The input signal to an audio amplifier is described by $v_I = (0.5 + 0.25 \sin 1200\pi t)$ V, and the output is described by $v_O = (2 + 4 \sin 1200\pi t + 0.8 \sin 2400\pi t + 0.4 \sin 3600\pi t)$ V. What is the voltage gain of the amplifier? What order harmonics are present in the signal? What is the total harmonic distortion in the output signal?

6.36. The input signal to an audio amplifier is a 1-kHz sine wave with an amplitude of 4 mV, and the output is described by $v_O = (5 \sin 2000\pi t + 0.5 \sin 6000\pi t + 0.20 \sin 10{,}000\pi t)$ V. What is the voltage gain of the amplifier? What order harmonics are present in the signal? What is the total harmonic distortion in the output signal?

6.37. (a) Use the FFT capability of MATLAB to find the Fourier series representation of $v_{o2}(t)$ in Fig. 6.15. (b) Use MATLAB to find the coefficients of the first three terms of the Fourier series for $v_{o2}(t)$ by evaluating the integral expression for the coefficients.

6.38. MATLAB limits the output of a sound signal to unity (1 V). Any signal value above this limit will be clipped (set to 1). (a) Use MATLAB to plot the following waveform: $y = \max(-1, \min(1, 1.5 \sin(1400\pi t)))$. (b) Use MATLAB to find the total harmonic distortion in waveform y. (c) Use the sound output on your computer to listen to and compare the following signals: $y = 1 \sin 1400\pi t$, $y = 1.5*\sin 1400\pi t$, and $y = \max(-1, \min(1, 1.5\sin(1400\pi t)))$. Describe what you hear.

6.39. Suppose the VTC for an amplifier is described by $v_O = 10 \tanh(2v_I)$ volts. (a) Plot the VTC. (b) Calculate the distortion based upon the first three frequency components in the output if $v_I = 0.75 \sin 2000\pi t$ volts.

6.7 Differential Amplifier Model

6.40. A differential amplifier connected in the circuit in Fig. P6.40 has the parameters listed below with $R_I = 5$ kΩ and $R_L = 600$ Ω. (a) Find the overall voltage gain A_v, current gain A_i, and power gain A_P for the amplifier, and express the results in dB. (b) What is the amplitude V_I of the sinusoidal input signal needed to develop a 20-V peak-to-peak signal at v_o?

Input resistance $R_{id} = 1$ MΩ
Output resistance $R_o = 25$ Ω
$A = 66$ dB
$v_i = V_I \sin \omega t$

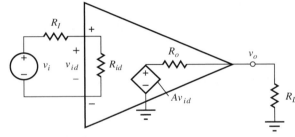

Figure P6.40

6.41. Suppose that the amplifier in Fig. P6.40 has been designed to match the source and load resistances in Prob. 6.40 with the parameters below. (a) What is the amplitude of the input signal v_i needed to develop a 15-V peak-to-peak signal at v_o? (b) How much power is dissipated in the amplifier when 0.5 W is delivered to the load resistor?

Input resistance $R_{id} = 5$ kΩ
Output resistance $R_o = 1$ kΩ
$A = 30$ dB

6.42. An amplifier has a sinusoidal output signal and is delivering 100 W to a 50-Ω load resistor. What output resistance is required if the amplifier is to dissipate no more than 2 W in its own output resistance?

6.43. The input to an amplifier comes from a transducer that can be represented by a 1-mV voltage source in series with a 50-kΩ resistor. What input resistance is required of the amplifier for $v_{id} \geq 0.99$ mV?

6.44. Suppose a differential amplifier has $A = 120$ dB, and it is operating in a circuit with an open-circuit output voltage $v_o = 15$ V. What is the input voltage v_{id}? How large must the voltage gain be for $v_{id} \leq 1$ μV? What is the input current i_+ if $R_{id} = 1$ MΩ?

6.45. Draw a graph of the gain of the amplifier in Fig. 6.13 versus v_{ID}.

6.46. (a) Draw a graph of the gain for a differential amplifier with a gain of 5 and ± 5 V power supplies. (b) Repeat for a gain of 4 and power supplies of $+5$ and -3 V. (c) Repeat for a gain of 10 and power supplies of $+5$ and 0 V.

6.8 Amplifier Frequency Response

*6.47. Find the poles and zeros of the following transfer functions:

(a) $A_i(s) = -\dfrac{5 \times 10^9 s^2}{(s^2 + 51s + 50)(s^2 + 13{,}000s + 3 \times 1}$

*(b) $A_v(s)$

$$= -\dfrac{2 \times 10^5(s^2 + 51s + 50)}{s^5 + 1000s^4 + 50{,}000s^3 + 20{,}000s^2 + 13{,}000s + 3 \times 10^7}$$

6.48. What are A_{mid} in dB, f_H, f_L, and the BW in Hz for the amplifier described by

$$A_v(s) = \dfrac{3 \times 10^4 s}{s + 200\pi}$$

What type of amplifier is this?

6.49. What are A_{mid} in dB, f_H, f_L, and the BW in Hz for the amplifier described by

$$A_v(s) = \frac{2\pi \times 10^6}{s + 200\pi}$$

What type of amplifier is this?

6.50. What are A_{mid} in dB, f_H, f_L, and the BW in Hz for the amplifier described by

$$A_v(s) = \frac{5\pi \times 10^7 s}{(s + 20\pi)(s + 2\pi \times 10^4)}$$

What type of amplifier is this?

6.51. What are A_{mid} in dB, f_H, f_L, and the BW in Hz for the amplifier described by

$$A_v(s) = 30 \frac{s^2 + 10^{12}}{s^2 + 10^4 s + 10^{12}}$$

What type of amplifier is this?

6.52. What are A_{mid} in dB, f_H, f_L, the BW in Hz, and the Q for the amplifier described by

$$A_v(s) = -\frac{3 \times 10^6 s}{s^2 + 10^5 s + 10^{14}}$$

What type of amplifier is this?

*6.53. What are A_{mid} in dB, f_H, f_L, and the BW in Hz for the amplifier described by

$$A_v(s) = \frac{5\pi^2 \times 10^{14} s^2}{(s + 20\pi)(s + 50\pi)(s + 2\pi \times 10^5)(s + 2\pi \times 10^6)}$$

What type of amplifier is this?

6.54. (a) Use MATLAB, a spreadsheet, or other computer program to generate a Bode plot of the magnitude and phase of the transfer function in Prob. 6.48. (b) Repeat for Prob. 6.49.

6.55. (a) Use MATLAB, a spreadsheet, or other computer program to generate a Bode plot of the magnitude and phase of the transfer function in Prob. 6.50. (b) Repeat for the transfer function in Prob. 6.51.

6.56. (a) Use MATLAB, a spreadsheet, or other computer program to generate a Bode plot of the magnitude and phase of the transfer function in Prob. 6.52. (b) Repeat for the transfer function in Prob. 6.53.

6.57. The voltage gain of an amplifier is described by the transfer function in Prob. 6.48 and has an input $v_i = 0.003 \sin \omega t$ V. Write an expression for the amplifier's output voltage at a frequency of (a) 5 Hz, (b) 500 Hz, (c) 50 kHz.

6.58. The voltage gain of an amplifier is described by the transfer function in Prob. 6.50 and has an input $v_i = 0.2 \sin \omega t$ mV. Write an expression for the amplifier's output voltage at a frequency of (a) 1 Hz, (b) 50 Hz, (c) 5 kHz.

6.59. The voltage gain of an amplifier is described by the transfer function in Prob. 6.49 and has an input $v_i = 20 \sin \omega t$ μV. Write an expression for the amplifier's output voltage at a frequency of (a) 2 Hz, (b) 2 kHz, (c) 200 kHz.

6.60. The voltage gain of an amplifier is described by the transfer function in Prob. 6.51 and has an input $v_i = 0.005 \sin \omega t$ V. Write an expression for the amplifier's output voltage at a frequency of (a) 1.59 MHz, (b) 1 MHz, (c) 5 MHz.

6.61. The voltage gain of an amplifier is described by the transfer function in Prob. 6.52 and has an input $v_i = 0.275 \sin \omega t$ V. Write an expression for the amplifier's output voltage at a frequency of (a) 159 kHz, (b) 50 kHz, (c) 200 kHz.

6.62. The voltage gain of an amplifier is described by the transfer function in Prob. 6.53 and has an input $v_i = 0.002 \sin \omega t$ V. Write an expression for the amplifier's output voltage at a frequency of (a) 5 Hz, (b) 500 Hz, (c) 50 kHz.

6.63. (a) Write an expression for the transfer function of a low-pass voltage amplifier with a gain of 26 dB and $f_H = 5$ MHz. (b) Repeat if the amplifier exhibits a phase shift of 180° at $f = 0$.

6.64. (a) Write an expression for the transfer function of a voltage amplifier with a gain of 40 dB, $f_L = 400$ Hz, and $f_H = 100$ kHz. (b) Repeat if the amplifier exhibits a phase shift of 180° at $f = 0$.

6.65. (a) What is the bandwidth of the low-pass amplifier described by

$$A_v(s) = A_o \left(\frac{\omega_1}{s + \omega_1}\right)^3$$

if $A_o = -2000$ and $\omega_1 = 50{,}000\pi$? (b) Make a Bode plot of this transfer function. What is the slope of the magnitude plot for $\omega \gg \omega_H$ in dB/dec?

*6.66. The input to a low-pass amplifier with a gain of 10 dB is

$$v_S = 1 \sin(1000\pi t) + 0.333 \sin(3000\pi t) + 0.200 \sin(5000\pi t) \text{ V}$$

(a) If the phase shift of the amplifier at 500 Hz is 10°, what must be the phase shift at the other two frequencies if the shape of the output waveform is to be the same as that of the input waveform? Write an expression for the output signal. (b) Use the computer to check your answer by plotting the input and output waveforms.

Low-Pass Filters

6.67. Find the midband gain in dB and the upper cutoff frequency for the low-pass filter in Ex. 6.4 if $R_1 = 10$ kΩ, $R_2 = 200$ kΩ, and $C = 0.01$ µF.

6.68. Find the midband gain in dB and the upper cutoff frequency for the low-pass filter in Ex. 6.4 if $R_1 = 1$ kΩ, $R_2 = 1.5$ kΩ, and $C = 0.02$ µF.

6.69. (a) Design a low-pass filter using the circuit in Ex. 6.4 to provide a loss of no more than 0.5 dB at low frequencies and a cutoff frequency of 20 kHz if $R_1 = 560$ Ω. (b) Pick standard values from the tables in Appendix A.

6.70. What are A_{mid} in dB, f_H, f_L, and the BW in Hz for the amplifier described by

$$A_v(s) = \frac{2\pi \times 10^5}{s + 200\pi}$$

What type of amplifier is this?

High-Pass Filters

6.71. Find the midband gain in dB and the upper cutoff frequency for the high-pass filter in Ex. 6.5 if $R_1 = 10$ kΩ, $R_2 = 82$ kΩ, and $C = 0.01$ µF.

6.72. Find the midband gain in dB and the upper cutoff frequency for the high-pass filter in Ex. 6.5 if $R_1 = 8.2$ kΩ, $R_2 = 20$ kΩ, and $C = 0.02$ µF.

6.73. (a) Design a high-pass filter using the circuit in Ex. 6.5 to provide a loss of no more than 0.5 dB at high frequencies and a cutoff frequency of 20 kHz if $R_1 = 390$ Ω. (b) Pick standard values from the tables in Appendix A.

6.74. (a) Design a high-pass filter (choose R_1, R_2, and C) to have a high frequency input resistance of 20 kΩ, a gain of 20 dB, and a lower cutoff frequency of 1 kHz. (b) Choose element values from the tables in Appendix A.

6.75. What are A_{mid} in dB, f_H, f_L, and the BW in Hz for the amplifier described by

$$A_v(s) = \frac{5 \times 10^3 s}{s + 200\pi}$$

What type of amplifier is this?

术语对照

Active filters	有源滤波器
Audio frequency(AF)	音频(AF)
Band-pass amplifier	带通放大器
Bandwidth	带宽
Bias	偏置
Bode plot	伯德图
Closed-loop amplifier	闭环放大器
Closed-loop gain	闭环增益
Comparator	比较器
Critical frequency	截止频率
Current amplifier	电流放大器
Current gain(A_i)	电流增益(A_i)
Current-to-voltage(i-v) converter	电流-电压(i-v)转换器
Decibel(dB)	分贝(dB)
Ac-coupled amplifier	交流耦合放大器
Dc-coupled amplifiers	直流耦合放大器
Digital-to-analog converter(DAC)	数/模转换器(DAC)
Dual-ramp (dual slope) ADC	双斜坡(双斜率)模/数转换器
Feedback amplifier	反馈放大器
Feedback network	反馈网络
Gain-bandwidth product(ω_T)	增益带宽积(ω_T)
g-parameters	g参数
High-pass amplifier	高通放大器
High-pass filter	高通滤波器
Input resistance(R_{in})	输入电阻(R_{in})
Input offset voltage	输入偏移电压
Integrator	积分器
Intermediate frequency(IF)	中频(IF)
Inverted R-$2R$ ladder	反相R-$2R$梯形网络
Inverting amplifier	反相放大器
Inverting input	反相输入
Inverting amplifier voltage gain	反相放大器电压增益
Least significant bit(LSB)	最低有效位(LSB)
Linear amplifier	线性放大器
Lower-cutoff frequency	下限截止频率
Lower half-power point	下半功率点
Lower -3-dB frequency	下3dB频率
Most significant b(MSB)	最高有效位(MSB)
Noninverting amplifier	同相放大器
Op amp	运算放大器
Open-circuit	开路
Open-circuit input conductance	开路输入电导
Open-circuit input resistance	开路输入电阻
Open-circuit termination	开路输入端
Open-circuit voltage gain	开路电压增益
Open-loop gain	开环增益
Operational amplifier(op amp)	运算放大器(op amp)
Output resistance	输出电阻
Offset voltage	偏移电压
Phase angle	相位角

Phasor	相量
Phase angle	相角
Phasor representation	相量表示
Power gain(A_P)	功率增益（A_P）
R-$2R$ ladder	R-$2R$梯形网络
Radio frequency(RF)	射频（RF）
Short-circuit output conductance	输入短路时的输出电导
Short-circuit output resistance	输入短路时的输出电阻
Short-circuit current gain	短路电流增益
Short-circuit termination	短路输出端
Short-circuit	短路
Transresistance amplifier	互阻放大器
Two-port parameters	二端口参数
Two-port model	二端口模型
Two-port network	二端口网络
Unity-gain buffer	单位增益缓冲器
Unity-gain	单位增益
Voltage gain(A_V)	电压增益（A_V）
Weighted-resistor DAC	权电阻DAC的设计

CHAPTER 7

THE TRANSISTOR AS AN AMPLIFIER

第7章 晶体管放大器

本章提纲

7.1 晶体管放大器
7.2 耦合电容和旁路电容
7.3 用直流和交流等效电路进行电路分析
7.4 小信号模型简介
7.5 双极型晶体管的小信号模型
7.6 共射极放大器
7.7 重要限制及模型简化
7.8 场效应晶体管的小信号模型
7.9 BJT和FET小信号模型的小结与对比
7.10 共源极放大器
7.11 共射极放大器和共源极放大器小结
7.12 放大器功率和信号范围

本章目标

- 了解晶体管如何作为线性放大器；
- 了解直流和交流等效电路；
- 了解用耦合和旁路电容、电感修正直流和交流等效电路；
- 了解小信号电压和电流概念；
- 掌握共发射极和共源极放大器的区别；
- 掌握放大器特性，包括电压增益、输入电阻、输出电阻、线性信号范围；
- 掌握共发射极和共源极放大器的电压增益；
- 理解交流小信号传输函数和SPICE瞬态分析的应用和区别。

本章导读

本章进一步介绍线性放大器设计的基本知识，学习设计复杂模拟单元和系统的基本放大电路，例如高性能运算放大器、数/模（D/A）和模/数（A/D）转换器、音频设备、光盘播放器、无线设备等。本章开始研究与设计更多种类的复杂电路，并学习用直流和交流分析简化工作。为了预测电路的详细行为，需要创建数学模型描述电路，并基于这些模型进行分析与计算。

本章首先介绍单个晶体管放大器的基础知识，包括双极型晶体管和场效应晶体管放大器，深入探讨共发射极晶体管放大电路的工作原理，详细分析采用MOSFET的共源放大器。比较包含上述器件的电路，推导不同放大器电压增益和输入、输出电阻的数学表达式，并详细讨论各种电路的优缺点。

为了简化电路的分析和设计，本章将电路分成了两个部分进行分析：一是直流等效电路，用于确定晶体管的Q点；二是交流等效电路，用于分析电路的信号响应。直流等效电路（dc equivalent circuit）适用于稳态直流分析电路，用于找电路的Q点。要构建直流等效电路，需要假设电容开路、电感短路。一旦找到Q点，就可以用交流等效电路（ac equivalent circuit）来确定电路对交流信号的响应。文中给出直流和交流分析步骤，首先通过直流分析找出Q点，然后通过交流分析确定放大器电路的特性，并以实例说明如何构建BJT放大器的交流和直流等效电路。

基于线性的交流分析需要应用小信号模型，它展示了电压和电流端点间的线性关系。本章将提出小信号的概念，并详细讨论二极管、双极型晶体管和场效应晶体管的小信号模型。在交流分析中，一般用性能比较完善的线性电路分析技术。为了达到这一目的，信号的电流和电压必须足够小，以确保交流电路呈线性特性。因此，必须假设时变信号分量是小信号。小信号的幅度取决于设备，因此必须为每个设备开发的小信号模型进行定义。

对小信号模型的定义首先从二极管开始。二极管的小信号模型表示二极管的电压和电流之间在Q点附近的微小变化关系，文中对二极管的小信号模型给出了详细的推导及讲解。本章也详细研究并推导了双极型晶体管和场效应晶体管的小信号模型，给出了小信号模型的等效形式以及化简形式。设计人员通常要根据目标来权衡放大器的直流和交流特性设计，在电路中采用耦合电容、旁路电容和电感来改变电路的拓扑结构。

交流分析全都是基于晶体管的线性小信号模型进行的。本章对于二极管、双极型晶体管（混合π模型）、MOSFET管和JFET的小信号模型都进行了详细讨论。通过对前面章节中推导出的大信号模型方程进行计算，本章也给出了与Q点相关的跨导g_m、输出电阻r_o和输入电阻r_π的表达式。

本章详细比较了双极型晶体管和场效应晶体管的小信号模型，目的是突显出两种器件的异同点。BJT的跨导直接与电流呈比例关系，而FET的跨导只是与电流的平方根呈比例关系。电阻r_π和r_o与Q点的电流呈反比关系。在FET中电阻r_π无限大，因此在其小信号模型中并不会出现。每种类型的器件对都有同样的小信号模型，如npn型及pnp型BJT，n沟道及p沟道FET。

本章详细介绍了与晶体管的本征电压增益相关的概念。晶体管的本征电压增益也称为晶体管的放大系数，定义为$\mu_f=g_m r_o$，表示的是在C-E和C-S放大器中晶体管所能获得的最大增益，文中给出了BJT和FET的本征电压增益表达式的推导。对于BJT而言，参数μ_f与Q点无关；但对于FET而言，放大系数会随着工作电流的增大而减小。对于普通的工作点而言，BJT的μ_f可高达数千，而FET只有几十到几百。

本章最后讨论了放大器工作点的设计、功耗和输出信号摆幅之间的关系。放大器的输出信号摆幅受限于以下两个因素中的较小者：一是晶体管Q点的集电极-基极或漏极-栅极的电压值；二是Q点的集电极偏置电阻R_C，或漏极偏置电阻R_D上的压降。

在SPICE中，了解交流分析和瞬态分析之间的差别是非常重要的。交流分析假设电路网络是线性的，采用的是晶体管和二极管的小信号模型。由于电路是线性的，因此信号源的幅值可以选取任意合适的值，因而通常会选择1V和1A。相比之下，瞬态仿真利用了晶体管的全大信号非线性模型。如果想在瞬态仿真中获得线性行为，则所有信号都必须满足小信号的约束。

Chapter Outline

7.1 The Transistor as an Amplifier
7.2 Coupling and Bypass Capacitors
7.3 Circuit Analysis Using dc and ac Equivalent Circuits
7.4 Introduction to Small-Signal Modeling
7.5 Small-Signal Models for Bipolar Junction Transistors
7.6 The Common-Emitter (C-E) Amplifier
7.7 Important Limits and Model Simplifications
7.8 Small-Signal Models for Field-Effect Transistors
7.9 Summary and Comparison of the Small-Signal Models of the BJT and FET
7.10 The Common-Source (C-S) Amplifier
7.11 Common-Emitter and Common-Source Amplifier Summary
7.12 Amplifier Power and Signal Range
Summary
Key Terms
Reference
Problems

Chapter Goals

In Chapter 7, we develop a basic understanding of the following concepts related to linear amplification:

- Transistors as linear amplifiers
- dc and ac equivalent circuits
- Use of coupling and bypass capacitors and inductors to modify the dc and ac equivalent circuits
- The concept of small-signal voltages and currents
- Small-signal models for diodes and transistors
- Identification of common-emitter and common-source amplifiers
- Amplifier characteristics including voltage gain, input resistance, output resistance, and linear signal range
- Rule-of-thumb estimates for voltage gain of common-emitter and common-source amplifiers
- Improvement of our understanding of the use and differences between the ac small-signal transfer function, and transient analysis capabilities of SPICE

Chapter 7 begins our study of basic amplifier circuits that are used in the design of complex analog components and systems such as high-performance operational amplifiers, analog-to-digital and digital-to-analog converters, audio equipment, compact disk players, wireless devices, cellular telephones, and so on. At first glance, the operational amplifier schematic in the figure here represents an overwhelming interconnection of transistors and passive components. With this chapter, we begin our quest to understand and design a wide variety of such circuits. We will learn to simplify our job by separating the dc and ac circuit analyses.

In order to predict the detailed behavior of the circuit, we must also be able to build mathematical models that describe the circuit. This chapter develops these models. Then over the next several chapters, we become familiar with the basic subcircuits that serve as our electronic tool kit for building more complicated electronic systems. With practice over time, we will be able to spot these basic building blocks in more complex electronic circuits and use our knowledge of the subcircuits to understand the full system.

This chapter introduces the general techniques for employing individual transistors as amplifiers and then studies in detail the operation of the common-emitter bipolar transistor circuit. This is followed by analysis of common-source amplifiers employing MOSFETs. Circuits containing these devices are compared, and expressions are developed for the voltage gain and input and output resistances of the various amplifiers. The advantages and disadvantages of each are discussed in detail.

To simplify the analysis and design processes, the circuits are split into two parts: a dc equivalent circuit used to find the operating-point of the transistor, and an ac equivalent circuit used for analysis of the circuit's response to signal sources. As a by-product of this approach, we discover how capacitors and inductors are used to change the ac and dc circuit topologies.

The ac analysis is based on linearity and requires the use of "small-signal" models that exhibit ships linear relation between their terminal voltages and currents. The concept of a small signal is developed, and small-signal models for

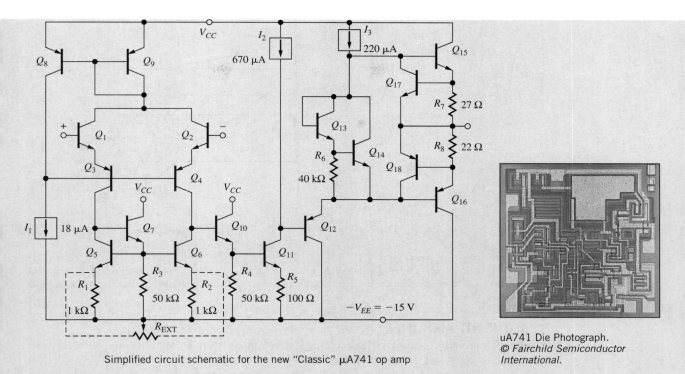

Simplified circuit schematic for the new "Classic" µA741 op amp

uA741 Die Photograph.
© Fairchild Semiconductor International.

the diode, bipolar transistor, and MOSFET are all discussed in detail.

Examples of the complete analysis of common-emitter and common-source amplifiers are included in this chapter. The relationships between the choice of operating point and the small-signal characteristics of the amplifier are developed, as is the relationship between operating-point design and output signal voltage swing.

7.1 THE TRANSISTOR AS AN AMPLIFIER

As mentioned in Part One, the bipolar junction transistor is an excellent amplifier when it is biased in the forward-active region of operation; field-effect transistors should be operated in the saturation or pinch-off region in order to be used as amplifiers. For simplicity, we will now refer to bipolar transistors operating in the forward-active region and FETs in the saturation region as simply being in the "**active region**" where they can be used as linear amplifiers. In these regions of operation, the transistors have the capacity to provide high voltage, current, and power gains. This chapter focuses on determining voltage gain, input resistance, and output resistance. We will find that we need the input and output resistance information in order to calculate the lower and upper cutoff frequencies of the amplifiers. Evaluation of current gain and power gain are addressed in later chapters.

We must provide bias to the transistor in order to stabilize the operating point (Q-point) in the active region of operation. Once the dc operating point has been established, we can then use the transistor as an amplifier. Choice of the Q-point controls many amplifier characteristics, including

- Small-signal parameters of the transistor
- Voltage gain, input resistance, and output resistance
- Maximum input and output signal amplitudes
- Power consumption

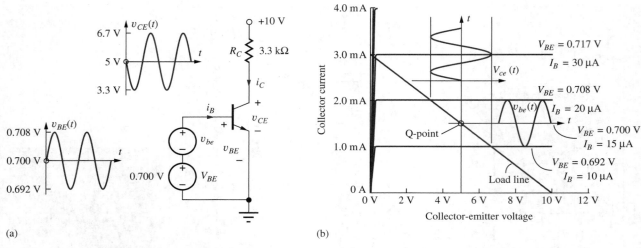

Figure 7.1 (a) BJT biased in the active region by the voltage source V_{BE}. A small sinusoidal signal voltage v_{be} is applied in series with V_{BE} and generates a similar but larger amplitude waveform at the collector; (b) load-line, Q-point, and signals for circuit of Fig. 7.1(a).

7.1.1 THE BJT AMPLIFIER

To get a clearer understanding of how a transistor can provide linear amplification, let us assume that a bipolar transistor is biased in the active region by the dc voltage source V_{BE} shown in Fig. 7.1. For this particular transistor, the fixed base-emitter voltage of 0.700 V sets the Q-point to be $(I_C, V_{CE}) = (1.5\text{ mA}, 5\text{ V})$ with $I_B = 15\ \mu\text{A}$, as indicated in Fig. 7.1(b). Both I_B and V_{BE} have been shown as parameters on the output characteristics in Fig. 7.1(b) (usually only I_B is given).

To provide amplification, a signal must be injected into the circuit in a manner that causes the transistor voltages and currents to vary with the applied signal. For the circuit in Fig. 7.1, the base-emitter voltage is forced to vary about its Q-point value by signal source v_{be} placed in series with dc bias source V_{BE}, so the total base-emitter voltage becomes

$$v_{BE} = V_{BE} + v_{be} \tag{7.1}$$

In Fig. 7.1(b), we see that the 8-mV peak change in base-emitter voltage produces a 5-μA change in base current and hence a 500-μA change in collector current ($i_c = \beta_F i_b$).

The collector-emitter voltage of the BJT in Fig. 7.1 can be expressed as

$$v_{CE} = 10 - i_C R_C = 10 - 3300 i_C \tag{7.2}$$

The change in collector current develops a time-varying voltage across load resistor R_C and at the collector terminal of the transistor. The 500 μA change in collector current develops a 1.65-V change in collector-emitter voltage.

If these changes in operating currents and voltages are all small enough ("small signals"), then the collector current and collector-emitter voltage waveforms will be undistorted replicas of the input signal. Small-signal operation is device-dependent; it will be precisely defined for the BJT when the small-signal model for the bipolar transistor is introduced.

In Fig. 7.1 we see that a small input voltage change at the base is causing a large voltage change at the collector. The voltage gain for this circuit is defined in terms of the frequency domain (phasor) representation of the signals:

$$A_v = \frac{\mathbf{V_{ce}}}{\mathbf{V_{be}}} = \frac{1.65 \angle 180°}{0.008 \angle 0°} = 206 \angle 180° = -206 \tag{7.3}$$

The magnitude of the collector-emitter voltage is 206 times larger than the base-emitter signal amplitude; this represents a voltage gain of 206. It is also important to note in Fig. 7.1 that the output signal voltage decreases as the input signal increases, indicating a 180° phase shift between the input and the output signals. Thus, this transistor circuit is an inverting amplifier. This 180° phase shift is often represented by the minus sign in Eq. (7.3). The transistor in Fig. 7.1 has its emitter connected to ground, and this circuit is known as a ***common-emitter amplifier***. Note that Eq. (7.2) represents the load line for this transistor.

> **EXERCISE:** The dc common-emitter current gain β_F of the bipolar transistor is defined by $\beta_F = I_C/I_B$. (a) What is the value of β_F for the transistor in Fig. 7.1? (b) The dc collector current of the BJT in the (forward) active region is given by $I_C = I_S \exp V_{BE}/V_T$. Use the Q-point data to find the saturation current I_S of the transistor in Fig. 7.1. (Remember $V_T = 0.025$ V.) (c) The ratio of v_{be}/i_b represents the small-signal input resistance R_{in} of the BJT. What is its value for the transistor in Fig. 7.1? (d) Does the BJT remain in the active region during the full range of signal voltages at the collector? Why? (e) Express the voltage gain in dB.
>
> **ANSWERS:** $\beta_F = 100$; $I_S = 1.04 \times 10^{-15}$ A; $R_{in} = 1.6$ kΩ; yes, $v_{CE}^{MIN} > v_{BE}$: 3.4 V > 0.708 V; 46.3 dB.

7.1.2 THE MOSFET AMPLIFIER

The amplifier circuit using a MOSFET in Fig. 7.2 is directly analogous to the BJT amplifier circuit in Fig. 7.1. Here the gate-to-source voltage is forced to vary about its Q-point value ($V_{GS} = 3.5$ V) by signal source v_{gs} placed in series with dc bias source V_{GS}. In this case, the total gate-source voltage is

$$v_{GS} = V_{GS} + v_{gs}$$

The resulting signal voltages are superimposed on the MOSFET output characteristics in Fig. 7.2(b). $V_{GS} = 3.5$ V sets the Q-point (I_D, V_{DS}) at (1.56 mA, 4.8 V), and the 1-V *p-p* change in v_{GS} causes a 1.25-mA *p-p* change in i_D.

In this circuit, the drain-source voltage of the MOSFET can be expressed as

$$v_{DS} = 10 - 3300 i_D \tag{7.4}$$

The 1.25 mA *p-p* change in drain current develops a 4.13-V change in the drain-source voltage of the MOSFET. If these changes in operating currents and voltages are again small enough to be considered "small signals," then the drain-source signal voltage waveform will be an undistorted

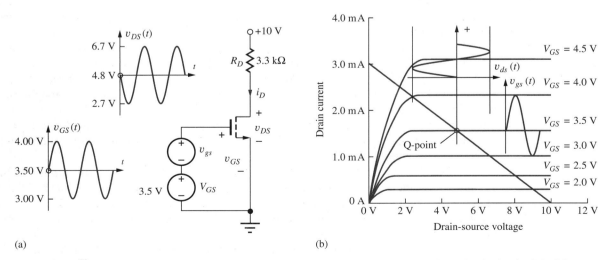

Figure 7.2 (a) A MOSFET common-source amplifier; (b) Q-point, load line, and signals for the circuit of Fig. 7.2(a).

replica of the input signal applied to the gate. The definition of a small-signal is different for the MOSFET than for the BJT, and it will be defined when MOSFET small-signal model is introduced.

In Fig. 7.2, the input voltage signal applied to the gate is causing a larger voltage change at the drain, and the voltage gain for this circuit is given by

$$A_v = \frac{\mathbf{v_{ds}}}{\mathbf{v_{gs}}} = \frac{4.13\angle 180°}{1\angle 0°} = 4.13\angle 180° = -4.13 \tag{7.5}$$

In this case, the source of the transistor is grounded, and this circuit is known as a ***common-source amplifier***. We see that the common-source configuration also forms an inverting amplifier, but its voltage gain is much smaller than that of the common-emitter amplifier operating at a similar Q-point. This is one of many differences we will explore in this and subsequent chapters.

> **EXERCISE:** (a) Does the MOSFET in Fig. 7.2 remain in the active region of operation during the full-output signal swing? (b) If the dc drain current of the MOSFET in the active region is given by $I_D = (K_n/2)(V_{GS} - V_{TN})^2$, what are the values of the parameter K_n and threshold voltage V_{TN} for the transistor in Fig. 7.2? (c) Express the amplifier voltage given in dB.
>
> **ANSWERS:** No, not near the positive peak of v_{GS}, corresponding to the peak negative excursion of v_{DS}; $K_n = 5 \times 10^4$ A/V^2, $V_{TN} = 1$ V; 12.3 dB

7.2 COUPLING AND BYPASS CAPACITORS

The constant base-emitter or gate-source voltage biasing techniques used in Figs. 7.1 and 7.2 are not very desirable methods of establishing the Q-point for a bipolar transistor or FET because the operating point is highly dependent on the transistor parameters. As discussed in detail in Chapters 4 and 5, the four-resistor bias network in Fig. 7.3, is much preferred for establishing a stable Q-point for the transistor.

However, to use the transistor as an amplifier, ac signals must be introduced into the circuit, but application of these ac signals must not disturb the dc Q-point that has been established by the bias network. One method of injecting an input signal and extracting an output signal without disturbing the Q-point is to use **ac coupling** through capacitors. The values of these capacitors are chosen to have negligible impedances in the frequency range of interest, but at the same time, the capacitors provide open circuits at dc so the Q-point is not disturbed. When power is first applied to the amplifier circuit, transient currents charge the capacitors, but the final steady-state operating point is not affected.

Figure 7.4 is an example of the use of capacitors; the transistor is biased by the same four-resistor network shown in Fig. 7.3. Input signal v_i is *coupled* onto the base node of the transistor through capacitor C_1, and the signal developed at the collector is coupled to load resistor R_3 through capacitor C_2. C_1 and C_2 are referred to as **coupling capacitors,** or **dc blocking capacitors.**

For now, the values of C_1 and C_2 are assumed to be very large, so their reactance $(1/\omega C)$ at the signal frequency ω will be negligible. This assumption is indicated in the figure by $C \to \infty$. Calculation of more exact values of the capacitors is left until the discussion of amplifier frequency response in Chapters 8 and 9.

Figure 7.4 also shows the use of a third capacitor C_3, called a **bypass capacitor.** In many circuits, we want to force signal currents to go around elements of the bias network. Capacitor C_3 provides a low-impedance path for ac current to "bypass" emitter resistor R_4. Thus R_4, which is required for good Q-point stability, can be effectively removed from the circuit when ac signals are considered.

Simulation results of the behavior of this circuit are shown in Fig. 7.5. A 5-mV sine wave signal at a frequency of 1 kHz is applied to the base terminal of transistor Q through coupling capacitor C_1;

7.2 Coupling and Bypass Capacitors   381

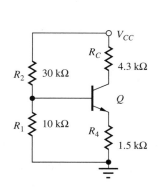

Figure 7.3 Transistor biased in the active region using the four-resistor bias network (see Sec. 4.11 for an example).

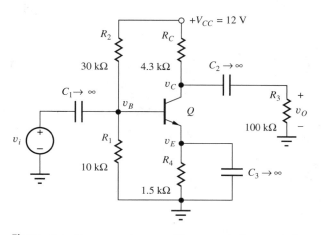

Figure 7.4 Common-emitter amplifier stage built around the four-resistor bias network. C_1 and C_2 function as coupling capacitors, and C_3 is a bypass capacitor.

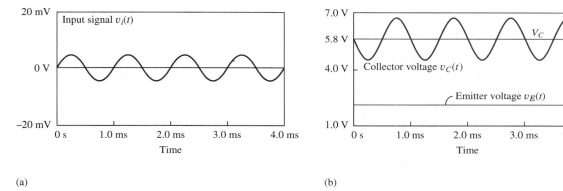

Figure 7.5 SPICE simulation results for v_S, v_C, and v_E for the amplifier in Fig. 7.4 with $v_I = 0.005 \sin 2000\pi t$ V.

this signal produces a sinusoidal signal at the collector node with an amplitude of approximately 1.1 V, centered on the Q-point value of $V_C \cong 5.8$ V. Note once again that there is a 180° phase shift between the input and output voltage signals. These values indicate that the amplifier is providing a voltage gain of

$$A_v = \frac{\mathbf{v_c}}{\mathbf{v_i}} = \frac{1.1\angle 180°}{.005\angle 0°} = 220\angle 180° = -220. \tag{7.6}$$

In Fig. 7.5, we should also observe that the voltage at the emitter node remains constant at its Q-point value of slightly more than 2 V. The very low impedance of the bypass capacitor prevents any signal voltage from being developed at the emitter. We say that the bypass capacitor maintains an "ac ground" at the emitter terminal. In other words, zero signal voltage appears at the emitter, so the emitter voltage remains constant at its dc Q-point value.

EXERCISE: Calculate the Q-point for the bipolar transistor in Fig. 7.3. Use $\beta_F = 100$, $V_{BE} = 0.7$ V, and $V_A = \infty$. What is the value of V_B? (See Sec. 4.11.1.)

ANSWERS: (1.45 mA, 3.57 V); 2.89 V

> **EXERCISE:** Write expressions for $v_C(t)$, $v_E(t)$, $i_c(t)$ and $v_B(t)$ based on the waveforms shown in Fig. 7.5.
>
> **ANSWERS:** $v_C(t) = (5.8 - 1.1 \sin 2000\pi t)$ V, $v_E(t) = 2.20$ V, $i_c(t) = -0.25 \sin 2000\pi t$ mA; $v_B(t) = (2.90 + 0.005 \sin 2000\pi t)$ V
>
> **EXERCISE:** Suppose capacitor C_3 is 500 μF. What is its reactance at a frequency of 1000 Hz?
>
> **ANSWER:** 0.318 Ω

7.3 CIRCUIT ANALYSIS USING dc AND ac EQUIVALENT CIRCUITS

To simplify the circuit analysis and design tasks, we break the amplifier into two parts, performing separate dc and ac circuit analyses. We find the Q-point of the circuit using the **dc equivalent circuit** — the circuit that is appropriate for steady-state dc analysis. To construct the dc equivalent circuit, we assume that capacitors are open circuits and inductors are short circuits. For example, Fig. 7.3 represents the dc equivalent circuit for the amplifier in Fig. 7.4.

Once we have found the Q-point, we determine the response of the circuit to the ac signals using an **ac equivalent circuit.** In constructing the ac equivalent circuit, we assume that the reactance of the coupling and bypass capacitors is negligible at the operating frequency ($|Z_C| = 1/\omega C = 0$), and we replace the capacitors by short circuits. Similarly, we assume the impedance of any inductors in the circuit is extremely large ($|Z_L| = \omega L \to \infty$), so we replace inductors by open circuits. Because the voltage at a node connected to a dc voltage source cannot change, these points represent grounds in the ac equivalent circuit (i.e., no ac voltage can appear at such a node: $v_{ac} = 0$). Furthermore, the current through a dc current source does not change even if the voltage across the source changes ($i_{ac} = 0$), so we replace dc current sources with open circuits in the ac equivalent circuit.

These ac equivalent circuits we are forming here are actually valid in the mid-band region of the amplifier as defined in Sec. 6.8.4, and the parameter values we calculate are those of the mid-band voltage gain, mid-band input resistance, mid-band output resistance and so on. In Chapter 8 and beyond, we explore how the upper and lower cutoff frequencies are related to the values of the capacitors and inductors in the circuit.

7.3.1 MENU FOR dc AND ac ANALYSIS

First we do a dc analysis to find the Q-point, and then we perform an ac analysis to determine the behavior of the circuit as an amplifier. The Q-point values must be found first because they ultimately determine the ac characteristics of the amplifier. To summarize, our analysis of amplifier circuits is performed using the two-part process listed here.

dc Analysis
1. Find the dc equivalent circuit by replacing all capacitors with open circuits and inductors by short circuits.
2. Find the Q-point from the dc equivalent circuit using the appropriate large-signal model for the transistor.

Midband ac Analysis
3. Find the midband ac equivalent circuit by replacing all capacitors by short circuits and all inductors by open circuits. dc voltage sources are replaced by short circuits, and dc current sources are replaced by open circuits in the ac equivalent circuit.
4. Replace the transistor by its small-signal model. (The small-signal model is Q-point dependent.)
5. Analyze the ac characteristics of the amplifier using the small-signal ac equivalent circuit from Step 4.

6. If desired, combine the results from Steps 2 and 5 to yield the total voltages and currents in the network.

Since we are most often interested in determining the ac behavior of the circuit, we seldom actually perform this final step of combining the dc and ac results.

DESIGN NOTE

dc equivalent circuits: capacitors are replaced with open circuits and inductors with short circuits.
ac equivalent circuits (midband): dc power supply nodes represent grounds and current sources represent open circuits; capacitors are replaced with short circuits and inductors with open circuits.

EXAMPLE 7.1 CONSTRUCTING ac AND dc EQUIVALENT CIRCUITS FOR A BJT AMPLIFIER

As has been stated several times, we usually split a circuit into its dc and ac equivalents in order to simplify the analysis or design problem. This is a critical step, since we cannot get the correct answer if the equivalent circuits are improperly constructed.

PROBLEM Draw the dc and ac equivalent circuits (menu steps 1 and 3) for the common-emitter amplifier in Fig. 7.6(a). The circuit topology is similar to that in Fig. 7.4 except resistor R_I, representing the Thévenin equivalent resistance of the signal source, has been added to the circuit. The resistor values have been changed to establish a new operating point.

SOLUTION **Known Information and Given Data:** The circuit with element values appears in Fig. 7.6(a).

Unknowns: dc equivalent circuit; ac equivalent circuit

Approach: Replace capacitors by open circuits to obtain the dc equivalent circuit. Replace capacitors and dc voltage sources by short circuits to obtain the ac equivalent circuit. Combine and simplify resistor combinations wherever possible.

Assumptions: Capacitor values are large enough that they can be treated as short circuits in the ac equivalent circuit.

ANALYSIS **dc Equivalent Circuit:** The dc equivalent circuit is found by open circuiting all the capacitors in the circuit. We find that the resulting dc equivalent circuit in Fig. 7.6(b) is identical to the four-resistor bias circuit of Fig. 7.3 (also see Sec. 4.11). Opening capacitors C_1 and C_2 disconnects v_I, R_I, and R_3 from the circuit.

ac Equivalent Circuit: To construct the ac equivalent circuit, we replace the capacitors by short circuits. Also, the dc voltage source becomes an ac ground in Fig. 7.6(c). In the ac equivalent circuit, source resistance R_I is now connected directly to the base node, and the external load resistor R_3 is connected directly to the collector node. Figure 7.6(d) is a redrawn version of Fig. 7.6(c). Although these two figures may look different, they are the same circuit! Note that resistor R_E is shorted out by the presence of bypass capacitor C_3 and has therefore been removed from Fig. 7.6(d). Because the power supply represents an "ac ground," bias resistors R_1 and R_2 appear in parallel between the base node and ground, and R_C and R_3 are in parallel at the collector. Do not overlook the fact that only the signal v_i is included in the ac equivalent circuit!

In Fig. 7.6(e), R_1 and R_2 have been combined into the resistor R_B, and R_C and R_3 have been combined into the resistor R_L:

$$R_B = R_1 \| R_2 = 160\,\text{k}\Omega \| 300\,\text{k}\Omega = 104\,\text{k}\Omega \quad \text{and} \quad R_L = R_C \| R_3 = 22\,\text{k}\Omega \| 100\,\text{k}\Omega = 18.0\,\text{k}\Omega$$

384 Chapter 7 The Transistor as an Amplifier

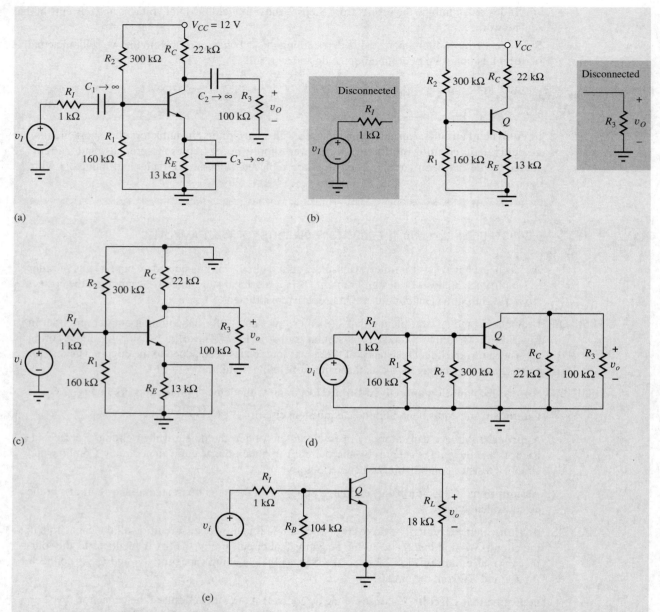

Figure 7.6 (a) Complete ac-coupled amplifier circuit; (b) simplified equivalent circuit for dc analysis; (c) circuit after step 3. Note that the input and output are now v_i and v_o; (d) redrawn version of Fig. 7.6(c); (e) continued simplification of the ac circuit.

Check of Results: In this case, the best way to verify our results is to double check our work—everything seems correct.

Discussion: Notice again how the capacitors have been used to modify the circuit topologies for dc and ac signals. C_1 and C_2 isolate the source and load from the bias circuit at dc. Capacitor C_3 causes the emitter node to be connected directly to ground in the ac circuit, effectively removing R_E from the circuit, and the dc power supply becomes an ac ground.

EXERCISE: What are the values of R_B and R_L in Fig. 7.6(e) if $R_1 = 20$ kΩ, $R_2 = 62$ kΩ, $R_C = 8.2$ kΩ, and $R_E = 2.7$ kΩ?

ANSWERS: 15.1 kΩ; 7.58 kΩ

EXERCISE: Redraw the dc and ac equivalent circuits in Fig. 7.6(b) and (e) if C_3 were eliminated from the circuit.

ANSWERS: (b) No change, (e) resistor R_E appears between the BJT emitter and ground.

EXAMPLE 7.2 CONSTRUCTING ac AND dc EQUIVALENTS FOR A MOSFET AMPLIFIER

This second example showing how to construct the dc and ac equivalent circuits of an overall amplifier circuit includes the use of a split-supply biasing technique, and an inductor has also been included in the circuit.

PROBLEM Draw the dc and ac equivalent circuits (menu steps 1 and 3) for the common-source amplifier in Fig. 7.7(a).

SOLUTION **Known Information and Given Data:** The circuit with labeled elements appears in Fig. 7.7(a).

Unknowns: dc equivalent circuit; ac equivalent circuit

Approach: Replace capacitors by open circuits and inductors by short circuits to obtain the dc equivalent circuit. To obtain the ac equivalent circuit, replace capacitors and dc voltage sources by short circuits, and current sources and inductors by open circuits. Combine and simplify resistor combinations wherever possible.

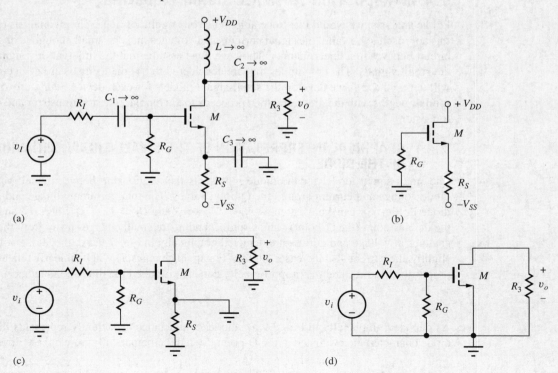

Figure 7.7 (a) An amplifier biased by two power supplies; (b) dc equivalent circuit for Q-point analysis; (c) first step in generating the ac equivalent circuit; (d) simplified ac equivalent circuit.

> **ANALYSIS**
>
> **dc Equivalent Circuit:** The dc equivalent circuit is found by replacing the capacitors by open circuits and the inductor by a short circuit, resulting in the circuit in Fig. 7.7(b). Capacitors C_1 and C_2 again disconnect v_I, R_I, and R_3 from the circuit, and the shorted inductor connects the drain of the transistor directly to V_{DD}.
>
> **ac Equivalent Circuit:** The ac equivalent circuit in Fig. 7.7(c) is obtained by replacing the capacitors by short circuits and the inductor by an open circuit. Figure 7.7(c) has been redrawn in final simplified form in Fig. 7.7(d). Only signal component v_i appears in the ac equivalent circuit.
>
> **Check of Results:** For this case, the best way to verify our results is to double check our work—all seems correct.
>
> **Discussion:** Here again, the designer has used the capacitors and inductor to achieve very different circuit topologies for the dc and ac equivalent circuits. Compare Figs. 7.7(b) and 7.7(d).

> **EXERCISE:** Redraw the dc and ac equivalent circuits in Fig. 7.7(b) and (d) if C_3 were eliminated from the circuit.
>
> **ANSWERS:** (b) No change, (d) resistor R_S appears between the MOSFET source and ground

7.4 INTRODUCTION TO SMALL-SIGNAL MODELING

For ac analysis, we would like to be able to use our wealth of linear circuit analysis techniques. For this approach to be valid, the signal currents and voltages must be small enough to ensure that the ac circuit behaves in a linear manner. Thus, we must assume that the time-varying signal components are **small signals.** The amplitudes that are considered to be small signals are device-dependent; we will define these as we develop the small-signal models for each device. Our study of **small-signal models** begins with the diode and then proceeds to the bipolar junction transistor and the field-effect transistor.

7.4.1 GRAPHICAL INTERPRETATION OF THE SMALL-SIGNAL BEHAVIOR OF THE DIODE

The small-signal model for the diode represents the relationship between small variations in the diode voltage and current around the Q-point values. The total terminal voltage and current for the diode in Fig. 7.8 can be written as $v_D = V_D + v_d$ and $i_D = I_D + i_d$ where I_D and V_D represent the dc bias point (the Q-point) values, and v_d and i_d are small changes away from the Q-point. The changes in voltage and current are depicted graphically in Fig. 7.9. As the diode voltage increases slightly, the current also increases slightly. For small changes, i_d will be linearly related (i.e., directly proportional) to v_d, and the proportionality constant is called the diode conductance g_d:

$$i_d = g_d v_d \tag{7.7}$$

As depicted graphically in Fig. 7.9(a), diode conductance g_d actually represents the slope of the diode characteristic evaluated at the Q-point. Stated mathematically, g_d can be written as

$$g_d = \left.\frac{\partial i_D}{\partial v_D}\right|_{\text{Q-point}} = \left.\frac{\partial}{\partial v_D}\left\{I_S\left[\exp\left(\frac{v_D}{V_T}\right)-1\right]\right\}\right|_{\text{Q-point}} = \frac{I_S}{V_T}\exp\left(\frac{V_D}{V_T}\right) = \frac{I_D + I_S}{V_T} \tag{7.8}$$

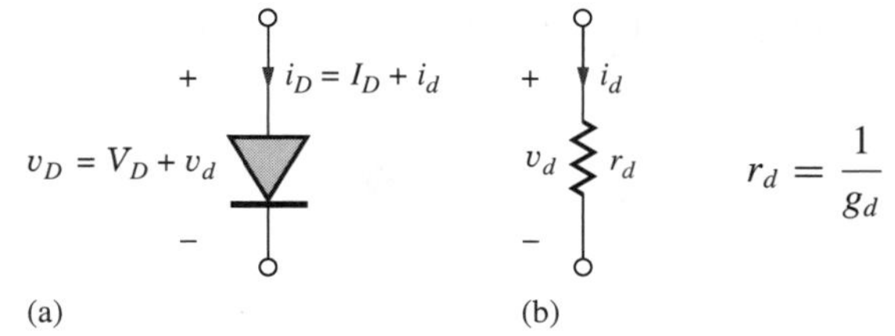

7.4 Introduction to Small-Signal Modeling

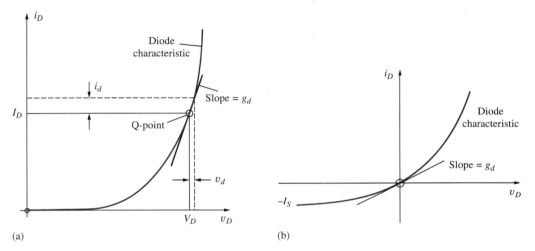

Figure 7.8 (a) Total diode terminal voltage and current; (b) small signal model for the divide.

Figure 7.9 (a) The relationship between small increases in voltage and current above the diode operating point (I_D, V_D). For small changes $i_d = g_d v_d$; (b) the diode conductance is not zero for $I_D = 0$.

where we have used our mathematical model for i_D. For forward bias with $I_D \gg I_S$, the diode conductance becomes

$$g_d \cong \frac{I_D}{V_T} \quad \text{or} \quad g_d \cong \frac{I_D}{0.025 \text{ V}} = 40 I_D / \text{V}^{\,1} \tag{7.9}$$

at room temperature. Note that g_d is small but not zero for $I_D = 0$ because the slope of the diode equation is nonzero at the origin, as depicted in Fig. 7.9(b).

7.4.2 SMALL-SIGNAL MODELING OF THE DIODE

Now we will use the diode equation to more fully explore the small-signal behavior of the diode and to actually define how large v_d and i_d can become before Eq. (7.7) breaks down. A relationship between the ac and dc quantities can be developed directly from the diode equation introduced in Chapter 3:

$$i_D = I_S \left[\exp\left(\frac{v_D}{V_T}\right) - 1 \right] \tag{7.10}$$

Substituting $v_D = V_D + v_d$ and $i_D = I_D + i_d$ into Eq. (7.10) yields

$$I_D + i_d = I_S \left[\exp\left(\frac{V_D + v_d}{V_T}\right) - 1 \right] = I_S \left[\exp\left(\frac{V_D}{V_T}\right) \exp\left(\frac{v_d}{V_T}\right) - 1 \right] \tag{7.11}$$

---

[1] Remember the more exact value is $V_T = 25.9$ mV and therefore $g_d = 38.6 I_D/$V.

Expanding the second exponential using Maclaurin's series and collecting the dc and signal terms together,

$$I_D + i_d = I_S \left[\exp\left(\frac{V_D}{V_T}\right) - 1\right] + I_S \exp\left(\frac{V_D}{V_T}\right) \left[\frac{v_d}{V_T} + \frac{1}{2}\left(\frac{v_d}{V_T}\right)^2 + \frac{1}{6}\left(\frac{v_d}{V_T}\right)^3 + \cdots\right] \quad (7.12)$$

We recognize the first term on the right-hand side of Eq. (7.12) as the dc diode current I_D:

$$I_D = I_S \left[\exp\left(\frac{V_D}{V_T}\right) - 1\right] \quad \text{and} \quad I_S \exp\left(\frac{V_D}{V_T}\right) = I_D + I_S \quad (7.13)$$

Subtracting I_D from both sides of the equation yields an expression for i_d in terms of v_d:

$$i_d = (I_D + I_S)\left[\frac{v_d}{V_T} + \frac{1}{2}\left(\frac{v_d}{V_T}\right)^2 + \frac{1}{6}\left(\frac{v_d}{V_T}\right)^3 + \cdots\right] \quad (7.14)$$

We want the signal current i_d to be a linear function of the signal voltage v_d. Using only the first two terms of Eq. (7.14), we find that linearity requires

$$\frac{v_d}{V_T} \gg \frac{1}{2}\left(\frac{v_d}{V_T}\right)^2 \quad \text{or} \quad \boxed{v_d \ll 2V_T = 0.05 \text{ V}} \quad (7.15)$$

If the relationship in Eq. (7.15) is met, then Eq. (7.14) can be written as

$$i_d = \frac{(I_D + I_S)}{V_T} v_d \quad \text{or} \quad i_d = g_d v_d \quad \text{and} \quad i_D = I_D + g_d v_d \quad (7.16)$$

in which g_d is the **small-signal conductance** of the diode originally given in Eq. (7.8). Equation (7.16) states that the total diode current is the dc current I_D (at the Q-point) plus a small change in current ($i_d = g_d v_d$) that is linearly related to the small voltage change v_d across the diode.

The values of the **diode conductance** g_d, or the equivalent **diode resistance** r_d, are determined by the operating point of the diode as defined in Eq. (7.9):

$$\boxed{g_d = \frac{I_D + I_S}{V_T} \cong \frac{I_D}{V_T} = 40 I_D \quad \text{and} \quad r_d = \frac{1}{g_d}} \quad (7.17)$$

The diode and its corresponding small-signal model, represented by resistor r_d, are given in Fig. 7.8.

Equation (7.15) defines the requirement for small-signal operation of the diode. The shift in diode voltage away from the Q-point value must be much less than 50 mV. Choosing a factor of 10 as adequate to satisfy the inequality yields $v_d \leq 0.005$ V for small-signal operation. This is indeed a small voltage change.

Note, however, that the maximum small-signal change in diode voltage represents a significant change in diode current:

$$i_d = g_d v_d = 0.005 \text{ V} \frac{I_D}{0.0025 \text{ V}} = 0.2 I_D \quad (7.18)$$

The 5-mV change in diode voltage corresponds to a 20 percent change in the diode current! This large change results from the exponential relationship between voltage and current in the diode.

DESIGN NOTE

Small changes in diode current and voltage are related by the small-signal conductance of the diode

$$i_d = g_d v_d \quad \text{where} \quad g_d \cong \frac{I_D}{V_T} \cong 40 I_D$$

at room temperature. For small-signal operation,

$$|v_d| \leq 0.005 \text{ V} \quad \text{and} \quad |i_d| \leq 0.20 I_D$$

EXERCISE: Calculate the values of the diode resistance r_d for a diode with $I_S = 1$ fA operating at $I_D = 0$, 50 μA, 2 mA, and 3 A.

ANSWERS: 25.0 TΩ, 500 Ω, 12.5 Ω, 8.33 mΩ

EXERCISE: What is the small-signal diode resistance r_d at room temperature for $I_D = 1.5$ mA? What is the small-signal resistance of this diode at $T = 100°C$?

ANSWERS: 16.7 Ω, 21.4 Ω

7.5 SMALL-SIGNAL MODELS FOR BIPOLAR JUNCTION TRANSISTORS

Now that the concept of small-signal modeling has been introduced, we shall develop the small-signal model for the more complicated bipolar transistor. The BJT is a three-terminal device, and its small-signal model is based on the two-port network representation[1] shown in Fig. 13.10 for which the input port variables are v_{be} and i_b, and the output port variables are v_{ce} and i_c. A set of two-port equations in terms of these variables can be written as

$$\begin{aligned} \mathbf{i_b} &= g_\pi \mathbf{v_{be}} + g_\mu \mathbf{v_{ce}} \\ \mathbf{i_c} &= g_m \mathbf{v_{be}} + g_o \mathbf{v_{ce}} \end{aligned} \quad (7.19)$$

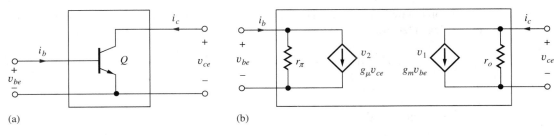

Figure 7.10 (a) Two-port representation of the *npn* transistor; (b) two-port representation for the transistor in Fig. 7.10(a).

---

[2] These equations actually represent a y-parameter two-port network. Remember our convention from Section 1.3, $R_x = 1/G_x$, $r_o = 1/g_o$, etc.

390 **Chapter 7** The Transistor as an Amplifier

The port variables in Fig. 7.10 can be considered to represent either the time-varying portion of the total voltages and currents or small changes in the total quantities away from the Q-point values.

or
$$v_{BE} = V_{BE} + v_{be} \qquad v_{CE} = V_{CE} + v_{ce}$$
$$i_B = I_B + i_b \qquad i_C = I_C + i_c \qquad (7.20)$$
$$v_{be} = \Delta v_{BE} = v_{BE} - V_{BE} \qquad v_{ce} = \Delta v_{CE} = v_{CE} - V_{CE}$$
$$i_b = \Delta i_B = i_B - I_B \qquad i_c = \Delta i_C = i_C - I_C$$

We can write the y-parameters in terms of small-signal voltages and currents or in terms of derivatives of the complete port variables, as in Eq. (7.21):

$$g_\pi = \left.\frac{i_b}{v_{be}}\right|_{v_{ce}=0} = \left.\frac{\partial i_B}{\partial v_{BE}}\right|_{\text{Q-point}} \qquad g_\mu = \left.\frac{i_b}{v_{ce}}\right|_{v_{be}=0} = \left.\frac{\partial i_B}{\partial v_{CE}}\right|_{\text{Q-point}}$$
$$\qquad\qquad\qquad\qquad\qquad\qquad\qquad\qquad\qquad\qquad\qquad (7.21)$$
$$g_m = \left.\frac{i_c}{v_{be}}\right|_{v_{ce}=0} = \left.\frac{\partial i_C}{\partial v_{BE}}\right|_{\text{Q-point}} \qquad g_o = \left.\frac{i_c}{v_{ce}}\right|_{v_{be}=0} = \left.\frac{\partial i_C}{\partial v_{CE}}\right|_{\text{Q-point}}$$

Because we have the transport model, Eq. (4.45), which expresses the BJT terminal currents in terms of the terminal voltages, as repeated in Eq. (7.22) for the active region, we use the derivative formulation to determine the g-parameters for the transistor:

$$i_C = I_S\left[\exp\left(\frac{v_{BE}}{V_T}\right)\right]\left[1 + \frac{v_{CE}}{V_A}\right] \qquad i_B = \frac{i_C}{\beta_F} = \frac{I_S}{\beta_{FO}}\left[\exp\left(\frac{v_{BE}}{V_T}\right)\right] \qquad (7.22)$$

$$\beta_F = \beta_{FO}\left[1 + \frac{v_{CE}}{V_A}\right]$$

Evaluating the various derivatives[3] of Eq. (7.22) yields the parameters for the BJT:

$$g_\mu = \left.\frac{\partial i_B}{\partial v_{CE}}\right|_{\text{Q-point}} = 0$$

$$g_m = \left.\frac{\partial i_C}{\partial v_{BE}}\right|_{\text{Q-point}} = \frac{I_S}{V_T}\left[\exp\left(\frac{v_{BE}}{V_T}\right)\right]\left[1 + \frac{v_{CE}}{V_A}\right]_{\text{Q-point}} = \frac{I_C}{V_T} \qquad (7.23)$$

$$g_o = \left.\frac{\partial i_C}{\partial v_{CE}}\right|_{\text{Q-point}} = \frac{I_S}{V_A}\left[\exp\left(\frac{V_{BE}}{V_T}\right)\right] = \frac{I_C}{V_A + V_{CE}}$$

Calculation of g_π has been saved until last because it requires a bit more effort and the use of some new information. The current gain of a BJT is actually operating point-dependent, and this dependence should be included when evaluating the derivative needed for g_π:

$$g_\pi = \left.\frac{\partial i_B}{\partial v_{BE}}\right|_{\text{Q-point}} = \left[\frac{1}{\beta_F}\frac{\partial i_C}{\partial v_{BE}} - \frac{i_C}{\beta_F^2}\frac{\partial \beta_F}{\partial v_{BE}}\right]_{\text{Q-point}} = \left[\frac{1}{\beta_F}\frac{\partial i_C}{\partial v_{BE}} - \frac{i_C}{\beta_F^2}\frac{\partial \beta_F}{\partial i_C}\frac{\partial i_C}{\partial v_{BE}}\right]_{\text{Q-point}}$$
$$\qquad\qquad\qquad\qquad\qquad\qquad\qquad\qquad\qquad\qquad\qquad (7.24)$$

Factoring out the first term:

$$g_\pi = \frac{1}{\beta_F}\frac{\partial i_C}{\partial v_{BE}}\left[1 - \frac{i_C}{\beta_F}\frac{\partial \beta_F}{\partial i_C}\right]_{\text{Q-point}} = \frac{I_C}{\beta_F V_T}\left[1 - \left(\frac{i_C}{\beta_F}\frac{\partial \beta_F}{\partial i_C}\right)_{\text{Q-point}}\right] \qquad (7.25)$$

---

[3] We could equally well have taken the direct approach used in analysis of the diode.

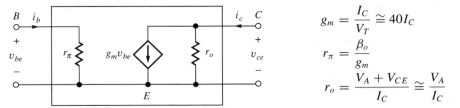

Figure 7.11 Hybrid-pi small-signal model for the intrinsic bipolar transistor.

Finally, Eq. (7.25) can be simplified by defining a new parameter β_o:

$$g_\pi = \frac{I_C}{\beta_o V_T} \quad \text{where} \quad \beta_o = \frac{\beta_F}{\left[1 - I_C \left(\frac{1}{\beta_F} \frac{\partial \beta_F}{\partial i_C}\right)_{Q\text{-point}}\right]} \quad (7.26)$$

β_o represents the **small-signal common-emitter current gain** of the bipolar transistor. Equation (7.26) will be discussed in more detail in Sec. 7.5.3.

Note that g_μ is zero and does not appear in the model, and this is a result of our transistor model which assumes that i_B is independent of v_{CE}. This assumption is good for most modern bipolar transistors. However, this assumption is not valid for every BJT, and a resistor r_μ can be added between the collector and base terminals of the hybrid-pi circuit to model the small changes in base current caused by changes in collector-emitter voltage for transistors in which the assumption breaks down. Resistor r_μ is typically much larger than r_o or r_π with a lower bound on r_μ given by $r_\mu \geq \beta_o r_o$. Throughout the rest of this text, we will assume r_μ is zero. One can always add it to the model if desired.

> **EXERCISE:** Redraw the hybrid-pi model in Fig. 7.11 to include resistor r_μ mentioned above.
>
> **ANSWERS:** Connect resistor r_μ between the top of resistor r_π and the top of current source $g_m v_{be}$.

7.5.1 THE HYBRID-PI MODEL

The **hybrid-pi model** is the most widely accepted small-signal model for the bipolar transistor. If one looks at the latest results in analog circuits as published in the *IEEE Journal of Solid-State Circuits*, for example, the analysis will most likely be cast in terms of the hybrid-pi model.[4]

The standard representation of the basic hybrid-pi small-signal model appears in Fig. 7.11, and the expressions for the model elements are given in Eq. (7.27). These results will be used throughout the rest of the text and should be committed to memory.

$$\begin{aligned}
\text{Transconductance:} \quad & g_m = \frac{I_C}{V_T} \cong 40 I_C \\
\text{Input resistance:} \quad & r_\pi = \frac{\beta_o V_T}{I_C} = \frac{\beta_o}{g_m} \\
\text{Output resistance:} \quad & r_o = \frac{1}{g_o} = \frac{V_A + V_{CE}}{I_C} \cong \frac{V_A}{I_C}
\end{aligned} \quad (7.27)$$

Arguably the most important small-signal parameter is transconductance g_m. The transconductance characterizes how the collector current changes in response to a change in the base-emitter

---

[4] An alternative model, called the T-model, appears in Prob. 7.64.

voltage, thereby modeling the forward gain of the device. Here again we see the fundamental voltage-controlled current behavior of the bipolar transistor, $i_c = g_m v_{be}$. At room temperature, $V_T \cong$ 0.025 mV, and transconductance $g_m \cong 40 I_C$. Also, collector-emitter voltage V_{CE} is typically much less than Early voltage V_A, so we can simplify the expression for the output resistance: $r_o \cong V_A / I_C$.

When we change the base-emitter voltage and hence the collector current, we must also supply a change in base current, and resistor r_π characterizes the relationship between changes in i_b and v_{be}. Similarly, when the collector-emitter voltage changes slightly, the collector current changes, and resistor r_o characterizes the relationship between changes in i_c and v_{ce}.

The two-port representation in Fig. 7.11 using these symbols shows the intrinsic low-frequency **hybrid-pi small-signal model** for the bipolar transistor. Additional elements will be added to model frequency dependencies in Chapter 17.

From Eq. (7.27) and Fig 7.11, we see that the values of the small-signal parameters are controlled explicitly by our choice of Q-point. Transconductance g_m is directly proportional to the collector current of the bipolar transistor, whereas input resistance r_π and output resistance r_o are both inversely proportional to the collector current. The output resistance exhibits a weak dependence on collector-emitter voltage (but generally $V_{CE} \ll V_A$). Note that these parameters are independent of the geometry of the BJT. For example, small high-frequency transistors or large-geometry power devices all have the same value of g_m for a given collector current.

7.5.2 GRAPHICAL INTERPRETATION OF THE TRANSCONDUCTANCE

Figure 7.12 depicts the diode-like exponential relation between total collector current i_C and total base-emitter voltage v_{BE} in the bipolar transistor. Transconductance g_m represents the slope of the $i_C - v_{BE}$ characteristic at the given operating point (Q-point). For a small increase v_{be} above the Q-point voltage V_{BE}, a small corresponding increase i_c occurs above the Q-point current I_C. When the small-signal condition $v_{be} \leq 5$ mV is met, these two changes are linearly related by the transconductance: $i_c = g_m v_{be}$ (see sec. 7.5.7).

7.5.3 SMALL-SIGNAL CURRENT GAIN

Two important auxiliary relationships also exist between the small-signal parameters. It can be seen in Eq. (7.27) that the parameters g_m and r_π are related by the small-signal current gain β_o:

$$\beta_o = g_m r_\pi \qquad (7.28)$$

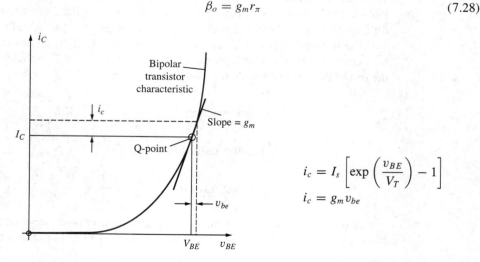

Figure 7.12 The relationship between small increases in base-emitter voltage and collector current above the BJT operating point (I_C, V_{CE}). For small changes $i_c = g_m v_{be}$.

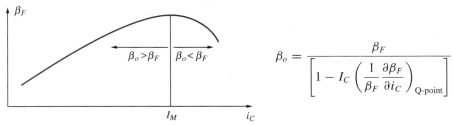

Figure 7.13 dc current gain versus current for the BJT.

As mentioned before, the dc current gain in a real transistor is not constant but is a function of operating current, as indicated in Fig. 7.13. From this figure, we see that

$$\frac{\partial \beta_F}{\partial i_C} > 0 \text{ for } i_C < I_M \quad \text{and} \quad \frac{\partial \beta_F}{\partial i_C} < 0 \text{ for } i_C > I_M$$

where I_M is the collector current at which β_F is maximum. Thus, for the **small-signal current gain** defined by

$$\beta_o = \frac{\beta_F}{\left[1 - I_C \left(\frac{1}{\beta_F} \frac{\partial \beta_F}{\partial i_C}\right)_{\text{Q-point}}\right]} \quad (7.29)$$

$\beta_o > \beta_F$ for $i_C < I_M$, and $\beta_o < \beta_F$ for $i_C > I_M$. That is, the ac current gain β_o exceeds the dc current gain β_F when the collector current is below I_M and is smaller than β_F when I_C exceeds I_M. In practice, the difference between β_F and β_o is usually ignored, and β_F and β_o are commonly assumed to be the same.

7.5.4 THE INTRINSIC VOLTAGE GAIN OF THE BJT

The second important auxiliary relationship is given by the **intrinsic voltage gain** μ_f[5], which is equal to the product of g_m and r_o:

$$\mu_f = g_m r_o = \frac{V_A + V_{CE}}{V_T} \cong \frac{V_A}{V_T} \quad \text{for} \quad V_{CS} \ll V_A \quad (7.30)$$

From Eq. (7.30), μ_f can be seen to be almost independent of operating point for $V_{CE} \ll V_A$. At room temperature $\mu_f \cong 40 V_A$.

We shall find that the intrinsic gain μ_f plays an important role in circuit design, and it appears often in the analysis of amplifier circuits. Parameter μ_f represents the maximum voltage gain that the individual transistor can provide and is also referred to as the **amplification factor** of the device. For V_A ranging from 25 V to 100 V, μ_f ranges from 1000 to 4000. Thus, if we are clever enough, we should be able to build a single transistor amplifier with a voltage gain of several thousand. In later chapters, we will explore how this can be achieved.

DESIGN NOTE

Remember, the voltage gain of a single transistor amplifier cannot exceed the transistor's intrinsic voltage gain μ_f, which ranges from 1000 to 4000 for the bipolar transistor.

$$\mu_f = \frac{V_A + V_{CE}}{V_T} = \frac{V_A}{V_T} \qquad \mu_f = 40 V_A \text{ at room temperature}$$

---

[5] or just intrinsic gain

TABLE 7.1
BJT Small-Signal Parameters versus Current: $\beta_o = 100$, $V_A = 75$ V, $V_{CE} = 10$ V

I_C	g_M	r_π	r_o	μ_f
1 μA	4×10^{-5} S	2.5 MΩ	85 MΩ	3400
10 μA	4×10^{-4} S	250 kΩ	8.5 MΩ	3400
100 μA	0.004 S	25.0 kΩ	850 kΩ	3400
1 mA	0.04 S	2.5 kΩ	85 kΩ	3400
10 mA	0.40 S	250 Ω	8.5 kΩ	3400

Table 7.1 displays examples of the variation of the small-signal parameters with operating point. The values of g_m, r_π, and r_o can each be varied over many orders of magnitude by changing the value of the dc collector current corresponding to the Q-point. Note that μ_f does not change with the choice of operating point. As we see later in this chapter, this is a very significant difference between the BJT and FET.

It is important to realize that although we developed the small-signal model of the BJT based on analysis of the transistor oriented in the common-emitter configuration in Fig. 7.10, the resulting hybrid-pi model can actually be used in the analysis of any circuit topology. This point will become clearer in Chapter 8.

EXERCISE: Calculate the values of g_m, r_π, r_o, and μ_f for a bipolar transistor with $\beta_o = 75$ and $V_A = 60$ V operating at a Q-point of (50 μA, 5 V).

ANSWERS: 2.00 mS, 37.5 kΩ, 1.30 MΩ, 2600

EXERCISE: Calculate the values of g_m, r_π, r_o, and μ_f for a bipolar transistor with $\beta_o = 50$ and $V_A = 75$ V operating at a Q-point of (250 μA, 15 V).

ANSWERS: 10.0 mS, 5.00 kΩ, 360 kΩ, 3600

EXERCISE: Use graphical analysis to find values of β_{FO}, g_m, β_o, and r_o at the Q-point for the transistor in Fig. 13.1(b). Calculate the value of r_π.

ANSWERS: 100, 62.5 mS, 100, ∞; 1.60 kΩ

EXERCISE: Suppose that we wanted to add r_μ to the hybrid-pi model. What values of r_μ would correspond to the Q-points in Table 13.1 if $r_\mu = \beta_o r_o$.

ANSWERS: 8.5 GΩ, 850 MΩ, 85 MΩ, 8.5 MΩ, 850 kΩ,

7.5.5 EQUIVALENT FORMS OF THE SMALL-SIGNAL MODEL

The small-signal model in Fig. 7.14 includes the voltage-controlled current source $g_m v_{be}$. It is often useful in circuit analysis to transform this model into a current-controlled source. Recognizing that the voltage v_{be} in Fig. 7.13 can be written in terms of the current i_b as $v_{be} = i_b r_\pi$, the voltage-controlled source can be rewritten as

$$g_m v_{be} = g_m r_\pi i_b = \beta_o i_b \quad \text{where} \quad \beta_o = g_m r_\pi \quad (7.31)$$

Figure 7.14 shows the two equivalent forms of the small-signal BJT model. The model in Fig. 7.14(a) recognizes the fundamental voltage-controlled current source nature of the transistor

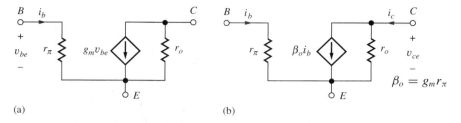

(a)　　　　　　　　　　　　　　　(b)

Figure 7.14 Two equivalent forms of the BJT small-signal model: (a) voltage-controlled current source model, and (b) current-controlled current source model.

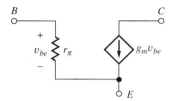

Figure 7.15 Simplified hybrid pi model in which r_o is neglected.

that is explicit in the transport model. From the second model, Fig. 7.14(b), we see that

$$i_c = \beta_o i_b + \frac{v_{ce}}{r_o} \cong \beta_o i_b \tag{7.32}$$

which demonstrates the auxiliary relationship that $i_c \cong \beta_o i_b$ in the active region of operation. For the most typical case, $v_{ce}/r_o \ll \beta_o i_b$. Thus, the basic relationship $i_C = \beta i_B$ is useful for both ac and dc analysis when the BJT is operating in the active region. We will find that sometimes circuit analysis is more easily performed using the model in Fig. 7.14(a), and at other times more easily performed using the model in Fig. 7.14(b).

7.5.6 SIMPLIFIED HYBRID-PI MODEL

As we investigate circuit behavior in more detail, we will find that output resistance r_o often has a relatively minor effect on circuit performance, especially on voltage gain, and we can often greatly simplify our circuit analysis if we neglect the output resistance in our model as shown in Fig. 7.15. Generally, we can make this simplification if the voltage gain of the circuit is much less than the intrinsic voltage gain μ_f. So our approach will be to neglect r_o, then calculate the voltage gain, and see if the result is consistent with the assumption that the voltage gain is much less than μ_f. However, r_o can have a much greater impact on amplifier output resistance calculations, and we must often add it back into the model in order to get nontrivial results for resistance calculations. We will see examples of this as we proceed through Chapters 7 and 8.

DESIGN NOTE

Output resistance r_o can be neglected in calculations of voltage gain A_v as long as $A_v \ll \mu_f$.

7.5.7 DEFINITION OF A SMALL SIGNAL FOR THE BIPOLAR TRANSISTOR

For small-signal operation, we want the relationship between changes in voltages and currents to be linear. We can find the constraints on the BJT corresponding to small-signal operation using the simplified transport model for the total collector current of the transistor in the active region:

$$i_C = I_S \left[\exp\left(\frac{v_{BE}}{V_T}\right)\right] = I_S \left[\exp\left(\frac{V_{BE} + v_{be}}{V_T}\right)\right] \qquad (7.33)$$

Rewriting the exponential as a product,

$$i_C = I_C + i_c = \left[I_S \exp\left(\frac{V_{BE}}{V_T}\right)\right]\left[\exp\left(\frac{v_{be}}{V_T}\right)\right] = I_C \left[\exp\left(\frac{v_{be}}{V_T}\right)\right] \qquad (7.34)$$

in which it has been recognized that the collector current I_C is given by

$$I_C = I_S \exp\left(\frac{V_{BE}}{V_T}\right) \qquad (7.35)$$

Now, expanding the remaining exponential in Eq. (7.34), its Maclaurin's series yields

$$i_C = I_C \left[1 + \frac{v_{be}}{V_T} + \frac{1}{2}\left(\frac{v_{be}}{V_T}\right)^2 + \frac{1}{6}\left(\frac{v_{be}}{V_T}\right)^3 + \cdots\right] \qquad (7.36)$$

Recognizing $i_c = i_C - I_C$ gives

$$i_c = I_C \left[\frac{v_{be}}{V_T} + \frac{1}{2}\left(\frac{v_{be}}{V_T}\right)^2 + \frac{1}{6}\left(\frac{v_{be}}{V_T}\right)^3 + \cdots\right] \qquad (7.37)$$

Linearity requires that i_c be proportional to v_{be}, so we must have

$$\frac{1}{2}\left(\frac{v_{be}}{V_T}\right)^2 \ll \frac{v_{be}}{V_T} \quad \text{or} \quad v_{be} \ll 2V_T \cong 0.05 \text{ V} \qquad (7.38)$$

where higher order terms have been neglected.

From Eq. (7.38), we see that small-signal operation requires the signal applied to the base-emitter junction to be much less than twice the thermal voltage, 50 mV at room temperature. In this book, we assume that a factor of 10 satisfies the condition in Eq. (7.38), and

$$|v_{be}| \leq 0.005 \text{ V} \qquad (7.39)$$

is our definition of a **small signal for the BJT**. If the condition in Eq. (7.39) is met, then Eq. (7.36) can be approximated as

$$i_C \cong I_C \left[1 + \frac{v_{be}}{V_T}\right] = I_C + \frac{I_C}{V_T} v_{be} = I_C + g_m v_{be} \qquad (7.40)$$

and the change in i_C is directly proportional to the change in v_{BE} (i.e., $i_c = g_m v_{be}$). The constant of proportionality is the transconductance g_m. Note that the quadratic, cubic, and higher-order powers of v_{be} in Eq. (7.37) are sources of the harmonic distortion that was discussed in Sec. 6.5.

From Eq. (7.39), the signal developed across the base-emitter junction must be no larger than 5 mV to qualify as a small signal! This is indeed small. But note well: We must not infer that signals at other points in the circuit need be small. Referring back to Fig. 7.1, we can see that a 16-mV p-p signal v_{be} generates a 3.3-V p-p signal at the collector. This is fortunate because we often want linear amplifiers to develop signals that are many volts in amplitude.

Let us now explore the change in collector current i_c that corresponds to small-signal operation. Using Eq. (7.40),

$$\frac{i_c}{I_C} = \frac{g_m v_{be}}{I_C} = \frac{v_{be}}{V_T} \leq \frac{0.005}{0.025} = 0.2 \qquad (7.41)$$

A 5-mV change in v_{BE} corresponds to a 20 percent deviation in i_C from its Q-point value as well as a 20 percent change in i_E since $\alpha_F \cong 1$. Some authors prefer to permit $|v_{be}| \leq 10$ mV, which corresponds to a 40 percent change in i_C from the Q-point value. In either case, relatively large changes in voltage can occur at the collector and/or emitter terminals of the transistor when the signal currents i_c and i_e flow through resistors external to the transistor.

The strict small-signal guidelines introduced above are frequently violated in practice. The designer must accept the trade-off between a larger signal amplitude and a higher level of distortion. As we move beyond the small-signal limit, our small-signal analysis becomes more approximate. However, our hand analysis still represents a useful estimate of circuit performance that we often refine with detailed transient simulation.

DESIGN NOTE

The small-signal limit for the bipolar transistor is set by

$$|v_{be}| \leq 0.005 \text{ V} \quad \text{and} \quad |i_c| \leq 0.2 I_C$$

EXERCISE: Does the amplitude of the signal in Fig. 7.1(a) and (b) satisfy the requirements for small-signal operation?

ANSWER: No, $|v_{be}| = 8$ mV exceeds our definition of a small signal.

7.5.8 SMALL-SIGNAL MODEL FOR THE *pnp* TRANSISTOR

The small-signal model for the *pnp* transistor is identical to that of the *npn* transistor. At first glance, this fact is surprising to most people because the dc currents flow in opposite directions. The circuits in Fig. 7.16 will be used to help explain this situation.

In Fig. 7.16, the *npn* and *pnp* transistors are each biased by dc current source I_B, establishing the Q-point current $I_C = \beta_F I_B$. In each case a signal current i_b is also injected *into* the base. For the *npn* transistor, the total base and collector currents are (for $\beta_o = \beta_F$):

$$i_B = I_B + i_b \quad \text{and} \quad i_C = I_C + i_c = \beta_F I_B + \beta_F i_b \quad (7.42)$$

An increase in base current of the *npn* transistor causes an increase in current entering the collector terminal.

For the *pnp* transistor,

$$i_B = I_B - i_b \quad \text{and} \quad i_C = I_C - i_c = \beta_F I_B - \beta_F i_b \quad (7.43)$$

The signal current injected into the base of the *pnp* transistor causes a decrease in the total collector current, which is again equivalent to an increase in the signal current entering the collector. Thus, for both the *npn* and *pnp* transistors, a signal current injected into the base causes a signal current to enter the collector, and the polarities of the current-controlled source in the small-signal model are identical, as in Fig. 7.17.

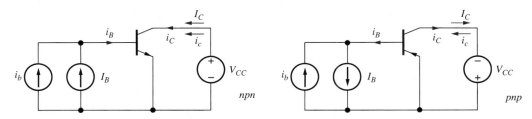

Figure 7.16 dc bias and signal currents for *npn* and *pnp* transistors.

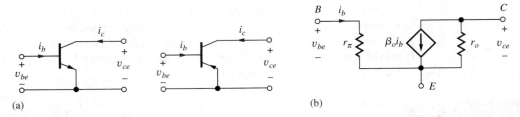

Figure 7.17 (a) Two-port notations for *npn* and *pnp* transistors; (b) the small-signal models are identical.

7.5.9 AC ANALYSIS VERSUS TRANSIENT ANALYSIS IN SPICE

Differences between the ac and transient analysis modes in SPICE are a constant source of confusion to new users of electronic simulation tools. ac analysis mirrors our hand calculations with small-signal models. In SPICE ac analysis, the transistors are automatically replaced with their small-signal models, and a linear circuit analysis is then performed. On the other hand, SPICE transient analysis provides a time-domain representation similar to what we see when we build the circuit and look at waveforms with an oscilloscope. The built-in models in SPICE attempt to fully account for the nonlinear behavior of the devices. If the small-signal limits are violated, distorted waveforms will result.

Once the circuit is converted to a linearized version, the magnitudes of the sources applied have no small-signal constraints. We typically use a value of 1 V or 1 A for convenience in ac analysis. On the other hand, this large a signal would cause significant distortion in many transient simulations.

7.6 THE COMMON-EMITTER (C-E) AMPLIFIER

Now we are in a position to analyze the small-signal characteristics of the complete **common-emitter (C-E) amplifier** shown in Fig. 7.18(a). The ac equivalent circuit of Fig. 7.18(b) was constructed earlier (Ex. 7.1) by assuming that the capacitors all have zero impedance at the signal frequency and the dc voltage source represents an ac ground. For simplicity, we assume that we have found the Q-point and know the values of I_C and V_{CE}. In Fig. 7.18(b), resistor R_B represents the parallel combination of the two base bias resistors R_1 and R_2,

$$R_B = R_1 \| R_2 \tag{7.44}$$

and R_E is eliminated by bypass capacitor C_3.

Before we can develop an expression for the voltage gain of the amplifier, the transistor must be replaced by its small-signal model as in Fig. 7.18(c). A final simplification appears in Fig. 7.18(d), in which the resistor R_L represents the total equivalent load resistance on the transistor, the parallel combination of r_o, R_C and R_3:

$$R_L = r_o \| R_C \| R_3 \tag{7.45}$$

In Fig. 7.18(b) through (d), the reason why this amplifier configuration is called a common-emitter amplifier is apparent. The emitter terminal represents the common connection between the amplifier input and output ports. The input signal is applied to the transistor's base, the output signal appears at the collector, and both the input and output signals are referenced to the (common) emitter terminal.

7.6.1 TERMINAL VOLTAGE GAIN

Now we are ready to develop an expression for the overall gain of the amplifier in Fig. 7.18 from signal source v_i to the output voltage across resistor R_3. The voltage gain can be written as

$$A_v^{CE} = \frac{\mathbf{v_o}}{\mathbf{v_i}} = \left(\frac{\mathbf{v_o}}{\mathbf{v_b}}\right)\left(\frac{\mathbf{v_b}}{\mathbf{v_i}}\right) = A_{vt}^{CE}\left(\frac{\mathbf{v_b}}{\mathbf{v_i}}\right) \qquad \text{where} \qquad A_{vt}^{CE} = \left(\frac{\mathbf{v_o}}{\mathbf{v_b}}\right) \tag{7.46}$$

7.6 The Common-Emitter (C-E) Amplifier

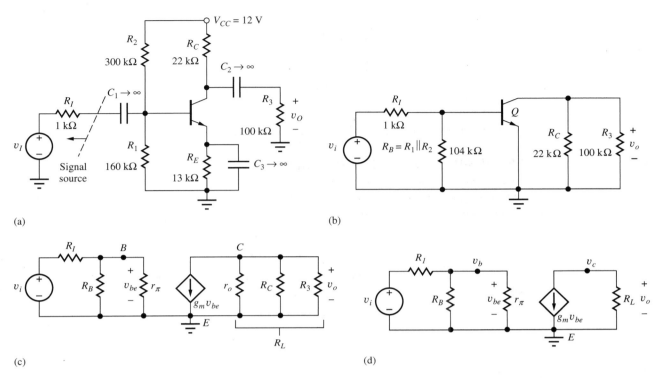

Figure 7.18 (a) Common-emitter amplifier circuit employing a bipolar transistor; (b) ac equivalent circuit for the common-emitter amplifier in part (a); the common-emitter connection should now be evident; (c) ac equivalent circuit with the bipolar transistor replaced by its small-signal model; (d) final equivalent circuit for ac analysis of the common-emitter amplifier.

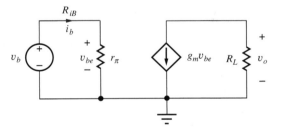

Figure 7.19 Simplified circuit model for finding the common-emitter terminal voltage gain A_{vt}^{CE} and input resistance R_{iB}.

A_{vt}^{CE} represents the voltage gain between the base and collector terminals of the transistor, the "**terminal gain.**" We will first find expressions for terminal gain A_{vt}^{CE} as well as the input resistance at the base of the transistor. Then we can relate v_b to v_i to find the overall voltage gain.

In Fig. 7.19, the BJT is replaced with its small-signal model, and the base terminal of the transistor is driven by test source v_b. Output voltage v_o is given by $\mathbf{v_o} = -g_m R_L \mathbf{v_b}$ and

$$A_{vt}^{CE} = \frac{\mathbf{v_o}}{\mathbf{v_b}} = -g_m R_L \qquad (7.47)$$

The minus sign indicates that the common-emitter stage is an inverting amplifier in which the input and output are 180° out of phase. The gain is proportional to the product of the transistor transconductance g_m and load resistor R_L. This product places an upper bound on the gain of the amplifier, and we will encounter the $g_m R_L$ product over and over again as we study transistor amplifiers. We will explore gain expression (Eq. 7.48) in more detail shortly.

7.6.2 INPUT RESISTANCE

The resistance looking into the base terminal of the transistor R_{iB} in Fig. 7.19 is simply the ratio of v_b and i_b,

$$R_{iB} = \frac{v_b}{i_b} = r_\pi \qquad (7.48)$$

The input resistance looking into the base of the transistor is equal to r_π.

7.6.3 SIGNAL SOURCE VOLTAGE GAIN

The overall voltage gain A_v^{CE} of the amplifier, including the effect of source resistance R_I, can now be found using the input resistance and terminal gain expressions. Voltage v_b at the base of the bipolar transistor in Fig. 7.18(d) is related to v_i by

$$v_b = v_i \frac{R_B \| R_{iB}}{R_I + (R_B \| R_{iB})} \qquad (7.49)$$

Combining Eqs. (7.46), (7.47), and (7.49), yields a general expression for the overall voltage gain of the common-emitter amplifier:

$$A_v^{CE} = A_{vt}^{CE} \left(\frac{v_b}{v_i}\right) = -g_m R_L \left[\frac{R_B \| r_\pi}{R_I + (R_B \| r_\pi)}\right] \qquad (7.50)$$

In this expression, we see that the overall voltage gain is equal to the terminal gain A_{vt}^{CE} reduced by the voltage division between R_I and the equivalent resistance at the base of the transistor. Terminal gain A_{vt}^{CE} places an upper limit on the voltage gain since the voltage division factor will be less than one.

7.7 IMPORTANT LIMITS AND MODEL SIMPLIFICATIONS

Let us now explore the limits to the voltage gain of common-emitter amplifiers. First, we will assume that the source resistance is small enough that $R_I \ll R_B \| R_{iB}$ so that

$$A_v^{CE} \cong A_{vt}^{CE} = -g_m R_L = -g_m (r_o \| R_C \| R_3) \qquad (7.51)$$

This approximation is equivalent to saying that the total input signal appears at the base of the transistor. Equation (7.51) states that the terminal voltage gain of the common-emitter stage is equal to the product of the transistor's transconductance g_m and load resistance R_L, and the minus sign indicates that the output voltage is "inverted" or 180° out of phase with respect to the input. Equation (7.51) places an upper limit on the gain we can achieve from a common-emitter amplifier with an external load resistor.

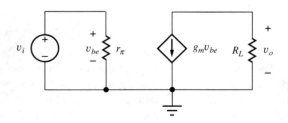

Figure 7.20 Simplified circuit corresponding to Eq. (7.51) with $R_E = 0$.

7.7.1 A DESIGN GUIDE FOR THE COMMON-EMITTER AMPLIFIER

In most amplifier designs, $r_o \gg R_C$ and we try to achieve $R_3 \gg R_C$. For these conditions, the load resistance on the collector of the transistor is approximately equal to R_C, and Eq. (7.51) can be reduced to

$$A_v^{CE} \cong A_{vt}^{CE} = -g_m R_C = -\frac{I_C R_C}{V_T} \qquad (7.52)$$

The $I_C R_C$ product represents the dc voltage dropped across the collector resistor R_C. Assuming $I_C R_C = \zeta V_{CC}$ with $0 \leq \zeta \leq 1$, and remembering that the reciprocal of V_T is 40 V^{-1}, Eq. (7.52) can be rewritten as

$$A_v^{CE} \cong -\frac{I_C R_C}{V_T} \cong -40\zeta V_{CC} \quad \text{with} \quad 0 \leq \zeta \leq 1 \qquad (7.53)$$

A common design allocates 1/3 of the power supply voltage across the collector resistor. For this case, $\zeta = 1/3$, $I_C R_C = V_{CC}/3$, and Eq. (7.53) becomes $A_v \cong -13 V_{CC}$. To further account for the approximations that led to this result and produce a number that is easy to remember, we will use this expression for our voltage gain estimate:

$$A_v^{CE} \cong -10 V_{CC} \quad \text{with the emitter at ac ground} \qquad (7.54)$$

Equation (7.54) represents our basic rule-of-thumb for the design of resistively loaded common-emitter amplifiers; that is, the magnitude of the voltage gain is approximately equal to 10 times the power supply voltage.[6] We need to know only the supply voltage to make a rough prediction of the gain of the common-emitter amplifier. For a C-E amplifier operating from a 15-V power supply, we estimate the gain to be -150 or 44 dB; a C-E amplifier with a 10-V supply would be expected to produce a gain of approximately -100 or 40 dB.

DESIGN NOTE

The magnitude of the voltage gain of a resistively loaded common-emitter amplifier *with emitter at ac ground* is approximately equal to 10 times the power supply voltage.

$$A_v^{CE} \cong -10 V_{CC}$$

This result represents an excellent way to quickly check the validity of more detailed calculations. Remember that the rule-of-thumb estimate is not going to be exact, but will predict the order of magnitude of the gain, typically within a factor of two or so.

A Comparison of r_o and R_C

Let us formally compare r_o to R_C by multiplying each by the collector current I_C:

$$I_C R_o = V_A + V_{CE} \cong V_A \quad \text{whereas} \quad I_C R_C \cong V_{CC}/3 \qquad (7.55)$$

For typical values, say $V_A = 75$ V and $V_{CC} = 15$ V, we see $I_C R_C \ll I_C r_o$ and $R_C \ll r_o$. Therefore we also have

$$g_m R_C \ll g_m r_o \quad \text{or} \quad g_m R_C \ll \mu_f \qquad (7.56)$$

From Eq. (7.56), we see we can neglect r_o any time the voltage gain is much less than μ_f, the intrinsic voltage gain.

---

[6] For dual power supplies, the corresponding estimate would be $A_V = -10(V_{CC} + V_{EE})$.

DESIGN NOTE

The transistor output resistance r_o can be neglected in voltage gain calculations whenever the voltage gain is much less than the intrinsic voltage gain μ_f.

7.7.2 UPPER BOUND ON THE COMMON-EMITTER GAIN

If we can somehow find a circuit in which both R_C and R_3 are much greater than r_o[7], then we achieve the most gain we can possibly get from the transistor:

$$A_v^{CE} \cong -g_m r_o = -\mu_f \quad \text{for} \quad R_C \| R_3 \gg r_o \quad (7.57)$$

The gain approaches the intrinsic gain of the transistor ($\mu_f \approx 40\, V_A$), typically several thousand. We will explore ways to realize such high gains in later chapters.

7.7.3 SMALL-SIGNAL LIMIT FOR THE COMMON-EMITTER AMPLIFIER

For small-signal operation, the magnitude of the base-emitter voltage v_{be}, developed across r_π in the small-signal model, must be less than 5 mV (you may wish to review Sec. 7.5.7). This voltage can be found using Eq. (7.49):

$$v_i = v_{be}\left(\frac{R_I + R_B \| r_\pi}{R_B \| r_\pi}\right) \quad (7.58)$$

Requiring $|v_{be}|$ in Eq. (7.58) to be less than 5 mV gives

$$|v_i| \leq 0.005\left(1 + \frac{R_I}{R_B \| r_\pi}\right)\, \text{V}$$

$$v_i \leq 0.005\left(1 + \frac{R_I}{R_B \| r_\pi}\right)\, \text{V} \cong 0.005\, \text{V} \quad \text{for} \quad R_B \| r_\pi \gg R_I \quad (7.59)$$

EXAMPLE 7.3 **VOLTAGE GAIN OF A COMMON-EMITTER AMPLIFIER**

In this example, we find the small-signal parameters of the bipolar transistor and then calculate the voltage gain of a common-emitter amplifier.

PROBLEM Calculate the voltage gain of the common-emitter amplifier in Fig. 7.18 if the transistor has $\beta_F = 100$, $V_A = 75$ V, and the Q-point is (0.245 mA, 3.39 V). What is the maximum value of v_i that satisfies the small-signal assumptions? Compare the voltage gain to the common-emitter "rule-of-thumb" gain estimate and the intrinsic gain (amplification factor) of the transistor.

SOLUTION **Known Information and Given Data:** Common-emitter amplifier with its ac equivalent circuit given in Fig. 13.18; $\beta_F = 100$ and $V_A = 75$ V; the Q-point is (0.245 mA, 3.39 V); $R_I = 1$ kΩ, $R_1 = 160$ kΩ, $R_2 = 300$ kΩ, $R_C = 22$ kΩ, $R_E = 13$ kΩ, and $R_3 = 100$ kΩ.

Unknowns: Small-signal parameters of the transistor; voltage gain A_v^{CE}; small-signal limit for the value of v_i; rule-of-thumb estimate; value of μ_f.

Approach: Use the Q-point information to find r_π. Use the calculated and given values to evaluate the voltage gain expression in Eq. (7.50).

---

[7] For example, if $R_3 = \infty$, and R_C is replaced with a large value of inductance.

Assumptions: The transistor is in the active region, and $\beta_o = \beta_F$. The signal amplitudes are low enough to be considered as small signals. Assume r_o can be neglected. $T = 300$ K.

Analysis: To evaluate Eq. (7.50),

$$A_v^{CE} = -g_m R_L \left[\frac{R_B \| R_{iB}}{R_I + (R_B \| R_{iB})} \right] \quad \text{with} \quad R_B = R_1 \| R_2 \quad \text{and} \quad R_{iB} = r_\pi$$

the values of the various resistors and small-signal model parameters are required. We have

$$g_m = 40 I_C = 40(0.245 \text{ mA}) = 9.80 \text{ mS} \quad r_\pi = \frac{\beta_o V_T}{I_C} = \frac{100(0.025 \text{ V})}{0.245 \text{ mA}} = 10.2 \text{ k}\Omega$$

$$r_o = \frac{V_A + V_{CE}}{I_C} = \frac{75 \text{ V} + 3.39 \text{ V}}{0.245 \text{ mA}} = 320 \text{ k}\Omega \quad R_{iB} = r_\pi = 10.2 \text{ k}\Omega$$

$$R_B = R_1 \| R_2 = 104 \text{ k}\Omega \quad R_L = R_c \| R_3 = 18.0 \text{ k}\Omega$$

Using these values,

$$A_v^{CE} = -9.80 \text{ mS} \, (18 \text{ k}\Omega) \frac{104 \text{ k}\Omega \| 10.2 \text{ k}\Omega}{1 \text{ k}\Omega + (104 \text{ k}\Omega \| 10.2 \text{ k}\Omega)} = -159 \text{ or } 44.0 \text{ dB}$$

With the emitter bypassed, v_{be} is given by

$$v_{be} = v_i \left[\frac{R_B \| R_{iB}}{R_I + (R_B \| R_{iB})} \right] = v_i \frac{R_B \| r_\pi}{R_I + (R_B \| r_\pi)} = v_i \frac{104 \text{ k}\Omega \| 10.2 \text{ k}\Omega}{1 \text{ k}\Omega + (104 \text{ k}\Omega \| 10.2 \text{ k}\Omega)} = 0.903 \mathbf{v}_i$$

and the small-signal limit for v_i is

$$|v_i| \leq \frac{0.005 \text{ V}}{0.903} = 5.53 \text{ mV}$$

The rule-of-thumb estimate and intrinsic gain are

$$A_v^{CE} \cong -10(12) = -120 \quad \text{and} \quad \mu_f = 9.80 \text{ mS} \, (320 \text{ k}\Omega) = 3140$$

Check of Results: The calculated voltage gain is similar to the rule-of-thumb estimate so our calculation appears correct. Remember, the rule-of-thumb formula is meant to only be a rough estimate; it will not be exact. The gain is much less than the amplification factor, so the neglect of r_o is valid.

Computer-Aided Analysis: SPICE simulation yields the Q-point (0.248 mA, 3.30 V) that is consistent with the assumed value. The small difference results from V_A being included in the SPICE simulation and not in our hand calculations. An ac sweep from 10 Hz to 100 kHz with

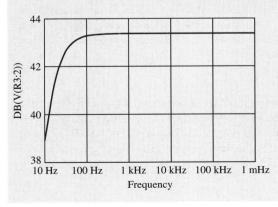
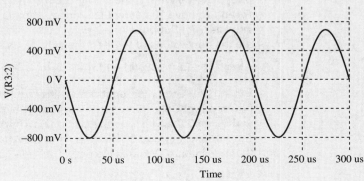

10 frequency points/decade is used to find the region in which the capacitors are acting as short circuits, and the gain is observed to be constant at 43.4 dB above a frequency of 1 kHz. The voltage gain is slightly less than our calculated value because r_o was neglected in our calculations. A transient simulation was performed with a 5-mV, 10-kHz sine wave. The output exhibits reasonably good linearity, but the positive and negative amplitudes are slightly different, indicating some waveform distortion. Enabling the Fourier analysis capability of SPICE yields THD = 3.6%.

Discussion: Let us complete our discussion of the common-emitter amplifier example by exploring the impact of tolerances on circuit performance. Here we assume that V_{CC} and all the resistors have 5 percent tolerances, and β_F has a 25 percent tolerance. Tolerances on V_{BE} and V_A are not included for simplicity.

The results of a 500-case Monte Carlo analysis appear in the table below. The collector current varies by approximately ±15 percent. Fortunately, the transistor's minimum collector-base voltage is ±1.11 V, so the transistor remains in the active region. If it were found to be saturated, the circuit would need to be redesigned. The gain varies from −125 to −169. Most of this variation can be traced to changes in the values of R_C, R_3, I_C, and β_F. So if each person in the class were to build this circuit in the lab, we should expect significant variations in Q-point and voltage gain from one individual's circuit to another.

Common-Emitter Amplifier 500-Case Monte Carlo Analysis Results

PARAMETER	NOMINAL VALUE	MAXIMUM VALUE	MINIMUM VALUE
I_C (μA)	245	285	211
V_{CE} (V)	3.40	4.36	2.52
V_{CB} (V)	2.44	3.60	1.11
A_v^{CE}	−146	−169	−125
r_π (kΩ)	10.6	14.2	7.36

EXERCISE: What is the terminal voltage gain A_{vt} ($-g_m R_L$) for the amplifier in Ex. 7.3? The actual gain of the amplifier was only −159. Where is most of this gain being lost?

ANSWER: −176; approximately 10 percent of the input signal is lost by voltage division between the source resistance R_I and the amplifier input resistance.

EXERCISE: (a) What is the voltage gain in the original circuit if $\beta_F = 125$? (b) Suppose resistors R_C and R_3 have 10 percent tolerances. What are the worst-case values of voltage gain for this amplifier? (c) Suppose the Q-point current in the original circuit increases to 0.275 mA. What are the new values of V_{CE} and the voltage gain?

ANSWERS: (a) −162; (b) −143, −175; (c) 2.34 V, −177

EXERCISE: A common-emitter amplifier similar to Fig. 7.18 is operating from a single +20-V power supply, and the emitter terminal is bypassed by capacitor C_3. The BJT has $\beta_F = 100$ and $V_A = 50$ V and is operating at a Q-point of (100 μA, 10 V). The amplifier has $R_I = 5$ kΩ, $R_B = 150$ kΩ, $R_C = 100$ kΩ, and $R_3 = \infty$. What is the voltage gain predicted from our rule of thumb estimate? What is the actual voltage gain? What is the value of μ_f for this transistor?

ANSWERS: −200; −278; 2400

DESIGN NOTE

Remember, the amplification factor μ_f places an upper bound on the voltage gain of a single-transistor amplifier. We can't do better than μ_f! For the BJT,

$$\mu_f \cong 40V_A$$

For $25\ V \leq V_A \leq 100\ V$, we have $1000 \leq \mu_f \leq 4000$.

EXERCISE: Verify the bias point values used in Ex. 7.3 by directly calculating the Q-point.

EXERCISE: What value of saturation current I_S must be used in SPICE to achieve $V_{BE} = 0.7$ V for $I_C = 245\ \mu A$? Assume a default temperature of 27°C.

ANSWER: 0.422 fA

7.8 SMALL-SIGNAL MODELS FOR FIELD-EFFECT TRANSISTORS

We now turn our attention to the small-signal model for the field-effect transistor and then use it in Sec. 7.9 to analyze the behavior of the common-source amplifier stage that is the FET version of the common-emitter amplifier. As mentioned earlier for the diode and bipolar transistor, we need to have a linearized model of the field-effect transistor that is valid for small changes in voltages and currents, in order to use our wealth of linear circuit analysis techniques to analyze the ac performance of the circuit. First, we consider the MOSFET as a three-terminal device; we then explore the changes necessary when the MOSFET is operated as a four-terminal device.

7.8.1 SMALL-SIGNAL MODEL FOR THE MOSFET

The small-signal model of the MOSFET is based on the two-port network representation in Fig. 7.21 with the input port variables defined as v_{gs} and i_g and the output port variables defined as v_{ds} and i_d. Rewriting Eq. (7.19) in terms of these variables yields

$$\mathbf{i_g} = g_\pi \mathbf{v_{gs}} + g_\mu \mathbf{v_{ds}}$$
$$\mathbf{i_d} = g_m \mathbf{v_{gs}} + g_o \mathbf{v_{ds}} \tag{7.60}$$

Remember that the port variables in Fig. 7.21(a) can be considered to represent either the time-varying portion of the total voltages and currents or small changes in the total quantities.

$$v_{GS} = V_{GS} + v_{gs} \qquad v_{DS} = V_{DS} + v_{ds}$$
$$i_G = I_G + i_g \qquad i_{DS} = I_D + i_d \tag{7.61}$$

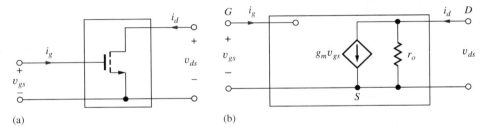

Figure 7.21 (a) The MOSFET represented as a two-port network; (b) small-signal model for the three-terminal MOSFET.

The parameters in Eq. (7.61) can be written in terms of the small-signal variations or in terms of derivatives of the complete port variables, as in Eq. (7.63):

$$g_\pi = \left.\frac{i_g}{v_{gs}}\right|_{v_{ds}=0} = \left.\frac{\partial i_G}{\partial v_{GS}}\right|_{\text{Q-point}} \qquad g_\mu = \left.\frac{i_g}{v_{ds}}\right|_{v_{gs}=0} = \left.\frac{\partial i_G}{\partial v_{DS}}\right|_{\text{Q-point}}$$

$$g_m = \left.\frac{i_d}{v_{gs}}\right|_{v_{ds}=0} = \left.\frac{\partial i_{DS}}{\partial v_{GS}}\right|_{\text{Q-point}} \qquad g_o = \left.\frac{i_d}{v_{ds}}\right|_{v_{gs}=0} = \left.\frac{\partial i_{DS}}{\partial v_{DS}}\right|_{\text{Q-point}} \qquad (7.62)$$

We can evaluate these parameters by taking appropriate derivatives of the large-signal model equations for the drain current of the active region MOS transistor, as developed in Chapter 5, and repeated here in Eq. (7.64).

$$i_D = \frac{K_n}{2}(v_{GS} - V_{TN})^2 (1 + \lambda v_{DS}) \qquad (7.63)$$

for $v_{DS} \geq v_{GS} - V_{TN}$, and $i_G = 0$, where $K_n = \mu_n C_{\text{ox}}(W/L)$.

$$g_\pi = \left.\frac{\partial i_G}{\partial v_{GS}}\right|_{v_{DS}} = 0 \quad \text{and} \quad g_\mu = \left.\frac{\partial i_G}{\partial v_{DS}}\right|_{v_{GS}} = 0$$

$$g_m = \left.\frac{\partial i_{DS}}{\partial v_{GS}}\right|_{\text{Q-point}} = K_n(V_{GS} - V_{TN})(1 + \lambda V_{DS}) = \frac{2I_D}{V_{GS} - V_{TN}} \qquad (7.64)$$

$$g_o = \left.\frac{\partial i_{DS}}{\partial v_{DS}}\right|_{\text{Q-point}} = \lambda \frac{K_n}{2}(V_{GS} - V_{TN})^2 = \frac{\lambda I_D}{1 + \lambda V_{DS}} = \frac{I_D}{\frac{1}{\lambda} + V_{DS}}$$

Because i_G is always zero and therefore independent of v_{GS} and v_{DS}, g_π and g_r are both zero. Remembering that the gate terminal is insulated from the channel by the gate oxide, we can reasonably expect that the input resistance $(1/g_\pi)$ of the transistor is infinite.

As for the bipolar transistor, g_m is called the transconductance, and $1/g_o$ represents the output resistance of the transistor.

$$\textbf{Transconductance:} \quad g_m = \frac{I_D}{\dfrac{V_{GS} - V_{TN}}{2}}$$

$$\textbf{Output resistance:} \quad r_o = \frac{1}{g_o} = \frac{\dfrac{1}{\lambda} + V_{DS}}{I_D} \cong \frac{1}{\lambda I_D} \qquad (7.65)$$

The small-signal circuit model for the MOSFET resulting from Eqs. (7.64) and (7.65) appears in Fig. 7.21(b) and contains only the voltage controlled current source and output resistance.

From Eq. (7.65), we see that the values of the small-signal parameters are directly controlled by the design of the Q-point. The form of the equations for g_m and r_o of the MOSFET directly mirrors that of the BJT. However, one-half the internal gate drive $(V_{GS} - V_{TN})/2$ replaces the thermal voltage V_T in the transconductance expression, and $1/\lambda$ replaces the Early voltage in the output resistance expression. The value of $V_{GS} - V_{TN}$ is often a volt or more in MOSFET circuits, whereas $V_T = 0.025$ V at room temperature. Thus, for a given operating current, the MOSFET can be expected to have a much smaller transconductance than the BJT. However, the value of $1/\lambda$ is similar to V_A, so the output resistances are similar for a given operating point $(I_D, V_{DS}) = (I_C, V_{CE})$. Here, and similar to the BJT case, drain-source voltage V_{DS} is typically much less than $1/\lambda$, so we can simplify the expression for the output resistance to $r_o \cong 1/\lambda I_D$.

The actual dependence of transconductance g_m on current is not shown explicitly by Eq. (7.65) because I_D is a function of $(V_{GS} - V_{TN})$. Rewriting the expression for g_m from Eq. (7.64) yields

$$g_m = K_n(V_{GS} - V_{TN})(1 + \lambda V_{DS}) = \sqrt{2K_n I_D (1 + \lambda V_{DS})}$$
$$g_m \cong K_n(V_{GS} - V_{TN}) \quad \text{or} \quad g_m \cong \sqrt{2K_n I_D} \tag{7.66}$$

where the simplifications require $\lambda V_{DS} \ll 1$.

Here we see two other important differences between the MOSFET and BJT. The MOSFET transconductance increases only as the square root of drain current, whereas the BJT transconductance is directly proportional to collector current. In addition, the MOSFET transconductance is dependent on the geometry of the transistor because $K_n \propto W/L$, whereas the transconductance of the BJT is geometry-independent. It is also worth noting that the current gain of the MOSFET is infinite. Since the value of $r_\pi = (1/g_\pi)$ is infinite for the MOSFET, the "current gain" $\beta_o = g_m r_\pi$ equals infinity as well. However, we will later see that this is only true at dc.

7.8.2 INTRINSIC VOLTAGE GAIN OF THE MOSFET

Another important difference between the BJT and MOSFET is the variation of the intrinsic voltage gain μ_f with operating point. Using Eq. (7.65) for g_m and r_o, we find that the intrinsic voltage gain becomes

$$\mu_f = g_m r_o = \frac{\frac{1}{\lambda} + V_{DS}}{\left(\frac{V_{GS} - V_{TN}}{2}\right)} \quad \text{and} \quad \mu_f \cong \frac{2}{\lambda(V_{GS} - V_{TN})} \cong \frac{1}{\lambda}\sqrt{\frac{2K_n}{I_D}} \tag{7.67}$$

for $\lambda V_{DS} \ll 1$.

The value of μ_f for the MOSFET decreases as the operating current increases. Thus the larger the operating current of the MOSFET, the smaller its voltage gain capability. In contrast, the intrinsic gain of the BJT is independent of operating point. This is an extremely important difference to keep in mind, particularly during the design process.

Table 7.2 displays examples of the values of the MOSFET small-signal parameters for a variety of operating points. Just as for the bipolar transistor, the values of g_m and r_o can each be varied over many orders of magnitude through the choice of Q-point. By comparing the results in Tables 7.1 and 7.2 we see that g_m, r_o, and μ_f of the MOSFET are all similar to those of the bipolar transistor at low currents. However, as the drain current increases, the value of g_m of the MOSFET does not grow as rapidly as for the bipolar transistor, and μ_f drops significantly. This particular MOSFET has a significantly lower intrinsic gain than the BJT for currents greater than a few tens of microamperes.

TABLE 7.2
MOSFET Small-Signal Parameters versus Current: $K_n = 1$ mA/V², $\lambda = 0.0133$ V⁻¹, $V_{DS} = 10$ V

I_D	g_m	r_π	r_o	μ_f
1 μA	4.76×10^{-5} S	∞	85.2 MΩ	4060
10 μA	1.51×10^{-4} S	∞	8.52 MΩ	1280
100 μA	4.76×10^{-4} S	∞	852 kΩ	406
1 mA	1.51×10^{-3} S	∞	85.2 kΩ	128
10 mA	4.76×10^{-3} S	∞	8.52 kΩ	40

EXERCISES: (a) Calculate the values of g_m, r_o, and μ_f for a MOSFET transistor with $K_n = 1$ mA/V² and $\lambda = 0.02$ V⁻¹ operating at Q-points of (250 μA, 5 V) and (5 mA, 10 V). (b) Use graphical analysis to find values of g_m and r_o at the Q-point for the transistor in Fig. 7.2(b).

ANSWERS: 7.42×10^{-4} S, 220 kΩ, 163; 3.46×10^{-3} S, 12.0 kΩ, 41.5; 1.3×10^{-3} S, ∞

DESIGN NOTE

The intrinsic voltage gain μ_f of the MOSFET is operating point dependent and decreases as drain current increases:

$$\mu_f \cong \frac{1}{\lambda}\sqrt{\frac{2K_n}{I_D}}$$

7.8.3 DEFINITION OF SMALL-SIGNAL OPERATION FOR THE MOSFET

The limits of linear operation of the MOSFET can be explored using the simplified drain-current expression ($\lambda = 0$) for the MOSFET in the active region:

$$i_D = \frac{K_n}{2}(v_{GS} - V_{TN})^2 \quad \text{for} \quad v_{DS} \geq v_{GS} - V_{TN} \tag{7.68}$$

Expanding this expression using $v_{GS} = V_{GS} + v_{gs}$ and $i_D = I_D + i_d$ gives

$$I_D + i_d = \frac{K_n}{2}\left[(V_{GS} - V_{TN})^2 + 2v_{gs}(V_{GS} - V_{TN}) + v_{gs}^2\right] \tag{7.69}$$

Recognizing that the dc drain current is equal to $I_D = (K_n/2)(V_{GS} - V_{TN})^2$ and subtracting this term from both sides of Eq. (7.69) yields an expression for signal current i_d:

$$i_d = \frac{K_n}{2}\left[2v_{gs}(V_{GS} - V_{TN}) + v_{gs}^2\right] \tag{7.70}$$

For linearity, i_d must be directly proportional to v_{gs}, which is achieved for

$$v_{gs}^2 \ll 2v_{gs}(V_{GS} - V_{TN}) \quad \text{or} \quad v_{gs} \ll 2(V_{GS} - V_{TN}) \tag{7.71}$$

Using a factor of 10 to satisfy the inequality gives

$$\boxed{v_{gs} \leq 0.2(V_{GS} - V_{TN})} \tag{7.72}$$

Because the MOSFET can easily be biased with $(V_{GS} - V_{TN})$ equal to several volts, we see that it can handle much larger values of v_{gs} than the values of v_{be} corresponding to the bipolar transistor. This is another fundamental difference between the MOSFET and BJT and can be very important in circuit design, particularly in RF amplifiers, for example.

Now let us explore the change in drain current that corresponds to small-signal operation. Using Eq. (7.73),

$$\frac{i_d}{I_D} = \frac{g_m v_{gs}}{I_D} = \frac{0.2(V_{GS} - V_{TN})}{\dfrac{V_{GS} - V_{TN}}{2}} \leq 0.4 \tag{7.73}$$

A 20% change of $(V_{GS} - V_{TN})$ in v_{GS} corresponds to a 40 percent deviation in the drain and source currents from the Q-point values.

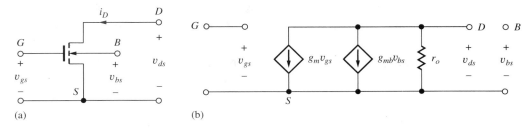

Figure 7.22 (a) MOSFET as a four-terminal device; (b) small-signal model for the four-terminal MOSFET.

EXERCISE: A MOSFET transistor with $K_n = 2.0$ mA/V² and $\lambda = 0$ is operating at a Q-point of (25 mA, 10 V). What is the largest value of v_{gs} that corresponds to a small signal? If a BJT is biased at the same Q-point, what is the largest value of v_{be} that corresponds to a small signal?

ANSWERS: 1 V; 0.005 V

7.8.4 BODY EFFECT IN THE FOUR-TERMINAL MOSFET

When the body terminal of the MOSFET is not connected to the source terminal, as in Fig. 7.22(a), an additional controlled source must be introduced into the small-signal model. Referring to the simplified drain-current expression for the MOSFET from Sec. 5.2.11:

$$i_D = \frac{K_n}{2}(v_{GS} - V_{TN})^2 \quad \text{and} \quad V_{TN} = V_{TO} + \gamma\left(\sqrt{v_{SB} + 2\phi_F} - \sqrt{2\phi_F}\right) \quad (7.74)$$

We recognize that the drain current is dependent on the threshold voltage, and the threshold voltage changes as v_{SB} changes. Thus, a **back-gate transconductance** can be defined:

$$g_{mb} = \left.\frac{\partial i_D}{\partial v_{BS}}\right|_{\text{Q-point}} = -\left.\frac{\partial i_D}{\partial v_{SB}}\right|_{\text{Q-point}} = -\left(\frac{\partial i_D}{\partial V_{TN}}\right)\left(\frac{\partial V_{TN}}{\partial v_{SB}}\right)\bigg|_{\text{Q-point}} \quad (7.75)$$

Evaluating the derivative terms in brackets,

$$\left.\frac{\partial i_D}{\partial V_{TN}}\right|_{\text{Q-point}} = -K_n(V_{GS} - V_{TN}) = -g_m \quad \text{and} \quad \left.\frac{\partial V_{TN}}{\partial v_{SB}}\right|_{\text{Q-point}} = \frac{\gamma}{2\sqrt{V_{SB} + 2\phi_F}} = \eta \quad (7.76)$$

in which η represents the **back-gate transconductance parameter.** Combining Eqs. (7.76) yields

$$g_{mb} = -(-g_m)\eta \quad \text{or} \quad g_{mb} = +\eta g_m \quad (7.77)$$

for typical values of γ and V_{SB}, $0 \leq \eta \leq 1$.

We also need to explore the question of whether there is a conductance connected from the bulk terminal to the other terminals. However, the bulk terminal represents a reverse-biased diode between the bulk and channel. Using our small-signal model for the diode, Eq. (7.15), we see that

$$\left.\frac{\partial i_B}{\partial v_{BS}}\right|_{\text{Q-point}} = \frac{I_D + I_S}{V_T} \cong 0 \quad (7.78)$$

because diode current $I_D \cong -I_S$ for the reverse-biased diode. Thus, there is no significant conductance indicated between the bulk and source or drain terminals in the small-signal model.

The resulting small-signal model for the four-terminal MOSFET is given in Fig. 7.22(b), in which a second voltage-controlled current source has been added to model the back-gate transconductance g_{mb}.

> **EXERCISE:** Calculate the values of η for a MOSFET transistor with $\gamma = 0.75$ $V^{0.5}$ and $2\phi_F = 0.6$ V for $V_{SB} = 0$ and $V_{SB} = 3$ V.
>
> **ANSWERS:** 0.48, 0.20

7.8.5 SMALL-SIGNAL MODEL FOR THE PMOS TRANSISTOR

Just as was the case for the *pnp* and *npn* transistors, the small-signal model for the PMOS transistor is identical to that of the NMOS device. The circuits in Fig. 7.23 should help reinforce this result.

In Fig. 7.23, the NMOS and PMOS transistors are each biased by the dc voltage source V_{GG}, establishing Q-point current I_D. In each case, a signal voltage v_{gg} is added in series with V_{GG} so that a positive value of v_{gg} causes the gate-to-source voltage of each transistor to increase. For the NMOS transistor, the total gate-to-source voltage and drain current are

$$v_{GS} = V_{GG} + v_{gg} \quad \text{and} \quad i_{DS} = I_D + i_d \qquad (7.79)$$

and an increase in v_{gg} causes an increase in current into the drain terminal. For the PMOS transistor,

$$v_{SG} = V_{GG} - v_{gg} \quad \text{and} \quad i_D = I_D - i_d \qquad (7.80)$$

A positive signal voltage v_{gg} reduces the source-to-gate voltage of the PMOS transistor and causes a decrease in the total current exiting the drain terminal. This reduction in total current is equivalent to an increase in the signal current entering the drain.

Thus, for both the NMOS and PMOS transistors, an increase in the value of v_{GS} causes an increase in current into the drain, and the polarities of the voltage-controlled current source in the small-signal model are identical, as depicted in Fig. 7.24.

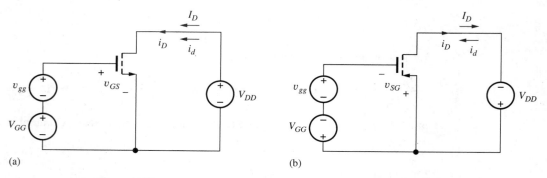

Figure 7.23 dc Bias and signal currents for (a) NMOS and (b) PMOS transistors.

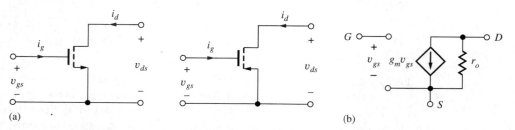

Figure 7.24 (a) NMOS and PMOS transistors. (b) The small-signal models are identical.

7.8.6 SMALL-SIGNAL MODEL FOR MOS TRANSISTORS IN WEAK INVERSION

In weak inversion, discussed in Sec. 5.11, the model for the MOS transistor becomes an exponential expression that is very similar to that of the bipolar transistor:

$$I_D = I_{DO} \exp\left(\frac{V_{GS} - V_{TN}}{nV_T}\right) \quad \text{with} \quad I_{DO} = 2nV_T^2 K_n \quad (7.81)$$

Following the same analyses as used in Sec. 7.5, the small-signal parameter parameters for the MOSFET become:

Transconductance: $g_m = \dfrac{I_D}{nV_T}$

Input resistance: ∞

Output resistance: $r_o = \dfrac{1}{\lambda I_D}$ (7.82)

Intrinsic gain: $\mu_f = \dfrac{1}{n\lambda V_T}$

Small-signal limit: $v_{gs} \leq 0.2nV_T \cong n(0.005 \text{ V})$

The main differences occur by replacement of V_T by nV_T and the Early voltage V_A by $1/\lambda$. For the small-signal limit, the MOS gate drive term $(V_{GS} - V_{TN})$ is replaced by nV_T. The above changes also appear in Table 7.3.

> **EXERCISE:** What is the value of the MOS transconductance efficiency (g_m/I_D) for the MOS transistor in weak inversion at room temperature with $n = 1.5$? What is the value for the BJT? What is the **weak inversion** for the MOSFET?
>
> **ANSWERS:** Using $V_T = 25.8$ mV: 25.8 for the MOSFET; 38.8 for the BJT; $S = 89.1$ mV/decade.

7.8.7 SMALL-SIGNAL MODEL FOR THE JUNCTION FIELD-EFFECT TRANSISTOR

The drain-current expressions for the JFET and MOSFET can be written in essentially identical form (see Prob. 7.147), so we should not be surprised that the small-signal models also have the same form. For small-signal analysis, we represent the JFET as the two-port network in Fig. 7.25. The small-signal parameters can be determined from the large-signal model given in Chapter 5 for the drain current of the JFET operating in the pinch-off region:

$$i_D = I_{DSS}\left[1 - \frac{v_{GS}}{V_P}\right]^2 [1 + \lambda v_{DS}] \quad \text{for } v_{DS} \geq v_{GS} - V_P \quad (7.83)$$

The total gate current i_G represents the current of the gate-to-channel diode, which we express in terms of the gate-to-source voltage v_{GS} and saturation current I_{SG}:

$$i_G = I_{SG}\left[\exp\left(\frac{v_{GS}}{V_T}\right) - 1\right] \quad (7.84)$$

Once again using the derivative formulation from Eq. (7.63):

$$g_\pi = \left.\frac{\partial i_G}{\partial v_{GS}}\right|_{\text{Q-point}} = \frac{I_G + I_{SG}}{V_T} \qquad g_\mu = \left.\frac{\partial i_G}{\partial v_{DS}}\right|_{\text{Q-point}} = 0$$

$$g_m = \left.\frac{\partial i_D}{\partial v_{GS}}\right|_{\text{Q-point}} = 2\frac{I_{DSS}}{-V_P}\left[1 - \frac{V_{GS}}{V_P}\right][1 + \lambda V_{DS}] = \frac{2I_D}{V_{GS} - V_P}$$

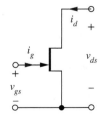

Figure 7.25 The JFET as a two-port network.

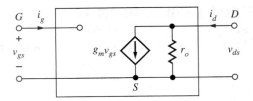

Figure 7.26 Small-signal model for the JFET.

Alternatively,

$$g_m = \frac{2}{|V_P|}\sqrt{I_{DSS}I_D(1+\lambda V_{DS})} \cong \frac{2}{|V_P|}\sqrt{I_{DSS}I_D} \cong 2\frac{I_{DSS}}{V_P^2}[V_{GS}-V_P] \qquad (7.85)$$

$$g_o = \left.\frac{\partial i_D}{\partial v_{DS}}\right|_{Q\text{-point}} = \lambda I_{DSS}\left[1-\frac{V_{GS}}{V_P}\right]^2 = \frac{\lambda I_D}{1+\lambda V_{DS}} = \frac{I_D}{\frac{1}{\lambda}+V_{DS}}$$

Because the JFET is normally operated with the gate junction reverse-biased,

$$I_G = -I_{SG} \qquad \text{and} \qquad r_\pi = \infty \qquad (7.86)$$

Thus, the small-signal model for the JFET in Fig. 7.26 is identical to that of the MOSFET, including the formulas used to express g_m and r_o when V_{TN} is replaced by V_P.

As a result, the definition of a small signal and the expression for the intrinsic gain μ_F are also similar to those of the MOSFET:

$$v_{gs} \leq 0.2(V_{GS}-V_P)$$

and $\qquad (7.87)$

$$\mu_f = g_m r_o = 2\frac{\frac{1}{\lambda}+V_{DS}}{V_{GS}-V_P} \cong \frac{2}{\lambda|V_P|}\sqrt{\frac{I_{DSS}}{I_D}}$$

> **EXERCISES:** Calculate the values of g_m, r_o, and μ_f for a JFET with $I_{DSS} = 5$ mA, $V_P = -2$ V, and $\lambda = 0.02$ V^{-1} if it is operating at a Q-point of (2 mA, 5 V). What is the largest value of v_{gs} that can be considered to be a small signal?
>
> **ANSWERS:** 3.32×10^{-3} S, 27.5 kΩ, 91; 0.253 V.

7.9 SUMMARY AND COMPARISON OF THE SMALL-SIGNAL MODELS OF THE BJT AND FET

Table 7.3 is a side-by-side comparison of the small-signal models of the bipolar junction transistor and the field-effect transistor; the table has been constructed to highlight the similarities and differences between the two types of devices.

The transconductance of the BJT is directly proportional to operating current, whereas that of the FET increases only with the square root of current. Both can be represented as the operating current divided by a characteristic voltage: V_T for the BJT and $(V_{GS}-V_{TN})/2$ for the MOSFET.

The input resistance of the bipolar transistor is set by the value of r_π, which is inversely proportional to the Q-point current and can be quite small at even moderate currents (1 to 10 mA). On the other hand, the input resistance of the FETs is extremely high, approaching infinity.

7.9 Summary and Comparison of the Small-Signal Models of the BJT and FET

TABLE 7.3
Small-Signal Parameter Comparison

PARAMETER	BIPOLAR TRANSISTOR	MOSFET STRONG INVERSION	MOSFET WEAK INVERSION	JFET
Transconductance g_m	$\dfrac{I_C}{V_T}$	$\dfrac{2I_D}{V_{GS}-V_{TN}} \cong \sqrt{2K_n I_D}$	$\dfrac{I_D}{nV_T}$	$\dfrac{2I_D}{V_{GS}-V_P} \cong \dfrac{2}{|V_P|}\sqrt{I_D I_{DSS}}$
Input resistance	$r_\pi = \dfrac{\beta_o}{g_m} = \dfrac{\beta_o V_T}{I_C}$	∞	∞	∞
Output resistance r_o	$\dfrac{V_A+V_{CE}}{I_C} \cong \dfrac{V_A}{I_C}$	$\dfrac{\frac{1}{\lambda}+V_{DS}}{I_D} \cong \dfrac{1}{\lambda I_D}$	$\dfrac{1}{\lambda I_D}$	$\dfrac{\frac{1}{\lambda}+V_{DS}}{I_D} \cong \dfrac{1}{\lambda I_D}$
Intrinsic voltage gain μ_f	$\dfrac{V_A+V_{CE}}{V_T} \cong \dfrac{V_A}{V_T}$	$\dfrac{2\left(\frac{1}{\lambda}+V_{DS}\right)}{V_{GS}-V_{TN}} \cong \dfrac{1}{\lambda}\sqrt{\dfrac{2K_n}{I_D}}$	$\dfrac{1}{n\lambda V_T}$	$\dfrac{2\left(\frac{1}{\lambda}+V_{DS}\right)}{V_{GS}-V_P} \cong \dfrac{2}{\lambda|V_P|}\sqrt{\dfrac{I_{DSS}}{I_D}}$
Small-signal requirement	$v_{be} \leq 0.005$ V	$v_{gs} \leq 0.2(V_{GS}-V_{TN})$	$v_{gs} \leq 0.2nV_T$	$v_{gs} \leq 0.2(V_{GS}-V_P)$

dc i-v active region expressions for use with Table 7.3:

BJT: $\quad I_C = I_S\left[\exp\left(\dfrac{V_{BE}}{V_T}\right)-1\right]\left[1+\dfrac{V_{CE}}{V_A}\right] \qquad V_T = \dfrac{kT}{q}$

MOSFET: $\quad I_D = \dfrac{K_n}{2}(V_{GS}-V_{TN})^2(1+\lambda V_{DS}) \qquad K_n = \mu_n C_{ox}\dfrac{W}{L}$

$I_D = I_{D0}\exp\left(\dfrac{V_{GS}-V_{TN}}{nV_T}\right) \qquad I_{D0} = 2nV_T^2 K_n$

JFET: $\quad I_D = I_{DSS}\left(1-\dfrac{V_{GS}}{V_P}\right)^2(1+\lambda V_{DS})$

The expressions for the output resistances of the transistors are almost identical, with the parameter $1/\lambda$ in the FET taking the place of the Early voltage V_A of the BJT. The value of $1/\lambda$ is similar to V_A, so the output resistances can be expected to be similar in value for comparable operating currents.

The intrinsic voltage gain of the BJT is nearly independent of operating current and has a typical value of several thousand at room temperature. In contrast, μ_f for the FETs is inversely proportional to the square root of operating current and decreases as the Q-point current is raised. At very low currents, μ_f of the FETs can be similar to that of the BJT, but in normal operation it is often much smaller and can even fall below 1 for high currents (see Prob. 7.85).

Small-signal operation is dependent on the size of the base-emitter voltage of the BJT or gate-source voltage of the FET. The magnitude of voltage that corresponds to small-signal operation can be significantly different for these two devices. For the BJT, v_{be} must be less than 5 mV. This value is indeed small, and it is independent of Q-point. In contrast, the FET requirement is $v_{gs} \leq 0.2(V_{GS}-V_{TN})$ or $0.2(V_{GS}-V_P)$, which is dependent on bias point and can be designed to be as much as a volt or more.

This discussion highlighted the similarities and differences between the bipolar and field-effect transistors. An understanding of Table 7.3 is extremely important to the design of analog circuits. As we study single and multistage amplifier design in the coming sections and chapters, we will note the effect of these differences and relate them to our circuit designs.

ELECTRONICS IN ACTION

Direct Measurement of Intrinsic Gain ($\mu_f = g_m r_o$)

Although direct measurement of transconductance g_m is relatively straightforward, the measurement of output resistance r_o can be more difficult, particularly at low currents where very high values of resistances are often involved. However, intrinsic gain μ_f can be directly measured [1] using the circuit in (a) of the following diagram. If desired, the output resistance can then be extracted using a measured value of g_m.

To make the measurement, the MOSFET is biased by current source I_S and the drain voltage is set by source v_{DD}. The drain voltage is swept over some range and the change in gate-source voltage is measured. The measurements are easily done utilizing a commercial semiconductor parameter analyzer.

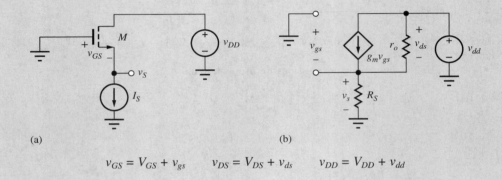

(a)   (b)

$$v_{GS} = V_{GS} + v_{gs} \qquad v_{DS} = V_{DS} + v_{ds} \qquad v_{DD} = V_{DD} + v_{dd}$$

Small-Signal Analysis: We desire to find the change in gate-source voltage (v_{gs}) in response to a change in drain-source voltage (v_{dd}) based upon the small-signal model in (b). Summing currents at the FET source in terms of conductances and noting that $v_s = -v_{gs}$ yields:

$$g_m v_{gs} + g_o v_{ds} - G_S v_s = 0 \qquad \text{or} \qquad g_o v_{ds} = -(g_m + G_S) v_{gs}$$

$$\frac{v_{ds}}{v_{gs}} = -r_o(g_m + G_S) = -\mu_f \left(1 + \frac{1}{g_m R_S}\right) \qquad \text{for} \qquad \mu_f = g_m r_o$$

$$\mu_f = -\left(\frac{v_{ds}}{v_{gs}}\right)\left(\frac{1}{1 + \frac{1}{g_m R_S}}\right) \qquad \text{and} \qquad \mu_f \cong -\left(\frac{v_{ds}}{v_{gs}}\right)$$

Noting that $v_{ds} = v_{dd} + v_{gs}$ also gives

$$\mu_f + 1 = -\left(\frac{v_{dd}}{v_{gs}}\right)$$

A similar analysis applies to the bipolar transistor with an additional term accounting for the additional current in r_π:

$$\mu_f = -\left(\frac{v_{ce}}{v_{bc}}\right)\left(\frac{1}{1 + \frac{1}{\beta_o} + \frac{1}{g_m R_S}}\right) \qquad \text{and} \qquad \mu_f \cong -\left(\frac{v_{ce}}{v_{be}}\right)$$

Measurement Example: Intrinsic gain μ_f is determined by sweeping the drain voltage v_{dd} over a range and measuring the resulting changes in v_{gs} and v_{ds}. The reciprocal of the slope of the graph of v_{ds} versus v_{gs} yields the value of the intrinsic gain as in the example graph below of a silicon carbide NMOS transistor. For this measurement, $\mu_f + 1 = 329$ with the transistor biased at 5 mA. Note that even though the change drain voltage is large, the gate-source voltage change is small (21 mV) and represents a small-signal change. Also, the slope is negative as expected.

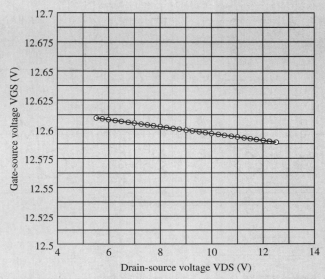

Source: R. C. Jaeger et al., "Direct measurement of the available voltage gain of bipolar and field-effect transistors," *IEEE Electron Device Letters,* vol. EDL-6, no. 5, pp. 219–220, May 1985.

7.10 THE COMMON-SOURCE (C-S) AMPLIFIER

Now we are in a position to analyze the small-signal characteristics of the **common-source (C-S) amplifier** shown in Fig. 7.27(a), which uses an enhancement-mode *n*-channel MOSFET ($V_{TN} > 0$) in a four-resistor bias network. The ac equivalent circuit of Fig. 7.27(b) is constructed by assuming that the capacitors all have zero impedance at the signal frequency and that the dc voltage sources represent ac grounds. In Fig. 7.27(c), the transistor has been replaced with its small-signal model. Bias resistors R_1 and R_2 appear in parallel and are combined into gate resistor R_G, and R_L represents the parallel combination of R_D, R_3 and r_o. In subsequent analysis, we will assume that the voltage gain is much less than the intrinsic voltage gain of the transistor so we can neglect transistor output resistance r_o. For simplicity at this point, we assume that we have found the Q-point and know the values of I_D and V_{DS}.

In Fig. 7.27(b) through (d), the common-source nature of this amplifier should be apparent. The input signal is applied to the transistor's gate terminal, the output signal appears at the drain, and both the input and output signals are referenced to the (common) source terminal. Note that the small-signal models for the MOSFET and BJT are virtually identical at this step, except that r_π is replaced by an open circuit for the MOSFET. Note again that only the signal portion of the input signal appears in the ac circuit model.

Our first goal is to develop an expression for the voltage gain A_v^{CS} of the circuit in Fig. 7.27(a) from the source v_i to the output v_o. As with the BJT, we will first find the terminal voltage gain A_{vt}^{CS} between the gate and drain terminals of the transistor. Then, we will use the terminal gain expression to find the gain of the overall amplifier.

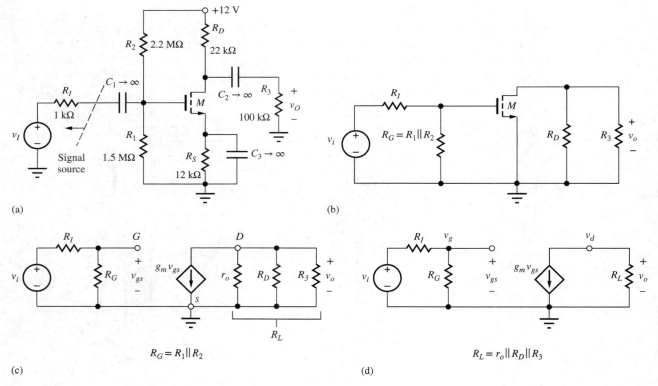

Figure 7.27 (a) Common-source amplifier circuit employing a MOSFET; (b) ac equivalent circuit for common-source amplifier in part (a); the common-source connection is now apparent; (c) ac equivalent circuit with the MOSFET replaced by its small-signal model; (d) final equivalent circuit for ac analysis of the common-source amplifier.

7.10.1 COMMON-SOURCE TERMINAL VOLTAGE GAIN

Starting with the circuit in Fig. 7.27(d), the terminal voltage gain is defined as

$$A_{vt}^{CS} = \frac{\mathbf{v_d}}{\mathbf{v_g}} = \frac{\mathbf{v_o}}{\mathbf{v_g}} \quad \text{where} \quad \mathbf{v_o} = -g_m \mathbf{v_{gs}} R_L \quad \text{and} \quad A_{vt}^{CS} = -g_m R_L \quad (7.88)$$

7.10.2 SIGNAL SOURCE VOLTAGE GAIN FOR THE COMMON-SOURCE AMPLIFIER

Now we can find the overall gain from source v_i to the voltage across R_L. The overall gain can be written as

$$A_v^{CS} = \frac{\mathbf{v_o}}{\mathbf{v_i}} = \left(\frac{\mathbf{v_o}}{\mathbf{v_g}}\right)\left(\frac{\mathbf{v_g}}{\mathbf{v_i}}\right) = A_{vt}^{CS}\left(\frac{\mathbf{v_g}}{\mathbf{v_i}}\right) \quad \text{where} \quad \mathbf{v_g} = \mathbf{v_i}\frac{R_G}{R_G + R_I} \quad (7.89)$$

v_g is related to v_i by the voltage divider formed by R_G and R_I. Combining Eqs. (7.88) and (7.89) yields a general expression for the voltage gain of the common-source amplifier:

$$A_v^{CS} = -g_m R_L \left(\frac{R_G}{R_G + R_I}\right) \quad (7.90)$$

We now explore the limits to the voltage gain of common-source amplifiers using model simplifications for zero and large values of resistance R_S. First, we will assume that the signal source resistance R_I is much less than R_G so that

$$A_v^{CS} \cong A_{vt}^{CS} = -g_m R_L \cong -g_m(R_D \parallel R_3 \parallel r_o) \quad \text{for} \quad R_I \ll R_G \quad (7.91)$$

This approximation is equivalent to saying that the total input signal appears at the gate terminal of the transistor.

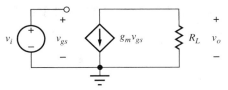

Figure 7.28 Simplified circuit for $R_G \gg R_I$.

Equation (7.88) places an upper limit on the gain we can achieve from a common-source amplifier with an external load resistor. Equation (7.88) states that the terminal voltage gain of the common-source stage is equal to the product of the transistor's transconductance g_m and load resistance R_L, and the minus sign indicates that the output voltage is "inverted" or 180° out of phase with respect to the input. The approximations that led to Eq. (7.88) are equivalent to saying that the total input signal appears across v_{gs} as shown in Fig. 7.28.

7.10.3 A DESIGN GUIDE FOR THE COMMON-SOURCE AMPLIFIER

When a resistive load is used with the common-source amplifier, we often try to achieve $R_3 \gg R_D$, and normally $r_o \gg R_D$. For these conditions, the total load resistance on the collector of the transistor is approximately equal to R_D, and Eq. (7.91) can be reduced to

$$A_v^{CS} \cong -g_m R_D = -\frac{2I_D R_D}{V_{GS} - V_{TN}} \tag{7.92}$$

using the expression for g_m from Eq. (7.64).

The product $I_D R_D$ represents the dc voltage drop across drain resistor R_D. This voltage is usually in the range of one-fourth to three-fourths the power supply voltage V_{DD}. Assuming $I_D R_D = V_{DD}/2$ and $V_{GS} - V_{TN} = 1$ V, we can rewrite Eq. (7.92) as

$$A_v^{CS} \cong -\frac{V_{DD}}{V_{GS} - V_{TN}} \cong -V_{DD} \tag{7.93}$$

Equation (7.93) is a basic rule of thumb for the design of the resistively loaded common-source amplifier; its form is very similar to that for the BJT in Eq. (7.51). The magnitude of the gain is approximately the power supply voltage divided by the internal gate drive ($V_{GS} - V_{TN}$) of the MOSFET. For a common-source amplifier operating from a 12-V power supply with a 1-V gate drive, Eq. (7.93) predicts the voltage gain to be -12.

Note that this estimate is an order of magnitude smaller than the gain for the BJT operating from the same power supply. Equation (7.92) should be carefully compared to the corresponding expression for the BJT, Eq. (7.52). Except in special circumstances, the denominator term $(V_{GS} - V_{TN})/2$ in Eq. (7.92) for the MOSFET is much greater than the corresponding term $V_T \cong 0.025$ V for the BJT, and the MOSFET voltage gain should be expected to be correspondingly lower.

DESIGN NOTE

The magnitude of the voltage gain of a resistively loaded common-source amplifier *with zero source resistance* is approximately equal to the power supply voltage:

$$A_v^{CS} \cong -V_{DD} \quad \text{for} \quad R_S = 0$$

This result represents an excellent way to quickly check the validity of more detailed calculations.

7.10.4 SMALL-SIGNAL LIMIT FOR THE COMMON-SOURCE AMPLIFIER

Using Eq. (7.89) and assuming $R_G \gg R_I$,

$$v_i = v_{gs}\frac{R_I + R_G}{R_G} \cong v_{gs} \quad \text{or} \quad v_i \leq 0.2(V_{GS} - V_{TN}) \tag{7.94}$$

The permissible input voltage is determined by the design of the bias point.

EXAMPLE 7.4   **VOLTAGE GAIN OF A COMMON-SOURCE AMPLIFIER**

In this example, we find the small-signal parameters of the MOSFET and then calculate the voltage gain of a common-source amplifier.

PROBLEM (a) Calculate the gain of the common-source amplifier in Fig. 7.27 if the transistor has $K_n = 0.500$ mA/V^2, $V_{TN} = 1$ V, and $\lambda = 0.0133$ V^{-1}, and the Q-point is (0.241 mA, 3.81 V). (b) Compare the result in (a) to the common-source "rule-of-thumb" gain estimate and the intrinsic gain of the transistor. (c) What is the largest value of v_i that can be considered to be a small-signal?

SOLUTION **Known Information and Given Data:** Common-source amplifier with its ac equivalent circuit given in Fig. 7.27; $K_n = 0.500$ mA/V^2, $V_{TN} = 1$ V, and $\lambda = 0.0133$ V^{-1}; the Q-point is (0.241 mA, 3.81 V); $R_I = 1$ kΩ, $R_1 = 1.5$ MΩ, $R_2 = 2.2$ MΩ, $R_D = 22$ kΩ, $R_3 = 100$ kΩ, $R_S = 12$ kΩ.

Unknowns: Small-signal parameters of the transistor; voltage gain A_v; small-signal limit for the value of v_i; rule-of-thumb estimate; intrinsic gain

Approach: Use the Q-point information to find g_m and r_o. Use the calculated and given values to evaluate the voltage gain expression in Eq. (7.90).

Assumptions: The transistor is in the active region of operation, and the signal amplitudes are below small-signal limit for the MOSFET. Transistor output resistance r_o can be neglected.

Analysis: (a) Calculating values of the various resistors and small-signal model parameters yields

$$g_m = \sqrt{2K_n I_{DS}(1 + \lambda V_{DS})}$$

$$= \sqrt{2\left(5 \times 10^{-4}\frac{A}{V^2}\right)(0.241 \times 10^{-3} \text{ A})\left(1 + \frac{0.0133}{V}3.81 \text{ V}\right)} = 0.503 \text{ mS}$$

$$r_o = \frac{\frac{1}{\lambda} + V_{DS}}{I_D} = \frac{\left(\frac{1}{0.0133} + 3.81\right) \text{V}}{0.241 \times 10^{-3} \text{ A}} = 328 \text{ k}\Omega$$

$$R_G = R_1 \| R_2 = 892 \text{ k}\Omega \quad\quad R_L \cong R_D \| R_3 = 18.0 \text{ k}\Omega$$

$$A_v^{CS} = -g_m R_L \frac{R_G}{R_G + R_I} = -0.503 \text{ mS}(18.0 \text{ k}\Omega)\frac{892 \text{ k}\Omega}{892 \text{ k}\Omega + 1 \text{ k}\Omega} = -9.04 \quad \text{or} \quad 19.1 \text{ dB}$$

(b) Our "rule-of-thumb" estimate for the voltage gain is $A_v = -V_{DD} = -12$, which somewhat overestimates the actual gain.

For the given Q-point,

$$V_{GS} - V_{TN} \cong \sqrt{\frac{2I_{DS}}{K_n}} = \sqrt{\frac{2 \times 0.241 \times 10^{-3} \text{ A}}{5 \times 10^{-4} \frac{\text{A}}{\text{V}^2}}} = 0.982 \text{ V}$$

and our simple estimate for the gain is

$$A_v^{CS} \cong -\frac{V_{DD}}{V_{GS} - V_{TN}} = -\frac{12 \text{ V}}{0.982 \text{ V}} = -12.2$$

which is similar to the simple rule-of-thumb estimate.

The intrinsic gain of the MOSFET is equal to

$$\mu_f = \frac{\frac{1}{\lambda} + V_{DS}}{\frac{V_{GS} - V_{TN}}{2}} = \frac{(75.2 + 3.71) \text{ V}}{0.491} = 161$$

With the source bypassed, essentially all of the input signal appears directly across the gate-source terminals of the transistor. The small-signal limit on the input signal is therefore

$$|v_{gs}| \le 0.2(V_{GS} - V_{TN}) = 0.2(0.982 \text{ V}) = 0.196 \text{ V} \quad \text{so} \quad |v_{gs}| \le 0.196 \text{ V}$$

Check of Results: The rule-of-thumb estimates are in reasonable agreement with the actual gains. The voltage gain is much less than the amplification factor, so neglect of r_o is valid.

Discussion: The rule-of-thumb produces a reasonable estimate for the gain of this amplifier. Although the amplification factor for this MOSFET is much smaller than that for the BJT, the gain of this resistively loaded amplifier circuit is still not limited by amplification factor μ_f.

Computer-Aided Analysis: SPICE simulation yields a Q-point of (0.242 mA, 3.77 V) that is consistent with the assumed value. An ac sweep from 0.1 Hz to 100 kHz with 10 frequency points/decade is used to find the region in which the capacitors are acting as short circuits, and the gain is observed to be constant at 18.7 dB above a frequency of 10 Hz. The voltage gain is slightly less than our calculated value because r_o was neglected in our calculations. A transient simulation was performed with a 0.15-V, 10-kHz sine wave. The output exhibits reasonably good linearity, but note that the positive and negative amplitudes are slightly different, indicating some waveform distortion.

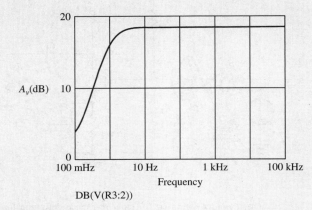

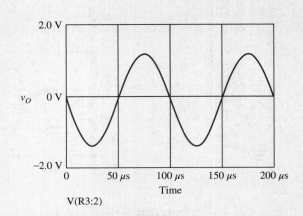

> **EXERCISE:** Calculate the Q-point for the transistor in Fig. 7.27.
>
> **EXERCISE:** Draw the small-signal ac equivalent circuit for the amplifier in Ex. 7.4 including the transistor output resistance. What is the total load resistance on the transistor? What is the new value of the voltage gain?
>
> **ANSWERS:** $R_L = r_o \| R_D \| R_3 = 328 \text{ k}\Omega \| 22 \text{ k}\Omega \| 100 \text{ k}\Omega = 17.1 \text{ k}\Omega$
>
> $A_v^{CS} = -0.503 \text{ mS } (17.1 \text{ k}\Omega) \dfrac{892 \text{ k}\Omega}{892 \text{ k}\Omega + 1 \text{ k}\Omega} = -8.59$ or 18.7 dB
>
> **EXERCISE:** Suppose we increase the transconductance parameter of the transistor to $K_n = 2 \times 10^{-3}$ A/V² by increasing the W/L ratio of the device. If the drain current is kept the same, find a new estimate for the voltage gain in Ex. 7.4. By what factor was the W/L ratio increased?
>
> **ANSWERS:** -24.4; 4

7.10.5 INPUT RESISTANCES OF THE COMMON-EMITTER AND COMMON-SOURCE AMPLIFIERS

If the voltage gain of the MOSFET amplifier is generally much lower than that of the BJT, there must be other reasons for using the MOSFET. One of the reasons was mentioned earlier: A small signal can be much larger for the MOSFET than for the BJT. Another important difference is in the relative size of the input impedance of the amplifiers. This section explores the input resistances of the common-emitter and common-source amplifiers.

The input resistance R_{in} to the common-emitter and common-source amplifiers is defined in Fig. 7.29(a) and (b) to be the total resistance looking into the amplifier at coupling capacitor C_1. R_{in} represents the total resistance presented to the signal source represented by v_I and R_I.

Common-Emitter Input Resistance

Let us first calculate the input resistance for the common-emitter stage. In Fig. 7.30, the BJT has been replaced by its small-signal model, and the input resistance is found to be

$$\mathbf{v_x} = \mathbf{i_x}(R_B \| r_\pi) \quad \text{and} \quad R_{\text{in}}^{CE} = \dfrac{\mathbf{v_x}}{\mathbf{i_x}} = R_B \| r_\pi = R_1 \| R_2 \| r_\pi \tag{7.95}$$

R_{in} is equal to the parallel combination of r_π and the two base-bias resistors R_1 and R_2.

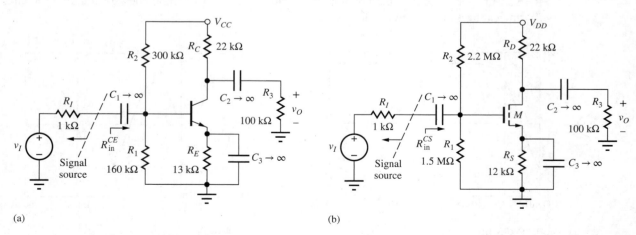

Figure 7.29 (a) Input resistance definition for the common-emitter amplifier; (b) input resistance definition for the common-source amplifier.

7.10 The Common-Source (C-S) Amplifier

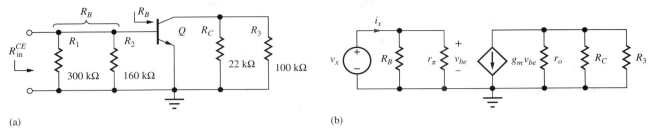

Figure 7.30 (a) ac Equivalent circuits for the input resistance for the common-emitter amplifier; (b) small-signal model.

EXAMPLE 7.5 **INPUT RESISTANCE OF THE COMMON-EMITTER AMPLIFIER**

Let us calculate R_{in} for the amplifier in Fig. 7.29 for a given Q-point.

PROBLEM Find the input resistance for the common-emitter amplifier in Figs. 7.29 and 7.30. The Q-point is (0.245 mA, 3.39 V).

SOLUTION **Known Information and Given Data:** The small-signal circuit topology appears in Fig. 7.30. The Q-point is given as (0.245 mA, 3.39 V). From Fig. 7.30, we have $R_1 = 160$ kΩ, $R_2 = 300$ kΩ, and $R_3 = 100$ kΩ.

Unknowns: Input resistance looking into the common-emitter amplifier.

Approach: Find r_π and use Eq. (7.95) to find the input resistance.

Assumptions: Small-signal conditions apply, $\beta_o = 100$, $V_T = 25$ mV

Analysis: The values of R_B and r_π are

$$R_B = R_1 \| R_2 = 160 \text{ k}\Omega \| 300 \text{ k}\Omega = 104 \text{ k}\Omega \quad \text{and} \quad r_\pi = \frac{\beta_o V_T}{I_C} = \frac{100(0.025)}{0.245 \text{ mA}} = 10.2 \text{ k}\Omega$$

$$R_{in}^{CE} = R_B \| R_{iB} = R_B \| r_\pi = 104 \text{ k}\Omega \| 10.2 \text{ k}\Omega = 9.29 \text{ k}\Omega$$

Check of Results: The input resistance must be smaller than any one of the resistors R_1, R_2, or r_π, since they all appear in parallel. The calculated value of input resistance is consistent with this observation.

Discussion: With the emitter terminal bypassed, the input resistance to the amplifier, 9.29 kΩ, is quite low and is dominated by r_π. In the next chapter we will discover how to increase the input resistance of the common-emitter amplifier.

Computer-Aided Analysis: (a) We may use an ac analysis of the circuit from Fig. 7.30(a) to determine R_{in} by finding the signal current in source v_I. (Note that a TF analysis cannot be used because of the presence of capacitors in the network.) The input resistance is equal to the base voltage divided by the current entering the base terminal through C_1. SPICE yields VB(Q1)I(C1) = 9.80 kΩ, which is 5 percent higher than our calculations. This discrepancy results from the values of ac current gain β_o and thermal voltage V_T used by SPICE, since both differ slightly from our hand calculations.

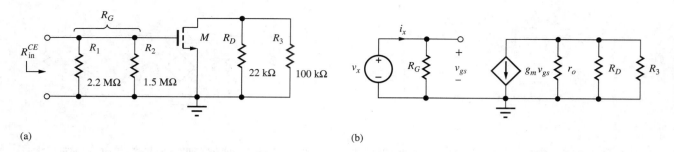

(a)   (b)

Figure 7.31 (a) ac Equivalent circuits for the input resistance for the common-source amplifiers; (b) small-signal model.

EXERCISE: What is the value of R_{in}^{CE} if the Q-point is changed to (0.725 mA, 3.86 V)?

ANSWER: 3.34 kΩ

Common-Source Input Resistance
Now let us compare the input resistance of the common-source amplifier to that of the common-emitter stage. In Fig. 7.31, the MOSFET in Fig. 7.29 has been replaced by its small-signal model. This circuit is similar to that in Fig. 7.30 except that $r_\pi \to \infty$. Because the gate terminal of the MOSFET itself represents an open circuit, the input resistance of the circuit is simply limited by our value of R_G:

$$\mathbf{v_x} = \mathbf{i_x} R_G \quad \text{and} \quad R_{in}^{CS} = R_G \tag{7.96}$$

In the C-S amplifier in Fig. 7.29, $R_G = 2.2\,\text{M}\Omega \parallel 1.5\,\text{M}\Omega = 892\,\text{k}\Omega$, so $R_{in}^{CS} = 892\,\text{k}\Omega$. We see that the input resistance of the C-S amplifier can easily be much larger than that of the corresponding C-E stage.

EXERCISE: What is the input resistance of the common-source amplifier in Fig. 7.29(b) if $R_2 = 1.0$ MΩ and $R_1 = 680$ kΩ? Is the Q-point of the amplifier changed?

ANSWERS: 405 kΩ; no, the Q-point remains the same because $I_G = 0$ and the dc voltage at the gate is unchanged.

7.10.6 COMMON-EMITTER AND COMMON-SOURCE OUTPUT RESISTANCES
The output resistances of the C-E and C-S amplifiers are defined in Fig. 7.32(a) and (b) as the total equivalent resistance looking into the output of the amplifier at coupling capacitor C_2. The definition of the output resistance is repeated in Fig. 7.33, in which the two amplifiers have been reduced to their ac equivalent circuits. For the output resistance calculation, independent input source v_I is set to zero.

Output Resistance of the Common-Emitter Amplifier
The transistors are replaced with their small-signal models in Fig. 7.34, and test source v_x is applied to the output in order to calculate the output resistance. For the BJT in Fig. 7.34(a), the current from v_x is equal to

$$\mathbf{i_x} = \frac{\mathbf{v_x}}{R_C} + \frac{\mathbf{v_x}}{r_o} + g_m \mathbf{v_{be}} \tag{7.97}$$

However, there is no excitation at the base node:

$$\frac{\mathbf{v_{be}}}{R_I} + \frac{\mathbf{v_{be}}}{R_B} + \frac{\mathbf{v_{be}}}{r_\pi} = 0 \quad \text{and} \quad v_{be} = 0 \tag{7.98}$$

7.10 The Common-Source (C-S) Amplifier   423

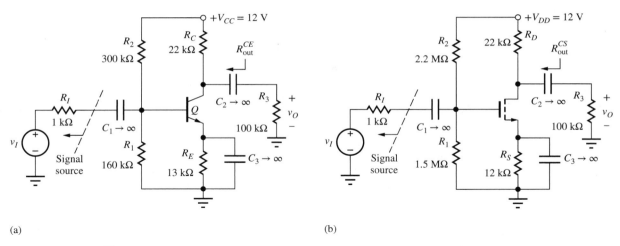

Figure 7.32 (a) Output resistance definition for the common-emitter amplifier; (b) output resistance definition for the common-source amplifier.

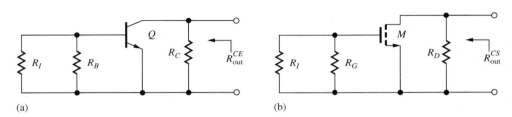

Figure 7.33 Output resistance for (a) common-emitter and (b) common-source amplifiers.

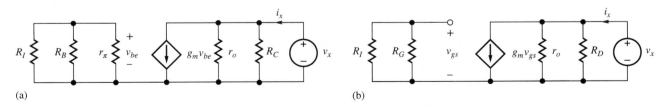

Figure 7.34 Small signal models for (a) C-E and (b) C-S amplifier output resistance.

Thus, $g_m v_{be} = 0$, and the output resistance is equivalent to the parallel combination of R_C and r_o given by

$$R_{out}^{CE} = \frac{v_x}{i_x} = r_o \| R_C \qquad (7.99)$$

For the common-emitter amplifier in Fig. 7.29(a), we have $R_{out}^{CE} = 320 \text{ k}\Omega \| 22 \text{ k}\Omega = 20.6 \text{ k}\Omega$ where the value of r_o was found earlier in Ex. 7.3.

Let us compare the values of r_o and R_C by multiplying each by I_C:

$$I_C r_o = I_C \frac{V_A + V_{CE}}{I_C} \cong V_A \quad \text{and} \quad I_C R_C \cong \frac{V_{CC}}{3} \qquad (7.100)$$

As discussed previously, the voltage developed across the collector resistor R_C is typically 0.25 to $0.75 V_{CC}$, but the apparent voltage across r_o is the Early voltage V_A. Thus, from the relations in Eq. (7.100), we expect $r_o \gg R_C$, and Eq. (7.99) yields $R_{out}^{CE} \cong R_C$.

Output Resistance of the Common-Source Amplifier

For the MOSFET in Fig. 7.34(b), the analysis is the same. Voltage v_{gs} will be zero, and R_{out} is equal to the parallel combination of r_o and R_D:

$$R_{\text{out}}^{CS} = \frac{\mathbf{v_x}}{\mathbf{i_x}} = r_o \| R_D \quad (7.101)$$

For the common-source amplifier in Fig. 7.29(b), we have $R_{\text{out}}^{CS} = 328 \text{ k}\Omega \| 22 \text{ k}\Omega = 20.6 \text{ k}\Omega$ where the value of r_o was found earlier in Ex. 7.4. Note that the output resistances of the common-emitter and common-source amplifier examples are essentially the same.

Comparing r_o and R_D in a manner similar to that for the BJT,

$$I_D r_o = I_D \frac{\frac{1}{\lambda} + V_{DS}}{I_D} \cong \frac{1}{\lambda} \quad \text{and} \quad I_D R_D \cong \frac{V_{DD}}{2} \quad (7.102)$$

where it is assumed that the voltage developed across the drain resistor R_D is $V_{DD}/2$. The effective voltage across r_o is equivalent to $1/\lambda$. Because the value of $1/\lambda$ is similar to the Early voltage V_A, we expect $r_o \gg R_D$, and Eq. (7.101) can be simplified to $R_{\text{out}}^{CS} \cong R_D$. We conclude that, for comparable bias points $(I_C, V_{CE}) = (I_{DS}, V_{DS})$, the output resistances of the C-E and C-S stages are similar and limited by the resistors R_C and R_D.

EXAMPLE 7.6 A COMMON-SOURCE AMPLIFIER USING A JFET

Our final example in this chapter is a common-source amplifier using an *n*-channel JFET as depicted in Fig. 7.35. Although not used as often as BJTs and MOSFETs, JFETs do play important roles in analog circuits, both discrete and integrated. Capacitors C_1 and C_2 are used to couple the signal into and out of the amplifier, and bypass capacitor C_3 provides an ac ground at the source of the JFET. The JFET is inherently a depletion-mode device; it requires only three resistors for proper biasing: R_G, R_4, and R_D.

PROBLEM Find the input resistance, output resistance, and voltage gain for the common-source amplifier in Fig. 7.35.

SOLUTION **Known Information and Given Data:** The circuit topology with element values appears in the Fig. 7.35. The transistor parameters are specified in the figure to be $I_{DSS} = 1$ mA, $V_P = -1$ V, and $\lambda = 0.02$ V^{-1}.

Unknowns: Q-point (I_D, V_{DS}); small-signal parameters, R_{in}, R_{out}, and A_v.

Approach: To analyze the circuit, we first draw the dc equivalent circuit and find the Q-point. Then we develop the ac equivalent circuit, find the small-signal model parameters, and characterize the small-signal properties of the amplifier.

Assumptions: Pinch-off region operation for the JFET; λ can be ignored in dc bias calculations; small-signal operating conditions apply.

Q-Point Analysis: The dc equivalent circuit, obtained from Fig. 7.35 by opening the capacitors, appears in Fig. 7.36. Assuming operation in the pinch-off region, the drain current of the JFET is expressed by [see Eq. (5.99)]

$$I_D = I_{DSS}\left(1 - \frac{V_{GS}}{V_P}\right)^2$$

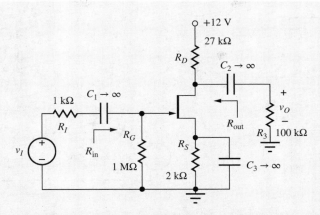

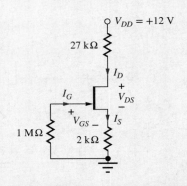

Figure 7.35 Common-source amplifier using a junction field-effect transistor. For the JFET, $I_{DSS} = 1$ mA, $V_P = -1$ V, $\lambda = 0.02$ V^{-1}.

Figure 7.36 Circuit for determining the Q-point of the JFET.

in which λ is neglected for dc analysis. The gate-source voltage may be related to the drain current by writing a loop equation including V_{GS}:

$$I_G(10^6) + V_{GS} + I_S(2000) = 0$$

However, the gate current is zero, so $I_S = I_D$ and $V_{GS} = -2000 I_D$. Substituting this result and the device parameters into the drain current expression yields a quadratic equation for V_{GS}:

$$V_{GS} = -(2 \times 10^3)(1 \times 10^{-3})\left[1 - \frac{V_{GS}}{(-1)}\right]^2$$

Rearranging this expression for V_{GS}, we get

$$2V_{GS}^2 + 5V_{GS} + 2 = 0 \quad \text{and} \quad V_{GS} = -0.50 \text{ V}, -2.0 \text{ V}$$

V_{GS} must be negative but less negative than the pinch-off voltage of the n-channel JFET, so the -0.50-V result must be the correct choice. The corresponding value of I_D becomes

$$I_D = 10^{-3} \text{A} \left(1 - \frac{-0.50 \text{ V}}{-1 \text{ V}}\right)^2 = 0.250 \text{ mA}$$

V_{DS} can be found by writing the load-line equation for the JFET,

$$12 = 27{,}000 I_D + V_{DS} + 2000 I_S$$

Substituting $I_S = I_D = 250$ µA gives the Q-point:

$$(250 \text{ µA}, 4.75 \text{ V})$$

Check of Results and Discussion: As always, we must check the region of operation to be sure our original assumption of pinch-off was correct:

$$V_{DS} \geq V_{GS} - V_P \quad 4.75 > -0.50 - (-1) \quad 4.75 > 0.50 \;\checkmark$$

In this dc analysis, we neglected the channel-length modulation term since we want to use the lowest complexity model that provides reasonable answers. For this problem, we see that λV_{DS} is $(0.02 \text{ V}^{-1})(5 \text{ V}) = 0.10$. Including the λV_{DS} term would change our answers by at most 10 percent but would considerably complicate the dc analysis. In addition, any differences in Q-point values would be less than the total uncertainty in the circuit and device parameter values.

426   Chapter 7   The Transistor as an Amplifier

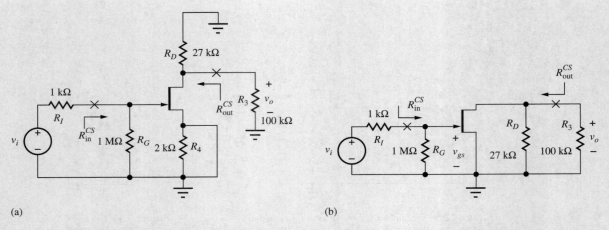

(a)   (b)

Figure 7.37 (a) Construction of the ac equivalent circuit; (b) redrawn version of the circuit in (a).

ac ANALYSIS As in Ex. 7.3 and Ex. 7.4, we begin the ac analysis by finding the ac equivalent circuit. In Fig. 7.37(a), the capacitors in Fig. 7.35 have been replaced with short circuits, and the dc voltage source has been replaced with a ground connection. The ac equivalent circuit is redrawn in Fig. 7.37(b) by eliminating resistor R_4 and indicating the parallel connection of R_D and R_3. We recognize this as a common-source circuit since the source of the JFET is clearly the terminal in common between the input and output ports.

Small-Signal Parameters and Voltage Gain: We wish to find the voltage gain from v_i to v_o for the amplifier in Fig. 7.37. The output voltage at the drain terminal is related to the voltage at the gate by the terminal gain in Eq. (7.91), $\mathbf{v_o} = -g_m R_L \mathbf{v_{gs}}$, where R_L is the total load resistance at the drain terminal. R_L is equal to R_{out} in parallel with external load resistor R_3, $R_L = R_{out} \| R_3$. Gate-source voltage v_{gs} is related to v_i through voltage division between the source resistance R_I and input resistance R_{in}. Combining these results yields an expression for the overall voltage gain:

$$A_v = \frac{\mathbf{v_o}}{\mathbf{v_i}} = -g_m(R_{out} \| R_3) \frac{R_{in}}{R_I + R_{in}} \tag{7.103}$$

The final step prior to mathematical analysis is to find the small-signal model parameters. Using the Q-point values,

$$g_m = \frac{2}{|V_P|} \sqrt{I_{DSS} I_{DS}(1 + \lambda V_{DS})} = \frac{2}{|-1\,\text{V}|} \sqrt{(0.001\,\text{A})(0.00025\,\text{A})\left(1 + \frac{0.02}{\text{V}} 4.75\,\text{V}\right)}$$

$$g_m = 1.05\,\text{mS}$$

$$r_o = \frac{\frac{1}{\lambda} + V_{DS}}{I_{DS}} = \frac{(50 + 4.75)\,\text{V}}{0.25 \times 10^{-3}\,\text{A}} = 219\,\text{k}\Omega \quad \text{and} \quad \mu_f = g_m r_o = 230$$

Input Resistance: The amplifier's input resistance is calculated looking into the position of coupling capacitor C_1 in Figs. 7.35 and 7.37, and the equivalent circuit for finding R_{in} is redrawn in Fig. 7.38. In Fig. 7.38(b), we see that the input resistance is set by gate-bias resistor R_G, because the input resistance of the JFET itself is infinite:

$$R_{in}^{CS} = R_G = 1\,\text{M}\Omega$$

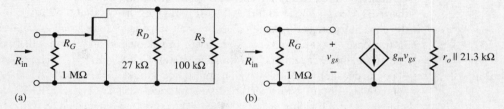

Figure 7.38 (a) ac Equivalent circuit for determining R_{in}; (b) small-signal model for the circuit in part (a).

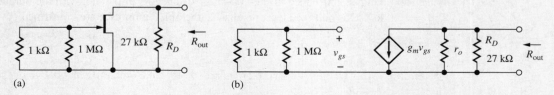

Figure 7.39 (a) ac Equivalent circuit for determining R_{out}; (b) small-signal model for circuit in (a).

Output Resistance: The amplifier's output resistance is calculated looking into the position of coupling capacitor C_2 in Figs. 7.35 and 7.37. The equivalent circuit for calculating R_{out}^{CS} is presented in the schematic in Fig. 7.39. In Fig. 7.39(b), the voltage $v_{gs} = 0$, and the output resistance is equal to the parallel combination of R_D and r_o:

$$R_{out}^{CS} = R_D \| r_o = 27 \text{ k}\Omega \| 219 \text{ k}\Omega = 24.0 \text{ k}\Omega$$

Voltage Gain: Substituting these values into Eq. (7.103) and solving for the voltage gain gives

$$A_v = \frac{v_o}{v_i} = -(1.05 \text{ mS})(24 \text{ k}\Omega \| 100 \text{ k}\Omega)\left(\frac{1 \text{ M}\Omega}{1 \text{ k}\Omega + 1 \text{ M}\Omega}\right) = -20.3$$

Thus, this particular common-source JFET amplifier is an inverting amplifier with a voltage gain of −20.3 or 26.2 dB.

Check of Results: We have found the answers requested in the problem. Our rule-of-thumb estimate for the voltage gain would be $A_v = -V_{DD} = -12$, so the calculated gain seems reasonable. Looking at the circuits in Fig. 7.37, we quickly see that the input and output resistances should not exceed 1 MΩ and 27 kΩ, respectively, which also agree with our more detailed calculations. In summary, our JFET amplifier provides the following characteristics:

$$A_v = -20.3 \qquad R_{in}^{CS} = 1.00 \text{ M}\Omega \qquad R_{out}^{CS} = 24.0 \text{ k}\Omega$$

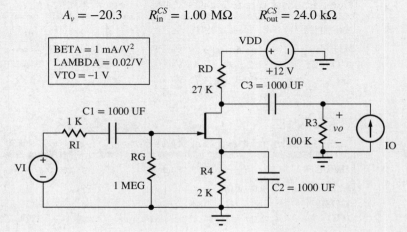

Computer-Aided Analysis: The JFET parameters must be correctly defined in SPICE. Remember that BETA = I_{DSS}/V_P^2 = 0.001 A/V². SPICE gives the Q-point (257 μA, 5.05 V). For ac analysis the capacitors are set to large values so their impedances are small at the frequencies of interest. In this case, 1000-μF capacitors are used. In Chapters 8 and 9, we will find how to choose the values for these capacitors. An ac analysis (DEC, FSTART = 1 kHz, FSTOP = 100 kHz, and 3 points/decade) with a 1-V value for source v_I and $i_o = 0$ yields $A_v = -20.4$. The current in source v_I is 999 nA corresponding to a total input resistance of 1.001 MΩ. Subtracting the 1-kΩ source resistance yields R_{in} = 1.00 MΩ. The output resistance can be found by driving the output with a 1-A ac current source with $v_I = 0$ yielding a total resistance of 19.3 kΩ at the output node. Removing the influence of the 100-kΩ resistance R_4 in parallel with the output node yields R_{out} = 23.9 kΩ. Our hand analysis results for the JFET amplifier are confirmed.

EXERCISE: What is the amplification factor of the JFET characterized by the parameters in Ex. 7.6? How does A_v compare to μ_f?

ANSWERS: 230; $|A_v| \ll \mu_f$

EXERCISE: What is the largest value of v_i that corresponds to a small signal for the JFET in this amplifier? What is the largest value of v_o that corresponds to a small signal in this amplifier?

ANSWERS: 100 mV; 2.04 V

EXERCISE: Verify the dc and ac analysis using SPICE. Compare the operating points with $\lambda = 0$ and $\lambda = 0.02$ V.

7.10.7 COMPARISON OF THE THREE AMPLIFIER EXAMPLES

Tables 7.4 and 7.5 compare the numerical results for the amplifiers analyzed in Exs. 7.3 through 7.5. The three amplifiers have all been designed to have similar Q-points, as indicated in Table 7.4. In this table, we see that the BJT yields a much higher voltage gain than either of the FET circuits. However, all the voltage gains are well below the value of the amplification factor, which is characteristic of amplifiers with resistive loads in which the gain is limited by the external resistors (for $r_o \gg R_C$ or R_D).

TABLE 7.4
Comparison of Three Amplifier Voltage Gains

AMPLIFIER	Q-POINT	A_v	μ_f	RULE-OF-THUMB ESTIMATES
BJT	(245 μA, 3.39 V)	−159	3140	−120
MOSFET	(241 μA, 3.81 V)	−9.04	161	−12
JFET	(250 μA, 4.75 V)	−20.4	230	−12

TABLE 7.5
Comparison of Input and Output Resistances

AMPLIFIER	R_{in}	R_B or R_G	r_π	R_{out}	R_C or R_D	r_o
BJT	9.29 kΩ	100 kΩ	10.2 kΩ	20.6 kΩ	22 kΩ	320 kΩ
MOSFET	892 kΩ	892 kΩ	∞	20.6 kΩ	22 kΩ	328 kΩ
JFET	1.00 MΩ	1.00 MΩ	∞	24.0 kΩ	27 kΩ	219 kΩ

Table 7.5 compares the input and output resistances. We see that the bipolar input resistance, in this case dominated by the value of r_π, is orders of magnitude smaller than that of the FETs. On the other hand, R_{in} of the FET stages is limited by the choice of gate-bias resistor R_G. All the output resistances are limited by the external resistors and are of similar magnitude.

7.11 COMMON-EMITTER AND COMMON-SOURCE AMPLIFIER SUMMARY

Table 7.6 presents a comparison of the ac small-signal characteristics of the common-emitter (C-E) and common-source (C-S) amplifiers based on the analyses presented in this chapter. The voltage gain expressions collapse to the same symbolic form, but the values will differ because the value of g_m for the BJT is usually much larger than that of the FET for a given operating current. The input resistance of the C-S stages is limited only by the design value of R_G and can be quite large, whereas the values of R_B and r_π limit the input resistance of the C-E amplifier to much smaller values. For a given operating point, the output resistances of the C-E and C-S stages are similar because R_{out} is limited by the collector- or drain-bias resistors R_C or R_D.

7.11.1 GUIDELINES FOR NEGLECTING THE TRANSISTOR OUTPUT RESISTANCE

In all these amplifier examples, we found that the transistor's own output impedance did not greatly affect the results of the various calculations. The following question naturally arises: Why not just neglect r_o altogether, which will simplify the analysis? The answer is: The resistance r_o must be included whenever it makes a difference. We use the following rule: The transistor output resistance r_o can be neglected in voltage gain calculations as long as the computed value of $A_v \ll \mu_f$. However, in Thévenin equivalent resistance calculations r_o can play a very important role and one must be careful not to overlook limitations due to r_o. If r_o is neglected, and an input or output resistance is calculated that is similar to or much larger than r_o, then the calculation should be rechecked with r_o included in the circuit. At this point, this procedure may sound mysterious, but in the next several chapters we shall find circuits in which r_o is very important.

DESIGN NOTE

You can neglect the transistor output resistance in voltage gain calculations as long as the computed value is much less than the transistor's intrinsic gain μ_f! When the output resistance is included in a calculation, we often do not know V_{CE} or V_{DS}, and it is perfectly acceptable to use the simplified expression for the output resistances:

$$r_o = \frac{V_A}{I_C} \quad \text{or} \quad r_o = \frac{1}{\lambda I_D}$$

TABLE 7.6
Common-Emitter/Common-Source Amplifier Characteristics

	COMMON-EMITTER AMPLIFIER	COMMON-SOURCE AMPLIFIERS
Terminal gain A_{vt}	$-g_m R_L$	$-g_m R_L$
Rule-of-thumb estimate for $g_m R_L$	$-10 V_{CC}$	$-V_{DD}$
Voltage gain A_v	$A_v = \dfrac{v_o}{v_i} = -g_m(R_{out} \| R_3)\left(\dfrac{R_{in}}{R_I + R_{in}}\right)$	
Input resistance R_{in}	$R_B \| r_\pi$	R_G
Output resistance R_{out}	$R_C \| r_o \cong R_C$	$R_D \| r_o \cong R_D$
Input small-signal limit	0.005 V	$0.2(V_{GS} - V_{TN})$ or $0.2(V_{GS} - V_P)$

7.12 AMPLIFIER POWER AND SIGNAL RANGE

We found in our examples how the selection of Q-point affects the value of the small-signal parameters of the transistors and hence affects the voltage gain, input resistance, and output resistance of common-emitter and common-source amplifiers. For the FET, the choice of Q-point also determines the value of v_{gs} that corresponds to small-signal operation. Two additional characteristics that are set by Q-point design are discussed in this section. The choice of operating point determines the level of power dissipation in the transistor and overall circuit, and it also determines the maximum linear signal range at the output of the amplifier.

7.12.1 POWER DISSIPATION

The static power dissipation of the amplifiers can be determined from the dc equivalent circuits used earlier. The power that is supplied by the dc sources is dissipated in both the resistors and transistors. For the amplifier in Fig. 7.40(a), for example, power P_D dissipated in the transistor is the sum of the power dissipation in the collector-base and emitter-base junctions:

$$P_D = V_{CB}I_C + V_{BE}(I_B + I_C) = (V_{CB} + V_{BE})I_C + V_{BE}I_B$$

or
$$P_D = V_{CE}I_C + V_{BE}I_B \quad \text{where} \quad V_{CE} = V_{CB} + V_{BE} \tag{7.104}$$

The total power P_S supplied to the amplifier is determined by the currents in the power supply:

$$P_S = V_{CC}(I_C + I_2) \tag{7.105}$$

Similarly for the MOSFET circuit in Fig. 7.40(b), the power dissipated in the transistor is given by

$$P_D = V_{DS}I_D + V_{GS}I_G = V_{DS}I_D \tag{7.106}$$

because the gate current is zero. The total power being supplied to the amplifier is equal to

$$P_S = V_{DD}(I_D + I_2) \tag{7.107}$$

> **EXERCISE:** What power is being dissipated by the bipolar transistor in Fig. 7.40(a)? Assume $\beta_F = 65$. What is the total power being supplied to the amplifier? Use the Q-point information given earlier (245 µA, 3.39 V).
>
> **ANSWERS:** 0.833 mW; 3.26 mW

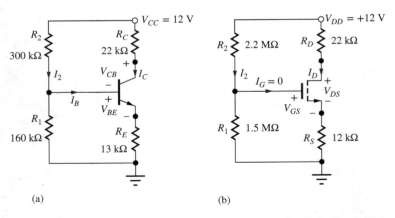

Figure 7.40 dc Equivalent circuits for the (a) BJT and (b) MOSFET amplifiers from Figs. 7.18(a) and 7.28(a).

EXERCISE: What power is being dissipated by the MOSFET in Fig. 7.40(b)? What is the total power being supplied to the amplifier? Use the Q-point information given earlier (241 µA, 3.81 V).

ANSWERS: 0.918 mW; 2.93 mW

7.12.2 SIGNAL RANGE

We next discuss the relationship between the Q-point and the amplitude of the signals that can be developed at the output of the amplifier. Consider the amplifier in Fig. 7.41 with $V_{CC} = 12$ V, and the corresponding waveforms, which are given in Fig. 7.42. The collector and emitter voltages at the operating point are 5.9 V and 2.10 V, respectively, and hence the value of V_{CE} at the Q-point is 3.8 V.

Because the bypass capacitor at the emitter forces the emitter voltage to remain constant, the total collector-emitter voltage can be expressed as

$$v_{CE} = V_{CE} - V_M \sin \omega t \tag{7.108}$$

in which $V_M \sin \omega t$ is the signal voltage being developed at the collector. The bipolar transistor must remain in the active region at all times, which requires that the collector-emitter voltage remain larger than base-emitter voltage V_{BE}:

$$v_{CE} \geq V_{BE} \quad \text{or} \quad v_{CE} \geq 0.7 \text{ V} \tag{7.109}$$

Thus the amplitude of the signal at the collector must satisfy

$$V_M \leq V_{CE} - V_{BE} \tag{7.110}$$

The positive power supply presents an additional limit to the signal swing. Writing an expression for the voltage across resistor R_C,

$$v_{R_C}(t) = I_C R_C + V_M \sin \omega t \geq 0 \tag{7.111}$$

In this circuit, the voltage across the resistor cannot become negative; that is, the voltage V_C at the transistor collector cannot exceed the power supply voltage V_{CC}. Equation (7.111) indicates

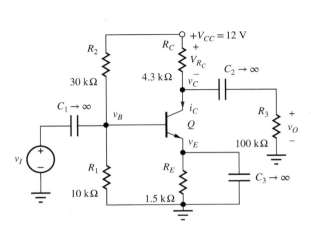

Figure 7.41 Common-emitter amplifier stage.

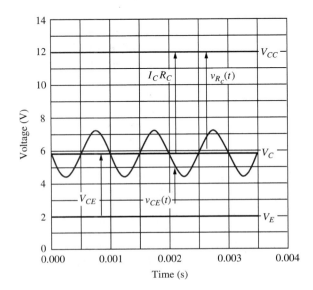

Figure 7.42 Waveforms for the amplifier in Fig. 7.41.

that the amplitude V_M of the ac signal developed at the collector must be smaller than the voltage drop across R_C at the Q-point:

$$V_M \leq I_C R_C \qquad (7.112)$$

Thus, the signal swing at the collector is limited by the smaller of the two limits expressed in Eq. (7.110) or (7.112):

$$V_M \leq \min[I_C R_C, (V_{CE} - V_{BE})] \qquad (7.113)$$

Similar expressions can be developed for field-effect transistor circuits. We must require that the MOSFET remains pinched off, or v_{DS} must always remain larger than $v_{GS} - V_{TN}$.

$$v_{DS} = V_{DS} - V_M \sin \omega t \geq V_{GS} - V_{TN} \qquad (7.114)$$

in which it has been assumed that $v_{gs} \ll V_{GS}$. In direct analogy to Eq. (7.112) for a BJT circuit, the signal amplitude in the FET case also cannot exceed the dc voltage drop across R_D:

$$V_M \leq I_D R_D \qquad (7.115)$$

So, for the case of the MOSFET, V_M must satisfy:

$$V_M \leq \min[I_D R_D, (V_{DS} - (V_{GS} - V_{TN}))] \qquad (7.116)$$
$$\leq \min[I_D R_D, (V_{DS} - V_{DSAT})]$$

EXERCISE: (a) What is V_M for the bipolar transistor amplifier in Ex. 7.3? (b) For the MOSFET amplifier in Ex. 7.4?

ANSWERS: 2.69 V; 2.83 V

ELECTRONICS IN ACTION

Electric Guitar Distortion Circuits

For most of this chapter we have focused on small-signal models and gain calculations. However, in some applications, it is desirable to intentionally violate small-signal constraints and generate a distorted waveform. In particular, electric guitars, the mainstay of contemporary music, intentionally use distortion to enrich the sound. The early Marshall and Fender tube amps, through substantial over-design and the natural characteristics of vacuum tube circuits, generated a rich soft-clipped sound when driven into overload. When excited with the right chords, the tube amplifier distortion can actually generate harmonics that are in-tune and add a great deal to the character of the electric guitar sound.

Modern guitar players use "pedal" boxes to produce distortion and other effects without the excessive power levels required to produce the overdrive sound. Typical forms of these circuits are shown below. The first is an op-amp circuit with a pair of diodes in the feedback network. R_2 is 50 to 200 times larger than R_1, so the circuit has a large gain. As the voltage across the amplifier exceeds the diode turn-on voltage, the diodes begin to conduct. Since the diode impedance is much less than R_2, the gain is reduced during diode conduction. The resulting "soft" clipped waveform is shown below.

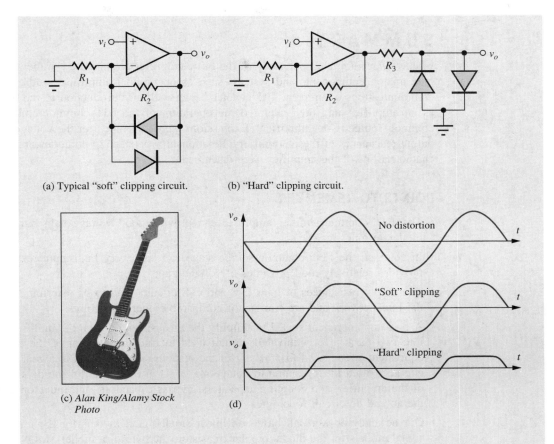

(a) Typical "soft" clipping circuit.
(b) "Hard" clipping circuit.
(c) *Alan King/Alamy Stock Photo*
(d)

Another form of distortion circuit is the "hard" clipping circuit. The amplifier gain is again set to be quite large, and resistor R_3 is typically a few kilohms. As v_o exceeds the diode turn-on voltage, the output is clipped to the diode voltage. In this case, the diode current is limited by R_3, so v_o changes very little once the diode turns on. This results in a "hard" clipped waveform as seen above. Typically, practical circuits also include some frequency shaping.

From Fourier analysis, we know that any cyclical waveform shape other than an ideal sine wave is composed of a possibly infinite set of harmonics or sine and cosine waves, each at frequencies which are multiples of the fundamental frequency. The sharper the transitions in a waveform, the more harmonic content it contains. The soft clipping circuit creates a waveform with smaller amplitude of harmonics than the hard clipping circuit.

There are also additional tones created by the intermodulation of the incoming frequencies. In these nonlinear clipping circuits, the incoming frequencies mix and give rise to sum and difference frequencies. This is an additional audible effect of the distortion circuits. There are many variations on these simple circuits which produce a wide range of sounds. The guitarist must select between a variety of different distortion and effects devices to create the sound that optimally presents their musical ideas.

SUMMARY

Chapter 7 has initiated our study of the basic amplifier circuits used in the design of more complex analog components and systems such as operational amplifiers, audio amplifiers, and RF communications equipment. The chapter began with an introduction to the use of the transistor as an amplifier, and then explored the operation of the BJT common-emitter (C-E) and FET common-source (C-S) amplifiers. Expressions were developed for the voltage gain and input and output resistances of these amplifiers. Relationships between Q-point design and the small-signal characteristics of the amplifier were discussed.

POINTS TO REMEMBER

- The BJT common-emitter amplifier can provide good voltage gain but has only a low-to-moderate input resistance.
- In contrast, the FET common-source stage can have very high-input resistance but typically provides relatively modest values of voltage gain.
- The output resistances of both C-E and C-S circuits tend to be determined by the resistors in the bias network and are similar for comparable operating points.
- A two-step approach is used to simplify the analysis and design of amplifiers. Circuits are split into two parts: a dc equivalent circuit used to find the Q-point of the transistor, and an ac equivalent circuit used for analysis of the response of the circuit to signal sources. The design engineer often must respond to competing goals in the design of the dc and ac characteristics of the amplifier, and coupling capacitors, bypass capacitors, and inductors are used to change the ac and dc circuit topologies.
- Our ac analyses were all based on linear small-signal models for the transistors. The small-signal models for the diode, bipolar transistor (the hybrid-pi model), MOSFET and JFET were all discussed in detail. The expressions relating the transconductance g_m, output resistance r_o, and input resistance r_π to the Q-point were all found by evaluating derivatives of the large-signal model equations developed in earlier chapters.
- The small-signal model for the diode is simply a resistor that has a value given by $r_d = V_T/I_D$.
- The results in Table 7.3 on page 389 for the three-terminal devices are extremely important. The structure of the models is similar. The transconductance of the BJT is directly proportional to current, whereas that of the FET increases only in proportion to the square root of current in strong inversion but becomes proportional to current in weak inversion. Resistances r_π and r_o are inversely proportional to Q-point current. Resistor r_π is infinite for the case of the FET, so it does not actually appear in the small-signal model. It was discovered that each device pair, the *npn* and *pnp* BJTs, and the *n*-channel and *p*-channel FETs, has the same structure in the small-signal model.
- The small-signal current gain of the BJT was defined as $\beta_o = g_m r_\pi$, and its value generally differs from that of the large-signal current gain β_F. The FET exhibits an infinite small-signal current gain at low frequency.
- The intrinsic voltage gain, also known as intrinsic gain or amplification factor of the transistor, is defined as $\mu_f = g_m r_o$ and represents the maximum gain available from the transistor in the C-E and C-S amplifiers. Expressions were evaluated for the intrinsic gain of the BJT and FETs. Parameter μ_f was found to be independent of Q-point for the BJT, but for the FET, the amplification factor decreases as operating current increases. For usual operating points, μ_f for the BJT will be several thousand, whereas that for the FET ranges between tens and hundreds.

- The definition of a small signal was found to be device-dependent. The signal voltage v_d developed across the diode must be less than 5 mV in order to satisfy the requirements of a small signal. Similarly, the base-emitter signal voltage v_{be} of the BJT must be less than 5 mV for small-signal operation. However, FETs can amplify much larger signals without distortion. For the MOSFET, $v_{gs} \leq 0.2(V_{GS} - V_{TN})$ represents the small-signal limit and can be designed to range from 100 mV to more than 1 V. For the JFET, $v_{gs} \leq 0.2(V_{GS} - V_P)$. In weak inversion, the FET becomes similar to the BJT, and the small-signal limit is less than 10 mV.

- Common-emitter and common-source amplifiers were analyzed in detail. Table 7.6 on page 405 is another extremely important table. It summarizes the overall characteristics of these amplifiers. The rule-of-thumb estimates in Table 7.6 were developed to provide quick predictions of the voltage gain of the C-E and C-S stages.

- The chapter closed with a discussion of the relationship between operating point design and the power dissipation and output signal swing of the amplifiers. The amplitude of the signal voltage at the output of the amplifier is limited by the smaller of the Q-point value of the collector-base or drain-gate voltage of the transistor, and by the Q-point value of the voltage across the collector or drain-bias resistors R_C or R_D.

- It is extremely important to understand the difference between ac analysis and transient analysis in SPICE. The ac analysis mode assumes that the network is linear and uses small-signal models for the transistors and diodes. Since the circuit is linear, any convenient value can be used for the signal source amplitudes, hence the common choice of 1-V and 1-A sources. In contrast, transient simulations utilize the full large-signal nonlinear models of the transistors. If we desire linear behavior in a transient simulation, all signals must satisfy the small-signal constraints.

KEY TERMS

ac coupling
ac equivalent circuit
Active region
Amplification factor
Back-gate transconductance
Back-gate transconductance parameter
Bypass capacitor
Common-emitter (C-E) amplifier
Common-source (C-S) amplifier
Coupling capacitor
dc blocking capacitor
dc equivalent circuit
Diode conductance
Diode resistance
Hybrid-pi model

Hybrid-pi small-signal model
Input resistance
Intrinsic voltage gain μ_f (or intrinsic gain)
Lambda λ
Output resistance r_o
Signal source voltage gain
Small signal
Small-signal conductance
Small-signal current gain
Small-signal models
Subthreshold operation
Terminal voltage
Terminal voltage gain
Transconductance g_m
Weak inversion

REFERENCE

1. R. C. Jaeger et al., "Direct Measurement of the Available Voltage Gain of Bipolar and Field-Effect Transistors," *IEEE Electron Device Letters*, vol. EDL-6, no. 5, pp. 219–220, May 1985.

PROBLEMS

Figures P7.3 through P7.13 are used in a variety of problems in this chapter. Assume all capacitors and inductors have infinite value unless otherwise noted. Assume $V_{BE} = 0.7$ V and $\beta_F = \beta_o$ unless otherwise specified.

7.1 The Transistor as an Amplifier

7.1. (a) Suppose $v_{be}(t) = 0.005 \sin 2000\pi t$ V in the bipolar amplifier in Fig. 7.1. Write expressions for $v_{BE}(t)$, $v_{ce}(t)$, and $v_{CE}(t)$. (b) What is the maximum value of I_C that corresponds to the active region of operation?

7.2. (a) Suppose $v_{gs}(t) = 0.25 \sin 2000\pi t$ V in the MOSFET amplifier in Fig. 7.2. Write expressions for $v_{GS}(t)$, $v_{ds}(t)$, and $v_{DS}(t)$. (b) What is the maximum value of I_D that corresponds to the active region of operation?

7.2 Coupling and Bypass Capacitors

7.3. (a) What are the functions of capacitors C_1, C_2, and C_3 in Fig. P7.3? (b) What is the magnitude of the signal voltage at the source of M_1?

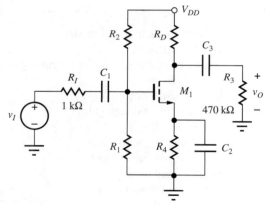

Figure P7.3

7.4. (a) What are the functions of capacitors C_1, C_2, and C_3 in Fig. P7.4? (b) What is the magnitude of the signal voltage at the top of C_3?

7.5. Repeat Prob. 7.4 if capacitor C_3 is connected between the transistor's emitter and ground.

7.6. (a) What are the functions of capacitors C_1, C_2, and C_3 in Fig. P7.6? (b) What is the magnitude of the signal voltage at the base of Q_1?

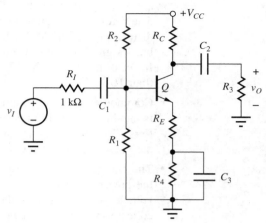

Figure P7.4

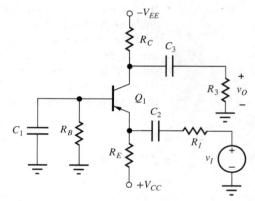

Figure P7.6

7.7. (a) What are the functions of capacitors C_1, C_2, and C_3 in Fig. P7.7? (b) What is the magnitude of the signal voltage at the emitter of Q_1?

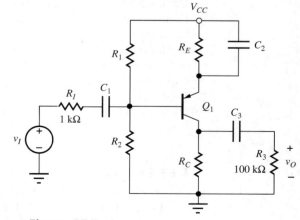

Figure P7.7

7.8. (a) What are the functions of capacitors C_1, C_2, and C_3, in Fig. P7.8? (b) What is the magnitude of the signal voltage at the source of M_1?

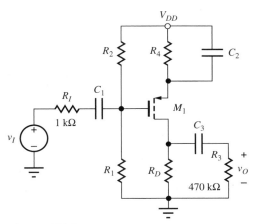

Figure P7.8

7.9. What are the functions of capacitors C_1 and C_2 in Fig. P7.9?

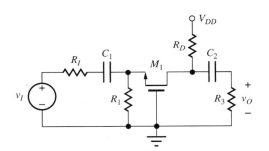

Figure P7.9

7.10. What are the functions of capacitors C_1, C_2, and C_3, in Fig. P7.10? What is the magnitude of the signal voltage at the emitter of Q_1?

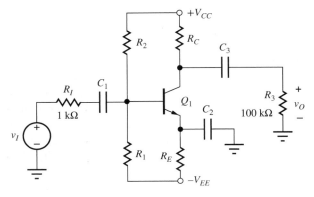

Figure P7.10

7.11. What are the functions of capacitors C_1, C_2, and C_3 in Fig. P7.11? What is the magnitude of the signal voltage at the collector of Q_1?

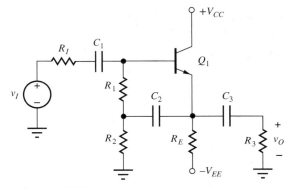

Figure P7.11

7.12. Describe the functions of capacitors C_1, C_2, and C_3 in Fig. P7.12. What is the magnitude of the signal voltage at the upper terminal of C_2?

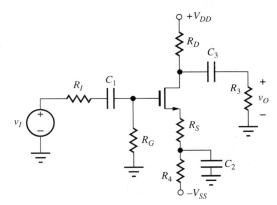

Figure P7.12

7.13. What are the functions of capacitors C_1 and C_2 in Fig. P7.13?

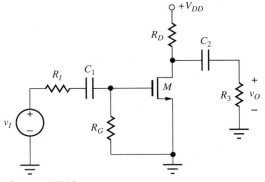

Figure P7.13

7.3 Circuit Analysis Using dc and ac Equivalent Circuits

BJT Q-Points

7.14. The phrase "dc voltage sources represent ac grounds" is used several times in the text. Use your own words to describe the meaning of this statement.

7.15. Draw the dc equivalent circuit and find the Q-point for the amplifier in Fig. P7.10. Assume $\beta_F = 50$, $V_{CC} = 9$ V, $-V_{EE} = -9$ V, $R_I = 1$ kΩ, $R_1 = 5$ kΩ, $R_2 = 10$ kΩ, $R_3 = 24$ kΩ, $R_E = 3$ kΩ, and $R_C = 6$ kΩ.

7.16. Use SPICE to find the Q-point for the circuit in Prob. 7.15. Compare the results to the hand calculations in Prob. 7.15.

7.17. Draw the dc equivalent circuit and find the Q-point for the amplifier in Fig. P7.4. Assume $\beta_F = 120$, $V_{CC} = 14$ V, $R_I = 2$ kΩ, $R_1 = 360$ kΩ, $R_2 = 750$ kΩ, $R_C = 270$ kΩ, $R_E = 7.5$ kΩ, $R_4 = 240$ kΩ, and $R_3 = 910$ kΩ.

7.18. (a) Use SPICE to find the Q-point for the circuit in Prob. 7.17. Assume $V_A = \infty$ and $I_S = 5$ fA. (b) Repeat with $V_A = 80$ V and $I_S = 5$ fA.

7.19. Draw the dc equivalent circuit and find the Q-point for the amplifier in Fig. P7.6. Assume $\beta_F = 90$, $V_{CC} = 3$ V, $-V_{EE} = -3$ V, $R_I = 0.47$ kΩ, $R_B = 3$ kΩ, $R_C = 36$ kΩ, $R_E = 75$ kΩ, and $R_3 = 120$ kΩ.

7.20. Use SPICE to find the Q-point for the circuit in Prob. 7.19. Compare the results to the hand calculations in Prob. 7.19.

7.21. Draw the dc equivalent circuit and find the Q-point for the amplifier in Fig. P7.7. Assume $\beta_F = 125$ and $V_{CC} = 12$ V, $R_1 = 36$ kΩ, $R_2 = 110$ kΩ, $R_C = 13$ kΩ, and $R_E = 3.9$ kΩ.

7.22. Use SPICE to find the Q-point for the circuit in Prob. 7.21. Compare the results to the hand calculations in Prob. 7.21.

7.23. Draw the dc equivalent circuit and find the Q-point for the amplifier in Fig. P7.11. Assume $\beta_F = 100$, $V_{CC} = 5$ V, $-V_{EE} = -5$ V, $R_I = 1$ kΩ, $R_1 = 43$ kΩ, $R_2 = 43$ kΩ, $R_3 = 24$ kΩ, and $R_E = 82$ kΩ.

7.24. Use SPICE to find the Q-point for the circuit in Prob. 7.23. Compare the results to the hand calculations in Prob. 7.23.

FET Q-Points

7.25. Draw the dc equivalent circuit and find the Q-point for the amplifier in Fig. P7.3. Assume $K_n = 250$ μA/V^2, $V_{TN} = 1$ V, $V_{DD} = 12$ V, $R_I = 1$ kΩ, $R_1 = 390$ kΩ, $R_2 = 1$ MΩ, $R_D = 82$ kΩ, and $R_4 = 27$ kΩ.

7.26. Use SPICE to find the Q-point for the circuit in Prob. 7.25. Compare the results to the hand calculations in Prob. 7.25.

7.27. Draw the dc equivalent circuit and find the Q-point for the amplifier in Fig. P7.9. Assume $K_n = 500$ μA/V^2, $V_{TN} = -2$ V, $V_{DD} = 22$ V, $R_I = 2$ kΩ, $R_1 = 6.2$ kΩ, $R_D = 7.5$ kΩ, and $R_3 = 51$ kΩ.

7.28. Use SPICE to find the Q-point for the circuit in Prob. 7.27. Compare the results to the hand calculations in Prob. 7.27.

7.29. Draw the dc equivalent circuit and find the Q-point for the amplifier in Fig. P7.8. Assume $K_p = 400$ μA/V^2, $V_{TP} = -1$ V, $V_{DD} = 15$ V, $R_1 = 2$ MΩ, $R_2 = 2$ MΩ, $R_D = 24$ kΩ, and $R_4 = 22$ kΩ.

7.30. Use SPICE to find the Q-point for the circuit in Prob. 7.29. Compare the results to the hand calculations in Prob. 7.29.

7.31. Draw the dc equivalent circuit and find the Q-point for the amplifier in Fig. P7.12. Assume $K_n = 400$ μA/V^2, $V_{TN} = -5$ V, $V_{DD} = 18$ V, $R_G = 10$ MΩ, $R_D = 5.6$ kΩ, $R_I = 10$ kΩ, $R_1 = 2$ kΩ, $R_S = 1.5$ kΩ, $R_4 = 1.5$ kΩ, and $R_3 = 36$ kΩ. Assume $V_{SS} = 0$.

7.32. Use SPICE to find the Q-point for the circuit in Prob. 7.31. Compare the results to the hand calculations in Prob. 7.31.

7.33. Draw the dc equivalent circuit and find the Q-point for the amplifier in Fig. P7.13. Assume $V_{DD} = 15$ V, $K_n = 225$ μA/V^2, $V_{TN} = -3$ V, $R_G = 2.2$ MΩ, $R_D = 8.2$ kΩ, $R_I = 10$ kΩ, and $R_3 = 220$ kΩ.

7.34. Use SPICE to find the Q-point for the circuit in Prob. 7.33. Compare the results to the hand calculations in Prob. 7.33.

ac Equivalent Circuits

7.35. (a) Draw the equivalent circuit used for ac analysis of the circuit in Fig. P7.4. (Use transistor symbols for this part.) Assume all capacitors have infinite value. (b) Redraw the ac equivalent circuit, replacing the transistor with its small-signal model. (c) Identify the function of each capacitor in the circuit (bypass or coupling).

7.36. (a) Repeat Prob. 7.35 for the circuit in Fig. P7.6. (b) Repeat Prob. 7.35 for the circuit in Fig. P7.7.

7.37. (a) Repeat Prob. 7.35 for the circuit in Fig. P7.11.
(b) Repeat Prob. 7.35 for the circuit in Fig. P7.13.

7.38. (a) Repeat Prob. 7.35 for the circuit in Fig. P7.3.
(b) Repeat Prob. 7.35 for the circuit in Fig. P7.9.

7.39. (a) Repeat Prob. 7.35 for the circuit in Fig. P7.8.
(b) Repeat Prob. 7.35 for the circuit in Fig. P7.10.

7.40. Describe the function of each of the resistors in the circuit in Fig. P7.4.

7.41. Describe the function of each of the resistors in the circuit in Fig. P7.6.

7.42. Describe the function of each of the resistors in the circuit in Fig. P7.7.

7.43. Describe the function of each of the resistors in Fig. P7.3.

7.44. Describe the function of each of the resistors in Fig. 7.13.

7.4 Introduction to Small-Signal Modeling

7.45. (a) Calculate r_d for a diode with $V_D = 0.6$ V if $I_S = 5$ fA. (b) What is the value of r_d for $V_D = 0$ V? (c) At what voltage does r_d exceed 10^{12} Ω?

7.46. What is the value of the small-signal diode resistance r_d of a diode operating at a dc current of 2 mA at temperatures of (a) 75 K, (b) 100 K, (c) 200 K, (d) 300 K, (e) 400 K, and (f) 4 K?

7.47. (a) Compare $[\exp(v_d/V_T) - 1]$ to v_d/V_T for $v_d = +5$ mV and -5 mV. How much error exists between the linear approximation and the exponential? (b) Repeat for $v_d = \pm 10$ mV.

7.5 Small-Signal Models for Bipolar Junction Transistors

7.48. (a) What collector current is required for a bipolar transistor to achieve a transconductance of 35 mS? (b) Repeat for a transconductance of 170 µS. (c) Repeat for a transconductance of 40 µS.

7.49. At what Q-point current will $r_\pi = 10$ kΩ for a bipolar transistor with $\beta_o = 75$? What are the approximate values of g_m and r_o if $V_A = 100$ V?

7.50. Repeat Prob. 7.49 for $r_\pi = 100$ MΩ with $\beta_o = 125$ and $V_A = 100$ V.

7.51. Repeat Prob. 7.49 for $r_\pi = 220$ kΩ with $\beta_o = 125$.

7.52. At what Q-point current will $r_\pi = 1$ MΩ for a bipolar transistor with $\beta_o = 75$? What are the values of g_m and r_o if $V_A = 100$ V?

7.53. The following table contains the small-signal parameters for a bipolar transistor. What are the values of β_F and V_A? Fill in the values of the missing entries in the table if $V_{CE} = 10$ V.

Bipolar Transistor Small-Signal Parameters					
I_C (A)	g_m (S)	r_π (Ω)	r_o (Ω)	μ_f	g_m/I_C
0.001			50,000		
	0.15	600	480,000		
0.0001		—		2000	

7.54. (a) Compare $[\exp(v_{be}/V_T) - 1]$ to v_{be}/V_T for $v_{be} = +5$ mV and -5 mV? How much error exists between the linear approximation and the exponential? (b) Repeat for $v_{be} = \pm 7.5$ mV. (c) Repeat for $v_{be} = \pm 2.5$ mV.

7.55. The output characteristics of a bipolar transistor appear in Fig. P7.144. (a) What are the values of β_F and β_o at $I_B = 4$ µA and $V_{CE} = 10$ V? (b) What are the values of β_F and β_o at $I_B = 8$ µA and $V_{CE} = 10$ V?

**7.56. (a) Suppose that a BJT is operating with a total collector current given by

$$i_C(t) = 0.001 \exp\left(\frac{v_{be}(t)}{V_T}\right) \text{ amp}$$

and $v_{be}(t) = V_M \sin 1500\pi t$ with $V_M = 5$ mV. What is the value of the dc collector current? Plot the collector current using MATLAB. Use the FFT capability of MATLAB to find the amplitude of i_c at 1500 Hz? At 3000 Hz? At 4500 Hz? (b) Repeat for $V_M = 50$ mV.

7.57. (a) Use SPICE to find the Q-point of the circuit in Fig. P7.10 using the element values in Prob. 7.15. Use $V_A = 75$ V. Use the Q-point information from SPICE to calculate the values of the small-signal parameters of transistor Q_1. Compare the values with those printed out by SPICE and discuss the source of any discrepancies. (b) Repeat part (a) for the circuit in Fig. P7.7 with the element values from Prob. 7.21.

*7.58. Another small-signal model, the "T-model" in Fig. P7.58, is of historical interest and quite useful in certain situations. Show that this model is equivalent to the hybrid-pi model if the emitter resistance $r_e = r_\pi/(\beta_o + 1) = \alpha_o/g_m = V_T/I_E$. (Hint: Calculate the short-circuit input admittance (y_{11}) for both models assuming $\beta_F = \beta_o$.)

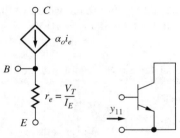

Figure P7.58

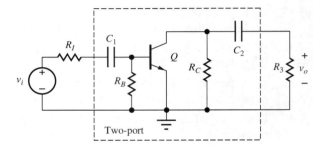

Figure P7.62

7.6 The BJT Common-Emitter (C-E) Amplifier

7.59. The ac equivalent circuit for an amplifier is shown in Fig. P7.59. Assume the capacitors have infinite value, $R_I = 750\ \Omega$, $R_B = 100\ \text{k}\Omega$, $R_C = 62\ \text{k}\Omega$, and $R_3 = 100\ \text{k}\Omega$. Calculate the voltage gain and input resistance for the amplifier if the BJT Q-point is $(40\ \mu\text{A},\ 10\ \text{V})$. Assume $\beta_o = 110$ and $V_A = 75\ \text{V}$.

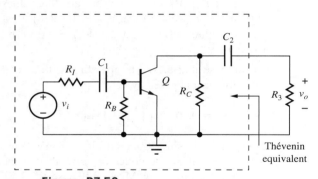

Figure P7.59

7.60. What are the worst-case values of voltage gain for the amplifier in Prob. 7.59 if β_o can range from 50 to 100? Assume that the Q-point is fixed.

7.61. The ac equivalent circuit for an amplifier is shown in Fig. P7.59. Assume the capacitors have infinite value, $R_I = 50\ \Omega$, $R_B = 4.7\ \text{k}\Omega$, $R_C = 4.7\ \text{k}\Omega$, and $R_3 = 10\ \text{k}\Omega$. Calculate the voltage gain for the amplifier if the BJT Q-point is $(2.0\ \text{mA},\ 7.5\ \text{V})$. Assume $\beta_o = 95$ and $V_A = 50\ \text{V}$.

7.62. The ac equivalent circuit for an amplifier is shown in Fig. P7.62. Assume the capacitors have infinite value, $R_I = 10\ \text{k}\Omega$, $R_B = 5\ \text{M}\Omega$, $R_C = 2\ \text{M}\Omega$, and $R_3 = 3.3\ \text{M}\Omega$. Calculate the voltage gain for the amplifier if the BJT Q-point is $(1\ \mu\text{A},\ 1.5\ \text{V})$. Assume $\beta_o = 40$ and $V_A = 50\ \text{V}$.

7.63. (a) Rework Prob. 7.62 if I_C is increased to 10 μA, and the values of R_C, R_B, and R_3 are all reduced by a factor of 10. (b) Rework Prob. 7.62 if I_C is increased to 100 μA, and the values of R_C, R_B, and R_3 are all reduced by a factor of 100.

7.64. Simulate the behavior of the BJT common-emitter amplifier in Fig. 7.18 and compare the results to the calculations in Ex. 7.3. Use 100 μF for all capacitor values and perform the ac analysis at a frequency of 1000 Hz.

7.65. (a) Use SPICE to simulate the dc and ac characteristics of the amplifier in Prob. 7.21. What is the Q-point? What is the value of the small-signal voltage gain? Use 100 μF for all capacitor values and perform the ac analysis at a frequency of 1000 Hz. (b) Compare the results to hand calculations.

7.7 Important Limits and Model Simplifications

7.66. A C-E amplifier is operating from a single 6-V supply. Estimate its voltage gain.

*7.67. A C-E amplifier is operating from symmetrical ±15-V power supplies. Estimate its voltage gain.

7.68. A battery-powered amplifier must be designed to provide a voltage gain of 50. Can a single-stage amplifier be designed to meet this goal if it must operate from two ±1.5-V batteries?

7.69. A battery-powered C-E amplifier is operating from a single 1.5-V battery. Estimate its voltage gain. What will the gain be if the battery voltage drops to 1 V?

*7.70. An amplifier is required with a voltage gain of 30,000 and will be designed using a cascade of several C-E amplifier stages operating from a single 7.5-V power supply. Estimate the minimum number of amplifier stages that will be required to achieve this gain.

*7.71. The common-emitter amplifier in Fig. P7.71 must develop a 5-V peak-to-peak sinusoidal signal across the 1.5-kΩ load resistor R_L. (a) What is the minimum collector current I_C that will satisfy the requirements of small-signal operation of the transistor? (b) What is the minimum power supply voltage V_{CC} if the transistor must remain in the forward-active region at all times?

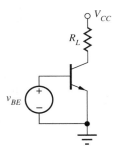

Figure P7.71

*7.72. The common-emitter amplifier in Fig. P7.71 has a voltage gain of 43 dB. What is the amplitude of the largest output signal voltage at the collector that corresponds to small-signal operation?

*7.73. A common-emitter amplifier has a gain of 47 dB and is developing a 15-V peak-to-peak ac signal at its output. Is this amplifier operating within its small-signal region? If the input signal to this amplifier is a sine wave, do you expect the output to be distorted? Why or why not?

7.74. What is the voltage gain of the common-emitter amplifier in Fig. P7.10? Assume $\beta_F = 135$, $V_{CC} = V_{EE} = 10$ V, $R_1 = 20$ kΩ, $R_2 = 62$ kΩ, $R_C = 13$ kΩ, and $R_E = 3.9$ kΩ.

7.75. What is the voltage gain of the amplifier in Fig. P7.7 if $V_{CC} = 20$ V? Use the resistor values from Prob. 7.74.

7.76. Resistor R_L in Fig. P7.71 is replaced with an inductor L. What is the voltage gain of the circuit at high frequencies for which $\omega L \gg r_o$ if $V_A = 75$ V?

7.8 Small-Signal Models for Field-Effect Transistors

7.77. The following table contains the small-signal parameters for a MOS transistor. What are the values of K_n and λ? Fill in the values of the missing entries in the table if $V_{DS} = 5$ V and $V_{TN} = 0.75$ V.

MOSFET Small-Signal Parameters

I_D	g_m (S)	r_o (Ω)	μ_f	SMALL-SIGNAL LIMIT V_{gs} (V)	g_m/I_D
0.8 mA		50,000			
50 μA	0.0002				
10 mA					

7.78. A MOSFET is needed with $g_m = 5$ mS at $V_{GS} - V_{TN} = 0.5$ V. What is W/L if $K'_n = 75$ μA/V^2?

7.79. What value of W/L is required to achieve $\mu_f = 200$ in a MOSFET operating at a drain current of 250 μA if $K'_n = 50$ μA/V^2 and $\lambda = 0.02$/V? What is the value of $V_{GS} - V_{TN}$?

7.80. An n-channel MOSFET has $K_n = 500$ μA/V^2, $V_{TN} = 1$ V, and $\lambda = 0.025$ V^{-1}. At what drain current will the MOSFET no longer be able to provide any voltage gain (that is, $\mu_F \leq 1$)?

7.81. (a) Compare $[1 + v_{gs}/(V_{GS} - V_{TN})]^2 - 1$ to $[2v_{gs}/(V_{GS} - V_{TN})]$ for $v_{gs} = 0.2(V_{GS} - V_{TN})$. How much error exists between the linear approximation and the quadratic expression? (b) Repeat for $v_{gs} = 0.4(V_{GS} - V_{TN})$. (c) Repeat for $v_{gs} = 0.1(V_{GS} - V_{TN})$.

7.82. Use SPICE to find the Q-point of the circuit in Prob. 7.25. Use the Q-point information from SPICE to calculate the values of the small-signal parameters of transistor M_1. Compare the values with those printed out by SPICE and discuss the source of any discrepancies.

7.83. Repeat Prob. 7.82 for the circuit in Prob. 7.29.

7.84. At approximately what Q-point will $R_{\text{out}}^{CS} = 100$ kΩ in a common-source amplifier if the transistor has $\lambda = 0.02$ V^{-1} and the power supply is 15 V?

*7.85. At approximately what Q-point can we achieve an input resistance of $R_{\text{in}}^{CS} = 2$ MΩ in a common-source amplifier if the transistor has $K_n = 500$ μA/V^2, $V_{TN} = 1$ V, $\lambda = 0.02$ V^{-1}, and the power supply is 12 V?

**7.86. Figure P7.86 gives the device characteristics and schematic of an amplifier circuit including a "new"[8] electronic device called a *triode* vacuum tube. (a) Write the equation for the load line for the circuit. (b) What is the Q-point (I_P, V_{PK})? Assume $i_G = 0$. (c) Using the following definitions, find

---

[8] New to us at least.

the values of g_m, r_o, and μ_f. (d) What is the voltage gain of the circuit?

$$g_m = \left.\frac{\Delta i_P}{\Delta v_{GK}}\right|_{\text{Q-point}}$$

$$r_o = \left(\left.\frac{\Delta i_P}{\Delta v_{PK}}\right|_{\text{Q-point}}\right)^{-1} \qquad \mu_f = g_m r_o$$

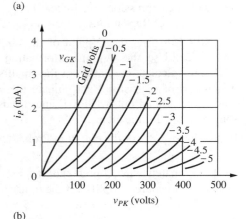

(a)

(b)

Figure P7.86 (a) "New" electron device—the triode vacuum tube; (b) triode output characteristics: G = grid, P = plate, K = cathode.

7.9 Summary and Comparison of the Small-Signal Models of the BJT and FET

7.87. A circuit is to be biased at a current of 7 mA and achieve an input resistance of at least 2 MΩ. Should a BJT or FET be chosen for this circuit and why?

7.88. A circuit requires the use of a transistor with a transconductance of 0.5 S. A bipolar transistor with $\beta_F = 60$ and a MOSFET with $K_n = 25$ mA/V² are available. Which transistor would be preferred and why?

7.89. A BJT has $V_A = 60$ V and a MOSFET has $K_n = 25$ mA/V² and $\lambda = .017$ V⁻¹. At what current level is the intrinsic gain of the MOSFET equal to that of the BJT if $V_{DS} = V_{CE} = 10$ V? What is μ_f for the BJT?

7.90. A BJT has $V_A = 50$ V, and a MOSFET has $\lambda = 0.02$/V with $V_{GS} - V_{TN} = 0.5$ V. What are the intrinsic gains of the two transistors? What are the transconductances if the transistors are both operating at a current of 200 μA?

7.91. An amplifier circuit is needed with an input resistance of 75 Ω. Should a BJT or MOSFET be chosen for this circuit? Discuss.

7.92. (a) We need to amplify a 0.25-V signal by 43 dB. Would a BJT or FET amplifier be preferred? Why? (b) RF amplifiers must often amplify microvolt signals in the presence of many other interfering signals with amplitudes of 100 mV or more. Does an FET or BJT seem most appropriate for this application? Why?

7.10 The Common-Source (C-S) Amplifier

7.93. A C-S amplifier is operating from a single 15-V supply with $V_{GS} - V_{TN} = 1$ V. Estimate its voltage gain.

7.94. A common-source amplifier has a gain of 16 dB and is developing a 15-V peak-to-peak ac signal at its output. Is this amplifier operating within its small-signal region? Discuss.

7.95. A C-S amplifier is operating from a single 18-V supply. The MOSFET has $K_n = 1$ mA/V². What is the Q-point current required for a voltage gain of 25?

7.96. A C-S amplifier is operating from a single 9-V supply. What is the maximum value of $V_{GS} - V_{TN}$ that can be used if the amplifier must have a gain of at least 35?

7.97. A MOSFET common-source amplifier must amplify a sinusoidal ac signal with a peak amplitude of 0.2 V. What is the minimum value of $V_{GS} - V_{TN}$ for the transistor? If a voltage gain of 36 dB is required, what is the minimum power supply voltage?

7.98. A MOSFET common-source amplifier must amplify a sinusoidal ac signal with a peak amplitude of 0.4 V. What is the minimum value of $V_{GS} - V_{TN}$ for the transistor? If a voltage gain of 26 dB is required, what is the minimum power supply voltage?

7.99. An amplifier is required with a voltage gain of 1500 and will be designed using a cascade of several C-S amplifier stages operating from a single 12-V power supply. Estimate the minimum number of amplifier stages required to achieve this gain.

7.100. What is the voltage gain of the amplifier in Fig. P7.100? Assume $K_n = 0.450$ mA/V^2, $V_{TN} = 1$ V, and $\lambda = 0.0133$ V^{-1}.

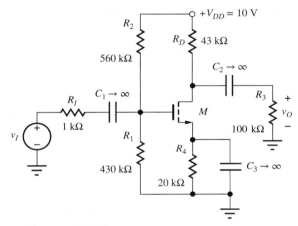

Figure P7.100

7.101. The ac equivalent circuit for an amplifier is shown in Fig. P7.101. Assume the capacitors have infinite value, $R_I = 100$ kΩ, $R_G = 6.8$ MΩ, $R_D = 50$ kΩ, and $R_3 = 120$ kΩ. Calculate the voltage gain for the amplifier if the MOSFET Q-point is (100 μA, 5 V). Assume $K_n = 450$ μA/V^2 and $\lambda = 0.02$ V^{-1}.

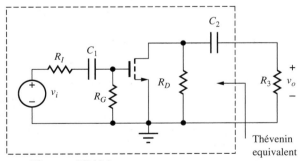

Figure P7.101

7.102. What are the worst-case values of voltage gain for the amplifier in Prob. 7.101 if K_n can range from 300 μA/V^2 to 700 μA/V^2? Assume the Q-point is fixed.

7.103. The ac equivalent circuit for an amplifier is shown in Fig. P7.101. Assume the capacitors have infinite value, $R_I = 100$ kΩ, $R_G = 10$ MΩ, $R_D = 560$ kΩ, and $R_3 = 1.5$ MΩ. Calculate the voltage gain for the amplifier if the MOSFET Q-point is (10 μA, 5 V). Assume $K_n = 100$ μA/V^2 and $\lambda = 0.02$ V^{-1}.

7.104. The ac equivalent circuit for an amplifier is shown in Fig. P7.104. Assume the capacitors have infinite value, $R_I = 10$ kΩ, $R_G = 1$ MΩ, $R_D = 3.9$ kΩ, and $R_3 = 33$ kΩ. Calculate the voltage gain for the amplifier if the MOSFET Q-point is (2 mA, 7.5 V). Assume $K_n = 1$ mA/V^2 and $\lambda = 0.015$ V^{-1}.

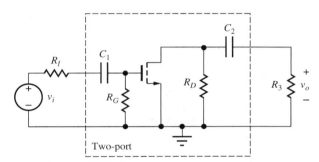

Figure P7.104

7.105. Use SPICE to simulate the dc and ac characteristics of the amplifier in Prob. 7.25. What is the Q-point? What are the values of the small-signal voltage gain, input resistance, and output resistance of the amplifier? Use 100 μF for all capacitor values and perform the ac analysis at a frequency of 1000 Hz.

7.106. Use SPICE to simulate the dc and ac characteristics of the amplifier in Prob. 7.29. What is the Q-point? What are the values of the small-signal voltage gain, input resistance, and output resistance of the amplifier? Use 100 μF for all capacitor values and perform the ac analysis at a frequency of 1000 Hz.

7.107. Use SPICE to simulate the dc and ac characteristics of the amplifier in Prob. 7.31. What is the Q-point? What are the values of the small-signal voltage gain, input resistance, and output resistance of the amplifier? Use 100 μF for all capacitor values and perform the ac analysis at a frequency of 1000 Hz.

7.108. Use SPICE to simulate the dc and ac characteristics of the amplifier in Prob. 7.33. What is the Q-point? What are the values of the small-signal voltage gain, input resistance, and output resistance of the amplifier? Use 100 μF for all capacitor values and perform the ac analysis at a frequency of 1000 Hz.

Input and Output Resistances of the Common-Emitter and Common-Source Amplifiers

7.109. The ac equivalent circuit for an amplifier is shown in Fig. P7.59. Assume the capacitors have infinite value, $R_I = 750$ Ω, $R_B = 100$ kΩ, $R_C = 100$ kΩ, and $R_3 = 100$ kΩ. Calculate the input resistance and output resistance for the amplifier if the BJT Q-point is (75 μA, 10 V). Assume $\beta_o = 125$ and $V_A = 75$ V.

7.110. The ac equivalent circuit for an amplifier is shown in Fig. P7.59. Assume the capacitors have infinite value, $R_I = 50$ Ω, $R_B = 4.7$ kΩ, $R_C = 4.3$ kΩ, and $R_3 = 10$ kΩ. Calculate the input resistance and output resistance for the amplifier if the BJT Q-point is (2.0 mA, 7.5 V). Assume $\beta_o = 75$ and $V_A = 50$ V.

7.111. What are the worst-case values of input resistance and output resistance for the amplifier in Prob. 7.59 if β_o can range from 60 to 100? Assume that the Q-point is fixed.

7.112. The ac equivalent circuit for an amplifier is shown in Fig. P7.62. Assume the capacitors have infinite value, $R_I = 10$ kΩ, $R_B = 5$ MΩ, $R_C = 1.5$ MΩ, and $R_3 = 3.3$ MΩ. Calculate the input resistance and output resistance for the amplifier if the BJT Q-point is (2 μA, 2 V). Assume $\beta_o = 60$ and $V_A = 50$ V.

7.113. (a) Rework Prob. 7.112 if I_C is increased to 10 μA, and the values of R_C, R_B, and R_3 are all reduced by a factor of 5. (b) Rework Prob. 7.112 if I_C is increased to 100 μA, and the values of R_C, R_B, and R_3 are all reduced by a factor of 50.

7.114. What are the input resistance and output resistance of the amplifier in Prob. 7.100?

7.115. Calculate the input and output resistances for the amplifier in Prob. 7.101.

7.116. What are the worst-case values of the input and output resistances for the amplifier in Prob. 7.101 if K_n can range from 300 to 700 μA/V²? Assume the Q-point is fixed.

7.117. Calculate the input and output resistances for the amplifier in Prob. 7.103.

7.118. Calculate the input and output resistances for the amplifier in Prob. 7.104.

7.119. Calculate the Thévenin equivalent representation for the amplifier in Prob. 7.59.

7.120. Calculate the Thévenin equivalent representation for the amplifier in Prob. 7.61.

7.121. Calculate the Thévenin equivalent representation for the amplifier in Prob. 7.103.

7.122. Calculate the Thévenin equivalent representation for the amplifier in Prob. 7.101.

7.11 Common-Emitter and Common-Source Amplifier Summary

7.123. (a) Find the voltage gain, input resistance and output resistance of the C-E stage in Fig. P7.123. Assume $\beta_F = 65$ and $V_A = 50$ V. (b) What is the Thévenin equivalent circuit for this amplifier? (c) What is the Norton equivalent circuit for this amplifier?

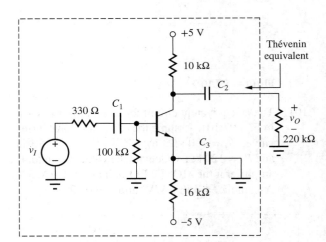

Figure P7.123

7.124. Simulate the behavior of the BJT common-emitter amplifier in Fig. P7.123 and compare the results to the calculations in Prob. 7.123. Use 100 μF for all capacitor values and perform the ac analysis at a frequency of 10,000 Hz.

7.125. The amplifier in Fig. P7.125 is the bipolar amplifier in Fig. P7.123 with currents reduced by a factor of approximately 10. What are the voltage gain and input resistance and output resistance of this amplifier? Compare to that in Fig. P7.123, and discuss the reasons for any differences in gain.

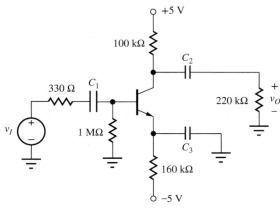

Figure P7.125

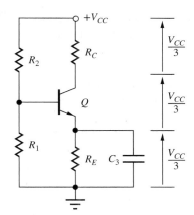

Figure P7.136

7.126. Simulate the behavior of the BJT common-emitter amplifier in Fig. P7.125 and compare the results to the calculations in Prob. 7.125. Use 100 μF for all capacitor values and perform the ac analysis at a frequency of 1000 Hz.

7.127. Use SPICE to simulate the behavior of the MOSFET common-source amplifier in Fig. 7.32(b) and compare the results to the calculations in the example. Use 100 μF for all capacitor values and perform the ac analysis at a frequency of 1000 Hz.

7.128. Use SPICE to simulate the voltage gain and input resistance and output resistance of the amplifier in Prob. 7.100. Use 100 μF for all capacitor values and perform the ac analysis at a frequency of 1000 Hz.

7.12 Amplifier Power and Signal Range

7.129. Calculate the dc power dissipation in each element in the circuit in Fig. 7.40(a) if $\beta_F = 65$. Compare the result to the total power delivered by the sources.

7.130. Calculate the dc power dissipation in each element in the circuit in Fig. 7.40(b). Compare the result to the total power delivered by the sources.

7.131. Calculate the dc power dissipation in each element in the circuit in Prob. 7.15. Compare the result to the total power delivered by the sources.

7.132. Repeat Prob. 7.131 for the circuit in Prob. 7.19.

7.133. Repeat Prob. 7.131 for the circuit in Prob. 7.25.

7.134. Repeat Prob. 7.131 for the circuit in Prob. 7.29.

7.135. Repeat Prob. 7.131 for the circuit in Prob. 7.31.

*7.136. A common bias point for a transistor is shown in Fig. P7.136. What is the maximum amplitude signal that can be developed at the collector terminal that will satisfy the small-signal assumptions (in terms of V_{CC})?

*7.137. The MOSFET in Fig. P7.137 has $K_n = 500\ \mu\text{A/V}^2$ and $V_{TN} = -1.25$ V. What is the largest permissible signal voltage at the drain that will satisfy the requirements for small-signal operation if $R_D = 15\ \text{k}\Omega$? What is the minimum value of V_{DD} if the transistor is to remain saturated at all times?

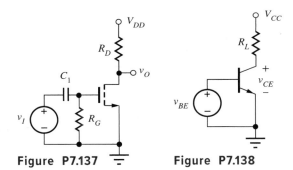

Figure P7.137   **Figure P7.138**

*7.138. The simple C-E amplifier in Fig. P7.138 is biased with $V_{CE} = V_{CC}/2$. Assume that the transistor can saturate with $V_{CESAT} = 0$ V and still be operating linearly. What is the amplitude of the largest sine wave that can appear at the output? What is the ac signal power P_{ac} being dissipated in the load resistor R_L? What is the total dc power P_S being supplied from the power supply? What is the efficiency ε of this amplifier if ε is defined as $\varepsilon = 100\% \times P_{ac}/P_S$?

7.139. What is the amplitude of the largest ac signal that can appear at the collector of the transistor in Fig. P7.7 that satisfies the small-signal limit? Use the parameter values from Prob. 7.21.

7.140. What is the amplitude of the largest ac signal that can appear at the drain of the transistor in Fig. P7.3 that satisfies the small-signal limit? Use the parameter values from Prob. 7.25.

7.141. What is the amplitude of the largest ac signal that can appear at the drain of the transistor in Fig. P7.8 that satisfies the small-signal limit? Use the parameter values from Prob. 7.29.

7.142. What is the amplitude of the largest ac signal that can appear at the collector of the transistor in Fig. P7.10 that satisfies the small-signal limit? Use the parameter values from Prob. 7.15.

7.143. What is the amplitude of the largest ac signal that can appear at the drain of the transistor in Fig. P7.13 that satisfies the small-signal limit? Use the parameter values from Prob. 7.33.

7.144. Draw the load line for the circuit in Fig. 7.1 on the output characteristics in Fig. P7.144 for $V_{CC} = 20$ V and $R_C = 20$ kΩ. Locate the Q-point for $I_B = 2$ µA. Estimate the maximum output voltage swing from the characteristics. Repeat for $I_B = 5$ µA.

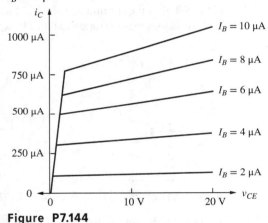

Figure P7.144

JFET Problems

7.145. Describe the functions of capacitors C_1, C_2, and C_3 in Fig. P7.145. What is the magnitude of the signal voltage at the source of J_1?

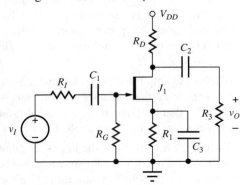

Figure P7.145

7.146. What are the functions of capacitors C_1 and C_2 in Fig. P7.146?

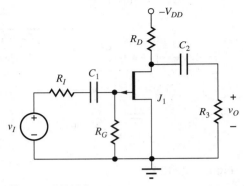

Figure P7.146

7.147. The JFET amplifier in Fig. P7.147 must develop a 10-V peak-to-peak sinusoidal signal across the 18-kΩ load resistor R_D. What is the minimum drain current I_D that will satisfy the requirements for small-signal operation of the transistor?

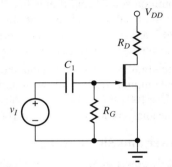

Figure P7.147

7.148. The ac equivalent circuit for an amplifier is shown in Fig. P7.148. Assume the capacitors have infinite value, $R_I = 10$ kΩ, $R_G = 1$ MΩ, $R_D = 7.5$ kΩ, and $R_3 = 120$ kΩ. Calculate the voltage gain, input resistance and output resistance for the amplifier if the JFET Q-point is (1.2 mA, 9 V). Assume $I_{DSS} = 1.2$ mA, $V_P = -3$ V, and $\lambda = 0.015$ V^{-1}.

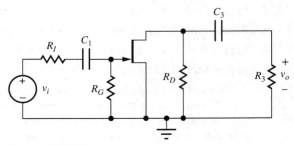

Figure P7.148

7.149. Show that the drain-current expression for the JFET can be represented in exactly the same form as that of the MOSFET using the substitutions $V_P = V_{TN}$ and $K_n = 2 I_{DSS}/V_P^2$.

7.150. What is the value of the maximum peak-to-peak output signal in the BJT amplifier in Fig. 7.1 if the transistor must remain in the forward-active region at all times?

7.151. (a) Asymmetry is clearly apparent in the SPICE waveform for the BJT circuit in Ex. 7.3 due to the change in operating current with signal. Calculate the values of collector current i_C and voltage gain of the circuit at the maximum and minimum of the "sinusoidal" waveform. (b) Estimate the total harmonic distortion in the waveform.

7.152. What is the value of the maximum peak-to-peak output signal in the MOS amplifier in Fig. 7.2 if the transistor must remain in the saturation region at all times? Assume $V_{TN} = 1$ V.

7.153. (a) Asymmetry is clearly apparent in the SPICE waveform for the MOSFET circuit in Ex. 7.4 due to the change in operating current with signal. Calculate the values of collector current i_D and voltage gain of the circuit at the maximum and minimum of the "sinusoidal" waveform. (b) Estimate the total harmonic distortion in the waveform.

Harmonic Distortion

7.154. (a) The signal voltage applied across the base-emitter terminals of a bipolar transistor is given by $v_{be} = V_M \sin 5000\pi t$, and the collector current is 1 mA. Calculate the total harmonic distortion in the collector current of the BJT based upon Eq. (7.37) if $V_M = 5$ mV. (b) repeat for $V_M = 10$ mV. (c) Repeat for $V_M = 2.5$ mV.

7.155. (a) The signal voltage applied across the gate-source terminals of a MOS transistor is given by $v_{gs} = V_M \sin 5000\pi t$, and $V_{GS} - V_{TN} = 0.75$ V. Calculate the total harmonic distortion in the drain current of the MOSFET based upon Eq. (7.70) if $V_M = 150$ mV. (b) Repeat for $V_M = 300$ mV. (c) Repeat for $V_M = 75$ mV.

7.156. (a) Perform a transient simulation of the amplifier in Ex. 7.3 for a sinusoidal input signal with an amplitude of 10 mV at a frequency of 10 kHz. Plot the input and output signals. What is the total harmonic distortion in the output signal? (b) Repeat for an amplitude of 15 mV.

7.157. (a) Perform a transient simulation of the amplifier in Ex. 7.4 for a sinusoidal input signal with an amplitude of 150 mV at a frequency of 10 kHz. Plot the input and output signals. What is the total harmonic distortion in the output signal? (b) Repeat for an amplitude of 300 mV.

Subthreshold Operation

7.158. The circuit in Fig. P7.100 is redesigned with the new values: $V_{DD} = 2.5$ V, $R_1 = 5.6$ MΩ, $R_2 = 10$ MΩ, $R_D = 12$ MΩ, $R_3 = 20$ MΩ, $R_4 = 0$, $V_{TN} = 1$ V, $I_{DO} = 100$ nA, and $n = 1.65$. Find the Q-point. What is the voltage gain of the circuit?

7.159. Repeat Prob. 7.158 if $R_1 = 5.1$ MΩ and V_{DD} is increased to 3 V.

7.160. Repeat Prob. 7.158 if $R_1 = 4.3$ MΩ and V_{DD} is increased to 4 V.

7.161. (a) Find the Q-point for the circuit in Fig. P7.161 assuming $V_{DD} = 1$ V, $R_G = 10$ MΩ, $R_D = 10$ MΩ, $V_{TN} = 0.5$ V, $I_{DO} = 200$ nA, and $n = 1.65$. (b) Redraw the circuit by adding an input source v_i, two capacitors, and a 20-MΩ load resistor to the circuit in Fig. P7.161 to turn it into a common-source amplifier. (c) Analyze the circuit and find the voltage gain.

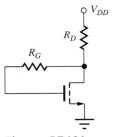

Figure P7.161

术语对照

Ac coupling	交流耦合
Ac equivalent circuit	交流等效电路
Active region	有源区域
Amplification factor	放大系数
Back-gate transconductance	背栅跨导
Back-gate transconductance parameter	背栅跨导参数
Bypass capacitor	旁路电容
Common-emitter (C-E) amplifier	共发射极放大器
Common-source (C-S) amplifier	共源极放大器
Coupling capacitor	耦合电容
Dc blocking	直流阻断
Dc blocking capacitor	直流阻断电容
Dc equivalent circuit	直流等效电路
Diode conductance	二极管电导
Diode resistance	二极管电阻
Hybrid-pi model	混合π模型
Hybrid-pi small-signal model	混合π小信号模型
Input resistance	输入电阻
Intrinsic voltage gain μ_f	本征电压增益 μ_f
Output resistance r_o	输出电阻 r_o
Signal source voltage gain	信号源电压增益
Small signal	小信号
Small-signal common-emitter current gain	小信号共发射极电流增益
Small-signal conductance	小信号电导
Small-signal current gain	小信号电流增益
Small-signal models	小信号模型
Terminal gain	端增益
Terminal voltage gain	端电压增益
Transconductance g_m	跨导 g_m

CHAPTER 8
TRANSISTOR AMPLIFIER BUILDING BLOCKS

第8章 采用单晶体管放大器构建块

本章提纲

8.1 放大器分类
8.2 反相放大器——共射极和共源极放大器电路
8.3 跟随电路——共集电极和共漏极放大器
8.4 同相放大器——共基极和共栅极电路
8.5 放大器原型回顾和比较
8.6 采用MOS反相器的共源极放大器
8.7 耦合和旁路电容设计
8.8 放大器设计实例
8.9 多级交流耦合放大器
8.10 直流耦合放大器简介

本章目标

- 了解三大类放大器:
  - 反相放大器(共发射极和共源极结构): 提供了具有180°相位差的高电压增益;
  - 跟随器(共集电极和共漏极结构): 提供了类似于运算放大器电压跟随器的单位增益;
  - 同相放大器(共基极和共栅极结构): 提供了无相位差的高电压增益。
- 对于每一类放大器,理解其设计相关的细节:
  - 电压增益和输入电压范围;
  - 电流增益;
  - 输入和输出电阻;
  - 耦合及旁路电容设计,低截止频率。
- 了解放大器的设计工具包(design kit),并将这些工具包用于一些设计问题实例。
- 继续加深对SPICE仿真和SPICE结果的解释的理解,同时也要初步理解SPICE交流(小信号)、瞬态(大信号)及传输函数分析模式的不同之处。

本章导读

本章将深入探讨三类单级放大器的小信号特性,并阐述为什么某些晶体管端口更适用于信号输入,而其他的则适用于信号输出,通过对晶体管的输入和输出信号分析将放大器分为三大类。对于BJT的三个可用端口,只有基极和发射极能用作信号输入端,而集电极和发射极可用作输出端。对于FET,源极和栅极可用作信号输入端,而漏极和源极可用作输出端。集电极会用作输入端,而基极或栅极不会用作输出端。因此,放大器分为基本的三大类: ①反相放大器——共发射极和共源极放大器; ②跟随器——共集电极和共漏极，也称为射极跟随器或源极跟随器；③同相放大器——共基极和共栅极放大器。

要区别这三类放大器的结构，在每种应用中需要将不同端点作为公共端或参考端。当采用双极型晶体管，该结构称为共发射极、共集电极和共基极放大器，对于采用场效应管的放大器，相应的名称为共源极、共漏极和共栅极放大器。每种类型的放大器都有各自不同的特性指标，包括电压增益、输入电阻、输出电阻和电流增益。

本章采用晶体管的小信号模型对三类放大器进行了详细的分析，得到了电压增益、电流增益、输入电阻、输出电阻和输入信号范围的表达式，并在给出的一系列重要表格中对这些结果进行了总结，从而构成了模拟电路设计人员的基本工具包，全面掌握这些结论是设计人员进行更为复杂模拟电路设计的先决条件。

本章采用的这些放大器是根据其交流等效电路的结构进行分类的，对每种放大器都进行了详细的分析。本章的电路实例中都采用如图 8.1 所示的四电阻偏置电路，目的是为不同放大器建立Q点。耦合和旁路电容用于改变交流等效电路，可以发现不同放大器的交流特性有显著的区别。

反相放大器（C-E和C-S放大器）可提供高电压和电流增益，同时还有高输入和输出电阻，本章对于共发射极和共源极放大器进行了详细的推导与分析。如果在晶体管的发射极或源极包含一个电阻，则电压增益会下降，但是可以与各个晶体管特性保持相对应关系。增益下降可以增加输入电阻、输出电阻和输入信号范围。由于具有更高跨导，BJT管相比FET管更容易获得高电压增益，而FET管无穷大的输入电阻可令其在获得高输入放大器方面占有优势。通常FET管比BJT管具有更大的输入信号范围。

射极跟随器和源极跟随器（C-C和C-D放大器）可提供近似为1的电压增益，并具有高输入及低输出电阻。跟随器可提供中等的电流增益，获得最高的输入信号范围。这些C-C和C-D放大器与第10章中介绍的电压跟随运算放大器电路的单晶体管等效电路相同。

同相放大器（C-B和C-G放大器）提供的电压增益信号范围和输出电阻与反相放大器非常类似，它们可提供相对较低的输入电阻，电流增益小于1。

所有类型的放大器都可提供至少中等的电压或电流（或二者都有）增益，因此如果设计合适，它们可提供较大的功率增益。本章给出了三种分别利用反相放大器、同相放大器及跟随器电路的放大器实例，同时也给出了利用蒙特卡洛分析来评估元件容限对电路性能影响的实例。

通过将电容电抗值设置成比电容两端的戴维南电阻小得多的值，可以选择耦合和旁边电容的值，电抗值在放大器频响中频带区域的最低频率处进行计算。当电容电抗值等于电容两端等效电阻值时的频率可决定截止频率下限f_L。本章所研究的放大器中，有两个或三个极点相互设定截止频率下限f_L的值，带宽缩减技术可将放大器的截止频率上推至超过每个电容单独作用时所设定的截止频率值。

在大多数情况下，单晶体管放大器无法满足给定放大器设计的所有要求。所需的电压增益通常会超出单个晶体管的放大系数，或者所要求的电压增益、输入电阻和输出电阻组合无法同时满足。根据本章对放大器的分析，单晶体管放大器无法同时满足这些要求，因此需要将多个放大器级联，构成一个新的放大器来满足所有的设计要求。本章以三级交流耦合放大器为例，通过确定该放大器的电压、输入和输出电阻、电流和功率增益以及输入信号的范围来描述它的特性，并估算放大器的下限截止频率。

Chapter Outline
8.1 Amplifier Classification
8.2 Inverting Amplifiers—Common-Emitter and Common-Source Circuits
8.3 Follower Circuits—Common-Collector and Common-Drain Amplifiers
8.4 Noninverting Amplifiers—Common-Base and Common-Gate Circuits
8.5 Amplifier Prototype Review and Comparison
8.6 Common-Source Amplifiers Using MOS Transistor Loads
8.7 Coupling and Bypass Capacitor Design
8.8 Amplifier Design Examples
8.9 Multistage ac-Coupled Amplifiers
8.10 Introduction to dc-Coupled Amplifiers
Summary
Key Terms
Additional Reading
Problems

Chapter Goals
In this chapter, we fully explore the small-signal characteristics of three families of single-stage amplifiers. We will discover why certain transistor terminals are preferred for signal input whereas others are used for signal outputs. The results define three broad classes of amplifiers.

- Inverting amplifiers—the common-emitter and common-source configurations—that provide high voltage gain with a 180° phase shift
- Followers—the common-collector and common-drain configurations—that provide nearly unity gain similar to the op amp voltage follower
- Noninverting amplifiers—the common-base and common-gate configurations—that provide high voltage gain with no phase shift

For each type of amplifier, we discuss the detailed design of

- Voltage gain and input voltage range
- Current gain
- Input and output resistances
- Coupling and bypass capacitor design and lower cutoff frequency

The results become our design toolkit and are used to solve a number of examples of design problems.

As in most chapters, we will continue to increase our understanding of SPICE simulation and interpretation of SPICE results. In particular, we try to understand the differences between SPICE ac (small signal), transient (large signal), and transfer function analysis modes

Chapter 7 introduced the common-emitter and common-source amplifiers, in which the input signal was applied to the base and gate terminals of the BJT and MOSFET, respectively, and the output signal was taken from the collector and drain. However, bipolar and field-effect transistors are three-terminal devices, and this chapter explores the use of other terminals for signal input and output. Three useful amplifier configurations are identified, each using a different terminal as the common or reference terminal. When implemented using bipolar transistors, these are called the common-emitter, common-collector, and common-base amplifiers; the corresponding names for the FET implementations are the common-source, common-drain, and common-gate amplifiers. Each amplifier category provides a unique set of characteristics in terms of voltage gain, input resistance, output resistance, and current gain.

The chapter expands the discussion of the characteristics of the common-emitter and common-source amplifiers, i.e., the inverting amplifiers that were developed in Chapter 7, and then looks in depth at the followers and noninverting amplifiers, focusing on the limits solid-state devices place on individual amplifier performance. Expressions are presented for the properties of each amplifier, and their similarities and differences are discussed in detail in order to build the understanding needed for the circuit design process. The transistor-level results are used throughout this book to analyze and design more complex single-stage and multi-stage amplifiers. We also explore amplifier frequency response at low frequencies and develop design equations useful for choosing coupling and bypass capacitors.

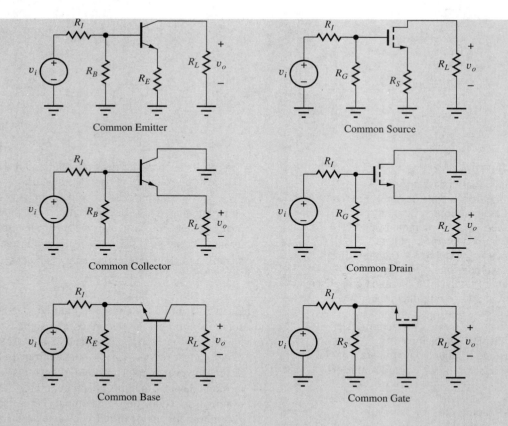

Common Emitter   Common Source
Common Collector   Common Drain
Common Base   Common Gate

Much discussion is devoted to single-transistor amplifiers because they are the heart of analog design. These single-stage amplifiers are an important part of the basic "tool set" of analog circuit designers, and a good understanding of their similarities and differences is a prerequisite for more complex amplifier design.

8.1 AMPLIFIER CLASSIFICATION

In Chapter 7, the input signal was applied to the base or gate of the transistor, and the output signal was taken from the collector or drain. However, the transistor has three separate terminals that may possibly be used to inject a signal for amplification: the base, emitter, and collector for the BJT; the gate, source, and drain for the FET. We will see shortly that only the base and emitter, or gate and source, are useful as signal insertion points; the collector and emitter, or drain and source, are useful points for signal removal. The examples we use in this chapter of the various amplifier configurations all use the same four-resistor bias circuits shown in Fig. 8.1. Coupling and bypass capacitors are then used to change the signal injection and extraction points and modify the ac characteristics of the amplifiers.

8.1.1 SIGNAL INJECTION AND EXTRACTION—THE BJT

For the BJT in Fig. 8.1(a), the large-signal transport model provides guidance for proper location of the input signal. In the forward-active region of the BJT,

$$i_C = I_S \left[\exp\left(\frac{v_{BE}}{V_T}\right) \right] \quad i_B = \frac{i_C}{\beta_F} = \frac{I_S}{\beta_{FO}} \left[\exp\left(\frac{v_{BE}}{V_T}\right) \right] \quad i_E = \frac{I_S}{\alpha_F} \left[\exp\left(\frac{v_{BE}}{V_T}\right) \right] \quad (8.1)$$

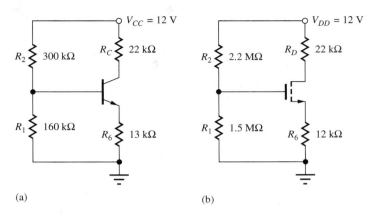

Figure 8.1 Four-resistor bias circuits for the (a) BJT and (b) MOSFET.

To cause i_C, i_E, and i_B to vary significantly, we need to change the base-emitter voltage v_{BE}, which appears in the exponential term. Because v_{BE} is equivalent to

$$v_{BE} = v_B - v_E \tag{8.2}$$

an input signal voltage can be injected into the circuit to vary the voltage at either the base or the emitter of the transistor. Note that the Early voltage has been omitted from Eq. (8.1), which indicates that varying the collector voltage has no effect on the terminal currents. Thus, the collector terminal is not an appropriate terminal for signal injection. Even for finite values of Early voltage, current variations with collector voltage are small, especially when compared to the exponential dependence of the currents on v_{BE}—again, the collector is not used as a signal injection point.

Substantial changes in the collector and emitter currents can create large voltage signals across the collector and emitter resistors R_C and R_6 in Fig. 8.1. Thus, signals can be removed from the amplifier at the collector or emitter terminals. However, because the base current i_B is a factor of β_F smaller than either i_C or i_E, the base terminal is not normally used as an output terminal.

DESIGN NOTE

The input signal can be applied to the base or emitter terminal of the bipolar transistor, and the output signal can be taken from the collector or emitter. The collector is not used as an input terminal, and the base is not used as an output.

8.1.2 SIGNAL INJECTION AND EXTRACTION—THE FET

A similar set of arguments can be used for the FET in Fig. 8.1(b), based on the expression for the n-channel MOSFET drain current in pinchoff:

$$i_S = i_D = \frac{K_n}{2}(v_{GS} - V_{TN})^2 \quad \text{and} \quad i_G = 0 \tag{8.3}$$

To cause i_D and i_S to vary significantly, we need to change the gate-source voltage v_{GS}. Because v_{GS} is equivalent to

$$v_{GS} = v_G - v_S \tag{8.4}$$

an input signal voltage can be injected so as to vary either the gate or source voltage of the FET. Varying the drain voltage has only a minor effect (for $\lambda \neq 0$) on the terminal currents, so the drain terminal is not an appropriate terminal for signal injection. As for the BJT, substantial changes in the drain or source currents can develop large voltage signals across resistors R_D and R_6 in Fig. 8.1(b). However, the gate terminal is not used as an output terminal because the gate current is zero.

454　Chapter 8　Transistor Amplifier Building Blocks

In summary, effective amplification requires a signal to be injected into either the base/emitter or gate/source terminals of the transistors in Fig. 8.1; the output signal can be taken from the collector/emitter or drain/source terminals. We do not inject a signal into the collector or drain or extract a signal from the base or gate terminals. These constraints yield three families of amplifiers: the **common-emitter/common-source (C-E/C-S)** circuits that we studied in Chapter 7, the **common-base/common-gate (C-B/C-G)** circuits, and the **common-collector/common-drain (C-C/C-D)** circuits.

These amplifiers are classified in terms of the structure of the ac equivalent circuit; each is discussed in detail in the next several sections. As noted earlier, the circuit examples all use the same four-resistor bias circuits in Fig. 8.1 in order to establish the Q-point of the various amplifiers. Coupling and bypass capacitors are then used to change the ac equivalent circuits. We will find that the ac characteristics of the various amplifiers are significantly different.

EXERCISE: Find the Q-points for the transistors in Fig. 8.1 and calculate the small-signal model parameters for the BJT and MOSFET. Use $\beta_F = 100$, $V_A = 50$ V, $K_n = 500$ μA/V^2, $V_{TN} = 1$ V, and $\lambda = 0.02$ V^{-1}. What are the values of μ_f? What is the value of $V_{GS} - V_{TN}$ for the MOSFET?

ANSWERS:

	I_C/I_D	V_{CE}/V_{DS}	$V_{GS} - V_{TN}$	g_m	r_π	r_o	μ_f
BJT	245 μA	3.39 V	...	9.80 mS	10.2 kΩ	218 kΩ	2140
FET	241 μA	3.81 V	0.982 V	0.491 mS	∞	223 kΩ	110

DESIGN NOTE

The input signal can be applied to the gate or source terminal of the FET, and the output signal can be taken from the drain or source. The drain is not used as an input terminal, and the gate is not used as an output.

8.1.3 COMMON-EMITTER (C-E) AND COMMON-SOURCE (C-S) AMPLIFIERS

The circuits in Fig. 8.2 are generalized versions of the common-emitter and common-source amplifiers discussed in Chapter 7. In these circuits, resistor R_6 in Fig. 8.1 has been split into two parts, with only resistor R_4 bypassed by capacitor C_2. We gain considerable flexibility in setting the voltage gain, input resistance, and output resistance of the amplifier by not bypassing all of the resistance in the transistor's emitter or source. In the C-E circuit in Fig. 8.2(a), the signal is injected into the base and taken out of the collector of the BJT. The emitter is the common terminal between the input and output ports. In the C-S circuit in Fig. 8.2(b), the signal is injected into the gate and taken out of the drain of the MOSFET; the source is the common terminal between the input and output ports.

The simplified ac equivalent circuits for these amplifiers appear in Fig. 8.2(c) and (d). We see that these network topologies are identical. Resistors R_E and R_S, connected between the emitter or source and ground, represent the unbypassed portion of the original bias resistor R_6. The presence of R_E and R_S in the ac equivalent circuits gives an added degree of freedom to the designer, and allows gain to be traded for increased input resistance, output resistance, and input signal range. Our comparative analysis will show that the C-E and C-S circuits can provide moderate-to-high values of voltage, current gain, input resistance, and output resistance.

8.1 Amplifier Classification   455

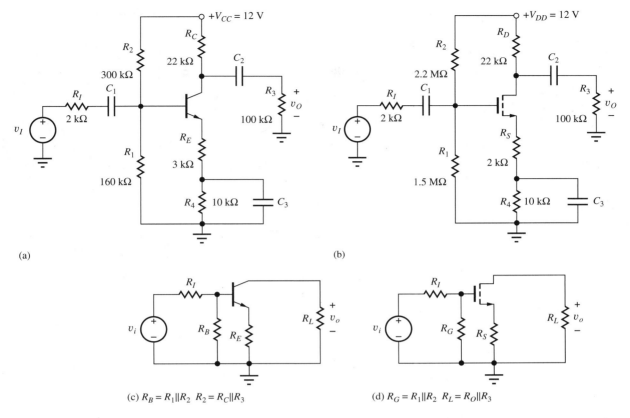

Figure 8.2 Generalized versions of the (a) common-emitter (C-E) and (b) common-source (C-S) amplifiers; (c) simplified ac equivalent circuit of the C-E amplifier in (a); (d) simplified ac equivalent circuit of the C-S amplifier in (b).

EXERCISE: Construct the ac equivalent circuit for the C-E and C-S amplifiers in Fig. 8.2, and show that the ac models are correct. What are the values of R_B or R_G, R_E or R_S, and R_L?

ANSWERS: 104 kΩ, 3.00 kΩ, 18.0 kΩ; 892 kΩ, 2.00 kΩ, 18.0 kΩ

8.1.4 COMMON-COLLECTOR (C-C) AND COMMON-DRAIN (C-D) TOPOLOGIES

The C-C and C-D circuits are shown in Fig. 8.3. Here the signal is injected into the base [Fig. 8.3(a)] or gate [Fig. 8.3(b)] and extracted from the emitter or source of the transistors. The collector and drain are bypassed directly to ground by the capacitors C_2 and represent the common terminals between the input and output ports. Once again, the ac equivalent circuits in Fig. 8.3(c) and (d) are identical in structure; the only differences are the resistor and transistor parameter values. Analysis will show that the C-C and C-D amplifiers provide a voltage gain of approximately 1, a high input resistance and a low output resistance. In addition, the input signals to the C-C and C-D amplifiers can be quite large without exceeding the small-signal limits. These amplifiers, often called emitter followers or source followers, are the single-transistor equivalents of the op amp voltage-follower circuit that we studied in Chapter 10.

EXERCISE: Construct the ac equivalent circuit for the C-C and C-D amplifiers in Fig. 8.3, and show that the ac models are correct. Verify the values of R_B, R_G, and R_L.

456　Chapter 8　Transistor Amplifier Building Blocks

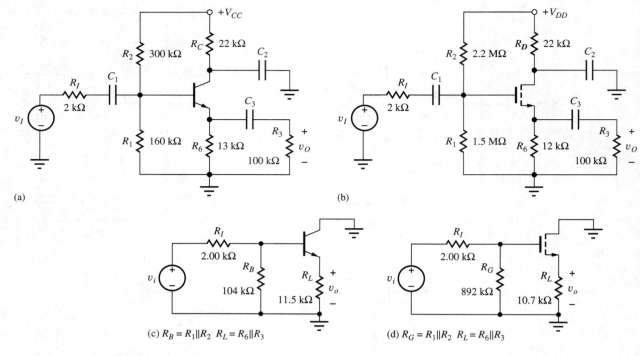

Figure 8.3 (a) Common-collector (C-C) amplifier; (b) common-drain (C-D) amplifier; (c) simplified ac equivalent circuit for the C-C amplifier; (d) simplified ac equivalent circuit for the C-D amplifier.

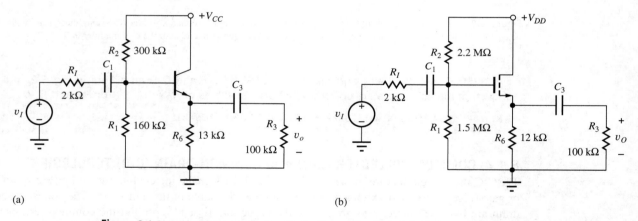

Figure 8.4 Simplified follower circuits with C_2, R_C, and R_D eliminated: (a) common-collector amplifier and (b) common-drain amplifier.

Circuit Simplification

For economy of design, we certainly do not want to include unneeded components, and the circuits in Fig. 8.3 can actually be simplified. The purpose of capacitor C_2 in the C-C and C-D amplifiers is to provide an ac ground at the collector or drain terminal of the transistor, and since we do not wish to develop a signal voltage at either of these terminals, there is no reason to have resistors R_C or R_D in the circuits. We can achieve the desired ac ground by simply connecting the collector and drain terminals directly to V_{CC} and V_{DD}, respectively, which eliminates components R_C, R_D, and C_2 from the circuits, as shown in Fig. 8.4.

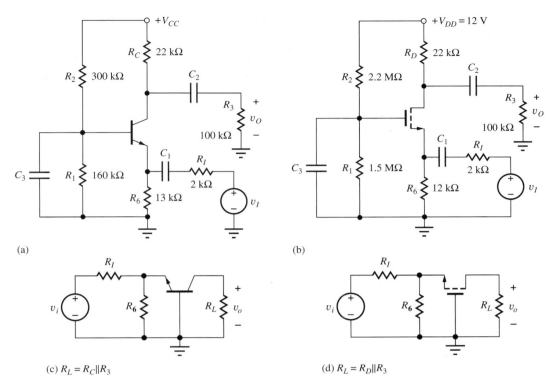

Figure 8.5 (a) Common-base (C-B) amplifier; (b) common-gate (C-G) amplifier; (c) simplified ac equivalent circuit for the C-B amplifier; (d) simplified ac equivalent circuit for the C-G amplifier.

8.1.5 COMMON-BASE (C-B) AND COMMON-GATE (C-G) AMPLIFIERS

The third class of amplifiers contains the C-B and C-G circuits in Fig. 8.5. ac signals are injected into the emitter or source and extracted from the collector or drain of the transistors. The base and gate terminals are connected to signal ground through bypass capacitors C_2; these terminals are the common connections between the input and output ports. The resulting ac equivalent circuits in Fig. 8.5(c) and (d) are again identical in structure. Analysis will show that the C-B and C-G amplifiers provide a voltage gain and output resistance very similar to those of the C-E and C-S amplifiers, but they have a much lower input resistance.

Analyses in the next several sections involve the simplified ac equivalent circuits given in Figs. 8.2(c), (d), 8.3(c), (d), and 8.5(c), (d). We assume for purposes of analysis that the circuits have been reduced to these "standard amplifier prototypes." These circuits are used to delineate the limits that the devices place on performance of the various circuit topologies. The results from these simplified circuits will then be used to analyze and design complete amplifiers.

The circuits in Figs. 8.2 to 8.5 showed only the BJT and MOSFET. The small-signal model of the JFET is identical to that of the three-terminal MOSFET, and the results obtained for the MOSFET amplifiers apply directly to those for the JFETs as well. JFETs can replace the MOSFETs in many circuits.

EXERCISE: Construct the ac equivalent circuit for the C-B and C-G amplifiers in Fig. 8.5, and show that the ac models are correct. What are the values of R_I, R_6, and R_L?

ANSWERS: 2 kΩ, 13.0 kΩ, 18.0 kΩ; 2 kΩ, 12.0 kΩ, 18.0 kΩ

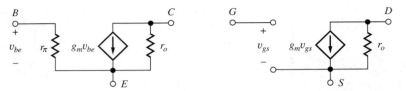

Figure 8.6 Small-signal models for the BJT and MOSFET.

TABLE 8.1
Small-Signal Transistor Models

SMALL-SIGNAL PARAMETER	BJT	MOSFET
g_m	$\dfrac{I_C}{V_T} \cong 40 I_C$	$\dfrac{2 I_D}{V_{GS} - V_{TN}} \cong \sqrt{2 K_n I_D}$
r_π	$\dfrac{\beta_o}{g_m}$	∞
r_o	$\dfrac{V_A + V_{CE}}{I_C} \cong \dfrac{V_A}{I_C}$	$\dfrac{(1/\lambda) + V_{DS}}{I_D} \cong \dfrac{1}{\lambda I_D}$
β_o	$g_m r_\pi$	∞
$\mu_f = g_m r_o$	$\dfrac{V_A + V_{CE}}{V_T} \cong 40 V_A$	$\dfrac{2}{\lambda(V_{GS} - V_{TN})} \cong \dfrac{1}{\lambda}\sqrt{\dfrac{2 K_n}{I_D}}$

8.1.6 SMALL-SIGNAL MODEL REVIEW

The small-signal models for the BJT and MOSFET appear in Fig. 8.6, and the formulae relating the small-signal model parameters to the Q-point are summarized again for reference in Table 8.1.

Again, we recognize that the topologies are very similar, except for the finite value of r_π for the BJT. Due to these similarities, we begin the analyses with that for the bipolar transistor because it has the more general small-signal model; we obtain results for the FET cases from the BJT expressions by taking limits as r_π and β_o approach infinity. In subsequent sections, expressions for the voltage gain, input resistance, output resistance, and current gain are developed for each of the single-transistor amplifiers based upon the small-signal models in Fig. 8.6.

8.2 INVERTING AMPLIFIERS—COMMON-EMITTER AND COMMON-SOURCE CIRCUITS

We begin our comparative analysis of the various amplifier families with the common-emitter and common-source amplifiers that we introduced in Chapter 7. The ac equivalent circuits are given in Fig. 8.2 and now include the addition of unbypassed resistors R_E and R_S. Here again we note that the topologies are identical. Performance differences arise because of the parametric differences between the transistors used in the circuits. As mentioned above we will analyze the common-emitter amplifier first and then simplify the C-E results for the common-source case.

8.2.1 THE COMMON-EMITTER (C-E) AMPLIFIER

Now we are in a position to analyze the small-signal characteristics of the complete **common-emitter (C-E) amplifier** shown in Fig. 8.7(a). The ac equivalent circuit of Fig. 8.7(b) is constructed by assuming that the capacitors all have zero impedance at the signal frequency and the dc voltage

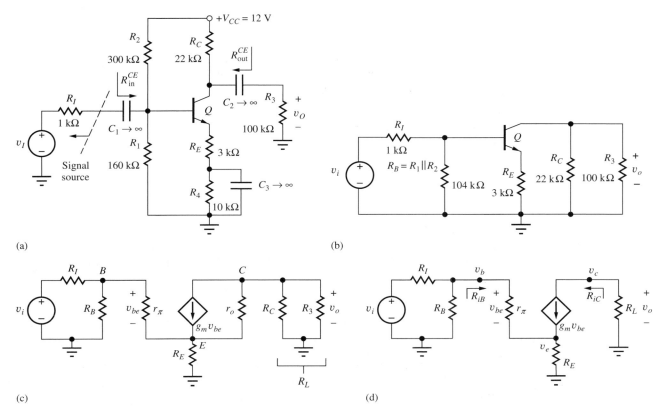

Figure 8.7 (a) Common-emitter amplifier circuit employing a bipolar transistor; (b) ac equivalent circuit for the common-emitter amplifier in part (a); the common-emitter connection should now be evident; (c) ac equivalent circuit with the bipolar transistor replaced by its small-signal model; (d) final equivalent circuit for ac analysis of the common-emitter amplifier in which r_o has been neglected.

source represents an ac ground. For simplicity, we assume that we have found the Q-point and know the values of I_C and V_{CE}. In Fig. 8.7(b), resistor R_B represents the parallel combination of the two base bias resistors R_1 and R_2,

$$R_B = R_1 \| R_2 \tag{8.5}$$

and R_4 is eliminated by bypass capacitor C_3.

Before we can develop an expression for the voltage gain of the amplifier, the transistor must be replaced by its small-signal model as in Fig. 8.7(c). A final simplification appears in Fig. 8.7(d), in which the resistor R_L represents the total equivalent external load resistance on the transistor, the parallel combination of R_C and R_3:

$$R_L = R_C \| R_3 \tag{8.6}$$

Note that transistor output resistance r_o has been neglected in the final circuit in Fig. 8.7(d) as discussed in Secs. 7.7 and 7.11.[1]

In Fig. 8.7(b) through (d), the reason why this amplifier configuration is called a common-emitter amplifier is apparent. The emitter portion of the circuit represents the common connection between the amplifier input and output ports. The input signal is applied to the transistor's base, the output signal appears at the collector, and both the input and output signals are referenced to the (common) emitter terminal (through R_E).

---

[1] We assume that the voltage gain is much less than μ_f.

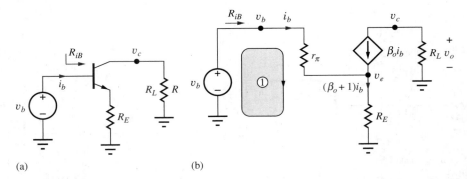

Figure 8.8 Simplified circuit and small-signal models for finding the common-emitter terminal voltage gain A_{vt}^{CE} and input resistance R_{iB}.

Terminal Voltage Gain

Now we are ready to develop an expression for the overall gain of the amplifier from signal source v_i to the output voltage across resistor R_3 ($R_L = R_3 \| R_C$). The voltage gain can be written as

$$A_v^{CE} = \frac{\mathbf{v}_o}{\mathbf{v}_i} = \left(\frac{\mathbf{v}_c}{\mathbf{v}_b}\right)\left(\frac{\mathbf{v}_b}{\mathbf{v}_i}\right) = A_{vt}^{CE}\left(\frac{\mathbf{v}_b}{\mathbf{v}_i}\right) \qquad \text{where} \qquad A_{vt}^{CE} = \left(\frac{\mathbf{v}_c}{\mathbf{v}_b}\right) \qquad (8.7)$$

A_{vt}^{CE} represents the voltage gain between the base and collector terminals of the transistor, the "**terminal gain.**" We will first find expressions for terminal gain A_{vt}^{CE} as well as the input resistance at the base of the transistor. Then we can relate v_b to v_i to find the overall voltage gain.

In Fig. 8.8, the BJT is replaced with its small-signal model, and the base terminal of the transistor is driven by test source v_b. Note that the small-signal model has been changed to its current-controlled form, and r_o is neglected as discussed before. Collector voltage v_c is given by

$$\mathbf{v_c} = -\beta_o \mathbf{i_b} R_L \qquad (8.8)$$

We can relate i_b to base voltage v_b by writing an equation around loop 1:

$$\mathbf{v_b} = \mathbf{i_b} r_\pi + (\mathbf{i_b} + \beta_o \mathbf{i_b}) R_E = \mathbf{i_b}[r_\pi + (\beta_o + 1) R_E] \qquad (8.9)$$

Solving for i_b and substituting the result in Eq. (8.8) yields

$$A_{vt}^{CE} = \frac{\mathbf{v_c}}{\mathbf{v_b}} = -\frac{\beta_o R_L}{r_\pi + (\beta_o + 1) R_E} \cong -\frac{g_m R_L}{1 + g_m R_E} \qquad (8.10)$$

in which the approximation assumes $\beta_o \gg 1$ and uses $\beta_o = g_m r_\pi$.

The minus sign indicates that the common-emitter stage is an inverting amplifier in which the input and output are 180° out of phase. The gain is proportional to the product of the transistor transconductance g_m and load resistor R_L. This product places an upper bound on the gain of the amplifier, and we will encounter the $g_m R_L$ product over and over again as we study transistor amplifiers. We will explore gain expression [Eq. (8.10)] in more detail shortly.

Input Resistance

The resistance looking into the base terminal R_{iB} in Fig. 8.8 can easily be found by rearranging Eq. (8.9). The input resistance is simply the ratio of v_b and i_b,

$$R_{iB} = \frac{\mathbf{v_b}}{\mathbf{i_b}} = r_\pi + (\beta_o + 1) R_E \cong r_\pi (1 + g_m R_E) \qquad (8.11)$$

in which the final approximation again assumes $\beta_o \gg 1$ and uses $\beta_o = g_m r_\pi$. The input resistance looking into the base of the transistor is equal to r_π plus the resistance reflected into the base by R_E. The effective value of R_E is increased by the current gain $(\beta_0 + 1)$.

DESIGN NOTE

The resistance R_{iB} looking into the base of the transistor itself is given by

$$R_{iB} = r_\pi + (\beta_o + 1)R_E \cong r_\pi(1 + g_m R_E)$$

The overall input resistance of the common-emitter amplifier, R_{in}^{CE}, is defined as the resistance looking into the amplifier at coupling capacitor C_1 in Fig. 8.7(a) and is equal to the parallel combination of R_{iB} and base bias resistance R_B:

$$R_{\text{in}}^{CE} = R_B \| R_{iB} \tag{8.12}$$

Signal Source Voltage Gain

The overall voltage gain A_v^{CE} of the amplifier, including the effect of source resistance R_1, can now be found using the input resistance and terminal gain expressions. Voltage v_b at the base of the bipolar transistor in Fig. 8.7(d) is related to v_i by

$$\mathbf{v_b} = \mathbf{v_i} \frac{R_B \| R_{iB}}{R_I + (R_B \| R_{iB})} \tag{8.13}$$

Combining Eqs. (8.7), (8.10), and (8.13), yields a general expression for the overall voltage gain of the common-emitter amplifier:

$$A_v^{CE} = A_{vt}^{CE} \left(\frac{\mathbf{v_b}}{\mathbf{v_i}} \right) = -\left(\frac{g_m R_L}{1 + g_m R_E} \right) \left[\frac{R_B \| R_{iB}}{R_I + (R_B \| R_{iB})} \right] \tag{8.14}$$

In this expression, we see that the overall voltage gain is equal to the terminal gain A_{vt}^{CE} reduced by the voltage division between R_I and the equivalent resistance at the base of the transistor. Terminal gain A_{vt}^{CE} places an upper limit on the voltage gain since the voltage division factor will be less than one.

Important Limits and Model Simplifications

We now explore the limits to the voltage gain of common-emitter amplifiers using model simplifications for large emitter resistance and zero emitter resistance. First, we will assume that the source resistance is small enough that $R_I \ll R_B \| R_{iB}$ so that

$$A_v^{CE} \cong A_{vt}^{CE} = -\frac{g_m R_L}{1 + g_m R_E} \qquad \text{for } R_I \ll R_B \| R_{iB} \tag{8.15}$$

This approximation is equivalent to saying that the total input signal appears at the base of the transistor.

Zero Resistance in the Emitter   In order to achieve as large a gain as possible, we need to make the denominator in Eq. (8.15) as small as possible, and this is achieved by setting $R_E = 0$. The gain is then

$$A_v^{CE} \cong -g_m R_L = -g_m(R_C \| R_3) \tag{8.16}$$

which is the expression we found for the basic common-emitter amplifier in Chapter 7. Equation (8.16) states that the terminal voltage gain of the common-emitter stage is equal to the product of the transistor's transconductance g_m and load resistance R_L, and the minus sign indicates that the output voltage is "inverted" or 180° out of phase with respect to the input. Equation (8.16) places an upper limit on the gain we can achieve from a common-emitter amplifier with an external load resistor. Remember that we already developed a simple rule-of-thumb estimate for the $g_m R_L$ product in Chapter 7:

$$g_m R_L \cong 10 \, V_{CC} \tag{8.17}$$

DESIGN NOTE

The magnitude of the voltage gain of a resistively loaded common-emitter amplifier *with zero emitter resistance* is approximately equal to 10 times the power supply voltage.

$$A_v^{CE} \cong -10 V_{CC} \quad \text{for} \quad R_E = 0$$

This result represents an excellent way to quickly check the validity of more detailed calculations. Remember that the rule-of-thumb estimate is not going to be exact, but will predict the order of magnitude of the gain, typically within a factor of two or so.

Large Emitter Resistance The presence of a nonzero value of emitter resistor R_E reduces the gain below that given by Eq. (8.16), and another very useful simplification occurs when the $g_m R_E$ product is much larger than one:

$$A_{vt}^{CE} = -\frac{g_m R_L}{1 + g_m R_E} \cong -\frac{R_L}{R_E} \quad \text{for} \quad g_m R_E \gg 1 \tag{8.18}$$

The gain is now set by the ratio of the load resistor R_L and emitter resistor R_E. This is an extremely useful result because the gain is now independent of the transistor characteristics that vary widely from device to device. The result in Eq. (8.18) is very similar to the one obtained for the op-amp inverting amplifier circuit and is a result of feedback introduced by resistor R_E.

Achieving the simplification in Eq. (8.18) requires $g_m R_E \gg 1$. We can relate this product to the dc bias voltage developed across R_E:

$$g_m R_E = \frac{I_C R_E}{V_T} = \alpha_F \frac{I_E R_E}{V_T} \cong \frac{I_E R_E}{V_T} \quad \text{and we need} \quad I_E R_E \gg V_T \tag{8.19}$$

$I_E R_E$ represents the dc voltage drop across emitter resistor R_E and must be much greater than 25 mV, for example 0.250 V, a value that is easily achieved.

Understanding Generalized Common-Emitter Amplifier Operation

Let us explore common-emitter operation a bit further by looking at the signal voltage developed at the emitter terminal with reference to Fig. 8.8 and Eq. (8.9):

$$\mathbf{v_e} = (\beta_o + 1)\mathbf{i_b} R_E = \frac{(\beta_o + 1) R_E}{r_\pi + (\beta_o + 1) R_E} \mathbf{v_b} \cong \frac{g_m R_E}{1 + g_m R_E} \mathbf{v_b} \cong \mathbf{v_b} \quad \text{for large } g_m R_E \tag{8.20}$$

The voltage v_b at the base of the transistor is transferred directly to the emitter, setting up an emitter current of v_b/R_E. Essentially all the emitter current must be supplied from the collector yielding a voltage gain equal to the ratio of R_L to R_E:

$$\mathbf{i_e} \cong \frac{\mathbf{v_b}}{R_E} \qquad \mathbf{v_o} = -\mathbf{i_c} R_L = -\alpha_o \mathbf{i_e} R_L \cong -\mathbf{i_e} R_L \quad \text{and} \quad A_{vt}^{CE} = \frac{\mathbf{v_o}}{\mathbf{v_b}} \cong -\frac{R_L}{R_E} \tag{8.21}$$

This unity signal voltage transfer from base to emitter should not be mysterious. We know that the base and emitter terminals are directly connected by a forward-biased diode whose voltage is virtually constant at 0.7 V. Thus the emitter signal voltage should be approximately the same as the base signal. The voltage transfer between the base and emitter terminals forms the basis of the emitter-follower operation to be discussed in detail in Sec. 8.3.

DESIGN NOTE

The gain of the generalized common-emitter amplifier is approximately equal to the ratio of the load and emitter resistors.

$$A_{vt}^{CE} = -\frac{g_m R_L}{1 + g_m R_E} \cong -\frac{R_L}{R_E} \quad \text{for} \quad g_m R_E \gg 1$$

Small-Signal Limit for the Common-Emitter Amplifier An important additional benefit of adding resistor R_E to the circuit is to increase the allowed size of the input signal v_b at the base. For small-signal operation, the magnitude of the base-emitter voltage v_{be}, developed across r_π in the small-signal model, must be less than 5 mV (you may wish to review Sec. 8.5.7). This voltage can be found using the input current i_b from Eq. (8.9):

$$\mathbf{v_{be}} = \mathbf{i_b} r_\pi = \mathbf{v_b} \frac{r_\pi}{r_\pi + (\beta_o + 1)R_E} \cong \frac{\mathbf{v_b}}{1 + g_m R_E} \tag{8.22}$$

The approximation requires $\beta_o \gg 1$. Requiring $|v_{be}|$ in Eq. (8.22) to be less than 5 mV gives

$$|v_b| \leq 0.005(1 + g_m R_E) \text{ V} \tag{8.23}$$

If $g_m R_E \gg 1$, then v_b can be increased well beyond the 5-mV limit.

DESIGN NOTE

Use of an emitter resistor in the common-emitter amplifier can significantly increase the input signal range of the amplifier.

$$|v_b| \leq 0.005 \text{ V}(1 + g_m R_E)$$

Resistance at the Collector of the Bipolar Transistor

The resistance looking into the collector terminal of the transistor, R_{iC}, can be found with the aid of the equivalent circuit in Fig. 8.9 in which input source v_i has been set to zero, and test source v_x is applied to the collector of the transistor. The Thévenin equivalent resistance on the base is then $R_{th} = R_B \| R_I$.

R_{iC} equals the ratio of $\mathbf{v_x}$ to $\mathbf{i_x}$, where i_x represents the current through independent source v_x. To find i, we write an expression for $\mathbf{v_e}$:

$$\mathbf{v_e} = (\beta_o + 1)\mathbf{i} R_E \quad \text{and} \quad \mathbf{i_x} = \beta_o \mathbf{i} \tag{8.24}$$

and realize that the current $\mathbf{i}$ can also be written directly in terms of $\mathbf{v_e}$:

$$\mathbf{i} = -\frac{\mathbf{v_e}}{R_{th} + r_\pi} \tag{8.25}$$

Combining Eqs. (8.24) and (8.25) yields

$$\mathbf{v_e}\left[1 + \frac{(\beta_o + 1)R_E}{r_\pi + R_{th}}\right] = 0 \quad \text{and} \quad \mathbf{v_e} = 0 \tag{8.26}$$

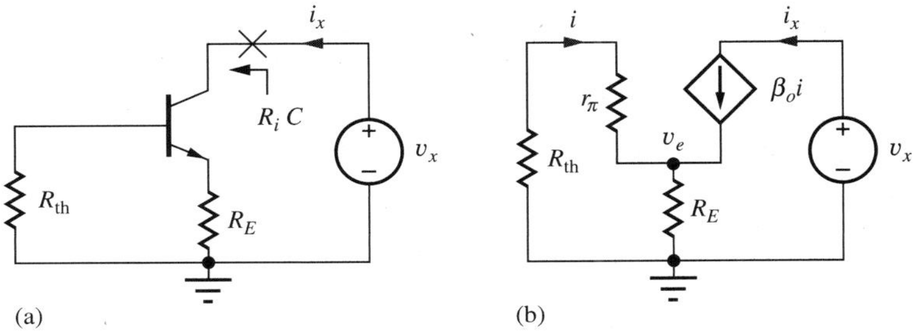

Figure 8.9 Circuits for calculating the resistance at the collector of the transistor.

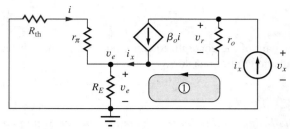

Figure 8.10 Collector resistance with r_o included.

Because $v_e = 0$, Eq. (8.25) requires that i equal zero as well. Hence, $i_x = 0$, and the output resistance of this circuit is infinite!

On the surface, this result may seem acceptable. However, a red flag should go up. We must be suspicious of the results that indicate resistances are infinite (or zero). Using the simplified circuit model in Fig. 8.9(b), in which r_o is neglected, has led to an unreasonable result!

We improve our analysis by moving to the next level of model complexity by including r_o, as shown in Fig. 8.10. For this analysis, the circuit is driven by the test current i_x, and the voltage v_x must be determined in order to find R_{out}.[2]

Summing the voltages around loop 1 and applying KCL at the output node,

$$v_x = v_r + v_e = (i_x - \beta_o i)r_o + v_e \tag{8.27}$$

Current i_x is forced through the parallel combination of $(R_{\text{th}} + r_\pi)$ and R_E, so that v_e can be expressed as

$$v_e = i_x[(R_{\text{th}} + r_\pi) \| R_E] = i_x \frac{(R_{\text{th}} + r_\pi)R_E}{R_{\text{th}} + r_\pi + R_E} \tag{8.28}$$

At the emitter node, current division can be used to find i in terms of i_x:

$$i = -i_x \frac{R_E}{R_{\text{th}} + r_\pi + R_E} \tag{8.29}$$

Combining Eqs. (8.27) through (8.29) yields a somewhat messy expression for the output resistance of the C-E amplifier:

$$R_{iC} = r_o \left(1 + \frac{\beta_o R_E}{R_{\text{th}} + r_\pi + R_E}\right) + (R_{\text{th}} + r_\pi) \| R_E \cong r_o \left(1 + \frac{\beta_o R_E}{R_{\text{th}} + r_\pi + R_E}\right) \tag{8.30}$$

If we now assume that $(r_\pi + R_E) \gg R_{\text{th}}$ and $r_o \gg R_E$ and remember that $\beta_o = g_m r_\pi$, we reach the approximate results that should be remembered:

$$R_{iC} \cong r_o[1 + g_m(R_E \| r_\pi)] = r_o + \mu_f(R_E \| r_\pi) \tag{8.31}$$

We should check to see that Eq. (8.31) reduces to the proper result for $R_E = 0$; that is, $R_{\text{out}} = r_o$. Since it does, we can feel comfortable that our level of modeling is sufficient to produce a meaningful result.

Equation (8.31) tells us that the output resistance of the common-emitter amplifier is equal to the output resistance r_o of the transistor itself plus the equivalent resistance $(R_E \| r_\pi)$ multiplied by the amplification factor of the transistor. For $g_m(R_E \| r_\pi) \gg 1$, $R_{\text{out}} \gg r_o$, we find that the resistance at the collector can be designed to be much greater than the output resistance of the transistor itself!

Important Limit for the Bipolar Transistor

The finite current gain of the bipolar transistor places an upper bound on the size of R_{iC} that can be achieved. Referring back to Fig. 8.10, we see that r_π appears in parallel with R_E when we neglect

---

[2] The upcoming sequence of equations has been developed by the author as an "easy" way to derive this result; this approach is not expected to be obvious. Alternatively, the circuit in Fig. 8.10 can be formulated as a two-node problem by combining R_{th} and r_π.

R_{th}. If we let $R_E \to \infty$ in Eq. (8.31), we find that the maximum value of output resistance is $R_{ic} \cong \mu_f r_\pi = \beta_o r_o$.

DESIGN NOTE

A quick design estimate for the resistance at the collector of a bipolar transistor with an unbypassed resistor R_E in the emitter is
$$R_{iC} \cong r_o[1 + g_m(r_\pi \| R_E)] \cong \mu_f(r_\pi \| R_E) < \beta_o r_o$$
However, remember R_{iC} can never exceed $\beta_o r_o$.

Output Resistance of the Overall Common-Emitter Amplifier

The output resistance of the overall common-emitter amplifier is defined as the resistance looking into the circuit at input coupling capacitor C_2 in Fig. 8.7(a). Thus R_{out}^{CE} equals the parallel combination of collector resistor R_C and the resistance looking into the collector of the transistor itself, R_{iC}, as defined in Fig. 8.7(c):

$$R_{out}^{CE} = R_C \| R_{iC} = R_C \| r_o \left(1 + \frac{\beta_o R_E}{R_{th} + r_\pi + R_E}\right) \quad (8.32)$$

Terminal Current Gain for the Common-Emitter Amplifier

The **terminal current gain** A_{it} is defined as the ratio of the current delivered to the load resistor R_L to the current being supplied to the base terminal. For the C-E amplifier in Fig. 8.11, the current in R_L is equal to i amplified by the current gain β_o, yielding a current gain equal to $-\beta_o$.

$$A_{it}^{CE} = -\beta_o \quad (8.33)$$

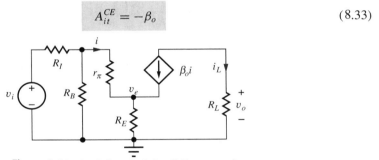

Figure 8.11 Circuit for calculating C-E current gain.

EXAMPLE 8.1 VOLTAGE GAIN OF A COMMON-EMITTER AMPLIFIER

In this example, we find the small-signal parameters of the bipolar transistor and then calculate the voltage gain of a common-emitter amplifier.

PROBLEM Calculate the voltage gain, input resistence, and output resistence of the common-emitter amplifier in Fig. 8.7 if the transistor has $\beta_F = 100$, $V_A = 75$V, $\lambda = 0.0133$ V^{-1}, and the Q-point is (0.245 mA, 3.39 V). What is the maximum value of v_i that satisfies the small-signal assumptions?

SOLUTION **Known Information and Given Data:** Common-emitter amplifier with its ac equivalent circuit given in Fig. 8.7; $\beta_F = 100$ and $V_A = 75$ V; the Q-point is (0.245 mA, 3.39 V); $R_I = 1$ kΩ, $R_1 = 160$ kΩ, $R_2 = 300$ kΩ, $R_C = 22$ kΩ, $R_E = 3$ kΩ, $R_4 = 10$ kΩ, and $R_3 = 100$ kΩ.

Unknowns: Small-signal parameters of the transistor; voltage gain A_v; R_{in}^{CE}; R_{out}^{CE}; small-signal limit for the value of v_i

Approach: Use the Q-point information to find g_m, r_π and r_o. Use the calculated and given values to evaluate the voltage gain expression in Eq. (8.14), and the expressions for R_{in}^{CE} and R_{out}^{CE}.

Assumptions: The transistor is in the active region, and $\beta_o = \beta_F$. The signal amplitudes are low enough to be considered as small signals. Assume r_o can be neglected.

Analysis: (a) To evaluate Eq. (8.14),

$$A_v^{CE} = -\left(\frac{g_m R_L}{1 + g_m R_E}\right)\left[\frac{R_B \| R_{iB}}{R_1 + (R_B \| R_{iB})}\right] \quad \text{with} \quad R_B = R_1 \| R_2 \quad \text{and} \quad R_{iB} = r_\pi + (\beta_o + 1) R_E$$

the values of the various resistors and small-signal model parameters are required. We have

$$g_m = 40 I_C = 40(0.245 \text{ mA}) = 9.80 \text{ mS} \qquad r_\pi = \frac{\beta_o V_T}{I_C} = \frac{100(0.025 \text{ V})}{0.245 \text{ mA}} = 10.2 \text{ k}\Omega$$

$$r_o = \frac{V_A + V_{CE}}{I_C} = \frac{75 \text{ V} + 3.39 \text{ V}}{0.245 \text{ mA}} = 320 \text{ k}\Omega \qquad R_{iB} = r_\pi + (\beta_o + 1) R_E = 313 \text{ k}\Omega$$

$$R_B = R_1 \| R_2 = 104 \text{ k}\Omega \qquad R_L = R_c \| R_3 = 18.0 \text{ k}\Omega$$

Using these values,

$$A_v^{CE} = -\left(\frac{9.80 \text{ mS}(18.0 \text{ k}\Omega)}{1 + 9.80 \text{ mS}(3.0 \text{ k}\Omega)}\right)\left[\frac{104 \text{ k}\Omega \| 313 \text{ k}\Omega}{1 \text{ k}\Omega + (104 \text{ k}\Omega \| 313 \text{ k}\Omega)}\right] = -5.80(0.987) = -5.72$$

Thus, the common-emitter amplifier in Fig. 8.7 provides a small-signal voltage gain $A_v = -5.72$ or 15.1 dB.

The common-emitter amplifier's input and output resistances are found as

$$R_{\text{in}}^{CE} = R_B \| R_{iB} = 104 \text{ k}\Omega \| 313 \text{ k}\Omega = 78.1 \text{ k}\Omega \qquad \text{and} \qquad R_{\text{out}}^{CE} = R_C \| R_{iC}$$

$$R_{iC} = R_C \| r_o \left(1 + \frac{\beta_o R_E}{R_{\text{th}} + r_\pi + R_E}\right) = 320 \text{ k}\Omega \left[1 + \frac{100(3 \text{ k}\Omega)}{0.99 \text{ k}\Omega + 10.2 \text{ k}\Omega + 3 \text{ k}\Omega}\right] = 7.09 \text{ M}\Omega$$

$$R_{\text{out}}^{CE} = 22 \text{ k}\Omega \| 7.09 \text{ M}\Omega = 21.9 \text{ k}\Omega$$

Small-signal operation requires $|v_{be}| \leq 0.005$ V. Based on Fig. 8.7, the base-emitter signal voltage can be related to v_i by

$$\mathbf{v_{be}} = \mathbf{v_b} \frac{r_\pi}{r_\pi + (\beta_o + 1) R_E} = \mathbf{v_i} \left[\frac{R_B \| R_{iB}}{R_I + R_B \| R_{iB}}\right]\left[\frac{r_\pi}{r_\pi + (\beta_o + 1) R_E}\right]$$

so that

$$|v_i| \leq (0.005 \text{ V}) \left[\frac{R_I + (R_B \| R_{iB})}{R_B \| R_{iB}}\right]\left[\frac{r_\pi + (\beta_o + 1) R_E}{r_\pi}\right]$$

$$|v_i| \leq (0.005 \text{ V}) \left[\frac{1 \text{ k}\Omega + (104 \text{ k}\Omega \| 313 \text{ k}\Omega)}{104 \text{ k}\Omega \| 313 \text{ k}\Omega}\right]\left[\frac{10.2 \text{ k}\Omega + 101(3 \text{ k}\Omega)}{10.2 \text{ k}\Omega}\right] = 0.155 \text{ V}$$

Check of Results: We have found the required information. The amplification factor is $\mu_f = g_m r_o = (9.80 \text{ mS})(320 \text{ k}\Omega) = 3140$. The magnitude of the voltage gain of a single-transistor amplifier cannot exceed this value. Using the result from Eq. (8.18), we estimate the gain to be $A_v^{CE} = -R_L / R_E = -18 \text{ k}\Omega / 3 \text{ k}\Omega = -6$. Our answer satisfies both these checks.

We can quickly check our R_{iC} calculation using the approximation $R_{iC} \approx \mu_f R_E$:

$$R_{iC} \cong \mu_f R_E = (g_m r_o) R_E = (9.80 \text{ mS})(320 \text{ k}\Omega)(3 \text{ k}\Omega) = 9.41 \text{ M}\Omega \quad \text{and} \quad 7.09 \text{ M}\Omega < 9.41 \text{ M}\Omega$$

The estimate of R_{iC} is somewhat high since R_{th} and r_π cannot be neglected relative to R_E.

Discussion: Note that the value of the voltage gain ($A_v = -5.72$) is much less than the intrinsic voltage gain ($\mu_f = 3140$), so neglecting r_o in the calculation should be valid. Note also that the value of r_o is much greater than the load resistance connected to the collector terminal of the amplifier (18 kΩ). This is also consistent with our being able to neglect r_o in the voltage gain calculation. The maximum allowed input signal is increased significantly to 0.155 V due to the presence of R_E.

We also see that we did a lot of work to find out that the overall output resistance is essentially equal to R_C. Finally we observe that the value of R_{iC} is less than 25 percent of the $\beta_o r_o$ limit of 32 MΩ.

Computer-Aided Analysis: Now, let us close up check our hand analysis using the SPICE circuit below in which we must set the transistor parameters to be consistent with our hand analysis in order to achieve a similar Q-point: BF = 100, VAF = 75 V, and IS = 1 fA. We can use an ac analysis to find the voltage gain and will sweep from 1000 Hz to 100 kHz with five frequency points per decade. Several decades are simulated so we can be sure we are in a region where the effects of the capacitors are negligible. The capacitor values must be set to a large number, say 100 μF, so that they will have very small reactance at the simulation frequencies.

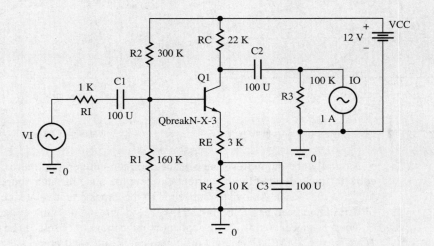

Source VI is the input source and has both an ac value (1∠0°) for small-signal analysis (ac sweep) and a sine wave component (0.15 sin 20,000 π t) for transient simulation. ac current source IO (1∠0°) is added to the output in order to find the output resistance. Note that only VI or IO should have a nonzero value at a given time.

The SPICE results are: Q-point = (0.248 mA, 3.30 V) and $A_v = -5.67$. [Note that an alternative method to check our calculations is to use SPICE to perform an ac analysis of the small-signal equivalent circuit in Fig. 8.7(d).]

The graph here shows the time-domain response of the amplifier output with an input of 0.15 V obtained with a transient simulation with TSTOP = 0.3 MS. In the graph, we observe good linearity with a gain of −5.7.

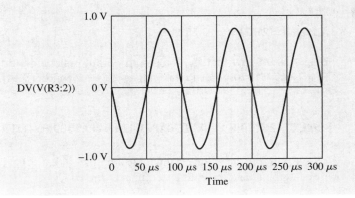

The (frequency dependent) input resistance of the common-emitter amplifier is equal to the voltage at the base node divided by the current entering the node through coupling capacitor C_1 : $R_{in}^{CE} = V(Q1:b)/I(C1)$. As frequency increases, bypass capacitor C_3 becomes effective and the input resistance drops. At frequencies above 10 Hz, the SPICE output becomes constant at 77.8 kΩ in agreement with our hand calculations.

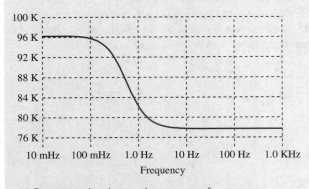

(a) Common-emitter input resistance versus frequency.

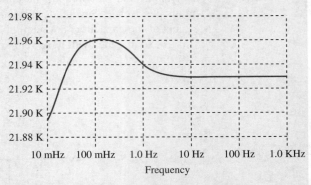

(b) Common-emitter output resistance versus frequency.

Similarly, the output resistance of the common-emitter amplifier is equal to the voltage at the collector node divided by the current entering the node through coupling capacitor C_2 : $R_{out}^{CE} = V(Q1:c)/I(C2)$. As bypass capacitor C_3 becomes effective at frequencies above 10 Hz, the SPICE output becomes constant at 21.93 kΩ in agreement with our hand calculations.

The resistance looking into the collector of the transistor itself can be found as $R_{iC} = VC(Q1)/IC(Q1) = 6.89$ MΩ. The slight disagreement is due to differences in the calculated Q-point and the small-signal parameters in SPICE.

EXERCISE: (a) Suppose resistors R_C, R_E, and R_3 have 10 percent tolerances. What are the worst-case values of voltage gain for this amplifier? (b) What is the voltage gain in the original circuit if $\beta_o = 125$? (c) Suppose the Q-point current in the original circuit increases to 0.275 mA. What are the new values of V_{CE} and the voltage gain?

ANSWERS: (a)−4.75, −6.99; (b) −5.74; (c) 2.34 V, −5.76

EXERCISE: What is the value of R_{out} for the common-emitter amplifier in Ex. 8.1 if R_E is changed to 2 kΩ? Assume the Q-point does not change. Compare the result to the new value of $\mu_f R_E$.

ANSWERS: 21.9 k$\Omega \ll$ 6.28 MΩ

EXERCISE: Show that the maximum output resistance for the common-emitter amplifier is $R_{out} \cong (\beta_o + 1)r_o$ by taking the limit as $R_E \to \infty$ in Eq. (8.31).

EXAMPLE 8.2 COMMON-EMITTER VOLTAGE GAIN WITH BYPASSED EMITTER

Now we will find the voltage gain of the amplifier in Ex. 8.1 with bypass capacitor C_3 connected between ground and the emitter terminal of the BJT.

PROBLEM (a) Find the voltage gain of the amplifier in Ex. 8.1 with bypass capacitor C_3 connected between ground and the emitter terminal of the BJT. (b) Compare the result in (a) to the common-emitter "rule-of-thumb" gain estimate and the amplification factor of the transistor. (c) Find the new value of the amplifier input and output resistances. (d) Find the value of v_i that corresponds to the small-signal limit.

SOLUTION **Known Information and Given Data:** Common-emitter amplifier in Fig. 8.7 with emitter terminal bypassed to ground. From Ex. 8.1, the Q-point = (0.245 mA, 3.39 V), g_m = 9.80 mS, r_π = 10.2 kΩ, and r_o = 320 kΩ.

Unknowns: Actual voltage gain, rule-of-thumb estimate, amplification factor of the transistor; R_{in}^{CE}; R_{out}^{CE}

Approach: (a) Evaluate the A_v^{CE} expression with $R_E = 0$ (see ac equivalent circuit on next page). (b) Estimate the voltage gain using $A_v^{CE} \cong -10V_{CC}$; calculate $\mu_f = g_m r_o$.

Assumptions: The bipolar transistor is operating in the active region. Signal amplitudes correspond to small-signal conditions. Transistor output resistance r_o can be neglected.

Analysis:

(a) With the emitter terminal bypassed to ground, $R_E = 0$:

$$R_{iB} = r_\pi + (\beta_o + 1)R_E = r_\pi \quad \text{and} \quad R_B = R_1 \| R_2$$

$$A_v^{CE} = -\left(\frac{g_m R_L}{1 + g_m R_E}\right)\left[\frac{R_B \| R_{iB}}{R_I + (R_B \| R_{iB})}\right] = -g_m R_L \frac{R_B \| r_\pi}{R_I + (R_B \| r_\pi)}$$

$$A_v^{CE} = -9.80 \text{ mS } (18 \text{ k}\Omega) \frac{104 \text{ k}\Omega \| 10.2 \text{ k}\Omega}{1 \text{ k}\Omega + (104 \text{ k}\Omega \| 10.2 \text{ k}\Omega)} = -159 \text{ or } 44.0 \text{ dB}$$

(b) $A_v^{CE} \cong -10(12) = -120$ (41.5 dB) and μ_f = 9.80 mS (320 kΩ) = 3140

(c) Evaluating the expressions of the C-E amplifier's input and output resistances gives

$$R_{in}^{CE} = R_B \| R_{iB} = 104 \text{ k}\Omega \| 10.2 \text{ k}\Omega = 9.29 \text{ k}\Omega$$

$$R_{out}^{CE} = R_C \| R_{iC} = R_C \| r_o \cong R_C = 22 \text{ k}\Omega$$

(d) With the emitter bypassed, v_{be} is given by

$$\mathbf{v_{be}} = \mathbf{v_i}\left[\frac{R_B \| R_{iB}}{R_I + (R_B \| R_{iB})}\right] = \mathbf{v_i}\frac{R_B \| r_\pi}{R_I + (R_B \| r_\pi)} = \mathbf{v_i}\frac{104 \text{ k}\Omega \| 10.2 \text{ k}\Omega}{1 \text{ k}\Omega + (104 \text{ k}\Omega \| 10.2 \text{ k}\Omega)} = 0.903 \mathbf{v_i}$$

and the small-signal limit for v_i is

$$|v_i| \leq \frac{0.005 \text{ V}}{0.903} = 5.53 \text{ mV}$$

Check of Results: The calculated voltage gain is similar to the rule-of-thumb estimate so our calculation appears correct. Remember, the rule-of-thumb formula is meant to only be a rough estimate; it will not be exact. The gain is much less than the amplification factor, so the neglect of r_o is valid.

Computer-Aided Analysis: SPICE simulation uses the circuit from Example 7.3 with bypass capacitor C_3 connected from the emitter to ground. Simulation yields the Q-point (0.248 mA, 3.30 V) that is consistent with the assumed value. The small difference results from V_A being included in the SPICE simulation and not in our hand calculations. An ac sweep from 10 Hz to 100 kHz with 10 frequency points/decade is used to find the region in which the capacitors are acting as short circuits, and the gain is observed to be constant at 43.4 dB above a frequency of 1 kHz. The voltage gain is slightly less than our calculated value because r_o was neglected in our calculations. A transient simulation was performed with a 5-mV, 10-kHz sine wave. The output exhibits reasonably good linearity, but the positive and negative amplitudes are slightly different, indicating some waveform distortion. Enabling the Fourier analysis capability of SPICE yields THD = 3.9%.

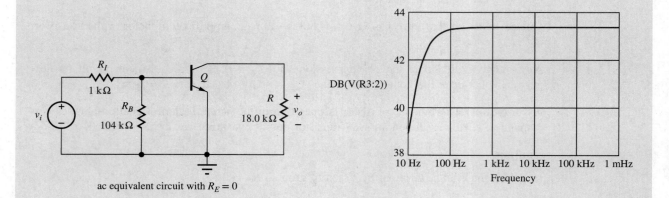

ac equivalent circuit with $R_E = 0$

The (frequency dependent) input resistance of the common-emitter amplifier is equal to the voltage at the base node divided by the current entering the node through coupling capacitor C_1: $R_{in}^{CE} = V(Q1:b)/I(C1)$. At frequencies above 10 Hz, the SPICE input becomes constant at 9.80 kΩ in agreement with our hand calculations. Similarly, the output resistance is given by $R_{out}^{CE} = V(Q1:c)/I(C2)$ which becomes constant at 20.6 kΩ for frequencies above 1 kHz. The discrepancies are due to differences in the SPICE values for the Q-point, temperature T, and the current gain.

8.2.2 COMMON-EMITTER EXAMPLE COMPARISON

The results from Examples 8.1 and 8.2 are listed in Table 8.2. Addition of the emitter resistor significantly reduces the voltage gain. This loss in gain is traded for a much higher input resistance and signal handling capability. The output resistances are both set by collector resistor R_C, so they are approximately the same.

TABLE 8.2
Common-Emitter Amplifier Comparison—SPICE Results

	BYPASSED EMITTER ($R_E = 0$)	$R_E = 3$ kΩ
A_v^{CE}	−159	−5.70
R_{in}^{CE}	9.29 kΩ	77.8 kΩ
R_{out}^{CE}	20.6 kΩ	21.9 kΩ
v_i^{max}, (THD)	5.53 mV (3.9 %)	155 mV (0.15 %)

EXERCISE: (a) What is the voltage gain A_v of the amplifier in Ex. 8.1 if R_E is changed to 1 kΩ? Assume the Q-point does not change. (b) What is the new value of R_4 required to maintain the Q-points unchanged in the amplifier?

ANSWERS: −16.0, 12 kΩ

EXERCISE: What value of saturation current I_S must be used in SPICE to achieve $V_{BE} = 0.7$ V for $I_C = 245$ μA? Assume a default temperature of 27°C.

ANSWER: 0.425 fA

EXERCISE: A common-emitter amplifier similar to Fig. 8.7 is operating from a single +20-V power supply, and the emitter terminal is bypassed by capacitor C_3. The BJT has $\beta_F = 100$ and $V_A = 50$ V and is operating at a Q-point of (100 μA, 10 V). The amplifier has $R_I = 5$ kΩ, $R_B = 150$ kΩ, $R_C = 100$ kΩ, and $R_3 = \infty$. What is the voltage gain predicted using our rule of thumb estimate? What is the actual voltage gain? What is the value of μ_f for this transistor?

ANSWERS: −200; −278; 2400

DESIGN NOTE

Remember, the intrinsic gain μ_f places an upper bound on the voltage gain of a single-transistor amplifier. We can't do better than μ_f! For the BJT,

$$\mu_f \cong 40V_A$$

For $25 \text{ V} \leq V_A \leq 100 \text{ V}$, we have $1000 \leq \mu_f \leq 4000$.

8.2.3 THE COMMON-SOURCE AMPLIFIER

Now we are in a position to analyze the small-signal characteristics of the **common-source (C-S) amplifier** shown in Fig. 8.12(a), which uses an enhancement-mode n-channel MOSFET ($V_{TN} > 0$) in a four-resistor bias network. The ac equivalent circuit of Fig. 8.12(b) is constructed by assuming that the capacitors all have zero impedance at the signal frequency and that the dc voltage sources represent ac grounds. Bias resistors R_1 and R_2 appear in parallel and are combined into gate resistor R_G, and R_L represents the parallel combination of R_D and R_3. In Fig. 8.12(c), the transistor has been replaced with its small-signal model. In subsequent analysis, we will assume that the voltage gain is much less than the intrinsic voltage gain of the transistor so we can neglect transistor output resistance r_o. For simplicity at this point, we assume that we have found the Q-point and know the values of I_D and V_{DS}.

In Fig. 8.12(b) through (d), the common-source nature of this amplifier should be apparent. The input signal is applied to the transistor's gate terminal, the output signal appears at the drain, and both the input and output signals are referenced to the (common) source terminal. Note that the small-signal models for the MOSFET and BJT are virtually identical at this step, except that r_π is replaced by an open circuit for the MOSFET.

Our first goal is to develop an expression for the voltage gain A_v^{CS} of the circuit in Fig. 8.12(a) from the source v_i to the output v_o. As with the BJT, we will first find the terminal voltage gain A_{vt}^{CS} between the gate and drain terminals of the transistor. Then, we will use the terminal gain expression to find the gain of the overall amplifier.

472 Chapter 8 Transistor Amplifier Building Blocks

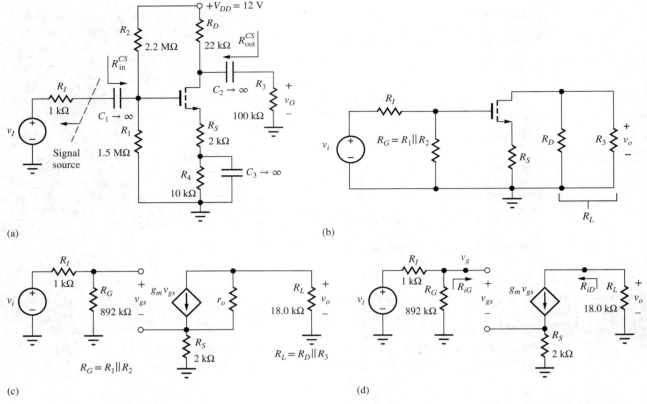

Figure 8.12 (a) Common-source amplifier circuit employing a MOSFET; (b) ac equivalent circuit for common-source amplifier in part (a); the common-source connection is now apparent; (c) ac equivalent circuit with the MOSFET replaced by its small-signal model; (d) final equivalent circuit for ac analysis of the common-source amplifier in which r_o is neglected.

Common-Source Terminal Voltage Gain

Starting with the circuit in Fig. 8.12(d), the terminal voltage gain is defined as

$$A_{vt}^{CS} = \frac{\mathbf{v_d}}{\mathbf{v_g}} = \frac{\mathbf{v_o}}{\mathbf{v_g}} \quad \text{where} \quad \mathbf{v_o} = -g_m \mathbf{v_{gs}} R_L \tag{8.34}$$

We can relate v_{gs} to v_g by applying KVL at the gate of the FET:

$$\mathbf{v_g} = \mathbf{v_{gs}} + g_m \mathbf{v_{gs}} R_S \quad \text{or} \quad \mathbf{v_{gs}} = \frac{\mathbf{v_g}}{1 + g_m R_S} \tag{8.35}$$

Combining Eqs. (8.34) and (8.35) yields an expression for the terminal gain.

$$A_{vt}^{CS} = -\frac{g_m R_L}{1 + g_m R_S} \tag{8.36}$$

Signal Source Voltage Gain for the Common-Source Amplifier

Now we can find the overall gain from source v_i to the voltage across R_L. The overall gain can be written as

$$A_v^{CS} = \frac{\mathbf{v_o}}{\mathbf{v_i}} = \left(\frac{\mathbf{v_o}}{\mathbf{v_g}}\right)\left(\frac{\mathbf{v_g}}{\mathbf{v_i}}\right) = A_{vt}^{CS}\left(\frac{\mathbf{v_g}}{\mathbf{v_i}}\right) \quad \text{where} \quad \mathbf{v_g} = \mathbf{v_i} \frac{R_G}{R_G + R_I} \tag{8.37}$$

in which v_g is related to v_i by the voltage divider formed by R_G and R_I. Combining Eqs. (8.36) and (8.37) yields a general expression for the voltage gain of the common-source amplifier:

$$A_v^{CS} = -\frac{g_m R_L}{1 + g_m R_S}\left(\frac{R_G}{R_G + R_I}\right) \tag{8.38}$$

We now explore the limits to the voltage gain of common-source amplifiers using model simplifications for zero and large values of resistance R_S. First, we will assume that the signal source resistance R_I is much less than R_G so that

$$A_v^{CS} \cong A_{vt}^{CS} = -\frac{g_m R_L}{1 + g_m R_S} \quad \text{for} \quad R_I \ll R_G \tag{8.39}$$

This approximation is equivalent to saying that the total input signal appears at the gate terminal of the transistor.

Common-Source Voltage Gain for Large Values of R_S

A very useful simplification occurs when R_S is large enough so that the $g_m R_S \gg 1$:

$$A_v^{CS} = -\frac{g_m R_L}{1 + g_m R_S} \cong -\frac{R_L}{R_S} \quad \text{for} \quad g_m R_S \gg 1 \quad \text{and} \quad R_G \gg R_I \tag{8.40}$$

The gain is now set by the ratio of the load resistor R_L and source resistor R_S. This is an extremely useful result because the gain is now independent of the transistor characteristics that vary widely from device to device. The result in Eq. (8.40) is very similar to the one that we obtained for the op-amp inverting amplifier circuit in Chapter 1 and is a result of feedback introduced by resistor R_S.

Achieving the simplification in Eq. (8.40) requires $g_m R_S \gg 1$. We can relate this product to the dc bias voltage developed across R_S:

$$g_m R_S = \frac{2}{(V_{GS} - V_{TN})} I_D R_S \quad \text{and we need} \quad I_D R_S \gg \frac{V_{GS} - V_{TN}}{2} \tag{8.41}$$

$I_D R_S$ represents the dc voltage drop across source resistor R_S and must be much greater than half the gate drive of the transistor. This inequality can be achieved, but not as easily as for the case of the BJT.

Understanding Generalized Common-Source Amplifier Operation

Let us explore common-source operation a bit further by looking at the signal voltage developed at the source terminal of the FET by referencing Fig. 8.12 and Eqs. (8.34) and (8.35):

$$\mathbf{v_s} = g_m \mathbf{v_{gs}} R_S = \frac{g_m R_S}{1 + g_m R_S} \mathbf{v_g} \cong \mathbf{v_g} \quad \text{for large} \quad g_m R_S \tag{8.42}$$

The voltage v_g at the gate of the transistor is transferred directly to the source, setting up a current of v_g/R_S. All the source current is supplied from the drain yielding a terminal voltage gain equal to the ratio of R_L and R_S:

$$\mathbf{i_s} \cong \frac{\mathbf{v_g}}{R_S} \qquad \mathbf{v_o} = -\mathbf{i_d} R_L = -\mathbf{i_s} R_L \quad \text{and} \quad A_{vt}^{CS} = \frac{\mathbf{v_o}}{\mathbf{v_g}} \cong -\frac{R_L}{R_S} \tag{8.43}$$

Unity signal voltage transfer from gate to source should not be mysterious. We know that the gate-source voltage has an approximately constant value of V_{GS}.[3] Thus the source signal

---

[3] Remember $v_{GS} = V_{GS} + v_{gs}$, and $v_{gs} \ll V_{GS}$ for small-signal operation.

voltage should be approximately the same as the gate signal. This voltage transfer between the gate and source terminals forms the basis of the source-follower operation to be discussed in detail in Sec. 8.3.

DESIGN NOTE

The gain of the generalized common-source amplifier is approximately equal to the ratio of the load and source resistors.

$$A_v^{CS} = -\frac{g_m R_L}{1 + g_m R_S} \cong -\frac{R_L}{R_S} \quad \text{for} \quad g_m R_S \gg 1$$

8.2.4 SMALL-SIGNAL LIMIT FOR THE COMMON-SOURCE AMPLIFIER

Equation (8.35) presents the general relation for the gate-source signal of the transistor that must be less than $0.2(V_{GS} - V_{TN})$ for small signal operation:

$$|v_g| = |v_{gs}|(1 + g_m R_S) < 0.2(V_{GS} - V_{TN})(1 + g_m R_S) \qquad (8.44)$$

The presence of a resistor in the source can substantially increase the signal handling capability of the common-source amplifier.

DESIGN NOTE

Use of a source resistor in the common-source amplifier can significantly increase the input signal range of the amplifier.

$$|v_g| \leq 0.2(V_{GS} - V_{TN})(1 + g_m R_S)$$

Zero Resistance in the Source

In order to achieve as large a gain as possible, we need to make the denominator in Eq. (8.39) as small as possible, and this is achieved by setting $R_S = 0$. The gain is then

$$A_v^{CS} \cong -g_m R_L = -g_m (R_D \| R_3) \quad \text{for} \quad R_S = 0 \qquad (8.45)$$

Equation (8.45) places an upper limit on the gain we can achieve from a common-source amplifier with an external load resistor. Equation (8.45) states that the terminal voltage gain of the common-source stage is equal to the product of the transistor's transconductance g_m and load resistance R_L, and the minus sign indicates that the output voltage is "inverted" or 180° out of phase with respect to the input. The approximations that led to Eq. (8.45) are equivalent to saying that the total input signal appears across v_{gs} as shown in Fig. 8.13.

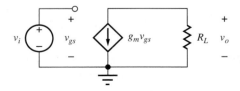

Figure 8.13 Simplified circuit for $R_G \gg R_I$ and $R_S = 0$.

8.2 Inverting Amplifiers—Common-Emitter and Common-Source Circuits

Common-Source Input Resistance

If we look in the gate terminal of the circuit in Fig. 8.12(d), we see an open circuit so $R_{iG} = \infty$. We can also find R_{iG} by taking the limit of the common-emitter input resistance as r_π approaches infinity with R_E replaced by R_S (and remembering $\beta_o = g_m r_\pi$):

$$R_{iG} = \lim_{r_\pi \to \infty} R_{iB} = \lim_{r_\pi \to \infty} [r_\pi + (\beta_o + 1)R_S] = \infty \tag{8.46}$$

The overall input resistance of the common-source amplifier R_{in}^{CS} is the resistance looking into the circuit at coupling capacitor C_1 in Fig. 8.12(a):

$$R_{in}^{CS} = R_G \| R_{iG} = R_G \| \infty = R_G \tag{8.47}$$

Common-Source Output Resistance

As in the BJT case, we must add r_o back into the small-signal model in order to achieve a unreasonable result for R_{iD}. The easiest way to find the resistance R_{iD} looking into the drain terminal of the transistor is to take the limit of the common-emitter output resistance as r_π approaches infinity with R_E replaced with R_S:

$$R_{iD} = \lim_{r_\pi \to \infty} R_{iC} = \lim_{r_\pi \to \infty} \left[r_o \left(1 + \frac{\beta_o R_S}{R_{th} + r_\pi + R_S} \right) \right] = r_o(1 + g_m R_S) = r_o + \mu_f R_S \tag{8.48}$$

The overall output resistance of the common-source amplifier R_{out}^{CS} is the resistance looking into the circuit at coupling capacitor C_2 in Fig. 8.12(a):

$$R_{out}^{CS} = R_D \| R_{iD} = R_D \| r_o(1 + g_m R_S) \cong R_D \tag{8.49}$$

The output resistance is approximately equal to the drain resistor R_D, since $r_o \gg R_D$.

DESIGN NOTE

The equations describing the behavior of the common-source amplifier are the same as those of the common-emitter amplifier in the limit as r_π and β_o approach infinity.

EXAMPLE 8.3   VOLTAGE GAIN OF A COMMON-SOURCE AMPLIFIER

In this example, we find the small-signal parameters of the MOSFET and then calculate the voltage gain of a common-source amplifier.

PROBLEM (a) Calculate the voltage gain, input resistance, and output resistance of the common-source amplifier in Fig. 8.12 if the transistor has $K_n = 0.500$ mA/V^2, $V_{TN} = 1$ V, and $\lambda = 0.0133$ V^{-1}, and the Q-point is (0.241 mA, 3.81 V). What is the largest value of v_i that does not violate the small-signal assumption? (b) Repeat part (a) if bypass capacitor C_3 is connected between the source terminal of the transistor and ground.

SOLUTION **Known Information and Given Data:** Common-source amplifier with its ac equivalent circuit given in Fig. 8.12; $K_n = 0.500$ mA/V^2, $V_{TN} = 1$ V, and $\lambda = 0.0133$ V^{-1}; the Q-point is (0.241 mA, 3.64 V); $R_I = 1$ kΩ, $R_1 = 1.5$ MΩ, $R_2 = 2.2$ MΩ, $R_D = 22$ kΩ, $R_3 = 100$ kΩ, $R_S = 2$ kΩ, $R_4 = 10$ kΩ.

Unknowns: Small-signal parameters of the transistor; voltage gain A_v^{CS}; input resistance R_{in}^{CS}; output resistance R_{out}^{CS}; small-signal limit for the value of v_i

Approach: Use the Q-point information to find g_m and r_o. Use the calculated and given values to evaluate the voltage gain and input and output resistance expression.

Assumptions: The transistor is in the active region of operation, and the signal amplitudes are below the small-signal limit for the MOSFET.

Analysis: We need to evaluate Eq. (8.38):

$$A_v^{CS} = -\frac{g_m R_L}{1 + g_m R_S}\left(\frac{R_G}{R_G + R_I}\right)$$

Calculating the values of the various resistors and small-signal model parameters yields

$$g_m = \sqrt{2 K_n I_{DS}(1 + \lambda V_{DS})}$$

$$= \sqrt{2\left(5 \times 10^{-4}\,\frac{A}{V^2}\right)(0.241 \times 10^{-3}\,A)\left(1 + \frac{0.0133}{V}3.81\,V\right)} = 0.503\,\text{mS}$$

$$r_o = \frac{\frac{1}{\lambda} + V_{DS}}{I_D} = \frac{\left(\frac{1}{0.0133} + 3.81\right)V}{0.241 \times 10^{-3}\,A} = 328\,\text{k}\Omega$$

$$R_G = R_1 \| R_2 = 892\,\text{k}\Omega \qquad R_L = R_D \| R_3 = 18.0\,\text{k}\Omega$$

$$g_m R_L = 9.05 \quad g_m R_S = 1.01 \quad A_v^{CS} = -\frac{9.05}{1 + 1.01}\left(\frac{892\,\text{k}\Omega}{892\,\text{k}\Omega + 1\,\text{k}\Omega}\right) = -4.50$$

Thus the common-source amplifier in Fig. 8.13 provides a small-signal voltage gain $A_v = -4.50$ or 13.1 dB.

Based on Eq. (7.87) for small-signal operation, we require

$$|v_i| \leq 0.2(V_{GS} - V_{TN})(1 + g_m R_S) = 0.2(0.982\,V)(2.01) = 0.395\,V$$

Thus, the input signal amplitude must not exceed 0.40 V for small-signal operation.

The overall input resistance of the common-source amplifier R_{in}^{CS} is set by gate bias resistor R_G:

$$R_{in}^{CS} = R_G = 892\,\text{k}\Omega$$

The overall output resistance of the common-source amplifier R_{out}^{CS} is approximately equal to the drain bias resistor R_D:

$$R_{out}^{CS} = R_D \| r_o(1 + g_m R_S) = 22\,\text{k}\Omega \| 328\,\text{k}\Omega[1 + (0.503\,\text{mS})(2\,\text{k}\Omega)] = 21.3\,\text{k}\Omega$$

(b) When the source is directly bypassed the results become

$$A_v^{CS} = -g_m R_L\left(\frac{R_G}{R_I + R_G}\right) = -9.04 \quad |v_i| \leq 0.2(V_{GS} - V_{TN}) = 0.2(0.982\,V) = 0.186\,V$$

$$R_{in}^{CS} = R_G = 892\,\text{k}\Omega \quad R_{out}^{CS} = R_D \| r_o = 22\,\text{k}\Omega \| 328\,\text{k}\Omega = 20.6\,\text{k}\Omega$$

Check of Results: We have found all the requested values. The intrinsic gain for this transistor is $\mu_f = g_m r_o = 165$. Our calculated voltage gain is much less than $\mu_f = 165$, so neglect of r_o is justified. With nonzero R_S, our estimate for the gain is $-R_L/R_S = -18\,\text{k}\Omega/2\,\text{k}\Omega = -9.00$. Our gain is lower than this prediction because the $g_m R_S$ product is not large compared to one. Checking the active region assumption: $V_{GS} - V_{TN} = 0.982\,V$ and $V_{DS} = 3.81\,V$. ✓

8.2 Inverting Amplifiers—Common-Emitter and Common-Source Circuits

Discussion: Note that this C-S amplifier has been designed to operate at nearly the same Q-point as the C-E amplifier in Fig. 8.7, and R_S has been chosen to give about the same gain.

Computer-Aided Analysis: (a) A SPICE operating point analysis (KP = 0.5 mA/V^2, VTO = 1 V, LAMBDA = 0.0133/V) yields the Q-point of (0.242 mA, 3.77 V). The slight variations result from including a nonzero value of λ. ac analysis yields a small-signal gain of -4.39. SPICE transient simulation results are given in the graphs below at a frequency of 10 kHz with TSTART = 0, TSTOP = 0.2 MS and TSTEP = 0.1 US. The first graph shows the results of an ac sweep from 0.1 Hz to 100 kHz with a 1-V input signal to identify the region (midband) where the capacitors are effectively short circuits. From the graph, we find that the gain is constant at -4.39 frequencies above 10 Hz. The second graph shows the result from the transient simulation with a 0.4-V, 10-kHz sine wave as the input. This amplitude equals our small-signal limit, and we observe some distortion since the positive and negative excursions of the sine wave have different amplitudes. SPICE gives the total harmonic distortion as 2.2 percent.

The (frequency dependent) input resistance of the common-source amplifier is equal to the voltage at the gate node divided by the current entering the node through coupling capacitor C_1: $R_{in}^{CE} = V(M1:g)/I(C1)$. At frequencies above 10 Hz, the SPICE input becomes constant at 892 kΩ in agreement with our hand calculations. Similarly, the output resistance is given by $R_{out}^{CS} = V(M1:d)/I(C2)$ which becomes constant at 21.3 kΩ for frequencies above 10 Hz. The discrepancies are due to differences in the SPICE values for the Q-point, temperature T, and the current gain.

(b) When the source terminal of the transistor is bypassed, SPICE yields the following results: $A_v^{CS} = -8.61$, $R_{in}^{CS} = 892$ kΩ, $R_{out}^{CS} = 20.6$ kΩ and the total harmonic distortion is 3.8 percent. Note that negative feedback from R_S reduces the harmonic distortion.

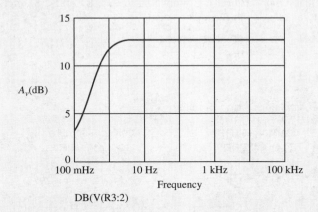

Frequency response (as sweep) for a 1-V ac input signal.

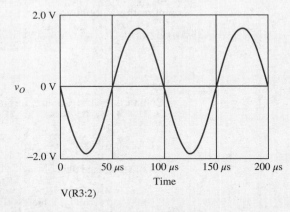

Transient response with $v_i = 0.4 \sin(20000\pi t)$ V.

EXERCISE: Calculate the Q-point for the transistor in Ex. 8.3.

EXERCISE: Convert the voltage gain in Ex. 8.3 to dB.

ANSWER: 13.1 dB

8.2.5 COMMON-EMITTER AND COMMON-SOURCE AMPLIFIER CHARACTERISTICS

Table 8.3 summarizes the results for the C-E and C-S amplifiers developed in this chapter. In the common-emitter circuit, resistor R_E adds feedback to the amplifier that reduces the voltage gain by the factor $(1 + g_m R_E)$, but increases the transistor's input resistance, output resistance, and input signal range by the same amount. Resistor R_S has a similar impact on the voltage gain, output resistance, and input signal range of the common-source amplifier. Since the resistance at the gate terminal of the FET is already infinite, the overall input resistance of the C-S amplifier is not affected by R_S. The presence of either R_E or R_S produces negative feedback in the circuit and reduces the harmonic distortion.

EXERCISE: (a) What is the voltage gain A_v of the two amplifiers in Fig. 8.2 if R_E and R_S are changed to 1 kΩ? Assume the Q-points do not change. (b) What are the new values of R_4 required to maintain the Q-points unchanged in the two amplifiers?

ANSWERS: −16.0, −6.02; 12 kΩ, 11 kΩ

EXERCISE: What is the voltage gain A_v of the two amplifiers in Fig. 8.2 if C_3 is removed from both circuits? What are the estimates for large emitter and large source resistances?

ANSWERS: −1.36, −1.29; −1.38, −1.50

EXERCISE: What value of saturation current I_S must be used in SPICE to achieve $V_{BE} = 0.7$ V for $I_C = 245$ μA? Assume a default temperature of 27°C.

ANSWER: 0.430 fA

EXERCISE: Evaluate $-g_m R_L$ and $-R_L/R$ for the C-E and C-S amplifiers in Exs. 8.1 and 8.3, and compare the magnitudes to the exact calculations in the examples. ($R = R_E$ or R_S)

ANSWERS: −176, −6.00; −8.84 < −9.00; 5.65 < 6.00; 4.46 < 8.84

TABLE 8.3
Common-Emitter/Common-Source Amplifier Design Summary

	COMMON-EMITTER (C-E) AMPLIFIER	COMMON-SOURCE (C-S) AMPLIFIER
Terminal voltage gain	$A_{vt}^{CE} = \dfrac{\mathbf{v_o}}{\mathbf{v_b}} = -\dfrac{g_m R_L}{1 + g_m R_E}$	$A_{vt}^{CS} = \dfrac{\mathbf{v_o}}{\mathbf{v_g}} = -\dfrac{g_m R_L}{1 + g_m R_S}$
Signal source voltage gain	$A_v^{CE} = \dfrac{\mathbf{v_o}}{\mathbf{v_i}} = A_{vt}^{CE} \dfrac{R_B \| R_{iB}}{R_I + R_B \| R_{iB}}$	$A_v^{CS} = \dfrac{\mathbf{v_o}}{\mathbf{v_i}} = A_{vt}^{CS} \dfrac{R_G}{R_I + R_G}$
Rule-of-thumb estimate for $g_m R_L$	$10(V_{CC} + V_{EE})$	$(V_{DD} + V_{SS})$
Input terminal resistance	$R_{iB} = r_\pi (1 + g_m R_E)$	$R_{iG} = \infty$
Output terminal resistance	$R_{iC} = r_o (1 + g_m R_E)$	$R_{iD} = r_o (1 + g_m R_S)$
Amplifier input resistance	$R_{in}^{CE} = R_B \| R_{iB}$	$R_{in}^{CS} = R_G$
Amplifier output resistance	$R_{out}^{CE} = R_C \| R_{iC}$	$R_{out}^{CS} = R_D \| R_{iD}$
Input signal range	$0.005(1 + g_m R_E)$ V	$0.2(V_{GS} - V_{TN})(1 + g_m R_S)$
Terminal current gain	β_o	∞

8.2 Inverting Amplifiers—Common-Emitter and Common-Source Circuits

TABLE 8.4
Common-Emitter/Common-Source Amplifier Comparison

	C-E AMPLIFIER	C-S AMPLIFIER
Voltage gain	−5.70	−4.39
Input resistance	77.8 kΩ	892 kΩ
Output resistance	21.9 kΩ	21.3 kΩ
Input signal range	155 mV	395 mV

8.2.6 C-E/C-S AMPLIFIER SUMMARY

The numeric results for the two specific amplifier examples are presented in Table 8.4. The common-emitter and common-source amplifiers have similar voltage gains. The C-E amplifier approaches the R_L/R_E limit (−6) more closely because $g_m R_E = 29.4$ for the BJT case, but only 0.982 for the MOSFET. The C-S amplifier provides high input resistance, but that of the BJT amplifier is also improved due to the $\beta_o R_E$ term. The output resistances of the C-E and C-S amplifiers are similar. The input signal levels have been increased above the R_S or $R_E = 0$ case—again by a substantial amount in the BJT case.

8.2.7 EQUIVALENT TRANSISTOR REPRESENTATION OF THE GENERALIZED C-E/C-S TRANSISTOR

The equations in Table 8.3 can actually provide us with a way to "absorb" resistor R_E into the transistor. This action can often simplify our circuit analysis or help provide insight into the operation of a circuit that we haven't seen before. The process is depicted in Fig. 8.14, in which the original transistor Q and resistor R are replaced by a new equivalent transistor Q'. The small-signal parameters of the new transistors are given by

$$g'_m = \frac{g_m}{1 + g_m R} \qquad r'_\pi = r_\pi (1 + g_m R) \qquad r'_o = r_o (1 + g_m R) \qquad (8.50)$$

Here we see the direct trade-off between reduced transconductance and increased input and output resistance. It is also important to note, however, that current gain and amplification factor of the transistor are conserved—we cannot exceed the limitations of the transistor itself!

$$\beta'_o = g'_m r'_\pi = \beta_o \qquad \text{and} \qquad \mu'_f = g'_m r'_o = \mu_f \qquad (8.51)$$

Similar results apply to the FET except that the current gain and input resistance are both infinite.

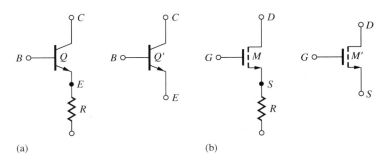

Figure 8.14 Composite transistor representation of (a) transistor Q and resistor R; (b) transistor M and resistor R.

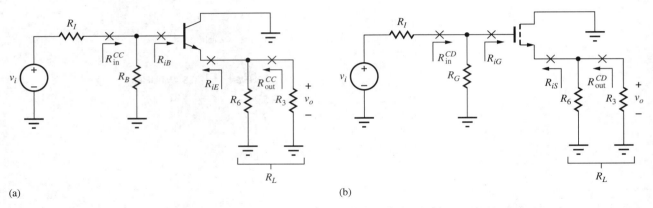

Figure 8.15 (a) ac equivalent circuit for the C-C amplifier; (b) ac equivalent circuit for the C-D amplifier. (See Fig. 8.4.)

8.3 FOLLOWER CIRCUITS—COMMON-COLLECTOR AND COMMON-DRAIN AMPLIFIERS

We now consider a second class of amplifiers, the common-collector (C-C) and common-drain (C-D) amplifiers, as represented by the ac equivalent circuits in Fig. 8.15. We will see that the follower circuits provide high input resistance and low output resistance with a gain of approximately one. The BJT circuit is analyzed first, and then the MOSFET circuit is treated as a special case with $r_\pi \to \infty$.

ELECTRONICS IN ACTION

Noise in Electronic Circuits

The linear signal-level limitations of transistors that we have introduced in this chapter may seem small, but we often deal with signal levels that are far below even the 5-mV v_{be} limit for the BJT. For example, the radio frequency signals from antennas on our cell phones can be in the microvolt range, and high-frequency communications receivers often have minimum detectable signals of less than 0.1 μV! The minimum detectable signals are often set by the noise in the RF amplifiers connected to the antennas. These amplifiers are referred to as low noise amplifiers, or LNAs, in which the noise actually comes from the transistors and resistors that make up the circuit.

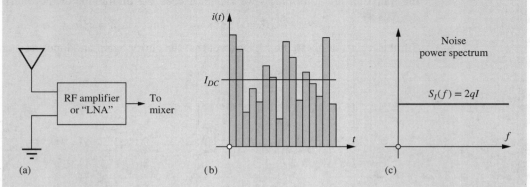

We often think that the dc voltages and currents that we calculate or measure with a dc voltmeter are constants, but they really represent averages of noisy signals. For example, the currents that we encounter in this text are made up of very large numbers of small current pulses due to individual electrons (e.g., 1 μA = 6.3 × 10^{12} electrons/sec). The current is constantly fluctuating or varying about the dc value as shown in the graph here, and these fluctuations represent one of the sources of noise in electronic devices. If we somehow listened to this current, it would sound much like rain on a tin roof. The background "din" from the rain is actually made up of the noise from a huge number of individual drops. This form of electronic noise is termed "shot" noise.

We model the noise in electron devices by adding noise voltage and current generators to our circuits. The noise generators represent random signals with zero mean and are therefore characterized by either their rms or mean square values. For example, both the base and collector currents in the bipolar transistor produce shot noise, and the noise is modeled by

$$\overline{i_{cn}^2} = \overline{[i_C(t) - I_C]^2} = 2qI_CB \quad \text{and} \quad \overline{i_{bn}^2} = \overline{[i_B(t) - I_B]^2} = 2qI_BB$$

These sources are referred to as "white noise" sources in which the noise power spectrum is independent of frequency. The mean square value of the noise current is directly proportional to the dc current and depends upon the bandwidth B (Hz) associated with the measurement. For instance, the rms value of the collector shot noise for $I_C = 1$ mA and $B = 1$ kHz would be

$$\sqrt{\overline{i_{cn}^2}} = \sqrt{2(1.6 \times 10^{-19})(10^{-3})(10^3)} = 0.566 \text{ nA}$$

In addition to shot noise, resistors and other resistive elements in electronic circuits exhibit noise due to the thermal agitation of electrons in the resistor. This "thermal" noise or "Johnson" noise is modeled by a noise voltage source in series with the resistor as shown for the base resistance of the BJT in the figure below (also see Chapter 14). The mean square value of the noise voltage associated with a resistor R is given by

$$\overline{v_{rn}^2} = 4kTRB$$

where k is Boltzmann's constant, T is absolute temperature, and B is the bandwidth of interest. For a 1-kΩ resistor operating at 300 K with $B = 1$ kHz,

$$\sqrt{\overline{v_{rn}^2}} = \sqrt{4(1.38 \times 10^{-23})(300)(10^3)(10^3)} = 0.129 \text{ μV}$$

Remembering that the channel region of the MOSFET is really a voltage-controlled resistor, we can model the thermal noise of the channel by a (Norton equivalent) current source whose mean square value is

$$\overline{i_{dn}^2} = \frac{8}{3}kTg_mB$$

The final figure presents our basic transistor noise models. For the BJT, current sources are added to model the shot noise of both the base and collector currents, and the thermal

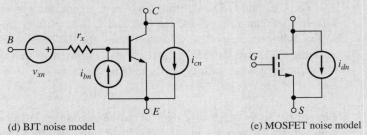

(d) BJT noise model　　　　　　　　(e) MOSFET noise model

noise of the base resistance is also included. For the MOSFET, the thermal noise of the channel is modeled by an equivalent noise current source. These noise models are built into SPICE, and NOISE is one of the analysis options available. For further information on how to make noise calculations and use the noise analysis capability in SPICE, see Connect.

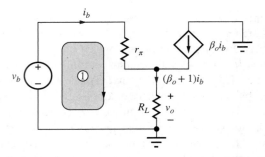

Figure 8.16 Simplified small-signal model for the C-C amplifier. $R_L = R_3 \| R_6$.

8.3.1 TERMINAL VOLTAGE GAIN

To find the terminal gain in Fig. 8.15(a), the bipolar transistor is replaced by its small-signal model in Fig. 8.16 (r_o is again neglected). The output voltage v_o now appears across load resistor R_L connected to the emitter of the transistor and is equal to

$$\mathbf{v_o} = +(\beta_o + 1)\mathbf{i_b}R_L \quad \text{where} \quad R_L = R_3 \| R_6 \tag{8.52}$$

The input current is related to applied voltage v_b by

$$\mathbf{v_b} = \mathbf{i_b}r_\pi + (\beta_o + 1)\mathbf{i_b}R_L = \mathbf{i_b}[r_\pi + (\beta_o + 1)R_L] \tag{8.53}$$

Combining Eqs. (8.52) and (8.53) yields an expression for the terminal gain of the common-collector amplifier:

$$A_{vt}^{CC} = \frac{\mathbf{v_e}}{\mathbf{v_b}} = +\frac{(\beta_o + 1)R_L}{r_\pi + (\beta_o + 1)R_L} \cong +\frac{g_m R_L}{1 + g_m R_L} \tag{8.54}$$

where the approximation holds for large β_o.

Letting r_π (and β_o) approach infinity in Eq. (8.54) yields the corresponding terminal gain for the FET follower in Fig. 8.15(b):

$$A_{vt}^{CD} = \frac{\mathbf{v_o}}{\mathbf{v_g}} = +\frac{g_m R_L}{1 + g_m R_L} \tag{8.55}$$

In most common-collector and common-drain designs, $g_m R_L \gg 1$, and Eqs. (8.54) and (8.55) reduce to

$$A_{vt}^{CC} \cong A_{vt}^{CD} \cong 1 \tag{8.56}$$

The C-C and C-D amplifiers both have a gain that approaches 1. That is, the output voltage follows the input voltage, and the C-C and C-D amplifiers are often called **emitter followers** and **source followers,** respectively. In most cases, the BJT does a better job of achieving $g_m R_L \gg 1$ than does the FET, and the BJT gain is closer to unity than that of the FET. However, in both cases the value of voltage gain typically falls in the range of

$$0.70 \leq A_{vt} \leq 1 \tag{8.57}$$

Obviously, A_{vt} is much less than μ_f, so neglecting r_o in the model of Fig. 8.16 is valid. Note, however, that r_o appears in parallel with R_L, and its effect can be included by replacing R_L with $(R_L \| r_o)$ in the equations.

Understanding Follower Operation

Unity signal transfer between the input and output of the follower circuits should not be mysterious. We know that the base and emitter terminals of the BJT are connected by a forward-biased diode whose voltage is virtually constant at 0.7 V. Thus the emitter signal voltage should be approximately the same as the base signal. (Remember that $v_{BE} = V_{BE} + v_{be}$, but $v_{be} \ll V_{BE}$.) FET followers behave in a similar manner. The gate-source voltage is approximately constant, so the signal voltage at the transistor source should be approximately the same as the applied gate signal. In this case, $v_{GS} = V_{GS} + v_{gs}$, but $v_{gs} \ll V_{GS}$. Thus the output of either follower should mirror the input with only a dc level shift between the two signals.

DESIGN NOTE

The terminal gain of single transistor voltage followers is given by

$$A_{vt}^{CC} \cong A_{vt}^{CD} = +\frac{g_m R_L}{1 + g_m R_L} \quad \text{and typically} \quad 0.70 < A_{vt}^{CD} < A_{vt}^{CC} < 1$$

8.3.2 INPUT RESISTANCE

The input resistance at the base terminal of the BJT is simply equal to the last term in brackets in Eq. (8.53):

$$R_{iB} = \frac{\mathbf{v_b}}{\mathbf{i_b}} = r_\pi + (\beta_o + 1)R_L \cong r_\pi(1 + g_m R_L) \quad \text{and} \quad R_{iG} = \infty \quad (8.58)$$

letting r_π approach infinity for the MOSFET. The input resistance of the emitter follower is equal to r_π plus an amplified replica of load resistor R_L, and can be made quite large. Of course, we see that the input resistance of the source follower is very large.

The overall input resistance R_{in}^{CC} to the common-collector amplifier in Fig. 8.15(a) is equal to the parallel combination of bias resistor and the equivalent resistance at the base of the BJT:

$$R_{\text{in}}^{CC} = R_B \parallel R_{iB} = R_B \parallel r_\pi(1 + g_m R_L) \quad \text{where} \quad R_L = R_6 \parallel R_3 \quad (8.59)$$

The overall input resistance R_{in}^{CD} to the common-drain amplifier in Fig. 8.15(b) is equal to the parallel combination of bias resistor and the equivalent resistance at the gate of the FET:

$$R_{\text{in}}^{CD} = R_G \parallel R_{iG} = R_G \parallel \infty = R_G \quad (8.60)$$

8.3.3 SIGNAL SOURCE VOLTAGE GAIN

The overall voltage gains from source v_i in Fig. 8.15 to the output are found using the terminal gain and input resistance expressions

$$A_v^{CC} = \frac{\mathbf{v_o}}{\mathbf{v_i}} = \left(\frac{\mathbf{v_o}}{\mathbf{v_b}}\right)\left(\frac{\mathbf{v_b}}{\mathbf{v_i}}\right) = A_{vt}^{CC}\left(\frac{\mathbf{v_b}}{\mathbf{v_i}}\right)$$

Voltage v_b at the base of the bipolar transistor in Fig. 8.15 is related to v_i by

$$\mathbf{v_b} = \mathbf{v_i}\frac{R_B \parallel R_{iB}}{R_I + (R_B \parallel R_{iB})}$$

for $R_B = R_1 \parallel R_2$. Combining these expressions,

$$A_v^{CC} = A_{vt}^{CC}\left[\frac{R_B \parallel R_{iB}}{R_I + (R_B \parallel R_{iB})}\right] \quad (8.61)$$

For the common-source case with infinite input resistance, Eq. (8.61) reduces to

$$A_v^{CD} = A_{vt}^{CD}\left(\frac{R_G}{R_I + R_G}\right) \quad (8.62)$$

8.3.4 FOLLOWER SIGNAL RANGE

Because the emitter- and source-follower circuits have a gain approaching unity, only a small portion of the input signal actually appears across the base-emitter or gate-source terminals. Thus, these circuits can be used with relatively large input signals without violating their respective small-signal limits.

The voltage developed across r_π in the small-signal model must be less than 5 mV for small-signal operation of the BJT. An expression for v_{be} is found in a manner identical to that used to derive Eq. (8.53):

$$\mathbf{v_{be}} = \mathbf{i_b} r_\pi = \mathbf{v_b} \frac{r_\pi}{r_\pi + (\beta_o + 1)R_L} \tag{8.63}$$

Requiring the amplitude of voltage v_{be} to be less than 5 mV gives

$$|v_b| \leq 0.005(1 + g_m R_L) \text{ V} \tag{8.64}$$

for large β_o. Normally, $g_m R_L \gg 1$, and the magnitude of v_b can be increased well beyond the 5-mV limit.

For the case of the FET (letting $r_\pi \to \infty$), the corresponding expression becomes

$$|v_{gs}| = \frac{|v_g|}{1 + g_m R_L} \leq 0.2(V_{GS} - V_{TN}) \tag{8.65}$$

and

$$|v_g| \leq 0.2(V_{GS} - V_{TN})(1 + g_m R_L) \tag{8.66}$$

which also increases the permissible range for v_i.

DESIGN NOTE

An unbypassed resistor R in series with the emitter or source of a transistor increases the signal handling capability of the amplifier by a factor of approximately $(1 + g_m R)$.

> **EXERCISE:** What are the largest values of v_i that correspond to small-signal operation of the amplifiers in Fig. 8.3 if the transistor currents are 0.25 mA and $V_{GS} - V_{TN} = 1$ V?
>
> **ANSWERS:** 0.592 V; 1.27 V

8.3.5 FOLLOWER OUTPUT RESISTANCE

The resistance looking into the emitter terminal R_{iE} can be calculated based on the circuit in Fig. 8.17, in which test source v_x is applied directly to the emitter terminal. Using KCL at the emitter node yields

$$\mathbf{i_x} = -\mathbf{i} - \beta_o \mathbf{i} = \frac{\mathbf{v_x}}{r_\pi + R_{\text{th}}} - \beta_o \left(-\frac{\mathbf{v_x}}{r_\pi + R_{\text{th}}} \right) \tag{8.67}$$

Collecting terms and rearranging gives

$$R_{iE} = \frac{r_\pi + R_{\text{th}}}{\beta_o + 1} \cong \frac{1}{g_m} + \frac{R_{\text{th}}}{\beta_o} \tag{8.68}$$

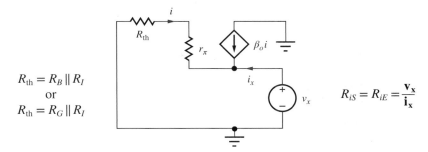

Figure 8.17 C-C/C-D output resistance calculation.

for $\beta_o \gg 1$. Because the current gain is infinite for the FET,

$$R_{iS} = \frac{1}{g_m} \qquad (8.69)$$

From Eqs. (8.68) and (8.69), it can be observed that the transistor's output resistance is primarily determined by the reciprocal of the transconductance of the transistor. This is an extremely important result to remember. For the BJT case, an additional term is added, but it is usually small, unless R_{th} is very large. The value of the output resistance for the C-C and C-D circuits can be quite low. For instance, at a current of 5 mA, the g_m of the bipolar transistor is $40 \times 0.005 = 0.2$ S, and $1/g_m$ is only 5 Ω whereas the value could be 100 Ω for the FET.

Using the results above, the overall output resistance of the follower circuits in Fig. 8.9 are also determined primarily by the transistor transconductances,

$$R_{\text{out}}^{CC} = R_6 \| R_{iE} \cong \frac{1}{g_m} \qquad \text{and} \qquad R_{\text{out}}^{CD} = R_6 \| R_{iS} \cong \frac{1}{g_m} \qquad (8.70)$$

and can be quite small in value.

EXERCISE: Redraw the small-signal equivalent circuit and derive a new expression for R_{iE} of the common-collector amplifier including r_o. Simplify the result.

ANSWER: $R_{iE} = r_o \| \left(\frac{1}{g_m} + \frac{R_{\text{th}}}{\beta_o} \right) \cong \frac{1}{g_m}$ since $r_o \gg \frac{1}{g_m}$

DESIGN NOTE

The equivalent resistance looking into the emitter or source of a transistor is approximately $1/g_m$!

Let us further interpret the two terms in Eq. (8.68) by injecting a current into the emitter of the BJT, as in Fig. 8.18. Multiplying **i** by the input resistance gives the voltage that must be developed at the emitter:

$$\mathbf{v_e} = \frac{\alpha_o \mathbf{i}}{g_m} + \frac{\mathbf{i}}{\beta_o + 1} R_{\text{th}} \qquad (8.71)$$

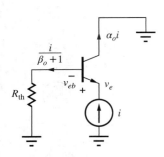
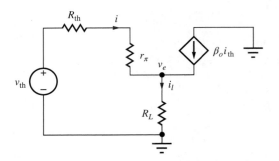

Figure 8.18 Circuit to aid in interpreting Eq. (8.71).

Figure 8.19 Circuit for calculating C-C/C-D terminal current gain.

Current ($\alpha_o \mathbf{i}$) comes out of the collector and must be supported by the emitter-base voltage $\mathbf{v_{eb}} = \alpha_o \mathbf{i}/g_m$, represented by the first term in Eq. (8.69). Base current $\mathbf{i_b} = -\mathbf{i}/(\beta_o + 1)$ creates a voltage drop in resistance R_{th} and yields the second term. In the FET case, only the first term exists because $i_g = 0$.

> **EXERCISE:** Drive the emitter node in Fig. 8.17 with a test current source i_x, and verify the output resistance results in Eq. (8.68).

8.3.6 CURRENT GAIN

Terminal current gain A_{it} is the ratio of the current delivered to the load element to the current being supplied from the Thévenin source. In Fig. 8.19, the current i plus its amplified replica ($\beta_o i$) are combined in load resistor R_L, yielding a current gain equal to ($\beta_o + 1$). For the FET, r_π is infinite, i is zero, and the current gain is infinite. Thus, for the C-C/C-D amplifiers,

$$A_{it}^{CC} = \frac{i_l}{i} = \beta_o + 1 \quad \text{and} \quad A_{it}^{CD} = \infty \tag{8.72}$$

8.3.7 C-C/C-D AMPLIFIER SUMMARY

Table 8.5 summarizes the results that have been derived for the common-collector and common-drain amplifiers in Fig. 8.20. As before, the FET results in the table can always be obtained from

TABLE 8.5
Common-Collector/Common-Drain Amplifier Design Summary

	COMMON-COLLECTOR (C-C) AMPLIFIER	COMMON-DRAIN (C-D) AMPLIFIER
Terminal voltage gain	$A_{vt}^{CC} = \dfrac{\mathbf{v_o}}{\mathbf{v_1}} = +\dfrac{g_m R_L}{1 + g_m R_L} \cong 1$	$A_{vt}^{CD} = \dfrac{\mathbf{v_o}}{\mathbf{v_1}} = +\dfrac{g_m R_L}{1 + g_m R_L} \cong 1$
Signal source voltage gain	$A_v^{CC} = \dfrac{\mathbf{v_o}}{\mathbf{v_i}} = A_{vt}^{CC} \dfrac{R_B \| R_{iB}}{R_I + R_B \| R_{iB}}$	$A_v^{CD} = \dfrac{\mathbf{v_o}}{\mathbf{v_i}} = A_{vt}^{CD} \dfrac{R_G}{R_I + R_G}$
Rule-of-thumb estimate for $g_m R_L$	$10(V_{CC} + V_{EE})$	$(V_{DD} + V_{SS})$
Input terminal resistance	$R_{iB} = r_\pi(1 + g_m R_L)$	$R_{iG} = \infty$
Output terminal resistance	$R_{iE} \cong \dfrac{1}{g_m} + \dfrac{R_{th}}{\beta_o}$	$R_{iS} = \dfrac{1}{g_m}$
Input signal range	$0.005(1 + g_m R_L)$ V	$0.2(V_{GS} - V_{TN})(1 + g_m R_L)$
Terminal current gain	$\beta_o + 1$	∞

8.3 Follower Circuits—Common-Collector and Common-Drain Amplifiers

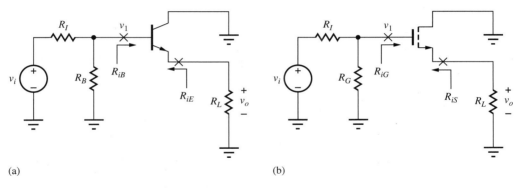

Figure 8.20 (a) Common-collector and (b) common-drain amplifiers for use with Table 8.5.

the BJT results by letting r_π and $\beta_o \rightarrow \infty$. The similarity between the characteristics of the C-C and C-D amplifiers should be readily apparent. Both amplifiers provide a gain approaching unity, a high input resistance, and a low output resistance. The differences arise because of the finite value of r_π and β_o of the BJT. The FET can more easily achieve very high values of input resistance because of the infinite resistance looking into its gate terminal, whereas the C-C amplifier can more easily reach very low levels of output resistance because of its higher transconductance for a given operating current. Both amplifiers can be designed to handle relatively large input signal levels. The current gain of the FET is inherently infinite, whereas that of the BJT is limited by its finite value of β_o.

EXAMPLE 8.4   FOLLOWER CALCULATIONS

The characteristics of the common-collector and common-drain amplifiers in Fig. 8.4 are calculated using the expressions derived in this section.

PROBLEM Calculate the overall gain A_v, input resistances, output resistances, and signal handling capability of the C-C and C-D amplifiers using the results from Sec. 8.3 and the parameter values from Exs. 8.1 and 8.3.

SOLUTION **Known Information and Given Data:** The equivalent circuits with element values are redrawn below. The Q-point and small-signal values appear in the table below from the previous examples.

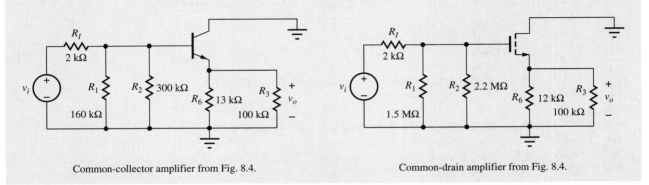

Common-collector amplifier from Fig. 8.4.   Common-drain amplifier from Fig. 8.4.

	I_C or I_D	V_{CE} or V_{DS}	$V_{GS} - V_{TN}$	g_m	r_π	r_o	μ_f
BJT	245 µA	3.64 V	—	9.80 mS	10.2 kΩ	219 kΩ	2150
FET	241 µA	3.81 V	0.982 V	0.491 mS	∞	223 kΩ	110

Unknowns: Voltage gains, input and output resistances, and maximum input signal levels for the C-C and C-D amplifiers

Approach: Substitute element values from the two circuits into the results in Table 8.5.

Assumptions: Use the parameter values tabulated above.

Analysis: For the C-C amplifier load resistor $R_L = R_6 \| R_3 = 11.5$ kΩ, and bias resistor $R_B = R_1 \| R_2 = 104$ kΩ. The resistances, terminal gain, and input signal level are

$$R_{iB} \cong r_\pi(1 + g_m R_L) = 10.2 \text{ kΩ}[1 + 9.80 \text{ mS}(11.5 \text{ kΩ})] = 1.16 \text{ MΩ}$$

$$R_{in}^{CC} = R_B \| R_{iB} = 104 \text{ kΩ} \| 1.16 \text{ MΩ} = 95.4 \text{ kΩ}$$

$$A_{vt}^{CC} \cong \frac{g_m R_L}{1 + g_m R_L} = \frac{9.80 \text{ mS}(11.5 \text{ kΩ})}{1 + 9.80 \text{ mS}(11.5 \text{ kΩ})} = 0.991$$

$$R_{th} = 2 \text{ kΩ} \| 160 \text{ kΩ} \| 300 \text{ kΩ} = 0.781 \text{ kΩ}$$

$$R_{iE} \cong \frac{1}{g_m} + \frac{R_{th}}{\beta_o} = \frac{1}{9.80 \text{ mS}} + \frac{781 \text{ Ω}}{100} = 110 \text{ Ω}$$

$$R_{out}^{CC} \cong R_6 \| R_{iE} = 13 \text{ kΩ} \| 110 \text{ Ω} = 109 \text{ Ω}$$

$$|v_i| \leq 0.005 \text{ V}(1 + g_m R_L)\left(\frac{R_I + R_{in}^{CC}}{R_{in}^{CC}}\right)$$

$$|v_i| \leq 0.005 \text{ V}[1 + 9.80 \text{ mS}(11.5 \text{ kΩ})]\frac{2 \text{ kΩ} + 95.4 \text{ kΩ}}{95.4 \text{ kΩ}} = 0.580 \text{ V}$$

Using Eq. (8.61), we find the overall gain to be

$$A_v^{CC} = A_{vt}^{CC}\left[\frac{R_{in}^{CC}}{R_I + R_{in}^{CC}}\right] = 0.991\left[\frac{95.4 \text{ kΩ}}{2.00 \text{ kΩ} + 95.4 \text{ kΩ}}\right] = 0.971$$

For the C-D amplifier, load resistor $R_L = R_6 \| R_3 = 10.7$ kΩ, and $R_G = R_1 \| R_2 = 892$ kΩ.

$$A_{vt}^{CD} = \frac{g_m R_L}{1 + g_m R_L} = \frac{0.491 \text{ mS}(10.7 \text{ kΩ})}{1 + 0.491 \text{ mS}(10.7 \text{ kΩ})} = 0.840$$

and

$$A_v^{CD} = A_{vt}^{CD}\left(\frac{R_G}{R_I + R_G}\right) = 0.840\left(\frac{892 \text{ kΩ}}{2 \text{ kΩ} + 892 \text{ kΩ}}\right) = 0.838$$

The overall input resistance for the common-drain amplifier is

$$R_{in}^{CD} = R_G \| R_{iG} = 892 \text{ kΩ} \| \infty = 892 \text{ kΩ}$$

The output resistance of the C-D transistor and the source follower are

$$R_{iS} \cong \frac{1}{g_m} = \frac{1}{0.491 \text{ mS}} = 2.04 \text{ kΩ} \qquad R_{out}^{CD} = R_6 \| R_{iS} = 12 \text{ kΩ} \| 2.04 \text{ kΩ} = 1.74 \text{ kΩ}$$

The input signal limit is

$$|v_i| \leq 0.2(V_{GS} - V_{TN})(1 + g_m R_L)\left(\frac{R_I + R_{in}^{CD}}{R_{in}^{CD}}\right)$$

$$|v_i| \leq 0.2(0.982 \text{ V})[1 + 0.491 \text{ mS}(10.7 \text{ k}\Omega)]\frac{2 \text{ k}\Omega + 892 \text{ k}\Omega}{892 \text{ k}\Omega} = 1.23 \text{ V}$$

Check of Results: Both voltage gains are approximately +1, as expected for a voltage follower. Both results are in the range specified in Eq. (8.57).

Discussion: The C-C amplifier has a gain much closer to 1 because $g_m R_L$ is much larger than it is for the C-D case. The C-C amplifier will normally have a gain closer to one than will the C-D stage.

Computer-Aided Analysis[4]: We can check the voltage gains using SPICE by performing an operating point analysis followed by an ac analysis with v_I as the input and v_O as the output voltage. Make all the capacitor values large, say 100 μF, and sweep the frequency to find the midband range of frequencies (e.g., FSTART = 1 Hz and FSTOP = 100 kHz with 10 frequency points per decade). Analysis of the two circuits yields +0.971 for the gain of the common-collector amplifier and +0.843 for the gain of the common-drain stage. Both agree well with hand calculations. The transistor output resistance r_o is included in the simulations (VAF = 50 V or LAMBDA = 0.02 V^{-1}) and appears to have only a small effect.

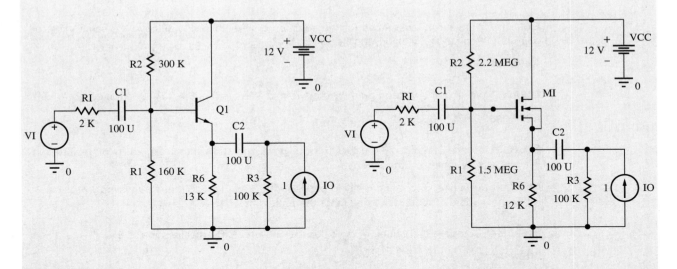

The input and resistances for the C-C circuit can be found as $R_{iB} = VB(Q1)/IB(Q1)$ and $R_{in}^{CC} = VB(Q1)/I(C1)$, and those of the C-D circuit are given by $R_{iG} = VG(M1)/IG(M1)$ and $R_{in}^{CD} = VG(M1)/I(C1)$. Similarly, the output resistances are given by $R_{iE} = VE(Q1)/IE(Q1)$, $R_{out}^{CC} = VE(Q1)/I(C2)$, $R_{iS} = VS(M1)/IS(M1)$ and $R_{out}^{CD} = VS(M1)/I(C2)$. The results appear in Table 8.6.

The input resistance of the common-drain amplifier is much higher than that of the common-collector stage because the lack of base current in the FET allows much larger

---

[4] See the MCD Connect site for help with this circuit.

TABLE 8.6
Follower Comparison—SPICE Results

	COMMON-COLLECTOR	COMMON-DRAIN
A_v^{CC}, A_v^{CD}	0.971	0.843
R_{iB}, R_{iG}	1.25 MΩ	∞
R_{in}^{CC}, R_{in}^{CD}	96.3 kΩ	892 kΩ
R_{iE}, R_{iS}	119 Ω	1.90 kΩ
R_{out}^{CC}, R_{out}^{CD}	119 Ω	1.64 kΩ
v_i^{max}, (THD)	580 mV (0.033 %)	1.23 V (0.73 %)

resistors to be used for R_1 and R_2. In contrast, the common-collector output resistance is much smaller than that of the common-drain stage because the transconductance of the BJT is much higher than that of the FET at a given operating current. The input signal capability and harmonic distortion of both stages are improved substantially by the presence of the resistances in the emitter and source of the transistors. The values all agree well with our hand calculations.

EXERCISE: How large must R_L be for the common-drain amplifier to achieve the same value of gain as the common-collector amplifier in Ex. 8.4?

ANSWER: 73.1 kΩ.

EXERCISE: What is the voltage gain for the two amplifiers in Fig. 8.4 if R_3 is removed ($R_3 \to \infty$)?

ANSWERS: 0.971, 0.853.

EXERCISE: Redraw the circuit in Fig. 8.16 including r_o and show that it can easily be included in the analysis by changing the value of R_L.

ANSWER: Resistor r_o appears directly in parallel with R_L in Fig. 8.16; hence we simply replace R_L with a new value of load resistance in all the equations: $R_L' = R_L \| r_o$.

EXERCISE: Compare the values of $g_m R_L$ for the C-C and C-D amplifiers in Ex. 8.4.

ANSWER: 113 ≫ 5.25.

8.4 NONINVERTING AMPLIFIERS—COMMON-BASE AND COMMON-GATE CIRCUITS

The final class of amplifiers to be analyzed consists of the common-base and common-gate amplifiers represented by the two ac equivalent circuits in Fig. 8.21. From our analyses, we will find that the noninverting amplifiers provide a voltage gain and output resistance similar to that of the C-E/C-S stages but with much lower input resistance. As in Sec. 8.3, we analyze the BJT circuit first and treat the MOSFET in Fig. 8.21(b) as a special case of Fig. 8.21(a).

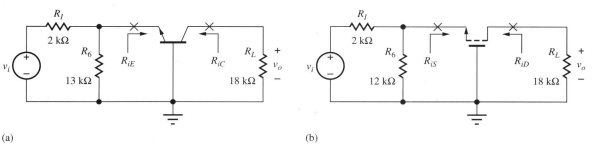

Figure 8.21 ac equivalent circuits for the (a) C-B and (b) C-G amplifiers. (See Fig. 8.5.)

ELECTRONICS IN ACTION

Revisiting the CMOS Imager Circuitry

In the first Electronics in Action feature in Chapter 5, we introduced a CMOS imager circuit. The chip contains 33 million pixels in the imaging array. A typical photodiode based imaging pixel consists of a photo diode with sensing and access circuitry. Let us revisit this sensor circuit in light of what we have learned about single transistor amplifiers.

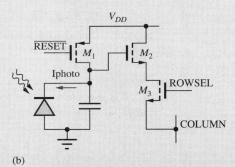

DALSA 33 MegaPixel CMOS image sensor.*
Teledyne DALSA. Reprinted by permission.

Typical photo diode pixel architecture.

M_1 is a reset switch, and after the $\overline{\text{RESET}}$ signal is asserted, the storage capacitor is fully charged to V_{DD}. The reset signal is then removed, and light incident on the photodiode generates a photo current that discharges the capacitor. Different light intensities produce different voltages on the capacitor at the end of the light integration time. Transistor M_2 is a source follower that buffers the photo-diode node. The source follower transfers the signal voltage at the photo-diode node to the output with nearly unity gain, and M_2 does not disturb the voltage at the photo diode output since it has an infinite dc input resistance. The voltage at the source of M_2 is then transferred to the output column via switch M_3. The source follower provides a low output resistance to drive the capacitance of the output column. The W/L ratio of switch M_3 must be chosen carefully so it does not significantly degrade the overall output resistance.

---

* The chip pictured above is a DALSA CMOS image sensor and is reprinted here with permission from Dalso Corporation.

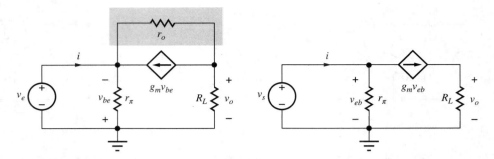

Figure 8.22 (a) Small-signal model for the common-base amplifier; (b) simplified model neglecting r_o and reversing the direction of the controlled source.

8.4.1 TERMINAL VOLTAGE GAIN AND INPUT RESISTANCE

The bipolar transistor is replaced by its small-signal model in Fig. 8.22(a). Because the amplifier has a resistor load, the circuit model is simplified by neglecting r_o, as redrawn in Fig. 8.22(b). In addition, the polarities of v_{be} and the dependent current source $g_m v_{be}$ have both been reversed.

For the common-base circuit, output voltage v_o appears at the collector across resistor R_L and is equal to

$$\mathbf{v_o} = +g_m \mathbf{v_{eb}} R_L = +g_m R_L \mathbf{v_e} \qquad (8.73)$$

and the terminal gain for the common-base transistor is

$$A_{vt}^{CB} = \frac{\mathbf{v_o}}{\mathbf{v_e}} = +g_m R_L \qquad (8.74)$$

which is the same as that for the common-emitter stage except for the sign. The input current i and input resistance at the emitter are given by

$$\mathbf{i} = \frac{\mathbf{v_e}}{r_\pi} + g_m \mathbf{v_e} \quad \text{and} \quad R_{iE} = \frac{\mathbf{v_e}}{\mathbf{i}} = \frac{r_\pi}{\beta_o + 1} \cong \frac{1}{g_m} \qquad (8.75)$$

assuming $\beta_o \gg 1$.

The corresponding expressions for the common-gate stage ($r_\pi \to \infty$) are

$$A_{vt}^{CG} = \frac{\mathbf{v_o}}{\mathbf{v_e}} = +g_m R_L \quad \text{and} \quad R_{iS} = \frac{1}{g_m} \qquad (8.76)$$

Understanding Common-Base and Common-Gate Amplifier Operation

When an input signal v_i is applied to the emitter of the C-B amplifier, or the source of the C-G amplifier, a current enters the transistor that is set by the transistor's input resistance ($R_{in} = 1/g_m$). This current exits the transistor from the collector or drain terminal and goes through the load resistor to produce output voltage v_o. The terminal voltage gain is equal to the ratio of the load resistance to the input resistance:

$$\mathbf{i_{in}} = \frac{\mathbf{v_i}}{R_{in}} = g_m \mathbf{v_i} \qquad \mathbf{v_o} = +\alpha_o \mathbf{i_{in}} R_L \cong +\mathbf{i_{in}} R_L \quad \text{and} \quad A_{vt}^{CB,CG} = \frac{\mathbf{v_o}}{\mathbf{v_i}} = +g_m R_L \qquad (8.77)$$

However, voltage division between the low-input resistance and the resistance associated with the signal source can often cause the signal source gain to be substantially less than the terminal gain of the noninverting amplifier.

8.4.2 SIGNAL SOURCE VOLTAGE GAIN

The overall gains for the amplifiers in Fig. 8.21 can now be expressed as

$$A_v^{CB} = \frac{\mathbf{V_o}}{\mathbf{V_i}} = \left(\frac{\mathbf{V_o}}{\mathbf{V_e}}\right)\left(\frac{\mathbf{V_e}}{\mathbf{V_i}}\right) = A_{vt}^{CB}\left[\frac{R_6 \| R_{iE}}{R_I + (R_6 \| R_{iE})}\right] \qquad (8.78)$$

and substituting $R_{iE} = 1/g_m$ yields

$$A_v^{CB} = \frac{g_m R_L}{1 + g_m(R_{\text{th}})}\left(\frac{R_6}{R_I + R_6}\right) \quad \text{and} \quad A_v^{CG} = \frac{g_m R_L}{1 + g_m(R_{\text{th}})}\left(\frac{R_6}{R_I + R_6}\right) \qquad (8.79)$$

where $R_{\text{th}} = R_6 \| R_I$. If we assume that $R_6 \gg R_I$, then the gain expressions in Eq. (8.79) become

$$A_v^{CB,CG} \cong \frac{g_m R_L}{1 + g_m R_I} \quad \text{for} \quad R_6 \gg R_I \qquad (8.80)$$

Because of the low input resistance of the common-base and common-gate amplifiers, the voltage gain A_v from the signal source to the output can be substantially less than the terminal gain. Note that the final expressions in Eq. (8.80) have a similar form to the overall gains for the inverting amplifiers and followers. We will explore this result more fully later in the chapter.

Note that the gain expressions in Eqs. (8.76) and (8.78) are positive, indicating that the output signal is in phase with the input signal. Thus, the C-B and C-G amplifiers are classified as noninverting amplifiers.

DESIGN NOTE

The terminal voltage gain of the noninverting amplifiers is given by

$$A_{vt}^{CB} = A_{vt}^{CG} \cong +g_m R_L$$

DESIGN NOTE

An estimate for the overall gain of the noninverting amplifiers is

$$A_v^{CB} = A_v^{CG} \cong +\frac{g_m R_L}{1 + g_m R_{\text{th}}} \cong +\frac{R_L}{R_{\text{th}}} \quad \text{for} \quad g_m R_{\text{th}} \gg 1 \quad \text{and} \quad R_{\text{th}} = R_6 \| R_I$$

DESIGN NOTE

The equivalent resistance R looking into the emitter or source of a transistor is approximately $R = 1/g_m$.

Important Limits

As for the C-E/C-S amplifiers, two limiting conditions are of particular importance (see Prob. 8.45). The upper bound occurs for $g_m R_I \ll 1$, for which Eq. (8.80) reduces to

$$A_v^{CB} \cong +g_m R_L \quad \text{and} \quad A_v^{CG} \cong +g_m R_L \qquad (8.81)$$

Equation (8.81) represents the upper bound on the gain of the C-B/C-G amplifiers and is the same as that for the C-E/C-S amplifiers, except the gain is noninverting.

However, if $g_m R_{th} \gg 1$, then Eq. (8.81) reduces to

$$A_v^{CB} = A_v^{CG} \cong +\frac{R_L}{R_{th}} \qquad (8.82)$$

For this case, the C-B and C-G amplifiers both have a gain that approaches the ratio of the value of the load resistor to that of the Thévenin source resistance ($R_{th} = R_6 \| R_I$) and is independent of the transistor parameters. For resistor loads, the limits in Eqs. (8.81) and (8.82) are much less than the amplification factor μ_f, so neglecting r_o is valid.

8.4.3 INPUT SIGNAL RANGE
The relationship between v_{eb} and v_i in Fig. 8.21(a) is given by

$$\mathbf{v_{eb}} = \mathbf{v_i}\frac{R_6 \| R_{iE}}{R_I + (R_6 \| R_{iE})} = \frac{\mathbf{v_i}}{1 + g_m(R_I \| R_6)}\left(\frac{R_6}{R_I + R_6}\right) \quad \text{and} \quad \mathbf{v_i} \cong \mathbf{v_{eb}}(1 + g_m R_I) \qquad (8.83)$$

for $R_6 \gg R_I$.

The small-signal limit for the BJT requires

$$|v_i| \le 0.005(1 + g_m R_I) \text{ V} \qquad (8.84)$$

For the FET case, replacing v_{eb} by v_{sg} yields $\mathbf{v_i} = \mathbf{v_{sg}}(1 + g_m R_I)$ and

$$|v_i| \le 0.2(V_{GS} - V_{TN})(1 + g_m R_I) \qquad (8.85)$$

The relative size of R_I and g_m will determine the signal-handling limits.

> **EXERCISE:** Calculate the maximum values of v_i for the C-B and C-G amplifiers in Ex. 8.4 and Fig. 8.21 based on Eqs. (8.84) and (8.85).
>
> **ANSWERS:** 103 mV; 389 mV

8.4.4 RESISTANCE AT THE COLLECTOR AND DRAIN TERMINALS
The resistance at the output terminal of the C-B/C-G transistors can be calculated for the circuit in Fig. 8.23, in which a test source v_x is applied to the collector terminal. The desired resistance is that looking into the collector with the base grounded and resistor R_{th} in the emitter. If the circuit is redrawn as shown in Fig. 8.23(b), we should recognize it to be the same as the C-E circuit in Fig. 8.9, repeated in Fig. 8.23(c), except that the equivalent resistance R_{th}^{CE} in the base is zero, and resistor R_E has been relabeled R_{th}.

Thus, the resistance at the output for the C-B device can be found using the results from the common-emitter amplifier, Eq. (8.30), without further detailed calculation, by substituting $R_{th}^{CE} = 0$ and replacing R_E with R_{th}:

$$R_{iC} = r_o\left(1 + \frac{\beta_o R_E}{R_{th}^{CE} + r_\pi + R_E}\right) = r_o\left(1 + \frac{\beta_o R_{th}}{r_\pi + R_{th}}\right) \qquad (8.86)$$

Using $\beta_o = g_m r_\pi$,

$$R_{iC} \cong r_o[1 + g_m(R_{th} \| r_\pi)] \quad \text{and} \quad R_{iD} = r_o(1 + g_m R_{th}) \qquad (8.87)$$

DESIGN NOTE

A quick design estimate for the output resistance of an inverting or noninverting amplifier with an unbypassed resistor R in the emitter or source is

$$R_o \cong r_o[1 + g_m(R \| r_\pi)] \cong \mu_f(R \| r_\pi)$$

8.4 Noninverting Amplifiers—Common-Base and Common-Gate Circuits 495

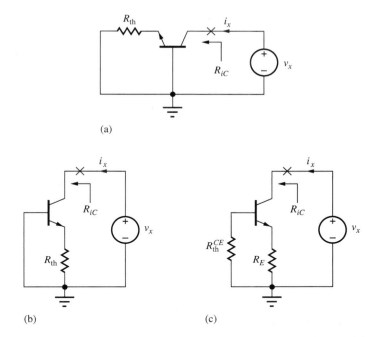

Figure 8.23 (a) Circuit for calculating the C-B output resistance; (b) redrawn version of the circuit in (a); (c) circuit used in common-emitter analysis (see Fig. 8.9).

EXERCISE: Calculate the output resistances of the C-B and C-G amplifiers in Fig. 8.21.

ANSWERS: 3.93 MΩ; 410 kΩ

8.4.5 CURRENT GAIN

The terminal current gain A_{it} is the ratio of the current through the load resistor to the current being supplied to the emitter. If a current i_e is injected into the emitter of the C-B transistor in Fig. 8.24, then the current $i_l = \alpha_o i_e$ comes out of the collector. Thus, the common-base current gain is simply α_o.

For the FET, α_o is exactly 1, and we have

$$A_{it}^{CB} = \frac{i_l}{i_e} = +\alpha_o \cong +1 \quad \text{and} \quad A_{it}^{CG} = +1 \tag{8.88}$$

8.4.6 OVERALL INPUT AND OUTPUT RESISTANCES FOR THE NONINVERTING AMPLIFIERS

The overall input and output resistances, R_{in}^{CB}, R_{in}^{CG}, R_{out}^{CB}, and R_{out}^{CG} of the common-base and common-gate amplifiers are defined looking into the input (C_1) and output (C_2) coupling capacitors in Fig. 8.5, as redrawn in the midband ac models in Fig. 8.25. The overall input resistance of

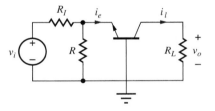

Figure 8.24 Common-base current gain.

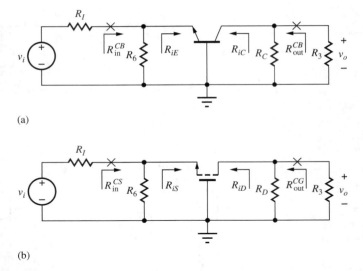

Figure 8.25 Midband ac equivalent circuits for the common-base and common gate amplifiers.

the common-base or common-gate amplifiers equals the parallel combination of resistor R_6 and the resistance looking into the emitter or source terminal of the transistor:

$$R_{in}^{CB} = R_6 \| R_{iE} \cong R_6 \| \frac{1}{g_m} \quad \text{and} \quad R_{in}^{CG} = R_6 \| R_{iS} = R_6 \| \frac{1}{g_m} \tag{8.89}$$

Similarly, the overall output resistance of the common-base or common-gate amplifiers equals the parallel combination of resistors R_C or R_D and the resistance looking into the collector or drain terminal of the transistor:

$$R_{out}^{CB} = R_C \| R_{iC} = R_C \| r_o[1 + g_m(R_6 \| R_I \| r_\pi)]$$
$$\text{and} \quad R_{out}^{CD} = R_D \| R_{iD} = R_D \| r_o[1 + g_m(R_6 \| R_1)] \tag{8.90}$$

EXAMPLE 8.5   NONINVERTING AMPLIFIER CHARACTERISTICS

A comparison of the characteristics of the common-base and common-gate amplifiers is provided by this example.

PROBLEM Calculate the signal-source voltage gains, input resistances, output resistances, and signal handling capability for the C-B and C-G amplifiers in Fig. 8.5.

SOLUTION **Known Information and Given Data:** The equivalent circuit with element values appear below. Q-point and small-signal values appear in the accompanying table.

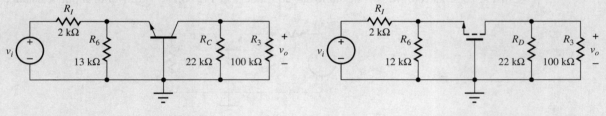

Common-base amplifier from Fig. 8.5.   Common-gate amplifier from Fig. 8.5.

8.4 Noninverting Amplifiers—Common-Base and Common-Gate Circuits

	I_C or I_D	V_{CE} or V_{DS}	$V_{GS} - V_{TN}$	g_m	r_π	r_o	μ_f
BJT	245 µA	3.64 V	—	9.80 mS	10.2 kΩ	219 kΩ	2150
FET	241 µA	3.81 V	0.982 V	0.491 mS	∞	223 kΩ	110

Unknowns: Voltage gains, input resistances, output resistances, and maximum input signal amplitudes for the common-base and common-gate amplifiers

Approach: Verify the value of R_L, and substitute element values from the two circuits the appropriate equations from Secs. 8.4.1–8.4.6.

Assumptions: Use the parameter values tabulated above.

Analysis: *For the C-B amplifier:* $R_I = 2$ kΩ, $R_6 = 13$ kΩ, $R_L = R_3 \| R_C = 18.0$ kΩ. The terminal input resistance and gain are

$$R_{iE} \cong \frac{1}{g_m} = \frac{1}{9.8 \text{ mS}} = 102 \, \Omega \quad \text{and} \quad A_{vt}^{CB} = +g_m R_L = 9.80 \text{ mS}(18.0 \text{ k}\Omega) = +176$$

and the overall voltage gain is

$$A_v^{CB} = \frac{A_{vt}^{CB}}{1 + g_m(R_6 \| R_I)} \left(\frac{R_6}{R_I + R_6}\right) = \frac{176}{1 + 9.8 \text{ mS}(1.73 \text{ k}\Omega)} \left(\frac{13 \text{ k}\Omega}{2 \text{ k}\Omega + 13 \text{ k}\Omega}\right) = +8.48$$

The input resistance of the common-base amplifier is found using Eq. (8.89)

$$R_{\text{in}}^{CB} = R_6 \| R_{iE} = 13 \text{ k}\Omega \| 102 \, \Omega = 101 \, \Omega$$

and the output resistance of the common-base amplifier is calculated using Eq. (8.90)

$$R_{iC} = r_o[1 + g_m(R_6 \| R_I \| r_\pi)] = 219 \text{ k}\Omega[1 + 9.80 \text{ mS}(13 \text{ k}\Omega \| 2 \text{ k}\Omega \| 10.2 \text{ k}\Omega)] = 3.40 \text{ M}\Omega$$

$$R_{\text{out}}^{CB} = R_C \| R_{iC} = 22 \text{ k}\Omega \| 3.40 \text{ M}\Omega = 21.9 \text{ k}\Omega$$

The maximum input signal amplitude is computed using Eq. (8.84).

$$|v_i| \leq 0.005V[1 + g_m(R_6 \| R_I)] \frac{R_I + R_6}{R_6}$$

$$|v_i| = 0.005V[1 + 9.80 \text{ mS}(13 \text{ k}\Omega \| 2 \text{ k}\Omega)] \frac{2 \text{ k}\Omega + 13 \text{ k}\Omega}{13 \text{ k}\Omega} = 104 \text{ mV}$$

For the C-G amplifier: $R_I = 2$ kΩ, $R_6 = 12$ kΩ, $R_L = R_3 \| R_D = 18.0$ kΩ. We have

$$R_{iS} = \frac{1}{g_m} = \frac{1}{0.491 \text{ mS}} = 2.04 \text{ k}\Omega \quad \text{and} \quad A_{vt}^{CG} = +g_m R_L = 0.491 \text{ mS}(18.0 \text{ k}\Omega) = +8.84$$

and

$$A_v^{CG} = \frac{A_{vt}^{CG}}{1 + g_m(R_6 \| R_I)} \left(\frac{R_6}{R_I + R_6}\right) = \frac{8.84}{1 + 0.491 \text{ mS}(1.71 \text{ k}\Omega)} \left(\frac{12 \text{ k}\Omega}{2 \text{ k}\Omega + 12 \text{ k}\Omega}\right) = +4.11$$

The input resistance of the common-gate amplifier is found using Eq. (8.89)

$$R_{\text{in}}^{CG} = R_6 \| R_{iS} = 12 \text{ k}\Omega \| 2.04 \text{ k}\Omega = 1.74 \text{ k}\Omega$$

and the output resistance of the common-gate amplifier is calculated using Eq. (8.90)

$$R_{iD} = r_o[1 + g_m(R_6 \| R_I)] = 223 \text{ k}\Omega[1 + 0.491 \text{ mS}(12 \text{ k}\Omega \| 2 \text{ k}\Omega)] = 411 \text{ k}\Omega$$

$$R_{\text{out}}^{CG} = R_D \| R_{iD} = 22 \text{ k}\Omega \| 411 \text{ k}\Omega = 20.9 \text{ k}\Omega$$

The maximum input signal amplitude is computed using Eq. (8.84).

$$|v_i| \leq 0.2(V_{GS} - V_{TN})[1 + g_m(R_6 \| R_I)]\frac{R_I + R_6}{R_6}$$

$$|v_i| = 0.2(0.982)[1 + 0.491 \text{ mS}(12 \text{ k}\Omega \| 2 \text{ k}\Omega)]\frac{2 \text{ k}\Omega + 12 \text{ k}\Omega}{12 \text{ k}\Omega} = 422 \text{ mV}$$

Check of Results: Both values are similar to and do not exceed the design estimate given by

$$A_v^{CB,CG} \cong +\frac{R_L}{R_I} = \frac{18 \text{ k}\Omega}{2 \text{ k}\Omega} = +9.00$$

Discussion: Note that the overall gain of the common-base amplifier is much less than its terminal gain because significant signal loss occurs due to the low input resistance of the transistor relative to the source resistance:

$$A_v^{CB} = A_{vt}^{CB}\left(\frac{R_6 \| R_{iE}}{R_I + R_6 \| R_{iE}}\right) = 176\left(\frac{13 \text{ k}\Omega \| 102 \text{ }\Omega}{2 \text{ k}\Omega + 13 \text{ k}\Omega \| 102 \text{ }\Omega}\right) = 176(0.0482) = +8.48$$

For the common-gate case, the loss factor is less,

$$A_v^{CG} = A_{vt}^{CG}\left(\frac{R_6 \| R_{iS}}{R_I + R_6 \| R_{iS}}\right) = 8.84\left(\frac{12 \text{ k}\Omega \| 2.04 \text{ k}\Omega}{2.00 \text{ k}\Omega + 12 \text{ k}\Omega \| 2.04 \text{ k}\Omega}\right)$$
$$= 8.84(0.466) = +4.12$$

Once again, we see that the overall C-G gain differs from the simple design estimate by more than that of the C-B stage because of the lower transconductance (and $g_m R_{th}$ product) of the FET. The gains are both well below the value of μ_f, so neglecting r_o in Fig. 8.22 is valid.

Computer-Aided Analysis[5]: We can check the characteristics of the noninverting amplifiers using SPICE by performing an operating point analysis followed by an ac analysis with v_I as the input and v_O as the output voltage. Make all the capacitor values large, say 100 μF, and sweep the frequency to find the midband region (e.g., FSTART = 1 Hz and FSTOP = 100 kHz with 10 frequency points per decade). Analysis of the two circuits yields +8.38 for the gain

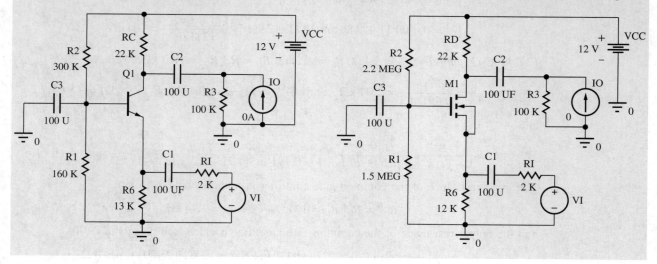

---

[5] See the MCD Connect site for help with this circuit.

TABLE 8.7
Noninverting Amplifier Comparison—SPICE Results

	COMMON-BASE	COMMON-GATE
A_v^{CB}, A_v^{CG}	+8.38	+4.05
R_{iE}, R_{iS}	112 Ω	2.08 kΩ
R_{in}^{CB}, R_{in}^{CG}	111 Ω	1.77 kΩ
R_{iC}, R_{iD}	3.26 MΩ	416 kΩ
R_{out}^{CB}, R_{out}^{CG}	21.9 kΩ	20.9 kΩ
v_i^{max}, (THD)	104 mV (0.27 %)	422 mV (2.1 %)

of the common-base amplifier and +4.05 for the gain of the common-gate stage. These values agree closely with our hand calculations. The transistor output resistance r_o is included in the simulations (VAF = 50 V or LAMBDA = 0.02 V^{-1}) and appears to have a negligible effect.

The input and resistances for the C-B circuit can be found as $R_{iE} = VE(Q1)/IE(Q1)$ and $R_{in}^{CB} = VE(Q1)/I(C1)$, and those of the C-G circuit are given by $R_{iS} = VS(M1)/IS(M1)$ and $R_{in}^{CG} = VS(M1)/I(C1)$. Similarly, the output resistances are given by $R_{iC} = VC(Q1)/IC(Q1)$, $R_{out}^{CB} = VC(Q1)/I(C2)$, $R_{iD} = VD(M1)/ID(M1)$, and $R_{out}^{CG} = VD(M1)/I(C2)$. The results appear in Table 8.7.

The input resistance of the common-base amplifier is much lower than that of the common-gate stage because the transconductance of the BJT is much higher than that of the FET at a given operating current. $R_{iC} \gg R_{iD}$ also because of the larger BJT transconductance, but the overall output resistances are the same since they are controlled by R_C and R_D. The input signal capability and harmonic distortion of both stages are increased substantially by the presence of the resistances in the emitter and source of the transistors. The values all agree well with our hand calculations.

EXERCISE: Show that Eq. (8.78) can be reduced to Eq. (8.79).

EXERCISE: What are the open circuit voltage gains ($R_3 = \infty$) for these two amplifiers?

ANSWERS: 10.4; 5.04

EXERCISE: Compare the gains of the C-B and C-G amplifiers calculated in Ex. 8.5 to the two limits developed in Eqs. (8.81) and (8.82).

ANSWERS: 8.98 < 10.4 ≪ 176; 4.11 < 8.48 < 10.5

EXERCISE: Estimate the ratio R_{iE}/R_{iS} for the BJT and FET assuming equal operating currents and $V_{GS} - V_{TN} = 1$ V.

ANSWER: 1/20

8.4.7 C-B/C-G AMPLIFIER SUMMARY

Table 8.8 summarizes the results derived for the common-base and common-gate amplifiers in Fig. 8.26, and the numeric results for the specific amplifiers in Fig. 8.4 are collected together in Table 8.7. The results show the symmetry between the various characteristics of the common-base

TABLE 8.8
Common-Base/Common-Gate Amplifier Summary

	C-B AMPLIFIER	C-G AMPLIFIER
Terminal voltage gain $A_{vt} = \dfrac{v_o}{v_1}$	$+g_m R_L$	$+g_m R_L$
Signal-source voltage gain $A_v = \dfrac{v_o}{v_i}$ $R_{th} = (R_I \| R_6)$	$\dfrac{g_m R_L}{1 + g_m R_{th}}\left(\dfrac{R_6}{R_I + R_6}\right)$	$\dfrac{g_m R_L}{1 + g_m R_{th}}\left(\dfrac{R_6}{R_I + R_6}\right)$
Input terminal resistance	$\dfrac{1}{g_m}$	$\dfrac{1}{g_m}$
Output terminal resistance	$r_o(1 + g_m R_{th}) = r_o + \mu_f R_{th}$	$r_o(1 + g_m R_{th}) = r_o + \mu_f R_{th}$
Input signal range	$0.005(1 + g_m R_{th})$	$0.2(V_{GS} - V_{TN})(1 + g_m R_{th})$
Terminal current gain	$\alpha_o \cong +1$	$+1$

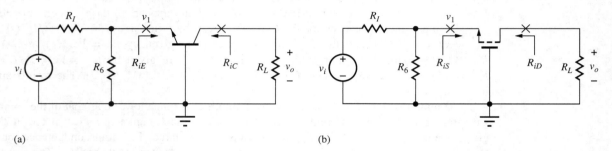

Figure 8.26 Circuits for use with summary Table 8.8; (a) common-base amplifier, (b) common-gate amplifier.

and common-gate amplifiers. The voltage gain and current gain are very similar. Numeric differences occur because of differences in the parameter values of the BJT and FET at similar operating points.

Both amplifiers can provide significant voltage gain, low input resistance, and high output resistance. The higher μ_F of the BJT gives it an advantage in achieving high output resistance; the C-B amplifier can more easily reach very low levels of input resistance because of the BJT's higher transconductance for a given operating current. The FET amplifier can inherently handle larger signal levels.

8.5 AMPLIFIER PROTOTYPE REVIEW AND COMPARISON

Sections 8.1 to 8.4 compared the three individual classes of BJT and FET circuits: the C-E/C-S, C-C/C-D, and C-B/C-G amplifiers. In this section, we review these results and compare the three BJT and FET amplifier configurations.

8.5.1 THE BJT AMPLIFIERS

Table 8.9 collects the results of analysis of the three BJT amplifiers in Fig. 8.27; Table 8.10 gives approximate results.

A very interesting and important observation can be made from review of Table 8.9. If we assume the voltage loss across the source resistance is small, the signal-source gains of the three amplifiers have exactly the same form:

$$|A_v| \cong \frac{g_m R_L}{1 + g_m R} \cong \frac{R_L}{\dfrac{1}{g_m} + R} \qquad (8.91)$$

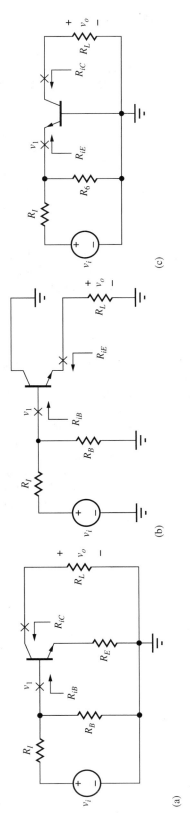

Figure 8.27 The three BJT amplifier configurations: (a) common-emitter amplifier, (b) common-collector amplifier, and (c) common-base amplifier.

TABLE 8.9
Single-Transistor Bipolar Amplifiers

	COMMON-EMITTER AMPLIFIER	COMMON-COLLECTOR AMPLIFIER	COMMON-BASE AMPLIFIER
Terminal voltage gain $A_{vt} = \dfrac{v_o}{v_1}$	$-\dfrac{g_m R_L}{1 + g_m R_E}$	$\dfrac{g_m R_L}{1 + g_m R_L} \cong +1$	$+g_m R_L$
Signal-source voltage gain $A_v = \dfrac{v_o}{v_i}$	$-\dfrac{g_m R_L}{1 + g_m R_E}\left[\dfrac{R_B \| R_{iB}}{R_I + (R_B \| R_{iB})}\right]$	$+\dfrac{g_m R_L}{1 + g_m R_L}\left[\dfrac{R_B \| R_{iB}}{R_I + (R_B \| R_{iB})}\right] \cong +1$	$+\dfrac{g_m R_L}{1 + g_m(R_I \| R_6)}\left(\dfrac{R_6}{R_I + R_6}\right)$
Input terminal resistance	$r_\pi + (\beta_o + 1) R_E$ $\cong r_\pi(1 + g_m R_E)$	$r_\pi + (\beta_o + 1) R_L$ $\cong r_\pi(1 + g_m R_L)$	$\dfrac{\alpha_o}{g_m} \cong \dfrac{1}{g_m}$
Output terminal resistance	$r_o(1 + g_m R_E)$	$\dfrac{\alpha_o}{g_m} + \dfrac{R_{\text{th}}}{\beta_o + 1}$	$r_o[1 + g_m(R_I \| R_6)]$
Input signal range	$\cong 0.005(1 + g_m R_E)$	$\cong 0.005(1 + g_m R_L)$	$\cong 0.005[1 + g_m(R_I \| R_6)]$
Terminal current gain	$-\beta_o$	$\beta_o + 1$	$\alpha_o \cong +1$

TABLE 8.10
Simplified Characteristics of Single BJT Amplifiers

	COMMON-EMITTER ($R_E = 0$)	COMMON-EMITTER WITH EMITTER RESISTOR R_E	COMMON-COLLECTOR	COMMON-BASE
Terminal voltage gain $A_{vt} = \dfrac{v_o}{v_i}$	$-g_m R_L \cong -10 V_{CC}$ (high)	$-\dfrac{R_L}{R_E}$ (moderate)	1 (low)	$+g_m R_L \cong +10 V_{CC}$ (high)
Input terminal resistance	r_π (moderate)	$\beta_o R_E$ (high)	$\beta_o R_L$ (high)	$1/g_m$ (low)
Output terminal resistance	r_o (moderate)	$\mu_f R_E$ (high)	$1/g_m$ (low)	$\mu_f (R_I \| R_4)$ (high)
Current gain	$-\beta_o$ (moderate)	$-\beta_o$ (moderate)	$\beta_o + 1$ (moderate)	1 (low)

in which R is the external resistance in the emitter of the transistor (R_E, R_L, or $R_I \| R_6$, respectively). We really only need to commit one formula to memory to get a good estimate of amplifier gain!

In addition, the same symmetry exists in the expressions for input signal range:

$$|v_{be}| \leq 0.005(1 + g_m R) \text{ V} \tag{8.92}$$

Note as well the similarity in the expressions for the input resistances of the C-E and C-C amplifiers, the input resistance of the C-B amplifier and the output resistance of the C-C amplifier, and the output resistances of the C-E and C-B amplifiers. Carefully review the three amplifier topologies in Fig. 8.27 to fully understand why these symmetries occur.

The second form of Eq. (8.91) deserves further discussion. The magnitude of the terminal gain of all three BJT stages can be expressed as the ratio of total resistance R_L at the output terminal to the total resistance in the emitter loop that equals the sum of the external resistance R [i.e., R_E, R_L, or $(R_I \| R_6)$, as appropriate] plus the resistance $(1/g_m)$ found looking back into the emitter of the transistor itself. This is an extremely important conceptual result.

Table 8.10 is a simplified comparison. The common-emitter amplifier provides moderate-to-high levels of voltage gain, and moderate values of input resistance, output resistance, and current gain. The addition of emitter resistor R_E to the common-emitter circuit gives added design flexibility and allows a designer to trade reduced voltage gain for increased input resistance, output resistance, and input signal range. The common-collector amplifier provides low voltage gain, high input resistance, low output resistance, and moderate current gain. Finally, the common-base amplifier provides moderate to high voltage gain, low input resistance, high output resistance, and low current gain.

8.5.2 THE FET AMPLIFIERS

Tables 8.11 and 8.12 are similar summaries for the three FET amplifiers shown in Fig. 8.28. The signal source voltage gain and signal range of all three amplifiers can again be expressed approximately as

$$|A_v| \cong \frac{g_m R_L}{1 + g_m R} = \frac{R_L}{\dfrac{1}{g_m} + R} \tag{8.93}$$

and

$$|v_{gs}| \leq 0.2(V_{GS} - V_{TN})(1 + g_m R) \text{ V} \tag{8.94}$$

in which R is the resistance in the source of the transistor (R_S, R_L, or $(R_I \| R_6)$, respectively). Note the symmetry between the output resistances of the C-S and C-G amplifiers. Also, the input resistance of the C-G amplifier and output resistance of the C-D amplifier are identical. Review

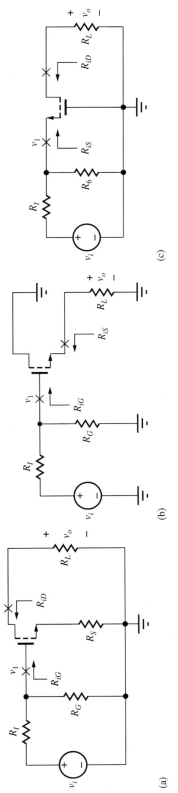

Figure 8.28 The three FET amplifier configurations: (a) common-source, (b) common-drain, and (c) common-gate.

TABLE 8.11
Single-Transistor FET Amplifiers

	COMMON-SOURCE AMPLIFIER	COMMON-DRAIN AMPLIFIER	COMMON-GATE AMPLIFIER
Terminal voltage gain $A_{vt} = \dfrac{v_o}{v_1}$	$-\dfrac{g_m R_L}{1 + g_m R_S}$	$+\dfrac{g_m R_L}{1 + g_m R_L} \cong +1$	$+g_m R_L$
Signal-source voltage gain $A_v = \dfrac{v_o}{v_i}$	$-\dfrac{g_m R_L}{1 + g_m R_S}\left(\dfrac{R_G}{R_I + R_G}\right)$	$+\dfrac{g_m R_L}{1 + g_m R_L}\left(\dfrac{R_G}{R_I + R_G}\right) \cong +1$	$+\dfrac{g_m R_L}{1 + g_m(R_I \| R_6)}\left(\dfrac{R_6}{R_I + R_6}\right)$
Input terminal resistance	∞	∞	$1/g_m$
Output terminal resistance	$r_o(1 + g_m R_S)$	$1/g_m$	$r_o[1 + g_m(R_I \| R_6)]$
Input signal range	$0.2(V_{GS} - V_{TN})(1 + g_m R_S)$	$0.2(V_{GS} - V_{TN})(1 + g_m R_L)$	$0.2(V_{GS} - V_{TN})[1 + g_m(R_I \| R_6)]$
Terminal current gain	∞	∞	$+1$

TABLE 8.12
Simplified Characteristics of Single FET Amplifiers

	COMMON-SOURCE ($R_S = 0$)	COMMON-SOURCE WITH SOURCE RESISTOR R_S	COMMON-DRAIN	COMMON-GATE
Terminal voltage gain $A_{vt} = \dfrac{v_o}{v_1}$	$-g_m R_L \cong -V_{DD}$ (moderate)	$-\dfrac{R_L}{R_S}$ (moderate)	1 (low)	$+g_m R_L \cong +V_{DD}$ (moderate)
Input terminal resistance	∞ (high)	∞ (high)	∞ (high)	$1/g_m$ (low)
Output terminal resistance	r_o (moderate)	$\mu_f R_S$ (high)	$1/g_m$ (low)	$\mu_f(R_I \| R_6)$ (high)
Current gain	∞ (high)	∞ (high)	∞ (high)	1 (low)

the three amplifier topologies in Fig. 8.28 carefully to fully understand why these symmetries occur. The addition of resistor R_S to the common-source circuit allows the designer to trade reduced voltage gain for increased output resistance and input signal range.

In a manner similar to the BJT amplifiers, the magnitude of the terminal gain of all three FET stages can be expressed as the ratio of total resistance R_L at the output terminal to the total resistance in the source loop which equals the sum of the external resistance R [i.e., R_S, R_L, or $(R_I \| R_6)$, as appropriate] plus the resistance $(1/g_m)$ found looking back into the source of the transistor itself. Thus, when properly interpreted, the form of the gain expressions for the single stage BJT and FET amplifier stages can all be considered as identical!

Table 8.12 is a relative comparison of the FET amplifiers. The common-source amplifier provides moderate voltage gain and output resistance but high values of input resistance and current gain. The common-drain amplifier provides low voltage gain and output resistance, and high input resistance and current gain. Finally, the common-gate amplifier provides moderate voltage gain, high output resistance, and low input resistance and current gain. Tables 8.9 to 8.12 are very useful in the initial phase of amplifier design, when the engineer must make a basic choice of amplifier configuration to meet the design specifications.

DESIGN NOTE

The magnitude of the overall gain of the single-stage amplifiers can all be expressed approximately by

$$|A_v| \cong \frac{g_m R_L}{1 + g_m R} = \frac{R_L}{\dfrac{1}{g_m} + R}$$

in which R is the external resistance in the emitter or source loop of the transistor.

Now we have a toolbox full of amplifier configurations that we can use to solve circuit design problems. Design Ex. 8.6 on page 483 demonstrates how to use our understanding to make design choices between the various configurations.

8.6 COMMON-SOURCE AMPLIFIERS USING MOS TRANSISTOR LOADS

Resistor loads are problematic in integrated circuits because they tend to take up a large amount of area relative to the size MOS transistors. However, we can use a transistor in place of the load resistor in a common-source amplifier as depicted in Fig. 8.29, where R_L is replaced as a transistor operating in the saturation region.[6] In digital logic circuits, this same circuit is called the "saturated load inverter."

Remember from Chapter 6 that the gain is equal to the slope of the amplifier's voltage transfer characteristic evaluated at the Q-point, $A_v = dv_O/dv_I|_{Q-pt}$, and the VTC in Fig. 8.29(b) has a region of high gain. In particular, if the circuit can be biased at a Q-point having $v_O = v_I$, then the inverter operates as a high-gain amplifier. It is actually easy to bias the MOS inverter into the high-gain region using negative feedback as in Fig. 8.29(c).[7] Since there is no dc current into the amplifier input, v_I and v_O must be equal, and the circuit operates in its high-gain region of the VTC with $v_o = v_i$.

8.6.1 VOLTAGE GAIN ESTIMATE

Let us estimate the gain of the circuit in Fig. 8.29(a) based upon the characteristics of the single-transistor amplifiers studied thus far. For now we will neglect the impact of resistor R_F. M_1 is connected as a common-source transistor, so the gain will be $A_v = -g_{m1}R_L$, where R_L is the overall load resistance connected to the drain of M_1. The load resistance consists of the parallel combination of the output resistance r_{o1} of M_1 and the resistance R_{iS2} looking into the source of M_2, which we now know is given by $R_{iS2} = 1/g_{m2}$:

$$R_L = r_{o1} \| R_{iS2} = r_{o1} \| \frac{1}{g_{m2}} \cong \frac{1}{g_{m2}} \tag{8.95}$$

Since the transistors must operate at the same drain current, we expect $r_{o1} \gg 1/g_{m2}$, and the voltage gain becomes

$$A_v^{CS} \cong -\frac{g_{m1}}{g_{m2}} = -\frac{\sqrt{2K_{n1}I_D}}{\sqrt{2K_{n2}I_D}} = -\sqrt{\frac{(W/L)_1}{(W/L)_2}} \tag{8.96}$$

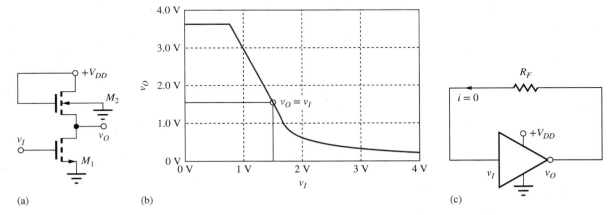

Figure 8.29 (a) Common-source amplifier with the load resistor replaced with a saturated transistor; (b) voltage transfer characteristic; (c) simple bias circuit for high gain operation.

---

[6] Saturated by connection (see Sec. S6.6).

[7] In Chapter 13, we will see how to eliminate R_F.

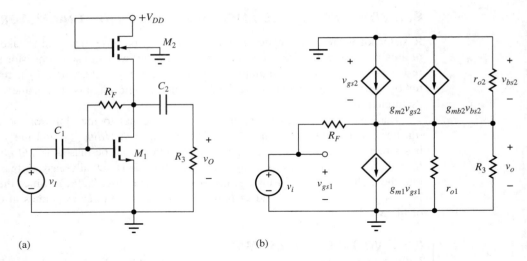

Figure 8.30 (a) Complete common-source amplifier; (b) small-signal model.

The gain of the amplifier with a saturated-load device is equal to the square root of the ratio of the (W/L) ratios of the input and load transistors. The gain is controlled by the designer's choice of the size of the transistors and is independent of the other transistor parameters. Unfortunately, even moderate gain requires large differences in the W/L ratios. For example, a 20 dB gain requires $(W/L)_1 = 100(W/L)_2$. So, although it works, this circuit still requires a relatively large amount of area to achieve much gain.

8.6.2 DETAILED ANALYSIS

Now let us explore the C-S amplifier in more detail in order to account for effects that have been neglected in our simplified analysis. The circuit in Fig. 8.30 includes bias resistor R_F, coupling capacitors C_1 and C_2, and external load resistor R_3, and the small-signal model includes the back-gate transconductance of transistor M_2 (see Sec. 7.8.4). An expression for the gain of the amplifier in Fig. 8.30 is found by writing a nodal equation at the output node:

$$G_F(\mathbf{v_o} - \mathbf{v_i}) + g_{m1}\mathbf{v_i} + \mathbf{v_o}(g_{o1} + g_{o2} + G_F + G_3) - g_{m2}\mathbf{v_{gs2}} - g_{mb2}\mathbf{v_{bs2}} = 0 \qquad (8.97)$$

Collecting terms, realizing that both v_{gs2} and v_{bs2} are equal to $-v_o$, and solving for the voltage gain yields

$$A_v^{CS} = \frac{\mathbf{v_o}}{\mathbf{v_i}} = -\frac{(g_{m1} - G_F)}{g_{m2}(1+\eta) + g_{o1} + g_{o2} + G_F + G_3} \qquad (8.98)$$

where $g_{mb2} = \eta g_{m2}$. This expression can be written in a more recognizable form as

$$A_v^{CS} \cong -g_{m1}R_L \qquad \text{where} \qquad R_L = R_3 \| R_F \| r_{o1} \| r_{o2} \| \frac{1}{g_{m2}(1+\eta)} \qquad (8.99)$$

is the total equivalent resistance on the output node. We already know that r_{o1} and r_{o2} will be much larger than $1/g_{m2}$, and R_F is normally designed to be much larger than R_3. In most cases, R_3 will also be much greater than $1/g_{m2}$, so the gain reduces to

$$A_v^{CS} \cong -\frac{g_{m1}}{g_{m2}(1+\eta)} = -\frac{1}{1+\eta}\sqrt{\frac{(W/L)_1}{(W/L)_2}} \qquad (8.100)$$

Equation (8.100) is the same as Eq. (8.96) except for the gain degradation caused by the back-gate transconductance. The effective load resistance is still limited by the relatively large conductance of the load transistor.

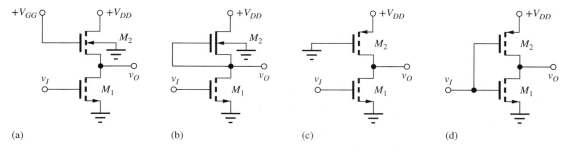

Figure 8.31 MOS inverting amplifiers: (a) linear load; (b) depletion-mode load; (c) pseudo NMOS; (d) CMOS.

EXERCISE: What is the W/L ratio of M_1 required to achieve a gain of 26 dB if $\eta = 0.2$ and $(W/L)_2 = 4/1$?

ANSWER: 2290/1

8.6.3 ALTERNATIVE LOADS

To improve the gain of the circuit, g_{m2} needs to be eliminated from the expression for the load resistance, Eq. (8.99), and there are a number of alternative transistor configurations for the load device as depicted in Fig. 8.31. NMOS transistors can be used as linear loads and depletion-mode loads, whereas PMOS transistors can be employed in psuedo NMOS and CMOS inverters. Any one of these circuits can be substituted for the saturated load inverter in Fig. 8.30.[8]

However, the linear load configuration achieves nothing, since the gate of M_2 is still at ac ground and R_{iS2} is still determined by $1/g_{m2}$. On the other hand, the depletion-mode load yields an improvement. Voltage v_{GS} is forced to be zero by connection, so the forward transconductance is eliminated and the gain is approximately

$$A_v^{CS} \cong -\frac{g_{m1}}{\eta g_{m2}} \qquad (8.101)$$

which improves the gain by a factor of $(1+\eta)/\eta$. For $\eta = 0.2$, the gain is improved by a factor of 6. In discrete circuits, v_{BS} can also be set to zero, and the back-gate transconductance is also eliminated. For this case the gain becomes

$$A_v^{CS} \cong -g_{m1} R_L = -g_{m1}(R_3 \| R_F \| r_{o1} \| r_{o2}) \cong -g_{m1} R_3 \qquad (8.102)$$

since we expect r_{o1} and r_{o2} to be much larger than R_3, and R_F can also be designed to be much larger than R_3. This configuration has more gain than our original C-S circuit because external load resistor R_3 is normally much larger than drain resistor R_D.

Circuits in Fig. 8.31(c) and (d) employ PMOS transistors and require CMOS technology. For the pseudo NMOS inverter, the load resistance on transistor M_1 is the same as that given in Eq. (8.102), and the gain is also the same. In the CMOS inverter case as depicted in Fig. 8.32, the transistors are connected in parallel: the gates are connected together, the drains are connected together, and the sources are both at ac ground potential. The input is applied to both gates so the gain expression becomes

$$A_v^{CS} \cong -(g_{m1} + g_{m2})R_L = -(g_{m1} + g_{m2})(R_3 \| R_F \| r_{o1} \| r_{o2}) \cong -(g_{m1} + g_{m2})R_3 \qquad (8.103)$$

which can be a factor of two improvement for a symmetrical inverter design ($K_p = K_n$).

---

[8] Logic applications of these circuits are discussed in detail in the supplemental chapters in the e-book.

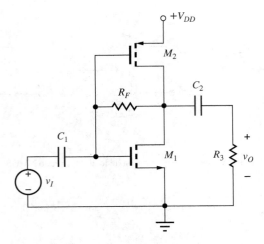

Figure 8.32 Common-source amplifier employing CMOS inverter.

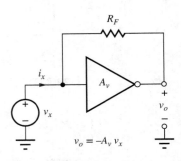

Figure 8.33 Circuit for determining input resistance.

Note that if we use a symmetrical CMOS inverter, eliminate R_3, and make R_F very large, the gain becomes approximately

$$A_v^{CS} \cong -(g_{m1} + g_{m2})(r_{o1} \| r_{o2}) = -2g_{m1}\frac{r_{o1}}{2} \cong -\mu_f \qquad (8.104)$$

We have discovered a circuit that achieves a gain equal to the intrinsic gain of the transistor, and we can't do better than that! Similar techniques will be used to design high-performance amplifiers in the next several chapters.

EXERCISE: Redraw the circuit in Fig. 8.32 using the amplifiers in Fig. 8.31(a), (b), and (c).

8.6.4 INPUT AND OUTPUT RESISTANCES

The input resistance of the amplifiers can be found with the assistance of the circuit in Fig. 8.33. R_{in} is calculated by finding an expression for i_x in terms of v_x:

$$i_x = \frac{v_x - v_o}{R_F} = \frac{v_x - (-A_v v_x)}{R_F} = v_x\left(\frac{1 + A_v}{R_F}\right) \quad \text{and} \quad R_{in} = \frac{v_x}{i_x} = \frac{R_F}{1 + A_v} \qquad (8.105)$$

For high gain, the input resistance is approximately equal to the feedback resistance divided by the amplifier's gain.

If input source v_i is set to zero for the circuit in Fig. 8.30(b), we immediately see that the output resistance there is given by

$$R_{out} = R_F \| r_{o1} \| r_{o2} \| \frac{1}{g_{m2}(1 + \eta)} \qquad \text{or} \qquad R_{out} = R_F \| r_{o1} \| r_{o2} \qquad (8.106)$$

depending upon the inverter configuration.

EXERCISE: Find the Q-point for M_1 in Fig. 8.30 if $R_F = 1\ M\Omega$, $K'_n = 100\ \mu A/V^2$, $V_{TN} = 1\ V$, $\lambda = 0.02$, $\eta = 0$, $(W/L)_1 = 8/1$, $(W/L)_2 = 2/1$ and $V_{DD} = 5\ V$. What is the amplifier voltage gain?

ANSWER: $V_o = 2.01\ V$, $I_o = 421\ \mu A$, $A_v = -1.85$.

EXERCISE: Find the Q-point for M_1 in Fig. 8.32 if $R_F = 560$ kΩ, $K'_n = 100$ µA/V^2, $K'_p = 40$ µA/V^2, $V_{TN} = 0.7$ V, $V_{TP} = -0.7$ V, $\lambda = 0.02$, $(W/L)_1 = 20/1$, $(W/L)_2 = 50/1$ and $V_{DD} = 3.3$ V. What is the amplifier's gain?

ANSWER: $V_o = 1.65$ V, $I_o = 932$ µA, $A_v = -104$.

DESIGN EXAMPLE 8.6

SELECTING AN AMPLIFIER CONFIGURATION

One of the first things we must do to solve a circuit design problem is to decide on the circuit topology to be used. A number of examples are given here.

PROBLEMS What is the preferred choice of amplifier configuration for each of these applications?

(a) A single-transistor amplifier is needed that has a gain of approximately 80 dB and an input resistance of 100 kΩ.

(b) A single-transistor amplifier is needed that has a gain of 52 dB and an input resistance of 250 kΩ.

(c) A single-transistor amplifier is needed that has a gain of 30 dB and an input resistance of at least 5 MΩ.

(d) A single-transistor amplifier is needed that has a gain of approximately 0 dB and an input resistance of 20 MΩ with a load resistor of 10 kΩ.

(e) A follower is needed that has a gain of at least 0.98 and an input resistance of at least 250 kΩ with a load resistance of 5 kΩ.

(f) A single-transistor amplifier is needed that has a gain of $+10$ and an input resistance of 2 kΩ.

(g) An amplifier is needed with an output resistance of 25 Ω.

SOLUTION **Known Information and Given Data:** In each case, we see that a minimum amount of information is provided, typically only a voltage gain and resistance specification.

Unknowns: Circuit topologies

Approach: Use our estimates of voltage gain, input resistance, and output resistance for the various configurations to make a selection.

Assumptions: Typical values of current gain, Early voltage, power supply voltage, and so on will be assumed as necessary: $\beta_o = 100$, 0.25 V $\leq V_{GS} - V_{TN} \leq 1$ V, $V_T = 0.025$ V, $V_A \leq 80$ V, $1/\lambda \leq 80$ V.

Analyses:

(a) The required voltage gain is $A_v = 10^{80/20} = 10{,}000$. This value of voltage gain exceeds the intrinsic voltage gain of even the best BJTs:

$$A_v \leq \mu_f = 40 V_A = 40(80) = 3200$$

An FET typically has a much lower value of intrinsic gain and is at an even worse disadvantage. Thus, such a large gain requirement cannot be met with a single-transistor amplifier.

(b) For the second set of specifications, we have $R_{in} = 250$ kΩ and $A_v = 10^{52/20} \cong 400$. We require both large gain and relatively large input resistance, which point us toward the common-emitter amplifier. For the C-E stage, $A_v = 10 V_{CC} \rightarrow V_{CC} = 40$ V, which is somewhat large. However, we know that the $10 V_{CC}$ estimate for the voltage gain is conservative and can easily be off by a factor of 2 or 3, so we can probably get by with a smaller power supply, say 20 V. Achieving the input of resistance requirement requires r_π to exceed 250 kΩ:

$$r_\pi = \frac{\beta_o V_T}{I_C} \geq 250 \text{ k}\Omega \rightarrow I_C \leq \frac{100(0.025 \text{ V})}{2.5 \times 10^5 \text{ }\Omega} = 10 \text{ }\mu\text{A}$$

which is small but acceptable. Achieving the gain specification with an FET would be much more difficult. For example, even with a small gate overdrive,

$$A_v = \frac{V_{DD}}{V_{GS} - V_{TN}} \cong \frac{V_{DD}}{0.25 \text{ V}} \rightarrow V_{DD} = 100 \text{ V}$$

which is unreasonably large for most solid-state designs. Note that the sign of the gain was not specified, so either positive or negative gain would be satisfactory, based on our limited specifications. However, the input resistance of the noninverting (C-B or C-G) amplifiers is low, not high.

(c) In this case, we require $R_{in} \geq 5 \text{ M}\Omega$ and $A_v = 10^{30/20} \cong 31.6$—large input resistance and moderate gain. These requirements can easily be met by a common-source amplifier:

$$A_v = \frac{V_{DD}}{V_{GS} - V_{TN}} = \frac{15 \text{ V}}{0.5 \text{ V}} = 30$$

The input resistance is set by our choice of gate bias resistors (R_1 and R_2 in Fig. 8.2), and 5 MΩ can be achieved with standard resistor values.

Since the gain is moderate, a C-E stage with emitter resistor could probably achieve the required high input resistance, although the values of the base bias resistor could become a limiting factor. For example, the input resistance and voltage gain could be met approximately with

$$R_{in} \cong \beta_o R_E \geq 5 \text{ M}\Omega \rightarrow R_E \geq \frac{5 \text{ M}\Omega}{100} = 50 \text{ k}\Omega \quad \text{and} \quad |A_v| = \frac{R_L}{R_E} \rightarrow R_L = 1.5 \text{ M}\Omega$$

(d) Zero-dB gain corresponds to a follower. For an emitter follower, $R_{in} \cong \beta_o R_L \cong 100(10 \text{ k}\Omega) = 1 \text{ M}\Omega$, so the BJT will not meet the input resistance requirement. On the other hand, a source follower provides a gain of approximately one and can easily achieve the required input resistance.

(e) A gain of 0.98 and an input resistance of 250 kΩ should be achievable with either a source follower or an emitter follower. For the MOSFET,

$$A_v = \frac{g_m R_L}{1 + g_m R_L} = 0.98 \quad \text{requires} \quad g_m R_L = \frac{2 I_D R_L}{V_{GS} - V_{TN}} = 49$$

which can be satisfied with $I_D R_L = 12.3$ V for $V_{GS} - V_{TN} = 0.5$ V.

The BJT can achieve the required gain with a much lower supply voltage and still meet the input resistance requirement: $R_{in} \cong \beta_o R_L \cong 100(5 \text{ k}\Omega) = 500 \text{ k}\Omega$.

$$g_m R_L = \frac{I_C R_L}{V_T} = 49 \rightarrow I_C R_L = 49(0.025 \text{ V}) = 1.23 \text{ V}$$

(f) A noninverting amplifier with a gain of 10 and an input resistance of 2 kΩ should be achievable with either a common-base or common-gate amplifier with proper choice of operating point. The gain of 10 is easily achieved with either the MOSFET or BJT design estimate: $A_v = V_{DD}/(V_{GS} - V_{TN})$ or $A_v = 10 V_{CC}$. $R_{in} \cong 1/g_m = 2 \text{ k}\Omega$ is within easy reach of either device.

(g) Twenty-five ohms represents a small value of output resistance. The follower stages are the only choices that provide low output resistances. For the followers, $R_\text{out} = 1/g_m$, and so we need $g_m = 40$ mS.

For the BJT: $\quad I_C = g_m V_T = 40 \text{ mS}(25 \text{ mV}) = 1 \text{ mA}$

For the MOSFET: $\quad I_D = \dfrac{g_m(V_{GS} - V_{TN})}{2} = \dfrac{40 \text{ mS}(0.5 \text{ V})}{2} = 10 \text{ mA}$

$$K_n = \dfrac{g_m^2}{2 I_D} = \dfrac{(40 \text{ mS})^2}{2(10 \text{ mA})} = 0.08 \dfrac{\text{A}}{\text{V}^2} \quad \text{and}$$

$$\dfrac{W}{L} = \dfrac{K_n}{K_n'} = \dfrac{80 \text{ mA/V}^2}{50 \text{ μA/V}^2} = \dfrac{1600}{1}$$

The 25-Ω requirement can be met with either device, but the BJT requires an order of magnitude less current. In addition, the MOSFET requires a large W/L ratio.

Discussion: The options developed here represent our first attempts, and there is no guarantee that we will actually be able to fully achieve the desired specifications. After attempting a full design, we may have to change the circuit choice or use more than one transistor in a more complex amplifier configuration.

EXERCISE: Suppose the BJT amplifier in part (b) of Design Ex. 8.6 will be designed with symmetric 15-V supplies using a circuit similar to the one in Figure P8.1(f). Choose a collector current.

ANSWER: 5 μA, (10 μA does not account for the effect of R_B)

EXERCISE: Estimate the collector current needed for a BJT to achieve the input resistance specification in part (f) of Design Ex. 8.6.

ANSWER: 12.5 μA

8.7 COUPLING AND BYPASS CAPACITOR DESIGN

Up to this point, we have assumed that the impedances of coupling and bypass capacitors are negligible, and have concentrated on understanding the properties of the single transistor building blocks in their "midband" region of operation. However, since the impedance of a capacitor increases with decreasing frequency, the coupling and bypass capacitors generally reduce amplifier gain at low frequencies. In this section, we discover how to pick the values of these capacitors to ensure that our midband assumption is valid. Each of the three classes of amplifiers will be considered in succession. The technique we use is related to the "short-circuit" time-constant (SCTC) method that we shall study in greater detail in Chapter 9. In this method, each capacitor is considered separately with all the others replaced by short circuits ($C \to \infty$).

8.7.1 COMMON-EMITTER AND COMMON-SOURCE AMPLIFIERS

Let us start by choosing values for the capacitors for the C-E and C-S amplifiers in Fig. 8.2. For the moment, assume that C_3 is still infinite in value, thus shorting the bottom of R_E and R_S to ground, as drawn in Fig. 8.34(a) and (b).

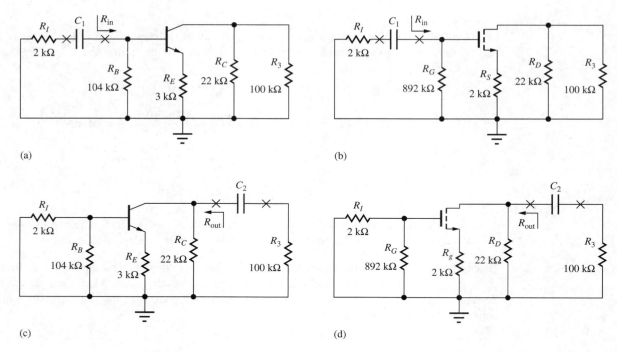

Figure 8.34 Coupling capacitors in the common-emitter and common-source amplifiers.

Coupling Capacitors C_1 and C_2

First, consider C_1. In order to be able to neglect C_1, we require the magnitude of the impedance of the capacitor (its capacitive reactance) to be much smaller than the equivalent resistance that appears at its terminals. Referring to Fig. 8.34, we see that the resistance looking to the left from capacitor C_1 (with $v_i = 0$) is R_I, and that looking to the right is R_{in}. Thus, design of C_1 requires

$$\frac{1}{\omega C_1} \ll (R_I + R_{\text{in}}) \qquad \text{or} \qquad C_1 \gg \frac{1}{\omega(R_I + R_{\text{in}})} \qquad (8.107)$$

Frequency ω is chosen to be the lowest frequency for which midband operation is required in the given application.

For the common-emitter stage, bias resistor R_B appears in parallel (shunt) with the input resistance of the transistor, so $R_{\text{in}} = R_B \| R_{iB}$. For the common-source stage, bias resistor R_G shunts the input resistance of the transistor, and $R_{\text{in}} = R_G \| R_{iG} = R_G$.

A similar analysis applies to C_2. We require the reactance of the capacitor to be much smaller than the equivalent resistance that appears at its terminals. Referring to Fig. 8.34(b), the resistance looking to the left from capacitor C_2 is R_{out}, and that looking to the right is R_3. Thus, C_2 must satisfy

$$\frac{1}{\omega C_2} \ll (R_{\text{out}} + R_3) \qquad \text{or} \qquad C_2 \gg \frac{1}{\omega(R_{\text{out}} + R_3)} \qquad (8.108)$$

For the common-emitter stage, collector resistor R_C appears in parallel with the output resistance of the transistor, and $R_{\text{out}} = R_C \| R_{iC}$. For the common-source stage, drain resistor R_D shunts the output resistance of the transistor, so $R_{\text{out}} = R_D \| R_{iD}$.

Bypass Capacitor C_3

The formula for C_3 is somewhat different. Figure 8.35 depicts the circuit assuming we can neglect the impedance of capacitors C_1 and C_2. At the terminals of C_3 in Fig. 8.35(a), the equivalent

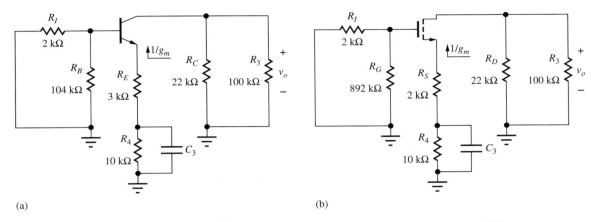

Figure 8.35 Bypass capacitors in the common-emitter and common-source amplifiers.

resistance is equal to R_4 in parallel with the sum $(R_E + 1/g_m)$,[9] the resistance looking up toward the emitter of the transistor. Thus, for the C-E and C-S amplifiers, C_3 must satisfy

$$C_3 \gg \frac{1}{\omega\left[R_4 \| \left(R_E + \frac{1}{g_m}\right)\right]} \quad \text{or} \quad C_3 \gg \frac{1}{\omega\left[R_4 \| \left(R_S + \frac{1}{g_m}\right)\right]} \qquad (8.109)$$

In order to satisfy the inequalities in Eqs. (8.107) through (8.109), we will set the capacitor value to be approximately 10 times that calculated in the equations.

DESIGN EXAMPLE 8.7 CAPACITOR DESIGN FOR THE C-E AND C-S AMPLIFIERS

In this example, we select capacitor values for the three capacitors in both inverting amplifiers in Figs. 8.2, 8.34, and 8.35.

PROBLEM Choose values for the coupling and bypass capacitors for the amplifiers in Fig. 8.2 so that the presence of the capacitors can be neglected at a frequency of 1 kHz (1 kHz represents an arbitrary choice in the audio frequency range).

SOLUTION **Known Information and Given Data:** Frequency $f = 1000$ Hz; for the C-E stage described in Fig. 8.2 and Table 8.3, $R_{iB} = 310$ kΩ, $R_{iC} = 4.55$ MΩ, $R_I = 2$ kΩ, $R_B = 104$ kΩ, $R_C = 22$ kΩ, $R_E = 3$ kΩ, $R_4 = 10$ kΩ, and $R_3 = 100$ kΩ; for the C-S stage from Table 8.4, $R_{iG} = \infty$, $R_{iD} = 442$ kΩ, $R_I = 2$ kΩ, $R_G = 892$ kΩ, $R_D = 22$ kΩ, $R_S = 2$ kΩ, $R_4 = 10$ kΩ, and $R_3 = 100$ kΩ

Unknowns: Values of capacitors C_1, C_2, and C_3 for the common-emitter and common-source amplifier stages.

Approach: Substitute known values in Eqs. (8.107) through (8.109). Choose nearest values from the appropriate table in Appendix A.

---

[9] For the BJT case, we are neglecting the $R_{th}/(\beta_o + 1)$ term. Since the additional term will increase the equivalent resistance, its neglect makes Eq. (8.109) a conservative estimate.

Assumptions: Small-signal operating conditions are valid, $V_T = 25$ mV.

Analysis: For the common-emitter amplifier,

$$R_{\text{in}} = R_B \| R_{iB} = 104 \text{ k}\Omega \| 310 \text{ k}\Omega = 77.9 \text{ k}\Omega$$

$$R_{\text{out}} = R_C \| R_{iC} = 22 \text{ k}\Omega \| 4.55 \text{ M}\Omega = 21.9 \text{ k}\Omega$$

$$C_1 \gg \frac{1}{\omega(R_I + R_{\text{in}})} = \frac{1}{2000\pi(2 \text{ k}\Omega + 77.9 \text{ k}\Omega)} = 1.99 \text{ nF} \rightarrow C_1 = 0.02 \text{ μF } (20 \text{ nF})^{10}$$

$$C_2 \gg \frac{1}{\omega(R_{\text{out}} + R_3)} = \frac{1}{2000\pi(21.9 \text{ k}\Omega + 100 \text{ k}\Omega)} = 1.31 \text{ nF} \rightarrow C_2 = 0.015 \text{ μF } (15 \text{ nF})$$

$$C_3 \gg \frac{1}{\omega\left[R_4 \|\left(R_E + \frac{1}{g_m}\right)\right]} = \frac{1}{2000\pi\left[10 \text{ k}\Omega \|\left(3 \text{ k}\Omega + \frac{1}{9.80 \text{ mS}}\right)\right]}$$

$$= 67.2 \text{ nF} \rightarrow C_3 = 0.68 \text{ μF}$$

For the common-source stage, $R_{\text{in}} = R_G$ since the input resistance at the gate of the transistor is infinite, and $R_{\text{out}} = R_D \| R_{iD}$.

$$C_1 \gg \frac{1}{\omega(R_I + R_{\text{in}})} = \frac{1}{2000\pi(2 \text{ k}\Omega + 892 \text{ k}\Omega)} = 178 \text{ pF} \rightarrow C_1 = 1800 \text{ pF}$$

$$C_2 \gg \frac{1}{\omega(R_{\text{out}} + R_3)} = \frac{1}{2000\pi(21.9 \text{ k}\Omega + 100 \text{ k}\Omega)} = 1.31 \text{ nF} \rightarrow C_2 = 0.015 \text{ μF } (15 \text{ nF})$$

$$C_3 \gg \frac{1}{\omega\left[R_4 \|\left(R_S + \frac{1}{g_m}\right)\right]} = \frac{1}{2000\pi\left[10 \text{ k}\Omega \|\left(2 \text{ k}\Omega + \frac{1}{0.491 \text{ mS}}\right)\right]}$$

$$= 55.3 \text{ nF} \rightarrow C_3 = 0.56 \text{ μF}$$

Check of Results: A double check of the calculations indicates they are correct. This would be a good place to check the analysis with simulation.

Discussion: We have chosen each capacitor to have negligible reactance at the frequency of 1 kHz and would expect the lower cutoff frequency of the amplifier to be well below this frequency. The choice of frequency in this example was arbitrary and depends upon the lowest frequency of interest in the application.

Computer-Aided Analysis: The graph below gives SPICE simulation results for the common-emitter amplifier with the capacitors as designed here. The midband gain is 15.0 dB and the lower cutoff frequency is 195 Hz. Note the two-pole roll-off at low frequencies indicated by the 40-dB/decade slope in the magnitude characteristic. The slope indicates that there are two zeros at dc, which are associated with capacitors C_1 and C_2. A signal cannot pass through either capacitor at dc, hence the frequency response exhibits a double zero at the origin. We have ended up with an amplifier that has three low frequency poles at approximately 100 Hz (1 kHz/10), and bandwidth shrinkage (Secs. 12.1.3 and 8.7.4) causes the resulting lower cutoff frequency f_L to increase to 195 Hz.

---

[10] We are using $C_1 = 10(1.99 \text{ nF})$ to satisfy the inequality.

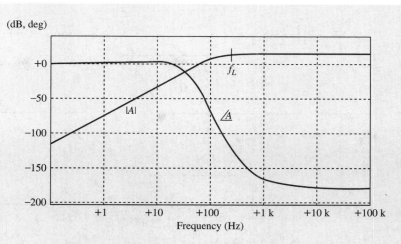

Frequency response for the common-emitter amplifier.

EXERCISE: Reevaluate the capacitor values for the two amplifiers in Ex. 8.7 if the frequency is 250 Hz and the values of R_I and R_3 are changed to 1 kΩ and 82 kΩ, respectively.

ANSWERS: 8.05 nF → 0.082 µF, 0.269 µF → 2.7 µF, 6.13 nF → 0.068 µF; 713 pF → 8200 pF, 6.40 nF → 0.068 µF, 0.221 µF → 2.2 µF

EXERCISE: Use SPICE to simulate the frequency response of the common-source amplifier and find the midband gain and lower cutoff frequency.

ANSWERS: 12.8 dB; 185 Hz

8.7.2 COMMON-COLLECTOR AND COMMON-DRAIN AMPLIFIERS

The simplified C-C and C-D amplifiers in Fig. 8.4 have only two coupling capacitors. In order to be able to neglect C_1, the reactance of the capacitor must be much smaller than the equivalent resistance that appears at its terminals. Referring to Fig. 8.36, we see that the resistance looking to the left from coupling capacitor C_1 is R_I, and that looking to the right is R_{in}. Thus, design of C_1 is the same as Eq. (8.107):

$$\frac{1}{\omega C_1} \ll (R_I + R_{in}) \qquad \text{or} \qquad C_1 \gg \frac{1}{\omega(R_I + R_{in})} \qquad (8.110)$$

Be sure to note that the values of the input and output resistances will be different in Eq. (8.110) from those in Eq. (8.107)! For the common-collector stage, bias resistor R_B shunts the input resistance of the transistor, so $R_{in} = R_B \| R_{iB}$. For the common-drain stage, gate bias resistor R_G appears in parallel with the input resistance of the transistor, and $R_{in} = R_G \| R_{iG}$.

For coupling capacitor C_2, the resistance looking to the left from capacitor C_2 is R_{out}, and that looking to the right is R_3. Thus, design of C_2 requires

$$\frac{1}{\omega C_2} \ll (R_{out} + R_3) \qquad \text{or} \qquad C_2 \gg \frac{1}{\omega(R_{out} + R_3)} \qquad (8.111)$$

where $R_{out} = R_6 \| R_{iE}$ or $R_6 \| R_{iS}$, because resistor R_6 appears in parallel with the output resistance of the transistor. Note again that the value of R_{out} in Eq. (8.111) differs from that in Eq. (8.108).

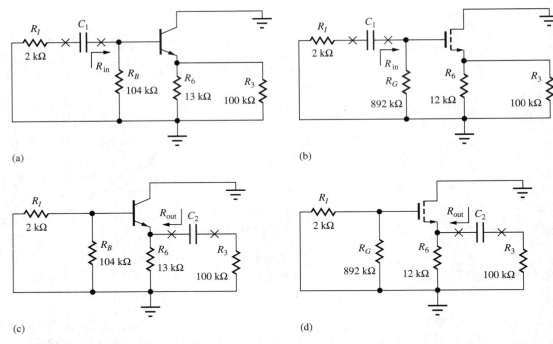

(a)　(b)　(c)　(d)

Figure 8.36 Coupling capacitors in the common-collector, (a) and (c), and common-drain amplifiers, (b) and (d).

DESIGN EXAMPLE 8.8   CAPACITOR DESIGN FOR THE C-C AND C-D AMPLIFIERS

This example selects capacitor values for the followers in Figs. 8.4 and 8.36.

PROBLEM Choose values for the coupling and bypass capacitors for the amplifiers in Figs. 8.4 and 8.36 so that the presence of the capacitors can be neglected at a frequency of 2 kHz.

SOLUTION **Known Information and Given Data:** Frequency $f = 2000$ Hz; for the C-C stage from Fig. 8.4 and Table 8.5, $R_{iB} = 1.17$ MΩ, $R_{iC} = 0.121$ kΩ, $R_6 = 13$ kΩ, $R_I = 2$ kΩ, $R_B = 104$ kΩ, and $R_3 = 100$ kΩ; for the C-S stage, $R_{iG} = \infty$, $R_{iS} = 2.04$ kΩ, $R_6 = 12$ kΩ, $R_I = 2$ kΩ, $R_G = 892$ kΩ, and $R_3 = 100$ kΩ

Unknowns: Values of capacitors C_1 and C_3 for the common-collector and common-drain amplifiers.

Approach: Substitute known values in Eqs. (8.110) and (8.111). Choose the nearest values from the capacitor table in Appendix A.

Assumptions: Small-signal operating conditions are valid.

Analysis: For the common-collector amplifier,

$$R_{\text{in}} = R_B \| R_{iB} = 104 \text{ k}\Omega \| 1.17 \text{ M}\Omega = 95.5 \text{ k}\Omega$$

$$C_1 \gg \frac{1}{\omega(R_I + R_{\text{in}})} = \frac{1}{4000\pi(2 \text{ k}\Omega + 95.5 \text{ k}\Omega)} = 816 \text{ pF} \rightarrow C_1 = 8200 \text{ pF}^{11}$$

---

[11] $C_1 = 10(816 \text{ pF})$ is used to satisfy the inequality.

$$R_{out} = R_6 \| R_{iC} = 13 \text{ k}\Omega \| 121 \text{ }\Omega = 120 \text{ }\Omega$$

$$C_2 \gg \frac{1}{\omega(R_{out} + R_3)} = \frac{1}{4000\pi(120 \text{ }\Omega + 100 \text{ k}\Omega)} = 795 \text{ pF} \rightarrow C_2 = 8200 \text{ pF}$$

and for the common-drain stage,

$$R_{in} = R_G \| R_{iG} = 892 \text{ k}\Omega \| \infty = 892 \text{ k}\Omega$$

$$C_1 \gg \frac{1}{\omega(R_I + R_{in})} = \frac{1}{4000\pi(2 \text{ k}\Omega + 892 \text{ k}\Omega)} = 89.0 \text{ pF} \rightarrow C_1 = 1000 \text{ pF}$$

$$R_{out} = R_6 \| R_{iS} = 12 \text{ k}\Omega \| 2.04 \text{ k}\Omega = 1.74 \text{ k}\Omega$$

$$C_2 \gg \frac{1}{\omega(R_{out} + R_3)} = \frac{1}{4000\pi(1.74 \text{ k}\Omega + 100 \text{ k}\Omega)} = 782 \text{ pF} \rightarrow C_2 = 8200 \text{ pF}$$

Check of Results: A double check of the calculations indicates they are correct. This represents a good place to check the analysis with simulation.

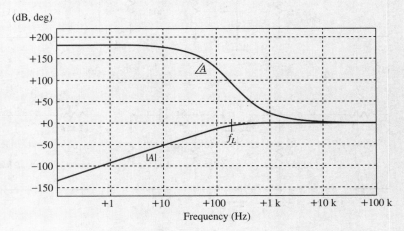

Emitter follower frequency response.

Discussion: We have chosen each capacitor to have negligible reactance at the frequency of 2 kHz and would expect the lower cutoff frequency of the amplifier to be well below this frequency. The choice of frequency in this example was arbitrary and depends upon the lowest frequency of interest in the application.

Computer-Aided Analysis: The graph on the previous page shows SPICE simulation results for the common-emitter amplifier with the capacitors as designed above. The midband gain is −0.262 dB (0.970) and the lower cutoff frequency is 310 Hz. Note the two-pole roll off at low frequencies indicated by the 40-dB/decade slope in the magnitude characteristic. As in Design Ex. 8.7, a dc signal cannot pass through capacitor C_1 or C_3, and the amplifier transfer function is characterized by a double zero at the origin.

EXERCISE: Reevaluate the capacitor values for the two amplifiers in Ex. 8.8 if the frequency is 250 Hz and the values of R_I and R_3 are changed to 1 kΩ and 82 kΩ, respectively.

ANSWERS: 6.79 nF → 0.068 µF, 8.16 nF → 0.082 µF; 713 pF → 8200 pF, 7.98 nF → 0.082 µF

518 Chapter 8 Transistor Amplifier Building Blocks

> **EXERCISE:** Use SPICE to simulate the frequency response of the common-drain amplifier and find the midband gain and lower cutoff frequency.
>
> **ANSWERS:** −1.54 dB; 293 Hz

8.7.3 COMMON-BASE AND COMMON-GATE AMPLIFIERS

For the C-B and C-G amplifiers, C_3 is first assumed to be infinite in value, thus shorting the base and gate of the transistors in Fig. 8.5 to ground as redrawn in Fig. 8.37. In order to neglect C_1 the magnitude of the impedance of the capacitor must be much smaller than the equivalent resistance

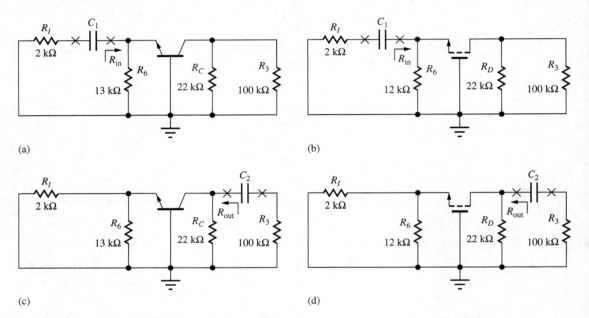

Figure 8.37 Coupling capacitors in the common-base and common-gate amplifiers.

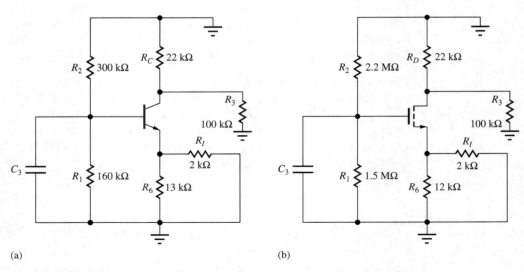

Figure 8.38 Bypass capacitors in the (a) common-collector and (b) common-drain amplifiers.

that appears at its terminals. Referring to Fig. 8.37, the resistance looking to the left from the capacitor is R_I, and that looking to the right is R_{in}. Thus, design of C_1 is the same as Eq. (8.107):

$$\frac{1}{\omega C_1} \ll (R_I + R_{in}) \quad \text{or} \quad C_1 \gg \frac{1}{\omega(R_I + R_{in})} \quad (8.112)$$

For the two amplifier stages, resistor R_6 appears in shunt with the input resistance of the transistor, so $R_{in} = R_6 \| R_{iE}$ or $R_{in} = R_6 \| R_{iS}$.

For C_2, we see that the resistance looking to the left from capacitor C_2 is R_{out}, and that looking to the right is R_3. Thus, design of C_2 requires

$$\frac{1}{\omega C_2} \ll (R_{out} + R_3) \quad \text{or} \quad C_2 \gg \frac{1}{\omega(R_{out} + R_3)} \quad (8.113)$$

For the amplifiers, resistor R_C or R_D appears in parallel with the output resistance of the transistor, so $R_{out} = R_C \| R_{iC}$ or $R_{out} = R_D \| R_{iD}$.

To be an effective bypass capacitor, the reactance of C_3 must be much smaller than the equivalent resistance at the base or gate terminal of the transistors in Fig. 8.5 with the other capacitors assumed to be infinite, as depicted in Fig. 8.38. The resistances at the base and gate nodes are

$$R_{eq}^{CB} = R_1 \| R_2 \| [r_\pi + (\beta_o + 1)(R_6 \| R_I)] \quad \text{and} \quad R_{eq}^{CG} = R_1 \| R_2 \quad (8.114)$$

respectively. The corresponding value of C_3 must satisfy

$$C_3 \gg \frac{1}{\omega R_{eq}^{CB, CG}}$$

DESIGN EXAMPLE 8.9 CAPACITOR DESIGN FOR THE C-B AND C-G AMPLIFIERS

This example selects capacitor values for the noninverting amplifiers in Fig. 8.5.

PROBLEM Choose values for the coupling and bypass capacitors for the amplifiers in Fig. 8.5 so that the presence of the capacitors can be neglected at a frequency of 1 kHz.

SOLUTION **Known Information and Given Data:** Frequency $f = 1000$ Hz; for the C-B stage from Fig. 8.5 and Table 8.6, $R_{iE} = 102$ Ω, $R_{iC} = 3.40$ MΩ, $R_I = 2$ kΩ, $R_1 = 160$ kΩ, $R_2 = 300$ kΩ, $R_C = 22$ kΩ, and $R_6 = 13$ kΩ; for the C-G amplifier, $R_{iS} = 2.04$ kΩ, $R_{iD} = 411$ kΩ, $R_I = 2$ kΩ, Ω, $R_1 = 1.5$ MΩ, $R_2 = 2.2$ MΩ, $R_6 = 12$ kΩ, and $R_D = 22$ kΩ

Unknowns: Values of capacitors C_1, C_2, and C_3

Approach: Substitute known values in Eqs. (8.112) through (8.114). Choose the nearest values from the capacitor table in Appendix A.

Assumptions: Small-signal operating conditions are valid. Use a factor of 10 to satisfy the inequalities.

Analysis: For the common-base amplifier,

$$R_{in} = R_6 \| R_{iE} = 13 \text{ kΩ} \| 102 \text{ Ω} = 100 \text{ Ω}$$

$$C_1 \gg \frac{1}{\omega(R_I + R_{in})} = \frac{1}{2000\pi(2 \text{ kΩ} + 100 \text{ Ω})} = 75.8 \text{ nF} \rightarrow C_1 = 0.82 \text{ μF}[12]$$

$$R_{out} = R_C \| R_{iC} = 22 \text{ kΩ} \| 3.40 \text{ MΩ} = 21.9 \text{ kΩ}$$

---

[12] $C_1 = 10(75.8$ nF) is used to satisfy the inequality.

$$C_2 \gg \frac{1}{\omega(R_{out} + R_3)} = \frac{1}{2000\pi(21.9 \text{ k}\Omega + 100 \text{ k}\Omega)} = 1.31 \text{ nF} \rightarrow C_2 = 0.015 \text{ µF} \quad (15 \text{ nF})$$

$$C_3 \gg \frac{1}{\omega(R_1 \| R_2 \| [r_\pi + (\beta_o + 1)(R_6 \| R_I)])}$$

$$= \frac{1}{2000\pi(160 \text{ k}\Omega \| 300 \text{ k}\Omega \| [10.2 \text{ k}\Omega + (101)(13 \text{ k}\Omega \| 2 \text{ k}\Omega)])}$$

$$= 2.38 \text{ nF} \rightarrow C_3 = 0.027 \text{ µF}$$

and for the common-gate stage,

$$R_{in} = R_6 \| R_{iS} = 12 \text{ k}\Omega \| 2.04 \Omega = 1.74 \text{ k}\Omega$$

$$C_1 \gg \frac{1}{\omega(R_I + R_{in})} = \frac{1}{2000\pi(2 \text{ k}\Omega + 1.74 \text{ k}\Omega)} = 42.6 \text{ nF} \rightarrow C_1 = 0.42 \text{ µF}$$

$$R_{out} = R_6 \| R_{iD} = 22 \text{ k}\Omega \| 411 \text{ k}\Omega = 20.9 \text{ k}\Omega$$

$$C_2 \gg \frac{1}{\omega(R_{out} + R_3)} = \frac{1}{2000\pi(20.9 \text{ k}\Omega + 100 \text{ k}\Omega)} = 1.31 \text{ nF} \rightarrow C_2 = 0.015 \text{ µF} \quad (15 \text{ nF})$$

$$C_3 \gg \frac{1}{\omega(R_1 \| R_2)} = \frac{1}{2000\pi(1.5 \text{ M}\Omega \| 2.2 \text{ M}\Omega)} = 178 \text{ pF} \rightarrow C_3 = 1800 \text{ pF}$$

Check of Results: A double check of the calculations indicates they are correct. This is a good place to check the analysis with simulation.

Discussion: We have chosen each capacitor to have negligible reactance at the frequency of 1 kHz and expect the lower cutoff frequency of the amplifier to be well below this frequency. The choice of frequency in this example was arbitrary and depends upon the lowest frequency of interest in the application.

Computer-Aided Analysis: The graph below shows SPICE simulation results for the common-base amplifier with the capacitors as just designed. The midband gain is 18.5 dB (8.41) and the lower cutoff frequency is 174 Hz. Note the two-pole roll-off at low frequencies indicated by the 40-dB/decade slope in the magnitude characteristic. Here again, since a dc signal cannot pass through capacitor C_1 or C_2, the amplifier transfer function exhibits a double zero at the origin.

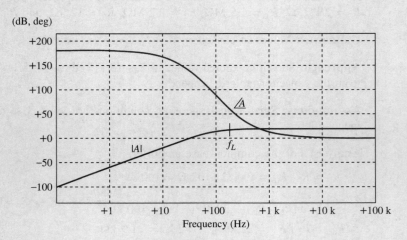

Common-base amplifier frequency response.

> **EXERCISE:** Recalculate the capacitor values for the two amplifiers in Design Ex. 8.9 if the frequency is 250 Hz and the values of R_I and R_3 are changed to 1 kΩ and 82 kΩ, respectively.
>
> **ANSWERS:** 0.579 µF → 6.8 µF, 6.13 nF → 0.068 µF; 12.2 nF → 0.12 µF; 0.232 µF → 2.2 µF, 6.19 nF → 0.068 µF, 714 pF → 8200 pF
>
> **EXERCISE:** Use SPICE to simulate the frequency response of the common-gate amplifier and find the midband gain and lower cutoff frequency.
>
> **ANSWERS:** 12.2 dB, 156 Hz

8.7.4 SETTING LOWER CUTOFF FREQUENCY f_L

In the previous sections, we have designed the coupling and bypass capacitors to have a negligible effect on the circuit at some particular frequency in the midband range of the amplifier. An alternative is to choose the capacitor values to set the lower cutoff frequency of the amplifier where we want it to be. Referring back to the high-pass filter analysis in Sec. 6.8.3, we see that the pole associated with the capacitor occurs at the frequency for which the capacitive reactance is equal to the resistance that appears at the capacitor terminals.

Multiple Poles and Bandwidth Shrinkage

In the circuits we have considered, there are several poles, and bandwidth shrinkage occurs at low frequencies in a manner similar to that which will be presented later in Chapter 12 for high frequencies. A transfer function which exhibits n identical poles at a low frequency ω_o can be written as

$$T(s) = A_{\text{mid}} \frac{s^n}{(s + \omega_o)^n} \tag{8.115}$$

$$|T(j\omega)| = A_{\text{mid}} \frac{\omega^n}{\left(\sqrt{\omega^2 + \omega_o^2}\right)^n} \tag{8.116}$$

$$|T(j\omega_L)| = \frac{A_{\text{mid}}}{\sqrt{2}} \rightarrow \omega_L = \frac{\omega_o}{\sqrt{2^{1/n} - 1}} \quad \text{or} \quad f_L = \frac{f_o}{\sqrt{2^{1/n} - 1}} \tag{8.117}$$

The factor in the denominator of Eq. (8.117) is less than 1, so that the lower cutoff frequency is higher than the frequency corresponding to the individual poles. Table 8.13 gives the relationship between ω_o and ω_L for various values of n.

In Design Exs. 8.7, 8.8, and 8.9, we have effectively located three poles of each amplifier at a frequency of 1/10 of the midband frequency specified in the problem. For three identical poles, $f_L = 1.96 f_o$. In Design Ex. 8.7, the three poles were placed at a frequency of approximately 100 Hz (1000 Hz/10), which should yield a cutoff of 196 Hz based on the numbers in Table 8.13. The simulation results yielded $f_L = 195$ Hz. In Design Ex. 8.9, the poles were also placed at a frequency of approximately 100 Hz, which should yield a cutoff of 195 Hz. The simulation results yielded a slightly smaller value of f_L, 174 Hz.

For the common-gate amplifier, no signal current exists in capacitor C_3 since the FET gate current is zero (without C_{GS} and C_{GD}). Therefore there are only two identical poles in the signal path yielding $f_L = 155$ Hz which agrees closely with simulation. (See the second exercise on top of the previous page.)

The situation in Design Ex. 8.8 is slightly different. With capacitor C_3 eliminated from the circuit, the C-C and C-D amplifiers exhibit two poles at low frequencies. In this example, the two poles are at 200 Hz, which should yield a cutoff frequency of 310 Hz, and the simulation results agree with $f_L = 310$ Hz.

TABLE 8.13 Bandwidth Shrinkage at Low Frequencies

n	f_L/f_o
1	1
2	1.55
3	1.96
4	2.30
5	2.59

Setting f_L with a Dominant Pole

It is often easy and preferable to have the pole associated with just one of the capacitors determine the lower cutoff frequency, rather than have f_L set by the interaction of several poles. In this case, we set f_L with one of the capacitors, and then choose the other capacitors to have their pole frequencies much below f_L. This is referred to as a dominant pole design. In Design Exs. 8.7, 8.8, and 8.9, we see that the capacitor associated with the emitter or source portion of the circuit tends to be the largest (C_3 in Fig. 8.35, C_2 in Fig. 8.36, and C_1 in Fig. 8.37) because of the low resistance presented by the emitter or source terminal of the transistor. It is common to use these capacitors to set f_L, and then increase the value of the other capacitors by a factor of 10 to push their corresponding poles to much lower frequencies.

For the C-E stage in Design Ex. 8.7, we could set $f_L = 1000$ Hz by choosing $C_3 = 0.067$ μF and leaving $C_1 = 0.02$ μF and $C_2 = 0.015$ μF. In the C-D amplifier in Fig. 8.36(b), using $C_2 = 780$ pF with $C_1 = 1000$ pF sets the lower cutoff frequency to 2000 Hz. Finally, for the C-B amplifier in Design Ex. 8.9, choosing $C_1 = 0.082$ μF, $C_2 = 0.027$ μF, and $C_3 = 0.015$ μF should set the cutoff frequency to approximately 1000 Hz.

> **EXERCISE:** Use SPICE to find the values of f_L for the three designs presented in the preceding paragraph.
>
> **ANSWERS:** C-E: 960 Hz; C-D: 2.04 kHz; C-B: 960 Hz.
>
> **EXERCISE:** (a) What value of capacitor C_3 should be used to set f_L to 1 kHz in the C-S amplifier in Design Ex. 8.7? (b) What value of capacitor C_2 should be used to set f_L to 2 kHz in the C-C amplifier in Design Ex. 8.8? (c) What value of capacitor C_1 should be used to set f_L to 1 kHz in the C-G amplifier in Design Ex. 8.9?
>
> **ANSWERS:** (a) 0.056 μF; (b) 820 pF; (c) 0.042 μF.

8.8 AMPLIFIER DESIGN EXAMPLES

Now that we have become "experts" in the characteristics of single-transistor amplifiers, we will use this knowledge to tackle several amplifier design problems. We should emphasize that no "cookbook" exists for design. Every design is a new, creative experience. Each design has its own unique set of constraints, and there may be more than one way to achieve the desired results. The examples presented here further illustrate the approach to design; they also underscore the interaction between the designer's choice of Q-point and the small-signal properties of the amplifiers.

DESIGN EXAMPLE 8.10 A FOLLOWER DESIGN

In this example, we will design a follower to meet a set of specifications.

PROBLEM Design an amplifier with a mid-band input resistance of at least 20 MΩ and a gain of at least 0.95 when driving an external load of at least 3 kΩ. Any capacitors present should not affect the performance of the circuit at frequencies above 50 Hz.

SOLUTION **Known Information and Given Data:** $A_v \geq 0.95$, $R_{in} \geq 20$ MΩ, $R_{out} \ll 3$ kΩ.

Unknowns: The circuit topology must be chosen, the Q-point must be selected, and the circuit element values must all be determined. The transistor parameters are unknown.

Approach: The gain is approximately one, a high input resistance is required, and the relatively small load resistance will require the amplifier to have a low output resistance. All three of these specifications lead us to consider a voltage follower. We must choose between the emitter-follower (C-C) and source-follower (C-D) configurations and then select the circuit values to meet the design specifications.

Assumptions: The transistors are operating in the active region. Small-signal operating conditions are satisfied, $V_T = 25$ mV.

Analysis: Reviewing Tables 8.10 and 8.12, we find that the input resistance of the C-D amplifier prototype is infinite, whereas that of the C-C amplifier is limited to $\beta_o R_L$. For a load resistance of 3 kΩ, a current gain β_o in excess of 6600 is required to meet the input-resistance specification. This current gain is beyond the range of normal bipolar transistors, so here we rule out the C-C amplifier. (However, be sure to watch for the Darlington circuit in Sec. 13.2.3.)

Figure 8.39 represents a basic source-follower circuit. In this amplifier, we recognize that R_{in} is set simply by the value of R_G, and we can pick $R_G = 22$ MΩ (± 5 percent) to meet the specification. The 22-MΩ value ensures that the design specifications are met when the effect of the tolerance is included.

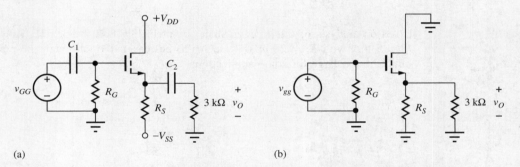

Figure 8.39 (a) Common-drain amplifier and (b) ac equivalent circuit.

The choices of source resistor R_S and power supply voltages are related to the voltage gain requirement:

$$\frac{g_m R_L}{1 + g_m R_L} \geq 0.95 \quad \text{or} \quad g_m R_L \geq 19 \quad \text{and} \quad R_L = R_S \| 3 \text{ k}\Omega \quad (8.118)$$

The $g_m R_L$ product can be related to the drain current and device parameter K_n by using $g_m \cong \sqrt{2 K_n I_D}$, and from Eq. (8.118),

$$\sqrt{2 K_n I_D}\, R_L \geq 19 \quad \text{or} \quad \sqrt{K_n I_D} \geq \frac{19}{\sqrt{2}\, R_L} \quad (8.119)$$

In Fig. 8.39(b), the equivalent load resistor $R_L = R_S \| 3 \text{ k}\Omega \leq 3$ kΩ. As is often the case in design, one equation—here, Eq. (8.119)—contains more than one unknown. We must make a design decision. Let us choose $R_L \geq 1.5$ kΩ (that is, $R_S \geq 3$ kΩ). Substituting this value into Eq. (8.119) yields

$$\sqrt{K_n I_D} \geq \frac{19/\sqrt{2}}{1.5 \text{ k}\Omega} = 8.96 \text{ mA} \quad (8.120)$$

Equation (8.120) indicates that the geometric mean of K_n and I_D must be at least 9 mA.

TABLE 8.14
Possible Solutions to Eq. (8.120)

I_D (mA)	K_n (mA/V²)	$(V_{GS} - V_{TN})$ (V)	V_{SS} (V)
3	10	0.78	$9.8 + V_{TN}$
5	10	1	$16 + V_{TN}$
8	10	1.27	$25.3 + V_{TN}$
5	20	0.71	$16.7 + V_{TN}$

We can now attempt to select an FET and Q-point current. Here again, Eq. (8.120) contains two unknowns. We must make another design decision. Table 8.14 presents some possible solution pairs for Eq. (8.121), as well as their impact on the values of $(V_{GS} - V_{TN})$ and negative supply voltage V_{SS} since

$$I_D = \frac{K_n}{2}(V_{GS} - V_{TN})^2 \tag{8.121}$$

and

$$V_{SS} = I_D R_S + V_{GS} \tag{8.122}$$

based on analysis of the dc equivalent circuit in Fig. 8.40 (remember $I_G = 0$). The choice of $I_D = 5$ mA with $K_n = 20$ mA/V² seems to be reasonable, although the power supply voltage might be too large for some applications.

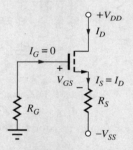

Figure 8.40 dc Equivalent circuit for the C-D amplifier.

Let us assume we have looked through our device catalogs and found a MOSFET with $V_{TN} = 1.5$ V and $K_n = 20$ mA/V². Evaluating Eq. (8.121) for this FET gives

$$V_{GS} = V_{TN} + \sqrt{\frac{2I_D}{K_n}} = 1.5 + \sqrt{\frac{2(0.005)}{0.02}} = 2.21 \text{ V} \tag{8.123}$$

Now we are finally in a position to find R_S using Eq. (8.123).

$$R_S = \frac{V_{SS} - V_{GS}}{I_D} = \frac{V_{SS} - 2.21}{0.005} \tag{8.124}$$

TABLE 8.15
Possible Solutions to Eq. (8.96)

V_{SS}	R_S
10 V	1.56 kΩ
15 V	2.56 kΩ
20 V	3.56 kΩ
25 V	4.56 kΩ

Values have been selected for V_{GS} and I_D but not for V_{SS}, and Eq. (8.124) is another equation with two unknowns. (The value in Table 8.14 represented only a lower bound.) Table 8.15 presents several possible solution pairs from which to make our design selection. Earlier in the design discussion, we assumed that $R_S \geq 3$ kΩ, so one acceptable choice is $V_{SS} = 20$ V and $R_S = 3.56$ kΩ.

Our final design decision is the choice of V_{DD}, which must be large enough to ensure that the MOSFET operates in the active region under all signal conditions:

$$v_{DS} \geq v_{GS} - V_{TN} \tag{8.125}$$

and

$$v_{DS} = v_D - v_S = V_{DD} + V_{GS} - v_s \tag{8.126}$$

for $v_S = V_S + v_s$ and $V_S = -V_{GS}$. Combining Eqs. (8.125) and (8.126) yields

$$V_{DD} + V_{GS} - v_s \geq V_{GS} - V_{TN} \quad \text{or} \quad V_{DD} \geq v_{gg} - V_{TN} = v_{gg} - 1.5 \text{ V} \tag{8.127}$$

The largest amplitude signal v_{gg} at the source that satisfies the small-signal requirements is

$$|v_{gg}| \leq 0.2(V_{GS} - V_{TN})(1 + g_m R_L)\frac{g_m R_L}{1 + g_m R_L} \leq 0.2(0.71)(19) = 2.70 \text{ V} \tag{8.128}$$

Thus, if we choose a V_{DD} of at least 1.2 V, then the MOSFET remains saturated for all signals that satisfy the small-signal criteria.

The final step in this design is to select values for the coupling capacitors. We desire the impedance of the capacitors at frequencies ≥ 50 Hz to be negligible with respect to the resistance that appears at their terminals. The resistance looking to the left from C_1 is zero, and that looking to the right is R_{in}^{CD}, which is 22 MΩ. Therefore,

$$\frac{1}{2\pi(50 \text{ Hz})C_1} \ll 22 \text{ MΩ} \quad \text{or} \quad C_1 \gg 145 \text{ pF}$$

For C_2, the resistance looking back toward the source is

$$R_{out} = R_S \left\| \frac{1}{g_m} = 3.6 \text{ kΩ} \right\| \frac{1}{\sqrt{2K_n I_D}} = 3.6 \text{ kΩ} \left\| \frac{1}{\sqrt{2(20 \text{ mS})(5 \text{ mA})}} = 69.4 \text{ Ω},\right.$$

and the resistance looking toward the right is 3 kΩ. Therefore,

$$\frac{1}{2\pi(50 \text{ Hz})C_2} \ll 3.069 \text{ kΩ} \quad \text{or} \quad C_2 \gg 1.04 \text{ μF}$$

Let us choose $C_1 = 1500$ pF and $C_2 = 10$ μF, which are standard values that exceed the minimum bound by a factor of approximately 10.

The final design appears in Fig. 8.41, in which the nearest 5 percent values have been used for the resistors and V_{DD} has been chosen to be a common power supply value of +5 V.

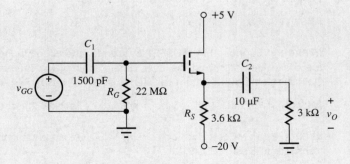

Figure 8.41 Completed source-follower design.

Check of Results: To check our design, we should now analyze the circuit and find the actual Q-point, input resistance, and voltage gain. This analysis is left as an exercise. Another approach at this point would be to check the analysis with SPICE.

Discussion: In this example, we see that even a problem that appears to be a relatively well specified problem takes considerable effort to achieve a design that meets the requirements and the design required a relatively large value of V_{SS}. Such is the situation in most real design situations. Most, if not all, real problems will be under-constrained with numerous choices to be made.

Computer-Aided Analysis: Simulation of the circuit in SPICE yields these results: Q-point: (4.94 mA, 7.20 V), $A_v = -0.369$ dB, and $f_L = 7.8$ Hz. With two poles at 5 Hz, the expected value of f_L is also 7.8 Hz.

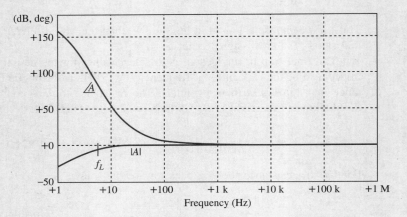

Source follower frequency response.

EXERCISE: Find the actual Q-point, input resistance, and voltage gain for the circuit in Fig. 8.41. ($K_n = 20$ mA/V^2, $V_{TN} = 1.5$ V)

ANSWERS: (4.94 mA, 7.20 V); 22 MΩ; +0.959.

EXERCISE: Find the output resistance of the amplifier in Fig. 8.41. What is the largest value of v_{gg} that satisfies the small-signal constraints?

ANSWERS: 69.1 Ω; 3.38 V.

EXERCISE: Suppose the MOSFET chosen for the circuit in Fig. 8.41 also had $\lambda = 0.015$ V^{-1}. What are the values of r_o and the new voltage gain? (Use the Q-point values from Design Ex. 8.10.) Does neglecting the output resistance seem a reasonable thing to do?

ANSWERS: 15.0 kΩ, 0.954; yes, r_o has little effect on the circuit.

EXERCISE: An MOS technology has $K'_n = 50$ μA/V^2. What is the W/L ratio required for the NMOS transistor in Design Ex. 8.10?

ANSWER: 400/1.

EXERCISE: (a) Create a Thévenin equivalent circuit for the midband region of the source follower in Fig. 8.41. (b) Use the model to calculate the voltage gain with the 3-kΩ load attached to the amplifier.

ANSWERS: (a) $R_{in} = 22$ MΩ, $A = +0.981$, $R_{out} = 69.4$ Ω; (b) 0.959

DESIGN EXAMPLE 8.11

A COMMON-BASE AMPLIFIER

The requirements of this design problem are even less specific than those in Design Ex. 8.10. A common-base amplifier will be found to be the most appropriate choice to meet the given design specifications.

PROBLEM Design an amplifier to match a 75-Ω source resistance (for example, a coaxial transmission line) and to provide a voltage gain of 34 dB. Design the capacitors to have negligible impact on the circuit for RF frequencies above 500 kHz.

SOLUTION **Known Information and Given Data:** Amplifier input resistance = 75 Ω; voltage gain = 50 (34 dB); capacitors should be negligible at a frequency of 500 kHz; source resistance = 75 Ω.

Unknowns: Amplifier topology; Q-point; circuit element values; transistor parameters

Approach: Use overall specifications to guide choice of circuit topology and transistor type; then choose circuit element values to meet numeric requirements

Assumptions: Active region operation; $V_{EB} = 0.7$ V; Small-signal conditions apply; $V_T = 25$ mV.

Analysis: Our first problem is to select a circuit configuration and transistor type. From the various examples in this and previous chapters, we realize that $A_v = 50$ (34 dB) is a moderate value of gain. At the same time, the required input resistance of 75 Ω is relatively low. Looking through our amplifier comparison charts in Tables 8.9 through 8.12, we find that the common-base and common-gate amplifiers most nearly meet these two requirements: good voltage gain and low input resistance. From past examples, we should recognize that it will probably be easier to achieve a gain of 50 with a BJT than with a FET, particularly since the matched input resistance requirement will increase the amplifier terminal gain requirement by a factor of 2! Thus, the common-base amplifier is the choice that seems to most nearly meet the problem specifications.

For simplicity, let us use the dual supply-bias circuit in Fig. 8.42, which requires only two bias resistors. In addition, to get some practice analyzing circuits using *pnp* devices, we have arbitrarily selected a *pnp* transistor. We happen to have a *pnp* transistor available with $\beta_F = 80$ and $V_A = 50$ V (e.g., a 2N3906—see MCD Web Resources).

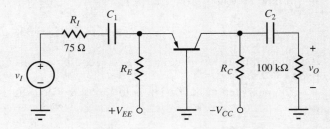

Figure 8.42 Common-base circuit topology.

Next, let us select the power supplies V_{CC} and V_{EE}. Remembering our rule-of-thumb from Chapter 7, $A_v = 10(V_{CC} + V_{EE})$. The matched input resistance situation causes a factor of two voltage loss between the signal source v_i and the emitter-base junction. Thus, an overall gain of 50 requires a value of $g_m R_L = 100$, and we estimate that a total supply voltage of 10 V is required. Using symmetrical supplies, we have $V_{CC} = V_{EE} = 5$ V.

Figures 8.43 and 8.44 are the dc and ac equivalent circuits needed to analyze the behavior of the amplifier in Fig. 8.42. Resistor R_E and the Q-point of the transistor can now be determined from the input resistance requirement. From Fig. 8.44, we recognize that the input resistance of the amplifier is equal to resistor R_E in parallel with the input resistance at the emitter of the transistor. From Table 8.8, $R_{iE} = 1/g_m$:

$$R_{in} = R_E \| R_{iE} = R_E \| \frac{1}{g_m} \tag{8.129}$$

Expanding Eq. (8.129) and using the expression for g_m yields

$$R_{in} = \frac{\frac{1}{g_m} R_E}{\frac{1}{g_m} + R_E} = \frac{R_E}{1 + g_m R_E} = \frac{R_E}{1 + 40 I_C R_E} \tag{8.130}$$

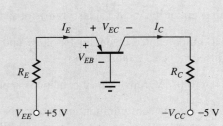

Figure 8.43 dc equivalent circuit for common-base amplifier.

Figure 8.44 ac equivalent circuits for the common-base amplifier.

Since $I_E \cong I_C$, the $I_C R_E$ product in Eq. (8.130) represents the dc voltage developed across the resistor R_E. Here again we see the direct coupling between the small-signal input resistance and the dc Q-point values. From the dc equivalent circuit in Fig. 8.43 and assuming $V_{EB} = 0.7$ V,

$$I_C R_E \cong I_E R_E = V_{EE} - V_{BE} = 5 - 0.7 = 4.3 \text{ V} \tag{8.131}$$

Combining Eqs. (8.130) and (8.131) with the input resistance specification,

$$75 = \frac{R_E}{1 + 40(4.3)} \quad \text{and} \quad R_E = 13.0 \text{ k}\Omega \tag{8.132}$$

I_C can now be found using Eq. (8.132):

$$I_C \cong I_E = \frac{4.3 \text{ V}}{13 \text{ k}\Omega} = 331 \text{ μA} \tag{8.133}$$

It is interesting to note that once V_{EE} was chosen for this circuit, R_E and I_C were both indirectly fixed.

The next step in the design is to choose collector resistor R_C. For the circuit in Fig. 8.44, the gain is

$$A_v^{CB} = g_m R_L \left(\frac{R_{in}}{R_I + R_{in}} \right) \tag{8.134}$$

For our circuit,

$$R_{in} = 75 \ \Omega \quad g_m = 40 I_C = 40(331 \ \mu A) = 13.2 \ mS \quad (8.135)$$
$$R_L = R_C \| 100 \ k\Omega$$

Solving for R_L in Eq. (8.134) yields

$$50 = (13.2 \ mS) \ R_L \left(\frac{75}{75 + 75} \right) \quad \text{and} \quad R_L = 7.58 \ k\Omega \quad (8.136)$$

Since $R_L = R_C \| 100 \ k\Omega$, $R_C = 8.20 \ k\Omega$.

The next step is to finish checking the Q-point of the transistor by calculating V_{EC}. Using the circuit in Fig. 8.43,

$$V_{EB} = V_{EC} + I_C R_C - 5 \quad (8.137)$$

and solving for V_{EC} yields

$$V_{EC} = 5 + V_{EB} - I_C R_C = 5 + 0.7 - (331 \ \mu A)(8.20 \ k\Omega) = 2.99 \ V \quad (8.138)$$

V_{EC} is positive and greater than 0.7 V, so the *pnp* transistor is operating in the active region, as required.

The final step in this design is to select values for the coupling capacitors. We desire the impedance of the capacitors for frequencies of 500 kHz and above to be negligible with respect to the resistance that appears at their terminals. The resistance looking to the left from C_1 is 75 Ω, and the resistance looking to the right is R_{in}, which is also 75 Ω. Therefore,

$$\frac{1}{2\pi(500 \ kHz) C_1} \ll 150 \ \Omega \quad \text{or} \quad C_1 \gg 2.12 \ nF$$

For C_2, the resistance looking back toward the collector is at most 8.2 kΩ, and the resistance looking toward the right is 100 kΩ. Therefore,

$$\frac{1}{2\pi(500 \ kHz) C_2} \ll 108 \ k\Omega \quad \text{or} \quad C_2 \gg 2.95 \ pF$$

Let us choose $C_1 = 0.022 \ \mu F$ and $C_2 = 33 \ pF$, which are standard values that are larger than the calculated values by a factor of at least 10.

The completed design is shown in Fig. 8.45, in which the nearest 5 percent values have been used for the resistors. This amplifier provides a gain of approximately 50 and an input resistance of approximately 75 Ω.

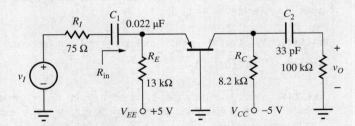

Figure 8.45 Final design for amplifier with $R_{in} = 75 \ \Omega$ and $A_v = 50$.

One serious limitation of this amplifier design is its signal-handling ability. Only 5 mV can appear across the emitter-base junction, which sets a limit on the signal v_i:

$$v_{eb} = v_i \frac{R_{in}}{R_I + R_{in}} = v_i \frac{75 \ \Omega}{75 \ \Omega + 75 \ \Omega} = \frac{v_i}{2} \quad (8.139)$$

Thus, for small-signal operation to be valid, the magnitude of the input signal v_i must not exceed 10 mV. However, input signals to RF amplifiers may often be in the microvolt range.

Check of Results: At this point, an excellent way to check the design is to simulate the circuit in SPICE, which yields a Q-point of (323 µA, 3.09 V). The frequency response results appear in the figure here.

Discussion: In this design, we were lucky that we remembered to account for the factor of 2 loss in Eq. (8.139) due to the matched resistance condition at the input. Otherwise, our initial choice of power supplies might not have been sufficient to meet the gain specification, and a second design iteration could have been required. The signal handling capability of this stage is small. If an FET were used in place of the bipolar transistor, a much higher input range could be achieved.

Computer-Aided Analysis: The frequency response generated by SPICE with v_I as the input appears below. The simulation parameters are FSTART = 1000 Hz and FSTOP = 10 MHz with 10 frequency points per decade. The midband voltage gain is found to be 33.5 dB, and f_L = 72 kHz.

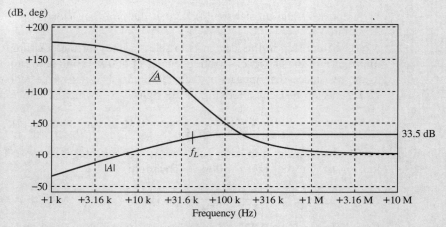

Common-base amplifier frequency response.

EXERCISE: Draw the *npn* version of the circuit in Fig. 8.45. Use the same circuit element values but change polarities as needed.

EXERCISE: What are the actual values of input resistance and gain for the amplifier in Fig. 8.45?

ANSWERS: 75.1 Ω, +50.4

EXERCISE: What is the largest sinusoidal signal voltage that can appear at the output of the amplifier in Fig. 8.45? What is the largest output signal consistent with the requirements for small-signal operation?

ANSWERS: $(2.90 \text{ V} - V_{EB}) \cong 2.29$ V, 0.500 V

EXERCISE: Suppose that both V_{EE} and V_{CC} were changed to 7.5 V. What are the new values of I_C, V_{EC}, R_E, and R_C required to meet the same specifications?

ANSWERS: 332 µA, 5.52 V, 20.5 kΩ, 8.06 kΩ

EXERCISE: Estimate the lower cutoff frequency of the circuit in Fig. 8.45.

ANSWER: 77.5 kHz

EXERCISE: Suppose the resistors and power supplies in the circuit in Fig. 8.45 all have 5 percent tolerances. Will the BJT remain in the active region in the worst-case situation? Repeat for tolerances of 10 percent. Do the values of current gain β_F or V_A have any significant effect on the design? Discuss.

ANSWERS: Yes; yes; no, not unless they become very small.

EXERCISE: (a) Create a Thévenin equivalent circuit for the midband region of the common-base amplifier in Fig. 8.45. (b) Use the model to calculate the voltage gain with the 100-kΩ load attached to the amplifier. $V_T = 25$ mV.

ANSWERS: (a) $v_{th} = 54.6\ v_i$, $R_{th} = 8200\ \Omega$; (b) +50.5

8.8.1 MONTE CARLO EVALUATION OF THE COMMON-BASE AMPLIFIER DESIGN

Before going on to the next design example, we carry out a statistical evaluation of the common-base design to see if it is a viable design for the mass production of large numbers of amplifiers. A spreadsheet is used here, although we could easily evaluate the same equation set using a simple computer program written in any high-level language or using the Monte Carlo option in some circuit simulation programs. SPICE can also do statistical analysis.

To perform a Monte Carlo analysis of the circuit in Fig. 8.45, we assign random values to V_{CC}, V_{EE}, R_C, R_E, and β_F; we then use these values to determine I_C and V_{EC}, R_{in}, and A_v. Referring back to Eq. (1.45) in Chapter 1, we write each parameter in the form

$$P = P_{nom}(1 + 2\varepsilon(\text{RAND}() - 0.5)) \quad (8.140)$$

where P_{nom} = nominal value of parameter

ε = parameter tolerance

RAND( ) = random-number generator in spreadsheet

For the design in Fig. 8.45, we assume that the resistors and power supplies have 5 percent tolerances and the current gain has a ±25 percent tolerance. As mentioned in Chapter 1 and Ex. 4.12, it is important that each variable invoke a separate evaluation of the random-number generator so that the random values are independent of each other. The random-element values are then used to characterize the Q-point, R_{in}, and A_v. The expressions for the Monte Carlo analysis are presented in a logical sequence for evaluation in Eq. (8.141):

$$
\begin{aligned}
&1.\ V_{CC} = 5(1 + 0.1(\text{RAND}() - 0.5)) \\
&2.\ V_{EE} = 5(1 + 0.1(\text{RAND}() - 0.5)) \\
&3.\ R_E = 13{,}000(1 + 0.1(\text{RAND}() - 0.5)) \\
&4.\ R_C = 8200(1 + 0.1(\text{RAND}() - 0.5)) \\
&5.\ \beta_F = 80(1 + 0.5(\text{RAND}() - 0.5)) \\
&6.\ I_C = \alpha_F I_E = \alpha_F \frac{V_{EE} - 0.7}{R_E} \\
&7.\ V_{EC} = 0.7 + V_{CC} - I_C R_C \\
&8.\ g_m = 40 I_C \\
&9.\ R_{in} = R_E \left\| \frac{\alpha_o}{g_m} \right. \\
&10.\ A_v = g_m R_L \frac{R_{in}}{R_I + R_{in}} \quad \text{where } R_L = R_C \| 100\text{ k}\Omega
\end{aligned}
\quad (8.141)
$$

Table 8.16 summarizes the results of a 1000-case analysis. The transistor is always in the active region. The mean collector current of 331 μA corresponds closely to the nominal value

TABLE 8.16
Monte Carlo Analysis of the Common-Base Amplifier Design

CASE #	V_{CC} (1)	V_{EE} (2)	R_E (3)	R_C (4)	β_F (5)	I_C (6)	V_{EC} (7)	g_m (8)	R_{in} (9)	A_v (10)
1	4.932	5.090	13602	8461	96.02	3.23E-04	2.902	1.29E-02	76.2	50.8
2	4.951	5.209	12844	8208	93.01	3.51E-04	2.769	1.40E-02	70.1	51.4
3	4.844	4.759	13418	8440	98.33	3.03E-04	2.990	1.21E-02	81.3	49.0
4	4.787	5.162	13193	8294	72.82	3.38E-04	2.682	1.35E-02	72.5	50.9
5	5.073	5.181	12358	8542	79.30	3.63E-04	2.676	1.45E-02	67.7	54.2
⋮										
996	4.863	5.058	12453	8134	68.56	3.50E-04	2.716	1.40E-02	70.0	50.8
997	5.157	5.016	12945	8225	98.03	3.33E-04	3.115	1.33E-02	73.8	50.3
998	4.932	5.183	12458	8211	78.17	3.60E-04	2.677	1.44E-02	68.2	52.0
999	5.034	4.940	13444	7969	76.71	3.15E-04	3.221	1.26E-02	77.8	47.4
1000	5.119	5.002	12948	7892	95.25	3.32E-04	3.196	1.33E-02	74.0	48.3
Mean	5.006	4.997	12992	8205	79.95	3.31E-04	2.990	1.32E-02	74.29	49.88
std. dev.	0.143	0.146	381	239	11.27	1.44E-05	0.199	5.75E-04	3.22	1.74
min.	4.750	4.751	12351	7792	60.04	2.97E-04	2.409	1.19E-02	66.85	45.36
max.	5.248	5.250	13650	8609	99.98	3.67E-04	3.613	1.47E-02	82.54	54.63

(X) = equation number in Equation Set (8.141).

for the standard 5 percent resistors that were selected for the final circuit. The mean values of R_{in} and A_v are 74.3 Ω and 49.9, respectively, and are also quite close to the design value. The 3σ limit corresponds to only slightly more than 10 percent deviation from the nominal design specification, and even the worst observed cases of R_{in} yield acceptable values of SWR (standing wave ratio) on the transmission line that the amplifier was designed to match. Overall, we should be able to mass produce this design and have few problems meeting the specifications.

DESIGN EXAMPLE 8.12 A COMMON-SOURCE AMPLIFIER

Let us now try to meet the requirements of the previous design using a C-E/C-S design.

PROBLEM Design an amplifier to match a 75-Ω source resistance (for example, a coaxial transmission line) and to provide a voltage gain of 34 dB. Design the capacitors to have negligible impact on the circuit for frequencies above 500 kHz.

SOLUTION **Known Information and Given Data:** Amplifier input resistance = 75 Ω; voltage gain = 50 (34 dB); frequency of application of amplifier is 500 kHz and above; source resistance = 75 Ω

Unknowns: Amplifier topology; Q-point; circuit element values; transistor parameters

Approach: Use overall specifications to guide choice of circuit topology and transistor type; then choose circuit element values to meet numeric requirements. Although the input resistance of the C-E and C-S amplifiers is usually considered in the moderate to high range, we can always limit it by reducing the size of the resistors in the bias network. For example, consider the common-source amplifier in Fig. 8.46. If the gate-bias resistor R_G is reduced to 75 Ω, then

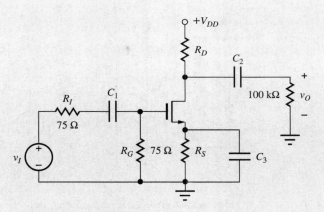

Figure 8.46 Common-source amplifier.

the input resistance of the amplifier will also be 75 Ω. (This design technique is sometimes referred to as **swamping** of the impedance level.) A BJT could also be used, but a depletion-mode MOSFET[13] has been chosen because it offers the potential of a higher signal-handling capability and simple bias circuit design.

Assumptions: The transistor is in the active region. Small-signal conditions apply.

Analysis: If resistor R_S is bypassed, this amplifier yields the full gain $-g_m R_L$, but the matched input causes a loss of input signal by a factor of 2:

$$\mathbf{v_{gs}} = \mathbf{v}_i \frac{R_G}{R_I + R_G} = \mathbf{v}_i \frac{75 \text{ Ω}}{75 \text{ Ω} + 75 \text{ Ω}} = \frac{\mathbf{v}_i}{2} \tag{8.142}$$

Thus, the prototype amplifier must deliver a gain of 100 for the overall amplifier to have a gain of 50. (This was also the case for the C-B amplifier designed in Design Ex. 8.11.) Referring back to Table 8.11 on page 477, we find that our design guide for the voltage gain of the common-source amplifier is

$$A_v = \frac{V_{DD}}{V_{GS} - V_{TN}} \tag{8.143}$$

Here again we have a single constraint equation with two variables; Table 8.17 presents some possible design choices. Let us choose the 20 V/0.2 V option.

Because $V_{GS} - V_{TN}$ must be small in order to achieve high gain, a MOSFET with a large K_n or K_p must be chosen if I_D is to be a reasonable current. Let us assume that we have found an n-channel depletion-mode MOSFET with $K_n = 10$ mS/V and $V_{TN} = -2$ V. With these parameters, the MOSFET drain current will be

$$I_D = \frac{K_n}{2}(V_{GS} - V_{TN})^2 = \frac{0.01}{2}(0.2)^2 = 0.200 \text{ mA} \tag{8.144}$$

With reference to the dc equivalent circuit in Fig 8.47(a), we can now calculate the value of R_S. Because the gate current is zero for the FET, the voltage developed across R_S equals $-V_{GS}$:

$$R_S = \frac{-V_{GS}}{I_D} = \frac{-(V_{TN} + 0.2 \text{ V})}{0.200 \text{ mA}} = \frac{1.8 \text{ V}}{0.200 \text{ mA}} = 9.00 \text{ kΩ} \tag{8.145}$$

TABLE 8.17 Possibilities for $A_v = 100$

V_{DD}	$V_{GS} - V_{TN}$
20 V	0.2 V
25 V	0.25 V
30 V	0.3 V

---

[13] This would also be a good place to use a JFET.

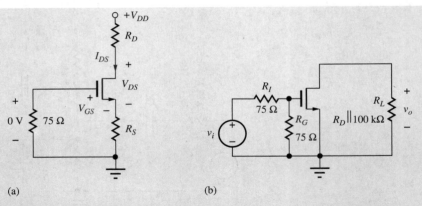

Figure 8.47 (a) dc and (b) ac equivalent circuits for the common-source amplifier.

The gain of the amplifier is

$$A_v = \frac{\mathbf{v_{gs}}}{\mathbf{v_i}}(-g_m R_L) = -\frac{g_m R_L}{2} \quad \text{where} \quad R_L = R_D \parallel 100 \text{ k}\Omega \quad (8.146)$$

Setting Eq. (8.87) equal to 50 and solving for R_L yields

$$R_L = \frac{2A_v}{g_m} = \frac{A_v(V_{GS} - V_{TN})}{I_D} = \frac{50(0.2 \text{ V})}{0.2 \text{ mA}} = 50 \text{ k}\Omega \quad (8.147)$$

For $R_L = 50$ kΩ, R_D must be 100 kΩ.

Now we have encountered a problem. A drain current of 0.2 mA in $R_D = 100$ kΩ requires a voltage drop equal to the total power supply voltage of 20 V. Thus, the power supply voltage must be increased. For active region operation, $V_{DS} \geq V_{GS} - V_{TN}$, where

$$V_{DS} = V_{DD} - I_D R_D - I_D R_S \quad (8.148)$$

Therefore,

$$V_{DD} - 20 - 1.8 \geq (-1.8) - (-2) \quad \text{or} \quad V_{DD} \geq 22 \text{ V} \quad (8.149)$$

is sufficient to ensure pinch-off operation. Let us choose $V_{DD} = 25$ V to provide additional design margin and room for additional signal voltage swing at the drain.

The final step in this design is to select values for the coupling capacitors. We desire the impedance of the capacitors for frequencies of 500 kHz and above to be negligible with respect to the resistance that appears at their terminals. The resistance looking to the left from C_1 is 75 Ω, and the input resistance looking to the right is also 75 Ω. Therefore,

$$\frac{1}{2\pi(500 \text{ kHz})C_1} \ll 150 \text{ }\Omega \quad \text{or} \quad C_1 \gg 2.12 \text{ nF}$$

For C_3, the resistance looking back toward the source is 9.1 kΩ in parallel with $(1/g_m)$ looking into the source of the transistor:

$$R_{eq} = 9.1 \text{ k}\Omega \parallel \frac{1}{g_m} = 9.1 \text{ k}\Omega \parallel \frac{1}{2 \text{ mS}} = 474 \text{ }\Omega$$

Therefore,

$$\frac{1}{2\pi(500 \text{ kHz})C_3} \ll 474 \text{ }\Omega \quad \text{or} \quad C_3 \gg 644 \text{ pF}$$

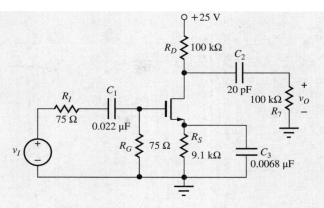

Figure 8.48 Final common-source amplifier design.

For C_2, the resistance looking back toward the drain is 100 kΩ, and the resistance looking toward the right is also 100 kΩ. Therefore,

$$\frac{1}{2\pi(500\text{ kHz})C_2} \ll 200\text{ k}\Omega \quad \text{or} \quad C_2 \gg 1.59\text{ pF}$$

Let us choose $C_1 = 0.022$ μF, $C_3 = 0.0068$ μF, and $C_2 = 20$ pF, which are standard values that are larger than the calculated values by a factor of approximately 10. The circuit corresponding to the final amplifier design is in Fig. 8.48, where standard 5 percent resistor values have once again been selected.

Check of Results: At this point, an excellent way to check the design is to simulate the circuit in SPICE, which yields a Q-point of (198 μA, 3.41 V) with a gain of 33.9 dB and $f_L = 91.5$ kHz. The frequency response appears on the next page.

Discussion: The designs in Design Exs. 8.11 and 8.12 demonstrate that usually more than one, often very different, design approaches can meet the specifications for a given problem. Choosing one design over another depends on many factors. For example, one criterion could be the use of power supply voltages that are already available in the rest of the system. Total power consumption might be an important issue. Our common-base design uses a power of approximately 3.3 mW and uses two power supplies, whereas the common-source design consumes 5 mW from a single 25-V supply. In actuality, the design is somewhat of a struggle using the FET. A large power supply voltage is combined with a FET operating very near cutoff. It may be difficult to find a FET with $K_n = 10$ mA with $V_P = -2$ V.

Another important factor could be amplifier cost. The core of the FET amplifier requires three resistors, R_D, R_G, and R_S; bypass capacitor C_3; and the JFET. The common-base amplifier core requires resistors R_E and R_C and the BJT. The cost of the additional parts, plus the expense of inserting them into a printed circuit board (often more expensive than the parts themselves), will probably tilt the economic decision away from the C-S design toward the C-B amplifier. However, the maximum input signal capability of the JFET amplifier, $|v_i| = 2 \times 0.2(V_{GS} - V_P) = 0.08$ V can be of overriding importance in certain applications. Obviously, the final decision involves many factors.

Computer-Aided Analysis: As already noted, the frequency response generated by SPICE with v_I as the input appears in the graph. The simulation parameters are FSTART = 1000 Hz and FSTOP = 10 MHz with 10 frequency points per decade. The midband voltage gain is 33.9 dB and $f_L = 91.5$ kHz. Based on Table 8.13, three poles at 50 kHz are expected to produce $f_L = 98.0$ kHz.

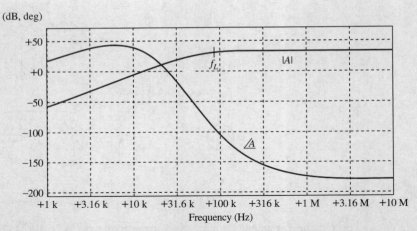

Frequency response of the common-source amplifier.

EXERCISE: Verify the results of the SPICE simulation of Design Ex. 8.12. What is the bandwidth predicted using the bandwidth shrinkage factor in Eq. (8.117)?

ANSWER: 89 Hz (using the average of the three pole frequencies).

EXERCISE: Suppose the FET chosen for the circuit in Fig. 8.48 also had $\lambda = 0.015$ V^{-1}. What are the values of r_o and the new voltage gain? (Use the Q-point values from the example.) Does neglecting the output resistance seem a reasonable thing to do?

ANSWERS: 333 kΩ, 43.5; No, r_o is important in this circuit! We will not achieve the desired gain with this FET.

EXERCISE: (a) Redesign the circuit using the 25 V/0.25 V case from Table 8.17. Use the same FET device parameters. (b) Verify your design with SPICE.

ANSWERS: 5.60 kΩ, 68 kΩ, $V_{DD} = 25$ V, 0.022 μF, 8200 pF, 20 pF.

EXERCISE: (a) Create a Thévenin equivalent circuit for the midband region of the common-source amplifier in Fig. 8.48. (b) Use the model to calculate the voltage gain with the 100-kΩ load attached to the amplifier. (c) Repeat for $r_o = 100$ kΩ.

ANSWERS: (a) $v_{th} = 100$ v_i, $R_{th} = 100$ kΩ; (b) 50.0; (c) $V_{TH} = 50$ v_i, $R_{TH} = 50$ kΩ, $A_V = 33.3$.

8.9 MULTISTAGE ac-COUPLED AMPLIFIERS

In most situations, a single-transistor amplifier cannot meet all the specifications of a given amplifier design. The required voltage gain often exceeds the amplification factor of a single transistor, or the required combination of voltage gain, input resistance, and output resistance cannot be met simultaneously. For example, consider the specifications of a good general purpose operational amplifier having an input resistance exceeding 1 MΩ, a voltage gain of 100,000, and an output resistance less than a few hundred ohms. It is clear from our investigation

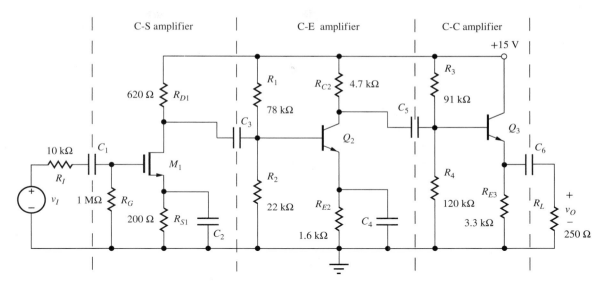

Figure 8.49 Three-stage ac-coupled amplifier.

of amplifiers in this chapter that these requirements cannot be met with a single-transistor amplifier. A number of stages must be cascaded in order to create an amplifier that can meet all the requirements.

8.9.1 A THREE-STAGE ac-COUPLED AMPLIFIER

In this section, we study the three-stage ac-coupled amplifier in Fig. 8.49. Signals are coupled from one stage to the next through the use of coupling capacitors C_1, C_3, C_5, and C_6, whereas the same capacitors provide dc isolation between stages that permits independent design of the bias circuitry of the individual stages.

The function of the various stages can more readily be seen in the midband ac equivalent circuit for this amplifier in Fig. 8.50(a) in which all the capacitors have been replaced with short circuits. MOSFET M_1, operating in the common-source configuration, provides a high input resistance with modest voltage gain. Bipolar transistor Q_2 in the common-emitter configuration provides a second stage with high voltage gain. Q_3, an emitter follower, provides a low output resistance and buffers the high-gain stage, Q_2, from the relatively small load resistance (250 Ω). In Fig. 8.50(a), the base bias resistors have been replaced by $R_{B2} = R_1 \| R_2$ and $R_{B3} = R_3 \| R_4$.

In the amplifier in Fig. 8.49, the input and output of the overall amplifier are ac-coupled through capacitors C_1 and C_6. Bypass capacitors C_2 and C_4 are used to obtain maximum voltage gain from the two inverting amplifier stages. Interstage coupling capacitors C_3 and C_5 transfer the ac signals between the amplifiers but provide isolation at dc. Thus, the individual Q-points of the transistors are not affected by connecting the stages together. Figure 8.50(b) gives the dc equivalent circuit for the amplifier in which the capacitors have all been removed. The isolation of the three individual transistor amplifier stages is apparent in this figure.

We want to characterize this amplifier by determining its voltage, input and output resistances, current and power gains, and input signal range using the transistor parameters in Table 8.18, and we will also estimate the lower cutoff frequency of the amplifier. First, the Q-points of the three transistors must be found. Each transistor stage in Fig. 8.50 is independently biased, and, for expediency, we assume that the Q-points listed in Table 8.19 have already been found using the dc analysis procedures developed in previous chapters. The details of these dc calculations are left for the next exercise.

538   Chapter 8   Transistor Amplifier Building Blocks

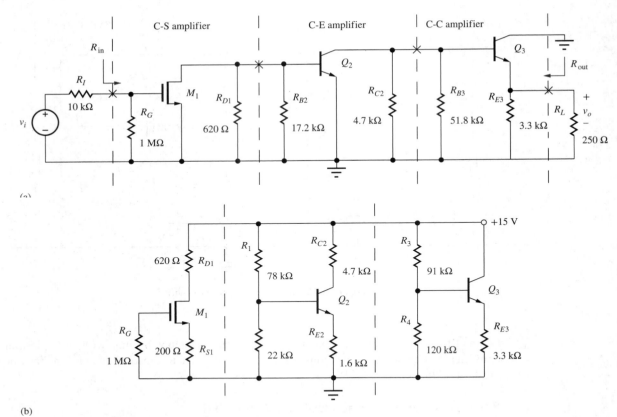

Figure 8.50 (a) Equivalent circuit for ac analysis. (b) dc equivalent circuit for the three-stage ac-coupled amplifier.

TABLE 8.18
Transistor Parameters for Figs. 8.49–8.54

M_1	$K_n = 10$ mA/V^2, $V_{TN} = -2$ V, $\lambda = 0.02$ V^{-1}
Q_2	$\beta_F = 150$, $V_A = 80$ V, $V_{BE} = 0.7$ V
Q_3	$\beta_F = 80$, $V_A = 60$ V, $V_{BE} = 0.7$ V

TABLE 8.19
Q-Points and Small-Signal Parameters for the Transistors in Fig. 8.50

	Q-POINT VALUES	SMALL-SIGNAL PARAMETERS
M_1	(5.00 mA, 10.9 V)	$g_{m1} = 10.0$ mS, $r_{o1} = 12.2$ kΩ
Q_2	(1.57 mA, 5.09 V)	$g_{m2} = 62.8$ mS, $r_{\pi 2} = 2.39$ kΩ, $r_{o2} = 54.2$ kΩ
Q_3	(1.99 mA, 8.36 V)	$g_{m3} = 79.6$ mS, $r_{\pi 3} = 1.00$ kΩ, $r_{o3} = 34.4$ kΩ

EXERCISE: Verify the values of the Q-points and small-signal parameters in Table 8.19.

EXERCISE: Why can't a single transistor amplifier meet the op amp specifications mentioned in the introduction to Section 8.9?

ANSWER: A single transistor cannot simultaneously meet the input resistance, output resistance and voltage gain of this amplifier.

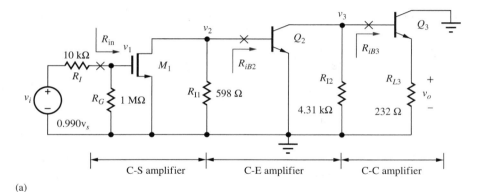

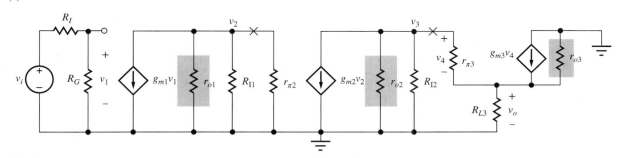

Figure 8.51 (a) Simplified ac equivalent circuit for the three-stage amplifier; (b) small-signal equivalent circuit for the three-stage amplifier. Resistances r_{o1}, r_{o2}, and r_{o3} are neglected in the calculations.

8.9.2 VOLTAGE GAIN

The ac equivalent circuit for the three-stage amplifier example has been redrawn and is shown in simplified form in Fig. 8.51, in which the three sets of parallel resistors have been combined into the following: $R_{I1} = 620\ \Omega\ \|\ 17.2\ \text{k}\Omega = 598\ \Omega$, $R_{I2} = 4.7\ \text{k}\Omega\ \|\ 51.8\ \text{k}\Omega = 4.31\ \text{k}\Omega$, and $R_{L3} = 3.3\ \text{k}\Omega\ \|\ 250\ \Omega = 232\ \Omega$. The voltage gain of the overall amplifier can be expressed as

$$A_v = \frac{\mathbf{v}_o}{\mathbf{v}_i} = \left(\frac{\mathbf{v}_o}{\mathbf{v}_3}\right)\left(\frac{\mathbf{v}_3}{\mathbf{v}_2}\right)\left(\frac{\mathbf{v}_2}{\mathbf{v}_1}\right)\left(\frac{\mathbf{v}_1}{\mathbf{v}_i}\right) = A_{vt1} A_{vt2} A_{vt3} \left(\frac{\mathbf{v}_1}{\mathbf{v}_i}\right) \tag{8.150}$$

where

$$\frac{\mathbf{v}_1}{\mathbf{v}_i} = \frac{R_{\text{in}}}{R_I + R_{\text{in}}} = \frac{R_G}{R_I + R_G} \tag{8.151}$$

Now it should be more clear why we developed expressions for the terminal gains earlier in this chapter. We see that the overall voltage gain is determined by the product of the individual terminal gains of the three amplifier stages, as well as the signal voltage loss across the source resistance.

We use our knowledge of single-transistor amplifiers, gained in Chapters 7 and 8, to determine expressions for the three voltage gains. The first stage is a common-source amplifier with a terminal gain

$$A_{vt1} = \frac{\mathbf{v}_2}{\mathbf{v}_1} = -g_{m1} R_{L1} \tag{8.152}$$

in which R_{L1} represents the total load resistance[14] connected to the drain of M_1. From the ac circuit in Fig. 8.51(a) and the small-signal version in (b), we can see that R_{L1} is equal to the

---

[14] The output resistances r_{o1}, r_{o2}, and r_{o3} are neglected because each amplifier has an external resistor as a load, and we expect $|A_v| \ll \mu_f$ for each stage.

parallel combination of R_{I1} and R_{iB2}, the input resistance at the base of Q_2. Because Q_2 is a common-emitter stage with zero emitter resistance, $R_{iB2} = r_{\pi 2}$,

$$R_{L1} = 598 \ \Omega \ \| \ r_{\pi 2} = 598 \ \Omega \ \| \ 2390 \ \Omega = 478 \ \Omega \tag{8.153}$$

and the gain of the first stage is

$$A_{vt1} = \frac{v_2}{v_1} = -0.01 \ \text{S} \times 478 \ \Omega = -4.78 \tag{8.154}$$

The terminal gain of the second stage is that of a common-emitter amplifier:

$$A_{vt2} = \frac{v_3}{v_2} = -g_{m2} R_{L2} \tag{8.155}$$

in which R_{L2} represents the total load resistance connected to the collector of Q_2. In Fig. 8.51, R_{L2} is equal to the parallel combination of R_{I2} and R_{iB3}, where R_{iB3} represents the input resistance of Q_3. Q_3 is an emitter follower with $R_{iB3} = r_{\pi 3}(1 + g_{m3} R_{L3})$. Thus, R_{L2} is equal to

$$R_{L2} = R_{I2} \ \| \ [r_{\pi 3} + (\beta_{o3} + 1) R_{L3}] = 4310 \ \Omega \ \| \ 1000 \ \Omega [1 + 79.6 \ \text{mS}(232 \ \Omega)] = 3.53 \ \text{k}\Omega \tag{8.156}$$

and the gain of the second stage is

$$A_{vt2} = -62.8 \ \text{mS} \times 3.53 \ \text{k}\Omega = -222 \tag{8.157}$$

Finally, the terminal gain of the emitter follower stage is

$$A_{vt3} = \frac{v_o}{v_3} = \frac{g_{m3} R_{L3}}{1 + g_{m3} R_{L3}} = \frac{(79.6 \ \text{mS})(232 \ \Omega)}{1 + 79.6 \ \text{mS}(232) \ \Omega} = 0.950 \tag{8.158}$$

Before we can complete the voltage gain calculation in Eq. (8.150), we must find input resistance R_{in} in order to evaluate the ratio v_1/v_i given in Eq. (8.151).

8.9.3 INPUT RESISTANCE

The input resistance R_{in} of this amplifier can be determined by referring to Figs. 8.50 through 8.52. Because the current i_g in Fig. 8.52 is zero, we see that the resistance presented to source v_i is simply $R_{in} = R_G = 1 \ \text{M}\Omega$. Note that this result is independent of the circuitry connected to the source or drain of M_1.

8.9.4 SIGNAL SOURCE VOLTAGE GAIN

Substituting the voltage gains and resistance values into Eqs. (8.150) and (8.151) gives the voltage gain for the overall amplifier:

$$A_v = A_{vt1} A_{vt2} A_{vt3} \frac{R_{in}^{CS}}{R_I + R_{in}^{CS}} = (-4.78)(-222)(0.95)\left(\frac{1 \ \text{M}\Omega}{10 \ \text{k}\Omega + 1 \ \text{M}\Omega}\right) = +998 \tag{8.159}$$

We find that the three-stage amplifier circuit realizes a noninverting amplifier with a voltage gain of approximately 60 dB and an input resistance of 1 MΩ. Because of the high input resistance, only a small portion (1 percent) of the input signal is lost across the source resistance.

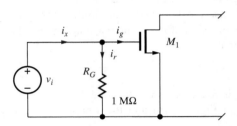

Figure 8.52 Input resistance of the three-stage amplifier.

EXERCISE: Recalculate A_v, including the influence of r_{o1}, r_{o2}, and r_{o3}.

ANSWER: 903 (59.1 dB)

EXERCISE: Estimate the gain of the amplifier in Fig. 8.49 using our simple design estimates if M_1 has $V_{GS} - V_{TN} = 1$ V. What is the origin of the discrepancy?

ANSWERS: $(-15)(-150)(1) = 2250$; only 3 V is dropped across R_{D1}, whereas the estimate assumes $V_{DD}/2 = 7.5$ V. Taking this difference into account, $(2250)(3/7.5) = 900$.

EXERCISE: What is the value of A_v if the interstage resistances R_{I1} and R_{I2} could be eliminated (made ∞)? Would r_{o1}, r_{o2}, and r_{o3} be required in this case?

ANSWERS: 28,200; r_{o2} would need to be included.

8.9.5 OUTPUT RESISTANCE

The output resistance R_{out} of the amplifier is defined looking back into the amplifier at the position of coupling capacitor C_6, as indicated in Figs. 8.49 and 8.50. To find R_{out}, test voltage v_x is applied to the amplifier output as in Fig. 8.53, and we see that the output resistance of the overall amplifier is determined by the output resistance of the emitter follower in parallel with the 3300-Ω resistor. Writing this mathematically gives,

$$i_x = i_r + i_e = \frac{v_x}{3300} + \frac{v_x}{R_{iE3}} \tag{8.160}$$

Using the results from Table 8.4, we find that the overall output resistance is

$$R_{out} = \frac{v_x}{i_x} = 3300 \| R_{iE3} \cong 3300 \left\| \left(\frac{1}{g_{m3}} + \frac{R_{th3}}{\beta_{o3}} \right) \right. \tag{8.161}$$

in which the Thévenin equivalent source resistance of stage 3, R_{th3}, must be found.

R_{th3} can be determined with the aid of Fig. 8.54. The third stage Q_3 is removed, and test voltage v_x is applied to node v_3. Current i_x from the test source v_x is equal to

$$i_x = \frac{v_x}{R_{I2}} + i_2 = \frac{v_x}{R_{I2}} + \frac{v_x}{R_{iC}} \quad \text{or} \quad R_{th3} = \frac{v_x}{i_x} = R_{I2} \| R_{iC} = R_{I2} \| r_{o2} \tag{8.162}$$

R_{th3} is equal to the parallel combination of interstage resistance R_{I2} and the resistance at the collector of Q_2, which we know is just equal to r_{o2}:

$$R_{th3} = 4310 \, \Omega \| 54200 \, \Omega = 3990 \, \Omega$$

Evaluating Eq. (8.161) for the output resistance of the overall amplifier yields

$$R_{out} = 3300 \, \Omega \left\| \left(\frac{1}{0.0796 \, \text{S}} + \frac{3990 \, \Omega}{80} \right) = 62.4 \, \Omega \right. \tag{8.163}$$

Note that R_{th3}/β_o is the most important term in the output resistance calculation.

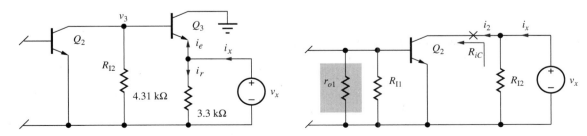

Figure 8.53 Output resistance of the three-stage amplifier.

Figure 8.54 Thévenin equivalent source resistance for stage 3.

8.9.6 CURRENT AND POWER GAIN

The input current delivered to the amplifier from source v_i in Fig. 8.50 is given by

$$\mathbf{i_i} = \frac{\mathbf{v_i}}{R_I + R_{in}} = \frac{\mathbf{v_i}}{10^4 + 10^6} = 9.90 \times 10^{-7} \mathbf{v_i} \quad (8.164)$$

and the current delivered to the load from the amplifier is

$$\mathbf{i_o} = \frac{\mathbf{v_o}}{250} = \frac{A_v \mathbf{v_i}}{250} = \frac{998 \, \mathbf{v_i}}{250} = 3.99 \, \mathbf{v_i} \quad (8.165)$$

Combining Eqs. (8.164) and (8.165) gives the current gain

$$A_i = \frac{\mathbf{i_o}}{\mathbf{i_i}} = \frac{3.99 \, v_i}{9.90 \times 10^{-7} v_i} = 4.03 \times 10^6 \quad (132 \text{ dB}) \quad (8.166)$$

Combining Eqs. (8.159) and (8.166) with the **power gain** expression from Chapter 6 yields a value for overall power gain of the amplifier:

$$A_P = \frac{P_o}{P_s} = \left|\frac{\mathbf{v_o} \, \mathbf{i_o}}{\mathbf{v_i} \, \mathbf{i_i}}\right| = |A_v A_i| = 998 \times 4.03 \times 10^6 = 4.02 \times 10^9 \quad (96.0 \text{ dB}) \quad (8.167)$$

Because input resistance to the common-source stage is large, only a small input current is required to develop a large output current. Thus, current gain is large. In addition, the voltage gain of the amplifier is significant, and combining a large voltage gain with a large current gain yields a very substantial power gain.

8.9.7 INPUT SIGNAL RANGE

Our final step in characterizing this amplifier is to determine the largest input signal that can be applied to the amplifier. In a multistage amplifier, the small-signal assumptions must not be violated anywhere in the amplifier chain. The first stage of the amplifier in Figs. 8.50 and 8.51 is easy to check. Voltage source v_1 appears directly across the gate-source terminals of the MOSFET, and to satisfy the small-signal limit, $v_1 (= 0.990 v_i)$ must satisfy

$$|\mathbf{v_i}| \leq 0.2(V_{GS1} - V_{TN}) \quad \text{or} \quad |v_i| \leq \frac{0.2(-1+2)}{0.990} = 0.202 \text{ V} \quad (8.168)$$

The first stage limits the input signal to 202 mV.

To satisfy the small-signal requirements, the base-emitter voltage of Q_2 must also be less than 5 mV. In this amplifier, $v_{be2} = v_2$, and we have

$$|\mathbf{v_2}| = |A_{vt1} v_1| \leq 5 \text{ mV}, \quad |\mathbf{v_1}| \leq \frac{5 \text{ mV}}{A_{vt1}} = \frac{0.005}{4.78} = 1.05 \text{ mV}$$

and
$$|\mathbf{v_i}| \leq \frac{1.05 \text{ mV}}{0.990} = 1.06 \text{ mV} \quad (8.169)$$

In this design, the small-signal requirements are violated at Q_2 if the amplitude of the input signal v_i exceeds 1.06 mV.

Finally, using Eq. (8.64) for the emitter-follower output stage (with $R_{th} = 0$),

$$\mathbf{v_{be3}} \cong \frac{\mathbf{v_3}}{1 + g_{m3} R_{L3}} = \frac{A_{vt1} A_{vt2} v_1}{1 + g_{m3} R_{L3}} = \frac{A_{vt1} A_{vt2}(0.990 v_i)}{1 + g_{m3} R_{L3}} \quad (8.170)$$

and requiring $|v_{be3}| \leq 5$ mV yields

$$|\mathbf{v_i}| \leq \frac{1 + g_{m3} R_{L3}}{A_{vt1} A_{vt2}(0.990)} 0.005 = \frac{1 + 0.0796 \text{ S}(232 \, \Omega)}{(-4.78)(-222)(0.99)} 0.005 \text{ V} = 92.7 \, \mu\text{V} \quad (8.171)$$

To satisfy all the small-signal limitations, the maximum amplitude of the input signal to the amplifier must be no greater than the smallest of the three values computed in Eqs. (8.168), (8.169), and (8.171):

$$|\mathbf{v_i}| \leq \min(202 \text{ mV}, 1.06 \text{ mV}, 92.7 \, \mu\text{V}) = 92.7 \, \mu\text{V} \quad (8.172)$$

In this design, output stage linearity limits the input signal amplitude to less than 93 μV. Note that the maximum output voltage that satisfies the small-signal limit is only

$$|\mathbf{v_o}| \leq A_v(92.7 \text{ μV}) = 998(92.7 \text{ μV}) = 92.5 \text{ mV} \qquad (8.173)$$

EXAMPLE 8.13 **THREE-STAGE AMPLIFIER SIMULATION**

Hand analysis of the three-stage amplifier is verified using SPICE simulation.

PROBLEM Use SPICE to find the midband voltage gain, input resistance, and output resistance of the amplifier in Fig. 8.49. Confirm the gain with both ac and transient analyses.

SOLUTION **Known Information and Given Data:** The original amplifier circuit appears in Fig. 8.49, and transistor parameters are given in Table 8.18.

Unknowns: A_v, R_{in}, and R_{out}

Approach: Use SPICE analysis to plot the frequency response and find the midband region. Then choose a midband frequency, and use ac analysis to find the voltage gain, input resistance, and output resistance. Assume large values for the capacitors. Verify the gain with a transient simulation.

Assumptions: The coupling and bypass capacitors are all arbitrarily set to 22 μF. Bipolar transistor parameters TF = 0.5 NS and CJC = 2 PF are added to the BJT models to cause the frequency response to roll off at high frequencies. These parameters are discussed in detail in Chapter 9.

Analysis: The circuit is created in SPICE using the schematic editor, as shown in the figure.

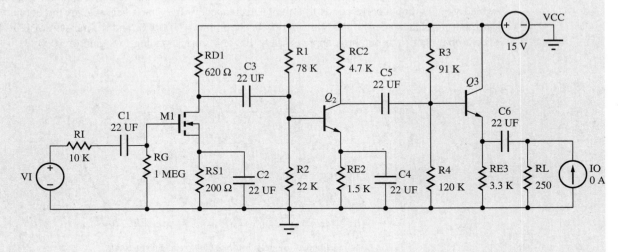

The MOSFET parameters are set to KP = 0.01 S/V, VTO = −2 V, and LAMBDA = 0.02 V^{-1}. The BJT parameters for Q_2 are set to BF = 150, VAF = 80 V, TF = 0.5 NS, and CJC = 2 PF. For Q_3, BF = 80, VAF = 60 V, TF = 0.5 NS, and CJC = 2 PF. As mentioned, TF and CJC are added to create a roll-off in the frequency response at high frequencies and will be discussed further in Chapter 9. Source VI is used for ac analysis of the voltage gain and input resistance. Source IO is an ac source used to find the output resistance.

First, we set VI = 1∠0° V and IO = 0∠0° A and perform an ac sweep from 10 Hz to 10 MHz with 20 points per decade in order to find the midband region. We obtain the response shown below.

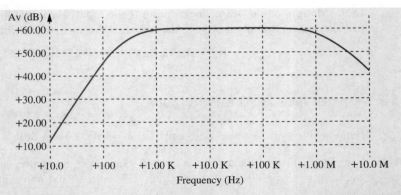

The midband region extends from approximately 500 Hz to 500 kHz. Choosing 20 kHz as a representative midband frequency, we find the gain is 60.1 dB ($A_v = 1010$), and the current in VI is -990 nA with a phase angle of 0°. The minus sign results from the sign convention in SPICE—positive current enters the positive terminal of an independent source. The input resistance presented to VI is 1 V/990 nA = 1.01 MΩ. Subtracting the 10-kΩ source resistance yields an amplifier input resistance of 1 MΩ. Both the gain and input resistance agree with our hand calculations.

The output resistance is found by setting VI = $0\angle 0°$ V and IO = $1\angle 0°$ A and finding the output voltage. The result yields $R = 45.6$ Ω. Removing the effect of the 250-Ω resistor in parallel with the output yields $R_{out} = 55.7$ Ω. The slight difference is caused by the value of current gain utilized in SPICE: $\beta_o = \text{BF}(1 + \text{VCB}/\text{VA}) = 80(1 + 7.6\,\text{V}/60\,\text{V}) = 90.1$.

Check of Results: As a second check on the gain, we can run a transient simulation at $f = 20$ kHz, which we now know corresponds to a midband frequency. The graph below gives the output with an input amplitude of 100 μV, Start time = 0, Stop time = 100 US, and a time step of 0.01 US. The amplitude of the output is approximately 100 mV corresponding to a gain of 1000.

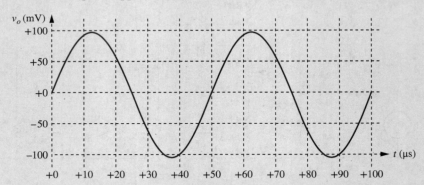

Simulation with slightly distorted output and gain of 1000.

Discussion: This input value in the transient simulation is just slightly above the small-signal limit that we calculated. The waveform looks like a slightly distorted[15] sine wave, and the Fourier analysis option in SPICE indicates that the total harmonic distortion in the waveform is less than 0.15 percent.

However, if one uses an input signal larger than about 650 μV in the transient solution, one discovers a new limitation. An example of the problem appears in the next figure. Because Q_3 is biased at a current of only 2 mA, the largest output signal that can be developed by the

---

[15] Note that the positive change is less than 100 mV and the negative change is greater than 100 mV.

emitter follower is approximately 2 mA × 250 Ω or 0.5 V. The output will begin to show substantial distortion before this value is reached. In the figure, the amplitude of input v_I is 750 μV. The bottom of the output waveform is "clipped off," and the total harmonic distortion has increased to 8.2 percent. This output waveform is not desirable.

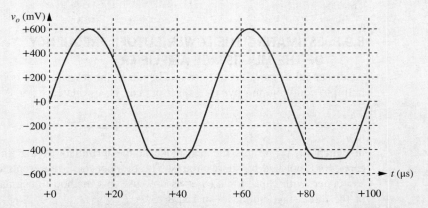

Highly distorted output with amplitude exceeding output voltage capability of amplifier.

EXERCISE: Reevaluate Eq. (8.163) using a current gain of 90.1.

ANSWER: 55.3 Ω

EXERCISE: Find the output waveform, voltage gain, and total harmonic distortion if the amplitude of VI is increased to (a) 400 μV, (b) 600 μV, and (c) 1 mV.

ANSWERS: (a) Looks like a sine wave, $A_v = 826$, THD = 0.28 percent; (b) looks like a sine wave, $A_v = 790$, THD = 2.4 percent; (c) bottom of the waveform is clipped off, $A_v = 760$, THD = 18.3 percent. Note that the overall voltage gain is dropping as the signal level increases.

Table 8.20 summarizes the characteristics for the three-stage amplifier in Fig. 8.49. The amplifier provides a noninverting voltage gain of approximately 60 dB, a high input resistance, and a low output resistance. The current and power gains are both quite large. The input signal must be kept below 92.7 μV in order to satisfy the small-signal limitations of the transistors.

TABLE 8.20
Three-Stage Amplifier Summary

	HAND ANALYSIS	SPICE RESULTS
Voltage gain	+998	+1010
Input signal range	92.7 μV	—
Input resistance	1 MΩ	1 MΩ
Output resistance	60.5 Ω	55.7 Ω
Current gain	$+4.03 \times 10^6$	—
Power gain	4.02×10^9	—

EXERCISE: (a) What would be the voltage gain of the amplifier if I_{D1} is reduced to 1 mA and R_{D1} is increased to 3 kΩ so that V_D is maintained constant? (b) The FET g_m decreases by $\sqrt{5}$. Why did the gain not increase by a factor of $\sqrt{5}$?

ANSWERS: 1150; although R_{D1} increases by a factor of 5, the total load resistance at the drain of M_1 does not.

8.9.8 ESTIMATING THE LOWER CUTOFF FREQUENCY OF THE MULTISTAGE AMPLIFIER

As discussed in more detail later in Chapter 9, the lower cutoff frequency for an amplifier having multiple coupling and bypass capacitors can be estimated from

$$\omega_L \cong \sum_{i=1}^{n} \frac{1}{R_{iS} C_i} \tag{8.174}$$

in which R_{iS} represents the resistance at the terminals of the ith capacitor with all the other capacitors replaced by short circuits. The product $R_{iS} C_i$ represents the short-circuit time constant associated with capacitor C_i. Let us now use this method to estimate the lower cutoff frequency of the three-stage amplifier in Ex. 8.13.

C_1: $R_{1S} = R_I + R_G = 1.01 \text{ M}\Omega$

C_2: $R_{2S} = R_{S1} \| R_{iS1} = R_{S1} \| \dfrac{1}{g_{m1}} = 200\,\Omega \| \dfrac{1}{0.01S} = 66.7\,\Omega$

C_3: $R_{3S} = R_{D1} + R_{I1} \| R_{iB2} = R_{D1} + R_{I1} \| r_{\pi 2} = 620\,\Omega + 17.2\,\text{k}\Omega \| 2.39\,\text{k}\Omega = 2.72\,\text{k}\Omega$

C_4: $R_{4S} = R_{E2} \| R_{iE2} = R_{E2} \| \dfrac{r_{\pi 2} + R_{th2}}{\beta_{o2} + 1} = 1.5\,\text{k}\Omega \| \dfrac{2.39\,\text{k}\Omega + (17.2\,\text{k}\Omega \| 620\,\Omega)}{151} = 19.2\,\Omega$

C_5: $R_{3S} = R_{C2} + R_{I2} \| R_{iB3} = R_{C2} + R_{I2} \| r_{\pi 3}(1 + g_{m3} R_{L3})$

$\qquad = 4.7\,\text{k}\Omega + 51.8\,\text{k}\Omega \| 1.0\,\text{k}\Omega [1 + 0.0796 S(232\,\Omega)] = 18.9\,\text{k}\Omega$

C_6: $R_{4S} = R_L + R_{E3} \| R_{iE3} = R_L + R_{E3} \| \dfrac{r_{\pi 3} + R_{th3}}{\beta_{o3} + 1}$

$\qquad = 250\,\Omega + 3.3\,\text{k}\Omega \| \dfrac{1.0\,\text{k}\Omega + (51.8\,\text{k}\Omega \| 4.7\,\text{k}\Omega)}{81} = 315\,\Omega$

$$f_L \cong \frac{1}{2\pi}\left[\frac{1}{1.01\,\text{M}\Omega(22\,\mu\text{F})} + \frac{1}{66.7\,\Omega(22\,\mu\text{F})} + \frac{1}{2.72\,\text{k}\Omega(22\,\mu\text{F})} + \frac{1}{19.2\,\Omega(22\,\mu\text{F})} \right.$$
$$\left. + \frac{1}{18.9\,\text{k}\Omega(22\,\mu\text{F})} + \frac{1}{315\,\Omega(22\,\mu\text{F})}\right]$$

$f_L \cong 511 \text{ Hz}$

The $f_L = 511$ Hz estimate obtained using the short-circuit time-constant approach agrees very well with the SPICE simulation results presented in Ex. 8.13.

8.10 INTRODUCTION TO dc-COUPLED AMPLIFIERS

The use of dc-coupled (or direct coupled) design can reduce the cost and complexity of circuit designs by eliminating unneeded components. For example, coupling capacitors between transistors can be eliminated by connecting the transistors directly together. Also, amplifiers are often needed that can amplify dc signals[16], and coupling capacitors must be eliminated from the design.

---

[16] In operational amplifiers for example.

ELECTRONICS IN ACTION

Humbucker Guitar Pickup

Electric guitar pickups are devices that convert motion of a steel string into electrical signals. They function through an interesting interaction of materials and magnetic fields. A basic schematic of a pickup is shown below. The magnet induces a magnetization (aligning of the magnetic domains) in the steel string. When the string vibrates, this creates a moving magnetic field, which we know from Faraday's law induces a current in the wire coil located beneath the string. The signal from the wire coil is then amplified and sent to the rest of the amplification system. The coil is typically composed of extremely thin wire, with several hundred to over a thousand turns. The choice of magnet, wire material, and number of turns in the coil generates a set of compromises in frequency response and sensitivity. Guitar players often use acoustic feedback to generate a sustained note by placing the guitar near the amplifier speakers. The acoustic energy couples into the guitar, causing the string to vibrate, which generates more signal through the amplifier. Unlike the highly undesirable feedback that we often hear with poorly configured public address systems, a skilled guitarist can use acoustic feedback to create intentionally sustained notes.

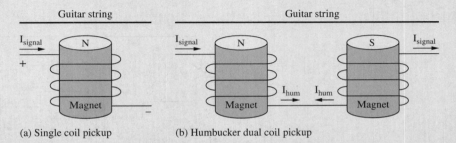

(a) Single coil pickup (b) Humbucker dual coil pickup

Inherent with the use of the single coil pickup is sensitivity to extraneous magnetic fields. In particular, the 60 Hz power moving through most buildings gives rise to magnetic fields at the same frequency. As a result, the guitar pickup coil will generate the desired string vibration signal as well as a 60 Hz signal commonly referred to as hum.

To eliminate hum, one has to make two important observations: First, the polarity of the string vibration signal is a function of both the magnetization polarity of the string and the orientation of the coil relative to the string. Second, the polarity of the undesired hum signal is a function of only the orientation of the coil to the hum-producing external magnetic field.

Making use of these two observations, the humbucker pickup shown above was created. A second pickup coil has been added in series with the first. In the second coil, the orientation of the magnet has been reversed, resulting in the reversing of the string magnetization in the area above the coil. Additionally, the orientation of the second coil with respect to the string has also been reversed. The result is a system where the string vibration signal of the two coils has the same polarity and is additive, while the hum signal, dependent only on the coil orientation, is of opposite sign in the two coils and is cancelled out.

The humbucker coil is an example of excellent sensor design. Recognizing the unique characteristics of the desired versus the undesired signals allowed the designers to implement a sensor that rejects everything but the signal of interest. Rejecting unwanted signals at the sensor is almost always preferred to attempting to reject undesired signals in post-processing after detection and amplification.

For more material refer to the excellent guitar building sites that can be found on the web.

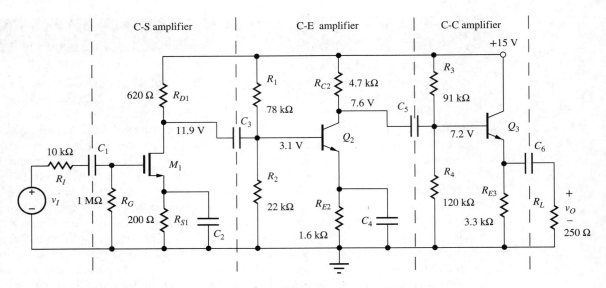

Figure 8.55 Original design from Fig. 8.49.

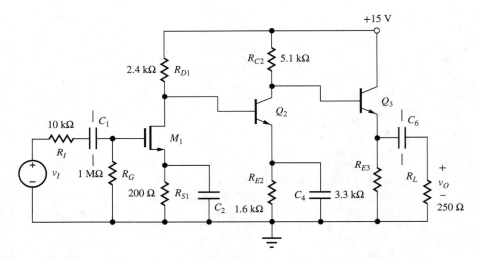

Figure 8.56 Modified design using dc coupling of the three interior amplifier stages.

8.10.1 A dc-COUPLED THREE-STAGE AMPLIFIER

The amplifier in Fig. 8.55 is the previous three-transistor design discussed in Section 8.9, and Fig. 8.56 is a modification of the amplifier in which direct coupling of the three transistors has been utilized to eliminate coupling capacitors C_3 and C_5 as well as the four base-bias resistors R_1 through R_4. The base of Q_2 is then connected directly to the drain of M_1, and the base of Q_3 is connected directly to the collector of Q_2. Since the voltage at the gate of M_1 is zero, input capacitor C_1 may also be eliminated if desired.

The voltages at the drain of M_1 (V_{D1}), the base (V_{B2}) and collector (V_{C2}) of Q_2 and the base of Q_3 (V_{B3}) were calculated using the transistor parameters and Q-points in Tables 8.18 and 8.19 and have been added to the schematic in Fig. 8.55.

There is significant mismatch between drain voltage V_{D1} and base voltage V_{B2}, so they cannot be connected directly together without greatly disturbing the Q-point of Q_2. Similarly, there is also a much lesser mismatch between voltages V_{C2} and V_{B3}. To solve this problem, the values of R_{D1} and R_{C2} have been changed to those shown in the **direct-coupled amplifier** in Fig. 8.56. A voltage reduction of 8.8 V is needed at the drain of M_1, and a shift of 0.4 V is required at the collector of Q_2. Based upon the Q-point currents in Table 8.19 and the 5 percent resistor values in Appendix A, the new values of R_{D1} and R_{C2} are 2.4 kΩ and 5.1 kΩ, respectively.

Note that a dc amplifier could be created by elimination of all the capacitors. However, removal of bypass capacitors C_2 and C_4 would significantly reduce the amplifier gain.

EXERCISE: (a) Verify the values of V_{D1}, V_{B2}, V_{C2}, and V_{B3} using the information in Tables 8.18 and 8.19. (b) Verify the new values of R_{D1} and R_{C2} in Fig. 8.56.

8.10.2 TWO TRANSISTOR dc-COUPLED AMPLIFIERS

There are several two-transistor dc-coupled amplifiers that are often treated as single-stage amplifiers. These amplifiers include bipolar and FET differential amplifiers; common-collector common-base (CC-CB) and common-drain common-gate (CD-CG) amplifiers; and common-emitter common-base (CE-CB) and common-source common-gate (CS-CG) amplifiers, usually referred to as cascode amplifiers. In addition, the current mirror is a dc-coupled circuit that is widely used in integrated circuit design. These configurations are introduced here and discussed in greater detail in later chapters.

Differential Amplifiers

Differential amplifiers, also known as **difference amplifiers,** consist of two transistors in the symmetrical circuits in Fig. 8.57 and are widely used in integrated circuit design. The differential amplifier amplifies the difference of the two input voltages, v_1 and v_2, and tends to reject any signal in common to the two inputs. The differential output v_{od} is the difference in the voltages between the two collectors or drains, and voltage gains for the two amplifiers can be expressed as

$$v_{od} = v_{C1} - v_{C2} = -g_m R_C(v_1 - v_2)$$
and (8.175)
$$v_{od} = v_{D1} - v_{D2} = -g_m R_D(v_1 - v_2)$$

By symmetry, the bias current in each transistor is equal to one-half of the current in the source at the bottom of the figure, so the pairs of transistor parameters have equal values (i.e., $g_{m2} = g_{m1} = g_m$, etc.). Differential amplifiers are discussed in great detail in Chapter 13.

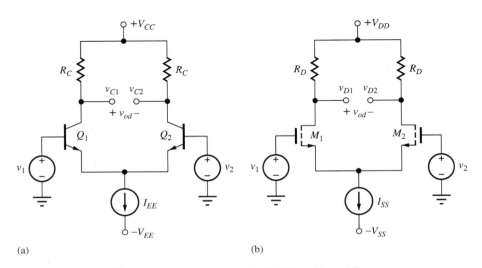

Fig. 8.57 (a) Bipolar and (b) NMOS differential amplifiers.

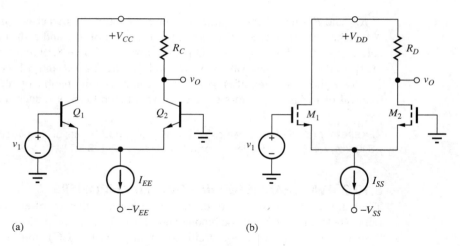

Fig. 8.58 (a) CC-CB and (b) CD-CG amplifiers.

Common-Collector, Common-Base (CD-CG) Amplifiers

If we remove the collector resistor from Q_1, connect the collector of Q_1 directly to V_{CC}, and connect v_2 to zero as in Fig. 8.58(a), we have a common-collector amplifier connected directly to a common-base amplifier as in Fig. 8.58(a). This amplifier is an asymmetrical form of the differential amplifier with one input and one output. The input signal is applied to the base of Q_1, and the output signal appears at the collector of Q_2. Here again the output of the ideal current source splits equally in the two transistors so that $g_{m1} = g_{m2} = g_m$. One half of the input signal appears at the top of the current source, and the resulting noninverting voltage gain is equal to

$$v_o \cong +\frac{g_m}{2} R_C v_1 \qquad (8.176)$$

The corresponding dc coupled common-drain, common-gate amplifier appears in Fig. 8.58(b).

Cascode Amplifiers

Figure 8.59(a) depicts a common-emitter transistor directly coupled to a common-base transistor with the input signal connected to the base of the Q_1 (M_1) and the output taken from the collector of the Q_2 (M_2). This circuit is commonly known as the **cascode amplifier.** The signal current from the collector of Q_1 goes directly through Q_2 to the load, so its gain is similar to the common-emitter stage by itself:

$$v_o \cong -g_m R_C v_1 \qquad (8.177)$$

However, the low resistance presented at the emitter of Q_2 ($1/g_{m2}$) substantially improves the bandwidth of the amplifier. The frequency response of the cascode amplifier is discussed in Chapter 9. The output resistance of the cascode is also higher than that of a single transistor common-emitter stage. The corresponding dc-coupled NMOS cascode gate amplifier appears in Fig. 8.59(b).

Current Mirrors

The circuits of NMOS and BJT current mirrors appear in Fig. 8.60. In the NMOS circuit, the drain current I_{D1} of M_1 must be equal to current source I_{REF} since the gate currents are zero, and gate-source voltage V_{GS} adjusts to the value required for $I_{D1} = I_{REF}$. V_{GS} also drives M_2, and I_O

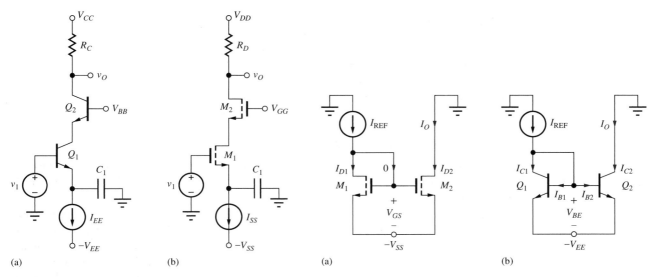

Fig. 8.59 (a) Bipolar and (b) NMOS cascode amplifiers.

Figure 8.60 (a) MOSFET current mirror and (b) BJT current mirror.

will equal I_{REF} if the transistors are identical (i.e., I_O "mirrors" I_{REF}). The currents can be scaled by changing the W/L ratios of the two transistors and the circuit can provide current gain:

$$I_O = I_{REF} \frac{(W/L)_2}{(W/L)_1} \tag{8.178}$$

The bipolar current mirror in Fig. 8.60(b) works in a similar manner. Assuming the current gain is very large, then base-emitter voltage V_{BE} adjusts to set $I_{C1} = I_{REF}$, and I_{C2} mirrors I_{C1}. If the transistor emitter areas are different, then

$$I_O = I_{REF} \frac{A_{E2}}{A_{E1}} \tag{8.179}$$

Current mirrors are widely used in integrated circuits design, and detailed analyses of current mirrors are developed at the beginning of Chapter 14.

EXERCISE: (a) Derive the expression for the current gain of the NMOS current mirror. (b) Repeat for the bipolar current mirror if β_F is not neglected.

SUMMARY

This chapter presented an in-depth investigation of the characteristics of amplifiers implemented using single transistors.

- Of the three available device terminals of the BJT, only the base and emitter are useful as signal input terminals, whereas the collector and emitter are acceptable as output terminals. For the FET, the source and gate are useful as signal input terminals, and the drain and source are acceptable as output terminals. Neither collector nor drain are used as input terminals, and the base or gate are not used as output terminals.

- There are three basic classifications of amplifiers: inverting amplifiers—the common-emitter and common-source amplifiers; followers—the common-collector and common-drain amplifiers (also known as emitter followers or source followers); and the noninverting amplifiers—common-base and common-gate amplifiers.

- Detailed analyses of these three amplifier classes were performed using the small-signal models for the transistors. These analyses produced expressions for the voltage gain, current gain, input resistance, output resistance, and input signal range, which are summarized in a group of important tables:

  Table 8.3 C-E/C-S Amplifier Summary
  Table 8.5 C-C/C-D Amplifier Summary
  Table 8.8 C-B/C-G Amplifier Summary
  Table 8.9 Single-Transistor Bipolar Amplifiers
  Table 8.11 Single-Transistor FET Amplifiers

  The results summarized in these tables form the basic toolkit of the analog circuit designer. A thorough understanding of these results is a prerequisite for design and for the analysis of more complex analog circuits.

- Inverting amplifiers (C-E and C-S amplifiers) can provide significant voltage and current gain, as well as high input and output resistance. If a resistor is included in the emitter or source of the transistor, the voltage gain is reduced but can be made relatively independent of the individual transistor characteristics. This reduction in gain is traded for increases in input resistance, output resistance, and input signal range. Because of its higher transconductance, the BJT more easily achieves higher values of voltage gain than the FET, whereas the infinite input resistance of the FET gives it the advantage in achieving high input resistance amplifiers. The FET also typically has a larger input signal range than the BJT.

- In MOS technology, a transistor can be used to replace the drain bias resistor in the common-source amplifiers, resulting in much more compact circuits suitable for IC realization. The resulting two-transistor amplifier circuits have the same circuit topology as the logic inverters studied in Chapters S6–S8. Similar circuits will be encountered in Chapters 13–14.

- Emitter and source followers (C-C and C-D amplifiers) provide a voltage gain of approximately 1, high input resistance, and low output resistance. The followers provide moderate levels of current gain and achieve the highest input signal range. These C-C and C-D amplifiers are the single-transistor equivalents of the voltage-follower operational-amplifier that will be discussed in Chapter 10.

- The noninverting amplifiers (C-B and C-G amplifiers) provide voltage gain, signal range, and output resistances very similar to those of the inverting amplifiers but have relatively low input resistance and a current gain of less than one.

- All the amplifier classes provide at least moderate levels of either voltage gain or current gain (or both) and are therefore capable of providing significant power gain with proper design.

- Table 8.21 presents a relative comparison of these three amplifier classes.

TABLE 8.21
Relative Comparison of Single-Transistor Amplifiers

	INVERTING AMPLIFIERS (C-E AND C-S)	FOLLOWERS (C-C AND C-D)	NONINVERTING AMPLIFIERS (C-B AND C-G)
Voltage gain	Moderate	Low ($\cong 1$)	Moderate
Input resistance	Moderate to high	High	Low
Output resistance	Moderate to high	Low	High
Input signal range	Low to moderate	High	Low to moderate
Current gain	Moderate	Moderate	Low ($\cong 1$)

- Design examples were presented for amplifiers using the inverting, noninverting, and follower configurations, and an example using Monte Carlo analysis to evaluate the effects of element tolerances on circuit performance was also given.
- The values of coupling and bypass capacitors can be chosen by setting the reactance of the capacitors to be much smaller than the Thévenin equivalent resistance that appears at the capacitor terminals. The reactance is calculated at the lowest frequency in the midband region of the amplifier's frequency response. The lower cutoff frequency f_L is determined by the frequency at which the capacitive reactance equals the equivalent resistance at the capacitor terminals. In the amplifiers in this chapter, there are two or three poles that interact to set f_L, and bandwidth shrinkage moves the cutoff frequency above that set by each individual capacitor acting alone.
- In direct-coupled amplifiers, transistor terminals are connected directly together and simplify design by reducing the number of coupling and bypass capacitors and resistors as much as possible.

KEY TERMS

Body effect
Cascode amplifier
Common-base (C-B) amplifier
Common-collector (C-C) amplifier
Common-emitter (C-E) amplifier
Common-drain (C-D) amplifier
Common-gate (C-G) amplifier
Common-source (C-S) amplifier
Current gain
Difference amplifier
Differential amplifier
Direct-coupled amplifier

Emitter follower
Input resistance
Output resistance
Power gain
Signal range
Source follower
Swamping
Terminal current gain
Terminal gain
Terminal voltage gain
Voltage gain

ADDITIONAL READING

P. R. Gray, P. J. Hurst, S. H. Lewis, and R. G. Meyer, *Analysis and Design of Analog Integrated Circuits,* 6th ed., John Wiley and Sons, New York: 2017.

M. N. Horenstein, *Microelectronic Circuits and Devices,* 2nd ed., Prentice-Hall, Englewood Cliffs, NJ: 1995.

C. J. Savant, M. S. Roden, and G. L. Carpenter, *Electronic Design—Circuits and Systems,* 2nd ed., Benjamin/Cummings, Redwood City, CA: 1990.

Adel S. Sedra, Kenneth C. Smith, Tony Chan Carusone, and Vincent Gaudet, *Microelectronic Circuits,* 8th ed., Oxford University Press, New York: 2019.

PROBLEMS

Assume all capacitors and inductors have infinite value unless otherwise indicated.

8.1 Amplifier Classification

8.1. (a) Draw the ac equivalent circuits for, and classify (that is, as C-S, C-G, C-D, C-E, C-B, C-C, and not useful), the amplifiers in Fig. P8.1 (a) to (q). (b) Redraw the circuits in Figs. P8.1(b) and P8.1(j) using *pnp* transistors instead of MOS devices. (c) Redraw the circuits in Figs. P8.1(n) and P8.1(o) using *pnp* and PMOS transistors, respectively.

554 Chapter 8 Transistor Amplifier Building Blocks

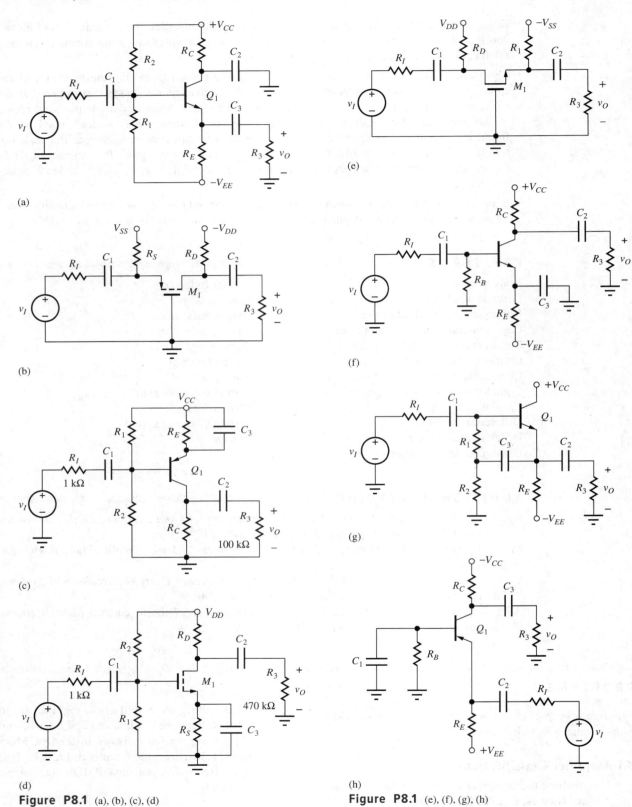

Figure P8.1 (a), (b), (c), (d)

Figure P8.1 (e), (f), (g), (h)

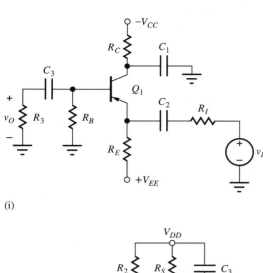

(i)

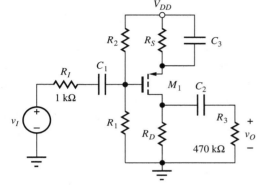

(j)

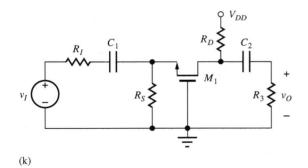

(k)

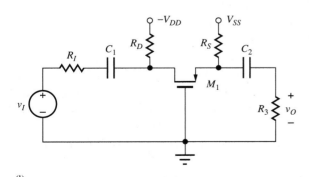

(l)

Figure P8.1 (i), (j), (k), (l)

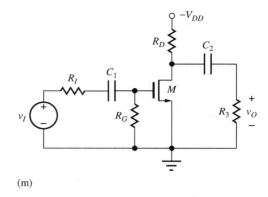

(m)

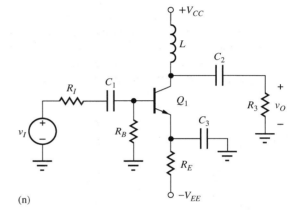

(n)

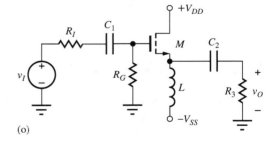

(o)

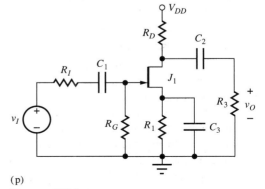

(p)

Figure P8.1 (m), (n), (o), (p)

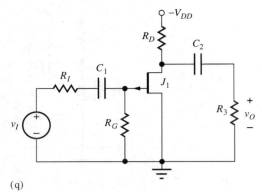

(q)
Figure P8.1 (q)

8.2. An *npn* transistor is biased by the circuit in Fig. P8.2. Using the external source and load configurations in the figure, add coupling and bypass capacitors to the circuit to turn the amplifier into a common-emitter amplifier with maximum gain.

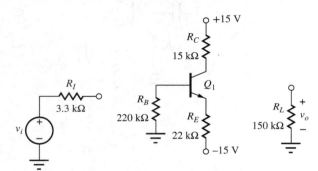

Figure P8.2

8.3. Modify the circuit in Problem 8.2 to give an inverting voltage gain of approximately 15 dB. (*Hint:* Consider splitting a resistor into two parts.)

8.4. (a) Repeat Prob. 8.2 to turn the amplifier into a common-base amplifier. (b) Eliminate R_B and any other unneeded components and draw the modified circuit.

8.5. (a) Repeat Prob. 8.2 to turn the amplifier into a common-collector amplifier. (b) Redesign the circuit by deleting any unneeded component(s). Draw the new circuit.

8.6. A *pnp* transistor is biased by the circuit in Fig. P8.6. Using the external source and load configurations in the figure, add coupling and bypass capacitors to the circuit to construct a common-collector amplifier.

8.7. Repeat Prob. 8.6 to construct a common-emitter amplifier with maximum gain.

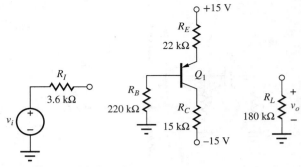

Figure P8.6

8.8. Modify the circuit in Problem 8.7 to give a voltage gain of approximately −10. (*Hint:* Consider splitting a resistor into two parts.)

8.9. Repeat Prob. 8.7 to construct a common-base amplifier.

8.10. An NMOS transistor is biased by the circuit in Fig. P8.10. Using the external source and load configurations in the figure, add coupling and bypass capacitors to the circuit to construct a common-source amplifier.

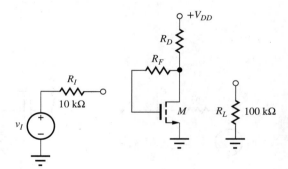

Figure P8.10

8.11. Repeat Prob. 8.10 to construct a common-gate amplifier.

8.12. Repeat Prob. 8.10 to construct a common-drain amplifier.

8.13. A PMOS transistor is biased by the circuit in Fig. P8.13. Using the external source and load configurations in the figure, add coupling and bypass capacitors to the circuit to turn the amplifier into a common-gate amplifier.

8.14. Repeat Prob. 8.13 to turn the amplifier into a common-drain amplifier with maximum gain.

8.15. Repeat Prob. 8.13 to turn the amplifier into a common-source amplifier.

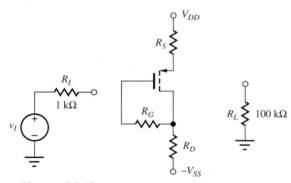

Figure P8.13

8.2 Inverting Amplifiers—Common-Emitter and Common-Source Circuits

8.16. (a) What are the values of A_v, R_{in}, R_{out}, and $A_i = i_o/i_i$ for the common-emitter stage in Fig. P8.16 if $g_m = 22$ mS, $\beta_o = 75$, $r_o = 100$ kΩ, $R_I = 500$ Ω, $R_B = 15$ kΩ, $R_L = 12$ kΩ, and $R_E = 300$ Ω? What is the small-signal limit for the value of v_i in this circuit? (b) What are the values if R_E is changed to 600 Ω?

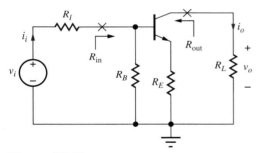

Figure P8.16

8.17. (a) What are the values of A_v, R_{in}, R_{out}, and $A_i = i_o/i_i$ for the common-source stage in Fig. P8.17 if $R_G = 2$ MΩ, $R_I = 75$ kΩ, $R_L = 3$ kΩ, and $R_S = 270$ Ω? Assume $g_m = 6$ mS and $r_o = 10$ kΩ. What is the small-signal limit for the value of v_i in this circuit? (b) What are the values of A_v, R_{in}, R_{out}, and A_i if R_S is bypassed by a capacitor?

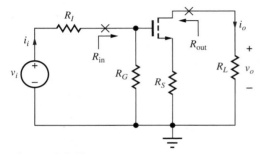

Figure P8.17

8.18. (a) Estimate the voltage gain of the inverting amplifier in Fig. P8.18. (b) Place a bypass capacitor in the circuit to change the gain to approximately -10. (c) Where should the bypass capacitor be placed to change the gain to approximately -20? (d) Where should the bypass capacitor be placed to achieve maximum gain? (e) Estimate this gain.

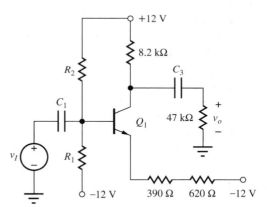

Figure P8.18

8.19. What values of R_E and R_L are required in the ac equivalent circuit in Fig. P8.19 to achieve $A_{vt} = -15$ and $R_{in} = 300$ kΩ? Assume $\beta_o = 75$.

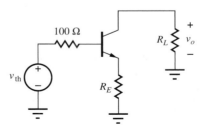

Figure P8.19

8.20. Assume that $R_E = 0$ in Fig. P8.19. What values of R_L and I_C are required to achieve $A_{vt} = 16$ dB and $R_{in} = 200$ kΩ? Assume $\beta_o = 95$.

8.21. Use nodal analysis to rederive the output resistance of the common-source circuit in Fig. P8.21, as expressed in Table 8.1.

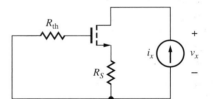

Figure P8.21

8.22. What are A_v, A_i, R_{in}, R_{out}, and the maximum amplitude of the signal source for the amplifier in Fig. P8.1(f) if $R_I = 500\ \Omega$, $R_E = 120\ k\Omega$, $R_B = 1\ M\Omega$, $R_3 = 500\ k\Omega$, $R_C = 56\ k\Omega$, $V_{CC} = 12\ V$, $-V_{EE} = -12\ V$? Use $\beta_F = 100$.

8.23. What are A_v, A_i, R_{in}, R_{out}, and the maximum amplitude of the signal source for the amplifier in Fig. P8.1(c) if $R_1 = 20\ k\Omega$, $R_2 = 62\ k\Omega$, $R_E = 3.9\ k\Omega$, $R_C = 10\ k\Omega$, and $V_{CC} = 10\ V$? Use $\beta_F = 75$. Compare A_v to our rule-of-thumb estimate and discuss the reasons for any discrepancy.

8.24. What are A_v, A_i, R_{in}, R_{out}, and the maximum amplitude of the signal source for the amplifier in Fig. P8.1(d) if $R_1 = 500\ k\Omega$, $R_2 = 1.4\ M\Omega$, $R_S = 33\ k\Omega$, $R_D = 82\ k\Omega$, and $V_{DD} = 15\ V$? Use $K_n = 250\ \mu A/V^2$ and $V_{TN} = 1.25\ V$. Compare A_V to our rule-of-thumb estimate and discuss the reasons for any discrepancy.

8.25. What are A_v, A_i, R_{in}, R_{out}, and the maximum amplitude of the signal source for the amplifier in Fig. P8.1(j) if $R_1 = 2.2\ M\Omega$, $R_2 = 2.2\ M\Omega$, $R_I = 22\ k\Omega$, $R_S = 22\ k\Omega$, $R_D = 18\ k\Omega$, and $V_{DD} = 18\ V$? Use $K_p = 400\ \mu A/V^2$ and $V_{TP} = -1.5\ V$.

8.26. What are A_v, A_i, R_{in}, R_{out}, and the maximum value of the signal source voltage for the amplifier in Fig. P8.1(m) if $R_I = 5\ k\Omega$, $R_G = 10\ M\Omega$, $R_3 = 36\ k\Omega$, $R_D = 1.8\ k\Omega$, and $V_{DD} = 12\ V$? Use $K_n = 0.4\ mS/V$ and $V_{TN} = -3\ V$.

8.27. What are A_v, A_i, R_{in}, R_{out}, and the maximum value of the signal source voltage for the amplifier in Fig. P8.1(n) if $R_I = 250\ \Omega$, $R_B = 20\ k\Omega$, $R_3 = 1\ M\Omega$, $R_E = 4.7\ k\Omega$, $V_{CC} = 10\ V$, and $V_{EE} = 10\ V$? Use $\beta_F = 80$ and $V_A = 100\ V$.

8.28. What are A_v, A_i, R_{in}, R_{out}, and the maximum value of the source voltage for the amplifier in Fig. P8.1(p) if $R_I = 5\ k\Omega$, $R_1 = 1\ k\Omega$, $R_G = 10\ M\Omega$, $R_3 = 36\ k\Omega$, $R_D = 1.8\ k\Omega$, $V_{DD} = 18\ V$? Use $I_{DSS} = 8\ mA$ and $V_P = -4.5\ V$.

8.3 Follower Circuits—Common-Collector and Common-Drain Amplifiers

8.29. What are the values of A_v, R_{in}, R_{out}, and A_i for the common-collector stage in Fig. P8.29 if $R_I = 10\ k\Omega$, $R_B = 56\ k\Omega$, $R_L = 1.2\ k\Omega$, $\beta_o = 75$, and $g_m = 0.5\ S$? ($A_i = i_o/i_i$). What is the small-signal limit for the value of v_i in this circuit?

8.30. What are the values of A_v, R_{in}, R_{out}, and A_i for the common-drain stage in Fig. P8.30 if $R_G = 1.5\ M\Omega$, $R_I = 100\ k\Omega$, $R_L = 3\ k\Omega$, and $g_m = 7.5\ mS$?

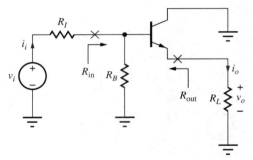

Figure P8.29

($A_i = i_o/i_i$). What is the small-signal limit for the value of v_i in this circuit?

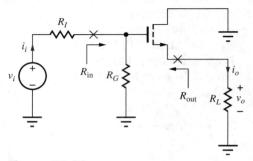

Figure P8.30

8.31. What are A_v, R_{in}, R_{out}, and maximum input signal amplitude for the amplifier in Fig. P8.1(a) if $R_I = 500\ \Omega$, $R_1 = 100\ k\Omega$, $R_2 = 100\ k\Omega$, $R_3 = 24\ k\Omega$, $R_E = 4.3\ k\Omega$, $R_C = 2\ k\Omega$, and $V_{CC} = V_{EE} = 12.5\ V$? Use $\beta_F = 100$ and $V_A = 58\ V$.

8.32. What are A_v, R_{in}, R_{out}, and maximum input signal for the amplifier in Fig. P8.1(o) if $R_I = 10\ k\Omega$, $R_G = 1\ M\Omega$, $R_3 = 120\ k\Omega$, and $V_{DD} = V_{SS} = 5\ V$? Use $K_n = 500\ \mu A/V^2$, $V_{TN} = 1.50\ V$, and $\lambda = 0.02\ V^{-1}$.

8.33. What are A_v, R_{in}, R_{out}, and the maximum input signal amplitude for the amplifier in Fig. P8.1(g) if $R_I = 500\ \Omega$, $R_1 = 470\ k\Omega$, $R_2 = 470\ k\Omega$, $R_3 = 500\ k\Omega$, $R_E = 430\ k\Omega$, and $V_{CC} = V_{EE} = 10\ V$? Use $\beta_F = 130$ and $V_A = 55\ V$.

*__8.34.__ The gate resistor R_G in Fig. P8.34 is said to be "bootstrapped" by the action of the source follower. (a) Assume that the FET is operating with $g_m = 3.54\ mS$ and r_o can be neglected. Draw the small-signal model and find A_v, R_{in}, and R_{out} for the amplifier. (b) What would R_{in} be if A_v were exactly $+1$?

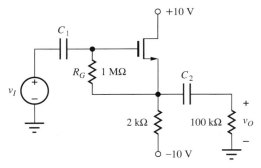

Figure P8.34

*8.35. Recast the signal-range formula for the common-collector amplifier in Table 8.5 in terms of the dc voltage developed across the emitter resistor R_E in Fig. 8.3(a). Assume $R_3 = \infty$.

8.36. Rework Prob. 8.34(a) by using the formulas for the bipolar transistor by "pretending" that R_G makes the FET equivalent to a BJT with $r_\pi = R_G$.

*8.37. The input to a common-collector amplifier is a triangular input signal with a peak-to-peak amplitude of 10 V. (a) What is the minimum gain required of the C-C amplifier to meet the small-signal limit? (b) What is the minimum dc voltage required across the emitter resistor in this amplifier to satisfy the limit in (a)?

*8.38. Design the emitter-follower circuit in Fig. P8.38 to meet the small-signal requirements when $v_o = 2.5 \sin 2000\pi t$ V. Assume $C_1 = C_2 = \infty$ and $\beta_F = 60$.

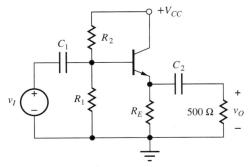

Figure P8.38

8.4 Noninverting Amplifiers—Common-Base and Common-Gate Circuits

8.39. What are the values of A_v, R_{in}, R_{out}, and A_i for the common-base stage in Fig. P8.39 operating with $I_C = 35$ μA, $\beta_o = 100$, $V_A = 60$ V, $R_I = 50$ Ω, $R_4 = 130$ kΩ and $R_L = 240$ kΩ? (b) What are the values if R_I is changed to 2.2 kΩ? ($A_i = i_o/i_i$). What is the small-signal limit for the value of v_i in this circuit?

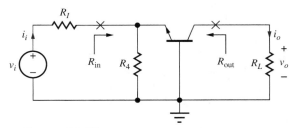

Figure P8.39

8.40. What are the values of A_v, R_{in}, R_{out}, and A_i for the common-gate stage in Fig. P8.40 operating with $g_m = 0.5$ mS, $R_I = 50$ Ω, $R_4 = 3$ kΩ and $R_L = 75$ kΩ? (b) What are the values if R_I is changed to 6.2 kΩ? ($A_i = i_o/i_i$).

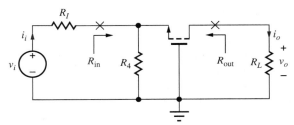

Figure P8.40

8.41. Find the voltage gain of the amplifier in Fig. P8.41 if $R_3 = 5.6$ kΩ, $R_2 = 10$ kΩ, $R_1 = 4.7$ kΩ, $V_A = 60$ V, $\beta_F = 100$, and $B_R = 2$.

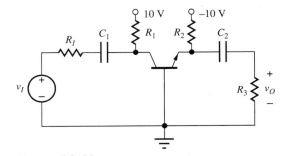

Figure P8.41

8.42. What are A_v, R_{in}, R_{out}, and the maximum input signal for the amplifier in Fig. P8.1(h) if $R_I = 500$ Ω, $R_B = 100$ kΩ, $R_3 = 100$ kΩ, $R_E = 82$ kΩ, $R_C = 39$ kΩ, and $V_{EE} = V_{CC} = 12$ V? Use $\beta_F = 75$ and $V_A = 50$ V.

8.43. What are A_v, R_{in}, R_{out}, and the maximum input signal for the amplifier in Fig. P8.1(h) if $R_I = 5$ kΩ, $R_B = 1$ MΩ, $R_3 = 1$ MΩ, $R_E = 820$ kΩ,

$R_C = 390$ kΩ, and $V_{EE} = V_{CC} = 9$ V? Use $\beta_F = 65$ and $V_A = 50$ V.

8.44. What are A_v, R_{in}, R_{out}, and the maximum input signal for the amplifier in Fig. P8.1(k) if $R_I = 1$ kΩ, $R_S = 3.9$ kΩ, $R_3 = 51$ kΩ, $R_D = 20$ kΩ, and $V_{DD} = 12$ V? Use $K_n = 450$ μA/V² and $V_{TN} = -2$ V.

8.45. What are A_v, R_{in}, R_{out}, and the maximum input signal for the amplifier in Fig. P8.1(b) if $R_I = 250$ Ω, $R_S = 68$ kΩ, $R_3 = 200$ kΩ, $R_D = 36$ kΩ, and $V_{DD} = V_{SS} = 16$ V? Use $K_p = 250$ μA/V² and $V_{TP} = -1$ V.

8.46. What are A_v, R_{in}, R_{out}, and the maximum input signal amplitude for the amplifier in Fig. P8.1(b) if $R_I = 500$ Ω, $R_S = 33$ kΩ, $R_3 = 100$ kΩ, $R_D = 24$ kΩ, and $V_{DD} = V_{SS} = 10$ V? Use $K_p = 200$ μA/V² and $V_{TP} = -1$ V.

8.47. The gain of the common-gate and common-base stages can be written as $A_v = R_L/[(1/g_m) + R_{th}]$. When $R_{th} \ll 1/g_m$, the circuit is said to be "voltage driven," and when $R_{th} \gg 1/g_m$, the circuit is said to be "current driven." What are the approximate voltage gain expressions for these two conditions? Discuss the reason for the use of these adjectives to describe the two circuit limits.

8.48. (a) What is the input resistance to the common-base stage in Fig. P8.48 if $I_C = 1$ mA and $\beta_F = 75$? (b) Repeat for $I_C = 120$ μA and $\beta_F = 125$. (c) Repeat for $V_A = 50$ V.

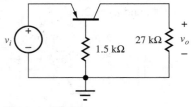

Figure P8.48

8.49. (a) What is the input resistance to the common-gate stage in Fig. P8.49 if $I_D = 1$ mA, $K_p = 1.25$ mA/V², and $V_{TP} = 2$ V? (b) Repeat for $I_D = 3$ mA and $V_{TP} = 2.5$ V. (c) Repeat with $\lambda = 0.02$/V.

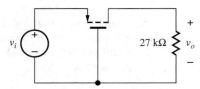

Figure P8.49

8.50. (a) Estimate the resistance looking into the collector of the transistor in Fig. P8.50 if $R_E = 330$ kΩ, $V_A = 75$ V, $\beta_F = 120$, and $V_{EE} = 12$ V? (b) What is the minimum value of V_{CC} required to ensure that Q_1 is operating in the forward-active region? (c) Repeat parts (a) and (b) if $R_E = 33$ kΩ.

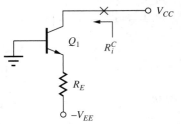

Figure P8.50

*8.51. What is the resistance looking into the collector terminal in Fig. P8.51 if $I_E = 40$ μA, $\beta_o = 110$, $V_A = 60$ V, and $V_{CC} = 10$ V? (*Hint:* r_o must be considered in this circuit. Otherwise $R_{out} = \infty$.)

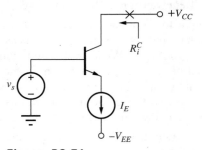

Figure P8.51

8.5 Amplifier Prototype Review and Comparison

8.52. A single-transistor amplifier is needed that has a gain of approximately 66 dB and an input resistance of 250 kΩ. What is the preferred choice of amplifier configuration, and why did you make this selection?

8.53. A single-transistor amplifier is needed that has a gain of approximately 0 dB and an input resistance of 20 MΩ with a load resistor of 20 kΩ. What is the preferred choice of amplifier configuration, and why did you make this selection?

8.54. A single-transistor amplifier is needed that has a gain of 20 dB and an input resistance of 10 MΩ. What is the preferred choice of amplifier configuration, and why did you make this selection?

8.55. A single-transistor amplifier is needed that has a gain of approximately +20 and an input resistance of 5 kΩ. What is the preferred choice of amplifier configuration? Discuss your reasons for making this selection.

8.56. A single-transistor amplifier is needed that has a gain of 56 dB and an input resistance of 50 kΩ. What is the preferred choice of amplifier configuration? Discuss your reasons for making this selection.

8.57. A single-transistor amplifier is needed that has a gain of 43 dB and an input resistance of 0.25 MΩ. What is the preferred choice of amplifier configuration? Discuss your reasons for making this selection.

8.58. A single-transistor amplifier is needed that has a gain of 46 dB and an input resistance of 500 Ω. What is the preferred choice of amplifier configuration? Discuss your reasons for your selection.

8.59. A single-transistor amplifier is needed that has a gain of -120 and an input resistance of 5 Ω. What is the preferred choice of amplifier configuration? Discuss your reasons for making this selection.

8.60. A follower is needed that has a gain of at least 0.97 and an input resistance of at least 250 kΩ with a load resistance of 5 kΩ. What is the preferred choice of amplifier configuration? Discuss your reasons for making this selection.

8.61. A common-collector amplifier is being driven from a source having a resistance of 250 Ω. (a) Estimate the output resistance of this amplifier if the transistor has $\beta_o = 150$, $V_A = 50$ V, and $I_C = 10$ mA. (b) Repeat for $I_C = 3.3$ mA.

8.62. An inverting amplifier is needed that has an output resistance of at least 1 GΩ. What is the preferred choice of amplifier configuration? Discuss your reasons for making this selection. Estimate the Q-point current and emitter or source resistance required to achieve this specification.

**8.63. Show that the emitter resistor R_E in Fig. P8.63 can be absorbed into the transistor by redefining the small-signal parameters of the transistor to be

$$g'_m \cong \frac{g_m}{1+g_m R_E} \qquad r'_\pi \cong r_\pi(1+g_m R_E)$$
$$r'_o \cong r_o(1+g_m R_E)$$

What is the expression for the common-emitter small-signal current gain β'_o for the new transistor? What is the expression for the amplification factor μ'_f for the new transistor?

*8.64. Perform a transient simulation of the behavior of the common-emitter amplifier in Fig. P8.64 for sinusoidal input voltages of 5 mV, 10 mV, and 15 mV at a frequency of 1 kHz. Use the Fourier analysis capability of SPICE to analyze the output waveforms. Compare the amplitudes of the 2-kHz and 3-kHz harmonics to the amplitude of the desired signal at 1 kHz. Assume $\beta_F = 100$ and $V_A = 70$ V.

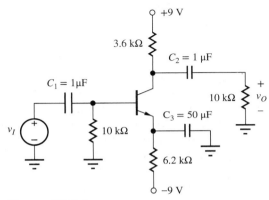

Figure P8.64

8.65. In the circuits in Fig. P8.65, $I_B = 10$ μA. Use SPICE to determine the output resistances of the two circuits by sweeping the voltage V_{CC} from 10 to 20 V. Use $\beta_F = 60$ and $V_A = 40$ V. Compare results to hand calculations using the small-signal parameter values from SPICE.

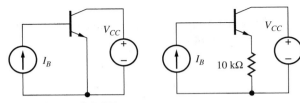

Figure P8.65

8.66. (a) What is the Thévenin equivalent representation for the amplifier in Fig. P8.66? (b) What are the values of v_{th} and R_{th} if $R_I = 270$ Ω, $\beta_o = 100$, $g_m = 4$ mS, and $r_o = 300$ kΩ?

8.67. What is the Thévenin equivalent representation for the amplifier in Fig. P8.67 if $R_I = 5$ kΩ, $R_L = 12$ kΩ, $\beta_o = 100$, and $g_m = 5$ mS?

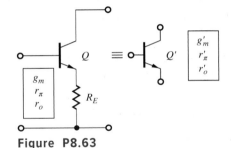

Figure P8.63

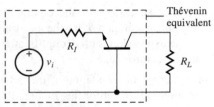

Figure P8.66

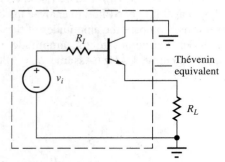

Figure P8.67

8.68. (a) What is the Thévenin equivalent representation for the amplifier in Fig. P8.68? (b) What are the values of v_{th} and R_{th} if $R_I = 100$ kΩ, $R_S = 27$ kΩ, $g_m = 500$ μS, and $r_o = 250$ kΩ?

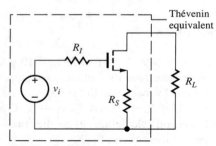

Figure P8.68

Assume the two-port description for Probs. 8.69 through 8.74 is

$$i_1 = G_\pi v_1 + G_r v_2$$
$$i_2 = G_m v_1 + G_o v_2$$

8.69. (a) An emitter follower is drawn as a two-port in Fig. P8.69. Calculate G_m and G_r for this amplifier in terms of the small-signal parameters. Compare the two results. (b) What are the values of G_m and G_r if $R_B = 150$ kΩ, $R_E = 2.4$ kΩ, $\beta_o = 125$, $g_m = 7.5$ mS, and $r_o = 250$ kΩ?

8.70. (a) A source follower is drawn as a two-port in Fig. P8.70. Calculate G_m and G_r for this amplifier in terms of the small-signal parameters. Compare

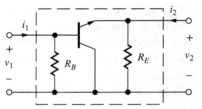

Figure P8.69

the two results. (b) What are the values of G_m and G_r if $R_G = 1$ MΩ, $R_D = 50$ kΩ, $g_m = 500$ μS, and $r_o = 475$ kΩ?

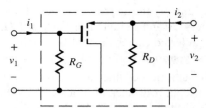

Figure P8.70

8.71. (a) A common-gate amplifier is drawn as a two-port in Fig. P8.71. Calculate G_m and G_r for this amplifier in terms of the small-signal parameters. Compare the two results. (b) What are the values of G_m and G_r if $R_S = 20$ kΩ, $R_D = 120$ kΩ, $g_m = 500$ μS, and $r_o = 500$ kΩ?

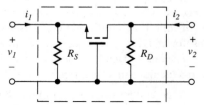

Figure P8.71

8.72. (a) A common-base amplifier is drawn as a two-port in Fig. P8.72. Calculate G_m and G_r for this amplifier in terms of the small-signal parameters. Compare the two results. (b) What are the values of G_m and G_r if $R_C = 18$ kΩ, $R_E = 3.6$ kΩ, $\beta_o = 120$, $g_m = 2.5$ mS, and $r_o = 750$ kΩ?

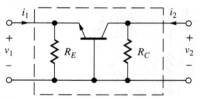

Figure P8.72

8.73. (a) A common-source amplifier is drawn as a two-port in Fig. P8.73. Calculate G_m and G_r for this amplifier in terms of the small-signal parameters. Compare the two results. (b) What are the values of G_m and G_r if $R_G = 1.5$ MΩ, $R_S = 12$ kΩ, $R_D = 140$ kΩ, $g_m = 750$ μS, and $r_o = 390$ kΩ?

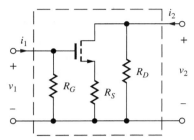

Figure P8.73

8.74. (a) A common-emitter amplifier is drawn as a two-port in Fig. P8.74. Calculate G_m and G_r for this amplifier in terms of the small-signal parameters. Compare the two results. (b) What are the values of G_m and G_r if $R_B = 180$ kΩ, $R_E = 13$ kΩ, $R_C = 120$ kΩ, $\beta_o = 125$, $g_m = 2.5$ mS, and $r_o = 1$ MΩ?

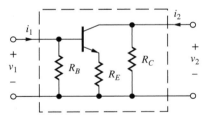

Figure P8.74

8.75. Our calculation of the input resistance of the common-gate and common-base amplifiers neglected r_o in the calculation. (a) Calculate an improved estimate for R_{in} for the common-gate stage in Fig. P8.75. (b) What is R_{in} if $R_L = r_o$?

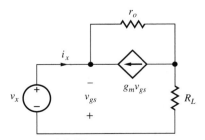

Figure P8.75

8.76. The circuit in Fig. P8.76 is called a phase inverter. Calculate the two gains $A_{v1} = \mathbf{v_{o1}}/\mathbf{v_i}$ and $A_{v2} = \mathbf{v_{o2}}/\mathbf{v_i}$. (a) What is the largest ac signal that can be developed at output v_{O1} in this particular circuit? Assume $\beta_F = 100$. (b) What is the largest value of v_i that satisfies the small signal limit?

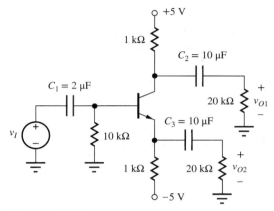

Figure P8.76

8.77. (a) Calculate the values of A_v, R_{in}, and R_{out} for the amplifier in Fig. P8.1(a) if $R_I = 600$ Ω, $R_1 = 100$ kΩ, $R_2 = 100$ kΩ, $R_3 = 24$ kΩ, $R_E = 4.7$ kΩ, $R_C = 2$ kΩ, and $V_{CC} = V_{EE} = 15$ V. Use $\beta_F = 125$ and $V_A = 50$ V. (b) Use SPICE to verify the results of your hand calculations. Assume $f = 10$ kHz and $C_1 = 10$ μF, $C_2 = 10$ μF, $C_3 = 47$ μF.

8.78. (a) Calculate the values of A_v, R_{in}, and R_{out} for the amplifier in Fig. P8.1(b) if $R_I = 500$ Ω, $R_S = 33$ kΩ, $R_3 = 100$ kΩ, $R_D = 24$ kΩ, and $V_{DD} = V_{SS} = 10$ V. Use $K_p = 250$ μA/V^2, $V_{TP} = -1$ V, and $\lambda = 0.02$ V^{-1}. (b) Use SPICE to verify the results of your hand calculations. Assume $f = 50$ kHz, $C_1 = 10$ μF, and $C_2 = 47$ μF.

8.79. (a) Calculate the values of A_v, R_{in}, and R_{out} for the amplifier in Fig. P8.1(c) if $R_1 = 20$ kΩ, $R_2 = 62$ kΩ, $R_E = 6.8$ kΩ, $R_C = 16$ kΩ, and $V_{CC} = 12$ V. Use $\beta_F = 75$ and $V_A = 60$ V. (b) Use SPICE to verify the results of your hand calculations. Assume $f = 5$ kHz and $C_1 = 2.2$ μF, $C_2 = 10$ μF, $C_3 = 47$ μF.

8.80. (a) Calculate the values of A_v, R_{in}, and R_{out} for the amplifier in Fig. P8.1(d) if $R_1 = 500$ kΩ, $R_2 = 1.4$ MΩ, $R_S = 27$ kΩ, $R_D = 75$ kΩ, and $V_{DD} = 18$ V. Use $K_n = 500$ μA/V^2, $\lambda = 0.02$ V^{-1}, and $V_{TN} = 1$ V. (b) Use SPICE to verify the results of your hand calculations. Assume $f = 5$ kHz and $C_1 = 2.2$ μF, $C_2 = 10$ μF, $C_3 = 47$ μF.

8.81. (a) Calculate the values of A_v, R_{in}, and R_{out} for the amplifier in Fig. P8.1(f) if $R_I = 500$ Ω, $R_E = 68$ kΩ, $R_B = 1$ MΩ, $R_3 = 500$ kΩ, $R_C = 39$ kΩ, $V_{EE} = -10$ V, and $V_{CC} = 10$ V. Use $\beta_F = 80$ and $V_A = 75$ V. (b) Use SPICE to verify the results of your hand calculations. Assume $f = 4$ kHz, $C_1 = C_2 = 2.2$ μF, and $C_3 = 47$ μF.

8.82. (a) Calculate the values of A_v, R_{in}, and R_{out} for the amplifier in Fig. P8.1(h) if $R_I = 500$ Ω, $R_B = 100$ kΩ, $R_3 = 100$ kΩ, $R_E = 82$ kΩ, $R_C = 39$ kΩ, and $V_{EE} = V_{CC} = 12$ V. Use $\beta_F = 50$ and $V_A = 50$ V. (b) Use SPICE to verify the results of your hand calculations. Assume $f = 12$ kHz and $C_1 = 4.7$ μF, $C_2 = 47$ μF, $C_3 = 10$ μF.

8.83. (a) Calculate the values of A_v, R_{in}, and R_{out} for the amplifier in Fig. P8.1(j) if $R_1 = 2.2$ MΩ, $R_2 = 2.2$ MΩ, $R_S = 110$ kΩ, $R_D = 90$ kΩ, and $V_{DD} = 18$ V. Use $K_p = 400$ μA/V^2, $\lambda = 0.02$ V^{-1}, and $V_{TP} = -1$ V. (b) Use SPICE to verify the results of your hand calculations. Assume $f = 7500$ Hz and $C_1 = 2.2$ μF, $C_2 = 10$ μF, $C_3 = 47$ μF.

8.84. (a) Calculate the values of A_v, R_{in}, and R_{out} for the amplifier in Fig. P8.1(k) if $R_I = 1$ kΩ, $R_3 = 10.0$ kΩ, $R_S = 51$ kΩ, $R_D = 20$ kΩ, and $V_{DD} = 15$ V. Use $K_n = 500$ μA/V^2, $\lambda = 0.02$ V^{-1}, and $V_{TN} = -2$ V. (b) Use SPICE to verify the results of your hand calculations. Assume $f = 20$ kHz and $C_1 = 47$ μF and $C_2 = 2.2$ μF.

8.85. (a) Calculate the values of A_v, R_{in}, and R_{out} for the amplifier in Fig. P8.1(m) if $R_I = 5$ kΩ, $R_G = 10$ MΩ, $R_3 = 36$ kΩ, $R_D = 1.8$ kΩ, and $V_{DD} = 16$ V. Use $K_n = 400$ μA/V^2, $V_{TN} = -5$ V, and $\lambda = 0.02$ V^{-1}. (b) Use SPICE to verify the results of your hand calculations. Assume $f = 3000$ Hz and $C_1 = 2.2$ μF, $C_2 = 10$ μF.

8.86. (a) Calculate the values of A_v, R_{in}, and R_{out} for the amplifier in Fig. P8.1(n) if $R_I = 250$ Ω, $R_B = 33$ kΩ, $R_3 = 1$ MΩ, $R_E = 7.8$ kΩ, $V_{CC} = 10$ V, and $V_{EE} = 10$ V. Use $\beta_F = 80$ and $V_A = 100$ V. (b) Use SPICE to verify the results of your hand calculations. Assume $f = 500$ kHz and $C_1 = 4.7$ μF, $C_2 = 1$ μF, $C_3 = 100$ μF, and $L = 1$ H.

8.87. (a) Calculate the values of A_v, R_{in}, and R_{out} for the amplifier in Fig. P8.1(o) if $R_I = 10$ kΩ, $R_G = 2$ MΩ, $R_3 = 100$ kΩ, and $V_{DD} = V_{SS} = 6$ V. Use $K_n = 400$ μA/V^2, $V_{TN} = 1$ V, and $\lambda = 0.02$ V^{-1}. (b) Use SPICE to verify the results of your hand calculations. Assume $f = 1$ MHz and $C_1 = 2.2$ μF, $C_2 = 4.7$ μF, and $L = 100$ mH.

8.88. (a) Calculate the values of A_v, R_{in}, and R_{out} for the amplifier in Fig. P8.1(p) if $R_I = 10$ kΩ, $R_1 = 10$ kΩ, $R_G = 500$ kΩ, $R_3 = 500$ kΩ, $R_D = 17$ kΩ, and $V_{DD} = 9$ V. Use $I_{DSS} = 1$ mA and $V_P = -3$ V. (b) Use SPICE to verify the results of your hand calculations. Assume $f = 10$ kHz and $C_1 = 10$ μF, $C_2 = 10$ μF, $C_3 = 47$ μF.

8.6 Common-Source Amplifiers Using MOS Transistor Loads

8.89. Find the Q-point, voltage gain, input resistance, and output resistance of the amplifier in Fig. P8.89 if $R_F = 750$ kΩ, $R_3 = 100$ kΩ, $K'_n = 100$ μA/V^2, $V_{TN} = 1$ V, $\lambda = 0.02$, $(W/L)_1 = 10/1$, $(W/L)_2 = 2/1$ and $V_{DD} = 5$ V.

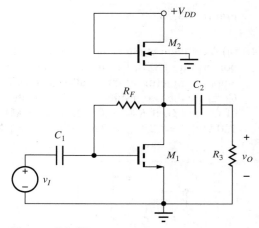

Figure P8.89

8.90. Find the Q-point, voltage gain, input resistance, and output resistance of the amplifier in Fig. P8.90 if $R_F = 750$ kΩ, $R_3 = 100$ kΩ, $K'_n = 100$ μA/V^2, $K'_p = 40$ μA/V^2, $V_{TN} = 1$ V, $V_{TP} = -1$ V, $\lambda = 0.02$, $(W/L)_1 = 40/1$, $(W/L)_2 = 100/1$ and $V_{DD} = 5$ V.

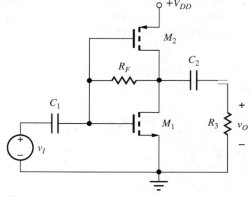

Figure P8.90

8.91. (a) Find the Q-point, voltage gain, input resistance, and output resistance of the amplifier in Fig. P8.91 if $R_1 = 240$ kΩ, $R_2 = 750$ kΩ, $R_3 = 100$ kΩ, $K'_n = 100$ μA/V^2, $V_{TN} = 1$ V, $\lambda = 0.02$, $(W/L)_1 = 5/1, (W/L)_2 = 5/1$ and $V_{DD} = 9$ V.

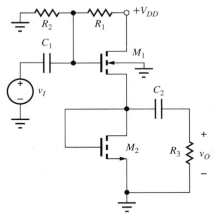

Figure P8.91

8.92. Redesign the W/L ratio for M$_2$ in Prob. 8.91 to achieve a voltage gain of 0.75.

8.93. The four amplifiers in Fig. 8.31 are used in the circuit in Fig. 8.33. What are the output resistance expressions for each of the four circuits?

8.7 Coupling and Bypass Capacitor Design

8.94. (a) The amplifier in Fig. P8.1(d) has $R_1 = 500$ kΩ, $R_2 = 1.4$ MΩ, $R_S = 33$ kΩ, $R_D = 75$ kΩ, and $V_{DD} = 15$ V. Use $K_n = 500$ μA/V^2, $V_{TN} = 1$ V, and $\lambda = 0.02$ V^{-1}. Choose values for C_1, C_2, and C_3 so that they can be neglected at a frequency of 100 Hz. (b) Choose C_3 to set the lower cutoff frequency to 4 kHz assuming C_1 and C_2 remain unchanged.

8.95. (a) The amplifier in Fig. P8.1(c) has $R_1 = 20$ kΩ, $R_2 = 62$ kΩ, $R_C = 8.2$ kΩ, $R_E = 5.1$ kΩ, and $V_{CC} = 15$ V. Choose values for C_1, C_2, and C_3 so that they can be neglected at a frequency of 400 Hz. Use $\beta_F = 85$ and $V_A = 60$ V. (b) Choose C_3 to set the lower cutoff frequency to 1000 Hz assuming C_1 and C_2 remain unchanged.

8.96. (a) Choose values of C_1 and C_2 in Fig. P8.96 so they will have negligible effect on the circuit at a frequency of 500 kHz. (b) Repeat for a frequency of 100 Hz.

8.97. Calculate the frequency for which each of the capacitors in Fig. P8.76 can be considered to have a negligible effect on the circuit.

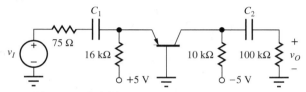

Figure P8.96

8.98. The amplifier in Fig. P8.1(a) has $R_I = 500$ Ω, $R_1 = 51$ kΩ, $R_2 = 100$ kΩ, $R_3 = 24$ kΩ, $R_E = 4.7$ kΩ, $R_C = 0$, and $V_{CC} = V_{EE} = 12$ V. Choose values for C_1, C_2, and C_3 so that they can be neglected at a frequency of 20 Hz. Use $\beta_F = 100$ and $V_A = 50$ V.

8.99. The amplifier in Fig. P8.1(k) has $R_I = 1$ kΩ, $R_s = 3.9$ kΩ, $R_3 = 100$ kΩ, $R_D = 20$ kΩ, and $V_{DD} = 15$ V. Choose values for C_1 and C_2 so that they can be neglected at a frequency of 1000 Hz. Use $K_n = 450$ μA/V^2, $V_{TN} = -2$ V, and $\lambda = 0.02$ V^{-1}.

8.100. (a) Use a dominant-pole approach to set the lower cutoff frequency of the C-S amplifier in Ex. 8.7 to 2500 Hz. Choose values for C_1, C_2, and C_3 based upon the values in the example. (b) Check your design with SPICE.

8.101. Use a dominant-pole approach to set the lower cutoff frequency of the C-C amplifier in Ex. 8.8 to 800 Hz. Choose values for C_1, and C_2 based upon the values in the example.

8.102. Use the dominant-pole approach to set the lower cutoff frequency of the C-G amplifier in Ex. 8.9 to 1500 Hz. Choose values for C_1, C_2, and C_3 based upon the values in the example. Check your design with SPICE.

8.8 Amplifier Design Examples

8.103. Repeat the source-follower design in Design Ex. 8.6 for a MOSFET with $K_n = 30$ mA/V^2 and $V_{TN} = 2.25$ V. Assume $V_{GS} - V_{TN} = 0.5$ V.

8.104. Rework Ex. 8.11 to achieve a 50-Ω input resistance with $R_I = 50$ Ω.

8.105. A common-base amplifier was used in the design problem in Ex. 8.11 to match the 75-Ω input resistance. One could conceivably match the input resistance with a common-emitter stage (with $R_E = 0$). What collector current is required to set $R_{in} = 75$ Ω for a BJT with $\beta_o = 100$?

*8.106. Redesign the bias network so that the common-base amplifier in Fig. 8.45 can operate from a single +10-V supply.

8.107. Redesign the amplifier in Fig. 8.45 to operate from symmetrical 9-V power supplies and achieve the same design specifications.

566   Chapter 8   Transistor Amplifier Building Blocks

8.108. $(1/g_m)$ is set to 50 Ω in a common-base design operating at 27°C. What are the values of $(1/g_m)$ at −40°C and +50°C?

*8.109. (a) Calculate worst-case estimates of the gain of the common-base amplifier in Fig. 8.45 if the resistors and power supplies all have 5 percent tolerances. (b) Compare your answers to the Monte Carlo results in Table 8.16.

**8.110. Use SPICE to perform a 1000-case Monte Carlo analysis of the common-base amplifier in Fig. 8.45 if the resistors and power supplies have 5 percent tolerances. Assume that the current gain β_F and V_A are uniformly distributed in the intervals [60, 100] and [50, 70], respectively. What are the mean and 3σ limits on the voltage gain predicted by these simulations? Compare the 3σ values to the worst-case calculations in Prob. 8.109. Compare your answers to the Monte Carlo results in Table 8.16. Use $C_1 = 47\ \mu F$, $C_2 = 4.7\ \mu F$, and $f = 10$ kHz.

8.111. A common-gate amplifier is needed with an input resistance of 10 Ω. Two n-channel MOSFETs are available: one with $K_n = 5$ mA/V² and the other with $K_n = 500$ mA/V². Both are capable of providing the desired value of R_{in}. Which one would be preferred and why? (*Hint:* Find the required Q-point current for each transistor.)

**8.112. The common-base amplifier in Fig. P8.96 is the implementation of the design from Design Ex. 8.11 using the nearest 1 percent resistor values. (a) What are the worst-case values of gain and input resistance if the power supplies have ±2 percent tolerances? (b) Use a computer program or spreadsheet to perform a 1000-case Monte Carlo analysis to find the mean and 3σ limits on the gain and input resistances. Compare these values to the worst-case estimates from part (a).

**8.113. Use SPICE to perform a 1000-case Monte Carlo analysis of the circuit in Fig. P8.96 assuming the resistors have 1 percent tolerances and the power supplies have ±2 percent tolerances. Find the mean and 3σ limits on the gain and input resistance at a frequency of 10 kHz. Assume that the current gain β_F and V_A are uniformly distributed in the intervals (60, 100) and (50, 70), respectively. Use $C_1 = 100\ \mu F$, $C_2 = 1\ \mu F$, and $f = 10$ kHz.

8.114. Suppose that we forgot about the factor of 2 loss in signal that occurs at the input of the common-base stage in Ex. 8.11 and selected $V_{CC} = V_{EE} = 2.5$ V. Repeat the design to see if the specifications can be met using these power supply values.

**8.115. (a) Use a spreadsheet or other computer tool to perform a Monte Carlo analysis of the design in Fig. 8.41. The resistors and power supplies have 5 percent tolerances. V_{TN} is uniformly distributed in the interval [1 V, 2 V], and K_n is uniformly distributed in the interval [10 mA/V², 30 mA/V²]. (b) Use the Monte Carlo option in PSPICE to perform the same analysis at a frequency of 10 kHz for $C_1 = 4.7\ \mu F$ and $C_2 = 68\ \mu F$. Compare the results.

Unless otherwise specified, use $\beta_F = 100$, $V_A = 70$ V, $K_p = K_n = 1$ mA/V², $V_{TN} = -V_{TP} = 1$ V, and $\lambda = 0.02$ V^{-1}.

8.9 Multistage ac-Coupled Amplifiers

8.116. What are the voltage gain, input resistance, and output resistance of the amplifier in Fig. 8.49 if bypass capacitors C_2 and C_4 are removed from the circuit?

8.117. Figure P8.117 is an "improved" version of the three-stage amplifier discussed in Sec. 8.9. Find the gain and input signal range for this amplifier. Was the performance actually improved?

8.118. Use SPICE to simulate the amplifier in Fig. P8.117 at a frequency of 2 kHz, and determine the voltage gain, input resistance, and output resistance. Assume the capacitors all have a value of 22 μF.

8.119. Find the midband voltage gain and input resistance of the amplifier in Fig. P8.117 if capacitors C_2 and C_4 are removed from the circuit.

8.120. Use SPICE to determine the gain of the amplifier in Fig. P8.117 if C_2 and C_4 are removed from the circuit. Assume the capacitors all have a value of 22 μF.

*8.121. Figure P8.121 shows another "improved" design of the three-stage amplifier discussed in Sec. 8.9. Find the gain and input signal range for this amplifier. Was the performance improved?

8.122. What is the gain of the amplifier in Fig. P8.121 if C_2 and C_4 are removed?

8.123. Use SPICE to simulate the amplifier in Fig. P8.121 at a frequency of 3 kHz and determine the voltage gain, input resistance, and output resistance. Assume the capacitors all have a value of 22 μF.

8.124. What are the midband voltage gain, input resistance, and output resistance of the amplifier in Fig. P8.124?

Problems 567

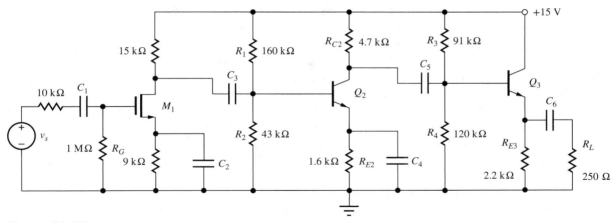

Figure P8.117

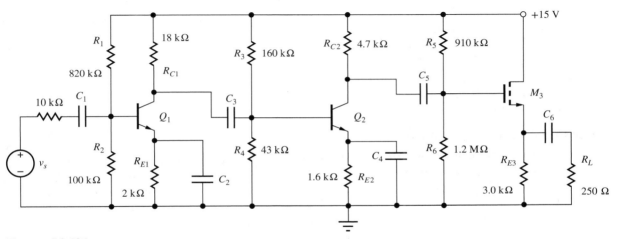

Figure P8.121

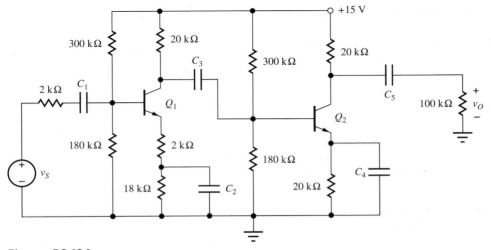

Figure P8.124

8.125. What are the voltage gain, input resistance, and output resistance of the amplifier in Fig. P8.124 if the bypass capacitors are removed?

8.126. Use SPICE to simulate the amplifier in Fig. P8.124 at a frequency of 5 kHz and determine the voltage gain, input resistance, and output resistance. Assume the capacitors all have a value of 10 µF.

8.127. Find the midband voltage gain, input resistance, and output resistance of the amplifier in Fig. P8.124 if capacitor C_2 is connected between the emitter of Q_1 and ground.

8.128. What are the midband voltage gain, input resistance, and output resistance of the amplifier in Fig. P8.128 if $K_n = 50$ mA/V^2 and $V_{TN} = -2$ V?

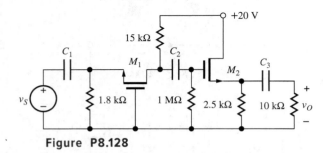

Figure P8.128

Lower Cutoff Frequency Estimates

8.129. Use the short-circuit time-constant technique to estimate the lower cutoff frequency for the amplifier in Prob. 8.77. Compare your answer to SPICE simulation.

8.130. Use the short-circuit time-constant technique to estimate the lower cutoff frequency for the amplifier in Prob. 8.78. Compare your answer to SPICE simulation.

8.131. Use the short-circuit time-constant technique to estimate the lower cutoff frequency for the amplifier in Prob. 8.79. Compare your answer to SPICE simulation.

8.132. Use the short-circuit time-constant technique to estimate the lower cutoff frequency for the amplifier in Prob. 8.80. Compare your answer to SPICE simulation.

8.133. Use the short-circuit time-constant technique to estimate the lower cutoff frequency for the amplifier in Prob. 8.81. Compare your answer to SPICE simulation.

8.134. Use the short-circuit time-constant technique to estimate the lower cutoff frequency for the amplifier in Prob. 8.82. Compare your answer to SPICE simulation.

8.135. Use the short-circuit time-constant technique to estimate the lower cutoff frequency for the amplifier in Prob. 8.84. Compare your answer to SPICE simulation.

8.136. Use the short-circuit time-constant technique to estimate the lower cutoff frequency for the amplifier in Prob. 8.85. Compare your answer to SPICE simulation.

8.137. Use the short-circuit time-constant technique to estimate the lower cutoff frequency for the amplifier in Prob. 8.128. Compare your answer to SPICE simulation. Use $C_1 = C_2 = C_3 = 1$ µF.

8.138. The MOS transistors in the circuit in Fig. 8.30(a) are replaced with *npn* transistors. What is the voltage gain of the new amplifier? Assume $g_m R_F \gg 1$.

Moderate and Weak Inversion

8.139. The circuit in Fig. P8.1(m) is operating in the weak inversion. (a) Find the Q-point for the circuit if $R_I = 100$ kΩ, $R_G = 2$ MΩ, $R_D = 1.25$ MΩ, $R_3 = 10$ MΩ, $V_{DD} = 0.75$ V, $K_n = 25$ mS/V, $V_{TN} = 0.4$ V, and $n = 1.5$. (b) What is the voltage gain of the amplifier? Assume that the capacitors have very large values.

8.140. The circuit in Fig. P8.34 is operating in the weak inversion with the 2-kΩ and 100-kΩ resistors changed to 1 MΩ. (a) Find the Q-point for the circuit if $R_G = 2$ MΩ, $K_n = 50$ mS/V, $V_{TN} = 0.35$ V, $n = 1.5$, and the power supply voltages are ± 1.5 V. (b) What is the voltage gain of the amplifier? Assume that the capacitors have very large values. (c) What are the minimum values of the two symmetrical power supplies?

8.141. (a) Find the Q-point of the circuit in Fig. P8.64 if $\beta_F = 100$ and $\beta_R = 5$. Assume $V_{BE} = 0.65$ V and ± 1.5 V supplies. What are the values of the voltage gain and input resistance? (b) An engineer accidentally inserts the *npn* transistor upside down in the same circuit. What is the new Q-point? What are the values of the new voltage gain and input resistance?

8.142. Capacitor C_1 and current source I_{SS} are removed from the circuit in Fig. 8.59(b), and the source of M_1 is connected directly to ground. (a) Draw the new circuit. (b) Find the Q-points if $R_D = 2$ MΩ, $K_n = 10$ mS/V, $V_{TN} = 0.3$ V, $n = 1.5$, $V_{DD} = 5$ V, and $V_{GG} = 2.5$ V. Assume that the dc value of v_1 is zero. (c) What is the voltage gain of the amplifier? (d) What is the minimum value of V_{GG} required to ensure saturation of M_1? What is the minimum value of V_{DD}?

8.143. Repeat Problem 8.90 if $V_{DD} = 1$ V, $V_{TN} = 0.5$ V, $V_{TP} = -0.5$ V, $n = 1.5$, $R_F = 20$ MΩ, and $R_3 = 2$ MΩ.

8.10 Introduction to dc-Coupled Amplifiers

8.144. (a) Verify the values of V_{D1}, V_{B2}, V_{C2}, and V_{B3} using the information in Tables 8.18 and 8.19. (b) Verify the new values of R_{D1} and R_{C2} in Fig. 8.56.

8.145. Find the new values of A_V, R_{in}, and R_{out} for the dc-coupled amplifier design in Fig. 8.56 and compare the results to the original design. (b) Estimate the new bandwidth of the amplifier using the short-circuit time-constant (SCTC) method. Use the capacitor values in Example 8.13. Compare the new bandwidth to the original ac-coupled design. (c) What is the new bandwidth if capacitor C_1 is eliminated from the amplifier?

8.146. Use SPICE to simulate the dc-coupled design in Fig. 8.56 and compare to the results of the original design. Use the capacitor values in Example 8.13.

8.147. A fully direct coupled amplifier can be achieved by removing all the capacitors from the circuit in Fig. 8.56. (a) Redraw the new circuit. (Be sure to connect the input source and load resistor to the amplifier.) (b) Find the new values of A_V, R_{in}, and R_{out}.

8.148. The amplifier in Fig. P8.128 can be direct coupled by removing capacitor C_2 and connecting the drain of M_1 directly to the gate of M_2. (a) Draw the new direct coupled version of the circuit. (b) Find the transistor Q-points and the voltage gain, input resistance, and output resistance of the amplifier. (c) Use the short-circuit time-constant technique to estimate the lower cutoff frequency if $C_1 = C_3 = 1$ μF. (d) Compare your result to SPICE simulation.

8.149. (a) Draw the small-signal circuit for the CD-CG amplifier in Fig. 8.58(b), assuming that r_o can be neglected in both transistors. (b) Derive the voltage gain expression in Eq. (8.176).

8.150. (a) Draw the small-signal circuit for the NMOS cascode amplifier in Fig. 8.59(b), assuming that r_o can be neglected in both transistors. (b) Derive the voltage gain expression in Eq. (8.176) assuming capacitor C_1 has a very large value.

8.151. (a) Draw the small-signal circuit for the CC-CB amplifier in Fig. 8.58(a), assuming that r_o can be neglected in both transistors. (b) Derive the voltage gain expression in Eq. (8.176).

8.152. (a) Draw the small-signal circuit for the BJT cascode amplifier in Fig. 8.59(a), assuming that r_o can be neglected in both transistors. (b) Derive the voltage gain expression in Eq. (8.176) assuming capacitor C_1 has a very large value.

8.153. For the cascode amplifier in Fig. 8.59(a), $I_{EE} = 25$ nA, $V_{BB} = 1$ V, $V_{DD} = 2.5$ V, $R_C = 50$ MΩ and the dc value of v_1 is zero. (a) Find the Q-points of the transistors. (b) What are the values of voltage gain, input resistance, and output resistance of the amplifier? (c) What are the minimum values of V_{BB} and V_{CC} if the BJTs must remain in the forward-active region? (d) How much power is being consumed by the amplifier if $V_{EE} = -0.5$ V?

8.154. (a) Calculate the mirror ratio $MR = I_O/I_{REF}$ for the NMOS current mirrors in Fig. P8.154 for $\lambda = 0$. (b) Repeat for $\lambda = 0.02$ V^{-1}, $V_{TN} = 1$ V, $K'_n = 25$ μA/V^2, and $I_{REF} = 50$ μA.

8.155. (a) Calculate the mirror ratio $MR = I_O/I_{REF}$ for the bipolar current mirrors in Fig. P8.155 for $V_A = \infty$ and $\beta_{FO} = \infty$. (b) Repeat for $V_A = 60$ V, $V_{BE} = 0.7$ V, and $\beta_{FO} = 75$.

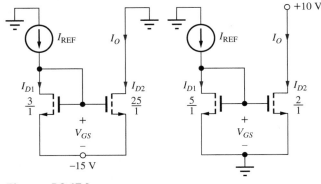

Figure P8.154

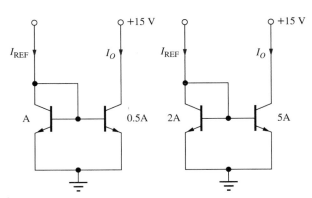

Figure P8.155

术语对照

Bandwidth Shrinkage	带宽缩减
Body effect	体效应
Bypass Capacitors	旁路电容
Common-base (C-B) amplifier	共基极放大器
Common-collector (C-C) amplifier	共集电极放大器
Common-drain (C-D) amplifier	共漏极放大器
Common-emitter (C-E) amplifier	共发射极放大器
Common-gate (C-G) amplifier	共栅极放大器
Common-source (C-S) amplifier	共源极放大器
Coupling capacitors	耦合电容
Current gain	电流增益
Cutoff frequency estimates	截止频率估计
Emitter follower	射极跟随器
Follower circuits	跟随器电路
Input resistance	输入电阻
Inverting amplifier	反相放大器
Multistage ac-coupled amplifier	多级交流耦合放大器
Noninverting amplifier	同相放大器
Output resistance	输出电阻
Power gain	功率增益
Signal range	信号范围
Signal source voltage gain	信号源电压增益
Source follower	源极跟随器
Swamping	阻抗压制
Terminal gain	端增益
Terminal current gain	端电流增益
Terminal voltage gain	端电压增益
Voltage gain	电压增益
Zero resistance	零电阻

CHAPTER 9
AMPLIFIER FREQUENCY RESPONSE

第 9 章 放大器频率响应

本章提纲

9.1 放大器频率响应
9.2 直接确定低频极点和零点——共源放大器
9.3 用短路时间常数法估算 ω_L 的值
9.4 高频晶体管模型
9.5 混合 π 模型中的基区电阻
9.6 共发射极和共源极放大器的高频响应
9.7 共基极和共栅极放大器的高频响应
9.8 共集电极和共漏极放大器的高频响应
9.9 单级放大器高频响应小结
9.10 多级放大器的频率响应
9.11 射频电路介绍
9.12 混频器和平衡调制器

本章目标

- 复习传输函数分析和确定截止频率;
- 理解放大器传输函数主极点的近似方法;
- 学会将交流电路分成低频和高频等效电路;
- 学会估计下限截止频率 f_L 的短路时间常数方法;
- 通过增加器件电容,完成双极型晶体管和 MOS 晶体管小信号模型的开发;
- 理解双极型晶体管和场效应晶体管的单位增益带宽积的限制;
- 学会估计上限截止频率 f_H 的开路时间常数方法;
- 推导反相、同相和跟随结构的上限截止频率的表达式;
- 证明反相、同相和跟随结构的增益带宽积的上限接近;
- 学会用两种时间常数方法分析多级放大器的频率响应;
- 掌握包括电流镜、级联放大器和差分对在内的双极型晶体管电路的带宽限制;
- 理解密勒效应;
- 掌握运放单位增益频率和放大器转换速率之间的关系;
- 了解简单的射频电路,包括调谐放大器、混频器和振荡器;
- 理解使用调谐电路生成宽带 (并联峰值) 和窄带 RF 放大器;
- 理解混频的基本概念;
- 掌握单平衡和双平衡混频电路,包括 Jones 混频器;
- 验证 SPICE 在交流分析中的应用;
- 验证 MATLAB 在显示频率响应信息的应用。

本章导读

本章主要讨论基本放大器的设计，并介绍分析模拟电路低频和高频频率响应的方法。作为讨论的一部分，将研究双极型晶体管和场效应晶体管器件的内部电容，并介绍与频率有关的晶体管小信号模型，采用小信号参数表达器件的单位带宽增益积。

频率响应是指放大器的增益与频率的关系。由于放大电路中存在电容，因此放大器对不同频率呈现的阻抗不同，对不同频率成分的放大倍数和相移也不相同。放大器的频率响应可通过将电路拆分成两个模型确定：一个模型在低频下有效，其中耦合电容和旁路电容最为重要；另一个模型在高频下有效，由器件电容控制电路的频率特性。

尽管对于单晶体管放大器而言，直接分析电路的频域特性通常是可行的，但对于多级放大器而言，直接分析法就显得不太现实。然而在绝大多数情况下，人们主要关注的是放大器的中频增益以及上限和下限截止频率，本章详细讲解了利用开路和短路时间常数法实现 f_H 和 f_L 的估算，并针对共发射极、共源极、共基极、共栅极、共集电极及共漏极等放大器形式分别运用短路时间常数法来计算下限截止频率 ω_L。

为研究放大器频率响应的上限，需要考虑晶体管的高频极限。所有的电子器件在其不同终端间都存在电容，这些电容可以限制器件提供有用的电压、电流或功率增益的频率。本章介绍了双极型晶体管与频率相关的混合π模型以及场效应晶体管的类似模型。通过在混合π模型中加入基极-发射极电容 C_π、基极-集电极电容 C_μ 和基极电阻 r_x，可以建立用于研究双极型晶体管频率特性的模型。C_x 的值与集电极电流 I_C 成正比，而 C_π 则与集电极-基极电压有着微弱的联系。$r_x C_\mu$ 的乘积是双极型晶体管频率限制的一个重要参数。通过在FET的π模型中加入栅源电容 C_{GS}、栅漏电容 C_{GD} 和栅极电阻 r_g，可以建立用于研究FET频率特性的模型。当FET工作在有源区时，C_{GS} 和 C_{GD} 的值与工作点无关。$r_g C_{GD}$ 的乘积是FET十分重要的参数。

在高频下，BJT和FET都具有有限电流增益，两种器件的单位增益带宽积都由器件电容和晶体管的跨导决定。在双极型晶体管中，截止频率 f_β 代表了电流增益低于其低频值3dB时所对应的频率。

本章详细研究了共发射极和共源极放大器、共基极和共栅极放大器、共集电极和共漏极放大器等三种基本单级放大器的高频响应。对于每一种基本放大器，都推导出其输入端和输出端的高频极点表达式，便于将分析结果扩展到多级放大器。本章介绍了密勒效应，并说明反相放大器相对较低的带宽是由放大器中晶体管的集电极-基极或栅极-漏极电容的密勒倍增(Miller multiplication)引起的。密勒效应使得跟随器的输入阻抗增加，跟随器主极点的表达式也可利用密勒倍增效应进行改写。

多级放大器的传递函数可能具有大量的极点和零点，本章详细介绍了采用短路和开路时间常数方法来估计上截止频率 ω_H 和下截止频率 ω_L。开路和短路时间常数法并不仅限于单晶体管放大器，还可直接运用到多级放大器当中。本章估算了两级直流耦合放大器的频率响应，其中包括差分放大器、Cascode放大器和电流镜放大器。因为这些放大器是直接耦合的，因此具有低通特性，所以只需确定 f_H。本章还分析了一个普通的三级放大器，并求出该放大器的 f_H 和 f_L。

本章还简要介绍了射频放大器和混频器等射频电路。射频电路的讨论包括宽带并联峰化和窄带（高Q值）调谐放大器。频率转换电路的介绍包括单平衡和双平衡混频器、无源和有源混频器电路。采用RLC电路的调谐放大器可被用来实现射频中的窄带放大器。设计中可采用单调谐或多级调谐电路。如果一个多级调谐放大器中的所有调谐电路都被设计成具有相同的中心频率，那么这一电路就被称为同步调谐。如果调谐电路被设计成具有不同的中心频率，那么该电路就被称为参差调谐。必须要确保调谐放大器不会变成振荡器，采用Cascode和C-C/C-B级联结构可以改进多级调谐电路之间的隔离性能。

混频器在通信电子电路中被广泛采用，用以转换信号的频率频谱。混频器要求对两个信号进行某种形式的相乘，从而产生两个输入频率的求和与求差。单平衡和双平衡架构可以从输出频谱中消除其中一个或两个输入信号。

Chapter Outline

9.1 Amplifier Frequency Response
9.2 Direct Determination of the Low-Frequency Poles and Zeros—The Common-Source Amplifier
9.3 Estimation of ω_L Using the Short-Circuit Time-Constant Method
9.4 Transistor Models at High Frequencies
9.5 Base and Gate Resistances in the Small-Signal Models
9.6 High-Frequency Common-Emitter and Common-Source Amplifier Analysis
9.7 Common-Base and Common-Gate Amplifier High-Frequency Response
9.8 Common-Collector and Common-Drain Amplifier High-Frequency Response
9.9 Single-Stage Amplifier High-Frequency Response Summary
9.10 Frequency Response of Multistage Amplifiers
9.11 Introduction to Radio Frequency Circuits
9.12 Mixers and Balanced Modulators
Summary
Key Terms
References
Problems

Chapter Goals

- Review transfer function analysis and determination of cutoff frequencies
- Understand dominant-pole approximations of amplifier transfer functions
- Learn to partition ac circuits into low-frequency and high-frequency equivalent circuits
- Learn the short-circuit time constant approach for estimating lower-cutoff frequency f_L
- Complete development of the small-signal models of both bipolar and MOS transistors with the addition of device capacitances
- Understand the unity-gain bandwidth product limitations of bipolar and field-effect transistors
- Learn the open-circuit time constant technique for estimating upper-cutoff frequency f_H
- Develop expressions for the upper-cutoff frequency of the inverting, noninverting, and follower configurations
- Demonstrate that the gain-bandwidth product limitations of the inverting, noninverting, and follower configurations approach the same upper limit
- Learn to apply the two time-constant approaches to the analysis of the frequency response of multistage amplifiers
- Explore bandwidth limitations of two-transistor circuits including current mirrors, cascode amplifiers, and differential pairs
- Understand the Miller effect
- Develop relationships between op amp unity-gain frequency and amplifier slew rate
- Introduce basic radio frequency (RF) circuits including tuned amplifiers, mixers, and oscillators
- Understand the use of tuned circuits to produce both broad-band (shunt-peaked) and narrow-band RF amplifiers
- Understand the basic concepts of mixing
- Explore single-balanced and double-balanced mixer circuits including the Jones Mixer
- Demonstrate the use of ac analysis in SPICE
- Demonstrate the use of MATLAB® to display frequency response information

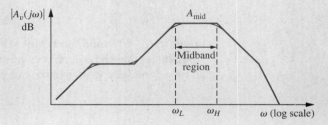

Chapters 6 to 8 discussed analysis and design of the midband characteristics of amplifiers. Low-frequency limitations due to coupling and bypass capacitors were also discussed, but the internal capacitances of electronic devices, which limit the response at high frequencies, were neglected. This chapter completes the discussion of basic amplifier design with the introduction of methods used to tailor the frequency response of analog circuits at both low and high frequencies. As part of this discussion, the internal device capacitances of bipolar and field-effect transistors are discussed, and frequency-dependent small-signal models of the transistors are introduced. The unity-gain bandwidth product of the devices is expressed in terms of the small-signal parameters.

In order to complete our basic circuit-building-block toolkit, expressions for the frequency responses of the single-stage inverting, noninverting, and follower configurations are each developed in detail. We show that the bandwidth of high-gain inverting and noninverting stages can be quite limited (although much wider than a typical op-amp stage of equal gain), whereas that of followers is normally very wide. Use of the cascode configuration is shown to significantly improve the frequency response of inverting amplifiers.

Transfer functions for multistage amplifiers may have large numbers of poles and zeros, and direct circuit analysis, although theoretically possible, can be complex and unwieldy. Therefore, approximation techniques — the short-circuit and open-circuit time-constant methods — have been developed to estimate the upper- and lower-cutoff frequencies ω_H and ω_L.

The Miller effect is introduced, and the relatively low bandwidth associated with inverting amplifiers is shown to be caused by Miller multiplication of the collector-base or gate-drain capacitance of the transistor in the amplifier.

This chapter also provides a brief introduction to radio frequency (RF) circuits including RF amplifiers and mixers. The RF circuit discussion includes both broad-band shunt-peaked and narrow-band (high-Q) tuned amplifiers. The presentation of frequency translation circuits includes single- and double-balanced mixers, including passive and active mixer circuits. High-frequency oscillators are discussed in Chapter 15.

9.1 AMPLIFIER FREQUENCY RESPONSE

Figure 9.1 is the Bode plot for the magnitude of the voltage gain of a hypothetical amplifier. Regardless of the number of poles and zeros, the voltage transfer function $A_v(s)$ can be written as the ratio of two polynomials in s:

$$A_v(s) = \frac{N(s)}{D(s)} = \frac{a_0 + a_1 s + a_2 s^2 + \cdots + a_m s^m}{b_0 + b_1 s + b_2 s^2 + \cdots + b_n s^n} \qquad (9.1)$$

In principle, the numerator and denominator polynomials of Eq. (9.1) can be written in factored form, and the poles and zeros can be separated into two groups. Those associated with the low-frequency response below the midband region of the amplifier can be combined into a function $F_L(s)$, and those associated with the high-frequency response above the midband region can be grouped into a function $F_H(s)$. Using F_L and F_H, $A_v(s)$ can be rewritten as

$$A_v(s) = A_{\text{mid}} F_L(s) F_H(s) \qquad (9.2)$$

in which A_{mid} is the **midband gain**[1] of the amplifier in the region between the **lower-** and **upper-cutoff frequencies** (ω_L and ω_H, respectively). For A_{mid} to appear explicitly as shown in Eq. (9.2), $F_H(s)$ and $F_L(s)$ must be written in the two particular standard forms defined by Eqs. (9.3) and (9.4):

$$F_L(s) = \frac{(s + \omega_{Z1}^L)(s + \omega_{Z2}^L) \cdots (s + \omega_{Zk}^L)}{(s + \omega_{P1}^L)(s + \omega_{P2}^L) \cdots (s + \omega_{Pk}^L)} \qquad (9.3)$$

$$F_H(s) = \frac{\left(1 + \dfrac{s}{\omega_{Z1}^H}\right)\left(1 + \dfrac{s}{\omega_{Z2}^H}\right) \cdots \left(1 + \dfrac{s}{\omega_{Zl}^H}\right)}{\left(1 + \dfrac{s}{\omega_{P1}^H}\right)\left(1 + \dfrac{s}{\omega_{P2}^H}\right) \cdots \left(1 + \dfrac{s}{\omega_{Pl}^H}\right)} \qquad (9.4)$$

The representation of $F_H(s)$ is chosen so that its magnitude approaches a value of 1 at frequencies well below the upper-cutoff frequency ω_H,

$$|F_H(j\omega)| \to 1 \quad \text{for} \quad \omega \ll \omega_{Zi}^H, \omega_{Pi}^H \quad \text{for } i = 1, 2, \cdots, l \qquad (9.5)$$

---

[1] You may wish to review some of the frequency response definitions in Chapter 6.

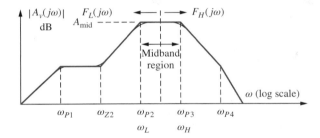

Figure 9.1 Bode plot for a general amplifier transfer function.

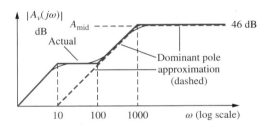

Figure 9.2 Bode plot for a complete transfer function and its dominant pole approximation.

Thus, at low frequencies, the transfer function $A_v(s)$ becomes

$$A_L(s) \cong A_{\text{mid}} F_L(s) \tag{9.6}$$

The form of $F_L(s)$ is chosen so its magnitude approaches a value of 1 at frequencies well above ω_L:

$$|F_L(j\omega)| \to 1 \quad \text{for} \quad \omega \gg \omega_{Zj}^L, \omega_{Pj}^L \quad \text{for } j = 1, 2, \cdots, k \tag{9.7}$$

Thus, at high frequencies, the transfer function $A_v(s)$ can be approximated by

$$A_H(s) \cong A_{\text{mid}} F_H(s) \tag{9.8}$$

9.1.1 LOW-FREQUENCY RESPONSE

In many designs, the zeros of $F_L(s)$ can be placed at frequencies low enough to not influence the lower-cutoff frequency ω_L. In addition, one of the low-frequency poles in Fig. 9.1, say ω_{P2}, can be designed to be much larger than the others. For these conditions, the low-frequency portion of the transfer function can be written approximately as

$$F_L(s) \cong \frac{s}{s + \omega_{P2}} \tag{9.9}$$

Pole ω_{P2} is referred to as the **dominant low-frequency pole** and the lower-cutoff frequency ω_L is approximately

$$\omega_L \cong \omega_{P2} \tag{9.10}$$

The Bode plot in Fig. 9.2 is an example of a transfer function and its dominant pole approximation. The overall transfer function $A_L(s)$ for this figure has two poles and two zeros.

9.1.2 ESTIMATING ω_L IN THE ABSENCE OF A DOMINANT POLE

If a **dominant pole** does not exist at low frequencies, then the poles and zeros interact to determine the lower-cutoff frequency, and a more complicated analysis must be used to find ω_L. As an example, consider the case of an amplifier having two zeros and two poles at low frequencies:

$$A_L(s) = A_{\text{mid}} F_L(s) = A_{\text{mid}} \frac{(s + \omega_{Z1})(s + \omega_{Z2})}{(s + \omega_{P1})(s + \omega_{P2})} \tag{9.11}$$

For $s = j\omega$,

$$|A_L(j\omega)| = A_{\text{mid}} |F_L(j\omega)| = A_{\text{mid}} \sqrt{\frac{(\omega^2 + \omega_{Z1}^2)(\omega^2 + \omega_{Z2}^2)}{(\omega^2 + \omega_{P1}^2)(\omega^2 + \omega_{P2}^2)}} \tag{9.12}$$

and remembering that ω_L is defined as the -3 dB frequency,

$$|A(j\omega_L)| = \frac{A_{\text{mid}}}{\sqrt{2}} \quad \text{and} \quad \frac{1}{\sqrt{2}} = \sqrt{\frac{(\omega_L^2 + \omega_{Z1}^2)(\omega_L^2 + \omega_{Z2}^2)}{(\omega_L^2 + \omega_{P1}^2)(\omega_L^2 + \omega_{P2}^2)}} \tag{9.13}$$

Squaring both sides and expanding Eq. (9.13),

$$\frac{1}{2} = \frac{\omega_L^4 + \omega_L^2(\omega_{Z1}^2 + \omega_{Z2}^2) + \omega_{Z1}^2 \omega_{Z2}^2}{\omega_L^4 + \omega_L^2(\omega_{P1}^2 + \omega_{P2}^2) + \omega_{P1}^2 \omega_{P2}^2} = \frac{1 + \frac{(\omega_{Z1}^2 + \omega_{Z2}^2)}{\omega_L^2} + \frac{\omega_{Z1}^2 \omega_{Z2}^2}{\omega_L^4}}{1 + \frac{(\omega_{P1}^2 + \omega_{P2}^2)}{\omega_L^2} + \frac{\omega_{P1}^2 \omega_{P2}^2}{\omega_L^4}} \quad (9.14)$$

If we assume that ω_L is larger than all the individual pole and zero frequencies, then the terms involving $1/\omega_L^4$ can be neglected, and the lower-cutoff frequency can be estimated from

$$\omega_L \cong \sqrt{\omega_{P1}^2 + \omega_{P2}^2 - 2\omega_{Z1}^2 - 2\omega_{Z2}^2} \quad (9.15)$$

For the more general case of n poles and n zeros, a similar analysis yields

$$\omega_L \cong \sqrt{\sum_n \omega_{Pn}^2 - 2\sum_n \omega_{Zn}^2} \quad (9.16)$$

EXERCISE: Use Eq. (9.15) to estimate f_L for the transfer functions given below:

$$A_v(s) = \frac{200s(s + 50)}{(s + 10)(s + 1000)} \quad \text{and} \quad A_v(s) = \frac{100s(s + 500)}{(s + 100)(s + 1000)}$$

ANSWERS: 159 Hz, 114 Hz

EXAMPLE 9.1 ANALYSIS OF A TRANSFER FUNCTION

The midband gain, poles, zeros, and cutoff frequency are identified from a specified transfer function.

PROBLEM Find the midband gain, $F_L(s)$, and lower-cutoff frequency f_L for

$$A_L(s) = 2000 \frac{s\left(\frac{s}{100} + 1\right)}{(0.1s + 1)(s + 1000)}$$

Identify the frequencies corresponding to the poles and zeros. Find a dominant pole approximation for the transfer function, if one exists.

SOLUTION **Known Information and Given Data:** The transfer function is specified.

Unknowns: A_{mid}, $F_L(s)$, f_L, poles, zeros, dominant-pole approximation

Approach: Rearrange $A_L(s)$ into the form of Eqs. (9.6) and (9.3). Identify the pole and zero frequencies. Find the midband region and A_{mid}. Since the poles and zeros can all be found, use Eq. (9.16) to find f_L. If the poles and zeros are widely separated, find the dominant pole representation.

Assumptions: None

Analysis: To begin, we need to rearrange the transfer function by factoring 0.01 out of the numerator and 0.1 out of the denominator in order to have all the poles and zeros written as in Eq. (9.3):

$$A_L(s) = 200 \frac{s(s + 100)}{(s + 10)(s + 1000)}$$

Now, $A_L(s) = A_{\text{mid}} F_L(s)$ with $A_{\text{mid}} = 200$ and

$$F_L(s) = \frac{s(s + 100)}{(s + 10)(s + 1000)}$$

Zeros occur at the values of s for which the numerator is zero: $s = 0$ and $s = -100$ rad/s. Poles occur at the frequencies s for which the denominator is zero: $s = -10$ rad/s and $s = -1000$ rad/s.

Substituting these values into Eq. (9.16) yields an estimate of f_L:

$$f_L = \frac{1}{2\pi}\sqrt{10^2 + 1000^2 - 2(0^2 + 100^2)^2} = \frac{990}{2\pi} = 158 \text{ Hz}$$

Note that these are all at low frequencies and are separated from one another by a decade of frequency. Thus, a dominant pole exists at $\omega = 1000$, and the lower-cutoff frequency is given approximately by $f_L \cong 1000/2\pi = 159$ Hz. For frequencies above a few hundred rad/s, the transfer function can be approximated by

$$A_L(s) \cong 200\frac{s}{s+1000} \quad \text{for} \quad \omega > 200 \text{ rad/s}$$

Check of Results: The requested unknowns have been found. For $\omega \gg 1000$, the transfer original function reaches its largest value and becomes constant—the midband region:

$$A_L(s)|_{s \gg 1000} \cong 200\frac{s^2}{s^2} = 200$$

Thus, $A_{\text{mid}} = 200$ or 46 dB. We also see that the value of f_L predicted by Eq. (9.16) is the same as that of the dominant-pole model, indicating correctness of the dominant-pole approximation.

Discussion: Figure 9.2 graphs the original transfer function and its dominant-pole approximation. The midband region is clearly viable for $\omega > 1000$ rad/s, and the single pole roll-off is valid for frequencies down to approximately 200 rad/s.

Computer-Aided Analysis: We can easily visualize the transfer function with the aid of MATLAB®: bode ([200 20000 0], [1 1010 10000]). The resulting graph of the magnitude and phase of $A_L(s)$ appears in the figure. The alternating sequence of zeros and poles is apparent in both the magnitude and phase plots, and the gain approaches 46 dB at high frequencies.

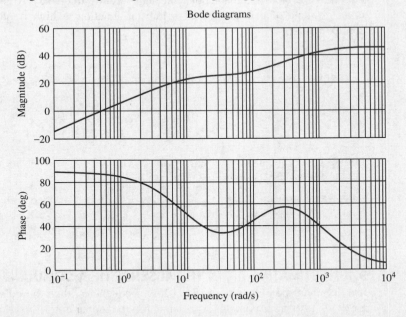

EXERCISE: For what range of frequencies does the approximation to $A_v(s)$ in Ex. 9.1 differ from the actual transfer function by less than 10 percent?

ANSWER: $\omega \geq 205$ rad/s

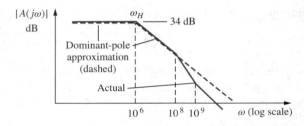

Figure 9.3 Bode plot for a complete transfer function and its dominant-pole approximation for midband and above.

9.1.3 HIGH-FREQUENCY RESPONSE

In the region above midband, $A_v(s)$ can be represented by its high-frequency approximation:

$$A_H(s) \cong A_{mid} F_H(s) \tag{9.17}$$

Many of the zeros of $F_H(s)$ are often at infinite frequency, or high enough in frequency that they do not influence the value of $F_H(s)$ near ω_H. If, in addition, one of the **pole frequencies**—for example, ω_{P3} in Fig. 9.1—is much smaller than all the others, then a **dominant high-frequency pole** exists in the high-frequency response, and $F_H(s)$ can be represented by the approximation:

$$F_H(s) \cong \frac{1}{1 + \dfrac{s}{\omega_{P3}}} \tag{9.18}$$

For the case of a dominant pole, the upper-cutoff frequency is given by $\omega_H \cong \omega_{P3}$. Figure 9.3 is an example of a Bode plot of a transfer function at high frequencies and its dominant-pole approximation.

> **EXERCISE:** The transfer function for the amplifier in Fig. 9.3 is
>
> $$A_H(s) = 50 \frac{\left(1 + \dfrac{s}{10^9}\right)}{\left(1 + \dfrac{s}{10^6}\right)\left(1 + \dfrac{s}{10^8}\right)}$$
>
> What are the locations of the poles and zeros of $A_H(s)$? What are A_{mid}, $F_H(s)$ for the dominant-pole approximation, and f_H?
>
> **ANSWERS:** $\omega_{Z1} = -10^9$ rad/s, $\omega_{P1} = -10^6$ rad/s, $\omega_{P2} = -10^8$ rad/s; 50, $F_H(s) = \dfrac{1}{\left(1 + \dfrac{s}{10^6}\right)}$, 159 kHz

9.1.4 ESTIMATING ω_H IN THE ABSENCE OF A DOMINANT POLE

If a dominant pole does not exist at high frequencies, then the poles and zeros interact to determine ω_H. An approximate expression for the upper-cutoff frequency can be found from the expression for F_H in a manner similar to that used to arrive at Eq. (9.16). Consider the case of an amplifier having two zeros and two poles at high frequencies:

$$A_H(s) = A_{mid} F_H(s) = A_{mid} \frac{\left(1 + \dfrac{s}{\omega_{Z1}}\right)\left(1 + \dfrac{s}{\omega_{Z2}}\right)}{\left(1 + \dfrac{s}{\omega_{P1}}\right)\left(1 + \dfrac{s}{\omega_{P2}}\right)} \tag{9.19}$$

and for $s = j\omega$,

$$|A_H(j\omega)| = A_{mid}|F_H(j\omega)| = A_{mid}\sqrt{\frac{\left(1 + \frac{\omega^2}{\omega_{Z1}^2}\right)\left(1 + \frac{\omega^2}{\omega_{Z2}^2}\right)}{\left(1 + \frac{\omega^2}{\omega_{P1}^2}\right)\left(1 + \frac{\omega^2}{\omega_{P2}^2}\right)}} \quad (9.20)$$

At the upper-cutoff frequency $\omega = \omega_H$,

$$|A(j\omega_H)| = \frac{A_{mid}}{\sqrt{2}} \quad \text{and} \quad \frac{1}{\sqrt{2}} = \sqrt{\frac{\left(1 + \frac{\omega_H^2}{\omega_{Z1}^2}\right)\left(1 + \frac{\omega_H^2}{\omega_{Z2}^2}\right)}{\left(1 + \frac{\omega_H^2}{\omega_{P1}^2}\right)\left(1 + \frac{\omega_H^2}{\omega_{P2}^2}\right)}} \quad (9.21)$$

By squaring both sides and expanding Eq. (9.21), and assuming ω_H is smaller than all the individual pole and zero frequencies, the upper-cutoff frequency can be found to be

$$\omega_H \cong \frac{1}{\sqrt{\frac{1}{\omega_{P1}^2} + \frac{1}{\omega_{P2}^2} - \frac{2}{\omega_{Z1}^2} - \frac{2}{\omega_{Z2}^2}}} \quad (9.22)$$

The expression for the general case of n poles and n zeros can be found in a manner similar to Eq. (9.22), and the resulting approximation for ω_H is

$$\omega_H \cong \frac{1}{\sqrt{\sum_n \frac{1}{\omega_{Pn}^2} - 2\sum_n \frac{1}{\omega_{zn}^2}}} \quad (9.23)$$

EXERCISE: Write the expression for the $A_H(s)$ below in standard form. What are the pole and zero frequencies? What are A_{mid}, $F_H(s)$, and f_H?

$$A_H(s) = \frac{2.5 \times 10^7 (s + 2 \times 10^5)}{(s + 10^5)(s + 5 \times 10^5)}$$

ANSWERS: $A_H(s) = 100 \dfrac{\left(1 + \dfrac{s}{2 \times 10^5}\right)}{\left(1 + \dfrac{s}{10^5}\right)\left(1 + \dfrac{s}{5 \times 10^5}\right)}$; -10^5 rad/s, -5×10^5 rad/s, -2×10^5 rad/s; ∞, 40 dB, 21.7 kHz

---

9.2 DIRECT DETERMINATION OF THE LOW-FREQUENCY POLES AND ZEROS—THE COMMON-SOURCE AMPLIFIER

To apply the theory in Sec. 9.1, we need to know the location of all the individual poles and zeros. In principle, the frequency response of an amplifier can always be calculated by direct analysis of the circuit in the frequency domain, so this section begins with an example of this form of analysis for the common-source amplifier. However, as circuit complexity grows, exact analysis by hand rapidly becomes intractable. Although SPICE analysis can always be used to study the characteristics of an amplifier for a given set of parameter values, a more general understanding of the factors that control the cutoff frequencies of the amplifier is needed for design. Because we are most often interested in the position of ω_L and ω_H, we subsequently develop approximation techniques that can be used to estimate ω_L and ω_H.

580 Chapter 9 Amplifier Frequency Response

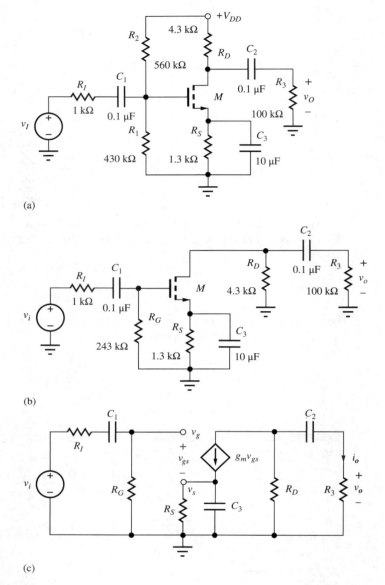

Figure 9.4 (a) A common-source amplifier, (b) low-frequency ac model, and (c) ac small-signal model.

The circuit for a common-source amplifier appears in Fig. 9.4(a) along with its ac equivalent circuit in Fig. 9.4(b). At low frequencies below midband, the impedance of the capacitors can no longer be assumed to be negligible, and they must be retained in the ac equivalent circuit. To determine circuit behavior at low frequencies, we replace transistor Q_1 by its low-frequency small-signal model, as in Fig. 9.4(c). Because the stage has an external load resistor, r_o is neglected in the circuit model.

In the frequency domain, output voltage $\mathbf{V_o}(s)$ can be found by applying current division at the drain of the transistor:

$$\mathbf{V_o}(s) = \mathbf{I_o}(s) R_3 \quad \text{where} \quad \mathbf{I_o}(s) = -g_m \mathbf{V_{gs}}(s) \frac{R_D}{R_D + \dfrac{1}{sC_2} + R_3}$$

and

$$\mathbf{V_o}(s) = -g_m(R_3 \| R_D) \frac{s}{s + \dfrac{1}{C_2(R_D + R_3)}} \mathbf{V_{gs}}(s) \quad (9.24)$$

Next, we must find $\mathbf{V_{gs}}(s) = \mathbf{V_g}(s) - \mathbf{V_s}(s)$. Because the gate terminal in Fig. 9.4(c) represents an open circuit, $\mathbf{V_g}(s)$ can be determined using voltage division:

$$\mathbf{V_g}(s) = \mathbf{V_i}(s) \frac{R_G}{R_I + \dfrac{1}{sC_1} + R_G} = \mathbf{V_i}(s) \frac{sC_1 R_G}{sC_1(R_I + R_G) + 1} \quad (9.25)$$

and the voltage at the source of the FET can be found by writing a nodal equation for $V_s(s)$:

$$g_m(\mathbf{V_g} - \mathbf{V_s}) - G_S \mathbf{V_s} - sC_3 \mathbf{V_s} = 0 \quad \text{or} \quad \mathbf{V_s} = \frac{g_m}{sC_3 + g_m + G_S} \mathbf{V_g} \quad (9.26)$$

and

$$\mathbf{V_{gs}}(s) = (\mathbf{V_g} - \mathbf{V_s}) = \mathbf{V_g}\left[1 - \frac{g_m}{sC_3 + g_m + G_S}\right] = \frac{sC_3 + G_S}{sC_3 + g_m + G_S} \mathbf{V_g} \quad (9.27)$$

By dividing through by C_3, Eq. (9.27) can be rewritten as

$$(\mathbf{V_g} - \mathbf{V_s}) = \frac{s + \dfrac{1}{C_3 R_S}}{s + \dfrac{1}{C_3\left(\dfrac{1}{g_m} \| R_S\right)}} \mathbf{V_g}(s) \quad (9.28)$$

Finally, combining Eqs. (9.24), (9.25), and (9.28) yields an overall expression for the voltage transfer function:

$$A_v(s) = \frac{\mathbf{V_o}(s)}{\mathbf{V_i}(s)} = A_{\text{mid}} F_L(s)$$

$$= \left[-g_m(R_3 \| R_D)\frac{R_G}{(R_I + R_G)}\right] \frac{s^2\left[s + \dfrac{1}{C_3 R_S}\right]}{\left[s + \dfrac{1}{C_1(R_I + R_G)}\right]\left[s + \dfrac{1}{C_3\left(\dfrac{1}{g_m}\| R_S\right)}\right]\left[s + \dfrac{1}{C_2(R_D + R_3)}\right]} \quad (9.29)$$

In Eq. (9.29), $A_v(s)$ has been written in the form that directly exposes the midband gain and $F_L(s)$:

$$A_v(s) = A_{\text{mid}} F_L(s) \quad \text{where} \quad A_{\text{mid}} = -g_m(R_D \| R_3)\frac{R_G}{R_G + R_I} \quad (9.30)$$

A_{mid} should be recognized as the voltage gain of the circuit with the capacitors all replaced by short circuits.

Although the analysis in Eqs. (9.24) to (9.30) may seem rather tedious, we nevertheless obtain a complete description of the frequency response. In this example, the poles and zeros of the transfer function appear in factored form in Eq. (9.29). Unfortunately, this is an artifact of this particular FET circuit and generally will not be the case. The infinite input resistance of the FET and absence of r_o in the circuit have decoupled the nodal equations for v_g, v_s, and v_o. In most cases, the mathematical analysis is even more complex. For example, if a bipolar transistor were used in which both r_π and r_o were included, the analysis would require the simultaneous solution of three equations in three unknowns.

EXERCISE: Draw the midband ac equivalent circuit for the amplifier in Fig. 9.2 and derive the expression for A_{mid} directly from this circuit.

ANSWER: Eq. (9.30)

Let us now explore the origin of the poles and zeros of the voltage transfer function. Equation (9.29) has three poles and three zeros, *one pole and one zero for each independent capacitor in the circuit*. Two of the zeros are at $s = 0$ (dc), corresponding to series capacitors C_1 and C_2, each of which blocks the propagation of dc signals through the amplifier. The third zero occurs at the frequency for which the impedance of the parallel combination of R_S and C_3 becomes infinite. At this frequency, propagation of signal current through the MOSFET is blocked, and the output voltage must be zero. Thus, the three zero locations are

$$s = 0, 0, -\frac{1}{R_S C_3} \qquad (9.31)$$

From the denominator of Eq. (9.29), the three poles are located at frequencies of

$$s = -\frac{1}{(R_I + R_G)C_1}, -\frac{1}{(R_D + R_3)C_2}, -\frac{1}{\left(R_S \left\| \frac{1}{g_m}\right.\right)C_3} \qquad (9.32)$$

These pole frequencies are determined by the time constants associated with the three individual capacitors. Because the input resistance of the FET is infinite, the resistance present at the terminals of capacitor C_1 is simply the series combination of R_I and R_G, and since the output resistance r_o of the FET has been neglected, the resistance associated with capacitor C_2 is the series combination of R_3 and R_D. The effective resistance in parallel with capacitor C_3 is the equivalent resistance present at the source terminal of the FET, which is equal to the parallel combination of resistor R_S and $1/g_m$. Section 9.3 has a more complete interpretation of these resistance expressions.

DESIGN NOTE

Each independent capacitor (or inductor) contributes one pole and one zero to the circuit transfer function. (Some poles or zeros may be at zero or infinite frequency.)

EXAMPLE 9.2   **DIRECT CALCULATION OF THE POLES AND ZEROS OF THE COMMON-SOURCE AMPLIFIER**

Analyze the low-frequency behavior of a common-source amplifier, including the effects of coupling and bypass capacitors.

PROBLEM Find the midband gain, poles, zeros, and cutoff frequency for the common-source amplifier in Fig. 9.4. Assume $g_m = 1.23$ mS. Write a complete expression for the amplifier transfer function. Write a dominant-pole representation for the amplifier transfer function.

SOLUTION **Known Information and Given Data:** The circuit with element values appears in Fig. 9.4, and $g_m = 1.23$ mS. Expressions for A_{mid} and the individual poles and zeros are given in Eqs. (9.29) and (9.30).

Unknowns: A_{mid}, poles, zeros, f_L, dominant-pole approximation, complete transfer function

Approach: Use the circuit element values to find A_{mid} and the poles and zeros from Eqs. (9.31) and (9.32). Use the pole and zero values to find f_L from Eq. (9.15).

Assumptions: Small-signal conditions apply; output resistance r_o can be neglected

Analysis: To begin, we will find A_{mid}:

$$A_{mid} = -(1.23 \text{ mS})(4.3 \text{ k}\Omega \parallel 100 \text{ k}\Omega)\frac{243 \text{ k}\Omega}{1.0 \text{ k}\Omega + 243 \text{ k}\Omega} = -5.05 \quad \text{or} \quad 14.1 \text{ dB}$$

From Eq. (9.29), the three zeros are

$$\omega_{Z1} = 0 \quad \omega_{Z2} = 0 \quad \omega_{Z3} = -\frac{1}{(10 \text{ μF})(1.3 \text{ k}\Omega)} = -76.9 \text{ rad/s}$$

and the three poles are

$$\omega_{P1} = -\frac{1}{(0.1 \text{ μF})(1 \text{ k}\Omega + 243 \text{ k}\Omega)} = -41.0 \text{ rad/s}$$

$$\omega_{P2} = -\frac{1}{(0.1 \text{ μF})(4.3 \text{ k}\Omega + 100 \text{ k}\Omega)} = -95.9 \text{ rad/s}$$

$$\omega_{P3} = -\frac{1}{(10 \text{ μF})\left(1.3 \text{ k}\Omega \parallel \frac{1}{1.23 \text{ mS}}\right)} = -200 \text{ rad/s}$$

The lower-cutoff frequency is given by

$$f_L = \frac{1}{2\pi}\sqrt{41.0^2 + 95.9^2 + 200^2 - 2(0^2 + 0^2 + 76.9^2)} = \frac{197}{2\pi} = 31.5 \text{ Hz}$$

and the complete transfer function is

$$A_v(s) = -5.05 \frac{s^2(s + 76.9)}{(s + 41.0)(s + 95.9)(s + 200)}$$

The dominant-pole estimate could be written using either the calculated value of f_L or the highest pole:

$$A_v(s) \cong -5.05 \frac{s}{s + 197} \quad \text{or} \quad A_v(s) \cong -5.05 \frac{s}{s + 200}$$

Check of Results: A double check of our math indicates the calculations are correct. We see that A_{mid} is small, so that neglecting r_o should be reasonable.

Discussion: Although the poles and zeros are not widely spaced, the lower-cutoff frequency is surprisingly close to ω_{P3}. This occurs because of an approximate pole-zero cancellation that is taking place between ω_{Z3} and ω_{P2}.

Computer-Aided Analysis: SPICE simulation results for the common-source amplifier appear in the figure here. The simulation used $V_{DD} = 12$ V, FSTART = 0.01 Hz, and FSTOP = 10 kHz with 10 frequency points per decade. The values of A_{mid} and f_L agree with our hand calculations.

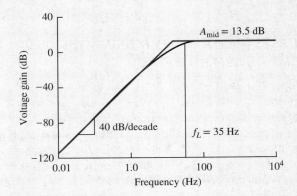

SPICE simulation results for the C-S amplifier in Fig. 9.4 ($V_{DD} = 12$ V).

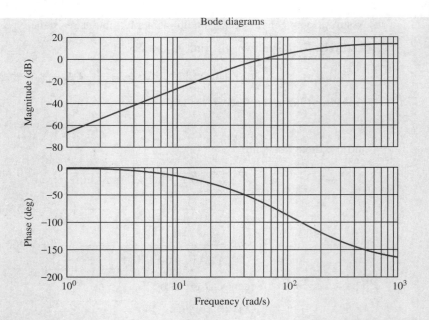

The small discrepancies are related to our neglect of r_o in the calculations. We can also plot $A_v(s)$ by multiplying out the numerator and denominator and then using MATLAB®: bode (−5.05*[1 76.9 0 0],[1 336.9 31311.9 786380]) or by using the convolution function to multiply the polynomials for us: bode(−5.05*[1 76.9 0 0],[conv([1 41],conv([1 95.9],[1 200]))]).

EXERCISE: Find the new values of A_{mid}, the poles and zeros, and f_L if the value of C_3 is reduced to 2 μF.

ANSWERS: −5.05; 0; 0; −385 rad/s; −41.0 rad/s; −95.9 rad/s; −1000 rad/s; 135 Hz

EXERCISE: What value of output resistance r_o is needed to account for the difference in A_{mid} between our hand calculations and the SPICE simulation results?

ANSWER: 57.5 kΩ

EXERCISE: Suppose that the output resistance in the previous exercise appears in parallel with R_D in the expressions for ω_{P2}. What are the new values of ω_{P2} and f_L?

ANSWERS: 96.2 rad/s, 31.5 Hz

9.3 ESTIMATION OF ω_L USING THE SHORT-CIRCUIT TIME-CONSTANT METHOD

To use Eq. (9.16) or Eq. (9.23), the location of all the poles and zeros of the amplifier must be known. In most cases, however, it is not easy to find the complete transfer function, let alone represent it in factored form. Fortunately, we are most often interested in the values of A_{mid}, and the upper- and lower-cutoff frequencies ω_H and ω_L that define the bandwidth of the amplifier, as indicated in Fig. 9.5. Knowledge of the exact position of all the poles and zeros is not necessary. Two techniques, the **short-circuit time-constant (SCTC) method** and the **open-circuit time-constant (OCTC) method,** (see Sec. 9.6) have been developed that yield good estimates of ω_L and ω_H, respectively, without having to find the complete transfer function.

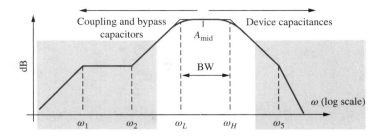

Figure 9.5 Midband region of primary interest in most amplifier transfer functions.

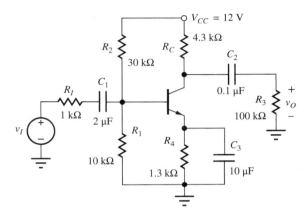

Figure 9.6 Common-emitter amplifier including finite capacitor values.

It can be shown theoretically [1] that the lower-cutoff frequency for a network having n coupling and bypass capacitors can be estimated from

$$\omega_L \cong \sum_{i=1}^{n} \frac{1}{R_{iS} C_i} \qquad (9.33)$$

in which R_{iS} represents the resistance at the terminals of the ith capacitor C_i with all the other capacitors replaced by short circuits. The product $R_{iS} C_i$ represents the short-circuit time constant associated with capacitor C_i. We now use the SCTC method to find ω_L for the three classes of single-stage amplifiers.

9.3.1 ESTIMATE OF ω_L FOR THE COMMON-EMITTER AMPLIFIER

We use the C-E amplifier in Fig. 9.6 that includes finite values for the capacitors as a first example of the SCTC method. The presence of r_π in the bipolar model causes direct calculation of the transfer function to be complex; including r_o leads to even further difficulty. Thus, the circuit is a good example of applying the method of short-circuit time constants to a network.

The ac model for the C-E amplifier in Fig. 9.7 contains three capacitors, and three short-circuit time constants must be determined in order to apply Eq. (9.33). The three analyses rely on the expressions for the midband input and output resistances of the BJT amplifier in Table 8.9 (page 475).

R_{1S}

For C_1, R_{1S} is found by replacing C_2 and C_3 by short circuits, yielding the network in Fig. 9.8. R_{1S} represents the equivalent resistance present at the terminals of capacitor C_1. Based on Fig. 9.8,

$$R_{1S} = R_I + (R_B \| R_{iB}) = R_I + (R_B \| r_\pi) \qquad (9.34)$$

R_{1S} is equal to the source resistance R_I in series with the parallel combination of the base bias resistor R_B and the input resistance r_π of the BJT.

The Q-point for this amplifier is found to be (1.66 mA, 2.70 V), and for $\beta_o = 100$ and $V_A = 75$ V,

$$r_\pi = 1.51 \text{ k}\Omega \qquad \text{and} \qquad r_o = 46.8 \text{ k}\Omega$$

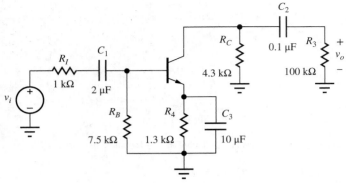

Figure 9.7 ac Model for the C-E amplifier in Fig. 9.6.

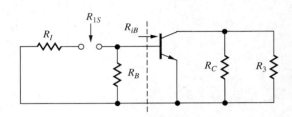

Figure 9.8 Circuit for finding R_{1S}.

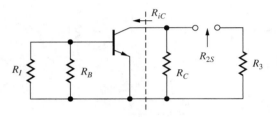

Figure 9.9 Circuit for finding R_{2S}.

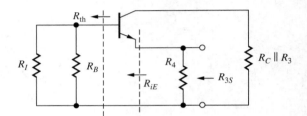

Figure 9.10 Circuit for finding R_{3S}.

Using these values and those of the other circuit elements,

$$R_{1S} = 1000\ \Omega + (7500\ \Omega \,\|\, 1510\ \Omega) = 2260\ \Omega$$

and

$$\frac{1}{R_{1S}C_1} = \frac{1}{(2.26\ \text{k}\Omega)(2.00\ \mu\text{F})} = 222\ \text{rad/s} \qquad (9.35)$$

R_{2S}

The network used to find R_{2S} is constructed by shorting capacitors C_1 and C_3, as in Fig. 9.9. For this network,

$$R_{2S} = R_3 + (R_C \,\|\, R_{iC}) = R_3 + (R_C \,\|\, r_o) \cong R_3 + R_C \qquad (9.36)$$

R_{2S} represents the combination of load resistance R_3 in series with the parallel combination of collector resistor R_C and the collector resistance r_o of the BJT. For the values in this particular circuit,

$$R_{2S} = 100\ \text{k}\Omega + (4.30\ \text{k}\Omega \,\|\, 46.8\ \text{k}\Omega) = 104\ \text{k}\Omega \qquad (9.37)$$

and

$$\frac{1}{R_{2S}C_2} = \frac{1}{(104\ \text{k}\Omega)(0.100\ \mu\text{F})} = 96.1\ \text{rad/s} \qquad (9.38)$$

R_{3S}

Finally, the network used to find R_{3S} is constructed by shorting capacitors C_1 and C_2, as in Fig. 9.10, and

$$R_{3S} = R_4 \,\|\, R_{iE} = R_4 \,\left\|\, \frac{r_\pi + R_\text{th}}{\beta_o + 1} \right. \qquad \text{where} \qquad R_\text{th} = R_I \,\|\, R_B \qquad (9.39)$$

and r_o has been neglected. R_{3S} represents the combination of emitter resistance R_4 in parallel with the equivalent resistance at the emitter terminal of the BJT. For the values in this particular circuit,

$$R_\text{th} = R_I \,\|\, R_B = 1000\ \Omega \,\|\, 7500\ \Omega = 882\ \Omega$$

$$R_{3S} = 1300\ \Omega \,\left\|\, \frac{1510\ \Omega + 882\ \Omega}{101} \right. = 23.3\ \Omega$$

and

$$\frac{1}{R_{3S}C_3} = \frac{1}{(23.3 \text{ }\Omega)(10 \text{ }\mu\text{F})} = 4300 \text{ rad/s} \tag{9.40}$$

The ω_L Estimate
Using the three time-constant values from Eqs. (9.35), (9.38), and (9.40) yields estimates for ω_L and f_L:

$$\omega_L \cong \sum_{i=1}^{3} \frac{1}{R_{iS}C_i} = 222 + 96.1 + 4300 = 4620 \text{ rad/s} \tag{9.41}$$

and

$$f_L = \frac{\omega_L}{2\pi} = 735 \text{ Hz}$$

The lower-cutoff frequency of the amplifier is approximately 735 Hz.

Note in this example that the time constant associated with emitter bypass capacitor C_3 is dominant; that is, the value of $R_{3S}C_3$ is more than an order of magnitude larger than the other two time constants so that $\omega_L \cong 1/R_{3S}C_3$ ($f_L \cong 4300/2\pi = 685$ Hz). This is a common situation and represents a practical approach to the design of ω_L. Because the resistance presented at the emitter or source of the transistor is low, the time constant associated with an emitter or source bypass capacitor is often dominant and can be used to set ω_L. The other two time constants can easily be designed to be much larger.

EXERCISE: Simulate the frequency response of the circuit in Fig. 9.6 using SPICE, and find the midband gain and lower-cutoff frequency. Use $\beta_o = 100$, $I_S = 1$ fA, and $V_A = 75$ V. What is the Q-point?

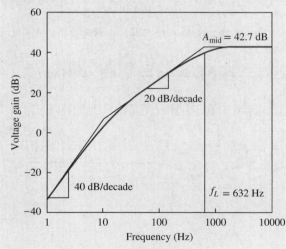

SPICE simulation results.

ANSWERS: 135, 635 Hz, (1.64 mA, 2.79 V)

EXERCISE: Find the short-circuit time constants and f_L for the common-emitter amplifier in Fig. 9.7 if $R_B = 75$ kΩ, $R_4 = 13$ kΩ, $R_C = 43$ kΩ, and $I_C = 175$ µA. Assume $\beta_o = 140$ and $V_A = 80$ V. The other values remain unchanged.

ANSWERS: 33.6 ms; 1.47 ms; 14.3 ms; 124 Hz

DESIGN EXAMPLE 9.3

LOWER-CUTOFF FREQUENCY DESIGN IN THE COMMON-EMITTER AMPLIFIER

Choose the coupling and bypass capacitors to set the value of f_L of the common-emitter amplifier to a specified value.

PROBLEM Choose C_1, C_2, and C_3 to set $f_L = 2000$ Hz in the amplifier in Fig. 9.6.

SOLUTION **Known Information and Given Data:** The circuit with resistor values appears in Fig. 9.6 with $\beta_o = 100$, $r_\pi = 1.51$ kΩ, and $r_o = 46.8$ kΩ. From Eqs. (9.34) through (9.40), we have $R_{1S} = 2.26$ kΩ, $R_{2S} = 23.3$ Ω, and $R_{3S} = 104$ kΩ.

Unknowns: C_1, C_2, and C_3

Approach: Because R_{3S} is much smaller than the other two resistors, its associated time constant can easily be designed to dominate the value of ω_L as occurred in Eq. (9.41). Thus, the approach taken here is to use C_3 to set f_L and to choose C_1 and C_2 so that their contributions are negligible.

Assumptions: Small-signal conditions apply. $V_T = 25.0$ mV.

Analysis: Choosing C_3 to set f_L yields

$$C_3 \cong \frac{1}{R_{3S}\omega_L} = \frac{1}{23.3 \; \Omega(2\pi)(2000 \; \text{Hz})} = 3.42 \; \mu\text{F}$$

Let us choose C_1 and C_2 so that their individual time constants are each 100 times larger than that associated with C_3—that is, each capacitor will contribute only a 1 percent error to f_L.

$$C_1 = 100\frac{R_{3S}C_3}{R_{1S}} = 100\frac{(23.2 \; \Omega)(3.42 \; \mu\text{F})}{2.26 \; \text{k}\Omega} = 3.51 \; \mu\text{F}$$

$$C_2 = 100\frac{R_{3S}C_3}{R_{2S}} = 100\frac{(23.2 \; \Omega)(3.42 \; \mu\text{F})}{104 \; \text{k}\Omega} = 0.0763 \; \mu\text{F}$$

Picking the nearest values from the capacitor table in Appendix A, we have $C_1 = 3.3$ μF, $C_2 = 0.082$ μF, and $C_3 = 3.3$ μF.

Check of Results: Let us check by calculating the actual values of f_L.

$$f_L = \frac{1}{2\pi}\left[\frac{1}{2.26 \; \text{k}\Omega(3.3 \; \mu\text{F})} + \frac{1}{104 \; \text{k}\Omega(0.082 \; \mu\text{F})} + \frac{1}{23.2 \; \Omega(3.3 \; \mu\text{F})}\right] = 2120 \; \text{Hz}$$

Discussion: The cutoff frequency is slightly higher than the design value because of the use of the 3.3-µF capacitor and the small contributions from C_1 and C_2. At additional cost, one could use two capacitors to make up the desired value. However, the tolerances on typical capacitors are relatively large, and one would need to use a precision capacitor (and resistors) if a more accurate value of f_L is required. (See simulation results on previous page.)

Computer-Aided Analysis: The frequency response with the new capacitor values can be simulated using SPICE ac analysis with FSTART = 100 Hz and FSTOP = 1 MHz with 20 frequency points per decade. The transistor parameters were set to IS = 3 fA, BF = 100, and VAF = 75 V. SPICE simulation results for the new common-emitter design results yields $A_{mid} = -138(42.8 \text{ dB})$ and $f_L = 2120$ Hz. The value of f_L is approximately 5 percent larger than our design value. This discrepancy is due to differences in V_T and the Q-point current as well as the smaller value for C_3.

EXERCISE: Estimate the midband gain for the circuit in Fig. 9.6. What is the source of the error between this value and SPICE?

ANSWER: −157; Neglect of r_o accounts for most of the difference.

9.3.2 ESTIMATE OF ω_L FOR THE COMMON-SOURCE AMPLIFIER

Equations (9.34), (9.36), and (9.39) can be applied directly to the C-S FET amplifier in Fig. 9.11 by substituting infinity for the values of the transistor's input resistance and current gain. These equations reduce directly to:

$$R_{1S} = R_I + (R_G \| R_{iG}) = R_I + R_G$$
$$R_{2S} = R_3 + (R_D \| R_{iD}) = R_3 + (R_D \| r_o) \cong R_3 + R_D \qquad (9.42)$$
$$R_{3S} = R_S \| R_{iS} = R_S \left\| \frac{1}{g_m} \right.$$

The three expressions in Eq. (9.42) represent the short-circuit resistances associated with the three capacitors in the circuit, as indicated in the ac circuit models in Figs. 9.12(a) to (c). Note that the three time constants are the same as those found by the direct approach that yielded Eq. (9.29).

EXERCISE: Find the short-circuit time constants and f_L for the common-source amplifier in Fig. 9.11 if $I_D = 1.5$ mA and $V_{GS} - V_{TN} = 0.5$ V. Assume $\lambda = 0.015/\text{V}$. The other values remain unchanged.

ANSWERS: 24.4 ms; 10.4 ms; 1.48 ms; 129 Hz

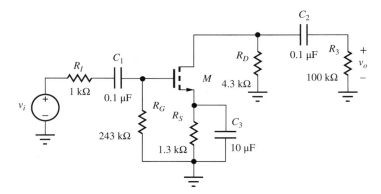

Figure 9.11 ac model for common-source amplifier.

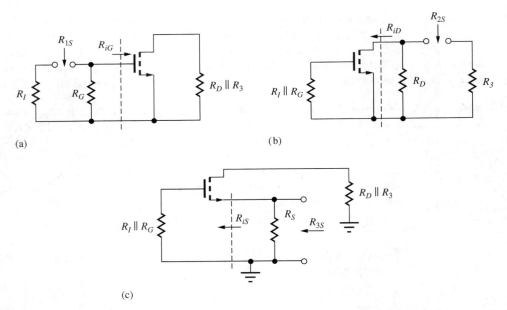

Figure 9.12 (a) Resistance R_{1S} at the terminals of C_1. (b) Resistance R_{2S} at the terminals of C_2. (c) Resistance R_{3S} at the terminals of C_3.

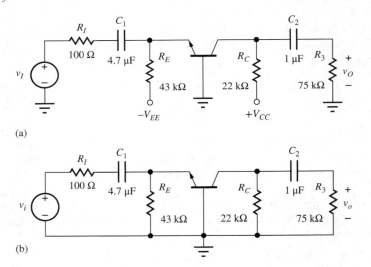

Figure 9.13 (a) Common-base amplifier. (b) Low-frequency ac equivalent circuit.

9.3.3 ESTIMATE OF ω_L FOR THE COMMON-BASE AMPLIFIER

Next, we apply the short-circuit time-constant technique to the common-base amplifier in Fig. 9.13. The results are also directly applicable to the common-gate case if β_o and r_π are set equal to infinity. Figure 9.13(b) is the low-frequency ac equivalent circuit for the common-base amplifier. In this particular circuit, coupling capacitors C_1 and C_2 are the only capacitors present, and expressions for R_{1S} and R_{2S} are needed.

R_{1S}

R_{1S} is found by shorting capacitor C_2, as indicated in the circuit in Fig. 9.14. Based on this figure,

$$R_{1S} = R_I + (R_E \| R_{iE}) \cong R_I + \left(R_E \left\| \frac{1}{g_m}\right.\right) \tag{9.43}$$

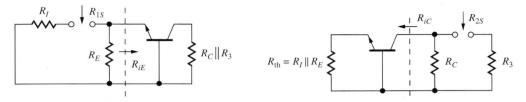

Figure 9.14 Equivalent circuit for determining R_{1S}. **Figure 9.15** Equivalent circuit for determining R_{2S}.

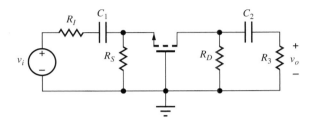

Figure 9.16 ac Circuit for common-gate amplifier.

R_{2S}
Shorting capacitor C_1 yields the circuit in Fig. 9.15, and the expression for R_{2S} is

$$R_{2S} = R_3 + (R_C \| R_{iC}) \cong R_3 + R_C \qquad (9.44)$$

because $R_{iC} \cong r_o(1 + g_m R_{th})$ is large.

EXERCISE: Find the short-circuit time constants and f_L for the common-base amplifier in Fig. 9.13 if $\beta_o = 100$, $V_A = 70$ V, and the Q-point is (0.1 mA, 5 V). What is A_{mid}?

ANSWERS: 1.64 ms, 97.0 ms, 98.7 Hz; 48.6

9.3.4 ESTIMATE OF ω_L FOR THE COMMON-GATE AMPLIFIER

The expressions for R_{1S} and R_{2S} for the common-gate amplifier in Fig. 9.16 are virtually identical to those of the common-base stage:

$$R_{1S} = R_I + (R_S \| R_{iS}) = R_I + \left(R_S \left\| \frac{1}{g_m}\right.\right) \qquad (9.45)$$
$$R_{2S} = R_3 + (R_D \| R_{iD}) \cong R_3 + R_D \quad \text{because} \quad R_{iD} \cong \mu_f(R_S \| R_I)$$

EXERCISE: Draw the circuits used to find R_{1S} and R_{2S} for the common-gate amplifier in Fig. 9.16 and verify the results presented in Eq. (9.45).

EXERCISE: Find the short-circuit time constants and f_L for the common-gate amplifier in Fig. 9.16 if $R_I = 100$ Ω, $R_S = 1.3$ kΩ, $R_D = 4.3$ kΩ, $R_3 = 75$ kΩ, $C_1 = 1$ μF, $C_2 = 0.1$ μF, $I_D = 1.5$ mA, and $V_{GS} - V_{TN} = 0.5$ V. Assume $\lambda = 0$.

ANSWERS: 0.248 ms; 7.93 ms; 663 Hz

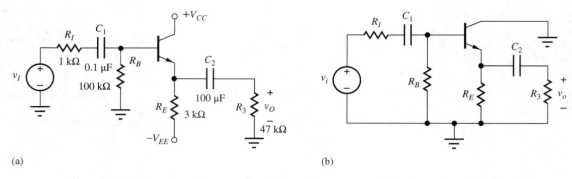

Figure 9.17 (a) Common-collector amplifier. (b) Low-frequency ac model for the common-collector amplifier.

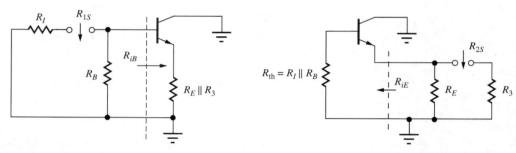

Figure 9.18 Circuit for finding R_{1S}. **Figure 9.19** Circuit for finding R_{2S}.

9.3.5 ESTIMATE OF ω_L FOR THE COMMON-COLLECTOR AMPLIFIER

Figures 9.17(a) and (b) are schematics of an emitter follower and its corresponding low-frequency ac model, respectively. This circuit has two coupling capacitors, C_1 and C_2. The circuit for R_{1S} in Fig. 9.18 is constructed by shorting C_2, and the expression for R_{1S} is

$$R_{1S} = R_I + (R_B \| R_{iB}) = R_I + (R_B \| [r_\pi + (\beta_o + 1)(R_E \| R_3)]) \qquad (9.46)$$

Similarly, the circuit used to find R_{2S} is found by shorting capacitor C_1, as in Fig. 9.19, and

$$R_{2S} = R_3 + (R_E \| R_{iE}) = R_3 + \left(R_E \left\| \frac{R_{th} + r_\pi}{\beta_o + 1} \right. \right) \qquad (9.47)$$

where R_{iE} neglects r_o in Eq. 9.47.

9.3.6 ESTIMATE OF ω_L FOR THE COMMON-DRAIN AMPLIFIER

The corresponding low-frequency ac model for the common-drain amplifier appears in Fig. 9.20. Taking the limits as β_o and r_π approach infinity, Eqs. (9.46) and (9.47) become

$$R_{1S} = R_I + (R_G \| R_{iG}) = R_I + R_G \quad \text{because} \quad R_{iG} = \infty$$

and (9.48)

$$R_{2S} = R_3 + (R_S \| R_{iS}) = R_3 + \left(R_S \left\| \frac{1}{g_m} \right. \right)$$

> **EXERCISE:** Find the short-circuit time constants and f_L for the common-collector amplifier in Fig. 9.17(a) if $\beta_o = 100$, $V_A = 70$ V, and the Q-point = (1 mA, 5 V). What is A_{mid}?
>
> **ANSWERS:** 7.52 ms, 4.70 s, 21.2 Hz; 0.978

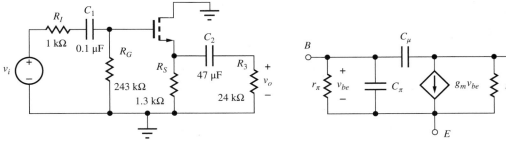

Figure 9.20 Low-frequency ac equivalent circuit for common-drain amplifier.

Figure 9.21 Capacitances in the hybrid-pi model of the BJT.

EXERCISE: Find the short-circuit time constants and f_L for the common-drain amplifier in Fig. 9.20 if $g_m = 1$ mS. What is A_{mid}?

ANSWERS: 24.4 ms, 1.16 ms, 6.66 Hz; 0.550

9.4 TRANSISTOR MODELS AT HIGH FREQUENCIES

To explore the upper limits of amplifier frequency response, the high-frequency limitations of the transistors, which we have ignored thus far, must be taken into account. All electronic devices have capacitances between their various terminals, and these capacitances limit the range of frequencies for which the devices can provide useful voltage, current, or power gain. This section develops the description of the frequency-dependent hybrid-pi model for the bipolar transistor, as well as a similar model for the field-effect transistor.

9.4.1 FREQUENCY-DEPENDENT HYBRID-PI MODEL FOR THE BIPOLAR TRANSISTOR

In the BJT, capacitances appear between the base-emitter and base-collector terminals of the transistor and are included in the small-signal hybrid-pi model in Fig. 9.21. C_μ connected between the base and collector terminals represents the capacitance of the reverse-biased collector-base junction of the bipolar transistor and is related to the Q-point through an expression equivalent to Eq. (3.21), Chapter 3:

$$C_\mu = \frac{C_{\mu o}}{\sqrt{1 + \frac{V_{CB}}{\phi_{jc}}}} \tag{9.49}$$

In Eq. (9.49), $C_{\mu o}$ represents the total collector-base junction capacitance at zero bias, and ϕ_{jc} is the built-in potential of the collector-base junction, typically 0.6 to 1.0 V.

The internal capacitance between the base and emitter terminals, denoted by C_π, represents the diffusion capacitance associated with the forward-biased base-emitter junction of the transistor. C_π is related to the Q-point through Eq. (4.42) in Sec. 4.8:

$$C_\pi = g_m \tau_F \tag{9.50}$$

in which τ_F is the forward transit-time of the bipolar transistor. In Fig. 9.21, C_π appears directly in parallel with r_π. For a given input signal current, the impedance of C_π causes the base-emitter voltage v_{be} to be reduced as frequency increases, thereby reducing the current in the controlled source at the output of the transistor.

Shunt capacitances such as C_π are always present in electronic devices and circuits. At low frequencies, the impedance of these capacitances is usually very large and so has negligible effect relative to the resistances such as r_π. However, as frequency increases, the impedance of C_π becomes smaller and smaller, and v_{be} eventually approaches zero. At very high frequencies, C_μ also shorts the base and collector terminals together. Thus, transistors cannot provide amplification at arbitrarily high frequencies.

9.4.2 MODELING C_π AND C_μ IN SPICE

In SPICE, the value of C_π is determined by the forward transit time TF and C_μ depends upon the zero-bias value of the collector-junction capacitance CJC, the built-in potential VJC of the collector-base junction, and the grading factor MJC of the collector-base junction. In SPICE, C_π and C_μ are referred to as C_{BE} and C_{BC}, respectively.

$$C_{BE} = g_m \text{TF} \quad \text{and} \quad C_{BC} = \frac{\text{CJC}}{\left(1 + \dfrac{\text{VCB}}{\text{VJC}}\right)^{\text{MJC}}} \quad (9.51)$$

VJC defaults to 0.75 V, and MJC defaults to 0.33.

9.4.3 UNITY-GAIN FREQUENCY f_T

A quantitative description of the behavior of the transistor at high frequencies can be found by calculating the frequency-dependent short-circuit current gain $\beta(s)$ from the circuit in Fig. 9.22. For a current $\mathbf{I_b}(s)$ injected into the base, the collector current $\mathbf{I_c}(s)$ consists of two components:

$$\mathbf{I_c}(s) = g_m \mathbf{V_{be}}(s) - \mathbf{I_\mu}(s) \quad (9.52)$$

Because the voltage at the collector is zero, v_{be} appears directly across C_μ and $\mathbf{I_\mu}(s) = sC_\mu \mathbf{V_{be}}(s)$. Therefore,

$$\mathbf{I_c}(s) = (g_m - sC_\mu)\mathbf{V_{be}}(s) \quad (9.53)$$

Because the collector is connected directly to ground, C_π and C_μ appear in parallel in this circuit, and the base current flows through the parallel combination of r_π and $(C_\pi + C_\mu)$ to develop the base-emitter voltage:

$$\mathbf{V_{be}}(s) = \mathbf{I_b}(s)\frac{r_\pi \dfrac{1}{s(C_\pi + C_\mu)}}{r_\pi + \dfrac{1}{s(C_\pi + C_\mu)}} = \mathbf{I_b}(s)\frac{r_\pi}{s(C_\pi + C_\mu)r_\pi + 1} \quad (9.54)$$

By combining Eqs. (9.53) and (9.54), we reach an expression for the frequency-dependent current gain:

$$\beta(s) = \frac{\mathbf{I_c}(s)}{\mathbf{I_b}(s)} = \frac{\beta_o\left(1 - \dfrac{sC_\mu}{g_m}\right)}{s(C_\pi + C_\mu)r_\pi + 1} \quad (9.55)$$

A right-half-plane transmission zero occurs in the current gain at an extremely high frequency, $\omega_Z = +g_m/C_\mu$, and can almost always be neglected. Neglecting ω_Z results in the following

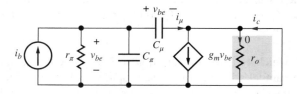

Figure 9.22 Finding the short-circuit current gain β of the BJT.

9.4 Transistor Models at High Frequencies

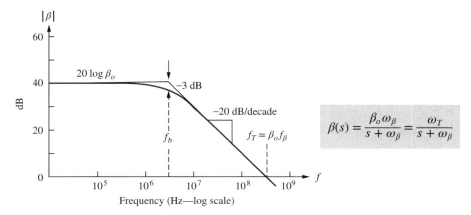

Figure 9.23 Common-emitter current gain versus frequency for the BJT.

simplified expression for $\beta(s)$:

$$\beta(s) \cong \frac{\beta_o}{s(C_\pi + C_\mu)r_\pi + 1} = \frac{\beta_o}{\frac{s}{\omega_\beta} + 1} \qquad (9.56)$$

in which ω_β represents the **beta-cutoff frequency,** defined by

$$\omega_\beta = \frac{1}{r_\pi(C_\pi + C_\mu)} \qquad \text{and} \qquad f_\beta = \frac{\omega_\beta}{2\pi} \qquad (9.57)$$

Figure 9.23 is a Bode plot for Eq. (9.56). From Eq. (9.56) and this graph, we see that the current gain has the value of $\beta_o = g_m r_\pi$ at low frequencies and exhibits a single-pole roll-off at frequencies above f_β, decreasing at a rate of 20 dB/decade and crossing through unity gain at $f = f_T$. The magnitude of the current gain is 3 dB below its low-frequency value at the beta-cutoff frequency, f_β.

Equation (9.56) can be recast in terms of $\omega_T = \beta_o \omega_\beta$ as

$$\beta(s) = \frac{\beta_o \omega_\beta}{s + \omega_\beta} = \frac{\omega_T}{s + \omega_\beta} \qquad (9.58)$$

where $\omega_T = 2\pi f_T$. Parameter f_T is referred to as the **unity gain-bandwidth product** of the transistor and characterizes one of the fundamental frequency limitations of the transistor. At frequencies above f_T, the transistor no longer offers any current gain and fails to be useful as an amplifier.

A relationship between the unity gain-bandwidth product and the small-signal parameters can be obtained from Eqs. (9.57) and (9.58):

$$\omega_T = \beta_o \omega_\beta = \frac{\beta_o}{r_\pi(C_\pi + C_\mu)} = \frac{g_m}{C_\pi + C_\mu} \qquad (9.59)$$

Note that the transmission zero occurs at a frequency beyond ω_T:

$$\omega_Z = \frac{g_m}{C_\mu} > \frac{g_m}{C_\pi + C_\mu} = \omega_T \qquad (9.60)$$

To perform numeric calculations, we determine the values of f_T and C_μ from a transistor's specification sheet and then calculate C_π by rearranging Eq. (9.59):

$$C_\pi = \frac{g_m}{\omega_T} - C_\mu \qquad (9.61)$$

596 Chapter 9 Amplifier Frequency Response

From Eq. (9.49) we can see that C_μ is only a weak function of operating point, but recasting g_m in Eq. (9.61) demonstrates that C_π is directly proportional to collector current:

$$C_\pi = \frac{40 I_C}{\omega_T} - C_\mu \qquad (9.62)$$

EXAMPLE 9.4   **BIPOLAR TRANSISTOR MODEL PARAMETERS**

Find a set of model parameters for a bipolar transistor from its specification sheet.

PROBLEM Find values of β_o, I_S, V_A, f_T, C_π, and C_μ for the CA-3096 *npn* transistors operating at a collector current of 1 mA using the specifications sheets on the MCD Connect site.

SOLUTION **Known Information and Given Data:** CA-3096 specification sheets; $I_C = 1$ mA

Unknowns: β_o, I_S, V_A, f_T, C_π, and C_μ

Approach: We will use our definitions of, and relationships between, the large-signal and small-signal parameters to find the unknown values.

Assumptions: $T = 25°C$ and $V_{CE} = 5$ V, corresponding to the electrical specification sheets; active region operation; $\beta_o \cong \beta_F$; the built-in potential of the collector-base junction is 0.75 V.

Analysis: Based on the typical values in the specification sheets, we find $\beta_F = h_{FE} = 390$, $V_{BE} = 0.69$ V, $f_T = 280$ MHz, and $C_{CB} = 0.46$ pF at $V_{CB} = 3$ V. From the graph of output resistance versus current, we find $r_o = 80$ kΩ for $I_c = 1$ mA. For $T = 25°C$, $V_T = 26.0$ mV.

We find the current gain and Early voltage using the values of h_{FE} and r_o,

$$\beta_o \cong h_{FE} = 390 \qquad V_A = I_C r_o - V_{CE} = 75 \text{ V}$$

and I_S is found from I_C, V_{BE}, and V_T:

$$I_S = \frac{I_C}{\exp\left(\dfrac{V_{BE}}{V_T}\right)} = \frac{1 \text{ mA}}{\exp\left(\dfrac{0.69 \text{ V}}{26.0 \text{ mV}}\right)} = 2.98 \text{ fA}$$

Capacitance C_μ is equal to the collector-base capacitance of the transistor, but it is specified at $V_{CB} = 3$ V. Using Eq. (9.49), we find $C_{\mu o}$, and then calculate C_μ for $V_{CB} = 5 - .69 = 4.31$ V.

$$C_{\mu o} \cong C_{CB} \sqrt{1 + \frac{V_{CB}}{\phi_{jc}}} = 0.46 \text{ pF} \sqrt{1 + \frac{3}{0.75}} = 1.03 \text{ pF}$$

$$C_\mu \cong \frac{C_{\mu o}}{\sqrt{1 + \dfrac{V_{CB}}{\phi_{jc}}}} = \frac{1.03 \text{ pF}}{\sqrt{1 + \dfrac{4.31}{0.75}}} = 0.397 \text{ pF}$$

Now we can find C_π:

$$C_\pi = \frac{g_m}{\omega_T} - C_\mu = \frac{1 \text{ mA}}{26.0 \text{ mV}} \frac{1}{2\pi(280 \text{ MHz})} - 0.40 \text{ pF} = 21.5 \text{ pF}$$

Check of Results: We have found the required values of β_o, I_S, V_A, f_T, C_π, and C_μ. The calculations appear correct and reasonable. The calculated value of C_μ agrees reasonably well with the graph of C_{CB} versus V_{CB} in the specification sheets.

Discussion: The values in the specification sheets must often be mapped into the parameters that we need, and the data supplied is often incomplete. Some may be presented in tabular form; others must be found from graphs. Note that the current gain peaks at a collector current of approximately 1 mA, whereas f_T peaks at approximately 4 mA.

Computer-Aided Analysis: Let us now attempt to create a SPICE model that has these parameters. We must set IS = 2.98 fA, BF = 390, and VAF = 75 V. Using Eq. (9.51), we also have TF = 559 ps, CJC = 1.03 pF, VJC = 0.75 V, and MJC = 0.5. Let us bias the transistor as in the circuit shown here, and request the device parameters as an output following an operating point analysis. The results are $I_C = 1$ mA, $V_{BE} = 0.685$ V, $V_{BC} = -5$ V, $g_m = 38.7$ mS, $\beta_o = g_m/g_\pi = 416$, $r_o = 1/g_o = 79.9$ kΩ, $C_\pi = 21.6$ pF, and $C_\mu = 0.372$ pF. Our set of device parameters appears to be correct. Note that $\beta_o = BF(1 + VCB/VAF) = 416$.

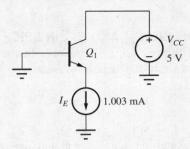

EXERCISE: A bipolar transistor has an $f_T = 500$ MHz and $C_{\mu o} = 2$ pF. What are the values of C_μ and C_π at Q-points of (100 μA, 8 V), (2 mA, 5 V), and (50 mA, 8 V)? Assume $V_{BE} = \phi_{jc} = 0.6$ V.

ANSWERS: 0.551 pF, 0.722 pF; 0.700 pF, 24.8 pF; 0.551 pF, 636 pF

9.4.4 HIGH-FREQUENCY MODEL FOR THE FET

To model the FET at high frequencies, gate-drain and gate-source capacitances C_{GD} and C_{GS} are added to the small-signal model, as shown in Fig. 9.24. For the MOSFET, these two capacitors represent the gate oxide and overlap capacitances discussed previously in Sec. 5.8. At high frequencies, currents through these two capacitors combine to produce a current in the gate terminal, and the signal current i_g can no longer be assumed to be zero. Thus, even the FET has a finite current gain at high frequencies!

The short-circuit current gain for the FET can be calculated in the same manner as for the BJT, as in Fig. 9.25:

$$\mathbf{I_d}(s) = (g_m - sC_{GD})\mathbf{V_{gs}}(s) = \mathbf{I_g}(s)\frac{(g_m - sC_{GD})}{s(C_{GS} + C_{GD})} \qquad (9.63)$$

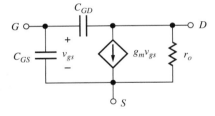

Figure 9.24 Pi model for the FET.

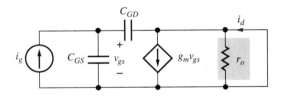

Figure 9.25 Circuit for calculating the short-circuit current gain of the FET.

598 Chapter 9 Amplifier Frequency Response

and

$$\beta(s) = \frac{\mathbf{I_d}(s)}{\mathbf{I_g}(s)} = \frac{g_m\left(1 - \frac{sC_{GD}}{g_m}\right)}{s(C_{GS} + C_{GD})} = \frac{\omega_T}{s}\left(1 - \frac{s}{\omega_T\left(1 + \frac{C_{GS}}{C_{GD}}\right)}\right) \quad (9.64)$$

At dc, the current gain is infinite but falls at a rate of 20 dB/decade as frequency increases. The unity gain-bandwidth product ω_T of the FET is defined in a manner identical to that of the BJT,

$$\omega_T = \frac{g_m}{C_{GS} + C_{GD}} \quad (9.65)$$

and the FET current gain falls below 1 for frequencies in excess of ω_T, just as for the case of the bipolar transistor. The transmission zero now occurs at $\omega_Z = \omega_T(1 + C_{GS}/C_{GD})$, a frequency above ω_T.

9.4.5 MODELING C_{GS} AND C_{GD} IN SPICE

As discussed in Sec. 5.5, we remember that the gate-source and gate-drain capacitances in the active region (pinch-off) are expressed as

$$C_{GS} = C'_{OL}W + \frac{2}{3}C''_{ox}WL \qquad C_{GD} = C'_{OL}W \qquad C''_{ox} = \frac{\varepsilon_{ox}}{T_{ox}} \quad (9.66)$$

The corresponding SPICE parameters are oxide thickness TOX, gate width W, gate length L, gate-source overlap capacitance per unit length CGSO, and gate-drain overlap capacitance per unit length CGDO. Note that SPICE permits definition of different values of overlap capacitance for the source and drain regions of the transistor.

9.4.6 CHANNEL LENGTH DEPENDENCE OF f_T

The unity-gain bandwidth product of the MOSFET is strongly dependent on the channel length, and this fact represents one of the reasons for continuing to scale the technology to smaller and smaller dimensions. The basic expression for the intrinsic $f_T(C'_{OL} = 0)$ of the MOSFET in terms of technology parameters can be found using Eqs. (9.65) and (9.66). If we remember that $g_m = K_n(V_{GS} - V_{TN})$, and assume $C'_{OL} = 0$:

$$f_T = \frac{\mu_n C''_{ox} \frac{W}{L}(V_{GS} - V_{TN})}{\frac{2}{3} C''_{ox} WL} = \frac{3}{2}\frac{\mu_n(V_{GS} - V_{TN})}{L^2} \quad (9.67)$$

The value of f_T is proportional to transistor mobility and inversely dependent upon the square of the channel length. Thus, an NMOS transistor will have a higher-cutoff frequency than a similar PMOS transistor for a given channel length and bias condition. Reducing the channel length by a factor of 10 results in an increase in f_T by a factor of 100!

EXAMPLE 9.5 **MOSFET MODEL PARAMETERS**

Find a set of model parameters for a MOSFET from its specification sheet.

PROBLEM Find values of V_{TN}, K_P, λ, C_{GS}, and C_{GD} for the NMOS transistors in the ALD-1116 transistor operating at a drain current of 10 mA using the specifications sheets on the MCD Connect site.

SOLUTION **Known Information and Given Data:** ALD-1116 specification sheets; $I_D = 10$ mA.

Unknowns: V_{TN}, K_n, λ, C_{GS}, and C_{GD}.

Approach: We will use our definitions of the relationships between the large-signal and small-signal parameters to find the unknown values.

Assumptions: $T = 25°C$ and $V_{DS} = 5$ V, corresponding to the electrical specification sheets; our square-law transistor model applies to the device; the transistor is symmetrical.

Analysis: Based on the typical values in the specification sheets, we find $V_{TN} = 0.7$ V and $I_D = 4.8$ mA for $V_{GS} = 5$ V, output conductance $g_o = 200$ μS at 10 mA, and $C_{ISS} = 1$ pF. First, we can find λ using the output conductance value:

$$\lambda = \left(\frac{I_D}{g_o} - V_{DS}\right)^{-1} = \left(\frac{10 \text{ mA}}{0.2 \text{ mS}} - 5 \text{ V}\right)^{-1} = 0.0222 \text{ V}^{-1}$$

Now we can find K_n using the MOS drain current expression in the active region:

$$K_n = \frac{2I_D}{(V_{GS} - V_{TN})^2(1 + \lambda V_{DS})} = \frac{2(4.8 \text{ mA})}{(5 - 0.7)^2\left(1 + \frac{5 \text{ V}}{45 \text{ V}}\right)} = 467 \frac{\mu A}{V^2}$$

From these results, we can set SPICE parameters to VTO = 0.7 V, KP = 467 μA/V^2, and LAMBDA = 0.0222 V^{-1}.

C_{ISS} is the short-circuit input capacitance of the transistor in the common-source configuration and is equal to the sum of C_{GS} and C_{GD}. Unfortunately, the test conditions are not specified, so we cannot be sure if the measurement was in the triode or active region of operation. If we assume active region operation and use Eq. (9.66), then

$$C_{GS} + C_{GD} = \frac{2}{3} C''_{ox} WL + 2C'_{OL} W = 1 \text{ pF}$$

However, we have no way of directly splitting the 1-pF capacitance between C_{GS} and C_{GD}. One approximation is to assume that the oxide capacitance term is approximately dominant. Then $C_{GS} \cong 1$ pF and $C_{GD} \cong 0$. As we shall see shortly, however, neglecting C_{GD} may cause significant errors in our calculations of the high-frequency response of amplifiers.

Check of Results: We have found the required values. The values of V_{TN} and λ appear reasonable. Let us see if the values of the amplification factor and f_T are reasonable.

$$g_m = \sqrt{2K_n I_D(1 + \lambda V_{DS})} = \sqrt{2(467 \text{ μA/V}^2)(10 \text{ mA})(1 + 5/45)} = 3.22 \text{ mS}$$

$$f_T = \frac{1}{2\pi} \frac{g_m}{C_{GS} + C_{GD}} = \frac{1}{2\pi} \frac{3.22 \text{ mS}}{1 \text{ pF}} = 513 \text{ MHz}$$

$$\mu_f = \frac{g_m}{g_o} = \frac{3.22}{0.2 \text{ mS}} = 16.1$$

The value of f_T is reasonable. The relatively low value of μ_f results from the 10-mA drain current condition. Although the value of g_m is larger than the typical value given in the table, it is reasonably consistent with the graph of the output characteristics: $g_m = \Delta I_D/\Delta V_{GS} = 4.5$ mA/2 V = 2.25 mS.

Discussion: The values in the specification sheets must often be mapped into the parameters that we need, and we often find that the data sheet information is incomplete and not necessarily self-consistent. We often must contact the manufacturer for more information. The manufacturer may be able to supply a SPICE model for the device. As a last resort, we can directly measure the parameters for ourselves, or choose another device.

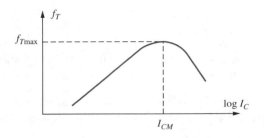

Figure 9.26 Current dependence of f_T.

> **EXERCISE:** What are the values of C_{GS} and C_{GD} if C_{ISS} in Ex. 9.5 had been measured in the triode region?
>
> **ANSWER:** 0.5 pF, 0.5 pF
>
> **EXERCISE:** An NMOSFET has $f_T = 200$ MHz and $K_n = 10$ mA/V^2 and is operating at a drain current of 10 mA. Assume that $C_{GS} = 5C_{GD}$ and find the values of these two capacitors.
>
> **ANSWERS:** $C_{GS} = 9.38$ pF, $C_{GD} = 1.88$ pF

9.4.7 LIMITATIONS OF THE HIGH-FREQUENCY MODELS

The pi-models of the transistor in Figs. 9.21 and 9.25 are good representations of the characteristics of the transistors for frequencies up to approximately $0.3\, f_T$. Above this frequency, the behavior of the simple pi-models begins to deviate significantly from that of the actual device. In addition, our discussion has tacitly assumed that ω_T is constant. However, this is only an approximation. In an actual BJT, ω_T depends on operating current, as shown in Fig. 9.26.

For a given BJT, there will be a collector current I_{CM}, which yields a maximum value of $f_T = f_{T_{max}}$. For the FET operating in the saturation region, C_{GS} and C_{GD} are independent of Q-point current so that $\omega_T \propto g_m \propto \sqrt{I_D}$. In the upcoming discussions, we assume that the specified value of f_T corresponds to the operating point being used.

> **EXERCISE:** As an example of the problem of using a constant value for the transistor f_T, repeat the calculation of C_π and C_μ for a Q-point of (20 μA, 8 V) if $f_T = 500$ MHz, $C_{\mu o} = 2$ pF, and $\phi_{jc} = 0.6$ V.
>
> **ANSWERS:** 0.551 pF, −0.296 pF. Impossible—C_π cannot have a negative value.

9.5 BASE AND GATE RESISTANCES IN THE SMALL-SIGNAL MODELS

One final circuit element, the **base resistance** r_x, completes the basic hybrid-pi description of the bipolar transistor. In the bipolar transistor cross-section in Fig. 9.27, base current i_b enters the transistor through the external base contact and traverses a relatively high resistance region before actually entering the active area of the transistor. Circuit element r_x models the voltage drop between the base contact and the active base region of the transistor and is included between the internal and external base nodes, B' and B, respectively, in the circuit model in Fig. 9.28(a). As discussed in the next section, the base resistance usually can be neglected at low frequencies. However, resistance r_x can represent an important limitation to the high frequency response of the transistor in low-source resistance applications. The thermal noise of r_x is also an important limitation in low-noise amplifier design. Typical values of r_x range from a few ohms to a thousand ohms or more. In SPICE, BJT base resistance is modeled by parameter RB.

9.5 Base and Gate Resistances in the Small-Signal Models

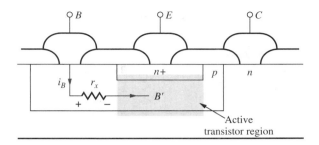

Figure 9.27 Base current in the BJT.

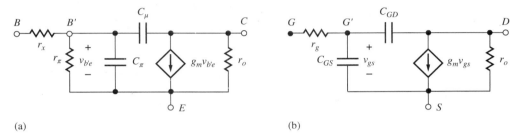

Figure 9.28 (a) Hybrid-pi model including base resistance r_x. (b) Small-signal MOSFET model including gate resistance R_G.

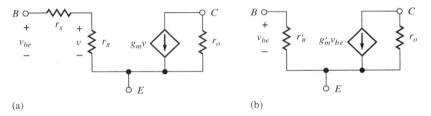

Figure 9.29 (a) Transistor model containing r_x. (b) Model transformation that "absorbs" r_x at midband.

A similar effect occurs in FETs. The gates of MOS transistors are typically formed from metals or polysilicon, all of which have finite conductivity, and there will also be contact resistance any time different layers of material are connected together. Thus, there is a gate resistance r_g in series with the internal and external gate nodes of the MOSFET as indicated in the small-signal model for the MOS transistor in Fig. 9.28(b). Resistance r_g can be important in limiting the bandwidth of very high-frequency amplifiers, and the thermal noise of r_g represents a potential limitation in low noise amplifier design.

9.5.1 EFFECT OF BASE AND GATE RESISTANCES ON MIDBAND AMPLIFIERS

Before considering the high-frequency response of single and multistage amplifiers, we explore the effect of base resistance on the midband gain expressions for single-stage amplifiers. Although the model used in deriving the midband voltage gain expressions in Chapters 7 and 8 did not include the effect of base resistance, the expressions can be easily modified to include r_x. A simple approach is to use the circuit transformation shown in Fig. 9.29, in which r_x is absorbed into an equivalent pi model. The current generator in the model in Fig. 9.29(a) is controlled by the voltage developed across r_π, which is related to the total base-emitter voltage through voltage division by

$$\mathbf{v} = \mathbf{v_{be}}\frac{r_\pi}{r_x + r_\pi} = \frac{\mathbf{v_{be}}}{1 + (r_x/r_\pi)} \tag{9.68}$$

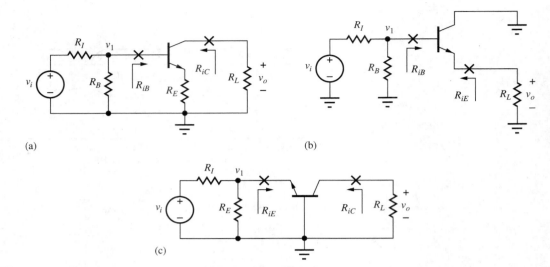

Figure 9.30 The three BJT amplifier configurations: (a) common-emitter; (b) common-collector; (c) common-base.

and the current in the controlled source is

$$\mathbf{i} = g_m \mathbf{v} = g_m \frac{r_\pi}{r_x + r_\pi} \mathbf{v_{be}} = g'_m \mathbf{v_{be}} \quad \text{where} \quad g'_m = \frac{\beta_o}{r_x + r_\pi} \quad (9.69)$$

From Eq. (9.68), we see that base resistance will only be important if it is a significant fraction of r_π, which occurs only for large values of r_x or small values of r_π (large collector current).

Equations (9.68) and (9.69) lead to the model in Fig. 9.29(b), in which the base resistance has been absorbed into r'_π and g'_m of an equivalent transistor Q' defined by

$$g'_m = g_m \frac{r_\pi}{r_x + r_\pi} = \frac{\beta_o}{r_x + r_\pi} \quad \text{and} \quad r'_\pi = r_x + r_\pi \quad (9.70)$$

Note that current gain is conserved during the transformation: $\beta'_o = \beta_o$.

Based on Eq. (9.70), the original expressions from Table 8.9 can be transformed to those in Table 9.1 for the three classes of amplifiers in Fig. 9.30 by simply substituting g'_m for g_m and r'_π for r_π. In many cases, particularly at bias points below a few hundred µA, $r_\pi \gg r_x$, and the expressions in Eq. (9.70) reduce to $g'_m \cong g_m$ and $r'_\pi \cong r_\pi$. The expressions in Table 9.1 then become identical to those in Table 8.9.

Gate resistance has little impact on the midband performance of FET amplifiers because, in the absence of C_{GS} and C_{GD}, the input resistance at node G' is infinite. Resistor r_G is of primary concern at high frequencies and low noise applications.

> **EXERCISE:** Recalculate the midband gain for the circuit in Fig. 9.6, including base resistance $r_x = 250\ \Omega$. What was the value of A_{mid} with $r_x = 0$?
>
> **ANSWERS:** −139; −157

9.6 HIGH-FREQUENCY COMMON-EMITTER AND COMMON-SOURCE AMPLIFIER ANALYSIS

Now that the complete hybrid-pi model has been described, we can explore the high-frequency limitations of the three basic single-stage amplifiers. For each of the basic stages, we will develop expressions for the high-frequency poles at the input and output of each stage. This approach will allow us to easily extend our analysis to multistage amplifiers.

TABLE 9.1
Single-Stage Bipolar Amplifiers, Including Base Resistance (See Fig. 9.30)

	COMMON-EMITTER AMPLIFIER	COMMON-COLLECTOR AMPLIFIER	COMMON-BASE AMPLIFIER
Terminal voltage gain $A_{vt} = \dfrac{\mathbf{v_o}}{\mathbf{v_i}}$ $r'_\pi = r_x + r_\pi$ $g'_m = \dfrac{\beta_o}{r'_\pi}$	$-\dfrac{\beta_o R_L}{r'_\pi + (\beta_o + 1)R_E}$ $\approx -\dfrac{g'_m R_L}{1 + g'_m R_E}$	$+\dfrac{\beta_o R_L}{r'_\pi + (\beta_o + 1)R_L}$ $\approx +\dfrac{g'_m R_L}{1 + g'_m R_L} \approx +1$	$+g'_m R_L$
Signal-source voltage gain $A_v = \dfrac{\mathbf{v_o}}{\mathbf{v_i}}$	$-\dfrac{g'_m R_L}{1 + g'_m R_E}\left(\dfrac{R_B \| R_{iB}}{R_I + R_B \| R_{iB}}\right)$	$+\dfrac{g'_m R_L}{1 + g'_m R_L}\left(\dfrac{R_B \| R_{iE}}{R_I + R_B \| R_{iE}}\right) \approx +1$	$+\dfrac{g'_m R_L}{1 + g'_m(R_I \| R_E)}\left(\dfrac{R_E}{R_I + R_E}\right)$
Input resistance	$r'_\pi + (\beta_o + 1)R_E$	$r'_\pi + (\beta_o + 1)R_L$	$\dfrac{1}{g'_m}$
Output resistance	$r_o(1 + g'_m R_E)$	$\dfrac{1}{g_m} + \dfrac{R_{th}}{\beta_o + 1}$	$r_o[1 + g'_m(R_I \| R_E)]$
Input signal range	$\cong 0.005(1 + g'_m R_E)$	$\cong 0.005(1 + g'_m R_L)$	$\cong 0.005[1 + g'_m(R_I \| R_E)]$
Current gain	$-\beta_o$	$\beta_o + 1$	$\alpha_o + 1 \approx +1$

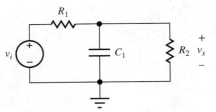

Figure 9.31 A two-resistor, one-capacitor circuit.

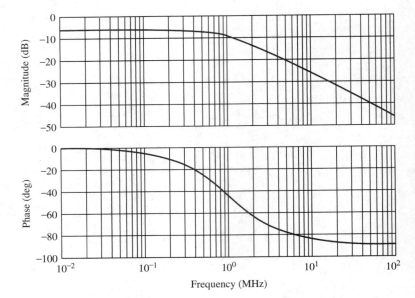

Figure 9.32 Magnitude and phase of single high-frequency pole with $R_1 = R_2$ and $f_p = 1$ MHz.

To begin our analysis, we will first review the high-frequency response of a single pole network as shown in Fig. 9.31. An expression for the high-frequency transfer characteristic for the RC circuit can be derived as

$$\frac{\mathbf{V_x}}{\mathbf{V_i}} = \frac{R_2 \left\| \frac{1}{sC_1} \right.}{R_1 + R_2 \left\| \frac{1}{sC_1} \right.} = \frac{\frac{R_2}{1+sR_2C_1}}{R_1 + \frac{R_2}{1+sR_2C_1}} = \frac{R_2}{R_1+R_2}\frac{1}{1+s(R_1 \| R_2)C_1} \qquad (9.71)$$

Substituting $s = j\omega$ and using $\omega_p = 1/(R_1 \| R_2)C_1$,

$$\frac{\mathbf{V_x}}{\mathbf{V_i}} = \frac{R_2}{R_1+R_2}\frac{1}{\left(1+j\frac{\omega}{\omega_p}\right)} = A_{\mathrm{mid}}F_H(s) \qquad (9.72)$$

This expression has two parts, the midband gain, $R_2/(R_2+R_1)$, and the high-frequency characteristic, $1/(1+j\omega/\omega_p)$. Notice that the equivalent resistance, $R_1 \| R_2$ is the total equivalent resistance to ground at the output of the example network. If other branch connections are present, the equivalent small-signal resistance of each branch will be added in parallel. Capacitance C_1 is the total equivalent capacitance to small-signal ground at the output. If other capacitors are present they too are simply added to find the total capacitance. The magnitude and phase of this single pole characteristic is shown in Fig. 9.32.

9.6.1 THE MILLER EFFECT

Figure 9.33(a) shows a typical variation on the simple network of Fig. 9.31. Here a capacitor connected at node X is connected across an amplifier with a gain A_{xy} from node X to node Y.

We would like to find a method to convert the physical capacitor C_{xy} across the amplifier to an equivalent capacitance, C_{eq}, to ground as shown in Fig. 9.33(b).

We can calculate the equivalent input capacitance by finding the current going into capacitor C_{xy} from node x. In the frequency domain,

$$\mathbf{I_c}(s) = sC_{xy}[\mathbf{V_x}(s) - \mathbf{V_y}(s)] \qquad (9.73)$$

9.6 High-Frequency Common-Emitter and Common-Source Amplifier Analysis

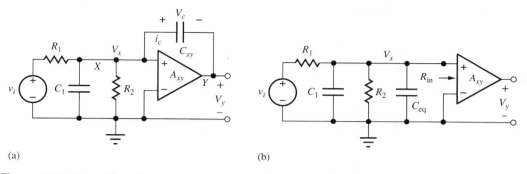

Figure 9.33 (a) Amplifier with a capacitance coupling its input and output. (b) The amplifier with the input-output capacitance replaced by an equivalent effective capacitance C_{eq} between the input and small-signal ground.

where $[\mathbf{V_x}(s) - \mathbf{V_y}(s)]$ represents the voltage across the capacitor, and the voltage at the output of the amplifier is a scaled replica of its input:

$$\mathbf{V_y}(s) = A_{xy}\mathbf{V_x}(s) \tag{9.74}$$

Combining the two equations above yields an expression for the input admittance at node x due to capacitor C_{xy}:

$$Y_s = \frac{\mathbf{I_c}(s)}{\mathbf{V_x}(s)} = sC_{xy}(1 - A_{xy}) = sC_{eq} \quad \text{and} \quad C_{eq} = C_{xy}(1 - A_{xy}) \tag{9.75}$$

The amplifier gain acts to produce an effective capacitance at its input which is scaled with respect to the physical capacitor by a factor $(1 - A_{xy})$. This is known as the **Miller effect,** or **Miller multiplication,** first described by John M. Miller in 1920.[2] Given our new equivalent capacitance, C_{eq}, based on our previous Eq. (9.70), we can now write an expression for the high-frequency transfer characteristic for the circuit in Fig. 9.33 as

$$\frac{\mathbf{V_x}}{\mathbf{V_i}} = \frac{R_2 \| R_{in}}{R_1 + R_2 \| R_{in}} \frac{1}{1 + s(R_1 \| R_2 \| R_{in})(C_1 + C_{eq})} \tag{9.76}$$

We see that the pole frequency is determined by a resistance of $R_1 \| R_2 \| R_{in}$, and a capacitance of $C_1 + C_{xy}(1 - A_{xy})$. As an example, consider the case with a gain $= -10$ V/V. The input capacitance due to C_{xy} will be $(1 - [-10])$ or eleven times larger than the physical capacitance C_{xy}. To understand this intuitively, consider $v_x = 10$ mV. With a gain of -10, v_y will be -100 mV, and the voltage v_c across capacitor C_{xy} will be 110 mV. In other words, as the voltage at the input of the capacitor is increasing, the other terminal is rapidly decreasing in voltage, causing the driving circuit to deliver much more current than would be expected given the actual value of the capacitor. On the other hand, for a gain of $+9$ V/V, the effective capacitance will be $(1 - 0.9)$, or 10 percent of the physical capacitance. For this gain value, the second terminal of the capacitor is approximately "following" the input terminal, leading to a much smaller delivery of current from the driving circuit. Using the Miller effect allows us to separate capacitively coupled sections of a circuit into simpler RC circuits which are more easily analyzed.

In the following sections, we will generalize this approach to develop the high-frequency response of an amplifier as the product of the midband gain we have developed in previous chapters and a high-frequency transfer characteristic representing the effects of the high-frequency time constant at each node along the signal path.

---

[2] J. M. Miller, "Dependence of the input impedance of a three-electrode vacuum tube upon the load in the plate circuit," *Scientific Papers of the Bureau of Standards*, vol. 15, no. 351, pp. 367–385, 1920.

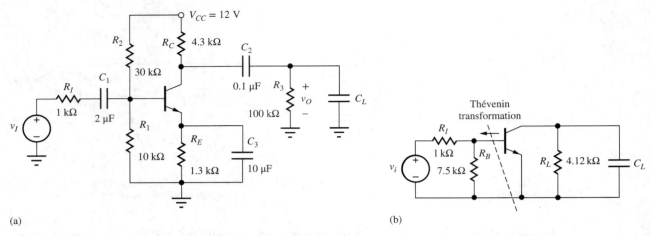

(a)   (b)

Figure 9.34 (a) Common-emitter amplifier. (b) High-frequency ac model for amplifier in (a).

9.6.2 COMMON-EMITTER AND COMMON-SOURCE AMPLIFIER HIGH-FREQUENCY RESPONSE

Figure 9.34 is a common-emitter amplifier with low-frequency coupling and bypass capacitors C_1, C_2, and C_3. In this section, we are concerned with the high-frequency response so we will consider the low-frequency capacitors to be open circuits at dc and short circuits at midband and high frequencies. C_L represents the high-frequency load capacitance. We will use a simplified analysis approach similar to that presented in the previous section. We calculate the midband gain and then calculate a time constant at the input and output signal nodes. The Miller effect will be used to calculate an equivalent capacitance at the input.

The ac small-signal equivalent circuit is shown in Fig. 9.35(a). The power supplies have been replaced with small-signal ground connections, and the low-frequency coupling and bypass capacitors are replaced with short circuits. The circuit has been further simplified in Fig. 9.35(c). The midband input gain is

$$A_i = \frac{\mathbf{v_b}}{\mathbf{v_i}} = \frac{R_{\text{in}}}{R_I + R_{\text{in}}} \cdot \frac{r_\pi}{r_x + r_\pi} = \frac{R_B \parallel (r_x + r_\pi)}{R_I + R_B \parallel (r_x + r_\pi)} \cdot \frac{r_\pi}{r_x + r_\pi} \quad (9.77)$$

where R_{in} is the parallel combination of R_1, R_2, and $(r_x + r_\pi)$, and R_L is the parallel combination of r_o, R_C, and R_3.

The terminal gain of the common-emitter amplifier (the effect of r_x was included in A_i) can be found as

$$A_{vt} = \frac{\mathbf{v_c}}{\mathbf{v_b}} = -g_m R_L \cong -g_m (R_C \parallel R_3) \quad (9.78)$$

We now use the Miller effect to calculate the input high-frequency pole at the base using the circuit in Fig. 9.35(a). The equivalent capacitance at node v_b is given by

$$\begin{aligned}C_{eqB} &= C_\mu(1 - A_{bc}) + C_\pi(1 - A_{be}) = C_\mu[1 - (-g_m R_L)] + C_\pi(1 - 0) \\ &= C_\pi + C_\mu(1 + g_m R_L)\end{aligned} \quad (9.79)$$

where $A_{bc} (= -g_m R_L)$ is the gain from the base to the collector and $A_{be} (= 0)$ is the gain from the base to the emitter. The equivalent small-signal resistance to ground at base node v_b is

$$R_{eqB} = r_\pi \parallel (r_x + R_B \parallel R_I) = r_{\pi o} \quad (9.80)$$

Remember the voltage source v_i has a small-signal impedance of zero. The resulting time constant at the input is

$$\tau_B = R_{eqB} C_{eqB} \quad (9.81)$$

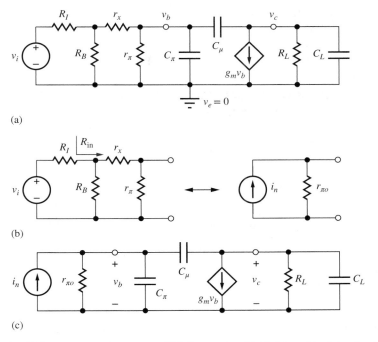

Figure 9.35 (a) Model for common-emitter stage at high frequencies. (b) Model used to determine the Norton source transformation for the CE amplifier. Resistor $r_{\pi 0}$ represents the equivalent resistance at the base node. (c) Simplified small-signal model for the high-frequency common-emitter amplifier.

At the collector output node, R_{eqC} is found as

$$R_{eqC} = R_L = r_o \| R_C \| R_3 \cong R_C \| R_3 \qquad (9.82)$$

We must now determine the equivalent capacitance at the collector, C_{eqC}. At first glance, we might expect to apply the Miller effect to model the equivalent capacitance at the output due to C_μ. However, the transistor does not operate in reverse: applying a signal at the collector does not result in a significant "output" signal at the base node. Therefore, the equivalent capacitance at the output only includes the physical capacitance C_μ as well as any additional load capacitance C_L.

$$C_{eqC} = C_\mu + C_L \qquad (9.83)$$

The resulting time constant at the output node is

$$\tau_C = R_{eqC} C_{eqC} \qquad (9.84)$$

If the input and output nodes are well isolated, we expect to calculate a separate pole frequency for the input and output time constants. However, in this case the input and output are coupled through C_μ. In addition, the input and output impedances are large, so we should expect the two time constants to interact. This interaction gives rise to a dominant pole equal to

$$\omega_{P1} = \frac{1}{R_{eqB} C_{eqB}} = \frac{1}{r_{\pi 0}[C_\pi + C_\mu(1 + g_m R_L)] + R_L[C_\mu + C_L]} \qquad (9.85)$$

We can rewrite this expression in terms of a single resistance and capacitance by factoring out resistance $r_{\pi 0}$. We will label the resulting capacitance as C_T.

$$C_T = [C_\pi + C_\mu(1 + g_m R_L)] + \frac{R_L}{r_{\pi 0}}[C_\mu + C_L] \qquad (9.86)$$

This substitution allows us to write the dominant pole frequency for the common-emitter as

$$\omega_{P1} = \frac{1}{r_{\pi o} C_T} = \frac{1}{r_{\pi o}\left([C_\pi + C_\mu(1 + g_m R_L)] + \frac{R_L}{r_{\pi o}}[C_\mu + C_L]\right)} \quad (9.87)$$

9.6.3 DIRECT ANALYSIS OF THE COMMON-EMITTER TRANSFER CHARACTERISTIC

At this point, it is desirable to check our simplified analysis approach via a direct analysis of the common-emitter transfer function. Writing and simplifying the nodal equations in the frequency domain for the circuit in Fig. 9.35(c) yields

$$\begin{bmatrix} \mathbf{I_n}(s) \\ 0 \end{bmatrix} = \begin{bmatrix} s(C_\pi + C_\mu) + g_{\pi o} & -s C_\mu \\ -(s C_\mu - g_m) & s(C_\mu + C_L) + g_L \end{bmatrix} \begin{bmatrix} \mathbf{V_b}(s) \\ \mathbf{V_c}(s) \end{bmatrix} \quad (9.88)$$

An expression for the output voltage, node voltage $\mathbf{V_c}(s)$, can be found using Cramer's rule:

$$\mathbf{V_c}(s) = \mathbf{I_n}(s) \frac{(s C_\mu - g_m)}{\Delta} \quad (9.89)$$

in which Δ represents the determinant of the system of equations given by

$$\Delta = s^2[C_\pi(C_\mu + C_L) + C_\mu C_L] + s[C_\pi g_L + C_\mu(g_m + g_{\pi o} + g_L) + C_L g_{\pi o}] + g_L g_{\pi o} \quad (9.90)$$

From Eqs. (9.89) and (9.90), we see that the high-frequency response is characterized by two poles, one finite zero, and one zero at infinity. The finite zero appears in the right-half of the s-plane at a frequency

$$\omega_Z = +\frac{g_m}{C_\mu} > \omega_T \quad (9.91)$$

The zero given by Eq. (9.91) can usually be neglected because it appears at a frequency above ω_T (for which the model itself is of questionable validity). Unfortunately, the denominator appears in unfactored polynomial form, and the positions of the poles are more difficult to find. However, good estimates for both pole positions can be found using the approximate factorization technique shown below. Note that even though there are three capacitors, the circuit only has two poles. The three capacitors are connected in a "pi" configuration, and only two of the capacitor voltages are independent. Once we know two of the voltages, the third is also defined.

Approximate Polynomial Factorization

We estimate the pole locations based on a technique for approximate factorization of polynomials. Let us assume that the polynomial has two real roots a and b:

$$(s + a)(s + b) = s^2 + (a + b)s + ab = s^2 + A_1 s + A_0 \quad (9.92)$$

If we assume that a dominant root exists—that is, $a \gg b$—then the two roots can be estimated directly from coefficients A_1 and A_0 using two approximations:

$$A_1 = a + b \cong a \quad \text{and} \quad \frac{A_0}{A_1} = \frac{ab}{a + b} \cong \frac{ab}{a} = b \quad (9.93)$$

so, $\qquad\qquad a \cong A_1 \quad \text{and} \quad b \cong \frac{A_0}{A_1}$

Note in Eq. (9.92) that the s^2 term is normalized to unity. Also note that the approximate factorization technique can be extended to polynomials having any number of widely spaced real roots.

9.6.4 POLES OF THE COMMON-EMITTER AMPLIFIER

For the case of the common-emitter amplifier, the smallest root is the most important because it is the one that limits the high-frequency response of the amplifier. From Eq. (9.93) we see that the smaller root is given by the ratio of coefficients A_0 and A_1, resulting in the following expression for the first pole:

$$\omega_{P1} = \frac{1}{r_{\pi o} C_T} = \frac{1}{r_{\pi o}\left([C_\pi + C_\mu(1 + g_m R_L)] + \frac{R_L}{r_{\pi o}}[C_\mu + C_L]\right)} \quad (9.94)$$

This result is identical to that of Eq. (9.87). The dominant pole is controlled by the combination of the input and output time constants set by the total equivalent capacitance and resistance at the input and output. Notice that if the driving resistance R_I is zero, $r_{\pi o}$ reduces to approximately r_x, and the bandwidth is primarily limited by r_x.

There is also a second pole resulting from the normalized version of coefficient A_1:

$$\omega_{P2} = \frac{C_\pi g_L + C_\mu(g_m + g_{\pi o} + g_L) + C_L g_{\pi o}}{C_\pi(C_\mu + C_L) + C_\mu C_L} \quad (9.95)$$

or

$$\omega_{P2} \cong \frac{g_m}{C_\pi\left(1 + \frac{C_L}{C_\mu}\right) + C_L} \cong \frac{g_m}{C_\pi + C_L} \quad (9.96)$$

in which the $C_\mu g_m$ term has been assumed to be the largest term in the numerator, as is most often the case for C-E stages with reasonably high gain. We can interpret the last approximation in Eq. (9.96) in this manner, particularly when C_μ is large. At high frequencies, capacitor C_μ effectively shorts the collector and base of the transistor together so that C_L and C_π appear in parallel, and the transistor behaves as a diode with a small-signal resistance of $1/g_m$. Recall also that there is a right-half plane zero equal to $+g_m/C_\mu$. While the zero is typically quite high in frequency and can be neglected, we will see in Chapter 15 that in FET amplifiers it can be an important aspect of the negative feedback amplifier stability analysis.

EXAMPLE 9.6   **HIGH-FREQUENCY ANALYSIS OF THE COMMON-EMITTER AMPLIFIER**

Find the midband gain and upper-cutoff frequency of a common-emitter amplifier.

PROBLEM Find the midband gain and upper-cutoff frequency of the common-emitter amplifier in Fig. 9.34 using the C_T approximation, assuming $\beta_o = 100$, $f_T = 500$ MHz, $C_\mu = 0.5$ pF, $r_x = 250$ Ω, and a Q-point of (1.60 mA, 3.00 V). Find the additional poles and zeros of the common-emitter amplifier. Assume $C_L = 0$, $C_1 = C_3 = 3.9$ µF, $C_2 = 0.082$ µF.

SOLUTION **Known Information and Given Data:** Common-emitter amplifier circuit in Fig. 9.34; Q-point = (1.60 mA, 3.00 V); $\beta_o = 100$, $f_T = 500$ MHz, $C_\mu = 0.5$ pF, and $r_x = 250$ Ω; expressions for the gain, poles, and zeros are given in Eqs. (9.77) through (9.96).

Unknowns: Values for A_mid, f_H, ω_{Z1}, ω_{P1}, and ω_{P2}

Approach: Find the small-signal parameters for the transistor. Find the unknowns by substituting the given and computed values into the expressions developed in the text.

Assumptions: Small-signal operation in the active region; $V_T = 25.0$ mV; $C_L = 0$

Chapter 9 Amplifier Frequency Response

Analysis: The common-emitter stage is characterized by Eqs. (9.77), (9.78), and (9.87).

$$A_{\text{mid}} = A_i A_{vt} \qquad A_i = \frac{R_{\text{in}}}{R_I + R_{\text{in}}}\left(\frac{r_\pi}{r_x + r_\pi}\right) \qquad A_{vt} = -g_m R_L$$

$$\omega_{P1} = \frac{1}{r_{\pi o} C_T} \qquad \omega_{P2} = \frac{g_m}{C_\pi + C_L} \qquad \omega_Z = \frac{g_m}{C_\mu}$$

$$r_{\pi o} = r_\pi \| (R_B \| R_I + r_x) \qquad C_T = C_\pi + C_\mu\left(1 + g_m R_L + \frac{R_L}{r_{\pi o}}\right)$$

The values of the various small-signal parameters must be found:

$$g_m = 40 I_C = 40(0.0016) = 64.0 \text{ mS} \qquad r_\pi = \frac{\beta_o}{g_m} = \frac{100}{0.064} = 1.56 \text{ k}\Omega$$

$$C_\pi = \frac{g_m}{2\pi f_T} - C_\mu = \frac{0.064}{2\pi(5 \times 10^8)} - 0.5 \times 10^{-12} = 19.9 \text{ pF}$$

$$R_{\text{in}} = 10 \text{ k}\Omega \| 30 \text{ k}\Omega \| 1.81 \text{ k}\Omega = 1.46 \text{ k}\Omega$$

$$R_L = R_C \| R_3 = 4.3 \text{ k}\Omega \| 100 \text{ k}\Omega = 4.12 \text{ k}\Omega$$

$$R_{\text{th}} = R_B \| R_I = 7.5 \text{ k}\Omega \| 1 \text{ k}\Omega = 882 \text{ }\Omega$$

$$r_{\pi o} = r_\pi \| (R_{\text{th}} + r_x) = 1.56 \text{ k}\Omega \| (882 \text{ }\Omega + 250 \text{ }\Omega) = 656 \text{ }\Omega$$

Substituting these values into the expression for C_T ($C_L = 0$) yields

$$C_T = C_\pi + C_\mu\left(1 + g_m R_L + \frac{R_L}{r_{\pi o}}\right)$$

$$= 19.9 \text{ pF} + 0.5 \text{ pF}\left[1 + 0.064(4120) + \frac{4120}{656}\right]$$

$$= 19.9 \text{ pF} + 0.5 \text{ pF}(1 + \underline{264} + 6.28) = 156 \text{ pF}$$

and

$$f_{P1} = \frac{1}{2\pi r_{\pi o} C_T} = \frac{1}{2\pi(656 \text{ }\Omega)(156 \text{ pF})} = 1.56 \text{ MHz}$$

$$\omega_{P2} \cong \frac{g_m}{C_\pi + C_L} = \frac{0.064}{19.9 \text{ pF}} = 3.22 \times 10^9 \text{ rad/sec}$$

$$f_{P2} = \frac{\omega_{P2}}{2\pi} = 512 \text{ MHz}$$

$$f_z = \frac{g_m}{2\pi C_\mu} = \frac{0.064}{2\pi(0.5 \text{ pF})} = 20.4 \text{ GHz}$$

$$A_i = \frac{1.46 \text{ k}\Omega}{1.00 \text{ k}\Omega + 1.46 \text{ k}\Omega}\left(\frac{1.56 \text{ k}\Omega}{250 \text{ }\Omega + 1.56 \text{ k}\Omega}\right) = 0.512 \qquad A_{vt} = -(0.064S)(4.12 \text{ k}\Omega) = -264$$

$$A_{\text{mid}} = 0.512(-264) = -135$$

Check of Results: We have found the desired information. By double checking, the calculations appear correct. Let us use the gain-bandwidth product as an additional check: $|A_{\text{mid}} f_{P1}| = 211$ MHz, which does not exceed f_T.

Discussion: The dominant pole is located at a frequency $f_{P1} = 1.56$ MHz, whereas f_{P2} and f_Z are estimated to be at frequencies above f_T (500 MHz). Thus, the upper-cutoff frequency f_H for this amplifier is determined solely by f_{P1}: $f_H \cong 1.56$ MHz. Note that this value of f_H is less than 1 percent of the transistor f_T and is consistent with the concept of GBW product. We should expect f_H to be no more than $f_T/A_{mid} = 3.3$ MHz for this amplifier. Note also that f_{P1} and f_{P2} are separated by a factor of almost 1000, clearly satisfying the requirement for widely spaced roots that was used in the approximate factorization.

It is important to keep in mind that the most important factor in determining the value of C_T is the term in which C_μ is multiplied by $g_m R_L$. To increase the upper-cutoff frequency f_H of this amplifier, the gain $(g_m R_L)$ must be reduced; a direct trade-off must occur between amplifier gain and bandwidth.

Computer-Aided Analysis: SPICE can be used to check our hand analysis, but we must define the device parameters that match our analysis. We set BF = 100 and IS = 5 fA, but let VAF default to infinity. The base resistance r_x must be added by setting SPICE parameter RB = 250 Ω. C_μ is determined in SPICE from the value of the zero-bias collector-junction capacitance CJC and the built-in potential ϕ_{jc}. The Q-point from SPICE gives $V_{CE} = 2.70$ V, which corresponds to $V_{BC} = 2.0$ V if $V_{BE} = 0.7$ V. In SPICE, VJC faults to 0.75 V, and MJC defaults to 0.33 (see Sec. 9.4). Therefore, to achieve $C_\mu = 0.5$ pF, CJC is specified as

$$\text{CJC} = 0.5 \text{ pF} \left(1 + \frac{20 \text{ V}}{0.75 \text{ V}}\right)^{0.33} = 0.768 \text{ pF}$$

C_π is determined by SPICE forward transit-time parameter TF, as defined by Eqs. (4.46) and (9.50):

$$\text{TF} = \frac{C_\pi}{g_m} = \frac{19.9 \text{ pF}}{64 \text{ mS}} = 0.311 \text{ ns}$$

After adding these values to the transistor model, we perform an ac analysis using FSTART = 100 Hz and FSTOP = 10 MHz with 20 frequency points per decade. The SPICE simulation results in the graph below yield $A_{mid} = -135$ (42.6 dB) and $f_H \cong 1.56$ MHz, which agree closely with our hand calculations. Checking the device parameters in SPICE, we also find r_x(RB) = 250 Ω, C_π(CBE) = 19.9 pF, and C_μ(CBC) = 0.499 pF, as desired.

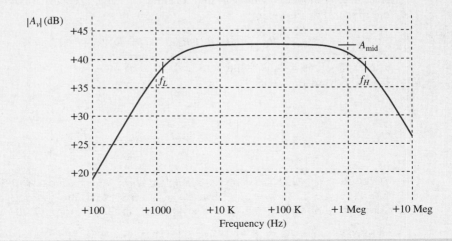

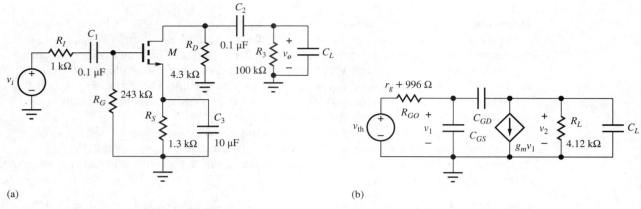

Figure 9.36 (a) Common-source amplifier. (b) The high-frequency small-signal model.

EXERCISE: Use SPICE to recalculate f_H for $V_A = 75$ V. For $V_A = 75$ V and $r_x = 0$?

ANSWERS: 1.67 MHz; 1.96 MHz

EXERCISE: Repeat the calculations in Ex. 9.6 if a load capacitance $C_L = 3$ pF is added to the circuit.

ANSWERS: 1.39 MHz, 445 MHz

EXERCISE: Find the midband gain and the frequencies of the poles and zeros of the common-emitter amplifier in Ex. 9.6 if the transistor has $f_T = 500$ MHz, but $C_\mu = 1$ pF.

ANSWERS: −135, 837 kHz, 525 MHz, 10.2 GHz

EXERCISE: Find the "exact" position of the two poles for the C-E amplifier by direct numerical evaluation of Eq. (9.90).

ANSWERS: $C_L = 0$, 602 MHz, 1.57 MHz; $C_L = 3$ pF, 93.2 MHz, 1.41 MHz

9.6.5 DOMINANT POLE FOR THE COMMON-SOURCE AMPLIFIER

Analysis of the C-S amplifier in Fig. 9.36 mirrors that of the common-emitter amplifier. The small-signal model is similar to that for the C-E stage, except that both r_x and r_π are absent from the model. For Fig. 9.36(b),

$$R_{GO} = r_g + R_G \| R_I \qquad R_L = R_D \| R_3 \qquad v_{th} = v_i \frac{R_G}{R_I + R_G} \qquad (9.97)$$

The expressions for the finite zero and poles of the C-S amplifier can be found by comparing Fig. 9.36(b) to Fig. 9.35:

$$\omega_{P1} = \frac{1}{R_{GO} C_T} \quad \text{and} \quad C_T = C_{GS} + C_{GD}\left(1 + g_m R_L + \frac{R_L}{R_{GO}}\right) + C_L \frac{R_L}{R_{GO}}$$

$$\omega_{P2} = \frac{g_m}{C_{GS} + C_L} \qquad \omega_Z = \frac{g_m}{C_{GD}} \qquad (9.98)$$

EXERCISE: (a) What is the upper-cutoff frequency for the amplifier in Fig. 9.36 if $r_g = 0$, $C_{GS} = 10$ pF, $C_{GD} = 2$ pF, $C_L = 0$ pF, and $g_m = 1.23$ mS? What are the positions of the second pole and the zero? What is the f_T of this transistor? (b) Repeat for $r_g = 300$ Ω.

ANSWERS: (a) 5.26 MHz; 19.6 MHz; 97.9 MHz; 16.3 MHz (b) 4.31 MHz, no change

ELECTRONICS IN ACTION

Graphic Equalizer

Graphic equalizers are used in audio applications to fine-tune the frequency response of an audio system. The equalizer is used to compensate for frequency-dependent absorption characteristics of a room, poor quality recordings, or just listener preferences. An example equalizer is shown in the figure below. This is the Ten/Series 2 analog equalizer marketed by Audio Control in 1983 and features total harmonic distortion of only 0.005 percent. The unit sold for $220 and weighed about 4 pounds. It has a set of slider controls that set boost or cut levels for different frequency bands within the audio frequency range. The term "graphic" is applied to equalizers where the physical position of the controls is representative of the boost or cut levels applied to the different bands.

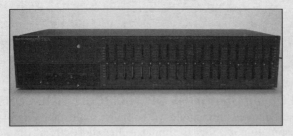

Audio Control/Ten Series 2
Mark Dierker/McGraw Hill

Typical graphic equalizer single band frequency response for different boost/cut settings.

A simplified schematic of a graphic equalizer is shown here. The circuit includes two summing amplifiers and a series of band-pass filters. The band-pass filters provide the frequency band selection. Resistor R_3 divides the output of the filters into two signals. The signal applied to the summing input of A_1 provides band reduction and the signal applied to the summing junction of A_2 provides boost for a particular frequency band.

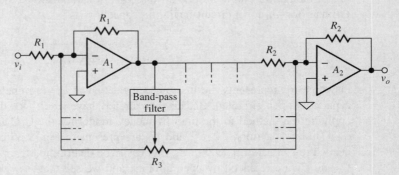

Typical graphic equalizer circuit.[3]

If potentiometer R_3 is set to the center point, the gain and boost signals are balanced, and no net signal is added or subtracted to the output. The slider controls in the picture above correspond to R_3 in the circuit diagram.

Graphic equalizers have been reduced greatly in size since the mid-1970s, and until recently, functioned similarly to the one shown above. However, with the advent of

---

[3] D. A. Bohn, "Constant-Q graphic equalizers." *J. Audio Eng. Soc.*, vol. 34, no. 9, September 1986.

high-precision, low-cost A/D converters and low-cost high-performance digital signal processing (DSP), graphic equalizers have moved into the digital domain. DSP based equalizers with excellent accuracy and controllability are now available. This new class of equalizer has an A/D converter, followed by DSP circuits, with a digital-to-analog (D/A) converter at the output to move the signal back into the analog domain. The DSP allows the designer to generate complex transfer functions that account for non-idealities such as channel-to-channel interactions. DSP equalizers are commonly found in digital music players, for example. As integrated circuit process technology advances, it is always important to reevaluate the appropriate boundaries between analog and digital signal processing.

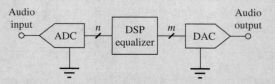

Graphic equalizer based upon digital signal processing.

9.6.6 ESTIMATION OF ω_H USING THE OPEN-CIRCUIT TIME-CONSTANT METHOD

Direct analysis by hand becomes intractable for complex multistage circuits. Fortunately, a technique also exists for estimating ω_H that is similar to the short-circuit time-constant method used to find ω_L. However, the upper-cutoff frequency ω_H is found by calculating the open-circuit time constants associated with the various device capacitances rather than the short-circuit time constants associated with the coupling and bypass capacitors. At high frequencies, the impedances of the coupling and bypass capacitors are negligibly small, and they effectively represent short circuits. The impedances of the device capacitances have now become small enough that they can no longer be neglected with respect to the internal resistances of the transistors. We will see shortly that the "C_T approximation" (Eq. 9.94) results can also be found using the OCTC method.

Although beyond the scope of this book, it can be shown theoretically[4] that the mathematical estimate for ω_H for a circuit having m capacitors is

$$\omega_H \cong \frac{1}{\sum_{i=1}^{m} R_{io} C_i} \quad (9.99)$$

in which R_{io} represents the resistance measured at the terminals of capacitor C_i with the other capacitors open circuited. Because we already have results for the C-E stages, let us practice by applying the method to the high-frequency model for the C-E amplifier in Fig. 9.35.

Three capacitors, C_π, C_μ, and C_L are present in Fig. 9.35(c), and $R_{\pi o}$, $R_{\mu o}$, and R_{Lo} will be needed to evaluate Eq. (9.99). $R_{\pi o}$ can easily be determined from the circuit in Fig. 9.37, in which C_μ and C_L are replaced by open circuits, and we see that

$$R_{\pi o} = r_{\pi o} = r_\pi \| (r_x + R_{\text{th}}) \quad (9.100)$$

R_{Lo} is found from Fig. 9.38. There is no current in $r_{\pi o}$, so $g_m v$ is zero, and

$$R_{Lo} = R_L \quad (9.101)$$

$R_{\mu o}$ can be determined from the circuit in Fig. 9.39, in which C_π is replaced by an open circuit. In this case, a bit more work is required. Test source i_x is applied to the network in

---

[4] See [1]. The OCTC and SCTC methods represent dominant root factorizations similar to Eq. (9.92).

9.6 High-Frequency Common-Emitter and Common-Source Amplifier Analysis

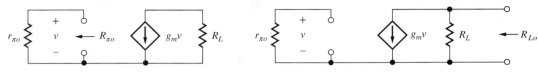

Figure 9.37 Circuit for finding $R_{\pi o}$.

Figure 9.38 Circuit for finding $R_{\pi o}$.

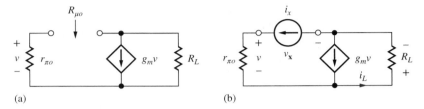

Figure 9.39 (a) Circuit defining $R_{\mu o}$. (b) Test source applied.

Fig. 9.39(b), and v_x can be found by applying KVL around the outside loop:

$$\mathbf{v_x} = \mathbf{i_x} r_{\pi o} + \mathbf{i_L} R_L = \mathbf{i_x} r_{\pi o} + (\mathbf{i_x} + g_m \mathbf{v}) R_L \qquad (9.102)$$

However, voltage $\mathbf{v}$ is equal to $\mathbf{i_x} r_{\pi o}$, and substituting this result into Eq. (9.102) yields

$$R_{\mu o} = \frac{\mathbf{v_x}}{\mathbf{i_x}} = r_{\pi o} + (1 + g_m r_{\pi o}) R_L = r_{\pi o}\left[1 + g_m R_L + \frac{R_L}{r_{\pi o}}\right] \qquad (9.103)$$

which should look familiar [see Eq. (9.86)]. Substituting Eqs. (9.100), (9.101), and (9.103) into Eq. (9.99) produces the estimate for ω_H:

$$\omega_H \cong \frac{1}{R_{\pi o} C_\pi + R_{\mu o} C_\mu + R_{Lo} C_L} = \frac{1}{r_{\pi o} C_\pi + r_{\pi o} C_\mu \left(1 + g_m R_L + \frac{R_L}{r_{\pi o}}\right) + R_L C_L} = \frac{1}{r_{\pi o} C_T} \qquad (9.104)$$

This is exactly the same result achieved from Eqs. (9.94) but with far less effort. (Remember, however, that this method does not produce an estimate for either the second pole or the zeros of the network.)

9.6.7 COMMON-SOURCE AMPLIFIER WITH SOURCE DEGENERATION RESISTANCE

Figure 9.40(a) shows a common-source amplifier with unbypassed source resistance R_S and Fig. 9.40(b) is the small-signal equivalent circuit. We find the input equivalent capacitance and resistance in the same manner as used for the common-emitter circuit. The midband gain is first calculated in two parts as before. The input gain expression is similar to that of the common-emitter except that r_π is not included since the impedance looking into the gate is infinite.

$$A_i = \frac{\mathbf{v_g}}{\mathbf{v_i}} = \frac{R_G}{R_I + R_G} = \frac{R_1 \| R_2}{R_I + (R_1 \| R_2)} \qquad (9.105)$$

The midband terminal gain of the common-source amplifier is found as

$$A_{gd} = \frac{\mathbf{v_d}}{\mathbf{v_g}} = \frac{-g_m R_L}{1 + g_m R_S} = \frac{-g_m (R_{iD} \| R_D \| R_3)}{1 + g_m R_S} \cong \frac{-g_m (R_D \| R_3)}{1 + g_m R_S} \qquad (9.106)$$

where

$$R_{iD} = r_o(1 + g_m R_S) \qquad (9.107)$$

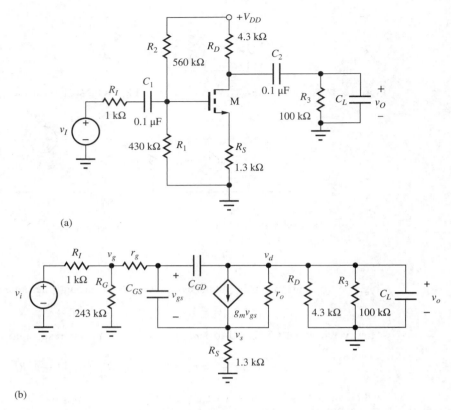

Figure 9.40 (a) Common-source amplifier with unbypassed source resistance. (b) Small-signal high-frequency equivalent circuit. Note that gate resistance r_g is included since it may be important at high frequencies.

As seen in Eq. (9.107), R_{iD} is typically quite large and can be neglected, and $R_L \cong R_D \| R_3$. We again use the Miller effect to calculate the input high-frequency time constant.

$$C_{eqG} = C_{GD}(1 - A_{gd}) + C_{GS}(1 - A_{gs})$$

$$= C_{GD}\left(1 - \frac{[-g_m R_L]}{1 + g_m R_S}\right) + C_{GS}\left(1 - \frac{g_m R_S}{1 + g_m R_S}\right) \quad (9.108)$$

$$= C_{GD}\left(1 + \frac{g_m(R_D \| R_3)}{1 + g_m R_S}\right) + \frac{C_{GS}}{1 + g_m R_S}$$

Note that we have used the expression for the gain of the common-drain amplifier to calculate the Miller multiplication of C_{GS}. Unlike the Miller effect with regard to C_{GD}, the effective capacitance of C_{GS} is reduced since A_{gs} will always be positive and less than 1. The unbypassed source resistance has also had the effect of reducing the effect of C_{GD} since the gate-to-drain gain has also been reduced by the $(1 + g_m R_S)$ term. The equivalent small-signal resistance to ground at the gate node is

$$R_{eqG} = r_g + R_G \| R_I = R_{GO} \quad (9.109)$$

The equivalent capacitance and resistance at the output is similar to that of the common-emitter amplifier.

$$R_{eqD} = R_{iD} \| R_D \| R_3 \cong R_D \| R_3 \quad \text{and} \quad C_{eqD} = C_{GD} + C_L \quad (9.110)$$

Combining these results and using Eq. (9.98), we find the following general form for the poles and right-half plane zero.

$$\omega_{P1} = \frac{1}{R_{GO}\left[\dfrac{C_{GS}}{1+g_m R_S} + C_{GD}\left(1 + \dfrac{g_m R_L}{1+g_m R_S}\right) + \dfrac{R_L}{R_{GO}}(C_{GD}+C_L)\right]} \quad (9.111)$$

$$\omega_{P2} = \frac{g_m}{(1+g_m R_S)(C_{GS}+C_L)} \quad (9.112)$$

$$\omega_z = \frac{+g_m}{(1+g_m R_S)(C_{GD})} \quad (9.113)$$

Notice the gain bandwidth trade-off indicated in Eqs. (9.106) and (9.111). The $(1+g_m R_S)$ term decreases gain while increasing the frequency of the dominant pole ω_{P1}. In our study of op amps, we will find that gain and bandwidth can be traded one for the other and the same relationship generally holds true in transistor circuits. Since the gain and bandwidth are inversely affected by the $(1+g_m R_S)$ term, the gain-bandwidth product is held relatively constant, similar to what we will also find in our study of the gain-bandwidth characteristics of op amp based amplifiers.

The second pole and zero equations are modified to account for the degeneration of the effective g_m by the source resistance. Notice that although the dominant pole increases in frequency, the frequencies of second pole and zero are decreased. Increasing the gain of the stage increases the frequency separation between ω_{P1} and ω_{P2}, resulting in what is often referred to as **pole-splitting.** Decreasing the gain moves the two poles closer in frequency, which can compromise the phase margin of feedback amplifiers.

If the small-signal resistance R_S in the source is reduced to zero, the equations for the common-source poles reduce to the simpler form of the common-source equations found previously. Likewise, if an unbypassed emitter resistance is included in the common-emitter amplifier, the pole equations can be modified in a manner similar to the common-source with source degeneration amplifier above, as discussed in the following section.

9.6.8 POLES OF THE COMMON-EMITTER WITH EMITTER DEGENERATION RESISTANCE

The equations for the common-emitter with unbypassed source resistance are modified in a manner similar to those of the common-source. In Fig. 9.41 a portion of the emitter resistance R_E is unbypassed, and the input stage gain A_i is modified due to the increased impedance looking into the base.

$$A_i = \frac{\mathbf{v_b}}{\mathbf{v_i}} = \frac{(R_B \| R_{iB})}{R_I + (R_B \| R_{iB})} \cdot \frac{r_\pi + (\beta_o+1)R_E}{r_x + r_\pi + (\beta_o+1)R_E} \cong \frac{R_B \| R_{iB}}{R_I + R_B \| R_{iB}} \quad (9.114)$$

Recall the impedance looking into the base is

$$R_{iB} = r_x + r_\pi + (\beta_o+1)R_E \quad (9.115)$$

The terminal gain of the common-emitter with unbypassed emitter resistance is found as

$$A_{bc} = \frac{\mathbf{v_c}}{\mathbf{v_b}} \cong \frac{-g_m(R_C \| R_3)}{1+g_m R_E} \cong \frac{-g_m R_L}{1+g_m R_E} \quad (9.116)$$

Including the effect of the emitter degeneration resistance R_E, the pole and zero equations are modified as follows:

$$\omega_{P1} = \frac{1}{r_{\pi o} C_T} = \frac{1}{r_{\pi o}\left(\left[\dfrac{C_\pi}{1+g_m R_E} + C_\mu\left(1 + \dfrac{g_m R_L}{1+g_m R_E}\right)\right] + \dfrac{R_L}{r_{\pi o}}[C_\mu + C_L]\right)} \quad (9.117)$$

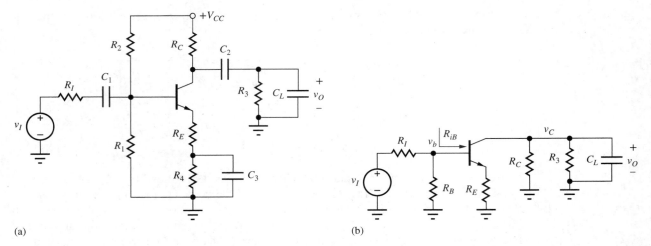

Figure 9.41 (a) Common-emitter amplifier with unbypassed emitter resistor R_E. (b) High-frequency ac equivalent circuit.

where

$$r_{\pi o} = R_{eqB} = (R_{th} + r_x) \| [r_\pi + (\beta_o + 1)R_E] \quad \text{with } R_{th} = R_B \| R_I \quad (9.118)$$

$$\omega_{P2} \cong \frac{g_m}{(1 + g_m R_E)(C_\pi + C_L)} \quad (9.119)$$

$$\omega_z = \frac{+g_m}{(1 + g_m R_E)(C_\mu)} \quad (9.120)$$

As with the common-source amplifier, the degeneration resistance causes a decrease in gain and an increase in the dominant pole frequency. The amplifier allows one to directly trade gain for bandwidth, approximately maintaining a constant gain-bandwidth product.

EXAMPLE 9.7 **COMMON-EMITTER AMPLIFIER WITH EMITTER DEGENERATION**

In this example, we explore the gain-bandwidth trade-off achieved by adding an unbypassed emitter resistor to the common-emitter amplifier from Ex. 9.6.

PROBLEM Find the midband gain, upper-cutoff frequency, and gain-bandwidth product for the common-emitter amplifier in Fig. 9.34 if a 300-Ω portion of the emitter resistor is not bypassed. Assume $\beta_o = 100$, $f_T = 500$ MHz, $C_\mu = 0.5$ pF, $r_x = 250$ Ω, and the Q-point = (1.6 mA, 3.0 V).

SOLUTION **Known Information and Given Data:** Common-emitter amplifier in Fig. 9.34 with a bypass capacitor placed around a 1000-Ω portion of the emitter resistor; $\beta_o = 100$, $f_T = 500$ MHz, $C_\mu = 0.5$ pF, $r_x = 250$ Ω; Q-point: (1.6 mA, 3.0 V).

Unknowns: A_{mid}, f_H, and GBW

Approach: Find A_{mid} and f_H using Eqs. (9.114)–(9.118). GBW = $A_{mid} \times f_H$.

Assumptions: $V_T = 25.0$ mV; small-signal operation in the active region

Analysis: Using the values from the analysis of Fig. 9.34 with $g_m = 40 I_C = 64$ mS:

$$r_\pi = \frac{\beta_o}{g_m} = \frac{100}{0.064} = 1.56 \text{ k}\Omega \quad R_{th} + r_x = 882\,\Omega + 250\,\Omega = 1130\,\Omega$$

$$R_{iB} = r_x + r_\pi + (\beta_o + 1)R_E = 250\,\Omega + 1560\,\Omega + (101)300\,\Omega = 32.1 \text{ k}\Omega$$

$$r_{\pi o} = R_{iB} \| (R_{th} + r_x) = 1.09 \text{ k}\Omega \quad 1 + g_m R_E = 1 + 0.064(300) = 20.2$$

9.6 High-Frequency Common-Emitter and Common-Source Amplifier Analysis

$$\omega_H \cong \frac{1}{r_{\pi o}\left[\dfrac{C_\pi}{1 + g_m R_E} + C_\mu\left(1 + \dfrac{g_m R_L}{1 + g_m R_E} + \dfrac{R_L}{r_{\pi 0}}\right)\right]}$$

$$\cong \frac{1}{1090\left[\dfrac{19.9 \text{ pF}}{20.2} + 0.5 \text{ pF}\left(1 + \dfrac{264}{20.2} + \dfrac{4120}{1090}\right)\right]}$$

$$f_H \cong \frac{1}{2\pi} \frac{1}{1090 \, \Omega (9.91 \text{ pF})} = 14.7 \text{ MHz}$$

$$A_i = \frac{R_1 \| R_2 \| R_{iB}}{R_1 + R_1 \| R_2 \| R_{iB}} = \frac{10 \text{ k}\Omega \| 30 \text{ k}\Omega \| 32.1 \text{ k}\Omega}{1 \text{ k}\Omega + 10 \text{ k}\Omega \| 30 \text{ k}\Omega \| 32.1 \text{ k}\Omega} = 0.859$$

$$A_{bc} = -\frac{g_m R_L}{1 + g_m R_E} = -\frac{0.064(4120 \, \Omega)}{1 + 0.064(300 \, \Omega)} = -13.0 \quad \text{or} \quad 22.3 \text{ dB}$$

$$A_{\text{mid}} = A_i A_{bc} = 0.859(-13.0) = -11.2 \qquad \text{GBW} = 11.2 \times 14.7 \text{ MHz} = 165 \text{ MHz}$$

Check of Results: A quick estimate for A_{mid} is $-R_L/R_E = -13.7$. Our more exact calculation is slightly less than this number, so it appears correct. The GBW product of the amplifier is 165 MHz, which is approximately $1/3$ of f_T, also a reasonable result.

Discussion: Remember, the original C-E stage with no emitter resistance had $A_{\text{mid}} = -153$ and $f_H = 1.56$ MHz for GBW = 239 MHz. With $R_E = 300 \, \Omega$, the gain has decreased by a factor of 14, and the bandwidth has increased by a factor of 8.9. The gain-bandwidth trade-off in the expression for ω_H is not exact because the effective resistance in the time constant only increases from 882 Ω to 1130 Ω, as well as the $R_L C_\mu$ term that is not scaled by the $(1 + g_m R_E)$ factor.

Computer-Aided Analysis: SPICE can be used to check our hand analysis, but we must define the device parameters that match our analysis. The base resistance, collector-base capacitance, and forward transit-time were calculated in Ex. 9.6: RB = 250 Ω, CJC = 0.768 pF, and TF = 0.311 ns. After adding these values to the transistor model, we perform an ac analysis using FSTART = 10 Hz and FSTOP = 100 MHz with 20 frequency points per decade. SPICE yields $A_{\text{mid}} = -11.0$ and $f_H \cong 15.0$ MHz, which agree well with our hand calculations. Note that the lower-cutoff frequency has changed to 158 Hz. The simulation used $C_1 = C_2 = 3.9 \, \mu\text{F}$, $C_3 = 0.082 \, \mu\text{F}$.

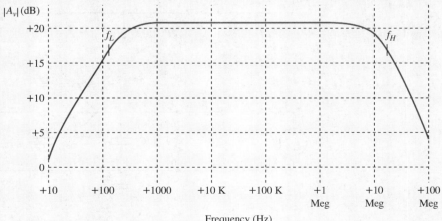

EXERCISE: Use SPICE to recalculate f_H for $V_A = 75$ V. For $V_A = 75$ V and $r_x = 0$?

ANSWERS: 14.8 MHz; 17.8 MHz

EXERCISE: Use the formulas to recalculate the midband gain, f_H, and GBW if the unbypassed portion of the emitter resistor is decreased to 100 Ω.

ANSWERS: −29.3; 6.70 MHz; 196 MHz

9.7 COMMON-BASE AND COMMON-GATE AMPLIFIER HIGH-FREQUENCY RESPONSE

We analyze the high-frequency response of the other single-stage amplifiers using the same approach we used in the previous section. At each node along the signal path, we determine an equivalent resistance to small-signal ground and an equivalent capacitance to small-signal ground. The resulting RC network gives rise to a high-frequency pole. We now apply this approach to the common-base amplifier shown in Fig. 9.42(a). The high-frequency ac equivalent circuit is shown in Fig. 9.42(b). Base resistance r_x has been neglected to simplify the analysis, as has output resistance r_o.

The input gain of the common base circuit is found as

$$A_i = \frac{v_e}{v_i} = \frac{R_{in}}{R_I + R_{in}} = \frac{R_E \| R_{iE}}{R_I + R_E \| R_{iE}} \tag{9.121}$$

where $R_{iE} = \dfrac{r_\pi}{\beta_o + 1} \cong \dfrac{1}{g_m}$. (To include the effect of r_x, we can add it to r_π.) Given that $R_{iE} \cong 1/g_m$ and if $1/g_m \ll R_E$, the input gain becomes

$$A_i \cong \frac{1}{1 + g_m R_I} \tag{9.122}$$

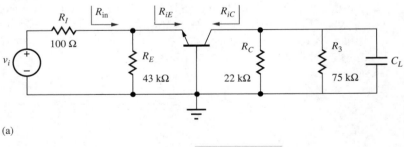

(a)

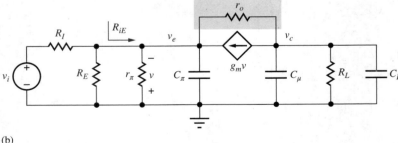

(b)

Figure 9.42 (a) High-frequency ac equivalent circuit for the common-base amplifier. (b) Small-signal model for the common-base amplifier neglecting r_x.

The terminal gain of the common-base amplifier is found as

$$A_{ec} = \frac{\mathbf{v_c}}{\mathbf{v_e}} = g_m(R_{iC} \| R_L) \cong +g_m R_L \tag{9.123}$$

where

$$R_{iC} = r_o[1 + g_m(r_\pi \| R_{th})] \quad \text{with} \quad R_{th} = R_E \| R_I \tag{9.124}$$

The expression for R_{iC} is the same as Eq. (8.90). Again, R_{iC} is typically much larger than the other resistances at the collector and can be neglected. The equivalent capacitance at the input is found as

$$C_{eqE} = C_\pi \tag{9.125}$$

An output capacitance associated with a driving stage would be added to C_π. To calculate the equivalent resistance at the emitter node, recall that due to the dependent generator, the resistance looking into the emitter is $R_{iE} \cong 1/g_m$. For the circuit in Fig. 9.42,

$$R_{eqE} = \frac{1}{g_m} \| R_E \| R_I \tag{9.126}$$

At the output, we determine the equivalent capacitance and resistance as

$$C_{eqC} = C_\mu + C_L \quad \text{and} \quad R_{eqC} = R_{iC} \| R_L \cong R_L \tag{9.127}$$

Since the input and output are well decoupled, we find the two poles for the common-base amplifier are

$$\omega_{P1} = \frac{1}{\left(\frac{1}{g_m} \| R_E \| R_I\right) C_\pi} \cong \frac{g_m}{C_\pi} \tag{9.128}$$

$$\omega_{P2} = \frac{1}{(R_{out} \| R_L)(C_\mu + C_L)} \cong \frac{1}{R_L(C_\mu + C_L)} \tag{9.129}$$

We should notice that the input pole of this stage has no Miller multiplication terms, and its equivalent resistance is dominated by the typically small $1/g_m$ term. As a result, the input pole of the common-base amplifier is typically a very high frequency, exceeding f_T. Therefore the bandwidth of the stage is dominated by the load resistance and capacitance, modeled with ω_{P2}.

EXERCISE: Find the midband gain and f_H using Eq. (9.129) for the common-base amplifier in Fig. 9.42 if the transistor has $\beta_o = 100, f_T = 500$ MHz, $r_x = 250\ \Omega, C_\mu = 0.5$ pF, and a Q-point (0.1 mA, 3.5 V). What is the gain-bandwidth product?

ANSWERS: +48.2; 18.7 MHz; 903 MHz

Figures 9.43(a) and (b) represent the high-frequency ac and small-signal equivalent circuits for a common-gate amplifier, and the analysis of the common-gate response is analogous to that of the common-base with R_4, C_{GS}, and C_{GD} replacing R_E, C_π, and C_μ with $r_g = 0$.

$$\omega_{P1} = \frac{1}{\left(\frac{1}{g_m} \| R_4 \| R_I\right) C_{GS}} \cong \frac{g_m}{C_{GS}} \tag{9.130}$$

$$\omega_{P2} = \frac{1}{[R_{out} \| R_L][C_{GD} + C_L]} \cong \frac{1}{R_L[C_{GD} + C_L]} \tag{9.131}$$

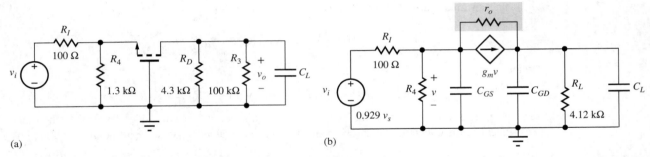

Figure 9.43 (a) High-frequency ac equivalent circuit for a common-gate amplifier. (b) Corresponding small-signal model ($r_g = 0$).

EXERCISE: Find the midband gain and f_H for the common-gate amplifier in Fig. 9.43 if the transistor $C_{GS} = 10$ pF, $C_{GD} = 1$ pF, $g_m = 3$ mS, and $C_L = 3$ pF. What are the gain-bandwidth product and f_T?

ANSWERS: +8.98, 9.65 MHz; 86.7 MHz; 43.4 MHz.

9.8 COMMON-COLLECTOR AND COMMON-DRAIN AMPLIFIER HIGH-FREQUENCY RESPONSE

The high-frequency responses of the common-collector and common-drain amplifiers are found in a manner similar to the other single-stage amplifiers. Figure 9.44 illustrates a typical common-collector amplifier and its small-signal equivalents. (Note that r_o is included in R_L.)

The midband input gain looks very similar to that of the common-emitter amplifier.

$$A_i = \frac{\mathbf{v_b}}{\mathbf{v_i}} = \frac{R_{\text{in}}}{R_I + R_{\text{in}}} = \frac{R_B \| R_{iB}}{R_I + R_B \| R_{iB}} = \frac{R_B \| [r_x + r_\pi + (\beta_o + 1) R_L]}{R_I + R_B \| [r_x + r_\pi + (\beta_o + 1) R_L]} \quad (9.132)$$

The base-to-emitter terminal voltage gain is

$$A_{be} = \frac{\mathbf{v_e}}{\mathbf{v_b}} = \frac{g_m R_L}{1 + g_m R_L} \quad (9.133)$$

Pole Estimation—ω_{P1}

To calculate the high-frequency poles, we first evaluate the equivalent small-signal resistance to ground, R_{eqB} at node v_b.

$$R_{eqB} = [(R_I \| R_B) + r_x] \| [r_\pi + (\beta_o + 1) R_L] = (R_{\text{th}} + r_x) \| [r_\pi + (\beta_o + 1) R_L] \quad (9.134)$$

The equivalent capacitance is found using Miller multiplication as

$$C_{eqB} = C_\mu (1 - A_{bc}) + C_\pi (1 - A_{be}) = C_\mu (1 - 0) + C_\pi \left(1 - \frac{g_m R_L}{1 + g_m R_L}\right) = C_\mu + \frac{C_\pi}{1 + g_m R_L} \quad (9.135)$$

Note that C_μ really appears directly between the base and ground, so Miller effect does not modify its value. On the other hand, the nearly unity gain between the transistor's base and emitter significantly reduces the effective size of C_π.

Pole Estimation—ω_{P2}

The equivalent small-signal resistance at the emitter can be found as

$$R_{eqE} = R_{iE} \| R_L = \left(\frac{r_\pi + R_{\text{th}} + r_x}{\beta_o + 1}\right) \| R_L \cong \frac{1}{g_m} + \frac{R_{\text{th}} + r_x}{\beta_o + 1} \quad (9.136)$$

9.8 Common-Collector and Common-Drain Amplifier High-Frequency Response

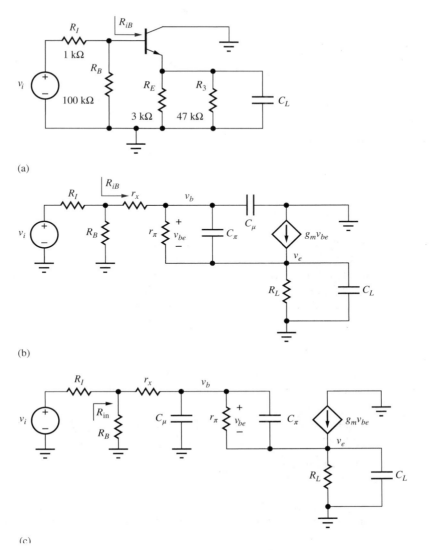

Figure 9.44 (a) Common-collector amplifier. (b) Small-signal model for the common-collector amplifier. (c) Simplification of the small-signal circuit for calculation of input and output high-frequency poles. Note $R_L = R_E \| R_3 \| r_o$.

where $R_{th} = R_B \| R_I$. The equivalent capacitance is found as the parallel combination of the load capacitance and the base-to-emitter capacitance.

$$C_{eqE} = C_\pi + C_L \tag{9.137}$$

Because of the low impedance at the output, the input and output time constants are relatively well decoupled, resulting in two poles for the common-collector amplifier.

$$\omega_{P1} = \frac{1}{([R_{th} + r_x] \| [r_\pi + (\beta_o + 1)R_L])\left(C_\mu + \dfrac{C_\pi}{1 + g_m R_L}\right)} \tag{9.138}$$

$$\omega_{P2} = \frac{1}{[R_{iE} \| R_L][C_\pi + C_L]} \cong \frac{1}{\left[\left(\dfrac{1}{g_m} + \dfrac{R_{th} + r_x}{\beta_o + 1}\right) \| R_L\right][C_\pi + C_L]} \tag{9.139}$$

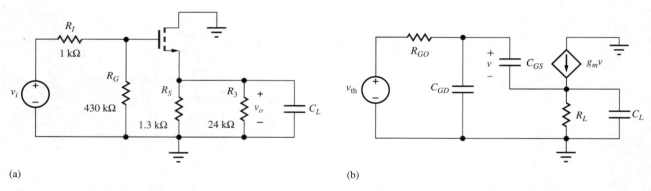

Figure 9.45 (a) High-frequency ac equivalent circuit for a source follower. (b) Corresponding high-frequency small-signal model. Note $R_L = R_S \| R_3 \| r_o$.

Notice that the output pole of this stage is dominated by the typically small $1/g_m$ term. As a result, the output pole of the common-collector amplifier is typically a very high frequency, approaching f_T. The bandwidth of the stage is dominated by f_{p1}, the pole associated with the input section equivalent resistance and capacitance. We will typically ignore the high-frequency pole at the emitter. Because of the feed-forward high-frequency path through C_π, a common-collector stage also includes a high-frequency zero.

$$\omega_z \cong \frac{g_m}{C_\pi} \qquad (9.140)$$

Notice that this zero is in the left-half plane. For low load capacitances, ω_z and ω_{P2} tend to cancel each other, so we should only include the effects of ω_{P2} when we also include ω_z.

> **EXERCISE:** Find A_{mid} and f_H for the common-collector amplifier in Fig. 9.44 if the Q-point is (1.5 mA, 5 V), $\beta_o = 100$, $r_x = 150\,\Omega$, $C_\mu = 0.5$ pF, and $f_T = 500$ MHz.
>
> **ANSWERS:** 0.980, 229 MHz.

A similar set of equations can be found for the common-drain amplifier of Fig. 9.45, making the appropriate changes for the different characteristics of the FET small-signal model.

$$\omega_{P1} = \frac{1}{R_{GO}\left(C_{GD} + \dfrac{C_{GS}}{1 + g_m R_L}\right)} \qquad R_{GO} = R_G \| R_I + r_g \qquad (9.141)$$

$$\omega_{P2} = \frac{1}{[R_{iS} \| R_L][C_{GS} + C_L]} \cong \frac{1}{\left[\dfrac{1}{g_m} \| R_L\right][C_{GS} + C_L]} \qquad (9.142)$$

$$\omega_z \cong \frac{g_m}{C_{GS}} \qquad (9.143)$$

Similar to the common-collector, the common-drain amplifier's high-frequency response is dominated by the input pole, f_{p1}, due to the small impedance associated with the output pole and zero.

> **EXERCISE:** (a) Find A_{mid} and f_H for the common-drain amplifier in Fig. 9.45 if $r_g = 0$, $C_{GS} = 10$ pF, $C_{GD} = 1$ pF, and $g_m = 3$ mS. (b) Repeat for $r_g = 250\,\Omega$.
>
> **ANSWERS:** 0.785, 51.0 MHz; 0.785, 40.8 MHz

9.9 SINGLE-STAGE AMPLIFIER HIGH-FREQUENCY RESPONSE SUMMARY

Table 9.2 collects the expressions for the dominant poles of the three classes of single-stage amplifiers. The inverting amplifiers provide high voltage gain but with the most limited bandwidth. The noninverting stages offer improved bandwidth with voltage gains similar to those of the inverting amplifiers. Remember, however, that the input resistance of the noninverting amplifiers is relatively low. The followers provide nearly unity gain with very wide bandwidth.

It is also worth noting at this point that both the C-E and C-B (or C-S and C-G) stages have a bandwidth that is always less than that set by the time constant of R_L and $(C_\mu + C_L)$ (or $C_{GD} + C_L$ and R_L) at the output node:

$$\omega_H < \frac{1}{R_L(C_\mu + C_L)} \quad \text{or} \quad \omega_H < \frac{1}{R_L(C_{GD} + C_L)}$$

9.9.1 AMPLIFIER GAIN-BANDWIDTH (GBW) LIMITATIONS

The importance of the base resistance r_x (and gate resistance in high-frequency FETs) in ultimately limiting the frequency response of amplifiers should not be overlooked. Consider first the common-emitter amplifiers described by Table 9.2. If the Thévenin equivalent source resistance R_I were reduced to zero in an attempt to increase the bandwidth, then $r_{\pi o}$ would not become zero, but would be limited approximately to the value of r_x. If one assumes that the gain is large and the $g_m R_L C_\mu$ term is dominant in determining ω_H, then the gain bandwidth product for the common-emitter stage becomes

$$\text{GBW} = A_{mid}\omega_H \le \frac{g_m R_L}{r_x(C_\mu g_m R_L)} \quad \text{and} \quad \text{GBW} \le \frac{1}{r_x C_\mu} \tag{9.144}$$

TABLE 9.2
Upper-Cutoff Frequency Estimates for the Single-Stage Amplifiers

	ω_H	
Common-emitter	$\dfrac{1}{r_{\pi o}C_T} = \dfrac{1}{r_{\pi o}\left[C_\pi + C_\mu(1+g_m R_L) + (C_u + C_L)\dfrac{R_L}{r_{\pi o}}\right]}$	$r_{\pi o} = r_\pi \| [r_x + (R_I \| R_B)]$
Common-source	$\dfrac{1}{R_{th}C_T} = \dfrac{1}{R_{th}\left[C_{GS} + C_{GD}(1+g_m R_L) + (C_{GD} + C_L)\dfrac{R_L}{R_{th}}\right]}$	$R_{th} = R_I \| R_G$
Common-emitter with emitter resistor R_E	$\dfrac{1}{r_{\pi o}\left[\dfrac{C_\pi}{1+g_m R_E} + C_\mu\left(1 + \dfrac{g_m R_L}{1+g_m R_E}\right) + (C_u + C_L)\dfrac{R_L}{r_{\pi o}}\right]}$	$r_{\pi o} = r_\pi \| [r_x + (R_1 \| R_B)]$
Common-source with source resistor R_S	$\dfrac{1}{R_{th}\left[\dfrac{C_{GS}}{1+g_m R_S} + C_{GD}\left(1 + \dfrac{g_m R_L}{1+g_m R_S}\right) + (C_{GD} + C_L)\dfrac{R_L}{R_{th}}\right]}$	$R_{th} = R_I \| R_G$
Common-base	$\dfrac{1}{R_L(C_\mu + C_L)}$	
Common-gate	$\dfrac{1}{R_L(C_{GD} + C_L)}$	
Common-collector	$\dfrac{1}{[(R_I \| R_B) + r_x]\left(\dfrac{C_\pi}{1+g_m R_L} + C_\mu\right)}$	
Common-drain	$\dfrac{1}{(R_I \| R_G)\left(\dfrac{C_{GS}}{1+g_m R_L} + C_{GD}\right)}$	

In the common-collector case, the gain is approximately one, and for $R_I = 0$ and large $g_m R_L$, the bandwidth becomes $1/r_x C_\mu$. Here again we find GBW $\leq 1/r_x C_\mu$. If we neglect C_π in Fig. 9.44, we can easily see that the bandwidth is set by r_x and C_μ, since the input resistance looking into r_π will be very large compared to r_x.

In order to simplify our analysis of the common-base amplifier, we neglected r_x. If r_x is included, it can be shown that the gain-bandwidth product is also limited by the $r_x C_\mu$ product in the C-B case as well. However, this limit is seldom reached since R_L is usually considerably larger than r_x.

Now we have found two important limits placed upon amplifier gain-bandwidth products by the characteristics of the transistor. The first was the unity-gain frequency for the current gain of the transistor, $f_T = g_m/(C_\pi + C_\mu)$. The second is set by GBW $\leq 1/r_x C_\mu$. However, in typical amplifier designs, the GBW product will reach less than 60 percent of either of these bounds.

For transistors designed for very high-frequency operation, minimization of the $r_x C_\mu$ product is one of the main goals guiding the choice of the physical structure and the impurity profiles of the devices. As is often the case in engineering, trade-offs are involved. The choices that minimize r_x increase C_μ, and vice-versa, and complex device designs are utilized to optimize the $r_x C_\mu$ (or $R_{iG} C_{GD}$) product.

9.10 FREQUENCY RESPONSE OF MULTISTAGE AMPLIFIERS

The open- and short-circuit time-constant methods are not limited to single-transistor amplifiers but are directly applicable to multistage circuits as well; the power of the technique becomes more obvious as circuit complexity grows. This section uses the techniques developed in the previous sections to estimate the frequency response of several important two-stage dc-coupled amplifiers, including the differential amplifier, the cascode stage, and the current mirror. Because these amplifiers are direct-coupled, they have low-pass characteristics and we only need to determine f_H. Following is an example of analysis of a general three-stage amplifier in which f_L and f_H are found.

9.10.1 DIFFERENTIAL AMPLIFIER

The differential amplifier is a key building block of analog circuits, and hence it is important to understand the frequency response of the differential pair. An important element, C_{EE}, has been included in the differential amplifier circuit in Fig. 9.46(a). C_{EE} represents the total capacitance at the emitter node of the differential pair. Analysis of the frequency response of the symmetrical amplifier in Fig. 9.46(a) is greatly simplified through the use of the half-circuits in (b) and (c) where differential-mode and common-mode input signals are given by $v_{id} = (v_1 - v_2)$ and $v_{ic} = (v_1 + v_2)/2$, respectively. Half-circuit techniques will be discussed in detail in Chapter 13.

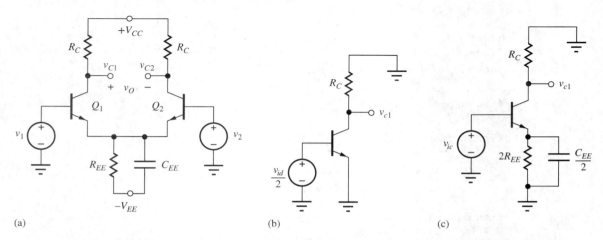

Figure 9.46 (a) Bipolar differential amplifier; (b) its differential-mode half-circuit; and (c) its common-mode half-circuit.

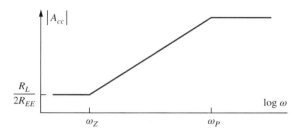

Figure 9.47 Bode plot for the common-mode gain of the differential pair.

Differential-Mode Signals

We recognize the differential-mode half-circuit in Fig. 9.46(b) as being equivalent to the standard common-emitter stage. Thus, the bandwidth for differential-mode signals is determined by the $r_{\pi o} C_T$ product that was developed in the analysis in Sec. 9.6, and we can expect amplifier gain-bandwidth products equal to a significant fraction of the f_T of the transistor. Because the emitter node is a virtual ground, C_{EE} has no effect on differential-mode signals.

Common-Mode Frequency Response

The important breakpoints in the Bode plot of the common-mode frequency response depicted in Fig. 9.47 can be determined from analysis of the common-mode half-circuit in Fig. 9.46(c). At very low frequencies, we know that the common-mode gain to either collector is small, given approximately by

$$|A_{cc}(0)| \cong \frac{R_C}{2R_{EE}} \ll 1 \qquad (9.145)$$

However, capacitance C_{EE} in parallel with emitter resistor R_{EE} introduces a transmission zero in the common-mode frequency response at the frequency for which the impedance of the parallel combination of R_{EE} and C_{EE} becomes infinite. This zero is given by

$$s = -\omega_z = -\frac{1}{R_{EE} C_{EE}} \qquad (9.146)$$

and typically occurs at relatively low frequencies. Although C_{EE} may be small, resistance R_{EE} is normally designed to be large, often the output resistance of a very high impedance current source. The presence of this zero causes the common-mode gain to increase at a rate of $+20$ dB/decade for frequencies above ω_Z. The common-mode gain continues to increase until the dominant pole of the pair is reached at relatively high frequencies.

The common-mode half-circuit is equivalent to a common-emitter stage with emitter resistor $2R_{EE}$. If we ignore base resistance r_x, we find that C_π and $C_{EE}/2$ appear in parallel, and the OCTC method yields

$$\omega_P = -\frac{1}{\left(C_\pi + \dfrac{C_{EE}}{2}\right) R_{EEO}} \qquad (9.147)$$

where R_{EEO} is the resistance at the terminals of $C_{EE}/2$:

$$R_{EEO} = \frac{1}{g_m} \parallel 2R_{EE} \parallel r_\pi \cong \frac{1}{g_m} \qquad (9.148)$$

The resulting pole position and common-mode gain are

$$\omega_P = -\frac{g_m}{C_\pi + \dfrac{C_{EE}}{2}} \quad \text{and} \quad A_{cc} = -\frac{g_m R_L}{1 + \dfrac{2C_\pi}{C_{EE}}} \qquad (9.149)$$

EXERCISE: Find f_Z and f_P for the common-mode response of the differential amplifier in Fig. 14.45 if $r_x = 250 \, \Omega$, $C_\mu = 0.5$ pF, $R_{EE} = 25 \, M\Omega$, $C_{EE} = 1$ pF, and $R_C = 50 \, k\Omega$.

ANSWERS: 6.37 kHz, 6.34 MHz

9.10.2 THE COMMON-COLLECTOR/COMMON-BASE CASCADE

Figure 9.48(a) is an unbalanced version of the differential amplifier. This circuit can also be represented as the cascade of a common-collector and common-base amplifier, as in Fig. 9.48(b). The poles of this two-stage amplifier are found by using the results of the previous single-stage amplifier analyses.

We assume that resistance R_{EE} is very large and neglect it in the analysis because the resistances presented at the emitters of Q_1 and Q_2 in Fig. 9.49 are both small:

$$R_{iE}^{CC1} = \frac{r_{x1} + r_{\pi1}}{\beta_{o1} + 1} \cong \frac{1}{g_{m1}} \quad \text{and} \quad R_{iE}^{CB2} = \frac{r_{x2} + r_{\pi2}}{\beta_{o2} + 1} \cong \frac{1}{g_{m2}} \quad (9.150)$$

The high-frequency response of the circuit of Fig. 9.49 is found by utilizing the results of the common-collector and common-base stages. We need to find estimates for the poles at the three nodes in the circuit: the base of Q_1, the emitter of Q_1, and the collector of Q_2.

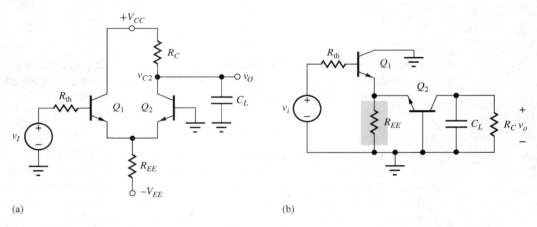

Figure 9.48 (a) The unbalanced differential amplifier and (b) its representation as a C-C/C-B cascade.

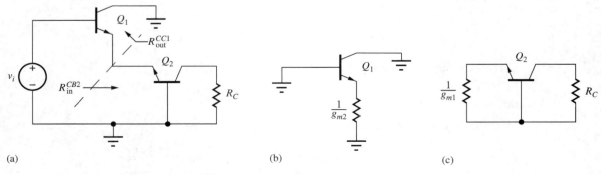

Figure 9.49 (a) Equivalent circuits for analysis of the poles of (b) Q_1 and (c) Q_2.

The pole at the input is that of a common-collector stage with $R_L = 1/g_{m2}$.

$$\omega_{PB1} = \frac{1}{([R_{th} + r_x] \| [r_{\pi 1} + (\beta_{o1} + 1)R_L])\left(C_\mu + \dfrac{C_\pi}{1 + g_{m1}R_L}\right)}$$

$$= \frac{1}{([R_{th} + r_{x1}] \| [2r_{\pi 1}])\left(C_{\mu 1} + \dfrac{C_{\pi 1}}{2}\right)} \qquad (9.151)$$

Notice that if the source impedance is zero, the input pole response is set by r_x. The second pole occurs at the emitter of Q_1 and Q_2 where the capacitance is $C_{\pi 1} + C_{\pi 2}$, and the resistance is $1/(g_{m1} + g_{m2})$. The resulting pole frequency is

$$\omega_{PE} = \frac{g_{m1} + g_{m2}}{C_{\pi 1} + C_{\pi 2}} \cong \frac{2g_m}{2C_\pi} = \frac{g_m}{C_\pi} > \omega_T \qquad (9.152)$$

and above the unity-gain frequency of the transistor. The pole at the collector of C_2 is

$$\omega_{PC2} \cong \frac{1}{R_C(C_{\mu 2} + C_L)} \qquad (9.153)$$

Depending on the impedances in the circuit, both the input and output poles could contribute significantly to the high-frequency response.

> **EXERCISE:** Compare the values of midband gain and f_H for the differential amplifier in Fig. 9.48 and the C-C/C-B cascade in Fig. 9.50 if $f_T = 500$ MHz, $C_\mu = 0.5$ pF, $I_{EE} = 200$ μA, $\beta_o = 100$, $r_x = 250$ Ω, and $R_C = 50$ kΩ.
>
> **ANSWERS:** −198, 3.16 MHz, 99.0, 6.27 MHz.

9.10.3 HIGH-FREQUENCY RESPONSE OF THE CASCODE AMPLIFIER

As mentioned in Chapter 8, the cascade of the common-emitter and common-base stages in Fig. 9.50 is referred to as the **cascode amplifier.** The cascode stage offers a midband gain and input resistance equal to that of the common-emitter amplifier but with a much improved upper-cutoff frequency f_H, as will be demonstrated by the forthcoming analysis.

The poles of the cascode stage follow directly from the analysis of the common-emitter and common-base stages using the model in Fig. 9.51. At the input to the cascode, we have the pole from our earlier analysis of this stage.

$$\omega_{PB1} = \frac{1}{r_{\pi o} C_T} = \frac{1}{r_{\pi o}\left([C_\pi + C_\mu(1 + g_{m1}R_L)] + \dfrac{R_L}{r_{\pi o}}[C_\mu + C_L]\right)} \qquad (9.154)$$

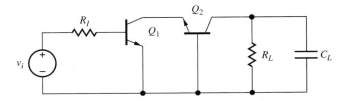

Figure 9.50 ac Model for the direct-coupled cascode amplifier.

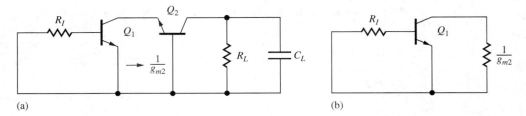

Figure 9.51 (a) Model for determining time constants associated with the two capacitances of Q_1. (b) Simplified model.

Since the load of the first stage is small $(1/g_m)$ we expect the second capacitive term in this expression to be insignificant, allowing us to simplify the expression. The bias current of the two transistors is the same, so the g_m values of the two transistors are also equal.

$$\omega_{PB1} = \frac{1}{r_{\pi o1}\left(\left[C_{\pi 1} + C_{\mu 1}\left(1 + \frac{g_{m1}}{g_{m2}}\right)\right] + \frac{1/g_{m2}}{r_{\pi o1}}[C_{\mu 1} + C_{\pi 2}]\right)} \cong \frac{1}{r_{\pi o1}(C_{\pi 1} + 2C_{\mu 1})} \quad (9.155)$$

The term due to Miller multiplication of C_μ has been reduced from a very large factor (264 in the C-E example in Sec. 9.6) to only 2, and the $R_L/r_{\pi o}$ term has also been essentially eliminated. These reductions are the primary advantage of the cascode amplifier and can greatly increase the bandwidth of the overall amplifier.

Similar to the previous circuit, the intermediate node impedance is quite low. At high frequencies, roughly equal to $1/g_{m2}$ (recall the high-frequency shunting of Q_1 due to C_μ). Because of this, we may expect the pole at the intermediate node to be quite high frequency at approximately $g_{m2}/(C_{\pi 2} + C_{\mu 1}) \approx \omega_T$.

The output pole is that of a common-base amplifier.

$$\omega_{PC2} \cong \frac{1}{R_L(C_{\mu 2} + C_L)} \quad (9.156)$$

Again, depending on the particular impedances in the circuit, both ω_{PB1} and ω_{PC2} could be significant in determining the overall high-frequency response.

EXERCISE: Find the midband value of A_v and the poles of the cascode amplifier in Fig. 9.48 assuming $\beta_o = 100$, $f_T = 500$ MHz, $C_\mu = 0.5$ pF, $r_x = 250$ Ω, $R_I = 882$ Ω, $R_L = 4.12$ kΩ, $C_L = 5$ pF, and a Q-point of (1.60 mA, 3.00 V) for Q_2.

ANSWERS: −151, 11.6 MHz, 7.02 MHz

9.10.4 CUTOFF FREQUENCY FOR THE CURRENT MIRROR

As a final example of the analysis of direct-coupled amplifiers, let us find ω_H for the current mirror configuration in Fig. 9.52 that was introduced in Chapter 8. The small-signal model in Fig. 9.52(b) represents the two-port model developed in Sec. 14.2 with the addition of the gate-source and gate-drain capacitances of M_1 and M_2. The gate-source capacitances of the two transistors appear in parallel, whereas the gate-drain capacitance of M_1 is shorted out by the circuit connection. The open-circuited load condition at the output represents a worst-case situation for estimating the current mirror bandwidth.

The circuit in Fig. 9.52(b) should be recognized as identical to the simplified model of the C-E stage in Fig. 9.35, and the results of the C_T approximation are directly applicable to the

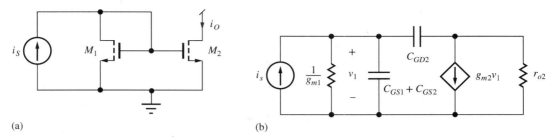

Figure 9.52 (a) MOS current mirror. (b) Small-signal model for the current mirror.

current mirror circuit with the following substitutions:

$$r_{\pi o} \to \frac{1}{g_{m1}} \quad R_L \to r_{o2} \quad C_\pi \to C_{GS1} + C_{GS2} \quad C_\mu \to C_{GD2} \quad (9.157)$$

Using the values from Eq. (9.157) in Eq. (9.94),

$$\omega_{P1} \cong \frac{1}{r_{\pi o} C_T} = \frac{1}{\frac{1}{g_{m1}}\left[C_{GS1} + C_{GS2} + C_{GD2}\left(1 + g_{m2}r_{o2} + \frac{r_{o2}}{\frac{1}{g_{m1}}}\right)\right]} \quad (9.158)$$

and for matched transistors with equal W/L ratios,

$$\omega_{P1} \cong \frac{1}{\frac{2C_{GS1}}{g_{m1}} + 2C_{GD2}r_{o2}} \cong \frac{1}{2C_{GD2}r_{o2}} \quad (9.159)$$

Equation (9.95) can be used to find an estimate for the second pole

$$\omega_{P2} \cong \frac{g_L C_\pi + 2g_m C_\mu}{C_\pi C_\mu} = \frac{1}{R_L C_\mu} + \frac{2g_m}{C_\pi} > \omega_T \quad (9.160)$$

which is again above ω_T.

The result in Eq. (9.158) indicates that the bandwidth of the current mirror is controlled by the time constant at the output of the mirror due to the output resistance and gate-drain capacitance of M_2. Note that the value of Eq. (9.159) is directly proportional to the Q-point current through the dependence of r_{o2}.

EXERCISE: (a) Find the bandwidth of the current mirror in Fig. 9.50 if $I_1 = 100$ μA, $C_{GD} = 1$ pF, and $\lambda = 0.02$ V^{-1}. (b) If $I_1 = 25$ μA.

ANSWERS: 159 kHz; 39.8 kHz

9.10.5 THREE-STAGE AMPLIFIER EXAMPLE

As an example of a more complex analysis, let us estimate the upper- and lower-cutoff frequencies for the multistage amplifier in Fig. 9.53 that was introduced in Chapter 8. We will use the method of short-circuit time constants to estimate the lower-cutoff frequency. In Chapters 11 and 15 we will need to know specific pole locations to accurately estimate feedback amplifier stability, so we will illustrate the calculation of high-frequency poles with this multistage example.

632 Chapter 9 Amplifier Frequency Response

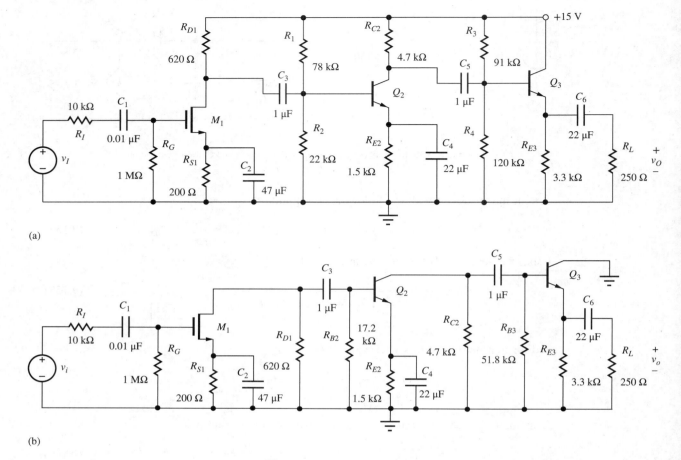

Figure 9.53 Three-stage amplifier and ac equivalent circuit.

EXAMPLE 9.8   MULTISTAGE AMPLIFIER FREQUENCY RESPONSE

The time-constant methods are used to find the upper- and lower-cutoff frequencies of a multistage amplifier.

PROBLEM Use direct calculation and short-circuit time-constant techniques to estimate the upper- and lower-cutoff frequencies of a multistage amplifier.

SOLUTION **Known Information and Given Data:** Three-stage amplifier circuit in Fig. 9.53; Q-points and small-signal parameters are given in Tables 8.19 and 9.3.

Unknowns: f_H, f_L, and bandwidth

Approach: The coupling and bypass capacitors determine the low-frequency response, whereas the device capacitances affect the high-frequency response. At low frequencies, the impedances of the internal device capacitances are very large and can be neglected. The coupling and bypass capacitors remain in the low-frequency ac equivalent circuit in Fig. 9.53(b), and an estimate for ω_L is calculated using the SCTC approach. An estimate for the upper-cutoff frequency is calculated based on the calculation of individual high-frequency poles from our single-stage analyses. At high frequencies, the impedances of the coupling and bypass capacitors are negligibly small, and we construct the circuit in Fig. 9.54 by replacing the coupling and bypass capacitors by short circuits.

TABLE 9.3
Transistor Parameters

	g_m	r_π	r_o	β_o	C_{GS}/C_π	C_{GD}/C_μ	r_x
M_1	10 mS	∞	12.2 kΩ	∞	5 pF	1 pF	0 Ω
Q_2	67.8 mS	2.39 kΩ	54.2 kΩ	150	39 pF	1 pF	250 Ω
Q_3	79.6 mS	1.00 kΩ	34.4 kΩ	80	50 pF	1 pF	250 Ω

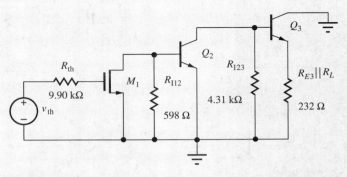

Figure 9.54 High-frequency ac model for three-stage amplifier in Fig. 9.53.

We develop expressions for the various time constants using our knowledge of input and output resistances of single-stage amplifiers. Finally, the expressions can be evaluated using known values of circuit elements and small-signal parameters.

Assumptions: Transistors are in the active region. Small-signal conditions apply. $V_T = 25$ mV.

ANALYSIS (a) **SCTC Estimate for the Lower-Cutoff Frequency ω_L:** The circuit in Fig. 9.53(b) has six independent coupling and bypass capacitors; Fig. 9.55 gives the circuits for finding the six short-circuit time constants. The analysis proceeds using the small-signal parameters in Table 9.3.

R_{1S}: Because the input resistance to M_1 is infinite in Fig. 9.55(a), R_{1S} is given by

$$R_{1S} = R_I + R_G \| R_{iG} = 10 \text{ k}\Omega + 1 \text{ M}\Omega \| \infty = 1.01 \text{ M}\Omega \qquad (9.161)$$

R_{2S}: R_{2S} represents the resistance present at the source terminal of M_1 in Fig. 9.55(b) and is equal to

$$R_{2S} = R_{S1} \left\| \frac{1}{g_{m1}} \right. = 200 \text{ }\Omega \left\| \frac{1}{0.01 \text{ S}} \right. = 66.7 \text{ }\Omega \qquad (9.162)$$

R_{3S}: Resistance R_{3S} is formed from a combination of four elements in Fig. 9.55(c). To the left, the output resistance of M_1 appears in parallel with the 620-Ω resistor R_{D1}, and on the right the 17.2-kΩ resistor R_{B2} is in parallel with the input resistance of Q_2:

$$R_{3S} = (R_{D1} \| R_{iD1}) + (R_{B2} \| R_{iB2}) = (R_{D1} \| r_{o1}) + (R_{B2} \| r_{\pi 2})$$

$$= (620 \text{ }\Omega \| 12.2 \text{ k}\Omega) + (17.2 \text{ k}\Omega \| 2.39 \text{ k}\Omega) = 2.69 \text{ k}\Omega \qquad (9.163)$$

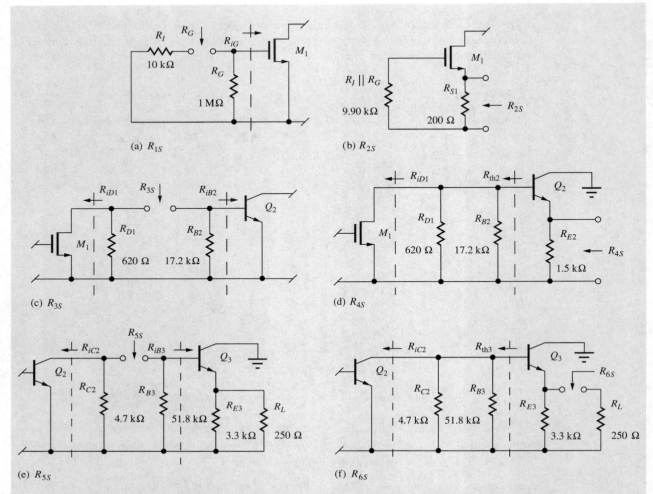

Figure 9.55 Subcircuits for finding the short-circuit time constants.

R_{4S}: R_{4S} represents the resistance present at the emitter terminal of Q_2 in Fig. 9.53(d) and is equal to

$$R_{4S} = R_{E2} \left\| \frac{R_{th2} + r_{\pi 2}}{(\beta_{o2} + 1)} \right. \quad \text{where} \quad R_{th2} = R_{B2} \| R_{D1} \| R_{iD1} = R_{B2} \| R_{D1} \| r_{o1}$$

$$R_{th2} = R_{B2} \| R_{D1} \| r_{o1} = 17.2 \text{ k}\Omega \| 620 \text{ }\Omega \| 12.2 \text{ k}\Omega = 571 \text{ }\Omega \quad (9.164)$$

$$R_{4S} = 1500 \text{ }\Omega \left\| \frac{571 \text{ }\Omega + 2390 \text{ }\Omega}{(150 + 1)} \right. = 19.4 \text{ }\Omega$$

R_{5S}: Resistance R_{5S} is also formed from a combination of four elements in Fig. 9.55(e). To the left, the output resistance of Q_2 appears in parallel with the 4.7-kΩ resistor R_{C2}, and to the right the 51.8-kΩ resistor R_{B3} is in parallel with the input resistance of Q_3:

$$R_{5S} = (R_{C2} \| R_{iC2}) + (R_{B3} \| R_{iB3}) = (R_{C2} \| r_{o2}) + (R_{B3} \| [r_{\pi 3} + (\beta_{o3} + 1)(R_{E3} \| R_L)])$$

$$= (4.7 \text{ k}\Omega \| 54.2) + 51.8 \text{ k}\Omega \| [1.00 \text{ k}\Omega + (80 + 1)(3.3 \text{ k}\Omega \| 250 \text{ }\Omega)]$$

$$= 18.4 \text{ k}\Omega \quad (9.165)$$

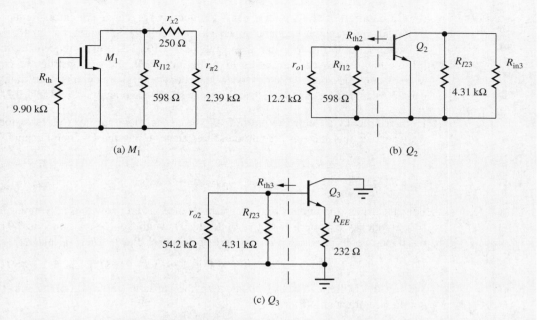

Figure 9.56 Subcircuits for evaluating the OCTC for each transistor.

R_{6S}: Finally, R_{6S} is the resistance present at the terminals of C_6 in Fig. 9.53(f):

$$R_{6S} = R_L + \left(R_{E3} \middle\| \frac{R_{th3} + r_{\pi3}}{\beta_{o3} + 1}\right) \quad \text{where} \quad R_{th3} = R_{B3} \| R_{C2} \| R_{iC2} = R_{B3} \| R_{C2} \| r_{o2}$$

$$R_{th3} = 51.8 \text{ k}\Omega \| 4.7 \text{ k}\Omega \| 54.2 \text{ k}\Omega = 3.99 \text{ k}\Omega \tag{9.166}$$

$$R_{6S} = 250 \text{ }\Omega + \left(3.3 \text{ k}\Omega \middle\| \frac{3.39 \text{ k}\Omega + 1.00 \text{ k}\Omega}{80 + 1}\right) = 311 \text{ }\Omega$$

An estimate for ω_L can now be constructed using Eq. (9.33) and the resistance values calculated in Eqs. (9.161) to (9.166):

$$\omega_L \cong \sum_{i=1}^{n} \frac{1}{R_{iS} C_i} = \frac{1}{R_{1S} C_1} + \frac{1}{R_{2S} C_2} + \frac{1}{R_{3S} C_3} + \frac{1}{R_{4S} C_4} + \frac{1}{R_{5S} C_5} + \frac{1}{R_{6S} C_6}$$

$$\cong \frac{1}{(1.01 \text{ M}\Omega)(0.01 \text{ }\mu\text{F})} + \frac{1}{(66.7 \text{ }\Omega)(47 \text{ }\mu\text{F})} + \frac{1}{(2.69 \text{ k}\Omega)(1 \text{ }\mu\text{F})}$$

$$+ \frac{1}{(19.4 \text{ }\Omega)(22 \text{ }\mu\text{F})} + \frac{1}{(18.4 \text{ k}\Omega)(1 \text{ }\mu\text{F})} + \frac{1}{(311 \text{ }\Omega)(22 \text{ }\mu\text{F})} \tag{9.167}$$

$$\cong 99.0 + 319 + 372 + \underline{2340} + 54.4 + 146 = 3330 \text{ rad/s}$$

$$f_L = \frac{\omega_L}{2\pi} = 530 \text{ Hz}$$

The estimate of the lower-cutoff frequency is 530 Hz. The dominant contributor is the fourth term, resulting from the time constant associated with emitter-bypass capacitor C_4, and for design purposes, this capacitor can be used to set the value of f_L. (Remember the design approach used in Design Ex. 9.3.)

(b) Calculation of the Upper-Cutoff Frequency f_H: The upper-cutoff frequency can be found by calculating the high-frequency poles at each of the nodes within the high-frequency ac model of the amplifier in Fig. 9.53 and then applying Eq. (9.23). At high frequencies, the impedances of the coupling and bypass capacitors are negligibly small, and we construct the circuit in Fig. 9.54 by replacing the coupling and bypass capacitors with short circuits. The high-frequency poles can be calculated at each node based on our single-stage analyses in Table 9.2.

High-frequency pole at the gate of M_1: From the subcircuit for the transistor in Fig. 9.56(a), we recognize this stage as a common-source stage. Using the C_T approximation from Table 9.2,

$$f_{p1} = \left(\frac{1}{2\pi}\right) \frac{1}{R_{\text{th}1}[C_{GS1} + C_{GD1}(1 + g_{m1}R_{L1}) + \frac{R_{L1}}{R_{\text{th}1}}(C_{GD1} + C_{L1})]} \quad (9.168)$$

For this circuit, the unbypassed source resistance is zero, so we use a simpler form of the input pole frequency equation. In Eq. (9.168), the Thévenin source resistance is 9.9 kΩ, and the load resistance is the parallel combination of resistances R_{I2}, $(r_{x2} + r_{\pi 2})$, and r_{o1}:

$$R_{L1} = R_{I12} \| r_{\pi 2} + r_x \| r_{o1} = 598\,\Omega \| (2.39\,\text{k}\Omega + 250\,\Omega) \| 12.2\,\text{k}\Omega = 469\,\Omega \quad (9.169)$$

We use the Miller effect to evaluate C_{L1}, the capacitance seen looking into the second stage common-emitter amplifier:

$$C_{L1} = C_{\pi 2} + C_{\mu 2}(1 + g_{m2}R_{L2}) \quad (9.170)$$

From Fig. 9.56(b), we evaluate R_{L2} as

$$\begin{aligned}R_{L2} &= R_{I23} \| R_{iB3} \| r_{o2} = R_{I23} \| [r_{x3} + r_{\pi 3} + (\beta_{o3} + 1)(R_{E3} \| R_L)] \| r_{o2} \\ &= 4.31\,\text{k}\Omega \| [250 + 1\,\text{k}\Omega + (80+1)(3.3\,\text{k}\Omega \| 250\,\Omega)] \| 54.2\,\text{k}\Omega \quad (9.171)\\ &= 3.33\,\text{k}\Omega\end{aligned}$$

Using this result we find C_{L1} as

$$C_{L1} = 39\,\text{pF} + 1\,\text{pF}[1 + 67.8\,\text{mS}(3.33\,\text{k}\Omega)] = 266\,\text{pF} \quad (9.172)$$

Combining these results, the pole at the input of M_1 becomes

$$f_{p1} = \left(\frac{1}{2\pi}\right)\frac{1}{(9.9\,\text{k}\Omega)[1\,\text{pF}(1 + 0.01\,\text{S}(469\,\Omega)] + 5\,\text{pF} + \frac{469\,\Omega}{9.9\,\text{k}\Omega}(1\,\text{pF} + 266\,\text{pF})]} = 689\,\text{kHz} \quad (9.173)$$

High-frequency pole at the base of Q_2: From the subcircuit for the transistor in Fig 9.56(b), we recognize this stage as a common-emitter stage. At first glance we might expect to use the C_T approximation for the pole at the output of stage 1 and the input of stage 2. However, if we recall the detailed analysis of the common-source and common-emitter stage, we find that the output pole of the common-source stage is described by Eq. (9.95), rewritten here for the common-source case:

$$f_{p2} = \left(\frac{1}{2\pi}\right)\frac{C_{GS1}g_{L1} + C_{GD1}(g_{m1} + g_{\text{th}1} + g_{L1}) + C_{L1}g_{\text{th}1}}{[C_{GS1}(C_{GD1} + C_{L1}) + C_{GD1}C_{L1}]} \quad (9.174)$$

In this particular case, C_{L1} is much larger than the other capacitances, so Eq. (9.173) simplifies to

$$f_{p2} \cong \left(\frac{1}{2\pi}\right)\frac{C_{L1}g_{\text{th}1}}{[C_{GS1}C_{L1} + C_{GD1}C_{L1}]} \cong \left(\frac{1}{2\pi}\right)\frac{1}{R_{\text{th}1}(C_{GS1} + C_{GD1})} \quad (9.175)$$

Substituting for the appropriate parameters, we calculate f_{p2} as

$$f_{p2} = \left(\frac{1}{2\pi}\right)\frac{1}{(9.9\,\text{k}\Omega)(5\,\text{pF} + 1\,\text{pF})} = 2.68\,\text{MHz} \quad (9.176)$$

High-frequency pole at the base of Q_3: From the subcircuit for the transistor in Fig. 9.56(c), we recognize the third stage as a common-collector stage. Again, due to the pole splitting effect of the common-emitter second stage, we expect that the pole at the base of Q_3 will be set by Eq. (9.95). In this case, due to the small load capacitance and high g_{m2} of the second stage, the $g_{m2}C_\mu$ term simplification of Eq. (9.95) will dominate the numerator. As a consequence, we can expect the pole at the interstage node between Q_2 and Q_3 to be governed by

$$f_{p3} \cong \left(\frac{1}{2\pi}\right) \frac{g_{m2}}{\left[C_{\pi 2}\left(1 + \frac{C_{L2}}{C_{\mu 2}}\right) + C_{L2}\right]} \tag{9.177}$$

The load capacitance of Q_2 is the input capacitance for the common-collector output stage. This is calculated as

$$C_{L2} = C_{\mu 3} + \frac{C_{\pi 3}}{1 + g_{m3}(R_{E3} \| R_L)} = 1\,\text{pF} + \frac{50\,\text{pF}}{1 + 79.6\,\text{mS}(3.3\,\text{k}\Omega \| 250\,\Omega)} = 3.55\,\text{pF} \tag{9.178}$$

To account for r_{x2}, we can use g_m as defined in Eq. (9.70) when evaluating f_{p3}.

$$f_{p3} \cong \left(\frac{1}{2\pi}\right) \frac{67.8\,\text{mS}[1\,\text{k}\Omega/(1\,\text{k}\Omega + 250\,\Omega)]}{\left[39\,\text{pF}\left(1 + \frac{3.55\,\text{pF}}{1\,\text{pF}}\right) + 3.55\,\text{pF}\right]} = 47.7\,\text{MHz} \tag{9.179}$$

There is an additional pole at the emitter of Q_3, but that will be at a very high frequency due to the relatively low equivalent resistance and capacitance at the output. The midband to high-frequency response can now be written as

$$\begin{aligned} A(f) &\cong \frac{A_{\text{mid}}}{\left(1 + j\frac{f}{f_{p1}}\right)\left(1 + j\frac{f}{f_{p2}}\right)\left(1 + j\frac{f}{f_{p3}}\right)} \\ &\cong \frac{998\,\text{V/V}}{\left(1 + j\frac{f}{689\,\text{kHz}}\right)\left(1 + j\frac{f}{2.68\,\text{MHz}}\right)\left(1 + j\frac{f}{47.7\,\text{MHz}}\right)} \end{aligned} \tag{9.180}$$

Applying Eq. (9.23), we estimate f_H as

$$f_H = \frac{1}{\sqrt{\frac{1}{f_{p1}^2} + \frac{1}{f_{p2}^2} + \frac{1}{f_{p3}^2}}} = 667\,\text{kHz} \tag{9.181}$$

Check of Results: SPICE is an excellent method to check an analysis of this complexity. After drawing the circuit with the schematic editor, we need to set the MOSFET and BJT parameters. We can set up the device parameters by referring back to Tables 8.18, 8.19, and 9.3. For the depletion-mode MOSFET, KP = 10 mA/V^2, VTO = -2 V, and LAMBDA = 0.02 V^{-1}. For this simulation, it is easiest to add external capacitors in parallel with the MOSFET to represent C_{GS} and C_{GD}. The values are 5 pF and 1 pF, respectively.

For the BJTs, RB = 250 Ω, BF = 150, and VAF = 80 V, and we can let IS take on its default value of 0.1 fA. The values of TF can also be found using the data in Table 9.3:

$$\text{TF}_2 = \frac{C_{\pi 2}}{g_{m2}} = \frac{39\,\text{pF}}{67.8\,\text{mS}} = 0.575\,\text{ns} \quad \text{and} \quad \text{TF}_3 = \frac{50\,\text{pF}}{79.6\,\text{mS}} = 0.628\,\text{ns}$$

The collector-emitter voltages from Table 8.19 are $V_{CE2} = 5.09$ V and $V_{CE3} = 8.36$ V. To achieve values of 1 pF for each C_μ, we must properly set the values of CJC:

$$\text{CJC2} = (1 \text{ pF})\left(1 + \frac{5.09 - 0.7}{0.75}\right)^{0.33} = 1.89 \text{ pF}$$

and

$$\text{CJC3} = (1 \text{ pF})\left(1 + \frac{8.36 - 0.7}{0.75}\right)^{0.33} = 2.22 \text{ pF}$$

Once the parameters are set, an ac analysis can be performed with FSTART = 10 Hz, FSTOP = 10 MHz, and 20 points per frequency decade. The resulting Bode magnitude plot appears next. We can also check the device parameters and see that the values of C_π and C_μ are approximately correct.

Discussion: Note that common-source stage M_1 and common-emitter stage Q_2 are both making contributions to f_H, whereas follower Q_3 represents a negligible contribution. Based on our calculated results, the midband region of the amplifier extends from $f_L = 530$ Hz to $f_H = 667$ kHz for a bandwidth BW = 666 kHz.

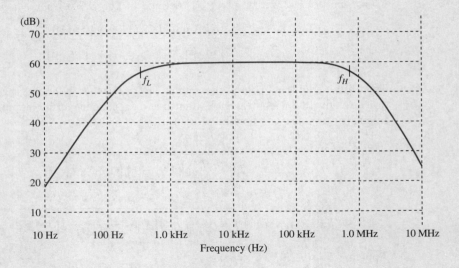

The SPICE results indicate that f_L and f_H are approximately 350 Hz and 675 kHz, respectively, and the midband gain is 60 dB. In this amplifier, we see that the SCTC method is overestimating the value of the lower-cutoff frequency. If we look at Eq. (9.167), we see that there is clearly a dominant time constant. If we use only this value, we get much better agreement with the SPICE results:

$$f_L \cong \frac{2340}{2\pi} = 372 \text{ Hz}$$

On the other hand, our estimate of f_H is in good agreement with simulation. We did have to be quite careful with our calculations to take into account the pole splitting behavior of common-emitter and common-source amplifiers. If we had not taken this into account, the estimate for f_H, based on dominant-pole calculations for each of the stages, would be less than 550 kHz. Of even more importance for feedback amplifier design, our analysis in this example also accurately characterizes the phase and magnitude response well beyond f_H.

EXERCISE: Calculate the reactance of $C_{\pi 2}$ at f_L and compare its value to $r_{\pi 2}$. Calculate the reactance of $C_{\mu 3}$ and compare it to $R_{B3} \| R_{in3}$ in Fig. 9.55(e).

ANSWERS: 7.7 M$\Omega \gg$ 2.39 kΩ; 300 M$\Omega \gg$ 14.3 kΩ

EXERCISE: Calculate the reactance of C_1, C_2, and C_3, in Fig. 9.51(b) at $f = f_H$, and compare the values to the midband resistances in the circuit at the terminals of the capacitors.

ANSWERS: 23.9 $\Omega \ll$ 1.01 MΩ; 5.08 m$\Omega \ll$ 66.7 Ω; 239 m$\Omega \ll$ 2.69 kΩ

9.11 INTRODUCTION TO RADIO FREQUENCY CIRCUITS

Since its inception, radio frequency (RF) communications has had a pervasive influence on our lives and the way we communicate with each other. There are several important circuits that appear over and over again in RF devices such as our cellular phones, radios, televisions, and so on. These include low-noise amplifiers, mixers, and oscillators.

For example, an architecture[5] for a hypothetical transceiver is shown in the block diagram in Fig. 9.57, and it could represent the RF portion of a device for a wireless local area network, or the transceiver for a cellular phone, depending upon the particular frequencies chosen for the design. In the 5-GHz digital radio system depicted here, the signal received from the antenna is amplified by a low-noise amplifier (LNA) and fed to two mixers, one for the in-phase (I) data channel and one for the quadrature (Q) data channel. Two quadrature[6] 5-GHz local oscillators (LOI and LOQ) are used to down-convert the incoming signals to low frequency base-band signals that are then amplified further by variable gain amplifiers and converted to digital form by the ADCs. Data is then recovered by the CMOS digital signal processor (DSP). On the transmit side, data is converted to analog form in the D/A converters and up-converted to the transmitting frequency by additional mixers and local oscillators. The signal level is increased by a power amplifier before being sent to the antenna. The next several sections will look at the basic building blocks of RF transceivers including RF amplifiers and mixers. High-frequency transistor oscillators are discussed in Chapter 15.

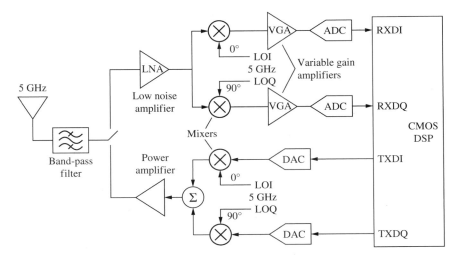

Figure 9.57 Example of an RF transceiver architecture.

---

[5] Known as a "direct-conversion" architecture.

[6] That is, sine and cosine.

9.11.1 RADIO FREQUENCY AMPLIFIERS

Sometimes we need a broad-band amplifier with a bandwidth extending from dc, or very low frequencies, well into the radio frequency range. A technique called **shunt peaking** utilizes an inductor to increase the bandwidth of the inverting amplifier with a capacitive load. However, RF amplifiers are more frequently needed with narrow bandwidths in order to select one signal from the large number that may be present (from an antenna, for example). The frequencies of interest are often above the unity-gain frequency of operational amplifiers so that the RC active filters discussed in Chapter 12 cannot be used. These amplifiers often have high Q; that is, f_H and f_L are very close together relative to the midband or center frequency of the amplifier. For example, a bandwidth of 20 kHz may be desired at a frequency of 1 MHz for an AM broadcast receiver application ($Q = 50$), or a bandwidth of 200 kHz could be needed at 100 MHz for an FM broadcast receiver ($Q = 500$). These applications often use resonant RLC circuits to form frequency selective **tuned amplifiers.**

9.11.2 THE SHUNT-PEAKED AMPLIFIER

In the shunt-peaked circuit [2], an inductor is added in series with the drain resistor as in Fig. 9.58(b). Inductor L forms a low Q resonant circuit with the circuit capacitance and enhances the bandwidth if the value of L is properly selected. As frequency goes up, the impedance of the inductor increases, enhancing the gain. The gain for the circuit in Fig. 9.58(a) is readily found to be

$$A_v(s) = \frac{\mathbf{V_o}(s)}{\mathbf{V_i}(s)} = -g_m Z_L = -g_m \frac{R}{1 + s(C_L + C_{GD})R} = -\frac{g_m R}{1 + sCR} \quad \text{for} \quad C = C_L + C_{GD} \quad (9.182)$$

and exhibits a single-pole roll-off with bandwidth $\omega_H = 1/RC$, where C is the total equivalent load capacitance at the output node. Replacing R by $(R + sL)$ in Eq. (9.182) yields the gain for the shunt-peaked stage:

$$A_{vsp} = -g_m \frac{R + sL}{1 + sCR + s^2 LC} = -g_m R \frac{1 + sL/R}{1 + sRC + s^2 LC} \quad (9.183)$$

Here the zero in the numerator tends to increase the gain as frequency goes up. Eventually the two poles in the denominator cause the gain to roll back off. The poles in the denominator can be real or complex depending upon the element values but are often complex to achieve bandwidth extension.

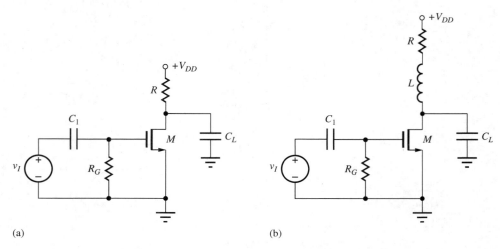

Figure 9.58 (a) Common-source amplifier with capacitive load; (b) Shunt-peaked amplifier.

ELECTRONICS IN ACTION

RF Network Transformations

The figure below provides a very useful set of series-to-parallel and parallel-to-series conversions. The impedances or admittances of the networks are equal at the frequency used to calculate the transformation.

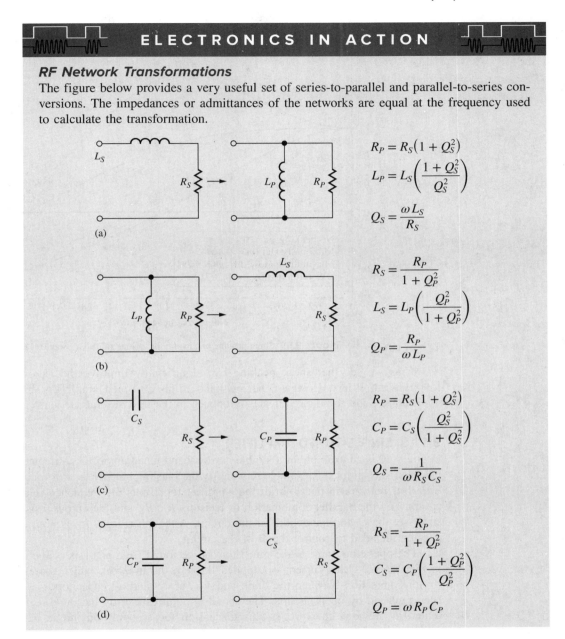

(a) $\quad R_P = R_S(1 + Q_S^2)$
$\quad L_P = L_S\left(\dfrac{1 + Q_S^2}{Q_S^2}\right)$
$\quad Q_S = \dfrac{\omega L_S}{R_S}$

(b) $\quad R_S = \dfrac{R_P}{1 + Q_P^2}$
$\quad L_S = L_P\left(\dfrac{Q_P^2}{1 + Q_P^2}\right)$
$\quad Q_P = \dfrac{R_P}{\omega L_P}$

(c) $\quad R_P = R_S(1 + Q_S^2)$
$\quad C_P = C_S\left(\dfrac{Q_S^2}{1 + Q_S^2}\right)$
$\quad Q_S = \dfrac{1}{\omega R_S C_S}$

(d) $\quad R_S = \dfrac{R_P}{1 + Q_P^2}$
$\quad C_S = C_P\left(\dfrac{1 + Q_P^2}{Q_P^2}\right)$
$\quad Q_P = \omega R_P C_P$

The improvement in bandwidth available through shunt peaking may be explored by normalizing the A_{vsp} expression. Setting $\omega_H = 1/RC = 1$ and defining parameter m as the ratio of the L/R and RC time constants $[m = (L/R)/(RC)]$, Eq. (9.183) can be rewritten as

$$A_{vn} = \left|\dfrac{A_{vsp}}{(-g_m R)}\right| = \dfrac{1 + ms}{1 + s + ms^2} \quad \text{where} \quad L = mR^2 C \qquad (9.184)$$

Equation (9.184) is plotted in Fig. 9.59 for several values of m. The $m = 0$ case corresponds to no shunt peaking, and the bandwidth ($|A_{vn}| = 0.707$) occurs for $\omega/\omega_H = 1$ as expected. The maximally flat, or Butterworth, response occurs for $m = 0.41$ and improves the bandwidth by a factor of 1.72. A maximum bandwidth of $1.85\omega_H$ is achieved with $m = 0.71$, but significant peaking of the gain can be observed in Fig. 9.59.

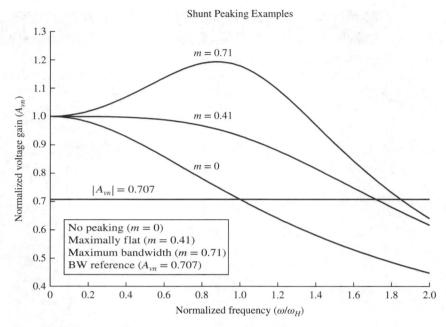

Figure 9.59 Shunt peaking bandwidth for various values of parameter m.

Thus we see that shunt peaking can significantly improve the bandwidth of wide-band low-pass amplifiers. However, narrow-band (high Q) tuned amplifiers are required in many applications, and these circuits are introduced in the next several sections.

9.11.3 SINGLE-TUNED AMPLIFIER

Figure 9.60 is an example of a simple narrow-band tuned amplifier. A depletion-mode MOSFET has been chosen for this example to simplify the biasing, but any type of transistor could be used. The *RLC* network in the drain of the amplifier represents the frequency-selective portion of the circuit, and the parallel combination of resistors R_D, R_3, and the output resistance r_o of the transistor set the Q and bandwidth of the circuit. Although resistor R_D is not needed for biasing, it is often included to control the Q of the circuit.

The operating point of the transistor can be found from analysis of the dc equivalent circuit in Fig. 9.60(b). Bias current is supplied through the inductor, which represents a direct short-circuit connection between the drain and V_{DD} at dc, and all capacitors C_1, C_2, C_S, and C have been replaced by open circuits. The actual Q-point can easily be found using the methods presented in previous chapters, so this discussion focuses only on the ac behavior of the tuned amplifier using the ac equivalent circuit in Fig. 9.61.

Writing a single nodal equation at the output node v_o of the circuit in Fig. 9.61(b) and observing that $v = v_i$ yields

$$(sC_{GD} - g_m)\mathbf{V_i}(s) = \mathbf{V_o}(s)\left[g_o + G_D + G_3 + s(C + C_{GD}) + \frac{1}{sL}\right] \quad (9.185)$$

Making the substitution $G_P = g_o + G_D + G_3$, and then solving for the voltage transfer function:

$$A_v(s) = \frac{\mathbf{V_o}(s)}{\mathbf{V_i}(s)} = (sC_{GD} - g_m)R_P \frac{\dfrac{s}{R_P(C + C_{GD})}}{s^2 + \dfrac{s}{R_P(C + C_{GD})} + \dfrac{1}{L(C + C_{GD})}} \quad (9.186)$$

9.11 Introduction to Radio Frequency Circuits

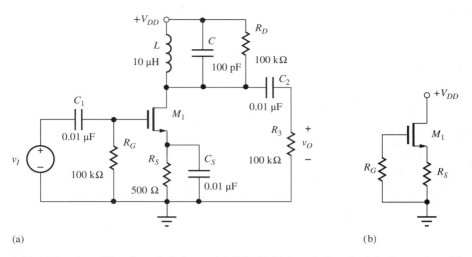

Figure 9.60 (a) Tuned amplifier using a depletion-mode MOSFET. (b) dc equivalent circuit for the tuned amplifier in (a).

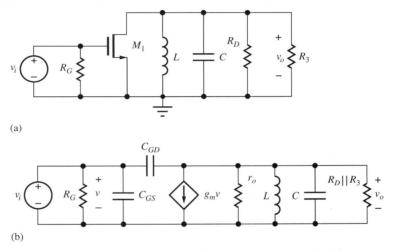

Figure 9.61 (a) High-frequency ac equivalent circuit and (b) small-signal model for the tuned amplifier in Fig. 9.60.

If we neglect the right-half-plane zero, then Eq. (9.186) can be rewritten as

$$A_v(s) \cong A_{\text{mid}} \frac{s\dfrac{\omega_o}{Q}}{s^2 + s\dfrac{\omega_o}{Q} + \omega_o^2} \quad \text{with} \quad \omega_o = \frac{1}{\sqrt{L(C + C_{GD})}} \quad (9.187)$$

In Eq. (9.186), ω_o is the **center frequency** of the amplifier, and the Q is given by

$$Q = \omega_o R_P (C + C_{GD}) = \frac{R_P}{\omega_o L}$$

The center or midband frequency of the amplifier is equal to the resonant frequency ω_o of the LC network. At the center frequency, $s = j\omega_o$, and Eq. (9.187) reduces to

$$A_v(j\omega_o) = A_{\text{mid}} \frac{j\omega_o \dfrac{\omega_o}{Q}}{(j\omega_o)^2 + j\omega_o \dfrac{\omega_o}{Q} + \omega_o^2} = A_{\text{mid}} \frac{j\omega_o \dfrac{\omega_o}{Q}}{-\omega_o^2 + j\omega_o \dfrac{\omega_o}{Q} + \omega_o^2} = A_{\text{mid}}$$

$$A_{\text{mid}} = -g_m R_P = -g_m(r_o \| R_D \| R_3) \quad (9.188)$$

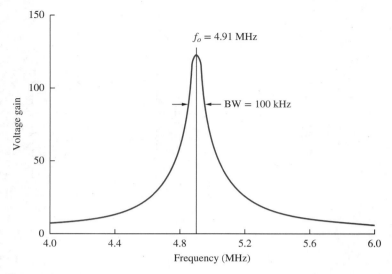

Figure 9.62 Simulated frequency response for the tuned amplifier in Fig. 9.58 with $C_{GS} = 50$ pF, $C_{GD} = 20$ pF, $V_{DD} = 15$ V, $K_n = 5$ mA/V^2, $V_{TN} = -2$ V, $\lambda = 0.02$ V^{-1}, and $R_3 = R_D = \infty$.

For narrow bandwidth circuits—that is, high-Q circuits—the bandwidth is equal to

$$\text{BW} = \frac{\omega_o}{Q} = \frac{1}{R_P(C + C_{GD})} = \frac{\omega_o^2 L}{R_P} \tag{9.189}$$

A narrow bandwidth requires a large value of equivalent parallel resistance R_P, large capacitance, and/or small inductance. In this circuit, the maximum value of $R_P = r_o$. For this case, the Q is limited by the output resistance of the transistor and thus the choice of operating point of the transistor, and the midband gain A_{mid} equals intrinsic gain μ_f.

An example of the frequency response of a tuned amplifier is presented in the SPICE simulation results in Fig. 9.62 for the amplifier in Fig. 9.60. This particular amplifier design has a center frequency of 4.91 MHz and a Q of approximately 50.

> **EXERCISE:** What is the impedance of the 0.01-µF coupling and bypass capacitors in Fig. 9.60 at a frequency of 5 MHz?
>
> **ANSWER:** $-j$ 3.18 Ω (note that $X_C \ll R_G$ and $X_C \ll R_3$)
>
> **EXERCISE:** Find the center frequency, bandwidth, Q, and midband gain for the amplifier in Fig. 9.60 using the parameters in Fig. 9.62, assuming $I_D = 3.20$ mA.
>
> **ANSWERS:** 4.59 MHz, 94.3 kHz, 49.2, −80.3
>
> **EXERCISE:** What are the new values of the center frequency and Q if V_{DD} is reduced to 10 V?
>
> **ANSWERS:** 4.59 MHz; 46.4

9.11.4 USE OF A TAPPED INDUCTOR—THE AUTO TRANSFORMER

The impedance of the gate-drain capacitance and output resistance of the transistor, C_{GD}, and r_o, can often be small enough in magnitude to degrade the characteristics of the tuned amplifier. The problem can be solved by connecting the transistor to a tap on the inductor instead of across the full inductor, as indicated in Fig. 9.63. In this case, the inductor functions as an auto transformer and changes the effective impedance reflected into the resonant circuit.

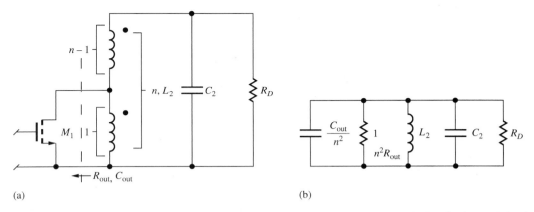

Figure 9.63 (a) Use of a tapped inductor as an impedance transformer. (b) Transformed equivalent for the tuned circuit elements in Fig. 9.63(a). This circuit can be used to find ω_o and Q.

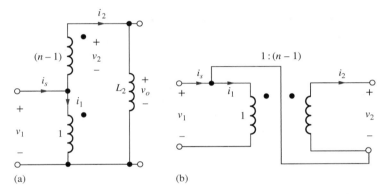

Figure 9.64 (a) Tapped inductor and (b) its representation by an ideal transformer.

The n-turn auto transformer can be modeled by its total magnetizing inductance L_2 in parallel with an ideal transformer having a turns ratio of $(n-1):1$. The ideal transformer has its primary and secondary windings interconnected, as in Fig. 9.64(b). Impedances are transformed by a factor of n^2 by the ideal transformer configuration:

$$\mathbf{V_o}(s) = \mathbf{V_2}(s) + \mathbf{V_1}(s) = (n-1)\mathbf{V_1}(s) + \mathbf{V_1}(s) = n\mathbf{V_1}(s)$$
$$\mathbf{I_s}(s) = \mathbf{I_1}(s) + \mathbf{I_2}(s) = (n-1)\mathbf{I_2}(s) + \mathbf{I_2}(s) = n\mathbf{I_2}(s) \tag{9.190}$$

and

$$\frac{\mathbf{V_o}(s)}{\mathbf{I_2}(s)} = \frac{n\mathbf{V_1}(s)}{\frac{\mathbf{I_s}(s)}{n}} = n^2\frac{\mathbf{V_1}(s)}{\mathbf{I_s}(s)} \qquad Z_s(s) = n^2 Z_p(s) \tag{9.191}$$

Thus, the impedance $Z_s(s)$ reflected into the secondary of the transformer is n^2 times larger than the impedance $Z_p(s)$ connected to the primary.

Using the result in Eq. (9.191), the resonant circuit in Fig. 9.63(a) can be transformed into the circuit representation in Fig. 9.63(b). L_2 represents the total inductance of the transformer. The equivalent output capacitance of the transistor is reduced by the factor of n^2, and the output resistance is increased by this same factor. Thus, a much higher Q can be obtained, and the center frequency is not shifted (detuned) significantly by changes in the value of C_{GD}.

A similar problem often occurs if the tuned circuit is placed at the input of the amplifier rather than the output, as in Fig. 9.65. For the case of the bipolar transistor in particular, the

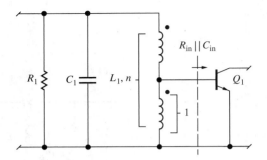

Figure 9.65 Use of an auto transformer at the input of transistor Q_1.

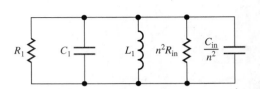

Figure 9.66 Transformed circuit model for the tuned circuit in Fig. 9.65.

equivalent input impedance of Q_1 represented by R_{in} and C_{in} can be quite low due to r_π and the large input capacitance resulting from the Miller effect. The tapped inductor increases the impedance to that in Fig. 9.66, in which L_1 now represents the total inductance of the transformer.

9.11.5 MULTIPLE TUNED CIRCUITS—SYNCHRONOUS AND STAGGER TUNING

Multiple RLC circuits are often needed to tailor the frequency response of tuned amplifiers, as in Fig. 9.67, which has tuned circuits at both the amplifier input and output. The high-frequency ac equivalent circuit for the double-tuned amplifier appears in Fig. 9.67(b). The source resistor is bypassed by capacitor C_S, and C_C is a coupling capacitor. The **radio frequency choke (RFC)** is used for biasing and is designed to represent a very high impedance (an open circuit) at the operating frequency of the amplifier.

Two tuned circuits can be used to achieve higher Q than that of a single LC circuit if both are tuned to the same center frequency (**synchronous tuning**), or a broader band amplifier can

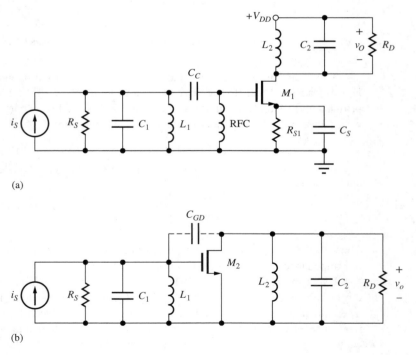

Figure 9.67 (a) Amplifier employing two tuned circuits. (b) High-frequency ac model for the amplifier employing two tuned circuits.

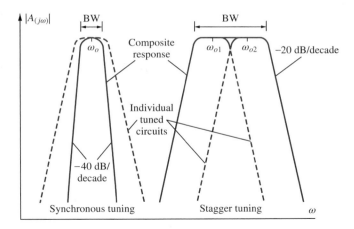

Figure 9.68 Examples of tuned amplifiers employing synchronous and stagger tuning of two tuned circuits.

be realized if the circuits are tuned to slightly different center frequencies (**stagger tuning**), as shown in Fig. 9.68. For the case of synchronous tuning, the overall bandwidth can be calculated using the bandwidth shrinkage factor that can be developed in a manner similar to that of Eq. 9.21:

$$\text{BW}_n = \text{BW}_1 \sqrt{2^{\frac{1}{n}} - 1} \qquad (9.192)$$

in which n is the number of synchronous tuned circuits, and BW_1 is the bandwidth for the case of a single-tuned circuit.

However, two significant problems can occur in the amplifier in Fig. 9.67, particularly for the case of synchronous tuning. First, alignment of the two tuned circuits is difficult because of interaction between the two tuned circuits due to the Miller multiplication of C_{GD}. Second, the amplifier can easily become an oscillator due to the coupling of signal energy from the output of the amplifier back to the input through C_{GD}.

A technique called **neutralization** can be used to solve this feedback problem but is beyond the scope of this discussion. However, two alternative approaches are shown in Fig. 9.69, in which the feedback path is eliminated. In Fig. 9.69(a), a cascode stage is used. Common-base transistor Q_2 effectively eliminates Miller multiplication and provides excellent isolation between the two tuned circuits. In Fig. 9.69(b), the C-C/C-B cascade is used to minimize the coupling between the output and input.

9.11.6 COMMON-SOURCE AMPLIFIER WITH INDUCTIVE DEGENERATION

In most RF systems, we desire to match the input resistance of the LNA to the resistance of the antenna, typically 50 or 75 Ω. In integrated circuits, the clever technique depicted in Fig. 9.70 creates an input match without the use of resistors that would degrade amplifier noise performance. The addition of inductor L_S in series with the source of transistor M_1 creates a positive real component in Z_{in}.

Input impedance Z_{in} can be found using our knowledge of the input resistance of the common-collector and common-drain amplifiers. For the moment, we ignore the gate-drain capacitance of the transistor. Based upon our follower analyses, the input impedance of the amplifier is the sum of impedances Z_{GS} and Z_S plus an amplified replica of Z_S:

$$Z_{iG}(s) = Z_{GS} + Z_S + (g_m Z_{GS})Z_S = \frac{1}{sC_{GS}} + sL_S + g_m \frac{L_S}{C_{GS}}$$

$$Z_{iG}(s) = \frac{1}{sC_{GS}} + sL_S + R_{eq} \quad \text{with} \quad R_{eq} = +g_m \frac{L_S}{C_{GS}} \qquad (9.193)$$

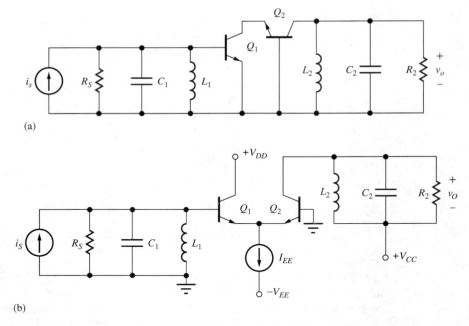

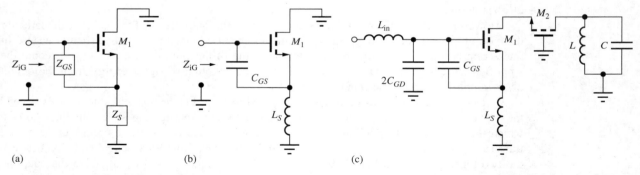

Figure 9.69 (a) Double-tuned cascode and (b) C-C/C-B cascade circuits that provide inherent isolation between input and output.

Figure 9.70 (a) Generalized input impedance circuit. (b) NMOS transistor with inductive source impedance. (c) Cascode LNA with inductor L_{in} added to cancel out the input capacitance part of the input impedance.

The input impedance is the series combination of the impedances of C_{GS} and L_S plus a real input resistance R_{eq} that can be adjusted to match 50 or 75 Ω with the proper choice of values for L_S and the W/L ratio and Q-point of the transistor.

Normally L_S and C_{GS} are not resonant at the desired operating frequency. In Fig. 9.70(c), a second inductor L_{in} is added in series with the input of the amplifier to resonate the input leaving a purely resistive input resistance. A cascode stage is often utilized to minimize the impact of the gate-drain capacitance that reflects an equivalent capacitance of approximately $2C_{GD}$ between the gate terminal and ground.

Let us now consider the general case of the circuit in Fig. 9.71(a) in which the small-signal model is redrawn in current controlled form. Above dc, the current gain of the MOSFET is no longer infinite but is given by

$$\beta(s) = g_m Z_{gs} = \frac{g_m}{s C_{GS}} \quad \text{and} \quad \beta(j\omega) = -j\frac{g_m}{\omega C_{GS}} \cong -j\frac{\omega_T}{\omega} \qquad (9.194)$$

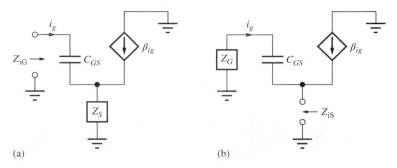

Figure 9.71 (a) Impedance Z_{iG} at the gate with impedance Z_S in the source. (b) Impedance Z_{iS} at the source with impedance Z_G at the gate.

The approximation assumes $C_{GS} \gg C_{GD}$. Using β, Z_{iG} in Eq. (9.193) can be rewritten in a familiar form, although the MOSFET now has a complex current gain with a $-90°$ phase shift!

$$Z_{iG} = Z_{GS} + (\beta + 1)Z_S = Z_{GS} + Z_S + \beta Z_S \qquad (9.195)$$

The complex current gain causes unexpected impedance transformations to occur that can ultimately lead to instability in certain circuits if one is not aware of the behavior. In Eq. (9.193) we observed that an inductor in the source is transformed to a resistance at the input. Table 9.4 gives the transformations that occur for a resistor, inductor or capacitor in place of Z_S. It is important to note that a load capacitance is transformed into a frequency-dependent negative input resistance term! This term has the potential to cause instability.

The impedance Z_{iS} looking into the source of the transistor in Fig. 9.71(b) with an impedance Z_G at the gate also sees unusual transformations. Assuming $\beta \gg 1$,

$$Z_{iS} = \frac{\frac{1}{sC_{GS}} + Z_G}{\beta + 1} \cong \frac{\frac{1}{sC_{GS}} + Z_G}{\beta} = \frac{1}{g_m} + \frac{Z_G}{\beta} = \frac{1}{g_m} + j\frac{\omega}{\omega_T}Z_G \qquad (9.196)$$

The complex current gain now causes a $+90°$ phase shift, and the impedances resulting from the second term in Eq. (9.196) appear in Table 9.4. For this case, an inductance at the gate is transformed into a frequency dependent negative resistance at the source of the transistor.

It is important to realize that a similar situation occurs in the bipolar transistor. For frequencies above ω_β, C_π becomes more important than r_π and the current gain of the BJT becomes

$$\beta(s) = g_m Z_{be} \cong \frac{g_m}{sC_\pi} \quad \text{and} \quad \beta(j\omega) = -j\frac{g_m}{\omega C_\pi} \cong -j\frac{\omega_T}{\omega} \quad \text{for} \quad \omega > \omega_\beta \qquad (9.197)$$

TABLE 9.4
Complex Current Gain Source and Gate Impedance Transformations $\left(\omega_T = \dfrac{g_m}{C_{GS}}\right)$

SOURCE FOLLOWER INPUT IMPEDANCE		SOURCE FOLLOWER OUTPUT IMPEDANCE	
Z_S	Resulting Impedance	Z_G	Resulting Impedance
R	$C_{EQ} = +\dfrac{1}{\omega_T R}$	R	$L_{EQ} = +\dfrac{R}{\omega_T}$
L	$R_{EQ} = +\omega_T L$	L	$R_{EQ} = -\dfrac{\omega^2}{\omega_T}L\,!$
C	$R_{EQ} = -\dfrac{\omega_T}{\omega^2 C}\,!$	C	$R_{EQ} = \dfrac{1}{\omega_T C}$

EXERCISE: An NMOS transistor has $\mu_n = 400$ cm^2/V-sec, $L = 0.5$ μm and is biased at 0.25 V above threshold. What value of L_S is required to achieve an input resistance of 75 Ω in Fig. 9.68(b)?

ANSWER: 1.88 nH

ELECTRONICS IN ACTION

Noise Factor, Noise Figure, and Minimum Detectable Signal

Resistors and transistors in amplifiers add thermal noise and shot noise to the signal during the amplification process (see the EIA on page 454). **Noise factor** F of an amplifier (or any electronic system) is a measure of the degradation of the **signal-to-noise ratio (SNR)** by these noisy elements where F is defined as the ratio of the total noise power at the output of the amplifier to the noise power at the output due to the noise of the source acting alone. F can also be expressed as the ratio of the SNR at the amplifier input to the SNR ratio at the output.

$$F = \frac{\text{Total noise power at the amplifier output}}{\text{Noise power at the output due to noise of the source}} = \frac{\text{SNR}_{in}}{\text{SNR}_{out}}$$

We can model the noise of the amplifier in terms of its noise factor as shown above where the "$F - 1$" noise source is added to model the internal noise of the amplifier. The quantity $F - 1$ indicates how much additional noise is added by the amplifier. If no noise were added, then F would be unity, and $F - 1$ would be zero.

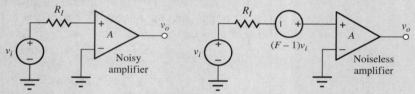

(a) Noisy amplifier and noise factor F.   (b) Noiseless amplifier model.

If the noise sources at the amplifier input are added up, the result is

$$\overline{v_{tot}^2} = \overline{v_i^2} + (F-1)\overline{v_i^2} = F\overline{v_i^2} \quad \text{and} \quad F = \frac{\overline{v_{tot}^2}}{\overline{v_i^2}}$$

where $\overline{v_i^2} = 4kTR_I B$ is the thermal noise of the source resistance in bandwidth B. **Noise figure** NF is the most often quoted quantity and is simply a conversion of the noise factor to dB: $NF = 10 \log F$. Note that the amplifier above will typically have BJTs or FETs connected to the external inputs, and r_x or r_g will appear directly in series with R_I, thereby becoming an important contributor to the noise figure of the amplifier.

The **minimum detectable signal (MDS)** is defined as the signal with a power equal to the equivalent input noise power of the amplifier. The total noise power available from the noise source in a matched system is

$$S_{mds} = \frac{\overline{v_{tot}^2}}{4R_I} = kTBF$$

The minimum detectable signal power is most often expressed in dBm ($10 \log S_{mds}/10^{-3}$), and for $T = 290$ K,

$$S_{mds} = -174 \text{ dBm} + 10 \log B + 10 \log F$$

For a bandwidth of 1 kHz and an NF of 3 dB, $S_{mds} = -142$ dBm.

9.12 MIXERS AND BALANCED MODULATORS

In radio frequency applications, we often need to translate signals from one frequency to another. This process includes both mixing and modulation, and generally requires some form of multiplication of two signals in order to generate sum and difference frequency components in the output spectrum. **Single-balanced mixers** eliminate one of the two input signals from the output, whereas the outputs of **double-balanced circuits** do not contain spectral components at either of the input frequencies.[7]

9.12.1 INTRODUCTION TO MIXER OPERATION

To achieve mixing, we need to multiply two signals together as indicated by the mixer symbol in Fig. 9.72. Suppose we form the product of two sine waves at frequencies ω_1 and ω_2 and expand the result using standard trigonometric identities:

$$s_o = s_2 \cdot s_1 = \sin\omega_2 t \cdot \sin\omega_1 t = \frac{\cos(\omega_2 - \omega_1)t - \cos(\omega_2 + \omega_1)t}{2} \tag{9.198}$$

The ideal mixer output contains signal components at frequencies $\omega_2 - \omega_1$ and $\omega_2 + \omega_1$. Usually filters or I and Q phasing techniques are used to select either the sum or difference output depending upon whether the application employs **up-conversion** ($\omega_2 + \omega_1$) or **down-conversion** ($\omega_2 - \omega_1$).

Figure 9.73 shows an FM receiver application in which a narrow-band VHF signal at 100 MHz is mixed with a **local oscillator** (LO) signal at 89.3 MHz. The narrow-band VHF spectrum is translated to both 10.7 MHz, which is selected by a band-pass filter, and 189.3 MHz, which is rejected by the same filter.

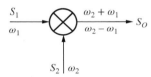

Figure 9.72 Basic mixer symbol indicating multiplication of signals s_1 and s_2.

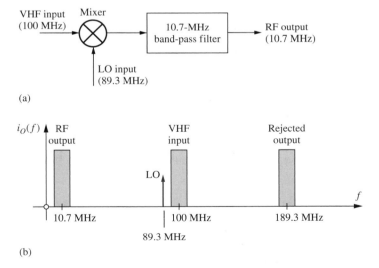

Figure 9.73 (a) Mixer block diagram and (b) spectrum in FM receiver application.

---

[7] The mixer is known as a linear time-varying (LTV) circuit.

EXERCISE: The LO signal in Fig. 9.73 could also be positioned above the VHF frequency. What would then be the local oscillator frequency and the center frequency of the unwanted output frequency signal?

ANSWERS: 110.7 MHz; 210.7 MHz

EXERCISE: (a) An FM receiver is to be tuned to receive a station at 104.7 MHz. What must be the local oscillator frequencies to set the output to the 10.7-MHz filter frequency? (b) Repeat for an input frequency of 88.1 MHz.

ANSWERS: 94.0 MHz or 115.4 MHz; 77.4 MHz or 98.8 MHz

Conversion Gain

In the amplifiers covered thus far in this text, gain expressions have generally involved signals at the same frequency. We have assumed that the amplifiers were linear and that the input and output signals were at the same frequency. In fact, a component at any other frequency was considered to be an undesirable distortion product. (Remember the definition of THD, total harmonic distortion.)

In contrast, the mixer is a circuit in which the output signals are at frequencies different from those at the input. A mixer's conversion gain is defined as the ratio of the phasor representation of the output signal to that of the input signal, and the fact that the signals are at two different frequencies is simply ignored. For example, the conversion gain of the mixer described by Eq. (9.193) is 0.5 or −6 dB for either output frequency.

Almost any nonlinear device can be used for mixing. For example, the $i - v$ characteristics of diodes, bipolar transistors, and field-effect transistors all contain quadratic (and higher) nonlinear terms in their mathematical representations and can therefore generate a wide range of product terms. However, we will focus in the next sections on switching mixers that have relatively high conversion gains (i.e., low conversion losses).

9.12.2 A SINGLE-BALANCED MIXER

There is actually no need for both signals to be sine waves in the mixer in Fig. 9.72. It is very convenient for one of the inputs to be a switching waveform, and the conversion gain is actually higher if a square wave is utilized. In its simplest form, the switching mixer consists of a signal source, a switch, and a load as depicted in Fig. 9.74(a). When the switch is closed, the output is equal to the input signal, and when the switch is open, the output is zero. Thus the output voltage is equal to input voltage v_1 multiplied by the square-wave switching function $s_s(t)$ in Fig. 9.74(b). If we assume that the input signal is a sine wave and represent the square wave by its Fourier series,

$$v_I(t) = A \sin \omega_1 t \quad \text{and} \quad s_S(t) = \frac{1}{2} + \frac{2}{\pi} \sum_{n \text{ odd}} \frac{1}{n} \sin n\omega_2 t \quad \text{with} \quad \omega_2 = \frac{2\pi}{T} \qquad (9.199)$$

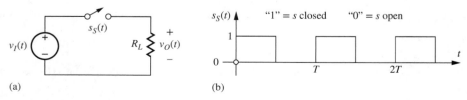

Figure 9.74 (a) Single-balanced mixer and (b) switching function $s_S(t)$.

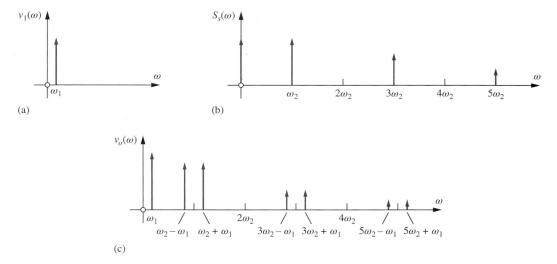

Figure 9.75 Single-balanced mixer spectra: (a) input 1; (b) input 2; (c) output.

then an expression for the output voltage becomes

$$v_O(t) = v_I \times s_S = \frac{A}{2}\sin \omega_1 t + \frac{2A}{\pi}\sum_{n\,\text{odd}} \frac{1}{n}\sin n\omega_2 t \sin \omega_1 t$$

or
(9.200)

$$v_O(t) = \frac{A}{2}\sin \omega_1 t + \frac{A}{\pi}\sum_{n\,\text{odd}} \frac{\cos(n\omega_2 - \omega_1)t - \cos(n\omega_2 + \omega_1)t}{n}$$

As a result of the mixing operation, the spectrum of the output signal has a component at the original input signal frequency ω_1, and copies of the input signal translated by odd multiples of switching frequency ω_2 as in Fig. 9.75(c). The terms corresponding to $n = 1$ are the most often utilized since they have the highest conversion gain.

Note that there are no components in the output at harmonics of the switching frequency ω_2, whereas there is a component at ω_1. This output is said to be single-balanced because only one of the fundamental input frequencies is eliminated from the output; the mixer in Fig. 9.75 is balanced with respect to s_S.

EXERCISE: What is the conversion gain of the single-balanced mixer in Fig. 9.74?

ANSWER: $1/\pi$ or -9.94 dB

9.12.3 THE DIFFERENTIAL PAIR AS A SINGLE-BALANCED MIXER

One concept for a single-balanced mixer appears in Fig. 9.76(a) with the switch implemented using a differential pair in Fig. 9.76(b). A signal v_1 at frequency ω_1 is used to vary the current supplied to the emitters of the pair:

$$i_{EE} = I_{EE} + I_1 \sin \omega_1 t \qquad (9.201)$$

The second input is driven by a large-signal square wave at frequency ω_2, which switches current i_{EE} back and forth between the two collectors and alternately multiplies the differential output voltage by $+1$ and -1. This multiplication can be represented by a unit amplitude square wave

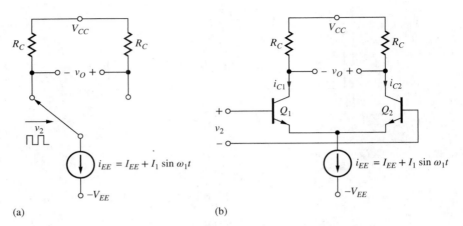

Figure 9.76 (a) Basic single-balanced mixer. (b) Differential pair implementation.

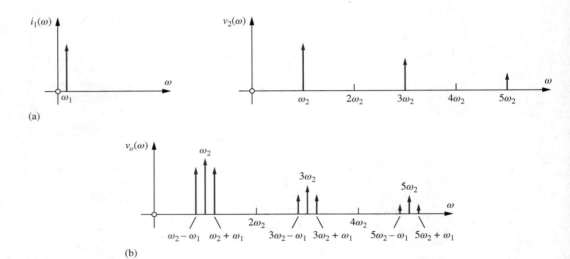

Figure 9.77 (a) Input and (b) output spectra for the mixer in Fig. 9.76.

with a Fourier series given by

$$v_2(t) = \sum_{n \text{ odd}} \frac{4}{n\pi} \sin n\omega_2 t \qquad (9.202)$$

Using Eqs. (9.201) and (9.202),

$$v_O(t) = [i_{C2}(t) - i_{C1}(t)]R_C = (I_{EE} + I_1 \sin \omega_1 t)R_C \sum_{n \text{ odd}} \frac{4}{n\pi} \sin n\omega_2 t$$

or $\qquad (9.203)$

$$V_O(t) = \sum_{n \text{ odd}} \frac{4}{n\pi}\left[I_{EE}R_C \sin n\omega_2 t + \frac{I_1 R_C}{2}\cos(n\omega_2 - \omega_1)t - \frac{I_1 R_C}{2}\cos(n\omega_2 + \omega_1)t\right]$$

The input and output spectra for the differential pair mixer appear in Fig. 9.77. The mixer in Fig. 9.76 is actually balanced relative to the input signal at frequency ω_1 rather than ω_2 as was the case in Fig. 9.74.

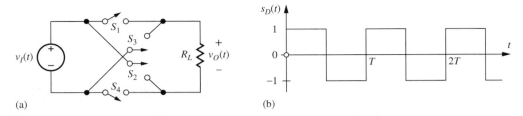

Figure 9.78 (a) Double-balanced mixers and (b) switching waveform $s_D(t)$.

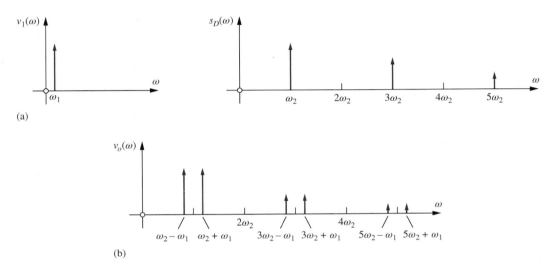

Figure 9.79 (a) Input and (b) output spectra for the mixer in Fig. 9.78.

9.12.4 A DOUBLE-BALANCED MIXER

In many cases, we prefer to eliminate both input signals from the output, and the double-balanced mixer solves this problem. If we study Eqs. (9.200) and (9.203), we see that the source of the balance problem is the dc term in the switching waveform, but the dc component can be eliminated by using four switches to modify the switching function as in Fig. 9.78. During the first half of the switching cycle, switches S_1 and S_4 are closed and the input source is connected directly to the output, but during the second half-cycle, switches S_2 and S_3 are closed reversing the polarity of the input signal. Thus the switching waveform alternates between $+1$ and -1 with zero average value!

The Fourier series for the switching waveform is now

$$s_D(t) = \frac{4}{\pi} \sum_{n \text{ odd}} \frac{1}{n} \sin n\omega_2 t \quad \text{where} \quad \omega_o = \frac{2\pi}{T} \qquad (9.204)$$

and the output signal becomes

$$v_O(t) = \frac{2A}{\pi} \sum_{n \text{ odd}} \frac{\cos(n\omega_2 - \omega_1)t - \cos(n\omega_2 + \omega_1)t}{n} \qquad (9.205)$$

Neither of the fundamental components of the input signals appears in the output in Fig. 9.79 Note however that the degree of balance depends upon the symmetry of the square wave, and any asymmetry between the two half cycles will produce a dc term that degrades the rejection of the undesired output signal.

EXERCISE: What is the conversion gain for the double-balanced mixer in Fig. 9.78?

ANSWER: $2/\pi$ or -3.92 dB

Passive Diode Mixers

Another popular form of passive double-balanced mixer appears in the figure below. The switches are implemented by a four-diode bridge and are driven by a high-level (10 dBm) local oscillator signal. Both the LO and RF inputs are transformer-coupled to the diode bridge, and the output signal appears at the IF port. Excellent balance can be achieved with well-matched diodes and carefully designed transformers.

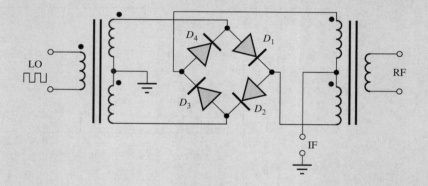

An example of such a mixer product produced by Mini-Circuits® appears in the photograph below. Similar mixer products cover a very wide range of frequencies and switching signal levels.

Mini-Circuits ZP-3LH+ Mixer: 0.15–400 MHz, 4.8 dB conversion loss, +10 dBm LO, 50 dB LO-RF isolation, 45 dB LO-IF isolation. Courtesy of Mini-Circuits® (www.minicircuits.com).
©2014 Scientific Components Corporation d/b/a Mini-Circuits. Used with permission.

A Passive MOS Double-Balanced Mixer

The circuit in Fig. 9.80 shows an implementation of the double-balanced mixer from Fig. 9.78 using four MOS transistors as switches in which the circuit is redrawn as a bridge. The circuit is considered to be a **passive mixer** because no power is required beyond that supplied from

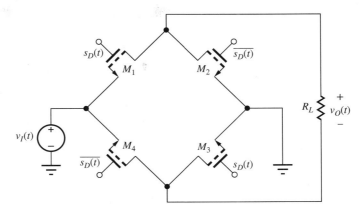

Figure 9.80 Passive NMOS double-balanced mixers.

input signal v_I and the switching signal applied to the gates of the MOSFETs. If the on-resistances of the MOSFETs are designed to be much smaller than load resistor R_L, then input signal v_I will appear across R_L when M_1 and M_3 are turned on, and the negative of v_I will appear across R_L when M_2 and M_4 are on. High levels of rejection can be achieved with well-matched transistors in integrated circuit realizations.

SPICE simulation results for the circuit in Fig. 9.80 appear in Fig. 9.81 for a 100-mV, 4-kHz sine wave and a $\pm$ 5-V, 50-kHz switching signal. The output waveform shows the signal polarity reversal that occurs at the switching signal rate and an amplitude loss caused by the on-resistance of the switches. The spectrum shows the mixer products 4 kHz above and below the odd harmonics of 50 kHz, whereas components at 4 kHz and the odd harmonics of 50 kHz are suppressed.

> **EXERCISE:** What is the actual conversion gain for the double-balanced mixer in Fig. 9.81?
>
> **ANSWER:** $0.7 \times 2/\pi$ or -7.02 dB

9.12.5 THE JONES MIXER—A DOUBLE-BALANCED MIXER/MODULATOR

The **Jones Mixer** of Fig. 9.82 operates as a **double-balanced modulator** or **mixer** when transistors Q_3–Q_6 in Fig. 9.82 are driven by the square-wave signal at input v_2 at carrier frequency ω_c. The second signal v_1 at modulating frequency ω_m is applied to the transconductance stage. For the circuit in Fig. 9.82, we have

$$i_{C1} = I_{BB} + \frac{V_m}{2R_1}\sin \omega_m t \quad \text{and} \quad i_{C2} = I_{BB} - \frac{V_m}{2R_1}\sin \omega_m t \qquad (9.206)$$

If we take a differential output, the dc current component cancels out, and the signal current at frequency ω_m is switched back and forth by the square-wave input and appears to be multiplied alternately by $+1$ and -1. Using Eqs. (9.204) and (9.206), the output signal between the collectors can be written as

$$v_O(t) = V_m \frac{R_C}{R_1} \sum_{n \text{ odd}} \frac{4}{n\pi} \sin n\omega_c t \, \sin \omega_m t$$

or $\hspace{12em}$ (9.207)

$$v_O(t) = V_m \frac{R_C}{R_1} \sum_{n \text{ odd}} \frac{2}{n\pi} [\cos(n\omega_c - \omega_m)t - \cos(n\omega_c + \omega_m)t]$$

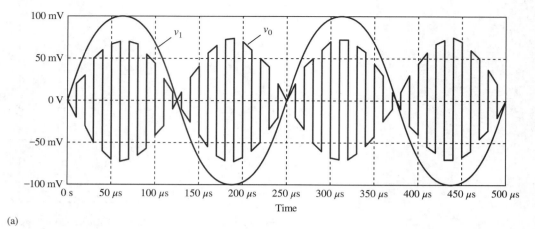

(a)

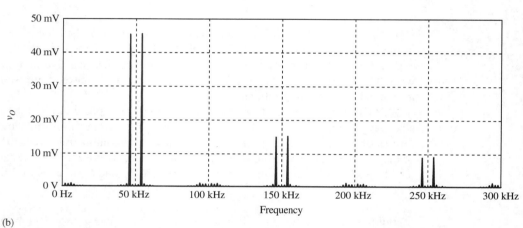

(b)

Figure 9.81 NMOS passive mixer. (a) SPICE output waveform and (b) spectrum.

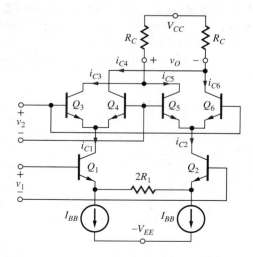

Figure 9.82 Classic double-balanced Jones mixer. Signal v_2 is a large-signal square-wave at the carrier frequency ω_c, and v_1 is the modulating signal at frequency ω_m.

9.12 Mixers and Balanced Modulators

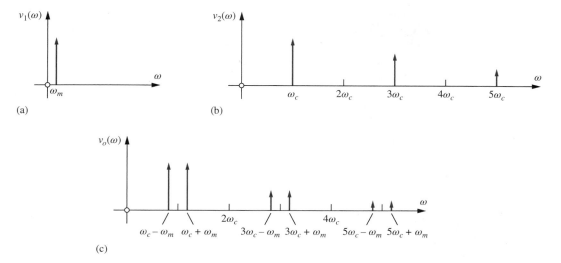

Figure 9.83 Spectra for double-balanced modulation. (a) Modulation input. (b) Switching input. (c) Output signal.

The output signal has spectral components at frequencies ω_m above and below each of the odd harmonics of the carrier frequency ω_c as in Fig. 9.83. Note that no signal energy at either the carrier or modulation signal frequencies ω_c or ω_m appears at the output, and the circuit operates as a double-balanced modulator or mixer. A bandpass filter or phasing techniques can be used to select the desired frequencies from the composite spectrum at the output.

In modulator applications, the circuit just described generates a double-sideband suppressed-carrier (DSBSC) output signal. An amplitude-modulated signal (with modulation index $M, 0 \leq M \leq 1$) can also be generated by adding a dc component to the modulating signal

$$v_1 = V_m(1 + M \sin \omega_m t) \tag{9.208}$$

The dc term unbalances the circuit relative to the carrier frequency thereby injecting a carrier frequency component into the output. (Note the same effect is caused by offset voltages due to mismatches in the transistors.) The output voltage becomes

$$v_O(t) = V_m \frac{R_C}{R_1} \sum_{n \text{ odd}} \frac{4}{n\pi} \left[\sin n\omega_c t + \frac{M}{2} \cos(n\omega_c - \omega_m)t - \frac{M}{2} \cos(n\omega_c + \omega_m)t \right] \tag{9.209}$$

In this case, the circuit remains balanced with respect to the modulation signal and operates as a single-balanced modulator.

EXERCISE: A 20-MHz carrier is modulated with a 10-kHz signal using the double-balanced modulator in Figs. 9.82 and 9.83. What are the frequencies of the spectral components in Fig. 9.83(c)?

ANSWERS: 19.99 MHz; 20.01 MHz; 59.99 MHz; 60.01 MHz; 99.99 MHz; 100.01 MHz

EXERCISE: The amplitude of the signal at 19.99 MHz in the previous exercise is 3 V. What are the amplitudes of the other components?

ANSWERS: 3 V; 1 V; 1 V; 0.6 V; 0.6 V

ELECTRONICS IN ACTION

The Jones Mixer

The patent for the classic Jones mixer circuit below[8] was filed in 1963, the very early days of transistor circuit design, and this innovative circuit is in pervasive use today in ICs developed for communication applications (see any recent issue of the *IEEE Journal of Solid-State Circuits*). The circuit appears in virtually every smart phone, satellite radio, communications receiver and software defined radio, to name a few. This circuit, however, has often been incorrectly attributed to Barrie Gilbert because of its similarity to his analog multiplier discussed in Sec. 14.10.

In the circuit from the Jones patent, transistors 121, 139, 83, 165, 157, and 103 correspond to transistors Q_1 to Q_6 in Fig. 9.82. Transistors 179 and 191 provide common-mode feedback to stabilize the operating points of the transistors in the circuit.

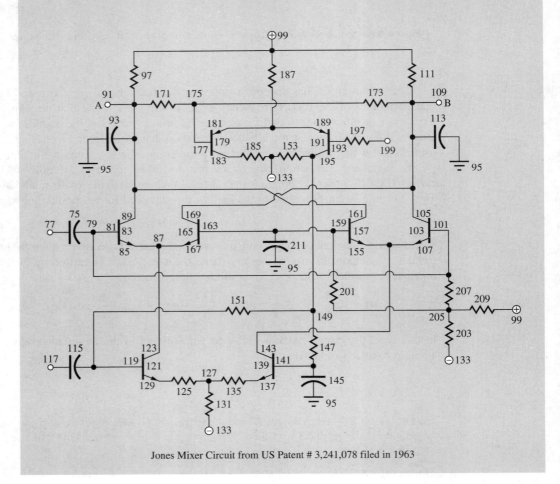

Jones Mixer Circuit from US Patent # 3,241,078 filed in 1963

---

[8] H. E. Jones, "Dual output synchronous detector utilizing differential amplifiers," US Patent # 3,241,078, filed June 18, 1963, issued March 15, 1966. Thanks to Professor Tom Lee of Stanford University for pointing out the original patent to the authors of this text.

SUMMARY

- Amplifier frequency response can be determined by splitting the circuit into two models, one valid at low frequencies where coupling and bypass capacitors are most important, and a second valid at high frequencies in which the internal device capacitances control the frequency-dependent behavior of the circuit.
- Direct analysis of these circuits in the frequency domain, although usually possible for single-transistor amplifiers, becomes impractical for multistage amplifiers. In most cases, however, we are primarily interested in the midband gain and the upper- and lower-cutoff frequencies of the amplifier, and estimates of f_H and f_L can be obtained using the open-circuit and short-circuit time-constant methods. More accurate results can be obtained using SPICE circuit simulation.
- The frequency-dependent characteristics of the bipolar transistor are modeled by adding the base-emitter and base-collector capacitors C_π and C_μ and base resistance r_x to the hybrid-pi model. The value of C_π is proportional to collector current I_C, whereas C_μ is weakly dependent on collector-base voltage. The $r_x C_\mu$ product is an important figure of merit for the frequency limitations of the bipolar transistor.
- The frequency dependence of the FET is modeled by adding gate-source and gate-drain capacitances, C_{GS} and C_{GD}, and gate resistance r_g to the pi-model of the FET. The values of C_{GS} and C_{GD} are independent of operating point when the FET is operating in the active region. The $r_g C_{GD}$ product represents an important figure of merit for the FET.
- Both the BJT and FET have finite current gain at high frequencies, and the unity gain-bandwidth product f_T for both devices is determined by the device capacitances and the transconductance of the transistor. In the bipolar transistor, the β-cutoff frequency f_β represents the frequency at which the current gain is 3 dB below its low-frequency value.
- In SPICE, the basic high-frequency behavior of the bipolar transistor is modeled using these parameters: forward transit-time TF, zero-bias collector-base junction capacitance CJC, collector junction built-in potential VJC, collector junction grading factor MJC, and base resistance RB.
- In SPICE, the high-frequency behavior of the MOSFET is modeled using the gate-source and gate-drain capacitances determined by the gate-source and gate-drain overlap capacitances CGSO and CGDO, as well as TOX, W, L, and RG.
- If all the poles and zeros of the transfer function can be found from the low- and high-frequency equivalent circuits, then f_H and f_L can be accurately estimated using Eqs. (9.16) and (9.23). In many cases, a dominant pole exists in the low- and/or high-frequency responses, and this pole controls f_H or f_L. Unfortunately, the complexity of most amplifiers precludes finding the exact locations of all the poles and zeros except through numerical means.
- For design purposes, however, one needs to understand the relationship between the device and circuit parameters and f_H and f_L. The short-circuit time constant (SCTC) and open-circuit time constant (OCTC) approaches, as well as Miller effect, provide the needed information and were used to find detailed, although approximate, expressions for f_H and f_L for the three classes of single-stage amplifiers, the inverting, noninverting, and follower stages.
- The input impedance of the inverting amplifier is decreased as a result of Miller multiplication, and the expression for the dominant pole of an inverting amplifier can be cast in terms of the Miller effect.
- In contrast, the input impedance of the followers is increased by the Miller effect, and the dominant pole of the follower can also be cast in terms of Miller multiplication.
- It was found that the inverting amplifiers provide high gain but the most limited bandwidth. Non-inverting amplifiers can provide improved bandwidth for a given voltage gain, but it is important to remember that these stages have a much lower input resistance. The follower

configurations provide nearly unity gain over a very wide bandwidth. The three basic classes of amplifiers show the direct trade-offs that occur between voltage gain and bandwidth.

- The SCTC approach is used to estimate the value of the lower-cutoff frequency in multistage amplifiers, whereas the OCTC method or Miller effect and equivalent time constant approach is applied to the nodes in the signal path to find the upper cutoff frequency. The frequency responses of the differential pair, cascode amplifier, C-C/C-B cascade stage, and current mirror were all evaluated, as well as an example of calculations for a three-stage amplifier. The frequency response of another multistage amplifier is calculated in Chapter 15.
- Shunt peaking utilizes an inductor to significantly extend the bandwidth of the inverting amplifier.
- Tuned amplifiers employing *RLC* circuits can be used to achieve narrow-band amplifiers at radio frequencies. Designs can use either single- or multiple-tuned circuits. If the circuits in a multiple-tuned amplifier are all designed to have the same center frequency, the circuit is referred to as synchronously tuned. If the tuned circuits are adjusted to different center frequencies, the circuit is referred to as stagger-tuned. Care must be taken to ensure that tuned amplifiers do not become oscillators, and the use of the cascode and C-C/C-B cascade configurations offers improved isolation between multiple-tuned circuits.
- Mixer circuits are widely used in communications electronics to translate the frequency spectrum of a signal. Mixing requires some form of multiplication of two signals, which generates sums and differences of the two input spectra. Single- and double-balanced configurations eliminate one or both of the input signals from the output spectrum. Single-balanced mixers can be designed using differential pairs. Double-balanced mixers often utilize circuits based on passive switching-type mixers employing FETs or diodes.
- The classic Jones mixer is an extremely important circuit that is widely employed in today's communications ICs.

KEY TERMS

Base resistance
Beta-cutoff frequency
Cascode amplifier
Center frequency
Dominant high-frequency pole
Dominant low-frequency pole
Dominant pole
Double-balanced circuits
Double-balanced mixers (modulators)
Down-conversion
Jones mixer
Local oscillator
Lower-cutoff frequency
Midband gain
Miller effect
Miller multiplication
Minimum detachable signal (MDS)
Neutralization

Noise factor F
Noise figure NF
Open-circuit time-constant (OCTC) method
Passive mixers
Pole frequencies
Pole-splitting
Q
Radio frequency choke (RFC)
Short-circuit time-constant (SCTC) method
Shunt peaking
Signal-to-noise ratio (SNR)
Single-balanced mixer
Stagger tuning
Synchronous tuning
Tuned amplifiers
Unity gain-bandwidth product
Up-conversion
Upper-cutoff frequency

REFERENCES

1. P. E. Gray and C. L. Searle, *Electronic Principles*, Wiley, New York: 1969.
2. S. S. Mohan, M. del Mar Hershenson, S. P. Boyd and T. H. Lee, "Bandwidth extension in CMOS with optimized on-chip inductors," *IEEE Journal of Solid-State Circuits*, vol. 35, no. 3, pp. 346–355, March 2000.

PROBLEMS

9.1 Amplifier Frequency Response

9.1. Find A_{mid} and $F_L(s)$ for this transfer function. Is there a dominant pole? If so, what is the dominant-pole approximation of $A_v(s)$? What is the cutoff frequency f_L of the dominant-pole approximation? What is the exact cutoff frequency using the complete transfer function?

$$A_v(s) = \frac{50s^2}{(s+3)(s+50)}$$

9.2. Find A_{mid} and $F_L(s)$ for this transfer function. Is there a dominant pole? If so, what is the dominant-pole approximation of $A_v(s)$? What is the cutoff frequency f_L of the dominant-pole approximation? What is the exact cutoff frequency using the complete transfer function?

$$A_v(s) = \frac{200s^2}{2s^2 + 1400s + 100{,}000}$$

9.3. Find A_{mid} and $F_L(s)$ for this transfer function. Is there a dominant pole? Use Eq. (9.16) to estimate f_L. Use the computer to find the exact cutoff frequency f_L.

$$A_v(s) = -\frac{150s(s+15)}{(s+11)(s+21)}$$

9.4. Find A_{mid} and $F_H(s)$ for this transfer function. Is there a dominant pole? If so, what is the dominant-pole approximation of $A_v(s)$? What is the cutoff frequency f_H of the dominant-pole approximation? What is the exact cutoff frequency using the complete transfer function?

$$A_v(s) = \frac{6 \times 10^{11}}{3s^2 + 3.3 \times 10^5 s + 3 \times 10^9}$$

9.5. Find A_{mid} and $F_H(s)$ for this transfer function. Is there a dominant pole? If so, what is the dominant-pole approximation of $A_v(s)$? What is the cutoff frequency f_H of the dominant-pole approximation? What is the exact cutoff frequency using the complete transfer function?

$$A_v(s) = \frac{(s + 2 \times 10^9)}{(s + 10^7)\left(1 + \dfrac{s}{7 \times 10^8}\right)}$$

9.6. Find A_{mid} and $F_H(s)$ for this transfer function. Is there a dominant pole? Use Eq. (9.16) to estimate f_H. Use the computer to find the exact cutoff frequency f_H.

$$A_v(s) = \frac{4 \times 10^9(s + 5 \times 10^5)}{(s + 1.3 \times 10^5)(s + 2 \times 10^6)}$$

9.7. Find A_{mid}, $F_L(s)$, and $F_H(s)$ for this transfer function. Is there a dominant pole at low frequencies? At high frequencies? Use Eqs. (9.16) and (9.23) to estimate f_L and f_H. Use the computer to find the exact cutoff frequencies and compare to the estimates.

$$A_v(s) = -\frac{2 \times 10^8 s^2}{(s+1)(s+2)(s+1000)(s+500)}$$

*__9.8.__ Find A_{mid}, $F_L(s)$ and $F_H(s)$ for this transfer function. Is there a dominant pole at low frequencies? At high frequencies? Use Eqs. (9.16) and (9.23) to estimate f_L and f_H. Use the computer to find the exact cutoff frequencies and compare to the estimates.

$$A_v(s) = \frac{2 \times 10^{10} s^2 (s+1)(s+200)}{(s+3)(s+5)(s+7)(s+100)^2(s+300)}$$

9.2 Direct Determination of the Low-Frequency Poles and Zeros—The Common-Source Amplifier

9.9. (a) Draw the low-frequency and midband equivalent circuits for the common-source amplifier in Fig. P9.9 if $R_I = 5\ \text{k}\Omega$, $R_1 = 430\ \text{k}\Omega$, $R_2 = 560\ \text{k}\Omega$, $R_S = 13\ \text{k}\Omega$, $R_D = 43\ \text{k}\Omega$, and $R_3 = 240\ \text{k}\Omega$. (b) What are the lower-cutoff frequency and midband gain of the amplifier if the Q-point = (0.2 mA, 5 V) and $V_{GS} - V_{TN} = 0.7$ V? (c) What is the value of V_{DD}?

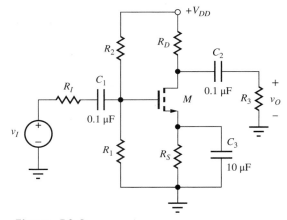

Figure P9.9

9.10. (a) Draw the low-frequency and midband equivalent circuits for the common-source amplifier in Fig. P9.9 if $R_I = 1$ kΩ, $R_1 = 4.3$ MΩ, $R_2 = 5.6$ MΩ, $R_S = 13$ kΩ, $R_D = 43$ kΩ, and $R_3 = 430$ kΩ. (b) What are the lower-cutoff frequency and midband gain of the amplifier if the Q-point = (0.2 mA, 5 V) and $V_{GS} - V_{TN} = 0.7$ V? (c) What is the value of V_{DD}?

9.11. (a) What is the value of C_3 required to set f_L to 50 Hz in the circuit in Prob. 9.9? (b) Choose the nearest standard value of capacitance from Appendix A. What is the value of f_L for this capacitor? (c) Repeat for the circuit in Prob. 9.10.

9.12. (a) Draw the low-frequency equivalent circuit for the common-gate amplifier in Fig. P9.12. (b) Write an expression for the transfer function of the amplifier and identify the location of the two low-frequency poles and two low-frequency zeros. Assume $r_o = \infty$ and $g_m = 4$ mS. (c) What are the lower-cutoff frequency and midband gain of the amplifier?

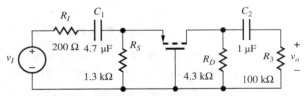

Figure P9.12

9.13. (a) What is the value of C_1 required to set f_L to 3000 Hz in the circuit in Prob. 9.12? (b) Choose the nearest standard value of capacitance from Appendix A. What is the value of f_L for this capacitor?

9.14. (a) Draw the low-frequency ac and midband equivalent circuits for the common-base amplifier in Fig. P9.14 if $R_I = 75$ Ω, $R_E = 4.3$ kΩ, $R_C = 2.2$ kΩ, $R_3 = 51$ kΩ, and $\beta_o = 100$. (b) Write an expression for the transfer function of the amplifier and identify the location of the two low-frequency poles and two low-frequency zeros. Assume $r_o = \infty$ and the Q-point = (1.5 mA, 5 V). (c) What are the midband gain and lower cutoff frequency of the amplifier? (d) What are the values of $-V_{EE}$ and V_{CC}? (e) What are the lower-cutoff frequency and midband gain of the amplifier if $R_E = 430$ kΩ, $R_C = 220$ kΩ, $R_3 = 510$ kΩ, and the Q-point is (15 μA, 5 V)? (f) What are the values of $-V_{EE}$ and V_{CC} in part (e)?

9.15. (a) What is the value of C_1 required to set f_L to 20 Hz in the circuit in Prob. 9.14(a)? (b) Choose the nearest standard value of capacitance from Appendix A. What is the value of f_L for this capacitor? (c) Repeat for the circuit in Prob. 9.14(e).

9.3 Estimation of ω_L Using the Short-Circuit Time-Constant Method

9.16. (a) The common-emitter circuit in Fig. 9.6 is redesigned with $R_1 = 100$ kΩ, $R_2 = 300$ kΩ, $R_E = 15$ kΩ, and $R_C = 43$ kΩ, and the Q-point is (175 μA, 2.3 V). The other values remain the same. Use the SCTC technique to find f_L. (b) Plot the frequency response of the amplifier with SPICE and find the value of f_L. (c) Calculate the Q-point for the transistor.

9.17. (a) What is the value of C_3 required to set f_L to 1500 Hz in the circuit in Fig. 9.6? (b) Choose the nearest standard value of capacitance from Appendix A. What is the actual value of f_L for this capacitor?

9.18. (a) Draw the low-frequency and midband equivalent circuits for the common-emitter amplifier in Fig. P9.18 if $R_I = 2$ kΩ, $R_1 = 110$ kΩ, $R_2 = 330$ kΩ, $R_E = 13$ kΩ, $R_C = 43$ kΩ, and $R_3 = 43$ kΩ. (b) What are the lower-cutoff frequency and midband gain of the amplifier assuming a Q-point of (0.164 mA, 2.79 V) and $\beta_o = 100$? (c) What is the value of V_{CC}?

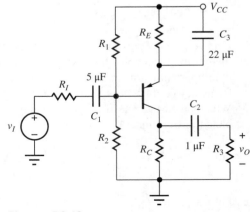

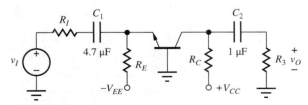

Figure P9.14

Figure P9.18

9.19. Use the SCTC technique to find the lower-cutoff frequency for the common-source amplifier in Fig. 9.11 if $R_G = 1$ MΩ, $R_3 = 68$ kΩ, $R_D = 22$ kΩ, $R_S = 6.8$ kΩ, and $g_m = 3$ mS. The other values remain unchanged.

9.20. Use the SCTC technique to find the lower-cutoff frequency for the common-source amplifier in Fig. 9.11 if $R_G = 500$ kΩ, $R_3 = 10$ kΩ, $R_D = 43$ kΩ, $R_S = 10$ kΩ and $g_m = 0.6$ mS. The other values remain unchanged.

9.21. (a) Draw the low-frequency and midband equivalent circuits for the common-gate amplifier in Fig. P9.21. (b) What are the lower-cutoff frequency and midband gain of the amplifier if the Q-point = (0.1 mA, 8.6 V), $V_{GS} - V_{TN} = 0.7$ V, $C_1 = 2.7$ μF, $C_2 = 0.2$ μF, and $C_3 = 0.1$ μF?

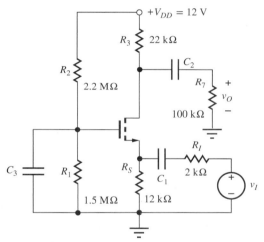

Figure P9.21

9.22. (a) Draw the low-frequency and midband equivalent circuits for the emitter follower in Fig. P9.22.

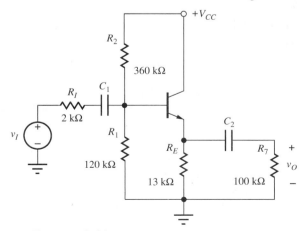

Figure P9.22

(b) What are the lower-cutoff frequency and midband gain of the amplifier if the Q-point is (0.25 mA, 10 V), $\beta_o = 100$, $C_1 = 3.3$ μF, and $C_2 = 8.2$ μF?

9.23. (a) Draw the low-frequency and midband equivalent circuits for the source follower in Fig. P9.23. (b) What are the lower-cutoff frequency and midband gain of the amplifier if the transistor is biased at 0.75 V above threshold with a Q-point = (0.1 mA, 6.3 V), $C_1 = 5.6$ μF, and $C_2 = 0.13$ μF? (c) What is the value of V_{DD}?

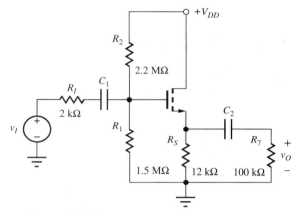

Figure P9.23

9.24. Redesign the value of C_3 in the C-S stage in Prob. 9.9 to set $f_L = 50$ Hz.

9.25. Redesign the value of C_1 in the C-G stage in Prob. 9.12 to set $f_L = 300$ Hz.

9.26. Redesign the value of C_3 in the C-E stage in Prob. 9.18 to set $f_L = 400$ Hz.

9.27. Redesign the value of C_1 in the C-G stage in Prob. 9.21 to set $f_L = 500$ Hz.

9.28. Redesign the value of C_2 in the C-C stage in Prob. 9.22 to set $f_L = 20$ Hz.

9.29. Redesign the value of C_2 in the C-D stage in Prob. 9.23 to set $f_L = 10$ Hz.

9.4 Transistor Models at High Frequencies

9.30. A bipolar transistor with $f_T = 500$ MHz and $C_{\mu o} = 2$ pF is biased at a Q-point of (2 mA, 5 V). What is the forward-transit time τ_F if $\phi_{jc} = 0.9$ V?

9.31. Fill in the missing parameter values for the BJT in the table if $r_x = 50$ Ω.

I_C	f_T	C_π	C_μ	$\frac{1}{2\pi r_x C_\mu}$
10 µA	40 MHz		0.50 pF	
200 µA	300 MHz	0.75 pF		
500 µA	2 GHz		0.25 pF	
5 mA		10 pF		1.59 GHz
1 µA		1 pF	1 pF	
	5 GHz	0.75 pF	0.25 pF	

9.32. Fill in the missing parameter values for the MOSFET in the table if $K_n = 2$ mA/V² and $r_g = 250$ Ω.

I_D	f_T	C_{GS}	C_{GD}	$\frac{1}{2\pi r_g C_{GD}}$
30 µA		1.5 pF	0.5 pF	
300 µA		1.5 pF	0.5 pF	
	3000 MHz	1.25 pF	0.25 pF	

9.33. (a) An n-channel MOSFET has a mobility of 500 cm²/V · s and a channel length of 1 µm. What is the transistor's f_T if $V_{GS} - V_{TN} = 0.25$ V? (b) Repeat for a PMOS device with a mobility of 200 cm²/V · s. (c) Repeat for transistors in a new technology with $L = 0.1$ µm. (d) Repeat for transistors in a technology with $L = 12$ nm.

9.34. (a) Suppose the polysilicon sheet resistance for the MOSFET in Fig 5.23 is 30 Ω/square, and the metal-to-polysilicon contact resistance is 12 Ω. Estimate the value of the gate resistance r_g for this transistor. (b) Suppose a second contact is made to the gate at the bottom end. What is the new value of r_g?

9.35. Suppose the polysilicon sheet resistance for the MOSFET in Design Example 14.4 is 25 Ω/square. Estimate the value of r_g for the gate stripes of the four transistors. Consider only the portions of the gates over the active regions of the transistors.

9.5 Base and Gate Resistances in the Small-Signal Models

9.36. (a) What is the midband gain for the common-emitter amplifier in Fig. P9.36 if $r_x = 400$ Ω, $I_C = 1$ mA, and $\beta_o = 110$? (b) If $r_x = 0$?

9.37. (a) What is the midband gain for the common-collector amplifier in Fig. P9.37 if $r_x = 250$ Ω, $I_C = 0.75$ mA, and $\beta_o = 165$? (b) If $r_x = 0$?

9.38. (a) What is the midband gain for the common-base amplifier in Fig. P9.38 if $r_x = 400$ Ω, $I_C = 0.125$ mA, and $\beta_o = 125$? (b) If $r_x = 0$?

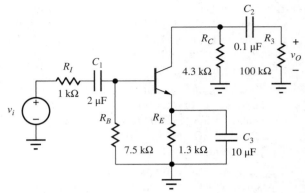

Figure P9.36

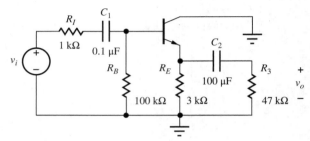

Figure P9.37

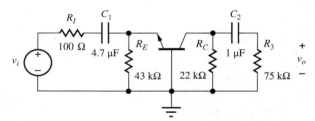

Figure P9.38

9.6 High-Frequency Response of the Common-Emitter and Common-Source Amplifiers

9.39. (a) What is the short-circuit current gain (magnitude and phase) at 75 MHz for a BJT with $r_x = 250$ Ω, $f_T = 400$ MHz, $C_\mu = 0.5$ pF, $\beta_o = 100$ and $I_C = 1.25$ mA? (b) What is the input impedance Z_{iB} at the base of a common-emitter amplifier at 50 MHz using this transistor if $g_m R_L = 20$?

9.40. (a) What is the short-circuit current gain (magnitude and phase) at 100 MHz for a MOSFET with $r_g = 250$ Ω, $C_{GS} = 1.25$ pF, $C_{GD} = 0.25$ pF, and $g_m = 50$ mS? (b) What is the input impedance Z_{iG} at the gate of a common-source amplifier at 20 MHz using this FET if $g_m R_L = 20$?

Factorization

9.41. Multiply out the transfer function coefficients to find the denominator polynomial in Problem 9.7. Estimate the four pole values utilizing the approximate factorization presented in Eqs. (9.92) and (9.93). Compare the approximate results to the exact pole positions.

9.42. Multiply out the transfer function coefficients to find the denominator polynomial in Problem 9.8. Estimate the four pole values utilizing the approximate factorization presented in Eqs. (9.92) and (9.93). Compare the approximate results to the exact pole positions.

9.43. Use dominant root factorization techniques to estimate the roots of these quadratic equations and compare the results to the exact roots: (a) $s^2 + 5000s + 500{,}000$; (b) $2s^2 + 500s + 30{,}000$; (c) $3s^2 + 3300s + 300{,}000$; (d) $1.5s^2 + 300s + 40{,}000$.

9.44. (a) Use dominant root factorization techniques to estimate the roots of this equation. (b) Compare the results to the exact roots.

$$s^3 + 1110s^2 + 111{,}000s + 1{,}000{,}000$$

**9.45. Use Newton's method to help find the roots of this polynomial. (*Hint:* Find the roots one at a time. Once a root is found, factor it out to reduce the order of the polynomial. Use approximate factorization to find starting points for iteration.)

$$s^6 + 142s^5 + 4757s^4 + 58{,}230s^3 + 256{,}950s^2 + 398{,}000s + 300{,}000$$

For Probs. 9.46 to 9.54, use $f_T = 500$ MHz, $r_x = 300\ \Omega$, $C_\mu = 0.75$ pF, $C_{GS} = C_{GD} = 2.5$ pF.

9.46. (a) What are the midband gain and upper-cutoff frequency for the common-emitter amplifier in Prob. 9.36(a) if $I_C = 1$ mA and $\beta_o = 135$? (b) What is the gain-bandwidth product for this amplifier? (c) What is the value of the current gain of the transistor at $f = f_H$? Make use of the C_T approximation.

9.47. Resistors R_1, R_2, R_E, and R_C in the common-emitter amplifier in Fig. 9.6 are all decreased in value by a factor of 3. (a) Draw the dc equivalent circuit for the amplifier, and find the new Q-point for the transistor. (b) Draw the ac small-signal equivalent circuit for the amplifier, and find the midband gain and upper-cutoff frequency for the amplifier. (c) What is the gain-bandwidth product for this amplifier?

9.48. The resistors in the common-emitter amplifier in Fig. 9.6 are all increased in value by a factor of 40. (a) Draw the dc equivalent circuit for the amplifier, and find the new Q-point for the transistor. (b) Draw the ac small-signal equivalent circuit for the amplifier, and find the midband gain and upper-cutoff frequency for the amplifier. (c) What is the gain-bandwidth product for this amplifier? (d) What are the values input impedance Z_{iB} and the current gain of the transistor at $f = f_H$?

9.49. (a) What are the midband gain and upper-cutoff frequency for the common-source amplifier in Prob. 9.9? (b) What are the values of input impedance Z_{iG} and the current gain of the transistor at $f = f_H$?

9.50. Simulate the frequency response of the amplifier in Prob. 9.9 and determine A_{mid}, f_L, and f_H.

9.51. In the common-source amplifier in Fig. 9.4, the value of R_S is changed to 3.9 kΩ and that of R_D to 10 kΩ. For the MOSFET, $K_n = 400$ μA/V^2, and $V_{TN} = 1$ V. (a) Draw the dc equivalent circuit for the amplifier, and find the new Q-point for the transistor if $V_{DD} = 15$ V. (b) Draw the ac small-signal equivalent circuit for the amplifier, and find the midband gain and upper-cutoff frequency for the amplifier. (c) What is the gain-bandwidth product for this amplifier?

9.52. (a) What are the midband gain and upper-cutoff frequency for the common-emitter amplifier in Prob. 9.18? (b) What are the values input impedance Z_{iB} and the current gain of the transistor at $f = f_H$?

9.53. Simulate the frequency response of the amplifier in Prob. 9.18 and determine A_{mid}, f_L, and f_H.

9.54. The network in Fig. P9.54 models a common emitter stage with a load capacitor[9] in parallel with R_L. (a) Write the two nodal equations and find the determinant of the system for the network in Fig. P9.54. (b) Use dominant root factorization to find the two poles. (c) There are three capacitors in the network. Why are there only two poles?

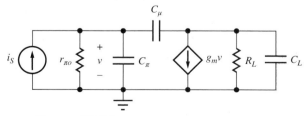

Figure P9.54

[9] For example, C_L could represent the collector-substrate capacitance of the BJT shown in Fig. 4.28.

The Miller Effect

9.55. (a) What is the total input capacitance in the circuit in Fig. 9.35(c) if $C_\pi = 25$ pF, $C_\mu = 1$ pF, $I_C = 5$ mA, and $R_L = 1$ kΩ? What is the f_T of this transistor? (b) Repeat for $I_C = 4$ mA and $R_L = 2$ kΩ.

***9.56.** (a) What is the input capacitance of the circuit in Fig. P9.56 if Z is a 120-pF capacitor and the amplifier is an op amp with a gain of $A = -100{,}000$? (**b) What is the input impedance of the circuit in Fig. P9.56 at $f = 1$ kHz if element Z is a 100-kΩ resistor and $A(s) = -10^6/(s + 25)$? (c) At 50 kHz? (d) At 1 MHz?

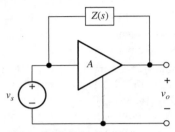

Figure P9.56

9.57. What is the input capacitance of the circuit in Fig. P9.56 if the amplifier gain is $A = +0.994$ and Z is a 45 pF capacitor?

****9.58.** (a) Find the transfer function of the Miller integrator in Fig. 10.34 if $A(s) = 20A_o/(s + 75)$. The transfer function is really that of a low-pass amplifier. What is the cutoff frequency if $A_o = 10^5$? (b) For $A_o = 10^6$? (c) Show that the transfer function approaches that of the ideal integrator if $A_o \to \infty$.

9.59. Use Miller multiplication to calculate the impedance presented to v_i by the circuit in Fig. P9.59 at $f = 1$ kHz if $r_x = 200$ Ω, $r_\pi = 2.5$ kΩ, $g_m = 0.04$ S, $R_L = 2.5$ kΩ, $C_\pi = 10$ pF, and $C_\mu = 1.25$ pF. (b) At 50 kHz. (c) At 1 MHz. (d) Compare your results to SPICE.

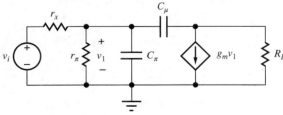

Figure P9.59

9.60. Use SPICE to find the midband gain, and upper- and lower-cutoff frequencies of the amplifier in Prob. 9.59.

9.61. (a) Estimate the upper-cutoff frequency for the common-emitter amplifier in Prob. 9.36(a) if $f_T = 500$ MHz and $C_\mu = 0.60$ pF. (b) Repeat for Prob. 9.36(b).

9.62. Resistors $R_1, R_2, R_E, R_B,$ and R_C in the common-emitter amplifier in Fig. P9.36 are all increased in value by a factor of 13, and the collector current is reduced to 100 µA. (a) Draw the ac small-signal equivalent circuit for the amplifier, and find the midband gain and upper-cutoff frequency for the amplifier if $\beta_o = 100$, $r_x = 250$ Ω, $C_\mu = 0.57$ pF, and $f_T = 575$ MHz. (b) What is the gain-bandwidth product for this amplifier? Calculate the upper bound on GBW given by the $r_x C_\mu$ product.

9.63. Estimate the upper-cutoff frequency for the common-source amplifier in Prob. 9.9 if $C_{GS} = 3$ pF and $C_{GD} = 1.8$ pF. What is the gain-bandwidth product for this amplifier?

Estimation of ω_H and ω_L for Inverting Amplifiers

9.64. What are the values of (a) A_{mid}, f_L, and f_H for the common-emitter amplifier in Fig. P9.64 if $C_1 = 1$ µF, $C_2 = 0.1$ µF, $C_3 = 2.2$ µF, $R_3 = 100$ kΩ, $\beta_o = 125$, $f_T = 350$ MHz, $r_x = 300$ Ω, $V_{CC} = 15$ V, and $C_\mu = 0.5$ pF? (b) What is the gain-bandwidth product?

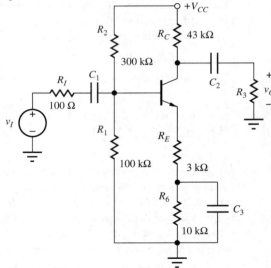

Figure P9.64

9.65. (a) Redesign the common-emitter amplifier in Fig. 9.34 and Ex. 9.6 to have an upper-cutoff frequency of 4.75 MHz by changing the value of the collector resistor R_C. What is the new value of the midband voltage gain? What is the gain-bandwidth product?

9.66. (a) Redesign the common-emitter amplifier in Fig. P9.64 to have an upper-cutoff frequency of 7.5 MHz by selecting new values for R_E and R_6. Maintain the sum $R_E + R_6 = 13$ kΩ. What is the new value of the midband voltage gain? What is the gain-bandwidth product?

9.67. Find (a) A_{mid}, (b) f_L, and (c) f_H for the amplifier in Fig. P9.67 if $\beta_o = 100, f_T = 225$ MHz, $C_\mu = 0.95$ pF, and $r_x = 200$ Ω.

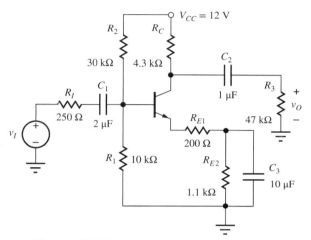

Figure P9.67

9.68. Redesign the values of R_{E1} and R_{E2} in the amplifier in Prob. 9.67 to achieve $f_H = 12.5$ MHz. Do not change the Q-point.

*9.69. The network in Fig. P9.69 has two poles. (a) Estimate the lower-pole frequency using the short-circuit time-constant technique if $C_1 = 1$ µF, $C_2 = 10$ µF, $R_1 = 12$ kΩ, $R_2 = 1.2$ kΩ, and $R_3 = 1.2$ kΩ. (b) Estimate the upper-pole frequency. (c) Why do the positions of the poles seem to be backward? (d) Find the system determinant and compare its exact roots to those in parts (a) and (b).

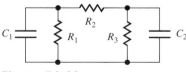

Figure P9.69

For Probs. 9.70–9.84, use $f_T = 550$ MHz, $r_x = 300$ Ω, $C_\mu = 0.60$ pF for the BJT, and $C_{GS} = 3$ pF and $C_{GD} = 0.55$ pF for the FET.

9.7 Common-Base and Common-Gate Amplifier High-Frequency Response

9.70. What are the midband gain and upper-cutoff frequency for the common-gate amplifier in Prob. 9.12?

9.71. Simulate the frequency response of the amplifier in Prob. 9.12 and determine A_{mid}, f_L, and f_H.

9.72. What are the midband gain and upper-cutoff frequency for the common-base amplifier in Prob. 9.14(e)?

9.73. Simulate the frequency response of the amplifier in Prob. 9.14 with $V_{CC} = V_{EE} = 5$ V and determine A_{mid}, f_L, and f_H.

9.74. What are the midband gain and upper-cutoff frequency for the common-base amplifier in Prob. 9.14 if $V_{CC} = -V_{EE} = 10$ V?

9.75. What are the midband gain and upper-cutoff frequency for the amplifier in Prob. 9.21?

9.76. What are the midband gain and upper- and lower-cutoff frequencies for the amplifier in Prob. 9.21 if V_{DD} is increased to 18 V?

*9.77. Find expressions for the open-circuit time constants and ω_H for the common-base stage including base resistance r_x. Assume source resistance $R_I = 0$, and R_E and r_π are very large. Show that the gain-bandwidth product is less than or equal to $1/r_x C_\mu$. What conditions are required to reach this limit?

9.78. Find the open-circuit time constants and ω_H for the common-gate stage including r_G. Assume that $R_I = 0$, and neglect R_4. Show that the gain-bandwidth product is less than or equal to $1/r_G C_{GD}$. What conditions are required to reach this limit?

9.8 Common-Collector and Common-Drain Amplifier High-Frequency Response

9.79. What are the midband gain and upper-cutoff frequency for the common-collector amplifier in Fig. P9.22 if V_{CC} is 12.5 V?

9.80. (a) What are the midband gain and upper-cutoff frequency for the emitter follower in Prob. 9.22? (b) Simulate the frequency response of the amplifier in Prob. 9.22 with $V_{CC} = 10$ V and determine A_{mid}, f_L, and f_H.

9.81. (a) What are the midband gain and upper-cutoff frequency for the source follower in Prob. 9.23? (b) Simulate the frequency response of the amplifier in Prob. 9.23 with $V_{DD} = 10$ V and determine A_{mid}, f_L, and f_H.

9.82. What are the midband gain and upper-cutoff frequency for the common-drain amplifier in Prob. 9.23 if V_{DD} is 15 V?

*9.83. Derive an expression for the total capacitance looking into the gate of the FET in Fig. 9.45(b). Use the expression to interpret Eq. (9.141).

**9.84. Derive an expression for the total input capacitance of the BJT in Fig. 9.44(c) looking into node v_b. Assume $C_L = 0$. Use it to interpret Eq. (9.138).

9.9 Single-Stage Amplifier High-Frequency Response Summary
Gain-Bandwidth Product

9.85. A bipolar transistor must be selected for use in a common-base amplifier with a gain of 42 dB and a bandwidth of 55 MHz. What should be the minimum specification for the transistor's f_T? What should be the minimum $r_x C_\mu$ product? (Use a factor of 5 safety margin for each estimate.)

9.86. A bipolar transistor must be selected for use in a common-emitter amplifier with a gain of 46 dB and a bandwidth of 5.5 MHz. What should be the minimum specification for the transistor's f_T? What should be the minimum $r_x C_\mu$ product? (Use a factor of 5 safety margin for each estimate.)

9.87. A BJT will be used in a differential amplifier with load resistors of 100 kΩ. What are the maximum values of r_x and C_μ that can be tolerated if the gain and bandwidth are to be 100 and 1.8 MHz, respectively?

*9.88. An FET with $C_{GS} = 10$ pF and $C_{GD} = 4$ pF will be used in a common-source amplifier with a source resistance of 100 Ω and a bandwidth of 25 MHz. Estimate the minimum Q-point current needed to achieve this bandwidth if $K_n = 25$ mA/V^2 and $V_{GS} - V_{TN} \geq 0.25$ V.

9.89. An FET with $C_{GS} = 8$ pF and $C_{GD} = 2$ pF will be used in a common-gate amplifier with a source resistance of 100 Ω, $A_{mid} = 20$, and a bandwidth of 25 MHz. Estimate the Q-point current needed to achieve these specifications if $K_n = 20$ mA/V^2 and $V_{DD} = 15$ V.

9.90. What is the upper bound on the bandwidth of the circuit in Fig. P9.14 if $R_C = 12$ kΩ, $R_3 = 47$ kΩ, and $C_\mu = 1.75$ pF?

**9.91. (a) Estimate the cutoff frequency of the C-C/C-E cascade in Fig. P9.91(a). (b) Estimate the cutoff frequency of the Darlington stage in Fig. P9.91(b).

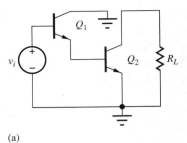

(a)

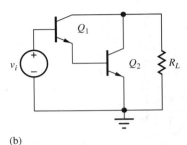

(b)

Figure P9.91

Assume $I_{C1} = 0.1$ mA, $I_{C2} = 1$ mA, $\beta_o = 100$, $f_T = 300$ MHz, $C_\mu = 0.5$ pF, $V_A = 50$ V, $r_x = 300$ Ω, and $R_L = \infty$. (c) Which configuration offers better bandwidth? (d) Which configuration is used in the second stage in the μA741 amplifier in Chapter 14? Why do you think it was used?

9.92. Draw a Bode plot for the common-mode rejection ratio for the differential amplifier in Fig. 9.48 if $I_C = 100$ μA, $R_{EE} = 10$ MΩ, $R_C = 6$ kΩ, $C_{EE} = 1$ pF, $\beta_o = 100$, $V_A = 50$ V, $f_T = 225$ MHz, $C_\mu = 0.3$ pF, $r_x = 175$ Ω, and $R_L = 100$ kΩ. R_L is connected between the collectors of transistors Q_1 and Q_2. Use ± 10-V supplies where needed.

9.93. Use SPICE to plot the graph for Prob. 9.92.

9.10 Frequency Response of Multistage Amplifiers

9.94. What is the minimum bandwidth of the MOS current mirror in Fig. P9.94 if $I_S = 250$ μA, $K'_n = 25$ μA/V^2, $\lambda = 0.02$ V^{-1}, $C_{GS1} = 3$ pF, $C_{GD1} = 1$ pF, $(W/L)_1 = 5/1$, and $(W/L)_2 = 25/1$? (b) What is the bandwidth if a parallel RC load of 800 Ω and 20 pF is connected to the drain of M_2?

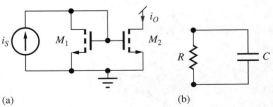

(a)   (b)

Figure P9.94

9.95. What is the minimum bandwidth of the NMOS current mirror in Fig. P9.94 if $I_S = 150$ µA, $K'_n = 25$ µA/V^2, $\lambda = 0.02$ V^{-1}, $C_{GS1} = 3$ pF, $C_{GD1} = 0.5$ pF, and $(W/L)_1 = 5/1 = (W/L)_2$?

*9.96. (a) What is the minimum bandwidth of the bipolar current mirror in Fig. P9.96 if $I_S = 200$ µA, $\beta_o = 100$, $V_A = 50$ V, $f_T = 500$ MHz, $C_\mu = 0.3$ pF, $r_x = 175$ Ω, and $A_{E2} = 4A_{E1}$? (b) What is the bandwidth if a parallel RC load of 1000 Ω and 18 pF is connected to the collector of Q_2?

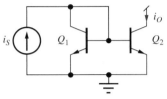

Figure P9.96

9.97. What is the minimum bandwidth of the *npn* current mirror in Fig. P9.96 if $I_S = 75$ µA, $\beta_o = 100$, $V_A = 60$ V, $f_T = 500$ MHz, $C_\mu = 0.5$ pF, and $A_{E2} = 10 A_{E1}$?

9.98. What is the minimum bandwidth of the *pnp* current mirror in Fig. P9.98 if $I_S = 80$ µA, $\beta_o = 50$, $V_A = 60$ V, $f_T = 65$ MHz, $C_\mu = 2.5$ pF, and $A_{E2} = A_{E1}$?

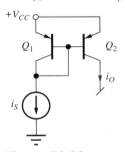

Figure P9.98

**9.99. Find the minimum bandwidth of the Wilson current mirror in Fig. P9.99 if $I_{REF} = 275$ µA, $K_n = 250$ µA/V^2, $V_{TN} = 0.75$ V, $\lambda = 0.02$ V^{-1}, $C_{GS} = 3$ pF, and $C_{GD} = 1$ pF.

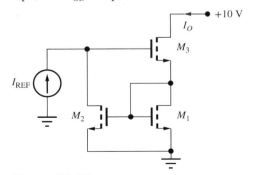

Figure P9.99

9.100. (a) The transistors in the differential amplifier in Fig. 9.48 are biased at a collector current of 18 µA, and $R_C = 390$ kΩ. The transistors have $f_T = 85$ MHz, $C_\mu = 0.5$ pF, and $r_x = 500$ Ω. What is the bandwidth of the differential amplifier? (b) Repeat if the collector current is increased to 50 µA and R_C is reduced to 150 kΩ.

9.101. (a) The transistors in the C-C/C-B cascade amplifier in Fig. 9.50 are biased with $I_{EE} = 225$ µA and $R_C = 75$ kΩ. The transistors have $f_T = 120$ MHz, $C_\mu = 1$ pF, and $r_x = 450$ Ω. What is the bandwidth of the amplifier? (b) Repeat if the current source is increased to 2 mA and R_C is reduced to 7.5 kΩ.

9.102. (a) The transistors in the cascode amplifier in Fig. 9.52 are biased at a collector current of 120 µA with $R_L = 82$ kΩ. The transistors have $f_T = 100$ MHz, $C_\mu = 0.9$ pF, and $r_x = 500$ Ω. What is the bandwidth of the amplifier if $R_{th} = 0$? (b) Repeat if the collector currents are increased to 1 mA and R_C is reduced to 7.5 kΩ.

9.103. The bias current in transistor Q_3 in Fig. 9.53(a) is doubled by reducing the value of resistors R_3, R_4, and R_{E3} by a factor of 2. What are the new values of midband gain, lower-cutoff frequency, and upper-cutoff frequency?

9.104. The bias current in transistor Q_2 in Fig. 9.53(a) is reduced by increasing the value of resistors R_1, R_2, R_{C2}, and R_{E2} by a factor of 2. What are the new values of midband gain, lower-cutoff frequency, and upper-cutoff frequency?

9.11 Introduction to Radio Frequency Circuits

Shunt-Peaked Amplifiers

9.105. The circuit in Fig. 9.58(a) has $C_L = 10$ pF and $R = 7.5$ kΩ, and the transistor parameters are $C_{GS} = 6$ pF, $C_{GD} = 2.5$ pF, and $g_m = 3$ mS. (a) What is the bandwidth of the amplifier? (b) Find the value of L required to extend the bandwidth to the maximally flat limit. What is the new bandwidth?

9.106. What value of L is required to increase the bandwidth of the amplifier in Prob. 9.105 by 45 percent?

9.107. What is the phase shift at the bandwidth frequency for each of the values of m in Fig. 9.59?

9.108. (a) The transistor in Fig. 9.58(a) is replaced with a bipolar transistor operating at 1 mA. What is the bandwidth of the amplifier if $C_L = 5$ pF and $R = 10$ kΩ, $f_T = 250$ MHz and $C_\mu = 2$ pF? (b) Find the value of L required to extend the bandwidth to the maximally flat limit. What is the new bandwidth?

Tuned Amplifiers

9.109. What are the center frequency, Q, and midband gain for the amplifier in Fig. P9.109 if the FET has $C_{GS} = 35$ pF, $C_{GD} = 4$ pF, $\lambda = 0.0167$ V^{-1}, and it is biased at 2 V above threshold with $I_D = 10$ mA and $V_{DS} = 10$ V?

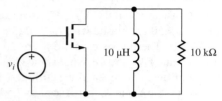

Figure P9.109

9.110. (a) What is the value of C required for $f_o = 10.7$ MHz in the circuit in Fig. P9.110 if $I_C = 10$ mA, $V_{CE} = 10$ V, $\beta_o = 125$, $C_\mu = 1.85$ pF, $f_T = 550$ MHz, and $V_A = 75$ V? (b) What is the Q of the amplifier? (c) Where should a tap be placed on the inductor to achieve a Q of 100? (d) What is the new value of C required to achieve $f_o = 10.7$ MHz?

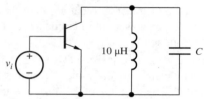

Figure P9.110

9.111. (a) Draw the dc and high-frequency ac equivalent circuits for the circuit in Fig. P9.111. (b) What is the resonant frequency of the circuit for $V_C = 0$ V if the diode is modeled by $C_{jo} = 15$ pF and $\phi_j = 0.87$ V? (c) For $V_C = 10$ V?

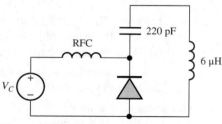

Figure P9.111

*9.112. (a) What are the center frequency, Q, and midband gain for the tuned amplifier in Fig. P9.112 if $L_1 = 5$ μH, $C_1 = 12$ pF, $C_2 = 12$ pF, $I_C = 1.2$ mA, $C_\pi = 5$ pF, $C_\mu = 1$ pF, $R_L = 5$ kΩ, $r_\pi = 2.5$ kΩ, and $r_x = 0$ Ω? (b) What would be the answers if the base

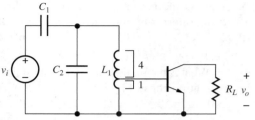

Figure P9.112

terminal of the transistor were connected to the top of the inductor?

9.113. (a) What are the midband gain, center frequency, bandwidth, and Q for the circuit in Fig. P9.113(a) if $I_D = 22.5$ mA, $\lambda = 0.02$ V^{-1}, $C_{GD} = 4$ pF, and $K_n = 5$ mA/V^2? (b) Repeat for the circuit in Fig. P9.113(b).

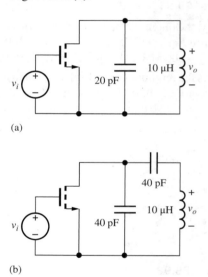

Figure P9.113

*9.114. Change the two capacitor values in the circuit in Fig. P9.113(b) to give the same center frequency as in Fig. P9.113(a). What are the Q and midband gain for the new circuit?

9.115. (a) Simulate the circuit in Prob. 9.113(a) and compare the results to the hand calculations in Prob. 9.113. (b) Simulate the circuit in Prob. 9.113(b) and compare the results to the hand calculations in Prob. 9.113. (c) Simulate the circuit in Prob. 9.114 and compare the results to the hand calculations in Prob. 9.114.

9.116. (a) What is the value of C_2 required to achieve synchronous tuning of the circuit in Fig. P9.116 if $L_1 = L_2 = 10$ μH, $C_1 = C_3 = 18$ pF, $C_{GS} = 22$ pF,

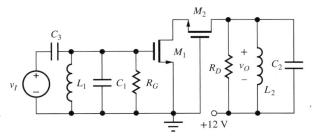

Figure P9.116

$C_{GD} = 5$ pF, $V_{TN1} = -1$ V, $K_{n1} = 10$ mA/V^2, $V_{TN2} = -4$ V, $K_{n2} = 10$ mA/V^2, and $R_G = R_D = 100$ kΩ? (b) What are the Q, midband gain, and bandwidth of your design?

9.117. Simulate the frequency response of the circuit design in Prob. 9.116 and find the midband gain, center frequency, Q, and bandwidth of the circuit. Did you achieve synchronous tuning of your design?

**9.118. (a) What is the value of C_2 required to adjust the resonant frequency of the tuned circuit connected to the drain of M_2 to a frequency 2 percent higher than that connected at the gate of M_1 in Fig. P9.116 if $L_1 = L_2 = 10$ μH, $C_1 = C_3 = 20$ pF, $C_{GS} = 18$ pF, $C_{GD} = 4.5$ pF, $V_{TN1} = -1$ V, $K_{n1} = 10$ mA/V^2, $V_{TN2} = -4$ V, $K_{n2} = 10$ mA/V^2, and $R_G = R_D = 100$ kΩ? (b) What are the Q and bandwidth of your design?

*9.119. Simulate the frequency response of the circuit design in Prob. 9.118 and find the midband gain, center frequency, Q and bandwidth, and the Q of the circuit. Was the desired stagger tuning achieved?

*9.120. (a) Derive an expression for the high frequency input admittance at the base of the common-emitter circuit in Fig. 9.34(b) and show that the input capacitance and input resistance can be represented by the expressions below for $\omega C_\mu R_L \ll 1$.

$$C_{in} = C_\pi + C_\mu(1 + g_m R_L)$$

$$R_{in} = r_\pi \left\| \frac{R_L}{(1 + g_m R_L)(\omega C_\mu R_L)^2} \right.$$

(b) A MOSFET has $C_{GS} = 5$ pF, $C_{GD} = 2.2$ pF, $g_m = 5$ mS, and $R_L = 10$ kΩ. What are the values of C_{in} and R_{in} at a frequency of 5 MHz?

9.121. (a) Find the equivalent input capacitance and resistance of the circuit in Fig. 9.70(b) if $L_S = 10$ nH, $C_{GS} = 112$ fF, $g_m = 1$ mS, $f = 1$ GHz. (b) Repeat for the circuit in Fig. 9.70(c) if $C_{GD} = 20$ fF, $L_{in} = 0$ and $f = 1$ GHz. (You may want to make use of the network transformations in the EIA on page 612.)

9.122. (a) Derive the expressions for the circuit transformation in part (a) of the RF Network Transformation EIA on page 612 by equating the admittances of the two networks. (b) Repeat for part (c) of the EIA. (c)

9.123. (a) Derive the expressions for the circuit transformation in part (b) of the RF Network Transformation EIA on page 612 by equating the impedances of the two networks. (b) Repeat for part (d) of the EIA.

9.124. Derive an expression for the high-frequency input impedance of the bipolar transistor with inductive degeneration L_E in the emitter. Assume $r_\pi \gg 1/\omega C_\pi$.

9.125. Perform an ac analysis of the circuit in Fig. P9.125 sweeping from 1 MHz to 100 MHz with 101 points per decade. Use $K_P = 100$ mA/V^2 and $V_{TO} = 1$ V. Plot the output voltage across C_L and discuss what you see. Calculate the low and high frequency asymptotes as well as the peak frequency. What is the maximum gain of the emitter follower?

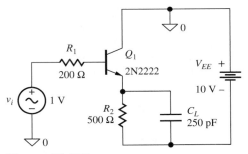

Figure P9.125

9.126. Perform an ac analysis of the circuit in Fig. P9.126 sweeping from 1 MHz to 100 MHz with 101 points per decade. Plot the output voltage across C_L and discuss what you see. Calculate the low and high frequency asymptotes as well as the peak frequency. What is the maximum gain of the source follower?

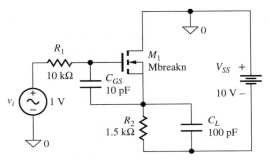

Figure P9.126

9.12 Mixers and Balanced Modulators

9.127. (a) A signal at 925 MHz is mixed with a local oscillator signal at 1.1 GHz. What is the frequency of the VHF output signal? The unwanted output signal? (b) Repeat for a local oscillator signal of 0.8 GHz.

9.128. A parallel LC circuit with a Q of 50 is used to select the VHF output signal in Prob. 9.127(a). What is the attenuation of the circuit at the unwanted signal frequency?

9.129. A parallel LC circuit with a Q of 75 is used to select the RF output signal at 10.7 MHz in Fig. 9.84. (a) Draw a possible circuit. (b) What is the attenuation of the circuit at the unwanted signal frequency?

9.130. (a) Cell phone signals in the range of 1.8 to 2.0 GHz are mixed with a local oscillator to produce an output signal at 70 MHz. What is the range of local oscillator (LO) frequencies required if the LO is below the cell phone signal frequency? What is the frequency range of the unwanted output signals? (b) Repeat if the local oscillator signal is above the cell phone frequency? (c) Which choice of LO frequency seems most desirable?

9.131. (a) Find the conversion gain for the single-balanced mixer in Fig. 9.76 for output frequencies centered around $3f_2$. (b) Repeat for output frequencies centered around $5f_2$.

9.132. Find the expression similar to Eq. (9.200) for the output voltage for the mixer in Fig. 9.74 if input $v_1 = A \cos\omega_1 t$.

9.133. Suppose the signal $s_S(t)$ driving the switch in Fig. 9.76 is not a perfect square wave. Instead, the switch spends 55 percent of the time in the closed position and 45 percent of the time in the open position. What is the amplitude of the output signal at frequency f_1?

9.134. Suppose that switching signal v_2 in the mixer in Fig. 9.76 is operating at a frequency of f_1, the same frequency as the signal part of i_{EE}. What are the amplitudes and frequencies of the first five spectral components of the output voltage if $I_{EE} = 2.5$ mA, $I_1 = 0.5$ mA, and $R_C = 2$ kΩ?

9.135. (a) Find the conversion gain for the double-balanced mixer in Fig. 9.78 for output frequencies centered around $3f_2$. (b) Repeat for output frequencies centered around $5f_2$.

9.136. Find the expression similar to Eq. (9.205) for the output voltage for the mixer in Fig. 9.78 if input $v_1 = A \cos\omega_1 t$.

9.137. Suppose the signal $s_D(t)$ driving the switch in Fig. 9.78 is not a perfect square wave. Instead, the switch spends 57.5 percent of the time in the closed position and 42.5 percent of the time in the open position. What is the amplitude of the output signal at frequency f_1?

9.138. (a) What is the on-resistance required of the NMOS transistors in the double-balanced mixer in Fig. 9.80 if the signal loss between the input and output voltages must be less than 5 percent with $R_L = 10$ kΩ? (b) What must be the W/L of the NMOS transistors if $K'_n = 50$ µA/V^2, $V_{TN} = 0.75$ V, and the amplitude of V_{GS} for the NMOS devices switches between 0 and +5 V? (b) Repeat for a signal loss of less than 1 percent.

9.139. Use SPICE to simulate the passive mixer in Fig. 9.80 and reproduce the results in Fig. 9.81. Use the default NMOS transistor model with $W/L = 10/1$ and $V_{TN} = 0.75$ V.

9.140. Suppose an AM signal is generated with the Jones mixer with $M = 1$. Compare the amplitudes of the carrier and each of the two sideband components.

9.141. (a) Write the expression for the output voltage for the Jones mixer in Fig. 9.82 for $I_{BB} = 2$ mA, $V_m = 10$ mV, $R_C = 5$ kΩ, $2R_1 = 1$ kΩ, $f_c = 90$ MHz, and $f_m = 10$ MHz. Include the terms for $n = 1$ and 2. (b) What is the largest value of V_1 that satisfies our small-signal assumption?

9.142. What is the conversion gain (for $n = 1$) for the doubly balanced mixer in Fig. 9.82 if $I_{BB} = 5$ mA, $R_C = 1$ kΩ, and $2R_1 = 200$ Ω. What is the largest value of V_m that satisfies our small-signal assumption?

9.143. The circuit in Fig. P9.143 provides the current i_{EE} for the mixer in Fig. 9.76(b) where $v_1 = V_1 \sin\omega_1 t$.

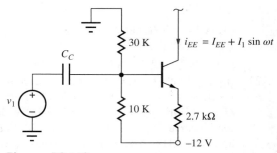

Figure P9.143

(a) If $V_1 = 0.25$ V, $R_C = 6.2$ kΩ, $f_1 = 2000$ Hz, and $f_2 = 1$ MHz, what are I_{EE} and I_1? (b) What are the amplitudes of the first five spectral components in the output signal? (c) What are the largest values of V_1 and I_1 that satisfy our small-signal assumptions?

*9.144. Suppose that signal v_2 driving the switch in Fig. 9.82 is not a perfect square wave. Instead, the switch spends 54 percent of the time in the left-hand position and 46 percent of the time in the right-hand position. What is the carrier suppression in dB (i.e., what is the gain at carrier frequency f_C)?

*9.145. Suppose that signal v_2 driving the switch in Fig. 9.76 is not a perfect square wave. Instead, the switch spends 40 percent of the time in the left-hand position and 60 percent of the time in the right-hand position. What is the amplitude of the output signal at frequency f_1?

术语对照

Amplitude stabilization	幅度稳定性
Base resistance	基极电阻
Beta-cutoff frequency	截止频率 β
Cascode amplifier	Cascode 放大器
Center frequency	中心频率
Dominant high-frequency pole	主高频极点
Dominant low-frequency pole	主低频极点
Dominant pole	主极点
Double-balanced mixers	双平衡混频器
Double-balanced circuits	双平衡电路
Double-balance modulator	双平衡调制器
Down-conversion	下变频
Gilbert mixer	Gilbert 混频器
Jones mixer	Jones 混频器
Lower-cutoff frequency	下限截止频率
Local oscillator	本地振荡器
Midband gain	中频增益
Miller compensation	密勒补偿
Miller effect	密勒效应
Miller integrator	密勒积分
Miller multiplication	密勒倍增
Mixer neutralization	混频中和
Minimum detectable signal	最小可测信号
npn	npn 管
Noise factor F	噪声因子 F
Noise figure(NF)	噪声系数（NF）
Open-circuit time-constant(OCTC)method	开路时域（OCTC）方法
One pole and one zero for each independent caoacitor	电路中每一个独立电容都具有一个极点和一个零点
Passive mixers	无源混频器
pnp	pnp 管
Pole frequencies	极点频率
Pole-splitting	极点分化
Radio frequency choke(RFC)	射频阻塞 (RFC)
Short-circuit time-constant(SCTC) method	短路时间常数（SCTC）方法
Short-peaked amplifier	短路峰化放大器
Signal-to-noise ratio	信噪比
Single-balanced mixer	单平衡混频器
Stagger tuning	交错调谐
Synchronous tuning	同步调谐
Tuned amplifiers	调谐放大器
Unity-gain-bandwidth product	单位增益带宽积
Up-conversion	上变频
Upper-cutoff frequency	上限截止频率

PART THREE
OPERATIONAL AMPLIFIERS AND FEEDBACK

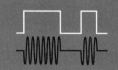

CHAPTER 10
IDEAL OPERATIONAL AMPLIFIERS

CHAPTER 11
NONIDEAL OPERATIONAL AMPLIFIERS AND FEEDBACK AMPLIFIER STABILITY

CHAPTER 12
OPERATIONAL AMPLIFIER APPLICATIONS

CHAPTER 13
DIFFERENTIAL AMPLIFIERS AND OPERATIONAL AMPLIFIER DESIGN

CHAPTER 14
ANALOG INTEGRATED CIRCUIT DESIGN TECHNIQUES

CHAPTER 15
TRANSISTOR FEEDBACK AMPLIFIERS AND OSCILLATORS

CHAPTER 10
IDEAL OPERATIONAL AMPLIFIERS

第10章 理想运算放大器

本章提纲

10.1 理想差分放大器和运算放大器
10.2 理想运算放大器电路的分析
10.3 反馈放大器的频率特性

本章目标

- 了解理想差分放大器和运算放大器（OP）的行为及特性；
- 了解用于分析电路（包括理想运算放大器）的技术方法；
- 了解用于确定通用放大器电路增益、输入阻抗和输出阻抗的方法；
- 熟悉典型的运算放大器电路，包括反相、非反相和求和放大器，电压跟随器和积分器；
- 了解设计运算放大器电路时必须考虑的因素。

本章导读

本章对于理想运算放大器进行了详细讲解，对理想预算放大器的假设进行了详细的分析。理想差分放大器的输出只与其两个输入端的电压差v_{ID}有关，而这个电压却与源和负载电阻无关。理想运算放大器具有两个基本假设，即增益为无穷大和输入电流为零，因此分析包含理想运算放大器的电路时常采用两个基本假设为：

(1) 差分输入电压为0，即$v_{ID}=0$。
(2) 输入电流都为0，即$i_+=0$和$i_-=0$。

运用假设(1)和假设(2)，并结合基尔霍夫电压定律和电流定律，可以分析基于运算放大器的电路模块的理想特性。在反相和同相放大器电路、电压跟随器、差分放大器和加法器中，常采用反馈系数为常数的电阻分压器，而在积分器、低通滤波器、高通滤波器和微分器中，常采用与频率相关的反馈电路。

线性放大器可用于调整正弦信号的幅值或相位,借助其频率响应可以描述其特性。在频域上反馈放大器的增益可用其传输函数进行表示。本章对低通和高通放大器的特性进行了讨论，利用伯德图可以方便地将放大器的特征表现出来，其中伯德图包括了传输函数的幅值(dB)和相位（度数）与对数频率刻度之间的关系。用 MATLAB可以轻松地绘制伯德图。在放大器中，中频增益A_{mid}表示的是放大器的最大增益。在上限截止频率f_H和下限截止频率f_L处，电压增益分别等于$A_{mid}/\sqrt{2}$，比中频增益值小3dB（$20\lg|A_{mid}|$）。放大器的带宽是指从f_L到f_H的频段，截止频率定义为BW $=f_H-f_L$。

本章还对有源低通和高通滤波器进行了讲解与举例。有源低通滤波器的传输函数表现出低通特性，带有单极点频率ω_H，即低通滤波器的上限截止频率（-3dB点）。当频率低于ω_H时，放大器表现为一个反相放大器，其增益为电阻R_2和R_1的比值；当频度高于ω_H时，放大器响应以-20dB/10倍频程的速率下降。有源高通滤波器的传输函数呈高通特性，并具有单个极点，极点频率为ω_L，为高通滤波器的下限截止频率。当频率高于ω_L时，放大器表现为一个反相放大器，当频率低于ω_L时，放大器以20dB/10倍频程的速率衰减。

Chapter Outline

10.1 Ideal Differential and Operational Amplifiers
10.2 Analysis of Circuits Containing Ideal Operational Amplifiers
10.3 Frequency Dependent Feedback
Summary
Key Terms
References
Additional Reading
Problems

741 operational amplifier (op amp) in dual-inline packages.
Richard Jaeger

Chapter Goals

Chapter 10 continues our study of the circuits used for analog electronic signal processing. We will develop an understanding of concepts used in the design of circuits containing ideal operational amplifiers.

- Behavior and characteristics of ideal differential and operational amplifiers (op amps)
- Techniques used to analyze circuits containing ideal op amps
- Techniques used to determine voltage gain, input resistance, and output resistance of general amplifier circuits
- Classic op amp circuits, including the inverting, noninverting, and summing amplifiers, the voltage follower, and the integrator
- Factors that must be considered in the design of circuits using operational amplifiers

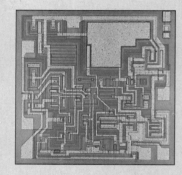

μA741 die photograph.
Fairchild Semiconductor International

The **operational amplifier (op amp)** is an essential building block used in millions of devices ranging from home appliances to automobiles to smartphones to robotic exploration vehicles on other planets! The IEEE Chip Hall of Fame describes the operational amplifier as follows:

> Operational amplifiers are the sliced bread of analog design. You can slap them together with almost anything and get something satisfying. Designers use them to make audio and video preamplifiers, voltage comparators, precision rectifiers, and many other subsystems essential to everyday electronics. [1].

Early op amps from the 1940s were made from vacuum tubes but quickly became transistor biased when those devices became available. Op amps were formally named in a paper by John Ragazzini [2] in 1947. The name "operational amplifier" originates from the use of this type of amplifier to perform specific electronic circuit functions or operations—such as scaling, summation, and integration—in analog computers.

Op amps became the ubiquitous components they are today after the first single-chip integrated circuit op amp (μA702) was designed by Bob Widlar at Fairchild Semiconductor in 1963. Widlar followed this with an improved version dubbed the μA709. Widlar then left Fairchild and joined National Semiconductor, where he designed the hugely successful LM101 op amp. Back at Fairchild, David Fullager (working for R&D head Gordon Moore) addressed some of the shortcomings of the LM101 with the design of the μA741 op amp (pictured above). The μA741, and many others, became workhorse components that found their way into countless designs. The internal circuit design of these op amps used 20 to 50 bipolar transistors. Integrated circuit operational amplifiers were in high demand within a competitive marketplace. The 702 design was initially $300 per op amp. A 741 op amp can now be purchased for about 50 cents due to the massive improvements in IC volume production and cost since 1963. Today there is an almost overwhelming array of operational amplifiers from which to choose.

The operational amplifier is a complete amplifier with a general set of characteristics that can be used as a component in the design of larger and more complex systems. Modern op amps are themselves quite complex in their details, but as in any good engineering design, we compartmentalize complexity within lower-level modules of a larger system. Op amps are an example of this modular approach, where details like biasing, input and output short-circuit protection, stability, gain, reduction of temperature sensitivity, bandwidth, and other details are handled by the op amp designer to generate an amplifier component that can be described with a simple set of parameters. The circuit board designer can then use the parameters to incorporate the op amps in a larger design to meet a specification.

Integrated circuits make it possible to easily include high-performance amplifiers in small, inexpensive electronics systems. Part One of this text provided the background for how the basic electronic building blocks—semiconductors, diodes, and transistors—were created and how they function, starting at the electron carrier level and building up to individual devices. In Part Two we learned how to characterize amplifiers and how to design basic amplifier structures out of the devices we analyzed in Part One. Part Three is where we put it all together. Although the reader is more likely to be a user of an op amp than a designer, a more advanced understanding of the internal op amp circuit techniques helps us understand potential system integration issues as well as use similar circuit techniques for many other applications. The first three chapters develop design techniques that utilize operational amplifiers to achieve higher-level design goals. In the second half of Part Three, we use the ideas and circuits we developed in Part Two to design and analyze the more capable circuit topologies that are used within operational amplifiers themselves. We begin with the ideal operational amplifier.

10.1 IDEAL DIFFERENTIAL AND OPERATIONAL AMPLIFIERS

An ideal **differential amplifier** would produce an output that depends only on the voltage difference v_{id} between its two input terminals, and this voltage would be independent of source and load resistances. Referring to Eq. (6.35), we see that this behavior can be achieved if the input resistance of the amplifier is infinite and the **output resistance** is zero (as pointed out previously in Sec. 6.4). For this case, Eq. (6.35) reduces to

$$\mathbf{v_o} = A\, \mathbf{v_{id}} \qquad \text{or} \qquad A_v = \frac{\mathbf{v_o}}{\mathbf{v_{id}}} = A \tag{10.1}$$

and the full differential amplifier gain is realized. A is referred to as either the **open-circuit voltage gain** or **open-loop gain** of the amplifier and represents the maximum voltage gain available from the device.

As introduced in earlier in Chapter 5, we often want to achieve a completely mismatched resistance condition in voltage amplifier applications ($R_{id} \gg R_I$ and $R_{out} \ll R_L$), so that maximum voltage gain in Eq. (10.1) can be achieved. For the mismatched case, the overall amplifier gain is independent of the source and load resistances, and multiple amplifier stages can be cascaded without concern for interaction between stages.

As noted earlier, the term "operational amplifier" grew from use of these high-performance amplifiers to perform specific electronic circuit functions or operations, such as scaling, summation, and integration, in analog computers. The operational amplifier used in these applications is an ideal differential amplifier with an additional property: infinite voltage gain. Although it is impossible to realize the **ideal operational amplifier,** its conceptual use allows us to understand the basic performance to be expected from a given analog circuit and serves as a model to help in circuit design. Once the properties of the ideal amplifier and its use in basic circuits are understood, then various ideal assumptions can be removed in order to understand their effect on circuit performance.

10.1.1 ASSUMPTIONS FOR IDEAL OPERATIONAL AMPLIFIER ANALYSIS

The ideal operational amplifier is a special case of the ideal difference amplifier in Fig. 6.16, in which $R_{id} = \infty$, $R_o = 0$, and, most importantly, voltage gain $A = \infty$. Infinite gain leads to the first of two assumptions used to analyze circuits containing ideal op amps. Solving for $\mathbf{v_{id}}$ in Eq. (10.1),

$$\mathbf{v_{id}} = \frac{\mathbf{v_o}}{A} \quad \text{and} \quad \lim_{A \to \infty} \mathbf{v_{id}} = 0 \qquad (10.2)$$

If A is infinite, then the input voltage v_{id} will be zero for any finite output voltage. We will refer to this condition as **Assumption 1** for ideal op-amp circuit analysis.

An infinite value for the input resistance R_{id} forces the two input currents i_+ and i_- to be zero, which will be **Assumption 2** for analysis of ideal op amp circuits. These two results, combined with Kirchhoff's voltage and current laws, form the basis for analysis of all ideal op amp circuits.

As just described, the two primary assumptions used for analysis of circuits containing ideal op amps are

1. Input voltage difference is zero: $v_{id} = 0$
2. Input currents are zero: $i_+ = 0$ and $i_- = 0$ \qquad (10.3)

Infinite gain and infinite input resistance are the explicit characteristics that lead to Assumptions 1 and 2. However, the ideal operational amplifier actually has quite a number of additional implicit properties, but these assumptions are seldom clearly stated. They are

- Infinite common-mode rejection
- Infinite power supply rejection
- Infinite output voltage range (not limited by $-V_{EE} \leq v_O \leq V_{CC}$)
- Infinite output current capability
- Infinite open-loop bandwidth
- Infinite slew rate
- Zero output resistance
- Zero input-bias currents and offset current
- Zero input-offset voltage

These terms may be unfamiliar at this point, but they will all be defined and discussed in detail in Chapter 11.

> **EXERCISE:** Suppose an amplifier is operating with $\mathbf{v_o} = +10$ V. What is the input voltage $\mathbf{v_{id}}$ if (a) $A = 100$? (b) $A = 5,000$? (c) $A = 120$ dB?
>
> **ANSWERS:** (a) 100 mV; (b) 2.00 mV; (c) 10.0 μV

DESIGN NOTE

Two assumptions are used for analysis of ideal operational amplifier circuits:

1. The differential input voltage of the op amp will be zero: $v_{id} = 0$.
2. The currents in both amplifier input terminals are zero: $i_+ = 0$ and $i_- = 0$.

10.2 ANALYSIS OF CIRCUITS CONTAINING IDEAL OPERATIONAL AMPLIFIERS

This section introduces a number of classic operational amplifier circuits, including the basic inverting and noninverting amplifiers; the unity-gain buffer, or voltage follower; the summing and difference amplifiers; the low-pass filter; the integrator; and the differentiator. Analysis of these various circuits demonstrates use of the two ideal op amp assumptions in combination with Kirchhoff's voltage and current laws (KVL and KCL, respectively). These classic op amp circuits are a fundamental part of our circuit design toolbox that we need to build more complex analog systems.

10.2.1 THE INVERTING AMPLIFIER

An **inverting-amplifier** circuit is built by grounding the positive input of the operational amplifier and connecting resistors R_1 and R_2, called the **feedback network**, between the inverting input and the signal source and amplifier output node, respectively, as in Fig. 10.1. We wish to find a set of two-port parameters that characterize the overall circuit, including the open-circuit voltage gain A_v, input resistance R_{in}, and output resistance R_{out}.

Inverting Amplifier Voltage Gain

We begin by determining the voltage gain. To find A_v, we need a relationship between v_i and v_o, which we can find by writing an equation for the single loop shown in Fig. 10.2.

$$\mathbf{v_i} - \mathbf{i_i} R_1 - \mathbf{i_2} R_2 - \mathbf{v_o} = 0 \tag{10.4}$$

Applying KCL at the inverting input to the amplifier yields a relationship between $\mathbf{i_i}$ and $\mathbf{i_2}$

$$\mathbf{i_i} = \mathbf{i_-} + \mathbf{i_2} \quad \text{or} \quad \mathbf{i_i} = \mathbf{i_2} \tag{10.5}$$

since Assumption 2 states that i_- must be zero. Equation (10.4) then becomes

$$\mathbf{v_i} - \mathbf{i_i} R_1 - \mathbf{i_i} R_2 - \mathbf{v_o} = 0 \tag{10.6}$$

Now, current i_i can be written in terms of v_i as

$$\mathbf{i_i} = \frac{\mathbf{v_i} - \mathbf{v_-}}{R_1} \tag{10.7}$$

where v_- is the voltage at the inverting input (negative input) of the op amp. But, Assumption 1 states that the input voltage v_{id} must be zero, so v_- must also be zero because the positive input is grounded:

$$\mathbf{v_{id}} = \mathbf{v_+} - \mathbf{v_-} = 0 \quad \text{but} \quad \mathbf{v_+} = 0 \quad \text{so} \quad \mathbf{v_-} = 0$$

Because $v_- = 0$, $\mathbf{i_i} = \mathbf{v_i}/R_1$, and Eq. (10.6) reduces to

$$-\mathbf{v_i}\frac{R_2}{R_1} - \mathbf{v_o} = 0 \quad \text{or} \quad \mathbf{v_o} = -\mathbf{v_i}\frac{R_2}{R_1} \tag{10.8}$$

The voltage gain is given by

$$A_v = \frac{\mathbf{v_o}}{\mathbf{v_i}} = -\frac{R_2}{R_1} \tag{10.9}$$

Referring to Eq. (10.9), we should note several things. The voltage gain is negative, indicative of an inverting amplifier with a 180° phase shift between dc or sinusoidal input and output signals. In addition, the magnitude of the gain can be greater than or equal to 1 if $R_2 \geq R_1$ (the most common case), but it can also be less than 1 for $R_1 > R_2$. The inverting amplifier in

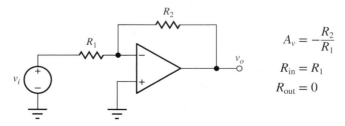

Figure 10.1 Inverting-amplifier circuit.

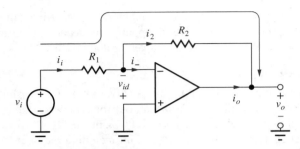

Figure 10.2 Inverting-amplifier circuit.

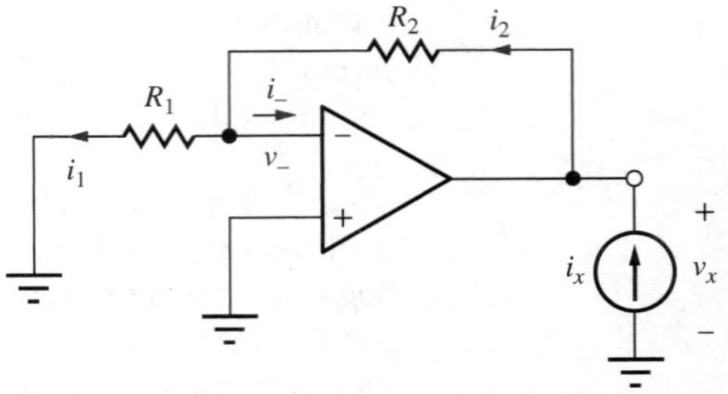

Figure 10.3 Test current applied to the amplifier to determine the output resistance: $R_{out} = v_x/i_x$.

Fig. 10.1 employs *negative feedback* in which a portion of the output signal is "fed back" to the op amp's negative input terminal through resistor R_2. Negative feedback is a requirement for stability of the feedback amplifier and will be discussed in more detail in Chapter 11.

Understanding Inverting Amplifier Operation

In the amplifier circuit in Figs. 10.1 and 10.2, the inverting-input terminal of the operational amplifier is at ground potential, 0 V, and is referred to as a **virtual ground.** The ideal operational amplifier adjusts its output to whatever voltage is necessary to force the differential input voltage to zero. Because of the virtual ground at the inverting input, input voltage v_i appears directly across resistor R_1 and establishes an input current v_i/R_1. The op amp forces this input current to flow through R_2 developing a voltage drop of $v_i \cdot (R_2/R_1)$. Thus $v_o = -v_i(R_2/R_1)$ and $A_v = -(R_2/R_1)$.

Note however, that although the inverting input represents a virtual ground, it is *not* connected directly to ground (there is no direct dc path for current to reach ground). Shorting this terminal to ground for analysis purposes is a common error that must be avoided!

EXERCISE: Find A_v, v_o, i_i, and i_o for the amplifier in Fig. 10.2 if $R_1 = 68$ kΩ, $R_2 = 360$ kΩ, and $v_i = 0.5$ V.

ANSWERS: −5.29, −2.65 V, 7.35 μA, −7.35 μA.

Input and Output Resistances of the Ideal Inverting Amplifier

The input resistance R_{in} of the overall amplifier is found directly from Eq. (10.7). Since $v_- = 0$ (virtual ground),

$$R_{in} = \frac{v_i}{i_i} = R_1 \tag{10.10}$$

The output resistance R_{out} is the Thévenin equivalent resistance at the output terminal; it is found by applying a test signal current source to the output of the amplifier circuit and determining the voltage, as in Fig. 10.3. All other *independent* voltage and current sources in the circuit must be turned off, and so v_i is set to zero in Fig. 10.3.

The output resistance of the overall amplifier is defined by

$$R_{out} = \frac{v_x}{i_x} \tag{10.11}$$

Writing a single-loop equation for Fig. 10.3 gives

$$v_x = i_2 R_2 + i_1 R_1 \tag{10.12}$$

but $i_1 = i_2$ because $i_- = 0$ based on op-amp Assumption 2. Therefore,

$$v_x = i_1(R_2 + R_1) \tag{10.13}$$

However, i_1 must be zero since the voltage is zero across R_1 because Assumption 1 tells us that $v_- = 0$. Thus $i_2 = 0$, and $v_x = 0$ independent of the value of i_x, and

$$R_\text{out} = 0 \tag{10.14}$$

DESIGN NOTE

For the **ideal inverting amplifier,** the closed-loop voltage gain A_v, input resistance R_in, and output resistance R_out are

$$A_v = -\frac{R_2}{R_1} \qquad R_\text{in} = R_1 \qquad R_\text{out} = 0$$

DESIGN EXAMPLE 10.1 INVERTING AMPLIFIER DESIGN

Design an op amp inverting amplifier to meet a pair of specifications.

PROBLEM Design an inverting amplifier (i.e., choose the values of R_1 and R_2) to have an input resistance of 20 kΩ and a gain of 40 dB.

SOLUTION **Known Information and Given Data:** In this case, we are given the values for the gain and input resistance, and the amplifier circuit configuration has also been specified: op amp inverting amplifier topology; voltage gain = 40 dB; R_in = 20 kΩ.

Unknowns: Values of R_1 and R_2 required to achieve the specifications

Approach: Based on Eqs. (10.9) and (10.10), we see that the input resistance is controlled by R_1, and the voltage gain is set by R_2/R_1. First find the value of R_1; then use it to find the value of R_2.

Assumptions: The op amp is ideal so that Eqs. (10.9) and (10.10) apply.

Analysis: We must convert the gain from dB before we use it in the calculations:

$$|A_v| = 10^{40\,\text{dB}/20\,\text{dB}} = 100 \qquad \text{so} \qquad A_v = -100$$

The minus sign is added since an inverting amplifier is specified. Using Eqs. (10.10) and (10.9):

$$R_1 = R_\text{in} = 20\ \text{k}\Omega \qquad \text{and} \qquad A_v = -\frac{R_2}{R_1} \rightarrow R_2 = 100 R_1 = 2\ \text{M}\Omega$$

Check of Results: We have found all the answers requested.

Evaluation and Discussion: Looking at Appendix A, we find that 20 kΩ and 2 MΩ represent standard 5 percent resistor values, and our design is complete. (Murphy has been on our side for a change.) Note in this example, that we have two design constraints and two resistors to choose.

Computer-Aided Analysis: In the SPICE circuit shown here in (a), the op amp is modeled by VCVS E1. In SPICE, we cannot set the gain of E1 to infinity. To approximate the ideal op amp, a value of -10^9 is assigned to E1. Remember that R2 = 2 MEG, not 2M = 0.002 Ω! A transfer function analysis from source VI to the output node is used to characterize the gain of the amplifier. A transient analysis gives the output voltage. VI is defined to have zero voltage offset, a 10-mV amplitude and a frequency of 1000 Hz (V = 0.01 sin 2000πt). The transient solution starts at $T = 0$, stops at $T = 0.003$ s and uses a time step of 1 μs.

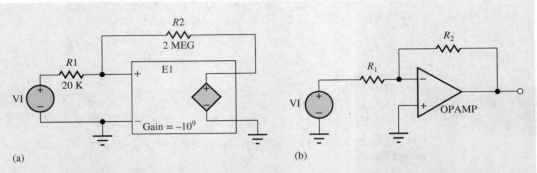

(a)   (b)

The SPICE results are: transfer function = -100, input resistance = $20\ k\Omega$, and output resistance = 0. These values confirm our design, and the output signal is an inverted 1-V, 1000-Hz sine wave, as expected. Note that the small input signal is actually present but hard to see on the graph because of the scale. Also, there is no sign of distortion in the waveform.

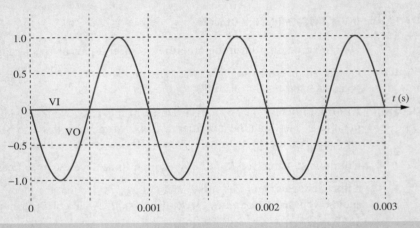

An alternate SPICE circuit appears in (b) in which a built-in OPAMP model is used. The adjustable parameters for this model are the voltage gain and two power supply voltages that need not be the same. Thus OPAMP models the voltage transfer characteristic presented in Fig. 6.13 in Chapter 6, but with zero offset voltage.

> **EXERCISE:** If $V_I = 2$ V, $R_1 = 3.9\ k\Omega$, and $R_2 = 22\ k\Omega$, find I_1, I_2, I_O, and V_O in Fig. 10.2. Why is the symbol V_I being used instead of v_i, and so on?
>
> **ANSWERS:** 0.513 mA, 0.513 mA, -0.513 mA, -11.3 V; the problem is stated specifically in terms of dc values.

10.2.2 THE TRANSRESISTANCE AMPLIFIER—A CURRENT-TO-VOLTAGE CONVERTER

In the inverting amplifier, the input voltage source injects a current into the summing junction through resistor R_1. If we instead inject the current directly from a current source as in Fig. 10.4, we form a **transresistance amplifier**, also called a **current-to-voltage (I-V) converter**. This circuit is widely used in receivers in fiber-optic communication systems for example.

Following the inverting amplifier analysis, we have

$$i_2 = i_1 \qquad i_2 = -\frac{v_o}{R_2} \quad \text{and} \quad A_{tr} = \frac{v_o}{i_i} = -R_2 \qquad (10.15)$$

10.2 Analysis of Circuits Containing Ideal Operational Amplifiers

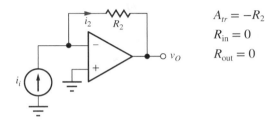

$$A_{tr} = -R_2$$
$$R_{in} = 0$$
$$R_{out} = 0$$

Figure 10.4 Transresistance amplifier.

The gain A_{tr} is the ratio of v_o to i_i and has units of resistance. Since the inverting input terminal is a virtual ground, the input resistance is zero, and zero output resistance appears at the output terminal of the ideal op amp.

ELECTRONICS IN ACTION

Fiber Optic Receiver

One of the important electronic blocks on the receiver side of a fiber optic communication link is the circuit that performs the optical-to-electrical (O/E) signal conversion, and a common approach is shown in the accompanying figure. Light exiting the optical fiber is incident upon a photodiode (see Sec. 3.18) that generates photocurrent i_{ph} as modeled by the current source in the figure. This photocurrent flows through feedback resistor R and generates a signal voltage at the output given by $v_o = i_{ph}R$. The voltage V_{BIAS} can be used to provide reverse bias to the photodiode. In this case, the total output voltage is $v_O = V_{BIAS} + i_{ph}R$.

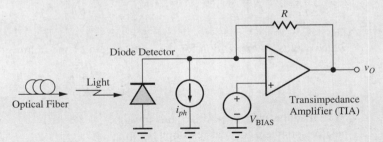

Optical-to-electrical interface for fiber optic data transmission.

Since the input to the amplifier is a current and the output is a voltage, the gain $A_{tr} = \mathbf{v_o}/\mathbf{i_{ph}}$ has the units of resistance, and the amplifier is referred to as a transresistance or (more generally) a transimpedance amplifier (TIA). The operational amplifier shown in the circuit must have an extremely wideband and linear design. The requirements are particularly stringent in communications systems in which multi-GHz signals coming from the optical fiber must be amplified without the addition of any significant phase distortion.

DESIGN NOTE

The gain of the *ideal transresistance amplifier* is set by feedback resistor R_2, and the input and output resistances are both zero:

$$A_{tr} = -R_2 \qquad R_{in} = 0 \qquad R_{out} = 0$$

> **EXERCISE:** We wish to convert a 25-µA sinusoidal current to a voltage with an amplitude of 5 V using a transresistance amplifier. What value of R_2 is required? If $i_i = 40 \sin 2000\pi t$ µA, what is the amplifier output voltage?
>
> **ANSWERS:** 200 kΩ; $v_o = -8 \sin 2000\pi t$ V

10.2.3 THE NONINVERTING AMPLIFIER

The operational amplifier can also be used to construct a **noninverting amplifier** utilizing the circuit schematic in Fig. 10.5. The input signal is applied to the positive or (noninverting) input terminal of the operational amplifier, and a portion of the output signal is fed back to the negative input terminal (negative feedback).

Analysis of the circuit is performed by relating the voltage at v_1 to both input voltage v_i and output voltage v_o. Because Assumption 2 states that input current i_- is zero, v_1 can be related to the output voltage through the voltage divider formed by R_1 and R_2:

$$\mathbf{v_1} = \mathbf{v_o}\frac{R_1}{R_1 + R_2} \tag{10.16}$$

Writing an equation around the loop including v_i, v_{id}, and v_1 yields a relationship between v_1 and v_i:

$$\mathbf{v_i} - \mathbf{v_{id}} = \mathbf{v_1} \tag{10.17}$$

However, Assumption 1 requires $v_{id} = 0$, so

$$\mathbf{v_i} = \mathbf{v_1} \tag{10.18}$$

Combining Eqs. (10.16) and (10.18) and solving for v_o in terms of v_i gives

$$\mathbf{v_o} = \mathbf{v_i}\frac{R_1 + R_2}{R_1} \tag{10.19}$$

which yields an expression for the voltage gain of the noninverting amplifier:

$$A_v = \frac{\mathbf{v_o}}{\mathbf{v_i}} = \frac{R_1 + R_2}{R_1} = 1 + \frac{R_2}{R_1} \tag{10.20}$$

Note that the gain is positive and must be greater than or equal to 1 because R_1 and R_2 are positive numbers for real resistors.

Understanding Noninverting Amplifier Operation

Since the voltage across the inputs of the op amp must be zero (Assumption 1), input voltage $\mathbf{v_i}$ appears directly across resistor R_1 and establishes a current $\mathbf{v_i}/R_1$. This current flows down through R_2 developing a scaled replica of v_i {i.e., $\mathbf{v_i}(R_2/R_1)$} across R_2. The output is the sum of the voltages across R_1 and R_2 yielding $\mathbf{v_o} = \mathbf{v_i} + \mathbf{v_i}(R_2/R_1)$ and $A_v = 1 + R_2/R_1$.

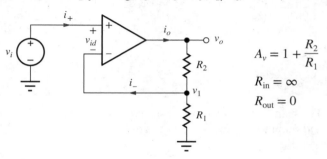

Figure 10.5 Noninverting amplifier configuration.

EXAMPLE 10.2 NONINVERTING AMPLIFIER ANALYSIS

Determine the characteristics of a noninverting amplifier with feedback resistors specified.

PROBLEM Find the voltage gain A_v, output voltage v_o, and output current i_o for the amplifier in Fig. 10.5 if $R_1 = 3$ kΩ, $R_2 = 43$ kΩ, and $v_i = +0.1$ V.

SOLUTION **Known Information and Given Data:** Noninverting amplifier circuit with $R_1 = 3$ kΩ, $R_2 = 43$ kΩ, and $v_i = +0.1$ V

Unknowns: Voltage gain A_v, output voltage v_o, and output current i_o

Approach: Use Eq. (10.20) to find the voltage gain. Use the gain to calculate the output voltage. Use the output voltage and KCL to find i_o.

Assumptions: The op amp is ideal.

Analysis: Using Eq. (10.20),

$$A_v = 1 + \frac{R_2}{R_1} = 1 + \frac{43 \text{ k}\Omega}{3 \text{ k}\Omega} = +15.3 \quad \text{and} \quad v_o = A_v v_i = (15.3)(0.1 \text{ V}) = 1.53 \text{ V}$$

Since the current $i_- = 0$,

$$i_o = \frac{v_o}{R_2 + R_1} = \frac{1.53 \text{ V}}{43 \text{ k}\Omega + 3 \text{ k}\Omega} = 33.3 \text{ }\mu\text{A}$$

Check of Results: We have found all the answers requested. SPICE is used to check the results.

Computer-Aided Analysis: The noninverting amplifier is characterized using a combination of an operating point analysis and a transfer function analysis. The gain of E1 of the op amp is set to 10^9 to model the ideal op amp. The transfer function analysis results are: transfer function = +15.3, input resistance = 10^{20} Ω, and output resistance = 0. The dc output voltage is 1.53 V, and the current in source E1 is -33.3 μA. These values agree with our hand analysis. Note that 10^{20} is the representation of infinity in this particular version of SPICE, and the current in E1 is negative because SPICE uses the passive sign convention which assumes that positive current enters the positive terminal of E1.

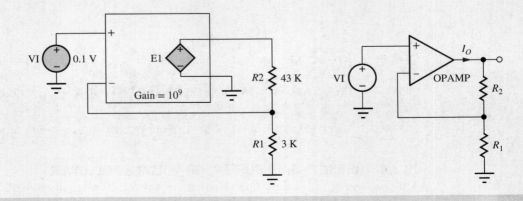

> **EXERCISE:** What are the voltage gain A_v, output voltage v_o, and output current i_o for the amplifier in Fig. 10.5 if $R_1 = 2$ kΩ, $R_2 = 36$ kΩ, and $v_I = -0.2$ V?
>
> **ANSWERS:** 19.0; −3.80 V; −100 µV

Input and Output Resistances of the Noninverting Amplifier

Using Assumption 2, $i_+ = 0$, we find that the input resistance of the noninverting amplifier is given by

$$R_{in} = \frac{v_i}{i_+} = \infty \tag{10.21}$$

To find the output resistance, a test current is applied to the output terminal, and the source v_i is set to 0 V. The resulting circuit is identical to that in Fig. 10.3, so the output resistance of the noninverting amplifier is also zero.

$$R_{out} = 0 \tag{10.22}$$

DESIGN NOTE

For the **ideal noninverting amplifier,** the closed-loop voltage gain A_v, input resistance R_{in}, and output resistance R_{out} are

$$A_v = 1 + \frac{R_2}{R_1} \qquad R_{in} = \infty \qquad R_{out} = 0$$

> **EXERCISE:** Draw the circuit used to determine the output resistance of the noninverting amplifier and convince yourself that it is indeed the same as Fig. 10.3.
>
> **EXERCISE:** What are the voltage gain in dB and the input resistance of the amplifier shown here? If $v_i = 0.25$ V, what are the values of v_o and i_o?
>
>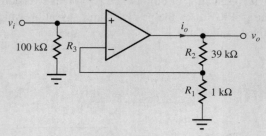
>
> **ANSWERS:** 32.0 dB, 100 kΩ; +10.0 V, 0.250 mA
>
> **EXERCISE:** Design a noninverting amplifier (choose R_1 and R_2 from Appendix A) to have a gain of 54 dB, and the current $i_o \leq 0.1$ mA when $v_o = 10$ V.
>
> **ANSWERS:** Two possibilities of many: (220 Ω and 110 kΩ) or (200 Ω and 100 kΩ)

10.2.4 THE UNITY-GAIN BUFFER, OR VOLTAGE FOLLOWER

A special case of the noninverting amplifier, known as the **unity-gain buffer,** or **voltage follower,** is shown in Fig. 10.6, in which the value of R_1 is infinite and that of R_2 is zero. Substituting

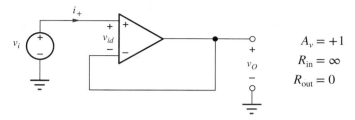

Figure 10.6 Unity-gain buffer (voltage follower).

these values in Eq. (10.20) yields $A_v = 1$. An alternative derivation can be obtained by writing a single-loop equation for Fig. 10.6:

$$v_i - v_{id} = v_o \quad \text{or} \quad v_o = v_i \quad \text{and} \quad A_v = 1 \quad (10.23)$$

because the ideal operational amplifier forces v_{id} to be zero.

Since i_+ is zero by Assumption 2, the input resistance to the voltage buffer is infinite. If we drive the output with a current source and set $v_i = 0$, the op amp will maintain the output voltage at zero. Thus the output resistance is zero.

Why is such an amplifier useful? The ideal unity-gain buffer provides a gain of 1 with infinite input resistance and zero output resistance and therefore provides a tremendous impedance-level transformation while maintaining the level of the signal voltage. Many transducers represent high-source impedances and cannot supply any significant current to drive a load. The ideal unity-gain buffer does not require any input current, yet can drive any desired load resistance without loss of signal voltage. Thus, the unity-gain buffer is found in many sensor and data acquisition applications.

This circuit is also often used as a building block within more complex circuits. It is used to transfer a voltage from one point in the circuit to another point without directly connecting the points together, thus buffering the first point from the loading of the second.

Understanding Voltage Follower Operation
The operation of this circuit is quite simple. The voltage across the inputs of the op amp must be zero (Assumption 1). Therefore the output voltage must equal (follow) the input voltage.

Summary of Ideal Inverting and Noninverting Amplifier Characteristics
Table 10.1 summarizes the properties of the ideal inverting and noninverting amplifiers; the properties are recapitulated here. The gain of the noninverting amplifier must be greater than or equal to 1, whereas the inverting amplifier can be designed with a gain magnitude greater than or less than unity (as well as exactly 1). The gain of the inverting amplifier is negative, indicating a 180° phase inversion between input and output voltages.

The input resistance represents an additional major difference between the two amplifiers. R_{in} is extremely large for the noninverting amplifier but is relatively low for the inverting amplifier, limited by the value of R_1. The output resistance of both ideal amplifiers is zero.

TABLE 10.1
Summary of the Ideal Inverting and Noninverting Amplifiers

	INVERTING AMPLIFIER	NONINVERTING AMPLIFIER
Voltage gain A_v	$-\dfrac{R_2}{R_1}$	$1 + \dfrac{R_2}{R_1}$
Input resistance R_{in}	R_1	∞
Output resistance R_{out}	0	0

EXAMPLE 10.3 INVERTING AND NONINVERTING AMPLIFIER COMPARISON

This example compares the characteristics of the inverting and noninverting amplifier configurations.

PROBLEM Explore the differences between the inverting and noninverting amplifiers in the figure here. Each amplifier is designed to have a gain of 40 dB.

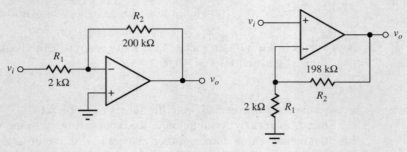

SOLUTION **Known Information and Given Data:** Inverting amplifier topology with $R_1 = 2$ kΩ and $R_2 = 200$ kΩ; noninverting amplifier topology with $R_1 = 2$ kΩ and $R_2 = 198$ kΩ.

Unknowns: Voltage gains, input resistances and output resistances for the two amplifier circuits

Approach: Use the given data to evaluate the amplifier formulas that have already been derived for the two topologies.

Assumptions: The operational amplifiers are ideal.

Analysis:

$$\text{Inverting amplifier: } A_v = -\frac{200 \text{ k}\Omega}{2 \text{ k}\Omega} = -100 \text{ or } 40 \text{ dB}$$

$$R_{\text{in}} = 2 \text{ k}\Omega \quad \text{and} \quad R_{\text{out}} = 0 \text{ }\Omega$$

$$\text{Noninverting amplifier: } A_v = 1 + \frac{198 \text{ k}\Omega}{2 \text{ k}\Omega} = +100 \text{ or } 40 \text{ dB}$$

$$R_{\text{in}} = \infty \quad \text{and} \quad R_{\text{out}} = 0 \text{ }\Omega$$

Check of Results: We have indeed found all the answers requested. A second check indicates the calculations are correct.

TABLE 10.2
Numeric Comparison of the Ideal Inverting and Noninverting Amplifiers

	INVERTING AMPLIFIER	NONINVERTING AMPLIFIER
Voltage gain A_v	−100 (40 dB)	+100 (40 dB)
Input resistance R_{in}	2 kΩ	∞
Output resistance R_{out}	0	0

Evaluation and Discussion: Table 10.2 lists the characteristics of the two amplifier designs. In addition to the sign difference in the gain of the two amplifiers, we see that the input resistance of the inverting amplifier is only 2 kΩ, whereas that of the noninverting amplifier is infinite. Note that the noninverting amplifier achieves our ideal voltage amplifier goals with $R_{\text{in}} = \infty$ and $R_{\text{out}} = 0$ Ω.

EXERCISE: What are the voltage gain A_v, input resistance R_{in}, output voltage v_o, and output current i_o for the amplifiers in Ex. 10.3 if $R_1 = 1.5$ kΩ, $R_2 = 30$ kΩ, and $v_i = 0.15$ V?

ANSWERS: -16.7, 1.8 kΩ, -2.50 V, -83.3 µA; 17.7, ∞, $+2.65$ V, $+83.3$ µA

EXERCISE: Use SPICE transfer function analysis to confirm the analyses in Ex. 10.3.

EXERCISE: Add a resistor to the noninverting amplifier circuit to change its input resistance to 2 kΩ.

ANSWER: Set resistor $R_3 = 2$ kΩ in the schematic on page 659.

10.2.5 THE SUMMING AMPLIFIER

Operational amplifiers can also be used to combine signals using the **summing-amplifier** circuit depicted in Fig. 10.7. Here, two input sources v_1 and v_2 are connected to the inverting input of the amplifier through resistors R_1 and R_2. Because the negative amplifier input represents a virtual ground,

$$i_1 = \frac{v_1}{R_1} \qquad i_2 = \frac{v_2}{R_2} \qquad i_3 = -\frac{v_o}{R_3} \tag{10.24}$$

Because $i_- = 0$, $i_3 = i_1 + i_2$, and substituting Eq. (10.24) into this expression yields

$$v_o = -\left(\frac{R_3}{R_1}v_1 + \frac{R_3}{R_2}v_2\right) \tag{10.25}$$

The output voltage sums the scaled replicas of the two input voltages, and the scale factors for the two inputs may be independently adjusted through the choice of resistors R_1 and R_2. These two inputs can be scaled independently because of the virtual ground maintained at the inverting-input terminal of the op amp.

The inverting-amplifier input node is also commonly called the **summing junction** because currents i_1 and i_2 are "summed" at this node and forced through the feedback resistor R_3. Although the amplifier in Fig. 10.7 has only two inputs, any number of inputs can be connected to the summing junction through additional resistors. A simple digital-to-analog converter can be formed in this way (see the EIA on the next page and Prob. 10.29).

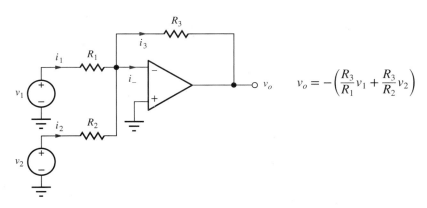

Figure 10.7 The summing amplifier.

EXERCISE: What is the summing amplifier output voltage v_o in Fig. 10.7 if $v_1 = 3 \sin 1000\pi t$ V, $v_2 = 4 \sin 2000\pi t$ V, $R_1 = 1$ kΩ, $R_2 = 2$ kΩ, and $R_3 = 3$ kΩ? What are the input resistances presented to sources v_1 and v_2? What is the current supplied by the op amp output terminal?

ANSWERS: $(-9 \sin 1000\pi t - 6 \sin 2000\pi t)$ V; 1 kΩ, 2 kΩ; $(-3 \sin 1000\pi t - 2 \sin 2000\pi t)$ mA

ELECTRONICS IN ACTION

Digital-to-Analog Converter (DAC) Circuits

One of the simplest **digital-to-analog converter (DAC)** circuits, known as the **weighted-resistor DAC,** is based upon the summing amplifier concept that we just encountered in Sec. 10.2.5. The DAC utilizes a binary-weighted resistor network, a reference voltage V_{REF}, and a group of single-pole, double-throw switches that are usually implemented using MOS transistors. Binary input data controls the switches, with a logical 1 indicating that the switch is connected to V_{REF} and a logical 0 corresponding to a switch connected to ground. Successive resistors are weighted progressively by a factor of 2, thereby producing the desired binary weighted contributions to the output:

$$v_O = (b_1 2^{-1} + b_2 2^{-2} + \cdots + b_n 2^{-n})V_{REF} \quad \text{for } b_i \in \{1, 0\}$$

Bit b_1 has the highest weight and is referred to as the **most significant bit (MSB),** whereas bit b_n has the smallest weight and is referred to as the **least significant bit (LSB).**

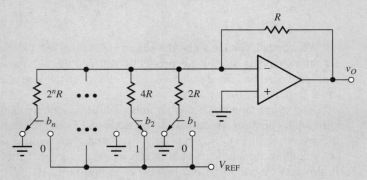

An n-bit weighted-resistor DAC.

Several problems arise in building a DAC using the weighted-resistor approach. The primary difficulty is the need to maintain accurate resistor ratios over a very wide range of resistor values (e.g., 4096 to 1 for a 12-bit DAC). Linearity and gain errors occur when the resistor ratios are not perfectly maintained. In addition, because the switches are in series with the resistors, their on-resistance must be very low, and they should have zero offset voltage. The designer can meet these last two requirements by using good MOSFETs (or JFETs) as switches, and the (W/L) ratios of the FETs can be scaled with bit position to equalize the resistance contributions of the switches. However, the wide range of resistor values is not suitable for monolithic converters of moderate to high resolution. We should also note that the current drawn from the voltage reference varies with the binary input pattern. This varying current causes a change in voltage drop in the Thévenin equivalent source resistance of the voltage reference and can lead to data-dependent errors sometimes called **superposition errors.**

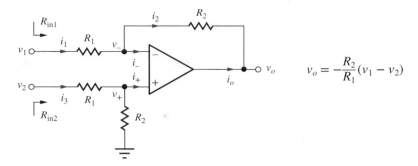

Figure 10.8 Circuit for the difference amplifier.

10.2.6 THE DIFFERENCE AMPLIFIER

Except for the summing amplifier, and D/A converter all the circuits thus far have utilized a single input. However, the operational amplifier may itself be used in a **difference amplifier** configuration, that amplifies the difference between two input signals as shown schematically in Fig. 10.8. Our analysis begins by relating the output voltage to the voltage at v_- as

$$v_o = v_- - i_2 R_2 = v_- - i_1 R_2 \tag{10.26}$$

because $i_2 = i_1$ since i_- must be zero. The current i_1 can be written as

$$i_1 = \frac{v_1 - v_-}{R_1} \tag{10.27}$$

Combining Eqs. (10.26) and (10.27) yields

$$v_o = v_- - \frac{R_2}{R_1}(v_1 - v_-) = \left(\frac{R_1 + R_2}{R_1}\right)v_- - \frac{R_2}{R_1}v_1 \tag{10.28}$$

Because the voltage between the op amp input terminals must be zero, $v_- = v_+$, and the current i_+ is also zero, v_- can be written using the voltage division formula as

$$v_- = v_+ = \frac{R_2}{R_1 + R_2}v_2 \tag{10.29}$$

Substituting Eq. (10.29) into Eq. (10.28) yields the final result

$$v_o = \left(-\frac{R_2}{R_1}\right)(v_1 - v_2) \tag{10.30}$$

Thus, the circuit in Fig. 10.8 amplifies the difference between v_1 and v_2 by a factor that is determined by the ratio of resistors R_2 and R_1. For the special case of $R_2 = R_1$,

$$v_o = -(v_1 - v_2) \tag{10.31}$$

This particular circuit is sometimes called a **differential subtractor**.

The input resistance of this circuit is limited by resistors R_2 and R_1. Input resistance R_{in2}, presented to source v_2, is simply the series combination of R_2 and R_1 because i_+ is zero. For $v_2 = 0$, input resistance R_{in1} equals R_1 because the circuit reduces to the inverting amplifier under this condition. However, for the general case, the input current i_1 is a function of both v_1 and v_2.

Understanding Differential Amplifier Circuit Operation
Probably the easiest way to understand how the differential amplifier operates is to employ superposition. If input v_2 is set to zero, then the circuit behaves as an inverting amplifier with gain $-R_2/R_1$. If v_1 is set to zero, then v_2 is attenuated by the voltage divider at the amplifier input formed by R_1 and R_2, and then amplified by a noninverting amplifier gain of $1 + R_2/R_1$. The total output is the sum of the outputs from the two individual inputs acting alone.

$$\text{For } v_2 = 0, \mathbf{v_{O1}} = -\frac{R_2}{R_1}\mathbf{v_1} \qquad \text{For } v_1 = 0, \mathbf{v_{O2}} = +\left(\frac{R_2}{R_1 + R_2}\right)\left(1 + \frac{R_2}{R_1}\right)\mathbf{v_2} = +\frac{R_2}{R_1}\mathbf{v_2}$$

Combining these results yields

$$\mathbf{v_O} = \mathbf{v_{O1}} + \mathbf{v_{O2}} = -\frac{R_2}{R_1}(\mathbf{v_1} - \mathbf{v_2})$$

EXAMPLE 10.4   DIFFERENCE AMPLIFIER ANALYSIS

Here we find the various voltages and currents within the single op amp difference amplifier circuit with a specific set of input voltages.

PROBLEM Find the values of V_O, V_+, V_-, I_1, I_2, and I_O for the difference amplifier in Fig. 10.8 with $V_1 = 5$ V, $V_2 = 3$ V, $R_1 = 10$ kΩ, and $R_2 = 100$ kΩ.

SOLUTION **Known Information and Given Data:** The input voltages, resistor values, and circuit topology are specified.

Unknowns: V_O, V_+, V_-, I_1, I_2, and I_O.

Approach: We must use circuit analysis (KCL and KVL) coupled with the ideal op amp assumptions to determine the various voltages and currents, but we must find the node voltages in order to find the currents.

Assumptions: Since the op amp is ideal, we know $I_+ = 0 = I_-$, and $V_+ = V_-$.

Analysis: Since $I_+ = 0$, V_+ can be found directly by voltage division,

$$V_+ = V_2 \frac{R_2}{R_1 + R_2} = 3 \text{ V} \frac{100 \text{ k}\Omega}{10 \text{ k}\Omega + 100 \text{ k}\Omega} = 2.73 \text{ V} \qquad \text{and} \qquad V_- = 2.73 \text{ V}$$

V_O can be related to V_1 using Kirchhoff's voltage law:

$$V_1 - I_1 R_1 - I_2 R_2 - V_O = 0$$

We know that $I_- = 0$, and $I_2 = I_1$. We can find I_1 since we know the values of V_1, V_-, and R_1:

$$I_1 = \frac{V_1 - V_-}{R_1} = \frac{5 \text{ V} - 2.73 \text{ V}}{10 \text{ k}\Omega} = 227 \text{ }\mu\text{A} \qquad \text{and} \qquad I_2 = 227 \text{ }\mu\text{A}$$

Then the output voltage can be found

$$V_O = V_1 - I_1 R_1 - I_2 R_2 = V_1 - I_1(R_1 + R_2)$$
$$V_O = 5 \text{ V} - (227 \text{ }\mu\text{A})(110 \text{ k}\Omega) = -20.0 \text{ V}$$

The op amp output current is $I_O = -I_2 = -227$ μA.

Check of Results: We have indeed found all the answers requested. The values of the voltages and currents all appear reasonable. This circuit is a difference amplifier that should amplify the difference in its inputs by the gain of $-R_2/R_1 = -10$. The output should be $-10(5 - 3) = -20$ V. ✓

Computer-Aided Analysis: SPICE can be used to check our calculations using the circuit below. The ideal op amp is modeled by OPAMP with a gain of 120 dB. An operating point analysis produces voltages that agree with our hand analysis: $V_+ = V_- = 2.73$ V, $V_O = -20$ V, $I_1 = 227$ μA, and $I_O = -227$ μA, (I(E1) = 227 μA).

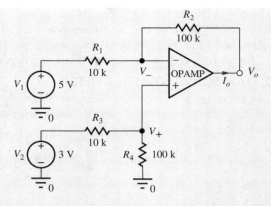

EXERCISE: What is the current exiting the positive terminal of source V_2 in Ex. 10.4?

ANSWER: 27.3 μA

EXERCISE: What are the voltage gain A_v, output voltage V_O, output current I_O, and the current in source V_2 for the amplifier in Ex. 10.4 if $V_1 = 3$ V and $V_2 = 5$ V?

ANSWERS: −10; 20.0 V; +154 μA, 45.5 μA

EXERCISE: What are the voltage gain A_v, output voltage V_O, and output current I_O for the amplifier in Fig. 10.8 if $R_1 = 2$ kΩ, $R_2 = 36$ kΩ, $V_1 = 8$ V, and $V_2 = 8.25$ V?

ANSWERS: A sawtooth waveform with a 10-V amplitude; −18.0; 4.50 V; −92.1 μA

10.3 FREQUENCY DEPENDENT FEEDBACK

Although the operational-amplifier circuit examples thus far have used only resistors in the feedback network, other passive elements or even solid-state devices can be part of the feedback path. The general case of the inverting configuration with passive feedback is shown in Fig. 10.9, in which resistors R_1 and R_2 have been replaced by general impedances $Z_1(s)$ and $Z_2(s)$, which may now be a function of frequency. (Note that resistive feedback is just a special case of the amplifier in Fig. 10.9.)

The gain of this amplifier can be described by its **transfer function** in the frequency domain in which $s = \sigma + j\omega$ represents the complex frequency variable:

$$A_v(s) = \frac{V_o(s)}{V_i(s)} \tag{10.32}$$

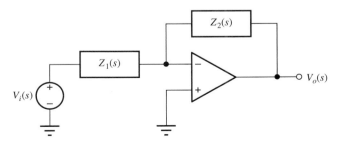

Figure 10.9 Generalized inverting-amplifier configuration.

General amplifier transfer functions can be quite complicated, having many poles and zeros, but their overall behavior can be broken down into a number of categories including low-pass, high-pass, and band-pass amplifiers to name a few. We will now extend our analysis techniques from Chapter 6 to understand several frequency dependent op amp circuits.

10.3.1 AN ACTIVE LOW-PASS FILTER

Now let us return to considering the generalized inverting-amplifier circuit in Fig. 10.10. The gain of the amplifier in this figure is obtained in a manner identical to that in the resistive-feedback case. Replacing R_1 by Z_1 and R_2 by Z_2 in Eq. (10.9) yields the transfer function $A_v(s)$:

$$A_v(s) = \frac{V_o(s)}{V_i(s)} = -\frac{Z_2(s)}{Z_1(s)} \tag{10.33}$$

One useful circuit involving frequency-dependent feedback is the single-pole, **low-pass filter** in Fig. 10.11, for which

$$Z_1(s) = R_1 \quad \text{and} \quad Z_2(s) = \frac{R_2 \frac{1}{sC}}{R_2 + \frac{1}{sC}} = \frac{R_2}{sCR_2 + 1} \tag{10.34}$$

Substituting the results from Eq. (10.34) into Eq. (10.33) yields an expression for the voltage transfer function for the low-pass filter.

$$A_v(s) = -\frac{R_2}{R_1}\frac{1}{(1 + sR_2C)} = -\frac{R_2}{R_1}\frac{1}{\left(1 + \frac{s}{\omega_H}\right)} \quad \text{where} \quad \omega_H = 2\pi f_H = \frac{1}{R_2C} \tag{10.35}$$

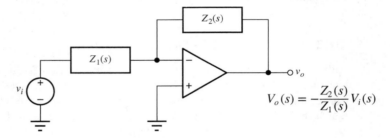

Figure 10.10 Generalized inverting-amplifier configuration.

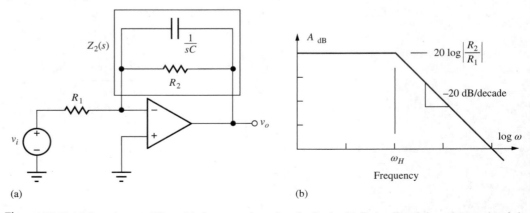

Figure 10.11 (a) Inverting amplifier with frequency-dependent feedback; (b) Bode plot of the voltage gain of the low-pass filter.

Figure 10.11(b) is the asymptotic Bode plot of the magnitude of the gain in Eq. (10.35). The transfer function exhibits a low-pass characteristic with a single pole at frequency ω_H, the upper-cutoff frequency (−3 dB point) of the low-pass filter. At frequencies below ω_H, the amplifier behaves as an inverting amplifier with gain set by the ratio of resistors R_2 and R_1; at frequencies above ω_H, the amplifier response rolls off at a rate of −20 dB/decade.

Note from Eq. (10.35) that the low-frequency gain and the cutoff frequency can be set independently in this **low-pass filter.** Indeed, because there are three elements—R_1, R_2, and C—the input resistance ($R_{in} = R_1$) can be a third design parameter. Since the inverting input terminal is at 0 V (remember it is a virtual ground), $R_{in} = R_1$. Filters that employ gain elements such as op amps or transistors are often referred to as **active filters.**

Understanding Low-Pass Filter Operation

The low-pass filter circuit functions in manner similar to the inverting amplifier. The virtual ground at the inverting input causes input voltage V_i to appear directly across resistor R_1 establishing an input current V_i/R_1 (using frequency domain notation). The op amp forces this input current to flow out through Z_2 developing a voltage drop of $(V_i/R_1)Z_2$ across Z_2. Because of the virtual ground, $V_o = -(V_i/R_1)Z_2$, and the gain is $-Z_2/R_1$ as expressed in Eq. (10.34).

At low frequencies (below ω_H), the amplifier operates as an inverting amplifier with gain of $-R_2/R_1$, since the capacitive reactance ($1/\omega C$) is much larger than that of R_2 and can be neglected. As frequencies increase, the magnitude of the impedance of the parallel combination of R_2 and C falls, reducing the gain. At high frequencies, the impedance of C becomes small, R_2 can be neglected, and the gain falls at a rate of 20 dB per decade as $(1/\omega C)$ decreases.

DESIGN EXAMPLE 10.5 ACTIVE LOW-PASS FILTER DESIGN

Design a single-pole low-pass filter using the single op-amp circuit in Fig. 10.11(a) to meet a given cutoff frequency specification.

PROBLEM Design an active low-pass filter (choose the values of R_1, R_2, and C) with $f_H = 2$ kHz, $R_{in} = 5$ kΩ, and $A_v = 40$ dB.

SOLUTION **Known Information and Given Data:** In this case, we are given the values for the bandwidth, gain and input resistance ($f_H = 2$ kHz, $R_{in} = 5$ kΩ, and $A_v = 40$ dB), and the amplifier circuit configuration has also been specified. However, we must convert the gain from dB to purely numeric form before we use it in the calculations:

$$|A_v| = 10^{40\,\text{dB}/20\,\text{dB}} = 100$$

Unknowns: Find the values of R_1, R_2, and C.

Approach: Use the single-pole low-pass filter circuit in Fig. 10.11(a). The three specifications should uniquely determine the three unknowns. We will use R_{in} to determine R_1, R_1 to find R_2, and R_2 to find C.

Assumptions: The op amp is ideal. Note that the specified gain actually represents the low-frequency gain of the amplifier and that a gain of either +100 or −100 will satisfy the gain specification.

Analysis: Since the inverting input represents a virtual ground, the input resistance is set directly by R_1 so that

$$R_1 = R_{in} = 5 \text{ k}\Omega$$

and

$$|A_v| = \frac{R_2}{R_1} \rightarrow R_2 = 100 R_1 = 500 \text{ k}\Omega$$

The value of C can now be determined from the f_H specification:

$$C = \frac{1}{2\pi f_H R_2} = \frac{1}{2\pi (2 \text{ kHz})(500 \text{ k}\Omega)} = 1.59 \times 10^{-10} \text{ F} = 159 \text{ pF}$$

Looking at Appendix A, we find the nearest values for R_1 and R_2 are 5.1 kΩ and 510 kΩ. In most applications, an input resistance of 5.1 kΩ (set by R_1) would be acceptable since it is only 2 percent higher than the design specification of 5 kΩ. Recalculating the value of C using the new value of R_2 yields

$$C = \frac{1}{2\pi f_H R_2} = \frac{1}{2\pi (2 \text{ kHz})(510 \text{ k}\Omega)} = 156 \text{ pF}$$

The closest capacitor value is 160 pF, which will lower f_H to 1.95 kHz. A second choice would be 150 pF for which $f_H = 2.08$ kHz.

Final Design: $R_1 = 5.1$ kΩ, $R_2 = 510$ kΩ, and $C = 160$ pF, yielding a slightly smaller bandwidth than the design specification.

Check of Results: We have found the three required values. The SPICE analysis here confirms the design.

Discussion: A third but more costly option would be to use a parallel combination of two capacitors, 100 pF and 56 pf. In a similar vein, R_1 and R_2 could be synthesized from the series combination of two resistors (e.g., $R_1 = 4.7$ k$\Omega + 300$ Ω). It might be preferable to just use more expensive 1 percent resistors with $R_1 = 4.99$ kΩ and $R_2 = 499$ kΩ. In order to select between these options, one would need to know more details about the application. Note that trying to use exact values of R and C doesn't provide much benefit if the resistor and capacitor tolerances are 5, 10, or 20 percent.

Computer-Aided Analysis: An ac analysis of the low-pass filter circuit is performed using the equivalent circuit below. The op amp gain is set to 10^6 (120 dB). The frequency response parameters are Start Frequency = 10 Hz, Stop Frequency = 100 KHz with 10 frequency points per decade. From the graph, $A_v = 40$ dB and $f_H = 1.95$ kHz as designed.

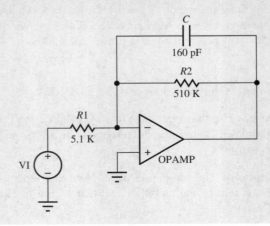

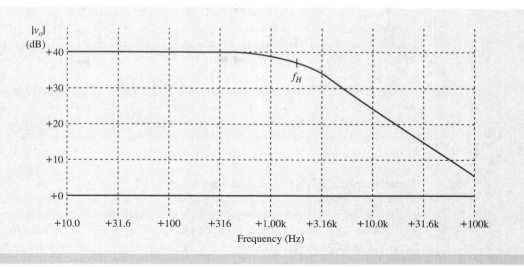

EXERCISE: Design an active low-pass filter (choose the values of R_1, R_2, and C) with $f_H = 3$ kHz, $R_{in} = 10$ kΩ, and $A_v = 26$ dB.

ANSWERS: Calculated values: 10 kΩ, 200 kΩ, 265 pF; Appendix A values: 10 kΩ, 200 kΩ, 270 pF

10.3.2 AN ACTIVE HIGH-PASS FILTER

We can also create an active **high-pass filter** by changing the form of impedances in the generalized inverting amplifier as in Fig. 10.12(a). Here Z_1 is replaced by the series combination of capacitor C and resistor R_1, and Z_2 is resistor R_2:

$$Z_1 = R_1 + \frac{1}{sC} = \frac{sCR_1 + 1}{sC} \quad \text{and} \quad Z_2 = R_2 \tag{10.36}$$

Substituting the results from Eq. (10.36) into Eq. (10.33) produces the voltage transfer function for the high-pass filter.

$$A_v(s) = -\frac{Z_2}{Z_1} = -\frac{R_2}{R_1}\frac{sCR_1}{sCR_1 + 1} = -\frac{R_2}{R_1}\frac{s}{s + \omega_L} = \frac{A_o}{1 + \frac{\omega_L}{s}} \tag{10.37}$$

$$\text{where} = A_o = -\frac{R_2}{R_1} \quad \text{and} \quad \omega_L = 2\pi f_L = \frac{1}{R_1 C}$$

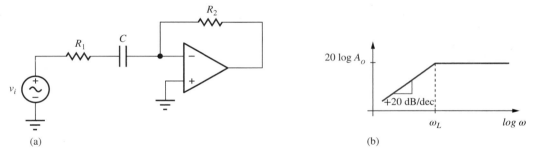

Figure 10.12 (a) Active high-pass filter circuit; (b) Bode plot for high-pass filter.

Figure 10.12(b) is the asymptotic Bode plot of the magnitude of the gain in Eq. (10.37). The transfer function exhibits a high-pass characteristic with a single pole at frequency ω_L, the lower cutoff frequency of the high-pass filter. At frequencies above ω_L, the amplifier behaves as an inverting amplifier with gain set by the ratio of resistors R_2 and R_1; at frequencies below ω_L, the amplifier rolls off at a rate of 20 dB/dec.

Realization of the transfer function in Eq. (10.39) and Fig. 10.12(b) is based upon use of an ideal operational amplifier with infinite bandwidth. In reality, we can only approximate the characteristic over a limited frequency range with a real op amp that has finite bandwidth. We will reexamine the behavior of a number of our basic building block circuits including low-pass and high-pass filters in Chapter 11 when we consider the frequency response limitations of op amps.

Understanding High-Pass Filter Operation

Once again, this filter circuit functions in manner similar to the inverting amplifier. The virtual ground at the inverting input causes input voltage V_i to appear directly across the series combination of resistor R_1 and C establishing an input current V_i/Z_1. The op amp forces this input current to flow out through R_2 developing a voltage drop of $(V_i/Z_1)R_2$ across R_2. Because of the virtual ground, $V_o = -(V_i/Z_1)R_2$, and the gain is $-R_2/Z_1$ as expressed in Eq. (10.37).

At high frequencies (above ω_L), the amplifier operates as an inverting amplifier with gain of $-R_2/R_1$, because the capacitive reactance $(1/\omega C)$ is much smaller than that of R_1 and can be neglected. At low frequencies, the impedance of C becomes large, R_1 can be neglected, and the gain falls at a rate of 20 dB/dec as the value of $(1/\omega C)$ increases.

> **EXERCISE:** Design an active high-pass filter (choose the values of R_1, R_2, and C) with $f_L = 5$ kHz, a high frequency gain of 20 dB, and an input resistance of 18 kΩ at high frequencies.
>
> **ANSWERS:** Calculated values: 18 kΩ, 180 kΩ, 1770 pF; Appendix A values: 18 kΩ, 180 kΩ, 1800 pF.

10.3.3 THE INTEGRATOR

The **integrator** is another highly useful building block formed from an operational amplifier with frequency-dependent feedback. In the integrator circuit, feedback-resistor R_2 is replaced by capacitor C, as in Fig. 10.13. This circuit provides an opportunity to explore op amp circuit analysis in the time domain (for frequency-domain analysis, see Prob. 10.50(a)).

Because the inverting-input terminal represents a virtual ground,

$$i_i = \frac{v_i}{R} \quad \text{and} \quad i_c = -C\frac{dv_o}{dt} \qquad (10.38)$$

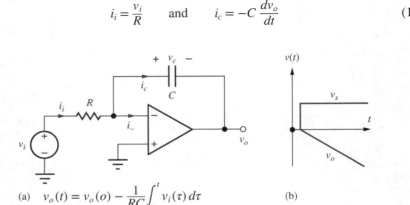

(a) $\quad v_o(t) = v_o(0) - \dfrac{1}{RC}\displaystyle\int_0^t v_i(\tau)\,d\tau$ \qquad (b)

Figure 10.13 (a) The integrator circuit; (b) output voltage for a step-function input with $v_C(0) = 0$.

For an ideal op amp, $i_- = 0$, so i_c must equal i_i. Equating the two expressions in Eq. (10.38) and integrating both sides yields

$$\int dv_o = \int -\frac{1}{RC} v_i \, d\tau \quad \text{or} \quad v_o(t) = v_o(0) - \frac{1}{RC} \int_0^t v_i(\tau) \, d\tau \tag{10.39}$$

in which the initial value of the output voltage is determined by the voltage on the capacitor at $t = 0$: $v_o(0) = -V_c(0)$. Thus the voltage at the output of this circuit at any time t represents the initial capacitor voltage minus the integral of the input signal from the start of the integration interval, chosen in this case to be $t = 0$.

Understanding Integrator Circuit Operation

The virtual ground at the inverting input of the op amp causes input voltage v_i to appear directly across resistor R establishing an input current v_i/R. As the input current flows out through C, the capacitor accumulates a charge equal to the integral of the current, $Q_C = \frac{1}{C} \int i \cdot dt$, and the overall scale factor becomes $-1/RC$.

EXERCISE: Suppose the input voltage $v_i(t)$ to an integrator is a 500-Hz square wave with a peak-to-peak amplitude of 10 V and 0 dc value. (a) What type of waveform appears at the output of the integrator? (b) Choose the values of R and C in the integrator so that the peak output voltage will be 10 V and $R_{in} = 10$ kΩ.

ANSWERS: 10 kΩ, 0.05 µF

ELECTRONICS IN ACTION

Dual-Ramp or Dual-Slope Analog-to-Digital Converters (ADCs)

The **dual-ramp** or **dual-slope analog-to-digital converter** is widely used as the ADC in data acquisition systems, digital multimeters, and other precision instruments. The heart of the dual-ramp converter is the integrator circuit discussed in Sec. 10.3.3. As illustrated in the circuit schematic below, the conversion cycle consists of two separate integration intervals.

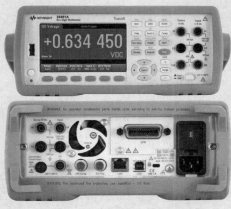

Keysight Digital Multimeter
© Keysight Technologies 2014 All Rights reserved

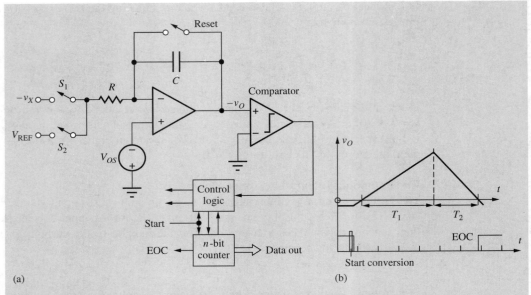

(a) Dual-ramp ADC and (b) timing diagram.

First, unknown voltage v_X is inverted and integrated for a known period of time T_1. The value of this integral is then compared to that of a known reference voltage V_{REF}, which is integrated for a variable length of time T_2.

At the start of conversion the counter is reset, and the integrator is reset to a slightly negative voltage. The unknown input is connected to the integrator input through switch S_1. Unknown voltage v_X is integrated for a fixed period of time $T_1 = 2^n T_C$, which begins when the integrator output crosses through zero where T_C is the period of the clock. At the end of time T_1, the counter overflows, causing S_1 to be opened and the reference input V_{REF} to be connected to the integrator input through S_2. The integrator output then decreases until it crosses back through zero, and the comparator changes state, indicating the end of the conversion. The counter continues to accumulate pulses during the down ramp, and the final number in the counter represents the quantized value of the unknown voltage v_X.

Circuit operation forces the integrals over the two time periods to be equal:

$$\frac{1}{RC}\int_0^{T_1} v_X(t)\,dt = \frac{1}{RC}\int_{T_1}^{T_1+T_2} V_{REF}\,dt$$

T_1 is set equal to $2^n T_C$ because the unknown voltage v_X was integrated over the amount of time needed for the n-bit counter to overflow. Time period T_2 is equal to NT_C, where N is the number accumulated in the counter during the second phase of operation.

Recalling the mean-value theorem from calculus, we have

$$\frac{1}{RC}\int_0^{T_1} v_X(t)\,dt = \frac{\langle v_X \rangle}{RC} T_1 \quad \text{and} \quad \frac{1}{RC}\int_{T_1}^{T_1+T_2} V_{REF}(t)\,dt = \frac{V_{REF}}{RC} T_2$$

because V_{REF} is a constant. Equating these last two results, we find the average value of the input $\langle v_X \rangle$ to be

$$\frac{\langle v_X \rangle}{V_{REF}} = \frac{T_2}{T_1} = \frac{N}{2^n}$$

assuming that the RC product remains constant throughout the complete conversion cycle. The absolute values of R and C do not enter directly into the relation between v_X and V_{FS}. The digital output word represents the average value of v_X during the first integration phase. Thus, v_X can change during the conversion cycle of this converter without destroying the validity of the quantized output value.

The conversion time T_T requires 2^n clock periods for the first integration period, and N clock periods for the second integration period. Thus the conversion time is variable and given by $T_T = (2^n + N)T_C \leq 2^{n+1}T_C$ because the maximum value of N is 2^n.

The dual ramp is a widely used converter. Although much slower than other forms of converters, the dual-ramp converter offers excellent linearity. By combining its integrating properties with careful design, one can obtain accurate conversion at resolutions exceeding 20 bits, but at relatively low conversion rates. In a number of recent converters and instruments, the basic dual-ramp converter has been modified to include extra integration phases for automatic offset voltage elimination. These devices are often called *quad-slope* or *quad-phase converters*. Another converter, the *triple ramp*, uses coarse and fine down ramps to greatly improve the speed of the integrating converter (by a factor of $2^{n/2}$ for an n-bit converter).

Normal-Mode Rejection

As mentioned before, the quantized output of the dual-ramp converter represents the average of the input during the first integration phase. The integrator operates as a low-pass filter with the normalized transfer function shown in the accompanying figure. Sinusoidal input signals, whose frequencies are exact multiples of the reciprocal of the integration time T_1, have integrals of zero value and do not appear at the integrator output. This property is used in many digital multimeters, which are equipped with dual-ramp converters having an integration time that is some multiple of the period of the 50- or 60-Hz power-line frequency. Noise sources with frequencies at multiples of the power-line frequency are therefore rejected by these integrating ADCs. This property is usually termed **normal-mode rejection.**

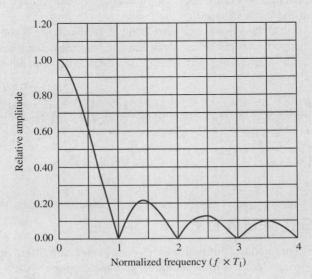

Normal-mode rejection for an integrating ADC.

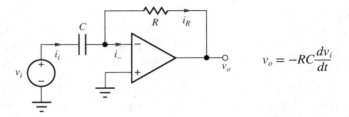

Figure 10.14 Differentiator circuit.

10.3.4 THE DIFFERENTIATOR

The derivative operation can also be provided by an op amp circuit. The **differentiator** is obtained by interchanging the resistor and capacitor of the integrator as drawn in Fig. 10.14. The circuit is less often used than the integrator because the derivative operation is an inherently "noisy" operation; that is, the high-frequency components of the input signal are emphasized. Analysis of the circuit is similar to that of the integrator. Since the inverting-input terminal represents a virtual ground,

$$i_i = C \frac{dv_i}{dt} \quad \text{and} \quad i_R = -\frac{v_o}{R} \tag{10.40}$$

Since $i_- = 0$, the currents i_i and i_R must be equal, and

$$v_o = -RC \frac{dv_i}{dt} \tag{10.41}$$

The output voltage is a scaled version of the derivative of the input voltage.

Understanding Differentiator Circuit Operation

The virtual ground at the inverting input of the op amp causes input voltage v_i to appear directly across capacitor C, establishing an input current that is proportional to the derivative of v_i. This current flows out through R yielding an output voltage that is a scaled version of the derivative of the input voltage.

Thinking in the frequency domain, the reactance of the capacitor $(1/\omega C)$ decreases as frequency increases. Therefore the input current, and the scaled output voltage, both increase directly with frequency, yielding a frequency dependence that corresponds to a differentiator.

EXERCISE: What is the output voltage of the circuit in Fig. 10.14 if $R = 20$ kΩ, $C = 0.02$ µF, and $v_i = 2.5 \sin 2000\pi t$ V?

ANSWER: $-6.28 \cos 2000\pi t$ V

SUMMARY

- This chapter introduced the operational amplifier, or op amp, and our circuit toolkit was expanded to include a number of classic op-amp-based building blocks. The op amp represents an extremely important tool for implementing basic amplifiers and more complex electronic circuits.

- Ideal operational amplifiers are assumed to have infinite gain and zero input current, and circuits containing these amplifiers were analyzed using two primary assumptions:
  1. The differential input voltage is zero: $v_{ID} = 0$.
  2. The input currents are zero: $i_{+} = 0$ and $i_{-} = 0$.
- Assumptions 1 and 2, combined with Kirchhoff's voltage and current laws, are used to analyze the ideal behavior of circuit building blocks based on operational amplifiers. Constant feedback with resistive voltage dividers is used in the inverting and noninverting amplifier configurations, the voltage follower, the difference amplifier, and the summing amplifier, whereas frequency-dependent feedback is used in the integrator, low-pass filter, high-pass filter, and differentiator circuits.
- Infinite gain and input resistance are the explicit characteristics that lead to Assumptions 1 and 2. However, many additional properties are implicit characteristics of ideal operational amplifiers; these assumptions are seldom clearly stated, though. They are
  - Infinite common-mode rejection
  - Infinite power supply rejection
  - Infinite output voltage range
  - Infinite output current capability
  - Infinite open-loop bandwidth
  - Infinite slew rate
  - Zero output resistance
  - Zero input-bias currents
  - Zero input-offset voltage

  These limitations will be explained in detail in the next two chapters.
- Unity gain bandwidth product (ω_T) is an important figure of merit for amplifiers and is the product of the amplifier's midband gain and its bandwidth.

KEY TERMS

Active filters
Current-to-voltage (I-V) converter
Difference amplifier
Differential amplifier
Differential subtractor
Differentiator
Digital-to-analog converter (DAC or D/A converter)
Dual-ramp (dual-slope) analog-to-digital converter
Feedback network
High-pass filter
Ideal inverting amplifier
Ideal noninverting amplifier
Ideal operational amplifier
Integrator
Inverting amplifier
Least significant bit (LSB)
Low-pass filter
Most significant bit (MSB)
Noninverting amplifier
Normal-mode rejection
Open-circuit voltage gain
Open-loop gain
Operational amplifier (op amp)
Output resistance
Summing amplifier
Summing junction
Superposition error
Transfer function
Transresistance amplifier
Unity-gain buffer
Virtual ground
Voltage follower
Weighted-resistor DAC

REFERENCES

1. https://spectrum.ieee.org/tech-history/silicon-revolution/chip-hall-of-fame-fairchild-semiconductor-a741-opamp
2. John R. Ragazzini, Robert H. Randall, and Frederick A. Russell, "Analysis of Problems in Dynamics by Electronic Circuits," *Proceedings of the IRE,* IEEE, vol. 35, no. 5 (May 1947), pp. 444–452.

ADDITIONAL READING

E. J. Kennedy, *Operational Amplifier Circuits—Theory and Applications.* Holt, Rinehart, and Winston, New York, NY: 1988.

Franco, Sergio, *Design with Operational Amplifiers and Analog Integrated Circuits,* Fourth Edition, McGraw-Hill, New York, NY: 2014.

P. R. Gray, P. J. Hurst, S. H. Lewis, and R. G. Meyer, *Analysis and Design of Analog Integrated Circuits,* Fifth Edition, John Wiley and Sons, New York, NY: 2009.

PROBLEMS

10.1 Ideal Differential and Operational Amplifiers

10.1. Suppose a differential amplifier has $A = 110$ dB, and it is operating in a circuit with an open-circuit output voltage $v_o = 15$ V. What is the input voltage v_{id}? How large must the voltage gain be for $v_{id} \leq 1$ µV? What is the input current i_+ if $R_{id} = 1$ MΩ?

10.2. An almost ideal op amp has an open-circuit output voltage $v_o = 10$ V and a gain $A = 100$ dB. What is the input voltage v_{id}? How large must the voltage gain be to make $v_{id} \leq 1$ µV?

10.2 Analysis of Circuits Containing Ideal Operational Amplifiers

Inverting Amplifiers

10.3. What are the voltage gain, input resistance, and output resistance of the amplifier in Fig. P10.3 if $R_1 = 10$ kΩ and $R_2 = 120$ kΩ? (b) Repeat for $R_1 = 160$ kΩ and $R_2 = 360$ kΩ. (c) Repeat for $R_1 = 4.3$ kΩ and $R_2 = 270$ kΩ.

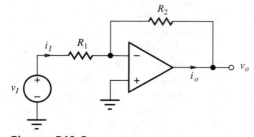

Figure P10.3

10.4. (a) What are the voltage gain, input resistance, and output resistance of the amplifier in Fig. P10.3 if $R_1 = 4.7$ kΩ and $R_2 = 200$ kΩ? Express the voltage gain in dB. (b) Repeat for $R_1 = 51$ kΩ and $R_2 = 2.2$ MΩ.

10.5. Write an expression for the output voltage $v_o(t)$ of the circuit in Fig. P10.3 if $R_1 = 680$ Ω, $R_2 = 9.1$ kΩ, and $v_i(t) = (0.05 \sin 4638t)$ V. Write an expression for the current $i_i(t)$.

10.6. $R_1 = 2.4$ kΩ and $R_2 = 200$ kΩ in the amplifier circuit in Fig. P10.3. (a) What is the output voltage if $v_i = 0$? (b) What is the output voltage if a dc signal $V_I = 0.22$ V is applied to the circuit? (c) What is the output voltage if an ac signal $v_I = 0.15 \sin 2500 \pi t$ V is applied to the circuit? (d) What is the output voltage if the input signal is $v_I = 0.22 - 0.15 \sin 2500 \pi t$ V? (e) What is the input current i_I for parts (b), (c), and (d)? (f) What is the op amp output current i_O for the input signals in parts (b), (c), and (d)? (g) What is the voltage at the inverting input of the op amp for the input signal in part (d)?

10.7. The amplifier in Fig. P10.3 has $R_1 = 7.5$ kΩ, $R_2 = 160$ kΩ and operates from ±12-V power supplies. (a) If $v_I = -0.2 + V_i \sin 2000\pi t$ volts, write an expression for the output voltage. (b) What is the maximum value of V_i for an undistorted output? (c) Repeat if $v_I = 0.6 + V_i \sin 2000\pi t$ volts.

10.8. The amplifier in Fig. P10.3 has $R_1 = 9.1$ kΩ, $R_2 = 160$ kΩ and operates from ±12-V power supplies. (a) What is the voltage gain $A_v = \mathbf{v_o}/\mathbf{v_i}$ of the circuit? (b) Suppose input source v_i is not ideal but actually has a 1.5 kΩ source resistance. What is the voltage gain $A_v = \mathbf{v_o}/\mathbf{v_i}$?

10.9. The amplifier in Fig. P10.3 has $R_1 = 10$ kΩ, $R_2 = 120$ kΩ and operates from ±10-V power supplies. (a) If $v_I = 0.5 + V_i \sin 5000\pi t$ volts, write an expression for the output voltage. (b) What is the maximum value of V_i for an undistorted output? (c) Repeat if $v_I = -0.25 + V_i \sin 2000\pi t$ volts.

10.10. The amplifier in Prob. 10.3(a) utilizes resistors with 10 percent tolerances. What are the nominal and worst-case values of the voltage gain and input resistance?

10.11. (a) The amplifier in Prob. 10.4(a) utilizes resistors with 5 percent tolerances. What are the nominal and worst-case values of the voltage gain and input resistance? (b) Repeat for Prob. 10.3(b). (c) Repeat for Prob. 10.3(c).

10.12. Design an inverting amplifier with an input resistance of 18 kΩ and a gain of 28 dB. Choose values from the 1 percent resistor table in Appendix A.

10.13. Design an inverting amplifier with an input resistance of 2 kΩ and a gain of 44 dB. Choose values from the 1 percent resistor table in Appendix A.

10.14. Design an inverting amplifier with an input resistance of 100 kΩ and a gain of 15 dB. Choose values from the 1 percent resistor table in Appendix A.

10.15. Find the voltage gain, input resistance, and output resistance for the circuit in Fig. P10.15.

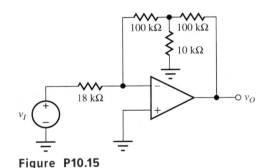

Figure P10.15

Transresistance Amplifiers

10.16. Find an expression for the output voltage v_O in Fig. P10.16.

10.17. Convert the inverting amplifier in Fig. 10.1 to the transresistance amplifier in Fig. P10.16 using a Norton transformation of v_i and R_1. What are the expressions for i_{TH} and R_{TH}? Write a expression for the gain $\mathbf{v_o}/\mathbf{v_i}$.

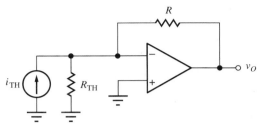

Figure P10.16

10.18. The current generated by some transducer falls in the range of ±2.5 μA, and its source resistance is 150 kΩ. A transresistance amplifier is needed to convert the current to a voltage between ±4 V. What value of R is required?

Noninverting Amplifiers

10.19. What are the voltage gain, input resistance, and output resistance of the amplifier in Fig. P10.19 if $R_1 = 6.8$ kΩ and $R_2 = 750$ kΩ? Express the voltage gain in dB.

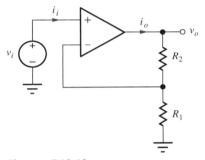

Figure P10.19

10.20. Write an expression for the output voltage $v_o(t)$ of the circuit in Fig. P10.19 if $R_1 = 910$ Ω, $R_2 = 7.5$ kΩ, and $v_i(t) = (0.04 \sin 9125t)$ V.

10.21. What are the voltage gain, input resistance, and output resistance of the amplifier in Fig. P10.19 if $R_1 = 33$ kΩ and $R_2 = 120$ kΩ? (b) Repeat for $R_1 = 18$ kΩ and $R_2 = 330$ kΩ. (c) Repeat for $R_1 = 3$ kΩ and $R_2 = 390$ kΩ.

10.22. $R_1 = 24$ kΩ and $R_2 = 360$ kΩ in the amplifier circuit in Fig. P10.19. (a) What is the output voltage if $v_I = 0$? (b) What is the output voltage if a dc signal $V_I = 0.35$ V is applied to the circuit? (c) What is the output voltage if an ac signal $v_i = 0.18 \sin 3250\pi t$ V is applied to the circuit? (d) What is the output voltage if the input signal is $v_I = 0.3 - 0.18 \sin 3250\pi t$ V? (e) Write an expression for the input current i_I for parts (b), (c), and (d). (f) Write an expression for the op amp output current i_O for the input signals in parts (b), (c) and (d). (g) What is the voltage at the inverting input of the op amp for the input signal in part (d)?

*10.23. (a) What are the gain, input resistance, and output resistance of the amplifier in Fig. P10.23 if $R_1 = 150\ \Omega$ and $R_2 = 51\ k\Omega$? Express the gain in dB. (b) If the resistors have 10 percent tolerances, what are the worst-case values (highest and lowest) of gain that could occur? (c) What are the resulting positive and negative tolerances on the voltage gain with respect to the ideal value? (d) What is the ratio of the largest to the smallest voltage gain? (e) Perform a 500-case Monte Carlo analysis of this circuit. What percentage of the circuits has a gain within ±5 percent of the nominal design value?

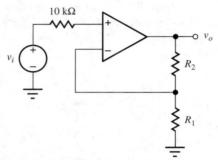

Figure P10.23

10.24. Design a noninverting amplifier with a gain of 20 dB. Choose values from the 1 percent resistor table in Appendix A, and use values that are no smaller than 2 kΩ.

10.25. Design a noninverting amplifier with an input resistance of 150 kΩ and a gain of 6 dB. Choose values from the 1 percent resistor table in Appendix A, and use values that are no smaller than 2 kΩ.

10.26. Design a noninverting amplifier with a gain of 40 dB. Choose values from the 1 percent resistor table in Appendix A, and use values that are no smaller than 1 kΩ.

10.27. What are the gain, input resistance, and output resistance for the circuits in Fig. P10.27.

Summing Amplifiers

10.28. Write an expression for the output voltage $v_o(t)$ of the circuit in Fig. P10.28 if $R_1 = 1\ k\Omega$, $R_2 = 2\ k\Omega$, $R_3 = 42\ k\Omega$, $v_2(t) = (0.01 \sin 3770t)$ V, and $v_1(t) = (0.04 \sin 10000t)$ V. Write an expression for the voltage appearing at the summing junction (v_-).

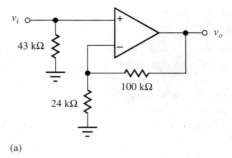

(a)

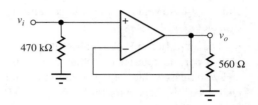

(b)

Figure P10.27

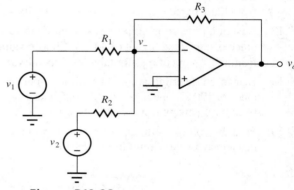

Figure P10.28

10.29. The summing amplifier can be used as a digitally controlled volume control using the circuit in Fig. P10.29. The individual bits of the 4-bit binary input word ($b_1 b_2 b_3 b_4$) are used to control the position of the switches with the resistor connected to 0 V if $b_i = 0$ and connected to the input signal v_I

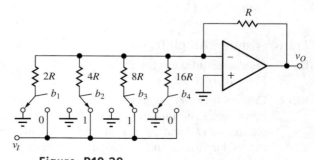

Figure P10.29

if $b_i = 1$. (a) What is the output voltage v_O with the data input 1001 if $v_I = 2 \sin 4000\pi t$ V? (b) Suppose the input changes to 1011. What will be the new output voltage? (c) Make a table giving the output voltages for all 16 possible input combinations.

10.30. The switches in Fig. P10.29 can be implemented using MOSFETs, as shown in Fig. P10.30. What are the W/L ratios of the transistors if the on-resistance of the transistor is to be less than 1 percent of the resistor $2R = 12$ kΩ? Assume that the voltage applied to the gate of the MOSFET is 5 V when $b_1 = 1$ and 0 V when $b_1 = 0$. For the MOSFET, $V_{TN} = 1$ V, $K'_n = 50$ µA/V^2, $2\phi_F = 0.6$ V, and $\gamma = 0.5 \sqrt{V}$.

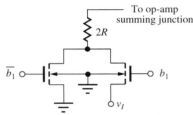

Figure P10.30

Difference Amplifier

10.31. (a) What is the gain of the circuit in Fig. P10.31 if $A_v = v_o/(v_1 - v_2)$ and $R = 6.8$ kΩ? (b) What is the input resistance presented to v_2? (c) What is the input resistance at terminal v_1? (d) What is the output voltage if $v_1 = 3$ V and $v_2 = 1.5 \cos 8300\pi t$ V? (e) What is the output voltage if $v_1 = [3 - 1.5 \cos 8300\pi t]$ V and $v_2 = 1.5 \sin 8300\pi t$ V?

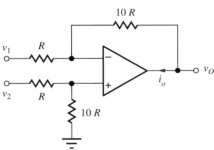

Figure P10.31

10.32. (a) What are the voltages at all the nodes in the difference amplifier in Fig. P10.31 if $V_2 = 3.1$ V, $V_1 = 3.4$ V, and $R = 100$ kΩ? (b) What is amplifier output current I_O? (c) What are the currents entering the circuit from v_1 and v_2?

10.33. Find v_O, i_1 and i_2 for the difference amplifier in Fig. P10.31 if $v_2 = 2 \sin 1000\pi t$ V, $v_1 = (2 \sin 1000\pi t + 2 \sin 2000\pi t)$ V, and $R = 15$ kΩ.

10.3 Frequency Dependent Feedback

Low-Pass Filters

10.34. Find the midband gain in dB and the upper cut-off frequency for the low-pass filter in Fig. 10.11 if $R_1 = 10$ kΩ, $R_2 = 100$ kΩ, and $C = 0.02$ µF. Find the midband gain in dB and the upper cutoff frequency for the low-pass filter in Fig. 10.11 if $R_1 = 1$ kΩ, $R_2 = 1.8$ kΩ, and $C = 0.02$ µF.

10.35. (a) Design a low-pass filter using the circuit in Fig. 10.11 to provide a loss of no more than 0.5 dB at low frequencies and a cutoff frequency of 24 kHz if $R_1 = 620$ Ω. (b) Pick standard values from the tables in Appendix A.

10.36. (a) What are the low-frequency voltage gain (in dB) and cutoff frequency f_H for the amplifier in Fig. 10.11 if $R_1 = 2.2$ kΩ, $R_2 = 10$ kΩ, and $C = 0.001$ µF? (b) Repeat for $R_1 = 3.3$ kΩ, $R_2 = 51$ kΩ, and $C = 100$ pF.

10.37. (a) Design a low-pass amplifier (i.e., choose R_1, R_2, and C) to have a low-frequency input resistance of 10 kΩ, a midband gain of 20 dB, and a bandwidth of 22 kHz. (b) Choose element values from the tables in Appendix A.

10.38. What are A_{mid} in dB, f_H, f_L, and the BW in Hz for the amplifier described by

$$A_v(s) = \frac{2\pi \times 10^5}{s + 200\pi}$$

What type of amplifier is this?

High-Pass Filters

10.39. Find the midband gain in dB and the upper cut-off frequency for the high-pass filter in Fig. 10.12 if $R_1 = 10$ kΩ, $R_2 = 75$ kΩ, and $C = 0.01$ µF.

10.40. Find the midband gain in dB and the upper cut-off frequency for the high-pass filter in Fig. 10.12 if $R_1 = 6.8$ kΩ, $R_2 = 20$ kΩ, and $C = 0.02$ µF.

10.41. (a) Design a high-pass filter using the circuit in Fig. 10.12 to provide a loss of no more than 0.5 dB at high frequencies and a cutoff frequency of 20 kHz if $R_1 = 330$ Ω. (b) Pick standard values from the tables in Appendix A.

10.42. What are the high-frequency voltage gain (in dB) and cutoff frequency f_L for the amplifier in Fig. 10.12 if $R_1 = 3.9$ kΩ, $R_2 = 20$ kΩ, and

$C = 560$ pF? (b) Repeat for $R_1 = 2.7$ kΩ, $R_2 = 51$ kΩ, and $C = 0.002$ µF.

10.43. (a) Design a high-pass filter (choose R_1, R_2, and C) to have a high frequency input resistance of 20 kΩ, a gain of 20 dB, and a lower cutoff frequency of 1 kHz. (b) Choose element values from the tables in Appendix A.

10.44. What are A_{mid} in dB, f_H, f_L, and the BW in Hz for the amplifier described by

$$A_v(s) = \frac{3 \times 10^3 s}{s + 200\pi}$$

What type of amplifier is this?

Integrator

10.45. The input voltage to the integrator circuit in Fig. 10.13 is given by $v_i = 0.1 \sin 2000\pi t$ V. What is the output voltage if $R = 15$ kΩ, $C = 0.005$ µF, and $v_o(0) = 0$?

10.46. The input voltage to the integrator circuit in Fig. 10.13 is a rectangular pulse with an amplitude of 5 V and a width 1 ms. Draw the waveform at the output of the integrator if the pulse starts at $t = 0$, $R = 12$ kΩ, and $C = 0.1$ µF. Assume $v_o = 0$ for $t \leq 0$. (b) Repeat for $v_o(0) = -2.5$ V.

*10.47. (a) What is the voltage transfer $V_o(s)/V_i(s)$ function for the integrator in Fig. 10.13. (b) What is the voltage transfer function for the circuit in Fig. P10.47?

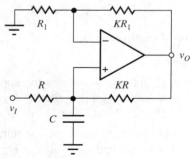

Figure P10.47

Differentiator

10.48. What is the transfer function $T(s) = V_O(s)/V_i(s)$ for the differentiator circuit in Fig. 10.14?

10.49. What is the output voltage of the differentiator circuit in Fig. 10.14 if $v_i(t) = 3 \cos 3000\pi t$ V with $C = 0.03$ µF and $R = 120$ kΩ?

10.50. What is the transfer function $A_v(s) = V_o(s)/V_i(s)$ for the circuit in Fig. P10.50? Draw a Bode plot for the transfer function.

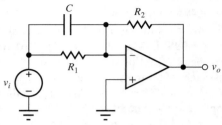

Figure P10.50

General Op Amp Problems

10.51. Find the voltage gain, input resistance, and output resistance for the circuits in Fig. P10.51.

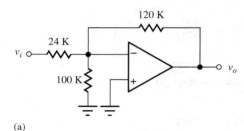

(a)

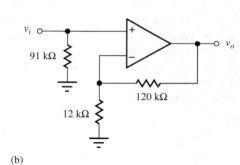

(b)

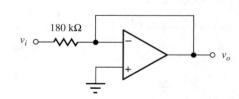

(c)

Figure P10.51

*10.52. (a) What is the output current I_O in the circuit of Fig. P10.52 if $-V_{EE} = -5$ V and $R = 10$ Ω? Assume that the MOSFET is saturated. (b) What is the minimum voltage V_{DD} needed to saturate the

MOSFET if $V_{TN} = 2.5$ V and $K_n = 0.25$ A/V^2? (c) What must be the power dissipation ratings of resistor R and the FET?

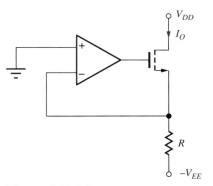

Figure P10.52

*10.53. (a) What is the output current I_O in the circuit in Fig. P10.53 if $-V_{EE} = -10$ V and $R = 15$ Ω? Assume that the BJT is in the forward-active region and $\beta_F = 50$. (b) What is the voltage at the output of the operational amplifier if the saturation current I_S of the BJT is 10^{-13} A? (c) What is the minimum voltage V_{CC} needed for forward-active region operation of the bipolar transistor? (d) Find the power dissipation rating of the resistor R. How much power is dissipated in the transistor if $V_{CC} = 15$ V?

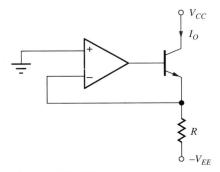

Figure P10.53

10.54. What is the transfer function for the voltage gain of the amplifier in Fig. P10.54?

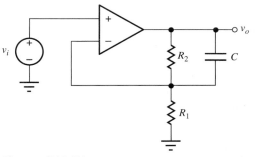

Figure P10.54

*10.55. The low-pass filter in Fig. P10.55 has $R_1 = 12$ kΩ, $R_2 = 300$ kΩ, and $C = 100$ pF. If the tolerances of the resistors are ±10 percent and that of the capacitor is +20 percent/−50 percent, what are the nominal and worst-case values of the low-frequency gain and cutoff frequency?

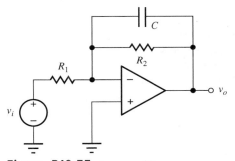

Figure P10.55 Low-pass filter

术语对照

Differential amplifier	差分放大器
Differential suntractor	差分减法器
Differential-mode gain	差模增益
Differential-mode input resistance	差模输入电阻
Differential-mode input voltage	差模输入电压
Differentiator	微分器
Ideal operational amplifier	理想运算放大器
Ideal voltage amplifier	理想电压放大器
Low-pass amplifier	低通放大器
Low-pass filter	低通滤波器
Magnitude	幅度
Midband gain	中频带增益
Noninverting input	同相输入
Normal mode rejection	常模抑制
Single-pole frequency response	单极点频率响应
Single Amplification	信号放大器
Source resistance(R_S)	电源内阻 (R_S)
Summing amplifier	求和放大器
Summing junction	求和点
Superposition errors	叠加误差
Total harmonic distortion	总谐波失真
Transfer function	传输函数
Upper-cutoff frequency	上限截止频率
Upper half-power point	上半功率点
Upper- 3-dB frequency	上3dB频率
Very high frequency	甚高频
Virtual ground	虚地
Voltage amplifier	电压放大器
Voltage follower	电压跟随器

CHAPTER 11

NONIDEAL OPERATIONAL AMPLIFIERS AND FEEDBACK AMPLIFIER STABILITY

第11章 非线性运算放大器和反馈放大器的稳定性

本章提纲
11.1 经典反馈系统
11.2 含有非理想运算放大器的电路分析
11.3 串联反馈和并联反馈电路
11.4 反馈放大器计算的统一方法
11.5 电压串联反馈放大器——电压放大器
11.6 电压并联反馈放大器——跨阻放大器
11.7 电流串联反馈放大器——跨导放大器
11.8 电流并联反馈放大器——电流放大器
11.9 使用持续电压和电流注入法计算回路增益
11.10 利用反馈减少失真
11.11 直流误差源和输出摆幅限制
11.12 共模抑制比和输入电阻
11.13 运算放大器的频率响应和带宽
11.14 反馈放大器的稳定性

本章目标
- 了解非理想运算放大器的工作原理；
- 了解用于分析包含非理想运算放大器电路的主要技术；
- 了解共模抑制的成因以及共模输入电阻的影响；
- 熟悉如何对直流误差进行建模，包括偏移电压、输入偏置电流和输入偏移电流；
- 了解电源电压和有限的电流输出能力对运算放大器带来的限制；
- 了解有限带宽和低转换速率对放大器建模的影响；
- 熟悉对非理想运算放大器的SPICE仿真；
- 理解电压串联反馈、电压并联反馈、电流串联反馈和电流并联反馈的拓扑结构及特征；
- 了解电流负载影响的反馈放大器电路分析方法；
- 理解反馈对频率响应的影响，并理解反馈放大器的稳定性；
- 理解相位裕度和增益裕度的概念；
- 理解根据奈奎斯特图和伯德图解释反馈放大器的稳定性；
- 能够使用SPICE的交流分析和传输函数分析方法确定反馈放大器的特性；
- 熟悉利用SPICE仿真或测量确定闭环放大器回路增益的方法。

本章导读
　　前面章节已经对理想运算放大器进行了介绍，本章主要讨论去除理想运算放大器假设之后的情况，并重点分析影响非理想运算放大器特性的情况。
　　为了能更好地理解非理想运算放大器对电路性能的影响，本章首先回顾了电子系统中经典的负反馈理论，这一理论最先是由贝尔电话公司的Harold Black提出的。他在1928年发明了反馈放大器，用以稳定电话中继器的增益。如今，几乎每个电子系统都会用到某种形式的反馈。本章将详细阐述反馈的概念，及其在电路系统的设计中所发挥的重大作用。通过将普通的电路重新设计为反馈放大器形式，可以进一步改进许多常见电子电路设计，提高电路设计性能。

目前，人们已经开发出数百种已经产品化的集成电路运算放大器，但如何选择和使用这些运算放大器，就必须对这些运算放大器有充分的了解，并理解实际运放的特性及其局限性。本章详细探讨了这些限制的影响，并给出了采用非理想运算放大器的电路的方法。一般来说，我们会对非理想运算放大器特性的每个因素进行单独研究，同时假设其他特征仍然是理想的，来讨论有限开环增益、有限输入电阻和非零输出电阻对同相和反相放大器整体特性的影响，看看如何通过改进设计使其接近理想目标。然后，将所有结论组合起来，以了解电路的一般行为。本章详细讨论了去除不同理想运放假设后的影响，并推导了闭环同相和反相放大器的增益、增益误差、输入电阻和输出电阻的表达式，发现回路增益$T=A\beta$对于确定闭环放大器的参数起着重要的作用。

串联和并联反馈类型在输入和输出上的组合可以得到四种反馈电路。当在一个放大器端口采用反馈时，采用串联反馈通常可以增加阻抗值，而采用并联反馈则相反，会使端口的阻抗值降低。电压放大器应该具有高输入电阻来测量所需电压，具有低输出电阻以驱动外部负载，对应的是电压串联反馈（Series-Shunt Feedback）电路。跨阻放大器将输入电流转换为输出电压，需要有较低输入阻抗来汇集所需的电流，有较低输出电阻来驱动外部负载，对应的是电压并联反馈（Shunt-Shunt Feedback）电路。电流放大器需要在输入端提供一个低阻电流源，在输出端提供一个高阻电流源，对应的是电流并联反馈（Shunt-Series Feedback）电路。跨导放大器是将输入电压转换为输出电流，需要有高的输入电阻和高的输出电阻，对应的是电流串联反馈（Series-Series Feedback）电路，其中放大器的输入端与输出端都与反馈网络串联。

在反馈放大器模型理解和使用方面，本章详细讲解了如何建立通用反馈放大器模型，并从电路中直接计算回路增益，不只在理论上，还要利用SPICE进行计算，并基于实际测量来验证。本章针对4种不同类型的放大器类型，分别从闭环增益计算、输入电阻计算、输出电阻计算等方面进行理论分析与计算，并分别给出了放大器电路闭环特性计算的实例，计算所示反馈放大器的闭环跨阻、输入电阻和输出电阻。

本章对非理想运算放大器（运放）特性产生的影响和限制进行了定量分析，书中对直流误差源和输出摆幅限制进行了讲解。直流误差源包括输入失调电压V_{OS}、输入偏置电流I_{B1}和I_{B2}及输出失调电流I_{OS}，所有这些都限制了运放电路的直流准确性，实际运放还存在输出电压、电流摆幅和摆率的限制。运算放大器内部电路所需偏置以及这些电路器件之间的失配是重要的误差源。为了让构成运算放大器的晶体管正常工作，必须要给放大器的每个输入端提供一个非零的直流偏置电流。对于采用双极型晶体管构成的放大器而言，这些电流就是基极电流，而对于采用MOSFET或JFET构成的放大器而言，这些电流就是栅极电流。尽管比较小，但偏置电流和失调电流都是额外的误差源。

本章对共模信号的概念进行了介绍。在数字系统中，高频信号的电容耦合会导致在多根信号线上感应出同一个信号。多数高速计算机总线采用差分信号，是为了采用具有良好CMRR的放大器来消除不希望出现的共模信号。电源抑制比（PSRR）是与CMRR密切相关的一个参数，由于长时间漂移或者电源中噪声的存在使得电源电压发生改变时，等效输入失调电压会发生轻微变化。PSRR用来衡量放大器对这些电源变化的抑制能力。

本章详细讲解了运算放大器的频率响应和带宽的概念及分析方法。基本单极点运算放大器的频率响应可通过开环增益A_o和增益带宽积ω_T两个参数描述。对同相放大器的增益和带宽的分析直接反映了这些放大器中闭环增益和闭环带宽之间的权衡关系，增益带宽积恒定，为了增大带宽必须降低闭环增益，反之亦然。

只要放大器电路中采用了反馈，就需要考虑稳定性问题。本章给出了带有放大器电路稳定性分析的方法，通过研究作为频率函数的反馈放大器的回路增益$T(s)=A(s)\beta(s)$的特性来确定，利用奈奎斯特图和伯德图两种方法研究放大器的稳定性。在奈奎斯特图中，稳定的条件是$T(j\omega)$曲线不包含$T=-1$的点。在伯德图中，$A(j\omega)$和$\beta(j\omega)$的幅值渐近线不能在超过20dB/10倍频程的区域相交。相位裕度和增益裕度可以通过奈奎斯特图或伯德图求得，也是衡量稳定性的重要标准。相位裕度确定了一个二阶反馈系统的阶跃响应中的过冲击大小，一般被设计成具有至少60°相位裕度。

本章给出了17个关于非线性运算放大器和反馈放大器的稳定性分析相关的设计实例，深入理解这些设计实例，有助于准确理解和掌握本章相关概念及设计分析方法。

Chapter Outline

11.1 Classic Feedback Systems
11.2 Analysis of Circuits Containing Nonideal Operational Amplifiers
11.3 Series and Shunt Feedback Circuits
11.4 Unified Approach to Feedback Amplifier Gain Calculation
11.5 Series-Shunt Feedback—Voltage Amplifiers
11.6 Shunt-Shunt Feedback—Transresistance Amplifiers
11.7 Series-Series Feedback—Transconductance Amplifiers
11.8 Shunt-Series Feedback—Current Amplifiers
11.9 Finding the Loop Gain Using Successive Voltage and Current Injection
11.10 Distortion Reduction Through the Use of Feedback
11.11 DC Error Sources and Output Range Limitations
11.12 Common-Mode Rejection and Input Resistance
11.13 Frequency Response and Bandwidth of Operational Amplifiers
11.14 Stability of Feedback Amplifiers
Summary
Key Terms
References
Problems

Chapter Goals

- Study nonideal operational amplifier behavior
- Demonstrate techniques used to analyze circuits containing nonideal op amps
- Determine the voltage gain, input resistance, and output resistance of general amplifier circuits
- Explore common-mode rejection limitations and the effect of common-mode input resistance
- Learn how to model dc errors including offset voltage, input bias current, and input offset current
- Explore limits imposed by power supply voltages and finite output current capability
- Model amplifier limitations due to limited bandwidth and slew rate of the op amp
- Perform SPICE simulation of nonideal op amp circuits
- Understand the topologies and characteristics of the series-shunt, shunt-shunt, series-series, and shunt-series feedback configurations
- Develop techniques for analysis of feedback amplifiers including the effects of circuit loading
- Understand the effects of feedback on frequency response and feedback amplifier stability
- Define phase and gain margins
- Learn to interpret feedback amplifier stability in terms of Nyquist and Bode plots
- Use SPICE ac and transfer function analyses to characterize feedback amplifiers
- Develop techniques to determine the loop-gain of closed-loop amplifiers using SPICE simulation or measurement

Chapter 10 explored the characteristics of circuits employing ideal operational amplifiers having infinite gain, zero input current, and zero output resistance. Real operational amplifiers, on the other hand, do not exhibit any of these

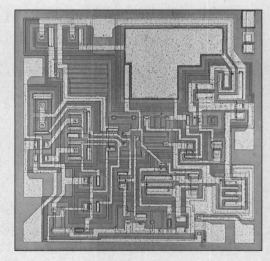

uA741 Die Photograph © *Fairchild Semiconductor International.*

ideal characteristics. In fact, they have a significant number of additional limitations as tabulated in the list below:

- Finite open loop gain
- Finite input resistance
- Nonzero output resistance
- Offset voltage
- Input bias and offset currents
- Limited output voltage range
- Limited output current capability
- Finite common-mode rejection
- Finite power supply rejection
- Limited bandwidth
- Limited slew rate

There are literally hundreds of commercial hybrid and integrated circuit operational amplifiers available to the engineer for use in circuit design. The only way to choose among this huge set of options is to fully understand the characteristics and limitations of real operational amplifiers. Thus, this chapter explores the impact of these limitations in detail and demonstrates the approaches used to analyze circuits employing nonideal op amps. Generally, we look at the effect of each of the nonideal characteristics independently, while assuming the others are still ideal. Then we can combine the results to understand how the circuits behave in general.

To better understand the impact of nonideal op amp characteristics on circuit performance, we will start by reviewing the classic theory of negative feedback in electronic systems that was first developed by Harold Black of the Bell Telephone System [1–3]. In 1928, he invented the feedback amplifier to stabilize the gain of early telephone repeaters. Today, some form of feedback is used in virtually every electronic system. This chapter formally develops the concept of feedback, which is an invaluable tool in the design of electronic circuits. Valuable insight into the operation of many common electronic circuits can be gained by recasting the circuits as feedback amplifiers.

Several of the advantages of negative feedback were actually uncovered during the discussion of ideal operational amplifier circuit design in Chapter 10. Generally, feedback can be used to achieve a trade-off between gain and many of the other properties of amplifiers:

- *Gain stability:* Feedback reduces the sensitivity of gain to variations in the values of transistor parameters and circuit elements.
- *Input and output impedances:* Feedback can increase or decrease the input and output resistances of an amplifier.
- *Bandwidth:* The bandwidth of an amplifier can be extended using feedback.
- *Nonlinear distortion:* Feedback reduces the effects of nonlinear distortion.

Feedback may also be **positive** (or **regenerative**), and we explore the use of positive feedback in sinusoidal **oscillator circuits** in Chapter 12. Positive feedback in amplifiers is usually undesirable. Excess phase shift in a feedback amplifier may cause the feedback to become regenerative and cause the feedback amplifier to break into oscillation, a situation that we must know how to avoid!

11.1 CLASSIC FEEDBACK SYSTEMS

Classic feedback systems are described by the block diagram in Fig. 11.1. This diagram may represent a simple feedback amplifier or a complex feedback control system. It consists of an amplifier with transfer function $A(s)$, referred to as the **open-loop amplifier**, a **feedback network** with transfer function $\beta(s)$, and a summing block indicated by Σ. The variables in this diagram are represented as voltages but could equally well be currents or even other physical quantities such as temperature, velocity, distance, and so on.

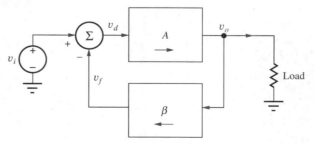

Figure 11.1 Classic block diagram for a feedback system.

11.1.1 CLOSED-LOOP GAIN ANALYSIS

In Fig. 11.1, the input to the open-loop amplifier A is provided by the summing block, that develops the difference between the input signal v_i and the feedback signal v_f:

$$\mathbf{v_d} = \mathbf{v_i} - \mathbf{v_f} \tag{11.1}$$

The output signal is equal to the product of the open-loop amplifier gain A and the input signal v_d:

$$\mathbf{v_o} = A\mathbf{v_d} \tag{11.2}$$

The signal fed back to the input is given by

$$\mathbf{v_f} = \beta \mathbf{v_o} \tag{11.3}$$

Combining Eqs. (11.1) to (11.3) and solving for the overall voltage gain of the system yields the classic expression for the **closed-loop gain** of a feedback amplifier:

$$A_v = \frac{\mathbf{v_o}}{\mathbf{v_i}} = \frac{A}{1 + A\beta} = \frac{1}{\beta}\left(\frac{A\beta}{1 + A\beta}\right) = A_v^{\text{Ideal}}\left(\frac{T}{1 + T}\right) \tag{11.4}$$

where A_v is the **closed-loop gain**, A is usually called the **open-loop gain**, and the product $T = A\beta$ is defined as the **loop gain** or **loop transmission**. A_v^{Ideal} is the **ideal gain** that would be achieved if the op amp were ideal $T = \infty$.

Remember in Chapter 10 that the linear amplifier circuits all assumed that the circuit was connected correctly with negative feedback. For the block diagram in Fig. 11.1, negative feedback requires $T > 0$, whereas $T < 0$ corresponds to a positive feedback condition. We will investigate some circuits called multivibrators that employ positive feedback in Chapter 12. We will explore the concepts of positive and negative feedback in more depth in this chapter and in Chapter 12.

A number of assumptions are implicit in this derivation. It is assumed that the blocks can be interconnected, as shown in Fig. 11.1, without affecting each other. That is, connecting the feedback network and the load to the output of the amplifier does not change the characteristics of the amplifier, nor does the interconnection of the summer, feedback network, and input of the open-loop amplifier modify the characteristics of either the amplifier or feedback network. In addition, it is tacitly assumed that signals flow only in the forward direction through the amplifier, and only in the reverse direction through the feedback network, as indicated by the arrows in Fig. 11.1.

Implementation of the block diagram in Fig. 11.1 with operational amplifiers having large input resistances, low output resistances, and essentially zero reverse-voltage gain is one method of satisfying these unstated assumptions. However, most general amplifiers and feedback networks do not necessarily satisfy these assumptions. In the next several sections we explore analysis and design of more general feedback systems that do not satisfy the implicit restrictions just outlined.

11.1.2 GAIN ERROR

In high precision applications it is important to know, or to control by design, just how far the actual gain in Eq. (11.4) deviates from the ideal value of the gain. The **gain error (GE)** is defined as the difference between the ideal gain and the actual gain:

$$\text{GE} = (\text{ideal gain}) - (\text{actual gain}) = A_v^{\text{Ideal}} - A_v = A_v^{\text{Ideal}}\left(1 - \frac{T}{1+T}\right) = \frac{A_v^{\text{Ideal}}}{1+T} \tag{11.5}$$

This error is more often expressed as a fractional error or percentage, and the **fractional gain error (FGE)** is defined as

$$\text{FGE} = \frac{(\text{ideal gain}) - (\text{actual gain})}{(\text{ideal gain})} = \frac{A_v^{\text{Ideal}} - A_v}{A_v^{\text{Ideal}}} = \frac{1}{1+T} \cong \frac{1}{T} \quad \text{for } T \gg 1 \tag{11.6}$$

For $T \gg 1$, we see that the value of FGE is determined by the reciprocal of the loop gain.

DESIGN NOTE

If the maximum fractional gain error is given as a design specification, then the value of the FGE places a lower bound on the value of the loop gain.

11.2 ANALYSIS OF CIRCUITS CONTAINING NONIDEAL OPERATIONAL AMPLIFIERS

In Chapter 10 we always assumed that the open-loop gain A of the op amp was infinite, which simplified the circuit analysis. If we take the limit of the expression in Eq. (11.4) as A, and therefore T, approach infinity

$$\lim_{A \to \infty} A_v = \lim_{T \to \infty} A_v^{\text{Ideal}} \left(\frac{T}{1+T} \right) = A_v^{\text{Ideal}} \qquad (11.7)$$

we see that the closed-loop voltage gain equals the ideal gain and is independent of the characteristics of the op amp! This independence is one goal of feedback amplifier design. From our work in Chapter 10 and Eq. (11.4), we recognize that $A_v^{\text{Ideal}} = 1/\beta$ represents the gain of the noninverting amplifier circuit employing an ideal amplifier.

In the next several sections, we remove various ideal assumptions as we explore the effects of finite open-loop gain, finite input resistance, and nonzero output resistance on the overall characteristics of the inverting and noninverting amplifiers that were introduced in Chapter 10, and see how close we can come to achieving our ideal goals.

11.2.1 FINITE OPEN-LOOP GAIN

A real operational amplifier provides a large but noninfinite gain. Op amps are commercially available with minimum open-loop gains of 80 dB (10,000) to over 120 dB (1,000,000). The finite open-loop gain contributes to deviations of the closed-loop gain, input resistance, and output resistance from those presented for the ideal amplifiers in Chapter 10.

Noninverting Amplifier

Evaluation of the closed-loop gain for the noninverting amplifier of Fig. 11.2 provides our first example of amplifier calculations involving nonideal amplifiers. In Fig. 11.2, the output voltage of the amplifier is given by

$$\mathbf{v_o} = A\mathbf{v_{id}} \quad \text{where} \quad \mathbf{v_{id}} = \mathbf{v_i} - \mathbf{v_1} \qquad (11.8)$$

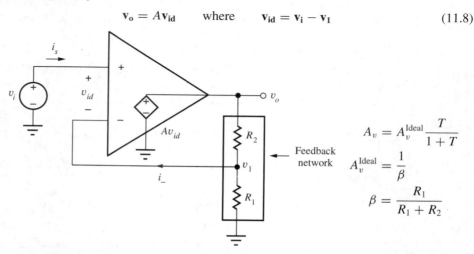

Figure 11.2 Operational amplifier with finite open-loop gain A.

Because $i_- = 0$ by ideal op-amp Assumption 2 (see Eq. (10.36)), v_1 is set by the voltage divider formed by resistors R_1 and R_2:

$$\mathbf{v_1} = \frac{R_1}{R_1 + R_2}\mathbf{v_o} = \beta\mathbf{v_o} \quad \text{where} \quad \beta = \frac{R_1}{R_1 + R_2} \quad (11.9)$$

The parameter β is called the **feedback factor** and represents the fraction of the output voltage that is fed back to the input from the output. Combining the last two equations gives

$$\mathbf{v_o} = A(\mathbf{v_i} - \beta\mathbf{v_o}) \quad (11.10)$$

and solving for v_o yields the classic **feedback amplifier** voltage-gain formula:

$$A_v = \frac{\mathbf{v_o}}{\mathbf{v_i}} = \frac{A}{1 + A\beta} = \frac{1}{\beta}\left(\frac{A\beta}{1 + A\beta}\right) = A_v^{\text{Ideal}}\left(\frac{T}{1+T}\right) \quad (11.11)$$

The product $A\beta$ is called the **loop gain** (or **loop transmission T**) and plays an important role in feedback amplifiers. For $T \gg 1$, A_v approaches the ideal gain expression found previously:

$$A_v^{\text{Ideal}} = \frac{1}{\beta} = 1 + \frac{R_2}{R_1} \quad (11.12)$$

The voltage v_{id} across the op amp input is given by

$$\mathbf{v_{id}} = \frac{\mathbf{v_o}}{A} = \frac{1}{A}\left(\frac{A}{1 + A\beta}\mathbf{v_i}\right) = \frac{\mathbf{v_i}}{1 + T} \quad (11.13)$$

Although v_{id} is no longer zero, it is small for large values of the loop gain T. Thus, when we apply an input voltage v_i, only a small portion of it appears across the input terminals.

We see we have the same result as in the "classic" feedback circuit in Fig. 11.1. So how does this circuit meet the implicit conditions mentioned in Sec. 11.1.1? An ideal op amp has zero output resistance, so attaching the resistive voltage divider between the output and ground does not affect the gain of the op amp, and the inverting input of the ideal op amp represents an infinite resistance, so connecting it to the feedback network does not change the value of feedback factor β. In addition, the op amp is unilateral only passing signals in the forward direction. There is also no signal injection from the input into the feedback network, so there is no reverse signal transmission through the feedback network to the output.

Inverting Amplifier

Evaluation of the closed-loop gain of the inverting amplifier in Fig. 11.3 is similar to that of the noninverting amplifier but yields a slightly different form of answer. In this case, the output voltage is

$$\mathbf{v_o} = A\mathbf{v_{id}} = -A\mathbf{v_-} \quad (11.14)$$

and the voltage at the inverting input terminal can be found using superposition:

$$\text{For } v_o = 0, \text{ then } \mathbf{v_-} = \mathbf{v_i}\frac{R_2}{R_1 + R_2} \quad \text{and for} \quad v_i = 0, \text{ then } \mathbf{v_-} = \mathbf{v_o}\frac{R_1}{R_1 + R_2} \quad (11.15)$$

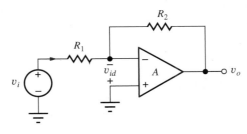

Figure 11.3 Inverting amplifier circuit.

Combining these results yields

$$\mathbf{v}_- = \mathbf{v}_i \frac{R_2}{R_1 + R_2} + \mathbf{v}_o \frac{R_1}{R_1 + R_2} \qquad (11.16)$$

After some additional algebra, the closed-loop gain can be written as

$$A_v = \frac{\mathbf{v}_o}{\mathbf{v}_i} = -\frac{R_2}{R_1}\left(\frac{A\beta}{1+A\beta}\right) = A_v^{\text{Ideal}}\left(\frac{T}{1+T}\right) \qquad (11.17)$$

where $A_v^{\text{Ideal}} = -R_2/R_1$, $T = A\beta$ and $\beta = R_1/(R_1 + R_2)$. First note that the expression for the feedback factor β, which represents the portion of the output voltage that is fed back to the input, is the same as we found for the noninverting amplifier, that is, β is independent of the configuration! In addition, as the loop-gain approaches infinity, we find that the ideal gain is again the same as we calculated in Chapter 10:

$$A_v^{\text{Ideal}} = \lim_{A \to \infty} A_v = \lim_{T \to \infty}\left(-\frac{R_2}{R_1}\right)\left(\frac{T}{1+T}\right) = -\frac{R_2}{R_1} \qquad (11.18)$$

The residual voltage across the op amp input terminals is

$$\mathbf{v}_{id} = \frac{\mathbf{v}_o}{A} = \frac{1}{A}\left(-\frac{R_2}{R_1}\frac{A\beta}{1+A\beta}\mathbf{v}_i\right) = -\frac{R_2}{R_1}\frac{\beta}{1+A\beta}\mathbf{v}_i \cong -\frac{R_2}{R_1}\frac{\mathbf{v}_i}{A} \qquad (11.19)$$

in which the approximation holds for large loop-gain T. Once again, only a very small portion of input signal v_i appears across the op amp input terminals.

> **EXERCISE:** Suppose $A = 10^5$, $\beta = 1/100$, and $v_i = 100$ mV. What are A_v^{Ideal}, T, A_v, v_o, and v_{id} for the noninverting amplifier.
>
> **ANSWERS:** 100, 99.9, 10.0 V, 100 μV (v_{id} is small but nonzero)
>
> **EXERCISE:** Repeat the previous exercise for the inverting amplifier.
>
> **ANSWERS:** −99, 1000, −98.9, −9.89 V, −98.9 μV
>
> **EXERCISE:** What are the nominal, minimum, and maximum values of the open-loop gain at 25°C for an OP-27 operational amplifier?
>
> **ANSWERS:** With 15-V supplies: 1,000,000; 1,800,000; no maximum value specified
>
> **EXERCISE:** What value of open-loop gain is guaranteed for the OP-27 op amp over the full temperature range with a load resistance of at least 2 kΩ?
>
> **ANSWER:** 600,000 with 15-V power supplies

EXAMPLE 11.1 GAIN ERROR ANALYSIS

Characterize the gain and gain error of a noninverting amplifier implemented with a finite gain operation amplifier.

PROBLEM A noninverting amplifier is designed to have a gain of 200 (46 dB) and is built using an operational amplifier with an open-loop gain of 80 dB. Find the values of the ideal gain, the actual gain, and the gain error. Express the gain error in percent.

SOLUTION **Known Information and Given Data:** Design a noninverting amplifier circuit with closed-loop gain of 46 dB. The open-loop gain of the op amp is 10,000 (80 dB).

Unknowns: Values of the ideal gain, the actual gain, and the gain error in percent

Approach: First, we need to clarify the meaning of some terminology. We normally design an amplifier to produce a given value of ideal gain, and then determine the deviations to be expected from the ideal case. So, when it is said that this amplifier is designed to have a gain of 200, we set $\beta = 1/200$. We do not normally try to adjust the design values of R_1 and R_2 to try to compensate for the finite open-loop gain of the amplifier. One reason is that we do not know the exact value of the gain A but generally only know its lower bound. Also, the resistors we use have tolerances, and their exact values are also unknown.

Assumptions: The op amp is ideal except for its finite open-loop gain.

Analysis: The ideal gain of the circuit is 200, so $\beta = 1/200$ and $T = 10^4/200 = 50$. The actual gain and FGE are given by

$$A_v = A_v^{\text{Ideal}} \frac{T}{1+T} = 200\left(\frac{50}{51}\right) = 196 \quad \text{and} \quad \text{FGE} = \frac{200 - 196}{200} = 0.02 \text{ or } 2\%$$

Check of Results: The three unknown values have been found. The value of A_v^{Ideal} is slightly less than A_v and therefore appears to be a reasonable result.

Discussion: The actual gain is 196, representing a 2 percent error from the ideal design gain of 200. Note that this gain error expression does not include the effects of resistor tolerances, which are an additional source of gain error in an actual circuit. If the gain must be more precise, a higher-gain op amp must be used, or the resistors can be replaced by a potentiometer so the gain can be adjusted manually. But note that the gain will still change with temperature since all the op amp parameters are temperature dependent.

Computer Aided Analysis: The circuit in Ex. 10.5 can be used to check the results of this example by setting R1 = 1 kΩ, R2 = 199 kΩ, and the gain of E1 to 10,000. A transfer function analysis gives $A_v = 196$, in agreement with our hand calculations.

EXERCISE: A noninverting amplifier is designed with $R_1 = 1$ kΩ, $R_2 = 39$ kΩ, and an op amp with an open-loop gain of 80 dB. What are the loop gain, closed-loop gain, ideal gain and fractional gain error of the amplifier?

ANSWERS: 250, +39.8, +40.0, 0.4 percent

EXERCISE: Repeat the previous exercise for the inverting amplifier.

ANSWERS: −38.8, −39.0, 0.398 percent

11.2.2 NONZERO OUTPUT RESISTANCE

The next effect we explore is the influence of a nonzero output resistance on the characteristics of the inverting and noninverting closed-loop amplifiers. In this case, we assume that the op amp has a nonzero output resistance R_o as well as a finite open-loop gain A. (As we shall see, finite gain must also be assumed; otherwise, we would get the same output resistance as for the ideal case.)

To determine the (Thévenin equivalent) output resistances of the two amplifiers in Fig. 11.4, each output terminal is driven with a test signal source v_x (a current source could also be used), and

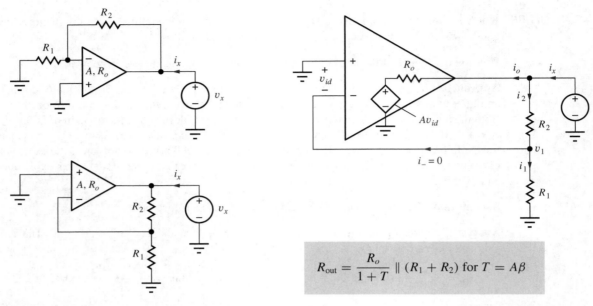

Figure 11.4 Circuits for determining output resistances of the inverting and noninverting amplifiers.

Figure 11.5 Circuit explicitly showing amplifier with A and R_o.

$$R_{\text{out}} = \frac{R_o}{1+T} \parallel (R_1 + R_2) \text{ for } T = A\beta$$

the current i_x is calculated; all other independent sources in the network must be turned off. The output resistance is then given by

$$R_{\text{out}} = \frac{\mathbf{v_x}}{\mathbf{i_x}} \tag{11.20}$$

By studying Fig. 11.4 we observe that the two amplifier circuits are identical for the output resistance calculation. Thus, analysis of the circuit in Fig. 11.5 gives the expression for R_{out} for both the inverting and noninverting amplifiers.

Analysis begins by expressing currents i_x and i_o as

$$\mathbf{i_x} = \mathbf{i_o} + \mathbf{i_2} \quad \text{and} \quad \mathbf{i_o} = \frac{\mathbf{v_x} - A\mathbf{v_{id}}}{R_o} \tag{11.21}$$

Current i_2 can be found from

$$\mathbf{v_x} = \mathbf{i_2} R_2 + \mathbf{i_1} R_1 \quad \text{or} \quad \mathbf{i_2} = \frac{\mathbf{v_x}}{R_1 + R_2} \tag{11.22}$$

because $i_1 = i_2$ due to op amp Assumption 2: $i_- = 0$. The input voltage v_{id} is equal to $-v_1$, and because $i_- = 0$,

$$\mathbf{v_1} = \frac{R_1}{R_1 + R_2} \mathbf{v_x} = \beta \mathbf{v_x} \tag{11.23}$$

Combining Eqs. (11.21) through (11.23) yields

$$\frac{1}{R_{\text{out}}} = \frac{\mathbf{i_x}}{\mathbf{v_x}} = \frac{1 + A\beta}{R_o} + \frac{1}{R_1 + R_2} \tag{11.24}$$

Equation (11.24) represents the output conductance of the amplifier and corresponds to the sum of the conductances of two parallel resistors. Thus, the output resistance can be expressed as

$$R_{\text{out}} = \frac{R_o}{1+T} \parallel (R_1 + R_2) \tag{11.25}$$

The output resistance in Eq. (11.25) represents the series combination of R_1 and R_2 in parallel with a resistance $R_o/(1 + A\beta)$ that represents the output resistance of the operational amplifier including the effects of feedback. In almost every practical situation, the value of $R_o/(1 + A\beta)$ is much less than that of $(R_1 + R_2)$, and the output resistance expression in Eq. (11.25) simplifies to

$$R_{\text{out}} \cong \frac{R_o}{1 + A\beta} = \frac{R_o}{1 + T} \qquad (11.26)$$

An example of the degree of dominance of the resistance term in Eq. (11.26) is given in Ex. 11.2.

Note that the output resistance would be zero if A were assumed to be infinite in Eq. (11.25) or (11.26). This is the reason why the analysis must simultaneously account for both finite A and nonzero R_o.

EXERCISE: What are the nominal, minimum, and maximum values of the open-loop gain and output resistance for an OP-77E operational amplifier (see MCD website)?

ANSWERS: 12,000,000 (142 dB); 5,000,000 (134 dB); no maximum value specified; 60 Ω; no minimum or maximum value specified.

EXAMPLE 11.2 OP AMP OUTPUT RESISTANCE

Perform a numeric calculation of the output resistance of a noninverting amplifier implemented using an op amp with a finite open-loop gain and nonzero output resistance.

PROBLEM A noninverting amplifier is constructed with $R_1 = 1$ kΩ and $R_2 = 39$ kΩ using an operational amplifier with an open-loop gain of 80 dB and an output resistance of 50 Ω. Find the output resistance of the noninverting amplifier.

SOLUTION **Known Information and Given Data:** Noninverting op amp amplifier circuit with $R_1 = 1$ kΩ, $R_2 = 39$ kΩ, $A = 10{,}000$, and $R_o = 50$ Ω.

Unknowns: Output resistance of the overall amplifier

Approach: Use known values to evaluate Eq. (11.25)

Assumptions: The op amp is ideal except for finite gain and nonzero output resistance.

Analysis: Evaluating Eq. (11.25):

$$1 + T = 1 + A\frac{R_1}{R_1 + R_2} = 1 + 10^4 \frac{1 \text{ k}\Omega}{1 \text{ k}\Omega + 39 \text{ k}\Omega} = 251$$

and

$$R_{\text{out}} = \frac{50 \text{ }\Omega}{251} \| (40 \text{ k}\Omega) = 0.199 \text{ }\Omega \| 40 \text{ k}\Omega = 0.198 \text{ }\Omega$$

Check of Results: The value of output resistance has been calculated. The value is much smaller than the value of R_o, as expected.

Evaluation and Discussion: We see that the effect of the feedback in the circuits in Fig. 11.4 is to reduce the output resistance of the closed-loop amplifier far below that of the individual op amp itself. In fact, the output resistance is quite small and represents a good practical approximation to that of an ideal amplifier ($R_\text{out} = 0$). This is a characteristic of shunt feedback at the output port, in which the feedback network is in parallel with the port. **Shunt feedback** tends to lower the resistance at a port, whereas feedback in series with a port, termed **series feedback**, tends to raise the resistance at that port. The properties of series and shunt feedback are explored in greater detail later in this chapter.

Computer-Aided Analysis: The output resistance of the noninverting amplifier can be simulated by adding the output resistance RO to the circuit of Ex. 10.5 as indicated in the figure. The gain of the OP AMP is set to 10,000. A transfer function analysis from VI to output node v_o gives a gain of $+39.8$ and an output resistance of $0.199\ \Omega$.

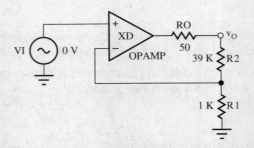

EXERCISE: What value of open-loop gain is required to achieve an output resistance of 0.1 Ω in the amplifier in Ex. 11.2?

ANSWER: 20,000

EXERCISE: Calculate the value of closed-loop gain in Ex. 11.2 and verify that the simulation result is correct.

EXERCISE: Suppose the resistors in Ex. 11.2 both have 5 percent tolerances. What are the worst-case (highest and lowest) values of gain that can be expected if the open-loop gain were infinite? What is the gain error for each of these two cases?

ANSWERS: 44.1, 36.2, 4.20 (10.5%), −3.70(−9.3%)

DESIGN EXAMPLE 11.3   OPEN-LOOP GAIN DESIGN

In this example, we find the value of open-loop gain required to meet an amplifier output resistance specification.

PROBLEM Design a noninverting amplifier to have a gain of 35 dB and an output resistance of no more than 0.2 Ω. The only op amp available has an output resistance of 250 Ω. What is the minimum open-loop gain of the op amp that will meet the design requirements?

SOLUTION **Known Information and Given Data:** For the noninverting amplifier: ideal gain = 35 dB, closed-loop output resistance = 0.2 Ω. For the operational amplifier used to realize the noninverting amplifier: open-loop output resistance = 250 Ω.

Unknowns: Open-loop gain A of the op amp

Approach: The required value of operational amplifier gain can be found using Eq. (11.26), in which all the variables are known except A.

Assumptions: The operational amplifier is ideal except for finite open-loop gain and nonzero output resistance.

Analysis: The closed-loop output resistance is given by

$$R_{\text{out}} = \frac{R_o}{1 + A\beta} \leq 0.2\ \Omega$$

R_o and R_{out} are given, and β is determined by the desired gain:

$$R_o = 250\ \Omega \qquad R_{\text{out}} = 0.2\ \Omega \qquad \beta = \frac{1}{|A_v|}$$

We must convert the gain from dB before we use it in the calculations:

$$|A_v| = 10^{35\,\text{dB}/20\,\text{dB}} = 56.2 \quad \text{and} \quad \beta = \frac{1}{|A_v|} = \frac{1}{56.2}$$

The minimum value of the open-loop gain A can now be determined from the R_{out} specification:

$$A \geq \frac{1}{\beta}\left(\frac{R_o}{R_{\text{out}}} - 1\right) = 56.2\left(\frac{250}{0.2} - 1\right) = 7.03 \times 10^4$$

$$A_{\text{dB}} = 20\log(7.03 \times 10^4) = 96.9\ \text{dB}$$

Check of Results: We have found the required unknown value.

Discussion: By exploring the world wide web, we see that op amps are available with 100 dB gain. So the value required by our design is achievable.

Computer-Aided Analysis: If we change the parameter values in the circuit in Ex. 11.2 and rerun the simulation, we will see if we meet the output resistance specification. Using R1 = 10 kΩ, R2 = 552 kΩ, and RO = 250 Ω with the OP AMP gain set to 7.03E4, the SPICE transfer function analysis yields $A_v = 56.2$ and $R_{\text{out}} = 0.200\ \Omega$. The values of R1 and R2 were deliberately chosen to be large so that they would not materially affect the output resistance. We see from the voltage gain result that we have chosen the correct value for the ratio R2/R1.

EXERCISE: A noninverting amplifier must have a closed-loop gain of 40 dB and an output resistance of less than 0.1 Ω. The only op amp available has an output resistance of 200 Ω. What is the minimum open-loop gain of the op amp that will meet the requirements?

ANSWER: 106 dB

11.2.3 FINITE INPUT RESISTANCE

Next we explore the effect of the finite input resistance of the operational amplifier on the open-loop input resistances of the noninverting and inverting amplifier configurations. In this case, we shall find that the results are greatly different for the two amplifiers.

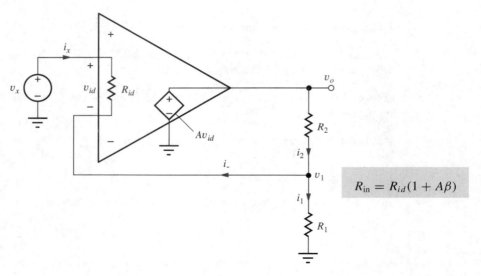

Figure 11.6 Input resistance of the noninverting amplifier.

Input Resistance for the Noninverting Amplifier

Let us first consider the noninverting amplifier circuit in Fig. 11.6, in which test source v_x is applied to the input and with input resistance R_{id} added across the input terminals of the op amp. Here again we must assume finite gain A in order to get a non trivial result. (Infinite gain would force zero volts across R_{id} so there would be no current in v_x.) To find R_{in}, we must calculate the current i_x give by

$$i_x = \frac{v_x - v_1}{R_{id}} \tag{11.27}$$

Voltage v_1 is equal to

$$v_1 = i_1 R_1 = (i_2 - i_-) R_1 \cong i_2 R_1 \tag{11.28}$$

which has been simplified by assuming that the input current i_- to the op amp can still be neglected with respect to i_2. We will check this assumption shortly. The assumption is equivalent to saying that $i_1 \cong i_2$ and permits the voltage v_1 to again be written in terms of the resistive voltage divider as

$$v_1 \cong \frac{R_1}{R_1 + R_2} v_o = \beta v_o = \beta(A v_{id}) = A\beta(v_x - v_1) \tag{11.29}$$

Solving for v_1 in terms of v_x yields

$$v_1 = \frac{A\beta}{1 + A\beta} v_x \tag{11.30}$$

and substituting this result into Eq. (11.27) yields an expression for R_{in}

$$i_x = \frac{v_x - \frac{A\beta}{1+A\beta} v_x}{R_{id}} = \frac{v_x}{(1+A\beta)R_{id}} \quad \text{and} \quad R_{in} = R_{id}(1+A\beta) = R_{id}(1+T) \tag{11.31}$$

Note from Eq. (11.31) that the input resistance can be very large—much larger than that of the op amp itself. R_{id} is often large (1 MΩ to 1 TΩ) to start with, and it is multiplied by the loop gain T, which is typically designed to be much greater than 1. If the loop-gain approaches infinity in Eq. (11.31) the input resistance also approaches its ideal value of infinity. Although the actual value of R_{in} cannot reach infinity in real circuits, it can be extremely large in value. This large input resistance occurs since only a small portion of the applied voltage v_x actually appears across R_{id} (see Eq. (11.13)). Thus the input current is very small, and the overall input resistance becomes very high.

11.2 Analysis of Circuits Containing Nonideal Operational Amplifiers

EXERCISE: What are the nominal, minimum, and maximum values of the open-loop gain and input resistance for an AD745 operational amplifier (see MCD website)? Repeat for the input resistance of the OP-27.

ANSWERS: 132 dB; 120 dB; no maximum value specified; 10^{10} Ω; minimum and maximum values not specified; 6 MΩ; 1.3 MΩ; no maximum specified.

EXAMPLE 11.4 NONINVERTING AMPLIFIER INPUT RESISTANCE

Find a numeric value for the input resistance of a noninverting feedback amplifier circuit.

PROBLEM The noninverting amplifier in Fig. 11.6 is built with an op amp having an input resistance of 2 MΩ and an open-loop gain of 90 dB. What is the amplifier input resistance if $R_1 = 20$ kΩ and $R_2 = 510$ kΩ?

SOLUTION **Known Information and Given Data:** A noninverting feedback amplifier circuit is built with feedback resistors $R_1 = 20$ kΩ and $R_2 = 510$ kΩ. For the op amp: $A = 90$ dB, $R_{id} = 2$ MΩ

Unknown: Closed-loop amplifier input resistance R_{in}

Approach: In this case, we are given the values necessary to directly evaluate Eq. (11.31) including A, R_{id}, and the two feedback resistors.

Assumptions: The operational amplifier is ideal except for finite open-loop gain and finite input resistance.

Analysis: In order to evaluate Eq. (11.31) we must find β, which is determined by the feedback resistors, since $R_{id} \gg R_1$:

$$\beta = \frac{R_1}{R_1 + R_2} = \frac{20 \text{ k}\Omega}{20 \text{ k}\Omega + 510 \text{ k}\Omega} = \frac{1}{26.5}$$

We must also convert the gain from dB before we use it in the calculations:

$$R_{id} = 2 \text{ M}\Omega \quad \text{and} \quad A = 10^{90 \text{ dB}/20 \text{ dB}} = 31{,}600$$

The closed-loop input resistance is given by

$$R_{\text{in}} = R_{id}(1 + A\beta) = 2 \text{ M}\Omega \left[1 + \frac{31{,}600}{26.5} \right] = 2.39 \times 10^9 \ \Omega = 2.39 \text{ G}\Omega$$

Check of Results: We have found the only unknown value. The value is large as expected from our analysis of the noninverting amplifier.

Discussion: The calculated input resistance of the noninverting amplifier is very large (although not infinite as for case of an ideal op amp). In fact, the calculated value of R_{in} is so large that we must consider other factors that may limit the actual input resistance. These include surface leakage of the printed circuit board in which the op amp is mounted as well as common-mode input resistance of the op amp itself, which we discuss in Sec. 11.12.4.

Computer-Aided Analysis: To check our result with SPICE, we add RID = 2 MEG to the noninverting amplifier circuit model as shown below (with the OP AMP gain set to 31,600) and perform a transfer function analysis. The results are $A_v = 26.5$ and $R_{\text{in}} = 2.39$ GΩ confirming our hand calculations.

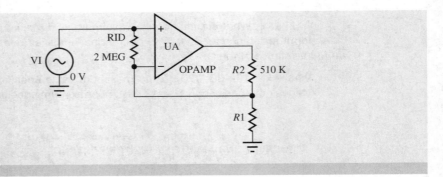

For the numbers in the preceding exercise, it is easy to see that the current i_-, which equals $-i_x$, is small compared to the current through R_2 and R_1 (see Prob. 11.16). Thus, our simplifying assumption that led to Eqs. (11.28) and (11.29) is well justified.

> **EXERCISE:** Suppose a noninverting amplifier has $R_{id} = 1$ MΩ, with $R_1 = 10$ kΩ, $R_2 = 390$ kΩ, and the open-loop gain is 80 dB. What is the input resistance of the overall amplifier? What are the currents I_- and I_1 for a dc input voltage $V_I = 1$ V? Is $I_- \ll I_1$?
>
> **ANSWERS:** 251 MΩ; −3.98 nA, 99.5 µA; yes

Input Resistance for the Inverting-Amplifier Configuration

The input resistance of the inverting amplifier can be determined using the circuit in Fig. 11.7(a) and is defined by

$$R_{\text{in}} = \frac{v_x}{i_x} \tag{11.32}$$

Test signal v_x can be expressed as

$$v_x = i_x R_1 + v_- \quad \text{and} \quad R_{\text{in}} = R_1 + \frac{v_-}{i_x} \tag{11.33}$$

The total input resistance R_{in} is equal to R_1 plus the resistance looking into the inverting terminal of the operational amplifier, which can be found using the circuit in Fig. 11.7(b). The input current in Fig. 11.7(b) is

$$i_1 = i_- + i_2 = \frac{v_1}{R_{id}} + \frac{v_1 - v_o}{R_2} = \frac{v_1}{R_{id}} + \frac{v_1 + A v_1}{R_2} \tag{11.34}$$

Using this result, the input conductance can be written as

$$G_1 = \frac{i_1}{v_1} = \frac{1}{R_{id}} + \frac{1 + A}{R_2} \tag{11.35}$$

which represents the sum of two conductances. Thus, the equivalent resistance looking into the inverting-input terminal is the parallel combination of two resistors,

$$R_{id} \parallel \left(\frac{R_2}{1 + A} \right) \tag{11.36}$$

and the overall input resistance of the inverting amplifier becomes

$$R_{\text{in}} = R_1 + R_{id} \parallel \left(\frac{R_2}{1 + A} \right) \tag{11.37}$$

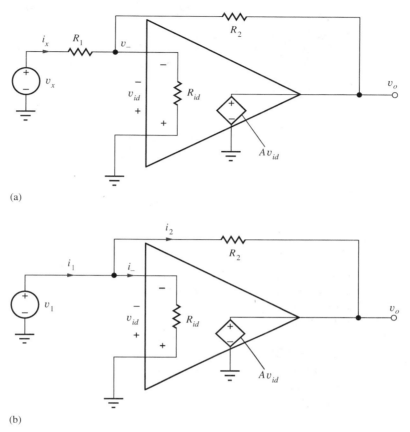

Figure 11.7 Inverting amplifier input-resistance calculation: (a) complete amplifier; (b) amplifier with R_1 removed.

Normally, R_{id} will be large and Eq. (11.37) can be approximated by

$$R_{in} \cong R_1 + \left(\frac{R_2}{1+A}\right) \tag{11.38}$$

For large A and common values of R_2, the input resistance approaches the ideal result $R_{in} \cong R_1$. In other words, we see that the input resistance is usually dominated by R_1 connected to the quasi virtual ground at the op amp input. (Remember, v_{id} is no longer exactly zero for a finite-gain amplifier.)

EXERCISE: Find the input resistance R_{in} of an inverting amplifier that has $R_1 = 1$ kΩ, $R_2 = 100$ kΩ, $R_{id} = 1$ MΩ, and $A = 100$ dB. What is the deviation of R_{in} from its ideal value?

ANSWERS: 1001 Ω; 1 Ω out of 1000 Ω or 0.1 percent

11.2.4 SUMMARY OF NONIDEAL INVERTING AND NONINVERTING AMPLIFIERS

Table 11.1 is a summary of the simplified expressions for the closed-loop voltage gain, input resistance, and output resistance of the inverting and noninverting amplifiers. These equations are most often used in the design of these basic amplifier circuits.

Op amp circuits are usually designed with large loop-gain $T = A\beta$; thus the simplified expressions in Table 11.1 normally apply. Except for very high precision circuits, the gain error caused by finite gain will be negligible, and resistor tolerances are much more likely to be the dominant source of gain error. Large values of T ensure low values of output resistance, although the output resistances do depend upon the value of T. The input resistance of the inverting amplifier is approximately equal to R_1, whereas that of the noninverting amplifier is large, but a direct function of T.

TABLE 11.1
Inverting and Noninverting Amplifier Summary

$\beta = \dfrac{R_1}{R_1 + R_2} \quad T = A\beta$	INVERTING AMPLIFIER	NONINVERTING AMPLIFIER
Voltage gain A_v	$-\dfrac{R_2}{R_1}\left(\dfrac{T}{1+T}\right) \cong -\dfrac{R_2}{R_1}$	$\left(1+\dfrac{R_2}{R_1}\right)\left(\dfrac{T}{1+T}\right) \cong 1+\dfrac{R_2}{R_1}$
Input resistance R_{in}	$R_1 + \left(R_{id} \Big\| \dfrac{R_2}{1+A}\right) \cong R_1$	$R_{id}(1+T)$
Output resistance R_{out}	$\dfrac{R_o}{1+T}$	$\dfrac{R_o}{1+T}$
Fractional gain error (FGE)	$\dfrac{1}{1+T}$	$\dfrac{1}{1+T}$

11.3 SERIES AND SHUNT FEEDBACK CIRCUITS

The properties of the feedback amplifier properties summarized earlier in Table 11.1 are characteristics of so-called **series feedback** and **shunt feedback**. When applied to an amplifier port, series feedback generally increases the impedance level. In contrast, shunt feedback decreases the impedance level at the amplifier port.

11.3.1 FEEDBACK AMPLIFIER CATEGORIES

The two types of feedback yield (2×2) or four possible circuit combinations of series and shunt feedback. These circuits are depicted in Fig. 11.8 and characterized in Table 11.2. Each type of feedback will be discussed in detail in the next several sections.

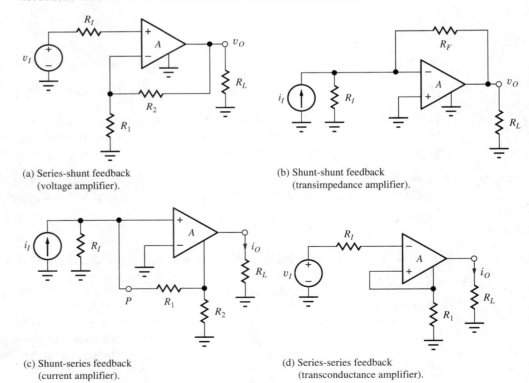

(a) Series-shunt feedback (voltage amplifier).

(b) Shunt-shunt feedback (transimpedance amplifier).

(c) Shunt-series feedback (current amplifier).

(d) Series-series feedback (transconductance amplifier).

Figure 11.8 Four types of feedback amplifiers.

TABLE 11.2
Feedback Amplifier Categories

FEEDBACK TYPE INPUT–OUTPUT	AMPLIFIER TYPE AND GAIN DEFINITION
Series-shunt	Voltage amplifier: $A_v = \dfrac{v_o}{v_i}$
Shunt-shunt	Transresistance amplifier: $A_{tr} = \dfrac{v_o}{i_i}$
Shunt-series	Current amplifier: $A_i = \dfrac{i_o}{i_i}$
Series-series	Transconductance amplifier: $A_{tc} = \dfrac{i_o}{v_i}$

11.3.2 VOLTAGE AMPLIFIERS—SERIES-SHUNT FEEDBACK

A voltage amplifier should have a high input resistance to measure the desired voltage and a low output resistance to drive the external load. These requirements correspond to the series-shunt feedback circuit shown in Fig. 11.8(a). To achieve the desired behavior, the input ports of the amplifier and feedback network are connected in series, and the output ports are connected in parallel (shunt).

11.3.3 TRANSIMPEDANCE AMPLIFIERS—SHUNT-SHUNT FEEDBACK

A transimpedance amplifier converts an input current to an output voltage. Thus it should have a low input resistance to sink the desired current and a low output resistance to drive the external load. These requirements correspond to the shunt-shunt feedback circuit shown in Fig. 11.8(b). To achieve the desired behavior, the input ports of the amplifier and feedback network are connected in parallel, and the output ports are connected in parallel.

11.3.4 CURRENT AMPLIFIERS—SHUNT-SERIES FEEDBACK

A current amplifier should provide a low resistance current sink at the input and a high resistance current source at its output. These attributes correspond to the shunt-series feedback as depicted in Fig. 11.8(c). The input ports of the amplifier and feedback network are connected in parallel, and the output ports are connected in series.

11.3.5 TRANSCONDUCTANCE AMPLIFIERS—SERIES-SERIES FEEDBACK

The last feedback configuration is the transconductance amplifier that converts an input voltage to an output current. This amplifier should have a high input resistance and a high output resistance and thus corresponds to the series-series feedback circuit in Fig. 11.8(d) in which both the input ports and the output ports of the amplifier and feedback networks are connected in series.

11.4 UNIFIED APPROACH TO FEEDBACK AMPLIFIER GAIN CALCULATION

We have already found that loop-gain T plays a very important role in determining the overall gain, input resistance, and output resistance of feedback amplifiers. We shall see shortly that T also determines feedback amplifier stability. Because of its importance, we need to understand how to model general feedback amplifiers and to calculate the loop gain directly from the circuit, not only theoretically, but also computationally using SPICE and experimentally based upon actual measurements. For the rest of this chapter, we will adopt a unified method for calculating the gain and terminal resistances associated with feedback amplifiers.

11.4.1 CLOSED-LOOP GAIN ANALYSIS

The closed-loop gain of all four feedback configurations in Fig. 11.8 can be written in the single form that we developed earlier in Sec. 11.2:

$$A_x = A_x^{\text{Ideal}}\left(\frac{T}{1+T}\right) \tag{11.39}$$

The ideal gain A_x^{Ideal} of each amplifier is set by its individual feedback network, and T is the amplifier's loop gain. In the next several sections, we will calculate T including the loading effects of R_{id}, R_o, R_I, R_L, and the feedback network. These loading effects were neglected in our earlier analysis, but they can be important in many real circuit cases.

11.4.2 RESISTANCE CALCULATIONS USING BLACKMAN'S THEOREM

R. B. Blackman was one of a group of individuals who first investigated the properties of feedback amplifiers at Bell Laboratories in the 1930s and 1940s [4], and Blackman's theorem provides a unified way to calculate impedances in feedback circuits. His highly useful result is stated in Eq. (11.40) and provides us with an alternate approach for calculating the input and output resistance [4] of a feedback amplifier.

$$R_X = R_X^D \frac{1+|T_{SC}|}{1+|T_{OC}|} \tag{11.40}$$

In this equation, R_X is the resistance of the closed-loop feedback amplifier looking into one of its ports (any terminal pair), R_X^D is the resistance looking into the same pair of terminals with the feedback loop disabled, T_{SC} is the loop-gain with a short circuit applied to the selected port, and T_{OC} is the loop gain with the same port open-circuited.

In order to apply Blackman's theorem, first we select the terminals where we desire to find the resistance. For example, we often wish to find the input resistance or the output resistance of a closed-loop feedback amplifier, and the resistance appears between one of the amplifier terminals and ground. Next, we select one of the controlled sources in the equivalent circuit of the amplifier. We use this source to disable the feedback loop, and the source is also used as the reference source for finding the two loop gains T_{SC} and T_{OC}. Resistance R_X^D represents the driving-point resistance at the port of interest calculated with the gain of the controlled source set to zero, whereas T_{SC} and T_{OC} are calculated with the port short-circuited and open-circuited, respectively. This procedure is best understood with the aid of several examples in the following section.

11.5 SERIES-SHUNT FEEDBACK—VOLTAGE AMPLIFIERS

The noninverting amplifier has been redrawn in Fig. 11.9 to more clearly delineate the series and shunt connections between the amplifier and feedback network F. The op amp has been drawn to explicitly include its own input resistance R_{id} and output resistance R_o. The feedback network consists of resistors R_1 and R_2, and its input and output port voltages, v_{if} and v_{of}, are defined in the figure. On the left side, applied input voltage v_i equals the sum of the op amp input voltage and the feedback network voltage: $v_i = v_{id} + v_{if}$. Thus there is **series feedback** at the input because the amplifier input and feedback network voltages are in series.

At the output, we see that the feedback network voltage equals the op amp output voltage: $v_{of} = v_o$. Thus the amplifier and feedback network are connected in parallel, or shunt, at the output, so we have **shunt feedback** at the output. We refer to this overall configuration as a *series-shunt feedback amplifier*. As summarized previously in Table 11.1, the gain is set by the feedback network, and we expect the input resistance to be increased over that of the op amp itself (R_{id}) by the series feedback at the input, and the output resistance is decreased below that of the op amp (R_o) by the shunt feedback at the output.

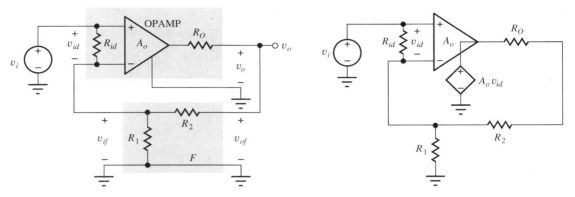

Figure 11.9 The noninverting amplifier as a series-shunt feedback amplifier.

Figure 11.10 Series-shunt feedback amplifier.

11.5.1 CLOSED-LOOP GAIN CALCULATION

In order to find the closed-loop gain for the series-shunt feedback amplifier, we need to evaluate the gain expression from Eq. (11.39) by finding the ideal gain and the loop gain:

$$A_v = A_v^{\text{Ideal}} \left(\frac{T}{1+T} \right) \quad \text{where} \quad A_v^{\text{Ideal}} = 1 + \frac{R_2}{R_1} \quad (11.41)$$

We already know from Chapter 10 that the ideal gain for the noninverting amplifier is given by $A_v^{\text{Ideal}} = 1 + (R_2/R_1)$.

Loop gain T represents the total gain through the amplifier and back around the feedback loop to the input, and we will use Fig. 11.10 to directly calculate the loop gain. To find T, we disable the feedback loop at some arbitrary point in the circuit, insert a test source into the loop, and calculate the gain around the loop. In the op amp circuit, it is convenient to use the source that already exists within the op amp model. We disable the feedback loop by assuming we know the value of v_{id} in source $A_o v_{id}$, (e.g., $\mathbf{v_{id}} = 1$ V and $A_o \mathbf{v_{id}} = A_o(1)$), and then calculate the value of v_{id} developed back at the op amp input. The loop gain is then equal to the negative of the ratio of the voltage returned through the loop to the op amp input. The negative sign accounts for the use of negative feedback.

Loop gain T can now be found by applying voltage division to the circuit in Figs. 11.10 or 11.11 (note that we must also turn off the independent input source v_i by setting its value to zero):

$$\mathbf{v_{id}} = -A_o(1) \frac{(R_1 \| R_{id})}{(R_1 \| R_{id}) + R_2 + R_o} \quad (11.42)$$

and

$$T = -\frac{\mathbf{v_{id}}}{1} = A_o \frac{R_1 \| R_{id}}{(R_1 \| R_{id}) + R_2 + R_o} \quad (11.43)$$

Note that the loop gain now incorporates all the nonideal resistance effects. R_{id} appears in parallel with R_1 of the feedback network, and R_o appears in series with R_2. If $R_{id} \| R_1 \gg (R_2 + R_o)$, then T approaches A_o. Otherwise, $T < A_o$.

11.5.2 INPUT RESISTANCE CALCULATIONS

Next let us find the input resistance of the series-shunt feedback amplifier by applying Blackman's theorem from Eq. (11.40):

$$R_{\text{in}} = R_{\text{in}}^D \frac{1 + |T_{SC}|}{1 + |T_{OC}|} \quad (11.44)$$

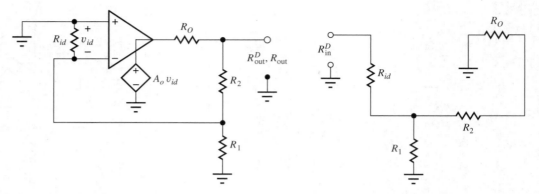

Figure 11.11 Circuit for finding T and R_{out}.

Figure 11.12 Circuit for finding R_{in}^D.

R_{in} is the resistance of the closed-loop feedback amplifier looking into any pair of terminals (a port), R_{in}^D is the resistance looking into the same terminals with the feedback loop disabled, T_{SC} is the loop gain with a short circuit applied to the port, and T_{OC} is the loop gain with the same port open.

In the series-shunt circuit, the input resistance is the resistance appearing between the noninverting input terminal and ground. R_{in}^D is found using the circuit in Fig. 11.12 in which the feedback loop has been disabled by setting $A_o = 0$, and R_{in}^D can then be written directly as

$$R_{in}^D = R_{id} + [R_1 || (R_2 + R_o)] \tag{11.45}$$

To find T_{SC}, the input terminals in the original circuit are shorted, and we recognize that the circuit is identical to the one used to find the loop gain T in Eq. (11.43). Therefore $|T_{SC}| = T$. To find T_{OC}, the input terminals are open-circuited. Then no current can flow through R_{id}, voltage v_{id} must be zero, and $T_{OC} = 0$ in Fig. 11.10. The final expression for the input resistance becomes

$$R_{in} = R_{in}^D \frac{1 + |T_{SC}|}{1 + |T_{OC}|} = R_{in}^D \frac{1 + T}{1 + 0} = [R_{id} + R_1 || (R_2 + R_o)](1 + T) \tag{11.46}$$

Compared to the results in Table 11.1, the input resistance now includes the additional influence of R_1, R_2, and R_o and the modified value of T.

11.5.3 OUTPUT RESISTANCE CALCULATIONS
Again, applying Blackman's theorem gives

$$R_{out} = R_{out}^D \frac{1 + |T_{SC}|}{1 + |T_{OC}|} \tag{11.47}$$

Be sure to note that the values of T_{SC} and T_{OC} depend upon the selected terminal pair, and the values in Eq. (11.47) will most likely differ from those just found in the previous section!

The output resistance is the resistance appearing between the amplifier output terminal and ground. We find R_{out}^D using the circuit in Fig. 11.11 in which we can disable the feedback by setting $A_o = 0$. The resistance can then written directly as

$$R_{out}^D = R_o || (R_2 + R_1 || R_{id}) \tag{11.48}$$

For T_{SC}, output terminal v_o is connected directly to ground, shorting out the feedback loop. Thus $T_{SC} = 0$. To find T_{OC}, the output terminals are open-circuited. Now we recognize that the circuit is identical to the one used to find the loop gain T in Eq. (11.43), and $|T_{OC}| = T$. The final expression for the output resistance becomes

$$R_{\text{out}} = R_{\text{out}}^D \frac{1+|T_{SC}|}{1+|T_{OC}|} = R_{\text{out}}^D \frac{1+0}{1+T} = \frac{R_o||(R_2+R_1||R_{id})}{1+T} \quad (11.49)$$

Compared to the results in Table 11.1, the input resistance now includes the additional influence of R_1, R_2, and R_{id} as well as the modified value of T. Note that the values of T_{SC} and T_{OC} are opposite from those found for the input resistance calculation.

During analysis of the impact of finite gain, finite input resistance, and nonzero output resistance on the characteristics of the closed-loop amplifiers at the beginning of this chapter, we assumed in several cases that the input current to the op amp was negligible with respect to the current in the feedback network, which is equivalent to assuming that R_{id} is much greater than R_1. However, our loop-gain analysis allows us to directly account for the impact of both R_{id} and R_o on the feedback amplifier's gain with no approximations, and it is directly extendable to include any number of additional resistances (see Ex. 11.5). Note that if $R_{id} \gg R_1$ and $R_1 + R_2 \gg R_o$, then $T \cong A_o\beta$.

11.5.4 SERIES-SHUNT FEEDBACK AMPLIFIER SUMMARY

The direct loop-gain analysis of the characteristics of the series-shunt feedback amplifier yielded the following results:

$$A_v = \left(1 + \frac{R_2}{R_1}\right)\frac{T}{1+T} \qquad R_{\text{in}} = R_{\text{in}}^D(1+T) \qquad R_{\text{out}} + \frac{R_{\text{out}}^D}{1+T} \quad (11.50)$$

The expressions in Eq. (11.50) are identical in form to those developed at the beginning of this chapter and summarized in Table 11.2. The overall input resistance is increased by the series feedback at the input, and the output resistance is decreased by the shunt feedback at the output. However, the equations now use the input resistance, output resistance, and loop gain of the amplifier including the effects of all the resistances in the circuit. Application of this theory appears in the next example.

EXAMPLE 11.5   **SERIES-SHUNT FEEDBACK AMPLIFIER ANALYSIS**

This example evaluates the closed-loop characteristics of an op-amp-based series-shunt feedback amplifier using the loop-gain approach. The analysis is extended to include a source resistance.

PROBLEM Find the closed-loop voltage gain, input resistance, and output resistance for the series-shunt feedback amplifier in Fig. 11.9, if the op amp has an open-loop gain of 80 dB, an input resistance of 25 kΩ, and an output resistance of 1 kΩ. Assume the amplifier is driven by a signal voltage with a 2-kΩ source resistance, and the feedback network is implemented with $R_2 = 91$ kΩ and $R_1 = 10$ kΩ.

SOLUTION **Known Information and Given Data:** The series-shunt feedback amplifier is drawn in the figure below with the source resistance added. For the op amp: $A_o = 80$ dB, $R_{id} = 25$ kΩ, and $R_o = 1$ kΩ.

Unknowns: Closed-loop gain A_v, closed-loop input resistance R_{in}, closed-loop output resistance R_{out}

Approach: Find A_v^{Ideal}, T, R_{in}^D, and R_{out}^D. Then calculate the values of the unknowns using the closed-loop feedback amplifier formulae derived in this section.

Assumptions: The op amp is ideal except for A_o, R_{id}, and R_o.

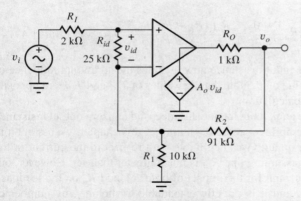
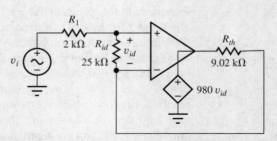

Figure 11.13 Series-shunt feedback amplifier with source resistance added.

Figure 11.14 Series-shunt feedback amplifier with Thévenin transformation.

Analysis: The amplifier circuit with the addition of R_I appears in Fig. 11.13.

Amplifier Analysis: Using Fig. 11.13,

$$A_v^{\text{Ideal}} = 1 + \frac{R_2}{R_1} = 1 + \frac{91\,\text{k}\Omega}{10\,\text{k}\Omega} = 10.1$$

$$R_{\text{in}}^D = R_I + R_{id} + R_1 \| (R_2 + R_O)$$

$$R_{\text{in}}^D = 2\,\text{k}\Omega + 25\,\text{k}\Omega + 10\,\text{k}\Omega \| (91\,\text{k}\Omega + 1\,\text{k}\Omega) = 36.0\,\text{k}\Omega$$

$$R_{\text{out}}^D = R_o \| [R_2 + R_1 \| (R_{id} + R_I)]$$

$$R_{\text{out}}^D = 1\,\text{k}\Omega \| [91\,\text{k}\Omega + 10\,\text{k}\Omega \| (25\,\text{k}\Omega + 2\,\text{k}\Omega)] = 990\,\Omega$$

The loop gain is most easily found by first taking a Thévenin equivalent of R_o, R_1, and the op amp output source yielding the circuit in Fig. 11.14:

$$\mathbf{v_{th}} = A_o \mathbf{v_{id}} \frac{R_1}{R_o + R_2 + R_1} = 10^4 \mathbf{v_{id}} \frac{10\,\text{k}\Omega}{1\,\text{k}\Omega + 91\,\text{k}\Omega + 10\,\text{k}\Omega} = 980\mathbf{v_{id}}$$

$$R_{th} = R_1 \| (R_2 + R_O) = 10\,\text{k}\Omega \| (91\,\text{k}\Omega + 1\,\text{k}\Omega) = 9.02\,\text{k}\Omega$$

Now we can assume $v_{id} = 1$ in v_{th} and solve for the op amp input voltage v_{id}:

$$\mathbf{v_{id}} = -\mathbf{v_{th}} \frac{R_{id}}{R_{th} + R_{id} + R_I} = -980(1)\frac{25\,\text{k}\Omega}{9.02\,\text{k}\Omega + 25\,\text{k}\Omega + 2\,\text{k}\Omega} = -680$$

$$T = -\frac{\mathbf{v_{id}}}{1} = 680$$

Closed-Loop Amplifier Results:

$$A_v = \left(1 + \frac{R_2}{R_1}\right)\frac{T}{1+T} = \left(1 + \frac{91\,\text{k}\Omega}{10\,\text{k}\Omega}\right)\frac{680}{1+680} = 10.1$$

$$R_{\text{in}} = R_{\text{in}}^D(1+T) = 36.0\,\text{k}\Omega(1+680) = 24.5\,\text{M}\Omega$$

$$R_{\text{out}} = \frac{R_{\text{out}}^D}{1+T} = \frac{990\,\Omega}{1+680} = 1.45\,\Omega$$

Check of Results: We have found the three unknowns. The loop gain is high, so we expect the closed-loop gain to be approximately 10.1. The calculated input resistance is much larger than that of the op amp itself, and the output resistance is much smaller than that of the op amp. These all agree with our expectations for the series-shunt feedback configuration.

Discussion: This analysis demonstrates a direct method for including the effects of nonideal op amp characteristics as well as loading effects of the feedback network and source (and/or load resistors) on the noninverting amplifier configuration. One of the following exercises looks at the impact of adding a load resistor to this amplifier. Note that the low value of R_{id} relative to R_1 and R_I in this problem reduces the loop gain substantially.

Computer-Aided Analysis: Our SPICE circuit mirrors the circuit in Fig. 11.13 and augments the built-in OPAMP model with the addition of R_{id} in parallel with its input and R_o in series with its output, and the gain is set to 10,000. A transfer function analysis from v_i to the output voltage yields: $A_v = 10.09$, $R_{\text{in}} = 24.54 \, \text{M}\Omega$, and $R_{\text{out}} = 1.453 \, \Omega$. These results agree closely with our hand calculations.

EXERCISE: Find the loop gain, closed-loop voltage gain, input resistance, and output resistance for the series-shunt feedback amplifier in Ex. 11.5 if the 2-kΩ source resistor is eliminated from the amplifier circuit.

ANSWERS: 720, 10.1, 24.5 MΩ, 1.57 Ω

EXERCISE: Find the loop gain, closed-loop voltage gain, input resistance, and output resistance for the series-shunt feedback amplifier in Ex. 11.5 if a 5-kΩ load resistor is connected to the output of the amplifier.

ANSWERS: 568, 10.1, 20.5 MΩ, 1.45 Ω. Note that the input resistance is a function of the load resistance R_L.

EXERCISE: What would the closed-loop gain, input resistance, and output resistance of the series-shunt feedback amplifier in the previous exercise have been if the loading effects of the feedback network and R_I and R_L had all been ignored?

ANSWERS: 10.1, 24.8 MΩ, 1.01 Ω

DISCUSSION: Note that the closed-loop gain is essentially unchanged because of the high value of loop gain, but the values of R_{in} and R_{out} differ by substantial percentages. Feedback stabilizes the value of the voltage gain but not those of the input and output resistances.

EXERCISE: What are the values of A_v, R_{in}, and R_{out} if the 2-kΩ source resistor is changed to 5kΩ in Fig. 11.14?

ANSWERS: 10.1; 15.7 MΩ; 1.59 Ω

ELECTRONICS IN ACTION

Three-Terminal IC Voltage Regulators

It is not easy to produce precise output voltages with rectifier circuits, particularly with changing load currents. Specially wound transformers may be required to produce the desired output voltages, and extremely large filter capacitances are required to reduce the output ripple voltage to very small values. A much better approach is to use integrated circuit voltage regulators to set the output voltage and remove the ripple. IC regulators are available with a wide range of fixed output voltages as well as adjustable output versions.

An example of a rectifier circuit with a three-terminal 5-V regulator is shown in the accompanying figure. The regulator uses feedback with high-gain amplifier circuitry to greatly reduce the ripple voltage at the output. **IC voltage regulators** also provide outstanding line and load regulation, maintaining a constant output voltage even though the output current may change by many orders of magnitude. Capacitor C is the normal rectifier filter capacitor, and C_{B1} and C_{B2} (typically 0.001–0.01 µf) are bypass capacitors that provide a low-impedance path for high-frequency signals and are needed to ensure proper operation of the voltage regulator.

The regulator can reduce the ripple voltage by a factor of 100 to 1000 or more. To minimize power dissipation in the regulator, the rectifier can be designed with a relatively large ripple voltage at the input to the regulator, thus reducing the average input voltage to the regulator. The main design constraint is set by the input-output voltage differential V_{REG} across the regulator, which must not fall below a minimum "dropout voltage" value specified for the regulator, typically a few volts. The current I_{REG} needed to operate the IC regulator is only a few mA and typically represents a small percentage of the total current supplied by the rectifier: $I_S = I_L + I_{REG}$. An example of a voltage regulator family can be found on the MCD website.

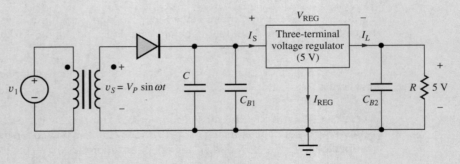

Half-wave rectifier and three-terminal IC voltage regulator.

A series-shunt feedback amplifier is one common implementation of the three-terminal voltage regulator as in the following figure. The op amp forces the attenuated output voltage to be equal to reference voltage:

$$V_O\left(\frac{R_1}{R_1 + R_2}\right) = V_{REF} \quad \text{or} \quad V_O = V_{REF}\left(1 + \frac{R_2}{R_1}\right)$$

Transistor Q_1, called a "pass transistor," is used to increase the regulator's output current capability far above that of the op amp alone.

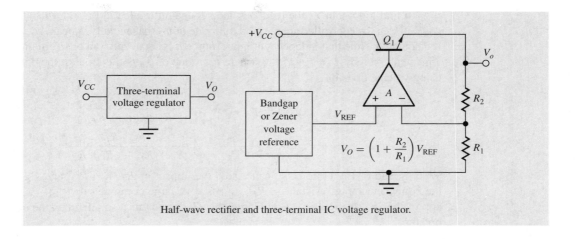

Half-wave rectifier and three-terminal IC voltage regulator.

11.6 SHUNT-SHUNT FEEDBACK—TRANSRESISTANCE AMPLIFIERS

The transresistance amplifier from Chapter 10 is redrawn as a shunt-shunt feedback amplifier in Fig. 11.15(a). In this circuit, the input source is represented by its Norton equivalent current i_i and resistance R_I, and R_{id} and R_o have again been included in the circuit so we can assess their impact on overall feedback amplifier performance. On the input side, the feedback network voltage equals the op amp input voltage: $v_{if} = -v_{id}$. Thus the amplifier and feedback network are connected in parallel, so we have **shunt feedback** at the input. At the output, $v_{of} = v_o$, so we also have **shunt feedback** at the output. We refer to this overall configuration as a *shunt-shunt feedback amplifier*. In this case, application of feedback results in both a low input resistance and a low output resistance, which is exactly what is required of a transimpedance amplifier (i.e., current in, voltage out).

11.6.1 CLOSED-LOOP GAIN CALCULATION

In order to find the closed-loop gain for the shunt-shunt feedback amplifier, we need to evaluate the gain expression from Eq. (11.51) by finding the ideal gain and the loop gain:

$$A_{tr} = A_{tr}^{\text{Ideal}} \left(\frac{T}{1+T} \right) \quad \text{where} \quad A_{tr}^{\text{Ideal}} = -R_F \quad (11.51)$$

We already know from Chapter 10 (Sec. 10.9.2) that the ideal gain for the transresistance amplifier is given by $A_{tr}^{\text{Ideal}} = -R_F$.

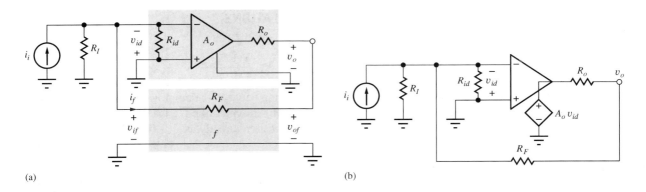

Figure 11.15 (a) Transresistance amplifier as a shunt-shunt feedback amplifier; (b) circuit for loop gain calculation.

To find loop-gain T, the feedback loop is again disabled in Fig. 11.15(b) by arbitrarily setting $v_{id} = 1$ V in the controlled source, and the value of v_{id} that actually appears back at the op amp input is calculated. Note that independent input source i_i is set to zero (an open circuit) in this calculation. T is then given by $T = -\mathbf{v_{id}}/1$ V, where the value of v_{id} at the input of the op amp is easily found using voltage division:

$$\mathbf{v_{id}} = -A_o(1\text{ V})\frac{R_{id}\|R_I}{R_o + R_F + (R_{id}\|R_I)} \tag{11.52}$$

$$T = -\frac{\mathbf{v_{id}}}{(1\text{V})} = A_o\frac{R_{id}\|R_I}{R_o + R_F + (R_{id}\|R_I)}$$

With the aid of Eq. (11.52) we can assess the impact of R_I, R_F, R_{id}, and R_o on the loop gain. If the equivalent input resistance $(R_I\|R_{id})$ is very large compared to $R_F + R_o$, then the loop gain approaches A_o. If $(R_I\|R_{id})$ is not large, then T can be much smaller than the open-loop gain of the op amp.

11.6.2 INPUT RESISTANCE CALCULATIONS

The input resistance of the series-shunt feedback amplifier is found by applying Blackman's theorem from Eq. (11.40):

$$R_{\text{in}} = R_{\text{in}}^D \frac{1 + |T_{SC}|}{1 + |T_{OC}|} \tag{11.53}$$

in which the input resistance is the resistance appearing between the inverting input terminal and ground. R_{in}^D is found from the circuit in Fig. 11.15(b) in which the feedback can be disabled by setting $A_o = 0$. The expression for R_{in}^D can then be written directly by inspection as

$$R_{\text{in}}^D = R_I\|R_{id}\|(R_F + R_o) \tag{11.54}$$

To find T_{SC}, the input terminals are shorted, thereby grounding the inverting input and forcing v_{id} and T_{SC} to be 0. To find T_{OC}, the input terminals are open-circuited, and we see that this is the same circuit used to find loop gain T. Therefore $|T_{OC}| = T$. The final expression for the input resistance becomes

$$R_{\text{in}} = R_{\text{in}}^D \frac{1+0}{1+T} = \frac{R_I\|R_{id}\|(R_F + R_o)}{1+T} \tag{11.55}$$

Compared to the results in Table 11.1, the input resistance now includes the additional influence of R_I, R_F, and R_o as well as the modified value of T. In the ideal case, T approaches infinity, and R_{in} approaches zero.

11.6.3 OUTPUT RESISTANCE CALCULATIONS

For the output resistance case, Blackman's theorem gives

$$R_{\text{out}} = R_{\text{out}}^D \frac{1 + |T_{SC}|}{1 + |T_{OC}|} \tag{11.56}$$

The output resistance is the resistance appearing between the amplifier output terminal and ground. R_{out}^D is found using the circuit in Fig. 11.15(b) in which the feedback loop is disabled by setting $A_o = 0$. The expression for R_{out}^D can then be written directly as

$$R_{\text{out}}^D = R_o\|(R_F + R_I\|R_{id}) \tag{11.57}$$

For T_{SC}, output terminal v_o is connected directly to ground, thereby shorting out the feedback loop. Thus $T_{SC} = 0$. To find T_{OC}, the output terminals are open-circuited. We again recognize that the circuit is identical to the one used to find the loop gain T in Eq. (11.52), and $|T_{OC}| = T$. The final expression for the output resistance becomes

$$R_{\text{out}} = R_{\text{out}}^D \frac{1+0}{1+T} = \frac{R_o \| (R_F + R_I \| R_{id})}{1+T} \quad (11.58)$$

Compared to the results in Table 11.1, the input resistance now includes the additional influence of R_I, R_F, and R_{id} as well as the modified value of T. If $(R_F + R_I \| R_{id}) \gg R_o$, then R_{out} approaches $R_o/(1+T)$. For infinite T, R_{out} becomes zero.

11.6.4 SHUNT-SHUNT FEEDBACK AMPLIFIER SUMMARY

The direct loop-gain analysis of the characteristics of the series-shunt feedback amplifier yielded the following results:

$$A_{tr} = (-R_F)\frac{T}{1+T} \qquad R_{\text{in}} = \frac{R_{\text{in}}^D}{1+T} \qquad R_{\text{out}} = \frac{R_{\text{out}}^D}{1+T} \quad (11.59)$$

The expressions in Eq. (11.59) are identical in form to those developed at the beginning of this chapter and summarized in Table 11.2. The ideal gain is determined by feedback resistor R_F. The overall input resistance is decreased by the shunt feedback at the input, and the output resistance is decreased by the shunt feedback at the output. However, the equations now use the input resistance, output resistance, and loop gain of the amplifier including the effects of all the resistances in the circuit (see the following example).

EXERCISE: Draw the simplified circuits used to find R_{in}^D and R_{out}^D that result from setting A_o to zero, and verify the expressions in Eqs. (11.54) and (11.57).

EXAMPLE 11.6 SHUNT-SHUNT FEEDBACK AMPLIFIER ANALYSIS

This example evaluates the closed-loop characteristics of an op-amp-based shunt-shunt feedback amplifier by finding the loop gain and applying Blackman's theorem.

PROBLEM Find T and the closed-loop transresistance, input resistance, and output resistance for the shunt-shunt feedback amplifier in Fig. 11.15 if the op amp has an open-loop gain of 80 dB, an input resistance of 25 kΩ, and an output resistance of 1 kΩ. Analyze the circuit with $R_I = 10$ kΩ, $R_F = 91$ kΩ, and a 5-kΩ load resistance connected to the output.

SOLUTION **Known Information and Given Data:** The shunt-shunt feedback amplifier is drawn in Fig. 11.16 with source and load resistances added. For the op amp: $A_o = 80$ dB, $R_{id} = 25$ kΩ and $R_o = 1$ kΩ, and $R_F = 91$ kΩ, $R_I = 10$ kΩ and $R_L = 5$ kΩ.

Unknowns: Closed-loop gain A_{tr}, closed-loop input resistance R_{in}, closed-loop output resistance R_{out}

Approach: Find A_{tr}^{Ideal}, T, R_{in}^D, and R_{out}^D. Then calculate the values of the unknowns using the closed-loop feedback amplifier formulae derived in this section.

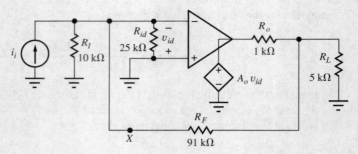

Figure 11.16 Shunt-shunt feedback amplifier with load resistor added.

Assumptions: The op amp is ideal except for A_o, R_{id}, and R_o.

Analysis: The ideal gain of the amplifier is $-R_F$, so

$$A_{tr}^{\text{Ideal}} = -R_F = -91\,\text{k}\Omega$$

We must modify the equations derived in the precious section to include the effect of the addition of R_L to the amplifier:

$$R_{\text{in}}^D = R_I \| R_{id} \| (R_F + R_L \| R_o) = 10\,\text{k}\Omega \| 25\,\text{k}\Omega \| (91\,\text{k}\Omega + 5\,\text{k}\Omega \| 1\,\text{k}\Omega) = 6.63\,\text{k}\Omega$$

$$R_{\text{out}}^D = R_L \| R_o \| (R_F + R_I \| R_{id}) = 5\,\text{k}\Omega \| 1\,\text{k}\Omega \| (91\,\text{k}\Omega + 10\,\text{k}\Omega \| 25\,\text{k}\Omega) = 826\,\Omega$$

The loop gain is most easily found by first taking a Thévenin equivalent (at point X) of R_F, R_o, R_L, and the op amp output source yielding the result shown in Fig. 11.17.

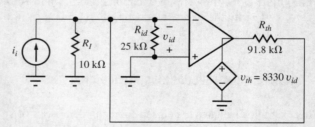

Figure 11.17 Amplifier following Thévenin transformation.

The Thévenin equivalent voltage and resistance looking back into R_F are

$$\mathbf{v_{th}} = A_o \mathbf{v_{id}} \frac{R_L}{R_o + R_L} = 10^4 \mathbf{v_{id}} \frac{5\,\text{k}\Omega}{1\,\text{k}\Omega + 5\,\text{k}\Omega} = 8330 \mathbf{v_{id}}$$

$$R_{th} = R_F + R_o \| R_L = 91\,\text{k}\Omega + 1\,\text{k}\Omega \| 5\,\text{k}\Omega = 91.8\,\text{k}\Omega$$

The loop gain is found by setting $A_o \mathbf{v_{id}} = A_o(1)$ and calculating v_{id}:

$$\mathbf{v_{id}} = -\mathbf{v_{th}} \frac{R_I \| R_{id}}{R_{th} + R_I \| R_{id}} = -A_o(1) \frac{R_L}{R_o + R_L} \frac{R_I \| R_{id}}{R_{th} + R_I \| R_{id}}$$

$$T = -\frac{\mathbf{v_{id}}}{1} = 10^4 \left(\frac{5\,\text{k}\Omega}{5\,\text{k}\Omega + 1\,\text{k}\Omega} \right) \left(\frac{10\,\text{k}\Omega \| 25\,\text{k}\Omega}{91.8\,\text{k}\Omega + 10\,\text{k}\Omega \| 25\,\text{k}\Omega} \right) = 602$$

Closed-Loop Amplifier Results:

$$A_{tr} = A_{tr}^{\text{Ideal}} \frac{T}{1+T} = -91\,\text{k}\Omega \frac{602}{1+602} = -90.9\,\text{k}\Omega$$

$$R_{\text{in}} = \frac{R_{\text{in}}^D}{1+T} = \frac{6.63\,\text{k}\Omega}{603} = 11.0\,\Omega \quad R_{\text{out}} = \frac{R_{\text{out}}^D}{1+T} = \frac{826\,\Omega}{603} = 1.37\,\Omega$$

Check of Results: We have found the unknowns. The loop gain is high, so we expect A_{tr} to be close to the ideal value of $-91\,\text{k}\Omega$. The calculated input and output resistances are both much smaller than those of the op amp itself. These all agree with our expectations for the shunt-shunt feedback configuration.

Computer-Aided Analysis: Our SPICE circuit mirrors the circuit in Fig. 11.16 and augments the built-in OPAMP model with the addition of R_{id} in parallel with its input and R_o in series with its output, and the gain is set to 10,000. A transfer function analysis from i_i to the voltage across R_L yields: $A_{tr} = -90.85\,\text{k}\Omega$, $R_{\text{in}} = 11.00\,\Omega$ and $R_{\text{out}} = 1.372\,\Omega$. These results agree closely with our hand calculations.

Discussion: This analysis illustrates the direct method for including the effects of nonideal op amp characteristics on the shunt-shunt feedback amplifier configuration, as well the accounting for loading effects of the feedback network and source and load resistors. The first exercise below looks at the errors that occur if these effects are neglected.

Note that the classic inverting amplifier can be transformed into a shunt-shunt feedback amplifier for analysis. The voltage gain, input resistance, and output resistance of the original inverting amplifier can then easily be found directly from the results above. The voltage gain equals the transresistance divided by R_I:

$$A_v = \frac{v_o}{v_i} = \frac{v_o}{i_i} \cdot \frac{i_i}{v_i} = A_{tr} \cdot \frac{1}{R_I} = -\frac{90.9\,\text{k}\Omega}{10\,\text{k}\Omega} = -9.09$$

$$R_{\text{in}}^{inv} = R_I + \left(\frac{1}{R_{\text{in}}} - \frac{1}{R_I}\right)^{-1} = 10.0\,\text{k}\Omega \quad \text{and} \quad R_{\text{out}}' = \left(\frac{1}{R_{\text{out}}} - \frac{1}{R_L}\right)^{-1} = 1.37\,\Omega$$

In the shunt-shunt analysis, R_I appears in parallel with the amplifier input, whereas it is in series with the input in the inverting amplifier. Thus R_{in}^{inv} is found by first removing R_I from the parallel combination and then adding it back as a series element. Similarly, the parallel effect of R_L can be removed directly from the output resistance.

EXERCISE: Calculate the loop gain, overall transresistance, input resistance, and output resistance of the transresistance amplifier if the loading effects of R_{id} and R_L are neglected.

ANSWERS: $T = 980$, $A_{tr} = -90.9\,\text{k}\Omega$, $R_{\text{in}} = 9.20\,\Omega$, $R_{\text{out}} = 1.01\,\Omega$

EXERCISE: Find the loop gain, closed-loop voltage gain, input resistance and output resistance for the series-shunt feedback amplifier in Ex. 11.6 if the 10-kΩ source resistor is eliminated from the amplifier circuit.

ANSWERS: 2140, $-91.0\,\text{k}\Omega$, $9.21\,\Omega$, $0.646\,\Omega$

ELECTRONICS IN ACTION

Fiber Optic Receiver

Interface circuits for optical communications were introduced in the Electronics in Action feature in Chapter 9. One of the important electronic blocks on the receiver side of such a fiber optic communication link is the circuit that performs the optical-to-electrical (O/E) signal conversion, and a common approach is shown in the accompanying figure. Light exiting an optical fiber is incident upon a photodiode (see Sec. 3.18) that generates photocurrent i_{ph} as modeled by the current source in the figure. This photocurrent flows through feedback resistor R and generates a signal voltage at the output given by $v_o = i_{ph}R$. The voltage V_{BIAS} can be used to provide reverse bias to the photodiode. In this case, the total output voltage is $v_O = V_{BIAS} + i_{ph}R$.

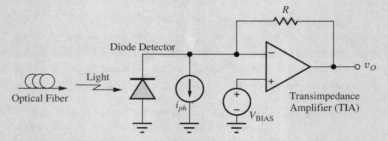

Optical-to-electrical interface for fiber optic data transmission.

Since the input to the amplifier is a current and the output is a voltage, the gain $A_{tr} = \mathbf{v_o}/\mathbf{i_{ph}}$ has the units of resistance, and the amplifier is referred to as a transresistance or (more generally) a transimpedance amplifier (TIA). The operational amplifier shown in the circuit must have an extremely wideband and linear design. The requirements are particularly stringent in OC-768 systems, for example, in which 40-GHz signals coming from the optical fiber must be amplified without the addition of any significant phase distortion.

11.7 SERIES-SERIES FEEDBACK—TRANSCONDUCTANCE AMPLIFIERS

In many circuits, we often require a high-performance transconductance amplifier that generates an output current that is proportional to the input voltage: $\mathbf{i_o} = A_{tc}\mathbf{v_i}$. In order to accurately measure the input voltage, series feedback is utilized to achieve a high input resistance, and to approximate a current source at the output, series feedback is again utilized to yield a high output resistance. Thus the transconductance amplifier is a **series-series feedback amplifier**. An op amp implementation is depicted in Fig. 11.18 in which the op amp has been drawn to explicitly include its own input resistance R_{id} and output resistance R_o and to show its four terminals.

The feedback network consists of only a single resistor R, and its input and output port voltages, v_{if} and v_{of}, are defined in the figure. On the left side, the applied input voltage v_i equals the sum of the op amp input voltage and the feedback network voltage: $v_i = v_{id} + v_{if}$. Thus we have **series feedback** at the input because the amplifier input and feedback network voltages are in series. At the output, we see that the output voltage v_o equals the sum of the op amp output voltage v_{op} and the feedback network voltage v_{of}. The amplifier and feedback network are connected in series at both the input and output ports, and this overall configuration represents a **series-series feedback amplifier**. In order to achieve negative feedback, the negative output terminal must be connected to the noninverting input terminal of the op amp as indicated in the figure. Source v_o indicates the location of the output terminals of the series-series feedback amplifier and the point where external circuitry would be connected. The analysis assumes that $v_o = 0$ as in Fig. 11.19.

11.7 Series-Series Feedback—Transconductance Amplifiers

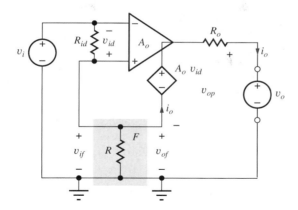

Figure 11.18 The four-terminal op amp as a series-series feedback amplifier.

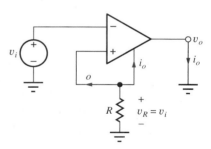

Figure 11.19 Ideal series-series feedback amplifier.

11.7.1 CLOSED-LOOP GAIN CALCULATION

In order to find the closed-loop gain for the series-shunt feedback amplifier, we need to evaluate the gain expression from Eq. (11.39) by finding the ideal gain A_{tc}^{Ideal} and the loop gain T:

$$A_{tc} = A_{tc}^{\text{Ideal}}\left(\frac{T}{1+T}\right) \tag{11.60}$$

The ideal case is depicted in Fig. 11.19. For infinite gain, the op amp input voltage is zero so that v_i appears directly across resistor R. Output current i_o must flow up through R since the input current to the ideal op amp is zero. Thus $\mathbf{i_o} = -\mathbf{v_i}/R$ and $A_{tc}^{\text{Ideal}} = \mathbf{i_o}/\mathbf{v_i} = -1/R$.

As a reminder, loop gain T represents the total gain through the amplifier and back around the feedback loop to the input. In the op amp circuit in Fig. 11.18, we can find T by assuming we know the value of $\mathbf{v_{id}}$ in source $A_o\mathbf{v_{id}}$, (e.g., $\mathbf{v_{id}} = 1$ V and $A_o\mathbf{v_{id}} = A_o(1)$), and then calculate the value of $\mathbf{v_{id}}$ developed at the op amp input. The loop gain is then equal to the negative of the ratio of the voltage returned through the loop to the op amp input.

Loop gain T can be found by applying voltage division to the circuit (remember the independent input source v_i must be turned off by setting its value to zero):

$$\mathbf{v_{id}} = -A_o(1)\frac{(R_{id}\|R)}{(R_{id}\|R)+R_o} \quad \text{and} \quad T = A_o\frac{(R_{id}\|R)}{(R_{id}\|R)+R_o} \tag{11.61}$$

The loop gain expression now incorporates all the nonideal op amp parameters. R_{id} appears in parallel with R of the feedback network, and R_o also affects the loop gain.

11.7.2 INPUT RESISTANCE CALCULATION

Next let us find the input resistance of the series-shunt feedback amplifier by applying Blackman's theorem from Eq. (11.40):

$$R_{\text{in}} = R_{\text{in}}^D \frac{1+|T_{SC}|}{1+|T_{OC}|} \tag{11.62}$$

R_{in} represents the resistance appearing between the noninverting input terminal and ground of the closed-loop feedback amplifier, R_{in}^D is the resistance looking into the same terminals with the feedback loop disabled, T_{SC} is the loop gain with a short circuit applied to the input port, and T_{OC} is the loop gain with the input port open.

We find R_{in}^D using the circuit in Fig. 11.18 in which the feedback is disabled by setting $A_o = 0$. The input resistance can then be written directly as

$$R_{\text{in}}^D = R_{id} + R\|R_o \tag{11.63}$$

To find T_{SC}, the input terminals are shorted, and we recognize that the circuit is identical to the one used to find the loop gain T in Eq. (11.61). Therefore $|T_{SC}| = T$. To find T_{OC}, the input terminals are open-circuited. Then no current can flow through R_{id}, so the voltage v_{id} must be zero, and $T_{OC} = 0$. The final expression for the input resistance becomes

$$R_{in} = R_{in}^D \frac{1+T}{1+0} = (R_{id} + R\| R_o)(1+T) \qquad (11.64)$$

We see that the input resistance is increased by T and can be very high.

11.7.3 OUTPUT RESISTANCE CALCULATION

Blackman's theorem gives

$$R_{out} = R_{out}^D \frac{1+|T_{SC}|}{1+|T_{OC}|} \qquad (11.65)$$

The output resistance represents the resistance appearing between the amplifier output terminal and ground. R_{out}^D is found using the circuit in Fig. 11.18 in which we set $v_i = 0$ and again disable the feedback by setting $A_o = 0$. The output resistance can then be written as

$$R_{out}^D = R_o + R\| R_{id} \qquad (11.66)$$

For T_{SC}, output terminal v_o is connected directly to ground, we recognize that the circuit is identical to the one used to find the loop gain T in Eq. (11.61), and $|T_{SC}| = T$. To find T_{OC}, the output terminals are open-circuited, and no current flows in the circuit. Thus $T_{OC} = 0$. The final expression for the output resistance becomes

$$R_{out} = R_{out}^D \frac{1+T}{1+0} = (R_o + R\| R_{id})(1+T) \qquad (11.67)$$

The output resistance of the op amp is multiplied by T, and the result can be very high.

11.7.4 SERIES-SERIES FEEDBACK AMPLIFIER SUMMARY

The analysis of the characteristics of the series-series feedback amplifier yielded the following results:

$$A_{tc} = \left(-\frac{1}{R}\right) \frac{T}{1+T} \qquad R_{in} = R_{in}^D(1+T) \qquad R_{out} = R_{out}^D(1+T) \qquad (11.68)$$

The transconductance amplifier has a high input resistance, a high output resistance, and the ideal transconductance equals the reciprocal of feedback resistance R.

EXERCISE: Draw the simplified circuits used to find R_{in}^D and R_{out}^D that result from setting A_o to zero and verify the expressions in Eqs. (11.63) and (11.66).

ANSWERS: R_{in}^D and R_{out}^D

EXAMPLE 11.7   **SERIES-SERIES FEEDBACK AMPLIFIER ANALYSIS**

This example evaluates the closed-loop characteristics of a op-amp-based series-series feedback amplifier utilizing the loop-gain approach and Blackman's theorem. The analysis is extended to include a source resistance.

PROBLEM Find the closed-loop transconductance, input resistance, and output resistance for the series-series feedback amplifier in Fig. 11.18 if the op amp has an open-loop gain of 80 dB, an input resistance of 25 kΩ, and an output resistance of 1 kΩ. Assume the amplifier is driven by a signal voltage with a 10-kΩ source resistance, and the feedback network is implemented with $R_2 = 91$ kΩ and $R_1 = 10$ kΩ.

SOLUTION **Known Information and Given Data:** The series-series feedback amplifier is drawn in the figure below with the 10 kΩ source resistance added. For the op amp: $A_o = 80$ dB, $R_{id} = 25$ kΩ and $R_o = 1$ kΩ.

Unknowns: Closed-loop transconductance A_{tc}, input resistance R_{in}, output resistance R_{out}

Approach: Find new expressions for A_{tc}^{Ideal}, T, R_{in}^D, and R_{out}^D. Then calculate the values of the unknowns using the closed-loop feedback amplifier formulae derived in this section.

Assumptions: The op amp is ideal except for A_o, R_{id}, and R_o. $v_o = 0$.

Analysis: The amplifier circuit with the addition of R_I appears in Fig. 11.20.

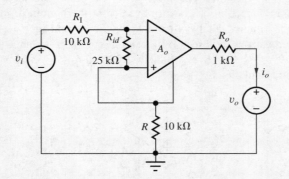

Figure 11.20 Series-series feedback amplifier with source resistance added.

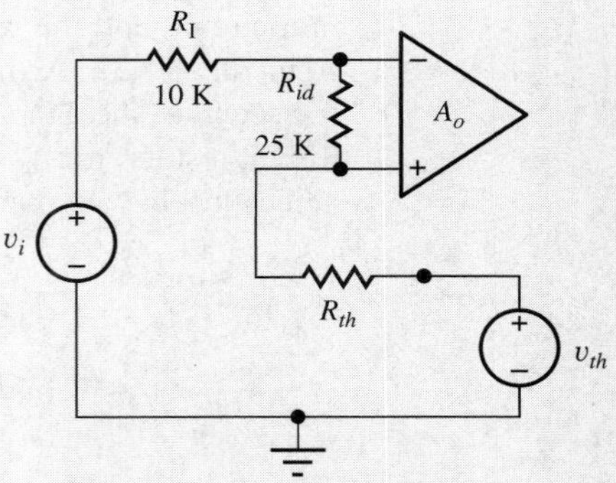

Figure 11.21 Series-shunt feedback amplifier with Thévinen transformation.

Amplifier Analysis:

$$A_{tc}^{\text{Ideal}} = -\frac{1}{R} = -\frac{1}{10\,\text{k}\Omega} = -10^{-4}\,S$$

$$R_{in}^D = R_I + R_{id} + R\|R_o = 10\,\text{k}\Omega + 25\,\text{k}\Omega + 10\,\text{k}\Omega\|1\,\text{k}\Omega = 35.9\,\text{k}\Omega$$

$$R_{out}^D = R_o + R\|(R_{id} + R_I) = 1\,\text{k}\Omega + 10\,\text{k}\Omega\|(25\,\text{k}\Omega + 10\,\text{k}\Omega) = 8.79\,\text{k}\Omega$$

The loop gain is most easily found by first taking a Thévinen equivalent of R_o, R, and the op amp output source with $v_o = 0$ as depicted in Fig. 11.21.

$$\mathbf{v_{th}} = -A_o\mathbf{v_{id}}\frac{R}{R + R_o} = -10^4\mathbf{v_{id}}\frac{10\,\text{k}\Omega}{10\,\text{k}\Omega + 1\,\text{k}\Omega} = -9090\mathbf{v_{id}}$$

$$R_{th} = R\|R_o = 10\,\text{k}\Omega\|1\,\text{k}\Omega = 909\,\Omega$$

Now we can assume $\mathbf{v_{id}} = 1$ in $\mathbf{v_{th}}$ and solve for the op amp input voltage $\mathbf{v_{id}}$:

$$\mathbf{v_{id}} = -\mathbf{v_{th}}\frac{R_{id}}{R_{th} + R_{id} + R_I} = 9090(1)\frac{25\,\text{k}\Omega}{0.909\,\text{k}\Omega + 25\,\text{k}\Omega + 10\,\text{k}\Omega} = 6330$$

$$T = 6330$$

Closed-Loop Amplifier Results:

$$A_{tc} = -\frac{1}{R}\left(\frac{T}{1+T}\right) = -\frac{1}{10\,\text{k}\Omega}\left(\frac{6330}{1+6330}\right) = -0.100\,\text{mS}$$

$$R_{in} = R_{in}^D(1+T) = 35.9\,\text{k}\Omega(1+6330) = 227\,\text{M}\Omega$$

$$R_{out} = R_{out}^D(1+T) = 8.79\,\text{k}\Omega(1+6330) = 55.7\,\text{M}\Omega$$

Check of Results: The three unknowns have been found. The loop gain is high, so we expect A_{tc} should be approximately -0.1 mS. The calculated input resistance is much larger than that of the op amp itself, and the output resistance is also much larger than that of the op amp. These all agree with our expectations for the series-series feedback configuration.

Computer-Aided Analysis: Our SPICE circuit mirrors the circuit in Fig. 11.20 and augments the built-in OPAMP model with the addition of R_{id} in parallel with its input and R_o in series with its output, and the gain is set to 10,000. Zero value voltage source v_o is added to the circuit in order to use the SPICE transfer function analysis. We must be careful in the description of the circuit for simulation. The op amp model that is being used assumes that its internal controlled source is connected to the reference node (ground). Fortunately, we are free to choose the reference node in a circuit, and the ground connection has been moved from the bottom to the top of resistor R in Fig. 11.22. A transfer function analysis from V_I to the current in V_O yields the overall transconductance, the input resistance as seen by V_I and the output resistance seen by V_O. The SPICE results are $A_{tc} = -9.998 \times 10^{-5}$ S, $R_{\text{in}} = 227.3$ MΩ and $R_{\text{out}} = 55.56$ MΩ. These results agree closely with our hand calculations.

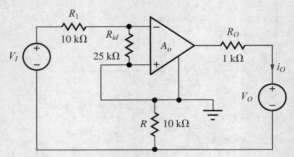

Figure 11.22 Circuit for simulation with new ground reference.

EXERCISE: Find the loop gain, closed-loop transconductance, input resistance and output resistance for the series-series feedback amplifier in Ex. 11.7 if the 10-kΩ source resistor is eliminated from the amplifier circuit.

ANSWERS: 8770, -1.00×10^{-4} S, 227 MΩ, 71.4 MΩ

11.8 SHUNT-SERIES FEEDBACK—CURRENT AMPLIFIERS

The last of the four feedback topologies is the current amplifier that is implemented using the shunt-series feedback configuration. For this case, we need to generate an output current that is proportional to the input current, $\mathbf{i_o} = A_i \mathbf{i_i}$. At the input we want to sense a current, so shunt feedback is utilized to achieve a low input resistance, and the output should approximate a current source, so series feedback is also used at the output to yield high output resistance. The op-amp-based shunt-series feedback amplifier is depicted in Fig. 11.23 in which the op amp input resistance R_{id} and output resistance R_o are explicitly shown. Voltage source v_o is included to identify the output terminals that would normally be connected to an external load.

The feedback network consists of resistors R_2 and R_1, and its input and output port voltages, v_{if} and v_{of}, are defined in the figure. On the left side, the $v_{id} = v_{if}$, so we have shunt feedback at the input. The output voltage equals the sum of the op amp output voltage and that of the feedback network, $v_o = v_{op} + v_{of}$, so there is a series connection at the output. Thus we refer to this overall configuration as a **shunt-series feedback amplifier.**

11.8 Shunt-Series Feedback—Current Amplifiers

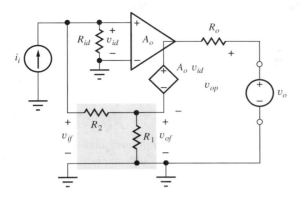

Figure 11.23 The four-terminal op amp as a shunt-series feedback amplifier.

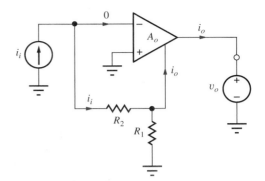

Figure 11.24 Ideal shunt-series feedback amplifier.

In order to achieve negative feedback, the negative output terminal must be connected back to the noninverting input terminal of the op amp through the feedback network as indicated in the figure. Source v_o indicates the location of the output terminals of the series-series feedback amplifier and the point where external circuitry would be connected. The analysis assumes that $v_o = 0$ as in Fig. 11.24.

11.8.1 CLOSED-LOOP GAIN CALCULATION

In order to find the closed-loop current gain A_i for the series-shunt feedback amplifier, we need to evaluate the gain expression from Eq. (11.39) by finding the ideal current gain A_i^{Ideal} and the loop gain T:

$$A_i = A_i^{\text{Ideal}} \left(\frac{T}{1+T} \right) \qquad (11.69)$$

The ideal case is depicted in Fig. 11.24. The op amp input current is zero, and input current i_i must go through resistor R_2. For infinite gain, the op amp input voltage must be zero.

Using these two ideal op amp assumptions, we can find the ideal current gain by writing a loop equation including R_2 and R_1:

$$\mathbf{i_i} R_2 + (\mathbf{i_i} - \mathbf{i_o}) R_1 = 0 \qquad \text{and} \qquad A_i^{\text{Ideal}} = \frac{\mathbf{i_o}}{\mathbf{i_i}} = 1 + \frac{R_2}{R_1} \qquad (11.70)$$

To find T, we set the value of $\mathbf{v_{id}}$ in source $A_o \mathbf{v_{id}}$ to 1 V ($A_o \mathbf{v_{id}} = A_o(1)$), and then calculate the value of v_{id} developed at the op amp input. The loop gain is then equal to the negative of the ratio of the voltage returned through the loop to the op amp input. In this case, T can be found by applying voltage division to the circuit in Fig. 11.23. (Remember independent input source i_i must be turned off by setting its value to zero.) First we find the voltage $\mathbf{v_1}$ across R_1, and then $\mathbf{v_{id}}$:

$$\mathbf{v_1} = -A_o(1) \frac{[R_1 \| (R_2 + R_{id})]}{R_1 \| (R_2 + R_{id}) + R_o} \qquad \text{and} \qquad \mathbf{v_{id}} = \mathbf{v_1} \left(\frac{R_{id}}{R_2 + R_{id}} \right)$$

$$T = A_o \frac{[R_1 \| (R_2 + R_{id})]}{R_1 \| (R_2 + R_{id}) + R_o} \left(\frac{R_{id}}{R_2 + R_{id}} \right)$$

(11.71)

The loop gain expression now incorporates the nonideal op amp parameters A_o, R_{id}, and R_o.

11.8.2 INPUT RESISTANCE CALCULATION

Next let us find the input resistance of the shunt-series feedback amplifier by applying Blackman's theorem from Eq. (11.40):

$$R_{\text{in}} = R_{\text{in}}^D \frac{1 + |T_{SC}|}{1 + |T_{OC}|} \qquad (11.72)$$

R_{in} represents the resistance appearing between the noninverting input terminal and ground of the closed-loop feedback amplifier, R_{in}^D is the resistance looking into the same terminals with the feedback loop disabled, T_{SC} is the loop gain with a short circuit applied to the input port, and T_{OC} is the loop gain with the input port open.

We find R_{in}^D using the circuit in Fig. 11.23 in which we disable the feedback by setting $A_o = 0$. The input resistance can then be written directly as

$$R_{in}^D = R_{id} \| (R_2 + R_1 \| R_o) \tag{11.73}$$

To find T_{SC}, the noninverting input terminal is connected to ground, and v_{id} is forced to be zero. Thus $T_{SC} = 0$. To find T_{OC}, the input terminals are open-circuited, and we recognize that the circuit is identical to the one used to find the loop gain T in Eq. (11.71). Therefore $|T_{oC}| = T$. The final expression for the input resistance becomes

$$R_{in} = R_{in}^D \frac{1+0}{1+T} = \frac{R_{id} \| (R_2 + R_1 \| R_o)}{1+T} \tag{11.74}$$

We see that the input resistance is decreased by T and can be small.

11.8.3 OUTPUT RESISTANCE CALCULATION
Blackman's theorem states

$$R_{out} = R_{out}^D \frac{1 + |T_{SC}|}{1 + |T_{OC}|} \tag{11.75}$$

The output resistance is the resistance appearing between the amplifier output terminal and ground, the resistance that is presented to source v_o. We find R_{out}^D using the circuit in Fig. 11.23 in which we set $i_i = 0$ and again disable the feedback by setting $A_o = 0$. The output resistance can then be written directly as

$$R_{out}^D = R_o + R_1 \| (R_2 + R_{id}) \tag{11.76}$$

For T_{SC}, output terminal v_o is connected directly to ground, and the circuit is identical to the one used to find the loop gain T in Eq. (11.71). Therefore $|T_{SC}| = T$. To find T_{OC}, the output terminals are open-circuited, and no current flows in the circuit. Thus $T_{OC} = 0$. The final expression for the output resistance becomes

$$R_{out} = R_{out}^D \frac{1+T}{1+0} = [R_o + R_1 \| (R_2 + R_{id})](1+T) \tag{11.77}$$

The output resistance is increased by T and can be very large.

11.8.4 SHUNT-SERIES FEEDBACK AMPLIFIER SUMMARY
The analysis of the characteristics of the shunt-series feedback amplifier are summarized in Eq. (11.78):

$$A_i = \left(1 + \frac{R_2}{R_1}\right)\frac{T}{1+T} \qquad R_{in} = \frac{R_{in}^D}{(1+T)} \qquad R_{out} = R_{out}^D(1+T) \tag{11.78}$$

The current amplifier has a small input resistance and a large output resistance, and its ideal current gain is $1 + R_2/R_1$.

EXERCISE: Draw the simplified circuits used to find R_{in}^D and R_{out}^D that result from setting A_o to zero and verify the expressions in Eqs. (11.73) and (11.76).

EXAMPLE 11.8 SHUNT-SERIES FEEDBACK AMPLIFIER ANALYSIS

This example evaluates the closed-loop characteristics of a op-amp-based shunt-series feedback amplifier by applying the loop-gain approach and Blackman's theorem. The analysis is extended to include a source resistance.

PROBLEM Find the closed-loop current gain, input resistance, and output resistance for the shunt-series feedback amplifier in Fig. 11.23 if the op amp has an open-loop gain of 80 dB, an input resistance of 25 kΩ, and an output resistance of 1 kΩ. Assume the amplifier is driven by a signal current with a 10-kΩ source resistance, and the feedback network is implemented with $R_2 = 27$ kΩ and $R_1 = 3$ kΩ.

SOLUTION **Known Information and Given Data:** The shunt-series feedback amplifier is drawn in Fig. 11.25 with the 10 kΩ source resistance added. For the op amp: $A_o = 80$ dB, $R_{id} = 25$ kΩ and $R_o = 1$ kΩ.

Unknowns: Closed-loop gain A_i, closed-loop input resistance R_{in}, closed-loop output resistance R_{out}.

Approach: Find new expressions for A_i^{Ideal}, T, R_{in}^D, and R_{out}^D incorporating the influence of R_I. Then calculate the values of the unknowns using the closed-loop feedback amplifier formulae derived in this section.

Assumptions: The op amp is ideal except for A_o, R_{id}, and R_o. $v_o = 0$.

Analysis: The amplifier circuit with the addition of R_I appears in Fig. 11.25.

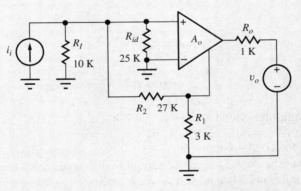

Figure 11.25 Shunt-series feedback amplifier with source resistance added.

Amplifier Analysis: In this circuit, we see that R_I is directly in parallel with R_{id}. Thus we can use the results from the previous section just by replacing R_{id} with

$$R'_{id} = R_I \| R_{id} = 10\,\text{k}\Omega \| 25\,\text{k}\Omega = 7.14\,\text{k}\Omega$$

$$A_i^{\text{Ideal}} = 1 + \frac{R_2}{R_1} = 1 + \frac{27\,\text{k}\Omega}{3\,\text{k}\Omega} = +10$$

$$R_{\text{in}}^D = R'_{id} \| (R_2 + R_1 \| R_o) = 7.14\,\text{k}\Omega \| (27\,\text{k}\Omega + 3\,\text{k}\Omega \| 1\,\text{k}\Omega) = 5.68\,\text{k}\Omega$$

$$R_{\text{out}}^D = R_o + R_1 \| (R_2 + R'_{id}) = 1\,\text{k}\Omega + 3\,\text{k}\Omega \| (27\,\text{k}\Omega + 7.14\,\text{k}\Omega) = 3.76\,\text{k}\Omega$$

The loop gain is found by modifying Eq. (11.71):

$$T = A_o \frac{[R_1|(R_2 + R'_{id})]}{R_1\|(R_2 + R'_{id}) + R_o}\left(\frac{R'_{id}}{R_2 + R'_{id}}\right)$$

$$T = 10^4 \frac{3\,\text{k}\Omega\|(27\,\text{k}\Omega + 7.14\,\text{k}\Omega)}{1\,\text{k}\Omega + 3\,\text{k}\Omega\|(27\,\text{k}\Omega + 7.14\,\text{k}\Omega)}\left(\frac{7.14\,\text{k}\Omega}{27\,\text{k}\Omega + 7.14\,\text{k}\Omega}\right) = 1535$$

Closed-Loop Amplifier Results

$$A_i = +10\left(\frac{T}{1+T}\right) = +10\left(\frac{1535}{1+1535}\right) = +9.99$$

$$R_{\text{in}} = \frac{R_{\text{in}}^D}{(1+T)} = \frac{5.68\,\text{k}\Omega}{1536} = 3.70\,\Omega$$

$$R_{\text{out}} = R_{\text{out}}^D(1+T) = 3.76\,\text{k}\Omega(1536) = 5.78\,\text{M}\Omega$$

Check of Results: We have found the three unknowns. The loop gain is high, so we expect the current gain to be approximately $+10$. The calculated input resistance is much smaller than that of the op amp itself, and the output resistance is also much larger than that of the op amp. These all agree with our expectations for the shunt-series feedback configuration.

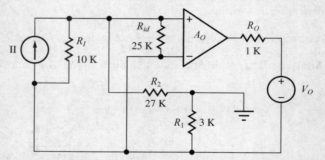

Figure 11.26 Circuit for shunt-series circuit simulation.

Computer-Aided Analysis: Our SPICE circuit in Fig. 11.26 augments the built-in OPAMP model with the addition of R_{id} in parallel with its input and R_o in series with its output, and the gain is set to 10,000. Zero value voltage source V_O is added to the circuit in order to use the transfer function analysis. We must again be careful in the simulation. The three-terminal op amp model assumes that its internal controlled source is connected to the reference node (ground). Fortunately, we are free to choose the reference node in a circuit, and the ground connection has been moved to the top of resistor R_1 in Fig. 11.26. A transfer function analysis from II to the current in V_O yields the overall current gain, the input resistance at the terminals of II and the output resistance at the terminals of V_O. The SPICE results are $A_i = 9.994$, $R_{\text{in}} = 3.698\,\Omega$ and $R_{\text{out}} = 5.773\,\text{M}\Omega$. These results agree well with our hand calculations.

EXERCISE: Find the loop gain, current gain, input resistance, and output resistance for the shunt-series feedback amplifier in Ex. 11.8 if the 10-kΩ source resistor is eliminated from the amplifier circuit.

ANSWERS: 3555, $+10.0$, 3.71 Ω, 13.7 MΩ

EXERCISE: Simulate the circuit in Fig. 11.25 and compare the results to those of Fig. 11.26.

EXERCISE: Find the loop gain, closed-loop current gain, input resistance, and output resistance for the shunt-series feedback amplifier in Ex. 11.8, if R_1 and R_2 are increased in value by a factor of 10.

ANSWERS: 248.5, +9.96, 27.9 Ω, 7.01 MΩ

11.9 FINDING THE LOOP GAIN USING SUCCESSIVE VOLTAGE AND CURRENT INJECTION

In many practical cases, particularly when the loop gain is large, the feedback loop cannot be opened to measure the loop gain because a closed loop is required to maintain a correct dc operating point. Another problem is electrical noise, which may cause an open-loop amplifier to saturate. A similar problem occurs in SPICE simulation of high-gain circuits, such as operational amplifiers, in which the circuit amplifies the numerical noise present in the calculations, and the open-loop analysis is unable to converge to a stable operating point. Fortunately, the method of **successive voltage and current injection** [5] can be used to measure the loop gain without opening the feedback loop.

Again consider the basic feedback amplifier in Fig. 11.27. To use the voltage and current injection method, an arbitrary point P within the feedback loop is selected, and a voltage source v_x is inserted into the loop, as in Fig. 11.27(a). The two voltages v_2 and v_1 on either side of the inserted source are measured, and T_v is calculated:

$$T_v = -\frac{v_2}{v_1} \tag{11.79}$$

Next, the voltage source is removed, a current i_x is injected into the same point P, and the ratio T_I of currents i_2 and i_1 is determined.

$$T_i = \frac{i_2}{i_1} \tag{11.80}$$

These two sets of measurements yield two equations in two unknowns: the loop gain T and the resistance ratio R_B/R_A. R_A represents the resistance seen looking to the left from test source v_x, and R_B represents the resistance seen looking to the right from the test source.

For the voltage injection case in Fig. 11.27(a),

$$\mathbf{v_1} = -\mathbf{i}R_A = (\mathbf{v_x} - A\mathbf{v_1})\frac{R_A}{R_A + R_B} \tag{11.81}$$

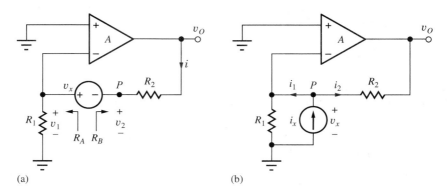

Figure 11.27 (a) Voltage injection at point P; and (b) current injection at point P. $R_{id} = \infty$.

756 Chapter 11 Nonideal Operational Amplifiers and Feedback Amplifier Stability

Solving for v_1 yields

$$\mathbf{v}_1 = \frac{\beta}{1+A\beta}\mathbf{v_x} \quad \text{where} \quad \beta = \frac{R_A}{R_A + R_B} \tag{11.82}$$

After some algebra, voltage v_2 is found to be

$$\mathbf{v}_2 = \mathbf{v}_1 - \mathbf{v_x} = \frac{\beta - (1+A\beta)}{1+A\beta}\mathbf{v_x} \tag{11.83}$$

and T_v is equal to

$$T_v = \frac{1 + A\beta - \beta}{\beta} \tag{11.84}$$

We recognize the $A\beta$ product as the loop gain T, and using $1/\beta = 1 + R_B/R_A$, T_v can be rewritten as

$$T_v = T\left(1 + \frac{R_B}{R_A}\right) + \frac{R_B}{R_A} \tag{11.85}$$

The current injection circuit in Fig. 11.27(b) provides the second equation in two unknowns. Injection of current i_x causes a voltage v_x to develop across the current generator; currents i_1 and i_2 can each be expressed in terms of this voltage:

$$\mathbf{i}_1 = \frac{\mathbf{v_x}}{R_A} \quad \text{and} \quad \mathbf{i}_2 = \frac{\mathbf{v_x} - (-A\mathbf{v_x})}{R_B} = \mathbf{v_x}\frac{1+A}{R_B} \tag{11.86}$$

Taking the ratio of these two expressions yields T_i

$$T_i = \frac{\mathbf{i}_2}{\mathbf{i}_1} = \frac{\frac{1+A}{R_B}}{\frac{1}{R_A}} = (1+A)\frac{R_A}{R_B} = \frac{R_A}{R_B} + A\frac{R_A}{R_B} \tag{11.87}$$

Multiplying the last term by β and again using $1/\beta = 1 + R_B/R_A$ yields

$$T_i = \frac{R_A}{R_B} + A\beta\frac{R_A}{R_B}\frac{1}{\beta} = \frac{R_A}{R_B} + T\left(1 + \frac{R_A}{R_B}\right) \tag{11.88}$$

Simultaneous solution of Eqs. (11.85) and (11.88) gives the desired result:

$$T = \frac{T_v T_i - 1}{2 + T_v + T_i} \quad \text{and} \quad \frac{R_B}{R_A} = \frac{1 + T_v}{1 + T_i} \tag{11.89}$$

Using this technique, we can find both the loop gain T and the resistance (or impedance) ratio at point P.

Although the resistance ratio would be dominated by R_2 and R_1 in the circuit in Fig. 11.27, R_B and R_A in the general case actually represent the two equivalent resistances that would be calculated looking to the right and left of the point P, where the loop is broken. This fact is illustrated more clearly by the SPICE analysis in Ex. 11.9.

EXAMPLE 11.9 **LOOP GAIN AND RESISTANCE RATIO CALCULATION USING SPICE**

We will use SPICE to find the loop gain for an amplifier using the successive voltage and current injection technique.

PROBLEM Find the loop gain T and the resistance ratio for the series-shunt feedback amplifier of Ex. 11.5 using the method of successive voltage and current injection at point P.

SOLUTION **Known Information and Given Data:** Series-shunt feedback amplifier with element values given in Ex. 11.5. Apply the voltage and current injection at point P between feedback resistors R_2 and R_1.

Unknowns: Loop gain T; resistance ratio R_B/R_A.

Approach: In the dc-coupled case, we can insert zero valued sources into the circuit and use the transfer function capability of SPICE to find the sensitivity of voltages v_1 and v_2 to changes in v_x and the sensitivity of i_1 and i_2 to changes in i_x.

Assumptions: From Ex. 11.5, $A_o = 80$ dB, $R_{id} = 25$ kΩ, $R_o = 1$ kΩ.

Analysis: The amplifier circuit is redrawn below with sources VX1, VX2, and IX added to the circuit. All three are zero-value sources, which do not affect the Q-point calculations. Source VX2 is added so that current I2 can be determined by SPICE.

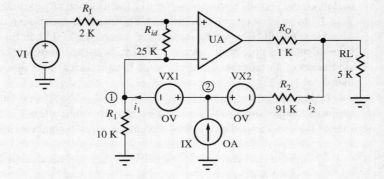

The results of the four SPICE transfer function analyses are

$$\frac{v_2}{v_{x1}} = 0.9999 \qquad \frac{v_1}{v_{x1}} = -1.294 \times 10^{-4}$$

$$\frac{i_2}{i_x} = 0.9984 \qquad \frac{i_1}{i_x} = 1.628 \times 10^{-3}$$

and the loop gain and resistance ratio calculated using these four values are

$$T_v = -\frac{-0.9999}{1.294 \times 10^{-4}} = 7730 \qquad T_i = \frac{0.9984}{1.628 \times 10^{-3}} = 613$$

$$T = \frac{T_v T_i - 1}{2 + T_v + T_i} = \frac{7730(613) - 1}{2 + 7730 + 613} = 568$$

$$\frac{R_B}{R_A} = \frac{1 + T_v}{1 + T_i} = \frac{1 + 7730}{1 + 613} = 12.6$$

Check of Results: The value of T computed by hand in the exercises following Ex. 11.5 was 568 which agrees with the result based on SPICE. Resistances R_A and R_B associated with the open feedback loop are identified in Fig. 11.27. Calculating these resistances and their ratio by hand gives

$$R_A = 10 \text{ k}\Omega \| (R_{id} + R_I) = 10 \text{ k}\Omega \| 27 \text{ k}\Omega = 7.30 \text{ k}\Omega$$

$$R_B = R_2 + (R_o \| R_L) = 91 \text{ k}\Omega + (1 \text{ k}\Omega + 5 \text{ k}\Omega) = 91.8 \text{ k}\Omega$$

$$\frac{R_B}{R_A} = 12.6$$

Again, we find good agreement with SPICE.

Discussion: The SPICE valuses and our hand calculations agree closely. As an alternative to using transfer function analyses, v_x and i_x can be made 1-V and 1-A ac sources, and two ac analyses can be performed. The ac source method has the advantage that it can find the loop gain and impedance ratio as a function of frequency. We must know the loop gain as a function of frequency in order to determine the stability of a feedback amplifier. This topic is discussed in detail in later in this chapter.

11.9.1 SIMPLIFICATIONS

Although analysis of the successive voltage and current injection method was performed using ideal sources, Middlebrook's analysis [5] shows that the technique is valid even if source resistances are included with both v_x and i_x. In addition, if point P is chosen at a position in the circuit where R_B is zero or R_A is infinite, then the equations can be simplified and T can be found from only one measurement. For example, if a point is found where R_A is infinite, then Eq. (11.85) reduces to $T = T_v$. In an ideal op amp circuit, such a point exists at the input of the op amp, as in Fig. 11.28(a).

Alternatively, if a point can be found where $R_B = 0$, then Eq. (11.85) also reduces to $T = T_v$. In an ideal op amp circuit, such a point exists at the output of the op amp, as in Fig. 11.28(b). A similar set of simplifications can be used for the current injection case. If $R_A = 0$ or R_B is infinite, then $T = T_I$.

In practice, the conditions $R_B \gg R_A$ or $R_A \gg R_B$ are sufficient to permit the use of the simplified expressions [6–8]. In the general case, where these conditions are not met, or we are not sure of the exact impedance levels, then the general method can always be applied.

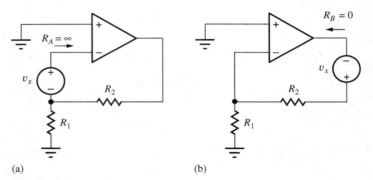

Figure 11.28 (a) Voltage injection at a point where $R_A = \infty$; and (b) voltage injection at a point where $R_B = 0$. (An ideal op amp is assumed.)

11.10 DISTORTION REDUCTION THROUGH THE USE OF FEEDBACK

Real amplifiers do not have the piece-wise linear voltage transfer functions depicted in Fig. 10.13. Actual VTCs have more of an "S" shape as in Fig. 11.29, and the nonlinearities introduce distortion into the output of the amplifier as discussed in Sec. 10.5. Fortunately, feedback can be used to significantly reduce distortion in amplifiers.

Consider the noninverting amplifier in Fig. 11.30 with an input of $v_i = V_i \sin \omega_o t$. Due to distortion, the op amp output will have both the desired output at frequency ω_o plus additional unwanted signal components at frequencies other than the input frequency [see Eq. (10.29)]:

$$v_o(t) = V_1 \sin(\omega_o t + \phi_1) + v_e(t)$$

where

$$v_e(t) = V_2 \sin(2\omega_o t + \phi_2) + V_3 \sin(3\omega_o t + \phi_3) + \ldots \quad (11.90)$$

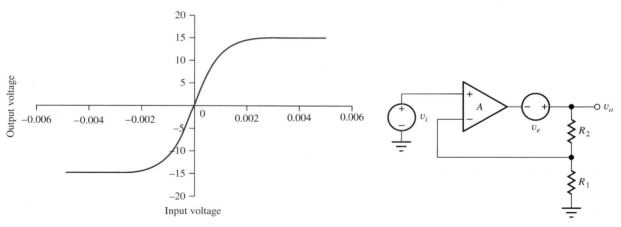

Figure 11.29 Realistic voltage transfer characteristic.

Figure 11.30 Noninverting amplifier with distortion source added.

We can model this behavior by inserting error source v_e in series with the output of the op amp as in Fig. 11.30. Now let us calculate the voltage at the output of the noninverting amplifier:

$$\mathbf{v_o} = A\mathbf{v_{id}} + \mathbf{v_e} \quad \text{and} \quad \mathbf{v_{id}} = \mathbf{v_i} - \beta \mathbf{v_o} \quad \text{where} \quad \beta = \frac{R_1}{R_1 + R_2} \quad (11.91)$$

and

$$\mathbf{v_o} = \frac{A}{1+A\beta}\mathbf{v_i} + \frac{\mathbf{v_e}}{1+A\beta} \quad (11.92)$$

In the first term, we see that the gain of the amplifier is unchanged, a result we should expect from our knowledge of superposition. In the second term, however, we find that the distortion terms are reduced by the feedback term $1 + A\beta$. In fact, in the ideal case distortion would be completely eliminated, since for $A = \infty$, the voltage across the op amp input must be zero. Since v_i contains no distortion terms, output voltage v_o cannot contain any distortion terms either, because βv_o must equal v_i. In the real case with finite gain, the distortion is still reduced by the factor $1 + A\beta$, which can be very large.

11.11 DC ERROR SOURCES AND OUTPUT RANGE LIMITATIONS

An important class of error sources results from the need to bias the internal circuits that form the operational amplifier and from mismatches between pairs of solid-state devices in these circuits. These dc error sources include input-offset voltage V_{OS}, input-bias currents I_{B1} and I_{B2}, and input-offset current I_{OS} which are each explored in the following subsections.

11.11.1 INPUT-OFFSET VOLTAGE

When the inputs of the amplifier in Fig. 11.31 are both zero, the output of the amplifier is not truly zero but is resting at some nonzero dc voltage level. A small dc voltage seems to have been applied to the input of the amplifier, which is being amplified by the gain.[1] The equivalent dc **input-offset**

---

[1] The voltage arises mainly from mismatches in the transistors in the input stage of the operational amplifier. We will explore this problem in Chapter 16.

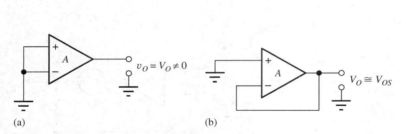

Figure 11.31 (a) Amplifier with zero input voltage but nonzero output voltage; (*Note:* The offset voltage cannot be measured in this manner.) (b) circuit for measuring offset voltage.

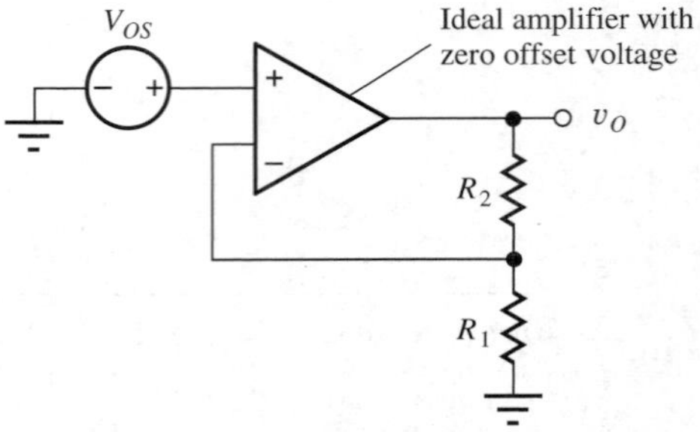

Figure 11.32 Offset voltage can be modeled by a voltage source V_{OS} in series with the amplifier input.

voltage V_{OS} is defined as

$$V_{OS} = \frac{V_O}{A}\bigg|_{v_+=0=v_-} \tag{11.93}$$

The op amp output voltage expression can be modified to include the effects of this offset voltage by adding the V_{OS} term:

$$v_O = A[v_{ID} + V_{OS}] \tag{11.94}$$

The first term in brackets represents the desired differential input signal to the amplifier, whereas the second term represents the offset-voltage error that corrupts the desired signal.

The offset voltage varies randomly from amplifier to amplifier so the actual sign of V_{OS} is not known, and only the magnitude of the worst-case offset voltage is typically specified. Most commercial operational amplifiers have offset-voltage specifications of less than 10 mV, and op amps can easily be purchased with V_{OS} specified to be less than a few mV. For additional cost, internally trimmed op amps are available with $V_{OS} < 50$ μV.

The offset voltage usually cannot actually be measured with the operational amplifier connected as depicted in Fig. 11.31(a) because of the high gain of the amplifier. However, the circuit in Fig. 11.31(b) can be used. Here the amplifier is connected as a voltage follower, and the output voltage is equal to the offset voltage of the amplifier (except for the small gain error of the amplifier since $A \neq \infty$.)

In Ex. 11.10, the effect of offset voltage is modeled as in Fig. 11.32, in which the offset voltage is represented by a source in series with the input to an otherwise ideal amplifier. V_{OS} is amplified just as any input signal source, and the dc output voltage of the amplifier in Fig. 11.32 is

$$v_O = \left(1 + \frac{R_2}{R_1}\right) V_{OS} \tag{11.95}$$

EXERCISE: (a) What is the actual output voltage in the circuit in Fig. 11.31(b) if $V_{os} = 1$ mV and $A = 60$ dB? (b) Calculate the output voltage for the circuit in Fig. 11.31(a).

ANSWERS: 0.999 mV; 1.00 V

> **EXERCISE:** What are the nominal, minimum, and maximum values of offset voltage for the AD745 operational amplifier at 25°C? (See MCD website for specification sheets.) Repeat for the OP77E.
>
> **ANSWERS:** 0.25 mV, no minimum value specified, 1.0 mV; 10 μV, no minimum value specified, 25 μV

EXAMPLE 11.10 OFFSET VOLTAGE ANALYSIS

This example calculates the output voltage of an op amp circuit caused by its offset voltage.

PROBLEM Suppose the amplifier in Fig. 11.32 has $|V_{OS}| \le 3$ mV and R_2 and R_1 are 99 kΩ and 1.2 kΩ, respectively. What is the quiescent dc voltage at the amplifier output?

SOLUTION **Known Information and Given Data:** Noninverting amplifier configuration with $R_1 = 1.2$ kΩ and $R_2 = 99$ kΩ. The amplifier has and equivalent input voltage of $|V_{OS}| \le 3$ mV.

Unknowns: Amplifier dc output voltage V_O

Approach: Use the known values to evaluate Eq. (11.95).

Assumptions: The op amp is ideal except for the specified value of nonzero offset voltage.

Analysis: Using Eq. (11.95), we find that the output voltage is

$$|V_O| \le \left(1 + \frac{99 \text{ k}\Omega}{1.2 \text{ k}\Omega}\right)(0.003) = 0.25 \text{ V}$$

Check of Results: We have found the value of the only unknown, and the value appears reasonable for standard IC power supplies.

Discussion: We do not actually know the sign of V_{OS} since the V_{OS} specification represents an upper bound. Therefore we actually know only that

$$-0.25 \text{ V} \le V_O \le 0.25 \text{ V}$$

> **EXERCISE:** Repeat the calculation in Ex. 11.10 if the noninverting amplifier gain is set to 50 and the offset voltage is 2 mV?
>
> **ANSWER:** -100 mV $\le V_O \le +100$ mV

11.11.2 OFFSET-VOLTAGE ADJUSTMENT

Addition of a potentiometer allows the **offset voltage** of most IC op amps to be manually adjusted to zero. Commercial amplifiers typically provide two terminals to which the potentiometer can be connected, as in Fig. 11.33. The third terminal of the potentiometer is connected to the positive or negative power supply voltage, and the potentiometer value depends on the internal design of the amplifier.

762   Chapter 11   Nonideal Operational Amplifiers and Feedback Amplifier Stability

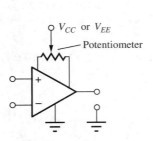

Figure 11.33 Offset-voltage adjustment of an operational amplifier.

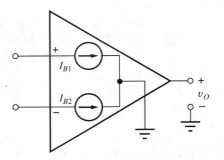

Figure 11.34 Operational amplifier with input-bias currents modeled by current sources I_{B1} and I_{B2}.

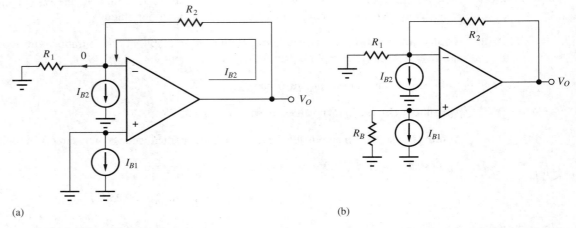

Figure 11.35 (a) Inverting amplifier with input-bias currents modeled by current sources I_{B1} and I_{B2}; (b) inverting amplifier with bias current compensation resistor R_B.

11.11.3 INPUT-BIAS AND OFFSET CURRENTS

For the transistors that form the operational amplifier to operate, a small but nonzero dc bias current must be supplied to each input terminal of the amplifier. These currents represent base currents in an amplifier built with bipolar transistors or gate currents in one designed with MOSFETs or JFETs. Although small, the bias and offset currents represent additional sources of error.

The bias currents can be modeled by two current sources I_{B1} and I_{B2} connected to the noninverting and inverting inputs of the amplifier, as in Fig. 11.34. The values of I_{B1} and I_{B2} are similar but not identical, and the actual direction of the currents depends on the details of the internal amplifier circuit (*npn*, *pnp*, NMOS, PMOS, and so on). The difference between the two bias currents is called the **offset current I_{OS}**.

$$I_{OS} = I_{B1} - I_{B2} \tag{11.96}$$

The offset-current specification for an op amp is normally expressed as an upper bound on the magnitude of I_{OS}, and the actual sign of I_{OS} for a given op amp is not known.

In an operational amplifier circuit, the **input-bias currents** produce an undesired voltage at the amplifier output. Consider the inverting amplifier in Fig. 11.35(a) as an example. In this circuit, I_{B1} is shorted out by the direct connection of the noninverting input to ground and does not affect the circuit. However, because the inverting input represents a virtual ground, the current in R_1 must be zero, forcing I_{B2} to be supplied by the amplifier output through R_2. Thus, the dc output voltage is equal to

$$V_O = I_{B2} R_2 \tag{11.97}$$

The output-voltage error in Eq. (11.97) can be reduced by placing a **bias current compensation resistor** R_B in series with the noninverting input of the amplifier, as in Fig. 11.35(b). Using analysis by superposition, the output due to I_{B1} acting alone is

$$V_O = -I_{B1} R_B \left(1 + \frac{R_2}{R_1}\right) \tag{11.98}$$

The total output voltage is the sum of Eqs. (11.97) and (11.98):

$$V_O^T = I_{B2} R_2 - I_{B1} R_B \left(1 + \frac{R_2}{R_1}\right) \tag{11.99}$$

If R_B is set equal to the parallel combination of R_1 and R_2, then the expression for the output-voltage error reduces to

$$V_O^T = (I_{B2} - I_{B1}) R_2 = -I_{OS} R_2 \quad \text{for} \quad R_B = \frac{R_1 R_2}{R_1 + R_2} \tag{11.100}$$

The value of the offset current is typically a factor of 5 to 10 times smaller than either of the individual bias currents, so the dc output-voltage error can be substantially reduced by using bias current compensation techniques.

Another example of the problems associated with offset-voltage and bias currents occurs in the integrator circuit in Fig. 11.36. A reset switch has been added to the integrator and is kept closed for $t < 0$. With the switch closed, the circuit is equivalent to a voltage follower, and the output voltage v_O is equal to the offset voltage V_{OS}. However, when the switch opens at $t = 0$, the circuit begins to integrate its own offset-voltage and bias current. Again using superposition analysis, it is easy to show (see Prob. 11.66) that the output voltage becomes

$$v_O(t) = V_{OS} + \frac{V_{OS}}{RC} t + \frac{I_{B2}}{C} t \quad \text{for} \quad t \geq 0 \tag{11.101}$$

The output voltage becomes a ramp with a constant slope determined by the values of V_{OS} and I_{B2}. Eventually, the integrator output saturates at a limit set by one of the two power supplies, as discussed in Chapter 10. If an integrator is built in the laboratory without a reset switch, the output is normally found to be resting near one of the power supply voltages.

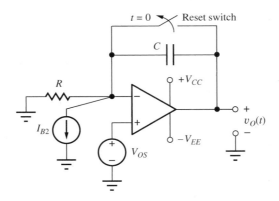

Figure 11.36 Example of dc offset-voltage and bias current errors in an integrator.

> **EXERCISE:** What are the nominal, minimum, and maximum values of the input bias and offset currents for the μA741C operational amplifier (see MCD website for specification sheets)? Repeat for the AD745J.
>
> **ANSWERS:** 80 nA, no minimum value, 500 nA; 20 nA, no minimum value, 200 nA; 150 pA, no minimum value, 400 pA; 40 pA, no minimum value, 150 pA
>
> **EXERCISE:** An inverting amplifier is designed with $R_1 = 1$ kΩ and $R_2 = 39$ kΩ. What value of resistance should be placed in series with the noninverting input terminal for bias current compensation?
>
> **ANSWERS:** 975 Ω. Note that 1 kΩ is the closest 5 percent resistor value.
>
> **EXERCISE:** An integrator has $R = 10$ kΩ, $C = 100$ pF, $V_{OS} = 1.5$ mV, and $I_{B2} = 100$ nA. How long will it take v_O to saturate (reach V_{CC} or V_{EE}) after the power supplies are turned on if $V_{CC} = V_{EE} = 15$ V?
>
> **ANSWER:** $t = 6.0$ ms

11.11.4 OUTPUT VOLTAGE AND CURRENT LIMITS

As discussed in Chapter 10, an actual operational amplifier has a limited range of voltage and current capability at its output. For example, the voltage at the output of the amplifier in Fig. 11.37 cannot exceed V_{CC} or be more negative than $-V_{EE}$. In fact, for many real op amps, the output-voltage range is limited to several volts less than the power supply span. For example, the output-voltage limits for a particular op amp might be specified as

$$(-V_{EE} + 1 \text{ V}) \leq v_O \leq (V_{CC} - 2 \text{ V}) \tag{11.102}$$

Commercial operational amplifiers also contain circuits that restrict the magnitude of the current in the output terminal in order to limit power dissipation in the amplifier to protect the amplifier from accidental short circuits. The current-limit specification is often given in terms of the minimum load resistance that an amplifier can drive with a given voltage swing. For example, an amplifier may be guaranteed to deliver an output of ± 10 V only for a total load resistance ≥ 5 kΩ. This is equivalent to saying that the total output current i_O is limited to

$$|i_O| \leq \frac{10 \text{ V}}{5 \text{ k}\Omega} = 2 \text{ mA} \tag{11.103}$$

The output-current specification not only affects the size of load resistor that can be connected to the amplifier, it also places lower limits on the value of the feedback resistors R_1 and R_2. The total

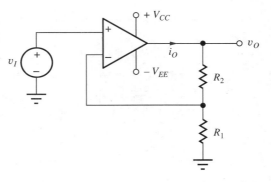

Figure 11.37 Amplifier with power supply voltages indicated.

11.11 DC Error Sources and Output Range Limitations

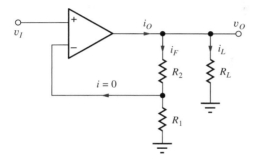

Figure 11.38 Output-current limit in the noninverting amplifier.

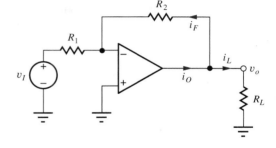

Figure 11.39 Output-current limit in the inverting-amplifier circuit.

output current i_O in Fig. 11.38 is given by $i_O = i_L + i_F$, and since the current into the ideal inverting input is zero,

$$i_O = \frac{v_O}{R_L} + \frac{v_O}{R_2 + R_1} = \frac{v_O}{R_{EQ}} \qquad (11.104)$$

The amplifier output must supply current not only to the load but also to its own feedback network! From Eq. (11.104), we see that the resistance that the noninverting amplifier must drive is equivalent to the parallel combination of the load resistance and the series combination of R_1 and R_2:

$$R_{EQ} = R_L \parallel (R_1 + R_2) \qquad (11.105)$$

For the case of the inverting amplifier in Fig. 11.39, R_{EQ} is given by

$$R_{EQ} = R_L \parallel R_2 \qquad (11.106)$$

since the inverting-input terminal of the amplifier represents a virtual ground.

The output-current constraint represented by Eqs. (11.105) and (11.106) often helps us choose the size of the feedback resistors during the design process.

EXERCISE: What is the maximum guaranteed value for the output current of the OP-27A operational amplifier (see MCD website for specification sheets)?

ANSWER: 12 V/2 kΩ = 6 mA

DESIGN EXAMPLE 11.11 INVERTING AMPLIFIER DESIGN WITH OUTPUT CURRENT LIMITS

Here we explore op amp circuit design including constraints on the output current capability of the op amp.

PROBLEM The amplifier in Fig. 11.39 is to be designed to have a gain of 20 dB and must develop a peak output voltage of at least 10 V when connected to a minimum load resistance of 5 kΩ. The op amp output current specification states that the output current must be less than 2.5 mA. Choose acceptable values of R_1 and R_2 from the table of 5 percent resistor values in Appendix A.

SOLUTION **Known Information and Given Data:** Inverting amplifier configuration; $A_v = 20$ dB, $|v_O| \leq 10$ V with $R_L \geq 5$ kΩ. The magnitude of the op amp output current must not exceed 2.5 mA.

Unknowns: Feedback resistors R_1 and R_2. Choose real values from the tables in Appendix A.

Approach: The op amp must supply current to both the load resistor and the feedback networks. We must account for both.

Assumptions: The op amp is ideal except for its limited output current capability.

Analysis: The equivalent load resistance on the amplifier must be greater than 4 kΩ:

$$R_{EQ} \geq \frac{10 \text{ V}}{2.5 \text{ mA}} = 4 \text{ k}\Omega \quad \text{or} \quad R_L \| R_2 \geq 4 \text{ k}\Omega$$

Because the minimum value of R_L is 5 kΩ, the feedback resistor R_2 must satisfy $R_2 \geq 20$ kΩ, and we also have $R_2/R_1 = 10$ because the gain was specified as 20 dB. We should allow some safety margin in the value of R_2. For example, a 27-kΩ resistor with a 5 percent tolerance will have a minimum value of 25.6 kΩ and would be satisfactory. A 22-kΩ resistor would have a minimum value of 20.9 kΩ and would also meet the specification. A wide range of choices still exists for R_1 and R_2. Several acceptable choices would be

$$R_2 = 22 \text{ k}\Omega \quad \text{and} \quad R_1 = 2.2 \text{ k}\Omega$$
$$R_2 = 27 \text{ k}\Omega \quad \text{and} \quad R_1 = 2.7 \text{ k}\Omega$$
$$R_2 = 47 \text{ k}\Omega \quad \text{and} \quad R_1 = 4.7 \text{ k}\Omega$$
$$R_2 = 100 \text{ k}\Omega \quad \text{and} \quad R_1 = 10 \text{ k}\Omega$$

Let us select the last choice: $R_1 = 10$ kΩ and $R_2 = 100$ kΩ to provide an input resistance of 10 kΩ.

Check of Results: The gain is $-R_2/R_1 = -10$, which is correct. The maximum output current will be

$$i_o \leq \frac{10 \text{ V}}{100 \text{ k}\Omega} + \frac{10 \text{ V}}{5 \text{ k}\Omega} = 2.1 \text{ mA}$$

which is less than 2.5 mA (2.2 mA if we include 5 percent tolerances).

Discussion: Note that an input resistance specification would help us decide on a value for R_1.

Computer-Aided Design: SPICE can be used to check our design using the circuit below. The gain of UA is set to 1E6 to approximate an ideal op amp. VI is set to -1 V to produce an output of $+10$ V. Operating point and transfer function analyses yield $V_O = 10$ V, $I_O = 2.1$ mA, $A_v = -10$, and $R_{in} = 10$ kΩ, all in agreement with our theory.

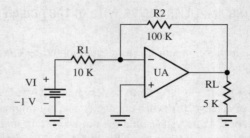

EXERCISE: What is the maximum guaranteed value for the output current of the AD745J operational amplifier (see MCD website for specification sheets)?

ANSWER: 12 V/2 kΩ = 6 mA

EXERCISE: Design a noninverting amplifier to have a gain of 20 dB and to develop a peak output voltage of at least 20 V when connected to a load resistance of at least 5 kΩ. The op amp output current specification states that the output current must be less than 5 mA. Choose acceptable values of R_1 and R_2 from the table of 5 percent resistor values in Appendix A.

ANSWER: Some possibilities: 27 kΩ and 3 kΩ; 270 kΩ and 30 kΩ; 180 kΩ and 20 kΩ; *but not* 18 kΩ and 2 kΩ because of tolerances.

EXERCISE: What is the largest output current for the design in Ex. 11.11 if the resistors have tolerances of 10 percent?

ANSWER: 2.33 mA

11.12 COMMON-MODE REJECTION AND INPUT RESISTANCE

11.12.1 FINITE COMMON-MODE REJECTION RATIO

Unfortunately, the output voltage of the real amplifier in Fig. 11.40 contains components in addition to the scaled replica of the input voltage ($A\mathbf{v_{id}}$). One of these was distortion that was discussed in Sec. 11.10. A real amplifier also responds to the signal that is in common to both inputs, called the **common-mode input voltage** v_{ic} defined as

$$v_{ic} = \left(\frac{v_1 + v_2}{2}\right) \tag{11.107}$$

The common-mode input signal is amplified by the **common-mode gain** A_{cm} to give an overall output voltage expressed by

$$\mathbf{v_o} = A(\mathbf{v_1} - \mathbf{v_2}) + A_{cm}\left(\frac{\mathbf{v_1} + \mathbf{v_2}}{2}\right) \quad \text{or} \quad \mathbf{v_o} = A\mathbf{v_{id}} + A_{cm}\mathbf{v_{ic}} \tag{11.108}$$

where A (or A_{dm}) = **differential-mode gain**

A_{cm} = common-mode gain

$v_{id} = (v_1 - v_2)$ = differential-mode input voltage

$v_{ic} = \left(\dfrac{v_1 + v_2}{2}\right)$ = common-mode input voltage

Simultaneous solution of these last two equations allows voltages v_1 and v_2 to be expressed in terms of v_{ic} and v_{id} as

$$v_1 = v_{ic} + \frac{v_{id}}{2} \quad \text{and} \quad v_2 = v_{ic} - \frac{v_{id}}{2} \tag{11.109}$$

and the amplifier in Fig. 11.40 can be redrawn in terms of v_{ic} and v_{id}, as in Fig. 11.41. This circuit is very useful for analysis by superposition that will be explored later.

An ideal amplifier would amplify the **differential-mode input voltage** v_{id} and totally reject the common-mode input signal ($A_{cm} = 0$), as has been tacitly assumed thus far. However, an actual

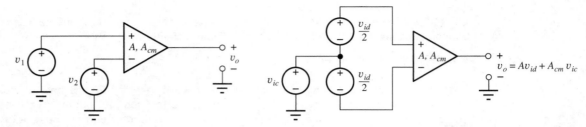

Figure 11.40 Operational amplifier with inputs v_1 and v_2.

Figure 11.41 Operational amplifier with common-mode and differential-mode in-puts shown explicitly.

amplifier has a nonzero value of A_{cm}, and Eq. (11.108) is often rewritten in a slightly different form by factoring out A:

$$\mathbf{v_o} = A\left[\mathbf{v_{id}} + \frac{A_{cm}\mathbf{v_{ic}}}{A}\right] = A\left[\mathbf{v_{id}} + \frac{\mathbf{v_{ic}}}{\text{CMRR}}\right] \quad (11.110)$$

In this equation, **CMRR** is the **common-mode rejection ratio,** defined by the ratio of A and A_{cm}

$$\text{CMRR} = \left|\frac{A}{A_{cm}}\right| \quad (11.111)$$

CMRR is often expressed in dB as

$$\text{CMRR}_{\text{dB}} = 20\log\left|\frac{A}{A_{cm}}\right| \text{ dB} \quad (11.112)$$

An ideal amplifier has $A_{cm} = 0$ and therefore infinite CMRR. Actual amplifiers usually have $A \gg A_{cm}$, and the CMRR typically falls in the range

$$60 \text{ dB} \leq \text{CMRR}_{\text{dB}} \leq 120 \text{ dB}$$

A value of 60 dB is a relatively poor level of common-mode rejection, whereas achieving 120 dB (or even higher) is possible but difficult. Generally, the sign of A_{cm} is unknown ahead of time. In addition, CMRR specifications represent a lower bound. An illustration of the problems that can be caused by finite common-mode rejection is given in the next section and Ex. 11.12.

>
> **EXERCISE:** What are the nominal, minimum, and maximum values of CMRR for the OP27 operational amplifier. (See MCD website for specification sheets.) Repeat for the AD745.
>
> **ANSWERS:** 126 dB, 114 dB, no maximum value specified; 95 dB, 80 dB, no maximum value specified

11.12.2 WHY IS CMRR IMPORTANT?

The common-mode signal concept may initially seem obscure, but we actually encounter common-mode signals quite often. In digital systems, capacitive coupling of high-frequency signals between signal lines on a bus or backplane can induce the same signal on more than one line. This induced signal often appears as a common-mode signal. Many high-speed computer buses utilize differential signaling so that the undesired common-mode signals can be eliminated by amplifiers with good CMRR.

ELECTRONICS IN ACTION

Low Voltage Differential Signaling (LVDS)

In Chapters 6 and 7, we found that one way to increase speed and/or reduce power in logic systems is to reduce the signal swing. In addition, coupling between digital signal lines (see Fig. 6.4) is a major source of noise in logic systems. The circuit below depicts a multi-drop low voltage differential signaling (LVDS) technique that has been developed to help address these problems.

Differential amplifiers are used as "receivers" ($R_{X1} - R_{X2} - R_{X3}$) to recover binary data represented by the polarity of the current driven down a transmission line by MOS (or bipolar) transistors. If data signal $D =$ "1," then transistors M_3 and M_2 are on, and M_1 and M_4 are off. The 3.5 mA current is sent through M_3, down the top conductor of the transmission line, through the 100-Ω resistor, and then returns through the lower transmission line via M_2 to ground. A +350 mV signal is developed across the 100-Ω resistor and the terminals of the differential receivers. For $D =$ "0," the current direction reverses via M_1 and M_4, and a -350 mV signal is developed at the receiver inputs. The differential-mode gain of the receiver amplifies the ±350 mV signals back up to the desired logic levels. At the same time, the amplifier's common-mode rejection capability is used to eliminate noise that is common to both conductors of the transmission line.

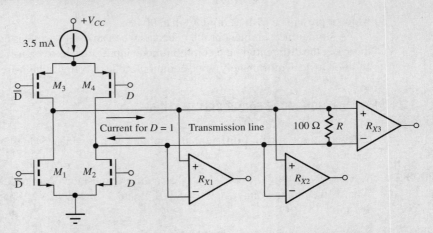

LVDS multi-drop digital logic circuitry employing the differential and common-mode characteristics of differential amplifiers.

Probably the most frequent time that we encounter common-mode signals is when we use instruments to make measurements. Consider the circuit in Fig. 11.42 in which we are trying to measure the voltage across the 100-Ω resistor with a digital multimeter (DMM). The dc voltage difference across the DMM input terminals (its differential-mode input V_{DM}) is easily found by voltage division:

$$V_{DM} = V_+ - V_- = 10 \text{ V} \left(\frac{100 \text{ }\Omega}{7300 \text{ }\Omega} \right) = 0.137 \text{ V}$$

However, there is also a dc common-mode input to the DMM:

$$V_{CM} = \frac{V_+ + V_-}{2} = \frac{1}{2} \left[10 \text{ V} \left(\frac{3700 \text{ }\Omega}{7300 \text{ }\Omega} \right) + 10 \text{ V} \left(\frac{3600 \text{ }\Omega}{7300 \text{ }\Omega} \right) \right] = 5.0 \text{ V}$$

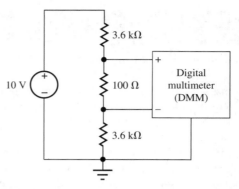

Figure 11.42 Common-mode input in a digital multimeter application.

Thus our DMM must accurately measure the 0.137 V differential input in the presence of a 5.0-V common-mode input, and this requires the digital multimeter to have good common-mode rejection capability. If we want the common-mode input to produce an error of less than 0.1 percent in the measurement of the 0.137-V input, we need

$$\frac{5.0}{\text{CMRR}} \leq 10^{-3} \times (0.137 \text{ V}) \quad \text{or} \quad \text{CMRR} \geq 3.65 \times 10^4$$

which represents a CMRR of more than 90 dB.

A similar measurement problem occurs when an oscilloscope is used in its differential mode. In this case, the differential- and common-mode inputs may have high-frequency signal components in addition to dc. Unfortunately, good common-mode rejection at high-frequencies is difficult to obtain.

EXAMPLE 11.12 COMMON-MODE ERROR CALCULATION

Calculate the error in a differential amplifier with nonideal values of gain and common-mode rejection.

PROBLEM Suppose the amplifier in Fig. 11.40 has a differential-mode gain of 2500 and a CMRR of 80 dB. What is the output voltage if $V_1 = 5.001$ V and $V_2 = 4.999$ V? What is the error introduced by the finite CMRR?

SOLUTION **Known Information and Given Data:** For the amplifier in Fig. 11.40: $A = 2500$, CMRR $= 80$ dB, $v_1 = 5.001$ V, and $v_2 = 4.999$ V.

Unknowns: Output voltage V_O; common-mode contribution to the error

Approach: Use the known values to evaluate Eq. (11.110).

Assumptions: The op amp is ideal except for finite gain and CMRR. The CMRR specification of 80 dB corresponds to CMRR $= \pm 10^4$. Let us assume CMRR $= +10^4$ for this example.

Analysis: The differential- and common-mode input voltages are

$$V_{ID} = 5.001 \text{ V} - 4.999 \text{ V} = 0.002 \text{ V} \quad \text{and} \quad V_{IC} = \frac{5.001 + 4.999}{2} \text{ V} = 5.000 \text{ V}$$

$$V_O = A\left[V_{ID} + \frac{V_{IC}}{\text{CMRR}}\right] = 2500\left[0.002 + \frac{5.000}{10^4}\right] \text{ V}$$

$$= 2500[0.002 + 0.0005] \text{ V} = 6.25 \text{ V}$$

The error introduced by the common-mode input is 25 percent of the differential input voltage.

Check for Results: We have found the required unknowns. The output voltage is a reasonable value for power supplies normally used with integrated circuit op amps.

Evaluation and Discussion: An ideal amplifier would amplify only v_{id} and produce an output voltage of 5.00 V. For this particular situation, the output voltage is in error by 25 percent due to the finite common-mode rejection of the amplifier. Common-mode rejection is important in measurements of small voltage differences in the presence of large common-mode voltages, as in the example shown here. Note in this case that

$$A_{cm} = \frac{A}{\text{CMRR}} = \frac{2500}{10{,}000} = 0.25 \text{ or } -12 \text{ dB}$$

Computer-Aided Analysis: Let's build a model to simulate this example. The output of the amplifier can be rewritten as

$$V_O = A_{dm}V_{ID} + \frac{A_{cm}}{2}V_1 + \frac{A_{cm}}{2}V_2 = 2500 V_{ID} + 0.125 V_1 + 0.125 V_2$$

which is implemented below using three voltage-controlled voltage sources. EDM depends on the voltage difference V1 – V2, ECM1 depends on the voltage V1, and ECM2 depends on the voltage V2. An operating point analysis confirms our hand analysis with $v_o = 6.25$ V. RL is added to have two connections at the output node and does not affect the calculation.

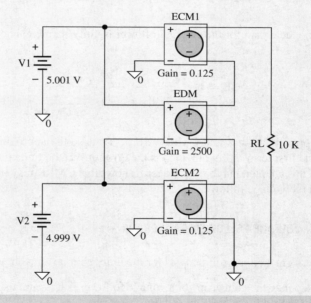

EXERCISE: The CMRR specification of 80 dB in Example 11.12 actually corresponds to $-10^4 \leq \text{CMRR} \leq +10^4$. What range of output voltages may occur?

ANSWER: $3.750 \text{ V} \leq V_O \leq 6.250 \text{ V}$

11.12.3 VOLTAGE-FOLLOWER GAIN ERROR DUE TO CMRR

Finite CMRR can also play an important role in determining the gain error in the voltage-follower circuit in Fig. 11.43, for which

$$v_{id} = v_i - v_o \quad \text{and} \quad v_{ic} = \frac{v_i + v_o}{2}$$

772 Chapter 11 Nonideal Operational Amplifiers and Feedback Amplifier Stability

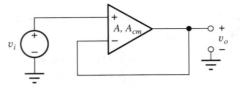

Figure 11.43 CMRR error in the voltage follower.

Using Eq. (11.108)

$$v_o = A\left[(v_i - v_o) + \frac{v_i + v_o}{2\,\text{CMRR}}\right] \quad (11.113)$$

Solving this equation for v_o yields

$$A_v = \frac{v_o}{v_i} = \frac{A\left[1 + \dfrac{1}{2\,\text{CMRR}}\right]}{1 + A\left[1 - \dfrac{1}{2\,\text{CMRR}}\right]} \quad (11.114)$$

The ideal gain for the voltage follower is unity, so the gain error is equal to

$$\text{GE} = 1 - A_v = \frac{1 - \dfrac{A}{\text{CMRR}}}{1 + A\left[1 - \dfrac{1}{2\,\text{CMRR}}\right]} \cong \frac{1}{A} - \frac{1}{\text{CMRR}} \quad (11.115)$$

Normally, both A and CMRR will be $\gg 1$, so the approximation in Eq. (11.115) is usually valid. The first term in Eq. (11.115) is the error due to the finite gain of the amplifier, as discussed earlier in this chapter, but the second term shows that CMRR may introduce an error of even greater import in the voltage follower.

EXAMPLE 11.13 VOLTAGE-FOLLOWER GAIN ERROR

Perform a gain error analysis for the unity gain op amp circuit.

PROBLEM Calculate the gain error for a voltage follower that is built using an op amp with an open-loop gain of 80 dB and a CMRR of 60 dB.

SOLUTION **Known Information and Given Data:** Operational amplifier configured as a voltage follower; $A = 80$ dB; CMRR $= 60$ dB

Unknowns: Gain error

Approach: Use the known values to evaluate Eq. (11.114).

Assumptions: The op amp is ideal except for finite open-loop gain and CMRR. The CMRR specification of 60 dB corresponds to CMRR $= \pm 1000$. Let us assume CMRR $= +1000$ for this example. Since both A and CMRR are much greater than one, we will use the approximate form of Eq. (11.115).

Analysis: Equation (12.38) gives a gain error of

$$\text{GE} \cong \frac{1}{10^4} - \frac{1}{10^3} = -9.00 \times 10^{-4} \quad \text{or} \quad -0.090 \text{ percent}$$

Check of Results: We have found the desired gain error. However, the sign is negative, which may seem a bit unusual. We better explore this result further.

Discussion: In this calculation, the error due to finite CMRR is ten times larger than that due to finite gain. As pointed out above, the gain error is negative, which corresponds to a gain that is greater than 1! Finite open-loop gain alone always causes A_v to be slightly less than 1. However, for this case,

$$A_v = \frac{A\left[1 + \dfrac{1}{2\,\text{CMRR}}\right]}{1 + A\left[1 - \dfrac{1}{2\,\text{CMRR}}\right]} = \frac{10^4\left[1 + \dfrac{1}{2(1000)}\right]}{1 + 10^4\left[1 - \dfrac{1}{2(1000)}\right]} = 1.001$$

Computer-Aided Analysis: The amplifier model from the previous example is reconnected in the circuit below as a voltage follower with V1 = 0. The gains of EDM, ECM1, and ECM2 are set to 10,000, 5, and 5, respectively. A SPICE transfer function analysis gives a voltage gain of +1.001.

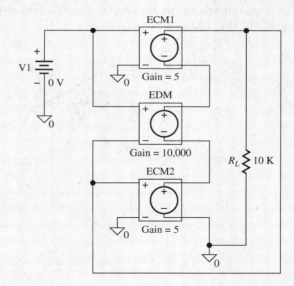

EXERCISE: What is the voltage gain in Ex. 11.13 if the CMRR is improved to 80 dB? If the differential-mode gain were only 60 dB?

ANSWERS: 1.000; 1.000

We must be aware of errors related to CMRR whenever we are trying to perform precision amplification and measurement. Discussion of CMRR often focuses on amplifier behavior at dc. However, CMRR can be an even greater problem at higher frequencies. **Common-mode rejection** decreases rapidly as frequency increases, typically with a slope of at least −20 dB/decade increase in frequency. This roll-off of the CMRR can begin at frequencies below 100 Hz. Thus, common-mode rejection at 60 or 120 Hz can be much worse than that specified for dc.

> **EXERCISE:** A voltage follower is to be designed to provide a gain error of less than 0.005 percent. Develop a set of minimum required specifications on open-loop gain and CMRR.
>
> **ANSWER:** Several possibilities: $A = 92$ dB, CMRR $= 92$ dB; $A = 100$ dB, CMRR $= 88$ dB; CMRR $= 100$ dB, $A = 88$ dB.

11.12.4 COMMON-MODE INPUT RESISTANCE

Up to now, the discussion of the input resistance of an op amp has been limited to the resistance R_{id}, which is actually the approximate resistance presented to a purely differential-mode input voltage v_{id}. In Fig. 11.44, two new resistors with value $2R_{ic}$ have been added to the circuit to model the finite common-mode input resistance of the amplifier.

When a purely common-mode signal v_{ic} is applied to the input of this amplifier, as depicted in Fig. 11.45 with $v_{id} = 0$, the input current is nonzero even though R_{id} is shorted out. In this situation, the total resistance presented to source v_{ic} is the parallel combination of the two resistors with value $2R_{ic}$, which thus equals R_{ic}. Therefore, R_{ic} is the equivalent resistance presented to the common-mode source; it is called the **common-mode input resistance** of the op amp. The value of R_{ic} is often much greater than that of the **differential-mode input resistance** R_{id}, typically in excess of 10^9 Ω (1 GΩ).

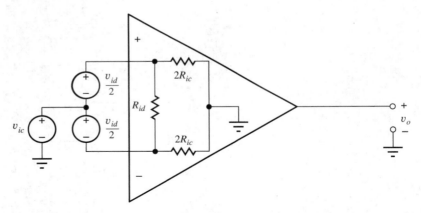

Figure 11.44 Op amp with common-mode input resistances added.

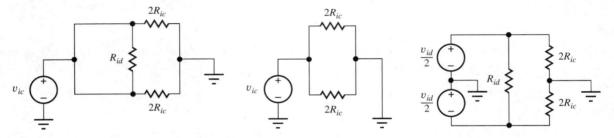

Figure 11.45 Amplifier with only a common-mode input signal present.

Figure 11.46 Amplifier input for a purely differential-mode input.

From Fig. 11.46, we see that a purely differential-mode input signal actually sees an input resistance equivalent to

$$R_{\text{in}} = R_{id} \parallel 4R_{ic} \tag{11.116}$$

As mentioned, however, R_{ic} is typically much greater than R_{id}, and the differential-mode input resistance is approximately equal to R_{id}.

11.12.5 AN ALTERNATE INTERPRETATION OF CMRR

If the differential input voltage v_{id} is set to zero in Eq. (11.111), then any residual output voltage is due to two equivalent input error voltage contributions:

$$\mathbf{v_O} = A\left(V_{OS} + \frac{\mathbf{v_{ic}}}{\text{CMRR}}\right) = A(\mathbf{v_{OS}}) \quad \text{where} \quad \mathbf{v_{OS}} = V_{OS} + \mathbf{v_{os}} \tag{11.117}$$

We can view the CMRR as being a measure of how the total offset voltage v_{OS} changes from its dc value V_{OS} when a common-mode voltage is applied. We may find CMRR as

$$\text{CMRR} = \frac{\mathbf{v_{os}}}{\mathbf{v_{ic}}} \quad \text{or} \quad \text{CMRR}_{\text{dB}} = 20 \log \left|\frac{\mathbf{v_{os}}}{\mathbf{v_{ic}}}\right|^{-1} \tag{11.118}$$

where the first form has units of μV/V.

11.12.6 POWER SUPPLY REJECTION RATIO

A parameter closely related to CMRR is the **power supply rejection ratio,** or **PSRR.** When power supply voltages change due to long-term drift or the existence of noise on the supplies, the equivalent input-offset voltage changes slightly. PSRR is a measure of the ability of the amplifier to reject these power supply variations.

In a manner similar to the CMRR, the power supply rejection ratio indicates how the offset voltage changes in response to a change in the power supply voltages.

$$\text{PSRR}_+ = \frac{\Delta V_{OS}}{\Delta V_{CC}} \quad \text{and} \quad \text{PSRR}_- = \frac{\Delta V_{OS}}{\Delta V_{EE}} \quad \text{usually expressed in } \frac{\mu\text{V}}{\text{V}} \tag{11.119}$$

PSRR is also often expressed in dB as $\text{PSRR}_{\text{dB}} = 20 \log |1/\text{PSRR}|$.

Generally the PSRR is different for changes in V_{CC} and V_{EE}, and the op amp PSRR specification usually represents the poorer of these two values. PSRR values are similar to those of CMRR with typical values ranging from 60 to 120 dB at dc. It is important to note that both CMRR and PSRR fall rapidly as frequency increases.

EXERCISE: What are the nominal, minimum, and maximum values of PSRR and CMRR for the OP77E operational amplifier (see MCD website for specification sheets)? Repeat for the AD741C.

ANSWERS: 123 dB, 120 dB, no maximum value specified; 140 dB, 120 dB, no maximum value specified; 90 dB, 76 dB, no maximum value specified; 90 dB, 70 dB, no maximum value specified.

ELECTRONICS IN ACTION

Offset Voltage, Bias Current, and CMRR Measurement

Conceptually, the three circuits in the figure below can be utilized to measure the offset voltage and bias currents of an operational amplifier. The output voltages of the three circuits are

$$V_o = V_{OS}\left(\frac{A}{1+A}\right) \qquad V_o = (V_{OS} - I_+ R_1)\left(\frac{A}{1+A}\right) \qquad V_o = (V_{OS} - I_- R_2)\left(\frac{A}{1+A}\right)$$

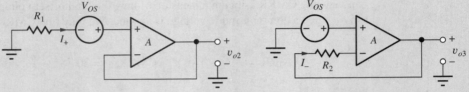

The first circuit produces a dc output voltage, which is approximately equal to the offset voltage. Adding R_1 to the circuit allows us to calculate bias current I_+, since the output voltage changes by approximately $\Delta V_O = -I_+ R_1$, where as adding R_2 to the original circuit permits calculation of bias current I_-, since the output voltage changes by approximately $\Delta V_o = +I_- R_2$. However, all of these measurements suffer from a low value of output voltage and a small gain error.

The next circuit addresses these issues by adding a second amplifier A_2. At dc, the overall open-loop gain is increased by the open-loop gain of A_2. The overall feedback amplifier can then be operated at a large closed-loop gain set by R_2 and R_1 and still have a large loop gain. The second amplifier is operated as an integrator to help stabilze the feedback loop. At dc, the integrator forces the output voltage of the device under test (DUT) to be zero, and the dc output voltage for the improved circuit is $V_o = V_{OS}(1 + R_2/R_1)$ (assuming $I_+ = 0 = I_-$). Resistor ratio R_2/R_1 can be set to increase the output voltage 10 to 1000 times V_{os} and still have a significantly reduced gain error. In this circuit, R_3 is chosen to provide bias current compensation. The bias currents can also be calculated by changing the values of R_1 and R_3.

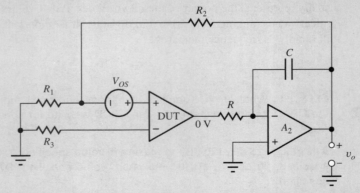

The final circuit extends the technique to include the calculation/measurement of CMRR. A common-mode input voltage is introduced by shifting the voltage at the connection between the two power supplies V_{CC} and V_{EE}. At dc, the integrator forces the output of the first op amp to equal the common-mode voltage V_{CM}, thereby shifting the operating points of the internal circuits of the op amp in the same manner that occurs for the application of a common-mode input voltage. The output voltage of the circuit including the effect of CMRR and assuming $T \gg 1$ is

$$V_O \cong \left(V_{OS} + \frac{V_{CM}}{\text{CMRR}}\right)\left(1 + \frac{R_2}{R_1}\right) \quad \text{and} \quad \frac{dV_O}{dV_{CM}} \cong \left(1 + \frac{R_2}{R_1}\right)\frac{1}{\text{CMRR}}$$

Resistor R_X has been added to the integrator to provide a zero as an additional variable that can be used to stabilize the feedback loop.

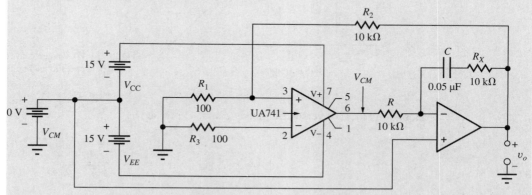

In SPICE, we can find the CMRR by using a transfer function analysis between V_{CM} and output voltage V_o. As a bonus, the power supply rejection ratios can be found in a similar manner:

$$\frac{dV_O}{dV_{CC}} \cong \left(1 + \frac{R_2}{R_1}\right)\frac{1}{\text{PSRR}_+} \quad \text{and} \quad \frac{dV_O}{dV_{EE}} \cong \left(1 + \frac{R_2}{R_1}\right)\frac{1}{\text{PSRR}_-}$$

Using the circuit above with the μA741 macro model built into SPICE yields the following values: $V_{OS} = 19.8\,\mu\text{V}$, CMRR $= 90.0$ dB, PSRR$_+$ $= 96$ dB, PSRR$_-$ $= 96$ dB.

The circuit above is convenient for use in SPICE and can be used in the laboratory. An easier way to implement the common-mode input is by shifting the two power supply voltages by an equal amount, for example, from ± 15 V to $+20$ V and -10 V. An example of this technique can be found in an Analog Devices application note.[2]

11.13 FREQUENCY RESPONSE AND BANDWIDTH OF OPERATIONAL AMPLIFIERS

Up to now we have assumed the op amp to have an ideal frequency response, that is infinite bandwidth. However, op amps are made with real electron devices that have internal capacitances, and every node in a real circuit has stray capacitance to ground. These capacitances all act to limit the bandwidth of the op amp. Most general-purpose operational amplifiers are low-pass amplifiers designed to have high gain at dc and a **single-pole frequency response** described by

$$A(s) = \frac{A_o \omega_B}{s + \omega_B} = \frac{\omega_T}{s + \omega_B} \tag{11.120}$$

in which A_o is the open-loop gain at dc, ω_B is the open-loop bandwidth of the op amp, and ω_T is called the **unity-gain frequency,** the frequency at which $|A(j\omega)| = 1$ (0 dB). The magnitude of

---

[2] "Op Amp common-Mode Rejection Ratio (CMRR)," Analog Devices Tutorial MT-042, see http://www.analog.com.

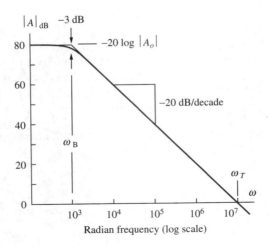

Figure 11.47 Voltage gain versus frequency for a single-pole operational amplifier.

Equation (11.120) versus frequency can be expressed as

$$|A(j\omega)| = \frac{A_o \omega_B}{\sqrt{\omega^2 + \omega_B^2}} = \frac{A_o}{\sqrt{1 + \frac{\omega^2}{\omega_B^2}}} \quad (11.121)$$

An example is depicted graphically in the Bode plot in Fig. 11.47. For $\omega \ll \omega_B$, the gain is constant at the dc value A_o. The bandwidth of the open-loop amplifier, the frequency at which the gain is 3 dB below A_o, is ω_B (or $f_B = \omega_B/2\pi$). In Fig. 11.47, $A_o = 10{,}000$ (80 dB) and $\omega_B = 1000$ rad/s (159 Hz).

At high frequencies, $\omega \gg \omega_B$, the transfer function can be approximated by

$$|A(j\omega)| \cong \frac{A_o \omega_B}{\omega} = \frac{\omega_T}{\omega} \quad (11.122)$$

Using Eq. (11.122), we see that the magnitude of the gain is indeed unity at $\omega = \omega_T$.

The amplifier in Fig. 11.47 has $\omega_T = 10^7$ rad/s, or $f_T = 1.59$ MHz.

The **gain-bandwidth product (GBW)** of the op amp is another figure of merit used to compare amplifiers (high values of GBW are usually preferred), and Eq. (11.123) defines GBW:

$$GBW = |A(j\omega)|\omega \cong \omega_T \quad (11.123)$$

Equation (11.123) states that, for any frequency $\omega \gg \omega_B$, the product of the magnitude of amplifier gain and frequency has a constant value equal to the unity-gain frequency ω_T. For this reason, the parameter ω_T (or f_T) is often referred to as the gain-bandwidth product of the amplifier. The important result in Eq. (11.123) is a property of *single-pole* amplifiers that can be represented by transfer functions of the form of Eq. (11.120).

EXAMPLE 11.14 OP AMP TRANSFER FUNCTION

Determine an op amp transfer function from a given Bode plot.

PROBLEM Write the transfer function that describes the frequency-dependent voltage gain of the amplifier in Fig. 11.47.

SOLUTION **Known Information and Given Data:** From the figure we see that the amplifier has a single-pole response as modeled by Eq. (11.120).

Unknowns: To evaluate the transfer function, we need to find A_o and ω_B.

Approach: The values must be found from the graph and converted into proper form before insertion into Eq. (11.120).

Assumptions: The amplifier can be modeled by the single-pole formula.

Analysis: At low frequencies, the gain asymptotically approaches 80 dB, which must be converted back from dB:

$$A_o = 10^{80\,\text{dB}/20\,\text{dB}} = 10^4$$

The cutoff frequency ω_B can also be read directly from the graph and is already in radian form: $\omega_B = 10^3$ rad/s. Substituting the values of A_o and ω_B into Eq. (11.120) yields the desired transfer function.

$$A_v(s) = \frac{A_o \omega_B}{s + \omega_B} = \frac{10^4(10^3)}{s + 10^3} = \frac{10^7}{s + 10^3} = \frac{10000}{1 + s/1000}$$

Check of Results: We have found the unknown transfer function. We can check the answer for consistency by observing that the numerator value represents the unity-gain frequency ω_T. From the graph we see that ω_T is indeed 10^7 rad/s.

Discussion: Note that we often express the frequency values in Hz and that $A_o f_B = f_T$.

$$f_B = \frac{\omega_B}{2\pi} = 159 \text{ Hz} \quad \text{and} \quad f_T = \frac{\omega_T}{2\pi} = 1.59 \text{ MHz}$$

EXERCISE: An op amp has a gain of 100 dB at dc and a unity-gain frequency of 5 MHz. What is f_B? Write the transfer function for the gain of the op amp.

ANSWERS: 50 Hz; $A(s) = \dfrac{10^7 \pi}{s + 100\pi}$

EXERCISE: What are the nominal values of open-loop gain and unity-gain frequency for the AD745 operational amplifier (see MCD website for specification sheets)? Write a transfer function for the op amp.

ANSWERS: 200,000; 1 MHz; $A_v(s) = \dfrac{A_o \omega_B}{s + \omega_B} = \dfrac{\omega_T}{s + \omega_B} = \dfrac{2\pi \times 10^6}{s + 10\pi}$

11.13.1 FREQUENCY RESPONSE OF THE NONINVERTING AMPLIFIER

We now use the frequency-dependent op-amp gain expression to study the closed-loop frequency response of the noninverting and inverting amplifiers. The closed-loop gain for the noninverting amplifier was found previously to be

$$A_v = \frac{A}{1 + A\beta} \quad \text{where} \quad \beta = \frac{R_1}{R_1 + R_2} \quad (11.124)$$

Our original derivation of this gain expression actually placed no restrictions on the functional form of A. Up to now, we have assumed A to be a constant, but we can explore the frequency response of the closed-loop feedback amplifier by replacing A in Eq. (11.124) by the frequency-dependent voltage-gain expression for the op amp, Eq. (11.120):

$$A_v(s) = \frac{A(s)}{1 + A(s)\beta} = \frac{\dfrac{A_o \omega_B}{s + \omega_B}}{1 + \dfrac{A_o \omega_B}{s + \omega_B}\beta} = \frac{A_o \omega_B}{s + \omega_B(1 + A_o \beta)} \quad (11.125)$$

Dividing by the factor $(1 + A_o\beta)\omega_B$, Eq. (11.125) can be written as

$$A_v(s) = \frac{\dfrac{A_o}{1+A_o\beta}}{\dfrac{s}{(1+A_o\beta)\omega_B} + 1} = \frac{A_v(0)}{\dfrac{s}{\omega_H} + 1} \qquad (11.126)$$

where the upper-cutoff frequency is

$$\omega_H = \omega_B(1 + A_o\beta) = \omega_T \frac{(1+A_o\beta)}{A_o} = \frac{\omega_T}{A_v(0)} \qquad (11.127)$$

The closed-loop amplifier also has a single-pole response of the same form as Eq. (11.120), but its dc gain and bandwidth are given by

$$A_v(0) = \frac{A_o}{1+A_o\beta} \quad \text{and} \quad \omega_H = \frac{\omega_T}{A_v(0)} \qquad (11.128)$$

For $A_o\beta \gg 1$, Eq. (11.127) reduces to

$$A_v(0) \cong \frac{1}{\beta} \quad \text{and} \quad \omega_H \cong \beta\omega_T \qquad (11.129)$$

Note that the gain-bandwidth product of the closed-loop amplifier is constant:

$$A_v(0)\,\omega_H = \omega_T$$

From Eq. (11.128), we see that the gain must be reduced in order to increase ω_H, or vice versa. We will explore this in more detail shortly.

The loop gain $A(s)\beta$ is now also a function of frequency depicted as in Fig. 11.48. At frequencies for which $|A(j\omega)\beta| \gg 1$, Eq. (11.124) reduces to $1/\beta$, the constant value derived previously for low frequencies. However, at frequencies for which $|A(j\omega)\beta| \ll 1$, Eq. (11.124) becomes $A_v \cong A(j\omega)$. At low frequencies, the gain is set by the feedback, but at high frequencies, we find that the gain must follow the gain of the amplifier. We should not expect a (negative) feedback amplifier to produce more gain than is available from the open-loop operational amplifier by itself.

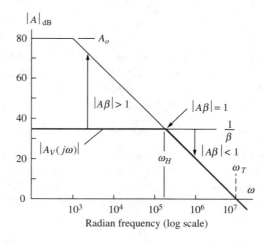

Figure 11.48 Graphical interpretation of operational amplifier with feedback.

11.13 Frequency Response and Bandwidth of Operational Amplifiers

These results are indicated graphically by the bold lines in Fig. 11.48 for an amplifier with $1/\beta = 35$ dB. Loop gain T can be expressed as

$$T = A\beta = \frac{A}{\left(\dfrac{1}{\beta}\right)} \quad \text{and (in dB)} \quad |A\beta|_{\text{dB}} = |A|_{\text{dB}} - \left|\frac{1}{\beta}\right|_{\text{dB}} \tag{11.130}$$

At any given frequency, the magnitude of the loop gain is equal to the difference between A_{dB} and $(1/\beta)_{\text{dB}}$ on the graph. The upper half-power frequency $\omega_H = \beta\omega_t$ corresponds to the frequency at which $(1/\beta)$ intersects $|A(j\omega)|$ corresponding to $|A\beta| = 1$ (actually $A\beta \cong -j1 = 1\angle{-90°}$). For the case in Fig. 11.48, $\beta = 0.0178$ (-35 dB) and $\omega_H = 0.0178 \times 10^7 = 1.78 \times 10^5$ rad/s.

EXAMPLE 11.15 NONINVERTING AMPLIFIER FREQUENCY RESPONSE

Characterize the frequency response of a noninverting amplifier built with a nonideal op amp having limited gain and bandwidth.

PROBLEM An op amp has a dc gain of 100 dB and a unity-gain frequency of 10 MHz. (a) What is the bandwidth of the op amp? (b) If the op amp is used to build a noninverting amplifier with a closed-loop gain of 60 dB, what is the bandwidth of the feedback amplifier? (c) Write an expression for the transfer function of the op amp. (d) Write an expression for the transfer function of the noninverting amplifier.

SOLUTION **Known Information and Given Data:** We are given $A_o = 10^5$ (100 dB) and $f_T = 10^7$ Hz. The desired closed-loop gain is $A_v = 1000$ (60 dB).

Unknowns: (a) Bandwidth f_B of the operational amplifier, (b) Bandwidth f_H of the closed-loop amplifier, (c) op amp transfer function, (d) noninverting amplifier transfer function

Approach: Evaluate Eqs. (11.124) to (11.129), which model the behavior of the noninverting amplifier.

Assumptions: Since we have been given values for A_o and f_T, we will assume that the amplifier is described by a single-pole transfer function. The op amp is ideal except for its single pole frequency response.

Analysis:

(a) The cutoff frequency of the op amp is f_B, its -3dB frequency:

$$f_B = \frac{f_T}{A_o} = \frac{10^7 \text{ Hz}}{10^5} = 100 \text{ Hz}$$

(b) Using Eq. (11.127), the bandwidth of the noninverting amplifier is

$$f_H = f_B(1 + A_o\beta) = 100(1 + 10^5 \cdot 10^{-3}) = 10.1 \text{ kHz}$$

in which the feedback factor β is determined by the desired closed-loop gain.

$$\beta = \frac{1}{A_v(0)} = \frac{1}{1000} = 10^{-3}$$

(c) Substituting the values of A_o and ω_B into Eq. (11.120) yields the op amp transfer function.

$$A_v(s) = \frac{A_o\omega_B}{s + \omega_B} = \frac{10^5(2\pi)(10^2)}{s + (2\pi)(10^2)} = \frac{2\pi \times 10^7}{s + 200\pi}$$

(d) Evaluating Eq. (11.125) yields the noninverting amplifier transfer function.

$$A_v(s) = \frac{A_o \omega_B}{s + \omega_B(1 + A_o\beta)} = \frac{10^5(2\pi)(10^2)}{s + (2\pi)(10^2)[1 + 10^5(10^{-3})]} = \frac{2\pi \times 10^7}{s + 2.02\pi \times 10^4}$$

Check of Results: We have found each of the requested answers. The numerators of the transfer functions should be equal to $\omega_T = 2\pi f_T$ and are correct. $A_v(0) = 990$ is also correct.

EXERCISE: An op amp has a dc gain of 90 dB and a unity-gain frequency of 5 MHz. What is the cutoff frequency of the op amp? If the op amp is used to build a noninverting amplifier with a closed-loop gain of 40 dB, what is the bandwidth of the feedback amplifier? Write an expression for the transfer function of the op amp. Write an expression for the transfer function of the noninverting amplifier.

ANSWERS: 158 Hz; 50 kHz; $A(s) = \dfrac{10^7 \pi}{s + 316\pi}$; $A_v(s) = \dfrac{10^7 \pi}{s + 10^5 \pi}$

EXERCISE: Show that $A\beta \cong -j1$ at the frequency at which $(1/\beta)$ intersects $|A(j\omega)|$.

11.13.2 INVERTING AMPLIFIER FREQUENCY RESPONSE

The frequency response for the inverting-amplifier configuration can be found in a manner similar to that for the noninverting case by substituting the frequency-dependent op amp gain expression, Eq. (11.120), into the equation for the closed-loop gain of the inverting amplifier.

$$A_v = \left(-\frac{R_2}{R_1}\right) \frac{A(s)\beta}{1 + A(s)\beta} \qquad \text{where} \qquad \beta = \frac{R_1}{R_1 + R_2}$$

or

(11.131)

$$A_v(s) = \left(-\frac{R_2}{R_1}\right) \frac{\dfrac{A_o \omega_B}{s + \omega_B}\beta}{1 + \dfrac{A_o \omega_B}{s + \omega_B}\beta} = \left(-\frac{R_2}{R_1}\right) \frac{\dfrac{A_o \beta}{1 + A_o \beta}}{\dfrac{s}{\omega_B(1 + A_o\beta)} + 1} = \left(-\frac{R_2}{R_1}\right) \frac{A_o \beta \omega_B}{s + \omega_B(1 + A_o\beta)}$$

For $A_o\beta \gg 1$, these equations reduce to

$$A_v = \frac{\left(-\dfrac{R_2}{R_1}\right)\dfrac{A_o\beta}{(1 + A_o\beta)}}{\dfrac{s}{\omega_H} + 1} \cong \frac{\left(-\dfrac{R_2}{R_1}\right)}{\dfrac{s}{\omega_H} + 1} \qquad \text{and} \qquad \omega_H = \frac{\omega_T}{\dfrac{A_o}{(1 + A_o\beta)}} \cong \beta\omega_T \qquad (11.132)$$

where the approximate values hold for $A_o\beta \gg 1$. This expression again represents a single-pole transfer function. The gain at low frequencies, $A_v(0)$, is set by the resistor ratio $(-R_2/R_1)$, and the bandwidth expression is identical to that of the noninverting amplifier, $\omega_H = \beta\omega_T$.

The frequency response characteristics of the inverting and noninverting amplifiers are summarized in Table 11.3, in which the expressions have been recast in terms of the ideal value of the gain at low frequencies. The expressions are quite similar. However, for a given value of dc gain, the noninverting amplifier will have slightly greater bandwidth than the inverting amplifier because of the difference in the relation between β and $A_v(0)$. The difference is significant only for amplifier stages designed with low values of closed-loop gain.

11.13 Frequency Response and Bandwidth of Operational Amplifiers

TABLE 11.3
Inverting and Noninverting Amplifier Frequency Response Comparison

$\beta = \dfrac{R_1}{R_1+R_2}$	NONINVERTING AMPLIFIER	INVERTING AMPLIFIER
dc gain	$A_v(0) = 1 + \dfrac{R_2}{R_1}$	$A_v(0) = -\dfrac{R_2}{R_1}$
Feedback factor	$\beta = \dfrac{1}{A_v(0)}$	$\beta = \dfrac{1}{1 + |A_v(0)|}$
Bandwidth	$f_B = \beta f_T$	$f_B = \beta f_T$
Input resistance	$R_{ic} \| R_{id}(1 + A\beta)$	$R_1 + \left(R_{ID} \| \dfrac{R_2}{1+A} \right)$
Output resistance	$\dfrac{R_o}{1 + A\beta}$	$\dfrac{R_o}{1 + A\beta}$

EXAMPLE 11.16 INVERTING AMPLIFIER FREQUENCY RESPONSE

Characterize the frequency response of an inverting amplifier built using a nonideal op amp having limited gain and bandwidth.

PROBLEM An op amp has a dc gain of 200,000 and a unity-gain frequency of 500 kHz. (a) What is the cutoff frequency of the op amp? (b) If the op amp is used to build an inverting amplifier with a closed-loop gain of 40 dB, what is the bandwidth of the feedback amplifier? (c) Write an expression for the transfer function of the op amp. (d) Write an expression for the transfer function of the inverting amplifier.

SOLUTION **Known Information and Given Data:** For the op amp; $A_o = 2 \times 10^5$, $f_T = 5 \times 10^5$ Hz; for the inverting amplifier, $A_v = -100$ (40 dB)

Unknowns: (a) Op amp cutoff frequency, (b) inverting amplifier bandwidth, (c) op amp transfer function, (d) inverting amplifier transfer function

Approach: Evaluate Eq. (11.120) for the op amp. Evaluate Eqs. (11.131) and (11.132), which model the behavior of the inverting amplifier.

Assumptions: The op amp has a single-pole frequency response. Otherwise it is ideal.

Analysis:

(a) The cutoff frequency of the op amp is f_B, its -3 dB frequency:

$$f_B = \frac{f_T}{A_o} = \frac{5 \times 10^5 \text{ Hz}}{2 \times 10^5} = 2.5 \text{ Hz}$$

(b) Using Eq. (11.132), the bandwidth of the inverting amplifier is

$$f_H = f_B(1 + A_o\beta) = 2.5 \text{ Hz} \left(1 + \frac{2 \times 10^5}{101} \right) = 4.95 \text{ kHz}$$

in which the feedback factor β is determined by the desired closed-loop gain (see Table 11.3).

$$\beta = \frac{1}{1 + |A_v(0)|} = \frac{1}{101}$$

(c) Substituting the values of A_o and ω_B into Eq. (11.120) yields the op amp transfer function.

$$A_v(s) = \frac{A_o \omega_B}{s + \omega_B} = \frac{\omega_T}{s + \omega_B} = \frac{(2\pi)(5 \times 10^5)}{s + (2\pi)(2.5)} = \frac{10^6 \pi}{s + 5\pi}$$

(d) Evaluating Eq. (11.131) yields the inverting amplifier transfer function:

$$A_v(s) = \left(-\frac{R_2}{R_1}\right) \frac{A_o \beta \omega_B}{s + \omega_B(1 + A_o \beta)} = (-100) \frac{(2 \times 10^5)\left(\frac{1}{101}\right)(2\pi)(2.5)}{s + (2\pi)(2.5)\left(1 + \frac{2 \times 10^5}{101}\right)}$$

$$= -\frac{9.90 \times 10^5 \pi}{s + 9.91 \times 10^3 \pi}$$

Check of Results: We have found the answers to all the unknowns. We can also double check the last transfer function by evaluating its dc gain and bandwidth.

$$A_v(0) = -\frac{9.90 \times 10^5 \pi}{9.91 \times 10^3 \pi} = -99.9 \quad \text{and} \quad f_H = \frac{9.91 \times 10^3 \pi}{2\pi} = 4.96 \text{ kHz}$$

The values agree within round-off error.

EXERCISE: An op amp has a dc gain of 90 dB and a unity-gain frequency of 5 MHz. What is the cutoff frequency of the op amp? If the op amp is used to build an inverting amplifier with a closed-loop gain of 50 dB, what is the bandwidth of the feedback amplifier? Write an expression for the transfer function of the op amp. Write an expression for the transfer function of the inverting amplifier.

ANSWERS: 158 Hz; 15.8 kHz; $A(s) = \dfrac{10^7 \pi}{s + 316\pi}$; $A(s) = \dfrac{10^7 \pi}{s + 3.16\pi \times 10^5}$

EXERCISE: If the amplifier in Ex. 11.15 is used in a voltage follower, what is its bandwidth? If the amplifier is used in an inverting amplifier with $A_v = -1$, what is its bandwidth?

ANSWERS: 10 MHz; 5 MHz

11.13.3 USING FEEDBACK TO CONTROL FREQUENCY RESPONSE

In the preceding sections, we found that feedback can be used to stabilize the gain and improve the input and output resistances of an amplifier, and that feedback can also be used to trade reduced gain for increased bandwidth in low-pass amplifiers. In this section, we extend the analysis to more general feedback amplifiers.

The closed-loop gain for all the feedback amplifiers in this chapter can be written as

$$A_v = \frac{A}{1 + A\beta} \quad \text{or} \quad A_v(s) = \frac{A(s)}{1 + A(s)\beta(s)} \tag{11.133}$$

Up to now, we have worked with the midband value of A and assumed it to be a constant. However, we can explore the frequency response of the general closed-loop feedback amplifier by substituting a frequency-dependent voltage gain expression for A into Eq. (11.133).

11.13 Frequency Response and Bandwidth of Operational Amplifiers

Suppose that amplifier A is an amplifier with cutoff frequencies of ω_H and ω_L and midband gain A_o as described by

$$A(s) = \frac{A_o \omega_H s}{(s + \omega_L)(s + \omega_H)} \tag{11.134}$$

Substituting Eq. (11.134) into Eq. (11.133) and simplifying the expression yields

$$A_v(s) = \frac{\frac{A_o \omega_H s}{(s + \omega_L)(s + \omega_H)}}{1 + \frac{A_o \omega_H s}{(s + \omega_L)(s + \omega_H)} \beta} = \frac{A_o \omega_H s}{s^2 + [\omega_L + \omega_H(1 + A_o \beta)]s + \omega_L \omega_H} \tag{11.135}$$

Assuming that $\omega_H(1 + A_o \beta) \gg \omega_L$, then dominant-root factorization (see Sec. 17.6.3) yields these estimates of the upper- and lower-cutoff frequencies and bandwidth of the closed-loop feedback amplifier:

$$\omega_L^F \cong \frac{\omega_L \omega_H}{\omega_L + \omega_H(1 + A_o \beta)} \cong \frac{\omega_L}{1 + A_o \beta}$$

$$\omega_H^F \cong \omega_L + \omega_H(1 + A_o \beta) \cong \omega_H(1 + A_o \beta) \tag{11.136}$$

$$\text{BW}_F = \omega_H^F - \omega_L^F \cong \omega_H(1 + A_o \beta)$$

The upper- and lower-cutoff frequencies and bandwidth of the feedback amplifier are all improved by the factor $(1 + A_o \beta)$. Using the approximations in Eq. (11.136), we find that the transfer function in Eq. (11.135) can be rewritten approximately as

$$A_v(s) \cong \frac{\frac{A_o}{(1 + A_o \beta)} s}{\left(s + \frac{\omega_L}{(1 + A_o \beta)}\right)\left[1 + \frac{s}{\omega_H(1 + A_o \beta)}\right]} \tag{11.137}$$

As expected, the midband gain is stabilized at

$$A_{\text{mid}} = \frac{A_o}{1 + A_o \beta} \cong \frac{1}{\beta} \tag{11.138}$$

It should once again be recognized that the gain-bandwidth product of the closed-loop amplifier remains constant:

$$\text{GBW} = A_{\text{mid}} \times \text{BW}_F \cong \frac{A_o}{1 + A_o \beta} \omega_H(1 + A_o \beta) = A_o \omega_H \tag{11.139}$$

These results are displayed graphically in Fig. 11.49 for an amplifier with $1/\beta = 20$ dB. The open-loop amplifier has $A_o = 40$ dB, $\omega_L = 100$ rad/s, and $\omega_H = 10{,}000$ rad/s, whereas the closed-loop amplifier has $A_v = 19.2$ dB, $\omega_L = 9.1$ rad/s, and $\omega_H = 110{,}000$ rad/s.

EXERCISE: An op amp has a dc gain of 100 dB and a unity-gain frequency of 10 MHz. What is the upper-cutoff frequency of the op amp itself? If the op amp is used to build a noninverting amplifier with a closed-loop gain of 60 dB, what is the bandwidth of the feedback amplifier? Write an expression for the transfer function of the op amp. Write an expression for the transfer function of the noninverting amplifier.

ANSWERS: 100 Hz; 10 kHz; $A(s) = 2\pi \times 10^7/(s + 200\pi)$; $A(s) = 2\pi \times 10^7/(s + 2\pi \times 10^4)$

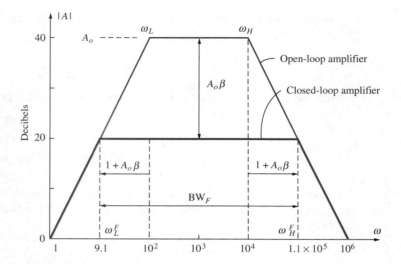

Figure 11.49 Graphical interpretation of feedback amplifier frequency response.

11.13.4 LARGE-SIGNAL LIMITATIONS—SLEW RATE AND FULL-POWER BANDWIDTH

Up to this point, we have tacitly assumed that the internal circuits that form the operational amplifier can respond instantaneously to changes in the input signal. However, the internal amplifier nodes all have an equivalent capacitance to ground, and only a finite amount of current is available to charge these capacitances. Thus, there will be some limit to the rate of change of voltage on the various nodes. This limit is described by the **slew-rate (SR)** specification of the operational amplifier. Slew rate defines the maximum rate of change of voltage at the output of the operational amplifier. Typical values of slew rate for general-purpose op amps fall into the range

$$0.1 \text{ V/}\mu\text{s} \leq \text{SR} \leq 10 \text{ V/}\mu\text{s}$$

although much higher values are possible in special designs. An example of a slew-rate limited signal at an amplifier output is shown schematically in Fig. 11.50.

For a given frequency, slew rate limits the maximum amplitude of a signal that can be amplified without distortion. Consider a sinusoidal output signal $v_o = V_M \sin \omega t$, for example. The maximum

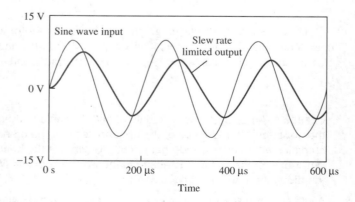

Figure 11.50 An example of a slew-rate limited output signal.

rate of change of this signal occurs at the zero crossings and is given by

$$\left.\frac{dv_O}{dt}\right|_{\max} = V_M \omega \cos \omega t |_{\max} = V_M \omega \quad (11.140)$$

For no signal distortion, this maximum rate of change must be less than the slew rate:

$$V_M \omega \leq \text{SR} \quad \text{or} \quad V_M \leq \frac{\text{SR}}{\omega} \quad (11.141)$$

The **full-power bandwidth** f_M is the highest frequency at which a full-scale signal can be developed. Denoting the amplitude of the full-scale output signal by V_{FS}, the full-power bandwidth can be written as

$$f_M \leq \frac{\text{SR}}{2\pi V_{FS}} \quad (11.142)$$

EXERCISE: Suppose that an op amp has a slew rate of 0.5 V/μs. What is the largest sinusoidal signal amplitude that can be reproduced without distortion at a frequency of 20 kHz? If the amplifier must deliver a signal with a 10-V maximum amplitude, what is the full-power bandwidth corresponding to this signal?

ANSWERS: 3.98 V; 7.96 kHz

11.13.5 MACRO MODEL FOR OPERATIONAL AMPLIFIER FREQUENCY RESPONSE

The actual internal circuit of an operational amplifier may contain from 20 to 100 bipolar and/or field-effect transistors. If the actual circuits were used for each op amp, simulations of complex circuits containing many op amps would be very slow. Simplified circuit representations, called **macro models,** have been developed to model the terminal behavior of the op amp. The two-port model that we used in this chapter is one simple form of macro model. This section introduces a model that can be used for SPICE simulation of the frequency response of circuits utilizing operational amplifiers.

To model the single-pole roll-off, an auxiliary loop consisting of the voltage-controlled voltage source with value v_1 in series with R and C is added to the interior of the original two-port, as depicted in Fig. 11.51. The product of R and C is chosen to give the desired −3 dB point for the open-loop amplifier. If a voltage source is applied to the input, then the open-circuit voltage gain ($R_L = \infty$) is

$$A_v(s) = \frac{V_o(s)}{V_1(s)} = \frac{A_o \omega_B}{s + \omega_B} \quad \text{where} \quad \omega_B = \frac{1}{RC} \quad (11.143)$$

This interior loop represents a "dummy" circuit added just to model the frequency response; the individual values of R and C are arbitrary. For example, $R = 1\,\Omega$ and $C = 0.0159\,\text{F}$, $R = 1000\,\Omega$ and $C = 15.9\,\mu\text{F}$, or $R = 1\,\text{M}\Omega$ and $C = 0.0159\,\mu\text{F}$ may all be used to model a cutoff frequency of 10 Hz. A similar form of single-pole macro model is utilized in a SPICE simulation in Ex. 11.17 in Sec. 11.14.

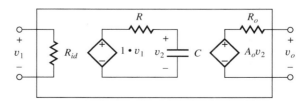

Figure 11.51 Simple macro model for an operational amplifier.

EXERCISE: Create a macro model for the OP27 based on Fig. 11.51 (see MCD website for specification sheets). Use the nominal specification values.

ANSWERS: $R_{id} = 6$ MΩ, $R = 2000$ Ω, $C = 17.9$ μF, $A_o = 1.8 \times 10^6$, $R_o = 70$ Ω. The individual values of R and C are arbitrary as long as $RC = (1/8.89\,\pi)$ s.

11.13.6 COMPLETE OP AMP MACRO MODELS IN SPICE

Most versions of SPICE contain sophisticated macro models for op amps including descriptions for many commercial op amps. These macro models include all of the nonideal limitations we have discussed in this chapter and contain a large number of parameters that can be adjusted to model op amp behavior. Both three- and five-terminal op amps may be included as shown in Fig. 11.52. An example parameter set is given in Table 11.4. In addition to those discussed previously in this chapter, we see parameters to model multiple poles and a zero in the op amp frequency response, to describe the input capacitance, and to set the input transistors to *npn* or *pnp* devices. This choice determines the direction of the input bias current; the bias current goes into the op amp terminals for an *n*-type input and comes out of the terminals for a *p*-type input.

11.13.7 EXAMPLES OF COMMERCIAL GENERAL-PURPOSE OPERATIONAL AMPLIFIERS

Now that we have explored the theory of circuits using ideal and nonideal operational amplifiers, let us look in more detail at the characteristics of a general-purpose operational amplifier. A portion of the specification sheets for one such commercial op amp, the AD745 series from Analog Devices Corporation, can be found on the MCD website. These amplifiers are fabricated with an IC technology that has both bipolar transistors and JFETs.

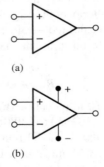

Figure 11.52 (a) Three-terminal op amp; (b) five-terminal op amp.

TABLE 11.4
Typical Op Amp Macro Model Parameter Set

PARAMETER	TYPICAL VALUE
Differential-mode gain (dc)	106 dB
Differential-mode input resistance	2 MΩ
Input capacitance	1.5 pF
Common-mode rejection ratio	90 dB
Common-mode input resistance	2 GΩ
Output resistance	50 Ω
Input offset voltage	1 mV
Input bias current	80 nA
Input offset current	20 nA
Positive slew rate	0.5 V/μs
Negative slew rate	0.5 V/μs
Maximum output source current	25 mA
Maximum output sink current	25 mA
Input type (*n*- or *p*-type)	*n*-type
Frequency of first pole	5 Hz
Frequency of zero	5 MHz
Frequency of second pole	2 MHz
Frequency of third pole	20 MHz
Frequency of fourth pole	100 MHz
Power supply voltage (3-pin model)	15 V

Note that many of the specifications are stated in terms of typical values plus either upper or lower bounds. For example, the voltage gain for the AD745J at $T = 25°C$ with ± 15-V power supplies is typically 132 dB but has a minimum value of 120 dB and no upper bound. The offset voltage is typically 0.25 mV with an upper bound of 1 mV, but the AD745K version is also available with a typical offset voltage of 0.1 mV and an upper bound of 0.5 mV. The input stage of this amplifier contains JFETs, so the input-bias current is very small at room temperature, and the nominal input resistance is very large.

The minimum common-mode rejection ratio (at dc) is 80 dB, and PSRR and CMRR specifications are the same. With ± 15-V power supplies, the amplifier can handle input signals with a common-mode range of $+13.3$ and -10.7 V, and the amplifier is guaranteed to develop an output-voltage swing of $+12$ V with a 2-kΩ load resistance.

The AD745J has a minimum gain-bandwidth product (unity-gain frequency f_T) of 20 MHz, and a slew rate of 12.5 V/μs. Considerable additional information is included concerning the performance of the amplifier family over a large range of power supply voltages and temperatures.

11.14 STABILITY OF FEEDBACK AMPLIFIERS

Whenever an amplifier is embedded within a feedback network, a question of **stability** arises. Up to this point, it has tacitly been assumed that the feedback is negative. However, as frequency increases, the phase of the loop gain changes, and it is possible for the feedback to become positive at some frequency. If the gain is also greater than or equal to 1 at this frequency, then instability occurs, typically in the form of oscillation.

The locations of the poles of a feedback amplifier can be found by analysis of the closed-loop transfer function described by

$$A_v(s) = \frac{A(s)}{1 + A(s)\beta(s)} = \frac{A(s)}{1 + T(s)} \quad (11.144)$$

The poles occur at the complex frequencies s for which the denominator becomes zero:

$$1 + T(s) = 0 \quad \text{or} \quad T(s) = -1 \quad (11.145)$$

The particular values of s that satisfy Eq. (11.145) represent the poles of $A_v(s)$. For amplifier stability, the poles must lie in the left half of the s-plane. Now we discuss two graphical approaches for studying stability using Nyquist and Bode plots.

11.14.1 THE NYQUIST PLOT

The **Nyquist plot** is a useful graphical method for qualitatively studying the locations of the poles of a feedback amplifier. The graph represents a mapping of the right half of the s-plane (RHP) onto the $T(s)$-plane, as in Fig. 11.53. Every value of s in the s-plane has a corresponding value of $T(s)$. The critical issue is whether any value of s in the RHP corresponds to $T(s) = -1$. However, checking every possible value of s would take a rather long time. Nyquist realized that to simplify the process, we need only plot $T(s)$ for values of s on the $j\omega$ axis

$$T(j\omega) = A(j\omega)\beta(j\omega) = |T(j\omega)| \angle T(j\omega) \quad (11.146)$$

which represents the boundary between the RHP and LHP. $T(j\omega)$ is normally graphed using the polar coordinate form of Eq. (11.146). If the -1 **point** is enclosed by this boundary, then there must be some value of s for which $T(s) = -1$, a pole exists in the RHP, and the amplifier is not stable.[3] However, if -1 lies outside the interior of the Nyquist plot, then the poles of the closed-loop amplifier are all in the left half-plane, and the amplifier is stable, although possibly only marginally so.

---

[3] If we mentally "walk" around the s-plane, keeping the shaded region on our right, then the corresponding region in the $T(s)$-plane will also be on our right as we "walk" in the $T(s)$-plane.

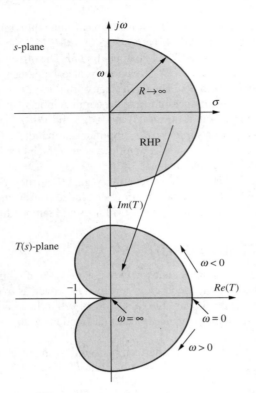

Figure 11.53 Nyquist plot as a mapping between the s-plane and the $T(s)$-plane.

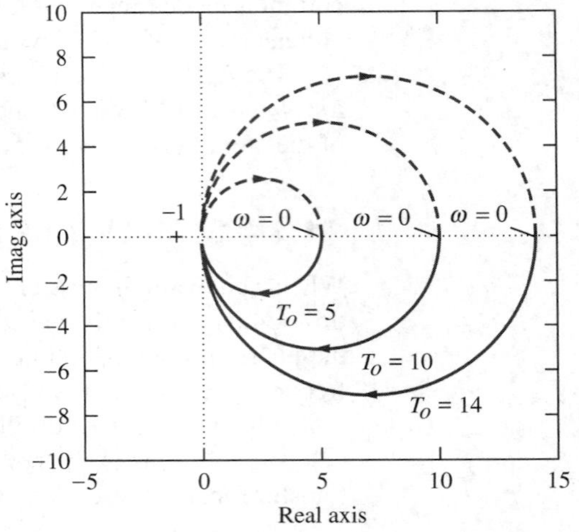

Figure 11.54 Nyquist plot for first-order $T(s)$ for $T_o = 5$, 10, and 14. (Nyquist plots are easily made using MATLAB. This figure is generated by three simple MATLAB statements: nyquist(14,[1 1]), nyquist(10,[1 1]), and nyquist(5,[1 1]).)

Today, we are fortunate to have computer tools such as MATLAB, which can quickly construct the Nyquist plot for us. These tools eliminate the tedious work involved in creating the graphs so that we can concentrate on interpretation of the information. Let us consider examples of basic first-, second-, and third-order systems.

11.14.2 FIRST-ORDER SYSTEMS

In most of the feedback amplifiers we have considered thus far, β was a constant and $A(s)$ was the frequency-dependent part of the loop gain $T(s)$. However, the important thing is the overall behavior of $T(s)$. The simplest case of $T(s)$ is that of a basic low-pass amplifier with a loop-gain described by

$$T(s) = \frac{A_o \omega_o}{s + \omega_o} \beta = \frac{T_o}{s + \omega_o} \tag{11.147}$$

For example, Eq. (11.147) might correspond to a single-pole operational amplifier with resistive feedback. The Nyquist plot for

$$T(j\omega) = \frac{T_o}{j\omega + 1} \tag{11.148}$$

is given in Fig. 11.54. At dc, $T(0) = T_o$, whereas for $\omega \gg 1$,

$$T(j\omega) \cong -j\frac{T_o}{\omega} \tag{11.149}$$

As frequency increases, the magnitude monotonically approaches zero, and the phase asymptotically approaches $-90°$.

From Eq. (11.149), we see that changing the feedback factor β scales the value of $T_o = T(0)$,

$$T(0) = A_o \omega_o \beta \qquad (11.150)$$

but changing $T(0)$ simply scales the radius of the circles in Fig. 11.54, as indicated by the curves for $T_o = 5$, 10, and 14. It is impossible for the graph in Fig. 11.54 to ever enclose the $T = -1$ point (indicated by the "+" symbol in Fig. 11.54), and the amplifier is stable regardless of the value of T_o. This is one reason why general-purpose op amps are often internally compensated to have a single-pole low-pass response. Single-pole op amps are stable for any fixed value of β.

11.14.3 SECOND-ORDER SYSTEMS AND PHASE MARGIN

A second-order loop-gain function can be described by

$$T(s) = \frac{A_o}{\left(1 + \dfrac{s}{\omega_1}\right)\left(1 + \dfrac{s}{\omega_2}\right)} \beta = \frac{T_o}{\left(1 + \dfrac{s}{\omega_1}\right)\left(1 + \dfrac{s}{\omega_2}\right)} \qquad (11.151)$$

An example appears in Fig. 11.55 for

$$T(s) = \frac{14}{(s+1)^2} \quad \text{and} \quad T(j\omega) = \frac{14}{(j\omega+1)^2} \qquad (11.152)$$

In this case, T_o is 14, but at high frequencies

$$T(j\omega) \cong (-j)^2 \frac{14}{\omega^2} = -\frac{14}{\omega^2} \qquad (11.153)$$

As frequency increases, the magnitude decreases monotonically from 14 toward 0, and the phase asymptotically approaches $-180°$. Again, it is theoretically impossible for this transfer function to encircle the -1 point since the curve never crosses the negative real axis. However, the second-order system can come arbitrarily close to this point, as indicated in Fig. 11.56, which is a blowup of the Nyquist plot in the region near the -1 point. The larger the value of T_o, the closer the curve will come to the -1 point. The curve in Fig. 11.56 is plotted for a T_o value of only 14, whereas an actual op amp circuit could easily have a T_o value of 1000 or more.

Although technically stable, the second-order system can have essentially zero **phase margin,** as defined in Fig. 11.57. Phase margin Φ_M represents the maximum increase in phase shift (phase lag)

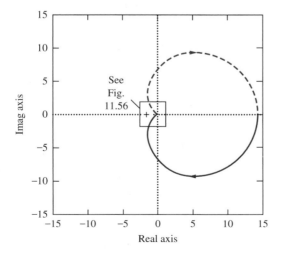

Figure 11.55 Nyquist plot for second-order $T(s)$. (Generated using MATLAB command: nyquist(14,[1 2 1]).)

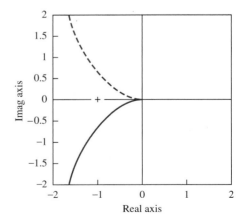

Figure 11.56 Blowup of Fig. 11.55 near the -1 point. The second-order system does not enclose the -1 point but may come arbitrarily close to doing so.

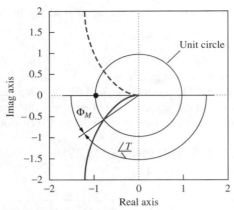

Figure 11.57 Definition of phase margin Φ_M.

that can be tolerated before the system becomes unstable. Φ_M is defined as

$$\Phi_M = \angle T(j\omega_1) - (-180°) = 180° + \angle T(j\omega_1) \quad \text{where} \quad |T(j\omega_1)| = 1 \quad (11.154)$$

To find Φ_M, we first must determine the frequency ω_1 for which the magnitude of the loop gain is unity, corresponding to the intersection of the Nyquist plot with the unit circle in Fig. 11.57, and then determine the phase shift of T at this frequency. The difference between this angle and $-180°$ is Φ_M.

Small phase margin leads to excessive peaking in the closed-loop frequency response and undesirable ringing in the step response. In addition, any rotation of the Nyquist plot due to additional phase shift (from poles that may have been neglected in the model, for example) can lead to instability.

11.14.4 STEP RESPONSE AND PHASE MARGIN

We are interested in phase margin not only because of stability concerns but also because phase margin is directly related to the time domain response of the feedback system and its overshoot and settling time. Consider an op amp with a transfer function containing the original low, frequency pole ω_B plus a second high-frequency pole at ω_2:

$$A(s) = \frac{A_o \omega_B}{(s + \omega_B)} \frac{\omega_2}{(s + \omega_2)} \quad (11.155)$$

If we assume $\omega_2 \gg \omega_B$, then the open-loop bandwidth of the op amp is approximately ω_B. For frequency independent feedback β, the closed-loop response is

$$A_{CL} = \frac{A(s)}{1 + A(s)\beta} = \frac{A_o \omega_B \omega_2}{s^2 + s(\omega_B + \omega_2) + \omega_B \omega_2(1 + A_o\beta)} = \frac{A_o}{1 + A_o\beta}\left(\frac{\omega_n^2}{s^2 + 2\zeta\omega_n s + \omega_n^2}\right)$$

$$\omega_n = \sqrt{\omega_B \omega_2 (1 + A_o\beta)} \quad \text{and} \quad \zeta = \frac{1}{2}\frac{(\omega_B + \omega_2)}{\omega_B \omega_2 (1 + A_o\beta)} = \frac{(\omega_B + \omega_2)}{2\omega_n^2} \quad (11.156)$$

in which two new parameters have been introduced: ζ is called the damping coefficient and ω_n is the pole frequency of the system. The natural frequencies of the system are given by

$$p_{1,2} = -\zeta\omega_n \pm j\omega_n\sqrt{1 - \zeta^2} \quad (11.157)$$

There are three cases to consider [9] based upon the value of ζ:

(i) Over damped $\zeta > 1$ (two real poles): $p_{1,2} = -\omega_n\left(\zeta \pm \sqrt{\zeta^2 - 1}\right)$

$$v_{od}(t) = V_F\left\{1 - \frac{1}{2\sqrt{\zeta^2 - 1}}\left[\frac{\varepsilon^{-\omega_n\left(\zeta - \sqrt{\zeta^2-1}\right)t}}{\zeta - \sqrt{\zeta^2 - 1}} - \frac{\varepsilon^{-\omega_n\left(\zeta + \sqrt{\zeta^2-1}\right)t}}{\zeta + \sqrt{\zeta^2 - 1}}\right]\right\} \quad \text{and} \quad V_F = \frac{A_o}{1 + A_o\beta}V_i$$

$$(11.158)$$

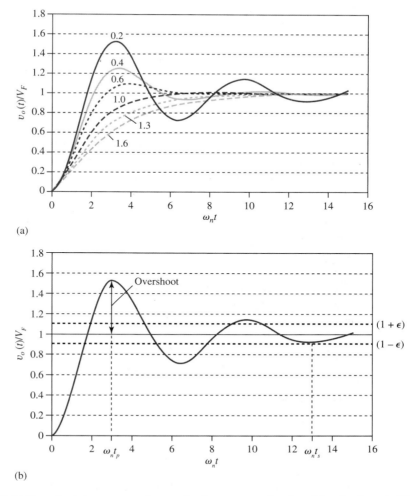

Figure 11.58 (a) Normalized step response of a second-order system as a function of damping factor ζ; (b) overshoot and settling time for $\zeta = 0.2$ and $\varepsilon = 0.1$.

(ii) Critically damped $\zeta = 1$ (two identical poles): $p_{1,2} = -\zeta \omega_n$

$$v_{cd}(t) = V_F \left[1 - (1+t)\varepsilon^{-\zeta \omega_n t} \right] \tag{11.159}$$

(iii) Under damped $\zeta < 1$ (two complex poles): $p_{1,2} = -\omega_n \zeta \pm j\omega_n \sqrt{1-\zeta^2}$

$$v_{ud}(t) = V_F \left[1 - \frac{1}{\sqrt{1-\zeta^2}} \varepsilon^{-\zeta \omega_n t} \sin\left(\sqrt{1-\zeta^2}\omega_n t + \phi\right) \right] \qquad \phi = \arctan\left(\frac{\sqrt{1-\zeta^2}}{\zeta}\right) \tag{11.160}$$

Graphs of the step response of the second-order system[4] as a function of damping factor ζ appear in Fig. 11.58(a) with definitions of overshoot and settling time in Fig. 11.58(b). *Overshoot* is a measure of how far the waveform initially exceeds its final value and can be specified as either a fraction or a percentage of final value. The overshoot peaks at time t_p. In Fig. 11.58(b) for which

---

[4] Note that for simulations, one must ensure that signals are small enough that the system is behaving linearly and nonlinear slew rate limiting is not occurring in the response.

$\zeta = 0.2$, the overshoot is 52.7 percent at $t_p = (3.21/\omega_n)$ seconds.

$$\text{Fractional overshoot} = \frac{\text{Peak value} - \text{Final value}}{\text{Final value}} = \exp\left(-\frac{\pi\zeta}{\sqrt{1-\zeta^2}}\right)$$

$$t_p = \frac{\pi}{\omega_n\sqrt{1-\zeta^2}} \qquad (11.161)$$

The time required for the waveform to fall within a given fractional error ε of final value is called the *settling time* t_s. In Fig. 11.58(b), the bold dashed lines represent 10 percent error bars ($\varepsilon = 0.1$), and the settling time t_s is slightly less than $13/\omega_n$ seconds.

An overdamped response ($\zeta > 1$) has no overshoot, but the settling time required to reach within a given error band of final value is the longest. The critically damped response ($\zeta = 1$) has the shortest settling time without overshoot and corresponds to the maximally flat frequency Butterworth response (see Sec. 12.3.1). However, shorter settling time can actually be obtained with an underdamped response ($\zeta < 1$), but the waveform exhibits overshoot and ringing that depend upon damping coefficient ζ.

Unity Gain Frequency and Phase Margin

The phase margin for the system can be found by calculating the unity-gain frequency for the loop gain based upon the transfer function in Eq. (11.155):

$$T(s) = A(s)\beta = \frac{A_o\beta\omega_B\omega_2}{s^2 + s(\omega_B + \omega_2) + \omega_B\omega_2} = \left(\frac{A_o\beta}{1 + A_o\beta}\right) \frac{\omega_n^2}{s^2 + 2\zeta\omega_n s + \frac{\omega_n^2}{1+A_o\beta}} \qquad (11.162)$$

In order to find the unity-gain frequency for the loop gain, we first must first construct the the expression for the magnitude of $T(j\omega)$ and set it equal to one which yields

$$|T(j\omega_T)| = 1 \quad \Rightarrow \quad \left(\frac{A_o\beta}{1 + A_o\beta}\right)^2 \omega_n^4 = \left(\frac{\omega_n^2}{1 + A_o\beta} - \omega_T^2\right)^2 + 4\zeta\omega_n^2\omega_T^2 \qquad (11.163)$$

After a considerable amount of algebra [9] and assuming $A_o\beta \gg 1$, it can be found that

$$\omega_T = \omega_n \left(\sqrt{4\zeta^4 + 1} - 2\zeta^2\right)^{\frac{1}{2}}$$

and

$$\phi_M = \tan^{-1} \frac{2\zeta}{\left(\sqrt{4\zeta^4 + 1} - 2\zeta^2\right)^{\frac{1}{2}}} = \cos^{-1}\left(\sqrt{4\zeta^4 + 1} - 2\zeta^2\right) \qquad (11.164)$$

The results in Eqs. (11.161) and (11.164) can be used to relate overshoot, phase margin, and damping coefficient. Sample results appear in Table 11.5. A common design specification is to achieve a minimum phase margin of 60° corresponding to an overshoot of less than 10 percent.

DESIGN NOTE

A common goal in feedback system design is to achieve a minimum phase margin of 60° that corresponds to an overshoot of less than 10 percent.

TABLE 11.5
Overshoot versus Phase Margin and Damping Coefficient

OVERSHOOT (%)	PHASE MARGIN (°)	DAMPING COEFFICIENT
1	71	0.83
2	69	0.78
3	67	0.75
5	65	0.69
7	62	0.65
10	59	0.59
20	48	0.46
30	39	0.36
50	24	0.22
70	13	0.11

EXERCISE: What damping factor and phase margin is required to achieve an overshoot of no more than 1 percent? How much overshoot will occur with a phase margin of 45°?

ANSWERS: 0.826, 70.9°; 23.4 percent

EXERCISE: Suppose we desire the amplifier in Fig. 11.58(b) to settle within 10 percent in 10 μsec. What value of ω_n is required? If the op amp has $f_T = 1$ MHz and $1/\beta = 10$, what is the value of $f_2 = \omega_2/2\pi$?

ANSWERS: ≥ 1.30 Mrad/s; ≥ 428 kHz

11.14.5 THIRD-ORDER SYSTEMS AND GAIN MARGIN

Third-order systems described by

$$T(s) = \frac{A_o}{\left(1 + \dfrac{s}{\omega_1}\right)\left(1 + \dfrac{s}{\omega_2}\right)\left(1 + \dfrac{s}{\omega_3}\right)} \beta = \frac{T_o}{\left(1 + \dfrac{s}{\omega_1}\right)\left(1 + \dfrac{s}{\omega_2}\right)\left(1 + \dfrac{s}{\omega_3}\right)} \quad (11.165)$$

can easily have stability problems. Consider the example in Fig. 11.59, for

$$T(s) = \frac{14}{s^3 + s^2 + 3s + 2} \quad (11.166)$$

For this case, $T(0) = 7$, and at high frequencies

$$T(j\omega) \cong (-j)^3 \frac{14}{\omega^3} = +j\frac{14}{\omega^3} \quad (11.167)$$

At high frequencies, the polar plot asymptotically approaches zero along the positive imaginary axis, and the plot can enclose the critical -1 point under many circumstances. The particular case in Fig. 11.59 represents an unstable closed-loop system.

Gain margin is another important concept and is defined as the reciprocal of the magnitude of $T(j\omega)$ evaluated at the frequency for which the phase shift is 180°:

$$\text{GM} = \frac{1}{|T(j\omega_{180})|} \quad \text{where} \quad \angle T(j\omega_{180}) = -180° \quad (11.168)$$

Gain margin is often expressed in dB as $\text{GM}_{\text{dB}} = 20\log(\text{GM})$.

Equation (11.168) is interpreted graphically in Fig. 11.60. If the magnitude of $T(s)$ is increased by a factor equal to or exceeding the gain margin, then the closed-loop system becomes unstable, because the Nyquist plot then encloses the -1 point.

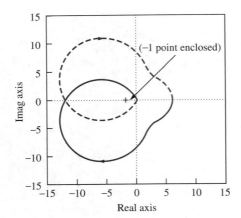

Figure 11.59 Nyquist plot for third-order $T(s)$. (Using MATLAB: nyquist(14,[1 1 3 2]).)

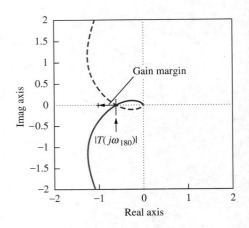

Figure 11.60 Nyquist plot showing gain margin of a third-order system. (Using MATLAB: nyquist(5,[1 3 3 1]).)

EXERCISE: Find the gain margin for the system in Fig. 11.60 described by

$$T(s) = \frac{5}{s^3 + 3s^2 + 3s + 1} = \frac{5}{(s+1)^3}$$

ANSWER: 4.08 dB

11.14.6 DETERMINING STABILITY FROM THE BODE PLOT

Phase and gain margin can also be determined directly from a **Bode plot** of the loop gain, as indicated in Fig. 11.61. This figure represents T for a third-order transfer function:

$$T = A\beta = \frac{2 \times 10^{19}}{(s + 10^5)(s + 10^6)(s + 10^7)} = \frac{2 \times 10^{19}}{s^3 + 11.1 \times 10^6 s^2 + 11.1 \times 10^{12} s + 10^{18}}$$

Phase margin is found by first identifying the frequency at which $|A\beta| = 1$ or 0 dB. For the case in Fig. 11.61, this frequency is approximately 1.2×10^6 rad/s. At this frequency, the phase shift is $-145°$, and the phase margin is $\Phi_M = 180° - 145° = 35°$. The amplifier can tolerate an additional phase shift of approximately $35°$ before it becomes unstable.

Gain margin is found by identifying the frequency at which the phase shift of the amplifier is exactly $180°$. In Fig. 11.61, this frequency is approximately 3.2×10^6 rad/s. The loop gain at this frequency is -17 dB, and the gain margin is therefore $+17$ dB. The gain must increase by 17 dB before the amplifier becomes unstable.

Using a tool like MATLAB, we can easily construct the Bode plot for the gain of the amplifier and use it to determine the range of closed-loop gains for which the amplifier will be stable. Stability can be determined by properly interpreting the Bode magnitude plot. We use this mathematical approach:

$$20 \log |A\beta| = 20 \log |A| - 20 \log \left|\frac{1}{\beta}\right| \tag{11.169}$$

Rather than plotting the loop gain $A\beta$ itself, the magnitude of the open-loop gain A and the reciprocal of the feedback factor β are plotted separately. (Remember, $A_v \cong 1/\beta$.) The frequency at which these two curves intersect is the point at which $|A\beta| = 1$, and the phase margin of the closed-loop amplifier can easily be determined from the phase plot.

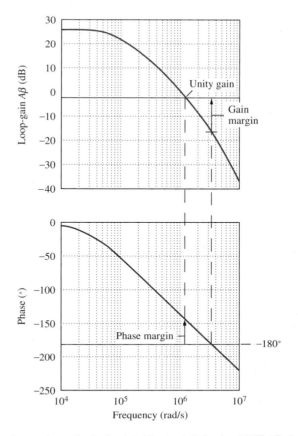

Figure 11.61 Phase and gain margin on the Bode plot. (Graph plotted using MATLAB: bode(2E19, [1 11.1E6 11.1 E12 1E18]).)

Let us use the Bode plot in Fig. 11.62 as an example. In this case,

$$A(s) = \frac{2 \times 10^{24}}{(s + 10^5)(s + 3 \times 10^6)(s + 10^8)} \qquad (11.170)$$

The asymptotes from Eq. (11.170) have also been included on the graph. For simplicity in this example, we assume that the feedback is independent of frequency (e.g., a resistive voltage divider) so that $1/\beta$ is a straight line.

Three closed-loop gains are indicated. For the largest closed-loop gain, $(1/\beta) = 80$ dB, the phase margin is approximately 85°, and stability is not a problem. The second case corresponds to a closed-loop gain of 50 dB and has a phase margin of only 15°. Although stable, the amplifier operating at a closed-loop gain of 50 dB exhibits significant overshoot and "ringing" in its step response. Finally, if an attempt is made to use the amplifier as a unity gain voltage follower, the amplifier will be unstable (negative phase margin). We see that the phase margin is zero for a closed-loop gain of approximately 35 dB.

Relative stability can be inferred directly from the magnitude plot. If the graphs of A and $1/\beta$ intersect at a "rate of closure" of 20 dB/decade, then the amplifier will be stable. However, if the two curves intersect in a region of 40 dB/decade, then the closed-loop amplifier will have poor phase margin (in the best case) or be unstable (in the worst case). Finally, if the rate of closure is 60 dB or greater, the closed-loop system will be unstable. The closure rate criterion is equally applicable to frequency-dependent feedback as well.

Chapter 11 Nonideal Operational Amplifiers and Feedback Amplifier Stability

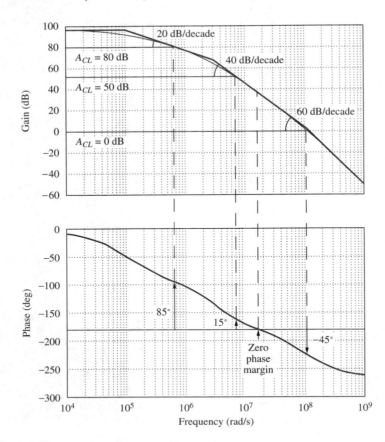

Figure 11.62 Determining stability from the Bode magnitude plot. Three values of closed-loop gain are indicated: 80 dB, 50 dB, and 0 dB. The corresponding phase margins are 85°, 15° (ringing and overshoot), and −45° (unstable). (Graph plotted using MATLAB: bode(2E24,[1   103.1E6   310.3E12   30E18]).)

EXAMPLE 11.17 PHASE MARGIN ANALYSIS

Even single-pole op amps can exhibit phase margin problems when driving heavy capacitive loads. This example evaluates the phase margin of the voltage follower with a large load capacitance connected to the output.

PROBLEM Find the phase margin for a voltage follower that is driving a 0.01 μF capacitive load. Assume the op amp has an open-loop gain of 100 dB, an f_T of 1 MHz, and an output resistance of 250 Ω.

SOLUTION **Known Information and Given Data:** The voltage follower with load capacitor C_L and output resistance R_O added is drawn in the figure. For the op amp circuit: $A_o = 100$ dB, $f_T = 1$ MHz, $C_L = 0.01$ μF, and $R_o = 250$ Ω. Feedback factor $\beta = \beta(s)$ is now a function of frequency.

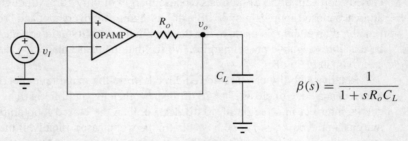

$$\beta(s) = \frac{1}{1 + sR_oC_L}$$

Voltage follower with output resistance and capacitive load.

Unknowns: Phase margin for the closed-loop amplifier

Approach: Find an expression for the loop gain T, find the frequency for which $|T| = 1$, evaluate the phase at this frequency, and compare it to $180°$.

Assumptions: The op amp is ideal except for A_o, f_T, and R_o.

Analysis: For the op amp, $f_B = f_T/A_o = 10^6/10^5 = 10$ Hz.

$$T(s) = A(s)\beta(s) = \frac{A_o}{1 + \dfrac{s}{\omega_B}} \cdot \frac{1}{1 + sR_oC_L} = \frac{10^5}{1 + \dfrac{s}{20\pi}} \cdot \frac{1}{1 + \dfrac{s}{4 \times 10^5}}$$

$$|T(j\omega)| = \frac{10^5}{\sqrt{1 + \left(\dfrac{\omega}{20\pi}\right)^2}\sqrt{1 + \left(\dfrac{\omega}{4\times 10^5}\right)^2}} \qquad \angle T(j\omega) = -\tan^{-1}\left(\frac{\omega}{20\pi}\right) - \tan^{-1}\left(\frac{\omega}{4 \times 10^5}\right)$$

Solving for $|T(j\omega)| = 1$ using a calculator or spreadsheet yields $f = 248$ kHz ($\omega = 1.57$ Mrad/s) and $\angle T = -166°$. Thus the phase margin is only $180° - 166° = 34°$.

Check of Results: We have found the phase margin to be only $34°$.

Discussion: A second pole is added to the system by the break between the load capacitance and op amp output resistance. Thus the overall feedback amplifier becomes a second-order system, and although it will not oscillate, the phase margin can be poor, and the step response may exhibit overshoot and ringing. Unfortunately, a real op amp has additional poles that can further degrade the phase margin and increase the possibility of oscillation. We can explore this issue further with SPICE simulation.

Computer-Aided Analysis: First we need to create a simulation model that reflects our analysis. One possibility that employs two op amps is given below. The first op amp is an ideal amplifier with 100 dB gain. R_1 and C_1 are added to model the single op amp pole at 10 Hz, and this "dummy network" is buffered from the output by a second amplifier with its gain set to 1. R_o and C_L complete the circuit. A small-signal pulse input with an amplitude of 5mV, a pulse width of 35 μs, a period of 70 μs, and rise and fall times of 10 ns is used so only a relatively short simulation time is needed.

In the simulated waveform, we observe more than 10 percent overshoot and ringing, and the output takes approximately 30 μs to settle to its final value. If we repeat the simulation with the original circuit utilizing a single μA741 op amp macro model, we find the real situation is much worse than our simple model predicts because the μA741 actually contains additional poles at high frequency.

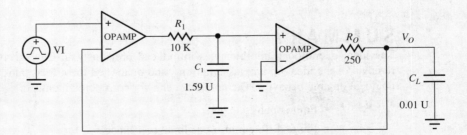

800 Chapter 11 Nonideal Operational Amplifiers and Feedback Amplifier Stability

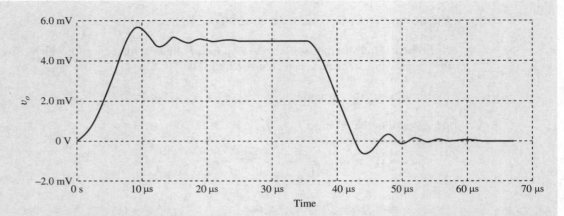

Step response of the circuit model that corresponds to our mathematical analysis.

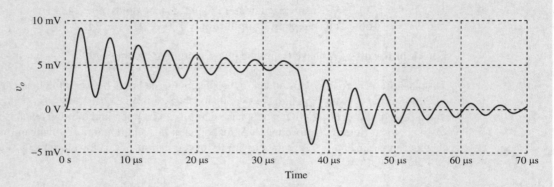

Simulation results of the original circuit using the μA741 op amp macro model in SPICE.

EXERCISE: Estimate the overshoot and calculate the phase margin for simulations in the two figures above.

ANSWERS: 14 percent, 54°; 85 percent, 6°

EXERCISE: Repeat the simulation to see what happens if the pulse amplitude is changed to 1 V. Explain what you see.

SUMMARY

The ideal operational amplifier was introduced previously in Chapter 10. Chapter 11 discussed removal of the ideal op amp assumptions, and quantified the effects and limitations caused by the nonideal op amp behavior. The nonideal behavior considered included:

- Finite open-loop gain
- Finite differential-mode input resistance
- Nonzero output resistance

- Offset voltage
- Input bias and offset currents
- Limited output voltage range
- Limited output current capability
- Finite common-mode rejection
- Finite common-mode input resistance
- Finite power supply rejection
- Limited bandwidth
- Limited slew rate

- The effect of removing the various ideal operational amplifier assumptions was explored in detail. Expressions were developed for the gain, gain error, input resistance, and output resistance of the closed-loop inverting and noninverting amplifiers, and it was found that the loop gain $T = A\beta$ plays an important role in determining the value of these closed-loop amplifier parameters.
- Series and shunt feedback are used to tailor and stabilize the characteristics of feedback amplifiers. Series feedback places the amplifier and feedback network in series and increases the overall impedance level at the series-connected port by $1 + T$. Shunt feedback is achieved by placing the amplifier and feedback network in parallel and decreases the impedance level at the shunt-connected port by $1 + T$.
- Feedback amplifiers are placed in four categories based upon the type of feedback utilized at the input and output of the amplifier:
  - Voltage amplifiers utilize series-shunt feedback to achieve high input impedance and low output impedance.
  - Current amplifiers utilize shunt-series feedback to achieve low input impedance and high output impedance.
  - Transresistance amplifiers utilize shunt-shunt feedback to achieve low input impedance and low output impedance.
  - Transconductance amplifiers utilize series-series feedback to achieve high input impedance and high output impedance.
- The loop gain $T(s)$ plays an important role in determining the characteristics of feedback amplifiers. For theoretical calculations, the loop gain can be found by breaking the feedback loop at some arbitrary point and directly calculating the voltage returned around the loop. However, both sides of the loop must be properly terminated before the loop-gain calculation is attempted.
- To find T in circuits employing op amps, it is convenient to disable the feedback loop by setting $v_{id} = 1$ V in the controlled source in the model for the op amp (i.e., $A_o v_{id} = A_o(1)$) and then calculating the value of the of v_{id} that appears at the op amp input.
- When using SPICE or making experimental measurements, it is often impossible to break the feedback loop. The method of successive voltage and current injection is a powerful technique for determining the loop gain without the need for opening the feedback loop.
- Whenever feedback is applied to an amplifier, stability becomes a concern. In most cases, a negative or degenerative feedback condition is desired. Stability can be determined by studying the characteristics of the loop gain $T(s) = A(s)\beta(s)$ of the feedback amplifier as a function of frequency, and stability criteria can be evaluated from either Nyquist diagrams or Bode plots.
- In the Nyquist case, stability requires that the plot of $T(j\omega)$ not enclose the $T = -1$ point.

- On the Bode plot, the asymptotes of the magnitudes of $A(j\omega)$ and $1/\beta(j\omega)$ should not intersect with a rate of closure exceeding 20 dB/decade.
- Phase margin and gain margin, which can be found from either the Nyquist or Bode plot, are important measures of stability.
- Phase margin determines the percentage overshoot in the step response of a second-order feedback system. Systems are typically designed to have at least a 60° phase margin.
- The dc error sources, including offset voltage, bias current, and offset current, all limit the dc accuracy of op amp circuits. Real op amps also have limited output voltage and current ranges as well as a finite rate of change of the output voltage called the slew rate. Circuit design options are constrained by these factors.
- The frequency response of basic single-pole operational amplifiers is characterized by two parameters: the open-loop gain A_o and the gain-bandwidth product ω_T. Analysis of the gain and bandwidth of the inverting and noninverting amplifier configurations demonstrated the direct tradeoff between the closed-loop gain and the closed-loop bandwidth of these amplifiers. The gain-bandwidth product is constant, and the closed-loop gain must be reduced in order to increase the bandwidth, or vice versa.
- Simplified macro models are often used for simulation of circuits containing op amps. Simple macro models can be constructed in SPICE using controlled sources, and most SPICE libraries contain comprehensive macro models for a wide range of commercial operational amplifiers.

KEY TERMS

Bandwidth (BW)
Bias
Bode plot
Closed-loop gain
Closed-loop input resistance
Closed-loop output resistance
Decibel (dB)
Feedback amplifier stability
Feedback network
Gain-bandwidth product
Gain margin (GM)
Ideal voltage amplifier
Input resistance (R_{in})
Inverting amplifier
Linear amplifier
Loop gain
Low-pass amplifier
Lower cutoff frequency
Midband gain
−1 Point
Negative feedback

Noninverting amplifier
Nyquist plot
Open-loop amplifier
Open-loop gain
Phase margin
Series-series feeback
Series-shunt feedback
Short-circuit termination
Shunt connection
Shunt-series feedback
Shunt-shunt feedback
Stability
Successive voltage and current injection
Total harmonic distortion
Transfer function
Transconductance amplifier
Transresistance amplifier
Two-port network
Upper cutoff frequency
Voltage amplifier

REFERENCES

1. H. S. Black, "Stabilized feed-back amplifiers," *Electrical Engineering*, vol. 53, pp. 114–120, January 1934.
2. Harold S. Black, "Inventing the negative feedback amplifier," *IEEE Spectrum*, vol. 14, pp. 54–60, December 1977. (50th anniversary of Black's invention of negative feedback amplifier.)
3. J. E. Brittain, "Scanning the past: Harold S. Black and the negative feedback amplifier," *Proceedings of the IEEE*, vol. 85, no. 8, pp. 1335–1336, August 1997.

4. R. B. Blackman, "Effect of feedback on impedance," *Bell System Technical Journal,* vol. 22, no. 3, 1943.
5. R. D. Middlebrook, "Measurement of loop gain in feedback systems," *International Journal of Electronics,* vol. 38, no. 4, pp. 485–512, April 1975. Middlebrook credits a 1965 Hewlett-Packard Application Note as the original source of this technique.
6. R. C. Jaeger, S. W. Director, and A. J. Brodersen, "Computer-aided characterization of differential amplifiers," *IEEE JSSC,* vol. SC-12, pp. 83–86, February 1977.
7. P. J. Hurst, "A comparison of two approaches to feedback circuit analysis," *IEEE Trans. on Education,* vol. 35, pp. 253–261, August 1992.
8. F. Corsi, C. Marzocca, and G. Matarrese, "On impedance evaluation in feedback circuits," *IEEE Trans. on Education,* vol. 45, no. 4, pp. 371–379, November 2002.
9. P. E. Allen and D. R. Holberg, *CMOS Analog Circuit Design,* 2nd ed., Oxford University Press, New York: 2002.

PROBLEMS

11.1 Classic Feedback Systems

11.1. The classic feedback amplifier in Fig. 11.1 has $\beta = 0.125$. What are the loop gain T, ideal closed-loop gain A_v^{Ideal}, actual closed-loop gain A_v, and the fractional gain error FGE if (a) $A = \infty$? (b) $A = 84$ dB? (c) $A = 20$?

11.2. A voltage follower's closed-loop voltage gain A_v is described by Eq. (11.4). (a) What is the minimum value of loop gain T required if the gain error is to be less than 0.02 percent for a voltage follower? (b) What value of open-loop gain A is required in the amplifier?

11.3. A amplifier's closed-loop voltage gain A_v is described by Eq. (11.4). What is the minimum value of loop gain T required if the gain error is to be less than 0.2 percent for an ideal gain of 50 dB?

11.4. (a) Calculate the sensitivity of the closed-loop gain A_v with respect to changes in open-loop gain A, $S_A^{A_v}$, using Eq. (11.4) and the definition of sensitivity given below:

$$S_A^{A_v} = \frac{A}{A_v} \frac{\partial A_v}{\partial A}$$

(b) Use this formula to estimate the percentage change in closed-loop gain if the open-loop gain A changes by 10 percent for an amplifier with $A = 100$ dB and $\beta = 0.01$.

11.2 Nonideal Op-Amp Circuits

11.5. A noninverting amplifier is built with $R_1 = 12$ kΩ and $R_2 = 150$ kΩ using an op amp with an open-loop gain of 86 dB. (a) What are the closed-loop gain, the gain error, and the fractional gain error for this amplifier? (b) Repeat if R_1 is changed to 1.2 kΩ.

11.6. A noninverting amplifier is built with $R_2 = 47$ kΩ and $R_1 = 6.2$ kΩ using an op amp with an open-loop gain of 94 dB. (a) What are the closed-loop gain, the gain error, and the fractional gain error for this amplifier? (b) Repeat if the open-loop gain is changed to 100 dB.

11.7. (a) What value of open-loop gain A is required of the amplifier in Prob. 11.3 if the amplifier is a noninverting amplifier? (b) If the amplifier is an inverting amplifier?

11.8. The feedback amplifier in Fig. P11.8(a) has $R_1 = 1$kΩ, $R_2 = 100$ kΩ, $R_I = 0$, and $R_L = 10$ kΩ. (a) What is β? (b) If $A = 92$ dB, what are the loop gain T and the closed loop gain A_v? (c) What are the values of GE and FGE?

11.9. An inverting amplifier is built with $R_1 = 11$ kΩ and $R_2 = 220$ kΩ using an op amp with an open-loop gain of 86 dB. (a) What are the closed-loop gain, the gain error, and fractional gain error for this amplifier? (b) Repeat if R_1 is changed to 1.1 kΩ.

11.10. The inverting amplifier in Fig. 11.3 is implemented with an op amp with finite gain $A = 80$ dB. If $R_1 = 1$ kΩ and $R_2 = 100$ kΩ, what are β, T, and A_v?

11.11. An inverting amplifier is built with $R_2 = 56$ kΩ and $R_1 = 5.6$kΩ using an op amp with an open-loop gain of 94 dB. (a) What are the closed-loop gain, the gain error, and fractional gain error for this

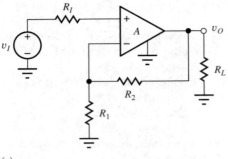

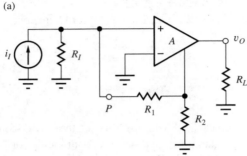

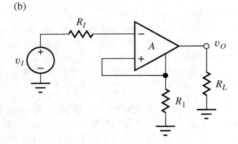

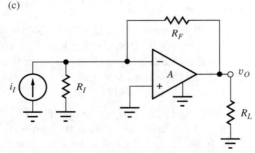

(d)
Figure P11.8 For each amplifier A: $A_o = 5000$, $R_{id} = 25$ kΩ, $R_o = 1$ kΩ.

amplifier? (b) Repeat if the open-loop gain is changed to 100 dB.

11.12. A noninverting amplifier is being designed to have a closed-loop gain of 36 dB. What op-amp gain is required to have the gain error less than 0.15 percent?

11.13. An inverting amplifier is being built with a closed-loop gain of 46 dB. What op amp gain is required to have the gain error below 0.1 percent?

11.14. A noninverting amplifier is built with 0.01 percent precision resistors and designed with $R_2 = 99 R_1$. What are the nominal and worst-case values of voltage gain if the op amp is ideal? What open-loop gain is required for the op amp if the gain-error due to finite op amp gain is to be less than 0.01 percent?

11.15. Calculate the currents i_1, i_2, and i_- for the amplifier in Fig. 11.6 if $v_x = 0.1$ V, $R_1 = 1$ kΩ, $R_2 = 47$ kΩ, $R_{id} = 1$ MΩ, and $A = 10^5$. What is i_+?

11.16. Repeat the derivation of the output resistance in Fig. 11.5 using a test current source rather than a test voltage source.

11.17. A noninverting amplifier is built with $R_2 = 75$ kΩ and $R_1 = 5.6$ kΩ using an op amp with an open-loop gain of 100 dB, an input resistance of 500 kΩ, and an output resistance of 300 Ω. (a) What are the closed-loop gain, input resistance, and output resistance for this amplifier? (b) Repeat if the open-loop gain is changed to 94 dB.

11.18. A noninverting amplifier is built with $R_1 = 15$ kΩ and $R_2 = 150$ kΩ using an op amp with an open-loop gain of 86 dB, an input resistance of 200 kΩ, and an output resistance of 200 Ω. (a) What are the closed-loop gain, input resistance, and output resistance for this amplifier? (b) Repeat if R_1 is changed to 1.2 kΩ.

11.19. An inverting amplifier is built with $R_2 = 47$ kΩ and $R_1 = 4.7$ kΩ using an op amp with an open-loop gain of 94 dB, an input resistance of 500 kΩ, and an output resistance of 200 Ω. (a) What are the closed-loop gain, input resistance, and output resistance for this amplifier? (b) Repeat if the open-loop gain is changed to 100 dB.

11.20. An inverting amplifier is built with $R_2 = 47$ kΩ and $R_1 = 2.4$ kΩ using an op amp with an open-loop gain of 100 dB, an input resistance of 300 kΩ, and an output resistance of 200 Ω. (a) What are the closed-loop gain, input resistance, and output resistance for this amplifier? (b) Repeat if the open-loop gain is changed to 94 dB.

11.21. An op amp has $R_{id} = 500$ kΩ, $R_o = 35$ Ω, and $A = 5 \times 10^4$. You must decide if a single-stage amplifier can be built that meets all of the specifications below. (a) Which configuration (inverting or noninverting) must be used and why? (b) Assume that the gain specification must be met and show which of the other specifications can or cannot be met.

$$|A_v| = 200 \quad R_{in} \geq 2 \times 10^8 \, \Omega \quad R_{out} \leq 0.2 \, \Omega$$

11.22. An op amp has $R_{id} = 1$ MΩ, $R_o = 100$ Ω, and $A = 1 \times 10^4$. Can a single-stage amplifier be built with this op amp that meets all of the following specifications? Show which specifications can be met and which cannot.

$$|A_v| = 200 \quad R_{in} \geq 10^8 \, \Omega \quad R_{out} \leq 0.2 \, \Omega$$

11.23. The overall amplifier circuit in Fig. P11.23 is a two-terminal network. $R_1 = 6.8$ kΩ and $R_2 = 120$ kΩ. What is its Thévenin equivalent circuit if the operational amplifier has $A = 7 \times 10^4$, $R_{id} = 1$ MΩ, and $R_o = 250$ Ω? What is its Norton equivalent?

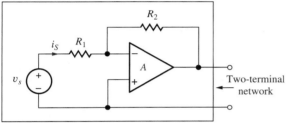

Figure P11.23

11.24. The circuit in Fig. P11.24 is a two-terminal network. $R_1 = 360$ Ω and $R_2 = 56$ kΩ. What is its Thévenin equivalent circuit if the operational amplifier has $A = 2 \times 10^4$, $R_{id} = 250$ kΩ, and $R_o = 250$ Ω?

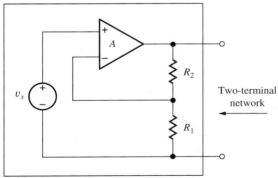

Figure P11.24

11.25. An inverting amplifier is to be designed to have a closed-loop gain of 60 dB. The only op amp that is available has an open-loop gain of 106 dB. What is the tolerance of the feedback resistors if the total gain error must be ≤ 1 percent? Assume that the resistors all have the same tolerances.

11.26. A noninverting amplifier is to be designed to have a closed-loop gain of 54 dB. The only op amp that is available has an open-loop gain of 40,000. What must be the tolerance of the feedback resistors if the total gain error must be ≤ 2 percent? Assume that the resistors all have the same tolerances.

11.3–11.4 Feedback Amplifier Characterization

11.27. Identify the type of negative feedback that should be used to achieve these design goals: (a) high input resistance and high output resistance, (b) low input resistance and high output resistance, (c) low input resistance and low output resistance, (d) high input resistance and low output resistance.

11.28. Identify the type of feedback being used in the four circuits in Fig. P11.8.

11.29. (a) Of the four circuits in Fig. P11.8, which tend to provide a high input resistance? (b) Which provide a relatively low input resistance?

11.30. (a) Of the four circuits in Fig. P11.8, which tend to provide a high output resistance? (b) Which provide a relatively low output resistance?

11.31. An amplifier has an open-loop voltage gain of 100 dB, $R_{id} = 40$ kΩ, and $R_o = 1000$ Ω. The amplifier is used in a feedback configuration with a resistive feedback network. (a) What is the largest value of input resistance that can be achieved in the feedback amplifier? (b) What is the smallest value of input resistance that can be achieved? (c) What is the largest value of output resistance that can be achieved? (d) What is the smallest value of output resistance that can be achieved?

11.32. An amplifier has an open-loop voltage gain of 90 dB, $R_{id} = 50$ kΩ, and $R_o = 5000$ Ω. The amplifier is used in a feedback configuration with a resistive feedback network. (a) What is the largest current gain that can be achieved with this feedback amplifier? (b) What is the largest transconductance that can be achieved with this feedback amplifier?

11.5 Voltage Amplifiers—Series-Shunt Feedback

11.33. (a) Find the closed-loop gain for the circuit in Fig. P11.8(a) if $R_L = 5.6$ kΩ, $R_1 = 4.3$ kΩ, $R_2 = 39$ kΩ, and $R_I = 1$ kΩ.

11.34. For the circuit in Fig. P11.8(a) find the voltage gain, input resistance, and output resistance of the feedback amplifier. Assume $R_I = 1$ kΩ, $R_L = 5$ kΩ, $R_1 = 5$ kΩ, and $R_2 = 45$ kΩ.

11.35. An op-amp-based noninverting amplifier has $R_1 = 15$ kΩ, $R_2 = 30$ kΩ, and $R_L = 20$ kΩ. The op amp has an open-loop gain of 94 dB, an input resistance of 30 kΩ, and an output resistance of 5 kΩ. Find the loop gain, ideal voltage gain, actual voltage gain, input resistance, and output resistance for the feedback amplifier.

11.36. Find the loop gain, ideal voltage gain, actual voltage gain, input resistance, and output resistance for the feedback amplifier in Ex. 11.5 if a 4-kΩ load resistor is connected to the output of the amplifier.

11.37. (a) Calculate the sensitivity of the closed-loop input resistance of the series-shunt feedback amplifier with respect to changes in open-loop gain A:

$$S_A^{R_{in}} = \frac{A}{R_{in}} \frac{\partial R_{in}}{\partial A}$$

(b) Use this formula to estimate the percentage change in closed-loop input resistance if the open-loop gain A changes by 10 percent for an amplifier with $A = 94$ dB and $\beta = 0.01$. (c) Calculate the sensitivity of the closed-loop output resistance of the series-shunt feedback amplifier with respect to changes in open-loop gain A:

$$S_A^{R_{out}} = \frac{A}{R_{out}} \frac{\partial R_{out}}{\partial A}$$

(d) Use this formula to estimate the percentage change in closed-loop output resistance if the open-loop gain A changes by 10 percent for an amplifier with $A = 100$ dB and $\beta = 0.01$.

11.6 Transresistance Amplifiers—Shunt-Shunt Feedback

11.38. Find the loop gain, ideal transresistance, actual transresistance, input resistance, and output resistance for the feedback amplifier in Fig. P11.8(d) with $R_I = 100$ kΩ, $R_L = 15$ kΩ, and $R_F = 10$ kΩ.

11.39. Simulate the amplifier in Prob. 11.38 using SPICE and compare the results to hand calculations.

11.40. Find the loop gain, ideal transresistance, actual transresistance, input resistance, and output resistance for the feedback amplifier in Fig. P11.8(d) with $R_I = 100$ kΩ, $R_L = 5$ kΩ, and $R_F = 24$ kΩ.

11.41. Simulate the amplifier in Prob. 11.40 using SPICE and compare the results to hand calculations.

11.42. An op-amp-based inverting amplifier has $R_1 = 15$ kΩ, $R_2 = 30$ kΩ, and $R_L = 20$ kΩ. The op amp has an open-loop gain of 92 dB, an input resistance of 30 kΩ, and an output resistance of 5 kΩ. Find the loop gain, ideal voltage gain, actual voltage gain, input resistance, and output resistance for the feedback amplifier.

11.43. Find an expression for the loop gain of the low-pass filter in Fig. 10.32 assuming the amplifier has a finite gain A_o. Show that the gain expression in Eq. (10.95) can be written in the form

$$A_v = A_v^{Ideal} \frac{T}{1+T}$$

11.44. The integrator in Fig. 10.34 is implemented with an op amp with finite gain $A = 86$ dB. If $R = 24$ kΩ and $C = 0.01$ µF, what are $T(s)$ and $A_v(s)$?

11.45. The differentiator in Fig. 10.35 is implemented with an op amp with finite gain $A = 80$ dB. If $R = 20$ kΩ and $C = 0.01$ µF, what are $T(s)$ and $A_v(s)$?

11.46. Find an expression for the loop gain of the integrator in Fig. 10.34 assuming the amplifier has a finite gain A_o. Show that the integrator voltage gain expression can be written in the form

$$A_v = A_v^{Ideal} \frac{T}{1+T}$$

11.47. Find an expression for the loop gain of the high-pass filter in Fig. 10.33 assuming the amplifier has a finite gain A_o. Show that the high-pass filter voltage gain expression can be written in the form

$$A_v = A_v^{Ideal} \frac{T}{1+T}$$

11.7 Transconductance Amplifiers—Series-Series Feedback

11.48. Find the loop gain, ideal transconductance, actual transconductance, input resistance, and output resistance for the feedback amplifier in Fig. P11.8(c) with $R_I = 15$ kΩ, $R_L = 10$ kΩ, and $R_1 = 3$ kΩ.

11.49. Simulate the amplifier in Prob. 11.43 using SPICE and compare the results to hand calculations.

11.50. Find the loop gain, ideal transconductance, actual transconductance, input resistance, and output resistance for the feedback amplifier in Fig. P11.8(c) with $R_I = 15$ kΩ, $R_L = 20$ kΩ, and $R_1 = 6$ kΩ.

11.51. Simulate the amplifier in Prob. 11.45 using SPICE and compare the results to hand calculations.

11.52. Find the loop gain, ideal transconductance, actual transconductance, input resistance, and output resistance for the feedback amplifier in Ex. 11.7 if a load resistance $R_L = 4$ kΩ is connected to the amplifier in place of v_o in Fig. 11.18.

11.8 Current Amplifiers—Shunt-Series Feedback

11.53. Find the loop gain, ideal current gain, actual current gain, input resistance, and output resistance for the feedback amplifier in Fig. P11.8(b) with $R_I = 100$ kΩ, $R_L = 10$ kΩ, $R_1 = 20$ kΩ and $R_2 = 2$ kΩ.

11.54. Simulate the amplifier in Prob. 11.43 using SPICE and compare the results to hand calculations.

11.55. Find the loop gain, ideal current gain, actual current gain, input resistance, and output resistance for the feedback amplifier in Fig. P11.8(b) with $R_I = 150$ kΩ, $R_L = 5$ kΩ, $R_1 = 10$ kΩ and $R_2 = 1$ kΩ.

11.56. Simulate the amplifier in Prob. 11.55 using SPICE and compare the results to hand calculations.

11.9 Loop Gain Calculations Using SPICE

11.57. Verify the value of the loop gain in Ex. 11.5 using successive voltage and current injection at the node connected to the inverting input of the op amp. What is the resistance ratio?

11.58. Verify the value of the loop gain in Ex. 11.6 using successive voltage and current injection at the right end of feedback resistor R_F. What is the resistance ratio?

11.59. Verify the value of the loop gain in Ex. 11.7 using successive voltage and current injection at the node connected to the noninverting input of the op amp. What is the resistance ratio?

11.60. Verify the value of the loop gain in Ex. 11.8 using successive voltage and current injection between the junction of R_2 and R_1. What is the resistance ratio?

11.10 Distortion Reduction Through the Use of Feedback

11.61. The VTC for an amplifier can be expressed as $v_o = 15 \tanh(1000 v_i)$ V. (a) Use MATLAB to find the total harmonic distortion for (a) $v_i = 0.001 \sin 2000\pi t$ V. (b) Repeat for $v_i = 0.002 \sin 2000\pi t$ V.

11.62. The VTC for the op amp in the circuit in Fig. 11.30 can be expressed as $v_o = 15 \tanh(1000 v_i)$ V and the closed-loop gain is set to 10. (a) Use MATLAB to find the total harmonic distortion for $v_i = 1 \sin 2000\pi t$ V. (b) Repeat for a closed-loop gain of 50 and $v_i = 0.2 \sin 2000\pi t$ V.

11.11 DC Error Sources and Output Range Limitations

11.63. Calculate the worst-case output voltage for the circuit in Fig. P11.63 if $V_{OS} = 2$ mV, $I_{B1} = 100$ nA, and $I_{B2} = 95$ nA. What is the ideal output voltage? What is the total error in this circuit? Is there a better choice for the value of R_1? If so, what is the value?

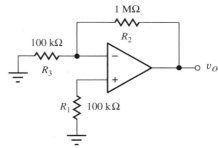

Figure P11.63

11.64. Repeat Prob. P11.63 if $V_{OS} = 8$ mV, $I_{B1} = 250$ nA, $I_{B2} = 200$ nA, and $R_2 = 510$ kΩ?

11.65. The voltage transfer characteristic for an operational amplifier is given in Fig. P11.65. (a) What are the values of gain and offset voltage for this op amp? (b) Plot voltage gain A versus v_{id} for the amplifier voltage transfer characteristic.

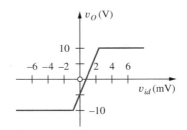

Figure P11.65

11.66. Use superposition to derive the result in Eq. (11.101).

11.67. The amplifier in Fig. P11.67 is to be designed to have a gain of 46 dB. What values of R_1 and R_2 should be used in order to meet the gain specification and minimize the effects of bias current errors?

*11.68. The op amp in the circuit of Fig. P11.68 has an open-loop gain of 10,000, an offset voltage of 1 mV, and an input-bias current of 100 nA. (a) What is

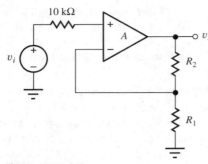

Figure P11.67

the output voltage for an ideal op amp? (b) What is the actual output voltage for the worst-case polarity of offset voltage? (c) What is the percentage error in the output voltage compared to the ideal output voltage?

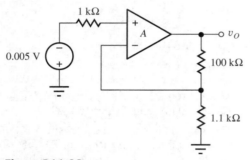

Figure P11.68

Voltage and Current Limits

11.69. The output-voltage range of the op amp in Fig. P11.69 is equal to the power supply voltages. What are the values of V_O and V_- for the amplifier if the dc input V_I is (a) -1 V and (b) $+2$ V?

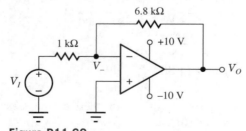

Figure P11.69

11.70. Plot a graph of the voltage transfer characteristic for the amplifier in Fig. P11.69.

11.71. The 6.8-kΩ resistor in Fig. P11.69 is replaced by a 10-kΩ resistor. What are the values of V_o and V_- if the dc input V_I is (a) 0.5 V and (b) 1.2 V?

11.72. Plot a graph of the voltage transfer characteristic for the amplifier in Prob. 11.71.

11.73. What are the voltages V_O and V_{ID} in the op amp circuit in Fig. P11.73 for dc input voltages of (a) $V_I = 300$ mV and (b) $V_I = 600$ mV if the output-voltage range of the op amp is limited to the power supply voltages.

11.74. Plot a graph of the voltage transfer characteristic for the amplifier in Fig. P11.73.

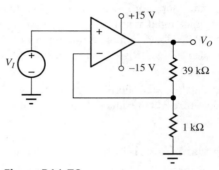

Figure P11.73

11.75. Repeat Prob. 11.73 if the 1-kΩ resistor in Fig. P11.73 is replaced by a 910-Ω resistor.

11.76. Design a noninverting amplifier with $A_v = 46$ dB that can deliver a ± 10-V signal to a 10-kΩ load resistor. Your op amp can supply only 1.5 mA of output current. Use standard resistor 5 percent values in your design. What is the gain of your design?

11.77. Design an inverting amplifier with $A_v = 43$ dB that can deliver ± 15 V to a 5-kΩ load resistor. Your op amp can supply only 4 mA of output current. Use standard resistor 5 percent values in your design. What is the gain of your design?

11.78. What is the minimum value of R in the circuit in Fig. P11.78 if the maximum op amp output current is 5 mA and the current gain of the transistor is $\beta_F \geq 60$?

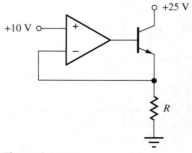

Figure P11.78

*11.79. (a) Design a single-stage inverting amplifier with a gain of 46 dB using an operational amplifier. The input resistance should be as low as possible while achieving the op amp output drive capability mentioned here. The amplifier must be able to produce the signal $v_o = (10 \sin 1000t)$ V at its output when an external load resistance $R_L \geq 5$ kΩ is connected to the output of the amplifier. You have an operational amplifier available whose output is guaranteed to deliver ± 10 V into a 4-kΩ load resistance. Otherwise, the amplifier is ideal. (b) If the amplifier input signal is $v_i = V \sin 1000t$, what is the largest acceptable value for the input signal amplitude V? (c) What is the input resistance of your amplifier?

11.12 Common-Mode Rejection and Input Resistance

11.80. The resistors in the difference amplifier in Fig. P11.80 are slightly mismatched due to their tolerances. (a) What are the differential-mode and common-mode input voltages if $v_1 = 3$ V and $v_2 = 3$ V? (b) What is the output voltage? (c) What would be the output voltage if the resistor pairs were matched? (d) What is the CMRR?

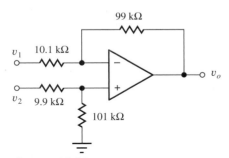

Figure P11.80

11.81. The resistors in the difference amplifier in Fig. P11.80 are slightly mismatched due to their tolerances. (a) What is the amplifier output voltage if $v_1 = 3.90$ V and $v_2 = 4.10$ V? (b) What would be the output voltage if the resistor pairs were matched? (c) What is the error in amplifying $(v_1 - v_2)$?

*11.82. The op amp in the amplifier circuit in Fig. P11.80 is ideal and

$$v_1 = (10 \sin 120\pi t + 0.25 \sin 5000\pi t) \text{ V}$$
$$v_2 = (10 \sin 120\pi t - 0.25 \sin 5000\pi t) \text{ V}$$

(a) What are the differential-mode and common-mode input voltages to this amplifier? (b) What are the differential-mode and common-mode gains of this amplifier? (c) What is the common-mode rejection ratio of this amplifier? (d) Find v_o.

11.83. The multimeter in Fig. P11.83 has a common-mode rejection specification of 86 dB. What possible range of voltages can be indicated by the meter?

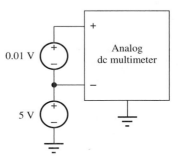

Figure P11.83

11.84. (a) We would like to measure the voltage $(V = V_1 - V_2)$ appearing across the 20-kΩ resistor in Fig. P11.84 with a voltmeter. What is the value of V? What is the common-mode voltage associated with $V (V_{CM} = (V_1 + V_2)/2)$? What CMRR is required of the voltmeter if we are to measure V with an error of less than 0.01 percent? (b) Repeat if the 10-kΩ resistor is changed to 100 Ω.

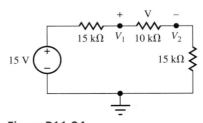

Figure P11.84

11.85. The common-mode rejection ratio of the difference amplifier in Fig. P11.85 is most often limited by the mismatch in the resistor pairs and not by the CMRR of the amplifier itself. Suppose that the nominal value of R is 10 kΩ and its tolerance is 0.025 percent. What is the worst-case value of the CMRR in dB?

11.86. What are the values of the common-mode and differential mode input resistances for the amplifier in Fig. P11.80?

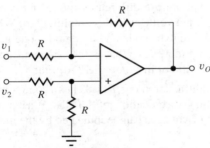

Figure P11.85

11.13 Frequency Response and Bandwidth of Operational Amplifier Circuits

11.87. (a) A single-pole op amp has an open-loop gain of 100 dB and a unity-gain frequency of 2 MHz. What is the open-loop bandwidth of the op amp? (b) A single-pole op amp has an open-loop gain of 100 dB and a bandwidth of 20 Hz. What is the unity-gain frequency of the op amp? (c) A single-pole op amp has unity-gain-bandwidth product of 30 MHz and a bandwidth of 200 Hz. What is the open-loop gain of the op amp?

11.88. A single-pole op amp has an open-loop gain of 92 dB and a unity-gain frequency of 1 MHz. What is the open-loop bandwidth of the op amp? (a) The op amp is used in a noninverting amplifier designed to have an ideal gain of 32 dB. What is the bandwidth of the noninverting amplifier? (b) Repeat for an inverting amplifier with ideal gain of 32 dB.

11.89. A single-pole op amp has an open-loop gain of 100 dB and a unity-gain frequency of 5 MHz. What is the open-loop bandwidth of the op amp? (a) The op amp is used in a voltage follower. What is the amplifier's bandwidth? (b) Repeat for a unity-gain inverting amplifier.

11.90. Repeat problem 11.89 if $f_T = 10$ MHz.

11.91. A single-pole op amp has an open-loop gain of 94 dB and a unity-gain frequency of 4 MHz. A noninverting amplifier is needed with bandwidth of at least 20 kHz. (a) What is the maximum closed loop-gain that a noninverting amplifier can have and still meet the frequency response specification? (b) What is the maximum closed loop-gain that an inverting amplifier can have and still meet the frequency response specification?

11.92. A single-pole op amp has an open-loop gain of 86 dB and a unity-gain frequency of 5 MHz. What is the open-loop bandwidth of the op amp? (a) The op amp is used in a noninverting amplifier with a gain of 3. What is the amplifier's bandwidth? (b) Repeat for an inverting amplifier with a gain of 3.

Using Feedback to Control Frequency Response

11.93. The voltage gain of an amplifier is described by
$$A(s) = \frac{2\pi \times 10^{10} s}{(s + 2000\pi)(s + 2\pi \times 10^6)}$$
(a) What are the mid-band gain and upper- and lower-cutoff frequencies of this amplifier? (b) If this amplifier is used in a feedback amplifier with a closed-loop gain of 100, what are the upper- and lower-cutoff frequencies of the closed-loop amplifier? (c) Repeat for a closed-loop gain of 40.

11.94. Repeat Prob. 11.93 if the voltage gain of the amplifier is given by
$$A(s) = \frac{2 \times 10^{14} \pi^2}{(s + 2\pi \times 10^3)(s + 2\pi \times 10^5)}$$

11.95. Repeat Prob. 11.93 if the voltage gain of the amplifier is given by
$$A(s) = \frac{4\pi^2 \times 10^{18} s^2}{(s + 200\pi)(s + 2000\pi)(s + 2\pi \times 10^6)(s + 2\pi \times 10^7)}$$

11.96. Derive an expression for the output impedance $Z_{\text{out}}(s)$ of the inverting and noninverting amplifiers assuming that the op amp has a transfer function given by Eq. (11.120) and an output resistance R_o.

11.97. A single-pole op amp has an open-loop gain of 86 dB, a unity-gain frequency of 5 MHz, and an output resistance of 250 Ω. The op amp is used in an inverting amplifier with an ideal gain of 20. (a) Find an expression for the output impedance of the amplifier. (b) Sketch a Bode plot for the output impedance as a function of frequency.

11.98. Use SPICE to make a Bode plot for the amplifier output resistance in Prob. 11.97.

11.99. Derive an expression for the input impedance $Z_{\text{in}}(s)$ of the noninverting amplifier assuming that the op amp has a transfer function given by Eq. (11.120) and an input resistance R_{id}.

11.100. A single-pole op amp has an open-loop gain of 86 dB, a unity-gain frequency of 2.5 MHz, and an input resistance of 100 kΩ. The op amp is used in a noninverting amplifier with an ideal gain of 20. (a) Find an expression for the input impedance of the amplifier. (b) Sketch a Bode plot for the input impedance as a function of frequency.

11.101. Use SPICE to make a Bode plot for the amplifier input resistance in Prob. 11.100.

11.102. A single-pole op amp has an open-loop gain of 97 dB and a unity-gain frequency of 2 MHz. Find an expression for the transfer function of the low-pass filter in Fig. 10.32 if $R_1 = 5.1$ kΩ, $R_2 = 51$ kΩ, and $C = 1600$ pF. Make a Bode plot comparing the ideal and the actual transfer functions.

11.103. A single-pole op amp has an open-loop gain of 100 dB and a unity-gain frequency of 2 MHz. Find an expression for the transfer function of the low-pass filter in Fig. 10.32 if $R_1 = 5.1$ kΩ, $R_2 = 100$ kΩ and $C = 750$ pF. Make a Bode plot comparing the ideal and the actual transfer functions.

11.104. A single-pole op amp has an open-loop gain of 92 dB and a unity-gain frequency of 1 MHz. Find an expression for the transfer function of the low-pass filter in Fig. 10.32 if $R_1 = 1.4$ kΩ, $R_2 = 27$ kΩ, and $C = 150$ pF. Make a Bode plot comparing the ideal and the actual transfer functions.

11.105. A single-pole op amp has an open-loop gain of 100 dB and a unity-gain frequency of 5 MHz. Find an expression for the transfer function of the integrator in Fig. 10.34 if $R = 10$ kΩ and $C = 0.05$ μF. Make a Bode plot comparing the ideal and the actual transfer functions.

11.106. A single-pole op amp has an open-loop gain of 94 dB and a unity-gain frequency of 1 MHz. Find an expression for the transfer function of the high-pass filter in Fig. 10.33 if $R_1 = 18$ kΩ, $R_2 = 180$ kΩ, and $C = 1800$ pF. Make a Bode plot comparing the ideal and the actual transfer functions.

11.107. What are the gain and bandwidth of the amplifier in Fig. P11.107 for the nominal and worst-case values of the feedback resistors if $A = 50,000$ and $f_T = 1$ MHz?

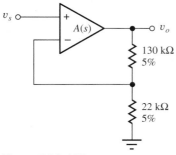

Figure P11.107

**11.108. Perform a Monte Carlo analysis of the circuit in Fig. P11.107. (a) What are the three sigma limits on the gain and bandwidth of the amplifier if $A_o = 50,000$ and $f_T = 1$ MHz? (b) Repeat if A_o is uniformly distributed in the interval $[5 \times 10^4, 1.5 \times 10^5]$ and f_T is uniformly distributed in the interval $[10^6, 3 \times 10^6]$.

Large Signal Limitations—Slew Rate and Full-Power Bandwidth

11.109. (a) An audio amplifier is to be designed to develop a 50-V peak-to-peak sinusoidal signal at a frequency of 20 kHz. What is the slew-rate specification of the amplifier? (b) Repeat for a frequency of 20 Hz.

11.110. An amplifier has a slew rate of 10 V/μs. What is the full-power bandwidth for signals having an amplitude of 18 V?

11.111. An amplifier must reproduce the output waveform in Fig. P11.111. What is its required slew rate?

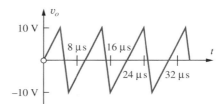

Figure P11.111

Macro Model for the Operational Amplifier Frequency Response

11.112. A single-pole op amp has these specifications:

$$A_o = 80,000 \qquad f_T = 5 \text{ MHz}$$
$$R_{id} = 250 \text{ k}\Omega \qquad R_o = 50 \text{ }\Omega$$

(a) Draw the circuit of a macro model for this operational amplifier. (b) Draw the circuit of a macro model for this operational amplifier if the op amp also has $R_{ic} = 500$ MΩ.

11.113. Draw a macro model for the amplifier in Prob. 11.112, including the additional elements necessary to model $R_{ic} = 100$ MΩ, $I_{B1} = 105$ nA, $I_{B2} = 95$ nA, and $V_{OS} = 1$ mV.

*11.114. A two-pole operational amplifier can be represented by the transfer function

$$A(s) = \frac{A_o \omega_1 \omega_2}{(s + \omega_1)(s + \omega_2)}$$

where

$$A_o = 80{,}000$$
$$f_1 = 1 \text{ kHz}$$
$$f_2 = 100 \text{ kHz}$$
$$R_{id} = 400 \text{ k}\Omega$$
$$R_o = 75 \text{ }\Omega$$

Create a macro model for this amplifier. (*Hint:* Consider using two "dummy" loops.)

Commercial General-Purpose Operational Amplifiers

11.115. (a) What are the element values for the macro model in Fig. 11.51 for the AD745J op amp? (b) Use $R = 1 \text{ k}\Omega$ and the nominal specifications.

11.116. (a) What are the worst-case values (minimum or maximum, as appropriate) of the following parameters of the AD745 op amp: open-loop gain, CMRR, PSRR, V_{OS}, I_{B1}, I_{B2}, I_{OS}, R_{ID}, slew rate, gain-bandwidth product, and power supply voltages? (b) Repeat for an LT1028 op amp.

11.14 Stability of Feedback Amplifiers

11.117. What are the phase and gain margins for the amplifier in Prob. 11.93?

11.118. What are the phase and gain margins for the amplifier in Prob. 11.94?

11.119. What are the phase and gain margins for the amplifier in Prob. 11.95?

11.120. The open-loop gain of an amplifier is described by

$$A(s) = \frac{4 \times 10^{19} \pi^3}{(s + 2\pi \times 10^4)(s + 2\pi \times 10^5)^2}$$

(a) If resistive feedback is used, find the frequency at which the loop gain will have a phase shift of 180°. (b) At what value of closed-loop gain will the amplifier break into oscillation? (c) Is the amplifier stable for larger or smaller values of closed-loop gain?

11.121. The open-loop gain of an amplifier is described by

$$A(s) = \frac{4 \times 10^{13} \pi^2}{(s + 2\pi \times 10^3)(s + 2\pi \times 10^4)}$$

(a) Will this amplifier be stable for a closed-loop gain of 4? (b) What is the phase margin?

11.122. A single-pole op amp has an open-loop gain of 106 dB and a unity-gain frequency of 1 MHz. (a) The op amp is used to build a noninverting amplifier with an ideal gain of 26 dB. What is the amplifier's phase margin? (b) Repeat for an ideal gain of 46 dB.

11.123. (a) Simulate the amplifier in Prob. 11.122(a) using the SPICE model for the 741 op amp and find the phase margin. Compare with hand calculations in Prob. 11.122. Discuss reasons for differences observed. (b) Repeat for the amplifier in Prob. 11.122(b).

11.124. A single-pole op amp has an open-loop gain of 95 dB and a unity-gain frequency of 2 MHz. (a) The op amp is used to build an inverting amplifier with an ideal gain of 20 dB. What is the amplifier's phase margin? (b) Repeat for an ideal gain of 46 dB.

11.125. (a) Simulate the amplifier in Prob. 11.124(a) using the SPICE model for the 741 op amp and find the phase margin. Compare with hand calculations in Prob. 11.124. Discuss reasons for differences observed. (b) Repeat for the amplifier in Prob. 11.124(b).

11.126. A single-pole op amp has an open-loop gain of 94 dB and a unity-gain frequency of 10 MHz. (a) If the op amp is used to build a voltage follower, what are its bandwidth and phase margin? (b) Repeat for an inverting amplifier with a gain of 0 dB.

11.127. A single-pole op amp has an open-loop gain of 86 dB and a unity-gain frequency of 4 MHz. Find an expression for the transfer function of the low-pass filter in Fig. 10.32 if $R_1 = 1.4 \text{ k}\Omega$, $R_2 = 27 \text{ k}\Omega$, and $C = 150 \text{ pF}$. What is the filter's phase margin?

11.128. A single-pole op amp has an open-loop gain of 100 dB and a unity-gain frequency of 3 MHz. Find an expression for the transfer function of the integrator in Fig. 10.34 if $R = 10 \text{ k}\Omega$ and $C = 0.05 \text{ }\mu\text{F}$. What is the integrator's phase margin?

11.129. A single-pole op amp has an open-loop gain of 94 dB and a unity-gain frequency of 4 MHz. Find an expression for the transfer function of the high-pass filter in Fig. 10.33 if $R_1 = 18 \text{ k}\Omega$, $R_2 = 180 \text{ k}\Omega$, and $C = 1800 \text{ pF}$. What is the filter's phase margin?

11.130. The voltage gain of an amplifier is described by

$$A(s) = \frac{2 \times 10^{14} \pi^2}{(s + 2\pi \times 10^3)(s + 2\pi \times 10^5)}$$

(a) Will this amplifier be stable for a closed-loop gain of 5? (b) What is the phase margin?

11.131. (a) Use MATLAB to make a Bode plot for the amplifier in Prob. 11.130 for a closed-loop gain of 5. Is the amplifier stable? What is the phase margin? (b) Repeat for the unity-gain case.

11.132. Find the loop gain for an integrator that uses a single-pole op amp with $A_o = 94$ dB and $f_T = 2$ MHz. Assume the integrator feedback elements are $R = 100$ kΩ and $C = 0.01$ μF. What is the phase margin of the integrator?

*11.133. Find the closed-loop transfer function of an integrator that uses a two-pole op amp with $A_o = 100$ dB, $f_{p1} = 1$ kHz, and $f_{p2} = 100$ kHz. Assume the integrator feedback elements are $R = 100$ kΩ and $C = 0.01$ μF. What is the phase margin of the integrator?

*11.134. (a) Write an expression for the loop gain $T(s)$ of the amplifier in Fig. P11.134 if $R_1 = 1$ kΩ, $R_2 = 20$ kΩ, $C_C = 0$, and the op amp transfer function is

$$A(s) = \frac{2 \times 10^{11} \pi^2}{(s + 2\pi \times 10^2)(s + 2\pi \times 10^4)}$$

(b) Use MATLAB to make a Bode plot of $T(s)$. What is the phase margin of this circuit? (c) Can compensation capacitor C_C be added to achieve a phase margin of 45°? If so, what is the value of C_C?

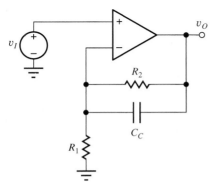

Figure P11.134

11.135. (a) Use MATLAB to make a Bode plot for the amplifier in Prob. 11.120. Find the frequency for which the phase shift is 180°. (b) At what value of closed-loop gain will the amplifier break into oscillation?

11.136. Repeat Prob. 11.135 for the amplifier in Prob. 11.93.

11.137. Repeat Prob. 11.135 for the amplifier in Prob. 11.94.

11.138. Repeat Prob. 11.135 for the amplifier in Prob. 11.95.

11.139. Use MATLAB to make a Bode plot for the op amp in Prob. 11.134 for a closed-loop gain of 100. Is the amplifier stable? What is the phase margin? What is the overshoot?

11.140. Use MATLAB to make a Bode plot for the integrator in Prob. 11.132. What is the phase margin of the integrator?

*11.141. Use MATLAB to make a Bode plot for the integrator in Prob. 11.133. What is the phase margin of the integrator?

11.142. The noninverting amplifier in Fig. P11.142 has $R_1 = 47$ kΩ, $R_2 = 390$ kΩ, and $C_S = 45$ pF. Find the phase margin and overshoot of the amplifier if amplifier voltage gain is described by the following transfer function:

$$A(s) = \frac{10^7}{(s + 50)}$$

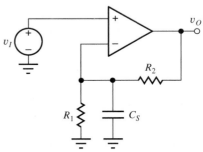

Figure P11.142

11.143. Use SPICE to draw a Bode plot for the low-pass filter circuit in Fig. 10.32 with $R_1 = 4.3$ kΩ, $R_2 = 82$ kΩ, and $C = 200$ pF if the op amp is not ideal but has an open-loop gain $A_o = 100$ dB and $f_T = 5$ MHz.

11.144. Use SPICE to draw a Bode plot for the integrator circuit in Fig. 10.34 with $R_1 = 10$ kΩ, and $C = 470$ pF if the op amp is not ideal but has an open-loop gain $A_o = 100$ dB and $f_T = 5$ MHz.

*11.145. What is the maximum load capacitance C_L that can be connected to the output of the voltage follower in Fig. P11.147 if the phase margin of the amplifier is to be 60°? Assume that the amplifier voltage gain is described by this transfer function, and that it has an output resistance of $R_o = 500\ \Omega$:

$$A(s) = \frac{10^7}{(s+50)}$$

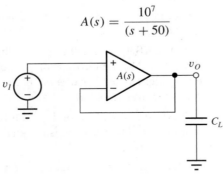

Figure P11.147

11.146. Suppose the op amp in Ex. 11.5 has an f_T of 1 MHz. What is the phase margin of the amplifier?

11.147. Suppose the op amp in Ex. 11.6 has an f_T of 3 MHz. What is the phase margin of the amplifier?

11.148. Suppose the op amp in Ex. 11.7 has an f_T of 0.5 MHz. What is the phase margin of the amplifier?

11.149. Suppose the op amp in Ex. 11.8 has an f_T of 2 MHz. What is the phase margin of the amplifier?

11.150. Calculate the phase margin and damping factor that correspond to an overshoot of (a) 0.5 percent, (b) 5 percent, (c) 25 percent.

11.151. A two-pole op amp has an open-loop gain of 100 dB and poles at 500 Hz and 1 MHz. Make a Bode plot for the gain of the op amp. What is its unity-gain frequency? (a) If the op amp is used to build a voltage follower, what are its bandwidth, phase margin, and overshoot? (b) Repeat for an inverting amplifier with a gain of 0 dB.

11.152. The op amp in Prob. 11.151 is used in a noninverting amplifier with a gain of 26 dB. What are the phase margin and overshoot for the amplifier?

11.153. An inverting amplifier utilizes the op amp in Prob. 11.151. (a) What closed-loop gain achieves a phase margin of 45°? What is the overshoot? (b) What closed-loop gain achieves a phase margin of 60°? What is the overshoot?

11.154. A two-pole op amp has an open-loop gain of 120 dB and its unity-gain frequency is 15 MHz. One of the op amp poles is at 1.5 MHz. What is the frequency of the second pole? (a) What are the bandwidth and phase margin if the op amp is used in an inverting amplifier with a gain 10? (b) What will be the overshoot if the op amp is used in a voltage follower?

11.155. A two-pole op amp has an open-loop gain of 94 dB and its first pole occurs at 500 Hz. (a) If the op amp is to be used in a noninverting amplifier with a gain of 3, what is the minimum frequency for the second pole if the phase margin is to be 60°? (b) Repeat for an inverting amplifier with a gain of 3.

11.156. A two-pole op amp has an open-loop gain of 100 dB and its first pole occurs at 100 Hz. The op amp is to be used in a noninverting amplifier with a gain of 5. What is the minimum frequency for the second pole, if the step response is to have an overshoot of less than 5 percent?

11.157. Use SPICE to simulate the amplifiers in problem 11.151.

11.158. Use SPICE to simulate the amplifier in problem 11.152.

11.159. Use SPICE to confirm the answers the problem 11.153.

11.160. Use SPICE to confirm the answers the problem 11.154.

11.161. Use SPICE to confirm the answers the problem 11.155.

11.162. Use SPICE to confirm the answers the problem 11.156.

术语对照

Bandwidth (BW)	带宽（BW）
Bias	偏置
Bias current compensation	偏置电流补偿
Bode plot	伯德图
Closed-loop gain	闭环增益
Closed-loop input resistance	闭环输入电阻
Closed-loop output resistance	闭环输出电阻
Common-mode input voltage	共模输入电压
Common-mode gain	共模增益
Common-mode rejection ratio	共模抑制比
Common-mode input resistance	共模输入电阻
Decibel (dB)	分贝（dB）
Differential-mode gain	差模增益
Differential-mode input voltage	差模输入电压
Differential-mode resistance	差模输入电阻
Feedback amplifier stability	反馈放大器的稳定性
Feedback network	反馈网络
Feedback factor	反馈系数
Fractional gain error	分数增益误差
Full-power bandwidth	满功率带宽
Gain-bandwidth product	增益带宽积
Gain margin(GM)	增益裕度
Gain error	增益误差
Input resistance（R_{in}）	输入电阻（R_{in}）
Input offset voltage	直流输入失调电压
Inverting amplifier	反相放大器
Ideal gain	理想增益
Loop gain	环路增益
Loop transmission	回路传输
Low-pass amplifier	低通放大器
Midband gain	中频带增益
Macro models	宏模型
−1 Point	−1点
Negative feedback	负反馈
Noninverting amplifier	同相放大器

Nyquist plot	奈奎斯特图
Open-loop amplifier	开环放大器
Open-loop gain	开环增益
Offset voltage	失调电压
Offset current	失调电流
Oscillator circuits	振荡器电路
Phase margin	相位裕度
Power supply rejection ratio	电源抑制比
Positive feedback	正反馈
IC voltage regulator	集成稳压器
Input bias current	输入偏置电流
Series feedback	串联反馈
Series-series feedback	电流串联反馈
Series-series feedback amplifier	电流串联反馈放大器
Series-shunt feedback	电压串联反馈
Series-shunt feedback amplifier	电压串联反馈放大器
Short-circuit termination	短路端
Shunt-series feedback	电流并联反馈
Shunt connection	并联连接
Shunt feedback	并联反馈
Shunt-shunt feedback	电压并联反馈
Stability	稳定性
Successive voltage and current injection	连续电压电流结
Single-pole frequency response	单极点频率响应
Slew-rate	摆率
Total harmonic distortion	总的谐波失真
Transfer function	传输函数
Transconductance amplifier	跨导放大器
Transresistance amplifier	跨阻放大器
Two-port network	二端口网络
Upper-cutoff frequency	上截止频率
Unity-gain frequency	单位增益频率
Voltage amplifier	电压放大器

CHAPTER 12
OPERATIONAL AMPLIFIER APPLICATIONS

第12章 运算放大器应用

本章提纲
12.1 级联放大器
12.2 仪表放大器
12.3 有源滤波器
12.4 开关电容电路
12.5 数/模转换
12.6 模/数转换
12.7 振荡器
12.8 非线性电路的应用
12.9 正反馈电路

本章目标
- 了解多级放大器的特性及其设计，包括增益、输入电阻、输出电阻等；
- 了解放大器级联的频率响应；
- 了解仪表放大器；
- 了解基于运放的有源滤波器，包括低通、高通、带通以及带阻电路；
- 了解滤波器的幅度和频率范围；
- 了解开关电容电路技术；
- 掌握模/数转换器和数/模转换器原理；
- 掌握数/模转换器和模/数转换器的基本结构；
- 掌握振荡器的巴克豪森准则；
- 理解基于运放的振荡器，包括文氏桥和相移电路；
- 理解振荡器的振幅稳定性；
- 理解精密半波和全波整流电路；
- 理解正反馈电路，包括施密特触发器和非稳态、单稳态多谐振荡器；
- 理解电压比较器。

本章导读

本章主要研究运放的应用，介绍了许多运算放大器的线性和非线性应用。通常来说，有时单级运放不能满足实际中所遇到的问题，需要多级运放才能实现。本章将通过一些实例进行多级运放的讨论，后面紧接着给出了用三个运放实现的仪表放大器。通常来说，单个放大器无法满足许多设计的要求，比如无法用一个放大器同时满足增益、输入阻抗和输出阻抗的要求，或者无法同时达到增益和带宽的要求，但是，这些设计要求常常可以通过将若干放大器进行级联来满足。级联放大器的二端口模型可用来简化整体放大器的表达形式，每一级放大器都可以简化成一个二端口网络。级联放大器的输入阻抗是由第一级放大器的输入阻抗决定的，而级联放大器的输出阻抗由最后一级放大器的输出阻抗决定。

当多个放大器级联时，总的传递函数可写为各级传递函数之积。多级放大器的带宽要小于每级放大器单独使用时的带宽。文中推导了一个由n个相同的放大器级联而成的放大器带宽表达式，并通过带宽缩减因子进行表示。本章给出了一个多级放大器设计的综合实例，并采用计算机电子制表软件来探寻设计的冗余度，并研究了电阻容限的影响。

仪表放大器是常用于数据采集系统的高性能电路。数据采集系统中经常需要放大两个信号的差异，由于输入电阻太低不能使用差分放大器，此时需要采用仪表放大器。本章给出了采用两个同相放大器与差分放大器组合形成的高性能复合仪表放大器，并以实例的形式给出了仪表放大器的分析与计算。

滤波器是运算放大器非常重要的一项应用。本章讲解了包括低通、高通和带通电路在内的有源RC滤波器。在这些电路中，RC反馈网络和运算用来替换庞大的电感，这些电感通常被设计在RLC滤波器中用以确定音频范围。单级放大器有源滤波器采用正负反馈相结合的形式，实现了二阶低通、高通和带通的传递函数。在滤波器电路设计中，需要着重考虑滤波器对无源元件和运算放大器参数容限的敏感度影响。多运算放大器和单独运算放大器相比敏感度较低，设计也相对简易。滤波器的阻抗和ω_0值可以通过幅度和频率的缩放来改变。

本章对开关电容（SC）技术进行了简单介绍。在CMOS工艺中该技术被广泛用来实现现代滤波器。在集成电路滤波器设计中，开关电容电路用电容器和开关的组合来替换电阻。这些滤波器给出连续RC滤波器的采样数据或离散时间的等价形式，可以与MOS集成电路工艺全面兼容，反向积分器和同相积分器都可用SC技术实现。

当前，数字信号处理技术日益发展，正逐步替代传统模拟电路的应用领域。模拟信号与数字信号的接口需要用到模/数转换器和数/模转换器。本章给出了模/数和数/模转换器的特性曲线和几种基本的电路实例。模/数转换器和数/模转换器的主要参数为增益、分辨率、误差、线性度和微分线性误差等。数/模转换器和模/数转换器的分辨率根据最低有效位（LSB）来测量。n位转换器的LSB等于$V_{FS}/2^n$，其中V_{FS}为转换器的满量程电压。转换器的最高有效位（MSB）等于$V_{FS}/2$。

权电阻数/模转换器电路是最简单的DAC电路之一，由权电阻网络、R-$2R$梯形网络和倒置R-$2R$梯形电路以及MOS晶体管开关构成。倒置R-$2R$梯形网络结构在梯形元件中保持恒定电流。在VLSI集成电路中，基于权电容和C-$2C$梯形网络形式的开关电容技术也广为应用。优质数/模转换器具有单调输入输出特性。

模/数转换器实现对连续模拟信号的采集，采集输入电压v_X，然后将其转换成易于被计算机处理的n位二进制数。这个n位的二进制数是一个二进制小数，表示的是未知输入电压v_X与转换器的满量程电压的比值：$V_{FS}=KV_{REF}$。计数转换器较为简单，但是速度相对较慢，逐次逼近转换器采用有效的二进制搜索算法来完成转换过程，仅需n个时钟周期，是一种十分流行的转换技术。

在单斜率和双斜率模/数转换器中，参考电压是带有明确斜率的模拟信号，通常由带有恒定输入电压的积分器产生，单斜率转换器的数字输出受到积分时间常数绝对值的影响，双斜率转换器很大程度上削弱了这一问题，能够达到较高的微分和积分线性度，但速度上只能达到每秒几次的转换速率，双斜率转换器广泛用于高精度仪表系统中。对周期为积分时间整数倍的正弦信号的抑制，称为常模抑制，它是积分转换器的重要特征之一。

速度最快的模/数转换器是并行转换器，也称作"快闪"转换器，这种转换器可同时将未知电压和所有可能的量化值进行比较。转换速度仅受到构成转换器的比较器和逻辑网络的速度限制。这种高速度要以提高硬件复杂性为代价。优质模/数转换器的线性和微分线性误差要小于1/2 LSB，并且没有丢码。

比较器是模/数转换器中比较重要的电路，用来将未知输入电压与精确参考电压进行比较。比较器可以被看作一个具有高增益、高速度且无反馈的运算放大器。

本章介绍了许多采用正反馈来实现的电路，包括用于产生信号的振荡器和多谐振荡器以及利用非线性反馈来实现的精密整流电路。文中还介绍了运算放大器的非线性电路的应用，包括几种精密整流电路。振荡器的巴克豪森准则要求满足两个准则，即在某一频率下，围绕反馈环路的相移必须是360°的整数倍，并且在这一频率下回路增益必须是1。振荡器用某种形式的选频反馈来决定振荡频率；RC和LC网络以及石英晶体都可用来设定振荡器的频率。文中给出了维恩电桥和相移振荡器用RC网络来设置振荡器频率的实例。对于真正的正弦振荡，振荡器的极点必须恰好落在s平面的$j\omega$轴上。否则，将会出现失真。为了得到正弦振荡，通常需要采用一些振幅稳定电路。这种稳定电路可能直接来自于电路中晶体管的固有非线性特性，或者来自额外的增益控制电路。

本章的最后介绍了利用正反馈实现的很多有用的非线性电路，包括比较器、施密特触发器和多谐振荡器电路。施密特触发器采用了一个比较器，其中比较器的参考电压通过一个跨越输出端的分压器获得。输入信号接反相输入端，参考电压接同相输入端（正反馈）。含正反馈的施密特触发器是一个带有两个稳定状态的电路实例。

多谐振荡器电路可用于产生不同形式的电脉冲。双稳态施密特触发器电路有两种稳定状态，常被用于处在噪声环境下的比较器中。单稳态多谐振荡器用于产生已知持续时间的单脉冲。非稳态多谐振荡器没有稳定状态，它持续振荡，可以产生方波输出。

Chapter Outline

12.1 Cascaded Amplifiers
12.2 The Instrumentation Amplifier
12.3 Active Filters
12.4 Switched-Capacitor Circuits
12.5 Digital-to-Analog Conversion
12.6 Analog-to-Digital Conversion
12.7 Oscillators
12.8 Nonlinear Circuit Applications
12.9 Circuits Using Positive Feedback
Summary
Key Terms
Additional Reading
Problems

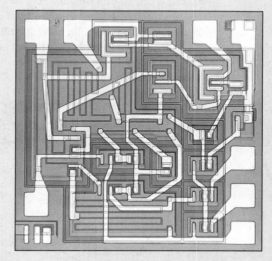

μA709 operational amplifier die photograph.
© Fairchild Semiconductor International.

Chapter Goals

In Chapter 12, we hope to achieve an understanding of the following topics:

- Characteristics and design of multistage amplifiers including gain, input resistance, and output resistance
- Frequency response of amplifier cascades
- The instrumentation amplifier
- Op-amp-based active filters including low-pass, high-pass, band-pass, and band-reject circuits
- Magnitude and frequency scaling of filters
- Switched capacitor circuit techniques
- Analog-to-digital (A/D) and digital-to-analog (D/A) converter specifications
- Basic forms of D/A and A/D converters
- The Barkhausen criteria for oscillation
- Op-amp-based oscillators including the Wein-bridge and phase-shift circuits
- Amplitude stabilization in oscillators
- Precision half-wave and full-wave rectifier circuits
- Circuits employing positive feedback including the Schmitt trigger and the astable and monostable multivibrators
- Voltage comparators

Chapter 12 continues our study of operational amplifier circuits by exploring a number of op amp applications. We frequently encounter a set of specifications that cannot be met with a single amplifier stage, and this chapter begins with discussion and an example of multistage amplifier design. This is followed with presentation of a precision instrumentation amplifier circuit that employs three op amps in its realization.

Filters represent an extremely important op amp application, and this chapter discusses op-amp-based active filters including low-pass, high-pass, and band-pass circuits. This is followed with a short introduction to switched capacitor techniques that are widely used to implement modern filters in CMOS technology.

Every day, more and more digital signal processing is utilized to enhance or replace traditional analog functions, and the interface between the analog and digital worlds requires an understanding of analog-to-digital (A/D) and digital-to-analog (D/A) converters. D/A and A/D converter characteristics are delineated, and a number of basic circuit implementations are presented.

In prior chapters, we generally assumed that the circuits utilized linear negative feedback. In this chapter, we introduce circuits that involve positive feedback including oscillators and multivibrators that are required for signal generation, as well as precision rectifier circuits that employ nonlinear feedback.

12.1 CASCADED AMPLIFIERS

Often, a set of design specifications cannot be met using a single amplifier. For example, we cannot simultaneously achieve the desired gain, input resistance, and output resistance, or the required gain and bandwidth at the same time with a just one amplifier.[1] However, the desired specifications can often be met by connecting several amplifiers in cascade as indicated by the three-stage cascade in Fig. 12.1. In this situation, the output of one amplifier stage is connected to the input of the next. If the output resistance of one amplifier is much less than the input resistance of the next, $R_{outA} \ll R_{inB}$ and $R_{outB} \ll R_{inC}$, then the loading of one amplifier on another can be neglected, and the overall voltage gain is simply the product of the open-circuit voltage gains of the individual stages. In order to understand this behavior more fully, we will represent the amplifiers using their simplified two-port models discussed next.

12.1.1 TWO-PORT REPRESENTATIONS

At each level in Fig. 12.1, we represent the "amplifier" as a **two-port model** with a value of voltage gain, input resistance, and output resistance, defined as in Fig. 12.1(b) and (c) (see Sec. 10.3).

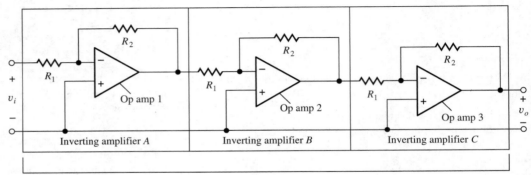

(a)

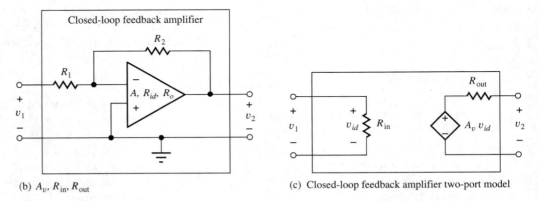

(b) A_v, R_{in}, R_{out} (c) Closed-loop feedback amplifier two-port model

Figure 12.1 (a) Three-stage amplifier cascade; (b) single stage inverting amplifier using an operational amplifier; (c) two-port representation for each feedback amplifier stage.

---

[1] See Probs. 11.21 to 11.22, for example.

12.1 Cascaded Amplifiers

TABLE 12.1
Feedback-Amplifier Terminology Comparison

	VOLTAGE GAIN	INPUT RESISTANCE	OUTPUT RESISTANCE
Open-Loop Amplifier	A	R_{id}	R_o
Closed-Loop Amplifier	A_v	R_{in}	R_{out}
Multi-Stage Amplifiers	A_v	R_{in}	R_{out}

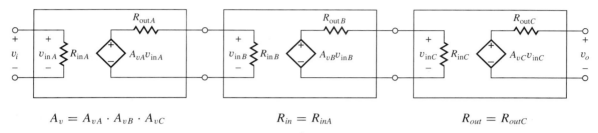

$$A_v = A_{vA} \cdot A_{vB} \cdot A_{vC} \qquad R_{in} = R_{inA} \qquad R_{out} = R_{outC}$$

Figure 12.2 Two-port representation for three-stage cascaded amplifier.

Each amplifier stage—A, B, and C—is built using an operational amplifier that has a gain A, input resistance R_{id}, and output resistance R_o. These quantities are usually called the open-loop parameters of the operational amplifier: *open-loop gain, open-loop input resistance,* and *open-loop output resistance.* They describe the op amp as a two-port by itself with no external elements connected.

Each single-stage amplifier built from an operational amplifier and the feedback network consisting of R_1 and R_2 is termed a **closed-loop amplifier.** We use A_v, R_{in}, and R_{out} for each closed-loop amplifier, as well as for the overall composite amplifier. Table 12.1 summarizes this terminology.

Two-Port Model for the Three-Stage Cascade Amplifier
In Fig. 12.2 each individual amplifier has been replaced by its two-port model. By proceeding through the amplifier from left to right, the overall gain expression can be written as

$$\mathbf{v_o} = A_{vA}\mathbf{v_i} \left(\frac{R_{inB}}{R_{outA} + R_{inB}} \right) A_{vB} \left(\frac{R_{inC}}{R_{outB} + R_{inC}} \right) A_{vC} \qquad (12.1)$$

For the voltage amplifiers considered so far, the output resistances are small (zero in the ideal case), so the impedance mismatch requirement is normally met (see Sec. 10.4), and the overall gain of the amplifier cascade is equal to the product of the open-loop gain of the three individual stages:

$$A_v = \frac{\mathbf{v_o}}{\mathbf{v_i}} = A_{vA} \cdot A_{vB} \cdot A_{vC} \qquad (12.2)$$

If a test source v_x is applied to the input and the input current i_x is calculated, we find that R_{in} of the overall amplifier is determined solely by the input resistance of the first amplifier. In this case, $R_{in} = \mathbf{v_x}/\mathbf{i_x} = R_{inA}$. Similarly, if we apply a test source v_x at the output and find the current i_x, we find that R_{out} of the overall amplifier is determined only by the output resistance of the last amplifier. In this case, $R_{out} = R_{outC}$.

Chapter 12 Operational Amplifier Applications

DESIGN NOTE

The input impedance of an amplifier cascade is set by the input impedance of the first amplifier, and the output impedance of an amplifier cascade is set by the output impedance of the final amplifier in the cascade. A very common mistake is to expect that the input resistance of a cascade of amplifiers results from some combination of the input resistances of all the individual amplifiers, or that the output resistance is a function of the output resistances of all the individual amplifiers.

EXERCISE: The amplifier in Fig. 12.1 has $R_2 = 68$ kΩ and $R_1 = 2.7$ kΩ. What are the values of A_{vA}, A_{vB}, A_{vC}, R_{inA}, R_{inB}, R_{inC}, and R_{outA}, R_{outB}, R_{outC}, for the amplifier equivalent circuit in Fig. 12.2?

ANSWERS: -25.2; -25.2; -25.2; 2.7 kΩ; 2.7 kΩ; 2.7 kΩ; 0; 0; 0.

EXERCISE: What are the gain A_v, input resistance, and output resistance of the three stage amplifier in Fig. 12.1(a) if $R_2 = 68$ kΩ and $R_1 = 2.7$ kΩ?

ANSWERS: $(-25.2)^3 = -1.60 \times 10^4$; 2.7 k$\Omega$; 0.

EXERCISE: Suppose the three output resistances in the amplifier in the previous exercise are not zero. What is the largest value of R_{out} that can be permitted if the gain is not to be reduced by more than 1 percent? Assume that the three output resistance values are the same.

ANSWER: 13.5 Ω.

12.1.2 AMPLIFIER TERMINOLOGY REVIEW

Now that we have analyzed a number of amplifier configurations, let us step back and review the terminology being used. Amplifier terminology is often a source of confusion because the portion of the circuit that is being called an *amplifier* must often be determined from the context of the discussion.

In Fig. 12.1, for example, an overall amplifier (the three-stage amplifier) is formed from the cascade connection of three inverting amplifiers (A, B, C), and each inverting amplifier has been implemented using an operational amplifier (op amp 1, 2, 3). Thus, we can identify at least seven different "amplifiers" in Fig. 12.1: operational amplifiers 1, 2, 3; inverting amplifiers A, B, C; and the composite three-stage amplifier ABC. Unfortunately, at any given time the "amplifier" that is being referenced must often be inferred from the context of the discussion.

EXAMPLE 12.1 CASCADED AMPLIFIER CALCULATIONS

This example characterizes a three-stage amplifier cascade and explores effects of power supply limits.

PROBLEM The op amps in the circuit on the next page operate from ± 12-V power supplies, but are ideal otherwise. (a) What are the voltage gain, input resistance, and output resistance of the overall amplifier? (b) If input voltage $v_I = 5$ mV, what are the voltages at each of the 10 nodes in the circuit? (c) Now suppose that $v_I = 10$ mV. What are the voltages at the outputs of the three op amps? (d) Suppose $v_I = V_i \sin 2000\pi t$. What is the largest value of input voltage V_i that corresponds to linear operation of the amplifier?

SOLUTION **Known Information and Given Data:** The three-stage amplifier circuit with resistor values appears below. The op amps are ideal except for power supplies of ±12 V.

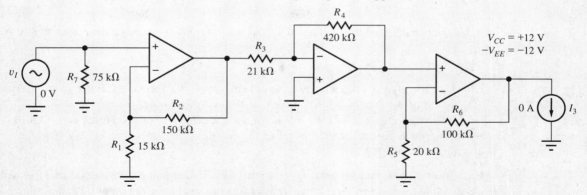

Unknowns: (a) Voltage gain, input resistance, and output resistance of the overall amplifier. (b) Voltages at each of the 10 nodes in the circuit with $v_I = 5$ mV. (c) Voltages at each of the 10 nodes in the circuit with $v_I = 10$ mV. (d) What is the largest amplitude of a sine wave input voltage that corresponds to linear operation of the amplifier?

Approach: Apply the inverting and noninverting amplifier formulas to the individual stages. The gain will be the product of the gains. The input resistance will be the input resistance of the first stage. The output resistance will equal that of the last amplifier.

Assumptions: The op amps are ideal except the power supplies are ±12 V. The interactions between R_{in} and R_{out} for the op amp circuits are negligible. Each op amp is using negative feedback and operating in its linear range.

Analysis: The three individual amplifier stages are recognized as noninverting, inverting, and noninverting amplifiers.

(a) Use the expressions in Table 10.3: $A_v = A_{v1}A_{v2}A_{v3}$

$$A_{v1} = 1 + \frac{R_2}{R_1} \qquad A_{v2} = -\frac{R_4}{R_3} \qquad A_{v1} = 1 + \frac{R_6}{R_5}$$

$$A_v = \left(1 + \frac{150\,k\Omega}{15\,k\Omega}\right)\left(-\frac{420\,k\Omega}{21\,k\Omega}\right)\left(1 + \frac{100\,k\Omega}{20\,k\Omega}\right) = -1320$$

The input resistance looking into the noninverting input of the first op amp is infinite, but the noninverting input is shunted by the 75 kΩ resistor. Thus, $R_{in} = 75\,k\Omega\|\infty = 75\,k\Omega$. The output resistance is equal to the the output resistance of the third amplifier. $R_{out} = R_{outC} = 0$

(b) $v_I = 5.00$ mV, $v_{OA} = +11v_I = 55.0$ mV, $v_{OB} = -20v_{OA} = -1.10$ V, $v_O = +6v_{OB} = -6.60$ V

Since the op amps are ideal, there must be zero volts across the input of each op amp:

$$v_{-A} = v_{+A} = +5.00\text{ mV}, v_{-B} = v_{+B} = 0\text{ V}, v_{-C} = v_{+C} = -6.60\text{ V}$$
$$V_+ = +12\text{ V}, V_- = -12\text{ V}, V_{gnd} = 0\text{ V}$$

(c) $v_I = 10.0$ mV, $v_{OA} = +11v_I = 110$ mV, $v_{OB} = -20v_{OA} = -2.2$ V, $v_O = +6v_{OB} = -13.2$ V < -12 V $\rightarrow v_O = -12$ V The output cannot exceed the power supply limits! The first two op amps are operating in their linear regions, so there will be zero volts across the input of the these two op amps: $v_{-A} = v_{+A} = +10.0$ mV. $v_{-B} = v_{+B} = 0$ V. However, the output of the third amplifier is saturated at -12 V, the gain of the third op amp is 0, and its feedback loop is "broken." Thus the inverting and noninverting inputs need no longer be equal.

$$v_{-C} = -12\,\text{V}\frac{20\,\text{k}\Omega}{20\,\text{k}\Omega + 100\,\text{k}\Omega} = -2\,\text{V} \quad V_{+} = +12\,\text{V} \quad V_{-} = -12\,\text{V} \quad V_{gnd} = 0\,\text{V}$$

(d) v_I must not exceed the voltage that causes the output to just reach the power supply voltages:

$$|v_I| \le \left|\frac{12\,\text{V}}{A_v}\right| = \frac{12}{1320} = 9.09\,\text{mV}$$

Check of Results: All the unknowns have been found. R_{in} should equal the 75 kΩ resistor, and R_{out} should be very small. For a 5-mV input, the expected output voltage $v_O = -1320(0.005) = -6.60$ V which checks. For a 10-mV input, the expected output voltage $v_O = -1320(0.005) = -13.2$ V which is less than the negative the power supply, so $v_O = -12$ V. The linear range assumption is violated.

Computer-Aided Analysis and Discussion: SPICE simulation utilizes a dc analysis with a dc input to find the node voltages and a transfer function analysis from input VI to the voltage across I3, V(I3) to find the gain, input resistance, and output resistance. The OPAMP power supplies are set to ± 12 V, and the gains default to 120 dB.

SPICE transfer function results are $A_v = -1320$, $R_{in} = 75$ kΩ, and $R_{out} = 0$, all in agreement with our hand calculations. The computed dc node voltages proceeding from left to right through the circuit are

dc Node Voltages from SPICE		
INPUT CASE	**5 mV**	**10 mV**
v_I	5.000 mV	10.00 mV
v_{-A}	5.000 mV	10.00 mV
v_{OA}	55.00 mV	110.0 mV
v_{-B}	1.100 μV	$-2.200\,\mu$V
v_{OB}	-1.100 V	-2.200 V
v_{-C}	-1.100 V	-2.000 V
v_{OC}	-6.600 V	-12.00 V
V_{+}	$+12.00$ V	$+12.00$ V
V_{-}	-12.00 V	-12.00 V
V_{gnd}	0.000 V	0.000 V

The computed node voltages agree with our hand calculations to four decimal places except for the voltage at the inverting input of the second amplifier where SPICE yields $v_{-B} = 1.100\,\mu$V. Let us try to understand the source of this discrepancy. The amplifiers in SPICE have finite gain of 120 dB. Thus the differential input voltage of the OPAMPs will not be zero. There must be a voltage across the input to generate the nonzero output voltage. The values of v_{ID} for each op amp will therefore be:

$$v_{IDA} = \frac{55.0\,\text{mV}}{10^6} = 55.0\,\text{nV} \quad v_{IDB} = \frac{-1.1\,\text{V}}{10^6} = -1.10\,\mu\text{V} \quad v_{IDC} = \frac{-6.6\,\text{V}}{10^6} = -6.60\,\mu\text{V}$$

The voltage at the inverting input of op amp B is the negative of v_{IDB} which agrees with SPICE. If we calculate v_{-A} and v_{-B}, the contributions of v_{ID} disappear in the round off:

$$v_{-A} = v_I - v_{IDA} = 5.000\,\text{mV} - 55.0\,\text{nV} = 55.00\,\text{mV}$$

$$v_{-C} = v_{OB} - v_{IDC} = -1.100\,\text{V} - (-6.600\,\mu\text{V}) = -1.100\,\text{V}$$

For the case with $v_I = 10$ mV, the node voltages are in agreement with our calculations except for v_{-B}. Note that the differential input voltage of the third op amp is not zero but equals -2.200 V $- (-2.000) = -0.200$ V! The output of the op amp cannot reach the value necessary to force $v_{ID} = 0$.

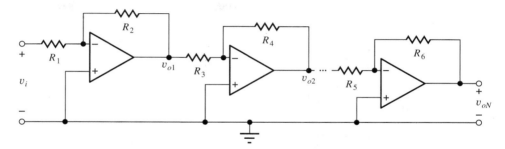

Figure 12.3 Multistage amplifier cascade.

12.1.3 FREQUENCY RESPONSE OF CASCADED AMPLIFIERS

When several amplifiers are connected in cascade, as in Fig. 12.3 for example, the overall transfer function can be written as the product of the transfer functions of the individual stages:

$$A_v(s) = \frac{V_{oN}(s)}{V_I(s)} = \frac{V_{o1}}{V_I} \frac{V_{o2}}{V_{o1}} \cdots \frac{V_{oN}}{V_{o(N-1)}} = A_{v1}(s) A_{v2}(s) \cdots A_{vN}(s) \quad (12.3)$$

It is extremely important to remember that this product representation implicitly assumes that the stages do not interact with each other, which can be achieved with $R_{\text{out}} = 0$ or $R_{\text{in}} = \infty$ (i.e., the interconnection of the various amplifiers must not alter the transfer function of any of the amplifiers).

In the general case, each amplifier has a different value of dc gain and bandwidth, and the overall transfer function becomes

$$A_v(s) = \frac{A_{v1}(0)}{\left(1 + \dfrac{s}{\omega_{H1}}\right)} \frac{A_{v2}(0)}{\left(1 + \dfrac{s}{\omega_{H2}}\right)} \cdots \frac{A_{vN}(0)}{\left(1 + \dfrac{s}{\omega_{HN}}\right)} \quad (12.4)$$

assuming single-pole amplifiers. The gain at dc ($s = 0$) is

$$A_v(0) = A_{v1}(0) A_{v2}(0) \cdots A_{vN}(0) \quad (12.5)$$

The overall bandwidth of the cascade amplifier is defined to be the frequency at which the voltage gain is reduced by a factor of $1/\sqrt{2}$ or -3 dB from its low-frequency value. Stated mathematically,

$$|A_v(j\omega_H)| = \frac{|A_{v1}(0) A_{v2}(0) \cdots A_{vN}(0)|}{\sqrt{2}} \quad (12.6)$$

In the general case, hand calculation of ω_H based on Eq. (12.6) can be quite tedious, and approximate techniques for estimating ω_H will be developed in Chapter 17. With the aid of a computer or calculator, solver routines or iterative trial-and-error can be used directly to find ω_H. Example 12.2 uses direct algebraic evaluation of Eq. (12.6) for the case of two amplifiers.

EXAMPLE 12.2   **TWO-AMPLIFIER CASCADE**

Calculate the gain and bandwidth of a two-stage amplifier.

PROBLEM   Two amplifiers with transfer functions $A_{v1}(s)$ and $A_{v2}(s)$ are connected in cascade. What are the dc gain and bandwidth of the overall two-stage amplifier?

$$A_{v1} = \frac{500}{1 + \dfrac{s}{2000}} \quad \text{and} \quad A_{v2} = \frac{250}{1 + \dfrac{s}{4000}}$$

SOLUTION Known Information and Given Data: A cascade connection of two amplifiers; the individual transfer functions are specified for the two amplifiers.

Unknowns: $A_v(0)$ and f_H for the overall two-stage amplifier

Approach: The transfer function for the cascade is given by $A_v = A_{v1} \times A_{v2}$. Find $A_v(0)$. Apply the definition of bandwidth to find f_H.

Assumptions: The amplifiers are ideal except for their frequency dependencies and can be cascaded without interaction—that is, the overall gain is equal to the product of the individual transfer functions.

Analysis: The overall transfer function is

$$A_v(s) = \left(\frac{500}{1 + \frac{s}{2000}}\right)\left(\frac{250}{1 + \frac{s}{4000}}\right) = \frac{125{,}000}{\left(1 + \frac{s}{2000}\right)\left(1 + \frac{s}{4000}\right)}$$

Calculating the dc gain $A_v(0)$:

$$A_v(0) = (500)(250) = 125{,}000 \text{ or } 102 \text{ dB}$$

Note that $A_v(0)$ is equal to the product of the dc gains of the two individual amplifiers.

The magnitude of the frequency response for $s = j\omega$ is

$$|A_v(j\omega)| = \frac{1.25 \times 10^5}{\sqrt{1 + \frac{\omega^2}{2000^2}}\sqrt{1 + \frac{\omega^2}{4000^2}}}$$

and we remember that ω_H is defined by

$$|A(j\omega_H)| = \frac{A_{\text{mid}}}{\sqrt{2}} = \frac{A_v(0)}{\sqrt{2}} = \frac{1.25 \times 10^5}{\sqrt{2}}$$

Equating the denominators of these two equations and squaring both sides yields

$$\left(1 + \frac{\omega_H^2}{2000^2}\right)\left(1 + \frac{\omega_H^2}{4000^2}\right) = 2$$

which can be rearranged into the following quadratic equation in terms of ω_H^2:

$$\left(\omega_H^2\right)^2 + 2.00 \times 10^7 \left(\omega_H^2\right) - 6.40 \times 10^{13} = 0$$

Using the quadratic formula or our calculator's root-finding routine gives these values for ω_H^2

$$\omega_H^2 = 2.81 \times 10^6 \quad \text{or} \quad -4.56 \times 10^7$$

The value of ω_H must be real, so the only acceptable answer is

$$\omega_H = 1.68 \times 10^3 \quad \text{or} \quad f_H = 267 \text{ Hz}$$

Check of Results: The bandwidth of the composite amplifier should be less than that of either individual amplifier:

$$f_{H1} = \frac{2000}{2\pi} = 318 \text{ Hz} \quad \text{and} \quad f_{H2} = \frac{4000}{2\pi} = 637 \text{ Hz}$$

The bandwidth we have calculated is indeed less than for either individual amplifier.

EXERCISE: An amplifier is formed by cascading two amplifiers with these transfer functions. What is the gain at low frequencies? What is the gain at f_H? What is f_H?

$$A_{v1}(s) = \frac{50}{1 + \frac{s}{10{,}000\pi}} \quad \text{and} \quad A_{v2}(s) = \frac{25}{1 + \frac{s}{20{,}000\pi}}$$

ANSWERS: 1250; 884; 4190 Hz

EXERCISE: An amplifier is formed by cascading three amplifiers with the following transfer functions. What is the gain at low frequencies? What is the gain at f_H? What is f_H?

$$A_{v1}(s) = \frac{-100}{1 + \frac{s}{10{,}000\pi}}, \quad A_{v2}(s) = \frac{66.7}{1 + \frac{s}{15{,}000\pi}}, \quad A_{v3}(s) = \frac{50}{1 + \frac{s}{20{,}000\pi}}$$

ANSWERS: -3.33×10^5; -2.36×10^5; 3450 Hz

Cascade of Identical Amplifier Stages

For the special case in which a cascade-amplifier configuration is composed of identical amplifiers, then a simple result can be obtained for the bandwidth of the overall amplifier. For N identical stages,

$$A_v(s) = \left[\frac{A_{v1}(0)}{1 + \frac{s}{\omega_{H1}}}\right]^N = \frac{[A_{v1}(0)]^N}{\left(1 + \frac{s}{\omega_{H1}}\right)^N} \quad \text{and} \quad A_v(0) = [A_{v1}(0)]^N \quad (12.7)$$

in which $A_{v1}(0)$ and ω_{H1} are the closed-loop gain and bandwidth of each individual amplifier stage. The bandwidth ω_H of the overall cascade amplifier is determined from

$$|A_v(j\omega_H)| = \frac{[A_{v1}(0)]^N}{\left(\sqrt{1 + \frac{\omega_H^2}{\omega_{H1}^2}}\right)^N} = \frac{[A_{v1}(0)]^N}{\sqrt{2}} \quad (12.8)$$

TABLE 12.2
Bandwidth Shrinkage Factor

N	$\sqrt{2^{1/N} - 1}$
1	1
2	0.644
3	0.510
4	0.435
5	0.386
6	0.350
7	0.323

Solving for ω_H in terms of ω_{H1} for the cascaded-amplifier bandwidth yields

$$\omega_H = \omega_{H1}\sqrt{2^{1/N} - 1} \quad \text{or} \quad f_H = f_{H1}\sqrt{2^{1/N} - 1} \quad (12.9)$$

The bandwidth of the cascade is less than that of the individual amplifiers. Sample values of the **bandwidth shrinkage factor** $\sqrt{2^{1/N} - 1}$ are given in Table 12.2.

Although most amplifier designs do not actually cascade identical amplifiers, Eq. (12.9) can be used to help guide the design of a multistage amplifier or, in some cases, to estimate the bandwidth of a portion of a more complex amplifier. (Additional results appear in Probs. 12.33 and 12.34.)

EXERCISE: Three identical amplifiers are connected in cascade as in Fig. 12.3. Each amplifier has $A_v = -30$ and $f_H = 33.3$ kHz. What are the gain and bandwidth of the composite three-stage amplifier?

ANSWERS: $-27{,}000$; 17.0 kHz

DESIGN EXAMPLE 12.3

A CASCADE AMPLIFIER DESIGN

In this example, a spreadsheet is used to assist in the design of a fairly complex multistage amplifier.

PROBLEM Design an amplifier to meet these specifications: $A_v \geq 100$ dB, bandwidth ≥ 50 kHz, $R_{out} \leq 0.1$ Ω, and $R_{in} \geq 20$ kΩ. Use an operational amplifier with these specifications: $A_o = 100$ dB, $f_T = 1$ MHz, $R_{id} = 1$ GΩ, and $R_o = 50$ Ω.

SOLUTION **Known Information and Given Data:** Op amp and overall amplifier specifications as already tabulated.

Unknowns: Choice between inverting and noninverting configurations; gain and bandwidth of each amplifier; feedback resistor values

Approach: Because the required value of R_{in} is relatively low and can be met by a resistor, both the inverting and noninverting amplifier stages should be considered. More than one stage will be required because a single op amp by itself cannot simultaneously meet the specifications for A_v, f_H, and R_{out}. For example, if we were to use the open-loop op amp by itself, it would provide a gain of 100 dB (10^5) but have a bandwidth of only $f_t/10^5 = 10$ Hz. Thus, we must reduce the gain of each stage in order to increase the bandwidth (i.e., we must trade gain for bandwidth).

For simplicity in the design, we assume that the amplifier will be built from a cascade of N identical amplifier stages. The design formulas will be set up in a logical order so that we can choose one design variable, and the rest of the equations can then be evaluated based on that single design choice. For this particular design, the gain and bandwidth are the most difficult specifications to achieve, whereas the required input and output resistance specifications are easily met. We can initially force our design to meet either the gain or the bandwidth specification and then find the number of stages that will be required to achieve the other specifications.

Assumptions: The design must have the minimum number of stages required to meet the specifications in order to achieve minimum cost.

Analysis: In this example, we force the cascade amplifier to meet the gain specification, and then find the number of stages needed to meet the bandwidth by repeated trial and error.

To meet the gain specification, we set the gain of each stage to

$$A_v(0) = \sqrt[N]{10^5}$$

Based on this choice, we can then calculate the other characteristics of the amplifier using this process:

1. Choose N.
2. Calculate the gain required of each stage $A_v(0) = \sqrt[N]{10^5}$.
3. Find β using the numerical result from step 2.
4. Calculate the bandwidth f_{H1} of each stage.
5. Calculate the bandwidth of N stages using Eq. (12.9).
6. Using $A\beta$, calculate R_{out} and R_{in}.
7. See if specifications are met. If not, go back to step 1 and try a new value of N.

The formulas for the noninverting and inverting amplifiers are slightly different, as summarized in Table 12.3. These formulas have been used for the results tabulated in the spreadsheet in Table 12.4.

From Table 12.4, we see that a cascade of six noninverting amplifiers meets all the specifications, whereas seven inverting amplifier stages are required. This occurs because the inverting

TABLE 12.3
N-Stage Cascades of Noninverting and Inverting Amplifiers

$\beta = \dfrac{R_1}{R_1 + R_2}$	**NONINVERTING AMPLIFIER**	**INVERTING AMPLIFIER**
1. Single-stage gain $A_v(0) = \sqrt[N]{10^5}$	$A_v(0) = 1 + \dfrac{R_2}{R_1}$	$A_v(0) = -\dfrac{R_2}{R_1}$
2. Feedback factor	$\beta = \dfrac{1}{A_v(0)}$	$\beta = \dfrac{1}{1 + |A_v(0)|}$
3. Bandwidth of each stage	$f_{H1} = \dfrac{f_T}{\dfrac{A_o}{1 + A_o\beta}}$	$f_{H1} = \dfrac{f_T}{\dfrac{A_o}{1 + A_o\beta}}$
4. N-stage bandwidth	$f_H = f_{H1}\sqrt{2^{1/N} - 1}$	$f_B = f_{H1}\sqrt{2^{1/N} - 1}$
5. Input resistance	$R_{id}(1 + A_o\beta)$	R_1
Output resistance	$\dfrac{R_o}{1 + A_o\beta}$	$\dfrac{R_o}{1 + A_o\beta}$

TABLE 12.4
Design of Cascade of N Identical Operational Amplifier Stages

CASCADE OF IDENTICAL NONINVERTING AMPLIFIERS

NUMBER OF STAGES	$A_V(0)$ GAIN PER STAGE $1/\beta$	f_{H1} SINGLE STAGE $\beta \times f_T$	f_H N STAGES	R_{in}	R_{out}
1	1.00E+05	1.00E+01	1.000E+01	2.00E+09	2.50E+01
2	3.16E+02	3.16E+03	2.035E+03	3.17E+11	1.58E−01
3	4.64E+01	2.15E+04	1.098E+04	2.16E+12	2.32E−02
4	1.78E+01	5.62E+04	2.446E+04	5.62E+12	8.89E−03
5	1.00E+01	1.00E+05	3.856E+04	1.00E+13	5.00E−03
6	**6.81E+00**	**1.47E+05**	**5.137E+04**	**1.47E+13**	**3.41E−03**
7	5.18E+00	1.93E+05	6.229E+04	1.93E+13	2.59E−03
8	4.22E+00	2.37E+05	7.134E+04	2.37E+13	2.11E−03

CASCADE OF IDENTICAL INVERTING AMPLIFIERS

NUMBER OF STAGES	$A_V(0)$ $(1/\beta) - 1$	f_{H1} SINGLE STAGE	f_H N STAGES	R_{in}	R_{out}
1	1.00E+05	1.00E+01	1.00E+01	R_1	2.50E+01
2	3.16E+02	3.15E+03	2.03E+03	R_1	1.58E−01
3	4.64E+01	2.11E+04	1.08E+04	R_1	2.32E−02
4	1.78E+01	5.32E+04	2.32E+04	R_1	8.89E−03
5	1.00E+01	9.09E+04	3.51E+04	R_1	5.00E−03
6	6.81E+00	1.28E+05	4.48E+04	R_1	3.41E−03
7	**5.18E+00**	**1.62E+05**	**5.22E+04**	R_1	**2.59E−03**
8	4.22E+00	1.92E+05	5.77E+04	R_1	2.11E−03

amplifier has a slightly smaller bandwidth than the noninverting amplifier for a given value of closed-loop gain. We are usually interested in the most economical design, so the six-stage amplifier will be chosen. Note that the R_{out} requirement is met with $N > 2$ for both amplifiers.

To complete the design, we must choose values for R_1 and R_2. From Table 12.4, the gain of each stage must be at least 6.81, requiring the resistor ratio R_2/R_1 to be 5.81. Because we will

TABLE 12.5
Cascade of Six Identical Noninverting Amplifiers

NUMBER OF STAGES	$A_V(0)$ GAIN PER STAGE $1/\beta$	N STAGE GAIN	f_{H1} SINGLE STAGE $\beta \times f_T$	f_H N STAGES	R_{in}	R_{out}
6	6.81E+00	1.00E+05	1.47E+05	5.137E+04	1.47E+13	3.41E−03
6	6.83E+00	1.02E+05	1.46E+05	5.121E+04	1.46E+13	3.42E−03
6	6.85E+00	1.04E+05	1.46E+05	5.107E+04	1.46E+13	3.43E−03
6	6.87E+00	1.05E+05	1.45E+05	5.092E+04	1.46E+13	3.44E−03
6	6.89E+00	1.07E+05	1.45E+05	5.077E+04	1.45E+13	3.45E−03
6	6.91E+00	1.09E+05	1.45E+05	5.062E+04	1.45E+13	3.46E−03
6	6.93E+00	1.11E+05	1.44E+05	5.048E+04	1.44E+13	3.47E−03
6	6.95E+00	1.13E+05	1.44E+05	5.033E+04	1.44E+13	3.48E−03
6	6.97E+00	1.15E+05	1.43E+05	5.019E+04	1.43E+13	3.49E−03
6	6.99E+00	1.17E+05	1.43E+05	5.004E+04	1.43E+13	3.50E−03
6	7.01E+00	1.19E+05	1.43E+05	4.990E+04	1.43E+13	3.51E−03

probably not be able to find two 5 percent resistors that give a ratio of exactly 5.81, we need to explore the acceptable range for the ratio now that we know we need six stages. In Table 12.5, a spreadsheet is again used to study six-stage amplifier designs having gains ranging from 6.81 to 7.01. As the single-stage gain is increased, the overall bandwidth decreases. From Table 12.5, we see that the specifications will be met for resistor ratios falling between 5.81 and 5.99.

Many acceptable resistor ratios can be found in the table of 5 percent resistors in Appendix C. Picking $A_v(0) = 6.91$, a value near the center of the range of acceptable gain and bandwidth, two possible resistors sets are

$$\text{(i)} \quad R_1 = 22 \text{ k}\Omega, \; R_2 = 130 \text{ k}\Omega$$

which gives

$$1 + \frac{R_2}{R_1} = 6.91, \; A(0) = 101 \text{ dB}, \; f_H = 50.6 \text{ kHz}, \; R_{out} = 3.46 \text{ m}\Omega$$

and

$$\text{(ii)} \quad R_1 = 5.6 \text{ k}\Omega, \; R_2 = 33 \text{ k}\Omega$$

which yields

$$1 + \frac{R_2}{R_1} = 6.89, \; A(0) = 101 \text{ dB}, \; f_H = 50.8 \text{ kHz}, \; R_{out} = 3.45 \text{ m}\Omega$$

The overall size of these resistors has been chosen so that the feedback resistors do not heavily load the output of the op amp. For example, the resistor pairs $R_1 = 220 \; \Omega$ and $R_2 = 1.3$ kΩ or $R_1 = 56 \; \Omega$ and $R_2 = 330 \; \Omega$, although providing acceptable resistor ratios, would not be desirable choices for a final design.

Check of Results: Based on the spreadsheet results, a design has been found that meets the specifications.

Discussion: This example has explored the design of a fairly complex multistage amplifier. Economical design requires the use of the minimum number of amplifier stages. In this case, spreadsheets were used to explore the design space, and the calculations indicated that the specifications could be met with a cascade of six identical noninverting amplifiers. The design was completed through the choice of feedback resistors from the set of available discrete resistor values.

Computer-Aided Analysis: With the level of complexity in this example, it would obviously be quite useful to use SPICE to check our final design, and this is done in Ex. 12.4.

TABLE 12.6
Cascade of Six Identical Noninverting Amplifiers — Worse-Case Analysis

R VALUES	ONE-STAGE GAIN	SIX-STAGE GAIN	f_{H1}	f_H	R_{in}	R_{out}
Nominal	6.91E + 00	1.09E + 05	1.45E + 05	5.065E + 04	1.45E + 13	3.45E − 03
Max	7.53E + 00	1.82E + 05	1.33E + 05	**4.647E + 04**	1.33E + 13	3.77E − 03
Min	6.35E + 00	**6.53E + 04**	1.58E + 05	5.514E + 04	1.58E + 13	3.17E − 03

The Influence of Tolerances on Design

Now that we have completed Ex. 12.3, let us explore the effects of the resistor tolerances on our design. We have chosen resistors that have 5 percent tolerances; Table 12.6 presents the results of calculating the worst-case specifications, in which

$$A_v^{\text{nom}} = 1 + \frac{130 \text{ k}\Omega}{22 \text{ k}\Omega} = 6.91$$

$$A_v^{\text{max}} = 1 + \frac{130 \text{ k}\Omega(1.05)}{22 \text{ k}\Omega(0.95)} = 7.53$$

$$A_v^{\text{min}} = 1 + \frac{130 \text{ k}\Omega(0.95)}{22 \text{ k}\Omega(1.05)} = 6.35$$

The nominal design values easily meet both specifications, with a margin of 9 percent for the gain but only 1.3 percent for the bandwidth. When the resistor tolerances are set to give the largest gain per stage, the gain specification is easily met, but the bandwidth shrinks below the specification limit. At the opposite extreme, the gain of the six stages fails to meet the required specification. This analysis gives us an indication that there may be a problem with the design. Of course, assuming that all the amplifiers reach the worst-case gain and bandwidth limits at the same time is an extreme conclusion. Nevertheless, the nominal bandwidth does not exceed the specification limit by very much.

A Monte Carlo analysis would be much more representative of the actual design results. Such an analysis for 10,000 cases of our six-stage amplifier indicates that if this circuit is built with 5 percent resistors, more than 30 percent of the amplifiers will fail to meet either the gain or bandwidth specification. (The details of this calculation and exact results are left for Prob. 12.33.)

DESIGN EXAMPLE 12.4 **MACRO MODEL APPLICATION**

Use a SPICE op amp macro model to simulate the frequency response of a multistage amplifier.

PROBLEM Use simulation to verify the frequency response of the six-stage amplifier designed in Ex. 12.3.

SOLUTION **Known Information and Given Data:** The six-stage noninverting amplifier cascade design from Ex. 12.3 with $R_1 = 22$ kΩ and $R_2 = 130$ kΩ. The op amp specifications are $A_o = 100$ dB, $f_T = 1$ MHz, $R_{id} = 1$ GΩ, and $R_o = 50$ Ω.

Unknowns: A Bode plot of the amplifier frequency response; the values of $A_v(0)$ and f_H

Approach: Use a SPICE macro model for the op amp and use it to simulate the frequency response of the six-stage amplifier. Use SPICE subcircuits to simplify the analysis.

Assumptions: The amplifier is a single-pole amplifier. Symmetrical 15-V power supplies are available.

Analysis: After drawing the circuit with the schematic editor, we need to set the parameters of the SPICE op amp model to agree with our specifications. The differential-mode gain and input resistance and the output resistance are given. We need to calculate the frequency of the first pole: $f_\beta = f_T/A_o = 10$ Hz.

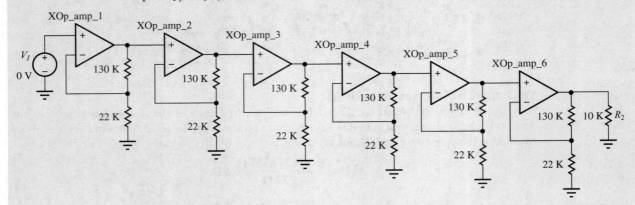

VI is an 1-V ac source with a dc value of 0 V. Since the amplifier is dc-coupled, a transfer function analysis from source VI to the output node will give the low-frequency gain, input resistance, and output resistance. An ac analysis using FSTART = 100 Hz, FSTOP = 1 MHz, and 10 frequency points per decade will produce the Bode plot needed to find the bandwidth. The simulation results yield a gain of 100.7 dB, $R_{in} = 28.9$ TΩ, $R_{out} = 3.52$ mΩ, and the bandwidth is 54.8 kHz.

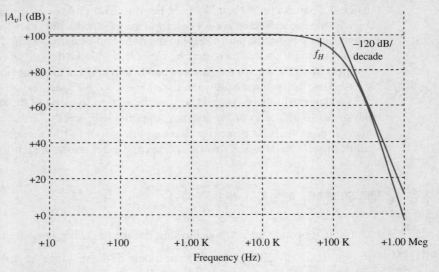

Check of Results: We see some discrepancies. The gain and output resistance agree with our calculations, but the bandwidth is larger than expected, and the input resistance is far too small. We should immediately be concerned about our simulation results. Indeed, a closer examination of the Bode plot also shows that the high-frequency roll-off is exceeding the 6(20 dB/decade) = 120 dB/decade slope that we should expect.

Discussion: The problems are buried in the macro model in which all of the unspecified parameters have default values. If we look at the op amp model in the version of SPICE used here, we find the default values are the same as given in Table 11.4: common-mode input resistance = 2 GΩ, second pole frequency = 2 MHz, offset voltage = 1 mV, input bias current = 80 nA, input offset current = 20 nA, etc. The input resistance cannot exceed the value set by R_{ic}, and the bandwidth and roll-off enhancements are actually caused by the second op amp pole at 2 MHz. If we change R_{ic} to 10^{15} Ω and set the higher-order pole frequencies all to 200 MHz, SPICE yields an input resistance of 28.9 TΩ and a bandwidth of 50.4 kHz, close to the expected values. In addition, the high-frequency roll-off rate is 120 dB/decade.

An additional problem was encountered in this simulation. In the initial simulation attempts, very small values of voltage gain were generated. An operating point analysis indicated that several of the op amp output voltages were at large values. Here again, the default parameter settings were causing the problem. This amplifier has very high overall gain, and a 1-mV offset voltage at the input of the first amplifier multiplied by the gain of 100,000 should produce 100 V at the output of the sixth amplifier! In order for the simulation to work, the offset voltage, input bias current, and input offset currents must all be set to zero in our op amp model!

The results discussed in the previous paragraph are also of significant practical interest! If we attempt to build this amplifier, we will encounter exactly the same problem. The offset voltages and input bias currents of the amplifier will cause the individual op amp outputs to saturate against the power supply levels.

EXERCISE: Simulate the amplifier with the dc value of VI set to 1 mV. What are the op amp output voltages?

ANSWERS: 6.91 mV, 47.7 mV, 330 mV, 2.28 V, 15 V, 15 V; the last two are saturated.

12.2 THE INSTRUMENTATION AMPLIFIER

We often need to amplify the difference in two signals but cannot use the difference amplifier in Fig. 12.3 because its input resistance is too low. In such a case, we can combine two noninverting amplifiers with a difference amplifier to form the high-performance composite **instrumentation amplifier** depicted in Fig. 12.4.

In this circuit, op amp 3, with resistors R_3 and R_4, forms a difference amplifier. Using Eq. (10.63), the output voltage v_o is

$$v_o = \left(-\frac{R_4}{R_3}\right)(v_a - v_b) \tag{12.10}$$

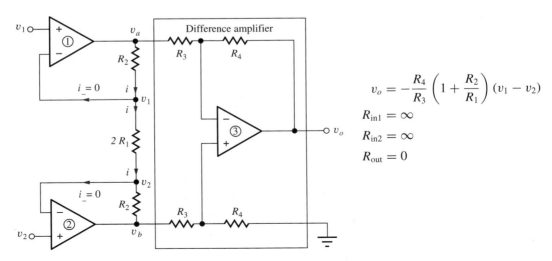

$$v_o = -\frac{R_4}{R_3}\left(1 + \frac{R_2}{R_1}\right)(v_1 - v_2)$$

$$R_{in1} = \infty$$
$$R_{in2} = \infty$$
$$R_{out} = 0$$

Figure 12.4 Circuit for the instrumentation amplifier.

834 Chapter 12 Operational Amplifier Applications

in which voltages v_a and v_b are the outputs of the first two amplifiers. Because the i_- input currents to amplifiers 1 and 2 must be zero, voltages v_a and v_b are related to each other by

$$v_a - iR_2 - i(2R_1) - iR_2 = v_b \quad \text{or} \quad v_a - v_b = 2i(R_1 + R_2) \quad (12.11)$$

Since the voltage across the inputs of op amps 1 and 2 must both be zero, the voltage difference $(v_1 - v_2)$ appears directly across the resistor $2R_1$, and

$$i = \frac{v_1 - v_2}{2R_1} \quad (12.12)$$

Combining Eqs. (12.10), (12.11), and (12.12) yields a final expression for the output voltage of the instrumentation amplifier:

$$v_o = -\frac{R_4}{R_3}\left(1 + \frac{R_2}{R_1}\right)(v_1 - v_2) \quad (12.13)$$

The ideal instrumentation amplifier amplifies the difference in the two input signals and provides a gain, that is, equivalent to the product of the gains of the noninverting and difference amplifiers. The input resistance presented to both input sources is infinite because the input current to both op amps is zero, and the output resistance is forced to zero by the difference amplifier.

EXAMPLE 12.5   **INSTRUMENTATION AMPLIFIER ANALYSIS**

The three op amp output voltages are calculated for a specific set of dc input voltages in this example.

PROBLEM Find the values of V_O, V_A, and V_B for the instrumentation amplifier in Fig. 12.4 if $V_1 = 2.5$ V, $V_2 = 2.25$ V, $R_1 = 15$ kΩ, $R_2 = 150$ kΩ, $R_3 = 15$ kΩ, and $R_4 = 30$ kΩ.

SOLUTION **Known Information and Given Data:** $V_1 = 2.5$ V, $V_2 = 2.25$ V, $R_1 = 15$ kΩ, $R_2 = 150$ kΩ, $R_3 = 15$ kΩ, and $R_4 = 30$ kΩ for the circuit configuration in Fig. 12.4.

Unknowns: The values of V_O, V_A, and V_B

Approach: All the values are specified to permit direct use of Eq. (12.13).

Assumptions: The op amps are ideal. Therefore $I_+ = 0 = I_-$ and $V_+ = V_-$ for each op amp.

Analysis: Using Eq. (12.13) with dc values, we find the output voltage is

$$V_O = -\frac{R_4}{R_3}\left(1 + \frac{R_2}{R_1}\right)(V_1 - V_2) = -\frac{30 \text{ k}\Omega}{15 \text{ k}\Omega}\left(1 + \frac{150 \text{ k}\Omega}{15 \text{ k}\Omega}\right)(2.5 - 2.25) = -5.50 \text{ V}$$

Since the op amp input currents are zero, V_A and V_B can be related directly to the two input voltages and current i

$$V_A = V_1 + IR_2 \quad \text{and} \quad V_B = V_2 - IR_2$$

$$I = \frac{V_1 - V_2}{2R_1} = \frac{2.5 \text{ V} - 2.25 \text{ V}}{2(15 \text{ k}\Omega)} = 8.33 \text{ }\mu\text{A}$$

$$V_A = 2.5 + (8.33 \text{ }\mu\text{A})(150 \text{ k}\Omega) = +3.75 \text{ V} \quad V_B = 2.25 - (8.33 \text{ }\mu\text{A})(150 \text{ k}\Omega) = 1.00 \text{ V}$$

Check of Results: The unknowns have all been determined. Let us check to see if these voltages are consistent with the difference amplifier that should amplify its input by a factor of -2:

$$V_O = -\frac{R_4}{R_3}(V_A - V_B) = -\frac{30 \text{ k}\Omega}{15 \text{ k}\Omega}(3.75 - 1.00) \text{ V} = -5.50 \text{ V} \checkmark$$

ELECTRONICS IN ACTION

CMOS Navigation Chip Prototype for Optical Mice

Agilent Technologies has sold over 100 million optical navigation mouse sensors, the devices at the core of most optical mice sold today. However, as is often the case in the engineering world, the navigation technology was originally developed for a different application. In 1993, a group of engineers at Hewlett-Packard Laboratories led by Ross Allen envisioned a handheld, battery powered document scanner that could be moved across a page in a freehand motion and still accurately recover the text. To help make this vision a reality, Travis Blalock and Dick Baumgartner at HP Labs designed a CMOS integrated circuit to optically measure movement of the scanner across the paper. The chip, known as "Magellan" within HP, is shown below.

Similar to digital cameras, the prototype contains a photo-receiver array to acquire images of the scanned surface, which is illuminated and positioned under the chip. The images are then transferred from the photo-receiver array to a computation array. The computation array always contains a reference image and a current image. Two-dimensional cross-correlations are then computed between the two images. The cross-correlation results can then be used to calculate the physical movement between the reference image and current image.

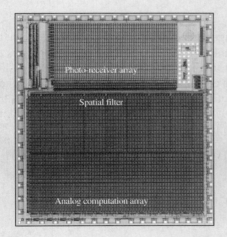
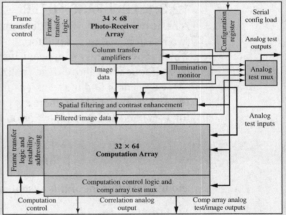

Optical navigation chip photo and block diagram.

The Magellan optical navigation chip contains over 6000 operational amplifiers and sample-and-hold circuits, over 2000 photo-transistor amplifiers, and acquires 25,000 images per second. The chip calculates the cross-correlations at a rate of over 1.5 billion computations per second.

After a successful technology demonstration, the prototype was transferred to a product division and modified to create a commercial product. At some point in this process, it was recognized that the optical navigation architecture could be used as the basis for an optical mouse. The navigation chip design was again modified and became the basis of an optical navigation module sold by Agilent Technologies (a spinoff of the Hewlett-Packard company) and is used as the original basis of most of the available optical mice on the market today.

EXERCISE: Suppose v_1 and v_2 are dc voltages with $V_1 = 5.001$ V, $V_2 = 4.999$ V, $R_1 = 1$ kΩ, $R_2 = 49$ kΩ, $R_3 = 10$ kΩ, and $R_4 = 10$ kΩ in Fig. 12.4. Write expressions for V_A and V_B. What are the values of V_O, V_A, V_B, and I?

ANSWERS: $V_A = V_1 + I R_2$, $V_B = V_2 - I R_2$; -0.100 V, 5.05 V, 4.95 V, 1.00 μA.

12.3 ACTIVE FILTERS

Filters come in many forms. We have looked at the characteristics of low-pass, high-pass, band-pass, and band-reject filters in Chapters 1 and 10. The simplest filter implementation uses passive components, resistors, capacitors, and inductors. In integrated circuits, however, inductors are difficult to fabricate, take up significant area, and can only be made with very small values of inductance. With the advent of low-cost high-performance op amps, new circuits were invented that could realize the desired filter characteristics without the use of inductors. These filters utilizing op amps are referred to as **active filters,** and this section discusses examples of active low-pass, high-pass, and band-pass filters. A simple active low-pass and high-pass filters were discussed in Secs. 10.10.5 and 10.10.6, but these circuits produced only a single pole. Many of the filters described in this section are more efficient in the sense that the circuits achieve two poles of filtering per op amp. The interested reader can explore the material further in many texts that deal exclusively with active-filter design.

12.3.1 LOW-PASS FILTER

A basic two-pole low-pass filter configuration is shown in Fig. 12.5 and is formed from an op amp with two resistors and two capacitors. In this particular circuit, the op amp operates as a voltage follower, which provides unity gain over a wide range of frequencies. The filter uses positive feedback through C_1 at frequencies above dc to realize complex poles without the need for inductors.

Let us now find the transfer function describing the voltage gain of this filter. The ideal op amp forces $V_o(s) = V_2(s)$, so there are only two independent nodes in the circuit. Writing nodal equations for $V_1(s)$ and $V_2(s)$ (with a Norton transformation of v_i and R_1) yields

$$\begin{bmatrix} G_1 V_I(s) \\ 0 \end{bmatrix} = \begin{bmatrix} sC_1 + G_1 + G_2 & -(sC_1 + G_2) \\ -G_2 & sC_2 + G_2 \end{bmatrix} \begin{bmatrix} V_1(s) \\ V_2(s) \end{bmatrix} \quad (12.14)$$

and the determinant of this system of equations is

$$\Delta = s^2 C_1 C_2 + s C_2 (G_1 + G_2) + G_1 G_2 \quad (12.15)$$

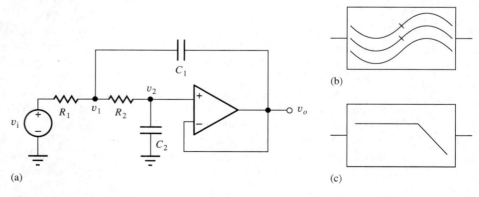

Figure 12.5 (a) A two-pole low-pass filter; (b) low-pass filter symbol; (c) alternate low-pass filter symbol.

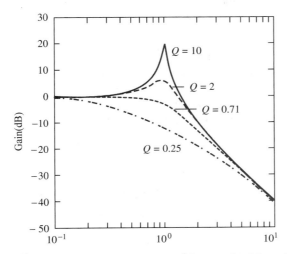

Figure 12.6 Low-pass filter response[2] for $\omega_o = 1$ and four values of Q.

Solving for $V_2(s)$ and remembering that $V_o(s) = V_2(s)$ yields

$$V_o(s) = V_2(s) = \frac{G_1 G_2}{\Delta} V_I(s) \tag{12.16}$$

which can be rearranged as

$$A_{LP}(s) = \frac{V_o(s)}{V_I(s)} = \frac{\frac{1}{R_1 R_2 C_1 C_2}}{s^2 + s \frac{1}{C_1}\left(\frac{1}{R_1} + \frac{1}{R_2}\right) + \frac{1}{R_1 R_2 C_1 C_2}} \tag{12.17}$$

Equation (12.17) is most often written in standard form as

$$A_{LP}(s) = \frac{\omega_o^2}{s^2 + s\dfrac{\omega_o}{Q} + \omega_o^2} \tag{12.18}$$

in which

$$\omega_o = \frac{1}{\sqrt{R_1 R_2 C_1 C_2}} \quad \text{and} \quad Q = \sqrt{\frac{C_1}{C_2}} \frac{\sqrt{R_1 R_2}}{R_1 + R_2} \tag{12.19}$$

The frequency ω_o is referred to as the cutoff frequency of the filter, although the exact value of the cutoff frequency, based on the strict definition of ω_H, is equal to ω_o only for $Q = 1/\sqrt{2}$. At low frequencies — that is, $\omega \ll \omega_o$ — the filter has unity gain, but for frequencies well above ω_o, the filter response exhibits a two-pole roll-off, falling at a rate of 40 dB/decade. At $\omega = \omega_o$, the gain of the filter is equal to Q.

Figure 12.6 shows the response of the filter for $\omega_o = 1$ and four values of Q: 0.25, $1/\sqrt{2}$, 2, and 10. $Q = 1/\sqrt{2}$ corresponds to the **maximally flat magnitude** response of a **Butterworth filter**, which gives the maximum bandwidth without a peaked response. For a Q larger than $1/\sqrt{2}$, the filter response exhibits a peaked response that is usually undesirable, whereas a Q below $1/\sqrt{2}$ does not take maximum advantage of the filter's bandwidth capability. Because the voltage follower must accurately provide a gain of 1, ω_o should be designed to be one to two decades below the unity-gain frequency of the op amp.

From a practical point of view, a much wider selection of resistor values than capacitor values exists, and the filters are often designed with $C_1 = C_2 = C$. Then ω_o and Q are adjusted by choosing

---

[2] Using MATLAB: Bode(1,[1 0.1 1]), for example.

different values of R_1 and R_2. For the equal capacitor design,

$$\omega_o = \frac{1}{C\sqrt{R_1 R_2}} \quad \text{and} \quad Q = \frac{\sqrt{R_1 R_2}}{R_1 + R_2} \quad (12.20)$$

Another practical consideration concerns op amp bias currents. In order to operate properly, the active filter circuits must provide dc paths for the op amp bias currents. In the circuit in Fig. 12.5, the dc current for the noninverting input is supplied from the dc-referenced signal source through R_1 and R_2. The dc current in the inverting input is supplied from the op amp output.

DESIGN NOTE

In order for an op amp circuit to operate properly, the feedback network must provide a dc path for the amplifier's input bias currents.

DESIGN EXAMPLE 12.6   LOW-PASS FILTER DESIGN

Determine the capacitor and resistor values required to meet a cutoff frequency specification in a two-pole active low-pass filter.

PROBLEM Design a low-pass filter using the circuit in Fig. 12.5 with an upper cutoff frequency of 5 kHz and a maximally flat response.

SOLUTION **Known Information and Given Data:** Second-order active low-pass filter circuit in Fig. 12.6; maximally flat design with $f_H = 5$ kHz

Unknowns: R_1, R_2, C_1, and C_2

Approach: As mentioned in Sec. 12.3, the maximally flat response for the transfer function in Eq. (12.18) is achieved for $Q = 1/\sqrt{2}$. For this case, we also find that $f_H = f_o$. Unfortunately, based on Eq. (12.19), the simple equal capacitor design cannot achieve this Q. We will need to explore another design option.

Assumptions: The operational amplifier is ideal.

Analysis: From Eq. (12.19), we see that one workable choice for the element values is $C_1 = 2C_2 = 2C$ and $R_1 = R_2 = R$. For these values,

$$R = \frac{1}{\sqrt{2}\omega_o C} \quad \text{and} \quad Q = \frac{1}{\sqrt{2}}$$

but we still have two values to select and only one design constraint. We must call on our engineering judgment to make the design choice. Note that $1/\omega_o C$ represents the reactance of C at the frequency ω_o, and R is 30 percent smaller than this value. Thus, the impedance level of the filter is set by the choice of C (or R). If the impedance level is too low, the op amp will not be able to supply the current needed to drive the feedback network.

At 5 kHz, a 0.01-μF capacitor has a reactance of 3.18 kΩ:

$$\frac{1}{\omega_o C} = \frac{1}{10^4 \pi (10^{-8})} = 3180 \ \Omega$$

This is a readily available value of capacitance, and so

$$R = \frac{3180 \ \Omega}{\sqrt{2}} = 2250 \ \Omega$$

Referring to the precision resistor table in Appendix A, we find that the nearest 1 percent resistor value is 2260 Ω. The completed design values are

$$R_1 = R_2 = 2.26 \text{ k}\Omega, C_1 = 0.02 \text{ μF}, C_2 = 0.01 \text{ μF}$$

Check of Results: Using the design values yields $f_o = 4980$ Hz and $Q = 0.707$.

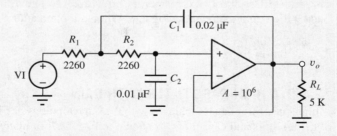

Discussion and Computer-Aided Analysis: The frequency response of the filter is simulated using the circuit above. The op amp gain is set to 10^6. An ac analysis is performed with VI as the source with FSTART = 10 Hz, FSTOP = 10 MHz, and 10 simulations points per frequency decade. The gain to the output node is 0 dB and $f_H = 5$ kHz, in agreement with the design specification. A second simulation result appears in the graph below giving the frequency response for a μA741 op amp. The differences are caused by the finite gain-bandwidth of the μA741.

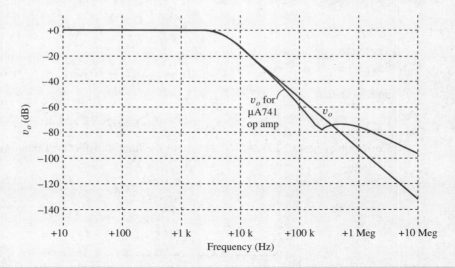

EXERCISE: What is $A_v(0)$ for the filter design in the above example? Show that $f_H = f_o$ for the maximally flat design with $Q = 1/\sqrt{2}$.

ANSWER: +1.00

EXERCISE: Redesign the filter in Ex. 12.6 to have an upper cutoff frequency of 10 kHz with a maximally flat response. Keep the impedance level of the filter the same.

ANSWERS: 0.01 μF, 0.005 μF, 2260 Ω, 2260 Ω.

EXERCISE: Starting with Eq. (12.18), show that $|A_{LP}(j\omega_o)| = Q$.

> **EXERCISE:** Change the cutoff frequency of this filter to 2 kHz by changing the values of R_1 and R_2. Do not change the Q.
>
> **ANSWERS:** $R_1 = R_2 = 5.62$ kΩ
>
> **EXERCISE:** Use the Q expression in Eq. (12.20) to show that $Q = 1/\sqrt{2}$ cannot be realized using the equal capacitance design. What is the maximum Q for $C_1 = C_2$?
>
> **ANSWER:** 0.5

12.3.2 A HIGH-PASS FILTER WITH GAIN

A **high-pass filter** can be achieved with the same topology as Fig. 12.5 by interchanging the position of the resistors and capacitors, as shown in Fig. 12.7. In many applications, filters with gain in the midband region are preferred, and the voltage follower in the low-pass filter has been replaced with a noninverting amplifier with a gain of K in the filter of Fig. 12.7. Gain K provides an additional degree of freedom in the design of the filter elements. Note that dc paths exist for both op amp input bias currents through resistor R_2 and the two feedback resistors.

The analysis is virtually identical to that of the low-pass filter. Nodes v_1 and v_2 are the only independent nodes because $v_o = +Kv_2$, and writing the two nodal equations yields this system of equations:

$$\begin{bmatrix} sC_1 V_I(s) \\ 0 \end{bmatrix} = \begin{bmatrix} s(C_1+C_2)+G_1 & -(sC_2+KG_1) \\ -sC_2 & sC_2+G_2 \end{bmatrix} \begin{bmatrix} V_1(s) \\ V_2(s) \end{bmatrix} \quad (12.21)$$

The system determinant is

$$\Delta = s^2 C_1 C_2 + s(C_1+C_2)G_2 + sC_2 G_1(1-K) + G_1 G_2 \quad (12.22)$$

and the output voltage is

$$V_o(s) = KV_2(s) = K \frac{s^2 C_1 C_2 V_I(s)}{\Delta} \quad (12.23)$$

Combining Eqs. (12.22) and (12.23) yields the filter transfer function that can be written in standard form as

$$A_{HP}(s) = \frac{V_o(s)}{V_I(s)} = K \frac{s^2}{s^2 + s\frac{\omega_o}{Q} + \omega_o^2} \quad (12.24)$$

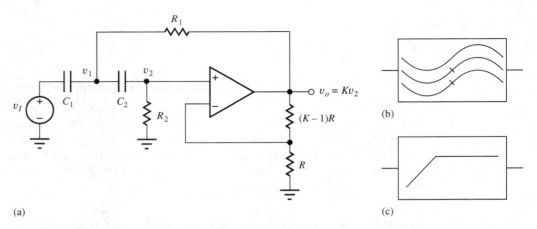

Figure 12.7 (a) A high-pass filter with gain; (b) high-pass filter symbol; (c) alternate high-pass filter symbol.

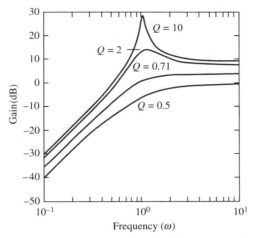

Figure 12.8 High-pass filter response[3] for $\omega_o = 1$ and four values of Q.

in which

$$\omega_o = \frac{1}{\sqrt{R_1 R_2 C_1 C_2}} \quad \text{and} \quad Q = \left[\sqrt{\frac{R_1}{R_2} \frac{C_1 + C_2}{\sqrt{C_1 C_2}}} + (1-K)\sqrt{\frac{R_2 C_2}{R_1 C_1}}\right]^{-1} \quad (12.25)$$

For the case $R_1 = R_2 = R$ and $C_1 = C_2 = C$, Eqs. (12.24) and (12.25) can be simplified to

$$A_{HP}(s) = K \frac{s^2}{s^2 + s\frac{3-K}{RC} + \frac{1}{R^2 C^2}} \qquad \omega_o = \frac{1}{RC} \quad \text{and} \quad Q = \frac{1}{3-K} \quad (12.26)$$

For this design choice, ω_o and Q can be adjusted independently.

Figure 12.8 shows the high-pass filter responses for a filter with $\omega_o = 1$ and four values of Q. The parameter ω_o corresponds approximately to the lower-cutoff frequency of the filter, and $Q = 1/\sqrt{2}$ again represents the maximally flat, or Butterworth, filter response.

The noninverting amplifier circuit in Fig. 12.7 must have $K \geq 1$. Note in Eq. (12.26) that $K = 3$ corresponds to infinite Q. This situation corresponds to the poles of the filter being exactly on the imaginary axis at $s = j\omega_o$ and results in sinusoidal oscillation. (Oscillators are discussed later in this chapter.) For $K > 3$, the filter poles will be in the right-half plane, and values of $K \geq 3$ correspond to unstable filters. Therefore, $1 \leq K < 3$.

> **EXERCISE:** What is the gain at $\omega = \omega_o$ for the filter described by Eq. (12.26)?
>
> **ANSWER:** $\frac{K}{3-K} < 90°$
>
> **EXERCISE:** The high-pass filter in Fig. 12.7 has been designed with $C_1 = 0.0047$ μF, $C_2 = 0.001$ μF, $R_1 = 10$ kΩ, and $R_2 = 20$ kΩ, and the amplifier gain is 2. What are f_o and Q for this filter?
>
> **ANSWERS:** 5.19 kHz, 0.828

---

[3] Using MATLAB: Bode([(3-sqrt(2)) 0 0],[1 sqrt(2) 1]), for example.

> **EXERCISE:** Derive an expression for the sensitivity of Q with respect to the closed-loop gain K for the high-pass filter in Fig. 12.7 (see Eq. 12.33). What is the value of sensitivity if $Q = 1/\sqrt{2}$?
> **ANSWERS:** $S_K^Q = (3 - Q)$; 1.12

12.3.3 BAND-PASS FILTER

A **band-pass filter** can be realized by combining the low-pass and high-pass characteristics of the previous two filters. Figure 12.9 is one possible circuit for such a band-pass filter. In this case, the op amp is used in its inverting configuration; this circuit is sometimes called an "infinite-gain" filter because the full open-loop gain of the op amp, ideally infinity, is utilized. Resistor R_3 is added to provide an extra degree of design freedom so that gain, center frequency, and Q can be set with a minimum of interaction. Note again that dc paths exist for both op amp input bias currents.

Analysis of the circuit in Fig. 12.9(b) can be reduced to a one-node problem by using op amp theory to relate $V_o(s)$ directly to $V_1(s)$:

$$sC_2 V_1(s) = -\frac{V_o(s)}{R_2} \quad \text{or} \quad V_1(s) = -\frac{V_o(s)}{sC_2 R_2} \tag{12.27}$$

Using KCL at node v_1,

$$G_{th} V_{th} = [s(C_1 + C_2) + G_{th}] V_1(s) - sC_1 V_o(s) \tag{12.28}$$

Combining Eqs. (12.27) and (12.28) yields

$$\frac{V_o(s)}{V_{th}(s)} = \frac{-\dfrac{s}{R_{th} C_1}}{s^2 + s\dfrac{1}{R_2}\left(\dfrac{1}{C_1} + \dfrac{1}{C_2}\right) + \dfrac{1}{R_{th} R_2 C_1 C_2}} \tag{12.29}$$

The band-pass output can now be expressed as

$$A_{BP}(s) = -\frac{V_o(s)}{V_I(s)} = -\sqrt{\frac{R_3}{R_1 + R_3} \frac{R_2 C_2}{R_1 C_1}} \frac{s\omega_o}{s^2 + s\dfrac{\omega_o}{Q} + \omega_o^2} \tag{12.30}$$

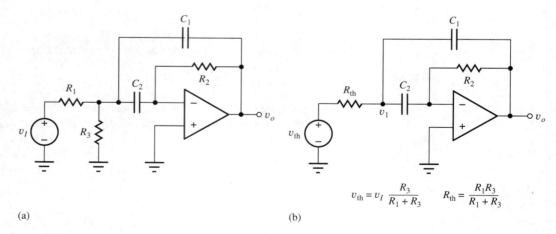

(a)   (b)

Figure 12.9 (a) Band-pass filter using inverting op amp configuration; (b) simplified band-pass filter circuit.

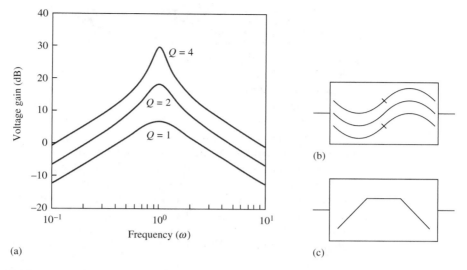

Figure 12.10 (a) Band-pass filter response[4] for $\omega_o = 1$ and three values of Q assuming $C_1 = C_2$ with $R_3 = \infty$; (b) band-pass filter symbol; (c) alternate band-pass filter symbol.

with

$$\omega_o = \frac{1}{\sqrt{R_{th}R_2C_1C_2}} \quad \text{and} \quad Q = \sqrt{\frac{R_2}{R_{th}}\frac{\sqrt{C_1C_2}}{C_1+C_2}} \quad (12.31)$$

If C_1 is set equal to $C_2 = C$, then

$$\omega_o = \frac{1}{C\sqrt{R_{th}R_2}} \quad Q = \frac{1}{2}\sqrt{\frac{R_2}{R_{th}}} \quad BW = \frac{2}{R_2C}$$

$$(12.32)$$

$$A_{BP}(s) = -\left(\frac{2Q}{1+\frac{R_1}{R_3}}\right)\left(\frac{s\omega_o}{s^2 + s\frac{\omega_o}{Q} + \omega_o^2}\right) \quad A_{BP}(\omega_o) = -\frac{1}{2}\left(\frac{R_2}{R_1}\right)$$

The response of the band-pass filter is shown in Fig. 12.10 for $\omega_o = 1$, $C_1 = C_2$, $R_3 = \infty$, and three values of Q. Parameter ω_o now represents the center frequency of the band-pass filter. The response peaks at ω_o, and the gain at the center frequency is equal to $2Q^2$. At frequencies much less than or much greater than ω_o, the filter response corresponds to a single-pole high- or low-pass filter, changing at a rate of 20 dB/decade.

EXERCISE: The filter in Fig. 12.9 is designed with $C_1 = C_2 = 0.02~\mu F$, $R_1 = 2~k\Omega$, $R_3 = 2~k\Omega$, and $R_2 = 82~k\Omega$. What are the values of f_o and Q?

ANSWERS: 879 Hz, 4.5

---

[4] Using MATLAB: Bode([4 0],[1 .5 1]), for example.

ELECTRONICS IN ACTION

Band-Pass Filters in BFSK Reception

Binary frequency shift keying (BFSK) is a basic form of modulation that is studied in communications classes and represents a type of communications that is commonly used for radio teletype transmissions in the high-frequency or "short wave" radio bands (3–30 MHz). The signal transmitting the data shifts back and forth between two closely spaced radio frequencies (e.g., 18,080,000 Hz and 18,080,170 Hz). In the block diagram in the accompanying figure, a communications receiver, that is receiving this transmission produces an audio signal at its output that shifts between 2125 Hz and 2295 Hz (or some other convenient frequency pair separated by a frequency shift of 170 Hz). In the analog signal processing circuit here, six-pole filters are used to separate these two audio tones. Each filter bank consists of a cascade of three, two-pole active band-pass filters as described in Sec. 12.3.3. The outputs of the two filter banks are rectified and filtered by an RC network to form a simple frequency discriminator. The output of the discriminator then drives circuitry that recovers the original data transmission.

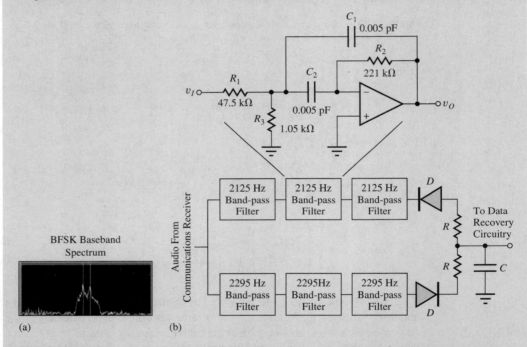

These same functions can be performed in the digital domain using digital signal processing (DSP) if the audio signal from the communications receiver is first digitized by an analog-to-digital (A/D) converter.

12.3.4 SENSITIVITY

An important concern in the design of active filters is the **sensitivity** of ω_o and Q to changes in passive element values and op amp parameters. The sensitivity of design parameter P to changes in circuit parameter Z is defined mathematically as

$$S_Z^P = \frac{\dfrac{\partial P}{P}}{\dfrac{\partial Z}{Z}} = \frac{Z}{P}\frac{\partial P}{\partial Z} \qquad (12.33)$$

Sensitivity S represents the fractional change in parameter P due to a given fractional change in the value of Z. For example, evaluating the sensitivity of ω_o with respect to the values of R and C using Eq. (12.19) yields

$$S_R^{\omega_o} = S_C^{\omega_o} = -\frac{1}{2} \qquad (12.34)$$

A 2 percent increase in the value of R or C will cause a 1 percent decrease in the frequency ω_o.

> **EXERCISE:** Calculate $S_{C_1}^Q$ and $S_{R_2}^Q$ for the low-pass filter using Eq. (12.19) and the values in the example.
>
> **ANSWERS:** +0.5; 0
>
> **EXERCISE:** Calculate $S_R^{\omega_o}$, $S_C^{\omega_o}$, and S_K^Q for the high-pass filter described by Eq. (12.26).
>
> **ANSWERS:** 1; 1; $K/(3-K)$
>
> **EXERCISE:** Calculate $S_{R_1}^{\omega_o}$, $S_{R_2}^{\omega_o}$, $S_{R_3}^{\omega_o}$, $S_C^{\omega_o}$, $S_{R_1}^Q$, $S_{R_2}^Q$, $S_{R_3}^Q$, S_C^Q and S_C^{BW} for the band-pass filter described by Eqs. (12.32).
>
> **ANSWERS:** $-\dfrac{1}{2}\dfrac{R_3}{R_1+R_3}$; $-\dfrac{1}{2}$; $-\dfrac{1}{2}\dfrac{R_1}{R_1+R_3}$; -1; $\dfrac{1}{2}\dfrac{R_3}{R_1+R_3}$; $+\dfrac{1}{2}$; $-\dfrac{1}{2}\dfrac{R_1}{R_1+R_3}$; 0; -1

12.3.5 MAGNITUDE AND FREQUENCY SCALING

The values of resistance and capacitance calculated for a given filter design may not always be convenient, or the values may not correspond closely to the standard values that are available. Magnitude scaling can be used to transform the values of the impedances of a filter without changing its frequency response. Frequency scaling, however, allows us to transform a filter design from one value of ω_o to another without changing the Q of the filter.

Magnitude Scaling

The magnitude of impedances of a filter may all be increased or decreased by a **magnitude scaling** factor K_M without changing ω_o or Q of the filter. To scale the magnitude of the impedance of the filter elements, the value of each resistor[5] is multiplied by K_M and the value of the capacitor is divided by K_M:

$$R' = K_M R \quad \text{and} \quad C' = \frac{C}{K_M} \quad \text{so that} \quad |Z'_C| = \frac{1}{\omega C'} = \frac{K_M}{\omega C} = K_M |Z_C| \quad (12.35)$$

In all the filters discussed in Sec. 12.3, Q is determined by ratios of capacitor values and/or resistor values whereas ω_o always has the form $\omega_o = 1/\sqrt{R_1 R_2 C_1 C_2}$. Applying magnitude scaling to the low-pass filter described by Eq. (12.19) yields

$$\omega'_o = \frac{1}{\sqrt{K_M R_1 (K_M R_2) \dfrac{C_1}{K_M} \dfrac{C_2}{K_M}}} = \frac{1}{\sqrt{R_1 R_2 C_1 C_2}} = \omega_o$$

and

$$Q' = \sqrt{\dfrac{\dfrac{C_1}{K_M}}{\dfrac{C_2}{K_M}}} \dfrac{\sqrt{K_M R_1 (K_M R_2)}}{K_M R_1 + K_M R_2} = \sqrt{\dfrac{C_1}{C_2}} \dfrac{\sqrt{R_1 R_2}}{R_1 + R_2} = Q \qquad (12.36)$$

Thus, both Q and ω_o are independent of the magnitude scaling factor K_M.

---

[5] In *RLC* filters, each inductor value is also increased by K_M: $L' = K_M L$ so $|Z'_L| = K_M |Z_L|$.

> **EXERCISE:** The filter in Fig. 12.9 is designed with $R_1 = R_2 = 2.26$ kΩ, $R_3 = \infty$, $C_1 = 0.02$ μF, and $C_2 = 0.01$ μF. What are the new values of C_1, C_2, R_1, R_2, f_o, and Q if the impedance magnitude is scaled by a factor of (a) 5 and (b) 0.885?
>
> **ANSWERS:** (a) 0.004 μF, 0.002 μF, 11.3 kΩ, 11.3 kΩ, 4980 Hz, 0.471; (b) 0.0226 μF, 0.0113 μF, 2.00 kΩ, 2.00 kΩ, 4980 Hz, 0.471

Frequency Scaling

The cutoff or center frequencies of a filter may be scaled by a **frequency scaling** factor K_F without changing the Q of the filter if each capacitor value is divided by K_F, and the resistor values are left unchanged.

$$R' = R \quad \text{and} \quad C' = \frac{C}{K_F}$$

Once again, using the low-pass filter as an example yields:

$$\omega'_o = \frac{1}{\sqrt{R_1 R_2 \frac{C_1}{K_F} \frac{C_2}{K_F}}} = \frac{K_F}{\sqrt{R_1 R_2 C_1 C_2}} = K_F \omega_o$$

and

$$Q' = \sqrt{\frac{\frac{C_1}{K_F}}{\frac{C_2}{K_F}}} \frac{\sqrt{R_1 R_2}}{R_1 + R_2} = \sqrt{\frac{C_1}{C_2}} \frac{\sqrt{R_1 R_2}}{R_1 + R_2} = Q \quad (12.37)$$

In this case, we see that the value of ω_o is increased by the factor K_F, but Q remains unaffected.

> **EXERCISE:** The filter in Fig. 12.9 is designed with $C_1 = C_2 = 0.02$ μF, $R_1 = 2$ kΩ, $R_3 = 2$ kΩ, and $R_2 = 82$ kΩ. (a) What are the values of f_o and Q? (b) What are the new values of C_1, C_2, R_1, R_2, f_o, and Q if the frequency is scaled by a factor of 4?
>
> **ANSWERS:** 880 Hz, 4.5; 0.005 μF, 0.005 μF, 1 kΩ, 82 kΩ, 3.5 kHz, 4.5.

12.4 SWITCHED-CAPACITOR CIRCUITS

As discussed in some detail in Chapter 6, resistors occupy inordinately large amounts of area in integrated circuits, particularly compared to MOS transistors. **Switched-capacitor (SC) circuits** are an elegant way to eliminate the resistors required in filters by replacing those elements with capacitors and switches. The filters become the discrete-time or sampled-data equivalents of the continuous-time filters discussed in Sec. 12.3, and the circuits then become compatible with high-density MOS IC processes. Switched capacitor circuits have become an extremely important and widely used approach to IC filter design. SC circuits provide low-power filters, and CMOS integrated circuits designed for signal processing and communications applications routinely include SC filters as well as SC analog-to-digital and digital-to-analog converters. These circuits will be discussed in Secs. 12.4 and 12.5.

12.4.1 A SWITCHED-CAPACITOR INTEGRATOR

A basic building block of SC circuits is the **switched-capacitor integrator** in Fig. 12.11. Resistor R of the continuous-time integrator is replaced by capacitor C_1 and MOSFET switches S_1 and S_2 in Fig. 12.11(b). The switches are driven by a **two-phase nonoverlapping clock** as depicted in Fig. 12.11(c). When phase Φ_1 is high, switch S_1 is on and S_2 is off, and when phase Φ_2 is high, switch S_2 is on and S_1 is off, assuming the switches are implemented using NMOS transistors.

12.4 Switched-Capacitor Circuits

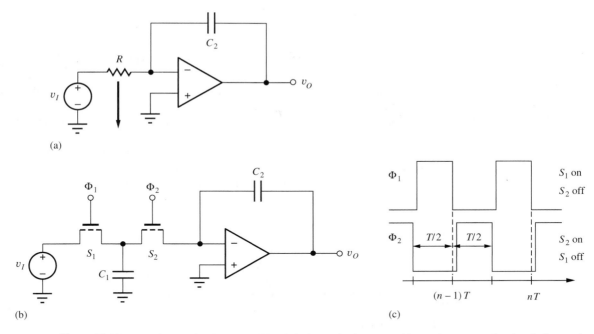

(a)

(b)

(c)

Figure 12.11 (a) Continuous-time integrator; (b) switched-capacitor integrator; (c) two-phase nonoverlapping clock controls the switches of the SC circuit. Switches S_1 and S_2 are implemented with NMOS transistors.

Figure 12.12 gives the (piecewise linear) equivalent circuits that can be used to analyze the circuit during the two individual phases of the clock. During phase 1, capacitor C_1 charges up to the value of source voltage v_I through switch S_1. At the same time, switch S_2 is open and the output voltage v_O stored on C_2 remains constant. During phase 2, capacitor C_1 becomes completely discharged because the op amp maintains a virtual ground at its input, and the charge stored on C_1 during the first phase is transferred directly to capacitor C_2 by the current that discharges C_1.

The charge stored on C_1 while phase 1 is positive (S_1 on) is

$$Q_1 = C_1 V_I \tag{12.38}$$

where $V_I = v_I[(n-1)T]$ is the voltage stored on C_1 when the switch opens at the end of the sampling interval. The change in charge stored on C_2 during phase 2 is

$$\Delta Q_2 = -C_2 \Delta v_O \tag{12.39}$$

Equating these two equations yields

$$\Delta v_O = -\frac{C_1}{C_2} V_I \tag{12.40}$$

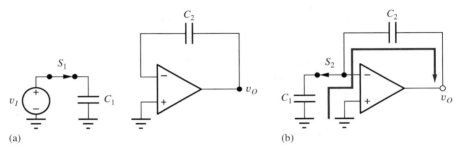

(a)   (b)

Figure 12.12 Equivalent circuits during (a) phase 1 and (b) phase 2.

The output voltage at the end of the nth clock cycle can be written as[6]

$$v_O[nT] = v_O[(n-1)T] - \frac{C_1}{C_2} v_I[(n-1)T] \quad (12.41)$$

During each clock period T, a packet of charge equal to Q_1 is transferred to storage capacitor C_2, and the output changes in discrete steps that are proportional to the input voltage with a gain determined by the ratio of capacitors C_1 and C_2. During phase 1, the input voltage is sampled and the output remains constant. During phase 2, the output changes to reflect the information sampled during phase 1.

An equivalence between the SC integrator and the continuous time integrator can be found by considering the total charge Q_I that flows from source v_I through resistor R during a time interval equal to the clock period T. Assuming a dc value of v_I for simplicity,

$$Q_I = IT = \frac{V_I}{R} T \quad (12.42)$$

Equating this charge to the charge stored on C_1 yields

$$\frac{V_I}{R} T = C_1 V_I \quad \text{and} \quad R = \frac{T}{C_1} = \frac{1}{f_C C_1} \quad (12.43)$$

in which f_C is the clock frequency. For a capacitance $C_1 = 1$ pF and a switching frequency of 100 kHz, the equivalent resistance $R = 10$ MΩ. This large value of R could not realistically be achieved in an integrated circuit realization of the continuous-time integrator.

> **EXERCISE:** The switched capacitor integrator in Fig. 12.11(b) has $V_I = 0.1$ V, $C_1 = 2$ pF, and $C_2 = 0.5$ pF. What are the output voltages at $t = T$, $t = 5T$, and $t = 9T$ if $V_O(0) = 0$?
>
> **ANSWERS:** −0.4 V; −2.0 V; −3.6 V

12.4.2 NONINVERTING SC INTEGRATOR

Switched-capacitor circuits also provide additional flexibility that is not readily available in continuous-time form. For example, the polarity of a signal can be inverted without the use of an amplifier. In Fig. 12.13, four switches and a floating capacitor are used to realize a **noninverting integrator**.

The circuits valid during the two individual phases appear in Fig. 12.14. During phase 1, switches S_1 are closed, a charge proportional to V_I is stored on C_1, and v_O remains constant. During phase 2, switches S_2 are closed, and a charge packet equal to $C_1 V_I$ is removed from C_2 instead of being added to C_2 as in the circuit in Fig. 12.12. For the circuit in Fig. 12.14, the output-voltage change at the end of one switch cycle is

$$\Delta v_O = + \frac{C_1}{C_2} V_I \quad (12.44)$$

The capacitances on the source-drain nodes of the MOSFET switches in Fig. 12.12 can cause undesirable errors in the inverting SC integrator circuit. By changing the phasing of the switches, as indicated in Fig. 12.15, the noninverting integrator of Fig. 12.14 can be changed to an inverting integrator. During phase 1 in Fig. 12.16(a), the source is connected through C_1 to the summing junction of the op amp, a charge equivalent to $C_1 V_I$ is delivered to C_2, and the output-voltage change

---

[6] Using z-transform notation, Eq. (12.47) can be written as

$$V_O(z) = z^{-1} V_O(z) - \frac{C_1}{C_2} z^{-1} V_S(z)$$

and the transfer function for the integrator is

$$T(z) = \frac{V_O(z)}{V_S(z)} = \frac{C_1}{C_2} \frac{z^{-1}}{1 - z^{-1}}$$

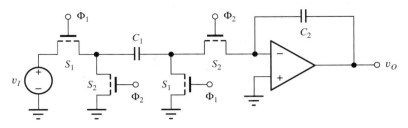

Figure 12.13 Noninverting SC integrator. (All transistors are NMOS devices.)

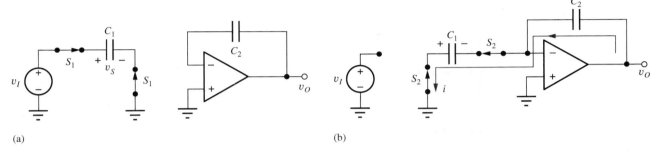

Figure 12.14 Equivalent circuits for the noninverting integrator during (a) phase 1 and (b) phase 2.

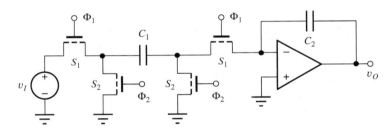

Figure 12.15 Inverting integrator achieved by changing clock phases of the switches.

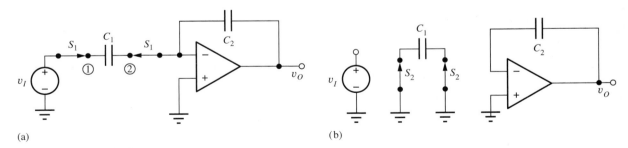

Figure 12.16 (a) Phase 1 of the stray-insensitive inverting integrator; (b) phase 2 of the stray-insensitive inverting integrator.

is given by Eq. (12.40). During phase 2, Fig. 12.16(b), the source is disconnected, v_O remains constant, and capacitor C_1 is completely discharged in preparation for the next cycle.

During phase 1, node 1 is driven by voltage source v_I and node 2 is maintained at zero by the virtual ground at the op amp input. During phase 2, both terminals of capacitor C_1 are forced to zero. Thus, any stray capacitances present at nodes 1 or 2 do not introduce errors into the charge transfer process. A similar set of conditions is true for the noninverting integrator. These two circuits are referred to as **stray-insensitive circuits** and are preferred for use in actual SC circuit implementations.

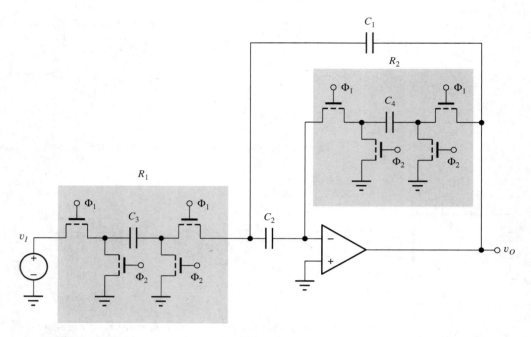

Figure 12.17 Switched-capacitor implementation of the second-order band-pass filter in Fig. 12.9.

12.4.3 SWITCHED-CAPACITOR FILTERS

Switched-capacitor circuit techniques have been developed to a high level of sophistication and are widely used as filters in audio applications as well as in RF and high-speed digital-to-analog and analog-to-digital converter designs. As an example, the SC implementation of the band-pass filter in Fig. 12.9 is shown in Fig. 12.17. For the continuous-time circuit, the center frequency and Q were described by

$$\omega_o = \frac{1}{\sqrt{R_{th} R_2 C_1 C_2}} \quad \text{and} \quad Q = \sqrt{\frac{R_2}{R_{th}} \frac{\sqrt{C_1 C_2}}{(C_1 + C_2)}} \quad (12.45)$$

In the SC version,

$$R_{th} = \frac{T}{C_3} \quad \text{and} \quad R_2 = \frac{T}{C_4} \quad (12.46)$$

in which T is the clock period. Substituting these values in Eq. (12.51) gives the equivalent values for the **switched-capacitor filter**:

$$\omega_o = \frac{1}{T}\sqrt{\frac{C_3 C_4}{C_1 C_2}} = f_C \sqrt{\frac{C_3 C_4}{C_1 C_2}} \quad \text{and} \quad Q = \sqrt{\frac{C_3}{C_4} \frac{\sqrt{C_1 C_2}}{(C_1 + C_2)}} \quad (12.47)$$

Note that the center frequency of this filter is tunable just by changing the clock frequency f_C, whereas the Q is independent of frequency. This property can be extremely useful in applications requiring tunable filters. However, since switched-capacitor filters are sampled-data systems, we must remember that the filter's input signal spectrum is limited to $f \leq f_C/2$ by the sampling theorem.

EXERCISE: What are the values of the center frequency, bandwidth, and voltage gain for the filter design in Fig. 12.17 for $C_1 = 3$ pF, $C_2 = 3$ pF, $C_3 = 4$ pF, $C_4 = 0.25$ pF, and a clock frequency of 200 kHz?

ANSWERS: 10.6 kHz; 5.31 kHz; −8.00.

ELECTRONICS IN ACTION

Body Sensor Networks

As low voltage integrated circuits have become more prevalent the opportunities to address ever lower power applications are expanding. It is now possible to combine high precision analog sensing circuits with ever more powerful digital computation capabilities at extremely low power levels.

An example of a new class of systems enabled by these low power integrated circuits are Body Area Sensor Networks (BSNs). BSNs comprise wearable or implanted sensors that collect, process, and communicate physiological information from the body. With health-care costs soaring, we need a more efficient and cost effective approach to medical diagnosis, treatment, and care. BSNs could revolutionize health care by offering miniaturized, unobtrusive, nearly continuous monitors to provide unprecedented levels of medical observation while simultaneously reducing the need for visits to the doctor. BSNs provide feedback to users that can illuminate potential health concerns earlier, encourage healthier lifestyles, and improve personal wellbeing. There is also the potential for using the collected data to provide real-time assessments of an individual's condition and need that could trigger a real-time assistance mechanism, such as a balance assistive device or a wearable defibrillator.[1]

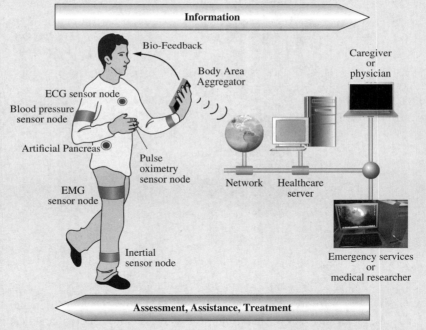

Globe: © *Design Pics/Design Pics Eye Traveller RF*; Laptop: © *D. Hurst/Alamy RF*; Monitor and CPU (Photo): © *Denise McCullough*; and Monitor and CPU (Line art): © *McGraw-Hill Education*.

A BSN consists of one or more sensor nodes that form a communication network. Usually, one of the nodes acts as a base station, which aggregates information from the body-distributed sensor nodes and ultimately conveys it across existing networks to other stakeholders like the wearer's caretakers and physicians. Smart phones are already supporting many useful health related apps, and future generation smart phones are the obvious choice to serve double duty as a BSN base station. The base station thus plays a different role from the other nodes in the system, and it has more resources in terms of bigger size, available energy, longer range

---

[1] M.A. Hanson, H.C. Powell Jr., A.T. Barth, K. Ringgenberg, B.H. Calhoun, J.H. Aylor, J. Lach, "Body Area Sensor Networks: Challenges and Opportunities," *IEEE Computer*, January 2009. Adapted from IEEE. Copyright 2009 IEEE.

radios, and more memory and computing power. The sensor nodes themselves each contain a set of sensors for particular physiological data, processing hardware for computation, and a wireless radio for communication (some BSN nodes may replace the radio with a transceiver that communicates across the body, using the surface of the skin as a communication channel[1]). Based on the application requirements, the sensed data may be streamed wirelessly or stored locally (typically in a flash memory) for later transmission or download.

The figure below illustrates a body sensor network and its environment. The sensors provide data to an aggregator such as a smart phone. The aggregators fuse data from multiple sensors, provide information to the user and communicate with health-care providers. The widespread adoption of smartphones has made such sensor/network integration more practical than just a few years ago.

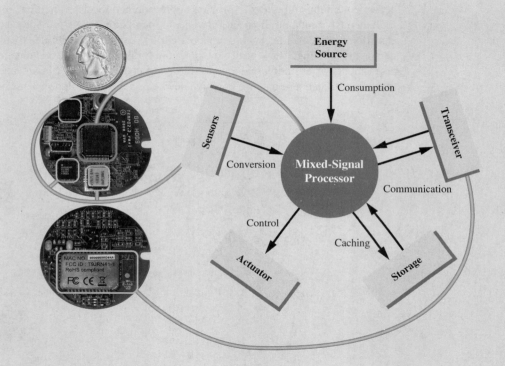

An example BSN node is shown above. It typically includes an energy source, analog sensing, A/D conversion, processing, storage, and communication. The node shown is the TEMPO inertial sensor node developed at the University of Virginia. It is used to aid elderly with balance issues and extract physiological data via gait dynamics. Current work includes dramatically reducing the power demands of the node to enable a new class of nodes powered entirely by power scavenged from the environment.

BSNs enable a diverse and powerful new direction in medical care. While technical and societal challenges remain, promising initial forays in BSN development show the potentially revolutionary possibilities for this technology.

---

[1] A.T. Barth, M.A. Hanson, H.C. Powell Jr., D. Unluer, S.G. Wilson, J. Lach, Adapted: "Body-Coupled Communication for Body Sensor Networks," International Conference on Body Area Networks, 2008.

12.5 DIGITAL-TO-ANALOG CONVERSION

As described briefly in Chapter 1, the **digital-to-analog converter**, often referred to as a **D/A converter** or **DAC**, provides an interface between the discrete signals of the digital domain and the continuous signals of the analog world. The D/A converter takes digital information, most often in binary form, as an input and generates an output voltage or current that may then be used for electronic control or information display.

12.5.1 D/A CONVERTER FUNDAMENTALS

In the DAC in Fig. 12.18, an n-bit binary input word $(b_1, b_2, \ldots b_n)$ is combined with the **dc reference voltage** V_{REF} to set the output of the D/A converter. The digital input is treated as a binary fraction with the binary point located to the left of the word. Assuming a voltage output, the behavior of the DAC can be expressed mathematically as

$$v_O = V_{FS}(b_1 2^{-1} + b_2 2^{-2} + \cdots + b_n 2^{-n}) + V_{OS} \quad \text{for } b_i \in \{1, 0\} \quad (12.48)$$

The DAC output may also be a current that can be represented as

$$i_O = I_{FS}(b_1 2^{-1} + b_2 2^{-2} + \cdots + b_n 2^{-n}) + I_{OS} \quad \text{for } b_i \in \{1, 0\} \quad (12.49)$$

The **full-scale voltage** V_{FS} or **full-scale current** I_{FS} is related to the internal reference voltage V_{REF} of the converter by

$$V_{FS} = K V_{\text{REF}} \quad \text{or} \quad I_{FS} = G V_{\text{REF}} \quad (12.50)$$

in which K and G determine the gain of the converter and are often set to a value of 1. Typical values of V_{FS} are 2.5, 5, 5.12, 10, and 10.24 V, whereas common values of I_{FS} are 2, 10, and 50 mA.

V_{OS} and I_{OS} represent the **offset voltage** or **offset current** of the converters, respectively, and characterize the converter output when the digital input code is equal to zero. The offset voltage is normally adjusted to zero, but the offset current of a current output DAC may be deliberately set to a nonzero value. For example, 2 to 10 mA and 10 to 50 mA ranges are used in some process control applications. The built-in offset current can be used to verify the integrity of the control circuit. For now, let us assume that the DAC output is a voltage.

EXERCISE: What are the decimal values of the following 8-bit binary fractions? (a) 0.01100001 (b) 0.10001000.

ANSWERS: (a) 0.37890625; (b) 0.5312500

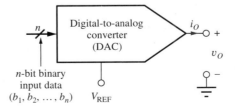

Figure 12.18 D/A converter with voltage output.

The smallest voltage change that can occur at the DAC output takes place when the **least significant bit (LSB)** b_n in the digital word changes from a 0 to a 1. This minimum voltage change is also referred to as the **resolution of the converter** and is given by

$$V_{\text{LSB}} = 2^{-n} V_{FS} \tag{12.51}$$

At the other extreme, b_1 is referred to as the **most significant bit (MSB)** and has a weight of one-half V_{FS}.

For example, a 12-bit converter with a full-scale voltage of 10.24 V has an LSB or resolution of 2.500 mV. However, resolution can be stated in different ways. A 12-bit DAC may be said to have 12-bit resolution, a resolution of 0.025 percent of full scale, or a resolution of 1 part in 4096. DACs are available with resolutions ranging from as few as 6 bits to 24 bits. Resolutions of 8 to 12 bits are quite common and economical. Above 12 bits, DACs become more and more expensive, and great care must be taken to truly realize their full precision.

EXERCISE: A 12-bit D/A converter has $V_{\text{REF}} = 5.12$ V. What is the output voltage for a binary input code of (101010101010)? What is V_{LSB}? What is the size of the MSB?

ANSWERS: 3.41250 V, 1.25 mV, 2.56 V

12.5.2 D/A CONVERTER ERRORS

Figure 12.19 and columns 1 and 2 in Table 12.7 present the relationship between the digital input code and the analog output voltage for an ideal three-bit DAC. The data points in the figure represent the eight possible output voltages, which range from 0 to $0.875 \times V_{FS}$. Note that the output voltage of the ideal DAC never reaches a value equal to V_{FS}. The maximum output is always 1 LSB smaller than V_{FS}. In this case, the maximum output code of 111 corresponds to 7/8 of full scale or $0.875\ V_{FS}$.

The ideal converter in Fig. 12.19 has been calibrated so that $V_{OS} = 0$ and 1 LSB is exactly $V_{FS}/8$. Figure 12.19 also shows the output of a converter with both gain and offset errors. The **gain error** of the D/A converter represents the deviation of the slope of the converter transfer function from that of the corresponding ideal DAC in Fig. 12.19, whereas the offset voltage is simply the output of the converter for a zero binary input code.

Although the outputs of both converters in Fig. 12.19 lie on a straight line, the output voltages of an actual DAC do not necessarily fall on a straight line. For example, the converter in Fig. 12.20

TABLE 12.7
D/A Converter Transfer Characteristics

BINARY INPUT	IDEAL DAC OUTPUT ($\times V_{FS}$)	DAC OF FIG. 12.21 ($\times V_{FS}$)	STEP SIZE (LSB)	DIFFERENTIAL LINEARITY ERROR (LSB)	INTEGRAL LINEARITY ERROR (LSB)
000	0.0000	0.0000			0.00
001	0.1250	0.1000	0.80	−0.20	−0.20
010	0.2500	0.2500	1.20	+0.20	0.00
011	0.3750	0.3125	0.50	−0.50	−0.50
100	0.5000	0.5625	2.00	+1.00	+0.50
101	0.6250	0.6250	0.50	−0.50	0.00
110	0.7500	0.8000	1.40	+0.40	+0.40
111	0.8750	0.8750	0.60	−0.40	0.00

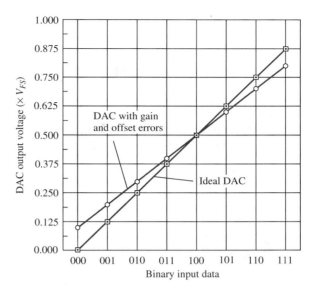

Figure 12.19 Transfer characteristic for an ideal DAC and a converter with both gain and offset errors.

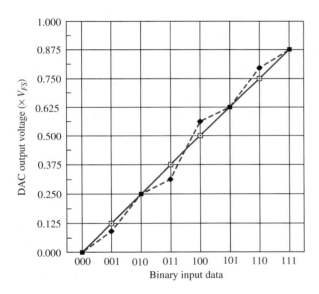

Figure 12.20 D/A converter with linearity errors.

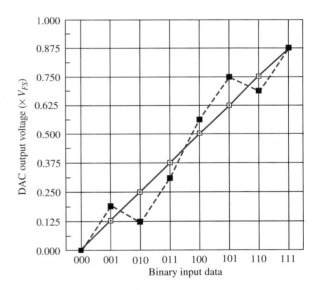

Figure 12.21 DAC with nonmonotonic output.

contains circuit mismatches that cause the output to no longer be perfectly linear. **Integral linearity error,** usually referred to as just **linearity error,** measures the deviation of the actual converter output from a straight line fitted to the converter output voltages. The error is usually specified as a fraction of an LSB or as a percentage of the full-scale voltage.

Table 12.7 lists the linearity errors for the nonlinear DAC in Fig. 12.20. This converter has linearity errors for input codes of 001, 011, 100, and 110. The overall linearity error for the DAC is specified as the magnitude of the largest error that occurs. Hence this converter will be specified as having a linearity error of either 0.5 LSB or 6.25 percent of full-scale voltage. A good converter exhibits a linearity error of less than 0.5 LSB.

A closely related measure of converter performance is the **differential linearity error.** When the binary input changes by 1 bit, the output voltage should change by 1 LSB. A converter's differential linearity error is the magnitude of the maximum difference between each output step of the converter and the ideal step size of 1 LSB. The size of each step and the differential linearity errors of the converter in Fig. 12.20 are also listed in Table 12.7. For instance, the DAC output changes by 0.8 LSB when the input code changes from 000 to 001. The differential linearity error represents the difference between this actual step size and 1 LSB. The integral linearity error for a given binary input represents the sum (integral) of the differential linearity errors for inputs up through the given input.

Another specification that can be important in many applications is **monotonicity.** As the input code to a DAC is increased, the output should increase in a monotonic manner. If this does not happen, then the DAC is said to be nonmonotonic. In the nonmonotonic DAC in Fig. 12.21, the output decreases from $\frac{3}{16}V_{FS}$ to $\frac{1}{8}V_{FS}$ when the input code changes from 001 to 010. A similar problem occurs for the 101 to 110 transition: In feedback systems, this behavior represents an unwanted 180° phase shift that effectively changes negative feedback to positive feedback and can lead to system instability.

In the upcoming exercise, we will find that this converter has a differential linearity error of 1.5 LSB, whereas the integral linearity error is 1 LSB. A tight linearity error specification does not necessarily guarantee good differential linearity. Although it is possible for a converter to have a differential linearity error of greater than 1 LSB and still be monotonic, a nonmonotonic converter always has a differential linearity error exceeding 1 LSB.

EXERCISE: Fill in the missing entries for step size, differential linearity error, and integral linearity error for the converter in Fig. 12.21.

BINARY INPUT	IDEAL DAC OUTPUT ($\times V_{FS}$)	ACTUAL DAC EXAMPLE	STEP SIZE (LSB)	DIFFERENTIAL LINEARITY ERROR (LSB)	INTEGRAL LINEARITY ERROR
000	0.0000	0.0000			0.00
001	0.1250	0.2000			
010	0.2500	0.1375			
011	0.3750	0.3125			
100	0.5000	0.5625			
101	0.6250	0.7500			
110	0.7500	0.6875			
111	0.8750	0.8750			0.00

ANSWERS: 1.5, −0.5, 1.5; 2.0, 1.5, −0.5, 1.5; 0.5, −1.5, 0.5, 1.0, 0.5, −1.5, 0.5; 0.5, −1.0, −0.5, 0.5, 1.0, −0.5, 0.0

EXERCISE: What are the offset voltage and step size for the nonideal converter in Fig. 12.21 if the endpoints are at 0.100 and 0.800V_{FS}?

ANSWERS: 0.100V_{FS}, 0.100V_{FS}

12.5.3 DIGITAL-TO-ANALOG CONVERTER CIRCUITS

One of the simplest DAC circuits, the **weighted-resistor DAC,** shown in Fig. 12.22, uses the summing amplifier that we encountered earlier in Chapter 10, the reference voltage V_{REF}, and a

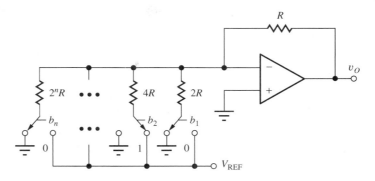

Figure 12.22 An n-bit weighted-resistor DAC.

weighted-resistor network. The binary input data controls the switches, with a logical 1 indicating that the switch is connected to V_{REF} and a logical 0 corresponding to a switch connected to ground. Successive resistors are weighted progressively by a factor of 2, thereby producing the desired binary weighted contributions to the output:

$$v_O = (b_1 2^{-1} + b_2 2^{-2} + \cdots + b_n 2^{-n}) V_{REF} \qquad \text{for } b_i \in \{1, 0\} \qquad (12.52)$$

Differential and integral linearity errors and gain error occur when the resistor ratios are not perfectly maintained. Any op amp offset voltage contributes directly to V_{OS} of the converter.

Several problems arise in building a DAC using the weighted-resistor approach. The primary difficulty is the need to maintain accurate resistor ratios over a very wide range of resistor values (e.g., 4096 to 1 for a 12-bit DAC). In addition, because the switches are in series with the resistors, their on-resistance must be very low and they should have zero offset voltage. The designer can meet these last two requirements by using good MOSFETs or JFETs as switches, and the (W/L) ratios of the FETs can be scaled with bit position to equalize the resistance contributions of the switches. However, the wide range of resistor values is not suitable for monolithic converters of moderate to high resolution. We should also note that the current drawn from the voltage reference varies with the binary input pattern. This varying current causes a change in voltage drop in the Thévenin equivalent source resistance of the voltage reference and can lead to data-dependent errors sometimes called **superposition errors**.

> **EXERCISE:** Suppose a 1-kΩ resistor is used for the MSB in an 8-bit converter similar to that in Fig. 12.24. What are the other resistor values?
>
> **ANSWERS:** 2 kΩ; 4 kΩ; 8 kΩ; 16 kΩ; 32 kΩ; 64 kΩ; 128 kΩ; 500 Ω

The R-2R Ladder

The **R-2R ladder** in Fig. 12.23 avoids the problem of a wide range of resistor values. It is well-suited to integrated circuit realization because it requires matching of only two resistor values, R and $2R$. The value of R typically ranges from 2 to 10 kΩ. By forming successive Thévenin equivalents proceeding from left to right at each node in the ladder, we can show that the contribution of each bit is reduced by a factor of 2 going from the MSB to LSB. Like the weighted-resistor DAC, this network requires switches with low on-resistance and zero offset voltage, and the current drawn from the reference still varies with the input data pattern.

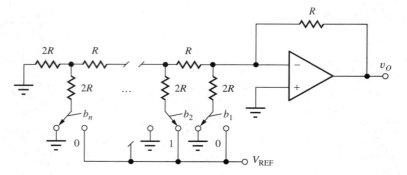

Figure 12.23 n-bit DAC using R-2R ladder.

> **EXERCISE:** What is the total resistance required to build an 8-bit R-2R ladder DAC if $R = 1$ kΩ? What is the total resistance required to build an 8-bit weighted resistor D/A converter if $R = 1$ kΩ?
>
> **ANSWERS:** 26 kΩ; 511 kΩ

Inverted R-2R Ladder

Because the currents in the resistor networks of the DACs in Figs. 12.22 and 12.23 change as the input data changes, power dissipation in the elements of the network changes, which can cause linearity errors in addition to superposition errors. Therefore some monolithic DACs use the configuration in Fig. 12.24, known as the **inverted R-2R ladder.** In this circuit, the currents in the ladder and reference are independent of the digital input because the input data cause the ladder currents to be switched either directly to ground or to the virtual ground input at the input of a current-to-voltage converter. Because both op amp inputs are at ground potential, the ladder currents are independent of switch position. Note that complementary currents, I and $\overline{I}$, are available at the output of the inverted ladder.

The inverted R-2R ladder is a popular DAC configuration, often implemented in CMOS technology. The switches still need to have low on-resistance to minimize errors within the converter. The R-2R ladder can be formed of diffused, implanted, or thin-film resistors; the choice depends on both the manufacturer's process technology and the required resolution of the D/A converter.

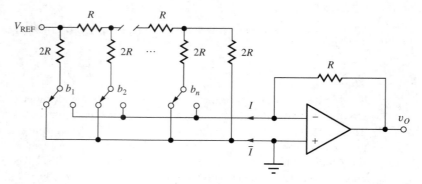

Figure 12.24 D/A converter using the inverted R-2R ladder.

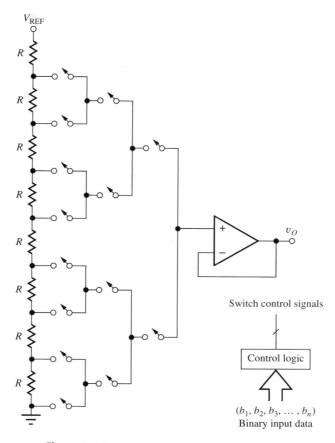

Figure 12.25 Inherently monotonic 3-bit D/A converter.

An Inherently Monotonic DAC

MOS IC technology has facilitated some alternate approaches to D/A converter design. Figure 12.25 shows a DAC whose output is inherently monotonic. A long resistor string forms a multi-output voltage divider connected between the voltage reference and ground. An analog switch tree connects the desired tap to the input of an operational amplifier operating as a voltage follower. The appropriate switches are closed by a logic network that decodes the binary input data.

Each tap on the resistor network is forced to produce a voltage greater than or equal to that of the taps below it, and the output must therefore increase monotonically as the digital input code increases. An 8-bit version of this converter requires 256 equal-valued resistors and 510 switches, plus the additional decoding logic. This DAC can be fabricated in NMOS or CMOS technology, in which the large number of MOSFET switches and the complex decoding logic are easily realized.

> **EXERCISE:** How many resistors and switches are required to implement a 10-bit DAC using the technique in Fig. 12.25?
>
> **ANSWERS:** 1024, 2046

Switched-Capacitor D/A Converters

D/A converters can be fabricated using only switches, capacitors, and operational amplifiers. Figure 12.26(a) is a **weighted-capacitor DAC**; Fig. 12.26(b) is a **C-2C ladder DAC**. Because these

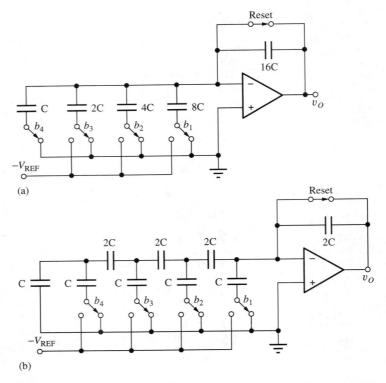

Figure 12.26 Switched-capacitor D/A converters: (a) weighted-capacitor DAC; (b) C-2C DAC.

circuits are composed only of switches and capacitors, the only static power dissipation in these circuits occurs in the op amps. However, dynamic switching losses occur just as in CMOS logic (see Sec. 7.4). These circuits represent the direct switched-capacitor (SC) analogs of the weighted-resistor and R-2R ladder techniques presented earlier.

When a switch changes state, current pulses charge or discharge the capacitors in the network. The current pulse is supplied by the output of the operational amplifier and changes the voltage on the feedback capacitor by an amount corresponding to the bit weight of the switch that changed state. These converters consume very little power, even when CMOS operational amplifiers are included on the same chip, and are widely used in VLSI systems.

> **EXERCISE:** (a) Suppose that an 8-bit weighted capacitor DAC is fabricated with the smallest unit of capacitance $C = 1.0$ pF. What is the total capacitance the DAC requires? (b) Repeat for a C-2C ladder DAC. (c) An IC process provides a thin oxide capacitor structure with a capacitance of 5 fF/μm^2. How much chip area is required for the C-2C ladder DAC?
>
> **ANSWERS:** 511 pF; 31 pF; 6200 μm^2

12.6 ANALOG-TO-DIGITAL CONVERSION

As described briefly in Chapter 1, the **analog-to-digital converter**, also known as an **A/D converter** or **ADC**, is used to transform analog information in electrical form into digital data. The ADC in Fig. 12.27 takes an unknown continuous analog input signal, most often a voltage v_X, and converts it into an n-bit binary number that can be readily manipulated by a digital computer. The n-bit number

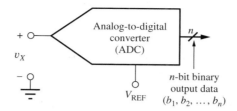

Figure 12.27 Block diagram representation for an A/D converter.

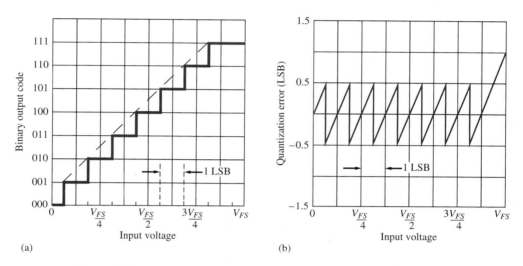

Figure 12.28 Ideal 3-bit ADC: (a) input-output relationship and (b) quantization error.

is a binary fraction representing the ratio between the unknown input voltage v_x and the converter's full-scale voltage $V_{FS} = KV_{\text{REF}}$.

12.6.1 A/D CONVERTER FUNDAMENTALS

Figure 12.28(a) is an example of the input-output relationship for an ideal 3-bit A/D converter. As the input increases from zero to full scale, the digital output code word stairsteps from 000 to 111. As the input voltage increases, the output code first underestimates the input voltage and then overestimates the input voltage. This error, called **quantization error,** is plotted against input voltage in Fig. 12.28(b).

For a given output code, we know only that the value of the input voltage v_X lies somewhere within a 1-LSB quantization interval. For example, if the output code of the 3-bit ADC is (101), then the input voltage can be anywhere between $\frac{9}{16}V_{FS}$ and $\frac{11}{16}V_{FS}$, a range of $V_{FS}/8$ V equivalent to 1 LSB of the 3-bit converter. From a mathematical point of view, the circuitry of an ideal ADC should be designed to pick the values of the bits in the binary word to minimize the magnitude of the quantization error v_ε between the unknown input voltage v_X and the nearest quantized voltage level:

$$v_\varepsilon = |v_X - (b_1 2^{-1} + b_2 2^{-2} + \cdots + b_n 2^{-n})V_{FS}| \qquad (12.53)$$

> **EXERCISE:** An 8-bit A/D converter has $V_{\text{REF}} = 5$ V. What is the binary output code word for an input of 1.2 V? What is the voltage range corresponding to 1 LSB of the converter?
>
> **ANSWERS:** (00111101); 19.5 mV

TABLE 12.8
A/D Converter Transfer Characteristics

BINARY OUTPUT CODE	IDEAL ADC TRANSITION POINT ($\times V_{FS}$)	ADC OF FIG. 12.29 ($\times V_{FS}$)	STEP SIZE (LSB)	DIFFERENTIAL LINEARITY ERROR (LSB)	INTEGRAL LINEARITY ERROR (LSB)
000	0.0000	0.0000	0.5	0	0
001	0.0625	0.0625	1.5	0.50	0.5
010	0.1875	0.2500	0.5	−0.50	0
011	0.3125	0.3125	1.0	0	0
100	0.4375	0.4375	1.0	0	0
101	0.5625	0.5625	1.50	0.50	0.5
110	0.6875	0.7500	0.5	−0.50	0
111	0.8125	0.8125	1.5	0	0

12.6.2 ANALOG-TO-DIGITAL CONVERTER ERRORS

As shown by the dashed line in Fig. 12.28(a), the code transition points of an ideal converter all fall on a straight line. However, an actual converter has integral and differential linearity errors similar to those of a digital-to-analog converter. Figure 12.29 is an example of the code transitions for a hypothetical nonideal converter. The converter is assumed to be calibrated so that the first and last code transitions occur at their ideal points.

In the ideal case, each code step, other than 000 and 111, would be the same width and should be equal to 1 LSB of the converter. Differential linearity error represents the difference between the actual code step width and 1 LSB, and integral linearity error is a measure of the deviation of the code transition points from their ideal positions. Table 12.8 lists the step size, differential linearity error, and integral linearity error for the converter in Fig. 12.29. Note that the ideal step sizes corresponding to codes 000 and 111 are 0.5 LSB and 1.5 LSB, respectively, because of the desired code transition points. As in D/A converters, the integral linearity error should equal the sum of the differential linearity errors for the individual steps.

Figure 12.30 is an uncalibrated converter with both offset and gain errors. The first code transition occurs at a voltage that is 0.5 LSB too high, representing a converter **offset error** of 0.5 LSB.

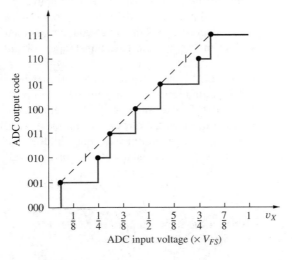

Figure 12.29 Example of code transitions in a nonideal 3-bit ADC.

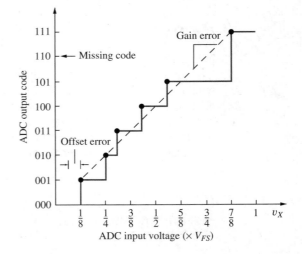

Figure 12.30 ADC with a missing code.

The slope of the fitted line does not give 1 LSB = $V_{FS}/8$, so the converter also exhibits a gain error.

A new type of error, which is specific to ADCs, can be observed in Fig. 12.30. The output code jumps directly from 101 to 111 as the input passes through $0.875 V_{FS}$. The output code 110 never occurs, so this converter is said to have a **missing code**. A converter with a differential linearity error of less than 1 LSB does not exhibit missing codes in its input-output function. An ADC can also be **nonmonotonic**. If the output code decreases as the input voltage increases, the converter has a nonmonotonic input–output relationship.

All these deviations from ideal A/D (or D/A) converter behavior are temperature-dependent; hence, converter specifications include temperature coefficients for gain, offset, and linearity. A good converter will be monotonic with less than 0.5 LSB linearity error and no missing codes over its full temperature range.

> **EXERCISE:** An A/D converter is used in a digital multimeter (DVM) that displays 6 decimal digits. How many bits are required in the ADC?
>
> **ANSWER:** 20 bits
>
> **EXERCISE:** What are the minimum and maximum code step widths in Fig. 12.30? What are the differential and integral linearity errors for this ADC based on the dashed line in the figure?
>
> **ANSWERS:** 0, 2.5 LSB; 1.5 LSB, 1 LSB

12.6.3 BASIC A/D CONVERSION TECHNIQUES

Figure 12.31 shows the basic conversion scheme for a number of analog-to-digital converters. The unknown input voltage v_X is connected to one input of an analog **comparator**, and a time-dependent reference voltage v_{REF} is connected to the other input of the comparator. If input voltage v_X exceeds input v_{REF}, then the output voltage will be high, corresponding to a logic 1. If input v_X is less than v_{REF}, then the output voltage will be low, corresponding to a logic 0.

In performing a conversion, the reference voltage is varied until the unknown input is determined within the quantization error of the converter. Ideally, the logic of the A/D converter will choose a set of binary coefficients b_i so that the difference between the unknown input voltage v_X and the final quantized value is less than or equal to 0.5 LSB. In other words, the b_i will be selected so that

$$\left| v_X - V_{FS} \sum_{i=1}^{n} b_i 2^{-i} \right| < \frac{V_{FS}}{2^{n+1}} \tag{12.54}$$

The basic difference among the operations of various converters is the strategy that is used to vary the reference signal V_{REF} to determine the set of binary coefficients $\{b_i, i = 1, 2, \cdots, n\}$.

Counting Converter

One of the simplest ways of generating the comparison voltage is to use a digital-to-analog converter. An n-bit DAC can be used to generate any one of 2^n discrete outputs simply by applying the appropriate digital input word. A direct way to determine the unknown input voltage v_X is to sequentially compare it to each possible DAC output. Connecting the digital input of the DAC to an n-bit binary counter enables a step-by-step comparison to the unknown input to be made, as shown in Fig. 12.32.

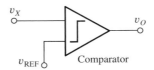

Figure 12.31 Block diagram representation for an A/D converter.

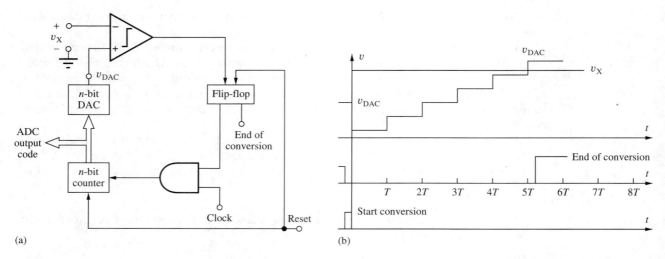

Figure 12.32 (a) Block diagram of the counting ADC; (b) timing diagram.

A/D conversion begins when a pulse resets the flip-flop and the counter output to zero. Each successive clock pulse increments the counter; the DAC output looks like a staircase during the conversion. When the output of the DAC exceeds the unknown input, the comparator output changes state, sets the flip-flop, and prevents any further clock pulses from reaching the counter. The change of state of the comparator output indicates that the conversion is complete. At this time, the contents of the binary counter represent the converted value of the input signal.

Several features of this converter should be noted. First, the length of the conversion cycle is variable and proportional to the unknown input voltage v_X. The maximum **conversion time** T_T occurs for a full-scale input signal and corresponds to 2^n clock periods or

$$T_T \leq \frac{2^n}{f_C} = 2^n T_C \qquad (12.55)$$

where $f_C = 1/T_C$ is the clock frequency. Second, the binary value in the counter represents the smallest DAC voltage that is larger than the unknown input; this value is not necessarily the DAC output which is closest to the unknown input, as was originally desired. Also, the example in Fig. 12.32(b) shows the case for an input that is constant during the conversion period. If the input varies, the binary output will be an accurate representation of the value of the input signal at the instant the comparator changes state.

The advantage of the counting A/D converter is that it requires a minimum amount of hardware and is inexpensive to implement. Some of the least expensive A/D converters have used this technique. The main disadvantage is the relatively low conversion rate for a given D/A converter speed. An n-bit converter requires 2^n clock periods for its longest conversion.

> **EXERCISE:** What is the maximum conversion time for a counting ADC using a 12-bit DAC and a 2-MHz clock frequency? What is the maximum possible number of conversions per second?
>
> **ANSWERS:** 2.05 ms; 488 conversions per second

Successive Approximation Converter

The **successive approximation converter** uses a much more efficient strategy for varying the reference input to the comparator, one that results in a converter requiring only n clock periods to complete an n-bit conversion. Figure 12.33 is a schematic of the operation of a 3-bit successive approximation

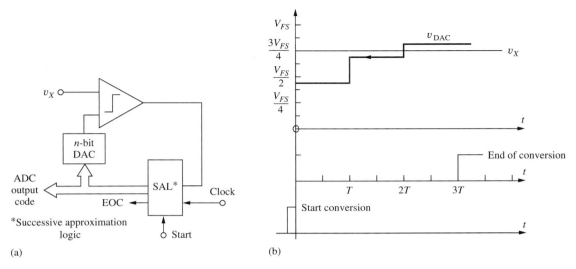

Figure 12.33 (a) Successive approximation ADC; (b) timing diagram.

converter. A "binary search" is used to determine the best approximation to v_X. After receiving a start signal, the successive approximation logic sets the DAC output to $(V_{FS}/2) - (V_{FS}/16)$ and, after waiting for the circuit to settle out, checks the comparator output. [The DAC output is offset by $(-\frac{1}{2}\text{LSB} = -V_{FS}/16)$ to yield the transfer function of Fig. 12.30.] At the next clock pulse, the DAC output is incremented by $V_{FS}/4$ if the comparator output was 1, and decremented by $V_{FS}/4$ if the comparator output was 0. The comparator output is again checked, and the next clock pulse causes the DAC output to be incremented or decremented by $V_{FS}/8$. A third comparison is made. The final binary output code remains unchanged if v_X is larger than the final DAC output or is decremented by 1 LSB if v_X is less than the DAC output. The conversion is completed following the logic decision at the end of the third clock period for the 3-bit converter, or at the end of n clock periods for an n-bit converter.

Figure 12.34 shows the possible code sequences for a 3-bit DAC and the sequence followed for the successive approximation conversion in Fig. 12.35. At the start of conversion, the DAC input is set to 100. At the end of the first clock period, the DAC voltage is found to be less than v_X, so the DAC code is increased to 110. At the end of the second clock period, the DAC voltage is still found to be too small, and the DAC code is increased to 111. After the third clock period, the DAC voltage is found to be too large, so the DAC code is decremented to yield a final converted value of 110.

Fast conversion rates are possible with a successive approximation ADC. This conversion technique is very popular and used in many 8 to 16-bit converters. The primary factors limiting the speed of this ADC are the time required for the D/A converter output to settle within a fraction of an LSB of V_{FS} and the time required for the comparator to respond to input signals that may differ by very small amounts.

EXERCISE: What is the conversion time for a successive approximation ADC using a 12-bit DAC and a 2-MHz clock frequency? What is the maximum possible number of conversions per second?

ANSWERS: 6.00 μs; 167,000 conversions per second.

In the discussion thus far, it has been tacitly assumed that the input remains constant during the full conversion period. A slowly varying input signal is acceptable as long as it does not change by more than 0.5 LSB ($V_{FS}/2^{n+1}$) during the conversion time ($T_T = n/f_C = nT_C$). The frequency of a

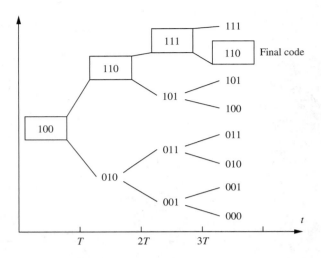

Figure 12.34 Code sequences for a 3-bit successive approximation ADC.

sinusoidal input signal with a peak-to-peak amplitude equal to the full-scale voltage of the converter must satisfy the following inequality:

$$T_T \left\{ \max\left[\frac{d}{dt}(V_{FS}\sin\omega_o t)\right]\right\} \leq \frac{V_{FS}}{2^{n+1}} \quad \text{or} \quad \frac{n}{f_C}(V_{FS}\omega_o) \leq \frac{V_{FS}}{2^{n+1}} \tag{12.56}$$

and

$$f_O \leq \frac{f_C}{2^{n+2}n\pi}$$

For a 12-bit converter using a 1-MHz clock frequency, f_O must be less than 1.62 Hz. If the input changes by more than 0.5 LSB during the conversion process, the digital output of the converter does not bear a precise relation to the value of the unknown input voltage v_X. To avoid this frequency limitation, a high-speed **sample-and-hold circuit**[7] that samples the signal amplitude and then holds its value constant is usually used ahead of successive approximation ADCs.

Single-Ramp (Single-Slope) ADC

The discrete output of the D/A converter in the counting ADC can be replaced by a continuously changing analog reference signal, as shown in Fig. 12.35. The reference voltage varies linearly with a well-defined slope from slightly below zero to above V_{FS}, and the converter is called a **single-ramp**, or **single-slope, ADC**. The length of time required for the reference signal to become equal to the unknown voltage is proportional to the unknown input.

Converter operation begins with a start conversion signal, which resets the binary counter and starts the ramp generator at a slightly negative voltage [see Fig. 12.35(b)]. As the ramp crosses through zero, the output of comparator 2 goes high and allows clock pulses to accumulate in the counter. The number in the counter increases until the ramp output voltage exceeds the unknown v_X. At this time, the output of comparator 1 goes high and prevents further clock pulses from reaching the counter. The number N in the counter at the end of the conversion is directly proportional to the input voltage because

$$v_X = KNT_C \tag{12.57}$$

---

[7] See Additional Reading or the Electronics in Action at the end of this section, for examples.

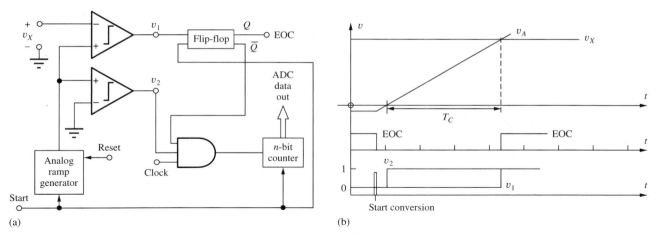

Figure 12.35 (a) Block diagram and (b) timing for a single-ramp ADC.

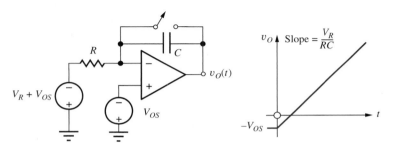

Figure 12.36 Ramp voltage generation using an integrator with constant input.

where K is the slope of the ramp in volts/second. If the slope of the ramp is chosen to be $K = V_{FS}/2^n T_C$, then the number in the counter directly represents the binary fraction equal to v_X/V_{FS}:

$$\frac{v_X}{V_{FS}} = \frac{N}{2^n} \qquad (12.58)$$

The conversion time T_T of the single-ramp converter is clearly variable and proportional to the unknown voltage v_X. Maximum conversion time occurs for $v_X = V_{FS}$, with

$$T_T \leq 2^n T_C \qquad (12.59)$$

As is the case for the **counter-ramp converter**, the counter output represents the value of v_X at the time that the end-of-conversion signal occurs.

The ramp voltage is usually generated by an integrator connected to a constant reference voltage, as shown in Fig. 12.36. When the reset switch is opened, the output increases with a constant slope given by V_R/RC:

$$v_O(t) = -V_{OS} + \frac{1}{RC} \int_0^t V_R \, dt \qquad (12.60)$$

The dependence of the ramp's slope on the RC product is one of the major limitations of the single-ramp A/D converter. The slope depends on the absolute values of R and C, which are difficult to maintain constant in the presence of temperature variations and over long periods of time. Because of this problem, the dual-ramp converter in the next section is preferred.

EXERCISE: What is the value of RC for an 8-bit single-ramp ADC with $V_{FS} = 5.12$ V, $V_R = 2.000$ V, and $f_C = 1$ MHz?

ANSWER: 0.1 ms

Dual-Ramp (Dual-Slope) ADC

The **dual-ramp**, or **dual-slope**, **ADC** solves the problems associated with the single-ramp converter and is commonly found in high-precision data acquisition and instrumentation systems. Figure 12.37 illustrates converter operation. The conversion cycle consists of two separate integration intervals. First, unknown voltage v_X is integrated for a known period of time T_1. The value of this integral is then compared to that of a known reference voltage V_{REF}, which is integrated for a variable length of time T_2.

At the start of conversion the counter is reset, and the integrator is reset to a slightly negative voltage. The unknown input v_X is connected to the integrator input through switch S_1. Unknown voltage v_X is integrated for a fixed period of time $T_1 = 2^n T_C$, which begins when the integrator output crosses through zero. At the end of time T_1, the counter overflows, causing S_1 to be opened and the reference input V_{REF} to be connected to the integrator input through S_2. The integrator output then decreases until it crosses back through zero, and the comparator changes state, indicating the end of the conversion. The counter continues to accumulate pulses during the down ramp, and the final number in the counter represents the quantized value of the unknown voltage v_X.

Circuit operation forces the integrals over the two time periods to be equal:

$$\frac{1}{RC}\int_0^{T_1} v_X(t)\,dt = \frac{1}{RC}\int_{T_1}^{T_1+T_2} V_{REF}\,dt \qquad (12.61)$$

T_1 is set equal to $2^n T_C$ because the unknown voltage v_X was integrated over the amount of time needed for the n-bit counter to overflow. Time period T_2 is equal to NT_C, where N is the number accumulated in the counter during the second phase of operation.

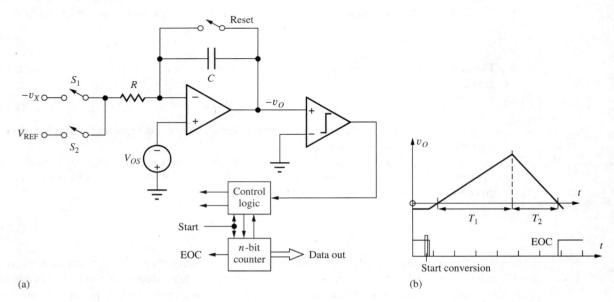

Figure 12.37 (a) Dual-ramp ADC and (b) timing diagram.

Recalling the mean-value theorem from calculus,

$$\frac{1}{RC} \int_0^{T_1} v_X(t)\, dt = \frac{\langle v_X \rangle}{RC} T_1 \qquad (12.62)$$

and

$$\frac{1}{RC} \int_{T_1}^{T_1+T_2} V_{\text{REF}}(t)\, dt = \frac{V_{\text{REF}}}{RC} T_2 \qquad (12.63)$$

because V_{REF} is a constant. Substituting these two results into Eq. (12.61), we find the average value of the input $\langle v_x \rangle$ to be

$$\frac{\langle v_X \rangle}{V_{\text{REF}}} = \frac{T_2}{T_1} = \frac{N}{2^n} \qquad (12.64)$$

assuming that the RC product remains constant throughout the complete conversion cycle. The absolute values of R and C no longer enter directly into the relation between v_X and V_{FS}, and the long-term stability problem associated with the single-ramp converter is overcome. Furthermore, the digital output word represents the average value of v_X during the first integration phase. Thus, v_X can change during the conversion cycle of this converter without destroying the validity of the quantized output value.

The conversion time T_T requires 2^n clock periods for the first integration period, and N clock periods for the second integration period. Thus the conversion time is variable and

$$T_T = (2^n + N) T_C \leq 2^{n+1} T_C \qquad (12.65)$$

because the maximum value of N is 2^n.

EXERCISE: What is the maximum conversion time for a 16-bit dual-ramp converter using a 1-MHz clock frequency? What is the maximum conversion rate?

ANSWERS: 0.131 s; 7.63 conversions per second

The dual ramp is a widely used converter. Although much slower than the successive approximation converter, the dual-ramp converter offers excellent differential and integral linearity. By combining its integrating properties with careful design, one can obtain accurate conversion at resolutions exceeding 20 bits, but at relatively low conversion rates. In a number of recent converters and instruments, the basic dual-ramp converter has been modified to include extra integration phases for automatic offset voltage elimination. These devices are often called *quad-slope* or *quad-phase converters*. Another converter, the *triple ramp*, uses coarse and fine down ramps to greatly improve the speed of the integrating converter (by a factor of $2^{n/2}$ for an n-bit converter).

Normal-Mode Rejection

As mentioned before, the quantized output of the dual-ramp converter represents the average of the input during the first integration phase. The integrator operates as a low-pass filter with the normalized transfer function shown in Fig. 12.38. Sinusoidal input signals, whose frequencies are exact multiples of the reciprocal of the integration time T_1, have integrals of zero value and do not appear at the integrator output. This property is used in many digital multimeters, which are equipped with dual-ramp converters having an integration time, that is some multiple of the period of the 50- or 60-Hz power-line frequency. Noise sources with frequencies at multiples of the power-line frequency are therefore rejected by these integrating ADCs. This property is usually termed **normal-mode rejection**.

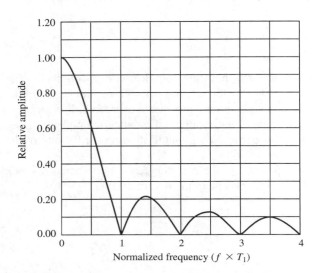

Figure 12.38 Normal-mode rejection for an integrating ADC.

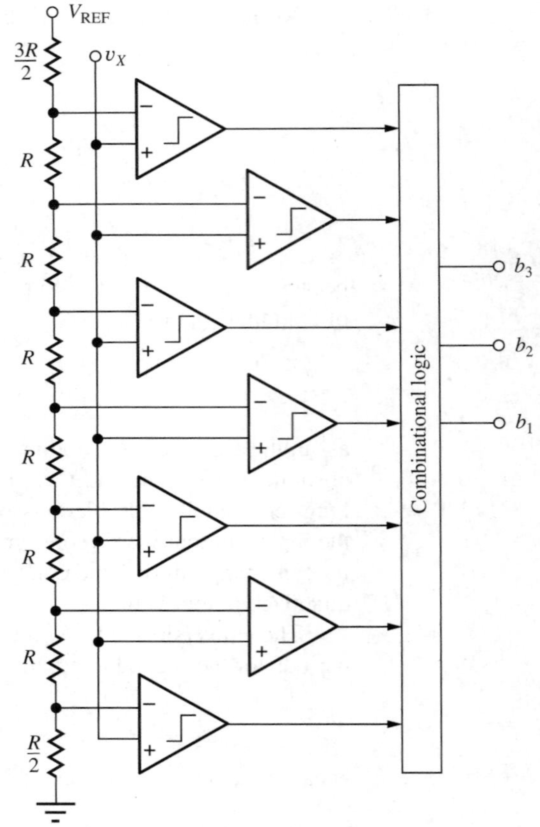

Figure 12.39 3-bit flash ADC.

The Parallel (Flash) Converter

The fastest converters employ substantially increased hardware complexity to perform a parallel rather than serial conversion. The term **flash converter** is sometimes used as the name of the parallel converter because of the device's inherent speed. Figure 12.39 shows a 3-bit parallel converter in which the unknown input v_X is simultaneously compared to seven different reference voltages. The logic network encodes the comparator outputs directly into three binary bits representing the quantized value of the input voltage. The speed of this converter is very fast, limited only by the time delays of the comparators and logic network. Also, the output continuously reflects the input signal delayed by the comparator and logic network.

The parallel A/D converter is used when maximum speed is needed and is usually found in converters with resolutions of 10 bits or less because $2^n - 1$ comparators and reference voltages are needed for an n-bit converter. Thus the cost of implementing such a converter grows rapidly with resolution. However, converters with 6-, 8-, and 10-bit resolutions have been realized in monolithic IC technology. These converters achieve effective conversion rates exceeding 10^9 conversions per second.

EXERCISE: How many resistors and comparators are required to implement a 10-bit flash ADC?

ANSWERS: 1024 resistors; 1023 comparators

12.6 Analog-to-Digital Conversion

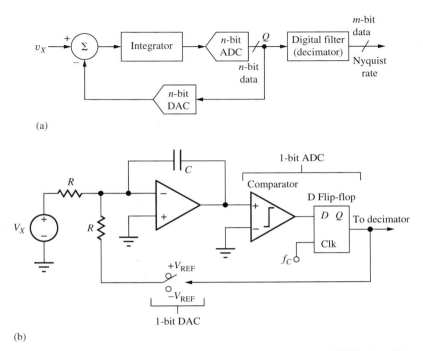

Figure 12.40 (a) Block diagram of a Delta-Sigma (Δ-Σ) ADC; (b) one-bit ($n = 1$) Δ-Σ ADC with utilizing a continuous time integrator.

Delta-Sigma A/D Converters

Delta-Sigma (Δ-Σ) converters are widely used in today's integrated circuits because they require a minimum of precision components and are easily implemented in switched capacitor form, making them ideal for use in digital signal processing applications. These converters are used in audio as well as high-frequency signal processing applications and mixed-signal integrated circuits.

The basic block diagram for the Δ-Σ ADC is shown in Fig. 12.40(a). The integrator accumulates the difference between unknown voltage v_X and the output of an n-bit D/A converter. The feedback loop forces the average value of the DAC output voltage to be equal to the unknown. In contrast to other types of converters, the internal ADC samples the integrator output at a rate that is much higher than the minimum required by the Nyquist theorem (remember that the sample rate must be at least twice the highest frequency present in the spectrum of the sampled signal). Typical sample rates for Δ-Σ ADCs range from 16 to 512 times the Nyquist rate, and the Δ-Σ converter is referred to as an "oversampled" A/D converter. Thus, the converter produces a high-rate stream of n-bit data words at output Q. This data stream is then processed by the digital filter to produce a higher resolution ($m > n$) representation of v_X at the Nyquist rate.

We can explore converter operation in more detail by referring to the implementation in Fig. 12.40(b). This most basic form of the Δ-Σ converter utilizes a continuous time integrator and 1-bit A/D and D/A converters. The integral of unknown dc voltage V_X is compared to the average of the D/A output that switches between $+V_{REF}$ and $-V_{REF}$. At the beginning of each clock interval, the 1-bit ADC decides if the output of the integrator is greater than zero ($Q = 1$) or less than zero ($Q = 0$), and the DAC output is set to force the integrator output back toward zero. If V_X is zero, for example, then the digital output alternates between 0 and 1, spending 50 percent of the time in each state. For other values of V_X, the switch will spend N clock periods connected to $-V_{REF}$ and $M - N$ clock periods connected to $+V_{REF}$, where the choice of the observation interval M depends on the desired resolution.

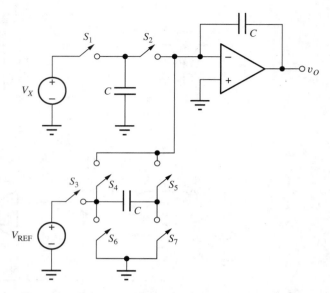

Figure 12.41 Switched capacitor integrator and reference switch.

We can get a quantitative representation of the output by using the fact that feedback loop attempts to force the integrator output to zero:

$$-V_X \left(\frac{MT_C}{RC} \right) - V_{REF} \left(\frac{NT_C}{RC} \right) + V_{REF} \left[\frac{(M-N)T_C}{RC} \right] = 0 \qquad (12.66)$$

or

$$V_X = V_{REF} \left(\frac{M - 2N}{M} \right) = V_{REF} \left(1 - 2\frac{N}{M} \right) \qquad (12.67)$$

where the ratio N/M represents the average value of the binary bit stream at the output. If we select $M = 2^m$, then

$$V_X = \left(\frac{V_{REF}}{2^m} \right) (2^m - 2N) \qquad (12.68)$$

and we see that the LSB is $V_{REF}/2^m$. The effective resolution is determined by how long we are willing to average the output. The simplest (although not necessarily the best) digital filter computes the average described here and converts the 1-bit data stream to m-bit parallel data words at the Nyquist sample rate. Converter operation is considerably more complex for a time-varying input signal, but the basic ideas are similar.

The circuit can be converted directly to switched capacitor form by replacing the continuous time integrator by the SC integrator in Fig. 12.41. Charge proportional to the input signal is added to the integrator output at each sample time, and a charge given by CV_{REF} is added or subtracted at each sample depending on the control sequence applied to the switches.

One of the advantages of the Δ-Σ converter is the inherent linearity of using a 1-bit DAC. Since there are only two levels, they must fall on a straight line, although an offset may be involved. For the case of the continuous time integrator, clock jitter still leads to errors. The SC integrator suffers less of a jitter problem as long as the clock interval is long enough for complete charge transfer to occur. The SC converter also offers the advantage of low-power operation.

ELECTRONICS IN ACTION

Sample-and-Hold Circuits

Sample-and-hold (S/H) circuits are used throughout sampled data systems and are needed ahead of many types of analog-to-digital converters in order to prevent the ADC input signal from changing during the conversion time. Several other op amp based S/H circuits are described here.[1]

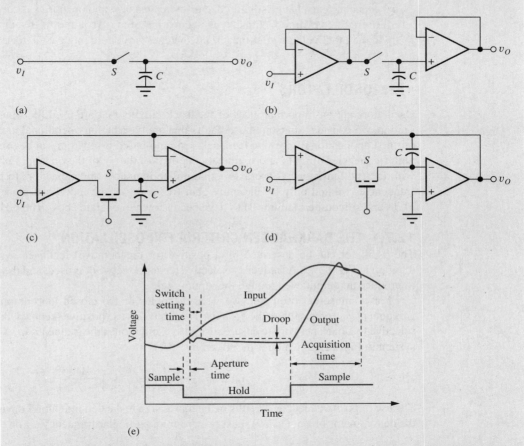

Sample-and-hold circuits: (a) basic (b) buffered (c) closed-loop (d) integrator (e) waveforms. *Copyright IEEE 1974. Reprinted with permission from [1].*

The basic sample-and-hold in (a) of the figure includes a sampling switch S and a capacitor C that stores the sampled voltage. However, this simple circuit can incur errors due to loading of the signal being sampled. Circuit (b) utilizes voltage followers to solve the problem by buffering both the input to, and the output from, sampling capacitor C. The closed-loop sample-and-hold circuits in (c) and (d) place C within a global feedback loop to improve circuit performance. The integrator circuit in (d) greatly increases the effective value of the sampling capacitor. If we apply our ideal op amp assumptions to each of the three S/H circuits, we find that both

---

[1] K. R. Stafford, P. R. Gray, and R. A. Blanchard, "A complete monolithic sample-and-hold," *IEEE Journal of Solid-State Circuits*, vol. SC-9, no. 6, pp. 381–387, December 1974.

the capacitor and output voltages are always forced to be equal to the input voltage v_I. It is worth noting that the switched-capacitor circuitry discussed in Sec. 11.2 utilize the basic sampling circuit in part (a) of the figure.

The graph in part (e) illustrates some of design issues associated with sample-and-hold operation. The aperture time represents the time required for the switching devices to change state between the sample and hold modes. A settling time is then required for the feedback circuits to recover from the switching transients. During the hold mode, the voltage stored on the capacitor can change slightly due to switch leakage and op amp bias currents. This change is referred to as "droop." Finally, an acquisition time is required for the circuit to catch back up to the input voltage after the circuit switches from hold mode back to sample mode.

12.7 OSCILLATORS

Oscillators are an important class of feedback circuits that are used for signal generation. In this section, we consider sinusoidal oscillators that are based upon operational amplifiers and represent our first application of positive feedback. Op-amp-based oscillators can be used to generate signals with frequencies of up to approximately one-half of the f_T of the op amp. Later, in Chapter 18, we will discuss transistor LC oscillators that utilize inductors and capacitors to generate signals with frequencies limited only by the unity-gain frequency of the individual devices. Oscillators using FETs and silicon-germanium BJTs have been shown to operate above 100 GHz!

12.7.1 THE BARKHAUSEN CRITERIA FOR OSCILLATION

The oscillator can be described by a positive (or regenerative) feedback system using the block diagram in Fig. 12.42. A frequency-selective feedback network is used, and the oscillator is designed to produce an output even though the input is zero.

For a sinusoidal oscillator, we want the poles of the closed-loop amplifier to be located at a frequency ω_o, precisely on the $j\omega$ axis. These circuits use positive feedback through the frequency-selective feedback network to ensure sustained oscillation at the frequency ω_o. Consider the feedback system in Fig. 12.42, which is described by

$$A_v(s) = \frac{A(s)}{1 - A(s)\beta(s)} = \frac{A(s)}{1 - T(s)} \quad (12.69)$$

The use of positive feedback results in the minus sign in the denominator. For sinusoidal oscillations, the denominator of Eq. (12.69) must be zero for a particular frequency ω_o on the $j\omega$ axis:

$$1 - T(j\omega_o) = 0 \quad \text{or} \quad T(j\omega_o) = +1 \quad (12.70)$$

The **Barkhausen criteria for oscillation** are a statement of the two conditions necessary to satisfy Eq. (12.70):

1. $\angle T(j\omega_o) = 0°$ or even multiples of $360° - 2n\pi$ rad
2. $|T(j\omega_o)| = 1$

$$(12.71)$$

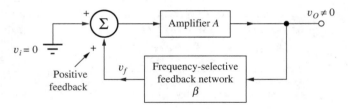

Figure 12.42 Block diagram for a positive feedback system.

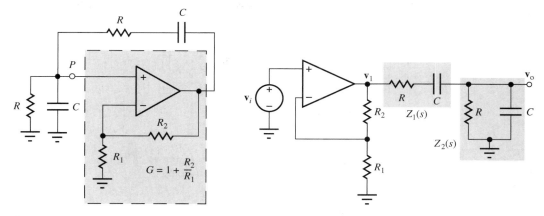

Figure 12.43 Wien-bridge oscillator circuit.

Figure 12.44 Circuit for finding the loop gain of the Wien-bridge oscillator.

These two criteria state that the phase shift around the feedback loop must be zero degrees, and the magnitude of the loop gain must be unity. Unity loop gain corresponds to a truly sinusoidal oscillator. A loop gain greater than 1 causes a distorted oscillation to occur.

In Sec. 12.7.2 we look at several RC oscillators that are useful at frequencies below a few megahertz. In Chapter 18, LC and crystal oscillators, both suitable for use at much higher frequencies, are presented.

12.7.2 OSCILLATORS EMPLOYING FREQUENCY-SELECTIVE RC NETWORKS

RC networks can be used to provide the required frequency-selective feedback at frequencies below a few megahertz. This section introduces two **RC oscillator** circuits: the Wien-bridge oscillator and the phase-shift oscillator. Other examples, a three-stage phase shift oscillator and the quadrature oscillator, appear in Probs. 12.95 and 12.96.

The Wien-Bridge Oscillator

The **Wien-bridge oscillator**[8] in Fig. 12.43 uses two RC networks to form the frequency-selective feedback network. The loop gain $T(s)$ for the Wien-bridge circuit can be found by breaking the loop at point P which represents a convenient point since the op amp represents an open circuit at its noninverting input and does not load the feedback network. The operational amplifier is operating as a noninverting amplifier with a gain $G = V_1(s)/V_I(s) = 1 + R_2/R_1$. The loop gain can be found from Fig. 12.44 using voltage division between $Z_1(s)$ and $Z_2(s)$:

$$V_o(s) = V_1(s)\frac{Z_2(s)}{Z_1(s) + Z_2(s)} \tag{12.72}$$

$$Z_1(s) = R + \frac{1}{sC} = \frac{sCR+1}{sC} \quad \text{and} \quad Z_2(s) = R\|\frac{1}{sC} = \frac{R}{sCR+1}$$

Simplifying Eq. (12.72) yields the transfer function for the loop gain:

$$V_o(s) = GV_I(s)\frac{sRC}{s^2R^2C^2 + 3sRC + 1}$$

$$T(s) = \frac{V_o(s)}{V_I(s)} = \frac{sRCG}{s^2R^2C^2 + 3sRC + 1} \tag{12.73}$$

For $s = j\omega$,

$$T(j\omega) = \frac{j\omega RCG}{(1 - \omega^2 R^2 C^2) + 3jwRC} \tag{12.74}$$

---

[8] A version of this oscillator was the product that launched the Hewlett-Packard Company.

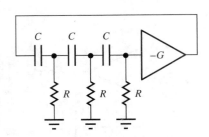

Figure 12.45 Basic concept for the phase-shift oscillator.

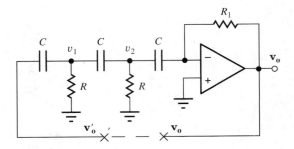

Figure 12.46 One possible realization of the phase-shift oscillator.

Applying the first Barkhausen criterion, we see that the phase shift will be zero if $(1 - \omega_o^2 R^2 C^2) = 0$. At the frequency $\omega_o = 1/RC$,

$$\angle T(j\omega_o) = 0° \quad \text{and} \quad |T(j\omega_o)| = \frac{G}{3} \quad (12.75)$$

At $\omega = \omega_o$, the phase shift is zero degrees. If the gain of the amplifier is set to $G = 3$, then $|T(j\omega_o)| = 1$, and sinusoidal oscillations will be achieved.

The Wien-bridge oscillator is useful up to frequencies of a few megahertz, limited primarily by the characteristics of the amplifier. In signal generator applications, capacitor values are often switched by decade values to achieve a wide range of oscillation frequencies. The resistors can be replaced with potentiometers to provide continuous frequency adjustment within a given range.

The Phase-Shift Oscillator

A second type of RC oscillator is the **phase-shift oscillator** depicted in Fig. 12.45. A three-section RC network is used to achieve a phase shift of 180°, which, added to the 180° phase shift of the inverting amplifier, results in a total phase shift of 360°.

The phase-shift oscillator has many practical implementations. One possible implementation combines a portion of the phase-shift function with an op amp gain block, as in Fig. 12.46. The loop gain can be found by breaking the feedback loop at x–x' and calculating $V_o(s)$ in terms of $V_o'(s)$.

Writing the nodal equations for voltages V_1 and V_2,

$$\begin{bmatrix} sCV_o'(s) \\ 0 \end{bmatrix} = \begin{bmatrix} (2sC + G) & -sC \\ -sC & (2sC + G) \end{bmatrix} \begin{bmatrix} V_1(s) \\ V_2(s) \end{bmatrix} \quad (12.76)$$

and using standard op amp theory:

$$\frac{V_o(s)}{V_2(s)} = -sCR_1 \quad (12.77)$$

Combining Eqs. (12.76) and (12.77) and solving for $V_o(s)$ in terms of $V_o'(s)$ yields

$$T(s) = \frac{V_o(s)}{V_o'(s)} = -\frac{s^3 C^3 R^2 R_1}{3s^2 R^2 C^2 + 4sRC + 1} \quad (12.78)$$

and

$$T(j\omega) = -\frac{(j\omega)^3 C^3 R^2 R_1}{(1 - 3\omega^2 R^2 C^2) + j4\omega RC} = \frac{j\omega^3 C^3 R^2 R_1}{(1 - 3\omega^2 R^2 C^2) + j4\omega RC} \quad (12.79)$$

We can see from Eq. (12.79) that the phase shift of $T(j\omega)$ will be zero if the real term in the denominator is zero:

$$1 - 3\omega_o^2 R^2 C^2 = 0 \quad \text{or} \quad \omega_o = \frac{1}{\sqrt{3}RC} \quad (12.80)$$

and
$$T(j\omega_o) = +\frac{\omega_o^2 C^2 R R_1}{4} = +\frac{1}{12}\frac{R_1}{R} \qquad (12.81)$$

For $R_1 = 12R$, the second Barkhausen criterion is met ($|T(j\omega_o)| = 1$).

Amplitude Stabilization in *RC* Oscillators

As power supply voltages, component values, and/or temperature change with time, the loop gain of an oscillator also changes. If the loop gain becomes too small, then the desired oscillation decays; if the loop gain is too large, waveform distortion occurs. Therefore, some form of **amplitude stabilization**, or gain control, is often used in oscillators to automatically control the loop gain and place the poles exactly on the $j\omega$ axis. Circuits will be designed so when power is first applied, the loop gain will be larger than the minimum needed for oscillation. As the amplitude of the oscillation grows, the gain control circuit reduces the gain to the minimum needed to sustain oscillation.

Two possible forms of amplitude stabilization are shown in Figs. 12.47 to 12.50. In the original Hewlett-Packard Wien-bridge oscillator, resistor R_1 was replaced by a nonlinear element, the lightbulb in Fig. 12.47. The resistance of the lamp is strongly dependent on the temperature of the filament of the bulb. If the amplitude is too high, the current is too large and the resistance of the lamp increases, thereby reducing the gain. If the amplitude is low, the lamp cools, the resistance decreases, and the loop gain increases. The thermal time constant of the bulb effectively averages the signal current, and the amplitude is stabilized using this clever technique.

In the Wien-bridge circuit in Fig. 12.48, diodes D_1 and D_2 and resistors R_1 to R_4 form an amplitude control network. For a positive output signal at node v_O, diode D_1 turns on as the voltage across R_3 exceeds the diode turn-on voltage. When the diode is on, resistor R_4 is switched in parallel with R_3, reducing the effective value of the loop gain. Diode D_2 functions in a similar manner on the negative peak of the signal. The values of the resistors should be chosen so that

$$\frac{R_2 + R_3}{R_1} > 2 \quad \text{and} \quad \frac{R_2 + R_3 \| R_4}{R_1} < 2 \qquad (12.82)$$

The first ratio should be set to be slightly greater than 2, and the second to slightly less than 2. Thus, when the diodes are off, the op amp gain is slightly greater than 3, ensuring oscillation, but when one of the diodes is on, the gain is reduced to slightly less than 3.

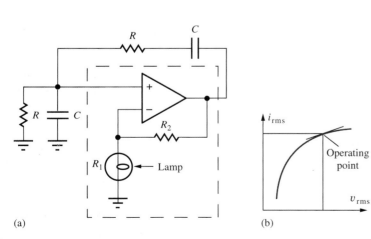

Figure 12.47 (a) Wien-bridge with amplitude stabilization; (b) bulb i-v characteristic.

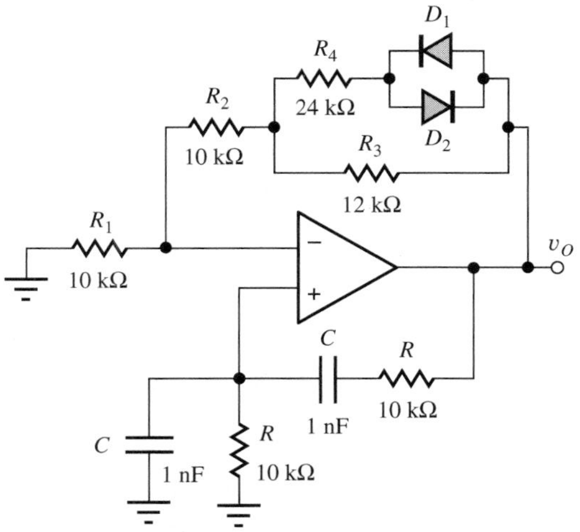

Figure 12.48 Diode amplitude stabilization of a Wien-bridge oscillator.

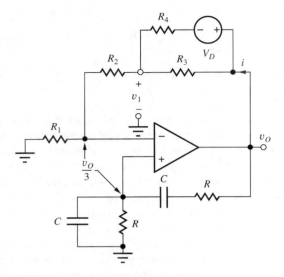

Figure 12.49 Equivalent circuit with diode D_1 on.

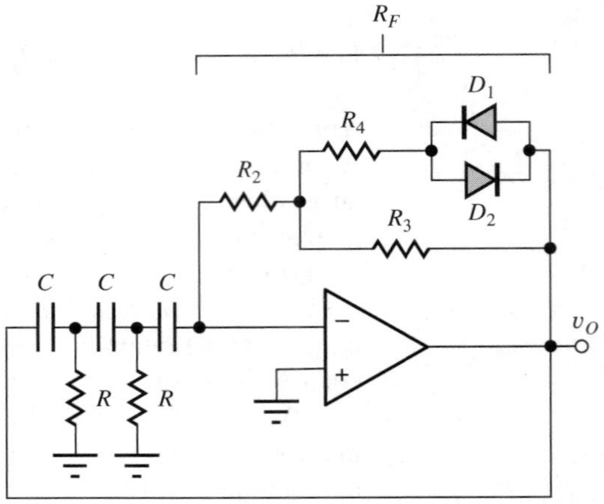

Figure 12.50 Diode amplitude stabilization of a phase-shift oscillator.

An estimate for the amplitude of oscillation can be determined from the circuit in Fig. 12.49, in which diode D_1 is assumed to be conducting with an on-voltage equal to V_D. The current i can be expressed as

$$i = \frac{v_O - v_1}{R_3} + \frac{v_O - v_1 - V_D}{R_4} \quad (12.83)$$

From Eq. (12.75) and ideal op amp behavior, we know that the voltages at both the inverting and noninverting input terminals are equal to one-third of the output voltage. Therefore,

$$v_1 = \frac{v_O}{3}\left(1 + \frac{R_2}{R_1}\right) \quad (12.84)$$

Combining Eqs. (12.83) and (12.84) and solving for v_O yields

$$v_O = \frac{3V_D}{\left(2 - \dfrac{R_2}{R_1}\right)\left(1 + \dfrac{R_4}{R_3}\right) - \dfrac{R_4}{R_1}} \quad \text{where} \quad \frac{R_2}{R_1} < 2 \quad (12.85)$$

Because the gain control circuit is actually a nonlinear circuit, Eq. (12.85) is only an estimate of the actual output amplitude; nevertheless, it does provide a good basis for circuit design.

A similar amplitude stabilization network is applied to the phase-shift oscillator in Fig. 12.50. In this case, conduction through the diodes adjusts the effective value of the total feedback resistance R_F, which determines the gain.

EXERCISE: What are the amplitude and frequency of oscillation for the Wien-bridge oscillator in Fig. 12.49? Assume $V_D = 0.6$ V.

ANSWERS: 15.9 kHz; 3.0 V.

EXERCISE: Simulate the Wein-bridge oscillator using SPICE and find the frequency and amplitude of oscillation. Model the op amp using a macromodel with a gain of 100,000.

ANSWERS: 15.9 kHz; 3.33 V.

12.8 NONLINEAR CIRCUIT APPLICATIONS

Up to this point, we have primarily considered operational amplifier circuits that use passive linear-circuit elements in the feedback network. But many interesting and useful circuits can be constructed using nonlinear elements, such as diodes and transistors in the feedback network. This section explores several examples of such circuits.

Except for oscillators, our op amp circuits thus far have involved only negative feedback configurations, but a number of other important nonlinear circuits employ positive feedback. Section 12.8 looks at this important class of circuits, including op amp implementations of rectifiers, astable and monostable multivibrators, and the Schmitt trigger circuit.

12.8.1 A PRECISION HALF-WAVE RECTIFIER

An op amp and diode are combined in Fig. 12.51 to form a **precision half-wave rectifier** circuit. Output v_O represents a rectified replica of the input signal v_I without loss of the voltage drop encountered with a normal diode rectifier circuit. The op amp tries to force the voltage across its input terminals to be zero. For $v_I > 0$, v_O equals v_I, and $i > 0$. Because current i_- must be zero, diode current i_D is equal to i, diode D is forward-biased, and the feedback loop is closed through the diode. However, for negative output voltages, currents i and i_D would be less than zero, but negative current cannot go through D_1. Thus, the diode cuts off ($i_D = 0$), the feedback loop is broken (inactive), and $v_O = 0$ because $i = 0$.

The resulting voltage transfer function for the precision rectifier is shown in Fig. 12.52. For $v_I \geq 0$, $v_O = v_I$, and for $v_I \leq 0$, $v_O = 0$. The rectification is precise; for $v_I \geq 0$, the operational amplifier adjusts its output v_1 to exactly absorb the forward voltage drop of the diode:

$$v_1 = v_O + v_D = v_I + v_D \tag{12.86}$$

This circuit provides accurate rectification even for very small input voltages and is sometimes called a **superdiode**. The primary sources of error are gain error due to the finite gain of the op amp, as well as an offset error due to the offset voltage of the amplifier. These errors were discussed in Chapter 11.

A practical problem occurs in this circuit for negative input voltages. Although the output voltage is zero, as desired for the rectifier, the voltage across the op amp input terminals is now negative, and the output voltage v_1 is saturated at the negative supply limit. Most modern op amps provide input voltage protection and will not be damaged by a large voltage across the input. However, unprotected op amps can be destroyed if the magnitude of the input voltage is larger than a few volts. The saturated output of the op amp is not usually harmful to protected amplifiers, but it does take time for the internal circuits to recover from the saturated condition, thus slowing down the response time of the circuit. It is preferable to prevent the op amp from saturating, if possible.

EXERCISE: Suppose diode D_1 in Fig. 12.51 has an "on-voltage" of 0.6 V, and the op amp is operating with ±10-V power supplies. What are the voltages v_O and v_1 for the circuit if $v_I = +1$ V? For $v_I = -1$ V? What is the minimum Zener breakdown voltage for the diode?

ANSWERS: +1 V, +1.6 V; 0 V, −10 V; 10 V

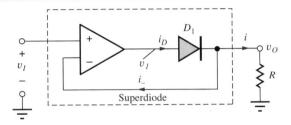

Figure 12.51 Precision half-wave rectifier circuit (or "superdiode").

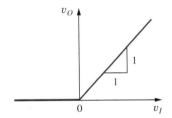

Figure 12.52 Voltage transfer characteristic for the precision rectifier.

12.8.2 NONSATURATING PRECISION-RECTIFIER CIRCUIT

The saturation problem can be solved using the circuit given in Fig. 12.53. An inverting-amplifier configuration is used instead of the noninverting configuration, and diode D_2 is added to keep the feedback loop closed when the output of the rectifier is zero.

For positive input voltages depicted in Fig. 12.53(b), the op amp output voltage v_1 becomes negative, forward-biasing diode D_2 so that current i_I passes through diode D_2 and into the output of the op amp. The inverting input is at virtual ground, the current in R_2 is zero, and the output remains at zero. Diode D_1 is reverse-biased.

For $v_I < 0$ in Fig. 12.53(c), diode D_1 turns on and supplies current i_I and load current i, and D_2 is off. The circuit behaves as an inverting amplifier with gain equal to $-R_2/R_1$. Thus, the overall voltage transfer characteristic can be described by

$$v_O = 0 \text{ for } v_I \geq 0 \quad \text{and} \quad v_O = -\frac{R_2}{R_1}v_I \text{ for } v_I \leq 0 \qquad (12.87)$$

as shown in Fig. 12.53(d). The output voltage of the op amp itself, v_1, is one diode-drop below zero for positive input voltages and one diode above the output voltage for negative input voltages. The inverting input is a virtual ground in both cases, and the negative feedback loop is always active: through D_1 and R_2 for $v_I < 0$ and through D_2 for $v_I > 0$.

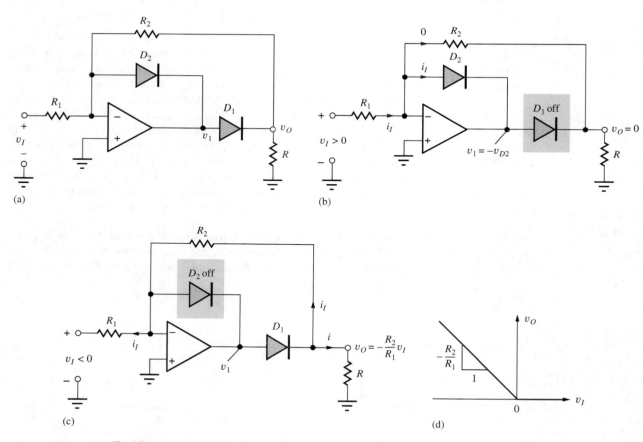

Figure 12.53 (a) Nonsaturating precision-rectifier circuit; (b) active feedback elements for $v_I \geq 0$; (c) active feedback elements for $v_I < 0$; (d) improved rectifier voltage transfer characteristic.

ELECTRONICS IN ACTION

An AC Voltmeter

The half-wave rectifier circuit can be combined with a low-pass filter to form a basic ac voltmeter circuit, as in the top figure. For a sinusoidal input signal with an amplitude V_M at a frequency ω_o, the output voltage v_1 is a rectified sine wave that can be described by its Fourier series as:

$$v_1(t) = -\left(\frac{R_2}{R_1}\right)\left(\frac{V_I}{\pi}\right)\left[1 + \frac{\pi}{2}\sin\omega_o t - \sum_{n=2}^{\infty}\frac{1+\cos n\pi}{(n^2-1)}\cos n\omega_o t\right]$$

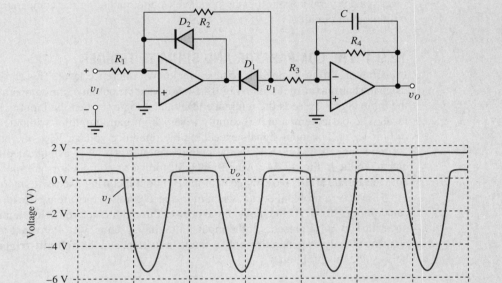

AC voltmeter circuit consisting of a half-wave rectifier and low-pass filter (top); voltage waveform at rectifier output v_1 for $v_I = (-5 \sin 120\pi t)$ V with $R_2 = R_1$, $R_4 = R_3$ and $f_C = 1.59$ Hz (bottom).

If the cutoff frequency of the low-pass filter is chosen such that $\omega_C \ll \omega_o$, then the output voltage v_O will consist primarily of the dc voltage component (see Prob. 12.107) given by

$$v_O = \frac{R_4}{R_3}\left[\frac{R_2}{R_1}\frac{V_I}{\pi}\right]$$

The voltmeter range (scale factor) can be adjusted through the choice of the four resistors.

EXERCISE: Suppose the diodes in Fig. 12.53 have "on-voltages" of 0.6 V, and the op amp is operating with ±15-V power supplies. What are the voltages v_O and v_1 for the circuit if $R_1 = 22$ kΩ, $R_2 = 68$ kΩ, and $v_I = +2$ V? For $v_I = -2$ V? Estimate the most negative input voltage for which the circuit will operate properly. What is the minimum Zener breakdown voltage specification for the diodes assuming they are both the same?

ANSWERS: 0 V, −0.6 V; +6.18 V, +6.78 V; −4.66 V; 15 V

> **EXERCISE:** What is the dc output voltage of the ac voltmeter circuit if $R_1 = 3.24$ kΩ, $R_2 = 10.2$ kΩ, $R_3 = 20$ kΩ, $R_4 = 20$ kΩ, and $V_I = 2$ V?
>
> **ANSWER:** 2.00 V

12.9 CIRCUITS USING POSITIVE FEEDBACK

Up to now, most of our circuits have used negative feedback: A voltage or current proportional to the output signal was returned to the inverting-input terminal of the operational amplifier. However, positive feedback can also be used to perform a number of useful nonlinear functions, and we investigate several possibilities in this final section, including the comparator, Schmitt trigger, and multivibrator circuits.

12.9.1 THE COMPARATOR AND SCHMITT TRIGGER

It is often useful to compare a voltage to a known reference level. This can be done electronically using the **comparator** circuit in Fig. 12.54. We want the output of the comparator to be a logic 1 when the input signal exceeds the reference level and a logic 0 when the input is less than the reference level. The basic comparator is simply a very high gain amplifier without feedback, as indicated in Fig. 12.54. For input signals exceeding the reference voltage V_{REF}, the output saturates at V_{CC}; for input signals less than V_{REF}, the output saturates at $-V_{EE}$, as indicated in the voltage transfer characteristic in Fig. 12.54.[9] Amplifiers built for use as comparators are specifically designed to be able to saturate at the two voltage extremes without incurring excessive internal time delays.

However, a problem occurs when high-speed comparators are used with noisy signals, as indicated in Fig. 12.55. As the input signal crosses through the reference level, multiple transitions may occur due to noise present on the input. In digital systems, we often want to detect this threshold crossing cleanly by generating only a single transition, and the **Schmitt-trigger** circuit in Fig. 12.56 helps solve this problem.

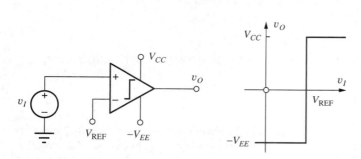

Figure 12.54 Comparator circuit using an infinite-gain amplifier.

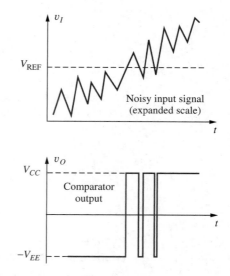

Figure 12.55 Comparator response to noisy input signal.

---

[9] In this section, we assume that the output can reach the supply voltages.

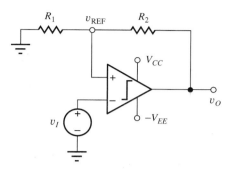

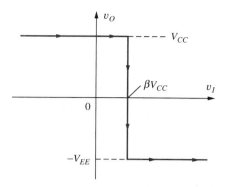

Figure 12.56 Schmitt-trigger circuit.

Figure 12.57 Voltage transfer characteristic for the Schmitt trigger as v_S increases from below $V_{\text{REF}} = +\beta V_{CC}$.

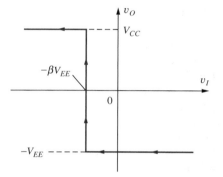

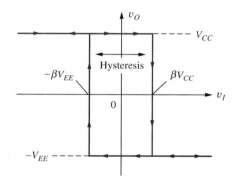

Figure 12.58 Voltage transfer characteristic for the Schmitt trigger as v_S decreases from above $V_{\text{REF}} = -\beta V_{EE}$.

Figure 12.59 Complete voltage transfer characteristic for the Schmitt trigger.

The Schmitt trigger uses a comparator whose reference voltage is derived from a voltage divider across the output. The input signal is applied to the inverting-input terminal, and the reference voltage is applied to the noninverting input (positive feedback). For positive output voltages, $V_{\text{REF}} = \beta V_{CC}$, but for negative output voltages, $V_{\text{REF}} = -\beta V_{EE}$, where $\beta = R_1/(R_1 + R_2)$. Thus, the reference voltage changes when the output switches state.

Consider the case for an input voltage increasing from below V_{REF}, as in Fig. 12.57. The output is at V_{CC} and $V_{\text{REF}} = \beta V_{CC}$. As the input voltage crosses through V_{REF}, the output switches state to $-V_{EE}$, and the reference voltage simultaneously drops, reinforcing the voltage across the comparator input. In order to cause the comparator to switch states a second time, the input must now drop below $V_{\text{REF}} = -\beta V_{EE}$, as depicted in Fig. 12.58.

Now consider the situation as v_I decreases from a high level, as in the voltage transfer characteristic in Fig. 12.58. The output is at $-V_{EE}$ and $V_{\text{REF}} = -\beta V_{EE}$. As the input voltage crosses through V_{REF}, the output switches state to V_{CC}, and the reference voltage simultaneously increases, again reinforcing the voltage across the comparator input.

The voltage transfer characteristics from Figs. 12.57 and 12.58 are combined to yield the overall voltage transfer characteristic for the Schmitt trigger given in Fig. 12.59. The arrows indicate the portion of the characteristic that is traversed for increasing and decreasing values of the input signal. The Schmitt trigger is said to exhibit **hysteresis** in its VTC, and will not respond to input noise that has a magnitude V_n smaller than the difference between the two threshold voltages:

$$V_n < \beta[V_{CC} - (-V_{EE})] = \beta(V_{CC} + V_{EE}) \qquad (12.88)$$

The Schmitt trigger with positive feedback is an example of a circuit with two stable states: a **bistable circuit,** or **bistable multivibrator.** Another example of a bistable circuit is the digital storage element usually called the flip-flop (see Chapter 8).

> **EXERCISE:** If $V_{CC} = +10$ V $= -V_{EE}$, $R_1 = 1$ kΩ, and $R_2 = 9.1$ kΩ, what are the values of the switching thresholds for the Schmitt-trigger circuit in Figs. 12.56 through 12.61 and the magnitude of the hysteresis?
>
> **ANSWERS:** +0.99 V; −0.99 V; 1.98 V

12.9.2 THE ASTABLE MULTIVIBRATOR

Another type of multivibrator circuit employs a combination of positive and negative feedback and is designed to oscillate and generate a rectangular output waveform. The output of the circuit in Fig. 12.60 has no stable state and is referred to as an **astable multivibrator.**

Operation of the astable multivibrator circuit can best be understood by referring to the waveforms in Fig. 12.61. The output voltage switches periodically (oscillates) between the two output voltages V_{CC} and $-V_{EE}$. Let us assume that the output has just switched to $v_O = V_{CC}$ at $t = 0$. The voltage at the inverting-input terminal of the op amp charges exponentially toward a final value of V_{CC} with a time constant $\tau = RC$. However, when the voltage on the comparator's inverting input exceeds that on the noninverting input, the output switches state. The voltage on the capacitor at the time of the output transition at $t = 0$ is $v_C = -\beta V_{EE}$. Thus, the expression for the voltage on the capacitor can be written as

$$v_C(t) = V_{CC} - (V_{CC} + \beta V_{EE}) \exp\left(-\frac{t}{RC}\right) \qquad (12.89)$$

The comparator changes state again at time T_1 when $v_C(t)$ just reaches βV_{CC}:

$$\beta V_{CC} = V_{CC} - (V_{CC} + \beta V_{EE}) \exp\left(-\frac{T_1}{RC}\right) \qquad (12.90)$$

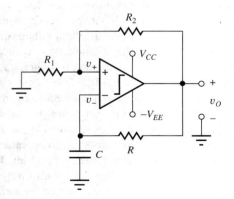

Figure 12.60 Operational amplifier in an astable multivibrator circuit.

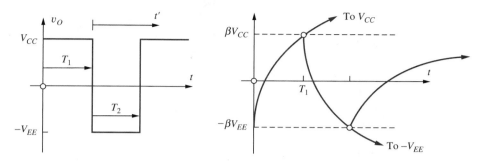

Figure 12.61 Waveforms for the astable multivibrator.

Solving for time T_1 yields

$$T_1 = RC \ln \frac{1 + \beta \left(\frac{V_{EE}}{V_{CC}}\right)}{1 - \beta} \qquad (12.91)$$

During time interval T_2, the output is low and the capacitor discharges from an initial voltage of βV_{CC} toward a final voltage of $-V_{EE}$. For this case, the capacitor voltage can be expressed as

$$v_C(t') = -V_{EE} + (V_{EE} + \beta V_{CC}) \exp\left(-\frac{t'}{RC}\right) \qquad (12.92)$$

in which $t' = 0$ at the beginning of the T_2 interval. At $t' = T_2$, $v_C = -\beta V_{EE}$,

$$-\beta V_{EE} = -V_{EE} + (V_{EE} + \beta V_{CC}) \exp\left(-\frac{T_2}{RC}\right) \qquad (12.93)$$

and T_2 is equal to

$$T_2 = RC \ln \frac{1 + \beta \left(\frac{V_{CC}}{V_{EE}}\right)}{1 - \beta} \qquad (12.94)$$

For the common case of symmetrical power supply voltages, $V_{CC} = V_{EE}$, and the output of the astable multivibrator represents a square wave with a period T given by

$$T = T_1 + T_2 = 2RC \ln \frac{1 + \beta}{1 - \beta} \qquad (12.95)$$

EXERCISE: What is the frequency of oscillation of the circuit in Fig. 12.62 if $V_{CC} = +5$ V, $-V_{EE} = -5$ V, $R_1 = 6.8$ kΩ, $R_2 = 6.8$ kΩ, $R = 10$ kΩ, and $C = 0.001$ μF?

ANSWER: 45.5 kHz

12.9.3 THE MONOSTABLE MULTIVIBRATOR OR ONE SHOT

A third type of multivibrator operates with one stable state and is used to generate a single pulse of known duration following application of a trigger signal. The circuit rests quiescently in its stable state, but can be "triggered" to generate a single transient pulse of fixed duration T. Once time T is past, the circuit returns to the stable state to await another **triggering** pulse. This **monostable circuit** is variously called a **monostable multivibrator,** a **single shot,** or a **one shot.**

An example of a comparator-based monostable multivibrator circuit is given in Fig. 12.62. Diode D_1 has been added to the astable multivibrator in Fig. 12.60 to couple the triggering signal v_T into the circuit, and clamping diode D_2 has been added to limit the negative voltage excursion on capacitor C.

ELECTRONICS IN ACTION

Numerically Controlled Oscillators and Direct Digital Synthesis

Modern D/A converter technology has advanced to the point that traditional analog feedback oscillators are being replaced with a direct digital synthesizer (DDS) that utilizes numerically controlled oscillators (NCOs) to synthesize the sinusoidal waveforms. The NCO can provide very small frequency step size and high-speed tuning. In the DDS, the signal waveform is constructed in the digital domain, and the analog output signal is produced using a digital-to-analog (DAC) converter followed by a low-pass filter.

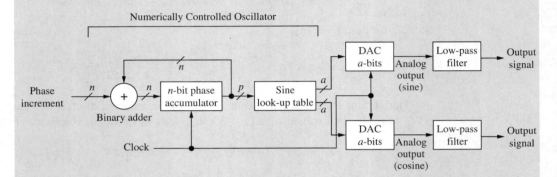

The digital NCO consists of an n-bit phase accumulator and a p-bit sine look-up table where $p \leq n$. To generate a sine wave, an n-bit phase increment is added to the accumulator during each clock cycle. A full counter (2^n counts) corresponds to 2π radians or 1 cycle of the output sine wave. If the counter is incremented by one at each clock interval, the maximum period $T_{\max}$ of the output waveform, corresponding to minimum output frequency $f_{\min}$, will be

$$T_{\max} = 2^n T_{\text{clk}} \quad \text{or} \quad f_{\min} = \frac{f_{\text{clk}}}{2^n}$$

where T_{clk} is the period of the clock, and f_{clk} is the clock frequency. This minimum output frequency also represents the frequency resolution of the DDS. To generate higher frequency signals, a larger phase increment N is added to the phase accumulator at each clock cycle, and $f_O = N f_{\min}$. For example, for $f_{\text{clk}} = 20$ MHz and $n = 24$, $f_{\min} \approx 1.192$ Hz. In order to generate a 10-kHz sine wave, a phase increment of 8389 (10,000/1.192) would be added to the counter at each clock cycle. Based upon the Nyquist sampling theorem, the highest frequency that can be generated is one-half of the clock frequency (using $N = 2^{n/2}$), since f_{clk} is the update rate of the D/A converter.

In order to reduce the size of the look-up table, only the upper p bits of the phase accumulator are used to address the sine table. A number of ROM compression techniques are utilized to further reduce the size of the ROM. The output of the sine table is an a-bit representation of the amplitude of the sine wave where "a" corresponds to the number of bits of resolution of the D/A converter. Finite resolution in the representation of both the signal phase and amplitude lead to distortion in the output waveform. The low-pass filter helps to remove distortion and the high-frequency content related to the update rate of the DAC (f_{clk}). Some DDS chips provide two D/A outputs, producing sine and cosine waves with very precise 90-degree phase relationships for the in-phase (I) and quadrature (Q) channels in RF transceivers.

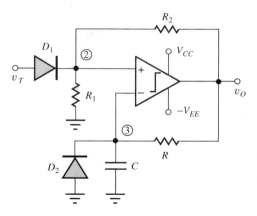

Figure 12.62 Example of an operational-amplifier monostable-multivibrator circuit.

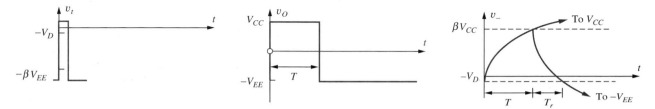

Figure 12.63 Monostable multivibrator waveforms.

The circuit rests in its quiescent state with $v_O = -V_{EE}$. If the trigger signal voltage v_T is less than the voltage at node 2,

$$v_T < -\frac{R_1}{R_1 + R_2}V_{EE} = -\beta V_{EE} \tag{12.96}$$

diode D_1 is cut off. Capacitor C discharges through R until diode D_2 turns on, clamping the capacitor voltage at one diode-drop V_D below ground potential. In this condition, the differential-input voltage v_{ID} to the comparator is given by

$$v_{ID} = -\beta V_{EE} - (-V_D) = -\beta V_{EE} + V_D \tag{12.97}$$

As long as the value of the voltage divider is chosen so that

$$v_{ID} < 0 \quad \text{or} \quad \beta V_{EE} > V_D \quad \text{where } \beta = \frac{R_1}{R_1 + R_2} \tag{12.98}$$

then the output of the circuit will have one stable state.

Triggering the Monostable Multivibrator

The monostable multivibrator can be triggered by applying a positive pulse to the trigger input v_t, as shown in the waveforms in Fig. 12.63. As the trigger pulse level exceeds a voltage of $-\beta V_{EE}$, diode D_1 turns on and subsequently pulls the voltage at node 2 above that at node 3. At this point, the comparator output changes state, and the voltage at the noninverting-input terminal rises abruptly to a voltage equal to $+\beta V_{CC}$. Diode D_1 cuts off, isolating the comparator from any further changes on the trigger input.

The voltage on the capacitor now begins to charge from its initial voltage $-V_D$ toward a final voltage of V_{CC} and can be expressed mathematically as

$$v_C(t) = V_{CC} - (V_{CC} + V_D)\exp\left(-\frac{t}{RC}\right) \tag{12.99}$$

where the time origin ($t = 0$) coincides with the start of the trigger pulse. However, the comparator changes state again when the capacitor voltage reaches $+\beta V_{CC}$. Thus, the pulse width T is given by

$$\beta V_{CC} = V_{CC} - (V_{CC} + V_D)\exp\left(-\frac{T}{RC}\right) \quad \text{or} \quad T = RC \ln \frac{1 + \left(\dfrac{V_D}{V_{CC}}\right)}{1 - \beta} \quad (12.100)$$

The output of the circuit consists of a positive pulse with a fixed duration T set by the values of R_1, R_2, R, and C.

ELECTRONICS IN ACTION

Function Generators
Analog Function Generators
The instrumentation in most introductory electronics laboratories includes some type of low frequency function generator that produces elementary waveforms including square, triangle, and sine wave outputs at frequencies up to a few MHz. For many years, inexpensive versions of these function generators utilized the astable multivibrator to generate the square wave signal as shown in the accompanying figure. The frequency of the multivibrator is varied by changing either R_3 or C_3. C_3 is often changed in decade steps; R_3 may be varied continuously using a potentiometer. The square wave output of the astable multivibrator drives an op amp integrator circuit to produce a triangular waveform. The output of the integrator can then be passed through a low-pass filter or piecewise linear shaping circuit to produce a low-distortion sine wave.

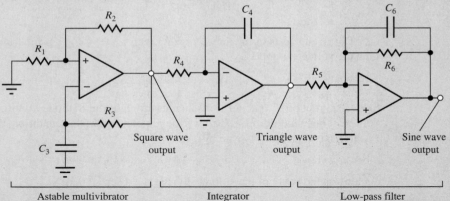

Simple function generator using an astable multivibrator, integrator, and low-pass filter.

Keysight Function Generator.
© Keysight Technologies 2014 All Rights reserved.

For a well-defined pulse width to be generated, this circuit should not be retriggered until the voltages on the various nodes have all returned to their quiescent steady-state values. Following the return of the output to $-V_{EE}$, the capacitor voltage charges from a value of βV_{CC} toward $-V_{EE}$, but reaches steady state when diode D_2 begins to conduct. Thus, the recovery time can be calculated from

$$-V_D = -V_{EE} + (V_{EE} + \beta V_{CC})\exp\left(-\frac{T_r}{RC}\right) \quad \text{and} \quad T_r = RC \ln \frac{1 + \beta\left(\dfrac{V_{CC}}{V_{EE}}\right)}{1 - \left(\dfrac{V_D}{V_{EE}}\right)} \quad (12.101)$$

EXERCISE: For the monostable multivibrator circuit in Fig. 12.64, $V_{CC} = +5$ V $= V_{EE}$, $R_1 = 22$ kΩ, $R_2 = 18$ kΩ, $R = 11$ kΩ, and $C = 0.002$ μF. What is the pulse width of the one shot? What is the minimum time between trigger pulses for this circuit?

ANSWERS: 20.4 μs; 33.4 μs

SUMMARY

Chapter 12 introduced a variety of linear and nonlinear applications of operational amplifiers. Key topics are outlined here.

- It is often impossible to realize a set of amplifier specifications utilizing a single amplifier stage, and we must cascade several stages in order to achieve the desired results.
- Two-port models for cascaded amplifiers can be used to simplify the representation of the overall amplifier.
- A comprehensive example of the design of a multistage amplifier was presented in which a computer spreadsheet was used to explore the design space. The influence of resistor tolerances on this design was also explored.
- The bandwidth of a multistage amplifier is less than the bandwidth of any of the single amplifiers operating alone. An expression was developed for the bandwidth of a cascade of N identical amplifiers and was cast in terms of the bandwidth shrinkage factor.
- The instrumentation amplifier is a high performance circuit often used in data acquisition systems.
- Active RC filters including low-pass, high-pass, and band-pass circuits were introduced. These designs use RC feedback networks and operational amplifiers to replace bulky inductors that would normally be required in RLC filters designed for the audio range. Single-amplifier active filters employ a combination of negative and positive feedback to realize second-order low-pass, high-pass, and band-pass transfer functions.
- Sensitivity of filter characteristics to passive component and op amp parameter tolerances is an important design consideration. Multiple op amp filters offer low sensitivity and ease of design, compared to their single op amp counterparts.
- Magnitude and frequency scaling can be used to change the impedance level and ω_o of a filter without affecting its Q.
- Switched-capacitor (SC) circuits use a combination of capacitors and switches to replace resistors in integrated circuit filter designs. These filters represent the sampled-data or discrete-time equivalents of the continuous-time RC filters and are fully compatible with MOS IC technology. Both inverting and noninverting integrators can be implemented using SC techniques.

- Digital-to-analog (D/A) and analog-to-digital (A/D) converters, also known as DACs and ADCs, provide the interface between the digital computer and the world of analog signals. Gain, offset, linearity, and differential linearity errors are important in both types of converters.
- The resolution of A/D and D/A converters is measured in terms of the least significant bit or LSB. The LSB of an n-bit converter is equal to $V_{FS}/2^n$, where V_{FS} is the full scale voltage range of the converter. The most significant bit or MSB of the converter is equal to $V_{FS}/2$.
- Simple MOS DACs can be formed using weighted-resistor, R-2R ladder and inverted R-2R ladder circuits, and MOS transistor switches. The inverted R-2R ladder configuration maintains a constant current within the ladder elements. Switched-capacitor techniques based on weighted-capacitor and C-2C ladder configurations are also widely used in VLSI ICs.
- Good-quality DACs have monotonic input-output characteristcs.
- Basic ADC circuits compare an unknown input voltage to a known time-varying reference signal. The reference signal is provided by a D/A converter in the counting and successive approximation converters. The counting converter sequentially compares the unknown to all possible outputs of the D/A converter; a conversion may take as many as 2^n clock periods to complete. The counting converter is simple but relatively slow. The successive approximation converter uses an efficient binary search algorithm to achieve a conversion in only n clock periods and is a very popular conversion technique.
- In the single- and dual-ramp ADCs, the reference voltage is an analog signal with a well-defined slope, usually generated by an integrator with a constant input voltage. The digital output of the single-ramp converter suffers from its dependence on the absolute values of the integrator time constant. The dual ramp greatly reduces this problem, and can achieve high differential and integral linearity, but with conversion rates of only a few conversions per second. The dual-ramp converter is widely used in high-precision instrumentation systems. Rejection of sinusoidal signals with periods that are integer multiples of the integration time, called normal-mode rejection, is an important feature of integrating converters.
- The fastest A/D conversion technique is the parallel or "flash" converter, which simultaneously compares the unknown voltage to all possible quantized values. Conversion speed is limited only by the speed of the comparators and logic network that form the converter. This high-speed is achieved at a cost of high hardware complexity.
- Good-quality ADCs exhibit linearity and differential linearity errors of less than 1/2 LSB and have no missing codes.
- A/D converters employ circuits called comparators to compare an unknown input voltage with a precision reference voltage. The comparator can be considered to be a high-gain, high-speed op amp designed to operate without feedback.
- In circuits called oscillators, feedback is actually designed to be positive or regenerative so that an output signal can be produced by the circuit without an input being present. The Barkhausen criteria for oscillation state that the phase shift around the feedback loop must be an even multiple of 360° at some frequency, and the loop gain at that frequency must be equal to 1.
- Oscillators use some form of frequency-selective feedback to determine the frequency of oscillation; RC and LC networks and quartz crystals can all be used to set the frequency.
- Wien-bridge and phase-shift oscillators are examples of oscillators employing RC networks to set the frequency of oscillation.
- For true sinusoidal oscillation, the poles of the oscillator must be located precisely on the $j\omega$ axis in the s-plane. Otherwise, distortion occurs. To achieve sinusoidal oscillation, some form of amplitude stabilization is normally required. Such stabilization may result simply from the inherent nonlinear characteristics of the transistors used in the circuit, or from explicitly added gain control circuitry.

- Nonlinear circuit applications of operational amplifiers were also introduced including several precision-rectifier circuits.
- Multivibrator circuits are used to develop various forms of electronic pulses. The bistable Schmitt-trigger circuit has two stable states and is often used in place of the comparator in noisy environments. The monostable multivibrator, or one shot, is used to generate a single pulse of known duration, whereas the astable multivibrator has no stable state and oscillates continuously, producing a square wave output.

KEY TERMS

Active filters
Amplitude stabilization
Analog-to-digital converter (ADC or A/D converter)
Astable circuit
Astable multivibrator
Band-pass filter
Barkhausen criteria for oscillation
Bistable circuit
Bistable multivibrator
Butterworth filter
C-2C ladder DAC
Comparator
Conversion time
Counter-ramp converter
Delta-Sigma ADC
Differential linearity error
Differential subtractor
Digital-to-analog converter (DAC or D/A converter)
Dual-ramp (dual-slope) ADC
Flash converter
Frequency scaling
Full-scale current
Full-scale voltage
Gain error
High-pass filter
Hysteresis
Instrumentation amplifier
Integral linearity error
Integrator
Inverted R-2R ladder
Inverting amplifier
Inverting input
Least significant bit (LSB)
Linearity error
Loop gain ($A\beta$)
Loop transmission (T)
Low-pass filter
Magnitude scaling

Maximally flat magnitude
-1 Point
Missing code
Monostable circuit
Monostable multivibrator
Monotonic converter
Most significant bit (MSB)
Negative feedback
Noninverting integrator
Nonmonotonic converter
Normal-mode rejection
Notch filter
One shot
Open-circuit voltage gain
Open-loop amplifier
Open-loop gain
Operational amplifier (op amp)
Oscillator circuits
Oscillators
Phase-shift oscillator
Positive feedback
Precision half-wave rectifier
Quantization error
R-2R ladder
RC oscillators
Reference current
Reference transistor
Reference voltage
Regenerative feedback
Resolution of the converter
Sample-and-hold circuit
Schmitt trigger
Sensitivity
Single-ramp (single-slope) ADC
Single shot
Sinusoidal oscillator
Stray-insensitive circuits
Successive approximation converter
Superdiode
Superposition errors

Switched-capacitor filters
Switched-capacitor integrator
Switched-capacitor (SC) circuits
Triggering
Two-phase nonoverlapping clock

Two-port model
Weighted-capacitor DAC
Weighted-resistor DAC
Wien-bridge oscillator

ADDITIONAL READING

Franco, Sergio, *Design with Operational Amplifiers and Analog Integrated Circuits*, Third Edition, McGraw-Hill, New York: 2001.

Ghausi, M. S. and K. R. Laker. *Modern Filter Design—Active RC and Switched Capacitor*. Prentice-Hall, Englewood Cliffs, NJ: 1981.

Gray, P. R., P. J. Hurst, S. H. Lewis, and R. G. Meyer, *Analysis and Design of Analog Integrated Circuits*, Fourth Edition, John Wiley and Sons, New York: 2001.

Huelsman, L. P. and P. E. Allen. *Introduction to Theory and Design of Active Filters*. McGraw-Hill, New York: 1980.

Kennedy, E. J. *Operational Amplifier Circuits—Theory and Applications*. Holt, Rinehart and Winston, New York: 1988.

PROBLEMS

12.1 Cascaded Amplifier

12.1. Seven amplifiers were identified in Fig. 12.1. Find two more possibilities.

12.2. An amplifier is formed by cascading two operational-amplifier stages, as shown in Fig. P12.2(a). (a) Replace each amplifier stage with its two-port representation. (b) Use the circuit model from part (a) to find the overall two-port representation (A_v, R_{in}, R_{out}) for the complete two-stage amplifier. (c) Draw the circuit of the two-port corresponding to the complete two-stage amplifier.

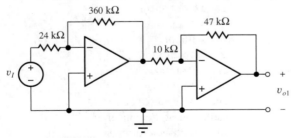

Figure P12.2

12.3. An amplifier is formed by cascading three identical operational-amplifier stages, as shown in Fig. P12.3. (a) Replace each op amp circuit with its two-port representation. (b) Use the circuit model from part (a) to find the overall two-port representation (A_v, R_{in}, R_{out}) for the complete three-stage amplifier. (c) Draw the two-port circuit corresponding to the complete three-stage amplifier.

12.4. An amplifier is formed by cascading the two operational amplifier stages shown in Fig. P12.2. What are the voltage gain, input resistance, and output resistance for this amplifier (a) if the op amps are ideal? (b) If the op amps have an open-loop gain of 10^5, an input resistance of 500 kΩ, and an output resistance of 200 Ω? (c) Draw the new circuit and repeat (a) and (b) if the two amplifier stages are interchanged.

12.5. An amplifier is formed by cascading the two operational amplifier stages shown in Fig. P12.5. What are the voltage gain, input resistance, and output resistance for this amplifier (a) if the op amps are ideal? (b) If the op amps have an open-loop gain of 86 dB, an input resistance of 250 kΩ, and an output resistance of 100 Ω? (c) Draw the new circuit and repeat (a) and (b) if the two amplifier stages are interchanged.

12.6. An amplifier is formed by cascading the two operational amplifier stages shown in Fig. P12.6. What are the voltage gain, input resistance, and output resistance for this amplifier (a) if the op amps are ideal? (b) If the op amps have an open-loop gain of 106 dB, an input resistance of 300 kΩ, and an output resistance of 200 Ω? (c) Draw the new circuit and repeat (a) and (b) if the two amplifier stages are interchanged.

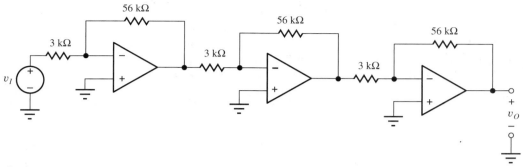

Figure P12.3

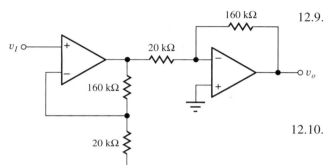

(c)
Figure P12.5

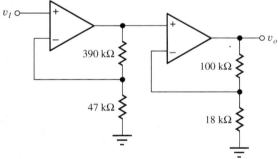

Figure P12.6

12.7. An amplifier is formed by cascading the three operational amplifier stages shown in Fig. P12.3. What are the voltage gain, input resistance, and output resistance for this amplifier (a) if the op amps are ideal? (b) If the op amps have an open-loop gain of 94 dB, an input resistance of 400 kΩ, and an output resistance of 250 Ω?

12.8. Assume the op amps in Fig. P12.3 are ideal except for power supply voltages of ± 12 V. (a) If the input voltage is 1 mV, what are the voltages at the 8 nodes in the circuit? (b) Repeat for an input voltage of 3 mV. (c) Repeat for an input voltage of 2 mV with an open-loop gain of 80 dB.

12.9. An amplifier is formed by cascading the three operational amplifier stages shown in Fig. P12.9. What are the voltage gain, input resistance, and output resistance for this amplifier (a) if the op amps are ideal? (b) If the op amps have an open-loop gain of 94 dB, an input resistance of 400 kΩ, and an output resistance of 250 Ω?

12.10. Assume the op amps in Fig. P12.9 are ideal except for power supply voltages of ± 12 V. (a) If the input voltage is 5 mV, what are the voltages at the 8 nodes in the circuit? (b) Repeat for an input voltage of 10 mV. (c) Repeat for an input voltage of 10 mV with an open-loop gain of 80 dB.

12.11. What are the values of A_v, R_{in}, and R_{out} for the overall three-stage amplifier in Fig. P12.3 if the 3-kΩ resistors are replaced with 3.9-kΩ resistors?

12.12. The 2-kΩ resistors in Fig. P12.3 are to be replaced with a value that gives an overall gain of 40 dB. What is the new resistor value? What is the new value of R_{in}?

12.13. The op amps in Fig. P12.9 are ideal. What are the nominal, minimum, and maximum values of the voltage gain, input resistance, and output resistance of the overall amplifier if all the resistors have 5 percent tolerances?

12.14. The op amps in Fig. P12.14 are ideal (a) What are the voltage gain, input resistance, and output resistance of the overall amplifier? (b) If the input voltage $v_I = 1$ mV, what are the voltages at each of the eight nodes in the amplifier circuit?

12.15. Repeat Prob. 12.14 if the 2-kΩ resistors are all replaced with 3-kΩ resistors, and the 1-MΩ resistor is replaced with a 470 kΩ resistor.

12.16. The op amps in Fig. P12.14 are ideal. What are the nominal, minimum, and maximum values of the voltage gain, input resistance, and output resistance

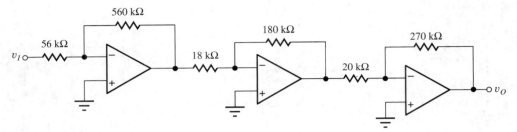

Figure P12.9

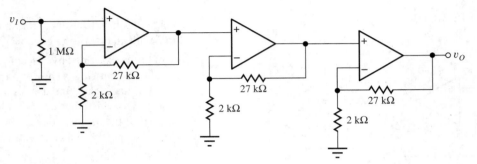

Figure P12.14

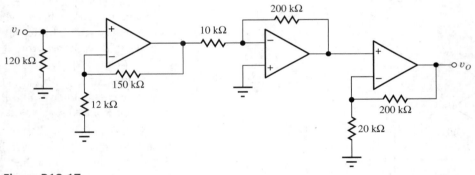

Figure P12.17

of the overall amplifier if all the resistors have 2 percent tolerances?

12.17. The op amps in Fig. P12.17 are ideal. (a) What are the voltage gain, input resistance, and output resistance of the overall amplifier? (b) If the input voltage $v_I = 0.004$ V, what are the voltages at each of the eight nodes in the amplifier circuit?

12.18. The op amps in Fig. P12.17 are ideal. What are the nominal, minimum, and maximum values of the voltage gain, input resistance, and output resistance of the overall amplifier if all the resistors have 1 percent tolerances?

12.19. What are the gain and bandwidth of the individual stages in the amplifier in Fig. P12.2 if the op amps have an $A_o = 86$ dB and $f_T = 3$ MHz? (a) What

are the overall gain and bandwidth of the two-stage amplifier? (b) Repeat for the amplifier in Fig. P12.5. (c) Repeat for the amplifier in Fig. P12.6.

12.20. (a) What are the gain and bandwidth of the individual amplifier stages in Fig. P12.3 if the op amps have $A_o = 10^5$ and $f_T = 3$ MHz? (b) What are the overall gain and bandwidth of the three-stage amplifier?

12.21. What are the gain and bandwidth of the individual stages in the amplifier in Fig. P12.14 if the op amps have an $A_o = 86$ dB and $f_T = 5$ MHz? What are the overall gain and bandwidth of the three-stage amplifier?

12.22. What are the gain and bandwidth of the individual stages in the amplifier in Fig. P12.17 if the op amps have an $A_o = 80$ dB and $f_T = 5$ MHz? What are the overall gain and bandwidth of the three-stage amplifier?

12.23. The op amps in Fig. P12.23 are described by $A_o = 86$ dB, $R_{id} = 250$ kΩ, $R_O = 200$ Ω, and $f_T = 4$ MHz, and the power supplies are ± 15 V. (a) What are the voltage gain, input resistance, output resistance, and bandwidth of the overall amplifier? (b) Assume the offset voltage of each op amp is equivalent to $+10$ mV at the positive input of the op amp. If the input voltage $v_I = 0$ V, what are the voltages (three significant digits) at each of the 10 nodes in the amplifier circuit?

12.24. What are the nominal, minimum, and maximum values of the voltage gain, input resistance, output resistance, and bandwidth of the overall amplifier in Prob. 12.23 if the resistors all have 10 percent tolerances?

**12.25. A cascade amplifier is to be designed to meet these specifications:

$$A_v = 5000 \quad R_{in} \geq 10 \text{ M}\Omega \quad R_{out} \leq 0.1 \ \Omega$$

How many amplifier stages will be required if the stages must use an op amp below? Because of bandwidth requirements, assume that no individual stage can have a gain greater than 50.

Op amp specifications: $A = 85$ dB
$$R_{id} = 1 \text{ M}\Omega$$
$$R_o = 100 \ \Omega$$
$$R_{ic} \geq 1 \text{ G}\Omega$$

**12.26. Use these op amp parameters to design a multistage amplifier that meets the specifications below.

$$A_v = 86 \text{ dB} \pm 1 \text{ dB} \quad R_{in} \geq 10 \text{ k}\Omega$$
$$R_{out} \leq 0.01 \ \Omega \quad f_H \geq 75 \text{ kHz}$$

The amplifier should use the minimum number of op amp stages that will meet the requirements. (A spreadsheet or simple computer program will be helpful in finding the solution.)

Op amp specifications: $A_o = 10^5$
$$R_{id} = 10^9 \ \Omega$$
$$R_o = 50 \ \Omega$$
$$\text{GBW} = 1 \text{ MHz}$$

**12.27. (a) Design the amplifier in Prob. 12.26, including values for the feedback resistors in each stage. (b) What is the bandwidth of your amplifier if the op amps have $f_T = 5$ MHz?

12.28. Simulate the frequency response of the nominal design of the six-stage cascade amplifier from Table 12.6. Use the macro model in Fig. 11.51 to represent the op amp.

12.29. Simulate the frequency response of the six-stage cascade amplifier from Table 12.6 using the μA741 op amp macro model in SPICE.

12.30. Use the Monte Carlo analysis capability in PSPICE to simulate 1000 cases of the behavior of the six-stage amplifier in Table 12.6. Assume that all resistors and capacitors have 5 percent tolerances and the open-loop gain and bandwidth of the op amps each has a 50 percent tolerance. Use uniform statistical distributions. What are the lowest and highest observed values of gain and bandwidth for the amplifier?

12.31. A cascade amplifier is to be designed to meet these specifications:

$$A_v = 60 \text{ dB} \pm 1 \text{ dB} \quad R_{in} = 27 \text{ k}\Omega$$
$$R_{out} \leq 0.1 \ \Omega \quad \text{Bandwidth} = 20 \text{ kHz}$$

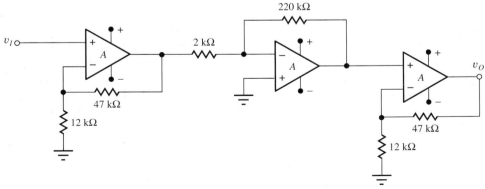

Figure P12.23

How many amplifier stages will be required if the stages must use these op amp specifications?

Op amp specifications: $A_o = 85$ dB
$f_T = 5$ MHz
$R_o = 100\ \Omega$
$R_{id} = 1\ M\Omega$
$R_{ic} \geq 1\ G\Omega$

12.32. Design the amplifier in Prob. 12.31, including values for the feedback resistors in each stage.

**12.33 (a) Perform a Monte Carlo analysis of the six-stage cascade amplifier design resulting from the example in Tables 12.4 and 12.5, and determine the fraction of the amplifiers that will not meet either the gain or bandwidth specifications. Assume the resistors are uniformly distributed between their limits.

$$A_v \geq 100 \text{ dB} \quad \text{and} \quad f_H \geq 50 \text{ kHz}$$

(b) What tolerance must be used to ensure that less than 0.1 percent of the amplifiers fail to meet both specifications?

The equation here can be used to estimate the location of the half-power frequency for N closely spaced poles, where $\overline{f_{H1}}$ is the average of the individual cutoff frequencies of the N stages and f_{H1}^i is the cutoff frequency of the ith individual stage.

$$f_H = \overline{f_{H1}}\sqrt{2^{1/N}-1}$$

where $\overline{f_{H1}} = \dfrac{1}{N}\sum_{i=1}^{N} f_{H1}^i$.

**12.34. (a) Show that the number of stages that optimizes the bandwidth of a cascade of identical noninverting amplifier stages having a total gain G is given by

$$N = \dfrac{\ln 2}{\ln\left[\dfrac{\ln G}{\ln G - \ln\sqrt{2}}\right]}$$

(b) Calculate N for the amplifier in Example 12.3.

12.2 Instrumentation Amplifier

12.35. What is the voltage gain of the instrumentation amplifier in Fig. 12.4 if $R_1 = 1.5$ kΩ, $R_2 = 75$ kΩ, $R_3 = 10$ kΩ, and $R_4 = 10$ kΩ. What is the output voltage if $v_1 = (2 + 0.1 \sin 2000\pi t)$ V and $v_2 = 2.1$ V?

12.36. What is the voltage gain of the instrumentation amplifier in Fig. 12.4 if $R_1 = 15$ kΩ, $R_2 = 75$ kΩ, $R_3 = 10$ kΩ, and $R_4 = 20$ kΩ. What is the output voltage if $v_1 = (4 - 0.2 \sin 4000\pi t)$ V and $v_2 = 3.5$ V?

12.37. What are the actual values of the two input resistances R_{in1} and R_{in2} and the output resistance R_{out} of the instrumentation amplifier in Fig. 12.4 if it is constructed using operational amplifiers with $A = 8 \times 10^4$, $R_{id} = 1\ M\Omega$, $R_{ic} = 800\ M\Omega$, and $R_o = 100\ \Omega$? Assume $R_1 = 2$ kΩ, $R_2 = 42$ kΩ, and $R_3 = R_4 = 10$ kΩ.

12.38. In the instrumentation amplifier in Fig. P12.38, $v_a = 5.02$ V and $v_b = 4.98$ V. Find the values of node voltages $v_1, v_2, v_3, v_4, v_5, v_6, v_o$, and currents $i_1, i_2,$ and i_3. What are the values of the common-mode gain, differential-mode gain, and CMRR of the amplifier? The op amps are ideal.

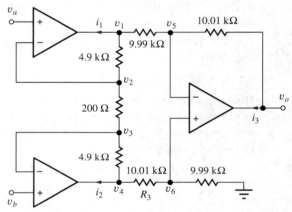

Figure P12.38

12.39. Find the values of $v_1, v_2, v_3, v_4, v_5, v_6, v_o, i_1, i_2,$ and i_3 in the instrumentation amplifier in Fig. P12.38 if $v_a = 3$ V and $v_b = 3$ V.

12.3 Active Filters

12.40. (a) Repeat Design Example 12.6 for a maximally flat second-order low-pass filter with $f_o = 25$ kHz, using the circuit in Fig. 12.5. Assume $C = 0.005\ \mu$F. What is the filter bandwidth? (b) Use frequency scaling to change f_o to 50 kHz.

*12.41. (a) Use MATLAB or other computer tool to make a Bode plot for the response of the filter in Prob. 12.40, assuming the op amp is ideal. (b) Use SPICE to simulate the characteristics of the filter in Prob. 12.40 using a 741 op amp. (c) Discuss any disagreement between the SPICE results and the ideal response.

12.42. Derive an expression for the input impedance of the filter in Fig. 12.5.

12.43. Use MATLAB or another computer tool to plot the input impedance of the low-pass filter in Fig. 12.5 versus frequency for $R_1 = R_2 = 2.26$ kΩ, $C_1 = 0.02$ μF, and $C_2 = 0.01$ μF.

*12.44. (a) What is the transfer function for the low-pass filter in Fig. P12.44? (b) What is S_K^Q for this filter if $R_1 = R_2$ and $C_1 = C_2$?

12.45. What are the expressions for $S_{R_1}^{\omega_o}$ and $S_{C_1}^{\omega_o}$ for the high-pass filter of Fig. 12.7?

12.46. What is $S_Q^{\omega_o}$ for the band-pass filter of Fig. 12.9 for $C_1 = C_2$? What is the value for $f_o = 12$ kHz and $Q = 10$?

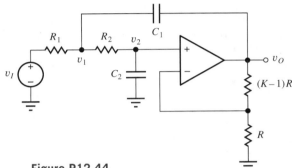

Figure P12.44

12.47. Design a maximally flat second-order low-pass filter with a bandwidth of 2 kHz using the circuit in Fig. 12.5.

12.48. Design a high-pass filter with a lower-half power frequency of 24 kHz and $Q = 1$ using the circuit in Fig. 12.7.

12.49. (a) Calculate f_o, Q, and the bandwidth for the band-pass filter in Fig. 12.9 if $R_{th} = 1$ kΩ, $R_2 = 200$ kΩ, and $C_1 = C_2 = 220$ pF. (b) Use magnitude scaling to change the element values so that $R_{th} = 3.3$ kΩ. (c) Use frequency scaling to double f_o for the filter in part (a).

12.50. (a) Design a band-pass filter with a center frequency of 600 Hz and $Q = 5$ using the circuit in Fig. 12.9 with $R_3 = \infty$. What is the filter bandwidth? (b) Use frequency scaling to change f_o to 2.25 kHz.

*12.51. (a) Use MATLAB or another computer tool to make a Bode plot for the response of the filter in Prob. 12.49(a), assuming the op amp is ideal. (b) Use SPICE to simulate the characteristics of the filter in Prob. 12.49(a) using a 741 op amp. (c) Discuss any disagreement between the SPICE results and the ideal response.

12.52. (a) Two identical band-pass filters having $\omega_o = 1$ and $Q = 3$ are designed using the circuit in Fig. 12.9 with $C_1 = C_2$ and $R_3 = \infty$. If the filters are cascaded, what are the center frequency, Q, and bandwidth of the overall filter? (b) Write the transfer function for the composite filter.

12.53. Use MATLAB or other computer tool to produce a Bode plot for the two-stage filter in Prob. 12.52.

*12.54. The first stage of a two-stage filter consists of a band-pass filter with $f_o = 5$ kHz and $Q = 5$. The second stage is also a band-pass filter, but it has $f_o = 6$ kHz and $Q = 5$. If the filters use Fig. 12.9 with $C_1 = C_2$ and $R_3 = \infty$, what are the center frequency, Q, and bandwidth of the overall filter?

12.55. Use MATLAB or another computer tool to produce a Bode plot for the two-stage filter in Prob. 12.54.

12.56. Find the expression for the loop gain of the filter in Fig. P12.44.

*12.57. Write an expression for the loop gain of the active low-pass filter in Fig. P12.57.

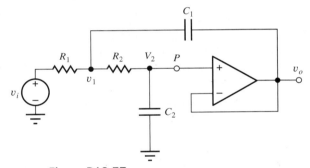

Figure P12.57

12.58. Write an expression for the loop gain of the active high-pass filter in Fig. P12.58.

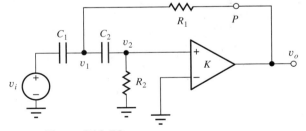

Figure P12.58

12.4 Switched-Capacitor Circuits

12.59. Draw a graph of the output voltage for the SC integrator in Fig. 12.13 for five clock periods if $C_1 = 4C_2$, $v_I = 1$ V, and $v_O(0) = 0$.

12.60. Draw a graph of the output voltage for the SC integrator in Fig. 12.15 for five clock periods if $C_1 = 4C_2$, $v_I = 1$ V, and $v_O(0) = 0$.

12.61. (a) Draw a graph of the output voltage for the SC integrator in Fig. 12.13 for five clock periods if $C_1 = 4C_2$ and v_I is the signal in Fig. P12.61. **(b) Repeat for the integrator in Fig. 12.15.

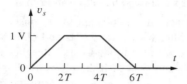

Figure P12.61

12.62. (a) What is the output voltage at the end of one clock cycle of the SC integrator in Fig. 12.11 if $C_1 = 1$ pF, $C_2 = 0.2$ pF, $v_I = 1$ V, and there is a stray capacitance $C_s = 0.1$ pF between each end of capacitor C_1 and ground? What are the gain and gain error of this circuit? (b) Repeat for the integrator in Fig. 12.15.

*12.63. (a) Simulate two clock cycles of the integrator of Prob. 12.62(a) using NMOS transistors with $W/L = 2/1$ and 100-kHz clock signals with rise and fall times of 0.5 μs. (b) Repeat for Prob. 12.59(b).

12.64. What are the center frequency, Q, and bandwidth of the switched capacitor band-pass filter in Fig. 12.17 if $C_1 = 0.4$ pF, $C_2 = 0.4$ pF, $C_3 = 1$ pF, $C_4 = 0.1$ pF, and $f_c = 100$ kHz?

12.65. Draw the circuit for a switched-capacitor implementation of the low-pass filter in Fig. 12.5.

12.5 Digital-to-Analog Conversion

12.66. Draw the transfer function, similar to Fig. 12.19, for a DAC with $V_{OS} = 0.5$ LSB and no gain error.

12.67. (a) What is the output voltage for the 4-bit DAC in Fig. P12.67, as shown with input data of 0110 if $V_{REF} = 2.56$ V? (b) Suppose the input data changes to 1001. What will be the new output voltage? (c) Make a table giving the output voltages for all 16 possible input data combinations.

12.68. The op amp in Fig. P12.67 has an offset voltage of +5 mV and the feedback resistor has a value of $1.05R$ instead of R. What are the offset and gain errors of this DAC?

12.69. Fill in the missing entries for step size, differential linearity error, and integral linearity error for the DAC in the accompanying table.

BINARY INPUT	DAC OUTPUT VOLTAGES	STEP SIZE (LSB)	DIFFERENTIAL LINEARITY ERROR (LSB)	INTEGRAL LINEARITY ERROR
000	0.0000			0.00
001	0.1000			
010	0.3000			
011	0.3500			
100	0.4750			
101	0.6300			
110	0.7250			
111	0.8750			0.00

12.70. Use Thévenin equivalent circuits for the R-2R ladder network in Fig. P12.70 to find the output voltage for the four input combinations 0001, 0010, 0100, and 1000 if $V_{REF} = 2.50$ V.

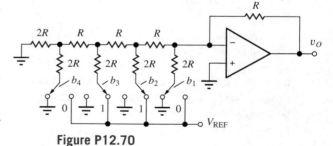

Figure P12.70

*12.71. The switches in Fig. P12.70 can be implemented using MOSFETs, as shown in Fig. P12.71. What must be the W/L ratios of the transistors if the on-resistance of the transistor is to be less than 1 percent of the resistor $2R = 12$ kΩ? Use $V_{REF} = 3.0$ V. Assume that the voltage applied to the gate of the MOSFET is 5 V when $b_1 = 1$ and 0 V when $b_1 = 0$. For the MOSFET, $V_{TN} = 1$ V, $K'_n = 50$ μA/V², $2\phi_F = 0.6$ V, and $\gamma = 0.5\sqrt{V}$.

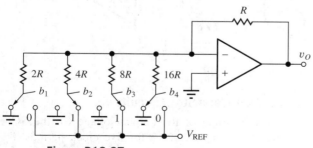

Figure P12.67

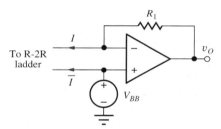

Figure P12.71

*12.72. A 3-bit weighted-resistor DAC similar to Fig. P12.67 is made using standard 5 percent resistors with $R = 1.2$ kΩ, $2R = 2.4$ kΩ, $4R = 4.8$ kΩ, and $8R = 9.1$ kΩ. (a) Tabulate the nominal output values of this converter in a manner similar to Table 12.8. What are the values of differential linearity and integral linearity errors for the nominal resistor values? (b) What are the worst-case values of linearity error that can occur with the 5 percent resistors? (*Note:* this converter has a gain error. You must recalculate the "ideal" step size. It is not 0.1250 V.)

**12.73. Perform a 200-case Monte Carlo analysis of the DAC in Prob. 12.67 and find the worst-case differential and integral linearity errors for the DAC. Use 5 percent resistor tolerances.

**12.74. The output voltage of a 3-bit weighted resistor DAC must have an error of no more than 5 percent of V_{REF} for any input combination. What can be the tolerances on the resistors R, $2R$, $4R$, and $8R$ if each resistor is allowed to contribute approximately the same error to the output voltage?

12.75. How many resistors are needed to realize an 11-bit weighted-resistor DAC? What is the ratio of the largest resistor to the smallest resistor?

*12.76. Tabulate the output voltages for the eight binary input words for the 3-bit DAC in Fig. P12.76, and find the differential and integral linearity errors if $V_{REF} = 5.00$ V, $R_{REF} = 250$ Ω, and $R = 1.2188$ kΩ.

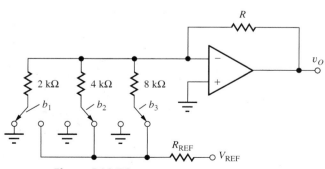

Figure P12.76

12.77. Suppose each switch in the DAC in Fig. P12.76 has an on-resistance of 200 Ω. (a) What value of R is required for zero gain error? (b) Find the differential and integral linearity errors if $V_{REF} = 5.00$ V, $R_{REF} = 0$ Ω. (c) Repeat for $R_{REF} = 250$ Ω.

**12.78. Perform a 200-case Monte Carlo analysis of the DAC in Prob. 12.76 and find the worst-case differential and integral linearity errors for the DAC. Use 10 percent resistor tolerances.

12.79. (a) Derive a formula for the total capacitance in an n-bit weighted-capacitor DAC. (b) In an n-bit C-2C DAC.

12.80. Perform a transient simulation of the C-2C DAC in Fig. 12.26(b) when b_3 switches from a 0 to a 1 and then back to a 0. Use $C = 0.5$ pF with $V_{REF} = 5$ V, and model the switches with NMOS transistors with $W/L = 10/1$, as in Fig. P12.71.

*12.81. A 3-bit inverted R-2R ladder with $R = 2.5$ kΩ and $V_{BB} = -2.5$ V is connected to the input of the op amp in Fig. P12.81. Draw the schematic of the complete D/A converter. Make a table of the output voltages versus input code if $R_1 = 5$ kΩ.

Figure P12.81

*12.82. Suppose a 10-bit DAC is built using the inherently monotonic circuit technique in Fig. 12.25. (a) If the resistor material has a sheet resistance of 50 Ω/square, and $R = 500$ Ω, estimate the number of squares that will be required for the resistor string. (b) If the minimum width of a resistor is 2.5 μm, what is the required length of the resistor string? Convert this length to inches.

12.6 Analog-to-Digital Conversion

12.83. A 14-bit ADC with $V_{FS} = 5.12$ V has an output code of 10101110110010. What is the possible range of input voltages?

12.84. A 20-bit ADC has $V_{FS} = 2$ V. (a) What is the value of the LSB? (b) What is the ADC output code for an input voltage of 1.630500 V? (c) What is the ADC output code for an input voltage of 0.997030 V?

12.85. Plot the transfer function and quantization error for an ideal 3-bit ADC that does not have the 0.5 LSB offset shown in Fig. 12.28. (That is, the first code transition occurs for $v_X = V_{FS}/8$.) Why is the design in Fig. 12.28 preferred?

12.86. A 12-bit counting converter with $V_{FS} = 5$ V and $f_c = 1$ MHz has an input voltage $V_X = 3.760$ V. (a) What is the output code? What is the conversion time T_T for this value of V_X if $f_c = 1$ MHz? (b) Repeat for $V_X = 4.333$ V.

12.87. A 10-bit counting converter with $V_{FS} = 5.12$ V uses a clock frequency of 1 MHz and has an input voltage $v_X(t) = 4\cos 5000\pi t$ V. What is the output code? What is the conversion time T_T for this input voltage?

12.88. (a) A 12-bit successive approximation ADC with $V_{FS} = 3.3$ V is designed using the circuit in Fig. 12.33. What is the maximum permissible offset voltage of the comparator if this offset error is to be less than 0.1 LSB? (b) Repeat for a 20-bit ADC.

12.89. A 16-bit successive approximation ADC is to be designed to operate at 50,000 conversions/second. What is the clock frequency? How rapidly must the unknown and reference voltage switches change state if the switch timing delay is to be equivalent to less than 0.1 LSB time?

*12.90. Figure P12.90 is the ramp generator for an integrating converter. (a) If the offset voltage of the op amp is 10 mV and $V = 3$ V, what is the effective reference voltage of this converter? (b) The integrator is used in a single-slope converter with an integration time of 1/30 s and a full-scale voltage of 5.12 V. What is the RC time constant? If $R = 50$ kΩ, what is the value of C?

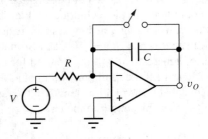

Figure P12.90

**12.91. A ramp generator using the op amp integrator circuit in Fig. P12.90 is built using an operational amplifier with an open-loop gain $A_o = 4 \times 10^4$. A 5-V step function is applied to the input of the integrator at $t = 0$. Write an expression for the output of the integrator in the time domain. What is the minimum RC product if the output ramp is to have an error of less than 1 mV at the end of a 200-ms integration interval?

*12.92. A 20-bit dual-ramp converter is to have an integration time $T_1 = 0.2$ s. How rapidly must the unknown and reference voltage switches change state if this timing uncertainty is to be equivalent to less than 0.1 LSB?

**12.93. Derive the transfer function for the integrator in the dual-ramp converter, and show that it has the functional form of $|\sin x/x|$.

12.94. Write formulas for the number of resistors and number of comparators that are required to implement an n-bit flash converter.

12.7 Oscillators

Frequency-Selective RC Networks

12.95. Derive an expression for the frequency of oscillation of the three-stage phase-shift oscillator in Fig. P12.95. What is the ratio R_2/R_1 required for oscillation?

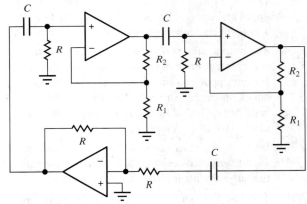

Figure P12.95

12.96. The circuit in Fig. P12.96 is called a quadrature oscillator. Derive an expression for its frequency of oscillation. What value of R_F is required for sinusoidal oscillation (in terms of R)?

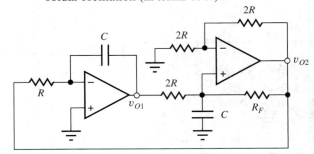

Figure P12.96

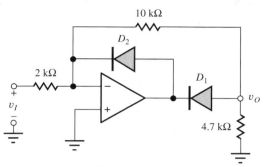
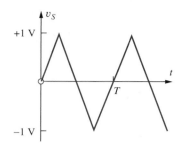

Figure P12.103

12.97. Calculate the frequency and amplitude of oscillation of the Wien-bridge oscillator in Fig. 12.48 if $R = 5.1$ kΩ, $C = 500$ pF, $R_1 = 10$ kΩ, $R_2 = 14$ kΩ, $R_3 = 6.8$ kΩ, and $R_4 = 10$ kΩ.

12.98. Use SPICE transient simulation to find the frequency and amplitude of the oscillator in Prob. 12.97. Start the simulation with a 1-V initial condition on the grounded capacitor C.

12.99. Calculate the frequency and amplitude of oscillation of the phase-shift oscillator in Fig. 12.50 if $R = 5$ kΩ, $C = 1000$ pF, $R_2 = 47$ kΩ, $R_3 = 15$ kΩ, and $R_4 = 68$ kΩ.

12.100. Use SPICE transient simulation to find the frequency and amplitude of the oscillator in Prob. 12.99. Start the simulation with a 1-V initial condition on the capacitor connected to the inverting input of the amplifier.

12.101. Four identical integrators from Fig. 10.34 are cascaded to form an oscillator. (a) Draw the circuit. (b) What is the frequency of oscillation if $R = 10$ kΩ, $C = 100$ pF, and the op amps are ideal? (c) If the output of the first integrator is $V_{o1} = 1\angle 0°$, what are the phasor representations of the other three output voltages? (d) Add an amplitude stabilization network to one of the amplifiers and design it to set the output voltage to approximately 2 V.

*12.102. (a) Repeat Prob. 12.101 if the op amps have an open-loop gain of 100 dB and a unity-gain frequency of 750 kHz.

12.8 Nonlinear Circuit Applications

Nonlinear Feedback

12.103. Draw the output voltage waveform for the circuit in Fig. P12.103 for the triangular input waveform shown. $T = 1$ ms.

*12.104. The signal v_I in Fig. P12.103 is used as the input voltage to the circuit in the figure in the EIA box on page 747. What will be the dc component of the voltage waveform at v_O if $R_1 = 2.7$ kΩ, $R_2 = 8.2$ kΩ, $R_3 = 10$ kΩ, $R_4 = 10$ kΩ, $C = 0.22$ μF, and $T = 1$ ms?

**12.105. What must be the cutoff frequency of the low-pass filter in the figure in the EIA box on page 747 if the rms value of the total ac component in the output voltage must be less than 1 percent of the dc voltage? Assume $R_1 = R_2$ and $v_I = -5\sin 120\pi t$ V.

12.106. The triangular waveform in Fig. P12.103 is applied to the circuit in Fig. P12.106. Draw the corresponding output waveform for $R_3 = R_2$.

12.107. The triangular waveform in Fig. P12.103 is applied to the circuit in Fig. P12.107. Draw the corresponding output waveform.

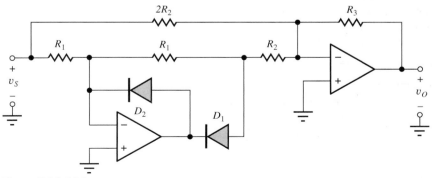

Figure P12.106

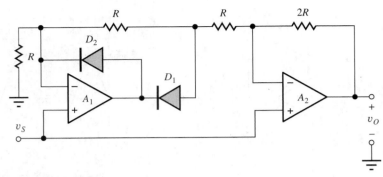

Figure P12.107

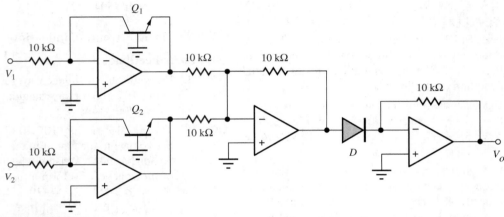

Figure P12.110

12.108. Simulate the circuit in Prob. 12.106 using $R_1 = 10$ kΩ, $R_2 = 10$ kΩ, $R_3 = 10$ kΩ, and $R_4 = 10$ kΩ. Use op amps with $A_o = 100$ dB.

12.109. Simulate the circuit in Prob. 12.107 for $R = 10$ kΩ. Use op amps with $A_o = 100$ dB.

*12.110. Write an expression for the output voltage in terms of the input voltage for the circuit in Fig. P12.110. Diode D is formed from a diode-connected transistor, and all three transistors are identical.

12.9 Circuits Using Positive Feedback

12.111. What are the values of the two switching thresholds and hysteresis in the Schmitt-trigger circuit in Fig. P12.111?

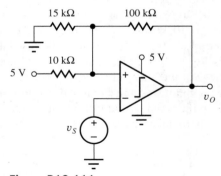

Figure P12.111

12.112. What are the switching thresholds and hysteresis for the Schmitt-trigger circuit in Fig. P12.112?

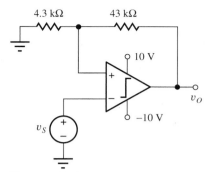

Figure P12.112

12.113. What are the switching thresholds and hysteresis for the Schmitt-trigger circuit in Fig. P12.113?

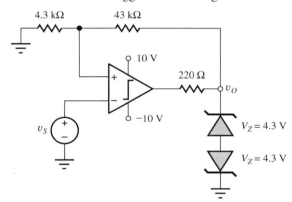

Figure P12.113

12.114. Design a Schmitt trigger to have its switching thresholds centered at 1 V with a hysteresis of ±0.05 V, using the circuit topology in Fig. P12.111.

12.115. What is the frequency of oscillation of the astable multivibrator in Fig. P12.115?

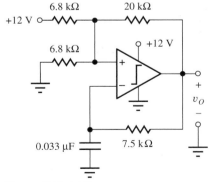

Figure P12.115

**12.116. Draw the waveforms for the astable multivibrator in Fig. P12.116. What is its frequency of oscillation? (Be careful — think before you calculate!)

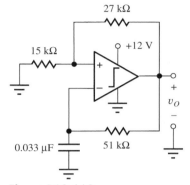

Figure P12.116

12.117. (a) Design an astable multivibrator to oscillate at a frequency of 1 kHz. Use the circuit in Fig. 12.60 with symmetric supplies of ±5 V. Assume that the total current from the op amp output must never exceed 1 mA. (b) If the resistors have ±5 percent tolerances and the capacitors have ±10 percent tolerances, what are the worst-case values of oscillation frequency? (c) If the power supplies are actually +4.75 and −5.25 V, what is the oscillation frequency for the nominal resistor and capacitor design values?

**12.118. The function generator circuit in the EIA on page 754 has been designed to generate a sine wave output voltage with an amplitude of 5 V at a frequency of 1 kHz. The low-pass filter has been designed to have a low-frequency gain of −1 and a cutoff frequency of 1.5 kHz. What are the magnitudes of the undesired frequency components in the output waveform at frequencies of 2, 3, and 5 kHz?

12.119. Two diodes are added to the circuit in Fig. P12.116 to convert it to a monostable multivibrator similar to the circuit in Fig. 12.62, and the power supplies are changed to ±7.5 V. What are the pulse width and recovery time of the monostable circuit?

12.120. Design a monostable multivibrator to have a pulse width of 20 μs and a recovery time of 5 μs. Use the circuit in Fig. 12.62 with ±5 V supplies.

术语对照

Active filter	有源滤波器
Amplitude stabilization	振幅稳定性
Amplitude stabilization in *RC* Oscillators	*RC*振荡器的振幅稳定电路
Analog function generators	模拟信号发生器
Analog-to-digital converter (ADC)	模/数转换器（ADC）
An inherently monotonic DAC	固有单调数/模转换器
Astable circuit	非稳态电路
Astable multivibrator	非稳态多谐振荡器
Band-pass filter	带通滤波器
Barkhausen criteria for oscillation	振荡器的巴克豪森准则
Bistable circuit	双稳态电路
Bistable multivibrator	双稳态多谐振荡器
Butterworth filter	巴特沃思滤波器
C-2C ladder DAC	C-2C梯形网络数/模转换器
Comparator	比较器
Conversion time	转换时间
Counter-ramp converter	计数斜坡转换器
Counting converter	计数转换器
Closed-loop amplifier	闭环放大器
Dc reference voltage	直流参考电压
Differential linearity error	差分线性误差
Digital-to-analog converter (DAC)	数/模转换器（DAC）
Differential subtractor	差分减法器
Dual-amp(dul-sape)ADC	双斜坡（双斜率）模/数转换器
Ounter-ramp converter	反斜坡转换器
Function generators	信号发生器
Flash converter	快闪转换器
Frequency scaling	频率缩放
Full-scale current	满度电流
Full-scale voltage	满度电压
High-pass filter	高通滤波器
Hysteresis	迟滞效应
Integral linearity error	积分线性误差
Instrumentation amplifier	两级级联放大器
Integral linearity error	积分线性误差
Integrator	积分器
Inverted R-2R ladder	反相R-2R梯形网络
Inverting amplifier	反相放大器
Inverting input	反相输入
Least significant bit(LSB)	最低有效位（LSB）
Instrumentation amplifier	仪表放大器
Linearity error	线性误差
Loop gain($A\beta$)	回路增益（$A\beta$）
Loop transmission(T)	传输回路（T）
Low-pass filter	低通滤波器
Magnitude scaling	幅度范围
Maximally flat magnitude	最大平坦性
−1 Point	−1点
Missing code	遗漏码
Monotonicity	单调性

Monostable circuit	单稳态电路
Monostable multivibrator	单稳态多谐振荡器
Monotonic converter	单调转换器
Most significant bit(MSB)	最高有效位(MSB)
Negative feedback	负反馈
Noninverting integrator	同相积分器
Nonmonotonic	非单调
Nonmonotonic converter	非单调转换器
Normal-mode rejection	常模抑制
Notch filter	陷波滤波器
One shot	单击器
Open-circuit voltage gain	开路电压增益
Open-loop amplifier	环路放大器
Open-loop gain	开环增益
Operational amplifier (op amp)	运算放大器(op amp)
Oscillator	振荡器
Oscillator circuits	振荡器电路
Offset current	失调电流
Offset error	偏移误差
Offset voltage	失调电压
Phase-shift oscillator	相移振荡器
Positive feedback	正反馈
Precision half-wave rectifier	精确半坡整流器
Quantization error	量化误差
R-2R ladder	R-2R梯形网络
RC oscillator	*RC*振荡器
Reference current	基准电流
Reference transistor	晶体管参考设计
Reference voltage	参考电压源
Regenerative feedback	再生反馈
Resolution of the converter	转换器的分辨率
Sample-and- hold circuit	采样保持电路
Schmitt trigger	施密特触发器
Sensitivity	灵敏度
Single shot	单击器
Single-ramp(single-slope)ADC	单斜坡ADC
Sinusoidal oscillator	正弦振荡器
Stray- sensitive circuits	杂散电容不敏感电路
Successive approximation converter	逐次逼近式转换器
Super diode	超级二极管
Superposition errors	叠加误差
Switched-capacitor integrator	开关电容积分器
Switched-capacitor filters	开关电容滤波器
Switched-capacitor D/A converters	开关电容D/A转换器
Switched-capacitor (SC) circuits	开关电容(SC)电路
Triggering	触发器
Triggering the monostable multivibrator	触发单稳态多谐振荡器
Two-phase nonoverlapping clock	双相非重叠时钟
Two- port model	二端口模型
The parallel (Flash) converter	并行(快闪)转换器
The Wien-bridge oscillator	维恩电桥振荡器
Weighted-capacitor DAC	权电容数/模转换器
Weighted-resistor DAC	权电阻数/模转换器
Wien-bridge oscillator	文氏桥振荡器

CHAPTER 13

DIFFERENTIAL AMPLIFIERS AND OPERATIONAL AMPLIFIER DESIGN

第13章　差分放大器和运算放大器设计

本章提纲
13.1　差分放大器
13.2　基本运算放大器的演进
13.3　输出级
13.4　电子电流源

本章目标
- 理解直流耦合多级放大器的分析和设计；
- 掌握差分放大器的直流和交流特性；
- 理解基本三级运算放大电路；
- 掌握A类、B类和AB类输出态的设计；
- 了解电子电流源的特性和设计。

本章导读
　　一般来说，单管放大器不能满足所有给定参数的设定要求，所需电压增益经常会超过单管放大器的放大系数，或者不能同时满足电压增益、输入电阻和输出电阻的组合要求。因此，单独使用单晶体管放大器不能同时满足这些指标，必须级联许多级以便创建能够满足这些所有要求的放大器。本章继续学习可实现更高性能的单管放大器级。

　　在多级放大器中，交流耦合和直流耦合方式都会用到，这取决于实际的应用。交流耦合允许每一级放大器的Q点设计可独立于其他级，而旁路电容可以用来消除放大器交流等效电路中的偏置元件。直流耦合可以去除部分电路元件，包括耦合电容器和偏置电阻，代表的是一种更加经济的电路设计方法。此外，直流耦合可得到一个低通放大器，能提供较高的直流增益，在大多数运算放大器设计中都会用到直流耦合。

　　最重要的直流耦合放大器是对称双晶体管差分放大器。差分放大器不仅是运算放大器设计中的关键电路，还是几乎所有模拟电路设计中的基本构建模块。本章详细研究了双极差分放大器和MOS差分放大器的晶体管级的实现。虽然差分放大器包含两个以对称方式放置的晶体管，但通常还是被看作单级放大器，其特性与共发射极放大器或共源极放大器的特性类似。差模增益、共模增益、共模抑制比（CMRR）、放大器的差模、共模输入和输出电阻都直接与晶体管的参数相关，从而与Q点设计相关。

　　集成电路（IC）工艺可以实现大量几乎相同的晶体管。尽管这些器件的绝对参数容限相对较差，但这些器件的实际特性偏差可控制在1%以内。这些大量的密切匹配的器件可以实现那些基于器件相似特征进行工作的特殊电路技术。匹配电路设计技术的使用贯穿了整个模拟电路的设计，可以用非常少的晶体管实现高性能电路。对于差分输入信号而言，差分放大器可以表现为反相放大器，也可以表现为同相放大器，但是会抑制两个输入端的共模信号。不过，只有当电路完全对称时，才可获得理想的差分放大器性能，最好的情况是采用集成电路工艺，这样晶体管特性可以做到近乎一致的匹配，只有在相同参数和特性的值都相等时才能称两个晶体管相匹配，即两个晶体管的参数（I_S, β_{FO}, V_A）或（K_n, V_{TN}, λ）Q点和温度都相同。

本章详细分析了双极型差分放大器和MOS管差分放大器的电路与性能，包括直流分析、传输特性、交流分析，分析了差模输入和单端输入的情况下的增益以及输入和输出电阻，评估了差分放大器的共模特性，研究了放大器对于抑制共模信号问题。对于差分放大器，给出了差分放大器的共模抑制比概念的定义，CMRR是抑制不希望出现的共模输入信号的能力，CMRR取决于设计者所选择的输出电压。为简化电路分析，本章进一步研究了通过半电路分析来实现差分放大器的简化分析，主要是利用差分放大器的对称性，将电路拆分为差模半电路，可以简化差分放大器电路的分析。一旦将电路画成对称的形式，就可利用两个基本的规则来构建电路，一个用于差模信号分析，另一个用于共模信号分析。

差分放大器的一个极为重要的应用就是作为运算放大器的输入级。运算放大器通常需要的电压增益比单级差分放大器所获得的要高很多，且多数运算放大器其增益通常至少有两级。为了获得更高的增益，本章给出了两级运算放大器的模型，将一个pnp共发射极放大器Q_3与由Q_1和Q_2组成的差分放大器的输出形成一个简单的两级运算放大器，详细研究了该放大器的直流分析、交流分析、输入输出电阻及CMRR分析等。本章研究了提高运算放大器性能的许多措施，采用附加电流源来改善电压增益，增加射极跟随器来降低输出电阻。在达灵顿电路中，两个dc耦合的npn或pnp晶体管可以实现更高的电流增益。

本章给出的运算放大器电路采用跟随器作为输出级，可以提供低输出电阻，同时又有相对较高的电流驱动能力。然而，放大器的输出级所消耗的功率很大，大约占总功率的2/3或者更多。

因为输出级需要经常给放大器的负载传递相对较大的功率，如此低的效率可能会导致放大器的高功耗。本章详细介绍了A类、B类以及AB类放大器的定义，分析了A类放大器的效率，然后引入了B类推挽输出级的概念。B类推挽输出级采用两个晶体管，每个晶体管仅在信号波形的一半导通，比A类的效率高得多。同时通过加入B类和AB类输出级进一步改善基本运放设计。在音频应用中，这些输出级常用于变压器耦合。

对于正弦信号而言，A类放大器的效率不会超过25%，而B类放大器的效率上限为78.5%。不过，B类放大器却受到因传输特性死区的影响而导致交越失真。AB类放大器以较小的静态功耗增加和效率消损消除了交越失真。选择合适的静态工作点，AB类放大器的效率可接近B类放大器的效率。通过用AB类输出级代替A类跟随输出级，可以大大改善基本运算电路的设计。在运算放大器中经常采用AB类输出级，且通常带有短路保护电路。

放大器还可以采用变压器耦合。变压器的阻抗转换特性可以用来简化那些需要驱动负载电阻值低的电路设计，如扬声器、耳机等。

直流电流源是非常有用的电流模块，本章举例介绍了多个电流源来为BJT和MOS运算放大器原型提供偏置，同时还能提高它们的性能。本章深入讨论了用于实现理想电流源的基本电子电路，通过研究适用于集成电路设计的专用技术对电流源设计进行更深入的探讨。

模拟电路的偏置一般由电流源提供。理想的电流源可以不依靠电源电压提供固定的电流输出，并具有无限大的输出电阻。电子电流源不能提供无限大的输出电阻，但可达到非常高的阻值。为实现高输入电阻，本章对多种基本电流电路进行介绍并进行了比较。

无论施加在电流源两端的电压如何变化，理想电流源可以提供恒定不变的输出电流，也就是说，电流源具有无穷大的输出电阻。尽管电子电流源无法达到无穷大的输出电阻，但是可以达到非常高的输出电阻值，多种基本电子电流源电路和技术可以用来实现高输出电阻。

对于一个电流源来说，电流源电流与输出电阻的乘积V_{CS}是一个用来进行电流源评价的品质因数。利用双极型晶体管可以建立一个单管电流源，其V_{CS}值可以接近BJT管$\beta_o V_A$的乘积值，品质很高的双极型晶体管，该乘积可以达到10 000V。对于FET，V_{CS}可达到$\mu_f V_{SS}$乘积值的很大部分，其中V_{SS}为电源电压。利用FET电流电源可以达到超过1000V的值。

在SPICE中可以用一个直流电流源和一个阻值等于电流输出值的电阻并联，来对电子电流源进行建模。为了达到最高的精确度，要对电流源的值进行调整，以满足输出电阻上流过的任何电流值。

Chapter Outline

13.1 Differential Amplifiers
13.2 Evolution to Basic Operational Amplifiers
13.3 Output Stages
13.4 Electronic Current Sources
Summary
Key Terms
References
Additional Reading
Problems

Chapter Goals

In this chapter, we learn to work with dc-coupled amplifiers that contain several interconnected stages, and important new amplifier concepts are introduced. Overall, we want to achieve these goals:

- Understand analysis and design of dc-coupled multistage amplifiers
- Explore the dc and ac properties of differential amplifiers
- Understand the basic three-stage operational amplifier circuit
- Explore the design of Class-A, Class-B, and Class AB output stages
- Discuss the characteristics and design of electronic current sources
- Learn how to analyze the effects of device and component mismatch on the performance of symmetrical amplifier circuits

In most situations, a single-transistor amplifier cannot meet all the given specifications. The required voltage gain often exceeds the amplification factor of a single transistor, or the combination of voltage gain, input resistance, and output resistance cannot be met simultaneously. For example, consider the specifications of a good general-purpose operational amplifier. Such an amplifier has an input resistance exceeding 1 MΩ, a voltage gain of 100,000, and an output resistance of less than 500 Ω. It should be clear from our investigation of amplifiers in Chapters 7 and 8 that these requirements cannot all be met simultaneously with a single-transistor amplifier. A number of stages must be cascaded in order to create an amplifier that can meet all these requirements.

Chapter 13 continues our study of combining single-transistor amplifier stages to achieve higher levels of overall performance. ac-coupled amplifiers discussed in Chapter 8 eliminate dc interactions between the various stages forming the amplifier, thus simplifying bias circuit design. On the other hand, in our work in Chapters 11 and 12, most of the operational amplifier circuits provided amplification of dc signals. To realize amplifiers of this type, coupling capacitors that block dc signal flow through the amplifier must be eliminated, which leads to the concept of direct-coupled or dc-coupled amplifiers that can satisfy the requirement for dc amplification. In the dc-coupled case, the operating point of one stage is dependent on the Q-point of the other stages, making the dc design somewhat more complex.

The most important dc-coupled amplifier is the symmetric two-transistor differential amplifier. Not only is the differential amplifier a key circuit in the design of operational amplifiers, it is also a fundamental building block in all analog IC design. In this chapter, we present the transistor-level implementation of BJT and FET differential amplifiers and explore how the differential-mode and common-mode gains, common-mode rejection ratio, differential-mode and common-mode input resistances, and output resistance of the amplifier are all related to transistor parameters.

Subsequently, a second gain stage and an output stage are added to the differential amplifier, creating the prototype for a basic operational amplifier. The definitions of class-A, class-B, and class-AB amplifiers are introduced, and the basic op amp design is further improved by adding class-B and class-AB output stages. In audio applications, these output stages often use transformer coupling.

Bias for analog circuits is most often provided by current sources. An ideal current source provides a fixed output current, independent of the voltage across the source; that is, the current source has an infinite output resistance. Electronic current sources cannot achieve infinite output resistance, but very high values are possible, and a number of

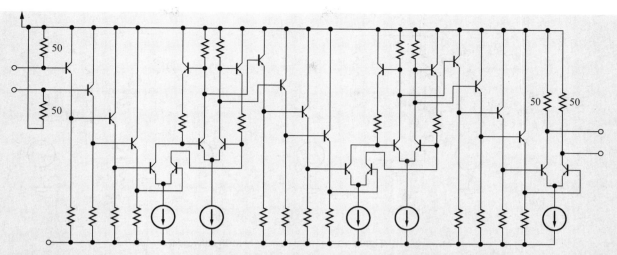

Schematic of a multistage dc-coupled amplifier in bipolar technology. Y. Baeyens et al., "InP D-HBT IC's for 40 Gb/s and higher bit rate lightwave transceivers," *IEEE J. Solid-State Circuits*, vol. 37, no. 9, September 2002, pp. 1152–1159. IEEE, 2002.

basic current source circuits and techniques for achieving high-output resistance are introduced and compared. Analysis of the various current sources uses the single-stage amplifier results from Chapters 7 and 8.

A circuit configuration is added to our circuit design tool box. In the Darlington circuit, two dc-coupled npn or pnp transistors create a much higher current gain composite replacement for a single npn or pnp device.

13.1 DIFFERENTIAL AMPLIFIERS

The coupling capacitors that were discussed in Chapter 8 limit the low-frequency response of the amplifiers and prevent their application as dc amplifiers. For an amplifier to provide gain at dc, capacitors in series with the signal path (e.g., C_1, C_3, C_5, and C_6 in Fig. 8.49) must be eliminated. Such an amplifier is called a **dc-coupled** or **direct-coupled amplifier.** Using a direct-coupled design can also eliminate additional resistors that are required to bias the individual stages in an ac-coupled amplifier, thus producing a less expensive amplifier.

The dc-coupled differential amplifier represents one of the most important additions to our "toolkit" of basic building blocks for analog design. Differential amplifiers appear in some form in almost every analog integrated circuit! This circuit forms the heart of operational amplifier design as well as of most dc-coupled analog circuits. Although the differential amplifier contains two transistors in a symmetrical configuration, it is usually thought of as a single-stage amplifier, and our analyses will show that it has characteristics similar to those of common-emitter or common-source amplifiers.

13.1.1 BIPOLAR AND MOS DIFFERENTIAL AMPLIFIERS

Figure 13.1 shows bipolar and MOS versions of the differential amplifier. Each circuit has two input terminals, v_1 and v_2, and the **differential-mode output voltage** v_{OD} is defined by the voltage difference between the collectors or drains of the two transistors. Ground-referenced outputs can also be taken between either collector or drain — v_{C1}, v_{C2}, or v_{D1}, v_{D2} — and ground.

The symmetrical nature of the amplifier provides useful dc and ac properties. We will find that the differential amplifier behaves as either an inverting or noninverting amplifier for differential input signals but tends to reject signals common to both inputs. However, ideal performance is obtained from the differential amplifier only when it is perfectly symmetrical, and the best versions are built using IC technology in which the transistor characteristics can be closely matched. Two transistors are

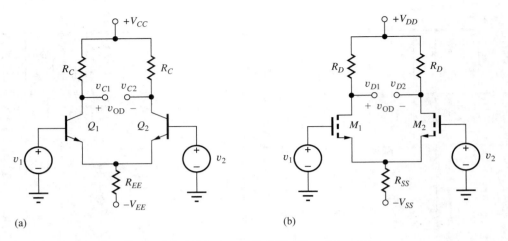

Figure 13.1 (a) Bipolar and (b) MOS differential amplifiers.

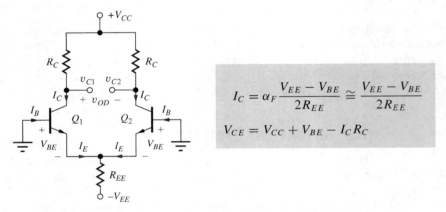

Figure 13.2 Circuit for dc analysis of the bipolar differential amplifier.

$$I_C = \alpha_F \frac{V_{EE} - V_{BE}}{2R_{EE}} \cong \frac{V_{EE} - V_{BE}}{2R_{EE}}$$

$$V_{CE} = V_{CC} + V_{BE} - I_C R_C$$

said to be **matched** if they have identical characteristics and parameter values; that is, the parameter sets (I_S, β_{FO}, V_A) or $(K_n, V_{TN}, \text{ and } \lambda)$, Q-points, and temperatures of the two transistors are identical.

13.1.2 dc ANALYSIS OF THE BIPOLAR DIFFERENTIAL AMPLIFIER

We begin our analysis of the differential amplifier by finding the transistor operating points. The quiescent operating points of the transistors in the bipolar differential amplifier can be found by setting both input signal voltages to zero, as in Fig. 13.2. In this circuit, both bases are grounded and the two emitters are connected together. Therefore, $V_{BE1} = V_{BE2} = V_{BE}$. If bipolar transistors Q_1 and Q_2 are assumed to be matched, then the terminal currents of the two transistors are identical; $I_{C1} = I_{C2} = I_C$, $I_{E1} = I_{E2} = I_E$, and $I_{B1} = I_{B2} = I_B$, and the symmetry of the circuit also forces $V_{C1} = V_{C2} = V_C$.

The emitter currents can be found by writing a loop equation starting at the base of Q_1,

$$V_{BE} + 2I_E R_{EE} - V_{EE} = 0 \quad \text{and} \quad I_C = \alpha_F I_E = \alpha_F \frac{V_{EE} - V_{BE}}{2R_{EE}} \qquad (13.1)$$

with $I_B = I_C/\beta_F$, and the voltages at the two collectors are equal to

$$V_{C1} = V_{C2} = V_{CC} - I_C R_C \qquad (13.2)$$

Also $V_{CE1} = V_{CE2} = V_{CC} + V_{BE} - I_C R_C$. For the symmetrical amplifier, the dc output voltage is zero:

$$V_{OD} = V_{C1} - V_{C2} = 0 \text{ V} \qquad (13.3)$$

EXAMPLE 13.1 DIFFERENTIAL AMPLIFIER Q-POINT ANALYSIS

In this example, we determine the Q-point for an "emitter-coupled pair" of bipolar transistors.

PROBLEM Find the Q-points plus V_C and I_B for the differential amplifier in Fig. 13.1(a) if $V_{CC} = V_{EE} = 15$ V, $R_{EE} = 75$ kΩ, $R_C = 75$ kΩ, and $\beta_F = 100$.

SOLUTION **Known Information and Given Data:** Circuit topology appears in Fig. 13.1(a); symmetrical 15-V power supplies are used to operate the circuit; $R_C = R_{EE} = 75$ kΩ; $\beta_F = 100$.

Unknowns: I_C, V_{CE}, V_C, I_B for Q_1 and Q_2

Approach: Use the circuit element values and follow the analysis presented in Eqs. (13.1) through (13.3).

Assumptions: Active region operation with $V_{BE} = 0.7$ V; $V_A = \infty$

Analysis: Using Eqs. (13.1) and (13.2):

$$I_E = \frac{V_{EE} - V_{BE}}{2R_{EE}} = \frac{(15 - 0.7) \text{ V}}{2(75 \times 10^3 \text{ }\Omega)} = 95.3 \text{ }\mu\text{A}$$

$$I_C = \alpha_F I_E = \frac{100}{101} I_E = 94.4 \text{ }\mu\text{A} \qquad I_B = \frac{I_C}{\beta_F} = \frac{94.4 \text{ }\mu\text{A}}{100} = 0.944 \text{ }\mu\text{A}$$

$$V_C = 15 - I_C R_C = 15 \text{ V} - (9.44 \times 10^{-5} \text{ A})(7.5 \times 10^4 \text{ }\Omega) = 7.92 \text{ V}$$

$$V_{CE} = V_C - V_E = 7.92 \text{ V} - (-0.7 \text{ V}) = 8.62 \text{ V}$$

Because of the circuit symmetry, both transistors in the differential amplifier are biased at a Q-point of (94.4 μA, 8.62 V) with $I_B = 0.944$ μA and $V_C = 7.92$ V.

Check of Results: A double check of results indicates the calculations are correct. Note that when R_C and R_{EE} are equal, the voltage drop across R_C should be approximately one half of the voltage across R_{EE}. Our calculations agree with this result. Also, $V_{CE} > V_{BE}$, so the assumption of forward-active region operation is correct.

Discussion: Note, that for $V_{EE} \gg V_{BE}$, I_E can be approximated by

$$I_E \cong \frac{V_{EE}}{2R_{EE}} = \frac{15 \text{ V}}{150 \text{ k}\Omega} = 100 \text{ }\mu\text{A}$$

This estimate represents only a 6 percent error compared to the more accurate calculation.

Computer-Aided Analysis: SPICE analysis with BF = 100 and IS = 5×10^{-16} A yields a Q-point of (94.6 μA, 8.57 V) with $V_{BE} = 0.672$ V. The collector voltage and base current values are 7.91 V and 0.946 μA, respectively, all in agreement with our hand calculations. We can also use SPICE to explore the effect of a nonzero Early voltage on the Q-point of the differential amplifier. A second simulation with VAF = 50 V yields Q-point values of (94.7 μA, 8.56 V). Almost no changes can be observed in the Q-point values! The collector voltage and base current values are now 7.90 V and 0.818 μA, respectively. Since the collector current has not changed,

V_C also has not changed. However, the base current has been reduced by 14 percent. We should wonder why this has occurred. Remember that the current gain of the transistor in our transport model is given by $\beta_F = \beta_{FO}(1 + V_{CE}/V_A)$. Also remember that a slightly different form is used in SPICE:

$$\beta_F = \beta_{FO}\left(1 + \frac{V_{CB}}{V_A}\right) = \beta_{FO}\left(1 + \frac{V_{CE} - V_{BE}}{V_A}\right) = 100\left(1 + \frac{7.90}{50}\right) = 116$$

and there is our discrepancy!

EXERCISE: What is the Q-point if β_F is 60 instead of 100?

ANSWER: (93.7 µA, 8.67 V)

EXERCISE: What are the actual values of I_C and V_{BE} for the transistor if the transistor saturation current is 0.5 fA? What value of I_S yields $V_{BE} = 0.7$ V if $V_T = 25.9$ mV?

ANSWERS: 0.649 V for $V_T = 25$ mV; 94.7 µA; 17.4 fA

EXERCISE: Draw a *pnp* version of the differential amplifier in Fig. 13.1(a).

ANSWER: See Fig. P13.15.

DESIGN NOTE

The voltage across the collector resistors will be approximately one half of that across R_{EE} when $R_C = R_{EE}$, since the current in R_{EE} splits in half.

13.1.3 TRANSFER CHARACTERISTIC FOR THE BIPOLAR DIFFERENTIAL AMPLIFIER

The differential amplifier provides advantages in terms of signal range and distortion characteristics over that of a single bipolar transistor. We can explore these advantages using results below in Eq. 13.4. For the symmetrical differential amplifier in Fig. 13.1 with a differential-mode input $v_1 = -v_2 = v_{id}/2$ as in Fig. 3.3, v_{BE1} and v_{BE2} become $v_{BE1} = V_{BE} + \frac{v_{id}}{2}$ and $v_{BE2} = V_{BE} - \frac{v_{id}}{2}$ and the individual collector currents are $i_{C1} = I_S\left[\exp\left(\frac{v_{BE1}}{V_T}\right) - 1\right]$ and $i_{C2} = I_S\left[\exp\left(\frac{v_{BE2}}{V_T}\right) - 1\right]$. With some effort, the collector current difference $(i_{C1} - i_{C2})$ and differential transconductance G_m are given by

$$\frac{i_{C1} - i_{C2}}{i_{C1} + i_{C2}} = \frac{\exp\left(\frac{v_{BE1}}{V_T}\right) - \exp\left(\frac{v_{BE2}}{V_T}\right)}{\exp\left(\frac{v_{BE1}}{V_T}\right) + \exp\left(\frac{v_{BE2}}{V_T}\right)} = \tanh\left(\frac{v_{BE1} - v_{BE2}}{2V_T}\right) \quad \text{and} \quad i_{C1} - i_{C2} = 2I_C \tanh\left(\frac{v_{id}}{2V_T}\right)$$

(13.4)

where $i_{C1} + i_{C2} = 2I_C$ and $G_m = \frac{d(i_{C1} - i_{C2})}{dv_{id}} = \frac{I_C}{V_T}\text{sech}^2\left(\frac{v_{id}}{2V_T}\right) = g_m \text{sech}^2\left(\frac{v_{id}}{2V_T}\right)$

Expanding the hyperbolic tangent using its Maclaurin series yields

$$I_{C1} - I_{C2} = 2I_C\left[\left(\frac{v_{id}}{2V_T}\right) - \frac{1}{3}\left(\frac{v_{id}}{2V_T}\right)^3 + \frac{2}{15}\left(\frac{v_{id}}{2V_T}\right)^5 - \frac{17}{315}\left(\frac{v_{id}}{2V_T}\right)^7 + \cdots\right]$$

(13.5)

First, we see that subtraction of the two collector currents eliminates the even order distortion terms in Eq. (13.5). Second, for small-signal operation, we desire the linear term to be dominant. Setting

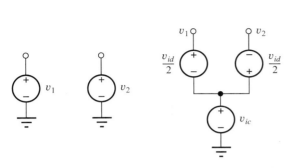

Figure 13.3 Definition of the differential-mode (v_{id}) and common-mode (v_{ic}) input voltages.

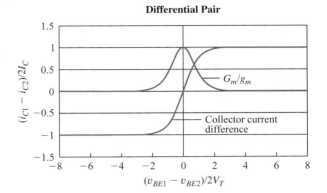

Figure 13.4 Large-signal transfer characteristic and transconductance for the bipolar differential pair.

the third-order term to be one tenth of the linear term requires $v_{id} \leq 2V_T\sqrt{0.3}$ or $v_{id} \leq 27$ mV. On the surface, one would expect an increase by a factor of 2 (to 10 mV) since the input signal is shared equally by the two transistors of the differential pair. However, cancellation of the even-order distortion terms further increases the signal-handling capability of the differential pair! This expanded linear region of the transfer function can clearly be seen at the center of the plot of Eq. (13.4) that appears in Fig. 13.4.

The transconductance (G_m) of the differential pair defined in Eq. (13.4) as the derivative of the transfer characteristic is also plotted in Fig. 13.4 as a function of the normalized input voltage. The value of G_m peaks at the transistor g_m when the pair is balanced with $i_{C1} = i_{C2}$ and falls to nearly zero for $|v_{id}| > 6V_T$ (150 mV). (i.e., the currents no longer change as v_{id} changes).

13.1.4 ac ANALYSIS OF THE BIPOLAR DIFFERENTIAL AMPLIFIER

Now that we have the Q-point information, we can proceed to use small-signal analysis to characterize the voltage gain and input and output resistances of the differential amplifier. The ac analysis of the differential amplifier can be simplified by breaking input sources v_1 and v_2 into their equivalent **differential-mode input (v_{id})** and **common-mode input (v_{ic})** signal components, shown in Fig. 13.3, and defined by

$$v_{id} = v_1 - v_2 \quad \text{and} \quad v_{ic} = \frac{v_1 + v_2}{2} \quad (13.6)$$

The total input voltages can be written in terms of v_{ic} and v_{id} as

$$v_1 = v_{ic} + \frac{v_{id}}{2} \quad \text{and} \quad v_2 = v_{ic} - \frac{v_{id}}{2} \quad (13.7)$$

The differential-mode input signal is the difference between inputs v_1 and v_2, whereas the common-mode input is the signal that is common to both inputs. Circuit analysis is performed using superposition of the differential-mode and common-mode input signal components. This technique was originally used in our study of operational amplifiers in Chapter 11.

The **differential-mode and common-mode output voltages**, v_{od} and v_{oc}, are defined in a similar manner:

$$v_{od} = v_{c1} - v_{c2} \quad \text{and} \quad v_{oc} = \frac{v_{c1} + v_{c2}}{2} \quad (13.8)$$

For the general amplifier case, voltages v_{od} and v_{oc} are functions of both v_{id} and v_{ic} and can be written as

$$\begin{bmatrix} v_{od} \\ v_{oc} \end{bmatrix} = \begin{bmatrix} A_{dd} & A_{cd} \\ A_{dc} & A_{cc} \end{bmatrix} = \begin{bmatrix} v_{id} \\ v_{ic} \end{bmatrix} \quad (13.9)$$

in which four gains are defined:

A_{dd} = **differential-mode gain**
A_{cd} = **common-mode** (to differential-mode) **conversion gain**
A_{cc} = **common-mode gain**
A_{dc} = **differential-mode** (to common-mode) **conversion gain**

For an ideal symmetrical amplifier with matched transistors, A_{cd} and A_{dc} are zero, and Eq. (13.9) reduces to

$$\begin{bmatrix} v_{od} \\ v_{oc} \end{bmatrix} = \begin{bmatrix} A_{dd} & 0 \\ 0 & A_{cc} \end{bmatrix} \begin{bmatrix} v_{id} \\ v_{ic} \end{bmatrix} \tag{13.10}$$

In this case, a differential-mode input signal produces a purely differential-mode output signal, and a common-mode input produces only a common-mode output.

However, when the differential amplifier is not completely balanced because of transistor or other circuit mismatches, A_{dc} or A_{cd} are no longer zero. In upcoming discussions, we assume that the transistors are identical unless stated otherwise.

> **EXERCISE:** Measurement of a differential amplifier yielded the following sets of values:
>
> $v_{od} = 2.2$ V and $v_{oc} = 1.002$ V for $v_1 = 1.01$ V and $v_2 = 0.990$ V
> $v_{od} = 0$ V and $v_{oc} = 5.001$ V for $v_1 = 4.995$ V and $v_2 = 5.005$ V
>
> What are v_{id} and v_{ic} for the two cases? What are the values of A_{dd}, A_{cd}, A_{dc}, and A_{cc} for the amplifier?
>
> **ANSWERS:** 0.02 V, 1.00 V; −0.01 V, 5.00 V; 100, 0.200, 0.0364, 1.00

Now we are in a position to fully characterize the signal properties of the differential amplifier. We want to find voltage gains A_{dd} and A_{cc}, and the input and output resistances of the amplifier. First, we will take a direct nodal analysis approach to the amplifier characterization. The results will subsequently lead us to a simplified analysis method called *half-circuit analysis* that is applicable to symmetric circuits.

13.1.5 DIFFERENTIAL-MODE GAIN AND INPUT AND OUTPUT RESISTANCES

Purely differential-mode input signals are applied to the differential amplifier in Fig. 13.5, and the two transistors are replaced with their small-signal models in Fig. 13.6. We will find the gain for both differential and single-ended outputs as well as the input and output resistances. Because the transistors have resistor loads, the output resistances will be neglected in the calculations.

Summing currents at the emitter node in Fig. 13.6:

$$g_\pi \mathbf{v}_3 + g_m \mathbf{v}_3 + g_m \mathbf{v}_4 + g_\pi \mathbf{v}_4 = G_{EE} \mathbf{v}_e \quad \text{or} \quad (g_m + g_\pi)(\mathbf{v}_3 + \mathbf{v}_4) = G_{EE} \mathbf{v}_e \tag{13.11}$$

These equations have been simplified by representing resistances r_π and R_{EE} with their equivalent conductances g_π and G_{EE}. The base-emitter voltages are

$$\mathbf{v}_3 = \frac{\mathbf{v}_{id}}{2} - \mathbf{v}_e \quad \text{and} \quad \mathbf{v}_4 = -\frac{\mathbf{v}_{id}}{2} - \mathbf{v}_e \tag{13.12}$$

giving $\mathbf{v}_3 + \mathbf{v}_4 = -2\mathbf{v}_e$. Combining Eq. (13.12) with Eq. (13.11) yields

$$\mathbf{v}_e(G_{EE} + 2g_\pi + 2g_m) = 0 \tag{13.13}$$

which requires $\mathbf{v}_e = 0$.

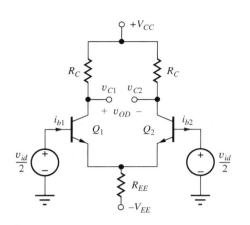
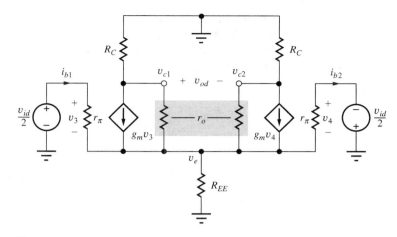

Figure 13.5 Differential amplifier with a differential-mode input signal.

Figure 13.6 Small-signal model for differential-mode inputs. The output resistances are neglected in the calculations.

For a purely differential-mode input voltage, the voltage at the emitter node is identically zero. This is an extremely important result. The "virtual ground" at the emitter node causes the differential amplifier to behave as a common-emitter (or common-source) amplifier.

DESIGN NOTE

The emitter node in the symmetrical differential amplifier represents a **virtual ground** for differential-mode input signals.

Because the voltage at the emitter node is zero, Eq. (13.12) yields

$$\mathbf{v_3} = \frac{\mathbf{v_{id}}}{2} \quad \text{and} \quad \mathbf{v_4} = -\frac{\mathbf{v_{id}}}{2} \tag{13.14}$$

and the output signal voltages are

$$\mathbf{v_{c1}} = -g_m R_C \frac{\mathbf{v_{id}}}{2} \qquad \mathbf{v_{c2}} = +g_m R_C \frac{\mathbf{v_{id}}}{2} \qquad \mathbf{v_{od}} = -g_m R_C \mathbf{v_{id}} \tag{13.15}$$

The differential-mode gain A_{dd} for a **balanced output**, $\mathbf{v_{od}} = \mathbf{v_{c1}} - \mathbf{v_{c2}}$, is

$$A_{dd} = \left. \frac{\mathbf{v_{od}}}{\mathbf{v_{id}}} \right|_{v_{ic}=0} = -g_m R_C \tag{13.16}$$

If either v_{c1} or v_{c2} alone is used as the output, referred to as a **single-ended** (or ground-referenced) **output,** then

$$A_{dd1} = \left. \frac{\mathbf{v_{c1}}}{\mathbf{v_{id}}} \right|_{v_{ic}=0} = -\frac{g_m R_C}{2} \quad \text{or} \quad A_{dd2} = \left. \frac{\mathbf{v_{c2}}}{\mathbf{v_{id}}} \right|_{v_{ic}=0} = +\frac{g_m R_C}{2} \tag{13.17}$$

depending on which output is selected.

The virtual ground condition at the emitter node causes the amplifier to behave as a single-stage common-emitter amplifier. The balanced differential output provides the full gain of a common-emitter stage, whereas the output at either collector provides a gain equal to one half that of the C-E stage.

The common-mode output voltage, defined by Eq. (13.8), is zero since $v_{c2} = -v_{c1}$, and therefore A_{dc} is indeed zero, as assumed in Eq. (13.10).

Differential-Mode Input Resistance

The **differential-mode input resistance** R_{id} represents the small-signal resistance presented to the full differential-mode input voltage appearing between the two bases of the transistors. R_{id} is defined as

$$R_{id} = \frac{v_{id}}{i_{b1}} = 2r_\pi \quad \text{because} \quad i_{b1} = \frac{\frac{v_{id}}{2}}{r_\pi} \tag{13.18}$$

If v_{id} is set to zero in Fig. 13.6, then $g_m v_3$ and $g_m v_4$ are zero, and the **differential-mode output resistance R_{od}** is equal to

$$R_{od} = 2(R_C \| r_o) \cong 2R_C \tag{13.19}$$

since node v_e represents a virtual ground. For single-ended outputs,

$$R_\text{out} \cong R_C \tag{13.20}$$

> **DESIGN NOTE**
>
> The differential pair behaves the same as a common-emitter or common-source amplifier for differential-mode input signals.

13.1.6 COMMON-MODE GAIN AND INPUT RESISTANCE

Next, we evaluate the common-mode characteristics of the differential amplifier and discover that it tends to reject common-mode input signals, a very useful property! Purely common-mode input signals are applied to the differential amplifier in Fig. 13.7. For this case, both sides of the amplifier are completely symmetrical. Thus, the two base currents, the emitter currents, the collector currents, and the two collector voltages must be equal. Using this symmetry as a basis, the output voltage can be developed by writing a loop equation including either base-emitter junction.

For the small-signal model in Fig. 13.8,

$$\mathbf{v_{ic}} = \mathbf{i_b} r_\pi + \mathbf{v_e} = \mathbf{i_b}[r_\pi + 2(\beta_o + 1)R_{EE}] \quad \text{and} \quad \mathbf{i_b} = \frac{\mathbf{v_{ic}}}{r_\pi + 2(\beta_o + 1)R_{EE}} \tag{13.21}$$

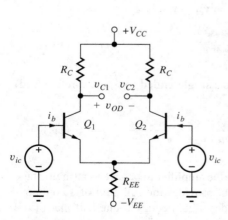

Figure 13.7 Differential amplifier with purely common-mode input.

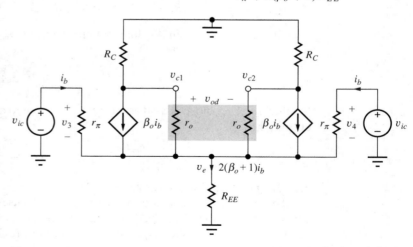

Figure 13.8 Small-signal model with common-mode input. The output resistances are neglected in the calculations. Note the change to the current-controlled form of the small-signal model.

The voltage at the emitter is

$$v_e = 2(\beta_o + 1)i_b R_{EE} = \frac{2(\beta_o + 1)R_{EE}}{r_\pi + 2(\beta_o + 1)R_{EE}} v_{ic} \cong v_{ic} \quad (13.22)$$

We recognize that Eq. (13.22) is identical to the gain of an emitter-follower with a resistance of $2R_{EE}$ in its emitter, and therefore emitter node voltage is approximately equal to the common-mode input signal. (Note that the circuit has been changed to the current-controlled form of the small-signal model, and the output resistances r_o are again neglected.)

The output voltage at either collector is given by

$$v_{C1} = v_{c2} = -\beta_o i_B R_C = -\frac{\beta_o R_C}{r_\pi + 2(\beta_o + 1)R_{EE}} v_{ic} \quad (13.23)$$

The common-mode output voltage v_{oc} is defined by Eq. (13.8), and the common-mode gain A_{cc} is given by

$$A_{cc} = \left. \frac{v_{oc}}{v_{ic}} \right|_{v_{id}=0} = -\frac{\beta_o R_C}{r_\pi + 2(\beta_o + 1)R_{EE}} \cong -\frac{R_C}{2R_{EE}} \quad (13.24)$$

for large β_o. Equation (13.24) is identical to the gain of an inverting amplifier with a resistance of $2R_{EE}$ in its emitter and a collector load resistance R_C. By multiplying and dividing Eq. (13.24) by collector current I_C, Eq. (13.24) can be rewritten as

$$A_{cc} = -\frac{I_C R_C}{2I_C R_{EE}} \cong \frac{\frac{V_{CC}}{2}}{2I_E R_{EE}} = \frac{V_{CC}}{2(V_{EE} - V_{BE})} \cong \frac{V_{CC}}{2V_{EE}} \quad (13.25)$$

where it is assumed that $\alpha_F = \alpha_o$ and $I_C R_C = V_{CC}/2$. In Eq. (13.25), we see that the common-mode gain A_{cc} is determined by the ratio of the two power supplies, and for symmetrical supplies, $A_{cc} = 0.5$. Note that the result in Eq. (13.25) only applies to the differential amplifier biased by resistor R_{EE}. We will shortly improve this result by replacing R_{EE} with an electronic current source.

Differential output voltage v_{od} is identically zero because the voltages at the two collectors are equal: $v_{od} = v_{c1} - v_{c2} = 0$. Therefore, the common-mode conversion gain for a differential output is also 0, as assumed in Eq. (13.10):

$$A_{cd} = \left. \frac{v_{od}}{v_{ic}} \right|_{v_{id}=0} = 0 \quad (13.26)$$

The result in Eq. (13.24) indicates that the common-mode output voltage and A_{cc} tend toward zero as R_{EE} approaches infinity. This is another suspicious result, and it is in fact a direct consequence of neglecting the output resistances in the circuit in Fig. 13.8. If r_o is included, a small current $v_{ic}/\beta_o r_o$ results from the finite current gain of the BJT and appears in the collector terminal. A more accurate expression for the common-mode gain is

$$A_{cc} \cong R_C \left(\frac{1}{\beta_o r_o} - \frac{1}{2R_{EE}} \right) \quad (13.27)$$

Now for infinite R_{EE}, we find that A_{cc} is limited to $R_C/\beta_o r_o \cong V_{CC}/2\beta_o V_A$. It is also interesting to note that the sign difference allows a theoretical cancellation to occur. (See Prob. 13.140.)

Common-Mode Input Resistance

The **common-mode input resistance** is determined by the total signal current $(2i_b)$ being supplied from the common-mode source and can be calculated using Eq. (13.21):

$$R_{ic} = \frac{v_{ic}}{2i_b} = \frac{r_\pi + 2(\beta_o + 1)R_{EE}}{2} = \frac{r_\pi}{2} + (\beta_o + 1)R_{EE} \quad (13.28)$$

As mentioned above, equations (13.21), (13.22), (13.23), and the numerator of Eq. (13.28) should be recognized as those of a common-emitter amplifier with a resistor of value $2R_{EE}$ in the emitter. This observation is discussed in detail shortly.

DESIGN NOTE

The characteristics of the differential pair with a common-mode input are similar to those of a common-emitter (common-source) amplifier with a large emitter (source) resistor.

13.1.7 COMMON-MODE REJECTION RATIO (CMRR)

As defined in Chapter 11, the **common-mode rejection ratio**, or **CMRR**, characterizes the ability of an amplifier to amplify the desired differential-mode input signal and reject the undesired common-mode input signal. For a general differential amplifier stage characterized by Eq. (11.110), CMRR is defined in Eq. (11.111) as

$$\text{CMRR} = \left| \frac{A_{dm}}{A_{cm}} \right| \qquad (13.29)$$

where A_{dm} and A_{cm} are the overall differential-mode and common-mode gains.[1]

For the differential amplifier, CMRR is dependent on the designer's choice of output voltage. For a differential output v_{od}, the common-mode gain of the balanced amplifier is zero, and the CMRR is infinite. However, if the output is taken from either collector, we have

$$\mathbf{v}_{c1} = \mathbf{v}_{oc} + \frac{\mathbf{v}_{od}}{2} = A_{cc}\mathbf{v}_{ic} + \frac{A_{dd}}{2}\mathbf{v}_{id} \quad \text{and} \quad \mathbf{v}_{c2} = \mathbf{v}_{oc} - \frac{\mathbf{v}_{od}}{2} = A_{cc}\mathbf{v}_{ic} - \frac{A_{dd}}{2}\mathbf{v}_{id} \qquad (13.30)$$

using Eqs. (13.8) and (13.10). Based upon Eqs. (13.29) with (13.17) and (13.27), the CMRR is given by

$$\text{CMMR} = \left| \frac{A_{dm}}{A_{cm}} \right| = \left| \frac{\frac{A_{dd}}{2}}{A_{cc}} \right| = \left| \frac{\frac{g_m R_c}{2}}{R_c \left(\frac{1}{\beta_o r_o} - \frac{1}{2R_{EE}} \right)} \right| = \left| \frac{1}{2 \left(\frac{1}{\beta_o \mu_f} - \frac{1}{2g_m R_{EE}} \right)} \right| \qquad (13.31)$$

For infinite R_{EE}, $\text{CMMR} \cong \beta_o \mu_f / 2$ and is limited by the $\beta_o \mu_f$ product of the transistor. On the other hand, if the term containing R_{EE} is dominant, we find the commonly quoted result:

$$\text{CMRR} \cong g_m R_{EE} \qquad (13.32)$$

Let us explore Eq. (13.32) a bit further by writing g_m in terms of the collector current.

$$\text{CMRR} = 40 I_C R_{EE} = 20(2 I_E R_{EE}) = 20(V_{EE} - V_{BE}) \cong 20 V_{EE} \qquad (13.33)$$

For the differential amplifier biased by resistor R_{EE}, CMRR is limited by the available negative power supply voltage V_{EE}. Also observe that the differential-mode gain is determined by the positive power supply voltage, that is $A_{dd} = -20 V_{CC}$ based on our design guide from Chapter 7 with $I_C R_C = V_{CC}/2$.

---

[1] A_{dm} and A_{cm} represent the differential-mode and common-mode gains of a general amplifier such as an op amp, whereas A_{dd}, A_{cc}, A_{dc}, and A_{cd} denote the characteristics of the differential amplifier stage by itself.

EXERCISE: Estimate the differential-mode gain, common-mode gain, and CMRR for a differential amplifier with V_{EE} and $V_{CC} = 15$ V if the differential output is used and if the output v_{C2} is used.

ANSWERS: $-300, 0, \infty$; $+150, -0.5, 49.5$ dB (a poor CMRR)

Effects of Mismatches

Although the CMRR for an ideal differential amplifier with differential output is infinite, an actual amplifier will not be perfectly symmetrical because of mismatches in the transistors, and the two conversion gains A_{cd} and A_{dc} will not be zero. For this case, many of the errors will be proportional to the result in Eq. (13.32) and will be of the form [1]:

$$\text{CMRR} \propto g_m R_{EE} \left(\frac{\Delta g}{g} \right) \qquad (13.34)$$

in which the $\Delta g/g = 2(g_1 - g_2)/(g_1 + g_2)$ factor represents the fractional mismatch between the small-signal device parameters on the two sides of the differential amplifier (see Probs. 13.21 and 13.23). Therefore, maximizing the $g_m R_{EE}$ product is equally important to improving the performance of differential amplifiers with differential outputs.

13.1.8 ANALYSIS USING DIFFERENTIAL—AND COMMON-MODE HALF-CIRCUITS

We noted that the differential amplifier behaves much as the single-transistor common-emitter amplifier. The analogy can be carried even further using the **half-circuit** method of analysis, in which the symmetry of the differential amplifier is used to simplify the circuit analysis by splitting the circuit into **differential-mode** and **common-mode half-circuits.**

Half-circuits are constructed by first drawing the differential amplifier in a fully symmetric form, as in Fig. 13.9. To achieve full symmetry, the power supplies have been split into two equal value sources in parallel, and the emitter resistor R_{EE} has been separated into two equal parallel resistors, each of value $2R_{EE}$. It is important to recognize from Fig. 13.9 that these modifications have not changed any of the currents or voltages in the circuit.

Once the circuit is drawn in symmetrical form, two basic rules are used to construct the half-circuits: one for differential-mode signal analysis and one for common-mode signal analysis:

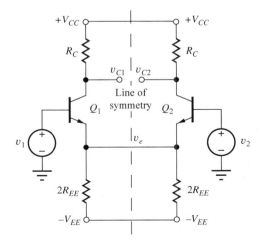

Figure 13.9 Circuit emphasizing symmetry of the differential amplifier.

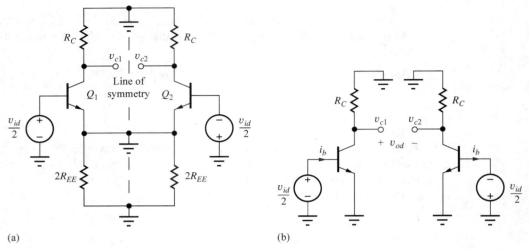

Figure 13.10 (a) ac Grounds for differential-mode inputs. (b) Differential-mode half-circuits.

DESIGN NOTE — RULES FOR CONSTRUCTING HALF-CIRCUITS

Differential-mode signals Points on the line of symmetry represent virtual grounds and can be connected to ground for ac analysis. (For example, remember that we found that $v_e = 0$ for differential-mode signals.)

Common-mode signals Points on the line of symmetry can be replaced by open circuits. (No current flows through these connections.)

Differential-Mode Half-Circuits

Applying the first rule to the circuit in Fig. 13.9 for differential-mode signals yields the circuit in Fig. 13.10(a). The two power supply lines and the emitter node all become ac grounds. (Of course, the power supply lines would become ac grounds in any case.) Simplifying the circuit yields the two differential-mode half-circuits in Fig. 13.10(b), each of which represents a common-emitter amplifier stage. The differential-mode behavior of the circuit, as described by Eqs. (13.15) to (13.20), can easily be found by direct analysis of the half-circuits:

$$\mathbf{v_{c1}} = -g_m R_C \frac{\mathbf{v_{id}}}{2} \qquad \mathbf{v_{c2}} = +g_m R_C \frac{\mathbf{v_{id}}}{2} \qquad \mathbf{v_o} = \mathbf{v_{c1}} - \mathbf{v_{c2}} = -g_m R_C \mathbf{v_{id}} = -A_{dd}\mathbf{v_{id}} \quad (13.35)$$

with

$$R_{id} = \frac{\mathbf{v_{id}}}{\mathbf{i_b}} = 2r_\pi \qquad \text{and} \qquad R_{od} = 2(R_C \| r_o) \tag{13.36}$$

Common-Mode Half-Circuits

If the second rule is applied to the circuit in Fig. 13.9, all points on the line of symmetry become open circuits, and we obtain the circuit in Fig. 13.11. The common-mode half-circuits obtained from Fig. 13.11 are redrawn in Fig. 13.12. The dc circuit with V_{IC} set to zero in Fig. 13.12(a) is used to find the Q-point of the amplifier. The circuit in Fig. 13.12(b) should be used to find the operating point when a dc common-mode input is applied, and the ac circuit of Fig. 13.12(c) is used for common-mode signal analysis.

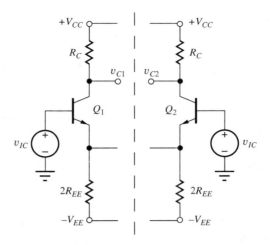

Figure 13.11 Construction of the common-mode half-circuit.

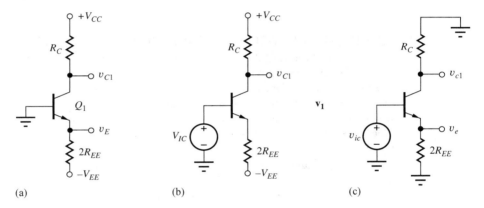

Figure 13.12 Common-mode half-circuits for (a) Q-point analysis; (b) dc common-mode input; (c) common-mode signal analysis.

The common-mode half-circuit in Fig. 13.12(c) simply represents the common-emitter amplifier with an emitter resistor $2R_{EE}$, which was studied in great detail in Chapter 8. In addition, Eqs. (13.24) and (13.28) could have been written down directly using the results of our analysis from Chapter 8.

We can see that use of the differential-mode and common-mode half-circuits can greatly simplify the analysis of symmetric circuits. Half-circuit techniques are used shortly to analyze the MOS differential amplifier from Fig. 13.1.

Common-Mode Input Voltage Range
Common-mode input voltage range is another important consideration in the design of differential amplifiers. The upper limit to the dc common-mode input voltage V_{IC} in the circuit in Fig. 13.12(b) is set by the requirement that Q_1 remains in the forward-active region of operation. Writing an expression for the collector-base voltage of Q_1,

$$V_{CB} = V_{CC} - I_C R_C - V_{IC} \geq 0 \quad \text{or} \quad V_{IC} \leq V_{CC} - I_C R_C \qquad (13.37)$$

in which

$$I_C = \alpha_F \frac{V_{IC} - V_{BE} + V_{EE}}{2R_{EE}} \qquad (13.38)$$

Solving the preceding two equations for V_{IC} yields

$$V_{IC} \leq V_{CC} \frac{1 - \alpha_F \dfrac{R_C}{2R_{EE}} \dfrac{(V_{EE} - V_{BE})}{V_{CC}}}{1 + \alpha_F \dfrac{R_C}{2R_{EE}}} \tag{13.39}$$

For symmetrical power supplies, $V_{EE} \gg V_{BE}$, and with $R_C = R_{EE}$, Eq. (13.39) yields $V_{IC} \leq V_{CC}/3$.

Note from Eq. (13.38) that I_C changes as V_{IC} changes, and the upper limit on V_{IC} is set by Eq. (13.39). As V_{IC} goes negative, the collector current reduces since $I_C \cong (V_{IC} - V_{BE} + V_{EE})/2R_{EE}$. The lower bound on V_{IC} is set by what reduction in bias current is deemed acceptable, and would probably be specified to be symmetrical, that is, $-V_{CC}/3 \leq V_{IC} \leq V_{CC}/3$.

> **EXERCISE:** Find the positive common-mode input voltage range for the differential amplifier in Fig. 13.7 if $V_{CC} = V_{EE} = 15$ V and $R_C = R_{EE}$.
>
> **ANSWER:** $\cong 5.30$ V

13.1.9 BIASING WITH ELECTRONIC CURRENT SOURCES

From Eqs. (13.1) and (13.2), we see that the Q-point of the differential amplifier is directly dependent on the value of the negative power supply, and from Eq. (13.31) we see that R_{EE} limits the CMRR. The Q-point is also dependent upon the dc common-mode voltage V_{IC} (see Eq. 13.38). In order to remove these limitations, most differential amplifiers are biased using electronic current sources, which both stabilize the operating point of the amplifier and increase the effective value of R_{EE}. Electronic current source biasing of both the BJT and MOSFET differential amplifiers is shown in Fig. 13.13. In these circuits, the current source replaces resistor R_{EE} or R_{SS}.

The rectangular symbols in Figs. 13.13 and 13.14 denote an electronic current source with a finite output resistance, as shown graphically in the i-v characteristic in Fig. 13.15. The electronic source has a Q-point current equal to I_{SS} and an output resistance equal to R_{SS}.

For hand analysis using the dc equivalent circuit, we will replace the electronic current source with a dc current source of value I_{SS}. For ac analysis, the ac equivalent circuit is constructed by replacing the source with its output resistance R_{SS}. These substitutions are depicted symbolically in Fig. 13.14.

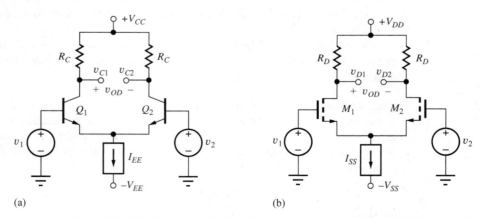

Figure 13.13 Differential amplifiers employing electronic current source bias.

13.1 Differential Amplifiers

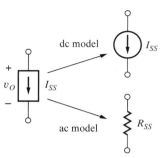
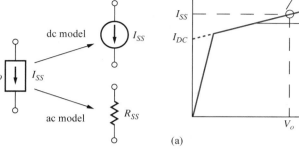
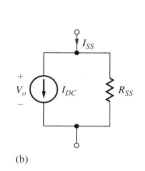

Figure 13.14 Electronic current source and models.

Figure 13.15 (a) i-v Characteristic for an electronic current source. (b) Proper SPICE representation of the electronic current source.

DESIGN NOTE

High common-mode rejection in differential amplifiers requires a large value of output resistance R_{SS} in the current source I_{SS} used for bias.

13.1.10 MODELING THE ELECTRONIC CURRENT SOURCE IN SPICE

Proper modeling of the electronic current source is slightly different in SPICE since the program creates its own dc and ac equivalent circuits. In order for SPICE to properly calculate the dc and ac behavior of a circuit, the network must contain both the dc current source and its output resistance R_{SS}. In the full circuit model in Fig. 13.15(b), a dc current will exist in the resistance R_{SS}, and the value of the current source in the SPICE circuit must be set to the value I_{DC} indicated in Fig. 13.15. I_{DC} represents the current in the equivalent circuit when voltage $V_o = 0$ and can be expressed as

$$I_{DC} = I_{SS} - \frac{V_o}{R_{SS}} \qquad (13.40)$$

The equivalent circuit to be used in SPICE appears in Fig. 13.15(b). In cases where R_{SS} is very large, I_{DC} is approximately equal to I_{SS}.

EXERCISE: Suppose an electronic current source has a current $I_{SS} = 100$ μA with an output resistance $R_{SS} = 750$ kΩ. (These values are representative of a single transistor current source operating at this current.) If $V_o = 15$ V, what is the value of I_{DC}?

ANSWER: 80 μA

13.1.11 dc ANALYSIS OF THE MOSFET DIFFERENTIAL AMPLIFIER

MOSFETS provide very high-input resistance and are often used in differential amplifiers implemented in CMOS and BiFET[2] technologies. In addition to high-input resistance, op amps with FET inputs typically have a much higher slew rate than those with bipolar input stages.

The MOS version of the differential amplifier circuit appears in Fig. 13.13(b). We will use the MOSFET differential amplifier as our first direct application of half-circuit analysis. For dc analysis

---

[2] BiFET technologies contain JFETs as well as bipolar transistors.

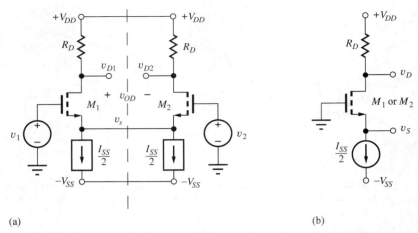

Figure 13.16 (a) Symmetric circuit representation of the MOS differential amplifier. (b) Half-circuit for dc analysis.

using half-circuits, the amplifier is redrawn in symmetrical form in Fig. 13.16(a). If the connections on the line of symmetry are replaced with open circuits, and the two input voltages are set to zero, we obtain in Fig. 13.16(b) the half-circuit needed for dc analysis.

It is immediately obvious from the dc half-circuit that the current in the source of the NMOS transistor must be equal to one half of the bias current I_{SS}:

$$I_S = \frac{I_{SS}}{2} \tag{13.41}$$

The gate-source voltage of the MOSFET can be determined directly from the drain-current expression for the transistor:

$$I_D = \frac{K_n}{2}(V_{GS} - V_{TN})^2 \quad \text{or} \quad V_{GS} = V_{TN} + \sqrt{\frac{2I_D}{K_n}} = V_{TN} + \sqrt{\frac{I_{SS}}{K_n}} \tag{13.42}$$

Note that $V_S = -V_{GS}$. The voltages at both MOSFET drains are

$$V_{D1} = V_{D2} = V_{DD} - I_D R_D \quad \text{and} \quad V_O = 0 \tag{13.43}$$

Thus, the drain-source voltages are

$$V_{DS} = V_{DD} - I_D R_D + V_{GS} \tag{13.44}$$

Analysis Using the Unified MOSFET Model

The Unified Model of Sect. 5.10.7 replaces $(V_{GS} - V_{TN})$ in the square law model with $V_{MIN} = \min\{(V_{GS} - V_{TN}), V_{DS}, V_{SAT}\}$. For values of internal gate drive $(V_{GS} - V_{TN})$ and V_{DS} exceeding the value of V_{SAT}, the drain current and transconductance become

$$i_D = K_n\left(v_{GS} - V_{TN} - \frac{V_{SAT}}{2}\right)V_{SAT} \quad \text{and} \quad g_m = K_n V_{SAT} \tag{13.45}$$

For a given level of drain current, the value of V_{GS} will change in order to support the current.

$$V_{GS} = V_{TN} + \frac{V_{SAT}}{2} + \frac{I_D}{K_n V_{SAT}} \tag{13.46}$$

The expressions for the drain and drain-source voltages remain the same.

EXAMPLE 13.2   MOSFET DIFFERENTIAL AMPLIFIER ANALYSIS

A dc Q-point analysis is provided for the MOSFET differential amplifier in this example.

PROBLEM (a) Find the Q-points for the MOSFETs in the differential amplifier in Fig. 13.13(b) if $V_{DD} = V_{SS} = 12$ V, $I_{SS} = 1$ mA, $R_{SS} = 100$ kΩ, $R_D = 13$ kΩ, $K_n = 500$ μA/V^2, $\lambda = 0.0133$ V^{-1}, and $V_{TN} = 0.7$ V. What is the maximum V_{IC} for which M_1 remains in the active region? (b) Repeat the analysis using the Unified Model from Sec. 4.7.7 with $V_{SAT} = 1$ V.

SOLUTION **Known Information and Given Data:** Circuit topology appears in Fig. 13.13(b); symmetrical 12-V power supplies are used to operate the circuit; $I_{SS} = 1000$ μA, $R_D = 13$ kΩ, $V_{TN} = 0.7$ V, and $K_n = 500$ μA/V^2

Unknowns: I_D, V_{DS}, for M_1 and M_2, and maximum dc common-mode input voltage V_{IC}

Approach: Use the circuit element values and follow the analysis presented in Eq. (13.41) through (13.44).

Assumptions: Active region operation; ignore λ and R_{SS} for hand bias calculations

Analysis: Using Eqs. (13.41) through (13.44):

$$I_D = \frac{I_{SS}}{2} = 500 \text{ μA} \qquad V_{GS} = 0.7 + \sqrt{\frac{1 \text{ mA}}{0.5 \text{ mA/V}^2}} = 2.11 \text{ V}$$

$$V_{DS} = 12 \text{ V} - (500 \text{ μA})(13 \text{ kΩ}) + 2.11 \text{ V} = 7.61 \text{ V}$$

Thus, both transistors in the differential amplifier are biased at a Q-point of (500 μA, 7.61 V). The voltages at the drain and source of the MOSFET are $V_D = 5.50$ V and $V_S = -2.11$ V.

Maintenance of pinch-off for M_1 for nonzero V_{IC} requires

$$V_{GD} = V_{IC} - (V_{DD} - I_D R_D) \leq V_{TN}$$

$$V_{IC} \leq V_{DD} - I_D R_D + V_{TN} = 6.2 \text{ V}$$

Check of Results: Checking for pinch-off, $V_{GS} - V_{TN} = 1.41$ V, and $V_{DS} \geq 1.41$. ✔

Discussion: Note that the drain currents are set by the current source and will be independent of device characteristics. This is demonstrated next using SPICE.

Computer-Aided Analysis: In our SPICE analysis, we can easily include λ and R_{SS} to see their impact on the Q-points of the transistors. Using Eq. (13.40) and the $V_{GS} = 1.2$ V as already calculated, the dc current source value for SPICE will be $200 - 21.6 = 178.4$ μA. We need to set up KP = 0.005 A/V^2, VTO = 1 V, and LAMBDA = 0.0133 V^{-1} in the SPICE device models. With these values, the Q-points from a SPICE operating point analysis are virtually the same as our hand calculations (100 μA, 6.99 V). Since the drain currents are locked by the current source, including λ causes only a small adjustment to occur in the value of gate-source voltage: $V_{GS} = 1.198$ V.

(b) From the analysis above, the value of $(V_{GS} - V_{TN})$ and V_{DS} are both greater than V_{SAT} of 1 V. Therefore for $I_D = 500$ μA,

$$V_{GS} = V_{TN} + \frac{V_{SAT}}{2} + \frac{I_D}{K_n V_{SAT}} = 0.7 \text{ V} + \frac{1 \text{ V}}{2} + \frac{500 \text{ μA}}{(500 \text{ μA/V}^2)(1 \text{ V})} = 2.20 \text{ V}$$

$$V_{DS} = 12 \text{ V} - 6.50 \text{ V} + 2.20 \text{ V} = 7.70 \text{ V} \quad \text{and} \quad V_{IC} \text{ doesn't change}$$

> **EXERCISE:** Draw a PMOS version of the NMOS differential amplifier in Fig. 13.13(b).
>
> **ANSWER:** See Figs. P13.40 and P13.42.
>
> **EXERCISE:** Replace the MOSFET in Ex. 13.2 by a four-terminal device with its substrate connected to $V_{SS} = -12$ V, and find the new Q-point for the transistor. Assume $V_{TO} = 1$ V, $\gamma = 0.75\sqrt{V}$, and $2\phi_F = 0.6$ V. What is the new value of V_{TN}?
>
> **ANSWERS:** (100 μA, 8.75 V); 2.75 V

13.1.12 DIFFERENTIAL-MODE INPUT SIGNALS

The differential-mode and common-mode half-circuits for the differential amplifier in Fig. 13.16 are given in Fig. 13.17. In the differential-mode half-circuit, the MOSFET sources represent a virtual ground. In the common-mode circuit, the electronic current source has been modeled by twice its small-signal output resistance R_{SS}, representing the finite output resistance of the current source.

The differential-mode half-circuit represents a common-source amplifier, and the output voltages are given by

$$v_{d1} = -g_m(R_D \| r_o)\frac{v_{id}}{2} \qquad v_{d2} = +g_m(R_D \| r_o)\frac{v_{id}}{2} \qquad v_{od} = -g_m(R_D \| r_o)v_{id} \qquad (13.47)$$

The differential-mode gain is

$$A_{dd} = \left.\frac{v_{od}}{v_{id}}\right|_{v_{ic}=0} = -g_m(R_D \| r_o) \cong -g_m R_D \quad \text{for } r_o \gg R_D \qquad (13.48)$$

whereas taking the single-ended output between either drain and ground provides a gain of one-half A_{dd}:

$$A_{dd1} = \left.\frac{v_{d1}}{v_{id}}\right|_{v_{ic}=0} \cong -\frac{g_m R_D}{2} = \frac{A_{dd}}{2} \quad \text{and} \quad A_{dd2} = \left.\frac{v_{d2}}{v_{id}}\right|_{v_{ic}=0} \cong +\frac{g_m R_D}{2} = -\frac{A_{dd}}{2} \qquad (13.49)$$

The differential-mode input and output resistances are infinite and $2R_D$, respectively:

$$R_{id} = \infty \quad \text{and} \quad R_{od} = 2(R_D \| r_o) \qquad (13.50)$$

The virtual ground at the source node causes the amplifier to again behave as a single-stage inverting amplifier. A differential output provides the full gain of the common-source stage, whereas using the single-ended output at either drain reduces the gain by a factor of 2.

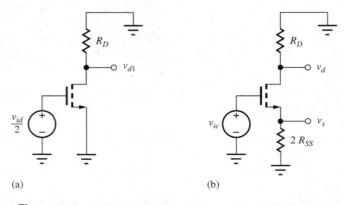

Figure 13.17 (a) Differential-mode and (b) common-mode half-circuits.

> **EXERCISE:** In a manner similar to the analysis of Fig. 13.8, derive the expressions for the differential-mode voltage gains of the MOS differential amplifier directly from the full small-signal model.

13.1.13 SMALL-SIGNAL TRANSFER CHARACTERISTIC FOR THE MOS DIFFERENTIAL AMPLIFIER

The MOS differential amplifier also provides improved linear input signal range and distortion characteristics over that of a single transistor. We can explore these advantages using the drain current expression for the MOSFET:

$$i_{D1} - i_{D2} = \frac{K_n}{2}[(v_{GS1} - V_{TN})^2 - (v_{GS2} - V_{TN})^2] \quad (13.51)$$

For the symmetrical differential amplifier with a purely differential-mode input, $v_{GS1} = V_{GS} + \frac{v_{id}}{2}$, $v_{GS2} = V_{GS} - \frac{v_{id}}{2}$, and

$$i_{D1} - i_{D2} = K_n(V_{GS} - V_{TN})v_{id} = g_m v_{id} \quad (13.52)$$

The second-order distortion product cancels out, and the output current expression is distortion free! As usual, we should question such a perfect result. In reality, MOSFETs are not perfect square-law devices, and some distortion will exist. There also will be distortion introduced through the voltage dependence of the output impedances of the transistors.

Impact of Velocity Saturation

When velocity saturation occurs, the drain current expression will be described by Eq. (13.47), and the difference in drain currents of the differential pair becomes

$$i_{D1} - i_{D2} = K_n V_{SAT} v_{id} = g_m v_{id} \quad (13.53)$$

Velocity saturation linearizes the drain current expressions, and the differential output current from the MOS pair is again distortion free. Note that the expression for transconductance g_m has changed [see Eq. (13.47)].

13.1.14 COMMON-MODE INPUT SIGNALS

The common-mode half-circuit in Fig. 13.17(b) is that of an inverting amplifier with a source resistor equal to $2R_{SS}$. Using the results from Chapter 8,

$$\mathbf{v_{d1}} = \mathbf{v_{d2}} = -\frac{g_m R_D}{1 + 2g_m R_{SS}} \mathbf{v_{ic}} \quad (13.54)$$

and the signal voltage at the source is

$$\mathbf{v_s} = \frac{2g_m R_{SS}}{1 + 2g_m R_{SS}} \mathbf{v_{ic}} \cong \mathbf{v_{ic}} \quad (13.55)$$

The differential output voltage is zero because the voltages are equal at the two drains:

$$\mathbf{v_{od}} = \mathbf{v_{d1}} - \mathbf{v_{d2}} = 0 \quad (13.56)$$

Thus, the common-mode conversion gain for a differential output is zero:

$$A_{cd} = \frac{\mathbf{v_{od}}}{\mathbf{v_{ic}}} = 0 \quad (13.57)$$

The common-mode gain is given by

$$A_{cc} = \frac{\mathbf{v_{oc}}}{\mathbf{v_{ic}}} = -\frac{g_m R_D}{1 + 2g_m R_{SS}} \cong -\frac{R_D}{2R_{SS}} \quad (13.58)$$

The common-mode input source is connected directly to the MOSFET gate. Thus, the input current is zero and

$$R_{ic} = \infty \tag{13.59}$$

Common-Mode Rejection Ratio (CMRR)

For a purely common-mode input signal, the output voltage of the balanced MOS amplifier is zero, and the CMRR is infinite. However, if a single-ended output is taken from either drain,

$$\text{CMRR} = \left|\frac{\frac{A_{dd}}{2}}{A_{cc}}\right| = \left|\frac{-\frac{g_m R_D}{2}}{-\frac{R_D}{2R_{SS}}}\right| = g_m R_{SS} \tag{13.60}$$

For high CMRR, a large value of R_{SS} is again desired. In Fig. 13.17, R_{SS} represents the output resistance of the current source in Fig. 13.13, and its value is much greater than resistor R_{EE}, which is used to bias the amplifier in Fig. 13.1. For this reason, as well as for Q-point stability, most differential amplifiers are biased by a current source, as in Fig. 13.13.

To compare the MOS amplifier more directly to the BJT analysis, however, let us assume for the moment that the MOS amplifier is biased by a resistor of value

$$R_{SS} = \frac{V_{SS} - V_{GS}}{I_{SS}} \tag{13.61}$$

Then, Eq. (13.60) can be rewritten in terms of the circuit voltages, as was done for Eq. (13.33):

$$\text{CMRR} = \frac{2I_D R_{SS}}{V_{GS} - V_{TN}} = \frac{I_{SS} R_{SS}}{V_{GS} - V_{TN}} = \frac{(V_{SS} - V_{GS})}{V_{GS} - V_{TN}} \tag{13.62}$$

Using the numbers from the example,

$$\text{CMRR} = \frac{(V_{SS} - V_{GS})}{V_{GS} - V_{TN}} = \frac{(12 - 1.2)}{0.20} = 54 \tag{13.63}$$

providing a paltry 35 dB. This is almost 10 dB worse than the result for the BJT amplifier. Because of the low values of CMRR in both the BJT and FET circuits when biased by only a resistor, the use of current sources with much higher effective values of R_{SS} or R_{EE} is common in all differential amplifiers.

13.1.15 MODEL FOR DIFFERENTIAL PAIRS

The ac analysis of circuits involving differential amplifiers can often be simplified by using the two-port small-signal model for the differential-pair appearing in Fig. 13.18. The model can be substituted directly for the differential pair, or it can be used as a conceptual aid in simplifying circuits. The two current sources represent the signal currents generated by the two transistors in the pair.

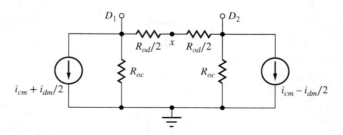

Figure 13.18 Two-port model for the differential pair.

ELECTRONICS IN ACTION

Limiting Amplifiers for Optical Communications

Interface circuits for optical communications were introduced in the Electronics in Action features in Chapter 11. Here, we discuss the limiting amplifier (LA), another of the important electronic blocks on the receiver side of the fiber optic communication link. The LA amplifies the low level output voltage (e.g., 10 mV) of the transimpedance amplifier up to a level that can drive the clock and data recovery circuits (e.g., 250 mV).

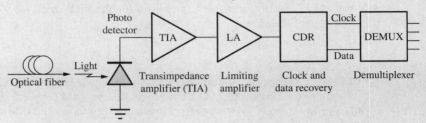

Optical fiber receiver block diagram.

A typical limiting amplifier consists of a wide-band multistage dc-coupled amplifier similar to the one in the circuit schematic here [2–4]. The input signal from the transimpedance amplifier is buffered and level-shifted by two stages of emitter followers (2EF). This is followed by a transadmittance amplifier (TAS) that converts the voltage to a current and then drives a transimpedance amplifier (TIS) that converts the current back to a voltage. This TAS-TIS cascade was developed by Cherry and Hooper [5] and represents an important technique for realizing amplifiers with very wide bandwidth. The output is level-shifted by two more emitter followers and amplified by a second Cherry-and-Hooper stage. A third pair of emitter followers drives a differential amplifier with load resistors chosen to match a transmission line impedance of 50 Ω. Note that 50-Ω matching is used at the LA input as well.

We see that differential pairs are used throughout the limiting amplifier in the TAS and TIS stages, and in the gain stage at the output. Since these optical-to-electrical interface circuits typically push the state-of-the-art in speed, only *npn* transistors are used in the design.

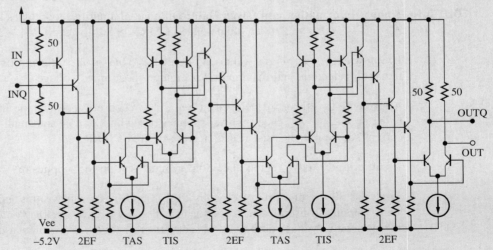

Schematic of a typical limiting amplifier in bipolar technology. (Note that this is a dc-coupled amplifier.)

Remember that *npn* transistors are inherently faster than *pnp* transistors because of the mobility advantage of electrons over that of holes.

Resistors R_{oc} are the common-mode output resistances appearing at each collector or drain, D_1 and D_2, and R_{od} is the differential output resistance that appears between the two collectors or drains. (Remember for symmetrical differential-mode circuits, the node "x" will be a virtual ground.) For the pairs in Fig. 13.13, approximate expressions for the elements are

$$\mathbf{i_{dm}} = g_m \mathbf{v_{dm}} \qquad \mathbf{i_{cm}} = \frac{g_m}{1 + 2g_m R_{EE}} \mathbf{v_{cm}} \cong \frac{\mathbf{v_{cm}}}{2R_{EE}} \qquad (13.64)$$

$$R_{od} = 2r_o \qquad R_{oc} \cong 2\mu_f R_{EE}$$

Substitute R_{SS} for R_{EE} in these expressions for the FET case. We will make use of this two-port in subsequent chapters.

EXERCISE: The bipolar differential amplifier in Fig. 13.13(a) is biased by a 75-µA current source with an output resistance of 1 MΩ. If the transistors have Early voltages of 60 V, estimate values of R_{od}, R_{oc}, i_{dm}, and i_{cm}. How many differential transistor pairs are used in the EIA on the previous page?

ANSWERS: 3.2 MΩ; 4.8 GΩ; $1.50 \times 10^{-3} v_{dm}$; $5.00 \times 10^{-7} v_{cm}$; 5

EXAMPLE 13.3 DIFFERENTIAL AMPLIFIER DESIGN

Design a differential amplifier to meet a given set of specifications.

PROBLEM Design a differential amplifier stage with $A_{dd} = 40$ dB, $R_{id} \geq 250$ kΩ, and an input common-mode input range of at least ±5 V. Specify a current source to give CMRR of at least 80 dB for a single-ended output. MOSFETs are available with $K_n' = 50$ µA/V^2, $\lambda = 0.0133$ V^{-1}, and $V_{TN} = 1$ V. BJTs are available with $I_S = 0.5$ fA, $\beta_F = 100$, and $V_A = 75$ V.

SOLUTION **Known Information and Given Data:** Differential amplifier topologies appear in Fig. 13.13; $A_{dd} = 40$ dB, $R_{id} \geq 250$ kΩ, single-ended CMRR ≥ 80 dB, and $|V_{IC}| \geq 5$ V.

Unknowns: Power supply values, Q-points, R_C, bias source current and output resistance, transistor selection, and maximum dc common-mode input voltage V_{IC}

Approach: Use theory developed in Sec. 13.1; choose transistor type and operating current based on A_{dm} and R_{id}; choose power supplies based on A_{dm}, V_{IC}, and small-signal range; choose current source output resistance to achieve desired CMRR.

Assumptions: Active region operation; symmetrical power supplies, $\beta_o = \beta_F$, $|v_{id}| \leq 30$ mV.

Analysis: 40 dB of gain corresponds to $A_{dd} = 100$. To achieve this gain with a resistively loaded amplifier, use of a BJT is indicated. For $A_{dd} = g_m R_C = 40 I_C R_C$, a gain of 100 can be achieved with a voltage drop of 2.5 V across the resistor R_C.[3] For a bipolar differential amplifier,

---

[3] Remember our rule-of-thumb for the FET: $g_m R_D \cong V_{DD}$ which would require very large V_{DD}.

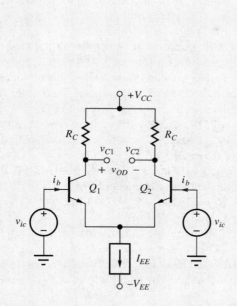
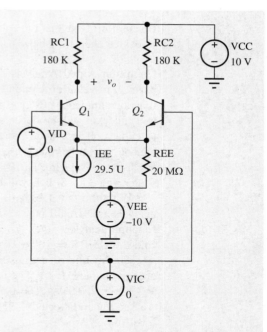

the input resistance $R_{id} = 2r_\pi$, therefore $r_\pi = 125$ kΩ, which requires

$$I_C \leq \frac{\beta_o V_T}{r_\pi} = \frac{100(0.025 \text{ V})}{125 \text{ k}\Omega} = 20 \text{ }\mu\text{A}$$

based on a current gain of 100. Let us choose $I_C = 15$ μA to provide some safety margin. Then, $R_C = 2.5$ V/15 μA $= 167$ kΩ. Choose $R_C = 180$ kΩ as the nearest value from the 5 percent resistor tables in Appendix A. (The larger value will also help compensate for our neglect of r_o in the gain calculation.)

A V_{IC} of 5 V requires the collector voltage of the BJT to be at least 5 V at all times. We do not know the signal level, but we know $|v_{id}| \leq 30$ mV for linearity in the differential pair. Thus, the ac component of the differential output voltage will be no greater than $100(0.03 \text{ V}) = 3$ V, half of which will appear at each collector. Thus, the dc + ac signal across R_C will not exceed 4 V (2.5-V dc + 1.5-V ac), and the positive power supply must satisfy

$$V_{CC} \geq V_{IC} + 4 \text{ V} = 5 + 4 = 9 \text{ V}$$

Choosing $V_{CC} = 10$ V provides a design margin of 1 V. For symmetrical supplies, $-V_{EE} = -10$ V.
The single-ended CMRR of 80 dB requires

$$R_{EE} \geq \frac{\text{CMRR}}{g_m} = \frac{10^4}{(40/\text{V})(15 \text{ }\mu\text{A})} = 16.7 \text{ M}\Omega$$

A current source with $I_{EE} = 30$ μA and $R_{EE} \geq 20$ MΩ will provide some design margin.

Check of Results: Using the design values, $A_{dd} = 40(15 \text{ }\mu\text{A})(180 \text{ k}\Omega) = 108$. CMRR $= 40(15 \text{ }\mu\text{A})(20 \text{ M}\Omega) = 12{,}000$ (81.6 dB), and $R_{id} = 2(2.5 \text{ V}/15 \text{ }\mu\text{A}) = 333$ kΩ. The bias voltages provide $V_{IC} = 6$ V. Thus, the amplifier design should meet the specifications. We will check it further shortly with SPICE.

Discussion: Note that the collector currents are set by the current source and will be independent of device characteristics.

Computer-Aided Analysis: The SPICE schematic input appears in the figure on the previous page. Zero-value differential- and common-mode sources VID and VIC are for use in transfer function simulations. Requesting the transfer function from VID to output voltage v_O between the two collectors will produce values for A_{dd}, R_{id}, and R_{od}. Requesting the transfer function from VIC to either collector node produces values for A_{cc} and R_{ic}. In our SPICE analysis, we can easily include the Early voltage (set VAF = 75 V) and R_{EE} to see their impact on the Q-points of the transistors. Using Eq. (13.40) with $V_{BE} = (0.025 \text{ V}) \ln(15 \text{ μA}/0.5 \text{ fA}) \cong 0.6$ V, the dc current source value for SPICE will be $30 - 0.5 = 29.5$ μA.

With these values and REE = 20 MEG, the Q-points from a SPICE operating point analysis are virtually the same as our hand calculations (14.9 μA, 7.33 V). Using two transfer function analyses gives $A_{dd} = 100$, $R_{id} = 382$ kΩ, $R_{od} = 349$ kΩ, and $A_{cc} = 0.00416$, and we find the CMRR = $100/0.00416 = 24{,}000$ or 87.6 dB.

The results of a transient simulation for the output voltage at the right-hand collector are given here for v_{id} equal to a 30-mV input sine wave at a frequency of 1 kHz and $V_{IC} = +5$ V. TSTART = 0, TSTOP = 0.002 s, and TSTEP = 0.001 ms (1 μs). As designed, we see an undistorted 1-kHz sine wave with an amplitude of 1.5 V biased at the Q-point level of 7.3 V.

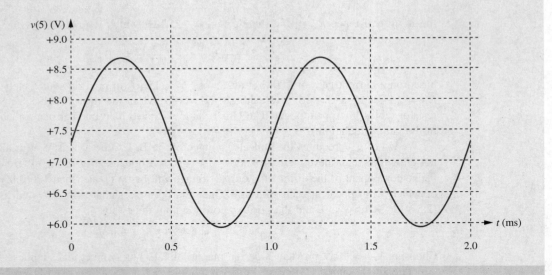

13.2 EVOLUTION TO BASIC OPERATIONAL AMPLIFIERS

One extremely important application of differential amplifiers is at the input stage of operational amplifiers. Differential amplifiers provide the desired differential input and common-mode rejection capabilities, and a ground-referenced signal is available at the output. However, an op amp usually requires higher voltage gain than is available from a single differential amplifier stage, and most op amps use at least two stages of gain. In addition, a third stage, the output stage, is added to provide low-output resistance and high-output current capability.

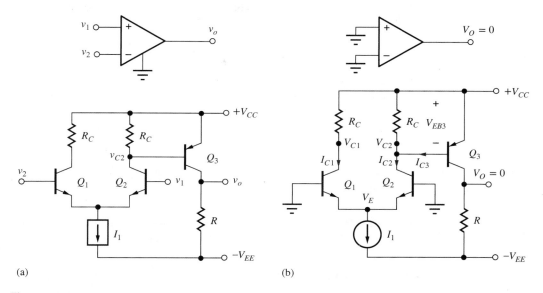

Figure 13.19 (a) A simple two-stage prototype for an operational amplifier. (b) dc equivalent circuit for the two-stage amplifier.

13.2.1 A TWO-STAGE PROTOTYPE FOR AN OPERATIONAL AMPLIFIER

To achieve a higher gain, a *pnp* common-emitter amplifier Q_3 has been connected to the output of differential amplifier, Q_1–Q_2, to form the simple two-stage op amp depicted in Fig. 13.19(a). Bias is provided by current source I_1. Note the dc coupling between Q_2 and Q_3. Also note that the positions of inputs v_1 and v_2 have been reversed to account for the additional phase inversion by Q_3.

dc Analysis

The dc equivalent circuit for the op amp is shown in Fig. 13.19(b) and will be used to find the Q-points of the three transistors. The emitter currents of Q_1 and Q_2 are each equal to one half of the bias current I_1: $I_{E1} = I_{E2} = I_1/2$. The voltage at the collector of Q_1 is equal to

$$V_{C1} = V_{CC} - I_{C1}R_C = V_{CC} - \alpha_{F1}\frac{I_1}{2}R_C \qquad (13.65)$$

and that at the collector of Q_2 is

$$V_{C2} = V_{CC} - (I_{C2} - I_{B3})R_C = V_{CC} - \left(\alpha_{F2}\frac{I_1}{2} - I_{B3}\right)R_C \qquad (13.66)$$

If the base current of Q_3 is neglected, then Eqs. (13.65) and (13.66) become

$$V_{C1} \cong V_{C2} \cong V_{CC} - \frac{I_1 R_C}{2} \qquad (13.67)$$

and because $V_E = -V_{BE}$,

$$V_{CE1} \cong V_{CE2} \cong V_{CC} - \frac{I_1 R_C}{2} + V_{BE} \qquad (13.68)$$

In this particular circuit, it is important to note that the voltage drop across R_C is constrained to be equal to the emitter-base voltage V_{EB3} of Q_3, or approximately 0.7 V.

The value of the collector current of Q_3 can be found by remembering that this circuit is going to represent an operational amplifier, and because both inputs in Fig. 13.19 are zero, V_O should also be zero. This is the situation that exists when the circuit is used in any of the negative feedback circuits discussed in Chapters 6, 11, and 12.

Since $V_O = 0$, I_{C3} must satisfy

$$I_{C3} = \frac{V_{EE}}{R} \qquad \text{and} \qquad V_{EC3} = V_{CC} \qquad (13.69)$$

We also know that V_{EB3} and I_{C3} are intimately related through our BJT model relationship,

$$V_{EB3} = V_T \ln\left(1 + \frac{I_{C3}}{I_{S3}}\right) \tag{13.70}$$

in which I_{S3} is the saturation current of Q_3. For the offset voltage of this amplifier to be zero, the value of R_C must be carefully selected, based on Eqs. (13.66) and (13.68):

$$R_C = \frac{V_T}{\left(\alpha_{F2}\dfrac{I_1}{2} - \dfrac{I_{C3}}{\beta_{F3}}\right)} \ln\left(1 + \frac{I_{C3}}{I_{S3}}\right) \tag{13.71}$$

Otherwise, a small input voltage (the offset voltage) will need to be applied to the op-amp input, to force the output to zero volts.

EXERCISE: Find the Q-points for the transistors in the amplifier in Fig. 13.19 if $V_{CC} = V_{CE1} = 15.0$ V, $I_1 = 150$ μA, $R_C = 10$ kΩ, $R = 20$ kΩ, and $\beta_F = 100$. What is the value of I_{S3} if the output voltage is zero? Assume $V_{CE1} = 15$ V.

ANSWERS: (74.3 μA, 14.9 V), (74.3 μA, 15.0 V), (750 μA, 15.0 V); 1.87×10^{-15} A

dc Bias Sensitivity—A Word of Caution

It should be noted that the circuit in Fig. 13.19 cannot be operated open-loop without some form of feedback to stabilize the operating point of transistor Q_3, because the collector current of Q_3 is exponentially dependent on the value of its emitter-base voltage. If one attempts to build this circuit, or even simulate it with the default values in SPICE, the output will be found to be saturated at one of the power supply "rails" due to our $V_{EB} = 0.7$ V approximation. This sensitivity could be reduced by putting a resistance in series with the emitter of Q_3, at the expense of a loss in voltage gain.

EXERCISE: Simulate the circuit in Fig. 13.19 with $V_{CC} = V_{EE} = 15$ V, $I_1 = 150$ μA, $R_C = 10$ kΩ, and $R = 20$ kΩ using the default transistor parameters in SPICE. What are the transistor Q-points and output voltage v_O?

ANSWERS: (74.4 μA, 14.9 V), (74.4 μA, 14.9 V), (164 μA, 26.7 V), −11.7 V—not quite saturated against the negative rail. (The exact values will depend on the default parameters in your version of SPICE.) The problem can be solved by connecting the base of Q_1 to the output. Try the simulation again.

ac Analysis

The ac equivalent circuit for the two-stage op amp is shown in Fig. 13.20, in which bias source I_1 has been replaced by its equivalent ac resistance R_1. Analysis of the differential-mode behavior of the op amp can be determined from the simplified equivalent circuit in Fig. 13.21 based on the differential-mode half-circuit for the input stage.

It is important to realize that the overall two-stage amplifier in Fig. 13.20 no longer represents a symmetrical circuit. Thus, half-circuit analysis is not theoretically justified. However, we know that voltage variations at the collector of Q_2 (or at the drain of an FET) do not substantially alter the current in the transistor when it is operating in the forward-active region (or saturation region for the FET). Thus, the emitters of the differential pair will remain approximately a virtual ground. One can also envision a fully symmetrical version of the amplifier with Q_3 and R replicated on the left-hand side of the circuit with the base of the additional transistor attached to the collector of Q_1. In fact, special op amps having both differential inputs and differential outputs are built in this manner. Thus, continuing to represent the differential amplifier by its half-circuit is a highly useful engineering approximation.

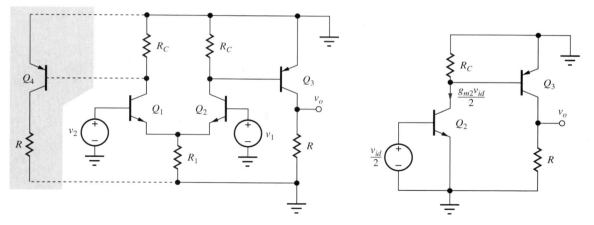

Figure 13.20 An ac equivalent circuit for the two-stage op amp.

Figure 13.21 Simplified model using differential-mode half-circuit.

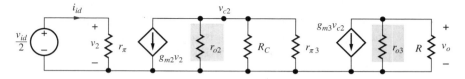

Figure 13.22 Small-signal model for Fig. 13.21.

The small-signal model corresponding to Fig. 13.21 appears in Fig. 13.22. In this analysis, output resistances r_{o2} and r_{o3} are neglected because they are in parallel with external resistors R_C and R. From Fig. 13.22, the overall differential-mode gain A_{dm} of this two-stage operational amplifier can be expressed as

$$A_{dm} = \frac{\mathbf{v_o}}{\mathbf{v_{id}}} = \frac{\mathbf{v_{c2}}}{\mathbf{v_{id}}}\frac{\mathbf{v_o}}{\mathbf{v_{c2}}} = A_{vt1}A_{vt2} \tag{13.72}$$

and the terminal gains A_{vt1} and A_{vt2} can be found from analysis of the circuit in the figure.

The first stage is a differential amplifier with the output taken from the inverting side,

$$A_{vt1} = \frac{\mathbf{v_{c2}}}{\mathbf{v_{id}}} = -\frac{g_{m2}}{2}R_{L1} = -\frac{g_{m2}}{2}\frac{R_C r_{\pi 3}}{R_C + r_{\pi 3}} \tag{13.73}$$

in which the load resistance R_{L1} is equal to the collector resistor R_C in parallel with the input resistance $r_{\pi 3}$ of the second stage.

The second stage is also a resistively loaded common-emitter amplifier with gain

$$A_{vt2} = \frac{\mathbf{v_o}}{\mathbf{v_{C2}}} = -g_{m3}R \tag{13.74}$$

Combining Eqs. (13.72) to (13.74) yields the overall voltage gain for the two-stage amplifier:

$$A_{dm} = A_{vt1}A_{vt2} = \left(-\frac{g_{m2}}{2}\frac{R_C r_{\pi 3}}{R_C + r_{\pi 3}}\right)(-g_{m3}R) = \frac{g_{m2}R_C}{2}\frac{\beta_{o3}R}{R_C + r_{\pi 3}} \tag{13.75}$$

Equation (13.75) appears to contain quite a number of parameters and is difficult to interpret. However, some thought and manipulation will help reduce this expression to its basic design parameters. Multiplying the numerator and denominator of Eq. (13.75) by g_{m3} and expanding the transconductances in terms of the collector currents yields

$$A_{dm} = \frac{1}{2} \frac{(40I_{C2}R_C)\beta_{o3}(40I_{C3}R)}{40\frac{I_{C3}}{I_{C2}}I_{C2}R_C + \beta_{o3}} \tag{13.76}$$

If the base current of Q_3 is neglected, then $I_{C2}R_C = V_{BE3} \cong 0.7$ V, and $I_{C3}R = V_{EE}$, as pointed out during the dc analysis. Substituting these results into Eq. (13.76) yields

$$A_{dm} = \frac{1}{2} \frac{(28)\beta_{o3}(40V_{EE})}{28\left(\frac{I_{C3}}{I_{C2}}\right) + \beta_{o3}} = \frac{560 V_{EE}}{1 + \frac{28}{\beta_{o3}}\left(\frac{I_{C3}}{I_{C2}}\right)} \tag{13.77}$$

In the final result in Eq. (13.77), A_{dm} is reduced to its basics. Once the power supply voltage V_{EE} and transistor Q_3 (that is, β_{o3}) are selected, the only remaining design parameter is the ratio of the collector currents in the first and second stages. An upper limit on I_{C2} and I_1 is usually set by the permissible dc bias current, I_{B2}, at the input of the amplifier, whereas the minimum value of I_{C3} is determined by the current needed to drive the total load impedance connected to the output node. Generally, I_{C3} is several times larger than I_{C1}.

Figure 13.23 is a graph of Eq. (13.77), showing the variation of amplifier gain versus the collector current ratio. Observe that the gain starts to drop rapidly as I_{C3}/I_{C2} exceeds approximately 5. Such a graph is useful as an aid in choosing the operating point during the design of the basic two-stage operational amplifier.

Input and Output Resistances

From the ac model of the amplifier in Figs. 13.21 and 13.22, the differential-mode input resistance and output resistance of the simple op amp are given by

$$R_{id} = \frac{v_{id}}{i_{id}} = 2r_{\pi 2} = 2r_{\pi 1} \quad \text{and} \quad R_{out} = R \,\|\, r_{o3} \cong R \tag{13.78}$$

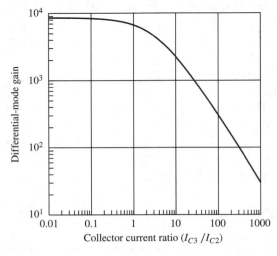

Figure 13.23 Differential-mode gain versus collector current ratio for $V_{EE} = 15$ V and $\beta_{o3} = 100$.

EXERCISE: What is the maximum possible gain of the amplifier described by Eq. (13.74) for $V_{CC} = V_{EE} = 15$ V, $\beta_{o1} = 50$, and $\beta_{o3} = 100$? What is the maximum voltage gain for the amplifier if the input bias current to the amplifier must not exceed 1 μA, and $I_{C3} = 500$ μA? Repeat if $I_{C3} = 5$ mA.

ANSWERS: 8400; 2210; 290.

EXERCISE: What are the input and output resistances for the two amplifier designs in the previous exercise?

ANSWERS: 50 kΩ, 30 kΩ; 50 kΩ, 3 kΩ

EXERCISE: What is the maximum possible gain of the amplifier described by Eq. (13.74) if $V_{CC} = V_{EE} = 1.5$ V?

ANSWER: 840

Before proceeding, we need to understand how the coupling and bypass capacitors have been eliminated from the two-stage op amp prototype. The virtual ground at the emitters of the differential amplifier allows the input stage to achieve the full inverting amplifier gain without the need for an emitter bypass capacitor. Use of the *pnp* transistor permits direct coupling between the first and second stages and allows the emitter of the *pnp* to be connected to the ac ground point at the positive power supply. In addition, the *pnp* provides the voltage **level shift** required to bring the output back to 0 V. Thus, the need for any bypass or coupling capacitors is entirely eliminated, and $v_O = 0$ for $v_1 = 0 = v_2$.

CMRR

The common-mode gain and CMRR of the two-stage amplifier can be determined from the ac circuit model with common-mode input that is shown in Fig. 13.24, in which the half-circuit has again been used to represent the differential input stage. If Fig. 13.24 is compared to Fig. 13.21, we see that the circuitry beyond the collector of Q_2 is identical in both figures. The only difference in output voltage is therefore due to the difference in the value of the collector signal current i_{c2}. In Fig. 13.24, i_{c2} is the collector current of a C-E stage with emitter resistor $2R_1$:

$$i_{c2} = \frac{\beta_{o2} \mathbf{v}_{ic}}{r_{\pi 2} + 2(\beta_{o2} + 1)R_1} \cong \frac{g_{m2} \mathbf{v}_{ic}}{1 + 2g_{m2} R_1} \quad (13.79)$$

whereas i_{c2} in Fig. 13.21 was

$$\mathbf{i}_{c2} = \frac{g_{m2}}{2} \mathbf{v}_{id} \quad (13.80)$$

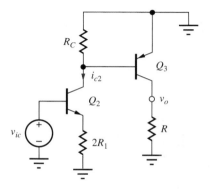

Figure 13.24 An ac equivalent circuit for common-mode inputs.

Thus, the common-mode gain A_{cm} of the op amp is found from Eq. (13.75) by replacing the quantity $g_{m2}/2$ by $g_{m2}/(1 + 2g_{m2}R_1)$:

$$A_{cm} = \frac{g_{m2}R_C}{1 + 2g_{m2}R_1} \frac{\beta_{o3}R}{R_C + r_{\pi 3}} = \frac{2A_{dm}}{1 + 2g_{m2}R_1} \qquad (13.81)$$

From Eq. (13.81), the CMRR of the simple op amp is

$$\text{CMRR} = \left|\frac{A_{dm}}{A_{cm}}\right| = \frac{1 + 2g_{m2}R_1}{2} \cong g_{m2}R_1 \qquad (13.82)$$

which is identical to the CMRR of the differential input stage alone.

EXERCISE: What is the CMRR of the amplifier in Fig. 13.19 if $I_1 = 100$ μA and $R_1 = 750$ kΩ?

ANSWER: 63.5 dB

13.2.2 IMPROVING THE OP AMP VOLTAGE GAIN

From the previous several exercises, we can see that the prototype op amp has a relatively low overall voltage gain and a higher output resistance than is normally associated with a true operational amplifier. This section explores the use of an additional current source to improve the voltage gain; the next section adds an emitter follower to reduce the output resistance.

Figure 13.23 indicates that the overall amplifier gain decreases rapidly as the quiescent current of the second stage increases. In the exercise, the overall gain is quite low when $I_{C3} = 5$ mA. One technique that can be used to improve the voltage gain is to replace resistor R by a second current source, as shown in Fig. 13.25. The modified ac model is in Fig. 13.25(b). The small-signal model is the same as Fig. 13.22 except R is replaced by output resistance R_2 of current source I_2. The load on Q_3 is now the output resistance R_2 of the current source in parallel with the output resistance of Q_3 itself. In Sec. 13.7, we shall discover that it is possible to design a current source with $R_2 \gg r_{o3}$, and, by neglecting R_2, the differential-mode gain expression for the overall amplifier becomes

$$A_{dm} = A_{vt1}A_{vt2} = \left(-\frac{g_{m2}}{2}\frac{R_C r_{\pi 3}}{R_C + r_{\pi 3}}\right)(-g_{m3}r_{o3}) \qquad (13.83)$$

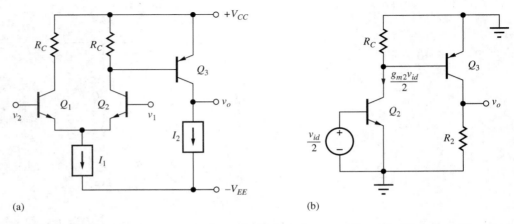

Figure 13.25 (a) Amplifier with improved voltage gain. (b) Approximate ac differential-mode equivalent for op amp.

We can reduce Eq. (13.81) to

$$A_{dm} = \frac{14\mu_{f3}}{1 + \frac{28}{\beta_{o3}}\left(\frac{I_{C3}}{I_{C2}}\right)} \cong \frac{560 V_{A3}}{1 + \frac{28}{\beta_{o3}}\left(\frac{I_{C3}}{I_{C2}}\right)} \qquad (13.84)$$

using the same steps that led to Eq. (13.77). This expression is similar to Eq. (13.77) except that power supply voltage V_{EE} has been replaced by the Early voltage of Q_3. For low values of the collector current ratio, excellent voltage gains, approaching $560 V_{A3}$, are possible from this simple two-stage amplifier. Also, note that the amplifier gain is no longer directly dependent on the choice of V_{CC} and V_{EE}.

Although adding the current source has improved the voltage gain, it also has degraded the output resistance. The output resistance of the amplifier is now determined by the characteristics of current source I_2 and transistor Q_3:

$$R_{\text{out}} = R_2 \| r_{o3} \cong r_{o3} \qquad (13.85)$$

Because of the relatively high-output resistance, this amplifier more nearly represents a transconductance amplifier with a current output ($A_{tc} = \mathbf{i_o}/\mathbf{v_{id}}$) rather than a true low-output resistance voltage amplifier.

> **EXERCISE:** Start with Eq. (13.83) and show that Eq. (13.84) is correct.
>
> **EXERCISE:** What is the maximum possible voltage gain for the amplifier described by Eq. (13.83) for $V_{CC} = 15$ V, $V_{EE} = 15$ V, $V_{A3} = 75$ V, $\beta_{o1} = 50$, and $\beta_{o3} = 100$? What is the voltage gain if the input bias current to the amplifier must not exceed 1 µA, and $I_{C3} = 500$ µA? Repeat if $I_{C3} = 5$ mA.
>
> **ANSWERS:** 42,000; 11,000; 1450
>
> **EXERCISE:** What are the input and output resistances for the last two amplifier designs?
>
> **ANSWERS:** 50 kΩ, 180 kΩ; 50 kΩ, 18 kΩ

13.2.3 DARLINGTON PAIRS

From the numerator of Eq. (13.84), we see that the Early voltage of Q_3 is very important in determining the amplifier voltage gain. However, the current gain of Q_3 is also an important part of the denominator. The **Darlington pairs** of *npn* and *pnp* transistor pairs in Fig. 13.26 provide an important technique for improving the effective current gain of the BJT. The Darlington pairs are used to replace a single BJT, but with much higher current gain.

Either circuit can be analyzed as a two port with the emitter of Q_2 as the common terminal. For example, collector current I_C of the *npn* composite transistor is the sum of the collector currents of the two transistors. The base current of Q_2 becomes the emitter current of Q_1. Thus, I_C can be expressed as

$$I_C = I_{C1} + I_{C2} = \beta_{F1} I_B + \beta_{F2} I_{E1} = \beta_{F1} I_B + \beta_{F2}(\beta_{F1} + 1) I_B \qquad (13.86)$$

and the current gain is found to be approximated by the product of the current gains of the two individual transistors:

$$\beta_F = \frac{I_C}{I_B} = \beta_{F1} + \beta_{F2}(\beta_{F1} + 1) \cong \beta_{F1}\beta_{F2} \qquad (13.87)$$

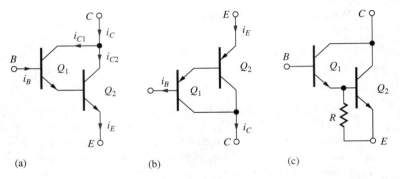

Figure 13.26 Darlington pairs: (a) *npn;* (b) *pnp;* and (c) resistor R added to increase bias current in Q_1.

The input resistance at the base terminal R_{iB} is easily found using our knowledge of common-emitter transistor circuits and is equal to $r_{\pi 1}$ of Q_1 plus the amplified value of the resistance in Q_1's emitter, $r_{\pi 2}$. The simplification occurs since $I_{C1} \cong I_{C2}/\beta_{F2}$.

$$R_{iB} = r_{\pi 1} + (\beta_{o1} + 1)r_{\pi 2} \cong 2\beta_{o1}r_{\pi 2} \quad \text{and} \quad R_{iC} \cong r_{o2} \| 2\frac{r_{o1}}{\beta_{o2}} \cong \frac{2}{3}r_{o2} \quad (13.88)$$

The resistance at the collector is dominated by r_{o2} reduced slightly by the presence of Q_1. This calculation is left for Prob. 13.65.

Issues do arise with this circuit technique however. Since Q_1 operates at 50 to 150 times smaller current than Q_2, its current gain may be much less than that of Q_2, and the overall current gain is less than β_{F2}^2 as one might hope. In addition the low current in Q_1 can lead to issues in the transient response of the Darlington circuit. A resistor is often added between the base and emitter of Q_2 to increase the bias current in Q_1 as in Fig. 13.26(c).

13.2.4 OUTPUT RESISTANCE REDUCTION

As just mentioned, the two-stage op amp prototype at this point more nearly represents a high-output resistance transconductance amplifier than a voltage amplifier with a low-output resistance. A third stage, that maintains the amplifier voltage gain but provides a low-output resistance, needs to be added to the amplifier. This sounds like the description of a follower circuit—unity voltage gain, high-input resistance, and low output resistance!

An emitter-follower (C-C) stage is added to the prototype amplifier in Fig. 13.27. In this case, the C-C amplifier is biased by a third current source I_3, and an external load resistance R_L has been connected to the output of the amplifier. The ac equivalent circuit is drawn in Fig. 13.27(b), in which the output resistances of I_2 and I_3 are assumed to be very large and will be neglected in the analysis. Based on the ac equivalent circuit, the overall gain of the three-stage operational amplifier can be expressed as

$$A_{dm} = \frac{v_2}{v_{id}} \frac{v_3}{v_2} \frac{v_o}{v_3} = A_{vt1} A_{vt2} A_{vt3} \quad (13.89)$$

The gain of the first stage is equal to the gain of the differential input pair (neglecting r_{o2}):

$$A_{vt1} = -\frac{g_{m2}}{2}(R_C \| r_{\pi 3}) \quad (13.90)$$

The second stage is a common-emitter amplifier with a load resistance equal to the output resistance of Q_3 in parallel with the input resistance of emitter follower Q_4:

$$A_{vt2} = -g_{m3}(r_{o3} \| R_{iB4}) \quad \text{where } R_{iB4} = r_{\pi 4}(1 + g_{m4}R_L) \quad (13.91)$$

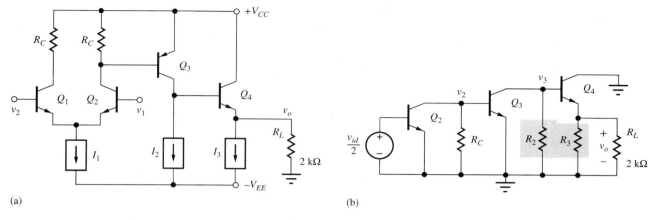

Figure 13.27 (a) Amplifier with common-collector stage Q_4 added. (b) Simplified ac equivalent circuit for the three-stage op amp. R_2 and R_3 are neglected in the analysis.

Finally, the gain of emitter follower Q_4 is (neglecting r_{o4}):

$$A_{vt3} = \frac{g_{m4} R_L}{1 + g_{m4} R_L} \cong 1 \qquad (13.92)$$

The input resistance is set by the differential pair, and the output resistance of the amplifier is now determined by the resistance looking back into the emitter of Q_4:

$$R_{id} = 2 r_{\pi 2} \quad \text{and} \quad R_{out} = \frac{1}{g_{m4}} + \frac{R_{th4}}{\beta_{o4} + 1} \qquad (13.93)$$

In this case, there is a relatively large Thévenin equivalent source resistance at the base of Q_4, $R_{th4} \cong r_{o3}$, and the overall output resistance is

$$R_{out} \cong \frac{1}{g_{m4}} + \frac{r_{o3}}{\beta_{o4}} = \frac{1}{g_{m4}} \left[1 + \frac{\mu_{f3} I_{C4}}{\beta_{o4} I_{C3}} \right] \qquad (13.94)$$

EXAMPLE 13.4   THREE-STAGE BIPOLAR OP AMP ANALYSIS

Let us now determine the characteristics of a specific implementation of the three-stage op amp implemented with bipolar transistors.

PROBLEM Find the differential-mode voltage gain, CMRR, input resistance, and output resistance for the amplifier in Fig. 13.27 if $V_{CC} = 15$ V, $V_{EE} = 15$ V, $V_{A3} = 75$ V, $\beta_{o1} = \beta_{o2} = \beta_{o3} = \beta_{o4} = 100$, $I_1 = 100$ μA, $I_2 = 500$ μA, $I_3 = 5$ mA, $R_1 = 750$ kΩ, and $R_L = 2$ kΩ. Assume R_2 and $R_3 = \infty$ and $T = 290$ K.

SOLUTION **Known Information and Given Data:** Three-stage prototype op amp in Fig. 13.27 with $V_{CC} = 15$ V, $V_{EE} = 15$ V, $V_{A3} = 75$ V, $\beta_{o1} = \beta_{o2} = \beta_{o3} = \beta_{o4} = 100$, $I_1 = 100$ μA, $I_2 = 500$ μA, $I_3 = 5$ mA, $R_1 = 750$ kΩ, and $R_L = 2$ kΩ. Assume R_2 and $R_3 = \infty$.

Unknowns: Q-point values, R_C, A_{dm}, CMRR, R_{id}, and R_{out}

Approach: We need to evaluate the expressions in Eqs. (13.89) through (13.94). First, we must find the Q-point and then use it to calculate the small-signal parameters including g_{m2}, $r_{\pi 2}$, $r_{\pi 3}$, g_{m2}, r_{o3}, and $r_{\pi 4}$. The required Q-point information can be found from Fig. 13.28, in which v_1 and v_2 equal zero.

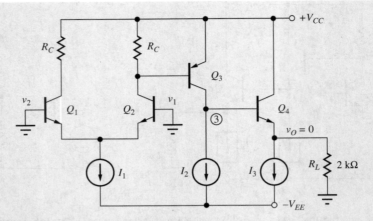

Figure 13.28 Operational amplifier dc equivalent circuit with $v_1 = 0 = v_2$.

Assumptions: The Q-point is found with v_1 and v_2 set to zero, and output voltage v_o is also assumed to be zero for this set of input voltages. The transistors are all in the active region with V_{BE} or V_{EB} equal to 0.7 V.

Analysis: The emitter current in the input stage is one half the bias current source I_1 and

$$g_{m2} = 40 I_{C2} = 40(\alpha_{F2} I_{E2}) = 40(0.99 \times 50\ \mu A) = 1.98\ mS$$

The collector of the second stage must supply the current I_2 plus the base current of Q_4:

$$I_{C3} = I_2 + I_{B4} = I_2 + \frac{I_{E4}}{\beta_{F4} + 1}$$

When the output voltage is zero, the current in load resistor R_L is zero, and the emitter current of Q_4 is equal to the current in source I_3. Therefore,

$$I_{C3} = I_2 + I_{B4} = I_2 + \frac{I_3}{\beta_{F4} + 1} = 5 \times 10^{-4}\ A + \frac{5 \times 10^{-3}\ A}{101} = 550\ \mu A^4$$

and

$$g_{m3} = 40 I_{C3} = \frac{40}{V}(5.5 \times 10^{-4}\ A) = 2.20 \times 10^{-2}\ S$$

$$r_{\pi 3} = \frac{\beta_{o3}}{g_{m3}} = \frac{100}{2.20 \times 10^{-2}\ S} = 4.55\ k\Omega$$

To find the output resistance of Q_3, V_{EC3} is needed. When properly designed, the dc output voltage of the amplifier will be zero when the input voltages are zero. Hence, the voltage at node 3 is one base-emitter voltage drop above zero, or +0.7 V, and $V_{EC3} = 15 - 0.7 = 14.3$ V. The output resistance of Q_3 is

$$r_{o3} = \frac{V_{A3} + V_{EC3}}{I_{C3}} = \frac{(75 + 14.3)\ V}{5.50 \times 10^{-4}\ A} = 162\ k\Omega$$

Remembering that $I_{E4} = I_3$

$$I_{C4} = \alpha_{F4} I_{E4} = 0.990 \times 5\ mA = 4.95\ mA$$

---

[4] Note that I_{B4} is becoming a significant part of I_{C3}.

and

$$g_{m4} = 40I_{C4} = 198 \text{ mS} \qquad r_{\pi 4} = \frac{\beta_{o4} V_T}{I_{C4}} = \frac{100 \times 0.025 \text{ V}}{4.95 \times 10^{-3} \text{ A}} = 505 \text{ }\Omega$$

Finally, the value of R_C is needed:

$$R_C = \frac{V_{EB3}}{I_{C2} - I_{B3}} = \frac{V_{EB3}}{I_{C2} - \frac{I_{C3}}{\beta_{F3}}} = \frac{0.7 \text{ V}}{\left(49.5 - \frac{550}{100}\right) \times 10^{-6} \text{ A}} = 15.9 \text{ k}\Omega$$

Now, the small-signal characteristics of the amplifier can be evaluated:

$$A_{vt1} = -\frac{g_{m2}(R_C \| r_{\pi 3})}{2} = -\frac{1.98 \text{ mS}(15.9 \text{ k}\Omega \| 4.55 \text{ k}\Omega)}{2} = -3.50$$

$$A_{vt2} = -g_{m3}[r_{o3} \| r_{\pi 4} + \beta_{o4} R_L] = -22 \text{ mS}(162 \text{ k}\Omega \| 203 \text{ k}\Omega) = -1980!$$

$$A_{vt3} = \frac{g_{m4} R_L}{r_{\pi 4}(1 + g_{m4} R_L)} = \frac{0.198\text{S}(2 \text{ k}\Omega)}{1 + 0.198\text{S}(2 \text{ k}\Omega)} = 0.998 \cong 1$$

$$A_{dm} = A_{vt1} A_{vt2} A_{vt3} = +6920$$

$$R_{id} = 2r_{\pi 2} = 2\frac{\beta_{o2}}{g_{m2}} = 2\frac{100}{(40/\text{V})(49.5 \text{ }\mu\text{A})} = 101 \text{ k}\Omega$$

$$R_{out} \cong \frac{1}{g_{m4}} + \frac{r_{o3}}{\beta_{o4}} = \frac{1}{(40/\text{V})(4.95 \text{ mA})} + \frac{162 \text{ k}\Omega}{100} = 1.62 \text{ k}\Omega$$

$$\text{CMRR} = g_{m2} R_1 = (40/\text{V})(49.5 \text{ }\mu\text{A})(750 \text{ k}\Omega) = 1490 \text{ or } 63.5 \text{ dB}$$

Check of Results: We can use our rule-of-thumb from Chapter 7 to estimate the voltage gain. The first stage should produce a gain of approximately $(1/2) \times 40 \times$ the voltage across the load resistor or $20(0.7) = 14$. The second stage should produce a gain of approximately $\mu_f = 40(75) = 3000$. The product is 42,000. Our detailed calculations give us about 1/6 of this value. Can we account for the discrepancies? We see the gain of the first stage is only 3.5 because $r_{\pi 3}$ is considerably smaller than R_C, and the gain of the second stage is approximately 2000 because the reflected loading of R_L is of the same order as r_{o3}. These two reductions account for the lower overall gain. The emitter follower produces a gain of one, as expected.

Discussion: This amplifier achieves a reasonable set of op-amp characteristics for a simple circuit: $A_v = 6920$, $R_{id} = 101$ kΩ, and $R_{out} = 1.62$ kΩ. Note that the second stage, loaded by current source I_2 and buffered from R_L by the emitter follower, is achieving a gain that is a substantial fraction of Q_3's intrinsic gain. However, even with the emitter follower, the reflected load resistance $\beta_{o4} R_L$ is similar to the value of r_{o3} and is reducing the overall voltage gain by a factor of almost 2. Also, note that the output resistance is dominated by r_{o3}, present at the base of Q_4, and not by the reciprocal of g_{m4}. These last two factors point to a way to increase the performance of the amplifier by replacing Q_4 with an *npn* Darlington stage (see Sec. 13.2.3).

Computer-Aided Analysis: Since this amplifier is dc coupled, a transfer function analysis from an input source to the output node will automatically yield the voltage gain, input resistance, and output resistance. In order to force the output to be nearly zero (the normal operating point), we must determine the offset voltage of the amplifier, and then apply it as a dc input to the amplifier. This is done by first connecting the amplifier as a voltage follower with the input grounded (see the figure below). For this amplifier, the SPICE yields $V_{OS} = 0.437$ mV. Note that a current of approximately 20 μA will exist in R_1, the output resistance of current source I_1. Be sure to choose $I_1 = 80$ μA so that the bias currents in Q_1 and Q_2 will each be approximately 50 μA.

Next, the offset voltage is applied to the amplifier input with the feedback connection removed, and a transfer function analysis is requested from source V_{OS} to the output (Fig. b). The computed values are $A_{dm} = 8280$, $R_{id} = 105$ kΩ, and $R_{out} = 960$ Ω. The values all differ from our hand calculations. Most of the differences can be traced to the higher temperature and hence higher value of V_T used in the simulations (T defaults to 27°C and $V_T = 25.9$ mV). R_{out} as calculated by SPICE includes the presence of R_L. Removing the 2-kΩ resistor from the SPICE result yields $R_{out} = [(1/960) - (1/2000)]^{-1} = 1.85$ kΩ, which agrees more closely with our hand calculations.

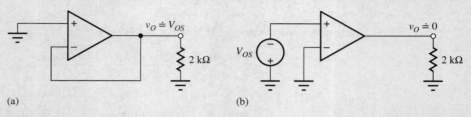

(a) (b)

EXERCISE: Suppose the output resistances of current sources R_2 and R_3 in the amplifier in Fig. 13.27 are 150 kΩ and 15 kΩ, respectively. (a) Recalculate the gain, input resistance, and output resistance. (b) Compare to SPICE simulation results. (c) What is the power consumption of the amplifier in Ex. 13.4?

ANSWERS: 4320, 101 kΩ, 776 Ω; 4480, 105 kΩ, 774 Ω; 168 mW

EXERCISE: Suppose the current gain β_F of all the transistors is 150 instead of 100. Recalculate the gain, input resistance, output resistance, and CMRR of the amplifier in Fig. 13.27.

ANSWERS: 11,000; 152 kΩ; 1.12 kΩ; 63.4 dB

EXERCISE: Suppose the Early voltage of Q_3 in the amplifier in Fig. 13.27 is 50 V instead of 75 V. Recalculate the gain, input resistance, output resistance, and CMRR.

ANSWERS: 5700; 101 kΩ; 1.16 kΩ; 63.5 dB

EXERCISE: The op amp in Ex. 13.4 is operated as a voltage follower. What are the closed-loop gain, input resistance, and output resistance?

ANSWERS: +0.99986, 699 MΩ, 0.233 Ω

13.2.5 A CMOS OPERATIONAL AMPLIFIER PROTOTYPE

Similar circuit design ideas have been used to develop the basic CMOS operational amplifier depicted in Fig. 13.29(a). A differential amplifier, formed by NMOS transistors M_1 and M_2, is followed by a PMOS common-source stage M_3 and NMOS source follower M_4. Current sources are again used to bias the differential input and source-follower stages and as a load for M_3. Referring to the ac equivalent circuit in Fig. 13.29(b), we see that the differential-mode gain is given by the product of the terminal gains of the three stages:

$$A_{dm} = A_{vt1} A_{vt2} A_{vt3} = \left(-\frac{g_{m2}}{2} R_D\right)[-g_{m3}(r_{o3} \| R_2)] \left[\frac{g_{m4}(R_3 \| R_L)}{1 + g_{m4}(R_3 \| R_L)}\right] \quad (13.95)$$

$$\cong \mu_{f3}\left(\frac{g_{m2}}{2} R_D\right)\left(\frac{g_{m4} R_L}{1 + g_{m4} R_L}\right) \quad (13.96)$$

in which we have assumed that $R_3 \gg R_L$ and $R_2 \gg r_{o3}$.

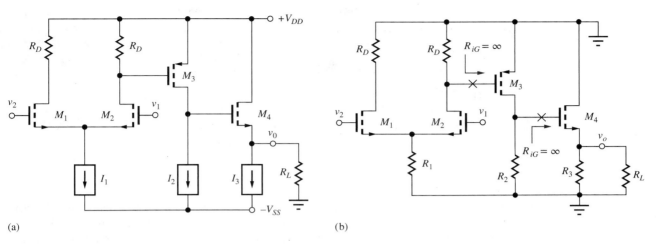

Figure 13.29 (a) A CMOS operational amplifier prototype. (b) ac equivalent circuit for the CMOS amplifier, in which the output resistances of current sources I_2 and I_3 have been neglected.

Equation (13.95) is relatively easy to construct using our single-stage amplifier formulas because the input resistance of each FET is infinite and the gain of one stage is not altered by the presence of the next. The overall differential-mode gain is approximately equal to the product of the voltage gain of the first stage and the amplification factor of the second stage.

Expanding g_{m2}, realizing that the product $I_{D2}R_D$ represents the voltage across R_D, which must equal V_{GS3}, and assuming that the source follower has a gain of nearly 1 yields

$$A_{dm} \cong A_{v1} A_{v2}(1) = \mu_{f3}\left(\frac{V_{SG3}}{V_{GS2} - V_{TN2}}\right) \tag{13.97}$$

Although Eq. (13.97) is a simple expression, we often prefer to have the gain expressed in terms of the various bias currents, and expanding μ_{f3}, V_{GS2}, and V_{SG3} yields

$$A_{dm} = \frac{1}{\lambda_3}\sqrt{\frac{K_{n2}}{I_{D2}}\frac{K_{p3}}{I_{D3}}}\left[\sqrt{\frac{2I_{D3}}{K_{p3}}} - V_{TP3}\right] \tag{13.98}$$

Because of the Q-point dependence of μ_f, there are more degrees of freedom in Eq. (13.98) than in the corresponding expression for the bipolar amplifier, Eq. (13.84). This is particularly true in the case of integrated circuits, in which the values of K_n and K_p can be easily changed by modifying the W/L ratios of the various transistors. However, the benefit of operating both gain stages of the amplifier at low currents is obvious from Eq. (13.98), and picking a transistor with a small value of λ for M_3 is also clearly important.

It is worth noting that because the gate currents of the MOS devices are zero, input-bias current does not place a restriction on I_{D1}, whereas it does place a practical upper bound on I_{C1} in the case of the bipolar amplifier. The input and output resistances of the op amp are determined by M_1, M_2, and M_4. From our knowledge of single-stage amplifiers,

$$R_{id} = \infty \qquad R_{out} = \frac{1}{g_{m4}} \bigg\| R_3 \qquad \text{CMRR} = g_{m2}R_1 \tag{13.99}$$

CMRR is once again determined by the differential input stage, where R_1 is the output resistance of current source I_1.

> **EXERCISE:** For the CMOS amplifier in Fig. 13.29(a), $\lambda_3 = 0.01$ V, $K_{n1} = K_{n4} = 5.0$ mA/V^2, $K_{p3} = 2.5$ mA/V^2, $I_1 = 200$ μA, $I_2 = 500$ μA, $I_3 = 5$ mA, $R_1 = 375$ kΩ, and $V_{TP3} = -1$ V. What is the actual gain of the source follower if $R_L = 2$ kΩ? What are the voltage gain, CMRR, input resistance, and output resistance of the amplifier?
>
> **ANSWERS:** 0.934; 2410, 51.5 dB, ∞, 141 Ω
>
> **EXERCISE:** What is the quiescent power consumption of this op amp if $V_{DD} = V_{SS} = 12$ V?
>
> **ANSWER:** 137 mW

13.2.6 BiCMOS AMPLIFIERS

A number of integrated circuit processes exist that offer the circuit designer a combination of bipolar and MOS transistors or bipolar transistors and JFETs. These are commonly referred to as BiCMOS and BiFET technologies, respectively. The combination of BJTs and FETs offers the designer the ability to use the best characteristics of both devices to enhance the performance of the circuit.

A simple BiCMOS op amp is shown in Fig. 13.30. In this case, a differential pair of PMOS transistors has been used as the input stage to demonstrate another design variation. The PMOS transistors at the input provide high-input resistance and can be biased at relatively high-input currents, since input current is not an issue. (We will discover later that this increased current improves the slew rate of the amplifier.) The second gain stage utilizes a bipolar transistor, which provides a superior amplification factor compared to the FET. Emitter resistor R_E increases the voltage across R_{D2} and hence, the voltage gain of the first stage without reducing the amplification factor of Q_1 (see Sec. 8.2.7). The follower stage uses another FET in order to maximize second-stage gain while maintaining a reasonable output resistance.

For the circuit shown, SPICE simulation uses VTO = −1 V, KP = 25 mA/V^2, VAF = 75 V, and BF = 100. SPICE is first used to find the offset voltage in the same manner as in Ex. 13.5. The value is found to be −11.37 mV, which is then applied to the input of the open-loop amplifier. A transfer function analysis from V_{OS} to the output yields infinite input resistance, a voltage gain of 13,200 and an output resistance of 61.4 Ω.

13.2.7 ALL TRANSISTOR IMPLEMENTATIONS

In NMOS and CMOS technology, it is often desirable to eliminate all the resistors wherever possible, and this can be done using the techniques introduced in Sec. 8.6. For example, we can replace the

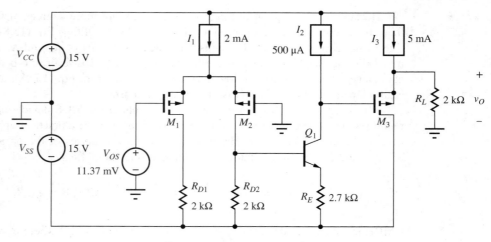

Figure 13.30 Basic BiCMOS op amp.

13.2 Evolution to Basic Operational Amplifiers

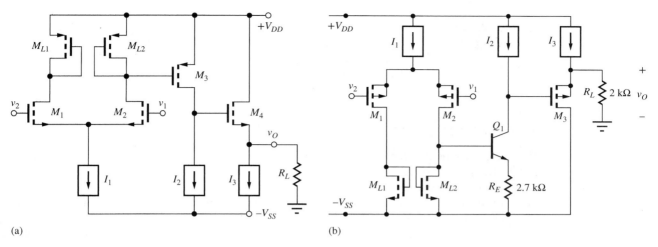

Figure 13.31 (a) A version of the CMOS amplifier in Fig. 13.29 with the drain resistors replaced with saturated PMOS transistors. (b) The CMOS amplifier from Fig. 13.30 with the drain resistors replaced with saturated NMOS transistors.

drain resistors in Fig. 13.29 with either NMOS or PMOS devices connected in saturation. Since the voltage across the transistor will provide the operating bias to M_3, it makes sense to use PMOS transistors as shown in Fig. 13.31(a) since the devices will then match to each other. The source-gate voltage of M_3 is set by the source-drain voltage of M_{L2}: $V_{SG3} = V_{SGL2}$.

The equivalent small-signal resistance for the "diode-connected" FET is approximately $R_{eq} = 1/g_m$, so the differential-mode gain of the input stage becomes $A_{dd} = -g_{m2}/g_{mL2}$. The expression for the voltage gain of the input stage is slightly different from that presented in Chapter 8 because the input and load transistors are not the same type.

$$A_{dd} = -\frac{g_{m2}}{g_{mL2}} = -\frac{\sqrt{2K_n I_{D2}}}{\sqrt{2K_p I_{DL2}}} = -\sqrt{\frac{K'_n}{K'_p}}\sqrt{\frac{(W/L)_2}{(W/L)_{L2}}} \qquad (13.100)$$

The difference between the transconductance parameters of the NMOS and PMOS transistors improves the gain for this case. Note that the PMOS transistors in Fig. 13.31(a) all have their sources tied to the power supply. The NMOS transistors could all be placed in individual p-wells in a p-well process to eliminate the body effect.

A similar technique is used to replace R_{D1} and R_{D2} in the BiCMOS amplifier in Fig. 13.31(b). NMOS transistors are used here so their sources can be connected to the negative power supply, and the PMOS transistors could be placed in separate p-wells if an n-well CMOS process were utilized. Resistor R_E is relatively small in value and would probably not be replaced with a transistor. In Chapter 14, we will find an even better way to configure load transistors M_{L1} and M_{L2} in both amplifiers in Fig. 13.31 by using "current mirror" circuits.

EXERCISE: Write an expression for upper bound on the gain of the amplifier in Fig. 13.31(a).

ANSWER: $A_{dm} = \frac{1}{2}\frac{g_{m2}}{g_{mL2}}\mu_{f3}\frac{g_{m4}R_L}{1 + g_{m4}R_L} \leq \frac{1}{2}\frac{g_{m2}}{g_{mL2}}\mu_{f3}$

EXERCISE: The input stage of the amplifier in Fig. 13.31(a) needs $A_{dd} = 10$ and $(W/L)_{L_2} = 4/1$. What value of $(W/L)_2$ is required?

ANSWER: 160/1

EXERCISE: Write an expression for the voltage gain of the input stage in Fig. 13.31(b), ignoring the loading of Q_1.

ANSWER: $A_{dd} = -\sqrt{\dfrac{K'_p}{K'_n}}\sqrt{\dfrac{(W/L)_2}{(W/L)_{L2}}}$

13.3 OUTPUT STAGES

The basic operational amplifier circuits discussed in Sec. 13.2 used followers for the output stages. The final stage of these amplifiers is designed to provide a low-output resistance as well as a relatively high current drive capability. However, because of this last requirement, the output stages of the amplifiers in the previous section consume approximately two-thirds or more of the total power of the amplifier.

Followers are **class-A amplifiers,** defined as circuits in which the transistors conduct during the full 360° of the signal waveform. The class-A amplifier is said to have a **conduction angle** $\theta_C = 360°$. Unfortunately, the maximum efficiency of the class-A stage is only 25 percent. Because the output stage must often deliver relatively large powers to the amplifier load, this low efficiency can cause high power dissipation in the amplifier. This section analyzes the efficiency of the class-A amplifier and then introduces the concept of the **class-B push-pull output stage.** The class-B push-pull stage uses two transistors, each of which conducts during only one half, or 180°, of the signal waveform ($\theta_C = 180°$) and can achieve much higher efficiency than the class-A stage. Characteristics of the class-A and class-B stages can also be combined into a third category, the class-AB amplifier, which forms the output stage of most operational amplifiers.

13.3.1 THE SOURCE FOLLOWER—A CLASS-A OUTPUT STAGE

We analyzed the small-signal behavior of follower circuits in detail in Chapter 8 and found that they provide high-input resistance, low-output resistance, and a voltage gain of approximately 1. Here we focus on the source-follower circuit in Fig. 13.32 biased by an **ideal current source.**

For $v_I \leq V_{DD} + V_{TN}$, M_1 will be operating in the saturation region (be sure to prove this to yourself). The current source forces a constant current I_{SS} to flow out of the transistor's source. Using Kirchhoff's voltage law, $v_O = v_I - v_{GS}$. Since the source current is constant, v_{GS} is also constant,

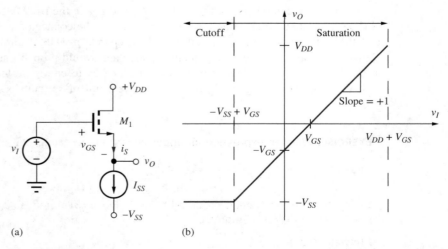

Figure 13.32 (a) Source-follower circuit. (b) Voltage transfer characteristic for the source follower.

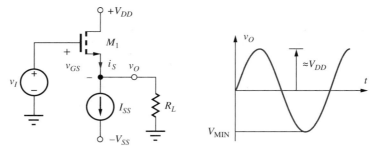

Figure 13.33 Source follower with external load resistor R_L.

and v_O is

$$v_O = v_I - V_{GS} = v_I - \left(V_{TN} + \sqrt{\frac{2I_{SS}}{K_n}}\right) \quad (13.101)$$

The difference between the input and output voltages is fixed. Thus, from a large-signal perspective (as well as from a small-signal perspective), we expect the source follower to provide a gain of approximately 1.

The voltage transfer characteristic for the source follower appears in Fig. 13.32(b). The output voltage at the source follows the input voltage with a slope of $+1$ and a fixed offset voltage equal to V_{GS}. For positive inputs, M_1 remains in saturation until $v_I = V_{DD} + V_{TN}$. The maximum output voltage is $v_o = V_{DD}$ for $v_I = V_{DD} + V_{GS}$. Note that to actually reach this output, the input voltage must exceed V_{DD}.

The minimum output voltage is set by the characteristics of the current source. An ideal current source will continue to operate even with $v_o < -V_{SS}$, but most electronic current sources require $v_o \geq -V_{SS}$. Thus, the minimum possible value of the input voltage is $v_I = -V_{SS} + V_{GS}$.

Source Follower with External Load Resistor
When a load resistor R_L is connected to the output, as in Fig. 13.33, the output voltage range is restricted by a new limit. The total source current of M_1 is equal to

$$i_S = I_{SS} + \frac{v_O}{R_L} \quad (13.102)$$

and must be greater than zero. In this circuit, current cannot go back into the MOSFET source, so the minimum output voltage occurs at the point at which transistor M_1 cuts off. In this situation, $i_S = 0$ and $v_{\text{MIN}} = -I_{SS}R_L$. M_1 cuts off when the input voltage falls to one threshold voltage drop above V_{MIN}: $v_I = -I_{SS}R_L + V_{TN}$.

13.3.2 EFFICIENCY OF CLASS-A AMPLIFIERS

Now consider the source follower in Fig. 13.33 biased with $I_{SS} = V_{SS}/R_L$ and using symmetrical power supplies $V_{DD} = V_{SS}$. Assuming that V_{GS} is much less than the amplitude of v_I, then a sinusoidal output signal can be developed with an amplitude approximately equal to V_{DD},

$$v_O \cong V_{DD} \sin \omega t \quad (13.103)$$

The efficiency ζ of the amplifier is defined as the power delivered to the load at the signal frequency ω, divided by the average power supplied to the amplifier.

The average power P_{av} supplied to the source follower is

$$P_{\text{av}} = \frac{1}{T}\int_0^T \left[I_{SS}(V_{DD} + V_{SS}) + \left(\frac{V_{DD}\sin \omega t}{R_L}\right)V_{DD}\right] dt$$
$$= I_{SS}(V_{DD} + V_{SS}) = 2I_{SS}V_{DD} \quad (13.104)$$

where T is the period of the sine wave. The first term in brackets in Eq. (13.104) is the power dissipation due to the dc current source; the second term results from the ac drain current of the transistor. The last simplification assumes symmetrical power supply voltages. The average of the sine wave current is zero, so the sinusoidal current does not contribute to the value of the integral in Eq. (13.104).

Because the output voltage is a sine wave, the power delivered to the load at the signal frequency is

$$P_{ac} = \frac{\left(\frac{V_{DD}}{\sqrt{2}}\right)^2}{R_L} = \frac{V_{DD}^2}{2R_L} \qquad (13.105)$$

Combining Eqs. (13.104) and (13.105) yields

$$\zeta = \frac{P_{ac}}{P_{av}} = \frac{\frac{V_{DD}^2}{2R_L}}{2I_{SS}V_{DD}} = \frac{1}{4} \quad \text{or} \quad 25\% \qquad (13.106)$$

because $I_{SS}R_L = V_{SS} = V_{DD}$. Thus, a follower, operating as a class-A amplifier, can achieve an efficiency of only 25 percent, at most, for sinusoidal signals (see Prob. 13.116). Equation (13.106) indicates that the low efficiency is caused by the Q-point current I_{SS} that flows continuously between the two power supplies.

13.3.3 CLASS-B PUSH-PULL OUTPUT STAGE

Class-B amplifiers improve the efficiency by operating the transistors at zero Q-point current, eliminating the quiescent power dissipation. A **complementary push-pull** (class-B) **output stage** using CMOS transistors is shown in Fig. 13.34, and the voltage and current waveforms for the composite output stage appear in Fig. 13.35. NMOS transistor M_1 operates as a source follower for positive input signals, and PMOS transistor M_2 operates as a source follower for negative inputs.

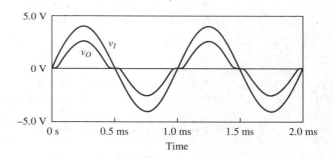

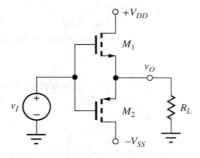

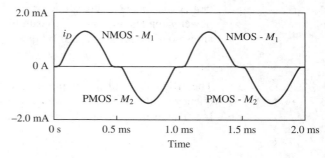

Figure 13.34 Complementary MOS class-B amplifier.

Figure 13.35 Cross-over distortion and drain currents in the class-B amplifier.

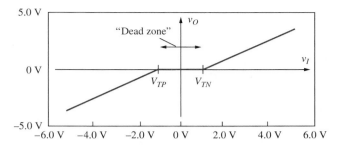

Figure 13.36 SPICE simulation of the voltage transfer characteristic for the complementary class-B amplifier.

Consider the sinusoidal input in Fig. 13.35, for example. As the input voltage v_I swings positive, M_1 turns on supplying current to the load, and the output follows the input on the positive swing. When the input becomes negative, M_2 turns on sinking current from the load, and the output follows the input on the negative swing.

Each transistor conducts current for approximately 180° of the signal waveform, as shown in Fig. 13.35. Because the n- and p-channel gate-source voltages are equal in Fig. 13.34, only one of the two transistors can be on at a time. Also, the Q-point current for $v_O = 0$ is zero, and the efficiency can be high.

However, although the efficiency is high, a distortion problem occurs in the class-B stage. Because V_{GS1} must exceed threshold voltage V_{TN} to turn on M_1, and V_{GS2} must be less than V_{TP} to turn on M_2, a "dead zone" appears in the push-pull class-B voltage transfer characteristic, shown in Fig. 13.36. Neither transistor is conducting for

$$V_{TP} \leq v_{GS} \leq V_{TN} \qquad (-1 \text{ V} \leq v_{GS} \leq 1 \text{ V in Fig. 13.36}) \tag{13.107}$$

This **dead zone,** or **cross-over region,** causes distortion of the output waveform, as shown in the simulation results in Fig. 13.35. As the sinusoidal input waveform crosses through zero, the output voltage waveform becomes distorted. The waveform distortion in Fig. 13.35 is called **cross-over distortion.**

Class-B Efficiency

Simulation results for the currents in the two transistors are also included in Fig. 13.35. If crossover distortion is neglected, then the current in each transistor can be approximated by a half-wave rectified sinusoid with an amplitude of approximately V_{DD}/R_L. Assuming $V_{DD} = V_{SS}$, the average power dissipated from each power supply is

$$P_{av} = \frac{1}{T} \int_0^{T/2} V_{DD} \frac{V_{DD}}{R_L} \sin \frac{2\pi}{T} t \, dt = \frac{V_{DD}^2}{\pi R_L} \tag{13.108}$$

The total ac power delivered to the load is still given by Eq. (13.105), and ζ for the class-B output stage is

$$\zeta = \frac{\frac{V_{DD}^2}{2R_L}}{2\frac{V_{DD}^2}{\pi R_L}} = \frac{\pi}{4} \cong 0.785 \tag{13.109}$$

By eliminating the quiescent bias current, the class-B amplifier can achieve an efficiency of 78.5 percent!

In closed-loop feedback amplifier applications such as those introduced in Chapters 10, 11, and 12, the effects of cross-over distortion are reduced by the loop gain $A\beta$. However, an even better solution is to eliminate the cross-over region by operating the output stage with a small nonzero quiescent current. Such an amplifier is termed a class-AB amplifier.

13.3.4 CLASS-AB AMPLIFIERS

The benefits of the class-B amplifier can be maintained, and cross-over distortion can be minimized by biasing the transistors into conduction, but at a relatively low quiescent current level. The basic technique is shown in Fig. 13.37. Bias voltage V_{GG} is used to establish a small quiescent current in both output transistors. This current is chosen to be much smaller than the peak ac current that will be delivered to the load. In Fig. 13.37, the bias source is split into two symmetrical parts so that $v_O = 0$ for $v_I = 0$.

Because both transistors are conducting for $v_I = 0$, the cross-over distortion can be eliminated, but the additional power dissipation can be kept small enough that the efficiency is not substantially degraded. The amplifier in Fig. 13.37 is classified as a **class-AB amplifier.** Each transistor conducts for more than the 180° of the class-B amplifier but less than the full 360° of the class-A amplifier.

Figure 13.37(b) shows the results of circuit simulation of the voltage transfer characteristic of the class-AB output stage with a quiescent bias current of approximately 60 μA. The distorted cross-over region has been eliminated, even for this small quiescent bias current.

Figure 13.38(a) shows one method for generating the needed bias voltage that is consistent with the CMOS operational amplifier circuit of Fig. 13.29. Bias current I_G develops the required bias voltage for the output stage across resistor R_G. If we assume that $K_p = K_n$ and $V_{TN} = -V_{TP}$ for the MOSFETs, and $v_O = 0$, then the bias voltage splits equally between the gate-source terminals

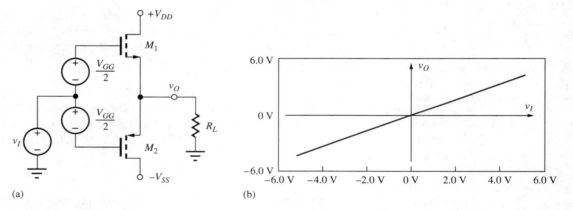

Figure 13.37 (a) Complementary output stage biased for class-AB operation. (b) SPICE simulation of voltage transfer characteristic for class-AB stage with $I_D \cong 60$ μA.

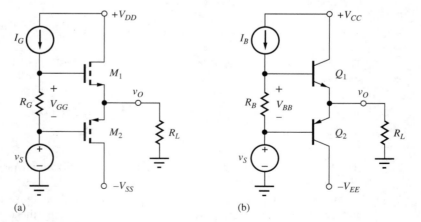

Figure 13.38 (a) Method for biasing the MOS class-AB amplifier. (b) Bipolar class-AB amplifier.

of the two transistors. The drain currents of the two transistors are both

$$I_D = \frac{K_n}{2}\left(\frac{V_{GG}}{2} - V_{TN}\right)^2 \quad (13.110)$$

The bipolar version of the class-AB push-pull output stage employs complementary *npn* and *pnp* transistors, as shown in Fig. 13.29(b). The principle of operation of the bipolar circuit is the same as that for the MOS case. Transistors Q_1 and Q_2 operate as emitter followers for the positive and negative excursions of the output signal, respectively. Current source I_B develops a bias voltage V_{BB} across resistor R_B, which is shared between the base-emitter junctions of the two BJTs.

For class-AB operation, voltage V_{BB} is designed to be approximately $2V_{BE} \cong 1.1$ V but the voltage is kept small enough that the transistors are conducting only a small collector current. If we assume the saturation currents of the two transistors are equal, then the bias voltage V_{BB} splits equally between the base-emitter junctions of the two transistors, and the two collector currents are

$$I_C = I_S \exp\left(\frac{I_B R_B}{2V_T}\right) \quad (13.111)$$

Each transistor is biased into conduction at a low level to eliminate cross-over distortion.

A simplified small-signal model for the class-AB stage is a single follower transistor with a current gain equal to the average of the gains of Q_1 and Q_2 or with a transconductance parameter equal to the average of the values for M_1 and M_2.

A class-B version of the bipolar push-pull output stage is obtained by setting V_{BB} to zero. For this case, the output stage exhibits cross-over distortion for an input voltage range of approximately $2V_{BE}$ (see Prob. 15.7).

EXERCISE: Find the bias current in the transistors in Fig. 13.38(a) for $v_O = 0$ if $K_n = K_p = 25$ mA/V^2, $V_{TN} = 1$ V, and $V_{TP} = -1$ V, $I_G = 500$ μA, and $R_G = 4.4$ kΩ.

ANSWER: 125 μA

EXERCISE: Find the bias current in the transistors in Fig. 13.38(b) for $v_O = 0$ if $I_S = 10$ fA, $I_B = 500$ μA, and $R_B = 2.4$ kΩ.

ANSWER: 265 μA

13.3.5 CLASS-AB OUTPUT STAGES FOR OPERATIONAL AMPLIFIERS

In Fig. 13.39(a) and (b), the follower output stages of the prototype CMOS and bipolar op amps have been replaced with complementary class-AB output stages. Current source I_2, which originally provided a high impedance load to transistors Q_3 and M_3, is also used to develop the dc bias voltage necessary for class-AB operation. The signal current is supplied by transistor M_3 or Q_3, respectively. The total quiescent power dissipation is greatly reduced in both these amplifiers.

13.3.6 SHORT-CIRCUIT PROTECTION

If the output of a follower circuit is accidentally shorted to ground, the transistor can be destroyed due to high current and high power dissipation, or, through direct destruction of the base-emitter junction of the BJT. To make op amps as "robust" as possible, circuitry is often added to the output stage to provide protection from short circuits but is a relatively small addition.

In Fig. 13.40, transistor Q_2 has been added to protect emitter follower Q_1. Under normal operating conditions, the voltage developed across R is less than 0.7 V, transistor Q_2 is cut off,

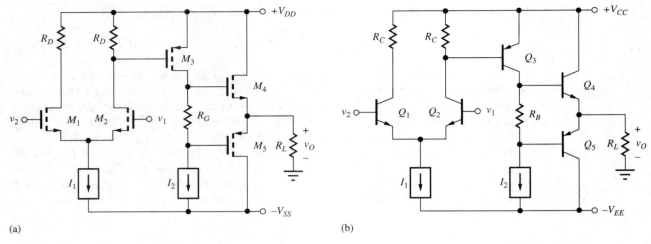

Figure 13.39 Class-AB output stages added to the (a) CMOS and (b) bipolar operational amplifiers.

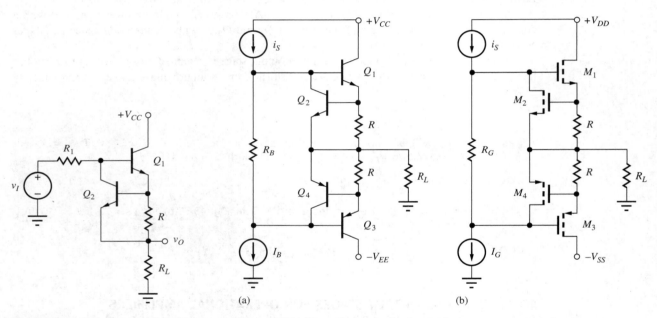

Figure 13.40 Short-circuit protection for an emitter follower.

Figure 13.41 Short-circuit protection for complementary output stages. ($i_S = I_B$ or I_G at the Q-point.)

and Q_1 functions as a normal follower. However, if emitter current I_{E1} exceeds a value of

$$I_{E1} = \frac{V_{BE2}}{R} = \frac{0.7 \text{ V}}{R} \tag{13.112}$$

then transistor Q_2 turns on and shunts any additional current from R_1 down through the collector of Q_2 and away from the base of Q_1. Thus, the output current is limited to approximately the value in Eq. (13.112). For example, $R = 25 \, \Omega$ will limit the maximum output current to 28 mA. Because R is directly in series with the output, however, the output resistance of the follower is increased by the value of R.

Figure 13.41(a) depicts the complementary class AB bipolar output stage including **short-circuit protection**. The *pnp* transistor Q_4 is used to limit the base current of Q_3 in a manner identical to that of Q_2 and Q_1. Similar **current-limiting circuits** can be applied to FET output stages, as shown in Fig. 13.41(b). Here, transistor M_2 steals the current needed to develop gate drive for M_1, and the output current is limited to

$$I_{S1} \cong \frac{V_{GS2}}{R} = \frac{V_{TN2} + \sqrt{\frac{2I_G}{K_{n2}}}}{R} \qquad (13.113)$$

Transistor M_4 provides similar protection to M_3.

13.3.7 TRANSFORMER COUPLING

Designing amplifiers to deliver power to low impedance loads can be difficult. For example, loudspeakers typically have only an 8- or 16-Ω impedance. To achieve good voltage gain and efficiency in this situation, the output resistance of the amplifier needs to be quite low. One approach would be to use a feedback amplifier to achieve a low-output resistance, as discussed in Chapter 12. An alternate approach to the problem is to use **transformer coupling.**

In Fig. 13.42, a follower circuit is coupled to load resistance R_L through an ideal transformer with a turns ratio of $n:1$. In this circuit, coupling capacitor C is required to block the dc path through the primary of the transformer. (See Prob. 13.125 for an alternate approach.)

As defined in network theory, the terminal voltages and currents of the ideal transformer are related by

$$\mathbf{v_1} = n\mathbf{v_2} \qquad \mathbf{i_2} = n\mathbf{i_1} \qquad \frac{\mathbf{v_1}}{\mathbf{i_1}} = n^2 \frac{\mathbf{v_2}}{\mathbf{i_2}} \qquad \text{or} \qquad Z_1 = n^2 Z_L \qquad (13.114)$$

The transformer provides an impedance transformation by the factor n^2. Based on these equations, the transformer and load resistor can be represented by the ac equivalent circuit in Fig. 13.41(b), in which the resistor has been moved to the primary side of the transformer and the secondary is now an open circuit. The effective resistance that the transistor must drive and the voltage at the transformer output are

$$R_{EQ} = n^2 R_L \qquad \text{and} \qquad \mathbf{v_o} = \frac{\mathbf{v_1}}{n} \qquad (13.115)$$

Transformer coupling can reduce the problems associated with driving very low impedance loads. However, the transformer obviously restricts operation to frequencies above dc.

Figure 13.43 is a second example of the use of a transformer, in which an inverting amplifier stage is coupled to the load R_L through the ideal transformer. The dc and ac equivalent circuits appear in Figs. 13.43(b) and (c), respectively. At dc, the transformer represents a short circuit, the

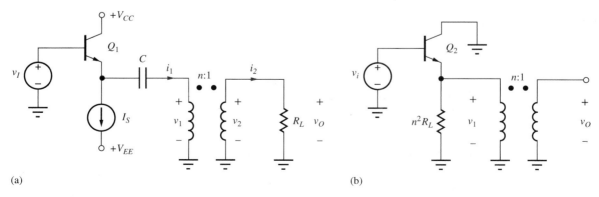

(a)   (b)

Figure 13.42 (a) Follower circuit using transformer coupling. (b) An ac equivalent circuit representation for the follower.

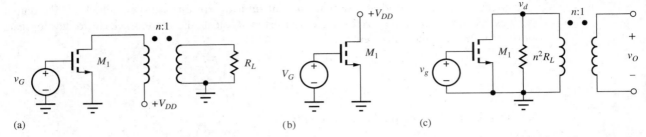

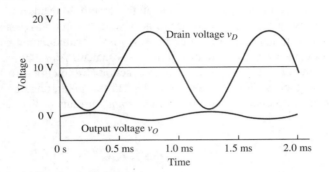

Figure 13.43 (a) Transformer-coupled inverting amplifier; (b) dc equivalent circuit; (c) ac equivalent circuit.

Figure 13.44 SPICE simulation of the transformer-coupled inverting amplifier stage for $n = 10$ with $V_{DD} = 10$ V.

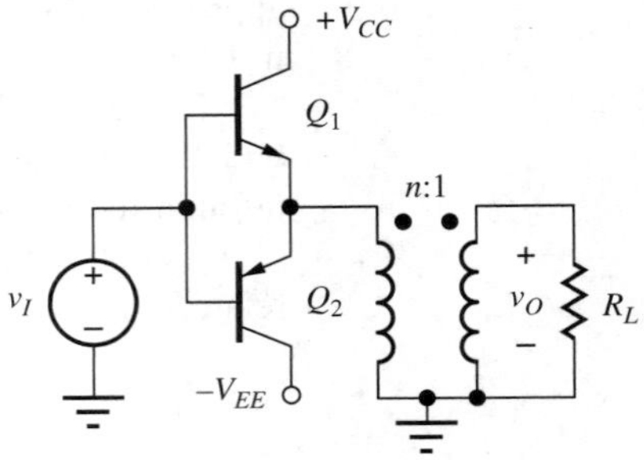

Figure 13.45 Transformer-coupled class-B output stage.

full dc power supply voltage appears across the transistor, and the quiescent operating current of the transistor is supplied through the primary of the transformer. At the signal frequency, a load resistance equal to $n^2 R_L$ is presented to the transistor.

Results of simulation of the circuit in Fig. 13.42 are in Fig. 13.44 for the case $R_L = 8 \, \Omega$, $V_{DD} = 10$ V, and $n = 10$. The behavior of this circuit is different from most that we have studied. The quiescent voltage at the drain of the MOSFET is equal to the full power supply voltage V_{DD}. The presence of the inductance of the transformer permits the signal voltage to swing symmetrically above and below V_{DD}, and the peak-to-peak amplitude of the signal at the drain can approach $2V_{DD}$.

Figure 13.45 is a final circuit example, which shows a transformer-coupled class-B output stage. Because the quiescent operating currents in Q_1 and Q_2 are zero, the emitters may be connected directly to the primary of the transformer.

> **EXERCISE:** Find the small-signal voltage gains
>
> $$A_{vi} = \frac{v_d}{v_g} \quad \text{and} \quad A_{vo} = \frac{v_o}{v_g}$$
>
> for the circuit in Fig. 13.43 if $V_{TN} = 1$ V, $K_n = 50$ mA/V^2, $V_G = 2$ V, $V_{DD} = 10$ V, $R_L = 8 \, \Omega$, and $n = 10$. What are the largest values of v_g, v_d, and v_o that satisfy the small-signal limitations?
>
> **ANSWERS:** -40, -4; 0.2 V, 8 V, 0.8 V

ELECTRONICS IN ACTION

Class-D Audio Amplifiers

As mentioned in the main body of the text, the efficiency of class-A, -B, and -AB amplifiers is limited to less than 80 percent. To achieve higher efficiencies, a number of forms of switching amplifiers have been developed for use in portable and other low-power electronic applications. One of these is the class-D amplifier shown here, in which the output is a pulse-width modulated (PWM) signal that switches rapidly between the positive and negative power supplies. High efficiency is achieved by using CMOS transistors as switches. In a manner similar to a CMOS inverter, the goal is to have only one transistor on at a given time.

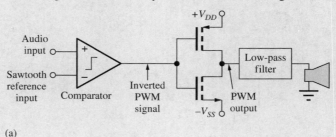

(a)

Conceptual implementation of a class-D audio amplifier.

A basic PWM signal can be generated by comparing the audio input signal to a sawtooth reference waveform. Referring to the sample waveforms, we see that the PWM output is switched high to V_{DD} when the audio input exceeds the reference waveform, and the output is switched to $-V_{SS}$ when the reference input exceeds the analog input. In the waveform illustration, the sawtooth reference input is operating at a frequency that is 10 times that of the sinusoidal input. For an audio signal with a bandwidth of 20 Hz to 20 kHz, the reference frequency may range from 250 kHz to more than 1 MHz. Before being fed to the speaker, the PWM signal is passed through a low-pass filter to remove the unwanted high frequency content.

In order to achieve higher power levels with a given supply voltage, the load is often driven in a differential fashion using a complementary "H-bridge." In the CMOS version shown here, output voltage v_O equals $(V_{DD} + V_{SS})$ when input V is high, and v_O equals $-(V_{DD} + V_{SS})$

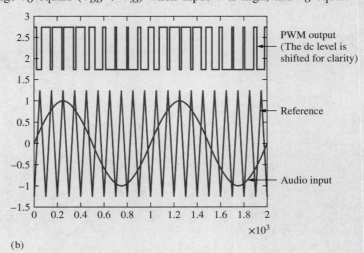

(b)

Illustration of PWM waveforms.

when input V is low. Thus, the total signal swing across the load is twice the sum of the power supply span, and the power that can be delivered to the load is four times that achieved without the H-bridge. A class-D amplifier using the H-bridge appears in the final figure in which the speaker is driven by the low-pass filtered output of the CMOS H-bridge.

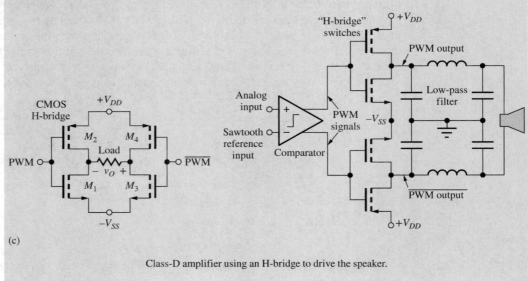

Class-D amplifier using an H-bridge to drive the speaker.

13.4 ELECTRONIC CURRENT SOURCES

The dc current source is clearly a fundamental and highly useful circuit component. In Sec. 13.3, we found that multiple current sources could be used to provide bias to the BJT and MOS op amp prototypes as well as to improve their performance. This section first explores the basic circuits used to realize electronic versions of ideal current sources and then explores current source design in more depth by looking at techniques specifically applicable to the design of integrated circuits.

In Fig. 13.46, the current-voltage characteristics of an ideal current source are compared with those of resistor and transistor current sources of Fig. 13.47. Current I_O through the ideal source

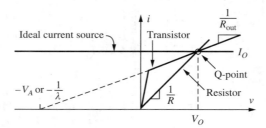

Figure 13.46 i-v Characteristics of basic electronic current sources.

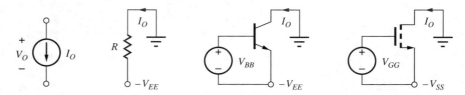

Figure 13.47 Ideal, resistor, BJT, and MOS current sources.

is independent of the voltage appearing across the source, and the output resistance of the ideal source is infinite, as indicated by the zero slope of the current source i-v characteristic.

For the ideal source, the voltage across the source can be positive or negative, and the current remains the same. However, **electronic current sources** must be implemented with resistors and transistors, and their operation is usually restricted to only one quadrant of the total i-v space. In addition, electronic sources have a finite output resistance, as indicated by the nonzero slope of the i-v characteristic. We will find that the output resistance of the transistor is much greater than a resistor for an equivalent Q-point.

In many cases, the circuit elements in Fig. 13.47 will actually be *sinking* current from the rest of the network, and some authors prefer to call these elements **current sinks**. In this book, we use the generic term *current source* to refer to both sinks and sources.

13.4.1 SINGLE-TRANSISTOR CURRENT SOURCES

The simplest forms of electronic current sources are shown in Fig. 13.47. A resistor is often used to establish bias currents in many circuits—differential amplifiers, for example—but it represents our poorest approximation to an ideal current source. Individual transistor implementations of current sources generally operate in only one quadrant because the transistors must be biased in the active regions in order to maintain high impedance operation. However, the transistor source can realize very high values of output resistance.

For simplicity, the transistors in Fig. 13.47 are biased into conduction by sources V_{BB} and V_{GG}. In these circuits, we assume that the collector-emitter and drain-source voltages are large enough to ensure operation in the forward-active or pinch-off (active) regions, as appropriate for each device.

13.4.2 FIGURE OF MERIT FOR CURRENT SOURCES

Resistor R in Fig. 13.47 will be used as a reference for comparing current sources. The resistor provides an output current and output resistance of

$$I_O = \frac{V_{EE}}{R} \quad \text{and} \quad R_{\text{out}} = R \quad (13.116)$$

The product of the dc current I_O and output resistance R_{out} is the effective voltage V_{CS} across the current source, and we will use it as a **figure of merit (FOM)** for comparing various current sources:

$$V_{CS} = I_O R_{\text{out}} \quad (13.117)$$

For a given Q-point current, V_{CS} represents the equivalent voltage that will be needed across a resistor for it to achieve the same output resistance as the given current source. The larger the value of V_{CS}, the higher the output resistance of the source. For the resistor itself, V_{CS} is simply equal to the power supply voltage V_{EE}.

If ac models are drawn for each transistor source in Fig. 13.47, the base, emitter, gate, and source of each transistor will be connected to ground, and each transistor will be considered operating in either the common-source or common-emitter configuration. The output resistance therefore will be equal to r_o in all cases, and the figures of merit for these sources will be

$$\begin{aligned} \text{BJT:} \quad & V_{CS} = I_O R_{\text{out}} = I_C r_o = I_C \frac{V_A + V_{CE}}{I_C} = V_A + V_{CE} \cong V_A \\ \text{FET:} \quad & V_{CS} = I_O R_{\text{out}} = I_D r_o = I_D \frac{\frac{1}{\lambda} + V_{DS}}{I_D} = \frac{1}{\lambda} + V_{DS} \cong \frac{1}{\lambda} \end{aligned} \quad (13.118)$$

V_{CS} for the C-E/C-S transistor current sources is approximately equal to either the Early voltage V_A or $1/\lambda$. We can expect that both these values generally will be at least several times the available power supply voltage. Therefore, any of the single transistor sources will provide an output resistance that is greater than that of a resistor.

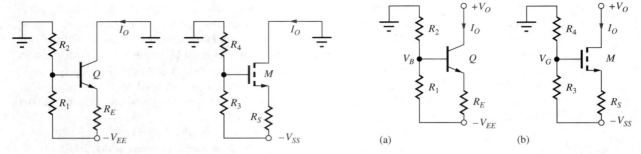

Figure 13.48 High-output resistance current sources.

Figure 13.49 (a) The *npn* and (b) NMOS current source circuits.

TABLE 13.1
Comparison of the Basic Current Sources $\beta_o = 100$, $V_A = 1/\lambda = 50$ V, $\mu_{f_{FET}} = 100$

TYPE OF SOURCE	R_{out}	V_{CS}	TYPICAL VALUES
Resistor	R	V_{EE}	15 V
Single transistor	r_o	V_A or $\frac{1}{\lambda}$	50–100 V
BJT with emitter resistor R_E	$\beta_o r_o$	$\cong \beta_o V_A$	5000 V
FET with source resistor R_S ($V_{SS} = 15$ V)	$\mu_f R_S$	$\cong \mu_f \dfrac{V_{SS}}{3}$	500 V or more

EXERCISE: Draw the small signal models for the two transistor sources in Fig. 13.47 and convince yourself that $R_{out} = r_o$.

13.4.3 HIGHER OUTPUT RESISTANCE SOURCES

From our study of single-stage amplifiers in Chapters 7 and 8, we know that placing a resistor in series with the emitter or source of the transistor, as in Fig. 13.48, increases the output resistance. Referring back to Eqs. (8.32) and (8.49), we find that the output resistances for the circuits in Fig. 13.48 are

$$\text{BJT:} \quad R_{out} = r_o\left[1 + \frac{\beta_o R_E}{R_1 \| R_2 + r_\pi + R_E}\right] \leq (\beta_o + 1) r_o \tag{13.119}$$

and

$$\text{FET:} \quad R_{out} = r_o(1 + g_m R_S) \cong \mu_f R_S \tag{13.120}$$

The figures of merit are

$$\text{BJT:} \ V_{CS} \cong \beta_o(V_A + V_{CE}) \cong \beta_o V_A \quad \text{and} \quad \text{FET:} \ V_{CS} \cong \mu_f \frac{V_{SS}}{3} \tag{13.121}$$

where it has been assumed that $I_o R_S \cong V_{SS}/3$. Based on these figures of merit, the output resistance of the current sources in Fig. 13.48 can be expected to reach very high values, particularly at low-current levels. Table 13.1 compares V_{CS} for the various sources for typical device parameter values.

DESIGN NOTE

Because of its importance in analog circuit design, the $\beta_o V_A$ product is often used as a basic figure of merit for the bipolar transistor.

13.4.4 CURRENT SOURCE DESIGN EXAMPLES

This section provides examples of the design of current sources using the three-resistor bias circuits in Fig. 13.49. The computer (via a spreadsheet) is used to help explore the design space. The current source requirements are provided in the following design specifications.

Design Specifications

Design a current source using the circuits in Fig. 13.49 with a nominal output current of 200 μA and an output resistance greater than 10 MΩ using a single −15-V power supply. The source must also meet the following additional constraints.

Output voltage (compliance) range should be as large as possible while meeting the output resistance specification.

The total current used by the source should be less than 250 μA.

Bipolar transistors are available with (β_o, V_A) of (80, 100 V) or (150, 75 V). FETs are available with $\lambda = 0.01$ V^{-1}; K_n can be chosen as necessary.

When used in an actual application, the collector and drain of the current sources in Fig. 13.49 will be connected to some other point in the overall circuit, as indicated by the voltage $+V_O$ in the figure. For the current source to provide a high-output resistance, the BJT must remain in the active region, with the collector-base junction reverse-biased ($V_O \geq V_B$), or the FET must remain in pinchoff ($V_O \geq V_G - V_{TN}$).

Specifications include the requirement that the output voltage range be as large as possible. Thus, the design goal is to achieve $I_O = 200$ μA and $R_{\text{out}} \geq 10$ MΩ with as low a voltage as possible at V_B or V_G. A range of designs is explored to see just how low a voltage can be used at V_B or V_G and still meet the I_O and R_{out} requirements. Investigating this design space is most easily done with the aid of the computer.

DESIGN EXAMPLE 13.5 — DESIGN OF A BIPOLAR TRANSISTOR CURRENT SOURCE

Here, we design a current source to meet a given set of design specifications using a bipolar transistor and the three resistor bias circuit. Example 13.6 explores the NMOS current source design.

PROBLEM Design a current source using the circuit in Fig. 13.49(a) with a nominal output current of 200 μA and an output resistance greater than 10 MΩ using a single −15-V power supply. The source must also meet the following additional constraints.

Output voltage (compliance) range should be as large as possible while meeting the output resistance specification.

The total current used by the source should be less than 250 μA.

Bipolar transistors are available with (β_o, V_A) of (80, 100 V) or (150, 75 V).

SOLUTION **Known Information and Given Data:** Current source circuit in Fig. 13.49(a); $I_O = 200$ μA; $V_{EE} = 15$ V; $I_{EE} < 250$ μA; $R_{\text{out}} > 10$ MΩ; V_B as low as possible; BJTs are available with (β_o, V_A) of (80, 100 V) and (150, 75 V)

Unknowns: Values of resistors R_1, R_2, and R_E

Approach: Set up equations for analysis; using a computer program or spreadsheet, search for a set of bias conditions that satisfy the requirements; choose nearest resistor values from 1 percent resistor table in Appendix A.

Assumptions: Active region and small-signal operating conditions apply; $V_{BE} = 0.7$ V; $V_T = 0.025$ V; choose $V_O = 0$ V as a representative value for the output voltage.

Analysis: We start the design of the bipolar version of the current source with the expression for the output resistance of the source. Because we will use a computer to help in the design, we use the most complete expression for the output resistance:

$$R_{\text{out}} = r_o\left[1 + \frac{\beta_o R_E}{R_E + r_\pi + R_1 \| R_2}\right] \leq \beta_o r_o \tag{13.122}$$

The figure of merit for this source is

$$V_{CS} = I_o R_{\text{out}} \leq \beta_o V_A \tag{13.123}$$

and the design specifications require

$$\beta_o V_A = I_o R_{\text{out}} \geq (200 \text{ μA})(10 \text{ MΩ}) = 2000 \text{ V} \tag{13.124}$$

Although both the specified transistors easily meet the requirement of Eq. (13.124), the denominator of Eq. (13.123) can substantially reduce the output resistance below that predicted by the $\beta_o r_o$ limit. Thus, it will be judicious to select the transistor with the higher $\beta_o V_A$ product—that is, (150, 75 V).

Having made this decision, the equations relating the dc Q-point design to the output resistance of the source can be developed. In Fig. 13.50, the three-resistor bias circuit is simplified using a $-V_{EE}$ referenced Thévenin transformation, for which

$$V_{BB} = 15\frac{R_1}{R_1 + R_2} = 15\frac{R_{BB}}{R_2} \quad \text{with} \quad R_{BB} = \frac{R_1 R_2}{R_1 + R_2} \tag{13.125}$$

The Q-point can be calculated using

$$I_B = \frac{V_{BB} - V_{BE}}{R_{BB} + (\beta_F + 1)R_E} \qquad I_O = I_C = \beta_F I_B$$

and

$$V_{CE} = V_O + V_{EE} - (V_{BB} - I_B R_{BB} - V_{BE}) \tag{13.126}$$

The small-signal parameters required for evaluating Eq. (13.122) are given by their usual formulas:

$$r_o = \frac{V_A + V_{CE}}{I_C} \quad \text{and} \quad r_\pi = \frac{\beta_o V_T}{I_C} \tag{13.127}$$

From Eq. (13.122), we can see that $R_{BB} = (R_1 \| R_2)$ should be made as small as possible in order to achieve maximum output resistance. From the design specifications, the complete current

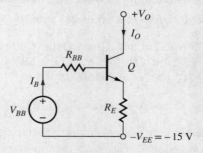

Figure 13.50 Equivalent circuit for the current source.

source must use no more than 250 µA. Because the output current is 200 µA, a maximum current of 50 µA can be used by the base bias network. The bias network current should be a factor of 5 to 10 times larger than the base current of the transistor which is 1.33 µA for the transistor with a current gain of 150. Thus, a bias network current of 20 µA is more than enough. However, in this case, we will trade increased operating current for a higher output resistance by picking a bias network current of 40 µA, which sets the sum of R_1 and R_2 to be (neglecting base current)

$$R_1 + R_2 \cong \frac{15 \text{ V}}{40 \text{ µA}} = 375 \text{ k}\Omega \tag{13.128}$$

Equations (13.122) to (13.128) provide the information necessary to explore the design space with the aid of a computer. These equations have been rearranged in order of evaluation in Eq. (13.129), with V_{BB} selected as the primary design variable.

Once V_{BB} is selected, R_1 and R_2 can be calculated. Then R_E and the Q-point can be determined, the small-signal parameters evaluated, and the output resistance determined from Eq. (13.119).

$$I_B = \frac{I_o}{\beta_F}$$

$$R_1 = (R_1 + R_2)\frac{V_{BB}}{15} = 375 \text{ k}\Omega\left(\frac{V_{BB}}{15}\right)$$

$$R_2 = (R_1 + R_2) - R_1 = 375 \text{ k}\Omega - R_1$$

$$R_{BB} = R_1 \parallel R_2$$

$$R_E = \alpha_F\left[\frac{V_{BB} - V_{BE} - I_B R_{BB}}{I_o}\right] \tag{13.129}$$

$$V_{CE} = V_{EE} - (V_{BB} - I_B R_{BB} - V_{BE})$$

$$r_o = \frac{V_A + V_{CE}}{I_o} \qquad r_\pi = \frac{\beta_o V_T}{I_o}$$

$$R_{\text{out}} = r_o\left[1 + \frac{\beta_o R_E}{R_{BB} + r_\pi + R_E}\right]$$

Table 13.2 presents the results of using a spreadsheet to assist in evaluating these equations for a range of V_{BB}. The smallest value of V_{BB} for which the output resistance exceeds 10 MΩ with some safety margin is 4.5 V, as indicated by the bold part in the table. Note that this value of output resistance is achieved as

$$R_{\text{out}} = 432 \text{ k}\Omega\left[1 + \frac{150(18.4 \text{ k}\Omega)}{(78.8 + 18.8 + 18.4) \text{ k}\Omega}\right] = 10.7 \text{ M}\Omega \tag{13.130}$$

TABLE 13.2
Spreadsheet Results for Current Source Design

V_{BB}	R_1	R_2	R_{BB}	R_E	r_o	R_{out}
1.0	2.50E + 04	3.50E + 05	2.33E + 04	1.34E + 03	4.49E + 05	2.52E + 06
2.0	5.00E + 04	3.25E + 05	4.33E + 04	6.17E + 03	4.44E + 05	6.46E + 06
3.0	7.50E + 04	3.00E + 05	6.00E + 04	1.10E + 04	4.39E + 05	8.52E + 06
3.5	8.75E + 04	2.88E + 05	6.71E + 04	1.35E + 04	4.36E + 05	9.31E + 06
4.0	1.00E + 05	2.75E + 05	7.33E + 04	1.59E + 04	4.34E + 05	1.00E + 07
4.5	**1.13E + 05**	**2.63E + 05**	**7.88E + 04**	**1.84E + 04**	**4.32E + 05**	**1.07E + 07**
5.0	1.25E + 05	2.50E + 05	8.33E + 04	2.08E + 04	4.29E + 05	1.13E + 07
5.5	1.38E + 05	2.38E + 05	8.71E + 04	2.33E + 04	4.27E + 05	1.20E + 07
6.0	1.50E + 05	2.25E + 05	9.00E + 04	2.57E + 04	4.24E + 05	1.26E + 07

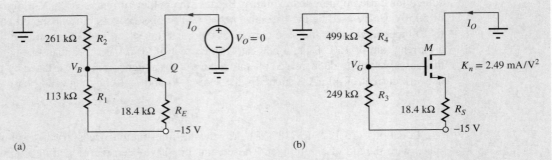

Figure 13.51 Final current source designs with $I_O = 200$ μA and $R_{out} \geq 10$ MΩ.

Check of Results: Analysis of the circuit with the 1 percent resistor values in Fig. 13.51(a) yields $I_O = 203$ μA, $R_{out} = 10.4$ MΩ, and the supply current is 244 μA.

Discussion: For this design, the denominator in Eq. (13.130) reduces the output resistance by a factor of 6.3 below the $\beta_o r_o$ limit. So, it was a wise decision to choose the transistor with the largest $\beta_o V_A$ product. The final design appears in Fig. 13.51 using the nearest values from the 1 percent table in Appendix A.

Computer-Aided Analysis: Now, we can check our hand design using SPICE with BF = 150, VAF = 75 V, and IS = 0.5 fA. (IS is selected to give $V_{BE} \cong 0.7$ V for a collector current or 200 μA.) In the circuit shown here, zero-value source V_O is added to directly measure the output current I_O as well as to provide a source that can be used to find R_{out} with a SPICE transfer function analysis. The results are $R_{out} = 11.4$ MΩ with $I_O = 205$ μA and $I_{EE} = 245$ μA, which meet all the design specifications. This could also be a good point to do a Monte Carlo analysis to explore the influence of tolerances on the design.

EXERCISE: What is the output resistance of the bipolar current source if the base were bypassed to ground with a capacitor?

ANSWER: 32.5 MΩ.

EXERCISE: The current source is to be implemented using the nearest 5 percent resistor values. What are the best values? Are resistors with a 1/4-W power dissipation rating adequate for use in this circuit? What are the actual output current and output resistance of your current source, based on these 5 percent resistor values?

ANSWERS: 110 kΩ, 270 kΩ, 18 kΩ; yes; 195 μA, 10.9 MΩ.

EXERCISE: Rework Design Ex. 13.5 using a bias network current of 20 μA. What are the new values of V_{BB}, R_1, R_2, R_E, and R_{out}?

ANSWERS: 9 V; 450 kΩ; 300 kΩ; 40.0 kΩ; 10.7 MΩ.

DESIGN EXAMPLE 13.6

DESIGN OF A MOSFET CURRENT SOURCE

Now, we design a current source to meet the same set of design specifications as in Ex. 13.5 but with a MOSFET replacing the BJT.

PROBLEM Design a current source using the circuit in Fig. 13.52 with a nominal output current of 200 μA and an output resistance greater than 10 MΩ using a single −15-V power supply. The source must also meet the following additional constraints.

- Output voltage (compliance) range should be as large as possible while meeting the output resistance specification.
- The total current used by the source should be less than 250 μA.
- MOS transistors are available with $\lambda = 0.01$ V^{-1}. K_n can be chosen as required.

SOLUTION **Known Information and Given Data:** Current source circuit in Fig. 13.52; $I_O = 200$ μA; $V_{SS} = 15$ V; $I_{SS} < 250$ μA; $R_{\text{out}} > 10$ MΩ; V_{GG} as low as possible; MOS transistors are available with $\lambda = 0.01$ V^{-1}. K_n can be chosen as required.

Unknowns: Values of resistors R_3, R_4, and R_S.

Approach: Use $R_S = R_E$ and $V_S = V_E$ from the bipolar design in the previous example so the two designs can be easily compared. Find the intrinsic gain and value of K_n required to meet the output resistance requirement. Find V_{GS} and V_{GG}, and then choose R_3 and R_4 from the 1 percent resistor table in Appendix A.

Assumptions: Active region and small-signal operating conditions apply; $V_{TN} = 1$ V; choose $V_O = 0$ V as a representative value for the output voltage.

Analysis: We begin the design of the MOSFET current source by writing the expression for the transistor's output resistance. Because of the infinite current gain of the MOSFET, the expression for the output resistance of the current source is much less complex than that of the BJT source and is given by

$$R_{\text{out}} = r_o(1 + g_m R_S) \cong \mu_f R_S$$

If values of R_S and V_S are selected that are the same as those of the BJT source, 18 kΩ and −11.4 V, respectively, then the MOSFET must have an intrinsic gain of

$$\mu_f \geq \frac{10 \text{ M}\Omega}{18 \text{ k}\Omega} = 556 \gg 1$$

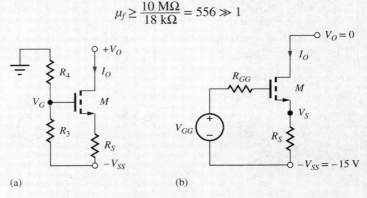

Figure 13.52 (a) MOSFET current source. (b) Equivalent circuit.

The intrinsic gain of the MOSFET is given by

$$\mu_f = \frac{1}{\lambda}\sqrt{\frac{2K_n}{I_D}}(1+\lambda V_{DS})$$

and solving for K_n yields

$$K_n = \frac{I_D}{2}\left(\frac{\lambda \mu_f}{1+\lambda V_{DS}}\right)^2 = 100\ \mu A \left(\frac{\frac{0.01}{V}(556)}{1+\frac{0.01}{V}(11.5\ V)}\right)^2 = 2.49\ \frac{mA}{V^2}$$

This value of K_n is achievable using either discrete components or integrated circuits.

In Fig. 13.52, the required gate voltage V_{GG} is

$$V_{GG} = I_D R_S + V_{GS} = 3.60 + V_{TN} + \sqrt{\frac{2I_D}{K_n}}$$

$$= 3.60\ V + 1\ V + \sqrt{\frac{2(0.2\ mA)}{2.49\ \frac{mA}{V^2}}} = 5.00\ V$$

If the current in the bias resistors is limited to 10 percent of the drain current, then

$$R_3 + R_4 = \frac{15\ V}{20\ \mu A} = 750\ k\Omega \quad \text{and} \quad R_3 = \frac{5.00\ V}{15\ V} 750\ k\Omega = 250\ k\Omega$$

The nearest 1 percent values from Appendix A are $R_3 = 249\ k\Omega$ and $R_4 = 499\ k\Omega$ with $R_S = 18.2\ k\Omega$. The final design appears in the figure below.

Check of Results: A recheck of the math indicates that our calculations are correct. SPICE can now be used to verify our design and the results appear below.

Discussion: For the MOS source, we can use a larger set of gate bias resistors, since the output resistance of the current source does not depend on R_{GG}.

Computer-Aided Analysis: Now, we can check our hand design using SPICE with VTO = 1 V, KP = 2.49 mA/V^2, and $\lambda = 0.01\ V^{-1}$. In the circuit shown here, zero-value source V_O is added to directly measure the output current I_O and to provide a source that can be used to find R_{out} with a SPICE Transfer Function analysis. The results using the 1 percent resistor values are $R_{out} = 11.3\ M\Omega$ with $I_O = 198\ \mu A$ and $I_{SS} = 219\ \mu A$, which meet all the design specifications. This could be a good point to do a Monte Carlo analysis to explore the influence of tolerances on the design. Also, more complex SPICE models can be used to double check the design.

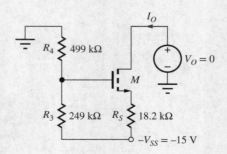

EXERCISE: What is the minimum drain voltage for which MOSFET *M* in the circuit on page 924 remains saturated?

ANSWER: −9.96 V

EXERCISE: What *W/L* ratio is required for the preceding FET if $K'_n = 25\ \mu A/V^2$?

ANSWER: 99.6/1

EXERCISE: What is the minimum collector voltage for which the BJT in Fig. 13.51(a) remains in the forward-active region?

ANSWER: −10.6 V

EXERCISE: The MOS current source is to be implemented using the nearest 5 percent resistor values. What are the best values? Are resistors with a 1/4-W power dissipation rating adequate for use in this circuit? What are the actual output current and output resistance of your current source based on these 5 percent resistor values?

ANSWERS: 510 kΩ, 240 kΩ, 18 kΩ; yes; 189 μA, 10.3 MΩ

ELECTRONICS IN ACTION

Medical Ultrasound Imaging

Medical ultrasound imaging systems are widely used in clinical applications for many diagnostic procedures such as characterization of tumors, measurement of cardiac function, and monitoring of prenatal development. Ultrasound systems work by sending 1 to 20 MHz acoustic pulses into the body and then measuring the acoustic echo. Different types of tissue absorb different amounts of acoustic energy, so the acoustic return varies with tissue type and characteristic. In order to measure tissue properties at specific points within the body, a phased array technique is used to focus the transmit and receive pulses. For example, a simplified view of the receive process is depicted below. The acoustic propagation time of the reflected wave varies with the distance from a particular transducer element to the focus point, resulting in a set of received waves separated in time. By introducing the appropriate delays to the received waves, they can then be summed. Random noise will average out, but the signal of interest adds coherently. The transmit process is also focused by time-varying the pulses driven onto each of the transducer elements.

A more detailed look at the electronics of an ultrasound system is shown in the accompanying figures. Because of the lossy nature of the transducers and body tissue, the received ultrasound signal is extremely small, often on the order of microvolts. As a consequence, the analog preamp must be a very low noise, multistage amplifier. Total gain is about 100 dB. With such a high gain, it is important that the amplifier be either ac coupled or have some form of offset-correction. For example, if the amplifier has an input offset of 5 mV, a gain of 100 dB would yield an output that is clipped. Another interesting aspect of ultrasound preamplifiers is the need for time gain control (TGC). As an ultrasound signal propagates through the body, it is heavily attenuated. The longer a signal propagates, the more it attenuates. This is compensated with a circuit that continuously varies the gain of the amplifier over a 60 to 80 dB range during the few microseconds required to receive an ultrasonic waveform.

After the preamp, the signal is sampled and then digitized. In a typical 128-channel system with 40 MSample/s 10-bit ADCs, the total data rate is 6.4 GB/s. This vast data pipeline is processed by several custom ASICs which digitally perform the real-time delay and summing operations, correcting for many non-idealities not described here.

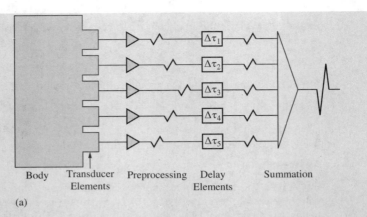

(a) Simplified view of ultrasound receive focusing.

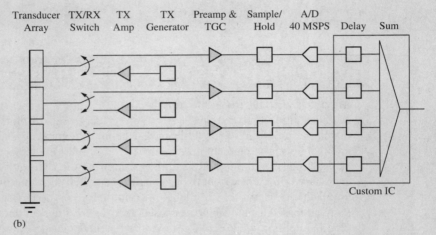

(b) Block diagram of typical commercial ultrasound system transmit/receive electronics.

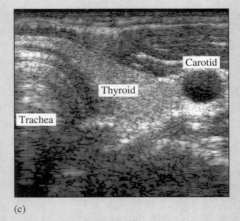

(c) Ultrasound image of trachea, thyroid, and carotid artery.*

---

*Ultrasound image appears courtesy of William F. Walker, University of Virginia.

Modern medical systems, such as the one shown here, are tremendous opportunities for innovative circuit design. As medical knowledge increases, it is increasingly important to accurately measure physiological responses and interactions to properly apply and utilize new understanding.

SUMMARY

In most situations, the single-stage amplifiers discussed in Chapters 7 and 8 cannot simultaneously meet all the requirements of an application (e.g., high-voltage gain, high-input resistance, and low-output resistance). Therefore, we must combine single-stage amplifiers in various ways to form multistage amplifiers that achieve higher levels of overall performance.

- Both ac- and dc-coupling (also called direct-coupling) methods are used in multistage amplifiers depending on the application. Ac coupling allows the Q-point design of each stage to be done independently of the other stages, and bypass capacitors can be utilized to eliminate bias elements from the ac equivalent circuit of the amplifier. However, dc coupling can eliminate circuit elements, including both coupling capacitors and bias resistors, and can represent a more economical approach to design. In addition, direct coupling achieves a low-pass amplifier that provides high gain at dc, and dc-coupling is utilized in most op amp designs.

- The most important dc-coupled amplifier is the symmetric two-transistor differential amplifier. Not only is the differential amplifier a key circuit in the design of operational amplifiers, but it is also a fundamental building block of all analog circuit design. In this chapter, we studied BJT and MOS differential amplifiers in detail. Differential-mode gain, common-mode gain, common-mode rejection ratio (CMRR), and differential- and common-mode input and output resistances of the amplifier are all directly related to transistor parameters and, hence, Q-point design.

- Either a balanced or a single-ended output is available from the differential amplifier. The balanced output provides a voltage gain that is twice that of the single-ended output, and the CMRR of the balanced output is inherently much higher (infinity for the ideal case).

- One of the most important applications of differential amplifiers is to form the input stage of the operational amplifier. By adding a second gain stage plus an output stage to the differential amplifier, a basic op amp is created. The performance of differential and operational amplifiers can be greatly enhanced by the use of electronic current sources. Op amp designs usually require a number of current sources, and, for economy of design, these multiple sources are often generated from a single-bias voltage.

- An ideal current source provides a constant output current, independent of the voltage across the source; that is, the current source has an infinite output resistance. Although electronic current sources cannot achieve infinite output resistance, very high values are possible, and there are a number of basic current source circuits and techniques for achieving high-output resistance.

- For a current source, the product of the source current and output resistance represents a figure of merit, V_{CS}, that can be used to compare current sources. A single-transistor current source can be built using the bipolar transistor in which V_{CS} can approach the $\beta_o V_A$ product of the BJT. For a very good bipolar transistor, this product can reach 10,000 V. For the FET case, V_{CS} can approach a significant fraction of $\mu_f V_{SS}$, in which V_{SS} represents the power supply voltage. Values well in excess of 1000 V are achievable with the FET source.

- The electronic current source can be modeled in SPICE as a dc current source in parallel with a resistor equal to the output resistance of the source. For greatest accuracy, the value of the dc source should be adjusted to account for any dc current existing in the output resistance.

- The Darlington connection of two *npn* or *pnp* transistors creates a much higher current gain replacement for a single *npn* or *pnp* device.
- Class-A, class-B, and class-AB amplifiers are defined in terms of their conduction angles: 360° for class-A, 180° for class-B, and between 180° and 360° for class-AB operation. The efficiency of the class-A amplifier cannot exceed 25 percent for sinusoidal signals, whereas that of the class-B amplifier has an upper limit of 78.5 percent. However, class-B amplifiers suffer from cross-over distortion caused by a dead zone in the transfer characteristic.
- The class-AB amplifier trades a small increase in quiescent power dissipation and a small loss in efficiency for elimination of the cross-over distortion. The efficiency of the class-AB amplifier can approach that of the class-B amplifier when the quiescent operating point is properly chosen. The basic op-amp design can be further improved by replacing the class-A follower output stage with a class-AB output stage. Class-AB output stages are often used in operational amplifiers and are usually provided with short-circuit protection circuitry.
- Amplifier stages may also employ transformer coupling. The impedance transformation properties of the transformer can be used to simplify the design of circuits that must drive low values of load resistances, such as loudspeakers, headphones, or earbuds.
- Integrated circuit (IC) technology permits the realization of large numbers of virtually identical transistors. Although the absolute parameter tolerances of these devices are relatively poor, device characteristics can actually be matched to within less than 1 percent. The availability of large numbers of such closely matched devices has led to the development of special circuit techniques that depend on the similarity of device characteristics for proper operation. These matched circuit design techniques are used throughout analog circuit design and produce high-performance circuits that require very few resistors.

KEY TERMS

Balanced output
Class-A, class-B, and class-AB amplifiers
Class-B push-pull output stage
Common-mode conversion gain
Common-mode gain
Common-mode half-circuit
Common-mode input resistance
Common-mode input voltage
Common-mode input voltage range
Common-mode output voltage
Common-mode rejection ratio (CMRR)
Complementary push-pull output stage
Conduction angle
Cross-over distortion
Cross-over region
Current-limiting circuit
Current sink
Darlington pair
dc-coupled (direct-coupled) amplifiers
Dead zone
Differential amplifier
Differential-mode conversion gain
Differential-mode gain
Differential-mode half-circuit
Differential-mode input resistance
Differential-mode input voltage
Differential-mode output resistance
Differential-mode output voltage
Electronic current source
Figure of merit (FOM)
Half-circuit analysis
Ideal current source
Level shift
Short-circuit protection
Single-ended output
Transformer coupling
Virtual ground

REFERENCES

1. R. D. Thornton et. al., *Multistage Transistor Circuits,* SEEC Volume 5, Wiley, New York: 1965.
2. H-M. Rein, "Multi-gigabit-per-second silicon bipolar IC's for future optical-fiber transmission systems," *IEEE Journal of Solid-State Circuits,* vol. 23, no. 3, pp. 664–675, June 1988.

3. R. Reimann and H-M. Rein, "Bipolar high-gain limiting amplifier IC for optical-fiber receivers operating up to 4 Gbits/s," *IEEE Journal of Solid-State Circuits,* vol. 22, no. 4, pp. 504–510, August 1987.
4. Y. Baeyens et al., "InP D-HBT IC's for 40-Gb/s and higher bit rate lightwave transceivers," *IEEE Journal of Solid-State Circuits,* vol. 37, no. 9, pp. 1152–1159, September 2002.
5. E. M. Cherry and D. E. Hooper, "The design of wide-band transistor feedback amplifiers," *Proceedings of the Institute of Electrical Engineers,* vol. 110, pp. 375–389, February 1963.

ADDITIONAL READING

P. R. Gray, P. J. Hurst, S. H. Lewis, and R. G. Meyer, *Analysis and Design of Analog Integrated Circuits,* 5th ed., John Wiley & Sons, New York: 2009.

R. C. Jaeger, "A high output resistance current source," *IEEE JSSC,* vol. SC-9, pp. 192–194, August 1974.

R. C. Jaeger, "Common-mode rejection limitations of differential amplifiers," *IEEE JSSC,* vol. SC-11, pp. 411–417, June 1976.

R. C. Jaeger, and G. A. Hellwarth. "On the performance of the differential cascode amplifier," *IEEE JSSC,* vol. SC-8, pp. 169–174, April 1973.

PROBLEMS

Unless otherwise specified, use $\beta_F = 100$, $V_A = 70$ V, $K_p = K_n = 1$ mA/V^2, $V_{TN} = -V_{TP} = 1$ V, $\lambda = 0.02$ V^{-1}, $V_T = 25.8$ mV, and $I_S = 10^{-15}$ A.

13.1 Differential Amplifiers

BJT Amplifiers

13.1. (a) What are the Q-points for the transistors in the amplifier in Fig. P13.1 if $V_{CC} = 12$ V, $V_{EE} = 12$ V, $R_{EE} = 240$ kΩ, $R_C = 270$ kΩ, and $\beta_F = 100$? (b) What are the differential-mode gain, and differential-mode input and output resistances? (c) What are the common-mode gain, CMRR, and common-mode input resistance for a single-ended output?

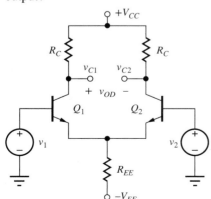

Figure P13.1

13.2. (a) What are the Q-points for the transistors in the amplifier in Fig. P13.1 if $V_{CC} = 1.5$ V, $V_{EE} = 1.5$ V, $\beta_F = 60$, $R_{EE} = 75$ kΩ, and $R_C = 100$ kΩ? (b) What are the differential-mode gain, common-mode gain, CMRR, and differential-mode and common-mode input and output resistances?

13.3. (a) Use SPICE to simulate the amplifier in Prob. 13.1 at a frequency of 1 kHz, and determine the differential-mode gain, common-mode gain, CMRR, and differential-mode and common-mode input resistances. (b) Apply a 25 mV, 1 kHz sine wave as an input signal, and plot the output signals using SPICE transient analysis. Use the SPICE distortion analysis capability to find the harmonic distortion in the output.

13.4. (a) What are the Q-points for the transistors in the amplifier in Fig. P13.1 if $V_{CC} = 12$ V, $V_{EE} = 12$ V, $R_{EE} = 100$ kΩ, $R_C = 100$ kΩ, and $\beta_F = 125$? (b) What are the differential- mode gain, common-mode gain, CMRR, and differential-mode and common-mode input and output resistances?

*13.5. (a) Use the common-mode gain to find voltages v_{C1}, v_{C2}, and v_{OD} for the differential amplifier in Fig. P13.1 if $V_{CC} = 15$ V, $V_{EE} = 15$ V, $R_{EE} = 270$ kΩ, $R_C = 240$ kΩ, $v_1 = 5.000$ V, and $v_2 = 5.000$ V. (b) Find the Q-points of the transistors directly with V_{IC} applied. Recalculate v_{c1} and v_{c2} and compare to the results in part (a). What is the origin of the discrepancy?

13.6. Design a differential amplifier to have a differential gain of 58 dB and $R_{id} = 100$ kΩ using the topology in Fig. P13.1, with $V_{CC} = V_{EE} = 10$ V and $\beta_F = 120$. (Be sure to check feasibility of the design using our rule-of-thumb estimates from Chapter 7 before you move deeper into the design calculations.)

13.7. Design a differential amplifier to have a differential gain of 46 dB and $R_{id} = 1$ MΩ using the topology in Fig. P13.1, with $V_{CC} = V_{EE} = 12$ V and $\beta_F = 100$. (Be sure to check feasibility of the design using our rule-of-thumb estimates from Chapter 7 before you move deeper into the design calculations.)

13.8. (a) What are the Q-points for the transistors in the amplifier in Fig. P13.8 if $V_{CC} = 15$ V, $V_{EE} = 15$ V, $I_{EE} = 400$ µA, $R_{EE} = 300$ kΩ, $R_C = 51$ kΩ, $V_A = \infty$, and $\beta_F = 100$? (b) What are the differential-mode gain, common-mode gain, CMRR, and differential-mode and common-mode input and output resistances? (c) Repeat part (b) for $V_A = 50$ V.

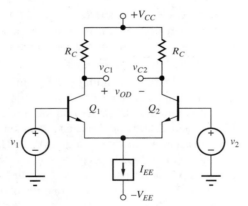

Figure P13.8

13.9. (a) Use SPICE to simulate the amplifier in Prob. 13.8 at a frequency of 1 kHz, and determine the differential-mode gain, common-mode gain, CMRR, and differential-mode and common-mode input resistances. Use $V_A = 60$ V. (b) Apply a 25-mV, 1-kHz sine wave as an input signal and plot the output signal using SPICE transient analysis. Use the SPICE distortion analysis capability to find the harmonic distortion in the output.

*13.10. What are the voltages v_{C1}, v_{C2}, and v_{OD} for the differential amplifier in Fig. P13.8 if $V_{CC} = 12$ V, $V_{EE} = 12$ V, $\beta_F = 75$, $I_{EE} = 300$ µA, $R_{EE} = 270$ kΩ, $R_C = 47$ kΩ, $v_1 = 1.995$ V, and $v_2 = 2.005$ V? What is the common-mode input range of this amplifier?

13.11. What is the value of the current I_{EE} required to achieve $R_{id} = 4$ MΩ in the circuit in Fig. P13.8 if $\beta_o = 150$? What output resistance R_{EE} is required for CMRR = 100 dB?

13.12. For the amplifier in Fig. P13.12, $V_{CC} = 7.5$ V, $V_{EE} = 7.5$ V, $\beta_F = 100$, $I_{EE} = 30$ µA, and $R_C = 220$ kΩ. (a) What are the output voltages v_o and V_O for the amplifier for $v_i = 0$ V and $v_i = 2$ mV? (b) What is the maximum value of v_i?

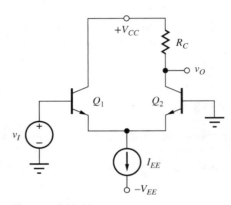

Figure P13.12

13.13. For the amplifier in Fig. P13.12, $V_{CC} = 12$ V, $V_{EE} = 12$ V, $\beta_F = 120$, $I_{EE} = 200$ µA, and $R_C = 100$ kΩ. (a) What are the output voltages V_O and v_o for the amplifier for $v_i = 0$ V and $v_i = 1$ mV? (b) What is the maximum value of v_i?

13.14. (a) Use SPICE to simulate the amplifier in Prob. 13.13 at a frequency of 1 kHz, and determine the differential-mode gain, common-mode gain, CMRR, and differential-mode and common-mode input resistances. Use $V_A = 60$ V. (b) Apply a 25-mV, 1-kHz sine wave as an input signal and plot the output signal using SPICE transient analysis. Use the SPICE distortion analysis capability to find the harmonic distortion in the output.

13.15. (a) What are the Q-points for the transistors in the amplifier in Fig. P13.15 if $V_{CC} = 10$ V, $V_{EE} = 10$ V, $\beta_F = 150$, $R_{EE} = 180$ kΩ, and $R_C = 200$ kΩ? (b) What are the differential-mode gain, common-mode gain, CMRR, and differential-mode and common-mode input resistances?

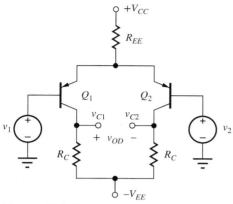

Figure P13.15

13.16. What are the voltages v_{C1}, v_{C2}, and v_{OD} for the differential amplifier in Fig. P13.15 if $V_{CC} = 12$ V, $V_{EE} = 12$ V, $\beta_F = 100$, $R_{EE} = 430$ kΩ, $R_C = 560$ kΩ, $v_1 = 1$ V, and $v_2 = 0.99$ V?

13.17. Use SPICE to simulate the amplifier in Prob. 13.15 at a frequency of 5 kHz, and determine the differential-mode gain, common-mode gain, CMRR, and differential-mode and common-mode input resistances.

13.18. What are the voltages v_{C1}, v_{C2}, and v_{OD} for the differential amplifier in Fig. P13.18 if $V_{CC} = 15$ V, $V_{EE} = 15$ V, $\beta_F = 100$, $I_{EE} = 1$ mA, $R_{EE} = 500$ kΩ, $R_C = 12$ kΩ, $v_1 = 0.01$ V, and $v_2 = 0$ V?

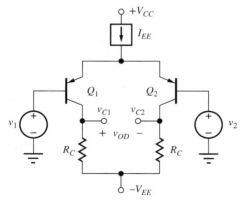

Figure P13.18

13.19. Use SPICE to simulate the amplifier in Prob. 13.18 at a frequency of 5 kHz, and determine the differential-mode gain, common-mode gain, CMRR, and differential-mode and common-mode input resistances.

13.20. (a) What are the Q-points for the transistors in the amplifier in Fig. P13.18 if $V_{CC} = 3$ V, $V_{EE} = 3$ V, $\beta_F = 80$, $I_{EE} = 10$ μA, $R_{EE} = 5$ MΩ, and $R_C = 390$ kΩ? (b) What are the differential-mode gain, common-mode gain, CMRR, differential-mode and common-mode input resistances, and common-mode input range?

*13.21. The differential amplifier in Fig. P13.21 has mismatched collector resistors. Calculate A_{dd}, A_{cd}, and the CMRR of the amplifier if the output is the differential output voltage v_{od}, and $R = 100$ kΩ, $\Delta R/R = 0.02$, $V_{CC} = V_{EE} = 12$ V, $R_{EE} = 100$ kΩ, and $\beta_F = 100$.

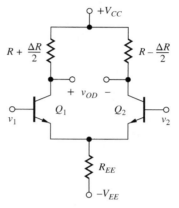

Figure P13.21

13.22. Use SPICE to simulate the amplifier in Fig. P13.21 at a frequency of 100 Hz, and determine the differential-mode gain, common-mode gain, and CMRR.

**13.23. The transistors in the differential amplifier in Fig. P13.23 have mismatched transconductances. Calculate A_{dd}, A_{cd}, and the CMRR of the amplifier if the output is the differential output voltage v_{OD}, and $R = 82$ kΩ, $g_m = 5$ mS, $\Delta g_m/g_m = 0.01$, $V_{CC} = V_{EE} = 15$ V, and $R_{EE} = 82$ kΩ.

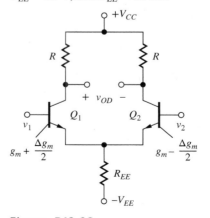

Figure P13.23

FET Differential Amplifiers

13.24. (a) What are the Q-points for the transistors in the amplifier in Fig. P13.24 if $V_{DD} = 15$ V, $V_{SS} = 15$ V, $R_{SS} = 12$ kΩ, and $R_D = 15$ kΩ? Assume $K_n = 400$ μA/V^2 and $V_{TN} = 0.8$ V. (b) What are the differential-mode gain, common-mode gain, CMRR, and differential-mode and common-mode input resistances?

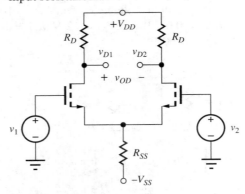

Figure P13.24

13.25. (a) What are the Q-points for the transistors in the amplifier in Fig. P13.24 if $V_{DD} = 5$ V, $V_{SS} = 5$ V, $R_{SS} = 2.4$ kΩ, and $R_D = 2.4$ kΩ? Assume $K_n = 400$ μA/V^2 and $V_{TN} = 0.7$ V. (b) What are the differential-mode gain, common-mode gain, CMRR, and differential-mode and common-mode input resistances?

13.26. Repeat Prob. 13.24 using the unified model from Sec. 5.10.7 assuming $V_{MIN} = 1$ V.

13.27. Repeat Prob. 13.25 using the unified model from Sec. 5.10.7 assuming $V_{MIN} = 1$ V.

13.28. (a) Use SPICE to simulate the amplifier in Prob. 13.25 at a frequency of 1 kHz, and determine the differential-mode gain, common-mode gain, CMRR, and differential-mode and common-mode input resistances. (b) Apply a 250-mV, 1-kHz sine wave as an input signal and plot the output signals using SPICE transient analysis. Use the SPICE distortion analysis capability to find the harmonic distortion in the output.

13.29. Design a differential amplifier to have a differential-mode output resistance of 10 kΩ and $A_{dm} = 20$ dB, using the circuit in Fig. P13.25 with $V_{DD} = V_{SS} = 5$ V. Assume $V_{TN} = 1$ V and $K_n = 25$ mA/V^2.

*13.30. (a) What are the Q-points for the transistors in the amplifier in Fig. P13.30 if $V_{DD} = 10$ V, $V_{SS} = 10$ V, $R_{SS} = 51$ kΩ, and $R_D = 51$ kΩ? Assume $K_n = 400$ μA/V^2, $\gamma = 0.75$ V$^{0.5}$, $2\phi_F = 0.6$ V, and $V_{TO} = 1$ V.
(b) What are the differential-mode gain, common-mode gain, CMRR, and differential-mode and common-mode input resistances? (c) What would the Q-points be if $\gamma = 0$?

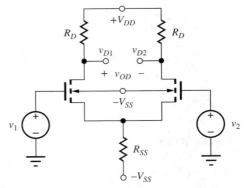

Figure P13.30

13.31. (a) Use SPICE to simulate the amplifier in Prob. 13.30 at a frequency of 1 kHz, and determine the differential-mode gain, common-mode gain, CMRR, and differential-mode and common-mode input resistances. (b) Apply a 250-mV, 1-kHz sine wave as an input signal and plot the output signal using SPICE transient analysis. Use the SPICE distortion analysis capability to find the harmonic distortion in the output.

*13.32. (a) What are the Q-points for the transistors in the amplifier in Fig. P13.30 if $V_{DD} = 12$ V, $V_{SS} = 12$ V, $R_{SS} = 270$ kΩ, and $R_D = 300$ kΩ? Assume $K_n = 400$ μA/V^2, $\gamma = 0.75$ V$^{0.5}$, $2\phi_F = 0.6$ V, and $V_{TO} = 1$ V. (b) What are the differential-mode gain, common-mode gain, CMRR, and differential-mode and common-mode input resistances? (c) What would the Q-points be if $\gamma = 0$?

13.33. (a) What are the Q-points for the transistors in the amplifier in Fig. P13.33 if $V_{DD} = 12$ V, $V_{SS} = 12$ V,

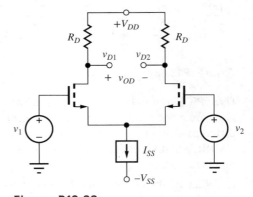

Figure P13.33

$I_{SS} = 1.5$ mA, $R_{SS} = 33$ kΩ, and $R_D = 15$ kΩ? Assume $K_n = 375$ µA/V^2 and $V_{TN} = 0.75$ V. (b) What are the differential-mode gain, common-mode gain, CMRR, and differential-mode and common-mode input resistances?

13.34. Repeat Prob. 13.33 using the unified model from Sec. 4.7.7 assuming $V_{SAT} = 1$ V.

13.35. (a) What are the Q-points for the transistors in the amplifier in Fig. P13.33 if $V_{DD} = 15$ V, $V_{SS} = 15$ V, $I_{SS} = 50$ µA, $R_{SS} = 1$ MΩ, and $R_D = 240$ kΩ? Assume $K_n = 400$ µA/V^2 and $V_{TN} = 1$ V. (b) What are the differential-mode gain, common-mode gain, CMRR, and differential-mode and common-mode input resistances?

*13.36. (a) What are the Q-points for the transistors in the amplifier in Fig. P13.36 if $V_{DD} = 12$ V, $V_{SS} = 12$ V, $I_{SS} = 300$ µA, $R_{SS} = 160$ kΩ, and $R_D = 75$ kΩ? Assume $K_n = 400$ µA/V^2, $\gamma = 0.75$ V$^{0.5}$, $2\phi_F = 0.6$ V, and $V_{TO} = 1$ V. (b) What are the differential-mode gain, common-mode gain, CMRR, and differential-mode and common-mode input resistances?

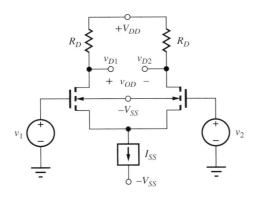

Figure P13.36

*13.37. (a) What are the Q-points for the transistors in the amplifier in Fig. P13.36 if $V_{DD} = 15$ V, $V_{SS} = 15$ V, $I_{SS} = 40$ µA, $R_{SS} = 1.25$ MΩ, and $R_D = 400$ kΩ? Assume $K_n = 400$ µA/V^2, $\gamma = 0.75$ V$^{0.5}$, $2\phi_F = 0.6$ V, and $V_{TO} = 1$ V. (b) What are the differential-mode gain, common-mode gain, CMRR, and differential-mode and common-mode input resistances?

13.38. Design a differential amplifier to have a differential-mode gain of 30 dB, using the circuit in Fig. P13.33 with $V_{DD} = V_{SS} = 8$ V. The circuit should have the maximum possible common-mode input range. Assume $V_{TN} = 1$ V and $K_n = 5$ mA/V^2.

13.39. Repeat Prob. 13.38 using the circuit in Fig. P13.36 with $2\phi_F = 0.6$ V and $\gamma = 0.75$ V$^{0.5}$.

13.40. (a) What are the Q-points for the transistors in the amplifier in Fig. P13.40 if $V_{DD} = 15$ V, $V_{SS} = 15$ V, $R_{SS} = 51$ kΩ, and $R_D = 82$ kΩ? Assume $K_p = 200$ µA/V^2 and $V_{TP} = -1$ V. (b) What are the differential-mode gain, common-mode gain, CMRR, and differential-mode and common-mode input resistances?

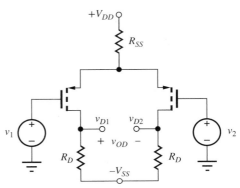

Figure P13.40

13.41. Use SPICE to simulate the amplifier in Prob. 13.40 at a frequency of 3 kHz, and determine the differential-mode gain, common-mode gain, CMRR, and differential-mode and common-mode input resistances.

*13.42. (a) What are the Q-points for the transistors in the amplifier in Fig. P13.42 if $V_{DD} = 9$ V, $V_{SS} = 9$ V, $I_{SS} = 40$ µA, $R_{SS} = 1.25$ MΩ, and $R_D = 300$ kΩ? Assume $K_p = 200$ µA/V^2, $\gamma = 0.6$ V$^{0.5}$, $2\phi_F = 0.6$ V, and $V_{TO} = -1$ V. (b) What are the differential-mode gain, common-mode gain, CMRR, and differential-mode and common-mode input resistances?

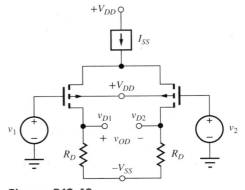

Figure P13.42

13.43. For the amplifier in Fig. P13.43, $V_{DD} = 15$ V, $V_{SS} = 15$ V, $I_{SS} = 20$ μA, and $R_D = 750$ kΩ. Assume $K_p = 1$ mA/V^2 and $V_{TP} = +1$ V. (a) What are the output voltages v_O for the amplifier for $v_1 = 0$ V and $v_1 = 20$ mV? (b) What is the maximum permissible value of v_i?

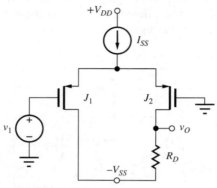

Figure P13.43

13.44. For the amplifier in Fig. P13.44, $V_{DD} = 12$ V, $V_{SS} = 12$ V, $I_{SS} = 20$ μA, and $R_D = 820$ kΩ. Assume $I_{DSS} = 1$ mA and $V_P = +2$ V. (a) What are the output voltages v_O for the amplifier for $v_1 = 0$ V and $v_1 = 20$ mV? (b) What is the maximum permissible value of v_s?

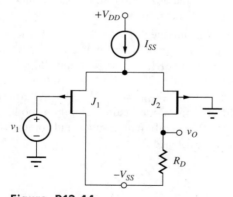

Figure P13.44

13.45. Redraw the circuit for the differential amplifier in Fig. P13.44 using n-channel JFETs.

Half-Circuit Analysis

*13.46. (a) Draw the differential-mode and common-mode half-circuits for the differential amplifier in Fig. P13.46. (b) Use the half-circuits to find the Q-points, differential-mode gain, common-mode gain, and differential-mode input resistance for the amplifier if $\beta_o = 150$, $V_{CC} = 20$ V, $V_{EE} = 20$ V, $R_{EE} = 100$ kΩ, $R_1 = 1.6$ kΩ, and $R_C = 100$ kΩ.

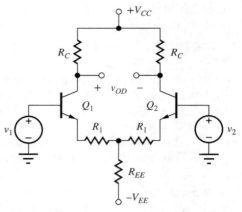

Figure P13.46

13.47. Use SPICE to simulate the amplifier in Prob. 13.46 at a frequency of 1 kHz, and determine the differential-mode gain, common-mode gain, and differential-mode input resistances.

*13.48. (a) Draw the differential-mode and common-mode half-circuits for the differential amplifier in Fig. P13.48. (b) Use the half-circuits to find the Q-points, differential-mode gain, common-mode gain, and differential-mode input resistance for the amplifier if $\beta_o = 120$, $V_{CC} = 15$ V, $V_{EE} = 15$ V, $I_{EE} = 100$ μA, and $R_{EE} = 600$ kΩ?

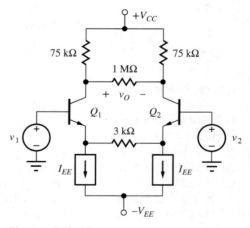

Figure P13.48

13.49. Use SPICE to simulate the amplifier in Prob. 13.48 at a frequency of 1 kHz, and determine the differential-mode gain, common-mode gain, and differential-mode input resistances.

*13.50. (a) Draw the differential-mode and common-mode half-circuits for the differential amplifier in Fig. P13.50. (b) Use the half-circuits to find the Q-points, differential-mode gain, common-mode gain, and differential-mode input resistance for the amplifier if $V_{CC} = 18$ V, $V_{EE} = 18$ V, $I_{EE} = 100$ μA, $R_D = 75$ kΩ, $R_{EE} = 600$ kΩ, $\beta_o = 100$, $K_n = 200$ μA/V², and $V_{TN} = -4$ V. (c) Show that Q_1 and Q_2 are in the active region.

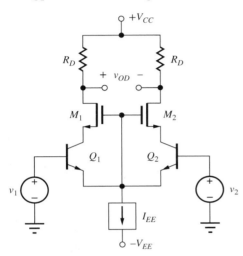

Figure P13.50

**13.51. (a) Draw the differential-mode and common-mode half-circuits for the differential amplifier in Fig. P13.51. (b) Use the half-circuits to find the Q-points, differential-mode gain, common-mode

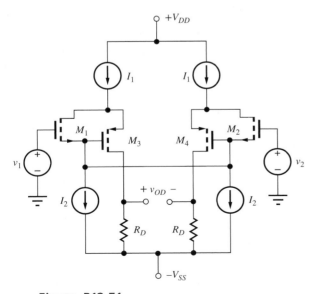

Figure P13.51

gain, and differential-mode input resistance for the amplifier if $K_n = 1000$ μA/V², $V_{TN} = 0.75$ V, $K_p = 500$ μA/V², $V_{TP} = -0.75$ V, $I_1 = 200$ μA, $I_2 = 100$ μA, $V_{DD} = 10$ V, $V_{SS} = 10$ V, and $R_D = 50$ kΩ.

13.52. (a) Repeat Prob. 13.51 for $V_{DD} = 2.5$ V, $-V_{SS} = -2.5$ V, and $R_D = 10$ kΩ. (b) What is the common-mode input range for this amplifier?

13.53. (a) Draw the differential-mode and common-mode half-circuits for the differential amplifier in Fig. P13.53. (b) Use the half-circuits to find the Q-points, differential-mode gain, common-mode gain, and differential-mode input resistance for the amplifier if $V_{CC} = 15$ V, $V_{EE} = 15$ V, $I_{EE} = 100$ μA, $R_D = 150$ kΩ, $R_{EE} = 500$ kΩ, $\beta_o = 100$, $I_{DSS} = 200$ μA, and $V_P = -4$ V. (c) Show that Q_1 and Q_2 are in the active region.

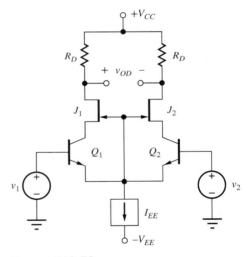

Figure P13.53

13.2 Evolution to Basic Operational Amplifiers

13.54. (a) What are the Q-points of the transistors in the amplifier in Fig. P13.54 if $V_{CC} = 18$ V, $V_{EE} = 18$ V, $I_1 = 50$ μA, $R = 24$ kΩ, $\beta_o = 100$, and $V_A = 60$ V? (b) What are the differential-mode voltage gain and input resistance? (c) What is the amplifier output resistance? (d) What is the common-mode input resistance? (e) Which terminal is the noninverting input?

13.55. What is the minimum collector-base breakdown voltage specification for transistor Q_3 in the amplifier in Prob. 13.54? Assume a large input signal so that the differential pair is acting as a current switch.

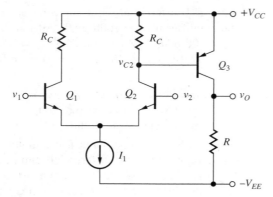

Figure P13.54

emitter is grounded and derive the expressions below.

$$I_{C1} = \beta_{F1} I_B \qquad I_{C2} = \beta_{F2}(\beta_{F1} + 1) I_B$$
$$I_C \cong \beta_{F1}\beta_{F2} I_B$$
$$g_{m2} = \beta_{o1} g_{m1} \qquad r_{\pi 1} = \beta_{o1} r_{\pi 2}$$
$$r_{o1} = \beta_{o1} r_{o2}$$
$$\beta_o = \frac{i_c}{i_B} \cong \beta_{o1}\beta_{o2}$$
$$G_m = \frac{i_c}{v_{be}} = \frac{g_{m1}}{2} + \frac{g_{m2}}{2} \cong \frac{g_{m2}}{2}$$
$$R_{iB} = \frac{v_{be}}{i_b} = r_{\pi 1} + (\beta_{o1} + 1) r_{\pi 2} \cong 2\beta_{o1} r_{\pi 2}$$
$$R_{iC} = \frac{v_{ce2}}{i_c} \cong r_{o2} \| 2 \frac{r_{o1}}{\beta_{o2}} \cong \frac{2}{3} r_{o2}$$

13.56. Use the technique in the EIA on pages 776–777 with SPICE to find the offset voltage, CMRR and PSRR for the amplifier in Prob. 13.54.

13.57. Use the technique in the EIA on pages 776–777 with SPICE to find the offset voltage, CMRR and PSRR for the amplifier in Prob. 13.67.

13.58. What is the common-mode input range for the amplifier in Prob. 13.54 if current source I_1 is replaced with an electronic current source that must have 0.75 V across it to operate properly?

13.59. Use SPICE to simulate the amplifier in Prob. 13.54 at a frequency of 1 kHz, and determine the differential-mode gain, CMRR, and differential-mode input resistance and output resistance.

13.60. (a) Repeat Prob. 13.54 with $V_{CC} = V_{EE} = 12$ V. (b) What is the new common-mode range as requested in Prob. 13.58?

13.61. Repeat Prob. 13.54 if I_1, R, and R_C are redesigned to increase the currents by a factor of 5.

13.62. Draw the two-port representation for the *npn* Darlington pair in Fig. 13.26(a) including the small-signal models of the transistors. Derive the expressions for R_{iB} and R_{iC} given in Sec. 13.2.3.

13.63. What are the values of β_F, I_B, I_{C1}, I_{C2}, R_{iB}, and R_{iC} for the *npn* Darlington circuit in Fig. 13.26 (a) if $I_C = 100$ μA, $\beta_{F1} = 40$, $\beta_{F2} = 110$, $V_{CE} = 6$ V and the Early voltages of both transistors are 60 V?

13.64. Use the technique in the EIA on pages 776–777 with SPICE to find the offset voltage, CMRR and PSRR for the amplifier in Prob. 13.66.

13.65. The circuit in Sec. 13.2.3 is called a Darlington connection of two transistors. Assume the

13.66. Transistor Q_3 in Fig. P13.54 is replaced with a *pnp* Darlington circuit. Draw the new amplifier and repeat Prob. 13.54. (See Fig. P13.99.)

13.67. (a) What are the Q-points of the transistors in the amplifier in Fig. P13.67 if $V_{CC} = 15$ V, $V_{EE} = 15$ V, $I_1 = 200$ μA, $R_E = 2$ kΩ, $R = 50$ kΩ, $\beta_o = 80$, and $V_A = 70$ V? (b) What are the differential-mode voltage gain and input resistance? (c) What is the amplifier output resistance? (d) What is the common-mode input resistance? (e) Which terminal is the noninverting input? (f) What is the common-mode input range for the amplifier if current source I_1 is replaced with an electronic current source that must have 0.75 V across it to operate properly?

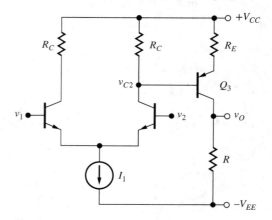

Figure P13.67

13.68. (a) What are the Q-points of the transistors in the amplifier in Fig. P13.67 if $V_{CC} = 18$ V, $V_{EE} = 18$ V, $I_1 = 200$ μA, $R_E = 0$, $R = 50$ kΩ, $\beta_o = 80$, and

$V_A = 70$ V? (b) What are the differential-mode voltage gain and input resistance? (c) What is the common-mode input resistance?

*13.69. Plot a graph of the differential-mode voltage gain of the amplifier in Prob. 13.68 versus the value of R_E. (The computer might be a useful tool.)

13.70. Design an amplifier to have $R_{out} = 2$ kΩ and $A_{dm} = 2000$, using the circuit in Fig. P13.54. Use $V_{CC} = V_{EE} = 10$ V, $V_A = 60$ V, and $\beta_F = 100$.

13.71. (a) What are the Q-points of the transistors in the amplifier in Fig. P13.71 if $V_{CC} = 15$ V, $V_{EE} = 15$ V, $I_1 = 200$ μA, $I_2 = 300$ μA, $R_E = 1.8$ kΩ, $\beta_o = 80$, and $V_A = 60$ V? (b) What are the differential-mode voltage gain, input resistance, and output resistance?

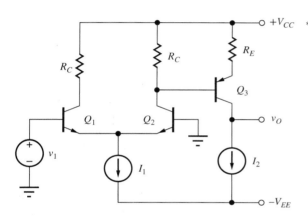

Figure P13.71

13.72. Use SPICE to simulate the amplifier in Prob. 13.71 and compare the results to hand calculations.

13.73. What are the Q-points of the transistors in the amplifier in Fig. P13.71 if $V_{CC} = V_{EE} = 16$ V, $I_1 = 300$ μA, $I_2 = 400$ μA, $R_E = 0$, $\beta_o = 100$, and $V_A = 70$ V?

*13.74. Plot a graph of the differential-mode voltage gain of the amplifier in Prob. P13.71 versus the value of R_E. (The computer might be a useful tool.)

13.75. (a) What are the Q-points of the transistors in the amplifier in Fig. P13.75 if $V_{CC} = V_{EE} = 15$ V, $I_1 = 500$ μA, $R_1 = 2$ MΩ, $I_2 = 500$ μA, $\lambda = 0.02$/V, and $R_2 = 2$ MΩ? Use $\beta_o = 80$, $V_A = 75$ V, $K_p = 5$ mA/V^2, and $V_{TP} = -1$ V. (b) What are the differential-mode voltage gain and input resistance and output resistance of the amplifier? (c) Which terminal is the noninverting input? (d) Which terminal is the inverting input? (e) What is the transconductance of this amplifier?

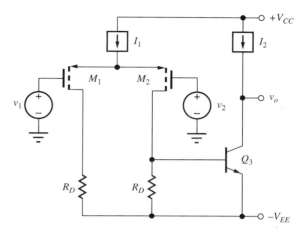

Figure P13.75

*13.76. Use SPICE to simulate the amplifier in Prob. 13.75 at a frequency of 1 kHz, and determine the differential-mode gain, CMRR, and differential-mode input resistance and output resistance.

13.77. What is the minimum collector-base breakdown voltage specification for transistor Q_3 in the amplifier in Prob. 13.75? Assume a large input signal so that the differential pair is acting as a current switch.

13.78. What is the voltage gain of the amplifier in Fig. P13.75 if $V_{CC} = V_{EE} = 5$ V, $I_1 = 400$ μA, $R_1 = 20$ MΩ, $I_2 = 200$ μA, $R_2 = 10$ MΩ, $\beta_o = 80$, $V_A = 75$ V, $K_p = 5$ mA/V^2, $\lambda = 0.02$/V, and $V_{TP} = -1$ V?

13.79. What is the common-mode input voltage range for the amplifier in Prob. 13.78 if current source I_1 must have a 0.75-V drop across it to operate properly?

13.80. (a) What are the Q-points of the transistors in the amplifier in Fig. P13.80 if $I_1 = 400$ μA, $R_1 = 1$ MΩ,

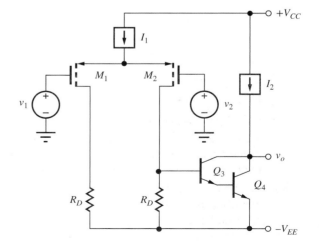

Figure P13.80

$I_2 = 400$ µA, $R_2 = 1$ MΩ, and $V_{CC} = V_{EE} = 7.5$ V. Use $\beta_o = 100$, $V_A = 75$ V, $K_p = 5$ mA/V^2, $\lambda = 0.02$/V, and $V_{TP} = -1$ V. (b) What are the differential-mode voltage gain and input resistance and output resistance of the amplifier? (See Sec. 13.2.3.)

*13.81. Use SPICE to simulate the amplifier in Prob. 13.80 at a frequency of 1 kHz, and determine the differential-mode gain, CMRR, and differential-mode input resistance and output resistance.

13.82. (a) Redraw the op amp circuit in Fig. 13.27(a) with Q_4 replaced by the *npn* Darlington configuration from Fig. 13.26. (b) What are the new values of voltage gain, CMRR, input resistance, and output resistance? Use the circuit element values from Ex. 13.4, and compare your results to those of the example.

13.83. Simulate the circuit in Prob. 13.82 using SPICE and compare the results of the two problems.

13.84. (a) What are the Q-points of the transistors in the amplifier in Fig. P13.84 if $V_{CC} = 20$ V, $V_{EE} = 20$ V, $I_1 = 200$ µA, $I_2 = 400$ µA, $I_3 = 1$ mA, $\beta_F = 100$, and $V_A = 50$ V? (b) What are the differential-mode voltage gain and input resistance? (c) What is the amplifier output resistance? (d) What is the common-mode input resistance? (e) Which terminal is the noninverting input?

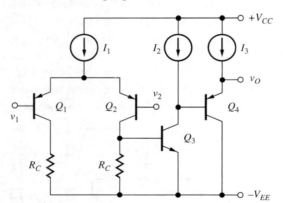

Figure P13.84

*13.85. Use SPICE to simulate the amplifier in Prob. 13.84 at a frequency of 1 kHz, and determine the differential-mode gain, CMRR, and differential-mode input resistance and output resistance.

13.86. (a) What are the Q-points of the transistors in the amplifier in Fig. P13.86 if $V_{DD} = 6$ V, $V_{SS} = 6$ V, $I_1 = 600$ µA, $I_2 = 500$ µA, $I_3 = 2$ mA, $K_n = 5$ mA/V^2, $V_{TN} = 0.70$ V, $\lambda_n = 0.02$ V^{-1}, $K_p = 2$ mA/V^2, $V_{TP} = -0.70$ V, and $\lambda_p = 0.015$ V^{-1}?

(b) What are the differential-mode voltage gain and input resistance and output resistance of the amplifier?

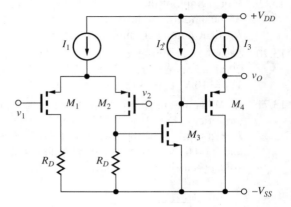

Figure P13.86

13.87. Simulate the circuit in Prob. 13.86 using SPICE and compare the results to hand calculations.

13.88. Transistor M_3 in Fig. P13.86 is replaced with an *npn* device with $\beta_o = 150$ and $V_A = 70$ V. What are the values of the differential-mode voltage gain and input resistance, and the output resistance of the new amplifier? Use the circuit element values from Prob. 13.86.

13.89. Simulate the circuit in Prob. 13.88 using SPICE and compare the results with hand calculations.

13.90. (a) What are the Q-points of the transistors in the amplifier in Fig. P13.90 if $V_{DD} = 15$ V, $V_{SS} = 15$ V, $I_1 = 800$ µA, $I_2 = 3$ mA, $I_3 = 6$ mA, $K_n = 5$ mA/V^2, $V_{TN} = 0.75$ V, $\lambda_n = 0.02$ V^{-1}, $K_p = 2$ mA/V^2, $V_{TP} = -0.75$ V, and $\lambda_p = 0.015$ V^{-1}? (b) What are the differential-mode voltage gain and input resistance and output resistance of the amplifier?

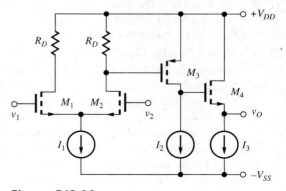

Figure P13.90

(c) Use SPICE to simulate the amplifier in Prob. 13.90 at a frequency of 1 kHz, and determine the differential-mode gain, CMRR, and differential-mode input resistance and output resistance.

13.91. Repeat the (a) and (b) parts of Prob. 13.90 using the unified model from Sec. 5.10.7 assuming $V_{SAT} = 1$ V.

13.92. ⓢ Use the technique in the EIA on page 776–777 with SPICE to find the offset voltage, CMRR and PSRR for the amplifier in Prob. 13.90.

13.93. ⓢ (a) What are the Q-points of the transistors in the amplifier in Fig. P13.93 if $V_{CC} = 5$ V, $V_{EE} = 5$ V, $I_1 = 200$ μA, $I_2 = 500$ μA, $I_3 = 2$ mA, $R_L = 2$ kΩ, $\beta_o = 100$, $V_A = 50$ V, $K_n = 5$ mA/V^2, and $V_{TN} = 0.70$ V? (b) What are the differential-mode voltage gain and input resistance and output resistance of the amplifier? (c) Use SPICE to simulate the amplifier in Fig. P13.93 at a frequency of 2 kHz, and determine the differential-mode gain, CMRR, and differential-mode input resistance and output resistance.

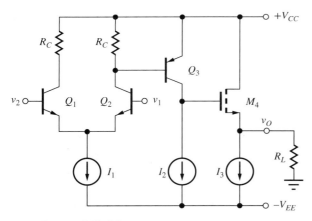

Figure P13.93

*13.94. (a) What are the Q-points of the transistors in the amplifier in Fig. P13.94 if $V_{CC} = 3$ V, $V_{EE} = 3$ V, $I_1 = 10$ μA, $I_2 = 60$ μA, $I_3 = 250$ μA, $R_{C1} = 300$ kΩ, $R_{C2} = 75$ kΩ, $R_L = 2$ kΩ, $\beta_{on} = 100$, $V_{AN} = 50$ V, $\beta_{op} = 50$, and $V_{AP} = 70$ V? (b) What are the differential-mode voltage gain and input resistance and output resistance of the amplifier? (c) Which terminal is the noninverting input? Which terminal is the inverting input? (d) What is the gain predicted by our rule-of-thumb estimate? What are the reasons for any discrepancy?

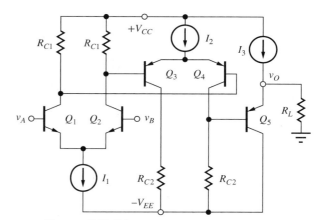

Figure P13.94

*13.95. (a) What are the Q-points of the transistors in the amplifier in Fig. P13.94 if $V_{CC} = V_{EE} = 15$ V, $I_1 = 100$ μA, $I_2 = 200$ μA, $I_3 = 750$ μA, $R_{C1} = 120$ kΩ, $R_{C2} = 170$ kΩ, $R_L = 2$ kΩ, $\beta_{on} = 100$, $V_{AN} = 50$ V, $\beta_{op} = 50$, and $V_{AP} = 70$ V? (b) What are the differential-mode voltage gain and input resistance and output resistance of the amplifier? (c) What is the common-mode input range? (d) Estimate the offset voltage of this amplifier.

13.96. (a) What are the Q-points of the transistors in the amplifier in Fig. P13.96 if $V_{CC} = V_{EE} = 15$ V, $I_1 = 250$ μA, $R = 10$ kΩ, $\beta_F = 100$, and $V_A = 70$ V? (b) What are the differential-mode voltage gain and input resistance and output resistance of the amplifier?

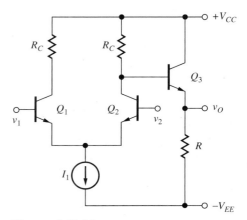

Figure P13.96

*13.97. 💡 Design an amplifier using the topology in Fig. P13.96 to have an input resistance of 250 kΩ and an output resistance of 100 Ω. Can these specifications all be met if $V_{CC} = V_{EE} = 12$ V,

$\beta_{FO} = 100$, and $V_A = 60$ V? If so, what are the values of I_1, R_C, and R, and the voltage gain of the amplifier? If not, what needs to be changed?

*13.98. Design an amplifier using the topology in Fig. P13.96 to have an input resistance of 1 MΩ and an output resistance ≤ 2 Ω. Can these specifications all be met if $V_{CC} = V_{EE} = 9$ V, $\beta_{FO} = 100$, and $V_A = 60$ V? If so, what are the values of I_1, R_C, and R, and the voltage gain of the amplifier? If not, what needs to be changed?

**13.99. (a) What are the Q-points of the transistors in the amplifier in Fig. P13.99 if $V_{CC} = V_{EE} = 20$ V, $I_1 = 100$ μA, $I_2 = 600$ μA, $I_3 = 6$ mA, $\beta_{on} = 100$, $V_{AN} = 50$ V, $\beta_{op} = 50$, and $V_{AP} = 70$ V? (b) What are the differential-mode voltage gain and input resistance and output resistance of the amplifier?

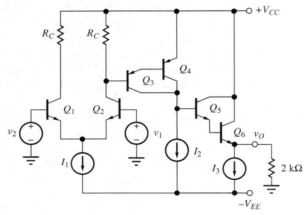

Figure P13.99

**13.100. (a) What are the Q-points of the transistors in the amplifier in Fig. P13.99 if $V_{CC} = V_{EE} = 15$ V, $I_1 = 250$ μA, $I_2 = 2$ mA, $I_3 = 7$ mA, $\beta_{on} = 100$, $V_{AN} = 50$ V, $\beta_{op} = 50$, and $V_{AP} = 70$ V? (b) What are the differential-mode voltage gain and input resistance and output resistance of the amplifier?

13.101. The drain resistors R_D in Prob. 13.90 are replaced with PMOS transistors as shown in Fig. 13.31(a). (a) What is the required value of K_p for these transistors? (b) What is the voltage gain of the new amplifier? (c) What was the original voltage gain?

13.102. The drain resistors R_D in Prob. 13.86 are replaced with NMOS transistors in a manner similar to that in Fig. 13.31(b). (a) What is the required value of K_n for these transistors? (b) What is the voltage gain of the new amplifier? (c) What was the original voltage gain?

13.103. The transconductance amplifier in Prob. 13.54 is connected as a voltage follower. What are the closed-loop voltage gain, input resistance, and output resistance of the voltage follower?

13.104. Resistor R in the transconductance amplifier in Prob. 13.54 is replaced with a 2-mA ideal current source. If the amplifier is connected as a voltage follower, what are the closed-loop voltage gain, input resistance, and output resistance of the voltage follower?

13.105. The transconductance amplifier in Prob. 13.71 is connected as a voltage follower. What are the closed-loop voltage gain, input resistance, and output resistance of the voltage follower?

13.106. (a) The op amp in Prob. 13.84 is connected as a noninverting amplifier with a gain of 8. What are the closed-loop voltage gain, input resistance, and output resistance of the amplifier? (b) Repeat for connection as a voltage follower.

13.107. (a) The op amp in Prob. 13.90 is connected as a noninverting amplifier with a gain of 5. What are the closed-loop voltage gain, input resistance, and output resistance of the amplifier? (b) Repeat for connection as a voltage follower.

13.3 Output Stages

13.108. What is the quiescent current in the class-AB stage in Fig. P13.108 if $K_p = K_n = 500$ μA/V^2 and $V_{TN} = -V_{TP} = 0.75$ V?

*13.109. What is the quiescent current in the class-AB stage in Fig. P13.108 if $K_p = 500$ μA/V^2, $K_n = 600$ μA/V^2, $V_{TP} = -0.7$ V, and $V_{TN} = 0.7$ V?

13.110. What is the quiescent current in the class-AB stage in Fig. P13.110 if both transistors have $I_S = 5 \times 10^{-15}$ A?

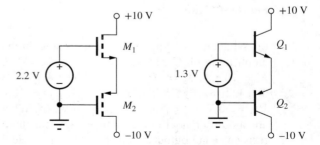

Figure P13.108   **Figure P13.110**

*13.111. What is the quiescent current in the class-AB stage in Fig. P13.110 if $I_S = 10^{-15}$ A for the *pnp* transistor and $I_S = 5 \times 10^{-15}$ A for the *npn* transistor?

13.112. Draw a sketch of the voltage transfer characteristic for the circuit in Fig. P13.112. Label important voltages on the characteristic.

13.113. Use SPICE to plot the voltage transfer characteristic for the class-AB stage in Fig. P13.112 if $I_S = 2 \times 10^{-15}$ A and $\beta_F = 50$ for the *pnp* transistor, $I_S = 6 \times 10^{-15}$ A and $\beta_F = 60$ for the *npn* transistor, $V_{BB} = 1.3$ V, and $R_L = 1$ kΩ.

13.114. What is the quiescent current in the class-AB stage in Fig. P13.114 if I_S for the *npn* transistor is 10^{-15} A, I_S for the *pnp* transistor is 2×10^{-16} A, $I_B = 250$ μA, and $R_B = 5$ kΩ? Assume $\beta_F = \infty$ and $v_O = 0$.

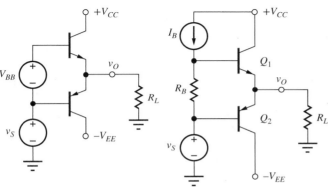

Figure P13.112   **Figure P13.114**

13.115. What is the quiescent current in the class-AB stage in Fig. P13.115 if $V_{TN} = 0.75$ V, $V_{TP} = -0.75$ V, $K_n = 500$ μA/V^2, $K_p = 300$ μA/V^2, $I_G = 500$ μA, and $R_G = 5$ kΩ?

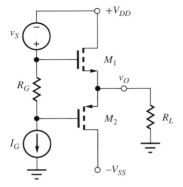

Figure P13.115

13.116. The source-follower in Fig. 13.33 has $V_{DD} = V_{SS} = 10$ V and $R_L = 1$ kΩ. If the amplifier is developing an output voltage of $4 \sin 2000\pi t$ V, what is the minimum value of I_{SS}? What are the maximum and minimum values of source current i_S that occur during the signal swing? What is the efficiency?

*13.117. An ideal complementary class-B output stage is generating a triangular output signal across a 40-kΩ load resistor with a peak value of 10 V from ±10-V supplies. What is the efficiency of the amplifier?

13.118. An ideal complementary class-B output stage is generating a square wave output signal across a 5-kΩ load resistor with a peak value of 5 V from ±5-V supplies. What is the efficiency of the amplifier?

**13.119. (a) Use the Fourier analysis capability of SPICE to find the amplitude of the first, second, third, fourth, and fifth harmonics of the input signal introduced by the cross-over region of the class-B amplifier in Fig. P13.112 if $V_{BB} = 0$, $V_{CC} = V_{EE} = 5$ V, $v_S = 4 \sin 2000\pi t$, and $R_L = 2$ kΩ. (b) Repeat for $V_{BB} = 1.3$ V.

Short-Circuit Protection

13.120. What is the current in R_L in Fig. 13.40 at the point when current just begins to limit ($V_{BE2} = 0.7$ V) if $\beta_o = 100$, $R = 15$ Ω, $R_1 = 1$ kΩ, and $R_L = 250$ Ω? For what value of v_I does the output begin to limit current?

13.121. Use SPICE to simulate the circuit in Prob. 13.120, and compare the results to your hand calculations. Discuss the reasons for any discrepancies.

13.122. What would be the Q-point currents in M_4 and M_5 in the amplifier in Fig. 13.39(a) if $V_{DD} = V_{SS} = 15$ V, $I_2 = 200$ μA, $R_G = 7$ kΩ, $R_L = 2$ kΩ, $R_G \to R_B$, $V_{TN} = 0.7$ V, $V_{TP} = -0.7$ V, $K_n = 5$ mA/V^2, and $K_p = 2$ mA/V^2?

13.123. What would be the currents in Q_4 and Q_5 in the amplifier in Fig. 13.39(b) if $V_{CC} = V_{EE} = 16$ V, $I_2 = 500$ μA, $R_B = 2.4$ kΩ, $R_L = 2$ kΩ, and Q_3 is modeled by a voltage of $V_{CESAT} = 0.2$ V in series with a resistance of 50 Ω when it is saturated?

Transformer Coupling

13.124. Calculate the output resistance of the follower circuit (as seen at R_L) in Fig. 13.42(a) if $n = 10$, $\beta_o = 100$, and $I_S = 10$ mA.

13.125. For the circuit in Fig. P13.125, $v_I = \sin 2000\pi t$, $R_E = 82$ kΩ, $R_B = 200$ kΩ, and $V_{CC} = V_{EE} = 9$ V. What value of n is required to deliver maximum power to R_L if $R_L = 10\ \Omega$? What is the power? Assume $C_1 = C_2 = \infty$.

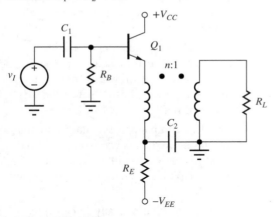

Figure P13.125

13.4 Electronic Current Sources

13.126. (a) What are the output current and output resistance of the current source in Fig. P13.126(a) if $V_{EE} = 10$ V, $R_1 = 2$ MΩ, $R_2 = 2$ MΩ, $R_E = 300$ kΩ, $\beta_o = 100$, and $V_A = 50$ V? (b) Repeat for the circuit in Fig. P13.126(b).

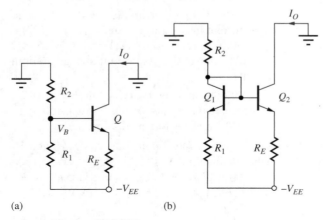

Figure P13.126

13.127. What are the output current and output resistance of the current source in Prob. 13.126(a) if node V_B is bypassed to ground with a capacitor?

13.128. (a) What are the output current and output resistance of the current source in Fig. P13.126(a) if $-V_{EE} = -6$ V, $R_1 = 100$ kΩ, $R_2 = 200$ kΩ, $R_E = 22$ kΩ, $\beta_o = 100$, and $V_A = 75$ V? (b) Repeat for the circuit in Fig. 13.126(b).

13.129. (a) What are the output current and output resistance of the current source in Fig. P13.126(a) if $-V_{EE} = -9$ V, $R_1 = 270$ kΩ, $R_2 = 470$ kΩ, $R_E = 18$ kΩ, $\beta_o = 150$, and $V_A = 75$ V? (b) Repeat for the circuit in Fig. P13.126(b).

13.130. Design a current source to provide an output current of 1mA using the topology of Fig. P13.126(a). The current source should use no more than 1.2 mA and have an output resistance of at least 500 kΩ. Assume $V_{EE} = 10$ V. (b) Repeat for Fig. 13.126(b).

13.131. What are the output current and output resistance of the current source in Fig. P13.131 if $V_O = V_{DD} = 10$ V, $R_4 = 680$ kΩ, $R_3 = 330$ kΩ, $R_S = 2$ kΩ, $K_n = 500\ \mu$A/V^2, $V_{TN} = 0.7$ V, and $\lambda = 0.01$ V^{-1}?

13.132. Repeat Prob. 13.131 using the unified model from Sec. 5.10.7 assuming $V_{SAT} = 1$ V.

13.133. What are the output current and output resistance of the current source in Fig. P13.131 if $V_O = V_{DD} = 5$ V, $R_4 = 200$ kΩ, $R_3 = 100$ kΩ, and $R_S = 15$ kΩ? Use the device parameters from Prob. 13.131.

13.134. Repeat Prob. 13.133 using the unified model from Sec. 5.10.7 assuming $V_{SAT} = 1$ V.

13.135. What are the output current and output resistance of the current source in Fig. P13.131 if $V_O = V_{DD} = 3$ V, $R_4 = 200$ kΩ, $R_3 = 68$ kΩ, and $R_S = 47$ kΩ? Use the device parameters from Prob. 13.131.

13.136. What are the output current and output resistance of the current source in Fig. P13.136 if $V_{CC} = 12$ V, $R_1 = 100$ kΩ, $R_2 = 200$ kΩ, $R_E = 47$ kΩ, $\beta_o = 75$, and $V_A = 50$ V?

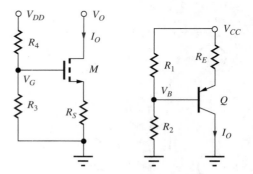

Figure P13.131   **Figure P13.136**

13.137. What are the output current and output resistance of the current source in Fig. P13.136 if $V_{CC} = 10$ V, $R_1 = 100$ kΩ, $R_2 = 270$ kΩ, $R_E = 16$ kΩ, $\beta_o = 90$, and $V_A = 60$ V?

13.138. What are the output current and output resistance of the current source in Fig. P13.136 if $V_{CC} = 3$ V, $R_1 = 10$ kΩ, $R_2 = 39$ kΩ, $R_E = 1.5$ kΩ, $\beta_o = 75$, and $V_A = 60$ V?

13.139. What are the output current and output resistance of the current source in Fig. P13.139 if $V_{DD} = 10$ V, $R_4 = 2$ MΩ, $R_3 = 1$ MΩ, $R_S = 130$ kΩ, $K_p = 750$ μA/V^2, $V_{TP} = -0.7$ V, and $\lambda = 0.01$ V^{-1}?

13.140. What are the output current and output resistance of the current source in Fig. P13.139 if $V_{DD} = 4$ V, $R_4 = 200$ kΩ, $R_3 = 100$ kΩ, and $R_S = 16$ kΩ? Use the device parameters from Prob. 13.139.

13.141. What are the output current and output resistance of the current source in Fig. P13.139 if $V_{DD} = 5$ V, $R_4 = 200$ kΩ, $R_3 = 47$ kΩ, and $R_S = 43$ kΩ? Use the device parameters from Prob. 13.139.

13.142. Design a current source to provide an output current of 175 μA using the topology in Fig. P13.139. The current source should use no more than 200 μA and have an output resistance of at least 2.5 MΩ. Assume $V_{DD} = 12$ V, $K_p = 200$ μA/V^2, $V_{TP} = -0.75$ V, and $\lambda = 0.02$ V^{-1}.

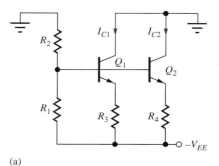

(a)

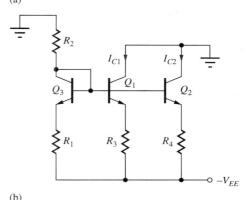

(b)

Figure P13.143

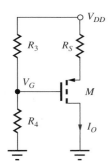

Figure P13.139

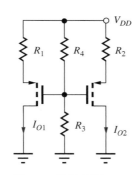

Figure P13.145

13.143. (a) What are the two output currents and output resistances of the current source in Fig. P13.143(a) if $V_{EE} = 10$ V, $\beta_o = 125$, $V_A = 50$ V, $R_1 = 33$ kΩ, $R_2 = 68$ kΩ, $R_3 = 20$ kΩ, and $R_4 = 30$ kΩ? (b) Repeat for the circuit in Fig. P13.143(b).

13.144. (a) Use SPICE to simulate the current source array in Prob. 13.143(a) and find the output currents and output resistances of the sources. Use transfer function analysis to find the output resistances. (b) Repeat for Prob. 13.143(b).

13.145. What are the two output currents and output resistances of the current sources in Fig. P13.145 if $V_{DD} = 10$ V, $K_p = 300$ μA/V^2, $V_{TP} = -0.6$ V, $\lambda = 0.02$ V^{-1}, $R_1 = 5$ kΩ, $R_2 = 24$ kΩ, $R_3 = 2$ MΩ, and $R_4 = 4$ MΩ?

13.146. Repeat Prob. 13.145 using the unified model from Sec. 5.10.7 assuming $V_{SAT} = 1$ V.

13.147. Use SPICE to simulate the current source array in Prob. 13.145, and find the output currents and output resistances of the source. Use transfer function analysis to find the output resistances.

*13.148. The op amp in Fig. P13.148 is used in an attempt to increase the overall output resistance of the current

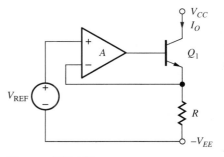

Figure P13.148

source circuit. If $V_{REF} = 5$ V, $V_{CC} = 0$ V, $V_{EE} = 12$ V, $R = 50$ kΩ, $\beta_o = 120$, $V_A = 70$ V, and $A = 75{,}000$, what are the output current I_O and output resistance of the current source? Did the op amp help increase the output resistance? Explain why or why not.

13.149. The op amp in Fig. P13.149 is used to increase the overall output resistance of current source M_1. If $V_{REF} = 5$ V, $V_{DD} = 0$ V, $V_{SS} = 12$ V, $R = 40$ kΩ, $K_n = 800$ μA/V^2, $V_{TN} = 0.8$ V, $\lambda = 0.02$ V^{-1}, and $A = 60{,}000$, what are the output current I_O and output resistance of the current source?

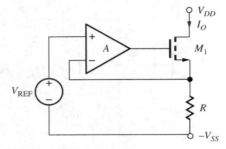

Figure P13.149

13.150. (a) What are the Q-points of the transistors in the amplifier in Fig. P13.150(a) if $\beta_o = 85$ and $V_A = 70$ V? (b) What are the differential-mode gain and CMRR of the amplifier? (c) Repeat for Fig. P13.150(b).

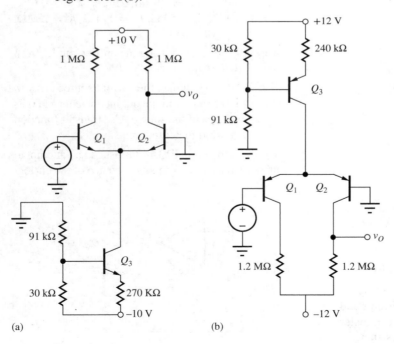

Figure P13.150

13.151. (a) What are the Q-points of the transistors in the amplifier in Fig. P13.151 if $K_n = 500$ μA/V^2, $V_{TN} = +1$ V, $\lambda = 0.02$ V^{-1}, $R_1 = 47$ kΩ, $R_2 = 100$ kΩ, $R_S = 6.8$ kΩ, and $R_D = 33$ kΩ? (b) What are the differential-mode gain and CMRR of the amplifier?

13.152. The output resistance of the MOS current source in Fig. P13.151 is given by $R_{out} = \mu_f R_S$. How much voltage must be developed across R_S to achieve an output resistance of 4 MΩ at a current of 150 μA if $K_n = 500$ μA/V^2 and $\lambda = 0.02$ V^{-1}?

13.153. (a) A current source with $R_{out} = \beta_o r_o$ is used to bias a standard bipolar differential amplifier. What is an expression for the CMRR of this amplifier for single-ended outputs?

**13.154. Use PSPICE to perform a Monte Carlo analysis of the circuits in Fig. 13.51. Assume 5 percent resistors and a 5 percent power-supply tolerance. Find the nominal and 3σ limits on I_O and R_{out}.

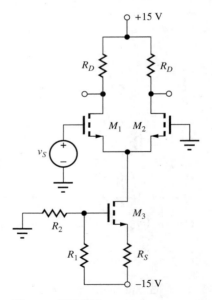

Figure P13.151

Low Voltage/Weak Inversion

(Use as needed: $V_T = 25.8$ mV, $V_{BE} = 0.55$ V, $\beta_F = 75$, $V_A = 60$ V, $I_S = 5 \times 10^{-15}$ A, $n = 1.5$, and $\lambda = 0.02$/V.)

13.155. A low-voltage version of the amplifier in Fig. P13.54 is to be designed to operate as a voltage-follower and have a sinusoidal output signal with an amplitude of 0.25 V. (a) What are the minimum values of the power supplies if the minimum value of V_{CE} is 0.2 V and the minimum voltage across the current source(s) is also 0.2 V? Assume $-V_{EE} = -V_{CC}$. (b) Estimate the open-loop gain of the amplifier. (c) What are the new values if the closed-loop amplifier is designed to have a gain of +5?

13.156. (a) Work Prob. 13.155 for the amplifier in Fig. P13.84 if the sinusoidal output signal is 0.5 V, the minimum value of V_{CE} is 0.25 V, and the minimum voltage across the current sources is 0.3 V? (b) What are the new values if the closed-loop amplifier is designed to have a gain of +4?

13.157. Work Prob. 13.155 for the amplifier in Fig. P13.96 if the sinusoidal output signal is 0.25 V, the minimum value of V_{CE} is 0.25 V, and the minimum voltage across the current sources is 0.2 V? The minimum open-loop gain is 60.

13.158. (a) A low-voltage version of the amplifier in Fig. P13.75 is to be designed to operate as a voltage-follower and have a sinusoidal output signal with an amplitude of 0.25 V. The MOSFETs are to be biased with $V_{SG} = 0$ V. What are the minimum values of the power supplies if the minimum values of V_{CE} and V_{SD} are 0.2 V and the minimum voltage across the current sources is also 0.2 V? Assume $-V_{EE} = -V_{CC}$. (b) Estimate the open-loop gain of the amplifier. (c) Repeat this problem for the amplifier in Fig. P13.80.

13.159. For the circuit in Fig. P13.90, M_1, M_2 and M_4 are biased with $V_{GS} = 0$ V and $V_{SG} = 0.2$ V for transistor M_3. What are the minimum values of the power supplies if V_{DS} and V_{SD} must be at least 0.2 V and the minimum voltage across the current sources is also 0.15 V? Assume $-V_{DD} = -V_{SS}$, and the amplifier is to be designed to operate as a voltage-follower with a sinusoidal output signal range of ± 0.35 V. (b) Estimate the open-loop gain of the amplifier.

13.160. (a) The 2.2-V source in Prob. 13.108 is reduced to 0.6 V so that the transistors are operating in weak inversion. What are the drain currents in the two transistors? Assume that the 10-V sources are changed to ± 1.5 V. (b) Repeat if the 2.2-V source is changed to 1.2 V.

13.161. (a) The 1.3-V source in Prob. 13.110 is reduced to 0.5 V. What are the collector currents in the two transistors? Assume that the 10-V sources are changed to ± 1.5 V. (b) Repeat if the 1.3-V source is changed to 0.7 V.

13.162. (a) Work Prob. 13.139 with V_{DD} changed to 1.5 V. (b) Repeat for $V_{DD} = 1.0$ V.

13.163. (a) Work Prob. 13.145 with V_{DD} changed to 1.5 V. (b) Repeat for $V_{DD} = 1.0$ V.

13.164. The power supplies in the amplifier in Fig. P13.150 are reduced to ± 2.5 V, the current-source resistors values are all increased by a factor of 20, and $R_C = 20$ MΩ. (a) Find the Q-points of the transistors. (b) What are the values of voltage gain and CMRR of the amplifier?

13.165. The power supplies in the amplifier in Fig. P13.151 are reduced to ± 2.5 V, the current-source resistors values are all increased by a factor of 10, and $R_D = 160$ MΩ. (a) Find the Q-points of the transistors. (b) What are the values of voltage gain and CMRR of the amplifier?

术语对照

ac-coupled amplifiers	交流耦合放大器
Balanced output	平衡输出
Cascode amplifier	Cascode放大器
Cascode current source	Cascode电流源
Class-A, class-B, and class-AB amplifiers	A类、B类和AB类放大器
Class-B push-pull output stage	B类推挽输出级
Common-mode conversion gain	共模转换增益
Common-mode gain	共模增益
Common-mode half-circuit	共模半电路
Common-code input resistance	共模输入电阻
Common-mode input voltage range	共模输入电压范围
Common-code rejection ratio(CMRR)	共模抑制比（CMRR）
Complementary push-pull output stage	互补推挽输出级
Conduction angle	导通角
Cross-over distortion	交越失真
Cross-over region	交越区
Current-limiting circuit	限流电路
Current sink	电流沉
Darlington circuit	达林顿电路
dc bias sensitivity	直流偏置灵敏度
Dead zone	死区
Differential amplifier	差分放大器
Differential-mode conversion gain	差模转换增益
Differential-mode gain	差模增益
Differential-mode half-circuit	差模半电路
Differential-mode input resistance	差模输入电阻
Differential-mode output resistance	差模输出电阻
Differential-mode output voltage	差模输出电压
Electronic current source	电子电流源
Figure of merit (FOM)	品质因数（FOM）
Half-circuit analysis	半电路分析
Ideal current source	理想电流源
Input resistances	输入电阻
Output resistances	输出电阻
Single-ended output	单端输出
Source follower	源极跟随器
Virtual ground	虚地
Voltage reference	参考电压
Velocity Saturation	速度饱和

CHAPTER 14

ANALOG INTEGRATED CIRCUIT DESIGN TECHNIQUES

第14章　模拟集成电路设计技术

本章提纲

14.1　电路元件匹配

14.2　电流镜

14.3　高输出电阻电流镜

14.4　参考电流的产生

14.5　与电源电压无关的偏置

14.6　带隙基准源

14.7　电流镜作为有源负载

14.8　运算放大器中的源负载

14.9　μA741运算放大器

14.10　Gilbert模拟乘法器

本章目标

　　本章的主要目标是了解基于紧密匹配器件特性的集成电路设计技术，并掌握运算放大器和其他IC设计中的一些关键构建模块。主要包括：

- 理解双极型和MOS电流镜工作原理和镜像比例错误；
- 分析高输出电阻电流源，包括级联和Wilson电流源电路；
- 学习用于分立和集成电路的电流源设计方法；
- 将参考电流电路添加到电路构建模块中，这些电路产生的电流与电源电压具有很大程度的独立性，包括基于V_{BE}的参考电压和Widlar电流；
- 研究带隙基准电路的工作原理和设计，该电路用于提供精确的参考电压（不依赖于电源和温度），是一类重要的电路；
- 用电流做差分放大器的有源负载，可以将单级放大器的电压增益提高到放大系数μ_f；
- 学习将器件失配的影响，如共模抑制比（CMRR），纳入放大器特性的计算中；
- 分析经典μA741运算放大器的设计；
- 学习实现大输入信号范围的四象限模拟器乘法设计技巧；
- 继续研究SPICE仿真。

本章导读

　　本章主要学习两位著名的集成电路设计师Robert Widlar和Barrie Gilbert的一些设计精巧的电路结构。Widlar

设计了μA702运放和μA741运放，也是μA723稳压器和带隙基准的设计者。Gilbert发明了四象限模拟乘法器，也就是现今的Gilbert乘法器。μA741电路技术包含了许多沿用至今的设计技巧。Brokaw的带隙参考电路版本被广泛使用，带隙参考和电压调节器电路也用作数字测温中的温度传感器。

集成电路能够集成更多参数相同的晶体管。虽然这些器件的绝对参数容差较差，但是器件参数的匹配度可达1%甚至更小，正是具有参数精确匹配的优势，开发了利用器件特性紧密匹配的特殊电路技术，本章介绍了电路元件匹配的概念及在集成设计中元器件匹配问题。大量高度匹配的器件应用在模拟电路设计中，实现了只采用少量电阻的高性能电路。

本章探讨了在MOS和双极技术中使用匹配的晶体管来设计电流源的方法。电流镜电路的输出电流是输入电流的镜像，该电路可产生多个镜像电流，并且电流镜的增益控制可以通过调整晶体管的发射极面积比例，或者FET的W/L比来实现。通过参数λ、V_A、β_F，电流镜的镜像比率系数的误差直接与晶体管的有限输出电阻和电流增益相关。

在双极型晶体管电流镜中，BJT晶体管的有限电流增益会导致电流镜比率误差，而缓冲电流镜电路的设计就是为了最小化这一误差。在FET和BJT电路中，电流镜的理想平衡受到镜像输入和输出部分之间的直流电压不匹配的干扰，失配程度由电流源的输出电阻决定。

基本电流镜的品质因数V_{CS}对于BJT电路约为V_A，对于MOS电路约为$1/\lambda$。但是，通过采用Cascode电流源或者Wilson电流源，V_{CS}的值可提升两个数量级。本章重点讨论了Widlar电源流和Cascode电流源，Widlar电源流可将V_{CS}增大到$\beta_0 V_A$或者μ_f/λ的数量级，而可调Cascode电流源可获得更高的V_{CS}值。

本章探讨了用于实现独立功率电压偏置的实现技术。电流镜可以用来产生与电源电压无关的电流。基于V_{BE}的参考源和Widlar参考源所产生的电流仅与电源电压的对数相关。将Widlar电流源与电流镜结合，可以实现一个独立于电源电压的一阶参考电源。

精准的电压基准源不仅要独立于电源电压，也要不受温度影响。本章详细介绍了带隙基准电路，这种电路主要采用pn结特性，能够产生准确的输出电压，而与电源和温度无关。带隙基准电路广泛用于参考电压源和稳压器中。由Widlar开发并由Brokaw优化的带隙电压电路，使用PTAT电路来消除BJT的基极-发射极结的负温度系数，从而形成具有非常低的TC和电源电压依赖性的、高度稳定的电压基准。该电路及其变体在模拟和数字集成电路中得到广泛使用。

电流镜通常用于偏置模拟电路中，或在差分和运算放大器中代替负载。这种有源负载电路能够显著增大放大器的电压增益，本章详细讲解了电流镜作为有源负载的CMOS差分放大器和双极型差分放大器，并给出了一些MOS和双极型电路相关实例。在差分放大器和运算放大器中替代负载电阻也是电流镜的一个十分重要的应用。这种有源负载电路在大大增强多数放大器的电压增益的同时，还保持了工作点的平衡，这是获得低偏移电压和良好的共模抑制能力的必要条件。带有有源负载的放大器可以实现单级电压增益，从而适应晶体管的放大系数。对采用电流镜的电路进行交流分析时，通常可利用应用于电流镜的双端口模型来简化。

本章详细介绍了μA741运算放大器的电路结构和工作原理，研究了输入级相关的直流分析、交流分析、电压增益、输出阻抗以及短路保护电路等内容。经典μA741运算放大器产生于20世纪60年代末，是第一个在输入级有击穿电压保护电路，在输出级有短路保护电路，且具有优异的整体性能的高鲁棒性放大器。在一个具有两级增益的放大器中，采用有源负载可获得超过100dB的电压增益。这一放大器迅速成为业界的标准运算放大器设计。

有源电流镜负载可用来增强双极型与MOS运算放大器的性能。本章最后给出了Gilbert的精密四象限模拟乘法器设计，两个输入电压是既可以为正也可以为负，解决了两个模拟信号精确相乘的难题。

Chapter Outline

14.1 Circuit Element Matching
14.2 Current Mirrors
14.3 High-Output-Resistance Current Mirrors
14.4 Reference Current Generation
14.5 Supply-Independent Biasing
14.6 The Bandgap Reference
14.7 The Current Mirror as an Active Load
14.8 Active Loads in Operational Amplifiers
14.9 The µA741 Operational Amplifier
14.10 The Gilbert Analog Multiplier
Summary
Key Terms
References
Additional Readings
Problems

Chapter Goals

In Chapter 14, we concentrate on understanding integrated circuit design techniques that are based upon the characteristics of closely matched devices and look at a number of key building blocks of operational amplifiers and other ICs. Our goals are to:

- Understand bipolar and MOS current mirror operation and mirror ratio errors
- Explore high-output resistance current sources including cascode, regulated cascode, and Wilson current source circuits
- Learn to design current sources for use in both discrete and integrated circuits
- Add reference current circuit techniques to our kit of circuit building blocks. These circuits produce currents that exhibit a substantial degree of independence from power supply voltage including the V_{BE}-based reference and the Widlar current source.
- Investigate the operation and design of bandgap reference circuits, one of the most important techniques for providing an accurate reference voltage that is independent of power supply voltages and temperature
- Use current mirrors as active loads in differential amplifiers to increase the voltage gain of single-stage amplifiers to the amplification factor μ_f
- Learn how to include the effects of device mismatch in the calculation of amplifier performance measures such as CMRR
- Analyze the design of the classic µA741 operational amplifier
- Understand the techniques used to realize four-quadrant analog multipliers with large input signal range
- Continue to increase our understanding of SPICE simulation techniques

In Chapter 14, we explore several extremely clever and exciting circuits designed by two of the legends of integrated circuit design, Paul Brokaw, Robert Widlar and Barrie Gilbert. Widlar designed the µA702 op amp and later developed the LM101 operational amplifier and many of the circuits that led to the design of the classic µA741 op amp. Widlar was also responsible for the µA723 voltage regulator and the original bandgap reference. Gilbert invented a four-quadrant analog multiplier circuit referred to today as the Gilbert multiplier. The µA741 circuit techniques spawned a broad range of follow-on designs that are still in use today. Brokaw's version of the bandgap reference circuit is widely used, and he is highly respected for mentoring many analog circuit designers. The bandgap reference forms the heart of most precision voltage references and voltage regulator circuits, and is also used as a temperature sensor in digital thermometry.

Integrated circuit (IC) technology allows the realization of large numbers of virtually identical transistors. Although the absolute parameter tolerances of these devices are relatively poor, device characteristics can be matched to within 1 percent or better. The ability to build devices with nearly identical characteristics has led to the development of special circuit techniques that take advantage of the tight matching of the device characteristics.

Chapter 14
Analog Integrated Circuit Design Techniques

(a)

(b)

(c)

Legends of analog design. (a) Paul Brokaw (b) Robert Widlar (c) Barrie Gilbert ((a) *Courtesy of Paul Brokaw* (b) *Courtesy Texas Instruments* (c) *Courtesy of Analog Devices*).

Chapter 14 begins by exploring the use of matched transistors in the design of current sources, called **current mirrors**, in both MOS and bipolar technology. Cascode and Wilson current sources are subsequently added to our repertoire of high-output-resistance current source circuits. Circuit techniques that can be used to achieve **power-supply-independent biasing** are introduced.

We will also study the bandgap reference circuit which uses the well-defined behavior of the pn junction to produce a precise output voltage that is highly independent of power supply voltage variations and temperature. The bandgap circuit is widely used in voltage references and voltage regulators.

The current mirror is often used to bias analog circuits and to replace load resistors in differential and operational amplifiers. This active-load circuit can substantially enhance the voltage gain capability of many amplifiers, and a number of MOS and bipolar circuit examples are presented. The chapter then discusses circuit techniques used in IC operational amplifiers, including the classic μA741 amplifier. This design provides a robust, high-performance, general-purpose operational amplifier with breakdown-voltage protection of the input stage and short-circuit protection of the output stage. The final section looks at the precision four-quadrant analog multiplier design of Gilbert.

14.1 CIRCUIT ELEMENT MATCHING

Integrated circuit (IC) technology allows the realization of large numbers of virtually identical transistors. Although the absolute parameter tolerances of these devices are relatively poor, device characteristics can be matched to within 1 percent or better. The ability to build devices with nearly identical characteristics has led to the development of special circuit techniques that take advantage of the tight matching of device characteristics. Transistors are said to be **matched** when they have identical sets of device parameters: (I_S, β_{FO}, V_A) for the BJT, (V_{TN}, K', λ) for the MOSFET, or (I_{DSS}, V_P, λ) for the JFET. The planar geometry of the devices can easily be changed in integrated designs, and so the emitter area A_E of the BJT and the W/L ratio of the MOSFET become important circuit design parameters. (Remember from our study of MOS digital circuits in Part II that W/L represents a fundamental circuit design parameter.)

In integrated circuits, absolute parameter values may vary widely from fabrication process run to process run, with ±25 to 30 percent tolerances not uncommon (see Table 14.1). However, the matching between nearby circuit elements on a given IC chip is typically within a fraction of a percent. Thus, IC design techniques have been invented that rely heavily on **matched device** characteristics and resistor ratios rather than absolute parameter values. The circuits described in this

TABLE 14.1
IC Tolerances and Matching [1]

	ABSOLUTE TOLERANCE, %	MISMATCH, %
Diffused resistors	30	≤ 2
Ion-implanted resistors	5	≤ 1
V_{BE}	10	≤ 1
I_S, β_F, V_A	30	≤ 1
V_{TN}, V_{TP}	15	≤ 1
K', λ	30	≤ 1

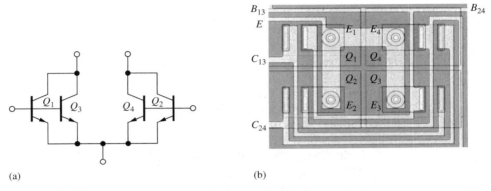

Figure 14.1 (a) Differential amplifier formed with a cross-connected quad of identical transistors. (b) Layout of the cross-coupled transistor quad in Fig. 14.1(a). Round emitters are used to improve device matching.

chapter depend, for proper operation, on the tight device matching that can be realized through IC fabrication processes, and many will not operate correctly if built with poorly mismatched discrete components. However, many of these circuits can be used in discrete circuit design if integrated transistor arrays are used in the implementation.

Figure 14.1 shows one example of the use of four matched transistors to improve the performance of the differential amplifier that we studied in the previous chapter. The four devices are cross-connected to further improve the overall parameter matching and temperature tracking of the circuit. In Sec. 14.2, we explore the use of matched bipolar and MOS transistors in the design of IC current sources called **current mirrors**. Circular emitters are used to further improve matching.

> **EXERCISE:** An IC resistor has a nominal value of 10 kΩ and a tolerance of ±30 percent. A particular process run has produced resistors with an average value 20 percent higher than the nominal value, and the resistors are found to be matched within 2 percent. What range of resistor values will occur in this process run?
>
> **ANSWER:** 11.88 kΩ ≤ R ≤ 12.12 kΩ.

14.2 CURRENT MIRRORS

Current mirror biasing is an extremely important technique in integrated circuit design. Not only is it heavily used in analog applications, it also appears routinely in digital circuit design as well. Figure 14.2 shows the circuits for basic MOS and bipolar current mirrors. In Fig. 14.2(a), MOSFETs M_1 and M_2 are assumed to have identical characteristics (V_{TN}, K'_n, λ) and W/L ratios; in Fig. 14.2(b),

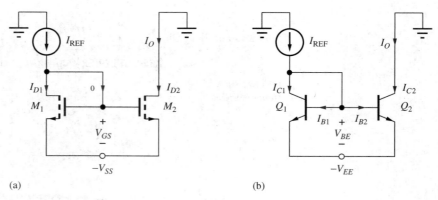

Figure 14.2 (a) MOS and (b) BJT current mirror circuits.

the characteristics of Q_1 and Q_2 are assumed to be identical (I_S, β_{FO}, V_A). In both circuits, a **reference current** I_{REF} provides operating bias to the mirror, and the output current is represented by current I_O. These basic circuits are designed to have $I_O = I_{REF}$; that is, the output current mirrors the reference current—hence, the name "current mirror." Note that the current mirror circuits do not utilize resistors that were required by the current sources studied in Chapter 13, a characteristic desired for integrated circuit realization.

14.2.1 dc ANALYSIS OF THE MOS TRANSISTOR CURRENT MIRROR

In the MOS current mirror in Fig. 14.2(a), reference current I_{REF} goes through "diode-connected" transistor M_1, establishing gate-source voltage V_{GS}. V_{GS} is applied to transistor M_2, developing an identical drain current $I_O = I_{D2} = I_{REF}$. Detailed analysis of the current mirror operation follows in the paragraphs below.

Because the gate currents are zero for the MOSFETs, reference current I_{REF} must flow into the drain of M_1, which is forced to operate in saturation (pinch-off) by the circuit connection because $V_{DS1} = V_{GS1} = V_{GS}$. V_{GS} will adjust itself to the value required for $I_{D1} = I_{REF}$. Assuming matched devices:[1]

$$I_{REF} = \frac{K_n}{2}(V_{GS1} - V_{TN})^2(1 + \lambda V_{DS1}) \quad \text{or} \quad V_{GS1} = V_{TN} + \sqrt{\frac{2 I_{REF}}{K_{n1}(1 + \lambda V_{DS1})}} \quad (14.1)$$

Current I_O is equal to the drain current of M_2:

$$I_O = I_{D2} = \frac{K_n}{2}(V_{GS2} - V_{TN})^2(1 + \lambda V_{DS2}) \quad (14.2)$$

However, the circuit connection forces $V_{GS2} = V_{GS1}$, and $V_{DS1} = V_{GS1}$. Substituting Eq. (14.1) into Eq. (14.2) yields

$$I_O = I_{REF}\frac{(1 + \lambda V_{DS2})}{(1 + \lambda V_{DS1})} \cong I_{REF} \quad (14.3)$$

For equal values of V_{DS}, the output current is identical to the reference current (that is, the output mirrors the reference current). Unfortunately, in most circuit applications, $V_{DS2} \neq V_{DS1}$, and there is a slight mismatch between the output current and the reference current, as demonstrated in Ex. 14.1.

---

[1] Matching between elements in the current mirror is very important; this is a case in which the $(1 + \lambda\ V_{DS})$ term is included in the dc, as well as ac, calculations.

For convenience, we define the ratio of I_O to I_{REF} to be the **mirror ratio** MR given by

$$\text{MR} = \frac{I_O}{I_{REF}} = \frac{(1 + \lambda V_{DS2})}{(1 + \lambda V_{DS1})} \qquad (14.4)$$

EXAMPLE 14.1 **OUTPUT CURRENT OF THE MOS CURRENT MIRROR**

In this example, we find the output current for the standard current mirror configuration.

PROBLEM Calculate the output current I_O for the MOS current mirror in Fig. 14.2(a) if $V_{SS} = 10$ V, $K_n = 250$ µA/V^2, $V_{TN} = 1$ V, $\lambda = 0.0133$ V^{-1}, and $I_{REF} = 150$ µA.

SOLUTION **Known Information and Given Data:** Current mirror circuit in Fig. 14.2(a); $V_{SS} = 10$ V; transistor parameters are given as $K_n = 250$ µA/V^2, $V_{TN} = 1$ V, $\lambda = 0.0133$ V^{-1}, and $I_{REF} = 150$ µA

Unknowns: Output current I_O

Approach: Find V_{GS1} and V_{DS2} and then evaluate Eq. (14.3) to give the output current.

Assumptions: Transistors are identical and operating in the active region of operation.

Analysis: We need to evaluate Eq. (14.3) and must find the value of V_{GS1} using Eq. (14.1). Since $V_{DS1} = V_{GS1}$, we can write

$$V_{DS1} = V_{TN} + \sqrt{\frac{2 I_{REF}}{K_n}} = 1 + \sqrt{\frac{2(150 \text{ µA})}{250 \frac{\text{µA}}{\text{V}^2}}} = 2.10 \text{ V}$$

in which we have neglected the $(1 + \lambda V_{DS1})$ term to simplify the dc bias calculation. Substituting this value and $V_{DS2} = 10$ V in Eq. (14.3):

$$I_O = (150 \text{ µA}) \frac{[1 + 0.0133(10)]}{[1 + 0.0133(2.10)]} = 165 \text{ µA}$$

The ideal output current would be 150 µA, whereas the actual currents are mismatched by approximately 10 percent.

Check of Results: A double check shows the calculations to be correct. M_1 is saturated by connection, and M_2 will also be active as long as its drain-source voltage exceeds $(V_{GS1} - V_{TN})$, which is easily met in Fig. 14.2(a) since $V_{DS2} = 10$ V.

Discussion: We could attempt to improve the precision of our answer slightly by including the $(1 + \lambda V_{DS1})$ term in the evaluation of V_{GS1}. The solution then requires an iterative analysis that barely changes the value of I_O. (See the exercise on next page.)

Computer-Aided Analysis: We can check our analysis directly with SPICE by setting the MOS transistor parameters to KP = 250 µA/V^2, VTO = 1 V, LEVEL = 1, and LAMBDA = 0. SPICE yields an output current of 150 µA with $V_{GS} = 2.095$ V. With nonzero λ, LAMBDA = 0.0133 V^{-1}, SPICE yields $I_O = 165$ µA with $V_{GS} = 2.081$ V. The values are in agreement with our hand calculations.

> **EXERCISE:** Suppose we include the $(1 + \lambda V_{DS1})$ term in the evaluation of V_{GS1}. Show that the equation to be solved is
>
> $$V_{DS1} = V_{TN} + \sqrt{\frac{2I_{REF}}{K_n(1 + \lambda V_{DS1})}}$$
>
> Find the new value of V_{DS1} using the numbers in Ex. 14.1. What is the new value of I_O?
>
> **ANSWERS:** 2.08 V; 165 μA
>
> **EXERCISE:** Based on the numbers in Ex. 14.1, what is the minimum value of the drain voltage required to keep M_2 saturated in Fig. 14.2(a)?
>
> **ANSWER:** −8.9 V

14.2.2 CHANGING THE MOS MIRROR RATIO

The power of the current mirror is greatly increased if the mirror ratio can be changed from unity. For the MOS current mirror, the ratio can easily be modified by changing the W/L ratios of the two transistors forming the mirror. In Fig. 14.3, for example, remembering that $K_n = K'_n(W/L)$ for the MOSFET, the K_n values of the two transistors are given by

$$K_{n1} = K'_n \left(\frac{W}{L}\right)_1 \quad \text{and} \quad K_{n2} = K'_n \left(\frac{W}{L}\right)_2 \tag{14.5}$$

Substituting these two different values of K_n in Eqs. (14.1) and (14.2) yields the **mirror ratio** (MR) given by

$$\text{MR} = \frac{\left(\frac{W}{L}\right)_2 (1 + \lambda V_{DS2})}{\left(\frac{W}{L}\right)_1 (1 + \lambda V_{DS1})} \tag{14.6}$$

In the ideal case ($\lambda = 0$) or for $V_{DS2} = V_{DS1}$, the mirror ratio is set by the ratio of the W/L values of the two transistors. For the particular values in Fig. 14.3, this design value of the mirror ratio would be 5, and the output current would be $I_O = 5I_{REF}$. However, the differences in V_{DS} will again create an error in the mirror ratio.

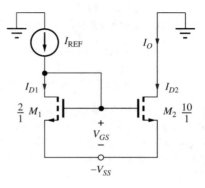

Figure 14.3 MOS current mirror with unequal (W/L) ratios.

EXERCISE: (a) Calculate the mirror ratio for the MOS current mirrors in the figure here for $\lambda = 0$.
(b) For $\lambda = 0.02$ V^{-1} if $V_{TN} = 1$ V, $K'_n = 25$ µA/V^2, and $I_{REF} = 50$ µA.

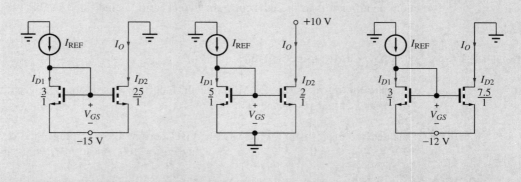

ANSWERS: 8.33, 0.400, 2.5; 10.4, 0.462, 2.97

14.2.3 dc ANALYSIS OF THE BIPOLAR TRANSISTOR CURRENT MIRROR

The operation of the bipolar current mirror in Fig. 14.2(b) is similar to that of the MOS circuit. Reference current I_{REF} goes through diode-connected transistor Q_1, establishing base-emitter voltage V_{BE}. V_{BE} also biases transistor Q_2, developing an almost identical collector current at its output: $I_O = I_{C2} \cong I_{REF}$. Detailed analysis of the current mirror operation follows in the paragraphs below.

Analysis of the BJT current mirror in Fig. 14.2(b) is similar to that of the FET. Applying KCL at the collector of diode-connected transistor Q_1 yields

$$I_{REF} = I_{C1} + I_{B1} + I_{B2} \quad \text{and} \quad I_O = I_{C2} \tag{14.7}$$

The currents needed to relate I_O to I_{REF} can be found using the transport model, noting that the circuit connection forces the two transistors to have the same base-emitter voltage V_{BE}:

$$I_{C1} = I_S \exp\left(\frac{V_{BE}}{V_T}\right)\left(1 + \frac{V_{CE1}}{V_A}\right) \quad I_{C2} = I_S \exp\left(\frac{V_{BE}}{V_T}\right)\left(1 + \frac{V_{CE2}}{V_A}\right)$$

$$\beta_{F1} = \beta_{FO}\left(1 + \frac{V_{CE1}}{V_A}\right) \quad\quad \beta_{F2} = \beta_{FO}\left(1 + \frac{V_{CE2}}{V_A}\right) \tag{14.8}$$

$$I_{B1} = \frac{I_S}{\beta_{FO}} \exp\left(\frac{V_{BE}}{V_T}\right) \quad\quad I_{B2} = \frac{I_S}{\beta_{FO}} \exp\left(\frac{V_{BE}}{V_T}\right)$$

Substituting Eq. (14.8) into Eq. (14.7) and solving for $I_O = I_{C2}$ yields

$$I_O = I_{REF} \frac{\left(1 + \frac{V_{CE2}}{V_A}\right)}{\left(1 + \frac{V_{CE1}}{V_A} + \frac{2}{\beta_{FO}}\right)} = I_{REF} \frac{\left(1 + \frac{V_{CE2}}{V_A}\right)}{\left(1 + \frac{V_{BE}}{V_A} + \frac{2}{\beta_{FO}}\right)} \tag{14.9}$$

If the Early voltage was infinite, Eq. (14.9) would give a mirror ratio of

$$\text{MR} = \frac{I_O}{I_{REF}} = \frac{1}{1 + \frac{2}{\beta_{FO}}} \tag{14.10}$$

and the output current would mirror the reference current, except for a small error due to the finite current gain of the BJT. For example, if $\beta_{FO} = 100$, the currents would match within 2 percent. As for the FET case, however, the collector-emitter voltage mismatch in Eq. (14.9) is generally more significant than the **current gain defect** term, as indicated in Ex. 14.2.

EXAMPLE 14.2 MIRROR RATIO CALCULATIONS

Compare the mirror ratios for MOS and BJT current mirrors operating with similar bias conditions and output resistances ($V_A = 1/\lambda$).

PROBLEM Calculate the mirror ratio for the MOS and BJT current mirrors in Fig. 14.2 for $V_{GS} = 2$ V, $V_{DS2} = 10$ V $= V_{CE2}$, $\lambda = 0.02$ V^{-1}, $V_A = 50$ V, and $\beta_{FO} = 100$. Assume $M_1 = M_2$ and $Q_1 = Q_2$.

SOLUTION **Known Information and Given Data:** Current mirror circuits in Fig. 14.2 with $M_2 = M_1$ and $Q_2 = Q_1$; $V_{SS} = 10$ V; operating voltages: $V_{GS} = 2$ V, $V_{DS2} = V_{CE2} = 10$ V and $V_{BE} = 0.7$ V; transistor parameters: $\lambda = 0.02$ V^{-1}, $V_A = 50$ V, and $\beta_{FO} = 100$

Unknowns: Mirror ratio MR for each current mirror

Approach: Use Eqs. (14.6) and (14.9) to determine the mirror ratios.

Assumptions: BJTs and MOSFETs are in the active region of operation, respectively. Assume $V_{BE} = 0.7$ V for the BJTs and the MOSFETs are enhancement-mode devices.

Analysis: For the MOS current mirror,

$$\text{MR} = \frac{(1 + \lambda V_{DS2})}{(1 + \lambda V_{DS1})} = \frac{\left[1 + \frac{0.02}{\text{V}}(10 \text{ V})\right]}{\left[1 + \frac{0.02}{\text{V}}(2 \text{ V})\right]} = 1.15$$

and for the BJT case

$$\text{MR} = \frac{\left(1 + \frac{V_{CE2}}{V_A}\right)}{\left(1 + \frac{2}{\beta_{FO}} + \frac{V_{CE1}}{V_A}\right)} = \frac{\left(1 + \frac{10 \text{ V}}{50 \text{ V}}\right)}{\left(1 + \frac{2}{100} + \frac{0.7 \text{ V}}{50 \text{ V}}\right)} = 1.16$$

Check of Results: A double check shows our calculations to be correct. M_1 is forced to be active by connection. M_2 has $V_{DS2} > V_{GS2}$ and will be pinched-off for $V_{TN} > 0$ (enhancement-mode transistor). Q_1 has $V_{CE} = V_{BE}$, so it is forced to be in the active region. Q_2 has $V_{CE2} > V_{BE2}$ and is also in the active region. The assumed regions of operation are valid.

Discussion: The FET and BJT mismatches are very similar—15 percent and 16 percent, respectively. The current gain error is a small contributor to the overall error in the BJT mirror ratio.

Computer-Aided Analysis: We can easily perform an analysis of the current mirrors using SPICE, which will be done shortly as part of Ex. 14.3.

> **EXERCISE:** What is the actual value of V_{BE} in the bipolar current mirror in Ex. 14.2 if $I_S = 0.1$ fA and $I_{REF} = 100$ μA? What is the minimum value of the collector voltage required to maintain Q_2 in the active region in Fig. 14.2(b)?
>
> **ANSWERS:** 0.691 V; $-V_{EE} + 0.691$ V

14.2.4 ALTERING THE BJT CURRENT MIRROR RATIO

In bipolar IC technology, the designer is free to modify the emitter area of the transistors, just as the W/L ratio can be chosen in MOS design. To alter the BJT mirror ratio, we use the fact that the saturation current of the bipolar transistor is proportional to its emitter area A_E and can be written as

$$I_S = I_{SO}\frac{A_E}{A} \tag{14.11}$$

I_{SO} represents the saturation current of a bipolar transistor with one unit of emitter area: $A_E = 1 \times A$. The actual dimensions associated with A are technology-dependent.

By changing the relative sizes of the emitters (**emitter area scaling**) of the BJTs in the current mirror, the IC designer can modify the mirror ratio. For the modified mirror in Fig. 14.4,

$$I_{C1} = I_{SO}\frac{A_{E1}}{A}\exp\left(\frac{V_{BE}}{V_T}\right)\left(1 + \frac{V_{CE1}}{V_A}\right) \quad I_{C2} = I_{SO}\frac{A_{E2}}{A}\exp\left(\frac{V_{BE}}{V_T}\right)\left(1 + \frac{V_{CE2}}{V_A}\right)$$

$$I_{B1} = \frac{I_{SO}}{\beta_{FO}}\frac{A_{E1}}{A}\exp\left(\frac{V_{BE}}{V_T}\right) \quad I_{B2} = \frac{I_{SO}}{\beta_{FO}}\frac{A_{E2}}{A}\exp\left(\frac{V_{BE}}{V_T}\right) \tag{14.12}$$

Substituting these equations in Eq. (14.7) and then solving for I_O yields

$$I_O = nI_{REF}\frac{1 + \dfrac{V_{CE2}}{V_A}}{1 + \dfrac{V_{BE}}{V_A} + \dfrac{1+n}{\beta_{FO}}} \quad \text{where} \quad n = \frac{A_{E2}}{A_{E1}} \tag{14.13}$$

In the ideal case of infinite current gain and identical collector-emitter voltages, the mirror ratio would be determined only by the ratio of the two emitter areas: MR $= n$.

However, for finite current gain,

$$\text{MR} = \frac{n}{1 + \dfrac{1+n}{\beta_{FO}}} \quad \text{where} \quad n = \frac{A_{E2}}{A_{E1}} \tag{14.14}$$

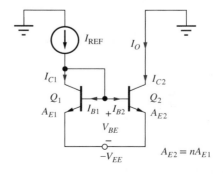

Figure 14.4 BJT current mirror with unequal emitter area.

For example, suppose $A_{E2}/A_{E1} = 10$ and $\beta_{FO} = 100$; then the mirror ratio would be 9.01. A relatively large error (10 percent) is occurring even though the effect of collector-emitter voltage mismatch has been ignored. For high mirror ratios, the current gain error term can become quite important because the total number of units of base current increases directly with the mirror ratio.

EXERCISE: (a) Calculate the ideal mirror ratio for the BJT current mirrors in the figure below if $V_A = \infty$ and $\beta_{FO} = \infty$. (b) If $V_A = \infty$ and $\beta_{FO} = 75$. (c) If $V_A = 60$ V, $\beta_{FO} = 75$, and $V_{BE} = 0.7$ V.

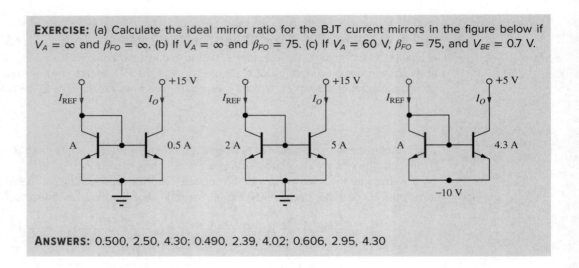

ANSWERS: 0.500, 2.50, 4.30; 0.490, 2.39, 4.02; 0.606, 2.95, 4.30

14.2.5 MULTIPLE CURRENT SOURCES

Analog circuits often require a number of different current sources to bias the various stages of the design. A single reference transistor, M_1 or Q_1, can be used to generate multiple output currents using the circuits in Fig. 14.5. In Fig. 14.5(a), the unusual connection of the gate terminals through the MOSFETs is being used as a "short-hand" method to indicate that all the gates are connected together. Circuit operation is similar to that of the basic current mirror. The reference current enters the diode-connected transistor—here, MOSFET M_1—establishing gate-source voltage V_{GS}, that is then used to bias transistors M_2 through M_5, each having a different W/L ratio. Because there is no current gain defect in MOS technology, a large number of output transistors can be driven from one reference transistor.

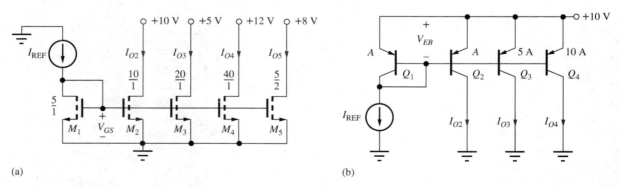

Figure 14.5 (a) Multiple MOS current sources generated from one reference voltage. (b) Multiple bipolar sources biased by one reference device.

EXERCISE: What are the four output currents in the circuit in Fig. 14.5(a) if $I_{REF} = 100$ μA and $\lambda = 0$ for all the FETs?

ANSWERS: 200 μA; 400 μA; 800 μA; 50.0 μA

EXERCISE: Recalculate the four output currents in the circuit in Fig. 14.5(a) if $\lambda = 0.02$ for all the FETs. Assume $V_{GS} = 2$ V.

ANSWERS: 231 μA; 423 μA; 954 μA; 55.8 μA

The situation is very similar in the *pnp* bipolar mirror in Fig. 14.5(b). Here again, the base terminals of the BJTs are extended through the transistors to simplify the drawing. In this circuit, reference current I_{REF} flows through diode-connected BJT Q_1 to establish the emitter-base reference voltage V_{EB}. V_{EB} is then used to bias transistors Q_2 to Q_4, each having a different emitter area relative to that of the reference transistor. Because the total base current increases with the addition of each output transistor, the base current error term gets worse as more transistors are added, which limits the number of outputs that can be used with the basic bipolar current mirror. The buffered current mirror in Sec. 14.2.6 was invented to solve this problem.

An expression for the output current from a given collector can be derived following the steps that led to Eq. (14.13):

$$I_{Oi} = n_i I_{REF} \frac{1 + \dfrac{V_{ECi}}{V_A}}{1 + \dfrac{V_{EB}}{V_A} + \dfrac{1 + \sum_{i=2}^{m} n_i}{\beta_{FO}}} \quad \text{where} \quad n_i = \frac{A_{Ei}}{A_{E1}} \quad (14.15)$$

EXERCISE: (a) What are the three output currents in the circuit in Fig. 14.5(b) if $I_{REF} = 10$ μA, $\beta_{FO} = 50$, and $V_A = \infty$ for all the BJTs? (b) Repeat for $V_A = 50$ V and $V_{EB} = 0.7$ V. Use Eq. (14.15).

ANSWERS: 7.46 μA, 37.3 μA, 74.6 μA; 8.86 μA, 44.3 μA, 88.6 μA

14.2.6 BUFFERED CURRENT MIRROR

The current gain defect in the bipolar current mirror can become substantial when a large mirror ratio is used or if many source currents are generated from one reference transistor. However, this error can be reduced greatly by using the circuit in Fig. 14.6, called a **buffered current mirror.** The current gain of transistor Q_3 is used to reduce the base current that is subtracted from the reference

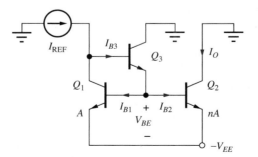

Figure 14.6 Buffered current mirror.

current. Applying KCL at the collector of transistor Q_1, and assuming that $V_A = \infty$ for simplicity, I_{C1} is expressed as

$$I_{C1} = I_{\text{REF}} - I_{B3} = I_{\text{REF}} - \frac{(1+n)\frac{I_{C1}}{\beta_{FO1}}}{\beta_{FO3}+1} \qquad (14.16)$$

and solving for the collector current yields

$$I_O = n I_{C1} = n I_{\text{REF}} \frac{1}{1+\frac{(1+n)}{\beta_{FO1}(\beta_{FO3}+1)}} \qquad (14.17)$$

The current gain error term in the denominator has been reduced by a factor of $(\beta_{FO3} + 1)$ from the error in Eq. (14.13).

> **EXERCISE:** What is the mirror ratio and the percent error for the buffered current mirror in Fig. 14.6 if $\beta_{FO} = 50$, $n = 10$, and $V_A = \infty$ for all the BJTs? (b) What is that value of V_{CE2} required to balance the mirror if $\beta_{FO} = \infty$?
>
> **ANSWERS:** 9.96, 0.430 percent; $V_{BE1} + V_{BE3} = 1.4$ V

14.2.7 OUTPUT RESISTANCE OF THE CURRENT MIRRORS

Now that we have found the dc output current of the current mirror, we will focus on the second important parameter that characterizes the **electronic current source**—the output resistance. The output resistance of the basic current mirror can be found by referring to the ac model of Fig. 14.7. Diode-connected bipolar transistor Q_1 represents a simple two-terminal device, and its small-signal model is easily found using nodal analysis of Fig. 14.8:

$$\mathbf{i} = g_\pi \mathbf{v} + g_m \mathbf{v} + g_o \mathbf{v} = (g_m + g_\pi + g_o)\mathbf{v} \qquad (14.18)$$

By factoring out g_m, an approximate result for the diode conductance is

$$g_D = \frac{\mathbf{i}}{\mathbf{v}} = g_m\left[1 + \frac{1}{\beta_o} + \frac{1}{\mu_f}\right] \cong g_m \quad \text{and} \quad r_D \cong \frac{1}{g_m} \qquad (14.19)$$

for β_o and $\mu_f \gg 1$. The small-signal model for the diode-connected BJT is simply a resistor of value $1/g_m$. Note that this result is the same as the small-signal resistance r_d of an actual diode that was developed in Sec. 7.4.

Using this diode model simplifies the ac model for the current mirror to that shown in Fig. 14.9. This circuit should be recognized as a common-emitter transistor with a Thévenin equivalent resistance $R_{\text{th}} = 1/g_m$ connected to its base; the output resistance just equals the output resistance r_{o2} of transistor Q_2.

The equation describing the small-signal model for the two-terminal diode-connected MOSFET is similar to that in Eq. (14.19) except that the current gain is infinite. Therefore, the two-terminal MOSFET is also represented by a resistor of value $1/g_m$, as in Fig. 14.10; the output resistance of the MOS current mirror is equal to r_{o2} of MOSFET M_2.

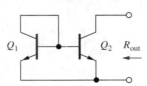

Figure 14.7 ac Model for the output resistance of the bipolar current mirror.

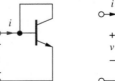

Figure 14.8 Model for diode-connected transistor.

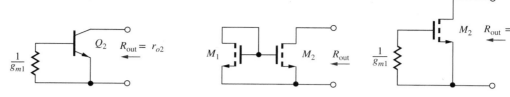

Figure 14.9 Simplified small-signal model for the bipolar current mirror.

Figure 14.10 Output resistance of the MOS current mirror.

Thus, the output resistance and **figure of merit (FOM)** V_{CS} (see Sec. 13.4.2) for the basic current mirror circuits are determined by output transistors Q_2 and M_2:

$$R_{\text{out}} = r_{o2} \quad \text{and} \quad V_{CS} \cong V_{A2} \quad \text{or} \quad V_{CS} = \frac{1}{\lambda_2} \quad (14.20)$$

EXERCISE: What are the output resistances of sources I_{O2} and I_{O3} in Fig. 14.5(a) for $I_{REF} = 100\ \mu A$ and Fig. 14.5(b) for $I_{REF} = 10\ \mu A$ if $V_A = 1/\lambda = 50$ V and $\beta_F = 100$?

ANSWERS: 260 kΩ, 130 kΩ; 5.94 MΩ, 1.19 MΩ

14.2.8 TWO-PORT MODEL FOR THE CURRENT MIRROR

We shall see shortly that the current mirror can be used not only as a dc current source but, in more complex circuits, as a current amplifier and active load. It will be useful to understand the small-signal behavior of the current mirror, redrawn as a two-port in Fig. 14.11.

The small-signal model for the current mirror is in Fig. 14.12, in which diode-connected transistor Q_1 is represented in its simplified form by $1/g_{m1}$. From the circuit in Fig. 14.12(a),

$$R_{\text{in}} = \frac{v_1}{i_1}\bigg|_{v_2=0} = \frac{1}{(g_{m1} + g_{\pi 2})} = \frac{1}{g_{m1}\left(1 + \dfrac{n}{\beta_{o2}}\right)} \cong \frac{1}{g_{m1}}$$

$$n = \frac{i_2}{i_1}\bigg|_{v_2=0} = \frac{g_{m2} r_{\pi 2}}{1 + g_{m1} r_{\pi 2}} \cong \frac{g_{m2}}{g_{m1}} = \frac{I_{C2}}{I_{C1}} = \frac{A_{E2}}{A_{E1}} \quad (14.21)$$

$$R_{\text{out}} = \frac{v_2}{i_2}\bigg|_{i_1=0} = r_{o2}$$

Figure 14.12(b) shows the final two-port model representation. The bipolar current mirror has an input resistance of $1/g_{m1}$, determined by diode Q_1 and an output resistance equal to r_{o2} of Q_2. The current gain is determined approximately by the emitter-area ratio $n = A_{E2}/A_{E1}$. Be sure to remember to use the correct values of I_{C1} and I_{C2} when calculating the values of the small-signal parameters.

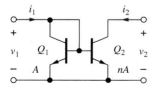

Figure 14.11 Current mirror as a two-port.

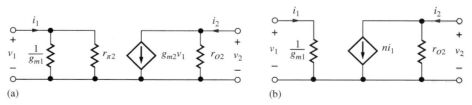

Figure 14.12 (a) Small-signal model for the current mirror. (b) Simplified small-signal model for the current mirror.

Analysis of the MOS current mirror yields similar results [or by simply setting $r_{\pi 2} = \infty$ in Eq. (14.21)]:

$$R_{in} = \frac{1}{g_{m1}} \qquad \beta = \frac{g_{m2}}{g_{m1}} \cong \frac{\left(\frac{W}{L}\right)_2}{\left(\frac{W}{L}\right)_1} \cong n \qquad R_{out} = r_{o2} \qquad (14.22)$$

In this case, the current gain n is determined by the W/L ratios of the two FETs rather than by the bipolar emitter-area ratio.

> **EXERCISE:** What are the values of I_{C1} and I_{C2} and the small-signal parameters for the current mirror in Fig. 14.4 if $I_{REF} = 100$ µA, $\beta_{FO} = 50$, $V_A = 50$ V, $V_{BE} = 0.7$ V, $V_{CE2} = 10$ V, and $n = 5$?
>
> **ANSWERS:** 89.4 µA; 529 µA; 280 Ω; 0; 5.92; 113 kΩ

EXAMPLE 14.3 CALCULATING THE TWO-PORT PARAMETERS OF A CURRENT MIRROR USING SPICE

Transfer function analysis is used to find the two-port parameters of the BJT current mirror.

PROBLEM Use the transfer function capability of SPICE to find the two-port parameters of the BJT current mirror biased by a reference current of 100 µA and a power supply of +10 V.

SOLUTION **Known Information and Given Data:** A current mirror using bipolar transistors; $I_{REF} = 100$ µA and $V_{CC} = 10$ V

Unknowns: Output current I_O, V_{BE}, R_{in}, n, and R_{out} for the current mirror

Approach: Construct the circuit using the schematic editor in SPICE. Use the transfer function analysis to find the forward transfer function from I_{REF} to $I(V_{CC})$ and reverse transfer function from V_{CC} to node 1. The SPICE transfer function analysis automatically calculates three values: the requested transfer function, the resistance at the input source node, and the resistance at the output source node. However, since the output node is connected to V_{CC}, the output resistance calculated at that node will be zero, and two analyses will be required to find the two-port parameters.

Assumptions: Use the current mirror with a single positive supply V_{CC} biased by current source I_{REF}, as shown in the figure here. $V_A = 50$ V, $\beta_{FO} = 100$, and $I_S = 0.1$ fA.

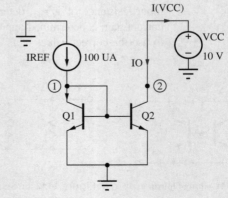

```
.op
.model NPN NPN BF=100 VAF=50V IS=0.1fA
*.TF I(VCC) IREF
*.TF V(N001) VCC
```

Analysis: First, we must set the BJT parameters to the desired values: BF = 100, VAF = 50 V, and IS = 0.1 fA. An operating point and two transfer function analyses are used in this example. The first asks for the transfer function from input source I_{REF} to output variable $I(V_{CC})$. The operating point analysis yields $V(1) = 0.719$ V and $I_O = 116$ μA. The transfer function analysis gives input resistance $R_{in} = 259$ Ω and current gain $n = +1.16$. The second analysis requests the transfer function from voltage source V_{CC} to node 1. SPICE analysis gives $R_{out} = 510$ kΩ.

Check of Results: Based on Eq. (14.21) and the operating point results, we expect

$$R_{in} = 250 \text{ Ω} \qquad n = +1.16 \qquad R_{out} = 517 \text{ kΩ}$$

and we see that agreement with theory is very good.

Discussion: One should always try to understand and account for the differences between our theory and SPICE. In this example, the input resistance difference can be traced to the use of $V_T = 25.9$ mV. Be careful not to make a sign error in interpreting the data for n. A negative sign appears in the SPICE output because of the assumed polarity of V_{CC} and $I(V_{CC})$. Finally, the SPICE model uses $r_o = (V_A + V_{CB})/I_C = 511$ kΩ, accounting for the small difference in the values of R_{out}.

EXERCISE: Use the transfer function capability of SPICE to find the two-port parameters for a MOS current mirror biased by a reference current of 100 μA and a power supply of +10 V. Assume $K_n = 1$ mA/V², $V_{TN} = 0.75$ V, and $\lambda = 0.02$/V.

ANSWERS: $I_O = 117$ μA, $V_{GS} = 1.19$ V; 220 Ω, 1.17, 512 kΩ

EXERCISE: Compare the answers in the previous exercise to hand calculations.

ANSWERS: $I_O = 117$ μA with $V_{GS} = 1.20$ V; 2.24 kΩ, 1.17, 513 kΩ

14.2.9 THE WIDLAR CURRENT SOURCE

Resistor R in the **Widlar**[2] **current source** circuit shown in the schematic in Fig. 14.13 gives the designer an additional degree of freedom in adjusting the mirror ratio of the current mirror. In this circuit, the difference in the base-emitter voltages of transistors Q_1 and Q_2 appears across resistor R and determines output current I_O. Transistor Q_3 buffers the mirror reference transistor in Fig. 14.13(b) to minimize the effect of finite current gain.

An expression for the output current may be determined from the standard expressions for the base-emitter voltage of the two bipolar transistors. In this analysis, we must accurately calculate the individual values of V_{BE1} and V_{BE2} because the behavior of the circuit depends on small differences in the values of these two voltages.

Assuming high current gain,

$$V_{BE1} = V_T \ln\left(1 + \frac{I_{REF}}{I_{S1}}\right) \cong V_T \ln\frac{I_{REF}}{I_{S1}} \quad \text{and} \quad V_{BE2} = V_T \ln\left(1 + \frac{I_O}{I_{S2}}\right) \cong V_T \ln\frac{I_O}{I_{S2}} \quad (14.23)$$

---

[2] Robert Widlar was a famous IC designer who made many lasting contributions to analog IC design. For examples, see [2, 3].

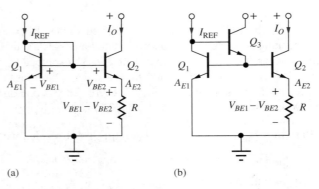
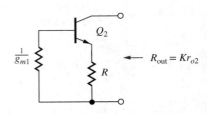

Figure 14.13 (a) Basic Widlar current source and (b) buffered Widlar source.

Figure 14.14 Widlar source output resistance – $K = 1 + \ln[(I_{REF}/I_{C2})(A_{E2}/A_{E1})]$.

The current in resistor R is equal to

$$I_{E2} = \frac{V_{BE1} - V_{BE2}}{R} = \frac{V_T}{R}\ln\left(\frac{I_{REF}}{I_O}\frac{I_{S2}}{I_{S1}}\right) \quad (14.24)$$

If the transistors are matched, then $I_{S1} = (A_{E1}/A)I_{SO}$ and $I_{S2} = (A_{E2}/A)I_{SO}$, and Eq. (14.24) can be rewritten as

$$I_O = \alpha_F I_{E2} \cong \frac{V_T}{R}\ln\left(\frac{I_{REF}}{I_O}\frac{A_{E2}}{A_{E1}}\right) \quad (14.25)$$

If I_{REF}, R, and the emitter-area ratio are all known, then Eq. (14.25) represents a transcendental equation that must be solved for I_O. The solution can be obtained by iterative trial and error or utilizing solvers in our calculators or on the internet.

Widlar Source Output Resistance
The ac model for the Widlar source in Fig. 14.13(a) represents a common-emitter transistor with resistor R in its emitter and a small value of R_{th} ($= 1/g_{m1}$) from diode Q_1 in its base, as indicated in Fig. 14.14. In normal operation, the voltage developed across resistor R is usually small ($\leq 10V_T$). By simplifying Eq. (13.114) for this case, we can reduce the output resistance of the source to

$$R_{out} \cong r_{o2}[1 + g_{m2}R] = r_{o2}\left[1 + \frac{I_O R}{V_T}\right] \quad (14.26)$$

in which $I_O R$ can be found from Eq. (14.25):

$$R_{out} \cong r_{o2}\left[1 + \ln\frac{I_{REF}}{I_O}\frac{A_{E2}}{A_{E1}}\right] = Kr_{o2} \quad \text{and} \quad V_{CS} \cong KV_{A2} \quad (14.27)$$

for

$$K = \left[1 + \ln\frac{I_{REF}}{I_O}\frac{A_{E2}}{A_{E1}}\right]$$

Using typical values, $1 < K < 10$.

> **EXERCISE:** What value of R is required to set $I_O = 25$ μA if $I_{REF} = 100$ μA and $A_{E2}/A_{E1} = 5$? What are the values of K and the output resistance in Eq. (14.27) for this source if $V_A + V_{CE} = 75$ V?
>
> **ANSWERS:** 3000 Ω; 4, 12 MΩ.

EXERCISE: Find the output current in the Widlar source if $I_{REF} = 100\ \mu A$, $R = 100\ \Omega$, and $A_{E2} = 10A_{E1}$. What are the values of K and the output resistance in Eq. (14.27) for this source if $V_A + V_{CE} = 75\ V$?

ANSWERS: 301 μA; 2.20, 551 kΩ.

ELECTRONICS IN ACTION

The PTAT Voltage

The voltage developed across resistor R in Fig. 14.13 represents an extremely useful quantity because it is directly **p**roportional **t**o **a**bsolute **t**emperature (referred to as PTAT). V_{PTAT} is equal to the difference in the two base-emitter voltages described by Eq. (14.23):

$$V_{\text{PTAT}} = V_{BE1} - V_{BE2} = V_T \ln\left(\frac{I_{C1} A_{E2}}{I_{C2} A_{E1}}\right) = \frac{kT}{q} \ln\left(\frac{I_{C1} A_{E2}}{I_{C2} A_{E1}}\right)$$

and the change of V_{PTAT} with temperature is

$$\frac{\partial V_{\text{PTAT}}}{\partial T} = +\frac{k}{q} \ln\left(\frac{I_{C1} A_{E2}}{I_{C2} A_{E1}}\right) = +\frac{V_{\text{PTAT}}}{T}$$

For example, suppose $T = 300$ K, $I_{C1} = I_{C2}$ and $A_{E2} = 10A_{E1}$. Then $V_{\text{PTAT}} = 59.6$ mV with a temperature coefficient of slightly less than +0.2 mV/K.

The PTAT voltage developed in the Widlar cell, combined with an analog-to-digital converter, forms the heart of all of today's highly accurate electronic thermometers.

PTAT Voltage Based Digital Thermometry

The PTAT generator produces a well-defined output voltage that is used in many of today's digital thermometers. One example is shown in the block diagram below that was produced as part of a Senior Design Project at Auburn University. The PTAT output voltage of the LM34DM reference IC is scaled directly to degrees Fahrenheit. This voltage is converted to digital form by the A/D converter in the ICL7136CMM that also contains its own reference generator and circuitry to directly interface with a liquid crystal digital display.

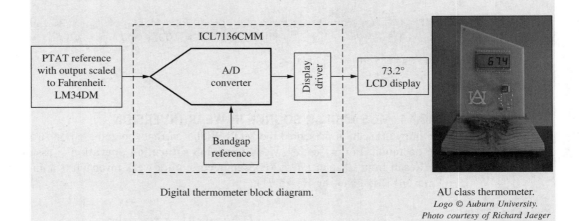

Digital thermometer block diagram.

AU class thermometer.
Logo © Auburn University.
Photo courtesy of Richard Jaeger

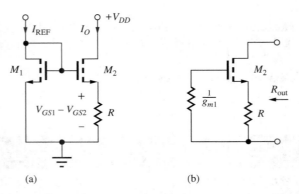

Figure 14.15 (a) MOS Widlar source and (b) small-signal model.

14.2.10 THE MOS VERSION OF THE WIDLAR SOURCE

Figure 14.15 is the MOS version of the Widlar source. In this circuit, the difference between the gate-source voltages of transistors M_1 and M_2 appears across resistor R, and I_O can be expressed as

$$I_O = \frac{V_{GS1} - V_{GS2}}{R} = \frac{\sqrt{\frac{2I_{REF}}{K_{n1}}} - \sqrt{\frac{2I_O}{K_{n2}}}}{R} = \frac{1}{R}\sqrt{\frac{2I_{REF}}{K_{n1}}}\left(1 - \sqrt{\frac{I_O}{I_{REF}}\frac{(W/L)_1}{(W/L)_2}}\right) \quad (14.28)$$

If I_O is known, then I_{REF} can be calculated directly from Eq. (14.28). If I_{REF}, R, and the W/L ratios are known, then Eq. (14.28) can be written as a quadratic equation in terms of $\sqrt{I_O/I_{REF}}$:

$$\left(\sqrt{\frac{I_O}{I_{REF}}}\right)^2 + \frac{1}{R}\sqrt{\frac{2}{K_{n1}I_{REF}}}\sqrt{\frac{(W/L)_1}{(W/L)_2}}\left(\sqrt{\frac{I_O}{I_{REF}}}\right) - \frac{1}{R}\sqrt{\frac{2}{K_{n1}I_{REF}}} = 0 \quad (14.29)$$

MOS Widlar Source Output Resistance

In Fig. 14.15(b), the small-signal model for the MOS Widlar source is recognized as a common-source stage with resistor R in its source. Therefore, from Table 8.9,

$$R_{out} = r_{o2}(1 + g_{m2}R) \cong \mu_{f2}R \quad (14.30)$$

> **EXERCISE:** (a) Find the output current in Fig. 14.15(a) if $I_{REF} = 200$ µA, $R = 2$ kΩ, and $K_{n2} = 10 K_{n1} = 250$ µA/V². (b) What is R_{out} if $\lambda = 0.02$/V and $V_{DS} = 10$ V?
>
> **ANSWERS:** 764 µA; 176 kΩ

14.2.11 MOS WIDLAR SOURCE IN WEAK INVERSION

The previous section presented the MOS Widlar current source based on strong inversion saturation operation. In this section, **weak inversion saturation operation** is assumed.

Recall from Chapter 5 that MOS drain current for weak inversion (subthreshold) in saturation ($V_{DS} > 3nV_T$) is given by Eq. (5.71),

$$I_D \cong I_{DO}\exp\left(\frac{V_{GS} - V_{TN}}{nV_T}\right) \quad (14.31)$$

where $I_{DO} = 2nV_T^2 K_n$.[3] Solving for V_{GS},

$$V_{GS} \cong nV_T\left[\ln\left(\frac{I_D}{I_{DO}}\right)\right] + V_{TN} \qquad (14.32)$$

Revisiting the schematic of Figure 14.15(a), the core of the Widlar current source, and once again solving for the voltage drop across the resistor R,

$$V_{GS1} - V_{GS2} \cong nV_T\left[\ln\left(\frac{I_{D1}}{I_{DO1}}\frac{I_{DO2}}{I_{D2}}\right)\right] \qquad (14.33)$$

This result neglects body effect. For $I_{D1} = I_{REF}$ and $I_{D2} = I_O$, then

$$V_{GS1} - V_{GS2} \cong nV_T\left[\ln\left(\frac{I_{REF}}{I_O}\frac{K_{n2}}{K_{n1}}\right)\right] \qquad (14.34)$$

In practical design, $L_1 = L_2$ for device matching purposes, thus $K_{n2}/K_{n1} = W_2/W_1$. Output current I_O is then given by

$$I_O = \frac{V_{GS1} - V_{GS2}}{R} \cong \frac{nV_T}{R}\ln\left(\frac{I_{REF}}{I_O}\frac{W_2}{W_1}\right) \qquad (14.35)$$

At a glance, this transcendental equation (with I_O on both sides) appears cumbersome to solve. However, using a current mirror circuit (with input current = I_O and output current = I_{REF}) to set the I_{REF}/I_O ratio, then the weak inversion MOS Widlar source's output current is readily determined. For example, using a 1:1 current mirror to set $I_{REF} = I_O$, then

$$I_O \cong \frac{nV_T}{R}\ln\left(\frac{W_2}{W_1}\right) \qquad (14.36)$$

A similar circuit implementation is described later in this chapter (see Sec. 14.5.4) and prior to that, note the mention of start-up circuit requirement.

> **EXERCISE:** (a) Neglecting body effect, find the output current at room temperature (300 K) in Fig. 14.15(a) for weak inversion saturation operation with $I_{REF}/I_O = 1$ and $K_{n2}/K_{n1} = 55$. (b) What is the output current at 350 K? (c) At 400 K? Assume $R = 10$ MΩ and $n = 1.4$.
>
> **ANSWERS:** 14.5 nA; 16.9 nA; 19.3 nA

14.3 HIGH-OUTPUT-RESISTANCE CURRENT MIRRORS

In our introductory discussion of differential amplifiers in Sec. 13.2, we found that current sources with very high output resistances are needed to achieve good CMRR. The basic current mirrors discussed in the previous sections have a figure of merit V_{CS} equal to V_A or $1/\lambda$; that for the Widlar source is typically a few times higher. This section continues our introduction to current mirrors by discussing three additional circuits, the Wilson and cascode current sources, which enhance the value of V_{CS} to the order of $\beta_o V_A$ or μ_f/λ, and the regulated cascode source that can achieve an even higher value of V_{CS}.

---

[3] Be careful not to confuse usage of "n" in I_{DO} with n in the mirror ratio.

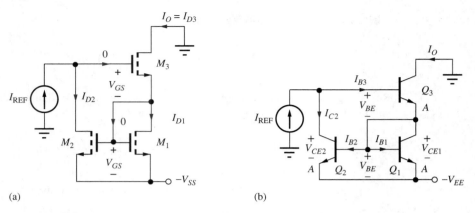

Figure 14.16 (a) MOS Wilson current source. (b) Original Wilson current source circuit using BJTs.

14.3.1 THE WILSON CURRENT SOURCES

The **Wilson current sources** [4] depicted in Fig. 14.16 use the same number of transistors as the buffered current mirror but achieve much higher output resistance; they are often used in applications requiring precisely matched current sources. In the MOS version, the output current is taken from the drain of M_3, and M_1 and M_2 form a current mirror. During circuit operation, the three transistors are all pinched-off and in the active region.

Because the gate current of M_3 is zero, I_{D2} must equal reference current I_{REF}. If the transistors all have the same W/L ratios, then

$$V_{GS3} = V_{GS1} = V_{GS} \quad \text{because} \quad I_{D3} = I_{D1}$$

The current mirror requires

$$I_{D2} = I_{D1} \frac{1 + 2\lambda V_{GS}}{1 + \lambda V_{GS}}$$

and because $I_O = I_{D3}$ and $I_{D3} = I_{D1}$, the output current is given by

$$I_O = I_{REF} \frac{1 + \lambda V_{GS}}{1 + 2\lambda V_{GS}} \quad \text{where} \quad V_{GS} \cong V_{TN} + \sqrt{\frac{2 I_{REF}}{K_n}} \tag{14.37}$$

For small λ, $I_O \cong I_{REF}$. For example, if $\lambda = 0.02/\text{V}$ and $V_{GS} = 2$ V, then I_O and I_{REF} differ by 3.7 percent.

The Wilson source actually appeared first in bipolar form as drawn in Fig. 14.16(b). The circuit operates in a manner similar to the MOS source, except for the loss of current from I_{REF} to the base of Q_3 and the current gain error in the mirror formed by Q_1 and Q_2. Applying KCL at the base of Q_3, $I_{REF} = I_{C2} + I_{B3}$ in which I_{C2} and I_{B3} are related through the current mirror formed by Q_1 and Q_2:

$$I_{C2} = \frac{1 + \dfrac{2V_{BE}}{V_A}}{1 + \dfrac{V_{BE}}{V_A} + \dfrac{2}{\beta_{FO}}} I_{E3} = \frac{1 + \dfrac{2V_{BE}}{V_A}}{1 + \dfrac{V_{BE}}{V_A} + \dfrac{2}{\beta_{FO}}} (\beta_{FO} + 1) I_{B3} \tag{14.38}$$

Note in Fig. 14.16(b) that $V_{CE1} = V_{BE}$ and $V_{CE2} = 2V_{BE}$.

Solving directly for $I_{C3} = \beta_F I_{B3}$ yields a messy expression that is difficult to interpret. However, if we assume the error terms are small, then we can eventually reduce (with considerable

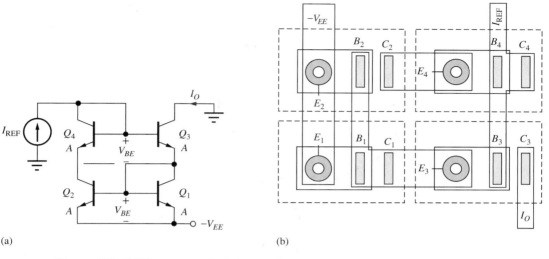

Figure 14.17 (a) Wilson source using balanced collector-emitter voltages. (b) Layout of Wilson source.

effort) the expression to the following approximate result:

$$I_O \cong I_{REF} \frac{1 + \dfrac{V_{BE}}{V_A}}{1 + \dfrac{2}{\beta_{FO}(\beta_{FO} + 2)} + \dfrac{2V_{BE}}{V_A}} \tag{14.39}$$

For $\beta_{FO} = 50$, $V_A = 60$ V, and $V_{BE} = 0.7$ V, the mirror ratio is 0.988. The primary source of error results from the collector-emitter voltage mismatch between transistors Q_1 and Q_2. The base current error has been reduced to less than 0.1 percent of I_{REF}.

The errors due to drain-source voltage mismatch in Fig. 14.16(a), or collector-emitter voltage mismatch in Fig. 14.16(b), may still be too large for use in precision circuits, but this problem can be significantly reduced by adding one more transistor to balance the circuit as in Fig. 14.17. Transistor Q_4 reduces the collector-emitter voltage of Q_2 by one V_{BE} drop and balances the collector-emitter voltages of Q_1 and Q_2:

$$V_{CE2} = V_{BE1} + V_{BE3} - V_{BE4} \cong V_{BE}$$

All four transistors are operating at approximately the same value of collector current, and the values of V_{BE} are all the same if the devices are matched with equal emitter areas.

EXERCISE: Draw a voltage-balanced version of the MOS Wilson source by adding one additional transistor to the circuit in Fig. 14.16(a).

ANSWER: See Fig. P14.42.

14.3.2 OUTPUT RESISTANCE OF THE WILSON SOURCE

The primary advantage of the Wilson source over the standard current mirror is its greatly increased output resistance. The small-signal model for the MOS version of the Wilson source is given in Fig. 14.18, in which test current i_x is applied to determine the output resistance.

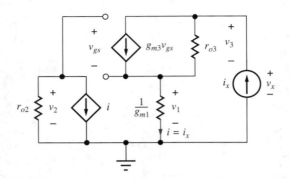

Figure 14.18 Small-signal model for the MOS version of the Wilson source.

The current mirror formed by transistors M_1 and M_2 is represented by its simplified two-port model assuming $n = 1$. Voltage v_x is determined from

$$\mathbf{v}_x = \mathbf{v}_3 + \mathbf{v}_1 = [\mathbf{i}_x - g_{m3}\mathbf{v}_{gs}]r_{o3} + \mathbf{v}_1 \qquad (14.40)$$

where

$$\mathbf{v}_{gs} = \mathbf{v}_2 - \mathbf{v}_1 \quad \text{with} \quad \mathbf{v}_1 = \frac{\mathbf{i}_x}{g_{m1}} \quad \text{and} \quad \mathbf{v}_2 = -\mu_{f2}\mathbf{v}_1$$

Combining these equations and recognizing that $g_{m1} = g_{m2}$ for $n = 1$ yields

$$R_{\text{out}} = \frac{\mathbf{v}_x}{\mathbf{i}_x} = r_{o3}\left[\mu_{f2} + 2 + \frac{1}{\mu_{f2}}\right] \cong \mu_{f2} r_{o3} \qquad (14.41)$$

and

$$V_{CS} = I_{D3}\mu_{f2}\frac{1 + \lambda_3 V_{DS3}}{\lambda_3 I_{D3}} \cong \frac{\mu_{f2}}{\lambda_3} \qquad (14.42)$$

Analysis of the bipolar source is somewhat more complex because of the finite current gain of the BJT and yields the following result:

$$R_{\text{out}} \cong \frac{\beta_{o3} r_{o3}}{2} \quad \text{and} \quad V_{CS} \cong \frac{\beta_o V_A}{2} \qquad (14.43)$$

Derivation of this equation is left for Prob. 14.39.

> **EXERCISE:** Calculate R_{out} for the Wilson source in Fig. 14.16(b) if $\beta_F = 150$, $V_A = 50$ V, $V_{EE} = 15$ V, and $I_O = I_{\text{REF}} = 50$ μA. What is the output resistance of a standard current mirror operating at the same current?
>
> **ANSWER:** 96.6 MΩ versus 1.30 MΩ
>
> **EXERCISE:** Use SPICE to find the output current and output resistance of the Wilson source in the previous exercise.
>
> **ANSWERS:** $I_O = 49.5$ μA; 118 MΩ

14.3.3 CASCODE CURRENT SOURCES

In this section, we learn that the output resistance of the cascode connection (C-E/C-B cascade) of two transistors is very high, approaching $\mu_f r_o$ for the FET case and $\beta_o r_o/2$ for the BJT circuit. Figure 14.19 shows the implementation of the MOS and BJT **cascode current sources** using current mirrors.

In the MOS circuit, $I_{D1} = I_{D3} = I_{\text{REF}}$. The current mirror formed by M_1 and M_2 forces the output current to be approximately equal to the reference current because $I_O = I_{D4} = I_{D2}$.

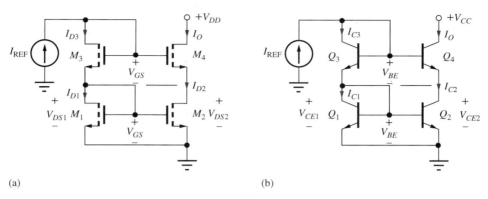

Figure 14.19 (a) MOS and (b) BJT cascode current sources.

Diode-connected transistor M_3 provides a dc bias voltage to the gate of M_4 and balances V_{DS1} and V_{DS2}. If all transistors are matched with the same W/L ratios, then the values of V_{GS} are all the same, and V_{DS2} equals V_{DS1}:

$$V_{DS2} = V_{GS1} + V_{GS3} - V_{GS4} = V_{GS} \quad \text{and} \quad V_{DS1} = V_{GS}$$

Thus, the M_1-M_2 current mirror is precisely balanced, and $I_O = I_{REF}$.

The BJT source in Fig. 14.19(b) operates in the same manner. For $\beta_F = \infty$, $I_{REF} = I_{C3} = I_{C1}$ on the reference side of the source. Q_1 and Q_2 form a current mirror, which sets $I_O = I_{C4} = I_{C2} = I_{C1} = I_{REF}$. Diode Q_3 provides the bias voltage at the base of Q_4 needed to keep Q_2 in the active region and balances the collector-emitter voltages of the current mirror:

$$V_{CE2} = V_{BE1} + V_{BE3} - V_{BE4} = 2V_{BE} - V_{BE} = V_{BE} = V_{CE1}$$

14.3.4 OUTPUT RESISTANCE OF THE CASCODE SOURCES

Figure 14.20 shows the small-signal model for the MOS cascode source; the two-port model has been used for the current mirror formed of transistors M_1 and M_2. Because current i represents the gate current of M_4, which is zero, the circuit can be reduced to that on the right in Fig. 14.20, which should be recognized as a common-source stage with resistor r_{o2} in its source. Thus, its output resistance is

$$R_{out} = r_{o4}(1 + g_{m4} r_{o2}) \cong \mu_{f4} r_{o2} \quad \text{and} \quad V_{CS} \cong \frac{\mu_{f4}}{\lambda_2} \cong \frac{\mu_{f4}}{\lambda_4} \quad (14.44)$$

Analysis of the output resistance of the BJT source in Fig. 14.21 is again more complex because of the finite current gain of the BJT. If the base of Q_4 were grounded, then the output

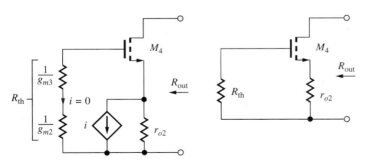

Figure 14.20 Small-signal model for the MOS cascode source.

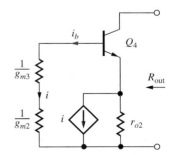

Figure 14.21 Small-signal model for the BJT cascode source.

resistance would be just equal to that of the cascode stage, $\beta_o r_o$. However, the base current i_b of Q_4 enters the current mirror, doubles the output current, and causes the overall output resistance to be reduced by a factor of 2:

$$R_{out} \cong \frac{\beta_{o4} r_{o4}}{2} \quad \text{and} \quad V_{CS} \cong \frac{\beta_{o4} V_{A4}}{2} \quad (14.45)$$

Detailed calculation of this result is left as Prob. 14.65.

> **EXERCISE:** Calculate the output resistance of the MOS cascode current source in Fig. 14.19(a) and compare it to that of a standard current mirror if $I_O = I_{REF} = 50$ µA, $V_{DD} = 15$ V, $K_n = 250$ µA/V^2, $V_{TN} = 0.8$ V, and $\lambda = 0.015$ V^{-1}.
>
> **ANSWER:** 379 MΩ versus 1.63 MΩ including all λV_{DS} terms
>
>
> **EXERCISE:** Use SPICE to find the output current and output resistance of the cascode current source in the previous exercise.
>
> **ANSWERS:** $I_O = 50.0$ µA; 382 MΩ
>
> **EXERCISE:** Calculate the output resistance of the BJT cascode current source in Fig. 14.19(b) and compare it to that of a standard current mirror if $I_O = I_{REF} = 50$ µA, $V_{CC} = 15$ V, $\beta_o = 100$, and $V_A = 67$ V.
>
> **ANSWER:** 81.3 MΩ versus 1.63 MΩ

14.3.5 REGULATED CASCODE CURRENT SOURCE

Another step up in current mirror output resistance can be achieved with the "regulated cascode" current source in Fig. 14.22 in which feedback through op amp A is used to further increase the output resistance. Output current I_O is set by the basic current mirror formed by M_1 and M_2. At dc, op amp A forces the voltage at the source of transistor M_3 to equal V_{REF}, whereas variations in the voltage at the source are reduced by the added loop-gain of A, thereby increasing the output resistance.

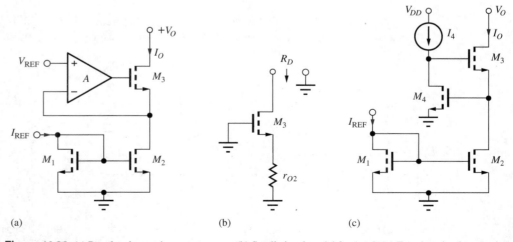

Figure 14.22 (a) Regulated cascode current source. (b) Small-signal model for $A = 0$. (c) Transistor implementation.

We can quickly find the regulated cascode output resistance by applying Blackman's Theorem (see Sec. 11.4.2):

$$R_{\text{out}} = R_D \frac{1 + |T_{SC}|}{1 + |T_{OC}|} \quad (14.46)$$

in which R_D is the resistance with the feedback loop disabled, and T_{OC} and T_{SC} represent the loop-gain with the output terminal open and shorted to ground, respectively. Setting gain A of the op amp to zero as in Fig. 14.22(b) yields $R_D = \mu_{f3} r_{o2}$. With the output terminal open, the drain-source signal current of M_2 will be zero, so $T_{OC} = 0$. With the output terminal connected to ac ground, T_{SC} equals the product of A and the gain of M_3 acting as a source follower:

$$T_{SC} = A \frac{g_{m3}(r_{o2} \| r_{o3})}{1 + g_{m3}(r_{o2} \| r_{o3})} \cong A \frac{\mu_{f3}/2}{1 + (\mu_{f3}/2)} \cong A \quad \text{and} \quad R_{\text{out}} \cong A \mu_{f3} r_{o2} \quad (14.47)$$

for $A \gg 1$. The output resistance is increased by the gain of the amplifier A.

A common implementation appears in Fig. 14.22(c), where amplifier A is realized by C-S transistor M_4 with current source load I_4. In this case $A = \mu_{f4}$, $R_{\text{out}} = \mu_{f4} \mu_{f3} r_{o2}$ and $V_{CS} \cong \mu_f^2/\lambda$!

14.3.6 CURRENT MIRROR SUMMARY

Table 14.2 is a summary of the current mirror circuits discussed in this chapter. The cascode and Wilson sources can achieve very high values of figure of merit V_{CS} and often find use in the design of differential and operational amplifiers, as well as in many other analog circuits. In the MOS case, it is possible to continue to stack cascode transistors (by adding M_5 and M_6 to the circuit in Fig. 14.19(a)) to further increase the current source output resistance. For instance, a stack of three MOS transistors will give $R_{\text{out}} = \mu_{f3} \mu_{f2} r_{o1}$. This does not work in the BJT case because the base current defect is always present in the uppermost transistor. The regulated cascode current source uses additional feedback to increase the output resistance to $\mu_f^2 r_o$.

TABLE 14.2
Comparison of the Basic Current Mirrors

TYPE OF SOURCE	R_{out}	V_{CS}	TYPICAL VALUES OF V_{CS}
Resistor	R	V_{EE}	15 V
Two-transistor mirror	r_o	V_A or $\frac{1}{\lambda}$	75 V
Cascode BJT	$\frac{\beta_o r_o}{2}$	$\frac{\beta_o V_A}{2}$	3750 V
Cascode FET	$\mu_f r_o$	$\frac{\mu_f}{\lambda}$	10,000 V
BJT Wilson	$\frac{\beta_o r_o}{2}$	$\frac{\beta_o V_A}{2}$	3750 V
FET Wilson	$\mu_f r_o$	$\frac{\mu_f}{\lambda}$	10,000 V
Regulated cascode	$\mu_f^2 r_o$	$\frac{\mu_f^2}{\lambda}$	1,000,000 V

DESIGN EXAMPLE 14.4

ELECTRONIC CURRENT SOURCE DESIGN

Design an IC current source to meet a given set of specifications.

PROBLEM Design a 1:1 current mirror with a reference current of 25 µA and a mirror ratio error of less than 0.1 percent when the output is operating from a 20-V supply. Devices with these parameters are available: $\beta_{FO} = 100$, $V_A = 75$ V, $I_{SO} = 0.5$ fA; $K'_n = 50$ µA/V^2, $V_{TN} = 0.75$ V, and $\lambda = 0.02$/V.

SOLUTION **Known Information and Given Data:** $I_{REF} = 25$ µA. A mirror ratio error of less than 0.1 percent requires an output current of 25 µA $\pm$ 25 nA when the output voltage is 20 V. Either a bipolar or MOS realization is acceptable.

Unknowns: Current source configuration; transistor sizes

Approach: The specifications define the required values of R_{out} and V_{CS}. Use this information to choose a circuit topology. Complete the design by choosing device sizes based on the output resistance expressions for the selected circuit topology.

Assumptions: Room temperature operation; devices are in the active region of operation.

Analysis: The output resistance of the current source must be large enough that 20 V applied across the output does not change (increase) the current by more than 25 nA. Thus, the output resistance must satisfy $R_{\text{out}} \geq 20$ V/25 nA = 800 MΩ. Let us choose $R_{\text{out}} = 1$ GΩ to provide some safety margin. The effective current source voltage is then $V_{CS} = 25$ µA (1 GΩ) = 25,000 V! From Table 14.2, we see that either a cascode or Wilson source will be required to meet this value of V_{CS}. In fact, the source must be an MOS version, since our BJTs can at best reach $V_{CS} = 100(75$ V$)/2 = 3750$ V.

The choice between the Wilson and cascode sources is arbitrary at this point. Let us pick the cascode source, which does not involve an internal feedback loop. In order to achieve the small mirror error, a voltage-balanced version is required. Our final circuit choice is therefore the circuit shown in Fig. 14.19(a). Now we must choose the device sizes. In this case, the W/L ratios are all the same since we require MR = 1.

Again referring to Table 14.2, the required intrinsic gain for the transistor is

$$\mu_f = \lambda V_{CS} = \left(\frac{0.02}{\text{V}}\right)(25{,}000 \text{ V}) = 500$$

The MOS transistor's intrinsic gain is given approximately by

$$\mu_f = g_m r_o \cong \sqrt{2 K_n I_D} \frac{1}{\lambda I_D}$$

Using $\mu_f = 500$, $\lambda = 0.02$/V, and $I_D = 25$ µA gives a value of $K_n = 1.25$ mS/V. Since $K_n = K'_n(W/L)$, we need a W/L ratio of 25/1 for the given technology. (This W/L ratio is easy to achieve in integrated circuit form.) In this circuit, all the transistors are operating at the same current, so the W/L ratios should all be the same size in order to maintain the required voltage balance.

Check of Results: Let us check the calculations by directly calculating the output resistance of the source.

$$R_{\text{out}} \cong g_{m4} r_{o4} r_{o2} \qquad g_{m4} = \sqrt{2 K_n I_D (1 + \lambda V_{DS4})} \qquad r_o = \frac{(1/\lambda) + V_{DS}}{I_D}$$

We can either neglect the values of V_{DS} in these expressions, or we can calculate them. In order to best compare with simulation, let us find V_{DS} and the corresponding values of g_m and r_o.

$$V_{DS2} = V_{GS2} = V_{TN} + \sqrt{\frac{2I_D}{K_n}} = 0.75 + \sqrt{\frac{50\ \mu\text{A}}{1.25\ \text{mS}}} = 0.95\ \text{V}$$

$$V_{DS4} = 20 - V_{DS2} = 19.0\ \text{V}$$

$$g_{m4} = \sqrt{2K_n I_D (1 + \lambda V_{DS4})} = \sqrt{2(1.25\ \text{mA/V}^2)(25\ \mu\text{A})[1 + .02(19)]} = 0.294\ \text{mS}$$

$$r_{o2} = \frac{(1/\lambda) + V_{DS2}}{I_D} = \frac{51.0\ \text{V}}{25\ \mu\text{A}} = 2.04\ \text{M}\Omega$$

$$r_{o4} = \frac{(1/\lambda) + V_{DS4}}{I_D} = \frac{69.0\ \text{V}}{25\ \mu\text{A}} = 2.76\ \text{M}\Omega$$

Multiplying the small-signal parameters together produces an output resistance estimate of 1.65 GΩ, which exceeds the design requirement that we originally calculated from the design specifications.

Discussion: Note that our ability to set the intrinsic gain of the MOS transistor was very important in achieving the design goals. In this case, $\mu_{f4} = 811$. A possible layout for the cascode current source is presented in the figure. The four 25/1 NMOS transistors are stacked vertically. G_1 and G_2 are the gates of the current mirror transistors. Gates G_1 and G_3 are connected directly to their respective drains. The drain of M_1 and the source of M_3 are merged as are those of M_2 and M_4. However, there are no contacts required to the connection between the drain of M_2 and the source of M_4.

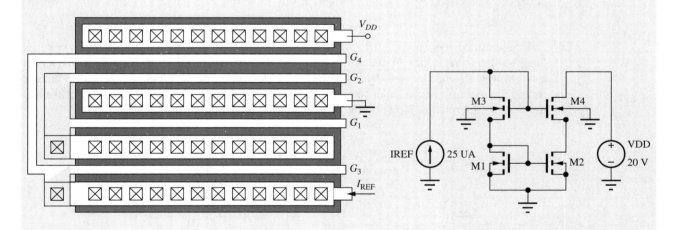

Computer-Aided Analysis: SPICE represents a good way to double check the results. First, we must set the MOS device parameters: KP = 50 μA/V², VTO = 0.75 V, LAMBDA = 0.02/V, $W = 25\ \mu$m, and $L = 1\ \mu$m. A dc simulation of the final circuit with the given device parameters yields an output current of 25.014 μA. In addition, the voltages at the drains of M_1 and M_2 are 0.948 V and 0.976 V, respectively, indicating that the voltage balancing is working as desired.

A transfer function analysis from source V_{DD} to the output node yields an output resistance of 1.66 GΩ, easily meeting the specifications with a satisfactory safety margin. We also have good agreement with the value of R_{out} that we calculated by hand.

> **EXERCISE:** In the SPICE results in Design Ex. 14.4, $I_O = 25.014$ μA at $V_{DD} = 20$ V. If $R_{out} = 1.66$ GΩ, what will be the output current at $V_{DD} = 10$ V?
>
> **ANSWER:** 25.008 μA
>
> **EXERCISE:** What is the minimum value of V_{DD} for which M_4 remains in the active region of operation?
>
> **ANSWER:** 1.15 V
>
> **EXERCISE:** Repeat the design in Design Ex. 14.4 for a current source with a mirror ratio of 2 ± 0.1 percent.
>
> **ANSWERS:** $(W/L)_3 = (W/L)_1 = 25/1$; $(W/L)_4 = (W/L)_2 = 50/1$

14.4 REFERENCE CURRENT GENERATION

A reference current is required by all the current mirrors that have been discussed. The least complicated method for establishing this reference current is to use resistor R, as shown in Fig. 14.23(a).

However, the source's output current is directly proportional to the supply voltage V_{EE}:

$$I_{\text{REF}} = \frac{V_{EE} - V_{BE}}{R} \qquad (14.48)$$

In MOS technology, the gate-source voltages of MOSFETs can be designed to be large, and several MOS devices can be connected in series between the power supplies to eliminate the need for large-value resistors. An example of this technique is given in Fig. 14.23(b), in which

$$V_{DD} + V_{SS} = V_{SG4} + V_{GS3} + V_{GS1}$$

and the drain currents must satisfy $I_{D1} = I_{D3} = I_{D4}$. However, any change in the supply voltages directly alters the values of the gate-source voltages of the three MOS transistors and again changes the reference current. Note that the series device technique is not as practical in bipolar technology because of the small fixed voltage ($\cong 0.7$ V) developed across each diode, as well as the exponential relationship between voltage and current in the diode.

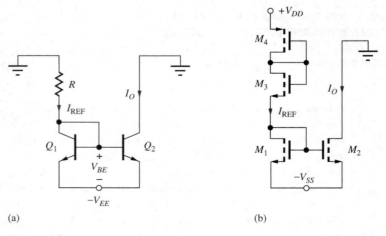

Figure 14.23 Reference current generation for current mirrors: (a) resistor reference and (b) series-connected MOSFETs.

EXERCISE: What is the reference current in Fig. 14.23(a) if $R = 43$ kΩ and $V_{EE} = -5$ V? (b) If $V_{EE} = -7.5$ V?

ANSWERS: 100 µA; 158 µA

EXERCISE: What is the reference current in Fig. 14.23(b) if $K_n = K_p = 400$ µA/V^2, $V_{TN} = -V_{TP} = 1$ V, $V_{DD} = 0$, and $V_{SS} = -5$ V? (b) If $V_{SS} = -7.5$ V?

ANSWERS: 88.9 µA; 450 µA. (*Note:* the variation is worse than in the resistor bias case because of the square-law MOSFET characteristic.)

14.5 SUPPLY-INDEPENDENT BIASING

In most cases, a supply voltage dependence of I_{REF} is undesirable. For example, we would like to fix the bias points of the devices in general-purpose op amps, even though they must operate from power supply voltages ranging from ± 3 V to ± 22 V. In addition, Eq. (14.48) indicates that relatively large values of resistance are required to achieve small operating currents, and these resistors use significant area in integrated circuits, as is discussed in detail in Sec. S6.5.9. Thus, a number of circuit techniques that yield currents relatively independent of the power supply voltages have been invented.

14.5.1 A V_{BE}-BASED REFERENCE

One possibility is the V_{BE}-**based reference,** shown in Fig. 14.24, in which the output current is determined by the base-emitter voltage of Q_1. For high current gain, the collector current of Q_1 is equal to the current through resistor R_1,

$$I_{C1} = \frac{V_{EE} - V_{BE1} - V_{BE2}}{R_1} \cong \frac{V_{EE} - 1.4 \text{ V}}{R_1} \qquad (14.49)$$

and the output current I_O is approximately equal to the current in R_2:

$$I_O = \alpha_{F2} I_{E2} = \alpha_{F2}\left(\frac{V_{BE1}}{R_2} + I_{B1}\right) \cong \frac{V_{BE1}}{R_2} \cong \frac{0.7 \text{ V}}{R_2} \qquad (14.50)$$

Rewriting V_{BE1} in terms of V_{EE},

$$I_O \cong \frac{V_T}{R_2} \ln \frac{V_{EE} - 1.4 \text{ V}}{I_{S1} R_1} \qquad (14.51)$$

A substantial degree of supply-voltage independence has been achieved because the output current is only logarithmically dependent on changes in the supply voltage V_{EE}. However, the output current is temperature dependent due to the temperature coefficients of V_{BE} and the resistors.

EXERCISE: (a) Calculate I_O in Fig. 14.24 for $I_S = 10^{-16}$ A, $R_1 = 39$ kΩ, $R_2 = 6.8$ kΩ, and $V_{EE} = 5$ V. Assume infinite current gains. (b) Repeat for $V_{EE} = -7.5$ V. (c) Calculate the sensitivity of I_O to changes in V_{EE}. [$S^{I_O}_{V_{EE}}$; see Eq. (12.33).]

ANSWERS: 101 µA; 103 µA; 0.009

14.5.2 THE WIDLAR SOURCE

Actually, we already discussed another source that achieves a similar independence from power supply voltage variations. The expression for the output current of the Widlar source given in Fig. 14.13 and Eq. (14.25) is

$$I_O = \alpha_F I_{E2} \cong \frac{V_T}{R} \ln\left(\frac{I_{REF}}{I_O} \frac{A_{E2}}{A_{E1}}\right) \qquad (14.52)$$

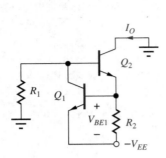

Figure 14.24 V_{BE}-based current source.

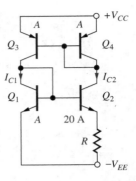

Figure 14.25 Power-supply-independent bias circuit using the Widlar source and a current mirror.

Here again, the output current is only logarithmically dependent on the reference current I_{REF} (which may be proportional to V_{CC}).

14.5.3 POWER-SUPPLY-INDEPENDENT BIAS CELL

Bias circuits with an even greater degree of power supply voltage independence can be obtained by combining the Widlar source with a standard current mirror, as indicated in the circuit in Fig. 14.25. Assuming high current gain, the *pnp* current mirror forces the currents on the two sides of the reference cell to be equal—that is, $I_{C1} = I_{C2}$. In addition, the emitter-area ratio of the Widlar source in Fig. 14.25 is equal to 20.

With these constraints, Eq. (14.52) can be satisfied by an operating point of

$$I_{C2} \cong \frac{V_T}{R}\ln(20) = \frac{0.0749 \text{ V}}{R} \quad (14.53)$$

In this example, a fixed voltage of approximately 75 mV is developed across resistor R, and this voltage is independent of the power supply voltages. Resistor R can then be chosen to yield the desired operating current.

Obviously, a wide range of mirror ratios and emitter-area ratios can be used in the design of the circuit in Fig. 14.25. Although the current, once established, is independent of supply voltage, the actual value of I_C still depends on temperature, as well as the absolute value of R and varies with run-to-run process variations.

Unfortunately, $I_{C1} = I_{C2} = 0$ is also a stable operating point for the circuit in Fig. 14.25. **Start-up circuits** must be included in IC realizations of this reference to ensure that the circuit reaches the desired operating point. A simple form of start-up circuit appears in Des. Ex. 14.6.

EXERCISE: Find the output current in the current source in Fig. 14.25 if $A_{E3} = 10A_{E4}$, $A_{E2} = 10A_{E1}$, and $R = 1\ k\Omega$.

ANSWER: 115 μA

EXERCISE: What is the minimum power supply voltage for proper operation of the supply-independent bias circuit in Fig. 14.25?

ANSWER: $V_{BE4} + V_{CE3} + V_R \cong 0.7 + 0.3 + 0.075 = 1.08$ V

EXERCISE: What is the temperature coefficient (TC) of current I_{C2} described by Eq. (14.53) if resistor R is assumed to be constant? (b) Repeat if the resistor has a TC of -2000 ppm/°C.

ANSWERS: 3300 ppm/°C; 5300 ppm/°C

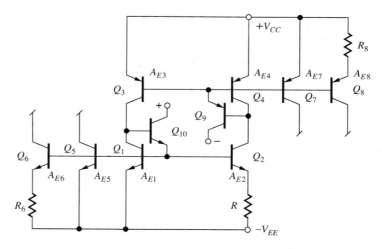

Figure 14.26 Multiple source currents generated from the supply-independent cell.

Once the current has been established in the reference cell consisting of Q_1–Q_4 in Fig. 14.25, the base-emitter voltages of Q_1 and Q_4 can be used as reference voltages for other current mirrors, as shown in Fig. 14.26. In this figure, buffered current mirrors have been used in the reference cell to minimize errors associated with finite current gains of the *npn* and *pnp* transistors. Output currents are shown generated from basic mirror transistors Q_5 and Q_7 and from Widlar sources, Q_6 and Q_8.

14.5.4 A SUPPLY-INDEPENDENT MOS REFERENCE CELL

The MOS analog of the circuit in Fig. 14.25 appears in Fig. 14.27. In this circuit, the PMOS current mirror forces a fixed relationship between drain currents I_{D3} and I_{D4}. For the particular case in Fig. 14.27, $I_{D3} = I_{D4}$, and so $I_{D1} = I_{D2}$. Substituting this constraint into Eq. (14.28) yields an equation for the value of R required to establish a given current I_{D2}:

$$R = \sqrt{\frac{2}{K_{n1}I_{D2}}}\left(1 - \sqrt{\frac{(W/L)_1}{(W/L)_2}}\right) \qquad (14.54)$$

Based on Eq. (14.54), we see that the MOS source is independent of supply voltage but is a function of the absolute values of R and K'_n.

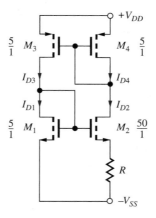

Figure 14.27 Supply-independent current source using MOS transistors.

EXERCISE: What value of R is required in the current source in Fig. 14.27 if I_{D2} is to be designed to be 100 µA and $K'_n = 25$ µA/V^2?

ANSWER: 8.65 kΩ

DESIGN EXAMPLE 14.5

REFERENCE CURRENT DESIGN

Design a supply-independent current source using bipolar technology.

PROBLEM Design a supply-independent current source to provide an output current of 45 µA at $T = 300$ K using the circuit topology in Fig. 14.25 with symmetrical 5-V power supplies. The circuit should use no more than 1 kΩ of resistance or 60 µA of total current. Use SPICE to determine the sensitivity of the design current to power supply voltage variations. Assume that a unit-area BJT has the following parameters: $\beta_{FO} = 100$, $V_A = 75$ V, and $I_{SO} = 0.1$ fA for both *npn* and *pnp* transistors.

SOLUTION **Known Information and Given Data:** Circuit topology in Fig. 14.25, $\beta_{FO} = 100$, $V_A = 75$ V, $I_{SO} = 0.1$ fA. Total current ≤ 60 µA.

Unknowns: R and the area ratio between Q_1 and Q_2; power supply sensitivity

Approach: The current in the circuit is described by Eq. (14.52). Use the maximum resistance values to select the area ratio. Select a current ratio in the sides of the reference to satisfy the total supply current requirement.

Assumptions: Transistors operate in the active region. $I_{C2} = 45$ µA.

Analysis: At $T = 300$ K and $V_T = 25.88$ mV, and from Eq. (14.52), we have

$$\ln\left(\frac{I_{C1} A_{E2}}{I_{C2} A_{E1}}\right) = \frac{I_{C2} R}{V_T} \leq \frac{(45 \text{ µA})(1 \text{ k}\Omega)}{25.88 \text{ mV}} = 1.739 \quad \text{or} \quad \frac{I_{C1} A_{E2}}{I_{C2} A_{E1}} \leq 5.69$$

In addition, the maximum current specification requires

$$\frac{I_{C2}}{I_{C1}} \geq \frac{45 \text{ µA}}{15 \text{ µA}} = \frac{3}{1}$$

Let us choose $I_{C2} = 5 I_{C1}$. Then $A_{E2}/A_{E1} \leq 28.5$. Choosing $A_{E2}/A_{E1} = 20$, we obtain

$$R = \frac{25.88 \text{ mV} \ln(4)}{45 \text{ µA}} = 797 \text{ }\Omega$$

The final design is $R = 797$ Ω, $A_{E1} = A$, $A_{E2} = 20$ A, $A_{E3} = A$, $A_{E4} = 5$ A with 35.88 mV across resistor R.

Check of Results: Since we need to use SPICE to find the power supply sensitivity, let us use it to also check our design.

Computer-Aided Analysis: The circuit shown is drawn using the schematic editor. Zero-valued sources V_{IC2} and V_{IC3} function as ammeters to measure the collector currents of transistors Q_2 and Q_3. First we must remember to set the *npn* and *pnp* BJT parameters to BF = 100, VAF = 75 V, IS = 0.1 fA, and TEMP = 27 C. We must also specify AREA = 1, AREA = 20, AREA = 1 and AREA = 5 for Q_1 through Q_4, respectively. SPICE then gives $I_{C2} = 49.6$ µA and $I_{C3} = 10.94$ µA with 39.89 mV across R. The currents and voltage are slightly higher than predicted, and this is primarily due to having neglected the mirror ratio error due to the different values of V_{EC4} and V_{EC3}. (Try the exercise after this example.) We can correct this error by modifying the emitter area ratio:

$$A_{E4} = 5\left(1 + \frac{V_{EC3} - V_{EC4}}{V_A}\right) = 5\left(1 + \frac{9.34 - 0.65}{75}\right) = 5.58$$

SPICE now yields $I_{C2} = 45.9$ μA, $I_{C3} = 9.08$ μA, and $V_{E2} = 36.9$ mV. A transfer function analysis from V_{CC} to V_{IC2} gives a total output resistance of 928 kΩ for the current source, and the sensitivity of I_{C2} to changes in V_{CC} is 0.808 μA/V.

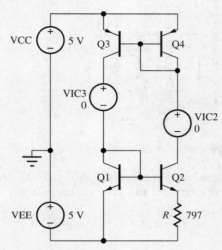

Discussion: The current source meets the specifications.

EXERCISE: Explore the errors caused by finite current gain and Early voltage by simulating the circuit with BF = 10,000 and VAF = 10,000 V. What are the new values of I_{C2}, I_{C3}, and the voltage developed across R?

ANSWERS: 45.0 μA; 9.01 μA; 35.88 mV

EXERCISE: What are the new design values if we choose $A_{E2}/A_{E1} = 25$?

ANSWERS: $R = 925$ Ω; $A_{E1} = A$; $A_{E2} = 25$ A; $A_{E3} = A$; $A_{E4} = 5.58$ A

14.6 THE BANDGAP REFERENCE

Precision **voltage references** need to not only be independent of power supply voltage, but also be independent of temperature. Although the circuits described in Sec. 14.5 can produce reference currents and voltages that are substantially independent of power supply voltage, they all still vary with temperature. Robert Widlar solved this problem with his invention of the elegant bandgap reference circuit, and today, the **bandgap reference** is the most common technique used to generate a precision voltage. It has supplanted Zener reference diodes in the majority of applications.

Based on his detailed understanding of bipolar transistor characteristics, Widlar realized that the negative temperature coefficient associated with the base-emitter junction could be canceled out by the positive temperature dependence of a voltage that is **proportional to absolute temperature**[4]

---

[4] see page 1007.

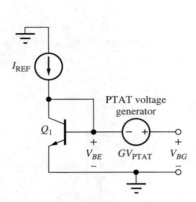

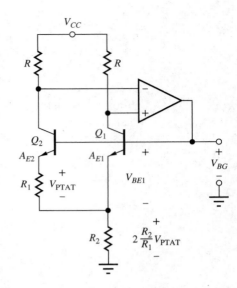

Figure 14.28 Concept for the bandgap reference.

Figure 14.29 Brokaw version of the bandgap reference.

(**PTAT**) as depicted conceptually in Fig. 14.28. He knew that a **PTAT voltage** is available from the difference between two base-emitter voltages:

$$V_{\text{PTAT}} = V_{BE1} - V_{BE2} = V_T \ln\left(\frac{I_{C1}}{I_{C2}}\frac{A_{E2}}{A_{E1}}\right) = \frac{kT}{q} \ln\left(\frac{I_{C1}}{I_{C2}}\frac{A_{E2}}{A_{E1}}\right) \tag{14.55}$$

The output voltage of the circuit in Fig. 14.28 can be written as

$$V_{BG} = V_{BE} + GV_{\text{PTAT}} \tag{14.56}$$

We desire this output voltage to have a zero temperature coefficient:

$$\frac{\partial V_{BG}}{\partial T} = \frac{\partial V_{BE}}{\partial T} + G\frac{\partial V_{\text{PTAT}}}{\partial T} = 0 \tag{14.57}$$

The dependence of V_{BE} on temperature was developed previously in Eqs. (3.14–3.15) and $\partial V_{\text{PTAT}}/\partial T = V_{\text{PTAT}}/T$. Substituting these values into Eq. (14.57) gives

$$\frac{\partial V_{BG}}{\partial T} = \frac{V_{BE} - V_{GO} - 3V_T}{T} + G\frac{V_{\text{PTAT}}}{T} = 0 \quad \text{or} \quad GV_{\text{PTAT}} = V_{GO} + 3V_T - V_{BE} \tag{14.58}$$

where V_{GO} is the silicon bandgap voltage at 0 K (1.12 V). Substituting this result into Eq. (14.58) reduces the output voltage to

$$V_{BG} = V_{GO} + 3V_T \tag{14.59}$$

The output voltage at which zero temperature coefficient is achieved is slightly above the bandgap voltage of silicon. Hence, this circuit is referred to as a "bandgap reference." At room temperature, the output voltage is approximately 1.20 V.

A circuit realization of the bandgap reference is shown in Fig. 14.29. This circuit is attributed to another talented designer, Paul Brokaw of Analog Devices [5], and is easier to understand than the original circuit of Widlar. In this case, the output voltage is equal to the sum of the base-emitter voltage of Q_1 plus the voltage across resistor R_2, which is a scaled replica of the PTAT voltage being developed across resistor R_1. The scaling factor is controlled by the op amp and resistors R.

The ideal op amp forces the voltage across the two matched collector resistors to be the same, thereby setting $I_{C2} = I_{C1}$ and $I_{E2} = I_{E1}$. Thus the PTAT voltage is equal to $V_T \ln(A_{E2}/A_{E1})$, and the emitter current of Q_2 equals V_{PTAT}/R_1. The current in R_2 is twice that in R_1, since $I_{E2} = I_{E1}$.

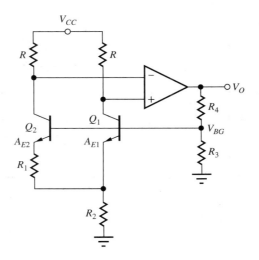

Figure 14.30 Bandgap reference with $V_O > V_{BG}$.

Combining these results yields an expression for the output voltage V_{BG}:

$$V_{BG} = V_{BE1} + 2\frac{R_2}{R_1} V_T \ln \frac{A_{E2}}{A_{E1}} \qquad (14.60)$$

For this circuit, the gain $G = 2R_2/R_1$, and based on Eq. (14.58), the resistor ratio is given by

$$\frac{R_2}{R_1} = -\frac{1}{2}\frac{\dfrac{\partial V_{BE1}}{\partial T}}{\dfrac{\partial V_{PTAT}}{\partial T}} = \frac{V_{GO} + 3V_T - V_{BE1}}{2V_{PTAT}} \qquad (14.61)$$

Often we want an output voltage that is not equal to 1.2 V, and other voltages are easy to achieve by adding a two-resistor voltage divider to the Brokaw circuit as in Fig. 14.30. In this case, the op-amp output voltage becomes

$$V_O = \left(1 + \frac{R_4}{R_3}\right) V_{BG} \qquad (14.62)$$

which can be scaled up to any desired value (e.g., 2.5 or 5 V).

A word of caution is needed here. In most bandgap reference designs, zero-output voltage is a valid operating point, and some additional circuitry must be added to ensure that the circuit "starts up" and reaches the desired operating point. In many simple circuit cases, SPICE will have considerable difficulty converging to the desired operating point (see Design Ex. 14.6).

DESIGN EXAMPLE 14.6 BANDGAP REFERENCE DESIGN

Design of the bandgap reference requires a slightly different sequence of calculations than the analysis in the previous example.

PROBLEM Design the bandgap reference in Fig. 14.30 to produce an output voltage of 5.000 V with zero temperature coefficient at a temperature of 47°C. Design for a collector current of 25 μA, and assume $I_S = 0.5$ fA.

SOLUTION **Known Information and Given Data:** The circuit is the Brokaw reference with amplified output given in Fig. 14.30. $V_O = 5.000$ V with a zero temperature coefficient (TC) at $T = 320$ K. Collector currents are to be 25 μA, and the transistor saturation current is 0.5 fA.

Unknowns: Values of resistors R, R_1, R_2, R_3, and R_4.

Approach: Find V_T and V_{PTAT}. Then use I_C to determine R_1. Use I_C to find V_{BE1}. Determine R_2 using Eq. (14.61). Choose R_4 and R_3 to set $V_O = 5$ V. Choose R to provide operating voltage to the op amp.

Assumptions: BJTs are in the active region of operation. $\beta_{FO} = \infty$ and $V_A = \infty$. $A_{E2} = 10 A_{E1}$ represents a reasonable emitter area ratio. Drop 2V across R.

Analysis: Because of the precision involved, we will carry four digits in our calculations.

$$V_T = \frac{kT}{q} = \frac{1.380 \times 10^{-23}(320)}{1.602 \times 10^{-19}} = 27.57 \text{ mV}$$

$$V_{PTAT} = V_T \ln\left(\frac{A_{E2}}{A_{E1}}\right) = V_T \ln(10) = 63.47 \text{ mV}$$

$$R_1 = \frac{V_{PTAT}}{I_E} = \frac{63.47 \text{ mV}}{25 \text{ μA}} = 2.539 \text{ k}\Omega$$

$$V_{BE1} = V_T \ln\left(\frac{I_{C1}}{I_{S1}}\right) = (27.57 \text{ mV}) \ln\left(\frac{25 \text{ μA}}{0.5 \text{ fA}}\right) = 0.6792 \text{ V}$$

$$\frac{R_2}{R_1} = \frac{V_{GO} + 3V_T - V_{BE1}}{2V_{PTAT}} = \frac{1.12 + 3(0.02757) - 0.6792}{2(0.06347)} = 4.124$$

$$R_2 = 4.124 R_1 = 10.47 \text{ k}\Omega$$

$$V_{BG} = V_{BE1} + 2\frac{R_2}{R_1} V_{PTAT} = 0.6792 + 2(4.124)(63.47 \text{ mV}) = 1.203 \text{ V}$$

$$\frac{R_4}{R_3} = \frac{V_O}{V_{BG}} - 1 = 3.157$$

We should not waste an excessive amount of current in the output voltage divider, so let us choose $I_3 = I_4 = 50$ μA. Also, set the voltage drop across R to 2 V.

$$R_3 = \frac{V_{BG}}{I_3} = \frac{1.203 \text{ V}}{50 \text{ μA}} = 24.0 \text{ k}\Omega \quad \text{and} \quad R_4 = \frac{V_O - V_{BG}}{I_3} = \frac{3.797 \text{ V}}{50 \text{ μA}} = 75.9 \text{ k}\Omega$$

$$R = \frac{2 \text{ V}}{25 \text{ μA}} = 80 \text{ k}\Omega$$

Check of Results: V_{BG} is approximately 1.20 V so our calculation appears correct. Our analysis showed that the output voltage should also be $V_{GO} + 3V_T = 1.203$ V, which also checks.

Discussion: Note that the voltage drop across the collector resistors must be enough to bring the inputs of the op amp into its common-mode operating range. In this circuit, the drop across the collector resistors is designed to be 2 V. Also note the total resistance required is probably too large for a realistic IC implementation, and a redesign would be needed.

Computer-Aided Analysis: We first set the *npn* parameters to BF = 10,000 and IS = 0.5 fA and let VAF default to infinity. Set AREA = 1 for Q_1, AREA = 10 for Q_2, and TEMP = 47°C. In the circuit shown here, the ideal op amp is modeled by EOPAMP whose controlling voltage appears across zero-value current source IOP. The gain is set to 10^6. Source VSTART may be needed in some versions of SPICE to help the circuit start up, although it introduces a small offset voltage across IOP[5]. (Remember that $V_O = 0$ is a valid operating point.) SPICE simulation produces $V_{BG} = 1.204$ V and $V_{PTAT} = 63.52$ mV and $V_O = 5.01$ V. With BF = 100 and VAF = 75 V, the values are $V_{BG} = 1.201$ V, $V_{PTAT} = 63.52$ mV, and $V_O = 5.03$ V.

.op
.model NPN NPN BF=100 VAF=75V IS=0.5fA
.TEMP 47
*.TEMP 0 25 50 75 100 125 150

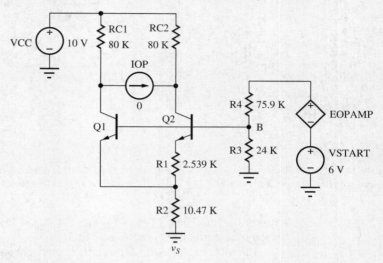

EXERCISE: Redesign the reference in Ex. 14.6 using $A_{E2} = 20A_{E1}$.

ANSWER: 3.30 kΩ, 10.5 kΩ, 24.0 kΩ, 75.9 kΩ, 80 kΩ

14.7 THE CURRENT MIRROR AS AN ACTIVE LOAD

We encountered use of transistors as replacements for the load resistors in amplifiers in Chapters 8 and 13. In this section, we find that one of the most important applications of the current mirror is as a replacement for the load resistors of differential amplifier stages in IC operational amplifiers. This elegant application of the current mirror can greatly improve amplifier voltage gain while maintaining the operating-point balance necessary for good common-mode rejection and low offset voltage. When used in this manner, the current mirror is referred to as an **active load** because the passive load resistors have been replaced with active transistor circuit elements.

14.7.1 CMOS DIFFERENTIAL AMPLIFIER WITH ACTIVE LOAD

Figure 14.31 shows a CMOS differential amplifier with an active load; the load resistors have been replaced by a PMOS current mirror. Let us first study the quiescent operating point of this circuit and then look at its small-signal characteristics.

dc Analysis
Assume for the moment that the amplifier is voltage balanced (in fact, it will turn out that it *is* balanced). Then, bias current I_{SS} divides equally between transistors M_1 and M_2, and I_{D1} and I_{D2} are each equal to $I_{SS}/2$. Current I_{D3} must equal I_{D1} and is mirrored as I_{D4} at the output of the PMOS

---
[5] Another alternative is to use the NODESET command to establish an initial condition on the bases of Q1–Q2 (node B) without using VSTART. For example: .NODESET V(B)=1.5V

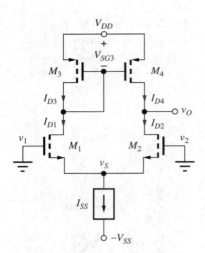

Figure 14.31 CMOS differential amplifier with PMOS active load.

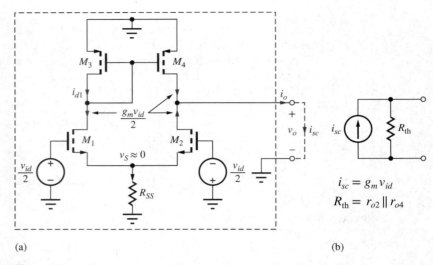

Figure 14.32 (a) CMOS differential amplifier with differential-mode input. (b) The circuit is a one port and can be represented by its Norton equivalent circuit.

current mirror. Thus, I_{D3} and I_{D4} are also equal to $I_{SS}/2$, and the current in the drain of M_4 is exactly the current required to satisfy M_2.

The mirror ratio set by M_3 and M_4 is precisely unity when $V_{SD4} = V_{SD3}$ and hence $V_{DS1} = V_{DS2}$. Thus, the differential amplifier is completely balanced at dc when the quiescent output voltage is

$$V_O = V_{DD} - V_{SD4} = V_{DD} - V_{SG3} = V_{DD} - \left(\sqrt{\frac{I_{SS}}{K_p}} - V_{TP}\right) \tag{14.63}$$

Q-Points

The drain-source voltages of M_1 and M_2 are

$$V_{DS1} = V_O - V_S = V_{DD} - \left(\sqrt{\frac{I_{SS}}{K_p}} - V_{TP}\right) + \left(V_{TN} + \sqrt{\frac{I_{SS}}{K_n}}\right)$$

or

$$V_{DS1} = V_{DD} + V_{TN} + V_{TP} + \sqrt{\frac{I_{SS}}{K_n}} - \sqrt{\frac{I_{SS}}{K_p}} \cong V_{DD} \tag{14.64}$$

and those of M_3 and M_4 are

$$V_{SD3} = V_{SG3} = \sqrt{\frac{I_{SS}}{K_p}} - V_{TP} \tag{14.65}$$

(Remember that $V_{TP} < 0$ for p-channel enhancement-mode devices.)

The drain currents of all the transistors are equal:

$$I_{DS1} = I_{DS2} = I_{SD3} = I_{SD4} = \frac{I_{SS}}{2} \tag{14.66}$$

Small-Signal Analysis

Now that we have found the operating points of the transistors, we can proceed to analyze the small-signal characteristics of the amplifier, including **differential-mode gain, differential-mode input** and **output resistances, common-mode gain, common-mode rejection ratio (CMRR),** and **common-mode input** and **output resistances.**

Differential-Mode Signal Analysis

Analysis of the ac behavior of the differential amplifier begins with the differential-mode input applied in the ac circuit model in Fig. 14.32. Upon studying the circuit in Fig. 14.32, we realize that it is a two-terminal network and can be represented by its Norton equivalent circuit consisting of the short-circuit output current and Thévenin equivalent output resistance. With the output terminals short circuited, the NMOS differential pair produces equal and opposite currents with amplitude $g_{m2}v_{id}/2$ at the drains of M_1 and M_2. Drain current i_{d1} is supplied through current mirror transistor M_3 and is replicated at the output of M_4. Thus, the total short-circuit output current is

$$\mathbf{i_o} = 2\frac{g_{m2}\mathbf{v_{id}}}{2} = g_{m2}\mathbf{v_{id}} \tag{14.67}$$

The current mirror provides a single-ended output but with a transconductance equal to the full value of the C-S amplifier!

The Thévenin equivalent output resistance will be found using the circuit in Fig. 14.33 in which the internal output resistances of M_2 and M_4 are shown next to their respective transistors. In the next sub section, we will show that R_{th} is equal to the parallel combination of r_{o2} and r_{o4}:

$$R_{th} = r_{o2} \| r_{o4} \tag{14.68}$$

The differential-mode voltage gain of the open-circuited differential amplifier is simply the product of i_{sc} and R_{th}:

$$A_{dm} = g_{m2}(r_{o2} \| r_{o4}) = \frac{\mu_{f2}}{1 + \frac{r_{o2}}{r_{o4}}} \cong \frac{\mu_{f2}}{2} \tag{14.69}$$

Equation (14.69) indicates that the gain of the input stage of the amplifier approaches one half the intrinsic gain of the transistors forming the differential pair. We are now within a factor of 2 of the theoretical voltage gain limit for the individual transistors!

Output Resistance of the Differential Amplifier

The origin of the output resistance expression in Eq. (14.69) can be thought of conceptually in the following (although technically incorrect) manner. At node 1 in Fig. 14.33, r_{o4} is connected directly to ac ground at the positive power supply, whereas r_{o2} appears connected to virtual ground at the sources of M_2 and M_1. Thus, r_{o2} and r_{o4} are effectively in parallel. Although this argument gives the correct answer, it is not precisely correct. Because the differential amplifier with active load no longer represents a symmetric circuit relative to the test signal applied to the output, the node at the sources of M_1 and M_2 is not truly a virtual ground.

Exact Analysis

A more precise analysis can be obtained from the circuit in Fig. 14.34. The output resistance r_{o4} of M_4 is indeed connected directly to ac ground and represents one component of the output resistance.

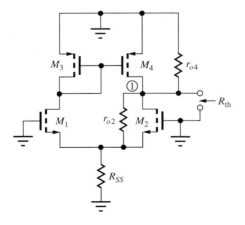

Figure 14.33 Simple CMOS op amp with active load in the first stage.

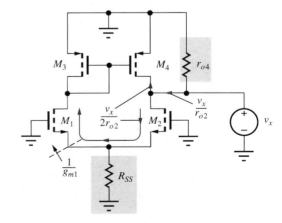

Figure 14.34 Output resistance component due to r_{o2}.

However, the current from v_x due to r_{o2} is more complicated. The actual behavior can be determined from Fig. 14.34, in which R_{SS} is assumed to be negligible with respect to $1/g_{m1}$ ($R_{SS} \gg 1/g_{m1}$).

Transistor M_2 is operating as a common-source transistor with an effective resistance in its source of $R_S = 1/g_{m1}$. Based on the results in Table 8.3, the resistance looking into the drain of M_2 is

$$R_{o2} = r_{o2}(1 + g_{m2}R_S) = r_{o2}\left(1 + g_{m2}\frac{1}{g_{m1}}\right) = 2r_{o2} \tag{14.70}$$

Therefore, the drain current of M_2 is equal to $v_x/2r_{o2}$. However, the current goes around the differential pair and into the current mirror at M_3. The current is replicated by the mirror to become the drain current of M_4. The total current from source v_x becomes $2(v_x/2r_{o2}) = v_x/r_{o2}$.

Combining this current with the current through r_{o4} yields a total current of

$$\mathbf{i_x^T} = \frac{\mathbf{v_x}}{r_{o2}} + \frac{\mathbf{v_x}}{r_{o4}} \quad \text{and} \quad R_{od} = r_{o2} \| r_{o4} \tag{14.71}$$

The equivalent resistance at the output node is, in fact, equal to the parallel combination of the output resistances of M_2 and M_4.

> **EXERCISE:** Find the Q-points of the transistors in Fig. 14.31 if $I_{SS} = 250$ μA, $K_n = 250$ μA/V², $K_p = 200$ μA/V², $V_{TN} = -V_{TP} = 0.75$ V, and $V_{DD} = V_{SS} = 5$ V. What are the transconductance, output resistance, and voltage gain of the amplifier if $\lambda = 0.0133$ V⁻¹?
>
> **ANSWERS:** (125 μA, 4.88 V), (125 μA, 1.87 V); 250 μS, 314 kΩ, 78.5

Common-Mode Input Signals

Figure 14.35 is the CMOS differential amplifier with a common-mode input signal. The common-mode input voltage causes a common-mode current i_{oc} in both sides of the differential pair consisting of M_1 and M_2. The common-mode current (i_{oc}) in M_1 is mirrored at the output of M_4 with a small error since no current can appear in r_{o4} with the output shorted. In addition, the small voltage difference developed between the drains of M_1 and M_2 causes a current in the differential output resistance ($2r_{o2}$) of the pair that is then doubled by the action of the current mirror.

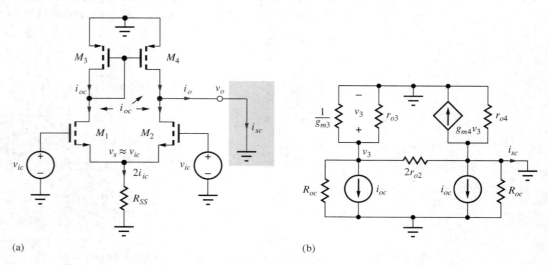

Figure 14.35 (a) CMOS differential amplifier with common-mode input. (b) Small-signal model.

An expression for the short-circuit output current can be found using the small-signal model for the circuit in Fig. 14.35(b). The differential pair with common-mode input is represented by the model from Sec. 13.1.15 with

$$\mathbf{i}_{oc} \cong \frac{\mathbf{v}_{ic}}{2R_{SS}} \qquad R_{od} = 2r_{o2} \qquad R_{oc} = 2\mu_f R_{SS} \qquad (14.72)$$

With the output short circuited, we have a one-node problem. Solving for v_3,

$$\mathbf{v}_3 = \frac{-\mathbf{i}_{oc}}{g_{m3} + g_{o3} + \frac{g_{o2}}{2} + G_{oc}} \quad \text{and} \quad \mathbf{i}_{sc} = -\left(\mathbf{i}_{oc} + g_{m4}\mathbf{v}_3 - \frac{g_{o2}}{2}\mathbf{v}_3\right) \qquad (14.73)$$

which together with Eq. (14.72) yields

$$\mathbf{i}_{sc} = -\frac{g_{o3} + g_{o2}}{g_{m3} + g_{o3} + \frac{g_{o2}}{2} + G_{oc}} \mathbf{i}_{oc} \cong -\frac{1 + \frac{r_{o3}}{r_{o2}}}{\mu_{f3}}\left(\frac{\mathbf{v}_{ic}}{2R_{SS}}\right) \qquad (14.74)$$

where it is assumed that $g_{m4} = g_{m3}$ and $G_{oc} \ll g_{o3}$. The Thévenin equivalent output resistance is exactly the same as found in the previous section, $R_{th} = r_{o2} \| r_{o4}$. Thus, the common-mode gain is

$$A_{cm} = \frac{\mathbf{i}_{sc} R_{th}}{\mathbf{v}_{ic}} = -\frac{\left(1 + \frac{r_{o3}}{r_{o2}}\right)}{2\mu_{f3}R_{SS}}(r_{o2} \| r_{o4}) \qquad (14.75)$$

where $\mu_{f3} \gg 1$ has been assumed. The common-mode rejection ratio is

$$\text{CMRR} = \left|\frac{A_{dm}}{A_{cm}}\right| = \frac{2\mu_{f3}g_{m2}R_{SS}}{\left(1 + \frac{r_{o3}}{r_{o2}}\right)} \cong \mu_{f3}g_{m2}R_{SS} \quad \text{for} \quad r_{o3} \cong r_{o2} \qquad (14.76)$$

which is improved by a factor of approximately μ_{f3} over that of the pair with a resistor load!

EXERCISE: Evaluate Eq. (14.76) for $K_p = K_n = 5$ mA/V^2, $\lambda = 0.0167$ V^{-1}, $I_{SS} = 200$ μA, and $R_{SS} = 10$ MΩ.

ANSWER: 6.00×10^6 or 136 dB

In the last exercise, we find that the CMRR predicted by Eq. (14.76) is quite large, whereas typical op-amp specs are 80 to 100 dB. We need to look deeper. In reality, this predicted level will not be achieved, but will be limited by mismatches between the devices in the circuit.

Mismatch Contributions to CMRR

In this section, we explore the techniques used to calculate the effects of device mismatches on CMRR. Figure 14.36 presents the small-signal model for the differential amplifier with mismatches in transistors M_1 and M_2 in which we assume

$$g_{m1} = g_m + \frac{\Delta g_m}{2} \qquad g_{m2} = g_m - \frac{\Delta g_m}{2} \qquad g_{o1} = g_o + \frac{\Delta g_o}{2} \qquad g_{o2} = g_o - \frac{\Delta g_o}{2} \qquad (14.77)$$

In this analysis, M_3 and M_4 are still identical. We desire to find the short-circuit output current $i_{sc} = (i_{d1} - i_{d2})$ in which i_{d1} is replicated by the current mirror. Let us use our knowledge of the gross behavior of the circuit to simplify the analysis. We have $v_{d2} = 0$, since we are finding the short-circuit output current, and based on previous common-mode analyses, we expect the signal at v_{d1} to be small. So let us assume that $v_{d1} \cong 0$. With this assumption, and noting that the two gate-source voltages are identical,

$$\mathbf{i}_{sc} = \mathbf{i}_{d1} - \mathbf{i}_{d2} = (g_{m1} - g_{m2})\mathbf{v}_{gs} - (g_{o1} - g_{o2})\mathbf{v}_s = \Delta g_m \mathbf{v}_{gs} - \Delta g_o \mathbf{v}_s \qquad (14.78)$$

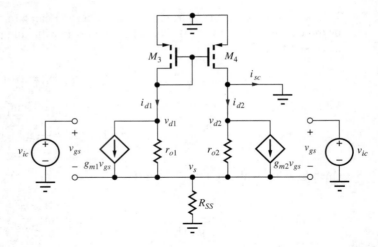

Figure 14.36 CMOS differential amplifier in which M_1 and M_2 are no longer matched.

To evaluate this expression, we need to find source voltage v_s and gate-source voltage v_{gs}. Writing a nodal equation for v_s with $v_{gs} = v_{ic} - v_s$, $v_{d1} = 0$ and $v_{d2} = 0$, yields

$$\left(g_m + \frac{\Delta g_m}{2} + g_m - \frac{\Delta g_m}{2}\right)(\mathbf{v_{ic}} - \mathbf{v_s}) = \left(g_o + \frac{\Delta g_o}{2} + g_o - \frac{\Delta g_o}{2} + G_{SS}\right)\mathbf{v_s}$$

in which we may be surprised to see all the mismatch terms cancel out! Thus, for common-mode inputs, v_s and v_{gs} are not affected by the transistor mismatches:[6]

$$\mathbf{v_s} \cong \frac{2g_m R_{SS}}{1 + 2g_m R_{SS}}\mathbf{v_{ic}} \cong \mathbf{v_{ic}} \quad \text{and} \quad \mathbf{v_{gs}} \cong \frac{1 + 2g_o R_{SS}}{1 + 2g_m R_{SS}}\mathbf{v_{ic}} \cong \left(\frac{1}{2g_m R_{SS}} + \frac{1}{\mu_f}\right)\mathbf{v_{ic}} \quad (14.79)$$

since $2g_m R_{SS} \gg 1$. The short-circuit output current goes through the Thévenin output resistance $R_{th} = r_{o2} \| r_{o4}$ to produce the output voltage, and

$$A_{cm} = \frac{\mathbf{i_{sc}} R_{th}}{\mathbf{v_{ic}}} = \left[\Delta g_m\left(\frac{1}{2g_m R_{SS}} + \frac{1}{\mu_f}\right) - \Delta g_o\right](r_{o2} \| r_{o4}) \quad (14.80)$$

The CMRR is then

$$\text{CMRR}^{-1} = \left|\frac{A_{cm}}{A_{dm}}\right| = \left|\frac{A_{cm}}{g_m(r_{o2} \| r_{o4})}\right|$$

$$= \left[\frac{\Delta g_m}{g_m}\left(\frac{1}{2g_m R_{SS}} + \frac{1}{\mu_f}\right) - \frac{\Delta g_o}{g_o}\frac{1}{\mu_f}\right] \quad (14.81)$$

For very large R_{SS}, we see that CMRR is now limited by the transistor mismatches and value of the intrinsic gain. For example, a 1 percent mismatch with an intrinsic gain of 500 limits the individual terms in Eq. (14.81) to 2×10^{-5}. Since we cannot predict the signs on the $\Delta g/g$ terms, the expected CMRR is 2.5×10^4 or 88 dB. This is much more consistent with observed values of CMRR.

---

[6] An exact analysis without assuming that $v_{d1} = 0$ shows that a negligibly small change actually occurs.

ELECTRONICS IN ACTION

G_m-C Integrated Filters

The design of integrated circuit filters is complicated by the lack of well-controlled resistive components in most mainstream CMOS processes. One approach to overcome this is the use of G_m-C filter topologies based on the operational transconductance amplifier (OTA). The OTA is characterized by both a high-input and high-output impedance. A simple form of an OTA is the CMOS differential amplifier from Fig. 14.31. The high-impedance output ($R_{out} = r_{o2} \| r_{o4}$) is a small-signal current given by the product of the differential pair g_m and the differential input voltage v_{id}. Typically, commercial OTA designs include additional devices to improve output resistance and voltage swing.

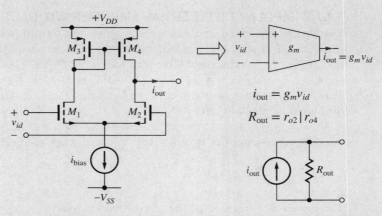

Equivalent schematic symbol for operational transconductance amplifier (OTA).

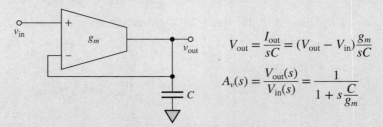

$$V_{out} = \frac{I_{out}}{sC} = (V_{out} - V_{in})\frac{g_m}{sC}$$

$$A_v(s) = \frac{V_{out}(s)}{V_{in}(s)} = \frac{1}{1 + s\frac{C}{g_m}}$$

Single pole G_m-C low-pass filter.

A simple low-pass filter formed with an OTA and a capacitor is shown above. The transfer characteristic is also included and indicates that the upper cutoff frequency occurs at $f_H = g_m/2\pi C$. One of the more useful characteristics of the G_m-C filter approach is the ease with which the characteristics can be tuned. Recalling that g_m is a function of the differential pair current, we see that the cutoff frequency of the filter is easily modified by adjusting the bias current. It is also very important to note that this circuit is a continuous-time filter that requires no resistors!

A second-order version (a biquad topology) is shown on the next page. This circuit permits the adjustment of cutoff frequency with constant Q, and still requires no resistors. High-pass, band-pass, and band-reject are also readily derived from this basic form. Because of their compatibility with standard CMOS processes and excellent power efficiency, G_m-C filters are frequently used in low power in communication circuits, A/D converter anti-alias filters, noise shaping, and many other applications.

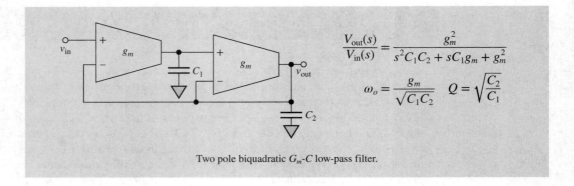

Two pole biquadratic G_m-C low-pass filter.

14.7.2 BIPOLAR DIFFERENTIAL AMPLIFIER WITH ACTIVE LOAD

The bipolar differential amplifier with an active load formed from a *pnp* current mirror is depicted in Fig. 14.37(a) with $v_1 = 0 = v_2$. If we assume that the circuit is balanced with $\beta_{FO} = \infty$, then the bias current I_{EE} divides equally between transistors Q_1 and Q_2, and I_{C1} and I_{C2} are equal to $I_{EE}/2$. Current I_{C1} is supplied by transistor Q_3 and is mirrored as I_{C4} at the output of *pnp* transistor Q_4. Thus, I_{C3} and I_{C4} are both also equal to $I_{EE}/2$, and the dc current in the collector of Q_4 is exactly the current required to satisfy Q_2.

If β_{FO} is very large, then the current mirror ratio is exactly 1 when $V_{EC4} = V_{EC3} = V_{EB}$, and the differential amplifier is completely balanced when the quiescent output voltage is

$$V_O = V_{CC} - V_{EB} \tag{14.82}$$

Q-Points

The collector currents of all the transistors are equal:

$$I_{C1} = I_{C2} = I_{C3} = I_{C4} = \frac{I_{EE}}{2} \tag{14.83}$$

The collector-emitter voltages of Q_1 and Q_2 are

$$V_{CE1} = V_{CE2} = V_C - V_E = (V_{CC} - V_{EB}) - (-V_{BE}) \cong V_{CC} \tag{14.84}$$

and for Q_3 and Q_4,

$$V_{EC3} = V_{EC4} = V_{EB} \tag{14.85}$$

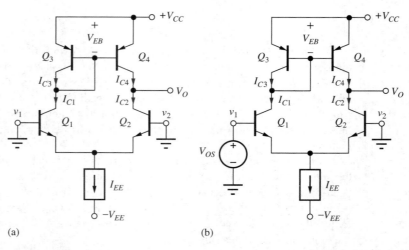

Figure 14.37 (a) Bipolar differential amplifier with active load. (b) Amplifier with offset voltage applied.

Finite Current Gain

The current gain defect in the current mirror upsets the balance of the circuit. However, the collector current of Q_4 must equal the collector current of Q_2, and the output voltage V_O adjusts itself to make up for the current mirror error. In Fig. 14.38(b), an offset voltage is applied to the amplifier to bring the amplifier back into voltage balance with $V_O = V_{C1}$, so that $V_{CE2} = V_{CE1}$, $V_{EC4} = V_{EC3}$. There is no longer any imbalance due to collector-emitter voltage mismatches, and the value of the offset voltage V_{OS} can be found directly from the ratio of the collector currents in Q_1 and Q_2 based upon Eq. (14.86):

$$\frac{I_{C1}}{I_{C2}} = \frac{I_S \exp\left(\frac{V_{BE} + V_{OS}}{V_T}\right)}{I_S \exp\left(\frac{V_{BE}}{V_T}\right)} = \exp\left(\frac{V_{OS}}{V_T}\right) \quad \text{or} \ldots \quad (14.86)$$

From Fig. 14.38(b), we have $I_{C3} = I_{C4} = I_{C2}$, and therefore

$$I_{C1} = I_{C3} + I_{B3} + I_{B4} = I_{C2}\left(1 + \frac{2}{\beta_{FO3}}\right) \quad (14.87)$$

The offset voltage is

$$V_{OS} = V_T \ln\left(1 + \frac{2}{\beta_{FO3}}\right) \cong V_T \frac{2}{\beta_{FO3}} \quad (14.88)$$

using $\ln(1 + x) \cong x$ for small values of x. V_{OS} in Eq. (14.88) represents the input voltage needed to force the differential output voltage v_{OD} to be zero. For *pnp* transistors with $\beta_{FO} = 80$, $V_{OS} \cong 0.625$ mV. To eliminate this error, a buffered current mirror is usually used as the active load, as shown in Fig. 14.38.

> **EXERCISE:** (a) What is the expression for V_{OS} if the buffered current mirror is utilized as the active load as in Fig. 14.38? (b) Calculate the value of V_{OS} if $\beta_{FO} = 80$ for the *pnp* transistors.
>
> **ANSWERS:** $V_{OS} \cong V_T \dfrac{2}{\beta_{FO3}(\beta_{FO11} + 1)}$; 7.72 μV

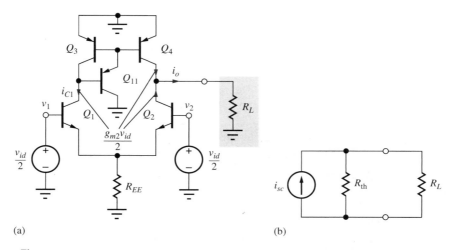

Figure 14.38 (a) BJT differential amplifier with differential-mode input. (b) Equivalent circuit.

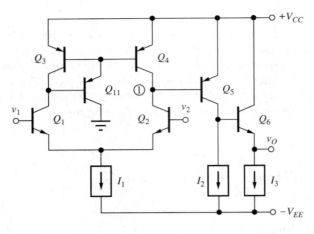

Figure 14.39 Bipolar op-amp prototype with active load in first stage.

Differential-Mode Signal Analysis

Analysis of the ac behavior of the differential amplifier begins with the differential-mode input applied in the ac circuit model in Fig. 14.38. The differential input pair produces equal and opposite currents with amplitude $g_{m2}v_{id}/2$ at the collectors of Q_1 and Q_2. Collector current i_{c1} is supplied by Q_3 and is replicated at the output of Q_4. Thus, the total short-circuit output current is equal to

$$\mathbf{i}_{sc} = 2\frac{g_{m2}\mathbf{v}_{id}}{2} = g_{m2}\mathbf{v}_{id} \tag{14.89}$$

The output resistance is identical to Eq. (14.71) $R_{th} = r_{o2} \| r_{o4}$ and

$$A_{dd} = \frac{\mathbf{i}_{sc}(R_L \| R_{th})}{\mathbf{v}_{dm}} = g_{m2}(R_L \| r_{o2} \| r_{o4}) = -g_{m2}R_L \tag{14.90}$$

The current mirror provides a single-ended output but with a voltage equal to the full gain of the C-E amplifier, just as for the FET case. Here, we have included R_L, which models the loading of the next stage in a multistage amplifier.

The power of the current mirror is again most apparent when additional stages are added, as in the prototype operational amplifier in Fig. 14.39. The resistance at the output of the differential input stage, node 1, is now equivalent to the parallel combination of the output resistances of transistors Q_2 and Q_4 and the input resistance of Q_5 ($R_L = r_{\pi5}$):

$$R_{eq} = r_{o2} \| r_{o4} \| r_{\pi5} \cong r_{\pi5} \tag{14.91}$$

The gain of the differential input stages becomes

$$A_{dm} = g_{m2}R_{eq} \cong g_{m2}r_{\pi5} = \beta_{o5}\frac{I_{C2}}{I_{C5}} \tag{14.92}$$

EXERCISE: What is the approximate differential-mode voltage gain of the differential input stage of the amplifier in Fig. 14.39 if $\beta_{FO} = 150$, $V_A = 75$ V, and $I_{C5} = 3\,I_{C2}$?

ANSWER: 50

Common-Mode Input Signals

The circuits in Fig. 14.40 represent the bipolar differential amplifier with current mirror load and a buffered current mirror load. The detailed analysis is quite involved and tedious, particularly

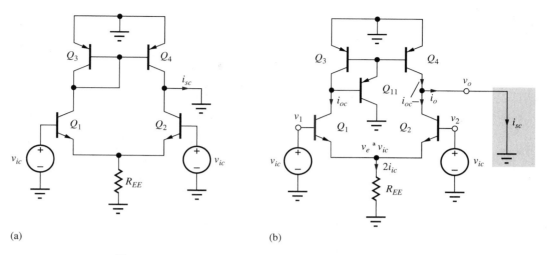

Figure 14.40 Bipolar differential amplifiers with common-mode input.

for the buffered mirror, so here we will argue the result based on earlier analyses. The common-mode current i_{oc} in Q_1 and Q_2 is found with the help of Eq. (13.27):

$$\mathbf{i}_{oc} = \frac{A_{cc}\mathbf{v}_{ic}}{R_C} = \mathbf{v}_{ic}\left(\frac{1}{\beta_o r_o} - \frac{1}{2R_{EE}}\right) \tag{14.93}$$

The current from Q_1 is mirrored at the output of Q_4 with a mirror error of $2/\beta_o$. Thus, the short-circuit output current is the error current

$$\mathbf{i}_{sc} = \mathbf{v}_{ic}\frac{2}{\beta_o}\left(\frac{1}{\beta_o r_o} - \frac{1}{2R_{EE}}\right) \tag{14.94}$$

The CMRR is

$$\text{CMRR} = \left|\frac{g_{m2}R_{th}}{i_{sc}R_{th}/v_{ic}}\right| \cong \left[\frac{2}{\beta_{o3}}\left(\frac{1}{\beta_{o2}\mu_{f2}} - \frac{1}{2g_{m2}R_{EE}}\right)\right]^{-1} \tag{14.95}$$

where R_{th} is the equivalent resistance at the amplifier output.

EXERCISE: Evaluate Eq. (14.96) for $\beta_F = 100$, $V_A = 75$ V, $I_{EE} = 200$ μA, and $R_{EE} = 10$ MΩ.

ANSWER: 5.45×10^6 or 135 dB

The expression in Eq. (14.95) yields a very large CMRR that is almost impossible to achieve. The CMRR predicted for the buffered current mirror is even larger, since the mirror error is approximately $2/\beta_{o11}\beta_{o3}$. In both these circuits, however, the CMRR will actually be limited to much smaller levels by small mismatches between the various transistors:

$$\text{CMRR}^{-1} = \left[\left(\frac{\Delta g_m}{g_m} + \frac{\Delta g_\pi}{g_\pi}\right)\left(\frac{1}{2g_m R_{SS}} + \frac{1}{\mu_f}\right) - \frac{\Delta g_o}{g_o}\frac{1}{\mu_f}\right] \tag{14.96}$$

Equation (14.96) is similar to the results for the FET from Eq. (14.81) with the addition of the $\Delta g_\pi/g_\pi$ term. In an actual amplifier, the common-mode gain is determined by small imbalances in the bipolar transistors and overall asymmetry of the amplifier.

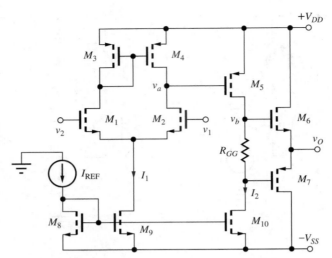

Figure 14.41 Complete CMOS op amp with current mirror bias.

14.8 ACTIVE LOADS IN OPERATIONAL AMPLIFIERS

Let us now explore more fully the use of active loads in MOS and bipolar operational amplifiers. Figure 14.41 shows a complete three-stage MOS operational amplifier. The input stage consists of NMOS differential pair M_1 and M_2 with PMOS current mirror load, M_3 and M_4, followed by a second common-source gain stage M_5 loaded by current source M_{10}. The output stage is a **class-AB amplifier** consisting of transistors M_6 and M_7. Bias currents I_1 and I_2 for the two gain stages are set by the current mirrors formed by transistors M_8, M_9, and M_{10}, and class-AB bias for the output stage is set by the voltage developed across resistor R_{GG}. At most, only two low value resistors are required: R_{GG} and one for the current mirror reference current.

14.8.1 CMOS OP-AMP VOLTAGE GAIN

Assuming that the gain of the output stage is approximately 1, then the overall differential-mode gain A_{dm} of the three-stage operational amplifier is approximately equal to the product of the terminal gains of the first two stages:

$$A_{dm} = \frac{\mathbf{v_o}}{\mathbf{v_{id}}} = \frac{\mathbf{v_a}}{\mathbf{v_{id}}}\frac{\mathbf{v_b}}{\mathbf{v_a}}\frac{\mathbf{v_o}}{\mathbf{v_b}} = A_{vt1}A_{vt2}(1) \cong A_{vt1}A_{vt2} \tag{14.97}$$

As discussed earlier, the input stage provides a gain of

$$A_{vt1} = g_{m2}(r_{o2} \| r_{o4}) \cong \frac{\mu_{f2}}{2} \tag{14.98}$$

The terminal gain of the second stage is equal to

$$A_{vt2} = g_{m5}(r_{o5} \| (R_{GG} + r_{o10})) \cong g_{m5}(r_{o5} \| r_{o10}) \cong g_{m5}(r_{o5} \| r_{o5}) = \frac{\mu_{f5}}{2} \tag{14.99}$$

assuming that the output resistances of M_5 and M_{10} are similar in value and $R_{GG} \ll r_{o10}$. Combining the three equations above yields

$$A_{dm} \cong \frac{\mu_{f2}\mu_{f5}}{4} \tag{14.100}$$

The gain approaches one quarter of the product of the intrinsic gains of the two gain stages.

The factor of 4 in the denominator of Eq. (14.100) can be eliminated by improved design. If a Wilson source is used in the first-stage active load, then the output resistance of the current mirror is much greater than r_{o2}, and A_{v1} becomes equal to μ_{f2}. The gain of the second stage can also be increased to the full amplification factor of M_5 if the current source M_{10} is replaced by

14.8 Active Loads in Operational Amplifiers

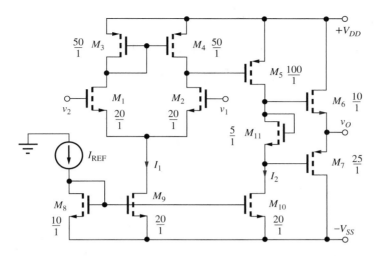

Figure 14.42 Op amp with current mirror bias of the class-AB output stage.

a Wilson or cascode source. If both these circuit changes are used (see Prob. 14.131), then the gain of the op amp can be increased to

$$A_{dm} \cong \mu_{f2}\mu_{f5} \qquad (14.101)$$

This discussion has only scratched the surface of the many techniques available for increasing the gain of the CMOS op amp. Several examples appear in the problems at the end of this chapter; further discussion can be found in the bibliography.

14.8.2 dc DESIGN CONSIDERATIONS

When the circuit in Fig. 14.41 is operating in a closed-loop op-amp configuration, the drain current of M_5 must be equal to the output current I_2 of current source transistor M_{10}. For the amplifier to have a minimum offset voltage, the (W/L) ratio of M_5 must be carefully selected so the source-gate bias of M_5, $V_{SG5} = V_{SD4} = V_{SG3}$, is precisely the proper voltage to set $I_{D5} = I_2$. Usually the W/L ratio of M_5 is also adjusted to account for V_{DS} and λ differences between M_5 and M_{10}. R_{GG} and the (W/L) ratios of M_6 and M_7 determine the quiescent current in the class-AB output stage.

Even resistor R_{GG} has been eliminated from the op amp in Fig. 14.42 by using the gate-source voltage of FET M_{11} to bias the output stage. The current in the class-AB stage is determined by the W/L ratios of the output transistors and the diode-connected MOSFET M_{11}.

EXAMPLE 14.7   CMOS OP-AMP ANALYSIS

Find the small-signal characteristics of a CMOS operational amplifier.

PROBLEM Find the voltage gain, input resistance, and output resistance of the amplifier in Fig. 14.42 if $K'_n = 25$ µA/V², $K'_p = 10$ µA/V², $V_{TN} = 0.75$ V, $V_{TP} = -0.75$ V, $\lambda = 0.0125$ V^{-1}, $V_{DD} = V_{SS} = 5$ V, and $I_{REF} = 100$ µA.

SOLUTION **Known Information and Given Data:** The schematic for the operational amplifier appears in Fig. 14.42; $V_{DD} = V_{SS} = 5$ V, and $I_{REF} = 100$ µA; device parameters are given as $K'_n = 25$ µA/V², $K'_p = 10$ µA/V², $V_{TN} = 0.75$ V, $V_{TP} = -0.75$ V, $\lambda = 0.0125$ V^{-1}.

Unknowns: Q-points, A_{dm}, R_{id}, and R_{out}

Approach: Find the Q-point currents and use the device parameters to evaluate Eq. (14.100) for A_{dm}. Since we have MOSFETs at the input, $R_{id} = R_{ic} = \infty$. R_{out} is set by M_6 and M_7: $R_{out} = (1/g_{m6}) \| (1/g_{m7})$.

Assumptions: MOSFETs operate in the active region.

Analysis: The gain can be estimated using Eq. (14.100).

$$A_{dm} \cong \frac{\mu_{f2}\mu_{f5}}{4} = \frac{1}{4}\left(\frac{1}{\lambda_2}\sqrt{\frac{2K_{n2}}{I_{D2}}}\right)\left(\frac{1}{\lambda_5}\sqrt{\frac{2K_{p5}}{I_{D5}}}\right)$$

For the amplifier in Fig. 14.42,

$$I_{D2} = \frac{I_1}{2} = \frac{2I_{REF}}{2} = 100 \text{ μA} \qquad I_{D5} = I_2 = 2I_{REF} = 200 \text{ μA}$$

$$K_{n2} = 20K'_n = 500 \frac{\mu A}{V^2} \qquad K_{p5} = 100K'_p = 1000 \frac{\mu A}{V^2}$$

and

$$A_{dm} \cong \frac{\mu_{f2}\mu_{f5}}{4} = \frac{1}{4}\left(\frac{1}{0.0125}\right)^2 V^2 \sqrt{\frac{2\left(500\frac{\mu A}{V^2}\right)}{100 \text{ μA}}}\sqrt{\frac{2\left(1000\frac{\mu A}{V^2}\right)}{200 \text{ μA}}} = 16{,}000$$

The input resistance is twice the input resistance of M_1, which is infinite: $R_{id} = \infty$. The output resistance is determined by the parallel combination of the output resistances of M_6 and M_7, which act as two source followers operating in parallel:

$$R_{out} = \frac{1}{g_{m6}}\bigg\|\frac{1}{g_{m7}} = \frac{1}{\sqrt{2K_{n6}I_{D6}}}\bigg\|\frac{1}{\sqrt{2K_{p7}I_{D7}}}$$

To evaluate this expression, the current in the output stage must be found. The gate-source voltage of M_{11} is

$$V_{GS11} = V_{TN11} + \sqrt{\frac{2I_{D11}}{K_{n11}}} = 0.75 \text{ V} + \sqrt{\frac{2(200 \text{ μA})}{125\left(\frac{\mu A}{V^2}\right)}} = 2.54 \text{ V}$$

In this design, $V_{TP} = -V_{TN}$ and the W/L ratios of M_6 and M_7 have been chosen so that $K_{p7} = K_{n6}$. Because I_{D6} must equal I_{D7}, $V_{GS6} = V_{SG7}$. Thus, both V_{GS6} and V_{SG7} are equal to one-half V_{GS11}, and

$$I_{D7} = I_{D6} = \frac{250}{2}\frac{\mu A}{V^2}(1.27 \text{ V} - 0.75 \text{ V})^2 = 33.7 \text{ μA}$$

The transconductances of M_6 and M_7 are also equal,

$$g_{m7} = g_{m6} = \sqrt{2\left(2.50 \times 10^{-4}\frac{\mu A}{V^2}\right)(33.7 \times 10^{-6} \text{ μA})} = 1.30 \times 10^{-4} \text{ S}$$

and the output resistance at the Q-point is $R_{out} = 3.85$ kΩ.

Check of Results: A double check of our hand calculations indicates they are correct. Because of the complexity of the circuit, SPICE simulation represents an excellent check of hand calculations. The simulation results appear in the next exercise.

Discussion: Simulation of the open-loop characteristics of high-gain amplifiers in SPICE can be difficult. The open-loop gain will amplify the offset voltage of the amplifier and may saturate the output. One approach is to first determine the offset voltage and then to apply a compensating voltage to the amplifier input to bring the output near zero. The steps are outlined next. In very high gain cases, SPICE may still be unable to converge because numerical "noise" during the simulation steps is amplified just as an input voltage. The successive voltage and current injection method discussed previously in Chapter 11 solves this problem.

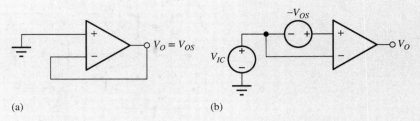

Figure 14.43 Op-amp setups for SPICE simulation. (a) Offset voltage determination. (b) Circuit for open-loop analysis using SPICE transfer functions.

Computer-Aided Analysis: After drawing the circuit of Fig. 14.42 with the schematic editor, be sure to set the device parameters to the desired values. For the NMOS devices, KP = 25 μA/V^2, VTO = 0.75 V, and LAMBDA = 0.0125 V^{-1}. For the PMOS devices, KP = 10 μA/V^2, VTO = −0.75 V, and LAMBDA = 0.0125 V^{-1}. W and L must be specified for each individual transistor. For example, use W = 5 μm and L = 1 μm for a 5/1 device.

The next step in the simulation is to find the offset voltage by operating the op amp in a voltage-follower configuration for which $V_O = V_{OS}$, as in Fig. 14.43(a). V_{OS} is then applied as a differential input to the amplifier in Fig. 14.43(b) with a common-mode input $V_{IC} = 0$. If the value of V_{OS} is correct, an operating point analysis should yield a value of approximately 0 for V_O. A transfer function analysis from V_{OS} to the output will give values of A_{dm}, R_{id}, and R_{out}. A transfer function analysis from V_{IC} to the output will give A_{cm}, R_{ic}, and R_{out}. The SPICE results are given as the answers to the next exercise.

EXERCISE: Simulate the amplifier in Fig. 14.42 using SPICE and compare the results to the answers in Ex. 14.7. Which terminal is the noninverting input? What are the offset voltage, common-mode and differential-mode gains, CMRR, common-mode and differential-mode input resistances, and output resistance?

ANSWERS: v_1; 64.164 μV; 17,800; 0.52; 90.7 dB; ∞; ∞; 3.63 kΩ

14.8.3 BIPOLAR OPERATIONAL AMPLIFIERS

As discussed earlier, active-load techniques can be applied equally well to bipolar op amps. In fact, most of the techniques discussed thus far were developed first for bipolar amplifiers and later applied to MOS circuits as NMOS and CMOS technologies matured. In the circuit in Fig. 14.44, a differential input stage with active load is formed by transistors Q_1 to Q_4. The first stage is followed by a high-gain C-E amplifier formed of Q_5 and its current source load Q_8. Load resistance R_L is driven by the class-AB output stage, consisting of transistors Q_6 and Q_7 biased by current I_2 and diodes Q_{11} and Q_{12}. (The diodes will actually be implemented with BJTs, in this case with emitter areas five times those of Q_6 and Q_7.) The circuit in Fig. 14.44 represents a complete implementation of the prototype in Fig. 14.39 with the emitter follower replaced with a Class AB output stage.

Based on our understanding of multistage amplifiers, the gain of this circuit is approximately $A_{dm} = A_{vt1}A_{vt2}A_{vt3}$ and

$$A_{dm} \cong [g_{m2}r_{\pi5}][g_{m5}(r_{o5} \| r_{o8} \| (\beta_{o6} + 1)R_L)][1] \cong \frac{g_{m2}}{g_{m5}}(g_{m5}r_{\pi5})\left(g_{m5}\frac{r_{o5}}{2}\right) = \frac{I_{C2}}{I_{C5}}\beta_{o5}\frac{\mu_{f5}}{2} \quad (14.102)$$

in which it has been assumed that the input resistance of the class-AB output stage is much larger than the parallel combination of r_{o5} and r_{o8}. Note that the upper limit to Eq. (14.102) is set by the $\beta_o V_A$ product of Q_5.

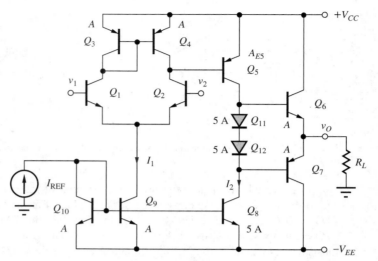

Figure 14.44 Complete bipolar operational amplifier.

> **EXERCISE:** Estimate the voltage gain of the amplifier in Fig. 14.44 using Eq. (14.102) if I_{REF} = 100 μA, V_{A5} = 60 V, β_{o1} = 150, β_{o5} = 50, R_L = 2 MΩ, and $V_{CC} = V_{EE}$ = 15 V. What is the gain of the first stage? The second stage? What should be the emitter area of Q_5? What is R_{ID}? Which terminal is the inverting input?
>
> **ANSWERS:** 6000; 5; 1200; 10 A; 150 kΩ; v_1.
>
>
>
> **EXERCISE:** Simulate the amplifier in the previous exercise using SPICE and determine the offset voltage, voltage gain, differential-mode input resistance, CMRR, and common-mode input resistance.
>
> **ANSWERS:** 3.28 mV; 8440; 165 kΩ; 84.7 dB; 59.1 MΩ.

14.8.4 INPUT STAGE BREAKDOWN

Although the bipolar amplifier designs discussed thus far have provided excellent voltage gain, input resistance, and output resistance, the amplifiers all have a significant flaw. The input stage does not offer **overvoltage protection** and can easily be destroyed by the large input voltage differences that can occur, not only under fault conditions but also during unavoidable transients during normal use of the amplifier. For example, the voltage across the input of an op amp can temporarily be equal to the total supply voltage span during slew-rate limited overload recovery.

Consider the worst-case fault condition applied to the differential pair in Fig. 14.45 where one input is connected to the positive power supply voltage and the other is connected to the negative

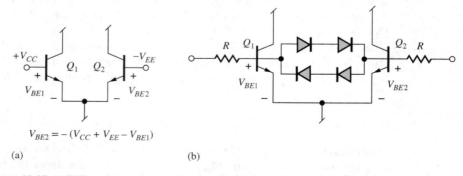

Figure 14.45 (a) Differential input stage voltages under a fault condition. (b) Simple diode input protection circuit.

supply. Under the conditions shown, the base-emitter junction of Q_1 will be forward-biased, and that of Q_2 reverse-biased by a voltage of $(V_{CC} + V_{EE} - V_{BE1})$. If $V_{CC} = V_{EE} = 22$ V, the reverse voltage exceeds 41 V. Because of heavy doping in the emitter, the typical Zener breakdown voltage of the base-emitter junction of an *npn* transistor is only 5 to 7 V. Thus, any voltage exceeding this value by more than one diode drop may destroy at least one of the transistors in the differential input pair.

Early IC op amps required circuit designers to add external diode protection across the input terminals, as shown Fig. 14.45(b). The diodes prevent the differential input voltage from exceeding approximately 1.4 V, but this technique adds extra components and cost to the design as well as extra capacitance at the inputs. The two resistors limit the current through the diodes. The µA741 described in the next section was the first commercial IC op amp to solve this problem by providing fully protected input, as well as output, stages.

14.9 THE µA741 OPERATIONAL AMPLIFIER

The classic Fairchild **µA741 operational-amplifier** design was the first to provide a highly robust amplifier from the application engineer's point of view. The amplifier provides excellent overall characteristics (high gain, input resistance and CMRR, low output resistance, and good frequency response) while providing overvoltage protection for the input stage and short-circuit current limiting of the output stage. The 741 style of amplifier design quickly became the industry standard and spawned many related designs. By studying the 741 design, we will find a number of new amplifier circuit design and bias techniques.

14.9.1 OVERALL CIRCUIT OPERATION

Figure 14.46 is a simplified schematic of the µA741 operational amplifier. The three bias sources shown in symbolic form are discussed in more detail following a description of the overall circuit. The op amp has two stages of voltage gain followed by a class-AB output stage. In the first stage, transistors Q_1 to Q_4 form a differential amplifier with a buffered current mirror active load, Q_5 to Q_7.

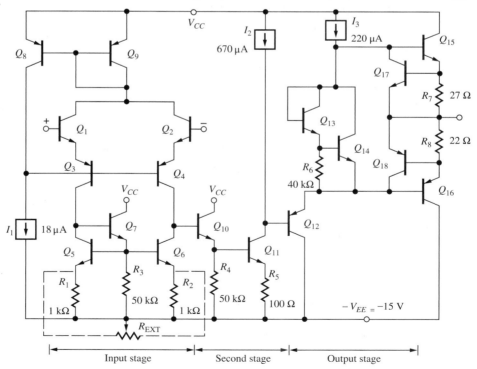

Figure 14.46 Overall schematic of the classic Fairchild µA741 operational amplifier (the bias network appears in Fig. 14.47).

The second stage consists of emitter follower Q_{10} driving common-emitter amplifier Q_{11} with current source I_2 and emitter-follower Q_{12} as load. Transistors Q_{13} to Q_{18} form a short-circuit protected class-AB push-pull output stage that is buffered from the second gain stage by Q_{12}. Practical operational amplifiers offer an offset voltage adjustment port, which is provided in the 741 through the addition of 1-kΩ resistors R_1 and R_2 and an external potentiometer R_{EXT}.

> **EXERCISE:** Reread this section and be sure you understand the function of each individual transistor in Fig. 14.46. Make a table listing the function of each transistor.

14.9.2 BIAS CIRCUITRY

The three current sources shown symbolically in Fig. 14.46 are generated by the bias circuitry in Fig. 14.47. The value of the current in the two diode-connected reference transistors Q_{20} and Q_{22} is determined by the power supply voltage and resistor R_5:

$$I_{\text{REF}} = \frac{V_{CC} + V_{EE} - 2V_{BE}}{R_5} = \frac{15 + 15 - 1.4}{39 \text{ k}\Omega} = 0.733 \text{ mA} \quad (14.103)$$

assuming ± 15-V supplies. Current I_1 is derived from the Widlar source formed of Q_{20} and Q_{21}. The output current for this design is

$$I_1 = \frac{V_T}{5000} \ln\left[\frac{I_{\text{REF}}}{I_1}\right] \quad (14.104)$$

Using the reference current calculated in Eq. (14.104) and iteratively solving for I_1 in Eq. (14.105) yields $I_1 = 18.4$ μA.

The currents in mirror transistors Q_{23} and Q_{24} are related to the reference current I_{REF} by their emitter areas using Eq. (14.13). Assuming $V_O = 0$ and $V_{CC} = 15$ V, and neglecting the voltage drop across R_7 and R_8 in Fig. 14.46, $V_{EC23} = 15 + 1.4 = 16.4$ V and $V_{EC24} = 15 - 0.7 = 14.3$ V. Using these values with $\beta_F = 50$ and $V_A = 60$ V, the two source currents are

$$I_2 = 0.75(733 \text{ μA}) \frac{1 + \dfrac{16.4 \text{ V}}{60 \text{ V}}}{1 + \dfrac{0.7 \text{ V}}{60 \text{ V}} + \dfrac{2}{50}} = 666 \text{ μA}$$

$$I_3 = 0.25(733 \text{ μA}) \frac{1 + \dfrac{14.4 \text{ V}}{60 \text{ V}}}{1 + \dfrac{0.7 \text{ V}}{60 \text{ V}} + \dfrac{2}{50}} = 216 \text{ μA}$$

$$(14.105)$$

and the two output resistances are

$$R_2 = \frac{V_{A23} + V_{EC23}}{I_2} = \frac{60 \text{ V} + 16.4 \text{ V}}{0.666 \text{ mA}} = 115 \text{ k}\Omega$$

$$R_3 = \frac{V_{A24} + V_{EC24}}{I_3} = \frac{60 \text{ V} + 14.3 \text{ V}}{0.216 \text{ mA}} = 344 \text{ k}\Omega$$

$$(14.106)$$

> **EXERCISE:** What are the values of I_{REF}, I_1, I_2, and I_3 in the circuit in Fig. 14.47 for $V_{CC} = V_{EE} = 22$ V?
>
> **ANSWERS:** 1.09 mA, 20.0 μA, 1.08 mA, 351 μA.

> **EXERCISE:** What is the output resistance of the Widlar source in Fig. 14.47 operating at 18.4 μA for $V_A = 60$ V and $V_{EE} = 15$ V?
>
> **ANSWER:** 18.8 MΩ.

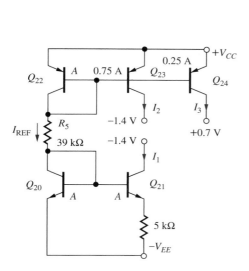

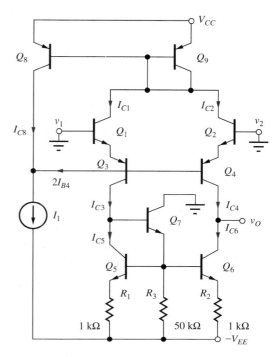

Figure 14.47 741 bias circuitry with voltages corresponding to $V_O = 0$ V.

Figure 14.48 μA741 input stage.

14.9.3 dc ANALYSIS OF THE 741 INPUT STAGE

The input stage of the μA741 amplifier is redrawn in the schematic in Fig. 14.48. As noted earlier, Q_1, Q_2, Q_3, and Q_4 form a differential input stage with an active load consisting of the buffered current mirror formed by Q_5, Q_6, and Q_7. In this input stage, there are four base-emitter junctions between inputs v_1 and v_2, two from the *npn* transistors and, more importantly, two from the *pnp* transistors. Therefore, $(v_1 - v_2) = (V_{BE1} + V_{EB3} - V_{EB4} - V_{BE2})$.

In standard bipolar IC processes, *pnp* transistors are formed from lateral structures in which both junctions exhibit breakdown voltages equal to that of the collector-base junction of the *npn* transistor. This breakdown voltage typically exceeds 50 V. Because most general-purpose op-amp specifications limit the power supply voltages to less than ±22 V, the emitter-base junctions of Q_3 and Q_4 provide sufficient breakdown voltage to fully protect the input stage of the amplifier, even under a worst-case fault condition, such as that depicted in Fig. 14.45(a).

Q-Point Analysis

In the 741 input stage in Fig. 14.48, the current mirror formed by transistors Q_8 and Q_9 operates with transistors Q_1 to Q_4 to establish the bias currents for the input stage. Bias current I_1 represents the output of the Widlar source discussed previously (18 μA) and must be equal to the collector current of Q_8 plus the base currents of matched transistors Q_3 and Q_4:

$$I_1 = I_{C8} + I_{B3} + I_{B4} = I_{C8} + 2I_{B4} \qquad (14.107)$$

For high current gain, the base currents are small and $I_{C8} \cong I_1$.

The collector current of Q_8 mirrors the collector currents of Q_1 and Q_2, which are summed together in mirror reference transistor Q_9. Assuming high current gain and ignoring the collector-voltage mismatch between Q_7 and Q_8,

$$I_{C8} = I_{C1} + I_{C2} = 2I_{C2} \qquad (14.108)$$

Combining Eqs. (14.108) and (14.109) yields the ideal bias relationships for the input stage

$$I_{C1} = I_{C2} \cong \frac{I_1}{2} \quad \text{and} \quad I_{C3} = I_{C4} \cong \frac{I_1}{2} \tag{14.109}$$

because the emitter currents of Q_1 and Q_3 and Q_2 and Q_4 must be equal. The collector current of Q_3 establishes a current equal to $I_1/2$ in current mirror transistors Q_5 and Q_6. Thus, transistors Q_1 to Q_6 all operate at a nominal collector current equal to one-half the value of source I_1.

Now that we understand the basic ideas behind the input stage bias circuit, let us perform a more exact analysis. Expanding Eq. (14.108) using the current mirror expression from Eq. (14.13),

$$I_1 = 2I_{C2} \frac{1 + \dfrac{V_{EC8}}{V_{A8}}}{1 + \dfrac{2}{\beta_{FO8}} + \dfrac{V_{EB8}}{V_{A8}}} + 2I_{B4} \tag{14.110}$$

I_{C2} is related to I_{B4} through the current gains of Q_2 and Q_4:

$$I_{C2} = \alpha_{F2} I_{E2} = \alpha_{F2}(\beta_{FO4} + 1)I_{B4} = \frac{\beta_{FO2}}{\beta_{FO2} + 1}(\beta_{FO4} + 1)I_{B4} \tag{14.111}$$

Combining Eqs. (14.111) and (14.112), and solving for I_{C2} yields

$$I_{C1} = \frac{I_1}{2} \times \left[\frac{1 + \dfrac{V_{EC8}}{V_{A8}}}{1 + \dfrac{2}{\beta_{FO8}} + \dfrac{V_{EB8}}{V_{A8}}} + \frac{1}{\dfrac{\beta_{FO2}}{\beta_{FO2} + 1}(\beta_{FO4} + 1)} \right]^{-1} \tag{14.112}$$

which is equal to the ideal value of $I_1/2$ but reduced by the nonideal current mirror effects from finite current gain and Early voltage.

The emitter current of Q_4 must equal the emitter current of Q_2, and so the collector current of Q_4 is

$$I_{C4} = \alpha_{F4} I_{E4} = \alpha_{F4} \frac{I_{C2}}{\alpha_{F2}} = \frac{\beta_{FO4}}{\beta_{FO4} + 1} \frac{\beta_{FO2} + 1}{\beta_{FO2}} I_{C2} \tag{14.113}$$

The use of buffer transistor Q_7 essentially eliminates the current gain defect in the current mirror. Note from the full amplifier circuit in Fig. 14.46 that the base current of transistor Q_{10}, with its 50-kΩ emitter resistor R_4, is designed to be approximately equal to the base current of Q_7, and $V_{CE6} \cong V_{CE5}$ as well. Thus, the current mirror ratio is quite accurate and $I_{C5} = I_{C6} = I_{C3} \cong I_1/2$.

If 50-kΩ resistor R_3 were omitted, then the emitter current of Q_7 would be equal only to the sum of the base currents of transistors Q_5 and Q_6 and would be quite small. Because of the Q-point dependence of β_F, the current gain of Q_7 would be poor. R_3 increases the operating current of Q_7 to improve its current gain, as well as to improve the dc balance and transient response of the amplifier. The value of R_3 is chosen to approximately match I_{B7} to I_{B10}.

To complete the Q-point analysis, the various collector-emitter voltages must be determined. The collectors of Q_1 and Q_2 are one V_{EB} below the positive power supply, whereas the emitters are one V_{BE} below ground potential. Hence,

$$V_{CE1} = V_{CE2} = V_{CC} - V_{EB9} + V_{BE2} \cong V_{CC} \tag{14.114}$$

The collector and emitter of Q_3 are approximately $2V_{BE}$ above the negative power supply voltage and one V_{BE} below ground, respectively:

$$V_{EC3} = V_{E3} - V_{C3} = -0.7 \text{ V} - (-V_{EE} + 1.4 \text{ V}) = V_{EE} - 2.1 \text{ V} \tag{14.115}$$

14.9 The µA741 Operational Amplifier   **1047**

The buffered current mirror effectively minimizes the error due to the finite current gain of the transistors, and $V_{CE6} = V_{CE5} \cong 2V_{BE} = 1.4$ V, neglecting the small voltage drop (<10 mV) across R_1 and R_2. Finally, the collector of Q_8 is $2V_{BE}$ below zero, and the emitter of Q_7 is one V_{BE} above $-V_{EE}$:

$$V_{EC8} = V_{CC} + 1.4 \text{ V} \qquad V_{CE7} = V_{EE} - 0.7 \text{ V} \tag{14.116}$$

EXAMPLE 14.8   **µA741 INPUT STAGE BIAS CURRENTS**

Find the currents in the 741 input stage.

PROBLEM Calculate the bias currents in the 741 input stage if $I_1 = 18$ µA, $\beta_{FOnpn} = 150$, $V_{Anpn} = 75$ V, $\beta_{FOpnp} = 60$, $V_{Apnp} = 60$ V, and $V_{CC} = V_{EE} = 15$ V.

SOLUTION **Known Information and Given Data:** µA741 input stage depicted in Fig. 14.48. $I_1 = 18$ µA, $\beta_{FOnpn} = 150$, $V_{Anpn} = 75$ V, $\beta_{FOpnp} = 60$, $V_{Apnp} = 60$ V, and $V_{CC} = V_{EE} = 15$ V.

Unknowns: I_{C1}, I_{C2}, I_{C3}, I_{C4}, I_{C5}, and I_{C6}

Approach: Use given data to evaluate Eqs. (14.112) through (14.116).

Assumptions: Transistors are in the active region; use default values of I_S.

Analysis: From Fig. 14.48, we find that the emitter-collector voltage of Q_8 is equal to $V_{CC} + V_{BE1} + V_{EB3} \cong 16.4$ V. Substituting the known values into Eq. (14.113) gives

$$I_{C1} = I_{C2} = \frac{18 \text{ µA}}{2} \cdot \frac{1}{\dfrac{1 + \dfrac{16.4 \text{ V}}{60 \text{ V}}}{1 + \dfrac{2}{50} + \dfrac{0.7 \text{ V}}{60 \text{ V}}} + \dfrac{1}{\dfrac{150}{150+1}(60+1)}} = 7.32 \text{ µA}$$

Equation (14.113) yields

$$I_{C3} = I_{C4} = \alpha_{F4}\frac{I_{C2}}{\alpha_{F2}} = \frac{\beta_{FO4}}{\beta_{FO4}+1}\left(\frac{\beta_{FO2}+1}{\beta_{FO2}}\right)I_{C2} = \frac{60}{61}\left(\frac{151}{150}\right)I_{C2} = 7.25 \text{ µA}$$

$$I_{C5} \cong I_{C3} = 7.25 \text{ µA} \qquad \text{and} \qquad I_{C6} = I_{C4} = 7.25 \text{ µA}$$

Check of Results: The basic objective of the bias circuit would be to set all currents to 18 µA/2 or 9 µA. Our calculations are close to this value and appear correct.

Discussion: The actual bias currents are slightly greater than 7 µA, whereas the ideal value would be 9 µA. The dominant source of error arises from the collector-emitter voltage mismatch of the *pnp* current mirror.

Computer-Aided Analysis: We draw the circuit using the schematic editor and set the BJT parameters. For the *npn* devices, BF = 150 and VAF = 75 V. For the *pnp* transistors, BF = 60 and VAF = 60 V. Source V_O is added to balance the circuit by forcing the output voltage to the same voltage as that which will appear at the collector of Q_5. Otherwise, the voltage at the collectors of Q_4 and Q_6 will float to a value determined by the difference in overall output resistances of transistors Q_4 and Q_6. When balance is achieved, the current in source V_O will be nearly zero. Table 14.3 summarizes the Q-points based on these calculations and Eqs. (14.109) to (14.116) and compares them with the SPICE operating point simulation results.

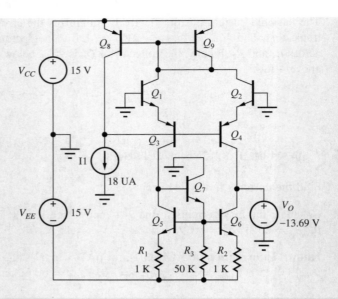

TABLE 14.3
Q-points of 741 Input Stage Transistors for $I_1 = 18$ μA and $V_{CC} = V_{EE} = 15$ V

TRANSISTORS	Q-POINT	SPICE RESULTS
Q_1 and Q_2	7.32 μA, 15 V	7.30 μA, 15.0 V
Q_3 and Q_4	7.25 μA, 12.9 V	7.24 μA, 13.0 V
Q_5 and Q_6	7.25 μA, 1.4 V	7.16 μA, 1.30 V
Q_7	12.2 μA, 14.3 V	13.1 μA, 14.3 V
Q_8	17.7 μA, 16.4 V	17.8 μA, 16.3 V
Q_9	14.0 μA, 0.7 V	14.1 μA, 0.66 V

EXERCISE: Remove V_O and simulate the 741 input stage amplifier. What are the new collector currents? What are the voltages at the collectors or Q_5 and Q_6?

ANSWERS: 7.31 μA, 7.28 μA, 7.25 μA, 7.22 μA, 7.18 μA, 7.22 μA, 13.1 μA, 17.8 μA, 14.1 μA; −13.7 V, −13.1 V

EXERCISE: Suppose buffer transistor Q_7 and resistor R_3 are eliminated from the amplifier in Fig. 14.48 and Q_5 and Q_6 were connected as a standard current mirror. What would be the collector-emitter voltage of Q_6 if $V_{BE6} = 0.7$ V, $\beta_{FO6} = 100$, and $V_{A6} = 60$ V?

ANSWER: 1.90 V

14.9.4 ac ANALYSIS OF THE 741 INPUT STAGE

The 741 input stage is redrawn in symmetric form in Fig. 14.49, with its active load temporarily replaced by two resistors. From Fig. 14.49, we see that the collectors of Q_1 and Q_2 as well as the bases of Q_3 and Q_4, lie on the line of symmetry of the amplifier and represent virtual grounds for differential-mode input signals.

The corresponding **differential-mode half-circuit** shown in Fig. 14.50 is a common-collector stage followed by a common-base stage, a C-C/C-B cascade. The characteristics of the C-C/C-B cascade can be determined from Fig. 14.50 and our knowledge of single-stage amplifiers.

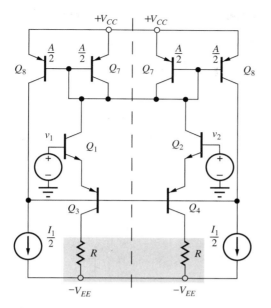

Figure 14.49 Symmetry in the 741 input stage.

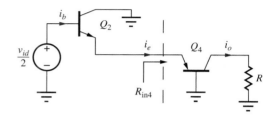

Figure 14.50 Differential-mode half-circuit for the 741 input stage.

The emitter current of Q_2 is equal to its base current i_b multiplied by $(\beta_{o2} + 1)$, and the collector current of Q_4 is α_{o4} times the emitter current. Thus, the output current can be written as

$$\mathbf{i_o} = \alpha_{o4}\mathbf{i_e} = \alpha_{o4}(\beta_{o2} + 1)\mathbf{i_b} \cong \beta_{o2}\mathbf{i_b} \qquad (14.117)$$

The base current is determined by the input resistance to Q_2:

$$\mathbf{i_b} = \frac{\dfrac{\mathbf{v_{id}}}{2}}{r_{\pi 2} + (\beta_{o2} + 1)R_{in4}} = \frac{\dfrac{\mathbf{v_{id}}}{2}}{r_{\pi 2} + (\beta_{o2} + 1)\left(\dfrac{r_{\pi 4}}{\beta_{o4} + 1}\right)} = \frac{\dfrac{\mathbf{v_{id}}}{2}}{r_{\pi 2} + r_{\pi 4}} \cong \frac{\mathbf{v_{id}}}{4r_{\pi 2}} \qquad (14.118)$$

in which $R_{in4} = r_{\pi 4}/(\beta_{o4} + 1)$ represents the input resistance of the common-base stage. Combining Eqs. (14.117) and (14.118) yields

$$\mathbf{i_o} \cong \beta_{o2}\frac{\mathbf{v_{id}}}{4r_{\pi 2}} = \frac{g_{m2}}{4}\mathbf{v_{id}} \qquad (14.119)$$

Each side of the C-C/C-B input stage has a transconductance equal to one-half of the transconductance of the standard differential pair. From Eq. (14.118) we can also see that the differential-mode input resistance is twice the value of the corresponding C-E stage:

$$R_{id} = \frac{\mathbf{v_{id}}}{\mathbf{i_b}} = 4r_{\pi 2} \qquad (14.120)$$

From Fig. 14.51, we can see that the output resistance is equivalent to that of a common-base stage with a resistor of value $1/g_{m2}$ in its emitter:

$$R_{out4} \cong r_{o4}(1 + g_{m4}R) = r_{o4}\left(1 + g_{m4}\frac{1}{g_{m2}}\right) = 2r_{o4} \qquad (14.121)$$

14.9.5 VOLTAGE GAIN OF THE COMPLETE AMPLIFIER

We now use the results from the previous section to analyze the overall ac performance of the op amp. We find a Norton equivalent circuit for the input stage and then couple it with a two-port model for the second stage.

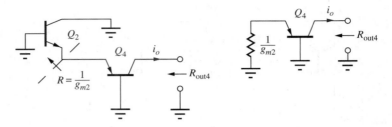

Figure 14.51 Output resistance of C-C/C-B cascade.

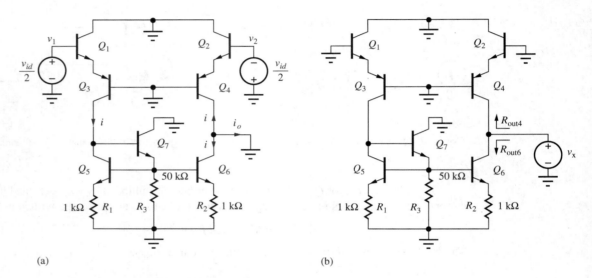

Figure 14.52 Circuits for finding the Norton equivalent of the input stage.

Norton Equivalent of the Input Stage

Figure 14.52 is the simplified differential-mode ac equivalent circuit for the input stage. We use Fig. 14.52(a) to find the short-circuit output current of the first stage. Based on our analysis of Fig. 14.50, the differential-mode input signal establishes equal and opposite currents in the two sides of the differential amplifier where $i = (g_{m2}/4)v_{id}$. Current i, exiting the collector of Q_3, is mirrored by the buffered current mirror so that a total signal current equal to $2i$ flows in the output terminal:

$$\mathbf{i_o} = -2\mathbf{i} = -\frac{g_{m2}\mathbf{v_{id}}}{2} = (-20I_{C2})\mathbf{v_{id}} = (-1.46 \times 10^{-4}\text{ S})\mathbf{v_{id}} \tag{14.122}$$

The Thévenin equivalent resistance at the output is found using the circuit in Fig. 14.52(b) and is equal to

$$R_{\text{th}} = R_{\text{out6}} \| R_{\text{out4}} \tag{14.123}$$

Because only a small dc voltage is developed across R_2, the output resistance of Q_6 can be calculated from

$$R_{\text{out6}} \cong r_{o6}[1 + g_{m6}R_2] \cong r_{o6}\left[1 + \frac{I_{C6}R_2}{V_T}\right] = r_{o6}\left[1 + \frac{0.0073\text{ V}}{0.025\text{ V}}\right] = 1.3r_{o6} \tag{14.124}$$

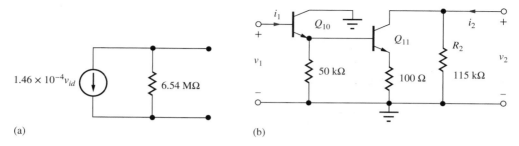

Figure 14.53 (a) Norton equivalent of the 741 input stage. (b) Two-port representation for the second stage.

The output resistance of Q_4 was already found in Eq. (14.121) to be $2r_{o4}$. Using these results,

$$R_{\text{th}} = 2r_{o4} \| 1.3r_{o6} = 0.79r_{o4} \cong 0.79 \frac{60 \text{ V}}{7.25 \times 10^{-6} \text{ A}} = 6.54 \text{ M}\Omega \qquad (14.125)$$

in which $r_{o4} = r_{o2}$ has been assumed for simplicity with $V_A + V_{CE} = 60$ V.

The resulting Norton equivalent circuit for the input stage appears in Fig. 14.53(a). Based on the values in this figure, the open-circuit voltage gain of the first stage is -955. SPICE simulations yield values very similar to those in Fig. 14.53(a): $(1.40 \times 10^{-4} \text{ S})v_{id}$, 6.95 M$\Omega$, and $A_{dm} = -973$.

> **EXERCISE:** Improve the estimate of R_{th} using the actual values of V_{CE6} and V_{CE4} if $V_{CC} = V_{EE} = 15$ V and $V_A = 60$ V. What are the values of R_{out4} and R_{out6}?
>
> **ANSWERS:** 7.12 MΩ; 20.2 MΩ, 11.0 MΩ

Model for the Second Stage

Figure 14.53(b) is a two-port representation for the second stage of the amplifier. Q_{10} is an emitter follower that provides high input resistance and drives a common-emitter amplifier consisting of Q_{11} and its current source load represented by output resistance R_2. A y-parameter model is constructed for this network.

From Fig. 14.46 and the bias current analysis, we can see that the collector current of Q_{11} is approximately equal to I_2 or 666 µA. Calculating the collector current of Q_{10} yields

$$I_{C10} \cong I_{E10} = \frac{I_{C11}}{\beta_{F11}} + \frac{V_{B11}}{50 \text{ k}\Omega} = \frac{666 \text{ µA}}{150} + \frac{0.7 + (0.67 \text{ mA})(0.1 \text{ k}\Omega)}{50 \text{ k}\Omega} = 19.8 \text{ µA} \qquad (14.126)$$

Using these values to find the small-signal parameters with ($\beta_{\text{on}} = 150$) gives

$$r_{\pi 10} = \frac{\beta_{o10} V_T}{I_{C10}} = \frac{3.75 \text{ V}}{19.8 \text{ µA}} = 189 \text{ k}\Omega \quad \text{and} \quad r_{\pi 11} = \frac{3.75 \text{ V}}{0.666 \text{ mA}} = 5.63 \text{ k}\Omega \qquad (14.127)$$

Parameters y_{11} and y_{21} are calculated by applying a voltage v_1 to the input port and setting $v_2 = 0$, as in Fig. 14.54. The input resistance to Q_{11} is that of a common-emitter stage with a 100-Ω emitter resistor:

$$R_{\text{in}11} = y_{11} = r_{\pi 11} + (\beta_{o11} + 1)100 \cong 5630 + (151)100 = 20.7 \text{ k}\Omega \qquad (14.128)$$

This value is used to simplify the circuit, as in Fig. 14.54(b), and the input resistance to Q_{10} is

$$R_{\text{in}10} = r_{\pi 10} + (\beta_{o10} + 1)(50 \text{ k}\Omega \| R_{\text{in}11}) = 189 \text{ k}\Omega + (151)(50 \text{ k}\Omega \| 20.7 \text{ k}\Omega) = 2.40 \text{ M}\Omega \qquad (14.129)$$

The gain of emitter follower Q_{10} is

$$\mathbf{v_e} = \mathbf{v_1} \frac{(\beta_{o10} + 1)(50 \text{ k}\Omega \| R_{\text{in}11})}{r_{\pi 10} + (\beta_{o10} + 1)(50 \text{ k}\Omega \| R_{\text{in}11})} = \frac{(151)(50 \text{ k}\Omega \| 20.7 \text{ k}\Omega)}{189 \text{ k}\Omega + (151)(50 \text{ k}\Omega \| 20.7 \text{ k}\Omega)} = 0.921 \mathbf{v_1} \qquad (14.130)$$

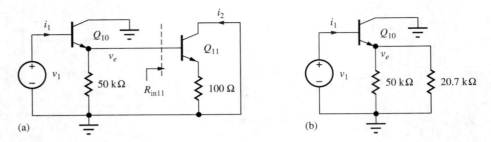

Figure 14.54 Network for finding y_{11} and y_{21}.

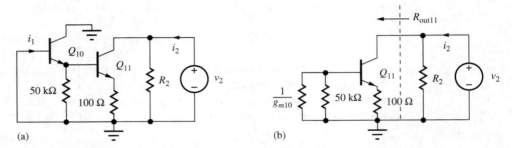

Figure 14.55 Network for finding y_{12} and y_{22}.

The output current i_2 in Fig. 14.53(a) is given by

$$i_2 = \frac{v_e}{\frac{1}{g_{m11}} + 100\ \Omega} = \frac{0.921 v_1}{\frac{1}{\frac{40}{V}(0.666\ \text{mA})} + 100\ \Omega} = 0.00670 v_1 \tag{14.131}$$

yielding a forward transconductance of

$$G_m = y_{21} = 6.70\ \text{mS} \tag{14.132}$$

Parameters y_{12} and y_{22} can be found from the network in Fig. 14.55. We assume that the reverse transconductance y_{12} is negligible and reserve its calculation for Prob. 14.146. The output conductance can be determined from Fig. 14.55(b).

$$G_o = y_{22} = i_2/v_2 = [R_2 \parallel R_{out11}]^{-1} \tag{14.133}$$

where $R_2 = 115\ \text{k}\Omega$ was calculated during the analysis of the bias circuit.

Because the voltage drop across the 100-Ω resistor is small, the output resistance of Q_{11} is approximately

$$R_{out11} = r_{o11}[1 + g_{m11} R_E] = \frac{V_{A11} + V_{CE11}}{I_{C11}} \left[1 + \frac{I_{C11} R_E}{V_T}\right]$$

$$= \frac{60\ \text{V} + 13.6\ \text{V}}{0.666\ \text{mA}} \left[1 + \frac{0.067\ \text{V}}{0.025\ \text{V}}\right] = 407\ \text{k}\Omega \tag{14.134}$$

and

$$R_o = 115\ \text{k}\Omega \parallel 407\ \text{k}\Omega = 89.1\ \text{k}\Omega \tag{14.135}$$

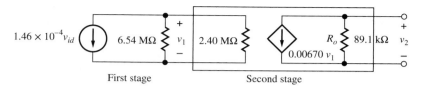

Figure 14.56 Combined model for first and second stages.

Figure 14.56 depicts the completed two-port model for the second stage, driven by the Norton equivalent of the input stage. Using this model, the open-circuit voltage gain for the first two stages of the amplifier is

$$\mathbf{v}_2 = -0.00670(89.1 \text{ k}\Omega)\mathbf{v}_1 = -597\mathbf{v}_1$$
$$\mathbf{v}_1 = -1.46 \times 10^{-4}(6.54 \text{ M}\Omega \| 2.40 \text{ M}\Omega)\mathbf{v}_{id} = -256\mathbf{v}_{id} \tag{14.136}$$
$$\mathbf{v}_2 = -597(-256\mathbf{v}_{id}) = 153{,}000\mathbf{v}_{id}$$

Note from Eq. (14.136) that the 2.42-MΩ input resistance of Q_{10} reduces the voltage gain of the first stage by a factor of almost 4.

> **EXERCISE:** What is the voltage gain of the input stage if transistor Q_{10} and its 50-kΩ emitter resistor are omitted so that the output of the first stage is connected directly to the base of Q_{11}? Use the small-signal element values already calculated.
>
> **ANSWER:** −3.00

14.9.6 THE 741 OUTPUT STAGE

Figure 14.57 shows simplified models for the 741 output stage. Transistor Q_{12} is the emitter follower that buffers the high impedance node at the output of the second stage and drives the push-pull output stage composed of transistors Q_{15} and Q_{16}. Class-AB bias is provided by the sum of the

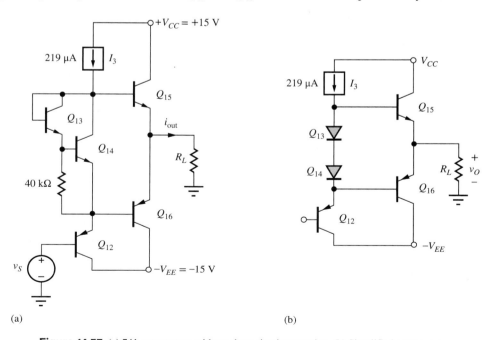

Figure 14.57 (a) 741 output stage without short-circuit protection. (b) Simplified output stage.

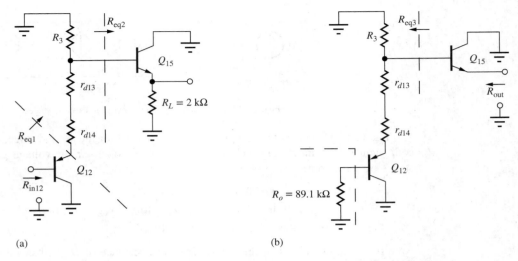

Figure 14.58 Circuits for determining input and output resistance of the output stage.

base-emitter voltages of Q_{13} and Q_{14}, represented as diodes in Fig. 14.57(b). The 40-kΩ resistor is used to increase the value of I_{C13}. Without this resistor, I_{C13} would only be equal to the base current of Q_{14}. The short-circuit protection circuitry in Fig. 14.46 is not shown in Fig. 14.57 in order to simplify the diagram.

The input and output resistances of the class-AB output stage are actually complicated functions of the signal voltage because the operating current in Q_{15} and Q_{16} changes greatly as the output voltage changes. However, because only one transistor conducts strongly at any given time in the class-AB stage, separate circuit models can be used for positive and negative output signals. The model for positive signal voltages is shown in Fig. 14.58. (The model for negative signal swings is similar except *npn* transistor Q_{15} is replaced by *pnp* transistor Q_{16} connected to the emitter of Q_{12}.)

Let us first determine the input resistance of transistor Q_{12}. If R_{in12} is much larger than the 89-kΩ output resistance of the two-port in Fig. 14.58, then it does not significantly affect the overall voltage gain of the amplifier. Using single-stage amplifier theory,

$$R_{in12} = r_{\pi 12} + (\beta_{o12} + 1)R_{eq1} \qquad (14.137)$$

where

$$R_{eq1} = r_{d14} + r_{d13} + R_3 \| R_{eq2} \quad \text{and} \quad R_{eq2} = r_{\pi 15} + (\beta_{o15} + 1)R_L \cong (\beta_{o15} + 1)R_L \qquad (14.138)$$

The value of R_3 (344 kΩ) was calculated in the bias circuit section. For $I_{C12} = 216$ µA, and assuming a representative collector current in Q_{15} of 2 mA,

$$R_{eq2} = r_{\pi 15} + (\beta_{o15} + 1)R_L = \frac{3.75 \text{ V}}{2 \text{ mA}} + (151)2 \text{ k}\Omega = 304 \text{ k}\Omega \qquad (14.139)$$

Note that the value of R_{eq2} is dominated by the reflected load resistance $\beta_{o15}R_L$. Resistor $r_{\pi 15}$ represents a small part of R_{eq2}, so knowing the exact value of I_{C15} is not critical.

$$R_{eq1} = r_{d14} + r_{d13} + R_3 \| R_{eq2} = 2\frac{0.025 \text{ V}}{0.216 \text{ mA}} + 344 \text{ k}\Omega \| 304 \text{ k}\Omega = 162 \text{ k}\Omega \qquad (14.140)$$

and

$$R_{in12} = r_{\pi 12} + (\beta_{o12} + 1)R_{eq1} = \frac{0.025 \text{ V}}{0.216 \text{ mA}} + (51)162 \text{ k}\Omega = 8.27 \text{ M}\Omega \qquad (14.141)$$

Because R_{in12} is approximately 100 times the output resistance R_o of the second stage, R_{in12} has little effect on the gain of the second stage. Although the value of R_{in12} changes for different values of load resistance, the overall op-amp gain is not affected because the value of R_{in12} is so much larger than the value of R_o in Fig. 14.56.

Similar results are obtained for negative signal voltages. The values are slightly different because the current gain of the *pnp* transistor Q_{16} differs from that of the *npn* transistor Q_{15}.

14.9.7 OUTPUT RESISTANCE

The output resistance of the amplifier for positive output voltages is determined by transistor Q_{15}

$$R_o = \frac{r_{\pi 15} + R_{eq3}}{\beta_{o15} + 1} \tag{14.142}$$

in which

$$R_{eq3} = R_3 \left\| \left[r_{d13} + r_{d14} + \frac{r_{\pi 12} + R_o}{\beta_{o12} + 1} \right] \right.$$

$$= 304 \text{ k}\Omega \left\| \left[2\frac{0.025 \text{ V}}{0.219 \text{ mA}} + \frac{5.17 \text{ k}\Omega + 89.1 \text{ k}\Omega}{51} \right] = 2.08 \text{ k}\Omega \tag{14.143}$$

From Fig. 14.49, we can see that the 27-Ω resistor R_7, which determines the short-circuit current limit, adds directly to the overall output resistance of the amplifier so that actual op-amp output resistance is

$$R_{out} = R_o + R_7 = \frac{1.88 \text{ k}\Omega + 2.08 \text{ k}\Omega}{151} + 27 \text{ }\Omega = 53 \text{ }\Omega \tag{14.144}$$

> **EXERCISE:** Repeat the calculation of R_{in12} and R_{out} if *pnp* transistor Q_{15} has a current gain of 50, $I_{C16} = 2$ mA, and $I_{C15} = 0$. Be sure to draw the new equivalent circuit of the output stage for negative output voltages.
>
> **ANSWERS:** 4.06 MΩ ($\gg$89.1 kΩ), 51 Ω + 27 Ω = 78 Ω

14.9.8 SHORT-CIRCUIT PROTECTION

For simplicity, the output short-circuit protection circuitry was not shown in Fig. 14.57. Referring back to the complete op-amp schematic in Fig. 14.46, we see that **short-circuit protection** is provided by resistors R_7 and R_8 and transistors Q_{17} and Q_{18}. The circuit is identical to the one presented in Fig. 13.41(a). Transistors Q_{17} and Q_{18} are normally off, but if the current in resistor R_7 becomes too high, then transistor Q_{17} turns on and steals the base current from Q_{15}. Likewise, if the current in resistor R_8 becomes too large, then transistor Q_{18} turns on and removes the base current from Q_{16}. The positive and negative short-circuit current levels will be limited to approximately V_{BE17}/R_7 and $-V_{EB18}/R_8$, respectively. As already mentioned, resistors R_7 and R_8 increase the output resistance of the amplifier since they appear directly in series with the output terminal.

> **EXERCISE:** Estimate the positive and negative short-circuit output current in the 741 op amp in Fig. 14.46.
>
> **ANSWERS:** 26 mA; −32 mA

14.9.9 SUMMARY OF THE μA741 OPERATIONAL AMPLIFIER CHARACTERISTICS

Table 14.4 is a summary of the characteristics of the μA741 operational amplifier. Column 2 gives our calculated values; column 3 presents values typically specified for the actual commercial product.

TABLE 14.4
µA741 Characteristics

	CALCULATION	TYPICAL VALUES
Voltage gain	153,000	200,000
Input resistance	2.05 MΩ	2 MΩ
Output resistance	53 Ω	75 Ω
Input bias current	49 nA	80 nA
Input offset voltage	—	2 mV

The observed values depend on the exact values of current gain and Early voltage for the *npn* and *pnp* transistors and vary from process run to process run.

14.10 THE GILBERT ANALOG MULTIPLIER

In Chapters 10, and 12, we saw how operational amplifiers could be used to perform scaling, addition, subtraction, integration, and differentiation of electronic signals. However, one of the more difficult operations to realize is accurate multiplication of two analog signals. Barrie Gilbert, another of the "legends" of integrated circuit design, discovered a solution to this problem using the characteristics of the bipolar transistor. The basic multiplier "core" in Fig. 14.59 consists of three differential pairs. The Q_1–Q_2 pair has significant emitter degeneration so that the transconductance of the pair is approximately[7] $1/R_1$. Under this assumption, the collector currents of the lower pair can be written as

$$i_{c1} \cong \frac{I_{BB}}{2} + \frac{v_1}{2R_1} \qquad i_{c2} \cong \frac{I_{BB}}{2} - \frac{v_1}{2R_1} \qquad \text{for} \qquad |v_1| \leq I_{BB}R_1 \qquad (14.145)$$

The bound on v_1 is determined by the requirement that neither collector current can be negative.

Multiplier output current v_o is taken from the upper two differential pairs and can be written as

$$v_o = [(i_{c3} + i_{c5}) - (i_{c4} + i_{c6})]R = [(i_{c3} - i_{c4}) + (i_{c5} - i_{c6})]R \qquad (14.146)$$

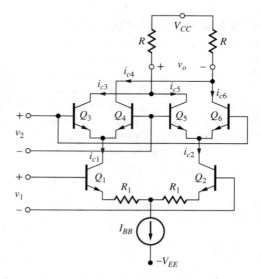

Figure 14.59 Gilbert multiplier core.

---

[7] More sophisticated voltage-to-current converters (transconductance amplifiers) can also be used.

Using Eq. (13.4), we can write expressions for the collector current differences in this equation:

$$i_{c3} - i_{c4} = i_{c1} \tanh\left(\frac{v_2}{2V_T}\right) \quad \text{and} \quad i_{c5} - i_{c6} = -i_{c2} \tanh\left(\frac{v_2}{2V_T}\right) \quad (14.147)$$

Using these equations, the output voltage can be reduced to

$$v_o = (i_{c1} - i_{c2})R \tanh\left(\frac{v_2}{2V_T}\right) = v_1\left(\frac{R}{R_1}\right) \tanh\left(\frac{v_2}{2V_T}\right) \quad (14.148)$$

At this point, one approach to multiplication is to expand the hyperbolic tangent as a series, and then keep only the first term:

$$\tanh(x) = x - \frac{x^3}{3} + \cdots \quad \text{and} \quad v_o \cong v_1\left(\frac{R}{R_1}\right)\left(\frac{v_2}{2V_T}\right) \quad \text{for} \quad \frac{x^3}{3} \ll x \quad (14.149)$$

where $x = v_2/2V_T$. However, this approach greatly restricts the input signal range of v_2 to only a few tens of mV [see discussion following Eq. (13.5)].

The key to the full range **Gilbert multiplier** is to use another pair of *pn* junctions to "predistort" the input signal as in Fig. 14.60. Diode connected transistors Q_9 and Q_{10} are driven by a second transconductance stage formed by Q_7 and Q_8 for which

$$i_{c7} \cong \frac{I_{EE}}{2} + \frac{v_3}{2R_3} \quad i_{c8} \cong \frac{I_{EE}}{2} - \frac{v_3}{2R_3} \quad \text{for} \quad |v_3| \leq I_{EE}R_3 \quad (14.150)$$

to develop voltage v_2:

$$v_2 = (V_{BB} - v_{BE10}) - (V_{BB} - v_{BE9}) = v_{BE9} - v_{BE10} \quad (14.151)$$

Using the standard expressions for the base-emitter voltages and assuming the two transistors are matched gives

$$v_2 = V_T \ln\left(\frac{\frac{I_{EE}}{2} + \frac{v_3}{2R_3}}{I_S}\right) - V_T \ln\left(\frac{\frac{I_{EE}}{2} - \frac{v_3}{2R_3}}{I_S}\right) = V_T \ln\left(\frac{1 + \frac{v_3}{I_{EE}R_3}}{1 - \frac{v_3}{I_{EE}R_3}}\right) \quad (14.152)$$

Searching our math tables, we might stumble on this identity:

$$\ln\left(\frac{1+x}{1-x}\right) = 2 \tanh^{-1}(x)$$

which can be used to rewrite the expression for v_2 as

$$v_2 = 2V_T \tanh^{-1}\left(\frac{v_3}{I_{EE}R_3}\right) \quad (14.153)$$

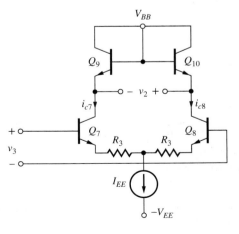

Figure 14.60 Inverse hyperbolic tangent "predistortion" circuit.

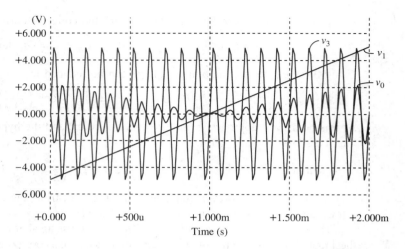

Figure 14.61 Gilbert multiplier simulation results for $v_3 = 5 \sin 20000\pi t$ and v_1 ramping between -5 and $+5$ V with a scale factor of 0.1.

Combining Eq. (14.153) with (14.148) gives the final result for the analog multiplier

$$v_o = \left(\frac{R}{I_{EE}R_1R_3}\right)v_1v_3 \qquad (14.154)$$

The circuit described by Eq. (14.154) is known as a **four-quadrant multiplier** since both input voltages are permitted to take on both positive and negative values. One common design sets the scaling constant to be 0.1 so that the input and output signals can all have a 10-V range.

An example of operation of an analog multiplier appears in Fig. 14.61. Input $v_3 = 5 \sin 20,000\pi t$ V. Signal v_1 is a ramp starting at -5 V at $t = 0$ and reaching $+5$ V at $t = 2$ ms. The product of the two waveforms appears in the figure. The product is zero and changes sign as v_1 crosses through 0 V at $t = 1$ ms.

EXERCISE: What should the scale factor be in Eq. (14.155) if all voltages are to have a 5-V range? A 1-V range?

ANSWERS: 0.2; 1.0.

EXERCISE: Simulate the full Gilbert multiplier with a 5-V, 1-kHz sine wave for v_1 with $v_3 = 5 \sin 20000\pi t$ V.

SUMMARY

Integrated circuit (IC) technology permits the realization of large numbers of virtually identical transistors. Although the absolute parameter tolerances of these devices are relatively poor, device characteristics can actually be matched to within less than 1 percent. The availability of large numbers of such closely matched devices has led to the development of special circuit techniques that depend on the similarity of device characteristics for proper operation. These matched circuit design techniques are used throughout analog circuit design and produce high-performance circuits that require very few resistors.

- One of the most important of the IC techniques is the current mirror circuit, in which the output current replicates, or mirrors, the input current. Multiple copies of the replicated current can be generated, and the gain of the current mirror can be controlled by scaling the emitter areas of bipolar transistors or the W/L ratios of FETs. Errors in the mirror ratio of current mirrors are related directly to the finite output resistance and/or current gain of the transistors through the parameters λ, V_A, and β_F.

- In bipolar current mirrors, the finite current gain of the BJT causes an error in the mirror ratio, which the buffered current mirror circuit is designed to minimize. In both FET and BJT circuits, the ideal balance of the current mirror is disturbed by the mismatch in dc voltages between the input and output sections of the mirror. The degree of mismatch is determined by the output resistance of the current sources.

- The figure of merit V_{CS} for the basic current mirror is approximately equal to V_A for the BJT or $1/\lambda$ for the MOS version. However, the value of V_{CS} can be improved by two orders of magnitude or more through the use of either the cascode, Wilson or regulated cascode current sources.

- Current mirrors can also be used to generate currents that are independent of the power supply voltages. The V_{BE}-based reference and the Widlar reference produce currents that depend only on the logarithm of the supply voltage. By combining a Widlar source with a current mirror, a reference is realized that exhibits first-order independence of the power supply voltages. The only variation is due to the finite output resistance of the current mirror and Widlar source used in the supply-independent cell.

- The Widlar cell produces a PTAT voltage (proportional to absolute temperature) which is used as the basic sensing element in most electronic thermometers and is a key part of the bandgap reference.

- The bandgap voltage circuit developed by Widlar and optimized by Brokaw uses a PTAT circuit to cancel the negative temperature coefficient of the base-emitter junction of the BJT thereby forming a highly stable voltage reference with a very low TC and supply voltage dependence. This circuit and its variations find pervasive use in both analog and digital ICs.

- An extremely important application of the current mirror is as a replacement for the load resistors in differential and operational amplifiers. This active-load circuit can substantially enhance the voltage gain capability of most amplifiers while maintaining the operating-point balance necessary for low offset voltage and good common-mode rejection. Amplifiers with active loads can achieve single-stage voltage gains that approach the amplification factor of the transistor. Analysis of the ac behavior of circuits employing current mirrors can often be simplified using a two-port model for the mirror.

- Active current mirror loads are used to enhance the performance of both bipolar and MOS operational amplifiers. The classic μA741 operational amplifier, introduced in the late 1960s, was the first highly robust design combining excellent overall amplifier performance with input-stage breakdown-voltage protection and short-circuit protection of the output stage. Active loads are used to achieve a voltage gain in excess of 100 dB in an amplifier with two stages of gain. This operational amplifier design immediately became the industry standard op amp and spawned many similar designs.

- Four-quadrant multiplication of analog signals can be accurately obtained using the Gilbert multiplier circuit.

KEY TERMS

Active load
Bandgap reference
Buffered current mirror
Cascode current source
Class-AB amplifiers
Common-mode gain
Common-mode input resistance
Common-mode output resistance
Common-mode rejection ratio (CMRR)
Current gain defect
Current mirror
Current mirror biasing
Differential-mode gain
Differential-mode half-circuit

Differential-mode input resistance
Differential-mode output resistance
Diode-connected transistor
Electronic current source
Emitter area scaling
Figure of merit (FOM)
Four-quadrant multiplier
Gilbert multiplier
Matched (devices)
Matched transistors
μA741 operational amplifier
Mirror ratio

Overvoltage protection
Power-supply-independent biasing
Proportional to absolute temperature (PTAT)
PTAT voltage
Reference current
Short-circuit protection
Startup circuit
V_{BE}-based reference
Voltage reference
Weak inversion saturation operation
Widlar current source
Wilson current source

REFERENCES

1. R. D. Thornton et. al., *Multistage Transistor Circuits*, SEEC Volume 5, Wiley, New York: 1965.
2. R. J. Widlar, "Some circuit design techniques for linear integrated circuits," *IEEE Transactions on Circuit Theory*, vol. CT-12, no. 12, pp. 586–590, December 1965.
3. R. J. Widlar, "Design techniques for monolithic operational amplifiers," *IEEE Journal of Solid-State Circuits*, vol. SC-4, no. 4, pp. 184–191, August 1969.
4. G. R. Wilson, "A monolithic junction FET-NPN operational amplifier," *IEEE Journal of Solid-State Circuits*, vol. SC-3, no. 6, pp. 341–348, December 1968.
5. A. Paul Brokaw, "A simple three-terminal IC bandgap reference," *IEEE Journal of Solid-State Circuits*, vol. SC-9, no. 6, pp. 388–393, December 1994.

ADDITIONAL READINGS

Barrie Gilbert, "The gears of genius," *IEEE Solid-State Circuits Society News*, vol. 12, no. 4, pp. 10–27, Fall 2007.

P. R. Gray, P. J. Hurst, S. H. Lewis, and R. G. Meyer, *Analysis and Design of Analog Integrated Circuits*, 5th ed., John Wiley & Sons, New York: 2009.

Robert J. Widlar, "New developments in IC voltage regulators," *IEEE Journal of Solid-State Circuits*, vol. SC-6, no. 1, pp. 2–7, January 1991.

PROBLEMS

14.1 Circuit Element Matching

14.1. An integrated circuit resistor has a nominal value of 3.95 kΩ. A given process run has produced resistors with a mean value 19 percent higher than the nominal value, and the resistors are found to be matched within 1.5 percent. What are the maximum and minimum resistor values that will occur?

14.2. (a) The emitter areas of two bipolar transistors are mismatched by 8 percent. What will be the base-emitter voltage difference between these two transistors when their collector currents are identical? (Assume $V_A = \infty$.) (b) Repeat for a 15 percent area mismatch. (c) What degree of matching is required for a base-emitter voltage difference of less than 0.5 mV?

14.3. The bipolar transistors in a differential pair are mismatched. (a) What will be the offset voltage if the current gains are mismatched by 7 percent? (b) If the saturation currents are mismatched by 7 percent? (c) If the Early voltages are mismatched by 7 percent and $V_{CE} = 0.1\ V_A$? (d) If the collector resistors are mismatched by 7 percent? (Remember, the offset voltage is the input voltage required to force the differential output voltage to be zero.)

*14.4. The collector currents of two BJTs are equal when the base-emitter voltages differ by 2 mV. (a) What is the fractional mismatch $\Delta I_S/I_S$ in the saturation current of the two transistors if $I_{S1} = I_S + \Delta I_S/2$ and $I_{S2} = I_S - \Delta I_S/2$? Assume that the collector-emitter voltages and Early voltages are matched.

(b) If $\Delta\beta_{FO}/\beta_{FO} = 5$ percent, what are the values of I_{B1} and I_{B2} for the transistors at a Q-point of (100 μA, 10 V)? Assume $\beta_{FO} = 150$ and $V_A = 70$ V.

14.5. What is the worst-case fractional mismatch $\Delta I_D/I_D$ in drain currents in two MOSFETs if $K_n = 225$ μA/V$^2 \pm 5$ percent and $V_{TN} = 0.7$ V $\pm$ 25 mV for (a) $V_{GS} = 2$ V? (b) $V_{GS} = 4$ V? Assume $I_{D1} = I_D + \Delta I_D/2$ and $I_{D2} = I_D - \Delta I_D/2$.

14.6. The MOS transistors in the differential pair are mismatched. The nominal value of $(V_{GS} - V_{TN}) = 0.7$ V. (a) What will be the offset voltage if the (W/L) ratios are mismatched by 7 percent? (b) If the threshold voltages are mismatched by 7 percent? (c) If the values of λ are mismatched by 7 percent and $\lambda V_{DS} = 0.1$? (d) If the drain resistors are mismatched by 7 percent? (Remember, the offset voltage is the input voltage required to force the differential output voltage to be zero.)

14.7. (a) A layout design error causes the W/L ratios of the two NMOSFETs in a differential amplifier to differ by 10 percent. What will be the gate-source voltage difference between these two transistors when their drain currents are identical if the nominal value of $(V_{GS} - V_{TN}) = 0.5$ V? (Assume $V_{TN} = 0.7$ V, $\lambda = 0$ and identical values of K'_n.) (b) What degree of matching is required for a gate-source voltage difference of less than 3 mV? (c) For 1 mV?

14.2 Current Mirrors

14.8. (a) What are the output currents and output resistances for the current sources in Fig. P14.8 if $I_{REF} = 50$ μA, $K'_n = 25$ μA/V^2, $V_{TN} = 0.7$ V and $\lambda = 0.01$ V^{-1}? (b) What are the currents if I_{REF} is changed to 40 μA? (c) What would be the values if $\lambda = 0$?

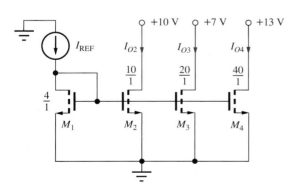

Figure P14.8

14.9. Simulate the current source array in Fig. P14.8 and compare the results to the hand calculations in Prob. 14.8.

14.10. (a) What are the output currents for the circuit in Prob. 14.8 if the W/L ratio of M_1 is changed to 2.5/1? (b) If $I_{REF} = 25$ μA and $(W/L)_1 = 5/1$?

14.11. The current sources in Prob. 14.8 could represent the binary weighted currents needed for a 3-bit D/A converter. (a) What are the ideal values of the three output currents (i.e., $\lambda = 0$)? (b) Express the current errors from Prob. 14.8 in terms of LSBs.

*14.12. What are the output currents and output resistances for the current sources in Fig. P14.12 if $R = 30$ kΩ, $K'_p = 15$ μA/V^2, $V_{TP} = -0.90$ V, and $\lambda = 0.01$ V^{-1}?

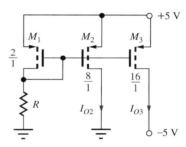

Figure P14.12

14.13. Simulate the current source array in Fig. P14.12 and compare the results to the hand calculations in Prob. 14.12.

14.14. (a) What are the output currents for the circuit in Prob. 14.12 if the W/L ratio of M_1 is changed to 3.3/1? (b) $R = 50$ kΩ and $(W/L)_1 = 4/1$?

14.15. What value of R is required in Fig. P14.12 to have $I_{O2} = 47$ μA? Use device data from Prob. 14.12.

14.16. (a) What are the output currents and output resistances for the current sources in Fig. P14.16(a) if $R = 60$ kΩ, $\beta_{FO} = 90$, and $V_A = 60$ V? (b) Repeat part (a) if the emitter areas of all the transistors are doubled. (c) Repeat for Fig. P14.16(b).

14.17. Simulate the current source array in Fig. P14.16(a) and compare the results to the hand calculations in Prob. 14.16. (b) Repeat for Fig. P14.16(b).

14.18. What value of R is required in Fig. P14.16(a) to have $I_{O3} = 200$ μA? What is the value of I_{O2}? Assume $\beta_{FO} = 80$ and $V_A = 60$ V. (b) Repeat for the circuit in Fig. P14.16(b).

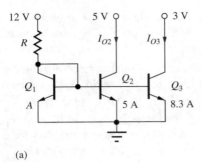

(a)

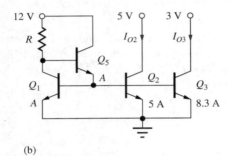

(b)

Figure P14.16

14.19. (a) What are the output currents in the circuit in Fig. P14.16(a) if $R = 100$ kΩ? Use $\beta_{FO} = 100$ and $V_A = 75$ V. (b) What value of R is required to produce the same output currents in Fig. P14.16(b)?

14.20. (a) What are the output currents in the circuit in Fig. P14.16(a) if the area of transistor Q_1 is changed to 2A, and $R = 60$ kΩ? Use $\beta_{FO} = 120$ and $V_A = 75$ V. (b) Repeat for Fig. P14.16(b).

14.21. (a) What are the output currents in the circuit in Fig. P14.16(b) if the area of transistor Q_1 is changed to 2.5A, and $R = 82$ kΩ? Use $\beta_{FO} = 100$ and $V_A = 75$ V. (b) Repeat if the area of Q_5 is also changed to 3 A.

14.22. (a) What are the output currents in Fig. P14.16(a) if $R = 110$ kΩ? (b) What are the output currents if the 5-V supply increases to 7 V? (c) What are the output currents if the 12-V supply decreases to 10 V? (d) Show that the change in I_{O2} in part (b) is equal to $g_{o2}\Delta V$. Use $\beta_{FO} = 80$ and $V_A = 60$ V.

14.23. What are the output currents and output resistances for the current sources in Fig. P14.23 if $R = 110$ kΩ, $\beta_{FO} = 75$, and $V_A = 60$ V?

14.24. What value of R is required in the circuit in Fig. P14.23 to set $I_{O3} = 75$ µA? What are the values of I_{O2} and I_{O4}?

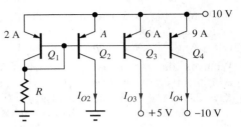

Figure P14.23

*14.25. Draw a buffered current mirror version of the source in Fig. P14.23 and find the value of R required to set $I_{REF} = 15$ µA if $\beta_{FO} = 70$ and $V_A = 60$ V. What are the values of the three output currents? What is the collector current of the additional transistor?

14.26. In Fig. P14.26, $R_2 = 5R_3$. What value of n is required to set I_{E3} to be equal to exactly $5I_{E2}$?

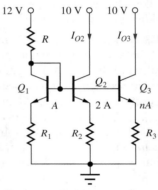

Figure P14.26

*14.27. What are the output currents and output resistances for the current sources in Fig. P14.26 if $R = 27$ kΩ, $R_1 = 10$ kΩ, $R_2 = 5$ kΩ, $R_3 = 2.5$ kΩ, $n = 4$, $\beta_{FO} = 75$, and $V_A = 75$ V?

*14.28. What values of n and R_3 would be required in Prob. 14.27 so that $I_{O2} = 3I_{O3}$?

14.29. Repeat Prob. 14.27 if the area of transistor Q_1 is changed to 0.5 A and R_1 is changed to 20 kΩ.

14.30. What are the output current I_O and output resistance in the circuit in Fig. P14.30 if $-V_{EE} = -8$ V, $n = 7.2$, $K_n = 50$ µA/V^2, $V_{TN} = 0.75$ V, $I_{REF} = 20$ µA, $\beta_{FO} = 100$, and $V_A = 60$ V?

14.31. Use SPICE to simulate the circuit in Prob. 14.30 and compare the results to hand calculations.

14.32. (a) What is the input resistance presented to source I_{REF} at the gate of transistor M_3 in Fig. P14.30 if $n = 1$? Use the other parameters from Prob. 14.30. (b) Use transfer function analysis in SPICE to verify your result.

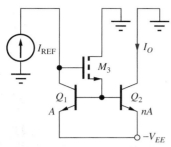

Figure P14.30

Widlar Sources

14.33. (a) What are the output current and output resistance for the Widlar current source I_{O2} in Fig. P14.33 if $R = R_2 = 15$ kΩ and $V_A = 60$ V? (b) For I_{O3} if $R_3 = 5$ kΩ and $n = 11$?

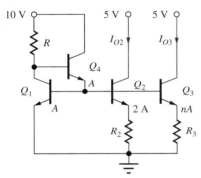

Figure P14.33

14.34. What value of R is required to set $I_{REF} = 50$ μA in Fig. P14.33? If $I_{REF} = 50$ μA, what value of R_2 is needed to set $I_{O2} = 5$ μA? If $R_3 = 2$ kΩ, what value of n is required to set $I_{O3} = 10$ μA?

14.35. Simulate the source of Prob. 14.34 and compare the results to hand calculations.

14.36. (a) What are the output current and output resistance for the Widlar current source I_{O2} in Fig. P14.36 if $R = 50$ kΩ and $R_2 = 5$ kΩ? Use $V_A = 70$ V and $\beta_F = 75$. (b) For I_{O3} if $R_3 = 2.5$ kΩ and $n = 18$?

14.37. What value of R is required to set $I_{REF} = 60$ μA in Fig. P14.36? If $I_{REF} = 60$ μA, what value of R_2 is needed to set $I_{O2} = 10$ μA? If $R_3 = 2$ kΩ, what value of n is required to set $I_{O3} = 10$ μA?

14.3 High-Output-Resistance Current Mirrors

Wilson Sources

14.38. $I_{REF} = 25$ μA, $-V_{EE} = -5$ V, $\beta_{FO} = 110$, and $V_A = 60$ V in the Wilson source in Fig. P14.38.

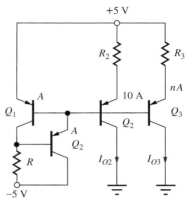

Figure P14.36

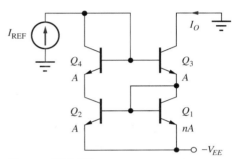

Figure P14.38

(a) What are the output current and output resistance for $n = 1$? (b) For $n = 4$? (c) What is the value of V_{CS} for the current source in (b)? (d) What is the minimum value of V_{EE}?

**14.39. Derive an expression for the output resistance of the BJT Wilson source in Fig. P14.38 and show that it can be reduced to Eq. (14.45), use $n = 1$. What assumptions were used in this simplification?

*14.40. Derive an expression for the output resistance of the Wilson source in Fig. P14.38 as a function of the area ratio n. Find an expression for the output resistance if $n = 5$.

14.41. What is the minimum voltage that can be applied to the collector of Q_3 in Fig. P14.38 and have the transistor remain in the active region if $I_{REF} = 40$ μA, $n = 5$, $\beta_{FO} = 110$, and $I_{SO} = 5$ fA? Calculate an exact value based on the value of I_{SO}.

14.42. $R = 50$ kΩ in the Wilson source in Fig. P14.42. (a) What is the output current if $(W/L)_1 = 5/1$, $(W/L)_2 = 20/1$, $(W/L)_3 = 20/1$, $K'_n = 25$ μA/V², $V_{TN} = 0.75$ V, $\lambda = 0$ V^{-1}, and $V_{SS} = -5$ V? What value of $(W/L)_4$ is required to balance the drain voltages of M_1 and M_2? (b) What is the output resistance if $\lambda = 0.015$ V^{-1}? Use the dc values from part (a). (c) Check your results in (b) with SPICE simulation.

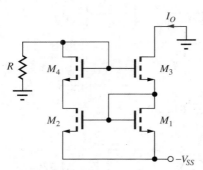

Figure P14.42

*14.43. Derive an expression for the output resistance of the Wilson source in Fig. P14.42 as a function of $(W/L)_1$, $(W/L)_2$, $(W/L)_3$, $(W/L)_4$, and the reference current I_{REF}. Assume $R = \infty$.

14.44. (a) Derive an expression for the equivalent resistance presented to I_{REF} in the Wilson source in Fig. 14.16(a). (b) Derive an expression for the equivalent resistance presented to I_{REF} in the Wilson source in Fig. 14.16(b).

14.45. What is the minimum voltage required on the drain of M_3 to maintain it in pinch-off in the circuit in Fig. P14.42 if $I_{REF} = 150$ μA, $(W/L)_{1,3} = 5/1$, $(W/L)_{2,4} = 20/1$, $K'_n = 25$ μA/V^2, $V_{TN} = 0.75$ V, $\lambda = 0$ V^{-1}, and $-V_{SS} = -10$ V?

14.46. In Fig. P14.42, $(W/L)_3 = 5/1$, $(W/L)_4 = 5/1$, and $I_{REF} = 50$ μA. What value of $(W/L)_2$ is required for $R_{out} = 300$ MΩ if $K'_n = 25$ μA/V^2, $V_{TN} = 0.75$ V, $\lambda = 0.0125$ V^{-1}? Assume $(W/L)_2 = (W/L)_1$, $R = \infty$, and $V_{SS} = 5$ V. Neglect V_{DS}.

**14.47. Redraw the equivalent circuit used to calculate the output resistance of the MOS Wilson source in Figs. 14.16(a) and 14.18 including a finite output resistance R_{REF} for the reference source. Based on this circuit, how large must R_{REF} be to keep from degrading the output resistance of the Wilson source? What type of current source could be used to implement I_{REF} to meet this requirement?

14.48. Use Blackman's theorem to find the expression for the output resistance of the MOS Wilson source in Fig. 14.16(a).

14.49. Find the output resistance of the MOS Wilson source in Prob. 14.42 by applying Blackman's theorem. What are the values of R_D, T_{OC}, T_{SC} and R_{out}?

14.50. Use Blackman's theorem to find the expression for the output resistance of the BJT Wilson source in Fig. 14.16(b).

14.51. Find the output resistance of the BJT Wilson source in Prob. 14.38 by applying Blackman's theorem. What are the values of R_D, T_{OC}, T_{SC}, and R_{out}?

Cascode Current Sources

14.52. (a) What are the output current and output resistance for the cascode current source in Fig. P14.52 if $I_{REF} = 50$ μA, $V_{DD} = 5$ V, $K_n = 80$ μA/V^2, $V_{TN} = 0.8$ V, and $\lambda = 0.0125$ V^{-1}? (b) What is the value of V_{CS} for this current source? (c) What is the minimum value of V_{DD}?

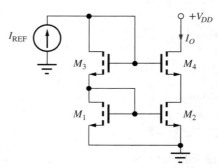

Figure P14.52

14.53. Use SPICE to simulate the current source in Prob. 14.52 and compare the results to your calculations.

14.54. (a) A layout error causes the W/L ratio of M_2 to be 5 percent larger than that of M_1 in Prob. 14.52. What is the error in the output current I_O? (a) Repeat if $M_1 = M_2$, but M_4 is 5 percent larger than that of M_3.

14.55. What is the output resistance of the source in Prob. 14.52 if the body terminals of M_3 and M_4 are connected to ground and $V_{TO} = 0.7$ V and $\gamma = 0.6$ V$^{0.5}$? Assume $\gamma = 0$ for the Q-point calculations.

14.56. In Fig. P14.52, $(W/L)_1 = 5/1$, $(W/L)_2 = 5/1$, $(W/L)_3 = 5/1$, and $I_{REF} = 60$ μA. What value of $(W/L)_4$ is required for $R_{out} = 250$ MΩ if $K'_n = 25$ μA/V^2, $V_{TN} = 0.75$ V, and $\lambda = 0.0125$ V^{-1}?

14.57. (a) Repeat Prob. 14.52 for $I_{REF} = 25$ μA. (b) Repeat Prob. 14.52 for $I_{REF} = 60$ μA.

14.58. (a) Find the output resistance of the cascode current source in Fig. P14.58. Use the parameters from Prob. 14.52. (b) What is the output resistance for the source in Fig. P14.52?

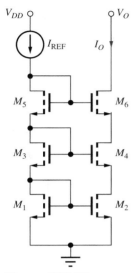

Figure P14.58

14.59. What is the equivalent resistance presented to I_{REF} in the cascode current source in Prob. 14.52?

14.60. (a) What are the output current and output resistance for the cascode current source in Fig. P14.60 if $I_{REF} = 20$ μA, $\beta_{FO} = 110$, and $V_A = 60$ V? (b) What is the value of V_{CS} for this current source? (c) What is the minimum value of V_{CC}?

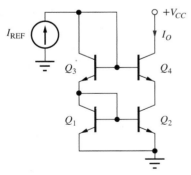

Figure P14.60

14.61. Simulate the current source in Prob. 14.60 and compare the results to hand calculations.

14.62. Repeat Prob. 14.60 if $I_{REF} = 150$ μA, $\beta_{FO} = 120$ and $V_A = 70$ V.

14.63. What is the equivalent resistance presented to I_{REF} in the cascode current source in Prob. 14.60?

*14.64. (a) Transistor Q_4 in Fig. P14.60 is replaced with a Darlington pair of BJTs (see Sec. 13.2.3). Find the output resistance of the new current source. Assume the circuit is voltage balanced and use the values in Prob. 14.60. (b) If Q_1, Q_2 and Q_4 all have an area of 4A, what must be the area of Q_3 to cause the circuit to be voltage balanced?

*14.65. Derive the expression for the output resistance of the cascode current source in Figs. 14.19(b) and 14.21.

Regulated Cascode

14.66. (a) Find the output resistance of the regulated cascode source in Fig. 14.22(c) if $I_{REF} = I_4 = 40$ μA, $K_n = 100$ μA/V², $V_{TN} = 0.75$ V and $\lambda = 0.015$ V⁻¹. (b) Repeat if I_4 is 80 μA.

14.67. Repeat Prob. 14.66 if I_4 is supplied by a standard PMOS current mirror whose transistors have $\lambda = 0.025$ V⁻¹.

14.68. Find the expression for the output resistance of the regulated cascode source in Fig. 14.22(c) if current source I_4 has an output resistance R_4. How large must R_4 be to not degrade the output resistance of the source?

14.69. Repeat the Des. Ex. 14.4 using a regulated cascode source.

*14.70. (a) Replace the NMOS transistor M_3 in Fig. 14.22(a) with an *npn* device, and find an new expression for the output resistance of the current source. What is limiting the output resistance? Why is the result different from the MOS version? (b) Replace transistor M_3 with a Darlington pair of BJTs (see Sec. 13.2.3) and find the new output resistance.

*14.71. (a) Replace the NMOS transistors M_3 and M_4 in Fig. 14.22(c) with *npn* devices, and find a new expression for the output resistance of the current source. What is limiting the output resistance? Why is the result different from the MOS version? (b) Now replace transistor M_3 with a Darlington pair of BJTs (see Sec. 13.2.3) and find the new output resistance.

14.4–14.5 Reference Current Generation and Supply-Independent Biasing (Use $\beta_F = 100$ and $V_A = 60$ V.)

14.72. What are the output current and output resistance for the Widlar source in Fig. P14.72 if $I_{REF} = 100$ μA and $R_2 = 500$ Ω? (b) What is the new value of the output current if a layout error causes the area of Q_2 to be 7 percent larger than desired? (c) What are the new values or output current and resistance if the emitter area of Q_2 is reduced to 14 A?

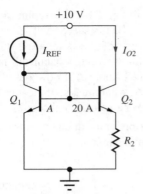

Figure P14.72

14.73. What are the output current and output resistance for the Widlar source in Fig. P14.72 if $I_{REF} = 40$ μA and $R_2 = 935$ Ω? (b) What are the new values if the emitter area of Q_1 is increased to 2 A?

14.74. $I_{REF} = 72$ μA in Fig. P14.72. (a) What value of R_2 is required to set $I_{O2} = 40$ μA? (b) To set $I_{O2} = 10$ μA? (c) To set $I_{O2} = 10$ μA if the area of Q_2 is changed to 10 A?

14.75. $I_{REF} = 68$ μA in Fig. P14.72. (a) What value of R_2 is required to set $I_{O2} = 16$ μA if the area Q_1 is changed to 2 A? (b) If the area of Q_2 is changed to 10 A?

14.76. Plot the variation of the output current vs. I_{REF} for the Widlar source in Fig. P14.72 for 50 μA ≤ I_{REF} ≤ 5 mA if $R_2 = 4$ kΩ and $\beta_{FO} = 100$.

14.77. (a) What is the output current of the V_{BE}-based reference in Fig. P14.77(a) if $I_S = 10^{-15}$ A, $\beta_F = \infty$, $R_1 = 8$ kΩ, $R_2 = 5$ kΩ, and $V_{EE} = 15$ V? (b) For $V_{EE} = 3.3$ V? (c) What is the output current of the V_{BE}-based reference in Fig. P14.77(b) if $R_1 = 15$ kΩ, $R_2 = 12$ kΩ, and $V_{CC} = 10$ V?

14.78. (a) Design the reference in Fig. P14.77(a) to produce an output current $I_O = 25$ μA. Assume $-V_{EE} = -3.3$ V and $I_S = 0.1$ fA and $\beta_{FO} = 130$ for both transistors. (b) Repeat for the circuit in Fig. P14.77(b) if $V_{CC} = 3.3$ V.

*14.79. What is the output current of the NMOS reference in Fig. P14.79(a) if $R_1 = 10$ kΩ, $R_2 = 20$ kΩ, $K_n = 250$ μA/V^2, $V_{TN} = 0.75$ V, $\lambda = 0.017$ V^{-1}, and $V_{DD} = 10$ V?

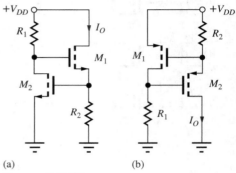

Figure P14.79

14.80. Design the reference in Fig. P14.79(a) to produce an output current $I_O = 100$ μA. Assume $V_{DD} = 5$ V and use the transistor parameters from Prob. 14.79.

*14.81. What is the output current of the PMOS reference in Fig. P14.79(b) if $R_1 = 10$ kΩ, $R_2 = 18$ kΩ, $K_p = 100$ μA/V^2, $V_{TP} = -0.75$ V, $\lambda = 0.02$ V^{-1}, and $V_{DD} = 3.3$ V?

14.82. Design the reference in Fig. P14.79(b) to produce an output current $I_O = 100$ μA. Assume $V_{DD} = 10$ V and use the transistor parameters from Prob. 14.81.

14.83. (a) What are the collector currents in Q_1 and Q_2 in the reference in Fig. P14.83 if $V_{CC} = V_{EE} = 1.5$ V,

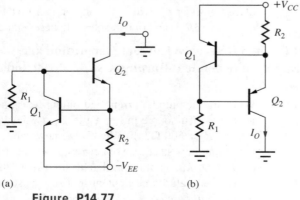

Figure P14.77

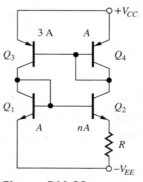

Figure P14.83

$n = 18$, and $R = 2.7$ kΩ? Assume $\beta_{FO} = \infty$ and $V_A = \infty$. (b) What are the temperature coefficients (TC) of currents I_{C2} and I_{C3} in the reference if resistor R has a TC of -2000 ppm/°C?

14.84. Simulate the reference in Prob. 14.83 using SPICE, assuming $\beta_{FO} = 100$ and $V_A = 50$ V. Compare the currents to hand calculations and discuss the source of any discrepancies. Use SPICE to determine the sensitivity of the reference currents to power supply voltage changes.

14.85. What are the collector currents in the four transistors in Fig. P14.83 if $V_{CC} = V_{EE} = 3.3$ V, $n = 8$, and $R = 4.7$ kΩ?

14.86. What is the smallest value of n required for the circuit in Fig. P14.83 to operate properly (i.e., $V_{PTAT} > 0$)?

14.87. (a) What value of R is required to set $I_{C2} = 40$ μA in Fig. P14.83 if $n = 5$ and $T = 50°C$? (b) For $n = 10$ and $T = 0°C$?

14.88. (a) What are the drain currents in M_1 and M_2 in the reference in Fig. P14.88 if $R = 6.8$ kΩ and $V_{DD} = V_{SS} = 5$ V? Use $K'_n = 25$ μA/V², $V_{TN} = 0.7$ V, $K'_p = 10$ μA/V², and $V_{TP} = -0.7$ V. Assume $\gamma = 0$ and $\lambda = 0$ for both transistor types. (b) Repeat for $\gamma_n = 0.6$ V$^{0.5}$ and $\gamma_p = 0.5$ V$^{0.5}$. (c) What is the temperature coefficient (TC) of current I_{D2} in the reference in Fig. 14.27 if mobility varies as $T^{-2.4}$ and resistor R has a TC of -2000 ppm/°C?

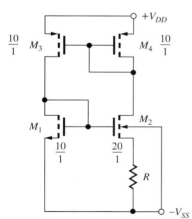

Figure P14.88

14.89. (a) Find the currents in both sides of the reference cell in Fig. P14.88 if $R = 15$ kΩ and $V_{DD} = V_{SS} = 5$ V, using $K'_n = 25$ μA/V², $V_{TON} = 0.7$ V, $K'_p = 10$ μA/V², $V_{TOP} = -0.7$ V, $\gamma_n = 0$, and $\gamma_p = 0$. Use $2\phi_F = 0.6$ V and $\lambda = 0$ for both transistor types.

(b) Repeat for $\gamma_n = 0.5$ V$^{0.5}$ and $\gamma_p = 0.75$ V$^{0.5}$ and compare the results.

14.90. Simulate the references in Prob. 14.89(a) and (b) using SPICE with $\lambda = 0.017$ V^{-1}. Compare the currents to hand calculations (with $\gamma = 0$ and $\lambda = 0$) and discuss the source of any discrepancies. Use SPICE to determine the sensitivity of the reference currents to power supply voltage changes.

14.91. What are the collector currents in Q_1 to Q_8 in the reference in Fig. P14.91 if $V_{CC} = 0$ V, $V_{EE} = 3.3$ V, $R = 15$ kΩ, $R_6 = 5$ kΩ, $R_8 = 5$ kΩ, and $A_{E2} = 5$ A, $A_{E3} = 2$ A, $A_{E4} = $ A, $A_{E5} = 2.5$ A, $A_{E6} = $ A, $A_{E7} = 5$ A, and $A_{E8} = 4$ A?

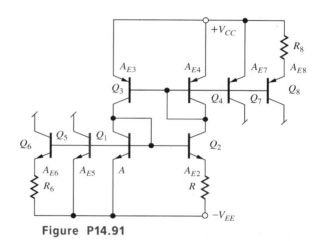

Figure P14.91

14.92. Repeat Prob. 14.91 if $A_{E2} = 10$A and $A_{E3} = $ A.

*14.93. (a) What are the collector currents in Q_1 to Q_7 in the reference in Fig. P14.93 if $V_{CC} = 5$ V and $R = 3600$ Ω? Assume $\beta_F = \infty = V_A$. (b) Repeat part (a) if the emitter areas of transistors Q_5, Q_6, and Q_7 are all changed to 2A.

*14.94. (a) Simulate the reference in Prob. 14.93 using SPICE. Assume $\beta_{FOn} = 100$, $\beta_{FOp} = 50$, and both Early voltages = 50 V. Compare the currents to hand calculations and discuss the source of any discrepancies. Use SPICE to determine the sensitivity of the reference currents to power supply voltage changes.

14.95. Repeat Prob. 14.93 assuming the emitter area of transistor Q_3 is changed to 2A. What is the minimum supply voltage required for proper operation of the circuit in Fig. P14.93?

*14.96. (a) What are the drain currents in M_1 and M_2 in the reference in Fig. P14.96 if $R = 3900$ Ω, $V_{DD} = 15$ V, $K'_n = 25$ μA/V², $V_{TN} = 0.7$ V, $K'_p = 10$ μA/V²,

$V_{TP} = -0.7$ V, and $\lambda = 0$ for both transistor types? (b) Repeat part (a) if the W/L ratios of transistors M_5, M_6, and M_7 are all increased to 16/1.

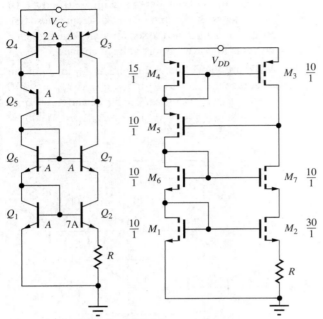

Figure P14.93   **Figure P14.96**

14.97. Simulate the reference in Prob. 14.96 with SPICE using $\lambda = 0.017$ V^{-1} for both transistor types. Compare the currents to those in Prob. 14.96 and discuss the source of any discrepancies. Use SPICE to determine the sensitivity of the reference currents to power supply voltage changes.

14.98. Repeat Prob. 14.96 assuming the W/L ratio of transistor M_3 is changed to 12.5/1.

14.6 The Bandgap Reference

14.99. Find I_C, V_{PTAT}, V_{BE} and V_{BG} for the bandgap reference in Fig. 14.29 if $R = 36$ kΩ, $R_1 = 1$ kΩ and $R_2 = 4.16$ kΩ. Assume $I_S = 0.2$ fA and $A_{E2} = 10A_{E1}$. What temperature corresponds to zero TC?

14.100. What are the bandgap reference output voltage and temperature coefficient of the reference in Prob. 14.99 if I_S changes to 0.6 fA?

14.101. (a) Process variations cause the value of the two collector resistors in the circuit in Prob. 14.99 to decrease to 32 kΩ. What is the new value of V_{BG}? What temperature corresponds to zero TC? (b) Repeat for $R = 24$ kΩ.

14.102. Find I_C, V_{PTAT}, V_{BE} and V_{BG} for the bandgap reference in Fig. 14.29 if $R = 50$ kΩ, $R_1 = 1$ kΩ and $R_2 = 4$ kΩ. Assume $I_S = 0.1$ fA and $A_{E2} = 8 A_{E1}$. What temperature corresponds to zero TC?

14.103. A layout error caused $A_{E2} = 9A_{E1}$ in the bandgap reference in Des. Ex. 14.6 (page 980). What is the new output voltage? What temperature corresponds to zero TC?

14.104. What are the bandgap reference output voltage and the temperature coefficient of the reference in Des. Ex. 14.6 at 320 K if I_S changes to 0.3 fA?

14.105. Process variations cause the values of the two collector resistors in the circuit in Des. Ex. 14.6 to be mismatched. If $R_1 = 83$ kΩ and $R_2 = 77$ kΩ, what is the new value of V_{BG}? What temperature corresponds to zero TC?

14.106. The bandgap reference in Des. Ex. 14.6 was designed to have zero temperature coefficient at 320 K. What will be the temperature coefficient at 280 K? At 300 K?

14.107. Redesign the bandgap reference in Des. Ex. 14.6 to use $A_{E2} = 8A_{E1}$.

14.108. Find the collector currents for each transistor and the base voltage of the *npn* transistors in the bandgap reference in Fig. P14.108. Assume $I_S = 8$ fA, *npn* current gain $= 150$, and *pnp* current gain $= 90$. B $= 8$.

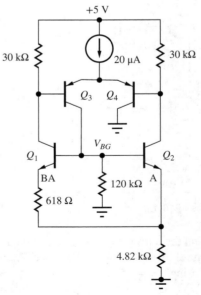

Figure P14.108

14.7 The Current Mirror as an Active Load

14.109. What are the values of A_{dd}, A_{cd}, and CMRR for the amplifier in Fig. 14.31 if $I_{SS} = 500$ µA, $R_{SS} = 10$ MΩ, $K_n = K_p = 500$ µA/V^2, $V_{TN} = -V_{TP} = 0.8$ V, and $\lambda = 0.015$/V for both transistors? What are the minimum power supply voltages if the common-mode input range must be ±5 V? Assume symmetrical supply voltages.

14.110. Use SPICE to simulate the amplifier in Prob. 14.109 and compare the results to hand calculations. Use symmetrical 12-V power supplies.

14.111. What are the values of A_{dd}, A_{cd}, and CMRR for the amplifier in Fig. 14.31 if $I_{SS} = 150$ µA, $R_{SS} = 25$ MΩ, $K_n = K_p = 500$ µA/V^2, $V_{TN} = 1$ V, and $V_{TP} = -1$ V and $\lambda = 0.02$ V^{-1} for both transistors?

14.112. Use SPICE to simulate the amplifier in Prob. 14.111 and compare the results to the hand calculations. Use symmetrical 12-V supplies.

**14.113. (a) What are A_{dd} and A_{cd} for the bipolar differential amplifier in Fig. 14.37 ($R_L = \infty$) if $\beta_{op} = 70$, $\beta_{on} = 125$, $I_{EE} = 180$ µA, $R_{EE} = 22$ MΩ, and the Early voltages for both transistors are 70 V? What is the CMRR for $v_{C1} = v_{C2}$? (b) What are the minimum power supply voltages if the common-mode input range must be ±1.5 V? Assume symmetrical supply voltages.

14.114. Use SPICE to calculate A_{dd} and A_{cd} for the differential amplifier in Prob. 14.113. Compare the results to hand calculations.

14.115. (a) Repeat Prob. 14.113 if I_{EE} is changed to 60 µA, $R_{EE} = 90$ MΩ, and $V_A = 65$ V. (b) Repeat part (a) for $V_A = 90$ V.

14.116. Use SPICE to simulate the amplifier in Prob. 14.115 and compare the results to hand calculations. Use symmetrical 3-V power supplies.

*14.117. (a) Find the Q-points of the transistors in the CMOS differential amplifier in Fig. P14.117 if $V_{DD} = V_{SS} = 10$ V, $I_{SS} = 200$ µA, and $R_{SS} = 25$ MΩ. Assume $K'_n = 25$ µA/V^2, $V_{TN} = 0.75$ V, $K'_p = 10$ µA/V^2, $V_{TP} = -0.75$ V, and $\lambda = 0.017$ V^{-1} for both transistor types. (b) What is the voltage gain A_{dd} of the amplifier? (c) Compare this result to the gain of the amplifier in Fig. 14.31 if the Q-point and W/L ratios of M_1 to M_4 are the same. (d) What is the offset voltage of the amplifier?

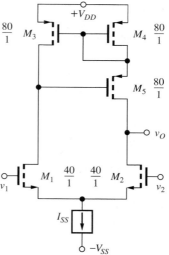

Figure P14.117

14.118. Use SPICE to simulate the amplifier in Prob. 14.117(a,b) and compare the results to hand calculations.

*14.119. Find the Q-points of the transistors in the "folded-cascode" CMOS differential amplifier in Fig. P14.119 if $V_{DD} = V_{SS} = 5$ V, $I_1 = 200$ µA, $I_2 = 200$ µA, $(W/L) = 30/1$ for all transistors, $K'_n = 25$ µA/V^2, $V_{TN} = 0.7$ V, $K'_p = 10$ µA/V^2, $V_{TP} = -0.7$ V, and $\lambda = 0.017$ V^{-1} for both transistor types. Draw the differential-mode half-circuit for transistors M_1 to M_4 and show that the circuit is in

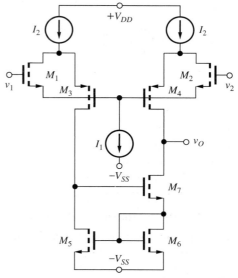

Figure P14.119

fact a cascode amplifier. What is the differential-mode voltage gain of the amplifier? Estimate the offset voltage of the amplifier.

14.120. Use SPICE to simulate the amplifier in Prob. 14.119 and determine its voltage gain, output resistance, and CMRR. Compare to hand calculations.

14.121. (a) Use the technique in the EIA on pages 742–743 with SPICE to find the offset voltage, CMRR and PSRR for the amplifier in Prob. 14.119. (b) Add a transistor to the active load to create a voltage-balanced version and repeat the simulation.

*14.122. Design a current mirror bias network to supply the three currents needed by the amplifier in Prob. 14.119.

Output Stages

14.123. What are the currents in Q_3 and Q_4 in the class-AB output stage in Fig. P14.123 if $R_1 = 12$ kΩ, $R_2 = 12$ kΩ, and $I_{S4} = I_{S3} = I_{S2} = 10^{-14}$ A? Assume $\beta_F = \infty$.

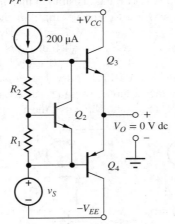

Figure P14.123

*14.124. (a) Show that the currents in Q_3 and Q_4 in the class-AB output stage in Fig. P14.124 are equal to $I_o = I_2 \sqrt{(A_{E3}A_{E4})/(A_{E1}A_{E2})}$. (b) What are the currents in Q_3 and Q_4 if $A_{E1} = 3A_{E3}$, $A_{E2} = 3A_{E4}$, $I_2 = 300$ μA, $I_{SOpnp} = 4$ fA, and $I_{SOnpn} = 10$ fA?

14.8 Active Loads in Operational Amplifiers

14.125. (a) Find the Q-points of the transistors in the CMOS op amp in Fig. 14.42 if $V_{DD} = V_{SS} = 5$ V, $I_{REF} = 300$ μA, $K'_n = 50$ μA/V^2, $V_{TN} = 0.7$ V, $K'_p = 20$ μA/V^2, and $V_{TP} = -0.7$ V. (b) What is the voltage gain of the op amp assuming the output

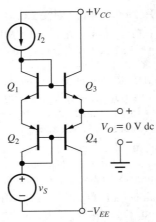

Figure P14.124

stage has unity gain and $\lambda = 0.017$ V^{-1} for both transistor types? (c) What is the voltage gain if I_{REF} is changed to 400 μA? (d) What is the offset voltage of the amplifier?

14.126. Based on the example calculations and your knowledge of MOSFET characteristics, what will be the gain of the op amp in Ex. 14.7 if I_{REF} is set to (a) 250 μA? (b) 20 μA? (*Note:* These should be short calculations.)

*14.127. Find the Q-points of the transistors in Fig. P14.127 if $V_{DD} = V_{SS} = 7.5$ V, $I_{REF} = 250$ μA, $(W/L)_{12} = 40/1$, $K'_n = 50$ μA/V^2, $V_{TN} = 0.75$ V, $K'_p = 20$ μA/V^2, and $V_{TP} = -0.75$ V. What is the differential-mode voltage gain of the op amp if $\lambda = 0.017$ V^{-1} for both transistor types? What is the offset voltage of the amplifier?

*14.128. What is the differential-mode gain of the amplifier in Fig. P14.127 if $V_{DD} = V_{SS} = 10$ V, $I_{REF} = 120$ μA, $K'_n = 50$ μA/V^2, $V_{TON} = 0.7$ V, $K'_p = 20$ μA/V^2, $V_{TOP} = -0.7$ V, $\gamma_n = 0$, and $\gamma_p = 0$? Use $\lambda = 0.017$ V^{-1} for both transistor types. What is the offset voltage of the amplifier?

14.129. (a) Use SPICE to find the Q-points of the transistors of the amplifier in Prob. 14.128. (b) Repeat with $2\phi_F = 0.8$ V, $\gamma_n = 0.60$ V$^{0.5}$, and $\gamma_p = 0.75$ V$^{0.5}$, and compare the results to part (a).

*14.130. (a) Estimate the minimum values of V_{DD} and V_{SS} needed for proper operation of the amplifier in Prob. 14.127. Use $K'_n = 25$ μA/V^2, $V_{TN} = 0.7$ V, $K'_p = 10$ μA/V^2, and $V_{TP} = -0.7$ V. (b) What are the minimum values of V_{DD} and V_{SS} needed to have at least a ±6-V common-mode input range in the amplifier?

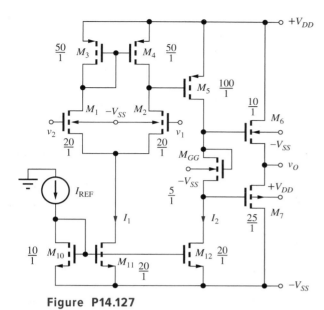

Figure P14.127

*14.131. (a) Find the Q-points of the transistors in Fig. P14.131 if $V_{DD} = V_{SS} = 10$ V, $I_{REF} = 200$ μA, $K'_n = 50$ μA/V², $V_{TN} = 0.7$ V, $K'_p = 20$ μA/V², and $V_{TP} = -0.7$ V. (b) What is the approximate value of the W/L ratio for M_6 of the CMOS op amp in order for the offset voltage to be zero? What is the differential-mode voltage gain of the op amp if $\lambda = 0.017$ V^{-1} for both transistor types?

*14.132. (a) Simulate the amplifier in Prob. 14.131 and compare its differential-mode voltage gain to the hand calculations in Prob. 14.131. (b) Use SPICE to calculate the offset voltage and CMRR of the amplifier.

14.133. What are the minimum supply voltages required for proper operation of the circuit in Fig. P14.131?

14.134. Draw the amplifier that represents the mirror image of Fig. 14.42 by interchanging NMOS and PMOS transistors. Choose the W/L ratios of the NMOS and PMOS transistors so the voltage gain of the new amplifier is the same as the gain of the amplifier in Fig. 14.42. Maintain the operating currents the same and use the device parameter values from Ex. 14.7.

14.135. Draw the amplifier that represents the mirror image of Fig. 14.44 by interchanging npn and pnp transistors. If $\beta_{on} = 150$, $\beta_{op} = 70$, and $V_{AN} = V_{AP} = 70$ V, which of the two amplifiers will have the highest voltage gain? Why?

*14.136. What is the approximate emitter area of Q_{16} needed to achieve zero offset voltage in the amplifier in Fig. P14.136 if $I_B = 200$ μA and $V_{CC} = V_{EE} = 5$ V? What is the value of R_{BB} needed to set the quiescent current in the output stage to 75 μA?

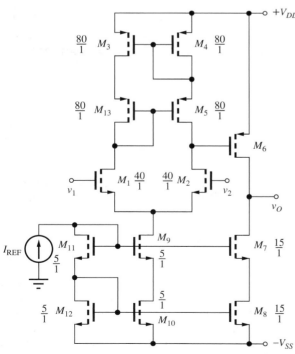

Figure P14.131

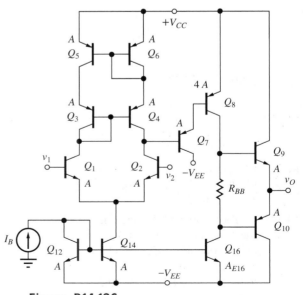

Figure P14.136

What are the voltage gain and input resistance of this amplifier? Assume $\beta_{on} = 120$, $\beta_{op} = 70$, $V_{AN} = V_{AP} = 50$ V, and $I_{SOnpn} = I_{SOpnp} = 15$ fA.

14.137. Use SPICE to simulate the characteristics of the amplifier in Prob. 14.136. Determine the offset voltage, voltage gain, input resistance, output resistance, and CMRR of the amplifier.

14.138. (a) What are the minimum values of V_{CC} and V_{EE} needed for proper operation of the amplifier in Fig. P14.136? (b) What are the minimum values of V_{CC} and V_{EE} needed to have at least a ± 1-V common-mode input range in the amplifier?

14.9 The µA741 Operational Amplifier

14.139. (a) What are the three bias currents in the source in Fig. P14.139 if $R_1 = 90$ kΩ, $R_2 = 5$ kΩ, and $V_{CC} = V_{EE} = 3$ V? (b) Repeat for $V_{CC} = V_{EE} = 20$ V. (c) Why is it important that I_1 in the µA741 be independent of power supply voltage whereas I_2 and I_3 are proportional to the supply voltages?

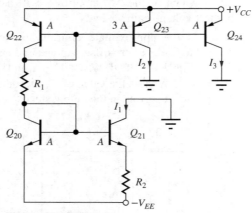

Figure P14.139

14.140. Choose the values of R_1 and R_2 in Fig. P14.139 to set $I_2 = 250$ µA and $I_1 = 50$ µA if $V_{CC} = V_{EE} = 12$ V. What is I_3?

14.141. Choose the values of R_1 and R_2 in Fig. P14.139 to set $I_3 = 125$ µA and $I_1 = 25$ µA if $V_{CC} = V_{EE} = 10$ V. What is I_2?

*14.142. (a) Based on the schematic in Fig. 14.46, what are the minimum values of V_{CC} and V_{EE} needed for proper operation of µA741 amplifier? (b) What are the minimum values of V_{CC} and V_{EE} needed to have at least a ± 1-V common-mode input range in the amplifier?

14.143. What are the values of the elements in the Norton equivalent circuit in Fig. 14.53(a) if I_1 in Fig. 14.46 is increased to 75 µA?

14.144. Suppose Q_{23} in Fig. 14.47 is replaced by a cascode current source. (a) What is the new value of output resistance R_2? (b) What are the new values of the y-parameters of Fig. 14.53(b)? (c) What is the new value of A_{dm} for the op amp?

14.145. Draw a schematic for the cascode current source in Prob. 14.144.

14.146. Create a small-signal SPICE model for the circuit in Fig. 14.53(b) and verify the values of R_{in10}, G_m, and G_o. What is the value of y_{12} for the circuit?

**14.147. Figure P14.147 represents an op-amp input stage that was developed following the introduction of the µA741. (a) Find the Q-points for all the transistors in the differential amplifier in Fig. P14.147 if $V_{CC} = V_{EE} = 15$ V and $I_{REF} = 100$ µA. (b) Discuss how this bias network operates to establish the Q-points. (c) Label the inverting and noninverting input terminals. (d) What are the transconductance and output resistance of this amplifier? Use $V_A = 70$ V.

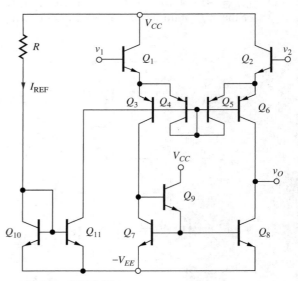

Figure P14.147

**14.148. Figure P14.148 represents an op-amp input stage that was developed following the introduction of the µA741. Find the Q-points for all the transistors in the differential amplifier in Fig. P14.148 if $V_{CC} = V_{EE} = 15$ V and $I_{REF} = 75$ µA. (b) Discuss how this bias network operates to establish the

Q-points. (c) Label the inverting and noninverting input terminals. (d) What are the transconductance and output resistance of this amplifier? Use $V_A = 60$ V.

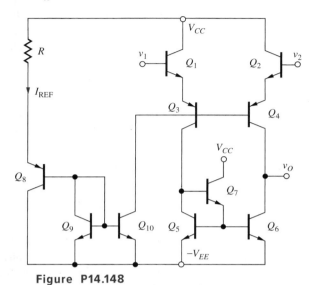

Figure P14.148

14.149. Use the technique in the EIA on pages 776–777 with SPICE to find the offset voltage, CMRR and PSRR for the amplifier in Prob. 14.148.

14.10 The Gilbert Analog Multiplier

14.150. Find the Q-points of the six transistors in Fig. 14.59 if $V_{CC} = -V_{EE} = 5$ V, $I_{BB} = 100$ µA, $R_1 = 10$ kΩ, and $R = 50$ kΩ. Draw the circuit assuming the bases of Q_1 and Q_2 are biased at a common-mode voltage of -2.5 V with $v_1 = 0$. Assume the bases of Q_3 through Q_6 are biased at a common-mode voltage of 0 V with $v_2 = 0$.

14.151. (a) Find the collector currents of the six transistors in Fig. 14.59 if $V_{CC} = -V_{EE} = 7.5$ V, $I_{BB} = 200$ µA, $R_1 = 8$ kΩ, and $R = 42$ kΩ. Draw the circuit assuming the bases of Q_1 and Q_2 are biased at a common-mode voltage of -3 V with $v_1 = 0.5$ V. Assume the bases of Q_3 through Q_6 are biased at a common-mode voltage of 0 V with $v_2 = 0$. (b) Repeat with $v_2 = 1$ V. (c) Repeat with $v_2 = -1$ V.

14.152. Write an expression for the output voltage for the circuit in Fig. 14.59 if $v_1 = 0.6 \sin 1000\pi t$, and v_2 is generated by the circuit in Fig. 14.60 with $v_3 = 0.6 \sin 12{,}000\pi t$? Assume $V_{CC} = -V_{EE} = 10$ V, $I_{EE} = 400$ µA, $R_1 = R_3 = 3$ kΩ, and $R = 12$ kΩ.

14.153. (a) Write expressions for the total collector currents i_{C1} and i_{C2} in Fig. 14.59 if $I_{BB} = 0.5$ mA, $R_1 = 4$ kΩ,

and $v_1 = 0.4 \sin 5000\pi t$ V. Assume the transistors are operating in the active region. (b) What is the transconductance G_m of the voltage-to-current converter formed by Q_1 and Q_2? [$G_m = \Delta(i_{C1} - i_{C2})/\Delta v_1$]

14.154. Use SPICE to plot the VTC for the circuit in Fig. 14.60 with $V_{BB} = 3$ V, $-V_{EE} = -5$ V, $I_{EE} = 300$ µA, and $R_3 = 3.3$ kΩ.

Low Voltage/Weak Inversion

Note: In the following problems you need to calculate the values of V_{BE} and V_{GS}. Do not assume a value! Use $V_T = 25.9$ mV unless otherwise specified. Newton's method, or "fzero," might be useful.

14.155. What are the gate-source voltage and output currents for the current mirror in Fig. P14.8 if $I_{REF} = 1$ nA, $K'_n = 25$ µA/V², $V_{TN} = 0.4$ V, $n = 1.5$, $\lambda = 0$, and all the power supply voltages are 1 V?

14.156. What are the source-gate voltage and output currents for the current mirror in Fig. P14.12 if $R = 150$ MΩ, $K'_p = 15$ µA/V², $V_{TP} = -0.45$ V, $n = 1.6$, $\lambda = 0$, and the positive power supply voltage is 1 V?

14.157. (a) What are the output currents in the circuit in Fig. P14.16(a) if $R = 12$ MΩ, $I_S = 10$ fA, $\beta_{FO} = 50$, and the supply voltages are all $+1$ V? Assume the Early voltage can be neglected. (b) What value of R is required to produce the same output currents in Fig. P14.16(b)?

14.158. What are the output currents in the circuit in Fig. P14.23 if $R = 10$ MΩ, $I_S = 3$ fA, $\beta_{FO} = 50$, and the 15-V supply is changed to 1.5 V? The collectors of Q_2, Q_3, and Q_4 are all connected to ground. Assume the Early voltage can be neglected.

14.159. The transistors in Fig. P14.38 have $I_S = 75$ fA, $\beta_{FO} = 60$, and the Early voltage can be neglected. (a) What is the output current for $I_{REF} = 1$ µA and $n = 1$? What is the value of V_{BE}? (b) What is the minimum value of $-V_{EE}$ if V_{CE} must be at least 120 mV? (c) Repeat parts (a) and (b) for $n = 5$.

14.160. Work Prob. 14.42 if $R = 100$ MΩ, $V_{TN} = 0.42$ V, $n = 1.5$, all transistors have a $W/L = 20/1$, and $-V_{SS} = -1.25$ V in Fig. P14.42. What is the minimum value of $-V_{SS}$ if V_{DS} must be at least 120 mV?

14.161. In Fig. P14.52, $I_{REF} = 100$ nA, V_{DD} is large enough to saturate the transistors, $V_{TN} = 0.38$ V,

$K_n = 75$ μA/V^2, $n = 1.5$, $\lambda = 0$, and all transistors $W/L = 10/1$. (a) What are the values of I_O and the V_{GS} of all the transistors? (b) What is the minimum value of V_{DD} if all V_{DS} values must be at least 100 mV?

14.162. In the interpolation model in Eq. (5.70), assume $I_{DO} = 100$ μA, $V_{GS} = 0.3$ V, $V_{TN} = 0.4$ V, and $n = 1.5$. What is the percentage error caused by the right-hand model term compared to Eq. (5.71) if $V_{DS} = 100$ mV?

14.163. The transistors in Fig. P14.60 have $I_S = 55$ fA, $\beta_{FO} = 60$, and the Early voltage can be neglected. (a) What is the output current for $I_{REF} = 150$ nA? What is the value of V_{BE}? (b) What is the minimum value of V_{CC} if V_{CE} must be at least 120 mV?

14.164. The transistors in the Widlar source in Fig. P14.72 have $I_S = 45$ fA, $\beta_{FO} = 60$, and the Early voltage can be neglected. (a) What is the output current with $I_{REF} = 150$ nA, $V_{CC} = 1.5$ V, and $R_2 = 300$ kΩ? What are the values of V_{BE}? (b) What is the minimum value of V_{CC} if V_{CE} must be at least 200 mV?

14.165. What is the output current I_O of the NMOS reference in Fig. P14.79(a) if $V_{DD} = 1.5$ V, $V_{TN} = 0.5$ V, $K'_n = 21$ μA/V^2, $n = 1.5$, $\lambda = 0$, and both transistor's $W/L = 10/1$? What are the values of I_O and V_{GS} of both the transistors if $R_1 = R_2 = 8.2$ MΩ?

14.166. What is the output current I_O of the PMOS reference in Fig. P14.79(b) if $V_{DD} = 1.25$ V, $V_{TP} = -0.45$ V, $K'_p = 10$ μA/V^2, $n = 1.6$, $\lambda = 0$, and both transistor's $W/L = 10/1$? What are the values of I_O and V_{SG} of both the transistors if $R_1 = R_2 = 10$ MΩ?

14.167. For the reference circuit in Fig. P14.88, $V_{DD} = 1.5$ V, $V_{TN} = 0.4$ V, $V_{TP} = -0.5$ V, $K'_n = 25$ μA/V^2, $K'_p = 10$ μA/V^2, $n = 1.6$, and $\lambda = 0$. Neglect the body effect in M_2. (a) What are the values of the drain currents and gate-source voltages for each of the transistors if $R = 820$ kΩ? (b) If $-V_{SS} = 0$, what is the minimum value of V_{DD} for proper operation of the circuit if the magnitudes of the drain-source voltages must be at least 100 mV?

14.168. Transistors Q_1, Q_5, Q_7, and Q_8 are removed from the circuit in Fig. P14.91. (a) Redraw the circuit with $V_{EE} = 0$. (b) If $I_S = 10$ fA, $\beta_F = $ infinity for all the transistors, and $R = 100$ kΩ, what are the values of the four collector currents? (c) If $-V_{EE} = 0$, what is minimum value of V_{CC} for proper operation of the circuit if the magnitudes of the collector-emitter voltages must be at least 100 mV?

14.169. (a) What are the collector currents in the reference circuit in Fig. P14.93 if $V_{CC} = 3.3$ V and $R = 1$ MΩ? (b) If $I_S = 33$ fA for a transistor with area A and the minimum magnitude of the collector-emitter voltages is 0.1 V, what is the minimum value of V_{CC} for proper operation of this reference circuit?

14.170. (a) What are the drain currents in the reference circuit in Fig. P14.96 if $V_{DD} = 3.3$ V and $R = 360$ kΩ? Assume $V_{TN} = 0.4$ V, $V_{TP} = -0.5$ V, $K'_n = 25$ μA/V^2, $K'_p = 10$ μA/V^2, $n = 1.5$, and $\lambda = 0$. (b) What is minimum value of V_{DD} for proper operation of the circuit if the magnitudes of the drain-source voltages must be at least 125 mV?

14.171. (a) Rework Prob. 14.108 with the value of the current source changed to 2 μA, the values of all the resistors are increased by a factor of 10, the emitter area of Q_1 is 8A, and $I_S = 20$ fA for transistors Q_2, Q_3 and Q_4. (b) What is the minimum value of the positive power supply voltage for proper operation of the circuit if the magnitudes of the collector-emitter voltages must be at least 125 mV?

14.172. (a) Rework Prob. 14.139(a) with power supplies changed to +1.5 V and −1.5 V. (b) What is the minimum value of the symmetrical power supply voltages for proper operation of the circuit if the magnitudes of the collector-emitter voltages must be at least 125 mV and $I_S = 20$ fA for transistors with area A?

14.173. (a) Work Prob. 14.147(a) with I_{REF} changed to 10 μA. (b) What is the minimum value of the symmetrical power supply voltages for proper operation of the circuit if the magnitudes of the collector-emitter voltages must be at least 120 mV, the output signal v_O must be able to range between ±0.25 V, and $I_S = 25$ fA for all transistors?

14.174. (a) Work Prob. 14.148(a) with I_{REF} changed to 10 μA. (b) What is the minimum value of the symmetrical power supply voltages for proper operation of the circuit if the magnitudes of the collector-emitter voltages must be at least 120 mV, the output signal v_O must be able to range between ±0.25 V, and $I_S = 25$ fA for all transistors?

术语对照

Active load	有源负载
Ac analysis	交流分析
Bandgap reference	带隙基准
Buffered current mirror	驱动电流镜
Bipolar transistor	双极型晶体管
Cascode current source	Cascode电流源
Class-AB amplifiers	AB类放大器
CMOS operational amplifier	CMOS运算放大器
CMOS differential amplifier	CMOS差分放大器
Common-mode gain	共模增益
Common-mode half-circuit	共模半电路
Common-mode input resistance	共模输入电阻
Common-mode input voltage range	共模输入电压范围
Common-mode rejection ratio(CMRR)	共模抑制比（CMRR）
Common-mode input signal	共模输入信号
Complementary push-pull output stage	互补推挽输出级
Current gain defect	电流增益缺陷
Current-limiting circuit	限流电路
Current mirror	电流镜
Dc analysis	直流分析
Differential-mode gain	差模增益
Differential-mode half-circuit	差模半电路
Differential-mode input resistance	差模输入电路
Differential-mode output resistance	差模输出电阻
Differential-mode output voltage	差模输出电压
Differential-mode signal	差模信号
Diode-connected transistor	二极管连接晶体管
Digital thermometry	数字测温
Electronic current source	电子电流源
Emitter area scaling	发射极面积比例
Figure of merit(FOM)	品质因数(FOM)
Four-quadrant multiplier	四象限乘法器

Finite current gain	有限电流增益
Gilbert multiplier	Gilbert乘法器
Half-circuit analysis	半电路分析
Matched(devices)	匹配（器件）
Matched transistors	匹配晶体管
µA741	µA741运放
Mirror ratio	镜像比
MOS Transistor	MOS晶体管
Multiple current sources	多级电流源
Norton equivalent circuit	诺顿等效电路
Overvoltage protection	过电压保护
Output stages	输出级
Power-supply-independent biasing	电源独立偏置
Proportional to Absolute Temperature (PTAT)	与绝对温度成正比
PTAT voltage	PTAT电压源
Reference current	参考电流，基准电流
Small-signal	小信号
Short-circuit protection	短路保护
Startup circuit	启动电路
Two-port model	二端口模型
VBE-based reference	基于VBE的参考源
Voltage reference	参考电压，基准电压
Widlar current source	Widlar 电流源
Wilson current source	Wilson电流源

CHAPTER 15

TRANSISTOR FEEDBACK AMPLIFIERS AND OSCILLATORS

第15章 晶体管反馈放大器与振荡器

本章提纲
- 15.1 基本反馈系统回顾
- 15.2 反馈放大器的中频分析
- 15.3 反馈放大电路举例
- 15.4 反馈放大器稳定性回顾
- 15.5 单极点运算放大器补偿
- 15.6 高频振荡器

本章目标
- 复习正反馈和负反馈的概念；
- 复习反馈传输回路分析技术；
- 复习反馈放大器的 Blackman 定理；
- 理解串-并、并-并、并-串和串-串反馈电路的结构和特性；
- 能用传输理论和 Blackman 定理分析各种反馈设置的中频特性；
- 理解反馈对频率响应和反馈放大器稳定性的影响；
- 学会用奈奎斯特理论和伯德图分析反馈放大器稳定性；
- 能采用 SPICE AC 分析法和传输函数分析法来分析反馈放大器特性；
- 能采用 SPICE 仿真或测量技术确定闭环放大器的环路增益；
- 学会采用密勒倍增设计运算放大器的频率补偿；
- 理解运放单位增益频率和电压转换速率（slew rate）之间的关系；
- 理解振荡器的 Barkhausen 准则；
- 理解高频 LC 和晶体振荡器电路；
- 理解振荡电路中的负阻；
- 理解石英晶体的 LCR 模型。

本章导读
　　本章首先给出了反馈系统的概念，介绍了反馈系统的主要系统结构，给出了传递函数、闭环增益、闭环阻抗等概念的定义及说明，并以实例说明了负反馈放大器的作用，负反馈可以最大限度地减小输入之间的差异，可以显著改变电路的特性，降低对电路参数变化的敏感度，利用反馈可以设计出对器件参数和其他参数具有鲁棒性的电路。通过重新从反馈的角度认识电路，可对许多普通电子电路有更加深入的理解。

　　本章介绍反馈的分析时，采用一个负反馈电路实例进行说明，对一个带有电流镜负载的差分放大器电路进行分析，通过从输出到反相输入的直接连接，得到单位增益放大器。通过该例的分析，给出了闭环增益、输入电阻、输出电阻以及偏移电压的计算方法。回路增益对确定反馈放大器特性有着重要的作用。在理论计算中，可以通过在某个任意点处断开反馈环路，并直接计算环路周围返回的电压来找到环路增益。但是，在计算循环增益之前，必须保证回路两侧都有适合的终端。利用Blackman定理来计算从负反馈放大器任意两端看过去的电阻，因此负反馈的输入电阻和输出电阻都可以借助Blackman定理进行计算。

根据放大器输入和输出端所采用反馈的类型，一般可将反馈放大器分为四类。电压放大器采用串-并反馈，跨阻放大器采用并-并反馈，跨导放大器采用串-串反馈，电流放大器采用并-串反馈。本章采用实例的形式，对这几种负反馈放大器进行计算和分析，给出了计算不同反馈结构的中频带闭环增益、闭环输入阻抗和闭环输出阻抗的步骤：

（1）判定反馈的输出连接是串联还是并联。如果是并联连接，则反馈网络感测的是输出电压。如果是串联连接，则反馈网络感测的是输出电流。

（2）判定反馈的输入连接是串联还是并联。如果是并联连接，则有电流反馈到电路输入端。如果是串联连接，则有电压反馈到电路输入端。

（3）如果已知反馈连接的类型，确定反馈系数β的单位，并且根据表15.4确定放大器的类型。

（4）计算理想化放大器的反馈系数β。

（5）考虑放大器和反馈网络负载的影响，计算回路增益。

（6）利用式（15.4）计算放大器的闭环增益。

（7）利用Blackman定理计算放大器的输入和输出阻抗，或电路中所需的、其他的任意阻抗值，同时计算包含非理想负载的开路和短路闭环增益。

（8）计算包含输入和输出负载影响的总增益。

本章依据这些步骤对多个不同拓扑结构的放大器进行了中频带分析。

本章在回顾负反馈稳定性设计问题的基础上，给出了反馈放大器稳定性的分析方法。通过研究反馈放大器回路增益$T(s)=A(s)\beta(s)$的特性，将其看作频率的函数，就可以确定放大器的稳定性。稳定性标准可通过奈奎斯特图或伯德图来评估。为了研究放大器的稳定性，需要深入分析回路增益T、回路增益的相频响应和幅频响应及相位裕度等方面的影响。为获得稳定的性能，需要相位裕度至少为45°，一般都要大于60°。相位裕度修正的基本方法是在放大器的某一个节点上增加一个电容，目的是创建一个主极点，以迫使在更低的频率处得到0dB的幅值响应，从而减小由于放大器较高频率处的极点引起的0dB相移。

本章通过实例讲解了单极点运算放大器补偿问题。反馈放大器利用内部频率补偿来迫使放大器具有单极点频率响应。对于普通的运算放大器而言，可以对单位增益缓冲器的稳定性进行补偿。密勒倍增是设置补偿运算放大器单位增益频率的一种有效方法，这种技术通常被称为密勒补偿。在这类运算放大器中，摆率与单位增益频率有密切关系。

振荡器电路中的反馈实际被设计成正反馈，在不需要输入的情况下，电路也可以产生输出信号。本章主要介绍了高频振荡器，给出了多种形式的高频振荡电路，并进行了深入的分析。在高频振荡器设计中，采用了特殊晶体管，其选频反馈网络由高Q值LC网络或石英晶体谐振元件组成。文中介绍了两种典型的LC振荡结构：一种是利用电容分压控制反馈的Colpitts振荡器，另一种是用电感分压的Hartley振荡器。集成振荡器主要利用晶体管差分对的负G_m单元来实现振荡的。晶体管振荡器采用石英晶体来代替LC振荡器中的电感。晶体在电学上可用一个超高Q值的谐振电路来建模，当将其用于振荡器时，晶体能准确地控制振荡频率。晶体振荡器是一种可以实现超高频率精度和高稳定性的振荡器。

为产生真正的正弦振荡，振荡器的极点必须严格位于s平面的$j\omega$轴上，否则会产生失真。为了实现正弦振荡，通常需要某种形式的幅值稳定技术。这种稳定技术可在电路中使用晶体管的固有非线性特性来实现，也可增加增益控制电路来实现。

Chapter Outline

15.1 Basic Feedback System Review
15.2 Feedback Amplifier Analysis at Midband
15.3 Feedback Amplifier Circuit Examples
15.4 Review of Feedback Amplifier Stability
15.5 Single-Pole Operational Amplifier Compensation
15.6 High-Frequency Oscillators
Summary
Key Terms
Additional Readings
Problems

Chapter Goals

- Review the concepts of negative and positive feedback
- Review loop transmission feedback analysis techniques
- Review the application of Blackman's theorem to feedback amplifiers
- Understand the topologies and characteristics of the series-shunt, shunt-shunt, shunt-series, and series-series feedback configurations
- Analyze midband characteristics of each feedback configuration with loop transmission theory and Blackman's theorem
- Understand the effects of feedback on frequency response and feedback amplifier stability
- Practice interpreting feedback amplifier stability in terms of Nyquist and Bode plots
- Use SPICE ac and transfer function analyses to characterize feedback amplifiers
- Develop techniques to determine the loop gain of closed-loop amplifiers using SPICE simulation or measurement
- Learn to design operational amplifier frequency compensation using Miller multiplication
- Develop relationships between op-amp unity gain frequency and slew rate.
- Discuss the Barkhausen criteria for oscillation
- Develop various types of oscillator circuits
- Understand high-frequency LC and crystal oscillator circuits
- Explore negative resistance in oscillator circuits
- Present the LCR model of the quartz crystal
- Understand ring oscillators

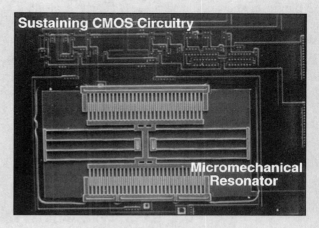

An oscillator employing a MEMS[1] frequency selective resonator.
©Nguyen, C. T-C. and R. T. Howe, "An integrated Micromechanical Resonator High-Q Oscillator." IEEE Journal of Solid-State Circuits 34, no. 4 (April 1999): 440–445. Copyright ©1999 IEEE. Reprinted with permission.

Examples of feedback systems abound in daily life. The thermostat that senses the temperature of a room and turns the air-conditioning system on and off is one example. Another is the remote control that we use to select a channel on the television or set the volume at an acceptable level. The heating and cooling system uses a simple temperature transducer to compare the temperature with a fixed set point. However, we are part of the TV remote control feedback system; we operate the control until our senses tell us that the audio and optical information is what we want.

The theory of negative feedback in electronic systems was first developed by Harold Black of the Bell Telephone System. In 1928, he invented the feedback amplifier to stabilize the gain of early telephone repeaters. Today, some form of feedback is used in virtually every

---

[1] Micro-Electro-Mechanical System. C. T-C. Nguyen and R. T. Howe, "An integrated micromechanical resonator high-Q oscillator," IEEE Journal of Solid-State Circuits, vol. 34, no. 4, pp. 440–445, April 1999.

electronic system. This chapter formally reviews the concept of feedback, which is an invaluable tool in the design of electronic systems. Valuable insight into the operation of many common electronic circuits can be gained by recasting the circuits as feedback amplifiers.

We already encountered **negative** (or **degenerative**) **feedback** in several forms. The four-resistor bias network uses negative feedback to achieve an operating point that is independent of variations in device characteristics. We also found that a source or emitter resistor can be used in an inverting amplifier to control the gain and bandwidth of the stage. Many of the advantages of negative feedback were actually uncovered during the discussion of operational amplifier circuit design. Generally, feedback can be used to achieve a tradeoff between gain and many of the other properties of amplifiers:

- *Gain stability:* Feedback reduces the sensitivity of gain to variations in the values of transistor parameters and circuit elements.
- *Input and output impedances:* Feedback can increase or decrease the input and output resistances of an amplifier.
- *Bandwidth:* The bandwidth of an amplifier can be extended using feedback.
- *Nonlinear distortion:* Feedback reduces the effects of nonlinear distortion. (For example, feedback can be used to minimize the effects of the dead zone in a class-B amplifier stage.)

Feedback may also be **positive** (or **regenerative**), and we explore the use of positive feedback in **oscillator circuits** in this chapter. We encountered the use of a combination of negative and positive feedback in the discussion of *RC* active filters and multivibrator circuits in Chapter 12. **Sinusoidal oscillators** use positive feedback to generate signals at specific desired frequencies; they use negative feedback to stabilize the amplitude of the oscillations. Ring oscillators are often utilized to generate high-frequency clock signals in digital circuits.

Positive feedback in amplifiers is usually undesirable. Excess phase shift in a feedback amplifier may cause the feedback to become regenerative and cause the feedback amplifier to break into oscillation. Remember that positive feedback was identified in Chapter 9 as a potential source of oscillation problems in tuned amplifiers.

15.1 BASIC FEEDBACK SYSTEM REVIEW

Let's review the feedback system introduced in Chapter 11. The diagram in Fig. 15.1 represents a simple feedback amplifier. It consists of an amplifier with transfer function $A(s)$, referred to as the **open-loop amplifier,** a **feedback network** with transfer function $\beta(s)$, and a summing block indicated by Σ.

15.1.1 CLOSED-LOOP GAIN

In Fig. 15.1, the input to the open-loop amplifier A is provided by the summing block, which actually develops the difference between the input signal v_i and the feedback signal v_f:

$$\mathbf{v_d} = \mathbf{v_i} - \mathbf{v_f} \tag{15.1}$$

The output signal is equal to the product of the open-loop amplifier gain and the input signal to the amplifier:

$$\mathbf{v_o} = A\mathbf{v_d} \tag{15.2}$$

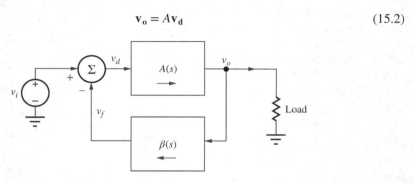

Figure 15.1 Classic block diagram for a feedback system.

and the signal fed back to the input is given by

$$\mathbf{v}_f = \beta \mathbf{v}_o \tag{15.3}$$

Combining these, as we did in Chapter 11, results in the core equations which describe the **closed-loop gain** of a negative feedback amplifier:

$$A_v = \frac{\mathbf{v_o}}{\mathbf{v_i}} = \frac{A}{1+A\beta} = \frac{1}{\beta}\left(\frac{A\beta}{1+A\beta}\right) = A_v^{\text{Ideal}} \frac{T}{1+T} \tag{15.4}$$

where A_v is the closed-loop gain, A is the **open-loop gain** of the amplifier, and the product $T = A\beta$ is defined as the **loop gain** or **loop transmission**. A_v^{Ideal} is the **ideal gain** that would be achieved if the amplifier were ideal. β is the feedback factor that describes how much of the output is fed back to the input of the amplifier. As in Chapter 11, we will need to ensure that our feedback is connected as negative feedback to match our basic topology defined in Fig. 15.1 and ensure **stability**. Also, Eq. (15.4) still holds if each of the terms are complex frequency-dependent terms instead of simple midband small-signal terms.

15.1.2 CLOSED-LOOP IMPEDANCES

Recall from Chapter 11 that we use **Blackman's theorem** to calculate the resistance (or impedance) looking into an arbitrary pair of terminals in a negative feedback amplifier:

$$R_X = R_X^D \frac{1+|T_{SC}|}{1+|T_{OC}|} \tag{15.5}$$

where R_X^D is the resistance seen with the feedback disabled, T_{SC} is the loop transmission with a short circuit across the selected terminal pair, and T_{OC} is the loop transmission with an open circuit across the selected terminal pair.

15.1.3 FEEDBACK EFFECTS

We now turn to an example negative feedback circuit to motivate our analyses. The circuit in Fig. 15.2 is a differential amplifier with current-mirror load. The only difference between this circuit and what we have previously analyzed is the negative feedback connection between the output and the inverting input of the differential pair. As we learned in our analysis of op-amp circuits, negative feedback works to minimize the difference between the inputs. With a direct connection from output to inverting input, the output is made to track the noninverting input, creating a unity-gain amplifier. Note that the differential pair provides the summation (subtraction) operation shown in Fig. 15.1.

To illustrate some of the effects of the feedback, let's explore some simulations of the circuit. We'll use BF = 100, VAF = 50 V, and IS = 1 fA for both the *npn* and *pnp* models. Simulations show that the midband gain $\mathbf{v_o}/\mathbf{v_i} = 0.996$, so the circuit is indeed a unity-gain amplifier. Without any understanding of feedback, we would expect the input resistance presented to source v_i is $2r_\pi + R_i$, or about 5.1 kΩ. If this is the correct value, increasing R_i to 5 kΩ should decrease the gain to about 0.5 due to voltage division at the input. However, Table 15.1 shows that the gain decreases less than 4 percent as the source impedance is increased well beyond our apparently erroneous calculation of the input impedance. So, feedback has increased the effective input resistance rather dramatically. Looking at Blackman's theorem, Eq. (15.5), we can see that the T_{OC} term is zero[2] when the noninverting input is left open-circuited.

On the output side, without the use of Blackman's theorem, we might calculate the output resistance as[3]

$$R_{\text{out}} = R_{iB2} \| R_{iC2} \| R_{iC4} \cong 2r_\pi \| r_{o2} \| r_{o4} \approx 4.2 \text{ k}\Omega \tag{15.6}$$

---

[2] There is actually a negligibly small value of T_{OC} because of conduction through r_o.

[3] $R_{iB2} = r_{\pi 2} + (\beta_{o2} + 1)\frac{r_{\pi 1} + R_i}{\beta_{o1}+1} = r_{\pi 2} + r_{\pi 1} + R_i \cong R_i + 2r_\pi = 5.1 \text{ k}\Omega$

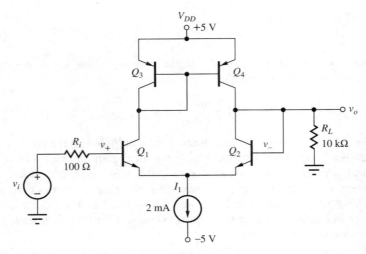

Figure 15.2 Single-stage differential feedback amplifier ($g_m = 0.04$ S, $r_\pi = 2.5$ kΩ, $r_o = 55$ kΩ).

TABLE 15.1 Gain Sensitivity to Source Resistance Variation

R_i (kΩ)	v_o/v_i
0.1	0.996
0.5	0.996
1	0.996
5	0.993
10	0.990
50	0.964

TABLE 15.2 Gain Sensitivity to Load Resistance Variation

R_L (kΩ)	v_o/v_i
10	0.996
5	0.994
1	0.974
0.5	0.950

TABLE 15.3 Gain Sensitivity to Parameter Variation

PARAMETER	v_o/v_i
BF = 100	0.996
BF = 200	0.997
BF = 50	0.996
VAF = 50	0.996
VAF = 100	0.997
VAF = 25	0.996

Note that we neglected the small R_i in this calculation. Given this result, we expect that the gain will decrease if we reduce the load resistance R_L. Table 15.2 shows the results from a series of simulations as the load resistance is changed, but the gain is only reduced by 5 percent when the load resistance has been reduced to 500 Ω. This indicates that the amplifier output resistance must be much less than our estimate of 4.2 kΩ. Here again we find that the negative feedback has significantly changed the circuit characteristics. In this case, the output resistance has been reduced by the feedback.

Another important characteristic of feedback amplifiers is reduced sensitivity to circuit parameter variations. For example, Table 15.3 shows how gain changes with changes in transistor forward current gain and Early Voltage. The doubling or halving of these parameters causes less than 0.1 percent change in the simulated closed-loop gain! The results in this section are explained by examination of Eq. (15.4). As long as T is large, $T/(1 + T)$ is nearly 1 and the gain will remain close to its ideal value.

Feedback allows us to create circuits that are robust to changes in device and other parameters. This explains how electronic systems are built with reliable characteristics despite large manufacturing tolerances for many of the parameters of the individual components that are used to build systems. Feedback is essential for reproducible and accurate behavior of amplifiers.

EXERCISE: Find the Q-points for the four transistors in the amplifier in Fig. 15.2. What is the Q-point value for the output voltage v_O?

ANSWERS: (1 mA, 5 V), (1 mA, 0.7 V), (1 mA, 0.7 V), (1 mA, 5 V); 0 V

15.2 FEEDBACK AMPLIFIER ANALYSIS AT MIDBAND

Referring to what we learned in Chapter 11, the amplifier in Fig. 15.2 is configured in a series-shunt topology. The feedback connection is directly sampling the output voltage and is therefore shunting the output. The feedback signal is a voltage applied in series (across the differential pair) with the input signal.

15.2.1 CLOSED-LOOP GAIN

We will now calculate the closed-loop gain using our feedback equation with the help of the ac equivalent circuit in Fig. 15.3. First, the ideal feedback factor β is unity, so the ideal gain A_v^{Ideal} is 1.0. Loop transmission (loop gain) T is calculated including the loading effects of the feedback connection to the output. If the signal voltage at the base of Q_2 is v_o, then the output voltage will be $(i_{c2} + i_{c4})$ times the resistance at the output node:

$$\mathbf{v_o} = (\mathbf{i_{c2}} + \mathbf{i_{c4}}) R_{\text{out}} = -\left(g_{m2}\frac{\mathbf{v_{b2}}}{2} + g_{m2}\frac{\mathbf{v_{b2}}}{2}\right)(r_{o2} \| r_{o4} \| R_{iB2} \| R_L) \tag{15.7}$$

Since the output is connected directly to the base of Q_2, the loop transmission is

$$T = -\frac{\mathbf{v_o}}{\mathbf{v_{b2}}} = A\beta = g_{m2}(r_{o2} \| r_{o4} \| R_{iB2} \| R_L) = (0.04S)(55 \| 55 \| 5.1 \| 10) \text{ k}\Omega = 120 \tag{15.8}$$

where

$$R_{iB2} = r_{\pi 2} + (\beta_{o2} + 1)\frac{r_{\pi 1} + R_i}{\beta_{o1} + 1} = r_{\pi 2} + r_{\pi 1} + R_i \cong R_i + 2r_\pi = 5.1 \text{ k}\Omega \tag{15.9}$$

Therefore, the closed-loop gain not including source attenuation is

$$A_v = (1)\left(\frac{120}{1 + 120}\right) = 0.992 \tag{15.10}$$

Recall that the simulated value of the closed-loop gain is 0.996, quite close to our calculated value.

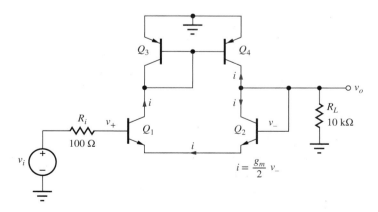

Figure 15.3 An ac equivalent circuit with currents for loop-gain calculation.

15.2.2 INPUT RESISTANCE

The **closed-loop input resistance** can be calculated from Blackman's theorem, and T_{SC}, T_{OC}, and R_{in}^D must be found in order to evaluate Eq. (15.5). To find T_{SC}, the input at v_i is shorted to ground, and we see that T_{SC} is the same as the loop transmission we calculated for the closed-loop gain. To find T_{OC}, we open the circuit at the base of Q_1. T_{OC} is approximately zero because the amplifier gain (with respect to an input at the base of Q_2) is zero with an ac open circuit at the base of Q_1 since no current can go through either Q_1 or Q_2 (neglecting the transistor output resistances). This condition is equivalent to an infinite resistance in series with the equivalent resistance of the differential pair.

Finally we need to find the input resistance with the feedback disabled, R_{in}^D. If we mentally ignore the presence of the feedback, then the input resistance, including the effects of the equivalent resistance on the base of Q_2, is $R_{in}^D = R_i + r_{\pi 1} + r_{\pi 2} + (r_{o4} \| r_{o2} \| R_L) = 12.2\,\text{k}\Omega$. These values enable the direct calculation of the input resistance from Blackman's theorem:

$$R_{in} = R_{in}^D \frac{1 + |T_{SC}|}{1 + |T_{OC}|} = 12.2\,\text{k}\Omega \left(\frac{1 + 120}{1 + 0}\right) = 1.48\,\text{M}\Omega \tag{15.11}$$

The simulated value of R_{in} is 1.43 MΩ. Note that although the closed-loop input resistance is increased by feedback, it is directly proportional to the load impedance in this simple feedback configuration, so it is clear that this topology would not work as a buffer to small load resistances. It is important to recognize that we do not actually disconnect the feedback network when we mentally disable the feedback for calculating the open-loop R_{in}. We still include the loading effects of the elements associated with the feedback connection.

15.2.3 OUTPUT RESISTANCE

Blackman's theorem is also used to calculate the **closed-loop output resistance.** In this case we need to find the loop transmission with the output open and the output shorted to ground. For this case, T_{OC} is the loop transmission we calculated earlier when calculating the voltage gain excluding the effect of R_L, since we are looking into the amplifier from the load. With this change to Eq. (15.7), we find $T_{OC} = 171$. T_{SC} is zero since the amplifier gain is zero when the output is shorted to small-signal ground. The output impedance with the feedback disabled, R_{out}^D, is simply the resistance looking into the amplifier output ignoring the effect of feedback and is given by

$$R_{out}^D = R_{iB2} \| r_{o2} \| r_{o4} = \left[r_{\pi 2} + (\beta_{o2} + 1)\left(\frac{r_{\pi 1} + R_i}{\beta_{o1} + 1}\right)\right] \| r_{o2} \| r_{o4} = (r_{\pi 2} + r_{\pi 1} + R_i) \| r_{o2} \| r_{o4} = 4.30\,\text{k}\Omega \tag{15.12}$$

R_{out} can now be calculated as

$$R_{out} = R_{out}^D \frac{1 + T_{SC}}{1 + T_{OC}} = 4.30\,\text{k}\Omega \left(\frac{1 + 0}{1 + 171}\right) = 25.0\,\Omega \tag{15.13}$$

The simulated value of R_{out} is 25.6 Ω. This is clearly much lower than our earlier estimate and explains why the amplifier gain is so insensitive to the changes in load resistance we simulated in the previous section.

We calculate the overall gain including source attenuation and output loading as

$$A = \left(\frac{R_{in}}{R_{in} + R_i}\right) A_v = 0.992\,\text{V/V} \tag{15.14}$$

Because the closed-loop input resistance is so large, the input attenuation due to source resistance is small. Notice in the above equations that we included R_L when calculating closed-loop gain and input resistance. As we can see from the above equations, this amplifier has a high open-loop output resistance and the output load plays a direct role in the gain and input impedance calculations. We therefore need to include R_L for accurate results. We will see later that amplifiers with low open-loop output resistance or very high loop gain can be analyzed independent of R_L with minimal impact on accuracy.

EXERCISE: For the output resistance calculation, convince yourself that $T_{SC} = 0$ when the output is short circuited.

EXERCISE: For the input resistance calculation, convince yourself that $T_{OC} = 0$ when the input is open circuited.

EXERCISE: Calculate T_{OC} as required for the output resistance calculation and confirm that its value is 171.

15.2.4 OFFSET VOLTAGE CALCULATION

An offset voltage arises in the amplifier in Fig. 15.2 because of the base current error in the *pnp* current mirror as well as the imbalances in collector-emitter voltages of the transistor pairs. Both these errors are small, so we will simplify the calculation by treating each error individually. First assume that the Early voltage is infinite so that the voltage mismatches do not cause an error. Then $I_{C4} = I_{C3}$, and

$$I_{C1} = I_{C3} + I_{B3} + I_{B4} = I_{C3}\left(1 + \frac{2}{\beta_{Fp}}\right) \quad \text{and} \quad I_{C2} = I_{C4} - I_{B2} \quad \text{yields} \quad I_{C2} = \frac{I_{C3}}{1 + \frac{2}{\beta_{Fn}}} \quad (15.15)$$

$$V_{OS}^{\beta} = V_T \ln\frac{I_{C2}}{I_{C1}} = -V_T \ln\left[\left(1 + \frac{2}{\beta_{Fp}}\right)\left(1 + \frac{1}{\beta_{Fn}}\right)\right] \cong -V_T\left(\frac{2}{\beta_{Fp}} + \frac{1}{\beta_{Fn}}\right)$$

since $\ln(1 + x) \cong x$ for small x. Now assume that the Early voltage is finite and β_F is infinite. The mismatches in currents are now caused by the collector-emitter voltage differences. For this circuit, we have $V_{EC3} = V_{CE2} = 0.7$ V and $V_{EC1} = V_{CE4} = 5$ V, and the circuit forces $I_{C1} = I_{C3}$ and $I_{C2} = I_{C4}$. Thus

$$I_S^n \exp\left(\frac{V_{BE1}}{V_T}\right)\left(1 + \frac{V_{CE1}}{V_{An}}\right) = I_{C3} \quad \text{and} \quad I_S^n \exp\left(\frac{V_{BE2}}{V_T}\right)\left(1 + \frac{V_{CE2}}{V_{An}}\right) = I_{C4}$$

$$\exp\left(\frac{V_{BE2} - V_{BE1}}{V_T}\right)\frac{\left(1 + \frac{V_{CE2}}{V_{An}}\right)}{\left(1 + \frac{V_{CE1}}{V_{An}}\right)} = \frac{I_{C4}}{I_{C3}} \quad \text{and} \quad \frac{I_{C4}}{I_{C3}} = \frac{\left(1 + \frac{V_{CE4}}{V_{Ap}}\right)}{\left(1 + \frac{V_{CE3}}{V_{Ap}}\right)}$$

$$V_{OS}^{VA} = (V_{BE2} - V_{BE1}) = V_T \ln\left[\frac{\left(1 + \frac{V_{CE4}}{V_{Ap}}\right)\left(1 + \frac{V_{CE1}}{V_{An}}\right)}{\left(1 + \frac{V_{CE3}}{V_{Ap}}\right)\left(1 + \frac{V_{CE2}}{V_{An}}\right)}\right] \quad (15.16)$$

$$\cong V_T\left[\frac{V_{CE4}}{V_{Ap}} + \frac{V_{CE1}}{V_{An}} - \frac{V_{CE3}}{V_{Ap}} - \frac{V_{CE2}}{V_{An}}\right]$$

For $\beta_{Fn} = \beta_{Fp} = 100$ and $V_{An} = V_{Ap} = 50$ V, the offset voltage is

$$V_{OS} = V_{OS}^{\beta} + V_{OS}^{VA} \cong -0.025\left(\frac{3}{100}\right) + 0.025\left[\frac{5}{50} + \frac{5}{50} - \frac{0.7}{50} - \frac{0.7}{50}\right] \quad (15.17)$$

$$V_{OS} = -0.75 \text{ mV} + 4.3 \text{ mV} = 3.55 \text{ mV}$$

EXERCISE: Simulate the circuit in Fig. 15.2 using the parameters above and verify the offset voltage calculations above.

15.3 FEEDBACK AMPLIFIER CIRCUIT EXAMPLES

In the following sections we will use this same approach to calculate midband closed-loop gain and the closed-loop input and output impedances of a variety of feedback topologies. The midband analysis can be summarized with the following steps:

1. Determine if the feedback output connection is shunt or series. If it is a shunt connection, the network is sensing output voltage. If series, the feedback is sensing the output current.
2. Determine if the feedback input connection is shunt or series. If shunt, current is being fed back to the input, and if series, voltage is being fed back to the input.
3. Given the type of feedback connections, the units for the feedback factor β are determined, and the type of amplifier can be found from Table 15.4.
4. Calculate feedback factor β for the idealized version of the amplifier. For example, with a series-shunt configuration, the idealized amplifier input impedance is infinite, and the output impedance is assumed to be zero. Ideal gain is then calculated as the reciprocal of the ideal feedback factor.
5. Calculate the loop transmission including the amplifier and feedback network loading effects.
6. Use Eq. (15.4) to calculate the closed-loop gain of the amplifier.
7. Use Blackman's theorem to calculate the amplifier input and output impedances or any other desired impedances in the circuit. The open-circuit and short-circuit loop gains are calculated including nonideal loading effects.
8. Calculate the overall gain including input and output loading.

We will now use these steps to perform midband analysis of a number of amplifier topologies.

15.3.1 SERIES-SHUNT FEEDBACK—VOLTAGE AMPLIFIERS

Figure 15.4 illustrates a two-stage feedback amplifier known as a **series-shunt feedback** pair. At first glance, the topology may appear a bit confusing. It appears to have two paths from input to output, one through the collector of Q_1 and another through the emitter of Q_2. While this is true, the dominant forward path is the high gain path through the two common-emitter stages of Q_1 and Q_2. Given this, the feedback path is apparently from the output through R_2 back to the emitter of Q_1. Since the feedback is connected directly to the output, it is a shunt connection and the feedback network is sampling voltage.

The feedback network does not seem to be summing a current into the input network, so it is apparently a series connection at the input. Recall that the small-signal output current at the collector of a transistor is $g_m \mathbf{v_{be}} = g_m(\mathbf{v_b} - \mathbf{v_e})$. So, the transistor is acting as a differential amplifier, generating an output that is proportional to the difference between the small-signal input voltage at the base and the small-signal voltage fed back to the emitter. This leads us to Fig. 15.5, a simplified small-signal equivalent of the series-shunt voltage amplifier shown in Fig. 15.4. Let's now calculate the midband gain, input, and output impedances in Ex. 15.1.

TABLE 15.4
Determining Amplifier Type Based on Feedback Connections

FEEDBACK CONNECTION	SENSED SIGNAL	FED BACK SIGNAL	FEEDBACK FACTOR, β	GAIN RATIO	AMPLIFIER GAIN
Series-shunt	Voltage	Voltage	V/V	V/V	Voltage
Shunt-shunt	Voltage	Current	I/V	V/I	Transresistance
Series-series	Current	Voltage	V/I	I/V	Transconductance
Shunt-series	Current	Current	I/I	I/I	Current

15.3 Feedback Amplifier Circuit Examples

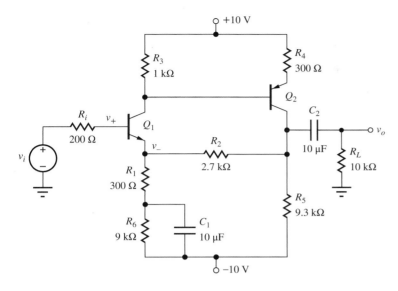

Figure 15.4 Two-stage feedback voltage amplifier–the series-shunt feedback pair.

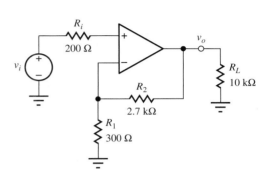

Figure 15.5 Ideal small-signal version of amplifier in Fig. 15.4.

EXAMPLE 15.1   TWO-STAGE SERIES-SHUNT VOLTAGE AMPLIFIER

Perform an analysis of a two-stage series-shunt feedback amplifier.

PROBLEM The amplifier of Fig. 15.4 has been constructed. Find the small-signal gain and input and output resistances. Assume dc base currents are negligible, $V_A = \infty$, and β_0 is 100.

SOLUTION **Known Information and Given Data:** The circuit diagram appears in Fig. 15.4 and $\beta_0 = 100$. Transistor output resistances are infinite.

Unknowns: Ideal gain, open-loop gain, loop transmission, and Blackman terms for the input and output impedances.

Approach: Use amplifier gain analysis from previous chapters, feedback analysis procedure, and Blackman's theorem.

Assumptions: $V_{BE} = 0.7$ V, $V_T = 25$ mV, and small-signal midband conditions apply; dc base currents are negligible and r_o is infinite.

Analysis: First, we must draw the dc equivalent circuit in (a) below and find the dc solution. Neglecting the dc base current, $V_{B1} = 0$ V, so $V_{E1} = -0.7$ V. If we assume the dc current through R_2 is negligible

$$I_{E1} \cong \frac{-0.7 - (-10)}{9.3 \text{ K}} = 1 \text{ mA} \qquad V_{C1} = 10 - I_{C1}R_3 = 9 \text{ V}$$

$$I_{E2} = \frac{10 - (9 - V_{BE2})}{R_4} = 1 \text{ mA} \qquad V_{C2} = I_{C2}R_5 + (-10V) = -0.7 \text{ V}$$

Now our assumption that the dc current in R_2 is zero needs to be checked:

$$I_{R2} = \frac{V_{C2} - V_{E1}}{R_2} = 0 \checkmark$$

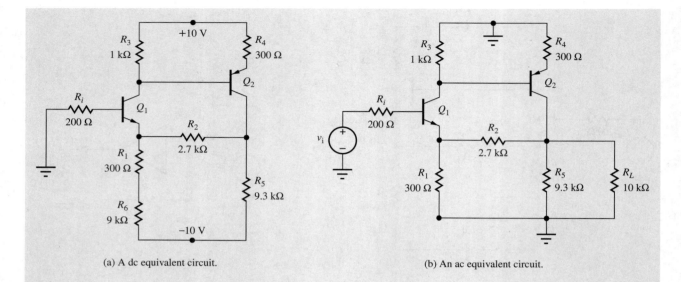

(a) A dc equivalent circuit. (b) An ac equivalent circuit.

Next, we draw the ac equivalent circuit by shorting all the capacitors and placing ac grounds at the two power supplies as in (b) above. The small-signal parameters are

$$g_{m1} = g_{m2} = 40(0.001) = 0.04\,\text{S}, r_\pi = 100/g_m = 2500\,\Omega, r_o = \infty$$

Now we turn to our feedback analysis procedure.

Step 1: As discussed above, the feedback network is composed of R_2 and R_1. The output voltage is directly sampled by R_2, so it is a shunt connection.

Step 2: The signal fed back is the small-signal voltage at the emitter of Q_1. This is in series with the input voltage at the base of Q_1, so this is a **series feedback connection.**

Step 3: Since voltage is sampled and voltage is fed back to the input, the feedback factor has units of voltage/voltage and the amplifier is a series-shunt voltage amplifier.

Step 4: As is apparent in Fig. 15.5, the amplifier type is a series-shunt configuration, so the ideal feedback factor is just the voltage division across the feedback network, $\beta = R_1/(R_1 + R_2)$, and the ideal gain is

$$A_v^{\text{Ideal}} = \frac{R_2 + R_1}{R_1} = 10$$

Step 5: The loop transmission is found by injecting a signal at a point in the circuit and calculating how much signal is returned to that point through the feedback path. Another approach is to calculate the gain around the loop, making sure to include all of the loading effects. We will start at the emitter of Q_1 and calculate the gain back around the loop to the same point. This requires calculating the common-base gain of Q_1, the common-emitter gain of Q_2, and finally the nonideal voltage division at the feedback network.

$$T = [g_{m1}(R_3 \| R_{iB2})]\left[-\frac{g_{m2}}{1+g_{m2}R_4}([R_2 + R_1 \| R_{iE1}] \| R_5 \| R_L)\right]\left(\frac{R_{iE1} \| R_1}{R_{iE1} \| R_1 + R_2}\right)$$

$$R_{iB2} = r_{\pi 2} + (\beta_o + 1)R_4 = 32.8\,\text{k}\Omega \quad R_{iE1} = \frac{r_{\pi 1} + R_i}{(\beta_{o1} + 1)} = \frac{200 + 2500}{101}\,\Omega = 26.7\,\Omega$$

$$T = 0.04\text{S}(970\,\Omega)\left[-\frac{0.04\text{S}(1720\,\Omega)}{1+0.04\text{S}(300\,\Omega)}\right]\left(\frac{26.7\,\Omega}{26.7\,\Omega + 2700\,\Omega}\right) = -2.01$$

Notice that T is quite low. As a consequence, the gain error will be large. We also see that T is negative. Remember we must always check that we have negative feedback when we are building a feedback amplifier.

Step 6: The closed-loop gain of the feedback amplifier is calculated according to Eq. (15.4). However, remember that the negative sign on T is already included in our high-level feedback description in Fig. 15.1, so T will be positive when we evaluate Eq. (15.4):

$$A_v = A_v^{\text{Ideal}} \frac{T}{1+T} = 10 \frac{2.01}{1+2.01} = 6.68$$

This expression does not include attenuation at the input due to voltage division. We expect this factor to be fairly insignificant since the source impedance is low. The actual value can be calculated after the input impedance is calculated.

Step 7: Since the loop transmission is low, the input and output impedances will not be changed much by feedback.

Input Resistance: First we calculate the open-loop input resistance looking into the base of Q_1 ignoring the effect of feedback.

$$R_{\text{in}}^D = R_{\text{in}B1} = r_{\pi 1} + (\beta_0 + 1)(R_1 \| [R_2 + R_5 \| R_L \| R_{iC2}]) \cong 31.7 \text{ k}\Omega$$

T_{OC} is zero for the input resistance calculation since the gain through Q_1 is reduced to zero if the impedance looking out of the base of Q_1 is infinite (i.e., for an open circuit, zero base current yields zero collector and emitter currents).[4] T_{SC} is very close to the value calculated above except that the R_{iE1} term is reduced slightly since the impedance looking out of the base is zero instead of 200 Ω. With this change, $T_{SC} = 1.86$ and we calculate the closed-loop resistance looking into the input (the base of Q_1):

$$R_{\text{in}} = R_{\text{in}}^D \frac{1 + T_{SC}}{1 + T_{OC}} = 31.7 \text{ k}\Omega \left(\frac{1 + 1.86}{1 + 0} \right) = 90.7 \text{ k}\Omega$$

Output Resistance: For the output resistance, we mentally disable the feedback and calculate the impedance looking into the output of the amplifier, not including the load resistance.

$$R_{\text{out}}^D = R_{iC2} \| R_5 \| (R_2 + R_{iE1} \| R_1) \cong 2.11 \text{ k}\Omega$$

where $R_{iC2} = r_{o2}(1 + g_{m2} R_4)$ is negligible. T_{SC} is now zero since the amplifier gain is zero when the output is shorted to small-signal ground. T_{OC} is nearly identical to the loop transmission calculated earlier except that R_L is not included:

$$T_{OC} = [g_{m1}(R_3 \| R_{iB2})] \left(-\frac{g_{m2}}{1 + g_{m2} R_4} [(R_2 + R_1 \| R_{\text{in}E1}) \| R_5] \right) \left(\frac{R_{iE1} \| R_1}{R_{iE1} \| R_1 + R_2} \right)$$

$$|T_{OC}| = 0.04S(970 \,\Omega) \left[\frac{0.04S(2107 \,\Omega)}{1 + 0.04S(300 \,\Omega)} \right] \left(\frac{26.7 \,\Omega}{26.7 \,\Omega + 2.7 \text{ k}\Omega} \right) = 2.46$$

---

[4] Again, r_{o1} is being neglected.

The output resistance is now calculated with Blackman's theorem.

$$R_{\text{out}} = R_{\text{out}}^D \frac{1 + |T_{SC}|}{1 + |T_{OC}|} = 2.11 \text{ k}\Omega \left(\frac{1+0}{1+2.46}\right) = 610 \text{ }\Omega$$

Step 8: We calculate the overall gain including source attenuation and output loading as

$$A = \left(\frac{R_{\text{in}}}{R_{\text{in}} + R_i}\right) A_v = \frac{90.7 \text{ k}\Omega}{90.7 \text{ k}\Omega + 200 \text{ }\Omega} 6.68 = 6.67$$

We again see that due to the low ratio of source to input resistance, the overall gain is nearly identical to the amplifier gain. We should also note that we included R_L directly in the amplifier gain calculations so there is no need to account for signal attenuation from output resistance to load resistance in this equation.

An Alternate Approach: This solution uses a slightly different approach to the calculations in which R_i and R_L are considered to be part of the amplifier and are included in all the calculations. The closed-loop gain now represents the gain from source v_i to the output. The input resistance is the total resistance presented to source v_i, and the output resistance includes the shunting effect of R_L. The effects of R_i and R_L can easily be eliminated at the end of the calculations if desired.

Closed-Loop Gain: The loop gain was originally calculated including the effects of R_i and R_L, so $T = -2.01$, and

$$A_v = A_v^{\text{Ideal}} \frac{T}{1+T} = 10 \frac{2.01}{1+2.01} = 6.68$$

The input resistance without feedback now includes R_i:

$$R_{\text{in}}^D = R_i + R_{iB1} = 31.9 \text{ k}\Omega$$

The loop gain T_{OC} with v_i open is zero, and the loop gain with v_i set to zero is $T_{SC} = T$. Thus

$$R_{\text{in}} = 31.9 \text{ k}\Omega \frac{1 + 2.01}{1 + 0} = 96.0 \text{ k}\Omega$$

The input resistance at the base of Q_1 would then be $R_{\text{in}B1} = R_{\text{in}} - 200 \text{ }\Omega = 95.8 \text{ k}\Omega$. The output resistance now includes R_L:

$$R_{\text{out}}^D = R_L \| R_{iC2} \| R_5 \| (R_2 + R_{iE1} \| R_1) \cong 1.74 \text{ k}\Omega$$

The loop gain T_{OC} with the output open is equal to T, and the loop gain with a short at the output is zero. Therefore

$$R_{\text{out}} = 1.74 \text{ k}\Omega \left(\frac{1+0}{1+2.01}\right) = 578 \text{ }\Omega$$

Removing R_L from the output resistance yields

$$R_{\text{out}}' = \left[\frac{1}{R_{\text{out}}} - \frac{1}{R_L}\right]^{-1} = 614 \text{ }\Omega$$

Discussion: Our first conclusion from this analysis is that this is not a particularly "good" amplifier design. The loop gain is quite low, so we are not taking advantage of many of the characteristics of negative feedback. In particular, our gain error (reciprocal of $1 + T$) will be high and the input and output impedances are not significantly enhanced by the negative feedback.

On a more general note, from our equations we see that the output load directly impacts the input impedance, so the design is also not well buffered. Increased loop gain would increase the input to output impedance ratio and therefore also improve this characteristic of the amplifier.

Computer-Aided Analysis: Simulations of this amplifier with BF = 100, VAF = 1000, and IS = 1 fA show a gain of 6.52, an input resistance of 87.0 kΩ, and an output impedance of 644 Ω. These results confirm our hand calculations. The discrepancies are due to the different values of T, g_m, and r_π that are used in SPICE. Note that SPICE transfer function analysis *should not* be used on this problem because of the presence of bypass and coupling capacitors!

EXERCISE: Calculate midband loop gain, R_{in}, R_{out}, and overall gain of the previous circuit if a 10 μF bypass capacitor is placed across R_4.

ANSWERS: −19.2, 596 kΩ, 86.8 Ω, 9.50; SPICE: −16.5, 533 kΩ, 105 Ω, 9.43

15.3.2 DIFFERENTIAL INPUT SERIES-SHUNT VOLTAGE AMPLIFIER

The amplifier in Fig. 15.6 is a more traditional **series-shunt voltage amplifier.** It is a simple op-amp structure with a FET differential input stage, a common-source gain stage, and a common-drain output buffer stage. The feedback network is composed of R_2 and R_1. In the following example we will analyze the characteristics of this negative feedback amplifier. Be aware that without additional modification, this amplifier will likely be unstable. Later we will learn how to predict and compensate for feedback instability.

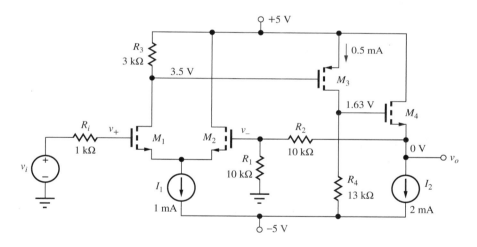

Figure 15.6 Three-stage MOSFET amplifier with negative feedback (K_n = 10 mA/V², K_p = 4 mA/V², V_{TN} = 1 V, V_{TP} = −1 V).

EXAMPLE 15.2   **DIFFERENTIAL INPUT SERIES-SHUNT VOLTAGE AMPLIFIER**

Perform an analysis of the three-stage differential input series-shunt feedback amplifier in Fig. 15.6.

PROBLEM The amplifier of Fig. 15.6 has been designed. Find the small-signal gain, input resistance, and output resistance. K_n = 10 mA/V², K_p = 4 mA/V², V_{TN} = 1 V, V_{TP} = −1 V. The dc bias currents and voltages are shown on the schematic.

SOLUTION **Known Information and Given Data:** The circuit diagram is presented in Fig. 15.6 with the indicated dc bias values.

Unknowns: Ideal gain, open-loop gain, loop transmission, and Blackman terms for the input and output impedances.

Approach: Use amplifier gain analysis from previous chapters, feedback analysis procedure, and Blackman's theorem.

Assumptions: Since λ is unspecified, assume transistor output impedances r_o are infinite; small-signal midband conditions apply. $T = 300$ K.

Analysis: The dc bias currents are easily found from the schematic (0.5 mA, 0.5 mA, 0.5 mA, 2 mA), and the small-signal parameters can be directly calculated:

$$g_{m1} = g_{m2} = \sqrt{2K_n I_D} = 3.16 \text{ mS},$$

$$g_{m3} = 2.00 \text{ mS}, \quad g_{m4} = 6.33 \text{ mS}, \quad r_o = \infty$$

Now we turn to our feedback analysis procedure. First we draw the ac equivalent circuit.

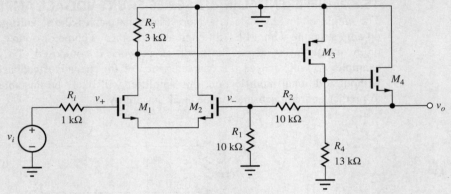

An ac equivalent circuit.

Step 1: As discussed above, the feedback network is composed of R_2 and R_1. The output voltage is directly sampled by R_2, so it is a shunt connection.

Step 2: The signal fed back is the small-signal voltage at the gate of M_2. This voltage is in series with the input voltage at the gate of M_1 (across the differential pair), so this is a series feedback connection.

Step 3: Since voltage is sampled and voltage is fed back to the input, the feedback factor has units of voltage/voltage and the amplifier is a series-shunt voltage amplifier.

Step 4: The amplifier type is series-shunt, so the ideal feedback factor is just the voltage division across the feedback network, $\beta = R_1/(R_1 + R_2)$. The ideal gain is therefore

$$A_v^{\text{Ideal}} = \frac{1}{\beta} = \frac{R_2 + R_1}{R_1} = +2$$

Step 5: We calculate the gain around the loop starting at the gate of M_2 and work our way around the loop back to our starting point. This requires calculating the differential pair gain, the common-source gain, the common-drain gain, and finally the attenuation of the feedback voltage divider. Recall that R_{iG} and the small-signal resistance of an ideal current source are infinite at midband and $r_o = \infty$ was assumed for this problem. These conditions simplify our equations considerably.

$$T = \left(+\frac{g_{m2}}{2}R_3\right)(-g_{m3}R_4)\left[\frac{g_{m4}(R_2+R_1)}{1+g_{m4}(R_2+R_1)}\right]\left(\frac{R_1}{R_1+R_2}\right)$$

$$T = (4.74)(-26.0)(.992)(0.5) = -61.1$$

Notice that T is much larger for this three-stage topology. We also see that T is again negative, satisfying our requirement for negative feedback.

Step 6: The closed-loop gain of the feedback amplifier is calculated according to Eq. (15.4). Remember that the negative sign on T is already included in our high-level feedback description in Fig. 15.1, so T will be positive when we evaluate Eq. (15.4):

$$A_v = A_v^{\text{Ideal}} \frac{T}{1+T} = 2\frac{61.1}{1+61.1} = 1.97$$

Due to the high midband resistance looking into the gate of M_1, there should be no signal loss at the input due to source resistance.

Step 7: Input Resistance: For this topology, input resistance is straightforward since the open-loop resistance looking into the gate of M_1 is approximately infinite. If we needed to calculate it, we would find that $T_{OC} = 0$, and T_{SC} is equal to the loop transmission we found for the closed-loop gain calculation.

Output Resistance: For the output resistance, we mentally disable the feedback[5] and calculate the impedance looking into the output of the amplifier, not including the load resistance.

$$R_{\text{out}}^D = R_{iS4} \| (R_2 + R_1) = (1/g_{m4}) \| (R_2 + R_1) = 157\,\Omega$$

T_{SC} is zero since the amplifier gain is zero when the output is shorted to small-signal ground. T_{OC} is identical to the loop gain calculated earlier since there is no R_L in the problem.

$$T_{OC} = T(\text{loop gain}) = 61.1$$

The output resistance is now calculated with Blackman's theorem.

$$R_{\text{out}} = R_{\text{out}}^D \frac{1 + |T_{SC}|}{1 + |T_{OC}|} = 157\left(\frac{1+0}{1+61.1}\right) = 2.53\,\Omega$$

Step 8: Due to our high-input impedance and low output impedance, our overall gain will be approximately equal to the amplifier gain.

Discussion: This topology is well suited to high forward gain feedback amplifiers. It can be augmented with active loads and a more efficient output stage to produce a true operational amplifier. As mentioned earlier, feedback stability issues will be addressed later in this chapter.

Computer-Aided Analysis: Here we can use a SPICE dc analysis followed by a transfer function analysis from input v_i to the voltage across I_2. The results show a gain of 1.98, an extremely high-input resistance, and an output impedance of 2.50 Ω. These results agree well with our hand calculations.

EXERCISE: Use Blackman's theorem to calculate the midband resistance between the drain of M_1 and small-signal ground. What are R_X^D, T_{SC}, T_{OC}, and R_X?

ANSWERS: 3 kΩ, 0, 61.1, 48.3 Ω.

EXERCISE: What are the Q-points of the four transistors in the amplifier in Fig. 15.6?

ANSWERS: (0.5 mA, 4.82 V), (0.5 mA, 6.32 V), (0.5 mA, 3.37 V), (2 mA, 5.00 V)

---

[5] Set g_{m2} or g_{m3} to zero.

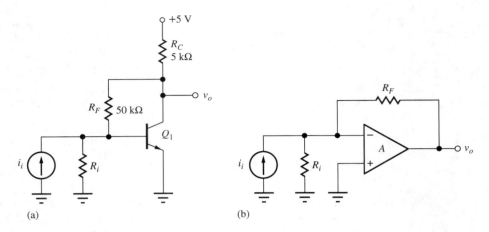

Figure 15.7 (a) Single-transistor transresistance amplifier; (b) idealized transresistance amplifier.

15.3.3 SHUNT-SHUNT FEEDBACK—TRANSRESISTANCE AMPLIFIERS

Figure 15.7 illustrates a simple single transistor **shunt-shunt feedback amplifier**. The amplifier itself is considered to be from the base of Q_1 to the collector. Resistor R_i and current source i_i represent the Norton equivalent of the signal source. The amplifier converts input current i_i to a voltage at the output. We will see that the feedback allows the circuit to present a low impedance to the source network to act as an efficient current sink and generate a voltage at the output. The gain is expressed as a voltage/current ratio that leads to the amplifier classification as transresistance. The feedback network is simply R_F. The output voltage is sampled, and a current is fed back to the input node at the base of Q_1. The input source also delivers a current to the base of Q_1, so the input current and the feedback current are summed at the base node.

EXAMPLE 15.3   **SHUNT-SHUNT FEEDBACK AMPLIFIER ANALYSIS**

Use our feedback analysis procedure to understand the operation of a single transistor **transresistance amplifier**.

PROBLEM Find the small-signal gain, input and output resistance of the amplifier in Fig. 15.7 for the idealized case with $R_i = \infty$. Use $\beta_F = 150$ and $V_A = 50$ V.

SOLUTION **Known Information and Given Data:** The circuit schematic appears in Fig. 15.7; transistor parameters: $\beta_F = 150$ and $V_A = 50$ V.

Unknowns: Ideal gain, open-loop gain, loop transmission, and Blackman terms for the input and output impedances.

Approach: Find the dc operating point; use amplifier gain analysis from previous chapters, the feedback analysis procedure, and Blackman's theorem.

Assumptions: $V_{BE} = 0.7$ V, $V_T = 25$ mV, and small-signal midband conditions apply.

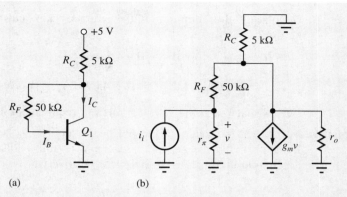

(a) A dc equivalent circuit. (b) An ac equivalent circuit ($g_m = 32.0$ mS, $r_\pi = 4.69$ kΩ, $r_o = 62.4$ kΩ).

Analysis: We first find the dc operating point from the dc equivalent circuit. Knowing the relationship between base and collector current ($I_B = I_C/\beta_F$), we can sum the voltage drops with a loop equation,

$$5V = (I_C + I_B)R_C + I_B R_F + V_{BE} \quad \text{or} \quad 5V - V_{BE} = I_C\left(R_C + \frac{R_C}{\beta_F} + \frac{R_F}{\beta_F}\right)$$

Solving for the collector current yields

$$I_C = \frac{5V - V_{BE}}{R_C + \frac{R_C + R_F}{\beta_F}} = 0.801 \text{ mA}$$

and the collector-emitter voltage is

$$V_{CE} = 5V - (I_C + I_B)R_C = I_C\left(1 + \frac{1}{\beta_F}\right)R_C = 0.968 \text{ V}$$

The corresponding small-signal parameters are

$$g_m = 40(0.801) = 32.0 \text{ mS} \quad r_\pi = \frac{150}{g_m} = 4.69 \text{ k}\Omega \quad r_o \cong \frac{50 \text{ V}}{0.801 \text{ mA}} = 62.4 \text{ k}\Omega$$

Now we turn to our feedback analysis procedure.

Step 1: As discussed above, the feedback network is the resistor R_F. R_F directly samples the output voltage, so it is a shunt connection.

Step 2: The signal fed back is a current to the base of Q_1 and this current is summed directly with the input current i_i. This is therefore a **shunt feedback connection.**

Step 3: Since voltage is sampled and current is fed back to the input, the feedback factor has units of current/voltage and the amplifier is a shunt-shunt transresistance amplifier.

Step 4: The amplifier type is shunt-shunt, so the ideal feedback factor is just the reciprocal of the resistance of the feedback network. The ideal gain is therefore

$$A_{tr}^{\text{Ideal}} = -\frac{1}{\beta} = -R_F = -50{,}000 \, \Omega \text{ (V/A)}$$

The negative sign accounts for the polarity of the voltage drop across R_F when i_i is positive.

Step 5: We calculate the gain around the loop starting at the base of Q_1 in the ac equivalent circuit above and work our way around the loop back to our starting point. This requires calculating the common-emitter voltage gain and the attenuation of the feedback network.

$$T = [-g_m(R_C \| (R_F + r_\pi) \| r_o)]\left(\frac{r_\pi}{r_\pi + R_F}\right)$$

$$T = -0.032S(5\,\text{k}\Omega \| (50\,\text{k}\Omega + 4.69\,\text{k}\Omega) \| 62.4\,\text{k}\Omega)\left(\frac{4.69\,\text{k}\Omega}{4.69\,\text{k}\Omega + 50\,\text{k}\Omega}\right) = -11.7$$

We see that T is negative, satisfying our requirement for negative feedback.

Step 6: The closed-loop gain of the feedback amplifier is calculated according to Eq. (15.4). Remember that the negative sign on T is already included in our high-level feedback description in Fig. 15.1, so T will be positive when we evaluate Eq. (15.4):

$$A_{tr} = A_{tr}^{\text{Ideal}} \frac{T}{1+T} = -50\,\text{k}\Omega\,\frac{11.7}{1+11.7} = -46.1\,\text{k}\Omega$$

The relatively low loop transmission results in a significant reduction (8%) of our gain from the ideal value.

Step 7: Input Resistance: To calculate input resistance, we start with the open-loop resistance with the feedback disabled (e.g., set $g_m v = 0$).

$$R_{\text{in}}^D = r_\pi \| (R_F + R_C \| r_o) = 4.32\,\text{k}\Omega$$

T_{SC} is zero since the signal gain is zero when the base of Q_1 is shorted to ground, whereas T_{OC} is the same as we calculated for the gain, $T_{SC} = 11.7$. Combining these results yields the midband input resistance

$$R_{\text{in}} = R_{\text{in}}^D \frac{1 + |T_{SC}|}{1 + |T_{OC}|} = 4.32\,\text{k}\Omega\,\frac{1+0}{1+11.7} = 340\,\Omega$$

The negative feedback has significantly reduced the input resistance, improving its suitability for sinking input currents.

Output Resistance: For the output resistance, we again mentally disable the feedback (by setting $g_m v = 0$) and calculate the impedance looking into the output of the amp.

$$R_{\text{out}}^D = R_C \| r_o \| (R_F + r_\pi) = 4.27\,\text{k}\Omega$$

T_{SC} is zero since the amplifier gain is zero when the output is shorted to signal ground. T_{OC} equals the previously calculated loop gain, $T_{OC} = 11.7$. The output resistance is now calculated with Blackman's theorem.

$$R_{\text{out}} = R_{\text{out}}^D \frac{1 + |T_{SC}|}{1 + |T_{OC}|} = 4.27\,\text{K}\,\frac{1+0}{1+11.7} = 336\,\Omega$$

Step 8: Since the signal source is an ideal current source and there is no external load in the circuit, the overall transresistance gain is as calculated earlier.

$$A_{tr} = \frac{\mathbf{v}_{\text{out}}}{\mathbf{i}_i} = -46.1\,\text{k}\Omega$$

Discussion: The transresistance amplifier is widely used for amplification of signals from current-mode detectors such as photodiodes. The low input impedance shunts the stray capacitance of the detector to maintain fast response times. We should recognize that the node at the base of Q_1 is equivalent to the virtual ground in op-amp circuits we studied earlier. However, to decrease R_{in} to levels similar to the op-amp version, the loop gain must be significantly increased.

Computer-Aided Analysis: Here we can use a SPICE dc analysis followed by a transfer function analysis from input i_i to the output voltage. The results yield a transresistance gain of $-46.0\,\text{k}\Omega$, an input resistance of 352 Ω, and an output impedance of 335 Ω. These results agree well with our hand calculations.

EXERCISE: Repeat the calculations in the example above if $R_i = 10$ kΩ. What are the new values of T, R_{in}, and R_{out}?

ANSWERS: 8.17, 329 Ω, 464 Ω

Impact of Source and Load Resistances

Now that we have looked at the behavior of the basic single-transistor transresistance amplifier, we will look at the impact of including source and load resistances on the performance of the amplifier, as illustrated in Fig. 15.8. Here we use the results of the previous example to create a model for the amplifier.

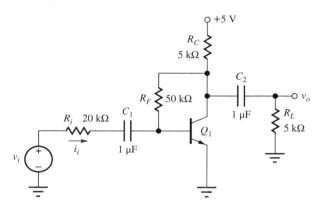

Figure 15.8 Transresistance amplifier with signal source and load resistance.

EXAMPLE 15.4   SHUNT-SHUNT FEEDBACK WITH NEW SOURCE AND LOAD IMPEDANCES

Understand the interaction of a transresistance amplifier with different source and load impedances. We will also explore the relationship between an inverting voltage amplifier and a transresistance amplifier.

PROBLEM Find the small-signal gain of the amplifier in Fig. 15.8. Use $\beta_F = 150$, $V_A = 50$ V, and the results from Ex. 15.3.

SOLUTION **Known Information and Given Data:** The circuit diagram appears in Fig. 15.8; transistor parameters are $\beta_F = 150$ and $V_A = 50$ V; results from Ex. 15.3.

Unknowns: Find the gain based on results from the previous exercise.

Approach: Include input and output loading effects to adjust previous results to a new source and load impedance. This will be an approximation, so our goal is to understand the validity of the approach.

Assumptions: $V_{BE} = 0.7$ V, $V_T = 25$ mV, and small-signal midband conditions apply.

Analysis: We have previously found the input impedance, output impedance, and transresistance of the circuit in Fig. 15.7.

$$A_{tr} = -46.1 \text{ k}\Omega \qquad R_{in} = 340 \text{ }\Omega \qquad R_{out} = 336 \text{ }\Omega$$

Figure 15.9 shows the equivalent circuit we will use to find the gain of the amplifier in Fig. 15.8. The results from Ex. 15.3 are used to generate a model of the transresistance amplifier. The amplifier itself is modeled with an ideal op amp with the calculated input and output resistances pulled out of the amplifier to account for loading effects. Notice that when the circuit is driven with a voltage source, the circuit becomes equivalent to an inverting amplifier.

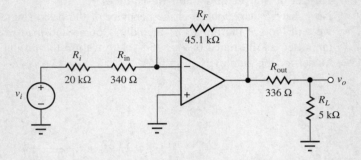

Figure 15.9 Approximate small-signal equivalent circuit for Fig. 15.8.

We can now calculate v_o/v_i using our knowledge of the op-amp inverting amplifier and voltage division.

$$A_v = \frac{v_o}{v_i} = \left(\frac{-R_F}{R_i + R_{in}}\right)\left(\frac{R_L}{R_{out} + R_L}\right) = \left(\frac{-45.1\,k\Omega}{20\,k\Omega + 340\,\Omega}\right)\left(\frac{5\,k\Omega}{336 + 5\,k\Omega}\right) = -2.08$$

The first portion of the equation is the basic inverting amplifier equation and the second term reflects the voltage division due to the output impedance of the circuit in Fig. 15.9.

Computer-Aided Analysis: Simulations of the amplifier produce a small-signal gain of -2.09 V/V. This is quite close to our hand calculation.

Discussion: We find that we can use results from an unloaded amplifier to approximate the response for the loaded situation. In this particular case it is quite accurate, but it is less accurate for other combinations of source and load resistance. Predictably, as the source or load impedances approach the input or output resistance the accuracy is compromised. While this is an approximation, it is much more efficient than reevaluating loop transmission for each case and should be one of the tools we apply to design problems.

EXERCISE: For the circuit in Fig. 15.8, compare the calculated and simulated gain values for $R_i = 2\,k\Omega$, $R_L = 10\,k\Omega$, and $R_i = 10\,k\Omega$, $R_L = 2\,k\Omega$. Also find the error for the two cases.

ANSWERS: Calculated: -18.6, -3.73; Simulated: -17.9, -3.60; Error: 4.2 percent, 3.6 percent

EXERCISE: A simulation of the TIA circuit in Fig. (b) in the EIA feature on the next page yields $A_{tr} = -48.5\,k\Omega$ and $R_{out} = 12\,\Omega$. What are the values of T, g_{m3} and R_{in}?

ANSWERS: 32.3, 2.50 mS, 1.51 kΩ

ELECTRONICS IN ACTION

A Transresistance Amplifier Implementation

The application of transresistance amplifiers in optical communications was introduced in Electronics in Action features in Chapters S9 and 6. The transresistance amplifier, Fig. (a), that converts the photodiode current i_{ph} into an output voltage $v_o = -i_{ph}R_F$ is often realized by a shunt-shunt feedback amplifier, and a basic CMOS implementation of such an amplifier appears in Fig. (b). M_1 and M_2 form a high-gain CMOS inverter that is connected to source follower M_3 to achieve a lower output resistance. Current source I_{bias} and the W/L ratios of the transistors determine the Q-points of the three devices. Reverse bias for photodiode D_1 is provided by the gate-source voltage of M_1.

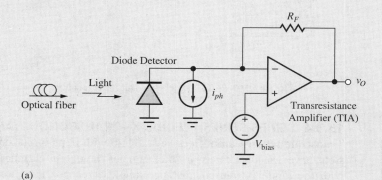

(a)

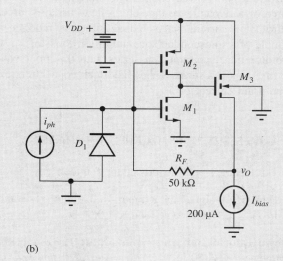

(b)

The transresistance, input resistance, and output resistance of the shunt-shunt feedback amplifier can be found using the theory just presented in Sec. 15.2:

$$A_{tr} = -R_F\left(\frac{T}{1+T}\right) \qquad R_{in} = \left(R_F + \frac{1}{g_{m3}}\right)\left(\frac{1}{1+T}\right) \qquad R_{out} = \left(\frac{1}{g_{m3}}\right)\left(\frac{1}{1+T}\right)$$

(The output resistance of the reverse-biased diode has been assumed to be extremely large.)

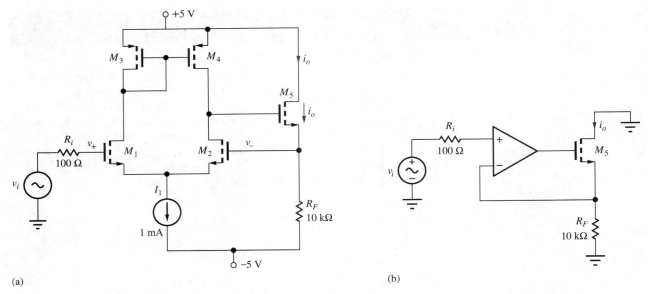

Figure 15.10 (a) Two-stage transconductance amplifier; (b) idealized transconductance amplifier.

15.3.4 SERIES-SERIES FEEDBACK—TRANSCONDUCTANCE AMPLIFIERS

Transconductance amplifiers have a gain with units of current/voltage. Negative feedback versions have a feedback network that samples the output current and feeds back a voltage in series with the input. A transconductance amplifier can be used to create a high-impedance current source or a dynamic voltage-controlled current source, and in other applications where precise control of a current via a voltage signal is required. The circuit in Fig. 15.10 is a two-stage example of a transconductance amplifier. Resistor R_F senses the M_5 output current and generates a voltage that is summed in series with the input voltage across the input differential pair, M_1 and M_2. This is a **series-series feedback** connection. Remember that the amplifier output current is the drain current of M_5.

EXAMPLE 15.5   **SERIES-SERIES FEEDBACK AMPLIFIER ANALYSIS**

Use our feedback analysis procedure to understand the operation of a transconductance amplifier.

PROBLEM Find the small-signal gain, input, and output resistance of the amplifier in Fig. 15.10. $K_n = 10$ mA/V^2, $K_p = 4$ mA/V^2, and $\lambda = 0.01$/V.

SOLUTION **Known Information and Given Data:** The circuit diagram appears in Fig. 15.10; transistor parameters: $K_n = 10$ mA/V^2, $K_p = 4$ mA/V^2, and $\lambda = 0.01$/V. In a typical application, the drain of M_5 would connect to some functional circuit which accepts the output current.

Unknowns: Ideal gain, open-loop gain, loop transmission, and Blackman terms for the input and output impedances.

Approach: Find the dc operating point; use amplifier gain analysis from previous chapters, the feedback analysis procedure, and Blackman's theorem.

Assumptions: Small-signal midband conditions apply.

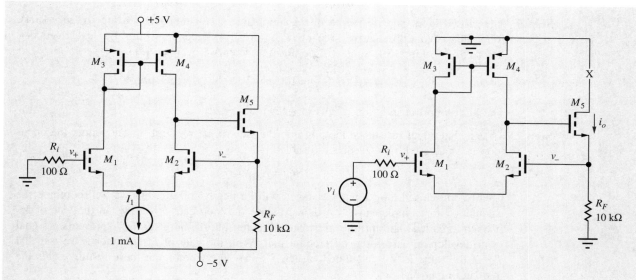

(a) dc equivalent circuit   (b) ac equivalent circuit

Analysis: We first draw the dc equivalent circuit in (a) and find the dc operating point. By inspection we see that M_1–M_4 are all biased at 0.5 mA. Negative feedback works to keep v_- equal to v_+; therefore, the M_5 channel current is $[0-(-5)]/10\,\text{k}\Omega = 0.5\,\text{mA}$. The corresponding small-signal parameters are

$$g_{m1} = g_{m2} = g_{m5} = \sqrt{2K_n I_D} = 3.16\,\text{mS}$$

$$g_{m3} = g_{m4} = 2.00\,\text{mS} \quad r_o \approx \frac{1}{\lambda I_D} = 200\,\text{k}\Omega$$

Now we turn to our feedback analysis procedure.

Step 1: As discussed above, the feedback network is the resistor R_F. R_F samples the output current, so it is a series connection.

Step 2: The signal fed back to the gate of M_2 is the voltage developed across R_F and summed in series with the input voltage signal. This is therefore a series feedback connection.

Step 3: Since current is sampled and voltage is fed back to the input, the feedback factor has units of voltage/current and the amplifier is a series-series transconductance amplifier.

Step 4: The amplifier employs series-series feedback that forces v_- to be equal to v_i, and the output signal current is then $i_o = v_i/R_F$. Thus, the ideal feedback factor is just the resistance of the feedback network, R_F, and the ideal gain is

$$A_{tc}^{\text{Ideal}} = \frac{1}{\beta} = \frac{1}{R_F} = 100\,\mu\text{A/V}$$

Step 5: We calculate the gain around the loop in the ac equivalent circuit in (b) starting at the gate of M_2 and work our way around the loop back to our starting point. This requires calculating the differential pair gain and the M_5 common-drain gain.

$$T = -g_{m2}(r_{o2}\|r_{o4})\left(\frac{g_{m5}(R_F\|r_{o5})}{1+g_{m5}(R_F\|r_{o5})}\right)$$

$$T = -3.16\,\text{mS}(100\,\text{k}\Omega)(0.968) = -306$$

We see that T is negative. Note again that we must always have negative feedback when we are building a feedback amplifier.

Step 6: The closed-loop gain of the feedback amplifier is calculated according to Eq. (15.4). Remember that the negative sign on T is already included in our high-level feedback description in Fig. 15.1, so T will be positive when we evaluate Eq. (15.4):

$$A_{tc} = A_{tc}^{\text{Ideal}} \frac{T}{1+T} = 100 \, \mu\text{A/V} \frac{306}{1+306} = 99.7 \, \mu\text{A/V}$$

The high loop transmission results in a low gain error.

Step 7: To calculate input resistance looking into the gate of M_1 we start by calculating the open-loop resistance with the feedback disabled.

$$R_{\text{in}}^D = R_{iG1} \approx \infty$$

Clearly, the closed-loop input resistance will be nearly infinite, but we will continue the calculation for completeness. T_{OC} is zero since the loop gain is zero when the gate of M_1 is open circuited. The infinite impedance looking out of the gate of M_1 prevents a signal from developing across the differential pair. With the gate of M_1 grounded, we see that T_{SC} is the same as we calculated earlier, $T_{SC} = -306$. Combining these results yields the midband input resistance

$$R_{\text{in}} = R_{\text{in}}^D \frac{1+|T_{SC}|}{1+|T_{OC}|} = \infty \left(\frac{1+0}{1+306} \right) = \infty \, \Omega$$

For the output resistance, we again mentally disable the feedback by setting $g_{m2} = 0$ and calculate the resistance looking into the output of the amp.

$$R_{\text{out}}^D = r_{o5}(1 + g_{m5}R_F) = 6.52 \, \text{M}\Omega$$

T_{OC} is zero since the amplifier gain is zero when the drain of M_5 is an open circuit (zero drain and source current in M_5). For T_{SC} the drain of M_5 is connected to ac ground, and T_{SC} is equal to the previously calculated loop gain: $T_{SC} = -306$. The output resistance is now calculated with Blackman's theorem.

$$R_{\text{out}} = R_{\text{out}}^D \frac{1+|T_{SC}|}{1+|T_{OC}|} = 6.52 \, \text{M}\Omega \frac{1+306}{1+0} = 2.00 \, \text{G}\Omega$$

Step 8: Since there is no appreciable signal loss across the low source resistance, the overall transconductance is the value we calculated earlier.

$$\frac{i_o}{v_i} = 99.7 \, \mu\text{A/V}$$

Discussion: The high-output impedance of this circuit confirms that a transconductance amplifier can be used to build a nearly ideal current source. Notice that if we took our output at the source of M_5 we have a unity-gain voltage amplifier.

Computer-Aided Analysis: SPICE simulation of the amplifier uses a dc analysis and a transfer function analysis from source v_i to the drain current of M_5. The TF results show a transconductance gain of 99.69 μA/V, extremely high-input resistance, and an output impedance of 2.216 GΩ. These results agree well with our hand calculations.

EXERCISE: Calculate the closed-loop gain if a voltage output is taken from the source of M_5. What are A_v^{Ideal}, T, and A_v? What is the closed-loop output resistance? Find R_{out}^D, T_{SC}, T_{OC}, and R_{out}.

ANSWERS: 1.00 V/V, 306, 0.997 V/V; 307 Ω, 0, −306, 1.00 Ω

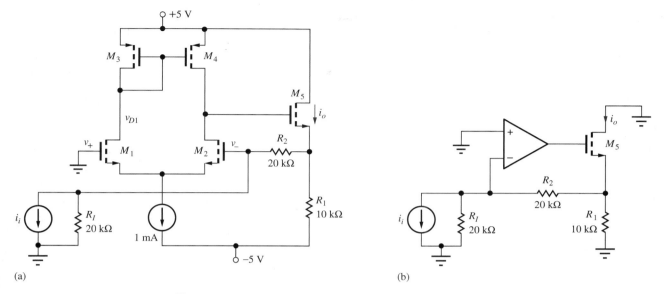

Figure 15.11 (a) Two-stage shunt-series current amplifier; (b) idealized current amplifier.

EXERCISE: Calculate the midband resistance between the drain node of M_1 and small-signal ground in Fig. (b) of Ex. 15.5. What are R_x^D, T_{SC}, T_{OC}, and R_x?

ANSWERS: 500 Ω, −204, −306, 334 Ω

15.3.5 SHUNT-SERIES FEEDBACK—CURRENT AMPLIFIERS

Negative feedback **current amplifiers** are used to produce a precise scaled current. Such current amplifiers have a feedback network that samples the output current and feeds back a portion of the sampled current to a current summing node at the input. Like the transconductance amplifier, a current amplifier can be used to create a high-impedance current source, but one that is controlled by an input current in this topology. The circuit in Fig. 15.11 is a two-stage example of a current amplifier. Feedback resistors R_2 and R_1 sense the M_5 output current and act as a current divider to feed back a portion of the output current to the input summing node. This is a **shunt-series feedback** connection. Note that the amplifier output current is the drain current of M_5.

EXAMPLE 15.6   SHUNT-SERIES FEEDBACK AMPLIFIER ANALYSIS

Use our feedback analysis procedure to understand the operation of a current amplifier.

PROBLEM Find the small-signal gain, and input and output resistances of the amplifier in Fig. 15.11. $K_n = 10$ mA/V^2, $K_p = 4$ mA/V^2, and $\lambda = 0.01$/V.

SOLUTION **Known Information and Given Data:** The circuit diagram appears in Fig. 15.11; transistor parameters: $K_n = 10$ mA/V^2, $K_p = 4$ mA/V, and $\lambda = 0.01$/V. In a typical application, a functional circuit would connect to the drain of M_6.

Unknowns: Ideal gain, open-loop gain, loop transmission, and Blackman terms for the input and output resistances.

Approach: Find the DC operating point; use amplifier gain analyses from previous chapters, the feedback analysis procedure, and Blackman's theorem.

Assumptions: Small-signal midband conditions apply.

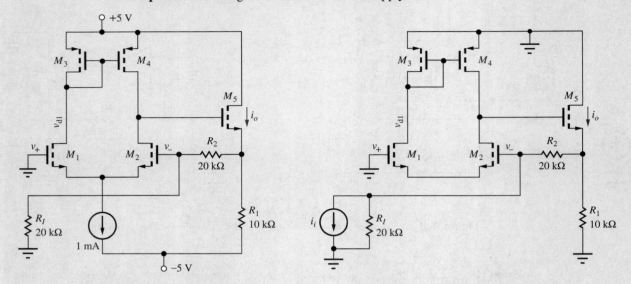

(a) dc equivalent circuit.   (b) ac equivalent circuit.

Analysis: We first draw the dc equivalent circuit in Fig. (a) and find the dc operating point. By inspection we see that M_1–M_4 are all biased at 0.5 mA. Negative feedback works to keep v_- equal to v_+. So there is no dc current flow through R_2, and the M_5 channel current is $[0 - (-5)]/10$ kΩ = 0.5 mA. The corresponding small-signal parameters are

$$g_{m1} = g_{m2} = g_{m5} = \sqrt{2K_n I_D} = 3.16 \text{ mS}$$

$$g_{m3} = g_{m4} = 2.00 \text{ mS} \quad r_o \approx \frac{1}{\lambda I_D} = 200 \text{ k}\Omega$$

We now use our feedback procedure to analyze the performance of the current amplifier.

Step 1: As discussed above, the feedback network is made up of resistors R_2 and R_1. In the ideal case, the feedback keeps v_- equal to v_+, so R_2 and R_1 act as a current divider to feed back a portion of i_o. Since we are sampling current and not voltage, this is a series connection.

Step 2: The signal fed back to the summing node at gate of M_2 is the fraction of output current sampled by the R_2 and R_1 current divider. This is a shunt connection of the feedback signal.

Step 3: Since current is sampled and current is fed back to the input, the feedback factor has units of current/current and the amplifier is a shunt-series current amplifier.

Step 4: The amplifier type uses shunt-series feedback which forces current i_i through R_2.[6] The output current becomes i_i plus the current through R_1 given by $i_i R_2/R_1$. The total current and ideal gain are therefore

$$\mathbf{i_o} = \mathbf{i_i}\left(1 + \frac{R_2}{R_1}\right) \quad \text{and} \quad A_c^{\text{Ideal}} = \frac{i_o}{i_i} = \frac{1}{\beta} = \frac{R_2 + R_1}{R_1} = 3$$

If the sign of the input current is changed we will need to change the sign of our gain. This is often confusing for a number of current amplifier topologies.

---

[6] Note that an ideal differential amplifier forces v_- to be zero volts, and there would be no current in R_I.

Step 5: We calculate the loop gain starting at the gate of M_2 in the ac equivalent circuit in Fig. (b) and work our way around the loop back to our starting point. This requires calculating the differential pair gain, the M_5 common-drain gain, and the attenuation through R_2 and R_I. Assume $r_{o5} \gg R_1 \| R_2$.

$$T = -g_{m2}(r_{o2} \| r_{o4}) \left(\frac{g_{m5}[R_1 \| (R_2 + R_I)]}{1 + g_{m5}[R_1 \| (R_2 + R_I)]} \right) \left(\frac{R_I}{R_I + R_2} \right)$$

$$T = -3.16 \text{ mS } (100 \text{ k}\Omega)(0.962)(0.5) = -152$$

We must always check that we have negative feedback and not positive when we are building a feedback amplifier, and we see that T is negative.

Step 6: The closed-loop gain of the feedback amplifier is calculated according to Eq. (15.4). Remember that the negative sign on T is already included in our high-level feedback description in Fig. 15.1, so T will be positive when we evaluate Eq. (15.4):

$$A_c = A_c^{\text{Ideal}} \frac{T}{1+T} = 3\left(\frac{152}{1+152}\right) = 2.98$$

High loop transmission results in a low gain error.

Step 7: To calculate input resistance looking into the gate of M_2 we start by calculating the open-loop resistance with the feedback disabled.

$$R_{\text{in}}^D = R_{iG2} \| R_I \| \left(R_2 + R_1 \| \frac{1}{g_{m5}} \right) = 10.1 \text{ k}\Omega$$

Loop gain T_{SC} is found with v_- shorted to ac ground, so $T_{SC} = 0$. With the input open, the loop gain will be the same as calculated above: $T_{OC} = -152$.

$$R_{\text{in}} = R_{\text{in}}^D \frac{1 + |T_{SC}|}{1 + |T_{OC}|} = 10.1 \text{ k}\Omega \left(\frac{1+0}{1+152}\right) = 66.0 \text{ }\Omega$$

For the output resistance, we again mentally disable the feedback and calculate the impedance looking into the output of the amp.

$$R_{\text{out}}^D = r_{o5}(1 + g_{m5}[R_1 \| (R_2 + R_I)]) = 5.26 \text{ M}\Omega$$

T_{OC} is zero since the amplifier gain is zero when the output at the drain of M_5 is open-circuited. T_{SC} is again equal to the previously calculated loop gain, so $T_{SC} = 152$. The output resistance is now calculated with Blackman's theorem.

$$R_{\text{out}} = R_{\text{out}}^D \frac{1 + |T_{SC}|}{1 + |T_{OC}|} = 5.26 \text{ M}\Omega \left(\frac{1+152}{1+0}\right) = 805 \text{ M}\Omega$$

Step 8: When feedback is active, there is a virtual ground at v_- and

$$A = A_c = 2.98$$

Discussion: The high-output impedance of this circuit confirms that like the transconductance case, a current amplifier can be used to build a nearly ideal current source. We also see that current sampling at the output leads to an output impedance scaled by $1 + T$.

Computer-Aided Analysis: SPICE transfer function analysis from source i_i to the current through M_5 yields a current gain of 2.984, extremely high-input resistance, an output impedance of 904.0 MΩ and an input resistance of 63.09 Ω with $V_{OS} = 6.571$ mV and $T = 159.3$. These results agree well with our hand calculations, although our output impedance is somewhat low due to our approximate calculation of r_o.

> **EXERCISE:** Calculate the closed-loop gain if a voltage output is taken from the source of M_5 and the input source is replaced with a voltage source v_i in series with a 20-kΩ resistance. What are A_v^{Ideal}, T, and A_v? What is the closed-loop output resistance? Find R_{out}^D, T_{SC}, T_{OC}, and R_{out}.
>
> **ANSWERS:** −1.00 V/V, 152, −0.994 V/V; 304 Ω, 0, 152, 1.99 Ω

ELECTRONICS IN ACTION

Fully Differential Design

Mixed-signal integrated circuits employing A/D and D/A converters, switched capacitor filters, and so on can be in a relatively noisy environment because of the high frequency clocking circuits, and these systems frequently utilize fully differential design throughout the circuitry in order to take advantage of the common-mode rejection capability of differential circuits. An example of a simple amplifier with differential input and differential output (*DI-DO*) was shown in Fig. 13.20, and a typical symbol for the *DI-DO* amplifier appears to the left. A similar symbol is used for a differential transconductance amplifier.

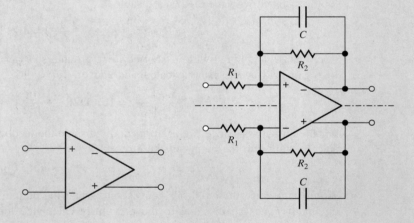

An example of the use of the fully differential amplifier in an active low-pass filter appears at the right. Note that both paths through the filter employ negative feedback and that the dashed line represents a line of symmetry for differential signals. Thus, the active low-pass filter discussed in Section 10.3.1, represents the differential-mode half-circuit for the differential filter.

Examples of full differential design can be observed in the EIA on page 887 and in the first two stages of Fig. P13.94. To complete the differential circuit in Fig. P13.94, a second emitter follower would be added to the collector of Q_3.

15.4 REVIEW OF FEEDBACK AMPLIFIER STABILITY

We will now review the negative feedback stability design issues with which we must contend to successfully design and implement feedback amplifiers. We will also discuss our approach to **feedback amplifier stability** analysis and review some important governing equations.

15.4 Review of Feedback Amplifier Stability

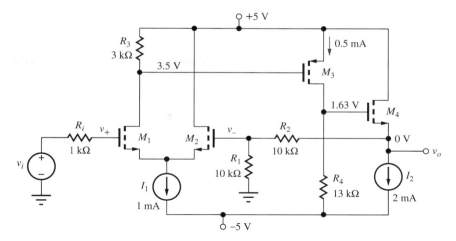

Figure 15.12 Three-stage MOSFET series-shunt feedback amplifier ($K_n = 10$ mA/V^2, $K_p = 4$ mA/V^2, $V_{TN} = 1$ V, $V_{TP} = -1$ V).

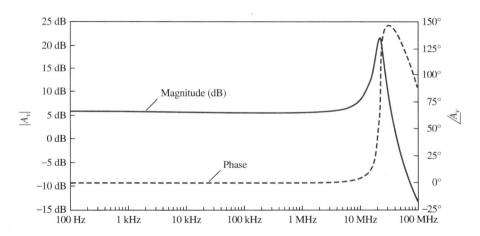

Figure 15.13 Small-signal gain versus frequency of the uncompensated three-stage amplifier.

15.4.1 CLOSED-LOOP RESPONSE OF THE UNCOMPENSATED AMPLIFIER

Figure 15.12 is the series-shunt three-stage amplifier that we analyzed at midband frequencies in Ex. 15.2. In our midband analysis of this circuit, we mentioned that this amplifier design wasn't complete since we had not yet addressed stability issues. Figure 15.13 presents simulation results for the small-signal gain $\mathbf{v_{out}}/\mathbf{v_i} = A_v$ over a wide range of frequencies after capacitances $C_{GS} = 5$ pF and $C_{GD} = 1$ pF have been added to each device model in order to extend our analysis to high frequencies. Note the excessive "peaking" of the response near 20 MHz. This is characteristic of a feedback amplifier with poor **phase margin.** Poor phase margin is typical of an "uncompensated" amplifier, an amplifier whose designer has not yet adjusted the design to address stability issues. A well "compensated" amplifier, an amplifier with good phase margin, will exhibit a smooth roll off from the midband response.

Recall Eq. (15.4), the closed-loop gain feedback equation:

$$A_v = \frac{\mathbf{v_o}}{\mathbf{v_i}} = A_v^{\text{Ideal}} \frac{T}{1+T}$$

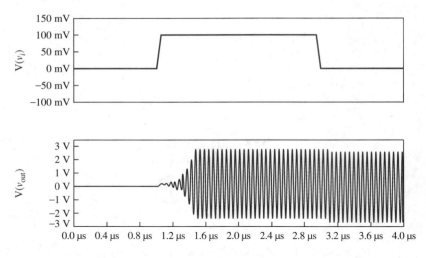

Figure 15.14 Input and output voltage of the uncompensated amplifier.

Loop gain T has a number of poles and perhaps zeros as well. The plot in Fig. 15.13 expresses the ratio $T/(1+T)$. This ratio results in a complicated relationship between gain and phase for the closed-loop amplifier. We do not have any tools to directly relate the response of this ratio to stability in a way that we can use to improve our design. As we will see shortly, we will instead examine loop gain T to gain insight into the amplifier stability.

Figure 15.14 is a transient simulation of the amplifier showing its response to a step input. The upper plot shows the input signal and the lower plot is the amplifier output. What is happening here? Recall from Chapter 11 that an amplifier with zero phase margin can oscillate. This occurs when the total phase shift around the loop reaches 360° at some frequency before the magnitude of the loop gain reaches 0 dB. The loop gain acquires 180° of phase shift from the inverting input to the amplifier, and gathers more phase shift as frequency increases due to poles associated with the parasitic capacitances of the transistors. In a physical circuit there will also be parasitic capacitances and perhaps inductances due to the physical structure and interconnect of the circuit. When the initial input transient reaches the circuit, enough energy has been added to the system to initiate the oscillation. In a real circuit, thermal noise and other signals would initiate the oscillation before the initial transient. Once the oscillation starts, the combination of loop gain and large phase shift around the loop sustains the oscillation. Clearly we will have to modify our design to make this circuit useful as an amplifier.

> **EXERCISE:** Estimate the frequency of oscillation in Fig. 15.14.
>
> **ANSWER:** 18.8 MHz (note that this is near the peak frequency in Fig. 15.13)
>
> **EXERCISE:** What are the values of f_T for the four transistors in Fig. 15.12 if $C_{GS} = 5$ pF and $C_{GD} = 1$ pF?
>
> **ANSWERS:** 83.8 MHz, 83.8 MHz, 53.1 MHz, 168 MHz

15.4.2 PHASE MARGIN

To guide the refinement of our design we must examine the phase and magnitude of the loop gain as a function of frequency. Based on equations from Chapter 9, we can find the pole frequencies at each of the nodes in the loop. At the gate of M_2, the resistance is R_1 in parallel with the resistance looking back through R_2. The capacitance is C_{GD2} added to the Miller multiplication applied to C_{GS2}. Combining these results gives

$$f_{P1} = \frac{1}{2\pi \left[R_1 \left\| \left(R_2 + \frac{1}{g_{m4}}\right)\right.\right] [C_{GD2} + C_{GS2}(1 - A_{vgs2})]}$$

$$f_{P1} = \frac{1}{2\pi \left[10\,\text{k}\Omega \left\| \left(10\,\text{k}\Omega + \frac{1}{3.16\,\text{mS}}\right)\right.\right] [1\,\text{pF} + 5\,\text{pF}(1 - 0.5)]} = 8.96\,\text{MHz}$$

At the source of M_2, the resistance is $1/g_{m1}$ in parallel with $1/g_{m2}$ and the capacitance is $C_{GS1} + C_{GS2}$:

$$f_{P2} = \frac{1}{2\pi}\left(\frac{g_{m1} + g_{m2}}{C_{GS1} + C_{GS2}}\right) = \frac{1}{2\pi}\frac{g_{m2}}{C_{GS2}} = \frac{1}{2\pi}\frac{3.16\,\text{mS}}{5\,\text{pF}} = 101\,\text{MHz}$$

At the gate of M_3, we use the C_T approximation for the dominant pole of the common-source amplifier including the load capacitance from M_4.

$$f_{P3} = \frac{1}{2\pi} \frac{1}{R_3[C_{GD1} + C_{GS3} + C_{GD3}(1 - A_{vgd3})] + R_4[C_{GD3} + C_{GS4}(1 - A_{vgs4}) + C_{GD4}]}$$

$$f_{P3} = \frac{1}{2\pi} \frac{1}{3\,\text{k}\Omega[1\,\text{pF} + 5\,\text{pF} + 1\,\text{pF}(1 + 26)] + 13\,\text{k}\Omega[1\,\text{pF} + 5\,\text{pF}(1 - 0.992) + 1\,\text{pF}]}$$

$$= 1.27\,\text{MHz}$$

At the drain of M_3 we have the second pole of the common-source stage:

$$f_{P4} = \frac{1}{2\pi}\left(\frac{g_{m3}}{C_{GS3} + C_{L3}}\right) = \frac{1}{2\pi}\left(\frac{g_{m3}}{C_{GS3} + C_{GD4} + C_{GS4}(1 - A_{vgs4})}\right)$$

$$f_{P4} = \frac{1}{2\pi}\left(\frac{2\,\text{mS}}{5\,\text{pF} + 1\,\text{pF} + 5\,\text{pF}(1 - 0.992)}\right) = 52.7\,\text{MHz}$$

Finally, the pole frequency estimate at the source of M_4 is

$$f_{P5} = \frac{1}{2\pi \left[\frac{1}{g_{m4}} \| (R_1 + R_2)\right] C_{GS4}} = \frac{1}{2\pi \left[\frac{1}{6.33\,\text{mS}} \| 20\,\text{k}\Omega\right] 5\,\text{pF}} = 203\,\text{MHz}$$

The lowest frequency pole is found at the gate of M_3 due to the Miller multiplication across M_3. To generate a single-pole response, the product of the lowest frequency pole and the midband loop gain should be at a lower frequency than the next highest pole frequency. Recall that this product is also f_T, the 0 dB frequency for a single-pole response. In this amplifier, $f_{p3} \times T = 1.27$ MHz $\times$ 61.1 = 77.6 MHz. If we know f_T, we can calculate phase margin as

$$\theta_m = 360 - 180 \text{ (inverting input)} - \tan^{-1}\left(\frac{f_T}{f_{p1}}\right) - \tan^{-1}\left(\frac{f_T}{f_{p2}}\right) - \cdots \quad (15.18)$$

However, for this circuit, our estimate of f_T is higher in frequency than several of the pole frequencies, so we can't use Eq. (15.18) because we don't have a good estimate of the actual f_T. As a result, we expect our phase margin to be zero or worse. For this situation, we can simulate or numerically evaluate the phase margin.

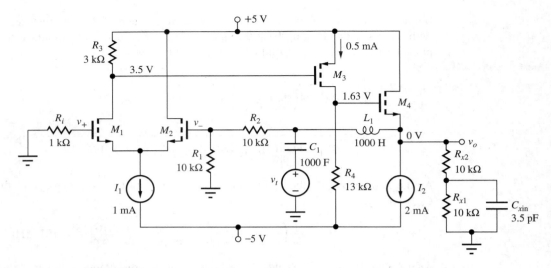

Figure 15.15 Circuit for simulating small-signal loop gain characteristics.

A circuit modification that facilitates the simulation of phase margin is shown in Fig. 15.15. This circuit provides negative feedback at dc but not at midband frequencies and beyond. Inductor L_1 blocks mid and high frequencies to effectively disconnect the feedback loop at those frequencies. Capacitor C_1 blocks dc so that the ac test voltage is only connected into the loop at mid and high frequencies. C_1 and L_1 are set to artificially high values so that the transition from the dc response to midband occurs at very low frequencies. Components R_{x1}, R_{x2}, and C_{xin} are used to model the output loading due to the feedback network and the inverting input of the amplifier. Note that since the output is nominally biased at 0 V, the resistors will have little impact on the operating point. If the amplifier is not biased at 0 V, a blocking capacitor can be added in series with R_{x2}. We might ask at this point, why not simply disconnect the feedback network and run our simulation without inserting L_1 and C_1? The negative feedback is serving to correct any bias errors and set the proper operating point. When an amplifier has large gain, even small bias errors can lead to large changes in the output dc voltage, and therefore the operating point of the amplifier transistors. Our approach allows us to get the benefits of feedback at dc for operating point stability while still effectively disconnecting the amplifier feedback connection at mid and high frequencies.

Figure 15.16 is the phase and magnitude response of the loop gain T of our uncompensated amplifier based upon simulation results for the circuit in Fig. 15.15. Phase margin is the difference between 360° and the phase of the loop gain at the frequency for which the loop gain magnitude is 0 dB, that is, f_T. Recall from Chapter 11 that an ideal single-pole loop gain will have a phase margin of 90°:

$$\theta_m = 90 = 360 - 180 \text{ (inverting input)} - 90 \text{ (max phase shift of single pole)}$$

Unfortunately, our amplifier phase margin is −9°, so the oscillation we saw earlier is not surprising. For stable performance, we desire a phase margin of at least +45°, but typically greater than +60°.

Now that we have assessed the phase margin, we need to correct the phase margin to make our design usable as an amplifier. The basic approach is to add capacitance to a node in the amplifier to create a dominant pole to force the magnitude response to 0 dB at a lower frequency, thereby reducing the phase shift at 0 dB due to higher frequency poles in the amplifier. Let's assume we would like to move our 0-dB frequency from its present value of about 22 MHz to

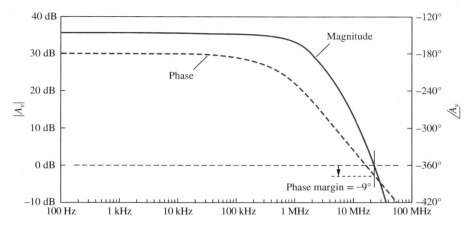

Figure 15.16 Loop transmission magnitude and phase of the three-stage amplifier.

2 MHz. If we successfully create a dominant pole, f_B, so that the magnitude response approximates a single-pole response, the 0-dB frequency can be found as

$$f_T = 2\,\text{MHz} = Tf_B = A_0\beta f_B$$

From Ex. 15.2 we know that $T(\text{midband}) = 61.1$. Therefore, we need to set our dominant pole as

$$f_B = f_T/T = f_T/A_0\beta = 32.8\,\text{kHz}$$

The equivalent capacitance at the gate of M_3 is dominated by the Miller capacitance due to high gain from the gate to the drain of M_3. This makes this node a good candidate for setting the dominant pole since the Miller multiplication allows us to use a relatively small physical capacitance to generate a large equivalent capacitance to place the dominant pole at a low frequency. Substituting 32.8 kHz for the pole frequency at the gate of M_3, we can calculate the expected phase margin with Eq. (15.14):

$$\theta_M = 360 - 180 - \tan^{-1}\left(\frac{2}{8.96}\right) - \tan^{-1}\left(\frac{2}{101}\right) - \tan^{-1}\left(\frac{2}{0.0328}\right) - \tan^{-1}\left(\frac{2}{52.7}\right) - \tan^{-1}\left(\frac{2}{203}\right)$$

$$\theta_M = 360 - 180 - 12.6 - 1.13 - 89.1 - 2.17 - 0.564 = 74.4°$$

This is an acceptable phase margin, so we will continue with the design. To achieve this phase margin we add capacitor C_C between the gate and drain of M_3 as in Fig. 15.19. If we assume the response at that node is dominated by Miller capacitance, our dominant pole frequency can be approximated as

$$f_B = 32.8\,\text{kHz} = \frac{1}{2\pi R_{eq}C_{eq}} = \frac{1}{2\pi(R_3 \| R_{iD1})([C_{gd} + C_c][1 - A_{vt3}])} = \frac{1}{2\pi(3\,\text{k}\Omega)([1\,\text{pF} + C_c][27])}$$

Solving for C_C yields

$$C_C = \frac{1}{2\pi(3\,\text{k}\Omega)(27)(32.8\,\text{kHz})} - 1\,\text{pF} \approx 60\,\text{pF}$$

Adding C_C to our simulation yields the loop transmission response shown in Fig. 15.17. We find the phase margin by identifying the phase at the frequency for a magnitude response of 0 dB, f_T, and calculate how far this is from 360°. Note that in the previous loop gain plot the phase axis on the right went from −420° to −120° while this plot goes from −150 to 180. This is an inconsistency in most SPICE circuit simulators. In one case the midband phase shift (due to the inverting input) is −180 while it is +180 in the plot on the next page. For a continuous sine wave these two values are identical. So, in Fig. 15.17, phase margin is measured relative to 0° rather than 360°.

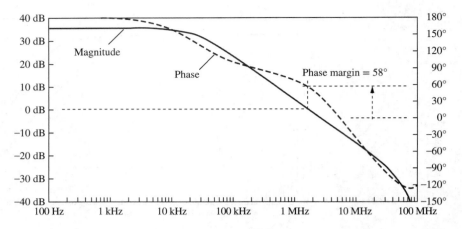

Figure 15.17 Loop gain response with a compensation capacitance of 60 pF added from gate-to-drain of M_3.

15.4.3 HIGHER-ORDER EFFECTS

Rather than the anticipated phase margin of 74°, our simulation yields 58°. This is due to an interesting aspect of the plot in Fig. 15.17 shown between 2 MHz and 20 MHz. The magnitude response is falling rapidly, but the phase shift is inflecting upward. This is surprising since a pole is of the form $1/(1 + jf/f_p)$ and contributes negative phase shift as frequency increases and a negative inflection in the magnitude response. A zero in the numerator is usually of the form $(1 + jf/f_z)$ and generates a positive inflection in the magnitude response and in the phase response.

What we see here is known as a right-half plane zero, or a zero that takes the form $(1 - jf/f_z)$. We found in the previous chapter that this occurs with common-emitter and common-source amplifier stages. This has a negative phase response and positive magnitude response. As we shall see in more detail later, this is due to a feed-forward path through the large compensation capacitor we added around M_3. At high frequencies signals can propagate through C_C instead of through M_3. Because of the positive inflection of the magnitude and related negative phase shift, a right-half plane zero can dramatically reduce phase margin. In this particular example, the effect is not pronounced (degrading ϕ_M by 14°), but in many situations it can have a major impact on stability.

The corresponding time domain response is shown in Fig. 15.18. The input pulse is plotted in the top plot and the output is seen in the bottom plot. This response illustrates the relationship

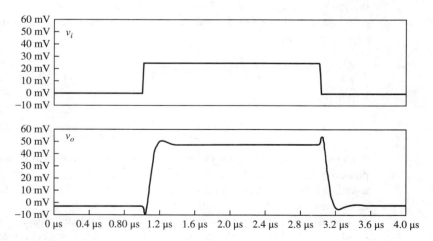

Figure 15.18 Compensated amplifier step response.

15.4 Review of Feedback Amplifier Stability

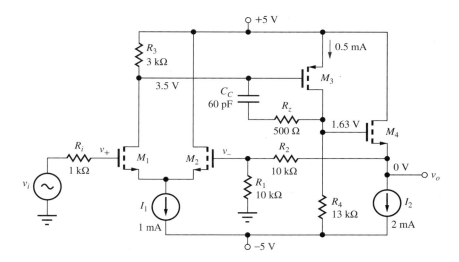

Figure 15.19 Frequency compensated three-stage feedback amplifier.

between overshoot and phase margin presented in Chapter 11. We see overshoot at the top of the rising edge of the output of about 7 percent. From Table 11.5, we expect a phase margin of 62° for an overshoot of 7 percent. This is in good agreement with our graphical estimate of 58° when we consider that Table 11.5 was generated for a second-order system and our circuit is higher order. We should also note the small negative transient at the beginning of each pulse transition. This is also due to the feed-forward signal path through the compensation capacitor. The leading edge of the input signal couples through the compensation capacitance faster than transistor M_3 can respond. (Recall that sharp transitions have significant high-frequency content.) After a short delay, the signal path through the transistor catches up and again dominates the output response.

As we will learn in the next section, adding an appropriately valued resistance in series with the compensation capacitor can mitigate the effects of the feed-forward path. The calculation for arriving at the value of R_Z is discussed in the following sections. Figure 15.19 shows the complete schematic with the compensation capacitor, C_C, and the feed-forward cancellation resistor R_Z.

15.4.4 RESPONSE OF THE COMPENSATED AMPLIFIER

A simulation of the loop gain of the compensated amplifier in Fig. 15.19 is shown in Fig. 15.20. The addition of R_Z has removed the effects of the right-half plane zero and added an additional

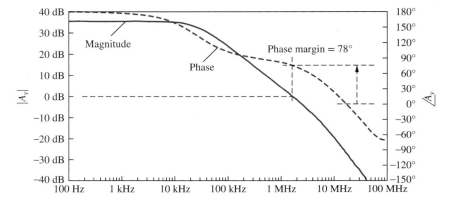

Figure 15.20 Loop gain characteristic with C_C and R_Z.

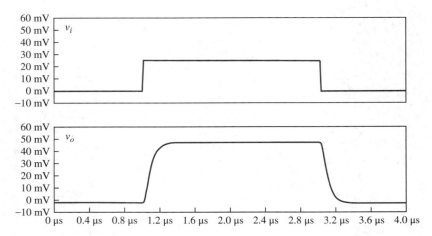

Figure 15.21 Amplifier transient response with C_C and R_Z.

20° to the phase margin. This is typical of MOSFET feedback amplifiers due to their low g_m compared to BJT amplifiers. The magnitude response no longer shows the positive inflection at 3 MHz, and the phase shift does not drop as rapidly as it did in Fig. 15.17. This improved phase margin agrees well with our calculated value of 74.5°.

With the improved phase margin we should expect a reduction in the overshoot of our pulse response simulation. According to Table 11.5, a phase margin of 78° should correspond to a pulse response with no overshoot, and this is confirmed in the simulation results shown in Fig. 15.21 for the fully compensated amplifier. Notice also that the transients at the start of each transition due to the feed-forward path have also been eliminated. The addition of R_Z increased the high-frequency impedance through the feed-forward path, resulting in less feed-forward signal. The extent to which the addition of R_Z mitigates this issue will vary with the specific conditions of a particular amplifier. Another important characteristic of the final amplifier is an increase in rise and fall time. The edge transitions in Fig. 15.21 are slower than the previous simulation. In some applications, a faster edge transition may be desirable even at the expense of some overshoot.

We started this section by looking at the small-signal closed-loop gain characteristic of our amplifier. Figure 15.22 shows our new closed-loop gain with the addition of C_C and R_Z. Clearly, this response is a more desirable characteristic than the results before compensation. Recall from Chapter 11 that the high-frequency corner, $f_{-3\text{dB}}$, is equal to the loop gain 0 dB frequency, f_T, for a single-pole response. The frequency compensation changes our design to create a dominant

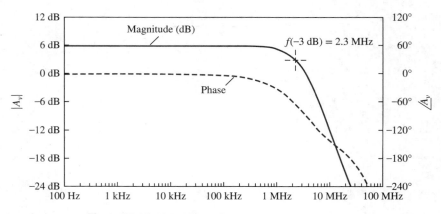

Figure 15.22 Closed-loop gain characteristic with C_C and R_Z.

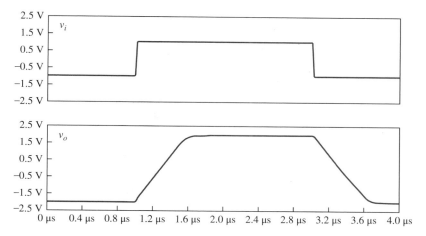

Figure 15.23 Large-signal response of the compensated amplifier.

pole and approximate a single-pole response. Our simulation shows that our high-frequency corner is 2.3 MHz, close to the f_T target of 2 MHz. The two values don't match precisely since we are only approximating a single-pole response.

15.4.5 SMALL-SIGNAL LIMITATIONS

Before we look into the analysis of feedback in transistor amplifiers in more detail, we need to recognize the small-signal limitations of our analysis to this point. To illustrate the issue, Fig. 15.23 shows a simulation of the amplifier pulse response to a 2-V pulse instead of the 25-mV pulse we simulated earlier. The rise and fall times of this output pulse are much slower than for the small-signal and are limited by amplifier slew rate. As we will see in a later section, slew rate is usually limited by how much current is available to charge the relatively large compensation capacitor. As we can see in Fig. 15.23, when an amplifier is slewing, the pulse edges deviate from an exponential RC settling characteristic to a linear charging shape. This is a large signal nonideality of a feedback amplifier, and we must be careful to recognize this when simulating or testing small-signal characteristics in the lab. In short, if we are attempting to measure a small-signal parameter and see a characteristic large signal slewing, we must reduce the amplitude of the input signal to ensure that we are operating in the small-signal regime.

In the following sections, we will analyze and design several feedback amplifiers for stable operation. While the example in this section is for an amplifier with a closed-loop gain of 2 and a β of 0.5, we typically design compensation for the worst-case unity-gain situation with gain of 1 and β of 1. Unless stated otherwise, we should assume this as the design goal.

15.5 SINGLE-POLE OPERATIONAL AMPLIFIER COMPENSATION

As discussed in the previous section and in Chapter 11, feedback amplifiers use internal frequency compensation to force the amplifier to have a single-pole frequency response. For general-purpose operational amplifiers, we will compensate the amplifiers for stable operation as unity-gain buffers, the worst-case situation for amplifier stability. The voltage transfer functions of these amplifiers can be represented by Eq. (15.19):

$$A_v(s) = \frac{A_o \omega_B}{s + \omega_B} = \frac{\omega_T}{s + \omega_B} \qquad (15.19)$$

This form of transfer function can be obtained by connecting a compensation capacitor C_C around the second gain stage of the basic operational amplifier, as depicted in Fig. 15.24.

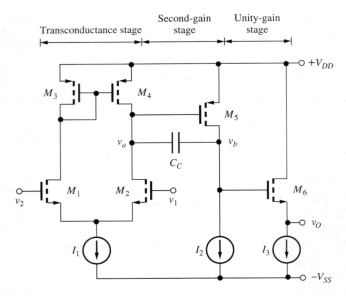

Figure 15.24 Frequency-compensation technique for single-pole operational amplifiers.

15.5.1 THREE-STAGE OP-AMP ANALYSIS

Figure 15.25 is a simplified representation for the three-stage op amp. The input stage is modeled by its Norton equivalent circuit, represented by current source $G_m v_{dm}$ and output resistance R_o. The second stage provides a voltage gain $A_{v2} = g_{m5} r_{o5} = \mu_{f5}$, and the follower output stage is represented as a unity-gain buffer.

The circuit in Fig. 15.25 can be further simplified using the **Miller effect** relations. Feedback capacitor C_C is multiplied by the factor $(1 + A_{v2})$ and placed in parallel with the input of the second-stage amplifier, as in Fig. 15.26, and an expression for the output voltage can now be

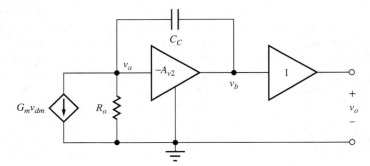

Figure 15.25 Simplified model for three-stage op amp.

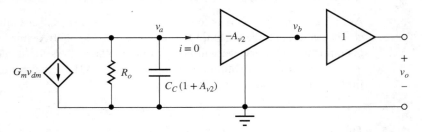

Figure 15.26 Equivalent circuit based on Miller multiplication.

obtained from analysis of this figure. The output voltage $V_o(s)$ must equal $V_b(s)$ because the output buffer has a gain of 1. Also, $V_b(s)$ equals $-A_{v2}V_a(s)$.

Writing the nodal equation for $V_a(s)$ assuming $i = 0$

$$-G_m V_{dm}(s) = V_a(s)[sC_C(1 + A_{v2}) + G_o] \tag{15.20}$$

and

$$\frac{V_a(s)}{V_{dm}(s)} = \frac{-G_m R_o}{sR_o C_C(1 + A_{v2}) + 1} \tag{15.21}$$

Combining these results gives the overall gain of the op amp:

$$A_v(s) = \frac{V_o(s)}{V_{dm}(s)} = \frac{V_b(s)}{V_{dm}(s)} = \frac{-A_{v2}V_a(s)}{V_{dm}(s)} = \frac{G_m R_o A_{v2}}{1 + sR_o C_C(1 + A_{v2})} \tag{15.22}$$

Rewriting Eq. (15.21) in the form of Eq. (15.19) yields

$$A_v(s) = \frac{\frac{G_m A_{v2}}{C_C(1 + A_{v2})}}{s + \frac{1}{R_o C_C(1 + A_{v2})}} = \frac{\omega_T}{s + \omega_B} = \frac{A_o \omega_B}{s + \omega_B} \tag{15.23}$$

Figure 15.27 is a **Bode plot** for this transfer function. At low frequencies the gain is $A_o = G_m R_o A_{v2}$, and the gain rolls off at 20 dB/decade above the frequency ω_B. Comparing Eq. (15.22) to (15.19)

$$\omega_B = \frac{1}{R_o C_C(1 + A_{v2})} \quad \text{and} \quad \omega_T = \frac{G_m A_{v2}}{C_C(1 + A_{v2})} \tag{15.24}$$

For large A_{v2}

$$\omega_T \cong \frac{G_m}{C_C} \tag{15.25}$$

Equation (15.24) is an extremely useful result. The unity gain frequency of the operational amplifier is set by the designer's choice of the values of the input stage transconductance and **compensation capacitor C_C**.

The single pole of the amplifier is at a relatively low frequency, as determined by the large values of the output resistance of the first stage and the Miller input capacitance of the second stage.

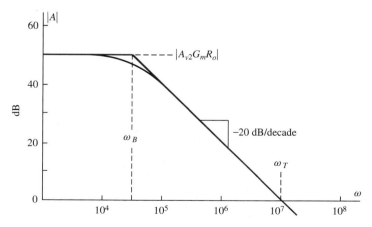

Figure 15.27 Gain magnitude plot for the ideal single-pole op amp.

> **EXERCISE:** What are the approximate values of G_m, R_o, f_T, and f_B for the op amp in Fig. 15.24 if $K_{n2} = 1$ mA/V^2, $K_{p5} = 1$ mA/V^2, $C_C = 20$ pF, $\lambda = 0.02$ V^{-1}, $I_1 = 100$ µA, and $I_2 = 500$ µA?
>
> **ANSWERS:** 0.316 mS, 500 kΩ, 2.52 MHz, 158 Hz

15.5.2 TRANSMISSION ZEROS IN FET OP AMPS

Equation (15.24) presents an excellent method for controlling the frequency response of the operational amplifier with two gain stages. Unfortunately, however, we have overlooked a potential problem in the analysis of this amplifier: The simplified Miller approach does not take into account the finite transconductance of the second-stage amplifier.

The source of the problem can be understood by using the complete small-signal model for transistor M_5, as incorporated in Fig. 15.28. The previous analysis overlooked the zero that is determined by g_{m5} and the total feedback capacitance between the drain and gate of M_5. The circuit in Fig. 15.28 should once again look familiar. It is the same topology as the circuit for the simplified C-E amplifier, and we can use the results of the analysis in Eq. (9.94) by making the appropriate symbolic substitutions identified in Eq. (15.25):

$$r_{\pi o} \to R_o \qquad R_L \to r_{o5} \qquad C_\pi \to C_{GS5} \qquad C_\mu \to C_C + C_{GD5} \qquad (15.26)$$

With these transformations, the transfer function becomes

$$A_{vth}(s) = (-g_{m5} r_{o5}) \frac{\left(1 - \dfrac{s}{\omega_Z}\right)}{\left(1 + \dfrac{s}{\omega_{P1}}\right)} \quad \text{in which} \quad \omega_Z = \frac{g_{m5}}{C_C + C_{GD5}} = \omega_T \frac{g_{m5}}{g_{m2}}$$

and $\qquad\qquad\qquad\qquad\qquad\qquad\qquad\qquad\qquad\qquad\qquad\qquad\qquad\qquad\qquad\qquad\qquad$ (15.27)

$$\omega_{P1} = \frac{1}{R_o C_T} \quad \text{where} \quad C_T = C_{GS5} + (C_C + C_{GD5})\left(1 + \mu_{f5} + \frac{r_{o5}}{R_o}\right)$$

In the case of many FET amplifier designs, ω_Z cannot be neglected because of the relatively low ratio of transconductances between M_5 and M_2. In bipolar designs, ω_Z can usually be neglected because of the much higher transconductance that is achieved for a given Q-point current. However, ω_z can also be a problem in common-emitter amplifiers with emitter resistors that reduce the overall transconductance of the amplifier stage.

The problem can be overcome in FET amplifiers, however, through the addition of resistor R_Z in Fig. 15.29(a), which cancels the zero in Eq. (15.26). If we assume that $C_C \gg C_{GD}$, then the location of ω_Z in the numerator of Eq. (15.26) becomes

$$\omega_Z = \frac{\left(\dfrac{1}{g_{m5}}\right) - R_Z}{C_C} \qquad (15.28)$$

and the zero can be eliminated by setting $R_Z = 1/g_{m5}$.

Figure 15.29(b) depicts an alternate approach. Source follower M_7 replicates the voltage at the drain of M_5 and drives compensation capacitor C_C providing the desired negative feedback, but at high frequencies, the feed-forward current is diverted through M_7 to the power supply (ac

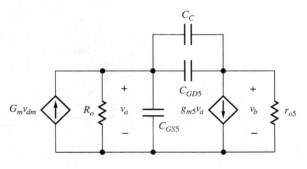

Figure 15.28 More complete model for op-amp compensation.

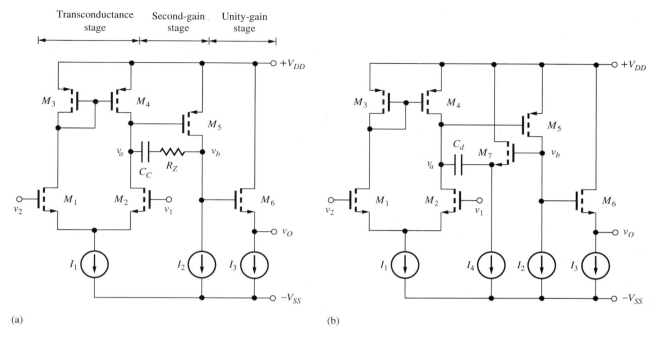

Figure 15.29 (a) Zero cancellation using resistor R_Z; (b) elimination of the feed-forward path with M_7.

ground) thereby eliminating the problematic right-half plane zero. M_7 does however introduce a high frequency left-half plane zero at $-g_{m7}/C_C$, but the effect of its phase shift is controllable with proper choice of I_4 and g_{m7}.

EXERCISE: Find the approximate location of f_Z for the op amp in Fig. 15.29 using the values from the previous exercise. What value of R_Z is needed to eliminate f_Z?

ANSWERS: 7.96 MHz; 1 kΩ

15.5.3 BIPOLAR AMPLIFIER COMPENSATION

The bipolar op amp shown in Fig. 15.30 is compensated in the same manner as the MOS amplifier. However, because the transconductance of the BJT is generally much higher than that of a FET for a given operating current, the transmission zero occurs at such a high frequency that it does not usually cause a problem. Applying Eq. (15.28) to the circuit in Fig. 15.30 yields an expression for the unity gain frequency of the two-stage bipolar amplifier:

$$\omega_T = \frac{g_{m2}}{C_C} = \frac{40 I_{C2}}{C_C} = \frac{20 I_1}{C_C} \quad \text{and} \quad \omega_Z = \frac{g_{m5}}{C_C} = \omega_T \left(\frac{I_{C5}}{I_{C2}}\right) \quad (15.29)$$

Because I_{C5} is 5 to 10 times I_{C2} in most designs, ω_Z is typically at a frequency of 5 to 10 times the unity gain frequency ω_T.

The simulated frequency response for the amplifier in Fig. 15.30(a) appears in Fig. 15.30(b) based on the values in the next exercise. The dominant pole, arising from the high resistance at the base of Q_5, occurs at approximately 565 Hz, and the unity-gain crossover occurs at 10 MHz. A second pole, due to the dominant pole of the *pnp* current mirror, causes the increased roll off beyond 10 MHz.

EXERCISE: Find the approximate locations of f_T, f_Z, and f_B for the bipolar op amp in Fig. 15.30 if $C_C = 30$ pF, $V_A = 50$, $I_1 = 100$ μA, $I_2 = 500$ μA, and $I_3 = 5$ mA.

ANSWERS: 10.6 MHz, 106 MHz, 584 Hz

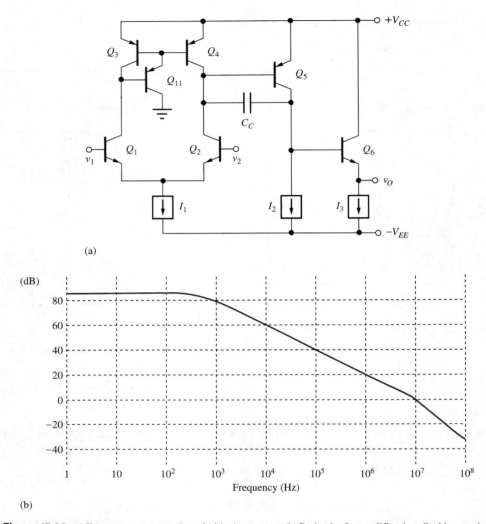

Figure 15.30 (a) Frequency compensation of a bipolar op amp; (b) Bode plot for amplifier described in exercise.

15.5.4 SLEW RATE OF THE OPERATIONAL AMPLIFIER

Errors caused by slew-rate limiting of the output voltage of the amplifier were discussed in Chapter 11. Slew-rate limiting occurs because there is a limited amount of current available to charge and discharge the internal capacitors of the amplifier. For an internally compensated amplifier, C_C typically determines the **slew rate.** Consider the example of the CMOS amplifier with the large input signal (no longer a small signal) in Fig. 15.31. In this case, the voltages applied to the differential input stage cause current I_1 to switch completely to one side of the differential pair.

Figure 15.32 is a simplified model for the amplifier in this condition. Because of the unity gain output buffer, output voltage v_O follows voltage v_B. Current I_1 must be supplied through compensation capacitor C_C, and the rate of change of the v_B, and hence v_O, must satisfy

$$I_1 = C_C \frac{d(v_B(t) - v_A(t))}{dt} = C_C \frac{d\left(v_B(t) + \frac{v_B(t)}{A_{v2}}\right)}{dt} \tag{15.30}$$

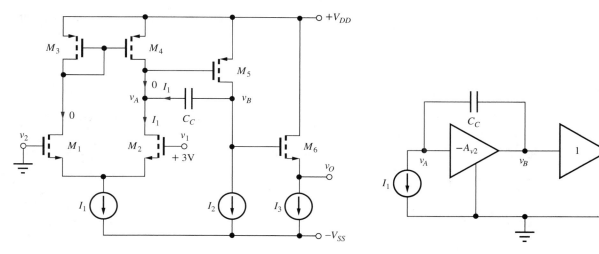

Figure 15.31 Operational amplifier with input stage overload.

Figure 15.32 Simplified model for three-stage op amp.

If A_{v2} is assumed to be very large, then the amplifier will behave in a manner similar to an ideal integrator; that is, node voltage v_A represents a virtual ground, and Eq. (15.26) becomes

$$I_1 \cong C_C \frac{dv_B(t)}{dt} = C_C \frac{dv_O(t)}{dt} \tag{15.31}$$

The slew rate is the maximum rate of change of the output signal, and

$$\text{SR} = \left.\frac{dv_O(t)}{dt}\right|_{\max} = \frac{I_1}{C_C} \tag{15.32}$$

The slew rate is determined by the total input stage bias current and the value of the compensation capacitor C_C. (It is seldom pointed out that this derivation tacitly assumes that the output of amplifier A_{v2} is capable of sourcing or sinking the current I_1. This requirement will be met as long as the amplifier is designed with $I_2 \geq I_1$.)

> **EXERCISE:** Show that the slew rate is symmetrical in the CMOS amplifier in Fig. 15.31; that is, what is the current in capacitor C_C if $v_1 = 0$ V and $v_2 = +3$ V?
>
> **ANSWER:** I_1

15.5.5 RELATIONSHIPS BETWEEN SLEW RATE AND GAIN-BANDWIDTH PRODUCT

Equation (15.31) can be related directly to the unity gain bandwidth of the amplifier using Eq. (15.25):

$$\text{SR} = \frac{I_1}{C_C} = \frac{I_1}{\left(\dfrac{G_m}{\omega_T}\right)} = \frac{\omega_T}{\left(\dfrac{G_m}{I_1}\right)} \tag{15.33}$$

For the simple CMOS amplifier in Fig. 15.24, the input stage transconductance is equal to that of transistors M_1 and M_2:

$$\left(\frac{G_m}{I_1}\right) = \frac{1}{I_1}\sqrt{2K_{n2}\frac{I_1}{2}} = \sqrt{\frac{2K_{n2}}{I_1}}$$

and

$$\text{SR} = \omega_T \sqrt{\frac{I_1}{K_{n2}}} \tag{15.34}$$

1122   Chapter 15   Transistor Feedback Amplifiers and Oscillators

For a given desired value of ω_T, the slew rate increases with the square root of the bias current in the input stage.

For the bipolar amplifier in Fig. 15.30

$$\left(\frac{G_m}{I_1}\right) = \left(\frac{40\frac{I_1}{2}}{I_1}\right) = 20 \quad \text{and} \quad SR = \frac{\omega_T}{20} \qquad (15.35)$$

In this case, the slew rate is related to the choice of unity gain frequency by a fixed factor.

EXERCISE: What is the slew rate of the CMOS amplifier in Fig. 15.24 if $K_{n2} = 1$ mA/V^2, $K_{p5} = 1$ mA/V^2, $C_C = 20$ pF, $\lambda = 0.02$ V^{-1}, $I_1 = 100$ µA, and $I_2 = 500$ µA?

ANSWER: 5.00 V/µS

EXERCISE: What is the slew rate of the bipolar amplifier in Fig. 15.30 if $C_C = 20$ pF, $I_1 = 100$ µA, and $I_2 = 500$ µA?

ANSWER: 5.00 V/µS

DESIGN EXAMPLE 15.7 — OPERATIONAL AMPLIFIER COMPENSATION

In this example, we will choose the value of the compensation capacitor in a BJT op amp to give a desired value of phase margin.

PROBLEM Design the compensation capacitor in the BJT op-amp circuit here to give a phase margin of 75°. Find the open-loop gain, bandwidth, and GBW product for the compensated op amp. For simplicity, assume that the *npn* and *pnp* transistors are described by the same set of SPICE parameters: BF = 100, VAF = 75 V, IS = 0.1 fA, RB = 250 Ω, TF = 0.75 ns, and CJC = 2 pF.

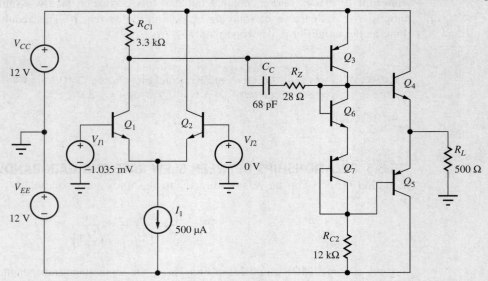

SOLUTION **Known Information and Given Data:** The three-stage op-amp circuit appears here and consists of an *npn* differential input stage driving a common-emitter *pnp* gain stage. $R_{C1} = 3.3$ kΩ and $R_{C2} = 12$ kΩ. The output stage is a complementary *npn-pnp* emitter-follower stage. Transistor parameters are given as BF = 100, VAF = 75 V, IS = 0.1 fA, RB = 250 Ω, TF = 0.75 ns, and CJC = 2 pF, $\Phi_M = 75°$.

15.5 Single-Pole Operational Amplifier Compensation

Unknowns: Value of C_C for 75° phase margin; the resulting open-loop gain and bandwidth, and unity-gain frequency; positions of the nondominant poles.

Approach: Find the Q-points of the transistors and the small-signal parameters of the transistors. Assume that the dominant pole of the amplifier is set by compensation capacitor C_C around the *pnp* common-emitter gain stage. Find the nondominant poles of the amplifier resulting from the differential input stage and the emitter follower. Then choose C_C to give the unity-gain frequency required to achieve the desired phase margin.

Assumptions: The dominant pole of the op amp is set by compensation capacitor C_C and the *pnp* common-emitter stage; R_Z is included to remove the zero associated with C_C; $T = 27°C$; the *pnp* and *npn* transistors are identical; $V_{BE} = V_{EB} = 0.75$ V. The quiescent value of $V_O = 0$. Neglect all base currents in the Q-point analyses. VJC = 0.75 V and MJC = 0.33. Transistors Q_4 and Q_5 operate in parallel. The small-signal resistances of diode-connected transistors Q_6 and Q_7 can be neglected.

ANALYSIS **Q-Point:** Bias current I_1 splits equally between Q_1 and Q_2 so that $I_{C1} = I_{C2} = 250$ μA. For $V_O = 0$, the voltage across $R_{C2} = 12 - 0.75 = 11.3$ V, and the current in Q_3 is $I_{C3} = 11.3$ V/12 kΩ $= 938$ μA. Q_4 and Q_5 mirror the currents in Q_7 and Q_8, so $I_{C4} = I_{C5} = 938$ μA. For $V_O = 0$, $V_{CE4} = 12$ V, $V_{EC5} = 12$ V, and $V_{EC3} = 11.3$ V. For $V_I = 0$, $V_{CE2} = 12.8$ V and $V_{CE1} = 12 - 3300(0.25$ mA$) + 0.75 = 11.9$ V.

Small-Signal Parameters: The small-signal parameters are found using these formulas cast in terms of the SPICE parameters:

$$\beta_o = \text{BF}\left(1 + \frac{V_{CE}}{\text{VAF}}\right) = 100\left(1 + \frac{V_{CE}}{75}\right) \qquad g_m = 40 I_C \qquad r_\pi = \frac{\beta_o}{g_m}$$

$$r_o = \frac{\text{VAF} + V_{CE}}{I_C} = \frac{75 + V_{CE}}{I_C}$$

$$C_\pi = g_m \text{TF} = g_m(0.75 \times 10^{-9}) \qquad C_\mu = \frac{\text{CJC}}{\left(1 + \frac{V_{CB}}{\text{VJC}}\right)^{\text{MJC}}} = \frac{2 \text{ pF}}{\left(1 + \frac{V_{CB}}{0.75}\right)^{0.33}}$$

	I_C (μA)	V_{CE} (V)	β_o	g_m (S)	r_π (kΩ)	r_o (kΩ)	C_π (pF)	C_μ (pF)
Q_1	250	11.9	116	0.01	11.6	348	7.50	0.803
Q_2	250	12.8	117	0.01	11.7	351	7.50	0.784
Q_3	938	11.3	115	0.0375	3.07	92.0	28.1	0.818
Q_4	938	12.0	116	0.0375	3.09	92.8	28.1	0.801
Q_5	938	12.0	116	0.0375	3.09	92.8	28.1	0.801

Open-Loop Gain: $A_o = A_{vt1} A_{vt2} A_{vt3}$

$$A_{vt1} = \frac{g_{m1}}{2}(2r_{o1} \| R_{C1} \| r_{\pi 3}) = \frac{0.01}{2}(696 \text{ kΩ} \| 3.3 \text{ kΩ} \| 3.07 \text{ kΩ}) = 7.93$$

$$A_{vt2} = g_{m3}\left[r_{o3} \| R_{C2} \| \left(\frac{r_{\pi 4}}{2} + (\beta_{o4} + 1)R_L\right)\right]$$

$$= 0.0375\left[92.0 \text{ kΩ} \| 12 \text{ kΩ} \| \left(\frac{3.09 \text{ kΩ}}{2} + (117)500 \text{ Ω}\right)\right] = 338$$

$$A_{vt3} = \frac{(\beta_o + 1)R_L}{\frac{r_{\pi 4}}{2} + (\beta_o + 1)R_L} = \frac{(117)500}{\frac{3090}{2} + (117)500} = 0.974$$

$$A_o = 2610$$

Compensation Capacitor Design: At the unity-gain frequency f_T, the dominant pole due to C_C will contribute a phase shift of 90°. The dominant poles of each of the other two stages will determine the phase margin. For a phase margin of 75°, the contributions of the additional poles can only be 15°. We expect these poles to be at frequencies above the op-amp unity-gain frequency, typically 50 to 200 MHz.

Input Stage Pole

We are interested in the transfer function for the loop gain. In the feedback path, the input stage appears as a C-C/C-B cascade. Thus, we will use the equation from Table 9.2 for the pole at the input of a common-collector stage with $R_{L2} = 1/g_{m1}$.

$$f_{pB2} = \left(\frac{1}{2\pi}\right) \frac{1}{([R_{th2} + r_{x2}] \| [r_{\pi 2} + (\beta_o + 1)R_{L2}])\left(C_{\mu 2} + \frac{C_{\pi 2}}{1 + g_{m2}R_{L2}}\right)}$$

$$= \left(\frac{1}{2\pi}\right) \frac{1}{([R_{th2} + r_{x2}] \| 2r_{\pi 2})\left(C_{\mu 2} + \frac{C_{\pi 2}}{2}\right)}$$

$$= \left(\frac{1}{2\pi}\right) \frac{1}{(250\,\Omega \| 2 \cdot 11.7\,\text{k}\Omega)\left(0.784\,\text{pF} + \frac{7.5\,\text{pF}}{2}\right)} = 142\,\text{MHz}$$

Gain Stage Pole

This pole will be dominated by the Miller effect capacitance associated with the compensation capacitance. The actual location of the pole will be calculated based on a desired phase margin.

Emitter-Follower Pole

The pole at the input to the emitter-follower stage will be affected by the pole-splitting action of the compensation capacitor placed across the gain stage. Assuming the compensation capacitor across the gain stage is much larger than the other capacitances in the circuit, the pole at the input to the follower stage is

$$f_{pB4} \cong \frac{g'_{m3}}{2\pi(C_{\pi 3} + C_{L3})}$$

To account for r_x, we will use g'_{m3} as defined in Eq. (9.70). The C_π term represents the total equivalent capacitance to small-signal ground at the input to the gain stage, including the output capacitance of the differential pair. C_{L3} is the capacitance looking into the complementary pair follower stage. Assuming only one of the two devices in the complementary pair is carrying a signal at any instant in time, the complementary pair device is represented by a device with the same g_m, a current gain equal to the average of the two devices, and TF roughly equal to the average TF. Since C_μ is a junction parasitic capacitance, we will see the cumulative capacitance due to the C_μ of both devices. Given these conditions, the pole is calculated as

$$f_{pB4} \cong \left(\frac{1}{2\pi}\right) \frac{0.0375\,\text{mS}(3.07\,\text{k}\Omega/3.32\,\text{k}\Omega)}{\left(0.8\,\text{pF} + 28.1\,\text{pF} + 2 \cdot 0.8\,\text{pF} + \frac{28.1\,\text{pF}}{1 + 0.0375\,\text{mS}\,(500\,\Omega)}\right)} = 173\,\text{MHz}$$

In addition to these terms, we should also expect to see a pole equal to approximately f_T, $[1/2\pi(\text{TF} + C_\mu/g_m)]$, at the emitter junction of the differential pair and at the output node since there is no additional output load capacitance. The f_T values for Q_1 and Q_4 are 192 MHz and 206 MHz, respectively.

We can now choose the unity-gain frequency, f_T, of the op amp to give the desired phase margin. At the unity-gain frequency, the primary pole of the op amp will contribute approximately

90° of phase shift. For a 75° phase margin, the remaining four poles can contribute an additional phase shift of 15°, which allows us to find the required value of f_T:

$$15° = \tan^{-1}\left(\frac{f_T}{142\,\text{MHz}}\right) + \tan^{-1}\left(\frac{f_T}{173\,\text{MHz}}\right) + \tan^{-1}\left(\frac{f_T}{192\,\text{MHz}}\right) + \tan^{-1}\left(\frac{f_T}{206\,\text{MHz}}\right)$$

Solving for the unity-gain frequency yields $f_T = 11.5$ MHz.

Using Eq. (15.25) from our op-amp analysis

$$(C_C + C_{\mu 3}) = \frac{G_{m1}}{\omega_T} = \left(\frac{g_{m1}}{2}\right)\left(\frac{1}{2\pi f_T}\right)$$

$$= \frac{0.005}{2\pi(11.5 \times 10^6)} = 69\,\text{pF}$$

since $C_{\mu 3}$ is approximately in parallel with C_C. To eliminate the unwanted zero associated with C_C, $R_Z = 1/g_{m3} = 27.5\,\Omega$. Now we can also find the open-loop bandwidth:

$$f_B = \frac{f_T}{A_o} = \frac{11.5\,\text{MHz}}{2610} = 4.41\,\text{kHz}$$

Thus, our design values are $A_o = 68.3$ dB, $f_T = 11.5$ MHz, $f_B = 4.41$ kHz, and $\Phi_M = 75°$.

Check of Results: We will check our analysis using SPICE as outlined below.

Computer-Aided Analysis: In order to simulate the gain with the feedback loop open, we must first find the offset voltage of the amplifier. With the amplifier connected as a voltage follower, the offset voltage was found to be 1.035 mV. The Q-point collector currents for transistors Q_1 through Q_7 are 242 μA, 254 μA, 936 μA, 1.05 mA, 1.05 mA, 917 μA, and 917 μA, respectively.

The offset voltage was then applied to the input of the open-loop amplifier to set the output voltage to zero, and an ac sweep was performed from 1 Hz to 100 MHz with 20 simulation points per decade. The resulting open-loop gain is plotted below. The open-loop gain is 67.2 dB, the open-loop bandwidth is 4.52 kHz, the unity-gain frequency is 10.7 MHz, and the phase margin is 74°. These values all agree well with our design calculations. The phase margin is being affected by a zero that can be seen in the magnitude response above 30 MHz and was not included in our analysis.

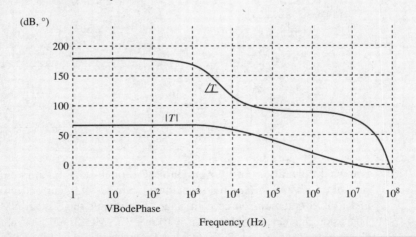

EXERCISE: Use the technique in Sec. 11.9 and Ex. 11.9 to verify the loop-gain plot in Ex. 15.7.

EXERCISE: Calculate the unity-gain frequency and phase margin for the amplifier in Ex. 15.7 if C_C is reduced to 50 pF.

ANSWERS: 15.9 MHz, 69.5°

DESIGN EXAMPLE 15.8

MOSFET OPERATIONAL AMPLIFIER COMPENSATION

In this example, we will choose the value of the compensation capacitor in a FET op amp to produce a desired value of phase margin for use in a unity-gain configuration.

PROBLEM Design the compensation capacitor in the FET op-amp circuit here to produce a phase margin of 70°. Find the open-loop gain, bandwidth, and GBW product for the compensated op amp. All of the NMOS FETs have SPICE parameters: KP = 10 mS/V, VTO = 1 V, LAMBDA = 0.01 V^{-1}. The PMOS FETs have SPICE parameters: KP = 4 mS/V, VTO = -1 V, LAMBDA = 0.01 V^{-1}. C_{GS} and C_{GD} are 5 pF and 1 pF, respectively, and will be added manually to the SPICE schematic. Consider M_5 to be the parallel combination of two PMOS FETs (or a PMOS with twice the W/L of the other PMOS FETs), KP = 8 mS/V, C_{GS} = 10 pF, and C_{GD} = 2 pF.

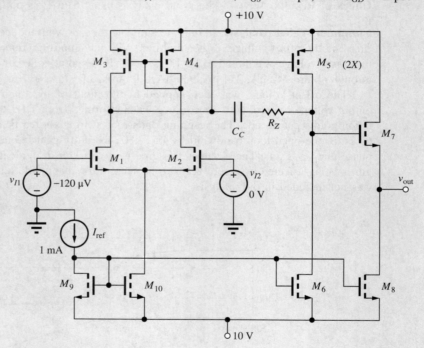

SOLUTION **Known Information and Given Data:** The three-stage op-amp circuit appears here and consists of an NMOS differential input stage with a PMOS current mirror load. The second stage is a PMOS common-source gain stage. The output stage is an NMOS source-follower stage. Transistor parameters given as KP = 10 mS/V (NMOS), KP = 4 mS/V (PMOS), VTO = -1 V, C_{GD} = 1 pF, and C_{GS} = 5 pF. Device M_5 is twice the width of the other PMOS FETs, so its KP, C_{GS}, and C_{GD} are doubled. $\Phi_M = 70°$.

Unknowns: Value of C_C for 70° phase margin; the resulting open-loop gain and bandwidth, and unity-gain frequency; positions of the nondominant poles.

Approach: Find the Q-points and small-signal parameters of the transistors. We initially assume that the dominant pole of the op amp is set by compensation capacitor C_C of the PMOS gain stage. Find the nondominant poles at the other nodes of the amplifier and use these to calculate the unity gain frequency required to achieve the desired phase margin.

Assumptions: The dominant pole is set by the compensation capacitor C_C and the C-S stage. We will include the appropriate value of R_Z to remove the right-half plane zero associated with the C-S gain stage. The circuit is operating at room temperature, and the circuit will be biased to produce a nominal output voltage of 0 V. We will neglect the finite output impedance effects on device currents when calculating the operating points.

ANALYSIS

Q-Point: The use of current mirror biasing and active loads greatly simplifies the calculation of the device operating currents. Given the reference current of 1 mA, we know that the bias currents for M_1–M_4 will all be 0.5 mA. M_6 and M_8 will nominally sink 1 mA. The V_{GS} of M_5 is of some interest. Because of the λ term in the FET current equation, we know that for I_{D3} and I_{D4} to be matched, they need to have the same V_{DS}. As a result, V_{GS} will nominally have the same value as the V_{GS} of M_3 and M_4. If M_5 is identical to M_4, their currents will therefore be approximately equal. However, M_6 is biased to sink twice the current of M_4, so the output voltage will be saturated near V_{SS} if M_5 is identical to M_4. This is why M_5 is specified as having twice the W/L of M_4, so it will produce twice the current of M_4, thus matching the current level of M_6. If the geometry of M_5 is not designed properly, the amplifier will have a nonzero offset voltage.

Small-Signal Parameters: The small-signal parameters are found using the following formulas:

$$r_o \cong \frac{1/\lambda + V_{DS}}{I_D} \qquad g_m = \sqrt{2K_n I_D(1 + \lambda V_{DS})}$$

	I_D (mA)	V_{DS} (V)	g_m (mS)	r_o (kΩ)	C_{GD} (pF)	C_{GS} (pF)
M1, M2	0.5	9.8	3.46	120	1	5
M3, M4	0.5	1.5	2.03	103	1	5
M5	1	8.6	4.33	58.6	2	10
M6	1	11.4	4.96	61.4	1	5
M7	1	10	4.90	60	1	5
M8	1	10	4.90	60	1	5
M9	1	1.45	4.50	101	1	5
M10	1	8.55	4.66	109	1	5

Open-Loop Gain: $A_o = A_{vt1} A_{vt2} A_{vt3}$

$$A_{vt1} = g_{m1,2}(r_{o1} \| r_{o3}) = 3.46 \text{ mS}(120 \text{ k}\Omega \| 103 \text{ k}\Omega) = 192 \text{ V/V}$$

$$A_{vt2} = -g_{m5}(r_{o5} \| r_{o6}) = 4.33 \text{ mS}(58.6 \text{ k}\Omega \| 61.4 \text{ k}\Omega) = -130 \text{ V/V}$$

$$A_{vt3} = \frac{g_{m7} R_{S7}}{1 + g_{m7} R_{S7}} = \frac{g_{m7}(r_{o7} \| r_{o8})}{1 + g_{m7}(r_{o7} \| r_{o8})} = \frac{4.90 \text{ mS}(60 \text{ k}\Omega \| 60 \text{ k}\Omega)}{1 + 4.90 \text{ mS}(60 \text{ k}\Omega \| 60 \text{ k}\Omega)} = 0.993 \text{ V/V}$$

$$A_o = -24{,}800 \text{ V/V} = 87.9 \text{ dB}$$

Compensation Capacitor Design: At f_T, the loop gain reaches 0 dB and the dominant pole will contribute approximately 90° of phase shift. To achieve a phase margin of 70°, the compensation capacitor is selected to set the unity-gain frequency such that the nondominant poles are contributing a total of 90 − 70 or 20° of phase shift (the inverting input contributes another 180°).

Input Stage Pole
We are interested in the transfer function for the loop gain, and in the feedback path, the input stage appears as a C-D/C-G cascade. Since we are driving our input with a zero impedance source, the pole at the gate of M_2 has infinite frequency. Since we are designing the op amp to be stable in a unity-gain configuration, we will include the M_2 input capacitance as an additional capacitive load at the output in our calculations to model the capacitive loading seen by the output when the negative feedback is connected.

$$C_{in} = C_{GD} + \frac{C_{GS}}{1 + g_{m2}R_{S2}} = C_{GD} + \frac{C_{GS}}{1 + \frac{g_{m2}}{g_{m1}}} = 1\,\text{pF} + \frac{5\,\text{pF}}{2} = 3.5\,\text{pF}$$

Differential Pair Source Node Pole
There is a high-frequency pole at the differential pair source node. This pole is found as

$$f_{pS1} \cong \left(\frac{1}{2\pi}\right)\frac{1}{\left(\frac{1}{g_{m1}} \parallel \frac{1}{g_{m2}}\right)(C_{GS1} + C_{GS2} + C_{GD10})}$$

$$= \left(\frac{1}{2\pi}\right)\frac{1}{\left(\frac{0.5}{3.46\,\text{mS}}\right)(5\,\text{pF} + 5\,\text{pF} + 1\,\text{pF})} = 100\,\text{MHz}$$

Gain Stage Pole
This pole will be dominated by the Miller effect capacitance at the input to the gain stage and associated with the compensation capacitance, C_C. The actual location of the pole will be calculated based on a desired phase margin.

Source Follower Input Pole
This pole at the input to the emitter-follower stage will be affected by the pole-splitting action of the compensation capacitor placed across the gain stage. Assuming the compensation capacitor across the gain stage is much larger than the other capacitances in the circuit, the pole at the input to the follower stage is

$$f_{pD5} \cong \frac{g_{m5}}{2\pi(C_{i5} + C_{L5})}$$

As with our bipolar example, the C_{GS} term above represents the total equivalent capacitance to small-signal ground at the input to the gain stage, including the output capacitance of the differential pair.

$$C_{i5} = C_{GD1} + C_{GD3} + C_{GS5} = (1 + 1 + 10)\,\text{pF} = 12\,\text{pF}$$

C_{L3} is the capacitance looking into the C-C output stage plus the capacitance seen looking into the current source.

$$C_{L5} = C_{GD6} + C_{GD7} + \frac{C_{GS7}}{1 + g_{m7}(r_{o7} \parallel r_{o8})} = 1 + 1 + \frac{5}{1 + 4.9\,\text{mS}(30\,\text{k}\Omega)} = 2.03\,\text{pF}$$

Given these results, the pole is calculated as

$$f_{pD5} \cong \left(\frac{1}{2\pi}\right)\frac{4.33\,\text{mS}}{(12\,\text{pF} + 2.03\,\text{pF})} = 49.2\,\text{MHz}$$

Output Pole
The pole at the output will be set by finding the equivalent resistance and capacitance at the output node. As mentioned earlier, we will include the capacitance at the gate of M_2 to model the loading of the output when the output is fed back to the input.

$$C_{eqS7} = C_{GS7} + C_{GD8} + C_{in} = 5 \text{ pF} + 1 \text{ pF} + 3.5 \text{ pF} = 9.5 \text{ pF}$$
$$R_{eqS7} \cong 1/g_{m7} = 204 \text{ }\Omega$$
$$f_{pS7} \cong \left(\frac{1}{2\pi}\right)\frac{1}{(204)(9.5 \text{ pF})} = 82.1 \text{ MHz}$$

We can now choose the unity-gain frequency, f_T, of the op amp to give the desired phase margin. At the unity-gain frequency, the primary pole of the op amp will contribute approximately 90° of phase shift. For a 70° phase margin, the remaining two poles can contribute an additional phase shift of 20°, which allows us to find the required value of f_T:

$$20° = \tan^{-1}\left(\frac{f_T}{49.2 \text{ MHz}}\right) + \tan^{-1}\left(\frac{f_T}{82.1 \text{ MHz}}\right) + \tan^{-1}\left(\frac{f_T}{100 \text{ MHz}}\right) \rightarrow f_T \cong 8.5 \text{ MHz}$$

Using our single-pole op-amp compensation result from the previous section, we calculate the compensation capacitor as

$$(C_C + C_{GD5}) = \frac{g_{m1}}{2\pi f_T} \rightarrow C_C = \frac{3.46 \text{ mS}}{2\pi(8.5 \text{ MHz})} - 2 \text{ pF} = 63 \text{ pF}$$

To eliminate the unwanted right-half plane zero associated with C_C, $R_Z = 1/g_{m5} = 230 \text{ }\Omega$. The open-loop bandwidth is now calculated as a function of the midband gain and the unity-gain frequency.

$$f_B = \frac{f_T}{A_o} = \frac{8.5 \text{ MHz}}{24,800} = 343 \text{ Hz}$$

Our final design values are $A_o = 87.9$ dB, $f_T = 8.5$ MHz, $f_B = 343$ Hz, and $\Phi_M = 70°$.

Check of Results: In this case, results will be verified by SPICE simulation.

Simulation: With such a high gain amplifier, we should expect to have a significant offset at the output when the amplifier is operated in open loop. Any bias error at the input gets multiplied by the gain of the amplifier. As with the previous BJT example, we connect the amplifier in a follower configuration and then apply the opposite offset to the input to cancel the offset and allow us to perform open loop ac simulations while maintaining a 0V bias at the output.

With the appropriate offset in place, the amplifier is simulated with an ac sweep from 1 Hz to 100 MHz. The first simulation is performed without R_Z to illustrate the stability problems created by the presence of the RHP zero in FET amplifiers.

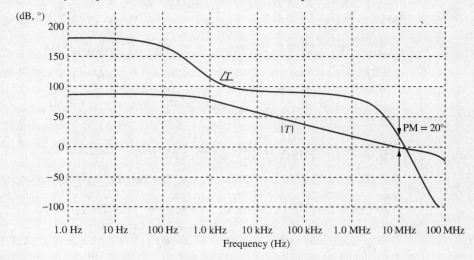

Loop gain without R_Z.

In the first simulation, the unity-gain frequency is 11 MHz but our phase margin is only 20°! The RHP zero is the worst of all conditions for stability. It simultaneously causes the magnitude response slope to decrease (increasing the 0 dB crossing frequency) while adding negative phase shift. If we resimulate with R_Z in place, we find the following loop gain response:

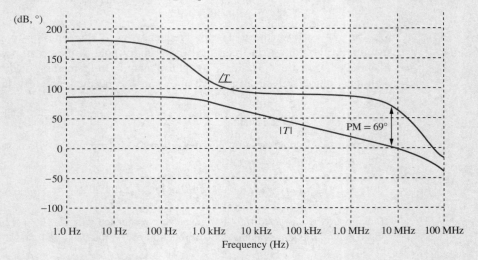

Loop gain with R_Z added to cancel RHP zero.

With the RHP zero cancelled by R_Z, our simulated result is quite close to the design values. The unity-gain frequency, f_T, is 8.4 MHz, and our phase margin is 69°. If we desire to increase the phase margin, R_Z can be increased to move the zero into the left-half plane and introduce some positive phase shift. The open-loop gain, A_o, is 86.5dB and the open-loop bandwidth, f_B, is approximately 410 Hz. These values are within the range of expected agreement with our design calculations.

EXERCISE: What is the slew rate of the amplifier in Ex. 15.8?

ANSWER: 15.4 V/μs.

EXERCISE: (a) What value of compensation capacitor is required to achieve a 60° phase margin in the amplifier in Ex. 15.8? (b) Verify your results with SPICE simulation.

ANSWER: 31.3 pF.

EXERCISE: (a) What is the offset voltage of the amplifier in Ex. 15.8 if $(W/L)_5 = (W/L)_4$? (b) If $(W/L)_5 = 3(W/L)_4$?

ANSWERS: +0.85 mV, −0.49 mV.

15.6 HIGH-FREQUENCY OSCILLATORS

Individual transistors are used in oscillators designed for high-frequency operation, and the frequency-selective feedback network is formed from a high-Q LC network or a quartz crystal resonant element. Two classic forms of **LC oscillator** are introduced here: The Colpitts oscillator uses capacitive voltage division to adjust the amount of feedback, and the Hartley oscillator employs an inductive voltage divider. Integrated circuit oscillators frequently utilize the negative G_m cell based upon a differential pair of transistors as presented in this section. Crystal oscillators are discussed in Sec. 15.6.6.

15.6.1 THE COLPITTS OSCILLATOR

Figure 15.33 shows the basic **Colpitts oscillator**. A resonant circuit is formed by inductor L and the series combination of C_1 and C_2; C_1, C_2, or L can be made variable elements in order to adjust the frequency of oscillation. The dc equivalent circuit is shown in Fig. 15.33(b). The gate of the FET is maintained at dc ground through inductor L, and the Q-point can be determined using standard techniques. In the small-signal model in Fig. 15.33(c), the gate-source capacitance C_{GS} appears in parallel with C_2, and the gate-drain capacitance C_{GD} appears in parallel with the inductor.

This circuit is used to illustrate an alternate approach to finding the conditions for oscillation to those discussed in Chapter 12. The algebra in the analysis can be simplified by defining $G = 1/(R_S \| r_o)$ and $C_3 = C_2 + C_{GS}$. Writing nodal equations for $\mathbf{V_g}(s)$ and $\mathbf{V_s}(s)$ yields

$$\begin{bmatrix} 0 \\ 0 \end{bmatrix} = \begin{bmatrix} \left(s(C_3 + C_{GD}) + \dfrac{1}{sL}\right) & -sC_3 \\ -(sC_3 + g_m) & (s(C_1 + C_3) + g_m + G) \end{bmatrix} \begin{bmatrix} \mathbf{V_g}(s) \\ \mathbf{V_s}(s) \end{bmatrix} \quad (15.36)$$

The determinant of this system of equations is

$$\Delta = s^2[C_1 C_3 + C_{GD}(C_1 + C_3)] + s[(C_3 + C_{GD})G + GC_3] + \frac{(g_m + G)}{sL} + \frac{(C_1 + C_3)}{L} \quad (15.37)$$

Because the oscillator circuit has no external excitation, we must require $\Delta = 0$ for a nonzero output voltage to exist. For $s = j\omega$, the determinant becomes

$$\Delta = \left(\frac{(C_1 + C_3)}{L} - \omega^2[C_1 C_3 + C_{GD}(C_1 + C_3)]\right) \\ + j\left(\omega[(g_m + G)C_{GD} + GC_3] - \frac{(g_m + G)}{\omega L}\right) = 0 \quad (15.38)$$

after collecting the real and imaginary parts. Setting the real part equal to zero defines the frequency of oscillation ω_o:

$$\omega_o = \frac{1}{\sqrt{L\left(C_{GD} + \dfrac{C_1 C_3}{C_1 + C_3}\right)}} = \frac{1}{\sqrt{LC_{TC}}} \quad \text{where} \quad C_{TC} = C_{GD} + \frac{C_1 C_3}{C_1 + C_3} \quad (15.39)$$

and setting the imaginary part equal to zero yields a constraint on the gain of the FET circuit:

$$\omega^2 L\left[C_{GD} + \frac{G}{(g_m + G)} C_3\right] = 1 \quad (15.40)$$

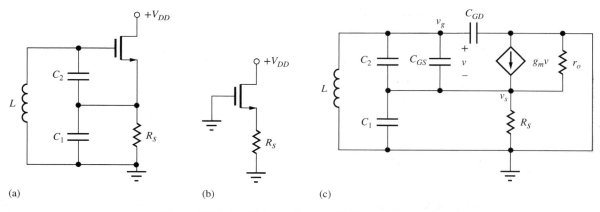

Figure 15.33 (a) Colpitts oscillator and (b) its dc and (c) small-signal models.

At $\omega = \omega_o$, the gain requirement expressed by Eq. (15.40) can be simplified to yield

$$g_m R = \frac{C_3}{C_1} \qquad \left(g_m R \geq \frac{C_3}{C_1}\right) \qquad (15.41)$$

From Eq. (15.39), we see that the frequency of oscillation is determined by the resonant frequency of the inductor L and the total capacitance C_{TC} in parallel with the inductor. The feedback is set by the capacitance ratio and must satisfy the condition in Eq. (15.41). A gain that satisfies the equality places the oscillator poles exactly on the $j\omega$ axis. However, normally, more gain is used to ensure oscillation, and some form of amplitude stabilization is used.

15.6.2 THE HARTLEY OSCILLATOR

Feedback in the **Hartley oscillator** circuit shown in Fig. 15.34 is set by the ratio of the two inductors L_1 and L_2. The dc circuit for this case appears in Fig. 15.34(b). The conditions for oscillation can be found in a manner similar to that used for the Colpitts oscillator. For simplicity, the gate-source and gate-drain capacitances have been neglected, and no mutual coupling appears between the inductors. Writing the nodal equations for the small-signal model shown in Fig. 15.34(c):

$$\begin{bmatrix} 0 \\ 0 \end{bmatrix} = \begin{bmatrix} sC + \dfrac{1}{sL_2} & -\dfrac{1}{sL_2} \\ -\left(\dfrac{1}{sL_2} + g_m\right) & \dfrac{1}{sL_1} + \dfrac{1}{sL_2} + g_m + g_o \end{bmatrix} \begin{bmatrix} \mathbf{V}_g(s) \\ \mathbf{V}_s(s) \end{bmatrix} \qquad (15.42)$$

The determinant of this system of equations is

$$\Delta = sC(g_m + g_o) + \frac{g_o}{sL_2} + \frac{1}{s^2 L_1 L_2} + C\left(\frac{1}{L_1} + \frac{1}{L_2}\right) \qquad (15.43)$$

For oscillation, we require $\Delta = 0$. After collecting the real and imaginary parts for $s = j\omega$, the determinant becomes

$$\Delta = \left[C\left(\frac{1}{L_1} + \frac{1}{L_2}\right) - \frac{1}{\omega^2 L_1 L_2}\right] + j\left(\omega C(g_m + g_o) - \frac{g_o}{\omega L_2}\right) = 0 \qquad (15.44)$$

Setting the real part equal to zero again defines the frequency of oscillation ω_o:

$$\omega_o = \frac{1}{\sqrt{C(L_1 + L_2)}} \qquad (15.45)$$

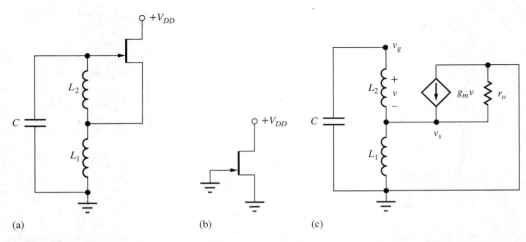

Figure 15.34 (a) Hartley oscillator using a JFET; (b) dc equivalent circuit; (c) small-signal model (C_{GS} and C_{GD} have been neglected for simplicity).

and setting the imaginary part equal to zero yields a constraint on the amplification factor of the FET:

$$1 + g_m r_o = \frac{1}{\omega C L_2} \tag{15.46}$$

At $\omega = \omega_o$, the gain requirement expressed by Eq. (15.46) becomes

$$\mu_f = \frac{L_1}{L_2} \quad \left(\mu_f \geq \frac{L_1}{L_2}\right) \tag{15.47}$$

The frequency of oscillation is set by the resonant frequency of the capacitor and the total inductance, $L_1 + L_2$. The feedback is set by the ratio of the two inductors and must satisfy the condition in Eq. (15.47). For poles on the $j\omega$ axis, the amplification factor must be large enough to satisfy the equality. Generally, more gain is used to ensure oscillation, and some form of **amplitude stabilization** is used.

15.6.3 AMPLITUDE STABILIZATION IN *LC* OSCILLATORS

The inherently nonlinear characteristics of the transistors are often used to limit oscillation amplitude. In JFET circuits, for example, the gate diode can be used to form a peak detector that limits amplitude (see Fig. 15.34). In bipolar circuits, rectification by the base-emitter diode often performs the same function. In the Colpitts oscillator shown in Fig. 15.35, a diode and resistor are added to provide the amplitude-limiting function. The diode and resistor R_G form a rectifier that establishes a negative dc bias on the gate. The capacitors in the circuit act as the rectifier filter. In practical circuits, the onset of oscillation is accompanied by a slight shift in the Q-point values as the oscillator adjusts its operating point to limit the amplitude.

15.6.4 NEGATIVE RESISTANCE IN OSCILLATORS

All oscillators need to have a negative input resistance in the oscillator in order for oscillation to occur. The **negative resistance** must be of the correct value to at least cancel the resistive losses in the circuit elements including bias resistors, the output resistance of the transistor, and the series resistance of the inductors.

As an example, let us calculate the resistance that appears at the terminals of the inductor in the Colpitts oscillator using the equivalent circuit in Fig. 15.36. We find the input resistance using the same technique that we applied to analysis of the common-source amplifier with inductive source degeneration. Based upon our knowledge of the input resistance of the common-collector and common-drain amplifiers, we have

$$\begin{aligned} Z_{in}(s) &= Z_{gs}(1 + g_m Z_s) + Z_s = \frac{1}{sC_1}\left(1 + g_m \frac{1}{sC_2}\right) + \frac{1}{sC_2} = \frac{1}{sC_1} + \frac{1}{sC_2} + \frac{g_m}{s^2 C_1 C_2} \\ Z_{in}(j\omega) &= \frac{1}{j\omega}\left(\frac{1}{C_1} + \frac{1}{C_2}\right) + R_{eq} \quad \text{with} \quad R_{eq} = -\frac{g_m}{\omega^2 C_1 C_2} \end{aligned} \tag{15.48}$$

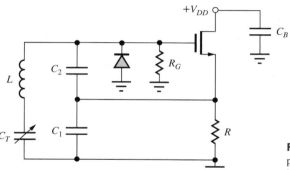

Figure 15.35 Tunable MOSFET version of the Colpitts oscillator with a diode rectifier for amplitude limiting.

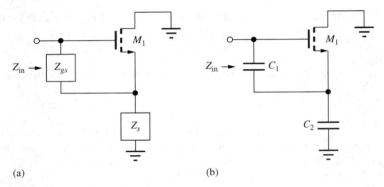

Figure 15.36 (a) Input impedance of common-drain transistor; (b) ac equivalent circuit at the inductor terminals of the Colpitts oscillator.

The input impedance is the series combination of the impedance of C_1 and C_2 plus a negative real input resistance R_{eq}.

EXERCISE: Show that the Hartley oscillator exhibits a negative input resistance using an analysis similar to that presented above.

ANSWER: $j\omega(L_1 + L_2) - \omega^2 g_m L_1 L_2$

15.6.5 NEGATIVE G_m OSCILLATOR

An oscillator that is widely utilized in integrated circuits employs a source-coupled pair of transistors biased by current source I_{SS} as in Fig. 15.37(a). The transistor pair is cross-coupled in a positive feedback configuration that causes a negative resistance to appear between the drains of the transistors. As long as the negative resistance of the cross-coupled transistors is sufficient to overcome the resistive loss in the inductors and output resistances of the transistors, then the circuit will oscillate.

Assuming symmetry, the bias current of each transistor will be $I_{SS}/2$. The desired mode of oscillation will be a differential signal appearing between the drains of the two transistors ($v_{d2} = -v_{d1}$), and the frequency of oscillation can be found from the ac equivalent circuit in Fig. 15.37(b). The resonant frequency for the circuit is determined by the equivalent capacitance that appears at

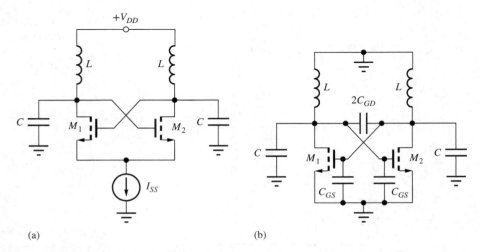

Figure 15.37 (a) Oscillator employing a negative resistance cell; (b) ac equivalent circuit.

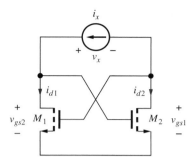

Figure 15.38 Circuit for finding the resistance of the cross-coupled transistor pair.

the terminals of the two inductors:

$$\omega_o = \frac{1}{\sqrt{2L\left(2C_{GD} + \frac{C+C_{GS}}{2}\right)}} = \frac{1}{\sqrt{LC_{eq}}} \quad \text{with} \quad C_{eq} = C + C_{GS} + 4C_{GD} \quad (15.49)$$

External capacitance C is usually designed to dominate the device capacitances and is often replaced with a varactor diode for electronic adjustment of the oscillator frequency. Note that the current source in Fig. 15.37 presents a high impedance for common-mode signals and prevents common-mode oscillations ($v_{d2} = v_{d1}$) from occurring.

The equivalent input resistance R_{in} of the **negative G_m cell** can be found with the aid of the circuit in Fig. 15.38 in which small-signal test current i_x is applied, and $R_{\text{in}} = \mathbf{v_x}/\mathbf{i_x}$. By applying KVL to the circuit, v_x is found to equal the difference in the gate-source voltages of transistors M_2 and M_1, and Kirchoff's current law indicates current i_x must enter the drain of M_1 and exit the drain of M_2:

$$\mathbf{v_x} = \mathbf{v_{gs2}} - \mathbf{v_{gs1}} \quad \text{with} \quad \mathbf{i_{d1}} = \mathbf{i_x} \quad \text{and} \quad \mathbf{i_{d2}} = -\mathbf{i_x} \quad (15.50)$$

The FET drain current and gate-source voltage are related by $\mathbf{i_d} = g_m \mathbf{v_{gs}}$ yielding

$$\mathbf{v_{gs1}} = \frac{\mathbf{i_{d1}}}{g_m} = +\frac{\mathbf{i_x}}{g_m} \quad \text{and} \quad \mathbf{v_{gs2}} = \frac{\mathbf{i_{d2}}}{g_m} = -\frac{\mathbf{i_x}}{g_m} \quad (15.51)$$

We find that the positive feedback loop results in a negative input resistance:

$$\mathbf{v_x} = -\frac{\mathbf{i_x}}{g_m} - \frac{\mathbf{i_x}}{g_m} \quad \text{and} \quad R_{\text{in}} = -\frac{2}{g_m} \quad (15.52)$$

Oscillation requires the overall conductance between the drains[7] of the transistors in Fig. 15.39 to be negative:

$$-\frac{g_m}{2} + \frac{g_o + G_P}{2} \leq 0 \quad \text{or} \quad g_m R_P \geq 1 \quad \text{for} \quad r_o \gg R_P \quad (15.53)$$

where R_P is the equivalent resistance in parallel with the inductor $R_P = (1 + Q_S^2) R_S \cong Q_S^2 R_S$. The requirement for oscillation can be written as

$$g_m > \frac{1}{Q_S^2 R_S} \quad \text{for} \quad Q_S = \frac{\omega L}{R_S} \quad (15.54)$$

Equation (15.54) places a lower bound on the transistor transconductance in terms of the inductor characteristics.

> **EXERCISE:** Draw a symmetric version of the oscillator in Fig. 15.37(b). Assume that points on the line of symmetry are virtual grounds and demonstrate that the frequency of oscillation is determined by L and C_{eq} as defined in Eq. (15.49).

---

[7] Or any other circuit port.

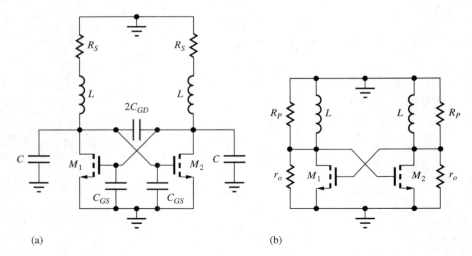

Figure 15.39 (a) Oscillator with finite Q inductors; (b) transformed equivalent circuit including output resistance of the transistors with capacitances removed.

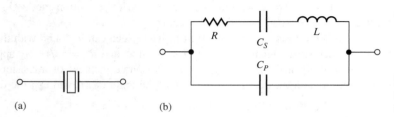

Figure 15.40 Symbol and electrical equivalent circuit for a quartz crystal.

15.6.6 CRYSTAL OSCILLATORS

Oscillators with very high-frequency accuracy and stability can be formed using quartz crystals as the frequency-determining element (**crystal oscillators**). The crystal is a piezoelectric device that vibrates in response to electrical stimulus. Although the frequency of vibration of the crystal is determined by its mechanical properties, the crystal can be modeled electrically by a very high Q ($>10{,}000$) resonant circuit, as shown in Fig. 15.40.

L, C_S, and R characterize the intrinsic series resonance path through the crystal element itself, whereas parallel capacitance C_P is dominated by the capacitance of the package containing the quartz element. The equivalent impedance of this network exhibits a series resonant frequency ω_S at which C_S resonates with L, and a parallel resonant frequency ω_P that is determined by L resonating with the series combination of C_S and C_P.

The impedance of the crystal versus frequency can easily be calculated using the circuit model in Fig. 15.40:

$$Z_C = \frac{Z_P Z_S}{Z_P + Z_S} = \frac{\frac{1}{sC_P}\left(sL + R + \frac{1}{sC_S}\right)}{\frac{1}{sC_P} + \left(sL + R + \frac{1}{sC_S}\right)} = \frac{1}{sC_P}\left(\frac{s^2 + s\frac{R}{L} + \frac{1}{LC_S}}{s^2 + s\frac{R}{L} + \frac{1}{LC_T}}\right) \qquad (15.55)$$

where $C_T = \dfrac{C_S C_P}{C_S + C_P}$

The figure accompanying Ex. 15.9 is an example of the variation of crystal reactance with frequency. Below ω_S and above ω_P, the crystal appears capacitive; between ω_S and ω_P, it exhibits an inductive reactance. As can be observed in the figure, the region between ω_S and ω_P is quite narrow. If the crystal is used to replace the inductor in the Colpitts oscillator, a well-defined frequency of oscillation will exist. In most crystal oscillators, the crystal operates between the two resonant points and represents an inductive reactance, replacing the inductor in the circuit.

EXAMPLE 15.9   QUARTZ CRYSTAL EQUIVALENT CIRCUIT

The values of L and C_S that represent the crystal have unusual magnitudes because of the extremely high Q of the crystal.

PROBLEM Calculate the equivalent circuit element values for a crystal with $f_S = 5$ MHz, $Q = 20{,}000$, $R = 50\ \Omega$, and $C_P = 5$ pF. What is the parallel resonant frequency?

SOLUTION **Known Information and Given Data:** The crystal parameters are specified as $f_S = 5$ MHz, $Q = 20{,}000$, $R = 50\ \Omega$, and $C_P = 5$ pF.

Unknowns: L and C_S

Approach: Use the definitions of Q and series resonant frequency to find the unknowns.

Assumptions: The equivalent circuit in Fig. 15.40 is adequate to model the crystal.

Analysis: Using Q, R, and f_S for a series resonant circuit

$$L = \frac{RQ}{\omega_S} = \frac{50(20{,}000)}{2\pi(5 \times 10^6)} = 31.8\ \text{mH} \qquad C_S = \frac{1}{\omega_S^2 L} = \frac{1}{(10^7\pi)^2(0.0318)} = 31.8\ \text{fF}$$

Typical values of C_P fall in the range of 5 to 20 pF. For $C_P = 5$ pF, the parallel resonant frequency will be

$$f_P = \frac{1}{2\pi\sqrt{L\dfrac{C_S C_P}{C_S + C_P}}} = \frac{1}{2\pi\sqrt{(31.8\ \text{mH})(31.6\ \text{fF})}} = 5.02\ \text{MHz}$$

whereas

$$f_S = 5.00\ \text{MHz}$$

Check of Results: Let us use our values of L and C_S to calculate f_S.

$$f_S = \frac{1}{2\pi\sqrt{31.8\ \text{mH}\ (31.8\ \text{fF})}} = 5.00\ \text{MHz}\ \checkmark$$

Discussion: Note that the two resonant frequencies differ by only 0.4 percent, and the high Q of the crystal results in a relatively large effective value for L and a small value for C_S.

Computer-Aided Analysis: The graph below presents results from a computer calculation of the reactance of the crystal versus frequency using the parameters calculated in Ex. 15.9.

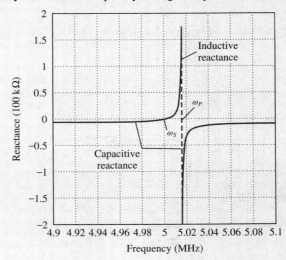

Reactance versus frequency for crystal parameters calculated in the example.

Below the series resonant frequency and above the parallel resonant frequency, the crystal exhibits capacitive reactance. Between f_S and f_P, the crystal appears inductive. In many oscillator circuits, the crystal behaves as an inductor and resonates with external capacitance. The oscillator frequency will therefore be between f_S and f_P.

EXERCISE: Calculate the parallel resonant frequency of the crystal if a 2-pF capacitor is placed in parallel with the crystal. Repeat for a 20-pF capacitor.

ANSWERS: 5.016 MHz; 5.008 MHz

Several examples of crystal oscillators are given in Figs. 15.41 to 15.44. Many variations are possible, but most of these oscillators are topological transformations of the Colpitts or Hartley oscillators. For example, the circuit in Fig. 15.41(a) represents a Colpitts oscillator with the source terminal chosen as the ground reference. The same circuit is drawn in a different form in Fig. 15.41(b). Figures 15.42 and 15.43 show Colpitts oscillators using bipolar and JFET devices.

The final crystal oscillator, shown in Fig. 15.44, represents a circuit that is often implemented using a CMOS inverting amplifier. The circuit forms yet another Colpitts oscillator, similar to Fig. 15.41(b). The inverter is initially biased into the middle of its operating region by feedback resistor R_F to ensure that the Q-point of the gate is in a region of high gain.

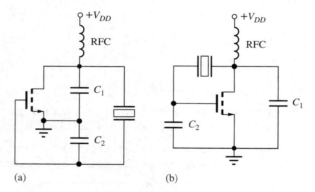

Figure 15.41 Two forms of the same Colpitts crystal oscillator.

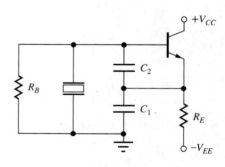

Figure 15.42 Crystal oscillator using a bipolar transistor.

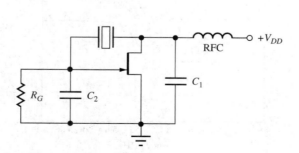

Figure 15.43 Crystal oscillator using a JFET.

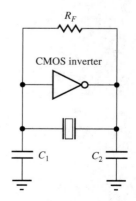

Figure 15.44 Crystal oscillator using a CMOS inverter as the gain element.

15.6.7 RING OSCILLATORS

The **ring oscillator,** depicted in Fig. 15.45, is frequently utilized to generate clock signals for logic and mixed signal circuits such as switched-capacitor filters; D/A and A/D converters; and VHF, UHF, or higher local oscillator signals for RF applications including mixers and modulators. In addition, ring oscillators are also used to measure and compare propagation delays of logic technologies (see Chapter S7 in the e-book).

The basic ring oscillator consists of an odd number of inverting amplifiers connected in a loop. The amplifiers can be as simple as a two-transistor CMOS inverter or a differential pair having differential input and differential output as in Fig. 15.46. If the voltage at the input to amplifier one in Fig. 15.45 starts to increase, then its output will decrease. Following the signal propagation through the rest of the ring, we find that the output of amplifier seven decreases. So, for an odd number of amplifiers in the ring, the feedback is negative at low frequencies. When the N-amplifier circuit is oscillating, its period T equals two times the total delay around the loop where τ_{P0} is the average time delay of the signal through one of the amplifiers as in Fig. 15.45(b) and the frequency of oscillation is given by

$$f = \frac{1}{T} = \frac{1}{2N\tau_{P0}} \tag{15.56}$$

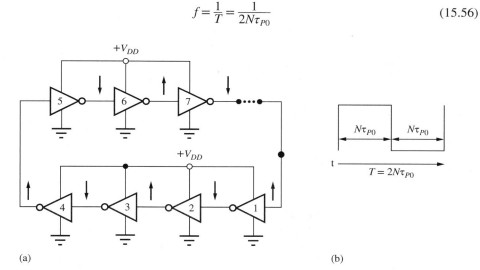

Figure 15.45 (a) A ring oscillator formed from an odd number of inverting amplifiers exhibits negative feedback at low frequencies. (b) Ring oscillator period $T = 2N\tau_{P0}$.

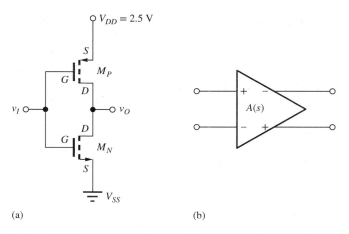

Figure 15.46 (a) CMOS inverting amplifier. (b) Differential amplifier with differential output.

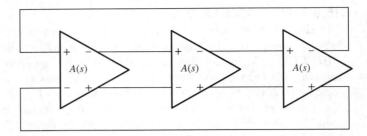

Figure 15.47 Three-stage ring oscillator utilizing fully differential amplifiers.

For oscillation, the **Barkhausen criteria** must be satisfied. N cascaded amplifiers will have a gain magnitude much greater than 1, and oscillation will occur at the frequency at which the overall phase shift will be multiples of $360°$ (or $2N\pi$ radians). For the loop of seven inverting amplifiers, an additional phase shift of $180°$ is required, and oscillation will occur at the frequency at which the additional phase is $180/7 = 25.7°$ per amplifier.

Similarly, Fig. 15.47 depicts a ring oscillator formed from three cascaded fully differential amplifiers connected in a negative feedback configuration.

One of the advantages of the ring oscillator is its ability to generate multiple signal outputs with different phases by taking outputs from various points in the ring. Even numbers of amplifiers can be used in the ring if additional circuitry is used to prevent latchup.[8]

EXERCISE: What is the total phase shift required through each of the individual amplifiers in the three-stage ring oscillator in Fig. 15.47?

ANSWER: $-240°$ or $+120°$

15.6.8 POSITIVE FEEDBACK AND LATCHUP

If a ring oscillator is constructed with an even number of amplifiers the ring will suffer from a condition termed **latchup.** Consider the cascade of two CMOS inverting amplifiers in Fig. 15.48.[9] If input voltage v_I increases, then the output of the first inverting amplifier decreases causing the output v_O of the second inverter to be high. Voltage v_O reinforces voltage v_I, creating positive feedback, and the two inverters will, "latch up" in a stable state. Similarly, a stable state also will occur if v_I decreases.

The voltage transfer characteristic (VTC) for a cascaded amplifier pair with $V_{DD} = 3$ V appears in Fig. 15.48(b). The diagonal line is a locus for $v_O = v_I$. Stable Q-points occur for $v_O = v_I$ at 0 V and 3 V. A third equilibrium point occurs at 1.5 V, however it is not stable. If any small shift away from 1.5 V occurs (noise for example), positive feedback will drive the voltages into latchup at one of the two stable operating points depending upon the polarity of the disturbance.

In digital logic, the circuit in Fig. 15.48(a) is actually called a latch and is the primary storage cell for binary data. The latch forms the heart of flip-flops, data registers, and static random access memories (SRAMs), to name a few. By adding two additional transistors, the latch of Fig. 15.48(a) becomes an R-S flip-flop, as drawn in Fig. 15.49. The additional R and S NMOS transistors can force the latch into either of the two stable states.

---

[8] For more information, see Behzad Razavi, "The ring oscillator—A circuit for all seasons," *IEEE Solid-State Circuits Magazine*, pp. 10–13, 81, Fall 2019.

[9] Or the output of amplifiers 2, 4, or 6 in Fig. 15.45(a).

15.6 High-Frequency Oscillators   1141

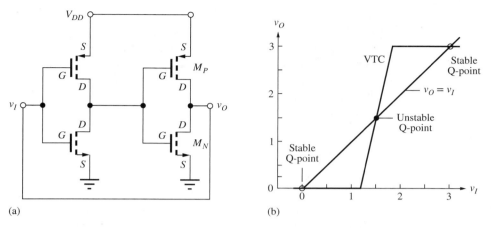

Figure 15.48 (a) Two inverting amplifiers connected in a ring. (b) The voltage transfer characteristic for two cascaded amplifiers having $V_{DD} = 3$ V and exhibiting three potential operating points with $v_O = v_I$ at 0, 1.5, and 3 V. Zero and 3 V are stable latchup points; the third point at 1.5 V is not stable.

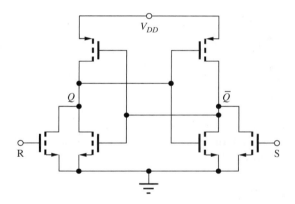

Figure 15.49 The latch becomes an R-S flip-flop with the addition of two NMOS transistors (R and S) that can force the latch into a known state.

ELECTRONICS IN ACTION

A MEMS Oscillator

Crystal oscillators have long been a mainstay for creating accurate, stable oscillators for clocks in watches and computer systems. Unlike oscillators based on integrated inductors and capacitors, crystal oscillators have very low equivalent series resistance, leading to low loss and high Q. However, conventional crystal oscillators are relatively bulky and are not easily integrated with CMOS processes. For this reason, researchers have developed microelectromechanical systems (MEMS) based resonant structures that can be integrated directly onto CMOS integrated circuits. Illustrated below is an early MEMS micromechanical resonator published in 1999 by Clark Nguyen and Roger Howe.[10] A photomicrograph of the device is shown on the next page. The structure is an electrostatic comb-drive constructed

---

[10] C. T-C. Nguyen and R. T. Howe, "An integrated CMOS micromechanical resonator high-Q oscillator," *IEEE Journal of Solid-State Circuits*, vol. 34, no. 4, pp. 440–445, April 1999.

from polysilicon material. A cross section of the MEMS post-processing is also shown. The large polysilicon structure to the left is an example of the structures used to make the resonator structure. The structure makes electrical contact to a metal layer through a thin deposited polysilicon layer. Note that the horizontal beam in the left of the figure is actually suspended above the substrate. The structural polysilicon is deposited over a sacrificial phosphosilicate glass (PSG) that had been previously deposited and patterned. After the structural polysilicon is deposited and patterned, the PSG is chemically etched away, leaving the polysilicon beams suspended above the substrate.

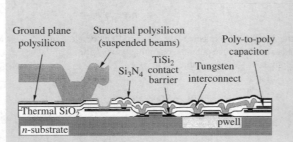

©Nguyen, C. T-C. and R. T. Howe, "An integrated Micromechanical Resonator High-Q Oscillator." IEEE Journal of Solid-State Circuits 34, no. 4 (April 1999): 440–445. Copyright ©1999 IEEE. Reprinted with permission.

The physical structure of the comb drive is more clearly seen in the block diagram below. By driving the leftmost finger structure with a voltage, the suspended structure in the middle is pulled to the left. When the voltage is removed, the structure is pulled back to the right by the suspension. When the frequency of the drive voltage approaches the resonant frequency of the structure, sustainable oscillation begins. Similar to a quartz oscillator, the micromechanical resonator has a series RLC and parallel capacitance model. As the center structure oscillates back and forth, a displacement current is generated on the output port comb structure due to the changing capacitance as the comb fingers move in and out. In this design, the displacement current is sensed by a transresistance amplifier which amplifies the signal and drives the input. At the resonant frequency, the Barkhausen criteria is satisfied and the oscillation is sustained.

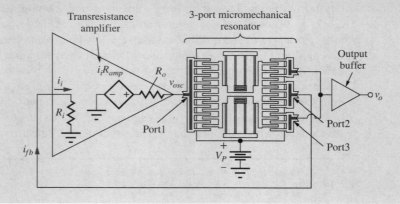

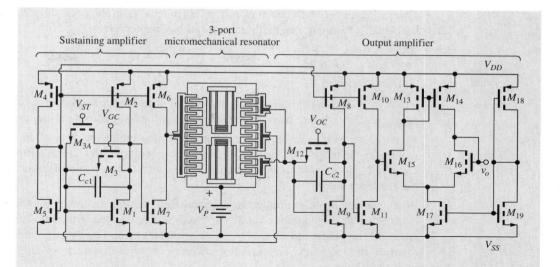

MEMS-based devices are enabling fully integrated mixers, filters, and other resonator-based structures. Because the structures are typically made from polysilicon material, they are compatible with conventional CMOS IC processing. The structure shown above has a resonant frequency in the tens of kilohertz, but researchers have developed other resonator forms with demonstrated resonant frequencies in the hundreds of megahertz. The combination of MEMS and CMOS enables highly efficient single-chip radio frequency transceivers that do not rely on the relatively lossy integrated capacitors and inductors used today. MEMS-based oscillators are currently in production at several companies.

SUMMARY

- General feedback amplifiers are separated into four classes depending on the type of feedback utilized at the input and output of the amplifier. Voltage amplifiers employ series-shunt feedback, transresistance amplifiers use shunt-shunt feedback, transconductance amplifiers utilize series-series feedback, and current amplifiers use shunt-series feedback.
- Series feedback places ports in series and increases the overall impedance level at the series-connected port. Shunt feedback is achieved by placing ports in parallel and reduces the overall impedance level at the shunt-connected port.
- The closed-loop gain of a feedback amplifier can be written as

$$A_{cl} = A_{cl}^{\text{Ideal}} \frac{T(s)}{1 + T(s)}$$

where A_{cl}^{Ideal} is the ideal closed-loop gain and $T(s)$ is the frequency-dependent loop gain or loop transmission.

- The loop gain $T(s)$ plays an important role in determining the characteristics of feedback amplifiers. For theoretical calculations, the loop gain can be found by breaking the feedback loop at some arbitrary point and directly calculating the voltage returned around the loop. However, both sides of the loop must be properly terminated before the loop-gain calculation is attempted.

- The resistance R_x between any pair of terminals in a feedback circuit can be found using Blackman's theorem, originally introduced in Chapter 11:

$$R_x = R_x^D \frac{1 + |T_{SC}|}{1 + |T_{OC}|}$$

where R_x^D is the resistance at the terminal pair with the feedback loop disabled, T_{SC} is the loop gain with the terminal pair shorted, and T_{OC} is the loop gain with the terminal pair open.

- When using SPICE or making experimental measurements, it is often impossible to break the feedback loop. The method of successive voltage and current injection, discussed in Chapter 11, is a powerful technique for determining the loop gain without the need for opening the feedback loop.

- Whenever feedback is applied to an amplifier, stability becomes a concern. In most cases, a negative or degenerative feedback condition is desired. Stability can be determined by studying the characteristics of the loop gain $T(s) = A(s)\beta(s)$ of the feedback amplifier as a function of frequency, and stability criteria can be evaluated from either Nyquist diagrams or Bode plots.

- In the Nyquist case, stability requires that the plot of $T(j\omega)$ not enclose the $T = -1$ point.

- On the Bode plot, the asymptotes of the magnitudes of $A(j\omega)$ and $1/\beta(j\omega)$ must not intersect with a rate of closure exceeding 20 dB/decade.

- Phase margin and gain margin, which can be found from either the Nyquist or Bode plot, are important measures of stability.

- Miller multiplication represents a useful method for setting the unity-gain frequency of internally compensated operational amplifiers. This technique is often called Miller compensation. In these op amps, slew rate is directly related to the unity-gain frequency.

- In circuits called oscillators, feedback is actually designed to be positive or regenerative so that an output signal can be produced by the circuit without an input being present. The Barkhausen criteria for oscillation, introduced in Chapter 12, state that the phase shift around the feedback loop must be an even multiple of 360° at some frequency, and the loop gain at that frequency must be equal to 1.

- Oscillators use some form of frequency-selective feedback to determine the frequency of oscillation; at high frequencies LC networks and quartz crystals are used to set the frequency.

- Most LC oscillators are versions of either the Colpitts or Hartley oscillators. In the Colpitts oscillator, the feedback factor is set by the ratio of two capacitors; in the Hartley case, a pair of inductors determines the feedback. Negative G_m cells are common in integrated circuit oscillators.

- Crystal oscillators use a quartz crystal to replace the inductor in LC oscillators. A crystal can be modeled electrically as a very high-Q resonant circuit, and when used in an oscillator, the crystal accurately controls the frequency of oscillation.

- Ring oscillators can generate clock signals for logic and mixed signal circuits such as switched-capacitor filters; D/A and A/D converters; and for VHF, UHF, or higher local oscillator signals for RF applications including mixers and modulators. In addition, ring oscillators are also used to measure and compare propagation delays of logic technologies.

- In order to oscillate, the circuit must develop a negative resistance to cancel losses in the circuit from bias resistors, transistor output resistances, and loss in the inductors and capacitors that form the resonant circuit.

- For true sinusoidal oscillation, the poles of the oscillator must be located precisely on the $j\omega$ axis in the s-plane. Otherwise, distortion occurs. To achieve sinusoidal oscillation, some form of amplitude stabilization is normally required. Such stabilization may result simply from the inherent nonlinear characteristics of the transistors used in the circuit, or from explicitly added gain control circuitry.

KEY TERMS

Amplitude stabilization
Barkhausen criteria for oscillation
Blackman's theorem
Bode plot
Closed-loop gain
Closed-loop input resistance
Closed-loop output resistance
Colpitts oscillator
Compensation capacitor
Crystal oscillator
Current amplifier
Degenerative feedback
Feedback amplifier stability
Feedback network
Hartley oscillator
Ideal gain
Latchup
LC oscillators
Loop gain
Loop transmission
Miller effect
Negative feedback
Negative G_m cell
Negative G_m oscillator
Negative resistance
Open-loop amplifier
Open-loop gain
Oscillator circuits
Phase margin
Positive feedback
Regenerative feedback
Ring oscillators
Series feedback connection
Series-series feedback
Series-shunt feedback
Series-shunt voltage amplifier
Shunt feedback connection
Shunt-series feedback
Shunt-shunt feedback amplifier
Sinusoidal oscillator
Slew rate
Stability
Transconductance amplifier
Transresistance amplifier
Voltage amplifier

ADDITIONAL READINGS

R. B. Blackman, "Effect of feedback on impedance," *Bell System Technical Journal,* vol. 22, no. 3, 1943.

F. Corsi, C. Marzocca, and G. Matarrese, "On impedance evaluation in feedback circuits," *IEEE Transaction on Education,* vol. 45, no. 4, pp. 371–379, November 2002.

P. J. Hurst, "A comparison of two approaches to feedback circuit analysis," *IEEE Transaction on Education,* vol. 35, pp. 253–261, August 1992.

R. C. Jaeger, S. W. Director, and A. J. Brodersen, "Computer-aided characterization of differential amplifiers," *IEEE JSSC,* vol. SC-12, pp. 83–86, February 1977.

R. D. Middlebrook, "Measurement of loop gain in feedback systems," *International Journal of Electronics,* vol. 38, no. 4, pp. 485–512, April 1975. Middlebrook credits a 1965 Hewlett-Packard Application Note as the original source of this technique.

Behzad Razavi, "The ring oscillator—A circuit for all seasons," *IEEE Solid-State Circuits Magazine,* pp. 10–13, 81, Fall 2019.

PROBLEMS

15.1 Basic Feedback System Review

15.1. The classic feedback amplifier in Fig. 15.1 has $\beta = 0.20$. What are the loop gain T, the closed-loop gain A_v, and the fractional gain error *FGE* (see Sec. 11.1.2) if $A = 120$ dB? (b) If $A = 60$ dB? (c) If $A = 15$?

15.2. The feedback amplifier in Fig. P15.2(a) has $R_1 = 1$ kΩ, $R_2 = 43$ kΩ, $R_I = 0$, and $R_L = 4.7$ kΩ. (a) What is β? (b) If $A = 80$ dB, what are the loop gain T and the closed-loop gain A_v?

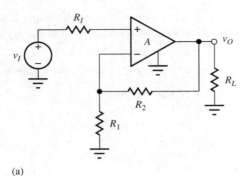

(a)

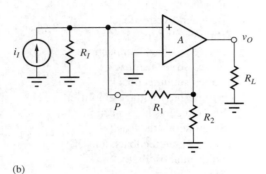

(b)

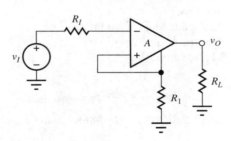

(c)

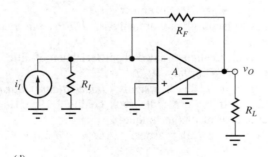

(d)

Figure P15.2 For each amplifier $A_o = 4000$, $R_{id} = 22$ kΩ, $R_o = 400$ Ω.

15.3. Find the gain, input resistance and output resistance for the circuit in Fig. 15.2 with power supplies of $+1$ V and -1 V, the current source of 2 μA, $R_i = 1$ kΩ, $R_L = 1$ MΩ, $\beta_F = 60$, $V_A = 75$ V, and $V_{BE} = 0.5$ V.

15.4. The inverting amplifier in Fig. P15.4 is implemented with an op amp with finite gain $A = 84$ dB. If $R_1 = 2$ kΩ and $R_2 = 78$ kΩ, what are β, T, and A_v?

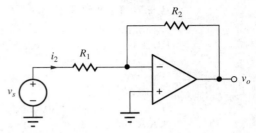

Figure P15.4

15.5. An amplifier's closed-loop voltage gain A_v is described by Eq. (15.4). What is the minimum value of open-loop gain needed if the gain error is to be less than 0.025 percent for a voltage follower ($A_v \approx 1$ with $\beta = 1$)?

15.6. An amplifier's closed-loop voltage gain is described by Eq. (15.4). What is the minimum value of open-loop gain needed if the gain error is to be less than 0.25 percent for an ideal gain of 150?

15.7. Use SPICE to simulate and compare the transfer characteristics of the two class-B output stages in Fig. P15.7 if the op amp is described by $A_o = 2500$, $R_{id} = 150$ kΩ, and $R_o = 150$ Ω. Assume $V_I = 0$.

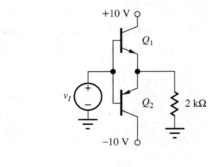

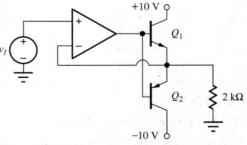

Figure P15.7

15.8. (a) Calculate the sensitivity of the closed-loop gain A_v with respect to changes in open-loop gain

A, $S_A^{A_v}$, using Eq. (15.4) and the definition of sensitivity originally presented in Chapter 12:

$$S_A^{A_v} = \frac{A}{A_v}\frac{\partial A_v}{\partial A}$$

(b) Use this formula to estimate the percentage change in closed-loop gain if the open-loop gain A changes by 20 percent for an amplifier with $A = 80$ dB and $\beta = 0.02$.

15.9. What is the maximum amplitude of a sinusoidal output in Fig. P15.7 if the power supplies are $+1.5$ V and -1.5 V if the transistors must stay in the active region as discussed in Sec. 4.4.1 of this text?

15.2 and 15.3 Feedback Amplifier Analysis at Midband and Feedback Amplifier Circuit Examples

15.10. Identify the type of feedback being used in the four circuits in Fig. P15.2.

15.11. Identify the type of negative feedback that should be used to achieve these design goals: (a) high-input resistance and high-output resistance, (b) low input resistance and low output resistance, (c) high input resistance and low output resistance, (d) low input resistance and high output resistance.

15.12. Of the four circuits in Fig. P15.2. (a) Which circuits use negative feedback to increase the output resistance? (b) Which tend to decrease the output resistance?

15.13. Consider the circuits in Fig. P15.2. (a) Which circuits use negative feedback to decrease the input resistance? (b) Which tend to increase the input resistance?

15.14. (a) Given the circuit in Fig. P15.14, use Blackman's theorem to find R_x. Assume the amplifier inputs and outputs are ideal. For $A = 400, R_1 = 680\,\Omega$, $R_2 = 2$ kΩ, and $R_3 = 2$ kΩ, find R_x^D, T_{SC}, T_{OC}, and R_x. (b) Repeat for R_y.

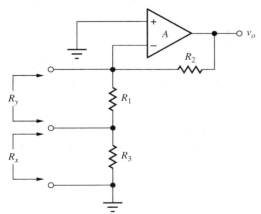

Figure P15.14

15.15. Rework Ex. 15.1 with power supplies of $+2.5$ and -2.5 V.

15.16. Repeat Problem 15.14 with $A = 200, R_1 = 4.3$ kΩ, $R_2 = 5.6$ kΩ, and $R_3 = 1$ kΩ.

15.17. For the circuit in Fig. 15.2, use Blackman's theorem to find the small-signal resistance, R_x, looking into the node at the collectors of Q_1 and Q_3. Find R_x^D, T_{SC}, T_{OC}, and R_x.

15.18. For the circuit in Fig. 15.2, use Blackman's theorem to find the small-signal resistance, R_x, looking into the node at the emitters of Q_1 and Q_2. Find R_x^D, T_{SC}, T_{OC}, and R_x.

15.19. An amplifier has an open-loop gain of 88 dB, $R_{id} = 70$ kΩ, and $R_o = 1250\,\Omega$. The amplifier is used in a feedback amplifier configuration with a resistive feedback network. (a) What is the largest value of input resistance that can be achieved in the feedback amplifier? (b) What is the smallest value of input resistance that can be achieved? (c) What is the largest value of output resistance that can be achieved in the feedback amplifier? (d) What is the smallest value of output resistance that can be achieved?

15.20. An amplifier has an open-loop gain of 100 dB, $R_{id} = 70$ kΩ, and $R_o = 1250\,\Omega$. The amplifier is used in a feedback amplifier configuration with a resistive feedback network. (a) What is the largest value of current gain that can be achieved with this feedback amplifier? (b) What is the largest value transconductance that can be achieved with this feedback amplifier?

15.21. (a) Calculate the offset voltage of the amplifier in Fig. 15.2 assuming $\beta_{Fn} = 175, \beta_{Fp} = 80, V_{An} = 75$ V, and $V_{Ap} = 60$ V. (b) Verify your calculation with SPICE.

15.22. (a) Develop an expression for the offset voltage of the amplifier in Fig. 15.2 if the current mirror is replaced with a buffered current mirror. (b) What is the offset voltage if $\beta_F = 100$ and $V_A = 50$ V? (c) Verify your result with SPICE.

Voltage Amplifiers—Series-Shunt Feedback

15.23. Find the closed-loop voltage gain, input resistance, and output resistance for the circuit in Fig. P15.2(a). Assume $R_I = 1$ kΩ, $R_1 = 5.6$ kΩ, $R_2 = 56$ kΩ, and $R_L = 10$ kΩ.

15.24. Find the closed-loop voltage gain, input resistance, and output resistance for the circuit in Fig. P15.2(a). Assume $R_I = 1$ kΩ, $R_1 = 3.6$ kΩ, $R_2 = 56$ kΩ, and $R_L = 5.6$ kΩ.

15.25. Rework Ex. 15.2 with $I_1 = 2$ μA, $I_2 = 10$ μA, $+1.5$ V and -1.5 V power supplies, $V_{TN} = -V_{TP} = 0.2$ V, and $K_n = K_p = 2$ mA/V². $R_1 = R_2 = R_4 = 130$ kΩ, $R_3 = 250$ kΩ, and $n = 1.5$. Use Eq 5.71.

15.26. Find the closed-loop gain, input resistance, and output resistance for the circuit in Fig. P15.26. Assume $R_1 = 2$ kΩ, $R_2 = 10$ kΩ, $\beta_0 = 150$, $V_A = 75$ V, $I = 100$ µA, $V_{CC} = 7.5$ V, $A = 40$ dB, $R_{id} = 75$ kΩ, and $R_o = 600$ Ω.

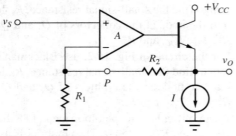

Figure P15.26

15.27. Rework Prob. 15.26 with $V_{CC} = 1.5$ V, $I = 1$ µA, $R_1 = 20$ kΩ and $R_2 = 200$ kΩ.

15.28. For the circuit in Fig. 15.4, use Blackman's theorem to find the small-signal resistance, R_x, looking into the node at the emitter of Q_2. Find R_x^D, T_{SC}, T_{OC}, and R_x.

15.29. Rework Ex. 15.1 with a bypass capacitor across R_4. Use 10 µF for the added capacitor.

15.30. Figure P15.30 is the circuit of Fig. 15.4 with an emitter-follower stage (Q_3) added to the feedback network. Use feedback analysis to find the small-signal midband gain, v_o/v_i, the input resistance R_{in} looking into the base of Q_1, and the output resistance, R_{out}. Assume $R_i = 100$ Ω, $R_1 = 200$ Ω, $R_2 = 2$ kΩ, $R_3 = 2$ kΩ, $R_4 = 300$ Ω, $R_5 = 8$ kΩ, $R_6 = 14.4$ kΩ, $R_7 = 10$ kΩ, $R_L = 10$ kΩ, $C_1 = 10$ µF. Do not treat Q_1 and Q_3 as a differential pair, you may assume that $V_O = V_I$, neglect dc base currents, and use $r_o = \infty$. Use ± 10-V power supplies.

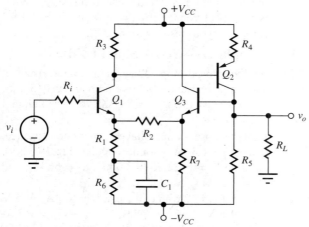

Figure P15.30

15.31. Simulate the circuit in Prob. 15.30 with SPICE and compare the results to those obtained in Prob. 15.30.

15.32. For the circuit in Prob. 15.30, use Blackman's theorem to find the small-signal resistance looking into the node at the emitter of Q_1.

15.33. Repeat Ex. 15.2 with $R_2 = 18$ kΩ, $I_1 = 200$ µA, and $R_3 = 15$ kΩ.

15.34. Use Blackman's theorem to find the small-signal impedance looking into the node at the drain of M_3 for the circuit in Prob. 15.33.

15.35. Use feedback analysis to find the voltage gain v_o/v_{ref}, input resistance, and output resistance for the circuit in Fig. P15.35. Use the results of these calculations to find the transconductance $A_{tc} = i_o/v_{\text{ref}}$. Assume $\beta_0 = 150$, $V_A = 75$ V, $I = 100$ µA, $V_{\text{ref}} = 0$ V, and $R = 7.5$ kΩ.

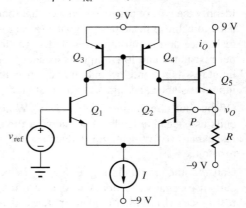

Figure P15.35

15.36. Simulate the circuit in Prob. 15.35 with SPICE and compare the results to those obtained in Prob. 15.35.

15.37. (a) Calculate the sensitivity of the closed-loop output resistance of the series-shunt feedback amplifier with respect to changes in open-loop gain A:

$$S_A^{R_{\text{out}}} = \frac{A}{R_{\text{out}}}\frac{\partial R_{\text{out}}}{\partial A}$$

(b) Use this formula to estimate the percentage change in closed-loop output resistance if the open-loop gain A changes by 5 percent for an amplifier with $A = 80$ dB and $\beta = 0.02$. (c) Calculate the sensitivity of the closed-loop input resistance of the series-shunt feedback amplifier with respect to changes in open-loop gain A:

$$S_A^{R_{\text{in}}} = \frac{A}{R_{\text{in}}}\frac{\partial R_{\text{in}}}{\partial A}$$

(Part (d) on next page.)

(d) Use this formula to estimate the percentage change in closed-loop input resistance if the open-loop gain A changes by 10 percent for an amplifier with $A = 80$ dB and $\beta = 0.02$.

Transresistance Amplifiers—Shunt-Shunt Feedback

15.38. Find the closed-loop transresistance, input resistance, and output resistance for the circuit in Fig. P15.2(d). Assume $R_I = 200$ kΩ, $R_L = 12$ kΩ, and $R_F = 36$ kΩ.

15.39. Find the closed-loop transresistance, input resistance, and output resistance for the circuit in Fig. P15.2(d). Assume $R_I = 62$ kΩ, $R_L = 12$ kΩ, and $R_F = 62$ kΩ.

15.40. The circuit in Fig. P15.40 is a shunt-shunt feedback amplifier. Use feedback analysis to find the midband input resistance, output resistance, and transresistance of the amplifier if $R_I = 500$ Ω, $R_E = 2$ kΩ, $\beta_0 = 100$, $V_A = 50$ V, $R_L = 5.6$ kΩ, and $R_F = 47$ kΩ, when v_i and R_I are replaced by a Norton equivalent circuit. What is the voltage gain for the circuit as drawn?

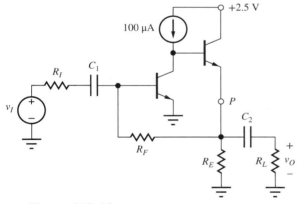

Figure P15.40

15.41. Use SPICE to find the midband input resistance, output resistance, and transresistance for the amplifier in Fig. P15.40. Compare the results to those in Prob. 15.38. $C_1 = 82$ µF and $C_2 = 47$ µF.

15.42. Use feedback analysis to find the midband transresistance, input resistance, and output resistance of the amplifier in Fig. P15.42 if $g_m = 4$ mS and $r_o = 60$ kΩ.

15.43. Use SPICE to find the midband input resistance, output resistance, and transresistance for the amplifier in Fig. P15.42. Compare the results to those in Prob. 15.42.

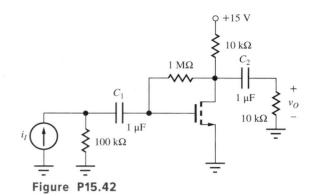

Figure P15.42

15.44. Find the gain, input resistance, and output resistance for the circuit in Fig. 15.7 with a power supply of $+1.5$ V, $R_F = 270$ kΩ, $R_i = 10$ kΩ, $R_C = 100$ kΩ, $\beta_F = 80$, $V_A = 75$ V, $V_{BE} = 0.65$ V, and neglect R_L.

15.45. Find the input resistance, output resistance, and transresistance of the TIA in page 1050 if $V_{DD} = 2.5$ V, $K_n = K_p = 500$ µA/V^2, $V_{TN} = 0.5$ V, and $V_{TP} = -0.5$ V. $\lambda_n = \lambda_p = 0.02$/V.

15.46. Rework Prob. 15.42 if the power supply in Fig. P15.42 is reduced to 1.0 V, the 10 kΩ resistors are changed to 470 kΩ, and $V_{TN} = 0.5$ V and $I_{DO} = 1$ µA for the MOSFET.

Transconductance Amplifiers—Series-Series Feedback

15.47. Find the closed-loop transconductance, input resistance, and output resistance for the circuit in Fig. P15.2(c). Assume $R_I = 2.7$ kΩ, $R_L = 8.2$ kΩ, and $R_1 = 7.5$ kΩ.

15.48. For the circuit in Fig. P15.35 with the output taken as the small-signal current i_o at the collector of Q_5, calculate the midband input resistance, output resistance looking into the collector of Q_5, and transconductance. Assume $\beta_0 = 150$, $V_A = 75$ V, $I = 100$ µA, $V_{\text{ref}} = 0$ V, and $R = 10$ kΩ.

15.49. Repeat Prob. 15.48 using SPICE to perform the analysis. Compare your results to those found in Prob. 15.48.

15.50. For the small-signal equivalent circuit in Fig. P15.50, find expressions for the midband transconductance $\mathbf{i_o}/\mathbf{v_i}$, input resistance, and output resistance looking into the collector of Q_3. Use $\beta_0 = 100$, $V_A = 50$ V, $g_m = 50$ mS, and $V_{CC} \ll V_A$. Use $R_{L1} = R_{L2} = R_{L3} = 4$ kΩ, $R_{E1} = R_{E2} = 1$ kΩ, $R_F = 10$ kΩ, and $R_I = 200$ Ω.

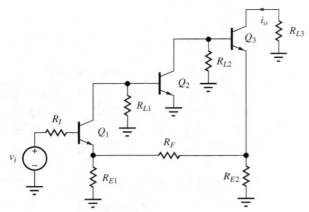

Figure P15.50

15.51. Taking the output of the circuit in Prob. 15.26 as the small-signal current flowing into the collector of the output transistor, find the transconductance and the output resistance looking into the output transistor collector.

Current Amplifiers—Shunt-Series Feedback

15.52. Find the closed-loop current gain, input resistance, and output resistance for the circuit in Fig. P15.2(b). Assume $R_I = 100$ kΩ, $R_L = 7.5$ kΩ, $R_1 = 9.1$ kΩ, and $R_2 = 1$ kΩ.

15.53. Find the input resistance, output resistance looking into the drain of M_3, and current gain $\mathbf{i_o}/\mathbf{i_{ref}}$ for the Wilson current source in Fig. P15.53. Use the small-signal two-port model in Fig. P15.53(b) for the current mirror. Assume $g_m = 4$ mS and $r_o = 60$ kΩ.

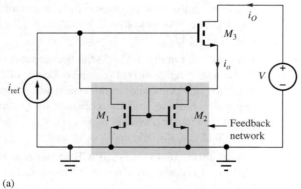

(a)

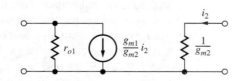

(b)

Figure P15.53

15.54. Use midband feedback analysis to find the current gain $\mathbf{i_o}/\mathbf{i_{ref}}$, input resistance, and output resistance of the Wilson BJT current source in Fig. P15.54. Assume all transistors have the same emitter area with $\beta_0 = 150$, $V_A = 75$ V, $g_m = 40$ mS, and $V_{CC} \ll V_A$.

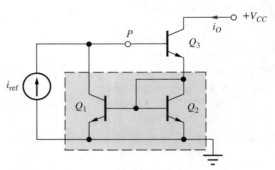

Figure P15.54

15.55. Use SPICE to simulate the Wilson BJT current source in Fig. P15.54 to find the output resistance. Use $I_{REF} = 200$ μA, $V_{CC} = 6$ V, and $V_A = 50$ V for current gains of 10^2, 10^4, and 10^6. Show that R_{out} goes from a limit of $\beta_o r_o/2$ to $\mu_f r_o$.

15.56. Find the input resistance, output resistance looking into the drain of M_3, and current gain $\mathbf{i_o}/\mathbf{i_{ref}}$ for the Wilson current source in Fig. P15.53(a) if the power supply V is 1.0 V, $V_{TN} = 0.5$ V, $K_n = 1$ μA/V^2, $I_{REF} = 100$ μA, and $\lambda = 0.02$/V for the MOSFET.

15.57. Repeat the hand calculation portions of Ex. 15.6 with all of the MOSFETS replaced by BJT transistors with $\beta_0 = 125$ and $V_A = 80$ V.

15.58. Use SPICE to perform the analysis in Prob. 15.57. Compare your results to those from Prob. 15.57.

15.59. Use Blackman's theorem to find the output resistance R_{out} looking into the drain of M_4, for the regulated cascode current source in Fig. P15.59. Use $I_1 = I_2 = 200$ μA, and $K_n = 500$ μA/V^2, $V_{TN} = 1$ V, $V_{DD} = 10$ V, and $\lambda = 0.01$/V.

15.60. Repeat Prob. 15.59 with SPICE and compare your results to hand calculations from Prob. 15.59.

15.61. Use Blackman's theorem to find the output resistance R_{out} for the regulated cascode current source in Fig. P15.59 if the MOSFETs are all replaced with BJTs with $\beta_F = 120$ and $V_A = 65$ V. Use the other element values from Prob. 15.59.

15.62. Repeat Prob. 15.61 with SPICE and compare your results to hand calculations from Prob. 15.61.

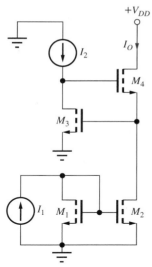

Figure P15.59

15.63. Use feedback theory to derive an expression for the input impedance of the "shunt-shunt" feedback amplifier in Fig. P15.63. Compare your result to the C_T approximation in Sec. 9.6.2.

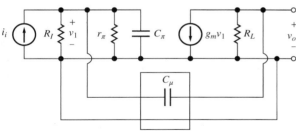

Figure P15.63

Successive Voltage and Current Injection

15.64. Use the successive voltage and current injection technique introduced in Chapter 11 at point P with SPICE to calculate the loop gain of the amplifier in Fig. P15.26. Assume $R_1 = 2$ kΩ, $R_2 = 12$ kΩ, $\beta_F = 180$, $V_A = 75$ V, $I = 125$ μA, $V_{CC} = 9$ V, $A = 40$ dB, $R_{id} = 60$ kΩ, and $R_o = 500$ Ω.

15.65. Use the successive voltage and current injection technique introduced in Chapter 11 at point P with SPICE to calculate the loop gain of the amplifier in Fig. P15.35. Assume $\beta_F = 150$, $V_A = 70$ V, $I = 100$ μA, $V_{ref} = 0$ V, and $R = 6$ kΩ.

15.66. Use the successive voltage and current injection technique introduced in Chapter 11 at point P with SPICE to calculate the loop gain of Wilson current source in Fig. P15.54. Assume all transistors have the same emitter area with $\beta_o = 145$, $V_A = 65$ V, $g_m = 60$ mS, and $V_{CC} \ll V_A$.

15.67. Use the successive voltage and current injection technique introduced in Chapter 11 at point P with SPICE to calculate the loop gain of the amplifier in Fig. P15.40. Assume all transistors have the same emitter area with $R_I = 500$ Ω, $R_E = 3$ kΩ, $\beta_F = 100$, $V_A = 50$ V, $R_L = 5.6$ kΩ, and $R_F = 39$ kΩ.

Additional problems appear at the end of the problem set.

15.4 Review of Feedback Amplifier Stability

15.68. Work Prob. 11.120.

15.69. Work Prob. 11.121.

15.70. What is the maximum load capacitance that can be connected to the voltage follower in Fig. P15.70 if the phase margin is to be 55°? Assume that the op-amp output resistance is 300 Ω, and $A(s)$ is given by

$$A(s) = \frac{2 \times 10^7}{s + 50}$$

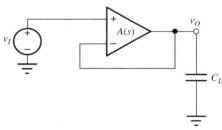

Figure P15.70

15.71. Use SPICE to find the phase margin of the shunt-series feedback pair in Ex. 15.1 if the transistors have $f_T = 300$ MHz and $C_\mu = 1$ pF.

15.72. For the circuit in Fig. P15.72 with the indicated gain characteristic, find β, T, A_v, and the phase margin if the amplifier inputs and output are ideal. Will the amplifier exhibit overshoot? If so, estimate how much overshoot. Use $R_2 = R_1 = 5$ kΩ.

$$A = \frac{575}{\left(1 + j\frac{f}{10\,\text{kHz}}\right)\left(1 + j\frac{f}{10\,\text{MHz}}\right)}$$

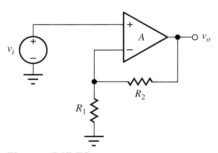

Figure P15.72

15.73. Repeat Prob. 15.72 if $R_2 = 0$, $R_1 = \infty$, and the voltage gain of the amplifier is given by
$$A = \frac{25{,}000}{\left(1 + j\frac{f}{100\,\text{Hz}}\right)\left(1 + j\frac{f}{10\,\text{MHz}}\right)}$$

15.74. Repeat Prob. 15.72 if $R_2 = 0$, $R_1 = \infty$, and the voltage gain of the amplifier is given by
$$A = \frac{25{,}000\left(1 - j\frac{f}{4\,\text{MHz}}\right)}{\left(1 + j\frac{f}{100\,\text{Hz}}\right)\left(1 + j\frac{f}{10\,\text{MHz}}\right)}$$

15.75. If the amplifier in Prob. 15.72 has an output resistance $R_o = 200\,\Omega$, what is the maximum load capacitance that can be connected at the output of the amplifier and still maintain a phase margin of 60°?

15.76. If $R_1 = 5\,\text{k}\Omega$, what must R_2 in Prob. 15.74 be to change the phase margin to 70°? What is the closed-loop gain for this value of R_2?

15.5 Single-Pole Operational Amplifier Compensation

15.77. (a) What are the unity-gain frequency and positive and negative slew rates for the CMOS amplifier in Fig. 15.24 if $I_1 = 500\,\mu\text{A}$, $I_2 = 600\,\mu\text{A}$, $K_{n1} = 1\,\text{mA/V}^2$, and $C_C = 10\,\text{pF}$? (b) If $I_1 = 400\,\mu\text{A}$, $I_2 = 400\,\mu\text{A}$, $K_{n1} = 1\,\text{mA/V}^2$, and $C_C = 5\,\text{pF}$?

15.78. Repeat Prob. 15.77 for $I_1 = 250\,\mu\text{A}$, $I_2 = 2\,\text{mA}$, $C_C = 12\,\text{pF}$.

15.79. Simulate the loop transmission frequency response of the CMOS amplifier in Fig. 15.29(a) for $R_Z = 0$ and $R_Z = 1.5\,\text{k}\Omega$. Compare the values of the unity gain frequency and phase shift of the amplifier at the unity-gain frequency. Use $I_1 = 200\,\mu\text{A}$, $I_2 = 500\,\mu\text{A}$, $I_3 = 2\,\text{mA}$, $(W/L)_1 = 30/1$, $(W/L)_3 = 40/1$, $(W/L)_5 = 80/1$, $(W/L)_6 = 60/1$, and $C_C = 10\,\text{pF}$. $V_{DD} = V_{SS} = 10\,\text{V}$. Use CMOS models from Appendix B.

15.80. Repeat Prob. 15.79 for the circuit in Fig. 15.29(b). Choose M_7 and I_4 to set $g_{m7} = 1/1500$.

15.81. What are the unity-gain frequency and slew rate of the bipolar amplifier in Fig. 15.30 if $I_1 = 100\,\mu\text{A}$, $I_2 = 400\,\mu\text{A}$, and $C_C = 10\,\text{pF}$? (b) If $I_1 = 300\,\mu\text{A}$, $I_2 = 350\,\mu\text{A}$, and $C_C = 10\,\text{pF}$?

15.82. Repeat Prob. 15.78(a) for $I_1 = 400\,\mu\text{A}$, $I_2 = 2\,\text{mA}$, and $C_C = 12\,\text{pF}$.

15.83. (a) What are the positive and negative slew rates of the amplifier in Fig. P15.83 just after a 1-V step function is applied to input v_2 if $I_1 = 50\,\mu\text{A}$, $I_2 = 400\,\mu\text{A}$, $I_3 = 500\,\mu\text{A}$, $V_{CC} = V_{EE} = 9\,\text{V}$, and $C_C = 7\,\text{pF}$? Assume v_1 is grounded. (b) Check your answers with SPICE.

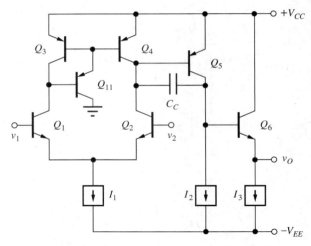

Figure P15.83

15.84. Repeat Prob. 15.83 for $I_1 = 200\,\mu\text{A}$, $I_2 = 500\,\mu\text{A}$, $I_3 = 1\,\text{mA}$, and $C_C = 10\,\text{pF}$.

15.85. (a) Calculate the poles at the base of Q_2, the base of Q_5, the collector of Q_5, and the emitter of Q_6 for the amplifier of Prob. 15.83(a) with $C_C = 0$. Use $f_T = 300\,\text{MHz}$ and $C_\mu = 1\,\text{pF}$ in addition to the bias parameters in Prob. 15.83. (b) Calculate the gain of the amplifier with $V_A = 50\,\text{V}$ and $\beta_o = 100$. (c) Assuming a unity-gain feedback connection, calculate the phase margin of the amplifier. (d) What value of C_C is necessary to set the phase margin at 75°?

15.86. Use SPICE to confirm the results from Prob. 15.85. (BF = 100, VAF = 50, TF = 530 PS, CJC = 1 pF.) Discuss the reasons for any discrepancy.

15.87. For the circuit in Fig. 15.12, find R_3 to set $I_{D3} = 0.8\,\text{mA}$ with $V_{DD} = V_{SS} = 10\,\text{V}$, and $I_{D4} = 4\,\text{mA}$. Calculate a new R_4 to maintain $V_O = 0\,\text{V}$. Calculate the poles of the amplifier and calculate R_Z to cancel the right-half plane zero. Find the unity-gain frequency such that the phase margin is set to 70°. For the phase margin calculation, assume the dominant pole contributes 90° of phase shift at the unity-gain frequency. Calculate the C_C required to achieve the desired unity-gain frequency and phase margin.

15.88. Use SPICE to simulate the results from Prob. 15.87. Discuss the reasons for any discrepancy. Use NMOS: KP = 0.01 A/V² and VTO = 1 V, PMOS: KP = 0.004 A/V² and VTO = −1 V, LAMBDA = 0.01/V, $C_{GS} = 5\,\text{pF}$, $C_{GD} = 1\,\text{pF}$. (You will need to manually add C_{GS} and C_{GD} into the circuit for this simplified model.)

15.89. (a) For the circuit discussed in Sec. 15.4, recalculate the phase margin for a unity-gain feedback connection. (b) Recalculate C_C to set the phase margin back to the value in the text when the feedback factor β was 0.5.

15.90. Use SPICE to simulate the results from Prob. 15.89. Discuss the reasons for any discrepancy. Use NMOS: KP = 0.01 A/V^2 and VTO = 1 V, PMOS: KP = 0.004 A/V^2 and VTO = -1 V, LAMBDA = 0.01/V, C_{GS} = 5 pF, C_{GD} = 1 pF. (You will need to manually add C_{GS} and C_{GD} into the circuit for this simplified model.)

15.91. Consider the compensation of the circuit in Ex. 15.1 by creating a dominant pole at the base of Q_2 with a compensation capacitor C_C from base to collector of Q_2. For this problem, assume the resistor R_4 is bypassed with a 50 µF capacitor. (a) Calculate the value of C_C required to set the unity-gain frequency of the circuit to 5 MHz. (b) Calculate the other poles in the loop transmission at the emitter of Q_1 and the collector of Q_2. (c) Calculate the phase margin of the circuit. Use $V_A = \infty$, $\beta_F = 100$, $f_T = 300$ MHz, and $C_\mu = 1$ pF.

15.92. Use SPICE to confirm the results from Prob. 15.91. BF = 100, VAF = 50, TF = 505 PS, CJC = 2.32 pF.) Discuss the reasons for any discrepancy.

15.93. Figure P15.93 presents another technique to minimize the effect of the RHP zero using a cascode amplifier at the input. Simulate the loop gain of the amplifier using SPICE and find the amplifier phase margin. Use the W/L values from Prob. 15.79. Assume that $M_{11} = M_{12} = M_1$ and $M_6 = M_9 = M_{10} = M_1$.

15.6 High-Frequency Oscillators

Colpitts Oscillators

**15.94. The ac equivalent circuit for a Colpitts oscillator is given in Fig. P15.94. (a) What is the frequency of oscillation if g_m = 10 mS, β_o = 100, R_E = 1 kΩ, L = 6 µH, C_1 = 22 pF, C_2 = 120 pF, and C_3 = infinity? Assume that the capacitances of the transistor can be neglected. (b) A variable capacitor C_3 is added to the circuit and has a range of 5–50 pF. What range of frequencies of oscillation can be achieved? (c) What is the minimum transconductance needed to ensure oscillation in part (a)? What is the minimum collector current required in the transistor?

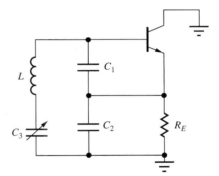

Figure P15.94

15.95. The ac equivalent circuit for a Colpitts oscillator is given in Fig. P15.94. (a) What is the frequency of oscillation if L = 15 µH, C_1 = 18 pF, C_2 = 100 pF, C_3 = infinity, f_T = 500 MHz, $r_\pi = \infty$, V_A = 50 V, r_x = 0, R_E = 1 kΩ, C_μ = 3 pF, and the transistor is operating at a Q-point of (2.5 mA, 5 V)? (b) What is the frequency of oscillation if the Q-point current is doubled?

*15.96. Design a Colpitts oscillator for operation at a frequency of 22.5 MHz using the circuit in Fig. 15.33(a). Assume L = 3 µH, K_n = 1.25 mA/V^2, and $V_{TN} = -4$ V. Ignore the device capacitances.

15.97. What is the frequency of oscillation of the MOSFET Colpitts oscillator in Fig. P15.97 if L = 7.5 µH, C_1 = 45 pF, C_2 = 45 pF, C_3 = 0 pF, C_{GS} = 10 pF,

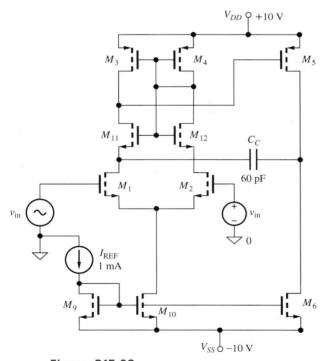

Figure P15.93

and $C_{GD} = 4$ pF? What is the minimum amplification factor of the transistor?

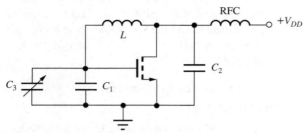

Figure P15.97

15.98. Capacitor C_3 is added to the Colpitts oscillator in Prob. 15.97 to allow tuning the oscillator. (a) Assume C_3 can vary from 5 to 50 pF and calculate the frequencies of oscillation for the two adjustment extremes. (b) What is the minimum value of amplification factor needed to ensure oscillation throughout the full tuning range?

15.99. A variable-capacitance diode is added to the Colpitts oscillator in Fig. P15.99 to form a voltage tunable oscillator. (a) The parameters of the diode are $C_{jo} = 18$ pF and $\phi_j = 0.8$ V [see Eq. (3.21)]. Calculate the frequencies of oscillation for $V_{TUNE} = 2$ V and 20 V if $L = 10$ μH, $C_1 = 75$ pF, and $C_2 = 75$ pF. Assume the RFC has infinite impedance and C_C has zero impedance. (b) What is the minimum value of voltage gain needed to ensure oscillation throughout the full tuning range?

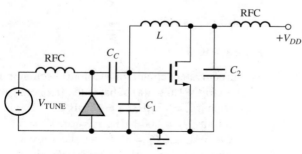

Figure P15.99

15.100. (a) Perform a SPICE transient simulation of the Colpitts oscillator in Fig. 15.33 and compare its frequency of oscillation to hand calculations if $V_{DD} = 10$ V, $K_n = 1.25$ mA/V^2, $V_{TN} = -4$ V, $R_S = 820$ Ω, $C_2 = 220$ pF, $C_1 = 470$ pF, and $L = 10$ μH. (b) Repeat if $C_2 = 470$ pF and $C_1 = 220$ pF.

15.101. Perform a SPICE transient simulation of the Colpitts oscillator in Fig. P15.97 if $L = 10$ μH, $C_1 = 50$ pF, $C_2 = 50$ pF, $C_3 = 0$ pF, RFC = 20 mH, $V_{DD} = 12$ V, $K_n = 10$ mA/V^2, $V_{TN} = 1$ V, $C_{GS} = 10$ pF, and $C_{GD} = 4$ pF. What are the amplitude and frequency of oscillation?

Hartley Oscillators

15.102. What is the frequency of oscillation of the Hartley oscillator in Fig. P15.102 if the diode is replaced by a short circuit and $L_1 = 8$ μH, $L_2 = 12$ μH, and $C = 20$ pF? Neglect C_{GS} and C_{GD}.

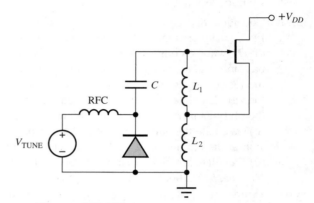

Figure P15.102

15.103. A variable-capacitance diode is added to the Hartley oscillator in Prob. 15.102 to form a voltage-tunable oscillator, and the value of C is changed to 220 pF. (a) If the parameters of the diode are $C_{jo} = 20$ pF and $\phi_j = 0.8$ V [see Eq. (3.21)], calculate the frequencies of oscillation for $V_{TUNE} = 2$ V and 20 V. Assume the RFC has infinite impedance. (b) What is the minimum value of amplification factor of the FET needed to ensure oscillation throughout the full tuning range?

15.104. Find the expression for the input impedance in the Hartley oscillator using the circuit in Fig. 15.36 with $Z_{gs} = L_1$ and $Z_s = L_2$ and demonstrate that the real part is negative.

15.105. Redraw the Hartley oscillator circuit in Fig. 15.34 with a depletion-mode NMOS transistor replacing the JFET.

15.106. Find the expression for the frequency of oscillation of the Hartley oscillator in Fig. 15.34 if the inductors have a mutual coupling M.

15.107. What is the expression for the frequency of oscillation of the Hartley oscillator in Fig. 15.34 if C_{GS} and C_{GD} of the FET are included in the circuit?

Negative G_m Oscillator

15.108. Write nodal equations for the negative G_m oscillator in Fig. 15.37(b) and directly derive the frequency of oscillation and gain required to sustain oscillation. Assume a differential-mode oscillation.

15.109. What are the Q-points of the transistors in the oscillator in Fig. 15.37(a) if $V_{DD} = 3.3$ V, $I_{EE} = 2$ mA, $V_{TN} = 0.75$ V, and $K_n = 2.5$ mA/V^2?

15.110. The oscillator in Fig. 15.37 has $L = 10$ nH and the transistor has $C_{GS} = 3$ pF and $C_{GD} = 0.5$ pF. (a) What value of C is required to achieve oscillation at 450 MHz? (b) At 1 GHz?

15.111. What are the Q-points of the transistors in the oscillators in Fig 15.37(a) if $V_{DD} = 0.75$ V, $V_{TN} = 0.75$ V, $I_{DO} = 1$ μA, the current source is removed, and the NMOS source terminals are grounded?

15.112. The oscillator in Fig. 15.37 has $L = 4$ nH with a Q of 15 at 1 GHz, and the transistor has $C_{GS} = 1$ pF and $C_{GD} = 0.25$ pF. (a) What value of C is required to achieve oscillation at 1 GHz? (b) If the transistor has $K_n = 2.5$ mA/V^2 and $\lambda = 0$, what is the minimum value of I_{SS} required for oscillation? (c) Repeat part (b) for $\lambda = 0.08$. (d) *Estimate the amplitude of the differential output signal from the oscillator.

15.113. Draw the ac equivalent circuit for the oscillator in Fig. P15.113 and find an expression for the frequency of oscillation. Include capacitances C_{GS} and C_{GD} of the FETs.

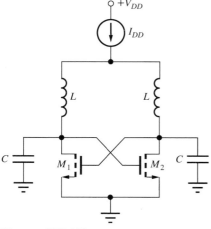

Figure P15.113

Crystal Oscillators

15.114. A crystal has a series resonant frequency of 10 MHz, series resistance of 40 Ω, Q of 25,000, and parallel capacitance of 12 pF. (a) What are the values of L and C_S for this crystal? (b) What is the parallel resonant frequency of the crystal? (c) The crystal is placed in an oscillator circuit in parallel with a total capacitance of 22 pF. What is the frequency of oscillation?

15.115. The crystal in the oscillator in Fig. P15.115 has $L = 17.5$ mH, $C_S = 20$ fF, and $R = 50$ Ω. (a) What is the frequency of oscillation if $R_E = 1$ kΩ, $R_B = 100$ kΩ, $V_{CC} = V_{EE} = 5$ V, $C_1 = 100$ pF, $C_2 = 470$ pF, and $C_3 = \infty$? Assume the transistor has $\beta_F = 100$, $V_A = 50$ V, and infinite f_T. (b) Repeat if $C_\mu = 5$ pF and $f_T = 275$ MHz.

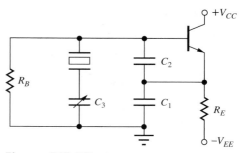

Figure P15.115

15.116. A variable capacitor C_3 is placed in series with the crystal in the oscillator in Prob. 15.115(a) to provide a calibration adjustment. Assume C_3 can vary from 1 pF to 50 pF and calculate the frequencies of oscillation for the two adjustment extremes.

15.117. Simulate the crystal oscillator in Fig. P15.115 and find the frequency of oscillation if $R_E = 1$ kΩ, $R_B = 100$ kΩ, $V_{CC} = V_{EE} = 5$ V, $C_1 = 100$ pF, $C_2 = 470$ pF, and $C_3 = \infty$. The crystal has $L = 15$ mH, $C_3 = 20$ fF, $R = 50$ Ω, and $C_P = 20$ pF. Assume the transistor has $\beta_F = 100$, $V_A = 50$ V, $C_\mu = 5$ pF, and $\tau_F = 1$ ns.

Ring Oscillators

15.118. A ring oscillator is constructed from five of the bipolar amplifiers described in Des. Ex. 15.7. What is the expected oscillation frequency based upon the Bode plot in the example?

15.119. A ring oscillator is constructed from three of the CMOS amplifiers described in Des. Ex. 15.8. (a) What is the expected oscillation frequency based upon the Bode plot for $R_Z = 0$ in the example? (b) What is the expected oscillation frequency based upon the Bode plot with R_Z added to the circuit?

15.120. (a) A ring oscillator is built from five amplifiers having $A(s) = -2/(s + 2 \times 10^6 \pi)$. What will be the frequency of operation? (b) Repeat for three amplifiers.

15.121. The operational amplifier in Fig. P15.121 is ideal and $R_1 = 2$ kΩ, $R_2 = 10$ kΩ, and $C = 270$ pF. What is the frequency of oscillator for a three-stage ring oscillator built using this amplifier circuit? (b) Repeat for a five-stage ring oscillator.

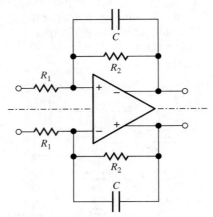

Figure P15.121

Additional Problems

*15.122. (a) Use SPICE and the method of successive voltage and current injection to find loop gain T as a function of frequency for the amplifier in Fig. 15.12. (b) Repeat for Fig. 15.19. (Don't overlook the simplifications in Sec. 11.9.1.)

*15.123. Use SPICE and the method of successive voltage and current injection to find loop gain T as a function of frequency for the amplifier in Des. Ex. 15.7. (Don't overlook the simplifications in Sec. 11.9.1.)

*15.124. Use SPICE and the method of successive voltage and current injection to find loop gain T as a function of frequency for the amplifier in Des. Ex. 15.8. (Don't overlook the simplifications in Sec. 11.9.1.)

**15.125. Use SPICE and the method of successive voltage and current injection to find loop gain T as a function of frequency for the oscillator in Prob. 15.94(a). What will be the frequency of oscillation? (Simulate only the ac equivalent circuit in SPICE.)

术语对照

Amplitude stabilization	幅度稳定性
Barkhausen criteria for oscillator	振荡器的Barkhausen准则
Blackman's theorem	Blackman定律
Bode plot	伯德图
Closed-loop gain	闭环增益
Closed-loop input resistance	闭环输入电阻
Closed-loop output resistance	闭环输出电阻
Colpitts oscillator	Colpitts振荡器
Compensation capacitor C_C	补偿电容C_C
Crystal oscillator	晶体振荡器
Current amplifier	电流放大器
Degenerative feedback	负反馈
Feedback amplifier stability	反馈放大器稳态
Feedback network	反馈网络
Gain margin(GM)	增益裕度(GM)
Gain stability	稳定增益
g-parameter	g参数
Hartley oscillator	Hartley振荡器
h-parameter	h参数
Input and output impedance	输入和输出阻抗
Ideal gain	理想增益
LC oscillators	LC振荡器
Loop gain	环路增益
Loop transmission	环路传输
−1 Point	−1 点
Miller effect	密勒效应
Negative feedback	负反馈
Negative Gm oscillator	负增益Gm振荡器
Negative resistance	负阻
Nonlinear distortion	非线性失真
Nyquist plot	奈奎斯特图
Open-loop amplifier	开环放大器

Chapter 15 Transistor Feedback Amplifiers and Oscillators

Open-loop gain	开环增益
Oscillator circuits	振荡电路
Oscillators	振荡器
Phase margin	相位裕度
Positive feedback	正反馈
Regenerative feedback	再生反馈
Series feedback connection	串联反馈连接
Sinusoidal oscillator	正弦振荡器
Stability	稳定性
Successive voltage and current injection technique	连续电压和电流注入技术
Transconductance amplifier	跨导放大器
Transresistance amplifier	跨阻放大器
Two-port network	二端口网络
Two-port parameters	二端口参数
Voltage amplifier	电压放大器
y-parameter	y 参数

APPENDICES

A: Standard Discrete Component Values

A.1 RESISTORS

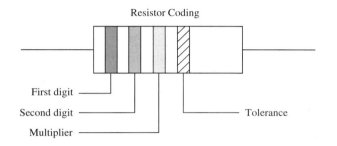

Resistor Coding
- First digit
- Second digit
- Multiplier
- Tolerance

TABLE A.1
Resistor Color Code

COLOR	DIGIT	MULTIPLIER	TOLERANCE, %
Silver	...	0.01	10
Gold	...	0.1	5
Black	0	1	
Brown	1	10	
Red	2	10^2	
Orange	3	10^3	
Yellow	4	10^4	
Green	5	10^5	
Blue	6	10^6	
Violet	7	10^7	
Gray	8	10^8	
White	9	10^9	

TABLE A.2
Standard resistor values (All values available with a 5 percent tolerance. Bold values are available with 10 percent tolerance.)

Ω								MΩ	
1.0	**5.6**	**33**	**180**	**1000**	**5600**	**33000**	**180000**	**1.0**	**5.6**
1.1	6.2	36	200	1100	6200	36000	200000	1.1	6.2
1.2	**6.8**	**39**	**220**	**1200**	**6800**	**39000**	**220000**	**1.2**	**6.8**
1.3	7.5	43	240	1300	7500	43000	240000	1.3	7.5
1.5	**8.2**	**47**	**270**	**1500**	**8200**	**47000**	**270000**	**1.5**	**8.2**
1.6	9.1	51	300	1600	9100	51000	300000	1.6	9.1
1.8	**10**	**56**	**330**	**1800**	**10000**	**56000**	**330000**	**1.8**	**10**
2.0	11	62	360	2000	11000	62000	360000	2.0	11
2.2	**12**	**68**	**390**	**2200**	**12000**	**68000**	**390000**	**2.2**	**12**
2.4	13	75	430	2400	13000	75000	430000	2.4	13
2.7	**15**	**82**	**470**	**2700**	**15000**	**82000**	**470000**	**2.7**	**15**
3.0	16	91	510	3000	16000	91000	510000	3.0	16
3.3	**18**	**100**	**560**	**3300**	**18000**	**100000**	**560000**	**3.3**	**18**
3.6	20	110	620	3600	20000	110000	620000	3.6	20
3.9	**22**	**120**	**680**	**3900**	**22000**	**120000**	**680000**	**3.9**	**22**
4.3	24	130	750	4300	24000	130000	750000	4.3	
4.7	**27**	**150**	**820**	**4700**	**27000**	**150000**	**820000**	**4.7**	
5.1	30	160	910	5100	30000	160000	910000	5.1	

TABLE A.3
Precision (1%) Resistors

						Ω						
10.0	19.1	36.5	69.8	133	255	487	931	1.78K	3.40K	6.49K	12.4K	23.7K	45.3K	84.5K	158K	294K	549K
10.2	19.6	37.4	71.5	137	261	499	953	1.82K	3.48K	6.65K	12.7K	24.3K	46.4K	86.6K	162K	301K	562K
10.5	20.0	38.3	73.2	140	267	511	976	1.87K	3.57K	6.81K	13.0K	24.9K	47.5K	88.7K	165K	309K	576K
10.7	20.5	39.2	75.0	143	274	523	1.00K	1.91K	3.65K	6.98K	13.3K	25.5K	48.7K	90.9K	169K	316K	590K
11.0	21.0	40.2	76.8	147	280	536	1.02K	1.96K	3.74K	7.15K	13.7K	26.1K	49.9K	93.1K	174K	324K	604K
11.3	21.5	41.2	78.7	150	287	549	1.05K	2.00K	3.83K	7.32K	14.0K	26.7K	51.1K	95.3K	178K	332K	619K
11.5	22.1	42.2	80.6	154	294	562	1.07K	2.05K	3.92K	7.50K	14.3K	27.4K	52.3K	97.6K	182K	340K	634K
11.8	22.6	43.2	82.5	158	301	576	1.10K	2.10K	4.02K	7.68K	14.7K	28.0K	53.6K	100K	187K	348K	649K
12.1	23.2	44.2	84.5	162	309	590	1.13K	2.15K	4.12K	7.87K	15.0K	28.7K	54.9K	102K	191K	357K	665K
12.4	23.7	45.3	86.6	165	316	604	1.15K	2.21K	4.22K	8.06K	15.4K	29.4K	56.2K	105K	196K	365K	681K
12.7	24.3	46.4	88.7	169	324	619	1.18K	2.26K	4.32K	8.25K	15.8K	30.1K	57.6K	107K	200K	374K	698K
13.0	24.9	47.5	90.9	174	332	634	1.21K	2.32K	4.42K	8.45K	16.2K	30.9K	59.0K	110K	205K	383K	715K
13.3	25.5	48.7	93.1	178	340	649	1.24K	2.37K	4.53K	8.66K	16.5K	31.6K	60.4K	113K	210K	392K	732K
13.7	26.1	49.9	95.3	182	348	665	1.27K	2.43K	4.64K	8.87K	16.9K	32.4K	61.9K	115K	215K	402K	750K
14.0	26.7	51.1	97.6	187	357	681	1.30K	2.49K	4.75K	9.09K	17.4K	33.2K	63.4K	118K	221K	412K	768K
14.3	27.4	52.3	100	191	365	698	1.33K	2.55K	4.87K	9.31K	17.8K	34.0K	64.9K	121K	226K	422K	787K
14.7	28.0	53.6	102	196	374	715	1.37K	2.61K	4.99K	9.53K	18.2K	34.8K	66.5K	124K	232K	432K	806K
15.0	28.8	54.9	105	200	383	732	1.40K	2.67K	5.11K	9.76K	18.7K	35.7K	68.1K	127K	237K	442K	825K
15.4	29.4	56.2	107	205	392	750	1.43K	2.74K	5.23K	10.0K	19.1K	36.5K	69.8K	130K	243K	453K	845K
15.8	30.1	57.6	110	210	402	768	1.47K	2.80K	5.36K	10.2K	19.6K	37.4K	71.5K	133K	249K	464K	866K
16.2	30.9	59.0	113	215	412	787	1.50K	2.87K	5.49K	10.5K	20.0K	38.3K	73.2K	137K	255K	475K	887K
16.5	31.6	60.4	115	221	422	806	1.54K	2.94K	5.62K	10.7K	20.5K	39.2K	75.0K	140K	261K	487K	909K
16.9	32.4	61.9	118	226	432	825	1.58K	3.01K	5.76K	11.0K	21.0K	40.2K	76.8K	143K	267K	499K	931K
17.4	33.2	63.4	121	232	443	845	1.62K	3.09K	5.90K	11.3K	21.5K	41.2K	78.7K	147K	274K	511K	953K
17.8	34.0	64.9	124	237	453	866	1.65K	3.16K	6.04K	11.5K	22.1K	42.2K	80.6K	150K	280K	523K	976K
18.2	34.8	66.5	127	243	464	887	1.69K	3.24K	6.19K	11.8K	22.6K	43.2K	82.5K	154K	287K	536K	1.00M
18.7	35.7	68.1	130	249	475	909	1.74K	3.32K	6.34K	12.1K	23.2K	44.2K					

A: Standard Discrete Component Values

A.2 CAPACITORS

TABLE A.4
Standard Capacitor Values (Larger values are also available)

pF	pF	pF	pF	µF	µF	µF	µF	µF	µF	µF
1	10	100	1000	0.01	0.1	1	10	100	1000	10000
	12	120	1200	0.012	0.12	1.2	12	120	1200	12000
1.5	15	150	1500	0.015	0.15	1.5	15	150	1500	15000
	18	180	1800	0.018	0.18	1.8	18	180	1800	
	20	200	2000	0.020	0.20				2000	20000
2.2	22	220	2200	0.022	0.22	2.2	22	220	2200	22000
	27	270	2700	0.027	0.27	2.7	27	270	2700	
3.3	33	330	3300	0.033	0.33	3.3	33	330	3300	33000
	39	390	3900	0.039	0.39	3.9	39	390	3900	
4.7	47	470	4700	0.047	0.47	4.7	47	470	4700	47000
5.0	50	500	5000	0.050	0.50					50000
5.6	56	560	5600	0.056	0.56	5.6	56	560	5600	
6.8	68	680	6800	0.068	0.68	6.8	68	680	6800	68000
8.2	82	820	8200	0.082	0.82	8.2	82	820	8200	

A.3 INDUCTORS

TABLE A.5
Standard Inductor Values

µH	µH	µH	µH	mH	mH	mH
0.10	1.0	10	100	1.0	10	100
	1.1	11	110			
	1.2	12	120	1.2	12	120
0.15	1.5	15	150	1.5	15	
0.18	1.8	18	180	1.8	18	
	2.0	20	200			
0.22	2.2	22	220	2.2	22	
	2.4	24	240			
0.27	2.7	27	270	2.7	27	
0.33	3.3	33	330	3.3	33	
0.39	3.9	39	390	3.9	39	
	4.3	43	430			
0.47	4.7	47	470	4.7	47	
0.56	5.6	56	560	5.6	56	
	6.2	62	620			
0.68	6.8	68	680	6.8	68	
	7.5	75	750			
0.82	8.2	82	820	8.2	82	
	9.1	91	910			

B: Solid-State Device Models and SPICE Simulation Parameters

B.1 pn JUNCTION DIODES (SECTION 3.3)

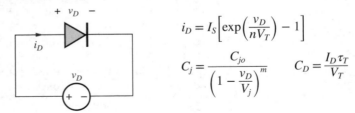

$$i_D = I_S\left[\exp\left(\frac{v_D}{nV_T}\right) - 1\right]$$

$$C_j = \frac{C_{jo}}{\left(1 - \frac{v_D}{V_j}\right)^m} \qquad C_D = \frac{I_D \tau_T}{V_T}$$

Figure B.1 Diode with applied voltage v_D.

TABLE B.1
Diode Parameters for Circuit Simulation

PARAMETER	NAME	DEFAULT	TYPICAL VALUE
Saturation current	IS	1×10^{-14} A	3×10^{-17} A
Emission coefficient (ideality factor — n)	N	1	1
Transit time (τ_T)	TT	0	0.15 nS
Series resistance	RS	0	10 Ω
Junction capacitance	CJO	0	1.0 pF
Junction potential (V_j)	VJ	1 V	0.8 V
Grading coefficient (m)	M	0.5	0.5

B.2 MOS FIELD-EFFECT TRANSISTORS (MOSFETs)

A summary of the mathematical models for both the NMOS and PMOS transistors follows. The terminal voltages and currents are defined in the accompanying figures.

NMOS TRANSISTOR MODEL SUMMARY (SECTION 5.2)

$$K_n = K'_n \frac{W}{L} = \mu_n C''_{ox} \frac{W}{L} \qquad i_G = 0 \text{ and } i_B = 0 \qquad \text{for all regions}$$

(a) NMOS transistor.

Threshold Voltage $\quad V_{TN} = V_{TO} + \gamma(\sqrt{v_{SB} + 2\phi_F} - \sqrt{2\phi_F})$

Cutoff Region $\quad i_D = 0 \quad$ for $v_{GS} \leq V_{TN}$

Triode Region $\quad i_D = K_n\left(v_{GS} - V_{TN} - \frac{v_{DS}}{2}\right)v_{DS} \quad$ for $v_{GS} - V_{TN} \geq v_{DS} \geq 0$

Saturation Region $\quad i_D = \frac{K_n}{2}(v_{GS} - V_{TN})^2(1 + \lambda v_{DS}) \quad$ for $v_{DS} \geq (v_{GS} - V_{TN}) \geq 0$

Unified Model $\quad i_D = K_n\left(v_{GS} - V_{TN} - \frac{V_{\text{MIN}}}{2}\right)V_{\text{MIN}}(1 + \lambda V_{DS}) \quad$ for $V_{GS} > V_{TN}$

$$V_{\text{MIN}} = \min\{(V_{GS} - V_{TN}), V_{DS}, V_{\text{SAT}}\}$$

Interpolation Model

$$I_D = I_{D0}\left\{\left[\ln\left(1 + \exp\left(\frac{V_{GS} - V_{TN}}{2nV_T}\right)\right)\right]^2 - \left[\ln\left(1 + \exp\left(\frac{V_{GS} - V_{TN} - nV_{DS}}{2nV_T}\right)\right)\right]^2\right\}$$

$$I_{D0} = 2nV_T^2 K_n$$

PMOS TRANSISTOR MODEL SUMMARY (SECTION 5.3)

$$K_p = K'_p \frac{W}{L} = \mu_p C''_{ox} \frac{W}{L} \qquad i_G = 0 \text{ and } i_B = 0 \qquad \text{for all regions}$$

Threshold Voltage $\quad V_{TP} = V_{TO} - \gamma(\sqrt{v_{BS} + 2\phi_F} - \sqrt{2\phi_F})$

Cutoff Region $\quad i_D = 0 \quad \text{for } v_{GS} \geq V_{TP}$

Triode Region $\quad i_D = K_p\left(v_{GS} - V_{TP} - \frac{v_{DS}}{2}\right)v_{DS}$
for $v_{GS} - V_{TP} \leq v_{DS} \leq 0$

Saturation Region $\quad i_D = \frac{K_p}{2}(v_{GS} - V_{TP})^2(1 + \lambda|v_{DS}|)$
for $v_{DS} \leq (v_{GS} - V_{TP}) \leq 0$

Unified Model $\quad i_D = K_p\left(v_{GS} - V_{TP} - \frac{V_{\text{MIN}}}{2}\right)V_{\text{MIN}}(1 + \lambda|v_{DS}|) \qquad \text{for } V_{GS} < V_{TP}$
$V_{\text{MIN}} = \max\{(V_{GS} - V_{TP}), V_{DS}, V_{\text{SAT}}\}$

(b) PMOS transistor.

Interpolation Model

$$I_{D0} = I_O\left\{\left[\ln\left(1 + \exp\left(\frac{V_{GS} - V_{TP}}{2nV_T}\right)\right)\right]^2 - \left[\ln\left(1 + \exp\left(\frac{V_{GS} - V_{TP} - nV_{DS}}{2nV_T}\right)\right)\right]^2\right\}$$

$$I_{D0} = 2nV_T^2 K_p$$

TABLE B.2
Types of MOSFET Transistors

NMOS DEVICE		PMOS DEVICE
Enhancement-mode	$V_{TN} > 0$	$V_{TP} < 0$
Depletion-mode	$V_{TN} \leq 0$	$V_{TP} \geq 0$

MOS TRANSISTOR PARAMETERS FOR CIRCUIT SIMULATION

For simulation purposes, use the LEVEL=1 models in SPICE with the following SPICE parameters in your NMOS and PMOS devices:

TABLE B.3
Representative MOS Device Parameters for SPICE Simulation (MOSIS 0.5-μm p-well process)

PARAMETER	SYMBOL	NMOS TRANSISTOR	PMOS TRANSISTOR
Threshold voltage	VTO	0.91 V	−0.77 V
Transconductance	KP	50 μA/V^2	20 μA/V^2
Body effect	GAMMA	0.99 $\sqrt{V}$	0.53 $\sqrt{V}$
Surface potential	PHI	0.7 V	0.7 V
Channel-length modulation	LAMBDA	0.02 V^{-1}	0.05 V^{-1}
Mobility	UO	615 cm^2	235 cm^2/s
Channel length	L	0.5 μm	0.5 μm
Channel width	W	0.5 μm	0.5 μm
Ohmic drain resistance	RD	0	0
Ohmic source resistance	RS	0	0
Junction saturation current	IS	0	0
Built-in potential	PB	0	0
Gate-drain capacitance per unit width	CGDO	330 pF/m	315 pF/m

(*Continued on next page*)

TABLE B.3
(Continued)

PARAMETER	SYMBOL	NMOS TRANSISTOR	PMOS TRANSISTOR
Gate-source capacitance per unit width	CGSO	330 pF/m	315 pF/m
Gate-bulk capacitance per unit length	CGBO	395 pF/m	415 pF/m
Junction bottom capacitance per unit area	CJ	3.9×10^{-4} F/m^2	2×10^{-4} F/m^2
Grading coefficient	MJ	0.45	0.47
Sidewall capacitance	CJSW	510 pF/m	180 pF/m
Sidewall grading coefficient	MJSW	0.36	0.09
Source-drain sheet resistance	RSH	22 Ω/square	70 Ω/square
Oxide thickness	TOX	4.15×10^{-6} cm	4.15×10^{-6} cm
Junction depth	XJ	0.23 µm	0.23 µm
Lateral diffusion	LD	0.26 µm	0.25 µm
Substrate doping	NSUB	2.1×10^{16}/cm^3	5.9×10^{16}/cm^3
Critical field	UCRIT	9.6×10^5 V/cm	6×10^5 V/cm
Critical field exponent	UEXP	0.18	0.28
Saturation velocity	VMAX	7.6×10^7 cm/s	6.5×10^7 cm/s
Fast surface state density	NFS	9×10^{11}/cm^2	3×10^{11}/cm^2
Surface state density	NSS	1×10^{10}/cm^2	1×10^{10}/cm^2

B.3 JUNCTION FIELD-EFFECT TRANSISTORS (JFETs)

CIRCUIT SYMBOLS AND JFET MODEL SUMMARY (SECTION 5.11)

The adjacent figures present the circuit symbols and terminal voltages and currents for n-channel and p-channel JFETs.

n-CHANNEL JFET $i_G \cong 0$ for $v_{GS} \leq 0$; $V_P < 0$

Cutoff Region $i_D = 0$ for $v_{GS} \leq V_P$

Linear Region $i_D = \dfrac{2I_{DSS}}{V_P^2}\left(v_{GS} - V_P - \dfrac{v_{DS}}{2}\right) v_{DS}$

for $v_{GS} - V_P \geq v_{DS} \geq 0$

Saturation Region $i_D = I_{DSS}\left(1 - \dfrac{v_{GS}}{V_P}\right)^2 (1 + \lambda v_{DS})$

for $v_{DS} \geq v_{GS} - V_P \geq 0$

p-CHANNEL JFET $i_G \cong 0$ for $v_{GS} \geq 0$; $V_P > 0$

Cutoff Region $i_D = 0$ for $v_{GS} \geq V_P$

Linear Region $i_D = \dfrac{2I_{DSS}}{V_P^2}\left(v_{SG} - V_P - \dfrac{v_{DS}}{2}\right) v_{DS}$

for $v_{GS} - V_P \leq v_{DS} \leq 0$

Saturation Region $i_D = I_{DSS}\left(1 - \dfrac{v_{GS}}{V_P}\right)^2 (1 + \lambda |v_{DS}|)$

for $v_{DS} \leq v_{GS} - V_P \leq 0$

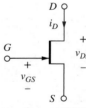

(a) n-channel JFET.

(b) p-channel JFET.

TABLE B.4
JFET Device Parameters for SPICE Simulation (NJF/PJF)

PARAMETER	SYMBOL	NJF DEFAULT	NJF EXAMPLE
Pinch-off voltage (V_P)	VTO	-2 V	-2 V ($+2$ V for PJF*)
Transconductance parameter	BETA $= \left(\dfrac{2I_{DSS}}{V_P^2}\right)$	100 µA/V^2	250 µA/V^2
Channel-length modulation	LAMBDA	0 V^{-1}	0.02 V^{-1}
Ohmic drain resistance	RD	0	100 Ω
Ohmic source resistance	RS	0	100 Ω
Zero-bias gate-source capacitance	CGS	0	10 pF
Zero-bias gate-drain capacitance	CGD	0	5 pF
Gate built-in potential	PB	1 V	0.75 V
Gate saturation current	IS	10^{-14} A	10^{-14} A

* There is an error in the implementation of many SPICE programs. The sign of VTO is entered as negative for both types of JFETs.

B.4 BIPOLAR-JUNCTION TRANSISTORS (BJTs) (SECTION 4.2)

TABLE B.5
Regions of Operation of the Bipolar Transistor

BASE-EMITTER JUNCTION	BASE-COLLECTOR JUNCTION	
	FORWARD BIAS	REVERSE BIAS
FORWARD BIAS	Saturation region (closed switch)	Forward active region (good amplifier)
REVERSE BIAS	Reverse active region (poor amplifier)	Cutoff region (open switch)

npn TRANSPORT MODEL EQUATIONS

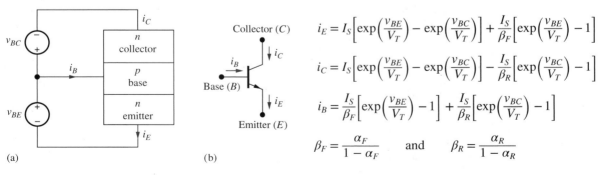

$$i_E = I_S\left[\exp\left(\frac{v_{BE}}{V_T}\right) - \exp\left(\frac{v_{BC}}{V_T}\right)\right] + \frac{I_S}{\beta_F}\left[\exp\left(\frac{v_{BE}}{V_T}\right) - 1\right]$$

$$i_C = I_S\left[\exp\left(\frac{v_{BE}}{V_T}\right) - \exp\left(\frac{v_{BC}}{V_T}\right)\right] - \frac{I_S}{\beta_R}\left[\exp\left(\frac{v_{BC}}{V_T}\right) - 1\right]$$

$$i_B = \frac{I_S}{\beta_F}\left[\exp\left(\frac{v_{BE}}{V_T}\right) - 1\right] + \frac{I_S}{\beta_R}\left[\exp\left(\frac{v_{BC}}{V_T}\right) - 1\right]$$

$$\beta_F = \frac{\alpha_F}{1 - \alpha_F} \quad \text{and} \quad \beta_R = \frac{\alpha_R}{1 - \alpha_R}$$

npn transistor.

npn FORWARD-ACTIVE REGION, INCLUDING EARLY EFFECT (SECTION 4.7.2)

$$i_C = I_S\left[\exp\left(\frac{v_{BE}}{V_T}\right)\right]\left[1 + \frac{v_{CE}}{V_A}\right]$$

$$\beta_F = \beta_{FO}\left[1 + \frac{v_{CE}}{V_A}\right]$$

$$i_B = \frac{I_S}{\beta_{FO}}\left[\exp\left(\frac{v_{BE}}{V_T}\right)\right]$$

pnp TRANSPORT MODEL EQUATIONS

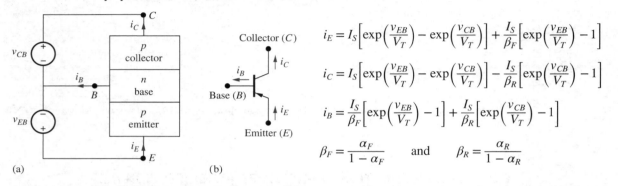

$$i_E = I_S\left[\exp\left(\frac{v_{EB}}{V_T}\right) - \exp\left(\frac{v_{CB}}{V_T}\right)\right] + \frac{I_S}{\beta_F}\left[\exp\left(\frac{v_{EB}}{V_T}\right) - 1\right]$$

$$i_C = I_S\left[\exp\left(\frac{v_{EB}}{V_T}\right) - \exp\left(\frac{v_{CB}}{V_T}\right)\right] - \frac{I_S}{\beta_R}\left[\exp\left(\frac{v_{CB}}{V_T}\right) - 1\right]$$

$$i_B = \frac{I_S}{\beta_F}\left[\exp\left(\frac{v_{EB}}{V_T}\right) - 1\right] + \frac{I_S}{\beta_R}\left[\exp\left(\frac{v_{CB}}{V_T}\right) - 1\right]$$

$$\beta_F = \frac{\alpha_F}{1 - \alpha_F} \quad \text{and} \quad \beta_R = \frac{\alpha_R}{1 - \alpha_R}$$

(a) (b)

pnp transistor.

pnp FORWARD-ACTIVE REGION, INCLUDING EARLY EFFECT

$$i_C = I_S\left[\exp\left(\frac{v_{EB}}{V_T}\right)\right]\left[1 + \frac{v_{EC}}{V_A}\right]$$

$$\beta_F = \beta_{FO}\left[1 + \frac{v_{EC}}{V_A}\right]$$

$$i_B = \frac{I_S}{\beta_{FO}}\left[\exp\left(\frac{v_{EB}}{V_T}\right)\right]$$

TABLE B.6
Bipolar Device Parameters for Circuit Simulation (*npn*/*pnp*)

PARAMETER	NAME	DEFAULT	TYPICAL *npn* VALUES
Saturation current	IS	10^{-16} A	3×10^{-17} A
Forward current gain	BF	100	100
Forward emission coefficient	NF	1	1.03
Forward Early voltage	VAF	∞	75 V
Reverse current gain	BR	1	0.5
Base resistance	RB	0	100 Ω
Collector resistance	RC	0	10 Ω
Emitter resistance	RE	0	1 Ω
Forward transit time	TF	0	0.15 nS
Reverse transit time	TR	0	15 nS
Base-emitter junction capacitance	CJE	0	0.5 pF
Base-emitter junction potential	PHIE	0.75 V	0.8 V
Base-emitter grading coefficient	ME	0.5	0.5
Base-collector junction capacitance	CJC	0	1 pF
Base-collector junction potential	PHIC	0.75 V	0.7 V
Base-collector grading coefficient	MC	0.33	0.33
Collector-substrate junction capacitance	CJS	0	3 pF

C: Two-Port Review (Section 6.3)

The **two-port network** in Fig. C.1(a) is very useful for modeling the behavior of amplifiers in complex systems. We can use the two-port to provide a relatively simple representation of a much more complicated circuit. Thus, the two-port helps us hide or encapsulate the complexity of the circuit, so we can more easily manage the overall analysis and design. One important limitation must be remembered, however. The two-ports we use are linear network models, and are valid under small-signal conditions that are fully discussed in Chapter 6.

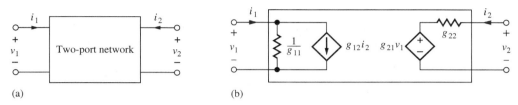

Figure C.1 (a) Two-port network representation; (b) two port g-parameter representation.

From network theory, we know that two-port networks can be represented in terms of **two-port parameters**. Four of these sets are often used as models for amplifiers: the g-, h-, y-, and z-parameters; the s- and *abcd*-parameters are not required in this text. Note in these two-port representations that (v_1, i_1) and (v_2, i_2) represent the signal components of the voltages and currents at the two ports of the network.

C.1 THE g-PARAMETERS

The g-**parameter** description is one of the most commonly used representations for a voltage amplifier:

$$\mathbf{i_1} = g_{11}\mathbf{v_1} + g_{12}\mathbf{i_2}$$
$$\mathbf{v_2} = g_{21}\mathbf{v_1} + g_{22}\mathbf{i_2}$$
(C.1)

Figure C.1(b) is a network representation of these equations.

The g-parameters are determined from a given network using a combination of **open-circuit** ($i = 0$) and **short-circuit** ($v = 0$) **termination** conditions by applying these parameter definitions:

$$g_{11} = \left.\frac{\mathbf{i_1}}{\mathbf{v_1}}\right|_{\mathbf{i_2}=0} = \text{open-circuit input conductance}$$

$$g_{12} = \left.\frac{\mathbf{i_1}}{\mathbf{i_2}}\right|_{\mathbf{v_1}=0} = \text{reverse short-circuit current gain}$$

$$g_{21} = \left.\frac{\mathbf{v_2}}{\mathbf{v_1}}\right|_{\mathbf{i_2}=0} = \text{forward open-circuit voltage gain}$$

$$g_{22} = \left.\frac{\mathbf{v_2}}{\mathbf{i_2}}\right|_{\mathbf{v_1}=0} = \text{short-circuit output resistance}$$
(C.2)

C.2 THE HYBRID OR h-PARAMETERS

The h-**parameter** description is also widely used in electronic circuits and is one convenient model for a current amplifier:

$$\mathbf{v_1} = h_{11}\mathbf{i_1} + h_{12}\mathbf{v_2}$$
$$\mathbf{i_2} = h_{21}\mathbf{i_1} + h_{22}\mathbf{v_2}$$
(C.3)

Figure C.2 is the network representation of these equations.

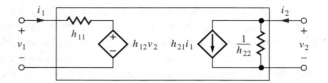

Figure C.2 Two-port h-parameter representation.

As with the g-parameters, the h-parameters are determined from a given network using a combination of open- and short-circuit measurement conditions:

$$h_{11} = \left.\frac{v_1}{i_1}\right|_{v_2=0} = \text{short-circuit input resistance}$$

$$h_{12} = \left.\frac{v_1}{v_2}\right|_{i_1=0} = \text{reverse open-circuit voltage gain}$$

$$h_{21} = \left.\frac{i_2}{i_1}\right|_{v_2=0} = \text{forward short-circuit current gain}$$

$$h_{22} = \left.\frac{i_2}{v_2}\right|_{i_1=0} = \text{open-circuit output conductance}$$

(C.4)

C.3 THE ADMITTANCE OR y-PARAMETERS

The admittance, or **y-parameter,** description is useful in modeling transconductance amplifiers.

$$i_1 = y_{11}v_1 + y_{12}v_2$$
$$i_2 = y_{21}v_1 + y_{22}v_2$$

(C.5)

Figure C.3 is a network representation of these equations.

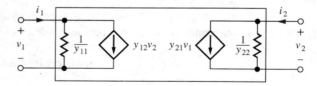

Figure C.3 Two-port y-parameter representation.

The y-parameters are often referred to as the short-circuit parameters because they are determined from a given network using only short-circuit terminations:

$$y_{11} = \left.\frac{i_1}{v_1}\right|_{v_2=0} = \text{short-circuit input conductance}$$

$$y_{12} = \left.\frac{i_1}{v_2}\right|_{v_1=0} = \text{reverse short-circuit transconductance}$$

$$y_{21} = \left.\frac{i_2}{v_1}\right|_{v_2=0} = \text{forward short-circuit transconductance}$$

$$y_{22} = \left.\frac{i_2}{v_2}\right|_{v_1=0} = \text{short-circuit output conductance}$$

(C.6)

C.4 THE IMPEDANCE OR z-PARAMETERS

The impedance, or **z-parameters,** can also be used for modeling voltage amplifiers.

$$v_1 = z_{11}i_1 + z_{12}i_2$$
$$v_2 = z_{21}i_1 + z_{22}i_2 \qquad (C.7)$$

Figure C.4 is a network representation of Eq. (C.7).

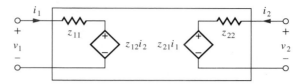

Figure C.4 Two-port z-parameter representation.

The z-parameters are determined from a given network using open-circuit measurement conditions and are often referred to as the open-circuit parameters:

$$z_{11} = \left.\frac{v_1}{i_1}\right|_{i_2=0} = \text{open-circuit input resistance}$$

$$z_{12} = \left.\frac{v_1}{i_2}\right|_{i_1=0} = \text{reverse open-circuit transresistance}$$

$$z_{21} = \left.\frac{v_2}{i_1}\right|_{i_2=0} = \text{forward open-circuit transresistance} \qquad (C.8)$$

$$z_{22} = \left.\frac{v_2}{i_2}\right|_{i_1=0} = \text{open-circuit output resistance}$$

Device parameters are often hard to determine from discrete transistor specification sheets because of historical use of a variety of two-port parameter definitions. Table C.1 has a number of synonyms for various BJT and FET parameters. Devices used in ICs are typically much more fully characterized.

TABLE C.1
Device Parameter Descriptions

PARAMETER	DESCRIPTION	SYNONYM
BJTs		
V_{CEO}	Collector-emitter breakdown voltage with base open	
V_{CBO}	Collector-base breakdown voltage with emitter open	
V_{EBO}	Emitter-base breakdown voltage with collector open	
β_F	Common-emitter (C-E) dc forward short circuit current gain	h_{FE}
β_o	Common-emitter forward short-circuit current gain	h_{fe}
$r_x + r_\pi$	C-E short-circuit input resistance	h_{ie}
$g_o = 1/r_o$	C-E open-circuit output conductance	h_{oe}
$r_\pi/(r_\mu + r_\pi)$	C-E reverse open-circuit voltage gain (Voltage feedback ratio)	h_{re}
C_π	Common-base input capacitance with collector open	C_{ibo}
C_μ	Common-base output capacitance with emitter open	C_{obo}
FETs		
g_m	Common-source (C-S) forward transadmittance or transconductance	$\|y_{fs}\|$ or g_{fs}
g_o	C-S output admittance or conductance	$\|y_{os}\|$ or g_{os}
$C_{GS} + C_{GD}$	C-S short-circuit input capacitance	C_{iss}
C_{GD}	C-S short-circuit reverse transfer capacitance	C_{rss}
$C_{GD} + C_{DS}$	C-S short-circuit output capacitance	C_{oss}

D: Physical Constants and Transistor Model Summary

PHYSICAL CONSTANTS		
SYMBOL	**QUANTITY**	**VALUE**
N_{AV}	Avogadro constant	6.022×10^{26}/kg·mole
c	Speed of light in a vacuum	2.998×10^{10} cm/s
ε_o	Permittivity of free space	8.854×10^{-14} F/cm
ε_s	Relative permittivity of silicon	11.7
ε_{ox}	Relative permittivity of silicon dioxide	3.9
E_G	Bandgap of silicon	1.12 eV
h	Planck's constant	6.625×10^{-34} J·s
		4.135×10^{-15} eV·s
k	Boltzmann's constant	1.381×10^{-23} J/K
		8.617×10^{-5} eV/K
$\frac{kT}{q}$	Thermal voltage at 300 K	0.0259 V
m_o	Electron rest mass	9.1095×10^{-31} kg
m_p	Proton rest mass	1.6726×10^{-27} kg
n_i	Silicon intrinsic carrier density at room temperature	10^{10}/cm^3
q	Electronic charge	1.602×10^{-19} C

CONVERSION FACTORS	
1 angstrom = 10^{-8} cm	$\mu = 10^{-6}$
1 µm = 10^{-4} cm	n = 10^{-9}
1 mil = 25.4 µm	p = 10^{-12}
1 eV = 1.602×10^{-19} J	f = 10^{-15}
	k = 10^{3}
	M = 10^{6}
	G = 10^{9}
	T = 10^{12}

D: Physical Constants and Transistor Model Summary

DIODE EQUATIONS

$$i_D = I_S\left[\exp\left(\frac{v_D}{nV_T}\right) - 1\right] \qquad V_T = \frac{kT}{q} \qquad C_j = \frac{C_{jo}A}{\sqrt{1 - \frac{v_D}{\phi_j}}} \qquad C_D = \frac{I_D}{V_T}\tau_T$$

(FORWARD) ACTIVE REGION EQUATIONS— npn TRANSISTOR ($V_{BE} > 0$ AND $V_{CE} \geq V_{BE}$)

$$i_C = I_S \exp\left(\frac{v_{BE}}{V_T}\right)\left(1 + \frac{v_{CE}}{V_A}\right) \qquad i_C = \beta_F i_B \qquad i_E = (\beta_F + 1)i_B \qquad \beta_F = \beta_{FO}\left(1 + \frac{v_{CE}}{V_A}\right)$$

BJT SMALL-SIGNAL MODEL PARAMETER RELATIONSHIPS ($\beta_o \cong \beta_F$)

$$g_m = \frac{I_C}{V_T} \cong 40 I_C \qquad \beta_o = g_m r_\pi \qquad r_o = \frac{V_A + V_{CE}}{I_C} \cong \frac{V_A}{I_C} \qquad \mu_f = g_m r_o \qquad C_\pi = g_m \tau_F \qquad \omega_T = \frac{g_m}{C_\pi + C_\mu}$$

LARGE SIGNAL MODEL EQUATIONS—NMOS TRANSISTOR

Triode (Linear) Region ($v_{GS} > V_{TN}$ and $v_{DS} \leq v_{GS} - V_{TN}$)

$$i_D = K_n\left(v_{GS} - V_{TN} - \frac{v_{DS}}{2}\right)^2 v_{DS} \qquad i_G = 0 \qquad i_S = i_D \qquad K_n = K_n'\frac{W}{L}$$

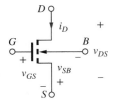

Active (Saturation) Region ($v_{GS} > V_{TN}$ and $v_{DS} \geq v_{GS} - V_{TN}$)

$$i_D = \frac{K_n}{2}(v_{GS} - V_{TN})^2(1 + \lambda v_{DS}) \qquad i_G = 0 \qquad i_S = i_D \qquad K_n = K_n'\frac{W}{L}$$

$$V_{TN} = V_{TO} + \gamma(\sqrt{v_{SB} + 2\phi_f} - \sqrt{2\phi_f})$$

All-region Interpolation Model

$$I_D = I_{D0}\left\{\left[\ln\left(1 + \exp\left(\frac{V_{GS} - V_{TN}}{2nV_T}\right)\right)\right]^2 - \left[\ln\left(1 + \exp\left(\frac{V_{GS} - V_{TN} - nV_{DS}}{2nV_T}\right)\right)\right]^2\right\}$$

$$I_{D0} = 2nV_T^2 K_n$$

FET SMALL-SIGNAL MODEL PARAMETER RELATIONSHIPS

$$g_m = \frac{2I_D}{V_{GS} - V_{TN}} \cong \sqrt{2K_n I_D} \qquad r_o = \frac{1 + \lambda V_{DS}}{\lambda I_D} \cong \frac{1}{\lambda I_D} \qquad \mu_f = g_m r_o \qquad \omega_T = \frac{g_m}{C_{GS} + C_{GD}}$$

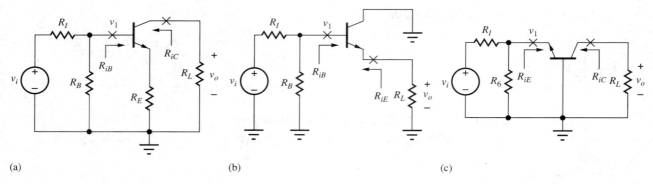

The three BJT amplifier configurations: (a) common-emitter, (b) common-collector, and (c) common-base.

SINGLE TRANSISTOR BJT AMPLIFIERS—APPROXIMATE EXPRESSIONS			
	COMMON-EMITTER AMPLIFIER	COMMON-COLLECTOR AMPLIFIER	COMMON-BASE AMPLIFIER
Terminal voltage gain A_{vt}	$\cong -\dfrac{g_m R_L}{1+g_m R_E}$	$\cong +\dfrac{g_m R_L}{1+g_m R_L} \cong +1$	$+g_m R_L$
Signal-source voltage gain A_v	$-\dfrac{g_m R_L}{1+g_m R_E}\left[\dfrac{R_B\|R_{iB}}{R_I+(R_B\|R_{iB})}\right]$	$+\dfrac{g_m R_L}{1+g_m R_L}\left[\dfrac{R_B\|R_{iB}}{R_I+(R_B\|R_{iB})}\right]\cong +1$	$+\dfrac{g_m R_L}{1+g_m(R_I\|R_6)}\left(\dfrac{R_6}{R_I+R_6}\right)$
Input terminal resistance	$r_\pi+(\beta_o+1)R_E$ $\cong r_\pi(1+g_m R_E)$	$r_\pi+(\beta_o+1)R_L$ $\cong r_\pi(1+g_m R_L)$	$\dfrac{\alpha_o}{g_m}\cong \dfrac{1}{g_m}$
Output terminal resistance	$r_o(1+g_m R_E)$	$\dfrac{\alpha_o}{g_m}+\dfrac{R_{th}}{\beta_o+1}$	$r_o[1+g_m(R_I\|R_6)]$
Input signal range	$\cong 0.005(1+g_m R_E)$	$\cong 0.005(1+g_m R_L)$	$\cong 0.005[1+g_m(R_I\|R_6)]$
Terminal current gain	$-\beta_o$	β_o+1	$\alpha_o\cong +1$

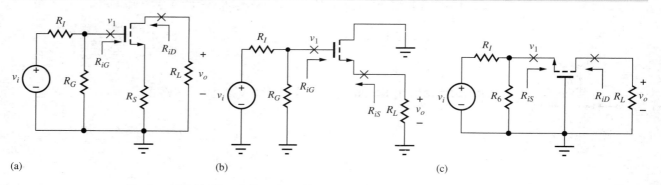

The three FET amplifier configurations: (a) common-source, (b) common-drain, and (c) common-gate.

SINGLE TRANSISTOR FET AMPLIFIERS—APPROXIMATE EXPRESSIONS			
	COMMON-SOURCE AMPLIFIER	COMMON-DRAIN AMPLIFIER	COMMON-GATE AMPLIFIER
Terminal voltage gain A_{vt}	$-\dfrac{g_m R_L}{1+g_m R_S}$	$+\dfrac{g_m R_L}{1+g_m R_L}\cong +1$	$+g_m R_L$
Signal-source voltage gain A_v	$-\dfrac{g_m R_L}{1+g_m R_S}\left(\dfrac{R_G}{R_I+R_G}\right)$	$+\dfrac{g_m R_L}{1+g_m R_L}\left(\dfrac{R_G}{R_I+R_G}\right)\cong +1$	$+\dfrac{g_m R_L}{1+g_m(R_I\|R_6)}\left(\dfrac{R_6}{R_I+R_6}\right)$
Input terminal resistance	∞	∞	$1/g_m$
Output terminal resistance	$r_o(1+g_m R_S)$	$1/g_m$	$r_o[1+g_m(R_I\|R_6)]$
Input signal range	$0.2(V_{GS}-V_{TN})(1+g_m R_S)$	$0.2(V_{GS}-V_{TN})(1+g_m R_L)$	$0.2(V_{GS}-V_{TN})[1+g_m(R_I\|R_6)]$
Terminal current gain	∞	∞	$+1$